FURNITURE

EXHIBITION

DESIGN

Furniture Exhibition & Design
Display & Exhibition of Furniture Products in the Space

Address: 5F Aikusu Building, 1-44-8, Jimbo-cho,
Kanda, Chiyoda-ku, Tokyo 101-0051 Japan
Tel: 81-3-3292-7601
Fax: 81-3-3292-7602
E-mail: azur@galaxy.ocn.ne.jp
Website: www.azurbook.co.jp

Distributed by Azur Corporation

ISBN978-4-903233-54-3

Design & Layout: Xiao Li / Bell
Staff: Joanna / Ray / Mark Gang

Printed and bound in China

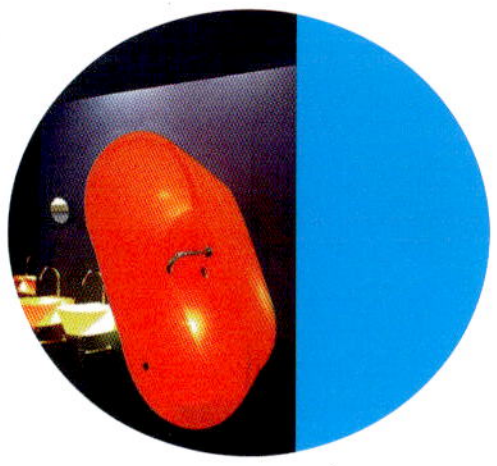

CONTENTS

PREFACE

Household items highlight their taste of life in the process of exhibiting & displaying.

The International Salon is known as one of the world's three major annual exhibitions. It is the pioneer of global fashion arena featuring the release of the world's top household brand new products as well as the indicator of international fashion trends. It attracts countless people involved in household design, fashion and media circle. The Salon is representative, because in there you can see the great works designed by masters. The furniture represents tendencies of the contemporary furniture design. The Salon has a complete system and is in a leading level in design, which makes it the indicator of the world's fashion trends. Today's furniture is not only functional, but also is embodied a concept of blending technological innovation, industrial experiment and the refined traditional techniques. Now the relationship between decoration and design is closer and closer. Decorative function has become an important part or even the core of design. International Salon is an incubator of global furniture design, for all the world's popular fashions start from there. Therefore, we can say that the annual International Salon succeeds both in design and commerce. People are all intoxicated by those exhibits that can express their thoughts

The materials used in the furniture and cabinets are getting closer and closer. Some organic polymer materials which have aesthetic feeling as well as fire resistance, moisture-proof will be found in the design of cabinets. In modern design concept for kitchen, it's not enough for the kitchen to just play to role of cooking. People begin to make the kitchen a place full of funs. The open design makes an unobstructed view of everything. There is no cover-up in the space by removing all elements that impede people's sight. Openness has become one of the popular key words in the International Salon.

The origin of the open design is the small space. Many things need to be removed due to space restriction, such as push down the wall and get rid of the cupboard door. Further action is taken to simplify the design and only a few the most basic lines will be kept. In order to get a better feeling of space openness, glass is a relatively ideal material because it is transparent and has no sense of oppression.

"The chair is not just a chair, but a pleasing artwork. Chair is the most anticipated part in the furniture. A few years ago the style of furniture design style was romantic. Now the trend becomes increasingly "cool."

"Simpleness" in the most admired concept in household industry. Simple lines and sophisticated techniques are integrated into one to achieve a perfect unity. That the integration of modern style and modern art makes chairs become a multi-angle artwork is the designer's favorite. The fun and satisfaction of life can come from a chair, as long as the chair has enough aesthetics, a certain degree of comfort and self-style. The sofa uses metal frame with a leather surface and satin cover, which shows a kind of special glaze. Going back to nature is the core that everyone focuses on, because nature is an eternal theme for human beings. Both flowers and trees, or breeze and clouds, have given a lot of inspiration to designers. It is a kind of grace, a feeling of freedom and a unique taste of life of returning to nature and searching for the nature of life in the hustle and bustle city. Advocating the simple natural style is still one of the mainstreams. Tension in modern life affects people's aesthetic taste; so they are desirous of luxury elegant quiet palace life.

Honorable, successful, tasteful life starts from the details at home. Luxury does not mean pursuing seemingly extravagant surface, but the perfection in the ordinary life. Low-key elegance and elaborately carved details are the essence of wisdom in the furniture design. This is a sophisticated-skill curved wooden chair design. The color of the special translucent plastic pad matches that of the roses. The sofa not only has nice appearance, but also is useful. All unimaginable things can be transformed into furniture. It is very surprising, isn't it?

"Open, transparent-Glass is a relatively ideal material to expand the open sense of space, because it is transparent and has no sense of oppression.

Natural, elegance-Horizontal, vertical, oblique, curved, its nature and elegance are fully shown through the reflection in the glass and the water. The bright orange, no matter combined with modern style, or with classical elements both has the special charm that can seize people's attention. Red, black or white is mysterious, rich, young, modern, open-minded, enthusiastic, self-style. People are always looking forward to the furniture salon. Space is a big topic. What will you think of when we talk about space? It can be as big as the universe and time or as small as a building. No matter it is still or changes slowly, it will always stimulate our imagination. In addition, it is a great source of inspiration for designers. Space not only changes our living habit and level, but also awakens our memory of primitive living demands. Maybe this is the power of space.

Fairy-like world of lamps

Designers won't abandon any aspect that can show self-style personalities. Some special materials can also be found in furniture design. For example, the classic clothing elements like tassels and folds can also be found in the furniture. This is the typical product of the environmental protection awareness and modern technology. People who have been living in the hustle and bustle city for a long time will naturally have the inborn affection for natural and primitive furniture, particular wooden furniture and environmental protection furniture. A lot of furniture design is also mixed with "behavior artistic" elements, such as the gauze wrapped sofa. Good design should nice but applicable and functional. This chandelier consists of tassels like lampshade which is beautiful and elegant. The sofa wears a black satin and is covered with black fur. It is just like a lady in the middle time of Europe with a bit of gold embellishment. The shape of the chandelier frame is like a cuttlefish, which is similar to Chinese lantern. The oval cut-out chandelier also shows some oriental charm. The world of lamps also provides designers with a broad design space, so we look forward to each exhibition.

The combination of classical art and modern industrial procedures makes the design and manufacture of benches, sofas, tables, beds and furniture reach to a higher level. That people's demand of improving living standards constantly gives designers ample room for development. Design is to achieve the perfect match of visible shapes and the geometry elements.

The Display & Exhibition of Furniture Products in the Space

Lights Design

DESIGN
Turn on the light it shines brilliantly and seems as an ancient ornament which brings comfort and uniqueness

Recently, the exhibition continues to introduce new works. Each product is a unique artwork. Different from traditional shapes, the work is full of humanity, showing the designer's concept of minimalism, and also reflecting his superb design.

The chandelier has the superb peculiarity to use the simplest element. Made in half cut-out plastic, it does not have complicated construction, but simple and elegant. It is heat resisting and light, producing light that is full of charm. The lighting gives a deep and vivid impression by its forward and aggressive style, attracting many well-known designers to the exhibition.

porada

The chandelier has the superb peculiarity to use the simplest element. Made in half cut-out plastic, it does not have complicated construction, but simple and elegant. It is heat resisting and very light, producing light that is full of charm. The lampshade is made from natural rattans, which shows the innovation of the lamp. It is very special.

Always BY CASA BELLA COLLECTION
mueble de españa
A06
Casa Bella Collection

m u e b l e d e e s p a n a

Classical chandeliers with exquisite techniques and the retro accessories show their magnificence and brilliance. The whole space is low-key and yet gorgeous. The various designs in the changes of the colors and materials as well as the highly creative visual languages have a strong impact on people's eye. The mort important is that this feeling has been integrated into life, which realizes values of the art. The golden trimmed stainless steel products have smooth surface and beautiful lines. Even the reflection of light and shade are taken consideration in the design, so the entire space glitters under the lights.

LIGHTS

White is and will be the leading color in furniture design. The main white body contrasts the black color, and stresses each other's shiny surface. In addition, the new color-grass green is playing an important role in the furniture design. Another tendency is the gray gradient color. So this year's red is not pure red, but contains a little bit purple. So does green. It must be with some blue so as to be call "pop color."

The feature of simple furniture is the simple and nice lines. This is the result of the designer's excellent craftsmanship. The furniture has simple style as well as advanced design. When placed in the room, though it is simple, it can enhance the atmosphere immediately. The designer's subtle creativity can be felt through the simple use of the materials and colors.

Fairy-like world of lamps

Human beings' progress depends on the invention. Invention is the primary achievement of mankind's ability in creation. The goal of the creative is that the material world should be controlled by the spiritual world. The lights design circle is still making progress. The progress does not rely on the countless forms, but through the expansion of each scope in design and production processes. Each element has to be useful.

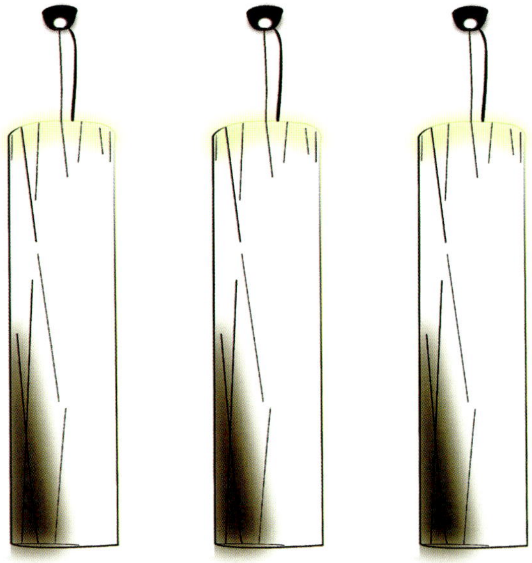

The design theory of lighting fixtures, in fact is the needs that designers understand and analyze people's life. It realizes different people's dreams, thus many excellent lighting fixtures come into being. The unique feature of the light is melted into various series of products. For instance, the sofas in the room are grouped according to their characteristics. Each series is unique in functions and concepts. At night, here is ablaze with light and filled with artistic feeling. The refraction of lights makes here warm and comfortable. With fashionable and modern design looking, the lighting fixtures are transparent and glazed.

LIGHTS

Classical chandeliers with exquisite techniques and the retro accessories show their magnificence and brilliance. The whole space is low-key and yet gorgeous. The various designs in the changes of the colors and materials as well as the highly creative visual languages have a strong impact on people's eye. The mort important is that this feeling has been integrated into life, which realizes values of the art. The golden trimmed stainless steel products have smooth surface and beautiful lines. Even the reflection of light and shade are taken consideration in the design, so the entire space glitters under the lights.

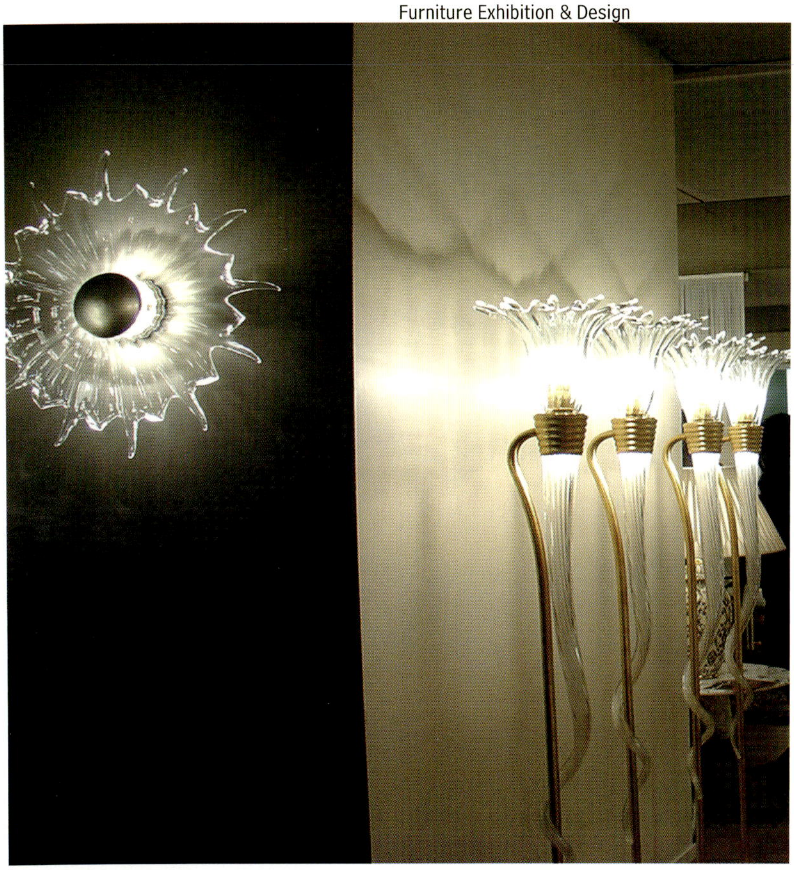

The light is designed in a method of using transparent glass. It has soft shape and elegant appearance. The lamplight is adjustable and full of modern appearance. Simple and fashionable makes it popular in the market. The furniture seems to realize their "flower" dreams. From the design, color, to shape, all have something to do with flowers. Designers find the inspiration from life, which perhaps is one the trends of the future furniture design.

With unique design, the lamp is implicit and accompanied with elegance, simple but fresh. It not only embodies the traditional dignity, but also reveals a peaceful and natural style. Now the market is gradually mature, and combines the function of both practical and decorative. People will begin to discuss when seeing the light. Does it look like a bird's nest hanging upside down? The shape is creative. What is inside is not a bird, but a lamp bulb.

The chandelier is a polygon sphere. The concept is not only innovative, but also makes the world of lights more attractive. The eggshell-shaped chandelier is like a broken egg under the yellow light, which embodies the unity of color and style. In addition to providing products that meet the demands of daily life, they also try to find the new direction of development through continuous innovation in order to invent more new products.

Charlotte
EDITION in BLACK
Charlotte
EDITION in BLACK
FARBEN FREUDE
FARBEN FREUDE
FARBEN FREUDE

The chandelier and floor lamp with different materials, delicate textures and ornaments reflect women's gentle beauty. The designer used white lights to highlight the romantic atmosphere. The dark brown sofa matches well with the wall in flower patterns.

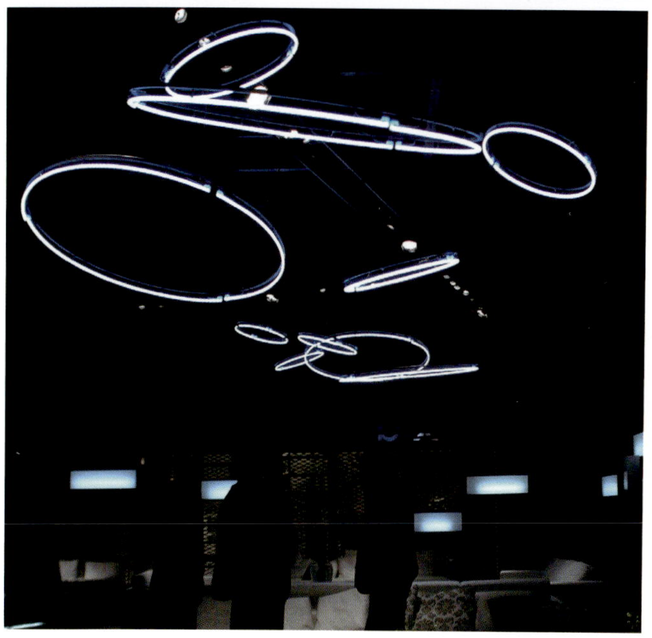

When you turn on the light, it shines brilliantly and seems as an ancient ornament, which brings comfort and uniqueness. The wave of simplicity has the impact on lighting design and expands the design scope.

The product effectively seizes the unchanged living ways in people's minds. With flexible design, it keeps the classic features and also catches modern aesthetic standards. So it is very suitable for those who love changing interior living environment. Plastic and glass are clear and transparent. They are shinning dazzlingly, brilliant but novel, which makes you feel like walking into the fairyland.

This year's exhibition got rid of the nostalgic and romantic trend and turned to pop "bio-mimetic" concept which pays attention to the life itself. Designers began to explore things in the most primitive state and use modern means to display them. For example, they try to find the passion through some simple and plain elements. Therefore some of the traditional raw materials such as timbers are re-used. What is different from the past is that designers have invented a new way to display the material's grains. The multifunctional furniture reflects the integrity and consistency of the architecture and interior design. Furniture, household fabrics, blankets, home accessories and household lamps, daily necessities will be exhibited together.

With the rise of new design power, lights designers inevitably adsorb new elements to their design. Take this lamp for example, designed in a fashionable way, it contains plenty of milk-white spheres. Therefore, it can be applied to various places according to people's needs. As daily durable products, the furniture not only must have the usage function, but also acts as the carrier of design culture.

With unique design, the lamp is implicit and accompanied with elegance, simple but fresh. It not only embodies the traditional dignity, but also reveals a peaceful and natural style. Now the market is gradually mature, and combines the function of both practical and decorative. People will begin to discuss when seeing the light. Does it look like a bird's nest hanging upside down? The shape is creative. What is inside is not a bird, but a lamp bulb.

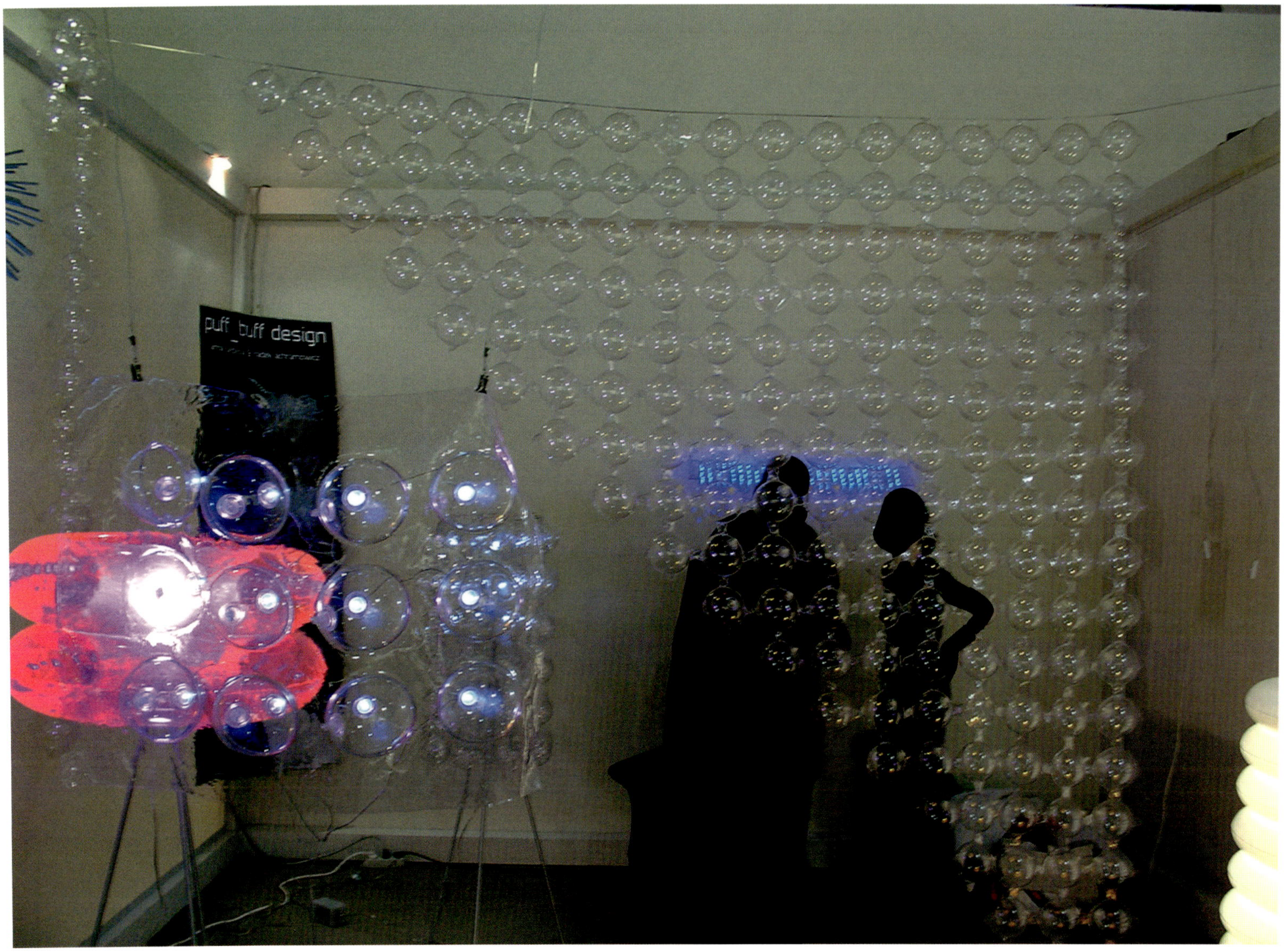

Plastic and glass are clear and transparent. They are shinning dazzlingly, brilliant but novel, which makes you feel like walking into the fairyland. Minimalism was in full flourish in the 90s of last century. European designers made the minimalism to the extreme by using natural wood, leather or other materials to meet the demands of modern people. Simplicity does not mean merely simplifying, but to simplify in the rich centralization and to achieve the most complicated artwork by the simplest method.

Today designers still act as the innovator in the technologies and materials of the lighting fittings. The current use materials include glass, metal and plastic, etc. Now the market of the lighting fittings contains Europe, the USA, Japan and Asia. The representative features of current lights and lamps are the high-tech industrial design, industrial materials, as well as industrial equipment. The study in the past had been focused on the design styles and works, and paid little attention to design schools and well-known brands. What the modern life needs is the style that has individual and the aesthetic demands of the details as well as the trace of the origin for the traditional culture and the humanities. The application of new technologies, elements, materials and colors brings a dazzling experience and allow you to feel the same things the other aspect, the real and mysterious aspect. From art to design works, the trend of molding "the dreaming world" has been triggering more and more ideas and concepts.

Kitchen & Bath DESIGN

nest!

Kitchen

Bath Design

DESIGN

Lighting will receive much attention because it has good at expressing the emotions of their switch novel chic very exciting.

BATH

LIGHTS

Imagine that oranges and candies, chocolate and butter, mixed with the cook flavor, you will get this strong color, rich feature prescription. There are various types of pots, bowls, all kinds of bottles for cooking oil, salt, soy sauce, vinegar, as well as all kinds of tableware items. How to use the space effectively to place these items neatly by using hardware pendants, multi-layer boards or wood storage racks? You will need to have more ideas. Kitchen as part of the living place, its function has been further developed. Home and office, rest and work are merged into one and coexist in the same space. This is a new living way

The plain color cabinet as well as the square ventilator as if is telling us the high taste of their master. The khaki-colored tile in the corner is the interpretation of kitchen space and the unlimited extension. Simple but with pink decoration, the charm of the room appears immediately. With the improvement of living standards and the pursuit of high quality life, people are not only more demanding with food, but also the dining atmosphere.

LIGHTS

Carefully selected decorative accessories not only make the kitchen more stylish and beautiful, but also increase the kitchen's accommodation space. Designers usually have many accessories programs. For example, we can put some glass clapboard or some stainless steel accessories on the wall. You can either place some flowers and dolls, or put the seasonings on them. The insider of the drawers can be decorated with plastic boxes to separate the space. These carefully selected small items not only make the kitchen more beautiful, but also increase the kitchen's containment capacity.

The wall between the kitchen and the living room is movable. The designer's purpose is to lay the base for the following magic. At the same time it makes both sides of the wall have different appearance compared with the rest. If you want to redo the wall, you will be able to choose the color of the materials based on the overall style of the house. The two sides of the wall face the hallway and living room respectively, so different methods are used to deal with them. The side facing the hallway is with a piece of glass clapboard, on which vases can be placed. The side facing the living room has a rectangular space to display the host's collections. The fixed side of the wall has a window for air ventilation of the kitchen. In this way, cooking fumes are completely cut off inside. Outside has a perfect image.

LIGHTS

The texture of materials affects the product's impression and the usage effect. The color, shading, texture of the chosen materials affect the style and the impression of the kitchen to a large extent. Moreover, the capacity of clean-resistance, pressure-resistance, corrosion-resistance determines the kitchen's long-term quality and your experience of using the kitchen. Therefore, all details should be paid attention--from the texture of the wood to the grinding pressure of the metal surface.

A good kitchen designer will optimize the space in accordance with the owner's individual needs, and make a variety of programs for reference. More rational use of space creates a sense of a family comfort. To expand the area of the kitchen in the visual sense brings a relaxing mood, which makes the process of meal preparation more enjoyable. The unique style makes the kitchen full of modern and fashionable aesthetic. "Different colors will give people different feelings." This kitchen is both nice and practical. The secret lies in the selection of color patterns. As a matter of fact, not all colors are suitable for kitchen. Some color is very brisk at first glance, but the visual fatigue will appears when you look at it too much. The reason is that the tone and saturation of different patterns brings people different psychological suggestion. Therefore, the warm color is better than cool colors. The graphics with strong contrast may bring negative stimulation. The same color is a little bit too dull. In addition, you can ask kitchen consultants for advice to adjust the color and light of your kitchen to make it more comfortable and wider to highlight your personality.

White is the eternal color in the minimalism. The use of white color in the kitchen makes narrow space to achieve the effect of a longitudinal perspective. Black is also a popular color used in the kitchen. But due to the black color is too dark, so there won't be large black area in the kitchen. We tend to use the wooden yellow which is very popular now as a decorative color. Wood, fire resistant board and stone are the basic materials for cabinets. The basic materials used in cardboards are wood and stone. In the minimalism, the new materials such as aluminum, carbon fiber, plastic, high-density glass, etc. are very popular, which enriches the combination of the kitchen world, such as waterproof, scratch resistance, refreshing, light-penetrable, etc.

The designer's creation can hide the entire kitchen: A vertical cabinet is placed respectively on both sides of the --shaped or L-shaped cardboard. The four sides of the cabinet can stretch to the four directions to combine long doors. If necessary, the four doors can fold to the middle to hide the table, counter, floor cupboard. The original --shaped kitchen now becomes a fashionable vertical cabinet. The original L-shaped kitchen turns into an L-shaped wardrobe. The light of the circle-shaped kitchen changes with different light. It not only looks modern, but also is a very good place for dining.

Highly innovative household products are from the International Salon. Popular colors, refreshing construction and fashionable concepts of life are gathered in this furniture feast. The world's top furniture designers and manufacturers took the opportunity to flourish their new spirit to the rest of the world. It is an artwork full of vitality. If you take a rest here, is just like to be familiar with a new concept. Its beauty is not limited to superficial beauty of the commodity, but the noble beauty. It uses raw materials such rough clay and stones to show the quality.

Smooth shiny metals and coated plastics. The flexible use of the white color gives the furniture a unique atmosphere as if we were on the moon. For old-fashioned kitchens, if there is no cabinet, we can only use the wall of the kitchen. The warm and harmonious wardrobe is like a perfect scene. The real protagonists here are the smoke exhauster, gas stove, disinfecting cabinet and dishwasher. The layout is in scientific way. People really fall in love with the kitchen. From now on, cooking seems to have become a kind of enjoyment.

LIGHTS

Home kitchen is the most important soul in the living space. Open kitchen becomes part of the living room. Refrigerator, oven, etc all become part of the kitchen cabinet. The way is reflected the performance of the design. It basically uses straight and simple lines to emphasize the sense of an open space. There are two colors used in the kitchen. One is the original wooden color; the other is dye the color according to the texture of the wood. Both kinds of solid woods are fireproofing. The wooden color can reflect the natural beauty of wood itself and highlight the design of furniture.

The three-storage box consists of an easy and simple space. At the same time it also increases the visual sense. The plain color cabinet as well as the square ventilator as if is telling us the high taste of their master. The khaki-colored tile in the corner is the interpretation of kitchen space and the unlimited extension. Simple but with pink decoration, the charm of the room appears immediately.

B24
DESIGN

ESDESIGN

No matter how small the house is, the kitchen is essential. Therefore, it really requires great efforts to create a nice and practical kitchen in a small space. Open kitchen is becoming more and more popular. It connects enclosure kitchen with the living space and makes the room more spacious and more modern looking. Kitchen in fact is a complicated place. If the design can simplify the procedure and let people do the house work in the clean bright environment, then it will be more efficient and the mood will be up. Do you have the desire to have beautiful music around you when cooking in the kitchen? Do you want to watch TV leisurely when cooking the soup? Do you look forward to working in the kitchen for hours and yet still not feeling tired? The answer of course is “Yes”. I believe this cabinet can liberate you completely from the boring kitchen life and make your kitchen with entertainment enjoyment.

The kitchen of the newly-wedding family is warm, which represents love and the wife to receive a new life. The whole color series have a sweet feeling. Each style has its own representative color. The warm color is the representative color of minimalism, which has a close relationship with the birthplace of minimalism. Because of the warm weather, the wood has light texture colors; there is less modified ingredients in household items. Therefore, simple and warm become the representative language of the minimalism.

Smooth shiny metals and coated plastics. The flexible use of the white color gives the furniture a unique atmosphere as if we were on the moon. For old-fashioned kitchens, if there is no cabinet, we can only use the wall of the kitchen. The warm and harmonious wardrobe is like a perfect scene. The real protagonists here are the smoke exhauster, gas stove, disinfecting cabinet and dishwasher. The layout is in scientific way. People really fall in love with the kitchen. From now on, cooking seems to have become a kind of enjoyment.

CUCINE
LUBE

newform

Simas
ACQUA SPACE

Listen carefully! There is no other noise except of the sound of flowing water. The home bathroom is generally not very large but the utilization rate is very high. Due to there are many toiletries and other objects in the bathroom, the design must focus on how to use limited effectively. For instance, we can put embedded frames, mirror cabinets, and other metal pendants, etc.

The design of this bath reflects the people's desire to enjoy the beauty of nature. It is just like drifting in the ocean. The design makes people relaxing. The whole design is novel and reflects strong individualism. The materials used in this bathroom are high-tech ones. The designer's inspiration comes from the demands of life in the twenty-first century.

These arc-type sanitary wares are more like the works of art. Kept in the spacious bathrooms, they are distinguished as well as graceful. Meanwhile the rectangular Hand-washing facility is the passionate crash of classical and modern. The designer applied an elegant and very humane way of design, which helps realize the perfect integration of the modern and the tradition. If the items in the bathroom are placed properly, you will find though the space is small, there is till somewhere that we can take use of. As long as we plan well, plenty of items can still find their place here.

redvision

Glass Agua

The bathroom is made with ecological materials. The decoration of the whole bathroom is in a solid and uniform color, which is quite novel. The entire decorative material is completely waterproof, and particularly suitable to use in the bathroom.

The ladder is placed in the bathroom. You can put those that are not frequently used on its higher place. It can also be used as the shelf for towels and books. If you want to read magazines in the bathroom, you can just get it from the ladder. You can place a basket which can contain towels or laundry near the bathtub. It will be very convenient for you. Do not put too many things on the dresser. Soapbox and hand sanitizer in small bottles are the better choice, which enables the easy maintenance of the table. The necessary items can be place on the top of the frame and the hand-wash platform frame, as well as the space under the cupboard. If you feel that this is inadequate to meet your requirements, you can post some stick hooks on the wall for hanging towels and clothes.

There are many standards about choosing sanitary products. Besides the factors of the brand and follow-up service, there are factors of materials, design, workmanship, the function of water-saving & water-proof, easy clean-up, comfort, appearance, etc. To sum up, there are two points. One is easy to use, that is utility; the other is nice looking, that is, appearance first. Based on this, the author searched several types of easy-use and nice sanitary items from large shopping centers to share with you all.

ZAZZERI
PRESS

ZAZZERI

The shapes of the furniture is subtly used the geometric forms. The color and materials are created dramatically. The appearance and design are adhered to Baroque and Rococo style. Meanwhile it also demonstrates simple and purist elements.

This basin is superior outstanding. The whole basin body and frame are with mosaic gold and look very grand. The lines of mosaic gold are very soft and flexible, which matches perfectly with the luxurious elegant and noble European furniture style. It is in ceramic structure, so it has advantage in waterproofing and cleaning.

totale - total impermeability

The interior design is regarded as the best way to show the activity of modern living environment. It not only reveals people's basic self-cultivation of dealing with the environment and the awareness of life, but also has an important impact on people's spiritual life. The space of bathrooms is a special space can reflect people's pursuit of the spiritual life. Dominated by this concept, in the design of bathroom many people pursue a personalized and comfortable style. The content of both senses and the spirit is each user's requirement. Modern bathroom is to let people enjoy the space inside the bathroom for more compensation. The wooden plate bathroom has a natural and warm feeling. The tones are attractive, which makes it match well with the whole decoration style.

Simpleness presents the formal simplicity which strikes the balance between functionality and aestheticism. For a designer, the rational affection of structure is no longer a game of metaphysical philosophy. We can clearly feel the attempt of getting a glimpse of the future design in these horizontal and vertical graphics. The design in the early 21st century differs from the one in the past several decades which just emphasized outside innovation in the shape of the minimalist design is that the application of new materials and techniques will re-gain attention. The use of new materials and technologies symbolizes the new are of logo design, which breaks the general boring rules in furniture design and gave people unlimited reverie. Leisure and relaxing design has become the byword of top design.

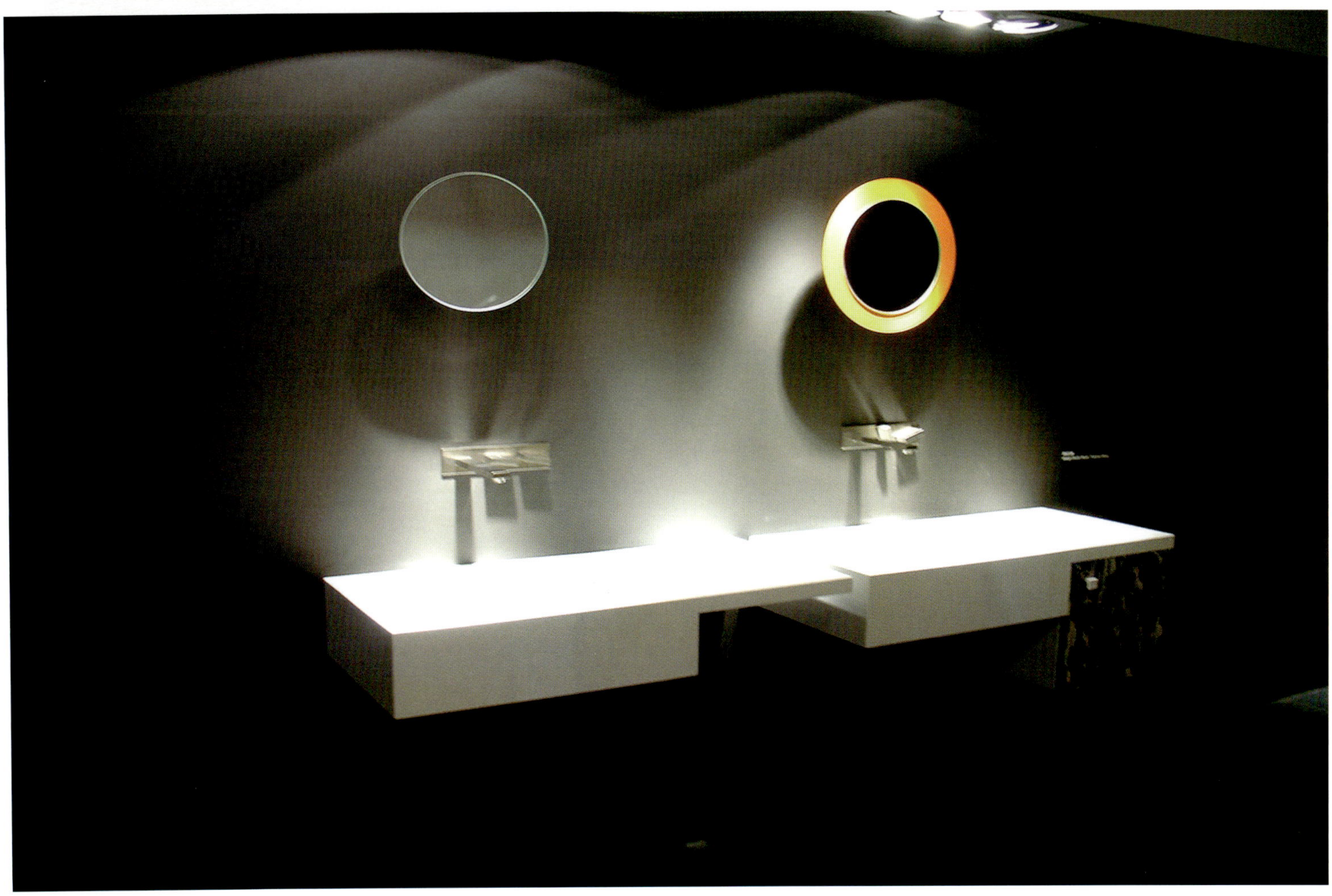

consolle
Talocci Design
young
design C&P

This is the shower limiter which restricts the direction of water flow. The pool also can be use to reserve water. It will be very convenient and fast when you use the small flush button. The design of the entire bathroom is very simple.

For some people, it will be extremely boring if it is merely a wall. Even if the wall is blue or yellow, we still will feel tired when time goes by. The floor mirror and the brushed aluminum furniture are dazzling, as if yelling that why not use the most pop color tiles to cover the whole wall? In this particular space of the bathroom, you can find some leisure time to take care of every inch of your skin by preparing a pleasant, enjoyable and comfortable SPA for yourself!

SAT
CELIA · FLASH

The three-door sliding screen belongs to bathroom series. The stretched brushed silver handle is the most special part. It uses the patented gear design to ensure the stability of the handle. This track groove is much easier to clean up because it adopts slotless push-pull tracks.

Now people become more demanding about the colors of the home, and gradually begin to enjoy the beauty brought by those colors. The large area of black, white and gray has become less and less in the bedroom. The decoration of bathrooms is mainly used bright colors. Pursuing the classic living, you may as well use the most primitive black-and-white color. Black and white bathroom design not just to calms your spirit, but also brings a more distillated spirit.

The mainstream of the design for toilet bowls is the siphon and the siamese one in recent years. First of all, there are no return water pipes in the cesspipe. The smell will be returned if we don't use the siphon one. Secondly, there are fewer bacteria bred by the toilet in the humid environment if we use the siamese toilet bowl. This siphon toilet bowl uses urea-formaldehyde and buffer land cover design. It is anti-aging, anti-scratching and easy to use. In the design, it uses straight lines, and the decorative effect is also very good.

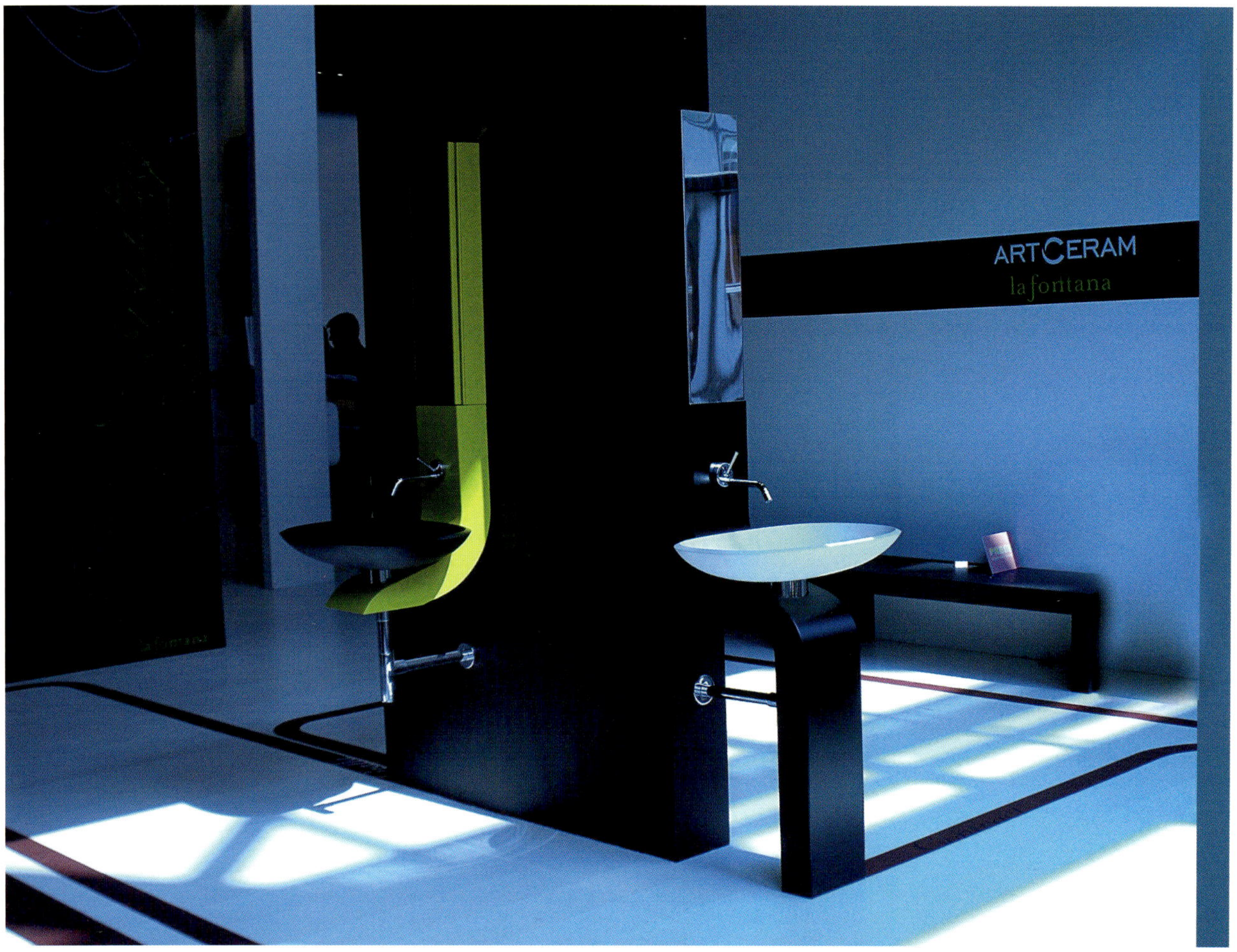

The modular multi-functional bathroom reflects a clear trend in the furniture design and foreign trade. Now, it is more and more important to provide consumers with all-round household items. People not only focus on furniture, but also pay attention to the accessories of bathrooms and other household products, in order to create a new way of life. The home decoration stressed on romantic nostalgia a few years ago. Now it turns to a new concept of "framework", which brings the focus to life itself. Designers began to explore the very beginning state of things and, and demonstrate it with modern means. Therefore some traditional materials such as wood have been re-used. What is different from the past is that designers have invented new ways to display the texture of the wood.

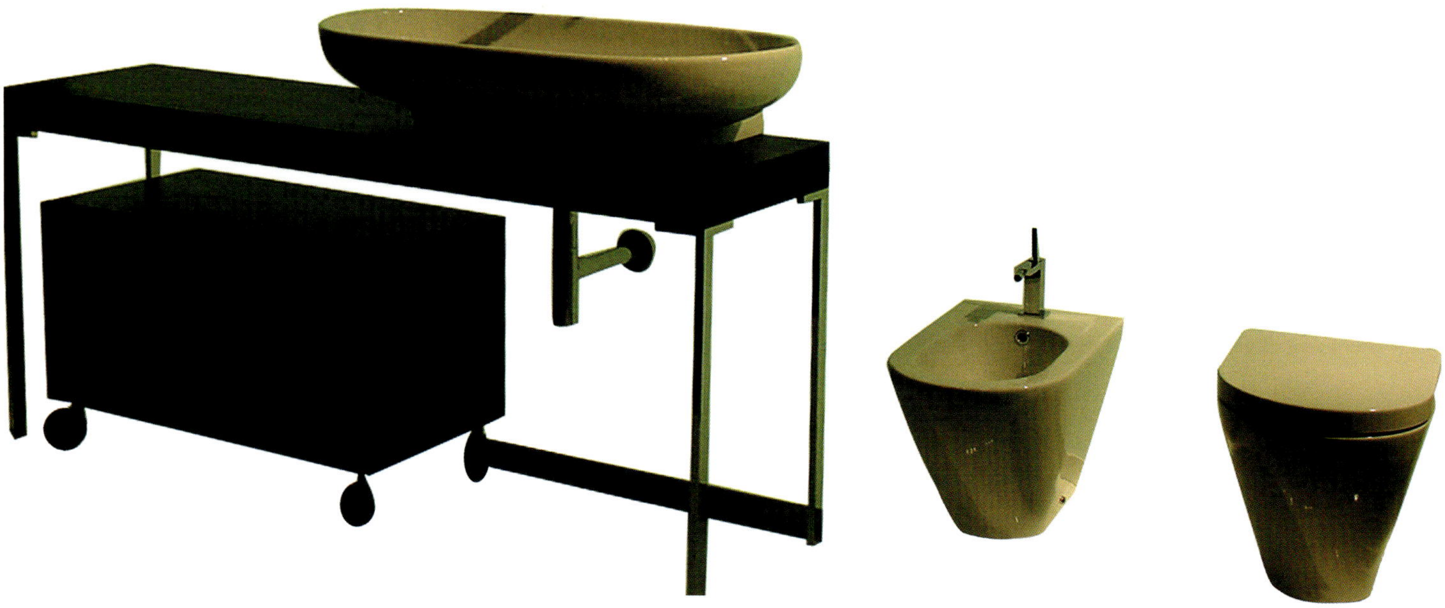

This set of bathroom design is full of colorful hue and simple floriation. The decoration emphasizes on space, which makes people have the impulse to go home. The color is mainly in black, red and white. They are bright but not messy.

PATTERN &
ACCESSORY DESIGN

nest!

Pattern
Accessory
Design

DESIGN

Fashion and design are two separate industries but they have been more frequent overlapping since this year They both can express the same theme and mood which provides the designer more creative space

Natural curly lines and decorative floral motifs continue to be prevalent in the Salon. The most popular design in this year is the fine and small pattern based on plants. Initially fashion and design are two separate industries, but they have been more frequent overlapping since this year. They both can express the same theme and mood, which provides the designer more creative space. Many top fashion designers have begun to set foot in the furniture design, thus the two industries have some common concerns. The elements that can represent the furniture's romantic sense are the graceful and smooth curves. There are also many new elements in the modern furniture design. For example, based on the ergonomic principles, we can mix wood, glass, plastics, leathers, rattans, textile fabrics and other ingenious to make the forms of the furniture more diversified.

The charm of the European and American furniture lays in their elegant shapes, textures, colors and accessories. Each detail conveys the ancient flavor. Complete respect for the history is the kind of furniture's original motivation. Retro is always the most classic sample. Initially fashion and design are two separate industries, but they have been more frequent overlapping since this year. They both can express the same theme and mood, which provides the designer more creative space. Many top fashion designers have begun to set foot in the furniture design, thus the two industries have some common concerns.

ESBACEAY

In the Italian Salon the various kinds of works which come from outstanding designers are really an eye-opener. From now on furniture is no longer plain, but becomes the perfect artwork! In fact, we need to appreciate the vanguard of the avant-garde design. But what we need badly is the nice and comfortable furniture. This is the relationship between art and life.

Charming
Salute
Elasticità
Living fashion

The mix and match trend in the fashion circle gradually comes to the furniture sector. It not only influences the furniture style, but also the colors. The rarely used or conflicting lines & colors in the past are now being applied to the furniture. Designers can make the iron full of passion, leathers and stones become sentimental. Italian designers even can make the ordinary plastic become sexy and enchanting, which completely upgrade the identity and taste of the plastic furniture. The furniture's beauty in the material and shapes is fully displayed through designers' design. Furniture becomes more and more beautiful and various, just like the fashion which always arouses people's desire to purchase them.

Superb Cutting Skills and Refined Beautiful Patterns: The charm of the European and American furniture lays in their elegant shapes, textures, colors and accessories. Each detail conveys the ancient flavor. Complete respect for the history is the kind of furniture's original motivation. Retro is always the most classic sample.

This is the series of furniture with flowers as the subject. There is petal furniture which can be combined at your disposal. The wooden petals can be placed together to constitute a flower in full bloom. Old design with new materials and techniques will express a quite different feeling. It is true that the sense of good taste comes from a person's inborn sensitivity to beauty. Besides this, it is also very important to cultivate one's good taste by widening his horizon. So, if you want to make progress in a certain area, the best and most effective way is to enhance the absorption of the updated information.

This year's exhibition got rid of the nostalgic and romantic trend and turned to pop "bio-mimetic" concept which pays attention to the life itself. Designers began to explore things in the most primitive state and use modern means to display them. For example, they try to find the passion through some simple and plain elements. Therefore some of the traditional raw materials such as timbers are re-used. What is different from the past is that designers have invented a new way to display the material's grains. The multifunctional furniture reflects the integrity and consistency of the architecture and interior design. Furniture, household fabrics, blankets, home accessories and household lamps, daily necessities will be exhibited together.

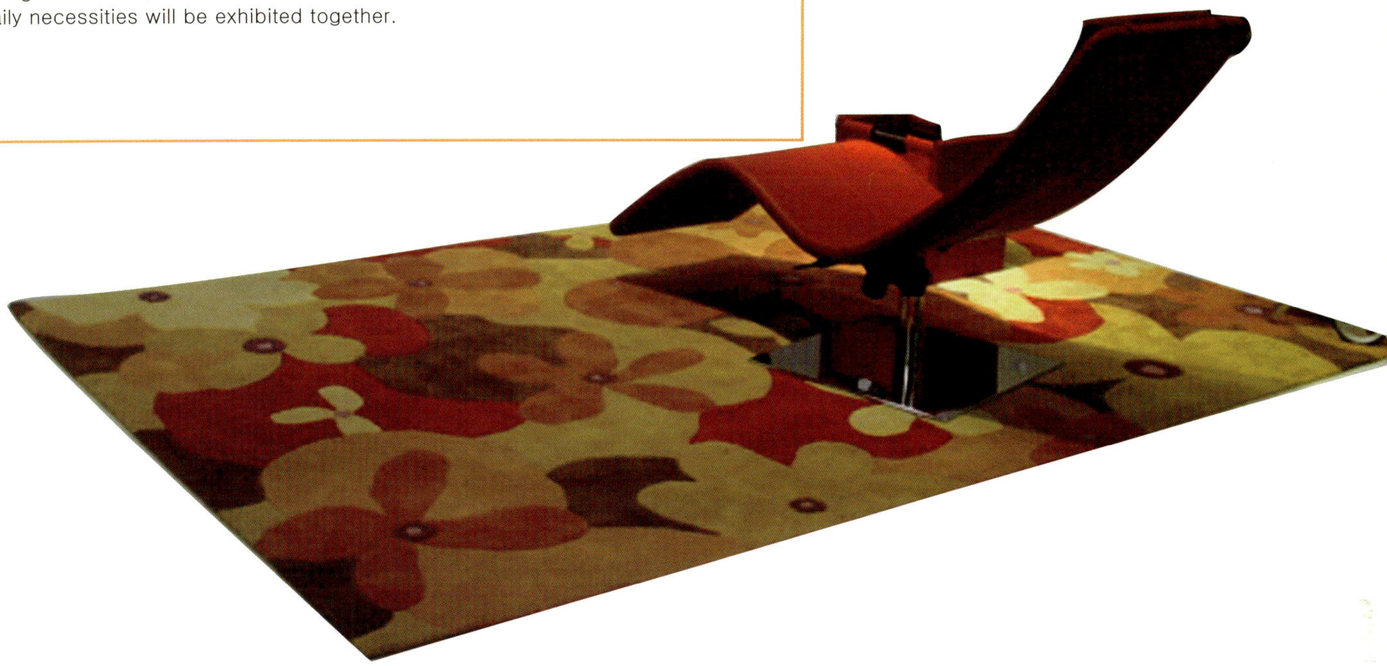

The application of new technologies, elements, materials and colors brings a dazzling experience and allow you to feel the same things the other aspect, the real and mysterious aspect. From art to design works, the trend of molding "the dreaming world" has been triggering more and more ideas and concepts. The shapes of the furniture is subtly used the geometric forms. The color and materials are created dramatically. The appearance and design are adhered to Baroque and Rococo style. Meanwhile it also demonstrates simple and purist elements.

The designer's works, say, the simplest sofas, are showing various kinds of appearances, elegant, conservative, or pure. People's life styles are taken into account by designers, as well as the spatial distribution of home and the use of different habits. Therefore they bring environmental protection into household item design. The design of the living room must consider the occasions to treat friends and relatives to express the host's hospitality, and also has to take into account that it is also a kind of personal space to shaking off the troubles so as to enjoy the world of private space. Designers must keep the two aspects in mind when designing the living room.

These innovative household items all come from this year's International Salon. Pop colors, refreshing and fashionable concepts of life are gathered in this furniture feast. The world's top furniture designers and manufacturers took the opportunity to spread the new spirits to the rest of the world. The fuzzy function in the product design makes the sole function in the past become multiple ones. It makes people no longer regard those products as simple tools. You can see more through their surface. They are embodied certain artistic through the meanings and symbols, which makes them more interesting and beautiful. Italian designers can combine pragmatism and technical knowledge with classical culture, so that their design from the post-war till now has their own brand, expressing of a design culture full of imagination and personality. Any item from life can be a showpiece. If you liked it, you can give it a different soul. The furniture is the complex of architecture, modern design, personal style, traditional crafts, modern manufacturing and so on. Designers do not want their works only be permanent collections. They really want them to enter into people's life so as to bring enjoyment to life.

MARTINI'S
DESIGN

Cooking

The application of new technologies, elements, materials and colors brings a dazzling experience and allow you to feel the same things the other aspect, the real and mysterious aspect. From art to design works, the trend of molding "the dreaming world" has been triggering more and more ideas and concepts. All bright colors such as bright red, orange red, grass green, bright yellow, bright purple, etc. are and will be the popular colors in this year and the coming years. Many designers have chosen a beautiful color to be the main color for the whole furniture design. Furniture becomes more and more beautiful, just like that of fashion, which always arouses people's desire to take them home. Black, white, red, green are the major colors. Much attention is paid to the combination of the colors, so the furniture seems more brilliant.

SOFT
LINE

The space created by a certain material that is refined through the design process can have a strong resonance with the mankind. Maybe it is a path, or one place for people to take a rest. The designer always uses the touching patterns to show the double connotation of his design works. They are both for the public and the history.

Just as years of n

The mix and match trend in the fashion circle gradually comes to the furniture sector. It not only influences the furniture style, but also the colors. The criteria of modern aesthetics are the pursuit of personality details and yet, keep the style. Therefore, fashionable, avant-garde and distinctive products are extremely popular in the exhibition. The mix-use of styles, materials, and elements gives people a new sensory feeling. The shapes of the furniture is subtly used the geometric forms. The color and materials are created dramatically. The appearance and design are adhered to Baroque and Rococo style. Meanwhile it also demonstrates simple and purist elements. Highly innovative household products are from the International Salon. Popular colors and refreshing and fashionable life concepts are gathered in this feast of the furniture. The world's top furniture designers and manufacturers take the opportunity to promote their new spirit and concepts to the rest of the world.

The multi-functional furniture reflects the integrity and consistency of architecture and interior design. Furniture, household fabrics, blankets, home accessories and household lamps, daily necessities will be exhibited together. European and American furniture is the best one in any space. Their brilliance makes the user have the feeling of nostalgia. The best state of life is in fact the semi-sufficient one, so does the design, which is to get the balance point among various kinds of elements.

The 21st century is an era of rational consumption. The taste is in no way reflected in thc increase of wealth, but the concern of the needs at the bottom of people's hearts. Really valuable things can stand the test of the time. When you use it, its proper function makes you satisfied. When it is placed in a place, its artistic feature immediately appears and makes the whole space have a romantic aura. Italian furniture design stresses more on artistry. The use of colors and lines is very lively and bold, which fully display this nation's romantic oceanic culture. It is also the most prominent feature of Italian modern furniture – It's not enough just to cope with the life's needs, the most important thing is the sense of beauty.

Veneta Cucine

Veneta Cucine
FORTI CARATTERI

EXHIBITION DESIGN

nest!

Exhibition Design

DESIGN
Beauty is the common language design works such as beauty does not have like safflower lack of leaves overshadowed

morretticompact

What exhibitors show is not just a business product, but designers' insight for the future. Visitors not only include buyers, but also designers and architects and design majors. They try to get some inspiration from the Salon for their upcoming career. The annual International Salon succeeds both in design and commerce. People are all intoxicated by those exhibits that can express their thoughts.

julia
arredamenti

colico
design
www.colico.com

AZCLASSICA
DESIGN FURNITURE

Furniture designers, manufacturers all over the world furniture are paying more and more attention to the practical function of the furniture. They place the practical function in the first place on the condition of guaranteeing the nice looking and innovative appearance. The exhibition has two main characteristics. It is not only an exhibition but also a kind of cultural activities. The furniture design circle currently stresses more on the nature and environmental protection. Therefore, there are a lot of natural materials used in the furniture.

B21

The International Salon is still showing a wide range of styles, such as the blending of Chinese and Western cultural elements. This year's exhibition got rid of the nostalgic and romantic trend and turned to pop "bio-mimetic" concept which pays attention to life itself. Designers began to explore things in the most primitive state and use modern means to display them. For example, they try to find the passion through some simple and plain elements. Therefore some of the traditional raw materials such as timbers are re-used. What is different from the past is that designers have invented a new way to display the material's grains. The design and decoration of the space is more closed the life but higher than life in that it pays more attention to the color and light of the space and emphasizes the integration, which is to blend the furniture, accessories with the design. Designers hold the belief that luxury is the perfect state of ordinary. The sense of luxury is not the surface one. It emphasizes more on the comfort to create a low-key magnificence through various kinds of materials such as leather, wood, glass, etc.

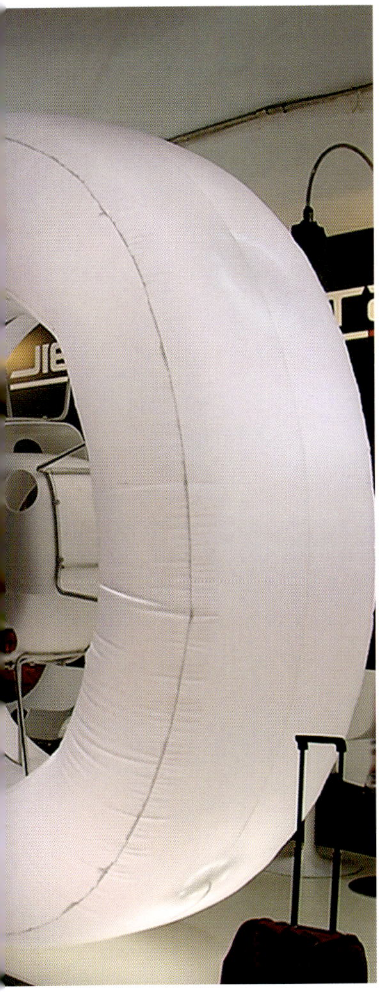

MAGIS

H01

Vibieffe

Gallery

In Milano veritas.

Initially fashion and design are two separate industries, but they have been more frequently overlapping since this year. They both can express the same theme and mood, which provides the designer more creative space. Many top fashion designers have begun to set foot in the furniture design, thus the two industries have some common concerns.

AXOLO
ATIKA
artdesigngroup

International Furniture Salon is the pioneer in modern furniture design. It is not only the platform of Italian furniture exportation but also is annual event that the world's furniture industry is eagerly waiting for. International Furniture Salon is like a fresh delicious feast which challengers our nerves in the sense of flavor, taste and smell.

ELLESALOTTI

B04
ELLESALOTTI

LUXFORM

The furniture design stresses more on artistry. The use of colors and lines is very lively and bold, which fully displays this nation's romantic oceanic culture. It is also the most prominent feature of Italian modern furniture – It's not enough just to cope with the life's needs, the most important thing is the sense of beauty.

The Salon has a complete system and is in a leading level in design, which makes it the indicator of the world's fashion trends. The most distinguishing feature of the Salon is that it opens a window for us and leads us to the nature. Fresh, positive and harmonious, people-oriented are still the main themes.

It's a kind of creation by converting exhibits into concepts. In the past, the exhibition and activities were both restricted to a certain area. The purpose is to break the restrictions and expand the displaying area to integrating the trends of design, art and other relevant aspects in order to explore their relationship.

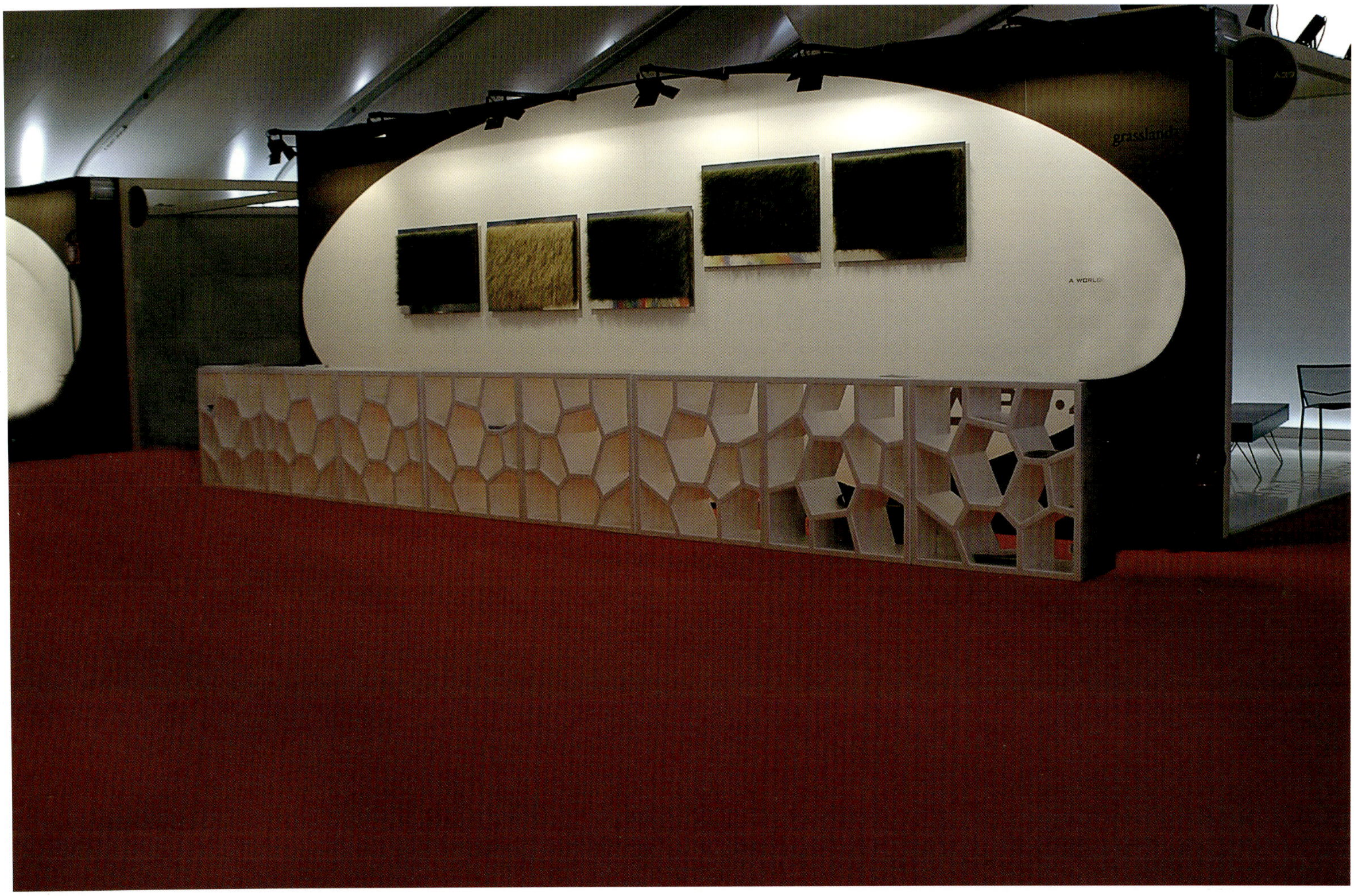

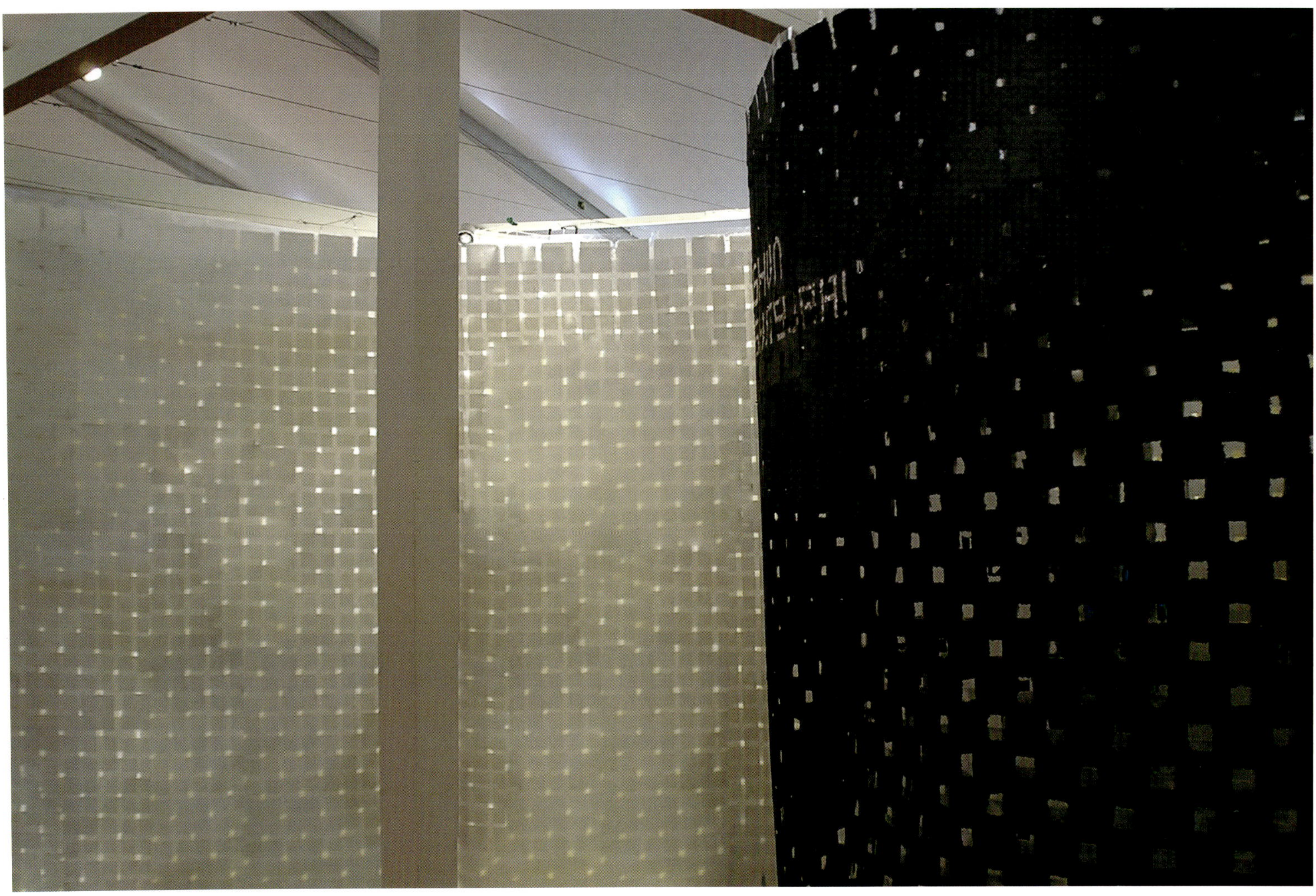

WITT
MANN

The International Salon is held annually in April. From the beginning of the 1980s, influenced by the post-modernism art of architecture, furniture design gradually flourished and was recognized as vanguard of fashion in the furniture design. The Salon has turned in an annual cultural event, not just the purely commercial exhibition, thus the exhibition has also become the world's furniture designers' "pilgrimage destination". If subject of the Paris Fashion Week is fashion, the one of the Salon is design.

The pattern glass table and large heavy table are made from glass. They are magnificent in style and give the sense of superiority. Modeling: the frame of the table, dining chair is made from the same material. When they appear alone, they are like an artwork. When appearing together, they constitute a warm and nice dining space.

You can enjoy the lighting booth while walking in the long corridors. The booth is simple design style, which matches well with the simple lighting fittings. With the decoration of the white and yellow light, different kinds of lights & lamps display in this place. Therefore, the place looks brighter and more enjoyable.

sicc
sic

EXHIBITION

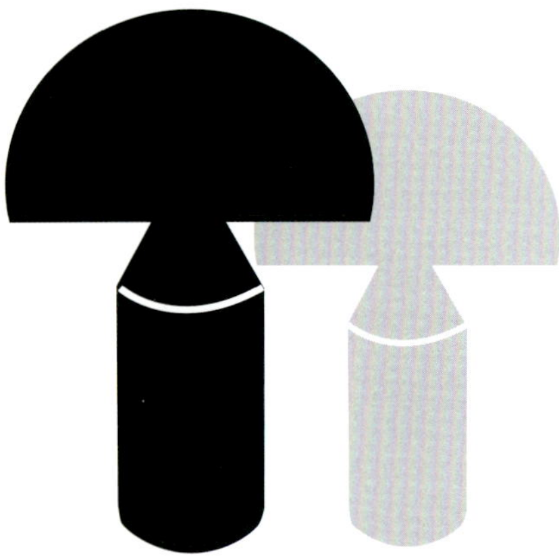

Alias

TOTAL-UMBAU
70%

This kind of design gives you a feeling of flourish. Futurism design, which is mainly based on black, red & white, has very strong color contrast. Therefore, the whole space gives you the impression that it is flourishing. The new ornamentation VS Trending-toward-nature is mainly mirrored in the close relationship between decoration and design. In the furniture design, decorative furniture has turned into a major part of the design and sometimes it is the most essential part.

en
Creative Front Line
魔

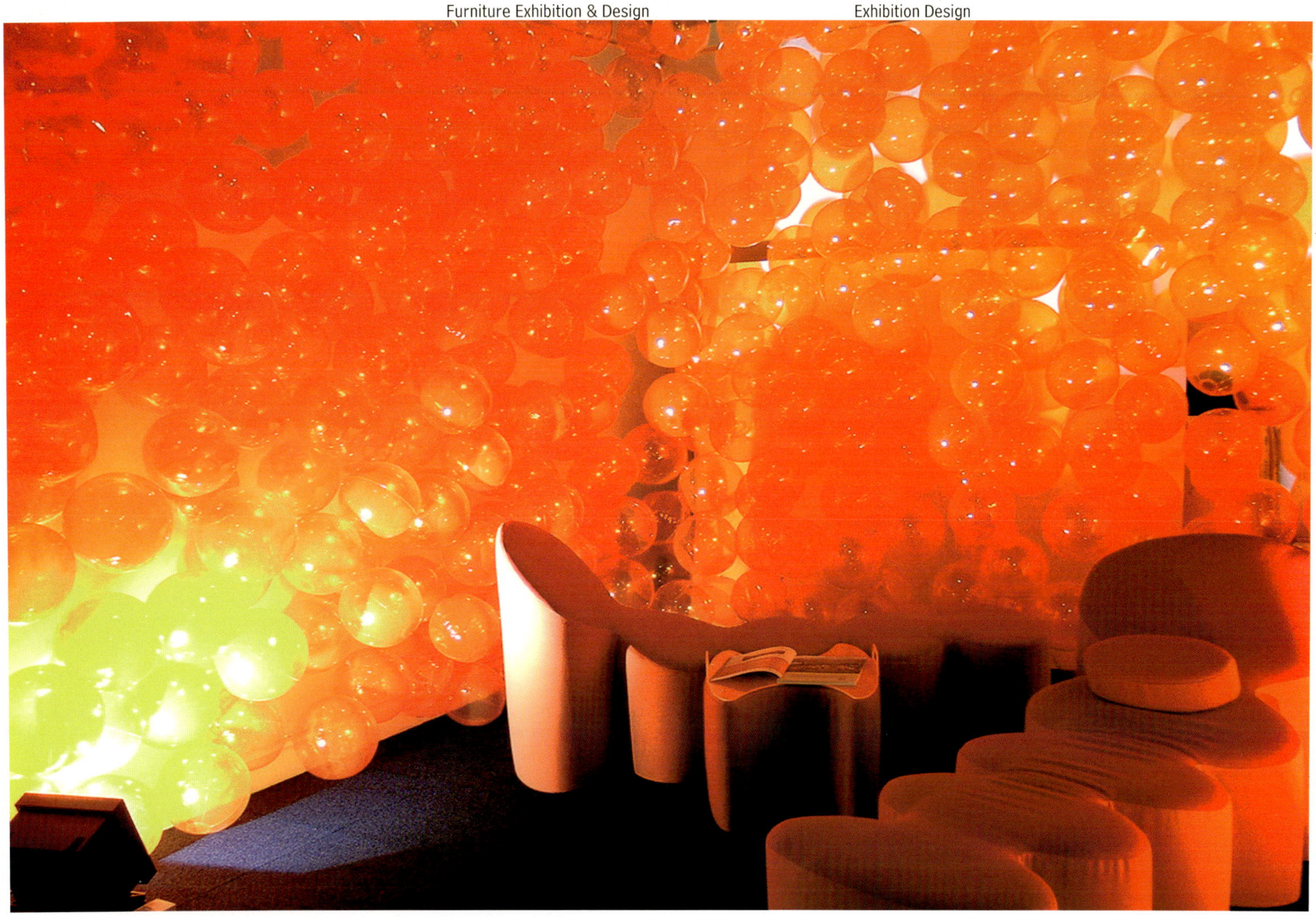

scialuppa
DRY • SPONGY • MUTABLE

CONCEPI
NIRMONN OFFICE

The design and decoration of the space is more closed the life but higher than life in that it pays more attention to the color and light of the space and emphasizes the integration, which is to blend the furniture, accessories with the design. Designers hold the belief that luxury is the perfect state of ordinary. The sense of luxury is not the surface one. It emphasizes more on the comfort to create a low-key magnificence through various kinds of materials such as leather, wood, glass, etc.

The interior design is regarded as the best way to show the activity of modern living environment. It not only reveals people's basic self-cultivation of dealing with the environment and the awareness of life, but also has an important impact on people's spiritual life.

Natural, elegance-Horizontal, vertical, oblique, curved, its nature and elegance are fully shown through the reflection in the glass and the water. The bright orange, no matter combined with modern style, or with classic embellishment, both has the special charm that can seize people's attention. Red, black or white is mysterious, rich, young, modern, open-minded, enthusiastic, self-style. People are always looking forward to the furniture salon. Space is a big topic. What will you think of when we talk about space? It can be as big as the universe and time or as small as a building. No matter it is still or changes slowly, it will always stimulate our imagination. In addition, it is a great source of inspiration for designers. Space not only changes our living habit and level, but also awakens our memory of primitive living demands. Maybe this is the power of space.

FACETT, CREATION RONAN ET ERWAN BOUROULLEC

Silver_new
FREZA

Silver_new
FREZA
Vega

Just as years of new celebrities open design, so that all the right under the nose exhaustive, Space there was no cover-up, remove all impede sight to the elements, Milan Salone became open on another pop Keywords. How can we open? One is the structural point of view, from a material point of view. The openness of the design origins in a small space, because space is restricted, so is not to dismiss many of the things So push the wall, Cupboard Door is not a further simplification, only the mgs So push the wall, Cupboard Door is not a further simplification, only the most basic of a few strokes to depict; Open space to be increased flu, glass is a relatively ideal material, transparent, no sense of oppression. ause space is restricted, so is not to dismiss many of the things So push the wall, Cupboard Door is not a further simplification, only the most basic of a few strokes to depict; Open space to be increased flu, glass is a relatively id

EXHIBITION

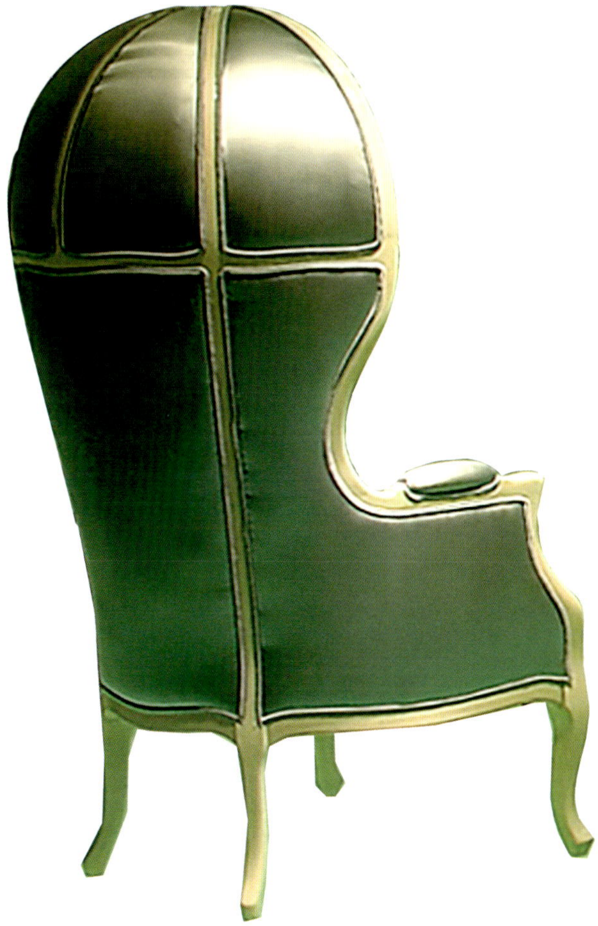

It's a kind of creation by converting exhibits into concepts. In the past, the exhibition and activities were both restricted to a certain area. The purpose is to break the restrictions and expand the displaying area to integrating the trends of design, art and other relevant aspects in order to explore their relationship.

94/9
97/00

92
94

EXHIBITION

design: Mirto Antonelli
ZEFIRO'S secrets:
Come rendere solida la leggerezza.
ZEFIRO

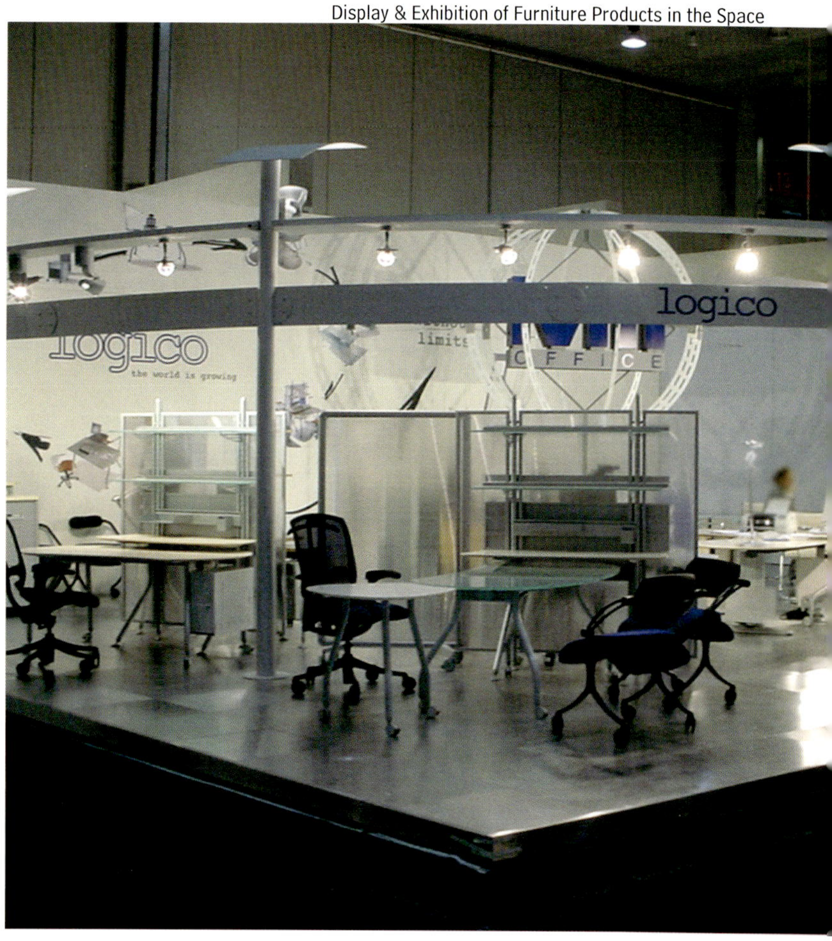
logico
logico
the world is growing
limits
OFFICE

B16

TE
MARTE

ZEFIRO'S secrets:
Come rendere solida la leggerezza.

Natural, elegance-Horizontal, vertical, oblique, curved, its nature and elegance are fully shown through the reflection in the glass and the water. The bright orange, no matter combined with modern style, or with classic embellishment, both has the special charm that can seize people's attention. Red, black or white is mysterious, rich, young, modern, open-minded, enthusiastic, self-style. People are always looking forward to the furniture salon. Space is a big topic. What will you think of when we talk about space? It can be as big as the universe and time or as small as a building. No matter it is still or changes slowly, it will always stimulate our imagination. In addition, it is a great source of inspiration for designers. Space not only changes our living habit and level, but also awakens our memory of primitive living demands. Maybe this is the power of space.

orseni

GIGLI

STEELMOBIL
flirt
STEELMOBIL

irt
UARE

GIORDA

FENDI

AXIL

These innovative household items all come from this year's International Salon. Pop colors, refreshing and fashionable concepts of life are gathered in this furniture feast. The world's top furniture designers and manufacturers took the opportunity to spread the new spirits to the rest of the world.

Mobi's

Italian Sedioliti

ERGO
PLUS
SEATING TODAY

A15
GEBRÜDER
THONET
VIENNA

GEBRÜDER
THONET
VIENNA

GEBRÜDER
THONET
VIENNA

Tom Dixonom.

The design of this exhibition hall is extremely creative. It is just like an outer space base which makes you feel like walking in the outer space. All exhibits are floating in the air.

BOSAL
VERSATILE COMFORT
www.eurobi.it

White is and will be the leading color in furniture design. The main white body contrasts the black color, and stresses each other's shiny surface. In addition, the new color-grass green is playing an important role in the furniture design. Another tendency is the gray gradient color. So this year's red is not pure red, but contains a little bit purple. So does green. It must be with some blue so as to be call "pop color."

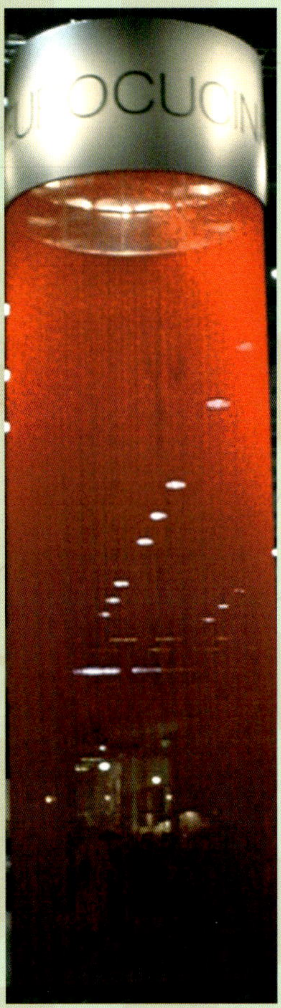

G09
Stile di Vita.
BRIANFORM

MARINELLI

The design and decoration of the space is more closed the life but higher than life in that it pays more attention to the color and light of the space and emphasizes the integration, which is to blend the furniture, accessories with the design. Designers hold the belief that luxury is the perfect state of ordinary. The sense of luxury is not the surface one. It emphasizes more on the comfort to create a low-key magnificence through various kinds of materials such as leather, wood, glass, etc.

MOBIL
Spider

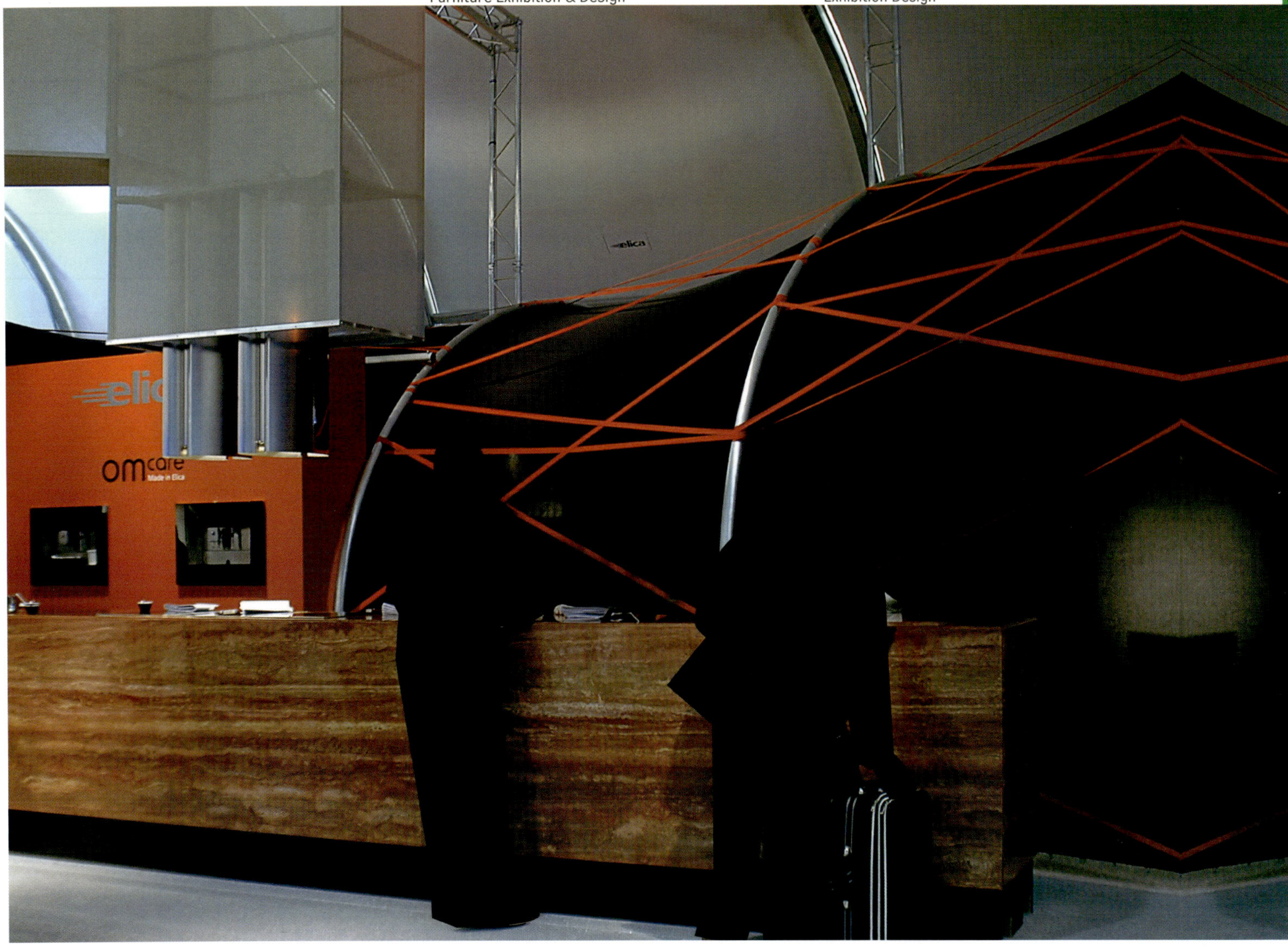
elica
om care
Made in Elica

afzimmer
PREIS HIT

Services
& Marques
ARTELANO

Brinkmann

Brinkmann
STUDIO

rosini
rosini
A07

A07
rosini

These innovative household items all come from this year's International Salon. Pop colors, refreshing and fashionable concepts of life are gathered in this furniture feast. The world's top furniture designers and manufacturers took the opportunity to spread the new spirits to the rest of the world.

GLOBAL
UPITALIA

Xten

COMPO

CLASSICAL
FURNITURE DESIGN

nest!

Classic Furniture Design

DESIGN

Life is not completely placed in a reasonable specification We need more from visible to intangible values from the material to the metaphysical

This kind of material uses wooden materials. They have the feeling of retro with the decorative silk or canvas. Metallic silver, bronze and other colors demonstrate the steadiness and prosperity of classical furniture. This is a kind of pursuit of gorgeous and elegant classical. In order to realize the grand style, the frame of furniture is decorated with golden thread or golden edge. The patterns of the wallpaper, carpets, curtains, bedspread, and valance are all the classical style.

The classical furniture demonstrates the essence of traditional culture through the combination and interaction between people and space. People can feel the charm of the classical. With the improvement of whole nation's qualities, more and more people are aware of the importance of family atmosphere. It has been the state's trend to highlight the individual's tastes and qualities through the furniture. Classical furniture still has the primitive feature by redesigning. You can enjoy the best life by just the simple combinations.

Life is not completely placed in a reasonable specification. We need more, from visible to intangible values, from the material to the metaphysical. We need to satisfy the demands to gain stability. The extravagant style tells us that life is not just the case, not just the yearning to quietness. Success does not need high-profile promotion. It always starts from the details to taste the flavors in life. This is a kind of pursuit of gorgeous and elegant classical. In order to realize the grand style, the frame of furniture is decorated with golden thread or golden edge. The patterns of the wallpaper, carpets, curtains, bedspread, and valance are all the classical style.

What European classical furniture brings to us is a calm and serenity living attitude. Modern European new aristocracy products have more advantages. Modern tensions of life more or less affect people's aesthetic taste, so the new direction of classical luxury and elegance appears. American furniture makes lots of European aristocracy furniture civilian. The design criteria are comfort and simplicity. Simple lines, bold looking, natural materials, more reserved colors and shapes, all these constitute to their characteristics.

COBEBRO

European historical panorama has been consisted of artistic architectures, sculptures, paintings and poetry since the beginning of the Renaissance 600 years ago. The classical furniture directly inherits European historical and cultural elements. The daintiness and luxury is far best to be caught by any of the furniture. They stress excellent manual cutting, carving and inlay work. In the aspect of the design in lines and proportions they fully display the rich artistic imagination, which makes them very popular in the middle class.

Classical furniture always gives a quiet and elegant feeling. Luxury is a kind of ordinary perfect, is the demand beyond the needs. The sense of luxury is not the pursuit of the surface luxury, but the low-key elegance, is the elaborately carved details. The cupboard is with exquisite patterns. It fully displays the artistic sense in view of the lines and the design proportion. Super techniques of solid wood, the royal family's luxurious and romantic flavor of life, and geometric shapes of the "modular" are all the representation of European classic style furniture. The trend of future development of the furniture is the simplicity of the overall shape and lines. However, for the neo-classical style furniture, the design tendency of this year is the partial exaggeration and the application of the Baroque or adult STUDIO elements for decorative embellishment.

The most distinguishing feature of the Salon is that it opens a window for us which leads us to the nature. Fresh, positive and harmonious, people-oriented are still the main themes. Metallic silver, bronze and other colors demonstrate the steadiness and prosperity of classical furniture. This is a kind of pursuit of gorgeous and elegant classical. In order to realize the grand style, the frame of furniture is decorated with golden thread or golden edge. The patterns of the wallpaper, carpets, curtains, bedspread, and valance are all the classical style.

Designers begin their trend of retro in a unique way. When anyone first sees the sofa, their eyes will become brighten. These sofas have a wealthy and elegant sense, which brings people the treatment like that of in the palace. The designer's retro concepts also include the theory of simplicity and luxury. Retro design makes the sofa looks very luxurious and has a noble feeling. It is the very right design that we want.

A15

Natural and man-made rattans, compared with those materials which either have bright colors or with high technologies, have their own natural features. There is also some highly refined craftsmanship in the traditional wooden furniture. The mixed design of cotton, flax, bamboo with glass is also very innovative, which shows the special design in the use of materials and also represents the future tendency of furniture design.

The European classical sofa has the elegant charm. A table lamp, a kitten chasing the ball everywhere… all these make us feel that as if we went back to the times when we played around our grandmother's knees. European furniture gives the feeling of luxurious elegance. The feature is the use of curves twists and turns to create special artistic effects, in order to present the mysterious atmosphere and the beautiful illusion. It also prefers the delicate and complex small lines such as the patterns of clam shells, vortex, water glass, etc., and is particular with rich colors and shiny gloss. Therefore they not only provide comfort but also offer the visual enjoyment to people.

LORENZON

Bedroom is a place that well displays Continental classical comfort. The board and the frame of the bed are decorated with gold lace. The gold embodied bed expresses the rich and extravagant feeling. Purely classical furniture often brings a heavy feeling and lacks energy. By adding some modern elements, they look more elegant and fashionable.

Classical and traditional furniture reminds people of the ancient traditional workshops. Classicality is a kind of trend and fashion in modern times. We can't use one sentence or two to explain the extensive and profound traditional culture. But we can carry forward these rich historical and cultural "precipitation" through the household item design. Moreover it enables our home to have the perfect quality. Blending the classical beauty into modern design makes neo-classical furniture more spiritual. It enables the classical beauty cut across the time and be active in our modern lives.

MODERN FASHIONABLE
FURNITURE DESIGN

nest!

The Display & Exhibition of Furniture Products in the Space

Modern Fashionable Furniture Design

DESIGN
What is beauty? Simplicity is beauty.

What is beauty? Simplicity is beauty. Life is not completely placed in a reasonable place. We need more, from the visible values to invisible ones, from the material to spiritual. We need to satisfaction. We need overflow to gain stability. The more extravagancy tells us that life is not just the case, not just the desire for quietness. Success doesn't need high-profile promotion. The taste of life always starts from details.

Selecting a product with classical style requires a series of sensory colors. The monotonous color will make the space looks too dull. The combination of dark and light shades of the same color can create the progressive depth of the space, which alleviates the discomfort in visual sense. Moreover, you can also find a total different color to make the space more eye-catching.

The feature of simple furniture is the simple but nice lines. This is the result of the designer's excellent craftsmanship. The furniture has simple style as well as advanced design. When placed in the room, though it is simple, it can enhance the atmosphere immediately. The designer's subtle creativity can be felt via the simple use of the materials and colors.

Minimalism was in full flourish in the 90s of last century. European designers made the minimalism to the extreme by using natural wood, leather or other materials to meet the demands of modern people. Simplicity does not mean merely simplification, but the simplifying in the rich centralization, and to achieve the most complicated artwork by the simplest method.

Bedroom is the "resort" in your spare time, so comfort and elegance should gain more attention in order to create a warm and romantic atmosphere that fits the exact mood of your work at home... In the colorful season, furniture design also can be free from scope. The living room, bedroom, kitchen, study, red, orange, yellow, green, they match each other perfectly.

This kind of design gives you a feeling of flourish. Futurism design, which is mainly based on black, red & white, has very strong color contrast. Therefore, the whole space gives you the impression that it is flourishing. The new ornamentation VS Trending-toward-nature is mainly mirrored in the close relationship between decoration and design. In the furniture design, decorative furniture has turned into a major part of the design and sometimes it is the most essential part.

Modern style is mainly based on the simple and modern Scandinavian style, which is very popular among the young people. The colors are still the fresh and clear light tones. Concise shape, less decoration and refined craftwork reflect the concision and sophistication of the highly industrialized products and has a strong personality. The materials used in this kind of furniture include solid wood, board, plastic, glass, metal, and so on. They stress on the match of the metal pieces and the board, so they are crystal-clear, simple, beautiful and very contemporary.

Leisure design reflects the nature and comfort of the furniture. Simple design is still the mainstream in the furniture design. The expression ways of design is various. Usually they come out unexpectedly, but on the other hand, reasonable.

White is and will be the leading color in furniture design. The main white body contrasts the black color, and stresses each other's shiny surface. In addition, the new color-grass green is playing an important role in the furniture design. Another tendency is the gray gradient color. So this year's red is not pure red, but contains a little bit purple. So does green. It must be with some blue so as to be call "pop color."

Simplicity is a living attitude that enables us to live more natural, simpler and quieter in the hustle and bustle city. Simplicity is the feeling of freedom and extraordinary taste. It is a higher level of life, other than the living attitude of simple, mean, perfunctory.

The design includes a series of plastic mats and chairs. They have different sizes and heights to meet the changes of circumstances. They can lean the wall or surround the sofa. Simplicity is to change the fixed thoughts in your mind. This is a four-storey cabinet with four wheels. Now only one element is changed, which simplifies the design. It not only reduces the storage space capacity, but to save people's living space and reserve enough room for your miscellaneous articles.

The designer used the colors have fresh senses. When a vibrant and vivid scene appears in front of you, you subconsciously will actively participate in this unexpected nature of interaction with the enthusiasm and forwarding momentum. It stresses and enjoys the fine details of life: love sunshine, like nature, and gain the pleasure from the relaxing exercises. High technical assistance enables designers to create some almost mystical products, and makes the emotional factors more and more obvious in the household items. For example, plastic, acrylic, and aluminum products are cleverly integrated into the new space.

Leisure furniture design is simple, natural, steady in style, and with a trace of a return to the traditional mood. They mainly use oak, maple, elm and other natural wood. These natural materials shorten the distance between human and nature, human and the furniture. It meets the modern citizens' desire to return to the nature. Besides natural wood, the leather with comfortable hand feeling and rich colors and other materials are also more and more popular in furniture production.

British and French country style: The furniture is mainly in fabrics and raw woods. Fresh flowers and grass as well as the lattice fabrics which never withdraw from the popular trends are always the leading actor in household items. Bright rose and elegant lavender are the "sweet flower clothes" for the furniture. Accompanied with natural color wooden furniture, it creates a kind of comfort and leisure that makes people feel like living in a fairy-tale land. They have brighter colors as well as a lot of tiny flowers. The whole style is simple, natural, romantic and elegant. In fact, we need the avant-garde design, but what we need badly is the nice and comfortable furniture. This is the relationship between art and life. What the new classic in the modern society pursue is not luxury, but the taste that has deep and inherent introversion. They are convinced that the success does not need high-profile publicity, and excitement always starts from details.

PARK PLUS

Now the circumstance is that the co-existence of simple and decorative, minimalism and stylism, modern and retro. Antithesis is always one of the most important sources for inspiration. Simplicity and minimalism almost reach the perfect level speaking from the notion. On the other hand, complex decoration can fully displayed extravagant and wealthy. The shapes of the furniture are diversifying. Straight lines are still popular. Moreover, curve lines, circles, ovals and fresh petals also bring fresh energy to the design.

Contemporary art of design has its own distinguish features. Keep customers' demands in mind, because this is the ultimate purpose of the design. Designers keep the constant innovation of the products. High quality materials, clear design and the outstanding logic are on the fly page of their dictionary. What modern people want is to live in a spacious, simple and comfortable place. The glimpse of urban white-collar class' life brings inspiration to the fashion design. Fashion designers express their preciseness through details, and create fashion highlights via contrast. If the language in fashion design is fashion, then the design of solid wood furniture makes fashion more practical.

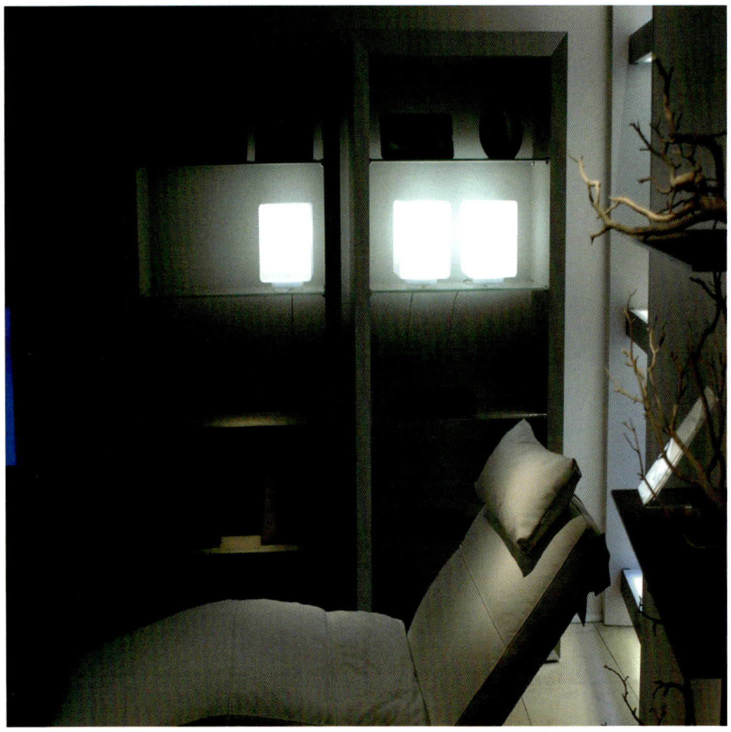

Furniture designers believe black and white is the most classic combined color. See, if you use black only, it will appear too dark. If you use white only, it will inevitably appear too dull. When the two are put together, is like the black and white piano keys. They mutually contrast with the ejection of the Intelligent. Futurism is always the most attaching topic for its imagination is endless and it always challenges human being's utmost limit. Most of the products in the satellite design belong to this group. Their success just indicates that modern people like to get in touch with these futuristic items and gain some kind of enjoyment. Stylish, pioneer…you can use any word that you like to describe them. The one that surprises us is the mirror. How awesome that the designer can demonstrate the classic in such a way!

The overlap between fashion and design becomes more and more frequent, say, the plenty use of metal elements. Some special materials are also applied to the furniture design. Standardization module demonstrates that furniture design pays more attention to the combination of the design in standardization and individualization. Through different combined colors and methods, various types of products are produced to meet people's demands.

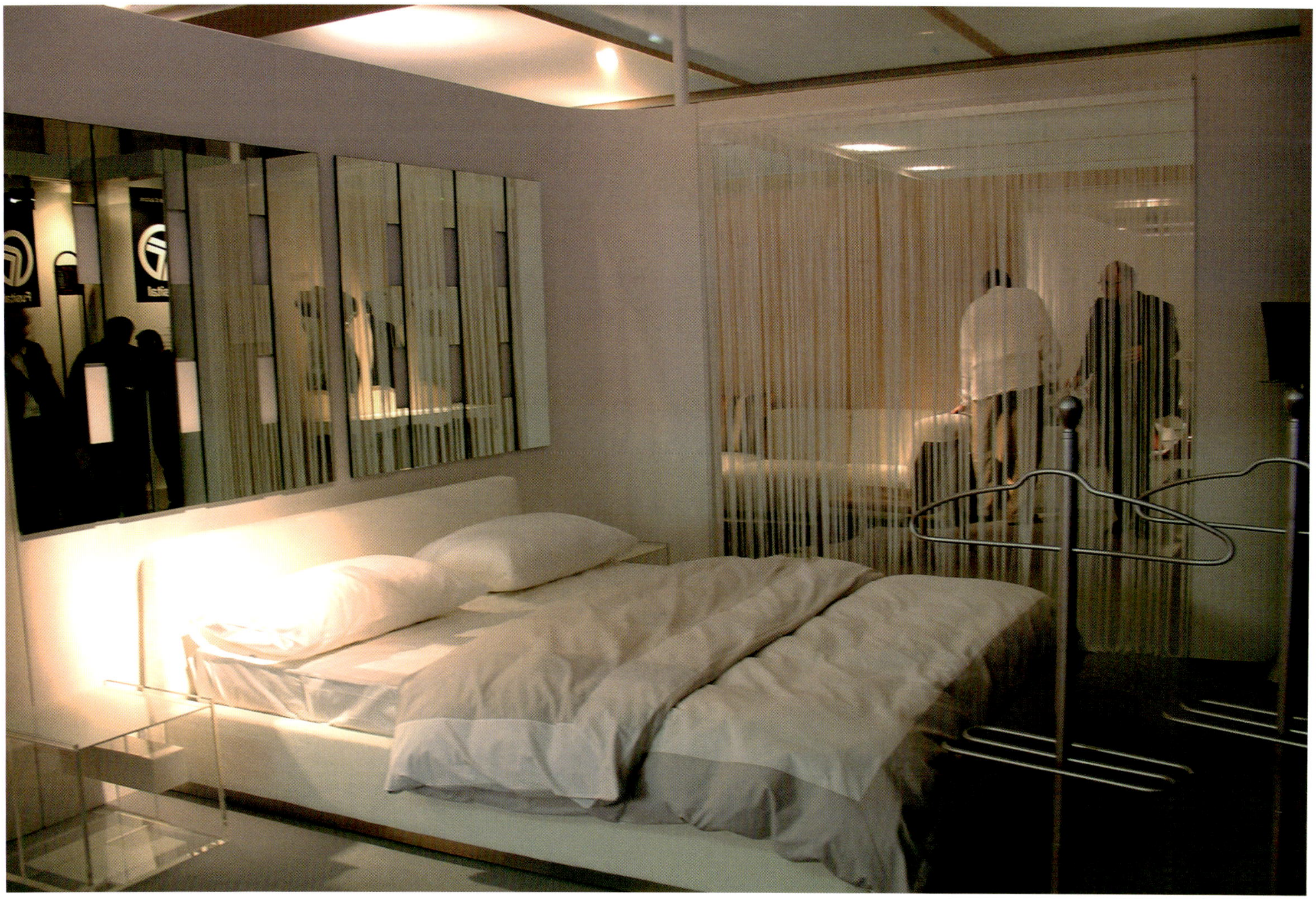

The advent of avant-garde design is always based on the application of new materials and technologies. With nowadays' rapid advances in technology, 3-D fiberglass, gel, foam aluminum, 3-D fabric and diving materials have emerged one after another. It is precisely because of them that the new products can be developed. Speaking from the aesthetic point of view, the future product design should be organic, portable, simple, functional and modular. Accompanied with the development of the world economy, the design of future products not only has to be functional, but also need to have rich connotation. Faced with the increasingly rising requirements from offices, the design work is far beyond the functional level. Constant update can achieve an unexpected and successful breakthrough. Design is a problem-solving process. After the issue is solved, then comes out the issue of the form. Because of this, design has been constantly changing and improving people's lives in the past few thousand years. Random mix is to meet different demands. Let your imagination free scope to find your own style. Natural, rugged, solid and durable not only maintains the material, color and, texture, but also the fresh lines of green leaf background. Elegance, tranquility as well as brilliance bring an impression of warm and natural pastoral atmosphere. Continental furniture design style is directly influenced by European architecture, literature, painting or even music. One of the most important impacts is that Continental furniture design pays more attention to the protection of the environment, so the design is not only environmental protection, but also fire protection, thermal insulation and durable.

When you enter the room, your eyes will be immediately attracted by the red furniture. The entire color series is red, with secondary color black/white to make it perfect. The designer originally uses black to intensify the austere decoration. It creates the color contrast of black, white and red, which makes the space filled with a warm feeling. In the meantime it keeps the simple and harmonious beauty of the living space. This is also one of the design ways of futurism. Furniture industry has been divided into two groups under the globalization-the new ultra-realism and the new rationalism. I believe many designers hold the concept that "Design comes from life". However, how should the furniture design industry face the volatile market? Due to the changes of people' s lifestyles, more and more classic furniture design products in the 20th century have been re-designed or re-packaged. They continue to be the popular and fashionable furniture products with the new popular fashion styles. This proves that classical furniture design products still have a very long life in view of the shapes and construction, the pre-condition is to combine with the needs of contemporary people's lifestyle and fashion tendencies.

The accessories of household items are mainly in streamlines and arcs, as well as leaves and flowers. They give the appearance of "Biology Imitation" to the furniture, and also offer a kind of poetic temptation and charm to the furniture. The application of emotional elements arouses people's desire to purchase, which is reflected in the decoration. Naturally curly lines and decorative floral motifs continue to be prevalent in the Salon. The most popular one is the small vein based on plant designs. The relationship between furniture and flowers in fact can be traced back to an earlier date. At present, the application of flowers in the furniture has reached an extremely high level, which makes the furniture more and more beautiful.

The two popular trends in this year are very interesting. One is the black-and-white furniture comes back. The other is the rare colors begin to take the lead in the design. Furniture designers believe that black and white are the most classic combination of colors -- you see, if using the black color , it will appear too dark; if using the white, it will be inevitably dull. After putting the two colors together, just like the black and white piano keys, they mutually contrast with each other to play the beautiful variations.

Futurism still has its own place in the furniture design. Black, white, red are still the main color. But designers pay more attention to the combination, so they come out more distinctive. The product not only has outstanding characteristics, but also has humane nature. They are fashionable, forwarding, stylish, and distinguishing. What the modern people desire is the kind of design style that can meet the demands of individual pursuit, and on the other hand it doesn't deviate from the origin of traditions and the humane innovation.

The futurism style round bed will draw your attention as soon as you enter the room. The whole series is mainly in black, which makes them look innovative and fashionable. The blanket, with plentiful accessories, shows a kind of luxurious style and contrasts with the room's simple decoration. The application of new technologies, elements and materials brings people dazzling experience and allow them to feel the same thing on the other side, the real and mysterious side. From art and design, the trend of modeling the "dreaming world" is triggering more and more new ideas. Black and white are very elegant colors. Although they belong to two different tones, they are well-integrated.

HOOPLA
ON/OFF

Moblex
DIAEZYONALI

The innovation of office modes directly influences the design of office furniture. It appears as that furniture design emphasizes the combination of standardization and personalization & flexible manufacture, with the combination of several basic standard modules through different ways and colors, so as to form divisible and flexible products to meet people's requirements. A lot of creation that comes from life is displayed in the design of furniture, jewelry and space, as well as the use of new materials, which surprises spectators to a great extent.

This is the perfect interpretation of contemporary office furniture. The nice proportions and well-selected materials appeal to be comfortable. Purely beautiful is meaningless unless you find the way to display its value. Design is a problem-solving process. After the issue is solved, then comes to the problem of formation. Just because of this, design in the past few thousand years has been changing and improving the people's life, as well as solving all sorts of problems that we encountered when using the tools. Silver materials give the furniture a strong personality and temperament royal characteristics.

mobilex
DIREZIONALI

OP/B
BABY/F
HOOP/B

A20
BABY/F
HOOP

Just as years of new celebrities open design, so that all the right under the nose exhaustive, Space there was no cover-up, remove all impede sight to the elements, Milan Salone became open on

FRANKE

The era of family office furniture has come. Not like the past it was just a tentative plan. It was no longer the reconstructed guest rooms. Today, most of the high-end residential buildings have office rooms. The boundary between work and personal life is increasingly blurred. "What can be more functional than a file cabinet? This file cabinet looks like a typical trencher cap. Why file cabinet must be metal and wood? It is a very old model. It is said that the cold metal is very helpful in work, but does not apply to family. The new concept believes that work is at home, and work is life. This is the whole life, which should be the best.

FRANKE

TEAM

Speaking from the space point of the view, the depth the 3-dimensional perspective of the narrow corridors, as well as the sense of space hierarchicy is the most beautiful part of the construction. It's not an easy thing for designers to show this artistic conception in the modern architecture, especially in the design of interior space. When entering the inside of the space we will see the horizontal and longitudinal space, which forms a very high and deep hallway in the visual sense to highlight the design. By comparison, the design is a more rational design and more pointedly one.

B11

Now very few of those who are engaged in the design will use this strictly- controlled ratio curves. Most of the designers have been accustomed to using the kind of careless or random lines. As long as it has the visual beauty it will be okay. But it takes more risks to use these lines that don't have golden section ratio or other aesthetic proportion ratio. Because it will be more likely to lose the aesthetics of the curves, so the application of the arc and curve should be more cautious and careful. No matter to design a building, or a piece of commodity, designers regard it as an artistic creation. They are very useful and can bring us enjoyment. The space design is the carrier that can make people experience this kind of design philosophy.

4xSvezia
4xSvezia
SÖDERBERGS
4x Svezia Inform

4xSvezia
4xSvezia
SÖDERBERGS
4x Svezia Informati

IDEAS

Designers do not have much ingenuity and creativity on the floor planning, which focuses on the rationalization of the combined skills on the use of space. On the other hand, designers pay more attention to the decoration. In fact, it is quite interesting to deal with such a small space. It not only needs to meet the functional design, but also has a certain degree of innovation and breakthrough. Therefore, it's very crucial to find the balance point among these mixed and complex factors. The design is clear and quiet from a comparative point of view. The main axis of the space is neat and clear. The entrance separates the space on both sides. The living room corresponds to restaurants, and the master bedroom corresponds with the study.

No matter to design a building, or a piece of commodity, designers regard it as an artistic creation. They are very useful and can bring us enjoyment. The space design is the carrier that can make people experience this kind of design philosophy. Leisure furniture design style is simple, natural, and with a trace of a return to the traditional mood. It mainly uses the oak maple, elm and other natural wood. These natural materials make the relationship between the nature and human beings closer, so does the distance between human beings and the furniture. It meets the modern people's demands of solid wood furniture complex and yearning pastoral countryside life. Apart from the natural wood, the leather that has soft hand feel and rich colors, as well as other materials used in the furniture are becoming more and more popular.

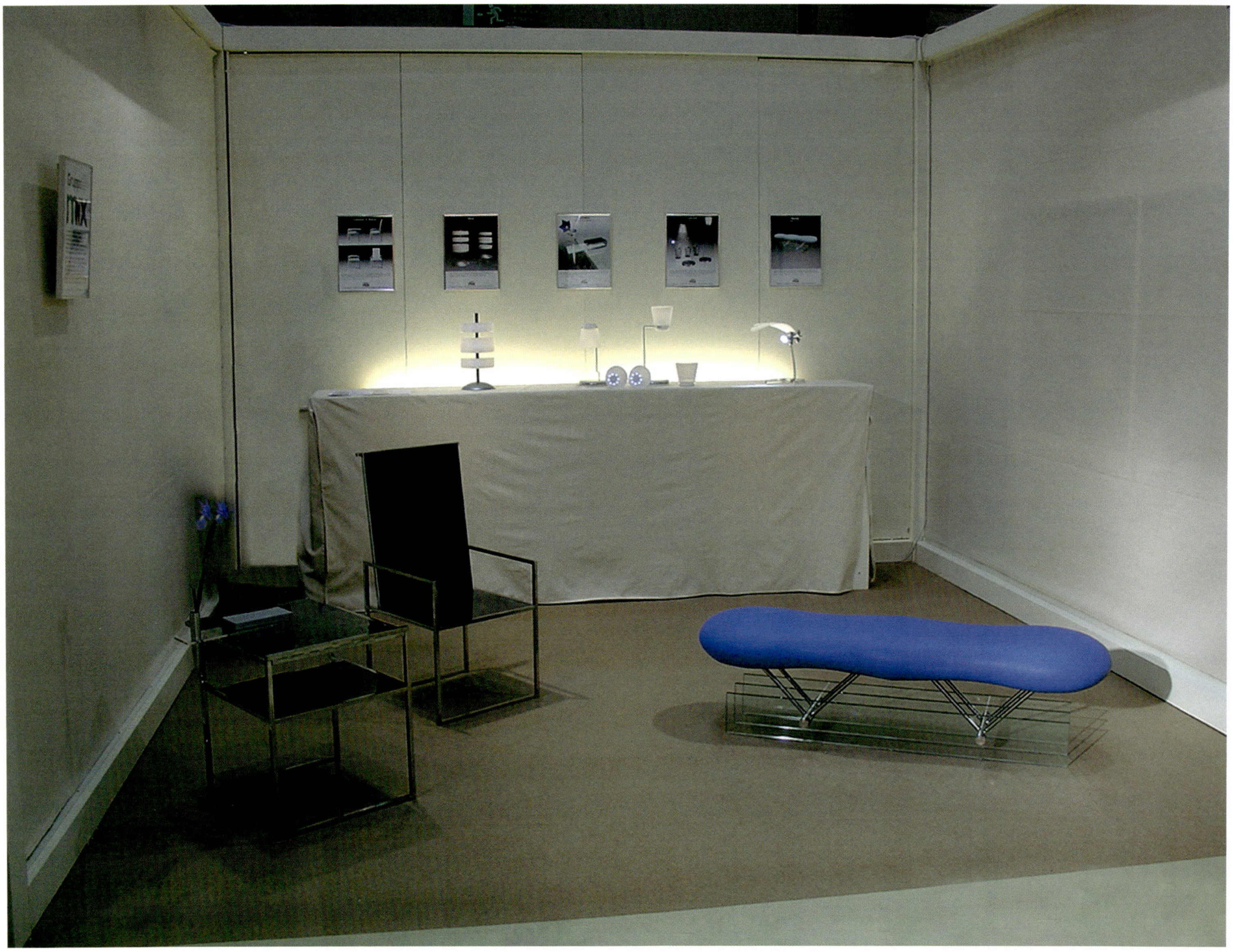

• Street Gallery
• Restaurant
Street Gallery

www.NadineMeisel.de

TO YOU. YOU. YOU. YOU. YOU AND YOU!
"THE LOST SON"
MERCI BEACOUP STATEN UND KAPITALET
KONSTNÄRSNÄMNDEN AND ALL THE REST...
KICK ME BEAUTIFUL PLEASE!
MERCI BEACOUP MADEMOISELLE!

ESTEL
ATELIER

Space is a large proposition. What will you think of when we talk about space? It can be as big as the universe and time or as small as a building. No matter it is still or changes slowly, it will always stimulate our imagination. In addition, it is a great source of inspiration for designers. Space not only changes our living habit and level, but also awakens our memory of primitive living demands. Maybe this is the power of space. Glass is a relatively ideal material to expand the open sense of space, because it is transparent and has no sense of oppression.

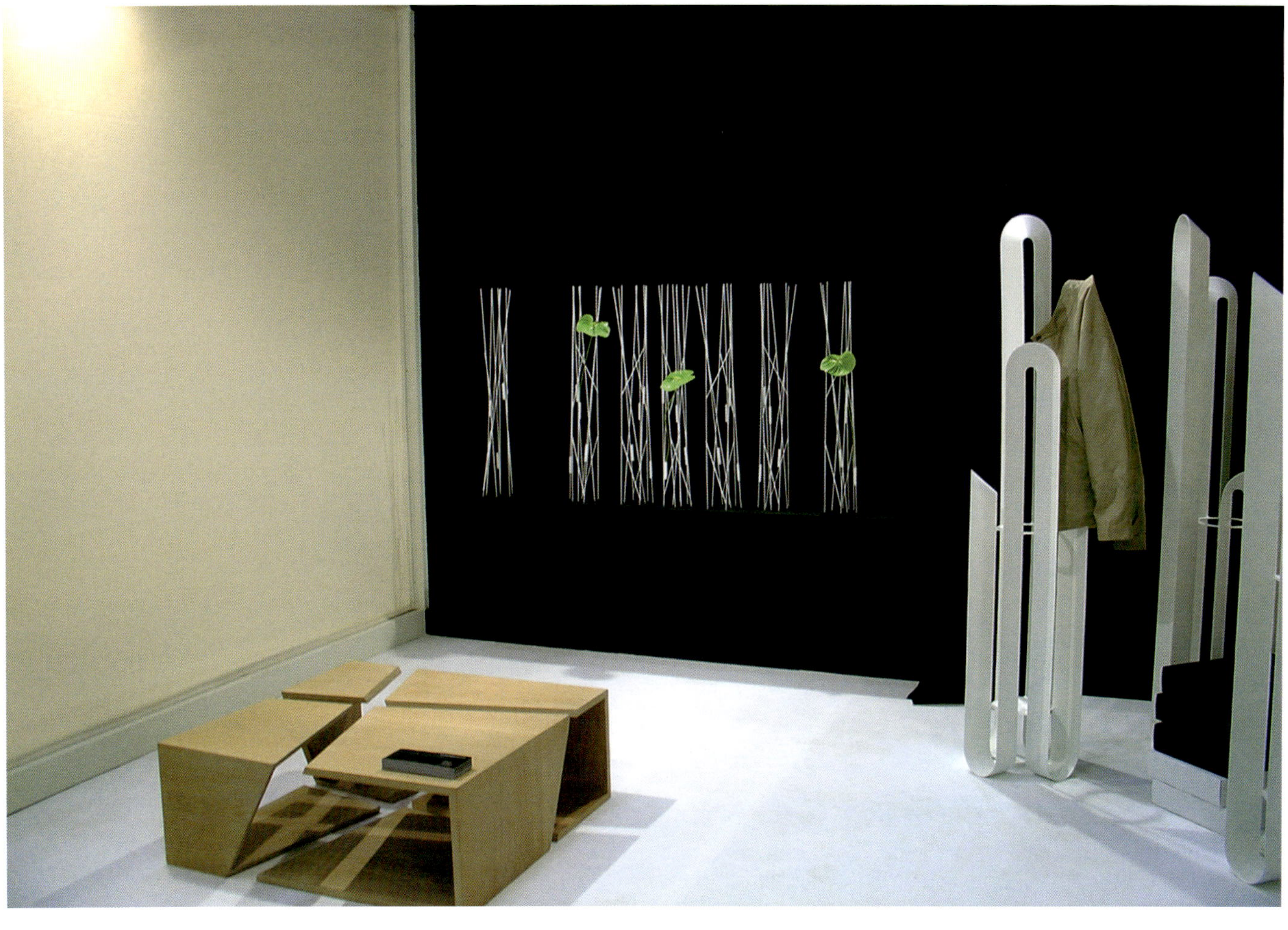

study space
play space

E 11-1
0 12-2

FLAVOUR
BOOGIE

The design and decoration of the space is more closed the life but higher than life in that it pays more attention to the color and light of the space and emphasizes the integration, which is to blend the furniture, accessories with the design. The furniture plays an essential role in any space. Human being's best state is the half-satisfied state. So does the furniture design. We need to find the balance point among those materials. Off-white and light colors make the space more obvious, which also is the interpretation of the whole clear and simple lines. All the materials are the cheap and easy-accessed ones. It's so amazing that the designer is able to use these materials to express the civilian noble life.

miniforms
miniforms

P A S T O E

guggenbichler design

Superfluous decorative lines were discarded as many as possible. Stainless steel made a point of contrast of materials and details. Large physical space use lines, corners and light of the ceiling to define the space boundary.

Designers create a vibrant and dynamic space, which is the kind that is full of blooming energy. When the basic concept is determined, designers will follow this logic to explain the language of all the design lines and decoration. These designs are more moderate. There are no creasy and mad curves. The language of decoration is also with a more modest graphics. To be more exactly, it is the kind of natural naughty.

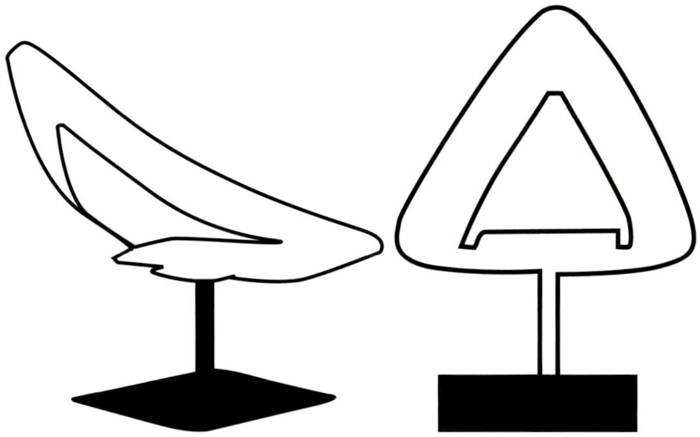

PARIS
ALITALIA
EUROPE
ALITALIA

Alitalia

Living Platform
Nelson.
Together Lazlo
Kite Jason.
Big blue.
Ribbon

Natural, elegance-Horizontal, vertical, oblique, curved, its nature and elegance are fully shown through the reflection in the glass and the water. The bright orange, no matter combined with modern style, or with classic embellishment, both has the special charm that can seize people's attention. Red, black or white is mysterious, rich, young, modern, open-minded, enthusiastic, self-style. People are always looking forward to the furniture salon. Space is a big topic. What will you think of when we talk about space? It can be as big as the universe and time or as small as a building. No matter it is still or changes slowly, it will always stimulate our imagination. In addition, it is a great source of inspiration for designers. Space not only changes our living habit and level, but also awakens our memory of primitive living demands. Maybe this is the power of space.

Generally speaking, economical and practical items are very common and owned by almost every family. But some luxurious and special items are quite rare. They are often owned by a very small group of people who either are very rich to buy whatever they want, or the ones who pursue personality so as to abandon economical, convenient and practical features to find or create some their favorite items. It is very difficult to understand the creation of those personalized seats designed by outstanding designers. You just can taste its meaning carefully so as to find the one that suits yourself. This multi-purpose outdoor lounge comes from Italy. Unique design unique and eye-catching bright colors will make you love it forever. Simple lines highlight the simplicity of the design style. Color contrasting of complementary mats makes the deck chair more dynamic and modern.

COMPASSO D'ORO 1954-2004

COLICO
design

Metal twisted frame chair displays the designer's minimalist style very well. Although the material is metal, but the technique has been improved. Laser cutting upgrades the decoration and comfort of the furniture. Simple lines, random technically twists, etc, make all these possible.

The metal-framed chair is the design that designer gets rid of the traditional design tactics, which enriches the forms of household products. The chair has simple form, and without colorful overcoats. You can just use a spanner to dismantle the metal-framed chair easily.

The curve lines of the portable chair were strictly designed in accordance with ergonomics. It doesn't affect the user's normal conversation, nor does it affect the rest time. These excellent designers use all their wisdom even it is just design an ordinary chair. All the details are displaying the designer's inspiration. The seemingly simple curves actually contain the principles of ergonomics, which meets the demands of modern fashion class who are pursuing for simple and comfortable life. The design perfectly demonstrates the "Italian style" design. It is worthy of domestic designers to learn from them.

by gerard v.d. berg.

The multi-used chairs made of plastic look strange though; in fact it can create different sitting postures with different ways of placement. It can either be used as lounge chair or as a double chair. It has analyzed the design according to human physical strictly, so it is very comfortable to sit.

Module-type stool, with comic-based caricature illustrations well expresses the modern art. The application of color modules has a very clear color contrast. What people take into consideration isn't whether it is comfortable to seat on these chairs. They are attracted by the art in front of their eyes. Even they are not very comfortable to sit, it is still a very nice display item.

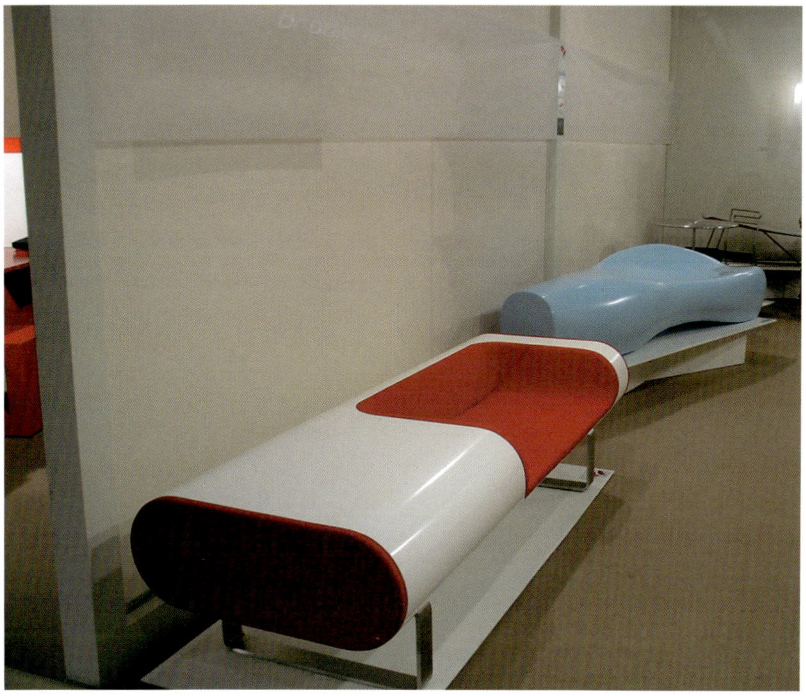

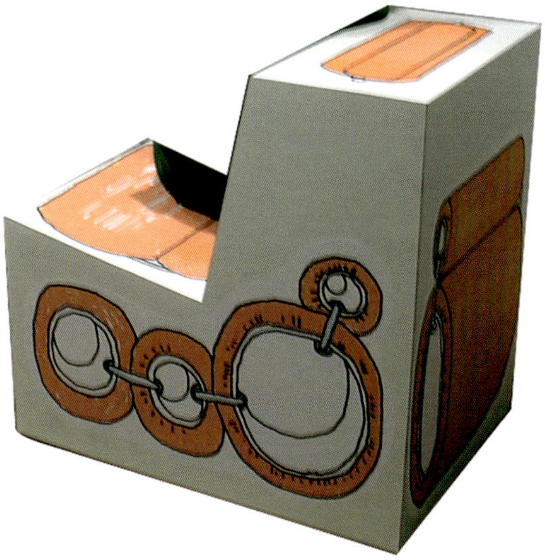

This chair is made of multi-layered lacquer wood. It has strong sculptural and very soft lines. In the furniture design circle, the use of curves is the alternative use of non-straight lines. The combination of accessories with different styles, colors of furniture brings a fresh feel to people.

The design of this chair is quite special. The designer used the traditional wooden furniture, so it has the traditional as well as the modern style. Young, enthusiastic, self-marked products are the main series for designers.

Armchair sofa with the prototype for the simple basic cube: This is the starting point of design, anti-design, none garish square, and highly ergonomic characteristics. In Furniture, streamline and arc are the main topic. Designers give the furniture the appearance of a "biological mimicry" decorative fabric, which forms a kind of poetic charm and temptation of the furniture. The application of emotional elements can arouse people's purchasing desire.

The Benson Collection of Corrupted Classics

flynn talbot

No need to worry about the colorful design of all colors will appear unpleasant. Furniture is used to lie on, and it is not a decoration for the artistic beauty.

There is no need to worry about it would be uncomfortable to sit in these colorfully-designed chairs. The function of the chair is to be seated. There is no chair just for the purpose of artistic beauty. The basic element of this design is the wood. People gasp with admiration at the marvelous design. The chair has majestic volumes, as well as the aesthetic simplicity which reflects the values of classic architecture. Designers make full use of the beauty of the furniture materials and the shapes through multi-level aesthetics to enrich people's life. When seeing the bracelet-shaped chair, we can't help longing for seating on it and wondering whether it is comfortable. Anyway, no one will refuse to place such an environmental friendly chair in the study, right?

No matter to design a product or a building, designers are all regard it as a kind of artistic creation. They are both useful and enjoyable. The furniture is the entity that helps people realize this kind of design philosophy. For designers, the temperament and content of life can come from a chair, as long as the chair has enough aesthetic feeling, suitable comfort and its own style. This is a new concept design. Wire twisted base together with the chairs makes it have strong characteristics and soft lines.

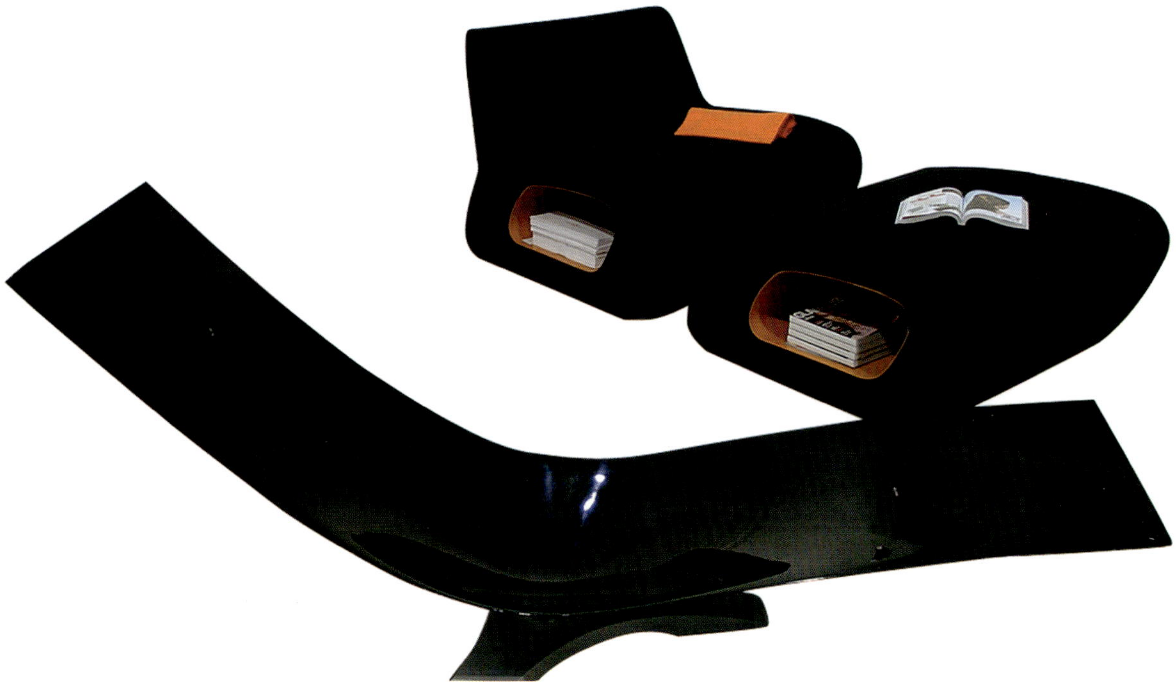

The chair has super-individual character. It uses very simple materials but with cool design. It is suitable for young avant-garde post-80. This series of chairs are very avant-garde in terms of design and materials. Therefore, we think this is more like new concept products from a concept designer. This chair is like a bullet type of creation. It can be split with four chairs and also can be put to form a new concept of bullet-type products. The chair is like a whirlpool-shaped sofa. With shiny materials, you won't think that it's vulgar but innovative.

4xSvezia

Italian style sofas, like most Italian men, have a romantic artistic characteristic. It always tempts you to "fall into its arms"—you are unwilling to get up after lying down. Even if it stands alone, it is still an artwork. When designing a chair, a sofa, we not only pursue the beauty of its shape, but also pay attention to the combination of the curved lines and the human body ' s structure. It breaks through the rigid concepts of the process and technology by incorporating them into human being's awareness and hereby becomes rational. A ridged chair, an elegant symbol makes the imagination played in our minds. The curves of the sofa are fascinating as if they had the inspiration of the Muses. The sofa is a two-way one in that it can be folded and unfolded in order to save the use of space.

The first impression that the armchair and sofa give is very beautiful, that it will be very comfortable to seat on them. The chair cover and sofa cover has a pleasant hand feeling and nice appearance.

These chairs are upgraded by adding 4 new elements. The high back seat is now accompanied by a chair, a low back seat, an easy chair, an armchair, and stools in two different sizes, which are designed for family easy use.

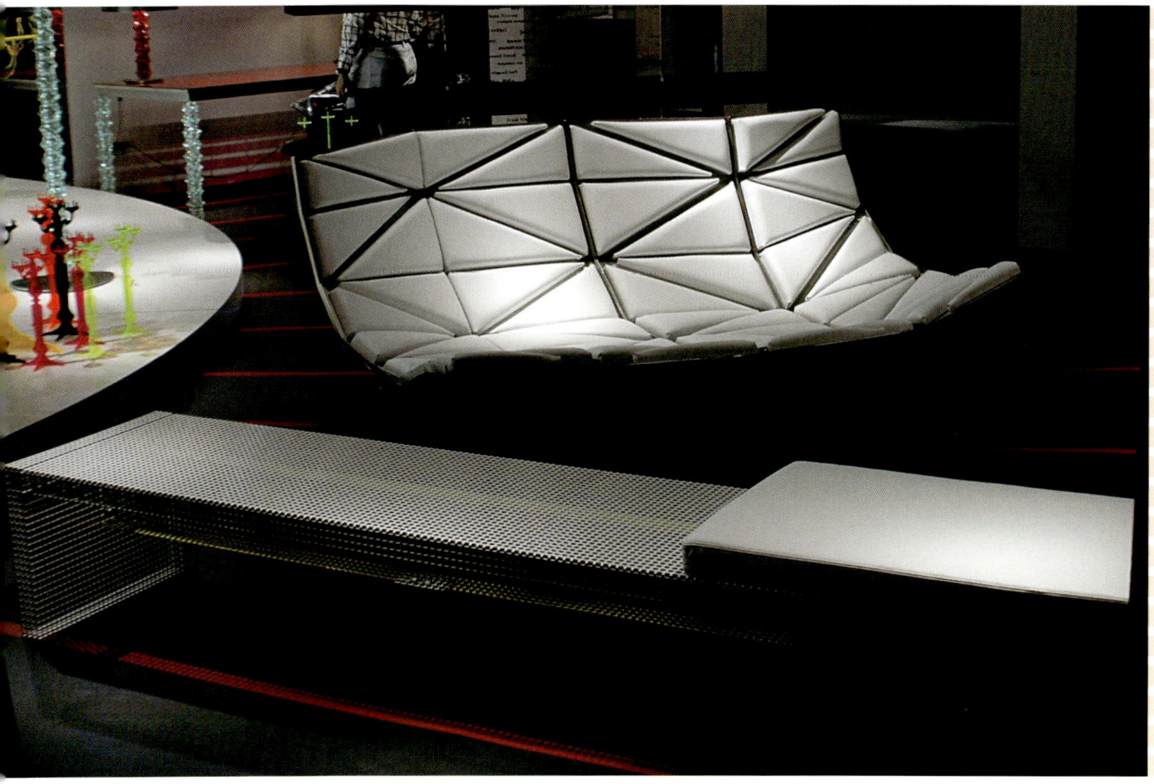

ligne roset
DÜSSEL
DORF

PRESS
STAMPA

ARCADE

The multi-used chairs made of plastic look strange though; in fact it can create different sitting postures with different ways of placement. It can either be used as lounge chair or as a double chair. It has analyzed the design according to human physical strictly, so it is very comfortable to sit.

Sofa is the necessary furniture for the living room. Well, are you fed up with those dull big wooden sofas? Why not choose a small sofa or chair for your living room? Small sofas and chairs not only occupy a small area, but also easy to move. Their styles are also simple and nice. In Italian style sofas, you would see cortex or movement with different diameters ranging from natural wrinkles or scars. This is the change of traditional aesthetics. No chemical pollution, no artificial odor, no artificial carving is the natural enjoyment that the Italian design pursues.

E04

design

Emmemobili

The materials used in this kind of furniture include solid wood, board, plastic, glass, metal, etc. It stresses the mix of metal and board, so the product comes out simple, beautiful and modern.

The above glass table is based on this theory. The arc table is supported by the raw round wood. There are two kinds of desktops: one is in frosted glass, the other mirror glass. Both types express the designer's concept of biological imitation. The special design makes us can not help taking these products home.

Around the Roses

Emmemobili

The most important feature is that each piece has the carven wood or metal as the frame. Texture desktops—Textures are largely used. You can choose the matching texture table in accordance to that of the wallpaper, sofas in your house, so that the whole space looks more unified.

The plain wooden table has very strong concepts. What it brings is not a rich dinner, but the new art with innovative concepts. The clear and smooth glass desktop is supported by the wood frame, and accompanied by wooden bar background. It has a kind of special style. The design is the simple design. It is extremely simple, so as to say.

The role of tables becomes increasingly important in the times of purchasing beauty for everything. Table has been the protagonist in restaurants. The table in the past was usually rectangular. People's position showed the status of each one in the family. In today's era of emphasis on individual style, the rules of the table can no longer restrict people's options.

These are stone furniture, chairs and tables. Stone is a kind of low-key and reserved material. The stone furniture becomes active and lively when the stone is given bright colors and creative modeling.

1983

SWEDESE

Three-dimensional and simple style: The table is the main character. It subtly connects the lines and planes to bring the fun into the space. One side of the table is decorated with cabinets, which prevents the lower part of the table becoming too heavy. Compared with the novel design, the model of the table matches the whole style well. Materials and colors: The table is made from plastics. Different styles fully display the beauty of the 3-D space. Although they use plastic materials, each type of plastic is used exclusively. The thick package of table increases the comfort in space and has been embodied with the flavor of life.

OSCOQUATTRO

The shape of tables has certain impact on the atmosphere of the living space. Rectangular dining table is more suitable for larger gatherings; and round table has a more democratic atmosphere. Rectangular tables are more suitable for family use. There are collapsible style tables which are more flexible than fixed ones.

The design of this chair is quite special. The designer used the traditional wooden furniture, so it has the traditional as well as the modern style. Young, enthusiastic, self-marked products are the main series for designers.

When designing a chair or a sofa, we not only pursue its shape of beauty, but also pay attention to the human body structure. It stresses the curve of the human body and how to achieve the perfect combination. It breaks through the rigid concepts in the process and technology and makes it very rational.

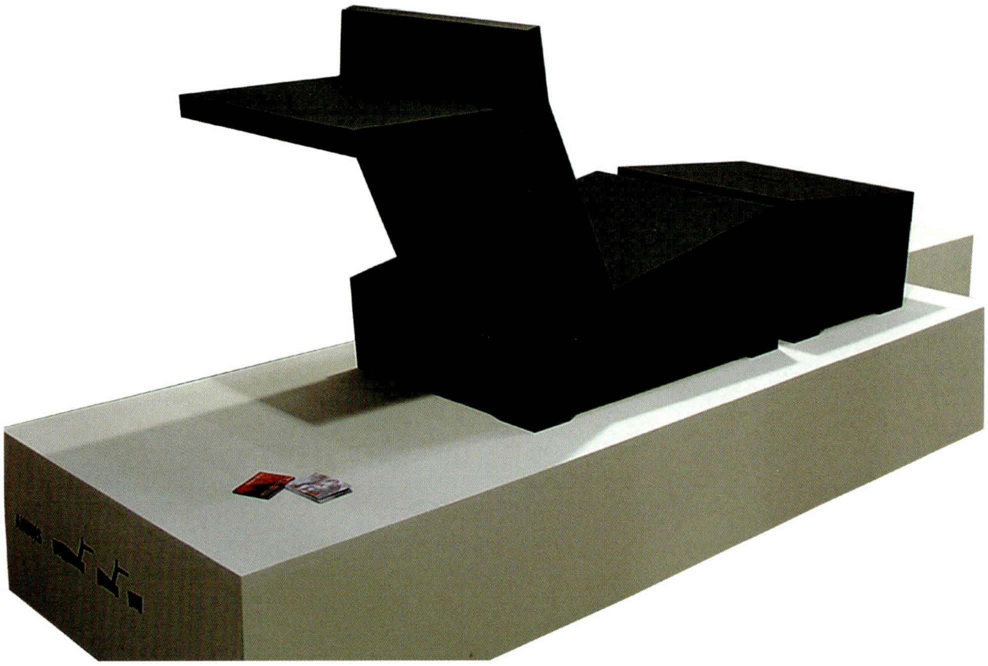

SOCA

They are magnificent in style and give the sense of superiority. Modeling: the frame of the table, dining chair is made from the same material. When they appear alone, they are like an artwork. When appearing together, they constitute a warm and nice dining space.

Metal twisted frame chair displays the designer's minimalist style very well. Although the material is metal, but the technique has been improved. Laser cutting upgrades the decoration and comfort of the furniture. Simple lines, random technically twists, etc, make all these possible.

softline

RICHARD LAMPERT

A beautiful table not only brings the possibility of a rich dinner, and sometimes also means that your friends like the warm atmosphere in your house, or envy the quality of your life ... The style is the typical Italian design. It is extremely simple in style. The main material is the raw wood. The chairs are in braided straw and very special.

It is said that a table is like a model that will let you dress up at your disposal. In order to show its unique style, you can choose different materials. Transparent glass has the flavor of traditional style. The clear and bright desktop conveys a noble and cheerful atmosphere. The matched tables are quite unique. People can enjoy the characteristics of the full set. It is such a wonderful enjoyment to have dinner with family or friends in such an enchanting atmosphere.

BURJ

FURNITURE EXHIBITION DESIGN

ACKNOWLEDGEMENTS

We would like to thank all the designers and companies who made a significant contribution to the compilation of this book. Without them, this project would not have been possible.
We would also like to thank all the producers for their invaluable assistance throughout. Its successful completion also owes a great deal to many professionals in the creative industry who have given us precious insights and comments. We are also very grateful to many other people whose names do not appear on the credits for making specific input and continuous support the whole time.

Neuroimmune Pharmacology and Therapeutics

Howard E. Gendelman • Tsuneya Ikezu
Editors

Neuroimmune Pharmacology and Therapeutics

Third Edition

Springer

Editors
Howard E. Gendelman
Department of Pharmacology and Experimental Neuroscience
University of Nebraska Medical Center Nebraska Medical Center
Omaha, NE, USA

Tsuneya Ikezu
Department of Neuroscience
Mayo Clinic Florida
Jacksonville, FL, USA

ISBN 978-3-031-68236-0 ISBN 978-3-031-68237-7 (eBook)
https://doi.org/10.1007/978-3-031-68237-7

This Springer imprint is published by the registered company Springer Nature Switzerland AG
The registered company address is: Gewerbestrasse 11, 6330 Cham, Switzerland

Preface

The preface was amended. We are honored to dedicate this book to Dr. Patrick Lucey McGeer who is acknowledged as a principal field originator. His life ended before the publication of the third edition. He was our sponsor, our role model, and our inspiration that birthed this field and much that was built on it. His life ended on August 29, 2022. Patrick was a Canadian physician, professor, and discoverer of the field of innate neuroimmunity and neuropharmacology. He was regarded without question as the leading authority on the causes and prevention of neurodegenerative immunity. It was his work as both an author and mentor that had given rise to the pathogenic concepts centering around the inflammatory hypothesis of Alzheimer's disease. In past days, he was also a "star" basketball player who had competed in the 1948 Olympics. He was a noted politician representing the Vancouver-Point Grey in the British Columbia legislator. He and his wife Edith were affirmed as members of the Order of Canada and inducted in the Royal Society of Canada as fellows. This edition of the textbook is therefore dedicated to Patrick's memory and in consideration of his influences on others and the many scientific and mentorship achievements. We have taken his prior preface and placed it anew as part of this new textbook to highlight the enduring scientific achievements he fashioned.

In the past four decades, enormous strides have been made in our understanding of the relationships between inflammation, immune responses, and degenerative human diseases. Neuroinflammation has grown from a tiny beginning to its current status as the largest field of brain research. Its growth continues to be rapid so this dominance will continue to expand. The field commenced with the obscure identification of activated microglia in the brains of Alzheimer's disease cases. A grant application of ours to the Canadian government to explore the ramifications of this finding was turned down with the terse comment: "This hypothesis is ridiculous!"

To grasp the dimensions of growth in this field, and to gain a perspective of future opportunities, today's neuroscientists only need to read this volume from cover to cover. The developing information has mostly appeared in specialty journals that have dealt only with isolated aspects of these tightly related fields. As a result, contemporary scientists have had a difficult time finding sources, even in review articles, that provide an integrated picture. This updated volume, by assembling chapters that demonstrate the relationship between these historically separated fields, overcomes that difficulty. Fifty-two chapters cover a broad spectrum of topics on the immunology of the nervous system. Included are diseases that result from immunological dysfunction, current therapeutic approaches, and prospects for the future. Overall, it integrates cutting-edge neuroscience, immunology, pharmacology, neurogenetics, neurogenesis, gene therapy, adjuvant therapy, nanomedicine, pharmacogenetics, biomarkers, proteomics, and magnetic resonance imaging. It is a rich harvest and readers will gain a perspective that has not previously been so readily available. Exposure to such a wealth of ideas is bound to inspire readers to undertake new and productive research initiatives.

The modern era of research into neuroinflammation and its relationship to neurodegenerative diseases began in the 1960s with the elaboration by Ralph van Furth of the monocyte phagocytic system. He injected labeled monocytes into animals and followed their migration and maturation into resident phagocytes in all body tissues. This provided a closure between Metchnikoff's 1882 discovery of mesodermal attack cells in starfish larvae, which he named phagocytes, and del Rio Hortega's 1919 discovery of phagocytic mesodermal cells entering the brain, which he named microglia. Hortega's results had always been questioned, and for more than two decades, the controversy continued as to whether microglia were truly phagocytes of mesodermal origin or were merely typical brain cells of epidermal origin. Resolving the controversy required the development of the techniques of immunohistochemistry and monoclonal antibody production. These tools for exploring brain biochemistry at the cellular level opened new vistas for understanding brain functioning and the pathogenesis of human disease. Using these tools, our laboratories and that of Joseph Rogers in Sun City at that time demonstrated that HLA-DR was strongly expressed in activated microglia. The identification of HLA-DR, a well-known leukocyte marker displayed by antigen-presenting cells, on these cells vindicated both Hortega and van Furth. The way was paved for many productive investigations exploring the properties of microglial cells and their relationship to inflammation and immune responses. This example of a conjunction between a fundamental concept and technical advances to establish its validity has been repeated many times since, as the chapters in this volume illustrate.

For a time, the concept that the brain is immunologically privileged held sway among neuroscientists. This was based on a narrow view that only the invasion of the brain by lymphocytes could be taken as evidence of an inflammatory response. But immunohistochemistry, coupled with newly developed molecular biological techniques, revealed that a spectrum of inflammatory mediators, including many known to cause tissue damage, was produced within the brain by resident brain cells. These discoveries required entirely new interpretations as to the nature of neuroinflammation and its relationship to immune responses. The innate immune system, operating at the local level in the brain, has proved to be the first line of defense. Indeed, the basic discoveries from studying the response of the brain in a variety of neurological diseases are causing a reevaluation of a number of peripheral degenerative disorders where innate immune responses, which had previously been ignored, have been shown to play a critical role in their pathogenesis. In other words, those studying the brain are providing immunologists with revolutionary new concepts regarding classical peripheral diseases. The insights of this volume need to be interpreted in this broader context.

Major neurologic disorders and details of their pathobiology are presented as individual chapters. They involve disorders where innate immune responses predominate, as in Alzheimer's disease, to others such as multiple sclerosis, where adaptive immune responses predominate, and others that seem to involve both. We have suggested that diseases involving self-damage generated by innate immune responses be defined as autotoxic to differentiate them from classical autoimmune diseases where self-damage is generated by adaptive immune responses. The common theme, however, is the involvement of microglia as the effector cells.

Genetics is well covered. It is a rapidly moving field. The methodology for linking familial disease to DNA mutations commenced in the late 1970s through the identification of restriction fragment-linked polymorphisms. By 1983, when James Gusella and his colleagues demonstrated a linkage of the G8 fragment to Huntington's disease, only about 18 markers were known. Now over 1,500,000 single-nucleotide polymorphisms have been localized so that every centimorgan of the human genome can be explored. This advance has been coupled with rapid methods for sequencing DNA. The report on genetics must be regarded as the tiny tip of a giant iceberg where much below the surface will soon be revealed.

The ultimate objective of neuroscientists studying human disease is to find more effective treatments. Part III covers the pharmacology of existing drugs, as well as describing approaches now in clinical evaluation, and those still at the bench level. Some of these include concepts that depart from established therapeutic approaches, giving the reader much food for thought.

There is an important chapter on the new field of biomarkers. Biomarkers have established that neurodegenerative disorders such as Alzheimer's disease commence at least a decade before clinical signs develop. A window of opportunity is opened up where early anti-inflammatory intervention can arrest disease progression and abort disease development. This can be combined with brain imaging to measure objectively the effects of therapeutic agents in diseases where progressive brain degeneration occurs. In summary, this is a volume not to be put on the shelf as a reference text, but to be read cover to cover by aspiring neuroscientists.

Vancouver, BC, Canada — Patrick L. McGeer

Introducing Neuroimmune Pharmacology and Therapeutics

The third edition of *Neuroimmune Pharmacology and Therapeutics* has broadened its focus for established and new investigators in the field. The book covers the molecular, biochemical, genetic, and pathogenetic parts of the disease complex to advance a deeper understanding of the pharmacologic and therapeutic parts of the field taking the backdrop of neuroimmunity. The book builds upon our now-established journal with the same title and aims to provide a two-pronged approach to advance this ever-broadening field. The overall intent of this edition is to merge laboratory, translational, and clinical research on a broad spectrum of topics that include the biology, pathophysiology, behavior, and structure-function relationships of innate and adaptive immunity for nervous system disease. The focus is on studies of the neurodegenerative, neuroinfectious, metabolic, inflammatory, behavioral, and infectious diseases. Each is covered in depth including those that continue to emerge. We present therapeutic works that focus on molecular, structural-biochemical, therapeutic, and biochemical approaches to attenuate, control, delay, or reverse disease severity and slow its onset and progression. The book serves to highlight new, interactive, and cross-disciplinary research that intersects divergent fields of drug delivery, neuroscience, nanomedicine, ultra-long-acting antivirals, and immunological therapies relevant to modulating, transforming, or controlling the neuroimmune axis in seeking therapeutic benefit. We cover divergent and prospective points of view of immediate interest designed to foster immediate information exchanges for neuroimmune therapeutics, disease prevention, and a broad range of treatment strategies. The body of work serves to align itself beyond the field and the journal with our society on NeuroImmune Pharmacology in order to support and encourage scientific exchange in and outside the field of immunotherapeutics that affect the nervous system. We trust that the included works can readily reach an international field of investigators.

To promote a field understanding we begin on the basics of etiology, prevention, and treatment of neuroimmune disorders and provide coverage from the basic aspects of laboratory research to translational efforts that bridge to the clinic. We have done this by defining where and how therapeutic intervention has proven most viable. This includes translational research in general and, in particular, a better understanding of the role played by neuroinfectious diseases, inflammation, and stress in disorders of the neuroimmune axis. Chapters serve to define neuroimmunomodulator properties of neuroimmune innervations and regeneration, and endogenous and exogenous neuroimmuno-modulators that include hormones, neuropeptides, neurotransmitters, cytokines, chemokines, and substance use disorders. Advances in the field have now included nanomedicines, improved drug delivery for therapeutics, and the advancement of technologies that include molecular interventional strategies and gene editing. The overarching goal for the textbook is to provide a platform for promoting a growing area of research and to bring that vision on a global scale in basic and clinical science.

Neuroscience, immunology, pharmacology, and related therapies are each of and by themselves broad disciplines that, without argument, impact upon a large component of what we come to know as biomedical science (Gendelman, 2002; Abdelmoaty et al., 2023; Holzer et al., 2015; McGeer & McGeer, 2004). Each, by itself and even more so when put together, is multidisciplinary and requires, for the student, both broad knowledge and deep understanding of molecular and cellular biology, neuroimmunity, the functional blood–brain barrier, neurochemistry, neuroinfectious and neurodegenerative disorders, cancer, and neurodevelopment. For many, it is considered a branch of immunology, but that is a starting point as the field bridges investigations of drug action and development with nervous system biology and disease pathogenesis. To those engaged in this field, linking the disciplines is ever more challenging as when they are joined, they become interactive. The bridges between disciplines are what we now call multidisciplinary science and require another level of insight. However, we posit that each need be first understood as a single entity. To this end, we forged chapters that cover the structure, function, and biology of each of the fields independent of one another. Then step-by-step they are each combined one with the other to form the basis of our engagement with the environment, to disease and a means to restore homeostasis by protective immune and pharmacologic means. Indeed, drugs that include immune modulators can certainly influence organ function, aging, and tissue homeostasis and improve clinical outcomes.

Indeed, a special feature of "humankind" among other species is the presence of an extraordinary complex immune system that can be used to protect against a plethora of harmful microbial pathogens, including viruses, bacteria, and parasites, as well as abnormal cells and proteins (Petranyi, 2002; Obermeier et al., 2016). This underlies the complexity of the human genome, which encodes expansive immune-related genes not found in lower species (Hughes, 2002). When the immune system is compromised, disease occurs and often does with ferocity; a wide range of clinical manifestations ensue that follows as a consequence of neurodegenerative, psychiatric, cancer, and infectious diseases, or those elicited by the immune system's attack on itself. The latter is commonly referred to as "destructive" autoimmunity (Christen & von Herrath, 2004; Khan et al., 2024). Interestingly enough, the immune system may sometimes be an impediment to therapy. Indeed, modulating its function is required for long-term and successful organ transplantation (Righi et al., 2024; Samaniego et al., 2006; Horst et al., 2016). On balance, modulation of the immune system can affect "neuroprotective" responses for certain diseases (Anderson et al., 2014).

Like the immune system, the nervous system contains surveillance functions and also possesses a number of functional roles that include mentation, movement, reasoning, sensation, vision, hearing, learning, breathing, and most behaviors. The nervous system includes defined tissue structures such as the brain, spinal cord, and peripheral nerves. On the cellular level, it includes networks of nerve cells with a variety of functional activities; complex networks and communications; supportive and regulatory cells, called glia; and a protective barrier that precludes the entry of a variety of macromolecules, cells, and proteins. It also possesses connections throughout the body that permits it function. Neuroscience is the discipline used to explore each of the nervous system regions and cells that include their networks and modes of communication in health and disease. As humans, we have ~100 billion neurons that are each functional units contained within the nervous system. Molecular and biochemical studies along with cell and animal systems were each used alone and together to explore and define neural biology. The task of neurosciences is to better understand the brain's function in the context of ontogeny, organism development, and aberrations during disease. Neuroscience is an interdisciplinary field and evolved as such during the past quarter-century. It includes neurobiology, neurochemistry, neurophysiology, mathematics, psychology, computer neuroscience, and learning and behavior. It also included the field of immunology (Sehgal & Berger, 2000; Zhang et al., 2024). In a historical context, the brain, for a long time, was considered an immune-privileged organ, meaning that its protective shield or barrier, commonly termed the blood–brain barrier, served to protect, defend, and as a consequence exclude ingress of toxins, cells, and pathogens (Streilein, 1993; Becher et al., 2000). This has undergone a reassessment of purpose (Louveau, 2015). Nonetheless, this is balanced by the fact that resident inflammatory cells do exist inside the brain and are capable of producing robust immune responses. In recent years, we have come to accept that it is the mononuclear phagocytes (MP; perivascular macrophage and microglia) that are disease perpetrators, while the astrocyte serves as "supportive" and "homeostatic" in nurturing neurons and in protection against the ravages of disease (Gendelman, 2002; Simard & Nedergaard, 2004; Trendelenburg & Dirnagl, 2005). The neuron in this neuroimmune model is the passive recipient of the battles that rage between the MP and the astrocyte. Findings that have emerged over the past half-decade have challenged this model. We now know that dependent upon environmental cues and disease, microglia, astrocytes, and other neural cell elements including endothelial cells and oligodendrocytes possess immunoregulatory functions. We also know that microglia and astrocytes dependent upon the environment and stimuli can be supportive, destructive, or both. Even more importantly, neurons can secrete immunoregulatory factors and engage directly into cell–cell–environmental cue stimulations. To make the system perhaps even more complex, local neuroimmune processes can result in the recruitment of T cells and enticement of the adaptive immune response, significantly affecting disease outcomes (Olson & Gendelman, 2016). All in all, things appear more complex than once thought even in the past decade.

These three disciplines and the complexities inherent in each academic field are perhaps the most multidisciplinary, serving to bring scientists and clinicians together with knowledge of neurobiology, immunology, pharmacology, biochemistry, cellular and molecular biology, virology, genetics, gene therapy, medicinal chemistry, nanomedicine, proteomics, pathology, and physiology. Even more than immunology and neuroscience, pharmacology integrates a broad knowledge in scientific disciplines, enabling the pharmacologist a unique perspective to tackle drug-, hormone-, immune-, and chemical-related pathways as they affect human health and behavior. Drug actions and therapeutic developments form the basis of such discoveries, but the central understanding of how they act provides vision for further research to improve human well-being and health.

This textbook is unique in scope by serving to investigate the intersection of this new discipline. Neuropharmacologists study drug actions including neurochemical disorders underlying a broad range of diseases such as psychiatric disorders (e.g., schizophrenia and depression) and neurodegenerative diseases (e.g., Alzheimer's and Parkinson's diseases). Drugs can also be used to examine neurophysiological or neurobiochemical changes as they affect brain, behavior, movement, and mental status. Immunopharmacology seeks to control the immune response in the treatment and prevention of disease. Research does include immunosuppressant agents used in organ transplant as well as developing agents that affect bone marrow function and cell differentiation in cancer therapies.

What then defines the field of *neuroimmune pharmacology and therapeutics*? Is it simply a field that intersects the three disciplines of neuroscience, immunology, and pharmacology in seeking to better define the epidemiology, prevention, and treatment of immune disorders of the nervous systems (Fig. 1)? Are these disorders limited in their scope in affecting behavior, cognition, motor, and sensory symptoms, or do they also involve developmental and degenerative disorders? It is clear that the immune system is linked, in whole or in part, to diseases that develop as a consequence of genetic abnormalities and a broad range of environmental cues (including microbial infections and abused drugs) and to toxins. So, where does neuroimmune pharmacology find its niche? These can occur, in part, as a consequence of neuropeptides, neurotransmitters, cytokines, chemokines, and abused drugs. Like much in science, we are left with more questions than answers. In the end, we seek avenues for translational research and better understanding of disease mechanisms. Diseases are together linked to microbial agents, by inflammatory processes, by emergence of cancerous cells or tumors, by stress, by environmental cues, and genetic disturbances. No matter the cause, harnessing the immune processes for pharmacological benefit will, at days end, provide "real" solutions to positively affect some of the most significant and feared disorders of our century.

What do we seek to accomplish by editing such a textbook? First, we would be remiss in not acknowledging the pivotal discoveries made by others when research fields intersect. These include the discovery and characterization of the guanosine triphosphate (GTP) binding and prion proteins (Gilman, 1995; Rodbell, 1995; Prusiner, 1998, 2001), neurotransmission and memory functions (Carlsson, 2001; Greengard, 2001; Kandel, 2001), and odorant receptors (Buck, 2000; Axel, 2005). We posit that new discoveries can and will be made through the intersections of neuroscience, immunology, and pharmacology and as such sought to define it

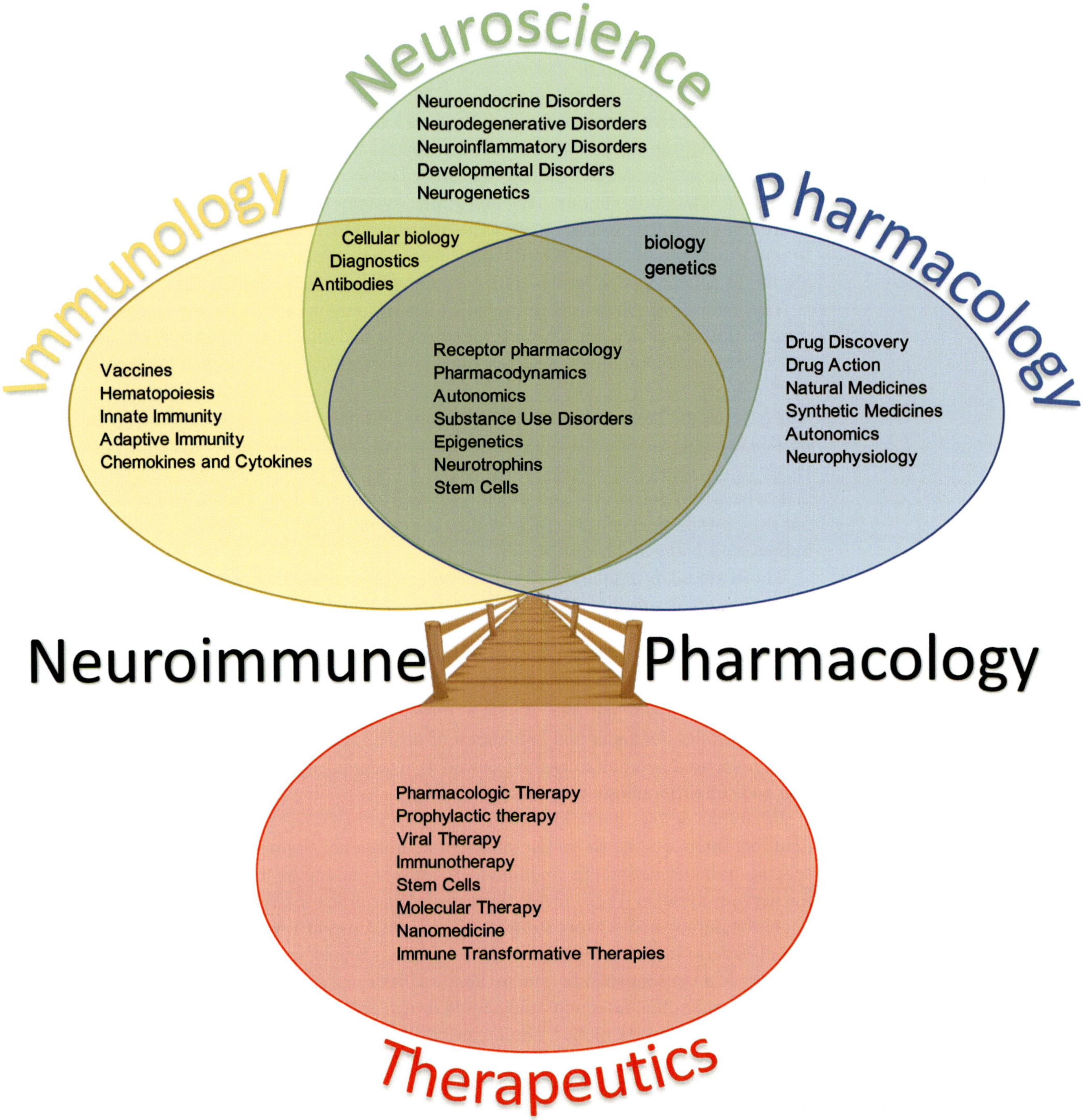

Fig. 1 This Venn diagram pictorially represents the fields of immunology, pharmacology, and neuroscience with common elements overlapping into the discipline of neuroimmune pharmacology. The blackened area depicts the diseases and research areas covered in this book and integral to the field of study

for the student. The notion that inflammation contributes in significant manner to neurodegeneration and significantly beyond autoimmune diseases is brought front and center and demonstrated without ambiguity for multiple sclerosis, peripheral neuropathies, Alzheimer's and Parkinson's disease, and amyotrophic lateral sclerosis as well as for microglial infections of the nervous system including NeuroAIDS where microglial activation is central to disease processes (Hooten et al., 2015; Appel et al., 1995; Toyka & Gold, 2003; McGeer & McGeer, 2004; Ercolini & Miller, 2005; Gendelman, 2002; Gendelman & Mosley, 2015). Perhaps most importantly, we have laid the groundwork for how the immune system can be harnessed either through its modulation, through altering blood–brain barrier integrity and function, or by drug-delivery strategies that target the brain. This third edition will add on to the success of the second edition with additional chapters on emerging topics that include therapeutics, mirobiota, innate and adaptive immunity, exosomes, microRNA, mental disorders, biomarkers and medicinal and polymer chemistry.

No doubt this textbook is an expansive read for the student and scholar alike. To this end, we are humbled by its realization and even more so during this third edition. These words lay only the beginnings to what we believe will be a significant future footprint into the integration between neuroscience, immunology, pharmacology, and therapeutics.

Department of Pharmacology and Experimental Neuroscience — Howard E. Gendelman
University of Nebraska Medical Center, Nebraska Medical Center
Omaha, NE, USA
Laboratory of Molecular Therapeutics, Mayo Clinic — Tsuneya Ikezu
Jacksonville, FL, USA

References

Abdelmoaty, M. M., Yeapuri, P., Machhi, J., Lu, Y., Namminga, K. L., Kadry, R., Lu, E., Bhattarai, S., Mosley, R. L., & Gendelman, H. E. (2023). Immune senescence in aged APP/PS1 mice. *NeuroImmune Pharmacology and Therapeutics, 2*(3), 317–330.

Anderson, K. M., Olson, K. E., Estes, K. A., Flanagan, K., Gendelman, H. E., & Mosley, R. L. (2014). Dual destructive and protective roles of adaptive immunity in neurodegenerative disorders. *Translational Neurodegeneration*, *3*(1), 25.

Appel, S. H., Smith, R. G., Alexianu, M., Siklos, L., Engelhardt, J., Colom, L. V., & Stefani, E. (1995). Increased intracellular calcium triggered by immune mechanisms in amyotrophic lateral sclerosis. *Clinical Neuroscience, 3*, 368–374.

Axel, R. (2005). Scents and sensibility: A molecular logic of olfactory perception (Nobel lecture). *Angewandte Chemie International Edition Engl*, *44*, 6110–6127.

Becher, B., Prat, A., & Antel, J. P. (2000). Brain-immune connection: Immuno-regulatory properties of CNS-resident cells. *Glia*, *29*, 293–304.

Buck, L. B. (2000). The molecular architecture of odor and pheromone sensing in mammals. *Cell*, *100*, 611–618.

Carlsson, A. (2001). A half-century of neurotransmitter research: Impact on neurology and psychiatry. Nobel lecture. *Bioscience Reports*, *21*, 691–710.

Christen, U., & von Herrath, M. G. (2004). Initiation of autoimmunity. *Current Opinion in Immunology*, *16*, 759–767.

Elenkov, I. J., Wilder, R. L., Chrousos, G. P., & Vizi, E. S. (2000). The sympathetic nerve—an integrative interface between two supersystems: The brain and the immune system. *Pharmacological Reviews*, *52*, 595–638.

Ercolini, A. M., & Miller, S. D. (2005). Role of immunologic cross-reactivity in neurological diseases. *Neurological Research*, *27*, 726–733.

Gendelman, H. E. (2002). Neural immunity: Friend or foe? *Journal of Neurovirology*, *8*, 474–479.

Gendelman, H. E., & Persidsky, Y. (2005). Infections of the nervous system. *The Lancet Neurology*, *4*, 12–13.

Gendelman, H. E., & Mosley, R. L. (2015). A perspective on roles played by innate and adaptive immunity in the pathobiology of neurodegenerative disorders. *Journal of Neuroimmune Pharmacology*, *10*(4), 645–650.

Gilman, A. G. (1995). Nobel lecture. G proteins and regulation of adenylyl cyclase. *Bioscience Reports*, *15*, 65–97.

Greengard, P. (2001). The neurobiology of dopamine signaling. *Bioscience Reports*, *21*, 247–269.

Holzer, P., Hassan, A. M., Jain, P., Reichmann, F., & Farzi, A. (2015). Neuroimmune pharmacological approaches. *Current Opinion in Pharmacology*, *25*, 13–22.

Hooten, K. G., Beers, D. R., Zhao, W., & Appel, S. H. (2015). Protective and toxic neuroinflammation in amyotrophic lateral sclerosis. *Neurotherapeutics, 12*(2), 364–375.

Horst, A. K., Neumann, K., Diehl, L., & Tiegs, G. (2016). Modulation of liver tolerance by conventional and nonconventional antigen-presenting cells and regulatory immune cells. *Cellular & Molecular Immunology*, *13*(3), 277–292.

Hughes, A. L. (2002). Natural selection and the diversification of vertebrate immune effectors. *Immunological Reviews*, *190*, 161–168.

Kandel, E. R. (2001). The molecular biology of memory storage: A dialogue between genes and synapses. *Science*, *294*, 1030–1038.

Khan, Z., Mehan, S., Gupta, G. D., & Narula, A. S. (2024). Immune system dysregulation in the progression of multiple sclerosis: Molecular insights and therapeutic implications. *Neuroscience*, *30*(548), 9–26.

Louveau, A., Smirnov, I., Keyes, T. J., Eccles, J. D., Rouhani, S. J., Peske, J. D., Derecki, N. C., Castle, D., Mandell, J. W., Lee, K. S., Harris, T. H., & Kipnis, J. (2015). Structural and functional features of central nervous system lymphatic vessels. *Nature*, *523*(7560), 337–341.

McGeer, P. L., McGeer, E. G. (2004). Inflammation and the degenerative diseases of aging. *Annals of the New York Academy of Sciences*, *1035*, 104–116.

Obermeier, B., Verma, A., & Ransohoff, R. M. (2016). The blood–brain barrier. *Handbook of Clinical Neurology*, *133*, 39–59.

Olson, K. E, & Gendelman, H. E. (2016). Immunomodulation as a neuroprotective and therapeutic strategy for Parkinson's disease. *Current Opinion in Pharmacology*, *26*, 87–95.

Petranyi, G. G. (2002). The complexity of immune and alloimmune response. *Transplant Immunology, 10*, 91–100.

Prusiner, S. B. (1998). Prions. *Proceedings of the National Academy of Sciences*, *95*, 13363–13383.

Prusiner, S. B. (2001). Shattuck lecture—neurodegenerative diseases and prions. *New England Journal of Medicine*, *344*, 1516–1526.

Righi, I., Trabattoni, D., Rosso, L., Vaira, V., & Clerici, M. (2024). Immune checkpoint molecules in solid organ transplantation: A promising way to prevent rejection. *Immunology Letters*, *25*(267), 106860.

Rodbell, M. (1995). Nobel lecture. Signal transduction: Evolution of an idea. *Bioscience Reports*, *15*, 117–133.

Samaniego, M., Becker, B. N., & Djamali, A. (2006). Drug insight: Maintenance immunosuppression in kidney transplant recipients. *Nature Clinical Practice Nephrology*, *2*, 688–699.

Sehgal, A., & Berger, M. S. (2000). Basic concepts of immunology and neuroimmunology. *Neurosurgical Focus, 9*, e1.

Simard, M., & Nedergaard, M. (2004). The neurobiology of glia in the context of water and ion homeostasis. *Neuroscience, 129*, 877–896.

Streilein, J. W. (1993). Immune privilege as the result of local tissue barriers and immunosuppressive microenvironments. *Current Opinion in Immunology*, *5*, 428–432.

Toyka, K. V., & Gold, R. (2003). The pathogenesis of CIDP: Rationale for treatment with immunomodulatory agents. *Neurology, 60*, S2–S7.

Trendelenburg, G., & Dirnagl, U. (2005). Neuroprotective role of astrocytes in cerebral ischemia: Focus on ischemic preconditioning. *Glia*, *50*, 307–320.

Zhang, W., Sun, H. S., Wang, X., Dumont, A. S., & Liu, Q. (2024). Cellular senescence, DNA damage, and neuroinflammation in the aging brain. *Trends in Neurosciences*, *9*, SO166-2236(00061-4).

Acknowledgments

This comprehensive volume would not have been possible without the tireless, dedicated, and determined effort of one extraordinary person, Amy Sather. Amy has proven to be a devoted, resourceful, and persistent leader. Any task was never too minor and never left unattended. The flow between myself and our associate editors, authors, administrators, and publishers was seamless. Amy remains more than a manager but is a conductor. Her attention to detail and organizational skills were vital to completing the text.

Our administrative staff, which includes Theresa Grutel, Kara Schmidt, Kim Morrison, and Reed Felderman at the University of Nebraska Medical Center who continues as the foundation for this text, our *Journal of Neuroimmune Pharmacology and Therapeutics*, and the function of our department of Pharmacology and Experimental Neuroscience. We are lucky to have them as partners in all our academic trials and appreciate being able to work with a team of talented individuals.

To our leaders and visionaries, Jeffrey Gold, Dele Davies, and Bradley Britigan, a simple thank you is inadequate for all you have done to ensure the success of our endeavors. To Carol Swarts, a friend, a colleague, a confidant, and a soul mate whose gifts of time and person ensure our sustained academic successes: Carol, you truthfully have no parallel. Your contribution to science is immeasurable. Thank you all for keeping the fire burning in all we do. To Robert Eisenberg, Howard Kooper, Milton Kleinberg, Peter Savas, Amy Volk, Tom Thompson, Jim Roubal, and the families of the late Frances, Louis Blumkin, and Harriet Singer, who have steadfastly supported our works for decades. Your faith in our research strengthens us each day.

To our chapter contributors and reviewers whose expertise, knowledge, intellect, and dedication proved invaluable in completing the tasks. To Bonnie Bloch and Seiko Ikezu, our partners in life, life journeys, and always best friends. To Yohei and Michiko Ikezu and the late Soffia Gendelman, our parents, navigators, and role models. To Yumiko Aoyama, for her continuous support of family and friendship, a simple thank you is nearly inadequate.

To our children and grandchildren, who are all our life treasures: Sierra, Jason, Samson and Sacha Tobias, Adam and Jen Wolf-Gendelman, Sidney Gendelman, Noa Sofia Gendelman, Lesley Gendelman, Eric, Simon, and Michael Greenfield, Emma Ehrenkranz, and Clark and David Ikezu.

We salute *the Journal of Neuroimmune Pharmacology and Therapeutics our editors and associate editors, which provided and continues to be the source of inspiration for this work. Thank you to a special life-long friend, Joel Alperson, for the gift of his time, years, knowledge, inspiration, and life pursuits and to our patients, students, mentors, and colleagues who all inspire and guide daily.*

Howard E. Gendelman
Tsuneya Ikezu

Contents

Part III Therapies and Diagnostics

About the Editors

Howard E. Gendelman is the Margaret R. Larson Professor of Internal Medicine and Infectious Diseases, Chairman of the Department of Pharmacology and Experimental Neuroscience, and Director of the Center for Neurodegenerative Disorders at the University of Nebraska Medical Center. Dr. Gendelman is credited in unraveling the functional immune alterations in brain and how they induce metabolic changes and lead to neural cell damage in a range of chronic infectious and neurodegenerative disorders. The discoveries led to immune transformative therapeutics aimed at preventing, slowing, or reversing neural maladies currently in clinical trials. He is also credited for the demonstration that AIDS dementia is a reversible metabolic encephalopathy. His work has led to regulatory T cell-based immunotherapies for Parkinson's and more recently Alzheimer's disease are being tested in early clinical trials as a result of intense translational investigations. He is a co-founder of Exavir Therapeutics, Inc., a biotechnology start-up focused on developing ultra-long acting therapies for chronic diseases. Dr. Gendelman obtained a Bachelor's degree in Natural Sciences and Russian Studies with honors from Muhlenberg College and his MD from the Pennsylvania State University-Hershey Medical Center where he was the 1999 Distinguished Alumnus. He completed a residency in Internal medicine at Montefiore Hospital, Albert Einstein College of Medicine and was a Clinical and Research Fellow in Neurology and Infectious Diseases at the Johns Hopkins University Medical Center. He occupied senior faculty and research positions at the Johns Hopkins Medical Institutions, the National Institute of Allergy and Infectious Diseases, the Uniformed Services University of the Health Sciences Center, the Walter Reed Army Institute of Research, and the Henry Jackson Foundation for the Advancement in Military Medicine before joining the University of Nebraska Medical Center faculty in March of 1993. He retired from the US Army with the rank of Lieutenant Colonel. Dr. Gendelman has authored over 580 peer-reviewed publications, edited 12 books and monographs, holds 112 patents, is Founder and the Editor-In-Chief of the *Journal of Neuroimmune Pharmacology* from 2004 to 2021 and currently *NeuroImmune Pharmacology and Therapy* along with service on numerous editorial boards, national and international scientific review, and federal and state committees. He has been an invited lecturer to more than 300 scientific seminars and symposia and the recipient of numerous local, national, and international honors. He is a true luminary of the medical center as the recipient of the University of Nebraska's IP Innovation and Commercialization Award based on his development of significant new intellectual property from concept to licensing/startup business. Dr. Gendelman has patented more than 200 new inventions and is moving to or already has begun several Phase I-II studies. In addition, he received the Henry L. Moses Award in Basic Science; the Carter-Wallace Fellow for Distinction in AIDS Research, the David T. Purtilo Distinguished Chair of Pathology and Microbiology; the Margaret R. Larson Professor of Infectious Diseases and Internal Medicine; being named a Nebraska Pioneer; a UNMC Scientist Laureate; an NU Outstanding Research and Creativity Award winner, the 2013 UNMC Innovator of the Year, the 2014 Outstanding Faculty Mentor of Graduate Students; the Life Saver Award from the Nebraska Coalition for Lifesaving Cures, the Intellectual Property, Innovation and Commercialization Award; and the Pioneer in Neurovirology Award by the International Society for NeuroVirology. He is also the President of the Society on NeuroImmune Pharmacology and the recipient of the humanitarian award given by the Jewish Federation of Omaha. The society granted him the Joseph Wybran, Herman Friedman, and Lifetime Service Awards. Dr. Gendelman was also named a J. William Fulbright Research Scholar at the Weizmann Institute of Science in Israel. He received the prestigious Jacob Javits Neuroscience Research Award from the National Institute of Neurological Disorders and Stroke and the Career Research Award in Medicine from the Department of Internal Medicine, UNMC. He enjoys a current Hirsch Index of 119 as listed on highly cited.com as one of the top-cited scientists in his field.

Tsuneya Ikezu is a Professor in the Department of Neuroscience and Director of Molecular Neurotherapeutics Laboratory at the Mayo Clinic in Florida. After receiving his MD and PhD at the University of Tokyo School of Medicine he devoted his academic pursuits to studies of gene therapy of Alzheimer's disease (AD) and precisely to study the therapeutic effect of the brain delivery of a spectrum of medicines designed to restore hippocampal neurogenesis and memory function and to the proteomic and biological characterization of EVs. Dr. Ikezu and his colleagues also contributed substantially to the classification of the activation phenotype of microglia, a cell type initially adapted from the M0, M1, and M2 classification of macrophages. His recent works characterizing the microglial gene expression phenotype in disease conditions show that microglia are essentially in homeostatic or disease-associated states. Prior to this, Dr. Ikezu was Professor of Pharmacology and Neurology at Boston University School of Medicine until 2020. He has authored more than 139 journal articles, which were cited over 23,000 times, and served on several editorial boards including the *Journal of Extracellular Vesicles*. Dr. Ikezu has received numerous awards including Inge Grundke-Iqbal Award from Alzheimer's Association (2016), Jack Spivack Excellence in Neuroscience Award from Boston University School of Medicine (2018), and Investigator of the Year Award from Mayo Clinic Florida (2023). Dr. Ikezu's educational background includes a medical degree from the Faculty of Medicine, University of Tokyo; a PhD from the University of Tokyo Graduate School of Medicine.

Abbreviations

^{1}H	Proton
^{1}H-MRSI	Proton magnetic resonance spectroscopic imaging
2D SDS-PAGE	Two-dimensional polyacrylamide gel electrophoresis
3'UTR	3'-Untranslated region
3HK	3-Hydroxykynurenine
5-ASA	5-Aminosalicylic acid
5-HIAA	5-Hydroxyindole acetic acid
5-HT	5-Hydroxy tryptophan
5-HT	5-Hydroxytryptamine
6-OHDA	6-Hydroxydopamine
8-OHdG	8-Hydroxy-2'-deoxyguanosine
γc	Common γ-chain
AAAD	Aromatic L-amino acid decarboxylase
AAV	Adeno-associated virus
Ab	Antibody
Aβ	Amyloid-β
ABD	Adamantiades-Behçet's disease
ABP	Actin-binding protein
AC	Anterior chamber
ACAID	Anterior chamber-associated immune deviation
αCamKII	α-Calcium/calmodulin-dependent protein kinase II
ACh	Acetylcholine
AChE	Acetylcholinesterase
ACTH	Adrenocorticotrophic hormone (pro-opiomelanocortin; POMC)
AD	Alzheimer's disease
ADHD	Attention deficit hypersensitivity disorder
ADI	Acceleration/deceleration injury
ADAM	A disintegrin and metalloprotease
ADCC	Antibody-dependent cellular cytotoxicity
ADEM	Acute disseminated encephalomyelitis
AF	Activation factor
Ag	Antigens
AGE	Advanced glycation end products
AGM	Aorta-gonad mesonephros
AHSCT	Autologous hematopoietic stem cell transplantation
AICA	Anterior inferior cerebellar artery
AID	Activation-induced (cytidine) deaminase
AIDP	Acute inflammatory demyelinating polyradiculoneuropathy
AIDS	Acquired immunodeficiency syndrome
AIF	Apoptosis inducing factor
AIR	Autoimmune retinopathy
AIRE	Autoimmune regulator

AIS	Anterior chamber-associated immune deviation (ACAID)-inducing signal
ALR	AIM2-like receptors
ALS	Amyotrophic lateral sclerosis
AMAN	Acute motor axonal neuropathy
AML	Acute myeloid leukemia
AMN	Adrenomyeloneuropathy
AMPA	α-Amino-3-hydroxy-5-methylisoxazole-4-propionic acid
AMSAN	Acute motor-sensory axonal neuropathy
ANG	Angiotensin II
ANI	Asymptomatic neurocognitive impairment
ANS	Autonomic nervous system
AP	Area postrema
AP-1	Activating protein-1
APAF 1	Apoptotic peptidase activating factor 1
APC	Antigen-presenting cells
aPKC	Atypical protein kinase C
aPL	Antiphospholipid antibody
APO1	Apoptosis antigen 1 (Fas/CD95)
apoE	Apolipoprotein E
APP	Amyloid precursor protein
APPs	Acute phase proteins
AQP4	Aquaporin-4
ARE	AU-rich response element
ARMD	Age-related macular degeneration
ART	Antiretroviral therapy
ASD	Autism spectrum disorders
ASL	Arterial spin-labeled
ASTIN	Acute stroke therapy by inhibition of neutrophils
AT	Adoptive transfer
ATL	Adult T cell leukemia
ATON	Atacicept in optic neuritis
ATP	Adenosine triphosphate
AVE	Anterior visceral endoderm
AVG	Anti-viral granule
AVP	Arginine vasopressin
AV3V	Anteroventral third ventricular
AZT	3'-Azidothymidine/zidovudine
BACE	Beta-site amyloid precursor protein cleaving enzyme
BBB	Blood–brain barrier
BBMEC	Bovine brain microvessel endothelial cells
BCR	B-cell receptor
BCRP	Breast cancer resistance protein
BCSFB	Blood-CSF barrier
BD	Bipolar disorder
BDNF	Brain-derived neurotrophic factor
bFGF	Basic fibroblast growth factor (FGF2)
bHLH	Basic helix-loop-helix
BHV-1	Bovine herpesvirus-1
BLBP	Brain lipid binding protein
BLIMP-1	B lymphocyte-induced maturation protein-1
BLV	Bovine leukemia virus
BM	Bone marrow

BMDM	Bone marrow-derived macrophages
BMP	Bone morphogenetic protein
BMVEC	Brain microvascular endothelial cell
BP	Biological processes
BrdU	Bromodeoxyuridine
BRMs	Biological response modifiers
BSE	Bovine spongiform encephalopathy
BTLA	B and T lymphocyte attenuator
C/EBP	CCAAT box/enhancer binding protein
C3d	Complement C3 fragment d
C4d	Complement C4 fragment d
CA	Cornu ammonis
CA II	Carbonic anhydrase II
CaMKII	Calcium/calmodulin-dependent protein kinase II
CAMs	Cell adhesion molecules
CAPS	Cryopyrin-associated periodic syndromes
CAR	Cancer-associated retinopathy
CARD	Caspase recruitment domain
CARD15	NOD2/caspase recruitment domain 15
CB	Cannabinoid
CB1	Cannabinoid receptor 1
CB2	Cannabinoid receptor 2
CBA	Cytokine bead arrays
CBF	Cerebral blood flow
CBF1	C promoter binding factor 1
CBP	cAMP-response element binding protein (CREB)-binding protein
CBV	Cerebral blood volume
CCI	Controlled cortical impact
CCK	Cholecystokinin
CCT	Central corneal thickness
CD	Cluster of differentiation
CD11b	Complement component 3 receptor 3 subunit (integrin alpha M; ITGAM)
CD40L	CD40 ligand (TNFSF5)
CDP	Common DC progenitor
CDR	Complementarity-determining region
CDV	Canine distemper virus
CEP	Carboxyethylpyrrole
CFT	2-β-Carbomethoxy-3β-(4-fluorophenyl) tropane
cGMP	Cyclic guanosine 5'-monophosphate
CGRP	Calcitonin gene-related peptide
CHAT	Choline acetyltransferase
CHN	Congenital hypomyelinating neuropathy
Cho	Choline
CIDP	Chronic inflammatory demyelinating polyradiculoneuropathy
CINC1	Cytokine-induced neutrophil chemoattractant-1
CIS	Clinically isolated syndrome
CJD	Creutzfeldt-Jakob disease
CLA	Cutaneous leukocyte antigen
CLL	Chronic lymphocytic leukemia
CLP	Common lymphoid progenitor
CMAP	Compound motor action potential
CMC	Critical micelle concentration

CMP	Common myeloid precursor
CMT	Charcot-Marie-Tooth (disease)
CMV	Cytomegalovirus
CNG	cGMP-gated
CNPase	Cyclic nucleotide 3' phospohydrolase
CNS	Central nervous system
CNTF	Ciliary neurotrophic factor
CO	Cytochrome oxidase
COMT	Catechol-*O*-methyltransferase
Con A	Concanavalin A
COP-1	Copolymer-1
COX	Cyclooxygenase (prostaglandin-endoperoxide synthase; PTGS)
CP	Choroid plexus
CR	Complement receptor
CRalBP	Cellular retinal binding protein (retinaldehyde binding protein 1; RLBP1)
CRD	Carbohydrate-recognition domains
CRE	cAMP-responsive element
Cre	Creatine
CREB	cAMP-response element binding protein
CRH	Corticotrophin-releasing hormone
CRID	Cytokine release inhibitory drugs
CRP	C-reactive protein
CRPM	Collapsing response mediator protein
CRVO	Central retinal vein occlusion
CSF	Cerebrospinal fluid
CSF-1R	Colony-stimulating factor receptor
CSPG	Chondroitin sulfate proteoglycans
CT	Computed tomography
CTA	Computed tomographic angiography
CTE	Chronic traumatic encephalopathy
cTEC	Cortical thymic epithelial cell (TEC)
CTL	Cytotoxic T lymphocyte
CTLA-4	Cytotoxic T lymphocyte antigen-4
CTP	Circulating T-cell progenitors
CVF	Cobra venom factor
CVO	Circumventricular organ
Cx	Connexin
cyto c	Cytochrome c
D1R	Type-1 family of dopamine receptors
D9-THC	D9-tetrahydrocannabinol
DA	Daniel's strain of Theiler's virus
DA	Dopamine
DAF	Decay-accelerating factor
DAG	Diacylglycerol
DAMP	Danger-associated molecular patterns
DAT	Dopamine transporter (solute carrier family 6A3; SLC6A3)
DβH	Dopamine-β-hydroxylase
DC	Dendritic cells
DCX	Doublecortin
ddC	Dideoxycytidine
ddI	Dideoxyinosine
DG	Dentate gyrus

dGTP	Deoxyguanosine triphosphate
DHA	Docosahexaenoic acid
Dhh	Desert hedgehog
DHP	Dihydropyridine
DIGE	Difference gel electrophoresis
DIRA	Deficiency of IL-1 receptor antagonist
DISC	Death-inducing signaling complex
DM	Diabetes mellitus
DN	Double/dominant negative
DNA	Deoxyribonucleic acids
Doc2	Double C2 protein
DOR	Delta-opioid receptor
Dox	Doxorubicin
Dox	Doxycycline
DR	Dopamine receptors
DRPLA	Dentatorubral-pallidoluysian atrophy
DSI	Depolarization induced suppression of inhibition
DSPN	Distal sensory peripheral neuropathy
DSS	Dejerine-Sottas syndrome
DTH	Delayed-type hypersensitivity
DTI	Diffusion tensor imaging
DTR	Diphtheria toxin receptor
DWI	Diffusion weighted imaging
E2F	Early-region-2 transcription factor
EAE	Experimental allergic/autoimmune encephalomyelitis
EAN	Experimental allergic/autoimmune neuritis
EAU	Experimental autoimmune uveitis
EBV	Epstein-Barr virus
ECT	Electroconvulsive therapy
EDSS	Expanded disability status scale
EEG	Electroencephalography
EGC	Embryonic germ cell
EGF	Epidermal growth factor
EGFR	Epidermal growth factor receptor
EIAV	Equine infectious anemia virus
ELAVL4	Embryonic lethal, abnormal vision, Drosophila-like 4
ELP	Early lymphoid progenitors
eLTP	Early long-term potentiation
ELVIS	Extravasation through Leaky Vasculature and subsequent Inflammatory cell-mediated Sequestration
EMG	Electromyography
EMP	Erythromyeloid progenitors
EndoG	Endonuclease G
EOAD	Early-onset Alzheimer's disease
EPI	Echo-planar imaging
EPR	Enhanced permeability and retention
EPS	Extrapyramidal symptoms
EPSC	Excitatory postsynaptic currents
EPSP	Excitatory postsynaptic potential
ER	Endoplasmic reticulum
ERK2	Extracellular signal-regulated kinase 2
ESC	Embryonic stem cell
ESCRT	Endosomal-sorting complex required for transport

ESI-MS/MS	Electrospray ionization-mass spectrometry/mass spectrometry
ET	Endothelin
ETP	Early T lineage progenitor
EV	Extracellular vesicles
FA	Fractional anisotropy
FAD	Familial Alzheimer's disease
fALS	Familial amyotrophic lateral sclerosis
FasL	Fas ligand (FASLG)
FcgR-1	Fc receptor IgG, high affinity-1
fCJD	Familial Creutzfeldt-Jakob disease
FcR	Fc receptor
FDA	Food and Drug Administration
FDC	Follicular dendritic cell
FDG	Fluorodeoxyglucose
FDOPA	6-[(18)F]fluoro-L-dopa
FFI	Fatal familial insomnia
FGF	Fibroblast growth factor
FID	Free induction decay
FIRE	Febrile infection-related epilepsy syndrome
FIV	Feline immunodeficiency virus
FKN	Fractalkine (CX3CL1)
fMRI	Functional magnetic resonance imaging
Foxp3	Forkhead box P3 transcription factor
FPI	Fluid percussion injury
FR	Folate receptor
FRC	Fibroblastic reticular cell
FRS2	Fibroblast growth factor receptor substrate 2
FS	Fisher syndrome
FSH	Follicle-stimulating hormone
FTD	Frontotemporal dementia
FT-ICR	Fourier transformed ion cyclotron resonance mass spectrometry
FTLD	Frontotemporal lobar dementia
FUS	Fused in sarcoma
Fz/PCP	Frizzled/planar cell polarity
GA	Glatiramer acetate
GABA	Gamma-aminobutyric acid
GAD	Glutamate decarboxylase
GalC	Galactocerebroside
GALT	Gut-associated lymphoid tissue
GAP	GTPase activating protein
GAPDH	Glyceraldehyde 3 phosphate dehydrogenase
GBM	Glioblastoma multiforme
GBS	Guillain-Barré syndrome
GC	Germinal center
GCDC	Germinal center dendritic cell
GCL	Ganglion cell layer
GEF	Guanine nucleotide exchange factor
GL	Granular cell layer
GC-MS	Gas chromatography combined with mass spectrometry
G-CSF	Granulocyte colony-stimulating factor
Gd	Gadolinium
GDF	Growth and differentiation factor
GDNF	Glial-derived neurotrophic factor

GDP	Guanosine diphosphate
GEF	GDP-GTP exchange factor
GFAP	Glial fibrillary acidic protein
GFP	Green fluorescent protein
GI	Gastrointestinal
GKAP	Guanylate kinase-associated protein
GLC1A	Chromosome 1 open-angle glaucoma gene
Gln	Glutamine
GLP-1	Glucagon-like peptide-1
GM-CSF	Granulocyte-macrophage colony-stimulating factor
GMP	Granulocyte monocyte precursor
GnRH	Gonadotropin-releasing hormone
GO	Gene ontology
GO	Graves' ophthalmopathy
GPCR	G-protein-coupled receptor
GPI	Glycosylphosphatidylinositol
GR	Glucocorticoid receptor
GRIP	Glucocorticoid receptor-interacting protein
GRO-α	Growth-related oncogene alpha
GSH	Glutathione
GSS	Gerstmann-Straussler-Scheinker disease
GSTO1	Glutathione *s*-transferase omega-1
GT	G-protein transducing
GTP	Guanosine triphosphate
GUCY	Guanylate cyclase
GWAS	Meta-genome-wide association study
HAART	Highly active antiretroviral therapy
HAD	HIV-associated dementia
HAM/TSP	HTLV-I associated myelopathy/tropical spastic paraparesis
HAND	HIV-associated neurocognitive disorder
HAT	Histone acetyltransferase
HAV	Hepatitis A virus
HBsAg	Hepatitis B surface antigen
HCV	Hepatitis C virus
HCMV	Human cytomegalovirus
HD	Huntington's disease
HDAC	Histone deacetylase
HDLS	Hereditary diffuse leukoencephalopathy with spheroids
HDR	Homology-dependent repair
HES	Hairy and enhancer of split homolog
HEV	High endothelial venule
HFS	High frequency stimulation
Hh	Hedgehog
HHH	Hypervolemic-hemodilution and hypertensive
HHV-6	Human herpes virus-6
HIV-1	Human immunodeficiency virus type 1
HIVE	Human immunodeficiency virus encephalitis
HLA	Human leukocyte antigen
HLH	Helix-loop-helix
HMG-CoA	3-Hydroxy-3-methylglutaryl coenzyme A
HNE	4-Hydroxy-2-nonenal
hnRNP-A1	Heterogeneous nuclear ribonuclear protein-A1
HPA	Hypothalamic-pituitary-adrenal

HPMA	*N*-(2-Hydroxypropyl) methacrylamide
HRP	Horseradish peroxidase
HSC	Hematopoietic stem cell
HSC70	Heat shock cognate protein 70
Hsp	Heat shock protein
HSV	Herpes simplex virus
HSVE	Herpes simplex virus-mediated encephalitis
HT	Huntington's disease
HTLV	Human T-cell lymphotropic virus type
HTRA2	High temperature requirement serine protease 2
Htt	Huntingtin
HUVEC	Human umbilical vein endothelial cells
HveA	Herpesvirus entry mediator A
I-1	Regulatory protein inhibitor-1
IBS	Inflammatory bowel syndrome
ICAM	Intracellular adhesion molecule
ICAT	Isotope-coded affinity tags
ICE	IL-1β-converting enzyme
ICGA	Indocyanine green angiography
ICH	Intracerebral hemorrhage
iCJD	Iatrogenic Creutzfeldt-Jakob disease
ICOS	Inducible co-stimulatory molecule
ICV	Intra-cerebro-ventricular
Id	Inhibitor of differentiation
IDE	Insulin degrading enzyme
IDO	Indoleamine 2,3-dioxygenase
IE	Immediate early
IF	Intermediate filament
IFN	Interferon
Ig	Immunoglobulin
IGF	Insulin-like growth factor
IgG	Immunoglobulin G
IGIV	Immunoglobulin intravenous therapy
Ihh	Indian hedgehog
IIDD	Idiopathic inflammatory demyelinating disease
IkB	Inhibitory kappa B
IKK	IkB kinase
IL	Interleukin
IL1RA	IL-1 receptor antagonist
ILBD	Incidental Lewy body disease
ILK	Integrin-linked kinase
ILM	Inner limiting membrane
ILV	Intraluminal vesicles
IM	Intramuscular
IMAC	Immobilized metal affinity chromatography
IMPDH	Inosine monophosphate dehydrogenase
iNKR	Inhibitory natural killer cell receptor
INL	Inner nuclear layer
INO	Internuclear ophthalmoplegia
iNOS	Inducible nitric oxide synthase (NOS2A)
IOP	Intraocular pressure
IP	Interferon-inducible protein
IPC	Intermediate progenitor cells

InsP3R	Inositol 1,4,5-triphosphate receptor
IPL	Inner plexiform layer
IRAK	IL-1 receptor-associated protein kinase
IRBP	Interphotoreceptor retinoid-binding protein
IRF	Interferon regulatory factor
ISCOM	Immunostimulating complexes
ISH	In situ hybridization
IT15	Interesting transcript 15
ITAM	Immunoreceptor tyrosine-based activation motif
ITGAM	Integrin, alpha M
ITR	Inverted terminal repeat
IVDU	Intravenous drug use
IVIg	Intravenous immunoglobulin
JAK	Janus kinase
JEV	Japanese encephalitis virus
JNK	c-Jun N-terminal kinase
KLH	Keyhole limpet hemocyanin
KO	Knockout
KOR	Kappa opioid peptide receptor
KYN	Kynurenine
KYNA	Kynurenic acid
LAK	Lymphokine-activated killer (cell)
L-AP4	L -2-Amino-4-phosphonobutyric acid
LAT	Latency-associated transcript
LAMP	Lysosome-associated membrane protein
LB	Lewy bodies
LC	Locus coeruleus
LCA	Leukocyte common antigen
LC-FTICR MS	Liquid chromatography Fourier transform ion cyclotron resonance mass
LC-MS	Liquid chromatography combined with mass spectrometry
LC-UV-SPE-NMR	Liquid chromatography, UV detection, solid phase extraction, and nuclear magnetic resonance
LD	Linkage disequilibrium
LDL	Low density lipoprotein
LFA-1	Leukocyte function-associated antigen-1 (integrin beta 2; ITGB2)
LGN	Lateral geniculate nucleus
LH	Luteinizing hormone
LIF	Leukemia inhibitory factor
Lingo-1	Leucine-rich repeat and Ig domain containing Nogo receptor-interacting protein-1
LMN	Lower motor neuron
LOAD	Late-onset Alzheimer's disease
LOS	Lipooligosaccharide
LPA	Lysophosphatidic acid
LPBN	Lateral parabrachial nucleus of the pons
LPS	Lipopolysaccharide
LRP-1	Lipoprotein receptor-related protein-1
LRR	Leucine-rich repeat
LRRK2	Leucine-rich repeat kinase 2
LT	Lymphotoxins
LTD	Long-term depression
LTNP	Long-term nonprogressors
LTP	Long-term potentiation

LTR	Long-terminal repeat
LT-bR	Lymphotoxin beta receptor
MCAO	Middle cerebral arterial occlusion
MCI	Mild cognitive impairment
MG	Myasthenia gravis
MHV	Mouse hepatitis virus
mI	Myoinositol
MIF	Migration inhibitory factor
Mint1	Munc-18 interacting protein 1
MIP	Macrophage inflammatory protein
MJO	Machado-Joseph disease
MME	Membrane metalloendopeptidase (neprilysin)
MMP	Matrix metalloproteinase
MMSE	Mini-Mental State Examination
MNGC	Multinucleated giant cell
MnPO	Median preoptic nucleus
Mn-SOD	Manganese superoxide dismutase
MOAT	Multispecific organic anion transporter
MOBP	Myelin-associated/oligodendrocyte basic protein
MOG	Myelin oligodendrocyte glycoprotein
MOI	Multiplicity of infection
MOR	mu opioid receptor
MOSP	Myelin/oligodendrocyte-specific protein
MP	Mononuclear phagocytes
MPA	Mycophenolic acid
MPL	Monophosphoryl lipid A
MPO	Myeloperoxidase
MPO	Medial preoptic area
MAC	Membrane attack complex
MPP	Multipotent progenitors
MPP+	1-Methyl-4-phenylpyridinium
MAdCAM-1	Mucosal addressin cell adhesion molecule-1
MAG	Myelin-associated glycoprotein
MAML	Mammalian mastermind-like
MAO	Monoamino-oxidase
MAP	Microtubule-associated protein
MAPK	Mitogen-activated protein kinases
Mash1	Mammalian achaete-scute homologue 1
MBGI	Myelin-based growth inhibitor
MBP	Myelin basic protein
MC	Mast cell
MC-1R	Melanocortin-1 receptor
MCMD	Minor cognitive motor disorder
MCP	Membrane cofactor protein (CD46)
MCP-1	Monocyte chemoattractant protein-1 (CCL2)
M-CSF	Macrophage colony-stimulating factor (CSF1)
MD	Major depression
MDA	Malondialdehyde
MDD	Major depressive disorder
MDM	Monocyte-derived macrophages
MDP	Muramyl-dipeptide
MDP	Monocytes and dendritic cell progenitor
MDR	Multidrug resistant

MDSC	Myeloid-derived suppressor cells
ME	Median eminence
MEG	Magnetoencephalography
MEPP	Miniature end-plate potential
MFS	Miller Fisher syndrome
MHC	Major histocompatibility complex
MHC-II	Class II major histocompatibility complex
MHPG	Methoxy-hydroxy-phenylethanolamine
MI	Myocardial infarction
MLR	Mixed lymphocyte reaction
MLV	Murine leukemia virus
MMEJ	Microhomology-mediated end-joining
MND	Mild neurocognitive disorder
MP	Mononuclear phagocyte
MPMV	Mason-Pfizer monkey virus
MPTP	1-Methyl-4-phenyl-1,2,3,6-tetrahydropyridine
MR	Mineralocorticoid receptors
MRA	Magnetic resonance angiography
MRI	Magnetic resonance imaging
MRP	Multidrug resistance protein
MRS	Magnetic resonance spectroscopy
MRSI	Magnetic resonance spectroscopic imaging
MS	Multiple sclerosis
MSA	Multisystem atrophy
MSC	Mesenchymal stem cell
MSCs	Myelinating Schwann cells
MSH	Melanocyte-stimulating hormone
MSN	Medium spiny neuron
mSOD1	Mutant Cu^{2+} /Zn^{2+} superoxide dismutase 1
MSRV	Multiple sclerosis retrovirus
MT	Magnetization transfer
Mtb	Mycobacterium tuberculosis
MTI	Magnetization transfer imaging
mTEC	Medullary thymic epithelial cell
mTOR	Mammalian target of rapamycin
MTR	Magnetization transfer ratio
MUC1	Mucin type 1 glycoprotein
MuLV	Murine leukemia virus
Munc-18	Mammalian homologue of unc-18
MV	Microvesicles
MVB	Multivesicular bodies
MVE	Murray Valley encephalitis virus
MW	Molecular weight
MZ	Marginal zone
NAA	*N*-Acetyl-aspartate
NAC	*N*-Acetyl cysteine
NADPH	Nicotinamide adenine dinucleotide phosphate
NB	Nucleotide-binding domain
NCAM	Neural cell adhesion molecule
NCC	Neural crest cell
NE	Norepinephrine
NEP	Neutral endopeptidase metalloendopeptidase
NET	Norepinephrine transporter

NeuN	Neuronal nuclei
NF	Neurofilament
NF-kB	Nuclear factor-k-B
NFAT	Nuclear factor of activated T lymphocytes
NFL	Nerve fiber layer
NFT	Neurofibrillary tangles
Ng-CAM	Neuronal-glial cell adhesion molecule (L1/NILE)
NGF	Nerve growth factor
NgR	Nogo-66 receptor
NHEJ	Non-homologous end-joining
NICD	Notch intracellular domain
NK	Natural killer (cells)
NKT	Natural killer T (cells)
NLR	Nod-like receptors
NMDA	*N*-methyl-D-aspartate
NMDAR	*N*-methyl-D-aspartate receptors
NMJ	Neuromuscular junction
NMO	Neuromyelitis optica
NMR	Nuclear magnetic resonance
NMSCs	Nonmyelinating Schwann cells
nNOS	Neuronal nitric oxide synthase
NNRTIs	Nonnucleoside analogue reverse transcriptase inhibitors
NO	Nitric oxide
NOD	Nucleotide oligomerization domain
NOD	Non-obese diabetic mice
NOS	Nitric oxide synthase
NOT	Nucleus of the optic tract
NP	Nanoparticle
NPC	Neural progenitor cell
NPY	Neuropeptide Y
NPZ-8	Neuropsychological Z score for 8 tests
NR	Nuclear receptor
NRG	Neuregulin
NRL	Nuclear receptor ligand
NRTI	Nucleoside analogue reverse transcriptase inhibitors
NSAID	Nonsteroidal anti-inflammatory drug
NSC	Neural stem cell
NSE	Neuron specific enolase
NSF	*N*-ethylmaleimide sensitive factor
NT	3-Nitrotyrosine
NTF	Neurotrophin
NVU	Neurovascular unit
OB	Olfactory bulb
OCB	Oligoclonal band
OCD	Obsessive compulsive disorder
ODN	Oligonucleotides
OE	Olfactory epithelium
OHT	Ocular hypertension
OL	Oligodendrocyte
OLM	Outer limiting membrane
OMgp	Oligodendrocyte-myelin glycoprotein
OMP	Olfactory marker protein
ONH	Optic nerve head

ONL	Outer nuclear layer
OP	Oligodendrocyte progenitors
OPC	Oligodendrocyte progenitor cell
OPCA	Olivopontocerebellar atrophy
OPL	Outer plexiform layer
ORF	Open reading frame
ORN	Olfactory response neuron
OSP	Oligodendrocyte-specific protein
OVA	Ovalbumin
OVLT	Organum vasculosum of the lamina terminalis
P0	Myelin protein zero
p75NTR	p75 neurotrophin receptor (nerve growth factor receptor; NGFR)
PACAP	Pituitary adenylate cyclase-activating polypeptide
PACT	Protein activator of the interferon-induced protein kinase
PAF	Platelet-activating factor
PAG	Periaqueductal gray
PAMP	Pathogen-associated molecular pattern
PANDAS	Pediatric autoimmune neuropsychiatric disorders associated with *Streptococcus*
PANSS	Positive and negative syndrome scale
PARP	Poly(ADP-ribose) polymerase
PASAT	Paced Auditory Serial Addition Test
PBBS	Peripheral benzodiazepine binding sites
PBL	Peripheral blood lymphocyte
PBMC	Peripheral blood mononuclear cell
PBR	Peripheral benzodiazepine receptor
PCP	Phencyclidine
PCR	Polymerase chain reaction
PD	Parkinson's disease
pDC	Plasmacytoid dendritic cells
PD1	Program death-1
PDE	Phosphodiesterase
PDGF	Platelet-derived growth factor
PDTC	Pyrrolidine dithiocarbamate
PE	Plasma exchange
Pe	Periventricular
PEG	Polyethylene glycol
PEI	Polyethyleneimine
PENK	Proenkephalin
PERG	Pattern electroretinogram
PET	Positron emission tomography
PFS	Periodic fever syndromes
PG	Prostaglandin
Pgp	P-glycoprotein
PHF	Paired helical filament
PI	Phosphatidylinositol
PI3K	Phosphatidylinositol-3-kinase
PICA	Posterior inferior cerebellar artery
PICK1	Protein interacting with C kinase 1
PKA	cAMP-dependent protein kinase
PKG	Protein kinase G
PLGA	Poly(D, L-lactide-co-glycolide)

PLOSL	Polycystic lipomembranous osteodysplasia with sclerosing leukoencephalopathy
PLP	Proteolipid protein
PMCA	Plasma membrane bound Ca^{2+}-ATPase
PMD	Pelizaeus-Merzbacher disease
PML	Progressive multifocal leukoencephalopathy
PMN	Polymorphonuclear (leukocyte)
PMP22	Peripheral myelin protein 22
PNS	Peripheral nervous system
POAG	Primary open-angle glaucoma
polyQ	Polyglutamine
POMC	Pro-opiomelanocortin
POU3F2	POU class 3 homeobox 2
PP	Protein phosphatase
PPAR	Peroxisome proliferator activated receptor
PPF	Paired-pulse facilitation
PPG	Poly(propylene glycol)
PP-MS	Primary progressive multiple sclerosis
PR	Photoreceptor
PrPc	Cellular prion protein
PrPres	Protease resistance prion
PrPsc	Disease-associated prion protein
PRR	Pattern recognition receptor
PS	Presenilin (PSEN)
PSA-NCAM	Poly-sialylated form of the neural cell adhesion molecule
PSCs	Perisynaptic Schwann cells
PSD	Postsynaptic density
PSP	Progressive supranuclear palsy
PSW	Periodic sharp wave
Ptc	Patched, a hedgehog receptor
PTM	Post-translational modifications
PTP	Post-tetanic potentiation
PTSD	Post-traumatic stress disorder
PTZ	Pentylenetetrazol
PVL	Periventricular leukomalacia
PVM	Perivascular macrophages
PVN	Paraventricular nucleus
PYD	Pyrin domain
PYY	Peptide YY
RA	Rheumatoid arthritis
Rag	Recombination-activating gene
RAGE	Receptor for advanced glycation end product
RANTES	Regulated upon activation normal T-cell expressed and secreted (CCL5)
RAS	Renin-angiotensin system
Rb	Retinoblastoma
RBP	RNA-binding proteins
REM	Rapid eye movement
RER	Rough endoplasmic reticulum
RF	Radiofrequency
RFLPs	Restriction fragment length polymorphisms
RGC	Retinal ganglion cell
RISC	RNA-induced silencing complex
RIG1	Retinoic acid-inducible gene 1

RIM	Rab3-interacting molecule
RIP	Receptor-interacting protein
RMS	Rostral migratory stream
RNAi	RNA interference
RNI	Reactive nitrogen intermediates
RNP	Ribonucleoprotein
RNS	Reactive nitrogen species
ROCK	Rho kinase
ROHHAD	Rapid-onset obesity, hypoventilation, hypothalamic dysfunction, and autonomic dysregulation
ROI	Reactive oxygen intermediate
RORγ	Retinoic-acid-receptor-related orphan receptor-γ
ROS	Reactive oxygen species
RP	Relapsing polychondritis
RPE	Retinal pigment epithelial (cells)
RRM	RNA recognition motifs
RRMS	Relapsing and remitting multiple sclerosis
RSV	Rous sarcoma virus
RTK	Receptor tyrosine kinase
rt-PA	Recombinant tissue plasminogen activator
RT-PCR	Reverse transcription polymerase chain reaction
RyR	Ryanodine receptor
sALS	Sporadic amyotrophic lateral sclerosis
SAP	Synapse-associated protein
SAPAP	SAP-associated protein (discs, large homolog-associated protein-1; DLGAP1)
SAPK	Stress-activated protein kinase (JNK, MAPK8)
sAPP	Secreted β-amyloid precursor protein
SBMA	Spinobulbar muscular atrophy
SCs	Schwann cells
SC	Superior colliculus
SCA-3	Spinocerebellar ataxia-3
scFv	Single-chain Fv antibodies
SCI	Spinal cord injury
SCID	Severe combined immunodeficiency
sCJD	Sporadic Creutzfeldt-Jakob disease
SCPs	Schwann cell precursor
sCrry	Soluble complement receptor-related protein y
SDF-1	Stromal cell-derived factor 1 (CXCL12)
SEC	Sinus endothelial cell
SELDI-TOF	Surface enhanced laser desorption ionization time-of-flight
SER	Smooth endoplasmic reticulum
SERCA	Sarco(endo)plasmic reticulum Ca^{2+}-ATPase
SERT	Serotonin transporter
sFI	Sporadic fatal insomnia
SFO	Subfornical organ
SG	Stress granule
SGLPG	Sulfated glucuronyl lactosaminyl paragloboside
SGZ	Subgranular zone
Shh	Sonic hedgehog
sIg	Surface immunoglobulin
sIL-2R	Soluble IL-2 receptor
SITA	Swedish interactive thresholding algorithm

SIV	Simian immunodeficiency virus
SIVE	Simian immunodeficiency virus encephalitis
SLE	Systemic lupus erythematosus
SMA	Spinal muscular atrophy
SMAC	Second mitochondrial-derived activator of caspase
SMase	Sphingomyelinase
SMN	Survival motor neuron gene
SN	Substantia nigra
SNAP	Sensory nerve action potential
SNAP-25	Synaptosome-associated protein of 25,000 daltons
SNARE	NSF attachment receptor
SNCA	Alpha-synuclein
SNpc	Substantia nigra pars compacta
SNPs	Single-nucleotide polymorphisms
SNS	Sympathetic nervous system
SOCS	Suppressors of cytokine signaling
SOD1	Superoxide dismutase 1
SON	Supraoptic nuclei
SP1	Specificity protein 1
SPARC	Secreted protein acidic and rich in cysteine
SPECT	Single photon emission computed tomography
SPG-II	Spastic paraplegia type II
SR-A	Scavenger receptor type A
sRAGE	Soluble Receptor for AGEs
SRBCs	Sheep red blood cells
SREBP	Sterol regulatory element-binding protein
SRF	Serum response factor
SSRI	Selective serotonin reuptake inhibitors
STAT	Signal transducers and activators of transcription
STP	Short-term potentiation
SV5	Simian virus 5
SVZ	Subventricular zone
SWAP	Short wavelength automated perimetry
SWATH	Sequential window acquisition of all theoretical mass spectra
SWI	Susceptibility weighted imaging
SYN	α-Synuclein
SZ	Schizophrenia
T	Tesla
TALENs	Transcription Activator-Like Effector Nucleases
Tat	Trans-activator of transcription
TBE	Tick-borne encephalitis virus
TBI	Traumatic brain injury
TBP	TATA-binding protein
TCA	Tricarboxylic acid
TCR	T-cell receptor
TCV	T-cell vaccination
TDO	Tryptophan 2,3-dioxygenase
TE	Echo time
TEC	Thymic epithelial cell
TES	Traumatic encephalopathy syndrome
Teff	T effector cells
TF	Transcription factor
TG	Trigeminal ganglia
TGF	Transforming growth factor

TH	Thyroid hormone
TH	Tyrosine hydroxylase
Th1	T helper type 1 cell
Th2	T helper type 2 cell
TIA	Transient ischemic attack
TIR	Toll/IL-1 receptor
TJ	Tight junction
TLR	Toll-like receptor
TMEV	Theiler's mouse encephalomyelitis virus
TMT	Trimethyltin
TMZ	Temozolomide
TN	Terminal nuclei
TNF	Tumor necrosis factor
TNFR	Tumor necrosis factor receptor
TOR1	Target of rapamycin 1
TRAF	TNF-receptor mediated factor
TRAIL	TNF-related apoptosis-inducing ligand (TNFSF10)
TRANCE	TNF-related activation-induced cytokine (TNFSF11)
TRANCER	TRANCE receptor (TNFRSF11A)
TRE	Tax responsive element
Treg	T regulatory cells
TREM	Triggering receptor expressed on myeloid cells
Trk	Receptor tyrosine kinase
TRP	Transient receptor potential
TRPC	Transient receptor potential canonical channels
TSA	Tissue-specific antigen
TSE	Transmissible spongiform encephalopathies
TSP	Thrombospondins
TULP-1	Tubby-lie protein 1
Tyk2	Protein tyrosine kinase 2
UACA	Uveal autoantigen with coiled domains and ankyrin repeats
UMN	Upper motor neuron
V1	Primary visual cortex
VAMP	Vesicle-associated membrane protein
VCAM	Vascular cell adhesion molecule
VEGF	Vascular endothelial growth factor
VEP	Visual evoked potential
VIP	Vasoactive intestinal peptide
VMAT-2	Vesicular monoamine transporter-2 (solute carrier family 18; SLC18A2)
VZV	Varicella-zoster virus
WDM	Welander distal myopathy
WKAH	Wistar-King-Aptekman-Hokudai
WNV	West Nile virus
X-ALD	X-adrenoleukodystrophy
XIAP	X-linked inhibitor of apoptosis protein
ZFNs	Zinc finger nucleases
ZO-1	Zonula occludens-1 (TJP1)

References

HUGO Gene Nomenclature Committee: http://www.genenames.org/index.html
"Abbreviations and Symbols for Chemical Names": www.blackwell-symergy.com/doi/pdf/10.1111/j.1432.1033.1967.tb00070.x

Part I

Immunology of the Nervous System

Defining Neuroinflammation

1

Howard E. Gendelman

Abstract

In steady-state conditions, neuroinflammation is defined solely as a central nervous system (CNS) innate immune response that serves to control infection through microbial elimination. This, together with scavenging cell debris and misfolded proteins, controls disease. To such ends, neuroinflammatory processes play a significant role in CNS homeostasis in health and disease control. The immune system affects the development, maintenance, and sustenance of brain cells and their connections. Linked to aging, epidemiological, animal, human, and therapeutic studies support the roles of neuroinflammatory cascades in sustaining neural function. This is highlighted by the neurotoxic microglial potential. In a steady state, microglia serve to protect the nervous system by acting as debris scavengers, killers of microbial pathogens, and regulators of innate and adaptive immune responses. In neurodegenerative diseases, activated microglia affect neuronal injury and death through the production of glutamate, pro-inflammatory factors, reactive oxygen species, and quinolinic acid, amongst others, and by mobilization of adaptive immune responses and cell chemotaxis, leading to transendothelial migration of immunocytes across the blood–brain barrier and perpetuation of neural damage. While the CNS is considered immunologically unique, glial–neuronal interactions occur under homeostatic and pathological conditions, with the mechanisms of immune privilege now reevaluated. One reevaluation changed the concept of immune privilege with the now-defined lymphatic drainage of CNS antigens. Notably, as the disease progresses, inflammatory secretions engage neighboring glial cells, including astrocytes and endothelial cells, resulting in a vicious cycle of autocrine and paracrine amplification of inflammation, perpetuating tissue injury. Such pathogenic processes contribute to neurodegeneration. The research seeks to harness such inflammatory processes to develop therapeutic interventions that positively affect the tempo and progression of human disease.

Keywords

Alzheimer's disease · Microglia · Neurodegenerative disorders · Neuroinflammatory processes · Parkinson's disease

H. E. Gendelman (✉)
Department of Pharmacology and Experimental Neuroscience, University of Nebraska Medical Center, Omaha, NE, USA
e-mail: hegendel@unmc.edu

Neuroinflammatory processes play a significant role in neural health and disease. These regulate the development, maintenance, and sustenance of brain cells and their connections. Linked to aging, epidemiological, animal, human, and therapeutic studies all support the presence of a neuroinflammatory cascade in disease. This is highlighted by the neurotoxic potential of microglia. In a steady state, microglia serve to protect the nervous system by acting as debris scavengers, killers of microbial pathogens, and regulators of innate and adaptive immune responses. In neurodegenerative diseases, activated microglia affect neuronal injury and death through the production of glutamate, pro-inflammatory factors, reactive oxygen species, and quinolinic acid, amongst others, and by mobilization of adaptive immune responses and cell chemotaxis, leading to transendothelial migration of immunocytes across the blood–brain barrier and perpetuation of neural damage. As the disease progresses, inflammatory secretions engage neighboring glial cells, including astrocytes and endothelial cells, resulting in a vicious cycle of autocrine and paracrine amplification of inflammation, perpetuating tissue injury. Such pathogenic processes contribute

H. E. Gendelman, T. Ikezu (eds.), *Neuroimmune Pharmacology and Therapeutics*,
https://doi.org/10.1007/978-3-031-68237-7_1

to neurodegeneration. Active research seeks to harness such inflammatory processes to develop therapeutic interventions that positively affect the tempo and progression of human disease.

As the life expectancy of the human population continues to increase, the possibility of developing neuroinflammatory and neurodegenerative diseases has increased considerably in the past 50 years. Of the neurodegenerative disorders, Alzheimer's disease has continued to be the leading cause of dementia in the aging population. Traditionally, neurodegenerative disorders have been defined as conditions where there is selective loss of neurons within specific regions of the brain accompanied by astrogliosis. However, in the past 20 years, we have learned that the pathological process leading to the dysfunction of selected circuitries in the brain initiates with damage to the synapses rather than the loss of neurons. Neuronal loss is a late event that is probably preceded by damage to axons and dendrites, followed by shrinkage of the neuronal cell body and abnormal accumulation of filamentous proteins.

Therefore, the revised concept of neurodegeneration suggests that neuronal injury initiates at the synaptic junction and propagates throughout selected circuitries, leading to neuronal dysfunction, which resolves in the classical clinical symptoms characteristic of each of the neurodegenerative disorders (Paglione et al., 2024). So, for example, in Alzheimer's disease, early damage to the synapses between the entorhinal cortex and the molecular layer of the dentate gyrus (perforant pathway) resolves in the short-term memory deficits characteristic of this dementing disorder. Later on, disconnection of the cortical-cortical fibers in the frontal, parietal, and temporal cortex resolves more severe memory deficits and alterations in executive functions and abstraction. The degeneration of connections between the nucleus basalis of Meynert and the neocortex resolves attention and memory deficits usually associated with the loss of cholinergic neurons. Other circuitries and neuronal populations are also affected in Alzheimer's disease, illustrating the complexity of these disorders and the fact that the concept of a single population being affected is limited. That is the case with several other disorders, including Parkinson's disease, where degeneration is not only limited to the dopaminergic system but also involves the limbic system, the raphe nucleus, the insula, and other systems.

In response to injury, neurons produce adhesion molecules and trophic factors that recruit astroglial and microglial cells to participate in the process of repair of the damage. In addition, the microvasculature and other glial systems might also participate in the process. Thus, neurodegeneration is accompanied by astrogliosis, microgliosis, and microvascular remodeling. While astroglial cells initially produce trophic factors and cytokines that aid in tissue repair, eventually these factors could amplify the inflammatory response, increase vascular permeability, and result in microglial activation, which, in turn, might lead to the production of more pro-inflammatory cytokines and chemokines. A critical balance between the repair and pro-inflammatory factors often determines the future rate and progression of the degenerative process.

The understanding of the mechanisms of neurodegeneration and neuroinflammation response in these neurological conditions has seen tremendous progress in the past 10 years. It is now recognized that small soluble misfolded protein aggregates denominated oligomers are responsible for injury. So, for example, in Alzheimer's disease, Aβ protein oligomers might damage the synapses in neocortical regions and the limbic system (Zeng et al., 2023). In Parkinson's disease, α-synuclein oligomers may damage the axons in the striatum, brain stem, and cortical areas. While significant progress has been made in understanding the fundamental mechanisms causing neuronal injury, less is known about the reasons for the selective neuronal vulnerability characteristic of these neurological conditions and the role of the innate immune system in this process. Considerable recent efforts have focused on immune transformation to restore immunological balance to clear misfolded aggregated proteins in disease and, at the same time, attenuate neuroinflammation. Such activities serve to support reparative and neurotrophic therapies that are developed in the chapters of this book. The overarching idea is to control innate neurotoxic activities and facilitate the clearance of protein aggregates. Maintaining the gut–immune–brain axis is an additional benefit in the control of bacterial products to limit neuroinflammation. The overarching goal in the control of chronic inflammation is to restore immunity and slow or prevent neurodegeneration. Each of these strategies can ensure the best disease outcomes (Saleh et al., 2022).

References

Paglione, M., Restivo, L., Zakhia, S., Llobet Rosell, A., Terenzio, M., & Neukomm, L. J. (2024). Local translatome sustains synaptic function in impaired Wallerian degeneration. *EMBO Reports*. https://doi.org/10.1038/s44319-024-00301-8

Saleh, M., Markovic, M., Olson, K. E., Gendelman, H. E., & Mosley, R. L. (2022). Therapeutic strategies for immune transformation in Parkinson's Disease. *Journal of Parkinsons Disease, 12*(s1), S201–S222.

Zeng, S. Y., Liu, Y. F., Liu, J. H., Zeng, Z. L., Xie, H., & Liu, J. H. (2023). *Aging Dis. 14*(6):2015–2027.

Defining Neuroimmune Pharmacology and Therapeutics

2

Howard E. Gendelman and Eric J. Benner

Abstract

Immunology, neuroscience, and pharmacology are each independent, broad, multidisciplinary fields well-developed for the biomedical sciences. The evolution of each at the molecular, biochemical, and cellular levels has been, by any measure, enormous in impact by their collective abilities to change the world as we know it for biomedical research. This book was fashioned to share with students and trained professionals a vision and integration of the three disciplines and to reach forward into their future.

Keywords

Adaptive immunity · Alzheimer's disease · Astrogliosis · HIV-associated neurocognitive disorders · Microglia · Neuroimmune pharmacology · Neuropharmacology · Parkinson's disease

2.1 Introduction

Notably, like other disciplines of science, the fields of immunology, pharmacology, and neuroscience intersect. How this has come about resides in history. That said, it did not begin in modern times but evolved over a century ago when Dr. Elie Metchnikoff discovered that an inflammatory cell type in starfish larvae could engulf foreign materials and, as a consequence, protect its host, thus birthing the field of immunology. It is in this field that some of the most significant medical advances have been realized. Indeed, immunology has become the most active and arguably prolific field of medicine. At its epicenter is the mononuclear phagocyte (MP) system (comprising monocytes, tissue macrophages, dendritic cells, and microglia) with a nonspecific ability to protect itself against pathogenic microbial and other injurious attacks. While the 1908 Nobel Prize in Medicine or Physiology was awarded to Dr. Metchnikoff, even he was uncertain of the exceedingly broad implications of his discovery in appreciating how these cells could affect the well-being of the human race. This realization came only decades later through many new developments in immune-based diagnostics and therapeutics. For example, MPs were subsequently found to induce intracellular killing of pathogenic microbes, process antigens for presentation, secrete biologically active factors, and, under certain circumstances, become instigators of pathological processes. It was considerably later when scientists discovered that the phagocytic functions of MPs were central to normal tissue homeostasis (Gardai et al., 2005; Fadok et al., 1998). So, how is MP function linked to disease? The obvious answer is through inflammation. On the one hand, inflammation is essential for life as it serves as the frontline defense against infection and tumor surveillance. Inflammation, for simplicity's sake, enriches the host's abilities to fend off disease-causing bacteria, viruses, and parasites. It comprises the first-line of immune defense to eliminate invaders and plays a critical role in the repair of damaged tissue. On the other hand, when these highly regulated processes go awry, inflammation evolves into pathological processes that are disadvantageous to the host as seen in cancers, infections, cardiovascular diseases, and autoimmune and neurodegenerative disorders.

Neuroscience is broadly defined as the investigation of networks and models of neural communication in health and disease amongst hundreds of billions of neurons and supporting glia (microglia, oligodendrocytes, and astrocytes). The intersection between the immune system and the nervous system is now considered significant. Indeed, within

H. E. Gendelman (✉)
Department of Pharmacology and Experimental Neuroscience, University of Nebraska Medical Center, Omaha, NE, USA
e-mail: hegendel@unmc.edu

E. J. Benner
Department of Pediatrics, Division of Neonatology, Duke University School of Medicine, Durham, NC, USA

H. E. Gendelman, T. Ikezu (eds.), *Neuroimmune Pharmacology and Therapeutics*,
https://doi.org/10.1007/978-3-031-68237-7_2

the central nervous system (CNS), immune activities abound. During disease, neuroinflammation perpetrated through the brain's MPs has been linked to widespread neurodegeneration in adult diseases that include Parkinson's, Alzheimer's, and Huntington's diseases (PD, AD, HD, respectively), HIV-1-associated neurocognitive disorder (HAND), stroke, multiple sclerosis (MS), and spongiform encephalopathies (prion-mediated neurodegeneration). We now understand that CNS injuries ranging from acute traumatic brain injury to chronic neurodegenerative disease lead to the recruitment and activation of the adaptive immune system (Benner et al., 2008; Appel et al., 2010). The polarization of this adaptive immune response appears to have a significant impact on local CNS MP or microglial secretory functions with the capacity for both deleterious and beneficial outcomes. A range of microglial-derived products have been linked to neuronal injury, including quinolinic acid, superoxide anions, matrix metalloproteinases (MMPs), nitric oxide (NO), arachidonic acid and its metabolites, chemokines, proinflammatory cytokines, and excitotoxins. Activated microglia commonly speed up disease progression, while anti-inflammatory therapies appear to either slow or halt it.

The ontogeny of the brain's immune system is still debated (Chan et al., 2007). However, recent lineage tracing experiments have revealed that microglia begin as yolk sac-derived Csflr$^+$ erythro-myeloid progenitor cells, which enter the brain during early embryogenesis and differentiate into a mature microglial pool (Gomez Perdiguero et al., 2015). Microglia serve as the resident CNS phagocytes capable of reacting promptly to brain insults that range from pathogens to aggregated proteins. For the most part, resident microglia are equipped to detect and rapidly remove infected or damaged cells and thus are critical to the brain's surveillance system. Both microglia and astrocytes were once thought of as simple support cells for neurons. However, new data suggest that these cells have distinct and direct functions in shaping neuronal synapse formation, suggesting far more important roles in neural function (Lui et al., 2016; Singh et al., 2016). The newer discovery, with regard to microglial function, shows that they have evolved into cells with distinct identities as compared to peripheral myeloid cells and as such have opened new areas of studies of cell biology.

Glial cells may elicit both neurotoxic and neuroprotective activities, these glial cells serve neuroprotective functions. Astrocytes serve as a primary supporter of neuronal and microglial activities made possible by the cell's abilities to produce neurotrophins, eliminate excitotoxins, and control tissue functions. However, it is the microglia that are primarily responsible for the removal of debris and clearance of microbes that have gained access to neural microenvironments. In particular, and under a broad range of diseased states and infectious processes, microglia serve with the astrocyte to promote neuronal survival. The activation of either cell type leads to profound morphological and functional changes. Activated microglia increase their numbers at the affected site and exhibit a "spider-like" ramified appearance with substantially enlarged cell bodies. Such cell signatures appear at the lesion close to injured or degenerating neurons. While such changes are clearly implicated in neurodegenerative processes, the cells serve the nervous system through their abilities to affect tissue and cell function, molecular trafficking, regeneration and elimination of lipids, and scavenging apoptotic cells. Such functions are coincident with innate cellular proliferation and surface expression of "activated" markers.

We can no longer propagate the perception that the CNS is an "immune-privileged" site. Recent disclosures have supported the presence of an anatomical lymphatic system within the CNS (Louveau et al., 2015). Such a unique system serves to traffic cells and macromolecules in and out of the CNS. Brain-derived antigens drain to the deep cervical lymph nodes following injury (Benner et al., 2008) by way of this system, and CNS-derived dendritic cells function in perivascular regions that may serve to maintain immune tolerance through regulatory T cells (Tregs). While naïve T cells are precluded from CNS entry, neuroinflammation aggressively recruits activated T cells to sites of active neurodegeneration (Karman et al., 2004). While the pathogenic role of T-cell autoimmunity has been extensively studied, more recent studies have suggested a potential beneficial role in some settings. Nonetheless, a far more complex relationship between neurons and immunity exists than previously thought. For example, numerous interleukins (ILs) and chemokines are readily expressed in neurons. Signaling between neurons, astrocytes, and microglia during neural injury incite inflammatory responses and leukocyte migration. Here is another example: the major histocompatibility complex (MHC) class I and CD3ζ molecules have a role in axonal guidance, activity-dependent remodeling, and plasticity in the developing nervous system (Huh et al., 2000). Within the neuronal synapse, MHC class I molecules may participate in synaptic connections. The cognate receptor for MHC class I peptide complexes is the αβ T-cell receptor (TCR) expressed on T cells. During neuroinflammatory responses, MHC class I molecules are upregulated on neurons without evidence of cytotoxic T lymphocyte (CTL)-mediated neural cell damage (Syken & Shatz, 2003; Nishiyori et al., 2004). Altered peptide profiles at the neuronal synapse may affect neuronal plasticity and brain remodeling. Additionally, a myriad of animal model systems of human disease have shown that deficiencies in adaptive immunity lead to neuronal loss, which are reversed following immune reconstitution. The relationship between T-cell responses following neural injury are now well-known. However, the specific identity of T-cell subpopulations and their divergent roles in neuroprotection and neurodegeneration are hotly debated.

How neurons are injured or die remains an important determinant in the ensuing microglial responses. This is of paramount importance as phagocytes clear cells undergoing programmed cell death and must do so in a manner that does not harm the neighboring cells (Sastry & Rao, 2000). The rapid clearance of such cells prevents the release of toxic or immunostimulatory molecules from dead neurons and is thought to play an important role in maintaining homeostatic functions (Magnus et al., 2001; Chan et al., 2001). Apoptosis leads to morphological and biochemical alterations at the cell surface, making them suitable targets for resident phagocytes. Such changes include the reversal of membrane polarity, yielding a net anionic charge, and the surface expression of modified lipid (phosphatidylserine) and carbohydrate motifs. Apoptosis proceeds without an inflammatory response, and phagocytosis of apoptotic cells inhibits innate immune activation (Savill et al., 2002). Actually, this process leads to anti-inflammatory cytokine production that includes transforming growth factor-beta (TGF-β) and interleukin-10 (IL-10). Alternatively, necrotic cell death alerts the immune system to danger through danger-associated molecular patterns (DAMPs) and elicits a significant inflammatory response (Krysko et al., 2011). Antigens from necrotic or stressed cells may trigger a class of pattern recognition receptors (PRRs) designed to recognize evolutionarily distant microorganisms and include toll-like receptors (TLRs). While minimal evidence exists for any infection etiology in parallel to neurodegenerative diseases, inflammation is nonetheless linked to neuronal loss. One example is the infusion of lipopolysaccharide (LPS), a pathogen-associated molecular pattern, into the brain, the eliciting of an inflammatory response, and its association with neuronal loss. LPS signals through TLR4 and CD14 and results in a robust activation of microglial cells (Elward & Gasque, 2003). Such immunostimulatory ligands include heat shock proteins (HSPs) (Asea et al., 2002), high mobility group 1 protein (HMG1) (Park et al., 2004), DNA (Collins et al., 2004), hyaluronan (Termeer et al., 2002), uric acid (Shi et al., 2003), and fibronectin (Okamura et al., 2001). Many of these target ligands are upregulated in stressed cells or liberated from the extracellular matrix (ECM) during tissue damage. As such, they make excellent candidates for molecular triggers of immune activation following tissue injury and may therefore contribute to the neurodegenerative process.

Interestingly, extracellular mitochondria, released during cell death, may also act as DAMPs to the innate immune system. Mitochondria likely evolved from a prokaryotic ancestor and have retained many phylogenic characteristics, including double membrane structures, primitive transcription and translation machinery, and stretches of unmethylated cytosine–phosphate–guanine (CpG) DNA motifs associated with microbial genomes. Both mitochondrial DNA (mtDNA) and DNA modified by oxidation (8-oxo-7,8-dihydro-2′-deoxyguanosine (8-Oxo-dG)) strongly activate cells of the MP lineage (Collins et al., 2004). This has important implications for neurodegenerative disease in which mitochondrial pathology and oxidative stress are associated with neuronal loss.

Aggregated proteins are also indicators of cellular stress and are potent activators of innate immunity. Most neurodegenerative diseases are associated with the accumulation of abnormal protein aggregates (proteopathies). Aggregated proteins can exert toxicity by disrupting intracellular transport, overwhelming the protein degradation machinery, or sequestering normal proteins away from their physiological roles. In addition to direct toxicity, mounting evidence suggests that abnormal protein conformations can activate innate immune cells through scavenger receptors, thus providing a mechanism for secondary tissue injury (Fischer et al., 2002; Yan et al., 1999; Bradt et al., 1998; Reynolds et al., 2008, 2009). Additionally, treatment of neurons with proteasome inhibitors and the subsequent accumulation of ubiquitinated proteins elicit prostaglandin synthesis, suggesting that intracellular accumulation of abnormal proteins can trigger the production of pro-inflammatory mediators (Rockwell et al., 2000). A major pathological hallmark of PD is the presence of eosinophilic, cytoplasmic inclusions of fibrillar, misfolded proteins called Lewy bodies (LBs), suggesting that derangements in proteostasis may be involved in disease pathogenesis (Dauer & Przedborski, 2003). While LBs are intracellular, the fate of associated protein aggregates following cell death is unknown and can be released following cell death. Moreover, neurons go to great lengths during times of cellular stress to prevent such aggregation by upregulating ATP-dependent chaperones such as HSPs (Muchowski, 2002). In addition to aggregated proteins, HSPs can be released during cell death and doing so may directly contribute to microglial activation (Kakimura et al., 2002).

Reactive oxygen species (ROS) include superoxide, hydrogen peroxide, and hydroxyl free radicals, as well as nitrogen intermediates (nitric oxide and peroxynitrite). These cause damage to neurons if produced in excess such as during prolonged neuroinflammatory responses. Much of the microglial-derived ROS such as superoxide cannot efficiently transverse cellular membranes, making it unlikely that these highly reactive molecules gain intracellular access to neurons and trigger intraneuronal toxic events (Ischiropoulos & Beckman, 2003). However, superoxide can rapidly react with NO in the extracellular space to form a more stable oxidant, peroxynitrite (Wu et al., 2003). The stability of peroxynitrite facilitates its crossing of cell membranes where it can damage intracellular components in neighboring neurons. Its toxicity has been linked to the disruption of the mitochondrial electron transport chain, lipid peroxidation, DNA damage, and the nitration of tyrosine residues in cellular proteins. NO plays an important role in innate immunity and is associated with tumoricidal and bactericidal activities of macrophages. In addition to the physiological roles of NO, under pathological settings with excessive production, it can act as a potent neurotoxin in a number of neurodegenerative models.

On a more general level, matrix metalloproteinases (MMPs) are a class of extracellular soluble or membrane-bound cysteine proteases involved in remodeling of the extracellular matrix (ECM). MMPs have also been implicated in a range of neurodegenerative diseases, including HD, AD, PD, and stoke. Gu et al. (2002) reported that S-nitrosylation of N-terminal cysteine residues within pro-MMP-9 leads to the subsequent activation of MMP-9 protease activity. These findings identify a potential extracellular proteolysis mechanism involved in neuronal cell death in which S-nitrosylation activates MMPs. In one study, an increase in MMP-9 expression was determined using an N-methyl-4-phenyl-1,2,3,6-tetrahydropyridine (MPTP) model and pharmacological inhibition of MMP-9 was shown to be neuroprotective (Lorenzl et al., 2004).

The complex role of cellular and molecular inflammatory mediators, particularly within the CNS, makes it difficult to define their specific beneficial or deleterious roles in pathophysiological states (Gendelman, 2002). Cytokines and chemokines are of particular importance in the initiation, maintenance, and resolution of inflammatory reactions within the CNS (Allan & Rothwell, 2001). Moreover, cytokines and chemokines, in particular, have important roles outside of inflammation, including chemotactic axonal guidance during the developmental stages of the CNS (Mennicken et al., 1999; Tran & Miller, 2003a). With regard to the immune system, cytokines and chemokines act as potent survival cues and their tissue levels directly modulate the activity of inflammatory cell types. Therefore, the function of these molecules within the immune system is reminiscent of that of neurotrophins within the nervous system (Tran & Miller, 2003b). In fact, many neurotrophins are structurally related to molecules within cytokine families, and these two classes of molecules appear to share functional overlap. IL-10 is a potent neuroprotective factor on purified cerebellar neurons and AD (Bachis et al., 2001; Kiyota et al., 2012), and Nerve Growth Factor (NGF) has emerged as a potent anti-inflammatory factor with beneficial effects in the mouse model of MS (Flugel et al., 2001). Indeed, neurons themselves can produce chemokines and express many of the their receptors on their surface, suggesting a direct dialogue between these two systems. Neuroimmune communication is not restricted to soluble mediators. The CD200 receptor (CD200R) that is expressed on myeloid cells, including microglia and its ligand CD200, is abundant on neuronal cell surfaces. Cognate CD200–CD200R interactions have potent inhibitory effects on the cells of the myeloid lineage (Barclay et al., 2002; Hoek et al., 2000; Varnum et al., 2015). Cells of the adaptive immune system can also communicate with neurons. Activated T cells express numerous neurotrophins in addition to the neurotrophic factor receptors, TrkB and TrkC, thus providing sufficient mechanistic means to establish direct T cell–neuron communications.

Studies on functional cellular immune responses can produce conflicting results. Virally or otherwise activated macrophages can have devastating consequences on neuronal function and survival. Depletion of peripheral monocyte/macrophage lineages in a spinal cord injury model can improve functional recovery (Popovich et al., 1999). Conversely, direct spinal injection of macrophages activated with damaged peripheral nerves ex vivo leads to enhanced functional recovery (Rapalino et al., 1998). Resident microglia, while contributing to noxious factors in the CNS disease detailed above, also produce a plethora of neurotrophic factors upon activation. Local production of these factors can establish trophic gradients, which stimulate and guide axonal sprouting following striatal injury (Batchelor et al., 1999, 2002). Thus, fine-tuning of cellular immune responses is critical for maximizing neuroprotection by immune cells.

The presence of T cells in CNS pathological settings has long been regarded as detrimental. One such case where T cells play a direct role in disease pathogenesis is multiple sclerosis (MS). MS is an inflammatory disease characterized by perivascular cuffs of mononuclear cells that include both lymphocyte and macrophage populations. Accumulation of these cells leads to significant damage to the myelin sheath and can involve the underlying axon (Kornek et al., 2000). However, provocative experimental data also implicate a role for adaptive immunity in otherwise non-autoimmune disorders of the CNS. Surprisingly, SCID (severe combined immunodeficiency) and $Rag2^{-/-}$ (recombination activating gene 2) mice, both deficient in adaptive immunity, show enhanced motor neuron loss within the facial nucleus following transection of the facial nerve. Regulatory T cells such as $CD4^{+}CD25^{+}$, which comprise 10% of the total $CD4^{+}$ population, play a vital role in inhibiting autoreactive T-cell activation and expansion and may be of paramount importance in maintaining peripheral tolerance against CNS antigens. Depletion of naturally occurring regulatory $CD4^{+}CD25^{+}$ T cells predisposes animals to development of organ-specific autoimmune diseases (Shevach, 2000). Depletion of $CD4^{+}CD25^{+}$ regulatory T cells enhances T cell-dependent protective response and hence improves post-injury neuronal survival following disease (Kipnis et al., 2002). Alternatively, immunodeficient mice have improved dopaminergic neuronal survival (Benner et al., 2008) and Treg cells can confer neuroprotection of the nigrostriatal axis in animal models of PD (Reynolds et al., 2010). Moreover, a recent study has demonstrated that the influence of gut microbiomes on ischemic stroke outcomes is mediated through Treg regulation of $\gamma\delta$ T-cell responses to injury (Benakis et al., 2016). In this study, Treg populations in the gut resulted in decreased $\gamma\delta$

T-cell-derived IL-17 in the brain following stroke, leading to improved outcomes. Thus, $CD4^+CD25^+$ regulatory T cells, at least for now, are both agonistic and antagonistic and require further investigation.

How can we harness these complex immune responses within the nervous system for therapeutic gain? This has given rise to the development of neuro- and immunopharmacology. In the context of the immune system, the drug acts on the biological system, which augments cellular responses against disease or helps maintain host homeostasis. Such cellular responses can be utilized for therapeutic gain. The ensuing events that occur within the nervous system ultimately is the field of neuroimmune pharmacology. For example, an unexpected role of T cells in neuronal function is immune-mediated neuroregeneration, differentiation, or protection. Surprisingly, for example, pro-inflammatory T_H1-cell responses generated through immunization with myelin-associated antigens can attenuate secondary neuronal loss following experimentally induced CNS injury (Weiner & Selkoe, 2002; Schwartz, 2001). It appears that such strategies can increase local production of pro-inflammatory cytokines as well as multiple neurotrophic factors within the injured tissue (Hammarberg et al., 2000; Moalem et al., 1999). These provocative observations have generated novel questions regarding neuroimmune interactions, and it has been suggested that the observed neuroprotection can be directly linked to the pro-inflammatory nature of the infiltrating T-cell population (Schwartz et al., 1999). However, detailed mechanistic data are lacking and the true mechanisms will likely prove as complicated as many other inflammatory mediators described above. Contrary to this, oral or nasal immunization that preferentially generates an anti-inflammatory T-cell response (T_H2/T_H3) also has a demonstrated neuroprotective element and may prove more clinically feasible (Frenkel et al., 2003; Monsonego et al., 2003).

Neurodegenerative disorders are linked to innate immune activation comprised of an astonishing array of cellular and molecular players. There is little evidence of an infectious process in most neurodegenerative diseases; however; multiple endogenous immune activators associated with cellular stress signals likely converge on resident glia, leading to an inflammatory phenotype. The intended purpose of this response is beneficial to the host. However, when regulatory controls are breached and inflammation is allowed to persist, tissue damage ensues. The identification of individual activation pathways operative in disease and the subsequent response they elicit may prove beneficial to the development of novel therapeutics. In addition to innate immunity, the adaptive immune system may also play a role (beneficial or toxic) in neurodegeneration, even in the absence of overt autoimmune pathology. Moreover, the adaptive immune system may be manipulated through peripheral immunization and/or immunotherapy in a way that modifies innate inflammatory mechanisms in the CNS. Strategies to maximize the benefit of protective innate and adaptive immunity against CNS disorders with emerging immunological, nanomedicinal, and molecular biological techniques is fundamental to these approaches. Careful pursuit of therapeutic strategies with an emphasis on safety may lead to novel treatment means to combat a broad range of neuropsychiatric, developmental, infectious, and degenerative disorders of the nervous system as highlighted in this second edition of the textbook. Here, we humbly dedicate this new book to the field's future.

2.2 Review Questions

1. Dr. Elie Metchnikoff's famous discovery that led to the birth of immunology was:
   (a) An inflammatory cell type in starfish larvae could engulf foreign materials and protect its host.
   (b) Cytotoxic T lymphocytes clear viruses and cancerous cells from the body during disease.
   (c) The gut is a central lymphoid organ responsible for a plethora of immune responses in the host, both during homeostasis and disease.
   (d) Macrophages and dendritic cells present antigens and mobilize adaptive immune responses.
   (e) All of the above.
2. Neuroscience is defined as:
   (a) Cell-to-cell interactions and communication networks of neurons and glia
   (b) Investigating networks and models of neural communication in health and disease amongst hundreds of billions of neurons and supporting glia (microglia, oligodendrocytes, and astrocytes)
   (c) The development of the nervous system
   (d) The reaction of the brain and spinal cord to injury and infections
   (e) All of the above
3. The ability to harness complex immune responses within the nervous system for therapeutic gain is defined by:
   (a) Neuroimmune pharmacology
   (b) How a drug acts on a specific biological system and modulates cellular responses against disease
   (c) Finding the means to boost the immune system to maintain host homeostasis
   (d) Engaging cellular immune responses for therapeutic gain
   (e) All of the above
4. Astrocytes are NOT defined by their ability:
   (a) To be the primary supporters of neuronal and microglial activities
   (b) To produce neurotrophins, eliminate excitotoxins, and control tissue functions

(c) To remove debris and clear microbes that have gained access to neural microenvironments
(d) To work in concert with microglia to promote neuronal survival
(e) To cause profound morphological and functional changes on activation

5. True or false.

While naïve T cells are precluded from CNS entry, neuroinflammation recruits activated T cells to sites of active neurodegeneration.

2.3 Answers

1. a
2. e
3. e
4. c
5. True

References

Allan, S. M., & Rothwell, N. J. (2001). Cytokines and acute neurodegeneration. *Nature Reviews. Neuroscience, 2*(10), 734–744.

Appel, S. H., Beers, D. R., & Henkel, J. S. (2010). T cell-microglial dialogue in Parkinson's disease and amyotrophic lateral sclerosis: Are we listening? *Trends in Immunology, 31*(1), 7–17. https://doi.org/10.1016/j.it.2009.09.003

Asea, A., Rehli, M., Kabingu, E., Boch, J. A., Bare, O., Auron, P. E., Stevenson, M. A., & Calderwood, S. K. (2002). Novel signal transduction pathway utilized by extracellular HSP70: Role of toll-like receptor (TLR) 2 and TLR4. *The Journal of Biological Chemistry, 277*(17), 15028–15034.

Bachis, A., Colangelo, A. M., Vicini, S., Doe, P. P., De Bernardi, M. A., Brooker, G., & Mocchetti, I. (2001). Interleukin-10 prevents glutamate-mediated cerebellar granule cell death by blocking caspase-3-like activity. *The Journal of Neuroscience, 21*(9), 3104–3112.

Barclay, A. N., Wright, G. J., Brooke, G., & Brown, M. H. (2002). CD200 and membrane protein interactions in the control of myeloid cells. *Trends in Immunology, 23*(6), 285–290.

Batchelor, P. E., Liberatore, G. T., Wong, J. Y., Porritt, M. J., Frerichs, F., Donnan, G. A., & Howells, D. W. (1999). Activated macrophages and microglia induce dopaminergic sprouting in the injured striatum and express brain-derived neurotrophic factor and glial cell line-derived neurotrophic factor. *The Journal of Neuroscience, 19*(5), 1708–1716.

Batchelor, P. E., Porritt, M. J., Martinello, P., Parish, C. L., Liberatore, G. T., Donnan, G. A., & Howells, D. W. (2002). Macrophages and microglia produce local trophic gradients that stimulate axonal sprouting toward but not beyond the wound edge. *Molecular and Cellular Neurosciences, 21*(3), 436–453.

Benakis, C., Brea, D., Caballero, S., Faraco, G., Moore, J., Murphy, M., Sita, G., Racchumi, G., Ling, L., Pamer, E. G., Iadecola, C., & Anrather, J. (2016). Commensal microbiota affects ischemic stroke outcome by regulating intestinal gammadelta T cells. *Nature Medicine, 22*(5), 516–523. https://doi.org/10.1038/nm.4068

Benner, E. J., Banerjee, R., Reynolds, A. D., Sherman, S., Pisarev, V. M., Tsiperson, V., Nemachek, C., Ciborowski, P., Przedborski, S., Mosley, R. L., & Gendelman, H. E. (2008). Nitrated alpha-synuclein immunity accelerates degeneration of nigral dopaminergic neurons. *PLoS One, 3*(1), e1376. https://doi.org/10.1371/journal.pone.0001376

Bradt, B. M., Kolb, W. P., & Cooper, N. R. (1998). Complement-dependent proinflammatory properties of the Alzheimer's disease beta-peptide. *The Journal of Experimental Medicine, 188*(3), 431–438.

Chan, A., Magnus, T., & Gold, R. (2001). Phagocytosis of apoptotic inflammatory cells by microglia and modulation by different cytokines: Mechanism for removal of apoptotic cells in the inflamed nervous system. *Glia, 33*(1), 87–95.

Chan, W. Y., Kohsaka, S., & Rezaie, P. (2007). The origin and cell lineage of microglia: New concepts. *Brain Research Reviews, 53*(2), 344–354. https://doi.org/10.1016/j.brainresrev.2006.11.002

Collins, L. V., Hajizadeh, S., Holme, E., Jonsson, I. M., & Tarkowski, A. (2004). Endogenously oxidized mitochondrial DNA induces in vivo and in vitro inflammatory responses. *Journal of Leukocyte Biology, 75*(6), 995–1000.

Dauer, W., & Przedborski, S. (2003). Parkinson's disease: Mechanisms and models. *Neuron, 39*(6), 889–909.

Elward, K., & Gasque, P. (2003). "Eat me" and "don't eat me" signals govern the innate immune response and tissue repair in the CNS: Emphasis on the critical role of the complement system. *Molecular Immunology, 40*(2–4), 85–94.

Fadok, V. A., Bratton, D. L., Konowal, A., Freed, P. W., Westcott, J. Y., & Henson, P. M. (1998). Macrophages that have ingested apoptotic cells in vitro inhibit proinflammatory cytokine production through autocrine/paracrine mechanisms involving TGF-beta, PGE2, and PAF. *The Journal of Clinical Investigation, 101*(4), 890–898. https://doi.org/10.1172/JCI1112

Fischer, B., von Knethen, A., & Brune, B. (2002). Dualism of oxidized lipoproteins in provoking and attenuating the oxidative burst in macrophages: Role of peroxisome proliferator-activated receptor-gamma. *Journal of Immunology, 168*(6), 2828–2834.

Flugel, A., Matsumuro, K., Neumann, H., Klinkert, W. E., Birnbacher, R., Lassmann, H., Otten, U., & Wekerle, H. (2001). Anti-inflammatory activity of nerve growth factor in experimental autoimmune encephalomyelitis: Inhibition of monocyte transendothelial migration. *European Journal of Immunology, 31*(1), 11–22.

Frenkel, D., Huang, Z., Maron, R., Koldzic, D. N., Hancock, W. W., Moskowitz, M. A., & Weiner, H. L. (2003). Nasal vaccination with myelin oligodendrocyte glycoprotein reduces stroke size by inducing IL-10-producing CD4+ T cells. *Journal of Immunology, 171*(12), 6549–6555.

Gardai, S. J., McPhillips, K. A., Frasch, S. C., Janssen, W. J., Starefeldt, A., Murphy-Ullrich, J. E., Bratton, D. L., Oldenborg, P. A., Michalak, M., & Henson, P. M. (2005). Cell-surface calreticulin initiates clearance of viable or apoptotic cells through trans-activation of LRP on the phagocyte. *Cell, 123*(2), 321–334. https://doi.org/10.1016/j.cell.2005.08.032

Gendelman, H. E. (2002). Neural immunity: Friend or foe? *Journal of Neurovirology, 8*(6), 474–479.

Gomez Perdiguero, E., Klapproth, K., Schulz, C., Busch, K., Azzoni, E., Crozet, L., Garner, H., Trouillet, C., de Bruijn, M. F., Geissmann, F., & Rodewald, H. R. (2015). Tissue-resident macrophages originate from yolk-sac-derived erythro-myeloid progenitors. *Nature, 518*(7540), 547–551. https://doi.org/10.1038/nature13989

Gu, Z., Kaul, M., Yan, B., Kridel, S. J., Cui, J., Strongin, A., Smith, J. W., Liddington, R. C., & Lipton, S. A. (2002). S-nitrosylation of matrix metalloproteinases: Signaling pathway to neuronal cell death. *Science, 297*(5584), 1186–1190.

Hammarberg, H., Lidman, O., Lundberg, C., Eltayeb, S. Y., Gielen, A. W., Muhallab, S., Svenningsson, A., Linda, H., van Der Meide,

P. H., Cullheim, S., Olsson, T., & Piehl, F. (2000). Neuroprotection by encephalomyelitis: Rescue of mechanically injured neurons and neurotrophin production by CNS-infiltrating T and natural killer cells. *The Journal of Neuroscience, 20*(14), 5283–5291.

Hoek, R. M., Ruuls, S. R., Murphy, C. A., Wright, G. J., Goddard, R., Zurawski, S. M., Blom, B., Homola, M. E., Streit, W. J., Brown, M. H., Barclay, A. N., & Sedgwick, J. D. (2000). Down-regulation of the macrophage lineage through interaction with OX2 (CD200). *Science, 290*(5497), 1768–1771.

Huh, G. S., Boulanger, L. M., Du, H., Riquelme, P. A., Brotz, T. M., & Shatz, C. J. (2000). Functional requirement for class I MHC in CNS development and plasticity. *Science, 290*(5499), 2155–2159.

Ischiropoulos, H., & Beckman, J. S. (2003). Oxidative stress and nitration in neurodegeneration: Cause, effect, or association? *The Journal of Clinical Investigation, 111*(2), 163–169.

Kakimura, J., Kitamura, Y., Takata, K., Umeki, M., Suzuki, S., Shibagaki, K., Taniguchi, T., Nomura, Y., Gebicke-Haerter, P. J., Smith, M. A., Perry, G., & Shimohama, S. (2002). Microglial activation and amyloid-beta clearance induced by exogenous heat-shock proteins. *The FASEB Journal, 16*(6), 601–603.

Karman, J., Ling, C., Sandor, M., & Fabry, Z. (2004). Initiation of immune responses in brain is promoted by local dendritic cells. *Journal of Immunology, 173*(4), 2353–2361.

Kipnis, J., Mizrahi, T., Hauben, E., Shaked, I., Shevach, E., & Schwartz, M. (2002). Neuroprotective autoimmunity: Naturally occurring CD4 + CD25+ regulatory T cells suppress the ability to withstand injury to the central nervous system. *Proceedings of the National Academy of Sciences of the United States of America, 99*(24), 15620–15625. https://doi.org/10.1073/pnas.232565399

Kiyota, T., Ingraham, K. L., Swan, R. J., Jacobsen, M. T., Andrews, S. J., & Ikezu, T. (2012). AAV serotype 2/1-mediated gene delivery of anti-inflammatory interleukin-10 enhances neurogenesis and cognitive function in APP + PS1 mice. *Gene Therapy, 19*(7), 724–733. https://doi.org/10.1038/gt.2011.126

Kornek, B., Storch, M. K., Weissert, R., Wallstroem, E., Stefferl, A., Olsson, T., Linington, C., Schmidbauer, M., & Lassmann, H. (2000). Multiple sclerosis and chronic autoimmune encephalomyelitis: A comparative quantitative study of axonal injury in active, inactive, and remyelinated lesions. *The American Journal of Pathology, 157*(1), 267–276.

Krysko, D. V., Agostinis, P., Krysko, O., Garg, A. D., Bachert, C., Lambrecht, B. N., & Vandenabeele, P. (2011). Emerging role of damage-associated molecular patterns derived from mitochondria in inflammation. *Trends in Immunology, 32*(4), 157–164. https://doi.org/10.1016/j.it.2011.01.005

Lorenzl, S., Calingasan, N., Yang, L., Albers, D. S., Shugama, S., Gregorio, J., Krell, H. W., Chirichigno, J., Joh, T., & Beal, M. F. (2004). Matrix metalloproteinase-9 is elevated in 1-methyl-4-phenyl-1,2,3,6-tetrahydropyridine-induced parkinsonism in mice. *Neuromolecular Medicine, 5*(2), 119–132.

Louveau, A., Smirnov, I., Keyes, T. J., Eccles, J. D., Rouhani, S. J., Peske, J. D., Derecki, N. C., Castle, D., Mandell, J. W., Lee, K. S., Harris, T. H., & Kipnis, J. (2015). Structural and functional features of central nervous system lymphatic vessels. *Nature, 523*(7560), 337–341. https://doi.org/10.1038/nature14432

Lui, H., Zhang, J., Makinson, S. R., Cahill, M. K., Kelley, K. W., Huang, H. Y., Shang, Y., Oldham, M. C., Martens, L. H., Gao, F., Coppola, G., Sloan, S. A., Hsieh, C. L., Kim, C. C., Bigio, E. H., Weintraub, S., Mesulam, M. M., Rademakers, R., Mackenzie, I. R., Seeley, W. W., Karydas, A., Miller, B. L., Borroni, B., Ghidoni, R., Farese, R. V., Jr., Paz, J. T., Barres, B. A., & Huang, E. J. (2016). Progranulin deficiency promotes circuit-specific synaptic pruning by microglia via complement activation. *Cell, 165*(4), 921–935. https://doi.org/10.1016/j.cell.2016.04.001

Magnus, T., Chan, A., Grauer, O., Toyka, K. V., & Gold, R. (2001). Microglial phagocytosis of apoptotic inflammatory T cells leads to down-regulation of microglial immune activation. *Journal of Immunology, 167*(9), 5004–5010.

Mennicken, F., Maki, R., de Souza, E. B., & Quirion, R. (1999). Chemokines and chemokine receptors in the CNS: A possible role in neuroinflammation and patterning. *Trends in Pharmacological Sciences, 20*(2), 73–78.

Moalem, G., Leibowitz-Amit, R., Yoles, E., Mor, F., Cohen, I. R., & Schwartz, M. (1999). Autoimmune T cells protect neurons from secondary degeneration after central nervous system axotomy. *Nature Medicine, 5*(1), 49–55.

Monsonego, A., Beserman, Z. P., Kipnis, J., Yoles, E., Weiner, H. L., & Schwartz, M. (2003). Beneficial effect of orally administered myelin basic protein in EAE-susceptible Lewis rats in a model of acute CNS degeneration. *Journal of Autoimmunity, 21*(2), 131–138.

Muchowski, P. J. (2002). Protein misfolding, amyloid formation, and neurodegeneration: A critical role for molecular chaperones? *Neuron, 35*(1), 9–12.

Nishiyori, A., Hanno, Y., Saito, M., & Yoshihara, Y. (2004). Aberrant transcription of unrearranged T-cell receptor beta gene in mouse brain. *The Journal of Comparative Neurology, 469*(2), 214–226.

Okamura, Y., Watari, M., Jerud, E. S., Young, D. W., Ishizaka, S. T., Rose, J., Chow, J. C., & Strauss, J. F., 3rd. (2001). The extra domain A of fibronectin activates Toll-like receptor 4. *The Journal of Biological Chemistry, 276*(13), 10229–10233.

Park, J. S., Svetkauskaite, D., He, Q., Kim, J. Y., Strassheim, D., Ishizaka, A., & Abraham, E. (2004). Involvement of toll-like receptors 2 and 4 in cellular activation by high mobility group box 1 protein. *The Journal of Biological Chemistry, 279*(9), 7370–7377.

Popovich, P. G., Guan, Z., Wei, P., Huitinga, I., van Rooijen, N., & Stokes, B. T. (1999). Depletion of hematogenous macrophages promotes partial hindlimb recovery and neuroanatomical repair after experimental spinal cord injury. *Experimental Neurology, 158*(2), 351–365.

Rapalino, O., Lazarov-Spiegler, O., Agranov, E., Velan, G. J., Yoles, E., Fraidakis, M., Solomon, A., Gepstein, R., Katz, A., Belkin, M., Hadani, M., & Schwartz, M. (1998). Implantation of stimulated homologous macrophages results in partial recovery of paraplegic rats. *Nature Medicine, 4*(7), 814–821.

Reynolds, A. D., Glanzer, J. G., Kadiu, I., Ricardo-Dukelow, M., Chaudhuri, A., Ciborowski, P., Cerny, R., Gelman, B., Thomas, M. P., Mosley, R. L., & Gendelman, H. E. (2008). Nitrated alpha-synuclein-activated microglial profiling for Parkinson's disease. *Journal of Neurochemistry, 104*(6), 1504–1525. https://doi.org/10.1111/j.1471-4159.2007.05087.x

Reynolds, A. D., Stone, D. K., Mosley, R. L., & Gendelman, H. E. (2009). Nitrated {alpha}-synuclein-induced alterations in microglial immunity are regulated by CD4+ T cell subsets. *Journal of Immunology, 182*(7), 4137–4149. https://doi.org/10.4049/jimmunol.0803982

Reynolds, A. D., Stone, D. K., Hutter, J. A., Benner, E. J., Mosley, R. L., & Gendelman, H. E. (2010). Regulatory T cells attenuate Th17 cell-mediated nigrostriatal dopaminergic neurodegeneration in a model of Parkinson's disease. *Journal of Immunology, 184*(5), 2261–2271. https://doi.org/10.4049/jimmunol.0901852

Rockwell, P., Yuan, H., Magnusson, R., & Figueiredo-Pereira, M. E. (2000). Proteasome inhibition in neuronal cells induces a proinflammatory response manifested by upregulation of cyclooxygenase-2, its accumulation as ubiquitin conjugates, and production of the prostaglandin PGE(2). *Archives of Biochemistry and Biophysics, 374*(2), 325–333.

Sastry, P. S., & Rao, K. S. (2000). Apoptosis and the nervous system. *Journal of Neurochemistry, 74*(1), 1–20.

Savill, J., Dransfield, I., Gregory, C., & Haslett, C. (2002). A blast from the past: Clearance of apoptotic cells regulates immune responses. *Nature Reviews. Immunology, 2*(12), 965–975. https://doi.org/10.1038/nri957

Schwartz, M. (2001). Harnessing the immune system for neuroprotection: Therapeutic vaccines for acute and chronic neurodegenerative disorders. *Cellular and Molecular Neurobiology, 21*(6), 617–627.

Schwartz, M., Moalem, G., Leibowitz-Amit, R., & Cohen, I. R. (1999). Innate and adaptive immune responses can be beneficial for CNS repair. *Trends in Neurosciences, 22*(7), 295–299.

Shevach, E. M. (2000). Regulatory T cells in autoimmmunity*. *Annual Review of Immunology, 18*, 423–449.

Shi, Y., Evans, J. E., & Rock, K. L. (2003). Molecular identification of a danger signal that alerts the immune system to dying cells. *Nature, 425*(6957), 516–521.

Singh, S. K., Stogsdill, J. A., Pulimood, N. S., Dingsdale, H., Kim, Y. H., Pilaz, L. J., Kim, I. H., Manhaes, A. C., Rodrigues, W. S., Jr., Pamukcu, A., Enustun, E., Ertuz, Z., Scheiffele, P., Soderling, S. H., Silver, D. L., Ji, R. R., Medina, A. E., & Eroglu, C. (2016). Astrocytes assemble thalamocortical synapses by bridging NRX1alpha and NL1 via Hevin. *Cell, 164*(1–2), 183–196. https://doi.org/10.1016/j.cell.2015.11.034

Syken, J., & Shatz, C. J. (2003). Expression of T cell receptor beta locus in central nervous system neurons. *Proceedings of the National Academy of Sciences of the United States of America, 100*(22), 13048–13053.

Termeer, C., Benedix, F., Sleeman, J., Fieber, C., Voith, U., Ahrens, T., Miyake, K., Freudenberg, M., Galanos, C., & Simon, J. C. (2002). Oligosaccharides of Hyaluronan activate dendritic cells via toll-like receptor 4. *The Journal of Experimental Medicine, 195*(1), 99–111.

Tran, P. B., & Miller, R. J. (2003a). Chemokine receptors in the brain: A developing story. *The Journal of Comparative Neurology, 457*(1), 1–6.

Tran, P. B., & Miller, R. J. (2003b). Chemokine receptors: Signposts to brain development and disease. *Nature Reviews. Neuroscience, 4*(6), 444–455.

Varnum, M. M., Kiyota, T., Ingraham, K. L., Ikezu, S., & Ikezu, T. (2015). The anti-inflammatory glycoprotein, CD200, restores neurogenesis and enhances amyloid phagocytosis in a mouse model of Alzheimer's disease. *Neurobiology of Aging, 36*(11), 2995–3007. https://doi.org/10.1016/j.neurobiolaging.2015.07.027

Weiner, H. L., & Selkoe, D. J. (2002). Inflammation and therapeutic vaccination in CNS diseases. *Nature, 420*(6917), 879–884.

Wu, D. C., Teismann, P., Tieu, K., Vila, M., Jackson-Lewis, V., Ischiropoulos, H., & Przedborski, S. (2003). NADPH oxidase mediates oxidative stress in the 1-methyl-4-phenyl-1,2,3,6-tetrahydropyridine model of Parkinson's disease. *Proceedings of the National Academy of Sciences of the United States of America, 100*(10), 6145–6150.

Yan, S. D., Roher, A., Schmidt, A. M., & Stern, D. M. (1999). Cellular cofactors for amyloid beta-peptide-induced cell stress. Moving from cell culture to in vivo. *The American Journal of Pathology, 155*(5), 1403–1411.

3 Immune Sensors, Regulators, and Effectors of Brain Health

Cigdem Acioglu and Stella Elkabes

Abstract

Traditionally, the brain was considered an immune-privileged organ. It is now well known that molecular and cellular mechanisms, classically described in the immune system, also occur in the brain. The brain mounts a robust and well-orchestrated innate immune response during infections, injuries, and neurodegenerative diseases. Resident cells of the brain play key roles in this process. Although peripheral immune cells are present at the meningeal borders of the brain, a relatively small number of immune cells infiltrate the parenchyma under physiological conditions. However, if the blood–brain barrier is perturbed in pathological conditions, immune cells infiltrate the brain in large numbers and participate in the immune reaction. Pattern recognition receptors (PRRs) and inflammasomes, sensors of danger signals, are expressed in glia and neurons and mediate the innate immunity of the brain, leading to neuroinflammation, which can be degenerative, protective, or reparative depending on the context. This chapter highlights the role of PRRs, especially toll-like receptors and inflammasomes, in the innate immune response of the brain and neuroinflammation. The danger signals that initiate the immune reaction, the effectors that mediate neuroinflammation, and the underlying mechanisms are discussed in the context of Alzheimer's and Parkinson's diseases and brain injuries (BIs).

Keywords

Innate immunity · Pattern recognition receptors · Inflammasome · Neurodegeneration · Neuroinflammation · Glia

C. Acioglu · S. Elkabes (✉)
Department of Neurosurgery, New Jersey Medical School, Rutgers, The State University of New Jersey, Newark, NJ, USA
e-mail: stella.elkabes@rutgers.edu

3.1 Overview of Pattern Recognition Receptors

Pattern recognition receptors (PRRs) are danger sensors that recognize highly conserved and specific molecular motifs in pathogens, collectively called pathogen-associated molecular patterns (PAMPs). They also bind endogenous ligands released by damaged cells in brain diseases or trauma, referred to as damage-associated molecular patterns (DAMPs). In neurodegenerative disorders, some pathogenic molecules such as amyloid beta (Aβ) or α-synuclein (α-syn) also bind PRRs. Ligand binding to PRRs activates intracellular cascades that lead to the release of effector molecules, including interleukin (IL)-1β, IL-18, and type I interferons (IFNs). Resident brain cells express functional PRRs, and activation of PRR signaling in microglia, astrocytes, and neurons results in a potent release of these effectors and neuroinflammation. In turn, effectors, including IL-1β, act on glia and neurons and promote the release of additional cytokines, chemokines, and other inflammatory mediators that chemoattract peripheral immune cells. These interactions fine-tune the immune response of the brain and neuroinflammation.

PRRs have been classified into several families: toll-like receptors (*TLRs*), nucleotide-binding and oligomerization domain (NOD)-like receptors (*NLRs*), retinoic acid-inducible gene (RIG)-like receptors (*RLRs*), absent in melanoma 2 (*AIM2*)-like receptors (*ALRs*), C-type lectin receptors (*CLRs*), and triggering receptors expressed on myeloid cells (*TREMs*). Some of these receptors are expressed in cells intrinsic to the central nervous system (CNS), in addition to immune cells that infiltrate the CNS in pathological conditions (Acioglu et al., 2022; Li et al., 2021a; Xu et al., 2019).

H. E. Gendelman, T. Ikezu (eds.), *Neuroimmune Pharmacology and Therapeutics*,
https://doi.org/10.1007/978-3-031-68237-7_3

3.1.1 Toll-Like Receptors

Toll-like receptors are the first and best-characterized PRR family (Gong et al, 2020). A total of 10 TLRs (TLR1–10) have been identified in humans. Rodents express TLR1–13, with the exemption of TLR10. TLR3, TLR7–9, and TLR13 are intracellular receptors located in the endosomal compartments. The other TLRs are found on the cell surface. In addition to immune cells, TLRs are expressed in both glia and neurons of the CNS (Acioglu et al., 2022; Li et al., 2021a).

Upon ligand binding, TLRs primarily form homodimers to initiate receptor signaling. TLR2 forms heterodimers with TLR1 or TLR6. Activation of TLR2/TLR1 and TLR2/TLR6 can lead to distinct outcomes. TLR4 also forms receptor signaling complexes with co-receptors upon binding a major TLR4 ligand, namely, lipopolysaccharide (LPS), from Gram-negative bacteria. The composition of such receptor complexes is distinct in different cell types. In myeloid cells, the LPS receptor complex consists of TLR4, cluster of differentiation 14 (CD14), and the myeloid differentiation protein 2 (MD-2). CD14 is a glycoprotein anchored to the cell surface, whereas MD-2 is a secreted glycoprotein. CD14 concentrates LPS, and MD-2 associates with the extracellular domain of TLR4, forming a stable complex (Miyake, 2004). However, in B lymphocytes, TLR4 interacts with MD-1, which associates with a TLR4 homolog, radioprotective 105 (RP105) (Miyake, 2004; Nagai et al., 2002). The RP105/MD-1 complex is necessary for TLR4/MD-2 signaling (Jennings & Kusler, 2009). The other cell surface TLRs recognize various bacterial components, including flagellin, peptidoglycans (PGNs), and lipoproteins. TLR2/TLR1 binds lipoproteins from bacteria, whereas diacylated lipopeptides, peptidoglycans, a fungal polysaccharide, and zymosan activate TLR2/TLR6. Endosomal TLRs recognize viral and bacterial nucleic acids. TLR3 is activated by double-stranded RNA (dsRNA) of a viral origin. TLR7 and TLR8 recognize single-stranded RNA (ssRNA) of a viral origin. TLR9 binds unmethylated cytosine–phosphate–guanine (CpG) dinucleotides derived from bacteria and viruses (Acioglu et al., 2022; Heiman et al., 2014; Li et al., 2021a).

There are also endogenous activators of TLRs. Self-mRNA released by necrotic cells, mitochondrial DNA (mtDNA) released by damaged mitochondria, high mobility group box 1 (HMGB1), a nuclear chromatin protein in neurons and activated glia, and heat shock proteins (HSPs) are among TLR ligands (Acioglu et al., 2022; Heiman et al., 2014; Li et al., 2021a). In neurodegenerative disorders, pathogenic mediators such as α-syn (Kim et al., 2013) and Aβ peptides (Liu et al., 2012) can bind and activate TLRs. Table 3.1 summarizes the pathogen-derived and endogenous ligands of different TLRs.

Toll-like receptors have a toll/IL-1 receptor (TIR) cytoplasmic domain, which initiates downstream signaling

Table 3.1 DAMPs and PAMPs are TLR ligands

Receptor	DAMPs found in the CNS	PAMPs	Origin of PAMPs
TLR2	Aβ	Triacyl lipopeptides, TLR2/TLR1	Bacteria
	Biglycan	Diacyl lipopeptides, TLR2/TLR6	Bacteria, viruses
	Fibronectin	Zymosan, TLR2/TLR6	Fungi
	HMGB1	Peptidoglycan	Gram-positive bacteria
	HSP60	Lipoteichoic acid	Gram-positive bacteria
	HSP70		
	HSP90B1		
	Hyaluronan		
	Monosodium urate crystals		
	α-Synuclein		
	Surfactant protein A		
	Versican		
TLR3	mRNA	dsRNA	Viruses
TLR4	Aβ	LPS	Gram-negative bacteria, viruses
	αA-crystallin		
	αB-crystallin		
	Fibronectin		
	Heparan sulfate		
	HSP60		
	HSP70		
	HSP72		
	HSP90B1		
	Hyaluronan		
	Lysozyme		
	Monosodium urate crystals		
	Peroxiredoxin 1		
	Resistin		
	S100 protein		
	Surfactant protein A		
	Tenascin C		
TLR5		Flagellin	Gram-positive or gram-negative bacteria
TLR7	Let-7b miRNA	ssRNA	Viruses
	miR-132-5p		
	miR-146a-5p		
	miR-340-3p		
TLR8	miR-21	ssRNA	Viruses
	miR-27		
	miR-132-5p		
	miR-340-3p		
TLR9	Chromatin–immunoglobulin (Ig)G complex	Unmethylated CpG motifs in DNA	Bacteria, viruses
	mtDNA		
	DNA		

Note that only DAMPs in the CNS are listed

using the adaptor proteins, myeloid differentiation primary response 88 (MyD88) or TIR-domain-containing adaptor-inducing interferon-β (TRIF). All TLRs, except TLR3, utilize the MyD88-dependent signaling pathway. TLR3 signals primarily through the TRIF-dependent pathway. TLR4 utilizes both the MyD88-dependent and the TRIF-dependent signaling pathways. Activation of TLR signaling prompts the translocation of nuclear factor kappa light chain enhancer of activated B cells (NF-κB), interferon regulatory factors (IRFs), and transcription factor activator protein 1 (AP-1) to the nucleus and transcription of inflammatory cytokines or type I IFNs (Heiman et al., 2014; Kawai & Akira, 2010) (Fig. 3.1). While this is the canonical pathway, noncanonical pathways have been described (Chen et al., 2019). Additional adaptor proteins, especially in neurons, have been reported (Mukherjee et al., 2015; Peng et al., 2010). However, information regarding noncanonical signaling pathways or other adaptor proteins is scarce and requires further investigation.

Ligand binding to TLRs triggers the formation of homodimers. TLR2 forms heterodimers with either TLR1 or TLR6 and interacts with co-receptors such as CD14. The binding of LPS to TLR4 promotes the formation of a complex containing MD-2 and CD14. TLR2 and TLR4 utilize the bridging adaptor TIRAP to interact with MyD88. TLR4 can also signal through the TRIF pathway and utilize the bridging adaptor TRAM. Activation of the MyD88 signaling pathway results in the activation of MAPKs, followed by induction of AP-1-dependent transcription, translocation of NF-κB to the nucleus, and initiation of NF-κB-mediated transcription or activation of IRF5-dependent transcription. Activation of the TRIF pathway induces either IRF3- and IRF7-dependent transcription or AP-1, NF-κB, and IRF5-dependent transcription via effects on TRAF6. The outcome is increased expression of pro-inflammatory cytokines and type 1 IFNs.

AP-1, activator protein 1; CD14, cluster of differentiation 14; IFN, interferon; IRAK, interleukin-1 receptor-associated kinase; IKK, IκB kinase; IRF, interferon regulatory factor; LPS, lipopolysaccharide; MyD88, myeloid differentiation primary response 88; MAPK, mitogen-activated protein kinase; NEMO, NF-κB essential modulator; NF-κB, nuclear factor kappa B; TLR, toll-like receptor; TIRAP, toll/interleukin-1 (IL-1) receptor (TIR)-domain-containing adaptor pro-

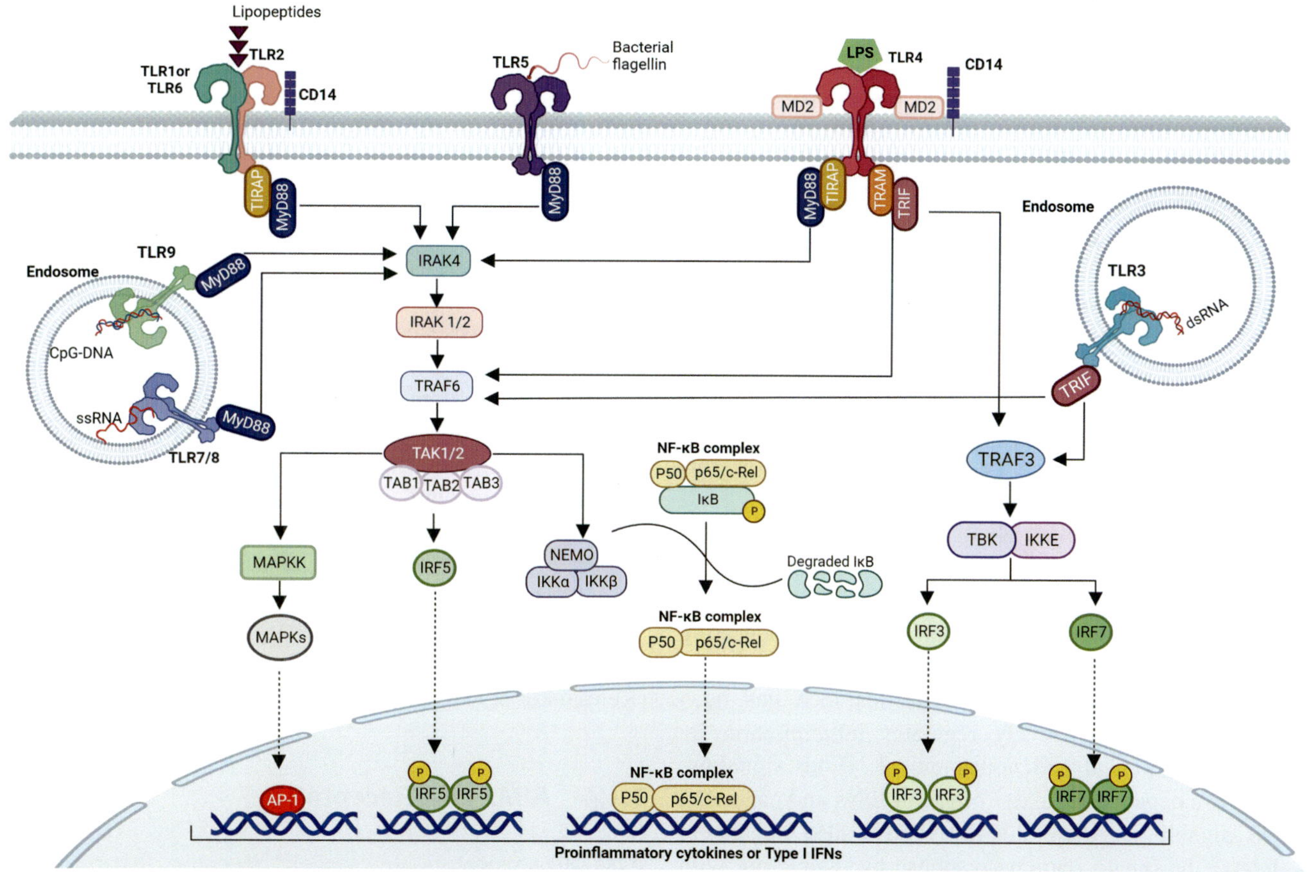

Fig. 3.1 Toll-like receptor signaling pathways. Schematic summarizing the signaling pathways utilized by the cell surface receptors TLR2, TLR4, and TLR5 and the intracellular receptors TLR3 and TLR7–9. Most TLRs utilize the MyD88-dependent signaling pathway, except for TLR3, which signals through TRIF

tein; TRAM, toll/IL-1R domain-containing adaptor-inducing IFN-β-related adaptor molecule; TRIF, toll/IL-1R domain-containing adaptor-inducing IFN-β. Created with BioRender.com

3.1.2 NOD-Like Receptors

NOD-like receptors are cytosolic PRRs consisting of three domains. The C-terminal leucine-rich repeat (LRR) domain facilitates interactions with ligands, the central nucleotide-binding and oligomerization (NACHT) domain mediates the self-oligomerization and activation of NLRs, and the variable N-terminal effector domain enables interactions with adaptor proteins and downstream effector proteins involved in signal transduction. NLRs are classified into subfamilies based on their N-terminal domains, which generally comprise a caspase activation and recruitment domain (CARD), a baculovirus inhibitor of apoptosis (BIR) domain, or a pyrin domain (PYD). *NLRA* contains one member, class II major histocompatibility complex transactivator (CIITA), which has an acidic transactivation domain. *NLRB* contains the BIR domain and includes neuronal apoptosis inhibitory proteins (NAIPs). *NLRC* contains a CARD and includes NLRC1–5. *NLRP* contains a PYD and NLRP1–14 (Looi et al., 2021). *NLRX* contains an N-terminal domain that does not have a strong analogy to the domains of all other NLRs. The X nomenclature indicates that the domain has not been well-characterized. NLRX1 is an unusual NLR because it is localized in mitochondria. Of note, NLRP1 contains both CARD and PYD, as well as the function to find domain (FIIND), a peptidase domain that has been implicated in autoproteolytic cleavage and is required for the activation of NLRP1. NLRC2 contains two CARDs (Fig. 3.2). Ligand binding to NLRs leads to auto-oligomerization and activation, followed by the recruitment of other proteins and the formation of multiprotein complexes. Among these protein complexes is the inflammasome, which will be discussed further in this chapter.

At the functional level, NLRs positively or negatively regulate the immune response. Some NLRs, such as NLRP12, NLRX1, and NLRC3, do not form inflammasomes and negatively regulate the signaling pathways underlying inflammation. Although NLRP12 was initially described as an inflammasome forming an NLR, this idea has been debated. NLRP12 negatively regulates inflammation by modulating canonical and noncanonical NF-κb signaling. NLRX1 is a negative regulator of type I IFNs and canonical NF-κB signaling (Allen, 2014). NLRC3 also negatively modulates the innate immune response by interacting with the stimulator of interferon genes (STING), a DNA-sensing signaling molecule that modulates the transcription of type I IFNs and other pro-inflammatory cytokines (Zhang et al., 2014). Most of the other NLRs function as positive regulators of innate immunity, primarily by forming inflammasomes.

3.1.3 RIG-Like Receptors

RIG-like receptors are cytosolic RNA sensors such as viral dsRNA. RIG-I, melanoma differentiation-associated protein 5 (MDA5), and laboratory of genetics and physiology 2 (LGP2) are among the members of this family. RIG-I recognizes short RNAs, whereas MDA5 senses long RNAs. RLRs have a helicase domain and a distinctive, regulatory C-terminal domain (CTD). Whereas RIG-I and MDA5 contain tandem CARDs at the N-terminal, LGP2 does not have a CARD. Downstream signaling of RIG-I and MDA5 is mediated through CARDs. In the nonactivated state, the RIG-I monomer is autoinhibited through interactions between its CARD, helicase domain, and an internal repressor domain (RD) at the C-terminus. Ligand binding initiates conformational changes that enable oligomerization and activation. RIG-I interacts with the adaptor protein membrane-bound mitochondrial antiviral signaling protein (MAVS, also named IPS-1 or CARDIF) through CARDs. These interactions are regulated by RD (Saito et al., 2007). RIG-I signaling initiates the transcriptional induction of type I IFNs and IFN-stimulated genes that collectively mount an antiviral response (Rehwinkel & Gack, 2020).

Evidence indicates that LGP2 regulates RIG-I and MDA5 activities positively or negatively (Rodriguez et al., 2014). LGP2 binds RNA with higher affinity than RIG-I and MDA5. Therefore, it has been postulated that it negatively regulates their activity through the sequestration of RNA (Rothenfusser et al., 2005). In addition, it has been hypothesized that LGP2, which contains an RD, interacts with the helicase domain of RIG-I and prevents RIG-I self-association and signaling (Saito et al., 2007). In contrast, studies investigating the response of LGP2-deficient mice to viral infection indicate that LGP2 positively regulates RIG-1 and MDA5-mediated antiviral signaling. These mice exhibit attenuated antiviral response and reduced IFN-β production, suggesting that LGP2 facilitates RIG-1 and MDA5 signaling (Satoh et al., 2010). Of note, in addition to viral RNA, RLRs recognize host-derived cellular RNAs (Rehwinkel & Gack, 2020).

3.1.4 AIM2-Like Receptors

AIM2-like receptors are cytosolic receptors that recognize dsDNA derived from pathogens or cytosolic self-DNA generated by endogenous retroviruses and retroelements. This family of PRRs belongs to the larger pyrin and hematopoi-

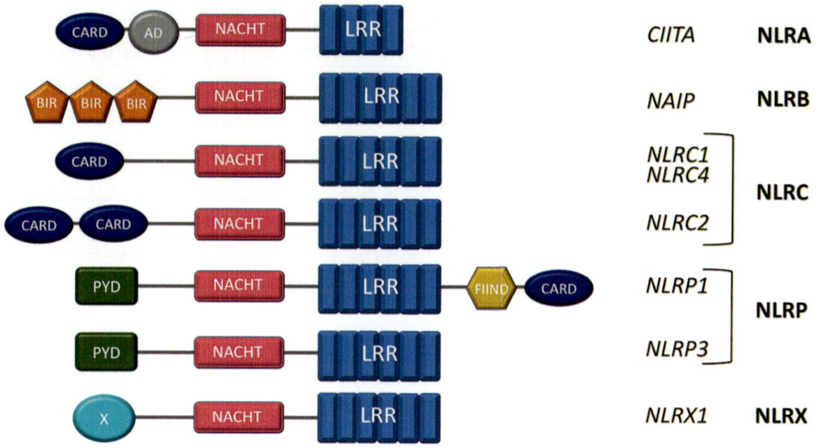

Fig. 3.2 Schematic illustrating the structure of NLRs. Only a few members of each subfamily are shown. AD acidic transactivation domain, BIR baculovirus inhibitor of apoptosis, CARD caspase activation and recruitment domain, CIITA major histocompatibility complex class II transactivator, FIIND function to find domain, LRR leucine-rich repeat, NACHT nucleotide-binding and oligomerization domain, NAIP neuronal apoptosis inhibitory protein, NLR nucleotide-binding and oligomerization domain (NOD)-like receptor, PYD pyrin domain, X uncharacterized domain

etic interferon-inducible nuclear protein (HIN) domain-containing (PYHIN) family and includes AIM2, IFN-inducible 16 (IFI16), and IFI250. A total of 4 ALR members have been characterized in humans and 13 in mice. Upon DNA binding, AIM2 and IFI16 activate the inflammasome pathway and induce IL-1β and IL-18 production and type I IFN transcription (Unterholzner et al., 2010). Stimulation of IFI250 by cytosolic retrotransposon DNA induces the type I IFN response via the stimulator of interferon genes (STING) pathway (Nakaya et al., 2017).

The interplay between ALRs is complex. An IFI16 isoform, IFI16β, binds AIM2 and impedes the formation of the AIM2 inflammasome by preventing interactions between AIM2 and ASC (apoptosis-associated speck-like protein containing a CARD). IFI16β also interferes with AIM2 inflammasome function through sequestration of dsDNA and reduction of its availability for AIM2. AIM2 prevents the interactions between IFI250 and STING and downregulates the IFI250-STING-mediated type I IFN response (Nakaya et al., 2017). In contrast, investigations on knockout mice with deletion of all ALRs have revealed that ALRs are not essential for the intracellular DNA-induced IFN response mediated by the STING pathway (Gray et al., 2016). Thus, further studies are necessary to better elucidate the interplay between different members of the ALR family and how this modulates the innate immune response.

3.1.5 C-Type Lectin Receptors

C-type lectin receptors are a large family that includes transmembrane and soluble receptors. A subgroup of these receptors that functions as PRRs recognizes a broad spectrum of pathogens and primarily belongs to the Dectin-1 or Dectin-2 family. These receptors are type II transmembrane proteins with an extracellular carbohydrate recognition domain, a transmembrane domain, and a cytoplasmic tail. In addition to pathogen-derived carbohydrates, DAMPs also activate CLRs (Nagata et al., 2017). Dendritic cell-associated C-type lectin-1 (Dectin-1 also named C-type lectin domain family 7A; CLEC7A), Dectin-2 (also named C-type lectin domain family 6A; CLEC6A), and macrophage-inducible C-type lectin (Mincle), a member of the Dectin-2 family, utilize the NF-κB and mitogen-activated protein kinase (MAPK) signaling pathways.

Dectin-1 recognizes fungal β-glucans and initiates an inflammatory response by inducing the secretion of tumor necrosis factor-α (TNF-α), IL-6, and IL-12. In addition to innate immune cells in the periphery, Dectin-1 is found in microglia. Dectin-1 signaling has been associated with neurotoxicity, neurodegeneration, and regeneration, depending on the context. When activated in macrophages by zymosan, a fungal cell wall extract, Dectin-1 confers neurotoxicity (Gensel et al., 2015). In contrast, intraocular zymosan injections in parallel with optic nerve crush injury promote axonal regeneration (Baldwin et al., 2015). There is also a crosstalk between Dectin-1 and TLRs to fine-tune inflammatory responses and cytokine release (Ferwerda et al., 2008).

Mincle mediates inflammatory responses to fungal and mycobacterial PAMPs. The best-characterized endogenous ligand of Mincle is splicing factor 3b subunit 3 (SF3B3) (Yamasaki et al., 2008), referred to as spliceosome-associated protein 130 (SAP130) in some reports (Metselaar et al., 2021). Studies on macrophages indicate that Mincle signals through CARD9 and via engagement of the spleen tyrosine kinase (syk), leading to TNF-α production (Drummond et al., 2011). There is controversy with respect to the expression of Mincle in CNS cells. Some reports show Mincle expression in microglia and neurons (He et al., 2015), whereas others exclude Mincle expression in microglia (Arumugam et al., 2017).

3.1.6 The Triggering Receptor Expressed on Myeloid Cells

The TREM family of PRRs includes TREM-1–4. TREM-1 is the best-characterized member of the TREM family. It is a cell surface receptor expressed in immune and nonimmune cells. TREM-1 is a transmembrane glycoprotein that does not have a cytoplasmic signaling motif and associates with the adaptor DNAX activating protein of 12 kDa (DAP12) to initiate signaling (Bouchon et al., 2000). Activation of TREM-1 leads to the phosphorylation of syk, which, in turn, activates the MAPK/extracellular signal-regulated kinase (ERK) pathway, resulting in the translocation of NF-κB to the nucleus and the transcription of cytokines. TREM-1 is best known for its role in amplifying the inflammatory reaction in response to bacterial infections (Bouchon et al., 2001). Two isoforms, membrane-bound and soluble (mTREM-1 and sTREM-1, respectively), have been described. mTREM-1 participates in the amplification of the inflammatory response, whereas sTREM-1, a glycosylated peptide, might negatively regulate mTREM-1 signaling by neutralizing its ligands. Interactions between TREM-1 and TLR4 have been reported, with TLR4 directly regulating the expression of TREM-1. TREM-1, in turn, regulates the expression of constituents in the TLR4 downstream signaling pathway (Dantas et al., 2020).

There is controversy with respect to the expression of TREM-1 in cells intrinsic to the CNS. Some reports showed expression in microglia but not in other glial subtypes or neurons (Xu et al., 2019). Other reports showed that infiltrating macrophages rather than microglia are the source of TREM-1 in the CNS.

TREM-2 is the most widely expressed member of the TREM family. TREM-2 signaling necessitates its association with DAP12 and activation of syk. TREM-2 is highly expressed in the hippocampus and spinal cord (Forabosco et al., 2013). Both human and mouse microglia express TREM-2 (Schmid et al., 2002).

TREM-2 is considered a negative regulator of inflammation. Abrogation of TREM-2 expression upregulates TLR-mediated release of pro-inflammatory cytokines by macrophages and increases the production of pro-inflammatory mediators by microglia. TREM-2 plays a protective role by promoting the rate of microglial phagocytosis of bacterial products, tissue debris, and Aβ. The expression of several anti-inflammatory genes in microglia and macrophages is TREM-2-dependent (Deczkowska et al., 2020). However, TREM-2 has also been linked to pro-inflammatory signaling and negative disease outcomes in some reports. Unraveling the tissue-, cell-, and disease-specific roles of TREM-2 and its soluble and membrane-bound forms can provide better insights into the way by which TREM-2 modulates inflammation in distinct pathological conditions.

3.2 Toll-like Receptors in Brain Pathology

Inflammation is a principal feature of neurodegenerative diseases and brain injury (BI). This prompted investigations on the role of PRRs in brain pathology to determine whether they could be targets for therapeutic interventions.

3.2.1 Alzheimer's Disease

Alzheimer's disease (AD) is the most common cause of dementia. The histopathology of AD is characterized by extracellular deposits of Aβ aggregates and the formation of plaques, intracellular accumulation of neurofibrillary tangles in neurons consisting primarily of hyperphosphorylated tau, and neuronal death. The hippocampus and entorhinal cortex are the brain regions most affected at early disease stages, followed by the cerebral cortex at later stages. Neuronal loss in these regions leads to deficits in memory, reasoning, and social behavior (Masters et al., 2015). Neuroinflammation has emerged as an essential player in the pathogenesis and progression of AD (Heppner et al., 2015).

Several TLRs have been implicated in AD pathology. By and large, investigations on TLRs have focused on animal models of AD with and without TLR deficiency and on in vitro models of neuronal degeneration and neuron–glia interactions with relevance to AD. At the cellular level, microglial TLRs have received the highest attention, followed by astrocytic TLRs. However, evidence indicates that neuronal TLRs might also contribute to AD pathogenesis.

Microglia play both beneficial and detrimental roles in AD. In the healthy CNS, microglia act as sentinels that patrol the environment for invading pathogens and tissue debris generated by cell death. In CNS pathologies, microglia undergo morphological, molecular, and functional changes in response to environmental stimuli and become activated. Activated microglia confer neuroprotection by clearing dead or infected cells, tissue debris, and other pathogenic substrates at the affected site via phagocytosis. Microglia also have the capacity to repair. However, persistent microglial activation leads to the release of neurotoxic factors and pro-inflammatory cytokines that cause damage and cell death.

Thus, the contribution of microglia to CNS pathology is often complex and context-dependent and changes during the course of the disease.

Studies on AD have mostly focused on TLR4. Expression of TLR4 is increased in the brain of individuals affected by AD. A link between TLR4 genetic polymorphism and a decreased risk of late-onset AD has been reported. This was attributed to the attenuation of neuroinflammation and IL-1β release because of impaired TLR4 signaling (Minoretti et al., 2006).

Fibrillary Aβ is an aggregated form of the Aβ peptide that causes neurotoxicity and activates the microglia surrounding plaques. TLR4, TLR2, and CD14 mediate Aβ-induced microglial activation. Although activated microglia exert a protective effect by clearing fibrillary Aβ, the accumulation of Aβ in microglia eventually causes the release of neurotoxic effectors, including reactive oxygen species (ROS) and pro-inflammatory cytokines, leading to neuronal death.

Studies on a transgenic mouse model of AD indicate that a loss-of-function mutation in TLR4 reduces microglial activation at an early disease stage. The TLR4 mutation does not affect amyloidogenesis, indicating that the receptor does not play a role in the initial deposition of Aβ. However, the Aβ load is higher in the brain of TLR4 mutant mice compared to wild-type (WT) controls at a later disease stage, suggesting that the loss of TLR4 function perturbs Aβ clearance by activated microglia. The higher Aβ load in the brain of TLR4 mutant mice is paralleled by the deterioration of cognitive function. In agreement with these findings, Aβ-induced TLR4 signaling mediates the activation and recruitment of microglia to the vicinity of plaques, which is paralleled by the demise of neurons in the affected brain regions (Song et al., 2011).

In addition to microglial TLR4, in vitro investigations raise the possibility that astrocytic TLR4 contributes to neurodegeneration and neuroinflammation in AD. In the healthy CNS, astrocytes regulate many key physiological mechanisms. During CNS pathology, astrocytes change morphology and function and become reactive. Reactive astrocytes proliferate and migrate to the affected sites. Astrocytes have emerged as key players in the innate immune response of the CNS, partly mediated by TLR signaling (Li et al., 2021a). In neuron–astrocyte co-cultures, stimulation of TLR4 by fibrillary Aβ induces TNF-α release and causes cell death, with neurons being the most vulnerable cell population. This effect of Aβ has been primarily attributed to astrocytic neurotoxicity, even though the direct action of Aβ on neurons could not be completely ruled out (Hughes et al., 2020).

In fact, Aβ activates neuronal TLR4 in vitro. It has been reported that TLR4 expression in cultured cortical neurons increases in response to Aβ(1–42). Exposure of cortical neurons to Aβ activated Jun N-terminal kinase (JNK) and caspase-3, resulting in apoptosis. This effect of Aβ(1–42) is significantly reduced in $TLR4^{-/-}$ cortical neuronal cultures compared to wild-type neurons (Tang et al., 2008). It has also been reported that Aβ enhances LPS-induced increase in intracellular Ca^{2+} and apoptosis of aged hippocampal neurons in vitro. Thus, LPS and Aβ act in synergy to promote the demise of hippocampal neurons (Calvo-Rodriguez et al., 2017).

High mobility group box 1 (HMGB1) is an endogenous TLR4 ligand that activates the TLR4/NF-κB signaling pathway in hippocampal neurons (Nan et al., 2019). The addition of Aβ to hippocampal neuronal cultures increases HMGB1, and TLR4 levels, activates NF-κB signaling, and upregulates IL-1β, IL-6, and TNF-α levels (Nan et al., 2019). Thus, HMGB1 and neuronal TLR4 participate in the mechanisms underlying Aβ-induced neuroinflammation. HMGB1 also promotes neurite degeneration through direct effects on cortical neurons and via TLR4-dependent phosphorylation of myristoylated alanine-rich C-kinase substrate (MARCKS), a protein that modulates the stability of the actin network. This has been linked to AD pathology because MARCKS phosphorylation is observed in a mouse AD model and the postmortem human brain. Anti-HMGB1 antibodies inhibit MARCKS phosphorylation and neurite degeneration while restoring cognitive function in the mouse model of AD. In addition, TLR4 antagonism or TLR4 knockdown significantly reduces HMGB1-induced MARCKS phosphorylation, confirming that the effects of HMGB1 are mediated through TLR4 (Fujita et al., 2016).

In summary, it is postulated that induction of microglial and astrocytic TLR4 signaling by Aβ aggregates causes the demise of neurons in the affected brain regions through the secretion of pro-inflammatory cytokines, release of neurotoxic factors, and interference with microglia-mediated Aβ clearance. Moreover, neuronal TLR4 increases the vulnerability of neurons to Aβ and participates in additional mechanisms that affect the integrity of neurons and cognitive function.

3.2.2 Parkinson's Disease

Parkinson's disease (PD) is a neurodegenerative disease characterized by motor dysfunction, including tremors, gait imbalance, and rigidity. Aberrant cognition, mood, and behavior are also among PD symptoms. PD presents in two different forms: familial early-onset PD and sporadic PD. Mutations in several genes, including *SNCA*, which encodes α-syn, have been associated with familial PD. Histopathological hallmarks of PD are the selective degeneration of dopaminergic (DA) neurons in the substantia nigra pars compacta (SNpc), which leads to abnormal dopaminergic signaling, the presence of α-syn containing cytoplasmic inclusions in neurons, also called Lewy body,

(Goedert et al., 2013), the presence of activated microglia in the vicinity of affected neurons, and neuroinflammation. Several animal models of PD and α-synucleinopathies have facilitated investigations on the mechanisms underlying α-syn pathology, degeneration of DA neurons, and neuroinflammation (Blandini & Armentero, 2012). Neurotoxin-induced degeneration of nigrostriatal neurons and transgenic mouse models overexpressing WT or mutated α-syn are among these animal models. Even though animal models mimic different, but not all pathological aspects of PD, they have been useful in advancing the understanding of PD pathogenesis and progression.

TLRs have been implicated in PD pathology. The role of TLRs in microglia- and astrocyte-mediated neuroinflammation and neurotoxicity and the contribution of neuronal TLRs to PD pathology have been investigated (Kouli et al., 2019). TLR2 is the receptor most frequently studied in the context of PD, followed by TLR4.

Studies on microglia have shown that neuron-derived oligomeric α-syn is a TLR2 ligand. α-Syn upregulates TLR2 expression in a microglia-like cell line, mediates microglial activation, and promotes the inflammatory response and neurotoxicity (Beraud et al., 2011; Kim et al., 2013). In agreement with these findings, treatment of primary microglial cultures or a microglia-like cell line with α-syn preformed fibrils (PFFs) enhances TLR2–MyD88 interactions, NF-κB activation, and IL-1β and inducible nitric oxide synthase (iNOS) expression. Pro-inflammatory agents released by PFF-treated microglia, in turn, increase α-syn expression in DA neurons in vitro. Similarly, in mice seeded with PFFs, treatment with an inhibitor of TLR2 signaling hinders microglial activation, prevents the spreading of α-syn pathology, protects DA neurons, and ameliorates functional deficits (Dutta et al., 2021).

The A53T mouse, which overexpresses the human α-syn that contains a mutation found in PD, is considered an animal model of PD and other α-synucleinopathies. The accumulation of α-syn in specific neuronal subpopulations and motor deficits are characteristics of these mice. TLR2 deficiency or knockdown in A53T mice reduces the accumulation of α-syn in neurons and neuronal loss and improves motor deficits. These investigations have also revealed that activation of TLR2 signaling in neuron-like cells perturbs the autophagy/lysosomal degradation pathway and is likely responsible for the accumulation of α-syn oligomers in neurons (Kim et al., 2015).

Propagation of α-syn pathology is a hallmark of PD progression. In vitro studies indicate that neuronal TLR2 mediates neuron-to-neuron α-syn transmission. In a model of donor and recipient neuron co-culture, activation of TLR2 signaling or overexpression of TLR2 in recipient cells enhances the internalization of α-syn released by donor neurons. The accumulation of α-syn in recipient neurons leads to neurotoxicity and activation of caspase-3, a marker of apoptotic cell death (Kim et al., 2018). Thus, neuronal TLR2 could play a key role in the progression of PD by propagating α-syn pathology and neuronal degeneration.

α-Syn-containing inclusions are often observed, not only in neurons but also in astrocytes. It has been hypothesized that α-syn in astrocytes is of a neuronal origin because the expression of α-syn in astrocytes is very low. In fact, direct transmission of α-syn from neurons to astrocytes has been demonstrated in vitro and by use of a mouse model of α-synucleinopathy exhibiting α-syn containing inclusions and motor deficits. The accumulation of α-syn in astrocytes is associated with pro-inflammatory gene expression and inflammatory mediators (Lee et al., 2010). In donor neuron-recipient human astrocyte co-cultures, overexpression of TLR2 increases neuron-to-astrocyte transmission of α-syn, whereas a TLR2-neutralizing antibody reduces the accumulation of α-syn in recipient astrocytes. In vivo, investigations concur with in vitro findings. Administration of TLR2-neutralizing antibodies reduces the accumulation of α-syn in neurons and astrocytes, attenuates neuroinflammation, mitigates neurodegeneration, and ameliorates functional deficits in a transgenic mouse model of PD (Kim et al., 2018). α-Syn-evoked activation of astrocytic TLR2 signaling leads to the production of pro-inflammatory cytokines and chemokines, including IL-1β, IL-6, TNF-α, CCL5, and CX3CL1, in vitro (Kim et al., 2018).

Analysis of postmortem human brains confirms the findings in mouse models of PD. TLR2 expression is increased in the postmortem brain of individuals with PD. In control brains, TLR2 is expressed exclusively in microglia, whereas in PD brains, TLR2 is primarily increased in neurons of the substantia nigra and anterior cingulate cortex. There is a strong correlation between the increase in neuronal TLR2 expression and the disease stage. The co-localization of TLR2 and α-syn in Lewy bodies further supports the notion of interactions between the two proteins. Activation of TLR2 signaling in human-induced pluripotent stem cell (iPSC)-derived neurons leads to the accumulation of α-syn due to inhibition of the autophagy/lysosomal degradation pathway (Dzamko et al., 2017), a mechanism by which α-syn is cleared (Chedid et al., 2022).

TLR4 signaling has also been implicated in α-syn-induced inflammatory responses in astrocytes. Genetic deletion of astrocytic TLR4 suppresses α-syn-induced activation of NF-κB and the transcription and release of pro-inflammatory cytokines, including TNF-α, CXCL1, and IL-6, without affecting α-syn uptake (Fellner et al., 2013; Rannikko et al., 2015).

It remains to be determined whether and how other TLRs contribute to PD pathology and whether TLR2 could be a therapeutic target.

3.2.3 Brain Injury

Studies on brain injuries (BIs) have been performed on various developmental or adult BI models, including ischemic-hypoxic BI, traumatic brain injury (TBI), and concussions. In vitro models that have relevance to BI have also been utilized. Investigations have primarily focused on the contribution of TLR4, followed by TLR2 to a lesser extent (Babcock et al., 2006; Lalancette-Hebert et al., 2009; Lehnardt et al., 2007) and TLR3 (Stridh et al., 2013). Most, but not all, investigations associate TLR2, TLR3, and TLR4 signaling with negative outcomes of BI and report that suppression of TLR signaling and expression, or genetic deletion of these TLRs, alleviates cellular loss, tissue damage, and functional deficits in a cell type-dependent manner. Microglia are the cells that receive the most attention, followed by astrocytes. Neuronal TLRs have not been investigated adequately, even though evidence indicates that they contribute to brain pathology following injury. However, some studies have shown that the contribution of TLRs to BI could also be beneficial depending on the milieu.

3.2.3.1 Ischemic and Hemorrhagic Brain Injury

TLR4 has been implicated in subarachnoid hemorrhage (Heinz & Schneider, 2022) and intracerebral hemorrhage (Sansing et al., 2011). Increased TLR2 and/or TLR4 expression has been observed in an animal model of ischemic BI (Caso et al., 2007; Hyakkoku et al., 2010; Tang et al., 2007; Ziegler et al., 2007). Neuronal and microglial TLR4 expression is upregulated in the ipsilateral but not in the contralateral ischemic brain. However, the increase in TLR4 expression in neurons as opposed to microglia occurs at distinct times post-injury. Microglial TLR4 expression is upregulated as early as 4 hours post-injury, whereas neuronal TLR4 expression increases 22 hours post-injury (Hyakkoku et al., 2010). The quick response of microglia is not surprising, given that they are the early sensors of perturbations in the environment. The upregulation of TLR4 is likely associated with the initiation of microglial activation, which occurs within 12–24 hours in rodent models of ischemia.

The importance of TLR2 and TLR4 in ischemic BI is demonstrated by findings indicating that TLR2 or TLR4 deficiency attenuates neurological deficits and decreases the infarction area, possibly due to reduced cell and tissue loss. Accordingly, TLR4 mutant and TLR2 knockout neurons are less vulnerable to death than wild-type neurons in an in vitro model of ischemic injury (Tang et al., 2007). Astrocytes subjected to ischemic damage also participate in neuronal degeneration through the engagement of TLR4. Activation of TLR4 signaling in ischemic brain astrocytes downregulates glutamate transporter expression and impairs glutamate clearance by astrocytes in vitro (Lin et al., 2020). Impairment of glutamate clearance can lead to excitotoxicity, resulting in cell death. Overall, the studies indicate that TLR4 signaling negatively impacts the outcomes of ischemic BI.

3.2.3.2 Traumatic Brain Injury

The role played by TLR4 in TBI is context-dependent. The detrimental effects of TLR4 in TBI are exemplified by a study showing that activation of TLR4 results in inflammation, synaptic loss, and functional deficits following TBI (Rosa et al., 2021). However, other investigations have highlighted the differential role of TLR4 in mice sustaining TBI as opposed to sham controls. Mild TBI in young rats has been used as a model of concussive BI, which is a risk factor for epilepsy and memory impairment. A transient increase in neuronal TLR4 expression is observed in the dentate gyrus (DG) following TBI. Local TLR4 antagonism differentially modulates DG excitability in injured mice and sham controls. The TLR4 antagonist decreases the excitability of DG, attenuates the vulnerability to seizures, and ameliorates memory impairment in injured mice, suggesting that TLR4 negatively impacts the outcomes of TBI. In contrast, TLR4 antagonism increases DG excitability, induces seizures, and impairs memory function in sham controls (Korgaonkar et al., 2020a, b). The cellular mechanisms underlying these distinct effects also appear to be different in sham controls as opposed to injured mice. It has been proposed that in sham controls, antagonism of glial TLR4 perturbs glia-mediated maintenance of homeostasis, leading to functional deficits. In contrast, in injured mice, the selective increase in neuronal TLR4 expression facilitates the negative effects of DAMPs and the TLR4 antagonist interferes with DAMP-induced TLR4 signaling, mitigating the detrimental outcomes.

Additional studies indicate that activation of TLR4 signaling can be either harmful or helpful depending on the time post-TBI when receptor signaling is induced. In a mild, repetitive TBI model in rats mimicking recurring head concussions, LPS-dependent activation of TLR4 1 day after the last injury reduces the inflammatory reaction, attenuates neuronal damage, and prevents depression and cognitive deficits. In contrast, when given 5 days following the last injury, LPS increases pro-inflammatory cytokines and exacerbates neuronal damage. A slight worsening of cognitive deficits and depression-like behavior is also observed (Corrigan et al., 2017). It remains to be determined whether and how temporal changes in the injury environment impact the outcomes of TLR4 signaling and which cellular and molecular mechanisms govern these differential effects of TLR4. It is also important to test whether endogenous activators of TLR4, rather than a bacterial component that binds a specific TLR4 receptor complex, produce similar results.

Currently, many investigations are focused on the identification of new pharmacotherapies that target TLR4 for the alleviation of functional deficits and cellular damage following ischemic BI or TBI. The goal is to prevent the detrimental effects of TLR4 without affecting the beneficial actions in BI.

3.3 Other Pattern Recognition Receptors in Brain Pathology

Although the contribution of TLRs to sterile inflammation in various brain pathologies has been extensively investigated, knowledge regarding the role of other PRRs in brain diseases and injury is relatively limited. Yet, evidence supports the notion that, in addition to TLRs, some of the other PRRs discussed in this chapter significantly contribute to neuroinflammation associated with brain disease and trauma. The following section will provide a few examples supporting this notion.

3.3.1 RIG-Like Receptors

In addition to infections, the involvement of RIG-I in reactive astrocyte function in neurodegenerative diseases has been reported by several laboratories.

3.3.1.1 RIG-I in Alzheimer's Disease

RIG-I levels are elevated in the temporal cortex of individuals with mild cognitive impairment with possible or probable AD and in the occipital cortex of individuals with definite AD. Aβ(3–42), the peptide most prevalent in the AD brain, increases RIG-I expression in human cortical astrocytes in vitro, and agonist-induced stimulation of RIG-I signaling elevates Aβ levels in astrocytes. Thus, RIG-I could contribute to AD pathology by modulating astrocyte function (de Rivero Vaccari et al., 2014).

3.3.1.2 RIG-I in Ischemic Brain Injury

Investigations on an animal model mimicking ischemic stroke have indicated an increase in astrocytic RIG-I followed by an elevation in IFN-α expression. Accordingly, when astrocytes were subjected to oxygen–glucose deprivation (OGD), an in vitro model of ischemia, RIG-I and IFN-α levels were increased. The studies suggested that astrocytic RIG-I could play a role in stroke and ischemic brain damage, although it is not yet determined whether this role is beneficial or detrimental (Brand et al., 2015).

3.3.2 C-Type Lectin Receptors

3.3.2.1 Mincle in Ischemic Brain Injury

Mincle has been implicated in ischemic brain injury and stroke (Suzuki et al., 2013). Mincle expression is increased in the brain and Cerebrospinal Fluid (CSF) of individuals with TBI and rodents sustaining a fluid percussion brain injury. Activation of Mincle in cortical neuronal cultures leads to the production of TNF-α, indicating that neurons can contribute to neuroinflammation via the engagement of Mincle (de Rivero Vaccari et al., 2015).

Investigations on Mincle knockout mice associate Mincle with the exacerbation of neuronal death, brain damage, and functional deficits in an animal model of ischemic BI. This was attributed to Mincle expressed in perivascular macrophages rather than microglia (Arumugam et al., 2017). Additional studies have shown the involvement of Mincle in neuroinflammation in subarachnoid hemorrhage. In these investigations, Mincle was associated with microglia and neurons but not astrocytes (He et al., 2015).

Even though these findings implicate a role for Mincle in animal models of stroke and ischemia, the precise cellular and molecular mechanisms by which Mincle signaling leads to damage in the brain require further investigation.

3.3.3 The Triggering Receptor Expressed on Myeloid Cells

3.3.3.1 TREM-2 in Alzheimer's Disease

An association between the genetic variants of *TREM-2* and an increased risk of AD has been documented (Guerreiro et al., 2013; Jonsson et al., 2013). TREM-2 deficiency has been linked to the reduction of AD-like pathology in mouse models (Jay et al., 2015). In contrast, *TREM-2* haplodeficiency in mice and humans impairs the ability of microglia to form a barrier around AD plaques without affecting their phagocytic activity and causes axonal dystrophy. However, microglial phagocytic activity is reduced in *TREM2*$^{-/-}$ mice. Thus, complete TREM-2 deficiency and a loss-of-function mutation in TREM-2 result in different outcomes (Yuan et al., 2016). Other reports indicate that a TREM-2-mediated increase in microglial phagocytosis plays a detrimental role at early AD stages by causing synaptic loss, whereas it plays a beneficial role at the middle or late stages of the disease by preventing amyloid deposition (Sheng et al., 2019).

Insights into the role of TREM-2 in AD have been provided in a comprehensive review by Deczkowska et al. (2020).

3.3.3.2 TREM-1 in Ischemic Brain Injury

TREM-1 is increased and remains high in the mouse brain following ischemia. Increased TREM-1 expression is found in microglia but not in other glial subtypes or neurons. Inhibition of TREM-1 reduces infarct progression, attenuates neuroinflammation and infiltration of immune cells, downregulates the expression of chemokines, including CXCL1, CXCL2, and MCP, prevents neuronal death, and improves behavioral deficits. In ischemic microglia, inhibition of TREM-1 signaling reduces IL-1β, IL-6, IL-18, CXCL1, CXCL2, and CCL2 expression in vitro. Collectively, these findings suggest that microglial TREM-1 contributes to neuroinflammation and neural damage in ischemia and highlight TREM-1 as a potential target for therapeutic interventions.

3.4 Inflammasomes

Inflammasomes are cytosolic multimeric protein complexes. Canonical inflammasomes consist of three main components: a sensor (PRR), the adaptor protein apoptosis-associated speck-like protein containing a CARD (ASC), and pro-caspase-1. The adaptor protein ASC that is common to most canonical inflammasomes has both a CARD and a PYD and facilitates the assembly of inflammasomes by interacting with both PRRs and pro-caspase-1. However, ASC is indispensable for assembling only some inflammasome subsets, as direct interactions between pro-caspase-1 and PRRs through CARD–CARD also occur. Caspases are aspartate-specific cysteine proteases. Their precursors, pro-caspases, are inactive. Autoproteolytic cleavage of the precursors yields active caspases. Caspases are classified as inflammatory or apoptotic caspases. Inflammatory caspases include caspase-1 and caspase-11 (Walsh et al., 2014).

Some members of the NLR and PYHIN families form inflammasomes. These inflammasomes are classified according to the name of the PRR that forms the backbone and include the NLRP, NLRC, and AIM2 inflammasomes. The assembly of the canonical NLR inflammasome protein complexes starts with the binding of PAMPs or DAMPs to the PRR and the formation of self-oligomers. Subsequently, ASC interacts with NLRs and recruits pro-caspase-1. Active caspase-1, generated by the autoproteolytic cleavage of pro-caspase-1, cleaves the precursor molecules pro-IL-1β and pro-IL-18, leading to the release of mature and biologically active IL-1β and IL-18. These cytokines initiate inflammation. Activation of caspase-1 can also cause pyroptosis, a form of programmed cell death, which is induced following the cleavage of the pyroptosis effector protein Gasdermin D (GSDMD). Gasdermin D is a pore-forming protein inserted into the cell membrane via its N-terminal and has a C-terminal, which contains an RD. Cleavage by caspase-1 separates the N-terminal from the C-terminal. Subsequently, the N-terminal oligomerizes, leading to large pores in the cell membrane and membrane rupture. Because pyroptosis is a type of lytic cell death paralleled by rapid membrane rupture, the contents of the cell, which include DAMPs and inflammatory cytokines, are released, resulting in inflammation. There are also noncanonical inflammasome activation pathways (Kayagaki et al., 2011), which will be discussed below.

3.4.1 Canonical Inflammasomes

In this section, we will focus on the NLRP1, NLRP3, NLRC4, and AIM2 inflammasomes, which are found in the central nervous system (CNS).

3.4.1.1 The NLRP1 Inflammasome

The NLRP1 inflammasome is the first inflammasome described in the CNS in relation to spinal cord injury. However, NLRP1 also mediates the innate immune response of the brain following traumatic injury and stroke. NLRP1 is unique among all NLRPs because it contains two signal transduction domains, a C-terminal CARD and an N-terminal PYD. The C-terminal CARD of NLRP1 can directly interact with the CARD of pro-caspase-1 and activate caspase-1 (Faustin et al., 2007). Consequently, NLRP1 does not necessitate the adaptor ASC for caspase-1 and inflammasome activation (Van Opdenbosch et al., 2014). However, ASC might enhance NLRP1 inflammasome activation. NLRP1 also contains a FIIND. Autoproteolytic processing of FIIND is necessary for the activation of the NLRP1 inflammasome.

3.4.1.2 The NLRP3 Inflammasome

The NLRP3 inflammasome is the most widely studied and the best-characterized inflammasome. It is activated by multiple PAMPs, including bacterial, fungal, and viral components, and DAMPs such as Aβ, extracellular adenosine triphosphate (ATP), uric acid crystals, and environmental microparticles such as silica crystals and asbestos. Thus, NLRP3 may act as a general cellular damage or stress sensor. The assembly of the NLRP3 inflammasome is presented in Fig. 3.3.

Unlike other inflammasomes, activation of the NLRP3 inflammasome requires a two-step process. A first signal, for example, a TLR ligand, primes cells to produce pro-IL-1β and increases the expression of NLRP3 and other components of the inflammasome complex. A second signal, such as ATP, triggers the assembly and activation of the NLRP3 inflammasome complex (Fig. 3.4).

3.4.1.3 The NLRC4 Inflammasome

NLRC4 forms an inflammasome complex with NAIPs. Bacterial flagellin and PrgJ, a type III secretion system of bacterial pathogens that are recognized by NAIPs, are activators of the NLRC4 inflammasome. NLRC4 and pro-caspase-1 can interact directly through their respective CARDs without the need for ASC. However, ASC potentiates NLRC4-induced activation of caspase-1. In addition to its classical role in bacterial infection, studies indicate that the NLRC4 inflammasome is also involved in sterile inflammation in CNS pathology (Denes et al., 2015; Poh et al., 2019; Ziegler et al., 2007). Although the DAMPs that bind NLRC4 have not been fully identified, lysophosphatidylcholine

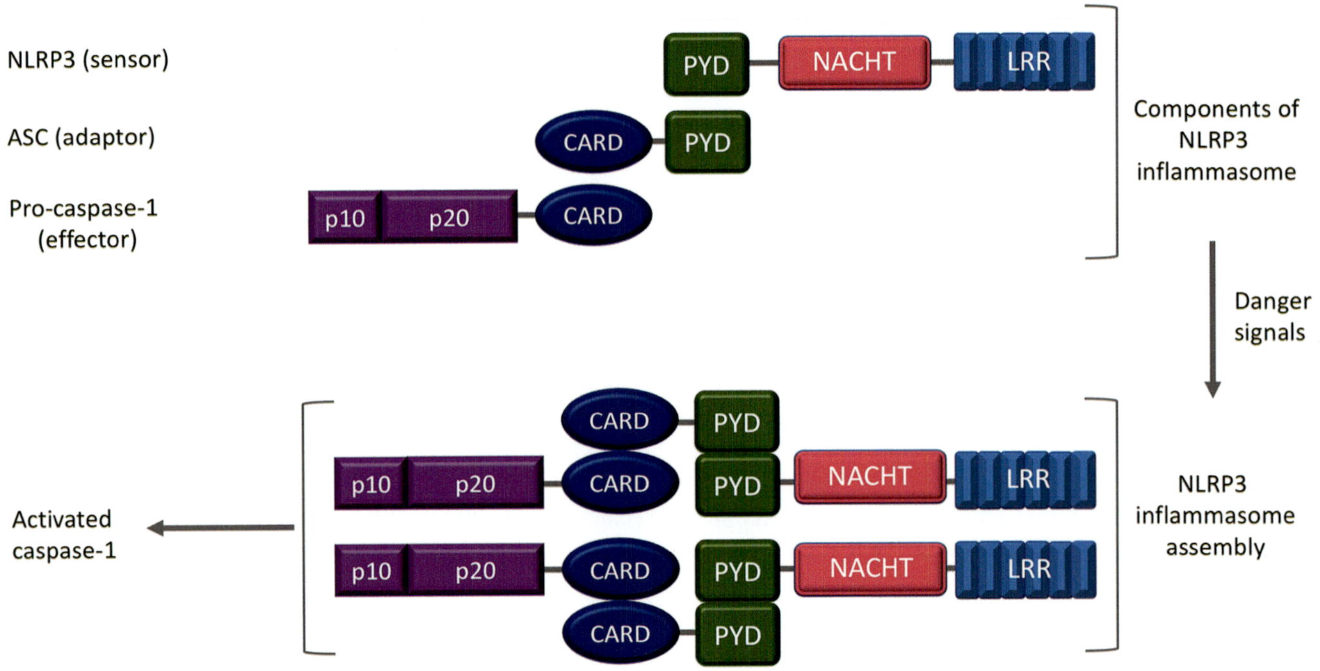

Fig. 3.3 Structure of the NLRP3 inflammasome. Domain organization of components forming the NLRP3 inflammasome (*upper panel*). NLRP3 contains a C-terminal LRR domain, a central NACHT domain, and an N-terminal PYD. Upon sensing danger signals (PAMPs or DAMPs), NLRP3 oligomers recruit the adaptor protein ASC through PYD–PYD interactions, and ASC recruits the effector pro-caspase-1 via CARD–CARD interactions (*lower panel*). The assembly of the inflammasome complex results in caspase-1 activation. ASC apoptosis-associated speck-like protein containing a CARD (ASC), CARD caspase activation and recruitment domain, DAMP danger-associated molecular pattern, LRR leucine-rich repeat, NACHT nucleotide-binding and oligomerization domain, NLR nucleotide-binding and oligomerization domain (NOD)-like receptor, PAMP pathogen-associated molecular pattern

(LPC), a major component of low-density protein, has been shown to induce NLRC4 inflammasome activation in microglia and astrocytes to mediate sterile inflammation.

3.4.1.4 The AIM2 Inflammasome

As stated above, ALRs are intracellular receptors that are primarily activated by viral dsDNA. However, dsDNA derived from dying cells has also been identified as an endogenous ligand for ALRs in the CNS. AIM2 has an N-terminal PYD and a 200-amino acid repeat HIN (HIN200) domain, which binds dsDNA at the C-terminus. The PYD enables interaction with ASC, which recruits pro-caspase-1, leading to the activation of caspase-1 and secretion of IL-1β and IL-18. ALRs also mediate pyroptosis. Upon ligand binding to the HIN domain, the inhibitory effect on PYD is lifted, resulting in the assembly and activation of the inflammasome, which triggers pyroptotic neuronal cell death.

3.4.2 Noncanonical Inflammasomes

Noncanonical inflammasomes have primarily been studied in immune system cells and, to a lesser extent, in microglia. Their signaling involves the activation of caspase-11 in mice (caspase-4 and caspase-5 in humans) in response to LPS (Kayagaki et al., 2011). These pro-inflammatory caspases act as PRRs and effector molecules in activating the noncanonical inflammasome pathway. This contrasts with canonical inflammasomes, which have distinct sensors and effectors. Caspase-11 directly binds LPS through interactions between the lipid A moiety of LPS and CARD of caspase-11. The binding of LPS initiates the oligomerization of the caspase-11–LPS complex and the autoproteolytic activation of caspase-11 (Shi et al., 2014). Activation of noncanonical inflammasomes induces pyroptosis through cleavage of GSDMD but does not directly mediate IL-1β and IL-18 release. Instead, caspase-11 cooperates with the NLRP3 inflammasome and, in doing so, indirectly promotes the secretion of IL-1β and IL-18 through the GSDMD pores (Yi, 2017). Thus, LPS promotes inflammation and pyroptotic cell death by activating noncanonical and canonical inflammasome signaling in a TLR4-independent manner (Hagar et al., 2013; Kayagaki et al., 2011).

Besides LPS, caspase-11 binds oxidized phospholipids such as 1-palmitoyl-2-arachidonoyl-sn-glycero-3-phosphorylcholine (ox-PAPC) released by dying cells through its catalytic domain. However, unlike LPS, ox-PAPC does not induce pyroptotic cell death (Zanoni et al., 2016), indicating that the biological outcomes of noncanonical inflammasome activation are context-dependent (Yi, 2020). In fact, in

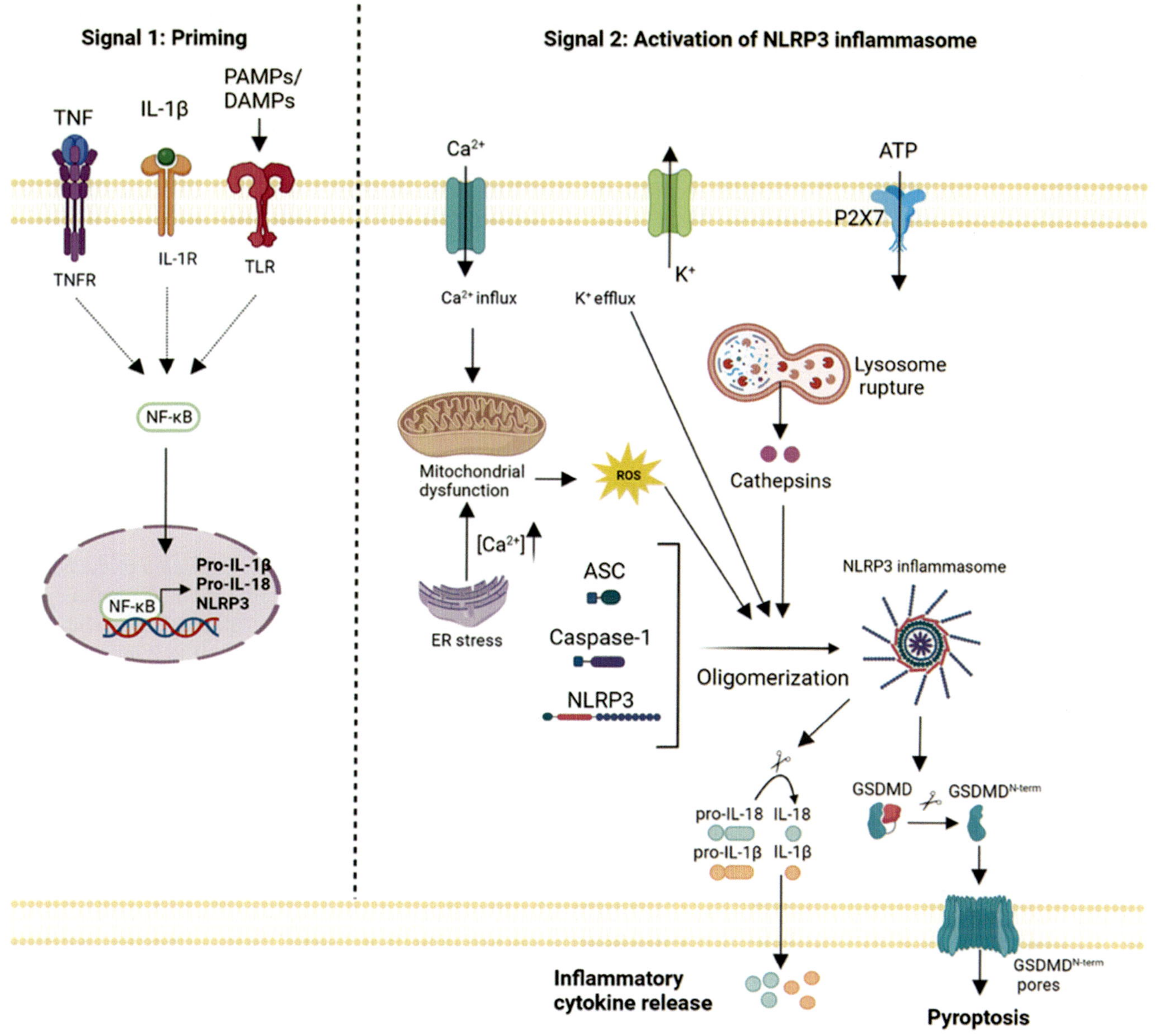

Fig. 3.4 Activation of the canonical NLRP3 inflammasome. A two-step process is necessary for the activation of the NLRP3 inflammasome. The first step is priming (*left panel*), which is triggered by cytokines or PAMPs/DAMPs. Activation of cytokine receptors or TLRs leads to pro-IL-1β, pro-IL-18, and NLRP3 expression. The second step (*right panel*) is NLRP3 inflammasome assembly and activation, which is induced by various signals, including K^+ efflux, Ca^{2+} influx, and extracellular ATP. Further, an increase in intracellular Ca^{2+} levels due to ER stress leads to mitochondrial damage. ROS released by damaged mitochondria and cathepsin released by ruptured lysosomes are additional triggers that promote the assembly and activation of the NLRP3 inflammasome. Activation of caspase-1 results in the proteolytic cleavage of pro-IL-1β and pro-IL-18 and the release of IL-1β and IL-18. The activation of the NLRP3 inflammasome also results in cleavage of GSDMD. The N-terminal fragment ($GSDMD^{N\text{-}term}$) generated incorporates into the cell membrane, forms pores, and induces pyroptosis. ASC apoptosis-associated speck-like protein containing a CARD, CARD9 caspase recruitment domain-containing protein 9, DAMP danger-associated molecular pattern, ER endoplasmic reticulum, GSDMD Gasdermin D, IL-1β interleukin-1β, IL-1R interleukin receptor 1B, IRF interferon regulatory factor, LPS lipopolysaccharide, NF-κB nuclear factor kappa B, NLRP3 NOD-, LRR-, and pyrin domain-containing protein 3, P2X purinoceptor 7, PAMP pathogen-associated molecular pattern, ROS reactive oxygen species, TLR toll-Like receptor, TNF tumor necrosis factor, TNFR TNF receptor. (Created with BioRender.com)

macrophages, ox-PAPC binds caspase-11 and averts the binding of LPS. This prevents LPS-induced pyroptosis, inflammatory cytokine release, and sepsis (Chu et al., 2018).

Noncanonical inflammasome pathways that involve caspase-8 have also been described. Although caspase-8 is well-known for its role in apoptosis, evidence indicates that it also plays non-apoptotic roles. In fact, caspase-8 modulates inflammation by processing pro-IL-1β and by regulating NLRP3 inflammasome activation. Recognition of extracellular fungal or mycobacterial PAMPs by Dectin-1 triggers the formation and activation of a noncanonical inflammasome complex that compromises caspase-8. Caspase-8 is associated with a protease that has homology to caspases and is named mucosa-associated lymphoid tissue lymphoma translocation protein 1 (MALT-1), which regulates NF-κB signaling. Upon Dectin-1 stimulation, the paracaspase MALT-1 forms a scaffold complex with two other proteins: caspase recruitment domain-containing protein 9 (CARD9) and B-cell lymphoma-10 (BCL-10). This complex then recruits caspase-8 and ASC, leading to the formation of the noncanonical MALT-1–caspase-8–ASC inflammasome and resulting in the activation of caspase-8, which processes pro-IL-1β (Gringhuis et al., 2012). Caspase-8 can also assemble with inflammasomes such as NLRC4 or NLRP3. In the absence of caspase-1 or caspase-11, stimulation of dendritic cells by LPS leads to the integration of caspase-8 into the NLRP3 inflammasome. Caspase-8 gets activated and processes pro-IL-1β. Similarly, in the absence of caspase-1, fungal pathogens induce the activation of the noncanonical NLRP3–ASC–caspase-8 inflammasome (Han et al., 2021).

Examples of noncanonical inflammasome activation are illustrated in Fig. 3.5.

3.5 Inflammasomes in Brain Pathologies

3.5.1 Alzheimer's Disease

Increasing evidence indicates that the microglial NLRP3 inflammasome plays a pivotal role in AD pathogenesis. Aβ fibrils bind microglia through receptor complexes that contain TLR2 or TLR4 and its co-receptor CD14. This leads to Aβ phagocytosis. It has been hypothesized that the accumulation of Aβ in microglia causes lysosomal damage and the release of cathepsin B, which acts as a DAMP to activate the NLRP3 inflammasome, resulting in IL-1β release (Halle et al., 2008).

In an animal model of AD, deficiency in the different components of the NLRP3 inflammasome reduces the deposition and spreading of Aβ, attenuates Aβ load in the brain, decreases IL-1β expression, and promotes neuroprotection (Yang et al., 2020).

As explained before, activation of the NLRP3 inflammasome requires two steps: a priming step and an assembly and activation step. Whereas TLR signaling can prime the NLRP3 inflammasome, stimuli associated with the perturbance of cellular functions such as K^+ efflux, ROS production, or aberrant Ca^{2+} signaling activate the NLRP3 inflammasome. Particularly, several mechanisms have been implicated in the activation of the NLRP3 inflammasome in AD. ATP released by dying neurons followed by the activation of the P2X7 purinergic receptor on microglia and astrocytes are among those mechanisms. K^+ efflux and engagement of pannexin 1 channels also contribute to NLRP3 inflammasome activation (Van Zeller et al., 2021). Even though mitochondrial damage, lysosomal dysfunction, and endoplasmic reticulum stress have been implicated in NLRP3 inflammasome activation or the regulation of its activity (Akbal et al., 2022), it is not yet known whether these stimuli contribute to the activation of the NLRP3 inflammasome in AD.

White matter pathology and oligodendrocyte (OL) degeneration have been described in AD, especially in Aβ plaques (Nasrabady et al., 2018), and have been linked to NLRP3 inflammasome activation (Zhang et al., 2020). The presence of GSDMD immunoreactivity in OL suggests damage by pyroptosis. Stimulation of mature OL cultures with Aβ(1–42) increases NLRP3 and causes GSDMD cleavage and release of IL-1β. Because rapid cell death is not observed, the authors suggested that Aβ(1–42) induces chronic, low-level NLRP3 inflammasome activation in OL, leading to GSDMD-mediated pyroptotic injury and resulting in myelin degeneration (Zhang et al., 2020).

In addition to Aβ pathology, the activation of the NLRP3 inflammasome plays an important role in tau pathology in AD (Ising et al., 2019). In an animal model that exhibits tau aggregation (Tau22 mice), the loss of NLRP3 inflammasome function decreases tau pathology and cognitive decline by regulating tau kinases and phosphatases. In vitro studies on microglia have revealed that tau monomers and oligomers induce IL-1β release in an ASC- and NLRP3-dependent manner. In turn, NLRP3-dependent IL-1β release increases tau phosphorylation in hippocampal neurons. The authors also showed that the NLRP3 inflammasome is a key mediator of Aβ-driven tau spreading. Consistent with these findings, NLRP3 deficiency reduces the seeding and propagation of tau pathology in the hippocampus and cortex, decreases hippocampal and cortical atrophy, and attenuates neurodegeneration (Stancu et al., 2019). Collectively, the studies suggest that NLRP3 plays a critical role in both Aβ and tau pathology in AD and should be further investigated as a potential target to develop therapeutic interventions.

3.5.2 Parkinson's Disease

Several studies have revealed that the NLRP3 inflammasome is involved in the key pathological features of PD, including α-syn aggregation, neuroinflammation, and DA neuron degeneration in the SNpc. Studies on human and animal models of PD have provided evidence supporting the involvement of NLRP3 inflammasome in the pathology of PD. Genetic polymorphisms that result in the abrogation of NLRP3 activity have been associated with reduced PD risk (von Herrmann et al., 2018). Pharmacological inhibition of NLRP3 inflammasome improves the behavioral deficits and neuroinflammation in mouse PD models.

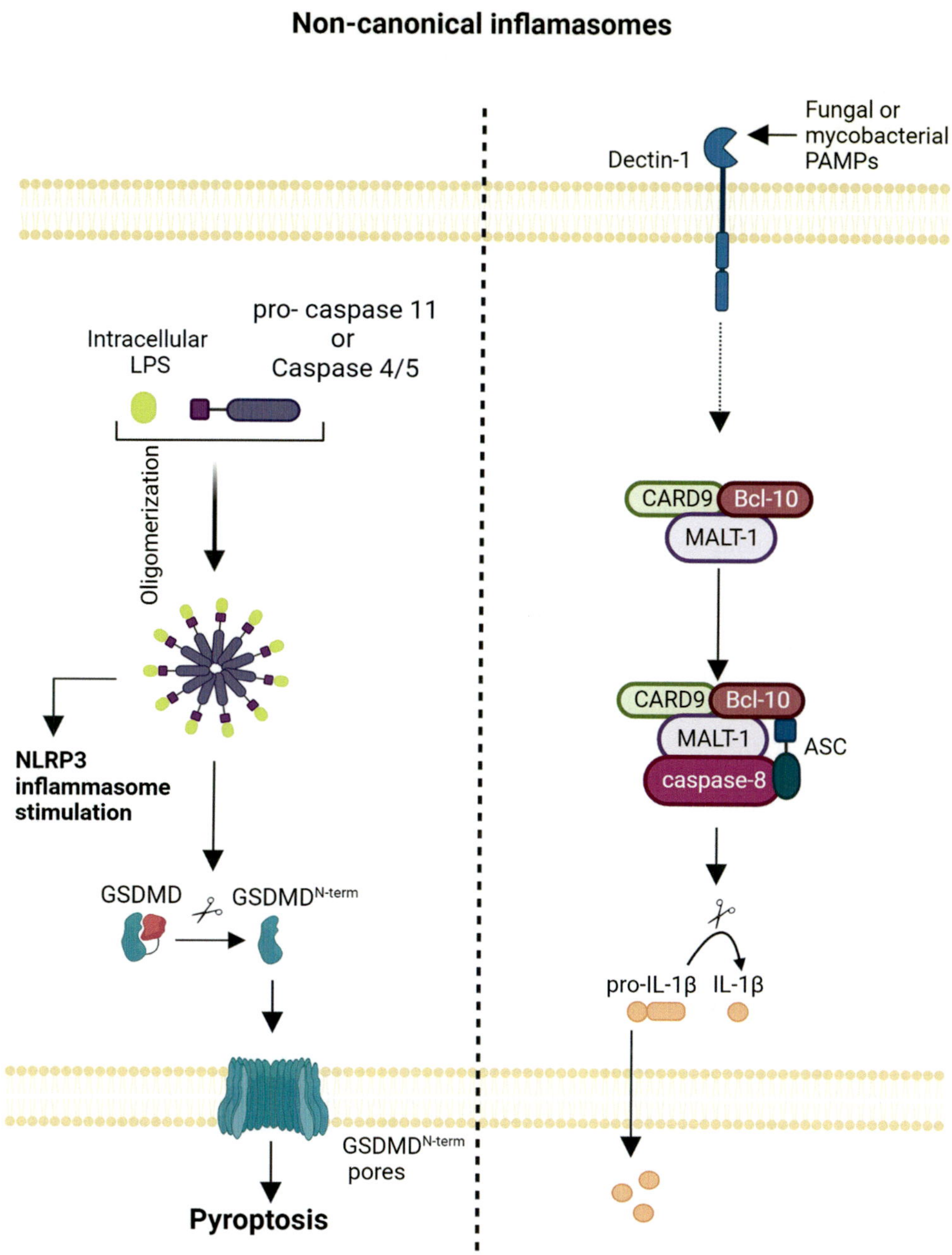

Fig. 3.5 Activation of noncanonical inflammasomes. Detection of intracellular LPS via caspase-11 (or caspase-4/5) activates noncanonical inflammasomes. Activated caspase-11 cleaves GSDMD, triggering pyroptosis. In addition, active caspase-11 further stimulates NLRP3 inflammasomes (*left panel*). Recognition of fungal pathogens by Dectin-1 triggers the formation and activation of another noncanonical inflammasome (*right panel*). Dectin-1 activation triggers CARD9–BCL-10–MALT-1 complex formation. Recruitment of ASC and caspase-8 to this complex induces caspase-8 activation, which then cleaves pro-IL-1β into its mature form. ASC apoptosis-associated speck-like protein containing a CARD, BCL-10 B-cell lymphoma-10, CARD9 caspase recruitment domain-containing protein 9, GSDMD Gasdermin D, IL-1β interleukin-1β, LPS lipopolysaccharide, MALT-1 mucosa-associated lymphoid tissue lymphoma translocation protein 1, NF-κB nuclear factor kappa B, NLRP3 NOD-, LRR-, and pyrin domain-containing protein 3, PAMP pathogen-associated molecular pattern. (Created with BioRender.com)

At the cellular level, the contribution of NLRP3 inflammasome to PD has been primarily studied in activated microglia (Gordon et al., 2018; Panicker et al., 2019). Microglial activation in the vicinity of DA neurons in the SNpc is one of the pathological hallmarks of PD and has been implicated in neuroinflammation that leads to DA neuronal loss. Microglia are the main source of NLRP3 inflammasome activity. The robust activation of NLRP3 inflammasome at the site of dopaminergic cell loss in the postmortem PD brain and in different rodent models of PD has been documented. NLRP3 and ASC are elevated in the affected areas of postmortem PD brains and predominantly localized in microglia, supporting the notion that the microglial NLRP3 inflammasome is a key player in neuronal degeneration in PD (Gordon et al., 2018). This is most likely through the release of IL-1β, which initiates neuroinflammation (Pike et al., 2022b; Scheiblich et al., 2021).

The priming and activation of NLRP3 inflammasome in microglia are often facilitated by TLRs through the engagement of α-syn. As mentioned before, neuron-derived α-syn fibrils and oligomers activate TLR2/NF-κB signaling in microglia, which provides the priming signal and increases the production of NLRP3 and pro-IL-1β. Subsequently, LPS-primed TLR4 facilitates the activation of the NLRP3 inflammasome by phagocytosed α-syn fibrils. Uptake of α-syn fibrils causes a variety of cellular perturbations, including K^+ efflux, lysosomal damage, and generation of mitochondrial ROS (mtROS) and mtDNA, triggers that activate the NLRP3 inflammasome, resulting in IL-1β release (Li et al., 2021b).

Although most studies indicate that LPS-dependent priming of NLRP3 inflammasome in microglia is indispensable for its activation by α-syn, other investigations have revealed alternate mechanisms independent of LPS. α-Syn aggregates prime and activate the NLRP3 inflammasome in the absence of LPS by interacting with TLR2 and C36, a scavenger receptor. Subsequently, CD36 recruits fyn kinase, which, in turn, activates NF-κb, resulting in an increased production of NLRP3 and pro-IL-1β. TLR2, CD36, and activated fyn kinase facilitate the uptake of α-syn aggregates, which induce mitochondrial damage. mtROS produced by damaged mitochondria activate NLRP3 inflammasome (Panicker et al., 2019).

Another report showed that α-syn monomers, and to a lesser extent, oligomers, but not fibrils, induce NLRP3 inflammasome priming and activation in primary microglia via TLR2 and TLR5 in the absence of LPS priming. At the same time, TLR2 ligation contributes to the priming of the NLRP3 inflammasome, and TLR5 ligation results in its activation. Activation of NLRP3 inflammasome interferes with the clearance of α-syn aggregates by microglia, contributing to α-syn pathology (Scheiblich et al., 2021). The studies suggest that α-syn/TLRs/NF-κB/NLRP3 inflammasome axis in microglia could play an important role in the pathogenesis of PD by modulating neuroinflammation and α-syn clearance.

Aberrant dopamine signaling, which is a hallmark of PD pathology, has been implicated in NLRP3-mediated neuroinflammation and neuroinflammation-induced DA neuron loss. Dopamine is a modulator of NLRP3 inflammasome activation in microglia. Dopamine inhibits canonical, noncanonical, and α-syn-mediated NLRP3 activation in primary human and mouse microglia. The inhibitory effects of dopamine on canonical NLRP3 activation are mediated by the dopamine D1 receptor (DRD1), whereas both DRD1 and DRD2 are involved in the inhibition of noncanonical and α-syn-mediated NLRP3 activation (Pike et al., 2022a). Accordingly, DRD1 agonism inhibits microglial inflammasome activation and IL-1β secretion in primary microglial cultures and a mouse PD model (Pike et al., 2022a). The exact mechanisms underlying the inhibitory effects of dopamine on α-syn-mediated NLRP3 activation have not been defined. However, it has been postulated that following microglial NLRP3 inflammasome activation by α-syn aggregates, dopamine released by healthy proximal neurons binds microglial dopamine receptors and counteracts α-syn-mediated NLRP3 activation. However, neuronal dopamine stores get depleted over time, and dopamine levels are no longer sufficient to restrict α-syn-mediated NLRP3 activation, resulting in neuroinflammation, which affects DA neuron integrity (Pike et al., 2022a).

In addition to microglia, a critical link between neuronal NLRP3 inflammasome activity and degeneration of DA neurons has been documented. This has been attributed to impaired Parkin function (Panicker et al. 2022). Parkin is a ubiquitin E3 ligase that ubiquitinates substrates that regulate the cellular processes involved in mitochondrial function. Loss-of-function mutations in the gene encoding Parkin cause familial, early-onset PD. Parkin has also been implicated in sporadic PD, as it can get inactivated by oxidative and nitrosative stress. Neuronal NLRP3 is a substrate of Parkin. In the absence of disease, Parkin mitigates neuronal NLRP3 inflammasome priming and activation because it causes the degradation of NLRP3 through ubiquitylation. However, in PD, loss of Parkin function impairs this process, leading to the priming and activation of NLRP3 inflammasome. Accordingly, NLRP3 inflammasome activation is observed in catecholaminergic neurons of the striatum in postmortem PD brains and a mouse model of α-synucleinopathy. The NLRP3 inflammasome is also activated in Parkin-deficient human DA neurons. Collectively, these findings suggest that loss of Parkin function in DA neurons leads to the priming and accumulation of NLRP3 in PD. This is followed by the activation of the NLRP3 inflammasome by triggers associated with damaged mitochondria. However, unlike microglia, the activation of NLRP3 inflammasome in neurons does not induce IL-1β or IL-18 release. Instead, it prompts GSDMD cleavage, resulting in pyroptotic cell death (Panicker et al., 2022). Thus, neuronal NLRP3 inflammasome activation following the loss of Parkin function could be a mechanism underlying the demise of DA neurons by pyroptosis.

It is worth noting that Parkin deficiency also increases NLRP3 activity in microglia and bone marrow-derived macrophages, resulting in excessive release of IL-1β and IL-18 in response to NLRP3 agonists. Parkin deficiency in microglia is associated with a significant decrease in A20 protein. A20 is a negative regulator of the NF-κB signaling pathway and prevents the activation of the NLRP3 inflammasome by suppressing NF-κB-mediated NLRP3 and pro-IL-1β production. Therefore, a reduction in A20 in Parkin-deficient microglia likely increases NF-κB-mediated NLRP3 production and NLRP3 inflammasome activation (Mouton-Liger et al., 2018). This mechanism may also be involved in the pathogenesis of PD caused by the loss of Parkin function.

In addition, inflammasome activation has been linked to α-syn accumulation in neuronal cultures overexpressing α-syn. Active caspase-1 cleaves α-syn at the Asp121 residue and generates a truncated form more prone to develop aggregates like those found in Lewy bodies (Wang et al., 2016). Since caspase-1 and α-syn are co-localized in the Lewy body, this mechanism is likely relevant to human disease pathology. However, the precise inflammasome involved in this mechanism has not been identified.

3.5.3 Brain Injury

3.5.3.1 Ischemic Brain Injury

The NLRP3 inflammasome has been studied in an animal model of ischemic stroke. Neuronal NLRP3 inflammasome expression and activity are robustly increased following ischemia. NLRP3 inflammasome inhibitors reduce the infarct volume, attenuate the infiltration of immune cells, and protect the integrity of the blood–brain barrier (BBB) (Franke et al., 2021). The NLRP3 inflammasome has also been implicated in ischemia-induced disruption of BBB integrity and functional deficits (Bellut et al., 2021; Palomino-Antolin et al., 2022).

In addition to NLRP3, the AIM2 inflammasome appears to contribute to ischemic stroke and poststroke cognitive impairment (PSCI) in animal models. The expression and activity of the AIM2 inflammasome are increased in the cortex and hippocampus following ischemia. In AIM2 knockout mice, PSCI is improved compared to controls. Therefore, studies have suggested that the AIM2 inflammasome could be a therapeutic target to ameliorate PSCI (Kim et al., 2020).

3.5.3.2 Traumatic Brain Injury

Investigations on animal models of TBI indicate that ASC levels, but not NLRP1, are upregulated in the cortex following injury, and an increase in active caspase-1 and caspase-11 parallels this. ASC co-immunoprecipitates with caspase-1 and caspase-11, and the association of ASC with NLRP1, caspase-1, and caspase-11 increases following TBI. Interference with inflammasome activation by the use of an ASC-neutralizing antibody reduces lesion volume following TBI (de Rivero Vaccari et al., 2009). Since NLRP1 inflammasome components are expressed in neurons, neuronal inflammasome activation may play a role in TBI pathology.

A role for microglial inflammasome activity in the induction of neuroinflammation and pyroptosis in animal models of penetrating ballistic-like brain injury has been documented (Lee et al., 2019; Lee et al., 2018). The expression of NLRP3 inflammasome components increases within hours post-TBI and remains high in subsequent days. Activation of NLRP3 inflammasome occurs within the same time frame. Therefore, NLRP3 inflammasome inhibitors have been used to investigate their effectiveness in modulating the inflammatory reaction to limit damage in animal models of TBI. Collectively, these preclinical studies indicate that NLRP3 inhibitors show promise in attenuating functional deficits associated with TBI (O'Brien et al., 2020).

3.6 Conclusions

The mechanisms that govern the innate immune response of the brain and neuroinflammation are complex and involve many different sensors that respond to specific motifs in pathogens or endogenous danger signals in a context-dependent manner. The outcomes of activating these sensors, which include PRRs and inflammasomes, are also cell-type-dependent. Whereas most studies focus on microglia and, to a lesser extent, astrocytes, the contribution of PRRs and inflammasomes to neurons and oligodendrocytes has received limited attention. A more comprehensive understanding of the way by which the various sensors and effectors mediate neuroinflammation requires a thorough analysis of PRR and inflammasome function in different neuronal and glial subtypes and their intercellular interactions. Additional investigations are necessary to unravel the cooperation or hindrance among PRRs, between PRRs and inflammasome signaling, and how interplay fine-tunes the neuroinflammatory response.

Because neuroinflammation has been increasingly implicated in many CNS diseases or injuries, PRRs and inflammasomes are now considered potential therapeutic targets. However, when designing therapies, it is important to remember that neuroinflammation can be beneficial and detrimental. The neuroinflammatory response and the PRRs or inflammasomes involved in this process could change during the course of the disease.

Despite all the challenges, CNS innate immunity, neuroinflammation, PRRs, and inflammasomes are excellent research directions to explore further, given their broad involvement, not only in neurodegenerative disorders but also in psychiatric diseases, mood and neurodevelopmental disorders, and normal aging.

3.7 Review Questions

1. Which of the following statements about TLR3 and TLR4 is false?
   (a) TLR3 and TLR4 form dimers upon ligand binding.
   (b) Both TLR3 and TLR4 utilize the TRIF-dependent signaling pathway.
   (c) *Both TLR3 and TLR4 utilize the MyD88-dependent signaling pathway.*
   (d) Lipopolysaccharide (LPS) binds TLR4.
   (e) TLR3 is an intracellular receptor.
2. The central nervous system cells that express TLRs are
   (a) Microglia
   (b) Astrocytes
   (c) Neurons
   (d) Oligodendrocytes
   (e) *All of the above*

3. Activation of the NLRP3 inflammasome differs from that of other inflammasomes. Which of the following is true?
   (a) Only virus-derived PAMPs activate NLRP3 inflammasome.
   (b) Activation of NLRP3 inflammasome is not dependent on complex formation.
   (c) Activation of NLRP3 inflammasome is not associated with pyroptosis.
   (d) *Activation of NLRP3 inflammasome is a two-step process.*
   (e) Activation of NLRP3 inhibits Gasdermin D cleavage.
4. TLRs and NLRP3 inflammasome cooperate to mediate neuroinflammation. Which of the following is true?
   (a) Endosomal TLRs physically interact with NLRP3 to facilitate inflammasome complex formation.
   (b) TLRs at the cell surface physically interact with NLRP3 to facilitate inflammasome complex formation.
   (c) *TLR signaling leads to pro-IL-1β synthesis, and activation of the NLRP3 inflammasome results in cleavage of pro-IL-1β and release of mature IL-1β.*
   (d) TLR signaling leads to IL-18 synthesis, which, in turn, activates NLRP3.
   (e) Both TLRs and the NLRP3 inflammasome induce caspase-1 activation.
5. Noncanonical inflammasomes differ from canonical inflammasomes. Which of the following is false?
   (a) Noncanonical inflammasomes utilize caspase-11 rather than caspase-1.
   (b) Noncanonical inflammasomes utilize caspase-8 rather than caspase-1.
   (c) Activated caspase-8 in noncanonical inflammasomes can cleave pro-IL-1β.
   (d) In noncanonical inflammasomes, sensors and effectors can be the same molecule.
   (e) *Dectin-1 activates only canonical inflammasomes.*

References

Acioglu, C., Heary, R. F., & Elkabes, S. (2022). Roles of neuronal toll-like receptors in neuropathic pain and central nervous system injuries and diseases. *Brain, Behavior, and Immunity, 102*, 163–178. https://doi.org/10.1016/j.bbi.2022.02.016

Akbal, A., Dernst, A., Lovotti, M., Mangan, M. S. J., McManus, R. M., & Latz, E. (2022). How location and cellular signaling combine to activate the NLRP3 inflammasome. *Cellular & Molecular Immunology, 19*, 1201–1214. https://doi.org/10.1038/s41423-022-00922-w

Allen, I. C. (2014). Non-inflammasome forming NLRs in inflammation and tumorigenesis. *Frontiers in Immunology, 5*, 169. https://doi.org/10.3389/fimmu.2014.00169

Arumugam, T. V., Manzanero, S., Furtado, M., Biggins, P. J., Hsieh, Y. H., Gelderblom, M., MacDonald, K. P., Salimova, E., Li, Y. I., Korn, O., Dewar, D., Macrae, I. M., Ashman, R. B., Tang, S. C., Rosenthal, N. A., Ruitenberg, M. J., Magnus, T., & Wells, C. A. (2017). An atypical role for the myeloid receptor Mincle in central nervous system injury. *Journal of Cerebral Blood Flow and Metabolism, 37*, 2098–2111. https://doi.org/10.1177/0271678X16661201

Babcock, A. A., Wirenfeldt, M., Holm, T., Nielsen, H. H., Dissing-Olesen, L., Toft-Hansen, H., Millward, J. M., Landmann, R., Rivest, S., Finsen, B., & Owens, T. (2006). Toll-like receptor 2 signaling in response to brain injury: An innate bridge to neuroinflammation. *The Journal of Neuroscience, 26*, 12826–12837. https://doi.org/10.1523/JNEUROSCI.4937-05.2006

Baldwin, K. T., Carbajal, K. S., Segal, B. M., & Giger, R. J. (2015). Neuroinflammation triggered by beta-glucan/dectin-1 signaling enables CNS axon regeneration. *Proceedings of the National Academy of Sciences of the United States of America, 112*, 2581–2586. https://doi.org/10.1073/pnas.1423221112

Bellut, M., Papp, L., Bieber, M., Kraft, P., Stoll, G., & Schuhmann, M. K. (2021). NLPR3 inflammasome inhibition alleviates hypoxic endothelial cell death in vitro and protects blood-brain barrier integrity in murine stroke. *Cell Death & Disease, 13*, 20. https://doi.org/10.1038/s41419-021-04379-z

Beraud, D., Twomey, M., Bloom, B., Mittereder, A., Ton, V., Neitzke, K., Chasovskikh, S., Mhyre, T. R., & Maguire-Zeiss, K. A. (2011). Alpha-synuclein alters Toll-like receptor expression. *Frontiers in Neuroscience, 5*, 80. https://doi.org/10.3389/fnins.2011.00080

Blandini, F., & Armentero, M. T. (2012). Animal models of Parkinson's disease. *The FEBS Journal, 279*, 1156–1166. https://doi.org/10.1111/j.1742-4658.2012.08491.x

Bouchon, A., Dietrich, J., & Colonna, M. (2000). Cutting edge: Inflammatory responses can be triggered by TREM-1, a novel receptor expressed on neutrophils and monocytes. *Journal of Immunology, 164*, 4991–4995. https://doi.org/10.4049/jimmunol.164.10.4991

Bouchon, A., Facchetti, F., Weigand, M. A., & Colonna, M. (2001). TREM-1 amplifies inflammation and is a crucial mediator of septic shock. *Nature, 410*, 1103–1107. https://doi.org/10.1038/35074114

Brand, F. J., 3rd, de Rivero Vaccari, J. C., Mejias, N. H., Alonso, O. F., & de Rivero Vaccari, J. P. (2015). RIG-I contributes to the innate immune response after cerebral ischemia. *Journal of Inflammation, 12*, 52. https://doi.org/10.1186/s12950-015-0101-4

Calvo-Rodriguez, M., de la Fuente, C., Garcia-Durillo, M., Garcia-Rodriguez, C., Villalobos, C., & Nunez, L. (2017). Aging and amyloid beta oligomers enhance TLR4 expression, LPS-induced Ca(2+) responses, and neuron cell death in cultured rat hippocampal neurons. *Journal of Neuroinflammation, 14*, 24. https://doi.org/10.1186/s12974-017-0802-0

Caso, J. R., Pradillo, J. M., Hurtado, O., Lorenzo, P., Moro, M. A., & Lizasoain, I. (2007). Toll-like receptor 4 is involved in brain damage and inflammation after experimental stroke. *Circulation, 115*, 1599–1608. https://doi.org/10.1161/CIRCULATIONAHA.106.603431

Chedid, J., Labrador-Garrido, A., Zhong, S., Gao, J., Zhao, Y., Perera, G., Kim, W. S., Halliday, G. M., & Dzamko, N. (2022). A small molecule toll-like receptor antagonist rescues α-synuclein fibril pathology. *Journal of Biological Chemistry, 298*, 102260. https://doi.org/10.1016/j.jbc.2022.102260

Chen, C. Y., Shih, Y. C., Hung, Y. F., & Hsueh, Y. P. (2019). Beyond defense: Regulation of neuronal morphogenesis and brain functions via Toll-like receptors. *Journal of Biomedical Science, 26*, 90. https://doi.org/10.1186/s12929-019-0584-z

Chu, L. H., Indramohan, M., Ratsimandresy, R. A., Gangopadhyay, A., Morris, E. P., Monack, D. M., Dorfleutner, A., & Stehlik, C. (2018). The oxidized phospholipid oxPAPC protects from septic shock by targeting the non-canonical inflammasome in macrophages. *Nature Communications, 9*, 996. https://doi.org/10.1038/s41467-018-03409-3

Corrigan, F., Arulsamy, A., Collins-Praino, L. E., Holmes, J. L., & Vink, R. (2017). Toll like receptor 4 activation can be either detrimental or beneficial following mild repetitive traumatic brain injury depending on timing of activation. *Brain, Behavior, and Immunity, 64*, 124–139. https://doi.org/10.1016/j.bbi.2017.04.006

Dantas, P., Matos, A. O., da Silva, F. E., Silva-Sales, M., & Sales-Campos, H. (2020). Triggering receptor expressed on myeloid cells-1 (TREM-1) as a therapeutic target in infectious and non-infectious disease: A critical review. *International Reviews of Immunology, 39*, 188–202. https://doi.org/10.1080/08830185.2020.1762597

de Rivero Vaccari, J. P., Lotocki, G., Alonso OF, Bramlett, H. M., Dietrich, W. D., & Keane, R. W. (2009). Therapeutic neutralization of the NLRP1 inflammasome reduces the innate immune response and improves histopathology after traumatic brain injury. *Journal of Cerebral Blood Flow and Metabolism, 29*, 1251–1261. https://doi.org/10.1038/jcbfm.2009.46

de Rivero Vaccari, J. P., Brand, F. J., 3rd, Sedaghat, C., Mash, D. C., Dietrich, W. D., & Keane, R. W. (2014). RIG-1 receptor expression in the pathology of Alzheimer's disease. *Journal of Neuroinflammation, 11*, 67. https://doi.org/10.1186/1742-2094-11-67

de Rivero Vaccari, J. C., Brand, F. J., 3rd, Berti, A. F., Alonso OF, Bullock, M. R., & de Rivero Vaccari, J. P. (2015). Mincle signaling in the innate immune response after traumatic brain injury. *Journal of Neurotrauma, 32*, 228–236. https://doi.org/10.1089/neu.2014.3436

Deczkowska, A., Weiner, A., & Amit, I. (2020). The physiology, pathology, and potential therapeutic applications of the TREM2 signaling pathway. *Cell, 181*, 1207–1217. https://doi.org/10.1016/j.cell.2020.05.003

Denes, A., Coutts, G., Lenart, N., Cruickshank, S. M., Pelegrin, P., Skinner, J., Rothwell, N., Allan, S. M., & Brough, D. (2015). AIM2 and NLRC4 inflammasomes contribute with ASC to acute brain injury independently of NLRP3. *Proceedings of the National Academy of Sciences of the United States of America, 112*, 4050–4055. https://doi.org/10.1073/pnas.1419090112

Drummond, R. A., Saijo, S., Iwakura, Y., & Brown, G. D. (2011). The role of Syk/CARD9 coupled C-type lectins in antifungal immunity. *European Journal of Immunology, 41*, 276–281. https://doi.org/10.1002/eji.201041252

Dutta, D., Jana, M., Majumder, M., Mondal, S., Roy, A., & Pahan, K. (2021). Selective targeting of the TLR2/MyD88/NF-kappaB pathway reduces alpha-synuclein spreading in vitro and in vivo. *Nature Communications, 12*, 5382. https://doi.org/10.1038/s41467-021-25767-1

Dzamko, N., Gysbers, A., Perera, G., Bahar, A., Shankar, A., Gao, J., Fu, Y., & Halliday, G. M. (2017). Toll-like receptor 2 is increased in neurons in Parkinson's disease brain and may contribute to alpha-synuclein pathology. *Acta Neuropathologica, 133*, 303–319. https://doi.org/10.1007/s00401-016-1648-8

Faustin, B., Lartigue, L., Bruey, J. M., Luciano, F., Sergienko, E., Bailly-Maitre, B., Volkmann, N., Hanein, D., Rouiller, I., & Reed, J. C. (2007). Reconstituted NALP1 inflammasome reveals two-step mechanism of caspase-1 activation. *Molecular Cell, 25*, 713–724. https://doi.org/10.1016/j.molcel.2007.01.032

Fellner, L., Irschick, R., Schanda, K., Reindl, M., Klimaschewski, L., Poewe, W., Wenning, G. K., & Stefanova, N. (2013). Toll-like receptor 4 is required for alpha-synuclein dependent activation of microglia and astroglia. *Glia, 61*, 349–360. https://doi.org/10.1002/glia.22437

Ferwerda, G., Meyer-Wentrup, F., Kullberg, B. J., Netea, M. G., & Adema, G. J. (2008). Dectin-1 synergizes with TLR2 and TLR4 for cytokine production in human primary monocytes and macrophages. *Cellular Microbiology, 10*, 2058–2066. https://doi.org/10.1111/j.1462-5822.2008.01188.x

Forabosco, P., Ramasamy, A., Trabzuni, D., Walker, R., Smith, C., Bras, J., Levine, A. P., Hardy, J., Pocock, J. M., Guerreiro, R., Weale, M. E., & Ryten, M. (2013). Insights into TREM2 biology by network analysis of human brain gene expression data. *Neurobiology of Aging, 34*, 2699–2714. https://doi.org/10.1016/j.neurobiolaging.2013.05.001

Franke, M., Bieber, M., Kraft, P., Weber, A. N. R., Stoll, G., & Schuhmann, M. K. (2021). The NLRP3 inflammasome drives inflammation in ischemia/reperfusion injury after transient middle cerebral artery occlusion in mice. *Brain, Behavior, and Immunity, 92*, 223–233. https://doi.org/10.1016/j.bbi.2020.12.009

Fujita, K., Motoki, K., Tagawa, K., Chen, X., Hama, H., Nakajima, K., Homma, H., Tamura, T., Watanabe, H., Katsuno, M., Matsumi, C., Kajikawa, M., Saito, T., Saido, T., Sobue, G., Miyawaki, A., & Okazawa, H. (2016). HMGB1, a pathogenic molecule that induces neurite degeneration via TLR4-MARCKS, is a potential therapeutic target for Alzheimer's disease. *Scientific Reports, 6*, 31895. https://doi.org/10.1038/srep31895

Gensel, J. C., Wang, Y., Guan, Z., Beckwith, K. A., Braun, K. J., Wei, P., McTigue, D. M., & Popovich, P. G. (2015). Toll-like receptors and dectin-1, a c-type lectin receptor, trigger divergent functions in CNS macrophages. *The Journal of Neuroscience, 35*, 9966–9976. https://doi.org/10.1523/JNEUROSCI.0337-15.2015

Goedert, M., Spillantini, M. G., Del Tredici, K., & Braak, H. (2013). 100 years of Lewy pathology. *Nature Reviews. Neurology, 9*, 13–24. https://doi.org/10.1038/nrneurol.2012.242

Gong, T., Liu, L., Jiang, W., Zhou, R. (2020). DAMP sensing receptors in sterile inflammation and inflammatory diseases. *Nature Reviews Immunology, 20*, 95–112. https://doi.org/10.1038/s41577-019-0215-7

Gordon, R., Albornoz, E. A., Christie, D. C., Langley, M. R., Kumar, V., Mantovani, S., Robertson, A. A. B., Butler, M. S., Rowe, D. B., O'Neill, L. A., Kanthasamy, A. G., Schroder, K., Cooper, M. A., & Woodruff, T. M. (2018). Inflammasome inhibition prevents alpha-synuclein pathology and dopaminergic neurodegeneration in mice. *Science Translational Medicine, 10*. https://doi.org/10.1126/scitranslmed.aah4066

Gray, E. E., Winship, D., Snyder, J. M., Child, S. J., Geballe, A. P., & Stetson, D. B. (2016). The AIM2-like receptors are dispensable for the interferon response to intracellular DNA. *Immunity, 45*, 255–266. https://doi.org/10.1016/j.immuni.2016.06.015

Gringhuis, S. I., Kaptein, T. M., Wevers, B. A., Theelen, B., van der Vlist, M., Boekhout, T., & Geijtenbeek, T. B. (2012). Dectin-1 is an extracellular pathogen sensor for the induction and processing of IL-1beta via a noncanonical caspase-8 inflammasome. *Nature Immunology, 13*, 246–254. https://doi.org/10.1038/ni.2222

Guerreiro, R., Wojtas, A., Bras, J., Carrasquillo, M., Rogaeva, E., Majounie, E., Cruchaga, C., Sassi, C., Kauwe, J. S., Younkin, S., Hazrati, L., Collinge, J., Pocock, J., Lashley, T., Williams, J., Lambert, J. C., Amouyel, P., Goate, A., Rademakers, R., et al. (2013). TREM2 variants in Alzheimer's disease. *The New England Journal of Medicine, 368*, 117–127. https://doi.org/10.1056/NEJMoa1211851

Hagar, J. A., Powell, D. A., Aachoui, Y., Ernst, R. K., & Miao, E. A. (2013). Cytoplasmic LPS activates caspase-11: Implications in TLR4-independent endotoxic shock. *Science, 341*, 1250–1253. https://doi.org/10.1126/science.1240988

Halle, A., Hornung, V., Petzold, G. C., Stewart, C. R., Monks, B. G., Reinheckel, T., Fitzgerald, K. A., Latz, E., Moore, K. J., & Golenbock, D. T. (2008). The NALP3 inflammasome is involved in the innate immune response to amyloid-beta. *Nature Immunology, 9*, 857–865. https://doi.org/10.1038/ni.1636

Han, J. H., Park, J., Kang, T. B., & Lee, K. H. (2021). Regulation of caspase-8 activity at the crossroads of pro-inflammation and anti-inflammation. *International Journal of Molecular Sciences, 22*. https://doi.org/10.3390/ijms22073318

He, Y., Xu, L., Li, B., Guo, Z. N., Hu, Q., Guo, Z., Tang, J., Chen, Y., Zhang, Y., Tang, J., & Zhang, J. H. (2015). Macrophage-inducible c-type lectin/spleen tyrosine kinase signaling pathway contributes to neuroinflammation after subarachnoid hemorrhage in rats. *Stroke, 46*, 2277–2286. https://doi.org/10.1161/STROKEAHA.115.010088

Heiman, A., Pallottie, A., Heary, R. F., & Elkabes, S. (2014). Toll-like receptors in central nervous system injury and disease: A focus on the spinal cord. *Brain, Behavior, and Immunity, 42*, 232–245. https://doi.org/10.1016/j.bbi.2014.06.203

Heinz, R., & Schneider, U. C. (2022). TLR4-pathway-associated biomarkers in subarachnoid hemorrhage (SAH): Potential targets for future anti-inflammatory therapies. *International Journal of Molecular Sciences, 23*. https://doi.org/10.3390/ijms232012618

Heppner, F. L., Ransohoff, R. M., & Becher, B. (2015). Immune attack: The role of inflammation in Alzheimer disease. *Nature Reviews. Neuroscience, 16*, 358–372. https://doi.org/10.1038/nrn3880

Hughes, C., Choi, M. L., Yi, J. H., Kim, S. C., Drews, A., George-Hyslop, P. S., Bryant, C., Gandhi, S., Cho, K., & Klenerman, D. (2020). Beta amyloid aggregates induce sensitised TLR4 signalling causing long-term potentiation deficit and rat neuronal cell death. *Communications Biology, 3*, 79. https://doi.org/10.1038/s42003-020-0792-9

Hyakkoku, K., Hamanaka, J., Tsuruma, K., Shimazawa, M., Tanaka, H., Uematsu, S., Akira, S., Inagaki, N., Nagai, H., & Hara, H. (2010). Toll-like receptor 4 (TLR4), but not TLR3 or TLR9, knock-out mice have neuroprotective effects against focal cerebral ischemia. *Neuroscience, 171*, 258–267. https://doi.org/10.1016/j.neuroscience.2010.08.054

Ising, C., Venegas, C., Zhang, S., Scheiblich, H., Schmidt, S. V., Vieira-Saecker, A., Schwartz, S., Albasset, S., McManus, R. M., Tejera, D., Griep, A., Santarelli, F., Brosseron, F., Opitz, S., Stunden, J., Merten, M., Kayed, R., Golenbock, D. T., Blum, D., et al. (2019). NLRP3 inflammasome activation drives tau pathology. *Nature, 575*, 669–673. https://doi.org/10.1038/s41586-019-1769-z

Jay, T. R., Miller, C. M., Cheng, P. J., Graham, L. C., Bemiller, S., Broihier, M. L., Xu, G., Margevicius, D., Karlo, J. C., Sousa, G. L., Cotleur, A. C., Butovsky, O., Bekris, L., Staugaitis, S. M., Leverenz, J. B., Pimplikar, S. W., Landreth, G. E., Howell, G. R., Ransohoff, R. M., & Lamb, B. T. (2015). TREM2 deficiency eliminates TREM2+ inflammatory macrophages and ameliorates pathology in Alzheimer's disease mouse models. *The Journal of Experimental Medicine, 212*, 287–295. https://doi.org/10.1084/jem.20142322

Jennings, C., Kusler, B., & Jones, P. P. (2009). Calcineurin inactivation leads to decreased responsiveness to LPS in macrophages and dendritic cells and protects against LPS-induced toxicity in vivo. *Innate Immunity, 15*, 109–120. https://doi.org/10.1177/1753425908100928

Jonsson, T., Stefansson, H., Steinberg, S., Jonsdottir, I., Jonsson, P. V., Snaedal, J., Bjornsson, S., Huttenlocher, J., Levey, A. I., Lah, J. J., Rujescu, D., Hampel, H., Giegling, I., Andreassen, O. A., Engedal, K., Ulstein, I., Djurovic, S., Ibrahim-Verbaas, C., Hofman, A., et al. (2013). Variant of TREM2 associated with the risk of Alzheimer's disease. *The New England Journal of Medicine, 368*, 107–116. https://doi.org/10.1056/NEJMoa1211103

Kawai, T., & Akira, S. (2010). The role of pattern-recognition receptors in innate immunity: Update on toll-like receptors. *Nature Immunology, 11*, 373–384. https://doi.org/10.1038/ni.1863

Kayagaki, N., Warming, S., Lamkanfi, M., Vande Walle, L., Louie, S., Dong, J., Newton, K., Qu, Y., Liu, J., Heldens, S., Zhang, J., Lee, W. P., Roose-Girma, M., & Dixit, V. M. (2011). Non-canonical inflammasome activation targets caspase-11. *Nature, 479*, 117–121. https://doi.org/10.1038/nature10558

Kim, C., Ho, D. H., Suk, J. E., You, S., Michael, S., Kang, J., Joong Lee, S., Masliah, E., Hwang, D., Lee, H. J., & Lee, S. J. (2013). Neuron-released oligomeric alpha-synuclein is an endogenous agonist of TLR2 for paracrine activation of microglia. *Nature Communications, 4*, 1562. https://doi.org/10.1038/ncomms2534

Kim, C., Rockenstein, E., Spencer, B., Kim, H. K., Adame, A., Trejo, M., Stafa, K., Lee, H. J., Lee, S. J., & Masliah, E. (2015). Antagonizing neuronal toll-like receptor 2 prevents synucleinopathy by activating autophagy. *Cell Reports, 13*, 771–782. https://doi.org/10.1016/j.celrep.2015.09.044

Kim, C., Spencer, B., Rockenstein, E., Yamakado, H., Mante, M., Adame, A., Fields, J. A., Masliah, D., Iba, M., Lee, H. J., Rissman, R. A., Lee, S. J., & Masliah, E. (2018). Immunotherapy targeting toll-like receptor 2 alleviates neurodegeneration in models of synucleinopathy by modulating alpha-synuclein transmission and neuroinflammation. *Molecular Neurodegeneration, 13*, 43. https://doi.org/10.1186/s13024-018-0276-2

Kim, H., Seo, J. S., Lee, S. Y., Ha, K. T., Choi, B. T., Shin, Y. I., Ju Yun, Y., & Shin, H. K. (2020). AIM2 inflammasome contributes to brain injury and chronic post-stroke cognitive impairment in mice. *Brain, Behavior, and Immunity, 87*, 765–776. https://doi.org/10.1016/j.bbi.2020.03.011

Korgaonkar, A. A., Li, Y., Sekhar, D., Subramanian, D., Guevarra, J., Swietek, B., Pallottie, A., Singh, S., Kella, K., Elkabes, S., & Santhakumar, V. (2020a). Toll-like receptor 4 signaling in neurons enhances calcium-permeable alpha-amino-3-hydroxy-5-methyl-4-isoxazolepropionic acid receptor currents and drives post-traumatic epileptogenesis. *Annals of Neurology, 87*, 497–515. https://doi.org/10.1002/ana.25698

Korgaonkar, A. A., Nguyen, S., Li, Y., Sekhar, D., Subramanian, D., Guevarra, J., Pang, K. C. H., & Santhakumar, V. (2020b). Distinct cellular mediators drive the Janus faces of toll-like receptor 4 regulation of network excitability which impacts working memory performance after brain injury. *Brain, Behavior, and Immunity, 88*, 381–395. https://doi.org/10.1016/j.bbi.2020.03.035

Kouli, A., Horne C. B., Williams-Gray C. H., (2019). Toll-like receptors and their therapeutic potential in Parkinson's disease and alpha-synucleinopathies. *Brain Behavior and Immunity 81*, 41–51. https://doi.org/10.1016/j.bbi.2019.06.042

Lalancette-Hebert, M., Phaneuf, D., Soucy, G., Weng, Y. C., & Kriz, J. (2009). Live imaging of Toll-like receptor 2 response in cerebral ischaemia reveals a role of olfactory bulb microglia as modulators of inflammation. *Brain, 132*, 940–954. https://doi.org/10.1093/brain/awn345

Lee, H. J., Suk, J. E., Patrick, C., Bae, E. J., Cho, J. H., Rho, S., Hwang, D., Masliah, E., & Lee, S. J. (2010). Direct transfer of α-synuclein from neuron to astroglia causes inflammatory responses in synucleinopathies. *The Journal of Biological Chemistry, 285*, 9262–9272. https://doi.org/10.1074/jbc.M109.081125

Lee, S. W., Gajavelli, S., Spurlock, M. S., Andreoni, C., de Rivero Vaccari, J. P., Bullock, M. R., Keane, R. W., & Dietrich, W. D. (2018). Microglial inflammasome activation in penetrating ballistic-like brain injury. *Journal of Neurotrauma, 35*, 1681–1693. https://doi.org/10.1089/neu.2017.5530

Lee, S. W., de Rivero Vaccari, J. P., Truettner, J. S., Dietrich, W. D., & Keane, R. W. (2019). The role of microglial inflammasome activation in pyroptotic cell death following penetrating traumatic brain injury. *Journal of Neuroinflammation, 16*, 27. https://doi.org/10.1186/s12974-019-1423-6

Lehnardt, S., Lehmann, S., Kaul, D., Tschimmel, K., Hoffmann, O., Cho, S., Krueger, C., Nitsch, R., Meisel, A., & Weber, J. R. (2007). Toll-like receptor 2 mediates CNS injury in focal cerebral ischemia. *Journal of Neuroimmunology, 190*, 28–33. https://doi.org/10.1016/j.jneuroim.2007.07.023

Li, L., Acioglu, C., Heary, R. F., & Elkabes, S. (2021a). Role of astroglial toll-like receptors (TLRs) in central nervous system infections, injury and neurodegenerative diseases. *Brain, Behavior, and Immunity, 91*, 740–755. https://doi.org/10.1016/j.bbi.2020.10.007

Li, Y., Xia, Y., Yin, S., Wan, F., Hu, J., Kou, L., Sun, Y., Wu, J., Zhou, Q., Huang, J., Xiong, N., & Wang, T. (2021b). Targeting microglial alpha-synuclein/TLRs/NF-kappaB/NLRP3 inflammasome axis in Parkinson's disease. *Frontiers in Immunology, 12*, 719807. https://doi.org/10.3389/fimmu.2021.719807

Lin, C. H., Chen, H. Y., & Wei, K. C. (2020). Role of HMGB1/TLR4 axis in ischemia/reperfusion-impaired extracellular glutamate clearance in primary astrocytes. *Cells, 9*. https://doi.org/10.3390/cells9122585

Liu, S., Liu, Y., Hao, W., Wolf, L., Kiliaan, A. J., Penke, B., Rube, C. E., Walter, J., Heneka, M. T., Hartmann, T., Menger, M. D., & Fassbender, K. (2012). TLR2 is a primary receptor for Alzheimer's amyloid beta peptide to trigger neuroinflammatory activation. *Journal of Immunology, 188*, 1098–1107. https://doi.org/10.4049/jimmunol.1101121

Looi, C. K., Hii, L.-W., Chung, F. F.-L., Mai, C.-W., Lim, W.-M., & Leong, C.-O. (2021). Roles of inflammasomes in Epstein–barr virus-associated nasopharyngeal cancer. *Cancers, 13*. https://doi.org/10.3390/cancers13081786

Masters, C. L., Bateman, R., Blennow, K., Rowe, C. C., Sperling, R. A., & Cummings, J. L. (2015). Alzheimer's disease. *Nature Reviews Disease Primers, 1*, 15056. https://doi.org/10.1038/nrdp.2015.56

Metselaar, P. I., Hos, C., Welting, O., Bosch, J. A., Kraneveld, A. D., de Jonge, W. J., & Te Velde, A. A. (2021). Ambiguity about splicing factor 3b subunit 3 (SF3B3) and Sin3A associated protein 130 (SAP130). *Cells, 10*. https://doi.org/10.3390/cells10030590

Minoretti, P., Gazzaruso, C., Vito, C. D., Emanuele, E., Bianchi, M., Coen, E., Reino, M., & Geroldi, D. (2006). Effect of the functional toll-like receptor 4 Asp299Gly polymorphism on susceptibility to late-onset Alzheimer's disease. *Neuroscience Letters, 391*, 147–149. https://doi.org/10.1016/j.neulet.2005.08.047

Miyake, K. (2004). Innate recognition of lipopolysaccharide by Toll-like receptor 4-MD-2. *Trends in Microbiology, 12*, 186–192. https://doi.org/10.1016/j.tim.2004.02.009

Mouton-Liger, F., Rosazza, T., Sepulveda-Diaz, J., Ieang, A., Hassoun, S. M., Claire, E., Mangone, G., Brice, A., Michel, P. P., Corvol, J. C., & Corti, O. (2018). Parkin deficiency modulates NLRP3 inflammasome activation by attenuating an A20-dependent negative feedback loop. *Glia, 66*, 1736–1751. https://doi.org/10.1002/glia.23337

Mukherjee, P., Winkler, C. W., Taylor, K. G., Woods, T. A., Nair, V., Khan, B. A., & Peterson, K. E. (2015). SARM1, not MyD88, mediates TLR7/TLR9-induced apoptosis in neurons. *Journal of Immunology, 195*, 4913–4921. https://doi.org/10.4049/jimmunol.1500953

Nagata, M., Izumi, Y., Ishikawa, E., Kiyotake, R., Doi, R., Iwai, S., Omahdi, Z., Yamaji, T., Miyamoto, T., Bamba, T., & Yamasaki, S. (2017). Intracellular metabolite beta-glucosylceramide is an endogenous Mincle ligand possessing immunostimulatory activity. *Proceedings of the National Academy of Sciences of the United States of America, 114*, E3285–E3294. https://doi.org/10.1073/pnas.1618133114

Nagai, Y., Shimazu, R., Ogata, H., Akashi, S., Sudo, K., Yamasaki, H., Hayashi, S., Iwakura, Y., Kimoto, M., Miyake, K. (2002). Requirement for MD-1 in cell surface expression of RP105/CD180 and B-cell responsiveness to lipopolysaccharide. *Blood 99*(5), 1699–1705. https://doi.org/10.1182/blood.v99.5.1699

Nakaya, Y., Lilue, J., Stavrou, S., Moran, E. A., & Ross, S. R. (2017). Aim2-like receptors positively and negatively regulate the interferon response induced by cytosolic DNA. *mBio, 8*. https://doi.org/10.1128/mBio.00944-17

Nan, K., Han, Y., Fang, Q., Huang, C., Yu, L., Ge, W., Xiang, F., Tao, Y. X., Cao, H., & Li, J. (2019). HMGB1 gene silencing inhibits neuroinflammation via down-regulation of NF-kappaB signaling in primary hippocampal neurons induced by Abeta(25-35). *International Immunopharmacology, 67*, 294–301. https://doi.org/10.1016/j.intimp.2018.12.027

Nasrabady, S. E., Rizvi, B., Goldman, J. E., & Brickman, A. M. (2018). White matter changes in Alzheimer's disease: A focus on myelin and oligodendrocytes. *Acta Neuropathologica Communications, 6*, 22. https://doi.org/10.1186/s40478-018-0515-3

O'Brien, W. T., Pham, L., Symons, G. F., Monif, M., Shultz, S. R., & McDonald, S. J. (2020). The NLRP3 inflammasome in traumatic brain injury: Potential as a biomarker and therapeutic target. *Journal of Neuroinflammation, 17*, 104. https://doi.org/10.1186/s12974-020-01778-5

Palomino-Antolin, A., Narros-Fernandez, P., Farre-Alins, V., Sevilla-Montero, J., Decouty-Perez, C., Lopez-Rodriguez, A. B., Fernandez, N., Monge, L., Casas, A. I., Calzada, M. J., & Egea, J. (2022). Time-dependent dual effect of NLRP3 inflammasome in brain ischaemia. *British Journal of Pharmacology, 179*, 1395–1410. https://doi.org/10.1111/bph.15732

Panicker, N., Sarkar, S., Harischandra, D. S., Neal, M., Kam, T. I., Jin, H., Saminathan, H., Langley, M., Charli, A., Samidurai, M., Rokad, D., Ghaisas, S., Pletnikova, O., Dawson, V. L., Dawson, T. M., Anantharam, V., Kanthasamy, A. G., & Kanthasamy, A. (2019). Fyn kinase regulates misfolded alpha-synuclein uptake and NLRP3 inflammasome activation in microglia. *The Journal of Experimental Medicine, 216*, 1411–1430. https://doi.org/10.1084/jem.20182191

Panicker, N., Kam, T. I., Wang, H., Neifert, S., Chou, S. C., Kumar, M., Brahmachari, S., Jhaldiyal, A., Hinkle, J. T., Akkentli, F., Mao, X., Xu, E., Karuppagounder, S. S., Hsu, E. T., Kang, S. U., Pletnikova, O., Troncoso, J., Dawson, V. L., & Dawson, T. M. (2022). Neuronal NLRP3 is a parkin substrate that drives neurodegeneration in Parkinson's disease. *Neuron, 110*, 2422–2437 e2429. https://doi.org/10.1016/j.neuron.2022.05.009

Peng, J., Yuan, Q., Lin, B., Panneerselvam, P., Wang, X., Luan, X. L., Lim, S. K., Leung, B. P., Ho, B., & Ding, J. L. (2010). SARM inhibits both TRIF- and MyD88-mediated AP-1 activation. *European Journal of Immunology, 40*, 1738–1747. https://doi.org/10.1002/eji.200940034

Pike, A. F., Longhena, F., Faustini, G., van Eik, J. M., Gombert, I., Herrebout, M. A. C., Fayed, M., Sandre, M., Varanita, T., Teunissen, C. E., Hoozemans, J. J. M., Bellucci, A., Veerhuis, R., & Bubacco, L. (2022a). Dopamine signaling modulates microglial NLRP3 inflammasome activation: Implications for Parkinson's disease. *Journal of Neuroinflammation, 19*, 50. https://doi.org/10.1186/s12974-022-02410-4

Pike, A. F., Szabo, I., Veerhuis, R., & Bubacco, L. (2022b). The potential convergence of NLRP3 inflammasome, potassium, and dopamine mechanisms in Parkinson's disease. *NPJ Parkinsons Disease, 8*, 32. https://doi.org/10.1038/s41531-022-00293-z

Poh, L., Kang, S. W., Baik, S. H., Ng, G. Y. Q., She, D. T., Balaganapathy, P., Dheen, S. T., Magnus, T., Gelderblom, M., Sobey, C. G., Koo, E. H., Fann, D. Y., & Arumugam, T. V. (2019). Evidence that NLRC4 inflammasome mediates apoptotic and pyroptotic microglial death following ischemic stroke. *Brain, Behavior, and Immunity, 75*, 34–47. https://doi.org/10.1016/j.bbi.2018.09.001

Rannikko, E. H., Weber, S. S., & Kahle, P. J. (2015). Exogenous alpha-synuclein induces toll-like receptor 4 dependent inflammatory responses in astrocytes. *BMC Neuroscience, 16*, 57. https://doi.org/10.1186/s12868-015-0192-0

Rehwinkel, J., & Gack, M. U. (2020). RIG-I-like receptors: Their regulation and roles in RNA sensing. *Nature Reviews. Immunology, 20*, 537–551. https://doi.org/10.1038/s41577-020-0288-3

Rodriguez, K. R., Bruns, A. M., & Horvath, C. M. (2014). MDA5 and LGP2: Accomplices and antagonists of antiviral signal transduction. *Journal of Virology, 88*, 8194–8200. https://doi.org/10.1128/JVI.00640-14

Rosa, J. M., Farre-Alins, V., Ortega, M. C., Navarrete, M., Lopez-Rodriguez, A. B., Palomino-Antolin, A., Fernandez-Lopez, E., Vila-Del Sol, V., Decouty, C., Narros-Fernandez, P., Clemente, D., & Egea, J. (2021). TLR4 pathway impairs synaptic number and cerebrovascular functions through astrocyte activation following traumatic brain injury. *British Journal of Pharmacology, 178*, 3395–3413. https://doi.org/10.1111/bph.15488

Rothenfusser, S., Goutagny, N., DiPerna, G., Gong, M., Monks, B. G., Schoenemeyer, A., Yamamoto, M., Akira, S., & Fitzgerald, K. A. (2005). The RNA helicase Lgp2 inhibits TLR-independent sensing of viral replication by retinoic acid-inducible gene-I. *Journal of Immunology, 175*, 5260–5268. https://doi.org/10.4049/jimmunol.175.8.5260

Saito, T., Hirai, R., Loo, Y. M., Owen, D., Johnson, C. L., Sinha, S. C., Akira, S., Fujita, T., & Gale, M., Jr. (2007). Regulation of innate antiviral defenses through a shared repressor domain in RIG-I and LGP2. *Proceedings of the National Academy of Sciences of the United States of America, 104*, 582–587. https://doi.org/10.1073/pnas.0606699104

Sansing, L. H., Harris, T. H., Welsh, F. A., Kasner, S. E., Hunter, C. A., & Kariko, K. (2011). Toll-like receptor 4 contributes to poor outcome after intracerebral hemorrhage. *Annals of Neurology, 70*, 646–656. https://doi.org/10.1002/ana.22528

Satoh, T., Kato, H., Kumagai, Y., Yoneyama, M., Sato, S., Matsushita, K., Tsujimura, T., Fujita, T., Akira, S., & Takeuchi, O. (2010). LGP2 is a positive regulator of RIG-I- and MDA5-mediated antiviral responses. *Proceedings of the National Academy of Sciences of the United States of America, 107*, 1512–1517. https://doi.org/10.1073/pnas.0912986107

Scheiblich, H., Bousset, L., Schwartz, S., Griep, A., Latz, E., Melki, R., & Heneka, M. T. (2021). Microglial NLRP3 inflammasome activation uponTLR2 and TLR5 ligation by distinct alpha-synuclein assemblies. *Journal of Immunology, 207*, 2143–2154. https://doi.org/10.4049/jimmunol.2100035

Schmid, C. D., Sautkulis, L. N., Danielson, P. E., Cooper, J., Hasel, K. W., Hilbush, B. S., Sutcliffe, J. G., & Carson, M. J. (2002). Heterogeneous expression of the triggering receptor expressed on myeloid cells-2 on adult murine microglia. *Journal of Neurochemistry, 83*, 1309–1320. https://doi.org/10.1046/j.1471-4159.2002.01243.x

Sheng, L., Chen, M., Cai, K., Song, Y., Yu, D., Zhang, H., & Xu, G. (2019). Microglial TREM2 induces synaptic impairment at early stage and prevents amyloidosis at late stage in APP/PS1 mice. *The FASEB Journal, 33*, 10425–10442. https://doi.org/10.1096/fj.201900527R

Shi, J., Zhao, Y., Wang, Y., Gao, W., Ding, J., Li, P., Hu, L., & Shao, F. (2014). Inflammatory caspases are innate immune receptors for intracellular LPS. *Nature, 514*, 187–192. https://doi.org/10.1038/nature13683

Song, M., Jin, J., Lim, J. E., Kou, J., Pattanayak, A., Rehman, J. A., Kim, H. D., Tahara, K., Lalonde, R., & Fukuchi, K. (2011). TLR4 mutation reduces microglial activation, increases Aβ deposits and exacerbates cognitive deficits in a mouse model of Alzheimer's disease. *Journal of Neuroinflammation, 8*, 92. https://doi.org/10.1186/1742-2094-8-92

Stancu, I. C., Cremers, N., Vanrusselt, H., Couturier, J., Vanoosthuyse, A., Kessels, S., Lodder, C., Brone, B., Huaux, F., Octave, J. N., Terwel, D., & Dewachter, I. (2019). Aggregated Tau activates NLRP3-ASC inflammasome exacerbating exogenously seeded and non-exogenously seeded Tau pathology in vivo. *Acta Neuropathologica, 137*, 599–617. https://doi.org/10.1007/s00401-018-01957-y

Stridh, L., Mottahedin, A., Johansson, M. E., Valdez, R. C., Northington, F., Wang, X., & Mallard, C. (2013). Toll-like receptor-3 activation increases the vulnerability of the neonatal brain to hypoxia-ischemia. *The Journal of Neuroscience, 33*, 12041–12051. https://doi.org/10.1523/JNEUROSCI.0673-13.2013

Suzuki, Y., Nakano, Y., Mishiro, K., Takagi, T., Tsuruma, K., Nakamura, M., Yoshimura, S., Shimazawa, M., & Hara, H. (2013). Involvement of Mincle and Syk in the changes to innate immunity after ischemic stroke. *Scientific Reports, 3*, 3177. https://doi.org/10.1038/srep03177

Tang, S. C., Arumugam, T. V., Xu, X., Cheng, A., Mughal, M. R., Jo, D. G., Lathia, J. D., Siler, D. A., Chigurupati, S., Ouyang, X., Magnus, T., Camandola, S., & Mattson, M. P. (2007). Pivotal role for neuronal Toll-like receptors in ischemic brain injury and functional deficits. *Proceedings of the National Academy of Sciences of the United States of America, 104*, 13798–13803. https://doi.org/10.1073/pnas.0702553104

Tang, S. C., Lathia, J. D., Selvaraj, P. K., Jo, D. G., Mughal, M. R., Cheng, A., Siler, D. A., Markesbery, W. R., Arumugam, T. V., & Mattson, M. P. (2008). Toll-like receptor-4 mediates neuronal apoptosis induced by amyloid beta-peptide and the membrane lipid peroxidation product 4-hydroxynonenal. *Experimental Neurology, 213*, 114–121. https://doi.org/10.1016/j.expneurol.2008.05.014

Unterholzner, L., Keating, S. E., Baran, M., Horan, K. A., Jensen, S. B., Sharma, S., Sirois, C. M., Jin, T., Latz, E., Xiao, T. S., Fitzgerald, K. A., Paludan, S. R., & Bowie, A. G. (2010). IFI16 is an innate immune sensor for intracellular DNA. *Nature Immunology, 11*, 997–1004. https://doi.org/10.1038/ni.1932

Van Opdenbosch, N., Gurung, P., Vande Walle, L., Fossoul, A., Kanneganti, T. D., & Lamkanfi, M. (2014). Activation of the NLRP1b inflammasome independently of ASC-mediated caspase-1 autoproteolysis and speck formation. *Nature Communications, 5*, 3209. https://doi.org/10.1038/ncomms4209

Van Zeller, M., Dias, D., Sebastião, A. M., Valente, C. A., & Wu, Z. (2021). NLRP3 Inflammasome: A starring role in smyloid-β- and Tau-driven pathological events in Alzheimer's disease. *Journal of Alzheimer's Disease, 83*, 939–961. https://doi.org/10.3233/jad-210268

von Herrmann, K. M., Salas, L. A., Martinez, E. M., Young, A. L., Howard, J. M., Feldman, M. S., Christensen, B. C., Wilkins, O. M., Lee, S. L., Hickey, W. F., & Havrda, M. C. (2018). NLRP3 expression in mesencephalic neurons and characterization of a rare NLRP3 polymorphism associated with decreased risk of Parkinson's disease. *NPJ Parkinsons Disease, 4*, 24. https://doi.org/10.1038/s41531-018-0061-5

Walsh, J. G., Muruve, D. A., & Power, C. (2014). Inflammasomes in the CNS. *Nature Reviews. Neuroscience, 15*, 84–97. https://doi.org/10.1038/nrn3638

Wang, W., Nguyen, L. T., Burlak, C., Chegini, F., Guo, F., Chataway, T., Ju, S., Fisher, O. S., Miller, D. W., Datta, D., Wu, F., Wu, C. X., Landeru, A., Wells, J. A., Cookson, M. R., Boxer, M. B., Thomas, C. J., Gai, W. P., Ringe, D., et al. (2016). Caspase-1 causes truncation and aggregation of the Parkinson's disease-associated protein alpha-synuclein. *Proceedings of the National Academy of Sciences of the United States of America, 113*, 9587–9592. https://doi.org/10.1073/pnas.1610099113

Xu, P., Zhang, X., Liu, Q., Xie, Y., Shi, X., Chen, J., Li, Y., Guo, H., Sun, R., Hong, Y., Liu, X., & Xu, G. (2019). Microglial TREM-1 receptor mediates neuroinflammatory injury via interaction with SYK in experimental ischemic stroke. *Cell Death & Disease, 10*, 555. https://doi.org/10.1038/s41419-019-1777-9

Yamasaki, S., Ishikawa, E., Sakuma, M., Hara, H., Ogata, K., & Saito, T. (2008). Mincle is an ITAM-coupled activating receptor that senses damaged cells. *Nature Immunology, 9*, 1179–1188. https://doi.org/10.1038/ni.1651

Yang, J., Wise, L., & Fukuchi, K. I. (2020). TLR4 cross-talk with NLRP3 Inflammasome and complement signaling pathways in Alzheimer's disease. *Frontiers in Immunology, 11*, 724. https://doi.org/10.3389/fimmu.2020.00724

Yi, Y. S. (2017). Caspase-11 non-canonical inflammasome: A critical sensor of intracellular lipopolysaccharide in macrophage-mediated inflammatory responses. *Immunology, 152*, 207–217. https://doi.org/10.1111/imm.12787

Yi, Y. S. (2020). Functional crosstalk between non-canonical caspase-11 and canonical NLRP3 inflammasomes during infection-mediated inflammation. *Immunology, 159*, 142–155. https://doi.org/10.1111/imm.13134

Yuan, P., Condello, C., Keene, C. D., Wang, Y., Bird, T. D., Paul, S. M., Luo, W., Colonna, M., Baddeley, D., & Grutzendler, J. (2016). TREM2 haplodeficiency in mice and humans impairs the microglia barrier function leading to decreased amyloid compaction and severe axonal dystrophy. *Neuron, 90*, 724–739. https://doi.org/10.1016/j.neuron.2016.05.003

Zanoni, I., Tan, Y., Di Gioia, M., Broggi, A., Ruan, J., Shi, J., Donado, C. A., Shao, F., Wu, H., Springstead, J. R., & Kagan, J. C. (2016). An endogenous caspase-11 ligand elicits interleukin-1 release from living dendritic cells. *Science, 352*, 1232–1236. https://doi.org/10.1126/science.aaf3036

Zhang, L., Mo, J., Swanson, K. V., Wen, H., Petrucelli, A., Gregory, S. M., Zhang, Z., Schneider, M., Jiang, Y., Fitzgerald, K. A., Ouyang, S., Liu, Z. J., Damania, B., Shu, H. B., Duncan, J. A., & Ting, J. P. (2014). NLRC3, a member of the NLR family of proteins, is a negative regulator of innate immune signaling induced by the DNA sensor STING. *Immunity, 40*, 329–341. https://doi.org/10.1016/j.immuni.2014.01.010

Zhang, X., Wang, R., Hu, D., Sun, X., Fujioka, H., Lundberg, K., Chan, E. R., Wang, Q., Xu, R., Flanagan, M. E., Pieper, A. A., & Qi, X. (2020). Oligodendroglial glycolytic stress triggers inflammasome activation and neuropathology in Alzheimer's disease. *Science Advances, 6*. https://doi.org/10.1126/sciadv.abb8680

Ziegler, G., Harhausen, D., Schepers, C., Hoffmann, O., Rohr, C., Prinz, V., Konig, J., Lehrach, H., Nietfeld, W., & Trendelenburg, G. (2007). TLR2 has a detrimental role in mouse transient focal cerebral ischemia. *Biochemical and Biophysical Research Communications, 359*, 574–579. https://doi.org/10.1016/j.bbrc.2007.05.157

B and T Cells

4

Taylor J. Dunning, Tiana Hua, Larisa Y. Poluektova, and R. Lee Mosley

Abstract

Neuroimmunology was an unforeseen discipline with impediments that, at times, forbade the use of immunology and neurology in the same paragraph much less the unconceivable emergence into its own discipline. However, once the notion of the brain as an immunologically privileged site was abandoned, the shared components and characteristics became clearer. Indeed, the immune and nervous systems share several associated properties as both share a functional basis of self and defense, both perform surveillance functions, both have a cellular basis of memory, and both have shared molecular components that, in many cases, function differentially in opposing compartments but seemingly maintain balance within each compartment under nonpathological conditions. This chapter is tasked with the overview of the adaptive immune system with its two primary lineages of effectors, B cells and T cells, in the context of development, interactions with the innate immune arm, different immune effector functions after activation by the innate system, and effector function in the central and peripheral nervous systems. Finally, a novel in vivo small animal platform is described in which the development and effector functions of the human immune system can be evaluated and the interactions of each immune arm with the central and peripheral nervous systems can be assessed for the missing pieces of the puzzle to form the entire picture of central nervous system immunity.

Authors Taylor J. Dunning and Tiana Hua have equally contributed to this chapter.

T. J. Dunning · T. Hua · L. Y. Poluektova · R. L. Mosley (✉)
Department of Pharmacology and Experimental Neuroscience, Center for Neurodegenerative Disorders, University of Nebraska Medical Center, Omaha, NE, USA
e-mail: rlmosley@unmc.edu

Keywords

Lymphocytes · T cells · B cells · Lymphocyte development · Neuroimmunity · Neurotransmitters · Humanized mice

4.1 Introduction

The immune system maintains the integrity of the human body by recognition of external but not host antigens. The control of immune reactions to infections, tumors, and degeneration in the brain is different from that in other tissues. The differences in immune control in the brain became clear with the development of transplantation technologies. In 1873, the Dutch ophthalmologist van Dooremaal observed prolonged mouse skin graft survival in the dog eye, and later, in 1948, Sir Peter Medawar, while trying to explain the success of corneal homograft transplantation, performed classic transplantation of the skin into the brain and anterior chamber of the eye of nonimmunized or immunized rabbits (previously transplanted with the same donor skin sample into the chest). He highlighted two mechanisms involved in the immunological control of the brain: existing systemic immunity to the donor skin and the status of the vasculature. Tissue grafted in the brain or eye survived longer than in any other place in the body without sensitized lymphocyte infiltration via new vessels of the graft. Several more places in the human body obtained the definition of "privileged" sites, including the testis, placenta, and anterior chamber of the eye, based upon immune tolerance to the graft. Only direct transplantation/injection allowed backward introduction of "unknown" antigens, and this process in the brain parenchyma, not meninges and choroids plexus, is also different than in nonprivileged sites due to the absence of endogenous professional antigen-presenting cells (APCs) capable of triggering an immune response (Hart & Fabre, 1981). This mechanism can be explained by the specific properties of nonprofessional antigen-presenting cells, such as microglial

H. E. Gendelman, T. Ikezu (eds.), *Neuroimmune Pharmacology and Therapeutics*,
https://doi.org/10.1007/978-3-031-68237-7_4

cells in parenchymal tissue (Matyszak & Perry, 1996). However, in situ, inflammation induced by pathogens or abnormal proteins with complementary peripheral adaptive immune responses can induce brain damage and the appearance of tertiary lymphoid structures (Aloisi & Pujol-Borrell, 2006). The role of professional antigen-presenting cells, microglial cells, and regulatory T cells (Tregs) in these events will be reviewed in other chapters. Currently, the regulation of immune responses within the brain is primarily attributed to specific cell membrane-bound molecules, soluble factors, and brain-associated immune deviation (BRAID) (Niederkorn, 2006). This chapter will review the development and function of lymphocytes, which include T cells (thymus-derived), B cells (Bursa- or bone marrow-derived), and natural killer (NK) cells (effectors of innate immunity) as the key cell types in the neuroimmunology of the brain.

The specialized functions of lymphocyte subsets are reflected in the differing patterns of molecules and genes they express. These molecules often underlie the unique function of the subsets they mark. For instance, CD3 defines T cells as essential components of the T-cell receptor (TCR) complex; CD4 is associated with major histocompatibility complex (MHC) class II recognition primarily for helper T-cell (Th) responses, whereas CD8 recognizes MHC class I and is associated with cytotoxic T-cell (Tc) responses. B cells are defined by their immunoglobulin (Ig) expression or production. Immunological research during much of the 1980–1990 decade was directed at identifying and characterizing leukocyte markers, first through monoclonal antibody (mAb) production and then through gene sequencing. This led to the discovery of most of the cluster of differentiation (CD) molecules and nomenclature used today to identify leukocyte subsets and define function. The final step of lymphocyte differentiation is transformation to antigen-specific memory cells. Only these cells are capable of surveillance in the brain. The mechanisms for surveillance of perivascular and parenchymal sites are reviewed elsewhere (Engelhardt, 2006; Ransohoff et al., 2003). Importantly, discussion of lymphopoiesis and lymphocyte function is necessary before lymphocyte–central nervous system (CNS) interactions during diseases of the nervous system.

4.2 Overview of Embryonic Lymphopoiesis

The common concept of lymphoid tissue ontogeny results from the understanding that circulating stem/progenitor cell populations migrate from the yolk sack into the aorta–gonad–mesonephros (AGM) region and then into organ anlages to initiate hematopoiesis in the fetal liver, bone marrow, bursa of Fabricius (avian), spleen, and thymus. The generation of lymphocytes from adult hematopoietic stem cells (HSCs) is different from embryonic development and will be discussed separately. Our knowledge about immune system ontogeny developed mainly from studies with rodents, wherein in vivo experiments tested whether precursor activities of certain cell populations were possible. By contrast, most information about human lymphopoiesis was derived from in vitro studies. The principles outlined in mouse models generally seem to be consistent with some underlying mechanisms of human lymphoid system development, but cell surface phenotypes of human transitional cell populations are often different from those in the mouse.

4.2.1 T Cells

The thymus is a primary lymphoid organ that is involved in the development of T cells (thymus-derived), which are only one lineage of effector cells for adaptive immune responses. The human thymus develops at approximately embryonic day 35 (E35) from the endoderm with subsequent epithelial–mesenchymal interactions. Again, information on T-cell ontogeny is mainly derived from studies in rodents. The thymic epithelium and surrounding mesenchyme, which are derived from the cephalic region of the neural crest, play important roles in T-cell development. During the embryonic phase, commitment to the lympho-myeloid cell lineage occurs before thymus development and as hematopoietic clusters expressing CD45, CD117, and CD34 within the aorta–gonad–mesonephros (AGM) region. In addition, multi-lineage progenitors were restricted to lineages that were identified in the AGM region, which included T cells, natural killer (NK) cells, and progenitor (p) lineages for T cells (pT), pT/myeloid (pTM) cells, progenitor B (pB) cells, and/or pB/myeloid (pBM) cells. At the time of entry into the thymic epithelium, immigrant cells express CD45, CD117, CD44, CD34, and α4 integrin but do not express CD62L, CD25, and Thy-1 in mice. One of the important cytokines that regulate survival of human lymphoid progenitors is interleukin (IL)-7. Before exposure to epithelial inductive influence by thymic stromal cells, including thymic epithelial cells (TECs), lympho-myeloid progenitors express IL-7 receptor alpha chain (IL-7Rα). Upon differentiation into functionally mature lymphocytes (CD4 and CD8), IL-7Rα is downregulated and cells lose their thymic homing capacity. In knockout models of mice, deletion of the *IL7* gene or related receptors abrogates development of lymphocytes and lymphoid organs (Liang et al., 2012; Ribeiro et al., 2013; Tsapogas et al., 2011). The transmembrane receptor Notch is essential during early T-lineage development. In the absence of Notch-1-mediated signals, T-cell development is arrested at an early stage, and B cells accumulate within the thymus. Fundamental studies in T-cell biology have enabled the discovery of T-cell memory, immunometabolism, exhaustion, and epigenetics, which has paved the way for development of precise therapies targeted toward T cells.

4.2.2 B Cells

B cells are continuously produced from hematopoietic stem progenitors throughout life. B lymphocytes are distinct from their T-cell counterparts in terms of the antigen receptor structure. The specific B-cell antigen receptor acts as a mechanism for detecting the presence of foreign molecules and aiding in pathogenic elimination by producing specialized proteins called immunoglobulins (Igs). B-cell proliferation into Ig-producing cells or plasma cells (PCs) occurs at the site of binding where these cells play a significant role in humoral immunity. While the exact location of B-cell commitment remains uncertain, B-cell precursors originate from the para-aortic region within the developing embryo. The process of B-cell commitment is believed to occur in the omentum and fetal liver. These precursor cells express a specific phenotype as $CD117^+AA4.1^+CD34^+CD45^+CD19^+Sca\text{-}1^-$cells. Historically, the fetal liver has been considered the initial site where B-cell commitment occurs. Subsequently, B-cell progenitors undergo a coordinated expansion resembling a wave-like pattern, with peak activity during the perinatal stage. B-cell development can also occur in the bone marrow after birth (Pieper et al., 2013). The transcription factor Pax-5 suppresses the development of myeloid cells and facilitates B-cell commitment. Additionally, B-cell development in the fetal liver is dependent on the C–X–C motif chemokine receptor 4 (CXCR4) and its associated ligand stromal cell-derived factor 1/C–X–C motif chemokine ligand 12 (SDF-1α/CXCL12). Alternatively, murine IL-7Rα expressing progenitors are derived from the fetal liver to play a role in B-cell development, along with their involvement in pre-lymphoid tissue inducers (pre-LTis) (van de Pavert & Mebius, 2010). Recent advancements in B-cell biology have led to a greater understanding of how these cells operate as therapeutic targets, not only for malignancies but also for autoimmune disorders and neurological diseases.

4.2.3 NK Cells and Other Innate Lymphoid Cells

Within the human immune system, 5–20% of all lymphocytes are natural killer (NK) cells (Abel et al., 2018). As an innate lymphocytic subset, NK cells are cytotoxic in nature and can process and present antigens, although antigen presentation is not required for NK-mediated clearance of viral infections or tumors. NK cells occupy secondary lymphoid tissues and areas of the lung, gut, liver, and skin and can be influenced by the microbiome environment.

NK cells are phenotypically identified by their expression of CD56 and CD16 (Fcγ receptor III or FcγRIII) and by the absence of CD3 expression. Two distinct populations of human NK cells can be examined by the density of CD56 expression on the cell surface. The majority (~90%) of human NK cells express a low density of CD56 ($CD56^{dim}$) and high levels of CD16, whereas approximately 10% of NK cells express higher CD56 density and lower or absent CD16 expression ($CD56^{bright}CD16^{dim}$ or $CD56^{bright}CD16^-$) (Caligiuri, 2008). One notable discovery is that the $CD56^{bright}$ NK cell subset expresses high-affinity IL-2 receptor (IL-2R) in a continuous manner. This subset is capable of producing interferon-gamma (IFN-γ) when exposed to extremely low concentrations of IL-2, in the picomolar range. This is significant as IL-2 is primarily produced by T cells and not by NK cells themselves. The presence of the high-affinity IL-2Rαβγ on $CD56^{bright}$ NK cells is crucial for enabling communication between NK cells and T cells through cytokine interactions in secondary lymphoid organs.

A key receptor found on the surface of NK cells is low-affinity FcγRIII, which binds to antibodies coated on target cells, enabling antibody-dependent cellular cytotoxicity (ADCC). This receptor is associated with subunits containing an immunoreceptor tyrosine-based activation motif (ITAM), which triggers cellular cytotoxicity (Miller, 2008). NK cells also express receptors for other cytokines produced by monocytes that include IFN-α/β, IL-1, IL-10, IL-12, IL-15, IL-18, IL-21, and IL-27. In most cases, a complete activation of the NK cell response requires the presence of a combination of two cytokines, IL-12 and IL-18 (Messlinger et al., 2018), and the earliest of these NK-activating cytokines most likely originate from monocytes during the innate immune response (Xu & Di Santo, 2013).

Recent evidence has categorized conventional NK cells as a subset of innate lymphoid cells (ILCs), which are devoid of TCRs. They are divided into three groups based on their characteristics and functions. Group 1 ILCs, (ILC1s) include NK cells, all of which produce IFN-γ upon activation. Group 2 ILCs (ILC2s) release IL-5 and IL-13 when stimulated by molecules such as IL-25 or IL-33, which are released by damaged epithelial cells or activated myeloid cells. Lastly, group 3 ILCs (ILC3s) comprise various subsets, including lymphoid tissue inducer (LTi) cells, as well as $CD4^+$ and $CD4^-$ subsets. The retinoic acid-related orphan receptor (ROR)γt is required by ILC3s to develop and produce cytokines such as IL-17A and/or IL-22 (Zhou & Sonnenberg, 2020). Functionally, the ILC family can be considered as "innate" equivalents of T effector cells (cytotoxic CD8 T cells and differentiated CD4 Th subsets), as they secret a similar array of cytokines and activate an analogous set of target cells (Xu & Di Santo, 2013).

4.2.4 Hematopoietic-Derived Immune Cells in the Brain

Within brain tissues, microglia are a singular type of specialized myeloid resident cell derived from the mononuclear phagocyte lineage. Over the past decade, advancements have expanded the comprehension of the development of the macrophage lineage in the brain. Various waves of population dynamics are operative in the brain starting with cells of the myeloid lineage and primitive macrophages originating in the yolk sac with a high proliferation capacity, then myeloid precursors from the fetal liver with the ability to proliferate, and, then, bone marrow-derived nondividing (or limited division) precursors in adults. Two types of brain phagocytic residents are found, namely, microglia and perivascular macrophages, both with nonselective but differently expressed human markers. These cells exhibit a common phenotype that includes $CD45^{lo}CD11c^{+}CD11b^{+}Ham56^{+}$ with variable expression of CD64, CD68, (*Ricinus communis* agglutinin-1) RCA-1, and MHC-II and serve as effector cells for innate immunity. These traits play a significant role in the establishment of adaptive immune reactions and immune privilege in situ within the parenchyma and the meninges. The nature of microglia and other brain myeloid residents is well-established in mice (Ginhoux et al., 2010; Perdiguero et al., 2015). These cells originate from yolk sac-derived erythromyeloid progenitors. The turnover and replacement with bone marrow-derived monocytes/tissue macrophages is exceptionally constrained, which indicates the unique and vital role of these cells in maintaining brain immune homeostasis.

4.3 Postnatal Development of Lymphocytes

4.3.1 Hematopoietic Stem Cells and Lymphoid Cell Lineages

The generation of lymphocytes from adult hematopoietic stem cells (HSCs) is different from the embryonic development of these same cells. Embryonically, HSCs develop through a three-wave process at different sites throughout the developing stages of the embryo. Following their initial differentiation into multipotent progenitor cells (MPPs), HSCs are subcategorized into two major commitment pathways: a lymphoid lineage and a myeloid lineage (Fig. 4.1). The lymphoid lineage focuses on the commitment of T, B, and NK cells, and, perhaps, dendritic cells (DCs), while the myeloid lineage is associated with formation of monocytes, macrophages, and other blood cell subtypes. HSCs are defined by their unique capacity to self-renew as well as their multilineage differentiation into all blood cell families (Gao et al., 2018; Lai & Kondo, 2008). Although representing approximately only 0.05–0.1% of total bone marrow cells, virtually all HSC activities have been shown to be contained within the lineage of progenitor cells that express little or no surface markers found on mature leukocytes ($Lin^{-/lo}$) and present a phenotype of ($Lin^{-/lo}Sca1^{+}c\text{-}Kit(CD117)^{lo/hi}CD34^{-/lo}$). The LSK ($Lin^{-/lo}Sca1^{+}c\text{-}Kit^{lo/hi}$) HSC pool can be further subdivided for short-term repopulating cells using additional markers (Peng et al., 2012). As HSCs acquire high levels of CD34, CD27, and Flk-2/Flt3, they lose long-term repopulating activity. HSCs are capable of self-renewal until their differentiation into MPPs where specific cell surface markers and transcription factors are expressed, thus leading to further differentiation into common lymphoid progenitors (CLPs) and the ILC-producing family. In the adult murine bone marrow, a population of $Lin^{+}IL\text{-}7R^{+}Thy\text{-}1^{-}Sca\text{-}1^{lo}CD117^{lo}$ cells contains CLPs based on their capacity to develop into lymphoid cells while being unable to support myeloid differentiation. The possibility remains that between the common myeloid precursors (CMPs) and the granulocyte monocyte precursors (GMPs), there exists a population of monocyte dendritic cell progenitors (MDPs), which is positive for CX3CR1 (fractalkine receptor) (Fogg et al., 2006)..

T and B cells arise after the differentiation of CLPs from HSCs and MPPs. Unlike their B-cell counterparts, T cells move through the blood from the bone marrow to ultimately mature in the thymus. The fate of the T-cell repertoire is dependent on environmental signals. Identification and characterization of thymus seeding progenitors (TSPs) have been challenging owing to the rarity of these cells in the bone marrow or blood at any given time, and, thus, their lineage potential and level of commitment are still largely debated. Potential candidates of TSPs include $Lin^{-}Sca\text{-}1^{+}cKit/CD117^{+}$ (LSK) cells; $Flt3^{+}$ LSK multipotent progenitors (MPPs) that are further subdivided based on the expression of C–C motif chemokine receptor 9 (CCR9) and VCAM-1; LSK $Flt3^{hi}Rag\text{-}1^{+}$ early lymphoid progenitors (ELPs) or lymphoid-primed multipotent progenitors (LMPPs); $Lin^{-}Sca\text{-}1^{lo}cKit^{-/lo}Flt3^{hi}IL\text{-}7R\alpha^{+}$ CLPs or their progeny $Lin^{-}cKit^{-/lo}B220^{+}IL\text{-}7R\alpha^{+}pre\text{-}T\alpha^{+}CLP\text{-}2^{+}$; and $Lin^{-}cKit^{lo}CD90^{+}CLP\text{-}2$-like circulating T-cell progenitors (CTPs). However, the identity of the cells that first enter the thymus is generally accepted to be a $Lin^{-}cKit^{+}Flt3^{+}CD24^{-/lo}PSGL1^{+}CCR9^{+}$, commonly referred to as early thymic progenitors (ETPs) (Luis et al., 2016; Thompson & Zúñiga-Pflücker, 2011).

The development of lymphoid cell lineages through hematopoietic stem cells is typically found to comprise the innate immune response in mucosal and epithelial spaces. Their capacity for self-renewal, which leads to the development of all blood cell lineages, renders HSCs crucial for therapeutic reconstitution of hematopoiesis in transplantation recipients undergoing bone marrow ablation.

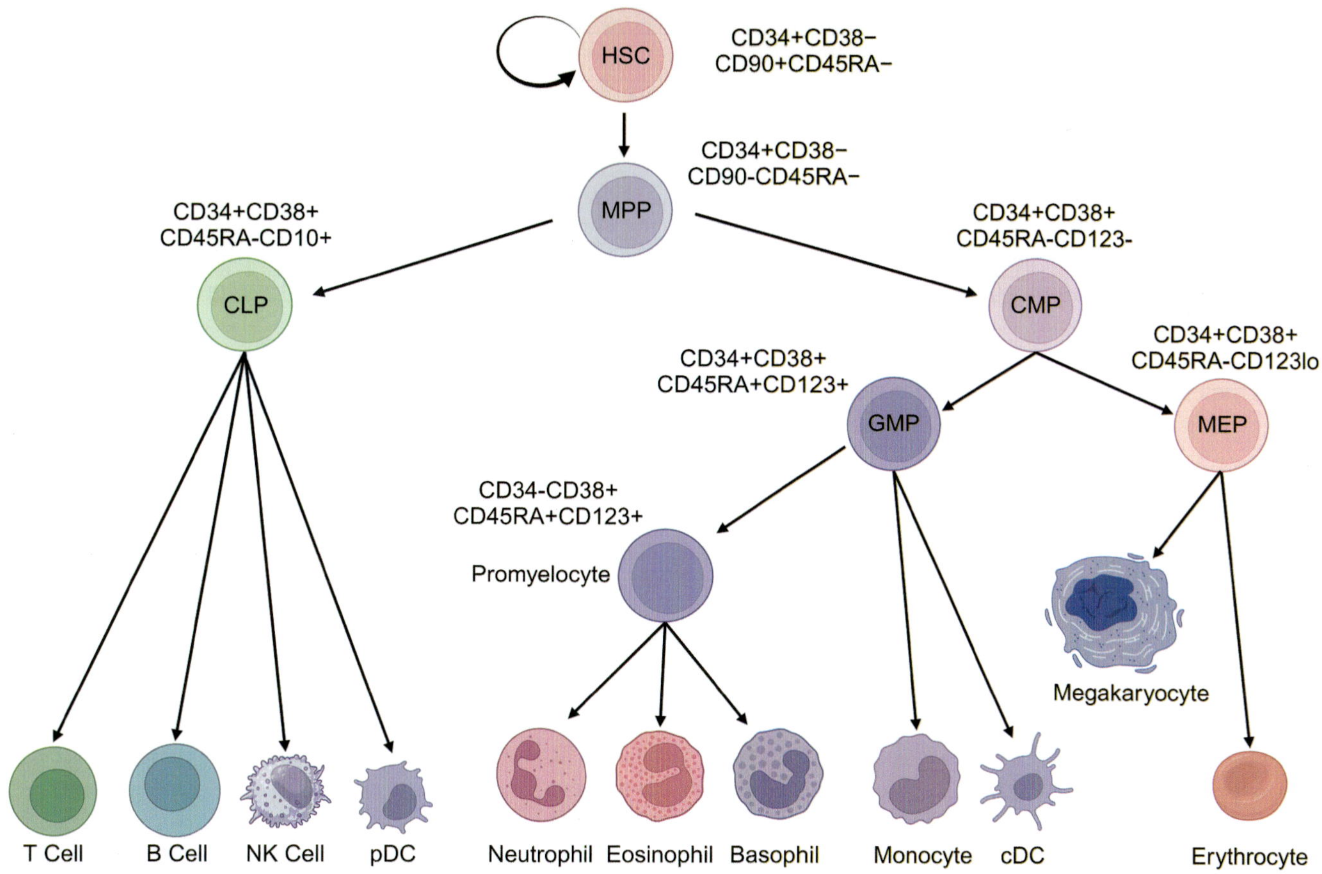

Fig. 4.1 Development of primary lineage commitment in hematopoiesis. Hematopoietic stem cells (HSCs) exhibit a self-renewing capacity that leads to the development of several multipotent progenitor cells (MPPs) and those develop into the lineage-committed common lymphoid progenitors (CLPs), common myeloid progenitors (CMPs), or megakaryocyte-erythroid progenitors (MEPs) from which develop lymphoid-type, myeloid-type, and erythroid mature blood cells, respectively. cDC, conventional dendritic cell; CLP, common lymphoid progenitor; CMP, common myeloid progenitor; GMP, granulocyte–monocyte progenitor; HSC, hematopoietic stem cell; MEP, megakaryocytic-erythroid progenitor; MPP, multipotent progenitor; pDC, plasmacytoid dendritic cell

4.3.2 Thymic Epithelial Cells

Although the exact nature of thymic epithelial cell (TEC) precursors and their maturation pathways have been debated since their discovery, recent advancements have led to a greater understanding of the molecular makeup of TECs and their role in T-cell development. Differentiation and self-tolerance of T cells occur in the thymus, a primary lymphoid organ. Major histocompatibility complex class I or II (MHC I or MHC II) act as critical checkpoints for differentiating between those cells that bind to MHC via T-cell receptor (TCR) interaction and those that cannot respond to MHC self-antigens. Differentiation between MHC I and MHC II occurs due in part to subtypes of TECs (Ishikawa et al., 2021), which are derived from self-renewing precursors of an endoderm origin in relation to these TECs (Anderson et al., 2014). These cells pass through a maturation pathway from cortical TECs (cTECs), without expression of a transcriptional regulator known as an autoimmune regulator (AIRE), through immature medullary TECs (mTECs), with low levels of AIRE expression, and then to mature AIRE$^+$ TECs to finally organize into Hassall's corpuscles. Expression of tissue-specific antigens (TSAs) is a physiological property of TECs, representing all parenchymal organs and including up to 5–10% of all currently known genes, including fetal stage-, tumor-, and sex-associated genes. TSA expression increases with TEC maturation. Brain-, intestine-, and liver-related TSAs have the highest presentation on mTECs. TECs do not form static scaffolds but, like the stratified epithelia of the skin and gut, have a steady-state turnover with a half-life of 3–4 weeks. Together with a turnover rate of 2 weeks for thymic dendritic cells (DCs), this makes the antigen-presenting cell (APC) population in the medulla a highly dynamic compartment (Derbinski & Kyewski, 2005; Perniola, 2012; Sun et al., 2013). Interaction with TECs is the major step in T-cell education in the thymus and the establishment of self-tolerance. Cross talk between derived early T-lineage progenitors (ETPs) and MHC II-positive cTECs gives rise to CD4$^-$CD8$^-$ double-negative (DN) thymocytes. These cells undergo proliferative expansion and lose B- and NK-cell potential. After that, T-cell lineage commitment and the onset of TCR β-chain rearrangement occur. Positive selection takes place through interaction of $CD4^+CD8^+$ double-positive cells with cTECs bearing MHC I and II. Then cross talk between mTECs and DCs expressing TSAs and circulating antigens leads to deletion (or induction of anergy) of self-reacting T cells, known as negative selection. These events take place at the corticomedullary junction and in the medulla; then, single positive $CD4^+$ or $CD8^+$ cells leave the thymus (Bommhardt et al., 2004; Yui & Rothenberg, 2014). Recently, the role of TECs under both pathological and healthy states has become better understood for their implications in immune system regulation.

4.3.2.1 T-Cell Development, TCR, MHC, and CD4/CD8 Commitment

Postnatal lympho-myeloid progenitors emerge within the bone marrow and circulate in the bloodstream, exhibiting an absence of IL-7Rα expression. However, during a subsequent phase, IL-7Rα reappears on common lymphoid progenitors (CLPs) within the mature bone marrow, before migration to the thymus for T-cell lineage commitment. The thymus, though not conducive to self-renewal, plays a key role in T-cell development (Fig. 4.2). To ensure the persistence of thymopoiesis during adulthood, bone marrow-derived progenitors are consistently introduced into the thymus, where Notch-1 and its corresponding ligand delta-like 4 (DLL4) play a critical role in the initial stages of T-cell development within the thymus. Without either of these components, the precursor cells of the thymus adopt fates other than T cells, such as B cells or other non-T-cell lineages.

The process of progenitor migration from the bloodstream to the thymus has been characterized by the expression profile of $CD25^-$ (IL-2Rα) and $CD117^+$ markers. This expression pattern delineates the trajectory of these progenitors as they transition from the bloodstream to the thymic microenvironment. Intrathymic T-cell development pivots on the orchestration of three fundamental gene clusters. First, TECs engage in promiscuous TSA expression. Subsequently, APCs, including cortical and medullary TECs, thymic DCs, and macrophages, express MHC class I and II molecules, thereby facilitating the presentation of diverse antigens. Lastly, mature T cells are characterized by the expression of T-cell receptor (TCR) molecules, which play a central role in antigen recognition and downstream signaling events.

TCR expression by T cells plays a pivotal role in governing the key stages of T-cell development and function. These developmental processes include maturation within the thymus from common lymphoid progenitors, survival of naive T cells after departure from the thymus, and specialization of T cells into distinct functional effector populations. The conserved CD3γ, δ, ε, and ζ subunits complex and form associations with highly diverse αβ- or γδ-chain heterodimers, whereby mature T cells express either TCRαβ or TCRγδ that recognize specific antigens presented by MHC. The expression of CD4 and CD8 co-receptors by TCRαβ lineage T cells influences the recognition of MHC II and MHC I molecules, respectively.

The genetic reorganization of segments contributes to the creation of genes that encode the distinct α-, β-, γ-, and δ-chains composing the TCR. At the pre-TCR expression stage, the preference between γδ- and αβ-chains is determined by the selection of cells with productive TCRβ rearrangements, irrespective of the Vβ gene segment employed. The reordering of TCRs leads to the development of receptors varying in their capacity to bind many peptides derived from self-proteins and foreign proteins. This binding occurs with MHC molecules found on thymic APCs and TECs. The destiny of thymocytes and the process of selection are intertwined with the extent to which TCRs are diminished and internalized after engaging with peptide/MHC combinations. In the thymus, full TCR downregulation corresponds with negative selection, signifying the removal of T cells with the highest affinity (agonism). Conversely, partial TCR downregulation is linked to positive selection, wherein T cells with lower affinity (antagonism) are retained.

Progression from the intermediate stages of early thymocytes as triple-negative ($CD3^-CD4^-CD8^-$) toward triple-positive phase ($CD3^+CD4^+CD8^+$) thymocytes is a pivotal process. This transformation occurs within the context of MHC II antigens, leading to the generation of $CD4^+$ T cells. During this developmental phase, the activity of Lck protein tyrosine kinase plays a significant role, particularly in response to TCR signaling strength. Robust TCR signaling with strong Lck activity is conducive to the development of $CD4^+$ cells, while weaker TCR signaling and decreased Lck activity promote the emergence of $CD8^+$ cells.

TECs and thymic hematopoietic DCs contribute substantially to inducing proliferation and differentiation of $CD4^+CD8^-CD25^-$ T cells into $CD4^+CD25^+FOXP3^+$ regulatory T cells (Tregs). This process relies on interactions between the TCR and MHC/peptide with high affinity, though not sufficient to induce apoptosis, and exposure to IL-2 and IL-15 for Treg development and emigration from the thymus (Apert et al., 2022). Lower-affinity interactions result in development of functional conventional T cells. After thymic education, these mature, yet naïve, T cells exit the peripheral blood and seed lymphoid organs in T-cell-specific zones and are functionally capable of acquiring encounter antigens in the context of MHC and becoming effector T cells (Teffs) necessary for elimination of infected or tumor cells, support for humoral immune responses, formation of immunological memory, prevention of excessive tissue damage, or facilitation of tissue regeneration.

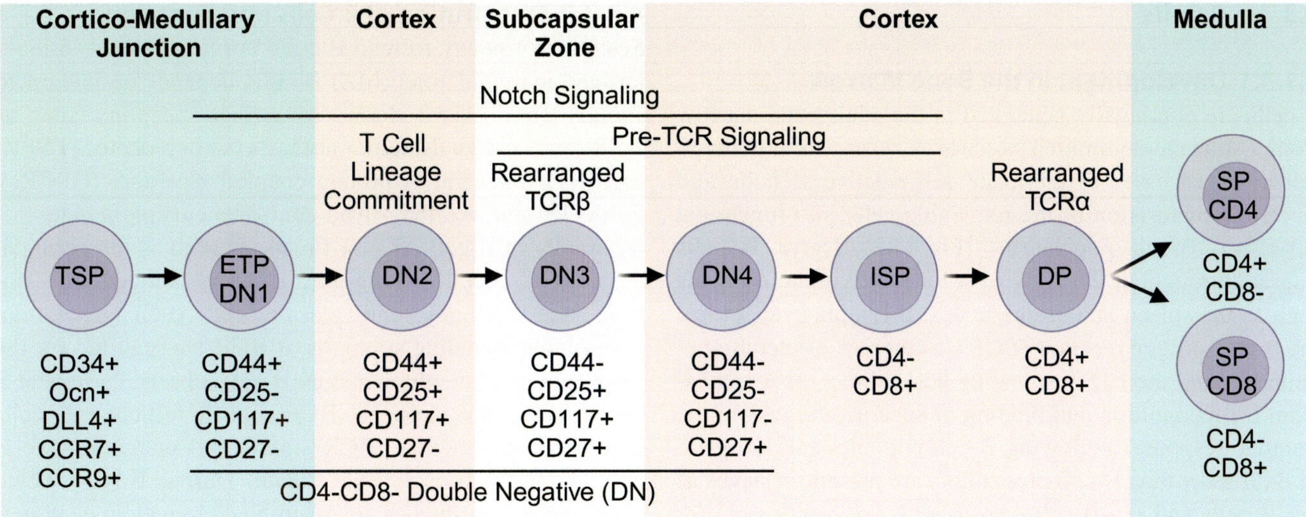

Fig. 4.2 Phenotypes and designated milestones of thymocyte development within the mouse thymus. In adults, thymic seeding progenitors (TSPs) are derived from the bone marrow and the common lymphoid progenitors (CLPs), but, due to the paucity of TSPs, a consensus phenotype is controversial. Nevertheless, TSPs enter the bloodstream, migrate to the thymus in response to chemotactic cues, and enter near the corticomedullary junction. Early thymic progenitors (ETPs) as $CD4^-CD8^-$ double-negative (DN) thymocytes progress through four stages of development (DN1–4) guided by Notch-1 signaling and defined by expression of CD44, CD25, and CD117 (c-Kit) within discrete microenvironments of the thymus. T-cell lineage commitment and initiation of TCR rearrangement begin at the DN2 stage with TCRγ, TCRδ (not shown), and TCRβ rearrangement, culminating in the DN3 stage, which marks the beginning of pre-TCR signaling. DN3 cells that successfully rearrange the *TCRγ* and *TCRδ* genes express TCRγδ and commit to the γδ-lineage (not shown). DN3 cells that successfully rearrange TCRβ are selected to become $CD44^-CD25^-CD117^-$ DN4 cells. TCRβ-lineage-committed DN4 thymocytes then mature through a cycling $CD4^-CD8^+TCR^-$ ISP stage and become $CD4^+CD8^+$ double-positive (DP) blasts. The DP thymocytes initiate V(D)J recombination of the *TCRα* locus, culminating in productive rearrangement and expression of a mature TCRαβ complex. DP thymocytes undergo a process of positive and negative selection in the cortex and further negative selection in the medulla as determined by TCR signaling in response to binding to self-peptide presented by self-MHC I and II. As a result, single positive (SP) thymocytes become MHC class I- or MHC class II-restricted $CD4^-CD8^+$ or $CD4^+CD8^-$ thymocytes, respectively, expressing functional TCRαβs, which emigrate from the thymus to peripheral lymphoid tissues

4.3.2.2 T Cells and the Brain

Neurons and synapses that provide long-term memory and movement must be strongly protected from damage by inflammatory/autoreactive $CD4^+$ and cytolytic/cytotoxic $CD8^+$ T cells. However, the brain parenchyma does not provide substantial support for the migration or survival of T cells. The highly specialized nature of the endothelial cell lining, with the cerebral vasculature and tight junctions, forms a physical and functional barrier to lymphocytes. For a short period of time, T cells can be found in perivascular spaces and make parenchymal foci as an inflammatory reaction to acute viral or bacterial infection. However, such physiological responses in the brain parenchyma are quickly terminated by downregulation of MHC class I and II and co-stimulatory molecules; by molecular expression of immunosuppressive mediators such as IL-10, TGF-β, and prostaglandins; or by apoptosis of T cells through Fas–Fas ligand (CD95/CD95L) interaction.

The activation of microglial cells, dendritic cells, and macrophages in the brain may also be inhibited by neuronal-derived substances. For instance, CD200 is a protein belonging to the Ig superfamily that can significantly counter the activation of myeloid cells via negative regulatory signaling through CD200 receptor (CD200R) interactions. The presence of T cells within the brain is an infrequent occurrence and is linked to persistent pathological conditions resembling tertiary lymphoid structures. Under these conditions, signals for sustenance and activation are provided by DCs, microglia, and macrophages and may lead to neurodegenerative diseases.

Activated microglia provide pro-inflammatory signals that induce vascular endothelial cells to upregulate adhesion molecules and stimulate T cells to infiltrate the brain tissue, leading to the generation of pro-inflammatory T-cell responses within the brain. This process is accompanied by the release of inflammatory molecules, intensifying blood–brain barrier (BBB) damage and amplifying the influx of T cells into the brain during neurodegenerative diseases such as amyotrophic lateral sclerosis and Alzheimer's, Parkinson's, and Huntington's diseases.

4.3.3 B Cells

4.3.3.1 Development in the Bone Marrow

B cells are continually generated in the adult bone marrow from B-lineage-committed precursors where they undergo selective processes to eliminate self-reactive B cells and direct the maturation of the remaining cells into functional peripheral B-cell populations (Fig. 4.3). Naïve B cells migrate throughout the body to secondary lymphoid organs such as the spleen and lymph nodes. B lymphocytes utilize the B-cell antigen receptor (BCR), a complex of membrane-bound Ig and the CD79 signaling heterodimer, as a mechanism for recognition and binding of specific antigens in the immune response. Following B-cell commitment, but prior to BCR assembly, B-cell progenitors are present in stages as pre-B cells and pro-B cells. Pre-B cells are responsible for completion of VDJ (variable–diversity–joining) recombination where heavy- and light-chain gene segments are rearranged to form various antigen-binding domains. Pre-B-cell niches serve as an important checkpoint in BCR signaling and B-lymphocyte development. FMS-like tyrosine kinase 3-expressing ($Flt3^+$) and $IL\text{-}7R^+$ populations in the bone marrow play vital roles in B-cell development as lymphopoiesis is diminished in $Flt3^{-/-}$ and $IL\text{-}7^{-/-}$ murine models. The generation of B lymphocytes from HSCs is controlled by two cytokine receptors (Flk-2/Flt3 and IL-7R) and six transcription factors (PU.1, Ikaros, E2A, Bcl11a, EBF, and Pax-5). Ikaros and PU.1 act in parallel pathways to control lymphoid progenitor development, in part, by regulating expression of essential signaling receptors Flt3, CD117, and IL-7Rα. The generation of the earliest B-cell progenitors depends on transcriptional factors, transcription factor 3 (TCF2/E2A), and EBF that coordinately activate the B-cell gene expression program and Ig heavy-chain gene rearrangements at the onset of B-cell lymphopoiesis. Pax-5 restricts the developmental options of lymphoid progenitors to the B-cell lineage by repressing transcription of lineage-inappropriate genes and simultaneously activating the expression of signaling molecules. Two other transcriptional factors, lymphoid enhancer-binding factor 1 (LEF1) and SRY-box transcription factor 4 (Sox4), contribute to the survival and proliferation of pro-B cells in response to extracellular signals. Finally, interferon regulatory factor (IRF)4 and IRF8 control the termination of pre-B-cell receptor signaling, thus promoting differentiation to small pre-B cells undergoing light-chain gene rearrangements (Singh et al., 2005). Productive variable region and joining-region rearrangements (VL/JL rearrangements) of the Ig light chain are completed in pre-B cells, and expression of the B-cell receptor (BCR) drives pre-B cells to the immature B-cell stage. Newly formed B cells are exported to peripheral lymphoid tissues as functionally immature or transitional intermediate cells (Clark et al., 2014).

4.3.3.2 Distribution of B Cells and Function

Selection of newly formed B cells into transitional, follicular, and marginal zone (MZ) B cells depends on integrated signals from several classes of surface receptors, such as BCRs and co-signals, tumor necrosis factor receptor (TNFR) family, and several G protein-coupled receptors (GPCRs). Follicular and MZ long-lived compartments participate differentially in B-cell effector functions such as the germinal center (GC) reaction, long-term memory, antigen presentation, and antibody and plasma cell (PC) generation. Differential signaling strengths of BCR are required for the development of the three mature B-cell subsets. Peritoneal B cells require the strongest BCR signal, follicular B cells require an intermediate BCR signal, and marginal zone B cells require a weaker BCR signal. During B-lymphocyte development, antibodies are assembled by random gene segment reassortment to produce a vast number of specificities. A potential disadvantage of this process is that some of the antibodies produced are self-reactive; in humans, the majority (55–75%) of all antibodies expressed by early immature B cells display self-reactivity, including poly-reactive and antinuclear specificities. Most autoantibodies are removed from the population at discrete checkpoints during B-cell development. Inefficient checkpoint regulation would lead to substantial increases in circulating autoantibodies. So, an important role of BCR at the immature B-cell stage is to induce efficient elimination of these potentially harmful cells. Immature self-reactive B cells can be eliminated by negative selection (BCR-induced cell death), inactivated by induction of anergy, or revision of BCR specificity (receptor editing). Finally, in general, secretion of IgM, IgG1, and IgG2c reflects Th1-dependent immune responses, whereas IgG2 and IgE production reflects Th2-dependent processes, and IgG3 production reflects T-cell-independent immune responses (Hua & Hou, 2013).

4.3.3.3 B Cells and the Brain

Presently, the role B cells in the brain is not as well-characterized as the connection with other immune cells. The past perception that the brain was immune-privileged is a contributing factor to this impediment. As the discipline of neuroimmunology continues to emerge, the roles of B cells in neurodegenerative diseases are becoming clearer. In contrast to T cells associated with different immunopathologies, the brain provides a more fostering environment for B cells. Although the role of B cells in the homeostatic brain is less clear, it remains an important topic for study in neuroimmunology. Primary lymphomas of the CNS are usually of a B-cell origin. The cerebrospinal fluid (CSF) of patients with chronic infectious and autoimmune diseases of the CNS typically contains a remarkably stable oligoclonal Ig repertoire. In the CNS of multiple sclerosis (MS) patients, clonally expanded B cells and plasma cells persist, while ectopic

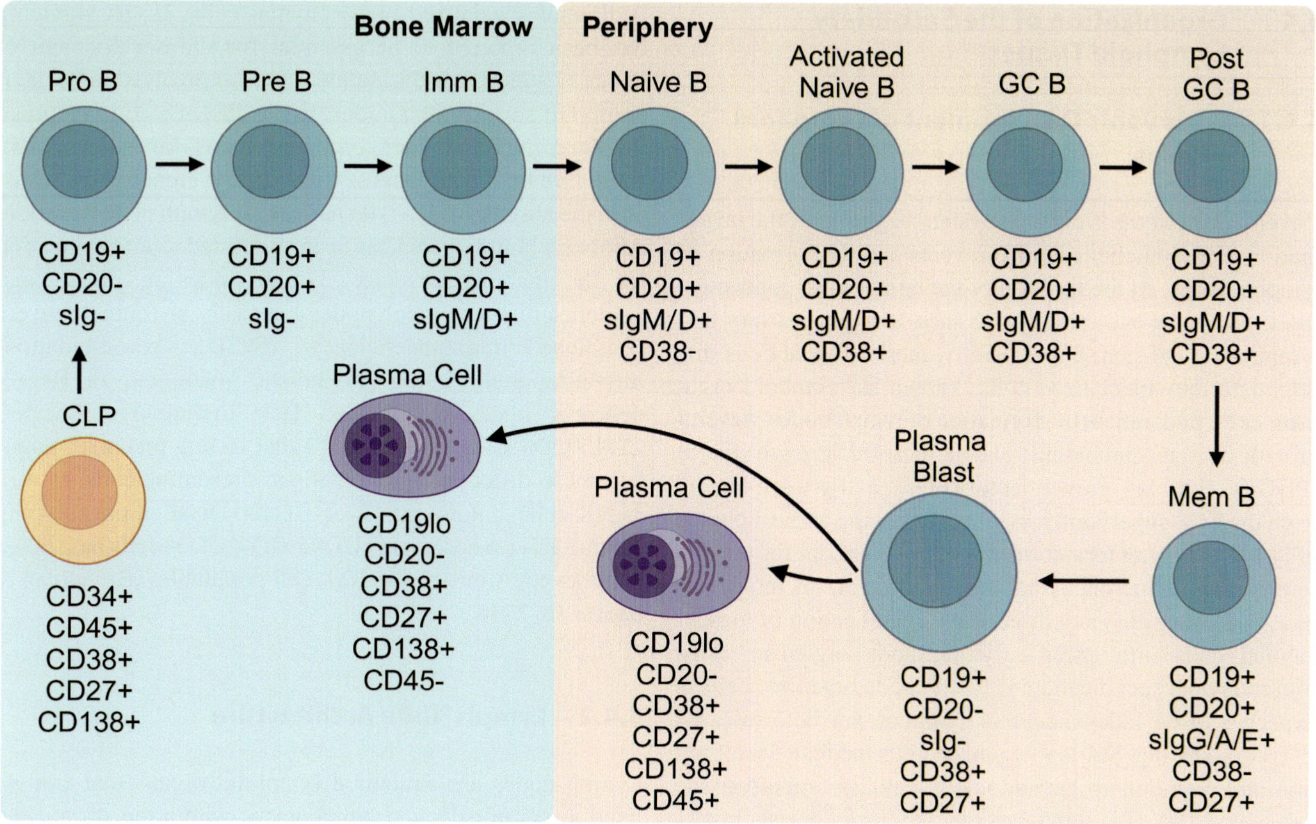

Fig. 4.3 Anatomical sites and cell surface markers associated with stages of B-cell development. B-cell development begins in the bone marrow from a population of hematopoietic stem cells, which give rise to common lymphoid progenitors (CLPs) with the potential to differentiate into various lymphoid cell types, including B cells. CLPs commit to the B-cell lineage and become pro-B cells (Pro B). Pro-B cells initiate rearrangement of immunoglobulin heavy-chain (*IgH*) genes through V(D)J recombination. Successful rearrangement of the *IgH* genes leads to the expression of the pre-B-cell (Pre B) receptor, consisting of a surrogate light chain and a pre-B-cell receptor heavy chain. Pre-B cells undergo further differentiation and continue to proliferate and rearrange their immunoglobulin light-chain (*IgL*) genes, both kappa and lambda. Production of a functional light chain leads to surface expression of the B-cell receptor (BCR or sIgM/D) on the surface of an immature B cell (Imm B). Immature B cells undergo a selection process in the bone marrow to ensure self-tolerance. Too strong ligation of BCR with self-antigens results in apoptosis of cell or receptor editing that modifies BCR specificity. Selected immature B cells progress to naïve B cells (Naïve B), exit the bone marrow, and enter the peripheral lymphoid organs. In the peripheral lymphoid organs, naive B cells encounter antigens that bind BCR and, with co-stimulatory signals, lead to an activated B cell (Activated Naïve B). Activated naive B cells enter germinal centers of secondary lymphoid tissues and become germinal center B cells (GC B). GC B cells undergo somatic hypermutation, which introduces point mutations into their immunoglobulin genes, leading to an increased affinity for antigens. Post-GC B cells (Post GC B) also undergo class switch recombination, changing the Ig constant region to produce different isotypes such as IgG, IgA, or IgE with the same antigen specificity. Following somatic hypermutation and class switching, some post-GC B cells differentiate into memory B cells (Mem B). Memory B cells are long-lived with the same antigen specificity for future encounters. An encounter with an antigen induces memory B cells to differentiate into plasma blasts, which are specialized for antibody production with a high rate of immunoglobulin synthesis and secrete a large number of antibodies, mostly IgG, IgA, or IgM. Plasma cells are fully differentiated Ig-producing cells and are responsible for continuous and high-level antibody secretion. Peripheral plasma cells have a limited lifespan and ultimately undergo apoptosis. Plasma cells also enter the bone marrow as a primary site for long-lived plasma cells that may reside for several years or decades

B-cell follicles develop in the meninges of patients with secondary progressive multiple sclerosis (MS), and B-cell differentiation may be recapitulated in the CNS of MS patients (Krumbholz et al., 2006). In recent years, the contribution of the BBB to the purported "immune-privileged" state of the brain has received substantial study. While the BBB serves as a selective barrier between the brain and the systemic circulation, low levels of B cells are identified in the CNS parenchyma of the homeostatic brain with the number of B cells significantly increased during inflammation or pathological conditions. Thus, B-cell depletion therapies have shown success in patients with MS and other neuroinflammatory conditions in neurodegenerative diseases (Sabatino et al., 2019).

4.4 Organization of the Secondary Lymphoid Tissues

4.4.1 Embryonic Development of Lymphoid Tissues

Development of the lymphatic system begins with the invagination of endothelial cells from veins and the formation of lymphoid sacs. At the location of the lymph sacs, connective tissue protrudes into these lymph sacs, forming the very first lymph node anlagen. Subsequently, mesenchymal cells differentiate into specialized cells, known as lymphoid organizer cells, and initiate the formation of lymph nodes. Several growth factors, including platelet-derived growth factor (PDGF), fibroblast growth factor (FGF), and growth factors from the TGF superfamily, are involved in the earliest phases of lymphoid organ formation and thus are crucial for the differentiation of mesenchymal cells. However, no molecules have been identified that direct early specification of mesenchymal cells into specific lymph node organizer cells. Mesenchymal specification of lymph node organizer cells is expected to be independent of lymphotoxin beta receptor (LTβR) signaling. Stromal organizer cells mediate the attraction and retention of hematopoietic cells (lymphoid tissue inducers, LTis), resulting in accumulation and initial clustering of cells (Bénézech et al., 2010; Cupedo & Mebius, 2005; Lu & Browning, 2014). For the generation of LTi cells, which are derived from IL-7R-expressing hematopoietic precursors, expression of retinoic acid—related orphan receptor-γ (RORγ) by LTi cells is necessary. LTi cells mediate triggering of LTβR on stromal cells through expression of LTα1β2, which is induced by ligation of the TNF-related activation-induced cytokine (TRANCE) that signals through the TRANCE receptor (TRANCER) and/or IL-7R. Functional LTi cells need to express CXCR5 as well as CCR7 to respond to C–X–C motif chemokine ligand 13 (CXCL13), C–C motif chemokine ligand 19 (CCL19), and CCL21 chemokines involved in lymph node formation. To generate functional lymphoid tissue organizer (LTo) cells, LTβR signaling leads to expression of mucosal addressin cell adhesion molecule 1 (MAdCAM-1), intercellular adhesion molecule 1 (ICAM-1), and vascular cell adhesion molecule 1 (VCAM-1) by LTo cells as well as the production of the chemokines CXCL13, CCL19, and CCL21. Mesenchymal cells produce more chemokines, express adhesion molecules, and establish the first positive feedback loop for recruitment of LTi cells. Days before birth, lymphocytes and dendritic cells are recruited to the lymph node and Peyer's patch anlagen. Because T and B cells themselves express LT-α1β2, an LTβR ligand, recruitment of lymphocytes is amplified by a second positive feedback loop and culminates in the formation of a mature lymph node or Peyer's patch, consisting of segregated T-cell zones and B-cell follicles, high endothelial venules (HEVs), and additional specialized mesenchymal cells. IL-7R signaling has been reported to be essential for the development of Peyer's patches, but this pathway is also involved in the generation of some lymph nodes. In the absence of a functional IL-2R common γ-chain (γc) and Janus kinase 3 (Jak3), which are both components of the IL-7R signaling pathway, or in the absence of IL-7, no peripheral lymph nodes develop. Although lymphopenia has been suggested as an explanation for the inability to detect lymph nodes in IL-7R- and γc-deficient mice, lymph nodes can easily be found in severe combined immunodeficiency (SCID), recombination-activating gene 1(Rag-1)-deficient mice, and in Rag-2-deficient mice, which also lack mature lymphocytes. $CD45^+CD4^+CD3^-$ inducer cells that form a precursor population can differentiate into antigen-presenting cells, as well as NK cells, but not into B or T cells. Detailed phenotyping of adult LTi (or accessory) $CD45^+CD4^+CD3^-$ cells has shown that they are similar to a $CD4^+$ cell population (Brendolan & Caamaño, 2012).

4.4.2 Lymph Node Architecture

Lymph nodes are structured lymphoid organs that contain fixed levels of cells and lymphocytes within the stroma and the parenchyma with its migrating cell populations. The structure of a lymph node includes the capsule, subcapsular sinus, cortex, and medulla. The primary function of lymph nodes is filtering interstitial fluid, which is collected from soft tissues, and eventually returning it to the vascular system. Filtering the interstitial fluid allows for T and B cells to be exposed to a range of antigens.

In contrast to nonlymphoid parenchymal organs, the stromal network has an indispensable functional role. Fibroblastic reticular cells (FRCs) constitute the main stromal population and form its internal three-dimensional network. FRCs express surface molecules and produce "homing" chemokines for T cells, B cells, and DCs and provide anatomical arrangements that influence the traffic patterns of lymphocytes and facilitate lymphocyte "crawling" along preformed "corridors." Moreover, FRCs typically wrap reticular fibers, mainly made up of type III and IV collagen, elastin, and laminin, to form a set of interconnected channels that comprise the FRC conduit system. This system allows efficient and rapid transfer of soluble molecules, such as antigens, chemokines, and immune complexes from the subcapsular sinus to the deeper cortex, high endothelial venules (HEVs), and, perhaps, the follicular compartment. Lymph nodes can undergo rapid and profound hypertrophy during an immune response, and FRCs are deeply involved in matrix reorganization and remodeling of the lymph node architecture. The conduit system and sinuses are lined by sinus endothelial cells (SECs). These are flattened cells that form a discontinuous layer, especially in the wall of the medullary sinus, which contain intercellular gaps or pores that are also found on the floor of the subcapsular sinus. FRCs and SECs are believed to be ontogenically related and serve to filter lymph fluid collected from all peripheral tissues, including brain interstitial tissue and cerebrospinal fluid. In the subcapsular sinus, macrophages and DCs sample the lymph and remove the microorganisms and debris contained therein. These cells also transport and/or process antigenic material for presentation to B and T cells. However, soluble lymph-derived material restricts access to the lymphocyte compartment and the cortex. Migrating naïve T and B cells, as well as monocytes, NK cells, and DCs, enter the lymph node via high endothelial venules (HEVs), migrate through the net of dendritic cells, and, if not acquired, exit back to the circulation via a single efferent lymph vessel for each nodule. The CCR7 chemokine receptor and two of its ligands, CCL19 and CCL21, have been attributed, but none are strongly confirmed to be expressed by human HEVs. Both ligands are possibly transported to the luminal surfaces of HEVs (Comerford et al., 2013; Pieper et al., 2013). All secondary lymphoid tissues have some structural similarity, and the major unit component is a lymphoid follicle with a germinal center (GC), a follicular mantle, and a marginal zone (van de Pavert & Mebius, 2010).

4.4.3 Antigen-Presenting Cells

The adaptive immune system is largely regulated by APCs as these cells present antigens to T cells. APCs can be classified into several subtypes that include, in order of decreasing antigen-presenting efficiency, dendritic cells, macrophages, and B cells. The lymph node harbors a composite population of DCs, both resident and emigrant, capable of selection, stimulation, and elimination or inactivation of effector lymphocytes, that is, those that trigger adaptive immunity and/or tolerance. In the human lymph node T-cell compartment, two primary subsets of DCs, compared to six in mice, have been identified: monocytic $CD11b^{-}CD11c^{+}$ DCs and plasmacytoid IFN-α-producing $CD123^{+}CD11b^{-}CD11c^{-}$ DCs. The latter subset clusters with T cells around HEVs. $CD11c^{+}$ DCs are present as stellate cells whose delicate processes or "veils" extend in many directions; these DCs are designated as interdigitating DCs. In the B-cell compartment, follicular DCs (FDCs) orchestrate B-cell function. Whether FDCs originate from hematopoietic progenitors or stromal elements is controversial. New evidence suggests the presence of two types of FDCs within human germinal centers: classical FDCs that express CD21 and fibroblast-related markers that represent stromal cells and the $CD3^{-}CD4^{-}CD11c^{-}$ germinal center dendritic cells (GCDCs), which represent hematopoietic cells that may be analogous to the antigen-transporting cells described in mice. Under specific in vitro conditions, human monocytes can downregulate expression of the hematopoietic marker CD45 and function as FDCs. An FDC is recognized as a bridging cell between innate and adaptive B-cell responses (Aguzzi et al., 2014). For example, FDCs are generally considered to be the major source of CXCL13 (also known as B-cell-attracting chemokine-1 or BCA-1) both in normal and immune-responding lymphoid tissues. Cells expressing the highest levels of CXCL13 in rheumatoid arthritis and ulcerative colitis are of a monocyte/macrophage lineage. These are not only found in irregular lymphoid aggregates within an FDC network but also within and near smaller collections of B cells in diseased tissue where no FDCs are detected. For the most part, these cells occupy the outer layers of primary and secondary lymph node follicles within the light zone and the follicular mantle. FDCs constitute a three-dimensional sponge-like network formed by the interdigitation of adjacent cells and, as shown by ultrastructural studies, present filiform or beaded dendrites with characteristic "ball-of-yarn" convolutions. FDCs play a pivotal role in promoting B-cell proliferation and differentiation in germinal centers. They are highly efficient in trapping and retaining antigen–antibody complexes through CD21–CD35 and FcγRIIB receptor interactions, which maintain expression for long periods and present as immune-complex-coated bodies (iccosomes) to memory B cells. Of interest, the spleen also contains lymph node-like organized structures; however, this organ is multifunctional (Kain & Owens, 2013).

4.4.4 Lymph Nodes and Primary Adaptive Immune Responses

4.4.4.1 Common Rules of Adaptive Immune Responses

To produce highly specific and effective immune responses against pathogens while maintaining immune tolerance to self-antigens, common rules govern the function of the adaptive immune system. The strength and quality of immune responses are based on highly polymorphic MHC molecules, varied recombination of TCRs and BCRs that comprise the respective antigen receptor repertoires, and regulatory T cells that dampen immune responses and maintain immunological tolerance to self-antigens. On the other hand, the specificity of proteins from viruses, bacteria, and parasites, as well as altered self-proteins due to chemical modifications, such as those produced by oxidation and nitration reactions, also play a crucial role. These antigens are transported or drained from affected tissues, including the brain, to regional draining lymph nodes via lymphatic vessels, and interact with antigen-processing and antigen-presenting machinery. Processing of antigens induces the production of cytokines and chemokines by capturing DCs and macrophages. In lymph nodes, the balance between triggering an immune response and promoting tolerance is carefully controlled. The initiation of antigen-specific immune responses or the establishment of tolerance relies on direct interactions between DCs and naïve T cells, which typically occur within the T-cell zones of lymph nodes, often near high endothelial venules (Batten et al., 2013; Siegert & Luther, 2012).

4.4.4.2 CD4$^+$ T-Cell Polarization

Naïve CD4$^+$CD25$^-$ T cells differentiate into various T-helper (Th) cell subsets, such as Th1, Th2, Th17, and T follicular helper (Tfh) cells, depending on the co-stimulatory signals and cytokine cues provided by DCs during the priming process. Upon activation of naïve CD4$^+$ T cells through TCR, a modest, but detectable transcription of both *IL-4* and *IFN-γ* genes is induced. T-box transcription factor 21 (Tbx21), or T-bet, plays a crucial role in Th1 development by promoting transcriptional competence at the *IFN-γ* gene locus and selective responsiveness to IL-12. With continued stimulation, in the presence of IL-12-induced signaling through the essential IL-12 receptor, mediated by the transcription factor STAT4 (signal transducer and activator of transcription 4), primed cells begin to express IFN-γ and differentiate into Th1-helper cells.

Conversely, Th2 differentiation involves the transcription factor GATA3, contributing to chromatin remodeling events that favor stable demethylation of the *IL-4* gene and production of IL-4. Signaling through the IL-4 receptor and downstream activation of STAT6 promote the stable production of not only IL-4 but also IL-5 and IL-13, leading to Th2 cell differentiation. It is unclear whether these cells differentiate as a distinct lineage during T-cell priming or emerge later from nonpolarized, primed T cells or even from polarized Th2 or Th1 cells.

The differentiation of naïve Th cells into the Th17 phenotype is facilitated by cytokines such as IL-1β, IL-12, IL-21, IL-23, and TGF-β. Th17 cells play a role in protecting against fungal and bacterial infections and are particularly abundant in the gut, where they help regulate the gut microbiota. Aberrant Th17 responses are strongly associated with autoimmune disorders.

Currently, at least four distinct Th cell subsets (Th1, Th2, Th17, and induced Tregs (iTregs)) have been characterized by the expression of unique transcription factors and hallmark cytokines. The cytokine environment, shaped by activated CD4$^+$ T cells themselves, dendritic cells, and other cell types during differentiation, plays a pivotal role in determining Th cell fate decisions (Yamane & Paul, 2012, 2013).

4.4.4.3 CD4$^+$ T-Cell–B-Cell Interaction

A proportion of newly primed T cells that have upregulated expression of CXCR5 migrate toward the T-cell zone–follicle boundary, where T-helper cells provide cognate help to antigen-specific B cells. In this location, T-helper cells interact with B cells and CD4$^+$CD3$^-$ accessory cells, i.e., LTi cell analogues for the development of lymph nodes. CD4$^+$CD3$^-$ accessory cells provide signals through CD134 (also known as OX40) and CD30 at the surface of follicular T cells, which direct localization of these cells to promote the maintenance of germinal center reactions and the generation of B-cell and CD4$^+$ T-cell memory responses.

These follicular T cells strongly express the CD28-related molecule-inducible T-cell co-stimulator (ICOS or CD278) and secrete moderate amounts of IL-10, which are co-stimulatory factors with crucial functions in T-helper cell and B-cell responses. After stimulation, T-helper cells express CD134 and rapidly upregulate the CD40 ligand (CD40L or CD154) from pre-formed stores. This is consistent with their capacity to induce germinal center B cells to survive, secrete antibodies, and express activation-induced cytidine deaminase (AICDA). AICDA is crucial for immunoglobulin class switching and somatic hypermutation, a process by which B cells adapt to new foreign elements. IL-21 co-stimulation potently induces the expression of both B lymphocyte-induced maturation protein-1 (BLIMP-1) and AICDA as well as the production of large amounts of IgG from B cells (Ettinger et al., 2005). Tfh, Th1, and Th2 cells can all participate in antibody responses. Th1 and Th2 cytokines and cytokine response elements then guide class switch recombination, resulting in immunoglobulin class and subclass production, thereby skewing antibody responses depending on the nature of the stimulus. For example, IFN-γ promotes switching to IgG2a during antiviral responses, and IL-4 promotes the production of IgE during antiparasitic responses. Both Th1 and Th2 cells are crucial helpers in the extrafollicular pathway of B-cell differentiation and are potent inducers of class switching to T-cell-dependent antibody classes and subclasses outside of germinal centers. CD4$^+$ Tregs suppressing Th1/Th2 cell function, expansion of CD8$^+$ cells, and B-cell antibody production have all also been described in control of immunoglobulin responses. Tregs are CD4$^+$CD25$^+$CD57$^-$ T cells that co-express CXCR5 and localize to T-cell zones and germinal centers. Also within the follicular areas are NK cells and T cells that express NK1.1, (i.e., CD4$^+$ NKT cells) and produce Th1 and Th2 cytokines but primarily Th1 cytokines.

4.4.4.4 B-Cell Differentiation and Humoral Immune Responses

Adaptive immune responses rely heavily on B-cell differentiation and humoral immune responses to defend against pathogens through selection of specific Igs. The maturation of B cells reflects a multistep process starting before differentiation into mature B cells with a unique cell surface BCR. As previously mentioned, V(D)J recombination of B cells allows a wide array of specificities for different antigens with unique BCRs that bind foreign antigens at the cell surface. The co-stimulatory factors of helper T cells allow for the production of high-affinity memory B cells and long-lived plasma cells that are specific for these foreign antigens. T-cell-independent immune responses usually follow direct activation of B cells by bacterial capsular polysaccharides or by microorganism-derived ligation of toll-like-receptors (TLRs). This is a rapid antibody response to pathogens through the generation of short-lived, low-affinity Ig production by extrafollicular plasma cells.

The essential structure for T-cell-dependent responses is the germinal center. Germinal centers arise in lymphoid tissues following antigenic stimulation and provide a milieu for proliferation of B cells, known as centroblasts. The genetic sequence of BCR is altered during affinity maturation, a selective process, which increases the affinity and avidity for antigens for a portion of germinal center B cells. Centroblasts exit the cell cycle to become a centrocyte and bind antigen that is associated with FDCs using newly expressed cell surface Igs; after processing, B cells present antigen in the context of MHC II to T-helper cells and help in Ig synthesis. Interaction with T-helper cells has several outcomes, including (1) survival and selection of high-affinity centrocytes, (2) class switch recombination of Ig, (3) differentiation into long-lived plasma cells and memory B cells, and (4) perpetuation of germinal center reactions by stimulating centrocytes to recycle back to centroblasts. In the absence of antigen-specific T cells, centrocytes are eliminated and germinal centers abort early after induction. Spontaneous ectopic germinal center formation is found in human autoimmune diseases and is a source of somatically mutated high-affinity autoantibodies.

The interplay between B-cell differentiation and humoral immune responses is evident in cellular processes that would not be possible without antibody-mediated immunity. Antibody production from plasma cells is essential for the elimination of pathogenic states through processes such as opsonization and activation of the complement cascade. Furthermore, immunity is maintained from previous infection or vaccination by antibody production from memory B cells.

4.4.4.5 CD8$^+$ T Cells

CD8$^+$ cells constitute a small fraction of follicular T cells within human germinal centers and comprise less than 5% in tonsils and 10–15% in lymph nodes. Unlike B cells, which heavily rely on CD4$^+$ T-cell assistance to achieve high specificity, magnitude, and stability in antibody production, a process central to humoral immune responses, CD8$^+$ cell-mediated cellular immune responses in the initial days after initiation are self-sufficient and do not depend totally on CD4$^+$ help, but CD4$^+$ T-cell assistance is primarily required for memory formation. This assistance hinges on CD40 ligation and production of IL-2 by CD4$^+$ Th1-helper cells, regulated by co-stimulatory molecules expressed by DCs.

Effector CD8$^+$ T cells possess a range of potent tools and are highly specialized in their ability to recognize target cells presenting short peptides on MHC I molecules. An encounter with a foreign peptide in the context of MHC I induces CD8$^+$ T cells to employ mechanisms that include both cytotoxic and noncytotoxic processes to eliminate pathogens from the host cell. These mechanisms include the release of pore-forming molecules and granzymes, the engagement of apoptosis-inducing ligands such as the Fas ligand, and the secretion of cytokines such as IFN-γ and TNF-α. This cellular arsenal is programmed with the lymph nodes during the priming processes and is deployed at the foci of cellular antigen engagement where CD8$^+$ T cells execute cytolytic functions. Cytotoxic T lymphocytes (CTLs) play a crucial role in eliminating virus- or bacteria-infected cells, including stromal cells, macrophages/DCs, or lymphocytes residing in lymph nodes. This action helps prevent conditions that induce lymphopenia and immune suppression, as seen in infections such as HIV/AIDS. Recent findings in immunology have continued to expand the intricate mechanisms of CD8$^+$ T-cell responses and their role in the maintenance of immune homeostasis.

4.4.4.6 Adaptive Immunity and the Brain

Like other tissues, the condition of post-capillary vein endothelial cells plays a crucial role in regulating the entry of activated lymphocytes into the brain parenchyma. Systemic activation of these cells, driven by exo- and endotoxins or infection, leads to increased production of pro-inflammatory cytokines like TNF-α and chemokines such as macrophage inflammatory protein-1 (MIP-1/CCL3), which upregulate endothelial adhesion molecules. This facilitates the migration of lymphocytes into the perivascular spaces within the meninges and choroid plexus, neither of which possess a well-defined BBB, thus allowing easier migration of immune cells into surrounding compartments. These compartments harbor both potential innate effector immune cells such as macrophages and DCs. Following viral and bacterial infections, APCs are engaged to process and present foreign antigens to elicit local immune responses and memory cell generation.

In the CSF of healthy individuals, low numbers of lymphocytes are typically present. Surprisingly, more than 80% of these lymphocytes are memory T cells characterized by a CD4$^+$CD45RO$^+$CD27$^+$CXCR3$^+$ phenotype. These memory cells engage in continuous recirculation between the bloodstream and the CSF. This is facilitated, in part, by the interaction of low levels of P-selectin (CD62P) on the endothelial cells of deep stromal veins within the choroid plexus and bridging meningeal veins with P-selectin glycoprotein ligand-1 (PSGL-1) on leukocytes that allow interchange between compartments. These memory cells are typically long-lived and respond to the reappearance of antigens, whether viral, bacterial, or tumor. However, memory recall in the brain is delayed due to the immunosuppressive environment within the brain and the activation of pro-apoptotic interactions between T cells and glia through molecules like CD95/CD95L (Fas/FasL), CD30/CD153, and other potential TNF-family related factors. Recent advancements in real-time imaging techniques for studying leukocyte behavior in cerebral infections have provided valuable insights into these complex processes (Pai et al., 2014).

4.5 Neuroimmune Interactions

4.5.1 Innervation of Lymphoid Tissues

The innervation of lymphoid tissues remains functionally operative for maintaining non-disease homeostasis at the intersection of the immune and nervous systems. Both peripheral and autonomic nervous systems have processes that impart profound implications in lymphoid tissues that connect neuroimmune interactions. Signals of pathogenic challenges from lymphoid tissues are communicated by the CNS through the peripheral nervous system. Meanwhile, the autonomic nervous system functions to release neurotransmitters, regulate cytokine production, and control inflammation. Clinical, epidemiological, and experimental data suggest the involvement of factors, hormones, and neural mediators that link the immune and nervous systems in the pathogenesis of multiple immune-mediated disorders of the nervous system. These molecules are members of the class 1 cytokine superfamily, which allow bidirectional neuroimmune interactions (Procaccini et al., 2014). In lymphoid organs, similar to blood vessels, sympathetic/noradrenergic and sympathetic/neuropeptide Y (NPY) innervations predominate over parasympathetic (cholinergic) innervations and are located primarily along blood vessels. In the thymus, at the ultrastructural level, noradrenergic fibers are seen in proximity to thymocytes, mast cells, and fibroblasts. Catecholaminergic nerve fibers also run in close contact with TECs. Since TECs form the blood–thymus barrier in the outer thymic cortex, these cells may be targets for circulating epinephrine from the adrenal gland or norepinephrine (NE) released from perivascular nerves. Sympathetic nerve fibers of the spleen are present among cells in the T-dependent area, and macrophages and B cells are found in the marginal zone and sinus where lymphocytes enter the spleen. In lymph nodes, noradrenergic fibers supply paracortical and cortical zones (T- and B-cell-rich regions, respectively) but are absent from nodular regions and B-cell-rich germinal centers. The gut, lung, and nasal mucosa have the highest intensity of innervation of associated lymphoid tissues. Nerves predominate in the T-cell zones of lymphoid aggregates that contain high levels of neuropeptides and norepinephrine, a sympathetic neurotransmitter.

4.5.2 Neurotransmitter/Neuropeptide Regulation of Immune Cell Function

As mentioned, APCs, starting with thymic TECs and extending to tissue-resident macrophages and DCs, play a vital role in regulating lymphocyte functions. This role spans from programming central tolerance and immunity to supporting effector functions and maintaining peripheral tolerance. To detect various primary signaling ligands of the nervous system, cells of a myeloid origin, such as macrophages and DCs, express functional receptors, including norepinephrine (NE) receptors ($\alpha1$, $\alpha2$, $\beta1$, $\beta2$), dopamine receptors (D1–5), neuropeptide Y receptor (Y1), opioid receptors (μ, δ, k), and parasympathetic nicotinic acetylcholine (Ach) receptors. The signaling pathways through these receptors have a significant impact on the activation status of APCs, such as regulating expression of co-stimulatory and co-inhibitory molecules, as well as the secreted cytokine and chemokine profiles of each cell. For instance, the ligation of ACh receptors on microglia exerts neuroprotective and anti-inflammatory actions under ischemic conditions in the brain by activation of nuclear factor erythroid-2-related factor 2 (Nrf2) that induces heme oxygenase-1 (HO-1). Both myeloid and lymphoid cells express various adrenergic receptors, which result in co-signaling during antigen presentation in the presence of increased concentrations of neurotransmitters such as NE or dopamine, and can influence T-cell polarization that facilitates Th2 responses.

Moreover, recent findings have revealed additional neurochemical factors involved in modulating immune responses. These include histamine, serotonin, neuropeptides such as substance P, vasoactive intestinal peptide (VIP), pituitary adenylate cyclase-activating polypeptide, calcitonin gene-related peptide, α-melanocyte-stimulating hormone, and leptin. These neurochemicals contribute to an intricate network of signaling that shapes immune responses mediated by antigen-presenting cells and lymphocytes (Elenkov et al., 2000; Scheiermann et al., 2013). More recently, vasoactive intestinal peptide (VIP) and VIPR2 agonists have been used to transform T-cell responses and induce neuroprotective Tregs in models of Parkinson's disease (Mosley et al., 2019; Olson et al., 2015). Together, these data underscore the multifaceted and dynamic interactions between the nervous and immune systems, revealing intricate regulatory mechanisms that can influence immune function and responses (Machhi et al., 2020; Olson et al., 2023).

4.6 Humanized Animal Models

4.6.1 Hematolymphoid Humanization of Mice

The field of humanized mouse models is advancing rapidly and has garnered significant attention as a valuable experimental platform for various areas of research. These models have proven crucial to the determination of the mechanisms associated with human hematopoietic and immune system development (Rongvaux et al., 2013); growth of human stem cells and normal tissues in the context of regenerative medicine (Isobe et al., 2014); and human tumor stem cells and tissues in cancer-related studies (Williams et al., 2013). To create such models, immunodeficient mice are typically utilized. Among the popular strains used are nonobese diabetic (NOD) mice carrying mutations in the DNA-dependent protein kinase gene (*Prkdc*) (Bosma et al., 1983) and lacking functional common cytokine receptor gamma chain (γc), resulting in either a complete loss of γc (γc^{null}) (Cao et al., 1995; DiSanto et al., 1995) in strains such as NOD.Cg-$Prkd^{cscid}$ $Il2rg^{tm1Wjl}$/SzJ (NSG) or a deletion of the intracytoplasmic domain for strains like NOD.Cg-$Prkdc^{scid}$ $Il2rg^{tm1Sug}$/JicTac (NOG) (Ikebe et al., 1997; Ohbo et al., 1996; Sugamura et al., 1996). These strains have been instrumental in studies involving human hematopoietic stem cells (HSCs) and in the development and function of the human immune system (Tanner et al., 2014).

A key factor contributing to the success of engraftment in these models is the low lytic activity of the C5 complement component due to a two base pair deletion in a 5′ exon of the murine *C5* gene (Baxter & Cooke, 1993). Additionally, defects in macrophage function, including impaired regulation of CSF-1 and IFN-γ receptors, as well as reduced secretion of IL-1 in response to LPS and decreased NK cell activity are correlated with improved engraftment (Shultz et al., 1995). Another critical factor is the murine polymorphism of signal regulatory protein alpha (SIRP-α, CD172α), which resides in the insulin-dependent diabetes 13 (*Idd13*) locus on chromosome 2 and contains a coding sequence polymorphism in the signal regulatory protein alpha (*Sirpa*) gene. This polymorphism has been shown to be responsible for the survival and development of human HSCs in a murine environment. SIRP-α interacts with CD47, a widely expressed cell surface glycoprotein, and interactions between these ligands on mouse macrophages and human HSCs provide a "stop" signal that prevents the destruction of xenogeneic human cells (Takenaka et al., 2007; Takizawa & Manz, 2007). For investigations requiring a functional human immune system when translational opportunities are unavailable, humanized mouse models provide an attractive alternative, whereby several key properties that lend this platform to investigators include:

1. High acceptance of human HSCs.
2. Stable populations of the mouse bone marrow.
3. Occupancy of the murine thymus with human thymocytes and their maturation into various T-cell subsets, including αβ, γδ, and regulatory T cells.
4. Complete reconstitution of residual murine lymph nodes with human T cells, B cells, and dendritic cells.
5. The formation of splenic white pulp.
6. The distribution of human cells of the macrophage lineage throughout the brain meninges and perivascular spaces.

Furthermore, humanized mice have proven valuable for studying chronic HIV-1 replication and recapitulating the course of HIV infection in humans, including loss of $CD4^+$ T cells, development of HIV-1-specific cytotoxic T lymphocytes (CTLs), and elicitation of humoral immune responses (Brainard et al., 2009; Gorantla et al., 2010a, b; Watanabe et al., 2007a, b).

Recent studies have introduced co-transplantation models in which a fragment of the human lung tissue is surgically implanted into an immunodeficient mouse humanized with progenitors from the bone marrow, liver, and thymus (BLT mouse), leading to the creation of a "BLT-L" model. When exposed to human cytomegalovirus (HCMV) infection, often administered through direct injection into the lung implant, the BLT-L model exhibited the capacity to initiate an antigen-specific adaptive immune response, effectively controlling viral replication (Chen et al., 2022). This approach broadens the susceptibility of the host mouse to respiratory pathogens with human tropism, including respiratory syncytial virus, Middle East respiratory syndrome coronavirus, and HCMV. Importantly, humanized mouse models are continually evolving, and ongoing works will likely yield further insights into and advancements in understanding human hematopoiesis, immunity, and disease modeling within these murine systems.

4.6.2 Mouse Brain Humanization

Recent advancements in stem cell differentiation techniques have facilitated the creation of chimeric mouse models containing human pluripotent stem cell-derived primitive macrophage progenitors (PMPs) through transplantation (Jiang & Alam, 2022). Xenografts of PMPs from human pluripotent stem cells undergo differentiation and maturation into microglial cells that maintain complex region-specific phenotypic and functional characteristics resembling those of the human microglia (Jiang et al., 2020). These chimeric models represent valuable tools for investigating the pathophysiology of the human microglia during brain development, neurodegenerative disorders, and novel functions associated with genes and variants expressed by human cells in in vivo models.

Additionally, transplantation and development of neurons and glial populations, such as astrocytes and oligodendrocytes, is possible and highly efficient in immunodeficient mice with NOD/scid or Rag-knockout backgrounds. For instance, the transplantation of human CD133$^+$ neuronal progenitor cells into irradiated neonatal mice results in the widespread distribution of mature neurons across various regions of the mouse brain with established connections within the mouse neuronal network (Benninger et al., 2000; Uchida et al., 2000). When human A2B5$^+$/PSA$^-$NCAM$^-$ glial progenitors from the fetal brain tissue are transplanted into immunodeficient mice, a significant portion of mouse astrocytes are replaced with human astrocytes, leading to improved cognitive abilities in the recipient mice (Han et al., 2013). Transplantation of the same phenotype progenitors into mice with immunodeficient backgrounds and deficiencies in myelinating cell development (such as shiverer or MBPshi/shi mice) give rise to oligodendrocytes (Windrem et al., 2014). Furthermore, the human glial cells that repopulate the mouse brain become susceptible to human-specific viral infections, such as the human glial tropic JC virus (Kondo et al., 2014).

While humanized mice provide important information about adaptive immune development and responses after reconstitution, those models were less conducive to development of the human microglia after HSC reconstitution, which proved to impede the assessment of some CNS viral infections. More recently, the human *IL-34* gene was microinjected into the germline of immunodeficient NOD.*Cg-Prkdc*^{scid}*il2g*^{tmlSug}/Jic (NOG) mice to yield novel NOG-hIL34 mice (Mathews et al., 2019). Transplantation of those mice with human HSCs not only provided reconstitution of a peripheral immune system but also human microglial-like cells in the brain expressing CD14, CD68, CD163, CD11b, ITGB2, CX3CR1, CSFR1, TREM2, and P2RY12. Moreover, infection of humanized NOG-hIL34 mice with macrophage-tropic strains of HIV resulted in persistent infection of brain microglia and recapitulated disease-specific innate immune-inflammatory and neurotoxic responses observed in HIV-infected human brains.

4.7 Conclusions

This chapter covers the development of the immune system, establishment of immune organs, and pivotal roles in gathering information about the status of various tissues, including the enigmatic realm of the brain. A profound interplay exists between the maturation of both nervous and immune systems, a relationship that undergoes significant transformation after birth. Innervation in lymphoid tissues and organs is crucial for regulating immune development and responses within those compartments. While the brain actively dispatches efferent signals, the development of effector immune responses in the brain faces constraints imposed by robust central tolerance mechanisms during T-cell education by TECs expressing brain antigens. Furthermore, peripheral tolerance is bolstered by the presence of soluble brain antigens, neurotransmitters, and neuropeptides, which are filtered through regional lymph nodes and presented by resident dendritic cells in the absence of essential co-stimulatory signals.

Humanized animal models, which enrich the understanding of the mechanisms of development and neurological interactions, provide a unique platform to dissect the complexities of human T- and B-cell immune responses in vivo, when and where translational models do not exist. Moreover, these platforms provide a valuable insight into therapeutics that target immune checkpoints prior to translation into the clinic. The human glial cell–mouse brain chimera has expanded our methodological toolkit, offering a novel means by which to study the intricate interactions between immune and nervous tissues while still in small animal platforms. Grasping the dynamic interplay between the two vital systems, which share so many components, adds to the knowledge necessary to apply mechanisms of immune development, responses to foreign or self-antigens, and the nature of the neurodegenerative disease and progression to be able to provide precise and efficacious immuno- or neurotherapeutics for beneficial outcomes.

4.8 Review Questions

1. What is the primary role of the thymus in lymphopoiesis?
   (a) Production of immunoglobulins
   (b) Development of T cells
   (c) Formation of B-cell precursors
   (d) Maturation of NK cells
2. What is the primary organ responsible for the development of B cells in humans?
   (a) Spleen
   (b) Bone marrow
   (c) Thymus
   (d) Liver
3. What is the main difference between embryonic and adult hematopoietic stem cell (HSC) development?
   (a) HSCs differentiate into the myeloid lineage in adults but not in embryos.
   (b) HSCs in embryos develop exclusively in the bone marrow.
   (c) HSCs in adults are primarily found in the yolk sac.

(d) HSCs in embryos have a unique capacity for self-renewal.
4. What is the significance of CD4 and CD8 co-receptors on TCRαβ lineage T cells?
   (a) CD4 promotes the development of $CD8^+$ T cells.
   (b) CD8 promotes the development of $CD4^+$ T cells.
   (c) CD4 and CD8 help T cells bind to TECs.
   (d) $CD4^+$ T cells recognize MHC I, and $CD8^+$ T cells recognize MHC II.
5. What determines the preference between γδ- and αβ-chains in the TCR of T cells?
   (a) TCRβ rearrangements and Vβ gene segments
   (b) MHC molecules on TECs
   (c) CD4 and CD8 co-receptors
   (d) The expression of CD25 and CD117 markers
6. Under what circumstances do T cells enter the brain parenchyma, and what impact do they have on neurodegenerative diseases?
   (a) T cells are always present in the brain, contributing to neurodegenerative diseases.
   (b) T cells are actively recruited by neurons to protect against neurodegenerative diseases.
   (c) T cells are restricted from entering the brain, ensuring immunity against neurodegenerative diseases.
   (d) T cells infiltrate the brain during acute infections and intensify inflammatory responses in neurodegenerative diseases.
7. Which type of B cells requires the strongest BCR signal?
   (a) Marginal zone B cells
   (b) Peritoneal B cells
   (c) Follicular B cells
   (d) Immature B cells
8. What is the primary function of pre-B cells in B-cell development?
   (a) Antigen recognition
   (b) Ig light-chain gene rearrangement
   (c) Formation of mature B cells
   (d) Generation of plasma cells
9. What is the primary role of BCR at the immature B-cell stage?
   (a) Inducing antibody production
   (b) Promoting anergy
   (c) Eliminating self-reactive B cells
   (d) Stimulating receptor editing
10. What is the primary function of inducible T-cell co-stimulator (ICOS or CD278) on follicular T cells?
   (a) To secrete IL-10
   (b) To induce somatic hypermutation
   (c) To suppress Tregs
   (d) To provide co-stimulation in T-helper cell and B-cell responses

4.9 Answers to the Questions

1. b
2. b
3. d
4. c
5. a
6. d
7. b
8. b
9. c
10. d

References

Abel, A. M., Yang, C., Thakar, M. S., & Malarkannan, S. (2018). Natural killer cells: Development, maturation, and clinical utilization. *Frontiers in Immunology, 9*, 1869. https://doi.org/10.3389/fimmu.2018.01869

Aguzzi, A., Kranich, J., & Krautler, N. J. (2014). Follicular dendritic cells: Origin, phenotype, and function in health and disease. *Trends in Immunology, 35*(3), 105–113. https://doi.org/10.1016/j.it.2013.11.001

Aloisi, F., & Pujol-Borrell, R. (2006). Lymphoid neogenesis in chronic inflammatory diseases. *Nature Reviews. Immunology, 6*(3), 205–217.

Anderson, G., Baik, S., Cowan, J. E., Holland, A. M., et al. (2014). Mechanisms of thymus medulla development and function. *Current Topics in Microbiology and Immunology, 373*, 19–47. https://doi.org/10.1007/82_2013_320

Apert, C., Galindo-Albarrán, A. O., Castan, S., Detraves, C., et al. (2022). IL-2 and IL-15 drive intrathymic development of distinct periphery-seeding CD4(+)Foxp3(+) regulatory T lymphocytes. *Frontiers in Immunology, 13*, 965303. https://doi.org/10.3389/fimmu.2022.965303

Batten, C. A., Edwards, L., & Oura, C. A. (2013). Evaluation of the humoral immune responses in adult cattle and sheep, 4 and 2.5 years post-vaccination with a bluetongue serotype 8 inactivated vaccine. *Vaccine, 31*(37), 3783–3785. https://doi.org/10.1016/j.vaccine.2013.06.033

Baxter, A. G., & Cooke, A. (1993). Complement lytic activity has no role in the pathogenesis of autoimmune diabetes in NOD mice. *Diabetes, 42*(11), 1574–1578. https://doi.org/10.2337/diab.42.11.1574

Bénézech, C., White, A., Mader, E., Serre, K., et al. (2010). Ontogeny of stromal organizer cells during lymph node development. *Journal of Immunology, 184*(8), 4521–4530. https://doi.org/10.4049/jimmunol.0903113

Benninger, Y., Marino, S., Hardegger, R., Weissmann, C., et al. (2000). Differentiation and histological analysis of embryonic stem cell-derived neural transplants in mice. *Brain Pathology, 10*(3), 330–341. https://doi.org/10.1111/j.1750-3639.2000.tb00265.x

Bommhardt, U., Beyer, M., Hunig, T., & Reichardt, H. M. (2004). Molecular and cellular mechanisms of T cell development. *Cellular and Molecular Life Sciences, 61*(3), 263–280.

Bosma, G. C., Custer, R. P., & Bosma, M. J. (1983). A severe combined immunodeficiency mutation in the mouse. *Nature, 301*(5900), 527–530. https://doi.org/10.1038/301527a0

Brainard, D. M., Seung, E., Frahm, N., Cariappa, A., et al. (2009). Induction of robust cellular and humoral virus-specific adaptive immune responses in human immunodeficiency virus-infected humanized BLT mice. *Journal of Virology, 83*(14), 7305–7321. https://doi.org/10.1128/jvi.02207-08

Brendolan, A., & Caamaño, J. H. (2012). Mesenchymal cell differentiation during lymph node organogenesis. *Frontiers in Immunology, 3*, 381. https://doi.org/10.3389/fimmu.2012.00381

Caligiuri, M. A. (2008). Human natural killer cells. *Blood, 112*(3), 461–469. https://doi.org/10.1182/blood-2007-09-077438

Cao, X., Shores, E. W., Hu-Li, J., Anver, M. R., et al. (1995). Defective lymphoid development in mice lacking expression of the common cytokine receptor gamma chain. *Immunity, 2*(3), 223–238. https://doi.org/10.1016/1074-7613(95)90047-0

Chen, J., Liao, S., Xiao, Z., Pan, Q., et al. (2022). The development and improvement of immunodeficient mice and humanized immune system mouse models. *Frontiers in Immunology, 13*, 1007579. https://doi.org/10.3389/fimmu.2022.1007579

Clark, M. R., Mandal, M., Ochiai, K., & Singh, H. (2014). Orchestrating B cell lymphopoiesis through interplay of IL-7 receptor and pre-B cell receptor signalling. *Nature Reviews. Immunology, 14*(2), 69–80. https://doi.org/10.1038/nri3570

Comerford, I., Harata-Lee, Y., Bunting, M. D., Gregor, C., et al. (2013). A myriad of functions and complex regulation of the CCR7/CCL19/CCL21 chemokine axis in the adaptive immune system. *Cytokine & Growth Factor Reviews, 24*(3), 269–283. https://doi.org/10.1016/j.cytogfr.2013.03.001

Cupedo, T., & Mebius, R. E. (2005). Cellular interactions in lymph node development. *Journal of Immunology, 174*(1), 21–25.

Derbinski, J., & Kyewski, B. (2005). Linking signalling pathways, thymic stroma integrity and autoimmunity. *Trends in Immunology, 26*(10), 503–506.

DiSanto, J. P., Müller, W., Guy-Grand, D., Fischer, A., et al. (1995). Lymphoid development in mice with a targeted deletion of the interleukin 2 receptor gamma chain. *Proceedings of the National Academy of Sciences of the United States of America, 92*(2), 377–381. https://doi.org/10.1073/pnas.92.2.377

Elenkov, I. J., Wilder, R. L., Chrousos, G. P., & Vizi, E. S. (2000). The sympathetic nerve—An integrative interface between two supersystems: the brain and the immune system. *Pharmacological Reviews, 52*(4), 595–638.

Engelhardt, B. (2006). Molecular mechanisms involved in T cell migration across the blood-brain barrier. *Journal of Neural Transmission, 113*(4), 477–485.

Ettinger, R., Sims, G. P., Fairhurst, A. M., Robbins, R., et al. (2005). IL-21 induces differentiation of human naive and memory B cells into antibody-secreting plasma cells. *Journal of Immunology, 175*(12), 7867–7879.

Fogg, D. K., Sibon, C., Miled, C., Jung, S., et al. (2006). A clonogenic bone marrow progenitor specific for macrophages and dendritic cells. *Science, 311*(5757), 83–87.

Gao, X., Xu, C., Asada, N., & Frenette, P. S. (2018). The hematopoietic stem cell niche: From embryo to adult. *Development, 145*(2), dev139691. https://doi.org/10.1242/dev.139691

Ginhoux, F., Greter, M., Leboeuf, M., Nandi, S., et al. (2010). Fate mapping analysis reveals that adult microglia derive from primitive macrophages. *Science, 330*(6005), 841–845. https://doi.org/10.1126/science.1194637

Gorantla, S., Makarov, E., Finke-Dwyer, J., Castanedo, A., et al. (2010a). Links between progressive HIV-1 infection of humanized mice and viral neuropathogenesis. *The American Journal of Pathology, 177*(6), 2938–2949. https://doi.org/10.2353/ajpath.2010.100536

Gorantla, S., Makarov, E., Finke-Dwyer, J., Gebhart, C. L., et al. (2010b). CD8+ cell depletion accelerates HIV-1 immunopathology in humanized mice. *Journal of Immunology, 184*(12), 7082–7091. https://doi.org/10.4049/jimmunol.1000438

Han, X., Chen, M., Wang, F., Windrem, M., et al. (2013). Forebrain engraftment by human glial progenitor cells enhances synaptic plasticity and learning in adult mice. *Cell Stem Cell, 12*(3), 342–353. https://doi.org/10.1016/j.stem.2012.12.015

Hart, D. N., & Fabre, J. W. (1981). Demonstration and characterization of Ia-positive dendritic cells in the interstitial connective tissues of rat heart and other tissues, but not brain. *The Journal of Experimental Medicine, 154*(2), 347–361.

Hua, Z., & Hou, B. (2013). TLR signaling in B-cell development and activation. *Cellular & Molecular Immunology, 10*(2), 103–106. https://doi.org/10.1038/cmi.2012.61

Ikebe, M., Miyakawa, K., Takahashi, K., Ohbo, K., et al. (1997). Lymphohaematopoietic abnormalities and systemic lymphoproliferative disorder in interleukin-2 receptor gamma chain-deficient mice. *International Journal of Experimental Pathology, 78*(3), 133–148. https://doi.org/10.1046/j.1365-2613.1997.230356.x

Ishikawa, T., Akiyama, N., & Akiyama, T. (2021). In pursuit of adult progenitors of thymic epithelial cells. *Frontiers in Immunology, 12*, 621824. https://doi.org/10.3389/fimmu.2021.621824

Isobe, T., Hisamori, S., Hogan, D. J., Zabala, M., et al. (2014). miR-142 regulates the tumorigenicity of human breast cancer stem cells through the canonical WNT signaling pathway. *eLife, 3*, e01977. https://doi.org/10.7554/eLife.01977

Jiang, P., & Alam, M. M. (2022). Rise of the human-mouse chimeric brain models. *Cell Regeneration, 11*(1), 32. https://doi.org/10.1186/s13619-022-00135-6

Jiang, P., Turkalj, L., & Xu, R. (2020). High-fidelity modeling of human microglia with pluripotent stem cells. *Cell Stem Cell, 26*(5), 629–631. https://doi.org/10.1016/j.stem.2020.04.004

Kain, M. J., & Owens, B. M. (2013). Stromal cell regulation of homeostatic and inflammatory lymphoid organogenesis. *Immunology, 140*(1), 12–21. https://doi.org/10.1111/imm.12119

Kondo, Y., Windrem, M. S., Zou, L., Chandler-Militello, D., et al. (2014). Human glial chimeric mice reveal astrocytic dependence of JC virus infection. *The Journal of Clinical Investigation, 124*(12), 5323–5336. https://doi.org/10.1172/jci76629

Krumbholz, M., Theil, D., Cepok, S., Hemmer, B., et al. (2006). Chemokines in multiple sclerosis: CXCL12 and CXCL13 up-regulation is differentially linked to CNS immune cell recruitment. *Brain, 129*(Pt 1), 200–211.

Lai, A. Y., & Kondo, M. (2008). T and B lymphocyte differentiation from hematopoietic stem cell. *Seminars in Immunology, 20*(4), 207–212. https://doi.org/10.1016/j.smim.2008.05.002

Liang, B., Hara, T., Wagatsuma, K., Zhang, J., et al. (2012). Role of hepatocyte-derived IL-7 in maintenance of intrahepatic NKT cells and T cells and development of B cells in fetal liver. *Journal of Immunology, 189*(9), 4444–4450. https://doi.org/10.4049/jimmunol.1201181

Lu, T. T., & Browning, J. L. (2014). Role of the lymphotoxin/LIGHT system in the development and maintenance of reticular networks and vasculature in lymphoid tissues. *Frontiers in Immunology, 5*, 47. https://doi.org/10.3389/fimmu.2014.00047

Luis, T. C., Luc, S., Mizukami, T., Boukarabila, H., et al. (2016). Initial seeding of the embryonic thymus by immune-restricted lympho-myeloid progenitors. *Nature Immunology, 17*(12), 1424–1435. https://doi.org/10.1038/ni.3576

Machhi, J., Kevadiya, B. D., Muhammad, I. K., Herskovitz, J., et al. (2020). Harnessing regulatory T cell neuroprotective activities for treatment of neurodegenerative disorders. *Molecular Neurodegeneration, 15*(1), 32. https://doi.org/10.1186/s13024-020-00375-7

Mathews, S., Branch Woods, A., Katano, I., Makarov, E., et al. (2019). Human Interleukin-34 facilitates microglia-like cell differentiation and persistent HIV-1 infection in humanized mice. *Molecular Neurodegeneration, 14*(1), 12. https://doi.org/10.1186/s13024-019-0311-y

Matyszak, M. K., & Perry, V. H. (1996). The potential role of dendritic cells in immune-mediated inflammatory diseases in the central nervous system. *Neuroscience, 74*(2), 599–608.

Messlinger, H., Sebald, H., Heger, L., Dudziak, D., et al. (2018). Monocyte-derived signals activate human natural killer cells in response to Leishmania parasites. *Frontiers in Immunology, 9*, 24. https://doi.org/10.3389/fimmu.2018.00024

Miller, J. S. (2008). How killers kill. *Blood, 112*(2), 213. https://doi.org/10.1182/blood-2008-05-152405

Mosley, R. L., Lu, Y., Olson, K. E., Machhi, J., et al. (2019). A synthetic agonist to vasoactive intestinal peptide receptor-2 induces regulatory T cell neuroprotective activities in models of Parkinson's disease. *Frontiers in Cellular Neuroscience, 13*, 421. https://doi.org/10.3389/fncel.2019.00421

Niederkorn, J. Y. (2006). See no evil, hear no evil, do no evil: The lessons of immune privilege. *Nature Immunology, 7*(4), 354–359.

Ohbo, K., Suda, T., Hashiyama, M., Mantani, A., et al. (1996). Modulation of hematopoiesis in mice with a truncated mutant of the interleukin-2 receptor gamma chain. *Blood, 87*(3), 956–967.

Olson, K. E., Kosloski-Bilek, L. M., Anderson, K. M., Diggs, B. J., et al. (2015). Selective VIP receptor agonists facilitate immune transformation for dopaminergic neuroprotection in MPTP-intoxicated mice. *The Journal of Neuroscience, 35*(50), 16463–16478. https://doi.org/10.1523/jneurosci.2131-15.2015

Olson, K. E., Mosley, R. L., & Gendelman, H. E. (2023). The potential for treg-enhancing therapies in nervous system pathologies. *Clinical and Experimental Immunology, 211*(2), 108–121. https://doi.org/10.1093/cei/uxac084

Pai, S., Qin, J., Cavanagh, L., Mitchell, A., et al. (2014). Real-time imaging reveals the dynamics of leukocyte behaviour during experimental cerebral malaria pathogenesis. *PLoS Pathogens, 10*(7), e1004236. https://doi.org/10.1371/journal.ppat.1004236

Peng, C., Chen, Y., Shan, Y., Zhang, H., et al. (2012). LSK derived LSK- cells have a high apoptotic rate related to survival regulation of hematopoietic and leukemic stem cells. *PLoS One, 7*(6), e38614. https://doi.org/10.1371/journal.pone.0038614

Perdiguero, E. G., Klapproth, K., Schulz, C., Busch, K., et al. (2015). The origin of tissue-resident macrophages: When an erythro-myeloid progenitor is an erythro-myeloid progenitor. *Immunity, 43*(6), 1023–1024. https://doi.org/10.1016/j.immuni.2015.11.022

Perniola, R. (2012). Expression of the autoimmune regulator gene and its relevance to the mechanisms of central and peripheral tolerance. *Clinical & Developmental Immunology, 2012*, 207403. https://doi.org/10.1155/2012/207403

Pieper, K., Grimbacher, B., & Eibel, H. (2013). B-cell biology and development. *The Journal of Allergy and Clinical Immunology, 131*(4), 959–971. https://doi.org/10.1016/j.jaci.2013.01.046

Procaccini, C., Pucino, V., De Rosa, V., Marone, G., et al. (2014). Neuro-endocrine networks controlling immune system in health and disease. *Frontiers in Immunology, 5*, 143. https://doi.org/10.3389/fimmu.2014.00143

Ransohoff, R. M., Kivisakk, P., & Kidd, G. (2003). Three or more routes for leukocyte migration into the central nervous system. *Nature Reviews. Immunology, 3*(7), 569–581.

Ribeiro, A. R., Rodrigues, P. M., Meireles, C., Di Santo, J. P., et al. (2013). Thymocyte selection regulates the homeostasis of IL-7-expressing thymic cortical epithelial cells in vivo. *Journal of Immunology, 191*(3), 1200–1209. https://doi.org/10.4049/jimmunol.1203042

Rongvaux, A., Takizawa, H., Strowig, T., Willinger, T., et al. (2013). Human hemato-lymphoid system mice: Current use and future potential for medicine. *Annual Review of Immunology, 31*, 635–674. https://doi.org/10.1146/annurev-immunol-032712-095921

Sabatino, J. J., Jr., Pröbstel, A. K., & Zamvil, S. S. (2019). B cells in autoimmune and neurodegenerative central nervous system diseases. *Nature Reviews. Neuroscience, 20*(12), 728–745. https://doi.org/10.1038/s41583-019-0233-2

Scheiermann, C., Kunisaki, Y., & Frenette, P. S. (2013). Circadian control of the immune system. *Nature Reviews. Immunology, 13*(3), 190–198. https://doi.org/10.1038/nri3386

Shultz, L. D., Schweitzer, P. A., Christianson, S. W., Gott, B., et al. (1995). Multiple defects in innate and adaptive immunologic function in NOD/LtSz-scid mice. *Journal of Immunology, 154*(1), 180–191.

Siegert, S., & Luther, S. A. (2012). Positive and negative regulation of T cell responses by fibroblastic reticular cells within paracortical regions of lymph nodes. *Frontiers in Immunology, 3*, 285. https://doi.org/10.3389/fimmu.2012.00285

Singh, H., Medina, K. L., & Pongubala, J. M. (2005). Contingent gene regulatory networks and B cell fate specification. *Proceedings of the National Academy of Sciences of the United States of America, 102*(14), 4949–4953.

Sugamura, K., Asao, H., Kondo, M., Tanaka, N., et al. (1996). The interleukin-2 receptor gamma chain: Its role in the multiple cytokine receptor complexes and T cell development in XSCID. *Annual Review of Immunology, 14*, 179–205. https://doi.org/10.1146/annurev.immunol.14.1.179

Sun, L., Li, H., Luo, H., Zhang, L., et al. (2013). Phosphatase Wip1 is essential for the maturation and homeostasis of medullary thymic epithelial cells in mice. *Journal of Immunology, 191*(6), 3210–3220. https://doi.org/10.4049/jimmunol.1300363

Takenaka, K., Prasolava, T. K., Wang, J. C., Mortin-Toth, S. M., et al. (2007). Polymorphism in Sirpa modulates engraftment of human hematopoietic stem cells. *Nature Immunology, 8*(12), 1313–1323. https://doi.org/10.1038/ni1527

Takizawa, H., & Manz, M. G. (2007). Macrophage tolerance: CD47-SIRP-alpha-mediated signals matter. *Nature Immunology, 8*(12), 1287–1289. https://doi.org/10.1038/ni1207-1287

Tanner, A., Taylor, S. E., Decottignies, W., & Berges, B. K. (2014). Humanized mice as a model to study human hematopoietic stem cell transplantation. *Stem Cells and Development, 23*(1), 76–82. https://doi.org/10.1089/scd.2013.0265

Thompson, P. K., & Zúñiga-Pflücker, J. C. (2011). On becoming a T cell, a convergence of factors kick it up a Notch along the way. *Seminars in Immunology, 23*(5), 350–359. https://doi.org/10.1016/j.smim.2011.08.007

Tsapogas, P., Zandi, S., Åhsberg, J., Zetterblad, J., et al. (2011). IL-7 mediates Ebf-1-dependent lineage restriction in early lymphoid progenitors. *Blood, 118*(5), 1283–1290. https://doi.org/10.1182/blood-2011-01-332189

Uchida, N., Buck, D. W., He, D., Reitsma, M. J., et al. (2000). Direct isolation of human central nervous system stem cells. *Proceedings of the National Academy of Sciences of the United States of America, 97*(26), 14720–14725. https://doi.org/10.1073/pnas.97.26.14720

van de Pavert, S. A., & Mebius, R. E. (2010). New insights into the development of lymphoid tissues. *Nature Reviews. Immunology, 10*(9), 664–674. https://doi.org/10.1038/nri2832

Watanabe, S., Ohta, S., Yajima, M., Terashima, K., et al. (2007a). Humanized NOD/SCID/IL2Rgamma(null) mice transplanted with hematopoietic stem cells under nonmyeloablative conditions show prolonged life spans and allow detailed analysis of human immunodeficiency virus type 1 pathogenesis. *Journal of Virology, 81*(23), 13259–13264. https://doi.org/10.1128/jvi.01353-07

Watanabe, S., Terashima, K., Ohta, S., Horibata, S., et al. (2007b). Hematopoietic stem cell-engrafted NOD/SCID/IL2Rgamma null mice develop human lymphoid systems and induce long-lasting HIV-1 infection with specific humoral immune responses. *Blood, 109*(1), 212–218. https://doi.org/10.1182/blood-2006-04-017681

Williams, S. A., Anderson, W. C., Santaguida, M. T., & Dylla, S. J. (2013). Patient-derived xenografts, the cancer stem cell paradigm, and cancer pathobiology in the 21st century. *Laboratory Investigation, 93*(9), 970–982. https://doi.org/10.1038/labinvest.2013.92

Windrem, M. S., Schanz, S. J., Morrow, C., Munir, J., et al. (2014). A competitive advantage by neonatally engrafted human glial progenitors yields mice whose brains are chimeric for human glia. *The Journal of Neuroscience, 34*(48), 16153–16161. https://doi.org/10.1523/jneurosci.1510-14.2014

Xu, W., & Di Santo, J. P. (2013). Taming the beast within: Regulation of innate lymphoid cell homeostasis and function. *Journal of Immunology, 191*(9), 4489–4496. https://doi.org/10.4049/jimmunol.1301759

Yamane, H., & Paul, W. E. (2012). Memory CD4+ T cells: Fate determination, positive feedback and plasticity. *Cellular and Molecular Life Sciences, 69*(10), 1577–1583. https://doi.org/10.1007/s00018-012-0966-9

Yamane, H., & Paul, W. E. (2013). Early signaling events that underlie fate decisions of naive CD4(+) T cells toward distinct T-helper cell subsets. *Immunological Reviews, 252*(1), 12–23. https://doi.org/10.1111/imr.12032

Yui, M. A., & Rothenberg, E. V. (2014). Developmental gene networks: A triathlon on the course to T cell identity. *Nature Reviews. Immunology, 14*(8), 529–545. https://doi.org/10.1038/nri3702

Zhou, W., & Sonnenberg, G. F. (2020). Activation and suppression of group 3 innate lymphoid cells in the gut. *Trends in Immunology, 41*(8), 721–733. https://doi.org/10.1016/j.it.2020.06.009

The Rise of Humanized Mouse Models

5

Larisa Y. Poluektova and Santhi Gorantla

Abstract

In this chapter, we present our view on immunity and approaches to studying interactions between the lymphatic and nervous systems. Cooperation between the two systems plays a fundamental role in health and disease. Recent advances in xenobiology have paved the way for new approaches to studying the relationships between immunity and nervous system homeostasis, response to injuries, and disease in animal models. Humanized animal models help explore the complexity of human immune system development and adaptive immune responses. Humanizing a mouse brain with human glial cells provides an additional tool to study hematolymphoid brain interactions in small animals. Humanized mouse models are the driving forces and workhorses of studying neuroimmune pharmacology and therapeutics.

Keywords

Innate immunity · Adaptive immunity · Hematopoietic stem cells · Humanized mice · Neural stem cells · Astrocytes · Oligodendrocytes

5.1 Introduction

According to the Cambridge Dictionary, immunity is *a situation in which you are protected against disease or from legal action.* Each living organism indeed has multiple protection systems that prevent damage by outside invaders. These protection systems are graded as innate immunity and adaptive immunity. Both systems are created by evolution, and they are species-specific (Duggal & Emerman, 2012; Mandujano-Tinoco et al., 2021). There are multiple restriction factors and interferons (IFNs) that protect the host from immediate infection. Similar immediate immune responses are provided by various innate lymphoid cells (ILCs), neutrophils, and macrophages. The adaptive immune system with specialized antigen-presenting cells (APCs), thymus-produced T cells, and bone marrow-produced B cells can "fight" invaders by developing short or lifelong memory to protect the host. The immune and nervous systems are in a close interplay during development and homeostasis. Despite complete segregation by the blood–brain barrier (BBB) and the privileged status of the mammalian central nervous system (CNS) after birth, neuroimmune interactions have shown common developmental paths and shared cell types and regulatory molecules. Similarities between the nervous and immune systems are apparent with the essential homeostatic and coordinated responses to danger. Both organs can establish direct and indirect connections to distant parts of the body and carry information from and to those parts using a range of chemical signaling molecules. Both communicate rapidly through the chemical and cellular networks by utilizing shared and conserved secondary messengers and signaling pathways. They form synapses to facilitate information transmission and have threshold sensitivity of cellular receptors for ligands. Nervous and immune organs acquire memory with maturation, training, and exposure to new signals (Chu et al., 2020). For a long time, we could not study the fundamentals of human immunity in response to human-specific viral infections, like HIV-1, in animal models.

The discovery of immunodeficient mice (Bosma et al., 1983) almost paralleled that of human immunodeficiency virus type 1 (HIV-1) as a cause of acquired immunodeficiency syndrome (AIDS) in humans. One of the first applications of severe combined immunodeficiency (*scid*) mice was the transplantation of human immune cells to study human-specific HIV-1 infection and the evaluation of antiretroviral drugs (Namikawa et al., 1988; McCune et al., 1990). Eventually, with the improvement of the mouse genetic background for its ability to engraft human hematopoietic

L. Y. Poluektova (✉) · S. Gorantla (✉)
Department of Pharmacology and Experimental Neuroscience, University of Nebraska Medical Center, Omaha, NE, USA
e-mail: lpoluekt@unmc.edu; sgorantla@unmc.edu

H. E. Gendelman, T. Ikezu (eds.), *Neuroimmune Pharmacology and Therapeutics*,
https://doi.org/10.1007/978-3-031-68237-7_5

stem cells (HSCs) efficiently (Ito et al., 2002), a mouse model with the full spectrum of human immune cells (except for neutrophils) became available for investigation. The discovery of neural stem cells and the improvement of their isolation techniques allow their transplantation into the brain and spinal cord of immunodeficient mice (Uchida et al., 2000). Depending on the environmental cues, neural stem cells generate neurons, astrocytes, and oligodendrocytes (Cummings et al., 2005; Uchida et al., 2012).

Finally, studying the human immune system and nervous system development and function in mice became possible in severely immunodeficient mice by transplantation of human hematopoietic and neural stem cells, and they are widely known as humanized mouse models.

5.2 Humanized Animal Models

5.2.1 History of Humanized Mice: Hematolymphoid and Mouse Brain Humanization

The humanized mouse field is continuously evolving and has received significant attention as an experimental platform to study human hematopoietic and immune system development (Martinov et al., 2021), human stem cells and human tumor stem cells, and tissue biology in cancer-related research (Cogels et al., 2021). For such models, scientists use immunodeficient mice, for example, nonobese diabetic mice (NOD) carrying a mutation of DNA-dependent protein kinase (*Prkdc*), called NOD/*Scid* mice (Bosma et al., 1983), where the *Prkdc* mutation leads to SCID (severe combined immunodeficiency) with a defective recombination mechanism that affects the normal development of T and B lymphocytes. The NOD background of these mice results in a partial impairment of mouse natural killer (NK) cells and macrophage function. Over the years, improvements in immunodeficient mouse strains have gradually enriched the human immune cell reconstitution in mice. Initial strains in the 1980s and 90s were *Scid* and NOD/*Scid* mice, and they were widely used to reconstitute human peripheral blood lymphocytes (huPBLs) or human peripheral blood mononuclear cells (huPBMCs). The first human PBMC transplantation in mice was described in C.B17-*Scid* (Mosier et al., 1988). Upon transplantation, T and B lymphocytes engraft in recipient mice in all the tissues. Most of the human T cells express CD45RO, indicating mature and activated phenotypes. Human-specific immunoglobulins (Igs) are also detected in the model, with human IgG levels. huPBL mice have been used to study the immune response to different human-specific viral infections, including EBV, HIV, SARS-CoV-1), and DENV-1. To recapitulate HIV infection in the brain, initially, human myeloid cells infected with HIV-1 were intracranially injected into the caudate putamen of CB17-*Scid* mice. Human monocyte-derived macrophages (MDMs) were used to represent the microglia, the major cellular targets of HIV in the brain. The mice develop HIV encephalitis (HIVE) pathology with HIV-positive multinucleated giant cells, myelin pallor, and dendritic loss at the injected site, and the model is called *Scid*-HIVE. However, the pathology is focal as it is only confined to the injected site and lacks adaptive immune cells. To overcome the deficiency of adaptive immune cells, the HIVE model was combined with huPBLs to study neuroimmune interactions during HIV infection. Both HIVE and huPBL-HIVE models were used to study acute HIV infection, and the highly activated phenotype of transplanted lymphocytes was a major caveat. A physiologically relevant humanized mouse model with a naïve human immune system that can lead to the natural progression of HIV infection and pathology in the periphery and brain was required.

Complete deletion of common cytokine receptor gamma chain (g_c null) on NOD/Scid (Cao et al., 1995; DiSanto et al., 1995) to develop NOD.Cg-Prkdc^{scid} Il2rgtm1Wjl/SzJ (NSG) or deletion of the intracytoplasmic domain of g_c for NOD.Cg-Prkdcscid Il2rgtm1Sug/JicTac (NOG) (Ohbo et al., 1996; Ikebe et al., 1997; Sugamura et al., 1996) became most popular for the ability to reconstitute the human immune system consisting of a wide range of human immune cell subtypes. The more recent addition has been the development of the NOD CRISPR *Prkdc IL2r gamma* (NCG) mouse model by Nanjing Biomedical Research Institute of Nanjing University and Nanjing Galaxy Biopharma in 2014 and transferred to Charles River in 2016. It was created by sequential CRISPR/Cas9 editing of the *Prkdc* and *Il2rg* loci in the NOD/Nju mouse, generating a mouse coisogenic to the NOD/Nju. Another immunodeficient strain is Rag2$^{-/-}$IL2R$g_c^{-/-}$, where Rag2 (recombination activating 2) knockout mice have deficiencies in mouse T- and B-lymphocyte receptor recombination and development. In all these new mouse strains, because of common cytokine receptor gamma chain (g_c) deficiency, mice have defective NK cells due to the lack of interleukin (IL)-15 signaling and CD11c$^+$ DCs, which reduces IFN production. These strains were used to study hematopoiesis from human CD34$^+$ hematopoietic stem cells (HSCs) and immune system development and function (Tanner et al., 2014). A marked increase in engraftment was attributed to the low lytic activity of the C5 component of complement [due to two base pair deletions in a 5′ exon of the murine C5 gene (Baxter & Cooke, 1993)], defects in macrophage function [defective regulation of colony-stimulating factor-1 and interferon-γ receptors and reduced secretion of IL-1 in response to lipopolysaccharide (LPS)], and NK cell activity (Shultz et al., 1995). Long-term engraftment of human immune cells was also attributed to the murine polymorphism of signal regulatory protein-α (SIRP-α,

CD172α)—the region at the insulin-dependent diabetes 13 (Idd13) locus on chromosome 2 containing a coding sequence polymorphism of Sirp-a. The polymorphism of Sirp-a was responsible for the survival and subsequent development of human HSCs in a murine environment. Interaction of Sirp-a on mouse macrophages and cell surface transmembrane glycoprotein CD47 (integrin-associated protein, IAP) on human HSC surfaces provides an "eat me" signal for destroying xenogeneic human cells. A polymorphism in mouse SIRP-α leads to the loss of the "eat me" signal, resulting in better survival of the engrafted human cells (Takenaka et al., 2007; Takizawa & Manz, 2007).

Transplantation of human neural progenitor cells (NPCs) in NOD/*Scid* (Uchida et al., 2000) and other types of immunodeficient mice (Kondo et al., 2014) gives rise to the development of human astrocytes and oligodendrocytes, which were named here as humanized astroglia mice (hAGL). Manipulation with human cytokine substitutions and knock-in of human IL-34 on NOG (Mathews et al., 2019) changed the distribution of human monocytes and a significant number of human microglia in mouse brains, which we named humanized microglia mice (hMGL).

We have summarized the history of the development of humanized mice in Fig. 5.1.

The major characteristics of chimeric humanized models on NSG and NOG mouse strains are as follows:

1. The high rate of acceptance of human HSCs.
2. The stable population of human HSCs in the mouse bone marrow and human-like hematopoiesis.
3. The occupancy of the murine thymus with human thymocytes and their maturation into αβ, γδ, and regulatory T cells.
4. The complete reconstitution of residual murine lymph nodes with human T, B, and dendritic cells.
5. The formation of splenic white pulp.
6. The distribution of human cells of the macrophage lineage throughout the brain meninges and perivascular spaces.

The significant deficiencies are as follows:

1. The absence of human erythrocytes in peripheral blood. This could be corrected by depletion of mouse phagocytic cells (Chen et al., 2017), introduction of Kit (Wv) mutation (Yurino et al., 2016), or depletion of the C3 component of complement (Yamaguchi et al., 2021).
2. The absence of human granulocytes.
3. The reduced functionality of human macrophages and natural killer cells. Functionality can be improved by the introduction of multiple human cytokines: macrophage colony-stimulating factor 1 (CSF1), granulocyte-macrophage colony-stimulating factor 2 (CSF2), IL-2, IL-3, IL-15, and others (Miller et al., 2013).
4. Limited secondary lymphoid organs—gut- and nasal-associated lymphatics, peripheral lymph nodes. The latest was improved by overexpression of mouse thymic stromal lymphopoietin (Li et al., 2018).

Despite these limitations, humanized mice support chronic HIV-1 replication and recapitulate the course of progressive loss of $CD4^+$, development of HIV-1-specific cytotoxic T lymphocytes (CTLs), and humoral immune responses (Watanabe et al., 2007; Gorantla et al., 2010a, b; Brainard et al., 2009). Moreover, chronic untreated HIV-1 infection induces damage to the central nervous system (CNS), detectable by magnetic resonance imaging (MRI) (Dash et al., 2011; Boska et al., 2014). The schematic representation of mouse humanization is shown in Fig. 5.2.

5.2.2 Hematolymphoid and Microglial "Humanization" of Mice

Humanized mice created by transplantation of HSCs at birth do not have or may have an insignificant number of human microglia-like cells, an important HIV-1 target in the central nervous system. Recently, we have developed a new strain of mice. Human interleukin-34 (hIL-34) was transgenically expressed in NOG (NOG-hIL-34). A sufficient population of the human microglia in the mouse brain upon humanization with HSCs was observed (Mathews et al., 2019). Figure 5.3 shows that transgenic overexpression of human IL-34 in NOG mice facilitates the reconstitution of the mouse brain with a significant number of human microglia compared to non-transgenic NSG.

Newly developed NOG-hIL-34 transgenic mice allow straightforward population of the mouse brain by human microglia derived from bone marrow HSCs. Human CD34-positive hematopoietic stem cells injected at birth into NOG-hIL-34 mice develop a human hematolymphoid system and HSC-derived human microglia in the mouse brain (Fig. 5.3). There was a clear difference in human microglial reconstitution of the mouse brain between the NOG-hIL-34 and NSG mouse strains. The human microglia were positive for P2Ry12, CD14, CD68, and Iba-1. They demonstrated typical microglial morphology. Human microglial reconstitution was nearly up to 80% of total Iba-1-positive cells in certain mouse brain regions. They also responded robustly to insult, like HIV infection with antiviral defense genes related to interferon (IFN) pathways, upregulated pattern recognition receptors and toll-like receptor (TLR), and increased expression of genes linked to the activation of the nuclear factor kappa B (NF-κB) transcription factor (Mathews et al., 2019).

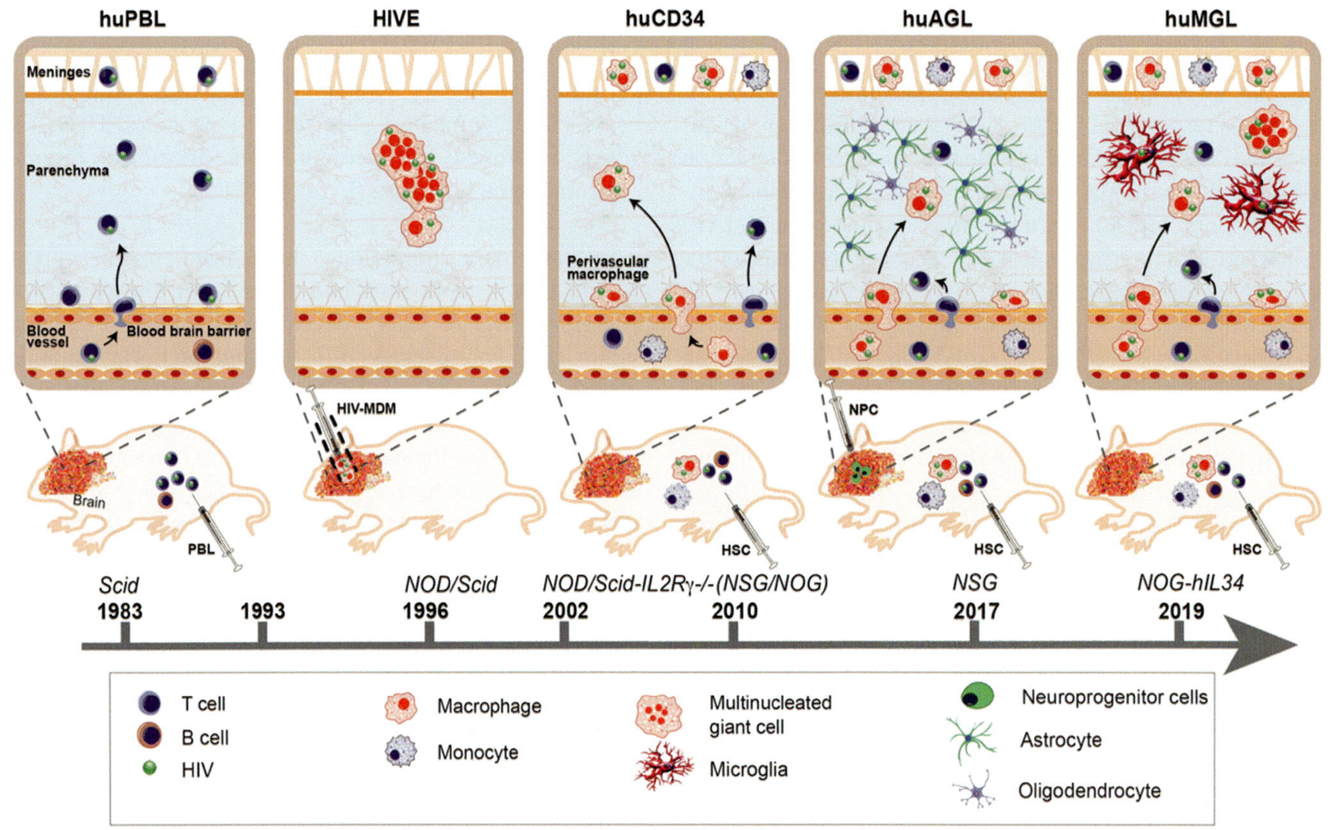

Fig. 5.1 The schematic diagram represents the history of humanized mice and human cells in the mouse brain in each model. huPBL mouse reconstitution in early years was in Scid and NOD/Scid mice, and the mice are engrafted mainly with activated lymphocytes in the periphery and brain. The HIVE model is generated by injecting HIV-infected human monocyte-derived macrophages into the caudate putamen of 4–5-week-old Scid mice, and the HIVE pathology was seen near the injected site. CD34-NSG mice reconstituted with human $CD34^+$ HSCs developed a human immune system consisting of different immune subsets. However, the brains had macrophages in the meninges and a few in the perivascular areas and parenchyma. hAGL are created by transplanting human NPCs into the left ventricle and HSCs intrahepatically into newborn NSG mice. hAGL reconstitute the human immune system from HSCs and astrocytes and oligodendrocytes from NPCs. hMGL are generated by transplanting human HSCs intrahepatically into newborn human IL-34 transgenic NOG mice. They develop the human immune system and human microglia from HSCs. Newly developed hMGL allow for the studies of systemic HIV infection and neuroimmune interactions. Abbreviations: huPBL Human Peripheral Blood Lymphocyte, HIVE HIV Encephalitis, HSC Hematopoietic Stem Cell, NSG NOD/Scid-IL2Rg$_c^{-/-}$, hAGL Humanized Astroglia Mice, hMGLx Humanized Microglia Mice

It was known that the cytokines involved in human monocyte–macrophage–microglia homeostasis are colony-stimulating factor-1 (CSF1) and IL-34, which signal via the common receptor CSF1R. Human IL-34 is specific in supporting and developing microglia. In CD34-NSG mice, it is obvious that mouse cytokines do not support human microglial development. Human cytokine support is essential for human cells. Moreover, transgenic expression of human CSF1 in NSG mice also supported the proliferation of human microglial precursors derived from inducible pluripotent stem cells (iPSCs) when transplanted into the mouse brain (Svoboda et al., 2019). The ability to support human microglial function in the mouse brain provides the opportunity to evaluate their role in pathologic conditions. There are several examples where human microglia and transgenic expression of human CSF1 are used to study neuropathology. Hasselmann et al. (2019) used iPSC-derived hematopoietic progenitor cells. Transplantation of these cells into NSG-hCSF1 transgenic newborn mice leads to the development and wide distribution of human microglia cells with transcriptional profiles similar to those of microglia from human brains. Moreover, human microglia responded to the brain parenchyma damage induced by focal laser ablation or mild closed-head injury with activation and phagocytosis (clearance) of neuronal components similar to murine microglia. Human microglia also responded robustly to a systemic inflammatory challenge with lipopolysaccharide (LPS), which differs from in vitro responses to LPS.

Another model was used to assess the response of human microglia to familial Alzheimer's disease (FAD), the mutant human amyloid beta (A4) precursor protein 695 (APP) (Swedish, Florida, and London), and the mutant FAD human presenilin 1 (PS1) (M146L and L286V) (Oakley et al., 2006). For this purpose, the BALB/c/129 model that includes two knockouts alleles, namely, recombination activating gene 2 ($Rag2^{tm1.1Flv}$) and γc ($Il2rg^{tm1.1Flv}$), and three humanized

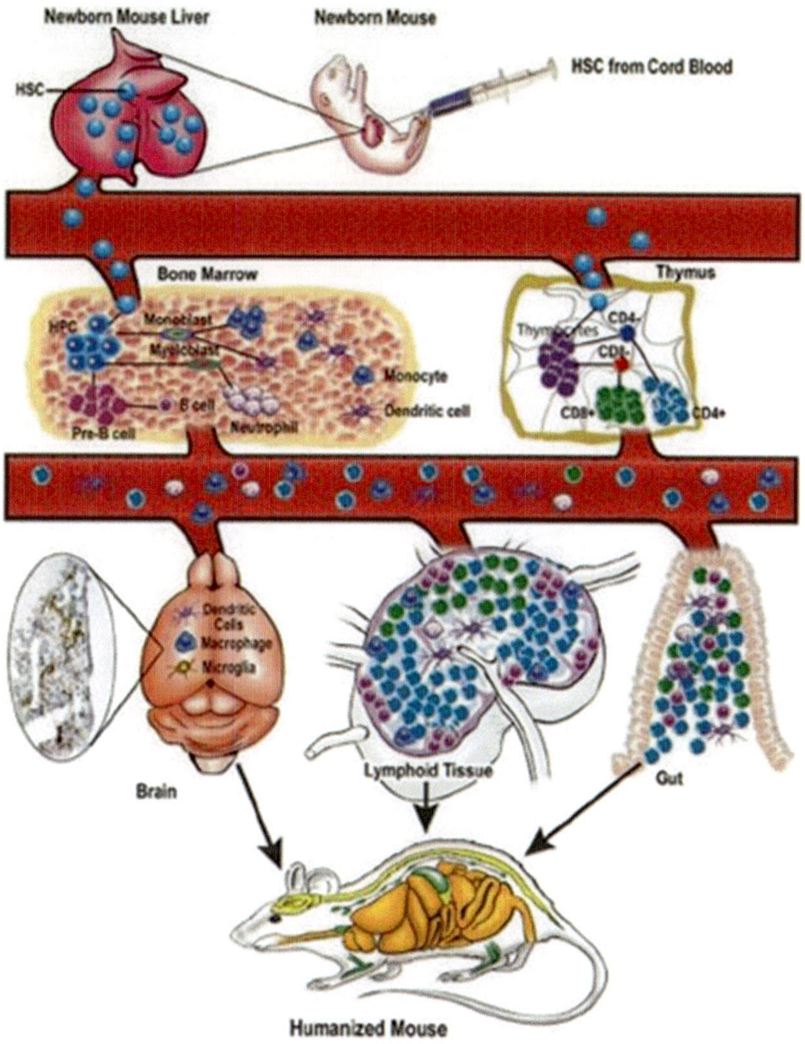

Fig. 5.2 A mouse model of humanized hematolymphoid organs. Human CD34⁺ stem cells (HSCs) isolated from the umbilical cord blood or fetal liver were injected intrahepatically into 1-day-old irradiated pups. A complete human immune system is developed in NSG/NOG mice. The injected HSCs reach the mouse lymphoid organs, including the bone marrow, spleen, lymph nodes, and gut, and develops into a broad range of immune cell lineages. The presence of human CD34⁺ stem cells, myeloblasts, B-cell precursors, erythroblasts, promyelocytes, granulocytes, and monocytes in the mouse bone marrow was detected. Human T-cell development occurred in the mouse thymus, as evidenced by the presence of CD4/CD8 T cells. In lymphoid tissue, distinct follicles filled with human T and B lymphocytes and macrophages were observed. Lymph nodes were also reconstituted with human lymphocytes, macrophages, and dendritic cells. A mature human immune system develops in the mouse by 20–22 weeks. (From Gorantla et al., 2012)

knock-in alleles, namely, CSF1 ($Csf1^{tm1(CSF1)Flv}$), IL-3/CSF2 ($Csf2/Il3^{tm1.1(CSF2,IL3)Flv}$), and thrombopoietin ($Thpo^{tm1.1(TPO)Flv}$) was crossed with APP/PS1 FAD mice. Transplantation of iPSC-derived human microglia into the brains of this mouse strain at birth and analysis at 9 months revealed the transformation of human microglia to the disease-associated microglia (DAM)-like phenotype. The expression of numerous DAM markers in microglia near Aβ plaques was observed. Human cells in contact with plaques downregulated the homeostatic marker P2RY12, adopted an activated ameboid morphology, and occasionally phagocytosed fibrillar Aβ. Plaque-associated human cells upregulated the DAM markers CD9, MERTK, APOE, CD11C, and TREM2. The advantage of using iPSCs is the possibility to manipulate genes and modify TREM2, where the AD-associated R47H mutation in TREM2 has been shown to impair the association between microglia and plaques in human AD subjects. Surprisingly, transplantation of genetically modified microglia, as expected, revealed significantly less clustering of microglia around Aβ plaques when compared to wild-type (WT), nonetheless, with no significant difference in the overall plaque burden across mice. Taken as a whole, these data demonstrate that human microglia derived from iPSCs, indeed, adopt relevant activation states in response to amyloid pathology and that the impact of disease-relevant microglial mutations can be accurately examined using this model. Unfortunately, these complicated models require thorough genotyping of mice and transplantation of cells via intracranial injection. Moreover, they lack a peripheral human hematolymphoid system to study brain and T-lymphocyte interactions.

In the near future, immunodeficient mouse models can be combined with mouse models of human pathology associated with genetic mutations in Alzheimer's disease. Amyotrophic lateral sclerosis (ALS), or toxic damage to specific brain regions (such as in Parkinson's disease), in combination with human hematolymphoid and brain microglia reconstitution, provide helpful information in the context of human cells and aid in the development of new therapeutics relevant to human translation.

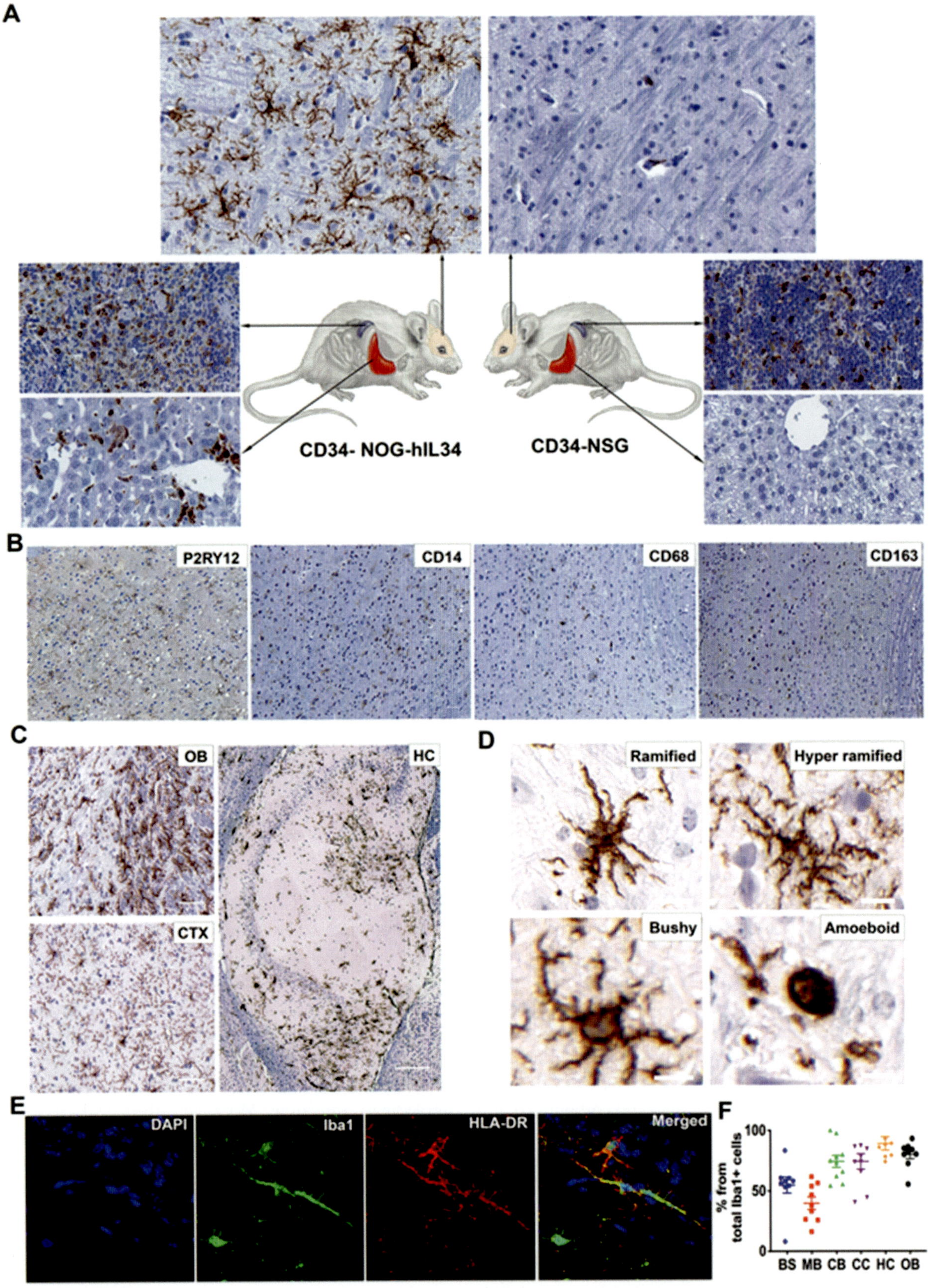

Fig. 5.3 Human microglia-like cells in CD34-NOG-hIL-34 mouse brains. (**a**) A comparison of tissue macrophage reconstitution between CD34-NOG-hIL-34 and CD34-NSG mice. Brain sections were stained for human leukocyte antigen DR (HLA-DR) and liver and spleen for human CD68 (scale bar: 10 μm). HLA-DR⁺ human cells were widely distributed in the brain parenchyma of CD34-NOG-hIL-34 mice. (**b**) Mouse brain sections showing human cells positive for P2Ry12, CD14, CD68, and CD163. (**c**) Magnified views of the olfactory bulb (OB, 20×), cortex (CTX, 20×), and hippocampus (HC, 10×) from CD34-NOG-hIL-34 mouse brain sections stained for HLA-DR. (**d**) Microglial morphology of human cells shown at a higher magnification (scale bar: 2 μm). (**e**) Confocal images of the brain stained for HLA-DR and Iba-1 show double-positive human microglia. (**f**) Quantification of HLA-DR/Iba-1 double-positive human microglial cells from total Iba-1-positive cells. (From Mathews et al., 2019)

5.2.3 Astroglia of Human Origin in the Mouse Brain

Since the discovery of neural stem cells in the 1960s (Takagi, 2016) and the development of techniques for their growth and differentiation in vitro into neurons, astrocytes, and oligodendrocytes (Galiakberova & Dashinimaev, 2020), it has become obvious that transplantation of neural stem cells or specific neural or glial progenitors into the brains of newborn immunodeficient mice can generate neurons and glial cells (Uchida et al., 2000, 2012). The successful humanization of the mouse brain with human glial cells—astrocytes and oligodendrocytes—was used to study the myelination (Uchida et al., 2012) and pathogenesis of John Cunningham virus (JCV), which is a human neurotropic polyomavirus, the etiologic agent of progressive multifocal leukoencephalopathy (PML) (Kondo et al., 2014). PML is a rare, fatal, demyelinating disease of the central nervous system, which almost always occurs in immunocompromised patient, including those with AIDS (Gordon & Khalili, 1998).

Humanization of the immune system and brain glial cells brought about the opportunity to investigate the interaction between immune and glial cells during HIV-1 infection. This approach was tested by Li et al. (2017) in double-humanized hAGL, and the transcriptomic changes associated with HIV-1 infection were analyzed (Fig. 5.4).

The authors successfully showed that human NPCs, when transplanted at an early postnatal age, demonstrate broad glial distribution throughout the brain. By 6 months of age, many human astrocytes were seen to have repopulated multiple brain regions along the white matter tracts. Moreover, considerable numbers of human precursors of oligodendrocytes were present. The majority of human immunocytes were present in the meninges. These were predominantly $CD163^+$ human macrophages. High levels of systemic HIV-1 infection with infected meningeal- and brain-infiltrating cells at lower numbers than those seen in peripheral tissues were detected. Thus, the model reflects three principal features of human brain disease: a functional immune system with mobile brain-penetrant monocyte macrophages, sustained high levels of systemic viral infection, and glial target cells that determine alterations in the brain microenvironment. Most importantly, each of the cell types was able to interact with the other physiologically. As a result, the transcriptomic analysis showed vigorous type 1 IFN signaling response and downregulation of genes that are crucial for oligodendrocyte differentiation and myelination. The transcriptional profiles from the chimeric brains overlap, in large measure, with the disease profile found in human HIV-1-infected patients with cognitive impairments, encephalitic brains, and acute SIV infection (Borjabad et al., 2011; Gelman et al., 2012; Sanfilippo et al., 2018). This model provides a platform to further investigate the pathobiology of HIV nervous system infection and viral cell reservoirs needed for novel disease-combating therapies and eradication measures.

5.3 Humanized Mice as Workhorses of Therapeutics Evaluation

Human-specific viral infections like HIV-1 can be studied in hematolymphoid humanized mice. Starting from the first success with atazanavir (AZT) evaluation in humanized mice (McCune et al., 1990; Kaneshima et al., 1991; Shih et al., 1991), broad panels of all classes of antiretroviral drugs and long-acting formulations were tested in these models. Both human PBL reconstituted mice (huPBL) and human CD34 HSC reconstituted mice (CD34-NSG) were used to evaluate HIV therapeutics. Other humanized mouse models of human hepatocyte reconstitution in thymidine kinase (TK) transgenic immunodeficient NOG mice (TK-NOG) were used to test therapeutics for hepatitis viruses, HBV and HCV (Table 5.1).

Humanized mice provide an opportunity to study therapeutic antibodies and vaccines. Several examples are shown in Table 5.2.

5.4 Humanized Mice for Assessment of Human Pathology

Chimeric models always face the question of how human immune cells and biomolecules will interact with the mouse environment. Is it possible to reproduce human pathology in the system where cross talk between the host and donor cells is compromised by species specificity? Could humanized mice reflect the differences in viral strains or human immune cell phenotypes during viral infections? Is it possible to evaluate the responses of human and mouse brain cells to the viral insult? Is it possible to address questions about viral latency and reservoirs? All these questions could be addressed in humanized mice, and several examples are shown in Table 5.3.

We have a long history of HIV research on humanized mice at the University of Nebraska Medical Center (UNMC), and we have performed evaluations of multiple drugs and their combinations in in vivo models of HIV, HBV, and HCV infections. Future directions will be based on the clinical need to address questions about immunology and nervous system interactions in pathologic conditions. Recently developed humanized mouse models with the human immune system, human microglia, and human glia will help fill the knowledge gap in understanding neuroimmune interactions in disease and in the development of therapeutics.

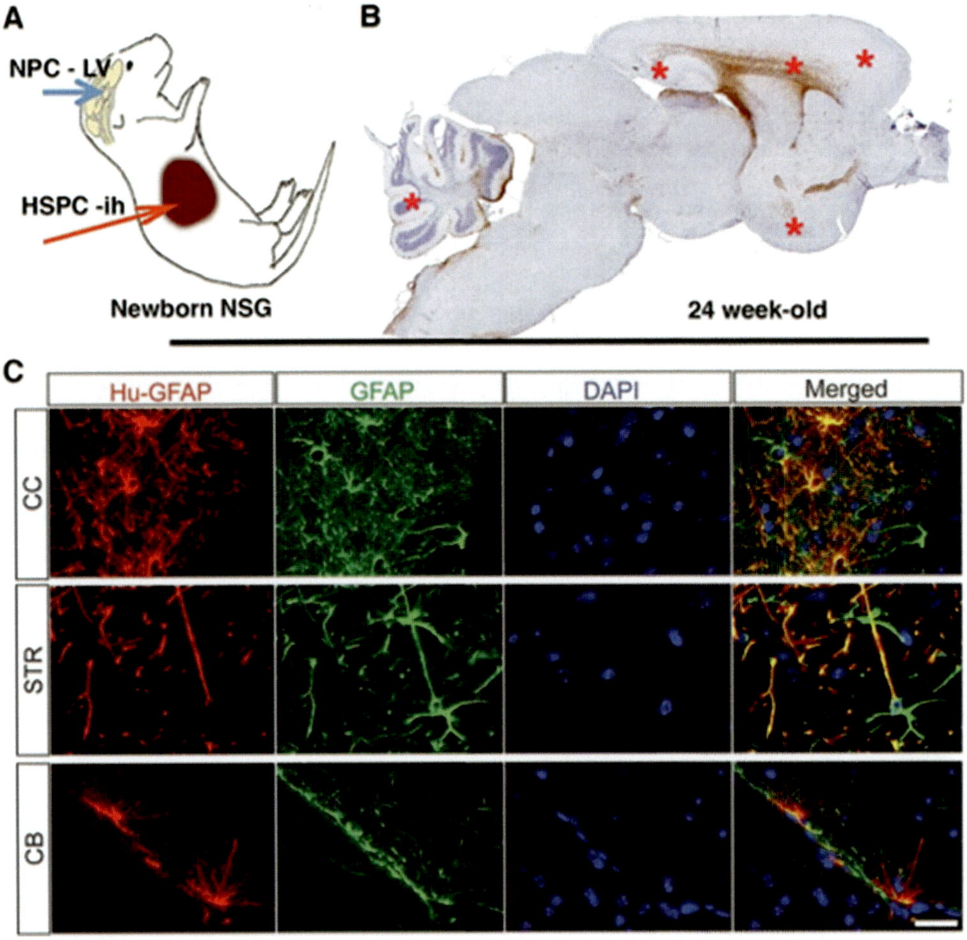

Fig. 5.4 Glial humanization of the mouse brain. (**a**) Human hematopoietic stem and brain progenitor cells were transplanted into mice at birth. The blue arrow indicates the injection of neural progenitor cells (NPCs) into the lateral ventricle (LV). The red arrow indicates intrahepatic (ih) injection of hematopoietic stem/progenitor cells (HSPCs). (**b**) Representative view under a 1× objective lens from a 5-μm paraffin-embedded sagittal section stained for human-specific GFAP (brown). Asterisks indicate fields where cell images were captured for display. (**c**) Staining with human astrocyte-specific antibodies (hu-GFAP, red) and total glial fibrillary acidic protein (GFAP; green) was performed. Representative images from different brain regions are shown. Original images were collected by a Leica DM6 system at 1000× magnification. Nuclei were stained with 4′,6-diamidino-2-phenylindole (blue). Scale bar: 20 μm. (From Li et al., 2017)

5.5 Review Questions

1. Common cytokine receptor gamma chain (g_c) knockout or truncation on NOD/*Scid* mice facilitates successful long-term engraftment of human HSC-derived immune system because g_c deficiency leads to:
   (a) Macrophage depletion
   (b) Reduced NK-cell function
   (c) Disruption of IL-15 signaling
   (d) All of the above
   (e) B and C
2. What statement is not true about humanized microglia mice generated on CSF1 and IL-34 transgenic immunodeficient mice?
   (a) Human microglia in IL-34 transgenic mice have a microglial phenotype and express microglial markers.
   (b) CSF1 transgenic mice support human iPSC-derived microglial reconstitution in the mouse brain.
   (c) In IL-34 transgenic mice, human microglia reconstitution in the mouse brain is from bone marrow HSCs.
   (d) CSF1 transgenic mice transplanted with iPSC-derived cells engraft the human immune system and human microglia.
   (e) IL-34 transgenic mice transplanted with HSCs engraft the human immune system and human microglia.
3. Human neuroprogenitor cells injected into the brains of newborn immunodeficient mice can differentiate into:
   (a) Astrocytes
   (b) Oligodendrocytes
   (c) Microglia
   (d) All of the above
   (e) A and B

Table 5.1 Evaluation of antiretroviral and antiviral therapeutics in humanized mice, hu-PBL,humanl lymphocytes transplanted immunodeficient mice; CD34-NSG, human CD34-positive hematopoietic stem cells transplanted NSG mice; AZT, zidovudine; 3TC, Lamivudine; ABC, Abacavir; PK/PD, pharmacokinetic/pharmacodynamic; HBV, hepatitis B virus; TK-NOG, Herpes simplex virus thymidine kinase (TK) transgenic NOG mice; AAV-9-CRISPR/Cas, adeno-associated viral (AAV) vector 9 to deliver CRISPR/Cas9 to cells.

Antiretrovirals	Animal model	References
Protease inhibitors		
Atazanavir, ritonavir poloxamer-coated, folic acid-linked poloxamer 407	huPBL, CD34-NSG mice	"Long-acting nanoformulated antiretroviral therapy elicits potent antiretroviral and neuroprotective responses in HIV-1-infected humanized mice" and others (Dash et al., 2012; Kanmogne et al., 2012; Roy et al., 2012; Gautam et al., 2013, 2014; Puligujja et al., 2015; Zhang et al., 2016)
Nucleoside reverse transcriptase inhibitors (NTRIs)		
AP peptide-decorated and cholesteryl-ε-poly-L-lysine nanogels with NRTI 5′-succinate derivatives AZT, 3TC, and ABC	HIV encephalitis mice (HIVE-*Scid*)	"Nano-NRTIs: efficient inhibitors of HIV type-1 in macrophages with a reduced mitochondrial toxicity" and others (Gerson et al., 2014; Senanayake et al., 2015)
Dual-acting anti-HBV/HIV NTRIs, anti-HCV peptide complexes, and immune modulators		
3TC ProTide	In vitro HIV, PK/PD on rats	"A long-acting nanoformulated lamivudine ProTide" and others (Guo et al., 2017; Gautam et al., 2018; Smith et al., 2019)
3TC ProTide	In vivo HBV in TK-NOG humanized liver mice	"A long-acting 3TC ProTide nanoformulation suppresses HBV replication in humanized mice" (Wang et al., 2020)
Antiviral peptide nanocomplex	In vitro HIV, HCV, and huPBL mice	"Antiviral peptide nanocomplexes as a potential therapeutic modality for HIV/HCV co-infection" and others (Zhang et al., 2013, 2015)
Long-acting tenofovir	In vitro HIV, PK/PD on mice In vivo HBV in TK-NOG humanized liver mice	"Transformation of tenofovir into stable ProTide nanocrystals with long-acting pharmacokinetic profiles" (Cobb et al., 2021; Das et al., 2022)
HIV integrase strand transfer inhibitor		
A nanoformulated myristoylated dolutegravir prodrug	huPBL and CD34-NSG	"Creation of a long-acting nanoformulated dolutegravir" and others (McMillan et al., 2018; Sillman et al., 2018)
Myristoylated CAB prodrug	Efficacy on CD34-NSG	"Creation of a nanoformulated cabotegravir prodrug with improved antiretroviral profiles" and others (McMillan et al., 2018; Zhou et al., 2018a, b)
MLK3 inhibitor URMC-099 and dolutegravir	CD34-NSG	"Autophagy facilitates macrophage depots of sustained release nanoformulated antiretroviral drugs" (Gnanadhas et al., 2017)
Combinatorial treatment for HIV eradication		
Dolutegravir, 3TC, abacavir, rilpivirine, and AAV-9-CRISPR/Cas excision	Anti-HIV efficacy on CD34-NSG	"Sequential LASER ART and CRISPR Treatments Eliminate HIV-1 in a Subset of Infected Humanized Mice" "CRISPR editing of CCR5 and HIV-1 facilitates viral elimination in antiretroviral drug-suppressed virus-infected humanized mice" (Dash et al., 2019, 2023)

hu-PBL human lymphocytes transplanted immunodeficient mice, *CD34-NSG* human CD34-positive hematopoietic stem cells transplanted NSG mice, *AZT* zidovudine, *3TC* Lamivudine, *ABC* Abacavir, *PK/PD* pharmacokinetic/pharmacodynamic, *HBV* hepatitis B virus, *TK-NOG* Herpes simplex virus thymidine kinase (TK) transgenic NOG mice, *AAV-9-CRISPR/Cas* adeno-associated viral (AAV) vector 9 to deliver CRISPR/Cas9 to cells

4. What are the deficiencies in the human hematolymphoid system in CD34-NSG humanized mice? Choose all the correct answers.
   (a) Low erythrocytes, granulocytes, NK cells
   (b) Low erythrocytes, lymphocytes, granulocytes
   (c) Low macrophages, lymphocytes, granulocytes
   (d) Low erythrocytes, microglia, granulocytes
   (e) Low leukocytes, erythrocytes, granulocytes

5.6 Correct Answers

1. e.
2. d.
3. e.
4. a and d.

Table 5.2 Evaluation of anti-HIV neutralizing antibodies and vaccines

Reagent	Animal model	References
Monoclonal IgG 3A5	huPBMC	"Deficient synthesis of class-switched, HIV-neutralizing antibodies to the CD4 binding site and correction by electrophilic gp120 immunogen" (Planque et al., 2014)
Vaccine candidate amplicon HSV-gp120$_{MN/LAI}$	NOD/scid transplanted with autologous huPBL and dendritic cells	"Human dendritic cells transduced with herpes simplex virus amplicons encoding human immunodeficiency virus type 1 (HIV-1) gp120 elicit adaptive immune responses from human cells engrafted into NOD/SCID mice and confer partial protection against HIV-1 challenge" (Gorantla et al., 2005)
Adenovirus serotype 5 vectors encoding two derivatives of the HIV-1 clade C envelope proteins	CD34-Rag2$^{-/-}$gc$^{-/-}$ and CD34-NSG	Gorantla S et al. (2008) immunization of humanized mice with adenoviral vector encoding HIV-1 envelope protein induces specific immune responses but *not viral protection*; unpublished

Table 5.3 Characterization of different aspects of human pathology in humanized mouse models. Rag2 gc , recombinase activated gene 2 and gamma chain knockout; CD34-NSG, human CD34-positive hematopoietic stem cells transplanted NSG mice; CAB, cabotegravir; RPV, Rilpivirine, 3TC, Lamivudine; ABC, Abacavir; PK/PD, pharmacokinetic/pharmacodynamic; HIVE, HIV encephalitis; hu-PBL,humanl lymphocytes transplanted immunodeficient mice; uPA-NOG, urokinase-type plasminogen activator transgenic NOG mice; TKNOG, Herpes simplex virus thymidine kinase (TK) transgenic NOG mice

Characterization	Model	References
HIV-1, clades B, C, and primary isolates	Rag2$^{-/-}$gc$^{-/-}$	"Human immunodeficiency virus type 1 pathobiology studied in humanized BALB/c-Rag2^{-}/$^{-}$gc^{-}/$^{-}$ mice" (Gorantla et al., 2007)
HIV lymphoid reservoir	CD34-NSG	"HIV-1 cellular and tissue replication patterns in infected humanized mice" (Arainga et al., 2016)
HIV reservoirs in long-acting ARV-suppressed mice	CD34-NSG treated with CAB and RPV, or CAB, 3TC, abacavir and RPV	"A mature macrophage is a principal HIV-1 cellular reservoir in humanized mice after treatment with long-acting antiretroviral therapy" (Arainga et al., 2017)
Role of CD8 in HIV immune pathogenesis	CD34-NSG	"CD8^{+} cell depletion accelerates HIV-1 immunopathology in humanized mice" (Gorantla et al., 2010b)
CNS comorbidity	HIVE, huPBL, CD34-NSG, CD34-NOG-IL34	"Links between progressive HIV-1 infection of humanized mice and viral neuropathogenesis" (Gorantla et al., 2010a; Epstein et al., 2013; Boska et al., 2014; Mathews et al., 2019)
	Astroglia and immune system dual reconstituted NSG mice	"Systemic HIV-1 infection produces a unique glial footprint in humanized mouse brains" (Li et al., 2017)
HIV and model-specific immune activation	CD34-NSG, huPBL	"Immune Activations and Viral Tissue Compartmentalization During Progressive HIV-1 Infection of Humanized Mice" (Su et al., 2019)
In vivo HIV viral outgrowth assay	CD34-NSG	"Amplification of Replication Competent HIV-1 by Adoptive Transfer of Human Cells from Infected Humanized mice" (Su et al., 2020)
Immune system and hepatocyte dual reconstitution	uPA-NOG	"Human hepatocytes and hematolymphoid dual reconstitution in treosulfan-conditioned uPA-NOG mice" (Gutti et al., 2014)
HIV effect on the immune system and hepatocytes	Dual reconstituted TK-NOG mice	"Human hepatocyte depletion in the presence of HIV-1 infection in dual reconstituted humanized mice" (Dagur et al., 2018)
HBV and HIV coinfection	TK-NOG mice with human hepatocytes and immune cells	"Small Animal Models for Human Immunodeficiency Virus (HIV), Hepatitis B, and Tuberculosis: Proceedings of an NIAID Workshop" (Akkina et al., 2020)

Rag2$^{-/-}$ gc$^{-/-}$ recombinase activated gene 2 and gamma chain knockout, *CD34-NSG* human CD34-positive hematopoietic stem cells transplanted NSG mice, *CAB* cabotegravir, *RPV* Rilpivirine, *3TC* Lamivudine, *ABC* Abacavir, *PK/PD* pharmacokinetic/pharmacodynamic, *HIVE* HIV encephalitis, *hu-PBL* human lymphocytes transplanted immunodeficient mice, *uPA-NOG* urokinase-type plasminogen activator transgenic NOG mice, *TKNOG* Herpes simplex virus thymidine kinase (TK) transgenic NOG mice

Acknowledgments We thank Dr. Jatin Machhi for the creation of Fig. 5.1.

References

Akkina, R., et al. (2020). Small animal models for human immunodeficiency virus (HIV), hepatitis B, and tuberculosis: Proceedings of an NIAID workshop. *Current HIV Research, 18*, 19–28.

Arainga, M., Su, H., Poluektova, L. Y., Gorantla, S., & Gendelman, H. E. (2016). HIV-1 cellular and tissue replication patterns in infected humanized mice. *Scientific Reports, 6*, 23513.

Arainga, M., Edagwa, B., Mosley, R. L., Poluektova, L. Y., Gorantla, S., & Gendelman, H. E. (2017). A mature macrophage is a principal HIV-1 cellular reservoir in humanized mice after treatment with long acting antiretroviral therapy. *Retrovirology, 14*, 17.

Baxter, A. G., & Cooke, A. (1993). Complement lytic activity has no role in the pathogenesis of autoimmune diabetes in NOD mice. *Diabetes, 42*(11), 1574–1578. https://doi.org/10.2337/diab.42.11.1574

Borjabad, A., Morgello, S., Chao, W., Kim, S. Y., Brooks, A. I., Murray, J., Potash, M. J., & Volsky, D. J. (2011). Significant effects of antiretroviral therapy on global gene expression in brain tissues of patients with HIV-1-associated neurocognitive disorders. *PLoS Pathogens, 7*, e1002213.

Boska, M. D., Dash, P. K., Knibbe, J., Epstein, A. A., Akhter, S. P., Fields, N., High, R., Makarov, E., Bonasera, S., Gelbard, H. A., Poluektova, L. Y., Gendelman, H. E., & Gorantla, S. (2014). Associations between brain microstructures, metabolites, and cognitive deficits during chronic HIV-1 infection of humanized mice. *Molecular Neurodegeneration, 9*, 58.

Bosma, G. C., Custer, R. P., & Bosma, M. J. (1983). A severe combined immunodeficiency mutation in the mouse. *Nature, 301*, 527–530.

Brainard, D. M., Seung, E., Frahm, N., Cariappa, A., Bailey, C. C., Hart, W. K., . . . Tager, A. M. (2009). Induction of robust cellular and humoral virus-specific adaptive immune responses in human immunodeficiency virus-infected humanized BLT mice. *Journal of Virology, 83*(14), 7305–7321. https://doi.org/10.1128/JVI.02207-08

Cao, X., Shores, E. W., Hu-Li, J., Anver, M. R., Kelsall, B. L., Russell, S. M., . . . et al. (1995). Defective lymphoid development in mice lacking expression of the common cytokine receptor gamma chain. *Immunity, 2*(3), 223–238. https://doi.org/10.1016/1074-7613(95)90047-0

Chen, B., Fan, W., Zou, J., Zhang, S., He, J., Shu, C., Zhao, G., Sun, T., Hu, Z., & Yang, Y. G. (2017). Complement depletion improves human red blood cell reconstitution in immunodeficient mice. *Stem Cell Reports, 9*, 1034–1042.

Chu, C., Artis, D., & Chiu, I. M. (2020). Neuro-immune interactions in the tissues. *Immunity, 52*, 464–474.

Cobb, D. A., Smith, N., Deodhar, S., Bade, A. N., Gautam, N., Shetty, B. L. D., McMillan, J., Alnouti, Y., Cohen, S. M., Gendelman, H. E., & Edagwa, B. (2021). Transformation of tenofovir into stable ProTide nanocrystals with long-acting pharmacokinetic profiles. *Nature Communications, 12*, 5458.

Cogels, M. M., Rouas, R., Ghanem, G. E., Martinive, P., Awada, A., Van Gestel, D., & Krayem, M. (2021). Humanized mice as a valuable pre-clinical model for cancer immunotherapy research. *Frontiers in Oncology, 11*, 784947.

Cummings, B. J., Uchida, N., Tamaki, S. J., Salazar, D. L., Hooshmand, M., Summers, R., Gage, F. H., & Anderson, A. J. (2005). Human neural stem cells differentiate and promote locomotor recovery in spinal cord-injured mice. *Proceedings of the National Academy of Sciences of the United States of America, 102*, 14069–14074.

Dagur, R. S., Wang, W., Cheng, Y., Makarov, E., Ganesan, M., Suemizu, H., Gebhart, C. L., Gorantla, S., Osna, N., & Poluektova, L. Y. (2018). Human hepatocyte depletion in the presence of HIV-1 infection in dual reconstituted humanized mice. *Biology Open, 7*.

Das, S., Wang, W., Ganesan, M., Fonseca-Lanza, F., Cobb, D. A., Bybee, G., . . . Poluektova, L. Y. (2022). An ultralong-acting tenofovir ProTide nanoformulation achieves monthslong HBV suppression. *Science Advances, 8*(51), eade9582. https://doi.org/10.1126/sciadv.ade9582

Dash, P. K., Gorantla, S., Gendelman, H. E., Knibbe, J., Casale, G. P., Makarov, E., Epstein, A. A., Gelbard, H. A., Boska, M. D., & Poluektova, L. Y. (2011). Loss of neuronal integrity during progressive HIV-1 infection of humanized mice. *The Journal of Neuroscience, 31*, 3148–3157.

Dash, P. K., Gendelman, H. E., Roy, U., Balkundi, S., Alnouti, Y., Mosley, R. L., Gelbard, H. A., McMillan, J., Gorantla, S., & Poluektova, L. Y. (2012). Long-acting nanoformulated antiretroviral therapy elicits potent antiretroviral and neuroprotective responses in HIV-1-infected humanized mice. *AIDS, 26*, 2135–2144.

Dash, P. K., et al. (2019). Sequential LASER ART and CRISPR treatments eliminate HIV-1 in a subset of infected humanized mice. *Nature Communications, 10*, 2753.

Dash, P. K., et al. (2023). CRISPR editing of CCR5 and HIV-1 facilitates viral elimination in antiretroviral drug-suppressed virus-infected humanized mice. *Proceedings of the National Academy of Sciences of the United States of America, 120*, e2217887120.

DiSanto, J. P., Muller, W., Guy-Grand, D., Fischer, A., & Rajewsky, K. (1995). Lymphoid development in mice with a targeted deletion of the interleukin 2 receptor gamma chain. *Proceedings of the National Academy of Sciences of the United States of America, 92*(2), 377–381. https://doi.org/10.1073/pnas.92.2.377

Duggal, N. K., & Emerman, M. (2012). Evolutionary conflicts between viruses and restriction factors shape immunity. *Nature Reviews. Immunology, 12*, 687–695.

Epstein, A. A., Narayanasamy, P., Dash, P. K., High, R., Bathena, S. P., Gorantla, S., Poluektova, L. Y., Alnouti, Y., Gendelman, H. E., & Boska, M. D. (2013). Combinatorial assessments of brain tissue metabolomics and histopathology in rodent models of human immunodeficiency virus infection. *Journal of Neuroimmune Pharmacology, 8*, 1224–1238.

Galiakberova, A. A., & Dashinimaev, E. B. (2020). Neural stem cells and methods for their generation from induced pluripotent stem cells in vitro. *Frontiers in Cell and Developmental Biology, 8*.

Gautam, N., Roy, U., Balkundi, S., Puligujja, P., Guo, D., Smith, N., Liu, X. M., Lamberty, B., Morsey, B., Fox, H. S., McMillan, J., Gendelman, H. E., & Alnouti, Y. (2013). Preclinical pharmacokinetics and tissue distribution of long-acting nanoformulated antiretroviral therapy. *Antimicrobial Agents and Chemotherapy, 57*, 3110–3120.

Gautam, N., Puligujja, P., Balkundi, S., Thakare, R., Liu, X. M., Fox, H. S., McMillan, J., Gendelman, H. E., & Alnouti, Y. (2014). Pharmacokinetics, biodistribution, and toxicity of folic acid-coated antiretroviral nanoformulations. *Antimicrobial Agents and Chemotherapy, 58*, 7510–7519.

Gautam, N., Lin, Z., Banoub, M. G., Smith, N. A., Maayah, A., McMillan, J., Gendelman, H. E., & Alnouti, Y. (2018). Simultaneous quantification of intracellular lamivudine and abacavir triphosphate metabolites by LC-MS/MS. *Journal of Pharmaceutical and Biomedical Analysis, 153*, 248–259.

Gelman, B. B., Chen, T., Lisinicchia, J. G., Soukup, V. M., Carmical, J. R., Starkey, J. M., Masliah, E., Commins, D. L., Brandt, D., Grant, I., Singer, E. J., Levine, A. J., Miller, J., Winkler, J. M., Fox, H. S., Luxon, B. A., Morgello, S., & National Neuro, A. T. C. (2012). The National NeuroAIDS tissue consortium brain gene array: Two types of HIV-associated neurocognitive impairment. *PLoS One, 7*, e46178.

Gerson, T., Makarov, E., Senanayake, T. H., Gorantla, S., Poluektova, L. Y., & Vinogradov, S. V. (2014). Nano-NRTIs demonstrate low neurotoxicity and high antiviral activity against HIV infection in the brain. *Nanomedicine, 10*, 177–185.

Gnanadhas, D. P., Dash, P. K., Sillman, B., Bade, A. N., Lin, Z., Palandri, D. L., Gautam, N., Alnouti, Y., Gelbard, H. A., McMillan, J., Mosley, R. L., Edagwa, B., Gendelman, H. E., & Gorantla, S. (2017). Autophagy facilitates macrophage depots of sustained-release nanoformulated antiretroviral drugs. *The Journal of Clinical Investigation, 127*, 857–873.

Gorantla, S., Santos, K., Meyer, V., Dewhurst, S., Bowers, W. J., Federoff, H. J., Gendelman, H. E., & Poluektova, L. (2005). Human dendritic cells transduced with herpes simplex virus amplicons encoding human immunodeficiency virus type 1 (HIV-1) gp120 elicit adaptive immune responses from human cells engrafted into NOD/SCID mice and confer partial protection against HIV-1 challenge. *Journal of Virology, 79*, 2124–2132.

Gorantla, S., Sneller, H., Walters, L., Sharp, J. G., Pirruccello, S. J., West, J. T., Wood, C., Dewhurst, S., Gendelman, H. E., & Poluektova, L. (2007). Human immunodeficiency virus type 1 pathobiology studied in humanized BALB/c–Rag2–/–gammac–/– mice. *Journal of Virology, 81*, 2700–2712.

Gorantla, S., Liu, J., Wang, T., Holguin, A., Sneller, H. M., Dou, H., . . . Gendelman, H. E. (2008). Modulation of innate immunity by copolymer-1 leads to neuroprotection in murine HIV-1 encephalitis. *Glia, 56*(2), 223–232. https://doi.org/10.1002/glia.20607

Gorantla, S., Makarov, E., Finke-Dwyer, J., Castanedo, A., Holguin, A., Gebhart, C. L., Gendelman, H. E., & Poluektova, L. (2010a). Links between progressive HIV-1 infection of humanized mice and viral neuropathogenesis. *The American Journal of Pathology, 177*, 2938–2949.

Gorantla, S., Makarov, E., Finke-Dwyer, J., Gebhart, C. L., Domm, W., Dewhurst, S., Gendelman, H. E., & Poluektova, L. Y. (2010b). CD8+ cell depletion accelerates HIV-1 immunopathology in humanized mice. *Journal of Immunology, 184*, 7082–7091.

Gorantla, S., Poluektova, L., & Gendelman, H. E. (2012). Rodent models for HIV-associated neurocognitive disorders. *Trends in Neurosciences, 35*(3), 197–208. https://doi.org/10.1016/j.tins.2011.12.006

Gordon, J., & Khalili, K. (1998). The human polyomavirus, JCV, and neurological diseases (review). *International Journal of Molecular Medicine, 1*, 647–655.

Guo, D., Zhou, T., Arainga, M., Palandri, D., Gautam, N., Bronich, T., Alnouti, Y., McMillan, J., Edagwa, B., & Gendelman, H. E. (2017). Creation of a long-acting nanoformulated 2′,3′-Dideoxy-3′-Thiacytidine. *Journal of Acquired Immune Deficiency Syndromes, 74*, e75–e83.

Gutti, T. L., Knibbe, J. S., Makarov, E., Zhang, J., Yannam, G. R., Gorantla, S., Sun, Y., Mercer, D. F., Suemizu, H., Wisecarver, J. L., Osna, N. A., Bronich, T. K., & Poluektova, L. Y. (2014). Human hepatocytes and hematolymphoid dual reconstitution in treosulfan-conditioned uPA-NOG mice. *The American Journal of Pathology, 184*, 101–109.

Hasselmann, J., Coburn, M. A., England, W., Figueroa Velez, D. X., Kiani Shabestari, S., Tu, C. H., . . . Blurton-Jones, M. (2019). Development of a Chimeric Model to Study and Manipulate Human Microglia In Vivo. *Neuron, 103*(6), 1016–1033 e1010. https://doi.org/10.1016/j.neuron.2019.07.002

Ikebe, M., Miyakawa, K., Takahashi, K., Ohbo, K., Nakamura, M., Sugamura, K., . . . Tomita, K. (1997). Lymphohaematopoietic abnormalities and systemic lymphoproliferative disorder in interleukin-2 receptor gamma chain-deficient mice. *International Journal of Experimental Pathology, 78*(3), 133–148. https://doi.org/10.1046/j.1365-2613.1997.230356.x

Ito, M., Hiramatsu, H., Kobayashi, K., Suzue, K., Kawahata, M., Hioki, K., Ueyama, Y., Koyanagi, Y., Sugamura, K., Tsuji, K., Heike, T., & Nakahata, T. (2002). NOD/SCID/gamma(c)(null) mouse: An excellent recipient mouse model for engraftment of human cells. *Blood, 100*, 3175–3182.

Kaneshima, H., Shih, C. C., Namikawa, R., Rabin, L., Outzen, H., Machado, S. G., & McCune, J. M. (1991). Human immunodeficiency virus infection of human lymph nodes in the SCID-hu mouse. *Proceedings of the National Academy of Sciences of the United States of America, 88*, 4523–4527.

Kanmogne, G. D., Singh, S., Roy, U., Liu, X., McMillan, J., Gorantla, S., Balkundi, S., Smith, N., Alnouti, Y., Gautam, N., Zhou, Y., Poluektova, L., Kabanov, A., Bronich, T., & Gendelman, H. E. (2012). Mononuclear phagocyte intercellular crosstalk facilitates transmission of cell-targeted nanoformulated antiretroviral drugs to human brain endothelial cells. *International Journal of Nanomedicine, 7*, 2373–2388.

Kondo, Y., Windrem, M. S., Zou, L., Chandler-Militello, D., Schanz, S. J., Auvergne, R. M., Betstadt, S. J., Harrington, A. R., Johnson, M., Kazarov, A., Gorelik, L., & Goldman, S. A. (2014). Human glial chimeric mice reveal astrocytic dependence of JC virus infection. *The Journal of Clinical Investigation, 124*, 5323–5336.

Li, W., Gorantla, S., Gendelman, H. E., & Poluektova, L. Y. (2017). Systemic HIV-1 infection produces a unique glial footprint in humanized mouse brains. *Disease Models & Mechanisms, 10*, 1489–1502.

Li, Y., Masse-Ranson, G., Garcia, Z., Bruel, T., Kök, A., Strick-Marchand, H., Jouvion, G., Serafini, N., Lim, A. I., Dusseaux, M., Hieu, T., Bourgade, F., Toubert, A., Finke, D., Schwartz, O., Bousso, P., Mouquet, H., & Di Santo, J. P. (2018). A human immune system mouse model with robust lymph node development. *Nature Methods, 15*, 623–630.

Mandujano-Tinoco, E. A., Sultan, E., Ottolenghi, A., Gershoni-Yahalom, O., & Rosental, B. (2021). Evolution of cellular immunity effector cells; perspective on cytotoxic and phagocytic cellular lineages. *Cells, 10*, 1853.

Martinov, T., McKenna, K. M., Tan, W. H., Collins, E. J., Kehret, A. R., Linton, J. D., Olsen, T. M., Shobaki, N., & Rongvaux, A. (2021). Building the next generation of humanized hemato-lymphoid system mice. *Frontiers in Immunology, 12*, 643852.

Mathews, S., Branch Woods, A., Katano, I., Makarov, E., Thomas, M. B., Gendelman, H. E., Poluektova, L. Y., Ito, M., & Gorantla, S. (2019). Human Interleukin-34 facilitates microglia-like cell differentiation and persistent HIV-1 infection in humanized mice. *Molecular Neurodegeneration, 14*, 12.

McCune, J. M., Namikawa, R., Shih, C. C., Rabin, L., & Kaneshima, H. (1990). Suppression of HIV infection in AZT-treated SCID-hu mice. *Science, 247*, 564–566.

McMillan, J., Szlachetka, A., Slack, L., Sillman, B., Lamberty, B., Morsey, B., Callen, S., Gautam, N., Alnouti, Y., Edagwa, B., Gendelman, H. E., & Fox, H. S. (2018). Pharmacokinetics of a long-acting nanoformulated dolutegravir prodrug in rhesus macaques. *Antimicrobial Agents and Chemotherapy, 62*, e01316–17.

Miller, P. H., Cheung, A. M., Beer, P. A., Knapp, D. J., Dhillon, K., Rabu, G., Rostamirad, S., Humphries, R. K., & Eaves, C. J. (2013). Enhanced normal short-term human myelopoiesis in mice engineered to express human-specific myeloid growth factors. *Blood, 121*, e1–e4.

Mosier, D. E., Gulizia, R. J., Baird, S. M., & Wilson, D. B. (1988). Transfer of a functional human immune system to mice with severe combined immunodeficiency. *Nature, 335*, 256–259.

Namikawa, R., Kaneshima, H., Lieberman, M., Weissman, I. L., & McCune, J. M. (1988). Infection of the SCID-hu mouse by HIV-1. *Science, 242*, 1684–1686.

Oakley, H., Cole, S. L., Logan, S., Maus, E., Shao, P., Craft, J., Guillozet-Bongaarts, A., Ohno, M., Disterhoft, J., Van Eldik, L., Berry, R., &

Vassar, R. (2006). Intraneuronal beta-amyloid aggregates, neurodegeneration, and neuron loss in transgenic mice with five familial Alzheimer's disease mutations: Potential factors in amyloid plaque formation. *The Journal of Neuroscience, 26*, 10129–10140.

Ohbo, K., Suda, T., Hashiyama, M., Mantani, A., Ikebe, M., Miyakawa, K., . . . Sugamura, K. (1996). Modulation of hematopoiesis in mice with a truncated mutant of the interleukin-2 receptor gamma chain. *Blood, 87*(3), 956–967. Retrieved from https://www.ncbi.nlm.nih.gov/pubmed/8562967

Planque, S. A., Mitsuda, Y., Chitsazzadeh, V., Gorantla, S., Poluektova, L., Nishiyama, Y., Ochsenbauer, C., Morris, M. K., Sapparapu, G., Hanson, C. V., Massey, R. J., & Paul, S. (2014). Deficient synthesis of class-switched, HIV-neutralizing antibodies to the CD4 binding site and correction by electrophilic gp120 immunogen. *AIDS, 28*, 2201–2211.

Puligujja, P., Arainga, M., Dash, P., Palandri, D., Mosley, R. L., Gorantla, S., Poluektova, L., McMillan, J., & Gendelman, H. E. (2015). Pharmacodynamics of folic acid receptor targeted antiretroviral nanotherapy in HIV-1-infected humanized mice. *Antiviral Research, 120*, 85–88.

Roy, U., McMillan, J., Alnouti, Y., Gautum, N., Smith, N., Balkundi, S., Dash, P., Gorantla, S., Martinez-Skinner, A., Meza, J., Kanmogne, G., Swindells, S., Cohen, S. M., Mosley, R. L., Poluektova, L., & Gendelman, H. E. (2012). Pharmacodynamic and antiretroviral activities of combination nanoformulated antiretrovirals in HIV-1-infected human peripheral blood lymphocyte-reconstituted mice. *The Journal of Infectious Diseases, 206*, 1577–1588.

Sanfilippo, C., Pinzone, M. R., Cambria, D., Longo, A., Palumbo, M., Di Marco, R., Condorelli, F., Nunnari, G., Malaguarnera, L., & Di Rosa, M. (2018). OAS gene family expression is associated with HIV-related neurocognitive disorders. *Molecular Neurobiology, 55*, 1905–1914.

Senanayake, T. H., Gorantla, S., Makarov, E., Lu, Y., Warren, G., & Vinogradov, S. V. (2015). Nanogel-conjugated reverse transcriptase inhibitors and their combinations as novel antiviral agents with increased efficacy against HIV-1 infection. *Molecular Pharmaceutics, 12*, 4226–4236.

Shih, C. C., Kaneshima, H., Rabin, L., Namikawa, R., Sager, P., McGowan, J., & McCune, J. M. (1991). Postexposure prophylaxis with zidovudine suppresses human immunodeficiency virus type 1 infection in SCID-hu mice in a time-dependent manner. *The Journal of Infectious Diseases, 163*, 625–627.

Shultz, L. D., Schweitzer, P. A., Christianson, S. W., Gott, B., Schweitzer, I. B., Tennent, B., . . . et al. (1995). Multiple defects in innate and adaptive immunologic function in NOD/LtSz-scid mice. *The Journal of Immunology, 154*(1), 180-191. Retrieved from https://www.ncbi.nlm.nih.gov/pubmed/7995938

Sillman, B., Bade, A. N., Dash, P. K., Bhargavan, B., Kocher, T., Mathews, S., Su, H., Kanmogne, G. D., Poluektova, L. Y., Gorantla, S., McMillan, J., Gautam, N., Alnouti, Y., Edagwa, B., & Gendelman, H. E. (2018). Creation of a long-acting nanoformulated dolutegravir. *Nature Communications, 9*, 443.

Smith, N., Bade, A. N., Soni, D., Gautam, N., Alnouti, Y., Herskovitz, J., Ibrahim, I. M., Wojtkiewicz, M. S., Dyavar Shetty, B. L., McMillan, J., Gendelman, H. E., & Edagwa, B. (2019). A long acting nanoformulated lamivudine ProTide. *Biomaterials, 223*, 119476.

Su, H., Cheng, Y., Sravanam, S., Mathews, S., Gorantla, S., Poluektova, L. Y., Dash, P. K., & Gendelman, H. E. (2019). Immune activations and viral tissue compartmentalization during progressive HIV-1 infection of humanized mice. *Frontiers in Immunology, 10*, 340.

Su, H., Sravanam, S., Gorantla, S., Kaminski, R., Khalili, K., Poluektova, L., Gendelman, H. E., & Dash, P. K. (2020). Amplification of replication competent HIV-1 by adoptive transfer of human cells from infected humanized mice. *Frontiers in Cellular and Infection Microbiology, 10*, 38.

Sugamura, K., Asao, H., Kondo, M., Tanaka, N., Ishii, N., Ohbo, K., . . . Takeshita, T. (1996). The interleukin-2 receptor gamma chain: its role in the multiple cytokine receptor complexes and T cell development in XSCID. *Annual Review of Immunology, 14*, 179–205. https://doi.org/10.1146/annurev.immunol.14.1.179

Svoboda, D. S., Barrasa, M. I., Shu, J., Rietjens, R., Zhang, S., Mitalipova, M., Berube, P., Fu, D., Shultz, L. D., Bell, G. W., & Jaenisch, R. (2019). Human iPSC-derived microglia assume a primary microglia-like state after transplantation into the neonatal mouse brain. *Proceedings of the National Academy of Sciences of the United States of America, 116*, 25293–25303.

Takagi, Y. (2016). History of neural stem cell research and its clinical application. *Neurologia Medico-Chirurgica (Tokyo), 56*, 110–124.

Takenaka, K., Prasolava, T. K., Wang, J. C., Mortin-Toth, S. M., Khalouei, S., Gan, O. I., . . . Danska, J. S. (2007). Polymorphism in Sirpa modulates engraftment of human hematopoietic stem cells. *Nature Immunology, 8*(12), 1313–1323. https://doi.org/10.1038/ni1527

Takizawa, H., & Manz, M. G. (2007). Macrophage tolerance: CD47-SIRP-alpha-mediated signals matter. *Nature Immunology, 8*(12), 1287–1289. https://doi.org/10.1038/ni1207-1287

Tanner, A., Taylor, S. E., Decottignies, W., & Berges, B. K. (2014). Humanized mice as a model to study human hematopoietic stem cell transplantation. *Stem Cells and Development, 23*(1), 76–82. https://doi.org/10.1089/scd.2013.0265

Uchida, N., Buck, D. W., He, D., Reitsma, M. J., Masek, M., Phan, T. V., Tsukamoto, A. S., Gage, F. H., & Weissman, I. L. (2000). Direct isolation of human central nervous system stem cells. *Proceedings of the National Academy of Sciences of the United States of America, 97*, 14720–14725.

Uchida, N., Chen, K., Dohse, M., Hansen, K. D., Dean, J., Buser, J. R., Riddle, A., Beardsley, D. J., Wan, Y., Gong, X., Nguyen, T., Cummings, B. J., Anderson, A. J., Tamaki, S. J., Tsukamoto, A., Weissman, I. L., Matsumoto, S. G., Sherman, L. S., Kroenke, C. D., & Back, S. A. (2012). Human neural stem cells induce functional myelination in mice with severe dysmyelination. *Science Translational Medicine, 4*, 155ra136.

Wang, W., Smith, N., Makarov, E., Sun, Y., Gebhart, C. L., Ganesan, M., Osna, N. A., Gendelman, H. E., Edagwa, B. J., & Poluektova, L. Y. (2020). A long-acting 3TC ProTide nanoformulation suppresses HBV replication in humanized mice. *Nanomedicine, 28*, 102185.

Watanabe, S., Ohta, S., Yajima, M., Terashima, K., Ito, M., Mugishima, H., . . . Yamamoto, N. (2007). Humanized NOD/SCID/IL2Rgamma(null) mice transplanted with hematopoietic stem cells under nonmyeloablative conditions show prolonged life spans and allow detailed analysis of human immunodeficiency virus type 1 pathogenesis. *Journal of Virology, 81*(23), 13259–13264. https://doi.org/10.1128/JVI.01353-07

Yamaguchi, T., Katano, I., Otsuka, I., Ito, R., Mochizuki, M., Goto, M., & Takahashi, T. (2021). Generation of novel human red blood cell-bearing humanized mouse models based on C3-deficient NOG mice. *Frontiers in Immunology, 12*, 671648.

Yurino, A., Takenaka, K., Yamauchi, T., Nunomura, T., Uehara, Y., Jinnouchi, F., Miyawaki, K., Kikushige, Y., Kato, K., Miyamoto, T., Iwasaki, H., Kunisaki, Y., & Akashi, K. (2016). Enhanced reconstitution of human erythropoiesis and thrombopoiesis in an immunodeficient mouse model with Kit(Wv) mutations. *Stem Cell Reports, 7*, 425–438.

Zhang, J., Mulvenon, A., Makarov, E., Wagoner, J., Knibbe, J., Kim, J. O., Osna, N., Bronich, T. K., & Poluektova, L. Y. (2013). Antiviral peptide nanocomplexes as a potential therapeutic modality for HIV/HCV co-infection. *Biomaterials, 34*, 3846–3857.

Zhang, J., Garrison, J. C., Poluektova, L. Y., Bronich, T. K., & Osna, N. A. (2015). Liver-targeted antiviral peptide nanocomplexes as potential anti-HCV therapeutics. *Biomaterials, 70*, 37–47.

Zhang, G., Guo, D., Dash, P. K., Arainga, M., Wiederin, J. L., Haverland, N. A., Knibbe-Hollinger, J., Martinez-Skinner, A., Ciborowski, P., Goodfellow, V. S., Wysocki, T. A., Wysocki, B. J., Poluektova, L. Y., Liu, X. M., McMillan, J. M., Gorantla, S., Gelbard, H. A., & Gendelman, H. E. (2016). The mixed lineage kinase-3 inhibitor URMC-099 improves therapeutic outcomes for long-acting antiretroviral therapy. *Nanomedicine, 12*, 109–122.

Zhou, T., Lin, Z., Puligujja, P., Palandri, D., Hilaire, J., Arainga, M., Smith, N., Gautam, N., McMillan, J., Alnouti, Y., Liu, X., Edagwa, B., & Gendelman, H. E. (2018a). Optimizing the preparation and stability of decorated antiretroviral drug nanocrystals. *Nanomedicine (London, England), 13*, 871–885.

Zhou, T., Su, H., Dash, P., Lin, Z., Dyavar Shetty, B. L., Kocher, T., Szlachetka, A., Lamberty, B., Fox, H. S., Poluektova, L., Gorantla, S., McMillan, J., Gautam, N., Mosley, R. L., Alnouti, Y., Edagwa, B., & Gendelman, H. E. (2018b). Creation of a nanoformulated cabotegravir prodrug with improved antiretroviral profiles. *Biomaterials, 151*, 53–65.

Rise of Macaque Models for Immunity and Infection Research

6

Arpan Acharya and Siddappa N. Byrareddy

Abstract

Nonhuman primates (NHPs) are crucial for immunity and infectious disease research due to their genetic similarities with humans and identical disease pathogenesis and immune responses against human pathogens. Multiple species of NHPs play a critical role in vaccine research and in understanding the transmission dynamics and disease pathogenesis against several viral infections, including human immunodeficiency virus-1 (HIV-1), Zika virus, hepatitis virus, and SARS-CoV-2, among others. NHPs are also extensively used as a preclinical model in HIV-1 cure research trials for screening the therapeutic efficacy of shock-and-kill strategies, broadly neutralizing antibodies (bNAbs), therapeutic vaccines, and gene editing technologies to remove HIV proviral DNA from host chromosomes. This chapter covers all these topics with a particular focus on the utility of rhesus macaques in immunity and infection research.

Keywords

Macaque models · SHIV · Eradication · Nonhuman primates · Immune system

6.1 Introduction

Animal models are an integral part of biomedical research. They are used to understand human infectious disease pathobiology and in the development of vaccines against invading microbes and therapeutics (Barre-Sinoussi & Montagutelli, 2015). Globally, small animal models are the workhorses of immunology and infectious disease research. The advantages of working with rodents includes (1) permitting adoptive transfer of immune cells, (2) comparative ease of developing transgenic models with selective deletion/insertion of the gene(s), (3) options to control for major histocompatibility complex (MHC) types, (4) short life cycles, and (5) relatively less expense, among others (da Silva Morais et al., 2018). Despite these advantages, the innate nature of experimental rodents, the inability of the species to replicate the corresponding human disease pathogenesis, and the absence of murine homologs of human pathogens impede the translation of immunological/therapeutic discoveries from murine models into clinical settings (Bracken, 2009). Nonhuman primate (NHP) models were introduced to overcome these limitations and to better understand human disease pathology, and to develop vaccines against emerging and reemerging viral diseases. NHPs also serve as an intermediate platform for translation of antiviral therapeutics from bench to bedside (Estes et al., 2018). NHP models have several advantages over murine models when it comes to studying the immune response and disease pathogenesis against an invading pathogen in humans; these include a significantly higher level of genetic homology, physiological and behavioral attributes as well as innate and adaptive immune responses against viral infections in a similar way to humans, the ability to cross-react with numerous reagents used to study the human immune response, the presence of multiple NHP genotypes that mimic the outbred human population, and the ability to better study vaccine efficacy and therapeutic interventions against fatal infections. In addition, NHPs share many similarities in central nervous system (CNS) neuroanatomy and morphology with humans, which makes them a suitable model for studying human brain function and disease. NHPs are widely used in neurobehavioral studies to understand the neuropathogenesis of an invading pathogen and associated comorbid conditions. The similarities between the human brain and NHPs include the comparable size, organization, and structures. NHPs and humans also have similar types of neurons, synapses, and neurotransmit-

A. Acharya · S. N. Byrareddy (✉)
Department of Pharmacology and Experimental Neuroscience, College of Medicine, University of Nebraska Medical Center, Omaha, NE, USA
e-mail: arpan.acharya@unmc.edu; sid.byrareddy@unmc.edu

ters. The white matter of humans and NHPs have a lot of similarities. The dopaminergic and serotonergic neurotransmitter systems of NHPs are like those of humans. Additionally, the functional organization of the NHP brains is comparable to those of humans regarding cognitive functions, motor controls, and sensory processing. Despite these similarities, the complexities of the brain regions of NHPs are quite low compared to those of humans. Further, humans have an extended prefrontal cortex with respect to NHPs, which is responsible for executive functions like decision-making and abstract reasoning. These factors must be considered when interpreting the outcomes of the NHP study.

6.2 Different NHP Models for Human Research

6.2.1 Old World NHPs

Old World macaques, belonging to the family Cercopithecidae, are the NHPs most used for research. These include *Macaca mulatta* (rhesus macaque), *Macaca fascicularis* (cynomolgus macaque), and *Macaca nemestrina* (pigtailed macaque) (Harding, 2017). *Macaca mulatta*, the most commonly used species, has a lifespan of around 25 years in captivity; males attain sexual and social maturity at 4–6 years of age, while females do so at 3–5 years and become adults at 5–7 years of age (Lewis & Prongay, 2015). Their ovarian cycle is 29 days, with a gestation length of 165–170 days. Their natural habitat extends from Afghanistan to eastern China, which covers the Indian subcontinent. In experimental laboratory settings, they are classified as Indian-origin and Chinese rhesus macaques (Hannibal et al., 2017). These two clads significantly differ in terms of morphology, physiology, temperament, and disease susceptibility. In biomedical research, rhesus macaques are mainly used in reproductive biology, neuroscience, immunology, and infectious disease research. Specifically, simian immunodeficiency virus (SIV)-infected rhesus macaques are widely used as a model for HIV infection in humans, as discussed in detail below. In recent times, they have also been used in aging research to understand the mechanism of the human aging process. The pharmaceutical industry uses them for drug development and biodistribution studies as they closely replicate the human blood–brain barrier (BBB) physiology. Viral epitope-loaded MHC tetramers that are available for rhesus macaques help in studying the functionality of virus-specific $CD8^+$ T cells. They express heterogeneous type I interferon-induced antiviral protein (tripartite motif-containing protein 5, alpha isoform (TRIM5α)) that differentially restricts SIV (Moretti et al., 2021; Wu et al., 2015).

Macaca fascicularis (cynomolgus monkeys) are known as long-tailed or crab-eating monkeys. Their lifespan is around 31 years, with a gestation length of 165 days. They are generally found in Southeast Asia and have also been introduced to Mauritius. Males attain sexual maturity at 7 years, while females do so at 4 years of age. Due to their similarity in physiology and genetic composition with humans, they are widely used in biomedical research. The limited diversity of MHC for these macaques makes them suitable for allogeneic responses and adoptive transfer experimental studies (Marsh et al., 2012).

The *Macaca nemestrina* or pigtailed macaque is native to the Malaysian and Thai rainforests. Their average lifespan is 26 years. They have a 32-day menstrual cycle, 172 days of gestational period, and breed throughout the year. Females reach sexual maturity at 3 years, while males do so at 4.5 years. They are extensively used in reproductive research, infectious disease research (for studying influenza, chlamyda, *Mycobacterium tuberculosis* etc), stem cell research, neuroscience studies, and immunology. They express only one tripartite motif-containing protein 5α (TRIM5α) genotype that does not restrict SIV infection (Lei et al., 2013) (Baskin et al., 2004; Ericsen et al., 2016). For pigtailed macaques, virus epitope-loaded MHC tetramers are available for studying the function of $CD8^+$ T cells.

Besides these Asian macaques, many other African NHP species are used in virology and immunology studies. The two most widely used African NHPs are sooty mangabeys (*Cercocebus atys*) and African green monkeys (*Chlorocebus pygerythrus and Chlorocebus sabaeus*). They are a natural host for SIV and are therefore a suitable model for understanding the mechanism of how the virus coevolves with its hosts without causing simian AIDS-like illness (Sodora et al., 2009). Baboons (*Papio* sp.) are also used to study host–virus interactions, but their availability and more significant body size restrict their use in research settings (Locher et al., 2001).

6.2.2 Apes

Chimpanzees are one of the ideal models for studying human viral infections as they are genetically most closely related to humans. A strain of SIV (SIVcpz) through cross-species transmission from chimpanzees to humans is responsible for the HIV-1 epidemic. This is the only NHP species in which HIV-1 replicates efficiently to develop productive infection (de Groot et al., 2017). Chimpanzees are the only species of NHPs in which hepatitis C virus (HCV) replicates and is used for developing vaccines and understanding the disease pathogenesis (Lemon & Thomas, 1997). In 2015, the US Fish and Wildlife Service (FWS) declared the captive chimpanzee an endangered species, which restricts all biomedical research studies involving this species. This regulation has considerably slowed the progress of vaccine and antiviral therapeutic development studies.

6.2.3 New World NHPs

New World monkeys (Platyrrhini) are small- and medium-sized primates. They are a diverse group that is local to forest habitats in Mexico, Brazil, Peru, and other regions of Central and South America. With the availability of the genome sequence of the common marmoset (*Callithrix jacchus*), the use of New World NHPs in biomedical research is growing rapidly. Other advantages of their use include less expense, less complex husbandry, and higher evolutionary proximity to humans than rodents. Due to limited diversity within their MHC loci, their use is limited in immunological studies and in understanding the immune response against invading pathogens (Fig. 6.1).

6.3 Immune Responses in NHPs

Innate and adaptive immune responses to antigens in NHPs follow a similar pattern to humans. The natural immune response against a viral infection is generally measured by the expression profile of interferons stimulating genes which is conserved among humans and NHPs (Aid et al., 2017; Harris et al., 2010; Jacquelin et al., 2009; Robinson et al., 2012). The functionality of virus-specific B and T cells is generally measured to understand the adaptive immune response in macaques. Major histocompatibility complex I (MHC-I) of rhesus macaques is more complex than that of humans (de Groot et al., 2014). While humans have one copy of *HLA-A*, *HLA-B*, and *HLA-C* genes per haplotype, macaques may have up to three copies of *MHC-A* genes per haplotype. Similarly, macaques have a more diverse pattern of MHC-II alleles when compared to human *HLA-DR*, *HLA-DP*, and *HLA-DQ* genes (Otting et al., 2017). The traditional methods used for measuring the virus-specific response and functionality of T cells in humans, including intracellular cytokine staining and measuring their expressions by flow cytometry, are also helpful in measuring virus-specific immune responses in macaques. Specifically, NHP T cells can be stimulated with viral antigen-derived overlapping peptide pools (tetramers, hexamers, and others). Then, intracellular production of effector cytokines is measured by the enzyme-linked immunospot (ELISPOT) assay or by intracellular cytokine staining and flow cytometry (Hansen et al., 2016; Harris et al., 2010). The individual MHC–peptide complex-specific T-cell diversity is comparable between humans and macaques, making them an ideal model for studying human virus-specific immune responses.

Generally, the B-cell response against an invading pathogen is measured by purifying B cells from peripheral blood mononuclear cells (PBMCs) and then looking at their antigen production capability after stimulating them with microbial antigens using ELISPOT assays. Similarly, the antibody response of NHP-specific B cells may be estimated by ELISPOT. In line with greater MHC diversity, the immunoglobulin (Ig) loci of macaques are more diverse compared to those of humans with higher sequence diversity and copy number variations (Ramesh et al., 2017). Overall, macaques' antibodies broadly complement those of humans with minute differences in IgG subtypes (Sundling et al., 2014) and represent a good model for vaccine development and immunotherapy validation studies against human viruses.

6.4 NHP Models of HIV Infection

Human immunodeficiency virus-1 (HIV-1) is the etiological agent responsible for the acquired immunodeficiency syndrome (AIDS) pandemic (Sharp & Hahn, 2011). HIV-1 originates from a cross-species transmission of the SIVcpz virus from Central African chimpanzees (*Pan troglodytes troglodytes*) to humans (de Groot et al., 2017). As mentioned earlier, chimpanzees are the only NHPs infected with HIV-1, and, due to federal restrictions, they cannot be used in biomedical research anymore (Aguilera et al., 2021). Because of this, the known simian immunodeficiency virus (SIVmac), isolated from a captive rhesus macaque having AIDS-like illness, is used as a surrogate for HIV (Letvin et al., 1983). SIVmac originated from a cross-species transmission of SIVsmn from sooty mangabeys (the natural host of SIVsmn) to rhesus macaques (Hirsch et al., 1989). As Asian macaques are not a natural host of lentiviruses, they developed simian immunodeficiency syndrome that resembles AIDS in humans (King et al., 1983). These features make SIV-infected rhesus macaques a suitable model for understanding the HIV disease pathogenesis, vaccine development, and in HIV cure research studies.

As macaques are not susceptible to HIV infection, chimers of SIVmac and HIV-1 were developed, known as simian-human immunodeficiency virus (SHIVs). Generally, in SHIVs the SIV envelopes are replaced with HIV-1 envelopes. Limited success is achieved in maintaining the virulence and antigenicity in engineered SHIVs compared to parent HIV strains. The binding of viral envelopes with the CD4 receptor of host cells is a rate-limiting step in HIV infectivity. It was observed that the restricted binding efficiency of HIV envelops with the rhesus CD4 receptor is the underlying mechanism of the limited infectivity of SHIVs in rhesus macaques. The amino acid residue at the 375 position of the SIV envelope is critical for binding with the CD4 receptor and exerts intense selection pressure in various primate lentiviruses. Therefore, in SHIVs generated using the envelope of transmitter/founder HIV-1 subtypes, a tyrosine or histidine residue is inserted at the amino acid position 375, replacing the serine found in the HIV-1 envelop to enhance the viral envelop affinity toward the rhesus CD4 receptor (Li

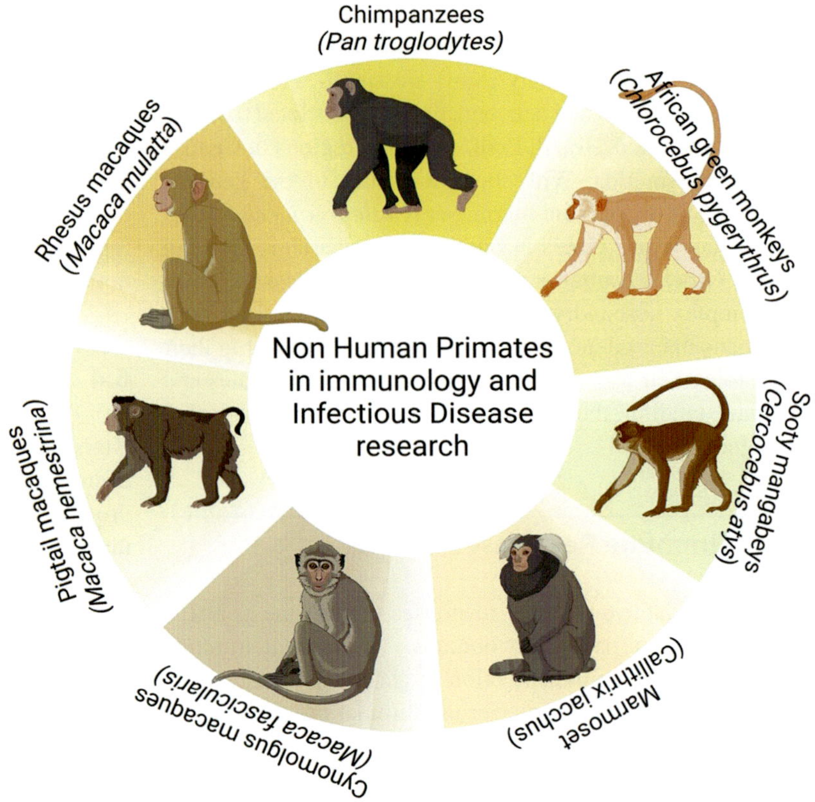

Fig. 6.1 Different Old- and New World nonhuman primates are used as a model in immunology and infectious disease research

et al., 2016). Other than this, in some chimeric SIVs, SIV's reverse transcriptase (RT) gene is replaced with HIV-1 RT gene, and is known as RSHIVmac239 (Policicchio et al., 2016).

The HIV-1 envelope must interact with the CD4 receptor and CCR5/CXCR4 co-receptors to invade the host cells. Based on their co-receptor usages, the viral strains are classified as CCR5 (R5) or CXCR4 (X4) tropic viruses or dual-tropic viruses (X4/R5) when they use both co-receptors for host cell entry. Almost exclusively, the transmitter/founder (T/F) HIV-1 strains are CCR5-tropic and usually evolve to CXCR4 or X4/R5 dual-tropic viruses during the chronic phase of the infection. Some SHIVs, like SHIV89.6P, are X4-tropic and have limited utility in HIV vaccine and cure research interventions.

Among all the clades of HIV-1, clade C viruses are the most prevalent globally, contributing to more than 50% of the total infections. They are rapidly transmissible, and animal models are required to understand their pathogenicity. Additionally, there is a need for a suitable animal model for the development and validation of a vaccine against HIV. In pursuit of this, several HIV-1 clade C T/F envelop-based SHIV clones were developed to understand these viruses' mucosal transmission and disease pathogenesis. Like SHIV-2873Nip, SHIV-1157ipd3N4, and SHIV-1157ipEL-p are developed, which are replication-competent R5-tropic SHIVs highly pathogenic, and transmitted through the mucosal route (Chenine et al., 2010; Siddappa et al., 2009; Siddappa et al., 2010) and thus have the potential to be used in HIV vaccine development studies using NHPs (Humbert et al., 2008) (Fig. 6.2).

6.4.1 NHPs in HIV Vaccine Research

Almost 40 years after the onset of the AIDS epidemic, a vaccine against HIV, the pathogenic agent responsible for the disease, remains elusive. Numerous vaccine candidates have been tested in humans with minimal success, most of which were validated in NHPs before translation into human trials. Much has been learned from the RV144 HIV vaccine trial in Thailand, one of the most prominent vaccine trials. This was the first HIV vaccine trial to demonstrate partial success in sterilizing protection against HIV transmission. In this trial, a modest 30% protection was reported. The protection was positively correlated with the functional HIV-specific antibodies generated against the V1/V2 loop of the HIV envelop protein (Haynes et al., 2012). This study also establishes the concept of protection against HIV by non-neutralizing antibodies. NHP models were proposed to understand the mechanism behind this phenomenon. In a pilot study, 11 rhesus macaques were immunized with SIV gp120 and then challenged with SIVmac251. After 5 challenges, 3 of the 11 macaques (27%) remained protected, mir-

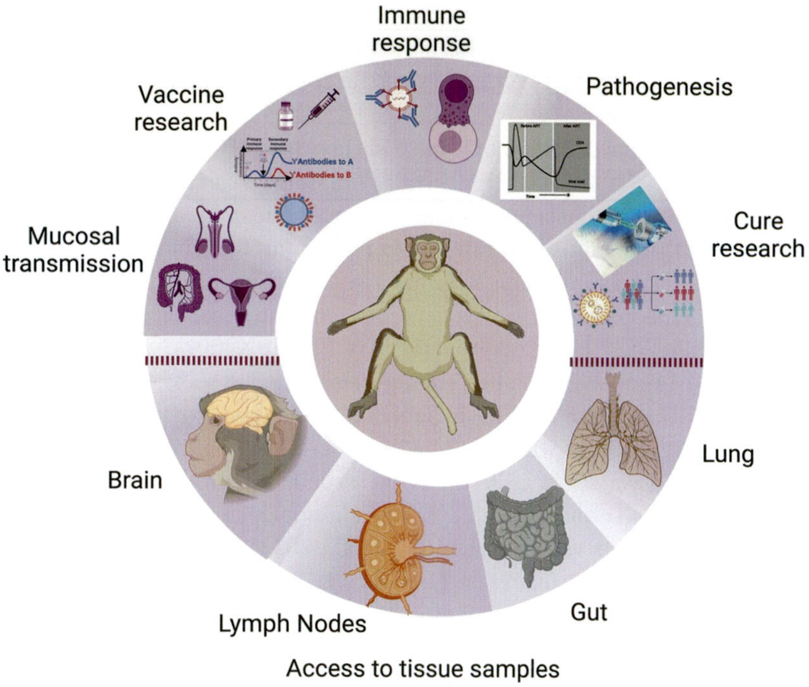

Fig. 6.2 Nonhuman primates help study various aspects of HIV infection, such as mucosal transmission, vaccine development, immune response, disease pathogenesis, and cure research. Additionally, the availability of different tissues in NHPs enables researchers to investigate tissue-level pathogenicity and potentially find cures for diseases primarily affecting immunologically restricted sites

roring the RV144 study. In functional analysis, it was found that there were no SIV-specific CD4 or CD8 T cells present, with strong avidity binding for the SIV envelop V1/V2 loop in the absence of antibody-dependent cellular cytotoxicity (ADCC) activity in the uninfected macaques (Page et al., 2012).

Although numerous vaccine development strategies have been tried, the goal of developing a vaccine with sterilizing protection against HIV remains unfulfilled. The more realistic approach is to develop a viral vaccine that may blunt systemic viremia, although it may not be able to provide complete protection. In NHP models, various immunogen, adjuvant, and prime-boost combinations were tested with variable efficacies (Virgin & Walker, 2010). Host genetics, like MHC alleles and restrictive TRIM5 alleles, add another layer of complexity to HIV vaccine studies. From these experiences, the minimal goal is to develop strategies to generate moderate cell-based immune responses in combination with modest antibody responses. The concept of protective antibody responses was developed from macaque studies incorporating passive immunization strategies (Mascola et al., 2000). It was observed that the antibodies generated exhibited strong ADCC activity. The cross-reactivity between HIV env gp41 and gut microbiota drifts away the expected immune responses from active immunization strategies (Williams et al., 2015).

During the first generation of HIV vaccine development strategies, the induction of a central memory T-cell response was significant. However, later on, in macaque studies using a rhesus cytomegalovirus vector, it was observed that the effector memory T-cell response can provide moderate protection, where half of the animals remain uninfected. In the rest, vaccine-induced immunity could suppress the plasma viremia to undetectable levels (Hansen et al., 2011). The lessons learned from these studies will help design better vaccine strategies against HIV.

From the essential knowledge gained from NHP models of HIV vaccine studies, we know that "the more, the better" is not always true for HIV vaccine-induced immune responses. Maintaining a delicate balance between the innate and adaptive immune response's quality, quantity, location, and timing is necessary for the best protection against the invading virus. As for example a robust immune activation and local inflammatory response may provide the virus with a better environment for establishing infection, replication, and systemic dissemination (Jensen et al., 2017; Qureshi et al., 2012). Human clinical trials mirrored these findings from macaque studies, for example, in the STEP study (a multicentre, double-blind, randomised, placebo-controlled, phase II test-of-concept study of the MRKAd5 HIV-1 gag/pol/nef vaccine), in which initially, even after generating a strong cell-based immune response, the vaccine failed to offer protection and, in some groups, even boosted viral replication (Buchbinder et al., 2008). In contrast, the Thai ALVAC/AIDSVAX trial (a recombinant canarypox vector vaccine (ALVAC-HIV) plus two booster injections of a recombinant HIV glycoprotein 120 subunit vaccine (AIDSVAX B/E) showed that despite a suboptimal immune response against the vaccine, it was able to protected one-fourth of the study population (Rerks-Ngarm et al., 2009). Overall, the knowledge gained from the NHP models of HIV vaccine research and their translation into human clinical trials offers insights into future vaccine development studies and serves as a launching pad for the rapid development of vaccines against SARS-CoV-2, the virus responsible for the COVID-19 pandemic.

In a significant development, the "Masico trial (HVTN 706/HPX3002)," a phase III HIV vaccine trial jointly carried out by Janssen pharmaceuticals and US government agencies (NIAID), failed to show efficacy in preventing the contraction of HIV among men who have sex with men and transgender people. Another HIV vaccine trial in African women, called the "Imbokodo trial (HVTN 705/HPX2008)" which used a similar vaccination strategy, also failed to show protection. This phase IIb trial included 2600 women from sub-Saharan Africa, of which 63 (4.32%) got infected in the placebo group versus 51 (3.61%) in the vaccinated group, scoring an overall efficacy of 25.2%, which is considered too low for an effective vaccine (Hargrave et al., 2021; Haynes et al., 2023). These setbacks in HIV vaccine development studies affirm the utility of NHP models in vaccine development, which will help study the underlying mechanism of these failures and make informed decisions for future studies.

6.4.2 The Mucosal Transmission Model

Due to their similarity to human physiology, genetic makeup, immune system development (Makori et al., 2003), and responses, NHPs are widely used for lentivirus transmission studies. To better understand the viral transmission across mucosal surfaces and the host–virus interaction during transmission and to know the timing of viral reservoirs seeding, macaques are used in intravaginal, intrarectal, and mother-to-child transmission studies of lentiviruses, including HIV (Abel, 2009; Keele et al., 2009; Miller, 1998; Miller et al., 1989).

Studies involving the vaginal transmission of SIV in macaques have informed us that the virus initially enters antigen-presenting cells, which carry them to the genital draining lymph nodes (LNs), followed by proximal lymph nodes before establishing the systemic infection (Miller, 1998). During the early days of vaginal transmissions, the effector memory T cells present in the vaginal mucosa are the primary targets of SIV. Dendritic cells and macrophages transport the virus to local lymphoid tissue (Zhang et al., 1999). It has also been revealed that at the mucosal port of entry, a solid innate inflammatory response against the virus is a prerequisite for the establishment of productive infection and this reconfirms the role of tissue-resident and trafficking $CD4^+$ T cells in the mucosal transmission of HIV (Haase, 2011; Zhang et al., 1999). These mucosal challenge studies have established a bottleneck for the virus entry. At the time of the establishment of infection, a limited number of viral particles are exposed to a small number of its target cells at the mucosal entry site. A single variant of the viral strain called the transmitted founder (T/F) virus initiates the productive infection (Miller et al., 2005; Parrish et al., 2013). These HIV mucosal transmission studies using NHP models were performed using both high- and low-dose viral stocks to mimic the HIV-1 transmission and help us understand the essential mechanisms of HIV-1 infection and the establishment of viral reservoirs during the acute phase of the disease (Liu et al., 2010; Roederer et al., 2014).

6.4.3 NHP Models in HIV Disease Pathogenesis

SIV-infected rhesus macaques do not identically mirror the disease pathogenesis of people living with HIV (PLWH). Nonetheless, they do recall several hallmarks of HIV infection that include the following: tropism for $CD4^+$ cells and the presence of CCR5/CXCR4 co-receptors for host cell entry, continuous depletion of $CD4^+$ T cells with the progression of the disease toward AIDS-like illness without antiretroviral therapy (ART), establishment of latent viral reservoirs immediately after productive infection, persistent chronic inflammatory responses in CNS and peripheral tissues, and development of neurocognitive disorders as reported in PLWH (Kumar et al., 2016). Chronic immune activation as reported in ART-treated PLWH was also observed in SIV-infected rhesus macaques. At the same time, this clinical features are absent in the sooty mangabeys and African green monkeys, the natural hosts of SIV (Deeks et al., 2004). These facts affirm the utility of differnt NHP models in comparative studies and their value in understanding the molecular pathobiology of chronic immune activation as observed in PLWH (Silvestri et al., 2007). Apart from this, gut-associated lymphoid tissue (GALT) is one of the important sites of HIV reservoirs. A rapid and profound loss of lymphocytes and damage to barrier integrity occur in GALT in the acute phase of HIV infection, which does not recover completely after initiating antiretroviral therapy (Estes, 2013). The microbial dysbiosis due to the leaky gut in PLWH is primarily responsable for the chronic immune activation observed in PLWH, as microbial products like lipopolysaccharide (LPS) and flagellin leach from the gut to the systemic circulation and activate the leucocytes (Brenchley et al., 2006). These damages in the gut and the associated microbial dysbiosis recapitulate in ART-treated SIV-infected rhesus macaques and confirm their utility in understanding the HIV pathogenesis in secondary lymphoid tissues.

Histopathological changes and lymphadenopathy are similar between PLWH and SIV-infected macaques (Chalifoux et al., 1987). Again, the collagen accumulation and associated lymphoid fibrosis observed in PLWH have similar broad lymph nodes to SIV-infected NHPs (Schacker et al., 2002). In both species, this is accompanied by overexpression of transforming growth factor-β (TGF-β) and a progressive increase in the number of regulatory T cells in the lymph nodes (Nilsson et al., 2006).

We have learned several important facts about host HIV interactions from NHP model studies and have subsequently used them in cure strategies. Like lymphoid tissues resting memory T cells are major viral reservoirs. A positive correlation exists between loss of central memory T cells and the development of AIDS-like illness, and a rapid loss of $CD4^+$ T cells and Th17 cells occurs in the gut early after infection (Brenchley et al., 2008; Li et al., 2005; Mattapallil et al., 2005; Veazey et al., 1998). NHP studies aid in understanding the mechanism of mucosal transmission of the virus. It has also been observed that after the initial encounter with the virus at the mucosal site of infection, $CD8^+$ T-cell response is delayed and insufficient to prevent the systemic dissemination of HIV (Reynolds et al., 2005). These studies focused on the early stages of HIV infection help understand how HIV breaks down the host immune defense to establish a productive infection that is impossible to perform in human clinical studies.

6.4.4 NHP Models in HIV Cure Research

With the systemic implementation of combined antiretroviral therapy (cART), long-term sustained virological suppression and reduction in immune activation are within reach of PLWH in well-developed regions like Europe and North America (Dragovic et al., 2016). However, cART cannot eradicate HIV from the body due to long-lived latent HIV reservoirs and suboptimal penetration of antiretroviral drugs (ARVs) in certain anatomical sanctuaries of HIV-1, like the brain. In patients with complete virological suppression, the viruses rebound after short- or long-term discontinuation of cART (Davey et al., 1999). This necessitates lifelong daily uptake of ARVs, and this ARV burden comes with certain caveats, including (A) pill fatigue, (B) stigma associated with antiretroviral therapy, (C) high cost of living associated with daily ART consumption, (D) ongoing and long-term drug toxicity, (E) premature aging, (F) long-term impact on neurocognitive function, and (G) emergence of resistance to the ongoing therapeutic regimen due to accrued mutations within the HIV-1 genome, which accelerate non-AIDS-related complications and pose a challenge to people wanting to live a disease progression-free life (Chawla et al., 2018). Due to these shortcomings of current therapy, novel therapeutic strategies to cure HIV-1 are urgently needed (Dinoso et al., 2009). In the absence of a broad neutralizing vaccine effective against the major subtypes of HIV-1 to prevent transmission of the virus (Haynes et al., 2016; Pegu et al., 2017), the current focus is on developing an effective cure strategy, which is classified as sterilization cure, to completely eradicate the virus from the body by gene therapy techniques or bone marrow transplantation, or to achieve a functional cure, where long-term complete suppression of viremia is achieved without any antiretroviral therapy using a combination of pharmacological agents and immunotherapies (Wu et al., 2018; Xu et al., 2017). Without any robust biomarker to assess the success of HIV-1 cure strategies, "analytical treatment interruption (ATI)" is considered as the benchmark to evaluate investigational therapeutic interventions in human clinical trials.

6.4.5 Use of Long-Acting Antiretroviral Therapy

Several long-acting formulations of currently approved antiretrovirals have been developed to overcome the limitations of the daily regimen of cART. The advantages of long-acting injectable ART include convenience, freedom from the daily pill burden, and lower toxicity. An ongoing effort is to use these long-acting formulations as a therapeutic vaccine (Soriano et al., 2022). Moreover, there is growing interest in using long-acting antiretrovirals (ARVs) in Pre-exposure prophylaxis (PrEP) (Thoueille et al., 2022). A combination formulation of long-acting cabotegravir and long-acting rilpivirine has been approved by the USFDA for PLWH, and this has been able to induce complete plasma viral suppression in people who are previously on oral daily cART (Slama et al., 2023). However, several concerns remain to be resolved in the widespread use of long-acting formulations, like the modalities of monthly/by-monthly injections, the relatively high volume of the injection, and the need for long-term monitoring of patients for possible development of drug resistance. To overcome these limitations, scientists develop once-injectable, long-acting formulations yearly and try to reduce the volume of the drug to be injected. Several preclinical developments used NHPs for pharmacokinetic/pharmacodynamic (PK/PD) studies (Deodhar et al., 2022; Kulkarni et al., 2020).

6.4.6 Risk of ATI

In the present era of HIV cure research, ATI is implemented to assess the efficacy of different therapeutic strategies to cure HIV. With the suspension of cART, there is a chance of a sudden burst of HIV plasma viral load and a rapid fall in the $CD4^+$ T-cell count. When the latter happens, the host immune system becomes weaker and the risk of acquiring opportunistic infections rises. Patients may develop "acute retroviral syndrome" with rapid viral rebound (Colven et al., 2000; Daar et al., 1998; Kilby et al., 2000; Knysz et al., 2005). The clinical symptoms related to this include febrile mononucleosis-like syndrome, fatigue, high-grade fever, enlarged peripheral lymph nodes, and others (Daar et al., 2008; Schacker et al., 1996). In some isolated cases, this may lead to multi-organ failure (Crespo et al., 2004). Most of the patients recover from the symptoms after reinitiation of ART. In some patients who underwent ATI and started to retake ART after meeting the therapy restart criteria, the plasma viral load and CD4$^+$ T-cell count did not revert to the baseline values. Moreover, there is a probability of the expansion of viral reservoirs (Montserrat et al., 2017). Contrary to that, some ongoing clinical trials involving ATI report that the expanded viral reservoirs during ATI return to their pre-ATI baseline values and diversity within a year of therapy restart (Clarridge et al., 2018; Kearney et al., 2016; Salantes et al., 2018). Minor quasi-species of HIV strains carrying drug resistance-associated mutations seeded in the viral reservoirs may be activated and proliferated during the ATI phase, which poses an additional challenge to controlling the rebounded virus pool after reintroduction of therapy (Arnedo-Valero et al., 2005; Palacios-Saucedo et al., 2011). This problem was more acute in patients on non-nucleoside reverse transcriptase inhibitor (NNRTI)-based ART before ATI was implemented (Henrich et al., 2014). Although the ATI protocol currently recommends intensive monitoring of the trial participants, there is always a chance of accruing superinfection and the emergence of recombinant viral strains in the absence of therapy, which will complicate the control of rebound viremia (Redd et al., 2013). Therapy interruption also increases the probability of HIV transmission to serodiscordant partners (Tubiana et al., 2002). Treatment interruption is also associated with chronic inflammation and immune activation (Klein et al., 2013; Tebas et al., 2008), which increases the risk of non-AIDS-related illnesses like premature aging, atherosclerosis, cardiovascular diseases, neurological disorders, and osteoporosis (Nasi et al., 2017; Sokoya et al., 2017; Thurman et al., 2020). Due to restrictions in the clinical trials of HIV cure research studies, NHP models of HIV-1 infection are widely used for preclinical validations of HIV cure strategies involving ATI. Indian-origin and Chinese rhesus macaques are the most used NHP species in HIV cure research. When infected with SIVmac239 (a molecular clone of SIV) or SIVmac251 (a swarm of SIVs), Indian-origin rhesus macaques show high peak plasma viremia that persists during chronic infection.

6.4.7 Use of NHPs in Current HIV Cure Research Approaches

6.4.7.1 Shock-and-Kill Strategies

Here, the goal is to reactivate the latent HIV reservoirs using latency-reversing agents (LRAs) and subsequently eliminate the cells expressing viral transcripts and proteins by the viral cytopathic effect or recognition and elimination by the host immune system (Gallo, 2016). Several LRAs, like histone deacetylase inhibitors (HDACs) and mitochondria-derived activator of caspases (SMAC) mimetics (AZD5582) are tested in SIV/SHIV-infected macaques to measure their efficacy in eliminating viral reservoirs with varying results (Nixon et al., 2020). Although LRAs show good latency reversal capacity in most studies, their killing effect is limited (Tanaka et al., 2022).

Several toll-like receptor (TLRs) agonists have been tested for their efficacy as LRAs. TLR7 and TLR9 agonists have been extensively studied alone or in combination with broadly neutralizing antibodies (bNAbs) as a part of the shock-and-kill strategy (Martinsen et al., 2020). TLR agonists also activate natural killer (NK) and memory B cells, potentiating antiviral immune responses (Lim et al., 2018). Along with this, using macaque models, it has been shown that immune checkpoint inhibitors like anti-cytotoxic T-lymphocyte-associated protein 4 (CTLA-4) or anti-programmed cell death protein 1 (PD-1) alone or in combination may act as LRAs (Harper et al., 2020).

6.4.7.2 Broadly Neutralizing Antibodies

Neutralizing and non-neutralizing antibody-based therapeutic strategies have been long tried to eliminate HIV-infected cells from the body with little success. With the technological advancements in single-cell antibody cloning modalities, numerous antibodies are isolated and characterized from PLWH, capable of neutralizing a wide variant of HIV-1 strains. They are called broadly neutralizing antibodies (bNAbs). Among the hundreds of bNAbs isolated and characterized, the ones with the highest breadth and the ability to neutralize maximum number of HIV-1 strains, including the variants resistant to traditional neutralizing antibodies, have been developed for use in HIV-1 prevention and cure research (Cohen & Corey, 2017). One of the limitations of using bNAbs is the development of viral quasi-species resistant to these antibodies. Similar to the combinational antiretroviral therapy approach, combinations of bNAbs are being tested to overcome this problem. SIV/SHIV-infected macaques are widely used as a preclinical model to assess the efficacy of these bNAbs (Nishimura et al., 2017).

As discussed earlier, during acute infection, a dramatic level of viral replication and depletion of CD4$^+$ T cells in gut-associated lymphoid tissue (GALT) is the hallmark of HIV infection. Long-term cART cannot ultimately rescue GALT from this initial damage. The CD4$^+$ T cells expressing the gut-homing molecule $\alpha_4\beta_7$ integrin are the principal targets for HIV/SIV to invade host cells (Cicala et al., 2009). Monoclonal antibodies against $\alpha_4\beta_7$ integrins were tested to block the trafficking of $\alpha_4\beta_7$ integrins expressing CD4$^+$ T cells to the gut to prevent gut damage and establish viral reservoirs in SIV-infected rhesus macaques and human clinical trials. Administration of monoclonal antibodies against $\alpha_4\beta_7$

integrins alone or combined with bNAbs protects against SIV transmission in repeated low-dose vaginal challenge studies in macaques (Byrareddy et al., 2014; Calenda et al., 2019; Frank et al., 2021). On the other hand, in functional cure studies, monoclonal antibodies against $\alpha_4\beta_7$ integrins delivered mixed results in delaying the rebound of plasma viremia based on the strains of SIV used in the study, viral titer, macaque genetic background, and development of anti-drug antibodies (Byrareddy et al., 2016; Iwamoto et al., 2019).

6.4.7.3 Therapeutic Vaccines

In this HIV cure strategy, vaccination is performed to induce adaptive immunity which is expected to provide a long-term delay in viral rebound after discontinuation of ART. In SIV-infected ART-treated rhesus macaques, a Sendai virus vector-based expression system, with SIV gag and vif inoculation that generates gag-specific CD8$^+$ T cells with enhanced poly-functionality without an impact on the viral rebound, describes the utility of these strategies in the generation of antiviral CD8$^+$ T-cell responses, which will be helpful in the functional cure of HIV when combined with other modalities (Nakamura-Hoshi et al., 2020). Similar strategies, including rhesus cytomegalovirus vectors expressing SIV gag, able to protect SIVmac239 infection in a subset of macaques in challenge studies, in which the efficacy of the vaccine in the immunized NHPs was tested by repeated exposure to the pathogen (Hansen et al., 2011, 2016).

6.4.7.4 Natural Killer (NK) Cell-Based Cure Strategies

NK cells play a significant role in recognizing and eliminating pathogen-infected cells. It has been shown that they also play a critical role in the depletion of HIV reservoirs in lymph nodes. Studies of SIVagm-infected with African Green Monkeys (AGMs), the natural host of the virus, facilitate the mechanism of NK cell-mediated control of SIV viremia. In AGMs, high CXCR5$^+$ NK cells in the lymph node follicles provide viral control. In contrast to the previous observations, NK cells are not detected in the lymph node follicles of SIV-infected rhesus macaques. A significant positive correlation has been observed between NK cell density and production of interleukin (IL)-15 in LN follicles. Therefore, a new strategy was formulated as part of HIV cure research, where IL-15 is administered to facilitate the trafficking of NK cells to LNs to achieve a functional cure for HIV. IL-15 and its super agonist N-803 were tested in SIV/SHIV-infected ART-treated macaques to test this hypothesis. Unfortunately, these molecules failed to reactivate viral reservoirs (Webb et al., 2020). N-803, when combined with CD8β depletion, can partially reactivate the latent reservoirs (McBrien et al., 2020).

6.4.7.5 Gene Editing/Therapy-Based HIV Cure Strategies

In the past couple of years, gene editing-based HIV cure research has gained momentum with the discovery of CRISPR-Cas9-based gene editing technology. As a part of cure research using CRISPR/Cas9, scientists are trying to excise either the HIV proviral genome from the host chromosome or the CCR5 co-receptor of HIV. One study involving SIV-infected macaques and CRISPR/Cas9 based gene editing strategy reports partial success in excising the proviral genome and eliminating SIV reservoirss (Mancuso et al., 2020). Chimeric antigen receptors (CARs) are engineered receptors for antigens designed to redirect the specificity and function of leucocytes. Recent advances in chimeric antigen receptor (CAR) T-cell-based therapy by genetically engineering autologous patient T cells for hematological malignancies have shown promising outcomes (Rust et al., 2020). This success has ignited interest in developing a CAR-based therapeutic approach for a functional cure of HIV that works independent of major histocompatibility complex (MHC) involvement. The majority of CAR-based therapeutic approaches target a conserved region of the HIV envelop glycoprotein that is protected from escape mutations. Another primary requirement of CAR-based therapy for HIV is to target significant viral reservoirs like CXCR5 expressing T follicular helper (Tfh) cells in lymph nodes (Haran et al., 2018). To this end, several CAR-based approaches have been developed and tested in SIV/SHIV-infected rhesus macaques for preclinical validation before translation into human trials. In one study, hematopoietic stem cell (HSC)-derived CD4CAR$^+$ cells transduced in SHIV-infected ART-treated macaques demonstrated these cells distribution and longtime persistence at all significant sites of HIV reservoirs (Barber-Axthelm et al., 2021). In another study, autologous CAR/CXCR5 T cells were transfused in SIV-infected ART-treated rhesus macaques. After ATI, the CAR T-cell-treated animals maintained a lower plasma viral load and lymph node follicular viral RNA level vs. controls without any observable adverse events, indicating the potential of these therapies in HIV cure research (Pampusch et al., 2022).

6.5 NHPs in Hepatitis Virus Research

There are five viruses responsible for viral hepatitis, namely, the hepatitis A, B, C, D, and E viruses. Hepatitis A and E cause acute infection, while B, C, and D cause acute and chronic liver infection. Although several different therapeutic options are now available for the treatment of hepatitis C virus, which includes direct-acting antivirals as well as interferon-based therapies, there is no vaccine available for the prevention of the spread of the virus (Lawitz et al., 2014). One of the primary bottlenecks of a comprehensive HCV vaccine research program is the unavailability of a rodent model for HCV infection. Chimpanzees are the only captive NHP species that can be infected with HCV. In the USA, they are no longer used for biomedical research due to federal restrictions (Bukh, 2004). In the past, chimpanzees played a vital role in discovering HCV and understanding the immune response against it during the acute and chronic phases of the infection. One of the limitations of the use of chimpanzees in HCV research was the difference in the disease pathogenesis between humans and chimpanzees. In chimpanzees, HCV infection resulted in milder hepatitis with no development of end-stage liver diseases, which is frequently observed in humans. One of the contributing factors to this may be the absence of other comorbid conditions like obesity, alcohol consumption, and others in chimpanzees. Occasionally, GB virus B (GBV-B)-infected marmosets are used as a surrogate host for HCV infection. Even though they recapitulate the HCV pathology, a chimeric version of GBV-B/HCV has been constructed for therapeutic development research against HCV (Manickam & Reeves, 2014).

Hepatitis E virus can infect different NHP species, like chimpanzees, marmosets, rhesus macaques, and cynomolgus macaques. Contrary to the direct impact of the hepatitis E virus (HEV) invasion, the damage observed in the liver is mainly caused by the host's immune response against the virus. Similar histopathological changes and inflammatory milieu in liver Kupffer cells have been reported in the liver of human patients and NHPs after HEV infection. Cynomolgus macaques are extensively used to develop vaccines against HEV; some are at different stages of clinical trials and waiting for final approval from the US FDA. One of the vaccines (HEV239) based on the ORF2 capsid protein of HEV was approved in China in 2012 (Li et al., 2020; Shrestha et al., 2007).

6.6 NHPs in Zika Virus Research

Zika virus infection in humans was first reported in 1952 in the Zika forest; hence the name. The Zika virus belongs to arboviruses and is transmitted through the Aedes mosquito (Rawal et al., 2016). Globally, two dominant strains of Zika persist. The African strain maintains its life cycle in macaques and apes with occasional cross-species transmissions to humans. For the Asian strain of Zika, humans are the dominant hosts (Haddow et al., 2012). Generally, Zika infection is not fatal, but some individuals develop neurological symptoms involving the peripheral and central nervous systems. Zika virus-infected macaque models are extensively used to understand the disease pathogenesis. Subcutaneous injection of the virus mimics the viral transmission through a mosquito bite. After the initial infection, macaques become viremic within 2–6 days and clear the systemic disease by 10 days. Zika RNA is detected in most body fluids and tissue organs, including the plasma, cerebrospinal fluid (CSF), urine, vaginal fluid, semen, lymph nodes, gastrointestinal (GI) tract, and brain (Dudley et al., 2016; Li et al., 2016). Zika-infected NHPs develop robust humoral and cellular immunity against subsequent homologous and heterologous challenges, which affirms the use of NHPs in developing vaccines against Zika (Dowd et al., 2016; Larocca et al., 2016). Two major concerns of Zika infection are its ability to cross the placenta and impact the neurological development of the growing fetus and how immunocompromised persons like PLWH respond to it. To understand the first concern, NHP models were developed. The fetus of pregnant macaques with Zika infection demonstrated more prolonged viremia and severe developmental defects, including delayed brain development, neuronal death, shrinkage in white matter volume, and loss of glial cells, all of which mirror the developmental defects reported in infants born to mothers infected with Zika virus during their pregnancy (Driggers et al., 2016; Koide et al., 2016). Similarly, NHP models of ZIKAV/HIV-1 co-infection were developed to understand the immune response and host–pathogen interaction during co-infection (Bidokhti et al., 2018).

6.7 NHP Models of SARS-CoV-2

At the onset of the COVID-19 pandemic, a suitable animal model was crucial to (a) understand the disease pathogenesis and (b) evaluate vaccines and develop antivirals/other therapeutics (Munoz-Fontela et al., 2020; Pandey et al., 2021). During earlier outbreaks of coronaviruses like SARS-CoV and MERS-CoV, several NHP species were used to understand the disease pathogenesis, especially with Old World NHPs like rhesus macaques, cynomolgus macaques, and African green monkeys as well as New World NHPs like marmosets, which demonstrated differential pathogenesis toward the coronaviruses (Rowe et al., 2004; Yao et al., 2014). Due to the close similarity between the angiotensin I-converting enzyme 2 (ACE-2) receptors of humans and NHPs, the primary receptors for SARS-CoV-2 to enter the host cells, NHPs serve as a good model of COVID-19 research (Pandey et al., 2021). SARS-CoV-2 also has a wide variation in the disease pathogenesis among different NHP species. SARS-CoV-2-infected rhesus and cynomolgus macaques recapitulate the histopathological changes, immune responses, and other clinical features observed in humans. They are widely used in vaccine studies and preclinical validation of therapeutics (Rockx et al., 2020; Singh et al., 2021). However, the COVID-19 disease pathogenesis in macaques is limited to mild and moderate cases observed in humans. After the initial transmission of the virus and establishment of productive infection in the lungs, SARS-CoV-2 can disseminate to multiple organs, primarily the GI track, liver, spleen, lymph nodes, kidney, and genital tracks (Salguero et al., 2021). New World NHP marmosets have limited susceptibility to SARS-CoV-2 compared to macaques, probably due to sequence mismatch with humans ACE-2 receptors at the positions where the receptor-binding domain of SARS-CoV-2 interacts with ACE-2 (Lu et al., 2020). It has also been observed that baboons and aged African green monkeys better represent the severe respiratory disease pathology and the cardiovascular and metabolic disorders marked in humans (Blair et al., 2021; Singh et al., 2021). Due to the similarity of their immune system with humans, NHPs are ideal for studying the humoral and cellular immune responses against SARS-CoV-2. In acute infection, infiltration of monocytes, neutrophils, NK cells, and dendritic cells has been observed in the lungs (Song et al., 2020). They release the pro-inflammatory cytokines responsible for inflammation in lung tissue. This is followed by severe lymphopenia in the blood. The B-cell count also drops, and it takes several weeks to recover to the pre-infection level. Like humans, in the plasma of SARS-CoV-2-infected macaques, many cytokine/chemokine levels become elevated, including IL-10, IL-1Ra, IL-4, IL-6, IL-15, and TNF-α), among others. In NHPs, it is also possible to monitor the level of these pro-inflammatory cytokines/chemokines within the lung at the local site of infection (Song et al., 2020).

Due to their close phylogenetic proximity to humans, NHPs are widely used in pharmacokinetics (PK) and pharmacodynamics (PD) studies and for preclinical development of small molecules and antibody-based therapeutics against SARS-CoV-2. For example, remdesivir is the most widely used antiviral for COVID-19 therapy. Its PK/PD study and preclinical validation were conducted in NHPs (Williamson et al., 2020). Similarly, preclinical validation studies of several monoclonal antibodies were performed in NHPs to block the interaction between the receptor-binding domain of SARS-CoV-2 and the host cells ACE-2 receptor (Jones et al., 2021). Several immune-modulatory therapeutic approaches were tested to document the cytokine storm observed in COVID-19 patients, like anti-IL-6 monoclonal antibody-based therapy and baricitinib (Janus kinase (JAK)1/2 inhibitor), having a strong anti-inflammatory effect. The macaques infected with SARS-CoV-2 and treated with baricitinib reduced the immune activation and blunted the production of pro-inflammatory cytokines/chemokines by alveolar macrophages without impacting the viral replication kinetics (Hoang et al., 2021). Now, baricitinib has been approved by the US FDA for the treatment of COVID-19 when used in combination with remdesivir. Building durable protective immunity against SARS-CoV-2 is the key to stopping the COVID-19 pandemic. It was observed that the protection developed after the viral infection and initial doses of vaccines were not long-lasting (Deng et al., 2020). The main reason for such short-term protection is the accumulation of mutations in the spike protein of the virus, which, with time, gives rise to different variants—these result in breakthrough infections in vaccinated people and reinfection in previously infected patients. NHPs, particularly rhesus macaques, were extensively used to understand the immune responses against SARS-CoV-2 and for developing preclinical evaluations of vaccines. Different NHP species were also used to assess the magnitude of the humoral response against SARS-CoV-2. In NHP, virus-specific antibodies that include IgM, IgG, and IgA are raised within 7–10 days of infection. A class switch between IgM and IgG is also observed 7/14 days post-infection (McMahan et al., 2021). IgGs purified from the plasma of convalescent rhesus macaques, when transferred to naive animals, show protection against SARS-CoV-2 in a dose-dependent manner, which affirms the role of neutralizing antibodies in the development of defense against invading viruses (Wang et al., 2020).

To reduce the devastating impact of the COVID-19 pandemic, a vaccine against SARS-CoV-2 was developed with electrifying speed and effort. Different species of NHPs, including rhesus macaques, cynomolgus macaques, and baboons, played a crucial role in these vaccine development and preclinical validation process. These studies helped to understand the different aspects of the vaccines, like the ideal delivery route, doses required for optimal response, antibody titers after vaccinations and their neutralization potential, and T-cell responses against the vaccine before translating into human clinical trials. Several different platforms were used to develop vaccines against SARS-CoV-2, including inactivated viral vaccines, nonreplicating viral vector-based vaccines, mRNA-based vaccines, protein subunit-based vaccines, and DNA-based vaccines. Various modalities were implemented in the preclinical validation of these vaccines using NHPs. The variables include the dose and route of inoculation, time of vaccination to challenge studies, the titer of the virus used in challenge studies, and techniques used to evaluate the neutralizing antibodies and T-cell responses (Klasse et al., 2021). Despite the variation in the vaccine development platforms and failure to generate sterilizing protection against SARS-CoV-2, the preclinical studies in NHPs indicate well-developed protection of the lower lung airways, dramatically lowering the disease severity and morbidity and mortality associated with COVID-19. These studies also confirm that vaccination's oral and intranasal routes develop comparable immune responses with subcutaneous/intramuscular inoculations (Albrecht et al., 2021).

6.8 The NHP Model of HIV and Substance Use Disorder Comorbidity

More than 50% of people living with HIV develop HIV-associated neurocognitive disorder (HAND). The symptoms of HAND include attention disorders, loss of memory, lack of motivation, and chronic depression. The current cART is not able to resolve HAND. Some antiretrovirals cannot penetrate the blood–brain barrier, are less effective in controlling viral replication in the CNS, and are known to aggravate the disease pathogenesis. The underlying mechanism of the development of HAND largely remains unknown due to the limited accessibility to the human CNS in longitudinal studies. PLWH are at a high risk of abusing substances compared to general populations. Most of these substances include drugs like opioids, methamphetamine, and cocaine. Cannabis is neuropathogenic and poses a significant comorbid condition for HAND. In human clinical cohorts, it is challenging to study the impact and the associated mechanism of the neuropathogenesis of these individual drugs due to polysubstance abuse, where most of the PLWH used multiple illicit substances. To overcome these issues, SIV-infected cART-treated NHPs are used as a model for studying the neuropathogenesis of substance use disorders among PLWH. These studies also have implications for HIV cure research, as it has been reported that abused substances negatively impact viral reservoirs in the brain (Acharya et al., 2021). This study says that chronic morphine use differentially regulates the SIV reservoirs in the brain and peripheral organs. In PBMCs and lymph nodes, morphine reduces the viral reservoir size, while in the brain, it potentiates the SIV reservoirs. It is also observed that morphine administration to SHIV-infected macaques promotes chronic inflammation in the brain by shifting the polarization of microglia/macrophages (Olwenyi et al., 2022). By single-cell gene expression analysis of brain-specific microglia/macrophages, morphine alters the gene expression pattern toward a neurodegenerative disease phenotype in SIV-infected ART-treated rhesus macaques (Fox et al., 2022). In the lungs, morphine aggravates SIV-induced fibrosis through induction of oxidative stress, TGF-β expression, and matrix degradation (Chemparathy et al., 2023).

6.9 NHP Models in Understanding the Role of the Microbiome in Human Health and Disease

The microbiome is a collection of microorganisms, including bacteria, viruses, fungi, and others, which reside in various parts of the human body. It is a diverse and unique ecosystem that varies among individuals and different body sites (Gilbert et al., 2018). The structure and composition of the microbiome also differ across species. The gut microbiome plays a vital role in digestion, nutrient absorption, and metabolism. Microbial metabolites, such as vitamins K and B, significantly impact human health. The breakdown of dietary fibers by gut microbes produces beneficial compounds called short-chain fatty acids (SCFAs), which contribute to overall host health (Rowland et al., 2018). Moreover, the gut microbiome plays a crucial role in the development and function of the immune system and in regulating inflammation throughout the body (Al Bander et al., 2020). Emerging research suggests that the gut microbiome significantly influences central nervous system (CNS) functions and maintains mental health conditions like anxiety and depression through the gut–brain axis (Diaz Heijtz et al., 2011). Microbial metabolites produced in the gut have been found to have essential implications in neurodegenerative disorders such as Alzheimer's and Parkinson's diseases (Szandruk-Bender et al., 2022). Dysbiosis, an imbalance in the gut microbiome, has been associated with various conditions, including inflammatory bowel disease (IBD), obesity, and allergies (Belizario & Faintuch, 2018). The vaginal microbiome plays a critical role in women's reproductive health by protecting against infections and maintaining a healthy pH. An imbalanced composition can lead to bacterial vaginosis and yeast infections (White et al., 2011). Similarly, the oral microbiome influences dental health and is involved in tooth decay, gum disease, and bad breath (Wade, 2013). Furthermore, the microbiome has been implicated in cancer development and can influence chemotherapy outcomes. It also plays a role in developing autoimmune diseases such as rheumatoid arthritis and multiple sclerosis, among many others. The microbiome has far-reaching effects on various aspects of human health and disease (Ogunrinola et al., 2020).

Rodent models are commonly used in scientific research to study the mechanisms of microbiome-associated diseases (Nguyen et al., 2015). However, it is crucial to acknowledge the significant variations in the specific microbial species and their relative abundance in humans and rodents (Krych et al., 2013). One key factor influencing these differences is the distinct dietary preferences of rodents and humans, which, in turn, affect their gut microbiomes. Additionally, the anatomical structure of the gut differs between humans and rodents, potentially influencing microbial colonization patterns and the interactions between the host and microbiota. Humans and rodents have evolved in separate environments, leading to differences in the coevolution of their respective host–microbiota relationships (Messer et al., 2017). The microbial communities in humans and rodents have adapted to their specific hosts over time, contributing to the observed dissimilarities. Genetic variations between humans and rodents also shape the composition and function of their respective microbiomes (Safari et al., 2020). Furthermore, environmental factors, such as living conditions, hygiene practices, and microbial exposures, differ between humans and rodents, and these factors can impact the acquisition and establishment of microbial communities. Therefore, it is essential to consider these variations when extrapolating findings from rodent studies to human health and disease (Hugenholtz & de Vos, 2018).

Contrary to rodents, NHPs exhibit several similarities in their microbial structure and composition to humans (Manara et al., 2019). These similarities include similar taxonomic groups such as bacteria, archaea, viruses, and fungi within their microbial communities (Nagpal et al., 2018). There is an overlap in core microbial species between humans and NHPs, indicating shared functional roles and potential coevolution. Both humans and NHPs have a dominant gut microbiota that significantly contributes to their overall microbial composition. Similar dietary patterns, particularly the consumption of plant-based foods, can result in comparable microbial communities in the gut (Mallott & Amato, 2022). The interaction between the host immune system and the gut microbiota also shows standard features, including immunomodulation and metabolic interactions. The early-life development of the microbiota in humans and NHPs undergoes similar changes influenced by host genetics and environmental factors. NHPs, like humans, live in social groups and display complex social behaviors, which can influence microbial transmission and community dynamics. Additionally, like humans, NHPs can experience dysbiosis, characterized by imbalances in the microbial composition, leading to potential health conditions (Brenchley & Ortiz, 2021). Considering these similarities, NHPs serve as valuable animal models for studying the impact of the gut microbiome on human health and diseases, particularly for understanding the role of the gut–brain axis in developing neurological disorders.

6.10 Conclusions

Nonhuman primate (NHP) models, particularly Old World NHPs like rhesus macaques, have proven invaluable in biomedical research due to their genetic homology, physiological similarities, and human immune response parallels. NHP models offer unique advantages for studying infectious diseases, vaccine development, therapeutic interventions, and HIV cure research, bridging bench discoveries and clinical applications. NHP models have played a critical role in advancing our understanding of HIV infection. By infecting rhesus macaques with simian immunodeficiency virus (SIV), researchers have created an effective model for studying HIV infection. To make the SIV infection model more relevant to HIV, chimeric simian–human immunodeficiency viruses (SHIVs) have been developed. SHIVs are created by replacing the SIV envelope with the envelope of HIV-1, the strain responsible for the AIDS pandemic. These modified viruses better mimic HIV infection in humans. NHP models have played a crucial role in HIV vaccine research. NHP studies have demonstrated the importance of both cellular and antibody responses in protecting against HIV. Lessons learned from NHP models have guided the development of strategies for generating moderate cell-based immune responses and antibody responses in HIV vaccine research. Mucosal transmission studies using NHP models have provided valuable insights into the early events of HIV infection. NHPs have been used to study viral transmission across mucosal surfaces, including intravaginal, intrarectal, and mother-to-child transmission. These studies have revealed the role of different cells and tissues in the initial stages of infection and the establishment of viral reservoirs. NHP models have also contributed to understanding HIV disease pathogenesis, including $CD4^+$ T-cell depletion, viral reservoirs, chronic immune activation, and neurocognitive disorders. Moreover, NHP models have played a vital role in studying the comorbidity of HIV and substance use disorder, particularly investigating HIV-associated neurocognitive disorder (HAND) and the impact of substance abuse on neuropathogenesis.

In addition to HIV research, NHP models have been extensively used in various other fields. They have contributed significantly to viral hepatitis research, including the study of hepatitis C virus (HCV) using GB virus B (GBV-B)-infected marmosets and chimeric versions of GBV-B/HCV. NHP models are also essential in Zika virus research, providing insights into disease pathogenesis, viral transmission, and the impact on fetal development. For SARS-CoV-2, NHPs have been crucial in understanding the disease pathogenesis, evaluating vaccines and therapeutics, and studying related respiratory and cardiovascular disorders. NHP models have also contributed to understanding the role of the microbiome in human health and disease, providing valuable insights into the complex interactions between the microbiome and various conditions, such as obesity, gastrointestinal diseases, and mental health disorders. Overall, NHP models have significantly advanced our knowledge in multiple areas of biomedical research. Their genetic homology, physiological similarities, and immune response parallels with humans make them invaluable resources for understanding immunity, infection, and disease research.

6.11 Review Questions

1. Which one is a New World nonhuman primate?
   (a) A pigtailed macaque
   (b) A marmoset
   (c) A cynomolgus macaque
   (d) A rhesus macaque

2. What cell surface co-receptor is required for host cell entry of HIV?
   (a) CXCR5
   (b) CD8
   (c) CCR5
   (d) CD16
3. Which one is not a neurotropic virus?
   (a) HIV
   (b) Zika
   (c) SARS-CoV-2
   (d) HBV
4. Which one acts as a restriction factor against SIV?
   (a) TRIM5α
   (b) IL-2
   (c) IL-4
   (d) IL-10
5. In which NHP species, can HIV replicate naturally?
   (a) Pigtailed macaques
   (b) Marmosets
   (c) Chimpanzees
   (d) Rhesus macaques

6.12 Answers

1. b
2. c
3. d
4. a
5. c

References

Abel, K. (2009). The rhesus macaque pediatric SIV infection model – a valuable tool in understanding infant HIV-1 pathogenesis and for designing pediatric HIV-1 prevention strategies. *Current HIV Research, 7*, 2–11.

Acharya, A., Olwenyi, O. A., Thurman, M., Pandey, K., Morsey, B. M., Lamberty, B., Ferguson, N., Callen, S., Fang, Q., Buch, S. J., Fox, H. S., & Byrareddy, S. N. (2021). Chronic morphine administration differentially modulates viral reservoirs in SIVmac251 infected rhesus macaque model. *Journal of Virology, 95*.

Aguilera, B., Perez Gomez, J., & DeGrazia, D. (2021). Should biomedical research with great apes be restricted? A systematic review of reasons. *BMC Medical Ethics, 22*, 15.

Aid, M., Abbink, P., Larocca, R. A., Boyd, M., Nityanandam, R., Nanayakkara, O., Martinot, A. J., Moseley, E. T., Blass, E., Borducchi, E. N., Chandrashekar, A., Brinkman, A. L., Molloy, K., Jetton, D., Tartaglia, L. J., Liu, J., Best, K., Perelson, A. S., De La Barrera, R. A., et al. (2017). Zika virus persistence in the central nervous system and lymph nodes of rhesus monkeys. *Cell, 169*, 610–620 e614.

Al Bander, Z., Nitert, M. D., Mousa, A., & Naderpoor, N. (2020). The gut microbiota and inflammation: An overview. *International Journal of Environmental Research and Public Health, 17*, 7618.

Albrecht, L., Bishop, E., Jay, B., Lafoux, B., Minoves, M., & Passaes, C. (2021). COVID-19 research: Lessons from non-human primate models. *Vaccines (Basel), 9*, 886.

Arnedo-Valero, M., Garcia, F., Gil, C., Guila, T., Fumero, E., Castro, P., Blanco, J. L., Miro, J. M., Pumarola, T., & Gatell, J. M. (2005). Risk of selecting de novo drug-resistance mutations during structured treatment interruptions in patients with chronic HIV infection. *Clinical Infectious Diseases, 41*, 883–890.

Barber-Axthelm, I. M., Barber-Axthelm, V., Sze, K. Y, Zhen, A, Suryawanshi, G. W., Chen, I. S., Zack, J. A., Kitchen, S. G., Kiem, H. P., & Peterson, C. W. (2021) Stem cell-derived CAR T cells traffic to HIV reservoirs in macaques. *JCI Insight, 6*(1), 141502. https://doi.org/10.1172/jci.insight.141502. PubMed PMID: 33427210; PMCID: PMC7821595.

Barre-Sinoussi, F., & Montagutelli, X. (2015). Animal models are essential to biological research: Issues and perspectives. *Future Science OA, 1*, FSO63.

Baskin, C. R., Garcia-Sastre, A., Tumpey, T. M., Bielefeldt-Ohmann, H., Carter, V. S., Nistal-Villan, E., & Katze, M. G. (2004). Integration of clinical data, pathology, and cDNA microarrays in influenza virus-infected pigtailed macaques (Macaca nemestrina). *Journal of Virology, 78*, 10420–10432.

Belizario, J. E., & Faintuch, J. (2018). Microbiome and gut dysbiosis. *Experientia. Supplementum, 109*, 459–476.

Bidokhti, M. R. M., Dutta, D., Madduri, L. S. V., Woollard, S. M., Norgren, R., Jr., Giavedoni, L., & Byrareddy, S. N. (2018). SIV/SHIV-Zika co-infection does not alter disease pathogenesis in adult non-pregnant rhesus macaque model. *PLoS Neglected Tropical Diseases, 12*, e0006811.

Blair, R. V., Vaccari, M., Doyle-Meyers, L. A., Roy, C. J., Russell-Lodrigue, K., Fahlberg, M., Monjure, C. J., Beddingfield, B., Plante, K. S., Plante, J. A., Weaver, S. C., Qin, X., Midkiff, C. C., Lehmicke, G., Golden, N., Threeton, B., Penney, T., Allers, C., Barnes, M. B., et al. (2021). Acute respiratory distress in aged, SARS-CoV-2-infected African Green Monkeys but not rhesus macaques. *The American Journal of Pathology, 191*, 274–282.

Bracken, M. B. (2009). Why animal studies are often poor predictors of human reactions to exposure. *Journal of the Royal Society of Medicine, 102*, 120–122.

Brenchley, J. M., & Ortiz, A. M. (2021). Microbiome studies in non-human primates. *Current HIV/AIDS Reports, 18*, 527–537.

Brenchley, J. M., Price, D. A., Schacker, T. W., Asher, T. E., Silvestri, G., Rao, S., Kazzaz, Z., Bornstein, E., Lambotte, O., Altmann, D., Blazar, B. R., Rodriguez, B., Teixeira-Johnson, L., Landay, A., Martin, J. N., Hecht, F. M., Picker, L. J., Lederman, M. M., Deeks, S. G., & Douek, D. C. (2006). Microbial translocation is a cause of systemic immune activation in chronic HIV infection. *Nature Medicine, 12*, 1365–1371.

Brenchley, J. M., Paiardini, M., Knox, K. S., Asher, A. I., Cervasi, B., Asher, T. E., Scheinberg, P., Price, D. A., Hage, C. A., Kholi, L. M., Khoruts, A., Frank, I., Else, J., Schacker, T., Silvestri, G., & Douek, D. C. (2008). Differential Th17 CD4 T-cell depletion in pathogenic and nonpathogenic lentiviral infections. *Blood, 112*, 2826–2835.

Buchbinder, S. P., Mehrotra, D. V., Duerr, A., Fitzgerald, D. W., Mogg, R., Li, D., Gilbert, P. B., Lama, J. R., Marmor, M., Del Rio, C., McElrath, M. J., Casimiro, D. R., Gottesdiener, K. M., Chodakewitz, J. A., Corey, L., Robertson, M. N., & Step Study Protocol, T. (2008). Efficacy assessment of a cell-mediated immunity HIV-1 vaccine (the step study): A double-blind, randomised, placebo-controlled, test-of-concept trial. *Lancet, 372*, 1881–1893.

Bukh, J. (2004). A critical role for the chimpanzee model in the study of hepatitis C. *Hepatology, 39*, 1469–1475.

Byrareddy, S. N., Kallam, B., Arthos, J., Cicala, C., Nawaz, F., Hiatt, J., Kersh, E. N., McNicholl, J. M., Hanson, D., Reimann, K. A., Brameier, M., Walter, L., Rogers, K., Mayne, A. E., Dunbar, P., Villinger, T., Little, D., Parslow, T. G., Santangelo, P. J., Villinger,

F., et al. (2014). Targeting alpha4beta7 integrin reduces mucosal transmission of simian immunodeficiency virus and protects gut-associated lymphoid tissue from infection. *Nature Medicine, 20*, 1397–1400.

Byrareddy, S. N., Arthos, J., Cicala, C., Villinger, F., Ortiz, K. T., Little, D., Sidell, N., Kane, M. A., Yu, J., Jones, J. W., Santangelo, P. J., Zurla, C., McKinnon, L. R., Arnold, K. B., Woody, C. E., Walter, L., Roos, C., Noll, A., Van Ryk, D., et al. (2016). Sustained virologic control in SIV+ macaques after antiretroviral and alpha4beta7 antibody therapy. *Science, 354*, 197–202.

Calenda, G., Frank, I., Arrode-Bruses, G., Pegu, A., Wang, K., Arthos, J., Cicala, C., Rogers, K. A., Shirreff, L., Grasperge, B., Blanchard, J. L., Maldonado, S., Roberts, K., Gettie, A., Villinger, F., Fauci, A. S., Mascola, J. R., & Martinelli, E. (2019). Delayed vaginal SHIV infection in VRC01 and anti-alpha4beta7 treated rhesus macaques. *PLoS Pathogens, 15*, e1007776.

Chalifoux, L. V., Ringler, D. J., King, N. W., Sehgal, P. K., Desrosiers, R. C., Daniel, M. D., & Letvin, N. L. (1987). Lymphadenopathy in macaques experimentally infected with the simian immunodeficiency virus (SIV). *The American Journal of Pathology, 128*, 104–110.

Chawla, A., Wang, C., Patton, C., Murray, M., Punekar, Y., de Ruiter, A., & Steinhart, C. (2018). A review of Long-term toxicity of antiretroviral treatment regimens and implications for an aging population. *Infectious Disease and Therapy, 7*, 183–195.

Chemparathy, D. T., Sil, S., Callen, S., Chand, H. S., Sopori, M., Wyatt, T. A., Acharya, A., Byrareddy, S. N., Fox, H. S., & Buch, S. (2023). Inflammation-associated lung tissue remodeling and fibrosis in morphine-dependent SIV-infected macaques. *The American Journal of Pathology, 193*, 380–391.

Chenine, A. L., Siddappa, N. B., Kramer, V. G., Sciaranghella, G., Rasmussen, R. A., Lee, S. J., Santosuosso, M., Poznansky, M. C., Velu, V., Amara, R. R., Souder, C., Anderson, D. C., Villinger, F., Else, J. G., Novembre, F. J., Strobert, E., O'Neil, S. P., Secor, W. E., & Ruprecht, R. M. (2010). Relative transmissibility of an R5 clade C simian-human immunodeficiency virus across different mucosae in macaques parallels the relative risks of sexual HIV-1 transmission in humans via different routes. *The Journal of Infectious Diseases, 201*, 1155–1163.

Cicala, C., Martinelli, E., McNally, J. P., Goode, D. J., Gopaul, R., Hiatt, J., Jelicic, K., Kottilil, S., Macleod, K., O'Shea, A., Patel, N., Van Ryk, D., Wei, D., Pascuccio, M., Yi, L., McKinnon, L., Izulla, P., Kimani, J., Kaul, R., et al. (2009). The integrin alpha4beta7 forms a complex with cell-surface CD4 and defines a T-cell subset that is highly susceptible to infection by HIV-1. *Proceedings of the National Academy of Sciences of the United States of America, 106*, 20877–20882.

Clarridge, K. E., Blazkova, J., Einkauf, K., Petrone, M., Refsland, E. W., Justement, J. S., Shi, V., Huiting, E. D., Seamon, C. A., Lee, G. Q., Yu, X. G., Moir, S., Sneller, M. C., Lichterfeld, M., & Chun, T. W. (2018). Effect of analytical treatment interruption and reinitiation of antiretroviral therapy on HIV reservoirs and immunologic parameters in infected individuals. *PLoS Pathogens, 14*, e1006792.

Cohen, M. S., & Corey, L. (2017). Broadly neutralizing antibodies to prevent HIV-1. *Science, 358*, 46–47.

Colven, R., Harrington, R. D., Spach, D. H., Cohen, C. J., & Hooton, T. M. (2000). Retroviral rebound syndrome after cessation of suppressive antiretroviral therapy in three patients with chronic HIV infection. *Annals of Internal Medicine, 133*, 430–434.

Crespo, M., Paradineiro, J. C., Ribera, E., Ruiz, I., Falco, V., Lopez-Quinones, J., Ocana, I., & Pahissa, A. (2004). A case of multiorgan failure following interruption of antiretroviral treatment. *European Journal of Clinical Microbiology & Infectious Diseases, 23*, 63–65.

da Silva Morais, A., Oliveira, J. M., & Reis, R. L. (2018). Small animal models. *Advances in Experimental Medicine and Biology, 1059*, 423–439.

Daar, E. S., Bai, J., Hausner, M. A., Majchrowicz, M., Tamaddon, M., & Giorgi, J. V. (1998). Acute HIV syndrome after discontinuation of antiretroviral therapy in a patient treated before seroconversion. *Annals of Internal Medicine, 128*, 827–829.

Daar, E. S., Pilcher, C. D., & Hecht, F. M. (2008). Clinical presentation and diagnosis of primary HIV-1 infection. *Current Opinion in HIV and AIDS, 3*, 10–15.

Davey, R. T., Jr., Bhat, N., Yoder, C., Chun, T. W., Metcalf, J. A., Dewar, R., Natarajan, V., Lempicki, R. A., Adelsberger, J. W., Miller, K. D., Kovacs, J. A., Polis, M. A., Walker, R. E., Falloon, J., Masur, H., Gee, D., Baseler, M., Dimitrov, D. S., et al. (1999). HIV-1 and T cell dynamics after interruption of highly active antiretroviral therapy (HAART) in patients with a history of sustained viral suppression. *Proceedings of the National Academy of Sciences of the United States of America, 96*, 15109–15114.

de Groot, N., Doxiadis, G. G., Otting, N., de Vos-Rouweler, A. J., & Bontrop, R. E. (2014). Differential recombination dynamics within the MHC of macaque species. *Immunogenetics, 66*, 535–544.

de Groot, N. G., Heijmans, C. M. C., & Bontrop, R. E. (2017). AIDS in chimpanzees: The role of MHC genes. *Immunogenetics, 69*, 499–509.

Deeks, S. G., Kitchen, C. M., Liu, L., Guo, H., Gascon, R., Narvaez, A. B., Hunt, P., Martin, J. N., Kahn, J. O., Levy, J., McGrath, M. S., & Hecht, F. M. (2004). Immune activation set point during early HIV infection predicts subsequent CD4+ T-cell changes independent of viral load. *Blood, 104*, 942–947.

Deng, W., Bao, L., Liu, J., Xiao, C., Liu, J., Xue, J., Lv, Q., Qi, F., Gao, H., Yu, P., Xu, Y., Qu, Y., Li, F., Xiang, Z., Yu, H., Gong, S., Liu, M., Wang, G., Wang, S., et al. (2020). Primary exposure to SARS-CoV-2 protects against reinfection in rhesus macaques. *Science, 369*, 818–823.

Deodhar, S., Sillman, B., Bade, A. N., Avedissian, S. N., Podany, A. T., McMillan, J. M., Gautam, N., Hanson, B., Dyavar Shetty, B. L., Szlachetka, A., Johnston, M., Thurman, M., Munt, D. J., Dash, A. K., Markovic, M., Dahan, A., Alnouti, Y., Yazdi, A., Kevadiya, B. D., et al. (2022). Transformation of dolutegravir into an ultra-long-acting parenteral prodrug formulation. *Nature Communications, 13*, 3226.

Diaz Heijtz, R., Wang, S., Anuar, F., Qian, Y., Bjorkholm, B., Samuelsson, A., Hibberd, M. L., Forssberg, H., & Pettersson, S. (2011). Normal gut microbiota modulates brain development and behavior. *Proceedings of the National Academy of Sciences of the United States of America, 108*, 3047–3052.

Dinoso, J. B., Kim, S. Y., Wiegand, A. M., Palmer, S. E., Gange, S. J., Cranmer, L., O'Shea, A., Callender, M., Spivak, A., Brennan, T., Kearney, M. F., Proschan, M. A., Mican, J. M., Rehm, C. A., Coffin, J. M., Mellors, J. W., Siliciano, R. F., & Maldarelli, F. (2009). Treatment intensification does not reduce residual HIV-1 viremia in patients on highly active antiretroviral therapy. *Proceedings of the National Academy of Sciences of the United States of America, 106*, 9403–9408.

Dowd, K. A., Ko, S. Y., Morabito, K. M., Yang, E. S., Pelc, R. S., DeMaso, C. R., Castilho, L. R., Abbink, P., Boyd, M., Nityanandam, R., Gordon, D. N., Gallagher, J. R., Chen, X., Todd, J. P., Tsybovsky, Y., Harris, A., Huang, Y. S., Higgs, S., Vanlandingham, D. L., et al. (2016). Rapid development of a DNA vaccine for Zika virus. *Science, 354*, 237–240.

Dragovic, G., Smith, C. J., Jevtovic, D., Dimitrijevic, B., Kusic, J., Youle, M., & Johnson, M. A. (2016). Choice of first-line antiretroviral therapy regimen and treatment outcomes for HIV in a middle income compared to a high income country: A cohort study. *BMC Infectious Diseases, 16*, 106.

Driggers, R. W., Ho, C. Y., Korhonen, E. M., Kuivanen, S., Jaaskelainen, A. J., Smura, T., Rosenberg, A., Hill, D. A., DeBiasi, R. L., Vezina, G., Timofeev, J., Rodriguez, F. J., Levanov, L., Razak, J., Iyengar, P., Hennenfent, A., Kennedy, R., Lanciotti, R., du Plessis, A., &

Vapalahti, O. (2016). Zika virus infection with prolonged maternal viremia and fetal brain abnormalities. *The New England Journal of Medicine, 374*, 2142–2151.

Dudley, D. M., Aliota, M. T., Mohr, E. L., Weiler, A. M., Lehrer-Brey, G., Weisgrau, K. L., Mohns, M. S., Breitbach, M. E., Rasheed, M. N., Newman, C. M., Gellerup, D. D., Moncla, L. H., Post, J., Schultz-Darken, N., Schotzko, M. L., Hayes, J. M., Eudailey, J. A., Moody, M. A., Permar, S. R., et al. (2016). A rhesus macaque model of Asian-lineage Zika virus infection. *Nature Communications, 7*, 12204.

Ericsen, A. J., Lauck, M., Mohns, M. S., DiNapoli, S. R., Mutschler, J. P., Greene, J. M., Weinfurter, J. T., Lehrer-Brey, G., Prall, T. M., Gieger, S. M., Buechler, C. R., Crosno, K. A., Peterson, E. J., Reynolds, M. R., Wiseman, R. W., Burwitz, B. J., Estes, J. D., Sacha, J. B., Friedrich, T. C., et al. (2016). Microbial translocation and inflammation occur in hyperacute immunodeficiency virus infection and compromise host control of virus replication. *PLoS Pathogens, 12*, e1006048.

Estes, J. D. (2013). Pathobiology of HIV/SIV-associated changes in secondary lymphoid tissues. *Immunological Reviews, 254*, 65–77.

Estes, J. D., Wong, S. W., & Brenchley, J. M. (2018). Nonhuman primate models of human viral infections. *Nature Reviews. Immunology, 18*, 390–404.

Fox, H. S., Niu, M., Morsey, B. M., Lamberty, B. G., Emanuel, K., Periyasamy, P., Callen, S., Acharya, A., Kubik, G., Eudy, J., Guda, C., Dyavar, S. R., Fletcher, C. V., Byrareddy, S. N., & Buch, S. (2022). Morphine suppresses peripheral responses and transforms brain myeloid gene expression to favor neuropathogenesis in SIV infection. *Frontiers in Immunology, 13*, 1012884.

Frank, I., Cigoli, M., Arif, M. S., Fahlberg, M. D., Maldonado, S., Calenda, G., Pegu, A., Yang, E. S., Rawi, R., Chuang, G. Y., Geng, H., Liu, C., Zhou, T., Kwong, P. D., Arthos, J., Cicala, C., Grasperge, B. F., Blanchard, J. L., Gettie, A., et al. (2021). Blocking alpha(4) beta(7) integrin delays viral rebound in SHIV(SF162P3)-infected macaques treated with anti-HIV broadly neutralizing antibodies. *Science Translational Medicine, 13*, eabf7201.

Gallo, R. C. (2016). Shock and kill with caution. *Science, 354*, 177–178.

Gilbert, J. A., Blaser, M. J., Caporaso, J. G., Jansson, J. K., Lynch, S. V., & Knight, R. (2018). Current understanding of the human microbiome. *Nature Medicine, 24*, 392–400.

Haase, A. T. (2011). Early events in sexual transmission of HIV and SIV and opportunities for interventions. *Annual Review of Medicine, 62*, 127–139.

Haddow, A. D., Schuh, A. J., Yasuda, C. Y., Kasper, M. R., Heang, V., Huy, R., Guzman, H., Tesh, R. B., & Weaver, S. C. (2012). Genetic characterization of Zika virus strains: Geographic expansion of the Asian lineage. *PLoS Neglected Tropical Diseases, 6*, e1477.

Hannibal, D. L., Bliss-Moreau, E., Vandeleest, J., McCowan, B., & Capitanio, J. (2017). Laboratory rhesus macaque social housing and social changes: Implications for research. *American Journal of Primatology, 79*, 1–14.

Hansen, S. G., Ford, J. C., Lewis, M. S., Ventura, A. B., Hughes, C. M., Coyne-Johnson, L., Whizin, N., Oswald, K., Shoemaker, R., Swanson, T., Legasse, A. W., Chiuchiolo, M. J., Parks, C. L., Axthelm, M. K., Nelson, J. A., Jarvis, M. A., Piatak, M., Jr., Lifson, J. D., & Picker, L. J. (2011). Profound early control of highly pathogenic SIV by an effector memory T-cell vaccine. *Nature, 473*, 523–527.

Hansen, S. G., Wu, H. L., Burwitz, B. J., Hughes, C. M., Hammond, K. B., Ventura, A. B., Reed, J. S., Gilbride, R. M., Ainslie, E., Morrow, D. W., Ford, J. C., Selseth, A. N., Pathak, R., Malouli, D., Legasse, A. W., Axthelm, M. K., Nelson, J. A., Gillespie, G. M., Walters, L. C., et al. (2016). Broadly targeted CD8(+) T cell responses restricted by major histocompatibility complex E. *Science, 351*, 714–720.

Haran, K. P., Hajduczki, A., Pampusch, M. S., Mwakalundwa, G., Vargas-Inchaustegui, D. A., Rakasz, E. G., Connick, E., Berger, E. A., & Skinner, P. J. (2018). Simian immunodeficiency virus (SIV)-specific chimeric antigen receptor-T cells engineered to target B cell follicles and suppress SIV replication. *Frontiers in Immunology, 9*, 492.

Harding, J. D. (2017). Nonhuman primates and translational research: Progress, opportunities, and challenges. *ILAR Journal, 58*, 141–150.

Hargrave, A., Mustafa, A. S., Hanif, A., Tunio, J. H., & Hanif, S. N. M. (2021). Current status of HIV-1 vaccines. *Vaccines (Basel), 9*, 1026.

Harper, J., Gordon, S., Chan, C. N., Wang, H., Lindemuth, E., Galardi, C., Falcinelli, S. D., Raines, S. L. M., Read, J. L., Nguyen, K., McGary, C. S., Nekorchuk, M., Busman-Sahay, K., Schawalder, J., King, C., Pino, M., Micci, L., Cervasi, B., Jean, S., et al. (2020). CTLA-4 and PD-1 dual blockade induces SIV reactivation without control of rebound after antiretroviral therapy interruption. *Nature Medicine, 26*, 519–528.

Harris, L. D., Tabb, B., Sodora, D. L., Paiardini, M., Klatt, N. R., Douek, D. C., Silvestri, G., Muller-Trutwin, M., Vasile-Pandrea, I., Apetrei, C., Hirsch, V., Lifson, J., Brenchley, J. M., & Estes, J. D. (2010). Downregulation of robust acute type I interferon responses distinguishes nonpathogenic simian immunodeficiency virus (SIV) infection of natural hosts from pathogenic SIV infection of rhesus macaques. *Journal of Virology, 84*, 7886–7891.

Haynes, B. F., Gilbert, P. B., McElrath, M. J., Zolla-Pazner, S., Tomaras, G. D., Alam, S. M., Evans, D. T., Montefiori, D. C., Karnasuta, C., Sutthent, R., Liao, H. X., DeVico, A. L., Lewis, G. K., Williams, C., Pinter, A., Fong, Y., Janes, H., DeCamp, A., Huang, Y., et al. (2012). Immune-correlates analysis of an HIV-1 vaccine efficacy trial. *The New England Journal of Medicine, 366*, 1275–1286.

Haynes, B. F., Shaw, G. M., Korber, B., Kelsoe, G., Sodroski, J., Hahn, B. H., Borrow, P., & McMichael, A. J. (2016). HIV-host interactions: Implications for vaccine design. *Cell Host & Microbe, 19*, 292–303.

Haynes, B. F., Wiehe, K., Borrow, P., Saunders, K. O., Korber, B., Wagh, K., McMichael, A. J., Kelsoe, G., Hahn, B. H., Alt, F., & Shaw, G. M. (2023). Strategies for HIV-1 vaccines that induce broadly neutralizing antibodies. *Nature Reviews. Immunology, 23*, 142–158.

Henrich, T. J., Hanhauser, E., Marty, F. M., Sirignano, M. N., Keating, S., Lee, T. H., Robles, Y. P., Davis, B. T., Li, J. Z., Heisey, A., Hill, A. L., Busch, M. P., Armand, P., Soiffer, R. J., Altfeld, M., & Kuritzkes, D. R. (2014). Antiretroviral-free HIV-1 remission and viral rebound after allogeneic stem cell transplantation: Report of 2 cases. *Annals of Internal Medicine, 161*, 319–327.

Hirsch, V. M., Dapolito, G., McGann, C., Olmsted, R. A., Purcell, R. H., & Johnson, P. R. (1989). Molecular cloning of SIV from sooty mangabey monkeys. *Journal of Medical Primatology, 18*, 279–285.

Hoang, T. N., Pino, M., Boddapati, A. K., Viox, E. G., Starke, C. E., Upadhyay, A. A., Gumber, S., Nekorchuk, M., Busman-Sahay, K., Strongin, Z., Harper, J. L., Tharp, G. K., Pellegrini, K. L., Kirejczyk, S., Zandi, K., Tao, S., Horton, T. R., Beagle, E. N., Mahar, E. A., et al. (2021). Baricitinib treatment resolves lower-airway macrophage inflammation and neutrophil recruitment in SARS-CoV-2-infected rhesus macaques. *Cell, 184*, 460–475 e421.

Hugenholtz, F., & de Vos, W. M. (2018). Mouse models for human intestinal microbiota research: A critical evaluation. *Cellular and Molecular Life Sciences, 75*, 149–160.

Humbert, M., Rasmussen, R. A., Song, R., Ong, H., Sharma, P., Chenine, A. L., Kramer, V. G., Siddappa, N. B., Xu, W., Else, J. G., Novembre, F. J., Strobert, E., O'Neil, S. P., & Ruprecht, R. M. (2008). SHIV-1157i and passaged progeny viruses encoding R5 HIV-1 clade C env cause AIDS in rhesus monkeys. *Retrovirology, 5*, 94.

Iwamoto, N., Mason, R. D., Song, K., Gorman, J., Welles, H. C., Arthos, J., Cicala, C., Min, S., King, H. A. D., Belli, A. J., Reimann, K. A.,

Foulds, K. E., Kwong, P. D., Lifson, J. D., Keele, B. F., & Roederer, M. (2019). Blocking alpha(4)beta(7) integrin binding to SIV does not improve virologic control. *Science, 365*, 1033–1036.

Jacquelin, B., Mayau, V., Targat, B., Liovat, A. S., Kunkel, D., Petitjean, G., Dillies, M. A., Roques, P., Butor, C., Silvestri, G., Giavedoni, L. D., Lebon, P., Barre-Sinoussi, F., Benecke, A., & Muller-Trutwin, M. C. (2009). Nonpathogenic SIV infection of African green monkeys induces a strong but rapidly controlled type I IFN response. *The Journal of Clinical Investigation, 119*, 3544–3555.

Jensen, K., Dela Pena-Ponce, M. G., Piatak, M., Jr., Shoemaker, R., Oswald, K., Jacobs, W. R., Jr., Fennelly, G., Lucero, C., Mollan, K. R., Hudgens, M. G., Amedee, A., Kozlowski, P. A., Estes, J. D., Lifson, J. D., Van Rompay, K. K., Larsen, M., & De Paris, K. (2017). Balancing trained immunity with persistent immune activation and the risk of simian immunodeficiency virus infection in infant macaques vaccinated with attenuated mycobacterium tuberculosis or Mycobacterium bovis BCG vaccine. *Clinical and Vaccine Immunology, 24*, e00360.

Jones, B. E., Brown-Augsburger, P. L., Corbett, K. S., Westendorf, K., Davies, J., Cujec, T. P., Wiethoff, C. M., Blackbourne, J. L., Heinz, B. A., Foster, D., Higgs, R. E., Balasubramaniam, D., Wang, L., Zhang, Y., Yang, E. S., Bidshahri, R., Kraft, L., Hwang, Y., Zentelis, S., et al. (2021). The neutralizing antibody, LY-CoV555, protects against SARS-CoV-2 infection in nonhuman primates. *Science Translational Medicine, 13*, eabf1906.

Kearney, M. F., Wiegand, A., Shao, W., Coffin, J. M., Mellors, J. W., Lederman, M., Gandhi, R. T., Keele, B. F., & Li, J. Z. (2016). Origin of rebound plasma HIV includes cells with identical proviruses that are transcriptionally active before stopping of antiretroviral therapy. *Journal of Virology, 90*, 1369–1376.

Keele, B. F., Li, H., Learn, G. H., Hraber, P., Giorgi, E. E., Grayson, T., Sun, C., Chen, Y., Yeh, W. W., Letvin, N. L., Mascola, J. R., Nabel, G. J., Haynes, B. F., Bhattacharya, T., Perelson, A. S., Korber, B. T., Hahn, B. H., & Shaw, G. M. (2009). Low-dose rectal inoculation of rhesus macaques by SIVsmE660 or SIVmac251 recapitulates human mucosal infection by HIV-1. *The Journal of Experimental Medicine, 206*, 1117–1134.

Kilby, J. M., Goepfert, P. A., Miller, A. P., Gnann, J. W., Jr., Sillers, M., Saag, M. S., & Bucy, R. P. (2000). Recurrence of the acute HIV syndrome after interruption of antiretroviral therapy in a patient with chronic HIV infection: A case report. *Annals of Internal Medicine, 133*, 435–438.

King, N. W., Hunt, R. D., & Letvin, N. L. (1983). Histopathologic changes in macaques with an acquired immunodeficiency syndrome (AIDS). *The American Journal of Pathology, 113*, 382–388.

Klasse, P. J., Nixon, D. F., & Moore, J. P. (2021). Immunogenicity of clinically relevant SARS-CoV-2 vaccines in nonhuman primates and humans. *Science Advances, 7*, eabe8065.

Klein, N., Sefe, D., Mosconi, I., Zanchetta, M., Castro, H., Jacobsen, M., Jones, H., Bernardi, S., Pillay, D., Giaquinto, C., Walker, A. S., Gibb, D. M., De Rossi, A., & Paediatric European Network for Treatment of, A.T.T. (2013). The immunological and virological consequences of planned treatment interruptions in children with HIV infection. *PLoS One, 8*, e76582.

Knysz, B., Gasiorowski, J., Czarnecki, M., & Gladysz, A. (2005). Viral rebound syndrome in two HIV-1-positive patients after structured treatment interruption. *Viral Immunology, 18*, 579–581.

Koide, F., Goebel, S., Snyder, B., Walters, K. B., Gast, A., Hagelin, K., Kalkeri, R., & Rayner, J. (2016). Development of a Zika virus infection model in cynomolgus macaques. *Frontiers in Microbiology, 7*, 2028.

Krych, L., Hansen, C. H., Hansen, A. K., van den Berg, F. W., & Nielsen, D. S. (2013). Quantitatively different, yet qualitatively alike: A meta-analysis of the mouse core gut microbiome with a view towards the human gut microbiome. *PLoS One, 8*, e62578.

Kulkarni, T. A., Bade, A. N., Sillman, B., Shetty, B. L. D., Wojtkiewicz, M. S., Gautam, N., Hilaire, J. R., Sravanam, S., Szlachetka, A., Lamberty, B. G., Morsey, B. M., Fox, H. S., Alnouti, Y., McMillan, J. M., Mosley, R. L., Meza, J., Domanico, P. L., Yue, T. Y., Moore, G., et al. (2020). A year-long extended release nanoformulated cabotegravir prodrug. *Nature Materials, 19*, 910–920.

Kumar, N., Chahroudi, A., & Silvestri, G. (2016). Animal models to achieve an HIV cure. *Current Opinion in HIV and AIDS, 11*, 432–441.

Larocca, R. A., Abbink, P., Peron, J. P., Zanotto, P. M., Iampietro, M. J., Badamchi-Zadeh, A., Boyd, M., Ng'ang'a, D., Kirilova, M., Nityanandam, R., Mercado, N. B., Li, Z., Moseley, E. T., Bricault, C. A., Borducchi, E. N., Giglio, P. B., Jetton, D., Neubauer, G., Nkolola, J. P., Maxfield, L. F., De La Barrera, R. A., Jarman, R. G., Eckels, K. H., Michael, N. L., Thomas, S. J., & Barouch, D. H. (2016). Vaccine protection against Zika virus from Brazil. *Nature, 536*, 474–478.

Lawitz, E., Poordad, F. F., Pang, P. S., Hyland, R. H., Ding, X., Mo, H., Symonds, W. T., McHutchison, J. G., & Membreno, F. E. (2014). Sofosbuvir and ledipasvir fixed-dose combination with and without ribavirin in treatment-naive and previously treated patients with genotype 1 hepatitis C virus infection (LONESTAR): An open-label, randomised, phase 2 trial. *Lancet, 383*, 515–523.

Lei, A. H., Pang, W., Zhang, G. H., & Zheng, Y. T. (2013). Use and research of pigtailed macaques in nonhuman primate HIV/AIDS models. *Dongwuxue Yanjiu, 34*, 77–88.

Lemon, S. M., & Thomas, D. L. (1997). Vaccines to prevent viral hepatitis. *The New England Journal of Medicine, 336*, 196–204.

Letvin, N. L., Eaton, K. A., Aldrich, W. R., Sehgal, P. K., Blake, B. J., Schlossman, S. F., King, N. W., & Hunt, R. D. (1983). Acquired immunodeficiency syndrome in a colony of macaque monkeys. *Proceedings of the National Academy of Sciences of the United States of America, 80*, 2718–2722.

Lewis, A. D., & Prongay, K. (2015). Chapter 6 – Basic physiology of Macaca mulatta. In J. Bluemel, S. Korte, E. Schenck, & G. F. Weinbauer (Eds.), *The nonhuman primate in nonclinical drug development and safety assessment* (pp. 87–113). Academic Press.

Li, Q., Duan, L., Estes, J. D., Ma, Z. M., Rourke, T., Wang, Y., Reilly, C., Carlis, J., Miller, C. J., & Haase, A. T. (2005). Peak SIV replication in resting memory CD4+ T cells depletes gut lamina propria CD4+ T cells. *Nature, 434*, 1148–1152.

Li, X. F., Dong, H. L., Huang, X. Y., Qiu, Y. F., Wang, H. J., Deng, Y. Q., Zhang, N. N., Ye, Q., Zhao, H., Liu, Z. Y., Fan, H., An, X. P., Sun, S. H., Gao, B., Fa, Y. Z., Tong, Y. G., Zhang, F. C., Gao, G. F., Cao, W. C., et al. (2016). Characterization of a 2016 clinical isolate of Zika virus in non-human primates. *eBioMedicine, 12*, 170–177.

Li, Y., Huang, X., Zhang, Z., Li, S., Zhang, J., Xia, N., & Zhao, Q. (2020). Prophylactic hepatitis E vaccines: Antigenic analysis and serological evaluation. *Viruses, 12*, 109.

Lim, S. Y., Osuna, C. E., Hraber, P. T., Hesselgesser, J., Gerold, J. M., Barnes, T. L., Sanisetty, S., Seaman, M. S., Lewis, M. G., Geleziunas, R., Miller, M. D., Cihlar, T., Lee, W. A., Hill, A. L., & Whitney, J. B. (2018). TLR7 agonists induce transient viremia and reduce the viral reservoir in SIV-infected rhesus macaques on antiretroviral therapy. *Science Translational Medicine, 10*, eaao4521.

Liu, J., Keele, B. F., Li, H., Keating, S., Norris, P. J., Carville, A., Mansfield, K. G., Tomaras, G. D., Haynes, B. F., Kolodkin-Gal, D., Letvin, N. L., Hahn, B. H., Shaw, G. M., & Barouch, D. H. (2010). Low-dose mucosal simian immunodeficiency virus infection restricts early replication kinetics and transmitted virus variants in rhesus monkeys. *Journal of Virology, 84*, 10406–10412.

Locher, C. P., Witt, S. A., Herndier, B. G., Tenner-Racz, K., Racz, P., & Levy, J. A. (2001). Baboons as an animal model for human immunodeficiency virus pathogenesis and vaccine development. *Immunological Reviews, 183*, 127–140.

Lu, S., Zhao, Y., Yu, W., Yang, Y., Gao, J., Wang, J., Kuang, D., Yang, M., Yang, J., Ma, C., Xu, J., Qian, X., Li, H., Zhao, S., Li, J., Wang, H., Long, H., Zhou, J., Luo, F., et al. (2020). Comparison of nonhuman primates identified the suitable model for COVID-19. *Signal Transduction and Targeted Therapy, 5*, 157.

Makori, N., Tarantal, A. F., Lu, F. X., Rourke, T., Marthas, M. L., McChesney, M. B., Hendrickx, A. G., & Miller, C. J. (2003). Functional and morphological development of lymphoid tissues and immune regulatory and effector function in rhesus monkeys: Cytokine-secreting cells, immunoglobulin-secreting cells, and CD5(+) B-1 cells appear early in fetal development. *Clinical and Diagnostic Laboratory Immunology, 10*, 140–153.

Mallott, E. K., & Amato, K. R. (2022). Butyrate production pathway abundances are similar in human and nonhuman primate gut microbiomes. *Molecular Biology and Evolution, 39*, msab279.

Manara, S., Asnicar, F., Beghini, F., Bazzani, D., Cumbo, F., Zolfo, M., Nigro, E., Karcher, N., Manghi, P., Metzger, M. I., Pasolli, E., & Segata, N. (2019). Microbial genomes from non-human primate gut metagenomes expand the primate-associated bacterial tree of life with over 1000 novel species. *Genome Biology, 20*, 299.

Mancuso, P., Chen, C., Kaminski, R., Gordon, J., Liao, S., Robinson, J. A., Smith, M. D., Liu, H., Sariyer, I. K., Sariyer, R., Peterson, T. A., Donadoni, M., Williams, J. B., Siddiqui, S., Bunnell, B. A., Ling, B., MacLean, A. G., Burdo, T. H., & Khalili, K. (2020). CRISPR based editing of SIV proviral DNA in ART treated non-human primates. *Nature Communications, 11*, 6065.

Manickam, C., & Reeves, R. K. (2014). Modeling HCV disease in animals: Virology, immunology and pathogenesis of HCV and GBV-B infections. *Frontiers in Microbiology, 5*, 690.

Marsh, A. K., Willer, D. O., Skokovets, O., Iwajomo, O. H., Chan, J. K., & MacDonald, K. S. (2012). Evaluation of cynomolgus macaque (Macaca fascicularis) endogenous retrovirus expression following simian immunodeficiency virus infection. *PLoS One, 7*, e40158.

Martinsen, J. T., Gunst, J. D., Hojen, J. F., Tolstrup, M., & Sogaard, O. S. (2020). The use of toll-like receptor agonists in HIV-1 cure strategies. *Frontiers in Immunology, 11*, 1112.

Mascola, J. R., Stiegler, G., VanCott, T. C., Katinger, H., Carpenter, C. B., Hanson, C. E., Beary, H., Hayes, D., Frankel, S. S., Birx, D. L., & Lewis, M. G. (2000). Protection of macaques against vaginal transmission of a pathogenic HIV-1/SIV chimeric virus by passive infusion of neutralizing antibodies. *Nature Medicine, 6*, 207–210.

Mattapallil, J. J., Douek, D. C., Hill, B., Nishimura, Y., Martin, M., & Roederer, M. (2005). Massive infection and loss of memory CD4+ T cells in multiple tissues during acute SIV infection. *Nature, 434*, 1093–1097.

McBrien, J. B., Mavigner, M., Franchitti, L., Smith, S. A., White, E., Tharp, G. K., Walum, H., Busman-Sahay, K., Aguilera-Sandoval, C. R., Thayer, W. O., Spagnuolo, R. A., Kovarova, M., Wahl, A., Cervasi, B., Margolis, D. M., Vanderford, T. H., Carnathan, D. G., Paiardini, M., Lifson, J. D., et al. (2020). Robust and persistent reactivation of SIV and HIV by N-803 and depletion of CD8(+) cells. *Nature, 578*, 154–159.

McMahan, K., Yu, J., Mercado, N. B., Loos, C., Tostanoski, L. H., Chandrashekar, A., Liu, J., Peter, L., Atyeo, C., Zhu, A., Bondzie, E. A., Dagotto, G., Gebre, M. S., Jacob-Dolan, C., Li, Z., Nampanya, F., Patel, S., Pessaint, L., Van Ry, A., et al. (2021). Correlates of protection against SARS-CoV-2 in rhesus macaques. *Nature, 590*, 630–634.

Messer, J. S., Liechty, E. R., Vogel, O. A., & Chang, E. B. (2017). Evolutionary and ecological forces that shape the bacterial communities of the human gut. *Mucosal Immunology, 10*, 567–579.

Miller, C. J. (1998). Localization of Simian immunodeficiency virus-infected cells in the genital tract of male and female Rhesus macaques. *Journal of Reproductive Immunology, 41*, 331–339.

Miller, C. J., Alexander, N. J., Sutjipto, S., Lackner, A. A., Gettie, A., Hendrickx, A. G., Lowenstine, L. J., Jennings, M., & Marx, P. A. (1989). Genital mucosal transmission of simian immunodeficiency virus: Animal model for heterosexual transmission of human immunodeficiency virus. *Journal of Virology, 63*, 4277–4284.

Miller, C. J., Li, Q., Abel, K., Kim, E. Y., Ma, Z. M., Wietgrefe, S., La Franco-Scheuch, L., Compton, L., Duan, L., Shore, M. D., Zupancic, M., Busch, M., Carlis, J., Wolinsky, S., & Haase, A. T. (2005). Propagation and dissemination of infection after vaginal transmission of simian immunodeficiency virus. *Journal of Virology, 79*, 9217–9227.

Montserrat, M., Plana, M., Guardo, A. C., Andres, C., Climent, N., Gallart, T., Leal, L., Gatell, J. M., Sanchez-Palomino, S., & Garcia, F. (2017). Impact of long-term antiretroviral therapy interruption and resumption on viral reservoir in HIV-1 infected patients. *AIDS, 31*, 1895–1897.

Moretti, S., Virtuoso, S., Sernicola, L., Farcomeni, S., Maggiorella, M. T., & Borsetti, A. (2021). Advances in SIV/SHIV non-human primate models of NeuroAIDS. *Pathogens, 10*, 1018.

Munoz-Fontela, C., Dowling, W. E., Funnell, S. G. P., Gsell, P. S., Riveros-Balta, A. X., Albrecht, R. A., Andersen, H., Baric, R. S., Carroll, M. W., Cavaleri, M., Qin, C., Crozier, I., Dallmeier, K., de Waal, L., de Wit, E., Delang, L., Dohm, E., Duprex, W. P., Falzarano, D., et al. (2020). Animal models for COVID-19. *Nature, 586*, 509–515.

Nagpal, R., Wang, S., Solberg Woods, L. C., Seshie, O., Chung, S. T., Shively, C. A., Register, T. C., Craft, S., McClain, D. A., & Yadav, H. (2018). Comparative microbiome signatures and short-chain fatty acids in mouse, rat, non-human primate, and human feces. *Frontiers in Microbiology, 9*, 2897.

Nakamura-Hoshi, M., Takahara, Y., Matsuoka, S., Ishii, H., Seki, S., Nomura, T., Yamamoto, H., Sakawaki, H., Miura, T., Tokusumi, T., Shu, T., & Matano, T. (2020). Therapeutic vaccine-mediated Gag-specific CD8(+) T-cell induction under anti-retroviral therapy augments anti-virus efficacy of CD8(+) cells in simian immunodeficiency virus-infected macaques. *Scientific Reports, 10*, 11394.

Nasi, M., De Biasi, S., Gibellini, L., Bianchini, E., Pecorini, S., Bacca, V., Guaraldi, G., Mussini, C., Pinti, M., & Cossarizza, A. (2017). Ageing and inflammation in patients with HIV infection. *Clinical and Experimental Immunology, 187*, 44–52.

Nguyen, T. L., Vieira-Silva, S., Liston, A., & Raes, J. (2015). How informative is the mouse for human gut microbiota research? *Disease Models & Mechanisms, 8*, 1–16.

Nilsson, J., Boasso, A., Velilla, P. A., Zhang, R., Vaccari, M., Franchini, G., Shearer, G. M., Andersson, J., & Chougnet, C. (2006). HIV-1-driven regulatory T-cell accumulation in lymphoid tissues is associated with disease progression in HIV/AIDS. *Blood, 108*, 3808–3817.

Nishimura, Y., Gautam, R., Chun, T. W., Sadjadpour, R., Foulds, K. E., Shingai, M., Klein, F., Gazumyan, A., Golijanin, J., Donaldson, M., Donau, O. K., Plishka, R. J., Buckler-White, A., Seaman, M. S., Lifson, J. D., Koup, R. A., Fauci, A. S., Nussenzweig, M. C., & Martin, M. A. (2017). Early antibody therapy can induce long-lasting immunity to SHIV. *Nature, 543*, 559–563.

Nixon, C. C., Mavigner, M., Sampey, G. C., Brooks, A. D., Spagnuolo, R. A., Irlbeck, D. M., Mattingly, C., Ho, P. T., Schoof, N., Cammon, C. G., Tharp, G. K., Kanke, M., Wang, Z., Cleary, R. A., Upadhyay, A. A., De, C., Wills, S. R., Falcinelli, S. D., Galardi, C., et al. (2020). Systemic HIV and SIV latency reversal via non-canonical NF-kappaB signalling in vivo. *Nature, 578*, 160–165.

Ogunrinola, G. A., Oyewale, J. O., Oshamika, O. O., & Olasehinde, G. I. (2020). The human microbiome and its impacts on health. *International Journal of Microbiology, 2020*, 8045646.

Olwenyi, O. A., Johnson, S. D., Bidokhti, M., Thakur, V., Pandey, K., Thurman, M., Acharya, A., Uppada, S., Callen, S., Giavedoni, L., Ranga, U., Buch, S. J., & Byrareddy, S. N. (2022). Systems biology analyses reveal enhanced chronic morphine distortion of gut-brain interrelationships in simian human immunodeficiency virus infected rhesus macaques. *Frontiers in Neuroscience, 16*, 1001544.

Otting, N., van der Wiel, M. K., de Groot, N., de Vos-Rouweler, A. J., de Groot, N. G., Doxiadis, G. G., Wiseman, R. W., O'Connor, D. H., & Bontrop, R. E. (2017). The orthologs of HLA-DQ and -DP genes display abundant levels of variability in macaque species. *Immunogenetics, 69*, 87–99.

Page, M., Stebbings, R., Berry, N., Hull, R., Ferguson, D., Davis, L., Duffy, L., Elsley, W., Hall, J., Ham, C., Hassall, M., Li, B., Mee, E. T., Quartey-Papafio, R., Rose, N. J., Mathy, N., Voss, G., Stott, E. J., & Almond, N. (2012). Heterologous protection elicited by candidate monomeric recombinant HIV-1 gp120 vaccine in the absence of cross neutralising antibodies in a macaque model. *Retrovirology, 9*, 56.

Palacios-Saucedo, G. C., Rivera-Morales, L. G., Vazquez-Guillen, J. M., Sanchez, L. M., Ramirez, T. J., Briones, E., Ortiz-Lopez, R., Vazquez, C. A., & Rodriguez-Padilla, C. (2011). Drug resistance mutations during structured interruptions of HAART in two HIV-1-infected children. *Current HIV Research, 9*, 154–159.

Pampusch, M. S., Abdelaal, H. M., Cartwright, E. K., Molden, J. S., Davey, B. C., Sauve, J. D., Sevcik, E. N., Rendahl, A. K., Rakasz, E. G., Connick, E., Berger, E. A., & Skinner, P. J. (2022). CAR/CXCR5-T cell immunotherapy is safe and potentially efficacious in promoting sustained remission of SIV infection. *PLoS Pathogens, 18*, e1009831.

Pandey, K., Acharya, A., Mohan, M., Ng, C. L., Reid, S. P., & Byrareddy, S. N. (2021). Animal models for SARS-CoV-2 research: A comprehensive literature review. *Transboundary and Emerging Diseases, 68*, 1868–1885.

Parrish, N. F., Gao, F., Li, H., Giorgi, E. E., Barbian, H. J., Parrish, E. H., Zajic, L., Iyer, S. S., Decker, J. M., Kumar, A., Hora, B., Berg, A., Cai, F., Hopper, J., Denny, T. N., Ding, H., Ochsenbauer, C., Kappes, J. C., Galimidi, R. P., et al. (2013). Phenotypic properties of transmitted founder HIV-1. *Proceedings of the National Academy of Sciences of the United States of America, 110*, 6626–6633.

Pegu, A., Hessell, A. J., Mascola, J. R., & Haigwood, N. L. (2017). Use of broadly neutralizing antibodies for HIV-1 prevention. *Immunological Reviews, 275*, 296–312.

Policicchio, B. B., Pandrea, I., & Apetrei, C. (2016). Animal models for HIV cure research. *Frontiers in Immunology, 7*, 12.

Qureshi, H., Ma, Z. M., Huang, Y., Hodge, G., Thomas, M. A., DiPasquale, J., DeSilva, V., Fritts, L., Bett, A. J., Casimiro, D. R., Shiver, J. W., Robert-Guroff, M., Robertson, M. N., McChesney, M. B., Gilbert, P. B., & Miller, C. J. (2012). Low-dose penile SIVmac251 exposure of rhesus macaques infected with adenovirus type 5 (Ad5) and then immunized with a replication-defective Ad5-based SIV gag/pol/nef vaccine recapitulates the results of the phase IIb step trial of a similar HIV-1 vaccine. *Journal of Virology, 86*, 2239–2250.

Ramesh, A., Darko, S., Hua, A., Overman, G., Ransier, A., Francica, J. R., Trama, A., Tomaras, G. D., Haynes, B. F., Douek, D. C., & Kepler, T. B. (2017). Structure and diversity of the rhesus macaque immunoglobulin loci through multiple De Novo genome assemblies. *Frontiers in Immunology, 8*, 1407.

Rawal, G., Yadav, S., & Kumar, R. (2016). Zika virus: An overview. *Journal of Family Medicine and Primary Care, 5*, 523–527.

Redd, A. D., Quinn, T. C., & Tobian, A. A. (2013). Frequency and implications of HIV superinfection. *The Lancet Infectious Diseases, 13*, 622–628.

Rerks-Ngarm, S., Pitisuttithum, P., Nitayaphan, S., Kaewkungwal, J., Chiu, J., Paris, R., Premsri, N., Namwat, C., de Souza, M., Adams, E., Benenson, M., Gurunathan, S., Tartaglia, J., McNeil, J. G., Francis, D. P., Stablein, D., Birx, D. L., Chunsuttiwat, S., Khamboonruang, C., et al. (2009). Vaccination with ALVAC and AIDSVAX to prevent HIV-1 infection in Thailand. *The New England Journal of Medicine, 361*, 2209–2220.

Reynolds, M. R., Rakasz, E., Skinner, P. J., White, C., Abel, K., Ma, Z. M., Compton, L., Napoe, G., Wilson, N., Miller, C. J., Haase, A., & Watkins, D. I. (2005). CD8+ T-lymphocyte response to major immunodominant epitopes after vaginal exposure to simian immunodeficiency virus: Too late and too little. *Journal of Virology, 79*, 9228–9235.

Robinson, B. A., Estep, R. D., Messaoudi, I., Rogers, K. S., & Wong, S. W. (2012). Viral interferon regulatory factors decrease the induction of type I and type II interferon during rhesus macaque rhadinovirus infection. *Journal of Virology, 86*, 2197–2211.

Rockx, B., Kuiken, T., Herfst, S., Bestebroer, T., Lamers, M. M., Oude Munnink, B. B., de Meulder, D., van Amerongen, G., van den Brand, J., Okba, N. M. A., Schipper, D., van Run, P., Leijten, L., Sikkema, R., Verschoor, E., Verstrepen, B., Bogers, W., Langermans, J., Drosten, C., et al. (2020). Comparative pathogenesis of COVID-19, MERS, and SARS in a nonhuman primate model. *Science, 368*, 1012–1015.

Roederer, M., Keele, B. F., Schmidt, S. D., Mason, R. D., Welles, H. C., Fischer, W., Labranche, C., Foulds, K. E., Louder, M. K., Yang, Z. Y., Todd, J. P., Buzby, A. P., Mach, L. V., Shen, L., Seaton, K. E., Ward, B. M., Bailer, R. T., Gottardo, R., Gu, W., et al. (2014). Immunological and virological mechanisms of vaccine-mediated protection against SIV and HIV. *Nature, 505*, 502–508.

Rowe, T., Gao, G., Hogan, R. J., Crystal, R. G., Voss, T. G., Grant, R. L., Bell, P., Kobinger, G. P., Wivel, N. A., & Wilson, J. M. (2004). Macaque model for severe acute respiratory syndrome. *Journal of Virology, 78*, 11401–11404.

Rowland, I., Gibson, G., Heinken, A., Scott, K., Swann, J., Thiele, I., & Tuohy, K. (2018). Gut microbiota functions: Metabolism of nutrients and other food components. *European Journal of Nutrition, 57*, 1–24.

Rust, B. J., Kean, L. S., Colonna, L., Brandenstein, K. E., Poole, N. H., Obenza, W., Enstrom, M. R., Maldini, C. R., Ellis, G. I., Fennessey, C. M., Huang, M. L., Keele, B. F., Jerome, K. R., Riley, J. L., Kiem, H. P., & Peterson, C. W. (2020). Robust expansion of HIV CAR T cells following antigen boosting in ART-suppressed nonhuman primates. *Blood, 136*, 1722–1734.

Safari, Z., Bruneau, A., Monnoye, M., Mariadassou, M., Philippe, C., Zatloukal, K., & Gerard, P. (2020). Murine genetic background overcomes gut microbiota changes to explain metabolic response to high-fat diet. *Nutrients, 12*, 287.

Salantes, D. B., Zheng, Y., Mampe, F., Srivastava, T., Beg, S., Lai, J., Li, J. Z., Tressler, R. L., Koup, R. A., Hoxie, J., Abdel-Mohsen, M., Sherrill-Mix, S., McCormick, K., Overton, E. T., Bushman, F. D., Learn, G. H., Siliciano, R. F., Siliciano, J. M., Tebas, P., & Bar, K. J. (2018). HIV-1 latent reservoir size and diversity are stable following brief treatment interruption. *The Journal of Clinical Investigation, 128*, 3102–3115.

Salguero, F. J., White, A. D., Slack, G. S., Fotheringham, S. A., Bewley, K. R., Gooch, K. E., Longet, S., Humphries, H. E., Watson, R. J., Hunter, L., Ryan, K. A., Hall, Y., Sibley, L., Sarfas, C., Allen, L., Aram, M., Brunt, E., Brown, P., et al. (2021). Comparison of rhesus and cynomolgus macaques as an infection model for COVID-19. *Nature Communications, 12*, 1260.

Schacker, T., Collier, A. C., Hughes, J., Shea, T., & Corey, L. (1996). Clinical and epidemiologic features of primary HIV infection. *Annals of Internal Medicine, 125*, 257–264.

Schacker, T. W., Nguyen, P. L., Beilman, G. J., Wolinsky, S., Larson, M., Reilly, C., & Haase, A. T. (2002). Collagen deposition in HIV-1 infected lymphatic tissues and T cell homeostasis. *The Journal of Clinical Investigation, 110*, 1133–1139.

Sharp, P. M., & Hahn, B. H. (2011). Origins of HIV and the AIDS pandemic. *Cold Spring Harbor Perspectives in Medicine, 1*, a006841.

Shrestha, M. P., Scott, R. M., Joshi, D. M., Mammen, M. P., Jr., Thapa, G. B., Thapa, N., Myint, K. S., Fourneau, M., Kuschner, R. A., Shrestha, S. K., David, M. P., Seriwatana, J., Vaughn, D. W., Safary, A., Endy, T. P., & Innis, B. L. (2007). Safety and efficacy of a recombinant hepatitis E vaccine. *The New England Journal of Medicine, 356*, 895–903.

Siddappa, N. B., Song, R., Kramer, V. G., Chenine, A. L., Velu, V., Ong, H., Rasmussen, R. A., Grisson, R. D., Wood, C., Zhang, H., Kankasa, C., Amara, R. R., Else, J. G., Novembre, F. J., Montefiori, D. C., & Ruprecht, R. M. (2009). Neutralization-sensitive R5-tropic simian-human immunodeficiency virus SHIV-2873Nip, which carries env isolated from an infant with a recent HIV clade C infection. *Journal of Virology, 83*, 1422–1432.

Siddappa, N. B., Watkins, J. D., Wassermann, K. J., Song, R., Wang, W., Kramer, V. G., Lakhashe, S., Santosuosso, M., Poznansky, M. C., Novembre, F. J., Villinger, F., Else, J. G., Montefiori, D. C., Rasmussen, R. A., & Ruprecht, R. M. (2010). R5 clade C SHIV strains with tier 1 or 2 neutralization sensitivity: Tools to dissect env evolution and to develop AIDS vaccines in primate models. *PLoS One, 5*, e11689.

Silvestri, G., Paiardini, M., Pandrea, I., Lederman, M. M., & Sodora, D. L. (2007). Understanding the benign nature of SIV infection in natural hosts. *The Journal of Clinical Investigation, 117*, 3148–3154.

Singh, D. K., Singh, B., Ganatra, S. R., Gazi, M., Cole, J., Thippeshappa, R., Alfson, K. J., Clemmons, E., Gonzalez, O., Escobedo, R., Lee, T. H., Chatterjee, A., Goez-Gazi, Y., Sharan, R., Gough, M., Alvarez, C., Blakley, A., Ferdin, J., Bartley, C., et al. (2021). Responses to acute infection with SARS-CoV-2 in the lungs of rhesus macaques, baboons and marmosets. *Nature Microbiology, 6*, 73–86.

Slama, L., Porcher, R., Linard, F., Chakvetadze, C., Cros, A., Carillon, S., Gallardo, L., Viard, J. P., & Molina, J. M. (2023). Injectable long acting antiretroviral for HIV treatment and prevention: Perspectives of potential users. *BMC Infectious Diseases, 23*, 98.

Sodora, D. L., Allan, J. S., Apetrei, C., Brenchley, J. M., Douek, D. C., Else, J. G., Estes, J. D., Hahn, B. H., Hirsch, V. M., Kaur, A., Kirchhoff, F., Muller-Trutwin, M., Pandrea, I., Schmitz, J. E., & Silvestri, G. (2009). Toward an AIDS vaccine: Lessons from natural simian immunodeficiency virus infections of African nonhuman primate hosts. *Nature Medicine, 15*, 861–865.

Sokoya, T., Steel, H. C., Nieuwoudt, M., & Rossouw, T. M. (2017). HIV as a cause of immune activation and immunosenescence. *Mediators of Inflammation, 2017*, 6825493.

Song, T. Z., Zheng, H. Y., Han, J. B., Jin, L., Yang, X., Liu, F. L., Luo, R. H., Tian, R. R., Cai, H. R., Feng, X. L., Liu, C., Li, M. H., & Zheng, Y. T. (2020). Delayed severe cytokine storm and immune cell infiltration in SARS-CoV-2-infected aged Chinese rhesus macaques. *Zoological Research, 41*, 503–516.

Soriano, V., Edagwa, B., Mendoza, C., Barreiro, P., Corral, O., Trevino, A., & Gendelman, H. E. (2022). Ultra-long-acting antivirals as chemical vaccines to prevent viral diseases. *Future Microbiology, 17*, 887–897.

Sundling, C., Zhang, Z., Phad, G. E., Sheng, Z., Wang, Y., Mascola, J. R., Li, Y., Wyatt, R. T., Shapiro, L., & Karlsson Hedestam, G. B. (2014). Single-cell and deep sequencing of IgG-switched macaque B cells reveal a diverse Ig repertoire following immunization. *Journal of Immunology, 192*, 3637–3644.

Szandruk-Bender, M., Wiatrak, B., & Szelag, A. (2022). The risk of developing Alzheimer's disease and Parkinson's disease in patients with inflammatory bowel disease: A meta-analysis. *Journal of Clinical Medicine, 11*, –3704.

Tanaka, K., Kim, Y., Roche, M., & Lewin, S. R. (2022). The role of latency reversal in HIV cure strategies. *Journal of Medical Primatology, 51*, 278–283.

Tebas, P., Henry, W. K., Matining, R., Weng-Cherng, D., Schmitz, J., Valdez, H., Jahed, N., Myers, L., Powderly, W. G., & Katzenstein, D. (2008). Metabolic and immune activation effects of treatment interruption in chronic HIV-1 infection: Implications for cardiovascular risk. *PLoS One, 3*, e2021.

Thoueille, P., Choong, E., Cavassini, M., Buclin, T., & Decosterd, L. A. (2022). Long-acting antiretrovirals: A new era for the management and prevention of HIV infection. *The Journal of Antimicrobial Chemotherapy, 77*, 290–302.

Thurman, M., Johnson, S., Acharya, A., Pallikkuth, S., Mahesh, M., & Byrareddy, S. N. (2020). Biomarkers of activation and inflammation to track disparity in chronological and physiological age of people living with HIV on combination antiretroviral therapy. *Frontiers in Immunology, 11*, 583934.

Tubiana, R., Ghosn, J., De-Sa, M., Wirden, M., Gautheret-Dejean, A., Bricaire, F., & Katlama, C. (2002). Warning: Antiretroviral treatment interruption could lead to an increased risk of HIV transmission. *AIDS, 16*, 1083–1084.

Veazey, R. S., DeMaria, M., Chalifoux, L. V., Shvetz, D. E., Pauley, D. R., Knight, H. L., Rosenzweig, M., Johnson, R. P., Desrosiers, R. C., & Lackner, A. A. (1998). Gastrointestinal tract as a major site of CD4+ T cell depletion and viral replication in SIV infection. *Science, 280*, 427–431.

Virgin, H. W., & Walker, B. D. (2010). Immunology and the elusive AIDS vaccine. *Nature, 464*, 224–231.

Wade, W. G. (2013). The oral microbiome in health and disease. *Pharmacological Research, 69*, 137–143.

Wang, S., Peng, Y., Wang, R., Jiao, S., Wang, M., Huang, W., Shan, C., Jiang, W., Li, Z., Gu, C., Chen, B., Hu, X., Yao, Y., Min, J., Zhang, H., Chen, Y., Gao, G., Tang, P., Li, G., et al. (2020). Characterization of neutralizing antibody with prophylactic and therapeutic efficacy against SARS-CoV-2 in rhesus monkeys. *Nature Communications, 11*, 5752.

Webb, G. M., Molden, J., Busman-Sahay, K., Abdulhaqq, S., Wu, H. L., Weber, W. C., Bateman, K. B., Reed, J. S., Northrup, M., Maier, N., Tanaka, S., Gao, L., Davey, B., Carpenter, B. L., Axthelm, M. K., Stanton, J. J., Smedley, J., Greene, J. M., Safrit, J. T., et al. (2020). The human IL-15 superagonist N-803 promotes migration of virus-specific CD8+ T and NK cells to B cell follicles but does not reverse latency in ART-suppressed, SHIV-infected macaques. *PLoS Pathogens, 16*, e1008339.

White, B. A., Creedon, D. J., Nelson, K. E., & Wilson, B. A. (2011). The vaginal microbiome in health and disease. *Trends in Endocrinology and Metabolism, 22*, 389–393.

Williams, W. B., Liao, H. X., Moody, M. A., Kepler, T. B., Alam, S. M., Gao, F., Wiehe, K., Trama, A. M., Jones, K., Zhang, R., Song, H., Marshall, D. J., Whitesides, J. F., Sawatzki, K., Hua, A., Liu, P., Tay, M. Z., Seaton, K. E., Shen, X., et al. (2015). HIV-1 VACCINES. Diversion of HIV-1 vaccine-induced immunity by gp41-microbiota cross-reactive antibodies. *Science, 349*, aab1253.

Williamson, B. N., Feldmann, F., Schwarz, B., Meade-White, K., Porter, D. P., Schulz, J., van Doremalen, N., Leighton, I., Yinda, C. K., Perez-Perez, L., Okumura, A., Lovaglio, J., Hanley, P. W., Saturday, G., Bosio, C. M., Anzick, S., Barbian, K., Cihlar, T., Martens, C., et al. (2020). Clinical benefit of remdesivir in rhesus macaques infected with SARS-CoV-2. *Nature, 585*, 273–276.

Wu, F., Ourmanov, I., Riddick, N., Matsuda, K., Whitted, S., Plishka, R. J., Buckler-White, A., Starost, M. F., & Hirsch, V. M. (2015). TRIM5alpha restriction affects clinical outcome and disease progression in simian immunodeficiency virus-infected rhesus macaques. *Journal of Virology, 89*, 2233–2240.

Wu, X., Guo, J., Niu, M., An, M., Liu, L., Wang, H., Jin, X., Zhang, Q., Lam, K. S., Wu, T., Wang, H., Wang, Q., Du, Y., Li, J., Cheng, L., Tang, H. Y., Shang, H., Zhang, L., Zhou, P., & Chen, Z. (2018). Tandem bispecific neutralizing antibody eliminates HIV-1 infection in humanized mice. *The Journal of Clinical Investigation, 128*, 2239–2251.

Xu, W., Li, H., Wang, Q., Hua, C., Zhang, H., Li, W., Jiang, S., & Lu, L. (2017). Advancements in developing strategies for sterilizing and functional HIV cures. *BioMed Research International, 2017*, 6096134.

Yao, Y., Bao, L., Deng, W., Xu, L., Li, F., Lv, Q., Yu, P., Chen, T., Xu, Y., Zhu, H., Yuan, J., Gu, S., Wei, Q., Chen, H., Yuen, K. Y., & Qin, C. (2014). An animal model of MERS produced by infection of rhesus macaques with MERS coronavirus. *The Journal of Infectious Diseases, 209*, 236–242.

Zhang, Z., Schuler, T., Zupancic, M., Wietgrefe, S., Staskus, K. A., Reimann, K. A., Reinhart, T. A., Rogan, M., Cavert, W., Miller, C. J., Veazey, R. S., Notermans, D., Little, S., Danner, S. A., Richman, D. D., Havlir, D., Wong, J., Jordan, H. L., Schacker, T. W., et al. (1999). Sexual transmission and propagation of SIV and HIV in resting and activated CD4+ T cells. *Science, 286*, 1353–1357.

Immunology of Nervous System Astrocytes, Oligodendrocytes, and Schwann Cells

7

Eliseo A. Eugenin

Abstract

Astrocytes are the most abundant cell type in the brain. They are involved in all the physiological functions of the nervous system. In pathological conditions such as Alzheimer's and Parkinson's diseases, multiple sclerosis, HIV-associated dementia, bacterial, fungal, parasitic, cancers, stroke, and traumatic brain injury, astrocytes play a key role in propagating disease and recovery. Other non-neuronal cell types, such as oligodendrocytes in the central nervous system and Schwann cells in the peripheral nervous system, are essential for the function of our bodies, but damage to them is irreversible. This chapter focuses on the historical and novel findings about the role of glial cells in healthy and disease-associated conditions.

Keywords

Astrocytes · Neurons · Oligodendrocytes · Schwann cells · Stroke · Regeneration · Stem cells · Cell-to-cell communication

7.1 General Introduction

The central and peripheral nervous systems (CNS and PNS, respectively) are composed of several cell types. These include endothelial cells (ECs), pericytes, microglia and brain macrophages, stem cells, neurons, ependymal cells, oligodendrocytes, and astrocytes. Glial cells include oligodendrocytes (Schwann cells in the periphery) and astrocytes, and they are categorized according to their shape, location, differentiation status, and function. Astrocytes are the most abundant cell type in the brain and create a unique intercommunicated network involved in neurogenesis, food supply, blood–brain barrier (BBB) integrity, ionic homeostasis, neurotransmitter metabolism, and synaptic activity/stability. The integration of these neuronal/glial functions results in complex tasks, including learning, memory, and behavior. Alterations in these glial-related mechanisms have been associated with aging and disease.

Neurons are classified according to their location, connectivity, and neurotransmitter identity. They could be myelinated or nonmyelinated. Myelinated axons enable better nerve impulse conduction due to the insulation of the action potential. Surprisingly, our knowledge about the interactions among these different cell types is still extremely limited, even though astrocytes and myelin-producing cells were described in the late 1800s. This chapter will describe the primary function and future challenges for the glial biology field in health and disease conditions.

7.1.1 The Ultrastructure of Glial Cells and Their Functional Consequences

Glia is a Greek word translating to "glue;" this name was suggested by Dr. Rudolph Virchow in the late 1800s to mean the connective tissue surrounding the neurons (Virchow, 1856). Dr. Virchow, a pathologist (at age 35 years), also provided the basis to disprove the "spontaneous generation theory" for cells and tissues. From his observations of cell division, he proposed that cells derive from other cells (*Omnis cellula e cellula*). Two years later, the first illustrations of glial cells were published (Virchow, 1858) and probably became one of the most outstanding books on glia/neuroglia biology (both publications are available at the Library of Congress, not in PubMed). In parallel, Dr. Heinrich Müller also published pictures of glial cells in the retina (now called Müller cells) in 1851 (Muller, 1851). Later, with the introduction of tissue staining, several new cell types were identified by Drs. Cajal, Golgi, and Bergmann

E. A. Eugenin (✉)
Department of Neurobiology, University of Texas Medical Branch (UTMB), Galveston, TX, USA
e-mail: eleugeni@utmb.edu

H. E. Gendelman, T. Ikezu (eds.), *Neuroimmune Pharmacology and Therapeutics*,
https://doi.org/10.1007/978-3-031-68237-7_7

(and several others) (see Fig. 7.1 for an example of Golgi's staining). Later, Drs. Rio-Hortega, Nissl, and Alzheimer suggested that some glial cells participate in pathological conditions, including dementia and other neurodegenerative diseases. These initial steps for identifying and characterizing reactive astrocytes and microglia remain the more beautiful examples of perseverance and discovery (Ramón y Cajal, 1995; Ramón y Cajal & Tello y Muñoz, 1933). We invite the new generations to read, learn, and challenge some of these early observations by these pioneers.

Despite the beauty of the original observations about neurons and glial cells, after decades of investigation, it has become evident that astrocytes are not just the "connective tissue or glue of the nervous system." They are critical brain regulators of an individual's life. Moreover, the highly heterogeneous nature of glial cells only confirms their plasticity and complexity.

7.1.2 Astrocytes

Astrocytes can be divided into multiple cell types based on their morphology, tissue location, cellular markers, neuronal interactions, and functional properties.

Astrocyte morphology or histological analysis As indicated above, several of the early histological definitions of glial cells are still valid, including their classification as protoplasmatic, fibrous, velate, inter-laminar, marginal, perivascular, radial astrocytes, pecten glial cells, Müller cells, and tanycytes depending on their interactions with neurons, the blood–brain barrier, the ventricle, or axons. Only in the last 10 years a new type of glial cell been identified, chondroitin sulfate proteoglycan 4 glial cells, also known as neuron glial antigen 2 (NG2), representing the fourth major population in the nervous system (5–10% of all brain cells), in addition to astrocytes, oligodendrocytes, and microglia (Dimou & Gotz, 2014). NG2 cells (also named oligodendrocyte precursor cells, OPCs) express the alpha receptor for platelet-derived growth factor (PDGFRalpha) and are distributed in the gray and the white matter. NG2 cells are the more active cycling population with a strong proliferative and differentiation capacity in the adult nervous system (Dimou & Gotz, 2014). They could differentiate into neurons (like brain stem cells) and other cell types, providing an excellent window for the next breakthrough in regenerative medicine. Moreover, we have to emphasize that most of these cell types are species-, development-, and location-dependent; thus, evolution has played a role in the development of these glial cells (Adams & Gallo, 2018; Bugiani et al., 2021; Dallerac et al., 2018).

Currently, it is accepted that most astrocytes are glial fibrillary acidic protein (GFAP)-positive. However, GFAP only stains the biggest processes. Fine astrocytic processes could reach up to 100,000 per astrocyte and interact with multiple synapse types, generating a unique network between

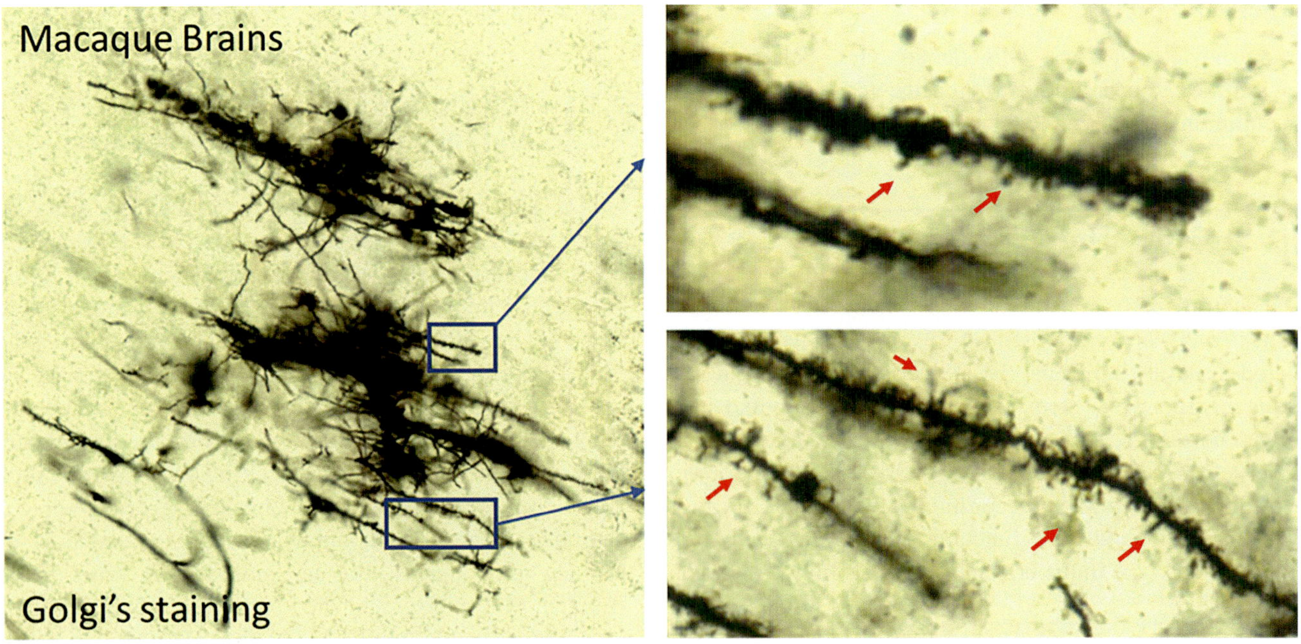

Fig. 7.1 Golgi's staining in the macaque cortex. An example of a group of neurons stained using Golgi's staining in the macaque cortex. The identification of multiple types of synapses on the dendrite or axons denoted by the red arrows should be noted. The blue boxes indicate the areas amplified on the right side. This staining corresponds to one of the first observations of cells and synapses

communication cells. Most groups named these kinds of interactions perisynaptic astrocytic processes (PVAPs). A substantial number of these cellular processes interact with other cell types, including blood vessels and other astrocytes, to form the larger syncytium within the CNS and PNS. Several groups propose that PVAPs are the main effectors of most astrocyte functions, including synaptic formation, collapse, and strength (Frizzo & Ohno, 2021; Lawal et al., 2022; Saint-Martin & Goda, 2023; Tan & Eroglu, 2021). Further, PVAPs modulate synaptic activity by controlling the baseline release of neurotransmitters into the synaptic cleft, including the neurotransmitters glutamate, GABA, acetylcholine, ATP, peptides, glycine, or adenosine. Any alterations in these functions (intracellular versus extracellular or recycling/storage) have been associated with neurodegenerative diseases (Frizzo & Ohno, 2021; Lawal et al., 2022; Saint-Martin & Goda, 2023; Tan & Eroglu, 2021). Lastly, PVAPs also surround arterioles to transduce neuronal activity to have better or restricted blood flow to support neuronal activity by altering the arteriolar diameter. However, overall, the function of these fine processes and their interactions are still poorly understood.

Astrocytes are functional units Several groups propose a different astrocyte classification, in addition to the histological one, based on functional anatomical units or specific tissue domains. This classification is based on data in rodents where a group of astrocytes poorly overlap (~10%) with a neighboring group of astrocytes on calcium signaling, generating functional anatomical domains as demonstrated in the rat hippocampus (Bushong et al., 2002; Ogata & Kosaka, 2002), but these functional units have not been proven in humans. This kind of glial specialization is believed to be calcium-, gap junction-, and adenosine-dependent (Iadecola & Nedergaard, 2007) and is organized as a Krogh tissue cylinder, which comprises around 50 μm of intercapillary distance (Auer et al., 1979). The cylinder organization is essential for calcium signaling, blood flow control, localized metabolism (glucose/lactate), integration of neuronal activity, K^+ buffering, and learning and memory (Scemes & Spray, 2012). Our data in uninfected and HIV-infected human brains support the existence of these functional units; however, each unit covers around 250–300 μm, suggesting a species specificity only present in humans so far (see example in Fig. 7.2 for a human brain, gray matter) (Eugenin & Berman, 2007; Eugenin et al., 2011; Malik et al., 2017, 2021; Valdebenito et al., 2021a). A critical complexity layer in astrocyte biology is their extensive communication by gap junction channels. Astrocytes form a syncytium connecting the cytoplasm of multiple cells by gap junction channels (Bennett et al., 2003). The interactions between glial and neuronal gap junctions and the tripartite synapses further expand the syncisium into extensive brain areas, contributing the the generation of brain waves such as gamma, beta, alpha, theta, and delta. This spatial complexity provides a unique function for these cells and a significant challenge for tissue repair and interventional approaches in all neurodegenerative diseases. All these areas are totally unknown, and future generations have a major challenge for cure efforts.

7.1.3 Oligodendrocytes and Schwann Cells

Both cell types are defined as myelin-producing cells. The main function of myelin is to insulate axons and enable efficient impulse transmission. Dr. Rio-Hortega (1928) characterized four types of oligodendrocytes based on their histology. He postulated that oligodendrocytes were involved in myelin formation, but these ideas were only demonstrated with the introduction of electron microscopy in the early 1960s (Peters, 1964a, b). The types of myelin-producing cells were defined according to the cell shape, numbers, the orientation of the cellular processes, and the distribution and size of the axons covered. Type I oligodendrocytes are located in the forebrain, cerebellum, and spinal cord. Type II are located only in the white matter. Type III is localized in the cerebral and cerebellar peduncles, the medulla oblongata, and the spinal cord. Type IV is restricted to large-diameter axons with a similar distribution to type III oligodendrocytes. Type IV oligodendrocytes are rarely present in small animals, suggesting that evolution has resulted in the generation or differentiation of these myelinic cells only in large animals. Interestingly, a single oligodendrocyte can establish 5–30 interactions with specific axons, suggesting a critical role of cell-to-cell communication between different axons and the surrounding oligodendrocytes (Anderson & Key, 1999; Hildebrand et al., 1993).

There are two types of Schwann cells derived from the neural crest: myelinating and nonmyelinating Schwann cells (both have myelin despite their names). Nonmyelinated cells collectively ensheath groups of small-caliber axons, whereas myelinating cells are always associated with a single axon of large caliber, especially in large animals. A third type of Schwann cell has been described at the neuromuscular junction (not discussed in the current chapter). The three types of Schwann cells form myelin, like oligodendrocytes within the CNS. Myelin is mostly composed of lipids, cholesterol, and a small protein/proteolipid fraction, including myelin basic protein (MBP) and proteolipid protein (PLP). Several groups have postulated that oligodendrocyte/Schwann cell maturation and axonal growth are interdependent; thus, both must occur concomitantly to define the molecular signature of each axon and its interactions. This point is essential because in pathological conditions such as multiple sclerosis (MS), stroke, and spinal cord injury (SCI), reestablishing these

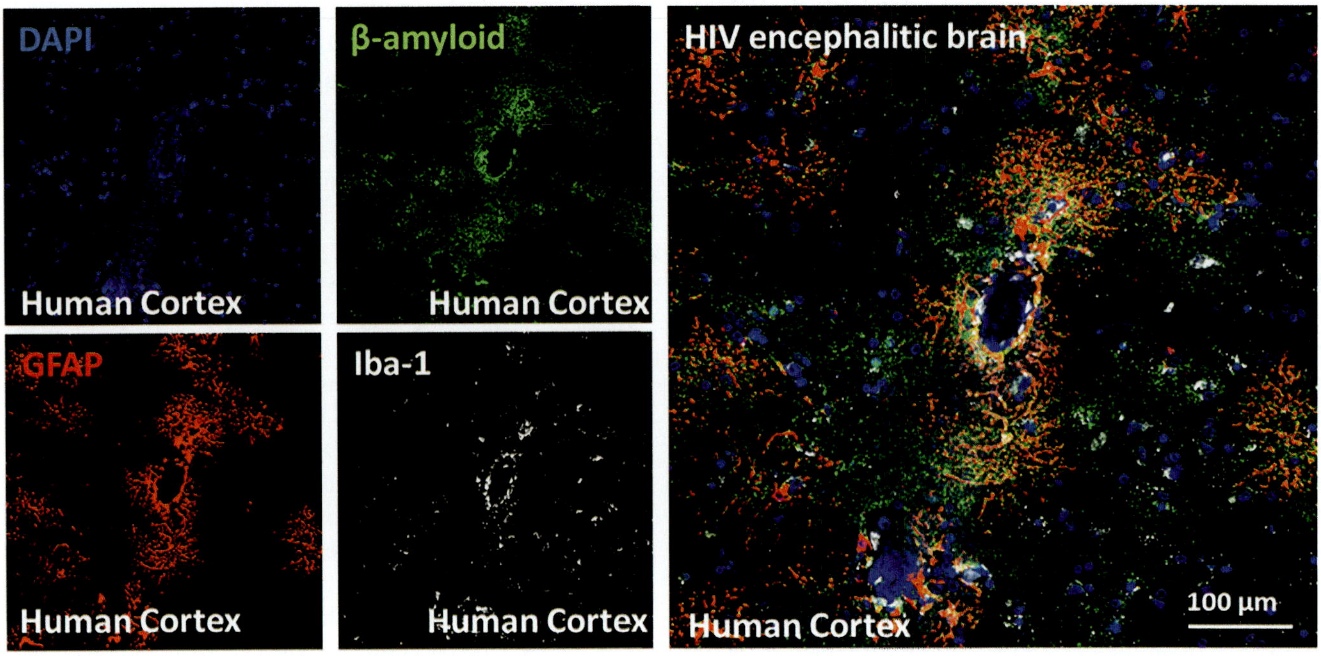

Fig. 7.2 The human cortex from an HIV-infected individual with encephalitis and stained for DAPI (to observe nuclei), β-amyloid, GFAP (to observe astrocytes), Iba-1 (to observe microglia/macrophages), and the merge of all colors. Note the interaction of the astrocytes with the blood vessels (center of the picture) and microglia/macrophages associated with the neovascular unit. In inflammatory conditions such as HIV encephalitis, astrocytes and microglia/macrophages become highly activated, as indicated in the figure

interactions and coordinating these mechanisms are essential for proper function recovery. Reprogramming or reestablishing these interactions is probably one of the biggest challenges in regenerative medicine.

7.1.4 Glial Lineage

Neuroepithelial cells (NECs) in the ventricular zone correspond to a unique group of multipotent stem cells with the capacity to generate neurons and macroglia (astrocytes and oligodendrocytes) during adulthood. For 130 years until 1980, most scientific communities accepted that these progenitor cells sequentially produce neurons and glial cells. This idea of development divergence between neurons and glial cells creates the belief that no new neurons were generated during adulthood despite the generation of new glial cells. However, glial lineage is an active area of investigation and controversy. The identification of new differentiation and proliferating markers demonstrated ongoing neurogenesis in the adult vertebrate brain and suggested a parallel development of both lineages upon differentiation (Avaro et al., 2022; Curtis et al., 2007; Huszar et al., 2022; Manfredi-Lozano et al., 2022).

The identification of Olig1/2 and O-2A$^+$ glial progenitor cells provided researchers with the tools or markers to identify and follow the differentiation of astrocyte and oligodendrocyte lineages in vitro and in vivo (Yakovlev et al., 1998). Several studies have identified that mutations of the *Olig2* gene compromised motor neuron generation and oligodendrocyte precursor cells (OPCs) but did not compromise the development of interneurons and astrocytes, supporting a different lineage in the brain and spinal cord (Marshall et al., 2005; Yang et al., 2022). These data suggest that the glial and neuronal lineage was dependent on location, differentiation, and species.

Currently, two germinal regions (we will not focus on the developmental brain or spinal cord in the current chapter) are described within the adult mammalian brain that contains neuronal progenitor cells: the subventricular zone (SVZ) and the subgranular zone (SGZ) within the dentate gyrus of the hippocampus (Bond et al., 2021; Feliciano et al., 2015; Gomes-Leal, 2021; Kalinina & Lagace, 2022; Nogueira et al., 2022; Ribeiro & Xapelli, 2021). The SVZ is proposed to be more active than the SGZ during adulthood due to the extensive cell-to-cell interactions, long-distance signaling from cerebrospinal fluid (CSF), and the surrounding blood vessels that promote cell division, differentiation, and migration into specific brain areas. The main signaling pathways and the environment associated with these stem cell niches will be discussed below.

A critical question for the following chapters is whether adult stem cells retain their capacity to generate early-born cell types and their complex interactions and whether we can induce or control them to repair a compromised nervous system. During development, various neuronal stem cells change morphology, location, and genetic profile/expression to generate different cell types. Later in adulthood, these mechanisms suffer a chronic "wear," preventing an effective differentiation compared to early developmental stages (Bond et al., 2021; Nogueira et al., 2022; Ribeiro & Xapelli, 2021). Even the transplantation of SVZ cells into the brain parenchyma, does not promote stem cell development, suggesting that disruption of the tissue cytoarchitecture compromises the germinal niche. These restriction mechanisms in adulthood have only recently been explored and could provide essential information to design new approaches for regenerative medicine.

7.1.5 Glia's Molecular Identity

A major difference between the SVZ and SGZ is the direct contact with the CSF secreted into the choroid plexus. CSF composition changes during corticogenesis, profoundly changing stem cells' differentiation and migratory properties. Currently, the timeline of brain differentiation and the composition of CSF at different stages of development are unknown. Also unknown is how to recapitulate specific developmental conditions to promote regeneration upon injury.

However, not all morphogens are soluble molecules; several of them are guided or targeted by specific long-range cell-to-cell interactions (named cytomenes or tunneling nanotubes) to "deliver" the right cargo cell into specific brain areas or cells (see review by Eugenin et al. (2009), Peretto and Paredes (2014), and Valdebenito et al. (2021b)). This area is so novel that early experiments during the development of the fly wings clearly indicated a highly orchestrated mechanism of targeted delivery of growth and differentiation factors from and to specific cell types (Chen et al., 2017; Huang et al., 2019; Patel et al., 2022; Sohr et al., 2019; Wood et al., 2021). However, most of these mechanisms are unknown. Below, we describe different variables involved in the formation and stability of the niche.

The extracellular matrix (ECM) influences stem cells' survival and differentiation As indicated above, there are two main sites for homing stem cells in the adult brain: SVZ and SGZ; both niches have strong cell-to-cell interactions with ECM and integrins and indirectly with CSF. ECM has a significant number of RGD sequences that enable the binding of integrins on the cell surface to promote not only adhesion but also differentiation. Integrins are a large family of heterodimeric transmembrane proteins (binding to a variety of molecules and working as anchors and receptors) composed of α- and β-subunits providing the binding specificity to ECM molecules such as fibronectin, laminin, and collagen. ECM–integrin interactions or the loss of their interactions induce profound changes in kinase activity to promote the proliferation and differentiation of glial stem cells at SVZ and SGZ. However, a detachment of the stem cells from the basal membrane could also result in apoptosis, suggesting that integrins can sense the environment and changes in CSF (Lathia et al., 2007; Stenzel et al., 2014). These ECM–integrin interactions are essential for tumor development and metastasis, suggesting that similar differentiation and migration mechanisms could be used to target cancer cells during adulthood (Deb et al., 2022; Kuznetsova et al., 2023; Oxburgh, 2022; Sirica, 2022). Some key examples involving stem cell differentiation and proliferating factors within the CNS are discussed below.

Fibroblast growth factors (FGFs) FGFs comprise a family of 22 soluble ligands that transduce into 4 tyrosine kinase receptors (R1–R4), resulting in Ras/Erk1/2 mitogen-activated protein (MAP) kinase activation and phosphoinositide 3-kinase (PI3 kinase), Akt, and protein kinase C (PKC) pathways (Woodbury & Ikezu, 2014). FGF signaling promotes and maintains the formation of self-renewal of progenitor cells. Deleting FGF receptors (FGFRs) in the dorsal telencephalon induces neurogenesis, resulting in reduced cortical areas. Alterations in FGF signaling also compromise Notch signaling, suggesting critical cross talk between these two pathways. Several groups have demonstrated that FGF signaling maintains the stemness of progenitor cells, resulting in controlled brain growth and differentiation (Allodi & Hedlund, 2014; Campos et al., 2020; Fares et al., 2019; Hagemann & Scholpp, 2012; Paradiso et al., 2013).

The Wnt/β-catenin signaling pathway Wnt ligands correspond to a family of secreted glycoproteins that bind to Frizzled and the lipoprotein receptor-related protein (LRP) 5/6 on the surface of stem cells during CNS development (Agirman et al., 2017). These interactions stabilize the adhesion/transcriptional factor, β-catenin, to avoid its degradation by phosphorylation by GSK3β. β-Catenin stabilization enables its translocation into the nucleus to associate with TCF/LEF transcription factors to promote transcription of the key genes associated with cell differentiation and proliferation (Mi & Johnson, 2007). The activation of β-catenin in progenitor cells and the nascent neuronal population results in an upregulation of N-myc and Ngn1 transcription, promoting the generation of oligodendroglial progenitors during late embryogenesis stages. Several groups have demonstrated that these pathways promote self-renewal of neuroepithelial cells (NECs), and radial glial cells (RGCs), favor neuronal differentiation as development proceeds, and finally induce oligodendrocyte differentiation from RGCs at the end of corticogenesis (Bhattacharya et al., 2021; Caron et al., 2022; Morante-Redolat & Porlan, 2019; Suciu & Caspary, 2021; Ventura et al., 2019). An important second function of the Wnt pathway is to regulate adherent junctions (AJs) to support stem niche formation and the associated differentiation/migration—all essential components for stem cell generation, differentiation, and migration.

The sonic hedgehog (Shh) signaling pathway Shh is a soluble protein secreted from the ventral telencephalon into CSF, but it is also produced by the choroid plexus in the marginal zone of the cerebral cortex. The Shh receptor Patched1 (Ptch1) and its related signaling components accumulate in the primary cilium of these cells. Upon Shh binding, Ptch1 translocates from the primary cilium, which relieves its inhibitory activity on Smoothened (Smo) to activate G-coupled protein and downstream Gli2 and Gli3 transcription factors, otherwise processed into truncated isoforms that act as transcriptional repressors (Embalabala et al., 2022). The molecular targets of Gli2 and Gli3 include Gli1, which enhances the signaling; Ptch1, which negatively regulates the signaling; Nkx2.1, which controls some aspects of the ventral telencephalon specification; FGF15; and cell cycle regulators, such as cyclin D_1, D_2, and E (Garcia, 2021; Hill et al., 2021; Nian & Hou, 2022; Tayyab et al., 2018). However, until now, Shh signaling has been proposed to be limited in the developing cortex but essential in the ventral telencephalon, suggesting region-specific signaling.

Notch signaling Notch is the receptor of a conserved signaling pathway mostly present during development and carcinogenesis (Nian & Hou, 2022). Notch signaling involves its proteolytic cleavage, releasing an intracellular fragment (Nicd) to regulate transcription and differentiation through an ADAM10 and γ-secretase mechanism. Nicd displaces a repression DNA-binding complex, DNA-binding CSL (CBF1, Su(H), and LAG-1), promoting the transcription of key genes such as the *Hairy enhancer of split* (*Hes*). *Hes* is a basic helix–loop–helix (bHLH) transcription factor that represses the expression of proneural genes, further ensuring RGC stemness maintenance and preventing neuronal generation (Kageyama et al., 2015; Kaise & Kageyama, 2021). *Hes* transcription follows an oscillatory pattern because *Hes* is a negative regulator of the his promoter and depends on the cell cycle. In the case of an asymmetric division of RGCs, the *Hes1* gene is epigenetically silenced, enabling some proneural genes, such as *Ngn2* or *Ascl1* or *DLL* or *MIB1*, to become dominants and promote the differentiation of daughter cells into a neuron or progenitor cells (Agirman et al., 2017). Overall, Notch regulates self-renewal and the survival and differentiation of daughter cells.

Bone morphogenetic proteins (BMPs) BMPs are members of the TGF-β superfamily that bind to heterotetrameric complexes composed of pair type I/II receptors and co-receptors, resulting in the phosphorylation of cytoplasmic R-Smads 1, 5, or 8, which bind to co-Smad4 and translocate into the nucleus to initiate transcriptional activity (Kashima & Hata, 2018; Yamashita et al., 2016). BMPs cooperate with Wnts to form the telencephalon and the choroid plexus. Similar to β-catenin and Shh, a BMP has an additional function of controlling the transition of multipolar to bipolar neuronal processes in conjunction with the microtubule-associated protein CRMP2 (collapsin response mediator protein 2), suggesting a dual role in postmitotic events. BMP-2 and BMP-4 signaling in the cortex favors neuronal differentiation by several mechanisms involving deacetylase HDAC6, a negative regulator of BMP-2/4 signaling.

Overall, several of the above examples denote exquisite signaling coordination for proper neuronal/glial development. However, we must consider that none of the pathways described above alone are sufficient to generate or recapitulate the stem cell niche described in the adult brain; thus, more research is needed to identify the proper combination of these molecules for regenerative medicine.

7.2 Glia in Health and Disease

Glial cells participate in most functions of the PNS and CNS, including controlling neuronal activity, axon potential, metabolism, immune activation, differentiation, learning/memory, and BBB integrity. However, all brain pathologies have also shown the participation of glial cells in the pathogenesis of the brain. Below, we describe the critical functions mediated by glial cells compromised during different diseases or aging conditions.

Astrocytes and the control of neuronal activity Here, we will focus on the mechanisms by which astrocytes regulate neurotransmitter metabolism. Astrocytes can regulate neuronal activity by altering ionic pools, cell polarization, metabolism, and expressing synaptic and extrasynaptic receptors/transporters. In addition, astrocytes form the more extensive syncytium within the CNS via gap junctions to establish long-range calcium, IP_3, electrotonic ionic, and BBB function to reach vast brain areas. Astrocytic gap junctions are essential to provide active or overactive synapses with an efficient K^+ buffering system to prevent neuronal overactivation or hyperpolarization upon repetitive action potential (Astiz et al., 2022; Durkee et al., 2021; Heuser & Enger, 2021; Oliveira & Araque, 2022; Saint-Martin & Goda, 2023; Takahashi, 2022).

In neurons, ion flux across the plasma membrane generates the action potential, resulting in extracellular K^+ concentrations $[K^+]_e$ from 2.5 mM to an additional 0.4 mM. However, in diseases such as epilepsy, $[K^+]_e$ could reach 10–12 mM; in animal models of spreading depression, the values can rise to 55 mM, representing a failure in $[K^+]_e$ regulation. The critical point is that neurons in a resting potential (no firing) are insensitive to changes in $[K^+]_e$; however, astrocytes are highly sensitive to those changes, making them the brain K^+ sensors (Orkand et al., 1966). This point is essential because glial cells correlated with the electrogenic transporter, Na^+-$HCO^{3}-$ cotransporter, ammonium and pH, and neurotransmitter uptake (driven by the Na^+ gradient) that can be reverted if any of these conditions are altered. Other chapters will explore the neuronal signaling described below in depth.

For example, glutamate is the most abundant neurotransmitter in the brain. Astrocytic-derived glutamate has been shown to enhance the probability of neurotransmitter release at the CA3–CA1 transient synapses through activation of type 1 glutamate receptors and to potentiate or depress inhibitory transmission through activation of kainate or group II/III metabotropic glutamate receptors, respectively (Escartin et al., 2021; Kastanenka et al., 2020; Perea & Araque, 2005, 2010; Perea et al., 2014).

ATP and adenosine are critical regulators of synaptic activity and depressed neurotransmission (Ben Achour & Pascual, 2012; Pascual et al., 2005). Astrocytic purinergic signaling also mediates heterosynaptic depression, which has been proposed to involve GABAergic neurons by a mechanism involving dysregulated astrocytic calcium signaling (Crosby et al., 2018; Jacob et al., 2014; Khan et al., 2019; Koizumi, 2015; Lalo et al., 2019; Lee et al., 2013; Wirkner et al., 2005). Dysregulation of these circuits induces changes in synaptic strength that could result in abnormal neuronal and glial activation stages, triggering loss of sensitivity, excitotoxicity, and compromised brain function. A critical mechanism of synaptic plasticity is long-term potentiation (LTP) or long-term depression (LTD), which is essential for long-term memory (Bazzari & Parri, 2019; Goncalves-Ribeiro et al., 2019; Klein, 2009; Sykova, 2001; Thalhammer & Cingolani, 2014). Astrocytes play a key role in neuronal function despite not being electrically excitable; they control local synaptic activity through indirect mechanisms such as calcium signaling, neurotransmitter metabolism, and networking via gap junctions. It is critical to understand that neurons and astrocytes have bidirectional signaling between both cell types, and equilibrium between excitatory and inhibitory synapses not only facilitates neuronal plasticity but also participates in critical events in the pathogenesis of several diseases, including Alzheimer's, Parkinson's, and multiple sclerosis.

The blood–brain barrier (BBB) The BBB is a complex barrier comprised of endothelial cells (ECs), pericytes, neurons, and glial cells. ECs at the BBB have unique morphological and functional properties (Rubin & Staddon, 1999). The BBB is characterized by interendothelial adherent junctions (AJs) and tight junctions (TJs) that form impermeable seals between ECs as well as the lack of fenestrations and several specific transport mechanisms in the luminal and abluminal sides of the BBB to regulate transport in and out of the nervous system. Astrocytes play a key role in supporting neurons and the endothelial layer's structure, differentiation, and permeability; in this way, astrocytes and pericytes are the only communicators between endothelial cells and neurons (see example in Fig. 7.3).

The primary function of the BBB is to protect the uniqueness of the nervous system's homeostasis through three specific mechanisms: first, by providing a physical barrier to isolate the nervous system from the blood using AJs and TJs. Astrocytes are essential to maintain the differentiation and localization of these AJs and TJs. AJs comprise transmembrane adhesion proteins of the cadherin family, which mediate homophilic adhesion at the cell borders. Brain ECs are unique because they expresses a specific cadherin, vascular endothelial (VE)-cadherin, which links AJs to actin (Bazzoni & Dejana, 2004). TJs comprise continuous, anastomosing intramembrane strands that inhibit paracellular exchange and maintain cell polarity (Mitic & Anderson, 1998; Prat et al., 2001; Rubin & Staddon, 1999). These strands comprise several proteins, including the integral membrane protein, occludin, the family of transmembrane proteins, claudins, and the peripheral membrane protein, zonula occludens-1 (ZO-1). The BBB also excludes most circulating leukocytes, although there is baseline trafficking of lymphocytes and monocytes. In pathological conditions, altered expression of chemoattractants and adhesion/tight junction molecules is associated with BBB disruption, enhanced leukocyte adhesion, and transmigration across CNS vessels (Sobel et al., 1990), resulting in subsequent neuronal/glia damage (Rubin & Staddon, 1999; Tsukita & Furuse, 1999). Second, the BBB is a selective transport system between the blood and neurons. Astrocytes are the most active cell type in the brain, taking up, releasing, and specifically regulating intercellular communication to control blood flow and associated metabolism in conjunction with pericytes. Some of these critical functions involve P-glycoprotein, which is a multidrug resistance-associated transporter for small compounds, including several drugs. ATP-binding cassette (ABC) transporters excrete xenobiotics such as antibiotics, dexamethasone, cancer drugs, morphine, and others, providing a significant barrier to drug delivery into the brain, making the CNS not only an immune sanctuary but also a hard tissue for disease treatment. In addition, at the BBB, there is also sodium uptake via luminal Na^+ channels and abluminal Na^+, K^+ ATPase with passive water transfer. In contrast, K^+ is excreted via basal K^+ channels and apical outward-directed K^+ channels. Transfer and accumulation of multiple hormones, metabolites, and drugs by endocytosis-mediated

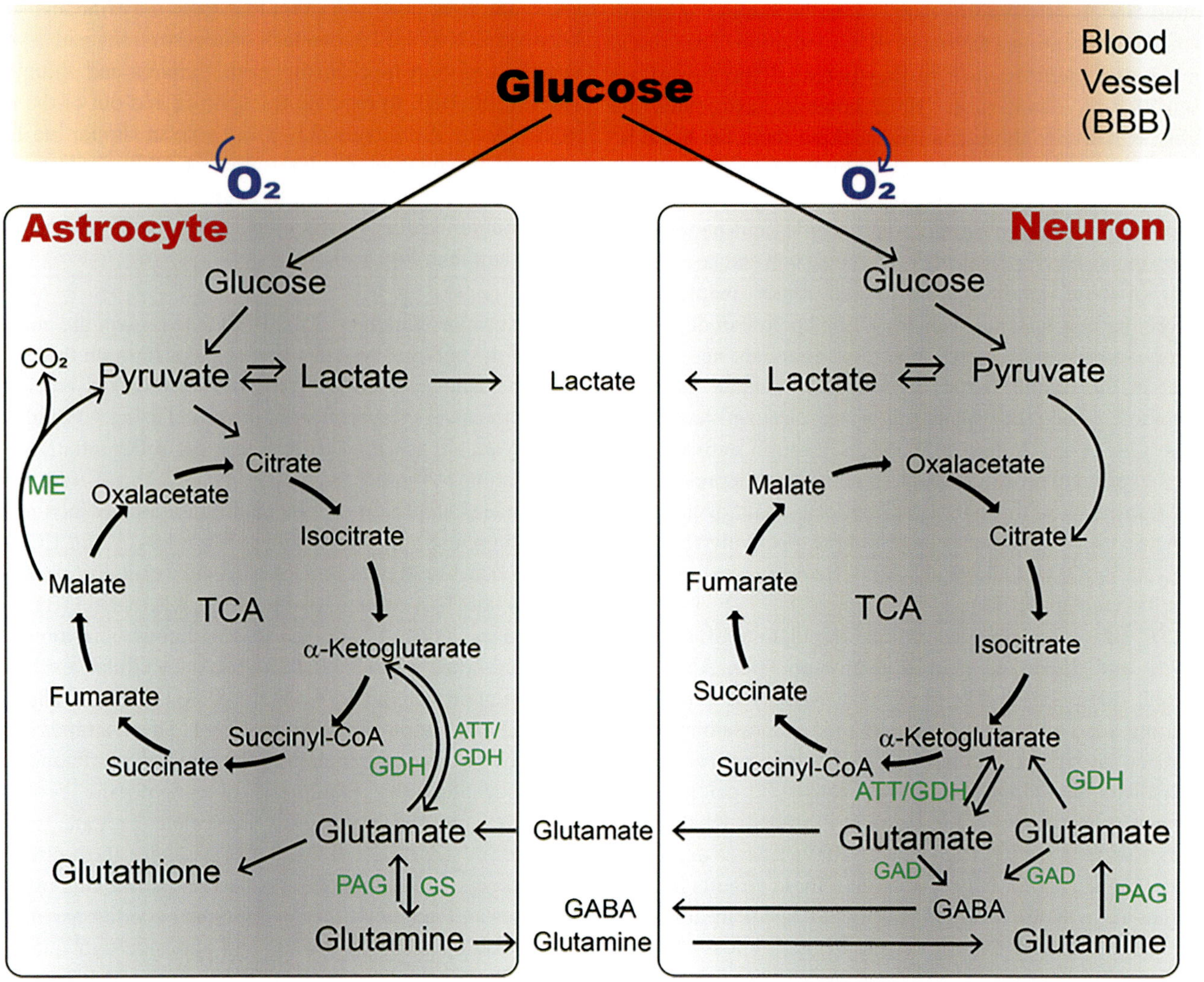

Fig. 7.3 A simplified scheme of glutamate and glutamine metabolism between astrocytes and neurons in the brain. The close interactions between the blood vessels (the blood–brain barrier, BBB) and the perivascular neurons and astrocytes process glucose into pyruvate to feed TCA. As indicated at the bottom of the cartoon, glutamate is released into the extracellular space and taken up by astrocytes to convert into glutamine to return to the neurons for glutamate and GABA storage. Several key enzymes involved in this process are indicated, including GDH, glutamate dehydrogenase; ATT/GDH, aspartate aminotransferase/GDH; PAG, phosphate-activated glutaminase; GS, glutamine synthetase; ME, mitochondrial malic enzyme; and GAD, glutamic acid decarboxylase. Several pathogens "learn" to take advantage of these synaptic mediators' high concentration and distribution (glutamate and metabolites) as a major energy source. Some of these are HIV viral reservoirs and glioblastoma stem cells

transport (soluble or receptor-mediated) also take place. Third, astrocytes also help control and regulate cell transmigration and immune patrol, including T cells, in healthy and inflammatory conditions. Fourth, astrocytes provide proper oxygenation and energy supply, including glycogen stores, to support neuronal activity. However, astrocytes can "overreact" and enhance the cytokine storm and amplify inflammation and CNS damage upon activation.

Neuro-energetics and neurotransmitter metabolism Astrocytes control both functions, and they are highly interrelated. First, the glutamate/glutamine (GABA, ATP, adenosine, and others) cycle to support glutamatergic synapses. Glutamate is the major excitatory neurotransmitter of the nervous system. Glutamate uptake is mediated by specific transporters, glutamate transporter-1 (GLT-1), and glutamate/aspartate transporter (GLAST). Upon uptake, it can be converted into glutamine by the enzyme glutamine synthetase or metabolized in the tricarboxylic acid cycle (TCA) into α-ketoglutarate by the sequential actions of the enzymes glutamate dehydrogenase (GDH), aspartate aminotransferase (AAT), or alanine aminotransferase (ALAT). The produced glutamine enables astrocytes to release large amounts of glutamine in areas with neuronal electrical activity or action potential. Then, glutamine is taken up by neurons and astrocytes by glutamine transporters. Glutamine in neurons is metabolized into glutamate by the phosphate-activated glutaminase concentrated mainly in neuronal mitochondria (abundant in the synapses). Then, glutamate is loaded into synaptic vesicles by the action of vesicular glutamate transporters to recycle the neurotransmitter into the synaptic cleft. A second possible pathway of neuronal glutamate synthesis is the use of α-ketoglutarate exchange between glia and neurons. In this case, glutamate can be formed by adding the amino group by conjugating free ammonia or leucine, both provided by the circulation. However, this second mechanism is minor compared to the first mechanism of glutamate recirculation because excessive use of α-ketoglutarate can compromise the TCA (Bowtell & Bruce, 2002; Corbet & Feron, 2015; Hertz et al., 1999; Maus & Peters, 2017; Shah & Chen, 2020; Shen et al., 2021).

Several diseases have "learned" to utilize these neurotransmitter pathways to survive and perpetuate disease, including HIV brain reservoirs (Castellano et al., 2019; Edara et al., 2020; Gorska & Eugenin, 2020) and aggressive glioblastoma stem cells (Tardito et al., 2015; Valdebenito et al., 2020, 2021b). Both diseases use glutamate/glutamine as a key energy source to survive within the CNS for extended periods because most of the brain has significant amounts of glutamate and glutamine for synaptic use. Clearly, dissecting these synaptic and metabolic mechanisms provides not only essential information about synaptic plasticity but also potential therapeutic approaches to reach a cure for HIV and glioblastoma.

Cytokine production Under healthy conditions, astrocytes and resident macrophages/microglia play a key role in CNS homeostasis, including the secretion of growth factors, neurotransmitters, chemokines, and vascular-related agents to maintain immune surveillance, BBB control, and synaptic function. However, in pathological conditions, the brain responds aggressively. For example, if systemic inflammation is detected, cells at the leptomeninges and choroid plexuses sense systemic inflammation to prepare the CNS for pathological conditions, including fever and vascular damage that could result in indirect CNS compromise as shown in sepsis, Reye's syndrome, syphilis (neurosyphilis), and ischemia/perfusion conditions. Another type of CNS damage that correlates with glial activation is a direct attack on the CNS, including meningitis, mechanical damage, encephalitis, cancers, bleeding, myelitis, and stroke. In both cases, these diseases are detected by clinical assessment and lumbar puncture to detect changes in CSF. As indicated above, CSF is key to the development and function of the brain. If the CNS damage is localized, astrocytes and macrophage/microglia become activated or hypertrophic in addition to secreting cytokines/chemokines to encapsulate the damage and prevent its amplification into unaffected areas. Glial activation generally produces specific chemokines or nucleotides, such as ATP, to recruit leukocytes to the injury site for resolution. However, full damage resolution is rare, especially in adults; most CNS damage leaves scars or calcifications that can be detected by different imaging techniques, including MRI/MRS. Upon damage, the neuronal function or survival is compromised according to the damage's extent and nature. Several cytokines have been associated with disease, including interferons, interleukins (−1, −6, −8, −10, −12, −15, −18, and TNF-α), inflammatory lipids (arachidonic acid and prostacyclin), chemokines (CCL2, CXCL8, CXCL12, CXCL10, CCL5, and CCL3), ATP/ADP/adenosine, and growth factors (MCSF, NGF, BDNF, PDGF, and GDNF). Most of these factors amplify inflammation and tissue compromise, but several are involved in tissue repair, including NGF, GDNF, LIF, CNTF, bFGF, and PDGF (Fig. 7.4).

However, we have to denote an efficient repair mechanism dependent on these factors' coordinated and equilibrated secretion. For example, ATP release is a major mechanism to find and clear apoptotic cells; however, in HIV-associated cognitive impairment or Alzheimer's disease, ATP raises, compromising the BBB and cognition, suggesting that the CNS equilibrium is delicate (Castillo et al., 2022; D'Amico et al., 2021; Hernandez & Eliseo, 2022; Khan et al., 2022; Singh et al., 2022; Velasquez et al., 2020). A similar mechanism has been described in severe cases of COVID-19, resulting in long-lasting NeuroCOVID (Tavcar et al., 2021). Another example is that CCL2 is normally expressed at low levels as a constitutive chemokine within the brain to regulate basal cell trafficking; however, in HIV conditions, infected cells upregulate the expression of CCR2, the receptor for CCL2. In these cases, HIV-infected monocytes become more sensitive to lower chemokine concentrations, resulting in their transmigration into the brain (Eugenin et al., 2006). In these conditions, pathogens such as HIV have evolved to use chemokine CNS gradients and the deleterious effects of cytokines at the BBB to induce damage to the central and peripheral nervous systems. In stroke conditions, the cytokine storm compromises most of the tissue under hypoxic conditions but simultaneously reduces the expression of growth factors required for neuronal/glial protection and proliferation of stem cells to repair the damage or help prevent later cell death in the penumbra area. Thus, cytokines/chemokines/growth factors could have a dual role in short- and long-term disease.

Antigen presentation and the generation of autoreactive immune cells The BBB protects the nervous system by excluding the passage of pathogens and immune cells and the diffusion of immune complexes, such as antibodies, into

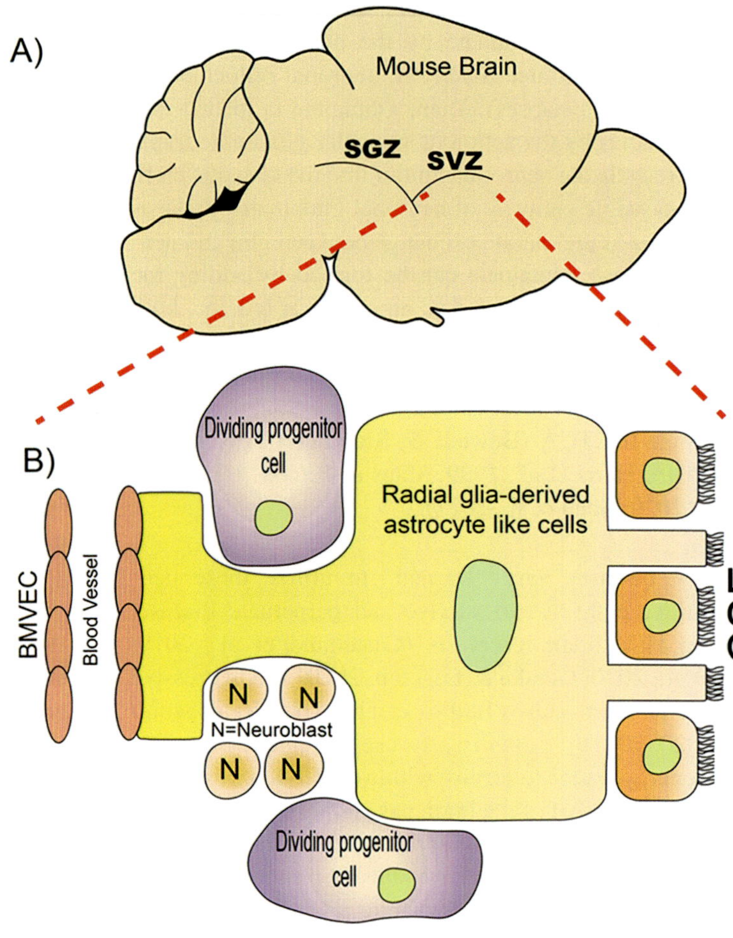

Fig. 7.4 Scheme of the location and basic structure of the SGZ and SVZ areas in the mouse brain. (**a**) The SGZ and SVZ areas produce new glial and neuronal cells in the adult brain, and the structure of the SVZ is indicated in panel (b). (**b**) A cartoon indicating the basic structure of the SVZ area, including direct access to the blood vessels and CSF with radial glia-derived astrocyte like cells, ependymal cells, neuroblasts, and dividing progenitor cells

the brain parenchyma. Under healthy conditions, no antigen presentation occurs within the CNS despite the constant transmigration of surveyed/patrol T cells into the CNS (Masthoff et al., 2021; Mundt et al., 2019; Ousman & Kubes, 2012). Normally, antigens cannot directly stimulate T cells. T cells need antigen-presenting cells that process and present the antigen in MHC class II (foreign antigens) or MHC class I (host antigens in the case of cancer and maybe MS) to recognize them and become activated.

Autoreactive T cells to, myelin proteins and phospholipids has been identified in, MS and other autoimmune diseases raising the possibility that not all CNS host antigens are presented to the immune system during development and small exposure during adulhood can contribute the generation of several auto-immune diseases. Upon disease, the immune system recognizes these host antigens, potentially contributing to the pathogenesis of several autoimmune diseases (D'Angelo et al., 2019; McFarland & Martin, 2007). To further support the presence of these myelin antigens, treatment of human microglia/macrophages with soluble MBP or PLP can induce inflammation (D'Aversa et al., 2013), suggesting that our immune system is not fully exposed to these brain antigens. These hypotheses are still under active investigation, especially in autoimmune diseases like MS and amyotrophic lateral sclerosis (ALS) and the long-term consequences of HIV, stroke, COVID-19, and spinal cord injury (SCI).

Inflammation and transmigration of immune components Leukocyte transmigration requires the opening of interendothelial cell junctions to allow for cell passage between tightly apposed ECs at the BBB, yet the barrier properties of the endothelium must be preserved to inhibit the uncontrolled transit of cellular and fluid components through the vessel wall into the parenchymal tissue. Leukocyte transmigration into the brain has been characterized as a dynamic multistep process involving the initial "rolling" of cells on the vessel endothelium in response to locally produced pro-inflammatory mediators and subsequent firm adhesion to and diapedesis across the BBB. The rolling of leukocytes along the endothelial surface is mediated by weak interactions between selectin molecules and their corresponding glycoprotein ligands, expressed by activated ECs and leukocytes. Rolling leukocytes are then stimulated by chemokines and other chemotactic molecules, activating the β_1-integrin, VLA-4, and the β_2-

integrin, LFA-1. Ultimately, leukocyte arrest occurs, mediated by strong interactions between LFA-1 and its EC counter-receptor, ICAM, and between VLA-4 and its EC counter-receptor, VCAM-1 (Liu et al., 2004; Muller, 2003; Springer, 1994; van Buul & Hordijk, 2004). The induction of active, cytoskeleton-driven migration culminates in leukocyte diapedesis across the blood vessel endothelium, a process mediated, in part, by homophilic and heterophilic binding molecules such as JAM (JAM-A, JAM-B, and JAM-C), platelet endothelial cell adhesion molecule 1 (PECAM-1; CD31), and CD99. Regulatory mechanisms to maintain low baseline levels of leukocyte extravasation across the BBB are important to minimize the deleterious effects of inflammatory events within the CNS parenchyma. Aberrant expression of adhesion molecules may be an early step in developing the pathology of several diseases, including genetic, environmental, stroke, infectious, and acquired. Astrocytes play a key role in all these processes because they are the main producers of several chemotactic agents.

Myelin and remyelination Myelinization is one of the key evolutionary events in vertebrates. Myelin is required for a much higher electrical conduction velocity for more extensive and more complex bodies despite rudimentary myelin elements being expressed in small bodies (Zalc et al., 2008). As described above, this key evolution point requires the divergence between astrocytes and oligodendrocytes/Schwann cells to enable the saltatory conduction and the clearing of high extracellular K^+ concentrations upon action potential by perisynaptic astrocytes. As described above, myelinated fibers consist of axons and myelin sheets. The myelin sheet had internodes, nodes of Ranvier, which are every 100 μm to 1 mm in distance. A single Schwann cell myelinates a single axon, and an oligodendrocyte myelinates 1–100 axons. The numbers are large and variable because conduction velocity depends on the axonal diameter and saltatory conduction. At the nodes of Ranvier, Na^+ channels cluster (1000/μm^2) in the axon membrane. Voltage-gated K^+ channels (fast, slow, and rectifier) are highly concentrated in the paranode section and decrease expression into the nodes of Ranvier. Both work together to maintain proper conduction and associated brain and muscle function. However, the examples provided here only denote the complexity of the system and the potential complications for functional recovery.

7.3 Summary

Astrocytes, neurons, progenitors, myelin-derived cells, and the BBB play a key role in brain development, synapse stability, and disease. In this chapter, we have tried to integrate the basic and early findings in glial biology with the recent developments in the field. Although astrocytes are one of the major cell types within the CNS, attention has been focused on neurons and oligodendrocytes. The relatively recent identification of neuronal/glial stem cells has provided a unique opportunity to prevent, repair, and enhance brain function for patients with brain dysfunctions, including acute, genetic, and long-term diseases such as Alzheimer's, stroke, Parkinson's, MS, and other diseases.

7.4 Review Questions/Problems

1. Name the two places within the CNS that generate neurons and glia during adulthood:
   (a) Subgranular zone and subventricular zone.
   (b) Subneuronal zone and subastrocyte zone.
   (c) Stem cell niches.
   (d) The hippocampus and third ventricle.

2. Which one is the main function of oligodendrocytes and Schwann cells?
   (a) Integrating the signaling of multiple axons.
   (b) Providing energy to the surrounding neurons.
   (c) Insulating the axons for better conduction.
   (d) Extracellular sodium and potassium clearing.
   (e) All of the above are true.

3. In the central nervous system, the major role of astrocytes is to:
   (a) Scavenge glutamine.
   (b) Scavenge cell debris.
   (c) Produce ATP.
   (d) All of the above are true.
   (e) None of the above.

4. In astrogliosis, astrocytes:
   (a) Become hyperreactive.
   (b) Form a bushy network surrounding the injury site.
   (c) Secrete cytokines and chemokines.
   (d) All of the above are true.

5. The functions of astrocytes involve:
   (a) Helping neurons for synaptic stability and collapse.
   (b) Generating a syncytium.
   (c) Helping maintain the BBB.
   (d) Responding to injury.
   (e) All of the above are true.

6. Is the species analyzed important for the role of oligodendrocytes and Schwann cells?
   (a) A human-specific type of myelin cell.
   (b) A human-specific oligodendrocyte to cover large-caliber axon type of myelinated cells.

(c) Myelin cells can integrate multiple axons.
(d) All of the above are true.

7. Which of the sequential stages is correct for oligodendroglial development?
   (a) Neural stem cells, progenitors, pre-oligodendrocytes, precursors, and mature oligodendrocytes.
   (b) Progenitors, precursors, pre-oligodendrocytes, mature oligodendrocytes, and neural stem cells.
   (c) Neural stem cells, precursors, progenitors, pre-oligodendrocytes, and mature oligodendrocytes.
   (d) Pre-oligodendrocytes, precursors, mature oligodendrocytes, progenitors, and neural stem cells.
   (e) None of the above.

8. Which of the following functions in the nervous system is not performed by oligodendrocytes?
   (a) Sheathing axons.
   (b) Supplying neurotrophic factors.
   (c) Forming the Ranvier's nodes.
   (d) Possessing phagocytic properties to remove debris.

9. A question without an answer: If you can regulate stem cell development, discuss the potential interventions to prevent or improve CNS dysfunction.

7.5 Answers to the Questions

1. a
2. e
3. d
4. d
5. e
6. d
7. c
8. d
9. Open for discussion

Acknowledgments This work was funded by the National Institute of Mental Health Grant Numbers MH128082 and MH134761 and the National Institute of Neurological Disorders and Stroke Grant Number NS105584 (to E.A.E).

References

Adams, K. L., & Gallo, V. (2018). The diversity and disparity of the glial scar. *Nature Neuroscience, 21*, 9–15.

Agirman, G., Broix, L., & Nguyen, L. (2017). Cerebral cortex development: An outside-in perspective. *FEBS Letters, 591*, 3978–3992.

Allodi, I., & Hedlund, E. (2014). Directed midbrain and spinal cord neurogenesis from pluripotent stem cells to model development and disease in a dish. *Frontiers in Neuroscience, 8*, 109.

Anderson, R. B., & Key, B. (1999). Novel guidance cues during neuronal pathfinding in the early scaffold of axon tracts in the rostral brain. *Development, 126*, 1859–1868.

Astiz, M., Delgado-Garcia, L. M., & Lopez-Mascaraque, L. (2022). Astrocytes as essential time-keepers of the central pacemaker. *Glia, 70*, 808–819.

Auer, L. M., Walter, G. F., & Mayer, G. (1979). Brain edema in acute arterial hypertension III: Fluorescence microscopic results. *Journal of Neurosurgical Sciences, 23*, 243–248.

Avaro, V., Hummel, T., & Calegari, F. (2022). Scent of stem cells: How can neurogenesis make us smell better? *Frontiers in Neuroscience, 16*, 964395.

Bazzari, A. H., & Parri, H. R. (2019). Neuromodulators and long-term synaptic plasticity in learning and memory: A steered-glutamatergic perspective. *Brain Sciences, 9*, 300.

Bazzoni, G., & Dejana, E. (2004). Endothelial cell-to-cell junctions: Molecular organization and role in vascular homeostasis. *Physiological Reviews, 84*, 869–901.

Ben Achour, S., & Pascual, O. (2012). Astrocyte-neuron communication: functional consequences. *Neurochemical Research, 37*, 2464–2473.

Bennett, M. V., Contreras, J. E., Bukauskas, F. F., & Saez, J. C. (2003). New roles for astrocytes: Gap junction hemichannels have something to communicate. *Trends in Neurosciences, 26*, 610–617.

Bhattacharya, D., Khan, B., & Simoes-Costa, M. (2021). Neural crest metabolism: At the crossroads of development and disease. *Developmental Biology, 475*, 245–255.

Bond, A. M., Ming, G. L., & Song, H. (2021). Ontogeny of adult neural stem cells in the mammalian brain. *Current Topics in Developmental Biology, 142*, 67–98.

Bowtell, J. L., & Bruce, M. (2002). Glutamine: an anaplerotic precursor. *Nutrition, 18*, 222–224.

Bugiani, M., Plug, B. C., Man, J. H. K., Breur, M., & van der Knaap, M. S. (2021). Heterogeneity of white matter astrocytes in the human brain. *Acta Neuropathologica*.

Bushong, E. A., Martone, M. E., Jones, Y. Z., & Ellisman, M. H. (2002). Protoplasmic astrocytes in CA1 stratum radiatum occupy separate anatomical domains. *The Journal of Neuroscience, 22*, 183–192.

Campos, J., Guerra-Gomes, S., Serra, S. C., Baltazar, G., Oliveira, J. F., Teixeira, F. G., & Salgado, A. J. (2020). Astrocyte signaling impacts the effects of human bone marrow mesenchymal stem cells secretome application into the hippocampus: A proliferation and morphometrical analysis on astrocytic cell populations. *Brain Research, 1732*, 146700.

Caron, A., Trzuskot, L., & Lindsey, B. W. (2022). Uncovering the spectrum of adult zebrafish neural stem cell cycle regulators. *Frontiers in Cell and Development Biology, 10*, 941893.

Castellano, P., Prevedel, L., Valdebenito, S., & Eugenin, E. A. (2019). HIV infection and latency induce a unique metabolic signature in human macrophages. *Scientific Reports, 9*, 3941.

Castillo, C., Saez-Orellana, F., Godoy, P. A., & Fuentealba, J. (2022). Microglial activation modulated by P2X4R in ischemia and repercussions in Alzheimer's disease. *Frontiers in Physiology, 13*, 814999.

Chen, W., Huang, H., Hatori, R., & Kornberg, T. B. (2017). Essential basal cytonemes take up hedgehog in the drosophila wing imaginal disc. *Development, 144*, 3134–3144.

Corbet, C., & Feron, O. (2015). Metabolic and mind shifts: From glucose to glutamine and acetate addictions in cancer. *Current Opinion in Clinical Nutrition and Metabolic Care, 18*, 346–353.

Crosby, K. M., Murphy-Royal, C., Wilson, S. A., Gordon, G. R., Bains, J. S., & Pittman, Q. J. (2018). Cholecystokinin switches the plasticity of GABA synapses in the dorsomedial hypothalamus via astrocytic ATP release. *The Journal of Neuroscience, 38*, 8515–8525.

Curtis, M. A., Eriksson, P. S., & Faull, R. L. (2007). Progenitor cells and adult neurogenesis in neurodegenerative diseases and injuries of the basal ganglia. *Clinical and Experimental Pharmacology & Physiology, 34*, 528–532.

Dallerac, G., Zapata, J., & Rouach, N. (2018). Versatile control of synaptic circuits by astrocytes: Where, when and how? *Nature Reviews Neuroscience, 19*, 729–743.

D'Amico, D., Valdebenito, S., & Eugenin, E. A. (2021). The role of Pannexin-1 channels and extracellular ATP in the pathogenesis of the human immunodeficiency virus. *Purinergic Signal, 17*, 563–576.

D'Angelo, C., Franch, O., Fernandez-Paredes, L., Oreja-Guevara, C., Nunez-Beltran, M., Comins-Boo, A., Reale, M., & Sanchez-Ramon, S. (2019). Antiphospholipid antibodies overlapping in isolated neurological syndrome and multiple sclerosis: Neurobiological insights and diagnostic challenges. *Frontiers in Cellular Neuroscience, 13*, 107.

D'Aversa, T. G., Eugenin, E. A., Lopez, L., & Berman, J. W. (2013). Myelin basic protein induces inflammatory mediators from primary human endothelial cells and blood-brain barrier disruption: Implications for the pathogenesis of multiple sclerosis. *Neuropathology and Applied Neurobiology, 39*, 270–283.

Deb, G., Cicala, A., Papadas, A., & Asimakopoulos, F. (2022). Matrix proteoglycans in tumor inflammation and immunity. *American Journal of Physiology. Cell Physiology, 323*, C678–C693.

Dimou, L., & Gotz, M. (2014). Glial cells as progenitors and stem cells: New roles in the healthy and diseased brain. *Physiological Reviews, 94*, 709–737.

Durkee, C., Kofuji, P., Navarrete, M., & Araque, A. (2021). Astrocyte and neuron cooperation in long-term depression. *Trends in Neurosciences, 44*, 837–848.

Edara, V. V., Ghorpade, A., & Borgmann, K. (2020). Insights into the gene expression profiles of active and restricted red/green-HIV(+) human astrocytes: Implications for shock or lock therapies in the brain. *Journal of Virology, 94*, 10–1128.

Embalabala, R. J., Brockman, A. A., Jurewicz, A. R., Kong, J. A., Ryan, K., Guinto, C. D., Alvarez-Buylla, A., Chiang, C., & Ihrie, R. A. (2022). GLI3 is required for OLIG2+ progeny production in adult dorsal neural stem cells. *Cells, 11*, 218.

Escartin, C., Galea, E., Lakatos, A., O'Callaghan, J. P., Petzold, G. C., Serrano-Pozo, A., Steinhauser, C., Volterra, A., Carmignoto, G., Agarwal, A., Allen, N. J., Araque, A., Barbeito, L., Barzilai, A., Bergles, D. E., Bonvento, G., Butt, A. M., Chen, W. T., Cohen-Salmon, M., Cunningham, C., Deneen, B., De Strooper, B., Diaz-Castro, B., Farina, C., Freeman, M., Gallo, V., Goldman, J. E., Goldman, S. A., Gotz, M., Gutierrez, A., Haydon, P. G., Heiland, D. H., Hol, E. M., Holt, M. G., Iino, M., Kastanenka, K. V., Kettenmann, H., Khakh, B. S., Koizumi, S., Lee, C. J., Liddelow, S. A., MacVicar, B. A., Magistretti, P., Messing, A., Mishra, A., Molofsky, A. V., Murai, K. K., Norris, C. M., Okada, S., Oliet, S. H. R., Oliveira, J. F., Panatier, A., Parpura, V., Pekna, M., Pekny, M., Pellerin, L., Perea, G., Perez-Nievas, B. G., Pfrieger, F. W., Poskanzer, K. E., Quintana, F. J., Ransohoff, R. M., Riquelme-Perez, M., Robel, S., Rose, C. R., Rothstein, J. D., Rouach, N., Rowitch, D. H., Semyanov, A., Sirko, S., Sontheimer, H., Swanson, R. A., Vitorica, J., Wanner, I. B., Wood, L. B., Wu, J., Zheng, B., Zimmer, E. R., Zorec, R., Sofroniew, M. V., & Verkhratsky, A. (2021). Reactive astrocyte nomenclature, definitions, and future directions. *Nature Neuroscience, 24*, 312–325.

Eugenin, E. A., & Berman, J. W. (2007). Gap junctions mediate human immunodeficiency virus-bystander killing in astrocytes. *The Journal of Neuroscience, 27*, 12844–12850.

Eugenin, E. A., Osiecki, K., Lopez, L., Goldstein, H., Calderon, T. M., & Berman, J. W. (2006). CCL2/monocyte chemoattractant protein-1 mediates enhanced transmigration of human immunodeficiency virus (HIV)-infected leukocytes across the blood-brain barrier: A potential mechanism of HIV-CNS invasion and NeuroAIDS. *The Journal of Neuroscience, 26*, 1098–1106.

Eugenin, E. A., Gaskill, P. J., & Berman, J. W. (2009). Tunneling nanotubes (TNT): A potential mechanism for intercellular HIV trafficking. *Communicative & Integrative Biology, 2*, 243–244.

Eugenin, E. A., Clements, J. E., Zink, M. C., & Berman, J. W. (2011). Human immunodeficiency virus infection of human astrocytes disrupts blood-brain barrier integrity by a gap junction-dependent mechanism. *The Journal of Neuroscience, 31*, 9456–9465.

Fares, J., Bou Diab, Z., Nabha, S., & Fares, Y. (2019). Neurogenesis in the adult hippocampus: History, regulation, and prospective roles. *The International Journal of Neuroscience, 129*, 598–611.

Feliciano, D. M., Bordey, A., & Bonfanti, L. (2015). Noncanonical sites of adult neurogenesis in the mammalian brain. *Cold Spring Harbor Perspectives in Biology, 7*, a018846.

Frizzo, M. E., & Ohno, Y. (2021). Perisynaptic astrocytes as a potential target for novel antidepressant drugs. *Journal of Pharmacological Sciences, 145*, 60–68.

Garcia, A. D. R. (2021). New tricks for an old (hedge)hog: Sonic hedgehog regulation of astrocyte function. *Cells, 10*, 1353.

Gomes-Leal, W. (2021). Adult hippocampal neurogenesis and affective disorders: New neurons for psychic well-being. *Frontiers in Neuroscience, 15*, 594448.

Goncalves-Ribeiro, J., Pina, C. C., Sebastiao, A. M., & Vaz, S. H. (2019). Glutamate transporters in hippocampal LTD/LTP: Not just prevention of excitotoxicity. *Frontiers in Cellular Neuroscience, 13*, 357.

Gorska, A. M., & Eugenin, E. A. (2020). The glutamate system as a crucial regulator of CNS toxicity and survival of HIV reservoirs. *Frontiers in Cellular and Infection Microbiology, 10*, 261.

Hagemann, A. I., & Scholpp, S. (2012). The tale of the three brothers - Shh, Wnt, and Fgf during development of the thalamus. *Frontiers in Neuroscience, 6*, 76.

Hernandez, C. A., & Eliseo, E. (2022). The role of pannexin-1 channels in HIV and NeuroHIV pathogenesis. *Cells, 11*, 2245.

Hertz, L., Dringen, R., Schousboe, A., & Robinson, S. R. (1999). Astrocytes: glutamate producers for neurons. *Journal of Neuroscience Research, 57*, 417–428.

Heuser, K., & Enger, R. (2021). Astrocytic ca(2+) signaling in epilepsy. *Frontiers in Cellular Neuroscience, 15*, 695380.

Hildebrand, C., Remahl, S., Persson, H., & Bjartmar, C. (1993). Myelinated nerve fibres in the CNS. *Progress in Neurobiology, 40*, 319–384.

Hill, S. A., Fu, M., & Garcia, A. D. R. (2021). Sonic hedgehog signaling in astrocytes. *Cellular and Molecular Life Sciences, 78*, 1393–1403.

Huang, H., Liu, S., & Kornberg, T. B. (2019). Glutamate signaling at cytoneme synapses. *Science, 363*, 948–955.

Huszar, R., Zhang, Y., Blockus, H., & Buzsaki, G. (2022). Preconfigured dynamics in the hippocampus are guided by embryonic birthdate and rate of neurogenesis. *Nature Neuroscience, 25*, 1201–1212.

Iadecola, C., & Nedergaard, M. (2007). Glial regulation of the cerebral microvasculature. *Nature Neuroscience, 10*, 1369–1376.

Jacob, P. F., Vaz, S. H., Ribeiro, J. A., & Sebastiao, A. M. (2014). P2Y1 receptor inhibits GABA transport through a calcium signalling-dependent mechanism in rat cortical astrocytes. *Glia, 62*, 1211–1226.

Kageyama, R., Shimojo, H., & Imayoshi, I. (2015). Dynamic expression and roles of Hes factors in neural development. *Cell and Tissue Research, 359*, 125–133.

Kaise, T., & Kageyama, R. (2021). Hes1 oscillation frequency correlates with activation of neural stem cells. *Gene Expression Patterns, 40*, 119170.

Kalinina, A., & Lagace, D. (2022). Single-cell and single-nucleus RNAseq analysis of adult neurogenesis. *Cells, 11*, 1633.

Kashima, R., & Hata, A. (2018). The role of TGF-beta superfamily signaling in neurological disorders. *Acta Biochimica et Biophysica Sinica (Shanghai), 50*, 106–120.

Kastanenka, K. V., Moreno-Bote, R., De Pitta, M., Perea, G., Eraso-Pichot, A., Masgrau, R., Poskanzer, K. E., & Galea, E. (2020). A roadmap to integrate astrocytes into systems neuroscience. *Glia, 68*, 5–26.

Khan, M. T., Deussing, J., Tang, Y., & Illes, P. (2019). Astrocytic rather than neuronal P2X7 receptors modulate the function of the tri-synaptic network in the rodent hippocampus. *Brain Research Bulletin, 151*, 164–173.

Khan, H., Kaur, P., Singh, T. G., Grewal, A. K., & Sood, S. (2022). Adenosine as a Key mediator of neuronal survival in cerebral ischemic injury. *Neurochemical Research, 47*(12), 3543–3555.

Klein, R. (2009). Bidirectional modulation of synaptic functions by Eph/ephrin signaling. *Nature Neuroscience, 12*, 15–20.

Koizumi, S. (2015). Gliaotransmission and brain functions. *Nihon Shinkei Seishin Yakurigaku Zasshi, 35*, 5–9.

Kuznetsova, A., Popova, O., Panchenkov, D., Dyuzheva, T., & Ivanov, A. (2023). Pancreatic ductal adenocarcinoma: Tumor microenvironment and problems in the development of novel therapeutic strategies. *Clinical and Experimental Medicine, 23*(3), 619–643.

Lalo, U., Bogdanov, A., & Pankratov, Y. (2019). Age- and experience-related plasticity of ATP-mediated signaling in the neocortex. *Frontiers in Cellular Neuroscience, 13*, 242.

Lathia, J. D., Patton, B., Eckley, D. M., Magnus, T., Mughal, M. R., Sasaki, T., Caldwell, M. A., Rao, M. S., Mattson, M. P., & C. ffrench-Constant. (2007). Patterns of laminins and integrins in the embryonic ventricular zone of the CNS. *The Journal of Comparative Neurology, 505*, 630–643.

Lawal, O., Ulloa Severino, F. P., & Eroglu, C. (2022). The role of astrocyte structural plasticity in regulating neural circuit function and behavior. *Glia, 70*, 1467–1483.

Lee, H. U., Yamazaki, Y., Tanaka, K. F., Furuya, K., Sokabe, M., Hida, H., Takao, K., Miyakawa, T., Fujii, S., & Ikenaka, K. (2013). Increased astrocytic ATP release results in enhanced excitability of the hippocampus. *Glia, 61*, 210–224.

Liu, Y., Shaw, S. K., Ma, S., Yang, L., Luscinskas, F. W., & Parkos, C. A. (2004). Regulation of leukocyte transmigration: Cell surface interactions and signaling events. *Journal of Immunology, 172*, 7–13.

Malik, S., Theis, M., & Eugenin, E. A. (2017). Connexin43 containing gap junction channels facilitate HIV bystander toxicity: Implications in NeuroHIV. *Frontiers in Molecular Neuroscience, 10*, 404.

Malik, S., Valdebenito, S., D'Amico, D., Prideaux, B., & Eugenin, E. A. (2021). HIV infection of astrocytes compromises inter-organelle interactions and inositol phosphate metabolism: A potential mechanism of bystander damage and viral reservoir survival. *Progress in Neurobiology, 206*, 102157.

Manfredi-Lozano, M., Leysen, V., Adamo, M., Paiva, I., Rovera, R., Pignat, J. M., Timzoura, F. E., Candlish, M., Eddarkaoui, S., Malone, S. A., Silva, M. S. B., Trova, S., Imbernon, M., Decoster, L., Cotellessa, L., Tena-Sempere, M., Claret, M., Paoloni-Giacobino, A., Plassard, D., Paccou, E., Vionnet, N., Acierno, J., Maceski, A. M., Lutti, A., Pfrieger, F., Rasika, S., Santoni, F., Boehm, U., Ciofi, P., Buee, L., Haddjeri, N., Boutillier, A. L., Kuhle, J., Messina, A., Draganski, B., Giacobini, P., Pitteloud, N., & Prevot, V. (2022). GnRH replacement rescues cognition in Down syndrome. *Science, 377*, eabq4515.

Marshall, C. A., Novitch, B. G., & Goldman, J. E. (2005). Olig2 directs astrocyte and oligodendrocyte formation in postnatal subventricular zone cells. *The Journal of Neuroscience, 25*, 7289–7298.

Masthoff, M., Freppon, F. N., Zondler, L., Wilken, E., Wachsmuth, L., Niemann, S., Schwarz, C., Fredrich, I., Havlas, A., Block, H., Gerwing, M., Helfen, A., Heindel, W., Zarbock, A., Wildgruber, M., & Faber, C. (2021). Resolving immune cells with patrolling behaviour by magnetic resonance time-lapse single cell tracking. *eBioMedicine, 73*, 103670.

Maus, A., & Peters, G. J. (2017). Glutamate and alpha-ketoglutarate: Key players in glioma metabolism. *Amino Acids, 49*, 21–32.

McFarland, H. F., & Martin, R. (2007). Multiple sclerosis: A complicated picture of autoimmunity. *Nature Immunology, 8*, 913–919.

Mi, K., & Johnson, G. V. (2007). Regulated proteolytic processing of LRP6 results in release of its intracellular domain. *Journal of Neurochemistry, 101*, 517–529.

Mitic, L. L., & Anderson, J. M. (1998). Molecular architecture of tight junctions. *Annual Review of Physiology, 60*, 121–142.

Morante-Redolat, J. M., & Porlan, E. (2019). Neural stem cell regulation by adhesion molecules within the subependymal niche. *Frontiers in Cell and Development Biology, 7*, 102.

Muller, H. (1851). Zur Histologie der Netzhaut. *Wissenschaftliche Zoologie, 3*, 234–237.

Muller, W. A. (2003). Leukocyte-endothelial-cell interactions in leukocyte transmigration and the inflammatory response. *Trends in Immunology, 24*, 327–334.

Mundt, S., Greter, M., Flugel, A., & Becher, B. (2019). The CNS immune landscape from the viewpoint of a T cell. *Trends in Neurosciences, 42*, 667–679.

Nian, F. S., & Hou, P. S. (2022). Evolving roles of notch signaling in cortical development. *Frontiers in Neuroscience, 16*, 844410.

Nogueira, A. B., Hoshino, H. S. R., Ortega, N. C., Dos Santos, B. G. S., & Teixeira, M. J. (2022). Adult human neurogenesis: Early studies clarify recent controversies and go further. *Metabolic Brain Disease, 37*, 153–172.

Ogata, K., & Kosaka, T. (2002). Structural and quantitative analysis of astrocytes in the mouse hippocampus. *Neuroscience, 113*, 221–233.

Oliveira, J. F., & Araque, A. (2022). Astrocyte regulation of neural circuit activity and network states. *Glia, 70*, 1455–1466.

Orkand, R. K., Nicholls, J. G., & Kuffler, S. W. (1966). Effect of nerve impulses on the membrane potential of glial cells in the central nervous system of amphibia. *Journal of Neurophysiology, 29*, 788–806.

Ousman, S. S., & Kubes, P. (2012). Immune surveillance in the central nervous system. *Nature Neuroscience, 15*, 1096–1101.

Oxburgh, L. (2022). The extracellular matrix environment of clear cell renal cell carcinoma. *Cancers (Basel), 14*, 4072.

Paradiso, B., Zucchini, S., & Simonato, M. (2013). Implication of fibroblast growth factors in epileptogenesis-associated circuit rearrangements. *Frontiers in Cellular Neuroscience, 7*, 152.

Pascual, O., Casper, K. B., Kubera, C., Zhang, J., Revilla-Sanchez, R., Sul, J. Y., Takano, H., Moss, S. J., McCarthy, K., & Haydon, P. G. (2005). Astrocytic purinergic signaling coordinates synaptic networks. *Science, 310*, 113–116.

Patel, A., Wu, Y., Han, X., Su, Y., Maugel, T., Shroff, H., & Roy, S. (2022). Cytonemes coordinate asymmetric signaling and organization in the Drosophila muscle progenitor niche. *Nature Communications, 13*, 1185.

Perea, G., & Araque, A. (2005). Glial calcium signaling and neuron-glia communication. *Cell Calcium, 38*, 375–382.

Perea, G., & Araque, A. (2010). GLIA modulates synaptic transmission. *Brain Research Reviews, 63*, 93–102.

Perea, G., Sur, M., & Araque, A. (2014). Neuron-glia networks: Integral gear of brain function. *Frontiers in Cellular Neuroscience, 8*, 378.

Peretto, P., & Paredes, R. G. (2014). Social cues, adult neurogenesis, and reproductive behavior. In C. Mucignat-Caretta (Ed.), *Neurobiology of chemical communication*.

Peters, A. (1964a). Further observations on the structure of myelin sheaths in the central nervous system. *The Journal of Cell Biology, 20*, 281–296.

Peters, A. (1964b). Observations on the connexions between myelin sheaths and glial cells in the optic nerves of young rats. *Journal of Anatomy, 98*, 125–134.

Prat, A., Biernacki, K., Wosik, K., & Antel, J. P. (2001). Glial cell influence on the human blood-brain barrier. *Glia, 36*, 145–155.

Ramón y Cajal, S. (1995). *Histology of the nervous system of man and vertebrates*. Oxford University Press.
Ramón y Cajal, S., & Tello y Muñoz, J. F. (1933). *Histology* (p. xiv, 738). Wood.
Ribeiro, F. F., & Xapelli, S. (2021). An overview of adult neurogenesis. *Advances in Experimental Medicine and Biology, 1331*, 77–94.
Rubin, L. L., & Staddon, J. M. (1999). The cell biology of the blood-brain barrier. *Annual Review of Neuroscience, 22*, 11–28.
Saint-Martin, M., & Goda, Y. (2023). Astrocyte-synapse interactions and cell adhesion molecules. *The FEBS Journal, 290*(14), 3512–3526.
Scemes, E., & Spray, D. C. (2012). Extracellular K(+) and astrocyte signaling via connexin and pannexin channels. *Neurochemical Research, 37*, 2310–2316.
Shah, R., & Chen, S. (2020). Metabolic signaling cascades prompted by glutaminolysis in cancer. *Cancers (Basel), 12*, 2624.
Shen, Y. A., Chen, C. L., Huang, Y. H., Evans, E. E., Cheng, C. C., Chuang, Y. J., Zhang, C., & Le, A. (2021). Inhibition of glutaminolysis in combination with other therapies to improve cancer treatment. *Current Opinion in Chemical Biology, 62*, 64–81.
Singh, S. A., Suresh, S., Singh, A., Chandran, L., & Vellapandian, C. (2022). Perspectives of ozone induced neuropathology and memory decline in Alzheimer's disease: A systematic review of preclinical evidences. *Environmental Pollution, 313*, 120136.
Sirica, A. E. (2022). Matricellular proteins in intrahepatic cholangiocarcinoma. *Advances in Cancer Research, 156*, 249–281.
Sobel, R. A., Mitchell, M. E., & Fondren, G. (1990). Intercellular adhesion molecule-1 (ICAM-1) in cellular immune reactions in the human central nervous system. *The American Journal of Pathology, 136*, 1309–1316.
Sohr, A., Du, L., Wang, R., Lin, L., & Roy, S. (2019). Drosophila FGF cleavage is required for efficient intracellular sorting and intercellular dispersal. *The Journal of Cell Biology, 218*, 1653–1669.
Springer, T. A. (1994). Traffic signals for lymphocyte recirculation and leukocyte emigration: The multistep paradigm. *Cell, 76*, 301–314.
Stenzel, D., Wilsch-Brauninger, M., Wong, F. K., Heuer, H., & Huttner, W. B. (2014). Integrin alphavbeta3 and thyroid hormones promote expansion of progenitors in embryonic neocortex. *Development, 141*, 795–806.
Suciu, S. K., & Caspary, T. (2021). Cilia, neural development and disease. *Seminars in Cell & Developmental Biology, 110*, 34–42.
Sykova, E. (2001). Glial diffusion barriers during aging and pathological states. *Progress in Brain Research, 132*, 339–363.
Takahashi, S. (2022). Metabolic contribution and cerebral blood flow regulation by astrocytes in the neurovascular unit. *Cells, 11*, 813.
Tan, C. X., & Eroglu, C. (2021). Cell adhesion molecules regulating astrocyte-neuron interactions. *Current Opinion in Neurobiology, 69*, 170–177.
Tardito, S., Oudin, A., Ahmed, S. U., Fack, F., Keunen, O., Zheng, L., Miletic, H., Sakariassen, P. O., Weinstock, A., Wagner, A., Lindsay, S. L., Hock, A. K., Barnett, S. C., Ruppin, E., Morkve, S. H., Lund-Johansen, M., Chalmers, A. J., Bjerkvig, R., Niclou, S. P., & Gottlieb, E. (2015). Glutamine synthetase activity fuels nucleotide biosynthesis and supports growth of glutamine-restricted glioblastoma. *Nature Cell Biology, 17*, 1556–1568.
Tavcar, P., Potokar, M., Kolenc, M., Korva, M., Avsic-Zupanc, T., Zorec, R., & Jorgacevski, J. (2021). Neurotropic viruses, astrocytes, and COVID-19. *Frontiers in Cellular Neuroscience, 15*, 662578.
Tayyab, M., Shahi, M. H., Farheen, S., Mariyath, M. P. M., Khanam, N., Castresana, J. S., & Hossain, M. M. (2018). Sonic hedgehog, Wnt, and brain-derived neurotrophic factor cell signaling pathway crosstalk: Potential therapy for depression. *Journal of Neuroscience Research, 96*, 53–62.
Thalhammer, A., & Cingolani, L. A. (2014). Cell adhesion and homeostatic synaptic plasticity. *Neuropharmacology, 78*, 23–30.
Tsukita, S., & Furuse, M. (1999). Occludin and claudins in tight-junction strands: Leading or supporting players? *Trends in Cell Biology, 9*, 268–273.
Valdebenito, S., Audia, A., Bhat, K. P. L., Okafo, G., & Eugenin, E. A. (2020). Tunneling nanotubes mediate adaptation of glioblastoma cells to temozolomide and ionizing radiation treatment. *iScience, 23*, 101450.
Valdebenito, S., Castellano, P., Ajasin, D., & Eugenin, E. A. (2021a). Astrocytes are HIV reservoirs in the brain: A cell type with poor HIV infectivity and replication but efficient cell-to-cell viral transfer. *Journal of Neurochemistry, 158*, 429–443.
Valdebenito, S., Malik, S., Luu, R., Loudig, O., Mitchell, M., Okafo, G., Bhat, K., Prideaux, B., & Eugenin, E. A. (2021b). Tunneling nanotubes, TNT, communicate glioblastoma with surrounding non-tumor astrocytes to adapt them to hypoxic and metabolic tumor conditions. *Scientific Reports, 11*, 14556.
van Buul, J. D., & Hordijk, P. L. (2004). Signaling in leukocyte transendothelial migration. *Arteriosclerosis, Thrombosis, and Vascular Biology, 24*, 824–833.
Velasquez, S., Prevedel, L., Valdebenito, S., Gorska, A. M., Golovko, M., Khan, N., Geiger, J., & Eugenin, E. A. (2020). Circulating levels of ATP is a biomarker of HIV cognitive impairment. *eBioMedicine, 51*, 102503.
Ventura, A. L. M., Dos Santos-Rodrigues, A., Mitchell, C. H., & Faillace, M. P. (2019). Purinergic signaling in the retina: From development to disease. *Brain Research Bulletin, 151*, 92–108.
Virchow, R. (1856). *Gesammelte Abhandlungen zur wissenschaftlichen Medicin* (p. 1024). Verlag von Meidinger Sohn.
Virchow, R. (1858). Die cellularpathologie in ihrer begründung auf physiologische und pathologische gewebelehre. Zwanzig vorlesungen gehalten während der monate februar, märz und april 1858 im Pathologischen institute zu Berlin. A. Hirschwald, Berlin,. xvi, 440 p.
Wirkner, K., Kofalvi, A., Fischer, W., Gunther, A., Franke, H., Groger-Arndt, H., Norenberg, W., Madarasz, E., Vizi, E. S., Schneider, D., Sperlagh, B., & Illes, P. (2005). Supersensitivity of P2X receptors in cerebrocortical cell cultures after in vitro ischemia. *Journal of Neurochemistry, 95*, 1421–1437.
Wood, B. M., Baena, V., Huang, H., Jorgens, D. M., Terasaki, M., & Kornberg, T. B. (2021). Cytonemes with complex geometries and composition extend into invaginations of target cells. *The Journal of Cell Biology, 220*.
Woodbury, M. E., & Ikezu, T. (2014). Fibroblast growth factor-2 signaling in neurogenesis and neurodegeneration. *Journal of Neuroimmune Pharmacology, 9*, 92–101.
Yakovlev, A. Y., Boucher, K., Mayer-Proschel, M., & Noble, M. (1998). Quantitative insight into proliferation and differentiation of oligodendrocyte type 2 astrocyte progenitor cells in vitro. *Proceedings of the National Academy of Sciences of the United States of America, 95*, 14164–14167.
Yamashita, K., Mikawa, S., & Sato, K. (2016). BMP3 expression in the adult rat CNS. *Brain Research, 1643*, 35–50.
Yang, L., Li, Z., Liu, G., Li, X., & Yang, Z. (2022). Developmental origins of human cortical oligodendrocytes and astrocytes. *Neuroscience Bulletin, 38*, 47–68.
Zalc, B., Goujet, D., & Colman, D. (2008). The origin of the myelination program in vertebrates. *Current Biology, 18*, R511–R512.

Microglial Signaling in Health and Disease

8

Mark Ryan Oppenheimer, Salma Singh, Danielle Antoine, Bridget Herlihy Truitt, and Sabita Roy

Abstract

Microglia are a key cell type in the central nervous system that play a role in a plethora of functions throughout development, in homeostasis, and in disease states. As the resident immune cells of the developed brain, microglia survey their environment for threats to the CNS and induce a cytokine and chemokine response to resolve the pathological insult. Activated microglia can polarize into pro-inflammatory and anti-inflammatory states to maintain homeostasis in the brain; however, the regulation of these activation phenotypes can become disordered in various pathological and disease states. Altered microglial signaling represents a critical pathway in neurodegenerative disease, psychiatric conditions, aging, and other areas of great clinical interest. Research into microglial signaling in health and disease pursues therapeutic interventions aimed at ameliorating these conditions.

Keywords

Synaptic plasticity · M1/M2 polarization · Neuroinflammation · Toll-like receptor

8.1 The Role of Microglia in Homeostasis

Microglia belong to the innate immune system and are the resident immune cells of the central nervous system (CNS). The term "microglia" was first used by Del Rio Hortega in the first half of the twentieth century (Rio-Hortega, 1939). Microglial cells play important roles in immune defense, regulation of CNS development, and maintenance of homeostasis. Activation of microglia and associated signaling pathways is implicated in a multitude of CNS pathologies, presenting targets for therapeutic intervention.

8.1.1 Microglial Morphology

Under physiological conditions, microglial cells have a small cell body and highly ramified processes that allow them to survey their local brain parenchyma for signs of pathogens or cellular damage (Leyh et al., 2021) (Fig. 8.1). In this stage, microglia were long believed to be "resting;" however, recent studies have demonstrated their dynamic nature (Kamei et al., 2022; Tremblay et al., 2011). Microglia perpetually change their morphology by extending and retracting highly motile processes in a matter of minutes in vivo (Nimmerjahn et al., 2005). Consequently, microglia under physiological conditions are now being described as "surveilling" rather than "resting" glial cells (Leyh et al., 2021).

Because microglia continually survey their cellular environment, they can immediately respond to injury. Once they encounter damage, akin to the immune cells of the periphery, microglia respond by phagocytosing—engulfing cellular debris and releasing cytokines—to generate significant innate and adaptive immune responses (Colonna & Butovsky, 2017). During this activated state, microglia undergo distinct morphological changes such as the retraction of their numerous fine processes and an increase in their cell body size (Anttila et al., 2017). These changes happen gradually and usually in a continuous process through a broad spectrum of microglial shapes. Microglia can process through several morphological states that indicate increasing levels of activation on the basis of which they present different morphologies, including highly ramified, rod-like, activated, and amoeboid shapes (Leyh et al., 2021)..

Ramified microglia scan their environment and are characterized by small cell bodies, large cell perimeters, and long processes. *Rod-like* microglial cells project similar skeleton lengths, branches, and end points to ramified microglia but

M. R. Oppenheimer · S. Singh · D. Antoine · B. H. Truitt · S. Roy (✉)
University of Miami Miller School of Medicine, Miami, FL, USA
e-mail: Sabita.Roy@miami.edu

H. E. Gendelman, T. Ikezu (eds.), *Neuroimmune Pharmacology and Therapeutics*,
https://doi.org/10.1007/978-3-031-68237-7_8

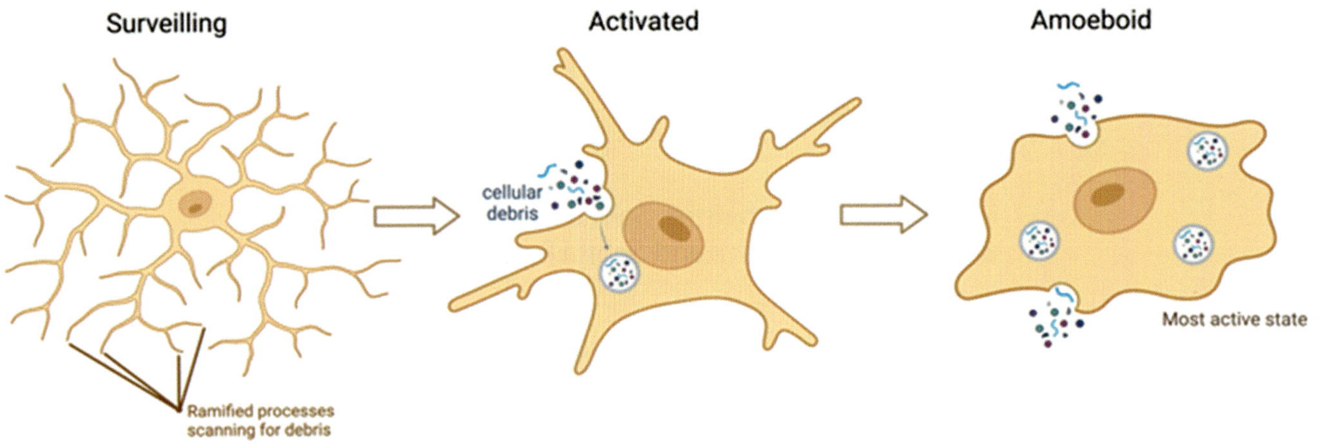

Fig. 8.1 Microglial activation: In their resting state, microglia survey the area with their ramified processes, searching for cellular debris. Once a site of injury or debris is found, microglia undergo a morphological change, bringing in the ramified processes and engulfing the debris. At their highest activation, microglia enter their amoeboid state, with their processes fully retracted and their focus on the phagocytosis of debris

exhibit a larger cell body size. Rod-like microglia are in an intermediate state of activation. The classically *activated* phenotype of microglia exhibits a smaller cell body and smaller cell perimeters as well as fewer skeleton lengths, branches, and end points than ramified and rod-like microglial cells. Lastly, *amoeboid* microglia exhibit the most activated morphology and are characterized by the smallest cell body, cell perimeter, and processes when compared to all the other morphological phenotypes (Fernández-Arjona et al., 2019; Leyh et al., 2021) (Fig. 8.1).

The function of microglial cells and their morphology are closely coupled during the resting/surveilling state or the activated state. In particular, the inflammatory activation of microglial cells is gradual and significantly correlates with morphological changes. Thus, microglia possess essential intrinsic pathways that allow them to exhibit different modes of motility. During the resting/surveilling state, microglia need to extend and retract mobile processes to survey the brain. When they encounter injury or cellular debris, they must enter the activated state and send out targeted processes to envelop sites of tissue damage. An important microglial-intrinsic process is the cAMP-regulated fine filopodia formation, which is distinct from large processes (Colonna & Butovsky, 2017). The fine filopodia allow for fast nanoscale sensing within the discrete regions of the brain parenchyma. Another intrinsic microglial process is the use of a two-pore domain potassium channel, THIK-1, which maintains a negative resting membrane potential to facilitate normal microglial ramification and surveillance of the brain (Madry et al., 2018). In summary, microglia have unique properties that allow them to patrol the adult brain and spinal cord, enter the activated state when needed, and be vigilant housekeepers of the CNS.

8.1.2 Microglial Role in Development

Prenatal development is a critical period for brain maturation, and microglia play an essential role in healthy CNS development. Microglia are one of the first tissue macrophages to be formed in the yolk sac at embryonic day 7.5, and they start to migrate to the developing brain at embryonic day 9.5 (Elmore et al., 2014; Ginhoux & Garel, 2018; Ginhoux et al., 2013; Prinz & Priller, 2014). This early colonization occurs prior to the differentiation of other resident nervous system cells, which allows microglia to be present at critical times to play essential roles in CNS development (Kriegstein & Alvarez-Buylla, 2009). In the developing CNS, microglia have several functions (Frost & Schafer, 2016; Lannes et al., 2017). They eliminate apoptotic cells and prevent the oversupply of neurons. They support the formation, migration, and differentiation of new neurons as well as the formation and growth of new synapses. Microglia also contribute substantially to the generation and maturation of astrocytes and oligodendrocytes (Lannes et al., 2017). The main mechanisms underlying these functions, which are still under investigation, include phagocytosis and cell-to-cell communication via direct and indirect cellular and molecular signaling.

During neurogenesis and positioning of the newly formed neurons, those that fail to assemble into functional circuits need to be eliminated. For the clearance of these apoptotic cells, microglia migrate to and accumulate in areas of cell death where they phagocytose and prevent the oversupply of neurons. Apoptotic neurons signal and attract microglial cells through nucleotide signaling, which facilitates microglial migration and colonization in the developing brain (Casano et al., 2016; Svahn et al., 2016). A study found the microglial colonization of the developing zebrafish midbrain to be triggered by apoptotic neuronal death, via releasing the chemical compound lysophosphatidylcholine (Xu et al., 2016). Interestingly, reduced neuronal cell death results in fewer microglia, whereas increased apoptosis enhances brain colonization, resulting in more microglia at later stages (Casano et al., 2016).

In addition to the clearance of apoptotic cells, microglial cells are essential to the formation, migration, and differentiation of new neurons during neurogenesis. The exact role of microglia in neural progenitor cells (NPCs) specifically remains controversial, as studies have come to differing conclusions, which may be due to differences in microglial activation state or experimental models. Microglia were found in vitro to promote the proliferation of NPCs by secreting osteopontin (Yamamiya et al., 2019). In co-cultures of both NPCs and microglia, there was an increase in survival and proliferation in both cell types compared with monocultures (Liu et al., 2013). Additionally, the depletion of microglia in mice lacking the receptor for colony-stimulating factor-1 (Csf-1R) reduced the pool of NPCs in the cerebral cortex (Arnò et al., 2014). In contrast, deactivating microglia in utero or eliminating microglia from the fetal cerebral cortex of macaques and rats was found to significantly increase the number of NPCs (Cunningham et al., 2013). Additionally, activating microglia in utero through maternal immune activation significantly decreased the number of NPCs by phagocytosis (Schafer & Stevens, 2015). Consequently, microglia may have the capacity to both promote the proliferation of NPCs by secreting factors and to reduce the number of NPCs by phagocytosis when activated. Without a doubt, normal microglial colonization is critical to the survival, proliferation, and maturation of NPCs in the developing brain.

Furthermore, elimination of excessive or unnecessary synaptic networks is essential to promote the refinement and maturation of the developing CNS. Microglia play an important role in refining synaptic circuits during development through the pruning of excess synapses. In culture, elevated synaptic density was observed in microglia-depleted brain slices. Additionally, the introduction of microglia to neuronal cultures decreased synaptic activity and reduced the density of synapses, spine numbers, and levels of synaptic adhesion molecules (Ji et al., 2013). Similarly, in vivo, the disruption or elimination of microglia in early life significantly altered synaptic structures, neural circuitry, and behavior during adulthood (Nelson & Lenz, 2017; Paolicelli et al., 2011). The mechanisms regulating glia-mediated synaptic elimination involve a complex process of many cues that promote and inhibit the removal of specific synaptic connections.

The main mechanism by which microglia eliminate excess synapses is through phagocytosis of pre- and/or postsynaptic elements (Wilton et al., 2019). In the developing visual system, microglia engulf and degrade presynaptic retinal inputs early in postnatal development when eye-specific segregation occurs (Schafer et al., 2012). Complement proteins, such as C1q and C3, and components of the innate immune system were identified to localize and bind to the unwanted material and promote their removal by microglial engulfment through interaction with the microglia-specific C3 receptor (CR3) (Schafer et al., 2012). In the hippocampus, other mechanisms of microglial synapse elimination have been identified. Neurons in need of pruning in the hippocampus release fractalkine, the ligand for the microglial receptor CX3CR1. The released fractalkine recruits microglia and signals for the removal of excessive synaptic connections. Indeed, CX3CR1 knockout mice display impaired synaptic pruning and disturbed social behavior (Paolicelli et al., 2011). Another mechanism of synaptic elimination is mediated by the triggering receptor expressed on myeloid cells-2, TREM2, an innate immune receptor on microglia that controls phagocytosis (Hsieh et al., 2009). TREM2 is essential for microglia-mediated synaptic refinement during the early stages of hippocampal circuitry, as the absence of *Trem2* results in impaired synapse elimination, enhanced excitatory neurotransmission, and reduced long-range functional connectivity (Filipello et al., 2018; Geloso & D'Ambrosi, 2021). Indeed, *Trem2*$^{-/-}$ mice display increased synaptic density, impaired connectivity, repetitive behavior, and altered sociability (Filipello et al., 2018). Additionally, in humans, TREM2 protein levels in the postmortem brains of patients affected by idiopathic autism were found to be negatively correlated with symptom severity (Filipello et al., 2018).

As with the different identification systems of supernumerary synapses, several mechanisms to spare normal synapses occur. Recognition molecules, such as those of the immunoglobin superfamily, play important roles in neuronal interaction, survival, and synaptic targeting (Maness & Schachner, 2007). Cluster of differentiation 47 (CD47), a protein belonging to the immunoglobulin superfamily, is widely expressed on the surface of neural membranes (Lehrman et al., 2018). CD47 provides a "do not eat" signal by binding to its receptor, signal regulatory protein alpha (SIRPα), expressed on microglia. Upon interaction, synaptic phagocytosis is inhibited and the neural synapse is spared from removal (Lehrman et al., 2018). Furthermore, microglial identification of synaptic function is based on the magnitude and timing of neural activity. Imaging studies have revealed that microglia dynamically interact with synapses by responding to changes in neural activity and neurotransmitter release (Schafer et al., 2013). Thus, to dissociate between active and weak synapses, microglia sense local neurotransmitter release and only phagocytose those exhibiting low neural activity. Taken together, nonactivated microglia play an essential role in modulating synaptic activity in the developing CNS by tightly regulating the number of new neurons and functional synapses through synaptic pruning.

In addition to new neurons and synapses, microglia also contribute substantially to the generation and maturation of other CNS cell types, including astrocytes and oligodendrocytes. Several studies in vitro have identified a significant role of microglia in nonneuronal cell development. Microglia-conditioned medium can induce astrocytic differentiation from neural stem and precursor cells (NSPCs) through microglia-derived interleukin (IL)-6 and leukemia inhibitory factor (LIF) (Nakanishi et al., 2007). Similarly, microglia-derived soluble factors, such as insulin-like growth factor 1 (IGF-1), nuclear factor kappa B (NF-κB), and interleukin-1 beta (IL-1β), regulate the behavior of oligodendrocyte precursor cells (OPCs). These secreted factors promote the survival and differentiation of cultured OPCs into mature oligodendrocytes (Nicholas et al., 2001; O'Kusky & Ye, 2012; Shigemoto-Mogami et al., 2014). Indeed, microglial suppression, which decreases the levels of microglia-derived cytokines, significantly inhibited neurogenesis and oligodendrogenesis in the forebrain subventricular zone (SVZ) of neonatal rats (Shigemoto-Mogami et al., 2014). Microglia also promote oligodendrogenesis by providing the main source of iron, which is a required cofactor for the enzymes involved in the myelination of oligodendrocytes (Todorich et al., 2009; Zhang et al., 2006). Microglia accumulate iron just before myelination, and a decrease in microglial iron load parallels the accumulation of iron in oligodendrocytes (Cheepsunthorn et al., 1998; Todorich et al., 2009). In conclusion, during development, microglia play important roles in the survival, proliferation, and maturation of several developing cell types of the CNS, including neurons, astrocytes, and oligodendrocytes.

8.1.3 Microglia in the Mature CNS

In adults, microglia are distributed throughout the entire CNS. Microglia are heterogeneous in the brain according to a number of variables, including region, sex, morphology, abundance, and functional gene expression (Tan et al., 2020). Microglia comprise an estimated 10% of the adult brain, which varies from 5% in the corpus callosum to 12% in the substantia nigra (Ochocka & Kaminska, 2021). Microglial phenotypic heterogeneity across brain regions may have functional implications; however, there is not sufficient evidence to establish a direct link (Tan et al., 2020).

The ontogeny of microglia and the nature of microglial self-renewal remain contentious issues. At this time, the prevailing consensus holds that microglia originate near-exclusively from the yolk sac in early development (Ginhoux et al., 2013; Zhang et al., 2022). Despite this, microglia have been shown to entirely repopulate the CNS with new cells, following a pharmacological depletion of microglia in the mouse brain (Najafi et al., 2018). This ostensible contradiction raises questions about the origin of these repopulated microglia, and a satisfactory answer has not yet been accepted. The most accepted explanation holds that microglia repopulate the brain by proliferation of a residual pool of microglia from development and possibly by migration of peripheral monocytes (Barnett et al., 2021; Huang et al., 2018). Despite the uncertainty, it is established that microglia maintain their numbers through some mechanism/s of self-renewal. Interestingly, repopulated microglia can have different functional profiles than the depleted microglia (Weber et al., 2019). This mechanism could potentially be therapeutically harnessed by depleting disease-associated microglia and allowing repopulation with microglia expressing transcriptional profiles closer to those of naive mice (Barnett et al., 2021).

8.1.4 Microglial Steady-State Functions

The role of microglia in the context of inflammation and disease has historically dominated the focus of microglial research so severely that microglia at homeostasis were long referred to as "resting." However, emerging studies have shed light on a plethora of functions that microglia carry out in homeostasis (Tremblay et al., 2011). Aside from CNS surveillance, steady-state microglia interact with almost all cell types to mediate synaptic plasticity and neurogenesis and regulate neuronal activity and other critical housekeeping measures (Woodburn et al., 2021) (Fig. 8.2).

Far from resting, steady-state microglia exist in a constant state of motion with elongated and motile processes that survey the brain parenchyma for pathogens or debris (Kierdorf & Prinz, 2017; Nimmerjahn et al., 2005). Primary microglial processes are further ramified by secondary branches and nanoscale filopodia, which enable rigorous monitoring of neighboring cells for any molecular abnormalities (Bernier et al., 2019; Petry et al., 2023). Filopodia extension is actin-dependent and driven by intracellular cyclic adenosine monophosphate (cAMP) concentration, whereas primary process extension is driven by decreasing intracellular cAMP concentration (Petry et al., 2023).

A major role of microglia in the steady state is the modulation of synaptic plasticity and long-term potentiation (LTP) (Kierdorf & Prinz, 2017). In addition to the removal of synapses from damaged neurons, microglia regulate synaptic and neuronal function in the adult brain (Bar & Barak, 2019). Microglia eliminate underused synaptic inputs via the complement system, a conserved family of the innate immune system involved in tagging weaker synapses and facilitating phagocytosis by microglia (Stevens et al., 2007). In the adult

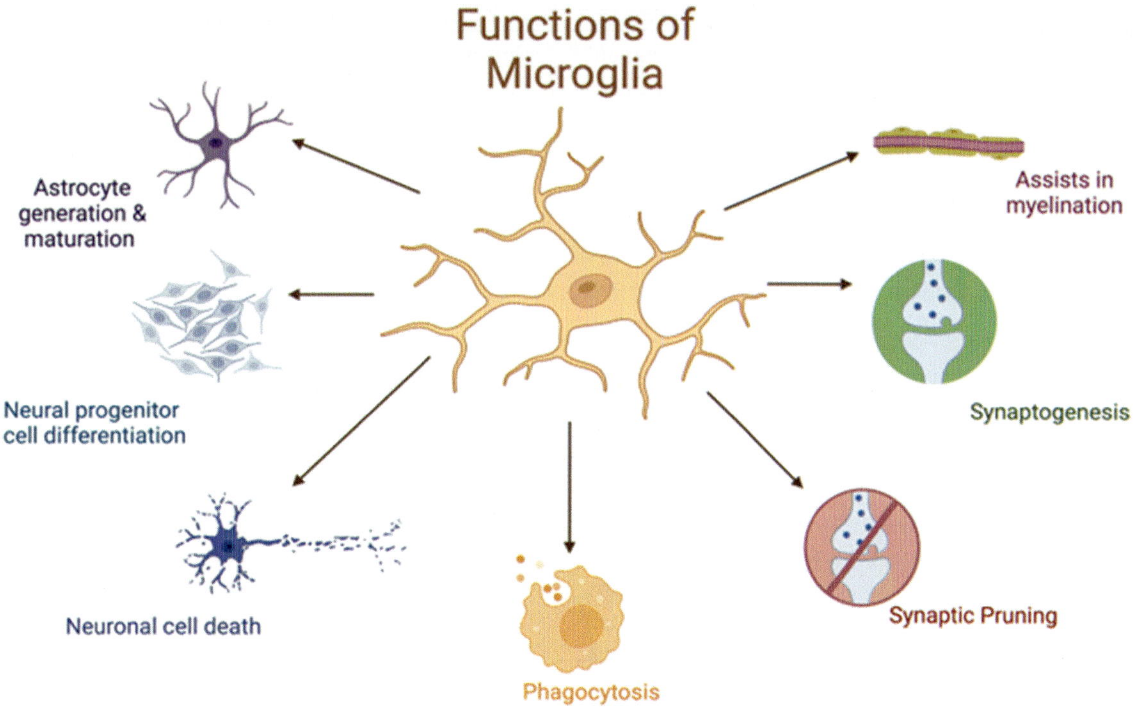

Fig. 8.2 Functions of microglia: Microglia aid in many functions of the brain. Some of the main functions of microglia include: (1) assisting in the generation and maturation of astrocytes, (2) aiding in the differentiation of neural progenitor cells, (3) death of neurons, (4) phagocytosis, (5) synaptic pruning, (6) synaptic plasticity, and (7) assisting in myelination

CNS, secretion of brain-derived neurotrophic factor (BDNF) by microglia promotes long-term potentiation (LTP) and neuronal survival (Deinhardt & Chao, 2014). The neuronal activity attracts microglia via adenosine triphosphate (ATP) signaling, which induces microglia to move their processes to the active synapses to both produce BDNF and suppress neuronal activity (Matejuk & Ransohoff, 2020). Depletion of microglia from adult mice has been shown to induce deficits in LTP and consequently impair learning and memory, highlighting the critical role of microglia in neural plasticity (Parkhurst et al., 2013).

In addition to modulating neuronal plasticity, microglia also promote adult neurogenesis (Kierdorf & Prinz, 2017). The generation of new neurons in the adult CNS takes place primarily in the subgranular zone of the dentate gyrus (part of the hippocampus) and the subventricular zone of the lateral ventricle (Cameron & McKay, 1998). Although the mechanisms are still debated, microglia have been shown to facilitate the migration and differentiation of neural precursor cells (NPCs) and ultimately promote NPC survival to become functional and integrated neurons (Aarum et al., 2003). Secretion of cytokines such as IL-1β, IFN-γ, and IGF-1 by microglia has also been shown to support adult neurogenesis (Shigemoto-Mogami et al., 2014).

Microglia are also necessary for the maintenance and growth of myelin by oligodendrocytes in the adult CNS (McNamara et al., 2023). Myelin provides insulation for neuronal electrical transmission; however, some CNS pathologies involve the deterioration of myelin integrity, which severely threatens neuronal functioning. Microglia have been shown to prevent both hypermyelination and demyelination, and microglial perturbations can have consequences for myelin health (McNamara et al., 2023). Phagocytosis of myelin debris and apoptotic neurons by microglia creates a favorable environment for myelin repair; however, oligodendrocytes can become unintended victims of the inflammatory signaling of activated microglia (Peferoen et al., 2014).

8.1.5 Protein Markers

To investigate microglia both in vivo and in vitro, protein markers are used for identification and visualization. Microglial markers include surface and intracellular proteins as well as released molecules. These markers can be detected at any phenotype as they can be expressed by both resting and activated microglial cells (Jurga et al., 2020). The distinction between the phenotypes can be achieved through the comparison of detectable protein levels as their expression changes with microglial activation (Jurga et al., 2020).

Some of the well-established protein markers include ionized calcium-binding adaptor molecule 1 (Iba-1) and cluster of differentiation (CD) molecules, including CD68, CD206,

CD45, and CD11b (Jeong et al., 2013; Rice et al., 2017; Walker & Lue, 2015; Yun et al., 2018). Under physiological conditions, microglia express some markers such as CD206 and CD45 at low levels. During the activated state, microglia upregulate some proteins that can be used as markers for microglial activation. Other cells like infiltrating macrophages in the CNS can be detected with many of these markers. Currently, the most specific general microglial markers are the transmembrane protein 119 (TMEM119) and the purinergic receptor (P2Y12R) (Amadio et al., 2014; Satoh et al., 2016).

8.2 Reactive Microglia

Microglia can become activated by various stimuli, including immunogenic molecules, neurodegenerative diseases, and pathological conditions. Upon activation, surveilling microglia retract their long processes and increase their cell soma size to take on an ameboid morphology and adopt new functions. Activated microglia classically respond to these neurophysiological threats by upregulation of inflammatory signaling, phagocytosis, and recruitment of peripheral immune cells. Although microglial activation is a critical function of host CNS defense, chronic and overactivation of microglia is a feature of various pathological conditions and neurodegenerative diseases.

8.2.1 Microglia and Neuroinflammation

Microglia are the primary affecters of neuroinflammation, including the induction of inflammatory cytokines and chemokines (Woodburn et al., 2021). It is important to understand the concept of "neuroinflammation" to understand the role of microglia in the CNS. Broadly, neuroinflammation refers to an inflammatory response in the brain or spinal cord through inflammatory cytokines and chemokines, reactive oxygen species (ROS), and secondary messengers. Neuroinflammation is an essential method of communication for the CNS that can have both adaptive and harmful effects, depending on the stimulus and context.

Activated microglia are typically classified by two polarization phenotypes: the classical pro-inflammatory M1 phenotype and the alternative anti-inflammatory M2 phenotype (Jurga et al., 2020). Generally, M1 microglia promote neuroinflammation and neurotoxicity, while M2 microglia promote neuroprotection (Fig. 8.3). More recently, however, microglia have been shown to exhibit mixed phenotypes as well as transition states and phenotypes unique to their microenvironment. A modern perspective recognizes a rich variety of microglial activation subtypes, but the M1/M2 classification remains a relevant model for discussing the functions of activated microglia (Ransohoff, 2016).

8.2.2 M1: Pro-inflammatory Microglia

When surveilling, when microglia recognize an injury or a threat, they polarize toward the M1 phenotype. M1 microglial activation, known as classical activation, is a pro-inflammatory and neurotoxic state typically induced by the binding of ligands such as IFN-γ or lipopolysaccharide (LPS) to toll-like receptors (TLRs) on the surface of microglial cells (Colonna & Butovsky, 2017). M1 activation ultimately results in the production of pro-inflammatory cytokines, chemokines, and reactive oxygen species (ROS). The objective of M1 activation is to kill foreign pathogens and recruit T cells to launch the adaptive immune response; however, pro-inflammatory activated microglia are an aspect of neurodegenerative disease pathology.

When steady-state microglia activate the M1 phenotype, they retract their ramified processes and assume the amoeboid morphology. Amoeboid microglia are similar to macrophages with a round shape, short pseudopodia, and phagocytotic vacuoles (Boche et al., 2013). In this state, M1 microglia are mobile and therefore capable of migrating to the site of the pathogen to express pro-inflammatory mediators and perform phagocytosis (Jurga et al., 2020).

Their general inflammatory profile is high IL-12 and low IL-10. Receptors associated with antigen presentation become upregulated in M1 microglia, such as MHC II, CD86, and Fcγ (Cherry et al., 2014). M1 microglia also induce oxidative stress by reactive oxygen species (ROS), leading to neuronal death (Simpson & Oliver, 2020). Other surface markers for M1 microglia include CD14, CD16, CD32, and CD40 (Jurga et al., 2020).

8.2.3 Toll-Like Receptor Signaling

Toll-like receptors (TLRs) are a highly conserved family of pattern recognition receptors (PRRs), usually expressed in the cell membrane of macrophages and dendritic cells, which recognize microbial-derived molecules and activate innate immune responses. More than 10 TLRs have been identified in mammals that are expressed in various cell types and bind one or more specific ligands. TLRs exist as dimers, with most functioning as homodimers, except for TLR2, which forms heterodimers with TLR1 or TLR6. TLR2 and TLR4 are highly expressed in the plasma membrane of microglia and play an important role in innate immune activation and microglial polarization (Cui et al., 2020). Microglial TLR signaling is summarized below in Fig. 8.4.

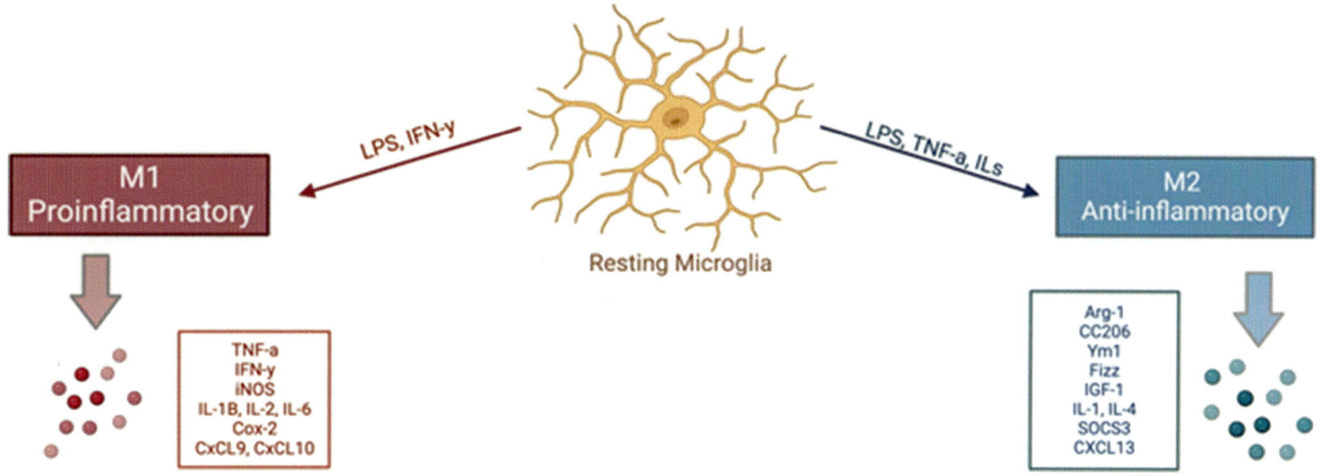

Fig. 8.3 Pro-inflammatory and anti-inflammatory microglia: Depending on the binding ligand that activates microglia, a pro-inflammatory or anti-inflammatory response can be activated. The ligands LPS and IFN-γ trigger the classical, pro-inflammatory response, pro-inflammatory cytokines, chemokines, and reactive oxygen species. The ligands TNF-α, ILs, and LPS activate the alternative anti-inflammatory response to downregulate inflammation and initiate tissue repair. This causes the release of anti-inflammatory cytokine and pro-inflammatory cytokine antagonists

TLR4 binding of its principal ligand lipopolysaccharide (LPS), a cell wall component of Gram-negative bacteria, promotes M1 microglial activation and a pro-inflammatory signaling cascade (Lu et al., 2008). Other TLR4 ligands include amyloid-β (Reed-Geaghan et al., 2009) and α-synuclein (Choi et al., 2020), making TLR4 activation a relevant pathway in neurodegenerative diseases. TLR2 heterodimerizes with multiple other TLRs as well as other non-TLR molecules to recognize a wide variety of pathogen-associated molecular patterns (PAMPs). The canonical ligand of TLR2 is lipoteichoic acid (LTA), a cell wall component of Gram-positive bacteria; however, TLR2 can recognize various bacterial, fungal, and viral molecules as well as other PAMPs due to its receptor "promiscuity" (Zähringer et al., 2008).

TLR activation recruits kinases and adapter proteins to promote inflammatory signaling through two potential pathways: a MyD88-dependent and a TRIF-dependent pathway.

All identified TLRs, except for TLR3, signal through recruitment of the MyD88 adapter protein. Mice with MyD88 genetically knocked out do not exhibit the production of inflammatory cytokines in response to TLR ligands (Takeda & Akira, 2004). Upon TLR ligand recognition, MyD88 binds their TIR domain and forms a complex with members of the IRAK family (Lin et al., 2010). IRAK1 then autophosphorylates, leading it to dissociate from the complex and bind TNFR-associated factor-6 (TRAF6) (Lye et al., 2004). TRAF6 facilitates the activation of TAK1, which activates two different pathways: the NF-κB signaling cascade and the MAPK signaling cascade. The result of both of these pathways is the production of pro-inflammatory cytokines, chemokines, and ROS.

One pathway that is activated downstream of the MyD88-dependent pathway is the IKK/NF-κB pathway. NF-κB naturally exists inactively in the cytoplasm due to an inhibitory molecule, IκBα. NF-κB regulation is dependent on deactivation of the inhibitory IκBα and subsequent translocation to the nucleus. Activation of NF-κB affects hundreds of gene targets, largely pro-inflammatory (Liu et al., 2017). Chief among the genes upregulated by NF-κB are the inflammatory cytokines TNF-α and IL-1β. Activated TAK1 from the final step of the MyD88-dependent pathway activates the IKK complex, which consists of IKKα, IKKβ, and IKK-γ (NEMO, short for NF-kappa-B essential modulator). The IKK complex catalyzes the phosphorylation of the inhibitory molecule IκBα, which is subsequently ubiquitinated and degraded by the 26 s proteasome, allowing the translocation of NF-κB into the nucleus and induction of inflammatory cytokines. This is termed the "canonical pathway" of NF-κB activation. A noncanonical pathway has been described involving phosphorylation of p100, another passive inhibitor of cytoplasmic NF-κB, by an IKKα homodimer. However, this pathway is clearly not associated with TLR signaling (Hayden et al., 2006; Kawai & Akira, 2007; Shabab et al., 2017).

The kinase cascade of the MAPK signaling pathway results in the activation of three mitogen-activated protein kinases (MAPKs): c-Jun N-terminal kinase (JNK), p38, and extracellular signal-regulated kinase (ERK). Activated TAK1 also phosphorylates two members of the MAPK family, MKK3 and MKK6, subsequently activating JNK and p38. ERK is activated by Tpl2, a MAP 3 K in the same family as TAK1. These MAPK family members mediate the activation of AP-1 transcription factors, which regulate pro-

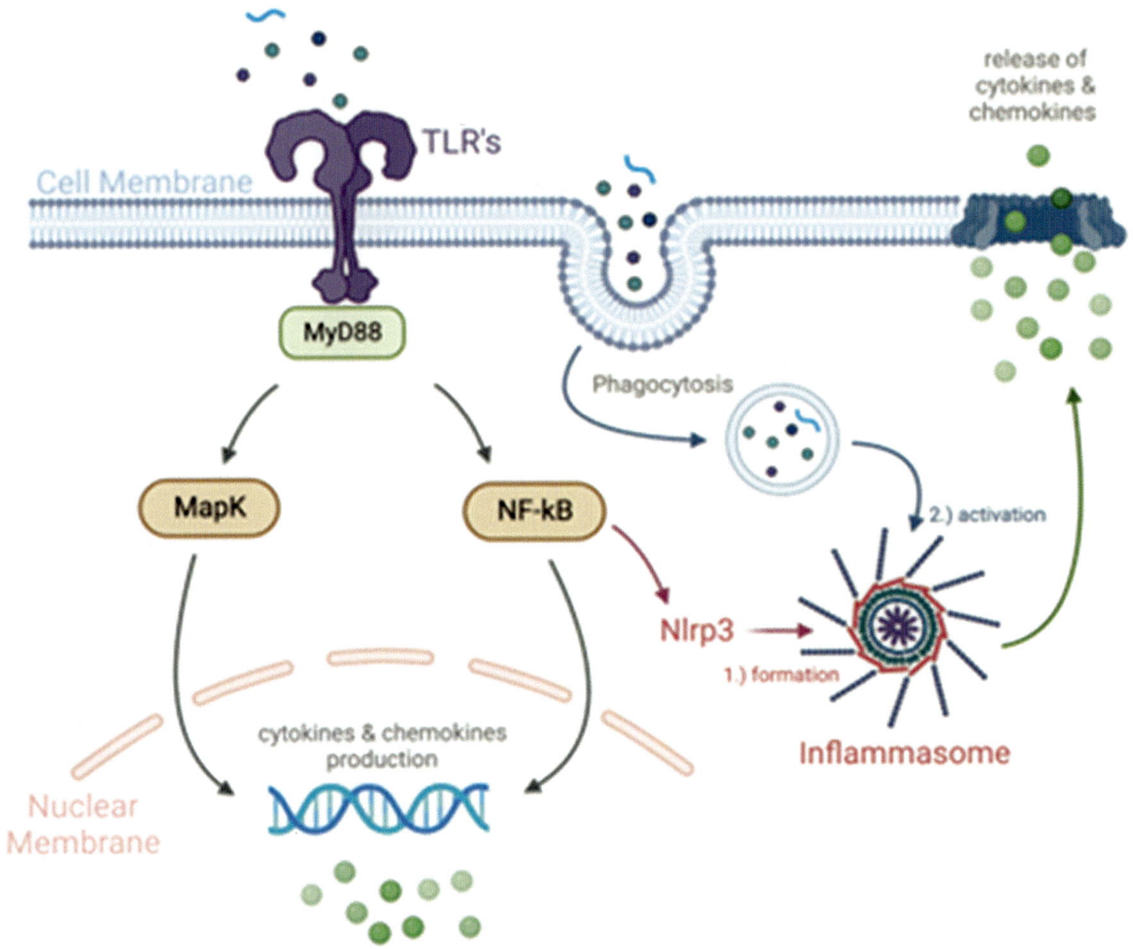

Fig. 8.4 TLR signaling: The binding of TLR ligands causes most TLRs (not TLR3) to recruit the MyD88 adaptor protein. This leads to activation of either the NF-κB or MAPK signaling cascade, both resulting in the production of pro-inflammatory cytokines, chemokines, and ROS. The NF-κB signaling pathway also signals for NLRP3 formation of the inflammasome. Inflammasome activation is triggered by the phagocytosis of DAMPs. This activation in microglia allows for the release of pro-inflammatory cytokines

inflammatory genes such as *COX2* and *NOS2* (Banerjee & Gerondakis, 2007; Barton & Medzhitov, 2003; Guha & Mackman, 2001; Rodríguez-Gómez et al., 2020; Shabab et al., 2017).

TLR3 and TLR4 function through the TRIF-dependent pathway independently of MyD88, leading to activation of NF-κB, MAPKs, and IRF3 to type I IFNs and inflammatory cytokines genes. TRIF-dependent signaling is responsible for TLR3-dependent production of IFN-β as well as TLR4-dependent activation of both NF-κB without the need for MyD88 and IFN-β.

Upon activation by dsDNA, TLR3 directly recruits TRIF, which interacts with TRAF3 and TRAF6. TRAF6 activates the TAK1 complex, which leads to the activation of NF-κB and MAPK signaling pathways and the production of inflammatory cytokines. TRAF3 recruits TBK1 and IKKi kinases and NEMO to accomplish the phosphorylation of IRF3. Dimerization and translocation of IRF3 from the cytoplasm into the nucleus leads to the induction of type I IFNs. TLR4 can also signal through the TRIF-dependent pathway in the absence of MyD88, but it requires the adapter protein TRAM to interact with TRIF (Chen et al., 2021; Kawai & Akira, 2006; Kawasaki & Kawai, 2014; Takeda & Akira, 2004; Ullah et al., 2016).

8.2.4 NOD-Like Receptor (NLR) Signaling

The more recently discovered phenomenon of inflammasome signaling has become a hot topic in biomedical research. Inflammasomes are multimeric protein complexes that have been shown to play a major role in a wide range of CNS pathologies, in particular the NLRP3 inflammasome (Hanslik & Ulland, 2020). Activation of the NF-κB pathway provides the first signal for NLRP3 inflammasome formation with expression of pro-IL-1β, pro-IL-18, and NLR family pyrin domain-containing 3 proteins (Nlrp3). Canonically, in response to an internalized cytosolic DAMP signal (the second signal), Nlrp3 recruits Asc and procaspase-1 to assemble the NLRP3 inflammasome.

The result of NLRP3 inflammasome activation is the expression of inflammatory cytokines IL-1β and IL-18 as well as a form of pro-inflammatory cell death called pyroptosis (Hanslik & Ulland, 2020). Pyroptosis serves to destroy the macrophages infected with pathogens by forming pores in the membrane, causing rupture and release of the pro-inflammatory contents of the cytosol into the extracellular space. This release of inflammatory cytokines differentiates pyroptosis from apoptosis (Wang et al., 2019).

M1 microglia and NLRP3 inflammasome activation are closely linked, in that inhibition of NLRP3 inflammasome activation promotes polarization to the M2 microglial phenotype (Li et al., 2021). For this reason, inhibition of NLRP3 activation has become an attractive target for treating the inflammatory aspects of neurological conditions. Other NLRP inflammasomes have been described but are far less studied.

8.2.5 Janus Kinase/Signal Transducer and Activator of Transcription (JAK/STAT) Signaling

The JAK/STAT signaling pathway regulates neuroinflammation and microglial phenotypic transition. This pathway is activated by multiple cytokines and interferons, including IFN-γ, and generally promotes M1 microglial polarization (Li et al., 2021).

Upon binding of a cytokine to its receptor on the membrane of microglia, associated JAKs activate and phosphorylate the receptor, allowing associated STATs to bind the cytokine receptor (Yan et al., 2018). STATs then translocate to the nucleus and modulate the expression of a wide array of inflammatory genes (Li et al., 2021). Individual cytokines utilize specific combinations of the multiple identified JAKs and STATs to produce unique regulation of neuroinflammation.

In the steady state, microglia express the suppressor of the cytokine signaling (SOCS) family of proteins, which serve as a negative regulator of the JAK/STAT pathway and promote the M2 microglial phenotype (Liau et al., 2018). Inhibition of the JAK/STAT pathway is a potential therapeutic target for reducing neuroinflammation.

8.2.6 Oxidative Stress

M1 microglial activation induces the production of reactive oxygen species (ROS), which can lead to oxidative stress when produced at levels that exceed the antioxidant capabilities of the CNS (Simpson & Oliver, 2020). ROS are important secondary messengers in homeostatic conditions but can lead to neuronal death when dysregulated. Production of ROS is largely mediated by the NADPH oxidase (NOX) enzymes NOX2 and NOX4 (Ma et al., 2017). The exact mechanisms of microglia-induced ROS production are yet to be elucidated; however, neuroinflammation as a consequence of PAMP/DAMP signaling is a likely culprit (Simpson & Oliver, 2020). Additionally, ROS such as H_2O_2 and $O_2^{\bullet-}$ can also induce neuroinflammation in microglia via the NF-κB and MAPK signaling pathways (Simpson & Oliver, 2020).

NOX-mediated oxidative stress is chronically elevated in disease-associated microglia, particularly associated with Alzheimer's disease (AD), which has been shown to contribute to neuronal loss (Multhaup et al., 1997).

8.2.7 M2: Anti-inflammatory Microglia

Although M1 microglial activation is critical to pathogen defense and injury response, the induced cytokines and chemokines result in neurotoxicity. To resolve the inflammatory state produced by M1 microglia, "alternative" M2 microglial activation carry out functions to downregulate inflammation and initiate tissue repair (Cherry et al., 2014). M2 microglia express anti-inflammatory cytokines such as IL-4, IL-10, IL-14, and TGF-β to antagonize pro-inflammatory cytokines (Tang & Le, 2016).

Anti-inflammatory M2 signaling in microglia has been a more recently discovered phenomenon than classical M1 signaling (Orihuela et al., 2016). Since then, multiple subtypes of M2 microglia have been identified, including M2a, M2b, and M2c. These subtypes are defined by the cytokines that stimulate them and their resulting signaling pathway. Surfaces of M2 microglial activation include CD163 and CD206 (Cherry et al., 2014). The shared phenotype between the subtypes of M2 microglia is high IL-10 and low IL-12 (Mosser & Edwards, 2008).

The first cytokine discovered to induce the "alternative" M2 phenotype in microglia was IL-4 (Stein et al., 1992). IL-4 and IL-13 signal through the JAK/STAT6 pathway to induce what is now termed "M2a activation," resulting in suppression of inflammatory cytokines such as IL-6 and TNF-α, inhibition of NF-κB, upregulation of Arg1, and reduction of NO release (Cherry et al., 2014; Gadani et al., 2012; Li et al., 2021; Tang & Le, 2016).

M2b microglia are stimulated by TLRs and immune complexes, and, unlike M2a and M2c, are associated with high levels of inflammatory cytokine production (Mosser, 2003). M2b microglia are classified under alternative activation rather than classical activation because, despite their pro-inflammatory phenotype, they maintain a ratio of low IL-12 and high IL-10 (Mantovani et al., 2004). Additionally, M2b microglia protect against LPS and promote Th2 differentiation (Mosser & Karp, 1999).

M2c microglia are induced following stimulation by IL-10, TGF-β, or glucocorticoids (Cherry et al., 2014; Martinez et al., 2008). This polarization has been referred to as "deactivated" due to the phenotype of downregulated pro-inflammatory cytokines, although M2c microglia carry out other functions to recover from inflammation such as cellular proliferation and tissue remodeling (Martinez et al., 2008; Mosser, 2003).

8.2.8 Beyond the M1/M2 Paradigm

As microglial research has progressed and matured, it has become increasingly apparent that the designations of M1/classical and M2/alternative activation are not accurate representations of microglia in vivo. These classifications of discrete actions are based on the outdated idea that microglia typically rest until they encounter a pathogen or diseased tissue, upon which they activate akin to an infiltrating macrophage. Our knowledge has evolved past this point; however, our nomenclature lags.

Two-photon imaging of microglia has since revealed the constant motion of microglia we once called "resting" (Nimmerjahn et al., 2005). We have since become aware of a multitude of functions that microglia carry out in homeostasis, with an equally rich signaling network to those associated with the M1/M2 classification.

Moreover, the distinction between M1 and M2 polarization is not as discrete as implied by the nomenclature. In fact, recent application of single-cell RNA sequencing technology has demonstrated that the canonical markers of polarization of both M1 and M2 can be highly expressed within the same cell (Kim et al., 2016). Genes that were found to be associated with the canonical polarization states in vitro did not hold up to single-cell analysis, implying either mixed-polarization states or indeed a lack of polarization as we understand it.

8.2.9 Summary

Microglial activation is commonly classified using the M1/pro-inflammatory and M2/anti-inflammatory framework. These signaling pathways are the major mechanisms in the CNS responsible for mediating inflammatory responses in the brain; however, this nomenclature fails to accurately capture the complexity of microglial signaling in various conditions and regions. Neuroinflammation is a native process of the brain that is not inherently harmful, but the dysregulation of these systems can have catastrophic cellular effects as seen in neurological disease.

8.3 Disease-Associated Microglia

Microglia, along with other immune cells, are key contributors to inflammatory responses of the central nervous system. Under homeostatic conditions, microglia have several protective roles in the brain, including promotion of neuronal health through various mechanisms such as the secretion of pro-inflammatory cytokines. In response to alterations in brain homeostasis, namely, pathological events, microglia can initiate neuroinflammatory responses that assist in the clearance process. The process of neuroinflammation can result from several reasons, most notably abnormal protein aggregation that is seen in several neurodegenerative diseases (Stelzmann et al., 1995). Microglia also largely contribute to the removal of foreign substances within the CNS via phagocytosis and lysosomal degradation. This clearance is promoted through the engulfment of synapses and sculpting of synaptic connections (Chung et al., 2015) and relies on effective communication amongst cells. Altogether, microglia are essential for maintaining homeostasis within the CNS, and alterations in their homeostatic functions can result in severe consequences.

Although microglia generally have a protective role, their activation can fail to cease upon neutralization of the pathological threat, becoming harmful to the health of the CNS. The failure of microglia to resolve their functions leads to continuous neuroinflammation, inducing neurotoxicity, further causing neurodegeneration, and contributing to the pathogenesis of neurodegenerative diseases, including Alzheimer's disease (AD) and Parkinson's disease (PD). Other factors that impact the role of microglia include the gut microbiome, gender, and age (Colonna & Butovsky, 2017).

8.3.1 Alzheimer's Disease

Currently, there are an estimated 6.7 million Americans, aged 65 years and older, living with dementia as a result of Alzheimer's disease (AD), a number that is expected to grow in future years (Rajan et al., 2021). There is currently no cure for AD, but treatments that alter disease progression and minimize symptoms are clinically available. Recent genetic studies have indicated that microglia play a significant role in the pathology of AD due to the abundance of microglia-specific genes found to be associated with AD. For this reason, research regarding the role of microglia in AD and many other neurodegenerative diseases has significantly expanded in pursuit of new therapeutic targets.

AD was identified in 1908 by Alois Alzheimer when he found patients exhibiting disorientation and progressive memory loss, key symptoms of what is now known as AD. Alzheimer also completed pathological assessments that demonstrated brain atrophy and other key pathologies, including morphological alterations in microglia, neurofibrillary tangles, and extracellular plaques containing cytotoxic levels of amyloid-beta (Aβ) (Stelzmann et al., 1995). These pathologies have also shown to be a result of aging due to its effect on "proteostasis," the collective term for mechanisms regarding protein synthesis, clearing, and post-translational modifications (Clayton et al., 2017).

Under homeostatic conditions, there is a balance between Aβ production and clearance. Microglia generally form a protective barrier around Aβ-containing plaques, such that new Aβ cannot be added to contribute to their cytotoxicity. Dysregulation in the production/clearance balance of Aβ can be a result of aging or genetic mutations, most notably in genes highly or solely expressed by microglia. As a result, abnormal Aβ aggregation occurs and becomes cytotoxic, thus triggering microglial activation. The continuous production of Aβ and its excessive interaction with microglia result in neurotoxic levels of reactive oxygen and nitrogen species. This increase in neurotoxicity causes a further increase in microglial activation and neuroinflammation, leading to neurodegeneration, and consequentially contributing to the development and progression of AD (Hickman et al., 2018). More specifically, microglial activation that is triggered by Aβ accumulation also promotes abnormal protein (tau) hyperphosphorylation (Hickman et al., 2018). Accumulation of phosphorylated tau has been found to cause neuronal dysfunction and synaptic impairment, resulting in neurofibrillary tangles, all key pathologies identified by Alzheimer's (Drummond et al., 2020).

While there is still no cure for AD, methods of treatment have been identified. The accumulation of β-amyloid plaques and its downstream effects represent one of the key pathologies identified in the development and progression of AD. For this reason, drugs that reduce these plaques have shown improvement in disease progression, but additional studies are currently being completed to elucidate their ability to slow cognitive decline.

8.3.2 Parkinson's Disease

PD is ranked as the second most common age-associated neurodegenerative disease in North America, predominantly affecting those over the age of 65 (Tansey & Goldberg, 2010). Symptoms of PD occur after the selective degeneration of dopaminergic neurons, which are responsible for cognitive functions such as motivation and habit learning, and the accumulation of Lewy bodies that consist of aggregated, misfolded neuronal protein α-synuclein within the substantia nigra (Willis et al., 2022).

Similar to AD, abnormal protein aggregation has been shown to greatly contribute to the pathogenesis of PD. The abnormal aggregation of neuronal protein α-synuclein causes microglial activation. Active microglia internalize and attempt to phagocytose α-synuclein, increasing inflammatory cytokines that are released by microglia. However, errors in the communication between neurons and glial cells can result in improper clearance of α-synuclein that is then able to spread to other parts of the brain. The excessive neuroinflammation that results from microglial activation due to abnormal α-synuclein aggregation contributes to toxicity in cerebrospinal fluid (CSF) in patients with PD, further causing neurodegeneration of dopaminergic neurons (Nagatsu et al., 2000; Vawter et al., 1996). Recent studies have found extensive levels of pro-inflammatory microglia in PD patients, but this did not longitudinally correlate with the clinical severity of PD, indicating that this activation may occur throughout the early stages of the disease and then remain relatively consistent (Gerhard et al., 2006). These findings suggest that microglia play a major role in the development and progression of PD, but their exact role is still unclear. PD is believed to be a result of hereditary and environmental factors, but there is still plenty to be investigated about the disease and its pathogenesis. Due to its complexity, there is currently no cure, but therapeutic measures exist that assist in replenishing dopamine levels.

8.3.3 Amyotrophic Lateral Sclerosis

Amyotrophic lateral sclerosis (ALS) is a neurodegenerative disease that is characterized by substantial loss of motor neurons and leads to paralysis and death. The majority of ALS cases are sporadically acquired, but other cases are a result of hereditary genetic mutations (Lall & Baloh, 2017). *SOD1* is a gene of interest, and mutations of this gene can be either inherited or sporadic, but other factors can contribute to the development and progression of ALS. In animal models, overexpression of this mutated gene has demonstrated a disease similar to ALS, although this alone is not sufficient to induce the motor neuron degeneration seen in ALS, suggesting other clinically relevant factors. Mutated SOD1 proteins do not display neurotoxicity without microglia, demonstrating the importance of microglia in the pathogenesis of ALS (Zhao et al., 2010). Its expression and interaction with microglia have been shown to accelerate the onset of the disease and exacerbate neurodegeneration. Mutations in SOD1 expression can cause consistent damage and result in increased neurotoxicity, thus chronically activating microglia. This chronic activation leads to neurotoxic levels of reactive oxygen and nitrogen species and the upregulation of pro-inflammatory cytokines, resulting in sustained inflammation that exacerbates neurodegeneration. While there is a clear association between microglia and ALS, the direct mechanism by which it results in motor neuron loss is still unknown.

As of date, there are no existing treatments capable of reversing neuronal damage or curing the disease. However, drugs that decrease glutamate levels can reduce subsequent motor neuron damage by decreasing the signaling between nerve cells and motor neurons.

8.3.4 Multiple Sclerosis/Experimental Autoimmune Encephalomyelitis

Multiple sclerosis (MS) is a condition that is among the most common disabling neurodegenerative diseases, characterized by inflammation in the CNS that results in motor, sensory, visual, and autonomic system deficits (Doshi & Chataway, 2017). The immune cells of patients with MS attack and compromise the integrity of myelin, the protective sheath of nerve fibers, causing errors in communication between the CNS and the rest of the body. As previously mentioned, these errors can result in chronic activation of microglia, ultimately contributing to disease progression and neurodegeneration.

Experimental autoimmune encephalomyelitis (EAE) is an animal model designed to mimic MS. These models have demonstrated that microglial activation induced by EAE onset can result in a release of proteases, reactive oxygen and nitrogen species, pro-inflammatory molecules, and reactive T lymphocytes, resulting in toxicity to neurons and cortical demyelination (Gao & Tsirka, 2011). As a result, neuroinflammation and the generation of demyelinating lesions occur, further leading to neurodegeneration. However, at the onset of EAE, microglia have been shown to demonstrate a protective role in axonal regeneration, the release of neurotrophic factors, and more (Napoli & Neumann, 2010). Collectively, studies have shown that the role of microglia in MS/EAE is dependent on the disease stage. It is clear that altered microglial functioning plays a role in the progression of EAE, most notably affecting their ability to generate a neuroprotective response, but further research must be conducted to determine the precise mechanisms. Current therapies that inhibit damaging microglial activity have shown progress in rescuing certain factors that contribute to the progression of MS.

8.3.5 Huntington's Disease

Huntington's disease (HD) is characterized by neurodegeneration that causes motor, cognitive, and psychiatric changes. This disease is inherited and caused by the expansion of the Huntington protein, in which more than 40 repeats result in HD (Bates et al., 2015). Studies have shown that levels of microglial activation are positively correlated with those of cell death and symptom severity (Sapp et al., 2001). Similar to other neurodegenerative diseases, mutations in the Huntington protein impact microglial activation in a manner that increases cytokine production and transcriptional abnormalities (Yang et al., 2017). There is currently no cure or treatment that exists to prevent disease development and progression, but there are drugs on the market to treat major symptoms. Microglia may promise a potential therapeutic target based on correlation with the key pathologies in HD; however, the mechanisms underlying microglia's role in HD remain unclear.

8.3.6 Summary

In many neurodegenerative diseases, microglia can shift from their homeostatic and protective role to a neurotoxic one, leading to neurodegeneration and progression of these diseases. Many studies have indicated that this change in roles is a result of abnormal protein activity or miscommunication within the body that triggers the chronic activation of microglia. Although microglia are generally protective under homeostatic conditions, their persistent activation has been found to ultimately contribute to neurodegeneration and, therefore, neurodegenerative diseases. Microglia have therefore been proposed as a therapeutic target for many of these diseases; however, our current mechanistic understanding has not yet yielded significant breakthroughs in the form of microglia-based therapies.

8.3.7 Peripheral Infection

Pathogenic exposure in peripheral tissues triggers microglial responses and neuroimmune signaling that differ from responses within the CNS (Passaro et al., 2021). Preclinical studies that mimic peripheral infections (i.e., lipopolysaccharide) have been used to understand the immune response elicited. These studies have found that peripheral infections result in pro-inflammatory signaling cascades in peripheral tissues that affect other components of an immune response, including microglia (Woodburn et al., 2021). Peripheral infections have been shown to indirectly impact microglial function because LPS and other similar pathogens are generally unable to reach the brain parenchyma, the area of the brain that is composed of neurons and glial cells. However, the mechanism by which microglia receive signals from neuroimmune pathways and neuroinflammation has yet to be clearly elucidated.

8.3.8 Psychiatric Disorders

In recent years, there have been an increasing number of studies regarding psychiatric disorders and the role of microglia. Patients who are affected by psychosis or affective disorders display increases in peripheral inflammatory biomarkers, but the mechanisms by which this occurs have not been clearly delineated (Brand et al., 2015). Microglia are important in the pathogenesis of neuropsychiatric and neurodevelopmental disorders, including autism, schizophrenia (SCZ), major depressive disorder (MDD), and substance abuse disorder (Mondelli et al., 2017). Psychological stress has been explicitly linked to microglial activation and neuroinflammation, creating a critical axis in various psychiatric conditions (Calcia et al., 2016). Early life stress has been postulated to "prime" microglia, potentiating future dysregulated microglial activation and contributing to psychiatric pathology (Cattane et al., 2022). Animal models have been widely used to study the role of microglia in psychiatric disorders, but research is limited by the inherent difficulties of studying microglia in living patients.

8.3.9 Major Depressive Disorder

Major depressive disorder (MDD) is a public health issue affecting many people across the world. To date, the biological mechanisms underlying MDD remain unclear due to the complexity of MDD pathology. Recent studies have shown that alterations to microglia are seen in several regions of the brain and contribute to the progression of depression, indicating that microglia can be a contributing factor (Jia et al., 2021). Specifically, microglia are substantially activated in the prefrontal cortex and anterior cingulate cortex throughout severe episodes of MDD (Wang et al., 2022). Excessive microglial activity has been linked to high levels of neuroinflammation, while microglial inhibition has been shown to alleviate depressive behavior (Yirmiya et al., 2015).

Flawed microglia–neuron communication has also been implicated in neurological disorders, including MDD. Dysfunctions in this communication axis promote defective synaptic pruning, a key process in the development of the nervous system that is characterized by the elimination of extra synapses and allows for effective signaling amongst neurons. Defects in this process are associated with ineffective neuron-to-neuron communication, synaptic transmission, and other pathologies seen in various psychiatric disorders, including MDD and schizophrenia (Rahimian et al., 2021).

Existing antidepressant medications have been shown to inhibit inflammatory microglial polarization; however, chronic use of these treatments can have a contradictory effect by increasing microglial activation (Kalkman & Feuerbach, 2016; Mariani et al., 2022). Future antidepressant therapies have been proposed based on switching microglial phenotypes away from the chronically pro-inflammatory state to a more anti-inflammatory phenotype (Zhang et al., 2018).

8.3.10 Autism Spectrum Disorder

Autism spectrum disorder (ASD) is a neurodevelopmental disorder characterized by deficiencies in social communication and restricted/repetitive patterns of behavior and interest. A study has shown that microglial activity was increased in the cortical regions of the brain of patients with ASD; however, this finding was contradicted by several studies indicating no significant changes in both ASD and schizophrenia (Kim et al., 2018). These differences could be related to genetics or due to evaluations being completed at different stages of these disorders. As mentioned before, the inflammatory role of microglia can be harmful, and excessive microglial activation has been shown to contribute to the development of ASD. When these dysfunctions are seen throughout the critical period of synaptic pruning, an association with autism is seen (Takano, 2015). More research is needed to uncover the involvement of microglia in the development of ASD.

8.3.11 Schizophrenia

Schizophrenia (SCZ) is a neurodevelopmental disorder in which patients display a range of psychotic symptoms and cognitive deficits. Recent evidence has indicated that abnormal changes in microglia throughout stages of neurodevelopment contribute to SCZ development—most notably increased microglial activation and, as a result, neuroinflammation (Monji et al., 2009). This neuroinflammation is associated with white matter pathology in SCZ patients and may result in structural and functional impairments (Chew et al., 2013). A loss of cortical gray matter is also seen as a result of the pruning of synapses, the phagocytosis of neurons, and the prevention of releasing neurotrophic factors, which support the overall health of neurons (Rački et al., 2016). Disruptions between cell communication further increase neuroinflammation and contribute to SCZ pathology. In contrast to several neurodegenerative diseases, genetic mutations that contribute to abnormal microglial activity have not been found related to SCZ susceptibility. Throughout long-term psychiatric disorders, increases in microglial density in patients with SCZ or depression have been associated with suicide (Steiner et al., 2008). Altogether, the evidence indicates that microglial regulation may be a therapeutic target.

Antipsychotic medications have been found to decrease microglial activation and associated neuroinflammation, which could point to an important mechanism for the resolution of symptoms in patients with SCZ (Kato et al., 2011).

8.3.12 Substance Abuse Disorder

Exposure to substances such as opioids or alcohol can trigger an inflammatory response, and the interaction between substance abuse and the neuroinflammatory process is a hot research topic (Namba et al., 2021). Drugs, including opioids, cause changes in neuronal behavior, including activation of the mesolimbic dopamine pathway, which is involved in mood and reward processing, causing rewarding properties to be associated with drug use (Taylor et al., 2016). In addition, the neuronal changes associated with drug use include alterations in synaptic plasticity that can contribute to an increased susceptibility to substance use disorder (Kauer & Malenka, 2007).

Human studies have suggested that microglia play a key role in alcohol use disorder; patients who were alcohol-dependent had higher glial cell densities than did those who were not dependent on alcohol (Hercher et al., 2009). Additionally, the release of ROS and pro-inflammatory molecules has been found in patients with alcohol use disorder, which is evidence of microglial activation (Tsermpini et al., 2022). Alcohol abuse can contribute to damage in the CNS, and microglia has been identified as a critical modulator of alcohol neurotoxicity (Henriques et al., 2018). Chronic exposure to psychostimulant substances like cocaine and methamphetamine similarly results in neurotoxicity associated with microglial dysregulation (Gonçalves et al., 2014). Opioid use disorder (OUD) results in high amounts of inflammatory cytokines released by microglia, most notably IL-1β, which causes impairments in neuronal synaptic plasticity and has shown to play a role in opioid withdrawal syndrome and analgesic tolerance (Herlihy & Roy, 2022). These changes in synaptic plasticity can lead to alterations in the reward pathways of the brain, contributing to adverse outcomes such as depression and addiction (Chen et al., 2010). The inflammatory response elicited by microglia also plays a role in the development of morphine tolerance due to the hyperactivity or central sensitization of neurons in the dorsal horn, which are responsible for processing sensory information (Wen et al., 2011). Collectively, microglia are key contributors to the development and progression of substance use disorder and sequelae, including tolerance and withdrawal.

8.3.13 Summary

Psychiatric disorders such as MDD are complex and require extensive investigation. Understanding the biological mechanisms of each disorder is essential for finding effective treatments or cures. Inflammatory markers as a result of microglial activation are increased in many patients with major depressive disorder, anxiety disorders, schizophrenia, and more. The neuroinflammation seen in these disorders has been associated with several symptoms, including reduced motivation, alterations in motor activity, and increased anxiety/alarm, contributing to their disorder progression. Neuroinflammation and psychiatric disorders have a clear association, suggesting that microglia may be an effective therapeutic target. Many studies demonstrate contradicting evidence regarding the role of microglia in psychiatric disorders, but it is clear that microglia are involved. The direct correlation has yet to be studied, as these findings have only been seen in a subset of patients. Antipsychotics and antidepressants have been shown to alter microglial activity and assist in slowing neurodegeneration; however, further research is required. This field is one of great interest, but, due to the complexity and heterogeneity of microglia and psychiatric disorders, there is much to be learned.

8.3.14 The Microbiome

The gut microbiome consists of many microorganisms that are key modulators of health and disease and play a significant role in the development of the brain, more specifically microglia (Sampson & Mazmanian, 2015). The microbiome exists in reciprocal communication with the CNS through various mechanisms, collectively termed the "gut–brain axis." Animal studies show that germ-free mice, which are raised in sterile conditions and theoretically contain no living microorganisms in their gut, differentially express genes that are associated with microglial maturation (Erny et al., 2015). More specifically, there was a downregulation of genes associated with cell activation and defense pathways, meaning their functions were significantly impaired. Supplementation with a limited microbiome was not effective in resolving this issue, but recolonization with a complex and diverse microbiome allowed for maturation of microglia, restoring their homeostatic development and function (Colonna & Butovsky, 2017). Additionally, drug-induced dysbiosis of the gut microbiome compromises the barrier of the gut and results in systemic neuroinflammation, indicating that alterations of the microbiome are correlated with altered microglial activity (Zhang et al., 2019). These findings show a clear relation between the microbiome and microglia, making these key targets in health and disease.

8.3.15 Sex

Recent studies have explored sex differences as they relate to microglia and their functional differences using animal models. During the early postnatal stages, male mice have higher levels of microglia in specific brain regions (hippocampus, amygdala, and cortex) than female mice (Han et al., 2021). This means that if an infection occurs during this period, male mice can produce higher quantities of inflammatory cytokines, which can cause long-term cognitive deficits (Colonna & Butovsky, 2017). However, throughout adulthood, females have more microglia in the same brain regions. This increase can be a result of the production of estradiol, which reduces the inflammatory characteristic of microglia, but further studies are required to elucidate the direct correlations between sex and microglia (Loram et al., 2012).

8.3.16 Age

Aging is one of the most prominent risk factors for the development of neurodegenerative diseases. As previously mentioned, aging impacts the process of "proteostasis," which is the collective term for mechanisms regarding protein synthesis, clearing, and posttranslational modifications (Clayton et al., 2017). As a result of these impairments, abnormal protein aggregation is observed, including Aβ seen in Alzheimer's disease, and produces cytotoxic conditions. In addition, aging also results in other impairments in the CNS, including a reduction in cognitive function and synaptic density as well as an increase in pro-inflammatory cytokines, increases in oxidative stress products, and a decrease in anti-inflammatory cytokines (Clayton et al., 2017).

Age-related alterations to microglial morphology, signaling profile, and dynamic behavior have been identified and hypothetically tied to neurodegeneration (Damani et al., 2011). As age increases, the homeostatic response elicited by microglia is delayed, yet enhanced, potentially contributing to neurodegeneration (Hefendehl et al., 2014). Microglial senescence in aging shifts the balance of microglial polarization toward a more pro-inflammatory state, even in the absence of disease, as part of a model termed "inflammaging" (Hu et al., 2019). The intersection of inflammation and neurodegenerative disease is an evolving field of study, with microglia at the center (Pizza et al., 2011).

8.4 Chapter Summary

Since their discovery just over a century ago, microglia have become a linchpin in our understanding of CNS mechanisms. As our technologies to isolate, sequence, image, and otherwise study microglia have developed, we have found their involvement in just about every corner of health and disease that research has considered. Our conception of microglial activation states is based on early research that explores the diversity of microglial signaling and the dynamic role of microglia in inflammatory systems. We have only recently started recognizing that microglia may express a plethora of "phenotypes" beyond just pro- or anti-inflammatory; however, it is possible that we may move toward a new framework entirely..

What has become clear from microglial research is that dysregulation of microglial signaling is a key element in most disease states in the brain. Because of the diversity in microglial functions, the solution to this dysregulation is not as simple as turning the microglia off or turning them back on. Despite these intricacies, the modulation of microglial activity represents an attractive therapeutic target for many conditions. Indeed, multiple companies are attempting to develop microglia-based therapies for use in neurodegenerative diseases and beyond (Table 8.1).

8.5 Review Questions

1. Which of the following statements is true?
   (a) Microglia are primarily neuroprotective
   (b) Microglia are primarily neurodestructive
   (c) A ramified microglial state is the active state
   (d) Microglia belong to the adaptive immune system
2. Describe the three morphological states of microglia and their function.
3. Name the main mechanisms by which microglia eliminate excess synapses.
4. What is the function of CD47 expression on microglial cells?
5. Name five main functions of microglia in the brain.
6. Name the most established surface marker of microglia.
7. Define neuroinflammation.
8. Which of the following is true? Microglial signaling pathways include:
   (a) TLR signaling pathways
   (b) NOD-like signaling pathway
   (c) JAK/STAT signaling pathway
   (d) All of the above
9. Name one of the key pathologies identified in the development and progression of AD.
10. Mutations in which genes, either inherited or sporadic, contribute to the development and progression of ALS.

8.6 Answers to the Questions

1. (a)
2. Ramified: Surveillance
   Rod-like: Intermediate state between surveillance function and activated state
   Activated: Mediate disposition of the injured tissues
   Amoeboid: Phagocytosis

Table 8.1 The activation status of microglia in different disease states and the consequences of microglia's involvement in disease progression

Disease/disorder	Microglial activation	Consequence	Sources
Alzheimer's disease	Increased activation	Increased Aβ cytotoxicity	8, 10
Parkinson's disease	Dysregulation, microglia communication	CSF toxicity, degeneration of dopamine neurons	9, 13, 19
Amyotrophic lateral sclerosis (ALS)	Chronic activation	Elevated oxygen and nitrogen species, pro-inflammatory cytokines released	24
Multiple sclerosis (MS)/experimental autoimmune encephalomyelitis (EAE)	Increased activation and glial communication errors	Reactive T lymphocytes, toxicity to neurons, cortical demyelination	7
Huntington's disease	Increased activation	Increased cytokine production, transcriptional abnormalities	2, 16
Infection	Increased pro-inflammatory activation	Signals immune response, increased inflammation	22
Major depression disorder	Excessive activity, impaired communication with neurons	Defective synaptic pruning, ineffective neural transmission	20
Autism	Increased activity in the cortical regions of the brain	Increased inflammation	11
Schizophrenia	Increased activation	White matter impairments, loss of gray matter	3, 17

3. Phagocytosis directed by complement components, fractalkine (CX3CL1), and TREM2.
4. CD47 provides a "do not eat" signal to microglia and diminishes synaptic pruning.
5. (a) Assist in generation and maturation of astrocytes
   (b) Differentiation of neuronal progenitor cells
   (c) Neuronal killing
   (d) Phagocytosis of foreign debris
   (e) Synaptic pruning
   (f) Synaptic plasticity
   (f) Myelination
6. While Iba-1 is the most common myeloid marker on the microglial surface, more specific markers that delineate the microglial lineage include TMEM119 and P2Y2R.
7. Neuroinflammation is an innate immune response in the brain or spinal cord mediated through pro-inflammatory cytokines and chemokines, reactive oxygen species (ROS), and secondary messengers.
8. (d)
9. Reactive microglia
   amyloid-beta plaques
   neurofibrillary tangles
10. Superoxide dismutase-1 (*SOD1*)

References

Aarum, J., Sandberg, K., Haeberlein, S. L. B., & Persson, M. A. (2003). Migration and differentiation of neural precursor cells can be directed by microglia. *Proceedings of the National Academy of Sciences, 100*(26), 15983–15988.

Amadio, S., Parisi, C., Montilli, C., Carrubba, A. S., Apolloni, S., & Volonté, C. (2014). P2Y(12) receptor on the verge of a neuroinflammatory breakdown. *Mediators of Inflammation, 2014*, 975849. https://doi.org/10.1155/2014/975849

Anttila, J. E., Whitaker, K. W., Wires, E. S., Harvey, B. K., & Airavaara, M. (2017). Role of microglia in ischemic focal stroke and recovery: Focus on toll-like receptors. *Progress in Neuro-Psychopharmacology & Biological Psychiatry, 79*(Pt A), 3–14. https://doi.org/10.1016/j.pnpbp.2016.07.003

Arnò, B., Grassivaro, F., Rossi, C., Bergamaschi, A., Castiglioni, V., Furlan, R., Greter, M., Favaro, R., Comi, G., Becher, B., Martino, G., & Muzio, L. (2014). Neural progenitor cells orchestrate microglia migration and positioning into the developing cortex. *Nature Communications, 5*(1), 5611. https://doi.org/10.1038/ncomms6611

Banerjee, A., & Gerondakis, S. (2007). Coordinating TLR-activated signaling pathways in cells of the immune system. *Immunology & Cell Biology, 85*(6), 420–424. https://doi.org/10.1038/sj.icb.7100098

Bar, E., & Barak, B. (2019). Microglia roles in synaptic plasticity and myelination in homeostatic conditions and neurodevelopmental disorders. *Glia, 67*(11), 2125–2141. https://doi.org/10.1002/glia.23637

Barnett, A. M., Crews, F. T., & Coleman, L. G. (2021, Jun). Microglial depletion and repopulation: A new era of regenerative medicine? *Neural Regeneration Research, 16*(6), 1204–1205. https://doi.org/10.4103/1673-5374.300439

Barton, G. M., & Medzhitov, R. (2003). Toll-like receptor signaling pathways. *Science, 300*(5625), 1524–1525. https://doi.org/10.1126/science.1085536

Bates, G. P., Dorsey, R., Gusella, J. F., Hayden, M. R., Kay, C., Leavitt, B. R., Nance, M., Ross, C. A., Scahill, R. I., Wetzel, R., Wild, E. J., & Tabrizi, S. J. (2015). Huntington disease. *Nature Reviews Disease Primers, 1*, 15005. https://doi.org/10.1038/nrdp.2015.5

Bernier, L.-P., Bohlen, C. J., York, E. M., Choi, H. B., Kamyabi, A., Dissing-Olesen, L., Hefendehl, J. K., Collins, H. Y., Stevens, B., Barres, B. A., & MacVicar, B. A. (2019). Nanoscale surveillance of the brain by microglia via cAMP-regulated Filopodia. *Cell Reports, 27*(10), 2895–2908.e2894. https://doi.org/10.1016/j.celrep.2019.05.010

Boche, D., Perry, V. H., & Nicoll, J. A. R. (2013). Review: Activation patterns of microglia and their identification in the human brain. *Neuropathology and Applied Neurobiology, 39*(1), 3–18. https://doi.org/10.1111/nan.12011

Brand, J. S., Moller, M., & Harvey, H. B. (2015). A review of biomarkers in mood and psychotic disorders: A dissection of clinical vs. preclinical correlates. *Current Neuropharmacology, 13*(3), 324–368.

Calcia, M. A., Bonsall, D. R., Bloomfield, P. S., Selvaraj, S., Barichello, T., & Howes, O. D. (2016). Stress and neuroinflammation: A systematic review of the effects of stress on microglia and the implications for mental illness. *Psychopharmacology, 233*(9), 1637–1650. https://doi.org/10.1007/s00213-016-4218-9

Cameron, H. A., & McKay, R. (1998). Discussion point stem cells and neurogenesis in the adult brain. *Current Opinion in Neurobiology, 8*(5), 677–680.

Casano, A. M., Albert, M., & Peri, F. (2016). Developmental apoptosis mediates entry and positioning of microglia in the zebrafish brain. *Cell Reports, 16*(4), 897–906. https://doi.org/10.1016/j.celrep.2016.06.033

Cattane, N., Vernon, A. C., Borsini, A., Scassellati, C., Endres, D., Capuron, L., Tamouza, R., Benros, M. E., Leza, J. C., Pariante, C. M., Riva, M. A., & Cattaneo, A. (2022). Preclinical animal models of mental illnesses to translate findings from the bench to the bedside: Molecular brain mechanisms and peripheral biomarkers associated to early life stress or immune challenges. *European Neuropsychopharmacology, 58*, 55–79. https://doi.org/10.1016/j.euroneuro.2022.02.002

Cheepsunthorn, P., Palmer, C., & Connor, J. R. (1998). Cellular distribution of ferritin subunits in postnatal rat brain. *Journal of Comparative Neurology, 400*(1), 73–86. https://doi.org/10.1002/(SICI)1096-9861(19981012)400:1<73::AID-CNE5>3.0.CO;2-Q

Chen, B. T., Hopf, F. W., & Bonci, A. (2010). Synaptic plasticity in the mesolimbic system: Therapeutic implications for substance abuse. *Annals of the New York Academy of Sciences, 1187*(1), 129–139.

Chen, Y., Lin, J., Zhao, Y., Ma, X., & Yi, H. (2021). Toll-like receptor 3 (TLR3) regulation mechanisms and roles in antiviral innate immune responses. *Journal of Zhejiang University. Science. B, 22*(8), 609–632. https://doi.org/10.1631/jzus.B2000808

Cherry, J. D., Olschowka, J. A., & O'Banion, M. K. (2014). Neuroinflammation and M2 microglia: The good, the bad, and the inflamed. *Journal of Neuroinflammation, 11*(1), 98. https://doi.org/10.1186/1742-2094-11-98

Chew, L.-J., Fusar-Poli, P., & Schmitz, T. (2013). Oligodendroglial alterations and the role of microglia in white matter injury: Relevance to schizophrenia. *Developmental Neuroscience, 35*(2–3), 102–129. https://doi.org/10.1159/000346157

Choi, I., Zhang, Y., Seegobin, S. P., Pruvost, M., Wang, Q., Purtell, K., Zhang, B., & Yue, Z. (2020). Microglia clear neuron-released α-synuclein via selective autophagy and prevent neurodegeneration. *Nature Communications, 11*(1), 1386.

Chung, W.-S., Welsh, C. A., Barres, B. A., & Stevens, B. (2015). Do glia drive synaptic and cognitive impairment in disease? *Nature Neuroscience, 18*(11), 1539–1545. https://doi.org/10.1038/nn.4142

Clayton, K. A., Van Enoo, A. A., & Ikezu, T. (2017). Alzheimer's disease: The role of microglia in brain homeostasis and Proteopathy [Review]. *Frontiers in Neuroscience, 11*. https://doi.org/10.3389/fnins.2017.00680

Colonna, M., & Butovsky, O. (2017). Microglia function in the central nervous system during health and neurodegeneration. *Annual Review of Immunology, 35*(1), 441–468. https://doi.org/10.1146/annurev-immunol-051116-052358

Cui, W., Sun, C., Ma, Y., Wang, S., Wang, X., & Zhang, Y. (2020). Inhibition of TLR4 induces M2 microglial polarization and provides neuroprotection via the NLRP3 inflammasome in Alzheimer's disease [Original research]. *Frontiers in Neuroscience, 14*. https://www.frontiersin.org/articles/10.3389/fnins.2020.00444

Cunningham, C. L., Martínez-Cerdeño, V., & Noctor, S. C. (2013). Microglia regulate the number of neural precursor cells in the developing cerebral cortex. *The Journal of Neuroscience, 33*(10), 4216–4233. https://doi.org/10.1523/jneurosci.3441-12.2013

Damani, M. R., Zhao, L., Fontainhas, A. M., Amaral, J., Fariss, R. N., & Wong, W. T. (2011). Age-related alterations in the dynamic behavior of microglia. *Aging Cell, 10*(2), 263–276. https://doi.org/10.1111/j.1474-9726.2010.00660.x

Deinhardt, K., & Chao, M. V. (2014). Shaping neurons: Long and short range effects of mature and proBDNF signalling upon neuronal structure. *Neuropharmacology, 76*, 603–609.

Doshi, A., & Chataway, J. (2017). Multiple sclerosis, a treatable disease. *Clinical Medicine (London, England), 17*(6), 530–536. https://doi.org/10.7861/clinmedicine.17-6-530

Drummond, E., Pires, G., MacMurray, C., Askenazi, M., Nayak, S., Bourdon, M., Safar, J., Ueberheide, B., & Wisniewski, T. (2020). Phosphorylated tau interactome in the human Alzheimer's disease brain. *Brain, 143*(9), 2803–2817. https://doi.org/10.1093/brain/awaa223

Elmore, M. R., Najafi, A. R., Koike, M. A., Dagher, N. N., Spangenberg, E. E., Rice, R. A., Kitazawa, M., Matusow, B., Nguyen, H., West, B. L., & Green, K. N. (2014). Colony-stimulating factor 1 receptor signaling is necessary for microglia viability, unmasking a microglia progenitor cell in the adult brain. *Neuron, 82*(2), 380–397. https://doi.org/10.1016/j.neuron.2014.02.040

Erny, D., Hrabě de Angelis, A. L., Jaitin, D., Wieghofer, P., Staszewski, O., David, E., Keren-Shaul, H., Mahlakoiv, T., Jakobshagen, K., & Buch, T. (2015). Host microbiota constantly control maturation and function of microglia in the CNS. *Nature Neuroscience, 18*(7), 965–977.

Fernández-Arjona, M. D. M., Grondona, J. M., Fernández-Llebrez, P., & López-Ávalos, M. D. (2019). Microglial morphometric parameters correlate with the expression level of IL-1β, and allow identifying different activated Morphotypes [Original research]. *Frontiers in Cellular Neuroscience, 13*. https://doi.org/10.3389/fncel.2019.00472

Filipello, F., Morini, R., Corradini, I., Zerbi, V., Canzi, A., Michalski, B., Erreni, M., Markicevic, M., Starvaggi-Cucuzza, C., Otero, K., Piccio, L., Cignarella, F., Perrucci, F., Tamborini, M., Genua, M., Rajendran, L., Menna, E., Vetrano, S., Fahnestock, M., Paolicelli, R. C., & Matteoli, M. (2018). The microglial innate immune receptor TREM2 is required for synapse elimination and Normal brain connectivity. *Immunity, 48*(5), 979–991.e978. https://doi.org/10.1016/j.immuni.2018.04.016

Frost, J. L., & Schafer, D. P. (2016). Microglia: Architects of the developing nervous system. *Trends in Cell Biology, 26*(8), 587–597. https://doi.org/10.1016/j.tcb.2016.02.006

Gadani, S. P., Cronk, J. C., Norris, G. T., & Kipnis, J. (2012). IL-4 in the brain: A cytokine to remember. *The Journal of Immunology, 189*(9), 4213–4219. https://doi.org/10.4049/jimmunol.1202246

Gao, Z., & Tsirka, S. E. (2011). Animal models of MS reveal multiple roles of microglia in disease pathogenesis. *Neurology Research International, 2011*, 383087.

Geloso, M. C., & D'Ambrosi, N. (2021). Microglial pruning: Relevance for synaptic dysfunction in multiple sclerosis and related experimental models. *Cells, 10*(3). https://doi.org/10.3390/cells10030686

Gerhard, A., Pavese, N., Hotton, G., Turkheimer, F., Es, M., Hammers, A., Eggert, K., Oertel, W., Banati, R. B., & Brooks, D. J. (2006). In vivo imaging of microglial activation with [11C](R)-PK11195 PET in idiopathic Parkinson's disease. *Neurobiology of Disease, 21*(2), 404–412. https://doi.org/10.1016/j.nbd.2005.08.002

Ginhoux, F., & Garel, S. (2018). The mysterious origins of microglia. *Nature Neuroscience, 21*(7), 897–899. https://doi.org/10.1038/s41593-018-0176-3

Ginhoux, F., Lim, S., Hoeffel, G., Low, D., & Huber, T. (2013). Origin and differentiation of microglia. *Frontiers in Cellular Neuroscience, 7*, 45. https://doi.org/10.3389/fncel.2013.00045

Gonçalves, J., Baptista, S., & Silva, A. P. (2014). Psychostimulants and brain dysfunction: A review of the relevant neurotoxic effects. *Neuropharmacology, 87*, 135–149.

Guha, M., & Mackman, N. (2001). LPS induction of gene expression in human monocytes. *Cellular Signalling, 13*(2), 85–94. https://doi.org/10.1016/S0898-6568(00)00149-2

Han, J., Fan, Y., Zhou, K., Blomgren, K., & Harris, R. A. (2021). Uncovering sex differences of rodent microglia. *Journal of Neuroinflammation, 18*(1), 1–11.

Hanslik, K. L., & Ulland, T. K. (2020). The role of microglia and the Nlrp3 Inflammasome in Alzheimer's disease [Mini review]. *Frontiers in Neurology, 11*. https://doi.org/10.3389/fneur.2020.570711

Hayden, M. S., West, A. P., & Ghosh, S. (2006). NF-κB and the immune response. *Oncogene, 25*(51), 6758–6780. https://doi.org/10.1038/sj.onc.1209943

Hefendehl, J. K., Neher, J. J., Sühs, R. B., Kohsaka, S., Skodras, A., & Jucker, M. (2014). Homeostatic and injury-induced microglia behavior in the aging brain. *Aging Cell, 13*(1), 60–69.

Henriques, J. F., Portugal, C. C., Canedo, T., Relvas, J. B., Summavielle, T., & Socodato, R. (2018). Microglia and alcohol meet at the crossroads: Microglia as critical modulators of alcohol neurotoxicity. *Toxicology Letters, 283*, 21–31.

Hercher, C., Parent, M., Flores, C., Canetti, L., Turecki, G., & Mechawar, N. (2009). Alcohol dependence-related increase of glial cell density in the anterior cingulate cortex of suicide completers. *Journal of Psychiatry & Neuroscience, 34*(4), 281–288.

Herlihy, B., & Roy, S. (2022). Gut-microbiome implications in opioid use disorder and related behaviors. *Advances in Drug and Alcohol Research, 2*, 10311.

Hickman, S., Izzy, S., Sen, P., Morsett, L., & El Khoury, J. (2018). Microglia in neurodegeneration. *Nature Neuroscience, 21*(10), 1359–1369. https://doi.org/10.1038/s41593-018-0242-x

Hsieh, C. L., Koike, M., Spusta, S. C., Niemi, E. C., Yenari, M., Nakamura, M. C., & Seaman, W. E. (2009). A role for TREM2 ligands in the phagocytosis of apoptotic neuronal cells by microglia. *Journal of Neurochemistry, 109*(4), 1144–1156. https://doi.org/10.1111/j.1471-4159.2009.06042.x

Hu, M.-Y., Lin, Y.-Y., Zhang, B.-J., Lu, D.-L., Lu, Z.-Q., & Cai, W. (2019). Update of inflammasome activation in microglia/macrophage in aging and aging-related disease. *CNS Neuroscience & Therapeutics, 25*(12), 1299–1307. https://doi.org/10.1111/cns.13262

Huang, Y., Xu, Z., Xiong, S., Qin, G., Sun, F., Yang, J., Yuan, T.-F., Zhao, L., Wang, K., Liang, Y.-X., Fu, L., Wu, T., So, K.-F., Rao, Y., & Peng, B. (2018). Dual extra-retinal origins of microglia in the model of retinal microglia repopulation. *Cell Discovery, 4*(1), 9. https://doi.org/10.1038/s41421-018-0011-8

Jeong, H. K., Ji, K., Min, K., & Joe, E. H. (2013). Brain inflammation and microglia: Facts and misconceptions. *Experimental Neurobiology, 22*(2), 59–67. https://doi.org/10.5607/en.2013.22.2.59

Ji, K., Akgul, G., Wollmuth, L. P., & Tsirka, S. E. (2013). Microglia actively regulate the number of functional synapses. *PLoS One, 8*(2), e56293. https://doi.org/10.1371/journal.pone.0056293

Jia, X., Gao, Z., & Hu, H. (2021). Microglia in depression: Current perspectives. *Science China Life Sciences, 64*, 911–925.

Jurga, A. M., Paleczna, M., & Kuter, K. Z. (2020). Overview of general and discriminating markers of differential microglia phenotypes [Review]. *Frontiers in Cellular Neuroscience, 14*. https://doi.org/10.3389/fncel.2020.00198

Kalkman, H. O., & Feuerbach, D. (2016). Antidepressant therapies inhibit inflammation and microglial M1-polarization. *Pharmacology & Therapeutics, 163*, 82–93.

Kamei, R., Urata, S., Maruoka, H., & Okabe, S. (2022). In vivo chronic two-photon imaging of microglia in the mouse hippocampus. *Journal of Visualized Experiments, 185*. https://doi.org/10.3791/64104

Kato, T. A., Monji, A., Mizoguchi, Y., Hashioka, S., Horikawa, H., Seki, Y., Kasai, M., Utsumi, H., & Kanba, S. (2011). Anti-inflammatory properties of antipsychotics via microglia modulations: Are antipsychotics a 'fire extinguisher' in the brain of schizophrenia? *Mini Reviews in Medicinal Chemistry, 11*(7), 565–574.

Kauer, J. A., & Malenka, R. C. (2007). Synaptic plasticity and addiction. *Nature Reviews Neuroscience, 8*(11), 844–858.

Kawai, T., & Akira, S. (2006). TLR signaling. *Cell Death & Differentiation, 13*(5), 816–825. https://doi.org/10.1038/sj.cdd.4401850

Kawai, T., & Akira, S. (2007). Signaling to NF-kappaB by toll-like receptors. *Trends in Molecular Medicine, 13*(11), 460–469. https://doi.org/10.1016/j.molmed.2007.09.002

Kawasaki, T., & Kawai, T. (2014). Toll-like receptor signaling pathways. *Frontiers in Immunology, 5*, 461. https://doi.org/10.3389/fimmu.2014.00461

Kierdorf, K., & Prinz, M. (2017). Microglia in steady state. *The Journal of Clinical Investigation, 127*(9), 3201–3209. https://doi.org/10.1172/jci90602

Kim, C. C., Nakamura, M. C., & Hsieh, C. L. (2016). Brain trauma elicits non-canonical macrophage activation states. *Journal of Neuroinflammation, 13*(1), 1–12.

Kim, J. W., Hong, J. Y., & Bae, S. M. (2018). Microglia and autism Spectrum disorder: Overview of current evidence and novel immunomodulatory treatment options. *Clinical Psychopharmacology and Neuroscience, 16*(3), 246–252. https://doi.org/10.9758/cpn.2018.16.3.246

Kriegstein, A., & Alvarez-Buylla, A. (2009). The glial nature of embryonic and adult neural stem cells. *Annual Review of Neuroscience, 32*, 149–184. https://doi.org/10.1146/annurev.neuro.051508.135600

Lall, D., & Baloh, R. H. (2017). Microglia and C9orf72 in neuroinflammation and ALS and frontotemporal dementia. *The Journal of Clinical Investigation, 127*(9), 3250–3258. https://doi.org/10.1172/jci90607

Lannes, N., Eppler, E., Etemad, S., Yotovski, P., & Filgueira, L. (2017). Microglia at center stage: A comprehensive review about the versatile and unique residential macrophages of the central nervous system. *Oncotarget, 8*(69), 114393–114413. https://doi.org/10.18632/oncotarget.23106

Lehrman, E. K., Wilton, D. K., Litvina, E. Y., Welsh, C. A., Chang, S. T., Frouin, A., Walker, A. J., Heller, M. D., Umemori, H., Chen, C., & Stevens, B. (2018). CD47 protects synapses from excess microglia-mediated pruning during development. *Neuron, 100*(1), 120–134.e126. https://doi.org/10.1016/j.neuron.2018.09.017

Leyh, J., Paeschke, S., Mages, B., Michalski, D., Nowicki, M., Bechmann, I., & Winter, K. (2021). Classification of microglial morphological phenotypes using machine learning [Methods]. *Frontiers in Cellular Neuroscience, 15*. https://doi.org/10.3389/fncel.2021.701673

Li, J., Shui, X., Sun, R., Wan, L., Zhang, B., Xiao, B., & Luo, Z. (2021). Microglial phenotypic transition: Signaling pathways and influencing modulators involved in regulation in central nervous system diseases [Review]. *Frontiers in Cellular Neuroscience, 15*. https://doi.org/10.3389/fncel.2021.736310

Liau, N. P. D., Laktyushin, A., Lucet, I. S., Murphy, J. M., Yao, S., Whitlock, E., Callaghan, K., Nicola, N. A., Kershaw, N. J., & Babon, J. J. (2018). The molecular basis of JAK/STAT inhibition by SOCS1. *Nature Communications, 9*(1), 1558. https://doi.org/10.1038/s41467-018-04013-1

Lin, S.-C., Lo, Y.-C., & Wu, H. (2010). Helical assembly in the MyD88–IRAK4–IRAK2 complex in TLR/IL-1R signalling. *Nature, 465*(7300), 885–890. https://doi.org/10.1038/nature09121

Liu, J., Hjorth, E., Zhu, M., Calzarossa, C., Samuelsson, E.-B., Schultzberg, M., & Åkesson, E. (2013). Interplay between human microglia and neural stem/progenitor cells in an allogeneic coculture model. *Journal of Cellular and Molecular Medicine, 17*(11), 1434–1443. https://doi.org/10.1111/jcmm.12123

Liu, T., Zhang, L., Joo, D., & Sun, S.-C. (2017). NF-κB signaling in inflammation. *Signal Transduction and Targeted Therapy, 2*(1), 1–9.

Loram, L. C., Sholar, P. W., Taylor, F. R., Wiesler, J. L., Babb, J. A., Strand, K. A., Berkelhammer, D., Day, H. E., Maier, S. F., & Watkins, L. R. (2012). Sex and estradiol influence glial pro-inflammatory responses to lipopolysaccharide in rats. *Psychoneuroendocrinology, 37*(10), 1688–1699.

Lu, Y.-C., Yeh, W.-C., & Ohashi, P. S. (2008). LPS/TLR4 signal transduction pathway. *Cytokine, 42*(2), 145–151. https://doi.org/10.1016/j.cyto.2008.01.006

Lye, E., Mirtsos, C., Suzuki, N., Suzuki, S., & Yeh, W.-C. (2004). The role of interleukin 1 receptor-associated kinase-4 (IRAK-4) kinase activity in IRAK-4-mediated signaling. *Journal of Biological Chemistry, 279*(39), 40653–40658.

Ma, M. W., Wang, J., Zhang, Q., Wang, R., Dhandapani, K. M., Vadlamudi, R. K., & Brann, D. W. (2017). NADPH oxidase in brain injury and neurodegenerative disorders. *Molecular Neurodegeneration, 12*, 1–28.

Madry, C., Kyrargyri, V., Arancibia-Cárcamo, I. L., Jolivet, R., Kohsaka, S., Bryan, R. M., & Attwell, D. (2018). Microglial ramification, surveillance, and interleukin-1β release are regulated by the two-pore domain K(+) channel THIK-1. *Neuron, 97*(2), 299–312. e296. https://doi.org/10.1016/j.neuron.2017.12.002

Maness, P. F., & Schachner, M. (2007). Neural recognition molecules of the immunoglobulin superfamily: Signaling transducers of axon guidance and neuronal migration. *Nature Neuroscience, 10*(1), 19–26. https://doi.org/10.1038/nn1827

Mantovani, A., Sica, A., Sozzani, S., Allavena, P., Vecchi, A., & Locati, M. (2004). The chemokine system in diverse forms of macrophage activation and polarization. *Trends in Immunology, 25*(12), 677–686. https://doi.org/10.1016/j.it.2004.09.015

Mariani, N., Everson, J., Pariante, C. M., & Borsini, A. (2022). Modulation of microglial activation by antidepressants. *Journal of Psychopharmacology, 36*(2), 131–150. https://doi.org/10.1177/02698811211069110

Martinez, F. O., Sica, A., Mantovani, A., & Locati, M. (2008). Macrophage activation and polarization. *Frontiers in Bioscience-Landmark, 13*(2), 453–461.

Matejuk, A., & Ransohoff, R. M. (2020). Crosstalk between astrocytes and microglia: An overview [Review]. *Frontiers in Immunology, 11*. https://www.frontiersin.org/articles/10.3389/fimmu.2020.01416

McNamara, N. B., Munro, D. A. D., Bestard-Cuche, N., Uyeda, A., Bogie, J. F. J., Hoffmann, A., Holloway, R. K., Molina-Gonzalez, I., Askew, K. E., Mitchell, S., Mungall, W., Dodds, M., Dittmayer, C., Moss, J., Rose, J., Szymkowiak, S., Amann, L., McColl, B. W., Prinz, M., Spires-Jones, T. L., Stenzel, W., Horsburgh, K., Hendriks, J. J. A., Pridans, C., Muramatsu, R., Williams, A., Priller, J., & Miron, V. E. (2023). Microglia regulate central nervous system myelin growth and integrity. *Nature, 613*(7942), 120–129. https://doi.org/10.1038/s41586-022-05534-y

Mondelli, V., Vernon, A. C., Turkheimer, F., Dazzan, P., & Pariante, C. M. (2017). Brain microglia in psychiatric disorders. *The Lancet Psychiatry, 4*(7), 563–572.

Monji, A., Kato, T., & Kanba, S. (2009). Cytokines and schizophrenia: Microglia hypothesis of schizophrenia. *Psychiatry and Clinical Neurosciences, 63*(3), 257–265.

Mosser, D. M. (2003). The many faces of macrophage activation. *Journal of Leukocyte Biology, 73*(2), 209–212. https://doi.org/10.1189/jlb.0602325

Mosser, D. M., & Edwards, J. P. (2008). Exploring the full spectrum of macrophage activation. *Nature Reviews Immunology, 8*(12), 958–969.

Mosser, D. M., & Karp, C. L. (1999). Receptor mediated subversion of macrophage cytokine production by intracellular pathogens. *Current Opinion in Immunology, 11*(4), 406–411.

Multhaup, G., Ruppert, T., Schlicksupp, A., Hesse, L., Beher, D., Masters, C. L., & Beyreuther, K. (1997). Reactive oxygen species and Alzheimer's disease. *Biochemical Pharmacology, 54*(5), 533–539.

Nagatsu, T., Mogi, M., Ichinose, H., & Togari, A. (2000). Cytokines in Parkinson's disease. *Journal of Neural Transmission. Supplementum, 2000*(58), 143–151.

Najafi, A. R., Crapser, J., Jiang, S., Ng, W., Mortazavi, A., West, B. L., & Green, K. N. (2018). A limited capacity for microglial repopulation in the adult brain. *Glia, 66*(11), 2385–2396. https://doi.org/10.1002/glia.23477

Nakanishi, M., Niidome, T., Matsuda, S., Akaike, A., Kihara, T., & Sugimoto, H. (2007). Microglia-derived interleukin-6 and leukaemia inhibitory factor promote astrocytic differentiation of neural stem/progenitor cells. *The European Journal of Neuroscience, 25*(3), 649–658. https://doi.org/10.1111/j.1460-9568.2007.05309.x

Namba, M. D., Leyrer-Jackson, J. M., Nagy, E. K., Olive, M. F., & Neisewander, J. L. (2021). Neuroimmune mechanisms as novel treatment targets for substance use disorders and associated comorbidities. *Frontiers in Neuroscience, 15*, 650785.

Napoli, I., & Neumann, H. (2010). Protective effects of microglia in multiple sclerosis. *Experimental Neurology, 225*(1), 24–28. https://doi.org/10.1016/j.expneurol.2009.04.024

Nelson, L. H., & Lenz, K. M. (2017). Microglia depletion in early life programs persistent changes in social, mood-related, and locomotor behavior in male and female rats. *Behavioural Brain Research, 316*, 279–293. https://doi.org/10.1016/j.bbr.2016.09.006

Nicholas, R. S., Wing, M. G., & Compston, A. (2001). Nonactivated microglia promote oligodendrocyte precursor survival and maturation through the transcription factor NF-kappa B. *The European Journal of Neuroscience, 13*(5), 959–967. https://doi.org/10.1046/j.0953-816x.2001.01470.x

Nimmerjahn, A., Kirchhoff, F., & Helmchen, F. (2005). Resting microglial cells are highly dynamic Surveillants of brain parenchyma in vivo. *Science, 308*(5726), 1314–1318. https://doi.org/10.1126/science.1110647

Ochocka, N., & Kaminska, B. (2021). Microglia diversity in healthy and diseased brain: Insights from single-cell omics. *International Journal of Molecular Sciences, 22*(6). https://doi.org/10.3390/ijms22063027

O'Kusky, J., & Ye, P. (2012). Neurodevelopmental effects of insulin-like growth factor signaling. *Frontiers in Neuroendocrinology, 33*(3), 230–251. https://doi.org/10.1016/j.yfrne.2012.06.002

Orihuela, R., McPherson, C. A., & Harry, G. J. (2016). Microglial M1/M2 polarization and metabolic states. *British Journal of Pharmacology, 173*(4), 649–665. https://doi.org/10.1111/bph.13139

Paolicelli, R. C., Bolasco, G., Pagani, F., Maggi, L., Scianni, M., Panzanelli, P., Giustetto, M., Ferreira, T. A., Guiducci, E., Dumas, L., Ragozzino, D., & Gross, C. T. (2011). Synaptic pruning by microglia is necessary for normal brain development. *Science, 333*(6048), 1456–1458. https://doi.org/10.1126/science.1202529

Parkhurst, C. N., Yang, G., Ninan, I., Savas, J. N., Yates, J. R., Lafaille, J. J., Hempstead, B. L., Littman, D. R., & Gan, W.-B. (2013). Microglia promote learning-dependent synapse formation through brain-derived neurotrophic factor. *Cell, 155*(7), 1596–1609.

Passaro, A. P., Lebos, A. L., Yao, Y., & Stice, S. L. (2021). Immune response in neurological pathology: Emerging role of central and peripheral immune crosstalk. *Frontiers in Immunology, 12*, 676621.

Peferoen, L., Kipp, M., van der Valk, P., van Noort, J. M., & Amor, S. (2014). Oligodendrocyte-microglia cross-talk in the central nervous system. *Immunology, 141*(3), 302–313. https://doi.org/10.1111/imm.12163

Petry, P., Oschwald, A., & Kierdorf, K. (2023). Microglial tissue surveillance: The never-resting gardener in the developing and adult CNS. *European Journal of Immunology, 53*(7), 2250232. https://doi.org/10.1002/eji.202250232

Pizza, V., Agresta, A., D'Acunto, W. C., Festa, M., & Capasso, A. (2011). Neuroinflamm-aging and neurodegenerative diseases: An overview. *CNS & Neurological Disorders-Drug Targets (Formerly*

Current Drug Targets-CNS & Neurological Disorders), 10(5), 621–634.

Prinz, M., & Priller, J. (2014). Microglia and brain macrophages in the molecular age: From origin to neuropsychiatric disease. *Nature Reviews. Neuroscience, 15*(5), 300–312. https://doi.org/10.1038/nrn3722

Rački, V., Petrić, D., Kučić, N., Gržeta, N., Jurdana, K., & Rončević-Gržeta, I. (2016). Cortical gray matter loss in schizophrenia: Could microglia be the culprit? *Medical Hypotheses, 88*, 18–21. https://doi.org/10.1016/j.mehy.2015.12.021

Rahimian, R., Wakid, M., O'Leary, L. A., & Mechawar, N. (2021). The emerging tale of microglia in psychiatric disorders. *Neuroscience & Biobehavioral Reviews, 131*, 1–29.

Rajan, K. B., Weuve, J., Barnes, L. L., McAninch, E. A., Wilson, R. S., & Evans, D. A. (2021). Population estimate of people with clinical Alzheimer's disease and mild cognitive impairment in the United States (2020–2060). *Alzheimers Dement, 17*(12), 1966–1975. https://doi.org/10.1002/alz.12362

Ransohoff, R. M. (2016). A polarizing question: Do M1 and M2 microglia exist? *Nature Neuroscience, 19*(8), 987–991. https://doi.org/10.1038/nn.4338

Reed-Geaghan, E. G., Savage, J. C., Hise, A. G., & Landreth, G. E. (2009). CD14 and toll-like receptors 2 and 4 are required for fibrillar A{beta}-stimulated microglial activation. *The Journal of Neuroscience, 29*(38), 11982–11992. https://doi.org/10.1523/jneurosci.3158-09.2009

Rice, R. A., Pham, J., Lee, R. J., Najafi, A. R., West, B. L., & Green, K. N. (2017, Jun). Microglial repopulation resolves inflammation and promotes brain recovery after injury. *Glia, 65*(6), 931–944. https://doi.org/10.1002/glia.23135

Rio-Hortega, P. (1939). The microglia. *The Lancet, 233*(6036), 1023–1026. https://doi.org/10.1016/S0140-6736(00)60571-8

Rodríguez-Gómez, J. A., Kavanagh, E., Engskog-Vlachos, P., Engskog, M. K. R., Herrera, A. J., Espinosa-Oliva, A. M., Joseph, B., Hajji, N., Venero, J. L., & Burguillos, M. A. (2020). Microglia: Agents of the CNS pro-inflammatory response. *Cells, 9*(7), 1717. https://www.mdpi.com/2073-4409/9/7/1717

Sampson, T. R., & Mazmanian, S. K. (2015). Control of brain development, function, and behavior by the microbiome. *Cell Host & Microbe, 17*(5), 565–576.

Sapp, E., Kegel, K. B., Aronin, N., Hashikawa, T., Uchiyama, Y., Tohyama, K., Bhide, P. G., Vonsattel, J. P., & DiFiglia, M. (2001). Early and progressive accumulation of reactive microglia in the Huntington disease brain. *Journal of Neuropathology and Experimental Neurology, 60*(2), 161–172. https://doi.org/10.1093/jnen/60.2.161

Satoh, J., Kino, Y., Asahina, N., Takitani, M., Miyoshi, J., Ishida, T., & Saito, Y. (2016). TMEM119 marks a subset of microglia in the human brain. *Neuropathology, 36*(1), 39–49. https://doi.org/10.1111/neup.12235

Schafer, D. P., & Stevens, B. (2015). Microglia function in central nervous system development and plasticity. *Cold Spring Harbor Perspectives in Biology, 7*(10), a020545. https://doi.org/10.1101/cshperspect.a020545

Schafer, D. P., Lehrman, E. K., Kautzman, A. G., Koyama, R., Mardinly, A. R., Yamasaki, R., Ransohoff, R. M., Greenberg, M. E., Barres, B. A., & Stevens, B. (2012). Microglia sculpt postnatal neural circuits in an activity and complement-dependent manner. *Neuron, 74*(4), 691–705. https://doi.org/10.1016/j.neuron.2012.03.026

Schafer, D. P., Lehrman, E. K., & Stevens, B. (2013). The "quad-partite" synapse: Microglia-synapse interactions in the developing and mature CNS. *Glia, 61*(1), 24–36. https://doi.org/10.1002/glia.22389

Shabab, T., Khanabdali, R., Moghadamtousi, S. Z., Kadir, H. A., & Mohan, G. (2017). Neuroinflammation pathways: A general review. *International Journal of Neuroscience, 127*(7), 624–633. https://doi.org/10.1080/00207454.2016.1212854

Shigemoto-Mogami, Y., Hoshikawa, K., Goldman, J. E., Sekino, Y., & Sato, K. (2014). Microglia enhance neurogenesis and oligodendrogenesis in the early postnatal subventricular zone. *The Journal of Neuroscience, 34*(6), 2231–2243. https://doi.org/10.1523/jneurosci.1619-13.2014

Simpson, D. S. A., & Oliver, P. L. (2020). ROS generation in microglia: Understanding oxidative stress and inflammation in neurodegenerative disease. *Antioxidants (Basel), 9*(8). https://doi.org/10.3390/antiox9080743

Stein, M., Keshav, S., Harris, N., & Gordon, S. (1992). Interleukin 4 potently enhances murine macrophage mannose receptor activity: A marker of alternative immunologic macrophage activation. *Journal of Experimental Medicine, 176*(1), 287–292. https://doi.org/10.1084/jem.176.1.287

Steiner, J., Bielau, H., Brisch, R., Danos, P., Ullrich, O., Mawrin, C., Bernstein, H. G., & Bogerts, B. (2008). Immunological aspects in the neurobiology of suicide: Elevated microglial density in schizophrenia and depression is associated with suicide. *Journal of Psychiatric Research, 42*(2), 151–157. https://doi.org/10.1016/j.jpsychires.2006.10.013

Stelzmann, R. A., Norman Schnitzlein, H., & Reed Murtagh, F. (1995). An english translation of alzheimer's 1907 paper, "über eine eigenartige erkankung der hirnrinde". *Clinical Anatomy, 8*(6), 429–431. https://doi.org/10.1002/ca.980080612

Stevens, B., Allen, N. J., Vazquez, L. E., Howell, G. R., Christopherson, K. S., Nouri, N., Micheva, K. D., Mehalow, A. K., Huberman, A. D., & Stafford, B. (2007). The classical complement cascade mediates CNS synapse elimination. *Cell, 131*(6), 1164–1178.

Svahn, A. J., Giacomotto, J., Graeber, M. B., Rinkwitz, S., & Becker, T. S. (2016). miR-124 contributes to the functional maturity of microglia. *Developmental Neurobiology, 76*(5), 507–518. https://doi.org/10.1002/dneu.22328

Takano, T. (2015). Role of microglia in autism: Recent advances. *Developmental Neuroscience, 37*(3), 195–202. https://doi.org/10.1159/000398791

Takeda, K., & Akira, S. (2004). TLR signaling pathways. *Seminars in Immunology, 16*(1), 3–9. https://doi.org/10.1016/j.smim.2003.10.003

Tan, Y.-L., Yuan, Y., & Tian, L. (2020). Microglial regional heterogeneity and its role in the brain. *Molecular Psychiatry, 25*(2), 351–367. https://doi.org/10.1038/s41380-019-0609-8

Tang, Y., & Le, W. (2016). Differential roles of M1 and M2 microglia in neurodegenerative diseases. *Molecular Neurobiology, 53*(2), 1181–1194. https://doi.org/10.1007/s12035-014-9070-5

Tansey, M. G., & Goldberg, M. S. (2010). Neuroinflammation in Parkinson's disease: Its role in neuronal death and implications for therapeutic intervention. *Neurobiology of Disease, 37*(3), 510–518. https://doi.org/10.1016/j.nbd.2009.11.004

Taylor, A. M., Becker, S., Schweinhardt, P., & Cahill, C. (2016). Mesolimbic dopamine signaling in acute and chronic pain: Implications for motivation, analgesia, and addiction. *Pain, 157*(6), 1194.

Todorich, B., Pasquini, J. M., Garcia, C. I., Paez, P. M., & Connor, J. R. (2009). Oligodendrocytes and myelination: The role of iron. *Glia, 57*(5), 467–478. https://doi.org/10.1002/glia.20784

Tremblay, M.-È., Stevens, B., Sierra, A., Wake, H., Bessis, A., & Nimmerjahn, A. (2011). The role of microglia in the healthy brain. *The Journal of Neuroscience, 31*(45), 16064–16069. https://doi.org/10.1523/jneurosci.4158-11.2011

Tsermpini, E. E., Plemenitaš Ilješ, A., & Dolžan, V. (2022). Alcohol-induced oxidative stress and the role of antioxidants in alcohol use disorder: A systematic review. *Antioxidants, 11*(7), 1374.

Ullah, M. O., Sweet, M. J., Mansell, A., Kellie, S., & Kobe, B. (2016). TRIF-dependent TLR signaling, its functions in host defense and inflammation, and its potential as a therapeutic target. *Journal of Leukocyte Biology, 100*(1), 27–45. https://doi.org/10.1189/jlb.2RI1115-531R

Vawter, M. P., Dillon-Carter, O., Tourtellotte, W. W., Carvey, P., & Freed, W. J. (1996). TGFbeta1 and TGFbeta2 concentrations are elevated in Parkinson's disease in ventricular cerebrospinal fluid. *Experimental Neurology, 142*(2), 313–322. https://doi.org/10.1006/exnr.1996.0200

Walker, D. G., & Lue, L. F. (2015). Immune phenotypes of microglia in human neurodegenerative disease: Challenges to detecting microglial polarization in human brains. *Alzheimer's Research & Therapy, 7*(1), 56. https://doi.org/10.1186/s13195-015-0139-9

Wang, S., Yuan, Y.-H., Chen, N.-H., & Wang, H.-B. (2019). The mechanisms of NLRP3 inflammasome/pyroptosis activation and their role in Parkinson's disease. *International Immunopharmacology, 67*, 458–464. https://doi.org/10.1016/j.intimp.2018.12.019

Wang, H., He, Y., Sun, Z., Ren, S., Liu, M., Wang, G., & Yang, J. (2022). Microglia in depression: An overview of microglia in the pathogenesis and treatment of depression. *Journal of Neuroinflammation, 19*(1), 132. https://doi.org/10.1186/s12974-022-02492-0

Weber, M. D., McKim, D. B., Niraula, A., Witcher, K. G., Yin, W., Sobol, C. G., Wang, Y., Sawicki, C. M., Sheridan, J. F., & Godbout, J. P. (2019). The influence of microglial elimination and repopulation on stress sensitization induced by repeated social defeat. *Biological Psychiatry, 85*(8), 667–678.

Wen, Y.-R., Tan, P.-H., Cheng, J.-K., Liu, Y.-C., & Ji, R.-R. (2011). Microglia: A promising target for treating neuropathic and postoperative pain, and morphine tolerance. *Journal of the Formosan Medical Association, 110*(8), 487–494.

Willis, A. W., Roberts, E., Beck, J. C., Fiske, B., Ross, W., Savica, R., Van Den Eeden, S. K., Tanner, C. M., & Marras, C. (2022). Incidence of Parkinson disease in North America. *npj Parkinson's Disease, 8*(1), 170. https://doi.org/10.1038/s41531-022-00410-y

Wilton, D. K., Dissing-Olesen, L., & Stevens, B. (2019). Neuron-glia signaling in synapse elimination. *Annual Review of Neuroscience, 42*, 107–127. https://doi.org/10.1146/annurev-neuro-070918-050306

Woodburn, S. C., Bollinger, J. L., & Wohleb, E. S. (2021). The semantics of microglia activation: Neuroinflammation, homeostasis, and stress. *Journal of Neuroinflammation, 18*(1), 258. https://doi.org/10.1186/s12974-021-02309-6

Xu, J., Wang, T., Wu, Y., Jin, W., & Wen, Z. (2016). Microglia colonization of developing zebrafish midbrain is promoted by apoptotic neuron and Lysophosphatidylcholine. *Developmental Cell, 38*(2), 214–222. https://doi.org/10.1016/j.devcel.2016.06.018

Yamamiya, M., Tanabe, S., & Muramatsu, R. (2019). Microglia promote the proliferation of neural precursor cells by secreting osteopontin. *Biochemical and Biophysical Research Communications, 513*(4), 841–845. https://doi.org/10.1016/j.bbrc.2019.04.076

Yan, Z., Gibson, S. A., Buckley, J. A., Qin, H., & Benveniste, E. N. (2018). Role of the JAK/STAT signaling pathway in regulation of innate immunity in neuroinflammatory diseases. *Clinical Immunology, 189*, 4–13. https://doi.org/10.1016/j.clim.2016.09.014

Yang, H.-M., Yang, S., Huang, S.-S., Tang, B.-S., & Guo, J.-F. (2017). Microglial activation in the pathogenesis of huntington's disease [Review]. *Frontiers in Aging Neuroscience, 9*. https://doi.org/10.3389/fnagi.2017.00193

Yirmiya, R., Rimmerman, N., & Reshef, R. (2015). Depression as a microglial disease. *Trends in Neurosciences, 38*(10), 637–658.

Yun, S. P., Kam, T.-I., Panicker, N., Kim, S., Oh, Y., Park, J.-S., Kwon, S.-H., Park, Y. J., Karuppagounder, S. S., Park, H., Kim, S., Oh, N., Kim, N. A., Lee, S., Brahmachari, S., Mao, X., Lee, J. H., Kumar, M., An, D., Kang, S.-U., Lee, Y., Lee, K. C., Na, D. H., Kim, D., Lee, S. H., Roschke, V. V., Liddelow, S. A., Mari, Z., Barres, B. A., Dawson, V. L., Lee, S., Dawson, T. M., & Ko, H. S. (2018). Block of A1 astrocyte conversion by microglia is neuroprotective in models of Parkinson's disease. *Nature Medicine, 24*(7), 931–938. https://doi.org/10.1038/s41591-018-0051-5

Zähringer, U., Lindner, B., Inamura, S., Heine, H., & Alexander, C. (2008). TLR2 – Promiscuous or specific? A critical re-evaluation of a receptor expressing apparent broad specificity. *Immunobiology, 213*(3), 205–224. https://doi.org/10.1016/j.imbio.2008.02.005

Zhang, X., Surguladze, N., Slagle-Webb, B., Cozzi, A., & Connor, J. R. (2006). Cellular iron status influences the functional relationship between microglia and oligodendrocytes. *Glia, 54*(8), 795–804. https://doi.org/10.1002/glia.20416

Zhang, L., Zhang, J., & You, Z. (2018). Switching of the microglial activation phenotype is a possible treatment for depression disorder. *Frontiers in Cellular Neuroscience, 12*, 306.

Zhang, L., Meng, J., Ban, Y., Jalodia, R., Chupikova, I., Fernandez, I., Brito, N., Sharma, U., Abreu, M. T., Ramakrishnan, S., & Roy, S. (2019). Morphine tolerance is attenuated in germfree mice and reversed by probiotics, implicating the role of gut microbiome. *Proceedings of the National Academy of Sciences, 116*(27), 13523. https://doi.org/10.1073/pnas.1901182116

Zhang, L., Cao, Y., Zhang, X., Gu, X., Mao, Y., & Peng, B. (2022). The origin and repopulation of microglia. *Developmental Neurobiology, 82*(1), 112–124. https://doi.org/10.1002/dneu.22862

Zhao, W., Beers, D. R., Henkel, J. S., Zhang, W., Urushitani, M., Julien, J. P., & Appel, S. H. (2010). Extracellular mutant SOD1 induces microglial-mediated motoneuron injury. *Glia, 58*(2), 231–243. https://doi.org/10.1002/glia.20919

Mitochondrial Autophagy in Brain

9

Jens O. Watzlawik, Xu Hou, Fabienne C. Fiesel, and Wolfdieter Springer

Abstract

Selective turnover of defective or worn-out mitochondria via autophagy is essential for maintaining a healthy mitochondrial network under both physiological and pathological conditions. This is important, especially in tissue with high-energy demand, such as the brain. As part of this chapter, we will give an update on mitochondrial autophagy with a focus on the PINK1–PRKN mitophagy pathway, which has been extensively studied since its discovery more than a decade ago. The rather low levels of patho−/physiological PINK1–PRKN signaling that occur in vivo and the general dynamic nature of mitophagy make it challenging to study. In addition to sensitive biochemical measurements, transgenic reporter mice that are available now allow analyses of mitophagy rates in a tissue-specific manner at baseline and under stress. We further describe the current status of knowledge on neuronal mitophagy and summarize findings in rodent models. A special emphasis is further placed on mitophagy in human neurological disorders including Parkinson's disease and Alzheimer's disease and the connection between defective mitophagy and neuroinflammation. The chapter concludes with current challenges and future therapeutic approaches to modulate mitophagy.

Keywords

Alzheimer's disease · CNS · Mitochondria · Mitophagy · Parkin · Parkinson's disease · Phosphorylated ubiquitin · PINK1 · PRKN · Selective autophagy

J. O. Watzlawik · X. Hou
Department of Neuroscience, Mayo Clinic, Jacksonville, FL, USA

F. C. Fiesel · W. Springer (✉)
Department of Neuroscience, Mayo Clinic, Jacksonville, FL, USA

Neuroscience PhD program, Mayo Clinic Graduate School of Biomedical Sciences, Jacksonville, FL, USA
e-mail: Springer.Wolfdieter@mayo.edu

9.1 Introduction

Neurons are specialized electrically excitable cells with high-energy demand and heavily rely on functional mitochondria and oxidative phosphorylation. With stress or during aging, mitochondrial function is compromised and mitochondria release deleterious substances including reactive oxygen species (ROS), which further damage mitochondrial proteins, lipids, and DNA in a vicious cycle (Wallace, 2005). There are a multitude of cellular and mitochondrial pathways infused in mitochondrial quality control (mitoQC) mechanisms to regulate and control mitochondrial health and function. Crosstalk between these pathways is essential for mitochondrial (and cellular) homeostasis including their biogenesis and degradation, fission and fusion events, transport, respiration, ATP production (metabolism in general), and the balance between mitochondrial and nuclear DNA expression. The extent and complexity of these pathways and their interaction have been summarized in recent review articles (Mandal & Drerup, 2019; Pearce et al., 2017; Pfanner et al., 2019; Vercellino & Sazanov, 2022; Youle & van der Bliek, 2012). The focus of this chapter is on mitophagy, a form of autophagy that selectively eliminates worn-out or damaged mitochondria.

During mitophagy, mitochondria become engulfed in double-membraned vesicles termed autophagosomes which then fuse with lysosomes to degrade their content. Basal mitophagy is active under normal, nondiseased conditions to maintain a healthy mitochondrial pool (McWilliams et al., 2018b; Watzlawik et al., 2021). Given its important regulatory role, mitophagy is tightly regulated and requires molecular sensors to determine organelle dysfunction in a sensitive and reliable manner. The sensors also serve as molecular docking stations for the autophagy machinery to initiate downstream tagging of defective mitochondrial units. The best-characterized mitophagy pathway so far is regulated by the ubiquitin (Ub) kinase PINK1 (PTEN-induced kinase 1) and the E3 Ub ligase PRKN (Parkin) (Pickles et al., 2018; Youle & Narendra,

H. E. Gendelman, T. Ikezu (eds.), *Neuroimmune Pharmacology and Therapeutics*,
https://doi.org/10.1007/978-3-031-68237-7_9

2011). Both enzymes work synergistically together to tag damaged mitochondria with Ub phosphorylated at serine 65 (p-S65-Ub), which is often used as a surrogate marker for mitophagy. Other parallel mitophagy pathways also exist, some of which are more developmental in nature and needed during cellular differentiation and maturation, while others respond to certain kinds of stress (Onishi et al., 2021).

Tremendous progress in better understanding mitophagy has been gained from simple cell model systems (stable cell lines that overexpress specific mitophagy markers) and in the case of PINK1–PRKN mitophagy also from nonphysiological, harsh conditions for example, high concentrations of mitochondrial depolarizers, respiratory chain inhibitors, or laser-induced ROS generation. Such an outcome clearly demonstrates the presence of active mitophagy pathways under maximal activation. However, the exact contribution of individual mitophagy pathways underin vivo conditions, especially in the vulnerable central nervous system (CNS), at baseline and under diseased conditions is still under investigation using disease human tissue or emerging mouse models. The impact of mitophagy on brain health and function is particularly important given its pathogenic role in many CNS diseases such as Parkinson's disease (PD) (Hou, et al., 2018a, Fiesel et al., 2015 ; Liu et al., 2019), Alzheimer's disease (AD) (Cai & Jeong, 2020; Watzlawik et al., 2021; Hou et al., 2020), traumatic brain injury (TBI) (Zhu et al., 2022), and stroke (Guan et al., 2018). While such studies are currently limited, available research will be covered and discussed in this chapter.

9.2 Mitophagy Pathways

9.2.1 PINK1–PRKN-Ubiquitin-Mediated Mitophagy

The Ub kinase PINK1 is the most upstream element and initiator of the pathway. Under physiological, nonstress conditions PINK1 is rapidly imported into mitochondria where it is N-terminally cleaved. The C-terminal PINK1 fragment is then released into the cytosol and subsequently eliminated by the Ub proteasome system (UPS) (Yamano & Youle, 2013). Under these conditions, the levels of full-length or cleaved PINK1 are very low and extremely difficult to detect, although the latter become visible upon inhibition of the UPS (Lin & Kang, 2008). However, mitochondrial damage disrupts PINK1's import into the mitochondrion and prevents subsequent cleavage by the protease PARL (Lazarou et al., 2012; Okatsu et al., 2013). Full-length PINK1 then accumulates on the mitochondrial surface by lateral insertion into the membrane (Hasson et al., 2013; Lazarou et al., 2012). Dimerization and autophosphorylation of this kinase at serine 228 (Rasool et al., 2022) are required for PINK1 activation but not for the accumulation of the PINK1 protein.

Stabilized PINK1 phosphorylates Ub at Ser65 (p-S65-Ub), and p-S65-Ub serves as the receptor (Ordureau et al., 2014) and as an allosteric activator of PRKN (Kane et al., 2014; Kazlauskaite et al., 2014; Koyano et al., 2014). Together with the direct phosphorylation of PRKN at Ser65 (p-S65-PRKN) (Kondapalli et al., 2012; Shiba-Fukushima et al., 2012), this renders PRKN fully active (Kazlauskaite et al., 2014; Koyano et al., 2014). Importantly, phosphorylation of PRKN at Ser65 has also been shown to be essential for its activation in vivo (McWilliams et al., 2018a). In addition to PRKN phosphorylation at Ser65, PRKN's E3 ligase activity is required for PRKN recruitment to mitochondria. It is of note that PRKN can utilize different E2-Ub conjugating enzymes during the process (Fiesel et al., 2014; Geisler et al., 2014; Wenzel et al., 2011). Once activated, PRKN decorates the outer mitochondrial membrane and cytosolic proteins with various Ub chain types (Chan et al., 2011; Narendra et al., 2012; Sarraf et al., 2013). A wealth of PRKN substrates has been identified now and further suggests that PRKN ubiquitylates substrates without particular sequence or structural specificity (Antico et al., 2021). However, certain Ub chain types and Ub density on mitochondria seem to act as main initiators of mitophagy (Sarraf et al., 2013). Indeed, additional Ub moieties ligated to mitochondrial proteins in turn then serve as further substrates for PINK1 and this leads to a positive feedback loop in which PINK1 and PRKN amplify the formation of the p-S65-Ub chains on damaged mitochondria (Ordureau et al., 2014).

The PINK1–PRKN pathway is antagonized by proteins that catalyze the dephosphorylation of mitophagy signals or the removal of Ub. Among the phosphatases that have been implicated in this process are phosphatase and tensin homolog (PTEN)-long (PTEN-L) (Wang et al., 2018a, b) and protein phosphatase with EF-hand domain 2 (PPEF2) (Wall et al., 2019). The ubiquitin-specific protease 30 (USP30) specifically removes Ub chains from mitochondrial proteins (Cunningham et al., 2015; Ordureau et al., 2014), and its overexpression or knockdown has been shown to prevent or enhance mitophagy, respectively. USP8 regulates endosomal trafficking (Clague & Urbe, 2006) and may also be involved in controlling PRKN recruitment to mitochondria (Durcan et al., 2014). The knockdown of USP8 prevents PRKN translocation to injured mitochondria and elevates PRKN expression (Durcan et al., 2014). The coordinated regulation of mitophagy by phosphatases and deubiquitinating enzymes in vivo is not fully uncovered and requires further investigation.

9.2.2 Other Mitophagy Pathways and Crosstalk

There are several receptors located at the outer mitochondrial membrane that can facilitate mitophagy. Each of them contains a LIR (Microtubule-associated *protein* 1A/1B-light chain 3 (LC3)-interacting region) motif, which enables the direct interaction with LC3 or Gamma-aminobutyric acid receptor-associated protein (GABARAP). These mitochondrial receptors are expressed in a tissue-specific manner and sometimes developmental-specific manner and respond to different triggers to activate the pathway. Mitochondrial proteins BCL2 and adenovirus E1B 19-kDa-interacting protein 3 (BNIP3) and BNIP3-like (BNIP3L/NIX) were the first mammalian mitophagy receptors identified (Hamacher-Brady et al., 2007; Sandoval et al., 2008; Schweers et al., 2007). During hypoxia, hypoxia-inducible factor 1 subunit alpha (HIF-1α) stimulates the expression of the pro-apoptotic BH3-only Bcl-2 family members BNIP3 and BNIP3L/NIX (Chinnadurai et al., 2008) to trigger mitophagy (Yussman et al., 2002). Another hypoxia-driven mitophagy receptor is the FUN14 domain-containing protein 1 (FUNDC1) (Liu et al., 2012). In analogy to BNIP3L/NIX and BNIP3, FUNDC1 functions independently from PINK1 and PRKN.

Cardiolipin is an anionic, dimeric phospholipid that contributes to 20% of total mitochondrial phospholipids and exists predominantly (65%) in the inner leaflet of the inner mitochondrial membrane (Ardail et al., 1990; Krebs et al., 1979). Translocated cardiolipin at the outside of the mitochondrial surface binds to LC3B via basic surface patches (Chu et al., 2013; Tanida et al., 2008) and results in autophagic destruction of damaged mitochondria. Mitochondrial stressors that induce cardiolipin-dependent mitophagy include rotenone, 6-hydroxy-dopamine (6-OHDA) as well as carbonyl cyanide m-chlorophenyl hydrazine (CCCP) (Chu et al., 2013). The cardiolipin–LC3 pathway does not depend on PINK1-mediated phosphorylation of Ub or PRKN translocation and does not require mitochondrial depolarization.

Other mitophagy receptors with LC3-binding motif are BCL2-like 13 (Bcl2L13) (Murakawa et al., 2015), the FKBP prolyl isomerase 8 (FKBP8) (Bhujabal et al., 2017; Saita et al., 2013), glyceraldehyde 3-phosphate dehydrogenase (GAPDH) (Hwang et al., 2015; Yogalingam et al., 2013), and the autophagy and Beclin 1 regulator 1 (AMBRA1) (Cianfanelli et al., 2015). AMBRA1 is an upstream autophagy regulator and can trigger both PRKN-dependent (Van Humbeeck et al., 2011a, b) and PRKN-independent mitophagy (Strappazzon et al., 2015). This is of interest because Beclin-1 also targets cardiolipin when located at the outer mitochondrial membrane and stimulates mitophagy by enhancing PRKN translocation to damaged mitochondria (Choubey et al., 2014). Therefore, AMBRA1 may bridge or co-regulate PINK1–PRKN- and cardiolipin-mediated mitophagy. Also, BNIP3 directly interacts with PINK1 at the mitochondrion to inhibit its proteolytic cleavage (Zhang et al., 2016).

While several of these different mitophagy pathways may be activated simultaneously by the same or different triggers and possibly interact with or regulate each other in a tissue and cell-type-specific manner, additional research is needed to uncover more mechanistic details and to determine their relevance and potential compensation in vitro and in vivo.

9.3 Mitophagy in Cellular and Neuronal Cultures

While most of the molecular mechanisms have been identified from nonneuronal cells, the importance of PINK1–PRKN mitophagy in cultured neurons has been confirmed in several studies, with PRKN being recruited to damaged mitochondria upon treatment with stressors such as CCCP, rotenone, and antimycin A (Ashrafi et al., 2014; Cai et al., 2012; Joselin et al., 2012; Koyano et al., 2014). A strong accumulation of p-S65-Ub has been observed as well (Fiesel et al., 2015). Of note, the p-S65-Ub response was particularly prominent in dopamine neurons compared to fibroblasts or induced pluripotent stem cells of the same individual likely due to the much greater PRKN protein levels in the respective cell types (Watzlawik et al., 2024). It is of interest that the extent of mitochondrial PRKN translocation was dependent on the trigger and the absence of antioxidants in the culture medium (Joselin et al., 2012). In contrast to studies using mitochondrial depolarizers, MitoKillerRed-induced axonal mitochondrial damage occurred within 1 h (Ashrafi et al., 2014). However, even with these more aggressive approaches, mitophagy was only induced in a subset of neuronal mitochondria, while the very same conditions were sufficient to completely disintegrate the mitochondrial network in non-neuronal cells (Ashrafi et al., 2014; Joselin et al., 2012; Koyano et al., 2014). These results suggest additional layers of mitoQC mechanisms in post-mitotic neurons. It should be noted that PINK1-PRKN signaling is constitutively active even at basal (non-stress) conditions both in cultured human cells and in rodent brain although at extremely low levels that requires ultrasensitive assays for detection (Watzlawik et al., 2024). Differences exist in enzymatic activity and expression levels for both PINK1 and PRKN across cell types, tissues and species (Watzlawik et al., 2024).

A quantitative approach to evaluate PINK1–PRKN signaling in embryonic stem cell-derived Neurogenin-2 (NGN2)-induced neurons (iNeurons) revealed a ~900-fold p-S65-Ub increase within 6 h of antimycin/oligomycin-induced mitochondrial depolarization (Ordureau et al., 2018). While the relative abundance of mitochondrial PRKN targets was comparable between iNeurons and HeLa cells overexpressing PRKN, the absolute abundance of Ub reflected the respective PRKN expression levels. Another recent proteomic analysis comparing mouse cortical neurons and human iNeurons revealed a conserved set of 49 PINK1 activation-dependent global di-Glycine ubiquitylation sites in 22 proteins (Antico et al., 2021). The authors also identified a subset of cytoplasmic proteins in mouse neurons recruited to mitochondria that undergo PINK1–PRKN-independent ubiquitination, indicating the presence of alternate Ub E3 ligase pathways that are activated by mitochondrial depolarization in neurons (Antico et al., 2021).

There is some debate about whether neuronal mitophagy occurs in the axon or in the soma. In vitro experiments with primary neurons have shown that mitochondrial fragments are engulfed within autophagosomes formed constitutively in the distal axon (Stavoe & Holzbaur, 2019); however, there is limited evidence that the engulfment of these fragments involves the machinery for selective mitophagy such as PINK1, PRKN, and Optineurin. Studies looking more specifically at mitophagy provide conflicting views. One group reported localized mitophagy along the axons of hippocampal neurons upon induction of cellular stress via mitochondrial depolarization or ROS generation (Ashrafi et al., 2014). In contrast, mitochondrial depolarization in cortical neurons resulted in PRKN-labeled mitochondria and accumulation in somatodendritic regions that were subsequently degraded locally by the autophagosomal–lysosomal pathway. Conversely, GFP-LC3 was rarely observed associated with axonal mitochondria in these experiments (Cai et al., 2012).

9.4 Mitophagy Studies in Animal Models

9.4.1 Mitophagy Reporter Mice

Within the last decade, fluorescent reporters were developed to monitor mitophagy. All reporters have in common that they are coupled to a mitochondrial targeting sequence and that they use pH-sensitive fluorescent protein(s) that behave differently in cytosolic (neutral) versus lysosomal (acidic) environments. The coral-derived red fluorescent protein Keima shifts its excitation spectrum in a pH-dependent manner and has been coupled to a mitochondrial-targeting sequence to direct it to the mitochondrial matrix to monitor mitophagy (Katayama et al., 2011; Sun et al., 2015). It is of note that Keima is (partially) resistant to lysosomal proteases and therefore protein degradation. Mito-QC mice express EGFP–mCherry tagged to the outer mitochondrial membrane of mitochondria. At neutral pH, both proteins, mCherry and EGFP, will fluoresce. Upon exposure to the acidic environment inside the lysosome, the green EGFP fluorescence is quenched, while the red mCherry fluorescence remains stable (McWilliams et al., 2016). One advantage of mito-QC over mt-Keima mice is the ability to fix tissues whereas mt-Keima mice/tissues require live imaging. Moreover, the mCherry fluorescence is an essential reference and indicates the presence of the fluorescent mCherry-GFP tag irrespective of mitophagy. The absence of mCherry fluorescence may either indicate a lack of EGFP-mCherry expression or protein degradation.

The characterization of both mouse models revealed a surprising heterogeneity of basal mitophagy levels between tissues (McWilliams et al., 2018b; Sun et al., 2015). In the brain of mt-Keima mice, basal mitophagy levels were lower in the cortex, substantia nigra, and striatum compared to dentate gyrus and lateral ventricles (Sun et al., 2015). Both mt-Keima and mito-QC mice showed high mitophagy rates in heart tissue with high metabolic demand (McWilliams et al., 2018b; Sun et al., 2015). However, conflicting results between the two mouse models were reported particularly for the detection of PINK1–PRKN mitophagy that suggested lower sensitivity of mito-QC over mt-Keima mice for PINK1–PRKN mitophagy (Liu et al., 2021b; McWilliams et al., 2018b). Observed differences between both models were based on different cellular localization and the nature of the fluorescent protein (outer mitochondrial membrane for tandem EGFP-mCherry (mito-QC) versus mitochondrial matrix for Keima (mt-Keima) (Liu et al., 2021b)). Alternatively, the different mouse strains used for both models, which are C57BL/6 j-ntac for mito-QC versus FVB/NJ for mt-Keima mice, may in part explain the observed differences (Ganley et al., 2021). Caution should be taken when using the mito-QC model to study PINK1–PRKN mitophagy (Liu et al., 2021b). Given the limitations of the Keima protein as a non-fixable reporter *Liu* et al., suggest how to improve the fixable mito-QC reporter by directing it to the mitochondrial matrix and by substituting mCherry with TOLLES, which has superior resistance to acidity (Liu et al., 2021a). These and other changes have already been implemented in the fixable mt-SRAI model with improved performance over mito-QC in cell culture models (Katayama et al., 2020).

9.4.2 Mouse Models Lacking Key Mitophagy Genes

Several groups created *pink1* or *prkn* knockout mice as potential new PD models to characterize the molecular function of PINK1 and PRKN in vivo and to validate the involvement of mitophagy in PD. Surprisingly, deletion of germline *Prkn* exons 2 (Perez & Palmiter, 2005), 3 (Goldberg et al., 2003; Itier et al., 2003; Palacino et al., 2004; Von Coelln et al., 2004), or 7 (Shin et al., 2011) resulted in only mild or no motor phenotypes, slightly changed dopamine metabolism, and no loss of dopaminergic (DA) neurons. Similar to *prkn* knockout mice, *pink1* knockout mouse models poorly recapitulate human disease. *Pink1* exon 4–7 or exon 4–5 deletion did not affect DA neurodegeneration or Lewy body formation (Akundi et al., 2011; Kitada et al., 2007). *Pink1* exon 4–5 deletion led to impaired DA release and decreased mitochondrial respiration in the striatum of mice (Akundi et al., 2011), and this was accompanied by elevated mitochondrial Ca^{2+} sensitivity and JNK activity that preceded reduced DA levels and abnormal DA homeostasis. While these observations are important, they are not equivalent to neurodegeneration seen in PD. Even 2-year-old *pink1−/−*, *prkn−/−*, and *dj1−/−* triple KO mice did not develop Lewy bodies or show loss of DA neurons in the SN (Kitada et al., 2009). Interestingly, while phenotypes were largely absent in the brain, tissue-specific mitophagy defects in other organs became apparent as *pink1−/−* mice developed progressive cardiomyopathy caused by mitochondrial dysfunction and oxidative stress (Billia et al., 2011). This might indicate alternative, compensatory mechanisms or the need for a second hit in some organs, such as the mouse brain.

Because *pink1−/−* and *prkn−/−* mice did not show overt phenotypes especially at younger age, different groups aged or crossed these strains with mouse models of mitochondrial dysfunction or neurodegenerative disease as additional metabolic stressors or as a second hit. Double knockouts of *prkn* and the mitochondrial transcription factor A (*TFAM*) to induce respiratory chain deficiency had no additional impact on the phenotype in DA neurons (Sterky et al., 2011). Similarly, *prkn* KO combined with Purkinje cell-specific *Drp-1* KO (Kageyama et al., 2012) or dopaminergic neuron-specific *Mfn2* KO (Lee et al., 2012) had also no impact on motor phenotypes. Crossing *prkn−/−* mice with mutant mitochondrial helicase Twinkle (Twinkle-TG mice; $Twinkle^{dup/+}$) led to very high mutant mtDNA concentration in the SN, very low mitochondrial function and low membrane potential, severe neurobehavioral deficits at 19 months and substantial loss of dopaminergic neurons in the SN combined with very low dopamine levels (Song et al., 2017).

Interestingly, *Sliter* et al. discovered an inflammatory phenotype in both *prkn* $^{-/-}$ and *pink1* $^{-/-}$ mice following exhaustive exercise as well as in *prkn* $^{-/-}$; *mutator* mice, which accumulate mutations in mitochondrial DNA (mtDNA) due to a mutation that affects the proof-reading ability of the mitochondrial polymerase *Polg*. The authors further observed loss of dopaminergic neurons in the substantia nigra pars compacta plus motor defect in aged *prkn* $^{-/-}$; *mutator* mice (Sliter et al., 2018). In *prkn* $^{-/-}$; *mutator* mice, dopaminergic neurons degenerate and cause a L-DOPA reversible motor deficit with other neuronal populations being unaffected (Pickrell et al., 2015). Phosphorylated Ub was increased in the brains of *mutator* mice, which indicates PINK1–PRKN pathway activation. Prkn loss caused mitochondrial dysfunction and affected the pathogenic phenotype but not the levels of mtDNA somatic mutations. A systemic loss of Parkin synergizes with mitochondrial dysfunction causing dopaminergic neuron death to model the PD pathogenic processes (Pickrell et al., 2015). However, a recent study was unable to reproduce this synergism (Scott et al., 2022). Nevertheless, in addition to the presence of mitochondrial stress, age appears to be an important factor for PRKN-mediated deficits in mice. *Prkn−/−* mice developed mitochondrial dysfunction with fragmented mitochondria and abnormal internal structures, motor locomotor impairment, and neuronal loss at 110 weeks of age (Noda et al., 2020).

9.5 Mitophagy Alteration in the Human Brain

Increasing evidence suggests that the accumulation of dysfunctional mitochondria combined with elevated oxidative stress levels are at least partially responsible for defective neurons during aging, in neurodegenerative diseases including PD and AD, and likely in other CNS disorders. However, the molecular mechanisms are still unclear and might vary between different conditions.

9.5.1 Mitophagy Alteration in Parkinson's Disease Brain

Dysfunctional mitophagy is intrinsically linked to PD as complete loss of PINK1 or PRKN function unequivocally cause early-onset PD. While the impact of heterozygous *PINK1* and *PRKN* mutations and their contribution to disease risk are still under debate (Lubbe et al., 2021), various other PD risk genes including *SNCA*, *GBA*, *LRRK2*, *PARK7*, *and VPS35*, all have been connected to mitochondrial and/or lysosomal biology, as shown in studies using fly, rodent, and human iPSC-derived model systems (Burchell et al., 2013; Chung et al., 2016; Clark et al., 2006; Hsieh et al., 2016; Lindstrom et al., 2017). This underscores the importance of functional mitochondria in PD and their mitophagic degradation in lysosomes (Pang et al., 2022; Xu et al., 2005).

The first indication for an important role of mitophagy in PD came from an ultrastructural study showing autophagic accumulation of neuronal mitochondria in patients with incidental Lewy body disease (LBD), PD, or dementia with Lewy bodies (DLB) (Zhu et al., 2003). In this study, the authors detected vesicular phospho-ERK-positive granules in midbrain sections that enveloped enlarged mitochondria by microscopy. Later, we demonstrated the presence and accumulation of the mitophagy tag p-S65-Ub in the human brain during normal aging and in disease (Fiesel et al., 2015;

Fiesel & Springer, 2015). In support of our findings, dopaminergic neurons with *PINK1* and *PRKN* mutations showed indeed compromised p-S65-Ub signaling and transport of axonal mitochondria upon mitochondrial stress with a similar p-S65-Ub pattern in CNS (Shiba-Fukushima et al., 2017). The identified p-S65-Ub-positive structures seem predominantly to accumulate in neurons and include cytoplasmic granules in the substantia nigra (SN) and hippocampus of sporadic LBD patients, which partially colocalize with mitochondrial, lysosomal, and total Ub markers (Hou, et al., 2018a, Fiesel et al., 2015).

It is currently not entirely clear whether increased levels of p-S65-Ub in sporadic LBD are primarily due to increased mitochondrial damage or are a result of impaired autophagic flux through lysosomes. This is different from familial PD cases with pathogenic *PINK1* or *PRKN* mutations causing complete or partial inactivation of the PINK1–PRKN-mediated mitophagy pathway with absent or very low p-S65-Ub levels (Hou, et al., 2018a Fiesel et al., 2015). However, it should further be noted that increased p-S65-Ub was associated with neuronal Lewy body-type α-synuclein pathology but not glial cytoplasmic inclusions of α-synuclein as seen in multiple system atrophy (Hou et al., 2023). In this context, it is worth mentioning that aside from α-synuclein, Lewy bodies consist of clusters of membranous compartments derived from mitochondria and lysosomes (Shahmoradian et al., 2019), and their formation may result from dysfunctions in either of these organelles (Erskine et al., 2021).

In summary, there is a strong genetic link between dysfunctional mitochondria, lysosomal impairment, and PD. While mitophagy may be altered at different steps of a sequential process, findings in both familial and sporadic cases indicate that a functional mitophagy pathway is crucial for DA neuron viability. Moreover, p-S65-Ub shows promise as a future biomarker for mitochondrial damage in aging and disease.

9.5.2 Mitophagy Alteration in Alzheimer's Disease Brain

The earliest AD studies describing alterations of mitochondrial morphology and metabolic changes in the mitochondrial tricarboxylic acid (TCA) cycle originate from 1980s showing swollen mitochondria with altered cristae organization and reduced activity of the pyruvate dehydrogenase complex (Saraiva et al., 1985; Sheu et al., 1985; Sorbi et al., 1983). More recently, fluoro-2-deoxy-D-glucose positron emission tomography (2DG-PET) brain imaging of individuals who later developed AD revealed that the brain energy metabolism is altered before the onset of cognitive deficits (Kapogiannis & Mattson, 2011) suggesting that mitochondrial dysfunction is a very early event in AD. Glucose hypometabolism was also observed in patients with mild cognitive impairment (MCI), a prodromal stage of AD (Croteau et al., 2018), and was linked to the patients' ApoE4 status (Paranjpe et al., 2019).

Fang et al. showed impaired mitophagy in the hippocampus of AD patients, induced pluripotent stem cell-derived human AD neurons, and animal AD models. Stimulation of mitophagy (through NAD^+ supplementation, urolithin A, and actinonin) reversed deficits through the PINK1–PRKN pathway in *C.elegans* and reduced levels of insoluble $A\beta_{1-42}$ and $A\beta_{1-40}$ and prevented cognitive impairment in APP/PS1 mice (Fang et al., 2019). Similarly, viral overexpression of PINK1 promoted the clearance of damaged mitochondria and alleviated amyloid-β-induced loss of synapses and cognitive decline in an AD mouse model (Du et al., 2017). *Ye* et al. found that while mitophagy is robustly induced in mutant APP neurons and the AD brain, the aberrant accumulation of dysfunctional mitochondria likely causes a reduced capacity to eliminate damaged mitochondria. Damaged mitochondria with altered morphology per neuronal cell body were significantly (threefold) increased in AD brains, which indicates insufficient or impaired clearance. In addition, also disease-associated tau was shown to sequester PRKN, thus impairing mitophagy as well (Cummins et al., 2019).

In support of the importance of PINK1-PRKN mitophagy in AD, our group demonstrated significant increases in the mitophagy tag p-S65-Ub in AD brain relative to control cases (Hou et al., 2020; Watzlawik et al., 2021). p-S65-Ub-positive structures strongly correlated with granulovacuolar degeneration (GVD), early phospho-tau deposits, and amyloid beta pathology. Single-cell analyses revealed differential co-localization of p-S65-Ub with mitochondria, GVD bodies, and/or lysosomes depending on the brain region analyzed. This outcome is indicative of selective organellar vulnerability in different brain regions in AD—more mitochondrial damage or subsequent accumulation of p-S65-Ub labeled mitochondria in autophagic–lysosomal structures (Hou et al., 2020).

In summary, strong evidence suggests successful initiation of mitophagy in AD, while the lysosomal digestion of damaged mitochondria and therefore also the autophagic flux is stalled. It is still unclear whether mitochondrial dysfunction precedes and potentially initiates Aβ and tau pathology and/or whether Aβ and tau pathology cause or exacerbate mitochondrial dysfunction with experimental evidence for both. Further live studies in iPSC-derived neurons from sporadic AD cases may be critical to further understand mechanistic details of stalled mitophagy and for intervention studies to evaluate future therapeutic approaches.

9.5.3 Mitophagy Alteration in Acute Diseases (Traumatic Brain Injury and Stroke)

Traumatic brain injury (TBI) results in increased brain mitophagy levels in humans that are associated with elevated outer mitochondrial membrane cardiolipin levels (Chao et al., 2019). The importance of cardiolipin-mediated mitophagy in TBI was further confirmed in rat models where knockdown of cardiolipin synthase or phospholipid scramblase-3 significantly decreased mitophagy (Chao et al., 2019). Inhibition of mitochondrial clearance by 3-methyladenine, mdivi-1, or phospholipid scramblase-3 knockdown after TBI led to a worse outcome, suggesting that mitophagy is neuroprotective (Chao et al., 2019). A different study found an association between BNIP3L/NIX expression levels, neuronal apoptosis, and neurological deficits in a rat TBI model. Neuronal BNIP3L/NIX expression was reduced after TBI, while BNIP3L/NIX upregulation stimulated autophagy and reduced neuronal apoptosis and neurological deficits (Ma et al., 2019). Activation of mitophagy pathways other than cardiolipin- and NIX-mediated mitophagy or their potential interaction in TBI remains elusive.

During cerebral ischemia in rats, hypoxia enhances *Bnip3* and *Bnip3l* expression, triggering the liberation of Beclin-1 from the Bcl-2/Beclin-1 complex, and ultimately induces mitophagy (Yu et al., 2013). BNIP3L/NIX-mediated mitophagy was also involved in cerebral ischemia–reperfusion (I–R)-induced injury in mice (Yuan et al., 2017). Knockout of *Bnip3l* impaired mitophagy and aggravated cerebral I–R injury in mice and was subsequently rescued by BNIP3L/NIX overexpression, which suppressed apoptosis. *Bnip3* and *prkn* double knockout mice showed a synergistic mitophagy deficiency upon I-R injury, suggesting that these are independent pathways. The loss of BNIP3 substantially lowers neuronal mitophagy and apoptosis but enhances nonselective autophagy following ischemic/hypoxic insults (Shi et al., 2014). Silenced BNIP3 cannot be functionally compensated for with BNIP3L/NIX overexpression and may cause excessive mitophagy (Shi et al., 2014). It has been concluded that BNIP3L/NIX regulates basal levels of mitophagy under physiological conditions. In contrast, BNIP3 can activate excessive mitophagy that results in detrimental cell death, which raises the question of how BNIP3 can be both neuroprotective and detrimental in stroke. Therefore, BNIP3 and BNIP3L/NIX are promising targets for the control of cell survival or death by regulating mitophagy after stroke.

In summary, mild to moderate stimulation of mitophagy is usually considered neuroprotective; however, stroke studies so far indicated that excessive mitophagy is highly detrimental with a severe impact on disease course. Therefore, mitophagy can be seen as a double-edged sword.

9.6 Defective Mitophagy and Inflammation in PD

Microglia-derived neuroinflammation in the nigrostriatal system has been reported as an early event in PD patients (Blum-Degen et al., 1995; Gerhard et al., 2006; Sanchez-Guajardo et al., 2013). The analysis of postmortem PD brains revealed NLRP3 inflammasome activation (Swanson et al., 2019) and active microgliosis in the SN (Gordon et al., 2018), which further supports the role of neuroinflammation in PD. Moreover, blood and CSF analysis from PD patients shows higher levels of classical innate immunity cytokines—IL-1β, IL-6, and TNFα compared to age-matched controls (Tansey et al., 2022).

9.6.1 Mitochondrial-Derived Danger-Associated Molecular Patterns Trigger Innate Immunity and Neuroinflammation in PD

Mitochondria have a circular genome and a unique translation machinery, with distinct ribosomes from the ones in the cytoplasm (Goncalves & Morais, 2021). When mitochondrial replication occurs, mitochondrial DNA fragments may escape to the cytosol, where they are sensed by Toll-like receptors (TLRs) as damage-associated molecular patterns (DAMPs) (Rodriguez-Nuevo & Zorzano, 2019). While this mechanism is usually beneficial for the clearance of damaged cells or the activation of a normal pro-inflammatory response, the cellular mitochondrial replication process in PD is more prone to fragmentation. This induces a continuous spilling of DAMPs of mitochondrial origin into the cytosol, which triggers a chronic immune response. When TLRs are activated by mitochondrial DAMPs, a signaling cascade, known as sterile inflammation, is elicited, following NF-kB translocation to the nucleus. As a result, ASC (Apoptosis-associated speck-like protein containing a CARD) and pro-IL1β expression increases, NLRP3 and ASC recruit pro-caspase-1 that in its active form cleaves and activate pro-inflammatory mediators like IL1β.

9.6.2 Mitochondrial-Derived Vesicles and PD

Mitochondrial-derived vesicles (MDVs) promote cellular communication for the maintenance of homeostasis or in disease for fast-acting responses. MDVs are important for the transport of damaged mitochondrial components, regulation of cell metabolism, or stimulation of the immune system. In the event of mitochondrial damage and increased ROS production, MDVs will be loaded with oxidized mitochondrial components and fuse with lysosomes for content degradation. This process is mediated by PINK1–PRKN and is Tollip-dependent (Ryan et al., 2020). Low expression of PRKN allows for mitochondrial cargo to be imported to extracellular vesicles (EVs) in a process mediated by OPA1 and sorting nexin 9 (Snx9) (Todkar et al., 2021). Given a stalled PINK1–PRKN mitophagy pathway in PD, this mechanism may facilitate the transport of oxidized/damaged mitochondrial content in EVs to other cells (Todkar et al., 2021).

9.6.3 Mitophagy and c-GAS-STING Pathway

The c-GAS-STING innate immunity pathway is a type I interferon production pathway induced by cyclic GMP-AMP synthase (c-GAS) and stimulator of interferon genes (STING), which is active in T lymphocytes, *NK cells*, *myeloid cells*, and monocytes when cells are infected with intracellular pathogens, including viruses and mycobacteria (Nakhaei et al., 2010). C-GAS-STING acts both as a cytosolic DNA sensor and adaptor protein in type I interferon signaling to activate downstream factors, which then mediate the antiviral and innate immune response against intracellular pathogens (Burdette & Vance, 2013). Translocation of STING from the ER to the Golgi apparatus and finally to the cytoplasm is required for double-stranded (ds) DNA-mediated immune responses where STING assembles with TBK1. STING also co-localizes with the autophagy proteins, LC3 and Atg9a after dsDNA stimulation. Interestingly, loss of Atg9a, but not Atg7, greatly enhances the assembly of STING and TBK1 by dsDNA, leading to aberrant activation of the innate immune response. It is of note that Atg9a functions both as a regulator of innate immunity following dsDNA stimulation as well as an essential autophagy protein (Saitoh et al., 2009). As mentioned earlier, *pink1−/−* and *prkn−/−* mice show no PD-relevant phenotypes. However, *Sliter* et al. demonstrated that exhaustive exercise in both knockout mouse strains causes a strong inflammatory phenotype which was completely rescued in the absence of STING (Sliter et al., 2018). Moreover, loss of dopaminergic neurons in the substantia nigra pars compacta and the motor defect observed in aged *prkn−/− mutator* mice are fully rescued by loss of STING as well, indicating that inflammation is the main cause of this mouse phenotype (Sliter et al., 2018).

9.6.4 Mitophagy and Adaptive Immunity

Both PINK1 and PRKN have been implicated in adaptive immunity via the presentation of mitochondrial antigens (MitAP) on MHC class I molecules in professional antigen-presenting cells (e.g., dendritic cells, macrophages) (Matheoud et al., 2016). MitAP relies on the generation and trafficking of MDVs to shuttle specific mitochondrial cargos to late endosomes (Sugiura et al., 2014). MDV formation and MitAP require the dynamin-binding partner Snx9, whose mitochondrial recruitment is inhibited by PRKN (Matheoud et al., 2016). Primary dendritic cells isolated from *pink1−/−* or *prkn−/−* mice showed increased MitAP of the mitochondrial matrix protein 2-oxoglutarate dehydrogenase relative to wild-type controls when stimulated with lipopolysaccharide (Matheoud et al., 2019). Intestinal infection of *pink1−/−* mice using gram-negative bacteria stimulated MitAP and autoimmune mechanisms that resulted in the establishment of cytotoxic mitochondria-specific CD8+ T cells in the periphery and the brain. It is of note that *pink1−/−* mice with intestinal infection revealed reduced levels of dopaminergic axons in the striatum, which was combined with increased CNS infiltrating monocytes and T cells and motor impairment (Matheoud et al., 2019). The outcome indicated an active role of both PINK1 and PRKN in MitAP repression in addition to their function as mitophagy regulators. The data also suggested that cytotoxic CD8+ T cells may contribute to an autoimmune mechanism exacerbating the pathology in PINK1- and PRKN-associated PD cases and highlights a potential role of intestinal infection as a trigger in PD and further emphasizes the relevance of the gut–brain axis in the disease.

9.7 Stimulating Mitophagy as a Therapeutic Goal

Mitochondrial dysfunction is a hallmark of several if not all neurodegenerative diseases. Stimulation of mitophagy in neurodegenerative diseases to clear out damaged mitochondria is a promising approach to achieve a significant therapeutic effect (Fang et al., 2017; Kerr et al., 2017; Lou et al., 2020). In addition to food restriction or rigorous exercise, different ways to stimulate mitophagy have been identified and tested in fruit flies, rodents, and cultured human cells and shown to stimulate mitophagy, act neuroprotective, improve overall lifespan and fitness, and prevent age-dependent memory decline (Gupta et al., 2013; Madeo et al., 2018).

NAD+ precursors protect against tau and Aβ pathology and prevent cognitive decline in an AD mouse model (Gong et al., 2013). NAD+ is essential for the activation of deacetylases sirtuin 1 and sirtuin 3 (SIRT1 and SIRT3), cAMP response element-binding (CREB) protein, and activating PI3K-Akt and MAPK/ERK1/2 (Gong et al., 2013; Hou, et al., 2018b; Liu et al., 2013; Turunc Bayrakdar et al., 2014). The NAD+ precursor nicotinamide riboside was also able to rescue mitochondrial dysfunction and neuronal loss in iPSC and PD *Drosophila* models (Schondorf et al., 2018). Urolithin A (UA) is an ellagitannins-derived metabolite shown to stimulate brain mitophagy and improve muscle health in old animals and in preclinical models of aging (Fang et al., 2019; Ryu et al., 2016). Mitophagy effects of both UA and the antibiotic actinonin were mediated by PINK1, PRKN, and NIX to weaken the detrimental outcome of AD on learning and memory deficits, and inflammation (Fang et al., 2019). In general, polyamines like spermidine stimulate autophagy-related gene expression and mitophagy via inhibition of the mammalian target of rapamycin (mTOR) and 5'-AMP-activated protein kinase (AMPK) activation (Fan et al., 2017; Madeo et al., 2018).

Approaches to directly stimulate PINK1-PRKN mitophagy are emerging too. The ATP analog kinetin triphosphate was shown to lead to higher PINK1 activity and PRKN recruitment to depolarized mitochondria, reduced mitochondrial motility in axons, and lower levels of apoptosis (Hertz et al., 2013). The nucleoside derivative, kinetin riboside, and other related compounds activate PINK1 in cells independent of mitochondrial depolarization (Lambourne et al., 2023; Osgerby et al., 2017). More recently, the N^6-substituted adenine, called MTK458 was shown to stabilize PINK1 in an active conformation and resolve stalled mitophagy (Chin et al., 2023). Interestingly, *Chin* et al. showed that oral dosing of MTK458 in an α-synuclein PD mouse model led to a dose-dependent decrease in α-synuclein pathology that was accompanied by improved motor activity and reduced levels of inflammatory markers. Likewise, the first positive allosteric modulators have been described enhanced PRKN enzymatic activity in vitro (Shlevkov et al., 2022). Another approach identified compounds that stimulate mitochondrial accumulation of the mitophagy-tag p-S65-Ub in PRKN haploinsufficiency human fibroblasts (Tufi et al., 2023). In further support of an earlier study (Tsefou et al., 2021), these molecules were shown to inhibit USP30, a negative regulator of PINK1-PRKN mitophagy.

It should be noted that stimulation of mitophagy is generally considered to be beneficial in health and disease as it would allow the removal of worn-out or defective mitochondria. However, it might be critical to simultaneously ensure proper flux through the autophagy system and efficient degradation in lysosomes to not exceed the cellular clearance capacity. In contrast, excessive mitophagy may occur in certain instances where mitochondrial damage is too severe or if mitochondrial biogenesis may not be able to keep up with the increased degradation leading to depletion of mitochondria.

9.8 Summary

Tremendous progress has been made in the field within the past 15 years, particularly for the PINK1–PRKN pathway with many new details discovered. However, molecular mechanisms have mostly been established in nonneuronal cells using PRKN overexpression and complete mitochondrial depolarization to induce the pathway. This paradigm provides maximal activation of PINK1–PRKN signaling and typically results in "wholesale" mitophagy where most mitochondria are cleared from the cells as they do not rely on oxidative phosphorylation. Yet, patho−/physiologically relevant PINK1–PRKN signaling seems much more localized and restricted to only mitochondrial subdomains that are damaged at any given time. Thus, mitophagy likely occurs at much lower levels overall and is more challenging to assess requiring more sophisticated and ultrasensitive detection methods (e.g., ELISAs for the detection of p-S65-Ub, PINK1, and PRKN). Moreover, the dynamic and transient nature of mitophagy makes the analysis of this pathway and the interpretation difficult, especially in vivo. Caution must be used when selecting a time point to investigate for example the induction of mitophagy upon a certain stressor. Low PINK1 or p-S65-Ub levels might be interpreted as weak mitophagy induction when in fact the peak of the response might have already passed, and levels are normalizing again after damaged mitochondria have been fed into the lysosomal machinery. Ideally, several time points should be used to scan the entire peak of the response. In the human postmortem brain, where analysis over time is not possible, the analysis of mitophagy remains difficult. Elevated p-S65-Ub levels, as observed in multiple neurodegenerative conditions, might reflect either increased mitochondrial stress or a decline of the lysosomal degradative capacity or they could stem from both. In either case, brain region-specific causes for elevated mitophagy markers are possible and very likely. The contribution of upstream damage versus downstream flux might depend also on the affected cell type and the type and extent of neuropathological lesion(s) that is present in certain regions. The comprehensive analysis of mitophagy in the human brain might further aid in addressing longstanding questions such as the selective vulnerability of very defined cell populations for certain neurodegenerative conditions.

9.9 Review Questions

1. Different mitophagy pathways are
   (a) *PINK1–PRKN mitophagy*
   (b) Beta-actin-mediated mitophagy
   (c) Sulfatide-mediated mitophagy
   (d) *BNIP3L/NIX-mediated mitophagy*
   (e) Src family kinase-mediated mitophagy
2. The PINK1–PRKN mitophagy pathway:
   (a) Cannot be activated by mitochondrial depolarizers, for example, CCCP
   (b) *Leads to the decoration of outer mitochondrial membrane proteins with p-S65-Ub*

(c) Is independent of p-S65-Ub
(d) Will not result in the degradation of damaged mitochondria
(e) *Is, so far, the best-documented mitophagy pathway*

3. The PINK1–PRKN positive feedback loop:
(a) *Is important to decorate mitochondria with p-S65-Ub*
(b) *Requires PINK1-mediated activation and recruitment of PRKN and Ub to mitochondria*
(c) Requires MUL-1 E3 Ub ligase activity
(d) Depends on healthy mitochondria
(e) Is based on LRRK2 phosphorylation

4. Detection of mitophagy at baseline:
(a) *Requires highly sensitive tools*
(b) Is easy to detect
(c) Is generally easier in vivo (mouse) compared to in vitro models
(d) *Is challenging due to low amounts of damaged mitochondria*
(e) Cannot be detected in mice

5. PINK1–PRKN mitophagy in PD:
(a) Is identical in sporadic and hereditary (genetic) cases
(b) Is independent of age
(c) *Can be best detected by measuring p-S65-Ub in the brain*
(d) Is not associated with Lewy-Body pathology
(e) Is indistinguishable between neurodegenerative and acute brain diseases

True/False

6. Mitophagy is an important cellular mechanism to eliminate damaged mitochondria.
7. Stimulation of mitophagy is not a double-edged sword.
8. Food restriction or rigorous exercise does not affect mitophagy in different animal models.
9. Mitophagy follows an all-or-nothing rule and cannot be stopped once initiated.
10. Mitophagy is a dynamic and transitory process.

9.10 Answers

1. a and d
2. b and e
3. a and b
4. a and d
5. c
6. T
7. F
8. F
9. F
10. T

Acknowledgments WS is partially supported by the National Institutes of Health (NIH)/National Institute of Neurological Disorders and Stroke (NINDS) [RF1 NS085070, R01 NS110085, and U54 NS110435], the NIH/National Institute of Aging (NIA) [R56 AG062556], the Department of Defense Congressionally Directed Medical Research Programs (CDMRP) [W81XWH-17-1-0248], the American Parkinson Disease Association (APDA) Center for Advanced Research at Mayo Clinic Florida, the Michael J. Fox Foundation for Parkinson's Research (MJFF), the Ted Nash Long Life Foundation, Mayo Clinic Foundation, the Center for Biomedical Discovery (CBD), and the Robert and Arlene Kogod Center on Aging. XH is supported by a pilot grant and a developmental project award from the Mayo Clinic Alzheimer Disease Research Center (ADRC, P30 AG062677) and fellowships awarded by the APDA and Alzheimer's Association [AARF-22-973152]. FCF is supported by the Florida Department of Health - Ed and Ethel Moore Alzheimer's Disease Research Program [22A07], the MJFF, and the Mayo Clinic Office of Research Equity, Inclusion and Diversity and the Mayo Clinic Center for Biomedical Discovery.

Competing Interest Mayo Clinic, WS, and FCF hold a patent related to PRKN activators (Small Molecule Activators of Parkin Enzyme Function, US patent, 11401255B2; August 02, 2022). Additional funding sources to disclose include grants from Amazentis SA [to WS]. All other authors declare they have no competing interests.

References

Akundi, R. S., Huang, Z., Eason, J., Pandya, J. D., Zhi, L., Cass, W. A., Sullivan, P. G., & Bueler, H. (2011). Increased mitochondrial calcium sensitivity and abnormal expression of innate immunity genes precede dopaminergic defects in Pink1-deficient mice. *PLoS One, 6*(1), e16038. https://doi.org/10.1371/journal.pone.0016038

Antico, O., Ordureau, A., Stevens, M., Singh, F., Nirujogi, R. S., Gierlinski, M., Barini, E., Rickwood, M. L., Prescott, A., Toth, R., Ganley, I. G., Harper, J. W., & Muqit, M. M. K. (2021). Global ubiquitylation analysis of mitochondria in primary neurons identifies endogenous Parkin targets following activation of PINK1. *Science Advances, 7*(46), eabj0722. https://doi.org/10.1126/sciadv.abj0722

Ardail, D., Privat, J. P., Egret-Charlier, M., Levrat, C., Lerme, F., & Louisot, P. (1990). Mitochondrial contact sites. Lipid composition and dynamics. *The Journal of Biological Chemistry, 265*(31), 18797–18802. Retrieved from https://www.ncbi.nlm.nih.gov/pubmed/2172233

Ashrafi, G., Schlehe, J. S., LaVoie, M. J., & Schwarz, T. L. (2014). Mitophagy of damaged mitochondria occurs locally in distal neuronal axons and requires PINK1 and Parkin. *The Journal of Cell Biology, 206*(5), 655–670. https://doi.org/10.1083/jcb.201401070

Bhujabal, Z., Birgisdottir, A. B., Sjottem, E., Brenne, H. B., Overvatn, A., Habisov, S., Kirkin, V., Lamark, T., & Johansen, T. (2017). FKBP8 recruits LC3A to mediate Parkin-independent mitophagy. *EMBO Reports, 18*(6), 947–961. https://doi.org/10.15252/embr.201643147

Billia, F., Hauck, L., Konecny, F., Rao, V., Shen, J., & Mak, T. W. (2011). PTEN-inducible kinase 1 (PINK1)/Park6 is indispensable for normal heart function. *Proceedings of the National Academy of Sciences of the United States of America, 108*(23), 9572–9577. https://doi.org/10.1073/pnas.1106291108

Blum-Degen, D., Muller, T., Kuhn, W., Gerlach, M., Przuntek, H., & Riederer, P. (1995). Interleukin-1 beta and interleukin-6 are elevated in the cerebrospinal fluid of Alzheimer's and de novo Parkinson's disease patients. *Neuroscience Letters, 202*(1–2), 17–20. https://doi.org/10.1016/0304-3940(95)12192-7

Burchell, V. S., Nelson, D. E., Sanchez-Martinez, A., Delgado-Camprubi, M., Ivatt, R. M., Pogson, J. H., Randle, S. J., Wray, S., Lewis, P. A., Houlden, H., Abramov, A. Y., Hardy, J., Wood, N. W., Whitworth, A. J., Laman, H., & Plun-Favreau, H. (2013). The Parkinson's disease-linked proteins Fbxo7 and Parkin interact to mediate mitophagy. *Nature Neuroscience, 16*(9), 1257–1265. https://doi.org/10.1038/nn.3489

Burdette, D. L., & Vance, R. E. (2013). STING and the innate immune response to nucleic acids in the cytosol. *Nature Immunology, 14*(1), 19–26. https://doi.org/10.1038/ni.2491

Cai, Q., & Jeong, Y. Y. (2020). Mitophagy in Alzheimer's disease and other age-related neurodegenerative diseases. *Cells, 9*(1). https://doi.org/10.3390/cells9010150

Cai, Q., Zakaria, H. M., Simone, A., & Sheng, Z. H. (2012). Spatial parkin translocation and degradation of damaged mitochondria via mitophagy in live cortical neurons. *Current Biology, 22*(6), 545–552. https://doi.org/10.1016/j.cub.2012.02.005

Chan, N. C., Salazar, A. M., Pham, A. H., Sweredoski, M. J., Kolawa, N. J., Graham, R. L., Hess, S., & Chan, D. C. (2011). Broad activation of the ubiquitin-proteasome system by Parkin is critical for mitophagy. *Human Molecular Genetics, 20*(9), 1726–1737. https://doi.org/10.1093/hmg/ddr048

Chao, H., Lin, C., Zuo, Q., Liu, Y., Xiao, M., Xu, X., Li, Z., Bao, Z., Chen, H., You, Y., Kochanek, P. M., Yin, H., Liu, N., Kagan, V. E., Bayir, H., & Ji, J. (2019). Cardiolipin-dependent mitophagy guides outcome after traumatic brain injury. *The Journal of Neuroscience, 39*(10), 1930–1943. https://doi.org/10.1523/JNEUROSCI.3415-17.2018

Chin, R. M., Rakhit, R., Ditsworth, D., Wang, C., Bartholomeus, J., Liu, S., Mody, A., Laishu, A., Eastes, A., Tai, C., Kim, R. Y., Li, J., Hansberry, S., Khasnavis, S., Rafalski, V., Herendeen, D., Garda, V., Phung, J., de Roulet, D., Ordureau, A., Harper, J. W., Johnstone, S., Stohr, J., & Hertz, N. T. (2023). Pharmacological PINK1 activation ameliorates pathology in Parkinson's disease models. *bioRxiv*. https://doi.org/10.1101/2023.02.14.528378

Chinnadurai, G., Vijayalingam, S., & Gibson, S. B. (2008). BNIP3 subfamily BH3-only proteins: Mitochondrial stress sensors in normal and pathological functions. *Oncogene, 27*(Suppl 1), S114–S127. https://doi.org/10.1038/onc.2009.49

Choubey, V., Cagalinec, M., Liiv, J., Safiulina, D., Hickey, M. A., Kuum, M., Liiv, M., Anwar, T., Eskelinen, E. L., & Kaasik, A. (2014). BECN1 is involved in the initiation of mitophagy: It facilitates PARK2 translocation to mitochondria. *Autophagy, 10*(6), 1105–1119. https://doi.org/10.4161/auto.28615

Chu, C. T., Ji, J., Dagda, R. K., Jiang, J. F., Tyurina, Y. Y., Kapralov, A. A., Tyurin, V. A., Yanamala, N., Shrivastava, I. H., Mohammadyani, D., Wang, K. Z. Q., Zhu, J., Klein-Seetharaman, J., Balasubramanian, K., Amoscato, A. A., Borisenko, G., Huang, Z., Gusdon, A. M., Cheikhi, A., Steer, E. K., Wang, R., Baty, C., Watkins, S., Bahar, I., Bayir, H., & Kagan, V. E. (2013). Cardiolipin externalization to the outer mitochondrial membrane acts as an elimination signal for mitophagy in neuronal cells. *Nature Cell Biology, 15*(10), 1197–1205. https://doi.org/10.1038/ncb2837

Chung, S. Y., Kishinevsky, S., Mazzulli, J. R., Graziotto, J., Mrejeru, A., Mosharov, E. V., Puspita, L., Valiulahi, P., Sulzer, D., Milner, T. A., Taldone, T., Krainc, D., Studer, L., & Shim, J. W. (2016). Parkin and PINK1 patient iPSC-derived midbrain dopamine neurons exhibit mitochondrial dysfunction and alpha-synuclein accumulation. *Stem Cell Reports, 7*(4), 664–677. https://doi.org/10.1016/j.stemcr.2016.08.012

Cianfanelli, V., Nazio, F., & Cecconi, F. (2015). Connecting autophagy: AMBRA1 and its network of regulation. *Molecular & Cellular Oncology, 2*(1), e970059. https://doi.org/10.4161/23723548.2014.970059

Clague, M. J., & Urbe, S. (2006). Endocytosis: The DUB version. *Trends in Cell Biology, 16*(11), 551–559. https://doi.org/10.1016/j.tcb.2006.09.002

Clark, I. E., Dodson, M. W., Jiang, C., Cao, J. H., Huh, J. R., Seol, J. H., Yoo, S. J., Hay, B. A., & Guo, M. (2006). Drosophila pink1 is required for mitochondrial function and interacts genetically with parkin. *Nature, 441*(7097), 1162–1166. https://doi.org/10.1038/nature04779

Croteau, E., Castellano, C. A., Fortier, M., Bocti, C., Fulop, T., Paquet, N., & Cunnane, S. C. (2018). A cross-sectional comparison of brain glucose and ketone metabolism in cognitively healthy older adults, mild cognitive impairment and early Alzheimer's disease. *Experimental Gerontology, 107*, 18–26. https://doi.org/10.1016/j.exger.2017.07.004

Cummins, N., Tweedie, A., Zuryn, S., Bertran-Gonzalez, J., & Gotz, J. (2019). Disease-associated tau impairs mitophagy by inhibiting Parkin translocation to mitochondria. *The EMBO Journal, 38*(3). https://doi.org/10.15252/embj.201899360

Cunningham, C. N., Baughman, J. M., Phu, L., Tea, J. S., Yu, C., Coons, M., Kirkpatrick, D. S., Bingol, B., & Corn, J. E. (2015). USP30 and parkin homeostatically regulate atypical ubiquitin chains on mitochondria. *Nature Cell Biology, 17*(2), 160–169. https://doi.org/10.1038/ncb3097

Du, F., Yu, Q., Yan, S., Hu, G., Lue, L. F., Walker, D. G., Wu, L., Yan, S. F., Tieu, K., & Yan, S. S. (2017). PINK1 signalling rescues amyloid pathology and mitochondrial dysfunction in Alzheimer's disease. *Brain, 140*(12), 3233–3251. https://doi.org/10.1093/brain/awx258

Durcan, T. M., Tang, M. Y., Perusse, J. R., Dashti, E. A., Aguileta, M. A., McLelland, G. L., Gros, P., Shaler, T. A., Faubert, D., Coulombe, B., & Fon, E. A. (2014). USP8 regulates mitophagy by removing K6-linked ubiquitin conjugates from parkin. *The EMBO Journal, 33*(21), 2473–2491. https://doi.org/10.15252/embj.201489729

Erskine, D., Koss, D., Korolchuk, V. I., Outeiro, T. F., Attems, J., & McKeith, I. (2021). Lipids, lysosomes and mitochondria: Insights into Lewy body formation from rare monogenic disorders. *Acta Neuropathologica, 141*(4), 511–526. https://doi.org/10.1007/s00401-021-02266-7

Fan, J., Yang, X., Li, J., Shu, Z., Dai, J., Liu, X., Li, B., Jia, S., Kou, X., Yang, Y., & Chen, N. (2017). Spermidine coupled with exercise rescues skeletal muscle atrophy from D-gal-induced aging rats through enhanced autophagy and reduced apoptosis via AMPK-FOXO3a signal pathway. *Oncotarget, 8*(11), 17475–17490. https://doi.org/10.18632/oncotarget.15728

Fang, E. F., Lautrup, S., Hou, Y., Demarest, T. G., Croteau, D. L., Mattson, M. P., & Bohr, V. A. (2017). NAD(+) in aging: Molecular mechanisms and translational implications. *Trends in Molecular Medicine, 23*(10), 899–916. https://doi.org/10.1016/j.molmed.2017.08.001

Fang, E. F., Hou, Y., Palikaras, K., Adriaanse, B. A., Kerr, J. S., Yang, B., Lautrup, S., Hasan-Olive, M. M., Caponio, D., Dan, X., Rocktaschel, P., Croteau, D. L., Akbari, M., Greig, N. H., Fladby, T., Nilsen, H., Cader, M. Z., Mattson, M. P., Tavernarakis, N., & Bohr, V. A. (2019). Mitophagy inhibits amyloid-beta and tau pathology and reverses cognitive deficits in models of Alzheimer's disease. *Nature Neuroscience, 22*(3), 401–412. https://doi.org/10.1038/s41593-018-0332-9

Fiesel, F. C., & Springer, W. (2015). Disease relevance of phosphorylated ubiquitin (p-S65-Ub). *Autophagy, 11*(11), 2125–2126. https://doi.org/10.1080/15548627.2015.1091912

Fiesel, F. C., Moussaud-Lamodiere, E. L., Ando, M., & Springer, W. (2014). A specific subset of E2 ubiquitin-conjugating enzymes regulate Parkin activation and mitophagy differently. *Journal of Cell Science, 127*(Pt 16), 3488–3504. https://doi.org/10.1242/jcs.147520

Fiesel, F. C., Ando, M., Hudec, R., Hill, A. R., Castanedes-Casey, M., Caulfield, T. R., Moussaud-Lamodiere, E. L., Stankowski, J. N., Bauer, P. O., Lorenzo-Betancor, O., Ferrer, I., Arbelo, J. M., Siuda, J., Chen, L., Dawson, V. L., Dawson, T. M., Wszolek, Z. K., Ross, O. A., Dickson, D. W., & Springer, W. (2015). (Patho-)physiological relevance of PINK1-dependent ubiquitin phosphorylation. *EMBO Reports, 16*(9), 1114–1130. https://doi.org/10.15252/embr.201540514

Ganley, I. G., Whitworth, A. J., & McWilliams, T. G. (2021). Comment on "mt-Keima detects PINK1-PRKN mitophagy in vivo with greater sensitivity than mito-QC". *Autophagy, 17*(12), 4477–4479. https://doi.org/10.1080/15548627.2021.1907269

Geisler, S., Vollmer, S., Golombek, S., & Kahle, P. J. (2014). The ubiquitin-conjugating enzymes UBE2N, UBE2L3 and UBE2D2/3 are essential for Parkin-dependent mitophagy. *Journal of Cell Science, 127*(Pt 15), 3280–3293. https://doi.org/10.1242/jcs.146035

Gerhard, A., Pavese, N., Hotton, G., Turkheimer, F., Es, M., Hammers, A., Eggert, K., Oertel, W., Banati, R. B., & Brooks, D. J. (2006). In vivo imaging of microglial activation with [11C](R)-PK11195 PET in idiopathic Parkinson's disease. *Neurobiology of Disease, 21*(2), 404–412. https://doi.org/10.1016/j.nbd.2005.08.002

Goldberg, M. S., Fleming, S. M., Palacino, J. J., Cepeda, C., Lam, H. A., Bhatnagar, A., Meloni, E. G., Wu, N., Ackerson, L. C., Klapstein, G. J., Gajendiran, M., Roth, B. L., Chesselet, M. F., Maidment, N. T., Levine, M. S., & Shen, J. (2003). Parkin-deficient mice exhibit nigrostriatal deficits but not loss of dopaminergic neurons. *The Journal of Biological Chemistry, 278*(44), 43628–43635. https://doi.org/10.1074/jbc.M308947200

Goncalves, F. B., & Morais, V. A. (2021). PINK1: A bridge between mitochondria and Parkinson's disease. *Life (Basel), 11*(5). https://doi.org/10.3390/life11050371

Gong, B., Pan, Y., Vempati, P., Zhao, W., Knable, L., Ho, L., Wang, J., Sastre, M., Ono, K., Sauve, A. A., & Pasinetti, G. M. (2013). Nicotinamide riboside restores cognition through an upregulation of proliferator-activated receptor-gamma coactivator 1alpha regulated beta-secretase 1 degradation and mitochondrial gene expression in Alzheimer's mouse models. *Neurobiology of Aging, 34*(6), 1581–1588. https://doi.org/10.1016/j.neurobiolaging.2012.12.005

Gordon, R., Albornoz, E. A., Christie, D. C., Langley, M. R., Kumar, V., Mantovani, S., Robertson, A. A. B., Butler, M. S., Rowe, D. B., O'Neill, L. A., Kanthasamy, A. G., Schroder, K., Cooper, M. A., & Woodruff, T. M. (2018). Inflammasome inhibition prevents alpha-synuclein pathology and dopaminergic neurodegeneration in mice. *Science Translational Medicine, 10*(465). https://doi.org/10.1126/scitranslmed.aah4066

Guan, R., Zou, W., Dai, X., Yu, X., Liu, H., Chen, Q., & Teng, W. (2018). Mitophagy, a potential therapeutic target for stroke. *Journal of Biomedical Science, 25*(1), 87. https://doi.org/10.1186/s12929-018-0487-4

Gupta, V. K., Scheunemann, L., Eisenberg, T., Mertel, S., Bhukel, A., Koemans, T. S., Kramer, J. M., Liu, K. S., Schroeder, S., Stunnenberg, H. G., Sinner, F., Magnes, C., Pieber, T. R., Dipt, S., Fiala, A., Schenck, A., Schwaerzel, M., Madeo, F., & Sigrist, S. J. (2013). Restoring polyamines protects from age-induced memory impairment in an autophagy-dependent manner. *Nature Neuroscience, 16*(10), 1453–1460. https://doi.org/10.1038/nn.3512

Hamacher-Brady, A., Brady, N. R., Logue, S. E., Sayen, M. R., Jinno, M., Kirshenbaum, L. A., Gottlieb, R. A., & Gustafsson, A. B. (2007). Response to myocardial ischemia/reperfusion injury involves Bnip3 and autophagy. *Cell Death and Differentiation, 14*(1), 146–157. https://doi.org/10.1038/sj.cdd.4401936

Hasson, S. A., Kane, L. A., Yamano, K., Huang, C. H., Sliter, D. A., Buehler, E., Wang, C., Heman-Ackah, S. M., Hessa, T., Guha, R., Martin, S. E., & Youle, R. J. (2013). High-content genome-wide RNAi screens identify regulators of parkin upstream of mitophagy. *Nature, 504*(7479), 291–295. https://doi.org/10.1038/nature12748

Hertz, N. T., Berthet, A., Sos, M. L., Thorn, K. S., Burlingame, A. L., Nakamura, K., & Shokat, K. M. (2013). A neo-substrate that amplifies catalytic activity of Parkinson's-disease-related kinase PINK1. *Cell, 154*(4), 737–747. https://doi.org/10.1016/j.cell.2013.07.030

Hou, X., Fiesel, F. C., Truban, D., Castanedes Casey, M., Lin, W. L., Soto, A. I., Tacik, P., Rousseau, L. G., Diehl, N. N., Heckman, M. G., Lorenzo-Betancor, O., Ferrer, I., Arbelo, J. M., Steele, J. C., Farrer, M. J., Cornejo-Olivas, M., Torres, L., Mata, I. F., Graff-Radford, N. R., Wszolek, Z. K., Ross, O. A., Murray, M. E., Dickson, D. W., & Springer, W. (2018a). Age- and disease-dependent increase of the mitophagy marker phospho-ubiquitin in normal aging and Lewy body disease. *Autophagy, 14*(8), 1404–1418. https://doi.org/10.1080/15548627.2018.1461294

Hou, Y., Lautrup, S., Cordonnier, S., Wang, Y., Croteau, D. L., Zavala, E., Zhang, Y., Moritoh, K., O'Connell, J. F., Baptiste, B. A., Stevnsner, T. V., Mattson, M. P., & Bohr, V. A. (2018b). NAD(+) supplementation normalizes key Alzheimer's features and DNA damage responses in a new AD mouse model with introduced DNA repair deficiency. *Proceedings of the National Academy of Sciences of the United States of America, 115*(8), E1876–E1885. https://doi.org/10.1073/pnas.1718819115

Hou, X., Watzlawik, J. O., Cook, C., Liu, C. C., Kang, S. S., Lin, W. L., DeTure, M., Heckman, M. G., Diehl, N. N., Al-Shaikh, F. S. H., Walton, R. L., Ross, O. A., Melrose, H. L., Ertekin-Taner, N., Bu, G., Petrucelli, L., Fryer, J. D., Murray, M. E., Dickson, D. W., Fiesel, F. C., & Springer, W. (2020). Mitophagy alterations in Alzheimer's disease are associated with granulovacuolar degeneration and early tau pathology. *Alzheimer's & Dementia*. https://doi.org/10.1002/alz.12198

Hou, X., Chen, T. H., Koga, S., Bredenberg, J. M., Faroqi, A. H., Delenclos, M., Bu, G., Wszolek, Z. K., Carr, J. A., Ross, O. A., McLean, P. J., Murray, M. E., Dickson, D. W., Fiesel, F. C., & Springer, W. (2023). Alpha-synuclein-associated changes in PINK1-PRKN-mediated mitophagy are disease context dependent. *Brain Pathology, 33*(5), e13175. https://doi.org/10.1111/bpa.13175

Hsieh, C. H., Shaltouki, A., Gonzalez, A. E., Bettencourt da Cruz, A., Burbulla, L. F., St Lawrence, E., Schule, B., Krainc, D., Palmer, T. D., & Wang, X. (2016). Functional impairment in miro degradation and mitophagy is a shared feature in familial and sporadic Parkinson's disease. *Cell Stem Cell, 19*(6), 709–724. https://doi.org/10.1016/j.stem.2016.08.002

Hwang, S., Disatnik, M. H., & Mochly-Rosen, D. (2015). Impaired GAPDH-induced mitophagy contributes to the pathology of Huntington's disease. *EMBO Molecular Medicine, 7*(10), 1307–1326. https://doi.org/10.15252/emmm.201505256

Itier, J. M., Ibanez, P., Mena, M. A., Abbas, N., Cohen-Salmon, C., Bohme, G. A., Laville, M., Pratt, J., Corti, O., Pradier, L., Ret, G., Joubert, C., Periquet, M., Araujo, F., Negroni, J., Casarejos, M. J., Canals, S., Solano, R., Serrano, A., Gallego, E., Sanchez, M., Denefle, P., Benavides, J., Tremp, G., Rooney, T. A., Brice, A., & Garcia de Yebenes, J. (2003). Parkin gene inactivation alters behaviour and dopamine neurotransmission in the mouse. *Human Molecular Genetics, 12*(18), 2277–2291. https://doi.org/10.1093/hmg/ddg239

Joselin, A. P., Hewitt, S. J., Callaghan, S. M., Kim, R. H., Chung, Y. H., Mak, T. W., Shen, J., Slack, R. S., & Park, D. S. (2012). ROS-dependent regulation of Parkin and DJ-1 localization during oxidative stress in neurons. *Human Molecular Genetics, 21*(22), 4888–4903. https://doi.org/10.1093/hmg/dds325

Kageyama, Y., Zhang, Z., Roda, R., Fukaya, M., Wakabayashi, J., Wakabayashi, N., Kensler, T. W., Reddy, P. H., Iijima, M., & Sesaki, H. (2012). Mitochondrial division ensures the survival of postmitotic neurons by suppressing oxidative damage. *The Journal of Cell Biology, 197*(4), 535–551. https://doi.org/10.1083/jcb.201110034

Kane, L. A., Lazarou, M., Fogel, A. I., Li, Y., Yamano, K., Sarraf, S. A., Banerjee, S., & Youle, R. J. (2014). PINK1 phosphorylates ubiquitin to activate Parkin E3 ubiquitin ligase activity. *The Journal of Cell Biology, 205*(2), 143–153. https://doi.org/10.1083/jcb.201402104

Kapogiannis, D., & Mattson, M. P. (2011). Disrupted energy metabolism and neuronal circuit dysfunction in cognitive impairment and Alzheimer's disease. *Lancet Neurology, 10*(2), 187–198. https://doi.org/10.1016/S1474-4422(10)70277-5

Katayama, H., Kogure, T., Mizushima, N., Yoshimori, T., & Miyawaki, A. (2011). A sensitive and quantitative technique for detecting autophagic events based on lysosomal delivery. *Chemistry & Biology, 18*(8), 1042–1052. https://doi.org/10.1016/j.chembiol.2011.05.013

Katayama, H., Hama, H., Nagasawa, K., Kurokawa, H., Sugiyama, M., Ando, R., Funata, M., Yoshida, N., Homma, M., Nishimura, T., Takahashi, M., Ishida, Y., Hioki, H., Tsujihata, Y., & Miyawaki, A. (2020). Visualizing and modulating mitophagy for therapeutic studies of neurodegeneration. *Cell, 181*(5), 1176–1187 e1116. https://doi.org/10.1016/j.cell.2020.04.025

Kazlauskaite, A., Kondapalli, C., Gourlay, R., Campbell, D. G., Ritorto, M. S., Hofmann, K., Alessi, D. R., Knebel, A., Trost, M., & Muqit, M. M. (2014). Parkin is activated by PINK1-dependent phosphorylation of ubiquitin at Ser65. *The Biochemical Journal, 460*(1), 127–139. https://doi.org/10.1042/BJ20140334

Kerr, J. S., Adriaanse, B. A., Greig, N. H., Mattson, M. P., Cader, M. Z., Bohr, V. A., & Fang, E. F. (2017). Mitophagy and Alzheimer's disease: Cellular and molecular mechanisms. *Trends in Neurosciences, 40*(3), 151–166. https://doi.org/10.1016/j.tins.2017.01.002

Kitada, T., Pisani, A., Porter, D. R., Yamaguchi, H., Tscherter, A., Martella, G., Bonsi, P., Zhang, C., Pothos, E. N., & Shen, J. (2007). Impaired dopamine release and synaptic plasticity in the striatum of PINK1-deficient mice. *Proceedings of the National Academy of Sciences of the United States of America, 104*(27), 11441–11446. https://doi.org/10.1073/pnas.0702717104

Kitada, T., Tong, Y., Gautier, C. A., & Shen, J. (2009). Absence of nigral degeneration in aged parkin/DJ-1/PINK1 triple knockout mice. *Journal of Neurochemistry, 111*(3), 696–702. https://doi.org/10.1111/j.1471-4159.2009.06350.x

Kondapalli, C., Kazlauskaite, A., Zhang, N., Woodroof, H. I., Campbell, D. G., Gourlay, R., Burchell, L., Walden, H., Macartney, T. J., Deak, M., Knebel, A., Alessi, D. R., & Muqit, M. M. (2012). PINK1 is activated by mitochondrial membrane potential depolarization and stimulates Parkin E3 ligase activity by phosphorylating Serine 65. *Open Biology, 2*(5), 120080. https://doi.org/10.1098/rsob.120080

Koyano, F., Okatsu, K., Kosako, H., Tamura, Y., Go, E., Kimura, M., Kimura, Y., Tsuchiya, H., Yoshihara, H., Hirokawa, T., Endo, T., Fon, E. A., Trempe, J. F., Saeki, Y., Tanaka, K., & Matsuda, N. (2014). Ubiquitin is phosphorylated by PINK1 to activate parkin. *Nature, 510*(7503), 162–166. https://doi.org/10.1038/nature13392

Krebs, J. J., Hauser, H., & Carafoli, E. (1979). Asymmetric distribution of phospholipids in the inner membrane of beef heart mitochondria. *The Journal of Biological Chemistry, 254*(12), 5308–5316. Retrieved from https://www.ncbi.nlm.nih.gov/pubmed/447651

Lambourne, O. A., Bell, S., Wilhelm, L. P., Yarbrough, E. B., Holly, G. G., Russell, O. M., Alghamdi, A. M., Fdel, A. M., Varricchio, C., Lane, E. L., Ganley, I. G., Jones, A. T., Goldberg, M. S., & Mehellou, Y. (2023). PINK1-Dependent mitophagy inhibits elevated ubiquitin phosphorylation caused by mitochondrial damage. *Journal of Medicinal Chemistry, 66*(11), 7645–7656. https://doi.org/10.1021/acs.jmedchem.3c00555

Lazarou, M., Jin, S. M., Kane, L. A., & Youle, R. J. (2012). Role of PINK1 binding to the TOM complex and alternate intracellular membranes in recruitment and activation of the E3 ligase Parkin. *Developmental Cell, 22*(2), 320–333. https://doi.org/10.1016/j.devcel.2011.12.014

Lee, S., Sterky, F. H., Mourier, A., Terzioglu, M., Cullheim, S., Olson, L., & Larsson, N. G. (2012). Mitofusin 2 is necessary for striatal axonal projections of midbrain dopamine neurons. *Human Molecular Genetics, 21*(22), 4827–4835. https://doi.org/10.1093/hmg/dds352

Lin, W., & Kang, U. J. (2008). Characterization of PINK1 processing, stability, and subcellular localization. *Journal of Neurochemistry, 106*(1), 464–474. https://doi.org/10.1111/j.1471-4159.2008.05398.x

Lindstrom, V., Gustafsson, G., Sanders, L. H., Howlett, E. H., Sigvardson, J., Kasrayan, A., Ingelsson, M., Bergstrom, J., & Erlandsson, A. (2017). Extensive uptake of alpha-synuclein oligomers in astrocytes results in sustained intracellular deposits and mitochondrial damage. *Molecular and Cellular Neurosciences, 82*, 143–156. https://doi.org/10.1016/j.mcn.2017.04.009

Liu, L., Feng, D., Chen, G., Chen, M., Zheng, Q., Song, P., Ma, Q., Zhu, C., Wang, R., Qi, W., Huang, L., Xue, P., Li, B., Wang, X., Jin, H., Wang, J., Yang, F., Liu, P., Zhu, Y., Sui, S., & Chen, Q. (2012). Mitochondrial outer-membrane protein FUNDC1 mediates hypoxia-induced mitophagy in mammalian cells. *Nature Cell Biology, 14*(2), 177–185. https://doi.org/10.1038/ncb2422

Liu, D., Pitta, M., Jiang, H., Lee, J. H., Zhang, G., Chen, X., Kawamoto, E. M., & Mattson, M. P. (2013). Nicotinamide forestalls pathology and cognitive decline in Alzheimer mice: Evidence for improved neuronal bioenergetics and autophagy procession. *Neurobiology of Aging, 34*(6), 1564–1580. https://doi.org/10.1016/j.neurobiolaging.2012.11.020

Liu, J., Liu, W., Li, R., & Yang, H. (2019). Mitophagy in Parkinson's disease: From pathogenesis to treatment. *Cells, 8*(7). https://doi.org/10.3390/cells8070712

Liu, Y. T., Sliter, D. A., Shammas, M. K., Huang, X., Wang, C., Calvelli, H., Maric, D. S., & Narendra, D. P. (2021a). Comment on "mt-Keima detects PINK1-PRKN mitophagy in vivo with greater sensitivity than mito-QC". *Autophagy, 17*(12), 4484–4485. https://doi.org/10.1080/15548627.2021.1909999

Liu, Y. T., Sliter, D. A., Shammas, M. K., Huang, X., Wang, C., Calvelli, H., Maric, D. S., & Narendra, D. P. (2021b). Mt-Keima detects PINK1-PRKN mitophagy in vivo with greater sensitivity than mito-QC. *Autophagy, 17*(11), 3753–3762. https://doi.org/10.1080/15548627.2021.1896924

Lou, G., Palikaras, K., Lautrup, S., Scheibye-Knudsen, M., Tavernarakis, N., & Fang, E. F. (2020). Mitophagy and neuroprotection. *Trends in Molecular Medicine, 26*(1), 8–20. https://doi.org/10.1016/j.molmed.2019.07.002

Lubbe, S. J., Bustos, B. I., Hu, J., Krainc, D., Joseph, T., Hehir, J., Tan, M., Zhang, W., Escott-Price, V., Williams, N. M., Blauwendraat, C., Singleton, A. B., Morris, H. R., & International Parkinson's Disease Genomics, C. (2021). Assessing the relationship between monoallelic PRKN mutations and Parkinson's risk. *Human Molecular Genetics, 30*(1), 78–86. https://doi.org/10.1093/hmg/ddaa273

Ma, J., Ni, H., Rui, Q., Liu, H., Jiang, F., Gao, R., Gao, Y., Li, D., & Chen, G. (2019). Potential roles of NIX/BNIP3L pathway in rat traumatic brain injury. *Cell Transplantation, 28*(5), 585–595. https://doi.org/10.1177/0963689719840353

Madeo, F., Eisenberg, T., Pietrocola, F., & Kroemer, G. (2018). Spermidine in health and disease. *Science, 359*(6374). https://doi.org/10.1126/science.aan2788

Mandal, A., & Drerup, C. M. (2019). Axonal transport and mitochondrial function in neurons. *Frontiers in Cellular Neuroscience, 13*, 373. https://doi.org/10.3389/fncel.2019.00373

Matheoud, D., Sugiura, A., Bellemare-Pelletier, A., Laplante, A., Rondeau, C., Chemali, M., Fazel, A., Bergeron, J. J., Trudeau, L. E., Burelle, Y., Gagnon, E., McBride, H. M., & Desjardins, M. (2016). Parkinson's disease-related proteins PINK1 and Parkin repress mitochondrial antigen presentation. *Cell, 166*(2), 314–327. https://doi.org/10.1016/j.cell.2016.05.039

Matheoud, D., Cannon, T., Voisin, A., Penttinen, A. M., Ramet, L., Fahmy, A. M., Ducrot, C., Laplante, A., Bourque, M. J., Zhu, L., Cayrol, R., Le Campion, A., McBride, H. M., Gruenheid, S., Trudeau, L. E., & Desjardins, M. (2019). Intestinal infection triggers Parkinson's disease-like symptoms in Pink1(−/−) mice. *Nature, 571*(7766), 565–569. https://doi.org/10.1038/s41586-019-1405-y

McWilliams, T. G., Prescott, A. R., Allen, G. F., Tamjar, J., Munson, M. J., Thomson, C., Muqit, M. M., & Ganley, I. G. (2016). mito-QC illuminates mitophagy and mitochondrial architecture in vivo. *The Journal of Cell Biology, 214*(3), 333–345. https://doi.org/10.1083/jcb.201603039

McWilliams, T. G., Barini, E., Pohjolan-Pirhonen, R., Brooks, S. P., Singh, F., Burel, S., Balk, K., Kumar, A., Montava-Garriga, L., Prescott, A. R., Hassoun, S. M., Mouton-Liger, F., Ball, G., Hills, R., Knebel, A., Ulusoy, A., Di Monte, D. A., Tamjar, J., Antico, O., Fears, K., Smith, L., Brambilla, R., Palin, E., Valori, M., Eerola-Rautio, J., Tienari, P., Corti, O., Dunnett, S. B., Ganley, I. G., Suomalainen, A., & Muqit, M. M. K. (2018a). Phosphorylation of Parkin at serine 65 is essential for its activation in vivo. *Open Biology, 8*(11). https://doi.org/10.1098/rsob.180108

McWilliams, T. G., Prescott, A. R., Montava-Garriga, L., Ball, G., Singh, F., Barini, E., Muqit, M. M. K., Brooks, S. P., & Ganley, I. G. (2018b). Basal mitophagy occurs independently of PINK1 in mouse tissues of high metabolic demand. *Cell Metabolism, 27*(2), 439–449 e435. https://doi.org/10.1016/j.cmet.2017.12.008

Murakawa, T., Yamaguchi, O., Hashimoto, A., Hikoso, S., Takeda, T., Oka, T., Yasui, H., Ueda, H., Akazawa, Y., Nakayama, H., Taneike, M., Misaka, T., Omiya, S., Shah, A. M., Yamamoto, A., Nishida, K., Ohsumi, Y., Okamoto, K., Sakata, Y., & Otsu, K. (2015). Bcl-2-like protein 13 is a mammalian Atg32 homologue that mediates mitophagy and mitochondrial fragmentation. *Nature Communications, 6*, 7527. https://doi.org/10.1038/ncomms8527

Nakhaei, P., Hiscott, J., & Lin, R. (2010). STING-ing the antiviral pathway. *Journal of Molecular Cell Biology, 2*(3), 110–112. https://doi.org/10.1093/jmcb/mjp048

Narendra, D., Walker, J. E., & Youle, R. (2012). Mitochondrial quality control mediated by PINK1 and Parkin: Links to parkinsonism. *Cold Spring Harbor Perspectives in Biology, 4*(11). https://doi.org/10.1101/cshperspect.a011338

Noda, S., Sato, S., Fukuda, T., Tada, N., Uchiyama, Y., Tanaka, K., & Hattori, N. (2020). Loss of Parkin contributes to mitochondrial turnover and dopaminergic neuronal loss in aged mice. *Neurobiology of Disease, 136*, 104717. https://doi.org/10.1016/j.nbd.2019.104717

Okatsu, K., Uno, M., Koyano, F., Go, E., Kimura, M., Oka, T., Tanaka, K., & Matsuda, N. (2013). A dimeric PINK1-containing complex on depolarized mitochondria stimulates Parkin recruitment. *The Journal of Biological Chemistry, 288*(51), 36372–36384. https://doi.org/10.1074/jbc.M113.509653

Onishi, M., Yamano, K., Sato, M., Matsuda, N., & Okamoto, K. (2021). Molecular mechanisms and physiological functions of mitophagy. *The EMBO Journal, 40*(3), e104705. https://doi.org/10.15252/embj.2020104705

Ordureau, A., Sarraf, S. A., Duda, D. M., Heo, J. M., Jedrychowski, M. P., Sviderskiy, V. O., Olszewski, J. L., Koerber, J. T., Xie, T., Beausoleil, S. A., Wells, J. A., Gygi, S. P., Schulman, B. A., & Harper, J. W. (2014). Quantitative proteomics reveal a feedforward mechanism for mitochondrial PARKIN translocation and ubiquitin chain synthesis. *Molecular Cell, 56*(3), 360–375. https://doi.org/10.1016/j.molcel.2014.09.007

Ordureau, A., Paulo, J. A., Zhang, W., Ahfeldt, T., Zhang, J., Cohn, E. F., Hou, Z., Heo, J. M., Rubin, L. L., Sidhu, S. S., Gygi, S. P., & Harper, J. W. (2018). Dynamics of PARKIN-Dependent mitochondrial ubiquitylation in induced neurons and model systems revealed by digital snapshot proteomics. *Molecular Cell, 70*(2), 211–227 e218. https://doi.org/10.1016/j.molcel.2018.03.012

Osgerby, L., Lai, Y. C., Thornton, P. J., Amalfitano, J., Le Duff, C. S., Jabeen, I., Kadri, H., Miccoli, A., Tucker, J. H. R., Muqit, M. M. K., & Mehellou, Y. (2017). Kinetin riboside and its ProTides activate the Parkinson's disease associated PTEN-Induced Putative Kinase 1 (PINK1) independent of mitochondrial depolarization. *Journal of Medicinal Chemistry, 60*(8), 3518–3524. https://doi.org/10.1021/acs.jmedchem.6b01897

Palacino, J. J., Sagi, D., Goldberg, M. S., Krauss, S., Motz, C., Wacker, M., Klose, J., & Shen, J. (2004). Mitochondrial dysfunction and oxidative damage in parkin-deficient mice. *The Journal of Biological Chemistry, 279*(18), 18614–18622. https://doi.org/10.1074/jbc.M401135200

Pang, S. Y., Lo, R. C. N., Ho, P. W., Liu, H. F., Chang, E. E. S., Leung, C. T., Malki, Y., Choi, Z. Y., Wong, W. Y., Kung, M. H., Ramsden, D. B., & Ho, S. L. (2022). LRRK2, GBA and their interaction in the regulation of autophagy: Implications on therapeutics in Parkinson's disease. *Translational Neurodegeneration, 11*(1), 5. https://doi.org/10.1186/s40035-022-00281-6

Paranjpe, M. D., Chen, X., Liu, M., Paranjpe, I., Leal, J. P., Wang, R., Pomper, M. G., Wong, D. F., Benzinger, T. L. S., Zhou, Y., & Alzheimer's Disease Neuroimaging, I. (2019). The effect of ApoE epsilon4 on longitudinal brain region-specific glucose metabolism in patients with mild cognitive impairment: A FDG-PET study. *Neuroimage Clinical, 22*, 101795. https://doi.org/10.1016/j.nicl.2019.101795

Pearce, S. F., Rebelo-Guiomar, P., D'Souza, A. R., Powell, C. A., Van Haute, L., & Minczuk, M. (2017). Regulation of mammalian mitochondrial gene expression: Recent advances. *Trends in Biochemical Sciences, 42*(8), 625–639. https://doi.org/10.1016/j.tibs.2017.02.003

Perez, F. A., & Palmiter, R. D. (2005). Parkin-deficient mice are not a robust model of parkinsonism. *Proceedings of the National Academy of Sciences of the United States of America, 102*(6), 2174–2179. https://doi.org/10.1073/pnas.0409598102

Pfanner, N., Warscheid, B., & Wiedemann, N. (2019). Mitochondrial proteins: From biogenesis to functional networks. *Nature Reviews. Molecular Cell Biology, 20*(5), 267–284. https://doi.org/10.1038/s41580-018-0092-0

Pickles, S., Vigie, P., & Youle, R. J. (2018). Mitophagy and quality control mechanisms in mitochondrial maintenance. *Current Biology, 28*(4), R170–R185. https://doi.org/10.1016/j.cub.2018.01.004

Pickrell, A. M., Huang, C. H., Kennedy, S. R., Ordureau, A., Sideris, D. P., Hoekstra, J. G., Harper, J. W., & Youle, R. J. (2015). Endogenous Parkin preserves dopaminergic substantia nigral neurons following mitochondrial DNA mutagenic stress. *Neuron, 87*(2), 371–381. https://doi.org/10.1016/j.neuron.2015.06.034

Rasool, S., Veyron, S., Soya, N., Eldeeb, M. A., Lukacs, G. L., Fon, E. A., & Trempe, J. F. (2022). Mechanism of PINK1 activation by autophosphorylation and insights into assembly on the TOM complex. *Molecular Cell, 82*(1), 44–59 e46. https://doi.org/10.1016/j.molcel.2021.11.012

Rodriguez-Nuevo, A., & Zorzano, A. (2019). The sensing of mitochondrial DAMPs by non-immune cells. *Cell Stress, 3*(6), 195–207. https://doi.org/10.15698/cst2019.06.190

Ryan, T. A., Phillips, E. O., Collier, C. L., Jb Robinson, A., Routledge, D., Wood, R. E., Assar, E. A., & Tumbarello, D. A. (2020). Tollip coordinates Parkin-dependent trafficking of mitochondrial-derived vesicles. *The EMBO Journal, 39*(11), e102539. https://doi.org/10.15252/embj.2019102539

Ryu, D., Mouchiroud, L., Andreux, P. A., Katsyuba, E., Moullan, N., Nicolet-Dit-Felix, A. A., Williams, E. G., Jha, P., Lo Sasso, G., Huzard, D., Aebischer, P., Sandi, C., Rinsch, C., & Auwerx, J. (2016). Urolithin A induces mitophagy and prolongs lifespan in C. elegans and increases muscle function in rodents. *Nature Medicine, 22*(8), 879–888. https://doi.org/10.1038/nm.4132

Saita, S., Shirane, M., & Nakayama, K. I. (2013). Selective escape of proteins from the mitochondria during mitophagy. *Nature Communications, 4*, 1410. https://doi.org/10.1038/ncomms2400

Saitoh, T., Fujita, N., Hayashi, T., Takahara, K., Satoh, T., Lee, H., Matsunaga, K., Kageyama, S., Omori, H., Noda, T., Yamamoto, N., Kawai, T., Ishii, K., Takeuchi, O., Yoshimori, T., & Akira, S. (2009). Atg9a controls dsDNA-driven dynamic translocation of STING and the innate immune response. *Proceedings of the National Academy of Sciences of the United States of America, 106*(49), 20842–20846. https://doi.org/10.1073/pnas.0911267106

Sanchez-Guajardo, V., Barnum, C. J., Tansey, M. G., & Romero-Ramos, M. (2013). Neuroimmunological processes in Parkinson's disease and their relation to alpha-synuclein: Microglia as the referee between neuronal processes and peripheral immunity. *ASN Neuro, 5*(2), 113–139. https://doi.org/10.1042/AN20120066

Sandoval, H., Thiagarajan, P., Dasgupta, S. K., Schumacher, A., Prchal, J. T., Chen, M., & Wang, J. (2008). Essential role for Nix in autophagic maturation of erythroid cells. *Nature, 454*(7201), 232–235. https://doi.org/10.1038/nature07006

Saraiva, A. A., Borges, M. M., Madeira, M. D., Tavares, M. A., & Paula-Barbosa, M. M. (1985). Mitochondrial abnormalities in cortical dendrites from patients with Alzheimer's disease. *Journal of Submicroscopic Cytology, 17*(3), 459–464. Retrieved from https://www.ncbi.nlm.nih.gov/pubmed/2991550

Sarraf, S. A., Raman, M., Guarani-Pereira, V., Sowa, M. E., Huttlin, E. L., Gygi, S. P., & Harper, J. W. (2013). Landscape of the PARKIN-dependent ubiquitylome in response to mitochondrial depolarization. *Nature, 496*(7445), 372–376. https://doi.org/10.1038/nature12043

Schondorf, D. C., Ivanyuk, D., Baden, P., Sanchez-Martinez, A., De Cicco, S., Yu, C., Giunta, I., Schwarz, L. K., Di Napoli, G., Panagiotakopoulou, V., Nestel, S., Keatinge, M., Pruszak, J., Bandmann, O., Heimrich, B., Gasser, T., Whitworth, A. J., & Deleidi, M. (2018). The NAD+ precursor nicotinamide riboside rescues mitochondrial defects and neuronal loss in iPSC and fly models of Parkinson's disease. *Cell Reports, 23*(10), 2976–2988. https://doi.org/10.1016/j.celrep.2018.05.009

Schweers, R. L., Zhang, J., Randall, M. S., Loyd, M. R., Li, W., Dorsey, F. C., Kundu, M., Opferman, J. T., Cleveland, J. L., Miller, J. L., & Ney, P. A. (2007). NIX is required for programmed mitochondrial clearance during reticulocyte maturation. *Proceedings of the National Academy of Sciences of the United States of America, 104*(49), 19500–19505. https://doi.org/10.1073/pnas.0708818104

Scott, L., Karuppagounder, S. S., Neifert, S., Kang, B. G., Wang, H., Dawson, V. L., & Dawson, T. M. (2022). The absence of Parkin does not promote dopamine or mitochondrial dysfunction in PolgA(D257A/D257A) mitochondrial mutator mice. *The Journal of Neuroscience, 42*(49), 9263–9277. https://doi.org/10.1523/JNEUROSCI.0545-22.2022

Shahmoradian, S. H., Lewis, A. J., Genoud, C., Hench, J., Moors, T. E., Navarro, P. P., Castano-Diez, D., Schweighauser, G., Graff-Meyer, A., Goldie, K. N., Sutterlin, R., Huisman, E., Ingrassia, A., Gier, Y., Rozemuller, A. J. M., Wang, J., Paepe, A., Erny, J., Staempfli, A., Hoernschemeyer, J., Grosseruschkamp, F., Niedieker, D., El-Mashtoly, S. F., Quadri, M., Van, I. W. F. J., Bonifati, V., Gerwert, K., Bohrmann, B., Frank, S., Britschgi, M., Stahlberg, H., Van de Berg, W. D. J., & Lauer, M. E. (2019). Lewy pathology in Parkinson's disease consists of crowded organelles and lipid membranes. *Nature Neuroscience, 22*(7), 1099–1109. https://doi.org/10.1038/s41593-019-0423-2

Sheu, K. F., Kim, Y. T., Blass, J. P., & Weksler, M. E. (1985). An immunochemical study of the pyruvate dehydrogenase deficit in Alzheimer's disease brain. *Annals of Neurology, 17*(5), 444–449. https://doi.org/10.1002/ana.410170505

Shi, R. Y., Zhu, S. H., Li, V., Gibson, S. B., Xu, X. S., & Kong, J. M. (2014). BNIP3 interacting with LC3 triggers excessive mitophagy in delayed neuronal death in stroke. *CNS Neuroscience & Therapeutics, 20*(12), 1045–1055. https://doi.org/10.1111/cns.12325

Shiba-Fukushima, K., Imai, Y., Yoshida, S., Ishihama, Y., Kanao, T., Sato, S., & Hattori, N. (2012). PINK1-mediated phosphorylation of the Parkin ubiquitin-like domain primes mitochondrial translocation of Parkin and regulates mitophagy. *Scientific Reports, 2*, 1002. https://doi.org/10.1038/srep01002

Shiba-Fukushima, K., Ishikawa, K. I., Inoshita, T., Izawa, N., Takanashi, M., Sato, S., Onodera, O., Akamatsu, W., Okano, H., Imai, Y., & Hattori, N. (2017). Evidence that phosphorylated ubiquitin signaling is involved in the etiology of Parkinson's disease. *Human Molecular Genetics, 26*(16), 3172–3185. https://doi.org/10.1093/hmg/ddx201

Shin, J. H., Ko, H. S., Kang, H., Lee, Y., Lee, Y. I., Pletinkova, O., Troconso, J. C., Dawson, V. L., & Dawson, T. M. (2011). PARIS (ZNF746) repression of PGC-1alpha contributes to neurodegeneration in Parkinson's disease. *Cell, 144*(5), 689–702. https://doi.org/10.1016/j.cell.2011.02.010

Shlevkov, E., Murugan, P., Montagna, D., Stefan, E., Hadzipasic, A., Harvey, J. S., Kumar, P. R., Entova, S., Bansal, N., Bickford, S., Wong, L. Y., Hirst, W. D., Weihofen, A., & Silvian, L. F. (2022). Discovery of small-molecule positive allosteric modulators of Parkin E3 ligase. *iScience, 25*(1), 103650. https://doi.org/10.1016/j.isci.2021.103650

Sliter, D. A., Martinez, J., Hao, L., Chen, X., Sun, N., Fischer, T. D., Burman, J. L., Li, Y., Zhang, Z., Narendra, D. P., Cai, H., Borsche, M., Klein, C., & Youle, R. J. (2018). Parkin and PINK1 mitigate STING-induced inflammation. *Nature, 561*(7722), 258–262. https://doi.org/10.1038/s41586-018-0448-9

Song, L., McMackin, M., Nguyen, A., & Cortopassi, G. (2017). Parkin deficiency accelerates consequences of mitochondrial DNA deletions and Parkinsonism. *Neurobiology of Disease, 100*, 30–38. https://doi.org/10.1016/j.nbd.2016.12.024

Sorbi, S., Bird, E. D., & Blass, J. P. (1983). Decreased pyruvate dehydrogenase complex activity in Huntington and Alzheimer brain. *Annals of Neurology, 13*(1), 72–78. https://doi.org/10.1002/ana.410130116

Stavoe, A. K. H., & Holzbaur, E. L. F. (2019). Autophagy in neurons. *Annual Review of Cell and Developmental Biology, 35*, 477–500. https://doi.org/10.1146/annurev-cellbio-100818-125242

Sterky, F. H., Lee, S., Wibom, R., Olson, L., & Larsson, N. G. (2011). Impaired mitochondrial transport and Parkin-independent degeneration of respiratory chain-deficient dopamine neurons in vivo. *Proceedings of the National Academy of Sciences of the United States of America, 108*(31), 12937–12942. https://doi.org/10.1073/pnas.1103295108

Strappazzon, F., Nazio, F., Corrado, M., Cianfanelli, V., Romagnoli, A., Fimia, G. M., Campello, S., Nardacci, R., Piacentini, M., Campanella, M., & Cecconi, F. (2015). AMBRA1 is able to induce mitophagy via LC3 binding, regardless of PARKIN and p62/SQSTM1. *Cell Death and Differentiation, 22*(3), 517. https://doi.org/10.1038/cdd.2014.190

Sugiura, A., McLelland, G. L., Fon, E. A., & McBride, H. M. (2014). A new pathway for mitochondrial quality control: Mitochondrial-derived vesicles. *The EMBO Journal, 33*(19), 2142–2156. https://doi.org/10.15252/embj.201488104

Sun, N., Yun, J., Liu, J., Malide, D., Liu, C., Rovira, I. I., Holmstrom, K. M., Fergusson, M. M., Yoo, Y. H., Combs, C. A., & Finkel, T. (2015). Measuring in vivo mitophagy. *Molecular Cell, 60*(4), 685–696. https://doi.org/10.1016/j.molcel.2015.10.009

Swanson, K. V., Deng, M., & Ting, J. P. (2019). The NLRP3 inflammasome: Molecular activation and regulation to therapeutics. *Nature Reviews. Immunology, 19*(8), 477–489. https://doi.org/10.1038/s41577-019-0165-0

Tanida, I., Ueno, T., & Kominami, E. (2008). LC3 and autophagy. *Methods in Molecular Biology, 445*, 77–88. https://doi.org/10.1007/978-1-59745-157-4_4

Tansey, M. G., Wallings, R. L., Houser, M. C., Herrick, M. K., Keating, C. E., & Joers, V. (2022). Inflammation and immune dysfunction in Parkinson disease. *Nature Reviews. Immunology, 22*(11), 657–673. https://doi.org/10.1038/s41577-022-00684-6

Todkar, K., Chikhi, L., Desjardins, V., El-Mortada, F., Pepin, G., & Germain, M. (2021). Selective packaging of mitochondrial proteins into extracellular vesicles prevents the release of mitochondrial DAMPs. *Nature Communications, 12*(1), 1971. https://doi.org/10.1038/s41467-021-21984-w

Tsefou, E., Walker, A. S., Clark, E. H., Hicks, A. R., Luft, C., Takeda, K., Watanabe, T., Ramazio, B., Staddon, J. M., Briston, T., & Ketteler, R. (2021). Investigation of USP30 inhibition to enhance Parkin-mediated mitophagy: Tools and approaches. *The Biochemical Journal, 478*(23), 4099–4118. https://doi.org/10.1042/BCJ20210508

Tufi, R., Clark, E. H., Hoshikawa, T., Tsagkaraki, C., Stanley, J., Takeda, K., Staddon, J. M., & Briston, T. (2023). High-content phenotypic screen to identify small molecule enhancers of Parkin-dependent ubiquitination and mitophagy. *SLAS Discovery, 28*(3), 73–87. https://doi.org/10.1016/j.slasd.2022.12.004

Turunc Bayrakdar, E., Uyanikgil, Y., Kanit, L., Koylu, E., & Yalcin, A. (2014). Nicotinamide treatment reduces the levels of oxidative stress, apoptosis, and PARP-1 activity in Abeta(1-42)-induced rat model of Alzheimer's disease. *Free Radical Research, 48*(2), 146–158. https://doi.org/10.3109/10715762.2013.857018

Van Humbeeck, C., Cornelissen, T., Hofkens, H., Mandemakers, W., Gevaert, K., De Strooper, B., & Vandenberghe, W. (2011a). Parkin interacts with Ambra1 to induce mitophagy. *The Journal of Neuroscience, 31*(28), 10249–10261. https://doi.org/10.1523/JNEUROSCI.1917-11.2011

Van Humbeeck, C., Cornelissen, T., & Vandenberghe, W. (2011b). Ambra1: A Parkin-binding protein involved in mitophagy. *Autophagy, 7*(12), 1555–1556. https://doi.org/10.4161/auto.7.12.17893

Vercellino, I., & Sazanov, L. A. (2022). The assembly, regulation and function of the mitochondrial respiratory chain. *Nature Reviews. Molecular Cell Biology, 23*(2), 141–161. https://doi.org/10.1038/s41580-021-00415-0

Von Coelln, R., Thomas, B., Savitt, J. M., Lim, K. L., Sasaki, M., Hess, E. J., Dawson, V. L., & Dawson, T. M. (2004). Loss of locus coeruleus neurons and reduced startle in parkin null mice. *Proceedings of the National Academy of Sciences of the United States of America, 101*(29), 10744–10749. https://doi.org/10.1073/pnas.0401297101

Wall, C. E., Rose, C. M., Adrian, M., Zeng, Y. J., Kirkpatrick, D. S., & Bingol, B. (2019). PPEF2 opposes PINK1-Mediated mitochondrial quality control by dephosphorylating ubiquitin. *Cell Reports, 29*(10), 3280–3292 e3287. https://doi.org/10.1016/j.celrep.2019.10.130

Wallace, D. C. (2005). A mitochondrial paradigm of metabolic and degenerative diseases, aging, and cancer: A dawn for evolutionary medicine. *Annual Review of Genetics, 39*, 359–407. https://doi.org/10.1146/annurev.genet.39.110304.095751

Wang, L., Cho, Y. L., Tang, Y., Wang, J., Park, J. E., Wu, Y., Wang, C., Tong, Y., Chawla, R., Zhang, J., Shi, Y., Deng, S., Lu, G., Wu, Y., Tan, H. W., Pawijit, P., Lim, G. G., Chan, H. Y., Zhang, J., Fang, L., Yu, H., Liou, Y. C., Karthik, M., Bay, B. H., Lim, K. L., Sze, S. K., Yap, C. T., & Shen, H. M. (2018a). PTEN-L is a novel protein phosphatase for ubiquitin dephosphorylation to inhibit PINK1-Parkin-mediated mitophagy. *Cell Research, 28*(8), 787–802. https://doi.org/10.1038/s41422-018-0056-0

Wang, L., Wang, J., Tang, Y., & Shen, H. M. (2018b). PTEN-L puts a brake on mitophagy. *Autophagy, 14*(11), 2023–2025. https://doi.org/10.1080/15548627.2018.1502565

Watzlawik, J. O., Hou, X., Fricova, D., Ramnarine, C., Barodia, S. K., Gendron, T. F., Heckman, M. G., DeTure, M., Siuda, J., Wszolek, Z. K., Scherzer, C. R., Ross, O. A., Bu, G., Dickson, D. W., Goldberg, M. S., Fiesel, F. C., & Springer, W. (2021). Sensitive ELISA-based detection method for the mitophagy marker p-S65-Ub in human cells, autopsy brain, and blood samples. *Autophagy, 17*(9), 2613–2628. https://doi.org/10.1080/15548627.2020.1834712

Watzlawik, J. O., Fiesel, F. C., Fiorino, G., Bustillos, B. A., Baninameh, Z., Markham, B. N., Hou, X., Hayes, C. S., Bredenberg, J. M., Kurchaba, N. W., Fričová, D., Siuda, J., Wszolek, Z. K., Noda, S., Sato, S., Hattori, N., Prasad, A. A., Kirik, D., Fox, H. S., Stauch, K. L., Goldberg, M. S., Springer, W. (2024). Basal activity of PINK1 and PRKN in cell models and rodent brain. *Autophagy, 20*(5): 1147–1158. https://doi.org/10.1080/15548627.2023.2286414.

Wenzel, D. M., Lissounov, A., Brzovic, P. S., & Klevit, R. E. (2011). UBCH7 reactivity profile reveals parkin and HHARI to be RING/HECT hybrids. *Nature, 474*(7349), 105–108. https://doi.org/10.1038/nature09966

Xu, J., Zhong, N., Wang, H., Elias, J. E., Kim, C. Y., Woldman, I., Pifl, C., Gygi, S. P., Geula, C., & Yankner, B. A. (2005). The Parkinson's disease-associated DJ-1 protein is a transcriptional co-activator that protects against neuronal apoptosis. *Human Molecular Genetics, 14*(9), 1231–1241. https://doi.org/10.1093/hmg/ddi134

Yamano, K., & Youle, R. J. (2013). PINK1 is degraded through the N-end rule pathway. *Autophagy, 9*(11), 1758–1769. https://doi.org/10.4161/auto.24633

Yogalingam, G., Hwang, S., Ferreira, J. C., & Mochly-Rosen, D. (2013). Glyceraldehyde-3-phosphate dehydrogenase (GAPDH) phosphorylation by protein kinase Cdelta (PKCdelta) inhibits mitochondria elimination by lysosomal-like structures following ischemia and reoxygenation-induced injury. *The Journal of Biological Chemistry, 288*(26), 18947–18960. https://doi.org/10.1074/jbc.M113.466870

Youle, R. J., & Narendra, D. P. (2011). Mechanisms of mitophagy. *Nature Reviews. Molecular Cell Biology, 12*(1), 9–14. https://doi.org/10.1038/nrm3028

Youle, R. J., & van der Bliek, A. M. (2012). Mitochondrial fission, fusion, and stress. *Science, 337*(6098), 1062–1065. https://doi.org/10.1126/science.1219855

Yu, D., Li, M., Ni, B., Kong, J., & Zhang, Z. (2013). Induction of neuronal mitophagy in acute spinal cord injury in rats. *Neurotoxicity Research, 24*(4), 512–522. https://doi.org/10.1007/s12640-013-9397-0

Yuan, Y., Zheng, Y., Zhang, X., Chen, Y., Wu, X., Wu, J., Shen, Z., Jiang, L., Wang, L., Yang, W., Luo, J., Qin, Z., Hu, W., & Chen, Z. (2017). BNIP3L/NIX-mediated mitophagy protects against ischemic brain injury independent of PARK2. *Autophagy, 13*(10), 1754–1766. https://doi.org/10.1080/15548627.2017.1357792

Yussman, M. G., Toyokawa, T., Odley, A., Lynch, R. A., Wu, G., Colbert, M. C., Aronow, B. J., Lorenz, J. N., & Dorn, G. W., 2nd. (2002). Mitochondrial death protein Nix is induced in cardiac hypertrophy and triggers apoptotic cardiomyopathy. *Nature Medicine, 8*(7), 725–730. https://doi.org/10.1038/nm719

Zhang, T., Xue, L., Li, L., Tang, C., Wan, Z., Wang, R., Tan, J., Tan, Y., Han, H., Tian, R., Billiar, T. R., Tao, W. A., & Zhang, Z. (2016). BNIP3 protein suppresses PINK1 kinase proteolytic cleavage to promote mitophagy. *The Journal of Biological Chemistry, 291*(41), 21616–21629. https://doi.org/10.1074/jbc.M116.733410

Zhu, J. H., Guo, F., Shelburne, J., Watkins, S., & Chu, C. T. (2003). Localization of phosphorylated ERK/MAP kinases to mitochondria and autophagosomes in Lewy body diseases. *Brain Pathology, 13*(4), 473–481. https://doi.org/10.1111/j.1750-3639.2003.tb00478.x

Zhu, M., Huang, X., Shan, H., & Zhang, M. (2022). Mitophagy in traumatic brain injury: A new target for therapeutic intervention. *Oxidative Medicine and Cellular Longevity, 2022*, 4906434. https://doi.org/10.1155/2022/4906434

Neurodevelopment and Drugs

10

Claire E. Wing, Emma G. Foster, Micah Summerlin, and Aditya N. Bade

Abstract

Neurodevelopment is a continuous, complex, and strictly regulated biological process that begins early after conception during embryonic life and continues into young adulthood. However, fetal brain development during gestation is a window of high vulnerability. Diverse coordinated biological processes occur during this critical period including neurulation, neurogenesis, microglial entry, synaptogenesis and pruning, programmed cell death, and myelination. Disturbances to these processes can produce short- or long-term detrimental effects on brain development. These include chronic forms of congenital malformations or long-term postnatal neurodevelopmental disorders. Congenital malformations include neural tube defects (NTDs) in which the development of the brain or the spinal cord structure is impaired. These malformations can be fatal. Moreover, impairment in biological processes can lead to a wide range of long-term functional disorders called neurodevelopmental disorders. These are a group of disorders characterized by impairments in developmental domains including cognition, communication, behavior, or motor skills. Genetic factors or environmental exposures during gestation have been linked with neurodevelopmental deficits. In this chapter, cellular and molecular pathways of neurogenesis are described in addition to characterizations of NTDs and postnatal neurodevelopmental disorders. Finally, an update on neurodevelopmental outcomes of prenatal exposure to drugs of abuse, prescribed medications, or environmental factors is provided.

Keywords

Pregnancy · Fetal neurodevelopment · Neural tube defects · Neurodevelopmental disorders · Drug-induced adverse events

10.1 Introduction

The fetal brain undergoes critical stages of development starting from the time of conception. Rapid structural changes occur over time during early brain development (Vasung et al., 2019). Biological, metabolic, and physiological processes critical for brain development are strictly regulated. Insults to these processes can have mild to severe adverse effects (Fig. 10.1). Neurodevelopment begins with the closure of the neural tube at 4 weeks post-fertilization. This mechanism is called neurulation (Schnoll et al., 2019; Tau & Peterson, 2010). After neurulation at approximately gestational week 4, neurogenesis begins. This process involves the differentiation of neural stem cells into neurons. Neurogenesis continues throughout gestation and is recognized in some brain regions into adulthood (Schnoll et al., 2019). Shortly after the beginning of neurogenesis, microglial cells enter the fetal central nervous system (CNS) between gestation weeks 4–16. In this process, microglia establish a resident immune cell population (Monier et al., 2006; Schafer & Stevens, 2015). At gestational week 12, synaptogenesis begins. In parallel, the pruning of excess synapses and dendritic spines is noted. At 18 weeks, cell populations undergo refinement through apoptosis to ensure the formation of proper synaptic connectivity (Schnoll et al., 2019; Tau & Peterson, 2010). Finally, at approximately gestational week 30, myelination of neurons begins and continues into adulthood (Budday et al., 2015; Schnoll et al., 2019).

These neurodevelopmental processes are highly susceptible to external modulators such as drugs, viruses, and inflammatory molecules. Even maternal health is important for healthy neurodevelopment. Any insults to the neurode-

C. E. Wing
Department of Biology, University of Nebraska, Omaha, NE, USA

E. G. Foster · M. Summerlin · A. N. Bade (✉)
Department of Pharmacology and Experimental Neuroscience, University of Nebraska Medical Center, Omaha, NE, USA
e-mail: aditya.bade@unmc.edu

H. E. Gendelman, T. Ikezu (eds.), *Neuroimmune Pharmacology and Therapeutics*,
https://doi.org/10.1007/978-3-031-68237-7_10

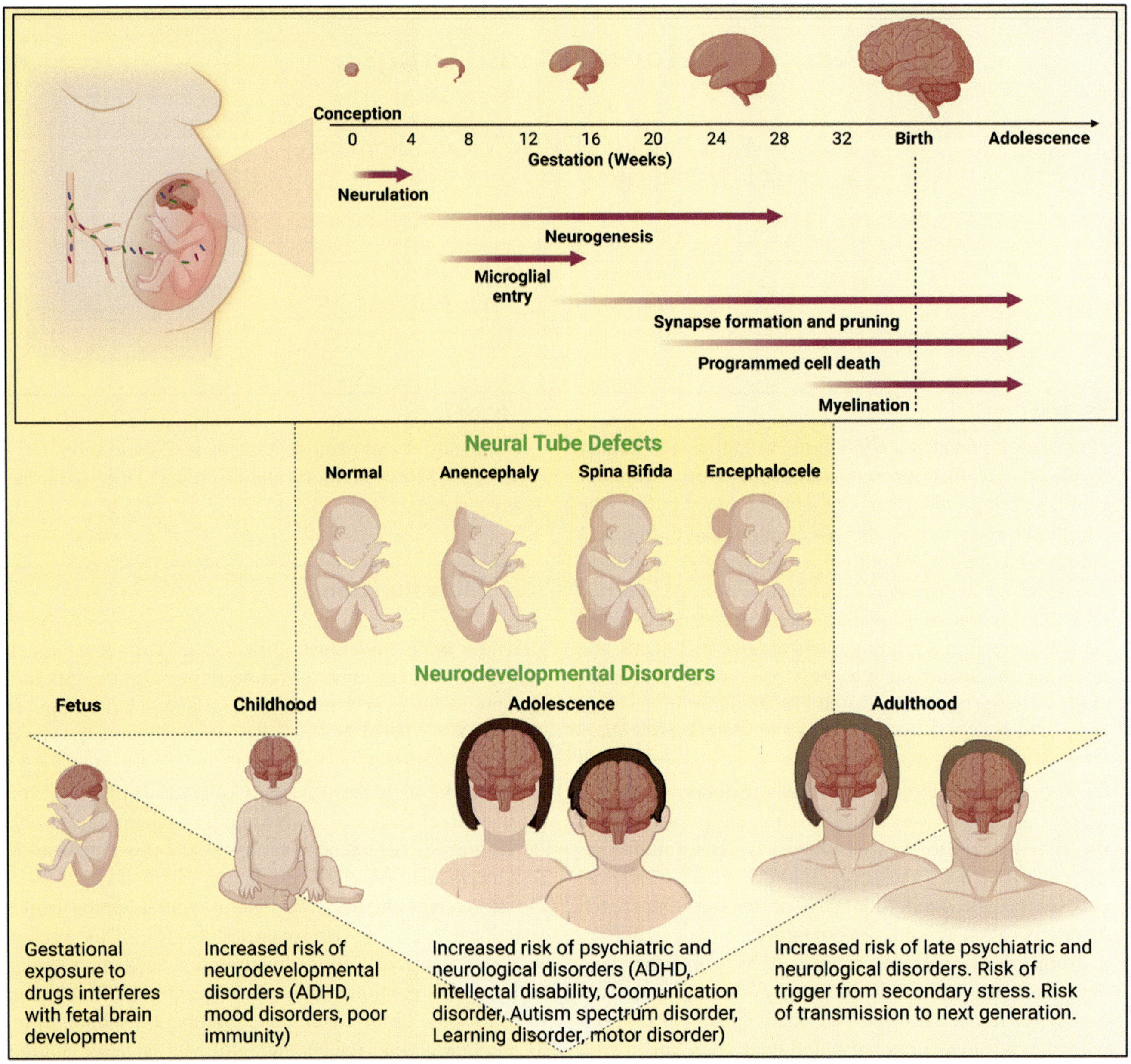

Fig. 10.1 Outline of human brain development and neurodevelopmental disorders. Neurodevelopment begins 4 weeks after conception with the closure of the neural tube. This process is called neurulation. Diverse biological processes occur after neurulation during gestation including neurogenesis, microglial entry, synaptogenesis and pruning, programmed cell death, and myelination. These neurodevelopmental processes are highly susceptible to external modulators such as drugs, viruses, and inflammatory proteins. Any insults to the neurodevelopmental processes can have serious adverse effects. Disturbances during the neurulation can result in neural tube defects (NTDs). Impairments in other processes can lead to long-term postnatal neurodevelopmental disorders

velopmental mechanisms can have detrimental effects. Disturbances during the neurulation process can result in neural tube defects (NTDs). These malformations occur due to incomplete neural tube closure and can be fatal. Moreover, changes in other biological processes can lead to a wide range of long-term postnatal neurodevelopmental disorders. Neurodevelopmental disorders in children include attention-deficit/hyperactivity disorder (ADHD), autism spectrum disorders, learning or intellectual disabilities, conduct disorders, cerebral palsy, etc. Children with neurodevelopmental disorders can experience difficulties with functions in language, motor skills, behavior, memory, or learning. Such disturbed brain development due to prenatal exposure to external modulators in combination with genetic susceptibility can increase the risk for postnatal neurodevelopmental disorders during childhood through adulthood. Subsequent postnatal challenges, such as drug abuse or trauma, may potentially prolong the risk for neurodevelopmental disorders by prenatal neurodevelopmental deficits. Disturbances in brain development may further instigate psychiatric and neurological disorders during adolescence and into adulthood. These neurological disturbances also carry the risk of getting passed on to the next generation.

While it can be difficult to assess cause–effect relationships in observational human gestational studies, the use of animal models in research has proven helpful in studying genetic or nongenetic neurodevelopmental disorders. In this chapter, we will focus on (1) cellular and molecular pathways of neurogenesis, (2) characterizations of NTDs and postnatal neurodevelopmental disorders, and (3) update on outcomes of prenatal exposure to drugs of abuse, prescribed medications, or environmental factors.

10.2 Neurogenesis

The development of a multicellular organism begins from a single-cell zygote. It undergoes rapid cell division to form a single layer of cells called the blastula (Gibb & Kovalchuk, 2017). Later, cells within the blastula rearrange themselves to form three layers. This process is called gastrulation (Gibb & Kovalchuk, 2017). Gastrulation is a key phase in embryonic development in which the blastula folds in on itself and cells migrate to form the endoderm, mesoderm, and ectoderm. Each of the layers is identified as a germ layer. These germ layers further differentiate into different tissues and organ systems. The endoderm gives rise to the linings of the digestive and respiratory systems as well as organs such as the stomach, intestines, lungs, liver, and pancreas. The mesoderm gives rise to the dermis, heart, connective tissue, muscle, vascular system, bone, and bone marrow. The ectoderm gives rise to the epidermis as well as the central nervous systems (CNS) and peripheral nervous systems (PNS). The CNS originates from the ectoderm's dorsal side and the epidermis originates from the ventral side (Monk et al., 2001). Cells within the ectoderm layer form neural tissue in a mechanism known as *Neural Induction* (Schoenwolf, 1982). The neuroepithelial cells (NECs) serve as the primary neural precursors of the CNS. Such mechanisms take place around gestation day 16 (O'Rahilly & Gardner, 1979). After neural induction, the *neurulation* process begins in which the induced neural region is transformed from a plate into a tube. During neural induction, the plate emerges thicker along the midline of the dorsal ectoderm. The plate also begins to elongate along the rostrocaudal axis, narrow on the lateral plane, and thicken on the apicobasal line (Sausedo et al., 1997). Such processes help the plate to continue to fold and ultimately fuse the ends of the neural fold to form the neural tube. The neural tube fusion starts around 22 days of gestation (Monk et al., 2001) and is completed around 26 days (Sidman & Rakic, 1982). The neural tube eventually forms the mature brain and spinal cord. The caudal portion becomes the spinal cord, while the rostral portion becomes the brain (Monk et al., 2001). The widest portion of the neural fold develops into the forebrain and the cranial flexure becomes the midbrain (Sidman & Rakic, 1982).

Neurogenesis is a complex process involving the generation of neurons in the brain and spinal cord. This process includes cell proliferation, production of migratory progeny, differentiation, and functional integration of cells into the neural circuitry of the CNS. Neurogenesis occurs after neural tube formation in temporally and spatially defined configurations along the anterior–posterior (AP), dorsal–ventral (DV), and lateral–medial body axes (Taverna et al., 2014). Morphogen gradients determine the patterning of NECs along these axes by governing where and how certain neural cell types develop. Morphogens found to play a role in this process include sonic hedgehog (SHH), bone morphogenetic proteins (BMPs), wingless/int (Wnt) proteins, and fibroblast growth factors (FGFs). In the patterning of the dorsal–ventral axis (DV), SHH is ventrally expressed in the floor plate and released toward the dorsal side, while BMPs and Wnts are dorsally expressed in the roof plate. Antagonism of BMP signaling by SHH signaling generates a counteracting gradient that allows for spatial differences in expression patterns of transcription factors (Pax6, Pax7, Irx3, Dbx2, Olig2, Nkx2.2, Nkx6.1, among others) (Fig. 10.2a). Therefore, these morphogens determine neural progenitor fates along the DV axis of the neural tube, which eventually develops into the spinal cord (Cohen et al., 2013; Dessaud et al., 2008; Ulloa & Briscoe, 2007). SHH signaling plays a critical role in DV patterning in both the anterior parts (brain region) and posterior parts (spinal cord) of the neural tube. However, mechanisms involved in the establishment of the DV axis differ from those for the posterior parts of the neuraxis. For example, SHH signaling antagonizes the activity of Gli3, which acts as a repressor of its target genes to determine DV patterning during the development of the telencephalon (Rallu et al., 2002).

The AP axis is established before and during gastrulation. During the gastrulation stage, nonneuronal tissues, the anterior visceral endoderm (AVE), and the prechordal plate are crucial for head induction and maintenance (Stern, 2001; Wilson & Houart, 2004). The AVE and prechordal plate act as signaling centers, specifying neural fate during early forebrain development. These centers produce morphogen antagonists, inhibiting diffusible posteriorizing factors like Wnts, FGFs, and Nodal and retinoic acid (RA) (Altmann & Brivanlou, 2001; Schier, 2001). Studies in fish and frogs suggest that secreted frizzled-related proteins (Frzb) and dickkopf (Dkk1-3 in mice) are expressed in the anterior endoderm (equivalent to the AVE) of embryos and can antagonize Wnts. Nodal and FGF signalings are inhibited by Lefty and Sprouty, respectively. Cyp26 proteins degrade RA into its

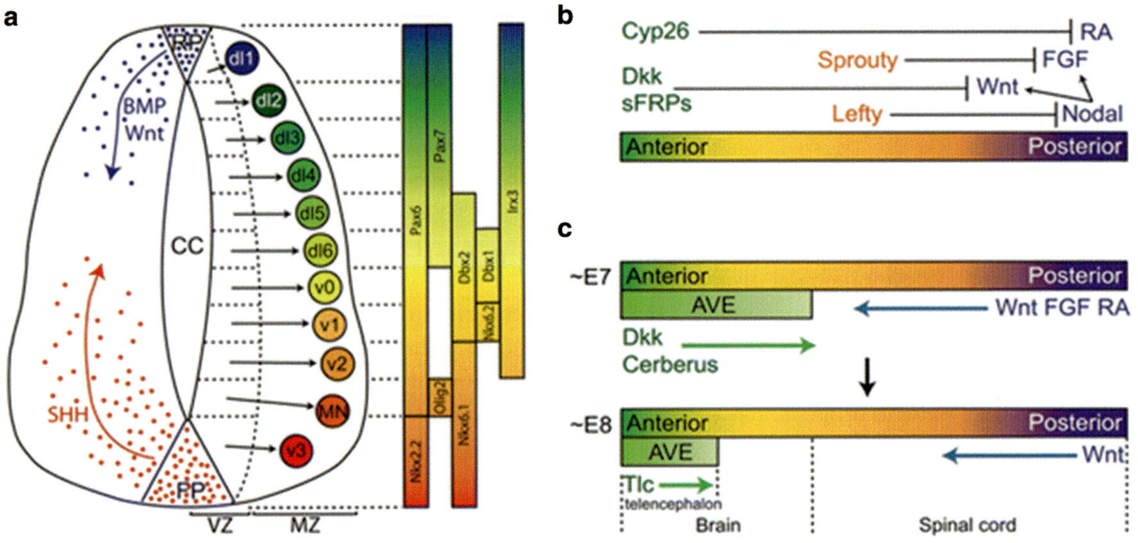

Fig. 10.2 (**a**) (Left) Diagram of a section of the spinal cord. The left half shows morphogen gradients of BMPs/Wnts from the roof plate (RF) and SHH from the floor plate (FP). The right half indicates the 11 neuronal progenitor domains located in the ventricular zone (VZ), medially adjacent to the central canal (CC). The morphogen gradients spatially control the position identity of the progenitor domains where distinct neuronal subtypes are generated. Postmitotic differentiated neurons are divided into six types of dorsal domains (dl1–dl6), four types of ventral interneuron domains (v0–v3), and motor neurons (MN) in the mantle zone (MZ). (Right) The spatial pattern of morphogen gradients regulates transcription factor expression. Different combinations of transcription factors define individual progenitor domains. (**b**) Anterior–posterior (AP) patterning during CNS development in fish and frogs. In neurulation, retinoic acid (RA) and Nodal, inducing Wnt and FGF, serve as diffusible posteriorizing factors. They are antagonized by Cyp26, Lefty, Frizzled-related proteins (Frzb: secreted FRPs) and dickkopf (Dkk), respectively, at the posterior part of embryos to specify the AP axis. (**c**) During mammalian development, Wnt, FGF, and RA characterize the AP axis in the neural plate as in fish and frog embryos. Dkk and Cerberus are expressed in the anterior visceral endoderm (AVE) and antagonize Wnt signaling, shaping the anterior neural plate domain. Subsequently, Tlc, a member of FRPs, further antagonizes Wnt signaling to subdivide the domain, designating future telencephalon region. E: embryonic day

polar metabolites, inhibiting discrete levels of RA signaling in the posterior part of the embryo (Fig. 10.2b) (Tuazon & Mullins, 2015). Studies on the removal or transplantation of the AVE or mutations in its genes support the role of the AVE in specifying the forebrain fate in unpatterned NECs (Rallu et al., 2002). Following anterior neural induction, TIc, a member of Frzb that serves as an antagonist of Wnt signaling, plays a role in mediating the establishment of the telencephalon at the anterior neural ridge and as well as partitioning the forebrain into the telencephalon, diencephalon, and optic territories (Fig. 10.2c) (Houart et al., 2002).

10.2.1 Cerebral Cortex Development and the VZ/SVZ

The telencephalon gives rise to the cerebral cortex, which is comprised of the neocortex and allocortex (also known as the heterogenetic cortex). The telencephalon territory is determined at approximately mouse embryonic day (E) 9.5. Initially, NECs undergo symmetric cell divisions leading to lateral and radial expansion of their cell populations during cerebral cortex development (Rakic, 1995). This occurs at the anterior edge as well as the area surrounding the lumen of the neuraxis. Meanwhile, NECs are polarized along the apical–basal axis. Mitosis occurs at the apical-most germinal layer, which is closest to the lumen, and a germinal niche called the ventricular zone (VZ). In this region, NECs form adherens junctions (AJ), connecting neighboring NECs (Marthiens & ffrench-Constant, 2009). NECs also undergo a process called interkinetic nuclear migration (INM), in which the NECs' nuclei migrate through the cytoplasm along the apical–basal axis of the neuroepithelium during the S-phase in mitosis, forming a pseudostratified epithelium (Taverna & Huttner, 2010). Cortical neurogenesis begins with the switch of NECs from symmetric to asymmetric differentiation, which can result in the generation of neural progenitor cells (NPCs) (Florio & Huttner, 2014).

The earliest cortical neurons are generated from NPCs at approximately E10.5 and settle within the preplate (PP) (De Carlos & O'Leary, 1992). The PP is comprised of the superficial marginal zone (MZ), Layer (L)1 of the postnatal cortex, the cortical plate (CP), postnatal L2–L6, and the subplate (SP) underneath the CP. Beginning around E11–12, NPCs transform into radial glial cells (RGCs) with their somata located at the VZ. Their short processes reach the ventricle (apical side), and their long processes reach the basal lamina. RGCs express astroglial markers, including glial fibrillary acidic protein (GFAP), glutamate/aspartate transporter (GLAST), and brain lipid-binding protein (BLBP) (Campbell

& Gotz, 2002). Similar to NPCs, RGCs undergo INM in the VZ, differentiating symmetrically and asymmetrically to produce daughter RGCs, intermediate progenitor cells (IPCs), or neurons (Gotz & Huttner, 2005). These newly formed neurons migrate radially from the VZ toward the pial surface through the intermediate zone. RGC basal processes are used as guiding scaffolds to help neurons reach their final destination (Rakic, 1988). Cajal-Retzius (CR) cells are generated as the earliest cortical neurons and play a major role in neuronal cell migration. These cells are found in the superficial MZ and serve as directors to regulate the identity of radial glia and migration of neurons during cerebral cortex development. They are considered transient, as most of the cells undergo cell death and are no longer present at postnatal stages (Derer & Derer, 1990; Soriano et al., 1997). IPCs migrate, accumulate, and line up above the VZ, forming a second germinal zone, called the subventricular zone (SVZ). Here, IPCs lose their processes and polarity, down-regulate astroglial markers, and undergo one round of symmetric neurogenic division, generating a pair of neurons that radially migrate from the SVZ to their final destination using RGC basal processes (Gal et al., 2006). New neurons generated from both RGCs and IPCs migrate to the cerebral cortex and reach the most superficial side of the CP and form the mantle layer, overlaying leading neurons allowing the CP to be formed in order of L6–L2. At approximately E17, RGCs retract their basal processes and become gliogenic to generate astrocytes and ependymal cells (ECs) that will line the ventricles (Fig. 10.3) (Bjornsson et al., 2015; Florio & Huttner, 2014; Kwan et al., 2012).

10.2.2 Hippocampal Formation During Development

The allocortex is comprised of three subtypes: the periallocortex, paleocortex, and archicortex. The hippocampal primordium is located within the archicortex. The hippocampus emerges from the caudomedial edge of the neuroepithelium after invagination of the roof in the dorsal telencephalon (Lee et al., 2000). The hippocampus, including the dentate gyrus (DG), originates from the VZ, but the formation is unique. At approximately E14.5, dentate NPCs migrate from the dentate neuroepithelium, or the primary matrix, which is a part of the VZ in the archicortex adjacent to the cortical hem (CH), toward the pial-midline side of the archicortex, forming the secondary matrix. CH-derived CR cells direct the migration of these cells (Del Rio et al., 1997). Dentate NPCs then migrate away from the secondary matrix toward the hippocampal fissure forming the tertiary matrix, a component of the DG. These cells migrate using glial scaffolds connecting the CH (the future hippocampal fimbria) to the pial surface and the hippocampal fissure and generate granule cells (GCs) that form the GC layer (GCL). In the VZ, asymmetric cell divisions of RGCs give rise to daughter RGCs and hippocampal neurons, which then migrate via RGC basal processes to the hippocampal cornu ammonis (CA) 1–3 and become pyramidal neurons. Dentate NPCs persist postnatally in the tertiary matrix, but only for a few weeks after birth. Eventually, the proliferation of NPCs is restricted to the subgranular zone (SGZ) of the DG (Urban & Guillemot, 2014; Yu et al., 2014).

10.3 Neural Tube Defects (NTDs)

NTDs are the second most common congenital malformations, in which the development of the fetal CNS is affected (Avagliano et al., 2019; Copp & Greene, 2013; Greene & Copp, 2014; Ravi et al., 2021). These are characterized as chronic disorders that affect the development of brain and spinal cord structures during gestation. Currently, the occurrence of NTDs is relatively low. The rate of NTD incidences varies across geographical regions, ranging from 0.2 to 11 occurrences per 1000 live births (Ravi et al., 2021). However, anencephaly and spina bifida are the most noted types of NTDs globally. Even with significant research efforts, a knowledge gap exists regarding the underlying mechanisms of these birth defects. These malformations occur due to incomplete neural tube closure (NTC) during the early gestation. As previously defined, the process of NTC is known as neurulation. This process occurs in human gestation between days 17 and 28 following conception. The mechanisms underlying the neurulation process are tightly regulated. These include the formation of an anterior (ANP) and posterior neuropore (PNP). These neuropores will progressively reduce in size to complete the process of NTC (Avagliano et al., 2019; Copp & Greene, 2013; Greene & Copp, 2014; Ravi et al., 2021).

Disturbances in the processes of NTC due to genetic or nongenetic factors can induce NTDs (Avagliano et al., 2019; Copp & Greene, 2013). The development of NTDs is recognized as multifactorial. It could be possible that an individual factor is unable to detrimentally affect NTC. However, the combined effect of genetic and nongenetic factors can potentially induce a synergistic effect that would cause failed NTC and thus the development of the abnormal embryonic phenotype. Nongenetic risk factors include exposure to a range of environmental factors such as air pollution, drugs such as valproate or phenobarbitol, or maternally toxic factors such as inadequate nutrition, maternal obesity, and diabetes mellitus (Avagliano et al., 2019; Copp & Greene, 2013).

Primary and secondary neurulation are essential steps in forming and closing the neural tube (Greene & Copp, 2014). In the primary neurulation process, neural tube formation involves bending the neuroepithelium at the midline

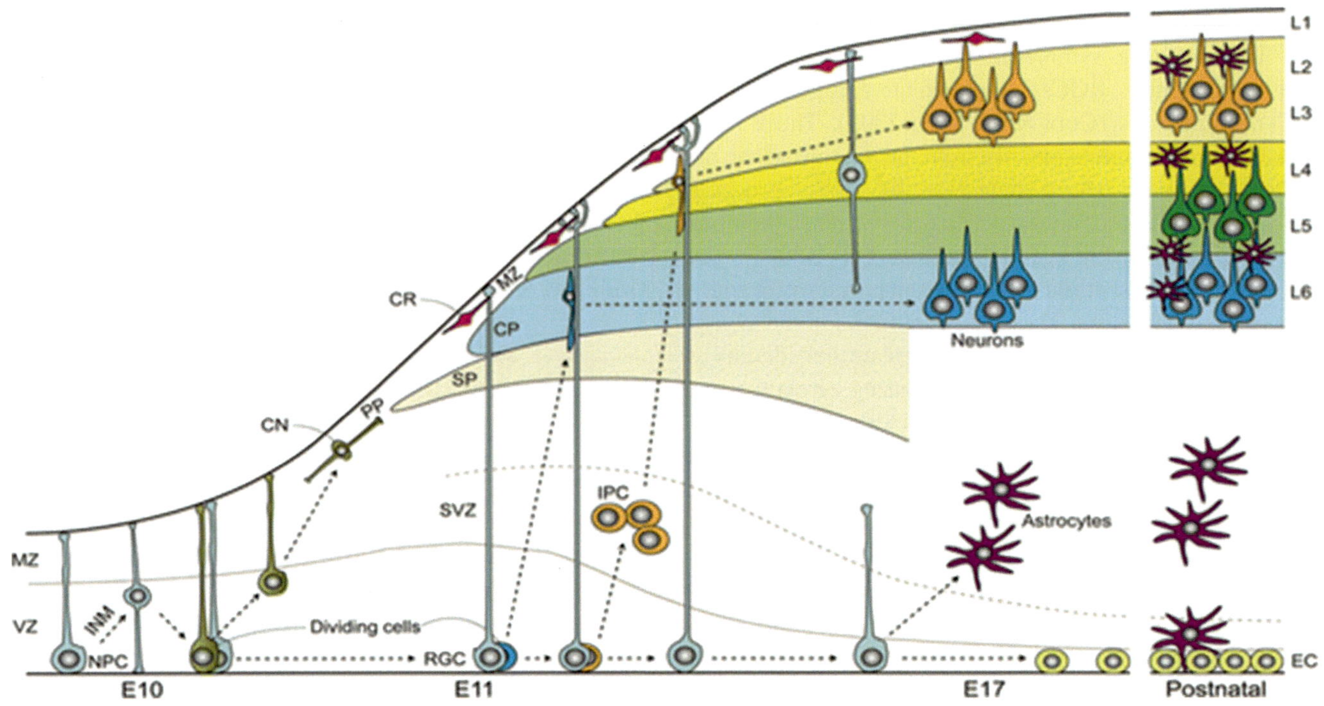

Fig. 10.3 Schematic of neuronal generation and migration in mouse neocortical development. The earliest cortical neurons (CNs) are generated from neural progenitor cells (NPCs) at around E10.5 and settle within the preplate (PP), which is split into the superficial marginal zone (MZ or Layer (L)1 of the postnatal cortex), the cortical plate (CP, the postnatal L2–L6) and the subplate (SP) at the later stages of development. At around E11–12, NPCs transform into radial glial cells (RGCs) with their somata positioned at the ventricular zone (VZ), and their long processes reaching the basal lamina. RGCs divide both symmetrically and asymmetrically to produce daughter RGCs and intermediate progenitor cells (IPCs) or neurons. Newly formed neurons radially migrate from the VZ toward the pial surface through the intermediate zone using RGC basal processes as guiding scaffolds and reach their final destination. The Cajal–Retzius (CR) cells lie in the superficial MZ, and serve as "directors," regulating the identity of radial glia and the migration of neurons. IPCs migrate above the VZ, form a second germinal zone called the subventricular zone (SVZ), and undergo one round of symmetric neurogenic division to generate pairs of neurons that also radially migrate from the SVZ to their final destination using RGC basal processes. Newly generated neurons from both RGCs and IPCs migrate into the cerebral cortex and reach the most superficial side of the CP to form the mantle layer, overlaying leading neurons to form the CP in order of L6–L2. At around E17, RGCs retract their basal processes and become gliogenic, generating astrocytes and giving rise to ependymal cells (ECs) that line the ventricles

to form neural folds. These folds further raise, connect, and fuse at the dorsal midline. Neural tube closure is a discontinuous process with initiation of closure at distinct sites located at axial levels, rather than simultaneous closing of folds along the rostrocaudal axis. Moreover, the morphological and molecular requirements for closure vary along the body axis. Thus, each NTD is the outcome of detrimental effects on only a defined portion of the neural tube. Altogether, NTDs can be ascribed to failure of a particular initiation process or impairment of closure between these sites. In the secondary neurulation process, the neural tube caudal to the mid-sacral region is continuous with the caudal end of the primary neural tube. Secondary neurulation involves cellular condensation and mesenchymal-to-epithelial transition to close the neural tube in the caudal region. Malformations resulting from disturbance of secondary neurulation are skin-covered. These defects usually involve tethering of the spinal cord and ectopic lipomatous material.

NTDs have been divided into two types, open NTDs and closed NTDs (Avagliano et al., 2019; Copp & Greene, 2013; Greene & Copp, 2014; Ravi et al., 2021). Commonly, open NTDs are characterized by the external protrusion and exposure of neural tissue to the surrounding environment. These open defects are the outcome of the failure of primary neurulation. Due to the nature of open NTDs and being exposed to the surrounding environment, degeneration of the open neural tissue takes place in utero. This leads to the loss of neurological functions. Open NTDs include craniorachischisis, exencephaly–anencephaly, myelomeningoceles, and lipomyelomeningoceles. In contrast, closed NTDs have an epithelial covering (skin covered) and present as defects in spinal cord structure. Thus, in these defects, neural tissue does not get exposed to the surrounding environment. These closed defects are commonly the outcome of the failure of secondary neurulation. Closed NTDs include encephalocele, split cord malformation, spina bifida occulta, dermal sinus, or dermal nevus.

10.3.1 Types and Characteristics of NTDs

10.3.1.1 Exencephaly and Anencephaly

Primarily, failed closure of the cranial neural tube leads to exencephaly (Avagliano et al., 2019; Copp & Greene, 2013; Greene & Copp, 2014; Ravi et al., 2021). Under this condition, the brain's neural folds remain open and exposed to the environment. Moreover, the neuroepithelium physically protrudes from the developing brain. The skull vault does not form over the open region. The degeneration of neural tissue due to exposure to the surrounding amniotic environment converts the exencephaly defect to the morphological appearance of anencephaly (Wilkins-Haug & Freedman, 1991; Wood & Smith, 1984). Human anencephaly can be subdivided into meroacrania and holoacrania (Seller, 1995). In meroacrania, defects are recognized in the rostral brain and skull, whereas in holoacrania, defects are recognized in the posterior brain and skull. Gross morphological assessments show the absence of the forebrain and the skull structures. The calvarium is generally absent. The parietal, frontal, and squama of the temporal and occipital bones physically appear as rudimentary fragments. The base of the skull is nearly normal (Goldstein & Filly, 1988) but appears thick and flattened. The sphenoidal bone is abnormally shaped. Generally, the face is normally structured; however, the eyes appear to protrude because of the shallowness of the orbits and the absence of a shortened forehead. Moreover, other associated facial abnormalities include low-set ears and a flattened nasal bridge (Welch, 1990). The remnants of the brain tissue are known as the area cerebrovasculosa (Avagliano et al., 2019). The area cerebrovasculosa appears as a dark brown undifferentiated mass. The content of the residual brain tissues is found to be variable. Generally, the structures of the forebrain including the thalamus, cerebrum, and midbrain are absent, whereas the brainstem may not be affected. Rodent models show effects on the forebrain, midbrain, hindbrain, or other neighboring regions, dependent on which part of development was disrupted (Copp & Greene, 2013). The pituitary is hypoplastic and without the intermediate and posterior lobes (Avagliano et al., 2019). Overall, anencephaly is a fatal condition, and even after early diagnosis, if the pregnancy is carried out to term, infants with anencephaly die within the first 2 days following birth (Avagliano et al., 2019).

10.3.1.2 Myelomeningocele

Disrupted development of the spinal neurulation along the body axis results in open spina bifida (Avagliano et al., 2019; Copp & Greene, 2013; Greene & Copp, 2014; Ravi et al., 2021). Similar to cranial NTDs, the morphological features vary substantially. These depend on the developmental stage of the examination. Myelomeningocele (spina bifida cystica) involves incomplete closure of the posterior spinal neural tube. The lumbar portion fails to fuse. In this NTD, neural tissue is contained within a meninges-covered sac that protrudes through a bony defect of the vertebral arch. However, in myelocele, neural tissue is exposed to the amniotic fluid without the extension of the meningeal sac. Therefore, a myelocele is an open defect and the cystic mass is absent. In normal development, the vertebral arches develop from the sclerotome of the axial mesoderm. These migrate dorsally to surround the closed neural tube before the beginning of cartilaginous and bony formation. However, when the neural folds fail to close, the sclerotome is unable to cover the neural tissue and a bifid vertebral column is the secondary outcome. In myelomeningocele, a cystic mass protruding though the vertebral arches may include cerebrospinal fluid drainage. The physical appearance of the neural tissue is translucent through the protruded meningeal sac. The neural placode, a part of nonneurulated embryonic neural tissue, is externally visible and extends above the skin surface. In myelocele, the neural tissue is clearly detectable though the vertebral cleft, and the neural placode is flush with the surface of the skin. The spinal cord above the myelomeningocele or myelocele deformation may appear to be irregular in position and shape, but it is not notably malformed (Emery & Lendon, 1973). Finally, in the absence of additional serious congenital malformations, infants with myelomeningocele or myelocele can survive with neurological impairment (Avagliano et al., 2019).

10.3.1.3 Craniorachischisis

Craniorachischisis is the most severe form of an open NTD (Avagliano et al., 2019; Copp & Greene, 2013; Greene & Copp, 2014; Ravi et al., 2021). It involves defects in both the cranial and spinal neural tube closure. The entire neural tube remains open from the midbrain to the low spine. Thus, defects in the development of both the skull and vertebral arch structures result in anencephaly and open spina bifida. The absence of the cranial vault and defective vertebrae at variable extents are noted in fetuses. Morphological characterization includes a short rostro-caudal body axis, short or absent neck, potential facial dysmorphisms, low-set and folded ears, and exophthalmos. In fetuses with craniorachischisis, both the brain and spinal cord are exposed to the intra-amniotic environment leading to the destruction of the nervous tissues. Histological characteristics of craniorachischisis are similar to anencephaly and myelomeningocele specimens. Abnormalities in the planar cell polarity (PCP) signaling pathway have been linked to craniorachischisis. In general, craniorachischisis is a lethal type of NTD, and treatments or surgical interventions are not available.

10.3.1.4 Encephalocele

Encephalocele is a sac-like protrusion of the brain or the meninges through an opening in the skull (Avagliano et al.,

2019; Copp & Greene, 2013; Greene & Copp, 2014; Ravi et al., 2021). Depending on the type of tissue (meninges, brain, or ventricles) involved in the sac-like protrusion, cephaloceles are classified as meningocele, encephalomeningocele, or encephalomeningocystocele (Avagliano et al., 2019; Copp & Greene, 2013). Meningocele is a herniation of the meninges, while encephalomeningocele is a herniation of the meninges and brain. Finally, encephalomeningocystocele is a herniation of the meninges, brain, and ventricles. Encephaloceles can happen through any part of the cranial vault and are classified by the location of the herniation. These include anterior, parietal, and occipital. In the anterior encephaloceles, the lesion is identified between the bregma and the anterior ethmoid bone. In the parietal encephaloceles, the lesion is identified between the bregma and the lambdoid suture. In occipital encephaloceles, the lesion is identified between the lambdoid suture and foramen magnum. The occipital encephalocele is the most common of the three. The anterior encephalocele is further sub-characterized according to the location of the defect. These subclassifications include frontal, sincipital, and basal anterior encephalocele. The frontal anterior encephaloceles are external lesions. They cause craniofacial abnormalities, which are commonly found near the root of the nose. Basal encephaloceles are internal lesions. These occur within the nose, pharynx, or orbit (Avagliano et al., 2019; Copp & Greene, 2013). The appearance of this encephalocele is characterized as a round, soft, compressible nodule with sub-scalp localization (Gao et al., 2014). The mass may appear solid according to the content of the herniation. The mass is commonly covered by alopecic skin which could be surrounded by a ring of hair (Gao et al., 2014). The surface of the encephalocele sometimes appears bluish, translucent, or glistening. Moreover, capillary malformations are notable (Sewell et al., 2015). In terms of prognosis, encephalocele treatment requires surgical correction.

10.3.1.5 Meningocele

Meningocele is a closed spina bifida. It consists of herniation of the meninges through the vertebral arch. However, the spinal cord remains in the normal position within the spinal canal (McComb, 2015). Spinal meningocele generally resembles encephalocele. Usually, the protruded mass is covered by skin (Avagliano et al., 2019) and occasionally, the skin may have a variable extent of dysplasia (McComb, 2015). The skin exhibits atrophic epidermis lacking skin appendages. This mass is developed from extremely thick meninges and contains vascularized stromal tissue. Children with meningocele have normal neurologic examination without deformity of lower extremities or sphincter dysfunction (McComb, 2015). Surgical procedures can be performed for meningocele treatment.

10.4 Neurodevelopmental Disorders

10.4.1 Autism Spectrum Disorders

Autism spectrum disorder (ASD) is a neurodevelopmental disorder that affects around 0.6% of the population and is more commonly diagnosed in boys (van Loo & Martens, 2007). ASD presents as deficits in communication and social interactions, and restrictive, repetitive behavior typically beginning before the age of 3 (Doernberg & Hollander, 2016; van Loo & Martens, 2007). Deficits in behavior and processing persist throughout the lifespan and severity varies greatly from mild to severe symptoms. ASD is characterized by perturbed signaling of various circuits in the brain, with an imbalance of excitation and inhibition a likely source of ASD-related behaviors (Golden et al., 2018). Understanding where and why this disruption occurs is key to understanding the etiology of ASD.

The brainstem hypothesis of ASD suggests that disruptions in signaling in the adolescent or adult brain are a result of improper development of the brainstem (Geva & Feldman, 2008; Inui et al., 2017). The brainstem not only supports basic physiological and behavioral functions but also plays a guiding role in hierarchical brain development, neuroarchitecture, and formation of neurocircuitry (Dadalko & Travers, 2018). Minor disruptions in brainstem development can have significant effects on brain connectivity and higher-order signaling (Inui et al., 2017; Thompson & Levitt, 2010). Particularly, impaired GABA signaling leading to an excitatory/inhibitory imbalance may contribute to the clinical manifestations of ASD (Inui et al., 2017). Brainstem dysfunction is known to affect the basal ganglia and limbic system, which can further affect higher cognitive function crucial for the regulation of arousal, attention, and emotional control (Inui et al., 2017). Evidence suggests that improper development of the brainstem due to insult after neural tube closing or impairments in crucial signaling pathways may be responsible for the development of ASD-like behaviors.

ASD is known to have multiple contributing factors such as genetic, epigenetic, and environmental factors (Thapar, 2018). Studies have shown that there is around a 90% heritability estimation, and these are shown in different twin, family, and adoption studies (Faraone & Khan, 2006). Yang and Gill et al. suggested that 7 chromosomal regions (7q21.2–q36.2, 16p12.1–p13.3, 6q14.3–q23.2, 2q24.1–q33.1, 17q11.1–q21.2, 1q21–q44 and 3q21.3–q29) are likely to contain risk genes for ASD (Yang & Gill, 2007). Other studies have explored phenotypes that depict particular chromosomal copy number changes and single-nucleotide changes to study autism, but predictions based on these findings are not common in patients (Badescu et al., 2016). Whole-genome sequence association studies have found that de

novo coding variants have a strong association with ASD (Kosmicki et al., 2017; Werling et al., 2018; Willsey et al., 2013). Multiple gene mutations in ASD patients are associated with the Wnt signaling pathway and both of its components, β-catenin and its regulators (Bae & Hong, 2018; Kalkman, 2012; Mulligan & Cheyette, 2017). Canonical and noncanonical Wnt signaling is crucial for proper neural development. Disruption in these signaling pathways is related to several neurodevelopmental disorders (Kwan et al., 2016; Mulligan & Cheyette, 2017; Okerlund & Cheyette, 2011; Tang, 2014; Wada & Okamoto, 2009). Other signaling pathways that have been proposed in ASD pathogenesis include BMP signaling (Li et al., 2016; Zhang et al., 2017), SHH signaling (Halepoto et al., 2015; Patel et al., 2017), and RA signaling (Niculae & Paval, 2016). These signaling pathways are crucial for proper brain development and function. Disruptions in these are likely to contribute to deficits in neuronal connectivity, impaired excitation and inhibition, and clinical manifestations of ASD. Other studies have identified miRNA dysregulation as an important contributor to ASD, as they are part of the epigenetic machinery and can influence gene expression of important neuronal genes (Badescu et al., 2016; Eshraghi et al., 2018; Vaishnavi et al., 2013). Finally, environmental factors like maternal viral infection, immune activation, or use of the anticonvulsant valproic acid (VPA) during pregnancy are associated with an increased risk of ASD (Chess, 1971; Moore et al., 2000). These could provide direct insults to the developing brain or affect pathways like the Wnt signaling pathway, leading to impaired neurodevelopment and ASD-like behaviors (Qin & Dai, 2015; Richetto et al., 2017; Zhou, 2015). Overall, ASD is accepted as a neurodevelopmental disorder, but the exact causes of structural and signaling abnormalities associated with ASD are not fully understood.

10.4.2 Schizophrenia

Schizophrenia is a mental illness characterized by disruptions in cognition and emotion with both positive (i.e., hallucinations) and negative (poor social functioning, apathy) symptoms (Fatemi & Folsom, 2009). Although serious symptoms manifest in early adulthood, schizophrenia is considered a neurodevelopmental disorder, as pathogenesis is thought to originate with brain insults occurring in the prenatal or early postnatal periods (Fatemi & Folsom, 2009; McGrath et al., 2003). Early proposals of the neurodevelopmental hypothesis suggest that a fixed brain lesion, regardless of brain region or cause, occurring early in life interacts with normal events of brain maturation, resulting in later manifestations of the disease (Weinberger, 1987). Similar interpretations suggest that following this insult, there is an inevitability that the disease will manifest and as the brain matures, manifestations will get progressively worse. This interpretation could explain minor cognitive deficiencies and memory impairments observed prior to the first episodes of psychosis (McGrath et al., 2003; Owen & O'Donovan, 2017). The "hit" model of schizophrenia takes the neurodevelopmental hypothesis further to account for multiple insults throughout the lifespan. This model suggests that multiple risk factors, occurring both during early neurodevelopment and later in life, are necessary for clinical manifestations (Keshavan, 1999; McGrath et al., 2003). Overall, current interpretations of schizophrenia etiology highlight the need to consider risk factors, both genetic and environmental, affecting the brain during neurodevelopment and preceding the onset of psychosis (Fatemi & Folsom, 2009; McGrath et al., 2003; Murray et al., 2017; Owen & O'Donovan, 2017).

A number of genetic and environmental risk factors have been proposed for the development of schizophrenia. It is widely accepted that schizophrenia has a genetic component and is highly hereditary. This genetic risk involves common, rare coding, and copy number variants (Owen & O'Donovan, 2017). Genome-wide association studies have identified schizophrenia-associated loci and allowed for polygenic risk score profiling (O'Donovan et al., 2008; Pardinas et al., 2018; Schizophrenia Psychiatric Genome-Wide Association Study, 2011; Schizophrenia Working Group of the Psychiatric Genomics, 2014; Stefansson et al., 2009). Schizophrenia is associated with genes involved in dopaminergic and glutaminergic neurotransmission, synaptic plasticity, calcium signaling, and immune functioning, among other genes and signaling pathways (O'Donovan et al., 2008; Pardinas et al., 2018; Schizophrenia Psychiatric Genome-Wide Association Study, 2011; Schizophrenia Working Group of the Psychiatric Genomics, 2014; Stefansson et al., 2009). Numerous nongenetic risk factors have also been identified. Pregnancy and birth complications, including prematurity, low birth weight, hypoxia–ischemia events, maternal inflammation, and maternal infections during pregnancy have been associated with an increased risk of psychosis (Brown et al., 2004; Byrne et al., 2007; Canetta et al., 2014; Cannon et al., 2002; Jones et al., 1998). Other environmental risk factors include but are not limited to childhood trauma (Popovic et al., 2019; Varese et al., 2012), migration (Bourque et al., 2011; Cantor-Graae & Selten, 2005), urbanicity (Bourque et al., 2011), and substance use (Gurillo et al., 2015; Koskinen et al., 2009; Marconi et al., 2016; Sara et al., 2015). However, it seems that none of these risk factors alone are sufficient to cause the onset of schizophrenia without gene–environment or environment–environment interactions (Stilo & Murray, 2019).

Numerous pathophysiological hypotheses have been proposed to explain the emergence of psychosis in schizophrenics. The dopamine hypothesis is prevalent and supported by the usage of dopamine inhibitors to improve psychosis; how-

ever, these inhibitors do not address other schizophrenia symptoms (Madras, 2013). This hypothesis attributes psychosis to the over-pruning of synapses and subsequent dopamine dysregulation (Howes & Shatalina, 2022). Aberrant synaptic pruning is also hypothesized to contribute to pathogenesis by resulting in lower dendritic spine density, possibly contributing to altered inhibitory mechanisms and disrupted glutaminergic signaling (Keshavan et al., 2020; Sellgren et al., 2019). As microglia play a major role in synaptic pruning, some have suggested that inhibiting microglia-mediated synapse uptake could be a target for delaying or preventing schizophrenia onset (Sellgren et al., 2019). An understanding of the mechanisms behind psychosis and other symptoms of schizophrenia is important for determining the etiology of schizophrenia and proposing disease-modifying treatments in the future.

10.4.3 Attention-Deficit/Hyperactivity Disorder

Attention-deficit/hyperactivity disorder (ADHD) impairs the ability to monitor behavior, caused by inattention, hyperactivity, and impulsivity (van Loo & Martens, 2007). It was first described in 1845 and affects around 1 in 20 children (Comings, 2001; Faraone et al., 2003). ADHD was recently recategorized as a neurodevelopmental disorder due to its early onset in life and diagnosis criteria in the DSM-5 of symptom manifestation before the age of 12 and persisting more than 6 months (Harris, 2014). ADHD often has comorbidities such as learning disabilities, oppositional defiant disorder, conduct disorder, Tourette's syndrome, or depressive illness (van Loo & Martens, 2007). The symptoms of ADHD are not concrete and have the ability to adapt over time(Nigg, 2006). However, when children are diagnosed with ADHD, they are known to not only exhibit similar symptoms but respond in a similar way to treatment and exhibit consistency from clinician to clinician, receiving good clinical validity (Nigg, 2006).

The risk factors that contribute to ADHD can be genetic or environmental, but ADHD is known to have a genetic component and is highly hereditary (Thapar, 2018). Studies performed on twins have shown a 70% chance of genetic heritability, suggesting genetic causation (Faraone & Khan, 2006). Many of these noted genes are related to neural transmission (Quist et al., 2003; Swanson et al., 2000; Ulke et al., 2019). Environmental factors that increase the risk of ADHD include maternal habits of smoking and alcohol consumption, preterm birth, low birth weight, psychosocial abnormalities, neurotoxicants, and lead (Agency, 2013). Specifically, lead is a prevalent environmental risk factor that has been studied thoroughly. Studies have shown that the combination of maternal lead exposure and cigarette smoking increases the risk of ADHD compared to lead exposure alone (Froehlich et al., 2009). This shows a compounding effect of different risk factors. The psychosocial adversity found to increase the likelihood of ADHD could be lower socioeconomic status or unstable home environment (Nigg et al., 2010). Animal models have shown supporting evidence (Eubig et al., 2010; Rice, 1996, 2000; Rossi-George et al., 2011; Virgolini et al., 2008). Genetic and environmental risk factors can affect development, leading to altered signaling and the behavioral deficits observed in ADHD.

Imaging studies have associated anatomical differences in dimensions of neural tissue with ADHD (Tripp & Wickens, 2009). Total brain size has been found to be reduced in the adolescent brain, especially the basal ganglia, cerebellum, and total gray matter (Castellanos et al., 2002; Hynd et al., 1993, 1990, 1991; Noordermeer et al., 2017). Structural and functional differences in the brains of those with ADHD could be responsible for observed behaviors. There are multiple different domains in which patients with ADHD can exhibit deficits including motivation, executive functioning, state regulation, and temporal information processing (Nigg et al., 2010). Mechanistically, many behaviors associated with ADHD are thought to be a result of dysfunction of the brain's natural reward cascade, mediated by dopamine (Blum et al., 2008). Multiple genes involved in dopamine neurotransmission have been associated with ADHD and may lead to dopamine imbalance in the neural networks associated with ADHD (DiMaio et al., 2003; Swanson et al., 2000; Veronneau-Veilleux et al., 2022; Volkow et al., 2009). Further, treatments eliciting an increase in dopamine in certain areas of the brain have been found to improve symptoms of ADHD (Volkow et al., 2012). Others have implicated norepinephrine (Fu et al., 2021; Ulke et al., 2019) and serotonin (Quist et al., 2003; Thapar, 2018) to play critical roles in ADHD. These hypotheses suggest that behaviors associated with ADHD are related to improper signaling, which could inform future treatments to remedy dysfunction.

10.5 Neurodevelopmental Consequences of Drug Exposures During Pregnancy

10.5.1 Alcohol

Prenatal alcohol exposure can produce a multitude of detrimental effects on fetal development. These developmental abnormalities are recognized and grouped as Fetal Alcohol Spectrum Disorders (FASDs). Children with FASDs may present a broad range of developmental defects including a combination of physical, neurodevelopmental, neurocognitive, and neurobehavioral deficits. It is noted that the manifestation of the severity of these effects varies and depends on the frequency, pattern, and timing of alcohol consumption

during pregnancy (Chung et al., 2021; Lipinski et al., 2012; Wozniak et al., 2019). Depending on the detectable clinical features, FASDs are divided into four categories. These are fetal alcohol syndrome (FAS), partial FAS (pFAS), alcohol-related neurodevelopmental disorder (ARND), and alcohol-related birth defects (ARBD). While FAS represents the most severe form of FASD, resulting from heavy maternal alcohol consumption during pregnancy, moderate alcohol consumption also produces significant neurobehavioral deficits. Such neurodevelopmental deficits can persist through young adulthood. The prevalence of FASD is variable and differs regionally depending on case definitions, methods of ascertainment, and examined population (Hymbaugh et al., 2002). FASD is commonly underdiagnosed due to the poor reliability of self-reported maternal drinking, an absence of definitive biomarkers, and the complexity involved in diagnosis (Wozniak et al., 2019). However, recent studies estimate that approximately 1–5% of US first-grade children have FASD.

Following are the clinical characterizations of Fetal Alcohol Spectrum Disorders (FASDs) (Young et al., 2014). FAS includes facial anomalies, growth deficit, and CNS dysfunction. Facial anomalies induced by gestational alcohol exposure include short palpebral fissures, broad philtrum, midface hypoplasia, thin upper lip and cleft palate, and epicanthal folds. Growth deficits during both pre-natal and post-natal periods are crucial characteristics of FAS. In addition to structural abnormalities, gestational alcohol exposure affects the developing fetal CNS. These include microcephaly, decreased intellectual ability, behavioral abnormalities, and impaired mental, and motor skills. pFAS is characterized by some facial components of FAS such as short palpebral fissure length, flattened philtrum, and thin upper lip. In addition to these, pFAS is usually diagnosed with at least one of the following impairments: growth restriction, CNS anomalies, or behavioral and cognitive deficiencies. Finally, ARND is diagnosed by the presence of CNS abnormalities and ARBD presents several congenital malformations.

A wide range of developmental neuropathological abnormalities are linked to FASD. Cognitive impairments in attention, executive function, learning and memory, motor function, and self- regulation have been reported (Mattson et al., 2011; Petrelli et al., 2018). Due to these cognitive and mental functional deficits, behavioral and emotional disturbances are common in FASD (Popova et al., 2016; Weyrauch et al., 2017). Moreover, a higher prevalence of mental disorders including ADHD, depression, and anxiety disorders is observed (Weyrauch et al., 2017). Hypoplasia of the corpus callosum and cerebellum, vascular anomalies, gliosis, perivascular space dilation, pituitary hypoplasia, ventriculomegaly, cavum septum pellucidum, and abnormal gyral patterns was identified in imaging studies performed on children with FASDs (Boronat et al., 2017). Likewise, in a retrospective study, heterotopia and brainstem anomalies, ventriculomegaly, microcephaly, NTDs, and holoprosencephaly were described in FASD patients (Jarmasz et al., 2017). Preclinical studies in animal models showed that exposure to alcohol during gastrulation produces craniofacial dysmorphology due to apoptotic cell death and reduced migration of neural crest cells (Eason et al., 2017). Whereas cell death and reduced proliferation of neural stem cells due to alcohol would result in microencephaly (Gangisetty et al., 2014; Riar et al., 2016). When prenatal alcohol exposure occurs after organogenesis, there is a reduced effect on craniofacial development; however, detrimental effects on brain development still occur (Wozniak et al., 2019). Decreases in spinal and cranial motor neuron production and size, neocortical and hippocampal dysgenesis, increased neuronal cell death, reduced neuronal migration, and a decrease in myelination were observed following alcohol exposure (Barrow Heaton et al., 1999; Ozer et al., 2000; Redila et al., 2006; Shetty & Phillips, 1992). In vitro studies show a disruption in neural cell migration and axon pathfinding by alcohol's ability to block cell adhesion and axon outgrowth (Dou et al., 2018). Alcohol can also induce immune and endocrine dysfunction (Wozniak et al., 2019). The direct effects of alcohol on the fetus can potentially affect the hypothalamic-pituitary-adrenal (HPA) axis. Such modulation of the HPA axis during fetal development can permanently affect its responsiveness to later stressors, setting off a cascade of behavioral, cognitive, and cellular alterations in the adult (Champagne & Meaney, 2001).

Alcohol and its metabolites can pass through the placenta (Heller & Burd, 2014; Petrelli et al., 2018). The initial concentration of alcohol in fetal blood is similar to maternal blood concentration, but the fetus's ability to clear alcohol and its metabolites is less efficient (Heller & Burd, 2014). This leads to a relative increase in levels of exposure and exposure duration for the fetus during pregnancy (Le Dare et al., 2019). Overall, alcohol exposure induces structural, cognitive, behavioral, and cellular alterations during CNS development. Such neurodevelopmental deficits can persist through adulthood and can potentially affect individuals social, health, school, and employment outcomes. Since FASD is not a genetic disorder, no treatment is available to reverse alcohol-induced CNS damage. However, intervention in alcohol usage during pregnancy may be a key to reducing the severity of FASD.

10.5.2 Methamphetamine

Methamphetamine (MA) is a psychostimulant classified as a noncatecholamine sympathomimetic substance (McCusker et al., 1997). MA has multiple mechanisms for targeting the sympathetic nervous system. It targets the monoaminergic

system by reversing the actions of monoamine transporters and increasing the release of dopamine, norepinephrine, and serotonin (Thompson et al., 2009). MA can block the reuptake and degradation of these neurotransmitters, which can lead to an increase in their concentration at the synaptic cleft (Thompson et al., 2009). MA inhibits the enzyme responsible for the oxidation of serotonin and norepinephrine, monoamine oxidase (Bennett et al., 1993). Increased concentration of these neurotransmitters stimulates motivation, movement, pleasure, and reward centers (Kish, 2008). In adults, MA is known to be neurotoxic to dopaminergic and serotonergic axons, axon terminal arbors, and potentially glutaminergic axons (Frost & Cadet, 2000; McCann & Ricaurte, 2004).

MA readily passes through the placental barrier due to its lipophilicity. During fetal neurodevelopment, the neurotransmitters affected by MA modulate multiple ontogenetic processes including neurogenesis, neuronal migration, and axon and dendritic growth (Frost & Cadet, 2000). The neurotoxicity of MA can also modify the dendritic structure in the nucleus accumbens and prefrontal cortex (Rodrigues Pdos & Dowling, 1990). Studies in animal models have shown that fetal MA exposure increases the production of ROS-mediated oxidative DNA damage, altering signal transduction (Jeng et al., 2005). ROS formation can oxidatively damage other cellular macromolecules including lipids and proteins (Wells et al., 2016). Thus, developing fetuses are substantially lacking antioxidative enzymes and are at risk of developmental deficits due to oxidative stress (Wells & Winn, 1996). Another study found that brain concentrations of MA in pups exposed prenatally were equivalent to 62.13% relative to the dam (Rambousek et al., 2014). The liver is responsible for the metabolism of MA. However, fetal livers are unable to process metabolic activity allowing the drug concentration to increase over time, eventually surpassing concentrations in the maternal system (Dattel, 1990). Prenatal MA exposure can cause morphological irregularities, including smaller subcortical structures, specifically the putamen, globus pallidus, caudate nucleus, and hippocampus (Chang et al., 2004). Children who were prenatally exposed to MA have a smaller striatum with fewer D2 dopamine receptors and decreased creatinine levels (Chang et al., 2007; Derauf et al., 2011). This could affect energy metabolism in the brain. Increased incidences of intraventricular hemorrhage and white matter densities have been demonstrated in human newborns exposed to MA (Dixon & Bejar, 1989). While volumetric assessments have shown whole brain volumes are comparable to the general population (Chang et al., 2004), MA exposed children have slight but significant size decreases and morphological differences in certain brain regions.

Prenatal MA exposure can affect behavioral and cognitive outcomes, but unlike alcohol, there are no identifiable patterns of neurobehavioral deficits following MA exposure. MA-exposed newborns have lower arousal scores, poor quality of movement, and increased stress signs. The most common MA-linked adverse effects observed in newborns are growth restriction, low birth weight and length, and head circumference (Smith et al., 2006). Neurocognitive testing has shown children with prenatal MA exposure score lower on sustained attention, long-term spatial and verbal memory, and visual motor integration (Chang et al., 2004). In rodents, the neurotoxic effect of MA is thought to reflect irreversible terminal damage (Acuff-Smith et al., 1996). The neurochemical alterations caused by the neurotoxic effects of MA are associated with learning impairment and behavioral deficits that have been seen in MA-exposed animals (Weissman & Caldecott-Hazard, 1995). Due to maternal health issues associated with MA usage, prenatal and postnatal care tends to be inadequate for MA-exposed infants (Tomaskova et al., 2020). Child development can be negatively impacted by stressful home environments, maternal polysubstance abuse, and comorbidities. Even though it is known that MA has been used by pregnant women for several years, there is a knowledge gap about the effects of MA during early fetal development and the long-term effects. A continued long-term investigation is required to fully understand the mechanistic and behavioral effects of MA on neurodevelopment.

10.5.3 Cocaine

Cocaine is a psychostimulant. It has comparable CNS cellular and behavioral effects to methamphetamine due to their similarity in pharmacological mode of action. Cocaine can easily cross both the placental and blood–brain barrier. Thus, it can induce teratogenic effects on the developing fetus (Ross et al., 2015). Mechanistically, the detrimental effects of neonatal cocaine exposure can arise from three different pathways. First, cocaine acts on the monoaminergic pathway by binding to monoaminergic transporters to prevent the uptake of monoamines into the presynaptic cleft (Thompson et al., 2009). This can lead to increased concentrations of dopamine, norepinephrine, and serotonin at the synaptic cleft. Alterations in fetal monoaminergic systems can affect short and long-term attention and cognitive development (Ross et al., 2015). Second, prenatal cocaine exposure has been attributed to indirect vasoconstrictive effects from maternal hypertension, vasoconstriction, and decreased uterine blood flow (Smith & Santos, 2016). This can cause impaired nutrient delivery and oxygen exchange to the fetus. Third, cocaine may also induce impairment in genetic programming during fetal development (Lester & Padbury, 2009).

Studies have demonstrated that fetal growth and neurogenesis can be affected by prenatal cocaine exposure. Cocaine use during pregnancy increases the risk of preterm birth, low birth weight and length, and smaller head circum-

ference (Gouin et al., 2011). Similar to MA, growth restriction is a common clinical feature in children exposed to cocaine during gestation (Minnes et al., 2006). It is also noted that such growth restriction may persist through adolescence (Covington et al., 2002; Richardson et al., 2013). Abnormal infant behaviors have been documented. These include lowered arousal, poor movement and self-regulation, and higher excitability (Lester et al., 2002). Prenatal cocaine exposure has not been linked to decreased global intelligence, but deficits in cognitive performance, language skills, and behavior have been observed in longitudinal studies (Ross et al., 2015). Moreover, adverse effects on executive function in cocaine-exposed children have been noted in long-term studies. Executive functions include attention, behavior, cognition, working memory, and information/problem-solving. Further, others found that cocaine exposure during the first trimester is linked to less sociability, withdrawn behavioral issues, and anxiety-related behaviors in the children. Such behaviors may serve as a precursor for psychiatric problems during later periods of life (Ross et al., 2015). Imaging studies showed morphological CNS irregularities including alterations in cortical and limbic regions (Rando et al., 2013) and reduced structural connectivity between the prefrontal cortex and the amygdala (Liu et al., 2013). Thin cortical thickness and reduced activation of portions of the prefrontal cortex were shown in adolescents (Liu et al., 2013). Several studies in animal models have been used to define the neuropathological effects of cocaine on developing brains. Disrupted cortical neurogenesis and migration, reduction of neuronal densities in the cortex, and altered dopaminergic functions have been modeled in animal studies (Crandall et al., 2004; Hamilton et al., 2010; McCarthy & Bhide, 2012; McCarthy et al., 2012). These studies further validated that prenatal cocaine exposure leads to long-lasting behavioral, cellular, and molecular changes. Overall, prenatal cocaine exposure can cause impaired fetal and long-term growth, deficits in language, behavior, and executive functions.

10.5.4 Methadone

For pregnant women with opioid dependence, opioid maintenance therapy (OMT) is used to decrease illicit drug use, preventing fetal exposure and maternal withdrawal. OMTs improve prenatal care, nutrition, and fetal health (Binder & Vavrinkova, 2008), but the opioid agonists used in treatment can put developing fetuses at risk. Methadone is the most widely known pharmacological treatment of opioid dependence and is recommended for pregnant women seeking OMT. Methadone is a long-acting mu-opioid receptor agonist that can cross the placental barrier and potentially alter fetal development. Clinical studies have shown that maternal methadone use can lead to premature birth, decreased birthweight (Cleary et al., 2012), and smaller head circumference (Welle-Strand et al., 2013). Morphological abnormalities including decreased brain and cerebellar weight, brain DNA content, and cortical thickness have been documented (Ford & Rhines, 1979). Moreover, motor, cognitive, and behavioral outcomes are also affected by prenatal methadone exposure. Methadone-exposed children experienced motor and cognitive delays, poorer communication and language skills, and higher levels of behavioral and emotional regulatory difficulties (Levine et al., 2021).

Opioid use or withdrawal comes with substantial risks and is not recommended for pregnant women. Opioid use increases the risk of spontaneous abortion, stillbirth, premature birth, low birth weight, and neural and birth defects (Tran et al., 2017). Abrupt withdrawal can lead to preterm labor, fetal distress, and fetal withdrawal symptoms (Tran et al., 2017). While OMT has associated risks, it is superior to the alternative of exposing developing fetuses to harmful drugs or maternal stress. OMT is an important means of treating at-risk mothers and supporting public health.

10.5.5 Valproic Acid

Epilepsy is a neurological disorder that requires maternal treatment throughout pregnancy. Studies have shown an increased risk of mental impairment in children of women who were on antiepileptic drugs (AED) during pregnancy (Kellogg & Meador, 2017). Prenatal AED exposure is associated with morphological anomalies and neurodevelopment impairments (Bromley et al., 2014). Valproic acid (VPA) is an anti-convulsant that blocks voltage-gated sodium and T-type calcium channels and inhibits the transamination of GABA (Landmark, 2007). VPA is also a potent histone deacetylase inhibitor, which has been proposed as a potential mechanism for teratogenicity (Parodi et al., 2021). VPA carries the highest risk for structural malformations and postnatal cognitive and behavioral abnormalities (Tomson et al., 2019). Fetal valproate spectrum disorder (FVSD) describes the multitude of anomalies including congenital malformations, developmental delay, intellectual disability, ASD, and craniofacial abnormalities, associated with prenatal exposure to VPA.

The mechanisms of VPA neuropathology are not fully established, but several effects of exposure have been documented. Animal models have shown apoptotic neurodegeneration (Bittigau et al., 2002), enhanced synaptic plasticity in the prefrontal cortex (Sui & Chen, 2012), and decreased folic acid levels (Wegner & Nau, 1992). These in turn can lead to reduced head circumference and impaired intellectual skills, autistic symptoms, and NTDs. Behavioral abnormalities were also noted in animal models. These include lowered sensitivity to pain, increased repetitive activity, anxiety, and decreased social interaction (Schneider & Przewlocki, 2005). Not all AEDs are comparable when evaluating neurodevelopmental outcomes. Neurodevelopmental deficiencies are also associated with phenobarbital and possibly phenytoin (Velez-Ruiz & Meador, 2015), while lamotrigine and levetiracetam may have low risk (Kellogg & Meador, 2017). Carbamazepine is another anticonvulsant that is believed to have similar mechanisms to VPA but relatively low teratogenic effects when compared to VPA (Maan et al., 2022; Ornoy et al., 2017). Current therapeutic interventions should continue to be evaluated and inform the development of new, safe, and effective antiepileptic drugs.

10.5.6 Dolutegravir

Dolutegravir (DTG) is a human immunodeficiency virus type-1 (HIV-1) integrase strand transfer inhibitor (INSTI). Due to its high potency and barrier to drug resistance, DTG is widely used for the treatment of HIV-1 infection (Smith et al., 2021). The United States Department of Health and Human Services (U.S. DHHS) Panel on Treatment of HIV During Pregnancy and Prevention of Perinatal Transmission currently designates DTG as a preferred antiretroviral (ARV) during pregnancy (recommendations: updated on January 31, 2024). However, concerns were raised for the usage of DTG by HIV-1-infected pregnant women or those of childbearing age in 2018. A report from the Botswana birth-outcomes surveillance study revealed a higher prevalence of neural tube defects (NTDs) among infants born to women on a DTG-based regimen at the time of conception compared to infants born to women on other antiretroviral (ART) regimens (Zash et al., 2018). Out of 426 infants born to women on a DTG-based regimen, 4 (0.94%) had NTDs, which included encephalocele, myelomeningocele, iniencephaly, and anencephaly. Later, a follow-up from the same study reported a lower incidence in 2019, with only 0.30% of infants born to women on a DTG-based regimen at conception having NTDs compared to 0.10% in infants born to women on any other ART at conception (Zash et al., 2019). While the prevalence was lower in the follow-up report, this observational study concluded that the incidence of NTDs is 2–3 times higher in association with DTG exposure at conception compared to other forms of ART. Since these reports, other studies have shown inconsistent results. Data from another surveillance study in Botswana and reports from Brazil found that DTG use at conception was associated with NTDs (Kreitchmann et al., 2021; Raesima et al., 2019). However, data from two large pregnancy registries (Vannappagari et al., 2019), the Canadian Perinatal HIV Surveillance Program (Money et al., 2019), and a retrospective study from Brazil (Pereira et al., 2021) did not find evidence for a higher prevalence of NTDs following DTG exposure at conception compared to other ART at conception. Similarly, animal and ex vivo studies have also provided conflicting results on DTG's association with NTDs (Bade et al., 2021; Cabrera et al., 2019; Kirkwood-Johnson et al., 2021; Mohan et al., 2020; Posobiec et al., 2021; Stanislaus et al., 2020).

While DTG's association with NTDs is a concern, there is a need to understand the effects of DTG exposure during pregnancy in infants born without structural malformations. With a relatively low (0.2–0.3%) risk of NTDs, the majority of children exposed to DTG during pregnancy are born without structural malformations but could suffer from postnatal neurodevelopmental adverse outcomes. Animal and in vitro studies have observed DTG-related neuronal toxicity (Montenegro-Burke et al., 2019; Smith et al., 2022). Moreover, DTG is known to cause neuropsychiatric adverse events in adults (Amusan et al., 2020; Yombi, 2018). Yet, few studies have investigated the effects of DTG on neurodevelopment. Notably, the Surveillance Monitoring for ART Toxicities (SMAART) study recently reported that DTG usage at conception and during the first trimester was associated with an increased risk of postnatal neurological abnormalities like febrile seizures, microcephaly, or epilepsy (Crowell et al., 2020). As DTG is known to show high transplacental transfer from mother to fetus (Bollen et al., 2021; Mulligan et al., 2018; Waitt et al., 2019) with drug accumulation and prolonged elimination of DTG from infants (Mulligan et al., 2018; Pain et al., 2015), it is important to consider the potential effects of DTG exposure during pregnancy, a critical period for brain development.

With reports highlighting DTG's effects on NTDs and neurodevelopmental toxicity, it is imperative to understand the potential mechanisms behind the adverse effects of DTG use during pregnancy. A potential mechanism behind DTG's association with NTDs could be the effects on folate levels, which is essential for neural tube development (Botto et al., 1999). Some studies suggest that DTG may affect the folate receptor or folate transport pathway, which could lead to NTDs (Gilmore et al., 2022; Matok et al., 2009). Potential mechanisms for DTG-related developmental neurotoxicity include induction of neuronal injury, neuroinflammation, and oxidative stress in the brains of mice exposed to DTG (Bade et al., 2021; Montenegro-Burke et al., 2019). Finally, an emerging mechanism for DTG developmental neurotoxicity is the inhibition of matrix-metalloproteinases, which play an important role in neurodevelopment (Bade et al., 2021). With the number of HIV-1-infected women of childbearing age on a DTG-based regimen increasing, the importance of understanding DTG's effects on neurodevelopment cannot be understated.

10.5.7 Efavirenz

Efavirenz (EFV) is a nonnucleoside reverse transcriptase inhibitor (NNRTI) used to treat HIV-1 infection. Concerns were raised in 1998 for the periconceptional use of EFV following a report of severe birth defects in a quarter of nonhuman primates (Cadman, 1998). This led to a recommendation by the Food and Drug Administration (FDA) for women receiving EFV to avoid pregnancy (Nightingale, 1998). Since then, findings on the safety of periconceptional EFV use have been inconsistent. In 2002, out of three women who conceived while on efavirenz, one NTD was reported (De Santis et al., 2002). However, reports from the Tsepamo study in Botswana (Zash et al., 2019), a pregnancy study in South Africa (Bera et al., 2010), and the European Pregnancy and Pediatric HIV Cohort Collaboration Study Group (Martinez de Tejada et al., 2019) suggest that EFV is not associated with an increase in the prevalence of birth defects following EFV exposure at conception or during the first trimester. With fewer concerns for severe birth defects, EFV is currently recommended by the US DHHS as an alternative for use during pregnancy for ART naïve women and women planning to become pregnant (recommendations: updated on January 31, 2024). It is also recommended that women on an EFV-containing regimen continue EFV use after becoming pregnant.

While the concern for severe birth defects has been reduced, EFV has now been linked to other adverse birth outcomes and neurodevelopmental issues. A multicenter cohort in the Republic of Congo found an increased incidence of adverse pregnancy outcomes, including low birth weight, premature delivery, stillbirth, and abortion in women treated with EFV compared to a non-EFV regimen (Bisio et al., 2015). In utero EFV exposure was associated with a higher risk of neurological case status, which included microcephaly, febrile seizures, seizure disorders, ophthalmologic disorders, and other neurologic disorders, which was stronger when comparing EFV exposure at conception to no exposure during pregnancy (Crowell et al., 2020). A recent report from the Surveillance Monitoring for ART Toxicity (SMARTT) study did find that in utero exposure to EFV was associated with an increased risk of microcephaly, which led to low scores on neurodevelopment assessments at ages 1 and 5 as well as a high prevalence of neurodevelopmental impairment (Williams et al., 2020). Further, an assessment of 24-month-old children from Botswana found that EFV-exposed children had lower scores on performance-based receptive language scales and were reported to have impairments in motor and social–emotional/self-regulatory skills compared to non-EFV-exposed peers (Cassidy et al., 2019).

These findings underscored the need for an understanding of the mechanisms of developmental toxicity. In rats, perinatal EFV exposure was associated with lower body weight, delayed reflex and motor development, and persistent structural changes in the motor cortex in adulthood (van de Wijer et al., 2019). Structural changes in the motor cortex were attributed to a reduction in mature neurons due to neuronal cell death, which was supported by a previous in vitro study revealing cell toxicity in response to EFV (Arend et al., 2016). It has also been found that EFV can affect the serotonergic system (Dalwadi et al., 2016; van de Wijer et al., 2019) and alter mitochondrial respiratory function, leading to mitochondrial dysfunction and oxidative stress (Funes et al., 2014, 2015). This could contribute to developmental toxicity. Further studies are necessary to understand the mechanisms of EFV-related detrimental effects on neurodevelopment.

10.5.8 Polycyclic Aromatic Hydrocarbons

Polycyclic aromatic hydrocarbons (PAHs) are persistent organic pollutants found in the environment. These are a product of incomplete combustion of organic material (ATSDR, 1995). Exposure of PAHs to humans occurs through air, dietary sources, drinking water, and soil (ATSDR, 1995). Benzo[a]pyrene (BaP) is a model pollutant commonly used to study PAHs. BaP and other PAHs are known to cause harmful effects in adults, including neurotoxicity. Determining the potential adverse effects of PAHs on the developing brain has gained research interest (Bukowska et al., 2022; Chepelev et al., 2015; Olasehinde & Olaniran, 2022). BaP exposure during pregnancy has been linked to adverse birth outcomes including pregnancy loss, reduced birth weights, and birth defects (Cao et al., 2021; Duarte-Salles et al., 2012; Legraverend et al., 1984; Ren et al., 2011; Ye et al., 2020). Further, studies in children have associated in utero exposure to PAHs, including BaP, with lower neurobehavioral development scores, slower processing speeds, and ADHD manifestations lasting into childhood (Cao et al., 2020; Peterson et al., 2015). Moreover, animal studies have found that developmental exposure to BaP is associated with learning and memory impairments, delays in normal reflex development, and hyperactivity (Bouayed et al., 2009; Chen et al., 2012; Gao et al., 2015; Honaker et al., 2022; Knecht et al., 2017). The mechanisms behind the fetal neurotoxicity of PAHs are incompletely understood. As the metabolism of BaP can result in the production of ROS and reactive metabolites, potential mechanisms include disturbances in neurotransmitter signaling, oxidative stress, changes in gene expression, and formation of DNA adducts, etc. (Bouayed et al., 2012, 2009; Elfawy et al., 2021; Gao et al., 2015; Gesto et al., 2009). Further, genetic polymorphisms in genes responsible for BaP metabolism, such as CYP1A1, CYP1A2, and CYP1B1, could lead to more susceptibility to the adverse effects of in utero BaP or PAH exposures (Honaker et al., 2022; Legraverend et al., 1984).

10.6 Review Questions

1. Which one of the following mechanisms defines closure of the neural tube?
   (a) Neurogenesis
   (b) Neurulation
   (c) Synaptogenesis
2. Which one of the following failed neurodevelopmental mechanisms leads to neural tube defects (NTDs)?
   (a) Neurogenesis
   (b) Neurulation
   (c) Synaptogenesis
3. Is autism spectrum disorder (ASD) a neurodevelopmental disorder?
   (a) Yes
   (b) No
4. Which growth factor contributes to maintaining neural progenitor cells?
   (a) FGF
   (b) EGF
   (c) Both a and b
   (d) None of the above
5. Would alcohol consumption during pregnancy cause birth defects and neurodevelopmental malformations?
   (a) Yes
   (b) No

10.7 Answers

1. b
2. b
3. a
4. c
5. a

Acknowledgments The authors thank Drs. Tomomi Kiyota and Howard E. Gendelman for their generosity in approving the use of Figs. 2 and 3. These figures were part of the chapter, "Neurogenesis and Brain Repair," in the 2nd edition of Neuroimmune Pharmacology.

References

Acuff-Smith, K. D., Schilling, M. A., Fisher, J. E., & Vorhees, C. V. (1996). Stage-specific effects of prenatal d-methamphetamine exposure on behavioral and eye development in rats. *Neurotoxicology and Teratology, 18*(2), 199–215. https://doi.org/10.1016/0892-0362(95)02015-2

Agency, U. S. E. P. (2013). Neurodevelopmental disorders. In *America's children and the enviroment* (3rd ed., pp. 233–253). CreateSpace Publishing.

Altmann, C. R., & Brivanlou, A. H. (2001). Neural patterning in the vertebrate embryo. *International Review of Cytology, 203*, 447–482. https://doi.org/10.1016/s0074-7696(01)03013-3

Amusan, P., Power, C., Gill, M. J., Gomez, D., Johnson, E., Rubin, L. H., & Fujiwara, E. (2020). Lifetime antiretroviral exposure and neurocognitive impairment in HIV. *Journal of Neurovirology, 26*(5), 743–753. https://doi.org/10.1007/s13365-020-00870-z

Arend, C., Rother, A., Stolte, S., & Dringen, R. (2016). Consequences of a chronic exposure of cultured brain astrocytes to the antiretroviral drug efavirenz and its primary metabolite 8-Hydroxy efavirenz. *Neurochemical Research, 41*(12), 3278–3288. https://doi.org/10.1007/s11064-016-2059-x

ATSDR. (1995). *Toxicological profile for polycyclic aromatic hydrocarbons*. Agency for Toxic Substances and Disease Registry (US).

Avagliano, L., Massa, V., George, T. M., Qureshy, S., Bulfamante, G. P., & Finnell, R. H. (2019). Overview on neural tube defects: From development to physical characteristics. *Birth Defects Research, 111*(19), 1455–1467. https://doi.org/10.1002/bdr2.1380

Bade, A. N., McMillan, J., Liu, Y., Edagwa, B. J., & Gendelman, H. E. (2021). Dolutegravir inhibition of matrix metalloproteinases affects mouse neurodevelopment. *Molecular Neurobiology*. https://doi.org/10.1007/s12035-021-02508-5

Badescu, G. M., Filfan, M., Sandu, R. E., Surugiu, R., Ciobanu, O., & Popa-Wagner, A. (2016). Molecular mechanisms underlying neurodevelopmental disorders, ADHD and autism. *Romanian Journal of Morphology and Embryology, 57*(2), 361–366. https://www.ncbi.nlm.nih.gov/pubmed/27516006

Bae, S. M., & Hong, J. Y. (2018). The Wnt signaling pathway and related therapeutic drugs in autism spectrum disorder. *Clinical Psychopharmacology and Neuroscience, 16*(2), 129–135. https://doi.org/10.9758/cpn.2018.16.2.129

Barrow Heaton, M. B., Kidd, K., Bradley, D., Paiva, M., Mitchell, J., & Walker, D. W. (1999). Prenatal ethanol exposure reduces spinal cord motoneuron number in the fetal rat but does not affect GDNF target tissue protein. *Developmental Neuroscience, 21*(6), 444–452. https://doi.org/10.1159/000017412

Bennett, B. A., Hyde, C. E., Pecora, J. R., & Clodfelter, J. E. (1993). Differing neurotoxic potencies of methamphetamine, mazindol, and cocaine in mesencephalic cultures. *Journal of Neurochemistry, 60*(4), 1444–1452. https://doi.org/10.1111/j.1471-4159.1993.tb03307.x

Bera, E., McCausland, K., Nonkwelo, R., Mgudlwa, B., Chacko, S., & Majeke, B. (2010). Birth defects following exposure to efavirenz-based antiretroviral therapy during pregnancy: A study at a regional South African hospital. *AIDS, 24*(2), 283–289. https://doi.org/10.1097/QAD.0b013e328333af32

Binder, T., & Vavrinkova, B. (2008). Prospective randomised comparative study of the effect of buprenorphine, methadone and heroin on the course of pregnancy, birthweight of newborns, early postpartum adaptation and course of the neonatal abstinence syndrome (NAS) in women followed up in the outpatient department. *Neuro*

Endocrinology Letters, 29(1), 80–86. https://www.ncbi.nlm.nih.gov/pubmed/18283247

Bisio, F., Nicco, E., Calzi, A., Giacobbe, D. R., Mesini, A., Banguissa, H., Vividila, N. E., Mahoungou, P., Boumba, J. D., Mboungou, F. A., Bruzzone, B., Ratto, S., Icardi, G., Viscoli, C., & Bruzzi, P. (2015). Pregnancy outcomes following exposure to efavirenz-based antiretroviral therapy in the Republic of Congo. *The New Microbiologica, 38*(2), 185–192. https://www.ncbi.nlm.nih.gov/pubmed/25938743

Bittigau, P., Sifringer, M., Genz, K., Reith, E., Pospischil, D., Govindarajalu, S., Dzietko, M., Pesditschek, S., Mai, I., Dikranian, K., Olney, J. W., & Ikonomidou, C. (2002). Antiepileptic drugs and apoptotic neurodegeneration in the developing brain. *Proceedings of the National Academy of Sciences of the United States of America, 99*(23), 15089–15094. https://doi.org/10.1073/pnas.222550499

Bjornsson, C. S., Apostolopoulou, M., Tian, Y., & Temple, S. (2015). It takes a village: Constructing the neurogenic niche. *Developmental Cell, 32*(4), 435–446. https://doi.org/10.1016/j.devcel.2015.01.010

Blum, K., Chen, A. L., Braverman, E. R., Comings, D. E., Chen, T. J., Arcuri, V., Blum, S. H., Downs, B. W., Waite, R. L., Notaro, A., Lubar, J., Williams, L., Prihoda, T. J., Palomo, T., & Oscar-Berman, M. (2008). Attention-deficit-hyperactivity disorder and reward deficiency syndrome. *Neuropsychiatric Disease and Treatment, 4*(5), 893–918. https://doi.org/10.2147/ndt.s2627

Bollen, P., Freriksen, J., Konopnicki, D., Weizsacker, K., Hidalgo Tenorio, C., Molto, J., Taylor, G., Alba-Alejandre, I., van Crevel, R., Colbers, A., Burger, D., & Pharmacokinetics of, A. a. i. H. I. V. i. p. w. N. (2021). The effect of pregnancy on the pharmacokinetics of total and unbound dolutegravir and its main metabolite in women living with human immunodeficiency virus. *Clinical Infectious Diseases, 72*(1), 121–127. https://doi.org/10.1093/cid/ciaa006

Boronat, S., Sanchez-Montanez, A., Gomez-Barros, N., Jacas, C., Martinez-Ribot, L., Vazquez, E., & Del Campo, M. (2017). Correlation between morphological MRI findings and specific diagnostic categories in fetal alcohol spectrum disorders. *European Journal of Medical Genetics, 60*(1), 65–71. https://doi.org/10.1016/j.ejmg.2016.09.003

Botto, L. D., Moore, C. A., Khoury, M. J., & Erickson, J. D. (1999). Neural-tube defects. *The New England Journal of Medicine, 341*(20), 1509–1519. https://doi.org/10.1056/NEJM199911113412006

Bouayed, J., Desor, F., Rammal, H., Kiemer, A. K., Tybl, E., Schroeder, H., Rychen, G., & Soulimani, R. (2009). Effects of lactational exposure to benzo[alpha]pyrene (B[alpha]P) on postnatal neurodevelopment, neuronal receptor gene expression and behaviour in mice. *Toxicology, 259*(3), 97–106. https://doi.org/10.1016/j.tox.2009.02.010

Bouayed, J., Bohn, T., Tybl, E., Kiemer, A. K., & Soulimani, R. (2012). Benzo[alpha]pyrene-induced anti-depressive-like behaviour in adult female mice: Role of monoaminergic systems. *Basic & Clinical Pharmacology & Toxicology, 110*(6), 544–550. https://doi.org/10.1111/j.1742-7843.2011.00853.x

Bourque, F., van der Ven, E., & Malla, A. (2011). A meta-analysis of the risk for psychotic disorders among first- and second-generation immigrants. *Psychological Medicine, 41*(5), 897–910. https://doi.org/10.1017/S0033291710001406

Bromley, R., Weston, J., Adab, N., Greenhalgh, J., Sanniti, A., McKay, A. J., Tudur Smith, C., & Marson, A. G. (2014). Treatment for epilepsy in pregnancy: Neurodevelopmental outcomes in the child. *Cochrane Database of Systematic Reviews, 2014*(10), CD010236. https://doi.org/10.1002/14651858.CD010236.pub2

Brown, A. S., Hooton, J., Schaefer, C. A., Zhang, H., Petkova, E., Babulas, V., Perrin, M., Gorman, J. M., & Susser, E. S. (2004). Elevated maternal interleukin-8 levels and risk of schizophrenia in adult offspring. *The American Journal of Psychiatry, 161*(5), 889–895. https://doi.org/10.1176/appi.ajp.161.5.889

Budday, S., Steinmann, P., & Kuhl, E. (2015). Physical biology of human brain development. *Frontiers in Cellular Neuroscience, 9*, 257. https://doi.org/10.3389/fncel.2015.00257

Bukowska, B., Mokra, K., & Michalowicz, J. (2022). Benzo[a]pyrene-environmental occurrence, human exposure, and mechanisms of toxicity. *International Journal of Molecular Sciences, 23*(11). https://doi.org/10.3390/ijms23116348

Byrne, M., Agerbo, E., Bennedsen, B., Eaton, W. W., & Mortensen, P. B. (2007). Obstetric conditions and risk of first admission with schizophrenia: A Danish national register based study. *Schizophrenia Research, 97*(1–3), 51–59. https://doi.org/10.1016/j.schres.2007.07.018

Cabrera, R. M., Souder, J. P., Steele, J. W., Yeo, L., Tukeman, G., Gorelick, D. A., & Finnell, R. H. (2019). The antagonism of folate receptor by dolutegravir: Developmental toxicity reduction by supplemental folic acid. *AIDS, 33*(13), 1967–1976. https://doi.org/10.1097/QAD.0000000000002289

Cadman, J. (1998). Efavirenz pregnancy warning. *GMHC Treatment Issues, 12*(3), 12. https://www.ncbi.nlm.nih.gov/pubmed/11365392

Campbell, K., & Gotz, M. (2002). Radial glia: Multi-purpose cells for vertebrate brain development. *Trends in Neurosciences, 25*(5), 235–238. https://doi.org/10.1016/s0166-2236(02)02156-2

Canetta, S., Sourander, A., Surcel, H. M., Hinkka-Yli-Salomaki, S., Leiviska, J., Kellendonk, C., McKeague, I. W., & Brown, A. S. (2014). Elevated maternal C-reactive protein and increased risk of schizophrenia in a national birth cohort. *The American Journal of Psychiatry, 171*(9), 960–968. https://doi.org/10.1176/appi.ajp.2014.13121579

Cannon, M., Jones, P. B., & Murray, R. M. (2002). Obstetric complications and schizophrenia: Historical and meta-analytic review. *The American Journal of Psychiatry, 159*(7), 1080–1092. https://doi.org/10.1176/appi.ajp.159.7.1080

Cantor-Graae, E., & Selten, J. P. (2005). Schizophrenia and migration: A meta-analysis and review. *The American Journal of Psychiatry, 162*(1), 12–24. https://doi.org/10.1176/appi.ajp.162.1.12

Cao, X., Li, J., Cheng, L., Deng, Y., Li, Y., Yan, Z., Duan, L., Yang, J., Niu, Q., Perera, F., Nie, J., & Tang, D. (2020). The associations between prenatal exposure to polycyclic aromatic hydrocarbon metabolites, umbilical cord blood mitochondrial DNA copy number, and children's neurobehavioral development. *Environmental Pollution, 265*(Pt B), 114594. https://doi.org/10.1016/j.envpol.2020.114594

Cao, C., Jia, Z., Shao, M., Li, R., Sun, Q., & Liu, D. (2021). Prenatal exposure to polycyclic aromatic hydrocarbons could increase the risk of low birth weight by affecting the DNA methylation states in a Chinese cohort. *Reproductive Biology, 21*(4), 100574. https://doi.org/10.1016/j.repbio.2021.100574

Cassidy, A. R., Williams, P. L., Leidner, J., Mayondi, G., Ajibola, G., Makhema, J., Holding, P. A., Powis, K. M., Batlang, O., Petlo, C., Shapiro, R., Kammerer, B., & Lockman, S. (2019). In utero efavirenz exposure and neurodevelopmental outcomes in HIV-exposed uninfected children in Botswana. *The Pediatric Infectious Disease Journal, 38*(8), 828–834. https://doi.org/10.1097/INF.0000000000002332

Castellanos, F. X., Lee, P. P., Sharp, W., Jeffries, N. O., Greenstein, D. K., Clasen, L. S., Blumenthal, J. D., James, R. S., Ebens, C. L., Walter, J. M., Zijdenbos, A., Evans, A. C., Giedd, J. N., & Rapoport, J. L. (2002). Developmental trajectories of brain volume abnormalities in children and adolescents with attention-deficit/hyperactivity disorder. *JAMA, 288*(14), 1740–1748. https://doi.org/10.1001/jama.288.14.1740

Champagne, F., & Meaney, M. J. (2001). Like mother, like daughter: Evidence for non-genomic transmission of parental behavior and stress responsivity. *Progress in Brain Research, 133*, 287–302. https://doi.org/10.1016/s0079-6123(01)33022-4

Chang, L., Smith, L. M., LoPresti, C., Yonekura, M. L., Kuo, J., Walot, I., & Ernst, T. (2004). Smaller subcortical volumes and cognitive deficits in children with prenatal methamphetamine exposure. *Psychiatry Research, 132*(2), 95–106. https://doi.org/10.1016/j.pscychresns.2004.06.004

Chang, L., Alicata, D., Ernst, T., & Volkow, N. (2007). Structural and metabolic brain changes in the striatum associated with methamphetamine abuse. *Addiction, 102*(Suppl 1), 16–32. https://doi.org/10.1111/j.1360-0443.2006.01782.x

Chen, C., Tang, Y., Jiang, X., Qi, Y., Cheng, S., Qiu, C., Peng, B., & Tu, B. (2012). Early postnatal benzo(a)pyrene exposure in Sprague-Dawley rats causes persistent neurobehavioral impairments that emerge postnatally and continue into adolescence and adulthood. *Toxicological Sciences, 125*(1), 248–261. https://doi.org/10.1093/toxsci/kfr265

Chepelev, N. L., Moffat, I. D., Bowers, W. J., & Yauk, C. L. (2015). Neurotoxicity may be an overlooked consequence of benzo[a]pyrene exposure that is relevant to human health risk assessment. *Mutation Research, Reviews in Mutation Research, 764*, 64–89. https://doi.org/10.1016/j.mrrev.2015.03.001

Chess, S. (1971). Autism in children with congenital rubella. *Journal of Autism and Childhood Schizophrenia, 1*(1), 33–47. https://doi.org/10.1007/BF01537741

Chung, D. D., Pinson, M. R., Bhenderu, L. S., Lai, M. S., Patel, R. A., & Miranda, R. C. (2021). Toxic and teratogenic effects of prenatal alcohol exposure on fetal development, adolescence, and adulthood. *International Journal of Molecular Sciences, 22*(16). https://doi.org/10.3390/ijms22168785

Cleary, B. J., Eogan, M., O'Connell, M. P., Fahey, T., Gallagher, P. J., Clarke, T., White, M. J., McDermott, C., O'Sullivan, A., Carmody, D., Gleeson, J., & Murphy, D. J. (2012). Methadone and perinatal outcomes: A prospective cohort study. *Addiction, 107*(8), 1482–1492. https://doi.org/10.1111/j.1360-0443.2012.03844.x

Cohen, M., Briscoe, J., & Blassberg, R. (2013). Morphogen interpretation: The transcriptional logic of neural tube patterning. *Current Opinion in Genetics & Development, 23*(4), 423–428. https://doi.org/10.1016/j.gde.2013.04.003

Comings, D. E. (2001). Clinical and molecular genetics of ADHD and Tourette syndrome. Two related polygenic disorders. *Annals of the New York Academy of Sciences, 931*, 50–83. https://doi.org/10.1111/j.1749-6632.2001.tb05773.x

Copp, A. J., & Greene, N. D. (2013). Neural tube defects--disorders of neurulation and related embryonic processes. *Wiley Interdisciplinary Reviews: Developmental Biology, 2*(2), 213–227. https://doi.org/10.1002/wdev.71

Covington, C. Y., Nordstrom-Klee, B., Ager, J., Sokol, R., & Delaney-Black, V. (2002). Birth to age 7 growth of children prenatally exposed to drugs: A prospective cohort study. *Neurotoxicology and Teratology, 24*(4), 489–496. https://doi.org/10.1016/s0892-0362(02)00233-7

Crandall, J. E., Hackett, H. E., Tobet, S. A., Kosofsky, B. E., & Bhide, P. G. (2004). Cocaine exposure decreases GABA neuron migration from the ganglionic eminence to the cerebral cortex in embryonic mice. *Cerebral Cortex, 14*(6), 665–675. https://doi.org/10.1093/cercor/bhh027

Crowell, C. S., Williams, P. L., Yildirim, C., Van Dyke, R. B., Smith, R., Chadwick, E. G., Seage, G. R., 3rd, Diperna, A., Hazra, R., & Pediatric, H. I. V. A. C. S. (2020). Safety of in-utero antiretroviral exposure: Neurologic outcomes in children who are HIV-exposed but uninfected. *AIDS, 34*(9), 1377–1387. https://doi.org/10.1097/QAD.0000000000002550

Dadalko, O. I., & Travers, B. G. (2018). Evidence for brainstem contributions to autism spectrum disorders. *Frontiers in Integrative Neuroscience, 12*, 47. https://doi.org/10.3389/fnint.2018.00047

Dalwadi, D. A., Kim, S., Amdani, S. M., Chen, Z., Huang, R. Q., & Schetz, J. A. (2016). Molecular mechanisms of serotonergic action of the HIV-1 antiretroviral efavirenz. *Pharmacological Research, 110*, 10–24. https://doi.org/10.1016/j.phrs.2016.04.028

Dattel, B. J. (1990). Substance abuse in pregnancy. *Seminars in Perinatology, 14*(2), 179–187. https://www.ncbi.nlm.nih.gov/pubmed/2187251

De Carlos, J. A., & O'Leary, D. D. (1992). Growth and targeting of subplate axons and establishment of major cortical pathways. *The Journal of Neuroscience, 12*(4), 1194–1211. https://doi.org/10.1523/JNEUROSCI.12-04-01194.1992

De Santis, M., Carducci, B., De Santis, L., Cavaliere, A. F., & Straface, G. (2002). Periconceptional exposure to efavirenz and neural tube defects. *Archives of Internal Medicine, 162*(3), 355. https://doi.org/10.1001/archinte.162.3.355

Del Rio, J. A., Heimrich, B., Borrell, V., Forster, E., Drakew, A., Alcantara, S., Nakajima, K., Miyata, T., Ogawa, M., Mikoshiba, K., Derer, P., Frotscher, M., & Soriano, E. (1997). A role for Cajal-Retzius cells and reelin in the development of hippocampal connections. *Nature, 385*(6611), 70–74. https://doi.org/10.1038/385070a0

Derauf, C., LaGasse, L., Smith, L., Newman, E., Shah, R., Arria, A., Huestis, M., Haning, W., Strauss, A., Della Grotta, S., Dansereau, L., Lin, H., & Lester, B. (2011). Infant temperament and high-risk environment relate to behavior problems and language in toddlers. *Journal of Developmental and Behavioral Pediatrics, 32*(2), 125–135. https://doi.org/10.1097/DBP.0b013e31820839d7

Derer, P., & Derer, M. (1990). Cajal-Retzius cell ontogenesis and death in mouse brain visualized with horseradish peroxidase and electron microscopy. *Neuroscience, 36*(3), 839–856. https://doi.org/10.1016/0306-4522(90)90027-2

Dessaud, E., McMahon, A. P., & Briscoe, J. (2008). Pattern formation in the vertebrate neural tube: A sonic hedgehog morphogen-regulated transcriptional network. *Development, 135*(15), 2489–2503. https://doi.org/10.1242/dev.009324

DiMaio, S., Grizenko, N., & Joober, R. (2003). Dopamine genes and attention-deficit hyperactivity disorder: A review. *Journal of Psychiatry & Neuroscience, 28*(1), 27–38. https://www.ncbi.nlm.nih.gov/pubmed/12587848

Dixon, S. D., & Bejar, R. (1989). Echoencephalographic findings in neonates associated with maternal cocaine and methamphetamine use: Incidence and clinical correlates. *The Journal of Pediatrics, 115*(5 Pt 1), 770–778. https://doi.org/10.1016/s0022-3476(89)80661-4

Doernberg, E., & Hollander, E. (2016). Neurodevelopmental disorders (ASD and ADHD): DSM-5, ICD-10, and ICD-11. *CNS Spectrums, 21*(4), 295–299. https://doi.org/10.1017/S1092852916000262

Dou, X., Menkari, C., Mitsuyama, R., Foroud, T., Wetherill, L., Hammond, P., Suttie, M., Chen, X., Chen, S. Y., Charness, M. E., & Collaborative Initiative on Fetal Alcohol Spectrum, D. (2018). L1 coupling to ankyrin and the spectrin-actin cytoskeleton modulates ethanol inhibition of L1 adhesion and ethanol teratogenesis. *The FASEB Journal, 32*(3), 1364–1374. https://doi.org/10.1096/fj.201700970

Duarte-Salles, T., Mendez, M. A., Morales, E., Bustamante, M., Rodriguez-Vicente, A., Kogevinas, M., & Sunyer, J. (2012). Dietary benzo(a)pyrene and fetal growth: Effect modification by vitamin C intake and glutathione S-transferase P1 polymorphism. *Environment International, 45*, 1–8. https://doi.org/10.1016/j.envint.2012.04.002

Eason, J., Williams, A. L., Chawla, B., Apsey, C., & Bohnsack, B. L. (2017). Differences in neural crest sensitivity to ethanol account for the infrequency of anterior segment defects in the eye compared with craniofacial anomalies in a zebrafish model of fetal alcohol syndrome. *Birth Defects Research, 109*(15), 1212–1227. https://doi.org/10.1002/bdr2.1069

Elfawy, H. A., Anupriya, S., Mohanty, S., Patel, P., Ghosal, S., Panda, P. K., Das, B., Verma, S. K., & Patnaik, S. (2021). Molecular toxicity of Benzo(a)pyrene mediated by elicited oxidative stress infer skeletal deformities and apoptosis in embryonic zebrafish. *Science of the Total Environment, 789*, 147989. https://doi.org/10.1016/j.scitotenv.2021.147989

Emery, J. L., & Lendon, R. G. (1973). The local cord lesion in neurospinal dysraphism (meningomyelocele). *The Journal of Pathology, 110*(1), 83–96. https://doi.org/10.1002/path.1711100110

Eshraghi, A. A., Liu, G., Kay, S. S., Eshraghi, R. S., Mittal, J., Moshiree, B., & Mittal, R. (2018). Epigenetics and autism spectrum disorder: Is there a correlation? *Frontiers in Cellular Neuroscience, 12*, 78. https://doi.org/10.3389/fncel.2018.00078

Eubig, P. A., Aguiar, A., & Schantz, S. L. (2010). Lead and PCBs as risk factors for attention deficit/hyperactivity disorder. *Environmental Health Perspectives, 118*(12), 1654–1667. https://doi.org/10.1289/ehp.0901852

Faraone, S. V., & Khan, S. A. (2006). Candidate gene studies of attention-deficit/hyperactivity disorder. *The Journal of Clinical Psychiatry, 67*(Suppl 8), 13–20. https://www.ncbi.nlm.nih.gov/pubmed/16961425

Faraone, S. V., Sergeant, J., Gillberg, C., & Biederman, J. (2003). The worldwide prevalence of ADHD: Is it an American condition? *World Psychiatry, 2*(2), 104–113. https://www.ncbi.nlm.nih.gov/pubmed/16946911

Fatemi, S. H., & Folsom, T. D. (2009). The neurodevelopmental hypothesis of schizophrenia, revisited. *Schizophrenia Bulletin, 35*(3), 528–548. https://doi.org/10.1093/schbul/sbn187

Florio, M., & Huttner, W. B. (2014). Neural progenitors, neurogenesis and the evolution of the neocortex. *Development, 141*(11), 2182–2194. https://doi.org/10.1242/dev.090571

Ford, D. H., & Rhines, R. K. (1979). Prenatal exposure to methadone HCL in relationship to body and brain growth in the rat. *Acta Neurologica Scandinavica, 59*(5), 248–262. https://doi.org/10.1111/j.1600-0404.1979.tb02935.x

Froehlich, T. E., Lanphear, B. P., Auinger, P., Hornung, R., Epstein, J. N., Braun, J., & Kahn, R. S. (2009). Association of tobacco and lead exposures with attention-deficit/hyperactivity disorder. *Pediatrics, 124*(6), e1054–e1063. https://doi.org/10.1542/peds.2009-0738

Frost, D. O., & Cadet, J. L. (2000). Effects of methamphetamine-induced neurotoxicity on the development of neural circuitry: A hypothesis. *Brain Research. Brain Research Reviews, 34*(3), 103–118. https://doi.org/10.1016/s0165-0173(00)00042-4

Fu, D., Wu, D. D., Guo, H. L., Hu, Y. H., Xia, Y., Ji, X., Fang, W. R., Li, Y. M., Xu, J., Chen, F., & Liu, Q. Q. (2021). The mechanism, clinical efficacy, safety, and dosage regimen of atomoxetine for ADHD therapy in children: A narrative review. *Frontiers in Psychiatry, 12*, 780921. https://doi.org/10.3389/fpsyt.2021.780921

Funes, H. A., Apostolova, N., Alegre, F., Blas-Garcia, A., Alvarez, A., Marti-Cabrera, M., & Esplugues, J. V. (2014). Neuronal bioenergetics and acute mitochondrial dysfunction: A clue to understanding the central nervous system side effects of efavirenz. *The Journal of Infectious Diseases, 210*(9), 1385–1395. https://doi.org/10.1093/infdis/jiu273

Funes, H. A., Blas-Garcia, A., Esplugues, J. V., & Apostolova, N. (2015). Efavirenz alters mitochondrial respiratory function in cultured neuron and glial cell lines. *The Journal of Antimicrobial Chemotherapy, 70*(8), 2249–2254. https://doi.org/10.1093/jac/dkv098

Gal, J. S., Morozov, Y. M., Ayoub, A. E., Chatterjee, M., Rakic, P., & Haydar, T. F. (2006). Molecular and morphological heterogeneity of neural precursors in the mouse neocortical proliferative zones. *The Journal of Neuroscience, 26*(3), 1045–1056. https://doi.org/10.1523/JNEUROSCI.4499-05.2006

Gangisetty, O., Bekdash, R., Maglakelidze, G., & Sarkar, D. K. (2014). Fetal alcohol exposure alters proopiomelanocortin gene expression and hypothalamic-pituitary-adrenal axis function via increasing MeCP2 expression in the hypothalamus. *PLoS One, 9*(11), e113228. https://doi.org/10.1371/journal.pone.0113228

Gao, Z., Massimi, L., Rogerio, S., Raybaud, C., & Di Rocco, C. (2014). Vertex cephaloceles: A review. *Child's Nervous System, 30*(1), 65–72. https://doi.org/10.1007/s00381-013-2249-7

Gao, D., Wu, M., Wang, C., Wang, Y., & Zuo, Z. (2015). Chronic exposure to low benzo[a]pyrene level causes neurodegenerative disease-like syndromes in zebrafish (Danio rerio). *Aquatic Toxicology, 167*, 200–208. https://doi.org/10.1016/j.aquatox.2015.08.013

Gesto, M., Tintos, A., Soengas, J. L., & Miguez, J. M. (2009). beta-Naphthoflavone and benzo(a)pyrene alter dopaminergic, noradrenergic, and serotonergic systems in brain and pituitary of rainbow trout (Oncorhynchus mykiss). *Ecotoxicology and Environmental Safety, 72*(1), 191–198. https://doi.org/10.1016/j.ecoenv.2008.03.007

Geva, R., & Feldman, R. (2008). A neurobiological model for the effects of early brainstem functioning on the development of behavior and emotion regulation in infants: Implications for prenatal and perinatal risk. *Journal of Child Psychology and Psychiatry, 49*(10), 1031–1041. https://doi.org/10.1111/j.1469-7610.2008.01918.x

Gibb, R., & Kovalchuk, A. (2017). Chapter 1 - Brain development. In R. Gibb & B. Kolb (Eds.), *The neurobiology of brain and behavioral development* (pp. 3–27). https://doi.org/10.1016/C2015-0-00695-5

Gilmore, J. C., Hoque, M. T., Dai, W., Mohan, H., Dunk, C., Serghides, L., & Bendayan, R. (2022). Interaction between dolutegravir and folate transporters and receptor in human and rodent placenta. *eBioMedicine, 75*, 103771. https://doi.org/10.1016/j.ebiom.2021.103771

Golden, C. E., Buxbaum, J. D., & De Rubeis, S. (2018). Disrupted circuits in mouse models of autism spectrum disorder and intellectual disability. *Current Opinion in Neurobiology, 48*, 106–112. https://doi.org/10.1016/j.conb.2017.11.006

Goldstein, R. B., & Filly, R. A. (1988). Prenatal diagnosis of anencephaly: Spectrum of sonographic appearances and distinction from the amniotic band syndrome. *AJR. American Journal of Roentgenology, 151*(3), 547–550. https://doi.org/10.2214/ajr.151.3.547

Gotz, M., & Huttner, W. B. (2005). The cell biology of neurogenesis. *Nature Reviews. Molecular Cell Biology, 6*(10), 777–788. https://doi.org/10.1038/nrm1739

Gouin, K., Murphy, K., Shah, P. S., Knowledge Synthesis group on Determinants of Low Birth, W, & Preterm, B. (2011). Effects of cocaine use during pregnancy on low birthweight and preterm birth: Systematic review and metaanalyses. *American Journal of Obstetrics and Gynecology, 204*(4), 340 e341–340 e312. https://doi.org/10.1016/j.ajog.2010.11.013

Greene, N. D., & Copp, A. J. (2014). Neural tube defects. *Annual Review of Neuroscience, 37*, 221–242. https://doi.org/10.1146/annurev-neuro-062012-170354

Gurillo, P., Jauhar, S., Murray, R. M., & MacCabe, J. H. (2015). Does tobacco use cause psychosis? Systematic review and meta-analysis. *Lancet Psychiatry, 2*(8), 718–725. https://doi.org/10.1016/S2215-0366(15)00152-2

Halepoto, D. M., Bashir, S., Zeina, R., & Al-Ayadhi, L. Y. (2015). Correlation between Hedgehog (Hh) protein family and Brain-Derived Neurotrophic Factor (BDNF) in Autism Spectrum Disorder (ASD). *Journal of the College of Physicians and Surgeons–Pakistan, 25*(12), 882–885. https://www.ncbi.nlm.nih.gov/pubmed/26691363

Hamilton, L. R., Czoty, P. W., Gage, H. D., & Nader, M. A. (2010). Characterization of the dopamine receptor system in adult rhesus monkeys exposed to cocaine throughout gestation. *Psychopharmacology, 210*(4), 481–488. https://doi.org/10.1007/s00213-010-1847-2

Harris, J. C. (2014). New classification for neurodevelopmental disorders in DSM-5. *Current Opinion in Psychiatry, 27*(2), 95–97. https://doi.org/10.1097/YCO.0000000000000042

Heller, M., & Burd, L. (2014). Review of ethanol dispersion, distribution, and elimination from the fetal compartment. *Birth Defects Research. Part A, Clinical and Molecular Teratology, 100*(4), 277–283. https://doi.org/10.1002/bdra.23232

Honaker, A., Kyntchev, A., Foster, E., Clough, K., Hawk, G., Asiedu, E., Berling, K., DeBurger, E., Feltner, M., Ferguson, V., Forrest, P. T., Jenkins, K., Massie, L., Mullaguru, J., Niang, M. D., Perry, C., Sene, Y., Towell, A., & Curran, C. P. (2022). The behavioral effects of gestational and lactational benzo[a]pyrene exposure vary by sex and genotype in mice with differences at the Ahr and Cyp1a2 loci. *Neurotoxicology and Teratology, 89*, 107056. https://doi.org/10.1016/j.ntt.2021.107056

Houart, C., Caneparo, L., Heisenberg, C., Barth, K., Take-Uchi, M., & Wilson, S. (2002). Establishment of the telencephalon during gastrulation by local antagonism of Wnt signaling. *Neuron, 35*(2), 255–265. https://doi.org/10.1016/s0896-6273(02)00751-1

Howes, O. D., & Shatalina, E. (2022). Integrating the neurodevelopmental and dopamine hypotheses of schizophrenia and the role of cortical excitation-inhibition balance. *Biological Psychiatry, 92*(6), 501–513. https://doi.org/10.1016/j.biopsych.2022.06.017

Hymbaugh, K., Miller, L. A., Druschel, C. M., Podvin, D. W., Meaney, F. J., Boyle, C. A., & Team, F. A. (2002). A multiple source methodology for the surveillance of fetal alcohol syndrome--The Fetal Alcohol Syndrome Surveillance Network (FASSNet). *Teratology, 66*(Suppl 1), S41–S49. https://doi.org/10.1002/tera.90010

Hynd, G. W., Semrud-Clikeman, M., Lorys, A. R., Novey, E. S., & Eliopulos, D. (1990). Brain morphology in developmental dyslexia and attention deficit disorder/hyperactivity. *Archives of Neurology, 47*(8), 919–926. https://doi.org/10.1001/archneur.1990.00530080107018

Hynd, G. W., Semrud-Clikeman, M., Lorys, A. R., Novey, E. S., Eliopulos, D., & Lyytinen, H. (1991). Corpus callosum morphology in attention deficit-hyperactivity disorder: Morphometric analysis of MRI. *Journal of Learning Disabilities, 24*(3), 141–146. https://doi.org/10.1177/002221949102400302

Hynd, G. W., Hern, K. L., Novey, E. S., Eliopulos, D., Marshall, R., Gonzalez, J. J., & Voeller, K. K. (1993). Attention deficit-hyperactivity disorder and asymmetry of the caudate nucleus. *Journal of Child Neurology, 8*(4), 339–347. https://doi.org/10.1177/088307389300800409

Inui, T., Kumagaya, S., & Myowa-Yamakoshi, M. (2017). Neurodevelopmental hypothesis about the etiology of autism spectrum disorders. *Frontiers in Human Neuroscience, 11*, 354. https://doi.org/10.3389/fnhum.2017.00354

Jarmasz, J. S., Basalah, D. A., Chudley, A. E., & Del Bigio, M. R. (2017). Human brain abnormalities associated with prenatal alcohol exposure and fetal alcohol spectrum disorder. *Journal of Neuropathology and Experimental Neurology, 76*(9), 813–833. https://doi.org/10.1093/jnen/nlx064

Jeng, W., Wong, A. W., Ting, A. K. R., & Wells, P. G. (2005). Methamphetamine-enhanced embryonic oxidative DNA damage and neurodevelopmental deficits. *Free Radical Biology & Medicine, 39*(3), 317–326. https://doi.org/10.1016/j.freeradbiomed.2005.03.015

Jones, P. B., Rantakallio, P., Hartikainen, A. L., Isohanni, M., & Sipila, P. (1998). Schizophrenia as a long-term outcome of pregnancy, delivery, and perinatal complications: A 28-year follow-up of the 1966 north Finland general population birth cohort. *The American Journal of Psychiatry, 155*(3), 355–364. https://doi.org/10.1176/ajp.155.3.355

Kalkman, H. O. (2012). A review of the evidence for the canonical Wnt pathway in autism spectrum disorders. *Molecular Autism, 3*(1), 10. https://doi.org/10.1186/2040-2392-3-10

Kellogg, M., & Meador, K. J. (2017). Neurodevelopmental effects of antiepileptic drugs. *Neurochemical Research, 42*(7), 2065–2070. https://doi.org/10.1007/s11064-017-2262-4

Keshavan, M. S. (1999). Development, disease and degeneration in schizophrenia: A unitary pathophysiological model. *Journal of Psychiatric Research, 33*(6), 513–521. https://doi.org/10.1016/s0022-3956(99)00033-3

Keshavan, M., Lizano, P., & Prasad, K. (2020). The synaptic pruning hypothesis of schizophrenia: Promises and challenges. *World Psychiatry, 19*(1), 110–111. https://doi.org/10.1002/wps.20725

Kirkwood-Johnson, L., Katayama, N., & Marikawa, Y. (2021). Dolutegravir impairs stem cell-based 3D morphogenesis models in a manner dependent on dose and timing of exposure: An implication for its developmental toxicity. *Toxicological Sciences, 184*(2), 191–203. https://doi.org/10.1093/toxsci/kfab112

Kish, S. J. (2008). Pharmacologic mechanisms of crystal meth. *CMAJ, 178*(13), 1679–1682. https://doi.org/10.1503/cmaj.071675

Knecht, A. L., Truong, L., Simonich, M. T., & Tanguay, R. L. (2017). Developmental benzo[a]pyrene (B[a]P) exposure impacts larval behavior and impairs adult learning in zebrafish. *Neurotoxicology and Teratology, 59*, 27–34. https://doi.org/10.1016/j.ntt.2016.10.006

Koskinen, J., Lohonen, J., Koponen, H., Isohanni, M., & Miettunen, J. (2009). Prevalence of alcohol use disorders in schizophrenia--a systematic review and meta-analysis. *Acta Psychiatrica Scandinavica, 120*(2), 85–96. https://doi.org/10.1111/j.1600-0447.2009.01385.x

Kosmicki, J. A., Samocha, K. E., Howrigan, D. P., Sanders, S. J., Slowikowski, K., Lek, M., Karczewski, K. J., Cutler, D. J., Devlin, B., Roeder, K., Buxbaum, J. D., Neale, B. M., MacArthur, D. G., Wall, D. P., Robinson, E. B., & Daly, M. J. (2017). Refining the role of de novo protein-truncating variants in neurodevelopmental disorders by using population reference samples. *Nature Genetics, 49*(4), 504–510. https://doi.org/10.1038/ng.3789

Kreitchmann, R., Oliveira, F. R., & Sprinz, E. (2021). Two cases of neural tube defects with dolutegravir use at conception in south Brazil. *The Brazilian Journal of Infectious Diseases, 25*(2), 101572. https://doi.org/10.1016/j.bjid.2021.101572

Kwan, K. Y., Sestan, N., & Anton, E. S. (2012). Transcriptional co-regulation of neuronal migration and laminar identity in the neocortex. *Development, 139*(9), 1535–1546. https://doi.org/10.1242/dev.069963

Kwan, V., Unda, B. K., & Singh, K. K. (2016). Wnt signaling networks in autism spectrum disorder and intellectual disability. *Journal of Neurodevelopmental Disorders, 8*, 45. https://doi.org/10.1186/s11689-016-9176-3

Landmark, C. J. (2007). Targets for antiepileptic drugs in the synapse. *Medical Science Monitor, 13*(1), RA1-7. https://www.ncbi.nlm.nih.gov/pubmed/17179916

Le Dare, B., Lagente, V., & Gicquel, T. (2019). Ethanol and its metabolites: Update on toxicity, benefits, and focus on immunomodulatory effects. *Drug Metabolism Reviews, 51*(4), 545–561. https://doi.org/10.1080/03602532.2019.1679169

Lee, S. M., Tole, S., Grove, E., & McMahon, A. P. (2000). A local Wnt-3a signal is required for development of the mammalian hippocampus. *Development, 127*(3), 457–467. https://doi.org/10.1242/dev.127.3.457

Legraverend, C., Guenthner, T. M., & Nebert, D. W. (1984). Importance of the route of administration for genetic differences in benzo[a]pyrene-induced in utero toxicity and teratogenicity. *Teratology, 29*(1), 35–47. https://doi.org/10.1002/tera.1420290106

Lester, B. M., & Padbury, J. F. (2009). Third pathophysiology of prenatal cocaine exposure. *Developmental Neuroscience, 31*(1–2), 23–35. https://doi.org/10.1159/000207491

Lester, B. M., Tronick, E. Z., LaGasse, L., Seifer, R., Bauer, C. R., Shankaran, S., Bada, H. S., Wright, L. L., Smeriglio, V. L., Lu, J., Finnegan, L. P., & Maza, P. L. (2002). The maternal lifestyle study: Effects of substance exposure during pregnancy on neurodevelopmental outcome in 1-month-old infants. *Pediatrics, 110*(6), 1182–1192. https://doi.org/10.1542/peds.110.6.1182

Levine, T. A., Davie-Gray, A., Kim, H. M., Lee, S. J., & Woodward, L. J. (2021). Prenatal methadone exposure and child developmental outcomes in 2-year-old children. *Developmental Medicine and Child Neurology, 63*(9), 1114–1122. https://doi.org/10.1111/dmcn.14808

Li, W., Yao, A., Zhi, H., Kaur, K., Zhu, Y. C., Jia, M., Zhao, H., Wang, Q., Jin, S., Zhao, G., Xiong, Z. Q., & Zhang, Y. Q. (2016). Angelman syndrome protein Ube3a regulates synaptic growth and endocytosis by inhibiting BMP signaling in drosophila. *PLoS Genetics, 12*(5), e1006062. https://doi.org/10.1371/journal.pgen.1006062

Lipinski, R. J., Hammond, P., O'Leary-Moore, S. K., Ament, J. J., Pecevich, S. J., Jiang, Y., Budin, F., Parnell, S. E., Suttie, M., Godin, E. A., Everson, J. L., Dehart, D. B., Oguz, I., Holloway, H. T., Styner, M. A., Johnson, G. A., & Sulik, K. K. (2012). Ethanol-induced face-brain dysmorphology patterns are correlative and exposure-stage dependent. *PLoS One, 7*(8), e43067. https://doi.org/10.1371/journal.pone.0043067

Liu, J., Lester, B. M., Neyzi, N., Sheinkopf, S. J., Gracia, L., Kekatpure, M., & Kosofsky, B. E. (2013). Regional brain morphometry and impulsivity in adolescents following prenatal exposure to cocaine and tobacco. *JAMA Pediatrics, 167*(4), 348–354. https://doi.org/10.1001/jamapediatrics.2013.550

Maan, J. S., Duong, T. V. H., & Saadabadi, A. (2022). Carbamazepine. *StatPearls*. https://www.ncbi.nlm.nih.gov/pubmed/29494062

Madras, B. K. (2013). History of the discovery of the antipsychotic dopamine D2 receptor: A basis for the dopamine hypothesis of schizophrenia. *Journal of the History of the Neurosciences, 22*(1), 62–78. https://doi.org/10.1080/0964704X.2012.678199

Marconi, A., Di Forti, M., Lewis, C. M., Murray, R. M., & Vassos, E. (2016). Meta-analysis of the association between the level of cannabis use and risk of psychosis. *Schizophrenia Bulletin, 42*(5), 1262–1269. https://doi.org/10.1093/schbul/sbw003

Marthiens, V., & ffrench-Constant, C. (2009). Adherens junction domains are split by asymmetric division of embryonic neural stem cells. *EMBO Reports, 10*(5), 515–520. https://doi.org/10.1038/embor.2009.36

Martinez de Tejada, B., Gayet-Ageron, A., Winterfeld, U., Thorne, C., Favarato, G., European, P., & Paediatric, H. I. V. C. C. S. G. (2019). Birth defects after exposure to Efavirenz-Based antiretroviral therapy at conception/first trimester of pregnancy: A multicohort analysis. *Journal of Acquired Immune Deficiency Syndromes, 80*(3), 316–324. https://doi.org/10.1097/QAI.0000000000001922

Matok, I., Gorodischer, R., Koren, G., Landau, D., Wiznitzer, A., & Levy, A. (2009). Exposure to folic acid antagonists during the first trimester of pregnancy and the risk of major malformations. *British Journal of Clinical Pharmacology, 68*(6), 956–962. https://doi.org/10.1111/j.1365-2125.2009.03544.x

Mattson, S. N., Crocker, N., & Nguyen, T. T. (2011). Fetal alcohol spectrum disorders: Neuropsychological and behavioral features. *Neuropsychology Review, 21*(2), 81–101. https://doi.org/10.1007/s11065-011-9167-9

McCann, U. D., & Ricaurte, G. A. (2004). Amphetamine neurotoxicity: Accomplishments and remaining challenges. *Neuroscience and Biobehavioral Reviews, 27*(8), 821–826. https://doi.org/10.1016/j.neubiorev.2003.11.003

McCarthy, D. M., & Bhide, P. G. (2012). Prenatal cocaine exposure decreases parvalbumin-immunoreactive neurons and GABA-to-projection neuron ratio in the medial prefrontal cortex. *Developmental Neuroscience, 34*(2–3), 174–183. https://doi.org/10.1159/000337172

McCarthy, D. M., Brown, A. N., & Bhide, P. G. (2012). Regulation of BDNF expression by cocaine. *The Yale Journal of Biology and Medicine, 85*(4), 437–446. https://www.ncbi.nlm.nih.gov/pubmed/23239946

McComb, J. G. (2015). A practical clinical classification of spinal neural tube defects. *Child's Nervous System, 31*(10), 1641–1657. https://doi.org/10.1007/s00381-015-2845-9

McCusker, J., Bigelow, C., Vickers-Lahti, M., Spotts, D., Garfield, F., & Frost, R. (1997). Planned duration of residential drug abuse treatment: Efficacy versus effectiveness. *Addiction, 92*(11), 1467–1478. https://www.ncbi.nlm.nih.gov/pubmed/9519490

McGrath, J. J., Feron, F. P., Burne, T. H., Mackay-Sim, A., & Eyles, D. W. (2003). The neurodevelopmental hypothesis of schizophrenia: A review of recent developments. *Annals of Medicine, 35*(2), 86–93. https://doi.org/10.1080/07853890310010005

Minnes, S., Robin, N. H., Alt, A. A., Kirchner, H. L., Satayathum, S., Salbert, B. A., Ellison, L., & Singer, L. T. (2006). Dysmorphic and anthropometric outcomes in 6-year-old prenatally cocaine-exposed children. *Neurotoxicology and Teratology, 28*(1), 28–38. https://doi.org/10.1016/j.ntt.2005.09.003

Mohan, H., Lenis, M. G., Laurette, E. Y., Tejada, O., Sanghvi, T., Leung, K. Y., Cahill, L. S., Sled, J. G., Delgado-Olguin, P., Greene, N. D. E., Copp, A. J., & Serghides, L. (2020). Dolutegravir in pregnant mice is associated with increased rates of fetal defects at therapeutic but not at supratherapeutic levels. *eBioMedicine, 63*, 103167. https://doi.org/10.1016/j.ebiom.2020.103167

Money, D., Lee, T., O'Brien, C., Brophy, J., Bitnun, A., Kakkar, F., Boucoiran, I., Alimenti, A., Vaudry, W., Singer, J., Sauve, L. J., & Canadian Perinatal, H. I. V. S. P. (2019). Congenital anomalies following antenatal exposure to dolutegravir: A Canadian surveillance study. *BJOG, 126*(11), 1338–1345. https://doi.org/10.1111/1471-0528.15838

Monier, A., Evrard, P., Gressens, P., & Verney, C. (2006). Distribution and differentiation of microglia in the human encephalon during the first two trimesters of gestation. *The Journal of Comparative Neurology, 499*(4), 565–582. https://doi.org/10.1002/cne.21123

Monk, C. S., Webb, S. J., & Nelson, C. A. (2001). Prenatal neurobiological development: Molecular mechanisms and anatomical change. *Developmental Neuropsychology, 19*(2), 211–236. https://doi.org/10.1207/S15326942DN1902_5

Montenegro-Burke, J. R., Woldstad, C. J., Fang, M., Bade, A. N., McMillan, J., Edagwa, B., Boska, M. D., Gendelman, H. E., & Siuzdak, G. (2019). Nanoformulated antiretroviral therapy attenuates brain metabolic oxidative stress. *Molecular Neurobiology, 56*(4), 2896–2907. https://doi.org/10.1007/s12035-018-1273-8

Moore, S. J., Turnpenny, P., Quinn, A., Glover, S., Lloyd, D. J., Montgomery, T., & Dean, J. C. (2000). A clinical study of 57 children with fetal anticonvulsant syndromes. *Journal of Medical Genetics, 37*(7), 489–497. https://doi.org/10.1136/jmg.37.7.489

Mulligan, K. A., & Cheyette, B. N. (2017). Neurodevelopmental perspectives on Wnt signaling in psychiatry. *Molecular Neuropsychiatry, 2*(4), 219–246. https://doi.org/10.1159/000453266

Mulligan, N., Best, B. M., Wang, J., Capparelli, E. V., Stek, A., Barr, E., Buschur, S. L., Acosta, E. P., Smith, E., Chakhtoura, N., Burchett, S., Mirochnick, M., & Team, I. P. P. (2018). Dolutegravir pharmacokinetics in pregnant and postpartum women living with HIV. *AIDS, 32*(6), 729–737. https://doi.org/10.1097/QAD.0000000000001755

Murray, R. M., Bhavsar, V., Tripoli, G., & Howes, O. (2017). 30 Years on: How the neurodevelopmental hypothesis of schizophrenia morphed into the developmental risk factor model of psychosis. *Schizophrenia Bulletin, 43*(6), 1190–1196. https://doi.org/10.1093/schbul/sbx121

Niculae, A. S., & Paval, D. (2016). From molecules to behavior: An integrative theory of autism spectrum disorder. *Medical Hypotheses, 97*, 74–84. https://doi.org/10.1016/j.mehy.2016.10.016

Nigg, J. T. (2006). *What causes ADHD? Understanding what goes wrong and why*. The Guilford Press.

Nigg, J., Nikolas, M., & Burt, S. A. (2010). Measured gene-by-environment interaction in relation to attention-deficit/hyperactivity disorder. *Journal of the American Academy of Child and Adolescent Psychiatry, 49*(9), 863–873. https://doi.org/10.1016/j.jaac.2010.01.025

Nightingale, S. L. (1998). From the food and drug administration. *JAMA, 280*(17), 1472. https://www.ncbi.nlm.nih.gov/pubmed/9809716

Noordermeer, S. D. S., Luman, M., Greven, C. U., Veroude, K., Faraone, S. V., Hartman, C. A., Hoekstra, P. J., Franke, B., Buitelaar, J. K., Heslenfeld, D. J., & Oosterlaan, J. (2017). Structural brain abnormalities of attention-deficit/hyperactivity disorder with oppositional defiant disorder. *Biological Psychiatry, 82*(9), 642–650. https://doi.org/10.1016/j.biopsych.2017.07.008

O'Donovan, M. C., Craddock, N., Norton, N., Williams, H., Peirce, T., Moskvina, V., Nikolov, I., Hamshere, M., Carroll, L., Georgieva, L., Dwyer, S., Holmans, P., Marchini, J. L., Spencer, C. C., Howie, B., Leung, H. T., Hartmann, A. M., Moller, H. J., Morris, D. W., et al. (2008). Identification of loci associated with schizophrenia by genome-wide association and follow-up. *Nature Genetics, 40*(9), 1053–1055. https://doi.org/10.1038/ng.201

O'Rahilly, R., & Gardner, E. (1979). The initial development of the human brain. *Acta Anatomica (Basel), 104*(2), 123–133. https://doi.org/10.1159/000145061

Okerlund, N. D., & Cheyette, B. N. (2011). Synaptic Wnt signaling-a contributor to major psychiatric disorders? *Journal of Neurodevelopmental Disorders, 3*(2), 162–174. https://doi.org/10.1007/s11689-011-9083-6

Olasehinde, T. A., & Olaniran, A. O. (2022). Neurotoxicity of polycyclic aromatic hydrocarbons: A systematic mapping and review of neuropathological mechanisms. *Toxics, 10*(8). https://doi.org/10.3390/toxics10080417

Ornoy, A., Weinstein-Fudim, L., & Ergaz, Z. (2017). Antidepressants, antipsychotics, and mood stabilizers in pregnancy: What do we know and how should we treat pregnant women with depression. *Birth Defects Research, 109*(12), 933–956. https://doi.org/10.1002/bdr2.1079

Owen, M. J., & O'Donovan, M. C. (2017). Schizophrenia and the neurodevelopmental continuum: Evidence from genomics. *World Psychiatry, 16*(3), 227–235. https://doi.org/10.1002/wps.20440

Ozer, E., Sarioglu, S., & Gure, A. (2000). Effects of prenatal ethanol exposure on neuronal migration, neuronogenesis and brain myelination in the mice brain. *Clinical Neuropathology, 19*(1), 21–25. https://www.ncbi.nlm.nih.gov/pubmed/10774947

Pain, J. B., Le, M. P., Caseris, M., Amiel, C., Lassel, L., Charpentier, C., Desnoyer, A., Farnoux, C., Pialoux, G., Descamps, D., & Peytavin, G. (2015). Pharmacokinetics of dolutegravir in a premature neonate after HIV treatment intensification during pregnancy. *Antimicrobial Agents and Chemotherapy, 59*(6), 3660–3662. https://doi.org/10.1128/AAC.00173-15

Pardinas, A. F., Holmans, P., Pocklington, A. J., Escott-Price, V., Ripke, S., Carrera, N., Legge, S. E., Bishop, S., Cameron, D., Hamshere, M. L., Han, J., Hubbard, L., Lynham, A., Mantripragada, K., Rees, E., MacCabe, J. H., McCarroll, S. A., Baune, B. T., Breen, G., et al. (2018). Common schizophrenia alleles are enriched in mutation-intolerant genes and in regions under strong background selection. *Nature Genetics, 50*(3), 381–389. https://doi.org/10.1038/s41588-018-0059-2

Parodi, C., Di Fede, E., Peron, A., Vigano, I., Grazioli, P., Castiglioni, S., Finnell, R. H., Gervasini, C., Vignoli, A., & Massa, V. (2021). Chromatin imbalance as the vertex between fetal valproate syndrome and chromatinopathies. *Frontiers in Cell and Development Biology, 9*, 654467. https://doi.org/10.3389/fcell.2021.654467

Patel, S. S., Tomar, S., Sharma, D., Mahindroo, N., & Udayabanu, M. (2017). Targeting sonic hedgehog signaling in neurological disorders. *Neuroscience and Biobehavioral Reviews, 74*(Pt A), 76–97. https://doi.org/10.1016/j.neubiorev.2017.01.008

Pereira, G. F. M., Kim, A., Jalil, E. M., Fernandes Fonseca, F., Shepherd, B. E., Veloso, V. G., Rick, F., Ribeiro, R., Pimenta, M. C., Beber, A., Correa, R. G., Lima, R., Maruri, F., McGowan, C. C., Schwartz Benzaken, A., Grinsztejn, B., Castilho, J. L., National Cohort Study of D, & Prica's children and the enviromregnancy Outcomes in, B. (2021). Dolutegravir and pregnancy outcomes in women on antiretroviral therapy in Brazil: A retrospective national cohort study. *Lancet HIV, 8*(1), e33–e41. https://doi.org/10.1016/S2352-3018(20)30268-X

Peterson, B. S., Rauh, V. A., Bansal, R., Hao, X., Toth, Z., Nati, G., Walsh, K., Miller, R. L., Arias, F., Semanek, D., & Perera, F. (2015). Effects of prenatal exposure to air pollutants (polycyclic aromatic hydrocarbons) on the development of brain white matter, cognition, and behavior in later childhood. *JAMA Psychiatry, 72*(6), 531–540. https://doi.org/10.1001/jamapsychiatry.2015.57

Petrelli, B., Weinberg, J., & Hicks, G. G. (2018). Effects of prenatal alcohol exposure (PAE): Insights into FASD using mouse models of PAE. *Biochemistry and Cell Biology, 96*(2), 131–147. https://doi.org/10.1139/bcb-2017-0280

Popova, S., Lange, S., Shield, K., Mihic, A., Chudley, A. E., Mukherjee, R. A. S., Bekmuradov, D., & Rehm, J. (2016). Comorbidity of fetal alcohol spectrum disorder: A systematic review and meta-analysis. *Lancet, 387*(10022), 978–987. https://doi.org/10.1016/S0140-6736(15)01345-8

Popovic, D., Schmitt, A., Kaurani, L., Senner, F., Papiol, S., Malchow, B., Fischer, A., Schulze, T. G., Koutsouleris, N., & Falkai, P. (2019). Childhood trauma in schizophrenia: Current findings and research perspectives. *Frontiers in Neuroscience, 13*, 274. https://doi.org/10.3389/fnins.2019.00274

Posobiec, L. M., Chapman, S. P., Murzyn, S. F., Rendemonti, J. E., Stanislaus, D. J., & Romach, E. H. (2021). No developmental toxicity observed with dolutegravir in rat whole embryo culture. *Birth Defects Research, 113*(16), 1190–1197. https://doi.org/10.1002/bdr2.1949

Qin, L., & Dai, X. (2015). Effect of sulindac on improving autistic behaviors in rats. *Nan Fang Yi Ke Da Xue Xue Bao, 35*(8), 1162–1165. https://www.ncbi.nlm.nih.gov/pubmed/26277514

Quist, J. F., Barr, C. L., Schachar, R., Roberts, W., Malone, M., Tannock, R., Basile, V. S., Beitchman, J., & Kennedy, J. L. (2003). The serotonin 5-HT1B receptor gene and attention deficit hyperactivity disorder. *Molecular Psychiatry, 8*(1), 98–102. https://doi.org/10.1038/sj.mp.4001244

Raesima, M. M., Ogbuabo, C. M., Thomas, V., Forhan, S. E., Gokatweng, G., Dintwa, E., Petlo, C., Motswere-Chirwa, C., Rabold, E. M., Tinker, S. C., Odunsi, S., Malima, S., Mmunyane, O., Modise, T., Kefitlhile, K., Dare, K., Letebele, M., Roland, M. E., Moore, C. A., et al. (2019). Dolutegravir use at conception - Additional surveillance data from Botswana. *The New England Journal of Medicine, 381*(9), 885–887. https://doi.org/10.1056/NEJMc1908155

Rakic, P. (1988). Specification of cerebral cortical areas. *Science, 241*(4862), 170–176. https://doi.org/10.1126/science.3291116

Rakic, P. (1995). A small step for the cell, a giant leap for mankind: A hypothesis of neocortical expansion during evolution. *Trends in Neurosciences, 18*(9), 383–388. https://doi.org/10.1016/0166-2236(95)93934-p

Rallu, M., Corbin, J. G., & Fishell, G. (2002). Parsing the prosencephalon. *Nature Reviews. Neuroscience, 3*(12), 943–951. https://doi.org/10.1038/nrn989

Rambousek, L., Kacer, P., Syslova, K., Bumba, J., Bubenikova-Valesova, V., & Slamberova, R. (2014). Sex differences in methamphetamine pharmacokinetics in adult rats and its transfer to pups through the placental membrane and breast milk. *Drug and Alcohol Dependence, 139*, 138–144. https://doi.org/10.1016/j.drugalcdep.2014.03.023

Rando, K., Chaplin, T. M., Potenza, M. N., Mayes, L., & Sinha, R. (2013). Prenatal cocaine exposure and gray matter volume in adolescent boys and girls: Relationship to substance use initiation. *Biological Psychiatry, 74*(7), 482–489. https://doi.org/10.1016/j.biopsych.2013.04.030

Ravi, K. S., Hassan, S. B., Pasi, R., Mittra, S., & Kumar, R. (2021). Neural tube defects: Different types and brief review of neurulation process and its clinical implication. *Journal of Family Medicine and Primary Care, 10*(12), 4383–4390. https://doi.org/10.4103/jfmpc.jfmpc_904_21

Redila, V. A., Olson, A. K., Swann, S. E., Mohades, G., Webber, A. J., Weinberg, J., & Christie, B. R. (2006). Hippocampal cell proliferation is reduced following prenatal ethanol exposure but can be rescued with voluntary exercise. *Hippocampus, 16*(3), 305–311. https://doi.org/10.1002/hipo.20164

Ren, A., Qiu, X., Jin, L., Ma, J., Li, Z., Zhang, L., Zhu, H., Finnell, R. H., & Zhu, T. (2011). Association of selected persistent organic pollutants in the placenta with the risk of neural tube defects. *Proceedings of the National Academy of Sciences of the United States of America, 108*(31), 12770–12775. https://doi.org/10.1073/pnas.1105209108

Riar, A. K., Narasimhan, M., Rathinam, M. L., Henderson, G. I., & Mahimainathan, L. (2016). Ethanol induces cytostasis of cortical basal progenitors. *Journal of Biomedical Science, 23*, 6. https://doi.org/10.1186/s12929-016-0225-8

Rice, D. C. (1996). Behavioral effects of lead: Commonalities between experimental and epidemiologic data. *Environmental Health Perspectives, 104*(Suppl 2), 337–351. https://doi.org/10.1289/ehp.96104s2337

Rice, D. C. (2000). Parallels between attention deficit hyperactivity disorder and behavioral deficits produced by neurotoxic exposure in monkeys. *Environmental Health Perspectives, 108*(Suppl 3), 405–408. https://doi.org/10.1289/ehp.00108s3405

Richardson, G. A., Goldschmidt, L., Larkby, C., & Day, N. L. (2013). Effects of prenatal cocaine exposure on child behavior and growth at 10 years of age. *Neurotoxicology and Teratology, 40*, 1–8. https://doi.org/10.1016/j.ntt.2013.08.001

Richetto, J., Massart, R., Weber-Stadlbauer, U., Szyf, M., Riva, M. A., & Meyer, U. (2017). Genome-wide DNA methylation changes in a mouse model of infection-mediated neurodevelopmental disorders. *Biological Psychiatry, 81*(3), 265–276. https://doi.org/10.1016/j.biopsych.2016.08.010

Rodrigues Pdos, S., & Dowling, J. E. (1990). Dopamine induces neurite retraction in retinal horizontal cells via diacylglycerol and protein kinase C. *Proceedings of the National Academy of Sciences of the United States of America, 87*(24), 9693–9697. https://doi.org/10.1073/pnas.87.24.9693

Ross, E. J., Graham, D. L., Money, K. M., & Stanwood, G. D. (2015). Developmental consequences of fetal exposure to drugs: What we know and what we still must learn. *Neuropsychopharmacology, 40*(1), 61–87. https://doi.org/10.1038/npp.2014.147

Rossi-George, A., Virgolini, M. B., Weston, D., Thiruchelvam, M., & Cory-Slechta, D. A. (2011). Interactions of lifetime lead exposure and stress: Behavioral, neurochemical and HPA axis effects. *Neurotoxicology, 32*(1), 83–99. https://doi.org/10.1016/j.neuro.2010.09.004

Sara, G. E., Large, M. M., Matheson, S. L., Burgess, P. M., Malhi, G. S., Whiteford, H. A., & Hall, W. D. (2015). Stimulant use disorders in people with psychosis: A meta-analysis of rate and factors affecting variation. *The Australian and New Zealand Journal of Psychiatry, 49*(2), 106–117. https://doi.org/10.1177/0004867414561526

Sausedo, R. A., Smith, J. L., & Schoenwolf, G. C. (1997). Role of nonrandomly oriented cell division in shaping and bending of the neural plate. *The Journal of Comparative Neurology, 381*(4), 473–488. https://www.ncbi.nlm.nih.gov/pubmed/9136804

Schafer, D. P., & Stevens, B. (2015). Microglia function in central nervous system development and plasticity. *Cold Spring Harbor Perspectives in Biology, 7*(10), a020545. https://doi.org/10.1101/cshperspect.a020545

Schier, A. F. (2001). Axis formation and patterning in zebrafish. *Current Opinion in Genetics & Development, 11*(4), 393–404. https://doi.org/10.1016/s0959-437x(00)00209-4

Schizophrenia Psychiatric Genome-Wide Association Study, C. (2011). Genome-wide association study identifies five new schizophrenia loci. *Nature Genetics, 43*(10), 969–976. https://doi.org/10.1038/ng.940

Schizophrenia Working Group of the Psychiatric Genomics, C. (2014). Biological insights from 108 schizophrenia-associated genetic loci. *Nature, 511*(7510), 421–427. https://doi.org/10.1038/nature13595

Schneider, T., & Przewlocki, R. (2005). Behavioral alterations in rats prenatally exposed to valproic acid: Animal model of autism. *Neuropsychopharmacology, 30*(1), 80–89. https://doi.org/10.1038/sj.npp.1300518

Schnoll, J. G., Temsamrit, B., Zhang, D., Song, H., Ming, G. L., & Christian, K. M. (2019). Evaluating neurodevelopmental consequences of perinatal exposure to antiretroviral drugs: Current challenges and new approaches. *Journal of Neuroimmune Pharmacology*. https://doi.org/10.1007/s11481-019-09880-z

Schoenwolf, G. C. (1982). On the morphogenesis of the early rudiments of the developing central nervous system. *Scanning Electron Microscopy, 1*, 289–308. https://www.ncbi.nlm.nih.gov/pubmed/7167749

Seller, M. J. (1995). Sex, neural tube defects, and multisite closure of the human neural tube. *American Journal of Medical Genetics, 58*(4), 332–336. https://doi.org/10.1002/ajmg.1320580406

Sellgren, C. M., Gracias, J., Watmuff, B., Biag, J. D., Thanos, J. M., Whittredge, P. B., Fu, T., Worringer, K., Brown, H. E., Wang, J., Kaykas, A., Karmacharya, R., Goold, C. P., Sheridan, S. D., & Perlis, R. H. (2019). Increased synapse elimination by microglia in schizophrenia patient-derived models of synaptic pruning. *Nature Neuroscience, 22*(3), 374–385. https://doi.org/10.1038/s41593-018-0334-7

Sewell, M. J., Chiu, Y. E., & Drolet, B. A. (2015). Neural tube dysraphism: Review of cutaneous markers and imaging. *Pediatric Dermatology, 32*(2), 161–170. https://doi.org/10.1111/pde.12485

Shetty, A. K., & Phillips, D. E. (1992). Effects of prenatal ethanol exposure on the development of Bergmann glia and astrocytes in the rat cerebellum: An immunohistochemical study. *The Journal of Comparative Neurology, 321*(1), 19–32. https://doi.org/10.1002/cne.903210103

Sidman, R., & Rakic, P. (1982). Histology and histopathology of the nervous system. In W. Haymaker & R. D. Adams (Eds.), *Development of the human central nervous system* (pp. 3–143). Thomas.

Smith, L. M., & Santos, L. S. (2016). Prenatal exposure: The effects of prenatal cocaine and methamphetamine exposure on the developing child. *Birth Defects Research. Part C, Embryo Today, 108*(2), 142–146. https://doi.org/10.1002/bdrc.21131

Smith, L. M., LaGasse, L. L., Derauf, C., Grant, P., Shah, R., Arria, A., Huestis, M., Haning, W., Strauss, A., Della Grotta, S., Liu, J., & Lester, B. M. (2006). The infant development, environment, and lifestyle study: Effects of prenatal methamphetamine exposure, polydrug exposure, and poverty on intrauterine growth. *Pediatrics, 118*(3), 1149–1156. https://doi.org/10.1542/peds.2005-2564

Smith, S. J., Zhao, X. Z., Passos, D. O., Lyumkis, D., Burke, T. R., Jr., & Hughes, S. H. (2021). Integrase strand transfer inhibitors are effective anti-HIV drugs. *Viruses, 13*(2). https://doi.org/10.3390/v13020205

Smith, A. S., Ankam, S., Farhy, C., Fiengo, L., Basa, R. C. B., Gordon, K. L., Martin, C. T., Terskikh, A. V., Jordan-Sciutto, K. L., Price, J. H., & McDonough, P. M. (2022). High-content analysis and kinetic image cytometry identify toxicity and epigenetic effects of HIV antiretrovirals on human iPSC-neurons and primary neural precursor cells. *Journal of Pharmacological and Toxicological Methods, 114*, 107157. https://doi.org/10.1016/j.vascn.2022.107157

Soriano, E., Alvarado-Mallart, R. M., Dumesnil, N., Del Rio, J. A., & Sotelo, C. (1997). Cajal-Retzius cells regulate the radial glia phenotype in the adult and developing cerebellum and alter granule cell migration. *Neuron, 18*(4), 563–577. https://doi.org/10.1016/s0896-6273(00)80298-6

Stanislaus, D. J., Posobiec, L. M., Laffan, S. B., Solomon, H. M., Ziejewski, M. K., & Romach, E. H. (2020). Absence of developmental and reproductive toxicity in animals exposed to dolutegravir. *Birth Defects Research, 112*(3), 245–261. https://doi.org/10.1002/bdr2.1635

Stefansson, H., Ophoff, R. A., Steinberg, S., Andreassen, O. A., Cichon, S., Rujescu, D., Werge, T., Pietilainen, O. P., Mors, O., Mortensen, P. B., Sigurdsson, E., Gustafsson, O., Nyegaard, M., Tuulio-Henriksson, A., Ingason, A., Hansen, T., Suvisaari, J., Lonnqvist, J., Paunio, T., et al. (2009). Common variants conferring risk of schizophrenia. *Nature, 460*(7256), 744–747. https://doi.org/10.1038/nature08186

Stern, C. D. (2001). Initial patterning of the central nervous system: How many organizers? *Nature Reviews. Neuroscience, 2*(2), 92–98. https://doi.org/10.1038/35053563

Stilo, S. A., & Murray, R. M. (2019). Non-genetic factors in schizophrenia. *Current Psychiatry Reports, 21*(10), 100. https://doi.org/10.1007/s11920-019-1091-3

Sui, L., & Chen, M. (2012). Prenatal exposure to valproic acid enhances synaptic plasticity in the medial prefrontal cortex and fear memories. *Brain Research Bulletin, 87*(6), 556–563. https://doi.org/10.1016/j.brainresbull.2012.01.011

Swanson, J. M., Flodman, P., Kennedy, J., Spence, M. A., Moyzis, R., Schuck, S., Murias, M., Moriarity, J., Barr, C., Smith, M., & Posner, M. (2000). Dopamine genes and ADHD. *Neuroscience and Biobehavioral Reviews, 24*(1), 21–25. https://doi.org/10.1016/s0149-7634(99)00062-7

Tang, S. J. (2014). Synaptic activity-regulated Wnt signaling in synaptic plasticity, glial function and chronic pain. *CNS & Neurological Disorders Drug Targets, 13*(5), 737–744. https://doi.org/10.2174/18715273126661312231114457

Tau, G. Z., & Peterson, B. S. (2010). Normal development of brain circuits. *Neuropsychopharmacology, 35*(1), 147–168. https://doi.org/10.1038/npp.2009.115

Taverna, E., & Huttner, W. B. (2010). Neural progenitor nuclei IN motion. *Neuron, 67*(6), 906–914. https://doi.org/10.1016/j.neuron.2010.08.027

Taverna, E., Gotz, M., & Huttner, W. B. (2014). The cell biology of neurogenesis: Toward an understanding of the development and evolution of the neocortex. *Annual Review of Cell and Developmental Biology, 30*, 465–502. https://doi.org/10.1146/annurev-cellbio-101011-155801

Thapar, A. (2018). Discoveries on the genetics of ADHD in the 21st century: New findings and their implications. *The American Journal of Psychiatry, 175*(10), 943–950. https://doi.org/10.1176/appi.ajp.2018.18040383

Thompson, B. L., & Levitt, P. (2010). The clinical-basic interface in defining pathogenesis in disorders of neurodevelopmental origin. *Neuron, 67*(5), 702–712. https://doi.org/10.1016/j.neuron.2010.08.037

Thompson, B. L., Levitt, P., & Stanwood, G. D. (2009). Prenatal exposure to drugs: Effects on brain development and implications for policy and education. *Nature Reviews. Neuroscience, 10*(4), 303–312. https://doi.org/10.1038/nrn2598

Tomaskova, A., Slamberova, R., & Cerna, M. (2020). Influence of prenatal methamphetamine abuse on the brain. *Epigenomes, 4*(3). https://doi.org/10.3390/epigenomes4030014

Tomson, T., Battino, D., & Perucca, E. (2019). Teratogenicity of antiepileptic drugs. *Current Opinion in Neurology, 32*(2), 246–252. https://doi.org/10.1097/WCO.0000000000000659

Tran, T. H., Griffin, B. L., Stone, R. H., Vest, K. M., & Todd, T. J. (2017). Methadone, buprenorphine, and naltrexone for the treatment of opioid use disorder in pregnant women. *Pharmacotherapy, 37*(7), 824–839. https://doi.org/10.1002/phar.1958

Tripp, G., & Wickens, J. R. (2009). Neurobiology of ADHD. *Neuropharmacology, 57*(7–8), 579–589. https://doi.org/10.1016/j.neuropharm.2009.07.026

Tuazon, F. B., & Mullins, M. C. (2015). Temporally coordinated signals progressively pattern the anteroposterior and dorsoventral body axes. *Seminars in Cell & Developmental Biology, 42*, 118–133. https://doi.org/10.1016/j.semcdb.2015.06.003

Ulke, C., Rullmann, M., Huang, J., Luthardt, J., Becker, G. A., Patt, M., Meyer, P. M., Tiepolt, S., Hesse, S., Sabri, O., & Strauss, M. (2019). Adult attention-deficit/hyperactivity disorder is associated with reduced norepinephrine transporter availability in right attention networks: A (S,S)-O-[(11)C]methylreboxetine positron emission tomography study. *Translational Psychiatry, 9*(1), 301. https://doi.org/10.1038/s41398-019-0619-y

Ulloa, F., & Briscoe, J. (2007). Morphogens and the control of cell proliferation and patterning in the spinal cord. *Cell Cycle, 6*(21), 2640–2649. https://doi.org/10.4161/cc.6.21.4822

Urban, N., & Guillemot, F. (2014). Neurogenesis in the embryonic and adult brain: Same regulators, different roles. *Frontiers in Cellular Neuroscience, 8*, 396. https://doi.org/10.3389/fncel.2014.00396

Vaishnavi, V., Manikandan, M., Tiwary, B. K., & Munirajan, A. K. (2013). Insights on the functional impact of microRNAs present in autism-associated copy number variants. *PLoS One, 8*(2), e56781. https://doi.org/10.1371/journal.pone.0056781

van de Wijer, L., Garcia, L. P., Hanswijk, S. I., Rando, J., Middelman, A., Ter Heine, R., Mast, Q., Martens, G. J. M., van der Ven, A., Kolk, S. M., Schellekens, A. F. A., & Homberg, J. R. (2019). Neurodevelopmental and behavioral consequences of perinatal exposure to the HIV drug efavirenz in a rodent model. *Translational Psychiatry, 9*(1), 84. https://doi.org/10.1038/s41398-019-0420-y

van Loo, K. M., & Martens, G. J. (2007). Genetic and environmental factors in complex neurodevelopmental disorders. *Current Genomics, 8*(7), 429–444. https://doi.org/10.2174/138920207783591717

Vannappagari, V., & Thorne, C. (2019). Pregnancy and neonatal outcomes following prenatal exposure to dolutegravir. *Journal of Acquired Immune Deficiency Syndromes, 81*(4), 371–378. https://doi.org/10.1097/QAI.0000000000002035

Varese, F., Smeets, F., Drukker, M., Lieverse, R., Lataster, T., Viechtbauer, W., Read, J., van Os, J., & Bentall, R. P. (2012). Childhood adversities increase the risk of psychosis: A meta-analysis of patient-control, prospective- and cross-sectional cohort studies. *Schizophrenia Bulletin, 38*(4), 661–671. https://doi.org/10.1093/schbul/sbs050

Vasung, L., Abaci Turk, E., Ferradal, S. L., Sutin, J., Stout, J. N., Ahtam, B., Lin, P. Y., & Grant, P. E. (2019). Exploring early human brain development with structural and physiological neuroimaging. *NeuroImage, 187*, 226–254. https://doi.org/10.1016/j.neuroimage.2018.07.041

Velez-Ruiz, N. J., & Meador, K. J. (2015). Neurodevelopmental effects of fetal antiepileptic drug exposure. *Drug Safety, 38*(3), 271–278. https://doi.org/10.1007/s40264-015-0269-9

Veronneau-Veilleux, F., Robaey, P., Ursino, M., & Nekka, F. (2022). A mechanistic model of ADHD as resulting from dopamine phasic/tonic imbalance during reinforcement learning. *Frontiers in Computational Neuroscience, 16*, 849323. https://doi.org/10.3389/fncom.2022.849323

Virgolini, M. B., Rossi-George, A., Lisek, R., Weston, D. D., Thiruchelvam, M., & Cory-Slechta, D. A. (2008). CNS effects of developmental Pb exposure are enhanced by combined maternal and offspring stress. *Neurotoxicology, 29*(5), 812–827. https://doi.org/10.1016/j.neuro.2008.03.003

Volkow, N. D., Wang, G. J., Kollins, S. H., Wigal, T. L., Newcorn, J. H., Telang, F., Fowler, J. S., Zhu, W., Logan, J., Ma, Y., Pradhan, K., Wong, C., & Swanson, J. M. (2009). Evaluating dopamine reward pathway in ADHD: Clinical implications. *JAMA, 302*(10), 1084–1091. https://doi.org/10.1001/jama.2009.1308

Volkow, N. D., Wang, G. J., Tomasi, D., Kollins, S. H., Wigal, T. L., Newcorn, J. H., Telang, F. W., Fowler, J. S., Logan, J., Wong, C. T., & Swanson, J. M. (2012). Methylphenidate-elicited dopamine increases in ventral striatum are associated with long-term symptom improvement in adults with attention deficit hyperactivity disorder. *The Journal of Neuroscience, 32*(3), 841–849. https://doi.org/10.1523/JNEUROSCI.4461-11.2012

Wada, H., & Okamoto, H. (2009). Roles of noncanonical Wnt/PCP pathway genes in neuronal migration and neurulation in zebrafish. *Zebrafish, 6*(1), 3–8. https://doi.org/10.1089/zeb.2008.0557

Waitt, C., Orrell, C., Walimbwa, S., Singh, Y., Kintu, K., Simmons, B., Kaboggoza, J., Sihlangu, M., Coombs, J. A., Malaba, T., Byamugisha, J., Amara, A., Gini, J., Else, L., Heiburg, C., Hodel, E. M., Reynolds, H., Mehta, U., Byakika-Kibwika, P., et al. (2019). Safety and pharmacokinetics of dolutegravir in pregnant mothers with HIV infection and their neonates: A randomised trial (DolPHIN-1 study). *PLoS Medicine, 16*(9), e1002895. https://doi.org/10.1371/journal.pmed.1002895

Wegner, C., & Nau, H. (1992). Alteration of embryonic folate metabolism by valproic acid during organogenesis: Implications for mechanism of teratogenesis. *Neurology, 42*(4 Suppl 5), 17–24. https://www.ncbi.nlm.nih.gov/pubmed/1574172

Weinberger, D. R. (1987). Implications of normal brain development for the pathogenesis of schizophrenia. *Archives of General Psychiatry, 44*(7), 660–669. https://doi.org/10.1001/archpsyc.1987.01800190080012

Weissman, A. D., & Caldecott-Hazard, S. (1995). Developmental neurotoxicity to methamphetamines. *Clinical and Experimental Pharmacology & Physiology, 22*(5), 372–374. https://doi.org/10.1111/j.1440-1681.1995.tb02022.x

Welch, G. W. (1990). The infant with anencephaly. *The New England Journal of Medicine, 323*(9), 615. https://doi.org/10.1056/nejm199008303230918

Welle-Strand, G. K., Skurtveit, S., Jones, H. E., Waal, H., Bakstad, B., Bjarko, L., & Ravndal, E. (2013). Neonatal outcomes following in utero exposure to methadone or buprenorphine: A National Cohort Study of opioid-agonist treatment of Pregnant Women in Norway from 1996 to 2009. *Drug and Alcohol Dependence, 127*(1–3), 200–206. https://doi.org/10.1016/j.drugalcdep.2012.07.001

Wells, P. G., & Winn, L. M. (1996). Biochemical toxicology of chemical teratogenesis. *Critical Reviews in Biochemistry and Molecular Biology, 31*(1), 1–40. https://doi.org/10.3109/10409239609110574

Wells, P. G., Bhatia, S., Drake, D. M., & Miller-Pinsler, L. (2016). Fetal oxidative stress mechanisms of neurodevelopmental deficits and exacerbation by ethanol and methamphetamine. *Birth Defects Research. Part C, Embryo Today, 108*(2), 108–130. https://doi.org/10.1002/bdrc.21134

Werling, D. M., Brand, H., An, J. Y., Stone, M. R., Zhu, L., Glessner, J. T., Collins, R. L., Dong, S., Layer, R. M., Markenscoff-Papadimitriou, E., Farrell, A., Schwartz, G. B., Wang, H. Z., Currall, B. B., Zhao, X., Dea, J., Duhn, C., Erdman, C. A., Gilson, M. C., et al. (2018). An analytical framework for whole-genome sequence association studies and its implications for autism spectrum disorder. *Nature Genetics, 50*(5), 727–736. https://doi.org/10.1038/s41588-018-0107-y

Weyrauch, D., Schwartz, M., Hart, B., Klug, M. G., & Burd, L. (2017). Comorbid mental disorders in fetal alcohol spectrum disorders: A systematic review. *Journal of Developmental and Behavioral Pediatrics, 38*(4), 283–291. https://doi.org/10.1097/DBP.0000000000000440

Wilkins-Haug, L., & Freedman, W. (1991). Progression of exencephaly to anencephaly in the human fetus--an ultrasound perspective. *Prenatal Diagnosis, 11*(4), 227–233. https://doi.org/10.1002/pd.1970110404

Williams, P. L., Yildirim, C., Chadwick, E. G., Van Dyke, R. B., Smith, R., Correia, K. F., DiPerna, A., Seage, G. R., 3rd, Hazra, R., Crowell, C. S., & Surveillance Monitoring for, A. R. T. T. s. o. t. P. H. I. V. A. C. S. (2020). Association of maternal antiretroviral use with microcephaly in children who are HIV-exposed but uninfected (SMARTT): A prospective cohort study. *Lancet HIV, 7*(1), e49–e58. https://doi.org/10.1016/S2352-3018(19)30340-6

Willsey, A. J., Sanders, S. J., Li, M., Dong, S., Tebbenkamp, A. T., Muhle, R. A., Reilly, S. K., Lin, L., Fertuzinhos, S., Miller, J. A., Murtha, M. T., Bichsel, C., Niu, W., Cotney, J., Ercan-Sencicek, A. G., Gockley, J., Gupta, A. R., Han, W., He, X., et al. (2013). Coexpression networks implicate human midfetal deep cortical projection neurons in the pathogenesis of autism. *Cell, 155*(5), 997–1007. https://doi.org/10.1016/j.cell.2013.10.020

Wilson, S. W., & Houart, C. (2004). Early steps in the development of the forebrain. *Developmental Cell, 6*(2), 167–181. https://doi.org/10.1016/s1534-5807(04)00027-9

Wood, L. R., & Smith, M. T. (1984). Generation of anencephaly: 1. Aberrant neurulation and 2. Conversion of exencephaly to anencephaly. *Journal of Neuropathology and Experimental Neurology, 43*(6), 620–633. https://www.ncbi.nlm.nih.gov/pubmed/6502191

Wozniak, J. R., Riley, E. P., & Charness, M. E. (2019). Clinical presentation, diagnosis, and management of fetal alcohol spectrum disorder. *Lancet Neurology, 18*(8), 760–770. https://doi.org/10.1016/S1474-4422(19)30150-4

Yang, M. S., & Gill, M. (2007). A review of gene linkage, association and expression studies in autism and an assessment of convergent evidence. *International Journal of Developmental Neuroscience, 25*(2), 69–85. https://doi.org/10.1016/j.ijdevneu.2006.12.002

Ye, Y., Jiang, S., Zhang, C., Cheng, Y., Zhong, H., Du, T., Xu, W., Azziz, R., Zhang, H., & Zhao, X. (2020). Environmental pollutant Benzo[a]pyrene induces recurrent pregnancy loss through promoting apoptosis and suppressing migration of extravillous trophoblast. *BioMed Research International, 2020*, 8983494. https://doi.org/10.1155/2020/8983494

Yombi, J. C. (2018). Dolutegravir neuropsychiatric adverse events: Specific drug effect or class effect. *AIDS Reviews, 20*(1), 14–26. https://www.ncbi.nlm.nih.gov/pubmed/29628511

Young, J. K., Giesbrecht, H. E., Eskin, M. N., Aliani, M., & Suh, M. (2014). Nutrition implications for fetal alcohol spectrum disorder. *Advances in Nutrition, 5*(6), 675–692. https://doi.org/10.3945/an.113.004846

Yu, D. X., Marchetto, M. C., & Gage, F. H. (2014). How to make a hippocampal dentate gyrus granule neuron. *Development, 141*(12), 2366–2375. https://doi.org/10.1242/dev.096776

Zash, R., Makhema, J., & Shapiro, R. L. (2018). Neural-tube defects with dolutegravir treatment from the time of conception. *The New England Journal of Medicine, 379*(10), 979–981. https://doi.org/10.1056/NEJMc1807653

Zash, R., Holmes, L., Diseko, M., Jacobson, D. L., Brummel, S., Mayondi, G., Isaacson, A., Davey, S., Mabuta, J., Mmalane, M., Gaolathe, T., Essex, M., Lockman, S., Makhema, J., & Shapiro, R. L. (2019). Neural-tube defects and antiretroviral treatment regimens in Botswana. *The New England Journal of Medicine, 381*(9), 827–840. https://doi.org/10.1056/NEJMoa1905230

Zhang, X., Rui, M., Gan, G., Huang, C., Yi, J., Lv, H., & Xie, W. (2017). Neuroligin 4 regulates synaptic growth via the bone morphogenetic protein (BMP) signaling pathway at the Drosophila neuromuscular junction. *The Journal of Biological Chemistry, 292*(44), 17991–18005. https://doi.org/10.1074/jbc.M117.810242

Zhou, H. (2015). Region specific effects of maternal immune activation on offspring neuroimmune function. *Open Journal of Immunology, 5*(2), 51–63. https://doi.org/10.4236/oji.2015.52006

Autophagic and Lysosomal Disorders

11

Sujyoti Chandra and Kalipada Pahan

Abstract

Lysosomes are acid hydrolase–containing membrane-enclosed organelles that play a central role in autophagy or self-eating. As a result, aberrant lysosomal function and autophagy have been associated with multiple lysosomal storage and neurodegenerative disorders. While mutations in genes that are important for normal lysosomal function lead to lysosomal storage disorders like Batten disease, Gaucher's disease, Tay–Sach's disease, Krabbe disease, Niemann–Pick disease, etc., age- and neurodegenerative stimuli-influenced decrease in lysosomal clearance is associated to the pathogenesis of neurodegenerative disorders like Alzheimer's disease, Parkinson's disease (PD), Huntington's disease. Recent studies have also identified an association between mutations of the glucocerebrosidase gene (for Gaucher's disease) and PD. Considering the pathologies of these autophagic and lysosomal diseases, enhancement of lysosomal clearance machinery by gene therapy, enzyme replacement therapy, and stimulation of TFEB, the master regulator of lysosomal biogenesis has emerged as attractive therapeutic strategies for these disorders.

Keywords

Lysosome · Autophagy · TFEB · Lysosomal storage disorders · Neurodegenerative disorders

S. Chandra
Department of Neurological Sciences, Rush University Medical Center, Chicago, IL, USA

Department of Neurosurgery, Stanford University Medical Center, Stanford, CA, USA

K. Pahan (✉)
Department of Neurological Sciences, Rush University Medical Center, Chicago, IL, USA

Division of Research and Development, Jesse Brown Veterans Affairs Medical Center, Chicago, IL, USA
e-mail: Kalipada_Pahan@rush.edu

11.1 Lysosome

11.1.1 Overview

Lysosomes, the central component of cellular degradative and recycling pathways, are acid hydrolase–containing membrane-enclosed dynamic organelles. They are regarded as the terminal compartment for degradation of macromolecules from the endocytic, autophagic, secretory, and phagocytic pathways (Ballabio, 2016; De Duve, 1963; Settembre et al., 2013a, b). A coordinated action of the lysosomal membrane proteins, acidification machinery, and enzyme activity is required for optimal lysosomal function. Lysosomal adaptation under different physiological and pathological scenarios is dependent on a regulatory gene network known as CLEAR, (coordinated lysosomal expression and regulation), with the master regulator transcription factor EB (TFEB) at the center (Martini-Stoica et al., 2016; Sardiello et al., 2009). TFEB is a basic helix-loop-helix transcription activator that belongs to the MiT family of transcription factors. By binding to the CLEAR element on the promoter of various lysosome-associated genes, it can positively regulate multiple cellular processes such as lysosomal biogenesis, autophagy, exocytosis, and endocytosis, making it possible to target this single central regulator molecule for designing innovative therapies for lysosomal storage diseases and neurodegenerative disorders (Martini-Stoica et al., 2016; Napolitano & Ballabio, 2016).

11.1.2 Historical Preview

Lysosomes were first described by Christian de Duve in 1955 when his team at the University of Louvain in Belgium was investigating hepatic glucose 6-phosphatase, which serendipitously led to the observation of a latent acid phosphatase in fresh rat liver homogenates (de Duve, 2005). Using differential centrifugation, De Duve and his team obtained

H. E. Gendelman, T. Ikezu (eds.), *Neuroimmune Pharmacology and Therapeutics*,
https://doi.org/10.1007/978-3-031-68237-7_11

the microsomal fraction from rat liver cells. Interestingly, when using fresh samples, they observed that the activity of the enzyme acid phosphatase, which they were originally using as a control for their glucose 6-phosphatase experiments, was decreased to 10% following differential centrifugation (Bainton, 1981; De Duve & Wattiaux, 1966). However, when they measured the enzyme activity in samples that were stored in the fridge for 5 days, they obtained much higher activity of the enzyme. These observations lead to the hypothesis that a membrane-bound "saclike structure" limited the availability of acid phosphatase to the substrate; when the samples were stored for a few days, the enzyme could diffuse and the activity was restored (Bainton, 1981; De Duve & Wattiaux, 1966). By 1955, several other hydrolases (β-glucuronidase, cathepsin, ribonuclease, and deoxyribonuclease) that exhibited similar latency characteristics as acid phosphatase were observed in the same particles, indicating a digestive function associated with it. Hence, the term "lysosome," meaning digestive body in Greek, was coined by De Duve (2005). The first electron micrograph of lysosome was obtained from cell fractions when Alex Novikoff visited De Duve's laboratory in 1955 (Bainton, 1981; Novikoff et al., 1956). De Duve was awarded the Nobel Prize in Physiology or Medicine in 1974 for the discovery of lysosomes.

11.1.3 Structure of Lysosome

Lysosomes are sac-like structures present ubiquitously in eukaryotic cells that are primarily known as the "waste disposal system" of the cell for degrading macromolecules like proteins, carbohydrates, lipids, nucleic acids, damaged organelles, and other cellular debris (De Duve & Wattiaux, 1966). At the ultrastructural level, they are identified as dense spherical or tubular bodies present in a perinuclear fashion inside the cell. The typical size of the lysosome is >1 um; however, it varies with different cell types as well as the amount of undigested substance within the lysosome (Appelqvist et al., 2013).

Lysosomes are enclosed by a single limiting phospholipid bilayer membrane (7–10 nm) containing highly glycosylated membrane proteins. On the internal side, a thick glycocalyx membrane protects the lysosomal membrane from degradation by the acid hydrolases present in the lumen (De Duve & Wattiaux, 1966). In addition to maintaining the luminal acidic pH and transport of molecules, the lysosomal membrane actively participates in the fusion of lysosomes with autophagosomes, late endosomes, and plasma membranes (Appelqvist et al., 2013; Settembre et al., 2013a, b).

The interior of the lysosome, the lysosomal lumen, is highly acidic with a pH of 4.5–5 which is required for activity of the lysosomal hydrolases. This internal acidic milieu is maintained by a large multicomponent membrane channel, vacuolar ATPase (vATPase), which actively (by utilizing the energy from ATP hydrolysis) transport protons through the lysosomal membrane (Ohkuma et al., 1982). This transmembrane proton gradient is required for bidirectional transport of molecules including metabolites, ions, substrates as well as for efficient delivery of lysosomal enzymes from the Golgi network to lysosomes. Disruption of the gradient can lead to defective cargo sorting, membrane traffic, and degradation of substrate, ultimately leading to metabolic aberrations (Lim & Zoncu, 2016; Saftig & Klumperman, 2009).

Until now, more than 100 lysosomal proteins have been recognized of which approximately 70 reside in the lysosomal matrix and approximately 50 belong to the lysosomal membrane (Schroder et al., 2010; Settembre et al., 2013a, b).

11.1.3.1 Membrane Proteins

- Membrane structural proteins include heavily glycosylated, abundantly present lysosome-associated membrane proteins (LAMPS), and lysosomal integral membrane proteins (LIMPS), which constitute approximately 50% of the lysosomal membrane proteins (Winchester, 2001).
- Lysosomal trafficking and fusion proteins include specific Rab GTPases and SNARE proteins (SNARE: N-ethylmaleimide-sensitive attachment protein (SNAP) receptor) (Rojas et al., 2008).
- The LYNUS (lysosomal nutrient sensing) machinery, composed of several proteins on the lysosomal surface, senses the nutritional status of the lysosome and sends signals to the nucleus (Settembre et al., 2013a, b; Settembre & Ballabio, 2014).
- Many ion channels on the lysosomal membrane mediate the transport of ions (H^+, Ca^{2+}, Na^+, K^+, Cl^-, Fe^{2+}, Zn^{2+}) across the membrane and help in lysosomal fusion and maintenance of the acidic interior of the lysosome (Xiong & Zhu, 2016; Xu & Ren, 2015). The TRPML1 or Mucolipin-1 participates in calcium signaling during fusion of lysosome with autophagosome or plasma membrane (Bach, 2005, Cantiello et al., 2005, Wong, Li et al., 2012). Another chloride channel, CLC-7 is involved in lysosomal acidification (Jentsch et al., 2005). K^+ channels on the lysosomal membrane include BK channel and TMEM175. Two pore channels or TPCs mediate Na^+ and Ca^{2+} export from endolysosome (Ishibashi et al., 2000).
- Membrane transporters mediate the influx and efflux of biomolecules across the membrane (Winchester, 2001; Settembre et al., 2013a, b; Xu & Ren, 2015). LAMP2A binds to cytosolic substrates for internalization into the lumen (Cuervo et al., 2000). Amino acid transporters on lysosomal membrane include cysteine transporter cystionosin, arginine/lysine exporter LAAT-1. These transporters, when mutated, result in intralysosomal

accumulation leading to lysosomal storage disorders (Winchester, 2001). For example, NPC1 mediates the export of cholesterol from lysosome, and mutations in NPC1 lead to Niemann-Pick type C disease (Infante et al., 2008; Lloyd-Evans et al., 2008).

11.1.3.2 Lysosomal Enzymes

There are approximately 60 different resident hydrolase enzymes that execute the catabolic processes in the acidic lumen of the lysosome. These enzymes, belonging to classes of proteases, phosphatases, peptidases, sulfatases, glycosidases, nucleases, and lipases, enable the degradation of a wide range of substrates in the lysosome (Appelqvist et al., 2013; Bainton, 1981). The lysosomal enzymes undergo a mannose-6-phosphate modification in the Golgi compartment which targets them to the lysosome (Settembre et al., 2013a, b). Following secretion, mannose-6-phosphate receptors on neighboring cells can recognize and uptake the enzymes inside the cell (Settembre et al., 2013a, b).

11.1.4 Functions of Lysosome

Apart from their classical role in cellular waste management, lysosomes play a central role in multiple cellular processes contributing to the maintenance of cellular homeostasis. The major functions of lysosomes can be broadly divided into three categories: degradation (and recycling), secretion, and signaling.

11.1.4.1 Lysosomal Clearance by Degradation of Macromolecules

Lysosomes are the terminal component for the degradation of extracellular and intracellular materials received via the processes of endocytosis and autophagy, respectively. The products generated through the catabolic processes in the lysosome are recycled for the formation of cellular components and utilized for generating energy depending on the cellular nutritional status, making the lysosome function as a nutritional reservoir of the cell.

Endocytosis

Endocytosis involves the internalization of extracellular materials, plasma membrane, and cell surface receptors via several mechanisms (phagocytosis, micropinocytosis, clathrin-mediated or independent endocytosis, caveolin-mediated or independent endocytosis) based on the type of the cargo (Conner & Schmid, 2003; Doherty & McMahon, 2009). Following internalization, the cargo is first sorted in the early endosome, from which the recycling receptors are delivered back to the plasma membrane via the recycling endosomes (Fig. 11.1). Early endosome (pH ~6), containing the cargo targeted for degradation, evolves into late endosome (pH 5–6) where, in response to a lower pH, the acid hydrolases are released from the mannose-6-phosphate receptors, following recycling of the receptors to the trans-Golgi network. Late endosome transfers the cargo to the lysosome (pH 4.5–5) (Luzio et al., 2007; Ohkuma & Poole, 1978). The formation of limiting membrane-containing intraluminal vesicles (ILVs) in the endosomal pathway helps in the delivery of lipids and membranes to the lysosome for degradation (Appelqvist et al., 2013; Woodman & Futter, 2008) (Fig. 11.1).

Maturation of early endosome to lysosome requires a series of changes which takes approximately 40 min and comprises of sequential decrease of the luminal pH, exchange of membrane components, Rab conversion (exchange of Rab5 on early endosome for Rab7 on late endosome), generation of ILVs, lysosomal component acquirement (hydrolases, membrane proteins), transport to the perinuclear region, and morphological changes (Appelqvist et al., 2013). Different markers are used to distinguish the highly dynamic intermediates of the endocytic pathway. While Rab5 and early endosomal antigen 1 (EEA1) selectively mark early endosome (Mu et al., 1995), late endosome and lysosome share multiple molecular features (LAMPs and luminal acid hydrolases), but they can be differentiated by the absence of mannose-6-phosphate receptors in the lysosome (Brown et al., 1986).

Autophagy

Intracellular cytoplasmic components (aggregated proteins, damaged organelles, intracellular pathogens, etc.) are transported to the lysosome for degradation via the process of autophagy (Griffey & Yamamoto, 2022), which can be classified into three categories: chaperone-mediated autophagy (CMA), microautophagy, and macroautophagy (Fig. 11.1) (Martini-Stoica et al., 2016). CMA is mediated by recognition of cytosolic proteins bearing a specific motif (KFERQ motif) by chaperone HSC 70 (heat-shock cognate protein of 70 kDa) followed by unfolding and internalization into the lysosome via active transport through the membrane receptor LAMP2A (Chiang et al., 1989; Cuervo & Dice, 1996; Dice, 1990; Kaushik & Cuervo, 2012). In microautophagy, cytoplasmic cargo is directly engulfed by invagination of the lysosomal or endosomal membrane. The major form of autophagy and macroautophagy involves the formation of autophagosome by sequestration of cytoplasmic material into the phagophore, a newly formed double membrane (Li et al., 2012; Settembre et al., 2013a, b). The process is divided into four phases: initiation, elongation, maturation, and fusion of autophagosome with lysosome to form autolysosome where the substrate is degraded by lysosomal hydrolases (Martini-Stoica et al., 2016; Parzych & Klionsky, 2014).

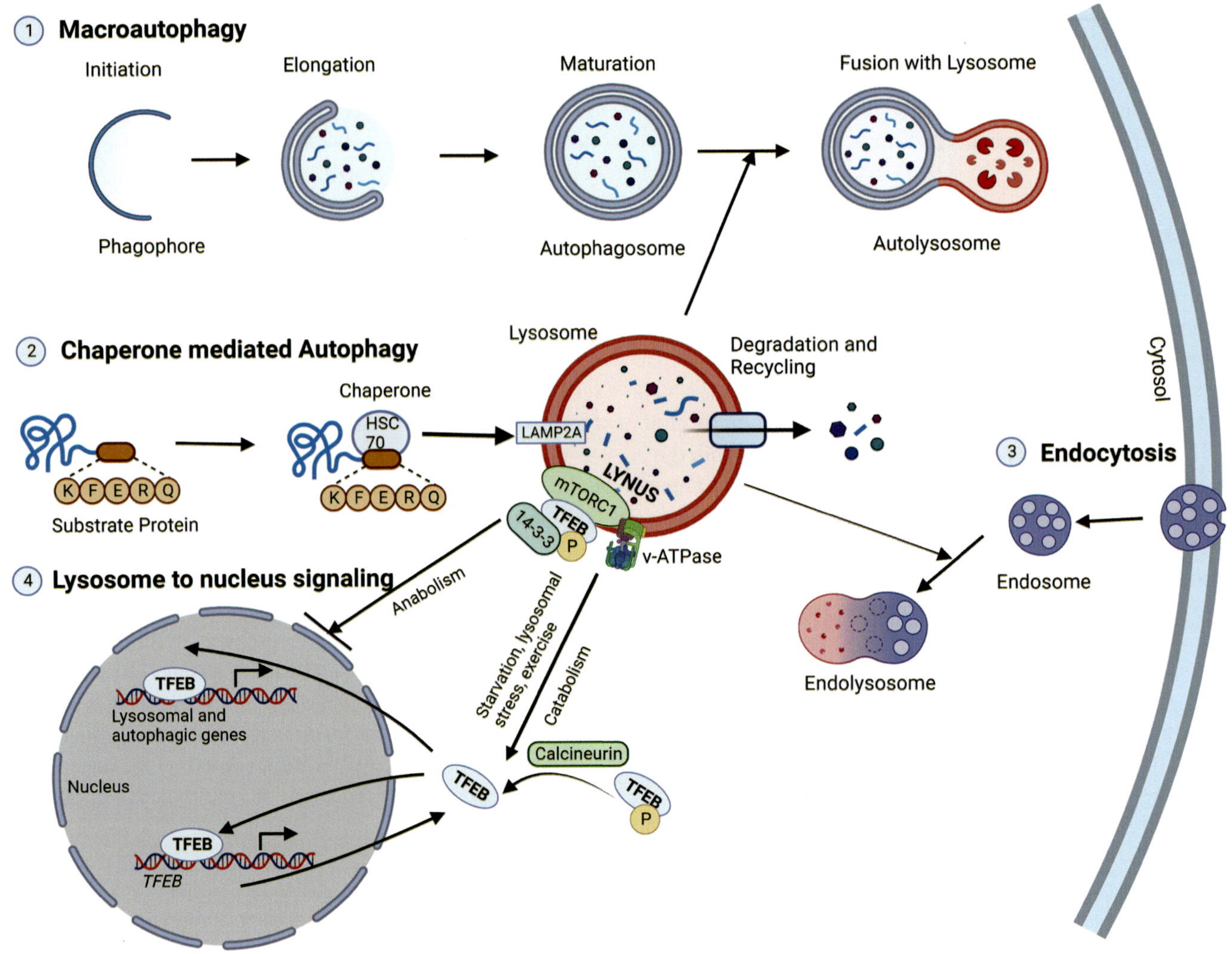

Fig. 11.1 Schematic presentation of lysosomal degradation of macromolecules and signaling. Lysosomes can degrade intracellular and extracellular materials received via autophagy (1, 2) and endocytosis, respectively (3). Lysosomal signaling is regulated by transcriptional program as well as feedback responses, with mTORC1 and TFEB as the central mediators. Under anabolic conditions, mTORC1 phosphorylates TFEB and sequesters it in the cytoplasm by binding to 14-3-3 proteins, thereby preventing it's translocation to the nucleus. During catabolic conditions (starvation, lysosomal stress, or exercise), mTORC1 is inactivated, and activation of the phosphatase calcineurin dephosphorylates TFEB which then can translocate to the nucleus and activate transcription of lysosomal and autophagic genes. TFEB also stimulates its own transcription via an autoregulatory loop (4). This model is created with BioRender.com

11.1.4.2 Lysosomal Secretion or Exocytosis

Lysosomal exocytosis is a Ca^{2+} regulated process by which lysosomes bulk release their content outside into the extracellular space (Andrews, 2000; Jaiswal et al., 2002; Rodriguez et al., 1997). It plays an important role in numerous physiological processes including immune responses like parasitic defense by mast cells, eosinophils, and platelet-mediated coagulations; bone resorption mediated by osteoclasts; hydrolase release from spermatozoa; melanocyte function during pigmentation; and repair of the plasma membrane (Gerasimenko et al., 2001, Reddy et al., 2001; Settembre et al., 2013a, b).

Lysosomal exocytosis involves a multistep mechanism initiating with repositioning of the lysosomes from the perinuclear region to space adjacent to the plasma membrane (Jaiswal et al., 2002) followed by the fusion of the lysosome with plasma membrane stimulated by enhanced intracellular calcium (Rodriguez et al., 1997) and finally, secretion of lysosomal content outside of the cell (Appelqvist et al., 2013). Several molecular components of the lysosomal and plasma membrane have been found to mediate the exocytosis including VAMP-7 (a v-SNARE), synaptotagmin VII (calcium sensor), multiple RAB proteins on the surface of the lysosome- and synaptosome-associated protein 23 (SNAP23), syntaxin 4 (t-SNARES) on the plasma membrane (Rao et al., 2004) as well as lysosomal membrane calcium channel mucolipin-1 (LaPlante et al., 2006).

11.1.4.3 Signaling

Lysosome functions as a signaling hub by acting as a nutrient sensor and metabolic regulator to coordinate multiple cellular pathways in response to different physiological conditions (Lim & Zoncu, 2016; Settembre & Ballabio, 2014). Lysosomal signaling is regulated by transcriptional programs as well as feedback responses, with the master cellular growth regulator mechanistic target of rapamycin complex 1 (mTORC1) (Fig. 11.1), and the basic helix-loop-helix transcription factor TFEB as the central mediators of lysosome-to-nucleus signaling. (Napolitano & Ballabio, 2016; Settembre et al., 2012). mTORC1, localized on the lysosomal surface, controls cellular biosynthesis and catabolism by inversely regulating autophagy. Under nutrient-rich conditions, a kinase complex (ULK1–ATG13–FIP200; ULK1: unc-51-like kinase 1; ATG13:mammalian autophagy-related gene 13; FIP200:focal adhesion kinase family-interacting protein of 200 kDa), that stimulates autophagosome biogenesis, is phosphorylated and inhibited by mTORC1 thus preventing autophagy initiation (Ganley et al., 2009; Hara et al., 2008; Hosokawa et al., 2009; Jung et al., 2009). Conversely, during starvation or mTORC1 inhibition, the ULK1–ATG13–FIP200 kinase complex is activated and induces autophagy (Martini-Stoica et al., 2016). Lysosomal signaling events establish this organelle as a regulator of cellular response to starvation and energy metabolism of the cell.

11.1.5 Mechanisms for the Regulation of Lysosomal Functions

A lysosome-to-nucleus signaling activates a transcriptional program and feedback mechanisms that regulate lysosomal adaptation (Settembre et al., 2012; Settembre & Ballabio, 2014). The lysosome nutrient sensing (LYNUS) machinery is comprised of several protein complexes that reside on the lysosome surface. Under normal conditions, vATPase on the lysosomal membrane can sense amino acid content and interact with the regulator that mediates the docking and activation of Rag GTPases. Active Rag GTPases then bind to mTORC1, recruit it to the lysosomal surface, and promote mTORC1 activation by Rheb GTPase (Sancak et al., 2008, 2010; Zoncu et al., 2011). Additionally, activated Rag GTPases bind to and transport TFEB to the lysosomal membrane (Martina & Puertollano, 2013). The mTORC1 complex, composed of mTOR and regulatory proteins Raptor (a positive regulatory subunit of mTOR complex 1), lethal with SEC13 protein 8 (LST8) and DEP-domain containing mTOR-interacting protein (DEPTOR), phosphorylates TFEB on residues Ser142 and Ser211. Phosphorylated TFEB is inactive and sequestered in the cytosol by binding to 14-3-3 family of proteins, which inhibits its nuclear transport by masking the nuclear translocation signal (Napolitano & Ballabio, 2016; Settembre et al., 2011).

During starvation or lysosomal stress, Rag GTPases are switched off, releasing mTORC1 from the LYNUS machinery on lysosomal membrane, thereby inactivating mTORC1 (Martini-Stoica et al., 2016; Sancak et al., 2010). Concurrently, Ca2+ is released through lysosomal membrane channel mucolipin1 (MCOLN1) generating higher Ca2+ concentrations outside the lysosomal surface that promotes the activation of the phosphatase calcineurin. Active calcineurin dephosphorylates and activates TFEB which, upon release from 14-3-3, translocates to the nucleus (Fig. 11.1) (Ballabio, 2016; Martini-Stoica et al., 2016; Medina et al., 2015). Inside the nucleus, TFEB transcriptionally regulates the expression of various genes of the lysosomal–autophagy pathway. Importantly, TFEB stimulates its own transcription via an autoregulatory loop further sustaining lysosomal response (Settembre et al., 2013a, b; Napolitano & Ballabio, 2016).

11.1.6 Mechanisms of Lysosomal Biogenesis: TFEB and CLEAR

Recent discoveries have shed light on the molecular mechanisms of lysosomal biogenesis. Using a microarray analysis approach, study by Sardiello et al. revealed that different lysosomal genes have a coordinated expression profile in diverse cell types under different cellular conditions (Sardiello et al., 2009; Settembre et al., 2013a, b). Furthermore, pattern discovery analysis of 96 lysosomal genes identified a palindromic motif, containing 10 base pairs (GTCACGTGAC), in the promoter (preferentially present within 200 base pairs of the transcription start site) of 68 genes. Owing to the enrichment of this motif in the promoter of genes involved in lysosomal biogenesis and function, the gene network and the motif were named Coordinated Lysosomal Expression and Regulation (CLEAR) and CLEAR element, respectively (Sardiello et al., 2009). Moreover, the CLEAR element was found to be overlapping with the E-box, target sequence for the basic helix-loop-helix (bHLH) family of transcription factors (Massari & Murre, 2000; Steingrimsson et al., 2004). Subsequent analyses revealed that transcription factor EB (TFEB), a member of the microphthalmia–transcription factor E (MiT/TFE) subfamily (comprised of MITF, TFE3, TFEB, and TFEC in humans) of bHLH, binds to CLEAR element and transcriptionally upregulate the lysosomal genes, identifying TFEB as the central regulator of CLEAR network. TFEB overexpression leads to enhancement in lysosomal numbers and levels of enzymes, resulting in increased lyososomal catabolism (Sardiello et al., 2009; Settembre et al., 2013a, b; Napolitano & Ballabio, 2016). Since its discovery, mounting evidence has established that TFEB coordinates the expression of several genes involved in lysosomal biogenesis, exocytosis, and autophagy (Settembre et al., 2011). Thus, TFEB-mediated induction of lysosomal genes, and thereby enhancement of cellular degradative capacity, is an attractive therapeutic strategy for disorders associated with lysosomal storage materials.

11.1.7 Mechanisms for the Regulation of TFEB

Since TFEB plays a key role in lysosomal biogenesis and autophagy, delineating mechanisms by which TFEB is regulated is an important area of research. Peroxisome proliferator-activated receptor alpha (PPARα) is a nuclear hormone receptor that participates in the proliferation of peroxisomes and β-oxidation of very long chain fatty acids (Seedorf, 1998; Pahan, 2006; Roy & Pahan, 2009, 2015). PPARα is known to bind peroxisome proliferator response element (PPRE) present in many genes that are ultimately involved in fatty acid metabolism via their protein products (Pahan, 2006; Pahan et al., 1994, 1998). However, analysis of *TFEB* gene promoter identifies the presence of a consensus PPRE (Fig. 11.2) (Ghosh et al., 2015). Accordingly, activation of PPARα increases the level of TFEB, and agonists of PPARα stimulate the recruitment of PPARα to the promoter of *TFEB* in brain cells (Fig. 11.2) (Ghosh et al., 2015; Ghosh & Pahan, 2016). In addition to PPARα, RXRα is also recruited to the TFEB gene promoter as a heterodimeric complex of PPARα and RXRα (Ghosh et al., 2015). Therefore, a combination of gemfibrozil (a prototype activator PPARα) and retinoic acid (a prototype activator of RXRα) is a potent inducer of TFEB and lysosomal biogenesis in brain cells (Ghosh et al., 2015). Other activators of PPARα (e.g., aspirin and cinnamic acid) have also been shown to upregulate TFEB and lysosomal biogenesis in brain cells (Fig. 11.2) (Chandra et al., 2019a, b). Recently, we have described that treadmill run increases PPARα in the nigra to ultimately stimulate TFEB and autophagic clearance of α-synuclein (α-syn; Fig. 11.2) (Dutta et al., 2022). It is interesting because until now, PPARα is the only known molecule to control the biogenesis of both peroxisomes and lysosomes.

11.1.8 Lysosomal Storage Disorders

11.1.8.1 General Introduction

LSDs are comprised of a group of recessively inherited inborn errors of metabolism characterized by progressive intralysosomal accumulation of undegraded macromolecules owing to lysosomal dysfunction. They are individually rare; however, the cumulative occurrence is estimated to be 1 in 5000 live births, relatively high compared to other rare diseases (Fuller et al., 2006). Approximately 60 different types of LSDs have been identified which are classified based on type of the storage material or the molecular defect (Boustany, 2013; Filocamo & Morrone, 2011). LSDs often involve multi-systematic symptoms and exhibit three forms: infantile (the most severe form), juvenile, and adult forms of the disease. LSDs are caused by mutations in genes encoding lysosomal enzymes involved in the degradation of macromolecules or lysosomal membrane proteins or proteins that contribute to normal lysosomal function (Platt et al., 2012; Settembre et al., 2013a, b; Raha et al., 2024).

11.1.8.2 Historical Preview of LSDs

The first case of LSDs was described by physicians Warren Tay in 1881 and by Bernard Sachs in 1887, which was later named Tay-Sachs disease (Filocamo & Morrone, 2011; Futerman & van Meer, 2004). Gaucher disease, the most common LSD at present, was described by Phillipe Gaucher in 1882. He reported a female patient having an enlarged spleen and unusually large cells in the spleen, which are now known as "Gaucher cells" (Futerman & van Meer, 2004). Various attempts have been made to understand the genetic, biochemical, and molecular bases of different LSDs that led to the development of therapeutic strategies. In 2008, the first clinical trial was organized for gene therapy for neuronal ceroid lipofuscinosis (NCL) (Eisenstein, 2016).

11.1.8.3 Major Biochemical and Cellular Events in LSD Pathogenesis

LSDs originate from mutations in gene encoding lysosomal enzymes, activators, transporters, membrane proteins, or non-lysosomal proteins required for normal lysosomal function (Parenti et al., 2015). These mutations often lead to

- An enzyme having reduced catalytic activity delivered to lysosome
- Impaired transport of an enzyme from ER due to
  - Mutations causing protein misfolding leading to degradation
  - Dysfunction of multi-enzyme complex that is required
- Aberrant glycosylation of an enzyme in Golgi resulting in:
  - Reduced catalytic activity
  - Inability to bind mannose-6-phosphate receptors and thereby defective targeting to the lysosome
- Deficiency of a protein that is required for optimal function of an enzyme
- Abnormal function of lysosomal membrane proteins including transporters defects in proteins necessary for regulation of lysosomal functions (Futerman & van Meer, 2004; Parenti et al., 2015)

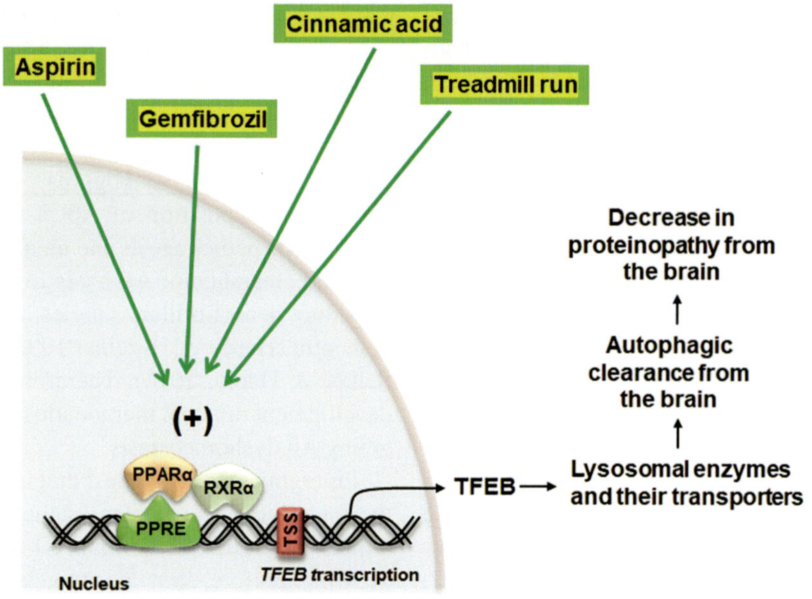

Fig. 11.2 Role of peroxisome proliferator-activated receptor α (PPARα) in lysosomal biogenesis and autophagy. Lipid-lowering drug gemfibrozil, nonsteroidal anti-inflammatory drug aspirin, naturally available aromatic acid cinnamic acid, and treadmill exercise activate PPARα and stimulate the recruitment of PPARα:RXRα complex to the peroxisome proliferator responsive element (PPRE) present in the promoter of TFEB gene to upregulate the transcription of TFEB. This leads to an increase in the lysosomal biogenesis and autophagic clearance of aggregated proteins from the brain. *TSS* transcription start site, *RXRα* retinoid X receptor alpha

These functional deficits lead to accumulation of the specific undigested substrate in the lysosome (primary storage) resulting in obstructed irregular lysosomal trafficking, further leading to accumulation of additional substrates (secondary storage). Abnormal lysosomal storage impacts the lysosomal function by aberrant calcium homeostasis, inflammation, impaired signaling, oxidative stress, and lysosomal membrane permeabilization leading to a cascade of pathogenic events (Vitner et al., 2010). Additionally, lysosomal dysfunction causes autophagy impairment and buildup of autophagy substrates (tertiary storage), which contributes to neurodegeneration and further drives the disease pathogenesis (Settembre et al., 2013a, b). These cellular perturbations ultimately lead to organ-specific manifestation of the disease.

11.1.8.4 Neuropathological Features of LSDs

Many of the LSDs exhibit neuropathological involvement and are associated with neurodegeneration in several brain regions, including the thalamus, cortex, hippocampus, and cerebellum (Pastores & Maegawa, 2013; Platt et al., 2012). The neurological components of LSDs, observed in about two-thirds of LSD patients, contribute significantly to the disease pathology and are vastly different ranging from progressive neurodegeneration and cognitive impairment to behavioral abnormalities, epileptic, and psychiatric disorders (Parenti et al., 2015; Walkley, 2009). Distinct spatial and temporal neuropathological characteristics are observed in LSDs which could be explained by diverse effects of specific material on different neuronal populations, synthesis of variable amounts of macromolecules by different types of neurons, and differential vulnerability of neurons to specific storage material (Platt et al., 2012). Astrogliosis along with activation of the innate immune system also contribute to the neuropathology in LSDs (Bosch & Kielian, 2015; Rama Rao & Kielian, 2016).

11.1.8.5 Therapeutic Approaches for Treating LSDs

Newborn screening, whenever applicable, is the best approach for managing the disease symptoms and early therapy for LSDs. Many innovative therapeutic approaches have been developed for treating LSDs, which are based on strategies focused on a specific stage of the disease pathogenesis (Parenti et al., 2015). The commonly used strategies include gene therapy, enzyme replacement therapy, and hematopoietic stem cell transplantation. Additionally, pharmacological chaperones and proteostasis regulators are used for stabilizing the mutant enzyme and therefore increasing its residual activity (Parenti, 2009; Parenti et al., 2015), whereas substrate reduction therapy involves reduction of the biosynthesis of the substrate (Coutinho et al., 2016). Multiple studies have established the vast application and potential therapeutic benefit of TFEB, the central regulator of lysosomal function, in several LSDs including Gaucher disease, Tay–Sachs disease (Song et al., 2013), Pompe disease (Spampanato et al., 2013), and cystinosis (Rega et al., 2016). Therefore, pharmacological stimulation of TFEB and thereby enhancement of lysosomal function are appealing therapeutic strategies for the treatment of LSDs.

11.2 Neurodegenerative Disorders

11.2.1 Alzheimer's Disease

The two pathological alterations described by Dr. Alzheimer are still considered the hallmarks of AD and are used as criteria for diagnosis. These major neuropathological characteristics of AD are the development of extracellular senile plaques containing amyloid beta and formation of intraneuronal neurofibrillary tangles (NFT) (LaFerla et al., 2007; Rangasamy et al., 2018; Chandra & Pahan, 2019; Chandra et al., 2019a, b). Abnormal accumulation of amyloid β (Aβ) (40 or 42) forms the senile plaques that can be broadly classified into two categories, diffuse and dense-core plaques, based on the morphological characteristic and staining using Thioflavin-S and Congo Red, dyes that specifically recognize the β-pleated sheet conformation (Chandra et al., 2018; Corbett et al., 2015; Serrano-Pozo et al., 2011). Diffuse plaques are Thioflavin-S-negative and non-neuritic that are frequently observed in the cognitively intact aged people. However, dense-core plaques, present in the brains of AD patients, are composed of fibrillar amyloid beta and stain positively for Thioflavin-S. These compact-core plaques are associated with dystrophic neurites or neuritic plaques, accumulation of reactive astrocytes, and microglia surrounding the plaque, neuronal degeneration, and synaptic loss (Knowles et al., 1999; Masliah et al., 1994; Modi et al., 2014, 2015; Rangasamy et al., 2015; Urbanc et al., 2002). While deposition of amyloid plaques is believed to start before the signs of cognitive decline, NFT development, neuronal and synaptic losses occur concomitantly with the cognitive impairment (Serrano-Pozo et al., 2011).

NFTs are formed by intraneuronal accumulation of abnormally hyperphosphorylated and aggregated tau. Tau is a microtubule-associated protein involved in stabilizing the microtubules and thereby facilitating axonal transport. Hyperphosphorylation of tau leads to misfolding and aggregation of the protein to generate paired helical fragments (PHFs), NFTs, neuropil threads, and ultimately extraneuronal ghost tangles following the death of the affected neuron (Kidd, 1963, 1964; Wisniewski et al., 1976). Accumulated evidence has established the burden and distribution of NFTs that can be correlated with the degree and duration of dementia and the neuropsychological hierarchy of AD–dementia (Serrano-Pozo et al., 2011).

11.2.1.1 Stimulation of Lysosomal Clearance as a Therapeutic Strategy in AD

Although AD involves a multifactorial etiology, the Amyloid cascade hypothesis, proposed by Hardy and Higgins, is widely established as the underlying model of AD pathophysiology (Hardy & Higgins, 1992). According to this hypothesis, deposition of Aβ is the initiating and driving event in AD pathogenesis and an imbalance between production and clearance of Aβ leads to other pathological events including neurofibrillary tangles, neuronal dysfunction, and dementia (Hardy & Higgins, 1992; Hardy & Selkoe, 2002; Selkoe & Hardy, 2016). Therefore, many strategies for the development of novel therapeutics for AD are focused on targeting Aβ dyshomeostasis.

Emerging studies suggest that impaired clearance of Aβ is the underlying mechanism of the widespread sporadic AD which constitutes 95% of all AD cases (Mawuenyega et al., 2010). Therefore, strategies targeted at effective clearance of Aβ from the brain have immense therapeutic implications for AD. Several studies have highlighted the potential of astrocytic lysosomal induction for efficient clearance of Aβ. Cultured adult astrocytes have been shown to migrate, bind, and degrade amyloid plaques in brain sections from an AD mouse model implicating inefficient astroglial clearance of Aβ in AD pathogenesis (Wyss-Coray et al., 2003). Further study indicated that apolipoprotein E plays an important role in astrocyte-mediated Aβ clearance (Koistinaho et al., 2004). Another study demonstrated that attenuating astrocytic activation by deletion of genes encoding intermediate filament proteins, glial fibrillary acidic protein (GFAP), and vimentin accelerates the amyloid plaque pathogenesis suggesting an important role of astrocyte activation in preventing AD pathogenesis (Kraft et al., 2013). Therefore, astrocytes play an important role in the clearance of Aβ plaques in AD.

Attempts to enhance the lysosomal function and restore normal autophagy by modulating TFEB, the essential regulator of the lysosome system, have generated promising therapeutic results in rescuing the amyloid pathogenesis in AD. Enhancing lysosomal function with TFEB leads to increased lysosomal degradation of holo-amyloid precursor protein (APP) in the neurons and thus reduces the amyloidogenic processing of APP and Aβ generation (Xiao et al., 2015). Recent studies have demonstrated that TFEB overexpression can alleviate AD pathology by regulation of the autophagy–lysosome pathway (Zhang & Zhao, 2015). Targeted TFEB expression in astrocytes mediated by viral gene transfer promotes attenuation of the amyloid pathology by enhancing lysosomal biogenesis and facilitating Aβ uptake and lysosomal degradation by astrocytes (Xiao et al., 2014). Given the immense importance of astrocytic lysosomal stimulation, in this study, we studied the effect of drug-mediated lysosomal biogenesis and its effect in an AD model. Recently, we have seen that activation of PPARα stimulates the uptake of Aβ by astrocytes and that a combination of gemfibrozil and retinoic acid reduces plaques and improves cognitive functions in a mouse model of AD via astrocytic PPARα (Raha et al., 2021a, b).

11.2.2 Parkinson's Disease

Parkinson's disease (PD) is the most common movement disorder and currently about six million people in the world are affected by PD (Roy & Pahan, 2011; Pahan & Pahan, 2015; Global, Regional, and National Burden of Neurological Disorders (1990–2016)). The disease is caused by the loss of dopamine in the brain due to the death of tyrosine hydroxylase (TH) positive dopaminergic (DAergic) neurons in the substantia nigra pars compacta (SNpc) (Benner et al., 2004; Ghosh et al., 2007, 2009). Another pathologic hallmark of PD is the presence of Lewy bodies (LBs) containing aggregated α-syn (Chandra et al., 2017; Dutta et al., 2021a, b, 2022; Raha et al., 2021a, b). In addition to PD, accumulation of α-syn is also seen in multiple brain regions of Lewy body dementia (LBD) and multiple system atrophy (MSA) (Goedert & Spillantini, 1998; Spillantini et al., 1997; Wakabayashi et al., 1998). While α-syn at the physiological level is responsible for vesicle trafficking and dopamine (DA) neurotransmission at the nerve terminals, overexpression of wild type α-syn or mutation of α-syn leads to the production of higher molecular weight oligomeric form of α-syn capable of evading cellular degradation machineries, ultimately causing neuronal demise (Hofer et al., 2005; Mueller et al., 2005). As a result, deficient cellular degradation machineries such as autophagy and lysosomal functions are important for the deposition of aggregated α-syn in the brain (Bellomo et al., 2020; Lee et al., 2013). Eventually neuron secreted forms of α-syn increasingly spread in selected brain regions of the CNS depending on the disease (Beach et al., 2009; Braak et al., 2003; Kantarci et al., 2010; Wenning & Jellinger, 2005).

11.2.2.1 Crosstalk Between Gaucher's Disease (GD) and PD

One of the most common genetic risk factors of PD is the presence of mutations in one copy of the glucocerebrosidase (GBA) gene (Bultron et al., 2010; Sidransky et al., 2009). Population studies have shown that about 9% of GBA carriers eventually develop PD (Sidransky et al., 2009). However, some reports indicate that about 30% of GD patients at the age of 80 years may be associated with PD (Anheim, et al., 2012). It has been also reported that patients with a homozygous mutation of *GBA* as seen in Gaucher's disease have a higher risk of developing PD even with an earlier age of onset (Thaler et al., 2017). However, it is not clear whether PD developed in GD patients exhibits severe phenotypes or not. It is becoming clear that PD develops due to a failure of the endolysosomal and the autophagic pathways (Moors et al., 2016). This is mainly due to the findings that the accumulation of α-syn in the dopaminergic neurons is one of the hallmarks of PD, that these clearance machineries are essential for the breakdown of α-syn and that the glucocerebrosidase in lysosomes plays a role in the degradation of α-syn (Mazzulli et al., 2011).

11.2.2.2 Upregulation of Autophagy as a Therapeutic Strategy for PD

The involvement of autophagy in the clearance of toxic protein species has generated interest in the development of therapeutic approaches for protein aggregation-associated neurodegenerative disorders (Rubinsztein et al., 2007; Wong & Cuervo, 2010). Decressac et al. have provided evidence that mechanistically connects defective TFEB function to α-syn–mediated toxicity in both rodent PD model and human PD (Decressac et al., 2013). Being a transcription factor, TFEB should be translocated to nucleus for optimum function and they have shown that nuclear translocation of TFEB is impaired in α-syn overexpressing dopaminergic neurons and in midbrain neurons in human PD (Decressac et al., 2013). However, overexpression of TFEB results in remarkable protection of nigral dopaminergic neurons and clearance of α-syn oligomers (Decressac et al., 2013). Therefore, drugs capable of upregulating the level of TFEB in brain cells may be beneficial for the survival of dopaminergic neurons in neurodegenerative brains.

Since the prion-like propagation of α-syn has also been recapitulated in animals by multiple studies (Mao et al., 2016; Rey et al., 2019; Sorrentino et al., 2017), we have studied the effect of a treadmill run on prion-like α-syn spreading. We have seen that regular treadmill exercise increases TFEB and autophagy in the brain via PPARα to stop α-syn spreading from striatum to nigra and motor cortex and protect dopaminergic neurons in the nigra (Dutta et al., 2022). It has been shown that gemfibrozil, an FDA-approved lipid-lowering drug, and cinnamic acid, a component of several plants including cinnamon bark, increase the expression of TFEB in brain cells via activation of PPARα (Ghosh et al., 2015; Chandra et al., 2019a, b). Prorok et al. have delineated that oral administration of cinnamic acid defends dopaminergic neurons in a mouse model of PD (Prorok et al., 2019). According to Gottschalk et al. (2021), oral gemfibrozil is also capable of protecting dopaminergic neurons in a mouse model of PD. It is likely that the TFEB-upregulating effect of cinnamic acid and gemfibrozil (Fig. 11.2) may play a role in helping the survival of dopaminergic neurons.

11.2.3 Huntington's Disease

Huntington's disease (HD) is an autosomal-dominant disorder caused by expansion of the polyQ tract at the N terminus of the protein huntingtin, which is encoded by CAG trinucleotide repeat in the first exon of the huntingtin (HTT) gene (Walker, 2007; Dutta et al., 2021a, b; Mondal et al., 2024). An expansion greater than 36Q results in the disease, and the number of the tandem repeat is positively correlated with the younger age of disease onset. The disease is characterized by loss of motor control and cognitive decline. The brain region most affected in HD is the striatum, which primarily regulates the movements of body, limbs, and eyes. At present, there is no disease-modifying therapy available for treating HD. Thus, modulation of the autophagy and lysosomal function has emerged as an attractive therapeutic strategy for treatment of the disease (Martin et al., 2015; Martini-Stoica et al., 2016).

Accumulation of mutant HTT (mHTT) aggregates in HD results in aberrant autophagic regulation at multiple steps, including defective cargo recognition, trafficking of autophagosomes, and decreased autophagosome-lysosome fusion (Martin et al., 2015). In particular, mHTT can induce autophagy by sequestering mTOR; however, deficit in cargo loading leads to increased abundance of autophagosomes that are devoid of contents ("empty autophagosomes"). A study by Martinez-Vicente et al. reported that a primary defect in cytosolic cargo recognition by autophagic vacuoles leads to inefficient autophagy in cellular and mouse models of HD and cells from HD patients (Martinez-Vicente et al., 2010). Furthermore, mHTT disrupts the motility of autophagic vesicles and prevents fusion of autophagosomes with lysosomes, which ultimately results in the accumulation of toxic materials in the cytoplasm (Martin et al., 2015; Martinez-Vicente et al., 2010).

11.2.3.1 Autophagy as a Therapeutic Target in HD

Induction of Chaperone-Mediated Autophagy

A possible crosstalk between different autophagic pathways has been widely reported in HD, indicating functional compensation by one autophagy pathway when others are compromised. Among many studies, a study by Koga et al. identified a compensatory upregulation of CMA in response to macroautophagic dysfunction in cellular and mouse models of HD (Koga et al., 2011).

Considering mutant HTT is an excellent CMA substrate (contains KFERQ motifs recognized by Hsc70) (Cortes & La Spada, 2014), induction of CMA, an alternative lysosome-mediated degradation pathway where cytosolic substrates could be directly translocated into the lysosomes, could be an appealing therapeutic strategy for the treatment of HD.

Modulation of Autophagy Using Autophagy Inducers

Multiple studies have revealed the efficacy of augmenting the autophagy system in countering the disease pathogenesis in different cellular and animal models of HD. Pharmacological compounds targeting the autophagy–lysosome pathway such as mTOR inhibitor rapamycin and its analog CCI-779 reduced the neurotoxicity in mouse, and fly models of HD highlighting the therapeutic benefit of autophagy induction in HD (Ravikumar et al., 2004). However, these mTOR-dependent inducer molecules present a host of side effects owing to a wide range of mTOR-regulated signaling pathways (Martin et al., 2015; Ravikumar et al., 2004).

Hence another class of compounds that are mTOR independent have also been studied and showed beneficial effects. For example, the FDA approved drug Rilmenidine was shown to stimulate autophagy and ameliorate the disease pathology in a HD mouse model (Rose et al., 2010). A combinatorial therapy with both mTOR-dependent and -independent compounds would achieve the greatest level of mHTT clearance and pre-symptomatic administration of such treatment would be most efficacious as the chances of consistent mHTT turnover and inhibition of aggregate overload are highest at this stage (Martin et al., 2015; Rose et al., 2010).

11.3 Conclusions

Most of the LSDs are monogenic, which are caused by the mutation of specific genes, ultimately resulting in the deposition of respective proteins in aggregated, folded, or complex forms. Although academically we are aware of the causes for these diseases, therapeutically we are far away from establishing viable therapeutic options for majority of these diseases (Bremova-Ertl & Schneider, 2023; Ellison et al., 2023). This is mainly because gene therapy, enzyme replacement therapy, and hematopoietic stem cell transplantation therapy are yet to be universally accepted in clinics due to the efficacy issue (Bremova-Ertl & Schneider, 2023; Ellison et al., 2023). For example, delivery of a particular gene into different parts of the brain has not been sorted out. Microinjection of viral vectors into different parts of the brain is definitely an option; however, sometimes it has fatal outcome. Although enzyme replacement therapy has been FDA-approved for some LSDs for the last several years, there are several side effects and shortcomings of this highly expensive therapy (Koto et al., 2021, 2022). Drug-mediated TFEB therapy has been proposed for LSDs and currently, these are at the preclinical stage (Ghosh et al., 2017; Jana et al., 2023; Kim et al., 2017). In contrast to LSDs, neurodegenerative disorders are mostly sporadic. Therefore, only boosting autophagy may be sufficient for the clearance of protein aggregates from the CNS of patients with neurodegenerative disorders, and this hypothesis is currently gaining momentum (Griffey & Yamamoto, 2022; Themistokleous, et al., 2023; Yang et al., 2023). It is known that physical exercise and running are beneficial for the brain (Palmer et al., 2023). Recent studies have shown that treadmill run stimulates autophagy and protects the brain from different neurodegenerative insults (Dutta et al., 2022; Rangasamy et al., 2023).

11.4 Review Questions

1. Which of the following is true about TFEB?
   (a) It regulates autophagy
   (b) It is involved in mitophagy
   (c) It stimulates the transcription of proinflammatory molecules
   (d) It controls the transcription of cell cycle genes
   (e) All of the above

2. Which of the following events could be controlled by autophagy?
   (a) Cerebral plaque level
   (b) Tangle formation in the brain
   (c) Removal of Lewy body pathology from the brain
   (d) Deposition of autofluorescent materials in the brain
   (e) All of the above

3. Which of the following is/are true about PPARα?
   (a) It is involved in peroxisome proliferation
   (b) It helps in fatty acid oxidation
   (c) It controls the transcription of TFEB gene
   (d) It forms heterodimer with RXRα
   (e) All of the above

4. Which of the following enzymes is not present in lysosomes?
   (a) Phosphatase
   (b) Peptidase
   (c) Nitric oxide synthase
   (d) Sulfatase
   (e) Glycosidase

5. Identify the procedures that may be therapeutically beneficial for Lysosomal storage disorders.
   (a) Gene therapy
   (b) Hematopoietic stem cell transplantation
   (c) Enzyme replacement therapy
   (d) All of the above
   (e) None of the above

11.5 Answer Key

1. a
2. e
3. e
4. c
5. d

Acknowledgments This study was supported by grants from NIH (AT10980 and AG050431) and merit awards (1I01BX005002 and I01BX005613) from the US Department of Veterans Affairs. Moreover, Dr. Pahan is the recipient of a Research Career Scientist Award (1IK6 BX004982) from the Department of Veterans Affairs. However, the views expressed in this article are those of the authors and do not necessarily reflect the position or policy of either the Department of Veterans Affairs or the United States government.

References

Andrews, N. W. (2000). Regulated secretion of conventional lysosomes. *Trends in Cell Biology, 10*(8), 316–321.

Anheim, M., Elbaz, A., Lesage, S., Durr, A., Condroyer, C., Viallet, F., Pollak, P., Bonaiti, B., Bonaiti-Pellie, C., Brice, A., French, G., & Parkinson Disease Genetic. (2012). Penetrance of Parkinson disease in glucocerebrosidase gene mutation carriers. *Neurology, 78*(6), 417–420.

Appelqvist, H., Waster, P., Kagedal, K., & Ollinger, K. (2013). The lysosome: From waste bag to potential therapeutic target. *Journal of Molecular Cell Biology, 5*(4), 214–226.

Bach, G. (2005). Mucolipin 1: endocytosis and cation channel—A review. *Pflügers Archiv, 451*(1), 313–317.

Bainton, D. F. (1981). The discovery of lysosomes. *The Journal of Cell Biology, 91*(3 Pt 2), 66s–76s.

Ballabio, A. (2016). The awesome lysosome. *EMBO Molecular Medicine, 8*(2), 73–76.

Beach, T. G., Adler, C. H., Lue, L., Sue, L. I., Bachalakuri, J., Henry-Watson, J., Sasse, J., Boyer, S., Shirohi, S., Brooks, R., Eschbacher, J., White, C. L., 3rd, Akiyama, H., Caviness, J., Shill, H. A., Connor, D. J., Sabbagh, M. N., & Walker, D. G. (2009). Unified staging system for Lewy body disorders: Correlation with nigrostriatal degeneration, cognitive impairment and motor dysfunction. *Acta Neuropathologica, 117*(6), 613–634.

Bellomo, G., Paciotti, S., Gatticchi, L., & Parnetti, L. (2020). The vicious cycle between alpha-synuclein aggregation and autophagic-lysosomal dysfunction. *Movement Disorders, 35*(1), 34–44.

Benner, E. J., Mosley, R. L., Destache, C. J., Lewis, T. B., Jackson-Lewis, V., Gorantla, S., Nemachek, C., Green, S. R., Przedborski, S., & Gendelman, H. E. (2004). Therapeutic immunization protects dopaminergic neurons in a mouse model of Parkinson's disease. *Proceedings of the National Academy of Sciences of the United States of America, 101*(25), 9435–9440.

Bosch, M. E., & Kielian, T. (2015). Neuroinflammatory paradigms in lysosomal storage diseases. *Frontiers in Neuroscience, 9*, 417.

Boustany, R. M. (2013). Lysosomal storage diseases—The horizon expands. *Nature Reviews. Neurology, 9*(10), 583–598.

Braak, H., Del Tredici, K., Rub, U., de Vos, R. A., Jansen Steur, E. N., & Braak, E. (2003). Staging of brain pathology related to sporadic Parkinson's disease. *Neurobiology of Aging, 24*(2), 197–211.

Bremova-Ertl, T., & Schneider, S. (2023). Current advancements in therapy for Niemann-pick disease: Progress and pitfalls. *Expert Opinion on Pharmacotherapy*, 1–19.

Brown, W. J., Goodhouse, J., & Farquhar, M. G. (1986). Mannose-6-phosphate receptors for lysosomal enzymes cycle between the Golgi complex and endosomes. *The Journal of Cell Biology, 103*(4), 1235–1247.
Bultron, G., Kacena, K., Pearson, D., Boxer, M., Yang, R., Sathe, S., Pastores, G., & Mistry, P. K. (2010). The risk of Parkinson's disease in type 1 Gaucher disease. *Journal of Inherited Metabolic Disease, 33*(2), 167–173.
Cantiello, H. F., Montalbetti, N., Goldmann, W. H., Raychowdhury, M. K., Gonzalez-Perrett, S., Timpanaro, G. A., & Chasan, B. (2005). Cation channel activity of mucolipin-1: The effect of calcium. *Pflügers Archiv, 451*(1), 304–312.
Chandra, S., & Pahan, K. (2019). Gemfibrozil, a lipid-lowering drug, lowers amyloid plaque pathology and enhances memory in a mouse model of Alzheimer's disease via peroxisome proliferator-activated receptor alpha. *Journal of Alzheimer's Disease Reports, 3*(1), 149–168.
Chandra, G., Roy, A., Rangasamy, S. B., & Pahan, K. (2017). Induction of adaptive immunity leads to nigrostriatal disease progression in MPTP mouse model of Parkinson's disease. *Journal of Immunology, 198*(11), 4312–4326.
Chandra, S., Jana, M., & Pahan, K. (2018). Aspirin induces lysosomal biogenesis and attenuates amyloid plaque pathology in a mouse model of Alzheimer's disease via PPARalpha. *The Journal of Neuroscience, 38*(30), 6682–6699.
Chandra, S., Roy, A., Jana, M., & Pahan, K. (2019a). Cinnamic acid activates PPARalpha to stimulate lysosomal biogenesis and lower amyloid plaque pathology in an Alzheimer's disease mouse model. *Neurobiology of Disease, 124*, 379–395.
Chandra, S., Roy, A., Patel, D. R., & Pahan, K. (2019b). PPARalpha between aspirin and plaque clearance. *Journal of Alzheimer's Disease, 71*(2), 389–397.
Chiang, H. L., Terlecky, S. R., Plant, C. P., & Dice, J. F. (1989). A role for a 70-kilodalton heat shock protein in lysosomal degradation of intracellular proteins. *Science, 246*(4928), 382–385.
Conner, S. D., & Schmid, S. L. (2003). Regulated portals of entry into the cell. *Nature, 422*(6927), 37–44.
Corbett, G. T., Gonzalez, F. J., & Pahan, K. (2015). Activation of peroxisome proliferator-activated receptor alpha stimulates ADAM10-mediated proteolysis of APP. *Proceedings of the National Academy of Sciences of the United States of America, 112*(27), 8445–8450.
Cortes, C. J., & La Spada, A. R. (2014). The many faces of autophagy dysfunction in Huntington's disease: From mechanism to therapy. *Drug Discovery Today, 19*(7), 963–971.
Coutinho, M. F., Santos, J. I., & Alves, S. (2016). Less is more: Substrate reduction therapy for lysosomal storage disorders. *International Journal of Molecular Sciences, 17*(7).
Cuervo, A. M., & Dice, J. F. (1996). A receptor for the selective uptake and degradation of proteins by lysosomes. *Science, 273*(5274), 501–503.
Cuervo, A. M., Gomes, A. V., Barnes, J. A., & Dice, J. F. (2000). Selective degradation of annexins by chaperone-mediated autophagy. *The Journal of Biological Chemistry, 275*(43), 33329–33335.
De Duve, C. (1963). The lysosome. *Scientific American, 208*, 64–72.
de Duve, C. (2005). The lysosome turns fifty. *Nature Cell Biology, 7*(9), 847–849.
De Duve, C., & Wattiaux, R. (1966). Functions of lysosomes. *Annual Review of Physiology, 28*, 435–492.
Decressac, M., Mattsson, B., Weikop, P., Lundblad, M., Jakobsson, J., & Bjorklund, A. (2013). TFEB-mediated autophagy rescues midbrain dopamine neurons from alpha-synuclein toxicity. *Proceedings of the National Academy of Sciences of the United States of America, 110*(19), E1817–E1826.
Dice, J. F. (1990). Peptide sequences that target cytosolic proteins for lysosomal proteolysis. *Trends in Biochemical Sciences, 15*(8), 305–309.
Doherty, G. J., & McMahon, H. T. (2009). Mechanisms of endocytosis. *Annual Review of Biochemistry, 78*, 857–902.
Dutta, D., Jana, M., Majumder, M., Mondal, S., Roy, A., & Pahan, K. (2021a). Selective targeting of the TLR2/MyD88/NF-kappaB pathway reduces alpha-synuclein spreading in vitro and in vivo. *Nature Communications, 12*(1), 5382.
Dutta, D., Majumder, M., Paidi, R. K., & Pahan, K. (2021b). Alleviation of Huntington pathology in mice by oral administration of food additive glyceryl tribenzoate. *Neurobiology of Disease, 153*, 105318.
Dutta, D., Paidi, R. K., Raha, S., Roy, A., Chandra, S., & Pahan, K. (2022). Treadmill exercise reduces alpha-synuclein spreading via PPARalpha. *Cell Reports, 40*(2), 111058.
Eisenstein, M. (2016). Myriad maladies. *Nature, 537*(7621), S146–S147.
Ellison, S., Parker, H., & Bigger, B. (2023). Advances in therapies for neurological lysosomal storage disorders. *Journal of Inherited Metabolic Disease*.
Filocamo, M., & Morrone, A. (2011). Lysosomal storage disorders: Molecular basis and laboratory testing. *Human Genomics, 5*(3), 156–169.
Fuller, M., Meikle, P. J., & Hopwood, J. J. (2006). In A. Mehta, M. Beck, & G. Sunder-Plassmann (Eds.), *Epidemiology of lysosomal storage diseases: An overview*. Fabry Disease: Perspectives from 5 Years of FOS.
Futerman, A. H., & van Meer, G. (2004). The cell biology of lysosomal storage disorders. *Nature Reviews. Molecular Cell Biology, 5*(7), 554–565.
Ganley, I. G., Lam Du, H., Wang, J., Ding, X., Chen, S., & Jiang, X. (2009). ULK1.ATG13.FIP200 complex mediates mTOR signaling and is essential for autophagy. *The Journal of Biological Chemistry, 284*(18), 12297–12305.
Gerasimenko, J. V., Gerasimenko, O. V., & Petersen, O. H. (2001). Membrane repair: Ca(2+)-elicited lysosomal exocytosis. *Current Biology, 11*(23), R971–R974.
Ghosh, A., & Pahan, K. (2016). PPARalpha in lysosomal biogenesis: A perspective. *Pharmacological Research, 103*, 144–148.
Ghosh, A., Roy, A., Liu, X., Kordower, J. H., Mufson, E. J., Hartley, D. M., Ghosh, S., Mosley, R. L., Gendelman, H. E., & Pahan, K. (2007). Selective inhibition of NF-kappaB activation prevents dopaminergic neuronal loss in a mouse model of Parkinson's disease. *Proceedings of the National Academy of Sciences of the United States of America, 104*(47), 18754–18759.
Ghosh, A., Roy, A., Matras, J., Brahmachari, S., Gendelman, H. E., & Pahan, K. (2009). Simvastatin inhibits the activation of p21ras and prevents the loss of dopaminergic neurons in a mouse model of Parkinson's disease. *The Journal of Neuroscience, 29*(43), 13543–13556.
Ghosh, A., Jana, M., Modi, K., Gonzalez, F. J., Sims, K. B., Berry-Kravis, E., & Pahan, K. (2015). Activation of peroxisome proliferator-activated receptor alpha induces lysosomal biogenesis in brain cells: Implications for lysosomal storage disorders. *The Journal of Biological Chemistry, 290*(16), 10309–10324.
Ghosh, A., Rangasamy, S. B., Modi, K. K., & Pahan, K. (2017). Gemfibrozil, food and drug administration-approved lipid-lowering drug, increases longevity in mouse model of late infantile neuronal ceroid lipofuscinosis. *Journal of Neurochemistry, 141*(3), 423–435.
Global, Regional, and National Burden of Neurological Disorders. (1990–2016). A systematic analysis for the Global Burden of Disease Study 2016. *Lancet Neurology, 18*(5), 459–480.
Goedert, M., & Spillantini, M. G. (1998). Lewy body diseases and multiple system atrophy as alpha-synucleinopathies. *Molecular Psychiatry, 3*(6), 462–465.
Gottschalk, C. G., Jana, M., Roy, A., Patel, D. R., & Pahan, K. (2021). Gemfibrozil protects dopaminergic neurons in a mouse model of Parkinson's disease via PPARalpha-dependent astrocytic GDNF pathway. *The Journal of Neuroscience, 41*(10), 2287–2300.

Griffey, C. J., & Yamamoto, A. (2022). Macroautophagy in CNS health and disease. *Nature Reviews. Neuroscience, 23*(7), 411–427.

Hara, T., Takamura, A., Kishi, C., Iemura, S., Natsume, T., Guan, J. L., & Mizushima, N. (2008). FIP200, a ULK-interacting protein, is required for autophagosome formation in mammalian cells. *The Journal of Cell Biology, 181*(3), 497–510.

Hardy, J. A., & Higgins, G. A. (1992). Alzheimer's disease: The amyloid cascade hypothesis. *Science, 256*(5054), 184–185.

Hardy, J., & Selkoe, D. J. (2002). The amyloid hypothesis of Alzheimer's disease: Progress and problems on the road to therapeutics. *Science, 297*(5580), 353–356.

Hofer, A., Berg, D., Asmus, F., Niwar, M., Ransmayr, G., Riemenschneider, M., Bonelli, S. B., Steffelbauer, M., Ceballos-Baumann, A., Haussermann, P., Behnke, S., Kruger, R., Prestel, J., Sharma, M., Zimprich, A., Riess, O., & Gasser, T. (2005). The role of alpha-synuclein gene multiplications in early-onset Parkinson's disease and dementia with Lewy bodies. *Journal of Neural Transmission (Vienna), 112*(9), 1249–1254.

Hosokawa, N., Hara, T., Kaizuka, T., Kishi, C., Takamura, A., Miura, Y., Iemura, S., Natsume, T., Takehana, K., Yamada, N., Guan, J. L., Oshiro, N., & Mizushima, N. (2009). Nutrient-dependent mTORC1 association with the ULK1-Atg13-FIP200 complex required for autophagy. *Molecular Biology of the Cell, 20*(7), 1981–1991.

Infante, R. E., Wang, M. L., Radhakrishnan, A., Kwon, H. J., Brown, M. S., & Goldstein, J. L. (2008). NPC2 facilitates bidirectional transfer of cholesterol between NPC1 and lipid bilayers, a step in cholesterol egress from lysosomes. *Proceedings of the National Academy of Sciences of the United States of America, 105*(40), 15287–15292.

Ishibashi, K., Suzuki, M., & Imai, M. (2000). Molecular cloning of a novel form (two-repeat) protein related to voltage-gated sodium and calcium channels. *Biochemical and Biophysical Research Communications, 270*(2), 370–376.

Jaiswal, J. K., Andrews, N. W., & Simon, S. M. (2002). Membrane proximal lysosomes are the major vesicles responsible for calcium-dependent exocytosis in nonsecretory cells. *The Journal of Cell Biology, 159*(4), 625–635.

Jana, M., Dutta, D., Poddar, J., & Pahan, K. (2023). Activation of PPARalpha exhibits therapeutic efficacy in a mouse model of juvenile neuronal ceroid Lipofuscinosis. *The Journal of Neuroscience, 43*(10), 1814–1829.

Jentsch, T. J., Poet, M., Fuhrmann, J. C., & Zdebik, A. A. (2005). Physiological functions of CLC Cl- channels gleaned from human genetic disease and mouse models. *Annual Review of Physiology, 67*, 779–807.

Jung, C. H., Jun, C. B., Ro, S. H., Kim, Y. M., Otto, N. M., Cao, J., Kundu, M., & Kim, D. H. (2009). ULK-Atg13-FIP200 complexes mediate mTOR signaling to the autophagy machinery. *Molecular Biology of the Cell, 20*(7), 1992–2003.

Kantarci, K., Avula, R., Senjem, M. L., Samikoglu, A. R., Zhang, B., Weigand, S. D., Przybelski, S. A., Edmonson, H. A., Vemuri, P., Knopman, D. S., Ferman, T. J., Boeve, B. F., Petersen, R. C., & Jack, C. R., Jr. (2010). Dementia with Lewy bodies and Alzheimer disease: Neurodegenerative patterns characterized by DTI. *Neurology, 74*(22), 1814–1821.

Kaushik, S., & Cuervo, A. M. (2012). Chaperone-mediated autophagy: A unique way to enter the lysosome world. *Trends in Cell Biology, 22*(8), 407–417.

Kidd, M. (1963). Paired helical filaments in electron microscopy of Alzheimer's disease. *Nature, 197*, 192–193.

Kidd, M. (1964). Alzheimer's disease—An electron microscopical study. *Brain, 87*, 307–320.

Kim, K., Kleinman, H. K., Lee, H. J., & Pahan, K. (2017). Safety and potential efficacy of gemfibrozil as a supportive treatment for children with late infantile neuronal ceroid lipofuscinosis and other lipid storage disorders. *Orphanet Journal of Rare Diseases, 12*(1), 113.

Knowles, R. B., Wyart, C., Buldyrev, S. V., Cruz, L., Urbanc, B., Hasselmo, M. E., Stanley, H. E., & Hyman, B. T. (1999). Plaque-induced neurite abnormalities: Implications for disruption of neural networks in Alzheimer's disease. *Proceedings of the National Academy of Sciences of the United States of America, 96*(9), 5274–5279.

Koga, H., Martinez-Vicente, M., Arias, E., Kaushik, S., Sulzer, D., & Cuervo, A. M. (2011). Constitutive upregulation of chaperone-mediated autophagy in Huntington's disease. *The Journal of Neuroscience, 31*(50), 18492–18505.

Koistinaho, M., Lin, S., Wu, X., Esterman, M., Koger, D., Hanson, J., Higgs, R., Liu, F., Malkani, S., Bales, K. R., & Paul, S. M. (2004). Apolipoprotein E promotes astrocyte colocalization and degradation of deposited amyloid-beta peptides. *Nature Medicine, 10*(7), 719–726.

Koto, Y., Ueki, S., Yamakawa, M., & Sakai, N. (2021). Experiences of patients with lysosomal storage disorders treated with enzyme replacement therapy: A qualitative systematic review protocol. *JBI Evid Synth, 19*(3), 702–708.

Koto, Y., Ueki, S., Yamakawa, M., & Sakai, N. (2022). Experiences of patients with lysosomal storage disorders who are receiving enzyme-replacement therapy and the experiences of their family members: A qualitative systematic review. *JBI Evid Synth, 20*(6), 1474–1510.

Kraft, A. W., Hu, X., Yoon, H., Yan, P., Xiao, Q., Wang, Y., Gil, S. C., Brown, J., Wilhelmsson, U., Restivo, J. L., Cirrito, J. R., Holtzman, D. M., Kim, J., Pekny, M., & Lee, J. M. (2013). Attenuating astrocyte activation accelerates plaque pathogenesis in APP/PS1 mice. *The FASEB Journal, 27*(1), 187–198.

LaFerla, F. M., Green, K. N., & Oddo, S. (2007). Intracellular amyloid-beta in Alzheimer's disease. *Nature Reviews. Neuroscience, 8*(7), 499–509.

LaPlante, J. M., Sun, M., Falardeau, J., Dai, D., Brown, E. M., Slaugenhaupt, S. A., & Vassilev, P. M. (2006). Lysosomal exocytosis is impaired in mucolipidosis type IV. *Molecular Genetics and Metabolism, 89*(4), 339–348.

Lee, H. J., Cho, E. D., Lee, K. W., Kim, J. H., Cho, S. G., & Lee, S. J. (2013). Autophagic failure promotes the exocytosis and intercellular transfer of alpha-synuclein. *Experimental & Molecular Medicine, 45*, e22.

Li, W. W., Li, J., & Bao, J. K. (2012). Microautophagy: Lesser-known self-eating. *Cellular and Molecular Life Sciences, 69*(7), 1125–1136.

Lim, C. Y., & Zoncu, R. (2016). The lysosome as a command-and-control center for cellular metabolism. *The Journal of Cell Biology, 214*(6), 653–664.

Lloyd-Evans, E., Morgan, A. J., He, X., Smith, D. A., Elliot-Smith, E., Sillence, D. J., Churchill, G. C., Schuchman, E. H., Galione, A., & Platt, F. M. (2008). Niemann-pick disease type C1 is a sphingosine storage disease that causes deregulation of lysosomal calcium. *Nature Medicine, 14*(11), 1247–1255.

Luzio, J. P., Pryor, P. R., & Bright, N. A. (2007). Lysosomes: Fusion and function. *Nature Reviews. Molecular Cell Biology, 8*(8), 622–632.

Mao, X., Ou, M. T., Karuppagounder, S. S., Kam, T. I., Yin, X., Xiong, Y., Ge, P., Umanah, G. E., Brahmachari, S., Shin, J. H., Kang, H. C., Zhang, J., Xu, J., Chen, R., Park, H., Andrabi, S. A., Kang, S. U., Goncalves, R. A., Liang, Y., Zhang, S., Qi, C., Lam, S., Keiler, J. A., Tyson, J., Kim, D., Panicker, N., Yun, S. P., Workman, C. J., Vignali, D. A., Dawson, V. L., Ko, H. S., & Dawson, T. M. (2016). Pathological alpha-synuclein transmission initiated by binding lymphocyte-activation gene 3. *Science, 353*(6307).

Martin, D. D., Ladha, S., Ehrnhoefer, D. E., & Hayden, M. R. (2015). Autophagy in Huntington disease and huntingtin in autophagy. *Trends in Neurosciences, 38*(1), 26–35.

Martina, J. A., & Puertollano, R. (2013). Rag GTPases mediate amino acid-dependent recruitment of TFEB and MITF to lysosomes. *The Journal of Cell Biology, 200*(4), 475–491.

Martinez-Vicente, M., Talloczy, Z., Wong, E., Tang, G., Koga, H., Kaushik, S., de Vries, R., Arias, E., Harris, S., Sulzer, D., & Cuervo, A. M. (2010). Cargo recognition failure is responsible for inefficient autophagy in Huntington's disease. *Nature Neuroscience, 13*(5), 567–576.

Martini-Stoica, H., Xu, Y., Ballabio, A., & Zheng, H. (2016). The autophagy-lysosomal pathway in neurodegeneration: A TFEB perspective. *Trends in Neurosciences, 39*(4), 221–234.

Masliah, E., Mallory, M., Hansen, L., DeTeresa, R., Alford, M., & Terry, R. (1994). Synaptic and neuritic alterations during the progression of Alzheimer's disease. *Neuroscience Letters, 174*(1), 67–72.

Massari, M. E., & Murre, C. (2000). Helix-loop-helix proteins: Regulators of transcription in eucaryotic organisms. *Molecular and Cellular Biology, 20*(2), 429–440.

Mawuenyega, K. G., Sigurdson, W., Ovod, V., Munsell, L., Kasten, T., Morris, J. C., Yarasheski, K. E., & Bateman, R. J. (2010). Decreased clearance of CNS beta-amyloid in Alzheimer's disease. *Science, 330*(6012), 1774.

Mazzulli, J. R., Xu, Y. H., Sun, Y., Knight, A. L., McLean, P. J., Caldwell, G. A., Sidransky, E., Grabowski, G. A., & Krainc, D. (2011). Gaucher disease glucocerebrosidase and alpha-synuclein form a bidirectional pathogenic loop in synucleinopathies. *Cell, 146*(1), 37–52.

Medina, D. L., Di Paola, S., Peluso, I., Armani, A., De Stefani, D., Venditti, R., Montefusco, S., Scotto-Rosato, A., Prezioso, C., Forrester, A., Settembre, C., Wang, W., Gao, Q., Xu, H., Sandri, M., Rizzuto, R., De Matteis, M. A., & Ballabio, A. (2015). Lysosomal calcium signalling regulates autophagy through calcineurin and TFEB. *Nature Cell Biology, 17*(3), 288–299.

Modi, K. K., Jana, A., Ghosh, S., Watson, R., & Pahan, K. (2014). A physically-modified saline suppresses neuronal apoptosis, attenuates tau phosphorylation and protects memory in an animal model of Alzheimer's disease. *PLoS One, 9*(8), e103606.

Modi, K. K., Roy, A., Brahmachari, S., Rangasamy, S. B., & Pahan, K. (2015). Cinnamon and its metabolite sodium benzoate attenuate the activation of p21rac and protect memory and learning in an animal model of Alzheimer's disease. *PLoS One, 10*(6), e0130398.

Mondal, S., Prieto, S., Rangasamy, S. B., Dutta, D., & Pahan, K. (2024). Nebulization of low-dose aspirin ameliorates Huntington's pathology in N171-82Q transgenic mice. *NeuroImmune Pharmacology and Therapeutics, 3*(1), 47–59.

Moors, T., Paciotti, S., Chiasserini, D., Calabresi, P., Parnetti, L., Beccari, T., & van de Berg, W. D. (2016). Lysosomal dysfunction and alpha-Synuclein aggregation in Parkinson's disease: Diagnostic links. *Movement Disorders, 31*(6), 791–801.

Mu, F. T., Callaghan, J. M., Steele-Mortimer, O., Stenmark, H., Parton, R. G., Campbell, P. L., McCluskey, J., Yeo, J. P., Tock, E. P., & Toh, B. H. (1995). EEA1, an early endosome-associated protein. EEA1 is a conserved alpha-helical peripheral membrane protein flanked by cysteine "fingers" and contains a calmodulin-binding IQ motif. *The Journal of Biological Chemistry, 270*(22), 13503–13511.

Mueller, J. C., Fuchs, J., Hofer, A., Zimprich, A., Lichtner, P., Illig, T., Berg, D., Wullner, U., Meitinger, T., & Gasser, T. (2005). Multiple regions of alpha-synuclein are associated with Parkinson's disease. *Annals of Neurology, 57*(4), 535–541.

Napolitano, G., & Ballabio, A. (2016). TFEB at a glance. *Journal of Cell Science, 129*(13), 2475–2481.

Novikoff, A. B., Beaufay, H., & De Duve, C. (1956). Electron microscopy of lysosomerich fractions from rat liver. *The Journal of Biophysical and Biochemical Cytology, 2*(4 Suppl), 179–184.

Ohkuma, S., & Poole, B. (1978). Fluorescence probe measurement of the intralysosomal pH in living cells and the perturbation of pH by various agents. *Proceedings of the National Academy of Sciences of the United States of America, 75*(7), 3327–3331.

Ohkuma, S., Moriyama, Y., & Takano, T. (1982). Identification and characterization of a proton pump on lysosomes by fluorescein-isothiocyanate-dextran fluorescence. *Proceedings of the National Academy of Sciences of the United States of America, 79*(9), 2758–2762.

Pahan, K. (2006). Lipid-lowering drugs. *Cellular and Molecular Life Sciences, 63*(10), 1165–1178.

Pahan, P., & Pahan, K. (2015). Can cinnamon bring aroma in Parkinson's disease treatment? *Neural Regeneration Research, 10*(1), 30–32.

Pahan, K., Gulati, S., & Singh, I. (1994). Phytanic acid alpha-oxidation in rat liver mitochondria. *Biochimica et Biophysica Acta, 1201*(3), 491–497.

Pahan, K., Khan, M., & Singh, I. (1998). Therapy for X-adrenoleukodystrophy: Normalization of very long chain fatty acids and inhibition of induction of cytokines by cAMP. *Journal of Lipid Research, 39*(5), 1091–1100.

Palmer, K. L., Shivgulam, M. E., Champod, A. S., Wilson, B. C., O'Brien, M. W., & Bray, N. W. (2023). Exercise training augments brain function and reduces pain perception in adults with chronic pain: A systematic review of intervention studies. *Neurobiology of Pain, 13*, 100129.

Parenti, G. (2009). Treating lysosomal storage diseases with pharmacological chaperones: From concept to clinics. *EMBO Molecular Medicine, 1*(5), 268–279.

Parenti, G., Andria, G., & Ballabio, A. (2015). Lysosomal storage diseases: From pathophysiology to therapy. *Annual Review of Medicine, 66*, 471–486.

Parzych, K. R., & Klionsky, D. J. (2014). An overview of autophagy: Morphology, mechanism, and regulation. *Antioxidants & Redox Signaling, 20*(3), 460–473.

Pastores, G. M., & Maegawa, G. H. (2013). Clinical neurogenetics: Neuropathic lysosomal storage disorders. *Neurologic Clinics, 31*(4), 1051–1071.

Platt, F. M., Boland, B., & van der Spoel, A. C. (2012). The cell biology of disease: Lysosomal storage disorders: The cellular impact of lysosomal dysfunction. *The Journal of Cell Biology, 199*(5), 723–734.

Prorok, T., Jana, M., Patel, D., & Pahan, K. (2019). Cinnamic acid protects the Nigrostriatum in a mouse model of Parkinson's disease via peroxisome proliferator-activated Receptoralpha. *Neurochemical Research, 44*(4), 751–762.

Raha, S., Dutta, D., Roy, A., & Pahan, K. (2021a). Reduction of Lewy body pathology by Oral cinnamon. *Journal of Neuroimmune Pharmacology, 16*(3), 592–608.

Raha, S., Ghosh, A., Dutta, D., Patel, D. R., & Pahan, K. (2021b). Activation of PPARalpha enhances astroglial uptake and degradation of beta-amyloid. *Science Signaling, 14*(706), eabg4747.

Raha, S., Paidi, R. K., Dutta, D., & Pahan, K. (2024). Cinnamic acid, a natural plant compound, exhibits neuroprotection in a mouse model of Sandhoff disease via PPARalpha. *NeuroImmune Pharmacology and Therapeutics, 3*(1), 17–32.

Rama Rao, K. V., & Kielian, T. (2016). Astrocytes and lysosomal storage diseases. *Neuroscience, 323*, 195–206.

Rangasamy, S. B., Corbett, G. T., Roy, A., Modi, K. K., Bennett, D. A., Mufson, E. J., Ghosh, S., & Pahan, K. (2015). Intranasal delivery of NEMO-binding domain peptide prevents memory loss in a mouse model of Alzheimer's disease. *Journal of Alzheimer's Disease, 47*(2), 385–402.

Rangasamy, S. B., Jana, M., Roy, A., Corbett, G. T., Kundu, M., Chandra, S., Mondal, S., Dasarathi, S., Mufson, E. J., Mishra, R. K., Luan, C. H., Bennett, D. A., & Pahan, K. (2018). Selective disruption of TLR2-MyD88 interaction inhibits inflammation and attenuates Alzheimer's pathology. *The Journal of Clinical Investigation, 128*(10), 4297–4312.

Rangasamy, S. B., Jana, M., Dasarathi, S., Kundu, M., & Pahan, K. (2023). Treadmill workout activates PPARalpha in the hippocampus to upregulate ADAM10, decrease plaques and improve cognitive functions in 5XFAD mouse model of Alzheimer's disease. *Brain, Behavior, and Immunity, 109*, 204–218.

Rao, S. K., Huynh, C., Proux-Gillardeaux, V., Galli, T., & Andrews, N. W. (2004). Identification of SNAREs involved in synaptotagmin VII-regulated lysosomal exocytosis. *The Journal of Biological Chemistry, 279*(19), 20471–20479.

Ravikumar, B., Vacher, C., Berger, Z., Davies, J. E., Luo, S., Oroz, L. G., Scaravilli, F., Easton, D. F., Duden, R., O'Kane, C. J., & Rubinsztein, D. C. (2004). Inhibition of mTOR induces autophagy and reduces toxicity of polyglutamine expansions in fly and mouse models of Huntington disease. *Nature Genetics, 36*(6), 585–595.

Reddy, A., Caler, E. V., & Andrews, N. W. (2001). Plasma membrane repair is mediated by Ca(2+)-regulated exocytosis of lysosomes. *Cell, 106*(2), 157–169.

Rega, L. R., Polishchuk, E., Montefusco, S., Napolitano, G., Tozzi, G., Zhang, J., Bellomo, F., Taranta, A., Pastore, A., Polishchuk, R., Piemonte, F., Medina, D. L., Catz, S. D., Ballabio, A., & Emma, F. (2016). Activation of the transcription factor EB rescues lysosomal abnormalities in cystinotic kidney cells. *Kidney International, 89*(4), 862–873.

Rey, N. L., Bousset, L., George, S., Madaj, Z., Meyerdirk, L., Schulz, E., Steiner, J. A., Melki, R., & Brundin, P. (2019). Alpha-Synuclein conformational strains spread, seed and target neuronal cells differentially after injection into the olfactory bulb. *Acta Neuropathologica Communications, 7*(1), 221.

Rodriguez, A., Webster, P., Ortego, J., & Andrews, N. W. (1997). Lysosomes behave as Ca2+-regulated exocytic vesicles in fibroblasts and epithelial cells. *The Journal of Cell Biology, 137*(1), 93–104.

Rojas, R., van Vlijmen, T., Mardones, G. A., Prabhu, Y., Rojas, A. L., Mohammed, S., Heck, A. J., Raposo, G., van der Sluijs, P., & Bonifacino, J. S. (2008). Regulation of retromer recruitment to endosomes by sequential action of Rab5 and Rab7. *The Journal of Cell Biology, 183*(3), 513–526.

Rose, C., Menzies, F. M., Renna, M., Acevedo-Arozena, A., Corrochano, S., Sadiq, O., Brown, S. D., & Rubinsztein, D. C. (2010). Rilmenidine attenuates toxicity of polyglutamine expansions in a mouse model of Huntington's disease. *Human Molecular Genetics, 19*(11), 2144–2153.

Roy, A., & Pahan, K. (2009). Gemfibrozil, stretching arms beyond lipid lowering. *Immunopharmacology and Immunotoxicology, 31*(3), 339–351.

Roy, A., & Pahan, K. (2011). Prospects of statins in Parkinson disease. *The Neuroscientist, 17*(3), 244–255.

Roy, A., & Pahan, K. (2015). PPARalpha signaling in the hippocampus: Crosstalk between fat and memory. *Journal of Neuroimmune Pharmacology, 10*(1), 30–34.

Rubinsztein, D. C., Gestwicki, J. E., Murphy, L. O., & Klionsky, D. J. (2007). Potential therapeutic applications of autophagy. *Nature Reviews. Drug Discovery, 6*(4), 304–312.

Saftig, P., & Klumperman, J. (2009). Lysosome biogenesis and lysosomal membrane proteins: Trafficking meets function. *Nature Reviews. Molecular Cell Biology, 10*(9), 623–635.

Sancak, Y., Peterson, T. R., Shaul, Y. D., Lindquist, R. A., Thoreen, C. C., Bar-Peled, L., & Sabatini, D. M. (2008). The rag GTPases bind raptor and mediate amino acid signaling to mTORC1. *Science, 320*(5882), 1496–1501.

Sancak, Y., Bar-Peled, L., Zoncu, R., Markhard, A. L., Nada, S., & Sabatini, D. M. (2010). Ragulator-rag complex targets mTORC1 to the lysosomal surface and is necessary for its activation by amino acids. *Cell, 141*(2), 290–303.

Sardiello, M., Palmieri, M., di Ronza, A., Medina, D. L., Valenza, M., Gennarino, V. A., Di Malta, C., Donaudy, F., Embrione, V., Polishchuk, R. S., Banfi, S., Parenti, G., Cattaneo, E., & Ballabio, A. (2009). A gene network regulating lysosomal biogenesis and function. *Science, 325*(5939), 473–477.

Schroder, B. A., Wrocklage, C., Hasilik, A., & Saftig, P. (2010). The proteome of lysosomes. *Proteomics, 10*(22), 4053–4076.

Seedorf, U. (1998). Peroxisomes in lipid metabolism. *Journal of Cellular Biochemistry. Supplement, 30-31*, 158–167.

Selkoe, D. J., & Hardy, J. (2016). The amyloid hypothesis of Alzheimer's disease at 25 years. *EMBO Molecular Medicine, 8*(6), 595–608.

Serrano-Pozo, A., Frosch, M. P., Masliah, E., & Hyman, B. T. (2011). Neuropathological alterations in Alzheimer disease. *Cold Spring Harbor Perspectives in Medicine, 1*(1), a006189.

Settembre, C., & Ballabio, A. (2014). Lysosomal adaptation: How the lysosome responds to external cues. *Cold Spring Harbor Perspectives in Biology, 6*(6).

Settembre, C., Di Malta, C., Polito, V. A., Garcia Arencibia, M., Vetrini, F., Erdin, S., Erdin, S. U., Huynh, T., Medina, D., Colella, P., Sardiello, M., Rubinsztein, D. C., & Ballabio, A. (2011). TFEB links autophagy to lysosomal biogenesis. *Science, 332*(6036), 1429–1433.

Settembre, C., Zoncu, R., Medina, D. L., Vetrini, F., Erdin, S., Erdin, S., Huynh, T., Ferron, M., Karsenty, G., Vellard, M. C., Facchinetti, V., Sabatini, D. M., & Ballabio, A. (2012). A lysosome-to-nucleus signalling mechanism senses and regulates the lysosome via mTOR and TFEB. *The EMBO Journal, 31*(5), 1095–1108.

Settembre, C., De Cegli, R., Mansueto, G., Saha, P. K., Vetrini, F., Visvikis, O., Huynh, T., Carissimo, A., Palmer, D., Klisch, T. J., Wollenberg, A. C., Di Bernardo, D., Chan, L., Irazoqui, J. E., & Ballabio, A. (2013a). TFEB controls cellular lipid metabolism through a starvation-induced autoregulatory loop. *Nature Cell Biology, 15*(6), 647–658.

Settembre, C., Fraldi, A., Medina, D. L., & Ballabio, A. (2013b). Signals from the lysosome: A control Centre for cellular clearance and energy metabolism. *Nature Reviews. Molecular Cell Biology, 14*(5), 283–296.

Sidransky, E., Nalls, M. A., Aasly, J. O., Aharon-Peretz, J., Annesi, G., Barbosa, E. R., Bar-Shira, A., Berg, D., Bras, J., Brice, A., Chen, C. M., Clark, L. N., Condroyer, C., De Marco, E. V., Durr, A., Eblan, M. J., Fahn, S., Farrer, M. J., Fung, H. C., Gan-Or, Z., Gasser, T., Gershoni-Baruch, R., Giladi, N., Griffith, A., Gurevich, T., Januario, C., Kropp, P., Lang, A. E., Lee-Chen, G. J., Lesage, S., Marder, K., Mata, I. F., Mirelman, A., Mitsui, J., Mizuta, I., Nicoletti, G., Oliveira, C., Ottman, R., Orr-Urtreger, A., Pereira, L. V., Quattrone, A., Rogaeva, E., Rolfs, A., Rosenbaum, H., Rozenberg, R., Samii, A., Samaddar, T., Schulte, C., Sharma, M., Singleton, A., Spitz, M., Tan, E. K., Tayebi, N., Toda, T., Troiano, A. R., Tsuji, S., Wittstock, M., Wolfsberg, T. G., Wu, Y. R., Zabetian, C. P., Zhao, Y., & Ziegler, S. G. (2009). Multicenter analysis of glucocerebrosidase mutations in Parkinson's disease. *The New England Journal of Medicine, 361*(17), 1651–1661.

Song, W., Wang, F., Savini, M., Ake, A., di Ronza, A., Sardiello, M., & Segatori, L. (2013). TFEB regulates lysosomal proteostasis. *Human Molecular Genetics, 22*(10), 1994–2009.

Sorrentino, Z. A., Brooks, M. M. T., Hudson, V., 3rd, Rutherford, N. J., Golde, T. E., Giasson, B. I., & Chakrabarty, P. (2017). Intrastriatal injection of alpha-synuclein can lead to widespread synucleinopathy independent of neuroanatomic connectivity. *Molecular Neurodegeneration, 12*(1), 40.

Spampanato, C., Feeney, E., Li, L., Cardone, M., Lim, J. A., Annunziata, F., Zare, H., Polishchuk, R., Puertollano, R., Parenti, G., Ballabio, A., & Raben, N. (2013). Transcription factor EB (TFEB) is a new therapeutic target for Pompe disease. *EMBO Molecular Medicine, 5*(5), 691–706.

Spillantini, M. G., Schmidt, M. L., Lee, V. M., Trojanowski, J. Q., Jakes, R., & Goedert, M. (1997). Alpha-synuclein in Lewy bodies. *Nature, 388*(6645), 839–840.

Steingrimsson, E., Copeland, N. G., & Jenkins, N. A. (2004). Melanocytes and the microphthalmia transcription factor network. *Annual Review of Genetics, 38*, 365–411.

Thaler, A., Gurevich, T., Bar Shira, A., Gana Weisz, M., Ash, E., Shiner, T., Orr-Urtreger, A., Giladi, N., & Mirelman, A. (2017). A "dose" effect of mutations in the GBA gene on Parkinson's disease phenotype. *Parkinsonism & Related Disorders, 36*, 47–51.

Themistokleous, C., Bagnoli, E., Parulekar, R., & M. K. M. M. (2023). Role of autophagy pathway in Parkinson's disease and related genetic neurological disorders. *Journal of Molecular Biology, 168144*.

Urbanc, B., Cruz, L., Le, R., Sanders, J., Ashe, K. H., Duff, K., Stanley, H. E., Irizarry, M. C., & Hyman, B. T. (2002). Neurotoxic effects of thioflavin S-positive amyloid deposits in transgenic mice and Alzheimer's disease. *Proceedings of the National Academy of Sciences of the United States of America, 99*(22), 13990–13995.

Vitner, E. B., Platt, F. M., & Futerman, A. H. (2010). Common and uncommon pathogenic cascades in lysosomal storage diseases. *The Journal of Biological Chemistry, 285*(27), 20423–20427.

Wakabayashi, K., Yoshimoto, M., Tsuji, S., & Takahashi, H. (1998). Alpha-synuclein immunoreactivity in glial cytoplasmic inclusions in multiple system atrophy. *Neuroscience Letters, 249*(2–3), 180–182.

Walker, F. O. (2007). Huntington's disease. *Lancet, 369*(9557), 218–228.

Walkley, S. U. (2009). Pathogenic cascades in lysosomal disease-why so complex? *Journal of Inherited Metabolic Disease, 32*(2), 181–189.

Wenning, G. K., & Jellinger, K. A. (2005). The role of alpha-synuclein in the pathogenesis of multiple system atrophy. *Acta Neuropathologica, 109*(2), 129–140.

Winchester, B. G. (2001). Lysosomal membrane proteins. *European Journal of Paediatric Neurology, 5 Suppl A*, 11–19.

Wisniewski, H. M., Narang, H. K., & Terry, R. D. (1976). Neurofibrillary tangles of paired helical filaments. *Journal of the Neurological Sciences, 27*(2), 173–181.

Wong, E., & Cuervo, A. M. (2010). Autophagy gone awry in neurodegenerative diseases. *Nature Neuroscience, 13*(7), 805–811.

Wong, C. O., Li, R., Montell, C., & Venkatachalam, K. (2012). Drosophila TRPML is required for TORC1 activation. *Current Biology, 22*(17), 1616–1621.

Woodman, P. G., & Futter, C. E. (2008). Multivesicular bodies: Co-ordinated progression to maturity. *Current Opinion in Cell Biology, 20*(4), 408–414.

Wyss-Coray, T., Loike, J. D., Brionne, T. C., Lu, E., Anankov, R., Yan, F., Silverstein, S. C., & Husemann, J. (2003). Adult mouse astrocytes degrade amyloid-beta in vitro and in situ. *Nature Medicine, 9*(4), 453–457.

Xiao, Q., Yan, P., Ma, X., Liu, H., Perez, R., Zhu, A., Gonzales, E., Burchett, J. M., Schuler, D. R., Cirrito, J. R., Diwan, A., & Lee, J. M. (2014). Enhancing astrocytic lysosome biogenesis facilitates Abeta clearance and attenuates amyloid plaque pathogenesis. *The Journal of Neuroscience, 34*(29), 9607–9620.

Xiao, Q., Yan, P., Ma, X., Liu, H., Perez, R., Zhu, A., Gonzales, E., Tripoli, D. L., Czerniewski, L., Ballabio, A., Cirrito, J. R., Diwan, A., & Lee, J. M. (2015). Neuronal-targeted TFEB accelerates lysosomal degradation of APP, reducing Abeta generation and amyloid plaque pathogenesis. *The Journal of Neuroscience, 35*(35), 12137–12151.

Xiong, J., & Zhu, M. X. (2016). Regulation of lysosomal ion homeostasis by channels and transporters. *Science China. Life Sciences, 59*(8), 777–791.

Xu, H., & Ren, D. (2015). Lysosomal physiology. *Annual Review of Physiology, 77*, 57–80.

Yang, J., Zhang, W., Zhang, S., Iyaswamy, A., Sun, J., Wang, J., & Yang, C. (2023). Novel insight into functions of transcription factor EB (TFEB) in Alzheimer's disease and Parkinson's disease. *Aging and Disease, 14*(3), 652–669.

Zhang, Y. D., & Zhao, J. J. (2015). TFEB participates in the Abeta-induced pathogenesis of Alzheimer's disease by regulating the autophagy-lysosome pathway. *DNA and Cell Biology, 34*(11), 661–668.

Zoncu, R., Bar-Peled, L., Efeyan, A., Wang, S., Sancak, Y., & Sabatini, D. M. (2011). mTORC1 senses lysosomal amino acids through an inside-out mechanism that requires the vacuolar H(+)-ATPase. *Science, 334*(6056), 678–683.

Cytokines in Neurodegenerative Disorders

12

Kalipada Pahan

Abstract

The pathogenesis of different neuroinflammatory and neurodegenerative diseases like multiple sclerosis, Parkinson's disease, Alzheimer's disease, Huntington's disease, and Amyotrophic lateral sclerosis is associated with elevated production of different proinflammatory cytokines (e.g. TNF-α, IL-1β, IL-6, etc.) within the central nervous system (CNS). It is found that resident glial cells in the CNS (microglia and astrocytes) produce high levels of proinflammatory cytokines in response to a wide variety of proinflammatory and neurodegenerative stimuli. Although pathways leading to the expression of different proinflammatory cytokines may vary in microglia and astrocytes, various signaling cascades are found to come together to activate several transcription factors for the transcription of proinflammatory cytokine genes in glial cells. Once produced in excess, proinflammatory cytokines cause nitrosative, oxidative, and inflammatory stress within the CNS leading to the loss of neurons and myelin-synthesizing oligodendrocytes. Here, we attempt to delineate our current familiarity with signaling mechanisms for the induction of proinflammatory cytokines in glial cells in response to neurodegenerative stimuli and how these proinflammatory cytokines may be involved in the disease process of neurodegenerative disorders. Such understanding may be helpful for possible intervention of neuroinflammation-mediated neurodegenerative disorders.

K. Pahan (✉)
Division of Research and Development, Jesse Brown Veterans Affairs Medical Center, Chicago, IL, USA

Department of Neurological Sciences, Rush University Medical Center, Chicago, IL, USA
e-mail: Kalipada_Pahan@rush.edu

Keywords

Proinflammatory cytokines · Microglia · Astrocytes · Signal transduction · Neurodegenerative disorders

12.1 Introduction

Proinflammatory cytokines (TNF-α, IL-1β, IL-6, etc.) are involved in various physiological and pathological processes in many organ systems including the CNS (Michaud et al., 2013; Colonna & Butovsky, 2017; Piancone et al., 2021). While at low concentrations, proinflammatory cytokines may play a role in neurotropism and neurotransmission, at higher concentrations, these molecules are implicated in having a role in the pathogenesis of demyelination and various neurodegenerative diseases (Carrieri et al., 1992; Pahan, 2011, 2013; Michaud et al., 2013; Colonna & Butovsky, 2017; Piancone et al., 2021). For example, TNF-α at low doses in the absence of other proinflammatory molecules stimulates the production of the neurotrophic factor BDNF from neurons and astrocytes (Saha et al., 2006). In the presence of other proinflammatory cytokines, the same TNFα stimulates the production of excessive nitric oxide causing nitrosative stress (Pahan et al., 1997a, b, 1998; Saha & Pahan, 2006a).

Microglia and astrocytes in the healthy brain do not express TNF-α, IL-1β, IL-6, etc. but following traumatic, neurotoxic, ischemic, or inflammatory insults, reactive microglia and astrocytes express different proinflammatory cytokines in the mouse, rat, and human (Jana et al., 2001, 2007, 2008; Xu et al., 2016; Colonna & Butovsky, 2017; Dutta et al., 2021a). Accordingly, proinflammatory cytokines derived from activated glial cells are presumed to participate in neuronal death during neurodegenerative diseases (e.g. Alzheimer's disease, Parkinson's disease, Huntington's disease, amyotrophic lateral sclerosis) and oligodendrocyte

H. E. Gendelman, T. Ikezu (eds.), *Neuroimmune Pharmacology and Therapeutics*,
https://doi.org/10.1007/978-3-031-68237-7_12

degeneration in demyelinating diseases (e.g. multiple sclerosis, X-Adrenoleukodystrophy) (Roy et al., 2006, 2008; Khasnavis et al., 2012; Sarlus & Heneka, 2017; Shabab et al., 2017; Rangasamy et al., 2018, 2021b; Chakrabarti et al., 2019). Therefore, characterization of intracellular pathways required for the expression of proinflammatory cytokines in activated glial cells and understanding mechanisms by which these molecules participate in neurodegeneration are active areas of investigation.

12.2 Induction of Proinflammatory Cytokines by Glial Cells

Similar to data in macrophages (MacMicking et al., 1997), different stimuli have been found to induce the expression of proinflammatory cytokines in astrocytes and microglia (Dauer & Przedborski, 2003; Dasgupta et al., 2005; Brahmachari et al., 2006, 2009; Brahmachari & Pahan, 2010; Chandra et al., 2017). In the subsequent lines, we are trying to describe and discuss these stimuli in major functional categories to highlight the diversity of inducers for the upregulation of proinflammatory cytokines in glial cells.

12.2.1 Bacterial Products

It is not known whether glial cells in patients with neurodegenerative disorders come in contact with any bacterial products such as lipopolysaccharide (LPS). However, it is probably the first stimulus shown to induce the production of TNFα from cultured glial cells (Sawada et al., 1989; Chung et al., 1991). Accordingly, many studies have used LPS to trigger the expression of proinflammatory cytokines in different glial cell lines and primary microglia (Jana et al., 2007, 2005; Roy et al., 2006; Brahmachari et al., 2009; Jana & Pahan, 2012; Roy & Pahan, 2013). Some studies have reported the induction of proinflammatory cytokines in primary astroglia by LPS in the presence of a co-stimulatory signal by IFN-γ (Bi et al., 2013). Toll-like receptor (TLR)4 serves as the primary receptor for LPS, and it has been shown that glial cells express TLR4 (Bowman et al., 2003; Lehnardt et al., 2003). Accordingly, several studies have reported the involvement of TLR4 in LPS-mediated upregulation of proinflammatory cytokines in glial cells (Qin et al., 2005; Zhou et al., 2006).

Similar to LPS, bacterial DNA has been shown to induce and stimulate the expression of proinflammatory cytokines in mouse glial cells (Dalpke et al., 2002; Rangasamy et al., 2018). TLR9, the receptor for bacterial DNA, is expressed in glia and thus permits sensitization of these cells against the foreign DNA containing unmethylated CpG dinucleotide (Dalpke et al., 2002). In addition to LPS and bacterial DNA, lipoteichoic acid (LTA) is also capable of triggering the expression of proinflammatory cytokines from glial cells (Rangasamy et al., 2018). However, in contrast to LPS and bacterial DNA, LTA requires the involvement of TLR2 to stimulate glial cells (Rangasamy et al., 2018; Dutta et al., 2021a).

12.2.2 Viral Products

Different viruses are capable of triggering the expression of proinflammatory cytokines in brain cells in a wide variety of species. For example, according to Salvador et al., 2021 (Salvador et al., 2021), virulent strains of Marek's Disease viruses induce proinflammatory cytokines in chicken brains. Sindbis virus and Theiler's murine encephalomyelitis virus also strongly induce the expression of proinflammatory cytokines in mouse brain cells (Mack et al., 2003). Human immunodeficiency virus (HIV) intrusion is also known to upregulate proinflammatory cytokines in the human brain (Uzasci et al., 2013). Although whole viruses are sometimes required for the upregulation of proinflammatory cytokines, isolated viral products such as viral nucleic acids or proteins, are also potent inducers of proinflammatory cytokines in glial cells. It has been shown that viral double-stranded RNA (dsRNA) induces proinflammatory cytokines in glial cells via the dsRNA-dependent protein kinase (PKR) pathway (Zhang et al., 2022). Accordingly, HIV coat protein gp41 and surface glycoprotein gp120 are capable of inducing different proinflammatory cytokines in glial cells (Guo et al., 2013). In addition, viral transcription factor HIV-1 Tat can also induce the expression of proinflammatory cytokines in glial cells (Mayne et al., 1998).

12.2.3 Neurodegenerative Toxins

Exposure to 1-methyl-4-phenyl-1, 2, 3, 6-tetrahydropyridine (MPTP) is neurotoxic as it causes PD in mice (Dauer & Przedborski, 2003; Ghosh et al., 2007, 2009; Mondal et al., 2019; Gottschalk et al., 2021). It has been found that MPP^+, an oxidation product of MPTP, stimulates the production of proinflammatory cytokines in glial cells (Ghosh et al., 2007, 2009). Accordingly, neuroinflammation is observed in the substantia nigra pars compacta (SNpc) of MPTP-intoxicated mice (Ghosh et al., 2007; Chandra et al., 2016, 2017). Accumulation of galactosylsphingosine (psychosine) is toxic for the brain as it is considered the etiological cause of Krabbe disease (Bradbury et al., 2021).

Several studies have reported the upregulation of proinflammatory cytokines by psychosine in astrocytes (Giri et al., 2002). Systemic administration of 3-nitropropionic acid (3-NP), a complex II inhibitor of the electron transport chain, is known to cause selective lesion in striatum in rodents, thus providing an animal model for HD (Borlongan et al., 1997). Interestingly, 3-NP treatment also causes an upregulation of proinflammatory cytokines like TNFα and IL-1β in glial cells (Kumar et al., 2014; Khan et al., 2015). Therefore, several neurodegenerative toxins are capable of inducing the expression of proinflammatory cytokines in glial cells.

12.2.4 Protein Aggregation

Certain proteins in the brain are aggregated and such protein aggregation is known to be stimulated during aging and various insults (Labbadia & Morimoto, 2015; Watanabe et al., 2020). During AD, toxic Aβ is formed by cleavage of amyloid precursor protein (APP) followed by aggregation of the cleaved peptides. These aggregated fibrillar Aβ peptides prompt the expression of proinflammatory cytokines in glial cells (Jana et al., 2008; Jana & Pahan, 2010a, b; Rangasamy et al., 2018). In addition to Aβ, the secreted derivative of APP also stimulates the expression of proinflammatory cytokines in microglia. Recently it has been shown that aggregated fibrillar tau, a component of neurofibrillary tangle, also induces the expression of proinflammatory cytokines in glial cells (Morales et al., 2013). A synthetic peptide consisting of amino acid residues 106-126 of the human prion protein, the endogenous misfolded villain of the Creutzfeldt-Jakob disease in humans, is known to induce the expression of proinflammatory cytokines in microglial cells (Veerhuis et al., 2005; Ye et al., 2016). A common pathological feature of the brains of patients with PD, multiple system atrophy (MSA), and Lewy body dementia (LBD) is the presence of Lewy bodies, which basically consist of aggregated α-synuclein (α-syn) (Dutta et al., 2021a, 2022; Raha et al., 2021). Similar to fibrillar Aβ and aggregated tau, α-syn fibrils also induce the activation of microglia leading to the production of proinflammatory cytokines (Dutta et al., 2021a). Recent studies have identified that aggregation of TDP-43 participates in the pathogenesis of amyotrophic lateral sclerosis (ALS) and aggregated TDP-43 has been shown to trigger the activation of astroglia (Lee et al., 2020; Peng et al., 2020). Accordingly, in ALS, strong immunoreactivity for proinflammatory cytokines is recorded in activated astrocytes (Ziff et al., 2022). In the brain of patients with Huntington's disease (HD), levels of proinflammatory cytokines are also elevated in glial cells in the region of the degenerating infarct (Birolini et al., 2021). However, until now, in HD, the exact inducer of glial proinflammatory cytokines has not been experimentally established.

12.2.5 Cell-to-Cell Contact

Although the brain is separated from our classical immune system by the blood-brain barrier (BBB), immune cells enter the brain, and this infiltration increases under disease conditions. For example, the leaky blood-brain barrier during multiple sclerosis (MS) and its animal model experimental autoimmune encephalomyelitis (EAE) permits the infiltration of many peripheral cells into the brain (Pahan, 2010, 2011, 2013, 2015).

T cell infiltration is also seen in the SNpc of PD patients, hemiparkinsonian monkeys, and the MPTP mouse model of PD (Chandra et al., 2016, 2017; Dutta et al., 2019; Mondal et al., 2019). It has been also found that motor neuron damage in the brain and spinal cord in ALS is associated with the infiltration of T cells and activation of antigen-presenting cells (Lewis et al., 2012). Infiltration of $CD8^+$ T cells into the brain parenchyma is also evident in animal models of AD (Unger et al., 2020). Although there are many mechanisms by which T cells can drive neurotoxicity, it has been found that activated T cells are capable of inducing the expression of proinflammatory cytokines in microglia via cell-to-cell contact (Dasgupta et al., 2002, 2005). Subsequently, it has been found that VLA4 integrin on the surface of activated T cells and its interaction with VCAM1 on glial cells (Fig. 12.1) play an important role in contact-mediated induction of proinflammatory cytokines in microglia and astroglia (Dasgupta et al., 2003a; Brahmachari & Pahan, 2010). Interestingly, VLA4-VCAM1 interaction-induced expression of glial proinflammatory cytokines is gender-sensitive (Fig. 12.1) in which male antigen-primed T cells are unable to induce the expression of proinflammatory cytokines in glial cells due to a defect in VLA4 in male T cells (Dasgupta et al., 2005; Brahmachari & Pahan, 2010). However, VLA4-VCAM1 interaction may not be the only mechanism representing cell-to-cell contact leading to microglial activation as ligation of microglial CD40 by either cross-linking antibodies or recombinant CD40 ligand (CD154) (Fig. 12.1) is also capable of triggering microglial proinflammatory cytokines (Jana et al., 2001, 2002).

12.3 Regulation of Glial Proinflammatory Cytokine Expression

Glial proinflammatory cytokines are mainly regulated at the transcriptional level.

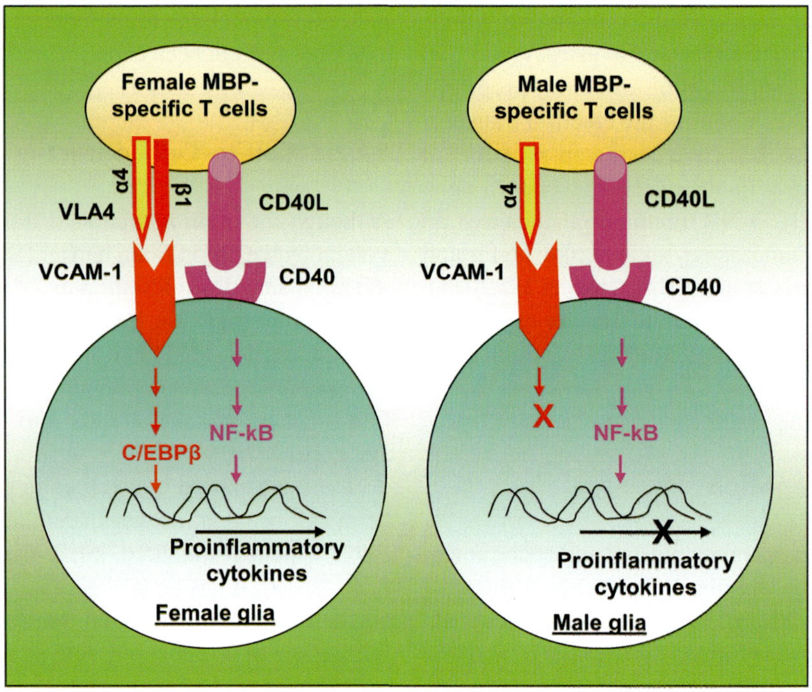

Fig 12.1 T-cell contact-mediated production of proinflammatory cytokines in glial cells is gender-sensitive. Very late antigen 4 (VLA4) is a heterodimer of α4 and β1 integrins. VLA4 and CD40 ligand (CD40L) of female neuroantigen-primed T cells interact with VCAM1 and CD40, respectively to upregulate the expression of proinflammatory cytokines in glial cells. While VLA4-VCAM1 interaction leads to the activation of C/EBPβ, CD40-CD40L ligation is responsible for the activation of NF-κB. Male T cells do not express β1 integrin and therefore, lack functional VLA4. respectively. Due to a defect in VLA4, male neuroantigen-primed T cells are unable to induce the expression of proinflammatory cytokines in glial cells

12.3.1 Involvement of Different Transcription Factors in the Synthesis of Proinflammatory Cytokine

Analysis of promoters of different cytokine genes such as TNF-α, IL-1β, IL-6, etc. reveals the presence of DNA-binding sites of multiple transcription factors (Ghosh et al., 2007; Dutta et al., 2021a). Among all the transcription factors working in concert to transactivate promoters of different proinflammatory cytokine genes, the most important one is definitely nuclear factor kappa B (NF-κB), a pleiotropic transcription factor involved in multiple cellular processes including inflammation (Pahan & Schmid, 2000; Hayden & Ghosh, 2004, 2008; Mondal et al., 2012b). Therefore, heterodimeric classical NF-κB p65:p50 is also termed as a proinflammatory transcription factor. The presence of multiple consensus sequences (kappaB elements) in the promoter region of TNF-α, IL-1β, IL-6, etc. for the binding of NF-κB and the reduction of proinflammatory cytokine gene expression in glial cells with the attenuation of NF-κB activation by N-acetyl cysteine, lovastatin, NF-κB essential modifier (NEMO)-binding domain (NBD) peptide, IκB kinase (IKK) inhibitor, or other inhibitors establishes an essential role of NF-κB activation in the induction of proinflammatory cytokines (Pahan et al., 1997b; Hayden & Ghosh, 2004; Ghosh et al., 2007, 2009; Jana et al., 2007; Brahmachari et al., 2009). It has been found that different signaling pathways transduced by stimuli of proinflammatory cytokine production ultimately converge into NF-κB activation (Fig. 12.2) (Hayden & Ghosh, 2004, 2008). More specifically, different signaling pathways activate IκB kinase (IKK) for the phosphorylation of NF-κB p50:p65 arresting protein, inhibitory kappa B (IκB) in the cytosol, thereby exposing it for ubiquitination and subsequent proteosomal degradation (Fig. 12.2). This releases the p50:p65 heterodimer to enter the nucleus, where the p65 subunit is further activated by phosphorylation and acetylation (Saha & Pahan, 2006a; Saha et al., 2007). Then the activated heterodimer binds to kappaB elements in the promoters of proinflammatory cytokine genes (Ghosh et al., 1998; Hayden & Ghosh, 2004; Hayden & Ghosh, 2008).

In addition to NF-κB, CCAAT/enhancer-binding protein (C/EBP) is another transcription factor that plays an important role in the induction of glial proinflammatory cytokines (Jana et al., 2002; Webb et al., 2002; Dasgupta et al., 2003a, 2005; Auch et al., 2004). Although the C/EBPs are a family of six basic leucine zipper transcription factors (Lekstrom-Himes & Xanthopoulos, 1998), C/EBPβ and C/EBPδ have

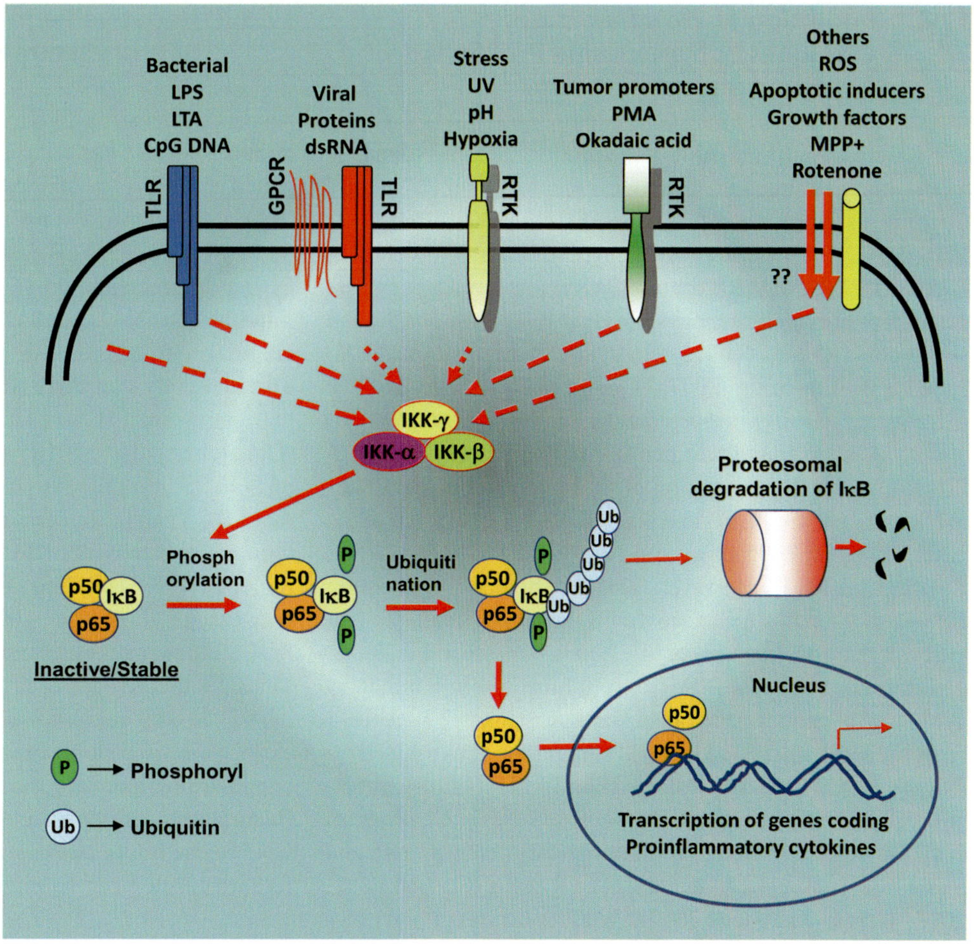

Fig. 12.2 Schematic presentation of how different inflammatory stimuli induce the activation of NF-κB to initiate the transcription of different proinflammatory cytokines in glial cells. Multiple signals converge at the activation of inhibitory kappa B (IκB) kinase (IKK), which leads to the phosphorylation of IκB. After phosphorylation, ubiquitination occurs leading to the proteosomal degradation of IκB. Since IκB arrests p65:p50 heterodimer in the cytoplasm keeping it at the inactive state, after the degradation of IκB, the classical p65:p50 heterodimer goes to the nucleus and initiates the transcription of proinflammatory cytokines including TNFα, IL-1β, and IL-6. LPS, lipopolysaccharides; LTA lipoteichoic acid; TLR toll-like receptor; RTK receptor tyrosine kinase; GPCR G protein-coupled receptor; ROS reactive oxygen species; MPP+ 1-methyl-4-phenylpyridinium

been reported to regulate the induction in glial proinflammatory cytokines (Jana et al., 2002; Webb et al., 2002; Dasgupta et al., 2003b; Won et al., 2003; Auch et al., 2004; Dasgupta et al., 2005). In our studies, we have found C/EBPβ to be a mandatory factor in the regulation of proinflammatory cytokine induction in microglial cells (Dasgupta et al., 2003a, 2005; Jana et al., 2005). How does C/EBP play such a critical role in the induction of proinflammatory cytokines? The answer perhaps lies in their ability to employ and phosphorylate histone acetyl-transferase (HAT) co-activators, like p300 and CREB binding protein (CBP). Activated C/EBPβ has been shown to recruit and phosphorylate p300-HAT (Schwartz et al., 2003), while C/EBPδ performs similar modifications on CBP-HAT for employing it to the active transcriptional complex (Kovacs et al., 2003). Eliminating this C/EBP-dependent phosphorylation of co-activators by site-directed mutagenesis reduces the transactivation of C/EBP contingent genes, probably by interfering with the formation of active transcriptosome.

In addition to kappaB and C/EBP binding elements, proinflammatory cytokine promoters also house the STAT-1α (signal transducers and activators of transcription-1α) binding sequences, called GAS elements (Rothhammer et al., 2016; Butturini et al., 2019). In some cases, Jak-STAT and NF-κB pathways cooperate with each other for the optimum induction of proinflammatory cytokines (Platanitis & Decker, 2018). Besides STAT-1α, IFN-γ is also known to activate interferon regulatory factor-1 (IRF-1) as a secondary response (Yang et al., 2022). Accordingly, interferon-stimulated responsive element (ISRE), the DNA binding sequence for IRF-1, has been identified in TNF-α as well as IL-1β gene promoters (Platanitis & Decker, 2018; Yang et al., 2022).

One more transcription factor that has been widely studied in the induction of proinflammatory cytokines is activator protein-1 (AP-1). Rodent and human TNF-α and IL-1 gene promoters harbor consensus AP-1 binding elements (Bailly et al., 1996; Falvo et al., 2010). However, in certain cases, the involvement of AP-1 is dependent on the type of stimulus used (Falvo et al., 2010).

Although the transcription factors (NF-κB, C/EBPβ, STAT1α, IRF1, AP-1, etc.) mentioned above regulate the transcription of proinflammatory cytokines positively, there are some transcription factors that can do the opposite controlling the synthesis of proinflammatory cytokines negatively. These transcription factors are outlined below:

12.3.2 Inhibition of Proinflammatory Cytokines by Peroxisome Proliferator-Activated Receptor (PPAR)

PPARs are ligand-inducible transcription factors belonging to the class of nuclear hormone receptors, which mainly function as lipid sensors (Keller et al., 1993; Reddy & Mannaerts, 1994). There are three different isoforms of PPARs (PPARα, PPARβ/δ, and PPARγ) that have distinct tissue expression patterns, physiological functions, and ligand specificity (Keller et al., 1993; Reddy & Mannaerts, 1994; Heneka & Landreth, 2007). Gemfibrozil is a ligand for peroxisome proliferator-activated receptor alpha (PPARα) and it has been found that gemfibrozil and other ligands of PPARα inhibit the expression of proinflammatory cytokines (TNFα, IL-1β, and IL-6) in activated glial cells (Pahan et al., 2002; Xu et al., 2005; Jana et al., 2007; Jana & Pahan, 2012; Kim et al., 2017). However, gemfibrozil does not require PPARα to suppress the expression of proinflammatory cytokines in microglia (Jana et al., 2007). On the other hand, gemfibrozil involves PPARβ to inhibit proinflammatory cytokines in human microglia (Jana & Pahan, 2012). IL-1 receptor antagonist (IL-1Ra) and suppressor of cytokine signaling (SOCS) are physiological anti-inflammatory molecules (Corbett et al., 2012; Ghosh et al., 2017; Chakrabarti et al., 2018, 2019). Gemfibrozil upregulates both IL-1Ra and SOCS3 in brain cells independent of PPARα (Corbett et al., 2012; Ghosh & Pahan, 2012). While gemfibrozil involves CREB for the upregulation of IL-1Ra in neurons (Corbett et al., 2012), it needs KLF4 for the stimulation of SOCS3 in astrocytes (Ghosh & Pahan, 2012). Along similar lines, 15-deoxy-12, 14-PGJ2 (15d-PGJ2), a ligand for PPARγ, attenuates the level of proinflammatory cytokines independent of the PPARγ itself (Zhao et al., 2006). However, in most instances, the activation of NF-κB is inhibited by activators of either PPARα or PPARγ (Glauert et al., 2008; Korbecki et al., 2019). Blocking of NF-κB and other transcription factors is undeniably a convenient way of shutting down inflammatory stimulus-induced responses in a short period of time. In addition, another mechanism has been suggested to explain the inhibitory effect of PPAR ligands on pro-inflammatory cytokines. PPARγ ligands 15d-PGJ2 and rosiglitazone are capable of reducing the phosphorylation and activation of JAK-STAT pathway in activated rat glial cells thereby leading to the attenuation of JAK-STAT-dependent inflammatory responses (Park et al., 2003). This inhibition is not reliant on PPARγ and is mediated by rapid transcription of suppressor of cytokine signaling (SOCS) 1 and 3 (Park et al., 2003). Moreover, SHP-2 is also involved in the anti-inflammatory action of many signaling receptors and it has been found that SHP2 is phosphorylated within minutes (Park et al., 2003) and as phosphorylated SHP2 dephosphorylates JAK, this creates yet another avenue of blocking the JAK-STAT pathway.

12.3.3 Negative Regulation of Proinflammatory Cytokines by SOCS

As part of human life, our cells are challenged with inflammation, we have been also endowed with a family of proteins called suppressors of cytokine signaling (SOCS) to challenge signaling pathways transduced by various proinflammatory cytokines. This family of proteins is now considered as important regulator of normal immune physiology and immune disease (Leroith & Nissley, 2005). Generally, SOCS are present in cells at very low levels. However, certain stimuli can rapidly transcribe SOCS. For example, we have found that gemfibrozil (Ghosh & Pahan, 2012) and cinnamic acid (Chakrabarti et al., 2018) are capable of increasing the level of SOCS3 in glial cells. As a result, several studies suggest that SOCS can inhibit the response of immune cells either by reducing the activity of JAK or by competing with signaling molecules for binding to the phosphorylated receptor (Alexander & Hilton, 2004). Such as, both SOCS1 and SOCS3 are capable of binding JAKs to inhibit their tyrosine kinase activity (Park et al., 2003). Therefore, up-regulation of SOCS may represent a critical step for suppressing the expression of proinflammatory cytokines in glial cells via negative regulation of the JAK-STAT pathway.Once proinflammatory cytokines are produced in excess within the CNS, these molecules participate in the pathogenesis of different neurological disorders via multiple mechanisms.

12.4 Pathways by Which Proinflammatory Cytokines May Promote Neurodegeneration

Once proinflammatory cytokines are produced in the CNS microenvironment, these can cause neurodegeneration via various mechanisms.

12.4.1 Oxidative Stress

Oxidative stress participates in the pathogenesis of many human diseases including neurodegenerative disorders (Jana & Pahan, 2007; Modi et al., 2015; Tonnies & Trushina, 2017). Although oxidative stress can be generated at multiple steps, whenever cells are challenged by any insults, injury, or infection, NADPH oxidase is quickly activated (Jana & Pahan, 2004a, b). In resting state, NADPH oxidase consists of both cytosolic and membrane-bound proteins. The two membrane-bound components (p22phox and gp91phox) form the heterodimeric flavocytochrome b558 that contains a putative NADPH binding site, FAD, and two hemes (Bokoch & Knaus, 2003; Ushio-Fukai, 2006). This part of NADPH oxidase is responsible for transferring two electrons from NADPH to two molecules of molecular oxygen. On the other hand, Rac-GTP and three phosphoproteins p47phox, p40phox, and p67phox are cytosolic components. During the activation of NADPH oxidase, all of the cytosolic components should be translocated to the membrane and assembled with flavocytochrome b558 to ensure electron flow (Bokoch & Knaus, 2003). Proinflammatory cytokines like TNFα and IL-1β are capable of activating NADPH oxidase and inducing the production of reactive oxygen species (ROS) in neurons (Lee et al., 2021) and microglia (Haslund-Vinding et al., 2017). It has been also shown that activation of NADPH oxidase is involved in neuronal death (Lee et al., 2021)). Accordingly, activation of NADPH oxidase is reported in postmortem AD brains and mouse models of AD (Ansari & Scheff, 2011; Ma et al., 2017). Ansari and Scheff (Ansari & Scheff, 2011) have reported a robust correlation between NOX activity and the individual's cognitive status such that as the enzyme activity increased, cognitive performance decreased. In 2003, Przedborski and colleagues (Wu et al., 2003) reported that postmortem substantia nigra samples from sporadic PD patients had higher levels of gp91phox than samples from control individuals. Several studies have also described an increase in microglial NADPH oxidase expression in experimental models of PD (Gao et al., 2003; Zhang et al., 2004).

12.4.2 Nitrosative Stress

Nitric oxide (NO), a short-lived gaseous molecule, is produced by the nitric oxide synthase (NOS). At low physiological concentrations, NO exhibits many beneficial functions including vasodilation and nerve transmission (Moncada & Bolanos, 2006; Saha & Pahan, 2006b, c). On the other hand, when NO is produced in excess, it is toxic for many cells including neurons and oligodendrocytes (Saha & Pahan, 2006b, c; Jana & Pahan, 2013). NOS enzyme family is comprised of three tissue-specific isoforms, neuronal NOS (nNOS; NOS1), inducible NOS (iNOS; NOS2), and endothelial NOS (eNOS; NOS3). While nNOS and eNOS are constitutive, iNOS is inducible and therefore, it requires a stimulus (Saha & Pahan, 2006b, c). Numerous studies have shown that proinflammatory cytokines like TNF-α, IL-1β, and IFNγ are potent inducers of iNOS and the production of NO in multiple cells including microglia and astrocytes in the brain (Jana et al., 2005, 2007; Saha & Pahan, 2006b, c).

This NO reacts with superoxide to produce peroxynitrite, a highly reactive derivative of NO (Beckman et al., 1993; Pacher et al., 2007). It exhibits damaging effects on cells, tissues, and organs (Beckman et al., 1993; Radi, 2013). Once peroxynitrite is produced, it is known to nitrate tyrosine residues of protein to form 3-nitro tyrosine (3-NT) (Beckman et al., 1993; Pacher et al., 2007; Radi, 2013). Now 3-NT is considered an important biomarker of oxidative and nitrosative stress (Bandookwala & Sengupta, 2020). The 3-NT formation has been reported in the CNS of patients with MS, AD, HD, PD, and ALS (Beckman et al., 2001; Bandookwala & Sengupta, 2020). In addition to nitrating proteins, peroxynitrite can also nitrate lipids and DNA (Pacher et al., 2007). Therefore, the alteration in the 3-NT profile occurs well before any symptoms appear and can be considered as a potential target for early diagnosis of neurodegenerative diseases (Bandookwala & Sengupta, 2020).

12.4.3 Inhibition of Mitochondrial Functions

Mitochondria play an essential role not only in maintaining normal energy homeostasis but also in the response to pathological conditions. According to Suliman et al (Suliman et al., 2004), electron transport chain (ETC) protein levels are decreased rapidly after insulting to endotoxin/TNFα exposure. It has been found that mitochondria are susceptible to damage from pro-inflammatory cytokines, like TNF-α and IL-1 generated by innate immune cells (Schulze-Osthoff et al., 1992). In turn, damaged mitochondria generate high levels of reactive oxygen species (ROS) and release intrinsic apoptosis-causing proteins to aggravate the outcome during inflammation (Taylor et al., 1995). According to Stommel et al (Stommel et al., 2007), TNF-α affects the neuritic transport of mitochondria in motor neurons, causing mitochondrial redistribution toward the cell soma. Ferri et al (Ferri et al., 2008) have reported significant alteration of morphological features and functionality of mitochondria in cells expressing mutant superoxide dismutase by combined treatment with TNFα and IFNγ. TNF-α may involve the autocrine stimulation of microglial glutamate release and the subsequent triggering of excitotoxic mechanisms to cause mitochondrial impairment (Takeuchi et al., 2006). Therefore, immunohistochemical, biochemical, and molecular studies demonstrate the coexistence of inflammatory processes and signs of mitochondrial impairment in AD brains (Querfurth & LaFerla, 2010). Mitochondrion is also considered the ultimate target leading to neurodegeneration in MS (Di Filippo et al., 2010). A crosstalk between neuroinflammation and mitochondrial dysfunction could potentially play a pathogenic role in PD as shown from postmortem description of complex I deficiency in the SN of patients with PD (Schapira, 1993) and evidence of proinflammatory cytokines (TNF-α, IL-1β, IL-6, etc.) in PD brains (Nagatsu et al., 2000).

12.4.4 Induction of Apoptosis

It has been reported that proinflammatory cytokine-mediated neuronal injury is accompanied by morphologic changes and a DNA fragmentation pattern consistent with apoptosis (Hu et al., 1997). Hu et al have described that the cytokine-mediated neuronal apoptosis is attenuated by the NOS inhibitor N-monomethyl-L-arginine (NMMA), indicating an important role of NO in neuronal apoptosis (Hu et al., 1997). However, according to Downen et al (Downen et al., 1999), IL-1β/IFNγ-induced apoptosis of neurons in human fetal CNS cultures is not significantly suppressed by NMMA. According to them (Downen et al., 1999), the cytokine combination of IL-1β and IFNγ kills human fetal neurons via TNF-α. Proinflammatory cytokine-mediated apoptosis was validated more when it was found that proinflammatory cytokines stimulated the production of ceramide, a lipid second messenger and an inducer of apoptosis, via induction of neutral sphingomyelinase (Jana et al., 2009; Jana & Pahan, 2010a, b). Several studies have established mechanisms behind TNF-α-mediated ceramide production. When TNF-α binds to TNFR1, the downstream adaptor protein FAN constitutively associates with a proximal membrane domain of TNFR1 called neutral sphingomyelinase activation domain (NSD), resulting in the activation of neutral sphingomyelinase, breakdown of sphingomyelin, and production of ceramide and phosphocholine (Adam-Klages et al., 1998). Accordingly, overexpression of full-length FAN stimulates N-SMase activity in TNF-treated cells, while truncated mutants of FAN inhibit the activity of neutral sphingomyelinase. Therefore, TNF-α induces apoptosis via TNFR1-FAN-neutral sphingomyelinase-ceramide pathway. Another proinflammatory cytokine IL-1β also induces the activation of neutral sphingomyelinase-ceramide pathway via IL-1R (Nalivaeva et al., 2000).

On the other hand, anti-inflammatory cytokines (IL-4, IL-10, IL-13, etc.) are known to inhibit apoptosis of brain cells. For example, it has been reported that IL-10 and IL-13 inhibit the induction of ceramide production in astrocytes via the activation of phosphatidylinositol (PI) 3-kinase and thereby protect cells from TNF-α-induced apoptosis (Pahan et al., 2000).

12.5 Neurodegenerative Disorders

As described above, by stimulating oxidative stress, inducing nitrosative stress, suppressing mitochondrial biogenesis, inhibiting autophagy, and promoting apoptosis, proinflammatory cytokines may participate in the pathogenesis of different neurodegenerative disorders.

12.5.1 Alzheimer's Disease (AD)

Although AD, the most common neurodegenerative disease in humans, is pathologically characterized by amyloid-beta (Aβ)-containing plaques and hyperphosphorylated tau-containing neurofibrillary tangles, recent studies indicate an important role of neuroinflammation in the pathogenesis of AD. Tarkowski et al have found higher levels of pro-inflammatory cytokine TNFα and lower levels of anti-inflammatory cytokine TNFβ in the CSF of patients with mild cognitive impairment who progressed to AD as compared to controls who did not progress to AD (Tarkowski et al., 2003). According to Brosseron et al, the level of some proinflammatory cytokines like IL-1β, TNF-α, and IL-6 increases in the early stage of AD (Brosseron et al., 2014). Upregulation of other cytokines such as IL-18, MCP-1, and IP-10 is also seen at a certain stage of AD (Brosseron et al., 2014). Nuclear factor-κB (NF-κB) is a proinflammatory transcription factor that is required for the transcription of most of the proinflammatory molecules (Ghosh et al., 1998, 2007). Accordingly, it has been found that NF-κB is activated in the hippocampus of AD patients and a mouse model of AD and that selective inhibition of NF-κB by NBD (NEMO-binding domain) peptides protect hippocampal neurons and improve spatial memory and learning in a mouse model of AD (Rangasamy et al., 2015). These results suggest that proinflammatory signaling pathways may play a role in the pathogenesis of AD.

However, despite the upregulation of proinflammatory molecules in AD, most clinical trials for anti-inflammatory drugs in AD patients, including prednisone, naproxen, diclofenac, indomethacin, and celecoxib, have failed to exhibit significant improvements (Aisen et al., 2000, 2003). Some of these treatments even revealed detrimental effects to some extents. Recently, upon comparing two different doses of minocycline and placebo in mild AD patients over 2 years, the Minocycline in Alzheimer's Disease Efficacy (MADE) trial also did not find significant beneficial effect (Howard et al., 2020). On the other hand, according to Pasqualetti et al, 12-month of ibuprofen treatment displays beneficial effects in a subgroup of mild-to-moderate AD patients with *APOE* ε4 carriers (Pasqualetti et al., 2009). Rogers et al have also described protective effects of indomethacin in mild-to-moderate AD patients (Rogers et al., 1993). However, significance of this study is minimum since many patients dropped out of this trial. The failure of different anti-inflammatory clinical trials could be explained by the possible influence of the anti-inflammatory medications on the beneficial effects of glial cells, the selection of participants, and the short-term follow-up period.

12.5.2 Parkinson's Disease (PD)

PD, the most common movement disorder in humans, is the second most frequent neurodegenerative disease after AD. While clinically, PD is categorized by tremor, bradykinesia, rigidity, and postural instability, pathologically, it is characterized by the presence of intracytoplasmic inclusions (Lewy bodies) and the progressive degeneration of the dopaminergic neurons in the substantia nigra pars compacta (SNpc) (Olanow & Tatton, 1999; Dauer & Przedborski, 2003). In addition, recent studies support the important role for microglial activation and associated inflammation in the death of nigral dopaminergic neurons and the loss of striatal dopamine (Dauer & Przedborski, 2003; Ghosh et al., 2007; Roy & Pahan, 2011). Accordingly, a number of proinflammatory cytokines such as tumor necrosis factor α (TNF-α), interleukin-1β (IL-1β), and IL-6 are detected in either CSF or SNpc in PD (Mogi et al., 1994; Bessler et al., 1999). It has been observed that transgenic mice with homozygous mutant alleles for both TNFR1 and TNFR2, but not the individual receptors, are completely protected against MPTP (Sriram et al., 2006), indicating an important role of TNFα – TNFR1/2 signaling pathway in nigrostriatal degeneration. Recently our laboratory has found that several chemokines including RANTES and eotaxin are also upregulated in the nigra of PD patients and mouse and monkey models of PD (Chandra et al., 2016, 2017). Because CCR5 is a common receptor of RANTES and eotaxin, CCR5 inhibitor maraviroc was tested in a primate model of PD and found to protect dopaminergic neurons in hemiparkinsonian monkeys (Mondal et al., 2019). Therefore, along with proinflammatory cytokines, proinflammatory chemokines also play a role in nigrostriatal degeneration.

Consistent with the upregulation of proinflammatory cytokines and chemokines in PD patients and animal models, it has been found that NF-κB is activated in the SNpc of PD patients and hemiparkinsonian monkeys and mice (Ghosh et al., 2007, 2009; Mondal et al., 2012b). Accordingly, selective inhibition of NF-κB by NBD (NEMO-binding domain) peptides protects dopaminergic neurons in hemiparkinsonian monkeys (Mondal et al., 2012b) and mice (Ghosh et al., 2007). Simvastatin, a cholesterol-lowering drug, inhibits the activation of NF-kB and the induction of proinflammatory molecules in microglia to exhibit protection of dopamine in the MPTP mouse model of PD (Ghosh et al., 2009). According to Hunot et al, the NF-κB-dependent signaling pathway in dopaminergic neurons may play a role in neuronal loss in PD (Hunot et al., 1997). Recently, it has been established that Nurr1, a nuclear orphan receptor, attenuates glial activation by docking to NF-κB-p65 on target inflammatory gene promoters and protects nigral neurons in vivo in the nigra (Saijo et al., 2009; Gottschalk et al., 2019). Therefore, targeting inflammation may be an important step to protect dopaminergic neurons in PD and its animal models.

12.5.3 Amyotropic Lateral Sclerosis (ALS)

ALS is an adult-onset progressive neurodegenerative disease that is also called Lou Gehrig's disease. Although ALS primarily involves motor neurons in the brain and spinal cord leading to progressive muscular atrophy, paralysis, speech and swallowing disturbances, and respiratory dysfunction (Pupillo et al., 2014), PET imaging indicates longitudinal changes in neuroinflammation in ALS patients (Corcia et al., 2012; Alshikho et al., 2018). Alshikho et al (2018) have reported increased neuroinflammation in the primary and supplementary motor and temporal cortices of ALS patients using PET imaging. Studies of post-mortem tissues have also shown microglial activation and associated inflammation in the CNS of ALS patients. Microglia are activated in both the motor cortex (Dols-Icardo et al., 2020) and spinal cord (D'Erchia et al., 2017) of sporadic ALS cases. However, in a subset of ALS cases, microglial proliferation is also found (Spiller et al., 2018). Accordingly, the level of the number of pro-inflammatory cytokines including TNF-α and IL-6 increases in plasma of ALS patients as compared to controls (Tortelli et al., 2020).

Proinflammatory cytokines are known to induce the expression of iNOS and accordingly, it is found that microglial iNOS is upregulated in the spinal cord of the symptomatic mutant SOD1 (mSOD1) mouse model of ALS (Almer et al., 1999) and that inhibition of iNOS extends lifespan in mSOD1 mice (Chen et al., 2010). The transcription factor NF-κB is considered a master regulator of proinflammatory molecules and activation of NF-κB involves phosphorylation of the p65 subunit (Ghosh et al., 2007; Saha et al., 2007) . Consistent with the presence of different proinflammatory molecules including cytokines in the CNS of ALS patients, an increase in phospho-p65 has been shown to correlate with the disease process in mSOD1 mice and inhibition of microglial-specific NF-κB increases the lifespan of mSOD1 mice (Frakes et al., 2014).

12.5.4 Huntington's Disease (HD)

Huntington's disease (HD) is a rare neurodegenerative disorder of the CNS characterized by unwanted choreatic movements, behavioral and psychiatric disturbances, and dementia (1993; van Duijn et al., 2007). Although HD is an autosomal dominantly inherited disease caused by an elongated CAG repeat on the short arm of chromosome 4p16.3 in the Huntingtin gene (1), it has been found that patients with HD exhibit increased circulating levels of inflammatory molecules, mainly related to innate immunity. Interestingly, it is seen that levels of inflammatory molecules increase with the advancement of the disease. According to Dalrymple et al (Dalrymple et al., 2007), plasma levels of IL-6 are higher in moderate HD in comparison with both controls and early HD patients. In an elegant study, Bjorkqvist et al (Bjorkqvist et al., 2008) have shown that IL-6 levels increase in HD gene carriers with a mean of 16 years before the predicted onset of clinical symptoms. In addition to IL-6, levels of TNFα, IL-10, and IL-4 are also increased in moderate-stage HD patients as compared to controls (Bjorkqvist et al., 2008). Data obtained in mouse models are also similar to those seen in HD patients, as plasma levels of several inflammatory molecules, including IL-6, IL-8, TNF-α, IL-10, IL-1β, and IL-12p70, are elevated in different mouse models of HD (R6/2, HdhQ150 expressing 150 CAG repeats inserted in the mouse endogenous HTT gene first exon, and YAC128) (Bjorkqvist et al., 2008; Crotti et al., 2014; Trager et al., 2015). Bouchard et al have shown that IL-6 is particularly important in HD as neutralization of IL-6 decreases weight loss at late stages and partially rescues motor deficits in R6/2 mice (Bouchard et al., 2012). Recently we have seen that oral administration of glyceryl tribenzoate, a food additive and a flavoring agent, inhibits microglia and astroglial activation and reduces the level of Huntingtin protein in the brain to protect locomotor behaviors in a transgenic mouse model of HD (Dutta et al., 2021b).

12.5.5 Multiple Sclerosis (MS)

Multiple sclerosis (MS) is one of the most common autoimmune demyelinating diseases of the CNS that affects mainly young adults (Jana & Pahan, 2010a, b; Pahan, 2010, 2013, 2015; Lassmann, 2018). MS is more prevalent in females than in males, with an incidence ratio of approximately 2:1 (Dasgupta et al., 2005; Brahmachari & Pahan, 2010). However, when a male gets MS, usually it leads to worse disability progression. Although the etiology of MS is poorly understood, it has been established that the observed demyelination in the CNS is a result of a T cell-mediated autoimmune response (Pahan, 2010, 2013, 2015; Lassmann, 2018). Experimental allergic encephalomyelitis (EAE) is an animal model of MS that is induced in rodents by immunization with myelin proteins or by adoptive transfer of myelin-reactive T cells (Pahan, 2010, 2013, 2015; Lassmann, 2018). The identification of wide range of proinflammatory cytokines (TNF-α, IL-1, IL-2, IL-6, and IFN-γ), cell adhesion molecules, inducible NO synthase (iNOS), and proinflammatory transcription factor (NF-κB) in CNS lesions of MS patients and EAE animals (Dasgupta et al., 2003b, 2004; Brahmachari & Pahan, 2007; Mondal & Pahan, 2015; Mondal et al., 2017, 2018, 2020) suggests that a broad-spectrum inflammatory process in the CNS plays an important role in the pathogenesis of these diseases. Accordingly, several studies in the 1990s were really upbeat about the role of TNF-α in MS pathogenesis. It was demonstrated that neutralizing antibodies against TNFα prevents EAE in an adoptive transfer model (Ruddle et al., 1990). Furthermore, increased expression of TNFα was seen in the CNS of TMEV-infected SJL/J mice and anti-TNFα antibody-protected mice from the disease (Inoue et al., 1996), indicating that TNF-α inhibitors might be beneficial for MS. However, clinical trials exhibited that recombinant TNF receptor lenercept (sTNFR-IgG p55) treatment exacerbated disease pathology in MS patients (1999), indicating a neuroprotective role of TNF-α in MS (Figiel, 2008).

On the other hand, recent studies highlight the involvement of the IL-23-Th17 axis in the pathogenesis of MS (Brahmachari & Pahan, 2008; Pahan, 2010, 2013; Mondal et al., 2012a). IL-23 is a heterodimer of p40 and p19 and accordingly, it has been found that IL-23p19-deficient mice are resistant to EAE (Cua et al., 2003). IL-23-driven $CD4^+$ T cells are also found to be highly pathogenic, which are characterized by the production of IL-17 (Chen et al., 2006). In addition to IL-17, Th17 cells are also characterized by the production of IL-17F, IL-21, and IL-22 (Chen et al., 2006).

12.6 Conclusion

The importance and contribution of any biomedical field should be judged from two fronts: academic and therapeutic. Academically, it is important to know which proinflammatory cytokines participate in the pathogenesis of different neurodegenerative disorders and how these molecules lead to the activation of glial cells and the demise of neurons and myelin-producing oligodendrocytes. Such intellectual expansion may help us to predict possible similarities among different neurodegenerative disorders and to look for pathological features of other CNS disorders to find a more coherent approach for a better understanding of physiological and pathological processes in the CNS. While therapeutically, one might expect direct application of neutralizing antibodies, shRNA/siRNA, and nontoxic small molecule drugs capable of blocking the proinflammatory cytokines, suppressing signaling pathways for glial production of proinflammatory cytokines and/or targeting neurotoxic functions of proinflammatory cytokines in patients with neurodegenerative diseases. From this aspect, it has, yet, achieved mixed success. The reason behind this lies in the fact that signaling pathways and transcription factors required for the generation of proinflammatory cytokines are being utilized for the transcription of other molecules as well. Several classes of drugs/supplements that reduce the expression of proinflammatory cytokines in other cells including glial cells such as salicylates, non-steroidal anti-inflammatory drugs, statins, and other mevalonate inhibitors, activators of PPAR, antioxidants, cAMP phosphodiesterase inhibitors, proteosome inhibitors, and natural compounds (cinnamon, cinnamein, curcumin, etc.) are being considered for various neurodegenerative disorders (Ghosh et al., 2009; Roy & Pahan, 2011; Khasnavis & Pahan, 2014; Modi et al., 2015; Chandra et al., 2018, 2019; Patel et al., 2018, 2019; Chandra & Pahan, 2019; Prorok et al., 2019; Gottschalk et al., 2021; Raha et al., 2021; Rangasamy et al., 2021a, 2022; Mondal et al., 2024; Raha et al., 2024). However, we must remember that in addition to suppressing the expression of proinflammatory cytokines, these compounds also modulate many other biological processes suggesting that therapeutic efficacy of these molecules cannot be credited only to the suppression of proinflammatory cytokines. Therefore, the challenge is to discover signaling pathways and transcription factors specific for the activation of different proinflammatory cytokine gene promoters, which may not be even possible. However, analysis of 5′-flanking regions of human proinflammatory cytokine genes may be an option in moving forward. Moreover, nothing is known about glial cell-specific signaling pathways for the induction of proinflammatory cytokines. It could be another important avenue for future investigation as specific targeting of glial proinflammatory cytokines may be a better option to achieve better therapeutic output.

12.7 Review Questions

1. Which of the following is true about neutral sphingomyelinase?
   (a) It breaks down sphingomyelin.
   (b) It is activated by TNF-α.
   (c) It generates ceramide.
   (d) It catalyzes the formation of phosphocholine.
   (e) All of the above
2. Which of the following cytokines inhibit the production of ceramide?
   (a) IL-1β and IL-6
   (b) IL-1β and IFNγ
   (c) IL-10 and IL-13
   (d) TNF-α and IL-2
   (e) IL-5 and IL-6
3. IL-23 is a heterodimeric cytokine. It is composed of
   (a) p40 and p28
   (b) p40 and p35
   (c) p19 and p35
   (d) p40 and p19
   (e) p35 and p23
4. Which of the following are anti-inflammatory cytokines?
   (a) IL-10 and IL-13
   (b) IL-6 and IL-8
   (c) TNF-α and IL-12
   (d) IL-12 and IL-23
   (e) IL-17 and IL-21
5. Identify the protein kinase responsible for the phosphorylation of IκB.
   (a) NEMO
   (b) MAP kinase
   (c) PKA
   (d) IKKβ
   (e) None of the above

12.8 Answer Key

(a) e
(b) c
(c) d
(d) a
(e) d

Acknowledgements This study was supported by grants from NIH (AT10980 and AG050431) and merit awards (1I01BX005002 and I01BX005613) from the US Department of Veterans Affairs. Moreover, Dr. Pahan is the recipient of a Research Career Scientist Award (1IK6 BX004982) from the Department of Veterans Affairs. However, the views expressed in this article are those of the authors and do not necessarily reflect the position or policy of the Department of Veterans Affairs or the United States government.

References

(1993). A novel gene containing a trinucleotide repeat that is expanded and unstable on Huntington's disease chromosomes. The Huntington's Disease Collaborative Research Group. *Cell, 72*(6), 971–983.

(1999). TNF neutralization in MS: results of a randomized, placebo-controlled multicenter study. The Lenercept Multiple Sclerosis Study Group and The University of British Columbia MS/MRI Analysis Group. *Neurology, 53*(3), 457–465.

Adam-Klages, S., Schwandner, R., Adam, D., Kreder, D., Bernardo, K., & Kronke, M. (1998). Distinct adapter proteins mediate acid versus neutral sphingomyelinase activation through the p55 receptor for tumor necrosis factor. *Journal of Leukocyte Biology, 63*(6), 678–682.

Aisen, P. S., Davis, K. L., Berg, J. D., Schafer, K., Campbell, K., Thomas, R. G., Weiner, M. F., Farlow, M. R., Sano, M., Grundman, M., & Thal, L. J. (2000). A randomized controlled trial of prednisone in Alzheimer's disease. Alzheimer's Disease Cooperative Study. *Neurology, 54*(3), 588–593.

Aisen, P. S., Schafer, K. A., Grundman, M., Pfeiffer, E., Sano, M., Davis, K. L., Farlow, M. R., Jin, S., Thomas, R. G., Thal, L. J., & Alzheimer's Disease Cooperative, S. (2003). Effects of rofecoxib or naproxen vs placebo on Alzheimer disease progression: A randomized controlled trial. *JAMA, 289*(21), 2819–2826.

Alexander, W. S., & Hilton, D. J. (2004). The role of suppressors of cytokine signaling (SOCS) proteins in regulation of the immune response. *Annual Review of Immunology, 22*, 503–529.

Almer, G., Vukosavic, S., Romero, N., & Przedborski, S. (1999). Inducible nitric oxide synthase up-regulation in a transgenic mouse model of familial amyotrophic lateral sclerosis. *Journal of Neurochemistry, 72*(6), 2415–2425.

Alshikho, M. J., Zurcher, N. R., Loggia, M. L., Cernasov, P., Reynolds, B., Pijanowski, O., Chonde, D. B., Izquierdo Garcia, D., Mainero, C., Catana, C., Chan, J., Babu, S., Paganoni, S., Hooker, J. M., & Atassi, N. (2018). Integrated magnetic resonance imaging and [(11) C]-PBR28 positron emission tomographic imaging in amyotrophic lateral sclerosis. *Annals of Neurology, 83*(6), 1186–1197.

Ansari, M. A., & Scheff, S. W. (2011). NADPH-oxidase activation and cognition in Alzheimer disease progression. *Free Radical Biology & Medicine, 51*(1), 171–178.

Auch, C. J., Saha, R. N., Sheikh, F. G., Liu, X., Jacobs, B. L., & Pahan, K. (2004). Role of protein kinase R in double-stranded RNA-induced expression of nitric oxide synthase in human astroglia. *FEBS Letters, 563*(1-3), 223–228.

Bailly, S., Fay, M., Israel, N., & Gougerot-Pocidalo, M. A. (1996). The transcription factor AP-1 binds to the human interleukin 1 alpha promoter. *European Cytokine Network, 7*(2), 125–128.

Bandookwala, M., & Sengupta, P. (2020). 3-Nitrotyrosine: A versatile oxidative stress biomarker for major neurodegenerative diseases. *The International Journal of Neuroscience, 130*(10), 1047–1062.

Beckman, J. S., Carson, M., Smith, C. D., & Koppenol, W. H. (1993). ALS, SOD and peroxynitrite. *Nature, 364*(6438), 584.

Beckman, J. S., Estevez, A. G., Crow, J. P., & Barbeito, L. (2001). Superoxide dismutase and the death of motoneurons in ALS. *Trends in Neurosciences, 24*(11 Suppl), S15–S20.

Bessler, H., Djaldetti, R., Salman, H., Bergman, M., & Djaldetti, M. (1999). IL-1 beta, IL-2, IL-6 and TNF-alpha production by peripheral blood mononuclear cells from patients with Parkinson's disease. *Biomedicine & Pharmacotherapy, 53*(3), 141–145.

Bi, J., Jiang, B., Zorn, A., Zhao, R. G., Liu, P., & An, L. J. (2013). Catalpol inhibits LPS plus IFN-gamma-induced inflammatory response in astrocytes primary cultures. *Toxicology In Vitro, 27*(2), 543–550.

Birolini, G., Verlengia, G., Talpo, F., Maniezzi, C., Zentilin, L., Giacca, M., Conforti, P., Cordiglieri, C., Caccia, C., Leoni, V., Taroni, F., Biella, G., Simonato, M., Cattaneo, E., & Valenza, M. (2021). SREBP2 gene therapy targeting striatal astrocytes ameliorates Huntington's disease phenotypes. *Brain, 144*(10), 3175–3190.

Bjorkqvist, M., Wild, E. J., Thiele, J., Silvestroni, A., Andre, R., Lahiri, N., Raibon, E., Lee, R. V., Benn, C. L., Soulet, D., Magnusson, A., Woodman, B., Landles, C., Pouladi, M. A., Hayden, M. R., Khalili-Shirazi, A., Lowdell, M. W., Brundin, P., Bates, G. P., Leavitt, B. R., Moller, T., & Tabrizi, S. J. (2008). A novel pathogenic pathway of immune activation detectable before clinical onset in Huntington's disease. *The Journal of Experimental Medicine, 205*(8), 1869–1877.

Bokoch, G. M., & Knaus, U. G. (2003). NADPH oxidases: not just for leukocytes anymore! *Trends in Biochemical Sciences, 28*(9), 502–508.

Borlongan, C. V., Koutouzis, T. K., & Sanberg, P. R. (1997). 3-Nitropropionic acid animal model and Huntington's disease. *Neuroscience and Biobehavioral Reviews, 21*(3), 289–293.

Bouchard, J., Truong, J., Bouchard, K., Dunkelberger, D., Desrayaud, S., Moussaoui, S., Tabrizi, S. J., Stella, N., & Muchowski, P. J. (2012). Cannabinoid receptor 2 signaling in peripheral immune cells modulates disease onset and severity in mouse models of Huntington's disease. *The Journal of Neuroscience, 32*(50), 18259–18268.

Bowman, C. C., Rasley, A., Tranguch, S. L., & Marriott, I. (2003). Cultured astrocytes express toll-like receptors for bacterial products. *GLIA, 43*(3), 281–291.

Bradbury, A. M., Bongarzone, E. R., & Sands, M. S. (2021). Krabbe disease: New hope for an old disease. *Neuroscience Letters, 752*, 135841.

Brahmachari, S., & Pahan, K. (2007). Sodium benzoate, a food additive and a metabolite of cinnamon, modifies T cells at multiple steps and inhibits adoptive transfer of experimental allergic encephalomyelitis. *Journal of Immunology, 179*(1), 275–283.

Brahmachari, S., & Pahan, K. (2008). Role of cytokine p40 family in multiple sclerosis. *Minerva Medica, 99*(2), 105–118.

Brahmachari, S., & Pahan, K. (2010). Gender-specific expression of beta1 integrin of VLA-4 in myelin basic protein-primed T cells: implications for gender bias in multiple sclerosis. *Journal of Immunology, 184*(11), 6103–6113.

Brahmachari, S., Fung, Y. K., & Pahan, K. (2006). Induction of glial fibrillary acidic protein expression in astrocytes by nitric oxide. *The Journal of Neuroscience, 26*(18), 4930–4939.

Brahmachari, S., Jana, A., & Pahan, K. (2009). Sodium benzoate, a metabolite of cinnamon and a food additive, reduces microglial and astroglial inflammatory responses. *Journal of Immunology, 183*(9), 5917–5927.

Brosseron, F., Krauthausen, M., Kummer, M., & Heneka, M. T. (2014). Body fluid cytokine levels in mild cognitive impairment and Alzheimer's disease: a comparative overview. *Molecular Neurobiology, 50*(2), 534–544.

Butturini, E., Boriero, D., Carcereri de Prati, A., & Mariotto, S. (2019). STAT1 drives M1 microglia activation and neuroinflammation under hypoxia. *Archives of Biochemistry and Biophysics, 669*, 22–30.

Carrieri, P. B., Maiorino, A., Provitera, V., Soscia, E., & Perrella, O. (1992). Cytokines in the pathogenesis of multiple sclerosis. *Acta Neurologica (Napoli), 14*(4-6), 333–341.

Chakrabarti, S., Jana, M., Roy, A., & Pahan, K. (2018). Upregulation of suppressor of cytokine signaling 3 in microglia by cinnamic acid. *Current Alzheimer Research, 15*(10), 894–904.

Chakrabarti, S., Roy, A., Prorok, T., Patel, D., Dasarathi, S., & Pahan, K. (2019). Aspirin up-regulates suppressor of cytokine signaling 3 in glial cells via PPARalpha. *Journal of Neurochemistry, 151*(1), 50–63.

Chandra, S., & Pahan, K. (2019). Gemfibrozil, a lipid-lowering drug, lowers amyloid plaque pathology and enhances memory in a mouse model of Alzheimer's disease via peroxisome proliferator-activated receptor αlpha. *Journal of Alzheimer's Disease Reports, 3*(1), 149–168.

Chandra, G., Rangasamy, S. B., Roy, A., Kordower, J. H., & Pahan, K. (2016). Neutralization of RANTES and eotaxin prevents the loss of dopaminergic neurons in a mouse model of Parkinson disease. *The Journal of Biological Chemistry, 291*(29), 15267–15281.

Chandra, G., Roy, A., Rangasamy, S. B., & Pahan, K. (2017). Induction of adaptive immunity leads to nigrostriatal disease progression in MPTP mouse model of Parkinson's disease. *Journal of Immunology, 198*(11), 4312–4326.

Chandra, S., Jana, M., & Pahan, K. (2018). Aspirin induces lysosomal biogenesis and attenuates amyloid plaque pathology in a mouse model of Alzheimer's disease via PPARa. *The Journal of Neuroscience, 38*(30), 6682–6699.

Chandra, S., Roy, A., Jana, M., & Pahan, K. (2019). Cinnamic acid activates PPARalpha to stimulate Lysosomal biogenesis and lower Amyloid plaque pathology in an Alzheimer's disease mouse model. *Neurobiology of Disease, 124*, 379–395.

Chen, Y., Langrish, C. L., McKenzie, B., Joyce-Shaikh, B., Stumhofer, J. S., McClanahan, T., Blumenschein, W., Churakovsa, T., Low, J., Presta, L., Hunter, C. A., Kastelein, R. A., & Cua, D. J. (2006). Anti-IL-23 therapy inhibits multiple inflammatory pathways and ameliorates autoimmune encephalomyelitis. *The Journal of Clinical Investigation, 116*(5), 1317–1326.

Chen, K., Northington, F. J., & Martin, L. J. (2010). Inducible nitric oxide synthase is present in motor neuron mitochondria and Schwann cells and contributes to disease mechanisms in ALS mice. *Brain Structure & Function, 214*(2-3), 219–234.

Chung, I. Y., Norris, J. G., & Benveniste, E. N. (1991). Differential tumor necrosis factor alpha expression by astrocytes from experimental allergic encephalomyelitis-susceptible and -resistant rat strains. *The Journal of Experimental Medicine, 173*(4), 801–811.

Colonna, M., & Butovsky, O. (2017). Microglia function in the central nervous system during health and neurodegeneration. *Annual Review of Immunology, 35*, 441–468.

Corbett, G. T., Roy, A., & Pahan, K. (2012). Gemfibrozil, a lipid-lowering drug, upregulates IL-1 receptor antagonist in mouse cortical neurons: Implications for neuronal self-defense. *Journal of Immunology, 189*(2), 1002–1013.

Corcia, P., Tauber, C., Vercoullie, J., Arlicot, N., Prunier, C., Praline, J., Nicolas, G., Venel, Y., Hommet, C., Baulieu, J. L., Cottier, J. P., Roussel, C., Kassiou, M., Guilloteau, D., & Ribeiro, M. J. (2012). Molecular imaging of microglial activation in amyotrophic lateral sclerosis. *PLoS One, 7*(12), e52941.

Crotti, A., Benner, C., Kerman, B. E., Gosselin, D., Lagier-Tourenne, C., Zuccato, C., Cattaneo, E., Gage, F. H., Cleveland, D. W., & Glass, C. K. (2014). Mutant Huntingtin promotes autonomous microglia activation via myeloid lineage-determining factors. *Nature Neuroscience, 17*(4), 513–521.

Cua, D. J., Sherlock, J., Chen, Y., Murphy, C. A., Joyce, B., Seymour, B., Lucian, L., W. To, Kwan, S., Churakova, T., Zurawski, S., Wiekowski, M., Lira, S. A., Gorman, D., Kastelein, R. A., & Sedgwick, J. D. (2003). Interleukin-23 rather than interleukin-12 is the critical cytokine for autoimmune inflammation of the brain. *Nature, 421*(6924), 744–748.

Dalpke, A. H., Schafer, M. K., Frey, M., Zimmermann, S., Tebbe, J., Weihe, E., & Heeg, K. (2002). Immunostimulatory CpG-DNA activates murine microglia. *Journal of Immunology, 168*(10), 4854–4863.

Dalrymple, A., Wild, E. J., Joubert, R., Sathasivam, K., Bjorkqvist, M., Petersen, A., Jackson, G. S., Isaacs, J. D., Kristiansen, M., Bates, G. P., Leavitt, B. R., Keir, G., Ward, M., & Tabrizi, S. J. (2007). Proteomic profiling of plasma in Huntington's disease reveals neuroinflammatory activation and biomarker candidates. *Journal of Proteome Research, 6*(7), 2833–2840.

Dasgupta, S., Jana, M., Liu, X., & Pahan, K. (2002). Myelin basic protein-primed T cells induce nitric oxide synthase in microglial cells. Implications for multiple sclerosis. *The Journal of Biological Chemistry, 277*(42), 39327–39333.

Dasgupta, S., Jana, M., Liu, X., & Pahan, K. (2003a). Role of very-late antigen-4 (VLA-4) in myelin basic protein-primed T cell contact-induced expression of proinflammatory cytokines in microglial cells. *The Journal of Biological Chemistry, 278*(25), 22424–22431.

Dasgupta, S., Zhou, Y., Jana, M., Banik, N. L., & Pahan, K. (2003b). Sodium phenylacetate inhibits adoptive transfer of experimental allergic encephalomyelitis in SJL/J mice at multiple steps. *Journal of Immunology, 170*(7), 3874–3882.

Dasgupta, S., Jana, M., Zhou, Y., Fung, Y. K., Ghosh, S., & Pahan, K. (2004). Antineuroinflammatory effect of NF-kappaB essential modifier-binding domain peptides in the adoptive transfer model of experimental allergic encephalomyelitis. *Journal of Immunology, 173*(2), 1344–1354.

Dasgupta, S., Jana, M., Liu, X., & Pahan, K. (2005). Myelin basic protein-primed T cells of female but not male mice induce nitric-oxide synthase and proinflammatory cytokines in microglia: Implications for gender bias in multiple sclerosis. *The Journal of Biological Chemistry, 280*(38), 32609–32617.

Dauer, W., & Przedborski, S. (2003). Parkinson's disease: Mechanisms and models. *Neuron, 39*(6), 889–909.

D'Erchia, A. M., Gallo, A., Manzari, C., Raho, S., Horner, D. S., Chiara, M., Valletti, A., Aiello, I., Mastropasqua, F., Ciaccia, L., Locatelli, F., Pisani, F., Nicchia, G. P., Svelto, M., Pesole, G., & Picardi, E. (2017). Massive transcriptome sequencing of human spinal cord tissues provides new insights into motor neuron degeneration in ALS. *Scientific Reports, 7*(1), 10046.

Di Filippo, M., Chiasserini, D., Tozzi, A., Picconi, B., & Calabresi, P. (2010). Mitochondria and the link between neuroinflammation and neurodegeneration. *Journal of Alzheimer's Disease, 20*(Suppl 2), S369–S379.

Dols-Icardo, O., Montal, V., Sirisi, S., Lopez-Pernas, G., Cervera-Carles, L., Querol-Vilaseca, M., Munoz, L., Belbin, O., Alcolea, D., Molina-Porcel, L., Pegueroles, J., Turon-Sans, J., Blesa, R., Lleo, A., Fortea, J., Rojas-Garcia, R., & Clarimon, J. (2020). Motor cortex transcriptome reveals microglial key events in amyotrophic lateral sclerosis. *Neurology-Neuroimmunology & Neuroinflammation, 7*(5).

Downen, M., Amaral, T. D., Hua, L. L., Zhao, M. L., & Lee, S. C. (1999). Neuronal death in cytokine-activated primary human brain cell culture: Role of tumor necrosis factor-alpha. *GLIA, 28*(2), 114–127.

Dutta, D., Kundu, M., Mondal, S., Roy, A., Ruehl, S., Hall, D. A., & Pahan, K. (2019). RANTES-induced invasion of Th17 cells into substantia nigra potentiates dopaminergic cell loss in MPTP mouse model of Parkinson's disease. *Neurobiology of Disease, 132*, 104575.

Dutta, D., Jana, M., Majumder, M., Mondal, S., Roy, A., & Pahan, K. (2021a). Selective targeting of the TLR2/MyD88/NF-kappaB pathway reduces alpha-synuclein spreading in vitro and in vivo. *Nature Communications, 12*(1), 5382.

Dutta, D., Majumder, M., Paidi, R. K., & Pahan, K. (2021b). Alleviation of Huntington pathology in mice by oral administration of food additive glyceryl tribenzoate. *Neurobiology of Disease, 153*, 105318.

Dutta, D., Paidi, R. K., Raha, S., Roy, A., Chandra, S., & Pahan, K. (2022). Treadmill exercise reduces alpha-synuclein spreading via PPARalpha. *Cell Reports, 40*(2), 111058.

Falvo, J. V., Tsytsykova, A. V., & Goldfeld, A. E. (2010). Transcriptional control of the TNF gene. *Current Directions in Autoimmunity, 11*, 27–60.

Ferri, A., Nencini, M., Cozzolino, M., Carrara, P., Moreno, S., & Carri, M. T. (2008). Inflammatory cytokines increase mitochondrial damage in motoneuronal cells expressing mutant SOD1. *Neurobiology of Disease, 32*(3), 454–460.

Figiel, I. (2008). Pro-inflammatory cytokine TNF-alpha as a neuroprotective agent in the brain. *Acta Neurobiologiae Experimentalis (Wars), 68*(4), 526–534.

Frakes, A. E., Ferraiuolo, L., Haidet-Phillips, A. M., Schmelzer, L., Braun, L., Miranda, C. J., Ladner, K. J., Bevan, A. K., Foust, K. D., Godbout, J. P., Popovich, P. G., Guttridge, D. C., & Kaspar, B. K. (2014). Microglia induce motor neuron death via the classical NF-kappaB pathway in amyotrophic lateral sclerosis. *Neuron, 81*(5), 1009–1023.

Gao, H. M., Liu, B., Zhang, W., & Hong, J. S. (2003). Critical role of microglial NADPH oxidase-derived free radicals in the in vitro MPTP model of Parkinson's disease. *The FASEB Journal, 17*(13), 1954–1956.

Ghosh, A., & Pahan, K. (2012). Gemfibrozil, a lipid-lowering drug, induces suppressor of cytokine signaling 3 in glial cells: Implications for neurodegenerative disorders. *The Journal of Biological Chemistry, 287*(32), 27189–27203.

Ghosh, S., May, M. J., & Kopp, E. B. (1998). NF-kappa B and Rel proteins: Evolutionarily conserved mediators of immune responses. *Annual Review of Immunology, 16*, 225–260.

Ghosh, A., Roy, A., Liu, X., Kordower, J. H., Mufson, E. J., Hartley, D. M., Ghosh, S., Mosley, R. L., Gendelman, H. E., & Pahan, K. (2007). Selective inhibition of NF-kappaB activation prevents dopaminergic neuronal loss in a mouse model of Parkinson's disease. *Proceedings of the National Academy of Sciences of the United States of America, 104*(47), 18754–18759.

Ghosh, A., Roy, A., Matras, J., Brahmachari, S., Gendelman, H. E., & Pahan, K. (2009). Simvastatin inhibits the activation of p21ras and prevents the loss of dopaminergic neurons in a mouse model of Parkinson's disease. *The Journal of Neuroscience, 29*(43), 13543–13556.

Ghosh, A., Rangasamy, S. B., Modi, K. K., & Pahan, K. (2017). Gemfibrozil, food and drug administration-approved lipid-lowering drug, increases longevity in mouse model of late infantile neuronal ceroid lipofuscinosis. *Journal of Neurochemistry, 141*(3), 423–435.

Giri, S., Jatana, M., Rattan, R., Won, J. S., Singh, I., & Singh, A. K. (2002). Galactosylsphingosine (psychosine)-induced expression of cytokine-mediated inducible nitric oxide synthases via AP-1 and C/EBP: Implications for Krabbe disease. *The FASEB Journal, 16*(7), 661–672.

Glauert, H. P., Calfee-Mason, K., Li, Y., Nilakantan, V., Twaroski, M. L., Tharappel, J., & Spear, B. T. (2008). The role of NF-kappaB in PPARalpha-mediated hepatocarcinogenesis. *PPAR Research, 2008*, 286249.

Gottschalk, C. G., Roy, A., Jana, M., Kundu, M., & Pahan, K. (2019). Activation of peroxisome proliferator-activated receptor-α increases the expression of nuclear receptor related 1 protein (Nurr1) in dopaminergic neurons. *Molecular Neurobiology, 56*(11), 7872–7887.

Gottschalk, C. G., Jana, M., Roy, A., Patel, D. R., & Pahan, K. (2021). Gemfibrozil protects dopaminergic neurons in a mouse model of Parkinson's disease via PPARα-dependent astrocytic GDNF pathway. *The Journal of Neuroscience, 41*(10), 2287–2300.

Guo, L., Xing, Y., Pan, R., Jiang, M., Gong, Z., Lin, L., Wang, J., Xiong, G., & Dong, J. (2013). Curcumin protects microglia and primary rat cortical neurons against HIV-1 gp120-mediated inflammation and apoptosis. *PLoS One, 8*(8), e70565.

Haslund-Vinding, J., McBean, G., Jaquet, V., & Vilhardt, F. (2017). NADPH oxidases in oxidant production by microglia: Activating receptors, pharmacology and association with disease. *British Journal of Pharmacology, 174*(12), 1733–1749.

Hayden, M. S., & Ghosh, S. (2004). Signaling to NF-kappaB. *Genes & Development, 18*(18), 2195–2224.

Hayden, M. S., & Ghosh, S. (2008). Shared principles in NF-kappaB signaling. *Cell, 132*(3), 344–362.

Heneka, M. T., & Landreth, G. E. (2007). PPARs in the brain. *Biochimica et Biophysica Acta, 1771*(8), 1031–1045.

Howard, R., Zubko, O., Bradley, R., Harper, E., Pank, L., O'Brien, J., Fox, C., Tabet, N., Livingston, G., Bentham, P., McShane, R., Burns, A., Ritchie, C., Reeves, S., Lovestone, S., Ballard, C., Noble, W., Nilforooshan, R., Wilcock, G., Gray, R., & Minocycline in Alzheimer Disease Efficacy Trialist, G. (2020). Minocycline at 2 different dosages vs placebo for patients with mild Alzheimer disease: A randomized clinical trial. *JAMA Neurology, 77*(2), 164–174.

Hu, S., Peterson, P. K., & Chao, C. C. (1997). Cytokine-mediated neuronal apoptosis. *Neurochemistry International, 30*(4-5), 427–431.

Hunot, S., Brugg, B., Ricard, D., Michel, P. P., Muriel, M. P., Ruberg, M., Faucheux, B. A., Agid, Y., & Hirsch, E. C. (1997). Nuclear translocation of NF-kappaB is increased in dopaminergic neurons of patients with parkinson disease. *Proceedings of the National Academy of Sciences of the United States of America, 94*(14), 7531–7536.

Inoue, A., Koh, C. S., Yahikozawa, H., Yanagisawa, N., Yagita, H., Ishihara, Y., & Kim, B. S. (1996). The level of tumor necrosis factor-alpha producing cells in the spinal cord correlates with the degree of Theiler's murine encephalomyelitis virus-induced demyelinating disease. *International Immunology, 8*(7), 1001–1008.

Jana, A., & Pahan, K. (2004a). Fibrillar amyloid-beta peptides kill human primary neurons via NADPH oxidase-mediated activation of neutral sphingomyelinase. Implications for Alzheimer's disease. *The Journal of Biological Chemistry, 279*(49), 51451–51459.

Jana, A., & Pahan, K. (2004b). Human immunodeficiency virus type 1 gp120 induces apoptosis in human primary neurons through redox-regulated activation of neutral sphingomyelinase. *The Journal of Neuroscience, 24*(43), 9531–9540.

Jana, A., & Pahan, K. (2007). Oxidative stress kills human primary oligodendrocytes via neutral sphingomyelinase: Implications for multiple sclerosis. *Journal of Neuroimmune Pharmacology, 2*(2), 184–193.

Jana, A., & Pahan, K. (2010a). Fibrillar amyloid-beta-activated human astroglia kill primary human neurons via neutral sphingomyelinase: Implications for Alzheimer's disease. *The Journal of Neuroscience, 30*(38), 12676–12689.

Jana, A., & Pahan, K. (2010b). Sphingolipids in multiple sclerosis. *Neuromolecular Medicine, 12*(4), 351–361.

Jana, M., & Pahan, K. (2012). Gemfibrozil, a lipid lowering drug, inhibits the activation of primary human microglia via peroxisome proliferator-activated receptor beta. *Neurochemical Research, 37*(8), 1718–1729.

Jana, M., & Pahan, K. (2013). Down-regulation of myelin gene expression in human oligodendrocytes by nitric oxide: implications for demyelination in multiple sclerosis. *Jounal of Clinical and Cellular Immunology, 4.*

Jana, M., Liu, X., Koka, S., Ghosh, S., Petro, T. M., & Pahan, K. (2001). Ligation of CD40 stimulates the induction of nitric-oxide synthase in microglial cells. *The Journal of Biological Chemistry, 276*(48), 44527–44533.

Jana, M., Dasgupta, S., Liu, X., & Pahan, K. (2002). Regulation of tumor necrosis factor-alpha expression by CD40 ligation in BV-2 microglial cells. *Journal of Neurochemistry, 80*(1), 197–206.

Jana, M., Anderson, J. A., Saha, R. N., Liu, X., & Pahan, K. (2005). Regulation of inducible nitric oxide synthase in proinflammatory cytokine-stimulated human primary astrocytes. *Free Radical Biology & Medicine, 38*(5), 655–664.

Jana, M., Jana, A., Liu, X., Ghosh, S., & Pahan, K. (2007). Involvement of phosphatidylinositol 3-kinase-mediated up-regulation of I kappa B alpha in anti-inflammatory effect of gemfibrozil in microglia. *Journal of Immunology, 179*(6), 4142–4152.

Jana, M., Palencia, C. A., & Pahan, K. (2008). Fibrillar amyloid-beta peptides activate microglia via TLR2: implications for Alzheimer's disease. *Journal of Immunology, 181*(10), 7254–7262.

Jana, A., Hogan, E. L., & Pahan, K. (2009). Ceramide and neurodegeneration: Susceptibility of neurons and oligodendrocytes to cell damage and death. *Journal of the Neurological Sciences, 278*(1-2), 5–15.

Keller, H., Mahfoudi, A., Dreyer, C., Hihi, A. K., Medin, J., Ozato, K., & Wahli, W. (1993). Peroxisome proliferator-activated receptors and lipid metabolism. *Annals of the New York Academy of Sciences, 684*, 157–173.

Khan, A., Jamwal, S., Bijjem, K. R., Prakash, A., & Kumar, P. (2015). Neuroprotective effect of hemeoxygenase-1/glycogen synthase kinase-3beta modulators in 3-nitropropionic acid-induced neurotoxicity in rats. *Neuroscience, 287*, 66–77.

Khasnavis, S., & Pahan, K. (2014). Cinnamon treatment upregulates neuroprotective proteins Parkin and DJ-1 and protects dopaminergic neurons in a mouse model of Parkinson's disease. *Journal of Neuroimmune Pharmacology, 9*(4), 569–581.

Khasnavis, S., Jana, A., Roy, A., Mazumder, M., Bhushan, B., Wood, T., Ghosh, S., Watson, R., & Pahan, K. (2012). Suppression of nuclear factor-kappaB activation and inflammation in microglia by physically modified saline. *The Journal of Biological Chemistry, 287*(35), 29529–29542.

Kim, K., Kleinman, H. K., Lee, H. J., & Pahan, K. (2017). Safety and potential efficacy of gemfibrozil as a supportive treatment for children with late infantile neuronal ceroid lipofuscinosis and other lipid storage disorders. *Orphanet Journal of Rare Diseases, 12*(1), 113.

Korbecki, J., Bobinski, R., & Dutka, M. (2019). Self-regulation of the inflammatory response by peroxisome proliferator-activated receptors. *Inflammation Research, 68*(6), 443–458.

Kovacs, K. A., Steinmann, M., Magistretti, P. J., Halfon, O., & Cardinaux, J. R. (2003). CCAAT/enhancer-binding protein family members recruit the coactivator CREB-binding protein and trigger its phosphorylation. *The Journal of Biological Chemistry, 278*(38), 36959–36965.

Kumar, P., Kumar, P., Khan, A., Deshmukh, R., & Lal Sharma, P. (2014). Role of neurosteroids in experimental 3-nitropropionic acid induced neurotoxicity in rats. *European Journal of Pharmacology, 723*, 38–45.

Labbadia, J., & Morimoto, R. I. (2015). The biology of proteostasis in aging and disease. *Annual Review of Biochemistry, 84*, 435–464.

Lassmann, H. (2018). Multiple sclerosis pathology. *Cold Spring Harbor Perspectives in Medicine, 8*(3).

Lee, S., Kim, S., Kang, H. Y., Lim, H. R., Kwon, Y., Jo, M., Jeon, Y. M., Kim, S. R., Kim, K., Ha, C. M., Lee, S., & Kim, H. J. (2020). The overexpression of TDP-43 in astrocytes causes neurodegeneration via a PTP1B-mediated inflammatory response. *Journal of Neuroinflammation, 17*(1), 299.

Lee, S. H., Lee, M., Ko, D. G., Choi, B. Y., & Suh, S. W. (2021). The role of NADPH oxidase in neuronal death and neurogenesis after acute neurological disorders. *Antioxidants (Basel), 10*(5).

Lehnardt, S., Massillon, L., Follett, P., Jensen, F. E., Ratan, R., Rosenberg, P. A., Volpe, J. J., & Vartanian, T. (2003). Activation of innate immunity in the CNS triggers neurodegeneration through a Toll-like receptor 4-dependent pathway. *Proceedings of the National Academy of Sciences of the United States of America, 100*(14), 8514–8519.

Lekstrom-Himes, J., & Xanthopoulos, K. G. (1998). Biological role of the CCAAT/enhancer-binding protein family of transcription factors. *The Journal of Biological Chemistry, 273*(44), 28545–28548.

Leroith, D., & Nissley, P. (2005). Knock your SOCS off! *The Journal of Clinical Investigation, 115*(2), 233–236.

Lewis, C. A., Manning, J., Rossi, F., & Krieger, C. (2012). The neuroinflammatory response in ALS: The roles of microglia and T cells. *Neurology Research International, 2012*, 803701.

Ma, M. W., Wang, J., Zhang, Q., Wang, R., Dhandapani, K. M., Vadlamudi, R. K., & Brann, D. W. (2017). NADPH oxidase in brain injury and neurodegenerative disorders. *Molecular Neurodegeneration, 12*(1), 7.

Mack, C. L., Vanderlugt-Castaneda, C. L., Neville, K. L., & Miller, S. D. (2003). Microglia are activated to become competent antigen presenting and effector cells in the inflammatory environment of the Theiler's virus model of multiple sclerosis. *Journal of Neuroimmunology, 144*(1-2), 68–79.

MacMicking, J., Xie, Q. W., & Nathan, C. (1997). Nitric oxide and macrophage function. *Annual Review of Immunology, 15*, 323–350.

Mayne, M., Bratanich, A. C., Chen, P., Rana, F., Nath, A., & Power, C. (1998). HIV-1 tat molecular diversity and induction of TNF-alpha: implications for HIV-induced neurological disease. *Neuroimmunomodulation, 5*(3-4), 184–192.

Michaud, M., Balardy, L., Moulis, G., Gaudin, C., Peyrot, C., Vellas, B., Cesari, M., & Nourhashemi, F. (2013). Proinflammatory cytokines, aging, and age-related diseases. *Journal of the American Medical Directors Association, 14*(12), 877–882.

Modi, K. K., Roy, A., Brahmachari, S., Rangasamy, S. B., & Pahan, K. (2015). Cinnamon and its metabolite sodium benzoate attenuate the activation of p21rac and protect memory and learning in an animal model of Alzheimer's disease. *PLoS One, 10*(6), e0130398.

Mogi, M., Harada, M., Kondo, T., Riederer, P., Inagaki, H., Minami, M., & Nagatsu, T. (1994). Interleukin-1 beta, interleukin-6, epidermal growth factor and transforming growth factor-alpha are elevated in the brain from parkinsonian patients. *Neuroscience Letters, 180*(2), 147–150.

Moncada, S., & Bolanos, J. P. (2006). Nitric oxide, cell bioenergetics and neurodegeneration. *Journal of Neurochemistry, 97*(6), 1676–1689.

Mondal, S., & Pahan, K. (2015). Cinnamon ameliorates experimental allergic encephalomyelitis in mice via regulatory T cells: implications for multiple sclerosis therapy. *PLoS One, 10*(1), e0116566.

Mondal, S., Martinson, J. A., Ghosh, S., Watson, R., & Pahan, K. (2012a). Protection of Tregs, suppression of Th1 and Th17 cells, and amelioration of experimental allergic encephalomyelitis by a physically-modified saline. *PLoS One, 7*(12), e51869.

Mondal, S., Roy, A., Jana, A., Ghosh, S., Kordower, J. H., & Pahan, K. (2012b). Testing NF-kappaB-based therapy in hemiparkinsonian monkeys. *Journal of Neuroimmune Pharmacology, 7*(3), 544–556.

Mondal, S., Dasarathi, S., & Pahan, K. (2017). Glyceryl tribenzoate: A flavoring ingredient, inhibits the adoptive transfer of experimental allergic encephalomyelitis via TGF-β: implications for multiple sclerosis therapy. *Journal of Clinical & Cellular Immunology, 8*(1).

Mondal, S., Jana, M., Dasarathi, S., Roy, A., & Pahan, K. (2018). Aspirin ameliorates experimental autoimmune encephalomyelitis through interleukin-11-mediated protection of regulatory T cells. *Science Signaling, 11*(558).

Mondal, S., Rangasamy, S. B., Roy, A., Dasarathy, S., Kordower, J. H., & Pahan, K. (2019). Low-dose maraviroc, an antiretroviral drug, attenuates the infiltration of T cells into the central nervous system and protects the nigrostriatum in hemiparkinsonian monkeys. *Journal of Immunology, 202*, 3412–3422.

Mondal, S., Kundu, M., Jana, M., Roy, A., Rangasamy, S. B., Modi, K. K., Wallace, J., Albalawi, Y. A., Balabanov, R., & Pahan, K. (2020). IL-12 p40 monomer is different from other IL-12 family members to selectively inhibit IL-12Rbeta1 internalization and suppress EAE. *Proceedings of the National Academy of Sciences of the United States of America, 117*(35), 21557–21567.

Mondal, S., Prieto, S., Rangasamy, S. B., Dutta, D., & Pahan, K. (2024). Nebulization of low-dose aspirin ameliorates Huntington's pathology in N171-82Q transgenic mice. *NeuroImmune Pharmacology and Therapeutics, 3*(1), 47–59.

Morales, I., Jimenez, J. M., Mancilla, M., & Maccioni, R. B. (2013). Tau oligomers and fibrils induce activation of microglial cells. *Journal of Alzheimer's Disease, 37*(4), 849–856.

Nagatsu, T., Mogi, M., Ichinose, H., & Togari, A. (2000). Changes in cytokines and neurotrophins in Parkinson's disease. *Journal of Neural Transmission, Suppl(60)*, 277–290.

Nalivaeva, N. N., Rybakina, E. G., Pivanovich, I., Kozinets, I. A., Shanin, S. N., & Bartfai, T. (2000). Activation of neutral sphingomyelinase by IL-1beta requires the type 1 interleukin 1 receptor. *Cytokine, 12*(3), 229–232.

Olanow, C. W., & Tatton, W. G. (1999). Etiology and pathogenesis of Parkinson's disease. *Annual Review of Neuroscience, 22*, 123–144.

Pacher, P., Beckman, J. S., & Liaudet, L. (2007). Nitric oxide and peroxynitrite in health and disease. *Physiological Reviews, 87*(1), 315–424.

Pahan, K. (2010). Neuroimmune pharmacological control of EAE. *Journal of Neuroimmune Pharmacology, 5*(2), 165–167.

Pahan, K. (2011). Immunomodulation of experimental allergic encephalomyelitis by cinnamon metabolite sodium benzoate. *Immunopharmacology and Immunotoxicology, 33*(4), 586–593.

Pahan, K. (2013). Multiple sclerosis and experimental allergic encephalomyelitis. *Jounal of Clinical and Cellular Immunology, 4*.

Pahan, K. (2015). Prospects of cinnamon in multiple sclerosis. *Journal of Multiple Sclerosis (Foster City), 2*(3), 1000149.

Pahan, K., & Schmid, M. (2000). Activation of nuclear factor-kB in the spinal cord of experimental allergic encephalomyelitis. *Neuroscience Letters, 287*(1), 17–20.

Pahan, K., Namboodiri, A. M., Sheikh, F. G., Smith, B. T., & Singh, I. (1997a). Increasing cAMP attenuates induction of inducible nitric-oxide synthase in rat primary astrocytes. *The Journal of Biological Chemistry, 272*(12), 7786–7791.

Pahan, K., Sheikh, F. G., Namboodiri, A. M., & Singh, I. (1997b). Lovastatin and phenylacetate inhibit the induction of nitric oxide synthase and cytokines in rat primary astrocytes, microglia, and macrophages. *The Journal of Clinical Investigation, 100*(11), 2671–2679.

Pahan, K., Sheikh, F. G., Namboodiri, A. M., & Singh, I. (1998). Inhibitors of protein phosphatase 1 and 2A differentially regulate the expression of inducible nitric-oxide synthase in rat astrocytes and macrophages. *The Journal of Biological Chemistry, 273*(20), 12219–12226.

Pahan, K., Khan, M., & Singh, I. (2000). Interleukin-10 and interleukin-13 inhibit proinflammatory cytokine-induced ceramide production through the activation of phosphatidylinositol 3-kinase. *Journal of Neurochemistry, 75*(2), 576–582.

Pahan, K., Jana, M., Liu, X., Taylor, B. S., Wood, C., & Fischer, S. M. (2002). Gemfibrozil, a lipid-lowering drug, inhibits the induction of nitric-oxide synthase in human astrocytes. *The Journal of Biological Chemistry, 277*(48), 45984–45991.

Park, E. J., Park, S. Y., Joe, E. H., & Jou, I. (2003). 15d-PGJ2 and rosiglitazone suppress Janus kinase-STAT inflammatory signaling through induction of suppressor of cytokine signaling 1 (SOCS1) and SOCS3 in glia. *The Journal of Biological Chemistry, 278*(17), 14747–14752.

Pasqualetti, P., Bonomini, C., Dal Forno, G., Paulon, L., Sinforiani, E., Marra, C., Zanetti, O., & Rossini, P. M. (2009). A randomized controlled study on effects of ibuprofen on cognitive progression of Alzheimer's disease. *Aging Clinical and Experimental Research, 21*(2), 102–110.

Patel, D., Roy, A., Kundu, M., Jana, M., Luan, C. H., Gonzalez, F. J., & Pahan, K. (2018). Aspirin binds to PPARalpha to stimulate hippocampal plasticity and protect memory. *Proceedings of the National Academy of Sciences of the United States of America, 115*(31), E7408–E7417.

Patel, D., Jana, A., Roy, A., & Pahan, K. (2019). Cinnamon and its metabolite protect the nigrostriatum in a mouse model of Parkinson's disease via astrocytic GDNF. *Journal of Neuroimmune Pharmacology, 14*(3), 503–518.

Peng, A. Y. T., Agrawal, I., Ho, W. Y., Yen, Y. C., Pinter, A. J., Liu, J., Phua, Q. X. C., Koh, K. B., Chang, J. C., Sanford, E., Man, J. H. K., Wong, P., Gutmann, D. H., Tucker-Kellogg, G., & Ling, S. C. (2020). Loss of TDP-43 in astrocytes leads to motor deficits by triggering A1-like reactive phenotype and triglial dysfunction. *Proceedings of the National Academy of Sciences of the United States of America, 117*(46), 29101–29112.

Piancone, F., La Rosa, F., Marventano, I., Saresella, M., & Clerici, M. (2021). The role of the inflammasome in neurodegenerative diseases. *Molecules, 26*(4).

Platanitis, E., & Decker, T. (2018). Regulatory networks involving STATs, IRFs, and NFkappaB in Inflammation. *Frontiers in Immunology, 9*, 2542.

Prorok, T., Jana, M., Patel, D., & Pahan, K. (2019). Cinnamic acid protects the nigrostriatum in a mouse model of Parkinson's disease via peroxisome proliferator-activated receptorα. *Neurochemical Research, 44*(4), 751–762.

Pupillo, E., Messina, P., Logroscino, G., Beghi, E., & Group, S. (2014). Long-term survival in amyotrophic lateral sclerosis: a population-based study. *Annals of Neurology, 75*(2), 287–297.

Qin, L., Li, G., Qian, X., Liu, Y., Wu, X., Liu, B., Hong, J. S., & Block, M. L. (2005). Interactive role of the toll-like receptor 4 and reactive oxygen species in LPS-induced microglia activation. *GLIA, 52*(1), 78–84.

Querfurth, H. W., & LaFerla, F. M. (2010). Alzheimer's disease. *The New England Journal of Medicine, 362*(4), 329–344.

Radi, R. (2013). Peroxynitrite, a stealthy biological oxidant. *The Journal of Biological Chemistry, 288*(37), 26464–26472.

Raha, S., Dutta, D., Roy, A., & Pahan, K. (2021). Reduction of Lewy body pathology by oral cinnamon. *Journal of Neuroimmune Pharmacology, 16*(3), 592–608.

Raha, S., Paidi, R. K., Dutta, D., & Pahan, K. (2024). Cinnamic acid, a natural plant compound, exhibits neuroprotection in a mouse model of Sandhoff disease via PPARalpha. *NeuroImmune Pharmacology and Therapeutics, 3*(1), 17–32.

Rangasamy, S. B., Corbett, G. T., Roy, A., Modi, K. K., Bennett, D. A., Mufson, E. J., Ghosh, S., & Pahan, K. (2015). Intranasal delivery of NEMO-binding domain peptide prevents memory loss in a mouse model of Alzheimer's disease. *Journal of Alzheimer's Disease, 47*(2), 385–402.

Rangasamy, S. B., Jana, M., Roy, A., Corbett, G. T., Kundu, M., Chandra, S., Mondal, S., Dasarathi, S., Mufson, E. J., Mishra, R. K., Luan, C. H., Bennett, D. A., & Pahan, K. (2018). Selective disruption of TLR2-MyD88 interaction inhibits inflammation and attenuates Alzheimer's pathology. *The Journal of Clinical Investigation, 128*(10), 4297–4312.

Rangasamy, S. B., Dasarathi, S., Nutakki, A., Mukherjee, S., Nellivalasa, R., & Pahan, K. (2021a). Stimulation of dopamine production by sodium benzoate, a metabolite of cinnamon and a food additive. *Journal of Alzheimer's Disease Reports, 5*(1), 295–310.

Rangasamy, S. B., Raha, S., Dasarathy, S., & Pahan, K. (2021b). Sodium benzoate, a metabolite of cinnamon and a food additive, improves cognitive functions in mice after controlled cortical impact injury. *International Journal of Molecular Sciences, 23*(1).

Rangasamy, S. B., Dutta, D., Mondal, S., Majumder, M., Dasarathy, S., Chandra, G., & Pahan, K. (2022). Protection of dopaminergic neurons in hemiparkinsonian monkeys by flavouring ingredient glyceryl tribenzoate. *NeuroImmune Pharmacology and Therapeutics, 1*(1), 7–22.

Reddy, J. K., & Mannaerts, G. P. (1994). Peroxisomal lipid metabolism. *Annual Review of Nutrition, 14*, 343–370.

Rogers, J., Kirby, L. C., Hempelman, S. R., Berry, D. L., McGeer, P. L., Kaszniak, A. W., Zalinski, J., Cofield, M., Mansukhani, L., Willson, P., et al. (1993). Clinical trial of indomethacin in Alzheimer's disease. *Neurology, 43*(8), 1609–1611.

Rothhammer, V., Mascanfroni, I. D., Bunse, L., Takenaka, M. C., Kenison, J. E., Mayo, L., Chao, C. C., Patel, B., Yan, R., Blain, M., Alvarez, J. I., Kebir, H., Anandasabapathy, N., Izquierdo, G., Jung, S., Obholzer, N., Pochet, N., Clish, C. B., Prinz, M., Prat, A., Antel, J., & Quintana, F. J. (2016). Type I interferons and microbial metabolites of tryptophan modulate astrocyte activity and central nervous system inflammation via the aryl hydrocarbon receptor. *Nature Medicine, 22*(6), 586–597.

Roy, A., & Pahan, K. (2011). Prospects of statins in Parkinson disease. *The Neuroscientist, 17*(3), 244–255.

Roy, A., & Pahan, K. (2013). Myelin basic protein-primed T helper 2 cells suppress microglial activation via AlphaVBeta3 integrin: implications for multiple sclerosis. *J Clin Cell Immunol, 7*, 158.

Roy, A., Fung, Y. K., Liu, X., & Pahan, K. (2006). Up-regulation of microglial CD11b expression by nitric oxide. *The Journal of Biological Chemistry, 281*(21), 14971–14980.

Roy, A., Jana, A., Yatish, K., Freidt, M. B., Fung, Y. K., Martinson, J. A., & Pahan, K. (2008). Reactive oxygen species up-regulate CD11b in microglia via nitric oxide: Implications for neurodegenerative diseases. *Free Radical Biology & Medicine, 45*(5), 686–699.

Ruddle, N. H., Bergman, C. M., McGrath, K. M., Lingenheld, E. G., Grunnet, M. L., Padula, S. J., & Clark, R. B. (1990). An antibody to lymphotoxin and tumor necrosis factor prevents transfer of experimental allergic encephalomyelitis. *The Journal of Experimental Medicine, 172*(4), 1193–1200.

Saha, R. N., & Pahan, K. (2006a). HATs and HDACs in neurodegeneration: A tale of disconcerted acetylation homeostasis. *Cell Death and Differentiation, 13*(4), 539–550.

Saha, R. N., & Pahan, K. (2006b). Regulation of inducible nitric oxide synthase gene in glial cells. *Antioxidants & Redox Signaling, 8*(5-6), 929–947.

Saha, R. N., & Pahan, K. (2006c). Signals for the induction of nitric oxide synthase in astrocytes. *Neurochemistry International, 49*(2), 154–163.

Saha, R. N., Liu, X., & Pahan, K. (2006). Up-regulation of BDNF in astrocytes by TNF-alpha: A case for the neuroprotective role of cytokine. *Journal of Neuroimmune Pharmacology, 1*(3), 212–222.

Saha, R. N., Jana, M., & Pahan, K. (2007). MAPK p38 regulates transcriptional activity of NF-kappaB in primary human astrocytes via acetylation of p65. *Journal of Immunology, 179*(10), 7101–7109.

Saijo, K., Winner, B., Carson, C. T., Collier, J. G., Boyer, L., Rosenfeld, M. G., Gage, F. H., & Glass, C. K. (2009). A Nurr1/CoREST pathway in microglia and astrocytes protects dopaminergic neurons from inflammation-induced death. *Cell, 137*(1), 47–59.

Salvador, C., Cortes, A. L., Pandiri, A. R., & Gimeno, I. M. (2021). Cytokine expression in the eye and brain of chickens following infection with a very virulent plus Marek's disease virus strain. *Veterinary Immunology and Immunopathology, 237*, 110277.

Sarlus, H., & Heneka, M. T. (2017). Microglia in Alzheimer's disease. *The Journal of Clinical Investigation, 127*(9), 3240–3249.

Sawada, M., Kondo, N., Suzumura, A., & Marunouchi, T. (1989). Production of tumor necrosis factor-alpha by microglia and astrocytes in culture. *Brain Research, 491*(2), 394–397.

Schapira, A. H. (1993). Mitochondrial complex I deficiency in Parkinson's disease. *Advances in Neurology, 60*, 288–291.

Schulze-Osthoff, K., Bakker, A. C., Vanhaesebroeck, B., Beyaert, R., Jacob, W. A., & Fiers, W. (1992). Cytotoxic activity of tumor necrosis factor is mediated by early damage of mitochondrial functions. Evidence for the involvement of mitochondrial radical generation. *The Journal of Biological Chemistry, 267*(8), 5317–5323.

Schwartz, C., Beck, K., Mink, S., Schmolke, M., Budde, B., Wenning, D., & Klempnauer, K. H. (2003). Recruitment of p300 by C/EBPbeta triggers phosphorylation of p300 and modulates coactivator activity. *The EMBO Journal, 22*(4), 882–892.

Shabab, T., Khanabdali, R., Moghadamtousi, S. Z., Kadir, H. A., & Mohan, G. (2017). Neuroinflammation pathways: A general review. *The International Journal of Neuroscience, 127*(7), 624–633.

Spiller, K. J., Restrepo, C. R., Khan, T., Dominique, M. A., Fang, T. C., Canter, R. G., Roberts, C. J., Miller, K. R., Ransohoff, R. M., Trojanowski, J. Q., & Lee, V. M. (2018). Microglia-mediated recovery from ALS-relevant motor neuron degeneration in a mouse model of TDP-43 proteinopathy. *Nature Neuroscience, 21*(3), 329–340.

Sriram, K., Matheson, J. M., Benkovic, S. A., Miller, D. B., Luster, M. I., & O'Callaghan, J. P. (2006). Deficiency of TNF receptors suppresses microglial activation and alters the susceptibility of brain regions to MPTP-induced neurotoxicity: role of TNF-alpha. *The FASEB Journal, 20*(6), 670–682.

Stommel, E. W., van Hoff, R. M., Graber, D. J., Bercury, K. K., Langford, G. M., & Harris, B. T. (2007). Tumor necrosis factor-alpha induces changes in mitochondrial cellular distribution in motor neurons. *Neuroscience, 146*(3), 1013–1019.

Suliman, H. B., Welty-Wolf, K. E., Carraway, M., Tatro, L., & Piantadosi, C. A. (2004). Lipopolysaccharide induces oxidative cardiac mitochondrial damage and biogenesis. *Cardiovascular Research, 64*(2), 279–288.

Takeuchi, H., Jin, S., Wang, J., Zhang, G., Kawanokuchi, J., Kuno, R., Sonobe, Y., Mizuno, T., & Suzumura, A. (2006). Tumor necrosis factor-alpha induces neurotoxicity via glutamate release from hemichannels of activated microglia in an autocrine manner. *The Journal of Biological Chemistry, 281*(30), 21362–21368.

Tarkowski, E., Andreasen, N., Tarkowski, A., & Blennow, K. (2003). Intrathecal inflammation precedes development of Alzheimer's disease. *Journal of Neurology, Neurosurgery, and Psychiatry, 74*(9), 1200–1205.

Taylor, D. E., Ghio, A. J., & Piantadosi, C. A. (1995). Reactive oxygen species produced by liver mitochondria of rats in sepsis. *Archives of Biochemistry and Biophysics, 316*(1), 70–76.

Tonnies, E., & Trushina, E. (2017). Oxidative stress, synaptic dysfunction, and Alzheimer's disease. *Journal of Alzheimer's Disease, 57*(4), 1105–1121.

Tortelli, R., Zecca, C., Piccininni, M., Benmahamed, S., Dell'Abate, M. T., Barulli, M. R., Capozzo, R., Battista, P., & Logroscino, G. (2020). Plasma inflammatory cytokines are elevated in ALS. *Frontiers in Neurology, 11*, 552295.

Trager, U., Andre, R., Magnusson-Lind, A., Miller, J. R., Connolly, C., Weiss, A., Grueninger, S., Silajdzic, E., Smith, D. L., Leavitt, B. R., Bates, G. P., Bjorkqvist, M., & Tabrizi, S. J. (2015). Characterisation of immune cell function in fragment and full-length Huntington's disease mouse models. *Neurobiology of Disease, 73*, 388–398.

Unger, M. S., Li, E., Scharnagl, L., Poupardin, R., Altendorfer, B., Mrowetz, H., Hutter-Paier, B., Weiger, T. M., Heneka, M. T., Attems, J., & Aigner, L. (2020). CD8(+) T-cells infiltrate Alzheimer's disease brains and regulate neuronal- and synapse-related gene expression in APP-PS1 transgenic mice. *Brain, Behavior, and Immunity, 89*, 67–86.

Ushio-Fukai, M. (2006). Localizing NADPH oxidase-derived ROS. *Science's STKE, 2006*(349), re8.

Uzasci, L., Nath, A., & Cotter, R. (2013). Oxidative stress and the HIV-infected brain proteome. *Journal of Neuroimmune Pharmacology, 8*(5), 1167–1180.

van Duijn, E., Kingma, E. M., & van der Mast, R. C. (2007). Psychopathology in verified Huntington's disease gene carriers. *The Journal of Neuropsychiatry and Clinical Neurosciences, 19*(4), 441–448.

Veerhuis, R., Boshuizen, R. S., Morbin, M., Mazzoleni, G., Hoozemans, J. J., Langedijk, J. P., Tagliavini, F., Langeveld, J. P., & Eikelenboom, P. (2005). Activation of human microglia by fibrillar prion protein-related peptides is enhanced by amyloid-associated factors SAP and C1q. *Neurobiology of Disease, 19*(1-2), 273–282.

Watanabe, Y., Taguchi, K., & Tanaka, M. (2020). Ubiquitin, autophagy and neurodegenerative diseases. *Cells, 9*(9).

Webb, S. J., McPherson, J. R., Pahan, K., & Koka, S. (2002). Regulation of TNF-alpha-induced IL-6 production in MG-63 human osteoblast-like cells. *Journal of Dental Research, 81*(1), 17–22.

Won, J. S., Im, Y. B., Key, L., Singh, I., & Singh, A. K. (2003). The involvement of glucose metabolism in the regulation of inducible nitric oxide synthase gene expression in glial cells: possible role of glucose-6-phosphate dehydrogenase and CCAAT/enhancing binding protein. *The Journal of Neuroscience, 23*(20), 7470–7478.

Wu, D. C., Teismann, P., Tieu, K., Vila, M., Jackson-Lewis, V., Ischiropoulos, H., & Przedborski, S. (2003). NADPH oxidase mediates oxidative stress in the 1-methyl-4-phenyl-1,2,3,6-tetrahydropyridine model of Parkinson's disease. *Proceedings of the National Academy of Sciences of the United States of America, 100*(10), 6145–6150.

Xu, J., Storer, P. D., Chavis, J. A., Racke, M. K., & Drew, P. D. (2005). Agonists for the peroxisome proliferator-activated receptor-alpha and the retinoid X receptor inhibit inflammatory responses of microglia. *Journal of Neuroscience Research, 81*(3), 403–411.

Xu, L., He, D., & Bai, Y. (2016). Microglia-mediated inflammation and neurodegenerative disease. *Molecular Neurobiology, 53*(10), 6709–6715.

Yang, X., Diaz, V., & Huang, H. (2022). The role of interferon regulatory factor 1 in regulating microglial activation and retinal inflammation. *International Journal of Molecular Sciences, 23*(23).

Ye, S., Zhou, T., Pan, D., Lai, Y., Yang, P., Chen, M., Wang, Y., Hou, Z., Ren, L., & Jiang, Y. (2016). Fullerene C(6)(0) derivatives attenuated microglia-mediated prion peptide neurotoxicity. *Journal of Biomedical Nanotechnology, 12*(9), 1820–1833.

Zhang, W., Wang, T., Qin, L., Gao, H. M., Wilson, B., Ali, S. F., Zhang, W., Hong, J. S., & Liu, B. (2004). Neuroprotective effect of dextromethorphan in the MPTP Parkinson's disease model: role of NADPH oxidase. *The FASEB Journal, 18*(3), 589–591.

Zhang, J., Zhang, X., Li, L., Bai, L., Gao, Y., Yang, Y., Wang, L., Qiao, Y., Wang, X., & Xu, J. T. (2022). Activation of double-stranded RNA-activated protein kinase in the Dorsal Root Ganglia and Spinal Dorsal Horn regulates neuropathic pain following peripheral nerve injury in rats. *Neurotherapeutics, 19*(4), 1381–1400.

Zhao, X., Zhang, Y., Strong, R., Grotta, J. C., & Aronowski, J. (2006). 15d-Prostaglandin J2 activates peroxisome proliferator-activated receptor-gamma, promotes expression of catalase, and reduces inflammation, behavioral dysfunction, and neuronal loss after intracerebral hemorrhage in rats. *Journal of Cerebral Blood Flow and Metabolism, 26*(6), 811–820.

Zhou, H., Lapointe, B. M., Clark, S. R., Zbytnuik, L., & Kubes, P. (2006). A requirement for microglial TLR4 in leukocyte recruitment into brain in response to lipopolysaccharide. *Journal of Immunology, 177*(11), 8103–8110.

Ziff, O. J., Clarke, B. E., Taha, D. M., Crerar, H., Luscombe, N. M., & Patani, R. (2022). Meta-analysis of human and mouse ALS astrocytes reveals multi-omic signatures of inflammatory reactive states. *Genome Research, 32*(1), 71–84.

The Blood–Brain Barriers

13

William A. Banks and Michelle A. Erickson

Abstract

The blood–brain barriers (BBBs) are mainly located at the levels of the vasculature, choroid plexus, circumventricular organs, and leptomeninges and play multiple roles in neuroimmunology. The ability of the BBBs to separate the blood and its contents from the central nervous system (CNS) is largely responsible for the CNS being an immune-privileged region. However, the BBB then revises this separation in a regulated way by a variety of mechanisms, including the ability to transport cytokines, control the trafficking of immune cells into the brain, and to itself secrete into the blood and the CNS immunoactive substances. The BBB thereby participates in a number of neuroimmune axes that allow communication between the CNS and the peripheral immune cells. Failure of the highly regulated activities of the BBBs can be both a cause and consequence of immune diseases.

Keywords

Blood-brain barrier · Neuroimmune · Cytokine · Opiate · Transport · Transcellular diffusion · Active transport · Facilitated diffusion · Virus · Adsorptive endocytosis · Adsorptive transcytosis · Neurovascular unit · Brain endothelial cell · Interleukin · Tumor necrosis factor · Endothelin · Brain · Central nervous system · Immune cell · AIDS

W. A. Banks (✉) · M. A. Erickson
Department of Medicine, Division of Gerontology and Geriatric Medicine, GRECC, Veterans Affairs Puget Sound Health Care System and University of Washington School of Medicine, Seattle, WA, USA
e-mail: wabanks1@uw.edu

13.1 Introduction

The roles played by the blood–brain barrier (BBB) in neuroimmunology are diverse and ever expanding. Conceptually, the BBB is those processes that restrict, control, or otherwise influence the exchange of substances between the peripheral circulation and the brain interstitial fluid and cerebrospinal fluid (CSF). Anatomically, there are multiple BBBs: the barrier formed by monolayers of endothelial cells (the vascular BBB), the barrier formed by ependymal/epithelial cells (the blood–CSF barrier), a barrier formed by tanycytes interposed between the circumventricular organs and the adjacent brain tissue, and a barrier formed by epithelial-like cells of the arachnoid layer of the leptomeninges. It is the restrictive properties of the BBB that limit and control the trafficking of immune cells and prevent the unrestricted leakage of immune active substances from the blood into the brain that renders the central nervous system (CNS) an immune-privileged area. But the BBB also transports immunoactive substances between the blood and CNS, responds to immunoactive substances secreted into the blood or CNS fluids, and secretes substances into those compartments. These last three processes of transport, responsiveness, and secretion allow the BBBs to be in constant cross-talk with other cells in both the CNS and periphery in a formation referred to as the neurovascular unit (NVU). The BBBs both respond to and influence the CNS and peripheral microenvironment and do so by reacting to immunoactive substances, including cytokines, chemokines, prostaglandins, and nitric oxide. Several consequences arise from this cross-talk, including that under physiological conditions the metabolic needs of the brain are met by the BBBs. The roles the BBBs play help to keep the CNS informed of peripheral events, including immune events, and are central to the formation and functioning of neuroimmune axes. When any of these physiological consequences are violated, diseases can arise and the BBB can be a target, a cause, or a conduit to treatment for those diseases.

H. E. Gendelman, T. Ikezu (eds.), *Neuroimmune Pharmacology and Therapeutics*,
https://doi.org/10.1007/978-3-031-68237-7_13

13.1.1 Structures and Functions of the BBB

Evidence for an interface between the circulation and the central nervous system (CNS) dates back to the end of the nineteenth century (Bradbury, 1979). The best known of those early studies were done by a young Paul Erlich who found that some dyes did not stain the brain after their peripheral injection. Erlich concluded erroneously that the lack of staining was because these dyes did not bind to brain tissue. Several decades later, Goldmann, a student of Erlich's, found that these dyes could stain the brain when injected intrathecally. Thus, these studies were reinterpreted as evidence in favor of some sort of barrier between the CNS and blood.

The location and nature of that barrier were controversial through much of the twentieth century. Elegant studies by Davson and colleagues identified the barrier at the vascular level (Davson, 1976). However, alternative opinions were held until Reese and coworkers conducted classic studies with the electron microscope in the late 1960s (Brightman & Reese, 1969; Reese & Karnovsky, 1967). Previous work had shown no difference between vascular beds of peripheral tissues and the CNS when studied grossly or at the light microscope level. However, Reese and coworkers found numerous differences at the ultrastructural level. These included a much reduced rate of pinocytosis, an absence of intracellular fenestrations, and the presence of intercellular tight junctions. The tight junctions, low rate of pinocytosis, and low number of intracellular fenestrations effectively eliminate intercellular gaps and pores. This, in turn, essentially eliminates the production of a plasma-derived ultrafiltrate and hence the leakage of serum proteins into the brain.

From this single change of the lack of a production of an ultrafiltrate, a large number of consequences for CNS function evolve. Obviously, it is the basis of the restriction of protein access that first defined the BBB in the late ninenteenth century. The need for a classic lymphatic system is eliminated, but this means that the CNS needs other methods to rid itself of the free water and wastes produced by metabolism and the secretions of the choroid plexus. Without the production of an ultrafiltrate, the CNS depends on other methods to extract nourishment from the blood. The BBB addresses this need with a large number of selective transporters for substances from electrolytes to regulatory proteins (Davson & Segal, 1996a, b, c). Because the CNS is not equipped to handle an ultrafiltrate, the reintroduction of a leaky BBB, as with a hypertensive crisis, can result in increased intracranial pressure and encephalopathy (Al-Sarraf & Phillip, 2003; Johansson, 1989; Mayhan & Heistad, 1985).

13.1.2 The Various BBBs

The BBB is not a single barrier, but several. In contrast with the testis–blood barrier, which consists of several barriers in series (Holash et al., 1993; Neaves, 1977), the BBBs exist in parallel. The most studied of these barriers are the vascular barrier and the choroid plexus. Often, the terms BBB and blood–cerebrospinal fluid (CSF) barrier are used to refer specifically to the vascular barrier and the barrier formed at the choroid plexus, respectively. The least studied barriers are the barriers formed by tanycytes at the circumventricular organs (CVO), other specialized neural barriers, such as the blood–retinal (Neuwelt et al., 2008), and the barrier of epithelial-like cells of the arachnoid that separates the dural and pial layers of the meninges (Derk et al., 2021).

13.1.2.1 Vascular BBB

The vascular BBB occurs because of the modifications noted by Reese and coworkers in the endothelial cells that comprise the capillary bed and line the venules and arterioles of the CNS (Fig. 13.1a). It is likely that these three regions are highly specialized. For example, immune cells primarily cross at the venules, and most of the classic transporters are located at the capillaries (Engelhardt & Wolburg, 2004). Recent single-cell RNAseq studies have defined distinct transcriptional profiles of endothelial cells from different segments of the brain vasculature further supporting that there are functional differences in endothelial cells from capillaries versus arterioles versus veins (Schaeffer & Iadecola, 2021) No CNS cell is more than about 40 microns from a capillary. This means that a substance that can cross the vascular BBB can immediately access the entire CNS. Substances that cross the vascular BBB can be either flow-dependent or not dependent on flow rate. A flow-dependent substance is one in which the BBB extracts the substance so rapidly that the concentration of the substance is lower at the venous end of the capillary than at the arteriolar end (Kety, 1987). In such a circumstance, increasing the flow rate of blood will increase the amount of the substance entering the brain. Glucose is an example of a flow-dependent substance (Rapoport et al., 1981). A particularly active brain region has its increased demand for glucose met by an increase in the regional blood flow. In contrast, the transport of a cytokine such as tumor necrosis factor-alpha (TNF) is not flow-dependent. Only a small percent of the TNF in the blood is extracted by the brain via the saturable transporter for TNF located at the BBB (Gutierrez et al., 1993). Alterations of blood flow within physiological limits do not alter the uptake of TNF from blood by the brain. However, extreme changes in the rate of blood flow or capillary tortuosity can result in rheological changes such as the loss of laminar flow. Such alterations likely occur in stroke, AIDS, and Alzheimer's dis-

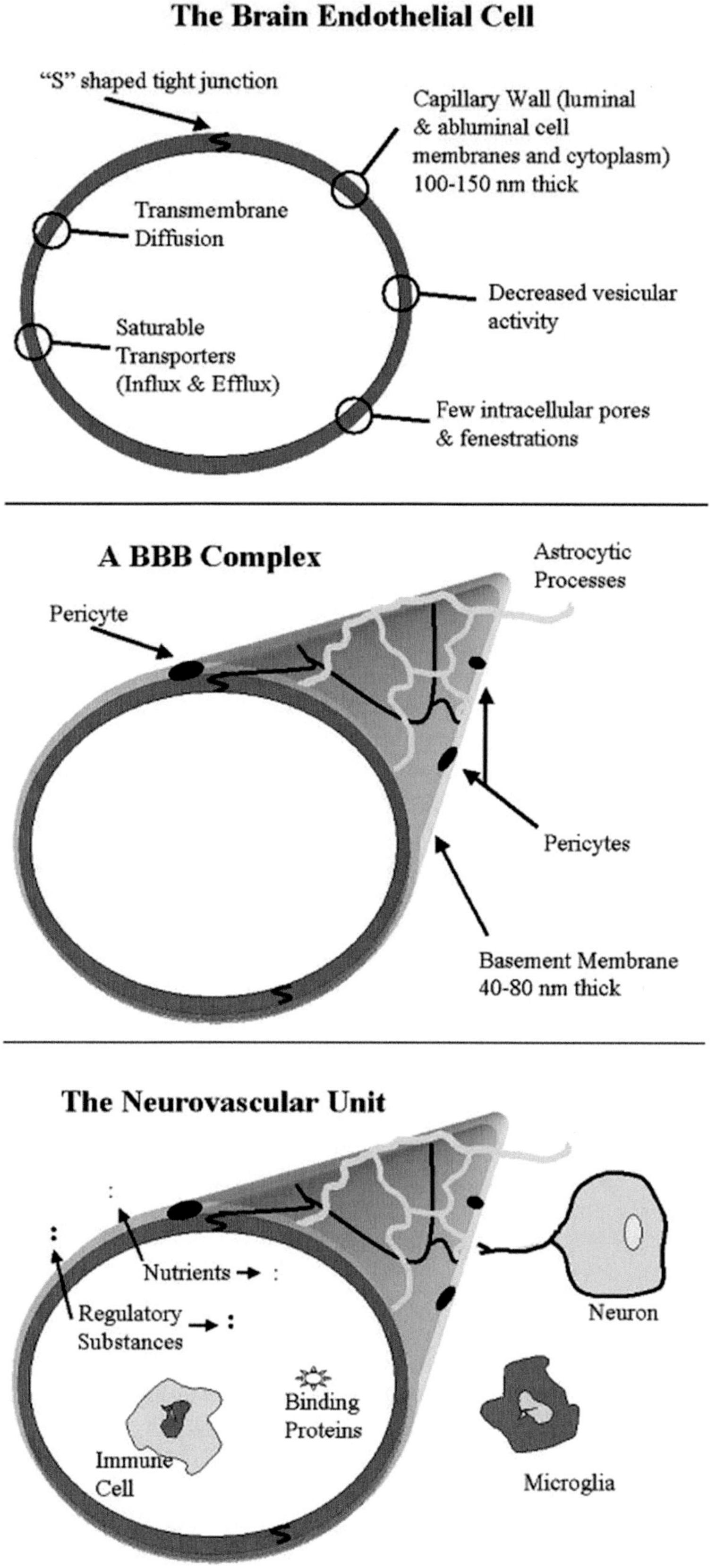

Fig. 13.1 The Vascular Blood–Brain Barrier: The upper panel illustrates the brain endothelial cell. This is the functional and anatomical site of both barrier function and saturable and non-saturable mechanisms of passage. The major modifications allowing both barrier function and selective penetration of substances are indicated. The middle panel illustrates other cell types and structures important in BBB function. Pericytes are embedded in a basement membrane and astrocytes form a structure over the capillary bed above the pericytes and basement membrane. Both cell types are in paracellular communication with the brain endothelial cells. The lower panel illustrates the neurovascular unit, a concept that emphasizes the integration of peripheral, BBB, and central interactions

ease (de la Torre & Mussivand, 1993; Nelson et al., 1999). This may result in impaired permeation of flow-dependent and non-flow dependent substances.

The vascular BBB has regional variations in terms of function and susceptibilities to disease. Huber et al. noted that some regions of the brain suffer larger and earlier disruptions to their barriers during diabetes mellitus than others (Huber, 2006, 2008). Many peptides and regulatory substances have unique regional variations in their transport rates across the BBB (Banks et al., 1996; Banks & Kastin, 1998). It is assumed that these reflect on brain function. For example, the brain region with the highest rate of transport of leptin is the region of the arcuate nucleus, an area important in leptin-mediated control of feeding (Banks et al., 1996; Schwartz et al., 2000).

13.1.2.2 Choroid Plexus and Meningeal Blood–CSF Barriers

The choroid plexus are bags composed of monolayers of epithelial cells that project into the ventricles and contain a capillary plexus (Johanson, 1988). The capillaries do not have a barrier function and so produce an ultrafiltrate that fills the bag. The epithelial cells have tight junctions and so prevent the ultrafiltrate from entering the ventricular space. Unlike the capillaries, the epithelial cells of the choroid plexus have a high rate of vesicular turnover, which is responsible for the production of CSF. However, the CSF is not an ultrafiltrate but a secreted substance. The choroid plexus also has many selective transport systems, some of which are specific to it or are enriched in comparison to the vascular BBB. In some cases, the choroid plexus seems not to be a system complementary to the vascular BBB, but one that is contrary to it. For example, efflux of amyloid beta peptide by the vascular BBB is impaired with aging but is increased with aging at the choroid plexus (Pascale et al., 2011). Thus, the choroid plexus should not be thought of as a secondary barrier but as an independent system with its unique characteristics. Another important blood–CSF barrier is formed by arachnoid barrier cells, which are epithelial-like cells that separate the permeable vasculature of the dura from the arachnoid CSF (Derk et al., 2021). The arachnoid barrier expresses tight junction proteins like claudin 11, and high levels of efflux transporters such as p-glycoprotein and breast cancer resistance protein at its apical membranes (Brochner et al., 2015; Yasuda et al., 2013), conferring a size-selective diffusion barrier and a barrier to endogenous and exogenous efflux transporter substrates.

13.1.2.3 Barriers at Circumventricular Organs

The CNS of mammals contains seven regions of the brain where the vasculature does not fully participate in a BBB (Gross et al., 1987). These regions have at least one side that faces a ventricle and so are termed circumventricular organs (CVOs). Together, they comprise about 0.5% of the brain by weight. Their capillaries allow the production of an ultrafiltrate and so their cells are in more intimate contact with the circulation. They are known to play vital roles as sensing organs for critical peripheral events; for example, they act as emetic centers and are important in blood pressure modulation (Ferguson, 1991; Johnson & Deckwerth, 1993). Some CVOs, such as the median eminence and neurohypophysis, are important sites of peptide secretion into the bloodstream and have a higher degree of leakage versus sensory CVOs (Morita & Miyata, 2012; Verheggen et al., 2020) CVOs can relay their signals to the rest of the brain by neurons that project from them to distant brain regions or project to them from other brain regions. However, the mixing of their interstitial fluids with that of adjacent brain tissue or CSF has been shown to be limited in most studies (Rethelyi, 1984, Plotkin et al., 1996, Peruzzo et al., 2000). Diffusion through brain tissue is poor and this alone would tend to produce a limit to mixing within a few hundred microns of the CVO (Cserr & Berman, 1978; de Lange et al., 1995). However, the major factor preventing leakage of substances from the CVO into the adjacent CSF and brain tissue is a physical diffusion barrier. The epithelial cells that line the ventricles form tight junctions when they are over CVOs thus limiting CVO-to-CSF diffusion. A functional barrier formed by bands of tanycytes also exists for the diffusion of substances from the CVO to the adjacent brain region (Peruzzo et al., 2000; Plotkin et al., 1996; Rethelyi, 1984). These tanycytes themselves can project to the cells and vessels within the CVO and can respond dynamically to relay information to adjacent regions of the brain that have BBBs (Langlet et al., 2013). A recent review by Rodriguez (Rodriguez et al., 2010) has explored the various aspects of the CVO barriers.

13.1.2.4 Other Specialized Neural Barriers

These barriers include the blood–retinal barrier, blood–spinal cord barrier, and blood–nerve barriers (for a fuller listing, see (Neuwelt et al., 2008)). These barriers generally include tight junctions participating in a vascular barrier but exhibit varying degrees of leakiness. They can also vary markedly in transporter activity from the vascular BBB during both health and in response to disease (Prockop et al., 1996, Pan & Kastin, 2003, Pan et al., 2008a, b).

13.1.3 Concept of the Neurovascular Unit and Comparison to Peripheral Vascular Beds

The endothelial cell is the anatomical unit forming the vascular BBB and containing its various saturable transporters (Fig. 13.1). Capillary beds from peripheral tissues have numerous intracellular and intercellular pores and fenestrations and high rates of pinocytosis that account for their leakiness. The brain endothelial cell engages in comparatively little pinocytosis, has few intracellular pores or fenestrations, and intercellular pores or gaps are eliminated because of tight junctions.

However, the brain endothelial cell does not function in isolation. The luminal (blood) side of the capillary is coated with a glycocalyx that can have barrier functions (Kutuzov et al., 2018) but can also be key to chemokine transport (Quaranta et al., 2022). The abluminal (brain) side of the capillary is encased in a basement membrane 40–80 nm thick. This membrane is not thought to act as a barrier to molecules but may influence the trafficking of viral-sized particles (Muldoon et al., 1999; Zhang et al., 2021) and immune cells (Engelhardt & Coisne, 2011). It also holds pericytes in close approximation to the endothelial cell (Balabanov & Dore-Duffy, 1998). The pericyte is a pluripotent cell and a modulator of BBB function (Deli et al., 2005; Dore-Duffy, 2008; Dore-Duffy et al., 2000). It, like astrocytes and microglia, secretes cytokines and other immune active substances both constitutively and when induced (Kovac et al., 2011). In the rat hippocampus, astrocytes project endfeet that surrounds the capillary in what looks at the ultrastructural level like a complete covering, albeit without intercellular tight junctions (Mathiisen et al., 2010). Astrocytes and pericytes both secrete substances that induce tight junction formation in endothelial cells (Daneman et al., 2010; Deli et al., 2005). All the major cell types of the NVU (pericytes, astrocytes, microglia, and endothelial cells) secrete a variety of substances, including cytokines, into their local environment (Banks, 2014; Fabry et al., 1993; Nath et al., 1999). Pericytes and astrocytes play interrelated but distinct roles in BBB function. For example, pericytes are the primary protectors of the blood–retinal and blood–brain barriers during glycemic stress (Nakaoke et al., 2007; Romeo et al., 2002).

It is clear that immune cell trafficking occurs across the normal BBB (Greenwood et al., 2011). One study found that about one in 5000 intravenously injected lymphocytes resided in the brain at any given time and that uptake was affected by the genetic strain of mouse used and immune activation (Banks et al., 2012). Microglia, which are yolk sac-derived cells that populate the brain during embryogenesis, are thought to be long-lived and self-renewing in the healthy brain. Monocytes can also enter the brain by crossing the BBB, especially following inflammatory insults or injuries. However, to what extent they can enter brain parenchyma to adopt a microglia-like phenotype versus that of a perivascular macrophage is controversial (Prinz & Priller, 2014) (Williams & Hickey, 1995). Other immune cells also enter and exit the CNS at unknown rates and frequencies as influenced by yet-to-be-determined factors. Clearly, secretions of prostaglandins, nitric oxide, and cytokines from each of these cells are important for intercellular communication and can influence endothelial cell permeability (Chao et al., 1994; Nath et al., 1999; Shafer & Murphy, 1997).

The concept of the neurovascular unit (NVU) emphasizes the interactive role that cells and events within the CNS and in the circulation play on BBB permeability as well as the consequences of the permeability itself. The NVU includes other factors long known to influence the penetration of substances across the BBB, such as degradation, sequestration, and serum protein binding. The encompassing concept of the NVU is particularly useful when considering the next section, the mechanisms of transport across the BBB.

13.2 Mechanisms of Transport Across the BBB

Substances can enter or exit the CNS by a variety of mechanisms. Some of these mechanisms are operational in both the blood-to-brain (influx) and the brain-to-blood (efflux) directions, whereas others are unidirectional.

13.2.1 Blood to CNS

Saturable and nonsaturable modes predominate influx. Within each of these categories are a diverse number of mechanisms. These different mechanisms tend to favor certain groups or types of substances.

13.2.1.1 Nonsaturable Passage

A hallmark of nonsaturable passage is that the percentage of material crossing into the CNS is not affected by the amount of material available for transport. The two main mechanisms of nonsaturable passage are transcellular diffusion and the extracellular pathways. The former is much better studied, and its principles are widely applied by the pharmaceutical industry for the development of CNS drugs; the latter has received much less attention.

Transcellular Diffusion

The most studied nonsaturable mechanism by which small molecules cross the BBB is by membrane diffusion, also sometimes termed transmembrane, transendothelial, or transcellular diffusion (Rapoport, 1976). The major determinant of passage is the degree to which the substance is lipid soluble. A substance that is too lipid soluble will be unable to repartition into the brain's interstitial fluid and so becomes trapped in the cell membranes of the BBB. A ratio of about 10:1 in favor of lipid versus aqueous solubility is near ideal for maximal passage across the BBB. The second most important determinant is molecular weight with passage being favored for smaller molecules. Other physicochemical determinants, such as charge, can occasionally become dominant for specific compounds. Work by Lipinksi in Caco-2 cells, an immortalized cell line derived from a gastrointestinal cancer, clearly shows that smaller, less charged, more lipid-soluble drugs are favored in transmembrane diffusion (Lipinski et al., 1997). Many exogenous substances, including many drugs with CNS activity, enter the brain predominantly by way of transcellular diffusion. Morphine and ethanol are prime examples of common substances that cross the BBB by this mechanism (Oldendorf, 1974).

Although higher molecular weight is an impedance to transcellular diffusion at the BBB, there seems to be no absolute molecular weight cut-off, despite many reviews stating that there is. A previous study that had thought to define such an absolute limit had discovered, in retrospect, early evidence for an efflux system (Levin, 1980). The largest substance to date noted to have a measurable uptake by the brain by transcellular diffusion is cytokine-induced neutrophil chemoattractant-1 (CINC1), with an MW of about 7.8 kDa (Pan & Kastin, 2001a, b). A surprisingly large number of small, lipid-soluble compounds cross the BBB at a rate considerably greater or lesser than that predicted by their physicochemical characteristics (Oldendorf, 1971, 1974). Binding to serum proteins and efflux systems are major factors decreasing influx, and the presence of a saturable blood-to-brain transporter is a major factor increasing influx.

Extracellular Pathways

Albumin derived from serum is present in small amounts in the CSF, showing that the BBB is not absolute. The amount of protein in CSF, however, is very small, being about 0.5%, or 1/200th, of that in plasma. The CSF is not an ultrafiltrate but a secreted fluid. This means that the relative and absolute concentrations of proteins, electrolytes, minerals, and other substances can differ tremendously from that of plasma. The extracellular pathways are another avenue by which substances can enter the CNS (Balin et al., 1986; Broadwell, 1993). These represent what has sometimes been termed "functional leaks" at discreet areas of the brain, including the large vessels of the pial surface and subarachnoid space, the circumventricular organs, the nasal epithelium, the sensory ganglia of spinal and cranial nerves, and some deep brain regions, such as the nucleus tractus solitarius (Broadwell & Banks, 1993).

The amount of a substance that enters the brain by the extracellular pathways is small. However, this route may be therapeutically relevant for compounds that have favorable peripheral pharmacokinetics, such as a long serum half-life and a small volume of distribution (Banks, 2004). Antibodies, erythropoietin, and enzymes can access the brain by way of the extracellular pathways (Kozlowski et al., 1992, Banks et al., 2004a, b, c, 2005a, b, 2007, Grubb et al., 2008), and this may underlie their therapeutic benefits when given in high doses (Alafaci et al., 2000; Ehrenreich et al., 2002; Erbyraktar et al., 2003; Farr et al., 2003; Grubb et al., 2008; Hock et al., 2003; Janus et al., 2000; Morgan et al., 2000).

13.2.1.2 Receptor-Mediated and Saturable Transport Systems

Saturable processes represent a diverse group of mechanisms. Here, we include diapedesis and adsorptive endocytosis/transcytosis, two other mechanisms of passage across the BBB that share some characteristics with more classical saturable systems.

Active Transport Versus Facilitated Diffusion

Saturable transporters (Yeagle, 1987) can be divided into those that require energy (active transport) and those that do not (facilitated diffusion). Both are dependent on a protein that acts as the transporter, may have co-factors, and be modulated by physiologic and disease processes. Energy-requiring systems can be unidirectional; that is, they may have only an influx or efflux component. Non-energy requiring saturable transport (facilitated diffusion) is bidirectional; that is, it transports substances in both directions with net flux being from the side of higher concentration to the side of lower concentration.

Most of the classic saturable transporters at the BBB are facilitated diffusion systems (Kaur et al., 1992). For example, GLUT-1, the transporter for glucose, is a facilitated diffusion transporter. If the level of glucose is artificially raised above that of serum (or if radioactive glucose is introduced into the CNS but not the serum), efflux of glucose can be shown.

Transcytotic Versus Transcellular Transport

Saturable transporters can also be categorized based on whether they use pores/channels or vesicles to transport their ligands across the BBB. In the pore system, the molecule crosses from one side of the cell membrane to the other by passing through a cavity in the transporter protein. The substance is thus transported either into or out of the cytoplasm of the BBB cell; a second set of transporters on the opposing cell membrane completes the transfer across the BBB or the substance can rely on membrane diffusion. With vesicular transport, often referred to as receptor-mediated transcytosis, the transported substance adheres to a binding site, usually a glycoprotein. Invagination then produces a vesicle that is then routed to the opposite membrane, and the contents of the vesicle are released from the cell surface. A specificity of transport distinguishes these vesicles from macropinocytosis whose reduction is a defining characteristic of the BBB (Reese & Karnovsky, 1967). Many modern CNS drug-delivery strategies have focused on using vesicular transport to deliver relatively BBB-impermeable substances, such as antibodies, to the brain. Such strategies usually entail the engineering and integration of a protein or antibody fragment that binds a vesicular transporter, an example of such a transporter is the transferrin receptor. While conceptually straightforward, optimization of this strategy is complicated, in that transporter binding must be optimized to permit both high-affinity binding in circulation and release of the intact ligand into the CNS. Binding conditions must also preserve the transcytotic routing of the vesicles and not induce rerouting to degradative cellular compartments such as the lysosome (Pulgar, 2018).

Most small molecules, such as glucose, electrolytes, and amino acids, use pores or channels. Pore systems may be either active or facilitated diffusion systems. Vesicular transporters, on the other hand, are energy-requiring and so are characterized by unidirectional transport. The best described of these vesicular-dependent systems is receptor-mediated transcytosis and is characterized by clathrin- and transglutaminase dependence (Davies et al., 1980). However, nonclathrin-dependent vesicles, such as podocytes, may also be present at the BBB.

It is reasonable to assume that very large molecules would be required to use vesicles rather than pores and channels to cross, but the molecular weight at which vesicles would be requisite is not known. It has been proposed that interleukin-2 (IL-2) is transported (Drach et al., 1996) by p-glycoprotein (P-gp). As P-gp is a pore system (Begley, 2004), IL-2 would be the largest substance currently known to be transported by a pore system. Peptides much smaller than IL-2 are known to cross by vesicular-dependent pathways (Shimura et al., 1991; Terasaki et al., 1992). It is clear, then, that a moderately large mass of the ligand alone does not dictate the need for vesicular transport.

Diapedesis of Immune Cells

A major shift in thinking about the relation of immune cells to the CNS and BBB has occurred over the last few years. The CNS was once viewed as separate from the immune system and sterile in terms of immune cell occupancy except under conditions of brain infection. As reviewed above, it is now clear that immune cells patrol the normal CNS, but the accessibility depends, in part, on the compartmentalization of CNS regions by brain barriers (Engelhardt et al., 2017). For example, activated T-cells can cross postcapillary venules of the vascular BBB, albeit at low efficiency. This process is inefficient because brain endothelial cells lack basal expression of components that facilitate rapid interactions with leukocytes (Mapunda et al., 2022). With neuroinflammation, brain endothelial cells upregulate the expression of selectins, cell adhesion molecules, and chemokine transporters that can facilitate the trafficking of many types of leukocytes into brain parenchyma. Immune surveillance also occurs readily at the brain borders, particularly in regions lacking a complete BBB such as the dura mater and the choroid plexus stroma (Engelhardt, 2008; Greenwood et al., 2011). The brain barriers that compartmentalize these regions can also participate in immune cell trafficking. For example, choroid plexus epithelial cells participate in immune cell trafficking to CSF and distal parenchymal sites of the brain and spinal cord (Kunis et al., 2013). Immune cells also survey the CSF and are mostly CD4+ central memory T-cells (Engelhardt & Ransohoff, 2005). The presence of CNS immune surveillance is likely important for protecting the brain from pathogens, as evidenced by progressive multifocal leukoencephalopathy (PML), a rare side effect in individuals taking the multiple sclerosis drug natalizumab, typically in combination with other immunosuppressive therapies and in individuals with HIV (Clifford et al., 2010). PML is caused by the reactivation of the JC virus, a very common infection that is typically asymptomatic (Paz et al., 2018). Natalizumab inhibits T-cell trafficking into the brain, which is protective in MS but may also inhibit CNS immune surveillance that is necessary for preventing JC virus-induced PML (Clifford et al., 2010). Many important questions remain on the origins, subtypes, and predominant anatomical routes of leukocyte trafficking into the CNS in health and disease.

Adsorptive Endo- and Trans-Cytosis

Adsorptive endocytosis occurs when a glycoprotein or proteoglycan on the brain's endothelial surface binds another glycoprotein, glycolipid, or cationic molecule in a ligand-like fashion (Broadwell, 1989; Broadwell et al., 1988). This second molecule (the ligand) may be free or attached to the surface of a virus or immune cell (Mellman et al., 1986). The binding can initiate endocytosis with the subsequent vesicle having several potential fates (Banks & Broadwell, 1994). In some cases, the vesicle is routed to lysosomes, the glycoprotein destroyed, and the vesicle rerouted to the endothelial cell surface for discharge of contents. In other cases, the vesicle can be routed to the golgi complex and endoplasmic reticulum. In other cases still, the vesicle can be discharged at the endothelial cell surface opposite to that of uptake. In this case, the vesicle has crossed the width of the endothelial cell in a transcytotic event and hence crossed the BBB. What determines the routing of these vesicles is largely unknown, but at least some vesicles can engage in more than one fate (Broadwell, 1993). It may be that binding of a large amount of glycoprotein to the endothelial cell can overwhelm the lysosomal pathway and result in the vesicles being routed to the transcytosis or golgi complex pathways.

Several principles of adsorptive endocytosis and transcytosis (also termed adsorptive-mediated transcytosis) are clear. Many of the glycoprotein ligands are toxic and endocytosis may represent a mechanism to rejuvenate or repair the membrane (Raub & Audus, 1990; Vorbrodt, 1994; Westergren & Johansson, 1993). Viruses and other pathogens that can infect or cross the BBBs have often co-opted adsorptive endocytosis/transcytosis mechanisms (Chou & Dix, 1989; Marsh, 1984; Schweighardt & Atwood, 2001). These processes may also be related to diapedesis as many of the events of immune cell passage across the BBB resemble these endocytic mechanisms. For example, both LFA-1 and ICAM, important to immune cell passage across the BBB, are glycoproteins. Although adsorptive endocytosis is in some sense saturable because of a finite amount of any single glycoprotein on a cell surface, it is not easy to demonstrate classical saturable kinetics for this process. In fact, excess glycoprotein can sometimes further stimulate endocytosis and so lead to a paradoxical increase, rather than decrease, in the rate of passage across the BBB (Banks et al., 1997). Glycoprotein distribution on brain endothelial cells is polarized; that is, a glycoprotein may be enriched on either the luminal or abluminal membranes (Vorbrodt, 1994; Zambenedetti et al., 1996). The tight junctions act as a "fence" to keep the glycoproteins confined to their respective sides of the endothelial cell (Deli et al., 2005). This means that the movement of a glycoprotein molecule (or a virus whose coat displays that glycoprotein) can be unidirectional as its transcytosis can only be initiated from the side of the brain endothelial cell that contains the ligand's complementary glycoprotein (Broadwell, 1989; Villegas & Broadwell, 1993). The possession and distribution of glycoproteins similarly dictate which viruses can invade the brain; neurovirulent viruses that invade the brain as free viruses (as opposed to entering in a Trojan horse fashion inside an infected immune cell) can do so because they possess a glycoprotein ligand capable of binding to the BBB. The degree to which the brain endothelial cell glycocalyx, which is the extracellular matrix present on the blood-facing side of endothelial cells, participates in adsorptive transcytosis is an active area of study. Heparan sulfate is a major glycosaminoglycan of the glycocalyx that is attached to the membrane via proteoglycans, which include syndecans and glypicans. These proteoglycans have been shown to mediate different modes of endocytosis (e.g., clathrin-dependent and independent) (Christianson & Belting, 2014) and may do so in brain endothelial cells as well. To date, it has been shown that heparan sulfate regulates the transcytosis of chemokines, exosomes, and HIV across the BBB (Banks et al., 2004a, b, c; Joshi & Zuhorn, 2021, Quaranta et al., 2022).

Other molecules besides glycoproteins can also induce adsorptive endocytosis/transcytosis-type mechanisms. A classic example is polycationic molecules such as the poly-L-lysines and the protamines. Protamines are peptides of about 30 amino acids that contain an abundance of arginine molecules. They can induce adsorptive transcytosis so vigorously as to result in BBB disruption (Hardebo & Kahrstrom, 1985; Vorbrodt et al., 1995). One of the proofs that viruses co-opt adsorptive transcytosis-like mechanisms is that they, like protamine, bind to heparins and heparans; indeed, protamine sulfate blocks viruses such as HIV-1 from binding to the BBB (Ramos-Kuri et al., 1996; Banks et al., 2004a, b, c; Bobardt et al., 2004). Many of the highly charged penetrating peptides, such as those derived from Tat, and many of the antibodies that target receptor-mediated transporters likely are taken up by adsorptive endocytosis/transcytosis-related mechanisms (Herve et al., 2008; Niewoehner et al., 2014; Weissmann, 1976).

13.2.2 CNS to Blood

Traditionally, the passage in the brain-to-blood direction (efflux) has been neglected. However, efflux often accounts for the inability of what would otherwise be effective drugs to accumulate in the CNS. Pharmacogenomic studies have suggested that the individual variation in efflux mechanisms may explain why some individuals are less sensitive to the CNS effects of drugs or more sensitive to their toxicities (Loscher & Potschka, 2002). Efflux mechanisms are important to the homeostasis of the CNS, ridding the brain of toxins (Taylor, 2002). The rate of efflux can be, in addition to synthesis and degradation, an important determinant of the level of a substance produced within the CNS (Chen et al., 1997; Chen & Reichlin, 1998; Maness et al., 1998).

13.2.2.1 Nonsaturable

Efflux, like influx, has both saturable and non-saturable mechanisms of entry. Transcellular diffusion occurs for both influx and efflux. Other mechanisms, such as bulk flow, are unique for efflux.

Transcellular Diffusion

Many of the principles that govern influx by transcellular diffusion are also important in efflux. The dramatic role that efflux by transcellular diffusion can play can be illustrated by comparing the fate of small, lipid-soluble molecules to that of a protein after intrathecal administration. Intrathecal application of small, lipid-soluble molecules, such as anesthetics, can have a local effect on spinal cord function but have little or no effect on the brain (Bernards, 1999). These substances readily cross the brain endothelial cell by transcellular diffusion and do this as easily in the brain-to-blood direction as in the blood-to-brain direction. Therefore, they are cleared from the CSF before they are able to reach the brain (McQuay et al., 1989). In contrast, proteins such as leptin and lysosomal enzymes are too large and water soluble to undergo significant transcellular diffusion (LeBel et al., 1999; McCarthy et al., 2002). Leptin, tetanus antitoxin, and the lysosomal enzyme idursulfase can reach the brain after intrathecal administration in amounts sufficient to produce CNS effects (LeBel et al., 1999; McCarthy et al., 2002; Kabura et al., 2006; Calias et al., 2012).

Efflux by transcellular diffusion can also contribute to the poor diffusion of substances within the brain parenchyma. Diffusion within the interstitial space of the brain is dependent on Brownian motion and the production of metabolic free water as driving forces and so is very slow (Cserr, 1984; Cserr & Berman, 1978). However, efflux by non-saturable (and saturable) mechanisms can further reduce the distance a substance will ultimately diffuse. For example, the less lipid-soluble drug atenolol can diffuse about 3 times further into brain tissue than the more lipid-soluble drug acetaminophen (de Lange et al., 1993).

Bulk Flow and the Glymphatic System
Bulk flow refers to the reabsorption of CSF into the blood, which occurs at the level of the arachnoid villi (Davson & Segal, 1996a, b, c) and cribriform plate (Widner et al., 1987; Yamada et al., 1991). Any substance dissolved in CSF will enter the blood by this mechanism (Jones & Robinson, 1982; Pollay & Davson, 1963). The glymphatic system provides an important mechanism for the mixing of the CSF and brain interstitial fluid, with aquaporin-dependent fluid production at the astrocytes providing the circulant and arteriole pulsations providing the directionality (Iliff et al., 2012). Thus, the glymphatics are important not only for bulk flow but also for the extracellular pathways, and possibly for the movement of CSF from the spinal cord into the cranium.

The characteristics of this system result in several surprising but important phenomena. For example, CSF drained at the cribriform plate, which is likely the dominant route for CSF drainage at normal CSF pressures (Boulton et al., 1999), can enter into the cervical lymphatic system. This can provide a direct route from the CNS to the cervical lymphatics (Oehmichen et al., 1979), as has been illustrated for gp120, the glycoprotein of the human immunodeficiency virus, HIV-1 (Cashion et al., 1999). This route to the lymphatics may explain why substances injected into the brain can produce a different immune response than when the substance is injected peripherally (Cserr & Knopf, 1992; Knopf et al., 1995). Another example is that in some cases, the levels of a substance in blood achieved after injection into the CSF can be sustained longer and at higher levels than after an intravenous bolus (Chen et al., 1997, Chen & Reichlin, 1998, Maness et al., 1998) This is because the central injection acts similarly to an intravenous infusion, slowly delivering drug to the blood. Impairment of the glymphatic circulation can result in decreased bulk flow and so may contribute to increased levels of protein and toxins in the CSF (Iliff et al., 2013).

13.2.2.2 Saturable Transport

The last decade has seen a huge increase in the interest of efflux by saturable mechanisms. Just as efflux by transcellular diffusion can limit the diffusion of a substance within the CNS, so can the presence of a saturable efflux transporter (Blasberg, 1977). Much of this interest centers on the multidrug efflux transport systems (Begley, 2004), most notably p-glycoprotein (P-gp). However, other efflux transporters for peptides, proteins, endogenous substances, and drugs are known to play important roles in physiology and disease (Drion et al., 1996; Martins et al., 1997; Mealey et al., 2001; Taylor, 2002). For example, peptide transport system-1 is a major regulator of brain levels of methionine enkephalin, an endogenous opiate that suppresses voluntary ethanol drinking (Plotkin et al., 1998). Depression and recovery of peptide transport systems-1 with ethanol drinking may relate to alcohol withdrawal seizures (Banks & Kastin, 1989; Banks & Kastin, 1994). IL-2 was the first cytokine known to be transported by a saturable efflux system (Banks et al., 2004a, b, c); some have postulated this transporter may be P-gp. Poor accumulation of protease inhibitors, antibiotics, AZT, anticancer drugs, and many other substances occurs because of efflux systems (Glynn & Yazdanian, 1998; King et al., 2001; Lee et al., 1998; Loscher & Potschka, 2002; Masereeuw et al., 1994; Spector & Lorenzo, 1974). P-gp plays a major role in the efflux of intrathecally administered opiate analgesics (Thompson et al., 2000). Brain-to-blood transport of corticotropin-releasing hormone is sufficient to influence splenic levels of beta-endorphin (Martins et al., 1997). Impaired efflux of amyloid ß peptide, the peptide believed to play a causal role in Alzheimer's disease, develops in mice that overexpress amyloid precursor protein; this loss of clearance further promotes accumulation within the brain of amyloid ß protein (Banks et al., 2003; Deane et al., 2004; Ghersi-Egea et al., 1996). Evidence suggests that impaired transport develops in humans as well and so may be a major mechanism for the induction of Alzheimer's disease (Shibata et al., 2000; Tanzi et al., 2004).

13.3 Neuroimmune Interactions

The above discussion of BBB fundamentals is tailored toward understanding the role of the BBB in neuroimmune interactions. Below are specific examples of how the BBB is involved in neuroimmune interactions.

13.3.1 Binding Sites at the BBB: Receptors and Transporters

An important distinction for understanding the function of the BBB is that of receptors versus transporters. The term "receptor" has undergone a transformation of its usage since its introduction in the late nineteenth century when it was first used to denote some physiological function. Eventually, the term receptor was used to denote a physical binding site through which a drug or hormone could exert its effects on a cell. In the 1980s, a distinction was made between "receptor" and "binding site," the former being coupled to intracellular machinery that translated its binding into a cellular effect. Binding sites on the brain endothelial cell can represent transporters, but they can also represent traditional receptors, that is, binding sites coupled to intracellular machinery. For example, brain endothelial cells have both insulin receptors and transporters. As a result, insulin is transported across the BBB to exert effects inside the CNS, but insulin also alters a number of functions of the brain endothelial cell. As examples of the latter, insulin alters the BBB transport of zidovudine (AZT, (Ayre et al., 1989), tryptophan (Cangiano et al., 1983), and leptin (Kastin & Akerstrom, 2001) and alters brain endothelial cell alkaline phosphatase activity (Catalan et al., 1988). Recent work has shown that the brain endothelial cell's insulin receptor and the insulin transporter are different proteins (Rhea et al., 2018).

Many BBB studies have assumed that a binding site represents transporter function and are so designed as to not consider whether receptors, as well as transporters, may exist at the BBB. However, a great deal of indirect evidence and some direct evidence indicates that the vascular BBB and the choroid plexus probably possess a large variety of receptors that can alter BBB functions. Besides insulin, substances that bind to and alter the function of brain endothelial cells include mu opiate receptor ligands (Baba et al., 1988; Chang et al., 2001; Vidal et al., 1998), cytokines (Ban et al., 1991; Cunningham et al., 1992; van Dam et al., 1996; Vidal et al., 1998; Khan et al., 2003; Moser et al., 2004), leptin (Bjorbaek et al., 1998; Kastin et al., 2000; Hsuchou et al., 2013a, b), acetylcholine (Grammas & Caspers, 1991), adrenergic (Kalaria & Harik, 1989; Walsh et al., 1987), glutamate (Koenig et al., 1992; Krizbai et al., 1998), and chemokines (Sanders et al., 1998).

13.3.2 Permeability to Cytokines and Related Substances

The BBB is known to transport several cytokines in the blood-to-brain direction. For example, the BBB transports the IL-1 s, IL-6, and TNF by three separate transport systems. Additionally, nerve growth factor, brain-derived neurotrophic factor, interferons, neurotrophins, eotaxin, fibroblast factor 19, leukemia inhibitory factor, and chemokines (Quaranta et al., 2022; Poduslo & Curran, 1996; Pan et al., 1997a, b, c, 1998a, b: Hsuchou et al., 2013a, b, Erickson et al., 2014) are also transported across the BBB. In some cases, the same gene that gives rise to a cytokine's receptor also produces the cytokine's transporter, whereas in other cases the receptor and transporter are different proteins (Banks et al., 2002; Pan & Kastin, 2002). For example, the BBB transporter for pituitary adenylate cyclase-activating polypeptide (PACAP) was found to be the same protein that acts as a neuronal receptor for enterostatin, but not for PACAP, and acts as a lipid transporter in the liver (Dogrukol-Ak et al., 2009; Martinez et al., 2003; Park et al., 2004). In general, BBB transporters occur throughout the CNS, including the spinal cord, although the transport rate across the BBB can vary greatly among CNS regions (Banks et al., 1994, McLay et al., 1997, Pan et al., 1997a, b, c, 1998a, b). Enough cytokine is transported into the brain to affect CNS function. For example, IL-1 alpha crosses the BBB at the posterior division of the septum where it mediates cognitive impairments (Banks et al., 2001). Similarly, serum TNF crosses the BBB to induce CNS release of TNF which in turn can induce apoptosis in the substantia nigra (Qin et al., 2007).

The cytokine transporters are not static but respond to physiological and pathological events. The transport rates of IL-1 and TNF each show diurnal variations (Banks et al., 1998a, b, Pan et al., 2002). The transport rate of TNF is altered in animals with experimental allergic encephalomyelitis (EAE), spinal cord injury, or blunt trauma to the brain (Pan et al., 1996; Pan et al., 1997a, b, c, 2003a, b, Pan & Kastin, 2001a, b, Pearse et al., 2003). The transport of CCL2 and CCL5 is increased by about 10fold following moderate systemic inflammation induced by bacterial lipopolysaccharide (Quaranta et al., 2022).

13.3.3 Permeability to Other Neuroimmune Substances

Other substances with neuroimmune actions are handled by the BBB in a variety of ways. Monoamines are largely excluded by the BBB (Hardebo & Owman, 1990; Kalaria et al., 1987), and opiates and opiate peptides as a rule enter the brain by transcellular diffusion but are transported by saturable systems in the brain-to-blood direction (Banks & Kastin, 1990; Elferink & Zadina, 2001; King et al., 2001). Pituitary adenylate cyclase-activating polypeptide, a member of the VIP/secretin/PACAP family, has immune functions (Arimura, 1992). Transport of its two major forms across the BBB is complex, involving both brain-to-blood and blood-to-brain components (Banks et al., 1993). Its blood-to-brain transport is altered with brain injury (Somogyvari-Vigh et al., 2000). Kynurenine is a tryptophan metabolite whose production is increased by inflammation and has immunomodulatory activities primarily through the aryl hydrocarbon receptor (Kaiser et al., 2020). Kynurenine and many of its bioactive metabolites are transported across the intact BBB via the large neutral amino acid transporter, LAT-1 (Fukui et al., 1991; Walker et al., 2019). Some of the other immune active substances whose passage across the

BBB has been investigated are melanocyte-stimulating hormone (Martins et al., 1996; Wilson et al., 1984), corticotrophin-releasing hormone (Martins et al., 1996), and enkephalins (Banks et al., 1986; Elferink & Zadina, 2001).

13.3.4 Permeability to Immune Cells

As discussed above, immune cells cross the BBB by the highly regulated process of diapedesis. The mechanism by which immune cells cross the BBB has also been greatly clarified in the past two decades. The review literature often states that immune cells enter the CNS by leaking across a disrupted BBB. However, immune cells cross by the highly regulated process of diapedesis and not by BBB disruption. Disruptions to the BBB are mediated by transcellular or paracellular processes. Transcellular processes are usually mediated by increased vesicular activity in the endothelial cells (Lossinsky et al., 1983; Mayhan & Heistad, 1985; Vorbrodt et al., 1995). These vesicles of 100 nm or so could not accommodate the passage of an immune cell 10,000 nm in diameter. Even in diseases where there is both increased immune cell trafficking into the CNS and a disrupted BBB, there is often a mismatch between the site of immune cell entry and BBB disruption (Engelhardt & Wolburg, 2004). The other route for disruption is the paracellular route. Although evidence suggests that immune cells can do this, a transcellular route is likely favored (Kivisakk et al., 2003; Wolburg et al., 2005). In brief, immune cells can tunnel through venular endothelial cells leaving the intercellular tight junctions intact (Engelhardt, 2008; Wolburg et al., 2005). Diapedesis is a complex process and is initiated when LFA-1 on an immune cell binds to ICAM on the brain endothelial cell. Other paracellular messengers, which likely include cytokines, are then released (Male, 1995; Persidsky et al., 1997). Protrusions and invaginations of the endothelial cell and protrusions of the immune cell occur, with the immune cell possibly using the tight junction as an initial anchoring site (Lossinsky et al., 1991). Other ligands that have been postulated to play a role in this transcytotic process include PECAM, VCAM, cadherin, members of the JAM family, and CD99 (Engelhardt & Wolburg, 2004). Some plasma inevitably accompanies the passage of the immune, which can give the appearance of a disrupted BBB (Avison et al., 2004; Greenwood et al., 1995; Persidsky et al., 2000).

The immune processes that induce an immune cell to cross the BBB are complex. Quan has shown that injection of interleukin-1 into brain tissue induces immune cell trafficking, but that such induction can be blocked by the injection of LPS into the periphery (Ching et al., 2005, 2006; Quan et al., 1994). Although trafficking is a product of a complex interaction between the immune cell and the BBB, the degree to which trafficking occurs can reside primarily with the immune cell or the BBB, and genetics or immune events can shift that dominance (Banks et al., 2012). Such events may underlie "euflammation," the phenomenon by which subclinical activation of the innate immune system renders it increasingly resistant to such activation by increasingly stronger stimuli (Tarr et al., 2014).

13.3.5 Permeability to Viruses

Whether a virus is neurovirulent or not depends largely on its ability to cross the BBB (Chou & Dix, 1989). However, viruses can induce neurotoxicity without themselves crossing the BBB by several mechanisms. For example, shed viral proteins might cross the BBB as could circulating cytokines whose release from peripheral sources is induced by the virus. However, most neurovirulent viruses seem to do their major damage directly after entering and replicating within the CNS. Some viruses can replicate within brain endothelial cells and are subsequently shed into the CNS (Cosby & Brankin, 1995). Other viruses invade the CNS by crossing the BBB (Nakaoke et al., 2005). Initial uptake by either route involves events reminiscent of adsorptive endocytosis as discussed above. Viral glycoproteins bind to brain endothelial cells (or choroid plexus) glycoproteins to initiate adsorptive endocytosis. As with adsorptive endocytosis, the virus-containing vesicle is subsequently routed to various membrane systems which can include discharge to the original side of uptake or transcytosis. Sialic acid and heparan sulfate are common components of the glycoproteins and proteoglycans involved in viral uptake by the BBB (Schweighardt & Atwood, 2001; Banks et al., 2004a, b, c). In some cases, the functional glycoprotein is known. For example, rabies can bind to acetylcholine and nerve growth factor receptors (Schweighardt & Atwood, 2001), and HIV-1 can bind to the mannose-6 phosphate receptor. Without an appropriate luminal or basal glycoprotein with which to bind to the BBB cell, the virus is largely excluded from the CNS.

Other principles reviewed above can play key roles in viral uptake by the brain. Activation of the immune system increases the transport of HIV-1 and other viruses across the BBB (Chou & Dix, 1989; Cosby & Brankin, 1995; Dohgu & Banks, 2008). In contrast, BBB transport of the S1 fragment of SARS-CoV-2 spike protein was not increased by inflammation (Rhea et al., 2021). In the case of HIV-1, this involves inducing secretion of GM-CSF and IL-6 from brain endothelial cells with the cytokines acting in an autocrine fashion to promote the transcytosis of the virus (Dohgu et al., 2011). This process is not dependent on pericytes but is enhanced by it (Dohgu & Banks, 2013). Eventually, the immune mechanisms combine to produce the cognitive and behavioral problems seen in AIDS (Hong & Banks, 2014).

13.3.6 Secretion of Neuroimmune-Active Substances

The brain endothelial cells and the epithelial cells of the choroid plexus are capable of secreting a large number of neuroimmune active substances. These include interleukins (Fabry et al., 1993; Hofman et al., 1999; Reyes et al., 1999), TNF (Lee et al., 2001), nerve growth factor, (Moser et al., 2004), endothelin (Didier et al., 2002), monocyte chemoattractant peptide (Chen et al., 2001), nitric oxide (Mandi et al., 1998), RANTES (Simpson et al., 1998), and prostacyclin (Faraci & Heistad, 1998). Some of these substances are secreted spontaneously, and many of them can be stimulated with immunoreactive substances such as lipopolysaccharide (LPS), bacteria, or viral proteins (Hofman et al., 1999; Reyes et al., 1999; Vadeboncoeur et al., 2003). The unique architecture of the BBB allows it to receive input from one of its surfaces and to secrete substances into the other. For example, LPS applied to the abluminal surface of brain endothelial cells in monolayer cultures will enhance release of IL-6 from the luminal surface (Verma et al., 2006).

13.3.7 Modulation of BBB Function by Neuroimmune Substances

Traditionally, neuroimmune modulation has been thought of in terms of disruption of the BBB. However, as the review above indicates, transporter functions are also vulnerable to manipulation by neuroimmune elements. Alterations in transport function are likely to be more common events than disruption, as the latter is likely seen only with extreme pathology, whereas the former is likely a physiological, as well as a pathological, aspect of neuroimmune regulation. Other functions of the BBB, such as brain endothelial cell secretions, are also clearly affected by neuroimmune events.

13.3.7.1 The BBB Coordinates Adaptive CNS Responses to Immune Stimuli

Mild and transient systemic inflammation, such as that which occurs during common infections, provokes a coordinated response by the CNS that adaptively alters physiology and behaviors that are beneficial for pathogen clearance and recovery of the host. Such adaptive changes include HPA axis activation, fever, and sickness behaviors. Brain barriers play important roles in these host responses, protecting the CNS from pathogen invasion and selectively relaying immune signals to the CNS from the periphery. Notably, widespread BBB disruption as measured by leakage of injected tracer molecules is generally not detectable unless the inflammatory response is severe, such as with high doses of LPS that mimic aspects of septic shock (Banks et al., 2015; Erickson et al., 2018; Quan et al., 1999), supporting that the BBB has mechanisms that preserve its integrity during mild systemic inflammation. Recent work has shown that endothelial cells of the BBB show fewer and less robust translational alterations in response to LPS versus endothelial cells from non-brain vascular beds (Cleuren et al., 2019); however specific responses of brain endothelial cells to systemic inflammatory stimuli regulate key CNS changes. For example, PGE2 in the hypothalamus is a critical mediator of fever, and brain endothelial cells are the main source of PGE2 that induces fever (Eskilsson et al., 2017; Inoue et al., 2002). Regional activation of the synthesis of inflammatory cytokines and I kappaB alpha in the brain was limited to circumventricular organs and brain endothelial cells at very low doses of LPS, whereas higher LPS doses that cause septic shock symptoms resulted in a more widespread activation of inflammatory markers throughout the brain (Quan et al., 1999). Therefore, brain endothelial cells and CVOs are important sensors of low-grade immune stimuli.

13.3.7.2 Agents That Modulate Permeability Through the BBB

Disruption was the first BBB function noted to be perturbed in neuroimmune disease. However, the review above makes it clear that an increase in BBB permeability can be induced for specific agents by increasing their blood-to-brain transport rate or inhibiting their brain-to-blood efflux rate. An additional observation is that the BBB does not respond to immune activation through a single pathway or even respond uniformly to an activated pathway. For example, indomethacin blocks the BBB disruption induced by LPS, suggesting that disruption involves prostaglandins (Minami et al., 1998a, b). In contrast, indomethacin enhances the LPS-induced increase in HIV-1 penetration across the BBB and does not affect the LPS-induced increase in insulin transport (Banks et al., 1999; Xaio et al., 2001).

13.3.7.3 Regulation of BBB Integrity and Tight Junction Function

The classic example of BBB disruption is that seen in multiple sclerosis and the animal model of that disease, EAE (Butter et al., 1991; Juhler et al., 1984; Pozzilli et al., 1988). LPS and treatment with cytokines such as TNF have also been shown to induce BBB disruption (Megyeri et al., 1992). As discussed above, paracellular (through tight junctions) and transcellular mechanisms of transport exist. Although either can underlie BBB disruption, classic studies have shown that the major cause of increased protein leakage across the BBB for almost every kind of insult to the BBB or the CNS is mediated by transcytotic mechanisms (Lossinsky et al., 1983; Vorbrodt et al., 1995). Nevertheless, recent advances in understanding tight junction assembly and regulation have encouraged many to investigate paracellular mechanisms. Both tight junction function and transcytosis are regulated events, although it is unclear to what extent protein leakage into the brain may be altered under physiological conditions. To some extent, paracellular and transcytotic routes likely involve some of the same cellular machinery, such as the cytoskeleton. TNF is known to induce rearrangements in cytoskeletal architecture. Additionally, cerebral ischemia, diabetes mellitus, and even intense pain are associated with alterations in the expression and cellular distribution of tight junction proteins and opening of the BBB (Brown & Davis, 2002; Chehade et al., 2002; Huber et al., 2002). The importance of regulatory processes in BBB disruption is vividly illustrated by the paradoxical finding that maximal disruption does not occur at the time of the CNS injury, even when the event is traumatic, but hours or days later (Baldwin et al., 1996). It is thought that it is the peripheral and central responses, such as cytokine release, to CNS injury rather than the CNS injury itself that results in BBB disruption.

13.3.7.4 Regulation of Saturable Transporters

Regulation of both influx and efflux transporters is influenced by neuroimmune events. Additionally, cytokine transporters are also affected by various CNS events. Much of what is known about the inflammatory regulation of BBB transport has been determined using bacterial lipopolysaccharide (LPS) as a stimulator of systemic inflammation.

LPS and Blood-to-Brain Transporters

LPS increases the transport across the BBB of cisplatin, insulin, leukemia inhibitory factor, amyloid beta, and the HIV-1 viral coat glycoprotein gp120, but not of TNF, pituitary adenylate cyclase-activating polypeptide, or the S1 subunit of SARS-CoV-2 spike protein (Minami et al., 1998a, b; Banks et al., 1999; Xaio et al., 2001; Osburg et al., 2002; Nonaka et al., 2005; Pan et al., 2008a, b; Jaeger et al., 2009a, b; Rhea et al., 2021). LPS affects leptin transport (Nonaka et al., 2004) through peripheral mechanisms and increases pituitary adenylate cyclase-activating polypeptide binding to receptors on the BBB but does not alter transport. CNS injuries such as ischemia or trauma to the spinal cord induce a cascade of events that can affect the transport of neuroimmune substances across the BBB as discussed below.

LPS, Cytokines, Prostaglandins, and Brain-to-Blood (Efflux) Transporters

Efflux systems are also altered by immune modulators. In vitro studies show that LPS, TNF, and interferon-gamma regulate P-gp (Stein et al., 1996; Theron et al., 2003; Hartz et al., 2006; Bauer et al., 2007; Yu et al., 2007a, b; Salkeni et al., 2009). More recently, in vivo studies have shown that LPS down-regulates P-gp function at the BBB (Salkeni et al., 2009). Interestingly, IL-2 appears to be both a substrate for P-gp and a modulator of its activity (Bonhomme-Faivre et al., 2002; Castagne et al., 2004; Drach et al., 1996). Because P-gp regulates the brain concentration of so many drugs and endogenous substances, immunomodulation could affect many other responses. For example, brain levels of exogenous opiate drugs such as morphine (King et al., 2001), endogenous opiates such as ß-endorphin (Kastin et al., 2002), and neurotoxins such as cyclosporine (Sakata et al., 1994) would all be expected to be increased in patients given IL-2. P-gp, along with the low-density lipoprotein receptor-related protein-1 (LRP-1), also contributes to the CNS clearance of amyloid beta, whose BBB efflux is inhibited by LPS (Jaeger et al., 2009a, b). Prostaglandins, including PGE2 (discussed in Sect. 13.3.7.1) and PGD2 are substrates for drug efflux transporters including MDR4 and Oat3 expressed at the BBB and BCSFB (Tachikawa et al., 2014). It has been proposed that these transporters have physiological functions such as regulating the duration of astrocyte-derived PGE2-mediated vasodilatory effects by rapidly eliminating local concentrations in the ISF, and diurnal rhythms of PDG2 which regulate sleep. Brain efflux of PGE2 is substantially reduced with inflammation (Tachikawa et al., 2014). Chemokine transport across the BBB in the brain-to-blood direction facilitates leukocyte recruitment to the CNS, and the duffy antigen chemokine receptor has been identified as one efflux transporter for chemokines that is active at the BBB and contributes to the progression of experimental autoimmune encephalomyelitis (Minten et al., 2014). Immunomodulation of efflux systems, therefore, could have a major effect on CNS metabolism and the response to drugs, as well as on neuroinflammation.

The activity of LRP-1 is also modulated by LPS. LRP-1 at the BBB acts as an efflux pump to amyloid beta protein (Deane et al., 2008), a substance associated causally with Alzheimer's disease. The neurovascular hypothesis states that decreases in LRP-1 activity at the BBB contribute to the loss of clearance of the amyloid beta protein from the brain and so promotes Alzheimer's disease (Zlokovic, 2005). Inhibition of LRP-1 activity leads to decreased efflux of amyloid beta protein, increases in levels of amyloid beta protein in the brain, and cognitive impairment (Jaeger et al., 2009a, b). Mice treated with LPS have decreased efflux of amyloid beta protein, providing a mechanism for the connection between inflammatory processes and Alzheimer's disease (Jaeger et al., 2009a, b).

13.4 Role of BBB in Neuroimmune Diseases

The above review has emphasized BBB/neuroimmune interactions under normal physiological conditions. However, the BBB is intimately involved in neuroimmune diseases as well. The BBB can be a target of such disease, its functions may be adaptive to the disease, or it can be a contributor to the disease process. Below are some examples of the ways in which the BBB is altered in diseases with neuroimmune processes.

13.4.1 TNF Transport and EAE

TNF has a biphasic effect on many neuroimmune processes, with too little or too much producing harmful effects (Pan et al., 1997a, b, c). TNF mediates many of its pathological effects through its central receptors and the transport of circulating TNF is one source of CNS TNF (Gutierrez et al., 1993; Osburg et al., 2002; Pan & Kastin, 2002). Induction of EAE is partially dependent on TNF and IL-1 (Schiffenbauer et al., 2000). Immune cell invasion in general and during EAE, in particular, is dependent on TNF-modulated expression of ICAM and VCAM on brain endothelial cells and of LFA-1 on immunocytes (Barten & Ruddle, 1994; Male, 1995). Finally, the saturable transport across the BBB of TNF itself is greatly increased in EAE (Pan et al., 1996).

13.4.2 CNS Injuries and Cytokine Transport

As noted above, CNS injuries can produce a disruption of the BBB, but this disruption is temporally dissociated from the injury (Baldwin et al., 1996; Pan et al., 1997a, b, c; Banks et al., 1998a, b). This dissociation is because the disruption is the consequence of the reactions to injury rather than to injury itself. Not surprisingly, then, CNS injuries can also produce complex alterations in the BBB transport of cytokines. Besides the example of TNF in EAE given above, TNF transport is also increased in spinal cord injury (Pan et al., 1997a, b, c, 2003a, b; Pan & Kastin, 2001a, b) and after blunt trauma to the brain (Pan et al., 2003a, b). This increase is neither confined to the site of injury, homogeneous through out the CNS, nor related to the disruption pattern of the BBB (Pan et al., 1997a, b, c; Pan & Kastin, 2001a, b; Xiang et al., 2005). It is also temporally and regionally independent of the changes in BBB transport rates of other cytokines and immunoactive substances whose transport rates are also altered with CNS injury (Banks et al., 1998a, b; Pan et al., 1998a, b). The pattern also is dependent on the type of CNS injury (Pan & Kastin, 2001a, b).

13.4.3 Anti-Retrovirals and the BBB

A major problem in treating viruses that can invade the CNS, such as HIV-1, is that antiretrovirals often cross the BBB poorly (Thomas, 2004). The major problem, however, is not that these substances are especially limited by their rate of transcellular diffusion, but that they are nearly all substrates for efflux systems. For example, AZT is 16 times more lipid soluble than sucrose and so should cross much more rapidly but crosses at the same rate (Wu et al., 1998). AZT is a ligand for at least two efflux systems (Masereeuw et al., 1994; Takasawa et al., 1997; Wang & Sawchuck, 1995), and the protease inhibitors are all substrates for P-gp (Lee et al., 1998). P-gp is expressed by immune and other cells as well with three major phenotypic clusters in humans. Those persons with higher expression of P-gp, and therefore less able to accumulate protease inhibitors in tissues, are more resistant to treatment for HIV-1.

Conditions associated with AIDS further complicate the interaction between efflux systems and the uptake of antiretrovirals by the brain. P-gp is modulated by inflammation (Yu et al., 2007a, b, 2008) and gp120 (Ronaldson & Bendayan, 2006). Many of the protease inhibitors induce P-gp expression and function thus likely further reducing their ability to cross the BBB (Chan et al., 2013).

Another therapeutic consideration regarding antiretroviral delivery to the CNS is that they may have some direct neurotoxic effects, albeit to a much lesser extent than those caused by untreated or poorly controlled HIV infection (Jordan-Sciutto, 2021). Pre-exposure prophylaxis (PREP) regimens of antiretrovirals, used to prevent new HIV infections in vulnerable populations, are also associated with increased biomarkers of systemic immune activation and BBB disruption; however, this study noted that other viral infections were more prevalent in HIV-negative subjects on PREP versus control volunteers who did not take PREP (Robertson et al., 2021). Recent work has also suggested that antiretroviral treatment reduces BBB disruption in HIV+ subjects with neurological symptoms, as measured by CSF/serum albumin ratios (Anesten et al., 2022); however, long-term ART in people with HIV has also been associated with cerebrovascular disease, which could be partly mediated through mitochondrial dysfunction (Bertrand et al., 2021).

13.4.4 Immune Cell Invasion and Other Interactions with the BBB

Immune cell trafficking into the CNS is important in mediating neuroimmune diseases. Immune cell invasion is an early event in multiple sclerosis and EAE (Wolburg et al., 2005). Infected immune cells are a mechanism by which HIV-1 (Koyanagi et al., 1997; Nottet et al., 1996) and perhaps prions (Klein et al., 1997) invade the CNS.

Immune cell passage across the BBB is, in turn, affected by immune modulators. LPS and the HIV-1 immunoreactive protein Tat increase expression by brain endothelial cells of ICAM and VCAM (Nottet et al., 1996; Pu et al., 2003), and monocytes treated with LPS have an increased rate of passage across the BBB (Persidsky et al., 1997). In vitro studies suggest that these events may be mediated through IL-1ß and IL-6 (De Vries et al., 1994).

Immune cells may also cause dysfunction through interactions with the vascular BBB in the absence of diapedesis. For example, increased interactions of leukocytes with the lumen of the BBB have been associated with changes in microvascular blood flow, termed capillary stalling which is increased in mouse models of CNS pericyte loss and Alzheimer's disease (Ali et al., 2022; Choe et al., 2022). Neutrophil obstruction of brain capillaries is predominant in the penumbral area following middle cerebral artery occlusion modeling of stroke in rodents, and neutrophil depletion in this model reduces infarct size and functional deficits (El Amki et al., 2020). Neutrophils may be especially adept at inducing capillary stalling due to their secretion of net-like extracellular traps upon activation (Vaibhav et al., 2020).

13.4.5 Efflux of HIV-1 Proteins and Cytokines

Because the BBB prevents the effective accumulation of many of the antivirals, the CNS can act as a reservoir of viruses. This reservoir could potentially reinfect the peripheral tissues. The CNS-to-blood movement of HIV-1 has not been investigated, but the movement of two of its proteins has been done. The coat glycoprotein HIV-1gp120 is cleared through bulk flow (Cashion et al., 1999), being reabsorbed predominately at the cribriform plate. As a result, it drains by way of lymphatic vessels directly to the cervical lymphatic nodes. If the whole virus also takes this route, then that means that lymph nodes could be directly reinfected without the virus having to enter the circulation where it could be exposed to antiviral agents.

A CNS reservoir of the virus could affect the peripheral immune system by a mechanism that does not involve the reinfection of peripheral tissues. HIV-1 Tat, like HIV-1gp120, is also reabsorbed with the CSF into the blood by a nonsaturable mechanism (Banks, et al. 2005). Proteins that are enzymatically resistant in blood, such as HIV-1 Tat, HIV-1gp120, and cytokines, can achieve high levels in blood even when their only source is the CSF. Production of these proteins within the CNS with subsequent reabsorption with the CSF into the blood could be a way in which the CNS virus produces toxic effects in peripheral tissues.

To date, IL-2 is the only cytokine known to be transported from the brain to the blood by a saturable transporter (Banks et al., 2004a, b, c). This transporter, along with binding to plasma proteins and robust degradation by the BBB or CNS, effectively prevents IL-2 from acutely entering the brain. Evidence suggests that this transporter is likely P-gp. P-gp activity is decreased with HIV encephalitis (Persidsky et al., 2000) and this could lead to blood-borne IL-2 entering the CNS. Chronically, however, the binding of IL-2 to plasma proteins extends its half-life, allowing time for it to accumulate in the brain by way of the extracellular pathways (Patel et al., 2012). Chronic IL-2 administration induces stereotypic behaviors and is used in an animal model of schizophrenia (Zalcman, 2001; Zalcman, 2002). Therefore, an enhanced entry of IL-2 is one mechanism by which HIV-1 could induce behavioral changes.

13.5 Summary

The BBB intimately interacts with cells and their secretions that are in both the CNS and periphery. Some neuroimmune substances, exemplified by cytokines, can cross the BBB directly and also have direct effects on the BBB. The BBB is itself a source of neuroimmune substances and can receive signals from one side, for example, the brain side, and release substances in response to that signal from its other side. The passage of immune cells across the BBB is a highly regulated event as is the passage of viruses and viral particles. Overall, the BBB is an important component of the neuroimmune axis and the only component that is simultaneously physically in both the peripheral and central compartments of the neuroimmune system.

13.6 Review Questions

1. The term diapedesis refers to:
   (a) Cells that have the histological appearance of having two feet
   (b) The mechanism used by immune cells to cross the blood-brain barrier
   (c) Production of fluids by the choroid plexus
   (d) A mechanism by which the brain can decrease inflammation
   (e) A vital part of adaptive immunity

2. Processes or agents shown to alter functions of the blood-brain barrier include (check all that apply):
   (a) Inflammation
   (b) Maturation
   (c) A liberal arts education
   (d) Lipopolysaccharide

3. What cell types form brain barriers (choose all that apply):
   (a) Brain endothelial cells
   (b) Epithelial cells of the choroid plexus

(c) Sertoli cells
(d) Tanycytes
(e) Microglia

13.7 Answers

1. b
2. a, b, d
3. a, b, d

References

Alafaci, C., Salpietro, G., Grasso, G., Sfacteria, A., Passalacqua, M., Morabito, A., Tripodo, E., Calapai, G., Buemi, M., & Tomasello, F. (2000). Effect of recombinant human erythropoietin on cerebral ischemia following experimental subarachnoid hemorrhage. *European Journal of Pharmacology, 406*, 219–225.

Ali, M., Falkenhain, K., Njiru, B. N., Murtaza-Ali, M., Ruiz-Uribe, N. E., Haft-Javaherian, M., Catchers, S., Nishimura, N., Schaffer, C. B., & Bracko, O. (2022). VEGF signalling causes stalls in brain capillaries and reduces cerebral blood flow in Alzheimer's mice. *Brain, 145*(4), 1449–1463.

Al-Sarraf, H., & Phillip, L. (2003). Effect of hypertension on the integrity of blood brain and blood CSF barriers, cerebral blood flow and CSF secretion in the rat. *Brain Research, 975*, 179–188.

Anesten, B., Zetterberg, H., Nilsson, S., Brew, B. J., Fuchs, D., Price, R. W., Gisslen, M., & Yilmaz, A. (2022). Correction: Effect of antiretroviral treatment on blood-brain barrier integrity in HIV-1 infection. *BMC Neurology, 22*(1), 197.

Arimura, A. (1992). Pituitary adenylate cyclase activating polypeptide (PACAP): Discovery and current status of research. *Regulatory Peptides, 37*, 287–303.

Avison, M. J., Nath, A., Greene-Avison, R., Schmitt, F. A., Bales, R. A., Ethisham, A., Greenberg, R. N., & Berger, J. R. (2004). Inflammatory changes and breakdown of microvascular integrity in early human immunodeficiency virus dementia. *Journal of Neurovirology, 10*, 223–232.

Ayre, S. G., Skaletski, B., & Mosnaim, A. D. (1989). Blood-brain barrier passage of azidothymidine in rats: Effect of insulin. *Research Communications in Chemical Pathology and Pharmacology, 63*, 45–52.

Baba, M., Oishi, R., & Saeki, K. (1988). Enhancement of blood-brain barrier permeability to sodium fluorescein by stimulation of mu opioid receptors in mice. *Naunyn-Schmiedeberg's Archives of Pharmacology, 37*, 423–428.

Balabanov, R., & Dore-Duffy, P. (1998). Role of the CNS microvascular pericyte in the blood-brain barrier. *Journal of Neuroscience Research, 53*, 637–644.

Baldwin, S. A., Fugaccia, I., Brown, D. R., Brown, L. V., & Scheff, S. W. (1996). Blood-brain barrier breach following cortical contusion in the rat. *Journal of Neurosurgery, 85*, 476–481.

Balin, B. J., Broadwell, R. D., Salcman, M., & El-Kalliny, M. (1986). Avenues for entry of peripherally administered protein to the central nervous system in mouse, rat, and squirrel monkey. *The Journal of Comparative Neurology, 251*(2), 260–280.

Ban, E., Milon, G., Prudhomme, N., Fillion, G., & Haour, F. (1991). Receptors for interleukin-1 (‡ and ·) in mouse brain: Mapping and neuronal localization in hippocampus. *Neuroscience, 43*, 21–30.

Banks, W. A. (2004). Are the extracellular pathways a conduit for the delivery of therapeutics to the brain? *Current Pharmaceutical Design, 10*, 1365–1370.

Banks, W. A. (2014). The blood-brain barrier in neuroimmunology: Tales of separation and assimilation. *Brain, Behavior, and Immunity, 14*.

Banks, W. A., & Broadwell, R. D. (1994). Blood to brain and brain to blood passage of native horseradish peroxidase, wheat germ agglutinin and albumin: Pharmacokinetic and morphological assessments. *Journal of Neurochemistry, 62*, 2404–2419.

Banks, W. A., & Kastin, A. J. (1989). Inhibition of the brain to blood transport system for enkephalins and Tyr-MIF-1 in mice addicted or genetically predisposed to drinking ethanol. *Alcohol, 6*, 53–57.

Banks, W. A., & Kastin, A. J. (1990). Editorial review: Peptide transport systems for opiates across the blood-brain barrier. *American Journal of Physiology, 259*, E1–E10.

Banks, W. A., & Kastin, A. J. (1994). In F. L. Strand, B. Beckwith, B. Chronwall, & C. A. Sandman (Eds.), *Brain-to-blood transport of peptides and the alcohol withdrawal syndrome. Models of neuropeptide action* (Vol. 739, pp. 108–118). New York Academy of Sciences.

Banks, W. A., & Kastin, A. J. (1998). Differential permeability of the blood-brain barrier to two pancreatic peptides: Insulin and amylin. *Peptides, 19*, 883–889.

Banks, W. A., Kastin, A. J., Fischman, A. J., Coy, D. H., & Strauss, S. L. (1986). Carrier-mediated transport of enkephalins and N-Tyr-MIF-1 across blood-brain barrier. *American Journal of Physiology, 251*, E477–E482.

Banks, W. A., Kastin, A. J., Komaki, G., & Arimura, A. (1993). Passage of pituitary adenylate cyclase activating polypeptide 1-27 and pituitary adenylate cyclase activating polypeptide 1-38 across the blood-brain barrier. *Journal of Pharmacology and Experimental Therapeutics, 267*, 690–696.

Banks, W. A., Kastin, A. J., & Ehrensing, C. A. (1994). Transport of blood-borne interleukin-1alpha across the endothelial blood-spinal cord barrier of mice. *Journal of Physiology (London), 479*, 257–264.

Banks, W. A., Kastin, A. J., Huang, W., Jaspan, J. B., & Maness, L. M. (1996). Leptin enters the brain by a saturable system independent of insulin. *Peptides, 17*, 305–311.

Banks, W. A., Kastin, A. J., & Akerstrom, V. (1997). HIV-1 protein gp120 crosses the blood-brain barrier: Role of adsorptive endocytosis. *Life Sciences, 61*, L119–L125.

Banks, W. A., Kastin, A. J., & Arimura, A. (1998a). Effect of spinal cord injury on the permeability of the blood-brain and blood-spinal cord barriers to the neurotropin PACAP. *Experimental Neurology, 151*, 116–123.

Banks, W. A., Kastin, A. J., & Ehrensing, C. A. (1998b). Diurnal uptake of circulating interleukin-1alpha by brain, spinal cord, testis and muscle. *Neuroimmunomodulation, 5*, 36–41.

Banks, W. A., Kastin, A. J., Brennan, J. M., & Vallance, K. L. (1999). Adsorptive endocytosis of HIV-1gp120 by blood-brain barrier is enhanced by lipopolysaccharide. *Experimental Neurology, 156*, 165–171.

Banks, W. A., Farr, S. A., La Scola, M. E., & Morley, J. E. (2001). Intravenous human interleukin-1 alpha impairs memory processing in mice: Dependence on blood-brain barrier transport into posterior division of the septum. *Journal of Pharmacology and Experimental Therapeutics, 299*, 536–541.

Banks, W. A., Niehoff, M. L., Martin, D., & Farrell, C. L. (2002). Leptin transport across the blood-brain barrier of the Koletsky rat is not mediated by a product of the leptin receptor gene. *Brain Research, 950*, 130–136.

Banks, W. A., Robinson, S. M., Verma, S., & Morley, J. E. (2003). Efflux of human and mouse amyloid · proteins 1-40 and 1-42 from brain: Impairment in a mouse model of Alzheimer's disease. *Neuroscience, 121*, 487–492.

Banks, W. A., Jumbe, N. L., Farrell, C. L., Niehoff, M. L., & Heatherington, A. (2004a). Passage of erythropoietic agents across the blood-brain barrier: A comparison of human and murine erythropoietin and the analog Darbopoetin alpha. *European Journal of Pharmacology, 505*, 93–101.

Banks, W. A., Niehoff, M. L., & Zalcman, S. (2004b). Permeability of the mouse blood-brain barrier to murine interleukin-2: Predominance of a saturable efflux system. *Brain, Behavior, and Immunity, 18*, 434–442.

Banks, W. A., Robinson, S. M., Wolf, K. M., Bess, J. W., Jr., & Arthur, L. O. (2004c). Binding, internalization, and membrane incorporation of human immunodeficiency virus-1 at the blood-brain barrier is differentially regulated. *Neuroscience, 128*(1), 143–153.

Banks, W. A., Pagliari, P., Nakaoke, R., & Morley, J. E. (2005a). Effects of a behaviorally active antibody on the brain uptake and clearance of amyloid beta proteins. *Peptides, 26*, 287–294.

Banks, W. A., Robinson, S. M., & Nath, A. (2005b). Permeability of the blood-brain barrier to HIV-1 tat. *Experimental Neurology, 193*, 218–227.

Banks, W. A., Farr, S. A., Morley, J. E., Wolf, K. M., Geylis, V., & Steinitz, M. (2007). Anti-amyloid beta protein antibody passage across the blood-brain barrier in the SAMP8 mouse model of Alzheimer's disease: An age related selective uptake with reversal of learning impairment. *Experimental Neurology, 206*, 248–256.

Banks, W. A., Niehoff, M. L., Ponzio, N. M., Erickson, M. A., & Zalcman, S. S. (2012). Pharmacokinetics and modeling of immune cell trafficking: Quantifying differential influences of target tissues versus lymphocytes in SJL and lippolysaccaride-treated mice. *Journal of Neuroinflammation, 9*, 231.

Banks, W. A., Gray, A. M., Erickson, M. A., Salameh, T. S., Damodarasamy, M., Sheibani, N., Meabon, J. S., Wing, E. E., Morofuji, Y., Cook, D. G., & Reed, M. J. (2015). Lipopolysaccharide-induced blood-brain barrier disruption: Roles of cyclooxygenase, oxidative stress, neuroinflammation, and elements of the neurovascular unit. *Journal of Neuroinflammation, 12*, 223.

Barten, D. M., & Ruddle, N. H. (1994). Vascular cell adhesion molecule-1 modulation by tumor necrosis factor in experimental allergic encephalomyelitis. *Journal of Neuroimmunology, 51*, 123–133.

Bauer, B., Hartz, A. M. S., & Miller, D. S. (2007). Tumor necorsis factor alpha and endothelin-1 increase P-glycoprotein expression and transport activity at the blood-brain barrier. *Molecular Pharmacology, 71*, 667–675.

Begley, D. J. (2004). ABC transporters and the blood-brain barrier. *Current Pharmaceutical Design, 10*, 1295–1312.

Bernards, C. M. (1999). Epidural and intrathecal drug movement. In T. L. Yaksh (Ed.), *Spinal Drug Delivery* (pp. 239–252). Elsevier.

Bertrand, L., Velichkovska, M., & Toborek, M. (2021). Cerebral vascular toxicity of antiretroviral therapy. *Journal of Neuroimmune Pharmacology, 16*(1), 74–89.

Bjorbaek, C., Elmquist, J. K., Michl, P., Ahima, R. S., van Beuren, A., McCall, A. L., & Flier, J. S. (1998). Expression of leptin receptor isoforms in rat brain microvessels. *Endocrinology, 139*, 3485–3491.

Blasberg, R. G. (1977). Methotrexate, cytosine arabinoside, and BCNU concentration in brain after ventriculocisternal perfusion. *Cancer Treatment Reports, 61*, 625–631.

Bobardt, M. D., Salmon, P., Wang, L., Esko, J. D., Gabuzda, D., Fiala, M., Trono, D., Van der Schueren, B., David, G., & Gallay, P. A. (2004). Contribution of proteoglycans to human immunodeficiency virus type 1 brain invasion. *Journal of Virology, 78*, 6567–6584.

Bonhomme-Faivre, L., Pelloquin, A., Tardivel, S., Urien, S., Mathieu, M. C., Castagne, V., Lacour, B., & Farinotti, R. (2002). Recombinant interleukin-2 treatment decreases P-glycoprotein activity and paclitaxel metabolism in mice. *Anit-Cancer Drugs, 13*, 51–57.

Boulton, M., Flessner, M., Armstrong, D., Mohamed, R., Hay, J., & Johnston, M. (1999). Contribution of extracranial lymphatics and arachnoid villi to the clearance of a CSF tracer in the rat. *The American Journal of Physiology, 276*, R818–R823.

Bradbury, M. (1979). *The concept of a blood-brain barrier*. John Wiley and Sons Ltd.

Brightman, M. W., & Reese, T. S. (1969). Junctions between intimately apposed cell membranes in the vertebrate brain. *The Journal of Cell Biology, 40*, 648–677.

Broadwell, R. D. (1989). Transcytosis of macromolecules through the blood-brain barrier: A cell biological perspective and critical appraisal. *ACTA Neuropathologica (Berlin), 79*, 117–128.

Broadwell, R. D. (1993). Endothelial cell biology and the enigma of transcytosis through the blood-brain barrier. *Advances in Experimental Medicine and Biology, 331*, 137–141.

Broadwell, R. D., & Banks, W. A. (1993). Cell biological perspective for the transcytosis of peptides and proteins through the mammalian blood-brain fluid barriers. In W. M. Pardridge (Ed.), *The Blood-Brain Barrier* (pp. 165–199). Raven Press, Ltd.

Broadwell, R. D., Balin, B. J., & Salcman, M. (1988). Transcytotic pathway for blood-borne protein through the blood-brain barrier. *Proceedings of the National Academy of Sciences USA, 85*, 632–636.

Brochner, C. B., Holst, C. B., & Mollgard, K. (2015). Outer brain barriers in rat and human development. *Frontiers in Neuroscience, 9*, 75.

Brown, R. C., & Davis, T. P. (2002). Calcium modulation of adherens tight junction function: A potential mechanism for blood-brain barrier disruption after stroke. *Stroke, 33*, 1706–1711.

Butter, C., Baker, D., O'Neill, J. K., & Turk, J. L. (1991). Mononuclear cell trafficking and plasma protein extravasation into the CNS during chronic relapsing experimental allergic encephalomyelitis in Biozzi AB/H mice. *Journal of Neurological Science, 104*, 9–12.

Calias, P., Papisov, M., Pan, J., Savioli, N., Belov, V., Huang, Y., Lotterhand, J., Alessandrini, M., Liu, N., Fischman, A. J., Powell, J. L., & Heartlein, M. W. (2012). CNS penetration of intrathecal-lumbar idursulfase in the monkey, dog and mouse: Implications for neurological outcomes of lysosomal storage disorder. *PLoS One, 7*, e30341.

Cangiano, C., Cardelli-Cangiano, P., Cascino, A., Patrizi, M. A., Barberini, F., Rossi, F., Capocaccia, L., & Strom, R. (1983). On the stimulation by insulin of tryptophan transport across the blood-brain barrier. *Biochemistry International, 7*, 617–627.

Cashion, M. F., Banks, W. A., Bost, K. L., & Kastin, A. J. (1999). Transmission routes of HIV-1 gp120 from brain to lymphoid tissues. *Brain Research, 822*, 26–33.

Castagne, V., Bonhomme-Faivre, L., Urien, S., Reguiga, M. D., Soursac, M., Gimenez, F., & Farinotti, R. (2004). Effect of recombinant interleukin-2 pretreatment on oral and intravenous digoxin pharmacokinetics and P-glycoprotein activity in mice. *Drug Metabolism and Disposition, 32*, 168–171.

Catalan, R. E., Martinez, A. M., Aragones, M. D., Miguel, B. G., & Robles, A. (1988). Insulin action on brain microvessels; effect on alkaline phosphatase. *Biochemical and Biophysical Research Communications, 150*, 583–590.

Chan, G. N., Patel, R., Cummins, C. L., & Bendayan, R. (2013). Induction of P-glycoprotein by antiretroviral drugs in human brain microvessel endothelial cells. *Antimicrobial Agents and Chemotherapy, 57*, 4481–4488.

Chang, S. L., Felix, B., Jiang, Y., & Fiala, M. (2001). Actions of endotoxin and morphine. *Advances in Experimental Medicine and Biology, 493*, 187–196.

Chao, C. C., Gekker, G., Sheng, W. S., Hu, S., Tsang, M., & Peterson, P. K. (1994). Priming effect of morphine on the production of tumor necrosis factor-alpha by microglia: Implications in respiratory burst activity and human immunodeficiency virus-1 expression. *Journal of Pharmacology and Experimental Therapeutics, 269*, 198–203.

Chehade, J. M., Hass, M. J., & Mooradian, A. D. (2002). Diabetes-related changes in rat cerebral occlusin and zonula occludens-1 (ZO-1) expression. *Neurochemical Research, 27*, 249–252.
Chen, G., & Reichlin, S. (1998). Clearance of [125 I]-tumor necrosis factor-alpha from the brain into the blood after intracerebroventricular injection into rats. *Neuroimmunomodulation, 5*, 261–269.
Chen, G., Castro, W. L., Chow, H. H., & Reichlin, S. (1997). Clearance of 125 I-labelled interleukin-6 from brain into blood following intracerebroventricular injection in rats. *Endocrinology, 138*, 4830–4836.
Chen, P., Shibata, M., Zidovetzki, R., Fisher, M., Zlokovic, B. V., & Hofman, F. M. (2001). Endothelin-1 and monocyte chemoattractant protein-1 modulation in ischemia and human brain-derived endothelial cell cultures. *Journal of Neuroimmunology, 116*, 62–73.
Ching, S., He, L., Lai, W., & Quan, N. (2005). IL-1 type I receptor plays a key role in mediating the recruitement of leukocytes into the central nervous system. *Brain, Behavior, and Immunity, 19*, 127–137.
Ching, S., Zhang, H., Lai, W., & Quan, N. (2006). Peripheral injection of lipopolysaccharide prevents brain recruitment of leukocytes induced by central injection of interleukin-1. *Neurosci, 137*, 717–726.
Choe, Y. G., Yoon, J. H., Joo, J., Kim, B., Hong, S. P., Koh, G. Y., Lee, D. S., Oh, W. Y., & Jeong, Y. (2022). Pericyte loss leads to capillary stalling through increased leukocyte-endothelial cell interaction in the brain. *Frontiers in Cellular Neuroscience, 16*, 848764.
Chou, S., & Dix, R. D. (1989). Viral infections and the blood-brain barrier. Implications of the blood-brain barrier and its manipulation. In E. A. Neuwelt (Ed.), *Volume 2: Clinical aspects* (pp. 449–468). Plenum Publishing Corporation.
Christianson, H. C., & Belting, M. (2014). Heparan sulfate proteoglycan as a cell-surface endocytosis receptor. *Matrix Biology, 35*, 51–55.
Cleuren, A. C. A., van der Ent, M. A., Jiang, H., Hunker, K. L., Yee, A., Siemieniak, D. R., Molema, G., Aird, W. C., Ganesh, S. K., & Ginsburg, D. (2019). The in vivo endothelial cell translatome is highly heterogeneous across vascular beds. *Proceedings of the National Academy of Sciences of the United States of America, 116*(47), 23618–23624.
Clifford, D. B., De Luca, A., Simpson, D. M., Arendt, G., Giovannoni, G., & Nath, A. (2010). Natalizumab-associated progressive multifocal leukoencephalopathy in patients with multiple sclerosis: Lessons from 28 cases. *Lancet Neurology, 9*(4), 438–446.
Cosby, S. L., & Brankin, B. (1995). Measles virus infection of cerebral endothelial cells and effect on their adhesive properties. *Veterinary Microbiology, 44*, 135–139.
Cserr, H. F. (1984). Convection of brain interstitial fluid. In K. Shapiro, A. Marmarou, & H. Portnoy (Eds.), *Hydrocephalus* (pp. 59–68). Raven Press.
Cserr, H. F., & Berman, B. J. (1978). Iodide and thiocyanate efflux from brain following injection into rat caudate nucleus. *American Journal of Physiology, 4*, F331–F337.
Cserr, H. F., & Knopf, P. M. (1992). Cervical lymphatics, the blood-brain barrier and the immunoreactivity of the brain: A new view. *Immunology Today, 13*, 507–512.
Cunningham, E. T., Jr., Wada, E., Carter, D. B., Tracey, D. E., Battey, J. F., & De Souza, E. B. (1992). In situ histochemical localization of type I interleukin-1 receptor messenger RNA in the central nervous system, pituitary, and adrenal gland of the mouse. *Journal of Neuroscience, 12*, 1101–1114.
Daneman, R., Zhou, L., Kebede, A. A., & Barres, B. A. (2010). Pericytes are required for blood-brain barrier integrity during embryogenesis. *Nature, 468*, 562–566.
Davies, P. J. A., Davies, D. R., Levitzki, A., Maxfield, F. R., Milhaud, P., Willingham, M. C., & Pastan, I. H. (1980). Transglutaminase is essential in receptor-mediated endocytosis of ‡-macroglobulin and polypeptide hormones. *Nature, 283*, 162–167.
Davson, H. (1976). Review lecture. The blood-brain barrier. *The Journal of Physiology, 255*(1), 1–28.
Davson, H., & Segal, M. B. (1996a). *Blood-brain-CSF relations. Physiology of the CSF and blood-brain barriers* (pp. 257–302). CRC Press.
Davson, H., & Segal, M. B. (1996b). *The proteins and other macromolecules of the CSF. Physiology of the CSF and the blood-brain barrier* (pp. 573–606). CRC Press.
Davson, H., & Segal, M. B. (1996c). Special aspects of the blood-brain barrier. In *Physiology of the CSF and Blood-Brain Barriers* (pp. 303–485). CRC Press.
de la Torre, J. C., & Mussivand, T. (1993). Can disturbed brain microcirculation cause Alzheimer's disease? *Neurological Research, 15*, 146–153.
de Lange, E. C. M., Bouw, M. R., Danhof, M., De Boer, A. G., & Breimer, D. D. (1993). Application of intracerebral microdialysis to study regional distribution kinetics of atenolol and acetaminophen in rat brain. In *The use of intracerebral microdialysis to study the blood-brain barrier transport characteristics of drugs* (pp. 93–106). (thesis, Leiden/Amsterdam Center for Drug Research). Leiden, Sinteur.
de Lange, E. C., Bouw, M. R., Mandema, J. W., Danhof, M., De Boer, A. G., & Breimer, D. D. (1995). Application of intracerebral microdialysis to study regional distribution kinetics of drug in rat brain. *British Journal of Pharmacology, 116*, 2538–2544.
De Vries, H. E., Moor, A. C., Blom-Roosemalen, M. C., De Boer, A. G., Breimer, D. D., van Berkel, T. J., & Kuiper, J. (1994). Lymphocyte adhesion to brain capillary endothelial cells in vitro. *Journal of Neuroimmunology, 52*, 1–8.
Deane, R., Wu, Z., Sagare, A., Davis, J., Du Yan, S., Hamm, K., Xu, F., Parisi, M., LaRue, B., Hu, H. W., Spijkers, P., Guo, H., Song, X., Lenting, P. J., Van Nostrand, W. E., & Zlokovic, B. (2004). LRP/amyloid beta-peptide interaction mediates differential brain efflux of Abeta isoforms. *Neuron, 43*(3), 333–344.
Deane, R., Sagare, A., & Zlokovic, B. (2008). The role of the cell surface LRP and soluble LRP in blood-brain barrier A· clearance in Alzheimer's disease. *Current Pharmaceutical Design, 14*, 1601–1605.
Deli, M. A., Abraham, C. R., Kataoka, Y., & Niwa, M. (2005). Permeability studies on in vitro blood-brain barrier models: Physiology, pathology, and pharmacology. *Cellular and Molecular Neurobiology, 25*, 59–127.
Derk, J., Jones, H. E., Como, C., Pawlikowski, B., & Siegenthaler, J. A. (2021). Living on the edge of the CNS: Meninges cell diversity in health and disease. *Frontiers in Cellular Neuroscience, 15*, 703944.
Didier, N., Banks, W. A., Creminon, C., Dereuddre-Bosquet, N., & Mabondzo, A. (2002). HIV-1-induced production of endothelin-1 in an in vitro model of the human blood-brain barrier. *Neuroreport, 13*(9), 1179–1183.
Dogrukol-Ak, D., Kumar, V. B., Ryerse, J. S., Farr, S. A., Verma, S., Nonaka, K., Nakamachi, T., Ohtaki, H., Niehoff, M. L., Edwards, J. C., Shioda, S., Morley, J. E., & Banks, W. A. (2009). Isolation of peptide transport system-6 from brain endothelial cells: Therapeutic effects with antisense inhibition in Alzheimer's and stroke models. *Journal of Cerebral Blood Flow and Metabolism, 29*, 411–422.
Dohgu, S., & Banks, W. A. (2008). Lipopolysaccharide-enhanced transcellular transport of HIV-1 across the blood-brain barrier is medited by the p38 mitogen-activated protein kinase pathway. *Experimental Neurology, 210*, 740–749.
Dohgu, S., & Banks, W. A. (2013). Brain pericytes increase the lipopolysaccharide-enhanced transcytosis of HIV-1 free virus across the in vitro blood-brain barrier: Evidence for cytokine-mediated pericyte-endothelial cell cross talk. *Fluids and Barriers of the CNS, 10*, 23.
Dohgu, S., Fleegal-DeMotta, M. A., & Banks, W. A. (2011). Lipopolysaccharide-enhanced transcellular transport of HIV-1

across the blood-brain barrier is mediated by luminal microvessel IL-6 and GM-CSF. *Journal of Neuroinflammation, 8*, 167.

Dore-Duffy, P. (2008). Pericytes: Pluripotent cells of the blood brain barrier. *Current Pharmaceutical Design, 14*, 1581–1593.

Dore-Duffy, P., Owen, C., Balabanov, R., Murphy, S., Beaumont, T., & Rafols, J. A. (2000). Pericyte migration from the vascular wall in response to traumatic brain injury. *Microvascular Research, 60*, 55–69.

Drach, J., Gsur, A., Hamilton, G., Zhao, S., Angerler, J., Fiegl, M., Zojer, N., Raderer, M., Haberl, I., Andreeff, M., & Huber, H. (1996). Involvement of P-glycoprotein in the transmembrane transport of interleukin-2 (IL-2), IL-4, and interferon-gamma in normal human T lymphocytes. *Blood, 88*, 1747–1754.

Drion, N., Lemaire, M., Lefauconnier, J. M., & Scherrmann, J. M. (1996). Role of p-glycoprotein in the blood-brain transport of colchicine and vinblastine. *Journal of Neurochemistry, 67*, 1688–1693.

Ehrenreich, H., Hasselblatt, M., Dembowski, C., Depek, L., Lewczuk, P., Stiefel, M., Rustenbeck, H. H., Breiter, N., Jacob, S., Knerlich, F., Bohn, M., Poser, W., Rither, E., Kochen, M., Gefeller, O., Gleiter, C., Wessel, T. C., De Ryck, M., Itri, L., Prange, H., Cerami, A., Brines, M., & Siren, A. L. (2002). Erythropoietin therapy for acute stroke is both safe and beneficial. *Molecular Medicine, 8*, 495–505.

El Amki, M., Gluck, C., Binder, N., Middleham, W., Wyss, M. T., Weiss, T., Meister, H., Luft, A., Weller, M., Weber, B., & Wegener, S. (2020). Neutrophils obstructing brain capillaries are a major cause of no-reflow in ischemic stroke. *Cell Reports, 33*(2), 108260.

Elferink, R. P. J. O., & Zadina, J. E. (2001). MDR1 P-glycoprotein transports endogenous opioid peptides. *Peptides, 22*, 2015–2020.

Engelhardt, B. (2008). The blood-central nervous system barriers actively control immune cell entry into the central nervous system. *Current Pharmaceutical Design, 14*, 1555–1565.

Engelhardt, B., & Coisne, C. (2011). Fluids and barriers of the CNS establish immune privilege by confining immune surveillance to a two-walled castle moat surrounding the CNS castle. *Fluids and Barriers of the CNS, 8*(1), 4.

Engelhardt, B., & Ransohoff, R. M. (2005). The ins and outs of T-lymphocyte trafficking to the CNS: Anatomical sites and molecular mechanisms. *Trends in Immunology, 26*(9), 485–495.

Engelhardt, B., & Wolburg, H. (2004). Minireview: Transendothelial migration of leukocytes: Through the front door or around the side of the house? *European Journal of Pharmacology, 34*, 2955–2963.

Engelhardt, B., Vajkoczy, P., & Weller, R. O. (2017). The movers and shapers in immune privilege of the CNS. *Nature Immunology, 18*(2), 123–131.

Erbyraktar, S., Grasso, G., Sfacteria, A., Xie, Q. W., Coleman, T., Kreilgaard, M., Torup, L., Sager, T., Erbayraktar, Z., Gokmen, N., Yilmaz, O., Ghezzi, P., Villa, P., Fratelli, M., Casagrande, S., Leist, M., Helboe, L., Gerwein, J., Christensen, S., Geist, M. A., Pedersen, L. O., Cerami-Hand, C., Wuerth, J. P., Cerami, A., & Brines, M. (2003). Asialoerythropoietin is a nonerythropoiectic cytokine with broad neuroprotective activity in vivo. *Proceedings of the National Academy of Sciences USA, 100*, 6741–6746.

Erickson, M. A., Morofuji, Y., Owen, J. B., & Banks, W. A. (2014). Rapid transport of CCL11 across the blood-brain barrier: Regional variation and importance of blood cells. *J Pharmacol Exp Therap, 349*, 497–507.

Erickson, M. A., Liang, W. S., Fernandez, E. G., Bullock, K. M., Thysell, J. A., & Banks, W. A. (2018). Genetics and sex influence peripheral and central innate immune responses and blood-brain barrier integrity. *PLoS One, 13*(10), e0205769.

Eskilsson, A., Matsuwaki, T., Shionoya, K., Mirrasekhian, E., Zajdel, J., Schwaninger, M., Engblom, D., & Blomqvist, A. (2017). Immune-induced fever is dependent on local but not generalized prostaglandin E(2) synthesis in the brain. *The Journal of Neuroscience, 37*(19), 5035–5044.

Fabry, Z., Fitzsimmons, K. M., Herlein, J. A., Moninger, T. O., Dobbs, M. B., & Hart, M. N. (1993). Production of the cytokines interleukin 1 and 6 by murine brain microvessel endothelium and smooth muscle pericytes. *Journal of Neuroimmunology, 47*, 23–34.

Faraci, F. M., & Heistad, D. D. (1998). Regulation of the cerebral circulation: Role of endothelium and potassium channels. *Physiological Reviews, 78*(1), 53–97.

Farr, S. A., Banks, W. A., Uezu, K., Sano, A., Gaskin, F. S., & Morley, J. E. (2003). Antibody to beta-amyloid protein increases acetylcholine in the hippocampus of 12 month SAMP8 male mice. *Life Sciences, 73*, 555–562.

Ferguson, A. V. (1991). The area postrema: A cardiovascular control Centre at the blood-brain interface? *Canadian Journal of Physiology & Pharmacology, 69*, 1026–1034.

Fukui, S., Schwarcz, R., Rapoport, S. I., Takada, Y., & Smith, Q. R. (1991). Blood-brain barrier transport of kynurenines: Implications for brain synthesis and metabolism. *Journal of Neurochemistry, 56*(6), 2007–2017.

Ghersi-Egea, J. F., Gorevic, P. D., Ghiso, J., Frangione, B., Patlak, C. S., & Fenstermacher, J. D. (1996). Fate of cerebrospinal fluid-borne amyloid ·-peptide: Rapid clearance into blood and appreciable accumulation by cerebral arteries. *Journal of Neurochemistry, 67*, 880–883.

Glynn, S. L., & Yazdanian, M. (1998). In vitro blood-brain barrier permeability of nevirapine compared to other HIV antiretroviral agents. *Journal of Pharmaceutical Sciences, 87*, 306–310.

Grammas, P., & Caspers, M. L. (1991). The effect of aluminum on muscarinic receptors in isolated cerebral microvessels. *Research Communications in Chemical Pathology and Pharmacology, 72*, 69–79.

Greenwood, J., Bamforth, S., Wang, Y., & Devine, L. (1995). In J. Greenwood, D. J. Begley, & M. B. Segal (Eds.), *The blood-retinal barrier in immune-mediated diseases of the retina. New concepts of a blood-brain barrier* (pp. 315–326). Plenum Press.

Greenwood, J., Heasman, S. J., Alvarez, J. I., Pratt, A., Lyck, R., & Engelhardt, B. (2011). Review: Leukocyte-endothelial cell crosstalk at the blood-brain barrier: A prerequisite for successful immune cell entry to the brain. *Neuropathology and Applied Neurobiology, 37*, 24–39.

Gross, P. M., Blasberg, R. G., Fenstermacher, J. D., & Patlak, C. S. (1987). The microcirculation of rat circumventricular organs and the pituitary gland. *Brain Research Bulletin, 18*, 73–85.

Grubb, J. H., Vogler, C., Levy, B., Galvin, N., Tan, Y., & Sly, W. S. (2008). Chemically modified beta-glucuronidase crosses blood-brain barrier and clears neuronal storage in murine mucopolysaccharidosis VII. *PNAS, 105*, 2616–2621.

Gutierrez, E. G., Banks, W. A., & Kastin, A. J. (1993). Murine tumor necrosis factor alpha is transported from blood to brain in the mouse. *Journal of Neuroimmunology, 47*, 169–176.

Hardebo, J. E., & Kahrstrom, J. (1985). Endothelial negative surface charge areas and blood-brain barrier function. *Acta Physiologica Scandinavica, 125*, 495–499.

Hardebo, J. E., & Owman, C. (1990). In B. B. Johansson, C. Owman, & H. Widner (Eds.), *Enzymatic barrier mechanisms for neurotransmitter monoamines and their precursors at the blood-brain barrier. Pathophysiology of the blood-brain barrier* (pp. 41–55). Elsevier.

Hartz, A. M. S., Bauer, B., Fricker, G., & Miller, D. S. (2006). Rapid modulation of P-glycoprotein-mediated transport at the blood-brain barrier by tumor necrosis factor-alpha and lipopolysaccharide. *Molecular Pharmacology, 69*, 462–470.

Herve, F., Ghinea, N., & Scherrmann, J. M. (2008). CNS delivery via adsorptive transcytosis. *The AAPS Journal, 10*, 455–472.

Hock, C., Konietzko, U., Streffer, J. R., Tracy, J., Signorell, A., Muller-Tillmanns, B., Lemke, U., Henke, K., Moritz, E., Garcia, E., Wollmer, M. A., Umbricht, D., de Quervain, D. J., Hofmann, M., Maddalena, A., Papassotiropoulos, A., & Nitsch, R. M.

(2003). Antibodies against beta-amyloid slow cognitive decline in Alzheimer's disease. *Neuron, 38*, 547–554.

Hofman, F., Chen, P., Incardona, F., Zidovetzki, R., & Hinton, D. R. (1999). HIV-tat protein induces the production of interleukin-8 by human brain-derived endothelial cells. *Journal of Neuroimmunology, 94*, 28–39.

Holash, J. A., Harik, S. I., Perry, G., & Stewart, P. A. (1993). Barrier properties of testis microvessels. *Proceedings of the National Academy of Sciences USA, 90*, 11069–11073.

Hong, S., & Banks, W. A. (2014). Role of the immune system in HIV-associated neuroinflammation and neurocognitive implications. *Brain, Behavior, and Immunity*.

Hsuchou, H., Jayaram, B., Kastin, A. J., Wang, Y., Ouyang, S., & Pan, W. (2013a). Endothelial cell leptin receptor mutant mice have hyperleptinemia and reduced tissue uptake. *Journal of Cellular Physiology, 228*, 1610–1616.

Hsuchou, H., Pan, W., & Kastin, A. J. (2013b). Fibroblast growth factor 19 entry into brain. *Fluids and Barriers of the CNS, 10*, 32.

Huber, J. D. (2008). Diabetes, cognitive function, and the blood-brain barrier. *Current Pharmaceutical Design, 14*, 1594–1600.

Huber, J. D., Hau, V. S., Borg, L., Campos, C. R., Egleton, R. D., & Davis, T. P. (2002). Blood-brain barrier tight junctions are altered during a 72-h exposure to lamba-carrageenan-induced inflammatory pain. *American Journal of Physiology, 283*, H1531–H1537.

Huber, J. D., Van Gilder, R. L., & Houser, K. A. (2006). Streptozotocin-induced diabetes progressively increases blood-brain barrier permeability in specific brain regions in rats. *American Journal of Physiology, 291*, H2660–H2668.

Iliff, J. J., Wang, M., Liao, Y., Plogg, B. A., Peng, W., Gundersen, G. A., Benveniste, H., Vates, G. E., Deane, R., Goldman, S. A., Nagelhus, E. A., & Nedergaard, M. (2012). A paravascular pathway facilitates CSF flow through the brain parenchyma and the clearance of interstitial solutes, including amyloid beta. *Science Translational Medicine, 4*, 147ra111.

Iliff, J. J., Lee, H. C., Yu, M., Feng, T., Logan, J., Nedergaard, M., & Benveniste, H. (2013). Brain-wide pathway for waste clearance captured by contrast-enhanced MRI. *The Journal of Clinical Investigation, 123*, 1299–1309.

Inoue, W., Matsumura, K., Yamagata, K., Takemiya, T., Shiraki, T., & Kobayashi, S. (2002). Brain-specific endothelial induction of prostaglandin E(2) synthesis enzymes and its temporal relation to fever. *Neuroscience Research, 44*(1), 51–61.

Jaeger, L. B., Dohgu, S., Hwang, M. C., Farr, S. A., Murphy, M. P., Fleegal-DeMotta, M. A., Lynch, J. L., Robinson, S. M., Niehoff, M. L., Johnson, S. N., Kumar, V. B., & Banks, W. A. (2009a). Testing the neurovascular hypothesis of Alzheimer's disease: LRP-1 antisense reduces blood-brain barrier clearance, increases brain levels of amyloid-beta protein, and impairs cognition. *Journal of Alzheimer's Disease, 17*, 553–570.

Jaeger, J. B., Dohgu, S., Lynch, J. L., Fleegal-DeMotta, M. A., & Banks, W. A. (2009b). Effects of lipopolysaccharide on the blood-brain barrier transport of amyloid beta protein: A mechanism for inflammation in the progression of Alzheimer's disease. *Brain, Behavior, and Immunity, 23*, 507–517.

Janus, C., Pearson, J., McLaurin, J., Mathews, P. M., Jiang, Y., Schmidt, S. D., Chishti, M. A., Horne, P., Heslin, D., French, J., Mount, H. T. J., Nixon, R. A., Mercken, M., Bergeron, C., Fraser, P. E., George-Hyslop, P., & Westaway, D. (2000). A· peptide immunization reduces behavioral impairment and plaques in a model of Alzheimer's disease. *Nature, 408*, 979–982.

Johanson, C. E. (1988). The choroid plexus-arachnoid membrane-cerebrospinal fluid system. In A. A. Boulton, G. B. Baker, & W. Walz (Eds.), *Neuromethods; the neuronal microenvironment* (pp. 33–104). The Humana Press.

Johansson, B. B. (1989). Hypertension and the blood-brain barrier. In E. A. Neuwelt (Ed.), *Implications of the blood-brain barrier and its manipulation. Vol 2. Clinical Aspects* (pp. 389–410). Plenum Publishing Co.

Johnson, E. M. J., & Deckwerth, T. L. (1993). Molecular mechanisms of developmental neuronal death. *Annual Review of Neuroscience, 16*, 31–46.

Jones, P. M., & Robinson, I. C. A. F. (1982). Differential clearance of neurophysin and neurohypophysial peptides from the cerebrospinal fluid in conscious Guinea pigs. *Neuroendocrinology, 34*, 297–302.

Jordan-Sciutto, K. L. (2021). Effects of antiretroviral therapy in the central nervous system: Beyond viral suppression. *Journal of Neuroimmune Pharmacology, 16*(1), 71–73.

Joshi, B. S., & Zuhorn, I. S. (2021). Heparan sulfate proteoglycan-mediated dynamin-dependent transport of neural stem cell exosomes in an in vitro blood-brain barrier model. *The European Journal of Neuroscience, 53*(3), 706–719.

Juhler, M., Barry, D. I., Offner, H., Konat, G., Klinken, L., & Paulson, O. B. (1984). Blood-brain and blood-spinal cord barrier permeability during the course of experimental allergic encephalomyelitis in the rat. *Brain Research, 302*, 347–355.

Kabura, L., Ilibagiza, D., Menten, J., & Van den Ende, J. (2006). Intrathecal vs intramuscular administration of human antitetanus immunoglobulin or equine tetanus antitoxin in the treatment of tetanus: A meta-analysis. *Tropical Medicine & International Health, 11*, 1075–1081.

Kaiser, H., Parker, E., & Hamrick, M. W. (2020). Kynurenine signaling through the aryl hydrocarbon receptor: Implications for aging and healthspan. *Experimental Gerontology, 130*, 110797.

Kalaria, R. N., & Harik, S. I. (1989). Increased alpha 2- and beta 2-adrenergic receptors in cerebral microvessels in Alzheimer disease. *Neuroscience Letters, 106*, 233–238.

Kalaria, R. N., Mitchell, M. J., & Harik, S. I. (1987). Correlation of 1-methyl-4-phenyl-1,2,3,6-tetrahydropyridine neurotoxicity with blood-brain barrier monoamine oxidase activity. *Proceedings of the National Academy of Sciences USA, 84*, 3521–3525.

Kastin, A. J., & Akerstrom, V. (2001). Glucose and insulin increase the transport of leptin through the blood-brain barrier in normal mice but not in streptozotocin-diabetic mice. *Neuroendocrinology, 73*, 237–242.

Kastin, A. J., Akerstrom, V., & Pan, W. (2000). Activation of urocortin transport into brain by leptin. *Peptides, 21*, 1811–1817.

Kastin, A. J., Fasold, M. B., & Zadina, J. E. (2002). Endomorphins, met-enkephalin, Tyr-MIF-1 and the P-glycoprotein efflux system. *Drug Metabolism and Disposition, 30*, 231–234.

Kaur, J., Jaswal, V. M., Nagpaul, J. P., & Mahmood, A. (1992). Chronic ethanol feeding and microvillus membrane glycosylation in normal and protein-malnourished rat intestine. *Nutrition, 8*, 338–342.

Kety, S. S. (1987). In G. Adelman (Ed.), *Cerebral circulation and its measurement by inert diffusible tracers. Encyclopedia of Neuroscience, Volume I* (pp. 206–208). Birkh, User.

Khan, N. A., DiCello, F., Nath, A., & Kim, K. S. (2003). Human immunodeficiency virus type 1 tat-mediated cytotoxicity of human brain microvascular endothelial cells. *Journal of Neurovirology, 9*, 584–593.

King, M., Su, W., Chang, A., Zuckerman, A., & Pasternak, G. W. (2001). Transport of opioids from the brain to the periphery by P-glycoprotein: Peripheral actions of central drugs. *Nature Neuroscience, 4*, 221–222.

Kivisakk, P., Mahad, D. J., Callahan, M. K., Trebst, C., Tucky, B., Wei, T., Wu, L., Baekkevold, E. S., Lassmann, H., Staugaitis, S. M., Campbell, J. J., & Ransohoff, R. M. (2003). Human cerebrospinal fluid central memory CD4+ T cells: Evidence for trafficking through the choroid plexus and meninges via P-selectin. *Proceedings of the National Academy of Sciences of USA, 100*, 8389–8394.

Klein, M. A., Frigg, R., Flechsig, E., Raeber, A. J., Kalinke, U., Bluethmann, H., Bootz, F., Suter, M., Zinkernagel, R. M., &

Aguzzi, A. (1997). A crucial role for B cells in neuroinvasive scrapie. *Nature, 390*, 687–690.
Knopf, P. M., Cserr, H. F., Nolan, S. C., Wu, T. Y., & Harling-Berg, C. J. (1995). Physiology and immunology of lymphatic drainage of interstitial and cerebrospinal fluid from the brain. *Neuropathology and Applied Neurobiology, 21*, 175–180.
Koenig, H., Trout, J. J., Goldstone, A. D., & Lu, C. Y. (1992). Capillary NMDA receptors regulate blood-brain barrier function and breakdown. *Brain Research, 588*, 297–303.
Kovac, A., Erickson, M. A., & Banks, W. A. (2011). Brain microvascular pericytes are immunoactive in culture: Cytokine, chemokine, nitric oxide, and LRP-1 expression in response to lipopolysaccharide. *Journal of Neuroinflammation, 8*, 139.
Koyanagi, Y., Tanaka, Y., Kira, J., Ito, M., Hioki, K., Misawa, N., Kawano, Y., Yamasaki, K., Tanaka, R., Suzuki, Y., Ueyama, Y., Terada, E., Tanaka, T., Myasaka, M., Kobayashi, T., Kumazawa, Y., & Yamamoto, N. (1997). Primary human immunodeficiency virus type 1 viremia and central nervous system invasion in a novel hu-PBL-immunodeficient mouse strain. *Journal of Virology, 71*, 2417–2424.
Kozlowski, G. P., Sterzl, I., & Nilaver, G. (1992). Localization patterns for immunoglobulins and albumins in the brain suggest diverse mechanisms for their transport across the blood-brain barrier (BBB). In A. Ermisch, R. Landgraf, & H. J. RÅhle (Eds.), *Progress in brain research* (Vol. 91, pp. 149–154). Elsevier.
Krizbai, I. A., Deli, M. A., Pestenacz, A., Siklose, L., Szabo, C. A., Andras, I., & Joo, F. (1998). Expression of glutamate receptors on cultured cerebral endothelial cells. *Journal of Neuroscience Research, 54*, 814–819.
Kunis, G., Baruch, K., Rosenzweig, N., Kertser, A., Miller, O., Berkutzki, T., & Schwartz, M. (2013). IFN-gamma-dependent activation of the brain's choroid plexus for CNS immune surveillance and repair. *Brain, 136*(Pt 11), 3427–3440.
Kutuzov, N., Flyvbjerg, H., & Lauritzen, M. (2018). Contributions of the glycocalyx, endothelium, and extravascular compartment to the blood-brain barrier. *Proceedings of the National Academy of Sciences of the United States of America, 115*(40), E9429–e9438.
Langlet, F., Mullier, A., Bouret, S. G., Prevot, V., & Dehouck, B. (2013). Tanycyte-like cells from a blood-cerebrospinal fluid barrier in the circumventricular organs of the mouse brain. *The Journal of Comparative Neurology, 521*, 3389–3405.
LeBel, C., Bourdeau, A., Lau, D., & Hunt, P. (1999). Biological response to peripheral and central administration of recombinant human leptin in dogs. *Obesity Research, 7*, 577–585.
Lee, C. G. L., Gottesman, M. M., Cardarelli, C. O., Ramachandra, M., Jeang, K. T., Ambudkar, S. V., Pastan, I., & Dey, S. (1998). HIV-1 protease inhibitors are substrates for the MDR1 multidrug transporter. *Biochemistry, 37*, 3594–3601.
Lee, Y. W., Hennig, B., Fiala, M., Kim, K. S., & Toborek, M. (2001). Cocaine activates redox-regulated transcription factors and induces TNF-alpha expression in human brain endothelial cells. *Brain Research, 920*, 125–133.
Levin, V. A. (1980). Relationship of octanol/water partition coefficient and molecular weight to rat brain capillary permeability. *Journal of Medicinal Chemistry, 23*, 682–684.
Lipinski, C. A., Lombardo, F., Dominy, B. W., & Feeney, P. J. (1997). Experimental and computational approaches to estimate solubility and permeability in drug discovery and developmental settings. *Advanced Drug Delivery Reviews, 23*, 3–25.
Loscher, W., & Potschka, H. (2002). Role of multidrug transporters in pharmacoresistance to antiepileptic drugs. *Journal of Pharmacology and Experimental Therapeutics, 30*, 7–14.
Lossinsky, A. S., Vorbrodt, A. W., & Wisniewski, H. M. (1983). Ultracytochemical studies of vesicular and canalicular transport structures in the injured mammalian blood-brain barrier. *ACTA Neuropathologica (Berlin), 61*, 239–245.
Lossinsky, A. S., Pluta, R., Song, M. J., Badmajew, V., Moretz, R. C., & Wisniewski, H. M. (1991). Mechanisms of inflammatory cell attachment in chronic relapsing experimental allergic encephalomyelitis: A scanning and high-voltage electron microscopic study of the injured mouse blood-brain barrier. *Microvascular Research, 41*, 299–310.
Male, D. (1995). The blood-brain barrier—no barrier to a determined lymphocyte. In J. Greenwood, D. J. Begley, & M. B. Segal (Eds.), *New concepts of a blood-brain barrier* (pp. 311–314). Press.
Mandi, Y., Ocsovszki, I., Szabo, D., Nagy, Z., Nelson, J., & Molnar, J. (1998). Nitric oxide production and MDR expression by human brain endothelial cells. *Anticancer Research, 18*(4C), 3049–3052.
Maness, L. M., Kastin, A. J., Farrell, C. L., & Banks, W. A. (1998). Fate of leptin after intracerebroventricular injection into the mouse brain. *Endocrinology, 139*, 4556–4562.
Mapunda, J. A., Tibar, H., Regragui, W., & Engelhardt, B. (2022). How does the immune system enter the brain? *Frontiers in Immunology, 13*, 805657.
Marsh, M. (1984). The entry of enveloped viruses into cells by endocytosis. *Biochemical Journal, 218*, 1–10.
Martinez, L. O., Jacquet, S., Esteve, J. P., Rolland, C., Cabezon, E., Champagne, E., Pineau, T., Georgeaud, V., Walker, J. E., Terce, F., Collet, X., Perret, B., & Barbaras, R. (2003). Ectopic-chain of ATP synthase is an apolipoprotein A-1 receptor in hepatic HDL endocytosis. *Nature, 421*, 75–79.
Martins, J. M., Kastin, A. J., & Banks, W. A. (1996). Unidirectional specific and modulated brain to blood transport of corticotropin-releasing hormone. *Neuroendocrinology, 63*, 338–348.
Martins, J. M., Banks, W. A., & Kastin, A. J. (1997). Transport of CRH from mouse brain directly affects peripheral production of ·beta-endorphin by the spleen. *American Journal of Physiology, 273*, E1083–E1089.
Masereeuw, R., Jaehde, U., Langemeijer, M. W. E., De Boer, A. G., & Breimer, D. D. (1994). In vivo and in vitro transport of zidovudine (AZT) across the blood-brain barrier and the effects of transport inhibitors. *Pharmaceutical Research, 11*, 324–330.
Mathiisen, T. M., Lehre, K. P., Danbolt, N. C., & Ottersen, O. P. (2010). The perivascular astroglial sheath prvides a complete covering of the brain microvessels: An electron microscopic 3D reconstruction. *Glia, 58*, 1094–1103.
Mayhan, W. G., & Heistad, D. D. (1985). Permeability of blood-brain barrier to various sized molecules. *The American Journal of Physiology, 248*, H712–H718.
McCarthy, T. J., Banks, W. A., Farrell, C. L., Adamu, S., Derdeyn, C. P., Snyder, A. Z., LaForest, R., Litzinger, D. C., Martin, D., LeBel, C. P., & Welch, M. J. (2002). Positron emission tomography shows that intrathecal leptin reaches the hypothalamus in baboons. *Journal of Pharmacology and Experimental Therapeutics, 307*, 878–883.
McLay, R. N., Kimura, M., Banks, W. A., & Kastin, A. J. (1997). Granulocyte-macrophage colony-stimulating factor crosses the blood-brain and blood-spinal cord barriers. *Brain, 120*, 2083–2091.
McQuay, H. J., Sullivan, A. F., Smallman, K., & Dickenson, A. H. (1989). Intrathecal opioids, potency and lipophilicity. *Pain, 36*, 111–115.
Mealey, K. L., Bentjen, S. A., Gay, J. M., & Cantor, G. H. (2001). Ivermectin sensitivity is associated with a deletion mutation of the mdr1 gene. *Pharmacogenetics, 11*, 727–733.
Megyeri, P., Abraham, C. S., Temesvari, P., Kovacs, J., Vas, T., & Speer, C. P. (1992). Recombinant human tumor necrosis factor alpha constricts pial arterioles and increases blood-brain barrier permeability in newborn piglets. *Neuroscience Letters, 148*, 137–140.
Mellman, I., Fuchs, R., & Helenius, A. (1986). Acidification of the endocytic and exocytic pathways. *Annual Reviews in Biochemistry, 55*, 663–700.
Minami, T., Okazaki, J., Kawabata, A., Kawaki, H., Okazaki, Y., & Tohno, Y. (1998a). Roles of nitric oxide and prostaglandins in the

increased permeability of the blood-brain barrier caused by lipopolysaccharide. *Environmental Toxicology and Pharmacology, 5*, 35–41.

Minami, T., Okazaki, J., Kawabata, A., Kuroda, R., & Okazaki, Y. (1998b). Penetration of cisplatin into mouse brain by lipopolyscharide. *Toxicology, 130*, 107–113.

Minten, C., Alt, C., Gentner, M., Frei, E., Deutsch, U., Lyck, R., Schaeren-Wiemers, N., Rot, A., & Engelhardt, B. (2014). DARC shuttles inflammatory chemokines across the blood-brain barrier during autoimmune central nervous system inflammation. *Brain, 137*(Pt 5), 1454–1469.

Morgan, D., Diamond, D. M., Gottschall, P. E., Ugen, K. E., Dickey, C., Hardy, J., Duff, K., Jantzen, P., DiCarlo, G., Wilcock, D., Connor, K., Hatcher, J., Hope, C., Gordon, M., & Arendash, G. W. (2000). A· peptide vaccination prevents memory loss in an animal model of Alzheimer's disease. *Nature, 408*, 982–985.

Morita, S., & Miyata, S. (2012). Different vascular permeability between the sensory and secretory circumventricular organs of adult mouse brain. *Cell and Tissue Research, 349*(2), 589–603.

Moser, K. V., Reindl, M., Blasig, I., & Humpel, C. (2004). Brain capillary endothelial cells proliferate in response to NGF, express NGF receptors and secrete NGF after inflammatiion. *Brain Research, 1017*, 53–60.

Muldoon, L., Pagel, M. A., Kroll, R. A., Roman-Goldstein, S., Jones, R. S., & Neuwelt, E. A. (1999). A physiological barrier distal to the anatomical blood-brain barrier in a model of transvascular delivery. *American Journal of Neuroradiology, 20*, 217–222.

Nakaoke, R., Ryerse, J. S., Niwa, M., & Banks, W. A. (2005). Human immunodeficiency virus type 1 transport across the in vitro mouse brain endothelial cell monolayer. *Experimental Neurology, 193*, 101–109.

Nakaoke, R., Verma, S., Niwa, M., Dohgu, S., & Banks, W. A. (2007). Glucose-regulated blood-brain barrier transport of insulin: Pericyte-astrocyte-endothelial cell cross talk. *International Journal of Neuroprotection and Neuroregeneration, 3*, 195–200.

Nath, A., Conant, K., Chen, P., Scott, C., & Major, E. O. (1999). Transient exposure to HIV-1 Tat protein results in cytokine production in macrophages and astrocytes: A hit and run phenomenon. *Journal of Biological Chemistry, 274*, 17098–17102.

Neaves, W. B. (1977). The blood-testis barrier. In A. D. Johnson & W. R. Gomes (Eds.), *The testis* (pp. 125–162). Academic Press.

Nelson, P. K., Masters, L. T., Zagzag, D., & Kelly, P. J. (1999). Angiographic abnormalities in progressive multifocal leukoencephalopathy: An explanation based on neuropathologic findings. *American Journal of Neuroradiology, 20*, 487–494.

Neuwelt, E., Abbott, N. J., Abrey, L., Banks, W. A., Blakley, B., Davis, T., Engelhardt, B., Grammas, P., Nedergaard, M., Nutt, J., Pardgridge, W., Rosenberg, G. A., Smith, Q., & Drewes, L. R. (2008). Strategies to advance translational research into brain barriers. *Lancet Neurology, 7*, 84–96.

Niewoehner, J., Borhrmann, B., Collin, L., Urich, E., Sade, H., Maier, P., Rueger, P., Stracke, J. O., Lau, W., Tissot, A. C. L., Ghosh, H. A., & Freskgard, P.-O. (2014). Increased brain penetration and potency of a therapeutic antibody using a monovalent molecular shuttle. *Neuron, 81*, 49–60.

Nonaka, N., Hileman, S. M., Shioda, S., Vo, P., & Banks, W. A. (2004). Effects of lipopolysaccharide on leptin transport across the blood-brain barrier. *Brain Research, 1016*, 58–65.

Nonaka, N., Shioda, S., & Banks, W. A. (2005). Effect of lipopolysaccharide on the transport of pituitary adenylate cyclase activating polypeptide across the blood-brain barrier. *Experimental Neurology, 191*, 137–144.

Nottet, H. S., Persidsky, Y., Sasseville, V. G., Nukuna, A. N., Bock, P., Zhai, Q. H., Sharer, L. R., McComb, R. D., Swindells, S., Soderland, C., & Gendelman, H. E. (1996). Mechanisms for the transendothelial migration of HIV-1-infected monocytes into brain. *The Journal of Immunology, 156*, 1284–1295.

Oehmichen, M., Gruninger, H., Wietholter, H., & Gencic, M. (1979). Lymphatic efflux of intracerebrally injected cells. *Acta Neuropathologica, 45*, 61–65.

Oldendorf, W. H. (1971). Brain uptake of radio-labelled amino acids, amines and hexoses after arterial injection. *American Journal of Physiology, 221*, 1629–1639.

Oldendorf, W. H. (1974). Lipid solubility and drug penetration of the blood-brain barrier. *Proceedings of the Society for Experimental Biology and Medicine, 147*, 813–816.

Osburg, B., Peiser, C., Domling, D., Schomburg, L., Ko, Y. T., Voight, K., & Bickel, U. (2002). Effect of endotoxin on expression of TNF receptors and transport of TNF-alpha at the blood-brain barrier of the rat. *American Journal of Physiology, 283*, E899–E908.

Pan, W., & Kastin, A. J. (2001a). Changing the chemokine gradient: CINC1 crosses the blood-brain barrier. *Journal of Neuroimmunology, 115*, 64–70.

Pan, W., & Kastin, A. J. (2001b). Increase in TNF alpha transport after SCI is specific for time, region, and type of lesion. *Experimental Neurology, 170*, 357–363.

Pan, W., & Kastin, A. J. (2002). TNF alpha transport across the blood-brain barrier is abolished in receptor knockout mice. *Experimental Neurology, 174*, 193–200.

Pan, W., & Kastin, A. J. (2003). Interactions of cytokines with blood-brain barrier: Implications for feeding. *Current Pharmaceutical Design, 9*, 827–831.

Pan, W., Banks, W. A., Kennedy, M. K., Gutierrez, E. G., & Kastin, A. J. (1996). Differential permeability of the BBB in acute EAE: Enhanced transport of TNF-alpha. *American Journal of Physiology, 271*, E636–E642.

Pan, W., Banks, W. A., & Kastin, A. J. (1997a). BBB permeability to ebiratide and TNF in acute spinal cord injury. *Experimental Neurology, 146*, 367–373.

Pan, W., Banks, W. A., & Kastin, A. J. (1997b). Permeability of the blood-brain barrier and blood-spinal cord barriers to interferons. *Journal of Neuroimmunology, 76*, 105–111.

Pan, W., Zadina, J. E., Harlan, R. E., Weber, J. T., Banks, W. A., & Kastin, A. J. (1997). Tumor necrosis factor-alpha: A neuromodulator in the CNS. *Neuroscience and Biobehavioral Reviews, 21*, 603–613.

Pan, W., Banks, W. A., Fasold, M. B., Bluth, J., & Kastin, A. J. (1998a). Transport of brain-derived neurotrophic factor across the blood-brain barrier. *Neuropharmacology, 37*, 1553–1561.

Pan, W., Banks, W. A., & Kastin, A. J. (1998b). Permeability of the blood-brain barrier to neurotrophins. *Brain Research, 788*, 87–94.

Pan, W., Cornelissen, G., Halberg, F., & Kastin, A. J. (2002). Selected contributions: Circadian rhythm of tumor necrosis factor-alpha uptake into mouse spinal cord. *Journal of Applied Physiology, 92*, 1357–1362.

Pan, W., Kastin, A. J., Rigai, T., McLay, R., & Pick, C. G. (2003a). Increased hippocampal uptake of tumor necrosis factor alpha and behavioral changes in mice. *Experimental Brain Research, 149*, 195–199.

Pan, W., Zhang, L., Liao, J., Csernus, B., & Kastin, A. J. (2003b). Selective increase in TNF alpha permeation across the blood-spinal cord barrier after SCI. *Journal of Neuroimmunology, 134*, 111–117.

Pan, W., Hsuchou, H., He, Y., Sakharkar, A., Cain, C., Yu, C., & Kastin, A. J. (2008a). Astrocyte leptin receptor (ObR) and leptin transport in adult-onset obese mice. *Endocrinology, 149*(6), 2798–2806.

Pan, W., Yu, C., Hsuchou, H., Zhang, Y., & Kastin, A. J. (2008b). Neuroinflammation facilitates LIF entry into brain: Role of TNF. *American Journal of Physiology. Cell Physiology, 294*, C1436–C1442.

Park, M., Lin, L., Thomas, S., Braymer, H. D., Smith, P. M., Harrison, D. H. T., & York, D. A. (2004). The F 1-ATPase—subunit is the putative enterostatin receptor. *Peptides, 25*, 2127–2133.

Pascale, C. L., Miller, M. C., Chiu, C., Boylan, M., Caralopoulos, I. N., Gonzalez, L., Johanson, C. E., & Silverberg, G. D. (2011). Amyloid-beta transporter expression at the blood-CSF barrier is age-dependent. *Fluids and Barriers of the CNS, 8*, 21.

Patel, A., Zhu, Y., Kuzhikandathil, E. V., Banks, W. A., Siegel, A., & Zalcman, S. S. (2012). Soluble interleukin-6 receptor induces motor stereotypies and co-localizes with gp130 in regions linked to cortico-striato-thalamo-cortical circuits. *PLoS One, 7*(7), e41623.

Paz, S. P. C., Branco, L., Pereira, M. A. C., Spessotto, C., & Fragoso, Y. D. (2018). Systematic review of the published data on the worldwide prevalence of John Cunningham virus in patients with multiple sclerosis and neuromyelitis optica. *Epidemiol Health, 40*, e2018001.

Pearse, D. D., Chatzipanteli, K., Marcillo, A. E., Bunge, M. B., & Dietrich, W. D. (2003). Comparison of iNOS inhibition by antisense and pharmacological inhibitors after spinal cord injury. *Journal of Neuropathology and Experimental Neurology, 62*, 1096–1107.

Persidsky, Y., Stins, M., Way, D., Witte, M. H., Weinand, M., Kim, K. S., Bock, P., Gendelman, H. E., & Fiala, M. (1997). A model for monocyte migration through the blood-brain barrier during HIV-1 encephalitis. *The Journal of Immunology, 158*, 3499–3510.

Persidsky, Y., Zheng, J., Miller, D., & Gendelman, H. E. (2000). Mononuclear phagocytes mediate blood-brain barrier compromise and neuronal injury during HIV-1-associated dementia. *Journal of Leukocyte Biology, 68*, 413–422.

Peruzzo, B., Pastor, F. E., Blazquez, J. L., Schobitz, K., Pelaez, B., Amat, P., & Rodriguez, E. M. (2000). A second look at the barriers of the medial basal hypothalamus. *Experimental Brain Research, 132*, 10–26.

Plotkin, S. R., Banks, W. A., & Kastin, A. J. (1996). Comparison of saturable transport and extracellular pathways in the passage of interleukin-1‡ across the blood-brain barrier. *Journal of Neuroimmunology, 67*, 41–47.

Plotkin, S. R., Banks, W. A., & Kastin, A. J. (1998). Enkephalin, PPE, mRNA, and PTS-1 in alcohol withdrawal seizure-prone and -resistant mice. *Alcohol, 15*, 25–31.

Poduslo, J. F., & Curran, G. L. (1996). Permeability at the blood-brain barrier and blood-nerve barriers of the neurotrophic factors: NGF, CNTF, NT-3, BDNF. *Molecular Brain Research, 36*, 280–286.

Pollay, M., & Davson, H. (1963). The passage of certain substances out of the cerebrospinal fluid. *Brain, 86*, 137–150.

Pozzilli, C., Bernardi, S., Mansi, L., Picozzi, P., Iannotti, F., Alfano, B., Bozzao, L., Lenzi, G. L., Salvatore, M., Conforti, P., & Fieschi, C. (1988). Quantitative assessment of the blood-brain barrier permeability in multiple sclerosis using 68-Ga-EDTA and positron emission tomography. *Journal of Neurology, Neurosurgery, and Psychiatry, 51*, 1058–1062.

Prinz, M., & Priller, J. (2014). Microglia and brain macrophages in the molecular age: From origin to neuropsychiatric disease. *Nature Reviews. Neuroscience, 15*(5), 300–312.

Prockop, L. D., Naidu, K. A., Binard, J. E., & Ransohoff, J. (1996). Selective permeability of [3 H]-D-mannitol and [14 C]-carboxyl-inulin across the blood-brain barrier and blood-spinal cord barrier in the rabbit. *The Journal of Spinal Cord Medicine, 18*, 221–226.

Pu, H., Tian, J., Flora, G., Lee, Y. W., Nath, A., Hennig, B., & Toborek, M. (2003). HIV-1 tat protein upregulates inflammatory mediators and induces monocyte invasion into the brain. *Molecular and Cellular Neurosciences, 24*, 224–237.

Pulgar, V. M. (2018). Transcytosis to cross the blood brain barrier, new advancements and challenges. *Frontiers in Neuroscience, 12*, 1019.

Qin, L., Wu, X., Block, M. L., Liu, Y., Breese, G. R., Hong, J. S., Knapp, D. J., & Crews, F. T. (2007). Systemic LPS causes chronic neuroinflammation and progressive neurodegeneration. *Glia, 55*, 453–462.

Quan, N., Sundar, S. K., & Weiss, J. M. (1994). Induction of interleukin-1 in various brain regions after peripheral and central injections of lipopolysaccharide. *Journal of Neuroimmunology, 49*, 125–134.

Quan, N., Stern, E. L., Whiteside, M. B., & Herkenham, M. (1999). Induction of pro-inflammatory cytokine mRNAs in the brain after peripheral injection of subseptic doses of lipopolysaccharide in the rat. *Journal of Neuroimmunology, 93*(1–2), 72–80.

Quaranta, D. V., Weaver, R. R., Baumann, K. K., Fujimoto, T., Williams, L. M., Kim, H. C., Logsdon, A. F., Omer, M., Reed, M. J., Banks, W. A., & Erickson, M. A. (2022). Transport of the pro-inflammatory chemokines CCL2 (MCP-1) and CCL5 (RANTES) across the intact mouse blood-brain barrier is inhibited by heparin and eprodisate and increased with systemic inflammation. *The Journal of Pharmacology and Experimental Therapeutics*.

Ramos-Kuri, M., Barron Romero, B. L., & Aguilar-Setien, A. (1996). Inhibition of three alphaherpesviruses (herpes simplex 1 and 2 and pseudorabies virus) by heparin, heparan and other sulfated polyelectrolytes. *Archives of Medical Research, 27*, 43–48.

Rapoport, S. I. (1976). *Blood brain barrier in physiology and medicine*. Raven Press.

Rapoport, S. I., Ohata, M., & London, E. D. (1981). Cerebral blood flow and glucose utilization following opening of the blood-brain barrier and during maturation of the rat brain. *Federation Proceedings, 40*, 2322–2325.

Raub, T. J., & Audus, K. L. (1990). Adsorptive endocytosis and membrane recycling by cultured primary bovine brain microvessel endothelial cell monolayers. *Journal of Cell Science, 97*, 127–138.

Reese, T. S., & Karnovsky, M. J. (1967). Fine structural localization of a blood-brain barrier to endogenous peroxidase. *The Journal of Cell Biology, 34*, 207–217.

Rethelyi, M. (1984). Diffusional barrier around the hypothalamic arcuate nucleus in the rat. *Brain Research, 307*, 355–358.

Reyes, T. M., Fabry, Z., & Coe, C. L. (1999). Brain endothelial cell production of a neuroprotective cytokine, interleukin-6, in response to noxious stimuli. *Brain Research, 851*, 215–220.

Rhea, E. M., Rask-Madsen, C., & Banks, W. A. (2018). Insulin transport across the blood-brain barrier can occur independently of the insulin receptor. *The Journal of Physiology, 596*, 4753–4765.

Rhea, E. M., Logsdon, A. F., Hansen, K. M., Williams, L. M., Reed, M. J., Baumann, K. K., Holden, S. J., Raber, J., Banks, W. A., & Erickson, M. A. (2021). The S1 protein of SARS-CoV-2 crosses the blood-brain barrier in mice. *Nature Neuroscience, 24*(3), 368–378.

Robertson, J., Eden, A., Nystrom, K., Hagberg, L., Yilmaz, A., Gostner, J. M., Fuchs, D., Nilsson, S., Blennow, K., Zetterberg, H., & Gisslen, M. (2021). Increased immune activation and signs of neuronal injury in HIV-negative people on preexposure prophylaxis. *AIDS, 35*(13), 2129–2136.

Rodriguez, E. M., Blazquez, J. L., & Guerra, M. (2010). The design of barriers in the hypothalamus allows the median eminence and the arcuate nucleus to enjoy private milieus: The former opens to the portal blood and the latter to the cerebrospinal fluid. *Peptides, 31*, 757–776.

Romeo, G., Liu, W. H., Asnaghi, V., Kern, T. S., & Lorenzi, M. (2002). Activation of nuclear factor-kappaB induced by diabetes and high glucose regulates a proapoptotic program in retinal pericytes. *Diabetes, 51*, 2241–2248.

Ronaldson, P. T., & Bendayan, R. (2006). HIV-1 viral envelope glycoprotein gp120 triggers an inflammatory response in cultured rat astrocytes and regulates the functional expression of P-glycoprotein. *Molecular Pharmacology, 70*, 1087–1098.

Sakata, A., Tamai, I., Kawazu, K., Deguchi, Y., Ohnishi, T., Saheki, A., & Tsuji, A. (1994). In vivo evidence for ATP-dependent and p-glycoprotein-mediated transport of cyclosporin A at the blood-brain barrier. *Biochemical Pharmacology, 48*, 1989–1992.

Salkeni, M. A., Lynch, J. L., Price, T. O., & Banks, W. A. (2009). Lipopolysaccharide impairs blood-brain barrier P-glycoprotein function in mice through prostaglandin- and nitric oxide-independent pathways and nitric oxide-independent pathways. *Journal of Neuroimmune Pharmacology, 4*, 276–282.

Sanders, V. J., Pittman, C. A., White, M. G., Wang, G., Wiley, C. A., & Achim, C. L. (1998). Chemokines and receptors in HIV encephalitis. *AIDS, 12*, 1021–1026.

Schaeffer, S., & Iadecola, C. (2021). Revisiting the neurovascular unit. *Nature Neuroscience, 24*(9), 1198–1209.

Schiffenbauer, J., Streit, W. J., Butfiloski, E., LaBow, M., Edward, C., & Moldawer, L. L., 3rd. (2000). The induction of EAE is only partially dependent on TNF receptor signaling but requires the IL-1 type 1 receptor. *Clinical Immunology, 95*, 117–125.

Schwartz, M. W., Woods, S. C., Porte, D., Jr., Seeley, R. J., & Baskin, D. G. (2000). Central nervous system control of food intake. *Nature, 404*, 661–671.

Schweighardt, B., & Atwood, W. J. (2001). Virus receptors in the human central nervous system. *Journal of Neurovirology, 7*, 187–195.

Shafer, R. A., & Murphy, S. (1997). Activated astrocytes induce nitric oxide synthase-2 in cerebral endothelium via tumor necrosis factor alpha. *Glia, 21*(4), 370–379.

Shibata, M., Yamada, S., Kumar, S. R., Calero, M., Bading, J., Frangione, B., Holtzman, D. M., Miller, C. A., Strickland, D. K., Ghiso, J., & Zlokovic, B. V. (2000). Clearance of Alzheimer's amyloid-· 1-40 peptide from brain by LDL receptor-related protein-1 at the blood-brain barrier. *Journal of Clinical Investigation, 106*, 1489–1499.

Shimura, T., Tabata, S., Ohnishi, T., Terasaki, T., & Tsuji, A. (1991). Transport mechanism of a new behaviorally highly potent adrenocorticotropic hormone (ACTH) analog, ebiratide, through the blood-brain barrier. *Journal of Pharmacology and Experimental Therapeutics, 258*, 459–465.

Simpson, J. E., Newcombe, J., Cuzner, M. L., & Woodrofe, M. N. (1998). Expression of monocyte chemoattractant protein-1 and other beta-chemokines by resident glia and inflammatory cells in multiple sclerosis. *Journal of Neuroimmunology, 84*, 238–249.

Somogyvari-Vigh, A., Pan, W., Reglodi, D., Kastin, A. J., & Arimura, A. (2000). Effect of middle cerebral artery occulsion on the passage of pituitary adenylate cyclase activating polypeptide across the blood-brain barrier in the rat. *Regulatory Peptides, 91*, 89–95.

Spector, R., & Lorenzo, A. V. (1974). The effects of salicylate and probenecid on the cerebrospinal fluid transport of penicillin, aminosalicylic acid and iodide. *Journal of Pharmacology and Experimental Therapeutics, 188*, 55–65.

Stein, U., Walther, W., & Shoemaker, R. H. (1996). Modulation of mdr1 expression by cytokines in human colon carcinoma cells: An approach for reversal of multidrug resistance. *British Journal of Cancer, 74*, 1384–1391.

Tachikawa, M., Hosoya, K., & Terasaki, T. (2014). Pharmacological significance of prostaglandin E2 and D2 transport at the brain barriers. *Advances in Pharmacology, 71*, 337–360.

Takasawa, M., Terasaki, T., Suzuki, H., & Sugiyama, Y. (1997). In vivo evidence for carrier-mediated efflux transport of 3'azido-3'deoxythymidine and 2′,3′-dideoxyinosine across the blood-brain barrier via a probenecid-sensitive transport system. *Journal of Pharmacology and Experimental Therapeutics, 281*, 369–375.

Tanzi, R. E., Moir, R. D., & Wagner, S. L. (2004). Clearance of Alzheimer's Abeta peptide: The many roads to perdition. *Neuron, 43*, 608–608.

Tarr, A. J., Liu, X., Reed, N. S., & Quan, N. (2014). Kinetic characteristics of euflammation: The induction of controlled inflammation without overt sickness behavior. *Brain, Behavior, and Immunity, 42*, 96–108.

Taylor, E. M. (2002). The impact of efflux transporters in the brain on the development of drugs for CNS disorders. *Clinical Pharmacokinetics, 41*, 81–92.

Terasaki, T., Takakuwa, S., Saheki, A., Moritani, S., Shimura, T., Tabata, S., & Tsuji, A. (1992). Absorptive-mediated endocytosis of an adrenocorticotropic hormone (ACTH) analogue, ebiratide, into the blood-brain barrier: Studies with monolayers of primary cultured bovine brain capillary endothelial cells. *Pharmaceutical Research, 9*, 529–534.

Theron, D., de Lagerie, S. B., Tardivel, S., Pelerin, H., Demeuse, P., Mercier, C., Mabondzo, A., Farinotti, R., Lacour, B., Roux, F., & Gimenez, F. (2003). Influence of tumor necrosis factor-alpha on the expression and function of P-glycoprotein in an immortalized rat brain capillary endothelial cell line, GPNT. *Biochemical Pharmacology, 66*, 579–587.

Thomas, S. A. (2004). Anti-HIV drug distribution to the central nervous system. *Current Pharmaceutical Design, 10*, 1313–1324.

Thompson, S. J., Koszdin, K., & Bernards, C. M. (2000). Opiate-induced analgesia is increased and prolonged in mice lacking P-glycoprotein. *Anesthesiology, 92*, 1392–1399.

Vadeboncoeur, N., Segura, M., Al-Numani, D., Vanier, G., & Gottschalk, M. (2003). Proinflammatory cytokine and chemokine release by human brain microvascular endothelial cells stimulated by Streptococcus suis serotype 2. *FEMS Immunology and Medical Microbiology, 35*, 49–58.

Vaibhav, K., Braun, M., Alverson, K., Khodadadi, H., Kutiyanawalla, A., Ward, A., Banerjee, C., Sparks, T., Malik, A., Rashid, M. H., Khan, M. B., Waters, M. F., Hess, D. C., Arbab, A. S., Vender, J. R., Hoda, N., Baban, B., & Dhandapani, K. M. (2020). Neutrophil extracellular traps exacerbate neurological deficits after traumatic brain injury. *Science Advances, 6*(22), eaax8847.

van Dam, A. M., De Vries, H. E., Kuiper, J., Zijlstra, F. J., De Boer, A. G., Tilders, F. J. H., & Berkenbosch, F. (1996). Interleukin-1 receptors on rat brain endothelial cells: A role in neuroimmune interaction? *FASEB, 10*, 351–356.

Verheggen, I. C. M., de Jong, J. J. A., van Boxtel, M. P. J., Postma, A. A., Verhey, F. R. J., Jansen, J. F. A., & Backes, W. H. (2020). Permeability of the windows of the brain: Feasibility of dynamic contrast-enhanced MRI of the circumventricular organs. *Fluids and Barriers of the CNS, 17*(1), 66.

Verma, S., Nakaoke, R., Dohgu, S., & Banks, W. A. (2006). Release of cytokines by brain endothelial cells: A polarized response to lipopolysaccharide. *Brain, Behavior, and Immunity, 20*, 449–455.

Vidal, E. L., Patel, N. A., Wu, G. D., Fiala, M., & Chang, S. L. (1998). Interleukin-1 induces the expression of Ê opioid receptors in endothelial cells. *Immunopharmacology, 38*, 261–266.

Villegas, J. C., & Broadwell, R. D. (1993). Transcytosis of protein through the mammalian cerebral epithelium and endothelium: II. Adsorptive transcytosis of WGA-HRP and the blood-brain and brain-blood barriers. *Journal of Neurocytology, 22*, 67–80.

Vorbrodt, A. W. (1994). Glycoconjugates and anionic sites in the blood-brain barrier. In M. Nicolini & P. F. Zatta (Eds.), *Glycobiology and the brain* (pp. 37–62). Press.

Vorbrodt, A. W., Dobrogowska, D. H., Ueno, M., & Lossinsky, A. S. (1995). Immunocytochemical studies of protamine-induced blood-brain barrier opening to endogenous albumin. *ACTA Neuropathologica (Berlin), 89*, 491–499.

Walker, A. K., Wing, E. E., Banks, W. A., & Dantzer, R. (2019). Leucine competes with kynurenine for blood-to-brain transport and prevents lipopolysaccharide-induced depression-like behavior in mice. *Molecular Psychiatry, 24*(10), 1523–1532.

Walsh, R. J., Slaby, F. J., & Posner, B. I. (1987). A receptor-mediated mechanism for the transport of prolactin from blood to cerebrospinal fluid. *Endocrinology, 120*, 1846–1850.

Wang, Y., & Sawchuck, R. J. (1995). Zidovudine transport in the rabbit brain during intravenous and intracerebroventricular infusion. *Journal of Pharmaceutical Sciences, 7*, 871–876.

Weissmann, G. (1976). Experimental enzyme replacment in genetic and other disorders. *Hosptial Practice, 11*, 49–58.

Westergren, I., & Johansson, B. B. (1993). Altering the blood-brain barrier in the rat by intracarotid infusion of polycations: A comparison between protamine, poly-L-lysine and poly-L-arginine. *Acta Physiologica Scandinavica, 149*, 99–104.

Widner, H., Jonsson, B. A., Hallstadius, L., Wingardh, K., Strand, S. E., & Johansson, B. B. (1987). Scintigraphic method to quantify the passage from brain parenchyma to the deep cervical lymph nodes in the rat. *European Journal of Nuclear Medicine, 13*, 456–461.

Williams, K. C., & Hickey, W. F. (1995). Traffic of hematogenous cells through the central nervous system. *Current Topics in Microbiology and Immunology, 202*, 221–245.

Wilson, J. F., Anderson, S., Snook, G., & Llewellyn, K. D. (1984). Quantification of the permeability of the blood-CSF barrier to ‡-MSH in the rat. *Peptides, 5*, 681–685.

Wolburg, H., Wolburg-Buchholz, K., & Engelhardt, B. (2005). Diapedesis of mononuclear cells across cerebral venules during experimental autoimmune encephalomyelitis leaves tight junctions intact. *Acta Neuropathologica, 109*, 181–190.

Wu, D., Clement, J. G., & Pardridge, W. M. (1998). Low blood-brain barrier permeability to azidothymidine (AZT), 3TC, and thymidine in the rat. *Brain Research, 791*, 313–316.

Xaio, H., Banks, W. A., Niehoff, M. L., & Morley, J. E. (2001). Effect of LPS on the permeability of the blood-brain barrier to insulin. *Brain Research, 896*, 36–42.

Xiang, S., Pan, W., & Kastin, A. J. (2005). Strategies to create a regenerating environment for the injured spinal cord. *Current Pharmaceutical Design, 11*, 1267–1277.

Yamada, S., DePasquale, M., Patlak, C. S., & Cserr, H. F. (1991). Albumin outflow into deep cervical lymph from different regions of rabbit brain. *American Journal of Physiology, 261*, H1197–H1204.

Yasuda, K., Cline, C., Vogel, P., Onciu, M., Fatima, S., Sorrentino, B. P., Thirumaran, R. K., Ekins, S., Urade, Y., Fujimori, K., & Schuetz, E. G. (2013). Drug transporters on arachnoid barrier cells contribute to the blood-cerebrospinal fluid barrier. *Drug Metabolism and Disposition, 41*(4), 923–931.

Yeagle, P. (1987). *Transport. The membranes of cells* (pp. 191–215). Academic Press, Inc.

Yu, C., Kastin, A. J., Tu, H., Waters, S., & Pan, W. (2007a). TNF activates P-glycoprotein in cerebral microvascular endothelial cells. *Cellular Physiology and Biochemistry, 20*, 853–858.

Yu, C., Pan, W., Tu, H., Waters, S., & Kastin, A. J. (2007b). TNF activates MDR1 (P-glycoprotein) in cerebral microvascular endothelial cells. *Cellular Physiology and Biochemistry, 20*, 853–858.

Yu, C., Argyropoulos, G., Zhang, Y., Kastin, A. J., Hsuchou, H., & Pan, W. (2008). Neuroinflammation activates Mdr1b efflux transport through NFkappaB: Promoter analysis in BBB endothelia. *Cellular Physiology and Biochemistry, 22*, 745–756.

Zalcman, S. S. (2001). Interleukin-2 potentiates novelty—and GBR 12909-induced exploratory activity. *Brain Research, 899*, 1–9.

Zalcman, S. S. (2002). Interleukin-2-induced increases in climbing behavior: Inhibition by dopamine D-1 and D-2 receptor antagonists. *Brain Research, 944*, 157–164.

Zambenedetti, P., Giordano, R., & Zatta, P. (1996). Indentification of lectin binding sites in the rat brain. *Glycoconjugate Journal, 13*, 341–346.

Zhang, L., Zhou, L., Bao, L., Liu, J., Zhu, H., Lv, Q., Liu, R., Chen, W., Tong, W., Wei, Q., Xu, Y., Deng, W., Gao, H., Xue, J., Song, Z., Yu, P., Han, Y., Zhang, Y., Sun, X., Yu, X., & Qin, C. (2021). SARS-CoV-2 crosses the blood-brain barrier accompanied with basement membrane disruption without tight junctions alteration. *Signal Transduction and Targeted Therapy, 6*(1), 337.

Zlokovic, B. V. (2005). Neurovascular mechanisms of Alzheimer's neurodegeneration. *Trends in Neurosciences, 28*, 202–208.

Neurotransmission of Immune Cells

14

Marco Consentino, Silvana Valdebenito,
and Eliseo Eugenin

Abstract

For decades, the interactions between the peripheral immune system and the brain have played a key role in disease. However, the molecular, biochemical, and cellular interactions that affect neuroimmune responses remain under investigation. Only in the last two decades has it become evident that the neuroimmune axis is reciprocal to maintaining health. When it becomes dysfunctional, it can underline disease pathobiology. This chapter focuses on what is known and what still remains to unravel the neuroimmune axis.

Keywords

Astrocytes · Neurons · Dopamine · Cytokines · Cell-to-cell communication

14.1 General Introduction

The brain was previously thought to have limited access to the immune system (Dave & Klein, 2022; Frederick et al., 2022; Mundt et al., 2022; Yoshida et al., 2022). However, recent findings demonstrated sustained neuro–immune interactions during health (Mundt et al., 2019; Ousman & Kubes, 2012) and disease (Rustenhoven, 2021). Only recently, local bone marrow from the skull and brain immunity has been associated with anxiety-like behaviors, inflammatory, cancer, and degenerative diseases (Alves de Lima et al., 2020a; Cugurra et al., 2021; Mazzitelli et al., 2022; Rustenhoven et al., 2021). This opened new theories of neuro–immune interactions.

Currently, most reports on the neuroimmune axis correspond to the immune system's role in infiltrating the central nervous system (CNS) in response to degeneration, inflammation, damage, or infections (Frederick et al., 2022; Padro & Sanders, 2014; Rustenhoven, 2021). This includes diseases such as Alzheimer's, Parkinson's, Multiple Sclerosis, stroke, spinal cord and traumatic brain injury, and bacterial and viral infections that include COVID-19 (Alluri et al., 2015; Buckner et al., 2006; Obermeier et al., 2016; Prat & Antel, 2005; Simoes Da Gama & Morin-Brureau, 2022). Overall, early brain injuries induce the local secretion of chemokines and cytokines by resident cells, resulting in the chemoattraction of peripheral immune cells (lymphocytes, dendritic cells, and monocytes) into compromised areas to clear the damaged areas or in eliminating infection or clearing cancerous cells. However, sustained inflammation and leukocyte transmigration result in microglial and astrocyte activation as well as neuronal synaptic compromise. These histopathological deficits commonly translate into cognitive, behavioral, and motor impairments. In general, brain damage and associated inflammation are responsible for several types of disabilities due to the loss of homeostatic glial–synaptic interaction as well as limitations that are known for brain cells to regenerate (Arino et al., 2022; Delage et al., 2021; Koyama & Shichita, 2022; Salvador & Kipnis, 2022; Tiedt et al., 2022).

In the brain, the main immune components are the glia, which consists of microglia oligodendrocytes, ependymal cells, and astrocytes. Normally, each of these cell types and in particular microglia and astrocytes, are the two essential cells responsible for proper brain development, synaptic pruning, energetics, blood–brain barrier (BBB), and immune surveillance (Ball et al., 2022; Chokr et al., 2022; Guedes et al., 2022; Rim et al., 2022). Any functional alterations have been associated with inflammatory and degenerative diseases (Bolton et al., 2022; Chithanathan et al., 2022;

M. Consentino (✉)
University of Insubria, Varese, Italy
e-mail: marco.cosentino@uninsubria.it

S. Valdebenito · E. Eugenin (✉)
Department of Neurobiology, The University of Texas Medical Branch (UTMB), Galveston, TX, USA
e-mail: eleugeni@utmb.edu

H. E. Gendelman, T. Ikezu (eds.), *Neuroimmune Pharmacology and Therapeutics*,
https://doi.org/10.1007/978-3-031-68237-7_14

Cornell et al., 2022; Dayananda et al., 2022; Dziabis & Bilbo, 2022; Gao et al., 2022; Kulkarni et al., 2022; Rim et al., 2022; Wang et al., 2022). Microglia are CD68, Iba-1, P_2Y12, CD45, and CD11b positive but normally remain quiescent during steady-state conditions. Astrocytes are mostly glial fibrillar acid protein-positive. Upon activation, microglia and astrocytes become highly reactive, resulting in morphological changes in cell shape, localization, and the cell's transcriptional profiles. This leads to a neuroinflammatory state by releasing cytokines and chemokines, increased clearance, phagocytosis, and cell proliferation. Depending on the injury type, there is also an associated complement-related activation and antigen presentation. The upregulation of CD68, the immune proteasome, toll-like receptors (TLR2 and 4), proteases/metalloproteases, and MHC class II expression ensue upon activation. All are common pathways toward cell activation and inflammation. Notably, both microglia and astrocytes secrete neurotrophins, such as nerve growth factor (NGF), to support neuronal and glia survival and promote their growth. It is clear that glial cells play a dual role in damage and CNS protection (Carlson et al., 2022; Dave & Klein, 2022; Hikosaka et al., 2022; Huang et al., 2022; Koyama & Shichita, 2022; Lu & Gao, 2022; Monje & Iwasaki, 2022). Thus, these cell types control brain immune activation and, indirectly, neuronal function. Another key to neuro–immune interactions is the vascular unit. The vascular unit is critical in neuronal–blood communication, metabolites/oxygen/CO_2 supply and clearance, antigen presentation, and leukocyte transmigration (Keaney & Campbell, 2015; McConnell et al., 2017; Torices et al., 2021). Recent data indicate that peripherical immune cells in the choroid plexus and meninges can exchange information between the immune system and brain, which modulates neuronal and glial activity (Alves de Lima et al., 2020b; Fernandez Zapata et al., 2022; Rodriguez-Lorenzo et al., 2022; Shipley et al., 2020; Zheng et al., 2022; Zhou et al., 2020). The focus on the recent developments in neuro–immune communication is highlighted in this chapter.

14.2 Brain and Immune Surveillance

The immune system regularly monitors CNS function (Humpel, 2008; Lehmann et al., 2022; Mundt et al., 2019; Ousman & Kubes, 2012). In healthy conditions, T cells enter the CNS or peripheral nervous system (PNS) through the BBB, enabling small immune cells to patrol the CNS parenchyma without triggering inflammation and neural compromise. Also, T cells are found in the CSF of healthy individuals and perform their function through a direct communication system from the choroid plexus and the meninges. These lymphocytes are $CD4^+CD45RA^-CD27^+CD69^+$-activated central memory T cells expressing high levels of CCR7 and CXCR3 (Giunti et al., 2003; Kivisakk et al., 2004; Schwab et al., 2012; Yokoseki et al., 2013). T-cell patrolling is not detrimental to brain function; immune responses are silenced under healthy conditions. These include the major histocompatibility complex expression, antigen presentation, and lymphatic drainage system that could carry antigens from the CNS into lymphoid compartments (Hsu et al., 2019; Louveau et al., 2018; Mitchell et al., 2015; Rickenbach & Gericke, 2021; Stern et al., 2014). However, most of these dogmas have evolved with the known glymphatic drainage during health and chronic diseases. Notably, the glymphatic system is a macroscopic waste clearance and distribution system that utilizes astrocyte perivascular channels to promote the elimination of CNS soluble proteins and metabolites but also can deliver brain nutrients glucose, lipids, amino acids, and neurotransmitters to brain subregions and the periphery.

A critical player in brain development, function, and response to disease is the BBB. The BBB is a highly selective cellular barrier composed of microvascular endothelial cells, pericytes, and astrocytes' end feet to separate or select the right exchange between the brain and the circulation. A continuous, highly selective, and impermeable barrier mediates most of the BBB function. This is due to high transendothelial electrical resistance mediated by the highly tight junction, adhesion molecules, and low pinocytic activity (Louveau et al., 2018; McConnell & Mishra, 2022; Simoes Da Gama & Morin-Brureau, 2022). Thus, leukocyte transmigration across the BBB is a coordinated process if the BBB is not compromised, including tethering, rolling, and adhesion of leukocytes into and from the CNS. The adhesion is mediated by endothelial cellular adhesion molecules (CAMs) and their ligands, chemokine gradients, and metalloproteases. Endothelial CAMs and selectins interact with leukocyte integrins such as very late antigen 4, VLA-4, and lymphocyte function-associated antigen-1, LFA-1, upon a chemotactic gradient to facilitate a directed rolling and leukocyte adhesion to the BBB. In healthy conditions, the BBB expresses low levels of CAMs, and the CNS chemokine gradients regulate circulating leukocyte movement in and out of the brain. For example, one of the first adhesion molecules described at the BBB is PECAM-1 (CD31), in addition to junctional adhesion molecules such as JAM-A, -B, and -C. ICAM-2, JAM-A, and PECAM-1 are essential for neutrophil diapedesis in the cremaster. CD99 is a glycoprotein that participates in monocyte diapedesis through human endothelial cells, including the BBB, forming homophilic interactions between endothelial and leukocytes. CD99 acts subsequently after PECAM-1 interactions during diapedesis. These findings suggest their participation at later stages post-initial binding (Jeong et al., 2022; Majerova et al., 2019; Muller, 2016; Pietronigro et al., 2016).

During inflammatory conditions, the expression of these adhesion molecules and tight junction proteins is strongly altered at the BBB cells and leukocytes, favoring the interaction of leukocytes with the BBB and facilitating the leukocyte transmigration across the BBB in conjunction with compromised endothelial permeability (Alluri et al., 2015; Buckner et al., 2006; Eugenin & Berman, 2003; Obermeier et al., 2016). For example, in human immunodeficiency virus-1 (HIV) infection, the virus early after primary infection invades the brain using low CCL2 chemotactic gradients present in the healthy brain to induce the initial recruitment. In addition, HIV infection of leukocytes (T cells and monocyte–macrophages) results in the upregulation of CCR2, the receptor for CCL2 (Eugenin, Osiecki, et al., 2006b). The leukocyte activation induced by viral infection increases the expression and plasma membrane localization of critical adhesion molecules such as ALCAM, PECAM-1, and JAMs (Calderon et al., 2017; Eugenin, Gamss, et al., 2006a; Eugenin, Osiecki, et al., 2006b; Leon-Rivera et al., 2021; Williams et al., 2012; Williams et al., 2013). Some of these proteins are proteolytically shed to promote BBB compromise (Buckner et al., 2006; Eugenin, Gamss, et al., 2006a; Eugenin, Osiecki, et al., 2006b). These mechanisms result in the enhanced transmigration of HIV-infected leukocytes across the BBB. Similar mechanisms of enhanced transmigration have been detected for most neurotropic viruses, bacteria, and fungi but also in neurodegenerative diseases such as multiple sclerosis, autoimmune diseases, vascular dementia, Alzheimer's, Parkinson's, stroke, and spinal cord injury. However, these mechanisms reveal the complex pathobiology nature of these diseases involving transmigration and invasion of the brain.

Recent studies have shown evidence that immune cells reside in the dura and meninges surrounding the brain and the dural lymphatic system that is connected to the cerebrospinal fluid (CSF) and out of the brain into the cervical lymph nodes (Albayram et al., 2022; Chachaj et al., 2022; Chen et al., 2020; Da Mesquita et al., 2018; Ding et al., 2021; Duque-Wilckens et al., 2022; Li et al., 2022; Song et al., 2020). This suggests that the brain has a local or specialized immunological environment (Fig. 14.1). These findings open new immune–neuronal interactions at this site that are currently being explored. In agreement, the CSF has direct access to bone marrow (Mazzitelli et al., 2022). For example, the injection of AMD3100, a CXCR4 antagonist, into the CSF resulted in the shift of monocytes and neutrophils from the skull bone marrow to the dura. Further, experiments with size-specific dyes indicated a parallel system next to the BBB to clear β-amyloid and other damaging agents (Achariyar et al., 2016; Iliff et al., 2012; Iliff et al., 2014; Ratner et al., 2017). All these data will have large implications in a broad range of infectious, inflammatory, traumatic, and degenerative disorders where the CXCR4 axis is compromised.

14.2.1 Neurons as Immunological Regulators

Neurons lack detectable MHC molecules and antigen presentation even after activation or infection including viruses (Chakraborty et al., 2010; Dachet et al., 2019; Hobson & Sulzer, 2022; Lucca et al., 2016; Moseman et al., 2020; Neumann, 2001). The lack of MHC class I by neurons prevents them from being eliminated by cytotoxic T lymphocytes (CTLs). However, it has been postulated that neuronal/glial apoptosis is regulated by the adaptive immune response to prevent significant brain infiltration and extensive tissue compromise. Moreover, neurons are not invisible to immune surveillance processes as specific inflammatory mediators do recognize these cells. In addition, the recent recognition that neurons can communicate with microglia, astrocytes, and immune cells by soluble mediators, and cell-to-cell interactions open the possibility of an indirect mechanism of immune activation.

Neurons secrete multiple soluble factors and neurotransmitters, including neuropeptides (substance P, vasoactive intestinal peptide, and neurokinin A), and ions that activate or repress the surrounding immune response by targeting astrocytes, microglia, monocytes and macrophages, and T cells. For example, dorsal root ganglia neurons can alter the secretion of immunoglobulins by B cells (Hikawa & Takenaka, 1996). Also, the denervation of lymph nodes with capsaicin compromised the cellular and humoral immune response at the site of damage (Beresford et al., 2004; Esquifino & Cardinali, 1994; Felten et al., 1993; Yamamoto et al., 2021). In addition, neurons express the CD200 receptor, and microglia express the CD200 ligand that prevents microglia activation and looks to play a key role in the pathogenesis of multiple sclerosis (Broderick et al., 2002; Hoek et al., 2000; Koning et al., 2007; Koning et al., 2009; Wright et al., 2000). Microglia obtained from CD200-deficient mice are activated, including increased expression of CD11b and CD45, and reactive microglia were detected upon experimental autoimmune encephalomyelitis and pain (Hernangomez et al., 2012; Manich et al., 2019; Zhao et al., 2020). Overall, these data indicate that expression of CD200 on the surface of the neurons helps to silence and downregulate neuroimmune responses.

Another example that neurons can regulate the function of neighboring immune cells is the expression of surface sialylated gangliosides on the plasma membrane (Belarbi et al., 2020; Gong et al., 2002; Lee et al., 2010; Sotnikov et al., 2013). If these neuronal gangliosides are added to activated T cells, the transcription of the IL-2 is inhibited by a mechanism that prevents the translocation of the nuclear factor kappa B into the nucleus (Hossain et al., 2012; Lu & Sharom, 1995). Also, gangliosides increased proliferation, cytotoxicity, and activation of T cells as well as dendritic cells. In monocytes, gangliosides prevent cytokine production and reduce activated immune cells' proliferation

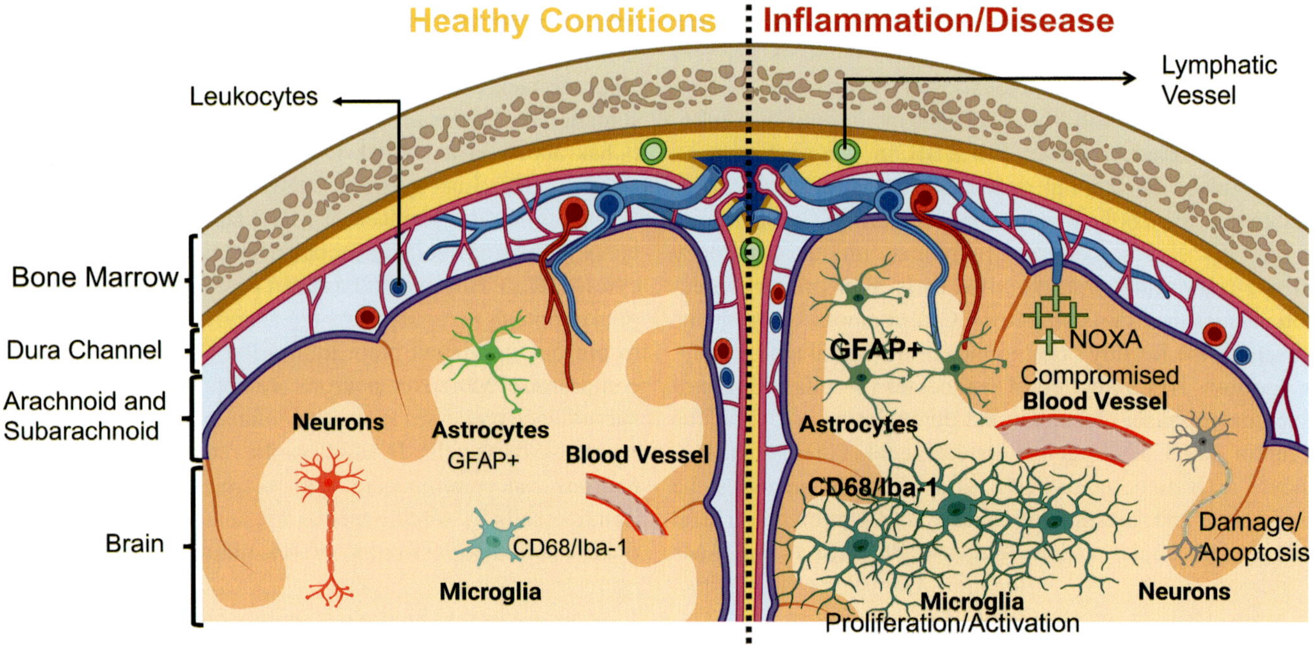

Fig. 14.1 The skull bone marrow supplies the dura and the brain with leukocytes. Under healthy conditions, the complex between the bone marrow and the dura meninges could provide an additional source of leukocytes and inflammatory cells. These studies are ongoing and the contribution of the skull bone marrow to most inflammation/diseases is unknown

(Claus et al., 1994; Hossain et al., 2012; Kimata, 1994; Lu & Sharom, 1995; Norihisa et al., 1994; Ortaldo et al., 1996; Schlaak & Claus, 1995). All these mechanisms indicate that sialylated gangliosides on the surface of the neurons or in a soluble manner serve as immune regulators of the surrounding immune system.

Another example of neuronal regulation of the immune system is related to the limited capacity of the brain to present antigens. The interactions of the T cell receptor and the MHC–peptide complex on the antigen-presenting cell surface initiate the antigen-specific immune responses. A healthy brain lacks expression of MHC molecules. However, T cells preactivated in peripherical immune organs circulate through the brain and transmigrate into the tissue parenchyma to patrol the brain. If these cells release inflammatory cytokines such as TFN-α or IFN-γ, they are induced to express MHC class I and II. Although all cells of the CNS can express MHCs, MHC class I in neurons is highly regulated and targeted by cytotoxic $CD8^+$ T lymphocytes (Medana et al., 2000; Medana et al., 2001; Neumann et al., 2002).

In contrast, microglia express MHC class I and II and co-stimulatory molecules such as CD80 and CD86 upon IFN-γ exposure (Gregerson & Yang, 2003; Wachholz et al., 2016). To a lesser extent, astrocytes are less competent in antigen presentation than microglia, but astrocytes lack the expression of co-stimulatory molecules (Blasko et al., 2000; Morga et al., 2000; Teige et al., 2006). Further, tetrodotoxin (TTX) to reduce neuronal electrical activity enables higher expression of MHC class II on microglia and astrocytes, suggesting that proper neuronal function could control immune activation within the brain (Goddard et al., 2007; Neumann et al., 1998). Also, we demonstrated that tumor necrosis factor-alpha (TNF-α) plus interferon-gamma (IFN-γ) induce gap junctional communication between adjacent microglia, while reduced gap junctional communication in neurons and astrocytes was observed (Eugenin et al., 2001; Eugenin et al., 2012). We and others later identified that TNF-α and IFN-γ not only induced gap junctional communication in microglia and monocytes but also generated tunneling nanotubes (TNTs), a long-distance cell-to-cell communication system, those share antigens, MHC-antigen complexes and organelles between connected cells (Ahani et al., 2022; Dupont et al., 2018; Rubio-Casillas et al., 2022; Schiller et al., 2013; Turos-Korgul et al., 2022; Valdebenito et al., 2019; Watkins & Salter, 2005). These novel communication systems have also been described in COVID-19, several neurodegenerative diseases, and during brain recovery, including glioblastoma (Khattar et al., 2022; Valdebenito et al., 2018, 2019, 2020, 2021). Several pathogens, such as SARS-CoV-2, HIV, and others, use TNTs and gap junctional communication to amplify infection and pathogenesis (Merolli et al., 2022; Okafo et al., 2017; Okura et al., 2021; Rubio-Casillas et al., 2022; Tiwari et al., 2021). This new field could provide new therapies for these diseases, especially because TNTs are not expressed in healthy conditions.

A critical regulation of the immune system is by neurotransmitters and neurotrophins in peripherical tissues and fluids (Chhatar & Lal, 2021; Daeron, 2022; De Virgiliis et al., 2022; Granero et al., 2021; Menees et al., 2022; Mravec, 2022; Mueller, 2022; Vasconcelos et al., 2022). Despite the potential of these mechanisms, few manuscripts have been published, about the consequences of stress, microbiota, and mental health in disease susceptibility and progression. Neurotransmitters such as glutamate, norepinephrine (NE), and dopamine generate intracellular signaling by cyclic nucleotides, kinases, and intracellular calcium changes. Although immune cells are nonexcitable like neurons and express low levels of neurotransmitter receptors/transporters or enzymes involved in producing or degrading neurotransmitters, immune cells respond to these neurotransmitters in multiple ways. For example, glutamate inhibits MHC class II induction in astrocytes but not in microglia (Lee et al., 1992; Neumann et al., 1998). NE and vasoactive intestinal polypeptide (VIP) prevented IFN-γ-mediated induction of MHC class II and Th1/Th2 polarization and dendritic cell maturation (Delgado et al., 2000; Frohman et al., 1988; Li et al., 2011; Lu et al., 2008). Also, neurotrophins such as brain-derived growth factor (BDNF), nerve growth factor (NGF), and neurotrophins can regulate MHC class II and co-stimulatory molecules in immune cells (Bracci-Laudiero et al., 2005; Gong et al., 2007; Neumann et al., 1998; Shinoda et al., 1996). Denote the few manuscripts available in the area of neuro–immune regulation and their dysregulation for external factors such as depression or drug use. In the context of drug abuse, most drugs are related to dopamine dysregulation and reward; however, most of these data are within the CNS, and there are few examples in the peripherical immune system. Our data and collaborators indicate that dopamine activates immune cells and can alter HIV entry and infection, suggesting a strong phenotype in healthy and disease conditions (Calderon et al., 2017; Coley et al., 2015; Gaskill et al., 2009; Gaskill et al., 2012, 2013, 2014; Matt et al., 2021; Matt & Gaskill, 2019; Nickoloff-Bybel et al., 2019, 2020; Nolan et al., 2019). Also, our data indicate that the neurotransmitter ATP continually circulates in the blood at low concentrations due to an effective degradation system into adenosine. However, in HIV conditions associated with cognitive impairment, extracellular ATP accumulates in the bloodstream, predicting the onset of declined cognition, suggesting a significant contribution of ATP to BBB compromise and leukocyte activation (D'Amico et al., 2021; Eugenin et al., 2011; Hazleton et al., 2012; Velasquez et al., 2016, 2020; Velasquez & Eugenin, 2014). Overall, neurotransmitters regulate the immune response effectively, altering T cell, dendritic, and monocyte activation. We expect that future generations will explore these exciting areas of research.

14.2.2 Inflammation and CNS Injuries

As indicated above, the brain can regulate the immune function through cell-to-cell contacts (gap junctions and TNTs), neurotransmitters, and several extracellular mediators, including cytokines-chemokines/growth factors, and exosomes. However, less is known about how the autonomic nervous system, through sympathetic and vagus nerve innervation, can also regulate the immune response. Afferent neurons express several receptors for inflammatory factors that activate the neural reflex circuits to regulate acute and chronic immune responses (Ordovas-Montanes et al., 2015; Shouman & Benarroch, 2021; Voisin et al., 2017). The vagus nerve activation results in NE release from splenic neurons, triggering β2 adrenergic receptor activation on splenic T lymphocytes (Bonaz et al., 2016; Guyot et al., 2019; Vida et al., 2011). The T cell activation results in acetylcholine production that binds to the α7 nACh receptors expressed on macrophages in the spleen's red pulp and marginal zone, suppressing NF-kB translocation into the nucleus and subsequent silencing cytokine production (Susarla et al., 2014; Wang et al., 2003). However, electrical stimulation of the vagus nerve improves systemic inflammation and the survival rate in different clinical conditions, including autoimmune diseases, synovitis, and obesity (Huston et al., 2006; Huston et al., 2009; Mina-Osorio et al., 2012; Reardon et al., 2013; Rosas-Ballina et al., 2011; Rosas-Ballina & Tracey, 2009).

Neurotransmitters such as catecholamines (NE, epinephrine, and dopamine) have been demonstrated to profoundly regulate immune activities such as chemotaxis, cytotoxic activity, and secretion of soluble factors, including antibodies in immune cells and immune organs. For example, dopamine reduces lymphocyte proliferation and cytotoxic activity (Gaskill et al., 2012, 2013, 2014; Nickoloff-Bybel et al., 2019, 2020; Rim et al., 2022; Shinoda et al., 1996). NE and adrenergic agents also reduce cytotoxic activity and increase lymphocyte proliferation (Edgar et al., 2002; Elenkov et al., 2000; Qiu et al., 1996; Zoukos et al., 1994). Another example of neuronal-immune interactions corresponds to the activation of nociceptors that leads to local axonal reflexes resulting in the release of several neuropeptides, including substance P (SP), neurokinins, vasoactive intestinal peptide (VIP), and neuropeptide Y (NPY), which induce the migration of innate and adaptive immune cells. More specifically, these neuropeptides can regulate chemotaxis, antigen presentation, and vascular permeability further enhancing their function in healthy and pathological conditions (Baral et al., 2019; Ganea & Delgado, 2001; Populin et al., 2021). Despite this exciting data, most of these manuscripts are sporadic, and few laboratories have examined the interactions between the immune and nervous systems.

Cytokines or neurotrophic factors mediate a more explored communication system between the immune and nervous systems (note: the differences between an immune and nervous system are due to the classification of the context identified, function, and group involved. Thus, an immune factor could be stronger in neuronal cells). For example, sensory neurons secrete and express several cytokine receptors, including IL-1, TNF-α, and NGF. Activation of these receptors results in the activation of danger-associated molecular patterns (DAMPs) receptors, Toll-like receptors (TLRs), pathogen-associated molecular patterns (PAMPs) to recognize environmental conditions (heat, cold, and pain), and pathogens (viruses, fungi, and bacteria) as well as soluble host danger signals such as extracellular ATP. All these factors enhance T cell function, APC, and cytotoxic-related activities. Extracellular ATP is interesting because it is one of the body's fastest and most powerful danger signals. However, ATP half-life is short and highly regulated by purinergic/adenosine receptors and the enzymes that degrade ATP (ecto-ATPases). Our group identified that ATP could be considered a systemic biomarker of cognitive impairment, suggesting that serum ATP levels play a key role in immune activation as well as a neuronal compromise in NeuroHIV and Alzheimer's disease (D'Amico et al., 2021; Hazleton et al., 2012; Velasquez et al., 2016, 2020).

In macrophages, VIP prevents the expression of inflammatory cytokines and chemokines, promoting the Th2 $CD4^+$ T lymphocyte proliferation and survival (Pozo & Delgado, 2004). NPY has been implicated in food consumption and circadian cycles but modulates lymphocyte proliferation, NK activity, and TNF-α release (Puerto et al., 2005). SP also stimulates lymphocyte migration, proliferation, IgA secretion, and phagocytosis (Bao et al., 2022; Calcagni & Elenkov, 2006). Another cytokine involved in food intake is leptin, which is also involved in autoimmune diseases, multiple sclerosis, and diabetes complications. Leptin is the product of the obese *(ob)* gene, mostly produced in the adipose tissue, and is linked to nutrition, metabolism, and immune homeostasis. However, the leptin-deficient and leptin receptor mice are not obese but have profound defects in reproduction, angiogenesis, bone metabolism, lipid/glucose metabolism, and innate and adaptive immunity (La Cava & Matarese, 2004; Procaccini et al., 2012).

Leptin upregulates differentiation marker expression in monocyte/macrophages (CD25, CD71, CD69, and CD38) in correlation with increased phagocytosis and cytokine secretion. In neutrophils, leptins promote chemotaxis and oxidative stress. In NK cells, leptin increases proliferation, differentiation, activation, and lytic activity. Most of these migratory phenotypes are associated with the enhanced expression of several adhesion molecules, including ICAM-1 and VLA-2. In tissues such as the thymus, leptin inhibits T cell apoptosis, enhancing their maturation and survival (Howard et al., 1999). Leptin deficiency protects against several diseases, including glomerulonephritis, multiple sclerosis, lupus, and type 1 diabetes, by preventing macrophage infiltration (Bluher et al., 2002; Holmes, 2016; Taleb et al., 2007; Valerio et al., 2009; Zheng et al., 2010). Intracerebroventricular leptin injection prevents activation of B cell development in the bone marrow of fasted mice, again providing in vivo evidence that leptin controls B cell development (Nagashima et al., 2010; Yoshida et al., 2010). Leptin, like NPY and other peptides, has profound immune regulation properties that must be examined (see Fig. 14.2).

14.2.3 Neural and Hormonal Neuroimmune Regulation

Sensory and autonomic nerves innervate all tissues. However, whether the local release of neurotransmitters can control the immune system is less known (Blake et al., 2019; Chiu et al., 2012; Liu et al., 2022; Ordovas-Montanes et al., 2015; Udit et al., 2022). For example, primary (thymus and the bone marrow, including the skull described above) and secondary (spleen and lymph nodes) lymphoid organs are innervated by sympathetic nerves (Cleypool, Lotgerink Bruinenberg, et al., 2021a; Cleypool, Mackaaij, et al., 2021b). In the spleen and lymph nodes, nervous fibers penetrate depth into T cell zones and toward the B cell follicles.

In contrast, sensory innervation is limited to the capsule and hilum (Popper et al., 1988). In both tissues, nerves interact with T cells, macrophages, and dendritic cells (Felten et al., 1987; Felten & Felten, 1987). These data strongly indicate a close interaction between the nervous and immune systems; however, the direct regulation of both systems is still poorly understood. Our data and others also suggest that immune cells may be capable of synthesizing some neurotransmitters due to their expression of tyrosine hydroxylase, an enzyme critical for catecholamine biosynthesis (Flierl et al., 2007). Myeloid-derived catecholamines amplify inflammation induced by endotoxin (Staedtke et al., 2018).

Our data in the context of HIV and drug use indicate that immune cells, including monocytes/macrophages and T cells, are highly responsive to catecholamines, resulting in increased HIV entry and replication and inflammation (Calderon et al., 2017; Coley et al., 2015). However, the role of endogenous concentrations of catecholamines in diseases such as HIV is unknown. For example, the adrenal medulla is a major source of systemic adrenaline, noradrenaline, and steroid hormones such as cortisol and mineralocorticoids. This point is important because several cytokines induce glucocorticoid production to promote resolution (Miller & Mayo, 1997). Currently, it is well accepted that β2 adrenergic activation alters most immune cells, except Th2 cells. NE

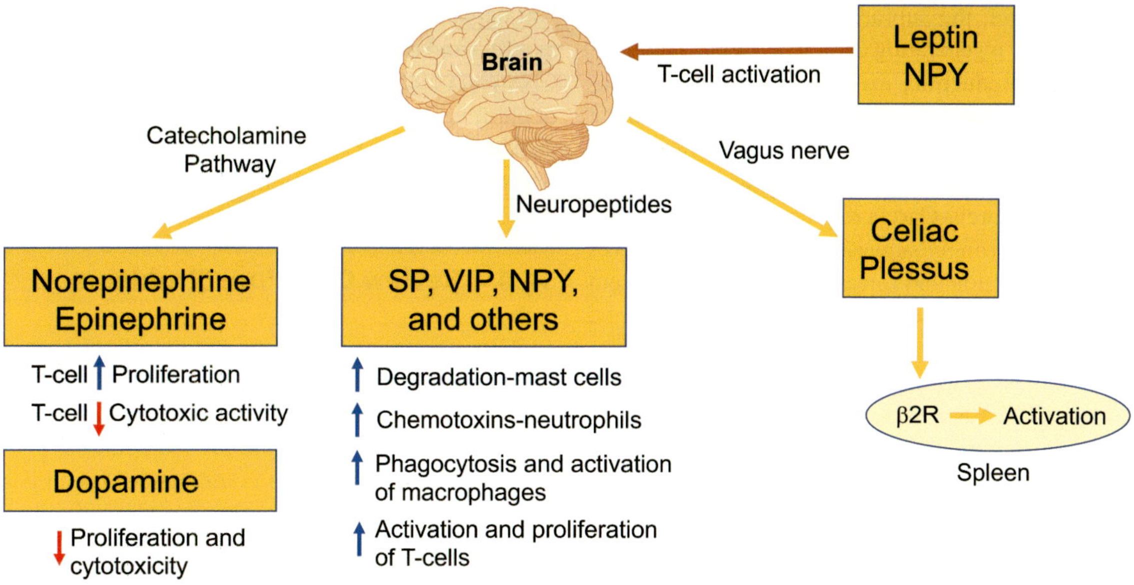

Fig. 14.2 The neuro-immune interactions between the brain and the immune system. The brain communicates with the immune system through catecholamines, neuropeptides, the vagus nerve, and leptin/NPY axis complexes. Activation of these pathways results in alterations in immune activation (phagocytosis, proliferation, activation, chemotaxis, degranulation, and cytotoxicity)

prevents IL-2 production in response to anti-CD3 antibody stimulation (Ramer-Quinn et al., 2000) and promotes naïve $CD4^+$ T lymphocyte differentiation into a Th1 phenotype (Swanson et al., 2001). β2-adrenergic receptor activation also enhanced immunoglobulin G1 response to CD86 ligation (Padro & Sanders, 2014). Also, monocyte/macrophages express both, α1 and α2 and β2 adrenergic receptors with diametral effects; β2 activation results in a reduction in cytokine production, whereas α1/α2 adrenergic activation results in cytokine release (Chhatar & Lal, 2021; Felten et al., 1993; Ramer-Quinn et al., 2000; Swanson et al., 2001).

Interestingly, adrenergic stimulation promotes myeloid cell recruitment into damaged areas. For example, β2-adrenergic receptor activation controls monocyte egress from the bone marrow but reduces lymphocyte egress of lymph nodes (Nakai et al., 2014). An interesting point is catecholamines regulate the innate and adaptive immune response during light–dark cycles. Several groups propose that high adrenergic tones are characterized by increased recruitment of myeloid cells into tissues and the sequestration of lymphocytes from the blood into lymphoid organs to be prepared for higher risk infections during the active phase of the light/dark cycle and keeping the adaptive immune response on hold (Scheiermann et al., 2012). In agreement with these ideas, T cells produce and release concentrations of epinephrine and NE to regulate local events in primary and secondary lymph tissues. Lymphocytes of the mesenteric lymph nodes regulate lymphocyte trafficking, cytokine production, activation, and blood vessel permeability (Peng et al., 2004; Qiu et al., 2004a, b). Further, macrophages and neutrophils release epinephrine and NE upon activation and blockage of α2-adrenoreceptors, but not β2-adrenoreceptors, reduced vascular leak in a lung injury model (Flierl et al., 2007). Also, blocking tyrosine hydroxylase or dopamine-β-hydroxylase diminished the inflammatory response (Flierl et al., 2007).

Thus, the immune–neuronal axis could affect behavior and increase susceptibility to several acute and chronic diseases. For example, stress and drug abuse influence the immune system. Short-term stress enhances immune response, but chronic stress is linked to immune dysfunction (Dhabhar et al., 2010) and diseases associated with a poor immune response, such as cancer and increased susceptibility to neuropsychological diseases. A critical example of immune–neuro interaction is the hypothalamus-pituitary-adrenal (HPA) axis through corticotropin-releasing hormone (CRH), adrenocorticotropic hormone (ACTH), and corticosterone in concert with nerve-derived NE. For example, acetylcholine (ACh)-producing choline acetyl transferase+ (ChAT+) T cells play a crucial role in plasma cell formation and development, but these studies also indicate a high communication between the spleen and the paraventricular nucleus (PVN) and the central nucleus of the amygdala (CeA) (Rosas-Ballina et al., 2011; Zhang et al., 2020). Also, suggestions have been suggested that the HPA axis, in association with stress, plays a role in B cell response, but most of these studies have been indirect. Most of these phenotypes are correlated with events of social defeat and enhanced release of adrenergic ligands, resulting in the recruitment and trafficking of monocytes into the brain, microglia activation, and the development of anxiety behaviors (Wohleb et al., 2014a, b, c).

As indicated above, the suprachiasmatic nucleus in the hypothalamus is key in generating circadian rhythms, including glucocorticoid production, as indicated by NE/adrenaline during diurnal hours. Nerve-derived NE in the bone marrow downregulates the expression of CXCL12 in stromal cells, enhancing the release of CXCR4-expressing stem cells into the circulation (Garcia-Garcia & Martin, 2019; Mendez-Ferrer et al., 2008), strongly suggesting that sensory nerves can control the production and release of bone marrow stem cells (Gao et al., 2021; Suekane et al., 2019). Interestingly, the adrenergic/circadian control system also regulates adhesion molecule expression, including ICAM-1, VCAM-1, P- and E-selectin, and MAdCAM-1, all required for activation, arrest, rolling, attachment, and transmigration of leukocytes into different tissues suggesting even during the day/night cycle an neuro-immune regulation. Also, we are not discussing the nervous system component that regulates blood flow and pressure here, including different degrees of activation in the morning for arteries and veins in the evening (Rhoads et al., 2020; Takeda & Maemura, 2011). Also, the sympathetic nervous system controls the lymph flow and leukocyte trafficking in and out of the primary and secondary lymph tissues (Bachmann et al., 2019).

In contrast, in stress or pharmacological stimulation of the adrenergic system, leukocyte trafficking is redirected to the circulation. Stressors like surgery that increases catecholamines and glucocorticoids had similar effects on leukocyte trafficking. The neuro–immune axis's role in cancer and metastasis is an exciting point. Clinical trials using β-blockers in cancer conditions reduce metastasis and improve the recruitment of immune cells into tumors (Erin et al., 2022; Sharawy, 2022; Shurin et al., 2021).

Although the tissue destination and associated functions of these cells induced to circulate by stress are unknown, their function probably provides another line of immune and brain defense against pathogens. Thus, these changes are highly dynamic and poorly examined but could have significant implications in acute, chronic, and vaccination strategies. However, despite the clear evidence of the role of stress on the immune system, strong direct neuronal control of the immune system by the brain remains controversial.

14.3 Conclusions

Overall, our chapter on neuro–immune modulation and its interactions indicates a complex and unexplored area of research. The immune–brain interactions are mediated by cytokines, hormones, and peptides released from the CNS and by activating afferent and efferent neurological pathways within the brain and lymph organs. Activation of these mechanisms can be immuno-suppressive or stimulant depending of the context including behavior conditions as stress and neuropsycological compromise. These immune/CNS responses are also regulated by external environmental or chronic conditions such as stress, cancer, and autoimmune diseases. But more importantly, it provides a unique opportunity for novel therapeutic interventions.

14.4 Review Questions/Problems

1. Neuro–immune interactions are characterized by the following:
   (a) Signals from neurons to immune cells
   (b) Signals from immune cells to neurons
   (c) External conditions could control both interactions
   (d) All of the above are true
2. Which neuronal receptors express immune cells?
   (a) Adrenergic
   (b) Neuropeptides
   (c) Cytokines/growth factors
   (d) Steroids
   (e) All of the above are true
3. Which sentences are true?
   (a) Neurotransmitters can alter cell trafficking
   (b) Neurotransmitters can alter bone marrow maturation and production of leukocytes
   (c) β2 blockers can be used to treat several neurological diseases
   (d) All of the above are true
   (e) None of the above
4. Can you discuss the significance of the skull bone marrow production of immune cells?
5. How could sympathetic nerves regulate the immune response?
6. Discuss with the class the potential therapeutic consequences of controlling short- and long-term stress in the patient population to promote an effective immune response.

14.5 Answers to the Questions

1. d
2. e
3. d
4. Open discussion
5. Open discussion
6. Open discussion

Acknowledgments This work was funded by the National Institute of Mental Health grant, MH128082 and MH134761, the National Institute of Neurological Disorders and Stroke, NS105584, and UTMB internal funding (to E.A.E).

References

Achariyar, T. M., Li, B., Peng, W., Verghese, P. B., Shi, Y., McConnell, E., Benraiss, A., Kasper, T., Song, W., Takano, T., Holtzman, D. M., Nedergaard, M., & Deane, R. (2016). Glymphatic distribution of CSF-derived apoE into brain is isoform specific and suppressed during sleep deprivation. *Molecular Neurodegeneration, 11*, 74.

Ahani, E., Fereydouni, M., Motaghed, M., & Kepley, C. L. (2022). Identification and characterization of tunneling nanotubes involved in human mast cell FcepsilonRI-mediated apoptosis of cancer cells. *Cancers (Basel), 14*, 14.

Albayram, M. S., Smith, G., Tufan, F., Tuna, I. S., Bostancikliogl u, M., Zile, M., & Albayram, O. (2022). Non-invasive MR imaging of human brain lymphatic networks with connections to cervical lymph nodes. *Nature Communications, 13*, 203.

Alluri, H., Wiggins-Dohlvik, K., Davis, M. L., Huang, J. H., & Tharakan, B. (2015). Blood-brain barrier dysfunction following traumatic brain injury. *Metabolic Brain Disease, 30*, 1093–1104.

Alves de Lima, K., Rustenhoven, J., Da Mesquita, S., Wall, M., Salvador, A. F., Smirnov, I., Martelossi Cebinelli, G., Mamuladze, T., Baker, W., Papadopoulos, Z., Lopes, M. B., Cao, W. S., Xie, X. S., Herz, J., & Kipnis, J. (2020a). Meningeal gammadelta T cells regulate anxiety-like behavior via IL-17a signaling in neurons. *Nature Immunology, 21*, 1421–1429.

Alves de Lima, K., Rustenhoven, J., & Kipnis, J. (2020b). Meningeal immunity and its function in maintenance of the central nervous system in health and disease. *Annual Review of Immunology, 38*, 597–620.

Arino, H., Heartshorne, R., Michael, B. D., Nicholson, T. R., Vincent, A., Pollak, T. A., & Vogrig, A. (2022). Neuroimmune disorders in COVID-19. *Journal of Neurology, 269*, 2827–2839.

Bachmann, S. B., Proulx, S. T., He, Y., Ries, M., & Detmar, M. (2019). Differential effects of anaesthesia on the contractility of lymphatic vessels in vivo. *The Journal of Physiology, 597*, 2841–2852.

Ball, J. B., Green-Fulgham, S. M., & Watkins, L. R. (2022). Mechanisms of microglia-mediated synapse turnover and synaptogenesis. *Progress in Neurobiology, 218*, 102336.

Bao, X., Qin, Y., Lu, L., & Zheng, M. (2022). Transcriptional regulation of early T-lymphocyte development in thymus. *Frontiers in Immunology, 13*, 884569.

Baral, P., Udit, S., & Chiu, I. M. (2019). Pain and immunity: Implications for host defence. *Nature Reviews. Immunology, 19*, 433–447.

Belarbi, K., Cuvelier, E., Bonte, M. A., Desplanque, M., Gressier, B., Devos, D., & Chartier-Harlin, M. C. (2020). Glycosphingolipids and neuroinflammation in Parkinson's disease. *Molecular Neurodegeneration, 15*, 59.

Beresford, L., Orange, O., Bell, E. B., & Miyan, J. A. (2004). Nerve fibres are required to evoke a contact sensitivity response in mice. *Immunology, 111*, 118–125.

Blake, K. J., Jiang, X. R., & Chiu, I. M. (2019). Neuronal regulation of immunity in the skin and lungs. *Trends in Neurosciences, 42*, 537–551.

Blasko, I., Veerhuis, R., Stampfer-Kountchev, M., Saurwein-Teissl, M., Eikelenboom, P., & Grubeck-Loebenstein, B. (2000). Costimulatory effects of interferon-gamma and interleukin-1beta or tumor necrosis factor alpha on the synthesis of Abeta1-40 and Abeta1-42 by human astrocytes. *Neurobiology of Disease, 7*, 682–689.

Bluher, M., Michael, M. D., Peroni, O. D., Ueki, K., Carter, N., Kahn, B. B., & Kahn, C. R. (2002). Adipose tissue selective insulin receptor knockout protects against obesity and obesity-related glucose intolerance. *Developmental Cell, 3*, 25–38.

Bolton, J. L., Short, A. K., Othy, S., Kooiker, C. L., Shao, M., Gunn, B. G., Beck, J., Bai, X., Law, S. M., Savage, J. C., Lambert, J. J., Belelli, D., Tremblay, M. E., Cahalan, M. D., & Baram, T. Z. (2022). Early stress-induced impaired microglial pruning of excitatory synapses on immature CRH-expressing neurons provokes aberrant adult stress responses. *Cell Reports, 38*, 110600.

Bonaz, B., Sinniger, V., & Pellissier, S. (2016). Anti-inflammatory properties of the vagus nerve: Potential therapeutic implications of vagus nerve stimulation. *The Journal of Physiology, 594*, 5781–5790.

Bracci-Laudiero, L., Aloe, L., Caroleo, M. C., Buanne, P., Costa, N., Starace, G., & Lundeberg, T. (2005). Endogenous NGF regulates CGRP expression in human monocytes, and affects HLA-DR and CD86 expression and IL-10 production. *Blood, 106*, 3507–3514.

Broderick, C., Hoek, R. M., Forrester, J. V., Liversidge, J., Sedgwick, J. D., & Dick, A. D. (2002). Constitutive retinal CD200 expression regulates resident microglia and activation state of inflammatory cells during experimental autoimmune uveoretinitis. *The American Journal of Pathology, 161*, 1669–1677.

Buckner, C. M., Luers, A. J., Calderon, T. M., Eugenin, E. A., & Berman, J. W. (2006). Neuroimmunity and the blood-brain barrier: Molecular regulation of leukocyte transmigration and viral entry into the nervous system with a focus on neuroAIDS. *Journal of Neuroimmune Pharmacology, 1*, 160–181.

Calcagni, E., & Elenkov, I. (2006). Stress system activity, innate and T helper cytokines, and susceptibility to immune-related diseases. *Annals of the New York Academy of Sciences, 1069*, 62–76.

Calderon, T. M., Williams, D. W., Lopez, L., Eugenin, E. A., Cheney, L., Gaskill, P. J., Veenstra, M., Anastos, K., Morgello, S., & Berman, J. W. (2017). Dopamine increases CD14(+)CD16(+) monocyte transmigration across the blood brain barrier: Implications for substance abuse and HIV neuropathogenesis. *Journal of Neuroimmune Pharmacology, 12*, 353–370.

Carlson, E. R., Guerin, S. P., Nixon, K., & Fonken, L. K. (2022). The neuroimmune system—Where aging and excess alcohol intersect. *Alcohol, 107*, 153.

Chachaj, A., Gasiorowski, K., Szuba, A., Sieradzki, A., & Leszek, J. (2022). Lymphatic system in the brain clearance mechanisms—New therapeutic perspectives for Alzheimer's disease. *Current Neuropharmacol, 21*, 380.

Chakraborty, S., Nazmi, A., Dutta, K., & Basu, A. (2010). Neurons under viral attack: Victims or warriors? *Neurochemistry International, 56*, 727–735.

Chen, J., Wang, L., Xu, H., Xing, L., Zhuang, Z., Zheng, Y., Li, X., Wang, C., Chen, S., Guo, Z., Liang, Q., & Wang, Y. (2020). Meningeal lymphatics clear erythrocytes that arise from subarachnoid hemorrhage. *Nature Communications, 11*, 3159.

Chhatar, S., & Lal, G. (2021). Role of adrenergic receptor signalling in neuroimmune communication. *Current Research in Immunology, 2*, 202–217.

Chithanathan, K., Xuan, F. L., Hickey, M. A., & Tian, L. (2022). Enhanced anxiety and olfactory microglial activation in early-stage familial Alzheimer's disease mouse model. *Biology (Basel), 11*, 11.

Chiu, I. M., von Hehn, C. A., & Woolf, C. J. (2012). Neurogenic inflammation and the peripheral nervous system in host defense and immunopathology. *Nature Neuroscience, 15*, 1063–1067.

Chokr, S. M., Milinkeviciute, G., Jimenez, G. A., Abubakr, H., & Cramer, K. S. (2022). Long-term microglia depletion impairs synapse elimination and auditory brainstem function. *Scientific Reports, 12*, 18521.

Claus, C., Schlaak, J., Dittmayer, M., K. Meyer zum Buschenfelde, & Dippold, W. (1994). Inhibition of anti-GD3-ganglioside antibody-induced proliferation of human CD8+ T cells by CD16+ natural killer cells. *European Journal of Immunology, 24*, 1208–1212.

Cleypool, C. G. J., Lotgerink Bruinenberg, D., Roeling, T., Irwin, E., & Bleys, R. (2021a). Splenic artery loops: Potential splenic plexus stimulation sites for neuroimmunomodulatory-based anti-inflammatory therapy? *Clinical Anatomy, 34*, 371–380.

Cleypool, C. G. J., Mackaaij, C., Lotgerink Bruinenberg, D., Schurink, B., & Bleys, R. (2021b). Sympathetic nerve distribution in human lymph nodes. *Journal of Anatomy, 239*, 282–289.

Coley, J. S., Calderon, T. M., Gaskill, P. J., Eugenin, E. A., & Berman, J. W. (2015). Dopamine increases CD14+CD16+ monocyte migration and adhesion in the context of substance abuse and HIV neuropathogenesis. *PLoS One, 10*, e0117450.

Cornell, J., Salinas, S., Huang, H. Y., & Zhou, M. (2022). Microglia regulation of synaptic plasticity and learning and memory. *Neural Regeneration Research, 17*, 705–716.

Cugurra, A., Mamuladze, T., Rustenhoven, J., Dykstra, T., Beroshvili, G., Greenberg, Z. J., Baker, W., Papadopoulos, Z., Drieu, A., Blackburn, S., Kanamori, M., Brioschi, S., Herz, J., Schuettpelz, L. G., Colonna, M., Smirnov, I., & Kipnis, J. (2021). Skull and vertebral bone marrow are myeloid cell reservoirs for the meninges and CNS parenchyma. *Science, 373*.

D'Amico, D., Valdebenito, S., & Eugenin, E. A. (2021). The role of Pannexin-1 channels and extracellular ATP in the pathogenesis of the human immunodeficiency virus. *Purinergic Signal, 17*, 563–576.

Da Mesquita, S., Louveau, A., Vaccari, A., Smirnov, I., Cornelison, R. C., Kingsmore, K. M., Contarino, C., Onengut-Gumuscu, S., Farber, E., Raper, D., Viar, K. E., Powell, R. D., Baker, W., Dabhi, N., Bai, R., Cao, R., Hu, S., Rich, S. S., Munson, J. M., Lopes, M. B., Overall, C. C., Acton, S. T., & Kipnis, J. (2018). Functional aspects of meningeal lymphatics in ageing and Alzheimer's disease. *Nature, 560*, 185–191.

Dachet, F., Liu, J., Ravits, J., & Song, F. (2019). Predicting disease specific spinal motor neurons and glia in sporadic ALS. *Neurobiology of Disease, 130*, 104523.

Daeron, M. (2022). The immune system as a system of relations. *Frontiers in Immunology, 13*, 984678.

Dave, V. A., & Klein, R. S. (2022). *The multitaskers of the brain: Glial responses to viral infections and associated post-infectious neurologic sequelae* (Vol. 71, p. 803). Glia.

Dayananda, K. K., Ahmed, S., Wang, D., Polis, B., Islam, R., & Kaffman, A. (2022). Early life stress impairs synaptic pruning in the developing hippocampus. *Brain, Behavior, and Immunity, 107*, 16–31.

De Virgiliis, F., Oliva, V. M., Kizil, B., & Scheiermann, C. (2022). Control of lymph node activity by direct local innervation. *Trends in Neurosciences, 45*, 704–712.

Delage, C., Taib, T., Mamma, C., Lerouet, D., & Besson, V. C. (2021). Traumatic brain injury: An age-dependent view of post-traumatic neuroinflammation and its treatment. *Pharmaceutics, 13*.

Delgado, M., Leceta, J., Sun, W., Gomariz, R. P., & Ganea, D. (2000). VIP and PACAP induce shift to a Th2 response by upregulating B7.2 expression. *Annals of the New York Academy of Sciences, 921*, 68–78.

Dhabhar, F. S., Saul, A. N., Daugherty, C., Holmes, T. H., Bouley, D. M., & Oberyszyn, T. M. (2010). Short-term stress enhances cellular immunity and increases early resistance to squamous cell carcinoma. *Brain, Behavior, and Immunity, 24*, 127–137.

Ding, X. B., Wang, X. X., Xia, D. H., Liu, H., Tian, H. Y., Fu, Y., Chen, Y. K., Qin, C., Wang, J. Q., Xiang, Z., Zhang, Z. X., Cao, Q. C., Wang, W., Li, J. Y., Wu, E., Tang, B. S., Ma, M. M., Teng, J. F., & Wang, X. J. (2021). Impaired meningeal lymphatic drainage in patients with idiopathic Parkinson's disease. *Nature Medicine, 27*, 411–418.

Dupont, M., Souriant, S., Lugo-Villarino, G., Maridonneau-Parini, I., & Verollet, C. (2018). Tunneling nanotubes: Intimate communication between myeloid cells. *Frontiers in Immunology, 9*, 43.

Duque-Wilckens, N., Teis, R., Sarno, E., Stoelting, F., Khalid, S., Dairi, Z., Douma, A., Maradiaga, N., Hench, S., Dharshika, C. D., Thelen, K. M., Gulbransen, B., Robison, A. J., & Moeser, A. J. (2022). Early life adversity drives sex-specific anhedonia and meningeal immune gene expression through mast cell activation. *Brain, Behavior, and Immunity, 103*, 73–84.

Dziabis, J. E., & Bilbo, S. D. (2022). Microglia and sensitive periods in brain development. *Current Topics in Behavioral Neurosciences, 53*, 55–78.

Edgar, V. A., Cremaschi, G. A., Sterin-Borda, L., & Genaro, A. M. (2002). Altered expression of autonomic neurotransmitter receptors and proliferative responses in lymphocytes from a chronic mild stress model of depression: Effects of fluoxetine. *Brain, Behavior, and Immunity, 16*, 333–350.

Elenkov, I. J., Wilder, R. L., Chrousos, G. P., & Vizi, E. S. (2000). The sympathetic nerve--an integrative interface between two supersystems: the brain and the immune system. *Pharmacological Reviews, 52*, 595–638.

Erin, N., Shurin, G. V., Baraldi, J. H., & Shurin, M. R. (2022). Regulation of carcinogenesis by sensory neurons and Neuromediators. *Cancers (Basel), 14*.

Esquifino, A. I., & Cardinali, D. P. (1994). Local regulation of the immune response by the autonomic nervous system. *Neuroimmunomodulation, 1*, 265–273.

Eugenin, E. A., & Berman, J. W. (2003). Chemokine-dependent mechanisms of leukocyte trafficking across a model of the blood-brain barrier. *Methods, 29*, 351–361.

Eugenin, E. A., Eckardt, D., Theis, M., Willecke, K., Bennett, M. V., & Saez, J. C. (2001). Microglia at brain stab wounds express connexin 43 and in vitro form functional gap junctions after treatment with interferon-gamma and tumor necrosis factor-alpha. *Proceedings of the National Academy of Sciences of the United States of America, 98*, 4190–4195.

Eugenin, E. A., Gamss, R., Buckner, C., Buono, D., Klein, R. S., Schoenbaum, E. E., Calderon, T. M., & Berman, J. W. (2006a). Shedding of PECAM-1 during HIV infection: A potential role for soluble PECAM-1 in the pathogenesis of NeuroAIDS. *Journal of Leukocyte Biology, 79*, 444–452.

Eugenin, E. A., Osiecki, K., Lopez, L., Goldstein, H., Calderon, T. M., & Berman, J. W. (2006b). CCL2/monocyte chemoattractant protein-1 mediates enhanced transmigration of human immunodeficiency virus (HIV)-infected leukocytes across the blood-brain barrier: A potential mechanism of HIV-CNS invasion and NeuroAIDS. *The Journal of Neuroscience, 26*, 1098–1106.

Eugenin, E. A., Clements, J. E., Zink, M. C., & Berman, J. W. (2011). Human immunodeficiency virus infection of human astrocytes disrupts blood-brain barrier integrity by a gap junction-dependent mechanism. *The Journal of Neuroscience, 31*, 9456–9465.

Eugenin, E. A., Basilio, D., Saez, J. C., Orellana, J. A., Raine, C. S., Bukauskas, F., Bennett, M. V., & Berman, J. W. (2012). The role of gap junction channels during physiologic and pathologic conditions of the human central nervous system. *Journal of Neuroimmune Pharmacology, 7*, 499–518.

Felten, D. L., & Felten, S. Y. (1987). Immune interactions with specific neural structures. *Brain, Behavior, and Immunity, 1*, 279–283.

Felten, D. L., Felten, S. Y., Bellinger, D. L., Carlson, S. L., Ackerman, K. D., Madden, K. S., Olschowki, J. A., & Livnat, S. (1987). Noradrenergic sympathetic neural interactions with the immune system: Structure and function. *Immunological Reviews, 100*, 225–260.

Felten, D. L., Felten, S. Y., Bellinger, D. L., & Madden, K. S. (1993). Fundamental aspects of neural-immune signaling. *Psychotherapy and Psychosomatics, 60*, 46–56.

Fernandez Zapata, C., Giacomello, G., Spruth, E. J., Middeldorp, J., Gallaccio, G., Dehlinger, A., Dames, C., Leman, J. K. H., van Dijk, R. E., Meisel, A., Schlickeiser, S., Kunkel, D., Hol, E. M., Paul, F., Parr, M. K., Priller, J., & Bottcher, C. (2022). Differential compartmentalization of myeloid cell phenotypes and responses towards the CNS in Alzheimer's disease. *Nature Communications, 13*, 7210.

Flierl, M. A., Rittirsch, D., Nadeau, B. A., Chen, A. J., Sarma, J. V., Zetoune, F. S., McGuire, S. R., List, R. P., Day, D. E., Hoesel, L. M., Gao, H., Van Rooijen, N., Huber-Lang, M. S., Neubig, R. R., & Ward, P. A. (2007). Phagocyte-derived catecholamines enhance acute inflammatory injury. *Nature, 449*, 721–725.

Frederick, N. M., Tavares, G. A., & Louveau, A. (2022). Neuroimmune signaling at the brain borders. *Immunological Reviews, 311*, 9–25.

Frohman, E. M., Frohman, T. C., Vayuvegula, B., Gupta, S., & van den Noort, S. (1988). Vasoactive intestinal polypeptide inhibits the

expression of the MHC class II antigens on astrocytes. *Journal of the Neurological Sciences, 88*, 339–346.
Ganea, D., & Delgado, M. (2001). Inhibitory neuropeptide receptors on macrophages. *Microbes and Infection, 3*, 141–147.
Gao, X., Zhang, D., Xu, C., Li, H., Caron, K. M., & Frenette, P. S. (2021). Nociceptive nerves regulate haematopoietic stem cell mobilization. *Nature, 589*, 591–596.
Gao, H., Huang, X., He, J., Zou, T., Chen, X., & Xu, H. (2022). The roles of microglia in neural remodeling during retinal degeneration. *Histology and Histopathology, 37*, 1–10.
Garcia-Garcia, A., & Martin, I. (2019). Extracellular matrices to modulate the innate immune response and enhance bone healing. *Frontiers in Immunology, 10*, 2256.
Gaskill, P. J., Calderon, T. M., Luers, A. J., Eugenin, E. A., Javitch, J. A., & Berman, J. W. (2009). Human immunodeficiency virus (HIV) infection of human macrophages is increased by dopamine: A bridge between HIV-associated neurologic disorders and drug abuse. *The American Journal of Pathology, 175*, 1148–1159.
Gaskill, P. J., Carvallo, L., Eugenin, E. A., & Berman, J. W. (2012). Characterization and function of the human macrophage dopaminergic system: Implications for CNS disease and drug abuse. *Journal of Neuroinflammation, 9*, 203.
Gaskill, P. J., Calderon, T. M., Coley, J. S., & Berman, J. W. (2013). Drug induced increases in CNS dopamine alter monocyte, macrophage and T cell functions: Implications for HAND. *Journal of Neuroimmune Pharmacology, 8*, 621–642.
Gaskill, P. J., Yano, H. H., Kalpana, G. V., Javitch, J. A., & Berman, J. W. (2014). Dopamine receptor activation increases HIV entry into primary human macrophages. *PLoS One, 9*, e108232.
Giunti, D., Borsellino, G., Benelli, R., Marchese, M., Capello, E., Valle, M. T., Pedemonte, E., Noonan, D., Albini, A., Bernardi, G., Mancardi, G. L., Battistini, L., & Uccelli, A. (2003). Phenotypic and functional analysis of T cells homing into the CSF of subjects with inflammatory diseases of the CNS. *Journal of Leukocyte Biology, 73*, 584–590.
Goddard, C. A., Butts, D. A., & Shatz, C. J. (2007). Regulation of CNS synapses by neuronal MHC class I. *Proceedings of the National Academy of Sciences of the United States of America, 104*, 6828–6833.
Gong, Y., Tagawa, Y., Lunn, M. P., Laroy, W., Heffer-Lauc, M., Li, C. Y., Griffin, J. W., Schnaar, R. L., & Sheikh, K. A. (2002). Localization of major gangliosides in the PNS: Implications for immune neuropathies. *Brain, 125*, 2491–2506.
Gong, N., Pleyer, U., Vogt, K., Anegon, I., Flugel, A., Volk, H. D., & Ritter, T. (2007). Local overexpression of nerve growth factor in rat corneal transplants improves allograft survival. *Investigative Ophthalmology & Visual Science, 48*, 1043–1052.
Granero, R., Pardo-Garrido, A., Carpio-Toro, I. L., Ramirez-Coronel, A. A., Martinez-Suarez, P. C., & Reivan-Ortiz, G. G. (2021). The role of iron and zinc in the treatment of ADHD among children and adolescents: A systematic review of randomized clinical trials. *Nutrients, 13*.
Gregerson, D. S., & Yang, J. (2003). CD45-positive cells of the retina and their responsiveness to in vivo and in vitro treatment with IFN-gamma or anti-CD40. *Investigative Ophthalmology & Visual Science, 44*, 3083–3093.
Guedes, J. R., Ferreira, P. A., Costa, J. M., Cardoso, A. L., & Peca, J. (2022). Microglia-dependent remodeling of neuronal circuits. *Journal of Neurochemistry, 163*, 74–93.
Guyot, M., Simon, T., Panzolini, C., Ceppo, F., Daoudlarian, D., Murris, E., Macia, E., Abelanet, S., Sridhar, A., Vervoordeldonk, M. J., Glaichenhaus, N., & Blancou, P. (2019). Apical splenic nerve electrical stimulation discloses an anti-inflammatory pathway relying on adrenergic and nicotinic receptors in myeloid cells. *Brain, Behavior, and Immunity, 80*, 238–246.
Hazleton, J. E., Berman, J. W., & Eugenin, E. A. (2012). Purinergic receptors are required for HIV-1 infection of primary human macrophages. *Journal of Immunology, 188*, 4488–4495.
Hernangomez, M., Mestre, L., Correa, F. G., Loria, F., Mecha, M., Inigo, P. M., Docagne, F., Williams, R. O., Borrell, J., & Guaza, C. (2012). CD200-CD200R1 interaction contributes to neuroprotective effects of anandamide on experimentally induced inflammation. *Glia, 60*, 1437–1450.
Hikawa, N., & Takenaka, T. (1996). Sensory neurons regulate immunoglobulin secretion of spleen cells: Cellular analysis of bidirectional communications between neurons and immune cells. *Journal of Neuroimmunology, 70*, 191–198.
Hikosaka, M., Kawano, T., Wada, Y., Maeda, T., Sakurai, T., & Ohtsuki, G. (2022). Immune-triggered forms of plasticity across brain regions. *Frontiers in Cellular Neuroscience, 16*, 925493.
Hobson, B. D., & Sulzer, D. (2022). Neuronal presentation of antigen and its possible role in Parkinson's disease. *Journal of Parkinson's Disease, 12*, S137–S147.
Hoek, R. M., Ruuls, S. R., Murphy, C. A., Wright, G. J., Goddard, R., Zurawski, S. M., Blom, B., Homola, M. E., Streit, W. J., Brown, M. H., Barclay, A. N., & Sedgwick, J. D. (2000). Down-regulation of the macrophage lineage through interaction with OX2 (CD200). *Science, 290*, 1768–1771.
Holmes, D. (2016). Inflammation: Leptin deficiency protects against SLE. *Nature Reviews. Endocrinology, 12*, 688.
Hossain, D. M., Mohanty, S., Ray, P., Das, T., & Sa, G. (2012). Tumor gangliosides and T cells: A deadly encounter. *Frontiers in Bioscience (Scholar Edition), 4*, 502–519.
Howard, J. K., Lord, G. M., Matarese, G., Vendetti, S., Ghatei, M. A., Ritter, M. A., Lechler, R. I., & Bloom, S. R. (1999). Leptin protects mice from starvation-induced lymphoid atrophy and increases thymic cellularity in ob/ob mice. *The Journal of Clinical Investigation, 104*, 1051–1059.
Hsu, M., Rayasam, A., Kijak, J. A., Choi, Y. H., Harding, J. S., Marcus, S. A., Karpus, W. J., Sandor, M., & Fabry, Z. (2019). Neuroinflammation-induced lymphangiogenesis near the cribriform plate contributes to drainage of CNS-derived antigens and immune cells. *Nature Communications, 10*, 229.
Huang, H., He, W., Tang, T., & Qiu, M. (2022). Immunological markers for central nervous system glia. *Neuroscience Bulletin, 39*, 379.
Humpel, C. (2008). Basolateral aggregated rat amyloidbeta(1-42) potentiates transmigration of primary rat monocytes through a rat blood-brain barrier. *Current Neurovascular Research, 5*, 185–192.
Huston, J. M., Ochani, M., Rosas-Ballina, M., Liao, H., Ochani, K., Pavlov, V. A., Gallowitsch-Puerta, M., Ashok, M., Czura, C. J., Foxwell, B., Tracey, K. J., & Ulloa, L. (2006). Splenectomy inactivates the cholinergic antiinflammatory pathway during lethal endotoxemia and polymicrobial sepsis. *The Journal of Experimental Medicine, 203*, 1623–1628.
Huston, J. M., Rosas-Ballina, M., Xue, X., Dowling, O., Ochani, K., Ochani, M., Yeboah, M. M., Chatterjee, P. K., Tracey, K. J., & Metz, C. N. (2009). Cholinergic neural signals to the spleen downregulate leukocyte trafficking via CD11b. *Journal of Immunology, 183*, 552–559.
Iliff, J. J., Wang, M., Liao, Y., Plogg, B. A., Peng, W., Gundersen, G. A., Benveniste, H., Vates, G. E., Deane, R., Goldman, S. A., Nagelhus, E. A., & Nedergaard, M. (2012). A paravascular pathway facilitates CSF flow through the brain parenchyma and the clearance of interstitial solutes, including amyloid beta. *Science Translational Medicine, 4*, 147ra111.
Iliff, J. J., Chen, M. J., Plog, B. A., Zeppenfeld, D. M., Soltero, M., Yang, L., Singh, I., Deane, R., & Nedergaard, M. (2014). Impairment of glymphatic pathway function promotes tau pathology after traumatic brain injury. *The Journal of Neuroscience, 34*, 16180–16193.
Jeong, H. W., Dieguez-Hurtado, R., Arf, H., Song, J., Park, H., Kruse, K., Sorokin, L., & Adams, R. H. (2022). Single-cell transcriptomics

reveals functionally specialized vascular endothelium in brain. *eLife, 11*.

Keaney, J., & Campbell, M. (2015). The dynamic blood-brain barrier. *The FEBS Journal, 282*, 4067–4079.

Khattar, K. E., Safi, J., Rodriguez, A. M., & Vignais, M. L. (2022). Intercellular communication in the brain through tunneling nanotubes. *Cancers (Basel), 14*.

Kimata, H. (1994). GM1, a ganglioside that specifically enhances immunoglobulin production and proliferation in human plasma cells. *European Journal of Immunology, 24*, 2910–2913.

Kivisakk, P., Mahad, D. J., Callahan, M. K., Sikora, K., Trebst, C., Tucky, B., Wujek, J., Ravid, R., Staugaitis, S. M., Lassmann, H., & Ransohoff, R. M. (2004). Expression of CCR7 in multiple sclerosis: Implications for CNS immunity. *Annals of Neurology, 55*, 627–638.

Koning, N., Bo, L., Hoek, R. M., & Huitinga, I. (2007). Downregulation of macrophage inhibitory molecules in multiple sclerosis lesions. *Annals of Neurology, 62*, 504–514.

Koning, N., Swaab, D. F., Hoek, R. M., & Huitinga, I. (2009). Distribution of the immune inhibitory molecules CD200 and CD200R in the normal central nervous system and multiple sclerosis lesions suggests neuron-glia and glia-glia interactions. *Journal of Neuropathology and Experimental Neurology, 68*, 159–167.

Koyama, R., & Shichita, T. (2022). Glial roles in sterile inflammation after ischemic stroke. *Neuroscience Research, 187, 67*.

Kulkarni, B., Cruz-Martins, N., & Kumar, D. (2022). Microglia in Alzheimer's disease: An unprecedented opportunity as prospective drug target. *Molecular Neurobiology, 59*, 2678–2693.

La Cava, A., & Matarese, G. (2004). The weight of leptin in immunity. *Nature Reviews. Immunology, 4*, 371–379.

Lee, S. C., Collins, M., Vanguri, P., & Shin, M. L. (1992). Glutamate differentially inhibits the expression of class II MHC antigens on astrocytes and microglia. *Journal of Immunology, 148*, 3391–3397.

Lee, S. Y., Kim, B., Yoon, S., Kim, Y. J., Liu, T., Woo, J. H., Chwae, Y. J., Joe, E. H., & Jou, I. (2010). Phosphatidylinositol 4-phosphate 5-kinase alpha is induced in ganglioside-stimulated brain astrocytes and contributes to inflammatory responses. *Experimental & Molecular Medicine, 42*, 662–673.

Lehmann, M. L., Samuels, J. D., Kigar, S. L., Poffenberger, C. N., Lotstein, M. L., & Herkenham, M. (2022). CCR2 monocytes repair cerebrovascular damage caused by chronic social defeat stress. *Brain, Behavior, and Immunity, 101*, 346–358.

Leon-Rivera, R., Veenstra, M., Donoso, M., Tell, E., Eugenin, E. A., Morgello, S., & Berman, J. W. (2021). Central Nervous System (CNS) viral seeding by mature monocytes and potential therapies to reduce CNS viral reservoirs in the cART era. *mBio, 12*, 12.

Li, J. M., Southerland, L., Hossain, M. S., Giver, C. R., Wang, Y., Darlak, K., Harris, W., Waschek, J., & Waller, E. K. (2011). Absence of vasoactive intestinal peptide expression in hematopoietic cells enhances Th1 polarization and antiviral immunity in mice. *Journal of Immunology, 187*, 1057–1065.

Li, X., Qi, L., Yang, D., Hao, S., Zhang, F., Zhu, X., Sun, Y., Chen, C., Ye, J., Yang, J., Zhao, L., Altmann, D. M., Cao, S., Wang, H., & Wei, B. (2022). Meningeal lymphatic vessels mediate neurotropic viral drainage from the central nervous system. *Nature Neuroscience, 25*, 577–587.

Liu, S. Q., Li, B., Li, J. J., Sun, S., Sun, S. R., & Wu, Q. (2022). Neuroendocrine regulations in tissue-specific immunity: From mechanism to applications in tumor. *Frontiers in Cell and Development Biology, 10*, 896147.

Louveau, A., Herz, J., Alme, M. N., Salvador, A. F., Dong, M. Q., Viar, K. E., Herod, S. G., Knopp, J., Setliff, J. C., Lupi, A. L., Da Mesquita, S., Frost, E. L., Gaultier, A., Harris, T. H., Cao, R., Hu, S., Lukens, J. R., Smirnov, I., Overall, C. C., Oliver, G., & Kipnis, J. (2018). CNS lymphatic drainage and neuroinflammation are regulated by meningeal lymphatic vasculature. *Nature Neuroscience, 21*, 1380–1391.

Lu, H. J., & Gao, Y. J. (2022). *Astrocytes in chronic pain: Cellular and molecular mechanisms* (Vol. 39, p. 425). Neuroscience Bulletin.

Lu, P., & Sharom, F. J. (1995). Gangliosides are potent immunosuppressors of IL-2-mediated T-cell proliferation in a low protein environment. *Immunology, 86*, 356–363.

Lu, J., Zheng, M. H., Yan, J., Chen, Y. P., & Pan, J. P. (2008). Effects of vasoactive intestinal peptide on phenotypic and functional maturation of dendritic cells. *International Immunopharmacology, 8*, 1449–1454.

Lucca, L. E., Axisa, P. P., Aloulou, M., Perals, C., Ramadan, A., Rufas, P., Kyewski, B., Derbinski, J., Fazilleau, N., Mars, L. T., & Liblau, R. S. (2016). Myelin oligodendrocyte glycoprotein induces incomplete tolerance of CD4(+) T cells specific for both a myelin and a neuronal self-antigen in mice. *European Journal of Immunology, 46*, 2247–2259.

Majerova, P., Michalicova, A., Cente, M., Hanes, J., Vegh, J., Kittel, A., Kosikova, N., Cigankova, V., Mihaljevic, S., Jadhav, S., & Kovac, A. (2019). Trafficking of immune cells across the blood-brain barrier is modulated by neurofibrillary pathology in tauopathies. *PLoS One, 14*, e0217216.

Manich, G., Recasens, M., Valente, T., Almolda, B., Gonzalez, B., & Castellano, B. (2019). Role of the CD200-CD200R Axis during homeostasis and Neuroinflammation. *Neuroscience, 405*, 118–136.

Matt, S. M., & Gaskill, P. J. (2019). Dopaminergic impact of cART and anti-depressants on HIV neuropathogenesis in older adults. *Brain Research, 1723*, 146398.

Matt, S. M., Nickoloff-Bybel, E. A., Rong, Y., Runner, K., Johnson, H., O'Connor, M. H., Haddad, E. K., & Gaskill, P. J. (2021). Dopamine levels induced by substance abuse alter efficacy of maraviroc and expression of CCR5 conformations on myeloid cells: Implications for neuroHIV. *Frontiers in Immunology, 12*, 663061.

Mazzitelli, J. A., Smyth, L. C. D., Cross, K. A., Dykstra, T., Sun, J., Du, S., Mamuladze, T., Smirnov, I., Rustenhoven, J., & Kipnis, J. (2022). Cerebrospinal fluid regulates skull bone marrow niches via direct access through dural channels. *Nature Neuroscience, 25*, 555–560.

McConnell, H. L., & Mishra, A. (2022). Cells of the blood-brain barrier: An overview of the neurovascular unit in health and disease. *Methods in Molecular Biology, 2492*, 3–24.

McConnell, H. L., Kersch, C. N., Woltjer, R. L., & Neuwelt, E. A. (2017). The translational significance of the neurovascular unit. *The Journal of Biological Chemistry, 292*, 762–770.

Medana, I. M., Gallimore, A., Oxenius, A., Martinic, M. M., Wekerle, H., & Neumann, H. (2000). MHC class I-restricted killing of neurons by virus-specific CD8+ T lymphocytes is effected through the Fas/FasL, but not the perforin pathway. *European Journal of Immunology, 30*, 3623–3633.

Medana, I., Martinic, M. A., Wekerle, H., & Neumann, H. (2001). Transection of major histocompatibility complex class I-induced neurites by cytotoxic T lymphocytes. *The American Journal of Pathology, 159*, 809–815.

Mendez-Ferrer, S., Lucas, D., Battista, M., & Frenette, P. S. (2008). Haematopoietic stem cell release is regulated by circadian oscillations. *Nature, 452*, 442–447.

Menees, K. B., Otero, B. A., & Tansey, M. G. (2022). Microbiome influences on neuro-immune interactions in neurodegenerative disease. *International Review of Neurobiology, 167*, 25–57.

Merolli, A., Kasaei, L., Ramasamy, S., Kolloli, A., Kumar, R., Subbian, S., & Feldman, L. C. (2022). An intra-cytoplasmic route for SARS-CoV-2 transmission unveiled by helium-ion microscopy. *Scientific Reports, 12*, 3794.

Miller, T. L., & Mayo, K. E. (1997). Glucocorticoids regulate pituitary growth hormone-releasing hormone receptor messenger ribonucleic acid expression. *Endocrinology, 138*, 2458–2465.

Mina-Osorio, P., Rosas-Ballina, M., Valdes-Ferrer, S. I., Al-Abed, Y., Tracey, K. J., & Diamond, B. (2012). Neural signaling in the spleen controls B-cell responses to blood-borne antigen. *Molecular Medicine, 18*, 618–627.

Mitchell, D. A., Batich, K. A., Gunn, M. D., Huang, M. N., Sanchez-Perez, L., Nair, S. K., Congdon, K. L., Reap, E. A., Archer, G. E., Desjardins, A., Friedman, A. H., Friedman, H. S., Herndon, J. E., 2nd, Coan, A., McLendon, R. E., Reardon, D. A., Vredenburgh, J. J., Bigner, D. D., & Sampson, J. H. (2015). Tetanus toxoid and CCL3 improve dendritic cell vaccines in mice and glioblastoma patients. *Nature, 519*, 366–369.

Monje, M., & Iwasaki, A. (2022). The neurobiology of long COVID. *Neuron, 110*, 3484–3496.

Morga, E., Faber, C., & Heuschling, P. (2000). Stimulation of endothelin B receptor modulates the inflammatory activation of rat astrocytes. *Journal of Neurochemistry, 74*, 603–612.

Moseman, E. A., Blanchard, A. C., Nayak, D., & McGavern, D. B. (2020). T cell engagement of cross-presenting microglia protects the brain from a nasal virus infection. *Science Immunology, 5*.

Mravec, B. (2022). Neurobiology of cancer: Definition, historical overview, and clinical implications. *Cancer Medicine, 11*, 903–921.

Mueller, S. N. (2022). Neural control of immune cell trafficking. *The Journal of Experimental Medicine, 219*.

Muller, W. A. (2016). Transendothelial migration: Unifying principles from the endothelial perspective. *Immunological Reviews, 273*, 61–75.

Mundt, S., Greter, M., Flugel, A., & Becher, B. (2019). The CNS immune landscape from the viewpoint of a T cell. *Trends in Neurosciences, 42*, 667–679.

Mundt, S., Greter, M., & Becher, B. (2022). The CNS mononuclear phagocyte system in health and disease. *Neuron, 110*, 3497–3512.

Nagashima, T., Shigematsu, N., Maruki, R., Urano, Y., Tanaka, H., Shimaya, A., Shimokawa, T., & Shibasaki, M. (2010). Discovery of novel forkhead box O1 inhibitors for treating type 2 diabetes: Improvement of fasting glycemia in diabetic db/db mice. *Molecular Pharmacology, 78*, 961–970.

Nakai, A., Hayano, Y., Furuta, F., Noda, M., & Suzuki, K. (2014). Control of lymphocyte egress from lymph nodes through beta2-adrenergic receptors. *The Journal of Experimental Medicine, 211*, 2583–2598.

Neumann, H. (2001). Control of glial immune function by neurons. *Glia, 36*, 191–199.

Neumann, H., Misgeld, T., Matsumuro, K., & Wekerle, H. (1998). Neurotrophins inhibit major histocompatibility class II inducibility of microglia: Involvement of the p75 neurotrophin receptor. *Proceedings of the National Academy of Sciences of the United States of America, 95*, 5779–5784.

Neumann, H., Medana, I. M., Bauer, J., & Lassmann, H. (2002). Cytotoxic T lymphocytes in autoimmune and degenerative CNS diseases. *Trends in Neurosciences, 25*, 313–319.

Nickoloff-Bybel, E. A., Mackie, P., Runner, K., Matt, S. M., Khoshbouei, H., & Gaskill, P. J. (2019). Dopamine increases HIV entry into macrophages by increasing calcium release via an alternative signaling pathway. *Brain, Behavior, and Immunity, 82*, 239–252.

Nickoloff-Bybel, E. A., Calderon, T. M., Gaskill, P. J., & Berman, J. W. (2020). HIV neuropathogenesis in the presence of a disrupted dopamine system. *Journal of Neuroimmune Pharmacology, 15*, 729–742.

Nolan, R. A., Muir, R., Runner, K., Haddad, E. K., & Gaskill, P. J. (2019). Role of macrophage dopamine receptors in mediating cytokine production: Implications for neuroinflammation in the context of HIV-associated neurocognitive disorders. *Journal of Neuroimmune Pharmacology, 14*, 134–156.

Norihisa, Y., McVicar, D. W., Ghosh, P., Houghton, A. N., Longo, D. L., Creekmore, S. P., Blake, T., Ortaldo, J. R., & Young, H. A. (1994). Increased proliferation, cytotoxicity, and gene expression after stimulation of human peripheral blood T lymphocytes through a surface ganglioside (GD3). *Journal of Immunology, 152*, 485–495.

Obermeier, B., Verma, A., & Ransohoff, R. M. (2016). The blood-brain barrier. *Handbook of Clinical Neurology, 133*, 39–59.

Okafo, G., Prevedel, L., & Eugenin, E. (2017). Tunneling nanotubes (TNT) mediate long-range gap junctional communication: Implications for HIV cell to cell spread. *Scientific Reports, 7*, 16660.

Okura, T., Taneno, A., & Oishi, E. (2021). Cell-to-cell transmission of Turkey herpesvirus in chicken embryo cells via tunneling nanotubes. *Avian Diseases, 65*, 335–339.

Ordovas-Montanes, J., Rakoff-Nahoum, S., Huang, S., Riol-Blanco, L., Barreiro, O., & von Andrian, U. H. (2015). The regulation of immunological processes by peripheral neurons in homeostasis and disease. *Trends in Immunology, 36*, 578–604.

Ortaldo, J. R., Mason, A. T., Longo, D. L., Beckwith, M., Creekmore, S. P., & McVicar, D. W. (1996). T cell activation via the disialoganglioside GD3: Analysis of signal transduction. *Journal of Leukocyte Biology, 60*, 533–539.

Ousman, S. S., & Kubes, P. (2012). Immune surveillance in the central nervous system. *Nature Neuroscience, 15*, 1096–1101.

Padro, C. J., & Sanders, V. M. (2014). Neuroendocrine regulation of inflammation. *Seminars in Immunology, 26*, 357–368.

Peng, Y. P., Qiu, Y. H., Jiang, J. L., & Wang, J. J. (2004). Effect of catecholamines on IL-2 production and NK cytotoxicity of rats in vitro. *Acta Pharmacologica Sinica, 25*, 1354–1360.

Pietronigro, E., Zenaro, E., & Constantin, G. (2016). Imaging of leukocyte trafficking in Alzheimer's disease. *Frontiers in Immunology, 7*, 33.

Popper, P., Mantyh, C. R., Vigna, S. R., Maggio, J. E., & Mantyh, P. W. (1988). The localization of sensory nerve fibers and receptor binding sites for sensory neuropeptides in canine mesenteric lymph nodes. *Peptides, 9*, 257–267.

Populin, L., Stebbing, M. J., & Furness, J. B. (2021). Neuronal regulation of the gut immune system and neuromodulation for treating inflammatory bowel disease. *FASEB Bioadvances, 3*, 953–966.

Pozo, D., & Delgado, M. (2004). The many faces of VIP in neuroimmunology: A cytokine rather a neuropeptide? *The FASEB Journal, 18*, 1325–1334.

Prat, A., & Antel, J. (2005). Pathogenesis of multiple sclerosis. *Current Opinion in Neurology, 18*, 225–230.

Procaccini, C., Jirillo, E., & Matarese, G. (2012). Leptin as an immunomodulator. *Molecular Aspects of Medicine, 33*, 35–45.

Puerto, M., Guayerbas, N., Alvarez, P., & De la Fuente, M. (2005). Modulation of neuropeptide Y and norepinephrine on several leucocyte functions in adult, old and very old mice. *Journal of Neuroimmunology, 165*, 33–40.

Qiu, Y., Peng, Y., & Wang, J. (1996). Immunoregulatory role of neurotransmitters. *Advances in Neuroimmunology, 6*, 223–231.

Qiu, Y. H., Peng, Y. P., Jiang, J. L., & Wang, J. J. (2004a). Effect of acetylcholine on in vitro IL-2 production and NK cell cytotoxicity of rats. *Lymphology, 37*, 31–38.

Qiu, Y. H., Peng, Y. P., Jiang, J. M., & Wang, J. J. (2004b). Expression of tyrosine hydroxylase in lymphocytes and effect of endogenous catecholamines on lymphocyte function. *Neuroimmunomodulation, 11*, 75–83.

Ramer-Quinn, D. S., Swanson, M. A., Lee, W. T., & Sanders, V. M. (2000). Cytokine production by naive and primary effector CD4+ T cells exposed to norepinephrine. *Brain, Behavior, and Immunity, 14*, 239–255.

Ratner, V., Gao, Y., Lee, H., Elkin, R., Nedergaard, M., Benveniste, H., & Tannenbaum, A. (2017). Cerebrospinal and interstitial fluid transport via the glymphatic pathway modeled by optimal mass transport. *NeuroImage, 152*, 530–537.

Reardon, C., Duncan, G. S., Brustle, A., Brenner, D., Tusche, M. W., Olofsson, P. S., Rosas-Ballina, M., Tracey, K. J., & Mak, T. W. (2013). Lymphocyte-derived ACh regulates local innate but not adaptive immunity. *Proceedings of the National Academy of Sciences of the United States of America, 110*, 1410–1415.

Rhoads, M. K., Balagee, V., & Thomas, S. J. (2020). Circadian regulation of blood pressure: Of mice and men. *Current Hypertension Reports, 22*, 40.

Rickenbach, C., & Gericke, C. (2021). Specificity of adaptive immune responses in central nervous system health, aging and diseases. *Front Neuroscience, 15*, 806260.

Rim, C., Park, H. S., You, M. J., Yang, B., Kim, H. J., Sung, S., & Kwon, M. S. (2022). Microglia involvement in sex-dependent behaviors and schizophrenia occurrence in offspring with maternal dexamethasone exposure. *Schizophrenia (Heidelb), 8*, 71.

Rodriguez-Lorenzo, S., van Olst, L., Rodriguez-Mogeda, C., Kamermans, A., van der Pol, S. M. A., Rodriguez, E., Kooij, G., & de Vries, H. E. (2022). Single-cell profiling reveals periventricular CD56(bright) NK cell accumulation in multiple sclerosis. *eLife, 11*.

Rosas-Ballina, M., & Tracey, K. J. (2009). Cholinergic control of inflammation. *Journal of Internal Medicine, 265*, 663–679.

Rosas-Ballina, M., Olofsson, P. S., Ochani, M., Valdes-Ferrer, S. I., Levine, Y. A., Reardon, C., Tusche, M. W., Pavlov, V. A., Andersson, U., Chavan, S., Mak, T. W., & Tracey, K. J. (2011). Acetylcholine-synthesizing T cells relay neural signals in a vagus nerve circuit. *Science, 334*, 98–101.

Rubio-Casillas, A., Redwan, E. M., & Uversky, V. N. (2022). SARS-CoV-2: A master of immune evasion. *Biomedicine, 10*.

Rustenhoven, J. (2021). A privileged brain. *Science, 374*, 548.

Rustenhoven, J., Drieu, A., Mamuladze, T., de Lima, K. A., Dykstra, T., Wall, M., Papadopoulos, Z., Kanamori, M., Salvador, A. F., Baker, W., Lemieux, M., Da Mesquita, S., Cugurra, A., Fitzpatrick, J., Sviben, S., Kossina, R., Bayguinov, P., Townsend, R. R., Zhang, Q., Erdmann-Gilmore, P., Smirnov, I., Lopes, M. B., Herz, J., & Kipnis, J. (2021). Functional characterization of the dural sinuses as a neuroimmune interface. *Cell, 184*, 1000–1016.

Salvador, A. F. M., & Kipnis, J. (2022). Immune response after central nervous system injury. *Seminars in Immunology, 101629*, 101629.

Scheiermann, C., Kunisaki, Y., Lucas, D., Chow, A., Jang, J. E., Zhang, D., Hashimoto, D., Merad, M., & Frenette, P. S. (2012). Adrenergic nerves govern circadian leukocyte recruitment to tissues. *Immunity, 37*, 290–301.

Schiller, C., Huber, J. E., Diakopoulos, K. N., & Weiss, E. H. (2013). Tunneling nanotubes enable intercellular transfer of MHC class I molecules. *Human Immunology, 74*, 412–416.

Schlaak, J. F., Claus, C., K.H. Meyer zum Buschenfelde, & Dippold, W. (1995). Anti-GD3 antibodies are potent activators of human gamma/delta and alpha/beta positive T cells. *Scandinavian Journal of Immunology, 41*, 475–480.

Schwab, N., Hohn, K. G., Schneider-Hohendorf, T., Metz, I., Stenner, M. P., Jilek, S., Du Pasquier, R. A., Gold, R., Meuth, S. G., Ransohoff, R. M., Bruck, W., & Wiendl, H. (2012). Immunological and clinical consequences of treating a patient with natalizumab. *Multiple Sclerosis, 18*, 335–344.

Sharawy, I. (2022). Neuroimmune crosstalk and its impact on cancer therapy and research. *Discover Oncology, 13*, 80.

Shinoda, M., Hoffer, B. J., & Olson, L. (1996). Interactions of neurotrophic factors GDNF and NT-3, but not BDNF, with the immune system following fetal spinal cord transplantation. *Brain Research, 722*, 153–167.

Shipley, F. B., Dani, N., Xu, H., Deister, C., Cui, J., Head, J. P., Sadegh, C., Fame, R. M., Shannon, M. L., Flores, V. I., Kishkovich, T., Jang, E., Klein, E. M., Goldey, G. J., He, K., Zhang, Y., Holtzman, M. J., Kirchhausen, T., Wyart, C., Moore, C. I., Andermann, M. L., & Lehtinen, M. K. (2020). Tracking calcium dynamics and immune surveillance at the choroid plexus blood-cerebrospinal fluid interface. *Neuron, 108*, 623–639.

Shouman, K., & Benarroch, E. E. (2021). Peripheral neuroimmune interactions: Selected review and some clinical implications. *Clinical Autonomic Research, 31*, 477–489.

Shurin, M. R., Baraldi, J. H., & Shurin, G. V. (2021). Neuroimmune regulation of surgery-associated metastases. *Cells, 10*.

Simoes Da Gama, C., & Morin-Brureau, M. (2022). Study of BBB dysregulation in neuropathogenicity using integrative human model of blood-brain barrier. *Frontiers in Cellular Neuroscience, 16*, 863836.

Song, E., Mao, T., Dong, H., Boisserand, L. S. B., Antila, S., Bosenberg, M., Alitalo, K., Thomas, J. L., & Iwasaki, A. (2020). VEGF-C-driven lymphatic drainage enables immunosurveillance of brain tumours. *Nature, 577*, 689–694.

Sotnikov, I., Veremeyko, T., Starossom, S. C., Barteneva, N., Weiner, H. L., & Ponomarev, E. D. (2013). Platelets recognize brain-specific glycolipid structures, respond to neurovascular damage and promote neuroinflammation. *PLoS One, 8*, e58979.

Staedtke, V., Bai, R. Y., Kim, K., Darvas, M., Davila, M. L., Riggins, G. J., Rothman, P. B., Papadopoulos, N., Kinzler, K. W., Vogelstein, B., & Zhou, S. (2018). Disruption of a self-amplifying catecholamine loop reduces cytokine release syndrome. *Nature, 564*, 273–277.

Stern, J. N., Yaari, G., Vander Heiden, J. A., Church, G., Donahue, W. F., Hintzen, R. Q., Huttner, A. J., Laman, J. D., Nagra, R. M., Nylander, A., Pitt, D., Ramanan, S., Siddiqui, B. A., Vigneault, F., Kleinstein, S. H., Hafler, D. A., & O'Connor, K. C. (2014). B cells populating the multiple sclerosis brain mature in the draining cervical lymph nodes. *Science Translational Medicine, 6*, 248ra107.

Suekane, A., Saito, Y., Nakahata, S., Ichikawa, T., Ogoh, H., Tsujikawa, K., & Morishita, K. (2019). CGRP-CRLR/RAMP1 signal is important for stress-induced hematopoiesis. *Scientific Reports, 9*, 429.

Susarla, R., Liu, L., Walker, E. A., Bujalska, I. J., Alsalem, J., Williams, G. P., Sreekantam, S., Taylor, A. E., Tallouzi, M., Southworth, H. S., Murray, P. I., Wallace, G. R., & Rauz, S. (2014). Cortisol biosynthesis in the human ocular surface innate immune response. *PLoS One, 9*, e94913.

Swanson, M. A., Lee, W. T., & Sanders, V. M. (2001). IFN-gamma production by Th1 cells generated from naive CD4+ T cells exposed to norepinephrine. *Journal of Immunology, 166*, 232–240.

Takeda, N., & Maemura, K. (2011). Circadian clock and cardiovascular disease. *Journal of Cardiology, 57*, 249–256.

Taleb, S., Herbin, O., Ait-Oufella, H., Verreth, W., Gourdy, P., Barateau, V., Merval, R., Esposito, B., Clement, K., Holvoet, P., Tedgui, A., & Mallat, Z. (2007). Defective leptin/leptin receptor signaling improves regulatory T cell immune response and protects mice from atherosclerosis. *Arteriosclerosis, Thrombosis, and Vascular Biology, 27*, 2691–2698.

Teige, I., Liu, Y., & Issazadeh-Navikas, S. (2006). IFN-beta inhibits T cell activation capacity of central nervous system APCs. *Journal of Immunology, 177*, 3542–3553.

Tiedt, S., Buchan, A. M., Dichgans, M., Lizasoain, I., Moro, M. A., & Lo, E. H. (2022). The neurovascular unit and systemic biology in stroke—Implications for translation and treatment. *Nature Reviews. Neurology, 18*, 597–612.

Tiwari, V., Koganti, R., Russell, G., Sharma, A., & Shukla, D. (2021). Role of tunneling nanotubes in viral infection, neurodegenerative disease, and cancer. *Frontiers in Immunology, 12*, 680891.

Torices, S., Cabrera, R., Stangis, M., Naranjo, O., Fattakhov, N., Teglas, T., Adesse, D., & Toborek, M. (2021). Expression of SARS-CoV-2-related receptors in cells of the neurovascular unit: Implications for HIV-1 infection. *Journal of Neuroinflammation, 18*, 167.

Turos-Korgul, L., Kolba, M. D., Chroscicki, P., Zieminska, A., & Piwocka, K. (2022). Tunneling nanotubes facilitate intercellular protein transfer and cell networks function. *Frontiers in Cell and Development Biology, 10*, 915117.

Udit, S., Blake, K., & Chiu, I. M. (2022). Somatosensory and autonomic neuronal regulation of the immune response. *Nature Reviews. Neuroscience, 23*, 157–171.

Valdebenito, S., Lou, E., Baldoni, J., Okafo, G., & Eugenin, E. (2018). The novel roles of Connexin channels and tunneling nanotubes in cancer pathogenesis. *International Journal of Molecular Sciences, 19*.

Valdebenito, S., D'Amico, D., & Eugenin, E. (2019). Novel approaches for glioblastoma treatment: Focus on tumor heterogeneity, treatment resistance, and computational tools. *Cancer Rep (Hoboken), 2*, e1220.

Valdebenito, S., Audia, A., Bhat, K. P. L., Okafo, G., & Eugenin, E. A. (2020). Tunneling nanotubes mediate adaptation of glioblastoma cells to temozolomide and ionizing radiation treatment. *iScience, 23*, 101450.

Valdebenito, S., Malik, S., Luu, R., Loudig, O., Mitchell, M., Okafo, G., Bhat, K., Prideaux, B., & Eugenin, E. A. (2021). Tunneling nanotubes, TNT, communicate glioblastoma with surrounding non-tumor astrocytes to adapt them to hypoxic and metabolic tumor conditions. *Scientific Reports, 11*, 14556.

Valerio, A., Dossena, M., Bertolotti, P., Boroni, F., Sarnico, I., Faraco, G., Chiarugi, A., Frontini, A., Giordano, A., Liou, H. C., De Simoni, M. G., Spano, P., Carruba, M. O., Pizzi, M., & Nisoli, E. (2009). Leptin is induced in the ischemic cerebral cortex and exerts neuroprotection through NF-kappaB/c-Rel-dependent transcription. *Stroke, 40*, 610–617.

Vasconcelos, D. P., Jabangwe, C., Lamghari, M., & Alves, C. J. (2022). The Neuroimmune interplay in joint pain: The role of macrophages. *Frontiers in Immunology, 13*, 812962.

Velasquez, S., & Eugenin, E. A. (2014). Role of Pannexin-1 hemichannels and purinergic receptors in the pathogenesis of human diseases. *Frontiers in Physiology, 5*, 96.

Velasquez, S., Malik, S., Lutz, S. E., Scemes, E., & Eugenin, E. A. (2016). Pannexin1 channels are required for chemokine-mediated migration of CD4+ T lymphocytes: Role in inflammation and experimental autoimmune encephalomyelitis. *Journal of Immunology, 196*, 4338–4347.

Velasquez, S., Prevedel, L., Valdebenito, S., Gorska, A. M., Golovko, M., Khan, N., Geiger, J., & Eugenin, E. A. (2020). Circulating levels of ATP is a biomarker of HIV cognitive impairment. *eBioMedicine, 51*, 102503.

Vida, G., Pena, G., Kanashiro, A., Thompson-Bonilla Mdel, R., Palange, D., Deitch, E. A., & Ulloa, L. (2011). beta2-Adrenoreceptors of regulatory lymphocytes are essential for vagal neuromodulation of the innate immune system. *The FASEB Journal, 25*, 4476–4485.

Voisin, T., Bouvier, A., & Chiu, I. M. (2017). Neuro-immune interactions in allergic diseases: Novel targets for therapeutics. *International Immunology, 29*, 247–261.

Wachholz, S., Esslinger, M., Plumper, J., Manitz, M. P., Juckel, G., & Friebe, A. (2016). Microglia activation is associated with IFN-alpha induced depressive-like behavior. *Brain, Behavior, and Immunity, 55*, 105–113.

Wang, H., Yu, M., Ochani, M., Amella, C. A., Tanovic, M., Susarla, S., Li, J. H., Wang, H., Yang, H., Ulloa, L., Al-Abed, Y., Czura, C. J., & Tracey, K. J. (2003). Nicotinic acetylcholine receptor alpha7 subunit is an essential regulator of inflammation. *Nature, 421*, 384–388.

Wang, Y. Y., Deng, Y. S., Dai, S. K., Mi, T. W., Li, R. Y., Liu, P. P., Liu, C., He, B. D., He, X. C., Du, H. Z., Yang, H. C., Tang, Y., Liu, C. M., & Teng, Z. Q. (2022). Loss of microglial EED impairs synapse density, learning, and memory. *Molecular Psychiatry, 27*, 2999–3009.

Watkins, S. C., & Salter, R. D. (2005). Functional connectivity between immune cells mediated by tunneling nanotubules. *Immunity, 23*, 309–318.

Williams, D. W., Eugenin, E. A., Calderon, T. M., & Berman, J. W. (2012). Monocyte maturation, HIV susceptibility, and transmigration across the blood brain barrier are critical in HIV neuropathogenesis. *Journal of Leukocyte Biology, 91*, 401–415.

Williams, D. W., Calderon, T. M., Lopez, L., Carvallo-Torres, L., Gaskill, P. J., Eugenin, E. A., Morgello, S., & Berman, J. W. (2013). Mechanisms of HIV entry into the CNS: Increased sensitivity of HIV infected CD14+CD16+ monocytes to CCL2 and key roles of CCR2, JAM-A, and ALCAM in diapedesis. *PLoS One, 8*, e69270.

Wohleb, E. S., McKim, D. B., Shea, D. T., Powell, N. D., Tarr, A. J., Sheridan, J. F., & Godbout, J. P. (2014a). Re-establishment of anxiety in stress-sensitized mice is caused by monocyte trafficking from the spleen to the brain. *Biological Psychiatry, 75*, 970–981.

Wohleb, E. S., McKim, D. B., Sheridan, J. F., & Godbout, J. P. (2014b). Monocyte trafficking to the brain with stress and inflammation: A novel axis of immune-to-brain communication that influences mood and behavior. *Frontiers in Neuroscience, 8*, 447.

Wohleb, E. S., Patterson, J. M., Sharma, V., Quan, N., Godbout, J. P., & Sheridan, J. F. (2014c). Knockdown of interleukin-1 receptor type-1 on endothelial cells attenuated stress-induced neuroinflammation and prevented anxiety-like behavior. *The Journal of Neuroscience, 34*, 2583–2591.

Wright, G. J., Puklavec, M. J., Willis, A. C., Hoek, R. M., Sedgwick, J. D., Brown, M. H., & Barclay, A. N. (2000). Lymphoid/neuronal cell surface OX2 glycoprotein recognizes a novel receptor on macrophages implicated in the control of their function. *Immunity, 13*, 233–242.

Yamamoto, Y., Otsuka, A., Ishida, Y., Wong, L. S., Seidel, J. A., Nonomura, Y., Nakashima, C., Nakajima, S., Kitoh, A., Nomura, T., Dainichi, T., Honda, T., Amano, W., Konishi, N., Hayashi, M., Matsushita, M., & Kabashima, K. (2021). Pituitary adenylate cyclase-activating polypeptide promotes cutaneous dendritic cell functions in contact hypersensitivity. *The Journal of Allergy and Clinical Immunology, 148*, 858–866.

Yokoseki, A., Saji, E., Arakawa, M., Hokari, M., Ishiguro, T., Yanagimura, F., Ishihara, T., Okamoto, K., Nishizawa, M., & Kawachi, I. (2013). Relapse of multiple sclerosis in a patient retaining CCR7-expressing T cells in CSF under fingolimod therapy. *Multiple Sclerosis, 19*, 1230–1233.

Yoshida, S., Tanaka, H., Oshima, H., Yamazaki, T., Yonetoku, Y., Ohishi, T., Matsui, T., & Shibasaki, M. (2010). AS1907417, a novel GPR119 agonist, AS an insulinotropic and beta-cell preservative agent for the treatment of type 2 diabetes. *Biochemical and Biophysical Research Communications, 400*, 745–751.

Yoshida, T. M., Wang, A., & Hafler, D. A. (2022). Basic principles of neuroimmunology. *Seminars in Immunopathology, 44*, 685–695.

Zhang, X., Lei, B., Yuan, Y., Zhang, L., Hu, L., Jin, S., Kang, B., Liao, X., Sun, W., Xu, F., Zhong, Y., Hu, J., & Qi, H. (2020). Brain control of humoral immune responses amenable to behavioural modulation. *Nature, 581*, 204–208.

Zhao, S. C., Heng, X., Ya-Ping, W., Di, L., Wen-Qian, W., Ling-Song, M., Chu, Z. H., & Xu, Y. (2020). CD200-CD200R1 signaling pathway regulates neuroinflammation after stroke. *Brain and Behavior: A Cognitive Neuroscience Perspective, 10*, e01882.

Zheng, J., Fang, J., Yin, Y. J., Wang, X. C., Ren, A. J., Bai, J., Sun, X. J., Yuan, W. J., & Lin, L. (2010). Leptin protects cardiomyocytes from serum-deprivation-induced apoptosis by increasing anti-oxidant defence. *Clinical and Experimental Pharmacology & Physiology, 37*, 955–962.

Zheng, W., Feng, Y., Zeng, Z., Ye, M., Wang, M., Liu, X., Tang, P., Shang, H., Sun, X., Lin, X., Wang, M., Li, Z., Weng, Y., Guo, W., Vakal, S., & Chen, J. F. (2022). Choroid plexus-selective inactivation of adenosine a(2A) receptors protects against T cell infiltration and experimental autoimmune encephalomyelitis. *Journal of Neuroinflammation, 19*, 52.

Zhou, Y. F., Huang, J. C., Zhang, P., Fan, F. M., Chen, S., Fan, H. Z., Cui, Y. M., Luo, X. G., Tan, S. P., Wang, Z. R., Feng, W., Yuan, Y., Yang, F. D., Savransky, A., Ryan, M., Goldwaser, E., Chiappelli, J., Rowland, L. M., Kochunov, P., Tan, Y. L., & Hong, L. E. (2020). Choroid plexus enlargement and Allostatic load in schizophrenia. *Schizophrenia Bulletin, 46*, 722–731.

Zoukos, Y., Kidd, D., Woodroofe, M. N., Kendall, B. E., Thompson, A. J., & Cuzner, M. L. (1994). Increased expression of high affinity IL-2 receptors and beta-adrenoceptors on peripheral blood mononuclear cells is associated with clinical and MRI activity in multiple sclerosis. *Brain, 117*(Pt 2), 307–315.

Growth and Neurotrophic Factors for HIV-Associated Neurocognitive Disorders

15

Uma Maheswari Deshetty, Shilpa Buch, and Palsamy Periyasamy

Abstract

The advent of combination antiretroviral therapy (cART) has achieved remarkable success in controlling the replication of the human immunodeficiency virus (HIV). This success has resulted in reduced morbidity and increased life expectancy for individuals living with the virus. However, paradoxically, as patients have extended lifespans, there has been a concerning increase in the prevalence of HIV-associated cognitive impairments. This rise can be attributed to the adverse effects of drug toxicity and the limited ability of these drugs to penetrate the central nervous system (CNS). Moreover, the presence of HIV proteins, which remain unaffected by antiretrovirals, contributes to neuroinflammation. Consequently, many individuals effectively managed by cART experience various forms of HIV-associated neurocognitive disorders (HAND). The mechanisms underlying these disorders encompass deficits in calcium flux, excitotoxicity, cell signaling, oxidative stress, and autophagy. Furthermore, disruptions in the processing and functionality of neurotrophic factors, including fibroblast growth factors (FGFs), brain-derived growth factor (BDNF), insulin-like growth factor (IGF), platelet-derived growth factor (PDGF), and glial cell-derived neurotrophic factor (GDNF), have been implicated in the disruption of neuronal function. Notably, these factors have demonstrated their capacity to interfere with the fundamental process of apoptotic cell death, a key feature of HAND and other neurodegenerative diseases. Additionally, neurotrophic factors can potentially promote neuronal differentiation, neurogenesis, and angiogenesis thereby facilitating neuronal repair. There is a conspicuous absence of effective adjunctive treatments for these diseases. Consequently, identifying novel neuroprotective compounds for addressing such conditions is paramount in this field. This chapter delves into the beneficial roles of various neurotrophic factors, including FGF, BDNF, IGF, PDGF, and GDNF, within the context of potential HIV therapies.

Keywords

HAND · HIV infection · BDNF · FGF · GDNF · PDGF · IGF · Neurodegeneration · Neuroinflammation

15.1 Introduction

At present, over 39 million individuals worldwide are living with HIV. The advent of combination antiretroviral therapy (cART) has undeniably achieved remarkable success in controlling viral replication, leading to an extended lifespan among those infected. Consequently, there has been a remarkable improvement in disease morbidities and mortality rates for HIV patients. However, this positive outcome has brought about an unforeseen and complex challenge—a surge in the prevalence of mild neurocognitive impairments collectively referred to as HIV-associated neurocognitive disorders (HAND). Clinically, HAND is associated with a diminished quality of life, memory loss, and social and occupational disabilities, affecting individuals' daily functioning and well-being. Pathologically, in its most severe forms, patients may exhibit prolonged manifestations of HIV-encephalitis (HIVE), while milder cases may show subtle evidence of gliosis, an indicator of central nervous system (CNS) damage. Notably, pathological findings don't always align with disease signs and symptoms, often presenting as common and subtle cognitive deficits rather than overt dementia. In cases where neurodegenerative pathology is evident, it typically involves cortical and subcortical regions

U. M. Deshetty · S. Buch (✉) · P. Periyasamy (✉)
Department of Pharmacology and Experimental Neuroscience, University of Nebraska Medical Center, Omaha, NE, USA
e-mail: sbuch@unmc.edu; palsamy.periyasamy@unmc.edu

H. E. Gendelman, T. Ikezu (eds.), *Neuroimmune Pharmacology and Therapeutics*,
https://doi.org/10.1007/978-3-031-68237-7_15

in milder phenotypes, accompanied by the simplification of the synaptodendritic structure within neuronal populations in the neocortex and hippocampus, which underscores the complex nature of HAND.

CNS homeostasis is an intricate equilibrium between the presence of neurotrophic and neurotoxic factors. While abundant research has been focused on neurotoxic factors and their detrimental effects on neuronal health, investigating neurotrophic factors has been relatively understudied despite their pivotal role in preserving neural integrity and function. Neurotrophic factors, including brain-derived neurotrophic factor (BDNF), fibroblast growth factor (FGF), nerve growth factor (NGF), and glial cell-derived neurotrophic factor (GDNF), have been implicated in safeguarding neurons against various neurotoxins, offering a potential avenue for therapeutic intervention (Almeida et al., 2005; Alzheimer & Werner, 2002; Colafrancesco & Villoslada, 2011; Deierborg et al., 2008). Following HIV infection, the downregulation of BDNF disrupts the balance between anti- and pro-apoptotic neurotrophins, setting the stage for neuronal vulnerability (Avdoshina et al., 2011; Bachis et al., 2012; Bachis & Mocchetti, 2005; Mocchetti et al., 2007; Nosheny et al., 2007). Remarkably, our research has uncovered yet another critical factor, platelet-derived growth factor (PDGF), which plays a pivotal role in mitigating neuronal toxicity induced by HIV proteins, specifically Trans-activator of transcription (Tat) and HIV gp120 (Yao et al., 2009). Neurotrophic factors exert their influence by preventing synaptic and axonal degeneration in the aftermath of neurotoxin exposure. These factors act as guardians of neuronal survival and interfere with the fundamental mechanism of apoptotic cell death, a hallmark feature of numerous neurodegenerative diseases, pointing toward their potential as therapeutic targets. Furthermore, neurotrophic factors can promote neuronal differentiation, neurogenesis, and angiogenesis, thus facilitating the process of repair and regeneration within the CNS, which is crucial for combating the effects of HIV-induced neurotoxicity (Avdoshina et al., 2011; Bachis et al., 2012; Bachis & Mocchetti, 2005; Mocchetti et al., 2007; Nosheny et al., 2007).

For HAND, a reduction in the expression of neurotrophic factors and/or their receptors can severely impair neural signaling and significantly contribute to the progression of the disease. Despite the efficacy of cART in inhibiting systemic virus replication, HIV proteins (which remain unaffected by cART) such as Tat or HIV gp120 persistently present in the CNS and lymph node tissues can induce neurotoxicity, creating a challenging scenario for clinical management. Viral proteins interfere with neuronal survival through various mechanisms, including generating free radicals, producing nitric oxide, and releasing excitatory neurotransmitters like glutamate, inflammatory cytokines, and other excitotoxins. This intricate interplay underscores the multifaceted nature of HIV-induced neurodegeneration (Kaul et al., 2001). Furthermore, neurotrophic factors, in addition to their protective roles, also play a pivotal role in promoting neuronal differentiation, neurogenesis, and angiogenesis, thereby contributing to the repair process within the CNS. Presently, effective treatments for most neurodegenerative diseases are lacking, underscoring the urgent need for innovative research to identify novel neuroprotective molecules. This chapter aims to comprehensively discuss the beneficial and multifaceted roles played by several neurotrophic factors, including FGF, BDNF, IGF, PDGF, and GDNF, within the complex and evolving context of HIV infection and its associated neurocognitive disorders.

15.2 Fibroblast Growth Factor (FGF)

The superfamily of FGF comprises an intricate network of 18 distinct family members, categorized into several subfamilies based on their sequence homology and phylogenetic relationships. Initially, this superfamily included four additional ligands, FGF11–14, which have since been reclassified as FGF homologous factors (FHF1–4). It is important to note that these FHF ligands, while structurally similar to other FGF family members, lack the ability to activate FGF receptors and have thus been excluded from the core FGF family (Goldfarb et al., 2007). These 18 FGF ligands engage with four distinct membrane-bound FGF receptor subtypes, each exhibiting varying affinities for specific ligands (Reuss et al., 2003).

Upon binding to their respective ligands, FGF receptors undergo dimerization and phosphorylation, ultimately culminating in the activation of three primary downstream signaling pathways: RAS/mitogen-activated protein kinase (MAPK), phosphatidylinositol 3-kinase (PI3K)/AKT, and phospholipase gamma (Fig. 15.1). These pathways orchestrate a symphony of cellular responses, dictating everything from cell growth and survival to differentiation and angiogenesis.

Ten out of the 18 FGFs are expressed within the brain, with FGF1 and FGF2 being the most abundant in the adult brain (Gomez-Pinilla et al., 1992; Turner et al., 2006; Wilcox & Unnerstall, 1991). FGF2, also known as basic FGF, is the first FGF cloned in rats (Kurokawa et al., 1988). Among the various brain regions, the hippocampus and cortical regions exhibit the highest expression of FGF2 (Gomez-Pinilla et al., 1994). Notably, FGF2 expression is found in both neurons and glial cells, with astrocytes displaying particularly elevated levels (Gonzalez et al., 1995). In contrast, FGF1, an acidic fibroblast growth factor, is predominantly expressed in neurons, with relatively limited expression outside the nervous system (Goodrich et al., 1989). FGFs have been extensively characterized for their multifaceted roles within

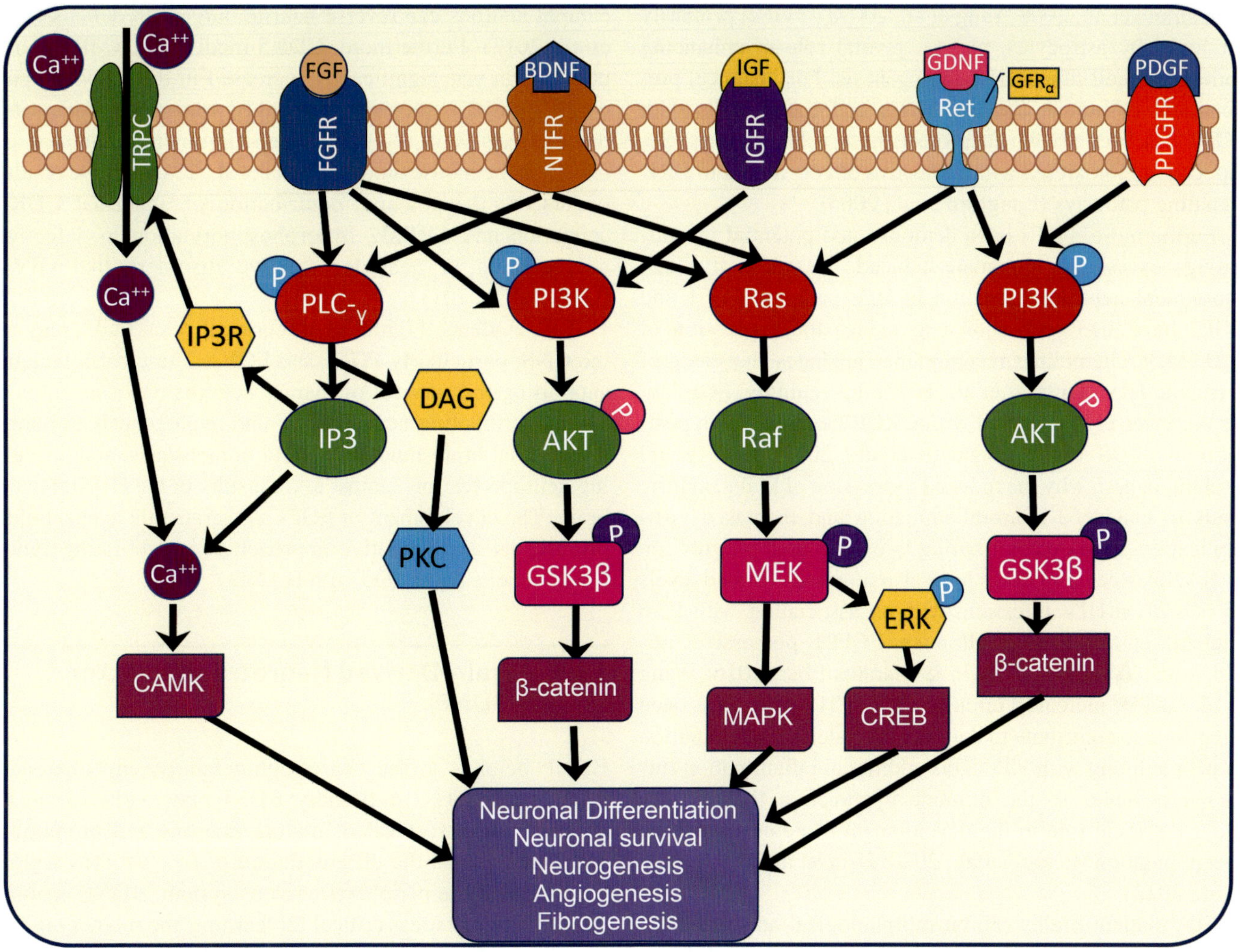

Fig. 15.1 Neuronal differentiation

the CNS. They promote neuronal differentiation, survival, and angiogenesis and have also demonstrated their ability to induce gliogenesis and neurogenesis when introduced into cultures of precursor cells from various brain regions (Qian et al., 1997; Vescovi et al., 1993) or expressed during brain development (Raballo et al., 2000). These effects also extend to adult brains in vivo (Shihabuddin et al., 1997; Yoshimura et al., 2001). FGFs also exhibit neuroprotective capabilities when neurons face toxic insults (Frim et al., 1993; Zechel et al., 2010). For example, FGF2 has effectively inhibited excitotoxicity from excessive glutamatergic stimulation (Brandoli et al., 1998; Fernandez-Sanchez & Novelli, 1993; Kirschner et al., 1995). Furthermore, FGF2 has been linked to reducing inflammatory responses and enhancing the recovery of neuronal function following mechanical injury or stroke (Kawamata et al., 1997; Peterson et al., 1996; Teng et al., 1999).

Given the protective effects of FGFs on mature neurons exposed to toxic stimuli, there is a compelling proposition that elevating the expression of these neurotrophic factors could serve as a therapeutic strategy to safeguard the CNS against toxicity mediated by HIV proteins like Tat and gp120 (Everall et al., 2001; Langford et al., 2002). FGF1 levels have shown an inverse correlation with neurotoxic effects observed in HIV patients (Everall et al., 2001). Archived brain tissues from HIV-positive patients with encephalitis but no apparent neurodegeneration displayed significantly elevated FGF1 levels compared to HIV-positive individuals exhibiting both encephalitis and neurodegenerative signs (Crews et al., 2009; Everall et al., 2001). In vitro studies utilizing primary human neuronal cultures have further demonstrated that FGF1 can protect neurons against gp120 toxicity in a dose-dependent manner (Everall et al., 2001). Notably, FGF1 overexpression has ameliorated neurodegeneration in gp120-expressing transgenic mice in vivo (Crews et al., 2009). Beyond neuronal protection, gp120 is also responsible for causing damage to the endothelial cells lining the blood–brain barrier (BBB). This damage results in breaches in BBB integrity, ultimately facilitating the enhanced migration of HIV-infected cells from the periphery into the brain

(Nakamuta et al., 2008; Yang et al., 2009a). FGF2, primarily produced by astrocytes, plays a pivotal role in enhancing endothelial cell fitness and angiogenesis. For instance, pretreatment of human umbilical vein endothelial cells with FGF2 has been shown to protect these cells from gp120-induced angiotoxicity, with protective effects linked to ERK signaling pathways (Langford et al., 2005).

Furthermore, FGFs have demonstrated potential in mitigating excessive glutamate-induced neurotoxicity and subsequent neuroinflammation in HIV patients. FGF1 and FGF2 have also been shown to reduce the expression of CXCR4, a chemokine receptor that mediates the entry of T-trophic HIV (Dittmar et al., 1997). Intriguingly, FGF levels are inversely correlated with CXCR4 expression in post-mortem AIDS brains (Sanders et al., 2000). This could explain, in part, why the reduced expression of FGFs, in turn, leads to enhanced neuroinflammation and increased virus replication, thereby contributing to exacerbated progression to HAND. Recent findings have also reported elevated levels of FGF21 in HIV infection, with an association with IL-6 (Eckard et al., 2020). Although FGF21 possesses anti-inflammatory effects (Fisher & Maratos-Flier, 2016; Wang et al., 2015), increased circulating FGF21 levels have been suggested to contribute to heightened systemic inflammation in people living with HIV. This additional inflammation further contributes to the immune dysfunction that already plays a critical role in the development of comorbidities in this population (Eckard et al., 2020; Hsue et al., 2012; Zhao et al., 2012).

HIV patient brains exhibit morphological and structural alterations in mature neurons, such as decreased synapse numbers, reduced spine density, neuronal loss, and significant defects in adult hippocampal neurogenesis (Avraham et al., 2014; Lee et al., 2013). These neurogenesis defects observed in HIV patient brains have also been replicated in a rodent HAND model—the gp120 transgenic mice (Okamoto et al., 2007). Interestingly, over-expression of FGF1 (Crews et al., 2009) ameliorates these neurogenesis defects, as indicated by increased doublecortin-expressing neurons (Langford et al., 2005). These findings suggest that impaired neurogenesis during HIV infection involves the neurotoxic effects of HIV proteins like gp120, vpr, and Tat on neuronal progenitor cells through the dysregulation of neurotrophic factor signaling pathways. In HIVE, abnormalities in GSK3β (Kehn-Hall et al., 2011; Maggirwar et al., 1999; Nguyen et al., 2009) and the aberrant activation of CDK5 (Lee et al., 2013; Patrick et al., 2011) have been implicated in the culmination of neurodegeneration and, ultimately, HAND. It has been demonstrated that over-activation of CDK5 results in hyperphosphorylation of collapsing-response mediator protein-2 (CRPM2) and decreased neurite outgrowth. Conversely, inhibiting CDK5 or the expression of a non-phosphorylatable (S522A) CRPM2 construct in adult hippocampal neurons can reverse neurite outgrowth deficits (Lee et al., 2013). Furthermore, CDK5-mediated CRMP2 phosphorylation was significantly increased in the hippocampus of HIV-positive subjects and gp120 transgenic mice, and these effects were rescued by genetic down-modulation of CDK5 in the mouse model (Crews et al., 2011). These results underscore the potential contribution of abnormal CDK5 activation and CRPM2 hyperphosphorylation to defective neurogenesis in neurodegenerative disorders like HIVE (Crews et al., 2011).

This evidence highlights the critical roles FGFs play in the CNS, particularly FGF1 and FGF2. These roles include supporting the survival of mature neurons exposed to toxic stimuli, promoting neurogenesis and angiogenesis to maintain normal brain functioning and homeostasis, and providing neuroprotection against toxic insults in the HIV-infected brain. The development of FGFs as therapeutic agents holds promise as an alternative approach to ameliorating functional deficits associated with HAND.

15.3 Brain-Derived Neurotrophic Factor (BDNF)

BDNF belongs to the neurotrophin family, encompassing NGF, NT-3, and NT-4. Broadly, BDNF plays a pivotal role in supporting the survival of mature neurons and promoting neurogenesis and the differentiation of new neurons within the CNS and the peripheral nervous system. BDNF is notably expressed in areas critical for learning and memory in the brain, including the hippocampus, cortex, and basal forebrain. Functionally, BDNF interacts with two key receptor types: tyrosine kinase receptor (TrkB) and low-affinity nerve growth factor receptor (LNGFR, also known as p75). It is important to emphasize that pro-BDNF and mature BDNF exhibit distinct binding affinities for TrkB and p75. Specifically, pro-BDNF prefers binding to p75, while mature BDNF predominantly interacts with TrkB. This distinct binding preference gives rise to differing cellular outcomes following exposure to pro- or mature BDNF. Pro-BDNF binding to p75 initiates pro-apoptotic signaling, whereas mature BDNF largely supports neuronal survival. Mechanistically, BDNF performs its functions by enhancing glutamatergic excitatory signaling, leading to increased activity of NMDA and AMPA receptors while simultaneously reducing the strength of GABAergic inhibitory signaling. The pivotal role of BDNF in neuronal survival and growth becomes especially significant when considering its implications in neurodegenerative disorders such as Alzheimer's disease (AD), drug addiction, and schizophrenia. In AD, alterations in BDNF transport have been identified, resulting in the degeneration of specific neuronal populations within the basal nucleus of Meynert and the hippocampus (Mufson et al., 1989; Mufson & Kordower, 1992; Poon et al., 2013). Reduced BDNF levels have also been closely associated with depression (Duman, 2004) and Huntington's disease (Zuccato & Cattaneo, 2009).

Recent evidence highlights the potential neuroprotective role of BDNF in the context of HIV. BDNF has demonstrated its capacity to protect neurons against gp120-mediated toxicity, both in vivo and in vitro, by modulating the neuronal expression of the HIV co-receptor, CXCR4. This receptor is abundantly found in neurons and CNS regions like the cortex, hippocampus, and striatum (Ahmed et al., 2008). Emerging data suggest that neuronal CXCR4 is a target of BDNF, with previous studies showing that BDNF reduces the expression of CXCR4 (Bachis et al., 2003; Nosheny et al., 2007). In contrast, heterozygous BDNF knockout animals exhibited increased CXCR4 mRNA levels in specific brain regions compared to wild-type controls, leading to a more pronounced neurotoxic effect of gp120 (Ahmed et al., 2008). BDNF also plays a critical role in reducing glial expression of CXCR4, effectively blocking virus infection in microglia and astrocytes within the CNS (Nosheny et al., 2004). Mechanistically, BDNF initiates this process by binding to the TrkB, which induces receptor dimerization. Subsequently, this leads to increased activation of ERK signaling, resulting in decreased CXCR4 expression and a subsequent reduction in gp120-mediated neuronal apoptosis (Bachis et al., 2003). CXCR4 internalization is essential to BDNF-mediated downregulation of CXCR4 levels (Bachis et al., 2003; Mocchetti et al., 2007, 2008; Mocchetti & Bachis, 2004). Interestingly, a study validating these findings showed that BDNF failed to prevent gp120 neurotoxicity in neurons lacking TrkB expression (Ahmed et al., 2008).

Beyond its role in CXCR4 regulation, BDNF supports the survival of T cells, which are depleted in AIDS patients. By maintaining the pool of $CD4^{+}$T cells, BDNF contributes to delaying the progression of AIDS, ultimately impacting CNS functioning. Notably, peripheral circulating monocytes infiltrate the brain during chronic inflammation induced in a non-human primate following simian immunodeficiency virus (SIV) infection, correlating with the severity of encephalitis (Burdo et al., 2010). This inflammation is associated with dendritic beading and loss of dendritic spines, culminating in a failure of long-term potentiation (LTP), a crucial process underlying learning and memory (Suzumura et al., 2006). BDNF plays a pivotal role in rescuing neuronal deficits by promoting dendritic branching and spine morphology, which are essential for BDNF-mediated LTP (Figurov et al., 1996; Horch & Katz, 2002; Tanaka et al., 2008).

HIV proteins are known to regulate BDNF levels in the brain. In the HIVE brain, for instance, gp120 and TNFα antagonize BDNF functions by inhibiting the anterograde transport of BDNF, reducing intracellular stores, and stimulating NMDA receptor activity (excitotoxicity), ultimately leading to reduced neuronal survival (Bachis et al., 2003; Nosheny et al., 2004). Furthermore, HIV indirectly inhibits BDNF activity by stimulating TNFα, which, in turn, prevents glucocorticoid receptor (GR) activity, resulting in the reduction of hippocampal neurogenesis. This indirect mechanism indicates how HIV-induced neuroinflammation silences GR activity (Suri & Vaidya, 2013). In HIV infection, the expression of TNF-α rises in tandem with reduced BDNF expression. Additionally, IL-1β, released in response to viral proteins (gp120 or Tat), decreases the retrograde transport of BDNF (Wenzel et al., 2019). At the same time, TNF-α interferes with its anterograde transport, further emphasizing the need to preserve BDNF transport along axons and dendrites as a potential therapeutic strategy for HAND (Adachi et al., 2014).

The pro-survival effects of BDNF are also diminished as HIV downregulates its levels in lymphocytes, leading to T-cell apoptosis and neurodegeneration (Avdoshina et al., 2011). The precise molecular mechanisms underlying HIV-induced downregulation of BDNF remain unclear. However, recent data suggest that gp120 reduces pro-BDNF processing in neurons by decreasing furin levels and altering the balance of anti- and pro-apoptotic neurotrophins (Bachis et al., 2012). Additionally, older HIV-positive patients with or without HIVE have exhibited decreased levels of BDNF and increased levels of pro-BDNF protein (Avdoshina et al., 2011). Also, the HIV Tat protein, produced by HIV-infected cells, decreases neurotrophic factor signaling. HIV Tat protein accomplishes this by reducing the production of BDNF through the suppression of the BDNF gene (Carvallo et al., 2017). Neurons exposed to HIV Tat for 24 h have demonstrated a 50% reduction in BDNF mRNA levels, resulting in the hindrance of LTP—a process vital for learning and memory. This effect arises from the diminished phosphorylation of a Camp response element-binding protein (CREB) transcription factor, ultimately leading to the downregulation of BDNF mRNA levels (Santerre et al., 2019).

Multiple cellular mechanisms participate in the processing of pro-BDNF into mature BDNF. Pro-BDNF is initially cleaved within the endoplasmic reticulum by furin and in regulated secretory vesicles by proconvertase enzymes. If pro-BDNF escapes to the extracellular environment, plasmin processes it, giving rise to mature BDNF. Mature BDNF can activate cell surface TrkB receptors or be recycled to the cell surface after cleavage (Pang et al., 2004). The site of BDNF translation within the neuron determines the form of BDNF released, with BDNF mRNA featuring a short 3′-UTR accumulating in the neuronal soma. On the other hand, BDNF mRNA with a long 3′-UTR traffics to dendrites. These neuronal processes support BDNF cleavage within the Golgi, while most dendrites lack Golgi elements necessary for pro-BDNF processing. Consequently, pro-BDNF predominates in the form released. The altered balance between mature

BDNF and pro-BDNF in HAND could compromise synaptic connections and neuronal survival (Barker, 2009; Yang et al., 2009b). These findings are underscored by a study demonstrating that exercise can reverse reduced neurite outgrowth and impaired neurogenesis in a gp120 transgenic mouse model through the production of BDNF (Lee et al., 2013). Moreover, the processing of pro-BDNF into mature BDNF appears to be diminished in HIV, contributing to synaptodendritic injury and synaptic dysfunction seen in HAND. Studies have shown that pro-BDNF levels are higher in HAND subjects than HIV-negative and HIV-positive individuals without dementia. The reduction in BDNF serum levels in HAND could also be attributed to thrombocytopenia, common in people living with HIV (Nascimento & Tanaka, 2012). Furthermore, genetic factors such as the rs6265 polymorphism in the BDNF gene may contribute to lower BDNF levels observed in HIV-positive individuals. This polymorphism results in a Val-Met amino acid substitution at codon 66 (Val66Met) (Avdoshina et al., 2011).

Taken together, accumulating evidence suggests that HIV-induced neuronal degeneration is associated with lower BDNF levels and that BDNF treatment could be a valuable therapeutic approach to mitigate complications of the CNS in HIV-infected patients. This notion is reinforced by observations of significantly lower BDNF levels in both the brain (Bachis et al., 2012) and serum (Avdoshina et al., 2011) of HIV-positive individuals compared to HIV-negative subjects. Furthermore, reduced BDNF levels are linked to impaired cortical innervation by serotonergic fibers (Lyons et al., 1999), highlighting how gp120-mediated reductions in BDNF levels can lead to neuronal degeneration even in the absence of inflammation (Nosheny et al., 2004). In conclusion, these findings emphasize the potential role of low BDNF levels and activity as a risk factor for HAND development and underscore the potential of BDNF delivery as an effective strategy to combat HIV-associated neurodegeneration.

15.4 Insulin-Like Growth Factors (IGFs)

IGFs are pivotal signaling protein molecules that intricately regulate critical cellular functions, encompassing cell proliferation, differentiation, and apoptosis. Recent reports have underscored the potential of IGFs as neurotherapeutic agents capable of modulating brain development and plasticity (Bondy & Cheng, 2002; Ciucci et al., 2007; Tropea et al., 2006). The IGF signaling system is a complex network consisting of two ligands, IGF-1 and IGF-2, two cell surface receptors (IGF1R and IGF2R), and a family of six high-affinity IGF-binding proteins (IGFBPs) and four low-affinity IGFBPs. Remarkably, the sequence homology of IGF-1 and IGF-2 closely resembles each other and proinsulin (Daughaday & Rotwein, 1989; Rotwein, 1991). IGF-1 and IGF-2 primarily bind to cell surface receptors within the insulin/IGF family, activating their intrinsic tyrosine kinase activities. IGF-1 exhibits a higher binding affinity to IGF1R than the insulin receptor, so IGF1R is often called the physiological receptor. Conversely, IGF2R functions as a clearance receptor, exclusively binding to IGF-2 and effectively sequestering it, thus preventing downstream IGF-2-mediated signaling events (Nakae et al., 2001). IGF signaling elicits a wide array of pleiotropic activities, including proliferation, maturation, survival, and growth across various neural cell types, such as neural stem cells, lineage-restricted neural precursor cells, post-mitotic neurons, oligodendrocytes, and astrocytes in the developing brain. However, the specific effects of IGFs can vary based on the unique characteristics of cell types, the local microenvironment, and the stage of development (O'Kusky & Ye, 2012).

IGF-1, a small polypeptide trophic hormone, is primarily synthesized in the liver and transported to the brain via IGFBPs in the plasma and cerebrospinal fluid (Jones & Clemmons, 1995). Additionally, IGF-1 is produced by various cell types within the brain, with its expression peaking perinatally in regions including the brainstem, cerebellum, cortex, hippocampus, hypothalamus, and spinal cord. The chief functions of IGF-1 encompass the regulation of development, cell differentiation, plasticity, and the promotion of nervous system development and survival (Benarroch, 2012). IGF-1 is also involved in brain development, neuroprotection following cellular injury, neurogenesis, myelination, and synaptogenesis. IGF-1 binds to the cell surface receptor IGFR1, which is expressed specifically during brain development or in response to injury, initiating growth-promoting signals. Canonical signaling through the IGF-1 receptor involves the activation of phosphoinositide-3-kinase-AKT-FOXO and RAS-MAPK pathways (Fig. 15.1). These pathways, in turn, modulate active gene transcription, subsequently activating numerous downstream kinases and phosphatases. These molecular events ultimately affect vital cellular functions, including protein synthesis, autophagy, apoptosis, and protection against oxidative stress associated with immunity and inflammation (Fernandez & Torres-Aleman, 2012). Furthermore, IGF-1 plays a crucial role in dorsal root ganglion neuronal development, peripheral axonal regeneration, and neuritogenesis (Jones et al., 2003; Kimpinski et al., 1997). It's important to note that the effects of IGF-1 are more pronounced in the adult brain compared to the developing brain (Bondy & Lee, 1993).

The relationship between the IGF axis and HIV is intricate. HIV infection is frequently associated with a noticeable decline in IGF-1, IGF-2, and IGFBP-3 levels and a significant elevation in IGFBP-1 and IGFBP-2 concentrations. Recently, IGF-1 has gained substantial attention due to its modulatory role in adult neurogenesis (Aberg et al., 2000,

2003; Lichtenwalner et al., 2001). Circulating IGF-1 plays a pivotal role in mediating exercise-induced enhancement of hippocampal neurogenesis in the adult brain (Trejo et al., 2001). Additionally, IGF-1 serves as a neuroprotective agent against neuronal death induced by supernatants from HIV-infected macrophages in vitro (Ying Wang et al., 2003). Observational studies have demonstrated an inverse relationship between IGF-1 and activation markers of macrophages/monocytes in individuals with HIV. While IGF-1 levels may be similar between HIV-positive and HIV-negative groups, those with IGF-1 levels below the average for their age and sex tend to exhibit higher levels of IL-6, soluble CD163, and soluble CD14 (markers of monocyte/macrophage activation) compared to those with IGF-1 levels at or above the average (Fourman et al., 2018). Furthermore, another study revealed that plasma IGF-1 levels were significantly lower in individuals receiving cART based on non-nucleoside reverse transcriptase inhibitors than those on protease inhibitors. This suggests that the choice of cART regimen may influence the reduction of IGF-1 levels, potentially contributing to the development of HAND. Additionally, the study found a positive correlation between plasma levels of IGF2R and viral load, while an inverse correlation was observed with current CD4+ T cell counts (Suh et al., 2015). Low levels of IGF-1, in conjunction with elevated levels of TNF-α, are known to exacerbate HIV-mediated neurodegeneration (Kumar et al., 2003). Furthermore, reduced peripheral IGF-1 levels have been reported in HIV patients with wasting syndrome and children with failure to thrive. Given the central role of IGF-1 in mediating growth hormone action, its involvement in anabolic effects has led to studies on IGF-1 levels in HIV-infected patients and the use of IGF-1 and growth hormone in treating cachectic patients (Mynarcik et al., 1999; Wilk et al., 2011). However, some studies suggest limited responsiveness to IGF-1 and growth hormone therapies in the context of HIV-1 wasting syndrome (Jain et al., 1998). Consequently, diminished levels of IGF-I in the CNS or the development of insulin resistance may compromise neuronal survival during HIV infection.

15.5 Glial-Derived Neurotrophic Factor (GDNF)

GDNF was initially discovered in the supernatant fluids of cultured rat glioma cells and has been demonstrated to exert profound effects on the survival and differentiation of various neuronal populations, including motor and dopaminergic neurons in the midbrain and striatum (Henderson et al., 1994; Yan et al., 1995). Beyond these, GDNF also exerts its neurotrophic influence on a diverse range of cells, such as sensory and autonomic ganglia (Buj-Bello et al., 1995; Ebendal et al., 1995), Purkinje cells in the cerebellum (Mount et al., 1995), hippocampal neurons (Martin et al., 1995), as well as noradrenergic (Arenas et al., 1995), serotoninergic (Ohta et al., 2002), and cholinergic neurons (Williams et al., 1996). GDNF belongs to a neurotrophic factor family that includes four distinct proteins: artemin, GDNF, neurturin, and persephin. This neurotrophic factor family falls under the umbrella of the transforming growth factor-β (TGF-β) superfamily. Notably, GDNF exhibits limited amino acid sequence homology with other members of the superfamily, but its conformational similarity is significantly conserved. All these neurotrophic factors function as homodimers, eliciting protective and regenerative signaling cascades in the developing and adult CNS (Airaksinen & Saarma, 2002; Allen et al., 2013). GDNF and its family members transmit their signals through the canonical signaling receptor Ret and a glycosyl-phosphatidylinositol-anchored co-receptor known as GDNF family receptor α (GFRα) (Fig. 15.1). This family of receptors comprises four isoforms: GFRα1, GFRα2, GFRα3, and GFRα4, each binding specifically to the ligands artemin, GDNF, neurturin, and persephin, respectively (Naveilhan et al., 1998).

Members of this neurotrophic factor family have been demonstrated to play pivotal roles in various biological processes, encompassing cell survival, differentiation, migration, and neurite outgrowth (Lin et al., 1993; Tomac et al., 1995). Notably, GDNF has exhibited the exclusive ability to promote the survival of dopaminergic neurons in a mouse model of Parkinson's disease (Hegarty et al., 2014; Zhao et al., 2014). These findings underscore GDNF as a promising therapeutic candidate for Parkinson's disease (Hegarty et al., 2014; Yasuda & Mochizuki, 2010). Furthermore, GDNF has been shown to protect neonatal and adult facial motor neurons from axotomy-induced cell death (Yan et al., 1995). Similar to GDNF, neurturin has also been found to shield nonhuman primate dopaminergic neurons from neurotoxicity induced by 1-methyl-4-phenyl-1,2,3,6-tetrahydropyridine (Grondin et al., 2008). Additionally, one study revealed that administering GDNF offered protective effects against AD-like alterations triggered by injecting aluminum complexes in rabbits (Ghribi et al., 2001). Interestingly, individuals in the early stages of AD exhibited elevated levels of GDNF in cerebrospinal fluid but reduced serum levels of GDNF. This suggests the presence of adaptive mechanisms within the impaired brain (Straten et al., 2011). Moreover, studies involving 10-month-old 3xTg-AD mice exposed to overexpression of GDNF for 6 months via recombinant lentiviral vectors demonstrated enhanced cognitive abilities related to learning and memory. This neuroprotective effect of GDNF was associated with a significant increase in the expression of BDNF, implying that the combined action of GDNF and BDNF may play a crucial role in countering neuronal atrophy and degeneration, potentially benefiting HAND pathology (Revilla et al., 2014).

The neurotrophic potential of GDNF is particularly relevant for HAND, especially in its late phase, where clinical features closely resemble those observed in Parkinson's disease, such as bradykinesia, loss of fine motor skills, involuntary movements, and postural instability (Berger & Nath, 1997; Mocchetti et al., 2014). Consequently, GDNF emerges as a potential therapeutic avenue for HAD. Interestingly, analogous to findings related to PDGF (as described earlier), intrastriatal microinjections of HIV envelope protein gp120 significantly reduce GDNF levels in the rat substantia nigra, leading to retrograde degeneration of nigrostriatal neurons (Nosheny et al., 2006). The same model has demonstrated chronic and progressive neurodegeneration mediated by gp120 (Nosheny et al., 2006). Intriguingly, GDNF expression was noticeably reduced in the substantia nigra of patients with dementia compared to those without dementia. Collectively, the neurotrophic potential of GDNF plays a crucial role in safeguarding the survival of substantia nigra neurons in HIV patients (Mocchetti et al., 2014).

15.6 Platelet-Derived Growth Factor (PDGF)

PDGF is a family of mesenchymal mitogens characterized by cysteine-knot-type structures. It is synthesized in various tissues, including brain cells such as neuronal progenitors, neurons, astrocytes, oligodendrocytes, and Schwann cells in the peripheral nervous system. Operating in an autocrine and paracrine manner, PDGF plays a pivotal role from gastrulation to maintaining neurons in adulthood. This growth factor family comprises four chains (yielding five isoforms) that form homo- or heterodimers connected by disulfide bonds, resulting in PDGF-AA, PDGF-AB, PDGF-BB, PDGF-CC, and PDGF-DD isoforms (Heldin & Westermark, 1999). These growth factors execute their biological functions by binding to their specific receptors, namely PDGF receptor (PDGFR)-α and PDGFR-β, activating their receptor tyrosine kinase activity. Upon PDGF binding, PDGFR dimerization occurs, initiating downstream signaling pathways. Depending on the PDGF isoform and receptor expression pattern, specific receptor dimers play crucial roles in cell-specific signaling. Typically, PDGFR-α binds to PDGF A-, B-, and C-chains, while PDGFR-β binds to PDGF B- and D-chains (Andrae et al., 2008; Chen et al., 2013).

While the mitogenic effects of PDGFs in various cell types are well documented, their significant roles in various CNS disorders are subjects of extensive research. For instance, PDGF-A serves as a potent mitogen for oligodendrocyte progenitors, and postnatally surviving PDGF-A null mice develop tremors due to severe myelination deficiency in the central nervous system (Betsholtz, 2004; Fruttiger et al., 1999). Additionally, PDGF-BB has been shown to protect primary hippocampal neurons from glutamate-induced neuronal injury by up-regulating PDGFR-β signaling, particularly targeting NR2B receptors (Beazely et al., 2009). Several studies have also demonstrated the neuroprotective potential of PDGFs by guarding neuronal and brain cells against in vitro glutamate-induced excitotoxicity and hypoxic-ischemic injury (Egawa-Tsuzuki et al., 2004; Ishii et al., 2006; Tseng & Dichter, 2005). Furthermore, PDGFs and PDGFR activations are essential for the differentiation of neuronal progenitor cells into neurons (Erlandsson et al., 2001; Williams et al., 1997), and they have been shown to protect hippocampal neurons from oxidative insults and energy deprivation (Cheng & Mattson, 1995). In AD pathology, PDGFR activation has demonstrated protective effects, potentially by inhibiting NMDA receptors. These studies collectively suggest a neuroprotective role for PDGF produced by neurons.

HIV infection primarily targets the basal ganglia region of the brain, resulting in the degeneration of dopaminergic neurons (Nath et al., 2000), ultimately leading to the development of HIV-associated dementia and other neurological complications in the later stages of the disease (Mocchetti et al., 2007). Recent investigations have unveiled a novel neuroprotective role of PDGFs, reversing neuronal toxicity induced by HIV proteins Tat and gp120 in primary midbrain neurons (Yao et al., 2009, 2014), as well as in SHSY-5Y neuroblastoma cells (Peng et al., 2008a, b; Zhu et al., 2009). Furthermore, these studies have shown that the neuroprotective potential of PDGF is mediated through transient receptor potential canonical channels (TRPC), Ca^{2+}-permeable, nonselective cationic channels that play crucial roles in various physiological functions, including neuronal survival (Fig. 15.1). In line with these cell culture findings, experiments validated that PDGF-mediated neuroprotection was compromised in mice pretreated with a TRPC blocker. Moreover, it was reported that the activation of the PDGF-BB/PDGF-R axis stimulates PI3K and PLC/IP3 receptor pathways, subsequently activating TRPC channels, resulting in the influx of calcium ions. This cascade ultimately leads to the activation of ERK/CREB and mTOR/4E-BP-p70S6K signaling pathways, highlighting TRPC1 as a potential novel target implicated in cell proliferation mediated by PDGF-BB. This has significant implications for potential therapeutic interventions aimed at restoring impaired neurogenesis due to the effects of HIV Tat, offering a potential target for cognitive impairment during the pathogenesis of HAND (Yao et al., 2012). Additional molecular mechanisms underlying PDGF-mediated neuroprotection against HIV protein toxicity involve activating PI3K/AKT and ERK/CREB pathways and the concurrent inactivation of pro-apoptotic mediators. Beyond its neuroprotective role, PDGF-BB also promotes the differentiation of rat hippocampal neural progenitor cells through the activation of TRPC channels, calcium influx, and the activation of ERK signaling, along with downstream transcription factor CREB, effectively reversing the impairments mediated by HIV Tat protein (Yang et al., 2016). Furthermore, PDGF exposure in neurons has decreased extracellular glutamate release and increased the upregulation of synaptic plasticity proteins (Peng et al., 2008a, b; Tseng & Dichter, 2005; Yao et al., 2009, 2014). Arc/Arg3.1 is one of the synaptic plasticity proteins whose neurite expression indicates synaptic activation. Interestingly, it has been demonstrated that PDGF-BB regulates the expression of the Arc/Arg3.1 gene in rat hippocampal neurons (Peng et al., 2010). Moreover, PDGF-CC has been shown to protect rat cortical neurons against HIV Tat protein-induced neurotoxicity by activating TRPC channels and AKT signaling (Peng et al., 2012).

15.7 PDGF Paradox in HAND

While members of the PDGF family play protective roles in supporting neuronal fitness and survival, their effects on endothelial cells and astrocytes tend to be more cytotoxic (Bethel-Brown et al., 2011, 2012; Mermis et al., 2011; Yao et al., 2011). For instance, in the context of pulmonary hypertension associated with HIV infection, exposure to HIV proteins has been found to induce oxidative stress in pulmonary endothelial cells. This, in turn, activates hypoxia-inducible factor-1α, leading to an increased expression of PDGF. PDGF then acts as a mitogen for neighboring smooth muscle cells, exacerbating the condition (Mermis et al., 2011). Similarly, HIV infection of human primary astrocytes results in an upregulation of PDGF expression, which subsequently leads to increased expression of the chemokine MCP-1 through the ERK1/2, JNK MAPK, and NFκB activation pathways. Elevated MCP-1 expression, in the end, results in increased monocyte transmigration and the disruption of endothelial barrier permeability in the brains of individuals infected with HIV (Bethel-Brown et al., 2012). In this context, the increased expression of PDGF in astrocytes proves detrimental to endothelial cells.

Among all the cells within the CNS, pericytes are known to express high constitutive levels of PDGF receptors. However, their role in the context of HIV infection remains relatively understudied (Nakagawa et al., 2012). It has been demonstrated that pericytes enhance and potentiate the transcytosis of HIV across brain microvascular endothelial cells through the lipopolysaccharide-mediated release of inflammatory cytokines (Dohgu & Banks, 2013). Recent findings have shed light on the role of PDGF-BB in pericyte functioning under both physiological and pathological conditions (Lindahl et al., 1997; McCarty et al., 2007). For instance, PDGF has been found to increase the migration of retinal microvascular pericytes and plays a crucial role in pericyte maintenance (Nadal et al., 2002). Conversely, exposure of pericytes to HIV Tat protein has been shown to induce the expression and release of PDGF-B. This, through an autocrine loop, results in an increased loss of pericytes from the endothelial barrier, ultimately leading to heightened neuroinflammation (Niu et al., 2014). The mechanism by which HIV Tat induces the expression of PDGF-BB in pericytes involves the activation of the ERK and JNK MAPK pathways and downstream activation of NF-κB (Niu et al., 2014).

15.8 Future Perspectives

One of the central challenges within the realm of growth factor therapy lies in the effective delivery of these therapeutic agents to the CNS. While some of these factors can minimally traverse the BBB through peripheral administration, direct administration into the CNS becomes imperative to ensure their reach to the neurons of the brain or spinal cord. However, a significant drawback arises when these growth and neurotrophic factors are directly introduced into the CNS, as their excessive presence in the nervous system can lead to intolerable adverse effects, especially when prolonged exposure occurs. Therefore, the optimal system for delivering growth factors to the nervous system must meet several criteria. It should be capable of achieving effective concentrations precisely at the sites where degenerating neurons are located while simultaneously limiting their distribution to distant regions, where they might produce unwanted side effects. Also, this delivery should be sustained over an extended period to ensure enduring therapeutic benefits. The absence of a safe and reliable delivery system for growth factor treatment in humans has hindered our ability to explore the full potential of these proteins in treating neurological diseases. Several potential strategies are being investigated to effectively overcome the challenge of delivering growth factors to the nervous system. These approaches include:

1. *Intraparenchymal protein infusion:* This involves the direct infusion of growth factors into specific brain or spinal cord regions. Careful targeting is essential to minimize off-target effects.
2. *Gene delivery* via *viral vectors:* Viral vectors can introduce genes encoding growth factors into the nervous system. This approach aims to achieve persistent, safe, and controllable expression of growth factors in the CNS.
3. *Drug-induced increases in growth factors:* Certain drugs may stimulate the natural production of growth factors within the CNS, enhancing their availability for therapeutic purposes.
4. *Peptide mimetics:* The development of peptide mimetics of growth factors specifically tailored to target degenerating cell systems is being explored. These mimetics can potentially replace invasive central administration and

circumvent the adverse effects associated with non-targeted distribution of growth factors.

Ongoing research in this field is dedicated to addressing these challenges comprehensively. Efforts are being made to improve gene therapy vectors to ensure persistent, safe, and regulatable expression of growth factors in the CNS. Additionally, there is a focus on enhancing the precision of vector and growth factor targeting within the brain, aided by real-time monitoring during treatment. The development of peptide mimetics represents a promising avenue to obviate the need for invasive central administration while minimizing the risks associated with the nonspecific distribution of growth factors. These advancements hold significant potential for advancing the field of growth factor therapy and improving the prospects for the treatment of neurological disorders.

15.9 Conclusions

In conclusion, neurotrophic factors are central to the complex landscape of HAND pathology. These multifaceted molecules offer a glimmer of hope in the quest for effective therapeutic strategies to counter the devastating effects of HIV/HIV proteins on the CNS. The treatment of HAND necessitates a comprehensive approach that combines the power of cART with adjunctive therapies featuring neuroprotective and neurodegenerative agents. Neurotrophic factors, renowned for their remarkable ability to bolster neuronal survival and drive neurogenesis, stand as promising adjuncts against HIV-mediated neurotoxicity. Neurotrophic factors have demonstrated their potential not only in protecting neurons from the toxic effects of HIV proteins but also in enhancing overall neuronal function. They achieve this by promoting synaptic plasticity and facilitating neurogenesis, which are crucial processes for repairing the neurological damage inflicted by the virus. While our exploration of the therapeutic properties of neurotrophic factors has primarily unfolded in cellular and animal models of HAND, these findings kindle optimism and motivate further research. The potential to translate these discoveries into clinical applications is an exciting prospect for individuals living with HIV-associated CNS complications. However, it is vital to emphasize that the journey toward their clinical use demands continued investigation. Deeper insights are needed to unlock the full potential of neurotrophic factors in the context of HAND management. Moreover, the paradoxical nature of some of these factors necessitates a cautious and precise approach to their development and targeted delivery to the CNS. This caution is pivotal in ensuring that the benefits of neurotrophic factor therapy are maximized while minimizing unintended off-target effects. In the evolving landscape of HIV research, neurotrophic factors shine as beacons of promise, illuminating the path toward a brighter future for those affected by HAND. As we delve deeper into the intricacies of these remarkable molecules, we hold the promise of not only ameliorating the impacts of HIV on the nervous system but also potentially reversing the course of neurocognitive decline in affected individuals. The continued pursuit of this research is a scientific endeavor and a bonfire of hope for the countless lives touched by HAND.

15.10 Review Questions

1. Loss of expression of neurotrophic factors and/or their receptors in HIV-associated neurocognitive disorders results in:
   (a) Impaired neuronal signaling
   (b) Axonal growth
   (c) Maintenance of normal neuronal morphology
   (d) Neuronal differentiation
2. Which downstream signaling pathways are activated following fibroblast growth factor binding to its receptors?
   (a) RAS/mitogen-activated protein kinase signaling
   (b) Phosphatidylinositol 3-kinase/AKT signaling
   (c) Phospholipase gamma signaling
   (d) All the above
3. Brain-derived neurotrophic factor primarily binds to:
   (a) TrkB receptor
   (b) Low-affinity nerve growth factor receptor
   (c) Dopamine receptor
   (d) a and b
4. In the context of HIV infection, brain-derived neurotrophic factor elicits neuroprotective effects through:
   (a) Superactivation of ERK signaling
   (b) Blocking of caspase-3 expression
   (c) Decreased expression of CXCR4
   (d) All the above
5. HIV mediated neurodegeneration is exacerbated by:
   (a) Low levels of IGF-1 with elevated levels of TNF-α
   (b) High levels of IGF-1 with low levels of TNF-α
   (c) Low levels of TNF-α
   (d) None of the above
6. GDNF belongs to a family of neurotrophic factors:
   (e) Belong to the transforming growth factor-β superfamily
   (f) Mediate their signaling through the canonical signaling receptor, Ret
   (g) Promote exclusively the survival of dopaminergic neuronal cells
   (h) All the above
7. Platelet-derived growth factors are produced by:
   (a) Neuronal progenitor cells
   (b) Neurons
   (c) Oligodendrocytes

   (d) All the above
8. Exposure of HIV proteins to pulmonary endothelial cells induces oxidative stress with subsequent activation of hypoxia-inducible factor-1α leading to:
   (a) Increased expression of PDGF
   (b) Increased expression of BDNF
   (c) Decreased expression of GDNF
   (d) None of the above
9. The paradoxical nature of neurotrophic factors necessitates caution in their development and delivery to the CNS to avoid what?
   (a) Off-target effects
   (b) Reducing their potency
   (c) Slowing down the therapeutic process
   (d) Making them ineffective
10. What is the primary challenge associated with effectively delivering growth factors to the CNS?
   (a) Ensuring rapid clearance of growth factors from the CNS
   (b) Achieving the maximum possible concentration of growth factors in the CNS
   (c) Avoiding adverse effects of non-targeted growth factor distribution
   (d) Increasing the frequency of growth factor administration

15.11 Answer to the Questions

1. a
2. d
3. d
4. d
5. a
6. d
7. d
8. a
9. a
10. c

References

Aberg, M. A., Aberg, N. D., Hedbacker, H., et al. (2000). Peripheral infusion of IGF-I selectively induces neurogenesis in the adult rat hippocampus. *The Journal of Neuroscience: The Official Journal of the Society for Neuroscience, 20*, 2896–2903. https://doi.org/10.1523/JNEUROSCI.20-08-02896.2000

Aberg, M. A., Aberg, N. D., Palmer, T. D., et al. (2003). IGF-I has a direct proliferative effect in adult hippocampal progenitor cells. *Molecular and Cellular Neurosciences, 24*, 23–40. https://doi.org/10.1016/s1044-7431(03)00082-4

Adachi, N., Numakawa, T., Richards, M., et al. (2014). New insight in expression, transport, and secretion of brain-derived neurotrophic factor: Implications in brain-related diseases. *World Journal of Biological Chemistry, 5*, 409–428. https://doi.org/10.4331/wjbc.v5.i4.409

Ahmed, F., Tessarollo, L., Thiele, C., et al. (2008). Brain-derived neurotrophic factor modulates expression of chemokine receptors in the brain. *Brain Research, 1227*, 1–11. https://doi.org/10.1016/j.brainres.2008.05.086

Airaksinen, M. S., & Saarma, M. (2002). The GDNF family: Signalling, biological functions and therapeutic value. *Nature Reviews. Neuroscience, 3*, 383–394. https://doi.org/10.1038/nrn812

Allen, S. J., Watson, J. J., Shoemark, D. K., et al. (2013). GDNF, NGF and BDNF as therapeutic options for neurodegeneration. *Pharmacology & Therapeutics, 138*, 155–175. https://doi.org/10.1016/j.pharmthera.2013.01.004

Almeida, R. D., Manadas, B. J., Melo, C. V., et al. (2005). Neuroprotection by BDNF against glutamate-induced apoptotic cell death is mediated by ERK and PI3-kinase pathways. *Cell Death and Differentiation, 12*, 1329–1343. https://doi.org/10.1038/sj.cdd.4401662

Alzheimer, C., & Werner, S. (2002). Fibroblast growth factors and neuroprotection. *Advances in Experimental Medicine and Biology, 513*, 335–351. https://doi.org/10.1007/978-1-4615-0123-7_12

Andrae, J., Gallini, R., & Betsholtz, C. (2008). Role of platelet-derived growth factors in physiology and medicine. *Genes & Development, 22*, 1276–1312. https://doi.org/10.1101/gad.1653708

Arenas, E., Trupp, M., Akerud, P., et al. (1995). GDNF prevents degeneration and promotes the phenotype of brain noradrenergic neurons in vivo. *Neuron, 15*, 1465–1473. https://doi.org/10.1016/0896-6273(95)90024-1

Avdoshina, V., Garzino-Demo, A., Bachis, A., et al. (2011). HIV-1 decreases the levels of neurotrophins in human lymphocytes. *AIDS, 25*, 1126–1128. https://doi.org/10.1097/QAD.0b013e32834671b3

Avraham, H. K., Jiang, S., Fu, Y., et al. (2014). The cannabinoid CB(2) receptor agonist AM1241 enhances neurogenesis in GFAP/Gp120 transgenic mice displaying deficits in neurogenesis. *British Journal of Pharmacology, 171*, 468–479. https://doi.org/10.1111/bph.12478

Bachis, A., Avdoshina, V., Zecca, L., et al. (2012). Human immunodeficiency virus type 1 alters brain-derived neurotrophic factor processing in neurons. *The Journal of Neuroscience: The Official Journal of the Society for Neuroscience, 32*, 9477–9484. https://doi.org/10.1523/JNEUROSCI.0865-12.2012

Bachis, A., Major, E. O., & Mocchetti, I. (2003). Brain-derived neurotrophic factor inhibits human immunodeficiency virus-1/gp120-mediated cerebellar granule cell death by preventing gp120 internalization. *The Journal of Neuroscience: The Official Journal of the Society for Neuroscience, 23*, 5715–5722. https://doi.org/10.1523/JNEUROSCI.23-13-05715.2003

Bachis, A., & Mocchetti, I. (2005). Brain-derived neurotrophic factor is neuroprotective against human immunodeficiency virus-1 envelope proteins. *Annals of the New York Academy of Sciences, 1053*, 247–257. https://doi.org/10.1196/annals.1344.022

Barker, P. A. (2009). Whither proBDNF? *Nature Neuroscience, 12*, 105–106. https://doi.org/10.1038/nn0209-105

Beazely, M. A., Lim, A., Li, H., et al. (2009). Platelet-derived growth factor selectively inhibits NR2B-containing N-methyl-D-aspartate receptors in CA1 hippocampal neurons. *The Journal of Biological Chemistry, 284*, 8054–8063. https://doi.org/10.1074/jbc.M805384200

Benarroch, E. E. (2012). Insulin-like growth factors in the brain and their potential clinical implications. *Neurology, 79*, 2148–2153. https://doi.org/10.1212/WNL.0b013e3182752eef

Berger, J. R., & Nath, A. (1997). HIV dementia and the basal ganglia. *Intervirology, 40*, 122–131. https://doi.org/10.1159/000150539

Bethel-Brown, C., Yao, H., Callen, S., et al. (2011). HIV-1 Tat-mediated induction of platelet-derived growth factor in astrocytes: Role of early growth response gene 1. *Journal of Immunology, 186*, 4119–4129. https://doi.org/10.4049/jimmunol.1002235

Bethel-Brown, C., Yao, H., Hu, G., et al. (2012). Platelet-derived growth factor (PDGF)-BB-mediated induction of monocyte chemoattractant protein 1 in human astrocytes: Implications for HIV-associated neuroinflammation. *Journal of Neuroinflammation, 9*, 262. https://doi.org/10.1186/1742-2094-9-262

Betsholtz, C. (2004). Insight into the physiological functions of PDGF through genetic studies in mice. *Cytokine & Growth Factor Reviews, 15*, 215–228. https://doi.org/10.1016/j.cytogfr.2004.03.005

Bondy, C. A., & Cheng, C. M. (2002). Insulin-like growth factor-1 promotes neuronal glucose utilization during brain development and repair processes. *International Review of Neurobiology, 51*, 189–217. https://doi.org/10.1016/s0074-7742(02)51006-0

Bondy, C. A., & Lee, W. H. (1993). Patterns of insulin-like growth factor and IGF receptor gene expression in the brain. Functional implications. *Annals of the New York Academy of Sciences, 692*, 33–43. https://doi.org/10.1111/j.1749-6632.1993.tb26203.x

Brandoli, C., Sanna, A., De Bernardi, M. A., et al. (1998). Brain-derived neurotrophic factor and basic fibroblast growth factor downregulate NMDA receptor function in cerebellar granule cells. *The Journal of Neuroscience: The Official Journal of the Society for Neuroscience, 18*, 7953–7961. https://doi.org/10.1523/JNEUROSCI.18-19-07953.1998

Buj-Bello, A., Buchman, V. L., Horton, A., et al. (1995). GDNF is an age-specific survival factor for sensory and autonomic neurons. *Neuron, 15*, 821–828. https://doi.org/10.1016/0896-6273(95)90173-6

Burdo, T. H., Soulas, C., Orzechowski, K., et al. (2010). Increased monocyte turnover from bone marrow correlates with severity of SIV encephalitis and CD163 levels in plasma. *PLoS Pathogens, 6*, e1000842. https://doi.org/10.1371/journal.ppat.1000842

Carvallo, L., Lopez, L., Fajardo, J. E., et al. (2017). HIV-tat regulates macrophage gene expression in the context of neuroAIDS. *PLoS One, 12*, e0179882. https://doi.org/10.1371/journal.pone.0179882

Chen, P. H., Chen, X., & He, X. (2013). Platelet-derived growth factors and their receptors: Structural and functional perspectives. *Biochimica et Biophysica Acta, 1834*, 2176–2186. https://doi.org/10.1016/j.bbapap.2012.10.015

Cheng, B., & Mattson, M. P. (1995). PDGFs protect hippocampal neurons against energy deprivation and oxidative injury: Evidence for induction of antioxidant pathways. *The Journal of Neuroscience: The Official Journal of the Society for Neuroscience, 15*, 7095–7104. https://doi.org/10.1523/JNEUROSCI.15-11-07095.1995

Ciucci, F., Putignano, E., Baroncelli, L., et al. (2007). Insulin-like growth factor 1 (IGF-1) mediates the effects of enriched environment (EE) on visual cortical development. *PLoS One, 2*, e475. https://doi.org/10.1371/journal.pone.0000475

Colafrancesco, V., & Villoslada, P. (2011). Targeting NGF pathway for developing neuroprotective therapies for multiple sclerosis and other neurological diseases. *Archives Italiennes de Biologie, 149*, 183–192. https://doi.org/10.4449/aib.v149i2.1376

Crews, L., Patrick, C., Achim, C. L., et al. (2009). Molecular pathology of neuro-AIDS (CNS-HIV). *International Journal of Molecular Sciences, 10*, 1045–1063. https://doi.org/10.3390/ijms10031045

Crews, L., Ruf, R., Patrick, C., et al. (2011). Phosphorylation of collapsin response mediator protein-2 disrupts neuronal maturation in a model of adult neurogenesis: Implications for neurodegenerative disorders. *Molecular Neurodegeneration, 6*, 67. https://doi.org/10.1186/1750-1326-6-67

Daughaday, W. H., & Rotwein, P. (1989). Insulin-like growth factors I and II. Peptide, messenger ribonucleic acid and gene structures, serum, and tissue concentrations. *Endocrine Reviews, 10*, 68–91. https://doi.org/10.1210/edrv-10-1-68

Deierborg, T., Soulet, D., Roybon, L., et al. (2008). Emerging restorative treatments for Parkinson's disease. *Progress in Neurobiology, 85*, 407–432. https://doi.org/10.1016/j.pneurobio.2008.05.001

Dittmar, M. T., McKnight, A., Simmons, G., et al. (1997). HIV-1 tropism and co-receptor use. *Nature, 385*, 495–496. https://doi.org/10.1038/385495a0

Dohgu, S., & Banks, W. A. (2013). Brain pericytes increase the lipopolysaccharide-enhanced transcytosis of HIV-1 free virus across the in vitro blood-brain barrier: Evidence for cytokine-mediated pericyte-endothelial cell crosstalk. *Fluids and Barriers of the CNS, 10*, 23. https://doi.org/10.1186/2045-8118-10-23

Duman, R. S. (2004). Role of neurotrophic factors in the etiology and treatment of mood disorders. *Neuromolecular Medicine, 5*, 11–25. https://doi.org/10.1385/NMM:5:1:011

Ebendal, T., Tomac, A., Hoffer, B. J., et al. (1995). Glial cell line-derived neurotrophic factor stimulates fiber formation and survival in cultured neurons from peripheral autonomic ganglia. *Journal of Neuroscience Research, 40*, 276–284. https://doi.org/10.1002/jnr.490400217

Eckard, A. R., Hughes, H. Y., Hagood, N. L., et al. (2020). Fibroblast growth factor 21 is elevated in HIV and associated with Interleukin-6. *Journal of Acquired Immune Deficiency Syndromes, 83*, e30–e33. https://doi.org/10.1097/QAI.0000000000002285

Egawa-Tsuzuki, T., Ohno, M., Tanaka, N., et al. (2004). The PDGF B-chain is involved in the ontogenic susceptibility of the developing rat brain to NMDA toxicity. *Experimental Neurology, 186*, 89–98. https://doi.org/10.1016/j.expneurol.2003.11.001

Erlandsson, A., Enarsson, M., & Forsberg-Nilsson, K. (2001). Immature neurons from CNS stem cells proliferate in response to platelet-derived growth factor. *The Journal of Neuroscience: The Official Journal of the Society for Neuroscience, 21*, 3483–3491. https://doi.org/10.1523/JNEUROSCI.21-10-03483.2001

Everall, I. P., Trillo-Pazos, G., Bell, C., et al. (2001). Amelioration of neurotoxic effects of HIV envelope protein gp120 by fibroblast growth factor: A strategy for neuroprotection. *Journal of Neuropathology and Experimental Neurology, 60*, 293–301. https://doi.org/10.1093/jnen/60.3.293

Fernandez-Sanchez, M. T., & Novelli, A. (1993). Basic fibroblast growth factor protects cerebellar neurons in primary culture from NMDA and non-NMDA receptor mediated neurotoxicity. *FEBS Letters, 335*, 124–131. https://doi.org/10.1016/0014-5793(93)80453-2

Fernandez, A. M., & Torres-Aleman, I. (2012). The many faces of insulin-like peptide signalling in the brain. *Nature Reviews. Neuroscience, 13*, 225–239. https://doi.org/10.1038/nrn3209

Figurov, A., Pozzo-Miller, L. D., Olafsson, P., et al. (1996). Regulation of synaptic responses to high-frequency stimulation and LTP by neurotrophins in the hippocampus. *Nature, 381*, 706–709. https://doi.org/10.1038/381706a0

Fisher, F. M., & Maratos-Flier, E. (2016). Understanding the physiology of FGF21. *Annual Review of Physiology, 78*, 223–241. https://doi.org/10.1146/annurev-physiol-021115-105339

Fourman, L. T., Czerwonka, N., Shaikh, S. D., et al. (2018). Insulin-like growth factor 1 inversely relates to monocyte/macrophage activation markers in HIV. *AIDS, 32*, 927–932. https://doi.org/10.1097/QAD.0000000000001775

Frim, D. M., Uhler, T. A., Short, M. P., et al. (1993). Effects of biologically delivered NGF, BDNF and bFGF on striatal excitotoxic lesions. *Neuroreport, 4*, 367–370. https://doi.org/10.1097/00001756-199304000-00006

Fruttiger, M., Karlsson, L., Hall, A. C., et al. (1999). Defective oligodendrocyte development and severe hypomyelination in PDGF-A knockout mice. *Development, 126*, 457–467. https://doi.org/10.1242/dev.126.3.457

Ghribi, O., Herman, M. M., Forbes, M. S., et al. (2001). GDNF protects against aluminum-induced apoptosis in rabbits by upregulating Bcl-2 and Bcl-XL and inhibiting mitochondrial Bax translocation. *Neurobiology of Disease, 8*, 764–773. https://doi.org/10.1006/nbdi.2001.0429

Goldfarb, M., Schoorlemmer, J., Williams, A., et al. (2007). Fibroblast growth factor homologous factors control neuronal excitability through modulation of voltage-gated sodium channels. *Neuron, 55*, 449–463. https://doi.org/10.1016/j.neuron.2007.07.006

Gomez-Pinilla, F., Lee, J. W., & Cotman, C. W. (1992). Basic FGF in adult rat brain: Cellular distribution and response to entorhinal lesion and fimbria-fornix transection. *The Journal of Neuroscience: The Official Journal of the Society for Neuroscience, 12*, 345–355. https://doi.org/10.1523/JNEUROSCI.12-01-00345.1992

Gomez-Pinilla, F., Lee, J. W., & Cotman, C. W. (1994). Distribution of basic fibroblast growth factor in the developing rat brain. *Neuroscience, 61*, 911–923. https://doi.org/10.1016/0306-4522(94)90412-x

Gonzalez, A. M., Berry, M., Maher, P. A., et al. (1995). A comprehensive analysis of the distribution of FGF-2 and FGFR1 in the rat brain. *Brain Research, 701*, 201–226. https://doi.org/10.1016/0006-8993(95)01002-x

Goodrich, S. P., Yan, G. C., Bahrenburg, K., et al. (1989). The nucleotide sequence of rat heparin binding growth factor 1 (HBGF-1). *Nucleic Acids Research, 17*, 2867. https://doi.org/10.1093/nar/17.7.2867

Grondin, R., Zhang, Z., Ai, Y., et al. (2008). Intraputamenal infusion of exogenous neurturin protein restores motor and dopaminergic function in the globus pallidus of MPTP-lesioned rhesus monkeys. *Cell Transplantation, 17*, 373–381.

Hegarty, S. V., O'Keeffe, G. W., & Sullivan, A. M. (2014). Neurotrophic factors: From neurodevelopmental regulators to novel therapies for Parkinson's disease. *Neural Regeneration Research, 9*, 1708–1711. https://doi.org/10.4103/1673-5374.143410

Heldin, C. H., & Westermark, B. (1999). Mechanism of action and in vivo role of platelet-derived growth factor. *Physiological Reviews, 79*, 1283–1316. https://doi.org/10.1152/physrev.1999.79.4.1283

Henderson, C. E., Phillips, H. S., Pollock, R. A., et al. (1994). GDNF: A potent survival factor for motoneurons present in peripheral nerve and muscle. *Science, 266*, 1062–1064. https://doi.org/10.1126/science.7973664

Horch, H. W., & Katz, L. C. (2002). BDNF release from single cells elicits local dendritic growth in nearby neurons. *Nature Neuroscience, 5*, 1177–1184. https://doi.org/10.1038/nn927

Hsue, P. Y., Deeks, S. G., & Hunt, P. W. (2012). Immunologic basis of cardiovascular disease in HIV-infected adults. *The Journal of Infectious Diseases, 205*(Suppl 3), S375–S382. https://doi.org/10.1093/infdis/jis200

Ishii, Y., Oya, T., Zheng, L., et al. (2006). Mouse brains deficient in neuronal PDGF receptor-beta develop normally but are vulnerable to injury. *Journal of Neurochemistry, 98*, 588–600. https://doi.org/10.1111/j.1471-4159.2006.03922.x

Jain, S., Golde, D. W., Bailey, R., et al. (1998). Insulin-like growth factor-I resistance. *Endocrine Reviews, 19*, 625–646. https://doi.org/10.1210/edrv.19.5.0348

Jones, D. M., Tucker, B. A., Rahimtula, M., et al. (2003). The synergistic effects of NGF and IGF-1 on neurite growth in adult sensory neurons: Convergence on the PI 3-kinase signaling pathway. *Journal of Neurochemistry, 86*, 1116–1128. https://doi.org/10.1046/j.1471-4159.2003.01925.x

Jones, J. I., & Clemmons, D. R. (1995). Insulin-like growth factors and their binding proteins: Biological actions. *Endocrine Reviews, 16*, 3–34. https://doi.org/10.1210/edrv-16-1-3

Kaul, M., Garden, G. A., & Lipton, S. A. (2001). Pathways to neuronal injury and apoptosis in HIV-associated dementia. *Nature, 410*, 988–994. https://doi.org/10.1038/35073667

Kawamata, T., Dietrich, W. D., Schallert, T., et al. (1997). Intracisternal basic fibroblast growth factor enhances functional recovery and up-regulates the expression of a molecular marker of neuronal sprouting following focal cerebral infarction. *Proceedings of the National Academy of Sciences of the United States of America, 94*, 8179–8184. https://doi.org/10.1073/pnas.94.15.8179

Kehn-Hall, K., Guendel, I., Carpio, L., et al. (2011). Inhibition of Tat-mediated HIV-1 replication and neurotoxicity by novel GSK3-beta inhibitors. *Virology, 415*, 56–68. https://doi.org/10.1016/j.virol.2011.03.025

Kimpinski, K., Campenot, R. B., & Mearow, K. (1997). Effects of the neurotrophins nerve growth factor, neurotrophin-3, and brain-derived neurotrophic factor (BDNF) on neurite growth from adult sensory neurons in compartmented cultures. *Journal of Neurobiology, 33*, 395–410. https://doi.org/10.1002/(sici)1097-4695(199710)33:4<395::aid-neu5>3.0.co;2-5

Kirschner, P. B., Henshaw, R., Weise, J., et al. (1995). Basic fibroblast growth factor protects against excitotoxicity and chemical hypoxia in both neonatal and adult rats. *Journal of Cerebral Blood Flow and Metabolism: Official Journal of the International Society of Cerebral Blood Flow and Metabolism, 15*, 619–623. https://doi.org/10.1038/jcbfm.1995.76

Kumar, M., Kumar, A. M., Waldrop, D., et al. (2003). HIV-1 infection and its impact on the HPA axis, cytokines, and cognition. *Stress, 6*, 167–172. https://doi.org/10.1080/10253890310001605376

Kurokawa, T., Seno, M., & Igarashi, K. (1988). Nucleotide sequence of rat basic fibroblast growth factor cDNA. *Nucleic Acids Research, 16*, 5201. https://doi.org/10.1093/nar/16.11.5201

Langford, D., Hurford, R., Hashimoto, M., et al. (2005). Signalling crosstalk in FGF2-mediated protection of endothelial cells from HIV-gp120. *BMC Neuroscience, 6*, 8. https://doi.org/10.1186/1471-2202-6-8

Langford, T. D., Letendre, S. L., Marcotte, T. D., et al. (2002). Severe, demyelinating leukoencephalopathy in AIDS patients on antiretroviral therapy. *AIDS, 16*, 1019–1029. https://doi.org/10.1097/00002030-200205030-00008

Lee, M. H., Amin, N. D., Venkatesan, A., et al. (2013). Impaired neurogenesis and neurite outgrowth in an HIV-gp120 transgenic model is reversed by exercise via BDNF production and Cdk5 regulation. *Journal of Neurovirology, 19*, 418–431. https://doi.org/10.1007/s13365-013-0194-6

Lichtenwalner, R. J., Forbes, M. E., Bennett, S. A., et al. (2001). Intracerebroventricular infusion of insulin-like growth factor-I ameliorates the age-related decline in hippocampal neurogenesis. *Neuroscience, 107*, 603–613. https://doi.org/10.1016/s0306-4522(01)00378-5

Lin, L. F., Doherty, D. H., Lile, J. D., et al. (1993). GDNF: A glial cell line-derived neurotrophic factor for midbrain dopaminergic neurons. *Science, 260*, 1130–1132. https://doi.org/10.1126/science.8493557

Lindahl, P., Johansson, B. R., Leveen, P., et al. (1997). Pericyte loss and microaneurysm formation in PDGF-B-deficient mice. *Science, 277*, 242–245. https://doi.org/10.1126/science.277.5323.242

Lyons, W. E., Mamounas, L. A., Ricaurte, G. A., et al. (1999). Brain-derived neurotrophic factor-deficient mice develop aggressiveness and hyperphagia in conjunction with brain serotonergic abnormalities. *Proceedings of the National Academy of Sciences of the United States of America, 96*, 15239–15244. https://doi.org/10.1073/pnas.96.26.15239

Maggirwar, S. B., Tong, N., Ramirez, S., et al. (1999). HIV-1 Tat-mediated activation of glycogen synthase kinase-3beta contributes to Tat-mediated neurotoxicity. *Journal of Neurochemistry, 73*, 578–586. https://doi.org/10.1046/j.1471-4159.1999.0730578.x

Martin, D., Miller, G., Rosendahl, M., et al. (1995). Potent inhibitory effects of glial derived neurotrophic factor against kainic acid mediated seizures in the rat. *Brain Research, 683*, 172–178. https://doi.org/10.1016/0006-8993(95)00369-2

McCarty, M. F., Somcio, R. J., Stoeltzing, O., et al. (2007). Overexpression of PDGF-BB decreases colorectal and pancreatic cancer growth by increasing tumor pericyte content. *The Journal of Clinical Investigation, 117*, 2114–2122. https://doi.org/10.1172/JCI31334

Mermis, J., Gu, H., Xue, B., et al. (2011). Hypoxia-inducible factor-1 alpha/platelet derived growth factor axis in HIV-associated pulmonary vascular remodeling. *Respiratory Research, 12*, 103. https://doi.org/10.1186/1465-9921-12-103

Mocchetti, I., & Bachis, A. (2004). Brain-derived neurotrophic factor activation of TrkB protects neurons from HIV-1/gp120-induced cell death. *Critical Reviews in Neurobiology, 16*, 51–57. https://doi.org/10.1615/critrevneurobiol.v16.i12.50

Mocchetti, I., Bachis, A., Campbell, L. A., et al. (2014). Implementing neuronal plasticity in NeuroAIDS: The experience of brain-derived neurotrophic factor and other neurotrophic factors. *Journal of Neuroimmune Pharmacology: The Official Journal of the Society on NeuroImmune Pharmacology, 9*, 80–91. https://doi.org/10.1007/s11481-013-9488-y

Mocchetti, I., Bachis, A., & Masliah, E. (2008). Chemokine receptors and neurotrophic factors: Potential therapy against aids dementia? *Journal of Neuroscience Research, 86*, 243–255. https://doi.org/10.1002/jnr.21492

Mocchetti, I., Nosheny, R. L., Tanda, G., et al. (2007). Brain-derived neurotrophic factor prevents human immunodeficiency virus type 1 protein gp120 neurotoxicity in the rat nigrostriatal system. *Annals of the New York Academy of Sciences, 1122*, 144–154. https://doi.org/10.1196/annals.1403.010

Mount, H. T., Dean, D. O., Alberch, J., et al. (1995). Glial cell line-derived neurotrophic factor promotes the survival and morphologic differentiation of Purkinje cells. *Proceedings of the National Academy of Sciences of the United States of America, 92*, 9092–9096. https://doi.org/10.1073/pnas.92.20.9092

Mufson, E. J., Bothwell, M., Hersh, L. B., et al. (1989). Nerve growth factor receptor immunoreactive profiles in the normal, aged human basal forebrain: Colocalization with cholinergic neurons. *The Journal of Comparative Neurology, 285*, 196–217. https://doi.org/10.1002/cne.902850204

Mufson, E. J., & Kordower, J. H. (1992). Cortical neurons express nerve growth factor receptors in advanced age and Alzheimer disease. *Proceedings of the National Academy of Sciences of the United States of America, 89*, 569–573. https://doi.org/10.1073/pnas.89.2.569

Mynarcik, D. C., Frost, R. A., Lang, C. H., et al. (1999). Insulin-like growth factor system in patients with HIV infection: Effect of exogenous growth hormone administration. *Journal of Acquired Immune Deficiency Syndromes, 22*, 49–55. https://doi.org/10.1097/00042560-199909010-00006

Nadal, J. A., Scicli, G. M., Carbini, L. A., et al. (2002). Angiotensin II stimulates migration of retinal microvascular pericytes: Involvement of TGF-beta and PDGF-BB. *American Journal of Physiology. Heart and Circulatory Physiology, 282*, H739–H748. https://doi.org/10.1152/ajpheart.00656.2001

Nakae, J., Kido, Y., & Accili, D. (2001). Distinct and overlapping functions of insulin and IGF-I receptors. *Endocrine Reviews, 22*, 818–835. https://doi.org/10.1210/edrv.22.6.0452

Nakagawa, S., Castro, V., & Toborek, M. (2012). Infection of human pericytes by HIV-1 disrupts the integrity of the blood-brain barrier. *Journal of Cellular and Molecular Medicine, 16*, 2950–2957. https://doi.org/10.1111/j.1582-4934.2012.01622.x

Nakamuta, S., Endo, H., Higashi, Y., et al. (2008). Human immunodeficiency virus type 1 gp120-mediated disruption of tight junction proteins by induction of proteasome-mediated degradation of zonula occludens-1 and -2 in human brain microvascular endothelial cells. *Journal of Neurovirology, 14*, 186–195. https://doi.org/10.1080/13550280801993630

Nascimento, F. G., & Tanaka, P. Y. (2012). Thrombocytopenia in HIV-infected patients. *Indian Journal of Hematology and Blood Transfusion, 28*, 109–111. https://doi.org/10.1007/s12288-011-0124-9

Nath, A., Anderson, C., Jones, M., et al. (2000). Neurotoxicity and dysfunction of dopaminergic systems associated with AIDS dementia. *Journal of Psychopharmacology, 14*, 222–227. https://doi.org/10.1177/026988110001400305

Naveilhan, P., Baudet, C., Mikaels, A., et al. (1998). Expression and regulation of GFRalpha3, a glial cell line-derived neurotrophic factor family receptor. *Proceedings of the National Academy of Sciences of the United States of America, 95*, 1295–1300. https://doi.org/10.1073/pnas.95.3.1295

Nguyen, T. B., Lucero, G. R., Chana, G., et al. (2009). Glycogen synthase kinase-3beta (GSK-3beta) inhibitors AR-A014418 and B6B3O prevent human immunodeficiency virus-mediated neurotoxicity in primary human neurons. *Journal of Neurovirology, 15*, 434–438. https://doi.org/10.1080/13550280903168131

Niu, F., Yao, H., Zhang, W., et al. (2014). Tat 101-mediated enhancement of brain pericyte migration involves platelet-derived growth factor subunit B homodimer: Implications for human immunodeficiency virus-associated neurocognitive disorders. *The Journal of Neuroscience: The Official Journal of the Society for Neuroscience, 34*, 11812–11825. https://doi.org/10.1523/JNEUROSCI.1139-14.2014

Nosheny, R. L., Ahmed, F., Yakovlev, A., et al. (2007). Brain-derived neurotrophic factor prevents the nigrostriatal degeneration induced by human immunodeficiency virus-1 glycoprotein 120 in vivo. *The European Journal of Neuroscience, 25*, 2275–2284. https://doi.org/10.1111/j.1460-9568.2007.05506.x

Nosheny, R. L., Bachis, A., Acquas, E., et al. (2004). Human immunodeficiency virus type 1 glycoprotein gp120 reduces the levels of brain-derived neurotrophic factor in vivo: Potential implication for neuronal cell death. *The European Journal of Neuroscience, 20*, 2857–2864. https://doi.org/10.1111/j.1460-9568.2004.03764.x

Nosheny, R. L., Bachis, A., Aden, S. A., et al. (2006). Intrastriatal administration of human immunodeficiency virus-1 glycoprotein 120 reduces glial cell-line derived neurotrophic factor levels and causes apoptosis in the substantia nigra. *Journal of Neurobiology, 66*, 1311–1321. https://doi.org/10.1002/neu.20288

O'Kusky, J., & Ye, P. (2012). Neurodevelopmental effects of insulin-like growth factor signaling. *Frontiers in Neuroendocrinology, 33*, 230–251. https://doi.org/10.1016/j.yfrne.2012.06.002

Ohta, K., Ohta, M., Mizuta, I., et al. (2002). The novel catecholaminergic and serotoninergic activity enhancer R-(−)-1-(benzofuran-2-yl)-2-propylaminopentane up-regulates neurotrophic factor synthesis in mouse astrocytes. *Neuroscience Letters, 328*, 205–208. https://doi.org/10.1016/s0304-3940(02)00461-5

Okamoto, S., Kang, Y. J., Brechtel, C. W., et al. (2007). HIV/gp120 decreases adult neural progenitor cell proliferation via checkpoint kinase-mediated cell-cycle withdrawal and G1 arrest. *Cell Stem Cell, 1*, 230–236. https://doi.org/10.1016/j.stem.2007.07.010

Pang, P. T., Teng, H. K., Zaitsev, E., et al. (2004). Cleavage of proBDNF by tPA/plasmin is essential for long-term hippocampal plasticity. *Science, 306*, 487–491. https://doi.org/10.1126/science.1100135

Patrick, C., Crews, L., Desplats, P., et al. (2011). Increased CDK5 expression in HIV encephalitis contributes to neurodegeneration via tau phosphorylation and is reversed with Roscovitine. *The American Journal of Pathology, 178*, 1646–1661. https://doi.org/10.1016/j.ajpath.2010.12.033

Peng, F., Dhillon, N., Callen, S., et al. (2008a). Platelet-derived growth factor protects neurons against gp120-mediated toxicity. *Journal of Neurovirology, 14*, 62–72. https://doi.org/10.1080/13550280701809084

Peng, F., Dhillon, N. K., Yao, H., et al. (2008b). Mechanisms of platelet-derived growth factor-mediated neuroprotection--implications in HIV dementia. *The European Journal of Neuroscience, 28*, 1255–1264. https://doi.org/10.1111/j.1460-9568.2008.06444.x

Peng, F., Yao, H., Akturk, H. K., et al. (2012). Platelet-derived growth factor CC-mediated neuroprotection against HIV tat involves

TRPC-mediated inactivation of GSK 3beta. *PLoS One, 7*, e47572. https://doi.org/10.1371/journal.pone.0047572

Peng, F., Yao, H., Bai, X., et al. (2010). Platelet-derived growth factor-mediated induction of the synaptic plasticity gene arc/Arg3.1. *The Journal of Biological Chemistry, 285*, 21615–21624. https://doi.org/10.1074/jbc.M110.107003

Peterson, D. A., Lucidi-Phillipi, C. A., Murphy, D. P., et al. (1996). Fibroblast growth factor-2 protects entorhinal layer II glutamatergic neurons from axotomy-induced death. *The Journal of Neuroscience: The Official Journal of the Society for Neuroscience, 16*, 886–898. https://doi.org/10.1523/JNEUROSCI.16-03-00886.1996

Poon, W. W., Carlos, A. J., Aguilar, B. L., et al. (2013). Beta-amyloid (Abeta) oligomers impair brain-derived neurotrophic factor retrograde trafficking by down-regulating ubiquitin C-terminal hydrolase, UCH-L1. *The Journal of Biological Chemistry, 288*, 16937–16948. https://doi.org/10.1074/jbc.M113.463711

Qian, X., Davis, A. A., Goderie, S. K., et al. (1997). FGF2 concentration regulates the generation of neurons and glia from multipotent cortical stem cells. *Neuron, 18*, 81–93. https://doi.org/10.1016/s0896-6273(01)80048-9

Raballo, R., Rhee, J., Lyn-Cook, R., et al. (2000). Basic fibroblast growth factor (Fgf2) is necessary for cell proliferation and neurogenesis in the developing cerebral cortex. *The Journal of Neuroscience: The Official Journal of the Society for Neuroscience, 20*, 5012–5023. https://doi.org/10.1523/JNEUROSCI.20-13-05012.2000

Reuss, B., Von Bohlen, & Halbach, O. (2003). Fibroblast growth factors and their receptors in the central nervous system. *Cell and Tissue Research, 313*, 139–157. https://doi.org/10.1007/s00441-003-0756-7

Revilla, S., Ursulet, S., Alvarez-Lopez, M. J., et al. (2014). Lenti-GDNF gene therapy protects against Alzheimer's disease-like neuropathology in 3xTg-AD mice and MC65 cells. *CNS Neuroscience & Therapeutics, 20*, 961–972. https://doi.org/10.1111/cns.12312

Rotwein, P. (1991). Structure, evolution, expression and regulation of insulin-like growth factors I and II. *Growth Factors, 5*, 3–18. https://doi.org/10.3109/08977199109000267

Sanders, V. J., Everall, I. P., Johnson, R. W., et al. (2000). Fibroblast growth factor modulates HIV coreceptor CXCR4 expression by neural cells. HNRC group. *Journal of Neuroscience Research, 59*, 671–679. https://doi.org/10.1002/(SICI)1097-4547(20000301)59:5<671::AID-JNR10>3.0.CO;2-B

Santerre, M., Bagashev, A., Gorecki, L., et al. (2019). HIV-1 Tat protein promotes neuronal dysregulation by inhibiting E2F transcription factor 3 (E2F3). *The Journal of Biological Chemistry, 294*, 3618–3633. https://doi.org/10.1074/jbc.RA118.003744

Shihabuddin, L. S., Ray, J., & Gage, F. H. (1997). FGF-2 is sufficient to isolate progenitors found in the adult mammalian spinal cord. *Experimental Neurology, 148*, 577–586. https://doi.org/10.1006/exnr.1997.6697

Straten, G., Saur, R., Laske, C., et al. (2011). Influence of lithium treatment on GDNF serum and CSF concentrations in patients with early Alzheimer's disease. *Current Alzheimer Research, 8*, 853–859. https://doi.org/10.2174/156720511798192754

Suh, H. S., Lo, Y., Choi, N., et al. (2015). Insulin-like growth factors and related proteins in plasma and cerebrospinal fluids of HIV-positive individuals. *Journal of Neuroinflammation, 12*, 72. https://doi.org/10.1186/s12974-015-0288-6

Suri, D., & Vaidya, V. A. (2013). Glucocorticoid regulation of brain-derived neurotrophic factor: Relevance to hippocampal structural and functional plasticity. *Neuroscience, 239*, 196–213. https://doi.org/10.1016/j.neuroscience.2012.08.065

Suzumura, A., Takeuchi, H., Zhang, G., et al. (2006). Roles of glia-derived cytokines on neuronal degeneration and regeneration. *Annals of the New York Academy of Sciences, 1088*, 219–229. https://doi.org/10.1196/annals.1366.012

Tanaka, J., Horiike, Y., Matsuzaki, M., et al. (2008). Protein synthesis and neurotrophin-dependent structural plasticity of single dendritic spines. *Science, 319*, 1683–1687. https://doi.org/10.1126/science.1152864

Teng, Y. D., Mocchetti, I., Taveira-DaSilva, A. M., et al. (1999). Basic fibroblast growth factor increases long-term survival of spinal motor neurons and improves respiratory function after experimental spinal cord injury. *The Journal of Neuroscience: The Official Journal of the Society for Neuroscience, 19*, 7037–7047. https://doi.org/10.1523/JNEUROSCI.19-16-07037.1999

Tomac, A., Lindqvist, E., Lin, L. F., et al. (1995). Protection and repair of the nigrostriatal dopaminergic system by GDNF in vivo. *Nature, 373*, 335–339. https://doi.org/10.1038/373335a0

Trejo, J. L., Carro, E., & Torres-Aleman, I. (2001). Circulating insulin-like growth factor I mediates exercise-induced increases in the number of new neurons in the adult hippocampus. *The Journal of Neuroscience: The Official Journal of the Society for Neuroscience, 21*, 1628–1634. https://doi.org/10.1523/JNEUROSCI.21-05-01628.2001

Tropea, D., Kreiman, G., Lyckman, A., et al. (2006). Gene expression changes and molecular pathways mediating activity-dependent plasticity in visual cortex. *Nature Neuroscience, 9*, 660–668. https://doi.org/10.1038/nn1689

Tseng, H. C., & Dichter, M. A. (2005). Platelet-derived growth factor-BB pretreatment attenuates excitotoxic death in cultured hippocampal neurons. *Neurobiology of Disease, 19*, 77–83. https://doi.org/10.1016/j.nbd.2004.11.007

Turner, C. A., Akil, H., Watson, S. J., et al. (2006). The fibroblast growth factor system and mood disorders. *Biological Psychiatry, 59*, 1128–1135. https://doi.org/10.1016/j.biopsych.2006.02.026

Vescovi, A. L., Reynolds, B. A., Fraser, D. D., et al. (1993). bFGF regulates the proliferative fate of unipotent (neuronal) and bipotent (neuronal/astroglial) EGF-generated CNS progenitor cells. *Neuron, 11*, 951–966. https://doi.org/10.1016/0896-6273(93)90124-a

Wang, W. F., Li, S. M., Ren, G. P., et al. (2015). Recombinant murine fibroblast growth factor 21 ameliorates obesity-related inflammation in monosodium glutamate-induced obesity rats. *Endocrine, 49*, 119–129. https://doi.org/10.1007/s12020-014-0433-5

Wenzel, E. D., Avdoshina, V., & Mocchetti, I. (2019). HIV-associated neurodegeneration: Exploitation of the neuronal cytoskeleton. *Journal of Neurovirology, 25*, 301–312. https://doi.org/10.1007/s13365-019-00737-y

Wilcox, B. J., & Unnerstall, J. R. (1991). Expression of acidic fibroblast growth factor mRNA in the developing and adult rat brain. *Neuron, 6*, 397–409. https://doi.org/10.1016/0896-6273(91)90248-x

Wilk, A., Urbanska, K., Yang, S., et al. (2011). Insulin-like growth factor-I-forkhead box O transcription factor 3a counteracts high glucose/tumor necrosis factor-alpha-mediated neuronal damage: Implications for human immunodeficiency virus encephalitis. *Journal of Neuroscience Research, 89*, 183–198. https://doi.org/10.1002/jnr.22542

Williams, B. P., Park, J. K., Alberta, J. A., et al. (1997). A PDGF-regulated immediate early gene response initiates neuronal differentiation in ventricular zone progenitor cells. *Neuron, 18*, 553–562. https://doi.org/10.1016/s0896-6273(00)80297-4

Williams, L. R., Inouye, G., Cummins, V., et al. (1996). Glial cell line-derived neurotrophic factor sustains axotomized basal forebrain cholinergic neurons in vivo: Dose-response comparison to nerve growth factor and brain-derived neurotrophic factor. *The Journal of Pharmacology and Experimental Therapeutics, 277*, 1140–1151.

Yan, Q., Matheson, C., & Lopez, O. T. (1995). In vivo neurotrophic effects of GDNF on neonatal and adult facial motor neurons. *Nature, 373*, 341–344. https://doi.org/10.1038/373341a0

Yang, B., Akhter, S., Chaudhuri, A., et al. (2009a). HIV-1 gp120 induces cytokine expression, leukocyte adhesion, and transmigration across the blood-brain barrier: Modulatory effects of STAT1 signaling.

Microvascular Research, 77, 212–219. https://doi.org/10.1016/j.mvr.2008.11.003
Yang, J., Siao, C. J., Nagappan, G., et al. (2009b). Neuronal release of proBDNF. *Nature Neuroscience, 12*, 113–115. https://doi.org/10.1038/nn.2244
Yang, L., Chen, X., Hu, G., et al. (2016). Mechanisms of platelet-derived growth factor-BB in restoring HIV tat-cocaine-mediated impairment of neuronal differentiation. *Molecular Neurobiology, 53*, 6377–6387. https://doi.org/10.1007/s12035-015-9536-0
Yao, H., Bethel-Brown, C., Niu, F., et al. (2014). Yin and Yang of PDGF-mediated signaling pathway in the context of HIV infection and drug abuse. *Journal of neuroimmune pharmacology: the official journal of the Society on NeuroImmune Pharmacology, 9*, 161–167. https://doi.org/10.1007/s11481-013-9481-5
Yao, H., Duan, M., & Buch, S. (2011). Cocaine-mediated induction of platelet-derived growth factor: Implication for increased vascular permeability. *Blood, 117*, 2538–2547. https://doi.org/10.1182/blood-2010-10-313593
Yao, H., Duan, M., Yang, L., et al. (2012). Platelet-derived growth factor-BB restores human immunodeficiency virus Tat-cocaine-mediated impairment of neurogenesis: Role of TRPC1 channels. *The Journal of Neuroscience: The Official Journal of the Society for Neuroscience, 32*, 9835–9847. https://doi.org/10.1523/JNEUROSCI.0638-12.2012
Yao, H., Peng, F., Fan, Y., et al. (2009). TRPC channel-mediated neuroprotection by PDGF involves Pyk2/ERK/CREB pathway. *Cell Death and Differentiation, 16*, 1681–1693. https://doi.org/10.1038/cdd.2009.108
Yasuda, T., & Mochizuki, H. (2010). Use of growth factors for the treatment of Parkinson's disease. *Expert Review of Neurotherapeutics, 10*, 915–924. https://doi.org/10.1586/ern.10.55
Ying Wang, J., Peruzzi, F., Lassak, A., et al. (2003). Neuroprotective effects of IGF-I against TNFalpha-induced neuronal damage in HIV-associated dementia. *Virology, 305*, 66–76. https://doi.org/10.1006/viro.2002.1690
Yoshimura, S., Takagi, Y., Harada, J., et al. (2001). FGF-2 regulation of neurogenesis in adult hippocampus after brain injury. *Proceedings of the National Academy of Sciences of the United States of America, 98*, 5874–5879. https://doi.org/10.1073/pnas.101034998
Zechel, S., Werner, S., Unsicker, K., et al. (2010). Expression and functions of fibroblast growth factor 2 (FGF-2) in hippocampal formation. *The Neuroscientist: A Review Journal Bringing Neurobiology, Neurology and Psychiatry, 16*, 357–373. https://doi.org/10.1177/1073858410371513
Zhao, L., Huang, Y., Tian, C., et al. (2012). Interferon-alpha regulates glutaminase 1 promoter through STAT1 phosphorylation: Relevance to HIV-1 associated neurocognitive disorders. *PLoS One, 7*, e32995. https://doi.org/10.1371/journal.pone.0032995
Zhao, Y., Haney, M. J., Gupta, R., et al. (2014). GDNF-transfected macrophages produce potent neuroprotective effects in Parkinson's disease mouse model. *PLoS One, 9*, e106867. https://doi.org/10.1371/journal.pone.0106867
Zhu, X., Yao, H., Peng, F., et al. (2009). PDGF-mediated protection of SH-SY5Y cells against Tat toxin involves regulation of extracellular glutamate and intracellular calcium. *Toxicology and Applied Pharmacology, 240*, 286–291. https://doi.org/10.1016/j.taap.2009.06.020
Zuccato, C., & Cattaneo, E. (2009). Brain-derived neurotrophic factor in neurodegenerative diseases. *Nature Reviews. Neurology, 5*, 311–322. https://doi.org/10.1038/nrneurol.2009.54

Specific Role of Extracellular Vesicles in Controlling Neuroinflammation in Alzheimer's and Parkinson's Disease: Is an EV-Based Therapy Possible?

16

Syamantak Ghosh, Sourav Hom Choudhury, Kamalika Mukherjee, and Suvendra N. Bhattacharyya

Abstract

Neurodegenerative diseases are associated with acute or chronic inflammation and death of neurons initiated by inflammatory cytokines and chemokines produced by brain resident glial or infiltrated immune cells in the affected brain parts. The cross-communications between neurons and glial cells are not static or mediated only by cell–cell interaction, neurotransmitters, or small peptides but are also controlled by small vesicles collectively known as extracellular vesicles or EVs secreted by neuronal and non-neuronal cells. While diverse types of cargo in EVs contribute to the well-being and survival of the neurons, a subset of EVs and their cargo were found to have detrimental effects on neuron–glia communication. EVs from virus-infected cells also determine the complexity of HIV1 infection-associated neurodegeneration by communicating viral proteins and RNAs, as part of EVs secreted by infected cells, to neighboring neurons and glial cells to affect the pathogenesis of HIV-associated neurocognitive dysfunction. We have explored the recent and relevant findings that have suggested significant and specific regulatory contributions of EVs and their cargo in neurodegenerative processes in the mammalian brain. Briefly, we have also pointed out how the future therapeutic applications of EVs may be able to reverse neurodegeneration after the successful resolution of specific challenges related to EV-based therapy.

Keywords

Extracellular vesicles · miRNA · Alzheimer's Disease · Parkinson's Disease · HIV-associated neurocognitive dysfunction (HAND) · HIV pathogenesis · Intercellular communication

Syamantak Ghosh and Sourav Hom Choudhury contributed equally with all other contributors.

S. Ghosh · S. Hom Choudhury
RNA Biology Research Laboratory, Molecular Genetics Division, CSIR-Indian Institute of Chemical Biology, Kolkata, India

K. Mukherjee (✉)
RNA Biology Research Laboratory, Molecular Genetics Division, CSIR-Indian Institute of Chemical Biology, Kolkata, India

Department of Anesthesiology, University of Nebraska Medical Center, Omaha, USA
e-mail: kmukherjee@unmc.edu

S. N. Bhattacharyya (✉)
RNA Biology Research Laboratory, Molecular Genetics Division, CSIR-Indian Institute of Chemical Biology, Kolkata, India

Department of Pharmacology and Experimental Neuroscience, University of Nebraska Medical Center, Omaha, NE, USA
e-mail: sbhattacharyya@unmc.edu

16.1 Introduction

16.1.1 Extracellular Vehicles (EVs) as Carriers of Different Cargo Between Cells and Tissues

The presence of vesicles in proximity to cell membranes in mammalian tissues or body fluids was first detected in the late 1960s (Anderson, 1969; Wolf, 1967), but it was only around 2011 that the umbrella term "extracellular vesicles" (EVs) was suggested to represent all lipid bilayer-encapsulated extracellular structures (Gyorgy et al., 2011). EVs can originate from direct budding out from the plasma membrane or from the end product of trafficking of the endocytic pathway where multivesicular late endosomes or multivesicular bodies (MVBs) fuse with the plasma membrane, as explained in several papers published in the 1980s (Harding et al., 1983; Pan et al., 1985). By this fusion process, intraluminal vesicles (ILVs) of MVBs get released in the extracellular space, leading to the formation of a class of EVs designated "exosomes" (Johnstone et al., 1987) that are similar in size to ILVs (<200 nm in diameter). Initially, it was

H. E. Gendelman, T. Ikezu (eds.), *Neuroimmune Pharmacology and Therapeutics*,
https://doi.org/10.1007/978-3-031-68237-7_16

suggested that the secretion of EVs is the mechanism by which cells eliminate unnecessary cellular proteins. Nevertheless, studies performed in the late 1990s showed that it could have been an intercellular communication mechanism, most notably in the context of immune responses and cancer (Raposo et al., 1996; Wolfers et al., 2001). This concept gained significant momentum in 2007 when it was shown that exosomes contained mRNAs and microRNAs, which retained their function and could alter recipient cells' behavior once taken up (Valadi et al., 2007). Subsequently, functional proteins and RNA have also been shown to be transferred between cells by other kinds of EVs, such as microvesicles, microparticles, or ectosomes, which are presumably released from the plasma membrane but without any endocytic pathway components involved in the process (Valadi et al., 2007) (Fig. 16.1).

Various cell types can engulf EVs, including macrophages, endothelial cells, and tumor cells. EVs are thus capable of transmitting their cargo to neighboring or distant cells once they have been released and can alter the gene expression, signal transduction, and overall functions of target cells (Barres et al., 2010; Skog et al., 2008; Tian et al., 2010). In the immune system, EVs are involved in signaling by which young dendritic cells deliver MHC peptide molecules to another dendritic cell so that they can trigger an immune response (Segura et al., 2005). Sorting and packaging specific mRNA and miRNA into exosomes can be a mode of intercellular communication (Valadi et al., 2007). In particular, it plays an essential role in the "exosomal" transport of immune synapses in the horizontal transfer of miRNA from activated T-lymphocytes toward antigen-presenting cells (Mittelbrunn et al., 2011). Within a recipient cell, the RNA species imported through exosomes may also be copied or translated into cDNA (Skog et al., 2008). EVs from cultured glioblastoma tumor cells contain a range of angiogenic peptides and RNA that can be exported and expressed into the recipient brain microvascular endothelial cells, generating proangiogenic properties, which in turn result in the dissemination of malignancy. In addition, serum EV of patients with Glioblastoma carries a tumor-specific mRNA called EGFRvIII (Skog et al., 2008) (Fig. 16.1).

Further studies have found that EVs from mast cells carry mRNA and miRNAs, and protein translation can occur from exported mRNA (Valadi et al., 2007). These crucial observations demonstrate the heterogeneous role of EVs as disease biomarkers, intercellular signal transducers, modulators, or potential drug delivery vectors (Alvarez-Erviti et al., 2011b). EVs can be capable of targeting cargo to a wide range of cells and sites, and their comparatively non-cytotoxic nature makes them efficient carriers or mediators for intercellular communication. miRNA transfer via EVs can rapidly modify gene expression and control essential processes, for example, growth, differentiation, cellular viability, or immunomodulatory activity (Zhang et al., 2015). In this chapter, we will discuss existing and recent findings that shed light on the role of EVs in the development, prognosis, and probable therapeutics of neurological disorders (Fig. 16.1).

16.2 EV-Mediated Cross-Communication Within Brain Cells

The brain is a very complex tissue to study in terms of anatomy and function. It is only at an early stage to study molecular mechanisms underpinning the development of brain disorders, which means that clinical treatments are limited. Brain diseases impose several social and economic burdens on society (Erkkinen et al., 2018). Due to their critical importance in disease progression and the potential for clinical use, EVs have attracted considerable interest in the study of brain diseases such as Alzheimer's disease, Parkinson's disease (PD), cerebrovascular accident, and traumatic brain injury in recent years. Hence, there has been an upsurge to explore the role and molecular mechanisms of EV-carried proteins such as amyloid precursor protein (APP), alpha-synuclein (α-syn), mutant huntingtin (mHtt), scrapie isoform of the prion protein (PrP^{sc}), which are related to the brain diseases (Hill, 2019; Wang et al., 2017b). In particular, the ability of EVs to move cargo has been a critical factor in disease spreading.

In contrast to conventional therapeutic agents, EVs-carrying drugs are more likely to pass the blood-brain-barrier (BBB), which helps them to reach the target tissues (Jarmalaviciute and Pivoriunas, 2016). EVs have been identified as potentially useful biomarkers of clinical diagnosis and disease prediction based on their widespread and easy availability in an organism and their role in a wide range of biological modulatory effects. Therefore, the specific mechanisms of involvement of EVs in brain disease have been investigated over the last few years by an increasing number of studies (Kalluri and LeBleu, 2020) (Fig. 16.1).

miRNAs, such as let-7c and let-7b, were reported to regulate the expression of the insulin-like growth factor 1(IGF1) gene and were observed to be downregulated significantly in patients having major depressive disorder (MDD) in comparison to healthy people. Moreover, the brain-derived neurotrophic factor (BDNF) signaling was promoted when IGF-1 was overexpressed by relaying an anti-depressant-like property (Gururajan et al., 2021). A higher level of miR-207 has been identified in EVs derived from unstressed mice, as revealed in the microarray analysis of the miRNA profile of EVs. miR-207 was confirmed to target TLR4 interactors with leucine-rich repeats (Tril) and also caused inhibition of NF-κB signaling in astrocytes. miR-207 may reduce the release of pro-inflammatory cytokines and inhibit the expression of Tril. In vivo studies have shown a decrease in the

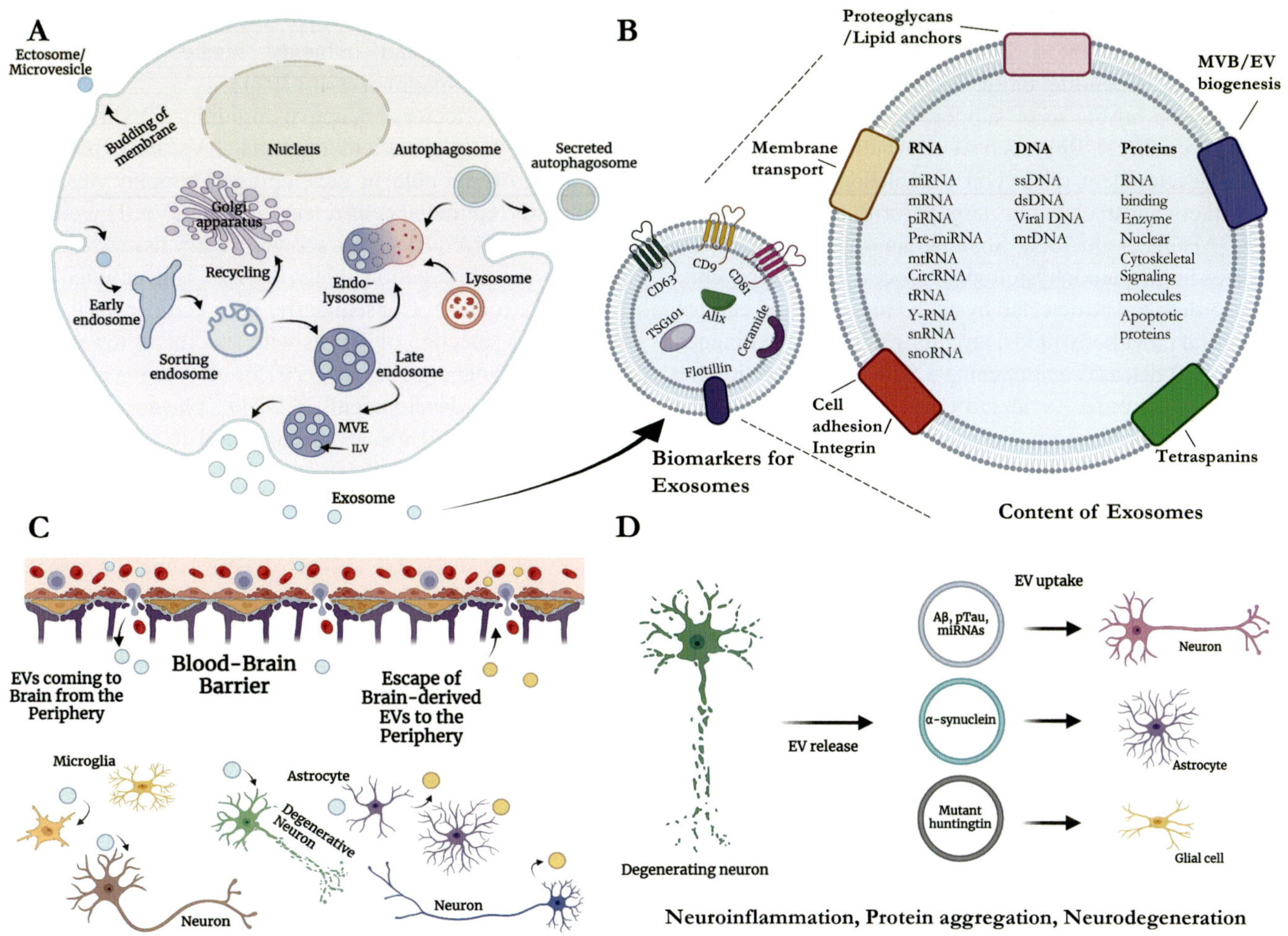

Fig. 16.1 Biogenesis of extracellular vesicles and their role in neurodegenerative diseases. (**a**) Exosomes and ectosomes (or microvesicles) are the two main subtypes of extracellular vesicles. Ectosomes are formed by directly budding off from the plasma membrane, and the exosomes are generated by the formation of multivesicular endosomes (MVEs) that, eventually, fuse with the plasma membrane, ultimately releasing their intraluminal vesicles (ILVs) as exosomes. (**b**) Several proteins can be used as exosome markers (CD9, CD81, CD63, flotillin, TSG101, and Alix). Tetraspanins, integrins, and membrane transporters are among the proteins residing on the exosome surface. Different types of cell surface proteins, intracellular proteins, RNA, DNA, and other molecules have been detected as the cargo of exosomes. (**c**) In physiological conditions, EVs may be released from neurons, astrocytes, oligodendrocytes, and microglia as a means of intercellular communication. However, neurotoxic proteins such as the infectious prion protein (PrP^{Sc}) or amyloid-β (Aβ) can also be transported through EVs, which can spread pathological proteins to proximal healthy neurons to spread the neurodegenerative diseases such as Alzheimer's disease (AD) and prion disease. Brain EVs contain proteins and miRNAs derived from the brain that could break down the blood–brain barrier, perhaps releasing them into the periphery where they can be isolated for use as biomarkers of neurodegenerative diseases. (**d**) Brain-derived exosomes, or EVs, are pivotal in escalating neurodegenerative diseases. Degenerating neurons can release EVs containing various neurotoxic proteins that healthy neuronal cells take up, leading to neuroinflammation, aggregated protein accumulation, and neurodegeneration (Image created with the help of BioRender)

antidepressant activity of EVs with low miR-207 levels (Li et al., 2020). After a cerebrovascular accident or stroke, glial scars are formed. In the M2 microglia, the formation of such glial scars was reduced by miR-124 present in EVs. The mechanism, signal transducer, and activator of transcription 3 (STAT3) level were downregulated by miR-124, which promotes recovery by proliferation and migration of astrocytes (Jiang et al., 2022; Xiong et al., 2022). In the model system of Parkinson's disease (PD), overexpression of the circular RNA–CircSV2b in EV was found to lower oxidative stress injury, maintain the function of the nigrostriatal pathway, protect the dopaminergic neurons, and recover motor defects by acting on miR-5107-5p-FOXK1 (fork-head box protein K1)-AKT1 (serine/threonine protein kinase) signaling pathways (Cheng et al., 2022). Apoptosis-related genes (SIRT1 and BCL2L11) are targeted by miR-215 and regulate the AMPK (adenosine 5′-monophosphate activated protein kinase)-SIRT1 (silent message regulatory factor 1)-TFEB (transcription factor EB) signaling pathway, thus preventings neuronal apoptosis. Upregulation of miR-215 is induced by exercise and probably has a preventive effect on Alzheimer's disease (AD) (Kaur et al., 2021).

However, depending on the biological signals they carry, EVs may be hostile to or supportive of the brain function (Kaur et al., 2021). Reports have shown that EVs play a role in the development of several encephalopathies, acting as biomarkers and therapeutic targets for diagnosis and treatment (Aheget et al., 2020; Shaimardanova et al., 2020). The changes in the pathophysiological process of neuropsychiatric disorders can be detected by analyzing EVs from cerebrospinal and other body fluids, providing new ways to diagnose and predict diseases and opening a new perspective in pharmacology. In order to identify and predict disease, EVs derived from damaged nerve cells that cross the blood-brain barrier and reach the peripheral circulation can be used as biomarkers. At the same time, the contents of the EVs can be modified, or therapeutic agents can be packaged into the EVs for targeted delivery. EVs play a role in the pathophysiological processes of neuropsychiatric disease, as they are a natural component of body fluids that are able to cross the blood-brain barrier. In conclusion, modified EVs in disease contexts may be used as possible biomarkers for diagnosing diseases or can act as a non-cytotoxic vector to deliver potential pharmaceutical compounds.

16.3 EV as a Modulator of the Immune System

EVs, which also play a role in the inflammatory process, are secreted by various immune cells that participate in inflammation. As a response by the innate immune system, EVs may have chemotactic effects and are also observed to carry arachidonic acid-derived bioactive lipid mediators such as eicosanoids and enzymes involved in their production. In addition, EVs derived from neutrophils transport arachidonic acid to platelets, which in turn use cyclooxygenase 1 (COX1) to produce thromboxane A2 from arachidonic acid and induce neutrophil extravasation (Rossaint et al., 2016). Furthermore, the enzyme that metabolizes arachidonic acid, platelet-type 12-lipoxygenase (12-LO), is exported by platelet-derived EVs. The EV-associated 12-LO, along with secreted phospholipase A1 type IIA, produces 12-hydroxyeicosatetraenoic acid from arachidonic acid that induces uptake of platelet-derived EVs by neutrophils, and this phenomenon is possibly involved in the inflammatory arthritis (Duchez et al., 2015). Wide arrays of reports show that EVs play pro-inflammatory roles by transferring various effector molecules (cytokines, bioactive lipids, etc.), stress signals, enzymes, and RNA involved in the cytokine production pathway, and several other mediators, as an important section of innate immunity. Moreover, EVs have also been observed to have anti-inflammatory effects in specific scenarios. So, EVs have a pleiotropic role depending on the context to modulate and influence innate immunity and inflammation (Burgelman et al., 2021).

As a major effector of adaptive immunity, EVs' roles have been highlighted in various contexts. EVs are expected to play a significant role in developing T-lymphocytes, with most thymic epithelial cells releasing their EVs. Thymic epithelial cell-derived EVs were shown to carry tissue-restricted antigens for presentation in the thymic conventional dendritic cells (cDCs). Consequently, EVs could contribute to the negative selection of T-cells with specificity toward self-antigens (Skogberg et al., 2015). In the context of B-lymphocyte development, CD24+ plasma-membrane-derived EVs (microvesicles) were found to be released by immature primary B-cells from bone marrow (Ayre et al., 2015). In 1996, it was observed that EVs are also capable of antigen-presentation, and this was the first milestone discovery that showed the possibility of EVs being an essential player in adaptive immunity (Raposo et al., 1996). Cross-presentation of exogenous antigens plays an intriguing role in immunity against viruses and tumors. Cross-presentation is done on MHC-I complexes to CD8+ T-cells, and it also generates an immune response upon vaccination and tolerance induction. EVs that carry peptide-MHC (pMHC) Class-I from pulsed cDCs could effectively activate naïve CD8+ T-cells (Andre et al., 2004). In addition, an association between the cross-priming of tumor-specific CD8+ T cells and the synaptic transfer of vesicles between donor and recipient DCs via close and sustained cell lines has been identified. Eventually, recent publications show that platelet-derived EVs (microvesicles) can work as an integral part of antigen presentation (Balan and Bhardwaj, 2020). EVs are also known to play a role in the functions of immune synapses, which form between lymphocytes and APC.

Additionally, the T-cells' surface microvilli are located at contact interfaces with the APCs and were initially assumed to be structures for screening purposes or attachment on surfaces of APCs. Recently, however, it has been demonstrated that these microvilli are fragmented into APC-enriched T-cell microvilli, known as T-cell microvilli particles (TMPs), deposited on the surface of APCs. TMPs were therefore proposed to constitute "immunological synaptosomes" and, to control the activation of DCs with minimum bystander effect, they have been suggested to allow for a specific and rapid transport of cargo molecules between physically interacting T-cells and DCs (Kim et al., 2018). It is also possible to establish antigen presentation synapses with T-cells by B-cells. EV-mediated transfer of miRNAs via synapses, such as miR-20a-5p, miR-25-3p, and miR-155-3p, was reported from T-cells to B-cells, which results in downregulation of phosphatase and tensin homolog (PTEN) and Bcl-2-interacting mediator of cell death (BIM) genes in B-cells. Apart from forming antigen-presentation synapses with

T-cells, B-cells have also been created in conjunction with follicular dendritic cells (FDCs) to capture and process antigens.

Several molecules known to play an important role in immunoregulation on the surface of EVs have been identified. Such molecules include immune-checkpoint molecules like cytotoxic T lymphocyte antigen-4 (CTLA4), programmed death ligand-1 (PDL1), and apoptosis-inducing ligand FASL. Regulatory T (T-reg) cells were reported to export EVs that play a part in the immunosuppressive activity of these cells by expression of CD73 on the surface of EVs as one of the mechanisms. Indeed, CD73 has been demonstrated to play a vital role in the immune function of EVs secreted by T-reg cells by producing adenosine (Smyth et al., 2013).

In this context, a highlight of some of the most significant recent developments in understanding how EVs regulate the immune system has been discussed. With the immune system's complexity in mind, it seems probable that there will also be many more discoveries related to EV and the development of novel biomarkers or therapeutic platforms based on EV.

16.4 Brain Inflammation and Possible Roles of EVs

In view of the ability of EVs to cross the blood-brain-barrier (BBB) and serve as cargo carriers for bidirectional communication between the CNS and the periphery, the function of EVs in the pathogenesis of central nervous system (CNS) disorders and the development of inflammation is of paramount importance for the diagnosis, prognosis, and treatment of CNS disorders.

Multiple sclerosis (MS) is a disease of the CNS affecting more than 2 million people globally, which causes chronic inflammation by demyelination of neuronal cells. EVs isolated from the serum/plasma and cerebrospinal fluid (CSF) have been shown to be a valuable source of biomarkers of disease prognosis and therapeutic effects in MS. The circulating EVs of patients with MS show a major change in myelin-related protein and miRNA profiles. miRNAs such as miR-122-5p, miR-196b-5p, miR-301a-3p, and miR-532-5p were specifically decreased; on the other hand, levels of miR-Let-7i were found to be increased in serum EVs of MS patients compared to healthy controls (Kimura et al., 2018). A widely used animal model to study MS is experimental autoimmune encephalomyelitis (EAE). As a matter of fact, microglia-secreted EVs abundantly carry pro-inflammatory cytokines such as interleukin (IL)-1b and miRNAs involved in the dysfunction of synapses. Intriguingly, EVs derived from microglia act in deleterious and protective roles in re-myelination, which the microglial phenotype decides. Pro-inflammatory microglia secreted EVs induce synaptic defects by hindering re-myelination, while myelin repair is promoted by EVs that were found to carry molecules like anandamide and sphingosine 1 phosphate (S1P) produced by pro-regenerative microglia (Lombardi et al., 2019). CNS infiltrating immune cells, particularly T-cells and myeloid cells, are considered to cause neuropathology in MS and EAE. EVs secreted by endothelial cells and platelets obtained from the blood serum of MS patients were reported to induce leakage of the BBB, which leads to T-cells and myeloid cells invading the CNS (Marcos-Ramiro et al., 2014).

A gradual irreversible loss of neurons and widespread neuroinflammation is characteristic of neurodegenerative disorders such as Alzheimer's disease (AD), amyotrophic lateral sclerosis (ALS), frontotemporal dementia (FTD), Creutzfeldt-Jakob disease (CJD), Huntington's disease (HD), and Parkinson's disease (PD). Neurodegenerative diseases are also known as proteinopathies because they show abnormal accumulation of proteins in the central nervous system. These misfolded and aggregated proteins lose their normal physiological functions, form insoluble aggregates or inclusions, and become neurotoxic. Such proteins are β-amyloid (Aβ), α-synuclein, huntingtin (Htt), tau, prion protein (PrP), fused in sarcoma (FUS), and transactive response DNA-binding protein 43 kDa (TDP-43). It is interesting that in patients with relevant neurodegenerative diseases, the presence of Aβ, α-synuclein, FUS, tau, and TDP-43 has been observed in brain exosomes, cerebral spinal fluid, or plasma, suggesting that EVs may have some role to play in neurodegenerative disorders. Useful information for diagnosing neurodegenerative disorders may be obtained from the analysis of blood exosomal Aβ and synaptic proteins, such as neurogranin, GAP43, or SNAP25 (Lim et al., 2019).

The distribution of misfolded proteins throughout specific brain regions, which indicates intercellular movement of these proteins contributing to disease processes, is also a common characteristic of neurodegenerative diseases. PrP^{Sc}, the misfolded and transmittable form of PrP, is the crucial constituent of the proteinaceous infectious particle designated as "prion" that leads to CJD, an infectious neurodegenerative disease in humans (Guo et al., 2015).

Local and distant neurons take in the exosomes containing toxic proteins that contribute to neuron loss, suggesting a mechanism for their spread into the neurodegenerative brain. Moreover, propagation of toxic proteins can also be mediated by microglia other than neurons, and activation of microglia is one of the hallmarks of neurodegenerative diseases, and they have been implicated as contributors to the spread of Aβ, tau, and α-synuclein via EVs. When chemical inhibitors block EV synthesis, a mouse model study lowered amyloid plaque formation and tau propagation. Notably, EVs enriched with tau protein have been observed to initiate aggregation of misfolded tau proteins and generate tauopathies in recipient mice, suggesting a central role of EVs in neurodegenerative disease propagation (Wang et al., 2017a). EVs can play an essential role in removing toxic cellular content, supporting neuronal function, and facilitating pathological protein propagation. Glycosphingolipids present on neurons allow the packaging of Aβ in EVs, and by phagocytosis, the Aβ content is eventually degraded by microglia, hinting that neurons and microglia carry out an interdependent mechanism to dispose of Aβ using EVs.

In neurodegenerative diseases, exosomes play both a protective and a detrimental role. Microglia cell-derived EVs are a critical switch and significant stimulator for neurodegenerative diseases. EVs facilitate the removal of protein aggregates by homeostatic microglia that can clear these proteins. Moreover, neurodegenerative microglia tend to incorporate aggregated proteins into EVs and spread them to support disease propagation under conditions where protein clearance capacity is impaired. Hence, enhancing the microglia's ability to clear protein could be a beneficial mechanism for treating neurodegenerative disease by altering the EV phenotype (Delpech et al., 2019).

16.5 Blood Brain Barrier and EVs

For the last decade or so, EVs have emerged as an important pharmacological and therapeutic agent in pre-clinical models of different brain diseases such as AD, PD, stroke, and other traumatic brain injuries. (Rufino-Ramos et al., 2017) However, whether these EVs can cross the biological blood-brain barrier (BBB) in a bidirectional manner to affect both the brain cells and peripheral tissue is still an underexplored area of research. BBB is a selective semi-permeable membrane, mainly composed of brain epithelial cells and tight junctions, that which acts as an interface to regulate the exchange of different micro and macro-molecules between the brain and blood circulation (Banks, 2016). Depending on their size and lipid solubility, different molecules can selectively pass through the BBB via passive diffusion or active transport. It has been reported that exosomes are able to cross BBB, but the mechanism behind it has not yet been fully understood. Evidence suggests that similar to immune cells and viral particles, exosomes also cross the BBB primarily by transcytosis, that is, vesicle-mediated transport across the cells (Banks et al., 2020). This transcytosis process takes place in three parts: first, the exosomes are taken up by the apical membrane of brain epithelial cells by endocytosis; secondly, the exosomal vesicles are subjected to intracellular transport, and finally, they are released from the basolateral side of the membrane. Studies reveal that larger exosomes accumulate on the apical/luminal side. In contrast, smaller exosomes accumulate on the basolateral/abluminal side of the brain epithelial cell membrane, suggesting that smaller the size of the exosome higher the efficiency of transport across the BBB (Chen et al., 2016).

Recent studies have also reported that exosomes can even affect the integrity of the BBB membrane. For example, exosomes released from breast cancer cells contain miR-105 that directly targets and represses the expression of tight junction protein ZO-1 to alter the permeability of the BBB (Zhou et al., 2014). A leaky BBB is one of the major reasons associated with neuroinflammation. As exosomes and exosome-derived factors alter the BBB membrane integrity, it may also indicate that exosomes have a role to play in neuroinflammation and the pathogenesis of neurodegenerative diseases and other mental disorders (Maes, 2008).

16.6 EVs and Alzheimer's Disease

Alzheimer's disease (AD) is a progressive age-related neurodegenerative disorder that is characterized by a variety of neuropsychiatric symptoms that finally lead to partial/complete cognitive impairment (Joe & Ringman, 2019). AD is the fundamental cause of 70% of cases of dementia, and between 1990 and 2016, the worldwide number of prevalent dementia cases showed an increase of 117% from 20·2 million to 43·8 million (Livingston et al., 2020). Genetically, AD is divided into two types: familial with Mendelian inheritance with an early onset of the disease and sporadic, where genetic and environmental risk factors contribute, and patients show a relatively late onset of the disease. Only about 5% of all global AD cases are familial due to a highly penetrant but rare mutation in one of these three genes: Amyloid Precursor Protein, Presenilin 1, and Presenilin 2. Other 95% of the AD cases are sporadic, where genetic factors and even environmental factors play a crucial role. The main genetic factor associated with sporadic AD is the epsilon allele of the ApoE gene (Piaceri et al., 2013). Though the real molecular basis behind the onset of Alzheimer's disease is still not completely understood, two main pathological hallmarks that have been identified are senile plaques, which consist of amyloid fibrils made of amyloid beta (Aβ) peptide and neurofibrillary tangles that are composed of hyperphosphorylated tau protein (Selkoe, 1991).

At present, there is no such ultimate cure for Alzheimer's disease. However, there are available treatments that just improve the symptoms by targeting different factors to delay the ongoing progression of the disease. The most common treatments include anti-amyloid therapy, anti-neuroinflammation therapy, anti-tau therapy, neuroprotection, and cognitive enhancers (Huang et al., 2020). As mentioned earlier, exosomes are increasingly recognized as a crucial communication vehicle involved in the pathogenesis of different neurodegenerative diseases such as Alzheimer's and Parkinson's. Evidence suggests exosomes can carry Aβ, tau proteins, α-synuclein, and other neuroinflammatory factors across the BBB. Exosomes spread these proteins in the interconnected neurons mainly by endosomal pathway or axonal transport (Polanco et al., 2018).

Moreover, in the senile plaques of patients with AD, significant enrichment of Aβ and exosomal protein (Alix) was found that in AD patients, neuronal cells-derived exosomes may contain Aβ peptide and transport them in the brain (Rajendran et al., 2006). In addition, recent studies also showed that EVs isolated from the cerebrospinal fluid (CSF) of advanced-stage AD patients are significantly more enriched with hyperphosphorylated tau (p-tau) protein as compared to exosomes isolated from early-stage AD patients. This finding suggests that exosome-mediated transport of p-tau proteins may play a key role in promoting neurofibrillary tangles formation in AD patients' brains (Saman et al., 2012). Accumulation of Aβ and p-tau proteins is associated with continuous activation of microglia and astrocytes which promotes neuroinflammatory response. Additionally, miRNAs are well-known exosomal/EV cargo that also induces neuroinflammation. Neuron-derived EVs contain miR-21-5p that can be engulfed by microglia and induce their M1 polarization, which promotes the release of neuroinflammatory factors, accumulation of p-tau protein, and apoptosis of PC12 cells and also inhibits neurite outgrowth (Yin et al., 2020).

On the contrary, neuron- and microglia-derived exosomes have also been associated with neuroprotective function. These exosomes were found to bind Aβ oligomers via their surface proteins and induce degradation or clearance of the beta-amyloid from the brain cells of AD patients to rescue long-term potentiation from Aβ-mediated impairment in vivo (Yuyama et al., 2012). Additionally, exosomes were found to be involved in the anti-inflammatory response. Under certain conditions, microglia-derived exosomes contain miR-124-3p that induces

M2 polarization of microglia, which, in turn, inhibits neuronal inflammation and promotes neurite outgrowth during traumatic brain injury (Fig. 16.2) (Huang et al., 2018).

It has already been mentioned that exosomes present in the bodily fluid can carry different factors such as Aβ, p-tau, and several miRNAs across the BBB bidirectionally under diseased conditions. Therefore, detecting specific protein factors and/or miRNAs by simple molecular biology techniques in the exosomes or EVs isolated from peripheral body fluids can be considered an excellent biomarker-based method in diagnosing early-stage AD and other neurodegenerative diseases. Combined detection of $A\beta_{1-42}$ and p-tau compared to individual detection in CSF-derived EVs significantly enhances the sensitivity and specificity and might also help in the early diagnosis of AD before the clinical onset (Schoonenboom et al., 2004). miRNAs are very easily detected as they remain stable in the exosomal vesicles found in the CSF and other extracellular body fluid, and reports suggest that as compared to the healthy individual, AD patients show alterations in the expression profiles of miRNAs and their target genes (Lugli et al., 2015). Therefore, detecting the altered miRNA expression pattern might provide better insight into preclinical and clinical studies. Compared to traditional candidate vectors for gene therapy, EVs, due to their RNA and protein-carrying capacity, stable presence in body fluids, and capability of crossing the BBB, show a more significant advantage in therapeutic effect against AD pathogenesis (Fig. 16.2) (Liu et al., 2013).

16.7 EVs and Parkinson's Disease

Parkinson's disease (PD) is the second most common neurodegenerative disorder worldwide after AD that is characterized by a progressive loss of motor function due to the loss of dopaminergic neurons in the substantia nigra pars compacta (SNpc) and accumulation of misfolded α-synuclein in the brain (Balestrino & Schapira, 2020). According to the data provided by Perkinson's Foundation, nearly 10 million people worldwide are suffering from this disease, and only the U.S. contributes for 1/10th of this number. As mentioned above, the main pathological hallmarks associated with PD are the progressive loss of pigmented dopaminergic neurons in the SNpc and the aggregation of misfolded α-synuclein in the cytoplasmic Lewy bodies (LBs). Patients suffering from PD experience primary motor symptoms such as tremors, rigidity, bradykinesia, and postural instability only after a significant amount of dopaminergic neurons have already been lost. Hence, early disease diagnosis is very difficult (El-Agnaf et al., 2006). Neurons of substantia nigra communicate with the neurons of the basal ganglia through the neurotransmitter Dopamine to control the motor functions. The loss of dopaminergic neurons leads to reduced production of dopamine in the brain. This lowering of dopamine results in gamma-aminobutyric acid (GABA) dysfunction, leading to inhibition of the thalamus. This inactive thalamus fails to communicate with the frontal cortex, resulting in PD's decreased motor activity characteristic (DeMaagd & Philip, 2015). LBs are characterized as intracellular cytoplasmic inclusions enriched with α-synuclein and ubiquitin. A pathogenic mutation in the α-synuclein coding gene–SNCA confers a greater tendency for the protein to misfold, become insoluble, and form β-sheet-rich amyloid aggregates in the brain that are assumed to be toxic for dopaminergic neurons (Spillantini et al., 1997). Currently, the treatment of PD is symptomatic and involves dopamine replacement therapy with Levodopa, using Dopamine agonists and others that are primarily focused on the dopaminergic pathway, but there are no disease-modifying treatments for PD until to date (Yuan et al., 2010).

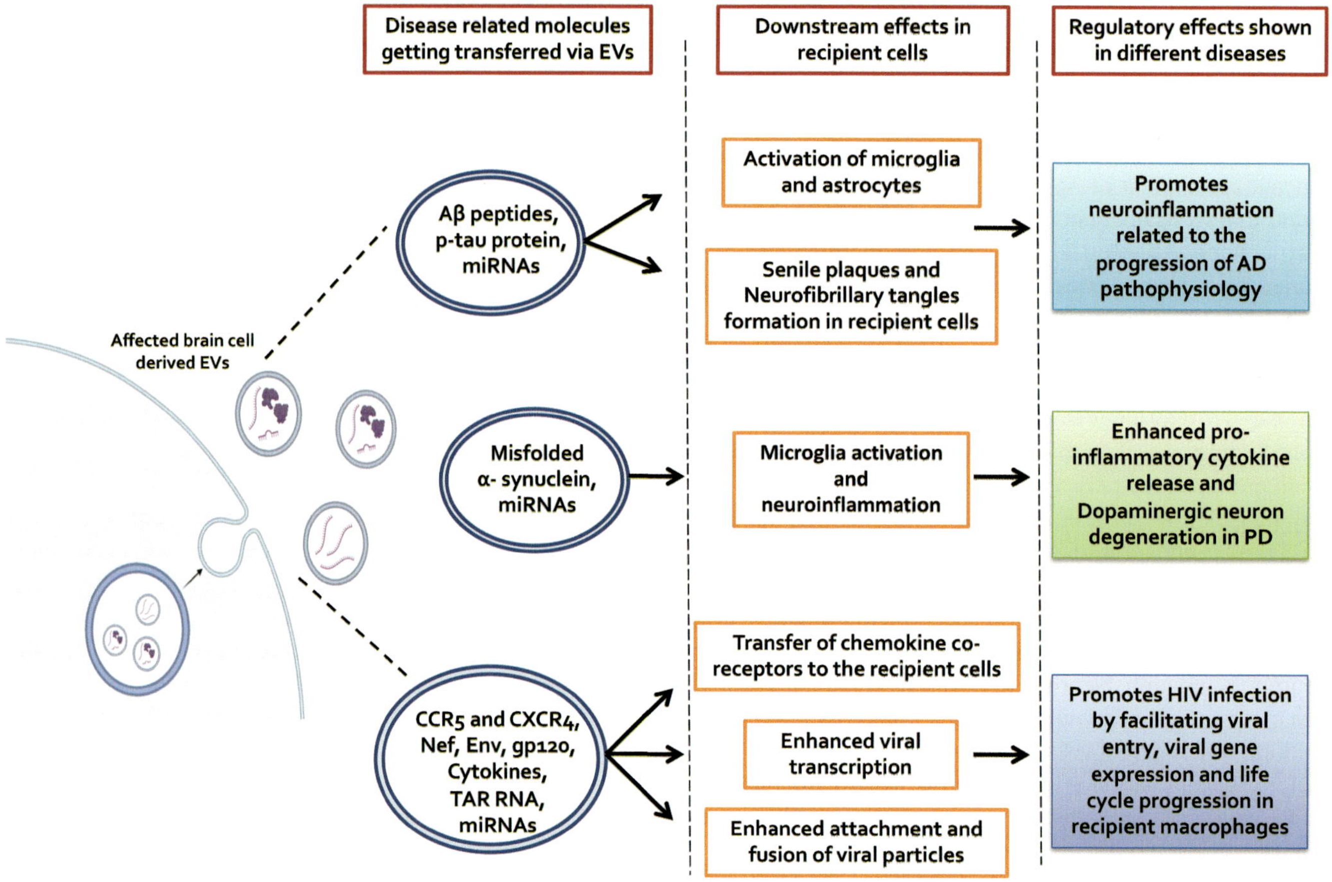

Fig. 16.2 Schematic outline to show the role of EVs in regulating the pathophysiology of AD, PD, and HIV infection. Exosomes released from the diseased cells alter the BBB membrane integrity, and different exosome-derived factors play a crucial role in neuroinflammation, neurodegenerative diseases, and other diseases. Moreover, infected cell-derived exosomes or EVs often transfer proteins, RNAs, miRNAs, and other factors to the noninfected recipient cells to control several downstream cellular signaling pathways, hence orchestrating the progress of the disease. EV-associated factors often activate different brain cells, such as microglia and astrocytes, leading to neuroinflammation. Apart from neurodegenerative diseases, even virus-infected cell-derived EVs carry host and viral proteins and RNAs, making the neighboring non-infected recipient cells more susceptible to viral entry and infection. EVs, due to their RNA and protein carrying capacity, stable presence in body fluids, and capability of crossing the BBB, show a great potential therapeutic effect against neurodegenerative disease pathogenesis

Studies have shown that α-synuclein is capable of intercellular transmission, which strengthens another hypothesis related to the pathophysiology of PD where the olfactory bulb and the gut are considered as the initial spreading sites for misfolded α-syn protein (Klingelhoefer & Reichmann, 2015). EVs in the extracellular fluids play a key role in the pathogenesis of PD by mediating cell-to-cell transmission of α-syn protein and intercellular transport of other PD-implicated miRNAs. Exosomes have been reported to carry α-syn protein from the damaged neuron to another nearby healthy neuron, which can form toxic aggregates and induce cell death. Evidence suggests that these EV-associated α-syn oligomers show higher cellular uptake and increased neurotoxicity than non-EV-associated oligomers (Alvarez-Erviti et al., 2011a). It has also been reported that EVs derived from the CSF of PD patients promote misfolded α-syn aggregation. When these α-syn-containing EVs were injected into the striatum of mice, they were found to induce dopaminergic neuron degeneration and loss of motor functions (Guo et al., 2020). Microglia play a significant role in degrading the α-syn and removing the aggregates from the affected part of the brain by phagocytosis. Both in vivo and in vitro studies have shown that α-syn containing exosomes can induce microglia activation and subsequent microglia-mediated neuroinflammation by releasing pro-inflammatory cytokines. Moreover, activated microglia-derived exosomes are responsible for the transport of these cytokine molecules from glia-to-glia or glia-to-neuron, which leads to dopaminergic neuron degeneration in PD patients. The vicious cycle continues as the damaged neurons release α-syn-containing EVs that activate microglia, repeating the process (Bruck et al., 2016) (Table 16.1).

Apart from mediating intercellular α-syn transmission in PD, EVs also contribute to PD pathogenesis as they transfer miRNAs between cells. A plethora of work has been done to identify a wide range of differentially expressed circulating

EV-miRNAs associated with PD. Some miRNAs target PD-related genes, such as miR-7, which binds to the 3′-untranslated region (UTR) of SNCA mRNA to inhibit its expression; thus, loss of miR-7 leads to increased α-syn production, aggregation, and dopaminergic neuron loss in PD-patient brains (McMillan et al., 2017). miR-133b was found to regulate the maturation and function of midbrain dopaminergic neurons, and in PD patients, there is a significant reduction in miR-133b expression level. In other studies, EV-associated miRNAs isolated from serum or CSF samples of PD patients showed specific differential expression compared to control healthy individuals. For example, miR-19b and miR-1 levels showed a reduction, whereas the levels of miR-195, miR-24, miR-153, miR-409-3p, miR-10a-5p, and let-7 g-3p were significantly upregulated in PD patients (Gui et al., 2015).

Studies have shown that EVs may have therapeutic potential as treatment vehicles and can also be considered a diagnostic tool for monitoring specific protein and miRNA levels in the diseased brain. The amount of α-synuclein contained within brain-derived EVs isolated from the body fluid can be a biomarker of early-stage Parkinson's disease (PD). The levels of EV-associated α-synuclein were significantly higher in PD patients than in the control individuals, and the abundance of α-synuclein in EV showed a gradual increase with disease progression (Niu et al., 2020). Similarly, specific miRNAs are differentially expressed in CSF-derived exosomes of PD patients that could also be evaluated as a potential diagnostic biomarker. However, despite the clinical potential of using EVs as drug delivery vehicles or diagnostic tools, some limitations and concerns must be considered before going to the next step.

16.8 EVs in Control of HIV Infection

The human immunodeficiency virus (HIV) is the causative agent behind the pathogenesis of acquired immunodeficiency syndrome (AIDS). According to the World Health Organization (WHO) data, a total of nearly 85 million people have been infected with HIV worldwide, and about 40 million people have died of HIV. At present, globally 39 million people are living with HIV at the end of 2022, with most of the disease concentrated in sub-Saharan Africa. HIV targets the host immune system by infecting and destroying the CD4+ T-cells, leading to compromised cellular immunity as the disease progresses (Chun & Fauci., 2012). HIV attaches to the CD4+ protein on the surface of the cells to gain entry. However, apart from the CD4+ molecules, HIV also uses different chemokine receptors, such as CCR5 (R5) and CXCR4 (X4 or fusin), to enter the different host macrophage cells (Naif., 2013). HIV-infected macrophages may be found in the brain, lungs, lymph nodes, skin, bone marrow, and blood, and these macrophages act as significant reservoirs for HIV-1 in tissues of the body. Continuous HIV replication inside the host causes a state of generalized activation of immune cells and enhanced release of pro-inflammatory cytokines (Brenchley et al., 2006). The progression of the disease is fairly slow, which means it might take several years from the primary viral infection to the development of HIV disease symptoms and immunosuppression. The symptomatic stage of the disease indicates the late phase of HIV infection (AIDS), where individuals have high plasma viral load and low CD4+ T-cells (below 200 cells per cubic millimeter of blood), and they may be very much susceptible to other opportunistic infections (El-Atrouni et al., 2006). Currently, there is no cure or any effective vaccine available against HIV. The management of HIV/AIDS usually involves the use of multiple antiretroviral drugs that are known as highly active antiretroviral therapy (HAART) as a strategy to control HIV infection. HAART slows the progression of the disease by lowering the total HIV load, maintaining a functional immune system, and preventing opportunistic infections that often lead to death (Moore & Chaisson, 1999).

Interestingly, evidence suggests that HIV infection affects the exosomal or EV content and utilizes the cellular exosome biogenesis and secretion pathway to enhance its reproduction and pathogenesis. It has been reported that, similar to EV biogenesis, the budding process of HIV also requires different ESCRT and associated proteins (Usami et al., 2009). During HIV infection, exosomes were also found to regulate the micro-environment of the host target cells by inducing a pro- or anti-viral immune response. Exosomes derived from monocytes and macrophages can transfer CCR5 and CXCR4 to the recipient cells lacking these co-receptors, making them more susceptible to viral entry (Mack et al., 2000). Exosomes can also carry different cytokines, such as IL-3, IL-4, IL-8, and IL-17, altering the viral life cycle and promoting HIV infection. Apart from cytokines, exosomes were also reported to contain specific HIV-derived proteins such as abundantly expressed HIV-associated protein, Nef, and envelope (Env) protein, gp120. Nef protein containing EVs leads to the lowering of the CD4+ cells by making the T-cells more susceptible to viral infection and also by promoting T-cell apoptosis.

Additionally, EVs containing gp120 enhance the attachment and fusion of the viral particles with the target cells to facilitate HIV infection (Chen et al., 2021; Arakelyan et al., 2017). On the contrary, T-cells release EVs with CD4 molecules displayed on their surface that can competitively bind to envelope proteins of HIV-1, inhibiting virus interaction with target cells and infection. Additionally, EVs reduce HIV infection in various ways, such as they contain the human cytidine deaminase APOBEC3G (A3G) that confers resistance to viral replication. EVs also promote transcriptional silencing of HIV by affecting the NF-κB/Sp1/Tat pathway (Khatua et al., 2009).

Table 16.1 List of miRNAs reported to act as regulators of Alzheimer's (AD), Parkinson's (PD), or HIV infection

miRNA	Disease affecting	Target(s)	Reference(s)
let-7e-5p	Alzheimer's disease	IL-6	Durur et al. (2022)
miR-9-5p	Alzheimer's disease	UBE4B, BACE1, SIRT1, GSK-3β	Subramanian et al. (2021), Liu et al. (2020), Turk et al. (2021)
miR-15a	Alzheimer's disease	ERK-1	Hebert et al. (2010)
miR-17-5p, miR-20a, miR-106b	Alzheimer's disease	APP	Hebert et al. (2009)
miR-29a, miR-29b-1	Alzheimer's disease	BACE1	Hebert et al. (2008)
miR-107	Alzheimer's disease	BACE1	Wang et al. (2008)
miR-132-3p, miR-212	Alzheimer's disease	PTEN, FOXO3a, P300	Wong et al. (2013)
miR-146a-5p	Alzheimer's disease	IRAK-1	Cui et al. (2010)
miR-193b	Alzheimer's disease	APP	Liu et al. (2014)
miR-339-5p	Alzheimer's disease	BACE1	Long et al. (2014)
miR-26a	Parkinson's disease	UBE4B,USP15, USP3	Martins et al. (2011)
miR-19b	Parkinson's disease	USP6, USP32,ST3GAL5	Fernandez-Santiago et al. (2015), Vallelunga et al. (2014)
miR-29b, miR-29c	Parkinson's disease	DBT	Vallelunga et al. (2014)
miR-17-5p, miR-20a	HIV	PCAF (Tat co-factor)	Triboulet et al. (2007)
miR-155	HIV	NF-κB pathway	Ruelas et al. (2015)
miR-34a	HIV	SIRT1	Zhang et al. (2012)
miR-101	HIV-associated neurocognitive disorders (HAND)	VE-cadherin	Mishra and Singh (2013)
mir-146a	Neuro-AIDS	CCL8/MCP-2	Rom et al. (2010)
miR-34c, miR-93, miR-381, miR-500a	HAND	PEX2, PEX7, PEX11B, PEX13	Xu et al. (2017)
miR-196a	HAND	p73 pathway	Bagashev et al. (2014)
miR-181a	Neuro-AIDS	SAMHD1	Pilakka-Kanthikeel et al. (2015)
miR-29a	HIV	Nef	Ahluwalia et al. (2008)
miR-32	HIV-associated neuroinflammation	TRAF3	Mishra et al. (2012)
Mir-128a	HIV-1 encephalopathy	SNAP25	Eletto et al. (2008)

Besides HIV-derived protein factors, EVs also contain several RNA molecules that are key in inducing inflammation. HIV-infected cell-derived EVs contain trans-activation response element (TAR) RNA that enhances viral transcription and regulates the gene expression of pro-inflammatory cytokines, such as IL-6 and TNF-β, through the TLR-3 pathway. (Sampey et al., 2016) EV-derived miRNAs also play an essential part in viral-mediated inflammation; miR-155, miR-223, and miR-29a induce inflammatory response in HIV-infected macrophages. Furthermore, miRNAs such as miR-29a, miR-128, miR-28, miR-125b, miR-150, miR-223, and miR-382–5p target the 3′-UTR of HIV mRNA to downregulate the expression of essential viral proteins thus its replication inside the host (Sadri Nahand et al., 2020). Studies have shown that miR-548a, miR-30e, miR-338, miR-454, miR-518f, miR-1243, miR-1247a, miR-150, miR-29a, miR-302c, miR-636, miR-872, and miR-875 were all upregulated, whereas the expression of miR-149, miR-382, miR-378, miR-324-5p, miR-223, miR-150, miR-125b, miR-29a, miR-29b, miR-29c, miR-92a, miR-28, miR-26a, miR-20a, miR-19a, miR-19b, and miR-17 were downregulated in HIV-infected cells (Roth et al., 2015; Aqil et al., 2014). Overall, the study of the altered patterns of EV-associated miRNAs could provide helpful insight to develop novel and prognostic biomarkers for predicting HIV disease progression toward AIDS and could open up new avenues in the field of anti-HIV drug discovery research.

16.9 EV-Mediated Communication to Virally Infected Cells

Extensive research has been going on worldwide for many years to understand extracellular vesicles' role in regulating viral pathogenesis. The involvement of EVs in viral spread and infection has become evident because EVs and viruses share common features in their size, structure, biogenesis, and uptake. Additionally, as mentioned earlier, EVs released from virus-infected cells often transfer viral proteins, RNAs, miRNAs, and other factors to the non-infected recipient cells to control the progress of the disease (Urbanelli et al., 2019). Epstein–Barr virus-infected cells release EVs that contain viral proteins, such as Latent Membrane Protein 1 (LMP-1), which induce the packaging of different growth factors inside these vesicles. When the recipient cells take up these vesicles, their LMP-1 content can upregulate adhesion molecules, such as ICAM-1, in non-EBV-infected cells thus promoting viral infectivity. EV-mediated transfer of growth factors in recipient cells can alter their immunosurveillance, inhibiting the immune system's destruction of the cells and favoring the oncogenic transformation (Ceccarelli et al., 2007). EVs released from EBV-infected cells also contain several viral miRNAs that target host anti-viral immune responses by regulating cytokine production and promoting cell apoptosis. Another herpesvirus, Kaposi's sarcoma-associated herpesvirus (KSHV) infected cell-derived EVs do not contain viral proteins; instead, they carry host protein factors that impair the host's immune response by affecting important cellular metabolic pathways in recipient cells. Viral miRNAs packaged inside the EVs released from KSHV-infected cells could play a role in the development of Primary Effusion Lymphoma (PEL), a KSHV-associated malignancy (Chugh et al., 2013; Meckes et al., 2013).

Studies on Cytomegalovirus (CMV) and Herpes simplex virus 1 (HSV-1) have shown that EVs may enhance viral infectivity by eliminating the host proteins relevant to anti-viral response. CMV-infected cells can increase the export of proteins, such as lectins, via EVs necessary for virus uptake (Plazolles et al., 2011). HSV-1-infected cells are found to package viral miRNAs miR-H28 and miR-H29 into EVs that play a crucial role in facilitating the infection of neighboring non-infected cells, perhaps by weakening their innate immune defenses (Han et al., 2016).

Hepatitis C virus (HCV) is a member of the Flaviviridae family, and it hijacks the cellular exosomal pathway for assembly and release. EV surface protein CD81 is the dominant co-receptor for HCV; therefore, HCV virions may be incorporated into EVs through their interaction with CD81 (Zhang et al., 2004). Thereby, CD81+ exosome-mediated infection has the advantage of masking the HCV virus with a host envelope, thus preventing an immune response. Additionally, nucleic acid host components have been retrieved in HCV-derived exosomes, which promotes viral replication. For example, the Ago2 protein and miR-122 have been shown to enhance HCV replication via binding to the 5′UTR of HCV RNA (Bukong et al., 2014).

Recently, the World Health Organization classified a novel coronavirus termed SARS-CoV-2 that causes a severe acute respiratory condition known as coronavirus disease 2019 (COVID-19) as a global pandemic in January 2020. SARS-CoV-2 infects cells by targeting angiotensin-converting enzyme 2(ACE2) receptor. Evidence suggests that EVs may facilitate the infection of SARS-CoV-2 using a similar strategy to that of HIV through an intercellular transfer of host molecules or viral materials (Zhang et al., 2020). Exosomes isolated from the infected cells carry ACE2 that, when transferred to the non-infected cells, can make them more susceptible to viral infection. EVs also help the virus to evade host immune recognition as they carry viral nucleocapsids and the spike protein inside them and transfer them to the target cells (Xia et al., 2021).

EVs not only facilitate the spreading of viral infections but can also promote the host's immune response against viruses. EVs from EBV-infected cells were found to carry viral antigens that stimulate B-cells to elicit an adaptive immune response (Gutzeit et al., 2014). HIV-derived EVs are responsible for the transport of host anti-viral protein APOBEC3G that inhibits viral replication. Similarly, the transfer of second messenger cyclic guanosine monophosphate–adenosine monophosphate (cGAMP) via EVs triggers anti-viral IFN responses in newly infected cells (Nolte-'t Hoen et al., 2016). EVs released from EBV-infected cells contain deoxyuridine triphosphatase (dUTPase) that induces NF-kB activation and cytokine secretion in primary DCs and peripheral mononuclear blood cells (PBMCs), thereby acting as a stimulator of host immune response.

Furthermore, EV-transformed B-cell-derived EVs can stimulate their proliferation and interfere with the fate of the human B cell's response (Urbanelli et al., 2019). CMV-infected endothelial cells release EVs carrying glycoprotein B that could indirectly activate $CD4^+$ T-cells, thereby inducing an adaptive immune response to control CMV infection. Studies have shown that exosomes released from HSV1-infected cells contain the host product Stimulator of INF genes (STING) protein in vesicles. Together with the packaged viral miRNAs within vesicles, STING inhibits the viral spread and increases host cell survival (Kalamvoki et al., 2014).

EVs can also transfer nucleic acids involved in anti-viral responses. Viruses and EVs transfer different RNA molecules that can be recognized as Pathogen-Associated Molecular Patterns (PAMPs) by Pathogen Recognition Receptors (PRRs) in target cells to stimulate innate anti-viral response. In the case of HIV, two viral miRNAs, miR-88 and miR-99, target the NF-kB signaling that facilitates the release of pro-inflammatory cytokine TNFα. EVs from infected cells transport these miRNAs to the bystander macrophages, thereby prompting an immune response against HIV (Bernard et al., 2014). Furthermore, the anti-viral response may not always be due to the transfer of viral mRNAs or miRNAs via EVs; instead, the transfer of host cell miRNAs via EVs can also contribute to the survival of the target cells. Studies have reported that cultured primary human placental trophoblasts show significant resistance against infection by HSV-1 and CMV and confer this resistance to nonplacental recipient cells by the EV-mediated delivery of miRNAs (Delorme-Axford et al., 2013).

To summarize, EVs produced by virus-infected cells are important mediators of anti-viral response and act as a vehicle to spread viral infection. EVs are shown to carry viral proteins or nucleic acids, which are involved in virus reactivation during latency or in broadening viral tropism.

Conversely, it has also been reported that EVs are involved in stimulating anti-viral mechanisms, but what cell conditions and virus types release EVs that favor or fight infection are still not clearly elucidated. Understanding the structure of EVs produced by infected cells, determining their cargo, and deciphering the fine mechanisms by which they affect viral infection are required for basic virology and translation into therapy. Since viral replication and transmission and EV biogenesis are intrinsically interlinked, understanding the interplay between viruses and extracellular vesicles can also provide a breakthrough for the therapeutic applications of EVs in the immunological field in order to exploit EVs for anti-viral strategies.

16.10 How EV May Affect Pharmacodynamics/Pharmacokinetics of Drugs: Possible Pathways

EVs can be developed as potential therapeutic agents for various diseases due to their biological functions. This field of research has been very active, with a number of researchers and companies looking at the use of native EV preparations as they are naturally produced by cells, as well as the potential for loading cargo molecules to EVs via biomimetic vesicles. There are inherent challenges for each of these approaches, which reflect technical competencies in characterizing and manipulating EVs. Moreover, EV therapeutics are currently not subject to a defined regulatory framework, although they may be part of the pharmaceutical class of biologicals. It is necessary to show pharmacokinetics and therapeutic effectiveness for a medicinal product to become clinically approved; currently, these are in the early stages of EV development. The therapeutic potential of EVs from different sources has been demonstrated in several in vivo studies. For example, EVs from mesenchymal, dendritic cells, plasma, and biologically inspired nanovesicles have been studied (Lener et al., 2015).

One of the advantages of using EVs as a therapeutic method is that they have low levels of immunogenicity when used in autologous therapy, which minimizes potential side effects. They are suitable for fusion with target cells due to their size and lipid membrane composition, allowing cargo delivery while protecting it from degradation. There are also advantages for these materials compared to inherently toxic nanomaterials, as is the case with liposomes. These advantages have led to the development of methods for packaging EVs with therapeutic cargo to treat a number of conditions. The method of loading EVs with cargo is altering an originating cell to express a specific cargo by electroporation or biochemical treatment of cells, which involves physically affixing small functional RNAs. In general, miRNA and siRNA have been packaged into EVs, which can be evaluated with luciferase gene expression-based assays to verify the recipients' delivery onto their cells. Also, native EVs prepared using plasma can be adapted to contain therapeutic cargo. An example is the incorporation of therapeutic antitumor miRNAs using electroporation to EVs isolated from human plasma, which promotes apoptosis in human tumor cell lines (Pomatto et al., 2019).

Other nanomaterials, including biomimetic vesicles, liposomes, carbon nanotubes, micelles, and dendrimers, can potentially be therapeutic agents in addition to cellular EVs. Techniques such as serial extrusion through filters of specific pore size can generate cell-derived biomimetic vesicles. In addition to their size and membrane composition, the vesicles formed through these methods possess characteristics similar to those of EVs. Several studies have shown that compared to small EVs from different parent cells can be used to produce functioning mimetic EVs over a 100 fold greater particle yield (Nasiri Kenari et al., 2019). So far, biomimetic vesicles have largely been housed in academic research laboratories, and pharmaceutical companies do not use them to develop therapeutic products. However, for the delivery of drugs like doxorubicin in order to treat cancer, a synthetic nanovesicle has long been used to utilize liposomes (Zagar et al., 2014).

Although EV has excellent potential for treating and diagnosing a wide range of diseases, several related challenges exist. A significant issue is the technological challenges associated with the isolation, characterization, and standardization of clinically appropriate EV preparations. Low-throughput techniques such as ultracentrifugation are used to isolate EV and its subtypes. The ability to prepare EVs from large volumes of culture media is demonstrated using new methods, such as tangential flow filtration or size exclusion chromatography. It is also a major technological problem to scale up cell culture so that enough EVs can be produced for clinical use. With this in mind, issues related to the use of bioreactors and media supplements, such as fetal bovine serum containing EV itself, have arisen, illustrating the relevance of characterization when scaling up production. It is necessary to define, first and foremost, the functional activity of EVs as part of the preparation to characterize an isolated product fully. It is also still challenging to determine effective dosage and even mechanism of action for Therapeutic Extracellular Vehicles, especially derived from stem cell sources. In order to develop an appropriate dose and functionality assessment, a greater mechanistic understanding of the mode of action for a particular EV preparation would be helpful. Increasing our knowledge of the heterogeneity of EVs as soon as these parameters are defined and improving isolation protocols to enrich vesicles with their most efficient function could be helpful in this regard. Another challenge is to develop potency assays that can measure the efficacy of a particular group of EVs. One approach that could help in this respect is to specify a particular biological action of an EV to measure drug efficacy. In order for EV therapies to be approved by the regulatory authorities, it is important to develop such assays (Gimona et al., 2021).

The question of whether or not the influx of external cargo disturbs and interferes with endogenous cargo, as well as creates issues related to off-target effects, should be investigated in relation to the use of EVs as drug delivery agents. The heterogeneity of EVs may prove to be an important obstacle when off-target effects exceed therapeutic objectives in a wide range of diseases, including cancer. In order to avoid immunogenicity problems and not deliver harmful toxic cargo, it is also necessary to determine a safe source of cells for EV isolation and therapy. In order to tackle these problems, it is necessary to establish appropriate preclinical models, and the selected EV preparations must satisfy the intended cargo as well as the characteristics of the individual molecules. Another aspect for which more study is needed is targeting EVs to sites where they will perform their therapeutic action. Reducing unintended adverse reactions is also a key consideration when designing treatment strategies for EVs targeting specific sites. Although the design of EVs has been demonstrated to improve tropism in specific cell types or enable delivery into organs like the brain via the BBB crossing, it is possible that technical difficulties associated with producing therapeutically relevant amounts of these vesicles may limit their further development. In particular, for diseases such as cancer and neurological conditions, the development of EV-based biomarkers has shown great promise, and they may consist of proteins, nucleic acids, lipids, or glycans. However, technological improvements will be needed to allow the discovery of EV Biomarkers for disease diagnostics in clinical laboratories, especially due to the development of high-throughput assays (Gimona et al., 2021).

In spite of ongoing challenges, important progress has been made in the area of EVs research over recent years. A limited number of EV-based biomarkers have now been applied for clinical and research purposes. Moreover, the stage for clinical development of EV therapies is being set by an increasing number of preclinical studies showing the potential to use EV therapy in a range of important diseases. Currently, small numbers are developing in the early stages. Technical advances to scale up production and the acquisition of appropriate regulatory authorizations for therapeutic use will be essential for further development. Although these technical and regulatory barriers are still to be solved, recent rapid growth in understanding EVs' physiological and pathological role has given rise to many potential diagnostics and therapeutic applications that have driven their immense clinical potential.

16.11 Where We Stand in EV-Based Therapy Development

Extracellular vesicles carry cargo proteins and miRNA from glial and infiltrating immune cells in the AD and PD brains, communicating the pathogenic proteins and RNA. These may affect the functionality and survival of neighboring neurons to spread the disease phenotype across the affected tissue and beyond. While some of the cargo within EVs may positively influence neuronal health, other cargo proteins (including amyloidogenic proteins) and specific miRNAs may have an opposite detrimental role on neurons in the affected part of the tissue. The non-brain-cells-derived EVs that cross the blood-brain barrier also have a role to play in disease progression (Ramos-Zaldivar et al., 2022). It may be possible that some of the immunomodulatory EVs are generated outside the brain and are even taken to our body from external sources such as milk. These "foreign" EVs may have a huge role to play in shaping the neuronal development and connectivity in the infant brain and may act as immunomodulators also in the adult brain due to their intrinsic capacity to cross blood-brain-barrier and the presence of anti-inflammatory miRNA and protein within milk EVs. Therefore, synthetic or engineered milk EVs with an arsenal on their surface can enable us to reach the proper target tissue. These EVs will be useful in delivering cargo to target cells in the brain specifically. We expect challenges in making the EV-mediated cargo delivery process specific and getting a high yield of immunomodulatory EVs. The longevity of EVs in ensuring a robust and long-term effect in the host tissue is another challenge.

One of the major obstacles associated with EV-based therapy is getting its production to a level useful for the therapeutic application of EVs. Spinner cultures of macrophage cells produce EVs with desired protein or miRNA cargo, and getting a substantial amount of EVs is now possible. Alternatively, the usage of EVs from milk would be an exciting alternative. However, the restriction of non-allogenic cross-species immunoreactivity and decoration of the milk-derived EVs with target molecules are additional challenges that must be addressed before the development of milk-EV-based therapy. Several new strategies, including loading EVs with miRNA and peptides by application of click-chemistry or electroporation, have been promising in recent times in loading EVs with desired cargo. Therefore, we may expect to find engineered EVs that can be used as immunomodulators in the therapy of neurodegenerative diseases in the near future.

16.12 Review Questions

1. Do EVs carry different classes of biomolecules, including the epigenetic material in the form of miRNA? It also carries:
   (a) MHC class of proteins
   (b) mRNA encoding cytokines and housekeeping genes
   (c) metabolites
   (d) All of the above.
2. HIV-infected cells release EVs that contain
   (a) HIV mRNAs
   (b) HIV specific proteins
   (c) Viral miRNAs
   (d) All of the above
3. Alzheimer's disease is never associated with
   (a) Cognitive impairment
   (b) Behavioral changes
   (c) Cardiovascular disease
   (d) Dysfunction in learning and memory
   (e) Motor disturbances
4. Can EV be engineered to have
   (a) Specific miRNA
   (b) Specific mRNA
   (c) Surface decorating proteins
   (d) All of the above

16.13 Answers to the Questions

1. d
2. d
3. c
4. d

Acknowledgments The authors would like to thank the UNMC Start-up fund to SNB.

References

Aheget, H., Tristan-Manzano, M., Mazini, L., Cortijo-Gutierrez, M., Galindo-Moreno, P., Herrera, C., Martin, F., Marchal, J. A., & Benabdellah, K. (2020).Exosome: A new player in translational nanomedicine. *Journal of Clinical Medicine, 9*, 2380.

Ahluwalia, J. K., Khan, S. Z., Soni, K., Rawat, P., Gupta, A., Hariharan, M., Scaria, V., Lalwani, M., Pillai, B., Mitra, D., & Brahmachari, S. K. (2008). Human cellular microRNA hsa-miR-29a interferes with viral nef protein expression and HIV-1 replication. *Retrovirology, 5*, 117.

Alvarez-Erviti, L., Seow, Y., Schapira, A. H., Gardiner, C., Sargent, I. L., Wood, M. J., & Cooper, J. M. (2011a). Lysosomal dysfunction increases exosome-mediated alpha-synuclein release and transmission. *Neurobiology of Disease, 42*, 360–367.

Alvarez-Erviti, L., Seow, Y., Yin, H., Betts, C., Lakhal, S., & Wood, M. J. (2011b). Delivery of siRNA to the mouse brain by systemic injection of targeted exosomes. *Nature Biotechnology, 29*, 341–34s.

Anderson, H. C. (1969). Vesicles associated with calcification in the matrix of epiphyseal cartilage. *The Journal of Cell Biology, 41*, 59–72.

Andre, F., Chaput, N., Schartz, N. E., Flament, C., Aubert, N., Bernard, J., Lemonnier, F., Raposo, G., Escudier, B., Hsu, D. H., Tursz, T., Amigorena, S., Angevin, E., & Zitvogel, L. (2004). Exosomes as potent cell-free peptide-based vaccine. I. Dendritic cell-derived exosomes transfer functional MHC class I/peptide complexes to dendritic cells. *Journal of Immunology, 172*, 2126–2136.

Aqil, M., Naqvi, A. R., Mallik, S., Bandyopadhyay, S., Maulik, U., & Jameel, S. (2014). The HIV Nef protein modulates cellular and exosomal miRNA profiles in human monocytic cells. *Journal of Extracellular Vesicles, 3*.

Arakelyan, A., Fitzgerald, W., Zicari, S., Vanpouille, C., & Margolis, L. (2017). Extracellular vesicles carry HIV Env and facilitate Hiv infection of human lymphoid tissue. *Scientific Reports, 7*, 1695.

Ayre, D. C., Elstner, M., Smith, N. C., Moores, E. S., Hogan, A. M., & Christian, S. L. (2015). Dynamic regulation of CD24 expression and release of CD24-containing microvesicles in immature B cells in response to CD24 engagement. *Immunology, 146*, 217–233.

Bagashev, A., Mukerjee, R., Santerre, M., Del Carpio-Cano, F. E., Shrestha, J., Wang, Y., He, J. J., & Sawaya, B. E. (2014). Involvement of miR-196a in HIV-associated neurocognitive disorders. *Apoptosis, 19*, 1202–1214.

Balan, S., & Bhardwaj, N. (2020). Cross-presentation of tumor antigens is ruled by synaptic transfer of vesicles among dendritic cell subsets. *Cancer Cell, 37*, 751–753.

Balestrino, R., & Schapira, A. H. V. (2020). Parkinson disease. *European Journal of Neurology, 27*, 27–42.

Banks, W. A. (2016). From blood-brain barrier to blood-brain interface: New opportunities for CNS drug delivery. *Nature Reviews. Drug Discovery, 15*, 275–292.

Banks, W. A., Sharma, P., Bullock, K. M., Hansen, K. M., Ludwig, N., & Whiteside, T. L. (2020). Transport of extracellular vesicles across the blood-brain barrier: Brain pharmacokinetics and effects of inflammation. *International Journal of Molecular Sciences, 21*, 4407.

Barres, C., Blanc, L., Bette-Bobillo, P., Andre, S., Mamoun, R., Gabius, H. J., & Vidal, M. (2010). Galectin-5 is bound onto the surface of rat reticulocyte exosomes and modulates vesicle uptake by macrophages. *Blood, 115*, 696–705.

Bernard, M. A., Zhao, H., Yue, S. C., Anandaiah, A., Koziel, H., & Tachado, S. D. (2014). Novel HIV-1 miRNAs stimulate TNFalpha release in human macrophages via TLR8 signaling pathway. *PLoS One, 9*, e106006.

Brenchley, J. M., Price, D. A., Schacker, T. W., Asher, T. E., Silvestri, G., Rao, S., Kazzaz, Z., Bornstein, E., Lambotte, O., Altmann, D., Blazar, B. R., Rodriguez, B., Teixeira-Johnson, L., Landay, A., Martin, J. N., Hecht, F. M., Picker, L. J., Lederman, M. M., Deeks, S. G., & Douek, D. C. (2006). Microbial translocation is a cause of systemic immune activation in chronic HIV infection. *Nature Medicine, 12*, 1365–1371.

Bruck, D., Wenning, G. K., Stefanova, N., & Fellner, L. (2016). Glia and alpha-synuclein in neurodegeneration: A complex interaction. *Neurobiology of Disease, 85*, 262–274.

Bukong, T. N., Momen-Heravi, F., Kodys, K., Bala, S., & Szabo, G. (2014). Exosomes from hepatitis C infected patients transmit HCV infection and contain replication competent viral RNA in complex with Ago2-miR122-HSP90. *PLoS Pathogens, 10*, e1004424.

Burgelman, M., Vandendriessche, C., & Vandenbroucke, R. E. (2021). Extracellular vesicles: A double-edged sword in sepsis. *Pharmaceuticals (Basel), 14*, 829.

Ceccarelli, S., Visco, V., Raffa, S., Wakisaka, N., Pagano, J. S., & Torrisi, M. R. (2007). Epstein-Barr virus latent membrane protein

1 promotes concentration in multivesicular bodies of fibroblast growth factor 2 and its release through exosomes. *International Journal of Cancer, 121*, 1494–1506.

Chen, C. C., Liu, L., Ma, F., Wong, C. W., Guo, X. E., Chacko, J. V., Farhoodi, H. P., Zhang, S. X., Zimak, J., Segaliny, A., Riazifar, M., Pham, V., Digman, M. A., Pone, E. J., & Zhao, W. (2016). Elucidation of exosome migration across the blood-brain barrier model in vitro. *Cellular and Molecular Bioengineering, 9*, 509–529.

Chen, J., Li, C., Li, R., Chen, H., Chen, D., & Li, W. (2021). Exosomes in HIV infection. *Current Opinion in HIV and AIDS, 16*, 262–270.

Cheng, Q., Wang, J., Li, M., Fang, J., Ding, H., Meng, J., Zhang, J., Fang, X., Liu, H., Ma, C., Chen, C., & Zhang, W. (2022). CircSV2b participates in oxidative stress regulation through miR-5107-5p-Foxk1-Akt1 axis in Parkinson's disease. *Redox Biology, 56*, 102430.

Chugh, P. E., Sin, S. H., Ozgur, S., Henry, D. H., Menezes, P., Griffith, J., Eron, J. J., Damania, B., & Dittmer, D. P. (2013). Systemically circulating viral and tumor-derived microRNAs in KSHV-associated malignancies. *PLoS Pathogens, 9*, e1003484.

Chun, T. W., & Fauci, A. S. (2012). HIV reservoirs: Pathogenesis and obstacles to viral eradication and cure. *AIDS, 26*, 1261–1268.

Cui, J. G., Li, Y. Y., Zhao, Y., Bhattacharjee, S., & Lukiw, W. J. (2010). Differential regulation of interleukin-1 receptor-associated kinase-1 (Irak-1) and Irak-2 by microRNA-146a and NF-kappaB in stressed human astroglial cells and in Alzheimer disease. *The Journal of Biological Chemistry, 285*, 38951–38960.

Delorme-Axford, E., Donker, R. B., Mouillet, J. F., Chu, T., Bayer, A., Ouyang, Y., Wang, T., Stolz, D. B., Sarkar, S. N., Morelli, A. E., Sadovsky, Y., & Coyne, C. B. (2013). Human placental trophoblasts confer viral resistance to recipient cells. *Proceedings of the National Academy of Sciences of the United States of America, 110*, 12048–12053.

Delpech, J. C., Herron, S., Botros, M. B., & Ikezu, T. (2019). Neuroimmune crosstalk through extracellular vesicles in health and disease. *Trends in Neurosciences, 42*, 361–372.

DeMaagd, G., & Philip, A. (2015). Parkinson's disease and its management: Part 1: Disease entity, risk factors, pathophysiology, clinical presentation, and diagnosis. *Pharmacy and Therapeutics, 40*, 504–532.

Duchez, A. C., Boudreau, L. H., Naika, G. S., Bollinger, J., Belleannee, C., Cloutier, N., Laffont, B., Mendoza-Villarroel, R. E., Levesque, T., Rollet-Labelle, E., Rousseau, M., Allaeys, I., Tremblay, J. J., Poubelle, P. E., Lambeau, G., Pouliot, M., Provost, P., Soulet, D., Gelb, M. H., & Boilard, E. (2015). Platelet microparticles are internalized in neutrophils via the concerted activity of 12-lipoxygenase and secreted phospholipase A2-IIA. *Proceedings of the National Academy of Sciences of the United States of America, 112*, E3564–E3573.

Durur, D. Y., Tastan, B., Ugur Tufekci, K., Olcum, M., Uzuner, H., Karakulah, G., Yener, G., & Genc, S. (2022). Alteration of miRNAs in small neuron-derived extracellular vesicles of Alzheimer's disease patients and the effect of extracellular vesicles on microglial immune responses. *Journal of Molecular Neuroscience, 72*, 1182–1194.

El-Agnaf, O. M., Salem, S. A., Paleologou, K. E., Curran, M. D., Gibson, M. J., Court, J. A., Schlossmacher, M. G., & Allsop, D. (2006). Detection of oligomeric forms of alpha-synuclein protein in human plasma as a potential biomarker for Parkinson's disease. *The FASEB Journal, 20*, 419–425.

El-Atrouni, W., Berbari, E., & Temesgen, Z. (2006). HIV-associated opportunistic infections. Bacterial infections. *The Lebanese Medical Journal, 54*, 80–83.

Eletto, D., Russo, G., Passiatore, G., Del Valle, L., Giordano, A., Khalili, K., Gualco, E., & Peruzzi, F. (2008). Inhibition of SNAP25 expression by HIV-1 Tat involves the activity of mir-128a. *Journal of Cellular Physiology, 216*, 764–770.

Erkkinen, M. G., Kim, M. O., & Geschwind, M. D. (2018). Clinical neurology and epidemiology of the major neurodegenerative diseases. *Cold Spring Harbor Perspectives in Biology, 10*, a033118.

Fernandez-Santiago, R., Iranzo, A., Gaig, C., Serradell, M., Fernandez, M., Tolosa, E., Santamaria, J., & Ezquerra, M. (2015). MicroRNA association with synucleinopathy conversion in rapid eye movement behavior disorder. *Annals of Neurology, 77*, 895–901.

Gimona, M., Brizzi, M. F., Choo, A. B. H., Dominici, M., Davidson, S. M., Grillari, J., Hermann, D. M., Hill, A. F., de Kleijn, D., Lai, R. C., Lai, C. P., Lim, R., Monguio-Tortajada, M., Muraca, M., Ochiya, T., Ortiz, L. A., Toh, W. S., Yi, Y. W., Witwer, K. W., Giebel, B., & Lim, S. K. (2021). Critical considerations for the development of potency tests for therapeutic applications of mesenchymal stromal cell-derived small extracellular vesicles. *Cytotherapy, 23*, 373–380.

Gui, Y., Liu, H., Zhang, L., Lv, W., & Hu, X. (2015). Altered microRNA profiles in cerebrospinal fluid exosome in Parkinson disease and Alzheimer disease. *Oncotarget, 6*, 37043–37053.

Guo, M., Wang, J., Zhao, Y., Feng, Y., Han, S., Dong, Q., Cui, M., & Tieu, K. (2020). Microglial exosomes facilitate alpha-synuclein transmission in Parkinson's disease. *Brain, 143*, 1476–1497.

Guo, B. B., Bellingham, S. A., & Hill, A. F. (2015). The neutral sphingomyelinase pathway regulates packaging of the prion protein into exosomes. *Journal of Biological Chemistry, 290*, 3455–3467.

Gururajan, A., Naughton, M. E., Scott, K. A., O'Connor, R. M., Moloney, G., Clarke, G., Dowling, J., Walsh, A., Ismail, F., Shorten, G., Scott, L., McLoughlin, D. M., Cryan, J. F., & Dinan, T. G. (2021). MicroRNAs as biomarkers for major depression: A role for let-7b and let-7c. *Translational Psychiatry, 6*, e862.

Gutzeit, C., Nagy, N., Gentile, M., Lyberg, K., Gumz, J., Vallhov, H., Puga, I., Klein, E., Gabrielsson, S., Cerutti, A., & Scheynius, A. (2014). Exosomes derived from Burkitt's lymphoma cell lines induce proliferation, differentiation, and class-switch recombination in B cells. *Journal of Immunology, 192*, 5852–5862.

Gyorgy, B., Szabo, T. G., Pasztoi, M., Pal, Z., Misjak, P., Aradi, B., Laszlo, V., Pallinger, E., Pap, E., Kittel, A., Nagy, G., Falus, A., & Buzas, E. I. (2011). Membrane vesicles, current state-of-the-art: Emerging role of extracellular vesicles. *Cellular and Molecular Life Sciences, 68*, 2667–2688.

Han, Z., Liu, X., Chen, X., Zhou, X., Du, T., Roizman, B., & Zhou, G. (2016). miR-H28 and miR-H29 expressed late in productive infection are exported and restrict HSV-1 replication and spread in recipient cells. *Proceedings of the National Academy of Sciences of the United States of America, 113*, E894–E901.

Harding, C., Heuser, J., & Stahl, P. (1983). Receptor-mediated endocytosis of transferrin and recycling of the transferrin receptor in rat reticulocytes. *The Journal of Cell Biology, 97*, 329–339.

Hebert, S. S., Horre, K., Nicolai, L., Papadopoulou, A. S., Mandemakers, W., Silahtaroglu, A. N., Kauppinen, S., Delacourte, A., & De Strooper, B. (2008). Loss of microRNA cluster miR-29a/b-1 in sporadic Alzheimer's disease correlates with increased BACE1/beta-secretase expression. *Proceedings of the National Academy of Sciences of the United States of America, 105*, 6415–6420.

Hebert, S. S., Horre, K., Nicolai, L., Bergmans, B., Papadopoulou, A. S., Delacourte, A., & De Strooper, B. (2009). MicroRNA regulation of Alzheimer's amyloid precursor protein expression. *Neurobiology of Disease, 33*, 422–428.

Hebert, S. S., Papadopoulou, A. S., Smith, P., Galas, M. C., Planel, E., Silahtaroglu, A. N., Sergeant, N., Buee, L., & De Strooper, B. (2010). Genetic ablation of Dicer in adult forebrain neurons results in abnormal tau hyperphosphorylation and neurodegeneration. *Human Molecular Genetics, 19*, 3959–3969.

Hill, A. F. (2019). Extracellular vesicles and neurodegenerative diseases. *Journal of Neuroscience, 39*, 9269–9273.

Huang, S., Ge, X., Yu, J., Han, Z., Yin, Z., Li, Y., Chen, F., Wang, H., Zhang, J., & Lei, P. (2018). Increased miR-124-3p in microglial exosomes following traumatic brain injury inhibits neuronal inflammation and contributes to neurite outgrowth via their transfer into neurons. *The FASEB Journal, 32*, 512–528.

Huang, L. K., Chao, S. P., & Hu, C. J. (2020). Clinical trials of new drugs for Alzheimer disease. *Journal of Biomedical Science, 27*, 18.

Jarmalaviciute, A., & Pivoriunas, A. (2016). Exosomes as a potential novel therapeutic tools against neurodegenerative diseases. *Pharmacological Research, 113*, 816–822.

Jiang, L., Chen, W., Ye, J., & Wang, Y. (2022). Potential role of exosomes in ischemic stroke treatment. *Biomolecules, 12*, 115.

Joe, E., & Ringman, J. M. (2019). Cognitive symptoms of Alzheimer's disease: Clinical management and prevention. *BMJ, 367*, l6217.

Johnstone, R. M., Adam, M., Hammond, J. R., Orr, L., & Turbide, C. (1987). Vesicle formation during reticulocyte maturation. Association of plasma membrane activities with released vesicles (exosomes). *The Journal of Biological Chemistry, 262*, 9412–9420.

Kalamvoki, M., Du, T., & Roizman, B. (2014). Cells infected with herpes simplex virus 1 export to uninfected cells exosomes containing STING, viral mRNAs, and microRNAs. *Proceedings of the National Academy of Sciences of the United States of America, 111*, E4991–E4996.

Kalluri, R., & LeBleu, V. S. The biology, function, and biomedical applications of exosomes. *(2020). Science, 367*, eaau6977.

Kaur, S., Verma, H., Dhiman, M., Tell, G., Gigli, G. L., Janes, F., & Mantha, A. K. (2021). Brain exosomes: Friend or foe in Alzheimer's disease? *Molecular Neurobiology, 58*, 6610–6624.

Khatua, A. K., Taylor, H. E., Hildreth, J. E., & Popik, W. (2009). Exosomes packaging APOBEC3G confer human immunodeficiency virus resistance to recipient cells. *Journal of Virology, 83*, 512–521.

Kim, H. R., Mun, Y., Lee, K. S., Park, Y. J., Park, J. S., Park, J. H., Jeon, B. N., Kim, C. H., Jun, Y., Hyun, Y. M., Kim, M., Lee, S. M., Park, C. S., Im, S. H., & Jun, C. D. (2018). T cell microvilli constitute immunological synaptosomes that carry messages to antigen-presenting cells. *Nature Communications, 9*, 3630.

Kimura, K., Hohjoh, H., Fukuoka, M., Sato, W., Oki, S., Tomi, C., Yamaguchi, H., Kondo, T., Takahashi, R., & Yamamura, T. (2018). Circulating exosomes suppress the induction of regulatory T cells via let-7i in multiple sclerosis. *Nature Communications, 9*, 17.

Klingelhoefer, L., & Reichmann H. (2015). Pathogenesis of Parkinson disease--the gut-brain axis and environmental factors. Nature Reviews. Neurology 11:625–636.

Lener, T., Gimona, M., Aigner, L., Borger, V., Buzas, E., Camussi, G., Chaput, N., Chatterjee, D., Court, F. A., Del Portillo, H. A., O'Driscoll, L., Fais, S., Falcon-Perez, J. M., Felderhoff-Mueser, U., Fraile, L., Gho, Y. S., Gorgens, A., Gupta, R. C., Hendrix, A., et al. (2015). Applying extracellular vesicles based therapeutics in clinical trials - An ISEV position paper. *Journal of Extracellular Vesicles, 4*, 30087.

Li, D., Wang, Y., Jin, X., Hu, D., Xia, C., Xu, H., & Hu, J. (2020). NK cell-derived exosomes carry miR-207 and alleviate depression-like symptoms in mice. *Journal of Neuroinflammation, 17*, 126.

Lim, C. Z. J., Zhang, Y., Chen, Y., Zhao, H., Stephenson, M. C., Ho, N. R. Y., Chung, J., Reilhac, A., Loh, T. P., Chen, C. L. H., & Shao, H.(2019). Subtyping of circulating exosome-bound amyloid beta reflects brain plaque deposition. *Nature Communications, 10*, 1144.

Liu, R., Liu, J., Ji, X., & Liu, Y. (2013). Synthetic nucleic acids delivered by exosomes: A potential therapeutic for generelated metabolic brain diseases. *Metabolic Brain Disease, 28*, 551–562.

Liu, C. G., Song, J., Zhang, Y. Q., & Wang, P. C. (2014). MicroRNA-193b is a regulator of amyloid precursor protein in the blood and cerebrospinal fluid derived exosomal microRNA-193b is a biomarker of Alzheimer's disease. *Molecular Medicine Reports, 10*, 2395–2400.

Liu, J., Zuo, X., Han, J., Dai, Q., Xu, H., Liu, Y., & Cui, S. (2020). MiR-9-5p inhibits mitochondrial damage and oxidative stress in AD cell models by targeting GSK-3beta. *Bioscience, Biotechnology, and Biochemistry, 84*, 2273–2280.

Livingston, G., Huntley, J., Sommerlad, A., Ames, D., Ballard, C., Banerjee, S., Brayne, C., Burns, A., Cohen-Mansfield, J., Cooper, C., Costafreda, S. G., Dias, A., Fox, N., Gitlin, L. N., Howard, R., Kales, H. C., Kivimaki, M., Larson, E. B., Ogunniyi, A., Orgeta, V., Ritchie, K., Rockwood, K., Sampson, E. L., Samus, Q., Schneider, L. S., Selbaek, G., Teri, L., & Mukadam, N. (2020). Dementia prevention, intervention, and care: 2020 report of the lancet commission. *Lancet, 396*, 413–446.

Lombardi, M., Parolisi, R., Scaroni, F., Bonfanti, E., Gualerzi, A., Gabrielli, M., de Rosbo, N. K., Uccelli, A., Giussani, P., Viani, P., Garlanda, C., Abbracchio, M. P., Chaabane, L., Buffo, A., Fumagalli, M., & Verderio, C. (2019). Detrimental and protective action of microglial extracellular vesicles on myelin lesions: Astrocyte involvement in remyelination failure. *Acta Neuropathology, 138*, 987–1012.

Long, J. M., Ray, B., & Lahiri, D. K. (2014). MicroRNA-339-5p down-regulates protein expression of beta-site amyloid precursor protein-cleaving enzyme 1 (BACE1) in human primary brain cultures and is reduced in brain tissue specimens of Alzheimer disease subjects. *The Journal of Biological Chemistry, 289*, 5184–5198.

Lugli, G., Cohen, A. M., Bennett, D. A., Shah, R. C., Fields, C. J., Hernandez, A. G., & Smalheiser, N. R. (2015). Plasma Exosomal miRNAs in persons with and without Alzheimer disease: Altered expression and prospects for biomarkers. *PLoS One, 10*, e0139233.

Mack, M., Kleinschmidt, A., Bruhl, H., Klier, C., Nelson, P. J., Cihak, J., Plachy, J., Stangassinger, M., Erfle, V., & Schlondorff, D. (2000). Transfer of the chemokine receptor CCR5 between cells by membrane-derived microparticles: A mechanism for cellular human immunodeficiency virus 1 infection. *Nature Medicine, 6*, 769–775.

Maes, M. (2008). The cytokine hypothesis of depression: Inflammation, oxidative & nitrosative stress (IO&NS) and leaky gut as new targets for adjunctive treatments in depression. *Neuro Endocrinology Letters, 29*, 287–291.

Marcos-Ramiro, B., Oliva Nacarino, P., Serrano-Pertierra, E., Blanco-Gelaz, M. A., Weksler, B. B., Romero, I. A., Couraud, P. O., Tunon, A., Lopez-Larrea, C., Millan, J., & Cernuda-Morollon, E. (2014). Microparticles in multiple sclerosis and clinically isolated syndrome: effect on endothelial barrier function. *BMC Neuroscience, 15*, 110.

Martins, M., Rosa, A., Guedes, L. C., Fonseca, B. V., Gotovac, K., Violante, S., Mestre, T., Coelho, M., Rosa, M. M., Martin, E. R., Vance, J. M., Outeiro, T. F., Wang, L., Borovecki, F., Ferreira, J. J., & Oliveira, S. A. (2011). Convergence of miRNA expression profiling, alpha-synuclein interacton and GWAS in Parkinson's disease. *PLoS One, 6*, e25443.

McMillan, K. J., Murray, T. K., Bengoa-Vergniory, N., Cordero-Llana, O., Cooper, J., Buckley, A., Wade-Martins, R., Uney, J. B., O'Neill, M. J., Wong, L. F., & Caldwell, M. A. (2017). Loss of microRNA-7 regulation leads to alpha-synuclein accumulation and dopaminergic neuronal loss in vivo. *Molecular Therapy, 25*, 2404–2414.

Meckes, D. G., Jr., Gunawardena, H. P., Dekroon, R. M., Heaton, P. R., Edwards, R. H., Ozgur, S., Griffith, J. D., Damania, B., & Raab-Traub, N. (2013). Modulation of B-cell exosome proteins by gamma herpesvirus infection. *Proceedings of the National Academy of Sciences of the United States of America, 110*, E2925–E2933.

Mishra, R., & Singh, S. K. (2013). HIV-1 tat C modulates expression of miRNA-101 to suppress VE-cadherin in human brain microvascular endothelial cells. *The Journal of Neuroscience, 33*, 5992–6000.

Mishra, R., Chhatbar, C., & Singh, S. K. (2012). HIV-1 tat C-mediated regulation of tumor necrosis factor receptor-associated factor-3 by microRNA 32 in human microglia. *Journal of Neuroinflammation, 9*, 131.

Mittelbrunn, M., Gutierrez-Vazquez, C., Villarroya-Beltri, C., Gonzalez, S., Sanchez-Cabo, F., Gonzalez, M. A., Bernad, A., & Sanchez-Madrid, F. (2011). Unidirectional transfer of microRNA-loaded exosomes from T cells to antigen-presenting cells. *Nature Communications, 2*, 282.

Moore, R. D., & Chaisson, R. E. (1999). Natural history of HIV infection in the era of combination antiretroviral therapy. *AIDS, 13*, 1933–1942.

Naif, H. M. (2013). Pathogenesis of HIV infection. *Infectious Disease Reports, 5*, e6.

Nasiri Kenari, A., Kastaniegaard, K., Greening, D. W., Shambrook, M., Stensballe, A., Cheng, L., & Hill, A. F. (2019). Proteomic and post-translational modification profiling of exosome-mimetic nanovesicles compared to exosomes. *Proteomics, 19*, e1800161.

Niu, M., Li, Y., Li, G., Zhou, L., Luo, N., Yao, M., Kang, W., & Liu, J. (2020). A longitudinal study on alpha-synuclein in plasma neuronal exosomes as a biomarker for Parkinson's disease development and progression. *European Journal of Neurology, 27*, 967–974.

Nolte-'t Hoen, E., Cremer, T., Gallo, R. C., & Margolis, L. B. (2016). Extracellular vesicles and viruses: Are they close relatives? *Proceedings of the National Academy of Sciences of the United States of America, 113*, 9155–9161.

Pan, B. T., Teng, K., Wu, C., Adam, M., & Johnstone, R. M. (1985). Electron microscopic evidence for externalization of the transferrin receptor in vesicular form in sheep reticulocytes. *The Journal of Cell Biology, 101*, 942–948.

Piaceri, I., Nacmias, B., & Sorbi, S. (2013). Genetics of familial and sporadic Alzheimer's disease. *Frontiers in Bioscience (Elite Edition), 5*, 167–177.

Pilakka-Kanthikeel, S., Raymond, A., Atluri, V. S., Sagar, V., Saxena, S. K., Diaz, P., Chevelon, S., Concepcion, M., & Nair, M. (2015). Sterile alpha motif and histidine/aspartic acid domain-containing protein 1 (SAMHD1)-facilitated HIV restriction in astrocytes is regulated by miRNA-181a. *Journal of Neuroinflammation, 12*, 66.

Plazolles, N., Humbert, J. M., Vachot, L., Verrier, B., Hocke, C., & Halary, F. (2011). Pivotal advance: The promotion of soluble DC-SIGN release by inflammatory signals and its enhancement of cytomegalovirus-mediated cis-infection of myeloid dendritic cells. *Journal of Leukocyte Biology, 89*, 329–342.

Polanco, J. C., Li, C., Durisic, N., Sullivan, R., & Gotz, J. (2018). Exosomes taken up by neurons hijack the endosomal pathway to spread to interconnected neurons. *Acta Neuropathologica Communications, 6*, 10.

Pomatto, M. A. C., Bussolati, B., D'Antico, S., Ghiotto, S., Tetta, C., Brizzi, M. F., & Camussi, G. (2019). Improved loading of plasma-derived extracellular vesicles to encapsulate antitumor miRNAs. *Molecular Therapy-Methods & Clinical Development, 13*, 133–144.

Rajendran, L., Honsho, M., Zahn, T. R., Keller, P., Geiger, K. D., Verkade, P., & Simons, K. (2006). Alzheimer's disease beta-amyloid peptides are released in association with exosomes. *Proceedings of the National Academy of Sciences of the United States of America, 103*, 11172–11177.

Ramos-Zaldivar, H. M., Polakovicova, I., Salas-Huenuleo, E., Corvalan, A. H., Kogan, M. J., Yefi, C. P., & Andia, M. E. (2022). Extracellular vesicles through the blood-brain barrier: A review. *Fluids and Barriers of the CNS, 19*, 60.

Raposo, G., Nijman, H. W., Stoorvogel, W., Liejendekker, R., Harding, C. V., Melief, C. J., & Geuze, H. J. (1996). B lymphocytes secrete antigen-presenting vesicles. *The Journal of Experimental Medicine, 183*, 1161–1172.

Rom, S., Rom, I., Passiatore, G., Pacifici, M., Radhakrishnan, S., Del Valle, L., Pina-Oviedo, S., Khalili, K., Eletto, D., & Peruzzi, F. (2010). CCL8/MCP-2 is a target for mir-146a in HIV-1-infected human microglial cells. *The FASEB Journal, 24*, 2292–2300.

Rossaint, J., Kuhne, K., Skupski, J., Van Aken, H., Looney, M. R., Hidalgo, A., & Zarbock, A. (2016). Directed transport of neutrophil-derived extracellular vesicles enables platelet-mediated innate immune response. *Nature Communications, 7*, 13464.

Roth, W. W., Huang, M. B., Addae Konadu, K., Powell, M. D., & Bond, V. C. (2015). Micro RNA in exosomes from HIV-infected macrophages. *International Journal of Environmental Research and Public Health, 13*, ijerph13010032.

Ruelas, D. S., Chan, J. K., Oh, E., Heidersbach, A. J., Hebbeler, A. M., Chavez, L., Verdin, E., Rape, M., & Greene, W. C. (2015). MicroRNA-155 reinforces HIV latency. *The Journal of Biological Chemistry, 290*, 13736–13748.

Rufino-Ramos, D., Albuquerque, P. R., Carmona, V., Perfeito, R., Nobre, R. J., & Pereira de Almeida, L. (2017). Extracellular vesicles: Novel promising delivery systems for therapy of brain diseases. *Journal of Controlled Release, 262*, 247–258.

Sadri Nahand, J., Bokharaei-Salim, F., Karimzadeh, M., Moghoofei, M., Karampoor, S., Mirzaei, H. R., Tabibzadeh, A., Jafari, A., Ghaderi, A., Asemi, Z., Mirzaei, H., & Hamblin, M. R. (2020). MicroRNAs and exosomes: Key players in HIV pathogenesis. *HIV Medicine, 21*, 246–278.

Saman, S., Kim, W., Raya, M., Visnick, Y., Miro, S., Saman, S., Jackson, B., McKee, A. C., Alvarez, V. E., Lee, N. C., & Hall, G. F. (2012). Exosome-associated tau is secreted in tauopathy models and is selectively phosphorylated in cerebrospinal fluid in early Alzheimer disease. *The Journal of Biological Chemistry, 287*, 3842–3849.

Sampey, G. C., Saifuddin, M., Schwab, A., Barclay, R., Punya, S., Chung, M. C., Hakami, R. M., Zadeh, M. A., Lepene, B., Klase, Z. A., El-Hage, N., Young, M., Iordanskiy, S., & Kashanchi, F. (2016). Exosomes from HIV-1-infected cells stimulate production of pro-inflammatory cytokines through Trans-activating Response (TAR) RNA. *The Journal of Biological Chemistry, 291*, 1251–1266.

Schoonenboom, N. S., Pijnenburg, Y. A., Mulder, C., Rosso, S. M., Van Elk, E. J., Van Kamp, G. J., Van Swieten, J. C., & Scheltens, P. (2004). Amyloid beta(1-42) and phosphorylated tau in CSF as markers for early-onset Alzheimer disease. *Neurology, 62*, 1580–1584.

Segura, E., Amigorena, S., & Thery, C. (2005). Mature dendritic cells secrete exosomes with strong ability to induce antigen-specific effector immune responses. *Blood Cells, Molecules & Diseases, 35*, 89–93.

Selkoe, D. J. (1991). The molecular pathology of Alzheimer's disease. *Neuron, 6*, 487–498.

Shaimardanova, A. A., Solovyeva, V. V., Chulpanova, D. S., James, V., Kitaeva, K. V., & Rizvanov, A. A. (2020). Extracellular vesicles in the diagnosis and treatment of central nervous system diseases. *Neural Regeneration Research, 15*, 586–596.

Skog, J., Wurdinger, T., van Rijn, S., Meijer, D. H., Gainche, L., Sena-Esteves, M., Curry, W. T., Jr., Carter, B. S., Krichevsky, A. M., & Breakefield, X. O. (2008). Glioblastoma microvesicles transport RNA and proteins that promote tumour growth and provide diagnostic biomarkers. *Nature Cell Biology, 10*, 1470–1476.

Skogberg, G., Lundberg, V., Berglund, M., Gudmundsdottir, J., Telemo, E., Lindgren, S., & Ekwall, O. (2015). Human thymic epithelial primary cells produce exosomes carrying tissue-restricted antigens. *Immunology and Cell Biology, 93*, 727–734.

Smyth, L. A., Ratnasothy, K., Tsang, J. Y., Boardman, D., Warley, A., Lechler, R., & Lombardi, G. (2013). CD73 expression on extracellular vesicles derived from CD4+ CD25+ Foxp3+ T cells contributes to their regulatory function. *European Journal of Immunology, 43*, 2430–2440.

Spillantini, M. G., Schmidt, M. L., Lee, V. M., Trojanowski, J. Q., Jakes, R., & Goedert, M. (1997). Alpha-synuclein in Lewy bodies. *Nature, 388*, 839–840.

Subramanian, M., Hyeon, S. J., Das, T., Suh, Y. S., Kim, Y. K., Lee, J. S., Song, E. J., Ryu, H., & Yu, K. (2021). UBE4B, a microRNA-9 target gene, promotes autophagy-mediated Tau degradation. *Nature Communications, 12*, 3291.

Tian, T., Wang, Y., Wang, H., Zhu, Z., & Xiao, Z. (2010). Visualizing of the cellular uptake and intracellular trafficking of exosomes by live-cell microscopy. *Journal of Cellular Biochemistry, 111*, 488–496.

Triboulet, R., Mari, B., Lin, Y. L., Chable-Bessia, C., Bennasser, Y., Lebrigand, K., Cardinaud, B., Maurin, T., Barbry, P., Baillat, V.,

Reynes, J., Corbeau, P., Jeang, K. T., & Benkirane, M. (2007). Suppression of microRNA-silencing pathway by HIV-1 during virus replication. *Science, 315*, 1579–1582.

Turk, A., Kunej, T., & Peterlin, B. (2021). MicroRNA-target interaction regulatory network in Alzheimer's disease. *Journal of Personalized Medicine, 11*, 1275.

Urbanelli, L., Buratta, S., Tancini, B., Sagini, K., Delo, F., Porcellati, S., & Emiliani, C. (2019). The role of extracellular vesicles in viral infection and transmission. *Vaccines (Basel), 7*.

Usami, Y., Popov, S., Popova, E., Inoue, M., Weissenhorn, W., & G.G. H. (2009). The ESCRT pathway and HIV-1 budding. *Biochemical Society Transactions, 37*, 181–184.

Valadi, H., Ekstrom, K., Bossios, A., Sjostrand, M., Lee, J. J., & Lotvall, J. O. (2007). Exosome-mediated transfer of mRNAs and microRNAs is a novel mechanism of genetic exchange between cells. *Nature Cell Biology, 9*, 654–659.

Vallelunga, A., Ragusa, M., Di Mauro, S., Iannitti, T., Pilleri, M., Biundo, R., Weis, L., Di Pietro, C., De Iuliis, A., Nicoletti, A., Zappia, M., Purrello, M., & Antonini, A. (2014). Identification of circulating microRNAs for the differential diagnosis of Parkinson's disease and multiple system atrophy. *Frontiers in Cellular Neuroscience, 8*, 156.

Wang, W. X., Rajeev, B. W., Stromberg, A. J., Ren, N., Tang, G., Huang, Q., Rigoutsos, I., & Nelson, P. T. (2008). The expression of microRNA miR-107 decreases early in Alzheimer's disease and may accelerate disease progression through regulation of beta-site amyloid precursor protein-cleaving enzyme 1. *The Journal of Neuroscience, 28*, 1213–1223.

Wang, Y., Balaji, V., Kaniyappan, S., Kruger, L., Irsen, S., Tepper, K., Chandupatla, R., Maetzler, W., Schneider, A., Mandelkow, E., & Mandelkow, E. M. (2017a). The release and trans-synaptic transmission of Tau via exosomes. *Molecular Neurodegeneration, 12*, 5.

Wang, J. K. T., Langfelder, P., Horvath, S., & Palazzolo, M. J. (2017b). Exosomes and homeostatic synaptic plasticity are linked to each other and to Huntington's, Parkinson's, and other neurodegenerative diseases by database-enabled analyses of comprehensively curated datasets. *Frontiers in Neuroscience, 11*, 149.

Wolf, P. (1967). The nature and significance of platelet products in human plasma. *British Journal of Haematology, 13*, 269–288.

Wolfers, J., Lozier, A., Raposo, G., Regnault, A., Thery, C., Masurier, C., Flament, C., Pouzieux, S., Faure, F., Tursz, T., Angevin, E., Amigorena, S., & Zitvogel, L. (2001). Tumor-derived exosomes are a source of shared tumor rejection antigens for CTL cross-priming. *Nature Medicine, 7*, 297–303.

Wong, H. K., Veremeyko, T., Patel, N., Lemere, C. A., Walsh, D. M., Esau, C., Vanderburg, C., & Krichevsky, A. M. (2013). De-repression of FOXO3a death axis by microRNA-132 and -212 causes neuronal apoptosis in Alzheimer's disease. *Human Molecular Genetics, 22*, 3077–3092.

Xia, X., Yuan, P., Liu, Y., Wang, Y., Cao, W., & Zheng, J. C. (2021). Emerging roles of extracellular vesicles in COVID-19, a double-edged sword? *Immunology, 163*, 416–430.

Xiong, Y., Song, J., Huang, X., Pan, Z., Goldbrunner, R., Stavrinou, L., Lin, S., Hu, W., Zheng, F., & Stavrinou, P. (2022). Exosomes derived from mesenchymal stem cells: novel effects in the treatment of ischemic stroke. *Frontiers in Neuroscience, 16*, 899887.

Xu, Z., Asahchop, E. L., Branton, W. G., Gelman, B. B., Power, C., & Hobman, T. C. (2017). MicroRNAs upregulated during HIV infection target peroxisome biogenesis factors: Implications for virus biology, disease mechanisms and neuropathology. *PLoS Pathogens, 13*, e1006360.

Yin, Z., Han, Z., Hu, T., Zhang, S., Ge, X., Huang, S., Wang, L., Yu, J., Li, W., Wang, Y., Li, D., Zhao, J., Wang, Y., Zuo, Y., Li, Y., Kong, X., Chen, F., & Lei, P. (2020). Neuron-derived exosomes with high miR-21-5p expression promoted polarization of M1 microglia in culture. *Brain, Behavior, and Immunity, 83*, 270–282.

Yuan, H., Zhang, Z. W., Liang, L. W., Shen, Q., Wang, X. D., Ren, S. M., Ma, H. J., Jiao, S. J., & Liu, P. (2010). Treatment strategies for Parkinson's disease. *Neuroscience Bulletin, 26*, 66–76.

Yuyama, K., Sun, H., Mitsutake, S., & Igarashi, Y. (2012). Sphingolipid-modulated exosome secretion promotes clearance of amyloid-beta by microglia. *The Journal of Biological Chemistry, 287*, 10977–10989.

Zagar, T. M., Vujaskovic, Z., Formenti, S., Rugo, H., Muggia, F., O'Connor, B., Myerson, R., Stauffer, P., Hsu, I. C., Diederich, C., Straube, W., Boss, M. K., Boico, A., Craciunescu, O., Maccarini, P., Needham, D., Borys, N., Blackwell, K. L., & Dewhirst, M. W. (2014). Two phase I dose-escalation/pharmacokinetics studies of low temperature liposomal doxorubicin (LTLD) and mild local hyperthermia in heavily pretreated patients with local regionally recurrent breast cancer. *International Journal of Hyperthermia, 30*, 285–294.

Zhang, J., Randall, G., Higginbottom, A., Monk, P., Rice, C. M., & McKeating, J. A. (2004). CD81 is required for hepatitis C virus glycoprotein-mediated viral infection. *Journal of Virology, 78*, 1448–1455.

Zhang, H. S., Chen, X. Y., Wu, T. C., Sang, W. W., & Ruan, Z. (2012). MiR-34a is involved in Tat-induced HIV-1 long terminal repeat (LTR) transactivation through the SIRT1/NFkappaB pathway. *FEBS Letters, 586*, 4203–4207.

Zhang, H., Penninger, J. M., Li, Y., Zhong, N., & Slutsky, A. S. (2020). Angiotensin-converting enzyme 2 (ACE2) as a SARS-CoV-2 receptor: Molecular mechanisms and potential therapeutic target. *Intensive Care Medicine, 46*, 586–590.

Zhang, J., Li, S., Li, L., Li, M., Guo, C., Yao, J., & Mi, S. (2015). Exosome and exosomal microRNA: Trafficking, sorting, and function. *Genomics Proteomics Bioinformatics, 13*, 17–24.

Zhou, W., Fong, M. Y., Min, Y., Somlo, G., Liu, L., Palomares, M. R., Yu, Y., Chow, A., O'Connor, S. T., Chin, A. R., Yen, Y., Wang, Y., Marcusson, E. G., Chu, P., Wu, J., Wu, X., Li, A. X., Li, Z., Gao, H., Ren, X., Boldin, M. P., Lin, P. C., & Wang, S. E. (2014). Cancer-secreted miR-105 destroys vascular endothelial barriers to promote metastasis. *Cancer Cell, 25*, 501–515.

Neurotrophins: Decades of Discoveries

17

Italo Mocchetti and Andrew Speidell

Abstract

Neuronal degeneration and loss of synapses are cardinal features of many neurological disorders. Thus, our great hope lies in our ability to develop therapies that prevent or reverse the neurodegenerative process. The central nervous system (CNS) expresses a variety of molecules with trophic activities, which could be used to rescue neurons from death. The most studied of these growth factors are neurotrophins, a family of diffusible polypeptides that promote the survival of various CNS cells. In this chapter, we will briefly present the old and the new emerging concepts of neurotrophic factors to highlight their potential as therapeutic agents to suppress neuronal degeneration.

Keywords

BDNF · Biological therapies · NGF · Neuroimmunology · Neuroprotection · NT-3 · Proneurotrophins · $p75^{NTR}$ · T-lymphocytes · Trk receptors

17.1 Introduction

The unprecedented growth of information on the structure, function, and genetics of the mammalian central nervous system (CNS) has generated high hopes for a more efficient management and treatment of neurological and psychiatric disorders. A critical element of this development is the realization that the mature CNS is much more capable of surviving and adapting to injury than formerly believed. Although there are multiple manifestations of the innate capability of the adult CNS to adjust to injury, a remarkable indication of its regenerative potential is the presence of neural stem cells in the telencephalon that can differentiate into any type of neuron when given the proper differentiation signals. More promising therapeutic avenues, however, seem to lay in our ability to develop substances that can support or augment the existing plasticity of the nervous system and consequently be employed as drugs. Many of these identified compounds to date have aimed to bolster the plastic capacity of the developed neuron. Plasticity in this chapter refers to the innate ability of the CNS to reorganize its structure and function in response to stimuli that could promote neuronal damage and/or synaptic pruning. In this way, identification of the pro-plastic mechanisms of action common to some or all of these pharmaceuticals is paramount to assessing their utility in addressing these diseases. The quintessential example of pro-plastic compounds are the neurotrophins (NTs). NT function is dependent on the neuroanatomical and developmental expression profiles, cell type-specific effects, and impact on higher-order neural functions. Still, a significant challenge for the modern clinical neurobiologist is the investigation of the fundamental mechanisms underlying neuronal injury in the adult CNS and how the nervous system may respond through NT to prevent permanent damage.

17.2 Neurotrophic Factors as Potential Biological Therapies

The search for drugs that can prevent or limit neural injury reflects the general concept of the so-called biological therapies, which are defined by two principal features: they presuppose some knowledge of pathogenetic mechanisms and address specific steps in the pathogenetic cascade. Ideally, biological therapies use physiological compounds employed by the organism for reasons related to the aims of the therapeutic intervention. In sharp contrast to biological therapeutics, the aim of traditional chemical pharmacology is to identify any type of molecule(s) that can alleviate the symptoms of a disease and is reasonably safe for chronic therapy,

I. Mocchetti (✉) · A. Speidell
Department of Neuroscience, Georgetown University Medical Center, Washington, DC, USA
e-mail: moccheti@georgetown.edu

H. E. Gendelman, T. Ikezu (eds.), *Neuroimmune Pharmacology and Therapeutics*,
https://doi.org/10.1007/978-3-031-68237-7_17

regardless of the mechanism of action. The original compounds in the pharmacopeia were discovered by serendipity, and more active compounds have been produced by analogical design, either based on a structural homology to the leading compound or a resemblance in their clinical effects.

Neurotrophic factors represent an example of a biological therapy primarily because they are naturally occurring peptides in mammalian organisms that promote the growth and survival of cells, including neurons, even in the face of pro-apoptotic stimuli. Equally important, neurotrophic factors are naturally employed to "nurture" and maintain mature neurons. Because loss of differentiation and death of neurons are hallmarks of many diseases of the CNS, especially neurodegenerative disorders, neurotrophic factors are natural candidates for the treatment of these illnesses. In addition, their efficacy in heterogeneous lesion paradigms indicates that neurotrophic factors can intervene in downstream steps in the pathogenetic cascade regardless of specific etiology or the nature of early ("upstream") events. With no exception, neurotrophic molecules act upon neurons by binding to specific membrane-associated receptors, a condition enabling them to target selected populations of neurons without direct interference with ubiquitous second messenger systems. These features are expected to minimize potential side effects stemming from nonspecific activation of signal transduction in many brain areas and organs outside the CNS.

Neurotrophic factors exert a variety of biological functions, providing both a trophic and tropic stimulus, and promoting synaptic plasticity. Tropic is the property of some neurotrophic factors to direct nerve fiber growth toward the cells that release them, whereas, the term trophism embraces the ability of neurotrophic factors to promote neuronal survival. Synaptic plasticity, in its broad formulation, encompasses multiple processes that serve to strengthen the structural and functional alterations occurring at the synapses; however, this phenomenon is not unique to neurotrophic factors. In fact, as described in the Hebbian model of trophism within neural networks, plastic homeostasis and augmented synaptic efficacy are induced by the repeated release of classical neurotransmitters from multiple presynaptic terminals. Therefore, an important issue to be considered is that the trophic function of neurotrophic factors may be disrupted in the presence of axonal and synaptic pathology, which commonly characterizes neurodegenerative disorders. This limitation is especially problematic for neurotrophic factors whose expression and signal transduction are regulated by neuronal activity. Moreover, unlike neurotransmitter receptors, a decrease in neurotrophic factor expression in the CNS may be followed by an increase in their receptors thereby maximizing the benefit of the exogenously delivered neurotrophic factors.

As mentioned above, neuronal degeneration is a common feature in many diseases of the CNS. These diseases were the first to be considered as targets of neurotrophic therapies. In conditions where degeneration occurs slowly, there is usually a considerable period between the cause of the disease, the pathological scenario, whether it contains neuronal or glial components, and the final cell death. It is also recognized that in some neurodegenerative diseases, the symptomatic phase of the illness does not necessarily involve cell death; rather, it derives from the degeneration and loss of synaptic connections. When a neuronal cell body is still present, although, under atrophy, there exists a window of opportunity to apply the neurotrophic factor therapy. This approach, therefore, aims at promoting the function of remaining synapses and potentially improving neuronal connectivity and axonal guidance. Alternatively, the delivery of recombinant neurotrophic factors may stimulate neuronal plasticity and promote neurogenesis which may help replace neurons and synapses that otherwise will be lost.

17.3 The Neurotrophins

A strategy to prevent cell death is to modify biological steps downstream of etiopathogenesis processes when neurons have already become afflicted with the initial injury but before they are committed to programmed cell death. A neurotrophic factor-based therapy would assist neurons in confronting the degenerative process from advancing to irreversible cell death. The neurotrophins (NTs) are one family of neurotrophic factors with therapeutic potential. In mammals, the NT family is composed of nerve growth factor (NGF), brain-derived neurotrophic factor (BDNF), and neurotrophins 3 (NT-3) and 4 (NT-4). NGF was discovered in the mid-1950s by Rita Levi-Montalcini, Victor Hamburg (1951), and Stanley Cohen (Cohen et al., 1954). The biological activity of NGF was initially believed to be confined to the sensory and autonomic systems. It was not until the late 1980s when NGF was discovered in the brain (Large et al., 1986), that several laboratories provided evidence that NGF is an important growth factor for the development, maintenance, and neuroprotection of cholinergic neurons (Gage et al., 1989; Koliatsos et al., 1990; Kromer, 1987; Li et al., 1995; Lucidi-Phillipi et al., 1996). However, the field of NTs and CNS plasticity exploded when Barde and coworkers characterized and cloned brain-derived neurotrophic factor (BDNF) (Leibrock et al., 1989). BDNF is the most abundant NT in the adult brain (Hofer et al., 1990) and exhibits neurotrophic activity in different populations of CNS neurons (Ghosh et al., 1994; Hyman et al., 1991; Koliatsos et al., 1993; Lyons et al., 1999). NT-3 was cloned and discovered based on sequence homology to NGF and BDNF (Maisonpierre et al., 1990). NT-3 trophic activity spans from modulating the function of central noradrenergic neurons (Akbarian et al., 2001), sympathetic neurons (Brennan et al.,

1999), and enteric neurons (Chalazonitis et al., 2001) thus expanding the potential use of NTs as biological therapies.

One mechanism used by NTs to be trophic is retrograde axonal transport (Bhattacharyya et al., 1997; Curtis et al., 1995; Kononenko et al., 2017; Zhao et al., 2016). According to this definition, a neuronal cell (whether it is a neuron or glia cell) releases trophic factors that influence an adjacent or distal neuron. By this mechanism, NTs promote the survival and maintain the differentiation of various neuronal cell types. Their effects could be long-term (with a single quantum of released trophic factors causing sustained effects), and their neurotrophic effects on survival and differentiation continue when a neuron is injured, regardless of the specific type of injury. NTs exert their biological function in a bidirectional manner. They can influence neuronal nuclei by being retrogradely transported by axons; such a mechanism requires an intact microtubular system and the motor proteins, kinesin and dynein. The function of neurotrophic factors as retrograde signals has inspired the construction of a powerful theory for the formation of the nervous system. However, it is now widely accepted that NTs are also transported anterogradely (Altar & DiStefano, 1998; Conner et al., 1998; Vaegter et al., 2011) and released from both axons and dendrites to support postsynaptic neurons that express their cognate receptors. Such transport can also be modulated by a variety of signals, including neurotransmitters and neuromodulators.

17.3.1 BDNF

Experimental and clinical evidence that NTs can be used as biological therapy is overwhelming and derive from the use of BDNF, the most abundant NT in the adult CNS. A reduction of BDNF secretion/activity is responsible for synaptic loss. This has been demonstrated not only in animals but also in humans. In fact, impaired learning and memory have been observed in both human subjects (Egan et al., 2003) and mice (Soliman et al., 2010) in which the regulated BDNF secretion is reduced due to a single nucleotide polymorphism in the BDNF gene that encodes a valine-(Val) to methionine (Met) substitution at codon 66 (Val66Met). In addition, correlative studies of human chronic neurodegenerative diseases characterized by reduced BDNF levels/activity have confirmed the role of BDNF in adult plasticity. For instance, a deficiency in BDNF synthesis has been described in postmortem brains (Phillips et al., 1991) or cerebrospinal fluid (Laske et al., 2007) of patients with Alzheimer's disease (AD), which is categorized as a disease with a reduced number of neurons and connections in the cortex and hippocampus as well as impaired short-term memory (Scheff & Price, 2006). Similarly, decreased BDNF synthesis has been suggested to be the main cause of the loss of neurons in Huntington's disease (Zuccato et al., 2001) and HIV-associated neurocognitive disorders (Bachis et al., 2012).

BDNF treatment has the potential to reduce or abolish neuronal death in several brain areas. Moreover, BDNF and NT-3 prevent the degeneration of motor descending pathways (Diener & Bregman, 1994; Schnell et al., 1994). Indeed, early experimental studies in animal models have shown a remarkable neuroprotective effect of BDNF against various neurotoxins that mimic human chronic neurodegenerative diseases such as Parkinson's and Huntington's disease (Alberch et al., 2002; Frim et al., 1994; Hyman et al., 1991). Similar neuroprotection has been observed in an animal model of AD in which the cortical input to the hippocampus contained in the perforant pathway is lesioned (Nagahara et al., 2009). Co-treatment with BDNF can also prevent synaptic loss in neurons exposed to HIV viral protein gp120 (Bachis et al., 2012). These and other experimental evidence have prompted the suggestion that BDNF could be a potential therapeutic agent for the treatment of these and other diseases characterized by loss of synaptic plasticity. Overall, BDNF remains one of the best-understood agents against neurodegeneration, and thus it deserves a critical analysis of its biological effect.

17.3.2 Neurotrophin Receptors: Dual Mechanism of Action

The biological activity of the NTs begins when they bind to a receptor complex composed of two different receptors (Fig. 17.1). One component of the neurotrophin receptor complex is the proto-oncogene Trk, a receptor tyrosine kinase which, like other tyrosine kinase receptors, is activated by ligand-induced formation of noncovalently associated receptor dimers (Cordon-Cardo et al., 1991; Kaplan et al., 1991). There are three structurally related Trks, and each neurotrophin binds selectively to a Trk: BDNF binds to TrkB, NGF to TrkA, and NT-3 to TrkC (Chao & Hempstead, 1995); however, at high concentrations, BDNF can also bind to TrkC (Klein et al., 1991). The second component of the receptor complex is the p75 neurotrophin receptor, a member of the tumor necrosis factor receptor family originally named p75 NGFR (Johnson et al., 1986; Radeke et al., 1987); however, p75 binds to all NTs with a similar affinity (Rodriguez-Tebar et al., 1990) and it is therefore named $p75^{NTR}$.

Both Trk and $p75^{NTR}$ subunits are necessary to confer high-affinity Trk binding to the NTs and to influence their biological activities (Hempstead et al., 1991; Barker, 1994; Brennan et al., 1999). When activation of $p75^{NTR}$ occurs without a concomitant activation of Trk, $p75^{NTR}$ promotes death of oligodendrocytes (Gu et al., 1999) as well as axonal degeneration in the peripheral (Kenchappa et al., 2006) and CNS (Park et al., 2010).

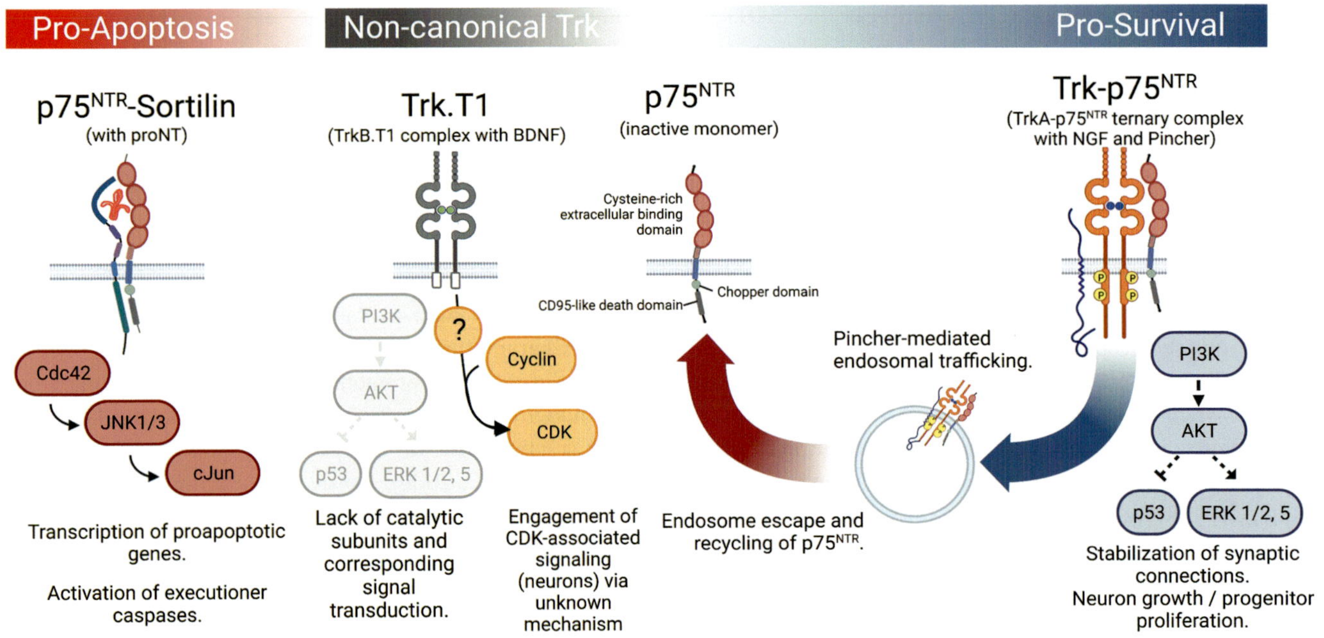

Fig. 17.1 General neurotrophin receptor stoichiometry and selected major signaling events. (Right) Functional high-affinity trk receptors are composed of both trk and p75^{NTR} subunits. These subunits are recycled back to the cell surface primarily via endosomal trafficking. (Center) Truncated trk receptors (trk.T1) lack catalytic domains but nonetheless activate distinct intracellular cascades through undetermined mechanisms. (Left) p75^{NTR} and co-receptors serve to engage major pro-apoptotic signaling pathways upon proNT binding

In addition to Trk, p75^{NTR} can also form a complex with the so-called truncated Trk (or Trk.T1), an isoform of Trk generated from an alternative splicing that does not contain the tyrosine kinase domain. TrkB.T1 can function as inhibitors of full-length Trk through a dominant-negative mechanism (Dorsey et al., 2006; Eide et al., 1996) or as an important regulator of the full-length TrkB signaling (Carim-Todd et al., 2009). A similar neurotrophic activity is seen with TrkC.T1 on neuronal crest proliferation and differentiation (Hapner et al., 1998). Nevertheless, neurons with high levels of TrkB.T1 receptor isoform, including motor neurons, appear to be less sensitive to BDNF. Indeed, the deletion of TrkB.T1 in an amyotrophic lateral sclerosis mouse model (Gurney et al., 1994) significantly slowed the onset of motor neuron degeneration (Yanpallewar et al., 2012). In addition, TrkB.T1 has been shown to promote the production of nitric oxide from astrocytes, which could cause neuronal degeneration (Colombo et al., 2012). Thus, the expression of Trk.T1 on neuronal and glial populations must be considered when approaching the use of BDNF for human neurological diseases.

17.3.3 p75^{NTR} and proNTs

p75^{NTR} can also bind the larger precursor protein for the NTs or proNTs (Hempstead, 2002). ProNTs, which are abundant in the mature brain (Fahnestock et al., 2001), can be released from neurons (Yang et al., 2009) and accumulate extracellularly unless astrocytes initiate their rapid uptake (Bergami et al., 2008). ProNTs can be cleaved in the endoplasmic reticulum by the proconvertase furin (Seidah et al., 1996) or extracellularly by proteases such as plasmin and matrix metalloproteases to be converted to mature NTs (Mowla et al., 1999). The conversion of proNTs to mature NTs is an important process for synaptic plasticity. In fact, binding of proNTs to p75^{NTR} in the absence of Trk activation leads to a constellation of molecular and cellular events that affect neuronal survival, including axonal retraction, cytoskeletal remodeling, and apoptosis (Harrington et al., 2004; Teng et al., 2005). While these effects are important during CNS development because they shape pre- and postsynaptic elements, in the adult CNS, they represent a major hurdle for neuronal survival. Several neuronal populations could be at risk because of low but constitutive p75^{NTR} expression, including GABAergic (Baho et al., 2019; Riffault et al., 2014), AMPA-expressing neurons (Rosch et al., 2005), and basal forebrain cholinergic neurons (Xie et al., 2019). Thus, reducing p75^{NTR} activation by the proNTs is paramount for an effective therapy.

The mechanisms of p75^{NTR}-mediated cell death include several "toxic" signaling pathways that include ceramide production (Dobrowsky et al., 1994), caspase-3 (Guo et al., 2016), and activation of nuclear factor-kB and Jun kinase (Bhakar et al., 2003; Carter et al., 1996) and proinflammatory responses (Luo et al., 2016). Apoptosis mediated by

$p75^{NTR}$ appears to require various membrane-bound and cytosolic adaptor proteins. These include NT receptor-interacting factor (NRIF) (Linggi et al., 2005; Volosin et al., 2008) and sortilin, a major $p75^{NTR}$ co-receptor located ubiquitously in neurons that serves to sort and transport proNTs (Nykjaer et al., 2004; Teng et al., 2005). Overall, $p75^{NTR}$ and Trk can promote survival or death depending upon the cellular context and the different adaptor proteins and signaling.

17.4 Neurotrophins and Glial Cells

A wealth of studies over the last three decades has illuminated an array of alternative roles for NTs in various immune cell types, which are not limited to the comparatively simple cell survival and proliferation functions typically associated with NT signaling. Moreover, sustained research into NTs and their receptors led to the observation that glial cells and immune cells synthetize NTs and express identical NT receptor isoforms as do their neuronal counterparts. Consequently, the effects of NT and NT receptor activation on these cell types have generated key insights into the wider neuroimmune axis and how it may sustain or restrain the immune response. Thus, it has been suggested and demonstrated that NTs may exert a broader span of immune-related functions than previously believed.

17.4.1 Neurotrophins and Microglia

Microglia, the primary immune effector cells in the CNS, respond to NT exposure and, in certain pathological conditions, release NTs. This special role as both a producer and target of NTs allows microglia to interact with neurons to enact an array of neuroimmune functions throughout the lifetime of the nervous system. In a developmental context, the microglial release of BDNF is thought to promote the outgrowth and stabilization of neurites to form functional synapses (Yang et al., 2012). Microglial-specific knockout of BDNF demonstrated that microglial-derived NTs assist in synapse formation and stabilization throughout development and adulthood in the rodent central nervous system (Parkhurst et al., 2013), which compares well with the "immune-like" role for microglia in eliminating synapses and synaptic debris through complement cascade proteins (Hong et al., 2016; Schafer et al., 2012). Intriguingly, the microglial release of NGF can also elicit programmed cell death in certain specialized nervous system structures, likely through the activation of the pan-proNT receptor $p75^{NTR}$ (Frade & Barde, 1998; Srinivasan et al., 2004).

Microglia may also respond to the acute release of NTs. In agreement with the broad trophic set of functions for the NT family of ligands in neurons, microglial proliferation and phagocytic capacity are known to be greatly enhanced following exposure to NT3, NGF, and BDNF (Elkabes et al., 1996). NTs and NT receptor signaling appear to modulate the microglial components of the neuroinflammatory response, as microglial co-exposure to NGF and LPS restrains the production of proinflammatory cytokines and corresponding activation of NF-κB and JNK pathways through TrkA (Fodelianaki et al., 2019). Co- or pre-treatment with some NTs, especially NGF, also appear to strongly upregulate microglial chemotaxis (De Simone et al., 2007), but intriguingly this treatment also biases microglial activation to prefer a neuroprotective phenotype of activation when exposed to inflammatory stimuli (Rizzi et al., 2018). Finally, microglia respond to certain inflammatory stimuli by increasing the synthesis and release of NTs, especially BDNF, which is likely to support the regeneration and return to the normal or resting functions of other CNS cell types (Nakajima et al., 2001).

17.4.2 NT and Astrocytes

The expression of NTs and their receptor in astrocytes has been a matter of debate. In vitro astrocytes obtained from several brain regions upregulate NTs in response to proinflammatory cytokines (Juric & Carman-Krzan, 2000; Saha et al., 2006; Yoshida et al., 1992). In vivo astrocytes are known to express low levels of NTs in a constitutive manner and thereby could contribute to the CNS pool of mature NTs, especially throughout development (Thompson et al., 2014). When astrocytes adopt an activated functional phenotype (i.e., "reactive astrocytes") in response to a traumatic and/or inflammatory stimulus, they strongly upregulate the production of NTs (Fulmer et al., 2014; Linker et al., 2010; Saha et al., 2006; Zhang et al., 2016). In parallel with the microglial release of NTs, it is also believed that the astrocytic release of NTs serves as a protective and counterbalancing factor against glial co-released neurotoxic factors central to the neuroinflammatory response. This hypothesis is supported by studies indicating that depletion of astrocyte-derived BDNF exacerbates outcomes in certain models of neurodegenerative diseases (Linker et al., 2010). However, it may be remiss to consider the NT effect on astrocytic immune function as a universal benefit for neural health, as astrocytes can also produce proNGF in response to kainic acid (Volosin et al., 2006). Some reports indicate that an increase in free BDNF and a corresponding upregulation of TrkB on astrocytes during CNS inflammation serve to potentiate the neurotoxic release of nitric oxide (NO) from these glial cells thereby aggravating inflammatory events (Colombo et al., 2012). Experiments employing culture systems have shown that exposure of astrocytes to NO indeed stimulates the release of proNGF from astrocytes and promotes apoptosis of cocultured motor neurons. Released proNGF could damage oligodendrocytes expressing $p75^{NTR}$ (Casaccia-Bonnefil et al., 1996), expanding the initial injury caused by neurotoxins.

Astrocytes are also expressors of truncated Trk receptors (Alderson et al., 2000; Aroeira et al., 2015; Saba et al., 2018) as well as $p75^{NTR}$, and are thereby poised to respond to the acute release of NTs in an autocrine or paracrine manner. Which immune-related functions of astrocytes are therefore activated or modulated by NT exposure? Glial scar formation, the physiological separation of healthy neural tissue from lesions, is likely facilitated by the activation of nearby astrocytes by NTs and wider members of the neurotrophic factor family, including glial cell line-derived (GDNF) and ciliary neurotrophic factors (CNTF) (Wang et al., 2018). There is also likely recruitment of activated microglia and oligodendrocyte precursors to sites of emerging glial scars through the astrocytic release of NTs. Ultimately, the general glial neuroimmune response in the presence of NT stimulation appears to favor a restrained yet activated state. Depending upon which NT species is released, glia can therefore adopt neuroprotective or neurotoxic roles.

17.5 Neurotrophins: Role in Immune Organs and the Systemic Immune Response

Biological therapies for the CNS should also consider potential effects in nonneuronal cells. This is particularly true for the NTs. Although NTs are most closely associated with their action in the CNS and PNS, NTs are known to have a complementary suite of functions in peripheral tissue, including within primary and secondary immune organs (Fig. 17.2). It has been established for over two decades that NGF and its receptors are expressed at sites of T-lymphocyte selection and maturation in the thymus (Aloe et al., 1997). In fact, studies with mice homozygous for targeted disruption of TrkA catalytic domain indicate that TrkA is necessary for normal differentiation and development of the thymus (Garcia-Suarez et al., 2000). In addition, TrkA-mutant mice exhibit dysregulation of Ig production and accumulation of B-cells (Coppola et al., 2004). Moreover, it is known that NGF acts as an autocrine survival factor for memory B lymphocytes (Torcia et al., 1996). Additional lines of investigation showed that NGF and TrkA are required for the development and regeneration of thymocytes (Garcia-Suarez et al., 2000; Vega et al., 2003). Remarkably, human and murine thymic tissue expresses multiple alternative semi- or nonfunctional splice variants of TrkA, which are hypothesized to throttle the maturation of thymocytes and T-lymphocytes in response to NGF throughout development (Tacconelli et al., 2007).

The BDNF–TrkB signaling axis is also heavily implicated in the differentiation of T-lymphocytes. Mice with functionally deficient TrkB display robust apoptosis of nascent lymphocytes in the thymic cortex (Garcia-Suarez et al., 2002). Immune reconstitution and the regeneration of immune organs following injury were recently found to be promoted through BDNF, likely derived from a population of circulating endothelial progenitors (Sharma et al., 2021). B-lymphocyte development also appears to rely heavily on BDNF exposure, as marrow stromal cells adjacent to sites of immature B-lymphocyte generation express and secrete BDNF (Paradisi et al., 2014). In support of this observation, B-lymphocytes, which express both TrkB and $p75^{NTR}$, are depleted in BDNF-deficient mice (Schuhmann et al., 2005). Although investigated with less attention, constituent cells within secondary lymphoid organs (SLOs) are known to synthesize or respond to NTs. A multitude of reports indicate that NTs and NTRs, or their transcripts, are expressed in the spleen (Laurenzi et al., 1994; Perez-Perez et al., 1999), tonsils (Artico et al., 2008; Bronzetti et al., 2006; Pezzati et al., 1992), and intestinal lymphoid tissue (Levanti et al., 1997; Pezzati et al., 1992) in animal models and in humans. Their exact function in these SLOs is unclear, but NTs and NTRs likely support the innervation of some of these structures by peripheral nervous pathways and assist in the maintenance of the pool of healthy leukocytes.

How might NTs regulate innate and acquired immune responses? Most mature leukocytes, including B- and T-lymphocytes, macrophages, mast cells, monocytes, and others, are known to express NTs or their receptors (Kozlov et al., 2020; Vega et al., 2003). Some NTs, especially NGF, are well connected to immune cell population differentiation, expansion and survival, albeit largely in culture systems (Bracci Laudiero & Manni, 2014). T- and B-lymphocytes appear to be particularly responsive to NT signaling both in the absence and presence of pathogenic stimuli. To this end, a plethora of studies involving these immune cells repeatedly describes the upregulation of IgG secretion (Brodie & Gelfand, 1994; Kimata et al., 1991), the release of immune-modulating peptides (Bracci-Laudiero et al., 1996; Bracci-Laudiero et al., 2002), enhanced chemoattraction (Samah et al., 2009), and augmented lymphocyte survival in the presence of NGF (Abram et al., 2009; Torcia et al., 1996). There are fewer reports describing BDNF-mediated modulation of immune responses. Nevertheless, these studies suggest that BDNF indeed promotes B-lymphocyte survival through TrkB (Fauchais et al., 2008; Talbot et al., 2020). Overall, NGF, BDNF, and other NTs serve to support the development, survival, and immune function of a variety of immune cell subtypes. Along with chemokines and other emerging immune regulators, these NTs, therefore, represent key mod-

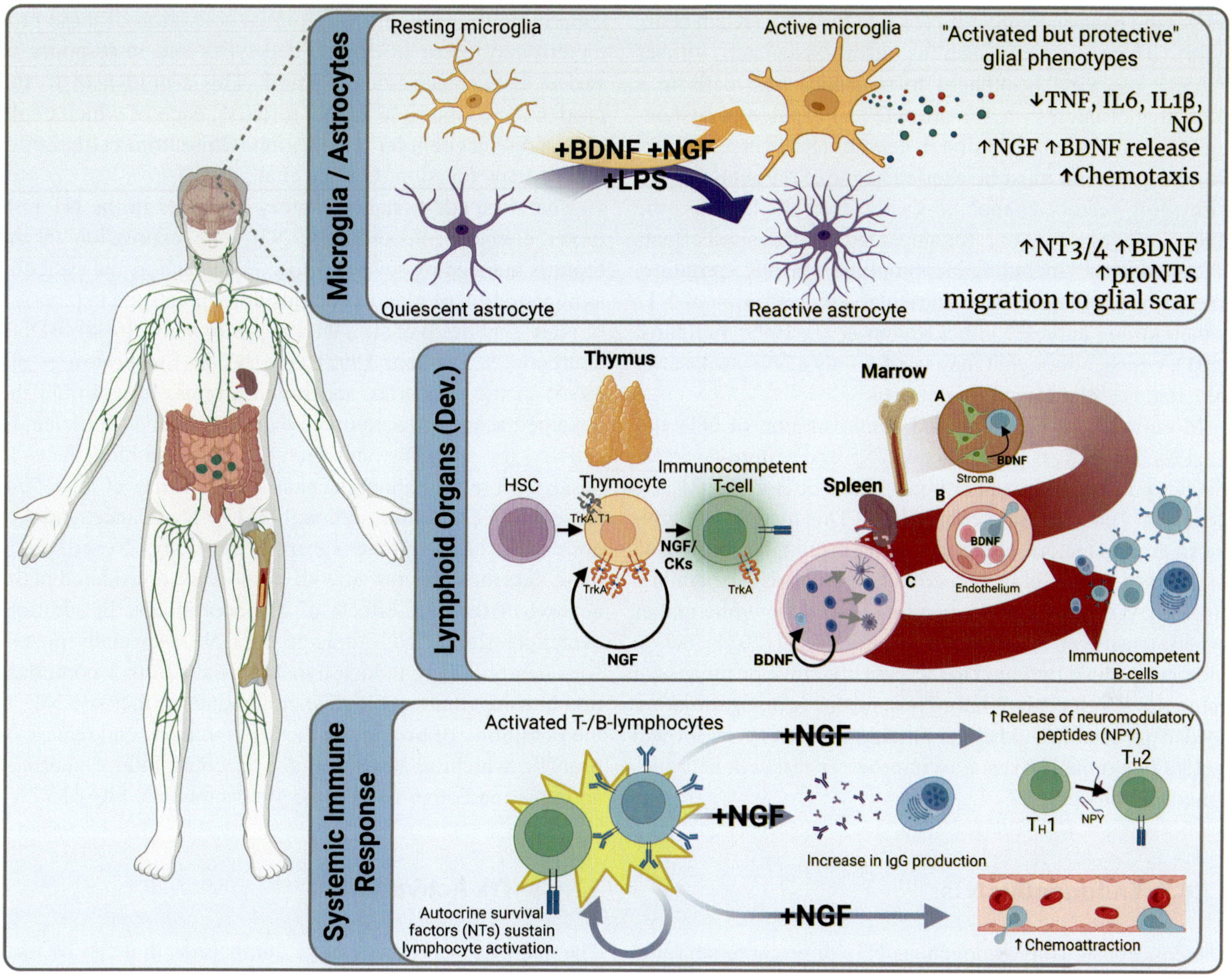

Fig. 17.2 Neurotrophin-mediated events in the immune system. (Top) BDNF and NGF affect glial modes of activation in the presence of pathogenic stimuli to favor protective phenotypes. (Center) Hematopoietic stem cells (HSC) are driven toward immunocompetency through exposure to cytokines (CKs) and action of NGF at TrkA in the thymus. Events in progenitor migration involving (**a**) exit from marrow stroma, (**b**) reseeding of lymphoid organs, and (**c**) B-cell maturation and differentiation in splenic germinal centers may be partially mediated through BDNF. (Bottom) Overview of proposed lymphocyte activation-modulating events in the presence of NTs

ulating factors to the broader immune response. Such responses are particularly crucial at the site of CNS injury, where NTs may initiate homological–heterological modulations of immune and neuronal cell function necessary for neuronal survival.

17.6 Challenges for the Clinical Use of NTs

17.6.1 Delivery of NTs

NTs have the ability to promote the re-establishment of synaptic connections and stop the progression of synaptic pathology. Thus, it would appear that NTs can, on the one hand, serve as ideal pharmacological tools for the treatment of neurological diseases. There are several issues to be considered in using NTs as a therapy for dying neurons. The first one is that NT receptors must be expressed or re-expressed in the brains of patients with neurological diseases, which are often characterized by loss of synapses. However, the principal aim of the trophic factor approach is neuronal regeneration not the replacement of missing axons. NTs, like other growth factors, have a short half-life in the blood due to active proteases; thus, the bioavailability of NTs after oral administration is negligible. Another challenge is that the polypeptidergic structure of NTs impairs their ability to cross the blood–brain barrier. This can be circumvented only by direct delivery of NTs into the ventricular system or brain parenchyma, which does not represent a viable nor practical treatment for humans. Viral vectors have been designed to deliver the gene encoding for NTs into the CNS (Klein et al., 1998). These vectors, such as adeno-associated viruses (AAVs), use viral backbones to transduce host cells in a brain area of interest. AAVs are safe and capable of sustaining long-term NTs expression in neurons (Klein et al., 1998). However, caution must be exerted as the expression of NTs with viral vectors cannot be easily controlled. Thus, the clinical efficacy could be accompanied by undesired effects of "overdosing," including inappropriate synaptic sprouting. Another side effect of intraventricular administration of NTs is their strong anorexic effect (Lyons et al., 1999; Williams, 1991), which, although it may last for only a few weeks after infusion, is still a reason for concern.

More recently, the intranasal administration of NTs has been used to deliver NTs into the CNS and restore neuronal plasticity in experimental animals (Braschi et al., 2021; De Rosa et al., 2005; Manni et al., 2023). This method of delivery of growth factors, and in particular insulin, has been used with mixed results in human conditions such as Alzheimer's disease (AD). In one study, insulin reduced the white matter hyperintensity of AD patients (Kellar et al., 2021) but in another one, insulin failed to increase cognitive or functional outcomes (Craft et al., 2020). Thus, further investigations are needed to establish whether intranasal delivery of growth factors is a reliable avenue to improve or rescue a neurodegenerative process.

17.6.2 Endogenous NTs

The concept of using endogenous NTs to promote neuronal survival originates from the evidence that the expression of NTs in certain parts of the CNS is regulated by neuronal activity. Thus, small molecules that change neuronal activity have the ability to modulate the synthesis and release of NTs. These phenomena require a basic knowledge of the gene regulatory elements, signal transduction, and transcriptional events underlying the neuronal-specific activation of the NT genes. For instance, NGF gene transcription is induced by increases in cAMP and activation of the protein kinase A (PKA) signaling pathway (Mocchetti & Colangelo, 2001). Among transcription factors that are cAMP-inducible, CCAAT/enhancer binding protein delta (C/EBPδ) is implicated in regulating NGF gene expression in the brain in response to beta-adrenergic receptor activation (Colangelo et al., 1998). The transcription repressor MeCP2 (Chen et al., 2003) and the calcium-responsive transcription factor CaRF modulate activity-dependent BDNF expression (Tao et al., 2002). The BDNF gene is also activated when cyclic AMP (cAMP) response element (CRE)-binding protein (CREB) binds to the corresponding regulatory elements in the BDNF promoter (Tao et al., 1998). However, the BDNF gene has a complex structure with nine promoters (Park & Poo, 2013), supporting the view that differential activation of these promoters may occur in diverse cell types and in response to varied activity-dependent stimuli. This would lead to the production of multiple BDNF mRNAs, each of which could have a distinct cellular localization (Chiaruttini et al., 2009) and perhaps function (Orefice et al., 2013).

There are additional regulatory elements in the NT promoter elements; accordingly, NT gene expression in the brain is induced by several drugs. For instance, the systemic administration of glucocorticoids upregulates the expression of NGF (Fabrazzo et al., 1991; Saporito et al., 1994), BDNF (Barbany & Persson, 1992), and NT-3 (Lauterborn et al., 1995) in the neocortex and hippocampus. In addition, the dexamethasone-induced upregulation of NGF is sufficient to activate trk signaling and thereby exert trophic effects in basal forebrain cholinergic neurons (Saporito et al., 1994; Shi et al., 1998). Thus, although in low abundance, endogenous NTs can improve synaptic plasticity. Nevertheless, these data must be viewed with caution and considered in the context of the side effects of glucocorticoids. In addition, overexpression of NGF outside the CNS, especially in sensory neurons, may induce hyperalgesia. Another complication in using pharmacological compounds to increase NTs is the possibility of promoting the accumulation and release of proNTs, which, as described above, could induce neuronal apoptosis and other neurotoxic effects mediated by $p75^{NTR}$.

17.6.3 Trk Activation

The need for neurotherapeutic compounds that can be used systemically to mimic NTs or antagonize $p75^{NTR}$ is critical. However, several considerations should be taken when developing these compounds. A neuropathological scenario may arise from a lack of Trk activation; in this case, Trk agonist or NT mimetics should be employed. Conversely, $p75^{NTR}$ and/or sortilin antagonists could function as a viable strategy to counteract pro-NT-mediated activation of neuronal apoptosis.

Importantly, given the fact that at high concentrations NTs can indiscriminately activate $p75^{NTR}$, caution must be exerted in delivering NTs into the mature injured brain where there is a possibility of an upregulation of $p75^{NTR}$ without a concomitant increase in Trk (Beattie et al., 2002; Brunello et al., 1990; Volosin et al., 2006).

Exogenous ganglioside GM1 was one of the first attempts to develop nonpeptidergic NT mimetics that activate Trk indirectly. Gangliosides are acidic glycosphingolipids containing sialic acid linked to an oligoglycosyl backbone attached to a ceramide base. They exhibit neurotrophic activity in different neuronal populations. Independent lines of investigation have provided evidence that GM1 activates TrkA via a classical receptor tyrosine kinase autophosphorylation mechanism. (Ferrari et al., 1995; Mutoh et al., 1995; Rabin & Mocchetti, 1995). This event is followed by Trk activation of several target proteins, including PLCγ and Erk1/2. Later studies that assessed the relative potency of gangliosides have established that gangliosides are robust activators of different Trks due to their ability to promote the release of NTs from neuronal cells. (Rabin et al., 2002). This approach could overcome a major hurdle for the clinical application of gangliosides because NT expression is down-regulated in most neurodegenerative diseases.

TrkB activation can be achieved either by the increased release of the ligand BDNF or through ligand-independent transactivation. A series of studies have shown that TrkB, and other tyrosine kinase receptors, can be transactivated by G-protein-coupled receptors, such as adenosine A2, even in the absence of ligands (Fischer et al., 2003; Wiese et al., 2007). This transactivation has a potential clinical relevance because, in rodents, it improves the survival of neurons after axotomy (Wiese et al., 2007). However, adenosine may promote unwanted side effects in the cardiovascular system and its use as an form of neurotherapy is likely limited (Sousa & Diniz, 2017).

17.6.4 NT Mimetics

Great emphasis is currently placed on the engineering of small peptides deriving from NT sequences that bind to Trks (O'Leary & Hughes, 2003; Zaccaro et al., 2005) or combine features of two or more NTs. The latter strategy presents the advantage of combined efficacy on more populations of neurons than single native NTs, an advantage confirmed in vitro for several "multifunctional" NTs generated through homolog-scanning or site-directed mutagenesis (Ilag et al., 1995; Suter et al., 1992; Urfer et al., 1994). Mutation analyses assessing the proteolytic cleavage of neurotrophin molecules confirm that sequences distant from the receptor-binding site (i.e. the first 10 N-terminal amino acids of NGF and BDNF) are crucial for receptor interaction, highlighting that protein structure is as important as sequence specificity for the neurotrophic effects of individual neurotrophins (Ibanez, 1995). These considerations have justified new research into small molecules with no ordered peptidergic structure that can be used systemically to mimic or antagonize NTs. Several considerations should be taken when developing these compounds. A neuropathological scenario may arise from a lack of Trk activation; in this case, Trk agonist or NT mimetics should be employed. Conversely, $p75^{NTR}$ and/or sortilin antagonists could function as a viable strategy to counteract pro-NT-mediated activation of neuronal apoptosis. Importantly, given the fact that at high concentrations, neurotrophins can indiscriminately activate $p75^{NTR}$, caution must be exerted in delivering NTs into the mature injured brain where there is a possibility of an upregulation of $p75^{NTR}$ without a concomitant increase in Trk (Beattie et al., 2002; Brunello et al., 1990; Volosin et al., 2006).

There are several small-molecule Trk agonists described in the literature; these include the flavone derivative tropoflavin (Obianyo & Ye, 2013). There are other structural classes reported in this field as well, such as LM22A-4 (Schmid et al., 2012) and the classic antidepressant, amitriptyline (Jang et al., 2009). Although all these compounds demonstrated trophic function, there is some skepticism about whether the compounds are truly Trk agonists because they may promote the release of the corresponding ligands or induce transactivation of Trk. Finding a small-molecule ligand for this receptor is indeed not an easy task. Conversely, $p75^{NTR}$ antagonists have been under recent development. These include LM11A-31, a nonpeptidergic small molecule that prevents proNGF-mediated death of neurons and oligodendrocytes (Massa et al., 2006) and reverses cholinergic neurite dystrophy observed in a mouse model of AD (Simmons et al., 2014). However, LM311A-31 has been shown to counteract the toxicity of anticancer drugs on neurons in culture (James et al., 2008). Thus, the trophic activity of LM311A-31 appears to go beyond its antagonist effect of $p75^{NTR}$. Clearly, more work needs to be done before using these approaches for human neurological diseases.

17.7 Conclusions

Besides their traditional neuronal role as molecules promoting regeneration, NTs have come to be increasingly recognized as potential mediators/modifiers of synaptic plasticity in the adult mammalian brain. The trophic effects of NTs broaden their role as mediators of both acute and chronic cellular changes that underlie adaptive functions of the forebrain, such as learning and memory. To this end, we recognized NTs and other neurotrophic factors as essential neuropeptides for CNS function. There are major concerns about using NTs in a clinical setting largely because NTs are large hydrophilic molecules that do not cross the BBB. However, this does not exclude NT-based therapies from additional consideration. For instance, physical exercise has been reported to increase BDNF levels in the hippocampus (Koob & Greenwell, 2004), which, in turn, promotes hippocampal-based learning, synaptic plasticity, and neurogenesis. Antidepressant drugs, which are safely used in humans, also increase BDNF synthesis as part of their range of neural effects (Hashimoto et al., 2002; Martin et al., 2004). Thus, the historical hurdle of BBB permeability may not be as formidable as once previously believed and more invasive routes of NT administration could soon be obviated. Overall, these results illustrate the difficult interface often encountered between experimental data and the potential for application in a clinical setting. Together, the lessons uncovered from decades of NT-focused research highlight an improving understanding of issues of the bioavailability of NTs to target nerve cells and the potential side effects of their long-term use.

17.8 Review Questions

1. The family of neurotrophins (NTs) is composed of (select the best answer)
   (a) BDNF, NGF, GM1, and NT-4
   (b) NT-3, NGF, GDNF, and BDNF
   (c) BDNF, CNTF, NGF, and NT-3
   (d) NGF, BDNF, NT-3, and NT-4
   (e) GDNF, NGF, CNTF, and BDNF

2. Which choice is not true regarding neurotrophin receptors (NTRs)?
   (a) Combinatorial arrangements of Trk and $p75^{NTR}$ subunits may form functional NTR complexes.
   (b) While $p75^{NTR}$ is a pan-proneurotrophin receptor, it may also bind NTs at lower affinity.
   (c) All NTRs follow a classic receptor tyrosine kinase autophosphorylation mode of activation upon ligand binding.
   (d) Truncated Trk receptors lack the catalytic domain of full-length Trks.
   (e) Activation of an NTR complex can be pro- or anti-apoptotic depending on ligand, receptor, and co-receptor composition.

3. Which choice best describes the role of neurotrophins in relation to immune system cells?
   (a) NTs have no identified neuroimmune role other than sustaining activation of existing lymphocytes.
   (b) NTs exacerbate immune responses in most cases by increasing proinflammatory cytokine release and upregulating expression of "eat me" signals on neighboring cells.
   (c) NTs inhibit the expansion of regulatory T cells while allowing for the proliferation of cytotoxic and helper variants of T lymphocytes.
   (d) Expression of NTRs on immune progenitor populations and within secondary lymphoid organs (SLOs) indicates that NTs likely serve as a trophic signal for further development and maturation of lymphocytes.
   (e) Expression of NTRs on cells of the SLOs assists in the developmental patterning of the SLO, but no further role for NTs is known in these organs.

4. Limitations to the use of NTs for neurological disorders include which of the following:
   (a) Poor penetrability of NTs through the blood–brain barrier.
   (b) Short half-life of NTs in the systemic circulation.
   (c) Significant off-target effects due to widespread expression of NTRs.
   (d) NTs alone are insufficient to restore or regenerate neural structures which may have already been lost.
   (e) All the above.

5. Which of the following approaches may have a detrimental effect as a therapy for CNS disorder?
   (a) Administration of proNTs into the CNS.
   (b) CNS-permeable, small-molecule inhibitors of $p75^{NTR}$.
   (c) Administration of gangliosides.
   (d) Injection of viral vectors encoding NTs.
   (e) Rigorous exercise.

17.9 Answers to Review Question

1. d
2. c
3. d
4. e
5. a

References

Abram, M., Wegmann, M., Fokuhl, V., Sonar, S., Luger, E. O., Kerzel, S., Radbruch, A., Renz, H., & Zemlin, M. (2009). Nerve growth factor and neurotrophin-3 mediate survival of pulmonary plasma cells during the allergic airway inflammation. *Journal of Immunology, 182*, 4705–4712.

Akbarian, S., Bates, B., Liu, R. J., Skirboll, S. L., Pejchal, T., Coppola, V., Sun, L. D., Fan, G., Kucera, J., Wilson, M. A., Tessarollo, L., Kosofsky, B. E., Taylor, J. R., Bothwell, M., Nestler, E. J., Aghajanian, G. K., & Jaenisch, R. (2001). Neurotrophin-3 modulates noradrenergic neuron function and opiate withdrawal. *Molecular Psychiatry, 6*, 593–604.

Alberch, J., Perez-Navarro, E., & Canals, J. M. (2002). Neuroprotection by neurotrophins and GDNF family members in the excitotoxic model of Huntington's disease. *Brain Research Bulletin, 57*, 817–822.

Alderson, R. F., Curtis, R., Alterman, A. L., Lindsay, R. M., & DiStefano, P. S. (2000). Truncated TrkB mediates the endocytosis and release of BDNF and neurotrophin-4/5 by rat astrocytes and schwann cells in vitro. *Brain Research, 871*, 210–222.

Aloe, L., Micera, A., Bracci-Laudiero, L., Vigneti, E., & Turrini, P. (1997). Presence of nerve growth factor in the thymus of prenatal, postnatal and pregnant rats. *Thymus, 24*, 221–231.

Altar, C. A., & DiStefano, P. S. (1998). Neurotrophin trafficking by anterograde transport. *Trends in Neurosciences, 21*, 433–437.

Aroeira, R. I., Sebastiao, A. M., & Valente, C. A. (2015). BDNF, via truncated TrkB receptor, modulates GlyT1 and GlyT2 in astrocytes. *Glia, 63*, 2181–2197.

Artico, M., Bronzetti, E., Felici, L. M., Alicino, V., Ionta, B., Bronzetti, B., Magliulo, G., Grande, C., Zamai, L., Pasquantonio, G., & De Vincentiis, M. (2008). Neurotrophins and their receptors in human lingual tonsil: An immunohistochemical analysis. *Oncology Reports, 20*, 1201–1206.

Bachis, A., Avdoshina, V., Zecca, L., Parsadanian, M., & Mocchetti, I. (2012). Human immunodeficiency virus type 1 alters brain-derived neurotrophic factor processing in neurons. *The Journal of Neuroscience, 32*, 9477–9484.

Baho, E., Chattopadhyaya, B., Lavertu-Jolin, M., Mazziotti, R., Awad, P. N., Chehrazi, P., Groleau, M., Jahannault-Talignani, C., Vaucher, E., Ango, F., Pizzorusso, T., Baroncelli, L., & Di Cristo, G. (2019). p75 Neurotrophin receptor activation regulates the timing of the maturation of cortical Parvalbumin interneuron connectivity and promotes juvenile-like plasticity in adult visual cortex. *The Journal of Neuroscience, 39*, 4489–4510.

Barbany, G., & Persson, H. (1992). Regulation of neurotrophin mRNA expression in the rat brain by glucocorticoids. *The European Journal of Neuroscience, 4*, 396–403.

Barker, P. E. S. (1994). Disruption of NGF binding to the low affinity neurotrophin receptor p75LNTR reduced NGF binding to trkA on PC12 cells. *Neuron, 13*, 203–215.

Beattie, M. S., Harrington, A. W., Lee, R., Kim, J. Y., Boyce, S. L., Longo, F. M., Bresnahan, J. C., Hempstead, B. L., & Yoon, S. O. (2002). ProNGF induces p75-mediated death of oligodendrocytes following spinal cord injury. *Neuron, 36*, 375–386.

Bergami, M., Santi, S., Formaggio, E., Cagnoli, C., Verderio, C., Blum, R., Berninger, B., Matteoli, M., & Canossa, M. (2008). Uptake and recycling of pro-BDNF for transmitter-induced secretion by cortical astrocytes. *The Journal of Cell Biology, 183*, 213–221.

Bhakar, A. L., Howell, J. L., Paul, C. E., Salehi, A. H., Becker, E. B., Said, F., Bonni, A., & Barker, P. A. (2003). Apoptosis induced by p75NTR overexpression requires Jun kinase-dependent phosphorylation of Bad. *The Journal of Neuroscience, 23*, 11373–11381.

Bhattacharyya, A., Watson, F. L., Bradlee, T. A., Pomeroy, S. L., Stiles, C. D., & Segal, R. A. (1997). Trk receptors function as rapid retrograde signal carriers in the adult nervous system. *The Journal of Neuroscience, 17*, 7007–7016.

Bracci Laudiero, L., & Manni, L. (2014). NGF and immune regulation. In R. Kostrzewa (Ed.), *Neurotoxicity handbook* (pp. 1849–1876). Springer.

Bracci-Laudiero, L., Aloe, L., Stenfors, C., Tirassa, P., Theodorsson, E., & Lundberg, T. (1996). Nerve growth factor stimulates production of neuropeptide Y in human lymphocytes. *Neuroreport, 7*, 485–488.

Bracci-Laudiero, L., Aloe, L., Buanne, P., Finn, A., Stenfors, C., Vigneti, E., Theodorsson, E., & Lundeberg, T. (2002). NGF modulates CGRP synthesis in human B-lymphocytes: A possible anti-inflammatory action of NGF? *Journal of Neuroimmunology, 123*, 58–65.

Braschi, C., Capsoni, S., Narducci, R., Poli, A., Sansevero, G., Brandi, R., Maffei, L., Cattaneo, A., & Berardi, N. (2021). Intranasal delivery of BDNF rescues memory deficits in AD11 mice and reduces brain microgliosis. *Aging Clinical and Experimental Research, 33*, 1223–1238.

Brennan, C., Rivas-Plata, K., & Landis, S. C. (1999). The p75 neurotrophin receptor influences NT-3 responsiveness of sympathetic neurons in vivo. *Nature Neuroscience, 2*, 699–705.

Brodie, C., & Gelfand, E. W. (1994). Regulation of immunoglobulin production by nerve growth factor: Comparison with anti-CD40. *Journal of Neuroimmunology, 52*, 87–96.

Bronzetti, E., Artico, M., Pompili, E., Felici, L. M., Stringaro, A., Bosco, S., Magliulo, G., Colone, M., Arancia, G., Vitale, M., & Fumagalli, L. (2006). Neurotrophins and neurotransmitters in human palatine tonsils: An immunohistochemical and RT-PCR analysis. *International Journal of Molecular Medicine, 18*, 49–58.

Brunello, N., Reynolds, M., Wrathall, J. R., & Mocchetti, I. (1990). Increased nerve growth factor receptor mRNA in contused rat spinal cord. *Neuroscience Letters, 118*, 238–240.

Carim-Todd, L., Bath, K. G., Fulgenzi, G., Yanpallewar, S., Jing, D., Barrick, C. A., Becker, J., Buckley, H., Dorsey, S. G., Lee, F. S., & Tessarollo, L. (2009). Endogenous truncated TrkB.T1 receptor regulates neuronal complexity and TrkB kinase receptor function in vivo. *The Journal of Neuroscience, 29*, 678–685.

Carter, B. D., Kaltschmidt, C., Kaltschmidt, B., Offenhauser, N., Bohm-Matthaei, R., Baeuerle, P. A., & Barde, Y. A. (1996). Selective activation of NF-kappa B by nerve growth factor through the neurotrophin receptor p75. *Science, 272*, 542–545.

Casaccia-Bonnefil, P., Carter, B. D., Dobrowsky, R. T., & Chao, M. V. (1996). Death of oligodendrocytes mediated by the interaction of nerve growth factor with its receptor p75. *Nature, 383*, 716–719.

Chalazonitis, A., Pham, T. D., Rothman, T. P., DiStefano, P. S., Bothwell, M., Blair-Flynn, J., Tessarollo, L., & Gershon, M. D. (2001). Neurotrophin-3 is required for the survival-differentiation of subsets of developing enteric neurons. *The Journal of Neuroscience, 21*, 5620–5636.

Chao, M. V., & Hempstead, B. L. (1995). p75 and Trk: A two-receptor system. *Trends in Neurosciences, 18*, 321–326.

Chen, W. G., Chang, Q., Lin, Y., Meissner, A., West, A. E., Griffith, E. C., Jaenisch, R., & Greenberg, M. E. (2003). Derepression of BDNF transcription involves calcium-dependent phosphorylation of MeCP2. *Science, 302*, 885–889.

Chiaruttini, C., Vicario, A., Li, Z., Baj, G., Braiuca, P., Wu, Y., Lee, F. S., Gardossi, L., Baraban, J. M., & Tongiorgi, E. (2009). Dendritic trafficking of BDNF mRNA is mediated by translin and blocked by the G196A (Val66Met) mutation. *Proceedings of the National Academy of Sciences of the United States of America, 106*, 16481–16486.

Cohen, S., Levi-Montalcini, R., & Hamburger, V. (1954). A nerve growth-stimulating factor isolated from Sarcom as 37 and 180. *Proceedings of the National Academy of Sciences of the United States of America, 40*, 1014–1018.

Colangelo, A. M., Johnson, P. F., & Mocchetti, I. (1998). Beta-adrenergic receptor-induced activation of nerve growth factor gene transcription in rat cerebral cortex involves CCAAT/enhancer-binding protein delta. *Proceedings of the National Academy of Sciences of the United States of America, 95*, 10920–10925.

Colombo, E., Cordiglieri, C., Melli, G., Newcombe, J., Krumbholz, M., Parada, L. F., Medico, E., Hohlfeld, R., Meinl, E., & Farina, C. (2012). Stimulation of the neurotrophin receptor TrkB on astrocytes drives nitric oxide production and neurodegeneration. *The Journal of Experimental Medicine, 209*, 521–535.

Conner, J. M., Lauterborn, J. C., & Gall, C. M. (1998). Anterograde transport of neurotrophin proteins in the CNS—A reassessment of the neurotrophic hypothesis. *Reviews in the Neurosciences, 9*, 91–103.

Coppola, V., Barrick, C. A., Southon, E. A., Celeste, A., Wang, K., Chen, B., Haddad el, B., Yin, J., Nussenzweig, A., Subramaniam, A., & Tessarollo, L. (2004). Ablation of TrkA function in the immune system causes B cell abnormalities. *Development, 131*, 5185–5195.

Cordon-Cardo, C., Tapley, P., Jing, S. Q., Nanduri, V., O'Rourke, E., Lamballe, F., Kovary, K., Klein, R., Jones, K. R., Reichardt, L. F., et al. (1991). The trk tyrosine protein kinase mediates the mitogenic properties of nerve growth factor and neurotrophin-3. *Cell, 66*, 173–183.

Craft, S., Raman, R., Chow, T. W., Rafii, M. S., Sun, C. K., Rissman, R. A., Donohue, M. C., Brewer, J. B., Jenkins, C., Harless, K., Gessert, D., & Aisen, P. S. (2020). Safety, efficacy, and feasibility of intranasal insulin for the treatment of mild cognitive impairment and Alzheimer disease dementia: A randomized clinical trial. *JAMA Neurology, 77*, 1099–1109.

Curtis, R., Adryan, K. M., Stark, J. L., Park, J. S., Compton, D. L., Weskamp, G., Huber, L. J., Chao, M. V., Jaenisch, R., Lee, K. F., et al. (1995). Differential role of the low affinity neurotrophin receptor (p75) in retrograde axonal transport of the neurotrophins. *Neuron, 14*, 1201–1211.

De Rosa, R., Garcia, A. A., Braschi, C., Capsoni, S., Maffei, L., Berardi, N., & Cattaneo, A. (2005). Intranasal administration of nerve growth factor (NGF) rescues recognition memory deficits in AD11 anti-NGF transgenic mice. *Proceedings of the National Academy of Sciences of the United States of America, 102*, 3811–3816.

De Simone, R., Ambrosini, E., Carnevale, D., Ajmone-Cat, M. A., & Minghetti, L. (2007). NGF promotes microglial migration through the activation of its high affinity receptor: Modulation by TGF-beta. *Journal of Neuroimmunology, 190*, 53–60.

Diener, P. S., & Bregman, B. S. (1994). Neurotrophic factors prevent the death of CNS neurons after spinal cord lesions in newborn rats. *Neuroreport, 5*, 1913–1917.

Dobrowsky, R., Werner, M., Castellino, A., Chao, M., & Hannun, Y. (1994). Activation of the sphingomyelin cycle through the low affinity neurotrophin receptor. *Science, 265*, 1596–1599.

Dorsey, S. G., Renn, C. L., Carim-Todd, L., Barrick, C. A., Bambrick, L., Krueger, B. K., Ward, C. W., & Tessarollo, L. (2006). In vivo restoration of physiological levels of truncated TrkB.T1 receptor rescues neuronal cell death in a trisomic mouse model. *Neuron, 51*, 21–28.

Egan, M. F., Kojima, M., Callicott, J. H., Goldberg, T. E., Kolachana, B. S., Bertolino, A., Zaitsev, E., Gold, B., Goldman, D., Dean, M., Lu, B., & Weinberger, D. R. (2003). The BDNF val66met polymorphism affects activity-dependent secretion of BDNF and human memory and hippocampal function. *Cell, 112*, 257–269.

Eide, F. F., Vining, E. R., Eide, B. L., Zang, K., Wang, X. Y., & Reichardt, L. F. (1996). Naturally occurring truncated trkB receptors have dominant inhibitory effects on brain-derived neurotrophic factor signaling. *The Journal of Neuroscience, 16*, 3123–3129.

Elkabes, S., DiCicco-Bloom, E. M., & Black, I. B. (1996). Brain microglia/macrophages express neurotrophins that selectively regulate microglial proliferation and function. *The Journal of Neuroscience, 16*, 2508–2521.

Fabrazzo, M., Costa, E., & Mocchetti, I. (1991). Stimulation of nerve growth factor biosynthesis in developing rat brain by reserpine: Steroids as potential mediators. *Molecular Pharmacology, 39*, 144–149.

Fahnestock, M., Michalski, B., Xu, B., & Coughlin, M. D. (2001). The precursor pro-nerve growth factor is the predominant form of nerve growth factor in brain and is increased in Alzheimer's disease. *Molecular and Cellular Neurosciences, 18*, 210–220.

Fauchais, A. L., Lalloue, F., Lise, M. C., Boumediene, A., Preud'homme, J. L., Vidal, E., & Jauberteau, M. O. (2008). Role of endogenous brain-derived neurotrophic factor and sortilin in B cell survival. *Journal of Immunology, 181*, 3027–3038.

Ferrari, G., Anderson, B. L., Stephens, R. M., Kaplan, D. R., & Greene, L. A. (1995). Prevention of apoptotic neuronal death by GM1 ganglioside. Involvement of Trk neurotrophin receptors. *The Journal of Biological Chemistry, 270*, 3074–3080.

Fischer, O. M., Hart, S., Gschwind, A., & Ullrich, A. (2003). EGFR signal transactivation in cancer cells. *Biochemical Society Transactions, 31*, 1203–1208.

Fodelianaki, G., Lansing, F., Bhattarai, P., Troullinaki, M., Zeballos, M. A., Charalampopoulos, I., Gravanis, A., Mirtschink, P., Chavakis, T., & Alexaki, V. I. (2019). Nerve growth factor modulates LPS—Induced microglial glycolysis and inflammatory responses. *Experimental Cell Research, 377*, 10–16.

Frade, J. M., & Barde, Y. A. (1998). Microglia-derived nerve growth factor causes cell death in the developing retina. *Neuron, 20*, 35–41.

Frim, D., Uhler, T., Galpern, W., Beal, M., Breakefield, X., & Isacson, O. (1994). Implanted fibroblasts genetically engineered to produce brain-derived neurotrophic factor prevent 1-methyl-4-phenylpyridinium toxicity to dopaminergic neurons in the rat. *Proceedings of the National Academy of Sciences USA, 24*, 5104–5108.

Fulmer, C. G., VonDran, M. W., Stillman, A. A., Huang, Y., Hempstead, B. L., & Dreyfus, C. F. (2014). Astrocyte-derived BDNF supports myelin protein synthesis after cuprizone-induced demyelination. *The Journal of Neuroscience, 34*, 8186–8196.

Gage, F. H., Batchelor, P., Chen, K. S., Chin, D., Higgins, G. A., Koh, S., Deputy, S., Rosenberg, M. B., Fischer, W., & Bjorklund, A. (1989). NGF receptor reexpression and NGF-mediated cholinergic neuronal hypertrophy in the damaged adult neostriatum. *Neuron, 2*, 1177–1184.

Garcia-Suarez, O., Germana, A., Hannestad, J., Ciriaco, E., Laura, R., Naves, J., Esteban, I., Silos-Santiago, I., & Vega, J. A. (2000). TrkA is necessary for the normal development of the murine thymus. *Journal of Neuroimmunology, 108*, 11–21.

Garcia-Suarez, O., Blanco-Gelaz, M. A., Lopez, M. L., Germana, A., Cabo, R., Diaz-Esnal, B., Silos-Santiago, I., Ciriaco, E., & Vega, J. A. (2002). Massive lymphocyte apoptosis in the thymus of functionally deficient TrkB mice. *Journal of Neuroimmunology, 129*, 25–34.

Ghosh, A., Carnahan, J., & Greenberg, M. E. (1994). Requirement for BDNF in activity-dependent survival of cortical neurons. *Science, 263*, 1618–1623.

Gu, C., Casaccia-Bonnefil, P., Srinivasan, A., & Chao, M. V. (1999). Oligodendrocyte apoptosis mediated by caspase activation. *The Journal of Neuroscience, 19*, 3043–3049.

Guo, J., Ji, Y., Ding, Y., Jiang, W., Sun, Y., Lu, B., & Nagappan, G. (2016). BDNF pro-peptide regulates dendritic spines via caspase-3. *Cell Death & Disease, 7*, e2264.

Gurney, M. E., Pu, H., Chiu, A. Y., Dal Canto, M. C., Polchow, C. Y., Alexander, D. D., Caliendo, J., Hentati, A., Kwon, Y. W., Deng, H. X., et al. (1994). Motor neuron degeneration in mice that express a human Cu,Zn superoxide dismutase mutation. *Science, 264*, 1772–1775.

Hapner, S. J., Boeshore, K. L., Large, T. H., & Lefcort, F. (1998). Neural differentiation promoted by truncated trkC receptors in collaboration with p75(NTR). *Developmental Biology, 201*, 90–100.

Harrington, A. W., Leiner, B., Blechschmitt, C., Arevalo, J. C., Lee, R., Morl, K., Meyer, M., Hempstead, B. L., Yoon, S. O., & Giehl, K. M. (2004). Secreted proNGF is a pathophysiological death-inducing ligand after adult CNS injury. *Proceedings of the National Academy of Sciences of the United States of America, 101*, 6226–6230.

Hashimoto, R., Takei, N., Shimazu, K., Christ, L., Lu, B., & Chuang, D.-M. (2002). Lithium induces brain-derived neurotrophic factor and activates TrkB in rodent cortical neurons: An essential step for neuroprotection against glutamate excitotoxicity. *Neuropharmacology, 43*, 1173.

Hempstead, B. L. (2002). The many faces of p75NTR. *Current Opinion in Neurobiology, 12*, 260–267.

Hempstead, B. L., Martin-Zanca, D., Kaplan, D. R., Parada, L. F., & Chao, M. V. (1991). High-affinity NGF binding requires coexpression of the trk proto-oncogene and the low-affinity NGF receptor. *Nature, 350*, 678–683.

Hofer, M., Pagliusi, S. R., Hohn, A., Leibrock, J., & Barde, Y. A. (1990). Regional distribution of brain-derived neurotrophic factor mRNA in the adult mouse brain. *The EMBO Journal, 9*, 2459–2464.

Hong, S., Beja-Glasser, V. F., Nfonoyim, B. M., Frouin, A., Li, S., Ramakrishnan, S., Merry, K. M., Shi, Q., Rosenthal, A., Barres, B. A., Lemere, C. A., Selkoe, D. J., & Stevens, B. (2016).

Complement and microglia mediate early synapse loss in Alzheimer mouse models. *Science, 352*, 712–716.

Hyman, C., Hofer, M., Barde, Y. A., Juhasz, M., Yancopoulos, G. D., Squinto, S. P., & Lindsay, R. M. (1991). BDNF is a neurotrophic factor for dopaminergic neurons of the substantia nigra. *Nature, 350*, 230–232.

Ibanez, C. F. (1995). Neurotrophic factors: From structure-function studies to designing effective therapeutics. *Trends in Biotechnology, 13*, 217–227.

Ilag, L. L., Curtis, R., Glass, D., Funakoshi, H., Tobkes, N. J., Ryan, T. E., Acheson, A., Lindsay, R. M., Persson, H., Yancopoulos, G. D., et al. (1995). Pan-neurotrophin 1: A genetically engineered neurotrophic factor displaying multiple specificities in peripheral neurons in vitro and in vivo. *Proceedings of the National Academy of Sciences of the United States of America, 92*, 607–611.

James, S. E., Burden, H., Burgess, R., Xie, Y., Yang, T., Massa, S. M., Longo, F. M., & Lu, Q. (2008). Anti-cancer drug induced neurotoxicity and identification of Rho pathway signaling modulators as potential neuroprotectants. *Neurotoxicology, 29*, 605–612.

Jang, S. W., Liu, X., Chan, C. B., Weinshenker, D., Hall, R. A., Xiao, G., & Ye, K. (2009). Amitriptyline is a TrkA and TrkB receptor agonist that promotes TrkA/TrkB heterodimerization and has potent neurotrophic activity. *Chemistry & Biology, 16*, 644–656.

Johnson, D., Lanahan, A., Buck, C. R., Sehgal, A., Morgan, C., Mercer, E., Bothwell, M., & Chao, M. (1986). Expression and structure of the human NGF receptor. *Cell, 47*, 545–554.

Juric, D. M., & Carman-Krzan, M. (2000). Cytokine-regulated secretion of nerve growth factor from cultured rat neonatal astrocytes. *Pflügers Archiv, 440*, R096–R098.

Kaplan, D. R., Hempstead, B. L., Martin-Zanca, D., Chao, M. V., & Parada, L. F. (1991). The trk proto-oncogene product: A signal transducing receptor for nerve growth factor. *Science, 252*, 554–558.

Kellar, D., Lockhart, S. N., Aisen, P., Raman, R., Rissman, R. A., Brewer, J., & Craft, S. (2021). Intranasal insulin reduces white matter hyperintensity progression in association with improvements in cognition and CSF biomarker profiles in mild cognitive impairment and Alzheimer's disease. *The Journal of Prevention of Alzheimer's Disease, 8*, 240–248.

Kenchappa, R. S., Zampieri, N., Chao, M. V., Barker, P. A., Teng, H. K., Hempstead, B. L., & Carter, B. D. (2006). Ligand-dependent cleavage of the P75 neurotrophin receptor is necessary for NRIF nuclear translocation and apoptosis in sympathetic neurons. *Neuron, 50*, 219–232.

Kimata, H., Yoshida, A., Ishioka, C., & Mikawa, H. (1991). Stimulation of Ig production and growth of human lymphoblastoid B-cell lines by nerve growth factor. *Immunology, 72*, 451–452.

Klein, R., Nanduri, V., Jing, S. A., Lamballe, F., Tapley, P., Bryant, S., Cordon-Cardo, C., Jones, K. R., Reichardt, L. F., & Barbacid, M. (1991). The trkB tyrosine protein kinase is a receptor for brain-derived neurotrophic factor and neurotrophin-3. *Cell, 66*, 395–403.

Klein, R. L., Meyer, E. M., Peel, A. L., Zolotukhin, S., Meyers, C., Muzyczka, N., & King, M. A. (1998). Neuron-specific transduction in the rat septohippocampal or nigrostriatal pathway by recombinant adeno-associated virus vectors. *Experimental Neurology, 150*, 183–194.

Koliatsos, V. E., Nauta, H. J., Clatterbuck, R. E., Holtzman, D. M., Mobley, W. C., & Price, D. L. (1990). Mouse nerve growth factor prevents degeneration of axotomized basal forebrain cholinergic neurons in the monkey. *The Journal of Neuroscience, 10*, 3801–3813.

Koliatsos, V. E., Clatterbuck, R. E., Winslow, J. W., Cayouette, M. H., & Price, D. L. (1993). Evidence that brain-derived neurotrophic factor is a trophic factor for motor neurons in vivo. *Neuron, 10*, 359–367.

Kononenko, N. L., Classen, G. A., Kuijpers, M., Puchkov, D., Maritzen, T., Tempes, A., Malik, A. R., Skalecka, A., Bera, S., Jaworski, J., & Haucke, V. (2017). Retrograde transport of TrkB-containing autophagosomes via the adaptor AP-2 mediates neuronal complexity and prevents neurodegeneration. *Nature Communications, 8*, 14819.

Koob, G. F., & Greenwell, T. N. (2004). Neuropeptide S: A novel activating anxiolytic? *Neuron, 43*, 441–442.

Kozlov, E. M., Grechko, A. V., Chegodaev, Y. S., Wu, W. K., & Orekhov, A. N. (2020). Contribution of neurotrophins to the immune system regulation and possible connection to alcohol addiction. *Biology (Basel), 9*.

Kromer, L. F. (1987). Nerve growth factor treatment after brain injury prevents neuronal death. *Science, 235*, 214–216.

Large, T. H., Bodary, S. C., Clegg, D. O., Weskamp, G., Otten, U., & Reichardt, L. F. (1986). Nerve growth factor gene expression in the developing rat brain. *Science, 234*, 352–355.

Laske, C., Stransky, E., Leyhe, T., Eschweiler, G. W., Maetzler, W., Wittorf, A., Soekadar, S., Richartz, E., Koehler, N., Bartels, M., Buchkremer, G., & Schott, K. (2007). BDNF serum and CSF concentrations in Alzheimer's disease, normal pressure hydrocephalus and healthy controls. *Journal of Psychiatric Research, 41*, 387–394.

Laurenzi, M. A., Barbany, G., Timmusk, T., Lindgren, J. A., & Persson, H. (1994). Expression of mRNA encoding neurotrophins and neurotrophin receptors in rat thymus, spleen tissue and immunocompetent cells. Regulation of neurotrophin-4 mRNA expression by mitogens and leukotriene B4. *European Journal of Biochemistry, 223*, 733–741.

Lauterborn, J., Berschauer, R., & Gall, C. (1995). Cell-specific modulation of basal and seizure-induced neurotrophin expression by adrenalectomy. *Neuroscience, 68*, 363–378.

Leibrock, J., Lottspeich, F., Hohn, A., Hofer, M., Hengerer, B., Masiakowski, P., Thoenen, H., & Barde, Y. A. (1989). Molecular cloning and expression of brain-derived neurotrophic factor. *Nature, 341*, 149–152.

Levanti, B., Hannestad, J., Esteban, I., Ciriaco, E., Germana, G., & Vega, J. A. (1997). Neurotrophin receptor-like proteins in Peyer's patches. *The Anatomical Record, 249*, 365–372.

Levi-Montalcini, R., & Hamburger, V. (1951). Selective growth stimulating effects of mouse sarcoma on the sensory and sympathetic nervous system of the chick embryo. *The Journal of Experimental Zoology, 116*, 321–361.

Li, Y., Holtzman, D. M., Kromer, L. F., Kaplan, D. R., Chua-Couzens, J., Clary, D. O., Knusel, B., & Mobley, W. C. (1995). Regulation of TrkA and ChAT expression in developing rat basal forebrain: Evidence that both exogenous and endogenous NGF regulate differentiation of cholinergic neurons. *The Journal of Neuroscience, 15*, 2888–2905.

Linggi, M. S., Burke, T. L., Williams, B. B., Harrington, A., Kraemer, R., Hempstead, B. L., Yoon, S. O., & Carter, B. D. (2005). Neurotrophin receptor interacting factor (NRIF) is an essential mediator of apoptotic signaling by the p75 neurotrophin receptor. *The Journal of Biological Chemistry, 280*, 13801–13808.

Linker, R. A., Lee, D. H., Demir, S., Wiese, S., Kruse, N., Siglienti, I., Gerhardt, E., Neumann, H., Sendtner, M., Luhder, F., & Gold, R. (2010). Functional role of brain-derived neurotrophic factor in neuroprotective autoimmunity: Therapeutic implications in a model of multiple sclerosis. *Brain, 133*, 2248–2263.

Lucidi-Phillipi, C. A., Clary, D. O., Reichardt, L. F., & Gage, F. H. (1996). TrkA activation is sufficient to rescue axotomized cholinergic neurons. *Neuron, 16*, 653–663.

Luo, C., Zhong, X. L., Zhou, F. H., Li, J. Y., Zhou, P., Xu, J. M., Song, B., Li, C. Q., Zhou, X. F., & Dai, R. P. (2016). Peripheral brain derived neurotrophic factor precursor regulates pain as an inflammatory mediator. *Scientific Reports, 6*, 27171.

Lyons, W. E., Mamounas, L. A., Ricaurte, G. A., Coppola, V., Reid, S. W., Bora, S. H., Wihler, C., Koliatsos, V. E., & Tessarollo, L. (1999). Brain-derived neurotrophic factor-deficient mice develop aggressiveness and hyperphagia in conjunction with brain sero-

tonergic abnormalities. *Proceedings of the National Academy of Sciences of the United States of America, 96*, 15239–15244.
Maisonpierre, P. C., Belluscio, L., Squinto, S., Ip, N. Y., Furth, M. E., Lindsay, R. M., & Yancopoulos, G. D. (1990). Neurotrophin-3: A neurotrophic factor related to NGF and BDNF. *Science, 247*, 1446–1451.
Manni, L., Conti, G., Chiaretti, A., & Soligo, M. (2023). Intranasal nerve growth factor for prevention and recovery of the outcomes of traumatic brain injury. *Neural Regeneration Research, 18*, 773–778.
Martin, G., Guadano-Ferraz, A., Morte, B., Ahmed, S., Koob, G. F., De Lecea, L., & Siggins, G. R. (2004). Chronic morphine treatment alters N-methyl-D-aspartate receptors in freshly isolated neurons from nucleus accumbens. *The Journal of Pharmacology and Experimental Therapeutics, 311*, 265–273.
Massa, S. M., Xie, Y., Yang, T., Harrington, A. W., Kim, M. L., Yoon, S. O., Kraemer, R., Moore, L. A., Hempstead, B. L., & Longo, F. M. (2006). Small, nonpeptide p75NTR ligands induce survival signaling and inhibit proNGF-induced death. *The Journal of Neuroscience, 26*, 5288–5300.
Mocchetti, I., & Colangelo, A. (2001). *In: (Mocchetti I, ed)* (pp. 631–654). F.P. Graham Publishing Co.
Mowla, S. J., Pareek, S., Farhadi, H. F., Petrecca, K., Fawcett, J. P., Seidah, N. G., Morris, S. J., Sossin, W. S., & Murphy, R. A. (1999). Differential sorting of nerve growth factor and brain-derived neurotrophic factor in hippocampal neurons. *The Journal of Neuroscience, 19*, 2069–2080.
Mutoh, T., Tokuda, A., Miyadai, T., Hamaguchi, M., & Fujiki, N. (1995). Ganglioside GM1 binds to the Trk protein and regulates receptor function. *Proceedings of the National Academy of Sciences of the United States of America, 92*, 5087–5091.
Nagahara, A. H., Merrill, D. A., Coppola, G., Tsukada, S., Schroeder, B. E., Shaked, G. M., Wang, L., Blesch, A., Kim, A., Conner, J. M., Rockenstein, E., Chao, M. V., Koo, E. H., Geschwind, D., Masliah, E., Chiba, A. A., & Tuszynski, M. H. (2009). Neuroprotective effects of brain-derived neurotrophic factor in rodent and primate models of Alzheimer's disease. *Nature Medicine, 15*, 331–337.
Nakajima, K., Honda, S., Tohyama, Y., Imai, Y., Kohsaka, S., & Kurihara, T. (2001). Neurotrophin secretion from cultured microglia. *Journal of Neuroscience Research, 65*, 322–331.
Nykjaer, A., Lee, R., Teng, K. K., Jansen, P., Madsen, P., Nielsen, M. S., Jacobsen, C., Kliemannel, M., Schwarz, E., Willnow, T. E., Hempstead, B. L., & Petersen, C. M. (2004). Sortilin is essential for proNGF-induced neuronal cell death. *Nature, 427*, 843–848.
Obianyo, O., & Ye, K. (2013). Novel small molecule activators of the Trk family of receptor tyrosine kinases. *Biochimica et Biophysica Acta, 1834*, 2213–2218.
O'Leary, P. D., & Hughes, R. A. (2003). Design of potent peptide mimetics of brain-derived neurotrophic factor. *The Journal of Biological Chemistry, 278*, 25738–25744.
Orefice, L. L., Waterhouse, E. G., Partridge, J. G., Lalchandani, R. R., Vicini, S., & Xu, B. (2013). Distinct roles for somatically and dendritically synthesized brain-derived neurotrophic factor in morphogenesis of dendritic spines. *The Journal of Neuroscience, 33*, 11618–11632.
Paradisi, M., Alviano, F., Pirondi, S., Lanzoni, G., Fernandez, M., Lizzo, G., Giardino, L., Giuliani, A., Costa, R., Marchionni, C., Bonsi, L., & Calza, L. (2014). Human mesenchymal stem cells produce bioactive neurotrophic factors: Source, individual variability and differentiation issues. *International Journal of Immunopathology and Pharmacology, 27*, 391–402.
Park, H., & Poo, M. M. (2013). Neurotrophin regulation of neural circuit development and function. *Nature Reviews. Neuroscience, 14*, 7–23.
Park, K. J., Grosso, C. A., Aubert, I., Kaplan, D. R., & Miller, F. D. (2010). p75NTR-dependent, myelin-mediated axonal degeneration regulates neural connectivity in the adult brain. *Nature Neuroscience, 13*, 559–566.
Parkhurst, C. N., Yang, G., Ninan, I., Savas, J. N., Yates, J. R., 3rd, Lafaille, J. J., Hempstead, B. L., Littman, D. R., & Gan, W. B. (2013). Microglia promote learning-dependent synapse formation through brain-derived neurotrophic factor. *Cell, 155*, 1596–1609.
Perez-Perez, M., Esteban, I., Garcia-Suarez, O., Hannestad, J., Naves, F. J., & Vega, J. A. (1999). Expression of the neurotrophin receptor TrkB in rat spleen macrophages. *Cell and Tissue Research, 298*, 75–84.
Pezzati, P., Stanisz, A. M., Marshall, J. S., Bienenstock, J., & Stead, R. H. (1992). Expression of nerve growth factor receptor immunoreactivity on follicular dendritic cells from human mucosa associated lymphoid tissues. *Immunology, 76*, 485–490.
Phillips, H. S., Hains, J. M., Armanini, M., Laramee, G. R., Johnson, S. A., & Winslow, J. W. (1991). BDNF mRNA is decreased in the hippocampus of individuals with Alzheimer's disease. *Neuron, 7*, 695–702.
Rabin, S. J., & Mocchetti, I. (1995). GM1 ganglioside activates the high-affinity nerve growth factor receptor trkA. *Journal of Neurochemistry, 65*, 347–354.
Rabin, S. J., Bachis, A., & Mocchetti, I. (2002). Gangliosides activate Trk receptors by inducing the release of neurotrophins. *The Journal of Biological Chemistry, 277*, 49466–49472.
Radeke, M. J., Misko, T. P., Hsu, C., Herzenberg, L. A., & Shooter, E. M. (1987). Gene transfer and molecular cloning of the rat nerve growth factor receptor. *Nature, 325*, 593–597.
Riffault, B., Medina, I., Dumon, C., Thalman, C., Ferrand, N., Friedel, P., Gaiarsa, J. L., & Porcher, C. (2014). Pro-brain-derived neurotrophic factor inhibits GABAergic neurotransmission by activating endocytosis and repression of GABAA receptors. *The Journal of Neuroscience, 34*, 13516–13534.
Rizzi, C., Tiberi, A., Giustizieri, M., Marrone, M. C., Gobbo, F., Carucci, N. M., Meli, G., Arisi, I., D'Onofrio, M., Marinelli, S., Capsoni, S., & Cattaneo, A. (2018). NGF steers microglia toward a neuroprotective phenotype. *Glia, 66*, 1395–1416.
Rodriguez-Tebar, A., Dechant, G., & Barder, Y. (1990). Binding of brain derived neurotrophic factor to the nerve growth factor receptor. *Neuron, 4*, 487–492.
Rosch, H., Schweigreiter, R., Bonhoeffer, T., Barde, Y. A., & Korte, M. (2005). The neurotrophin receptor p75NTR modulates long-term depression and regulates the expression of AMPA receptor subunits in the hippocampus. *Proceedings of the National Academy of Sciences of the United States of America, 102*, 7362–7367.
Saba, J., Turati, J., Ramirez, D., Carniglia, L., Durand, D., Lasaga, M., & Caruso, C. (2018). Astrocyte truncated tropomyosin receptor kinase B mediates brain-derived neurotrophic factor anti-apoptotic effect leading to neuroprotection. *Journal of Neurochemistry, 146*, 686–702.
Saha, R. N., Liu, X., & Pahan, K. (2006). Up-regulation of BDNF in astrocytes by TNF-alpha: A case for the neuroprotective role of cytokine. *Journal of Neuroimmune Pharmacology, 1*, 212–222.
Samah, B., Porcheray, F., Dereuddre-Bosquet, N., & Gras, G. (2009). Nerve growth factor stimulation promotes CXCL-12 attraction of monocytes but decreases human immunodeficiency virus replication in attracted population. *Journal of Neurovirology, 15*, 71–80.
Saporito, M. S., Brown, E. R., Hartpence, K. C., Wilcox, H. M., Robbins, E., Vaught, J. L., & Carswell, S. (1994). Systemic dexamethasone administration increases septal Trk autophosphorylation in adult rats via an induction of nerve growth factor. *Molecular Pharmacology, 45*, 395–401.
Schafer, D. P., Lehrman, E. K., Kautzman, A. G., Koyama, R., Mardinly, A. R., Yamasaki, R., Ransohoff, R. M., Greenberg, M. E., Barres, B. A., & Stevens, B. (2012). Microglia sculpt postnatal neural circuits in an activity and complement-dependent manner. *Neuron, 74*, 691–705.

Scheff, S. W., & Price, D. A. (2006). Alzheimer's disease-related alterations in synaptic density: Neocortex and hippocampus. *Journal of Alzheimer's Disease, 9*, 101–115.

Schmid, D. A., Yang, T., Ogier, M., Adams, I., Mirakhur, Y., Wang, Q., Massa, S. M., Longo, F. M., & Katz, D. M. (2012). A TrkB small molecule partial agonist rescues TrkB phosphorylation deficits and improves respiratory function in a mouse model of Rett syndrome. *The Journal of Neuroscience, 32*, 1803–1810.

Schnell, L., Schneider, R., Kolbeck, R., Barde, Y. A., & Schwab, M. E. (1994). Neurotrophin-3 enhances sprouting of corticospinal tract during development and after adult spinal cord lesion. *Nature, 367*, 170–173.

Schuhmann, B., Dietrich, A., Sel, S., Hahn, C., Klingenspor, M., Lommatzsch, M., Gudermann, T., Braun, A., Renz, H., & Nockher, W. A. (2005). A role for brain-derived neurotrophic factor in B cell development. *Journal of Neuroimmunology, 163*, 15–23.

Seidah, N. G., Benjannet, S., Pareek, S., Chretien, M., & Murphy, R. A. (1996). Cellular processing of the neurotrophin precursors of NT3 and BDNF by the mammalian proprotein convertases. *FEBS Letters, 379*, 247–250.

Sharma, G. P., Frei, A. C., Narayanan, J., Gasperetti, T., Veley, D., Amjad, A., Albano, K., Fish, B. L., & Himburg, H. A. (2021). Brain-derived neurotrophic factor promotes immune reconstitution following radiation injury via activation of bone marrow mesenchymal stem cells. *PLoS One, 16*, e0259042.

Shi, B., Rabin, S. J., Brandoli, C., & Mocchetti, I. (1998). Dexamethasone induces hypertrophy of developing medial septum cholinergic neurons: Potential role of nerve growth factor. *The Journal of Neuroscience, 18*, 9326–9334.

Simmons, D. A., Knowles, J. K., Belichenko, N. P., Banerjee, G., Finkle, C., Massa, S. M., & Longo, F. M. (2014). A small molecule p75NTR ligand, LM11A-31, reverses cholinergic neurite dystrophy in Alzheimer's disease mouse models with mid- to late-stage disease progression. *PLoS One, 9*, e102136.

Soliman, F., Glatt, C. E., Bath, K. G., Levita, L., Jones, R. M., Pattwell, S. S., Jing, D., Tottenham, N., Amso, D., Somerville, L. H., Voss, H. U., Glover, G., Ballon, D. J., Liston, C., Teslovich, T., Van Kempen, T., Lee, F. S., & Casey, B. J. (2010). A genetic variant BDNF polymorphism alters extinction learning in both mouse and human. *Science, 327*, 863–866.

Sousa, J. B., & Diniz, C. (2017). The adenosinergic system as a therapeutic target in the vasculature: New ligands and challenges. *Molecules, 22*.

Srinivasan, B., Roque, C. H., Hempstead, B. L., Al-Ubaidi, M. R., & Roque, R. S. (2004). Microglia-derived pronerve growth factor promotes photoreceptor cell death via p75 neurotrophin receptor. *The Journal of Biological Chemistry, 279*, 41839–41845.

Suter, U., Angst, C., Tien, C. L., Drinkwater, C. C., Lindsay, R. M., & Shooter, E. M. (1992). NGF/BDNF chimeric proteins: Analysis of neurotrophin specificity by homolog-scanning mutagenesis. *The Journal of Neuroscience, 12*, 306–318.

Tacconelli, A., Farina, A. R., Cappabianca, L., Cea, G., Panella, S., Chioda, A., Gallo, R., Cinque, B., Sferra, R., Vetuschi, A., Campese, A. F., Screpanti, I., Gulino, A., & Mackay, A. R. (2007). TrkAIII expression in the thymus. *Journal of Neuroimmunology, 183*, 151–161.

Talbot, H., Saada, S., Barthout, E., Gallet, P. F., Gachard, N., Abraham, J., Jaccard, A., Troutaud, D., Lalloue, F., Naves, T., Fauchais, A. L., & Jauberteau, M. O. (2020). BDNF belongs to the nurse-like cell secretome and supports survival of B chronic lymphocytic leukemia cells. *Scientific Reports, 10*, 12572.

Tao, X., Finkbeiner, S., Arnold, D. B., Shaywitz, A. J., & Greenberg, M. E. (1998). Ca2+ influx regulates BDNF transcription by a CREB family transcription factor-dependent mechanism. *Neuron, 20*, 709–726.

Tao, X., West, A. E., Chen, W. G., Corfas, G., & Greenberg, M. E. (2002). A calcium-responsive transcription factor, CaRF, that regulates neuronal activity-dependent expression of BDNF. *Neuron, 33*, 383–395.

Teng, H. K., Teng, K. K., Lee, R., Wright, S., Tevar, S., Almeida, R. D., Kermani, P., Torkin, R., Chen, Z. Y., Lee, F. S., Kraemer, R. T., Nykjaer, A., & Hempstead, B. L. (2005). ProBDNF induces neuronal apoptosis via activation of a receptor complex of p75NTR and sortilin. *The Journal of Neuroscience, 25*, 5455–5463.

Thompson, C. L., et al. (2014). A high-resolution spatiotemporal atlas of gene expression of the developing mouse brain. *Neuron, 83*, 309–323.

Torcia, M., Bracci-Laudiero, L., Lucibello, M., Nencioni, L., Labardi, D., Rubartelli, A., Cozzolino, F., Aloe, L., & Garaci, E. (1996). Nerve growth factor is an autocrine survival factor for memory B lymphocytes. *Cell, 85*, 345–356.

Urfer, R., Tsoulfas, P., Soppet, D., Escandon, E., Parada, L. F., & Presta, L. G. (1994). The binding epitopes of neurotrophin-3 to its receptors trkC and gp75 and the design of a multifunctional human neurotrophin. *The EMBO Journal, 13*, 5896–5909.

Vaegter, C. B., Jansen, P., Fjorback, A. W., Glerup, S., Skeldal, S., Kjolby, M., Richner, M., Erdmann, B., Nyengaard, J. R., Tessarollo, L., Lewin, G. R., Willnow, T. E., Chao, M. V., & Nykjaer, A. (2011). Sortilin associates with Trk receptors to enhance anterograde transport and neurotrophin signaling. *Nature Neuroscience, 14*, 54–61.

Vega, J. A., Garcia-Suarez, O., Hannestad, J., Perez-Perez, M., & Germana, A. (2003). Neurotrophins and the immune system. *Journal of Anatomy, 203*, 1–19.

Volosin, M., Song, W., Almeida, R. D., Kaplan, D. R., Hempstead, B. L., & Friedman, W. J. (2006). Interaction of survival and death signaling in basal forebrain neurons: Roles of neurotrophins and proneurotrophins. *The Journal of Neuroscience, 26*, 7756–7766.

Volosin, M., Trotter, C., Cragnolini, A., Kenchappa, R. S., Light, M., Hempstead, B. L., Carter, B. D., & Friedman, W. J. (2008). Induction of proneurotrophins and activation of p75NTR-mediated apoptosis via neurotrophin receptor-interacting factor in hippocampal neurons after seizures. *The Journal of Neuroscience, 28*, 9870–9879.

Wang, H., Song, G., Chuang, H., Chiu, C., Abdelmaksoud, A., Ye, Y., & Zhao, L. (2018). Portrait of glial scar in neurological diseases. *International Journal of Immunopathology and Pharmacology, 31*, 2058738418801406.

Wiese, S., Jablonka, S., Holtmann, B., Orel, N., Rajagopal, R., Chao, M. V., & Sendtner, M. (2007). Adenosine receptor A2A-R contributes to motoneuron survival by transactivating the tyrosine kinase receptor TrkB. *Proceedings of the National Academy of Sciences of the United States of America, 104*, 17210–17215.

Williams, L. R. (1991). Hypophagia is induced by intracerebroventricular administration of nerve growth factor. *Experimental Neurology, 113*, 31–37.

Xie, Y., Meeker, R. B., Massa, S. M., & Longo, F. M. (2019). Modulation of the p75 neurotrophin receptor suppresses age-related basal forebrain cholinergic neuron degeneration. *Scientific Reports, 9*, 5273.

Yang, J., Siao, C. J., Nagappan, G., Marinic, T., Jing, D., McGrath, K., Chen, Z. Y., Mark, W., Tessarollo, L., Lee, F. S., Lu, B., & Hempstead, B. L. (2009). Neuronal release of proBDNF. *Nature Neuroscience, 12*, 113–115.

Yang, H., Feng, G. D., Liang, Z., Vitale, A., Jiao, X. Y., Ju, G., & You, S. W. (2012). In vitro beneficial activation of microglial cells by mechanically-injured astrocytes enhances the synthesis and secretion of BDNF through p38MAPK. *Neurochemistry International, 61*, 175–186.

Yanpallewar, S. U., Barrick, C. A., Buckley, H., Becker, J., & Tessarollo, L. (2012). Deletion of the BDNF truncated receptor TrkB.T1 delays disease onset in a mouse model of amyotrophic lateral sclerosis. *PLoS One, 7*, e39946.

Yoshida, K., Kakihana, M., Chen, L. S., Ong, M., Baird, A., & Gage, F. H. (1992). Cytokine regulation of nerve growth factor-mediated cholinergic neurotrophic activity synthesized by astrocytes and fibroblasts. *Journal of Neurochemistry, 59*, 919–931.

Zaccaro, M. C., Lee, H. B., Pattarawarapan, M., Xia, Z., Caron, A., L'Heureux, P. J., Bengio, Y., Burgess, K., & Saragovi, H. U. (2005). Selective small molecule peptidomimetic ligands of TrkC and TrkA receptors afford discrete or complete neurotrophic activities. *Chemistry & Biology, 12*, 1015–1028.

Zhang, Y., Sloan, S. A., Clarke, L. E., Caneda, C., Plaza, C. A., Blumenthal, P. D., Vogel, H., Steinberg, G. K., Edwards, M. S., Li, G., Duncan, J. A., 3rd, Cheshier, S. H., Shuer, L. M., Chang, E. F., Grant, G. A., Gephart, M. G., & Barres, B. A. (2016). Purification and characterization of progenitor and mature human astrocytes reveals transcriptional and functional differences with mouse. *Neuron, 89*, 37–53.

Zhao, X., Chen, X. Q., Han, E., Hu, Y., Paik, P., Ding, Z., Overman, J., Lau, A. L., Shahmoradian, S. H., Chiu, W., Thompson, L. M., Wu, C., & Mobley, W. C. (2016). TRiC subunits enhance BDNF axonal transport and rescue striatal atrophy in Huntington's disease. *Proceedings of the National Academy of Sciences of the United States of America, 113*, E5655–E5664.

Zuccato, C., Ciammola, A., Rigamonti, D., Leavitt, B. R., Goffredo, D., Conti, L., MacDonald, M. E., Friedlander, R. M., Silani, V., Hayden, M. R., Timmusk, T., Sipione, S., & Cattaneo, E. (2001). Loss of huntingtin-mediated BDNF gene transcription in Huntington's disease. *Science, 293*, 493–498.

18 The Liver–Gut–Brain Axis and Hepatic Encephalopathy

Natalia A. Osna, Larisa Y. Poluektova,
and Kusum K. Kharbanda

Abstract

Hepatic encephalopathy (HE) is one of the severe complications of a broad complex of liver injuries, encircling an extensive range of reversible neurological and psychomotor dysfunctions varying from confusion to coma. Based on the etiology, it is categorized into three main categories: type A (resulting from acute liver failure), type B (with underlying portosystemic shunt), and type C (due to liver cirrhosis). A mortality rate could reach 65% depending on the type analyzed. HE affects patients' quality of life, and successful treatment starts with detecting the leading cause of precipitating events. It is unclear whether the neurological dysfunction is secondary to the etiology of liver disease or comorbidities. In some cases, it might be challenging to attribute neurological disturbances to a specific factor. This chapter elucidates the role of ammonia, altered microbiota, neuroinflammation, hepatitis viruses, and extracellular vesicles as possible mechanisms contributing to HE development.

N. A. Osna (✉)
Department of Pharmacology and Experimental Neuroscience, University of Nebraska Medical Center, Omaha, NE, USA

Department of Internal Medicine, University of University of Nebraska Medical Center, Omaha, NE, USA
e-mail: nosna@unmc.edu

L. Y. Poluektova
Department of Pharmacology and Experimental Neuroscience, University of Nebraska Medical Center, Omaha, NE, USA

K. K. Kharbanda
Department of Internal Medicine, University of University of Nebraska Medical Center, Omaha, NE, USA

Department of Biochemistry and Molecular Biology, University of University of Nebraska Medical Center, Omaha, NE, USA

Veterans Affairs Nebraska-Western Iowa Health Care System, Omaha, NE, USA

Keywords

Liver · Gut · Brain · Microbiota · Extracellular vesicles · Hepatic encephalopathy

18.1 Introduction

Hepatic encephalopathy (HE) represents a spectrum of reversible neuropsychiatric changes, which can occur in the course of acute or chronic liver disease. Most cases of HE are associated with liver cirrhosis and associated with a reduction of health-related quality of life (Rose et al., 2020). HE is the most common reason for hospitalization in patients with decompensated cirrhosis, accounting for more than 100,000 annual admissions with total hospital charges of seven billion dollars and increased mortality (Alsahhar & Rahimi, 2019). In addition, there is a significant family/caregiver burden given the need for high-level supportive care, with depression and anxiety reported in up to 20% of caregivers (Bajaj et al., 2011).

18.1.1 Types of HE

The 2014 AASLD and EASL clinical practice guidelines for managing HE recommend classifying according to the underlying liver disease, the severity of the manifestation, the time course, and the precipitating factors (Vilstrup et al., 2014; European Association for the Study of the Liver. Electronic address et al., 2017). HE is categorized as type A, B, or C, and graded between minimal to grades I–IV. HE in acute liver failure is HE type A, and its incidence rate is mostly the same as for acute liver failure (European Association for the Study of the Liver. Electronic address et al., 2017). Type B HE is caused exclusively by portosystemic shunts and is less rare than HE type A. Type C HE is

H. E. Gendelman, T. Ikezu (eds.), *Neuroimmune Pharmacology and Therapeutics*,
https://doi.org/10.1007/978-3-031-68237-7_18

caused primarily or exclusively by loss of functional liver mass due to cirrhosis and is the most common variant of HE with a prevalence rate of 40–60% in patients with cirrhosis.

18.1.2 Grades of HE

Grade 1 HE is the mildest clinically detectable but still non-disorientational cerebral involvement, with an estimated prevalence of 15–25% of cirrhotic patients. HE grade II is clinically manifest HE characterized by marked by disorientation and is present in 10–15% of cirrhosis patients at diagnosis. In the higher grade III, there is additional somnolence, and in grade IV, a coma-like state. The MELD (Model for End-stage Liver Disease) score, the disease severity indicator of patients with cirrhosis, does not correlate with HE severity. Other terms to describe HE are based on a wide range of manifestations, from minimal HE (MHE) and over overt HE (OHE) to frank coma. OHE is present in 10–14% of patients at the time of diagnosis of cirrhosis. Up to 40% of patients with cirrhosis will experience at least one episode; many will experience repeated episodes. MHE is diagnosed in 20–80% of patients with cirrhosis (Kornerup et al., 2018).

18.2 Pathogenesis of HE

It is generally accepted that ammonia, which is insufficiently eliminated in liver cirrhosis and inflammation, plays a major role in the pathogenesis of HE.

18.2.1 Role of Ammonia in HE

Although hyperammonemia is not the only mechanism in the development of HE, substantial research has proved that it plays a significant role and has multiple devastating neurotoxic effects as it can readily cross the blood–brain barrier (BBB) in its protonated and deprotonated form (Amodio, 2015).

Ammonia is a small nitrogenous metabolite that originates from the degradation of proteins and amino acids. Most of the research on the underlying pathophysiology of HE has been focused on the buildup of this nitrogenous waste product in the blood and brain of humans.

A variety of HE precipitants act through one or more of the three major pathways related to ammonia metabolism and toxicity: increased production, impaired excretion, and increased neurotoxicity (Jaffe et al., 2020).

18.2.1.1 Ammonia Production

The two main sites of ammonia production are the small/large intestine (50%) and the kidneys (40%). Within the intestine, two mechanisms exist: (1) urea from dietary protein is broken down into ammonia and carbon dioxide exclusively via urease-producing bacteria (i.e., Klebsiella, Proteus), predominantly in the large intestine and (2) direct amino acid degradation of glutamine under the effect of the glutaminase in the enterocytes generates glutamate and free ammonia. This unionized form of ammonia can then cross the intestinal epithelium freely and enter the portal circulation. The ionization of ammonia depends on the colonic luminal pH, with a more acidic pH leading to protonation and trapping of ammonium ($NH4+$) within the lumen.

Within the kidneys, the proximal tubular cells are capable of generating ammonia from glutamine and creating bicarbonate as a byproduct. Therefore, the body's acid–base status and potassium balance drive the renal production of ammonia. In the setting of metabolic acidosis or hypokalemia, glutamine is used to facilitate the disposal of acid or recovery of potassium. Through this mechanism, renal ammoniagenesis increases arterial ammonia levels (Han, 2011). In addition, intravascular hypovolemia in the setting of gastrointestinal bleeding and overdiuresis enhances renal ammonia production, mediated through angiotensin II (Nagami, 2002).

Smaller contributors to ammonia production include the release of ammonia from muscles via the adenylic acid metabolism, which plays a role in the purine nucleotide cycle, along with the release of glutamine from senescent red blood cells (Dasarathy et al., 2017).

In physiological conditions, the liver efficiently extracts (85%) the ammonia from portal blood and, through the urea cycle, converts it into urea, then excreted in the kidneys (75%) and the intestine (25%). Only 15% of the ammonia pool enters the systemic circulation (Rose et al., 2020). Defects in hepatic function, portal blood flow, and urea cycle enzymes or intermediates can result in hyperammonemia, as can excessive ammonia production in the gastrointestinal tract.

18.2.1.2 Ammonia Excretion

The major site of ammonia catabolism is the *urea cycle* (Krebs–Henseleit cycle), which exclusively takes place within the liver. Through this mechanism, two molecules of ammonia and one molecule of carbon dioxide generate urea and water. Urea is easily able to cross membranes given its small size and therefore is able to enter back into the intestines and the kidneys for excretion.

The ammonia concentration within the portal vein is 5–10 times higher than that of the systemic circulation (Dasarathy et al., 2017). When there is reduced capability of the liver to detoxify ammonia due to either hepatocellular damage or shunting around the liver from portal hypertension or spontaneous/iatrogenic portosystemic shunts, there is a resultant buildup within the systemic circulation.

The kidneys contribute to ammonia excretion through parallel H+ and ammonia secretion in the collecting duct, generating ammonium, which is eliminated through urine. Of note, almost all urinary ammonia is from renal ammoniagenesis. The proportion of ammonia generated that is excreted in the urine depends on several factors, most importantly acid–base and potassium balance, protein intake, and glucocorticoid hormones, but overall renal venous ammonia exceeds arterial ammonia, meaning that the kidneys actually increase systemic ammonia (Weiner et al., 2015). In the normal physiologic state, about 50% of ammonia gets excreted in the urine, with the remainder being returned to the systemic circulation via the renal vein. In acidosis, the kidneys excrete hydrogen ions in the urine and convert ammonia to ammonium. In alkalosis, there is a significant reduction in renal excretion of ammonia; similarly, when the renal perfusion is reduced (such as in dehydration and overdiuresis), there is a reduction in ammonia excretion (Dasarathy et al., 2017).

Muscles are capable of temporarily detoxifying ammonia by converting ammonia into glutamine via glutamine synthetase (GS). Impaired hepatic clearance of ammonia upregulates the gene coding for GS, and skeletal muscles become a significant site of ammonia metabolism for patients with cirrhosis. Therefore, sarcopenia is a risk factor for HE (Merli et al., 2013), and improving muscle mass is beneficial for HE treatment. Studies have shown an abnormal ratio in branched-chain amino acids (BCAAs), including isoleucine, leucine, and valine, along with aromatic amino acids in patients with cirrhosis, thought to be secondary to increased metabolism within skeletal muscles (Jawaro et al., 2016). Given the role of BCAAs to generate glutamate, BCAA deficiency leads to a reduced ability of the skeletal muscles to convert ammonia to glutamine.

18.2.1.3 Ammonia Neurotoxicity

Hallmarks of HE in acute and chronic liver failure are astrocyte swelling, cerebral oxidative stress, microglia activation, and altered neurotransmission.

Overwhelming evidence from both in vitro and in vivo studies indicates that astrocytes play a central role in the pathogenesis of ammonia toxicity and HE development. The ammonia passes freely into the brain and preferentially enters the astrocytes, as these cells are the only brain cells that contain GS, allowing for the detoxification of ammonia through the generation of glutamine (Norenberg, 1987). In fact, GS has a higher affinity for ammonia than it does for glutamate in astrocytes, and these cells are likely to take up to four times as much ammonia than any other cell type. Ammonia acts both as a substrate and product for a variety of biochemical reactions and is, therefore, capable of causing multiple cellular changes that alter cellular function. These detrimental changes range from altering pH, membrane potentials, and electrolyte balance to impairing oxidative metabolism via inhibition of the rate-limiting enzyme in the tricarboxylic acid cycle, altering calcium signaling, activating mitochondrial permeability transition leading to cell dysfunction (Felipo & Butterworth, 2002).

Several studies have demonstrated low-grade cerebral edema in cirrhotic patients. There are two mechanisms that contribute to cerebral edema in liver disease. Glutamine itself acts as an osmole, and increased intracellular concentrations of this molecule within astrocytes develop as these cells try and detoxify ammonia using GS. This creates an osmolar gradient resulting in astrocyte swelling (Haussinger et al., 1994; Haussinger et al., 2000; Lachmann et al., 2013; Dabrowska et al., 2018), known as an osmotic gliopathy (Dasarathy et al., 2017). Secondarily, glutamine enters mitochondria where special glutaminases catabolize it back into ammonia. This mitochondrial ammonia accumulation results in the generation of reactive oxygen species resulting in increased mitochondrial permeability and ultimately leads to astrocyte cell dysfunction and swelling.

According to current knowledge, HE can be seen as the clinical manifestation of a pathogenetic interplay between osmotic and oxidative stress in the astrocytes (Haussinger et al., 2000; Haussinger et al., 2021) that triggers posttranslational protein modifications, RNA oxidation, senescence, altered signaling and broad effects on gene expression. Further, the increased brain ammonia concentrations result in alterations in the expression of key astrocyte proteins, including glial fibrillary acidic protein, glutamate and glycine transporters, and "peripheral-type" (mitochondrial) benzodiazepine receptors. Such changes result in increased extracellular concentrations of excitatory and inhibitory substances. In addition, the ammonium ion has direct effects on excitatory–inhibitory transmission via distinct mechanisms involving cellular chloride extrusion and postsynaptic receptor function. Acute ammonia exposure leads to the activation of NMDA receptors and their signal transduction pathways. Chronic hyperammonemia also results in increased concentrations of neuroactive L-tryptophan metabolites, including serotonin and quinolinic acid (Felipo & Butterworth, 2002). All these alterations affect astrocytic/neuronal functions and synaptic plasticity, causing disturbances of oscillatory networks in the brain, which finally account for cognitive and motor HE symptoms (Butz et al., 2010; Haussinger et al., 2021; Nikolov et al., 2021).

In addition to the direct role of ammonia, systemic inflammation/oxidative stress and increased blood bile acids impact the blood-brain barrier (BBB) permeability, allowing an increased influx of molecules physiologically unable to cross it (Atluri et al., 2011). Consequently, the alterations of metabolites in cerebrospinal fluid and changes in neurotransmission, such as increased GABAergic tone, significantly modulate the onset of neurological decline (Keiding et al.,

2006; Riordan & Williams, 2010; Llansola et al., 2015). More recently, neurosteroids, endogenous benzodiazepines, and manganese have emerged as synergistic factors in the onset of HE (Rose et al., 2020). Furthermore, like a vicious circle, hyperammonaemia "per se" can induce neutrophil dysfunction and reactive oxygen species release, contributing to systemic oxidative stress and inflammation, and exacerbating its harmful effects in the brain (Shawcross et al., 2004).

18.2.2 Role of Inflammation in HE

In addition to astrocyte cell dysfunction, neuropathology studies have noted microglial cell activation (Butterworth, 2019), which results in a proinflammatory state leading to monocyte recruitment as well as increased cytokine gene expression within the brain (Butterworth, 2013; Schrimpf et al., 2022). The appearance of MHE is associated with a shift in the immune response, with the expansion and activation of Th17, Th22, and Tfh CD4+ lymphocytes and increased serum levels of pro-inflammatory cytokines, such as IL-6, IL-21, IL-17, IL-18, TNF-α, IL-1β, IL-15, IL-22- as well as of CCL20, CXCL13, and CX3CL1 (Montoliu et al., 2009; Mangas-Losada et al., 2017; Azhari & Swain, 2018). These alterations are transmitted to the brain and lead to an activation of the brain's immune system (neuroinflammation), which may alter neurotransmission and lead to cognitive and motor impairment (Cabrera-Pastor et al., 2019). Increased bile acids may also contribute to HE through sphingosine-1-phosphate receptor 2 (S1PR2) and TGR5 activation and increased C-C Motif Chemokine Ligand 2 (CCL2), which activates microglia and contributes to cognitive impairment (McMillin et al., 2017).

The several mechanisms, by which peripheral alterations may be transmitted to the brain to induce MHE, include (a) Infiltration of peripheral lymphocytes and monocytes into the brain (D'Mello et al., 2009; Balzano et al., 2018), (b) Infiltration of extracellular vesicles (EVs) from plasma (Izquierdo-Altarejos et al., 2022), as discussed in greater detail later, and (c) activation by peripheral cytokines of their receptors in endothelial cells and transmission of signals to the brain. All these mechanisms finally lead to neuroinflammation with the activation of microglia and astrocytes and increased synthesis of pro-inflammatory factors, which alter neurotransmission and trigger neurological impairment (Cabrera-Pastor et al., 2019).

Hyperammonemia and peripheral inflammation act synergistically to induce HE (Shawcross et al., 2004; Felipo et al., 2012; Coltart et al., 2013). In animal models of MHE (portacaval shunted rats) or chronic hyperammonemia, prevention of systemic inflammation by intravenous injection of infliximab, an anti-TNF-α antibody that does not cross the blood-brain barrier, also prevents the appearance of neurological impairment (Dadsetan et al., 2016; Balzano et al., 2020). TNF-α treatment of wild-type mice sensitizes the animals to the toxic effects of ammonia, whereas TNF-α- or TNF-α-receptor-1-deficient mice are protected in this respect (Pozdeev et al., 2017). The protection of TNFα-deficient mice against an ammonia load was paralleled by a decreased cerebral expression of NKCC1, which is expected to counteract ammonia-induced glia swelling. In line with this, TNF-α induces synergistically with ammonia astrocyte swelling and protein tyrosine nitration of cultured astrocytes (Haussinger & Schliess, 2008; Rama Rao et al., 2010; Lachmann et al., 2013). Hyperammonemia per se is enough to induce systemic inflammation, neuroinflammation, and neurological impairment (Rodrigo et al., 2010; Balzano et al., 2020).

18.3 Gut–Liver–Brain Axis: Role of Intestinal Microbiota Changes in HE

Liver disease is associated with changes in intestinal microbiota. Colonic mucosal microbiota is altered in cirrhotic patients, especially in those with HE, and this is linked to inflammation and impaired cognition (Bajaj et al., 2012; Dhiman, 2012; Iebba et al., 2018; Acharya & Bajaj, 2019; Luo et al., 2023). It is commonly accepted that neurological impairment and cognitive decline provoked by liver dysfunction result from blood-derived factors influencing the permeability and altering the integrity of the BBB. The gut–liver–brain axis refers to the bidirectional relationship between the gut and its microbiota, the liver, and the brain, resulting from integrating signals generated by dietary, genetic, and environmental factors (Mancini et al., 2018). In patients with liver cirrhosis, defective small intestinal motility, reduced gastric acid secretion, and weaker antimicrobial defense of intestinal mucosa determine small intestinal bacterial overgrowth (SIBO). The concomitant decrease in bile acids (BAs) synthesis are secondary to end-stage liver disease and synergistically act with SIBO to determine pathological changes in the intestinal microbiota composition, mainly characterized by a massive reduction in microbial diversity, a decline in autochthonous non-pathogenic bacteria (Bacteroidetes, Ruminococcus, Roseburia, Veillonellaceae, and Lachnospiraceae), and an overgrowth of potentially more pathogenic, urease-producing bacteria (Fusobacteria, Proteobacteria, Enterococcaceae, and Streptococcaceae) (Fukui, 2015; Acharya & Bajaj, 2017). Indeed, an increased risk of HE is associated with SIBO (Gupta et al., 2010), and a correlation with HE recurrence is observed with the relative abundances of *Alistipes*, *Bacteroides*, *Phascolarctobacterium* (Bajaj et al., 2014).

The paucity of bacteria involved in producing short-chain fatty acids (SCFAs) and converting primary into secondary BAs contributes to worsening gut dysbiosis and disrupting intestinal barrier integrity. Indeed, SCFAs (mainly butyrate, acetate, and propionate), produced in the colon by bacterial fermentation of dietary fibers and resistant starch, exert several beneficial functions: preserving intestinal barrier integrity, nourishing colonocytes, promoting mucus production, and reducing colonic inflammation (Nava & Stappenbeck, 2011; Rowland et al., 2018). Furthermore, the lower abundance of 7α-dehydroxylating bacteria in the colon (Lachonospiraceae, Ruminococcaceae, and Blautia) due to a reduction in primary BAs determines a change in secondary-to-primary BAs ratio that can favor the overgrowth of pathogenic taxa (Kakiyama et al., 2013; Ridlon et al., 2013). The impaired intestinal barrier integrity enhances bacterial translocation and the release of bacterial endotoxins in circulation, such as lipopolysaccharides (LPS), flagellin, peptidoglycan, and microbial nucleic acids, perpetuating liver damage, and contributing to systemic inflammation responsible for BBB dysfunction and neuroinflammation (Shawcross et al., 2004; Shawcross et al., 2007; Shawcross et al., 2011; Dhiman, 2012; Sole et al., 2021). Dysregulation of BBB permeability may also lead to a dramatic increase of certain BAs, such as lithocholic, taurocholic, and glycocholic acid, in the cerebrospinal fluid of patients with HE or brain tissue of rodent model of HE (Tripodi et al., 2012; McMillin et al., 2016; Weiss et al., 2016). Although a precise role for BAs in the pathogenesis of HE has not yet been completely defined, they are likely involved in aberrant neuronal signaling and the promotion of neuroinflammation through microglia activation (DeMorrow, 2019).

18.4 Role of Viral Hepatitis in the Induction of HE

In chronic inflammatory liver diseases, including viral hepatitis, altered mood states, and sick behavior are not rare events (Swain, 2006). HE is usually associated with advanced liver disease, but less advanced stages of chronic inflammatory liver disease may also induce changes within the CNS. Usually, the changes in neurotransmission found in cholestatic liver disease or viral hepatitis may promote behavioral alterations in the absence of visible organic CNS damage (Forton et al., 2004; Forton et al., 2008) and be observed in early-stage liver disease (Loftis et al., 2008).

In liver–brain communications accompanied by viral hepatitis, a product of gram-negative bacterial walls, LPS plays a significant role. LPS is known to be a potent inducer of three cytokines that are known to communicate between the periphery and brain, namely, TNF-α, IL-1β, and IL-6 (Capuron & Miller, 2011). The levels of these cytokines are elevated in chronic viral hepatitis, and neurotoxic events may be related to cytokine polymorphism. As an example, TNF-α polymorphism is linked with mood disorders, fatigue, and sleep abnormalities (Clerici et al., 2009). In the liver, these cytokines are mainly produced by macrophages (Kupffer cells), but hepatocytes also contribute to their production. Unlike liver endothelial cells, the CNS endothelial cells are not fenestrated, and large cytokines from the periphery cannot pass BBB. For communications with the brain, they use (1) the neural pathway, (2) signaling via cerebral endothelial cells (CECs), (3) signaling via the circumventricular organs (CVOs), and (4) immune cells. As reviewed by (D'Mello & Swain, 2011), the liver is innervated by vagal afferents responding to TNF-α, IL-1β, and IL-6. Vagal afferents project to the dorsal vagal complex, which includes the nucleus tractus solitarius (NTS), area postrema, and dorsal motor nucleus; from here, they then affect varied cerebral regions, including the paraventricular nucleus of the hypothalamus. The pro-inflammatory cytokines interact with their receptors expressed on cerebral endothelial cells to induce signal transduction. In addition, circulating cytokines can also access the brain via the circumventricular organs (CVOs) and the choroid plexus, regions of the brain that lack a blood-brain barrier. Furthermore, monocytes transmigrate to the brain due to microglia activation secreting MCP-1. Finally, these liver–brain communication pathways affect central neural activity and cause behavioral abnormalities.

Chronic viral hepatitis B and C (CHB and CHC), especially when potentiated by second hits, may progress to end-stage liver disease, like cirrhosis. Patients with cirrhosis frequently develop complications such as ascites, variceal bleeding, and hepatic encephalopathy (HE). The clinical manifestations of HE range from a mildly altered level of sensorium to severely altered consciousness levels, difficulty in judgment, the day–night reversal of sleep, flapping tremors of hands, and irrelevant talking or speech. (Younas et al., 2021). The exact mechanisms causing HE are still not clear, but some hypotheses include the effects of neurotoxins like ammonia, leading to a swelling of astrocytes, impairment of neurotransmission, and systemic response to infections and neuroinflammation (Ferenci, 2017). Hepatitis A, B, and E viruses represent the main cause of acute liver failure (ALF) in developing countries and the second leading cause in Western countries, which leads to HE (Caballol et al., 2019).

18.4.1 Hepatitis A

Hepatitis A virus (HAV) causes ALF in only 1% of cases, with a survival rate of 69%. These patients can be treated with plasma exchange. The removal of circulating damage-associated molecular patterns and the subsequent decrease in the production of proinflammatory cytokines could at least partially explain the positive effects of plasma exchange as these mediators have been involved in the pathogenesis of HE (Caballol et al., 2019). Also, Takotsubo syndrome is stressed-induced cardiomyopathy developing in response to a catecholamine storm causing reversible myocardial "stunning" during physical or emotional stress that complicates the course of HAV fulminant hepatitis and, thus, can be linked to HE (Caballol et al., 2019). Both Takotsubo syndrome and HE are reversible for HAV infection and require no liver transplantation.

18.4.2 Hepatitis B Virus (HBV) and Hepatitis C Virus (HCV)

HBV and HCV may cause fast-developed hepatic encephalopathy because of disease progression to liver cirrhosis or due to ALF. Neurological manifestations are linked to hyperammonemia and neuroinflammation. Immune activation in response to HCV infection results in increased IL-6, TNF-α, iNOS, COX-2, and IL-1 markers, which correlate with extrahepatic manifestations of this disease (Rahman et al., 2020). Serum levels of IL-6 and IL-18 correspond to the degree of encephalopathy in HBV infection as well (Tsai et al., 2015). As in other types of liver disease, in viral hepatitis, the gut–liver–brain axis is activated by gut microbiota, and six dominant phyla in gut microbiota: *Firmicutes* and *Bacteroidetes* (90%), *Proteobacteria, Actinobacteria, Fusobacteria,* and *Verrucomicrobia* (Catinean et al., 2018). Since intestinal dysbiosis has been demonstrated in HBV and HCV infections and leads to LPS leakage, it promotes an increase in TNF-α, IL-6, and C-reactive protein levels (Sandler et al., 2011), due to the activation of monocytes by LPS thereby contributing to neuroinflammation and HE. In addition, the microbiota-associated changes may be triggered by activation of the transformation of nondigestible carbohydrates to short-chain fatty acids (SCFA) or tryptophan to indole derivates, serotonin, and kynurenine. In addition, the microbiota's bacterial flagellin can be recognized by host TLR5/NFκB to induce the inflammation of systemic character. All these mechanisms work in HBV- and HCV infections (Neag et al., 2021). Gut microbiota studies showed the difference in the composition of intestinal microbiota in HCV and HBV patients with or without cirrhosis (Heidrich et al., 2018; Liu et al., 2019) and allowed identifying the targets for treatment with probiotics (Oo et al., 2016; Xia et al., 2018). Furthermore, regardless of changes in microbiota, HCV infection directly induces neuropsychiatric symptoms known as HCV brain syndrome (fatigue, depression, cognitive impairment) (Yarlott et al., 2017). In fact, these symptoms can be developed even in HCV patients with minimal liver disease. Some studies demonstrate evidence of CNS infection with HCV and compartmentalization of HCV within the cerebrospinal fluid (Yarlott et al., 2017). The postulated mechanism of HCV entry to CNS is via infected monocytes, but this data require further confirmation. Indirectly, the targeting of microglia by HCV proteins induces pro-inflammatory signal transduction via activation of ERK and STAT3 signaling (Paulino et al., 2011). Also, serotonin and dopamine transporters availability play a role in the impartment of cognitive function in HCV infection since this infection is accompanied by a decrease in glucose utilization in frontal and parietal cortices, indicating alteration of dopaminergic neurotransmission as a possible mechanism of cognitive decline (Heeren et al., 2011).

18.5 Induction of Hepatic Encephalopathy via EVs Transmission

The liver is one of the metabolic organs that plays a vital role in metabolizing toxic substances, synthesis of secretory proteins, and storing liver glucose. The development of encephalopathy is observed in many chronic liver diseases, indicating the distance crosstalk between the liver and brain (Wang et al., 2022). The incidence of neurological complications after liver transplantation is about 20% (Bronster et al., 2000). Also, hyperammonemia in liver cirrhotic patients induces moderate encephalopathy with cognitive and motor impairments (Balzano et al., 2020). Some biomolecules derived from liver cells and circulating with the blood cannot reach the brain since they are not crossing the BBB. EVs are one of the most effective vehicles for liver–brain communication because, size-wise, these vesicles are not limited to passing BBB.

18.5.1 What Are EVs?

EVs are small membrane-bound vesicles containing proteins, genes, lipids, and nuclear acids. These EVs are subdivided into exosomes, microvesicles, and apoptotic bodies based on their size. Apoptotic bodies formed from the dying apoptotic cells are big enough to serve for intracellular communication within the organ but not for the crosstalk between different organs and now, are only partially characterized in the context of some liver infections (like HIV) or alcohol-associated liver disease (Osna & Poluektova, 2023). The most well-studied EVs are exosomes. Exosomes have a size of 50–150 nm. They are formed as intraluminal vesicles by endosomal-sorting-complex-required-for transport (ESCRT) machinery components and are released when multivesicular bodies (MVB) fusing with the plasma membrane are not degraded by lysosomes (Rahman et al., 2020). Exosomes contain endosome-derived proteins, including Rab GTPases, SNARES, flotillins, annexins; ESCRT, membrane proteins, including tetraspanins (e.g., CD9, CD63, CD81, and CD82); lipid raft-associated proteins; heat shock proteins (HSP70 and HSP90β) as well as cell-type-specific proteins (van Niel et al., 2018). Exosomes and microvesicles also contain nucleic acids, such as DNA and RNA (mRNA and noncoding RNAs, ncRNAs); however, other characteristics of microvesicles (not exosomes) are very fuzzy. EV RNAs are of extreme importance because RNA cargo causes phenotypic alterations in recipient cells affecting gene expression (mRNA) or epigenetic regulation of this gene expression (ncRNAs). EV RNAs include those with already known functions (miRNAs, mRNAs, small interfering RNA), those with predicted functions (tRNA, small nuclear RNA, YRNA, vault RNA), and those whose functions are unknown (fragmented and degraded RNA) (Makrygianni & Chrousos, 2023).

18.5.2 Cell–Cell Communication by EVS

In liver disease, exosomes serve as a vehicle for both cell-to-cell communications within the liver (Osna et al., 2022) and for liver communications with other organs, including the brain, which is applied to healthy and pathological conditions (Colombo et al., 2014). Importantly, EVs play a role in bidirectional communications between the brain and other organs (Zhou et al., 2023). This is based on the functions of EV-carried mediators in the crosstalk between the brain, bone marrow, adipose tissue, liver, heart, lungs, and gut. As an example, an increase in soluble epoxide hydrolases in the brain and liver is related to the pathogenesis of major psychiatric disorders (Zhang et al., 2020). Furthermore, since the liver and gut are closely connected, the difference in gut bacterial composition causes mood and behavior impairment. Exposure to some nonpathogenic bacteria modulates brain chemistry and behavior (Nikolova et al., 2019). Thus, EVs are crucial for the brain–liver crosstalk (Fig. 18.1).

These interactions are supported by direct and indirect mechanisms. One of the indirect mechanisms is related to the brain–adipose tissue crosstalk, which may also affect the liver due to constant interactions between adipose and liver tissues linked to fatty liver disease development. Adipokines released from adipose tissue regulate neuroinflammation and oxidative stress, which is orchestrated by circulating noncoding RNAs transported by exosomes (Thomou et al., 2017). As shown, axonal sprouting and development, oligodendrocyte creation, tract connection, and remyelination after subcortical ischemic stroke are all mediated by exosomes of adipose origin (Otero-Ortega et al., 2020). However, adipose-derived exosomes play a dual role in the regulation of brain functions. On the one hand, exosomes carrying miR-30d-5p inhibit ischemic stroke-induced microglia, thereby promoting functional recovery and reducing brain damage; on the other hand, adipose exosomes provide cognitive impairment via synapsis loss (Zhou et al., 2023). Furthermore, in experimental settings of Alzheimer's Disease (AD), which is characterized by an imbalance between Aβ production and clearance leading to Aβ accumulation in the brain, exosomes containing enzymatically active neprilysin reduce Aβ40 and Aβ42 (Katsuda et al., 2013). In amyotrophic lateral sclerosis, exosomes derived from adipose stem cells prevent superoxide dismutase aggregation and mitochondrial dysfunction and, thus, have a therapeutic potential (Lee et al., 2016).

Hepatic encephalopathy is frequently developed in liver cirrhosis patients. These patients may have mild cognitive and motor impairments due to persistent hyperammonemia (Balzano et al., 2020), and EVs are involved in the transfer of these peripheral changes to the brain. In the experimental settings, when these EVs were injected into normal rats and reached the cerebellum, they induced neuroinflammation and motor incoordination (Izquierdo-Altarejos et al., 2022) and increased TNF-α levels, causing enhancement of cerebellar GABAergic transmitters and microglia activation. Moreover, apoprotein E4 (ApoE4) alleles are associated with age-dependent cognitive dysfunction and AD (Kanekiyo et al., 2014). This can be potentiated by age-related thyroid deficiency by enhancing EV-dependent transport of ApoE4 from the liver to the brain, activation of the NLRP3 inflammasome, and caspase 1-dependent pyroptosis of nerves (Zhang et al., 2022). While the role of EV-mediated communication between the liver and brain is not well investigated, it serves as an exciting area for future studies. However, currently, the origin, composition, biodistribution, and identity of EV targets found in systemic circulation are not clear.

The important liver-related aspect of EV-mediated crosstalk with the brain includes the role of EVs in communication between the gut and the brain, which may also affect the liver and is crucial for maintaining homeostasis (Zhou et al., 2023). Its disturbances are associated with AD. Multiple signaling pathways transmit signals to the CNS, and are related to the vagus nerve, circulating blood, and EVs (Zhao et al., 2021). EVs secreted from the gut might be of intestinal or microbial origin (Kim et al., 2021). By crossing the gut epithelial barrier, microbial products enter the blood circulation and become key mediators in altering brain physiology and behavior. Metabolomics profiling identified many metabolites in microbial EV cargo, including lipids, carbohydrates, amino acids, vitamins, organic acids, and neurotransmitter-related compounds, such as arachidonoyl-dopamine (NADA), gabapentin, glutamate, and N-acylethanolamines (Simpson & Redinbo, 2022). The majority of EVs are the products of the budding of the outer membrane of gram-negative bacteria and are designated as other membrane vesicles (OMV) (Perez-Cruz et al., 2013). In addition to gram-negative bacteria, other microorganisms like fungi may release EVs, which affect the brain (Rizzo et al., 2020).

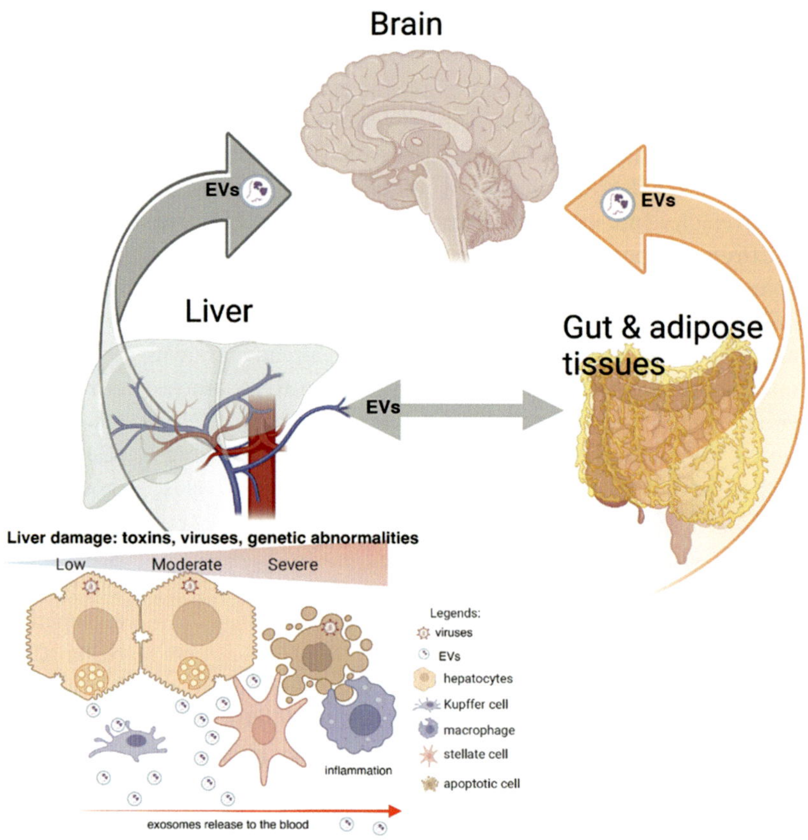

Fig. 18.1 Liver–brain communications via EVs. While apoptotic bodies (ABs) are mainly used for communication between various cells within the organ, the main crosstalk between the liver and brain is via exosomes carrying proteins, mRNAs, ncRNAs, DNAs, and lipids in their cargo. EVs from gut microbiota, adipose tissue, and other peripheral organs/tissue can also "program" the liver's and brain's dysfunctions. (Created with BioRender)

Bacterial EVs carry polysaccharides, proteins, nucleic acids, metabolites, and enzymes. Like other EVs, they can cross BBB to deliver cargo to the brain cells, and LPS is linked to neuroinflammation. Also, OMVs disrupt the integrity of the BBB. Promoting astrocyte and microglial activation, OMVs cause inflammation and tau hyperphosphorylation by activating the glycogen synthase kinase-3 beta (GSK-3β) pathway, which results in cognitive impairment (Wei et al., 2020). OMVs' effects are potentiated by EVs–vagus nerve interactions.

Many neurogenerative dysfunctions are related to microglia activation due to alterations in the composition of the microbiome. It has been demonstrated that *Actinomyces* OMVs can reach the brain in mice, and extracellular RNA (exRNA) transfer of OMVs to the brain causes neuroinflammatory diseases, such as AD, by activating the TLR8 and NF-κB inflammatory signaling pathways (Han et al., 2019). In addition, EVs from Paenalcaligenes hominis increase the amount of LPS-activated microglia in the hippocampus and contribute to cognitive impairments in the brain. Almost any type of infection causes transportation of OMVs to the brain. The inhibitory neurotransmitter gamma-aminobutyric acid (GABA) and its biosynthetic intermediates, α-ketoglutarate and glutamate, are present in EVs produced by *Bacillus fragilis* (Zakharzhevskaya et al., 2017). However, EVs from gut microorganisms may also provide not detrimental but beneficial effects. As an example, intraperitoneal injection of EVs from *Lactobacillus plantarum* increases BDNF and neurotrophin 4/5 (NT4/5) expression and blocks stress-induced depression-like behaviors (Choi et al., 2019). In addition, gut cells infected with *Escherichia coli LF*82 generate large exosomes enriched with miRNAs, including miR-130a, which affects BBB permeability (Wang et al., 2016). Since gut microbiota regulates intestinal miRNA profiles and these miRNAs are released with exosomes, this mechanism is of particular importance for the regulation of neuroinflammation via gut EVs (Zhao et al., 2021). In addition, the release of exosomes by intestine epithelial cells induces neuroinflammation via immune mechanisms activated by exosome-delivered exogenous peptides presented in the context of MHC class II (Kojima et al., 2018). These exosomes promote hippocampal neuronal damage, apoptosis, and excessive IL-1β in the circulation (Xi et al., 2021). Thus, gut dysfunctions with the involvement of microbiota, metabolic changes, and gut dysbiosis-induced inflammation developed in the settings of liver disease become crucial points for abnormalities in CNS functioning.

As already mentioned, the communication between the gut and brain is bidirectional. Stress not only modifies gut-devired EVs content but also increases the release of EVs by neural and neuroendocrine cells in the CNS (Makrygianni & Chrousos, 2023). Particularly, neuronal-derived EVs in acute stress prevent excitotoxicity in recipient neurons and program microglia for neuroprotection. However, under chronic stress, these EVs impair the survival of neurons, decrease synaptic efficiency, activate astrocytes and microglia, and promote neuroinflammation by moving immune cells to the CNS. Neuronal and neuroendocrine cell EVs are involved in neurotransmission and neuromodulation and change the phenotype of recipient cells by horizontal transfers. Thus, not only EVs are released from the cells of peripheral organs to target the CNS, but CNS exosomes could also affect other organs as well.

Here, we were able to overview major mechanisms of known EV-mediated communications between adipose tissue, gut, liver, and brain. These interplays are complex and may involve more endocrine, humoral, metabolic, and immune factors to bridge the liver-gut-adiposes axes and the CNS. There is no doubt that gut dysfunctions with the involvement of microbiota, metabolic changes, as well as gut dysbiosis and leakage-induced inflammation developed in the settings of liver disease may become crucial points for abnormalities in CNS functioning and require further investigations. EVs are used as biomarkers for neurological disorders, and EVs enriched for neuronal origin are more sensitive markers than just plasma EVs (Mustapic et al., 2017). Systemic inflammation can be induced by EVs carrying information about impaired host immune response, microbiome, and metabolic dysfunctions affecting CNS, which is partially controlled by probiotics and diet (Krakovski et al., 2022; Jian et al., 2023). Interestingly, exosomes derived from both human and cow milk may protect from neuroinflammation by preventing gut leakage (Tong et al., 2023).

18.6 Experimental Models for HE

18.6.1 Acute Liver Failure Models

ALF is marked by the rapid deterioration of liver functions in a previously well patient over days to weeks (Fig. 18.2). Though relatively rare, it is associated with high morbidity and mortality. This makes it a challenging disease to study clinically, necessitating reliance on preclinical models as a means to explore pathophysiology and novel therapies. Preclinical models of acute liver failure are artificial by nature and generally fall into one of the three categories: surgical, pharmacologic, or immunogenic. The preclinical models of acute liver failure and the consideration of their relevance in modeling clinical disease were reviewed by (Hefler et al., 2021).

Generally, clinically (except in cases of acute overdose), patients may present with progressive worsening over days and weeks (~26 weeks). The timeline for ALF in animal models is shorter than in humans. Laboratory animals have shorter lifespans, and in rodent models of ALF, liver failure develops over hours to days after a single inciting event. Additionally, most aspects of rodent biology, including metabolic rate and RNA and protein turnover, are several times faster than of humans. This is especially true for more complex biological processes. For instance, in rodent models of chronic liver injury, fibrosis develops in as little as 4 weeks, whereas cirrhosis in humans takes years (Hefler et al., 2021).

18.6.1.1 The Surgical Methods

The surgical approach is simple and reproducible. These methods are based on completely removing the liver by single or multi-stage procedures. These models are helpful for the testing of liver support devices. However, it is an irreversible procedure and lacks inflammatory responses, and is technically challenging in small animals. The development of HE is questionable. The method of partial hepatectomy with resection of 70–97% of liver tissues has a direct clinical correlation with HE and is helpful in studies of hepatic regeneration. Hepatic artery ligation and portocaval anastomosis procedures performed in one or two steps reliably have progressive encephalopathy and coma. The disadvantage of this model is the need for clinical correlation.

18.6.1.2 Pharmacological Models

The liver failure could be modeled with pharmacological agents. The well-known model that has clinical correlates is *acetaminophen-induced hepatic necrosis* (Mitchell et al., 1973). Acetaminophen (APAP) hepatotoxicity is mediated by the reactive metabolite N-acetyl-*p*-benzoquinonimine (NAPQI) (Moyer et al., 2011). This metabolic intermediate forms upon oxidation of APAP via the CYP2E1 enzyme. Typically, NAPQI is quickly conjugated with glutathione (GSH) via glutathione transferase to produce the nontoxic compound APAP-GSH. However, a rapid influx of NAPQI depletes GSH stores, causing NAPQI to build up in liver cells.

For this reason, patients who overdose on APAP are treated with N-acetyl cysteine (NAC), which stimulates GSH production to protect against liver damage. The toxic effects are dose-dependent and species-specific. Certain animals, like rats, are not sensitive to this treatment. In some animals, models may require cytochrome 450 induction or glutathione depletion. In large animals, the model is complicated by methemoglobinemia.

D-galactosamine (D-Gal) induces depletion of intracellular uridine stores necessary for RNA synthesis through metabolism in the Leloir pathway, which results in UDP-galactosamine accumulation. This also deprives the cell of uridine derivatives, UDP-glucose and UDP-galactose, necessary for glycogen synthesis. In this model, activation of Kupffer cells leads to the release of proinflammatory cytokines and neutrophil infiltration (Yang et al., 2016). D-Gal is

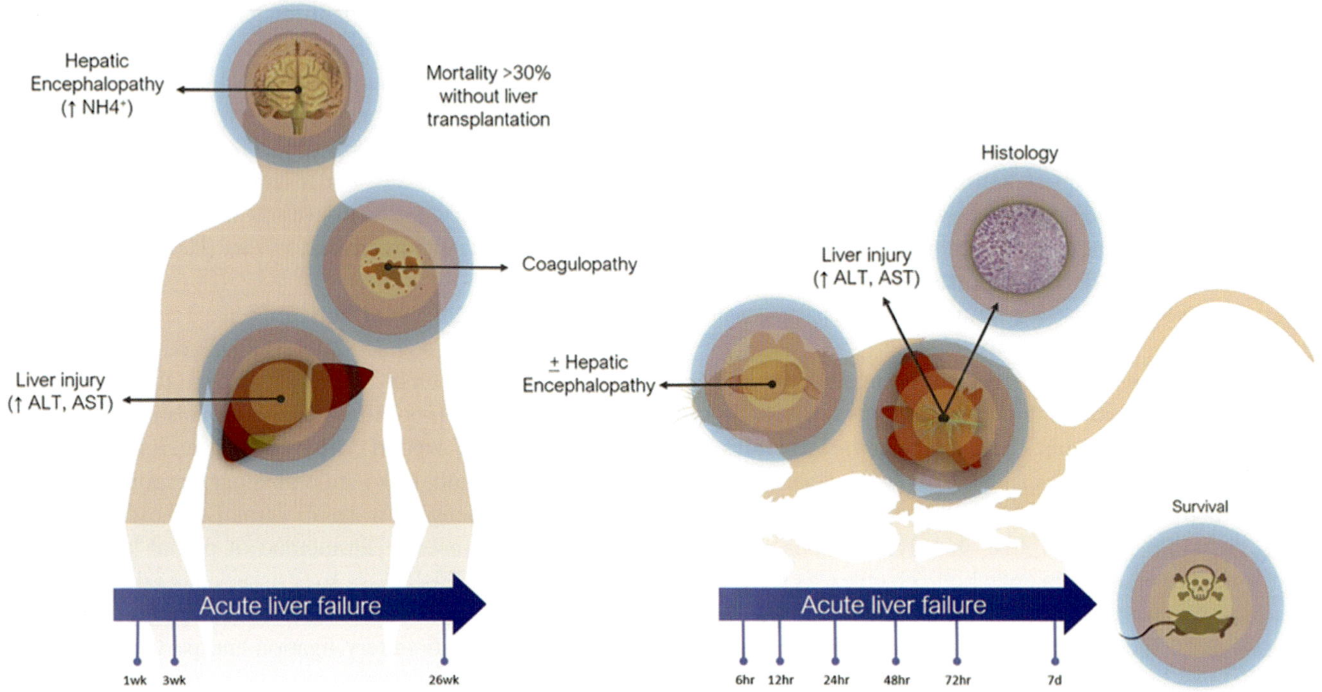

Fig. 18.2 Differences between clinical and preclinical models of acute liver failure

characterized by diffuse rather than zonal necrosis seen with most hepatotoxic drugs. This model (with and without the addition of LPS) has no direct clinical correlation.

Carbon tetrachloride (CCl_4) administration via cytochrome P450 metabolism produces highly reactive trichloromethyl and trichloromethyl peroxy radicals. This toxin is also used for the generation of chronic liver injury. Even though this model is efficient for generalized liver injury, it does not have a clinical correlation or reflect human HE development.

Several other hepatotoxic agents produce acute and chronic liver damage. These are *Thioacetamide, Azoxymethane, A-Amanitin* (from toxic mushrooms), and *A-Naphthyl Isothiocyanate* (cholestatic injury), which induce injury in mice, rats, pigs, and primates. In addition to the models mentioned above of hepatotoxicity, the administration of ammonia (either orally or intravenously) has been used for the acute induction of HE in isolation (Fick et al., 1989). These artificial models of hyperammonemia may be more appropriate as a control alongside acute or chronic models of liver injury, given the absence of other features of acute liver failure and the lack of a clinical correlation.

18.6.1.3 Immunogenic Models

Immunogenic lectin derived from jack bean *Concanavalin A* administration induces representative T cell-mediated hepatitis, relevant to autoimmune hepatitis clinically (Liu et al., 2022). The dosage of *Con A* needed to trigger hepatic injury varied significantly according to the genetic background of the mice. Generally, Th1-biased C57BL/6 and C3H mice are most susceptible to *Con A* and usually require lower doses of *Con A*, only around 15–20 mg/kg body weight. In contrast, mice with Th2-biased immune responses, such as BALB/c or NMRI mice, commonly need higher doses of this lectin up to >30 mg/kg body weight. Therefore, the genetic background of animals needs to be considered when establishing this hepatitis model because *Con A* is prone to induce a Th1-biased immune response. The development of HE is not well established in this model.

Fas antibody It is a monoclonal antibody (Jo-2 clone) produced by Pharmingen that binds to and activates the Fas receptor (CD95), inducing apoptosis. The Fas antibody affects hepatocytes, as well as nonparenchymal liver cells. The damage of liver sinusoidal endothelial cells generates rapid lethality within a short period (~8 hours). Lowering doses prolong the survival of animals with substantial liver injury and severe lesions on histology. However, there is no time for HE development.

Autoimmune hepatitis The knock-out of genes involved in immune tolerance also can lead to ALF. For example, PD-1 knockout BALB/c mice spontaneously developed AIH within several weeks when undergoing thymectomy at 3 days old. These mice showed the onset of AIH between 2–3 weeks, marked by peak AST > 3000 U/L and fatality of all mice by 4 weeks (Kido et al., 2008). The development of HE has not been investigated.

18.6.2 Acute on Chronic Liver Damage Models

Hepatic encephalopathy is commonly associated with cirrhosis, and portal hypertension is the most patients. There is no appropriate model to study hepatic encephalopathy associated with liver cirrhosis; nevertheless, some models have been developed (Lima et al., 2019; Gama et al., 2022; Zhai et al., 2023). The preclinical models have to show biochemical disturbances, with the elevation of serum ammonia levels and increased oxidative stress in the blood, alterations of neurotransmission systems, development of brain edema, astrocyte swelling, and inflammation. The first step is the establishment of cirrhosis, most often by chemical liver damage (Fig. 18.3). The second step is the induction of acute decompensation (Zhai et al., 2023).

Liver injury models applying chemical, biological, and surgery induction leading to cirrhosis and subsequent acute decompensation with LPS or Galactosamine hydrochloride or both associated, Klebsiella pneumonia i.p. injection or Ethyl alcohol (EtOH), and CS surgery to acute-on-chronic liver failure (ACLF) induction in rats or acute decompensation with EtOH in mice. Intense fibrosis and Aspartate aminotransferase, Alanine aminotransferase serum levels, inflammatory response, and impairment of mitochondrial function are observed. Unclear and nonreproducible data concerning the ACLF rabbit model induced by Carbon tetrachloride are described. SD: Sprague Dawley; HAS: Human albumin serum; PS: Porcine serum; TAA: Thioacetamide Administration; CS: Cecal slurry; BDL: Bile duct ligation; CCl4: Carbon tetrachloride; AST: Aspartate aminotransferase; ALT: Alanine aminotransferase; PT: Prothrombin time; ACLF: Acute-on-chronic liver failure (Gama et al., 2022).

Liver injury is highly complex, and the absence of adequate animal models able to faithfully reproduce human liver disease characteristics limits understanding the mechanisms underlying this condition and treatment. Rodent animal models are preferred in preclinical liver research, mainly in ACLF assessments, due to their easy maintenance and breeding in captivity, their genetic similarity to humans, similar liver morphology, and simple management concerning genetic manipulation tools. In contrast, large animal models, like pigs, previously used primarily for acute liver failure modeling, now play an essential role in assessing various acute and chronic liver diseases.

The significant challenge to developing an ACLF model is the ability to unite all the clinical characteristics observed in humans, as this is a multifactorial disease with multiple precipitators and complications and, therefore, varying disease phenotypes and organ failures, making it almost impossible to develop a single experimental model capable of triggering all of the most important clinical features. Chronic liver injury, acute hepatic insult,

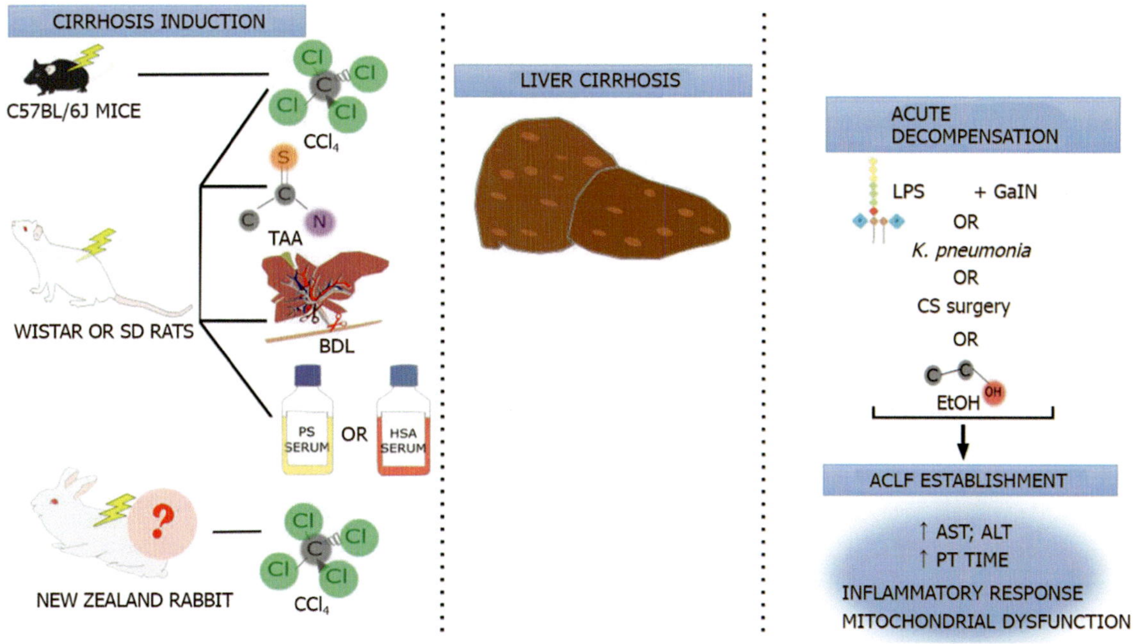

Fig. 18.3 Acute-on-chronic liver failure animal models based on a literature review

and bacterial infection induce severe liver injury with renal damage and high mortality, recapitulating some features of clinical ACLF. A severe liver injury model was prepared by combining chronic injury (CCl_4 injection), acute hepatic insult (injection of a CCl_4 double dose), and systemic bacterial infection (i.p., injection of bacteria *Klebsiella pneumonia*) (Xiang et al., 2020). The single bacterial infection step is crucial in inducing multi-organ failure in this model, as chronic-plus-acute liver injury did not drive the full course of ACLF in mice without bacterial infection.

There are three main patterns of ACLF experimental models worldwide, which were induced via hepatotoxic reagents, immune responses, and surgical procedures, respectively. Generally, each design of ACLF model always reflects a particular aspect of ACLF patients clinically, and it is tough to construct a single model to meet all the elements of the study of ACLF. Similar to the controversy in the definitions and criteria of ACLF among APASL, ESAL, and AASLD, animal models for ACLF would only be consistent once there is a universal agreement on the mechanism and definition of ACLF globally. At present, on the way to further reveal and elucidate the pathogenesis of ACLF, the optimal animal model of ACLF should be selected for the purpose of the study (Zhai et al., 2023).

18.7 Conclusion

- Hepatic encephalopathy (HE) is a nervous system disorder brought on by severe liver disease.
- Ammonia remains central to the pathophysiology of hepatic encephalopathy, and various HE precipitants affect ammonia production, excretion, and neurotoxicity.
- In addition to the liver and brain, other organs, such as the gut, kidneys, and muscles are critical for the metabolism of ammonia and glutamine, the key molecules in HE pathophysiology.
- New mechanisms related to inflammation, gut microbiome, and neurotransmitters have been identified.
- A better understanding of the intricate pathophysiologic mechanisms of HE is paramount for a comprehensive approach to the management of HE.
- Viral hepatitis leads to the activation of neuroinflammation in noncirrhotic patients and HE induction by ammonia in cirrhotic patients.
- EVs serve as an important tool for gut–adipose–liver–brain communications.
- Not all animal models of liver injury are suitable for experimental liver–brain studies.

18.8 Review Questions

1. Hepatic encephalopathy is the deterioration of brain function because toxic substances usually removed by the liver build up in the blood and reach the brain. Doctors recommend which of the following measures for people with hepatic encephalopathy?
   (a) Eating less meat
   (b) Avoiding acidic food
   (c) Reducing sodium intake
   (d) Taking Vitamin D

2. In older people, hepatic encephalopathy may be more difficult for doctors to recognize because its initial symptoms may resemble which of the following?
   (a) Delirium
   (b) Mild forgetfulness
   (c) Stroke
   (d) Parkinson disease

3. What is the role of the liver in ammonia turnover?
   (a) The liver is the organ that produces the highest levels of ammonia.
   (b) The liver extracts ammonia from portal blood and converts it into urea to be excreted in the kidney and intestine.
   (c) The liver promotes the release of ammonia into muscles.
   (d) The liver synthesizes ammonia to affect the brain.

4. How does the gut–liver axis contribute to HE development?
   (a) Via changes in microbiota and impaired intestinal barrier integrity
   (b) Via the production of antibodies
   (c) Via activation of intraepithelial lymphocytes
   (d) Via release of probiotics to the bloodstream

5. What is the main mechanism of HE development in viral hepatitis in patients without liver cirrhosis?
   (a) Hyperammonemia
   (b) Kidney dysfunctions
   (c) Increase in blood glucose
   (d) Cytokine storm and neuroinflammation

6. What are extracellular vesicles (EVs)?
   (a) EVs are always microbes.
   (b) EVs are round cells of a small size.
   (c) EVs are small membrane-bound vesicles containing proteins, genes, lipids, and nuclear acids.
   (d) EVs are the combination of lipids with nuclear acids.

7. What hepatotoxic reagents induce ACLF models, which are the most commonly used on rodent models and suitable for mimicking most clinical cases?
   (a) CCl4-induced cirrhosis
   (b) CCl4-induced cirrhosis in combination with D-galactosamine
   (c) CCl4-induced cirrhosis in combination with D-galactosamine and LPS injection
   (d) Bile duct ligation in combination with LPS injection

18.9 Answers to the Questions

1. a
2. a
3. b
4. a
5. d
6. c
7. c and d

References

Acharya, C., & Bajaj, J. S. (2017). Gut microbiota and complications of liver disease. *Gastroenterology Clinics of North America, 46*, 155–169.

Acharya, C., & Bajaj, J. S. (2019). Altered microbiome in patients with cirrhosis and complications. *Clinical Gastroenterology and Hepatology, 17*, 307–321.

Alsahhar, J. S., & Rahimi, R. S. (2019). Updates on the pathophysiology and therapeutic targets for hepatic encephalopathy. *Current Opinion in Gastroenterology, 35*, 145–154.

Amodio, P. (2015). Hepatic encephalopathy: Historical remarks. *Journal of Clinical and Experimental Hepatology, 5*, S4–S6.

Atluri, D. K., Prakash, R., & Mullen, K. D. (2011). Pathogenesis, diagnosis, and treatment of hepatic encephalopathy. *Journal of Clinical and Experimental Hepatology, 1*, 77–86.

Azhari, H., & Swain, M. G. (2018). Role of peripheral inflammation in hepatic encephalopathy. *Journal of Clinical and Experimental Hepatology, 8*, 281–285.

Bajaj, J. S., Wade, J. B., Gibson, D. P., Heuman, D. M., Thacker, L. R., Sterling, R. K., Stravitz, R. T., Luketic, V., Fuchs, M., White, M. B., Bell, D. E., Gilles, H., Morton, K., Noble, N., Puri, P., & Sanyal, A. J. (2011). The multi-dimensional burden of cirrhosis and hepatic encephalopathy on patients and caregivers. *The American Journal of Gastroenterology, 106*, 1646–1653.

Bajaj, J. S., Hylemon, P. B., Ridlon, J. M., Heuman, D. M., Daita, K., White, M. B., Monteith, P., Noble, N. A., Sikaroodi, M., & Gillevet, P. M. (2012). Colonic mucosal microbiome differs from stool microbiome in cirrhosis and hepatic encephalopathy and is linked to cognition and inflammation. *American Journal of Physiology. Gastrointestinal and Liver Physiology, 303*, G675–G685.

Bajaj, J. S., Heuman, D. M., Hylemon, P. B., Sanyal, A. J., White, M. B., Monteith, P., Noble, N. A., Unser, A. B., Daita, K., Fisher, A. R., Sikaroodi, M., & Gillevet, P. M. (2014). Altered profile of human gut microbiome is associated with cirrhosis and its complications. *Journal of Hepatology, 60*, 940–947.

Balzano, T., Forteza, J., Molina, P., Giner, J., Monzo, A., Sancho-Jimenez, J., Urios, A., Montoliu, C., & Felipo, V. (2018). The cerebellum of patients with steatohepatitis shows lymphocyte infiltration, microglial activation and loss of Purkinje and granular neurons. *Scientific Reports, 8*, 3004.

Balzano, T., Dadsetan, S., Forteza, J., Cabrera-Pastor, A., Taoro-Gonzalez, L., Malaguarnera, M., Gil-Perotin, S., Cubas-Nunez, L., Casanova, B., Castro-Quintas, A., Ponce-Mora, A., Arenas, Y. M., Leone, P., Erceg, S., Llansola, M., & Felipo, V. (2020). Chronic hyperammonemia induces peripheral inflammation that leads to cognitive impairment in rats: Reversed by anti-TNF-alpha treatment. *Journal of Hepatology, 73*, 582–592.

Bronster, D. J., Emre, S., Boccagni, P., Sheiner, P. A., Schwartz, M. E., & Miller, C. M. (2000). Central nervous system complications in liver transplant recipients--incidence, timing, and long-term follow-up. *Clinical Transplantation, 14*, 1–7.

Butterworth, R. F. (2013). The liver-brain axis in liver failure: Neuroinflammation and encephalopathy. *Nature Reviews. Gastroenterology & Hepatology, 10*, 522–528.

Butterworth, R. F. (2019). Hepatic encephalopathy in cirrhosis: Pathology and pathophysiology. *Drugs, 79*, 17–21.

Butz, M., Timmermann, L., Braun, M., Groiss, S. J., Wojtecki, L., Ostrowski, S., Krause, H., Pollok, B., Gross, J., Sudmeyer, M., Kircheis, G., Haussinger, D., & Schnitzler, A. (2010). Motor impairment in liver cirrhosis without and with minimal hepatic encephalopathy. *Acta Neurologica Scandinavica, 122*, 27–35.

Caballol, B., Reverter, E., Cid, J., Hernandez-Tejero, M., Triolo, M., Lozano, M., & Fernandez, J. (2019). Fulminant hepatitis A complicated by Takotsubo syndrome successfully treated with standard volume plasma exchange. *JHEP Reports, 1*, 445–448.

Cabrera-Pastor, A., Llansola, M., Montoliu, C., Malaguarnera, M., Balzano, T., Taoro-Gonzalez, L., Garcia-Garcia, R., Mangas-Losada, A., Izquierdo-Altarejos, P., Arenas, Y. M., Leone, P., & Felipo, V. (2019). Peripheral inflammation induces neuroinflammation that alters neurotransmission and cognitive and motor function in hepatic encephalopathy: Underlying mechanisms and therapeutic implications. *Acta Physiologica (Oxford, England), 226*, e13270.

Capuron, L., & Miller, A. H. (2011). Immune system to brain signaling: Neuropsychopharmacological implications. *Pharmacology & Therapeutics, 130*, 226–238.

Catinean, A., Neag, M. A., Muntean, D. M., Bocsan, I. C., & Buzoianu, A. D. (2018). An overview on the interplay between nutraceuticals and gut microbiota. *PeerJ, 6*, e4465.

Choi, J., Kim, Y. K., & Han, P. L. (2019). Extracellular vesicles derived from lactobacillus plantarum increase BDNF expression in cultured hippocampal neurons and produce antidepressant-like effects in mice. *Experimental Neurobiology, 28*, 158–171.

Clerici, M., Arosio, B., Mundo, E., Cattaneo, E., Pozzoli, S., Dell'osso, B., Vergani, C., Trabattoni, D., & Altamura, A. C. (2009). Cytokine polymorphisms in the pathophysiology of mood disorders. *CNS Spectrums, 14*, 419–425.

Colombo, M., Raposo, G., & Thery, C. (2014). Biogenesis, secretion, and intercellular interactions of exosomes and other extracellular vesicles. *Annual Review of Cell and Developmental Biology, 30*, 255–289.

Coltart, I., Tranah, T. H., & Shawcross, D. L. (2013). Inflammation and hepatic encephalopathy. *Archives of Biochemistry and Biophysics, 536*, 189–196.

Dabrowska, K., Skowronska, K., Popek, M., Obara-Michlewska, M., Albrecht, J., & Zielinska, M. (2018). Roles of glutamate and glutamine transport in ammonia neurotoxicity: State of the art and question marks. *Endocrine, Metabolic & Immune Disorders Drug Targets, 18*, 306–315.

Dadsetan, S., Balzano, T., Forteza, J., Cabrera-Pastor, A., Taoro-Gonzalez, L., Hernandez-Rabaza, V., Gil-Perotin, S., Cubas-Nunez, L., Garcia-Verdugo, J. M., Agusti, A., Llansola, M., & Felipo, V. (2016). Reducing peripheral inflammation with infliximab reduces neuroinflammation and improves cognition in rats with hepatic encephalopathy. *Frontiers in Molecular Neuroscience, 9*, 106.

Dasarathy, S., Mookerjee, R. P., Rackayova, V., Rangroo Thrane, V., Vairappan, B., Ott, P., & Rose, C. F. (2017). Ammonia toxicity: From head to toe? *Metabolic Brain Disease, 32*, 529–538.

DeMorrow, S. (2019). Bile acids in hepatic encephalopathy. *Journal of Clinical and Experimental Hepatology, 9*, 117–124.

Dhiman, R. K. (2012). Gut microbiota, inflammation and hepatic encephalopathy: A puzzle with a solution in sight. *Journal of Clinical and Experimental Hepatology, 2*, 207–210.

D'Mello, C., & Swain, M. G. (2011). Liver-brain inflammation axis. *American Journal of Physiology. Gastrointestinal and Liver Physiology, 301*, G749–G761.

D'Mello, C., Le, T., & Swain, M. G. (2009). Cerebral microglia recruit monocytes into the brain in response to tumor necrosis factoralpha signaling during peripheral organ inflammation. *The Journal of Neuroscience, 29*, 2089–2102.

European Association for the Study of the Liver. Electronic address eee, Clinical practice guidelines p, Panel m, representative EGB, Wendon, J., Cordoba, J., Dhawan, A., Larsen, F. S., Manns, M., Samuel, D., Simpson, K. J., Yaron, I., & Bernardi, M. (2017). EASL clinical practical guidelines on the management of acute (fulminant) liver failure. *Journal of Hepatology, 66*, 1047–1081.

Felipo, V., & Butterworth, R. F. (2002). Neurobiology of ammonia. *Progress in Neurobiology, 67*, 259–279.

Felipo, V., Urios, A., Montesinos, E., Molina, I., Garcia-Torres, M. L., Civera, M., Olmo, J. A., Ortega, J., Martinez-Valls, J., Serra, M. A., Cassinello, N., Wassel, A., Jorda, E., & Montoliu, C. (2012). Contribution of hyperammonemia and inflammatory factors to cognitive impairment in minimal hepatic encephalopathy. *Metabolic Brain Disease, 27*, 51–58.

Ferenci, P. (2017). Hepatic encephalopathy. *Gastroenterology Report (Oxford), 5*, 138–147.

Fick, T. E., Schalm, S. W., & de Vlieger, M. (1989). Continuous intravenous ammonia infusion as a model for the study of hepatic encephalopathy in rabbits. *The Journal of Surgical Research, 46*, 221–225.

Forton, D. M., Patel, N., Prince, M., Oatridge, A., Hamilton, G., Goldblatt, J., Allsop, J. M., Hajnal, J. V., Thomas, H. C., Bassendine, M., Jones, D. E., & Taylor-Robinson, S. D. (2004). Fatigue and primary biliary cirrhosis: Association of globus pallidus magnetisation transfer ratio measurements with fatigue severity and blood manganese levels. *Gut, 53*, 587–592.

Forton, D. M., Hamilton, G., Allsop, J. M., Grover, V. P., Wesnes, K., O'Sullivan, C., Thomas, H. C., & Taylor-Robinson, S. D. (2008). Cerebral immune activation in chronic hepatitis C infection: A magnetic resonance spectroscopy study. *Journal of Hepatology, 49*, 316–322.

Fukui, H. (2015). Gut microbiota and host reaction in liver diseases. *Microorganisms, 3*, 759–791.

Gama, J. F. G., Cardoso, L., Lagrota-Candido, J. M., & Alves, L. A. (2022). Animal models applied to acute-on-chronic liver failure: Are new models required to understand the human condition? *World Journal of Clinical Cases, 10*, 2687–2699.

Gupta, A., Dhiman, R. K., Kumari, S., Rana, S., Agarwal, R., Duseja, A., & Chawla, Y. (2010). Role of small intestinal bacterial overgrowth and delayed gastrointestinal transit time in cirrhotic patients with minimal hepatic encephalopathy. *Journal of Hepatology, 53*, 849–855.

Han, K. H. (2011). Mechanisms of the effects of acidosis and hypokalemia on renal ammonia metabolism. *Electrolyte Blood Press, 9*, 45–49.

Han, E. C., Choi, S. Y., Lee, Y., Park, J. W., Hong, S. H., & Lee, H. J. (2019). Extracellular RNAs in periodontopathogenic outer membrane vesicles promote TNF-alpha production in human macrophages and cross the blood-brain barrier in mice. *The FASEB Journal, 33*, 13412–13422.

Haussinger, D., & Schliess, F. (2008). Pathogenetic mechanisms of hepatic encephalopathy. *Gut, 57*, 1156–1165.

Haussinger, D., Laubenberger, J., vom Dahl, S., Ernst, T., Bayer, S., Langer, M., Gerok, W., & Hennig, J. (1994). Proton magnetic resonance spectroscopy studies on human brain myo-inositol in hypo-osmolarity and hepatic encephalopathy. *Gastroenterology, 107*, 1475–1480.

Haussinger, D., Kircheis, G., Fischer, R., Schliess, F., & vom Dahl, S. (2000). Hepatic encephalopathy in chronic liver disease: A clinical manifestation of astrocyte swelling and low-grade cerebral edema? *Journal of Hepatology, 32*, 1035–1038.

Haussinger, D., Butz, M., Schnitzler, A., & Gorg, B. (2021). Pathomechanisms in hepatic encephalopathy. *Biological Chemistry, 402*, 1087–1102.

Heeren, M., Weissenborn, K., Arvanitis, D., Bokemeyer, M., Goldbecker, A., Tountopoulou, A., Peschel, T., Grosskreutz, J., Hecker, H., Buchert, R., & Berding, G. (2011). Cerebral glucose utilisation in hepatitis C virus infection-associated encephalopathy. *Journal of Cerebral Blood Flow and Metabolism, 31*, 2199–2208.

Hefler, J., Marfil-Garza, B. A., Pawlick, R. L., Freed, D. H., Karvellas, C. J., Bigam, D. L., & Shapiro, A. M. J. (2021). Preclinical models of acute liver failure: A comprehensive review. *PeerJ, 9*, e12579.

Heidrich, B., Vital, M., Plumeier, I., Doscher, N., Kahl, S., Kirschner, J., Ziegert, S., Solbach, P., Lenzen, H., Potthoff, A., Manns, M. P., Wedemeyer, H., & Pieper, D. H. (2018). Intestinal microbiota in patients with chronic hepatitis C with and without cirrhosis compared with healthy controls. *Liver International, 38*, 50–58.

Iebba, V., Guerrieri, F., Di Gregorio, V., Levrero, M., Gagliardi, A., Santangelo, F., Sobolev, A. P., Circi, S., Giannelli, V., Mannina, L., Schippa, S., & Merli, M. (2018). Combining amplicon sequencing and metabolomics in cirrhotic patients highlights distinctive microbiota features involved in bacterial translocation, systemic inflammation and hepatic encephalopathy. *Scientific Reports, 8*, 8210.

Izquierdo-Altarejos, P., Martinez-Garcia, M., & Felipo, V. (2022). Extracellular vesicles from hyperammonemic rats induce neuroinflammation in cerebellum of normal rats: Role of increased TNFalpha content. *Frontiers in Immunology, 13*, 921947.

Jaffe, A., Lim, J. K., & Jakab, S. S. (2020). Pathophysiology of hepatic encephalopathy. *Clinics in Liver Disease, 24*, 175–188.

Jawaro, T., Yang, A., Dixit, D., & Bridgeman, M. B. (2016). Management of Hepatic Encephalopathy: A primer. *The Annals of Pharmacotherapy, 50*, 569–577.

Jian, H., Liu, Y., Wang, X., Dong, X., & Zou, X. (2023). Akkermansia muciniphila as a next-generation probiotic in modulating human metabolic homeostasis and disease progression: A role mediated by gut-liver-brain axes? *International Journal of Molecular Sciences, 24*, 3900.

Kakiyama, G., Pandak, W. M., Gillevet, P. M., Hylemon, P. B., Heuman, D. M., Daita, K., Takei, H., Muto, A., Nittono, H., Ridlon, J. M., White, M. B., Noble, N. A., Monteith, P., Fuchs, M., Thacker, L. R., Sikaroodi, M., & Bajaj, J. S. (2013). Modulation of the fecal bile acid profile by gut microbiota in cirrhosis. *Journal of Hepatology, 58*, 949–955.

Kanekiyo, T., Xu, H., & Bu, G. (2014). ApoE and Abeta in Alzheimer's disease: Accidental encounters or partners? *Neuron, 81*, 740–754.

Katsuda, T., Tsuchiya, R., Kosaka, N., Yoshioka, Y., Takagaki, K., Oki, K., Takeshita, F., Sakai, Y., Kuroda, M., & Ochiya, T. (2013). Human adipose tissue-derived mesenchymal stem cells secrete functional neprilysin-bound exosomes. *Scientific Reports, 3*, 1197.

Keiding, S., Sorensen, M., Bender, D., Munk, O. L., Ott, P., & Vilstrup, H. (2006). Brain metabolism of 13N-ammonia during acute hepatic encephalopathy in cirrhosis measured by positron emission tomography. *Hepatology, 43*, 42–50.

Kido, M., Watanabe, N., Okazaki, T., Akamatsu, T., Tanaka, J., Saga, K., Nishio, A., Honjo, T., & Chiba, T. (2008). Fatal autoimmune hepatitis induced by concurrent loss of naturally arising regulatory T cells and PD-1-mediated signaling. *Gastroenterology, 135*, 1333–1343.

Kim, M. H., van Noort, D., Sung, J. H., & Park, S. (2021). Organ-on-a-chip for studying gut-brain interaction mediated by extracellular vesicles in the gut microenvironment. *International Journal of Molecular Sciences, 22*, 13513.

Kojima, M., Costantini, T. W., Eliceiri, B. P., Chan, T. W., Baird, A., & Coimbra, R. (2018). Gut epithelial cell-derived exosomes trigger posttrauma immune dysfunction. *Journal of Trauma and Acute Care Surgery, 84*, 257–264.

Kornerup, L. S., Gluud, L. L., Vilstrup, H., & Dam, G. (2018). Update on the therapeutic management of hepatic encephalopathy. *Current Gastroenterology Reports, 20*, 21.

Krakovski, M. A., Arora, N., Jain, S., Glover, J., Dombrowski, K., Hernandez, B., Yadav, H., & Sarma, A. K. (2022). Diet-microbiome-gut-brain nexus in acute and chronic brain injury. *Frontiers in Neuroscience, 16*, 1002266.

Lachmann, V., Gorg, B., Bidmon, H. J., Keitel, V., & Haussinger, D. (2013). Precipitants of hepatic encephalopathy induce rapid astrocyte swelling in an oxidative stress dependent manner. *Archives of Biochemistry and Biophysics, 536*, 143–151.

Lee, M., Ban, J. J., Kim, K. Y., Jeon, G. S., Im, W., Sung, J. J., & Kim, M. (2016). Adipose-derived stem cell exosomes alleviate pathology of amyotrophic lateral sclerosis in vitro. *Biochemical and Biophysical Research Communications, 479*, 434–439.

Lima, L. C. D., Miranda, A. S., Ferreira, R. N., Rachid, M. A., & Simoes, E. S. A. C. (2019). Hepatic encephalopathy: Lessons from preclinical studies. *World Journal of Hepatology, 11*, 173–185.

Liu, Q., Li, F., Zhuang, Y., Xu, J., Wang, J., Mao, X., Zhang, Y., & Liu, X. (2019). Alteration in gut microbiota associated with hepatitis B and non-hepatitis virus related hepatocellular carcinoma. *Gut Pathogens, 11*, 1.

Liu, Y., Hao, H., & Hou, T. (2022). Concanavalin A-induced autoimmune hepatitis model in mice: Mechanisms and future outlook. *Open Life Sciences, 17*, 91–101.

Llansola, M., Montoliu, C., Agusti, A., Hernandez-Rabaza, V., Cabrera-Pastor, A., Gomez-Gimenez, B., Malaguarnera, M., Dadsetan, S., Belghiti, M., Garcia-Garcia, R., Balzano, T., Taoro, L., & Felipo, V. (2015). Interplay between glutamatergic and GABAergic neurotransmission alterations in cognitive and motor impairment in minimal hepatic encephalopathy. *Neurochemistry International, 88*, 15–19.

Loftis, J. M., Huckans, M., Ruimy, S., Hinrichs, D. J., & Hauser, P. (2008). Depressive symptoms in patients with chronic hepatitis C are correlated with elevated plasma levels of interleukin-1beta and tumor necrosis factor-alpha. *Neuroscience Letters, 430*, 264–268.

Luo, M., Xin, R. J., Hu, F. R., Yao, L., Hu, S. J., & Bai, F. H. (2023). Role of gut microbiota in the pathogenesis and therapeutics of minimal hepatic encephalopathy via the gut-liver-brain axis. *World Journal of Gastroenterology, 29*, 144–156.

Makrygianni, E. A., & Chrousos, G. P. (2023). Extracellular vesicles and the stress system. *Neuroendocrinology, 113*, 120–167.

Mancini, A., Campagna, F., Amodio, P., & Tuohy, K. M. (2018). Gut : Liver : Brain axis: The microbial challenge in the hepatic encephalopathy. *Food & Function, 9*, 1373–1388.

Mangas-Losada, A., Garcia-Garcia, R., Urios, A., Escudero-Garcia, D., Tosca, J., Giner-Duran, R., Serra, M. A., Montoliu, C., & Felipo, V. (2017). Minimal hepatic encephalopathy is associated with expansion and activation of CD(4+)CD28(−), Th22 and Tfh and B lymphocytes. *Scientific Reports, 7*, 6683.

McMillin, M., Frampton, G., Quinn, M., Ashfaq, S., de los Santos, M., 3rd, Grant, S., & DeMorrow, S. (2016). Bile acid signaling is involved in the neurological decline in a murine model of acute liver failure. *The American Journal of Pathology, 186*, 312–323.

McMillin, M., Frampton, G., Grant, S., Khan, S., Diocares, J., Petrescu, A., Wyatt, A., Kain, J., Jefferson, B., & DeMorrow, S. (2017). Bile acid-mediated sphingosine-1-phosphate receptor 2 signaling promotes neuroinflammation during hepatic encephalopathy in mice. *Frontiers in Cellular Neuroscience, 11*, 191.

Merli, M., Giusto, M., Lucidi, C., Giannelli, V., Pentassuglio, I., Di Gregorio, V., Lattanzi, B., & Riggio, O. (2013). Muscle depletion increases the risk of overt and minimal hepatic encephalopathy: Results of a prospective study. *Metabolic Brain Disease, 28*, 281–284.

Mitchell, J. R., Jollow, D. J., Potter, W. Z., Davis, D. C., Gillette, J. R., & Brodie, B. B. (1973). Acetaminophen-induced hepatic necrosis. I. Role of drug metabolism. *The Journal of Pharmacology and Experimental Therapeutics, 187*, 185–194.

Montoliu, C., Piedrafita, B., Serra, M. A., del Olmo, J. A., Urios, A., Rodrigo, J. M., & Felipo, V. (2009). IL-6 and IL-18 in blood may discriminate cirrhotic patients with and without minimal hepatic encephalopathy. *Journal of Clinical Gastroenterology, 43*, 272–279.

Moyer, A. M., Fridley, B. L., Jenkins, G. D., Batzler, A. J., Pelleymounter, L. L., Kalari, K. R., Ji, Y., Chai, Y., Nordgren, K. K., & Weinshilboum, R. M. (2011). Acetaminophen-NAPQI hepatotoxicity: A cell line model system genome-wide association study. *Toxicological Sciences, 120*, 33–41.

Mustapic, M., Eitan, E., Werner, J. K., Jr., Berkowitz, S. T., Lazaropoulos, M. P., Tran, J., Goetzl, E. J., & Kapogiannis, D. (2017). Plasma extracellular vesicles enriched for neuronal origin: A potential window into brain pathologic processes. *Frontiers in Neuroscience, 11*, 278.

Nagami, G. T. (2002). Enhanced ammonia secretion by proximal tubules from mice receiving NH(4)Cl: Role of angiotensin II. *American Journal of Physiology. Renal Physiology, 282*, F472–F477.

Nava, G. M., & Stappenbeck, T. S. (2011). Diversity of the autochthonous colonic microbiota. *Gut Microbes, 2*, 99–104.

Neag, M. A., Mitre, A. O., Catinean, A., & Buzoianu, A. D. (2021). Overview of the microbiota in the gut-liver axis in viral B and C hepatitis. *World Journal of Gastroenterology, 27*, 7446–7461.

Nikolov, P., Baumgarten, T. J., Hassan, S. S., Meissner, S. N., Fullenbach, N. D., Kircheis, G., Haussinger, D., Jordens, M. S., Butz, M., Schnitzler, A., & Groiss, S. J. (2021). Altered motor cortical plasticity in patients with hepatic encephalopathy: A paired associative stimulation study. *Clinical Neurophysiology, 132*, 2332–2341.

Nikolova, V., Zaidi, S. Y., Young, A. H., Cleare, A. J., & Stone, J. M. (2019). Gut feeling: Randomized controlled trials of probiotics for the treatment of clinical depression: Systematic review and meta-analysis. *Therapeutic Advances in Psychopharmacology, 9*, 2045125319859963.

Norenberg, M. D. (1987). The role of astrocytes in hepatic encephalopathy. *Neurochemical Pathology, 6*, 13–33.

Oo, K. M., Lwin, A. A., Kyaw, Y. Y., Tun, W. M., Fukada, K., Goshima, A., Shimada, T., & Okada, S. (2016). Safety and long-term effect of the probiotic FK-23 in patients with hepatitis C virus infection. *Bioscience of Microbiota, Food and Health, 35*, 123–128.

Osna, N. A., & Poluektova, L. Y. (2023). Elucidating the role of extracellular vesicles in liver injury induced by HIV. *Expert Review of Gastroenterology & Hepatology, 17*, 701–708.

Osna, N. A., Eguchi, A., Feldstein, A. E., Tsukamoto, H., Dagur, R. S., Ganesan, M., New-Aaron, M., Arumugam, M. K., Chava, S., Ribeiro, M., Szabo, G., Mueller, S., Wang, S., Chen, C., Weinman, S. A., & Kharbanda, K. K. (2022). Cell-to-cell communications in alcohol-associated liver disease. *Frontiers in Physiology, 13*, 831004.

Otero-Ortega, L., Alonso-Lopez, E., Perez-Mato, M., Laso-Garcia, F., Gomez-de Frutos, M. C., Diekhorst, L., Garcia-Bermejo, M. L., Conde-Moreno, E., Fuentes, B., Alonso de Lecinana, M., Armada, E., Buiza-Palomino, L., Diez-Tejedor, E., & Gutierrez-Fernandez, M. (2020). Similarities and differences in extracellular vesicle profiles between Ischaemic stroke and myocardial infarction. *Biomedicines, 9*, 8.

Paulino, A. D., Ubhi, K., Rockenstein, E., Adame, A., Crews, L., Letendre, S., Ellis, R., Everall, I. P., Grant, I., & Masliah, E. (2011). Neurotoxic effects of the HCV core protein are mediated by sustained activation of ERK via TLR2 signaling. *Journal of Neurovirology, 17*, 327–340.

Perez-Cruz, C., Carrion, O., Delgado, L., Martinez, G., Lopez-Iglesias, C., & Mercade, E. (2013). New type of outer membrane vesicle produced by the Gram-negative bacterium Shewanella vesiculosa M7T: Implications for DNA content. *Applied and Environmental Microbiology, 79*, 1874–1881.

Pozdeev, V. I., Lang, E., Gorg, B., Bidmon, H. J., Shinde, P. V., Kircheis, G., Herebian, D., Pfeffer, K., Lang, F., Haussinger, D., Lang, K. S., & Lang, P. A. (2017). TNFalpha induced up-regulation of Na(+),K(+),2Cl(−) cotransporter NKCC1 in hepatic ammonia clearance and cerebral ammonia toxicity. *Scientific Reports, 7*, 7938.

Rahman, M. A., Patters, B. J., Kodidela, S., & Kumar, S. (2020). Extracellular vesicles: Intercellular mediators in alcohol-induced pathologies. *Journal of Neuroimmune Pharmacology, 15*, 409–421.

Rama Rao, K. V., Jayakumar, A. R., Tong, X., Alvarez, V. M., & Norenberg, M. D. (2010). Marked potentiation of cell swelling by cytokines in ammonia-sensitized cultured astrocytes. *Journal of Neuroinflammation, 7*, 66.

Ridlon, J. M., Alves, J. M., Hylemon, P. B., & Bajaj, J. S. (2013). Cirrhosis, bile acids and gut microbiota: Unraveling a complex relationship. *Gut Microbes, 4*, 382–387.

Riordan, S. M., & Williams, R. (2010). Gut flora and hepatic encephalopathy in patients with cirrhosis. *The New England Journal of Medicine, 362*, 1140–1142.

Rizzo, J., Rodrigues, M. L., & Janbon, G. (2020). Extracellular vesicles in fungi: Past, present, and future perspectives. *Frontiers in Cellular and Infection Microbiology, 10*, 346.

Rodrigo, R., Cauli, O., Gomez-Pinedo, U., Agusti, A., Hernandez-Rabaza, V., Garcia-Verdugo, J. M., & Felipo, V. (2010). Hyperammonemia induces neuroinflammation that contributes to cognitive impairment in rats with hepatic encephalopathy. *Gastroenterology, 139*, 675–684.

Rose, C. F., Amodio, P., Bajaj, J. S., Dhiman, R. K., Montagnese, S., Taylor-Robinson, S. D., Vilstrup, H., & Jalan, R. (2020). Hepatic encephalopathy: Novel insights into classification, pathophysiology and therapy. *Journal of Hepatology, 73*, 1526–1547.

Rowland, I., Gibson, G., Heinken, A., Scott, K., Swann, J., Thiele, I., & Tuohy, K. (2018). Gut microbiota functions: Metabolism of nutrients and other food components. *European Journal of Nutrition, 57*, 1–24.

Sandler, N. G., Koh, C., Roque, A., Eccleston, J. L., Siegel, R. B., Demino, M., Kleiner, D. E., Deeks, S. G., Liang, T. J., Heller, T., & Douek, D. C. (2011). Host response to translocated microbial products predicts outcomes of patients with HBV or HCV infection. *Gastroenterology, 141*, 1220–1230, 1230 e1221–1223.

Schrimpf, A., Knappe, O., Qvartskhava, N., Poschmann, G., Stuhler, K., Bidmon, H. J., Luedde, T., Haussinger, D., & Gorg, B. (2022). Hyperammonemia-induced changes in the cerebral transcriptome and proteome. *Analytical Biochemistry, 641*, 114548.

Shawcross, D. L., Davies, N. A., Williams, R., & Jalan, R. (2004). Systemic inflammatory response exacerbates the neuropsychological effects of induced hyperammonemia in cirrhosis. *Journal of Hepatology, 40*, 247–254.

Shawcross, D. L., Wright, G., Olde Damink, S. W., & Jalan, R. (2007). Role of ammonia and inflammation in minimal hepatic encephalopathy. *Metabolic Brain Disease, 22*, 125–138.

Shawcross, D. L., Sharifi, Y., Canavan, J. B., Yeoman, A. D., Abeles, R. D., Taylor, N. J., Auzinger, G., Bernal, W., & Wendon, J. A. (2011). Infection and systemic inflammation, not ammonia, are associated with Grade 3/4 hepatic encephalopathy, but not mortality in cirrhosis. *Journal of Hepatology, 54*, 640–649.

Simpson, J. B., & Redinbo, M. R. (2022). Multi-omic analysis of host-microbial interactions central to the gut-brain axis. *Molecular Omics, 18*, 896–907.

Sole, C., et al. (2021). Alterations in gut microbiome in cirrhosis as assessed by quantitative metagenomics: Relationship with acute-on-chronic liver failure and prognosis. *Gastroenterology, 160*(206–218), e213.

Swain, M. G. (2006). Fatigue in liver disease: Pathophysiology and clinical management. *Canadian Journal of Gastroenterology, 20*, 181–188.

Thomou, T., Mori, M. A., Dreyfuss, J. M., Konishi, M., Sakaguchi, M., Wolfrum, C., Rao, T. N., Winnay, J. N., Garcia-Martin, R., Grinspoon, S. K., Gorden, P., & Kahn, C. R. (2017). Adipose-derived circulating miRNAs regulate gene expression in other tissues. *Nature, 542*, 450–455.

Tong, L., et al. (2023). Milk-derived extracellular vesicles protect intestinal barrier integrity in the gut-liver axis. *Science Advances, 9*, eade5041.

Tripodi, V., Contin, M., Fernandez, M. A., & Lemberg, A. (2012). Bile acids content in brain of common duct ligated rats. *Annals of Hepatology, 11*, 930–934.

Tsai, C. F., Chu, C. J., Huang, Y. H., Wang, Y. P., Liu, P. Y., Lin, H. C., Lee, F. Y., & Lu, C. L. (2015). Detecting minimal hepatic encephalopathy in an endemic country for hepatitis B: The role of psychometrics and serum IL-6. *PLoS One, 10*, e0128437.

van Niel, G., D'Angelo, G., & Raposo, G. (2018). Shedding light on the cell biology of extracellular vesicles. *Nature Reviews: Molecular Cell Biology, 19*, 213–228.

Vilstrup, H., Amodio, P., Bajaj, J., Cordoba, J., Ferenci, P., Mullen, K. D., Weissenborn, K., & Wong, P. (2014). Hepatic encephalopathy in chronic liver disease: 2014 practice guideline by the American Association for the Study of Liver Diseases and the European Association for the Study of the Liver. *Hepatology, 60*, 715–735.

Wang, M. D., Wang, Y., Xia, Y. P., Dai, J. W., Gao, L., Wang, S. Q., Wang, H. J., Mao, L., Li, M., Yu, S. M., Tu, Y., He, Q. W., Zhang, G. P., Wang, L., Xu, G. Z., Xu, H. B., Zhu, L. Q., & Hu, B. (2016). High serum MiR-130a levels are associated with severe perihematomal edema and predict adverse outcome in acute ICH. *Molecular Neurobiology, 53*, 1310–1321.

Wang, C., Liu, J., Yan, Y., & Tan, Y. (2022). Role of exosomes in chronic liver disease development and their potential clinical applications. *Journal of Immunology Research, 2022*, 1695802.

Wei, S., Peng, W., Mai, Y., Li, K., Wei, W., Hu, L., Zhu, S., Zhou, H., Jie, W., Wei, Z., Kang, C., Li, R., Liu, Z., Zhao, B., & Cai, Z. (2020). Outer membrane vesicles enhance tau phosphorylation and contribute to cognitive impairment. *Journal of Cellular Physiology, 235*, 4843–4855.

Weiner, I. D., Mitch, W. E., & Sands, J. M. (2015). Urea and ammonia metabolism and the control of renal nitrogen excretion. *Clinical Journal of the American Society of Nephrology, 10*, 1444–1458.

Weiss, N., Barbier Saint Hilaire, P., Colsch, B., Isnard, F., Attala, S., Schaefer, A., Amador, M. D., Rudler, M., Lamari, F., Sedel, F., Thabut, D., & Junot, C. (2016). Cerebrospinal fluid metabolomics highlights dysregulation of energy metabolism in overt hepatic encephalopathy. *Journal of Hepatology, 65*, 1120–1130.

Xi, S., Wang, Y., Wu, C., Peng, W., Zhu, Y., & Hu, W. (2021). Intestinal epithelial cell exosome launches IL-1beta-mediated neuron injury in sepsis-associated encephalopathy. *Frontiers in Cellular and Infection Microbiology, 11*, 783049.

Xia, X., Chen, J., Xia, J., Wang, B., Liu, H., Yang, L., Wang, Y., & Ling, Z. (2018). Role of probiotics in the treatment of minimal hepatic encephalopathy in patients with HBV-induced liver cirrhosis. *The Journal of International Medical Research, 46*, 3596–3604.

Xiang, X., Feng, D., Hwang, S., Ren, T., Wang, X., Trojnar, E., Matyas, C., Mo, R., Shang, D., He, Y., Seo, W., Shah, V. H., Pacher, P., Xie, Q., & Gao, B. (2020). Interleukin-22 ameliorates acute-on-chronic liver failure by reprogramming impaired regeneration pathways in mice. *Journal of Hepatology, 72*, 736–745.

Yang, P., Zhou, W., Li, C., Zhang, M., Jiang, Y., Jiang, R., Ba, H., Li, C., Wang, J., Yin, B., Gong, F., & Li, Z. (2016). Kupffer-cell-expressed transmembrane TNF-α is a major contributor to lipopolysaccharide and D-galactosamine-induced liver injury. *Cell and Tissue Research, 363*, 371–383.

Yarlott, L., Heald, E., & Forton, D. (2017). Hepatitis C virus infection, and neurological and psychiatric disorders – a review. *Journal of Advanced Research, 8*, 139–148.

Younas, A., Riaz, J., Chughtai, T., Maqsood, H., Saim, M., Qazi, S., Younus, S., Ghaffar, U., & Khaliq, M. (2021). Hyponatremia and its correlation with hepatic encephalopathy and severity of liver disease. *Cureus, 13*, e13175.

Zakharzhevskaya, N. B., Vanyushkina, A. A., Altukhov, I. A., Shavarda, A. L., Butenko, I. O., Rakitina, D. V., Nikitina, A. S., Manolov, A. I., Egorova, A. N., Kulikov, E. E., Vishnyakov, I. E., Fisunov, G. Y., & Govorun, V. M. (2017). Outer membrane vesicles secreted by pathogenic and nonpathogenic Bacteroides fragilis represent different metabolic activities. *Scientific Reports, 7*, 5008.

Zhai, H., Zhang, J., Shang, D., Zhu, C., & Xiang, X. (2023). The progress to establish optimal animal models for the study of acute-on-chronic liver failure. *Frontiers in Medicine (Lausanne), 10*, 1087274.

Zhang, J., Tan, Y., Chang, L., Hammock, B. D., & Hashimoto, K. (2020). Increased expression of soluble epoxide hydrolase in the brain and liver from patients with major psychiatric disorders: A role of brain – liver axis. *Journal of Affective Disorders, 270*, 131–134.

Zhang, M., Gong, W., Zhang, D., Ji, M., Chen, B., Chen, B., Li, X., Zhou, Y., Dong, C., Wen, G., Zhan, X., Wu, X., Cui, L., Feng, Y., Wang, S., Yuan, H., Xu, E., Xia, M., Verkhratsky, A., & Li, B. (2022). Ageing related thyroid deficiency increases brain-targeted transport of liver-derived ApoE4-laden exosomes leading to cognitive impairment. *Cell Death & Disease, 13*, 406.

Zhao, L., Ye, Y., Gu, L., Jian, Z., Stary, C. M., & Xiong, X. (2021). Extracellular vesicle-derived miRNA as a novel regulatory system for bi-directional communication in gut-brain-microbiota axis. *Journal of Translational Medicine, 19*, 202.

Zhou, W., Zhao, L., Mao, Z., Wang, Z., Zhang, Z., & Li, M. (2023). Bidirectional communication between the brain and other organs: The role of extracellular vesicles. *Cellular and Molecular Neurobiology, 43*, 2675–2696.

Hippocampus, Spatial Memory, and Neuroimmuomodulation

19

Huangui Xiong, Jingdong Zhang, and Jianuo Liu

Abstract

The hippocampus is a symmetrical structure located inside the medial temporal lobe on both sides of the human brain. In cross-sections, the hippocampus consists of two interlocking sheets of cortex with three distinct subregions: the dentate gyrus, the hippocampus proper (CA1–CA3), and the subiculum. The hippocampus has a much-defined laminar structure with layers visible where rows of pyramidal cells are arranged. A striking feature of the hippocampus is its connection circuitry. The connections within the hippocampus generally follow this laminar format and they are largely unidirectional. They form well-characterized closed loops that originate mainly in the adjacent entorhinal cortex. Thus, information flows through the hippocampus proceeds from the dentate gyrus to CA3–CA1 to the subiculum, forming the principal tri-synaptic circuit. Together with the adjacent amygdala and entorhinal cortex, the hippocampus forms the central axis of the limbic system and plays an important role in spatial learning and awareness, navigation, and formation/ retrieval of memories. In addition, the hippocampus is also involved in neuromodulation and neuroinflammation. Overall, the hippocampus plays a crucial role in both cognitive function and immune regulation, highlighting the interconnectedness of various systems in the body.

Keywords

CA1 · CA3 · Dentate gyrus · Hippocampus · Long-term potentiation · Mossy fibers · Neuroimmunomodulation · Neuroinflammation · Neurotransmitter · Perforant path · Schaffer collaterals · Spatial memory Synapse · Synaptic transmission

H. Xiong (✉) · J. Zhang · J. Liu
Department of Pharmacology and Experimental Neuroscience, University of Nebraska Medical Center, Omaha, NE, USA
e-mail: hxiong@unmc.edu

19.1 Introduction

The hippocampus is a brain structure located inside the temporal lobe. It forms a part of the limbic system and plays a crucial role in learning, memory, and spatial navigation (Chen & Wilson, 2023; Danieli et al., 2023). It has been shown that repetitive activation of excitatory synapses in the hippocampus causes an increase in synaptic strength that lasts for hours or days designated as long-term potentiation (LTP), an important key to understanding the cellular and molecular mechanisms by which memories are formed and stored. Sensory information enters the hippocampus mainly through the perforant pathway consisting of the axons of neurons in layers II and III of the entorhinal cortex. The perforant path axons terminate on the dendrites of the dentate gyrus granular cells. Then information flows through the hippocampus from the dentate gyrus to the CA3, the CA1, and the subiculum, forming a hippocampal intrinsic tri-synaptic circuit. In addition to its "traditional" role in learning and memory, the hippocampus has also been implicated in the regulation of the immune system through a complex network of interactions between neurons, glia, and immune cells. Studies have shown that the hippocampus can modulate immune responses by regulating the production of cytokines, small proteins that act as signaling molecules between cells of the immune system (Lathe, 2001). In addition, the hippocampus has been shown to be involved in the regulation of the stress response, which can also impact learning and memory as well as immune function (Fa et al., 2014). One of the ways in which the hippocampus may influence the immune system is through its interaction with the hypothalamic-pituitary-adrenal (HPA) axis, a neuroendocrine system that regulates the body''s response to stress and plays a role in the regulation of immune function (Bratt et al., 2001; Breivik et al., 2002). The hippocampus is known to regulate the HPA axis by controlling the release of corticotropin-releasing hormone (CRH), a key hormone involved in the stress response. Recent research has also sug-

H. E. Gendelman, T. Ikezu (eds.), *Neuroimmune Pharmacology and Therapeutics*,
https://doi.org/10.1007/978-3-031-68237-7_19

gested that the hippocampus may be involved in the regulation of neuroinflammation, a process that can contribute to the development of various neurological disorders (Shishkina et al., 2021). This is thought to be due to the hippocampus''s ability to regulate the activity of microglia, the immune cells of the central nervous system (Luo et al., 2021).

19.2 Anatomy of the Hippocampus

The hippocampus, so named because its shape vaguely resembles that of a seahorse, is a curved sheet of cortex folded into the medial surface of the temporal lobe. In transverse sections from rodent brain, the hippocampus has the appearance of two interlocking Cs with three distinct subfields: the dentate gyrus, the hippocampus proper (cornu ammonis, CA), and the subiculum (Fig. 19.1), these together are termed hippocampal formation. While the hippocampus proper generally includes only CA1–CA3, the CA4 is frequently called the hilus and considered part of the dentate gyrus (Fig. 19.1). The hippocampus of primates and humans occupies less of the telencephalon in proportion to the cerebral cortex than lower hierarchy animals do, reflecting it is among the phylogenetically oldest parts of the brain.

The anatomy of the hippocampus has been studied intensively in rodents since Ramón y Cajal (1911) published his famous drawing illustrating the main cells, connections, and flow of impulse traffic in the hippocampal formation (Fig. 19.1), followed by a consecutive of physiological, biochemical, and axonal tracing studies, making it one of the most studied and best-known structures in the brain (Frotscher, 1985; Raisman et al., 1966; Swanson et al., 1981; Swanson & Cowan, 1975). Its cytoarchitecture, neural circuit, main afferents, and efferents are summarized in Fig. 19.2. Its predominant intrinsic circuit is characterized as three-order projections from dentate granular cells to CA3 pyramidal cells and then from the later to the CA1 (also including CA2) and subiculum neurons (Fig. 19.2).

19.2.1 The Dentate Gyrus

19.2.1.1 Cytoarchitecture

The dentate gyrus is a sharply folded tri-layered cortex (molecular, granule cell, and polymorphic layers from outer gyrus to hilus) that forms a cap over the free edge of Ammon's horn (C or V shape in rodents). The granule cells are principal neurons in the dentate gyrus which are small (about 10 μm in diameter) and densely packed in the cell layer (Fig. 19.1). Their dendrites extend perpendicularly to the cell layer and into the molecular layer where they receive synaptic inputs mainly from perforant afferents (Figs. 19.1 and 19.2). The axons of granule cells are called mossy fibers and extend into the polymorphic layer in the hilus, which is a transition area between the dentate gyrus and the hippocampus proper. The mossy fibers synapse onto mossy cells and basket interneurons in the hilar area, the former give rise to excitatory recurrent projections to the granule cells and the latter may have more than five types mostly with inhibitory feedback to the granule cells (Fig. 19.2).

Interestingly, the dentate gyrus is one of the few brain regions where neurogenesis takes place. Neurogenesis is thought to play a role in the formation of new memories. It has been shown increased neurogenesis in response to both antidepressants and physical exercise, implying that neurogenesis may improve symptoms of depression (Yamashima et al., 2007).

19.2.1.2 Fiberarchitecture

Afferents to the Dentate Gyrus: The Perforant Path

The major afferents to the dentate gyrus are from the entorhinal cortex (but also perirhinal cortex, among others) by way

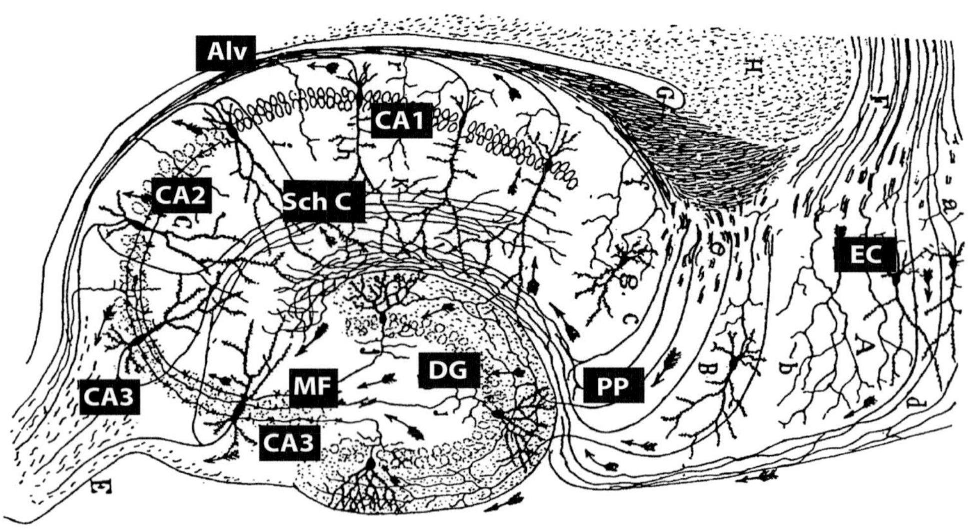

Fig. 19.1 Schematic drawing by Ramon y Cajal (1911) of the main cells, connections, and flow of impulse traffic in the hippocampus

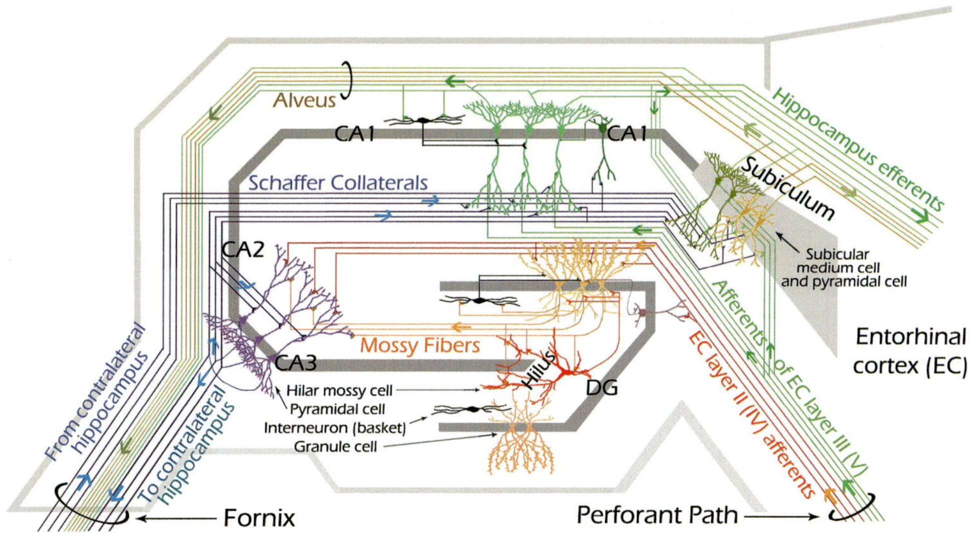

Fig. 19.2 Schematic drawing illustrating cytoarchitecture and fiberarchitecture of the hippocampus formation

of a fiber system called the perforant path and terminate in the outer two-thirds of the dentate molecular layer (equivalent to stratum lacunosum and moleculare) (Lorente de No, 1934). The dentate neurons also relay the massive projections from the entorhinal cortex to the CA3 pyramidal cells. Recurrent projections onto dentate neurons are conducted through hilar mossy cells and local basket interneurons (Figs. 19.1 and 19.2).

It was through the perforant pathway that LTP was first discovered (Bliss & Gardner-Medwin, 1973). Approximately 60–70% of the total dendritic fields of individual granule cells are taken up with the perforant input (Desmond & Levy, 1982). Other afferents to the dentate gyrus originate from brain stem raphe nuclei (Moore & Halaris, 1975), the locus coeruleus (Room et al., 1981), and the hypothalamic supramammillary nucleus (Wyss et al., 1979).

Efferents from the Dentate Gyrus: The Mossy Fibers

The dentate granule cells project predominantly to the CA3 region and less heavily to the hilus through mossy fibers. The mossy fiber terminals are large (3–6 μm) and make asymmetric synaptic contact on the dendritic shaft and dendritic spines (thorny excrescence) of the CA3 pyramidal cells. A single CA3 pyramidal cell can be innervated by more than one granule cell. This pathway has been studied extensively as a model for the functional roles of kainate receptors in synaptic plasticity. For instance, LTP is *N*-methyl-*D*-aspartate (NMDA) receptor-independent in this pathway, but it appears to involve pre-synaptic kainate receptors. The mossy fibers are glutamatergic; however, they have also been shown to be immunoreactive for γ-aminobutyric acid (GABA) and opiate peptides (dynorphin and enkephalin) (Commons & Milner, 1996; Gutierrez & Heinemann, 2006; Walker et al., 2002).

Neural tract tracing electron microscopy studies on mossy fibers have disclosed that fibers terminating onto the hilar mossy cells are branched from long projecting mossy fibers to the CA3 region (Blackstad & Kjaerheim, 1961).

19.2.2 The Hippocampus Proper (CA1–CA3 Fields)

19.2.2.1 Cytoarchitecture

The hippocampus proper is a U-shaped fold of cortex composed of CA1–CA3 fields (Fig. 19.1) extending from the subiculum to the hilus of the dentate gyrus. Adjacent to the subiculum is a field of tightly packed medium-sized cells, which Lorente de No (1934) named the CA1. Next to CA1 are the fields CA2 and CA3 containing large, less densely packed cells. CA2 pyramidal cells can be distinguished from CA3 pyramidal cells in golgi preparations by the absence of characteristic thorny spines on the proximal apical dendrites.

Hippocampal pyramidal cells generally have dendrite trees with apical and basal dendrites; the former heavily branches in the molecular layer (equivalent to stratum lacunosum and moleculare) and the latter gives off branches between polymorphic and pyramidal cell layers (equivalent to stratum oriens and pyramidal from outer border to center; Fig. 19.2). The axons generally originate from or near the basal dendrite and travel into the alveus where it may bifurcate. In the CA1 field, one branched axon travels from CA1 to the subicular area and the other toward the fimbria through the alveus bundle (Fig. 19.2). In the CA3 region, the pyramidal axons give rise to three branches in a polymorphic layer, and all of them are termed as Schaffer collaterals.

19.2.2.2 Fiberarchitecture

The CA3 pyramidal cells receive robust projections from the dentate gyrus through mossy fibers and accept less heavily afferent innervations from layer II (slightly from layer IV) of the entorhinal cortex. The CA3 neurons are also innervated by recurrent projections from themselves through Shaffer collaterals (Figs. 19.1 and 19.2).

The CA2 and CA1 pyramidal cells are predominantly innervated by bilateral CA3 neurons through Schaffer collaterals from both ipsi- and contralateral sides (Fig. 19.2). In the CA1 field, the pyramidal cells also receive projections from the layer III (slightly from layer V) of the entorhinal cortex (Fig. 19.2). The dendrite trees of CA1 neurons in both molecular and polymorphic layers are heavily innervated by CA3 neuronal axons. This Schaffer collateral to CA1 pathway is widely studied for NMDA receptor-dependent LTP and long-term depression (LTD).

Recurrent feedback innervations onto the CA1 pyramidal neurons are carried out through local interneurons and some small pyramidal cells (Fig. 19.2). The CA1 also receives a fairly substantial input from the amygdaloid complex, and a few noradrenergic and serotonergic projections.

The well-known efferents of the CA3 pyramidal cells are Schaffer collaterals projecting to both the ipsi- and contralateral hippocampus. These efferent collaterals also give rise to recurrent innervations to the CA3 neurons (Fig. 19.2). The efferent fibers from CA1 neurons travel through a polymorphic layer (equivalent to stratum oriens) and bifurcate into three bundles of branches, one group joins alveus toward fimbria, the other branches project onto the subiculum cells, and another group branches merge into the efferent bundle toward entorhinal cortex (Fig. 19.2). In addition, CA1 neuronal axons in the alveus again give rise to collaterals to local interneurons that feedback project to the CA1 pyramidal cells (Fig. 19.2).

The septal nucleus is the major extrahippocampal target of the pyramidal projections from all CA fields. Axons from both CA1 and CA3 terminate in the septal nucleus. The fibers from dorsal hippocampal project to dorsomedial areas of the lateral septal nucleus, while progressively more ventral parts of the hippocampus terminate in correspondingly more lateroventral bands in the lateral septal nucleus (Meibach & Siegel, 1977a). Some branches of the CA1 axons terminate sparsely in the olfactory bulb (de Olmos et al., 1978) and prefrontal cortex (Swanson, 1981).

19.2.3 The Subicular Complex

The subiculum curves anteriorly and laterally to wrap around the posterior extension of the dentate gyrus. It borders the medial entorhinal cortex and field of CA1 (Figs. 19.1 and 19.2). The subiculum can be divided into three distinct cytoarchitectural areas. The parasubiculum borders the medial entorhinal cortex and contains moderately packed medium-sized cells. The presubiculum is between the parasubiculum and the subiculum, which is characterized by densely packed small cells. The subiculum proper borders to the field of CA1 and has a loosely packed pyramidal cell layer and a wide molecular layer (Figs. 19.1 and 19.2).

Hippocampal intrinsic afferents to the subiculum originate from the CA1 neurons and bilateral CA3 pyramidal cells through the Schaffer collaterals (Swanson et al., 1978). The extra-hippocampal projections to the subiculum mostly originate from layer II (slightly from layer IV) of the entorhinal cortex and lessly from the raphe nuclei (Conrad et al., 1974), locus coeruleus (Haring & Davis, 1985; Jones & Moore, 1977), amygdaloid nuclei (Krettek & Price, 1977), septal nucleus (Meibach & Siegel, 1977b) and peririhnal cortex (Kosel et al., 1983).

The targeted regions for subicular projections include, but are not limited to, thalamic nuclei (Aggleton et al., 2005), reunions nucleus (Herkenham, 1978), the mammillary body of the hypothalamus(Aggleton et al., 2005), and amygdaloid nucleus (Kishi et al., 2006). The parasubiculum and presubiculum project heavily to the anterior thalamic nuclei (Cohen & Eichenbaum, 1993; Sikes et al., 1977).

19.3 Role of the Hippocampus in Learning and Memory

19.3.1 Memory Functions of the Hippocampus

The role of the hippocampus in learning and memory has been the focus of neuroscience studies since Scoville and Milner reported the case of Henry Molaison (HM) who had severe anterograde amnesia following bilateral medial temporal lobe resection (Scoville & Milner, 1957). The pattern of impaired and spared memory functions displayed by HM prompted a now commonly acknowledged characterization of the temporal lobe amnesic syndrome. It is widely believed that damage to the hippocampus disrupts declarative memory processes and more specifically episodic memory functions (Tulving, 2001; Tulving & Markowitsch, 1998). Declarative memories have been further divided into episodic and semantic memories. Episodic memory is concerned with conscious recall of specific episodes, and semantic memory with the storage of factual information. The memory functions that are spared in temporal lobe lesions have been classified as non-declarative or procedural memories (Cohen & Eichenbaum, 1993; Squire, 1992). Nondeclarative or procedural memory processes are thought to operate automatically and do not include information about where or when the learning experience took place. The pattern of memory deficits in humans with temporal lobe damage prompted the question, what role did the hippocampus play in memory?

The precise role of the hippocampus in learning memory remains to be determined. Accumulating evidence suggests that the hippocampus has an essential role in the formation of new memories about experienced events. Damage to the hippocampus usually results in profound difficulties in forming new memories (anterograde amnesia) and normally also affects access to memories prior to the damage (retrograde amnesia). Although the retrograde effect normally extends some years prior to the brain damage, in some cases older memories remain. This sparing of older memories leads to the idea that consolidation over time involves the transfer of memories out of the hippocampus to other parts of the brain. However, it is difficult to test the sparing of older memories experimentally. Also, in some cases of retrograde amnesia, the sparing appears to affect memories formed decades before the damage to the hippocampus occurred, so its role in maintaining these older memories remains controversial.

Animal studies indicate that the hippocampus plays a role in storing and processing spatial information. In rodents, the firing rate of hippocampal neurons was found to correlate to the location of the animal in a test environment, and these cells are referred to as place cells (O'Keefe & Conway, 1978; O'Keefe & Dostrovsky, 1971). There are many thousands of different place cells, which can be activated in response to a location in a particular environment. Different place cells have different place fields, which are not fixed in absolute space and are relative to spatial cues. The discovery of place cells led to the idea that the hippocampus might act as a cognitive map— a neural representation of the layout of the environment (O'Keefe & Nadel, 1978). There has been considerable support for the cognitive mapping theory from lesion and unit recording studies. It has been reported that rats with hippocampal lesions exhibit impairment in learning as detected in the radial arm (Jarrard, 1983), T-maze (Bannerman et al., 2001), and Morris water maze, where lesioned rats had poor performance in finding the hidden platform (Morris et al., 1982, 1986). Indeed, neuroimaging studies in humans revealed that the hippocampus becomes active during spatial navigation, suggesting that the hippocampus in humans contributes to the encoding and retrieval of spatial information. This finding corresponds well with the results in animal studies, which show a significant deficit in spatial navigation resulting from hippocampal damage.

19.3.2 Synaptic Mechanisms of Memory

The nature of the physiological basis of learning and memory remains an enigma in neurobiology. The assumption that information is stored in the brain as changes in synaptic efficacy was initially proposed by Ramon y Cajal and later refined by Hebb (Hebb, 1949). This assumption was not tested until the early 1970s when Timothy Bliss and Terje LØmo made an important discovery that brief high-frequency electrical stimulation of an excitatory pathway to the hippocampus produced a long-lasting enhancement in the strength of the stimulated synapses. This effect is now known as LTP.

LTP is expressed as a persistent increase in the size of the evoked synaptic response recorded from single cells or groups of cells. It can be induced by high-frequency stimulation (HFS, typically 100 Hz) or other types of stimulation, such as theta-burst stimulation. These stimulus paradigms resemble the synchronized firing patterns at similar frequencies that occur in the hippocampus during learning (Otto et al., 1991), making them useful experimental means to generate "learning activities" in the hippocampus. Over the past 30 years, LTP has been intensively studied because it is the leading experimental model for the synaptic changes that may underlie learning and memory.

19.3.2.1 Basic Properties of LTP

Although LTP was first demonstrated at the perforant path synapses on the granule cells in the dentate gyrus (Bliss & Gardner-Medwin, 1973), the majority of experiments on understanding the mechanisms of LTP have been performed on the Schaffer collateral/commissural synapses on the CA1 pyramidal cells in the hippocampus.

LTP in the hippocampus has three basic properties: cooperativity, associativity, and input-specificity (Bliss & Collingridge, 1993). Cooperativity refers to the fact that long-lasting synaptic enhancement following HFS increases with the number of stimulated afferents (McNaughton et al., 1978). A threshold stimulus intensity during HFS is required for synaptic enhancement. "Weak" HFS, which activates relatively few fibers, did not produce LTP, whereas strong stimulation at the same frequency and for the same duration produced LTP. Associativity means that a weak input (small number of stimulated afferents) can be potentiated if it is active at the same time as a strong tetanus (large number of stimulated afferents) to a separate but convergent input (Bliss & Collingridge, 1993). This associativity has often been viewed as a cellular analog of associative or classic conditioning (Malenka & Nicoll, 1999). Another basic property of LTP is its input-specificity. When LTP is induced by repetitive stimulation the increase in synaptic strength usually does not occur in other synapses (on the same cell) that are not active at the time of repetitive stimulation (Bliss & Collingridge, 1993). This property increases the storage capacity of individual neurons (Malenka & Nicoll, 1999).

One remarkable feature of LTP is that it can be induced by a brief HFS, lasting less than or equal to a second and consisting of stimulation frequency well within the range of normal axon discharging. Longevity is an additional feature of LTP. Once induced, LTP can persist for many hours in brain slices in vitro or in the anesthetized animal, and for days or weeks (also lasting a lifetime) in the freely moving animal.

19.3.2.2 Mechanisms of Hippocampal LTP

It is well accepted that activation of postsynaptic NMDA receptors, a subtype of glutamate receptors, is required for the induction of LTP in the hippocampus. The key role that the NMDA receptors play in LTP induction relies on the voltage-dependent block of its channel by Mg^{2+} (Ascher & Nowak, 1988). In this way, the NMDA receptor channel complex behaves as a molecular detector for LTP induction. To trigger the induction of LTP, two events must occur simultaneously: the cell membrane must be sufficiently depolarized to expel Mg^{2+} from NMDA channels at the same time L-glutamate binds to NMDA receptors and promotes the opening of these receptor—ligand-gated ion channels. The membrane depolarization can be achieved by repetitive tetanic stimulation of synapses or by directly depolarizing the cell while continuing low-frequency stimulation of synapses (Gustafsson et al., 1987). At Schaffer collateral—CA1 pyramidal cell synapses, Na^+ ions passing through the α-amino-3-hydroxy-5-methyl-4-isoxazolepropionic acid (AMPA) receptors are responsible for this membrane depolarization. When the membrane depolarization is sufficient and reaches a certain level, it expels Mg^{2+} from the NMDA receptor channel, allowing Ca^{2+} as well as Na^+ to enter the cell. The influx of Ca^{2+} through the NMDA receptor channel raises intracellular Ca^{2+} and triggers the induction of LTP.

Considerable evidence now links this rise in postsynaptic Ca^{2+} concentration to the induction of LTP. The most compelling evidence in support of this model comes from experimental results wherein LTP induction can be blocked by pharmacological inhibition of NMDA receptors (Collingridge et al., 1983) or prevented by the injection of a Ca^{2+} chelator into the postsynaptic neuron (Lynch et al., 1983; Yang et al., 1999). Ca^{2+} imaging studies have demonstrated that tetanic stimulation directly increases Ca^{2+} within dendrites and spines as a result of NMDA receptor activation (Perkel et al., 1993; Regehr & Tank, 1990; Yuste & Denk, 1995). Although NMDA receptors are the primary source of Ca^{2+} entry into the dendrites and spines, activation of dendritic voltage-gated Ca^{2+} channels and Ca^{2+} release from intracellular stores also elevate Ca^{2+} levels and contribute to the induction of LTP. However, the mechanisms underlying the Ca^{2+} channel-dependent LTP may differ from NMDA receptor-dependent LTP (Malenka & Nicoll, 1999).

19.3.2.3 Expression of LTP

It has long been a challenge for neurobiologists to identify whether the increase in synaptic strength is mediated primarily through a pre- or postsynaptic mechanism. Available experimental results indicate that the increase in the synaptic strength could be either presynaptic or postsynaptic or both, or through an extrasynaptic mechanism, such as a reduction in uptake of glutamate by glial cells resulting in an elevated concentration of glutamate at the synaptic cleft. Evidence for pre-synaptic mechanisms comes from experiments measuring the overflow of radiolabeled or endogenous L-glutamate in the hippocampus before and after the induction of LTP (Bliss et al., 1986). In addition, quanta analysis of synaptic transmission reveals the proportion of synaptic failures decreases after the induction of LTP (Kullmann & Siegelbaum, 1995; Malenka & Nicoll, 1999). As synaptic failures represent the failure of neurotransmitter release or silent synapses, it was concluded that the induction of LTP is the consequence of an increase in the probability of neurotransmitter release. Evidence supporting this conclusion came from the finding that the variation around the mean of excitatory postsynaptic currents (EPSCs) decreased during LTP (Kullmann & Siegelbaum, 1995; Malenka & Nicoll, 1999). If the probability of neurotransmitter release increases during LTP, then the quantal content will, on average, increase, and the coefficient of variation will decrease because the coefficient of variation (SD/mean) is inversely proportional to the quantal content (Malenka & Nicoll, 1999).

Experimental results from other studies have shown a postsynaptic mechanism for LTP induction. First, paired-pulse facilitation (PPF) is not altered after the induction of LTP, and this has been interpreted as evidence for a postsynaptic facilitation of LTP. PPF occurs when two pre-synaptic stimuli are delivered with a short interval (50–200 ms) and is thought to result from residual Ca2+ in the pre-synaptic terminal following the first stimulus, enhancing release during the second stimulus (Malenka & Nicoll, 1999). Various manipulations known to increase transmitter release do cause a decrease in the facilitation ratio because the release is already enhanced to near saturation during the first stimulus (Manabe et al., 1993). The failure in the alteration of the PPF ratio suggests a postsynaptic locus of LTP expression.

A number of studies have shown that during LTP, the AMPA receptor component of the EPSC is selectively enhanced, with little or no change in the component of NMDA receptors (Larkman & Jack, 1995), though both receptors are frequently co-localized at individual synapses. Pharmacological modulation of transmitter release affects AMPA and NMDA components equally, arguing for a selective postsynaptic alteration in either the density or properties of AMPA receptors (Kullmann & Siegelbaum, 1995).

19.4 Neuroimmunomodulation via Hippocampus

The central nervous system (CNS) has long been considered an immunologically privileged site which lacks an immune response to CNS antigens. Immune privilege in the CNS was believed to be the consequence of either poor immunosurveillance or modulation of the immune response. However, recent studies have demonstrated that the brain is an immunologically specialized organ with tight regulation of both innate and adaptive immune cells (Louveau et al., 2015). One of the important recent advances in understanding the biological basis of neurodegenerative disorders is the recognition that there is extensive communication between the CNS and the immune system. Initial evidence that the immune system may communicate with the CNS was provided by Besedovsky and colleagues (Besedovsky et al., 1977), who observed that activation of the immune system was accompanied by changes in hypothalamic, autonomic, and endocrine processes. The existence of neural—immune interactions is now supported by abundant evidence showing that the immune system communicates with the CNS through immunotransmitters (primarily cytokines), leading to direct CNS activation (Berkenbosch et al., 1987; Buckley & McGavern, 2022; Sapolsky et al., 1987; Zhu et al., 2022) or by the release of CNS-derived cytokines, and that the CNS regulates the immune system via neurotransmitters, hormones, and neuropeptides (Buckley & McGavern, 2022; Zhu et al., 2022). It has been shown that cells of the immune system can synthesize and release several immunomodulatory cytokines, hormones, neuropeptides, and catecholamines (Blalock, 1989, 1994; Zhu et al., 2022). For example, lymphocytes and macrophages produce the endogenous opioid peptides, norepinephrine, and epinephrine (Lolait et al., 1984; Harbour et al., 1987; Engler et al., 2005). An interesting study (Rivest, 2003) showed that the CNS responded to systemic bacterial infection with an innate immune reaction without the pathogen's direct access to the brain. Whether caused by a microbe, trauma, toxic metabolite, autoimmunity, or part of a wide degenerative process, activation of the immune system results in changes in the activity of discrete populations of brain neurons, including hippocampal neurons. Accumulating evidence indicates that this mutual regulation is relevant for the course of infectious, inflammatory, autoimmune, and neoplastic diseases.

As resident innate immune cells in the brain, microglia behave like a dual-edged sword. On one hand, they release neurotrophic factors to promote neuronal development and growth and, on the other hand, they produce neurotoxic substances resulting in neuronal dysfunction and injury (Block et al., 2007; Dermitzakis et al., 2023; Perry et al., 2010; Skaper et al., 2014). Microglia also functions as a CNS sensor and detects neural injury, inflammation, and/or infection (Arutyunov & Klein, 2023; Block et al., 2007; Perry et al., 2010; Perry & Holmes, 2014). Elevated levels of circulating pro-inflammatory cytokines and over-activation of microglia can result in CNS neuronal injury. It has been shown that microglia density in the hippocampus is relatively high than telencephalon cortex, diencephalon, brainstem, and cerebellum (Lawson et al., 1990), suggesting a potential occurrence of neuroimmunomodulatory activities in the hippocampus (Nebeling et al., 2023).

19.4.1 Immunomodulation of Neuronal Functions in Hippocampus

Hippocampal neuronal activities have been examined thoroughly in studies on neuroendocrine, autonomic, and cognitive function, as well as psychomotor behavior. Circumstantial evidence indicates that hippocampal physiology can be modulated by the immune system (Jankowsky et al., 2000; Jankowsky & Patterson, 1999; Yirmiya & Goshen, 2011). Immune cells such as microglia and astrocytes produce cytokines and other signaling molecules including, but not limited to, interleukin-1 beta (IL-1β), IL-2, IL-6, tumor necrosis factor-alpha (TNF-α), and interferon-gamma (INF-γ) that can influence the function of hippocampal neurons and synaptic plasticity (Levin et al., 2021; Wrona, 2006). Indeed, a number of cytokines, including those mentioned above, are expressed in the hippocampus, and alteration of their expression levels can affect hippocampal functions. Studies have shown that inflammatory cytokines, released in response to the detection of foreign substances (antigens), influence ion channel activities, intracellular Ca^{2+} homeostasis, and membrane potentials and suppress or enhance the induction of LTP in the hippocampus (Koller et al., 1997). Extensive experimental results have implicated IL-1 among the other cytokines as a likely candidate for a key immune transmitter, communicating immunological activation to the brain including the hippocampus (Besedovsky et al., 1975, 1986). It is worth pointing out that cytokines rarely work in isolation. For instance, the release of IL-1β is usually associated with the release of other proinflammatory cytokines, such as TNF-a and IL-6, which are indeed expressed in the hippocampus.

IL-1β and its receptors are expressed in the hippocampus (Farrar et al., 1987; Ban et al., 1991; Cunningham and De Souza, 1993). The high density of binding sites for IL-1β has been detected in the hippocampus, with the highest in the dentate gyrus (Takao et al., 1990). It has been shown that peripheral immune activation by lipopolysaccharide (LPS) up-regulates IL-1β mRNA expression and increases IL-1β protein in the hippocampus (Laye et al., 1994; Nguyen et al., 1998). This suggests that immune activation may modulate hippocampal function via releasing immune active molecules, such as IL-1β. As the hippocampus is a brain region involved in learning and memory, the immune-associated upregulation of IL-1β mRNA and protein expression may interrupt hippocampal functions such as learning and memory. Indeed, IL-1β suppressed the induction of LTP in the CA1 and CA3 areas of the hippocampus as well as in the dentate gyrus (Bellinger et al., 1993; Cunningham et al., 1996; Katsuki et al., 1990; O'Connor & Coogan, 1999; Xiong et al., 2000), while having no significant effects on excitatory postsynaptic potential (EPSP) evoked by low-frequency stimulation. The IL-1β-mediated suppression of LTP was antagonized by an IL-1β receptor antagonist, suggesting that IL-1β inhibits LTP through IL-1β receptors. In addition to IL-1, IL-2 had similar effects on LTP in the hippocampus. Application of recombinant IL-2 inhibited the induction of both short-term potentiation (STP) and LTP. It also inhibited post-tetanic potentiation (PTP) and LTP maintenance without affecting basal synaptic transmission (Tancredi et al., 1990). Moreover, IL-2 deficiency results in altered hippocampal cytoarchitecture (Beck et al., 2005). LTP was also suppressed by TNF-α in both the CA1 region (Tancredi et al., 1992) and the dentate gyrus (Cunningham et al., 1996). In the CA1, the induction of LTP was inhibited by TNF-α if the tetanic stimulation was given at least 50 min after TNF-α application. In contrast to IL-1 and IL-2, TNF-α increased basal synaptic transmission in the CA1 region of the slices acutely exposed to TNF-α (Cunningham et al., 1996) but not in the dentate gyrus. The underlying mechanisms, by which TNF-α increased basal synaptic transmission, have not been determined. Brief treatment of hippocampal slices with TNF-α did not influence LTP, while long-lasting application (>50 min) of TNF-α inhibited LTP.

Cytokines and chemokines are the messengers and regulators between neuron and glial cell communication (Harrison et al., 1998; Ransohoff, 2009). It has been shown that fractalkine (CX3CL1) is the cytokine expressed predominantly by neurons and its receptor—CX3CR1 is expressed in microglia (Ransohoff & Stevens, 2011). This CX3CL1 and CX3CR1 axis becomes a typical example of neural—immune communications and plays a pivotal role in neuronal modulation of microglia activity (Bachstetter et al., 2011). Studies by immunofluorescent staining have shown that CX3CL1 is highly expressed in pyramidal neurons in CA1 ~ CA3 regions of the hippocampus (Bachstetter et al., 2011; Sheridan et al., 2014), a brain structure associated with learning and memory. It has been demonstrated that high-frequency stimulation induces robust LTP in the hippocampus, and LTP has been considered as a cellular/synaptic mechanism for learning and memory (Otto et al., 1991; Tulving, 2001; Tulving & Markowitsch, 1998). Thus, alteration of the CX3CL1 and CX3CR1 axis may impact the induction of LTP and LTP-associated learning and memory (Maggi et al., 2009; Rogers et al., 2011; Sheridan et al., 2014).

19.4.2 Lesion of the Hippocampus Affects Neuroimmunity

The hippocampus plays an important role in the regulation of neuroimmunity through its interactions with immune cells in the CNS, regulation of BBB permeability, and modulation of astrocyte activity. It has been shown that during immune challenge, the hippocampus exhibits time-dependent changes in neurotransmitter levels and that an intact hippocampus is essential for the normal humoral immunity for the primary immune response in rats (Devi et al., 2004; Lathe, 2001). Lesions of the dorsal hippocampus were found to produce a transient increase in splenocytes and thymocytes, as well as increased T-cell mitogen responses (Brooks et al., 1982). Lesions of the hippocampus were also found to cause differential effects on humoral immunity depending on the lesions of different subfields of the hippocampus (Pan & Long, 1993). Axotomy of afferent fibers within the molecular layer of the dentate gyrus caused activation of neural—immune elements in the slice cultures (Coltman & Ide, 1996). In addition, electrical stimulation of the hippocampus increased the number of neutrophils and phagocytic index while also decreasing the number of lymphocytes and plasma corticosterone levels in rats (Devi et al., 1993). Lesions in the hippocampus induced by kainic acid resulted in elevated antibody production including IgM and IgG (Nance et al., 1987). Recent studies revealed that the hippocampus can regulate NLRP3 (Nucleotide-binding oligomerization domain, leucine-rich repeat, and pyrin-domain-containing 3) inflammasome activation through its control of microglia activity (Ghasemi et al., 2022; Liu et al., 2022), modulation of the stress response (Ghasemi et al., 2022), and regulation of mitochondrial function (Farris et al., 2019). Dysregulation of hippocampal function may contribute to the development of neurological disorders characterized by NLRP3 inflammasome activation and inflammation. Taken together, these studies have clearly shown that lesions (or stimulation) in the hippocampus affect immune functions. Dysregulation of hippocampal function may contribute to the development of neurological disorders characterized by neuroinflammation.

19.5 Review Questions

1. Where is the hippocampus located in the human brain?
2. What are the three major subfields in the hippocampus?
3. What are the principal cells in the dentate gyrus and the CA1 field?
4. The main input to the dentate gyrus originates from which part of the brain and via which fiber path?
5. Mossy fibers originate from which part of the hippocampus and synapse predominantly onto the neurons of which field(s)?
6. Describe sources of afferent fibers to the CA1 region.
7. Briefly describe synaptic information flow through the hippocampus.
8. Damage to the hippocampus disrupts what type(s) of memory processes?
9. Describe how long-term potentiation (LTP) is expressed) and why LTP is considered as the synaptic base for learning and memory?
10. Lesion of the hippocampus affects neuroimmunity?

19.6 Answers to Review Questions

1. It is located in the medial temporal lobe of the brain.
2. The three major subfields in the hippocampus are the dentate gyrus, CA3, and CA1.
3. The principal cells in the dentate gyrus and CA1 are granule cells and pyramidal cells, respectively.
4. The main input to the dentate gyrus originates from the entorhinal cortex via the performant pathway.
5. The Mossy fibers stem from granule cells in the dentate gyrus and their axonal fibers (the mossy fiber pathway) project predominantly to the CA3 neurons.
6. The major afferent fibers to the CA1 pyramidal cells come predominantly from bilateral CA3 neurons through Schaffer collaterals from both ipsi- and contralateral sides. The CA1 pyramidal cells also receive projections from layer III (slightly from layer V) of the entorhinal cortex.
7. Information enters the hippocampus mainly through the perforant pathway consisting of the axons of neurons in layers II and III of the entorhinal cortex. The perforant path axons terminate on the dendrites of the dentate gyrus granular cells. Then information flows through the hippocampus from the dentate gyrus to the CA3, the CA1, forming an intrinsic tri-synaptic circuit in the hippocampus.
8. Damage to the hippocampus disrupts declarative memory processes and more specifically episodic memory functions.
9. LTP is expressed as a persistent increase in the size of the evoked synaptic response recorded from single cells or groups of cells. It can be induced by high-frequency stimulation (HFS, typically 100 Hz) or other types of stimulation, such as theta-burst stimulation. These stimulus paradigms resemble the synchronized firing patterns at similar frequencies that occur in the hippocampus during learning, making them useful experimental means to generate "learning activities" in the hippocampus.
10. Increasing evidence indicates that the hippocampus is involved in neuroimmunity. Lesions of the hippocampus cause differential effects on neuroimmunity depending on the site of the lesion. Lesions of the dorsal hippocampus produce a transient increase in splenocytes and thymocytes, as well as increased T-cell mitogen responses. Lesions in the hippocampus induced by kainic acid result in elevated antibody production including IgM and IgG. In addition, electrical stimulation of the hippocampus increases the number of neutrophils and phagocytic index while also decreasing the number of lymphocytes and plasma corticosterone levels in rats. The aforementioned clearly shows that lesions (or stimulation) in the hippocampus affect immune functions.

Acknowledgments Supported by NIH grants R01 NS041862, R01 NS063878 and R01 DA050540 (HX).

References

Aggleton, J. P., Vann, S. D., & Saunders, R. C. (2005). Projections from the hippocampal region to the mammillary bodies in macaque monkeys. *The European Journal of Neuroscience, 22*, 2519–2530.

Arutyunov, A., & Klein, R. S. (2023). Microglia at the scene of the crime: What their transcriptomics reveal about brain health. *Current Opinion in Neurology, 36*, 207–213.

Ascher, P., & Nowak, L. (1988). The role of divalent cations in the N-methyl-D-aspartate responses of mouse central neurones in culture. *The Journal of Physiology, 399*, 247–266.

Bachstetter, A. D., Morganti, J. M., Jernberg, J., Schlunk, A., Mitchell, S. H., Brewster, K. W., Hudson, C. E., Cole, M. J., Harrison, J. K., Bickford, P. C., & Gemma, C. (2011). Fractalkine and CX 3 CR1 regulate hippocampal neurogenesis in adult and aged rats. *Neurobiology of Aging, 32*, 2030–2044.

Ban, E., Milon, G., Prudhomme, N., Fillion, G., & Haour, F. (1991). Receptors for interleukin-1 (alpha and beta) in mouse brain: Mapping and neuronal localization in hippocampus. *Neuroscience, 43*, 21–30.

Bannerman, D. M., Yee, B. K., Lemaire, M., Wilbrecht, L., Jarrard, L., Iversen, S. D., Rawlins, J. N., & Good, M. A. (2001). The role of the entorhinal cortex in two forms of spatial learning and memory. *Experimental Brain Research, 141*, 281–303.

Beck, R. D., Jr., King, M. A., Ha, G. K., Cushman, J. D., Huang, Z., & Petitto, J. M. (2005). IL-2 deficiency results in altered septal and hippocampal cytoarchitecture: Relation to development and neurotrophins. *Journal of Neuroimmunology, 160*, 146–153.

Bellinger, F. P., Madamba, S., & Siggins, G. R. (1993). Interleukin 1 beta inhibits synaptic strength and long-term potentiation in the rat CA1 hippocampus. *Brain Research, 628*, 227–234.

Berkenbosch, F., van Oers, J., del Rey, A., Tilders, F., & Besedovsky, H. (1987). Corticotropin-releasing factor-producing neurons in the rat activated by interleukin-1. *Science, 238*, 524–526.

Besedovsky, H., Sorkin, E., Keller, M., & Muller, J. (1975). Changes in blood hormone levels during the immune response. *Proceedings of the Society for Experimental Biology and Medicine, 150*, 466–470.

Besedovsky, H., Sorkin, E., Felix, D., & Haas, H. (1977). Hypothalamic changes during the immune response. *European Journal of Immunology, 7*, 323–325.

Besedovsky, H., del Rey, A., Sorkin, E., & Dinarello, C. A. (1986). Immunoregulatory feedback between interleukin-1 and glucocorticoid hormones. *Science, 233*, 652–654.

Blackstad, T. W., & Kjaerheim, A. (1961). Special axo-dendritic synapses in the hippocampal cortex: Electron and light microscopic studies on the layer of mossy fibers. *The Journal of Comparative Neurology, 117*, 133–159.

Blalock, J. E. (1989). A molecular basis for bidirectional communication between the immune and neuroendocrine systems. *Physiological Reviews, 69*, 1–32.

Blalock, J. E. (1994). Shared ligands and receptors as a molecular mechanism for communication between the immune and neuroendocrine systems. *Annals of the New York Academy of Sciences, 741*, 292–298.

Bliss, T. V., & Collingridge, G. L. (1993). A synaptic model of memory: Long-term potentiation in the hippocampus. *Nature, 361*, 31–39.

Bliss, T. V., & Gardner-Medwin, A. R. (1973). Long-lasting potentiation of synaptic transmission in the dentate area of the unanaestetized rabbit following stimulation of the perforant path. *The Journal of Physiology, 232*, 357–374.

Bliss, T. V., Douglas, R. M., Errington, M. L., & Lynch, M. A. (1986). Correlation between long-term potentiation and release of endogenous amino acids from dentate gyrus of anaesthetized rats. *The Journal of Physiology, 377*, 391–408.

Block, M. L., Zecca, L., & Hong, J. S. (2007). Microglia-mediated neurotoxicity: Uncovering the molecular mechanisms. *Nature Reviews. Neuroscience, 8*, 57–69.

Bratt, A. M., Kelley, S. P., Knowles, J. P., Barrett, J., Davis, K., Davis, M., & Mittleman, G. (2001). Long term modulation of the HPA axis by the hippocampus. Behavioral, biochemical and immunological endpoints in rats exposed to chronic mild stress. *Psychoneuroendocrinology, 26*, 121–145.

Breivik, T., Thrane, P. S., Gjermo, P., Cools, A., & Myhrer, T. (2002). Effects of hippocampal lesioning on experimental periodontitis in Wistar rats. *Journal of Periodontal Research, 37*, 360–365.

Brooks, W. H., Cross, R. J., Roszman, T. L., & Markesbery, W. R. (1982). Neuroimmunomodulation: Neural anatomical basis for impairment and facilitation. *Annals of Neurology, 12*, 56–61.

Buckley, M. W., & McGavern, D. B. (2022). Immune dynamics in the CNS and its barriers during homeostasis and disease. *Immunological Reviews, 306*, 58–75.

Cajal, R. (1911). *Histologie du systeme nerveux de l'homme et des vertebres* (Vol. 2). Instituto Ramon y Cajal.

Chen, Z. S., & Wilson, M. A. (2023). How our understanding of memory replay evolves. *Journal of Neurophysiology, 129*, 552–580.

Cohen, N. J., & Eichenbaum, H. (1993). *Memory, amnesia and the hippocampus system*. MIT Press.

Collingridge, G. L., Kehl, S. J., & McLennan, H. (1983). The antagonism of amino acid-induced excitations of rat hippocampal CA1 neurones in vitro. *The Journal of Physiology, 334*, 19–31.

Coltman, B. W., & Ide, C. F. (1996). Temporal characterization of microglia, IL-1 beta-like immunoreactivity and astrocytes in the dentate gyrus of hippocampal organotypic slice cultures. *International Journal of Developmental Neuroscience, 14*, 707–719.

Commons, K. G., & Milner, T. A. (1996). Ultrastructural relationships between leu-enkephalin- and GABA-containing neurons differ within the hippocampal formation. *Brain Research, 724*, 1–15.

Conrad, L. C., Leonard, C. M., & Pfaff, D. W. (1974). Connections of the median and dorsal raphe nuclei in the rat: An autoradiographic and degeneration study. *The Journal of Comparative Neurology, 156*, 179–205.

Cunningham, E. T., Jr., & De Souza, E. B. (1993). Interleukin 1 receptors in the brain and endocrine tissues. *Immunology Today, 14*, 171–176.

Cunningham, A. J., Murray, C. A., O'Neill, L. A., Lynch, M. A., & O'Connor, J. J. (1996). Interleukin-1 beta (IL-1 beta) and tumour necrosis factor (TNF) inhibit long-term potentiation in the rat dentate gyrus in vitro. *Neuroscience Letters, 203*, 17–20.

Danieli, K., Guyon, A., & Bethus, I. (2023). Episodic memory formation: A review of complex hippocampus input pathways. *Progress in Neuro-Psychopharmacology & Biological Psychiatry, 110757*.

de Olmos, J., Hardy, H., & Heimer, L. (1978). The afferent connections of the main and the accessory olfactory bulb formations in the rat: An experimental HRP-study. *The Journal of Comparative Neurology, 181*, 213–244.

Dermitzakis, I., Manthou, M. E., Meditskou, S., Tremblay, M. E., Petratos, S., Zoupi, L., Boziki, M., Kesidou, E., Simeonidou, C., & Theotokis, P. (2023). Origin and emergence of microglia in the CNS-an interesting (hi)story of an eccentric cell. *Current Issues in Molecular Biology, 45*, 2609–2628.

Desmond, N. L., & Levy, W. B. (1982). A quantitative anatomical study of the granule cell dendritic fields of the rat dentate gyrus using a novel probabilistic method. *The Journal of Comparative Neurology, 212*, 131–145.

Devi, R. S., Namasivayam, A., & Prabhakaran, K. (1993). Modulation of non-specific immunity by hippocampal stimulation. *Journal of Neuroimmunology, 42*, 193–197.

Devi, R. S., Sivaprakash, R. M., & Namasivayam, A. (2004). Rat hippocampus and primary immune response. *Indian Journal of Physiology and Pharmacology, 48*, 329–336.

Engler, K. L., Rudd, M. L., Ryan, J. J., Stewart, J. K., & Fischer-Stenger, K. (2005). Autocrine actions of macrophage-derived catecholamines on interleukin-1 beta. *Journal of Neuroimmunology, 160*, 87–91.

Fa, M., Xia, L., Anunu, R., Kehat, O., Kriebel, M., Volkmer, H., & Richter-Levin, G. (2014). Stress modulation of hippocampal activity—spotlight on the dentate gyrus. *Neurobiology of Learning and Memory, 112*, 53–60.

Farrar, W. L., Kilian, P. L., Ruff, M. R., Hill, J. M., & Pert, C. B. (1987). Visualization and characterization of interleukin 1 receptors in brain. *Journal of Immunology, 139*, 459–463.

Farris, S., Ward, J. M., Carstens, K. E., Samadi, M., Wang, Y., & Dudek, S. M. (2019). Hippocampal subregions express distinct dendritic transcriptomes that reveal differences in mitochondrial function in CA2. *Cell Reports, 29*(522–539), e526.

Frotscher, M. (1985). Mossy fibres form synapses with identified pyramidal basket cells in the CA3 region of the Guinea-pig hippocampus: A combined Golgi-electron microscope study. *Journal of Neurocytology, 14*, 245–259.

Ghasemi, M., Navidhamidi, M., Rezaei, F., Azizikia, A., & Mehranfard, N. (2022). Anxiety and hippocampal neuronal activity: Relationship and potential mechanisms. *Cognitive, Affective, & Behavioral Neuroscience, 22*, 431–449.

Gustafsson, B., Wigstrom, H., Abraham, W. C., & Huang, Y. Y. (1987). Long-term potentiation in the hippocampus using depolarizing current pulses as the conditioning stimulus to single volley synaptic potentials. *The Journal of Neuroscience, 7*, 774–780.

Gutierrez, R., & Heinemann, U. (2006). Co-existence of GABA and Glu in the hippocampal granule cells: Implications for epilepsy. *Current Topics in Medicinal Chemistry, 6*, 975–978.

Harbour, D. V., Smith, E. M., & Blalock, J. E. (1987). Splenic lymphocyte production of an endorphin during endotoxic shock. *Brain, Behavior, and Immunity, 1*, 123–133.

Haring, J. H., & Davis, J. N. (1985). Retrograde labeling of locus coeruleus neurons after lesion-induced sprouting of the coeruleohippocampal projection. *Brain Research, 360*, 384–388.

Harrison, J. K., Jiang, Y., Chen, S., Xia, Y., Maciejewski, D., McNamara, R. K., Streit, W. J., Salafranca, M. N., Adhikari, S., Thompson, D. A., Botti, P., Bacon, K. B., & Feng, L. (1998). Role for neuronally derived fractalkine in mediating interactions between neurons and CX3CR1-expressing microglia. *Proceedings of the National Academy of Sciences of the United States of America, 95*, 10896–10901.

Hebb, D. (1949). *The organization of behavior*. Johm Wiley and Sons.

Herkenham, M. (1978). The connections of the nucleus reuniens thalami: Evidence for a direct thalamo-hippocampal pathway in the rat. *The Journal of Comparative Neurology, 177*, 589–610.

Jankowsky, J. L., & Patterson, P. H. (1999). Cytokine and growth factor involvement in long-term potentiation. *Molecular and Cellular Neurosciences, 14*, 273–286.

Jankowsky, J. L., Derrick, B. E., & Patterson, P. H. (2000). Cytokine responses to LTP induction in the rat hippocampus: A comparison of in vitro and in vivo techniques. *Learning & Memory, 7*, 400–412.

Jarrard, L. E. (1983). Selective hippocampal lesions and behavior: Effects of kainic acid lesions on performance of place and cue tasks. *Behavioral Neuroscience, 97*, 873–889.

Jones, B. E., & Moore, R. Y. (1977). Ascending projections of the locus coeruleus in the rat. II. Autoradiographic study. *Brain Research, 127*, 25–53.

Katsuki, H., Nakai, S., Hirai, Y., Akaji, K., Kiso, Y., & Satoh, M. (1990). Interleukin-1 beta inhibits long-term potentiation in the CA3 region of mouse hippocampal slices. *European Journal of Pharmacology, 181*, 323–326.

Kishi, T., Tsumori, T., Yokota, S., & Yasui, Y. (2006). Topographical projection from the hippocampal formation to the amygdala: A combined anterograde and retrograde tracing study in the rat. *The Journal of Comparative Neurology, 496*, 349–368.

Koller, H., Siebler, M., & Hartung, H. P. (1997). Immunologically induced electrophysiological dysfunction: Implications for inflammatory diseases of the CNS and PNS. *Progress in Neurobiology, 52*, 1–26.

Kosel, K. C., Van Hoesen, G. W., & Rosene, D. L. (1983). A direct projection from the perirhinal cortex (area 35) to the subiculum in the rat. *Brain Research, 269*, 347–351.

Krettek, J. E., & Price, J. L. (1977). Projections from the amygdaloid complex and adjacent olfactory structures to the entorhinal cortex and to the subiculum in the rat and cat. *The Journal of Comparative Neurology, 172*, 723–752.

Kullmann, D. M., & Siegelbaum, S. A. (1995). The site of expression of NMDA receptor-dependent LTP: New fuel for an old fire. *Neuron, 15*, 997–1002.

Larkman, A. U., & Jack, J. J. (1995). Synaptic plasticity: Hippocampal LTP. *Current Opinion in Neurobiology, 5*, 324–334.

Lathe, R. (2001). Hormones and the hippocampus. *The Journal of Endocrinology, 169*, 205–231.

Lawson, L. J., Perry, V. H., Dri, P., & Gordon, S. (1990). Heterogeneity in the distribution and morphology of microglia in the normal adult mouse brain. *Neuroscience, 39*, 151–170.

Laye, S., Parnet, P., Goujon, E., & Dantzer, R. (1994). Peripheral administration of lipopolysaccharide induces the expression of cytokine transcripts in the brain and pituitary of mice. *Brain Research. Molecular Brain Research, 27*, 157–162.

Levin, S. G., Pershina, E. V., Bugaev-Makarovskiy, N. A., Chernomorets, I. Y., Konakov, M. V., & Arkhipov, V. I. (2021). Why do levels of anti-inflammatory cytokines increase during memory acquisition? *Neuroscience, 473*, 159–169.

Liu, D., Lv, F., Min, S., Yang, Y., & Chen, L. (2022). Inhibition of NLRP3 inflammasome-mediated neuroinflammation alleviates electroconvulsive shock-induced memory impairment via regulation of hippocampal synaptic plasticity in depressive rats. *Behavioural Brain Research, 428*, 113879.

Lolait, S. J., Lim, A. T., Toh, B. H., & Funder, J. W. (1984). Immunoreactive beta-endorphin in a subpopulation of mouse spleen macrophages. *The Journal of Clinical Investigation, 73*, 277–280.

Lorente de No, R. (1934). Studies on the structure of the cerebral cortex. II continuation of the study of ammonic system. *Journal für Psychologie und Neurologie, 46*, 113–177.

Louveau, A., Harris, T. H., & Kipnis, J. (2015). Revisiting the mechanisms of CNS immune privilege. *Trends in Immunology, 36*, 569–577.

Luo, G., Wang, X., Cui, Y., Cao, Y., Zhao, Z., & Zhang, J. (2021). Metabolic reprogramming mediates hippocampal microglial M1 polarization in response to surgical trauma causing perioperative neurocognitive disorders. *Journal of Neuroinflammation, 18*, 267.

Lynch, G., Larson, J., Kelso, S., Barrionuevo, G., & Schottler, F. (1983). Intracellular injections of EGTA block induction of hippocampal long-term potentiation. *Nature, 305*, 719–721.

Maggi, L., Trettel, F., Scianni, M., Bertollini, C., Eusebi, F., Fredholm, B. B., & Limatola, C. (2009). LTP impairment by fractalkine/CX3CL1 in mouse hippocampus is mediated through the activity of adenosine receptor type 3 (A3R). *Journal of Neuroimmunology, 215*, 36–42.

Malenka, R. C., & Nicoll, R. A. (1999). Long-term potentiation—A decade of progress? *Science, 285*, 1870–1874.

Manabe, T., Wyllie, D. J., Perkel, D. J., & Nicoll, R. A. (1993). Modulation of synaptic transmission and long-term potentiation: Effects on paired pulse facilitation and EPSC variance in the CA1 region of the hippocampus. *Journal of Neurophysiology, 70*, 1451–1459.

McNaughton, B. L., Douglas, R. M., & Goddard, G. V. (1978). Synaptic enhancement in fascia dentata: Cooperativity among coactive afferents. *Brain Research, 157*, 277–293.

Meibach, R. C., & Siegel, A. (1977a). Efferent connections of the hippocampal formation in the rat. *Brain Research, 124*, 197–224.

Meibach, R. C., & Siegel, A. (1977b). Efferent connections of the septal area in the rat: An analysis utilizing retrograde and anterograde transport methods. *Brain Research, 119*, 1–20.

Moore, R. Y., & Halaris, A. E. (1975). Hippocampal innervation by serotonin neurons of the midbrain raphe in the rat. *The Journal of Comparative Neurology, 164*, 171–183.

Morris, R. G., Garrud, P., Rawlins, J. N., & O'Keefe, J. (1982). Place navigation impaired in rats with hippocampal lesions. *Nature, 297*, 681–683.

Morris, R. G., Anderson, E., Lynch, G. S., & Baudry, M. (1986). Selective impairment of learning and blockade of long-term potentiation by an N-methyl-D-aspartate receptor antagonist, AP5. *Nature, 319*, 774–776.

Nance, D. M., Rayson, D., & Carr, R. I. (1987). The effects of lesions in the lateral septal and hippocampal areas on the humoral immune response of adult female rats. *Brain, Behavior, and Immunity, 1*, 292–305.

Nebeling, F. C., Poll, S., Justus, L. C., Steffen, J., Keppler, K., Mittag, M., & Fuhrmann, M. (2023). Microglial motility is modulated by neuronal activity and correlates with dendritic spine plasticity in the hippocampus of awake mice. *eLife, 12*.

Nguyen, K. T., Deak, T., Owens, S. M., Kohno, T., Fleshner, M., Watkins, L. R., & Maier, S. F. (1998). Exposure to acute stress induces brain interleukin-1beta protein in the rat. *The Journal of Neuroscience, 18*, 2239–2246.

O'Keefe, J., & Conway, D. H. (1978). Hippocampal place units in the freely moving rat: Why they fire where they fire. *Experimental Brain Research, 31*, 573–590.

O'Keefe, J., & Dostrovsky, J. (1971). The hippocampus as a spatial map. Preliminary evidence from unit activity in the freely-moving rat. *Brain Research, 34*, 171–175.

O'Keefe, J., & Nadel, L. (1978). *The hippocampua as a cognitive map*. Oxford University Press.

O'Connor, J. J., & Coogan, A. N. (1999). Actions of the pro-inflammatory cytokine IL-1 beta on central synaptic transmission. *Experimental Physiology, 84*, 601–614.

Otto, T., Eichenbaum, H., Wiener, S. I., & Wible, C. G. (1991). Learning-related patterns of CA1 spike trains parallel stimulation parameters optimal for inducing hippocampal long-term potentiation. *Hippocampus, 1*, 181–192.

Pan, Q., & Long, J. (1993). Lesions of the hippocampus enhance or depress humoral immunity in rats. *Neuroreport, 4*, 864–866.

Perkel, D. J., Petrozzino, J. J., Nicoll, R. A., & Connor, J. A. (1993). The role of Ca2+ entry via synaptically activated NMDA receptors in the induction of long-term potentiation. *Neuron, 11*, 817–823.

Perry, V. H., & Holmes, C. (2014). Microglial priming in neurodegenerative disease. *Nature Reviews. Neurology, 10*, 217–224.

Perry, V. H., Nicoll, J. A., & Holmes, C. (2010). Microglia in neurodegenerative disease. *Nature Reviews. Neurology, 6*, 193–201.

Raisman, G., Cowan, W. M., & Powell, T. P. (1966). An experimental analysis of the efferent projection of the hippocampus. *Brain, 89*, 83–108.

Ransohoff, R. M. (2009). Chemokines and chemokine receptors: Standing at the crossroads of immunobiology and neurobiology. *Immunity, 31*, 711–721.

Ransohoff, R. M., & Stevens, B. (2011). Neuroscience. How many cell types does it take to wire a brain? *Science, 333*, 1391–1392.

Regehr, W. G., & Tank, D. W. (1990). Postsynaptic NMDA receptor-mediated calcium accumulation in hippocampal CA1 pyramidal cell dendrites. *Nature, 345*, 807–810.

Rivest, S. (2003). Molecular insights on the cerebral innate immune system. *Brain, Behavior, and Immunity, 17*, 13–19.

Rogers, J. T., Morganti, J. M., Bachstetter, A. D., Hudson, C. E., Peters, M. M., Grimmig, B. A., Weeber, E. J., Bickford, P. C., & Gemma, C. (2011). CX3CR1 deficiency leads to impairment of hippocampal cognitive function and synaptic plasticity. *The Journal of Neuroscience, 31*, 16241–16250.

Room, P., Postema, F., & Korf, J. (1981). Divergent axon collaterals of rat locus coeruleus neurons: Demonstration by a fluorescent double labeling technique. *Brain Research, 221*, 219–230.

Sapolsky, R., Rivier, C., Yamamoto, G., Plotsky, P., & Vale, W. (1987). Interleukin-1 stimulates the secretion of hypothalamic corticotropin-releasing factor. *Science, 238*, 522–524.

Scoville, W. B., & Milner, B. (1957). Loss of recent memory after bilateral hippocampal lesions. *Journal of Neurology, Neurosurgery, and Psychiatry, 20*, 11–21.

Sheridan, G. K., Wdowicz, A., Pickering, M., Watters, O., Halley, P., O'Sullivan, N. C., Mooney, C., O'Connell, D. J., O'Connor, J. J., & Murphy, K. J. (2014). CX3CL1 is up-regulated in the rat hippocampus during memory-associated synaptic plasticity. *Frontiers in Cellular Neuroscience, 8*, 233.

Shishkina, G. T., Kalinina, T. S., Gulyaeva, N. V., Lanshakov, D. A., & Dygalo, N. N. (2021). Changes in gene expression and neuroinflammation in the hippocampus after focal brain ischemia: Involvement in the long-term cognitive and mental disorders. *Biochemistry (Mosc), 86*, 657–666.

Sikes, R. W., Chronister, R. B., & White, L. E., Jr. (1977). Origin of the direct hippocampus—anterior thalamic bundle in the rat: a combined horseradish peroxidase—Golgi analysis. *Experimental Neurology, 57*, 379–395.

Skaper, S. D., Facci, L., & Giusti, P. (2014). Neuroinflammation, microglia and mast cells in the pathophysiology of neurocognitive disorders: A review. *CNS & Neurological Disorders Drug Targets, 13*, 1654–1666.

Squire, L. R. (1992). Memory and the hippocampus: A synthesis from findings with rats, monkeys, and humans. *Psychological Review, 99*, 195–231.

Swanson, L. W. (1981). A direct projection from Ammon's horn to prefrontal cortex in the rat. *Brain Research, 217*, 150–154.

Swanson, L. W., & Cowan, W. M. (1975). Hippocampo-hypothalamic connections: Origin in subicular cortex, not ammon's horn. *Science, 189*, 303–304.

Swanson, L. W., Wyss, J. M., & Cowan, W. M. (1978). An autoradiographic study of the organization of intrahippocampal association pathways in the rat. *The Journal of Comparative Neurology, 181*, 681–715.

Swanson, L. W., Sawchenko, P. E., & Cowan, W. M. (1981). Evidence for collateral projections by neurons in Ammon's horn, the dentate gyrus, and the subiculum: A multiple retrograde labeling study in the rat. *The Journal of Neuroscience, 1*, 548–559.

Takao, T., Tracey, D. E., Mitchell, W. M., & De Souza, E. B. (1990). Interleukin-1 receptors in mouse brain: Characterization and neuronal localization. *Endocrinology, 127*, 3070–3078.

Tancredi, V., Zona, C., Velotti, F., Eusebi, F., & Santoni, A. (1990). Interleukin-2 suppresses established long-term potentiation and inhibits its induction in the rat hippocampus. *Brain Research, 525*, 149–151.

Tancredi, V., D'Arcangelo, G., Grassi, F., Tarroni, P., Palmieri, G., Santoni, A., & Eusebi, F. (1992). Tumor necrosis factor alters synaptic transmission in rat hippocampal slices. *Neuroscience Letters, 146*, 176–178.

Tulving, E. (2001). Episodic memory and common sense: How far apart? *Philosophical Transactions of the Royal Society of London. Series B, Biological Sciences, 356*, 1505–1515.

Tulving, E., & Markowitsch, H. J. (1998). Episodic and declarative memory: Role of the hippocampus. *Hippocampus, 8*, 198–204.

Walker, M. C., Ruiz, A., & Kullmann, D. M. (2002). Do mossy fibers release GABA? *Epilepsia, 43*(Suppl 5), 196–202.

Wrona, D. (2006). Neural-immune interactions: An integrative view of the bidirectional relationship between the brain and immune systems. *Journal of Neuroimmunology, 172*, 38–58.

Wyss, J. M., Swanson, L. W., & Cowan, W. M. (1979). Evidence for an input to the molecular layer and the stratum granulosum of the dentate gyrus from the supramammillary region of the hypothalamus. *Anatomy and Embryology (Berl), 156*, 165–176.

Xiong, H., Zeng, Y. C., Lewis, T., Zheng, J., Persidsky, Y., & Gendelman, H. E. (2000). HIV-1 infected mononuclear phagocyte secretory products affect neuronal physiology leading to cellular demise: Relevance for HIV-1-associated dementia. *Journal of Neurovirology, 6*(Suppl 1), S14–S23.

Yamashima, T., Tonchev, A. B., & Yukie, M. (2007). Adult hippocampal neurogenesis in rodents and primates: Endogenous, enhanced, and engrafted. *Reviews in the Neurosciences, 18*, 67–82.

Yang, S. N., Tang, Y. G., & Zucker, R. S. (1999). Selective induction of LTP and LTD by postsynaptic [Ca2+]i elevation. *Journal of Neurophysiology, 81*, 781–787.

Yirmiya, R., & Goshen, I. (2011). Immune modulation of learning, memory, neural plasticity and neurogenesis. *Brain, Behavior, and Immunity, 25*, 181–213.

Yuste, R., & Denk, W. (1995). Dendritic spines as basic functional units of neuronal integration. *Nature, 375*, 682–684.

Zhu, Y., Duan, S., Wang, M., Deng, Z., & Li, J. (2022). Neuroimmune interaction: A widespread mutual regulation and the weapons for barrier organs. *Frontiers in Cell and Development Biology, 10*, 906755.

20 The Vertebrate Retina

Wallace B. Thoreson

Abstract

The retina is an outpost of the central nervous system (CNS) with neuronal structures and proteins specialized for converting light into a neural code that the brain can interpret. Mutations of proteins involved in phototransduction, ion channels, and synaptic transmission in the retina lead to visual deficits ranging from subtle color vision defects to complete blindness. Although normally isolated from blood-mediated immune responses by the blood-retinal barrier, inflammatory responses, including gliosis and those mediated by the complement system, are important contributors to retinal degeneration, particularly age-related macular degeneration (ARMD). Studying immune responses in the retina and their pharmacological manipulation presents a promising avenue for treating and preventing retinal degeneration.

Keywords

Amacrine cell · Bipolar cell · Choroid · Cone · Fovea · Horizontal cell · Magnocellular ganglion cells · Melanopsin · mGluR6 · Müller cells · Opsin · Outer segment · Parvocellular ganglion cells · Photoreceptor · Retina · Retinal ganglion cell · Retinal pigment epithelium · Ribbon · Rod · Transducin

20.1 Introduction

Like other parts of the central nervous system (CNS), the retina derives embryologically from the neural tube. Retinal neurons and glia share many properties with other CNS cells but also exhibit specialized response properties and proteins that have evolved to serve the retina's key functions: converting light energy into nerve signals and analyzing the visual image. To encode small changes in light intensity, many retinal neurons respond to light with graded changes in membrane potential and do not exhibit sodium-dependent action potentials. In addition to proteins needed for phototransduction (e.g., rhodopsin and transducin), the retina uses a number of specialized proteins for the transmission and processing of visual information, including type 6 metabotropic glutamate receptors (mGluR6) in ON-type bipolar cells (Martemyanov & Sampath, 2017), type C γ-aminobutyric acid ($GABA_c$) receptors (Lukasiewicz et al., 2004), and CaV1.4 calcium channels in photoreceptors (Williams et al., 2022). This chapter summarizes some special features of the retina that help it transduce and process light.

20.2 Anatomy

The retina is a thin sheet of neural tissue (150–400 μm thick) at the back of the eye upon which the visual image is focused by the cornea and lens (Fig. 20.1a). A layer of darkly pigmented epithelial cells known as the retinal pigment epithelium (RPE, Fig. 20.1) lies directly behind the neurosensory retina. Beyond the RPE is a dense bed of capillaries known as the choroid. Surrounding the choroid is the sclera, composed of densely woven collagen fibers that help encase and protect the globe.

The retina has a highly organized laminar structure similar in all vertebrate species (Fig. 20.1b; Dowling, 2012; Rodieck, 1998). It is oriented so that photoreceptors lie at the back, adjacent to the RPE. Light must therefore pass through overlying retina layers to reach the photoreceptors (PR). The photoreceptor (PR) layer consists of the outer and inner segments of rod and cone photoreceptors. Separating the inner segments and photoreceptor cell bodies is the outer limiting membrane (OLM) formed from the apical processes of glial

W. B. Thoreson (✉)
Department of Ophthalmology and Visual Sciences, Truhlsen Eye Institute, University of Nebraska Medical Center, Omaha, NE, USA
e-mail: wbthores@unmc.edu

H. E. Gendelman, T. Ikezu (eds.), *Neuroimmune Pharmacology and Therapeutics*, https://doi.org/10.1007/978-3-031-68237-7_20

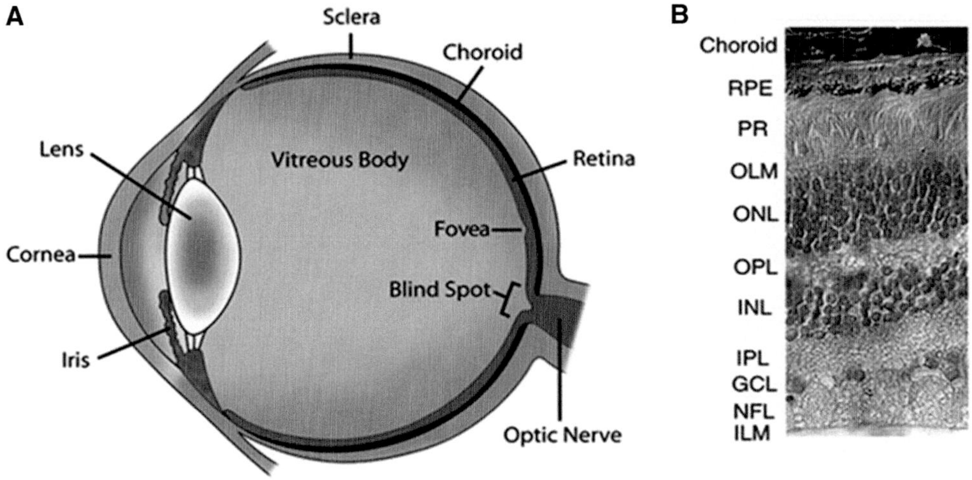

Fig. 20.1 (**a**) Schematic diagram of a primate eye. (**b**) Cross section showing the different layers in primate retina

Müller cells. The outer nuclear layer (ONL) contains the cell bodies of photoreceptors. Photoreceptors make synaptic contact with horizontal and bipolar cells in the outer plexiform layer (OPL). The inner nuclear layer (INL) contains cell bodies of horizontal, bipolar, and amacrine cells. Synaptic contacts among bipolar, amacrine, and retinal ganglion cells are made in the inner plexiform layer (IPL). Anterior to the IPL, closer to the front surface of the retina, is the ganglion cell layer (GCL), which contains cell bodies of retinal ganglion cells, along with cell bodies of some displaced amacrine cells. Axons from retinal ganglion cells create a nerve fiber layer (NFL) at the retina's inner surface. The axons bundle together as they exit the eye to form the optic nerve, which projects to higher visual centers. The absence of retina and photoreceptors at the optic nerve head creates a small blind spot (Fig. 20.1a). The vitreal surface of the retina is bounded by an inner limiting membrane (ILM) formed by Müller cell endfeet (Fig. 20.1b).

The human retina contains ~60 types of cells grouped into six major classes: photoreceptors (4 types), horizontal cells (2 types), bipolar cells (12 types), amacrine cells (25 types), ganglion cells (12 types), and nonneuronal cell types (Muller glia, astrocytes, microglia, and endothelial cells) (Yan et al., 2020b). Mouse retinas have nearly 140 distinct cell types, with 3 types of photoreceptors, 1 horizontal cell type, 15 bipolar cells, 63 amacrine cells, and 46 ganglion cells (Yan et al., 2020a). The different cell types vary considerably in their anatomy and physiology (Grunert & Martin, 2020; Van Hook et al., 2019). The larger number of amacrine and ganglion cells in the mouse retina compared to the human retina supports more extensive processing by retinal circuits in this species.

Photoreceptors release L-glutamate to stimulate bipolar cells, which in turn release L-glutamate onto ganglion cells. This forms a throughput pathway for transmitting light signals through the retina (Fig. 20.2). Horizontal and amacrine cells are predominantly inhibitory interneurons with processes extending laterally in the outer and inner plexiform layers, respectively. As light signals are transmitted from photoreceptors to bipolar cells to ganglion cells, they are modified by inhibitory synaptic feedback from horizontal cells onto photoreceptors and bipolar cells in the outer retina and by feedback from amacrine cells onto bipolar and ganglion cells in the inner retina. Photoreceptors and second-order retinal neurons (bipolar and horizontal cells) respond to light with graded changes in their membrane potential (Dowling, 2012). Fast, sodium-dependent action potentials are seen only in third-order neurons of the retina (amacrine and ganglion cells). The use of graded responses by cells early in the visual pathway is likely related to the fact that graded responses can contain more information than a spike code (Laughlin, 2001). However, ganglion cells must generate action potentials to propagate information over the distances needed to reach higher visual centers in the brain.

20.3 Cell Types

20.3.1 Rod and Cone Photoreceptor Cells

20.3.1.1 Outer Segments and Phototransduction

There are two major types of photoreceptors cells: rods and cones (Ebrey & Koutalos, 2001). Rods are specialized for transducing signals from dim lights. Cones are less sensitive than rods and provide high acuity and color vision at higher light levels. Rods and cones are named for their rod- and cone-shaped outer segments, respectively. The outer segments of both cell types contain the proteins necessary for phototransduction, packaged in disc-shaped organelles. Cone discs are formed from invaginations of the outer segment plasma membrane, whereas rod discs are completely sequestered within the outer segment.

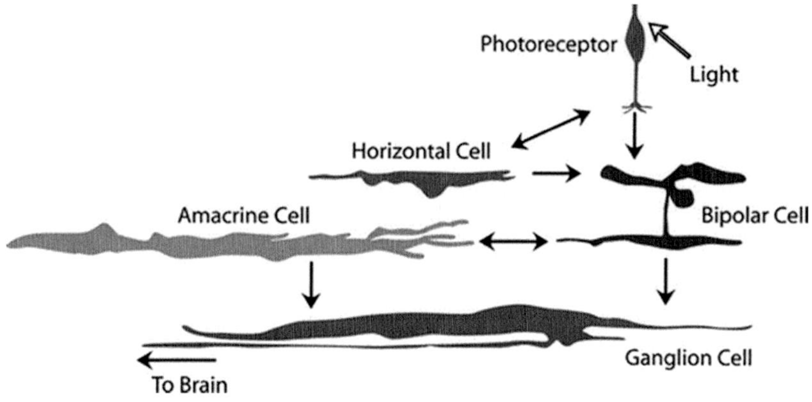

Fig. 20.2 The five major retinal cell types (photoreceptors, bipolar cells, horizontal cells, amacrine cells, and ganglion cells) and their synaptic connections

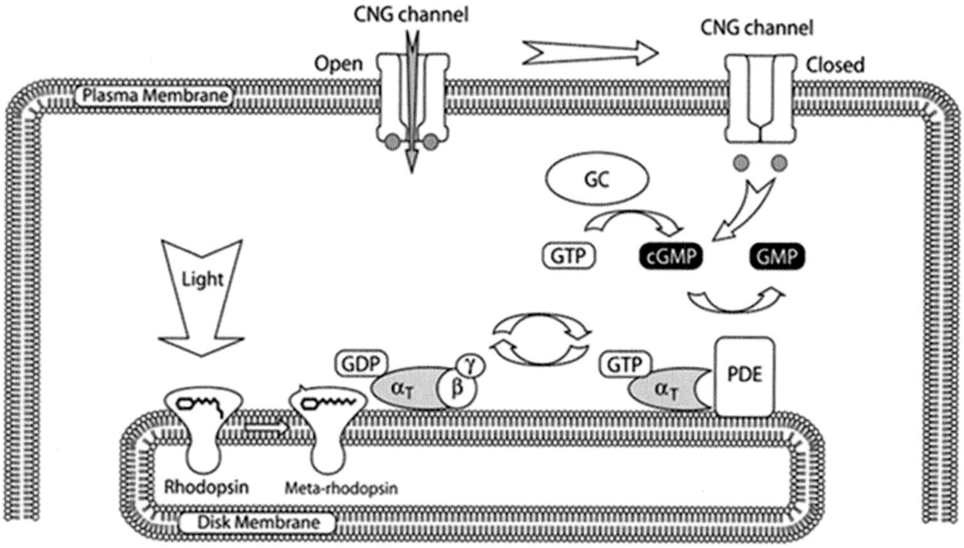

Fig. 20.3 The enzymatic cascade responsible for phototransduction

The principal job of photoreceptors is to transduce light into an electrical signal. Phototransduction involves a cascade of enzymes in the outer segment (Arshavsky & Burns, 2012) (Fig. 20.3). Phototransduction in rods is initiated by the capture of a photon by a G protein-coupled receptor (GPCR), rhodopsin. Cones use similar cone-specific opsins. Unlike other GPCRs that are activated by the binding of a neurotransmitter, the activating ligand for rhodopsin and cone opsins is a light-sensitive chromophore molecule, vitamin A aldehyde (or retinal), bound within a pocket of the opsin protein. Absorption of a photon by this chromophore initiates a conformational transition of 11-cis-retinal, which has a kink at the 11-cis carbon position, into all-trans-retinal, which has a straight chain configuration (Fig. 20.3). The conformational change in retinal produces conformational changes in the seven transmembrane domains of rhodopsin, causing it to assume an active configuration known as metarhodopsin. The activated GPCR, metarhodopsin, stimulates the exchange of guanosine triphosphate (GTP) for guanosine diphosphate (GDP) on the associated G protein, transducin (G_T). The activated alpha subunit of transducin (α_T) stimulates the activity of a cGMP-specific phosphodiesterase, which hydrolyzes cGMP. The plasma membrane of outer segments contains nonselective cation channels permeable to Na^+, K^+, and Ca^{2+} that are opened by the binding of cGMP to their intracellular face. Light-induced reduction in cytoplasmic [cGMP] causes cyclic nucleotide-gated cation (CNG) channels to close. Na^+ and Ca^{2+} influx through CNG channels causes photoreceptors to depolarize when open in the darkness. Conversely, light-induced closure of these channels causes photoreceptors to hyperpolarize. Because brighter lights cause a larger number of CNG channels to close, the amplitude of membrane hyperpolarization is graded with intensity.

The enzymatic transduction cascade initiated by photon absorption is terminated by two key mechanisms: (1) Light-activated metarhodopsin is inactivated in a process that begins with its phosphorylation by rhodopsin kinase. Following rhodopsin phosphorylation at three sites by rhodopsin kinase, the protein arrestin binds to rhodopsin and arrests further activity. The actions of rhodopsin kinase are analogous to the regulation of beta-adrenergic receptors by beta-adrenergic receptor kinase. (2) Light-activated transducin is also shut down by its intrinsic GTPase activity, which converts GTP into GDP and thereby de-activates the G protein. The intrinsic GTPase activity of transducin is accelerated by binding of the accessory proteins, RGS/Gβ5, to the transducin alpha subunit.

Tremendous amplification of the light signal by phototransduction in rods allows them to detect the absorption of a single photon (Field et al., 2005). To accomplish this feat, each activated metarhodopsin molecule catalyzes hundreds to thousands of transducin molecules, and although each transducin activates only a single phosphodiesterase (PDE) molecule, every PDE molecule hydrolyzes thousands of cGMP molecules. The net result is that activating a single rhodopsin molecule by a single photon of light causes the degradation of 10^5–10^6 cGMP molecules, causing many CNG channels to close and produce a small but detectable change in membrane potential.

Mutations in phototransduction proteins are a major cause of retinitis pigmentosa and other photoreceptor degenerations (Verbakel et al., 2018; Gill et al., 2019). Mutations in rhodopsin, phosphodiesterase, and CNG channels can all produce retinitis pigmentosa. Mutations in rod-specific transducin produce night blindness, whereas mutations in cone-specific transducin produce achromatopsia (rod monochromacy). Similarly, by preventing cones from responding to light, mutations in cone-specific CNG channels also lead to achromatopsia. For an updated list of genes involved in retinal diseases, consult the "Retinal Information Network" (https://web.sph.uth.edu/Retnet/).

20.3.1.2 Light Adaptation

Photoreceptors respond over a limited range of intensities before their responses are saturated. To maintain responsiveness over the entire range of intensities encountered in the natural world, the phototransduction apparatus adapts its sensitivity to increased light levels to maintain a constant relative response to increments in illumination ($\Delta I/I$ = constant response) (Fain et al., 2001; Burkhardt, 2001). This process of light adaptation is largely due to calcium-sensitive adjustments of the phototransduction enzyme cascade. As mentioned above, CNG channels are permeable to Ca^{2+}, allowing Ca^{2+} to enter the outer segments when these channels are open in the darkness. This steady Ca^{2+} influx is countered by the extrusion of Ca^{2+} from an outer segment by Na^+/Ca^{2+} exchangers. Closing CNG channels in response to light diminish the influx of Ca^{2+}, but its efflux via the Na^+/Ca^{2+} exchanger continues. The concentration of Ca^{2+} in the outer segment, therefore, diminishes in light. This decrease in $[Ca^{2+}]_i$ enhances the activity of guanylyl cyclase, stimulating cGMP production. The resulting increase in cGMP opens CNG channels leading to depolarization of the photoreceptor membrane potential. In cones, decreased $[Ca^{2+}]$ also increases the affinity of cGMP for CNG channels to promote their opening further. The membrane depolarization that results from the re-opening of CNG channels restores the operating range of the photoreceptor cell, allowing it to hyperpolarize in response to another flash of light.

While calcium-dependent adaptation is sufficient to maintain the sensitivity of cones over a large range of intensities, at extremely high intensities, there is also a contribution from photochemical adaptation in which a significant fraction of unbleached chromophore (11-cis-retinal) is bleached to all-trans-retinal (Burkhardt, 1994). At most light levels, there is a sufficient reservoir of 11-cis-retinal in the outer segments so that the bleaching of chromophore molecules does not appreciably limit the sensitivity of the opsin molecule as long as levels of 11-cis-retinal are soon restored. (As presented later, restoration of chromophore levels after bleaching involves the participation of RPE and Müller cells.) However, at extremely high light levels, the rate and amount of bleaching are sufficient to limit the availability of chromophore and thereby limit the sensitivity of opsin to light.

The combined effects of calcium-dependent and photochemical adaptation allow cones to maintain constant incremental sensitivity to light over 10^7-fold changes in intensity (Burkhardt, 1994). Rods, sensitive to dimmer lights, maintain relatively constant sensitivity over a 1000-fold change in intensity. Therefore, rod and cone systems allow the visual system to perform the impressive feat of maintaining relatively constant incremental sensitivity to light over a 10 billion-fold range of intensities. For comparison, pupillary constriction or dilation contributes a 16-fold change in sensitivity. However, it has the advantage of being relatively rapid, occurring within a couple of hundred ms compared to minutes for full light adaptation by photoreceptors.

20.3.1.3 Photoreceptor Inner Segment, Soma, and Synaptic Terminal

The inner segments of photoreceptors are packed with mitochondria to fuel their tremendous metabolic demands. Sodium ions that enter the outer segments through cGMP-gated cation channels in darkness are extruded by Na/K--ATPases in the inner segments. The continuous consumption of adenosine triphosphate (ATP) by these pumps makes photoreceptor cells among the most metabolically active in the body.

Below the inner segment is the soma and nucleus, which connects at its base to the axon and synaptic terminal. Photoreceptors release glutamate at ribbon synapses (Thoreson, 2019). Synaptic ribbons are specialized structures that regulate the sustained release of neurotransmitters. Ribbon synapses are also found in the terminals of retinal bipolar cells, as well as vestibular and cochlear hair cells (Moser et al., 2020). Synaptic ribbons received their name because of their planar structure in photoreceptor terminals.

The core of each ribbon is formed from tens of thousands of copies of the protein, Ribeye (Schmitz et al., 2000). This core is surrounded by various other proteins, including kinesin motor protein KIF3A, Rab3-interacting protein RIM, and piccolino. Eliminating KiF3A causes photoreceptors to degenerate, but this appears to be mostly due to disrupted ciliary transport (Jimeno et al., 2006). Beneath each ribbon is a trough-shaped structure known as the arciform density. The arciform density is a complex of proteins including bassoon, Munc13-2, Rim2, and ERC2/CAST. RIM protein mutations have been implicated in autosomal-dominant rod-cone dystrophy (Johnson et al., 2003). Ribbons help supply synaptic vesicles to release sites that line the base of each ribbon. Despite the presence of a motor protein, the movement of vesicles along the ribbon surface appears to be passive without the need for ATP-driven motors.

Glutamate release from synaptic terminals of photoreceptor and bipolar cells is regulated by calcium influx through L-type calcium channels (Pangrsic et al., 2018). The use of L-type channels at ribbon synapses contrasts with the reliance on N-, P-, and Q-type channels for neurotransmission at more conventional synapses used by spiking neurons. A retina-specific L-type channel, CaV1.4, is localized to rod terminals, and mutations to this channel can cause incomplete congenital stationary night blindness (Bech-Hansen et al., 1998; Zeitz et al., 2015).

20.3.1.4 The Fovea

In humans, the center of the visual image is focused on the fovea in the macula lutea, the region of highest acuity in the retina (Kolb et al., 2020). Primates are the only mammals with a fovea, although many birds and lizards also possess a fovea. Unlike the peripheral retina, where light must pass through overlying retinal layers to reach the photoreceptors, overlying neurons are displaced at the fovea to diminish light scattering. At the center of the resulting foveal pit, cone axons are the only structures that lie between the outer segments and vitreal surface. The presence of xanthophyll pigments in cone axons gives the macula lutea its characteristic yellow color. The fovea contains only cones. Visual acuity, the ability to resolve fine spatial details, is limited by the spacing between cones in the fovea and the density of cones parallels visual acuity in the retina. When perfectly corrected, acuity can be as high as 20/10 at the foveal center but falls off rapidly toward the retinal periphery. Although we are not typically conscious of these movements, the eye is constantly in motion, making small microsaccades that allow the high acuity fovea to scan various points and construct a high-resolution image at higher visual centers. Damage to the fovea (e.g., in macular degeneration) thus leads not only to a loss of central vision but also to a general loss of high acuity vision.

20.3.1.5 Cones and Cone Opsins

The retinas of Old World primates have three cone subtypes with different spectral sensitivities: short wavelength (S or blue-sensitive), middle wavelength (M or green-sensitive), and long-wavelength (L or red-sensitive) cones (Neitz & Neitz, 2011). Trichromacy evolved in Old World primates 40 million years ago with duplication of a single M/L ancestral pigment gene followed by a divergence into separate M and L pigments. Differences in spectral sensitivity among different cone types arise from the presence of different cone opsins. Cone opsins are 40% homologous to rhodopsin, S cone opsins are 40% homologous with M and L cone opsins, and M and L cone opsins are 97% homologous with one another (Nathans et al., 1986). The differences in spectral absorbance among different opsins result from differences in a few amino acid residues that alter the position of hydroxyl groups close to 11-cis-retinal. The 30 nm difference in spectral absorbance between primate M and L cones is primarily determined by only three amino acids (Asenjo et al., 1994).

Different photopigments are encoded by genes on different chromosomes (Nathans et al., 1986). The gene for rhodopsin is found on chromosome 3, the S cone pigment gene on chromosome 7, and M and L cone pigment genes on the X chromosome. M and L cone pigments are arranged in a tandem array on the X chromosome. Recombination between these genes on adjacent X chromosomes is the most frequent cause of color vision anomalies. For this reason, most color vision defects are X-linked recessive and involve changes to red-green color vision. Males with anomalous M or L cone pigments are most common (anomalous trichromats), with a frequency of ~6% among European males, while 2% are entirely lacking in one pigment or the other (dichromats).

20.3.2 Horizontal Cells

Horizontal cells are laterally arborizing interneurons in the outer retina that receive excitatory synaptic inputs from photoreceptors and make inhibitory synapses onto cones and bipolar cells. The main neurotransmitter released from horizontal cells is GABA. A light-evoked reduction of glutamate release from photoreceptors causes horizontal cells to hyperpolarize by reducing the activation of their α-amino-3-hydroxy-5-methyl-4-isoxazolepropionic acid (AMPA)-type glutamate receptors. Except for certain fish retinas, there appear to be no N-methyl-D-aspartate (NMDA) receptors in horizontal cells of the adult retina (Thoreson & Witkovsky, 1999). Rats and mice have only one type of horizontal cell, but most mammals have two types (Masland, 2012). Because both types hyperpolarize to light at all visible wavelengths, they are sometimes referred to as luminosity-type horizontal cells. Many non-mammalian vertebrates have additional chromaticity or color opponent horizontal cell types that depolarize to certain wavelengths and hyperpolarize to other wavelengths (Thoreson & Dacey, 2019).

Horizontal cells have large receptive fields due to extensive gap junction coupling among cells. Large receptive fields allow horizontal cells to measure illumination over a wide area. Inhibitory feedback from horizontal cells to cones and bipolar cells subtracts the mean luminance level measured over a wide area from signals transmitted to the inner retina about local luminance changes (Thoreson & Mangel, 2012). As discussed later, this negative feedback contributes to forming the center-surround arrangement of visual receptive fields, which is important for detecting edges as well as color vision.

20.3.3 Bipolar Cells

Bipolar cells transmit signals from photoreceptors to ganglion cells. They receive glutamatergic input from photoreceptors and inhibitory inputs—largely GABA and glycine—from amacrine cell contacts at their terminals in the inner retina. They may also receive inhibitory influences from horizontal cell contacts at their dendrites in the outer retina. Bipolar cells release L-glutamate at ribbon synapses that contact amacrine and ganglion cells in the inner plexiform layer. Depending on the species, there are 11–14 types of cone bipolar cells along with a single type of rod bipolar cell in mammalian retinas (Masland, 2012; Euler et al., 2014). Cone bipolar cells are grouped into two major physiological subtypes: ON cone bipolar cells that depolarize to light and OFF cone bipolar cells that hyperpolarize to light. Rod bipolar cells in the mammalian retina are all of the ON, depolarizing type. In lower vertebrates, ON and OFF bipolar cells receive a greater mixture of rod and cone inputs (Wu, 2010).

ON and OFF responses of bipolar cells result from the presence of different glutamate receptors in the two cell types. OFF bipolar cells possess AMPA and kainate receptors but not NMDA receptors (Thoreson & Witkovsky, 1999; Boccuni & Fairless, 2022). Mammalian bipolar cells largely use kainate receptors. Thus, like horizontal cells, the synapse from cones to OFF bipolar cells is sign-conserving, that is, light-evoked hyperpolarization of a cone reduces the depolarizing influence of AMPA/kainate receptors thereby causing the postsynaptic OFF bipolar cell to hyperpolarize.

ON bipolar cells do not possess ionotropic glutamate receptors and are activated by a metabotropic glutamate receptor, mGluR6 (Martemyanov & Sampath, 2017). mGluR6 is a G protein-coupled receptor that acts via the G protein, G_0, to close TRPM1 nonselective cation channels (Koike et al., 2010; Morgans et al., 2010). Light-evoked hyperpolarization of photoreceptors and the accompanying decline in glutamate release, therefore, causes TRPM1 channels in ON bipolar cells to open and depolarize the cell. By contrast with the sign-conserving synapse to OFF bipolar and horizontal cells, the synapse from photoreceptors to ON bipolar cells is, therefore, sign-inverting. Mutations in mGluR6, TRPM1, and other proteins associated with the mGluR6 receptor complex disrupt rod bipolar cell function and cause a complete form of congenital stationary night blindness (Zeitz et al., 2015). In addition to ON bipolar cells, TRPM1 ("melastatin") channels are expressed in melanocytes and their expression declines in metastatic melanoma. Circulating autoantibodies to TRPM1 can cause night blindness in melanoma-associated retinopathy by inhibiting rod bipolar cell function (Xiong et al., 2013).

ON and OFF bipolar cells excite ON- and OFF-type ganglion and amacrine cells, respectively. ON and OFF pathways remain segregated into the thalamus' lateral geniculate nucleus. While ON and OFF bipolar cells can respond equally well to positive and negative contrast steps (Burkhardt, 2001), the synaptic output of these cells is rectifying due to the voltage dependence of calcium channels that regulate release. Thus, the depolarizing responses of ON bipolar cells to light *increments* are transmitted more effectively than their responses to light decrements. Conversely, depolarizing responses of OFF bipolar cells to light *decrements* are transmitted more effectively than responses to light increments. Blocking ON bipolar cell responses by saturating mGluR6 with the selective agonist, L-2-amino-4-phosphonobutyric acid (L-AP4), produced an acute deficit in the perception of positive contrast (i.e., bright spots on a dark background) (Schiller et al., 1986). This supports the hypothesis that the ON pathway preferentially encodes information about positive contrast, and the OFF pathway preferentially encodes information about negative contrast.

Within the broad classification of ON and OFF bipolar cells, there are a half dozen cone ON cells and a half dozen cone OFF cells. Primate retinas possess a class of midget bipolar cells with tiny dendritic fields and receive input from very few cones. Near the fovea, single midget bipolar cells receive input from individual foveal cones, and each of these midget bipolar cells contacts a single small midget-type ganglion cell. This "private line" for individual cones maintains the highest possible resolution limited only by cone-to-cone spacing in the fovea. Other bipolar cells differ in their spectral sensitivity, spatial filtering, and kinetic properties, extracting different aspects of the visual image for transmission to ganglion cells.

20.3.4 Amacrine Cells

Amacrine cells are inhibitory interneurons with cell bodies in the INL (and occasionally in the GCL) and dendrites in the IPL. While most contain the inhibitory neurotransmitters GABA and/or glycine, almost every major neurotransmitter can be found in at least one type of amacrine cell. Amacrine cells are excited by glutamate released from bipolar cells. This glutamate acts principally on AMPA receptors, although many amacrine cells also possess NMDA receptors (Boccuni & Fairless, 2022). Amacrine cells can be classified by molecular criteria into 25 types in the human retina and 63 types in mice (Yan et al., 2020a, b). Physiological responses of amacrine cells include transient and sustained depolarization at light onset (ON cells), light offset (OFF cells), or both (ON/OFF cells). By contrast with the graded light responses of bipolar, horizontal and photoreceptor cells, many amacrine cells exhibit sodium-dependent action potentials making their responses more transient.

Many amacrine cells have relatively dedicated functions. Some examples include the following:

A. Large, radially symmetric starburst amacrine cells help create direction selectivity in directionally selective ganglion cells (Wei, 2018).
B. Dopaminergic amacrine cells with widespread dendritic arborizations increase dopamine release in response to increases in global illumination. Dopamine can diffuse through the retina to influence cells as far away as the RPE. The increased dopamine release by light modifies cell function to optimize retinal responses in daylight (Roy & Field, 2019).
C. AII amacrine cells transfer rod signals from rod bipolar cells cone bipolar cells and from there to ganglion cells (Grimes et al., 2018).

20.3.5 Ganglion Cells

Retinal ganglion cells are the output cells of the retina. Their axons course along the vitreal surface of the retina and bundle together to exit the eye as the optic nerve. Ganglion cells are excited by glutamate released from bipolar cells acting on both NMDA and non-NMDA glutamate receptors (Boccuni & Fairless, 2022).

Molecular criteria distinguish 46 types of ganglion cells in the mouse retina but only a dozen in the human retina (Yan et al., 2020a, b). In humans and other primates, the ganglion cell population is dominated by two types: magnocellular (M) cells and parvocellular (P) cells. These cell types project to the lateral geniculate nucleus (LGN) of the thalamus, whose cells project to the visual cortex. The larger number of ganglion cell types in the mouse retina appears to reflect the presence of more cell types with more specialized functions in reflex visual behavior, whereas more of the visual processing in primates occurs at higher levels in the LGN and cortex.

M cells receive their name from their large cell bodies and large dendritic arborizations resulting in large receptive fields (Rodieck, 1998). The large receptive fields of M cells limit their contribution to fine feature analysis. Instead, their output is primarily related to motion and other changes in illumination. M cells are classified anatomically as parasol ganglion cells.

P cells have small cell bodies with small dendritic arborizations resulting in small receptive fields.

As discussed further below, many P cells show spectral sensitivity, with some preferring red over green and others preferring green over red (Kolb, 2007; Wool et al., 2018). P cell thus contributes to fine feature analysis and color vision. P cells are classified anatomically as midget ganglion cells. As mentioned above, midget ganglion cells receive inputs from midget bipolar cells that can receive input from individual cones in the fovea.

M and P cells receive input from M and L cones and project to M and P cell layers in the LGN. Another population of ganglion cells, bistratified cells, receive selective input from blue-sensitive S cones and project predominantly to koniocellular interlaminar regions (Thoreson & Dacey, 2019).

A number of ganglion cells are intrinsically light sensitive and do not require photoreceptor inputs to respond to light (Berson et al., 2002; Fu et al., 2005; Lucas, 2013), although they nevertheless receive rod- and cone-driven synaptic inputs. The intrinsic light sensitivity of intrinsically photosensitive retinal ganglion cells is conferred by the presence of a photopigment, melanopsin. At least six types of ganglion cells (named M1-M6) have been shown to possess melanopsin. Melanopsin-driven, intrinsic light responses of these cells are slow and show minimal adaptation (Do, 2019). The mechanisms of melanopsin phototransduction vary among cell types. The most common mechanism involves a G protein-coupled receptor, G_q, acting via phospholipase C to open TRP channels, but there is also evidence for the regulation of other ion channels (Contreras et al., 2021). M1 ganglion cells were the first type of intrinsically photosensitive retinal ganglion cell to be identified and have the largest intrinsic light responses. M1 cells have large, sparse dendrites and project to areas regulating nonvisual responses to light, including the suprachiasmatic nucleus for circadian rhythm entrainment and the olivary pretectal nucleus for regulating pupil constriction. Melanopsin photosensitivity in other ganglion cells may contribute to image-forming vision (Schmidt et al., 2014; Estevez et al., 2012; Ecker et al., 2010).

20.4 Circuitry

In addition to converting photons into neural signals—that is, transducing the incoming light—the retina plays an important role in the initial process of analyzing visual information. In the following section, we consider the retinal circuitry employed to analyze edges, color, direction selectivity, and scotopic vision.

20.4.1 Edge Detection and Center-Surround Receptive Fields

The detection of contrast edges is enhanced by the center-surround arrangement of receptive fields in cones, bipolar cells, and ganglion cells (Kuffler, 1953; Baylor et al., 1971). How does this center-surround organization improve edge detection? Consider an ON-type ganglion cell (Fig. 20.4) in which the circular center of the receptive field is excited by light. Its center-surround arrangement is imparted by the presence of an annular inhibitory region flanking this central excitatory region (Fig. 20.4a). As illustrated by the diagram in Fig. 20.4b, illuminating the excitatory center with a small spot of light excites this cell but illuminating the surrounding inhibitory region with an annulus of light inhibits activity. Full-field illumination that stimulates both the excitatory center and surrounding inhibitory region thus generates weaker responses than a small centered spot. A bar or edge of light illuminating the entire excitatory center along with small portions of the inhibitory surround evokes a stronger response than full-field illumination but not as strong as a spot of light illuminating just the excitatory center. The net result of this center-surround receptive field arrangement is that cells respond most strongly to spots and more strongly to bars and edges than to full-field illumination. These same considerations can be extended to OFF-type cells that are excited by light decrements in the center and inhibited by light decrements in the surround. Another useful way to think about center-surround inhibition is that inhibitory feedback from the broad receptive fields of horizontal and amacrine cells subtracts the average surrounding luminance level from local changes in luminance. Machine vision uses the mathematical equivalent of a center-surround receptive field by spatially differentiating the image to enhance the salience of edges.

20.4.2 Color

Color vision is initiated in the retina by comparing the responses of cones with different spectral sensitivities (Thoreson & Dacey, 2019). The existence of three types of cones in the primate retina (L, M, and S cones) is the physiological basis for trichromatic vision (R, G, B). Responses of an individual cone are determined solely by the number of photons absorbed, irrespective of their wavelength. Thus, two different color lights can produce an identical response in a cone as long as one adjusts the intensities of the two lights to produce equal photon capture. Because a cone only reports how many photons it has captured, not the spectral nature of those photons, color discrimination requires the retina to compare the responses of spectrally different types of cones. Synaptic comparisons between different types of cones can produce responses of opposite polarity to different wavelengths (i.e., color opponent responses). Color opponency can arise at an individual cone synapse by inhibitory feedback from a horizontal cell that receives synaptic input from cones with different spectral sensitivity. Color opponency is also evident in some horizontal cells of non-mammalian vertebrates (Twig et al., 2003) but absent from horizontal cells of mammalian retinas (Thoreson & Dacey, 2019).

Color opponent ganglion cells in Old World primates can show red/green or blue/yellow opponent responses.

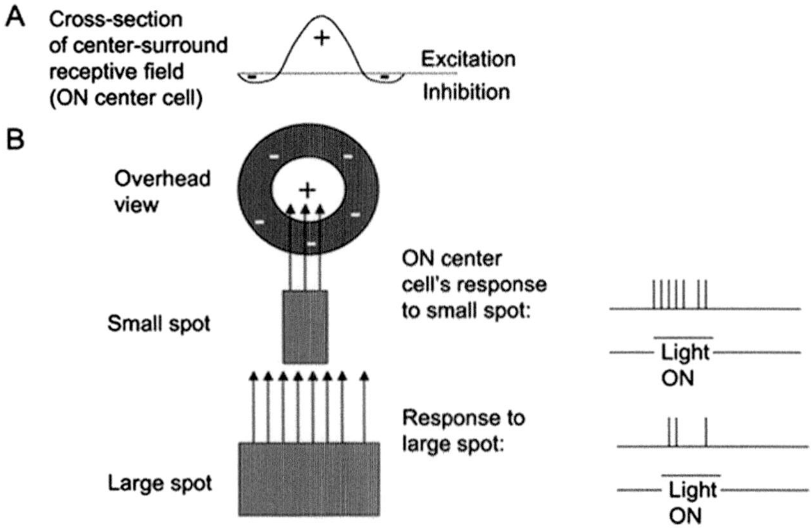

Fig. 20.4 Center-surround receptive fields in the retina enhance responses to small spots, edges, and bars of light. (**a**) Cross section of the center/surround receptive field of an ON center cell. Light falling on the center of the receptive field excites the cell, whereas light falling only in the periphery produces inhibition. (**b**) Overhead view of the center/surround arrangement of a receptive field of an ON center ganglion cell illustrating differences in the trains of action potentials evoked by a small spot vs. full-field illumination

Red/green opponency in primate P-type ganglion cells arises from midget bipolar cells that receive selective excitatory inputs from individual red or green cones along with inhibitory feedback from horizontal cells that receive a mixture of both red and green cone inputs yielding a spectrally distinct surround (Wool et al., 2018; Dacey & Packer, 2003). S cones transmit their responses to ganglion cells via specialized S cone bipolar cells, providing inputs to blue ON- and OFF-type ganglion cells. Responses to yellow light in the receptive field surrounding these cells are generated from the summed inputs of red and green cones onto horizontal cells. The presence of color opponency in retinal neurons contributes to the perception of colored after-images. Staring at a colored background can preferentially desensitize the responses of one cone population and thereby enhance the relative strength of the opponent cone population. Thus, the relative increase in green cone input strength produced by desensitizing red cones with a field of red generates a green percept when the red light is extinguished.

20.4.3 Direction Selectivity

Direction selective ganglion cells (DSGCs) are excited by stimuli moving along one axis of the receptive field but show little response to stimuli moving in the opposite direction. There are two types of DSGCs: ON DSGCs and ON-OFF DSGCs.

ON DSGCs cluster along three preferred axes of motion that correspond to the three axes of the semicircular canals (Wei & Feller, 2011; Borst & Euler, 2011). ON DSGCs project to the accessory optic system in the midbrain, where they drive optokinetic reflex eye movements that combine with vestibuloocular reflexes to correct for mismatches between head and eye motion ("retinal slip") that blur the image on the retina.

ON DSGCs receive synaptic inputs from bipolar cells and symmetrically radiating starburst amacrine cells (Vaney et al., 2012). Direction selectivity in ganglion cells arises from multiple mechanisms, including asymmetric inhibition from starburst amacrine cells involving differences in the number of GABAergic synapses in different parts of the cell (Briggman et al., 2011) and asymmetric excitation from bipolar cells (Wei, 2018).

ON-OFF DSGCs cluster along four axes matched to the major axes of the superior, inferior, medial, and lateral rectus extraocular muscles. These cells are important for the smooth pursuit eye movements that track moving objects. While direction selective cells were previously thought to contribute only to the control of eye reflexes, there is growing evidence that they also contribute to central motion processing (Rasmussen & Yonehara, 2020).

20.4.4 Rod Pathways

Scotopic or low-light vision is mediated by rod photoreceptors. The mammalian retina has a duplex organization in which rods communicate with ganglion cells using a largely separate circuit from cones. The primary rod circuit begins with rods communicating with ON-type rod bipolar cells. Unlike cone bipolar cells, rod bipolar cells do not contact ganglion cells directly but instead contact AII amacrine cells. AII amacrine cells form gap junctions with cone ON bipolar cell terminals as well as inhibitory, sign-inverting glycinergic synapses with cone OFF bipolar cells. Output from AII amacrine cells, therefore, feeds rod signals into both ON and OFF cone bipolar cells that in turn drive both ON and OFF ganglion cells. AII amacrine cells provide the primary pathway for rod signals at low light levels, but at mesopic light levels, there is also a contribution from direct rod inputs into OFF bipolar cells and the transmission of rod signals through gap junctions into neighboring cones (Grimes et al., 2018).

There is tremendous synaptic convergence in the rod pathway, with as many as 75,000 rods converging onto a single ganglion cell. By contrast, cones show much less convergence, with a 1:1 connection from cones to midget bipolar cells to midget ganglion cells in the foveal center. Limited convergence in the cone pathway is important for the high visual acuity mediated by cone circuits. On the other hand, the large convergence of signals in the rod pathway facilitates the detection of dim flashes of light. The ability of individual rods to respond to a single photon of light combined with convergence allows us to perceive the absorption of as few as a dozen photons within 100 ms (Field et al., 2005).

20.5 Retinal Pigment Epithelial Cells

The RPE is a monocellular epithelium of hexagonal cells whose name reflects the fact that it contains melanin pigment granules and forms the outermost layer of the retina. The apical processes of RPE cells ensheathe outer segments of rods and cones, with each RPE cell contacting 20–30 rods (Yang et al., 2021; Strauss, 2005; Marmor & Wolfensberger, 1998).

Melanin granules are concentrated in the apical processes and cytoplasm of RPE cells and are nearly absent from basal cytoplasm. By absorbing stray photons of light that have passed through outer photoreceptor segments, melanin granules improve the optical isolation of individual photoreceptors.

Neighboring RPE cells are connected by tight junctions that help to create the blood-retinal barrier separating the neurosensory retina from the fenestrated capillaries in the choroid (O'Leary & Campbell, 2021). The basement membrane of RPE, together with the adjacent basement membrane of the choroid, forms a structure known as Bruch's membrane. RPE cells possess a number of organic and ion transporters to move polar molecules across the blood-retinal barrier, including transporters for amino acids, folate, ascorbic acid, myo-inositol, organic anions, glucose, and lactate.

Photoreceptors continually synthesize new phototransduction proteins for incorporation into newly formed outer segment discs. As new discs are formed at the base of the outer segment, older discs are phagocytosed by RPE cells at the tip of the outer segment. Each outer segment disc is formed and shed in ~2 weeks. The phagocytosis of discs occurs in circadian bursts (rods at the end of the night, cones at the end of the day).

The bleached, all-trans form of the chromophore retinal shows little or no ability to convert into photosensitive 11-cis-retinal within rod photoreceptor cells. Instead, conversion to the light-sensitive 11-cis form in rods requires the participation of the RPE. As we discuss later, cones can use glial Müller cells for regenerating 11-cis-retinal (Lamb & Pugh, 2004; Wang & Kefalov, 2011). Following the conversion of 11-cis-retinal to all-trans-retinal by light, all-trans-retinal is rapidly converted to all-trans-retinol (all-trans-vitamin A) in the outer photoreceptor segment. All-trans-retinol is then transported out of the photoreceptor cell, through the interstitial space, and into adjacent RPE or Müller cells via a process involving interstitial retinoid-binding proteins (IRBP). Upon entering RPE cells, retinol is bound to cellular retinol-binding protein. All-trans-retinol is converted to 11-cis-retinol by retinol isomerase (RPE65), and 11-cis-retinol is converted back to 11-cis-retinal by 11-cis-retinol dehydrogenase. 11-cis-retinal is transferred to the cellular retinal binding protein (CRalBP) for transport to the RPE cell surface. Finally, regenerated 11-cis-retinal is transported back to photoreceptors via IRBP. By depriving photoreceptors of the chromophore, mutations of visual cycle proteins (e.g., RPE65, 11-cis-retinol dehydrogenase, CRalBP, and IRBP) can lead to photoreceptor degeneration. Restoring RPE65 function with gene therapy has been successful in treating certain forms of Leber's congenital amaurosis (Jacobson et al., 2012).

The interstitial space between RPE and photoreceptors contains a sticky interphotoreceptor matrix consisting of glycoproteins, proteoglycans, and hyaluronic acid that helps the retina adhere to the back of the eye. Retinal adhesion is also promoted by the extrusion of water from the RPE to the choroid. Basolateral Cl channels and apical Na/K/2Cl transporters are particularly important for water transport out of the RPE. Because photoreceptors are not physically bound to the RPE, the retina can detach from the RPE with a strong blow to the eye, fluid build-up behind the retina (rhegmatogenous detachment), or traction from overlying cells in proliferative vitreoretinopathy. Detachment of photoreceptor cells from the adjacent RPE prevents recycling of the photopigment, which blocks phototransduction by depriving opsin of the sufficient chromophore.

One by-product of photoisomerization in photoreceptor outer segments is N-retinylidene-N-retinylethanolamine (A2E). A2E is ingested by RPE cells when they phagocytose outer segments. However, A2E cannot be enzymatically degraded by RPE cells and thus accumulates in these cells, becoming a major component of lipofuscin granules (Sparrow & Boulton, 2005). Stargardt's macular degeneration involves a defect in the ABCR transporter that increases the accumulation of A2E, suggesting it may play a role in macular degeneration. The damage to the RPE associated with A2E accumulation may result from the stimulation of apoptosis in A2E-laden RPE cells produced by exposure to blue light.

A major risk factor for age-related macular degeneration (ARMD) is the presence of drusen deposits (Velez-Montoya et al., 2014). White drusen spots visible with an ophthalmoscope are formed by deposits between the RPE and Bruch's membrane. Drusen formation involves inflammatory reactions, and RPE cells overlying drusen often show signs of impending cell death. Studies on genetic factors contributing to ARMD have identified mutations in immune complement-related genes as important risk factors (Edwards et al., 2005; Haines et al., 2005; Klein et al., 2005; Gold et al., 2006). This has led to the suggestion that mutations that promote over-active inflammatory responses may contribute to drusen formation leading to RPE cell damage and ARMD (Wong et al., 2022). Transplantation of healthy RPE cells has shown promising initial results in treating macular degeneration (Van Gelder et al., 2022).

20.6 Glia

The predominant retinal glial cell is the Müller cell. Müller cells are radial glia that span the retina from the OLM to ILM and ensheathe virtually every cell in the retina. They play a number of important homeostatic roles in the retina (Bringmann et al., 2006). For example, Müller cells are a primary storage depot for glycogen in the retina that can provide metabolites (e.g., lactic acid) to neurons during metabolic stress. Müller cells also remove and redistribute metabolic waste products. They exhibit high levels of glutathione that help protect the retina from oxidative stress. Müller cells also possess Na/HCO_3 co-transport mechanisms and carbonic anhydrase to stabilize pH levels in the retina.

Another important role of Müller cells is the spatial redistribution of K^+ from regions of high concentration to regions of low concentration to maintain a stable, extracellular K^+ concentration of about 3 mM (Kofuji & Newman, 2004). When neuronal depolarization causes K^+ levels to increase in the IPL and OPL, excess K^+ ions enter Müller cells processes. A simultaneous efflux accompanies K^+ influx in the plexiform layers through K^+ channels clustered at the vitreal surface and alongside blood vessels. Conversely, a reduction in K^+ accompanying neuronal hyperpolarization is accompanied by an efflux of K^+ out of Müller cells into the plexiform layers. The current flowing through radially oriented Müller cells as a result of this spatial buffering of K^+ produces measurable trans-retinal potentials, such as slow PIII of the electroretinogram.

Müller cells are major sites for the uptake and removal of neurotransmitters, most notably glutamate and GABA. Glutamate uptake into Müller cells is thought to be the principal mechanism responsible for removing extracellular glutamate after synaptic release (Pow, 2001). In addition to neurotransmitter transporters, Müller cells possess neurotransmitter receptors and can release neuroactive substances (e.g., ATP) (Newman, 2004). Thus, the activity of retinal neurons can influence Müller cells, and Müller cell activity can, in turn, influence adjacent neurons.

As described in the previous section, RPE cells are required for regenerating 11-cis-retinal from all-trans-retinal derived from rods. However, regeneration of the bleached photopigment from cones is largely performed by Müller cells (Wang & Kefalov, 2011; Morshedian et al., 2019). All-trans-retinol is transported into Müller cells, where it is converted into the 11-cis isomer and then transported back into cones, where it is oxidized to 11-cis-retinal. Regeneration of cone pigment in Müller cells is generally similar to regeneration in RPE but uses a different isomerase (isomerase II in Müller cells vs. RPE65 in RPE cells), and Müller cells export the alcohol, 11-cis-retinol, rather than the aldehyde, 11-cis-retinal.

In addition to Müller cells, the retina contains three other types of glial cells: astrocytes, oligodendrocytes, and microglia. Retinal astrocytes are located primarily in the nerve fiber layer, and oligodendrocytes form the myelin sheath of axons in the optic nerve. Hematopoetically derived microglia are small stellate cells that, when quiescent, associate with inner retinal blood vessels.

Infection or damage to the retina stimulates Müller cell gliosis and migration of microglia to the injured area to assist in phagocytosing debris from dying cells. Microglia and Müller cells release cytokines in response to injury, and pro-inflammatory cytokines can sometimes exacerbate cell damage during retinal disease (Karlen et al., 2020; Chen et al., 2022). Cytokines released by Müller cells include vascular endothelial growth factor and transforming growth factor beta, which promotes neovascularization. In some species, Müller cells respond to injury and cytokines by dedifferentiating into progenitor cells that can give rise to neurons, suggesting possible strategies for repairing retina and treating retinal degeneration (Eastlake et al., 2021; Fischer & Bongini, 2010).

20.7 Blood Supply

Blood is supplied to the retina by the central retinal artery and choroidal blood vessels (Oyster, 1999). The central retinal artery arises from the ophthalmic artery, which in turn branches off the internal carotid artery. Upon entering the retina, the central retinal artery branches into deep capillary beds in the INL and superficial capillary beds in the ganglion cell layer. Endothelial cells of retinal capillaries are joined by tight junctions, contributing to the blood-retinal barrier. There is little or no autonomic regulation of the retinal circulation; blood flow through these capillaries is controlled primarily by autoregulation (Wangsa-Wirawan & Linsenmeier, 2003). Retinal capillaries drain into the central retinal vein.

Choroidal vessels derive from posterior and short posterior ciliary arteries that branch off from the ophthalmic artery like the central retinal artery. The choroidal circulation forms a dense bed of fenestrated capillaries, known as the choriocapillaris, just beneath the basolateral surface of the RPE. The flow rate through the choriocapillaris is among the highest in the body, and the arterio-venous drop in PO_2 is minimal. This high flow rate supplies the energetically demanding photoreceptors with large amounts of oxygen and maintains the retina at a constant temperature despite changing levels of radiant energy focused onto the back of the eye. Oxygen for photoreceptors comes primarily from the choroid. The high level of oxygen consumption by photoreceptor cells in darkness produces a PO_2 near zero at the level of the inner segments in the mammalian retina (Linsenmeier, 1986). Under these conditions, there is little or no oxygen available to the remainder of the retina from the choroid. Other retinal neurons instead receive their oxygen from the retinal circulation. By contrast with mammals, many cold-blood vertebrates lack retinal capillaries and rely on the choroid to supply the entire retina with oxygen and other nutrition. While capable of autoregulation (Kiel & Shepherd, 1992), blood flow in the choroicapillaris is largely regulated by autonomic inputs. Choroidal capillaries drain into four vortex veins, one in each quadrant of the eye.

20.8 Review Questions

1. Which of the following is the predominant glial cell type in the retina?
   (a) Astrocyte
   (b) *Müller cell*
   (c) Microglia
   (d) Schwann cell
   (e) RPE cell

2. In which layer of the retina do bipolar cell terminals contact ganglion cell dendrites?
   (a) Outer nuclear layer
   (b) Outer plexiform layer
   (c) Inner nuclear layer
   (d) *Inner plexiform layer*
   (e) Ganglion cell layer
3. Which of the following statements is true about the fovea?
   (a) The fovea contains both rods and cones.
   (b) The fovea is the region of the retina where ganglion cell axons exit the eye.
   (c) *The fovea is the region of the retina responsible for the highest acuity vision in humans and primates.*
   (d) The location of the fovea on the retina is not a fixed anatomical feature but varies with focus.
   (e) At the center of the fovea, light must first pass through ganglion, amacrine, horizontal and bipolar cells before reaching outer photoreceptor segments.
4. Upon which chromosome are genes for M cone pigments located?
   (a) *X chromosome*
   (b) Y chromosome
   (c) Chromosome 3
   (d) Chromosome 7
   (e) Chromosome 10
5. Which cell types are responsible for converting all-trans vitamin A into the 11-cis form needed for phototransduction?
   (a) Rods and cones
   (b) *Müller cells and RPE cells*
   (c) Choroid
   (d) Horizontal cells and bipolar cells
   (e) Astrocytes and microglia

References

Arshavsky, V. Y., & Burns, M. E. (2012). Photoreceptor signaling: Supporting vision across a wide range of light intensities. *The Journal of Biological Chemistry, 287*, 1620–1626.

Asenjo, A. B., Rim, J., & Oprian, D. D. (1994). Molecular determinants of human red/green color discrimination. *Neuron, 12*, 1131–1138.

Baylor, D. A., Fuortes, M. G., & O'Bryan, P. M. (1971). Receptive fields of cones in the retina of the turtle. *The Journal of Physiology, 214*, 265–294.

Bech-Hansen, N. T., Naylor, M. J., Maybaum, T. A., Pearce, W. G., Koop, B., Fishman, G. A., Mets, M., Musarella, M. A., & Boycott, K. M. (1998). Loss-of-function mutations in a calcium-channel alpha1-subunit gene in Xp11.23 cause incomplete X-linked congenital stationary night blindness. *Nature Genetics, 19*, 264–267.

Berson, D. M., Dunn, F. A., & Takao, M. (2002). Phototransduction by retinal ganglion cells that set the circadian clock. *Science, 295*, 1070–1073.

Boccuni, I., & Fairless, R. (2022). Retinal glutamate neurotransmission: From physiology to pathophysiological mechanisms of retinal ganglion cell degeneration. *Life, 12*, 638.

Borst, A., & Euler, T. (2011). Seeing things in motion: Models, circuits, and mechanisms. *Neuron, 71*, 9749–9794.

Briggman, K. L., Helmstaedter, M., & Denk, W. (2011). Wiring specificity in the direction-selectivity circuit of the retina. *Nature, 471*, 183–188.

Bringmann, A., Pannicke, T., Grosche, J., Francke, M., Wiedemann, P., Skatchkov, S. N., Osborne, N. N., & Reichenbach, A. (2006). Müller cells in the healthy and diseased retina. *Progress in Retinal and Eye Research, 25*, 397–424.

Burkhardt, D. A. (1994). Light adaptation and photopigment bleaching in cone photoreceptors in situ in the retina of the turtle. *The Journal of Neuroscience, 14*, 1091–1105.

Burkhardt, D. A. (2001). Light adaptation and contrast in the outer retina. *Progress in Brain Research, 131*, 407–418.

Chen, Y., Xia, Q., Zeng, Y., Zhang, Y., & Zhang, M. (2022). Regulations of retinal inflammation: Focusing on Müller Glia. *Frontiers in Cell and Development Biology, 10*, 898652.

Contreras, E., Nobleman, A. P., Robinson, P. R., & Schmidt, T. M. (2021). Melanopsin phototransduction: Beyond canonical cascades. *The Journal of Experimental Biology, 224*, 226522.

Dacey, D. M., & Packer, O. S. (2003). Colour coding in the primate retina: Diverse cell types and cone-specific circuitry. *Current Opinion in Neurobiology, 13*, 421–427.

Do, M. T. H. (2019). Melanopsin and the intrinsically photosensitive retinal ganglion cells: Biophysics to behavior. *Neuron, 104*, 205–226.

Dowling, J. E. (2012). *The retina: An approachable part of the brain.* Harvard University Press.

Eastlake, K., Lamb, W. D. B., Luis, J., Khaw, P. T., Jayaram, H., & Limb, G. A. (2021). Prospects for the application of Müller glia and their derivatives in retinal regenerative therapies. *Progress in Retinal and Eye Research, 85*, 100970.

Ebrey, T., & Koutalos, Y. (2001). Vertebrate photoreceptors. *Progress in Retinal and Eye Research, 20*, 49–94.

Ecker, J. L., Dumitrescu, O. N., Wong, K. Y., Alam, N. M., Chen, S. K., LeGates, T., Renna, J. M., Prusky, G. T., Berson, D. M., & Hattar, S. (2010). Melanopsin-expressing retinal ganglion-cell photoreceptors: Cellular diversity and role in pattern vision. *Neuron, 67*, 49–60.

Edwards, A. O., Ritter, R., 3rd, Abel, K. J., Manning, A., Panhuysen, C., & Farrer, L. A. (2005). Complement factor H polymorphism and age-related macular degeneration. *Science, 308*, 421–424.

Estevez, M. E., Fogerson, P. M., Ilardi, M. C., Borghuis, B. G., Chan, E., Weng, S., Auferkorte, O. N., Demb, J. B., & Berson, D. M. (2012). Form and function of the M4 cell, an intrinsically photosensitive retinal ganglion cell type contributing to geniculocortical vision. *The Journal of Neuroscience, 32*, 13608–13620.

Euler, T., Haverkamp, S., Schubert, T., & Baden, T. (2014). Retinal bipolar cells: Elementary building blocks of vision. *Nature Reviews Neuroscience, 15*, 507–519.

Fain, G. L., Matthews, H. R., Cornwall, M. C., & Koutalos, Y. (2001). Adaptation in vertebrate photoreceptors. *Physiological Reviews, 81*, 117–151.

Field, G. D., Sampath, A. P., & Rieke, F. (2005). Retinal processing near absolute threshold: From behavior to mechanism. *Annual Review of Physiology, 67*, 491–514.

Fischer, A. J., & Bongini, R. (2010). Turning Müller glia into neural progenitors in the retina. *Molecular Neurobiology, 42*, 199–209.

Fu, Y., Liao, H. W., Do, M. T., & Yau, K. W. (2005). Non-image-forming ocular photoreception in vertebrates. *Current Opinion in Neurobiology, 15*, 415–422.

Gill, J. S., Georgiou, M., Kalitzeos, A., Moore, A. T., & Michaelides, M. (2019). Progressive cone and cone-rod dystrophies: Clinical features, molecular genetics and prospects for therapy. *The British Journal of Ophthalmology, 103*, 711–720.

Gold, B., Merriam, J. E., Zernant, J., Hancox, L. S., Taiber, A. J., Gehrs, K., Cramer, K., Neel, J., Bergeron, J., Barile, G. R., Smith, R. T., AMD Genetics Clinical Study Group, Hageman, G. S., Dean, M., & Allikmets, R. (2006). Variation in factor B (BF) and complement component 2 (C2) genes is associated with age-related macular degeneration. *Nature Genetics, 38*, 458–462.

Grimes, W. N., Songco-Aguas, A., & Rieke, F. (2018). Parallel processing of rod and cone signals: Retinal function and human perception. *Annual Review of Vision Science, 4*, 123–141.

Grünert, U., & Martin, P. R. (2020). Cell types and cell circuits in human and non-human primate retina. *Progress in Retinal and Eye Research, 5*, 100844.

Haines, J. L., Hauser, M. A., Schmidt, S., Scott, W. K., Olson, L. M., Gallins, P., Spencer, K. L., Kwan, S. Y., Noureddine, M., Gilbert, J. R., Schnetz-Boutaud, N., Agarwal, A., Postel, E. A., & Pericak-Vance, M. A. (2005). Complement factor H variant increases the risk of age-related macular degeneration. *Science, 308*, 419–421.

Jacobson, S. G., Cideciyan, A. V., Ratnakaram, R., Heon, E., Schwartz, S. B., Roman, A. J., Peden, M. C., Aleman, T. S., Boye, S. L., Sumaroka, A., Conlon, T. J., Calcedo, R., Pang, J. J., Erger, K. E., Olivares, M. B., Mullins, C. L., Swider, M., Kaushal, S., Feuer, W. J., Iannaccone, A., Fishman, G. A., Stone, E. M., Byrne, B. J., & Hauswirth, W. W. (2012). Gene therapy for leber congenital amaurosis caused by RPE65 mutations: Safety and efficacy in 15 children and adults followed up to 3 years. *Archives of Ophthalmology, 130*, 9–24.

Jimeno, D., Lillo, C., Roberts, E. A., Goldstein, L. S., & Williams, D. S. (2006). Kinesin-2 and photoreceptor cell death: Requirement of motor subunits. *Experimental Eye Research, 82*, 351–353.

Johnson, S., Halford, S., Morris, A. G., Patel, R. J., Wilkie, S. E., Hardcastle, A. J., Moore, A. T., Zhang, K., & Hunt, D. M. (2003). Genomic organisation and alternative splicing of human RIM1, a gene implicated in autosomal dominant cone-rod dystrophy (CORD7). *Genomics, 81*, 304–314.

Karlen, S. J., Miller, E. B., & Burns, M. E. (2020). Microglia activation and inflammation during the death of mammalian photoreceptors. *Annual Review of Vision Science, 6*, 149–169.

Kiel, J. W., & Shepherd, A. P. (1992). Autoregulation of choroidal blood flow in the rabbit. *Investigative Ophthalmology & Visual Science, 33*, 2399–2410.

Klein, R. J., Zeiss, C., Chew, E. Y., Tsai, J. Y., Sackler, R. S., Haynes, C., Henning, A. K., SanGiovanni, J. P., Mane, S. M., Mayne, S. T., Bracken, M. B., Ferris, F. L., Ott, J., Barnstable, C., & Hoh, J. (2005). Complement factor H polymorphism in age-related macular degeneration. *Science, 308*, 385–389.

Kofuji, P., & Newman, E. A. (2004). Potassium buffering in the central nervous system. *Neuroscience, 129*, 1045–1056.

Koike, C., Numata, T., Ueda, H., Mori, Y., & Furukawa, T. (2010). TRPM1: A vertebrate TRP channel responsible for retinal ON bipolar function. *Cell Calcium, 48*, 95–101.

Kolb, H. (2007). Midget pathways of the primate retina underlie resolution and red green color opponency. In H. Kolb, E. Fernandez, & R. Nelson (Eds.), *Webvision: The Organization of the Retina and Visual System [Internet]*. University of Utah Health Sciences Center. http://webvision.med.utah.edu

Kolb, H., Nelson, R. F., Ahnelt, P. K., Ortuño-Lizarán, I., & Cuenca, N. (2020). The architecture of the human fovea. In H. Kolb, E. Fernandez, & R. Nelson (Eds.), *Webvision: The Organization of the Retina and Visual System [internet]*. University of Utah Health Sciences Center. http://webvision.med.utah.edu

Kuffler, S. W. (1953). Discharge patterns and functional organization of mammalian retina. *Journal of Neurophysiology, 16*, 37–68.

Lamb, T. D., & Pugh, E. N., Jr. (2004). Dark adaptation and the retinoid cycle of vision. *Progress in Retinal and Eye Research, 23*, 307–380.

Laughlin, S. B. (2001). Efficiency and complexity in neural coding. *Novartis Foundation Symposium, 239*, 177–192.

Linsenmeier, R. A. (1986). Effects of light and darkness on oxygen distribution and consumption in the cat retina. *The Journal of General Physiology, 88*, 521–542.

Lucas, R. J. (2013). Mammalian inner retinal photoreception. *Current Biology, 23*, R125–R133.

Lukasiewicz, P. D., Eggers, E. D., Sagdullaev, B. T., & McCall, M. A. (2004). GABAC receptor-mediated inhibition in the retina. *Vision Research, 44*, 3289–3296.

Marmor, M. F., & Wolfensberger, T. J. (1998). *The retinal pigment epithelium: Function and disease*. Oxford University Press.

Martemyanov, K. A., & Sampath, A. P. (2017). The transduction cascade in retinal on-bipolar cells: Signal processing and disease. *Annual Review of Vision Science, 3*, 25–51.

Masland, R. H. (2012). The neuronal organization of the retina. *Neuron, 76*(2), 266–80.

Morgans, C. W., Brown, R. L., & Duvoisin, R. M. (2010). TRPM1: The endpoint of the mGluR6 signal transduction cascade in retinal ON-bipolar cells. *BioEssays, 32*, 609–614.

Morshedian, A., Kaylor, J. J., Ng, S. Y., Tsan, A., Frederiksen, R., Xu, T., Yuan, L., Sampath, A. P., Radu, R. A., Fain, G. L., & Travis, G. H. (2019). Light-driven regeneration of cone visual pigments through a mechanism involving RGR opsin in Müller glial cells. *Neuron, 102*, 1172–1183.

Moser, T., Grabner, C. P., & Schmitz, F. (2020). Sensory processing at ribbon synapses in the retina and the cochlea. *Physiological Reviews, 100*, 103–144.

Nathans, J., Thomas, D., & Hogness, D. S. (1986). Molecular genetics of human color vision: The genes encoding blue, green, and red pigments. *Science, 232*, 193–202.

Neitz, J., & Neitz, M. (2011). The genetics of normal and defective color vision. *Vision Research, 51*, 633–651.

Newman, E. A. (2004). Glial modulation of synaptic transmission in the retina. *Glia, 47*, 268–274.

O'Leary, F., & Campbell, M. (2021). The blood-retina barrier in health and disease. *The FEBS Journal, 290*, 878. https://doi.org/10.1111/febs.16330

Oyster, C. W. (1999). *The human eye: Structure and function*. Sinauer Associates.

Pangrsic, T., Singer, J. H., & Koschak, A. (2018). Voltage-gated calcium channels: Key players in sensory coding in the retina and the inner ear. *Physiological Reviews, 98*, 2063–2096.

Pow, D. V. (2001). Amino acids and their transporters in the retina. *Neurochemistry International, 38*, 463–484.

Rasmussen, R., & Yonehara, K. (2020). Contributions of retinal direction selectivity to central visual processing. *Current Biology, 30*, R897–R903.

Rodieck, R. W. (1998). *The first steps in seeing*. Sinauer Associates.

Roy, S., & Field, G. D. (2019). Dopaminergic modulation of retinal processing from starlight to sunlight. *Journal of Pharmacological Sciences, 140*, 86–93.

Schiller, P. H., Sandell, J. H., & Maunsell, J. H. (1986). Functions of the ON and OFF channels of the visual system. *Nature, 322*, 824–825.

Schmidt, T. M., Alam, N. M., Chen, S., Kofuji, P., Li, W., Prusky, G. T., & Hattar, S. (2014). A role for melanopsin in alpha retinal ganglion cells and contrast detection. *Neuron, 82*, 781–788.

Schmitz, F., Königstorfer, A., & Südhof, T. C. (2000). RIBEYE, a component of synaptic ribbons: A protein's journey through evolution provides insight into synaptic ribbon function. *Neuron, 28*, 857–872.

Sparrow, J. R., & Boulton, M. (2005). RPE lipofuscin and its role in retinal pathobiology. *Experimental Eye Research, 80*, 595–606.

Strauss, O. (2005). The retinal pigment epithelium in visual function. *Physiological Reviews, 85*, 845–881.

Thoreson, W. B. (2019). Transmission at rod and cone ribbon synapses in the retina. *Pflügers Archiv, 473*, 1469–1491.

Thoreson, W. B., & Dacey, D. M. (2019). Diverse cell types, circuits, and mechanisms for color vision in the vertebrate retina. *Physiological Reviews, 99*, 1527–1573.

Thoreson, W. B., & Mangel, S. C. (2012). Lateral interactions in the outer retina. *Progress in Retinal and Eye Research, 31*, 407–441.

Thoreson, W. B., & Witkovsky, P. (1999). Glutamate receptors and circuits in the vertebrate retina. *Progress in Retinal and Eye Research, 18*, 765–810.

Twig, G., Levy, H., & Perlman, I. (2003). Color opponency in horizontal cells of the vertebrate retina. *Progress in Retinal and Eye Research, 22*, 31–68.

Van Gelder, R. N., Chiang, M. F., Dyer, M. A., Greenwell, T. N., Levin, L. A., Wong, R. O., & Svendsen, C. N. (2022). Regenerative and restorative medicine for eye disease. *Nature Medicine, 28*, 1149–1156.

Van Hook, M. J., Nawy, S., & Thoreson, W. B. (2019). Voltage- and calcium-gated ion channels of neurons in the vertebrate retina. *Progress in Retinal and Eye Research, 72*, 100760.

Vaney, D. I., Sivyer, B., & Taylor, W. R. (2012). Direction selectivity in the retina: Symmetry and asymmetry in structure and function. *Nature Reviews Neuroscience, 13*, 194–208.

Velez-Montoya, R., Oliver, S. C., Olson, J. L., Fine, S. L., Quiroz-Mercado, H., & Mandava, N. (2014). Current knowledge and trends in age-related macular degeneration: Genetics, epidemiology, and prevention. *Retina, 34*, 423–441.

Verbakel, S. K., van Huet, R. A. C., Boon, C. J. F., den Hollander, A. I., Collin, R. W. J., Klaver, C. C. W., Hoyng, C. B., Roepman, R., & Klevering, B. J. (2018). Non-syndromic retinitis pigmentosa. *Progress in Retinal and Eye Research, 66*, 157–186.

Wang, J. S., & Kefalov, V. J. (2011). The cone-specific visual cycle. *Progress in Retinal and Eye Research, 30*, 115–128.

Wangsa-Wirawan, N. D., & Linsenmeier, R. A. (2003). Retinal oxygen: Fundamental and clinical aspects. *Archives of Ophthalmology, 121*, 547–557.

Wei, W. (2018). Neural mechanisms of motion processing in the mammalian retina. *Annual Review of Vision Science, 4*, 165–192.

Wei, W., & Feller, M. B. (2011). Organization and development of direction-selective circuits in the retina. *Trends in Neurosciences, 34*, 638–645.

Williams, B., Maddox, J. W., & Lee, A. (2022). Calcium channels in retinal function and disease. *Annual Review of Vision Science, 8*, 53–77.

Wong, J. H. C., Ma, J. Y. W., Jobling, A. I., Brandli, A., Greferath, U., Fletcher, E. L., & Vessey, K. A. (2022). Exploring the pathogenesis of age-related macular degeneration: A review of the interplay between retinal pigment epithelium dysfunction and the innate immune system. *Frontiers in Neuroscience, 16*, 1009599.

Wool, L. E., Crook, J. D., Troy, J. B., Packer, O. S., Zaidi, Q., Dacey, D. M. (2018). Nonselective Wiring Accounts for Red-Green Opponency in Midget Ganglion Cells of the Primate Retina. *Journal of Neuroscience, 38*(6), 1520–1540.

Wu, S. M. (2010). Synaptic organization of the vertebrate retina: General principles and species-specific variations: The Friedenwald lecture. *Investigative Ophthalmology & Visual Science. 51*, 1263–1274.

Xiong, W. H., Duvoisin, R. M., Adamus, G., Jeffrey, B. G., Gellman, C., & Morgans, C. W. (2013). Serum TRPM1 autoantibodies from melanoma associated retinopathy patients enter retinal on-bipolar cells and attenuate the electroretinogram in mice. *PLoS One, 8*, e69506.

Yan, W., Laboulaye, M. A., Tran, N. M., Whitney, I. E., Benhar, I., & Sanes, J. R. (2020a). Mouse retinal cell atlas: Molecular identification of over sixty amacrine cell types. *The Journal of Neuroscience, 40*, 5177–5195.

Yan, W., Peng, Y. R., van Zyl, T., Regev, A., Shekhar, K., Juric, D., & Sanes, J. R. (2020b). Cell atlas of the human fovea and peripheral retina. *Scientific Reports, 10*, 9802.

Yang, S., Zhou, J., & Li, D. (2021). Functions and diseases of the retinal pigment epithelium. *Frontiers in Pharmacology, 12*, 727870.

Zeitz, C., Robson, A. G., & Audo, I. (2015). Congenital stationary night blindness: An analysis and update of genotype-phenotype correlations and pathogenic mechanisms. *Progress in Retinal and Eye Research, 45*, 58–110.

Part II

Immunology of Neurodegenerative, Neuroinflammatory, Neuroinfectious and Neuropsychiatric Disorders

Neurodegeneration

21

Kimberly Kwei and Serge Przedborski

Abstract

Neurodegeneration is a broad category of neurological illness that implies an underlying pathological process of primary neuronal destruction in the central nervous system, which furthermore does not seem to have an obvious external cause and is progressive. Because of their prevalence, this chapter focuses on Alzheimer's disease and Parkinson's disease. The epidemiology, classification, and clinical course of this group of illnesses are covered here. A discussion examining the potential pathogenesis of these illnesses and their related illnesses based on genetic, molecular, and pathological preclinical and clinical studies is also discussed.

Keywords

Alzheimer's disease · Amyotrophic lateral sclerosis · Asynchronous death · Basal ganglia · Cell autonomy · Cerebral cortex · Creutzfeldt-Jakob disease · Comorbidity · Dystonia · Etiology · Friedreich ataxia · Huntington's disease · Leucine-rich repeat kinase-2 (LRRK2) · Multiple sclerosis · Multisystem atrophy · Neurodegeneration · Neuropathology · Noncell autonomy · Olivopontocerebellar atrophy · Pathogenesis · Parkinson's disease · PD-ALS-dementia complex of Guam · Prion diseases · Progressive supranuclear palsy · Schizophrenia · Shy-Drager syndrome · Superoxide dismutase-1 · Striatonigral degeneration · Synucleinopathies · Tauopathies · Tourette's syndrome

21.1 Introduction

An examination of the literature indicates that since 1982, the term *neurodegeneration* appears in the title of over 9000 indexed publications. Neurodegeneration is a mentioned topic in virtually all neurological science textbooks. Thus, it is safe to assume that the meaning of the word *neurodegeneration* is universally understood. Most relevant textbooks however will not actually define neurodegeneration but will discuss the issue in bits and pieces as part of the discussion of various diseases of the nervous system rather than as a single chapter. As discussed previously (Przedborski et al., 2001, 2003) and in previous iterations of this chapter, the term neurodegeneration refers to a devolution or a process of decline involving a portion of the nervous system. Thus, neurodegeneration is any pathological condition in which the nervous system or nerve cell (i.e., neuron) loses its function, structure, or both. Neurodegeneration as a medical entity is an umbrella term for a large group of disorders of the nervous system that are heterogeneous both in terms of neuronal type and anatomical and functional territories affected. Neurodegenerative diseases frequently arise idiopathically (i.e., spontaneously and without cause) and progress in a relentless manner.

Notably, not all disorders of the nervous system are neurodegenerative, but rather they can be sorted into three broad categories: (i) pathologies that are restricted to the nervous system and which are *primary neuronal* diseases (referred henceforth as neurodegenerative diseases); (ii) pathologies that are restricted to the nervous system but are not *primary neuronal* diseases, such as brain neoplasm or cerebral edema and hemorrhage; and (iii) pathologies provoked by systemic causes that damage the nervous system, such cardiovascular arrest, carbon monoxide poison, or infections due to herpes simplex. Based on this simple categorization, hundreds of disorders of the nervous system, including Alzheimer's disease (AD), Parkinson's disease (PD), Huntington's disease (HD), and amyotrophic lateral sclerosis (ALS), clearly fulfill the criteria of neurodegenerative disorder and are unanimously regarded as such. Aside from these unambiguous neurodegenerative diseases, others such as torsion dystonia, Tourette's syndrome, or schizophrenia represent an interesting nosological challenge, as they do not show any distinct neuronal loss. Perhaps they have been traditionally included

K. Kwei (✉) · S. Przedborski
Neurological Institute, New York, NY, USA
e-mail: ktk2124@cumc.columbia.edu

H. E. Gendelman, T. Ikezu (eds.), *Neuroimmune Pharmacology and Therapeutics*,
https://doi.org/10.1007/978-3-031-68237-7_21

in this category because they are chronic diseases of the nervous system with an unknown cause. Finally, diseases such as multiple sclerosis (MS) also represent a perplexing situation. Conventionally, MS has been linked to pathology of the myelin that ensheath neuronal axons and not of the neuron per se. Several experts, however, have argued that MS is not only a demyelinating disease but also a disease where the neurons die due to the destruction of their axons. It is worth noting that long-term disability in MS has a higher correlation with axonal damage than with the degree of demyelination (Bjartmar et al., 2003). For these reasons, several scientific authorities (Trapp & Nave, 2008), are now wondering whether inflammatory demyelination in MS might not be secondary in the disease process and whether MS would be, in fact, a true neurodegenerative disease as defined above.

21.2 Frequency, Lifespan, and Comorbidity

The two most prevalent neurodegenerative diseases are AD and PD, and according to a report from the World Health Organization published about a decade ago, AD (together with other forms of dementias) and PD made up more than 14% of the global burden of neurological illnesses (Organization, 2006). Scientists estimate that up to 4.5 million Americans suffer from AD and 1.2 million from PD. This represents roughly 2.5% of the entire population in the United States. This estimate is likely valid for most, if not all, countries of the world as epidemiological studies have found roughly comparable incidence and prevalence rates of AD and PD around the globe.

Nevertheless, some geographic and temporal clusters of neurodegenerative conditions have been described. Two such examples include the marked increase in cases of parkinsonism in connection to the influenza pandemic of 1918 (Casals et al., 1998) and the high incidence of PD–ALS–dementia complex confined to the Chamorros Indians who live on the Western Pacific Island of Guam (Chen & Chase, 1986). These clusters while quite enlightening must be regarded as exceptions to the rule. The most consistent risk factor for developing a neurodegenerative disorder, especially in regards to AD and PD, is increasing age (Alzheimer's Association, 2016; Tanner, 1992; Simon et al., 2020; Scheltens et al., 2021), and this fact may have far-reaching implications for the generations to come. As previously noted (Przedborski et al., 2003), over the past century, the growth rate of the population ages 65 and above in industrialized countries has far exceeded that of the population as a whole. Therefore, it can be anticipated that, over the next generations, the proportion of elderly citizens will double. Consequently, unless effective preventive strategies are soon found, the number of persons suffering from a neurodegenerative disorder will rise dramatically and the epidemiological numbers provided above may have to be amended to higher percentages.

It is also important to emphasize the fact that nearly all neurodegenerative disorders shorten the life expectancy of affected patients, even in those settings where symptom-alleviating medications exist. This is well illustrated in the case of PD, in which symptoms worsen over time. As discussed elsewhere (Dauer & Przedborski, 2003), before the introduction of the potent symptomatic treatment levodopa, the mortality rate among PD patients was three times that of the normal age-matched subjects. Unexpectedly, population-based surveys suggest that PD patients continue to display a decreased longevity compared to the general population despite the fact that motor problems can be controlled by levodopa (Hely et al., 1989; Levy et al., 2002; Morgante et al., 2000). Yet, only a few neurodegenerative disorders are in and of themselves fatal. ALS is the hallmark example of a disease in which the affected neurological structures are implicated in controlling or driving vital physiological functions such as respiration or circulation. In ALS, the progressive loss of lower motor neurons innervating the intercostal and diaphragm muscles leads to fatal respiratory paralysis. In Friedreich ataxia, the association of neurodegeneration with heart disease (Harding, 1981) can also be a cause of death, although, in this case, death is not due to any neurodegenerative event but instead to cardio degenerative problems such as congestive heart failure. In most other neurodegenerative disorders, death is attributed neither to the damage of the nervous system nor to any associated extra-nervous system degeneration but rather to medical comorbidities that are a sequelae of the neurodegenerative process, such as increasing fall risk, aspiration pneumonia, pressure skin ulcers, malnutrition, and dehydration, which occur in relation to immobility, impaired balance, and cognitive decline. Given that the leading causes of death in much of the world include cancers and cardiovascular problems, it is important to note that comorbid illnesses such as atrial fibrillation and cancer double the chances of death of demented patients in nursing homes (Van Dijk et al., 1996; Vogelgsang et al., 2018). Cardiovascular problems such as high blood pressure have also been suggested to worsen dementing processes (Doraiswamy et al., 2002). Thus, medical comorbidities appear to aggravate the long-term prognosis of patients inflicted with a neurodegenerative disease. However, whether or not patients suffering from a neurodegenerative disease are at greater (or lower) risk for cancer, stroke, or heart attack, remains to be demonstrated (Potashkin et al., 2020).

21.3 Classification

The estimated number of different neurodegenerative diseases is a few hundred and, among these, many appear to overlap with one another clinically and pathologically. Classification of neurodegenerative diseases, while useful clinically, continues to remain an unfinished task while we do not fully understand for the most part why these illnesses develop. Most neurodegenerative diseases affect multiple brain areas. Yet, the degree with which these different brain areas are damaged often vary from one case to another, thus giving rise to different phenotypes. For example, in a disease such as multisystem atrophy where typically parkinsonism is a prominent feature, it may be accompanied with either severe ataxia, autonomic failure, or both depending on whether, in addition to the basal ganglia, the cerebellum, and the intermediolateral column of the thoracic spinal cord also degenerate (Wenning et al., 2004). Even in the case of a defined genetic defect, a similar mutation may also produce varied phenotypes. For instance, in familial parkinsonism linked to specific mutations in the leucine-rich repeat kinase-2 (LRRK2) gene, there is a pleomorphism in both the pattern of neuropathology that develops as well as clinical phenotypes, with some members of the affected family developing a motor neuron disease superimposed on dementia and/or parkinsonism (Zimprich et al., 2004). Despite these difficulties, categorization of neurodegenerative disorders by either the main clinical feature or the location of the predominant lesion, or often a combination of the two, seems to be the most succinct way to organize this disparate group of illnesses. This classification has been discussed previously (Przedborski et al., 2003), proposing that neurodegenerative disorders may be grouped into diseases by anatomical location within the central nervous system: cerebral cortex, basal ganglia, brainstem/cerebellum, and spinal cord. Within each anatomical group, diseases are further sub-grouped based on their main clinical features.

Based on this clinico-anatomical classification, diseases of the cerebral cortex may be divided into cortical diseases associated with dementia (e.g., AD) and without dementia. In theory, circumscribed degeneration of the cerebral cortex can occur in the absence of dementia as reported in some cases of progressive primary aphasia (Kirshner et al., 1987). However, because the cerebral cortex is so heavily involved in cognition, the progressive nature of neurodegenerative illness leads to spread beyond the initial locus of pathology, and as a result, the majority of these patients end up with dementia (Le Rhun et al., 2005). It is thus not surprising to find that neurodegenerative diseases of the cerebral cortex are usually equated to dementia. While AD is the most common illness in cortical neurodegenerative processes characterized clinically by dementia (Sulkava et al., 1983), about 50 other and less prominent dementing cortical diseases can be found (Tomlinson, 1977).

Diseases of the basal ganglia are essentially characterized by abnormal motor activity. Based on the type of the motor problem, diseases of the basal ganglia can be classified into hypokinetic or hyperkinetic groups. Hypokinetic basal ganglia disorders include PD, in which the amplitude and velocity of voluntary movements are diminished or, in extreme cases, nonexistent. Aside from PD, parkinsonism—which refers to a syndrome of at least two of the following clinical signs: resting tremor, slowness of movements, stiffness, and postural instability—is also found in a variety of other diseases of the basal ganglia (Dauer & Przedborski, 2003). In some, there is only parkinsonism (e.g., striatonigral degeneration), but in others, often called Parkinson-plus syndromes, there is parkinsonism plus signs of cerebellar ataxia (e.g., olivopontocerebellar atrophy), orthostatic hypotension (e.g., Shy-Drager syndrome), or paralysis of vertical eye movements (e.g., progressive supranuclear palsy). Because early on, parkinsonism may be the only clinical expression of Parkinson-plus syndromes, it is difficult to formulate a definitive and accurate diagnosis before the patient reaches a more advanced stage of the disease. This problem is brought into stark relief by the fact that more than 77% of patients with parkinsonism are diagnosed in life as having PD (Ballard et al., 2006; Irwin et al., 2012; Stacy & Jankovic, 1992; Walker et al., 2019), but as much as a quarter of the diagnosed patients are found at autopsy to have lesions incompatible with PD (Hughes et al., 1992). Of note, it is customary to classify the stiff-person syndrome among the hypokinetic diseases. However, available evidence would argue that the stiff-person syndrome while being a hypokinetic condition is an autoimmune disease involving the nervous system and not a neurodegenerative disease.

Found at the other end of the spectrum are the hyperkinetic basal ganglia disorders, such as HD and essential tremor. In these two conditions, excessive abnormal movements such as chorea or tremors are superimposed onto and interfere with normal voluntary movements. Although hyperkinetic basal ganglia disorders are probably as diverse as hypokinetic basal ganglia disorders, because there are more specific disease markers such as gene mutations for several of the hyperkinetic syndromes, classification of this group is less problematic.

All neurodegenerative diseases of the cerebellum and its connections are clinically associated with ataxia, and thus the main criteria of classification will rely not on the clinical presentation but on the loci of pathology. However, because of the breadth of cerebellar connections in the brain, a meaningful and functional classification of cerebellar neurodegenerative disease is a particularly challenging task. Some diseases of the cerebellum and its connections can easily be grouped into three main neuropathological types: cerebellar cortical atrophy (lesion confined to the Purkinje cells and the inferior olives); pontocerebellar atrophy (lesion affecting several cerebellar and brainstem structures); and Friedreich ataxia (lesion affecting the cerebellum and to a much greater extend the posterior column of the spinal cord, peripheral nerves, and the heart). However, several other diseases of the cerebellum and its connections fall somewhere in between these three well-delineated categories, such as dentorubral degeneration, in which the most conspicuous lesions are in the dentate and red nuclei, and Machado-Joseph disease, in which degeneration involves the lower and upper motor neurons, substantia nigra, and the dentate system (Koeppen, 2018; Samson & Claassen, 2017).

The main neurodegenerative disorders predominantly affecting the spinal cord can be split into those that affect the anterior and posterior portions of the spinal cord (Kanning et al., 2010). In ALS and spinal muscular atrophy, the most severe lesions are found in the anterior part of the spinal cord. In the already discussed Friedreich ataxia, the most severe lesions are found in the posterior part of the spinal cord.

Finally, there is one group of neurological diseases that are often, but not always, considered neurodegenerative because of their chronic and progressive course. However, they have an unknown etiopathogenesis as there are no apparent structural abnormalities on brain imaging or post-mortem pathology. These include torsion dystonia, Tourette syndrome, and schizophrenia. Notably, we are defining structural abnormalities in our discussion of neurodegeneration as a loss of neuronal cell bodies, but it is unknown whether or not the structural damage restricted to synaptic connections that is known to occur in torsion dystonia or schizophrenia could account for the disease phenotype (Balint et al., 2018; Bobilev et al., 2020). Moreover, in all these singular neurodegenerative disorders, various brain imaging studies and electrophysiological investigations have revealed significant functional abnormalities. Thus, could it be that some neurological diseases labeled as neurodegenerative disorders such as torsion dystonia may be caused by a loss of function of specific neural networks with minimal or no loss of neuronal viability?

Collectively, the above discussion was aimed at stressing some of the obtrusive shortcomings of the current classification of neurodegenerative diseases. To circumvent these problems, researchers have tried over the recent years to reconstitute and complete the clinical/pathological-based classifications by using molecular information obtained from various neurodegenerative diseases. Based on this novel approach, some neurodegenerative diseases, which used to belong to very distinct categories, are now brought together because of a common molecular alteration. For example, HD, spinal-cerebellar atrophy, and myotonic dystrophy now fall under the category of *trinucleotide repeat disorders* (Cummings & Zoghbi, 2000); Creutzfeldt-Jakob disease, Gerstmann–Straussler–Scheinker syndrome, and fetal familial insomnia fall under the category of Prion diseases (Prusiner, 1998); PD, progressive supranuclear palsy, and diffuse Lewy body dementia fall under the category of *synucleinopathies* (Galvin et al., 2001); and AD, corticobasal degeneration, frontotemporal dementia with parkinsonism linked to chromosome-17, and Pick's disease fall under the category of *tauopathies* (Goedert & Spillantini, 2001). Although the jury is still out on whether or not this new type of classification will alleviate the problems previously encountered, we strongly believe that a recasting of the conventional neuropathological-based classification to a molecular-based classification holds the promise of a more practical and less ambiguous organization for neurodegenerative illnesses from a clinical and therapeutic point-of-view.

21.4 Etiology of Neurodegenerative Diseases

The word *etiology* refers to the initiating factor of a disease. In some cases, a given clinical entity is linked to a single etiological factor such as in HD. Interestingly, however, several others can be caused by various and quite distinct factors. PD, for example, can be caused by overexpression of the small synaptic protein alpha-synuclein (Singleton et al., 2003), loss of parkin E3-ubiquitin ligase activity (Shimura et al., 2000), or by an increase in kinase activity of LRRK2 (West et al., 2005), to cite a few. These observations raise an important issue that plagues the field of neurodegeneration which is whether prominent disorders such as PD or AD are indeed diseases or rather syndromes, that is, very different diseases lumped together under the same name because they merely share the same clinical expression.

Regardless of this problem, it should be stated here that the causes of neurodegenerative diseases or syndromes are in most instances unknown. Even in some rare cases when the etiology has been identified, the mechanism by which the etiological factor leads to neuronal death remains, at best, speculative. For example, while the etiology of HD was identified more than 20 years ago (Group, 1993), we still do not know with certainty how mutant huntingtin kills striatal neurons. Similarly, it is not understood why mutations of superoxide dismutase-1 (SOD1) cause a familial form of ALS (Rosen et al., 1993).

Among the neurodegenerative disorders, only a few arise as a familial condition, which would support a purely genetic etiology for the disease. Within the affected members of a family, these genetic diseases can run as an autosomal dominant condition, which is the case for HD and dentatorubral-pallidoluysian atrophy. Less frequently, the disease can run as an autosomal recessive (e.g., familial spastic paraparesis), an X-linked (e.g., spinal and bulbar muscular atrophy), or even a maternally inherited trait (e.g., mitochondrial Leber optic neuropathy). In addition to these genetic-neurodegenerative diseases, others, while primarily sporadic, can also show a small contingent of patients in whom the illness is inherited. This is the case for PD, AD, and ALS, where less than 10% of all cases are generally familial.

For those in whom the disease is truly sporadic, which is the vast majority of patients, it appears that about 2% do carry known mutations, but in the about 98%, a genetic contribution to the neurodegenerative process would be minimal (Tanner et al., 1999). In these cases, environmental factors may be the prime suspects in initiating neurodegenerative processes. Relevant to this idea is the observation that some neurodegenerative conditions mentioned above might arise in geographic or temporal clusters. For instance, in the case of the PD–ALS–dementia complex of Guam which was introduced previously, it is believed that the pathology is caused by a toxic compound contained in the *Cycas circinalis*, an indigenous plant commonly ingested as food or medicine by the Chamorros Indians (Kurtland, 1988). Intoxication with 1-methyl-4-phenyl-1,2,3,6-tetrahydropyridine, a by-product of the synthesis of a meperidine compound, is also known to produce a severe and irreversible parkinsonian syndrome, which is almost identical to PD (Przedborski & Vila, 2001). Several large-scale epidemiological studies have failed, however, to link in a definitive manner environmental factors to diseases such as PD (Tanner, 1989). Furthermore, most of the toxic exposures known to cause neurological problems occur within a specific geographic, social, or professional context and in the absence of any significant association with an increased incidence of PD, AD, or other prominent neurodegenerative diseases. Collectively, the aforementioned findings argue that sporadic cases are neither clearly genetic nor clearly environmental. Conversely, this does not exclude the possibility that neurodegenerative disorders can result from a combination of both. Along this line, the demonstration of a nonsyndromic familial deafness linked to a mitochondrial point mutation (Prezant et al., 1993) provides compelling support for this view. In this study, family members who harbor the mitochondrial mutation develop a hearing impairment only upon exposure to the antibiotic aminoglycoside, thus illustrating the significant pathogenic interactions between genetics and environment. Also striking is the demonstration that the gene encoding for Ataxin-2, which is a known cause of spinocerebellar ataxia when its polyQ expansions is >34 glutamines, can also increase the risk of developing ALS when its polyQ expansions are between 27 and 33 (Elden et al., 2010). The latter work illustrates quite well the notion that genetic variants could predispose to environment factors.

21.5 Pathogenesis of Neurodegenerative Diseases

After considering the *etiology* or initiation of the disease process, it is also important to understand *pathogenesis* of disease. Pathogenesis refers to the cascade of cellular and molecular mechanisms set into motion by the etiological factor(s) which ultimately leads to the demise of the affected neurons. As mentioned above, only an exceedingly small group of so-called neurodegenerative diseases have no apparent neuropathological changes. While these clinically defined neurodegenerative processes may not have an apparent neuropathological correlate because we are constrained by using the loss of neuronal cell bodies as the cardinal feature used to define the neuropathology of a given neurodegenerative disorder. Yet it is increasingly recognized that in neurodegenerative disorders such as AD, PD, and ALS, especially early on, damage predominates not at the level of the cell body, but at the level of the synapse (Matuskey et al., 2020; Overk & Masliah, 2014). While synaptic lesions are more challenging to detect, it is critical to develop specific methods to survey the state of the synaptic fields as this may help to not only refine the classification of neurodegenerative disorders, including those which are associated with no apparent neuropathological changes but also to shed light into pathogenic mechanisms of neurodegeneration.

Most neurodegenerative illnesses do show overt neuropathology, mainly in the form of a focal loss of neurons with gliosis. Residual neurons may exhibit varying morphologies ranging from an almost normal appearance to a severe distortion with a combination of abnormal features such as process attrition, shape and size alterations of the cell body and nucleus, organelle fragmentation, dispersion of Nissl bodies, cytoplasmic vacuolization, and chromatin condensation. In several neurodegenerative disorders, spared neurons can also present with various types of intracellular proteinaceous inclusions, which, in the absence of any definite known pathogenic role, are quite useful in differentiating neurodegenerative disorders. As for the glial response it is comprised of innate resident immune cells including astrocytes and microglia. To a lesser extent, adaptive immune cells such as T-lymphocytes can also be found in the diseased areas of the brain.

The cell death seen in neurodegenerative disorders is usually limited to degeneration of specific subpopulations of neurons in specific structures of the nervous system. This means that at autopsy, areas of the brain that are damaged can be surrounded by healthy tissue and within the affected regions, degenerating and healthy neurons are intermingled. In some neurodegenerative diseases, such as olivopontocerebellar atrophy, multiple brain structures within the nervous system are affected. In these so called *system neurodegener-*

ative diseases, the spatial pattern of the lesions often becomes better defined as the disease progresses (Przedborski et al., 2003). Notably, the spatial pattern of the lesions is unrelated to the distribution of blood vessels (Sweeney et al., 2018), although dysfunction at the blood brain barrier is thought to be a contributor to pathogenesis in neurodegeneration. Yet, at least in some cases, the different lesions appear to be functionally and anatomically interconnected (Oppenheimer & Esiri, 1997). Such a linked degeneration is observed in ALS, where both the corticospinal tract and spinal cord lower motor neurons are affected, or in progressive supranuclear palsy, where both the globus pallidus and the subthalamic nucleus are damaged. Although such transneuronal degeneration is a well-recognized phenomenon (Saper et al., 1987), still very little is known about its molecular basis, except for the fact that this trans-synaptic demise occurs by programmed cell death (DeGiorgio et al., 1998; Ginsberg & Martin, 2002; Koga et al., 2021; Ramirez & Vonsattel, 2014). Some recent theories about the molecular basis of transneuronal degeneration hint at a self-propagating prion-like spread of toxic factors triggered by misfolded proteins such as synuclein or Tau (Goedert et al., 2010; Uchihara & Giasson, 2016). This prion hypothesis of neurodegenerations posits that abnormal protein in the synaptic milieu, such as synuclein, is prompted to adopt deleterious folding conformations. Once this toxic conformation is attained, the misfolded protein enters neighboring cells, further propagates more misfolded synuclein, perhaps by an alteration of free energy, and sets off a cascade of harmful protein stress.

Trans-neuronal degeneration cannot account for all the combinations of lesions found across regions of the nervous system in neurodegenerative diseases. For instance, in Friedreich ataxia, there is degeneration of the spinocerebellar tracts and the dentate nuclei, but not of the Purkinje cells, which would have been the link between these two lesions. It is thus more probable to consider that etiological factors are ubiquitous, but that the pathological threshold is attained by specific structures of the nervous system at various times. There are suggestions that the proteins tau and alpha-synuclein which are involved in Alzheimer's and Parkinson's diseases, respectively, may both be involved in a multitude of neurodegenerative illnesses, potentiating the destructive effects of each to cause neuronal destruction (Moussaud et al., 2014). In spinocerebellar ataxias, neurodegeneration is not only propagated by toxic proteins from the triplet repeat expansions that are pathognomic of the illness but also is propagated by deficiencies in transcription as a result of the pathologic genetic sequences (Matilla-Dueñas et al., 2014). In some neurodegenerative diseases, some areas of the nervous system may never reach this threshold and will thus remain unaffected throughout the span of the disease.

It is not an inevitability that neurodegenerative disorders will place large swathes of the nervous system at risk. In some specific neurodegenerative diseases, the lesions appear restricted to one or only a few brain regions. This is particularly well illustrated in spinal muscular atrophy, in which the degenerative process is limited to a loss of lower motor neurons, or in ALS, which is characterized by damage to the upper and lower motor neurons which may represent the sole neuropathological change. Nevertheless, almost all neurodegenerative disorders, which initially appeared *monosystemic,* will eventually become *multisystemic* over the progression of the degenerative process. In ALS, for example, it is typical to observe overt neuropathological changes mainly at the level of the spinal cord and motor cortex in patients who died after the short course of the disease, for unrelated causes (i.e., car accident or heart attack). However, in ALS patients with a more protracted course, the progressive degenerative changes have been shown to also affect the substantia nigra (Sasaki et al., 1992) and the oculomotor and trochlear nuclei (Hayashi et al., 1991). Recognition of this temporal spreading of the neurodegenerative process has led some neuropathologists (Braak et al., 2003) to propose that the neurodegenerative process starts on a specific circumscribed area of the nervous system where it begins a domino effect: The initial affected area is responsible for the disease occurring in the next region, whereby propagating the degenerative process. Although quite appealing, this model is speculative and would only be true if the different regions affected in the neurodegenerative conditions were anatomic neighbors, which, as mentioned above, is often not the case.

In most, if not all, neurodegenerative disorders, the locations of the principal lesions have been well established. However, it often remains difficult to determine the extent of degeneration affecting more than one group of neurons and, consequently, to define the exact neuropathological topography of certain diseases. This problem stems from at least three issues. First, lesions are often missed through incomplete examination of the brain and spinal cord. Second, quantitative morphology in postmortem samples seldom uses the rigorous counting methods necessary to generate reliable neuron numbers (Saper, 1996). The third issue pertains to sick neurons, which will not necessarily die. These compromised neurons often lose their phenotypic markers which are used for identification purposes in order to count them (Clarke & Oppenheim, 1995). Accordingly, the use of phenotypic markers in pathological conditions may lead to erroneous conclusions about the status of a given group of neurons. For these reasons, reported estimations regarding the distribution and magnitude of neuronal loss in neurodegenerative disorders may, perhaps with a few exceptions, be taken with caution.

21.6 Onset and Progression of the Disease

As indicated above, prominent neurodegenerative disorders are sporadic, and with the absence of pre-symptomatic markers for virtually all of them, patients are seeking medical attention only when the first symptoms of the diseases emerge. Because there is significant cellular redundancy in neuronal pathways, the onset of symptoms does not equate with the onset of the disease. Instead, the beginning of symptoms merely corresponds to a neurodegenerative stage at which the number of residual neurons in a given pathway drops below the number required to maintain its endowed function. It is thus clear that the onset of the disease occurs at some preceding time, which, depending on how fast the neurodegenerative process evolves, can range from a few months to several years. Our ability to determine the actual onset of the disease is at this point unfortunately undermined by the lack of pre-symptomatic markers and the little knowledge about the true kinetics of cell demise.

Occasionally there is a sudden worsening of a patient's condition. Although a sudden acceleration of the neurodegenerative process cannot be excluded, especially under the effect of intercurrent deleterious factors such as infection, it is more probable that the rate of neuronal death remains about the same throughout the natural course of the disease. However, the relationship between the clinical expression of a disease and the number of residual neurons does not have to be either linear or even constant. For instance, a patient may clinically, remain unchanged during a prolonged period of time despite a loss of many cells and then abruptly their condition may deteriorate as the number of neurons drop below a functional threshold. Another important aspect is that virtually all neurodegenerative disorders progress slowly over time, often taking several years to reach end-stage. It must be remembered that neuronal degeneration corresponds to an asynchronous death (Herrera et al., 2017; Pittman et al., 1999) in that cells within a population die at very different times. The correlate of this fact is that at any given time, only a small number of cells are actually dying and among these, they are at various stages along the cell death pathway. However, standard clinical, radiological, and biochemical measurements, which are critical to assessing the disease, generate information about the entire population of cells. Therefore, the rate of change in any of the usual clinical parameters reflects the decay of the entire population of affected cells and provides very little, if any insight into the pace at which the death of an individual cell occurs. Still, if one looks at the large body of in vitro data (Levy et al., 2009; Moujalled et al., 2021), it appears that once a cell gets sick, the entire process of death proceeds rapidly. Given these facts, it may be possible that the protracted clinical progression is the reflection of only a small number of neurons dying at any given time from a rapid demise.

21.7 Nonneuronal Factors in Neurodegeneration

As indicated above, neurodegeneration refers to pathology of neurons. It is thus not surprising to find from an examination of the literature that all theories about neurodegeneration pathogenesis revolve around neurons. This rather *neuronocentrist* view is increasingly at odds however with the current concept of neurodegeneration pathogenesis. As stressed above, in most prominent familial neurodegenerative diseases, the mutant proteins are ubiquitously expressed. This is the case, for example, with mutant alpha-synuclein in familial PD and of mutant SOD1 in ALS. While the underlying gene mutations are unquestionably pathogenic, ALS-linked SOD1 mutations expressed selectively in astrocytes (Gong et al., 2000) or in neurons (Lino et al., 2002; Pramatarova et al., 2001) fail to provoke the death of motor neurons in transgenic mice. Conversely, the expression of ALS-linked SOD1 mutations in all cells does cause an adult-onset paralytic phenotype in both transgenic mice and rats (Gurney et al., 1994; Nagai et al., 2001). Further support to the idea that many different cell types collaborate to achieve the demise of neurons in neurodegenerative diseases comes from the work of Clement and collaborators (Clement et al., 2003). In the study, the authors produced chimeric animals made of mixtures of normal and SOD1 mutant-expressing cells. They found that motor neurons that chronically expressed mutant SOD1 did not degenerate if they were surrounded by a sufficient number of normal nonneuronal cells. In addition, normal motor neurons surrounded by mutant-expressing nonneuronal cells showed to acquire intraneuronal ubiquitinated deposits consistent with the concept that mutant-expressing cells could transfer the disease phenotype to normal neighboring cells. These data suggest that the death of neurons in neurodegenerative diseases such as ALS may not be as neuron-autonomous as previously thought. Instead, the cellular environment surrounding neurons may play a critical role in determining the demise of specific neurons within these types of diseases of the nervous system. Among the prime candidates that mediate this nonneuronal autonomous mechanism of neurodegeneration are glial cells. Glial cells have been implicated in the pathogenesis of PD, both through the loss of neuronal homeostatic mechanisms and the subsequent neurotoxic downstream effects (Kam et al., 2020). Along this line, neuroinflammation has indeed emerged over the past decade as a key contributor to neurodegeneration in diseases such as AD, PD, and ALS (McGeer & McGeer, 2002; Przedborski & Goldman, 2004; Wyss-Coray & Mucke, 2002). Based on both epidemiological and pre-clinical studies in animal models of human diseases (Chen et al., 2003; McGeer & McGeer, 2002; Przedborski & Goldman, 2004; Wyss-Coray & Mucke, 2002), it is now proposed that neuroinflammation could stimulate neurodegeneration. Single-cell studies examining the blood and CSF of PD patients show that there are specific changes in T-cell populations that occur in PD (Wang et al., 2021). Similar changes in T-cell populations are seen in AD (Gate et al., 2020; Schafflick et al., 2020), which might point to an overall role of auto-immunity in neurodegeneration, similar to what has long been recognized in multiple sclerosis. Furthermore, the contribution of noncell autonomous mechanisms mediated by nonneuronal cells, such as astrocyte or oligodendrocytes, to the demise of neighboring neuronal cells are increasingly recognized (Garden & La Spada, 2012). Likewise, alterations in the neurovascular unit—which is comprised of neurons and glial cells as well as brain endothelial, pericytes and vascular smooth muscle cells—has received growing attention from the perspective of neurobiology of disease. For instance, Zlokovic (2010) is reminding us that the state of the blood—brain barrier, and more specifically, its nonneuronal cellular constituents, by regulating the exchange between the circulating blood and the brain extracellular fluid may be instrumental in driving the neurodegenerative process.

21.8 Summary

Neurodegenerative diseases represent a heterogeneous group of disorders affecting the nervous system. In most instances, they affect adults, their causes are unknown, and the progression is relentless. Some are genetic, but most are sporadic. They involve all parts of the nervous system, although the cerebral cortex and the basal ganglia are the most frequent loci of pathology. The historical classification of neurodegenerative diseases, based on clinical and pathological characteristics, is imperfect. New classifications are rather based on molecular determinants. Contrary to the common belief, it is now recognized that neurodegenerative disorders are multisystemic, even if specific neuronal pathways are more affected than others. The death of specific types of neurons in neurodegenerative diseases is provoked, not by a single pathogenic factor, but rather by a cascade of multiple deleterious molecular and cellular events.

21.9 Review Questions

1. Define neurodegeneration.
   (a) Aneurodegeneration is any pathological condition in which the nervous system or nerve cell (i.e., neuron) loses its function, structure, or both.
2. What are the three broad categories of diseases of the nervous system?
   (a) (i) pathologies which are restricted to the nervous system and which are primary neuronal diseases (referred henceforth as neurodegenerative diseases)
   (b) (ii) pathologies which are restricted to the nervous system but are not primary neuronal diseases
   (c) (iii) pathologies provoked by systemic causes which damage the nervous system
3. Provide examples of diseases from the three categories.
   (a) (i) Alzheimer's disease, Parkinson's disease
   (b) (ii) brain neoplasm or cerebral edema and hemorrhage
   (c) (iii) cardiovascular arrest, carbon monoxide poison, or infections due to herpes simplex.
4. What is expected to occur with the prevalence of neurodegenerative diseases in the forthcoming generations and why?
   (a) Prevalence is expected to increase both because of increased diagnosis and growth rate of the population in general.
5. What is the effect of neurodegeneration on life expectancy, and for what reason(s) typically patients with neurodegenerative disease die?
   (a) Will usually reduce life expectancy, in part because common medical co-morbidities aggravate symptoms in neurogenerative disease.
6. What is the common method used in classifying neurodegenerative diseases and what are the difficulties inherent with this type of classification?
   (a) A combination of the main clinical feature and the location of the predominant lesion.
7. Describe the different forms of genetic contributions among the neurodegenerative diseases.
   (a) Only a few are inherited genetically. Huntington's disease and dentatorubral-pallidoluysian atrophy are autosomally dominantly inherited. In PD, AD and ALS where less than 10% of all cases are generally familial.
8. What does multisystemic neurodegeneration refer to?
   (a) While most neurodegenerative disease start in one area of the central nervous system, most neurodegenerative illness eventually involve multiple CNS areas.
9. Why are neurodegenerative diseases progressive, and what does it mean at the level of the whole population of affected cells and a single affected cell?
   (a) Neurons are at different stages of cell death throughout the course of a neurodegenerative illness. Over time this means of course fewer and fewer neurons contributing to particular neuronal pathway, and it is notable that the rate of neuronal death is not linear to the onset of symptoms.

10. Define the notion of noncell autonomous neurodegeneration and provide an example.
    (a) Non-neuronal cells contribute to the demise of neurons, and more generally, many different cell types are "collaborating" to eventually lead to neuronal death. In the expression of ALS-linked SOD1 mutations, these mutations are needed in both astrocytes and neurons in order to lead to cell death, and mutations in one or the other dose not induce cell death pathways.
11. What are the two most prevalent neurodegenerative diseases?
    (a) Alzheimer's disease and stroke
    (b) Parkinson's disease and brain tumor
    (c) *Alzheimer's disease and Parkinson's disease*
    (d) Huntington's disease and amyotrophic lateral sclerosis
    (e) Mitochondrial encephalopathy and multiple sclerosis
12. What are typical neurodegenerative diseases without overt neuropathology?
    (a) Creutzfeldt-Jakob disease
    (b) Schizophrenia
    (c) Tourette's syndrome
    (d) Torsion dystonia
    (e) *All but a*
13. What is the most likely correct statement about the neuropathology of neurodegenerative diseases?
    (a) Neuropathologic changes in neurodegenerative diseases are always overt in at least one region of the nervous system.
    (b) Residual neurons look normal.
    (c) *It appears mainly as focal loss of neurons with gliosis.*
    (d) If different regions of the brain are affected, the type of neurons degenerating is identical.
    (e) The multisystemic nature of a neurodegenerative process is explained by a trans-synaptic phenomenon.
14. What is the most likely correct statement about the etiology of neurodegenerative diseases?
    (a) The etiology of most neurodegenerative diseases is known.
    (b) Most neurodegenerative diseases are genetic.
    (c) Neurodegenerative diseases are only linked to nuclear gene defects.
    (d) *Environmental toxic factors have been implicated in the occurrence of clusters of neurodegenerative diseases.*
    (e) When the etiology is known, the mechanism by which neurons degenerate is known as well.
15. Which of the following statements is most likely true about the pathogenesis?
    (a) The death of neurons is caused by a unique pathogenic mechanism triggered by an etiologic factor.
    (b) Neurodegenerative disorders do not become multisystemic upon progression of the disease.
    (c) It is established that the neurodegenerative process starts in one area of the nervous system and then spreads to others by a domino effect.
    (d) Before dying, compromised neurons never lose their functions and phenotypic markers.
    (e) *The distribution and magnitude of neuronal loss in neurodegenerative disorders is difficult to establish with certainty.*
16. Which of the following statements is most likely correct?
    (a) Onset of symptoms reflects the onset of the disease.
    (b) The prominent symptoms often reveal the main site of neuropathology.
    (c) Intercurrent infection can permanently exacerbate the symptoms.
    (d) Neurodegenerative diseases differ from nondegenerative diseases because the latter never progress in a stepwise manner.
    (e) *b and c*
17. Which of the following statements is most likely correct?
    (a) *Neurodegenerative diseases can involve either the central or the peripheral nervous systems, or both.*
    (b) A familial occurrence of a neurodegenerative disease is the signature of its genetic origin.
    (c) Neurodegenerative diseases are associated with increased risks for cancers.
    (d) Neurodegenerative diseases are associated with a shortened lifespan because they eventually affect brain areas regulating vital functions.
    (e) When available, symptomatic agents normalize lifespan in patients suffering from neurodegenerative diseases.
18. The transmissibility of a neurodegenerative phenotype can be?
    (a) Autosomal dominant
    (b) Autosomal recessive
    (c) X-linked
    (d) Mitochondrial
    (e) *All of the above*
19. Regarding the classification of neurodegenerative disorders which statement is not correct? But I think all the answers could be correct
    (a) *The clinico-anatomic classification is quite popular but is fraught with many shortcomings.*
    (b) The molecular classification groups together neurodegenerative diseases of quite different clinical expression.

(c) A parkinsonism can be observed in patients suffering from a wide variety of disorders, including Parkinson's disease.
(d) Members of the same family carrying an identical mutation always exhibit the same clinical and neuropathological phenotype.
(e) A similar clinical phenotype can be caused by distinct etiologic factors.

20. The concept of noncell autonomy in neurodegeneration implies.
 (a) *A disease phenotype can be transmitted from a mutant cell to a wild-type cell.*
 (b) The whole disease process does not rely solely on pathogenic events taking place in the degenerating cells.
 (c) The fate of neurons destined to die is at least in part, determined by other cells such as neighboring glia.
 (d) This concept applies to infectious diseases but not to neurodegenerative disorders.
 (e) All but d

21.10 Answers to the Questions

1. a
2. a
3. a, b, c
4. a
5. a
6. a
7. a
8. a
9. a
10. a
11. a
12. e
13. c
14. d
15. e
16. e
17. a
18. e
19. a
20. a

Acknowledgments The author is supported by NIH/NINDS (NS107442, NS117583), the US Department of Defense Grant (PR210956), the Parkinson's Foundation, the Chan Zuckerberg Initiative, and the ProjectALS

References

Alzheimer's Association. (2016). 2016 Alzheimer's disease facts and figures. *Alzheimer's & Dementia: The Journal of the Alzheimer's Association, 12*(4), 459–509. https://doi.org/10.1016/j.jalz.2016.03.001

Balint, B., Mencacci, N. E., Valente, E. M., Pisani, A., Rothwell, J., Jankovic, J., Vidailhet, M., & Bhatia, K. P. (2018). Dystonia. *Nature Reviews. Disease Primers, 4*(1), 25. https://doi.org/10.1038/s41572-018-0023-6

Ballard, C., Ziabreva, I., Perry, R., Larsen, J. P., O'Brien, J., McKeith, I., Perry, E., & Aarsland, D. (2006). Differences in neuropathologic characteristics across the Lewy body dementia spectrum. *Neurology, 67*(11), 1931–1934. https://doi.org/10.1212/01.wnl.0000249130.63615.cc

Bjartmar, C., Wujek, J. R., & Trapp, B. D. (2003). Axonal loss in the pathology of MS: Consequences for understanding the progressive phase of the disease. *Journal of the Neurological Sciences, 206*(2), 165–171. https://doi.org/10.1016/s0022-510x(02)00069-2

Bobilev, A. M., Perez, J. M., & Tamminga, C. A. (2020). Molecular alterations in the medial temporal lobe in schizophrenia. *Schizophrenia Research, 217*, 71–85. https://doi.org/10.1016/j.schres.2019.06.001

Braak, H., Del Tredici, K., Rüb, U., de Vos, R. A. I., Jansen Steur, E. N. H., & Braak, E. (2003). Staging of brain pathology related to sporadic Parkinson's disease. *Neurobiology of Aging, 24*(2), 197–211. https://doi.org/10.1016/s0197-4580(02)00065-9

Casals, J., Elizan, T. S., & Yahr, M. D. (1998). Postencephalitic parkinsonism—A review. *Journal of Neural Transmission (Vienna, Austria: 1996), 105*(6–7), 645–676. https://doi.org/10.1007/s007020050086

Chen, K.-M., & Chase, T. N. (1986). Parkinsonism-dementia. In P. J. Vinken, G. W. Bruyn, & H. L. Klawans (Eds.), *Handbook of clinical neurology. Extrapyramidal disorders* (Vol. 49, pp. 167–183). Elsevier.

Chen, H., Zhang, S. M., Hernán, M. A., Schwarzschild, M. A., Willett, W. C., Colditz, G. A., Speizer, F. E., & Ascherio, A. (2003). Nonsteroidal anti-inflammatory drugs and the risk of Parkinson disease. *Archives of Neurology, 60*(8), 1059–1064. https://doi.org/10.1001/archneur.60.8.1059

Clarke, P. G., & Oppenheim, R. W. (1995). Neuron death in vertebrate development: In vitro methods. *Methods in Cell Biology, 46*, 277–321.

Clement, A. M., Nguyen, M. D., Roberts, E. A., Garcia, M. L., Boillée, S., Rule, M., McMahon, A. P., Doucette, W., Siwek, D., Ferrante, R. J., Brown, R. H., Julien, J.-P., Goldstein, L. S. B., & Cleveland, D. W. (2003). Wild-type nonneuronal cells extend survival of SOD1 mutant motor neurons in ALS mice. *Science (New York, N.Y.), 302*(5642), 113–117. https://doi.org/10.1126/science.1086071

Cummings, C. J., & Zoghbi, H. Y. (2000). Trinucleotide repeats: Mechanisms and pathophysiology. *Annual Review of Genomics and Human Genetics, 1*, 281–328. https://doi.org/10.1146/annurev.genom.1.1.281

Dauer, W., & Przedborski, S. (2003). Parkinson's disease: Mechanisms and models. *Neuron, 39*(6), 889–909. https://doi.org/10.1016/s0896-6273(03)00568-3

DeGiorgio, L. A., Dibinis, C., Milner, T. A., Saji, M., & Volpe, B. T. (1998). Histological and temporal characteristics of nigral transneuronal degeneration after striatal injury. *Brain Research, 795*(1–2), 1–9. https://doi.org/10.1016/s0006-8993(98)00247-9

Doraiswamy, P. M., Leon, J., Cummings, J. L., Marin, D., & Neumann, P. J. (2002). Prevalence and impact of medical comorbidity in Alzheimer's disease. *The Journals of Gerontology. Series A, Biological Sciences and Medical Sciences, 57*(3), M173–M177. https://doi.org/10.1093/gerona/57.3.m173

Elden, A. C., Kim, H.-J., Hart, M. P., Chen-Plotkin, A. S., Johnson, B. S., Fang, X., Armakola, M., Geser, F., Greene, R., Lu, M. M., Padmanabhan, A., Clay-Falcone, D., McCluskey, L., Elman, L., Juhr, D., Gruber, P. J., Rüb, U., Auburger, G., Trojanowski, J. Q., et al. (2010). Ataxin-2 intermediate-length polyglutamine expansions are associated with increased risk for ALS. *Nature, 466*(7310), 1069–1075. https://doi.org/10.1038/nature09320

Galvin, J. E., Lee, V. M., & Trojanowski, J. Q. (2001). Synucleinopathies: Clinical and pathological implications. *Archives of Neurology, 58*(2), 186–190. https://doi.org/10.1001/archneur.58.2.186

Garden, G. A., & La Spada, A. R. (2012). Intercellular (mis)communication in neurodegenerative disease. *Neuron, 73*(5), 886–901. https://doi.org/10.1016/j.neuron.2012.02.017

Gate, D., Saligrama, N., Leventhal, O., Yang, A. C., Unger, M. S., Middeldorp, J., Chen, K., Lehallier, B., Channappa, D., De Los Santos, M. B., McBride, A., Pluvinage, J., Elahi, F., Tam, G. K.-Y., Kim, Y., Greicius, M., Wagner, A. D., Aigner, L., Galasko, D. R., et al. (2020). Clonally expanded CD8 T cells patrol the cerebrospinal fluid in Alzheimer's disease. *Nature, 577*(7790), 399–404. https://doi.org/10.1038/s41586-019-1895-7

Ginsberg, S. D., & Martin, L. J. (2002). Axonal transection in adult rat brain induces transsynaptic apoptosis and persistent atrophy of target neurons. *Journal of Neurotrauma, 19*(1), 99–109. https://doi.org/10.1089/089771502753460277

Goedert, M., & Spillantini, M. G. (2001). Tau gene mutations and neurodegeneration. *Biochemical Society Symposium, 67*, 59–71. https://doi.org/10.1042/bss0670059

Goedert, M., Clavaguera, F., & Tolnay, M. (2010). The propagation of prion-like protein inclusions in neurodegenerative diseases. *Trends in Neurosciences, 33*(7), 317–325. https://doi.org/10.1016/j.tins.2010.04.003

Gong, Y. H., Parsadanian, A. S., Andreeva, A., Snider, W. D., & Elliott, J. L. (2000). Restricted expression of G86R Cu/Zn superoxide dismutase in astrocytes results in astrocytosis but does not cause motoneuron degeneration. *The Journal of Neuroscience, 20*(2), 660–665.

Group, T. H. D. C. R. (1993). A novel gene containing a trinucleotide repeat that is expanded and unstable on Huntington's disease chromosomes. *Cell, 72*(6), 971–983.

Gurney, M. E., Pu, H., Chiu, A. Y., Dal Canto, M. C., Polchow, C. Y., Alexander, D. D., Caliendo, J., Hentati, A., Kwon, Y. W., & Deng, H. X. (1994). Motor neuron degeneration in mice that express a human Cu, Zn superoxide dismutase mutation. *Science (New York, N.Y.), 264*(5166), 1772–1775. https://doi.org/10.1126/science.8209258

Harding, A. E. (1981). Friedreich's ataxia: A clinical and genetic study of 90 families with an analysis of early diagnostic criteria and intrafamilial clustering of clinical features. *Brain: A Journal of Neurology, 104*(3), 589–620. https://doi.org/10.1093/brain/104.3.589

Hayashi, H., Kato, S., & Kawada, A. (1991). Amyotrophic lateral sclerosis patients living beyond respiratory failure. *Journal of the Neurological Sciences, 105*(1), 73–78. https://doi.org/10.1016/0022-510x(91)90121-m

Hely, M. A., Morris, J. G., Rail, D., Reid, W. G., O'Sullivan, D. J., Williamson, P. M., Genge, S., & Broe, G. A. (1989). The Sydney Multicentre Study of Parkinson's disease: A report on the first 3 years. *Journal of Neurology, Neurosurgery, and Psychiatry, 52*(3), 324–328. https://doi.org/10.1136/jnnp.52.3.324

Herrera, A., Muñoz, P., Steinbusch, H. W. M., & Segura-Aguilar, J. (2017). Are dopamine oxidation metabolites involved in the loss of dopaminergic neurons in the nigrostriatal system in Parkinson's disease? *ACS Chemical Neuroscience, 8*(4), 702–711. https://doi.org/10.1021/acschemneuro.7b00034

Hughes, A. J., Daniel, S. E., Kilford, L., & Lees, A. J. (1992). Accuracy of clinical diagnosis of idiopathic Parkinson's disease: A clinico-pathological study of 100 cases. *Journal of Neurology, Neurosurgery, and Psychiatry, 55*(3), 181–184. https://doi.org/10.1136/jnnp.55.3.181

Irwin, D. J., White, M. T., Toledo, J. B., Xie, S. X., Robinson, J. L., Van Deerlin, V., Lee, V. M.-Y., Leverenz, J. B., Montine, T. J., Duda, J. E., Hurtig, H. I., & Trojanowski, J. Q. (2012). Neuropathologic substrates of Parkinson disease dementia. *Annals of Neurology, 72*(4), 587–598. https://doi.org/10.1002/ana.23659

Kam, T.-I., Hinkle, J. T., Dawson, T. M., & Dawson, V. L. (2020). Microglia and astrocyte dysfunction in Parkinson's disease. *Neurobiology of Disease, 144*, 105028. https://doi.org/10.1016/j.nbd.2020.105028

Kanning, K. C., Kaplan, A., & Henderson, C. E. (2010). Motor neuron diversity in development and disease. *Annual Review of Neuroscience, 33*, 409–440. https://doi.org/10.1146/annurev.neuro.051508.135722

Kirshner, H. S., Tanridag, O., Thurman, L., & Whetsell, W. O. (1987). Progressive aphasia without dementia: Two cases with focal spongiform degeneration. *Annals of Neurology, 22*(4), 527–532. https://doi.org/10.1002/ana.410220413

Koeppen, A. H. (2018). The neuropathology of the adult cerebellum. *Handbook of Clinical Neurology, 154*, 129–149. https://doi.org/10.1016/B978-0-444-63956-1.00008-4

Koga, S., Sekiya, H., Kondru, N., Ross, O. A., & Dickson, D. W. (2021). Neuropathology and molecular diagnosis of synucleinopathies. *Molecular Neurodegeneration, 16*(1), 83. https://doi.org/10.1186/s13024-021-00501-z

Kurtland, L. T. (1988). Amyotrophic lateral sclerosis and Parkinson's disease complex on Guam linked to an environmental neurotoxin. *Trends in Neurosciences, 11*, 51–54.

Le Rhun, E., Richard, F., & Pasquier, F. (2005). Natural history of primary progressive aphasia. *Neurology, 65*(6), 887–891. https://doi.org/10.1212/01.wnl.0000175982.57472.84

Levy, G., Tang, M.-X., Louis, E. D., Côté, L. J., Alfaro, B., Mejia, H., Stern, Y., & Marder, K. (2002). The association of incident dementia with mortality in PD. *Neurology, 59*(11), 1708–1713. https://doi.org/10.1212/01.wnl.0000036610.36834.e0

Levy, O. A., Malagelada, C., & Greene, L. A. (2009). Cell death pathways in Parkinson's disease: Proximal triggers, distal effectors, and final steps. *Apoptosis: An International Journal on Programmed Cell Death, 14*(4), 478–500. https://doi.org/10.1007/s10495-008-0309-3

Lino, M. M., Schneider, C., & Caroni, P. (2002). Accumulation of SOD1 mutants in postnatal motoneurons does not cause motoneuron pathology or motoneuron disease. *The Journal of Neuroscience, 22*(12), 4825–4832.

Matilla-Dueñas, A., Ashizawa, T., Brice, A., Magri, S., McFarland, K. N., Pandolfo, M., Pulst, S. M., Riess, O., Rubinsztein, D. C., Schmidt, J., Schmidt, T., Scoles, D. R., Stevanin, G., Taroni, F., Underwood, B. R., & Sánchez, I. (2014). Consensus paper: Pathological mechanisms underlying neurodegeneration in spinocerebellar ataxias. *Cerebellum (London, England), 13*(2), 269–302. https://doi.org/10.1007/s12311-013-0539-y

Matuskey, D., Tinaz, S., Wilcox, K. C., Naganawa, M., Toyonaga, T., Dias, M., Henry, S., Pittman, B., Ropchan, J., Nabulsi, N., Suridjan, I., Comley, R. A., Huang, Y., Finnema, S. J., & Carson, R. E. (2020). Synaptic changes in Parkinson disease assessed with in vivo imaging. *Annals of Neurology, 87*(3), 329–338. https://doi.org/10.1002/ana.25682

McGeer, P. L., & McGeer, E. G. (2002). Inflammatory processes in amyotrophic lateral sclerosis. *Muscle & Nerve, 26*(4), 459–470. https://doi.org/10.1002/mus.10191

Morgante, L., Salemi, G., Meneghini, F., Di Rosa, A. E., Epifanio, A., Grigoletto, F., Ragonese, P., Patti, F., Reggio, A., Di Perri, R., & Savettieri, G. (2000). Parkinson disease survival: A population-based study. *Archives of Neurology, 57*(4), 507–512. https://doi.org/10.1001/archneur.57.4.507

Moujalled, D., Strasser, A., & Liddell, J. R. (2021). Molecular mechanisms of cell death in neurological diseases. *Cell Death & Differentiation, 28*(7), 2029–2044. https://doi.org/10.1038/s41418-021-00814-y

Moussaud, S., Jones, D. R., Moussaud-Lamodière, E. L., Delenclos, M., Ross, O. A., & McLean, P. J. (2014). Alpha-synuclein and tau: Teammates in neurodegeneration? *Molecular Neurodegeneration, 9*, 43. https://doi.org/10.1186/1750-1326-9-43

Nagai, M., Aoki, M., Miyoshi, I., Kato, M., Pasinelli, P., Kasai, N., Brown, R. H., & Itoyama, Y. (2001). Rats expressing human cytosolic copper-zinc superoxide dismutase transgenes with amyotrophic lateral sclerosis: Associated mutations develop motor neuron disease. *The Journal of Neuroscience, 21*(23), 9246–9254.

Oppenheimer, D. R., & Esiri, M. M. (1997). Diseases of the basal ganglia, cerebellum and motor neurons. In J. H. Adams, J. A. N. Corsellis, & L. W. Duchen (Eds.), *Greenfield's neuropathology* (6th ed., pp. 988–1045). Arnold.

Organization, W. H. (2006). *Neurological disorders. Public health challenges*. World Health Organization. http://www.columbia.edu/cgi-bin/cul/resolve?clio6099886

Overk, C. R., & Masliah, E. (2014). Pathogenesis of synaptic degeneration in Alzheimer's disease and Lewy body disease. *Biochemical Pharmacology, 88*(4), 508–516. https://doi.org/10.1016/j.bcp.2014.01.015

Pittman, R. N., Messam, C. A., & Mills, J. C. (1999). Asynchronous death as a characteristic feature of apoptosis. In V. E. Koliatsos & R. R. Ratan (Eds.), *Cell death and diseases of the nervous system* (pp. 29–43). Humana Press.

Potashkin, J., Huang, X., Becker, C., Chen, H., Foltynie, T., & Marras, C. (2020). Understanding the links between cardiovascular disease and Parkinson's disease. *Movement Disorders, 35*(1), 55–74. https://doi.org/10.1002/mds.27836

Pramatarova, A., Laganière, J., Roussel, J., Brisebois, K., & Rouleau, G. A. (2001). Neuron-specific expression of mutant superoxide dismutase 1 in transgenic mice does not lead to motor impairment. *The Journal of Neuroscience: The Official Journal of the Society for Neuroscience, 21*(10), 3369–3374.

Prezant, T. R., Agapian, J. V., Bohlman, M. C., Bu, X., Oztas, S., Qiu, W. Q., Arnos, K. S., Cortopassi, G. A., Jaber, L., & Rotter, J. I. (1993). Mitochondrial ribosomal RNA mutation associated with both antibiotic-induced and non-syndromic deafness. *Nature Genetics, 4*(3), 289–294. https://doi.org/10.1038/ng0793-289

Prusiner, S. B. (1998). Prions. *Proceedings of the National Academy of Sciences of the United States of America, 95*(23), 13363–13383. https://doi.org/10.1073/pnas.95.23.13363

Przedborski, S., & Goldman, J. E. (2004). Pathogenic role of glial cells in Parkinson's disease. In L. Hertz (Ed.), *Non-neuronal cells of the nervous system: Function and dysfunction* (Vol. 31, pp. 967–982). Elsevier.

Przedborski, S., & Vila, M. (2001). MPTP: A review of its mechanisms of neurotoxicity. *Clinical Neuroscience Research, 1*(6), 407–418. https://doi.org/10.1016/S1566-2772(01)00019-6

Przedborski, S., Jackson-Lewis, V., Naini, A. B., Jakowec, M., Petzinger, G., Miller, R., & Akram, M. (2001). The parkinsonian toxin 1-methyl-4-phenyl-1,2,3,6-tetrahydropyridine (MPTP): A technical review of its utility and safety. *Journal of Neurochemistry, 76*(5), 1265–1274. https://doi.org/10.1046/j.1471-4159.2001.00183.x

Przedborski, S., Vila, M., & Jackson-Lewis, V. (2003). Neurodegeneration: What is it and where are we? *The Journal of Clinical Investigation, 111*(1), 3–10. https://doi.org/10.1172/JCI17522

Ramirez, E. P. C., & Vonsattel, J. P. G. (2014). Neuropathologic changes of multiple system atrophy and diffuse Lewy body disease. *Seminars in Neurology, 34*(2), 210–216. https://doi.org/10.1055/s-0034-1381732

Rosen, D. R., Siddique, T., Patterson, D., Figlewicz, D. A., Sapp, P., Hentati, A., Donaldson, D., Goto, J., O'Regan, J. P., & Deng, H. X. (1993). Mutations in Cu/Zn superoxide dismutase gene are associated with familial amyotrophic lateral sclerosis. *Nature, 362*(6415), 59–62. https://doi.org/10.1038/362059a0

Samson, M., & Claassen, D. O. (2017). Neurodegeneration and the cerebellum. *Neuro-Degenerative Diseases, 17*(4–5), 155–165. https://doi.org/10.1159/000460818

Saper, C. B. (1996). Any way you cut it: A new journal policy for the use of unbiased counting methods. *The Journal of Comparative Neurology, 364*(1), 5. https://doi.org/10.1002/(SICI)1096-9861(19960101)364:1<5::AID-CNE1>3.0.CO;2-9

Saper, C. B., Wainer, B. H., & German, D. C. (1987). Axonal and transneuronal transport in the transmission of neurological disease: Potential role in system degenerations, including Alzheimer's disease. *Neuroscience, 23*(2), 389–398. https://doi.org/10.1016/0306-4522(87)90063-7

Sasaki, S., Tsutsumi, Y., Yamane, K., Sakuma, H., & Maruyama, S. (1992). Sporadic amyotrophic lateral sclerosis with extensive neurological involvement. *Acta Neuropathologica, 84*(2), 211–215. https://doi.org/10.1007/BF00311398

Schafflick, D., Xu, C. A., Hartlehnert, M., Cole, M., Schulte-Mecklenbeck, A., Lautwein, T., Wolbert, J., Heming, M., Meuth, S. G., Kuhlmann, T., Gross, C. C., Wiendl, H., Yosef, N., & Meyer Zu Horste, G. (2020). Integrated single cell analysis of blood and cerebrospinal fluid leukocytes in multiple sclerosis. *Nature Communications, 11*(1), 247. https://doi.org/10.1038/s41467-019-14118-w

Scheltens, P., De Strooper, B., Kivipelto, M., Holstege, H., Chételat, G., Teunissen, C. E., Cummings, J., & van der Flier, W. M. (2021). Alzheimer's disease. *The Lancet, 397*(10284), 1577–1590. https://doi.org/10.1016/S0140-6736(20)32205-4

Shimura, H., Hattori, N., Kubo, S. I., Mizuno, Y., Asakawa, S., Minoshima, S., Shimizu, N., Iwai, K., Chiba, T., Tanaka, K., & Suzuki, T. (2000). Familial Parkinson disease gene product, parkin, is a ubiquitin-protein ligase. *Nature Genetics, 25*(3), 302–305. https://doi.org/10.1038/77060

Simon, D. K., Tanner, C. M., & Brundin, P. (2020). Parkinson disease epidemiology, pathology, genetics, and pathophysiology. *Clinics in Geriatric Medicine, 36*(1), 1–12. https://doi.org/10.1016/j.cger.2019.08.002

Singleton, A. B., Farrer, M., Johnson, J., Singleton, A., Hague, S., Kachergus, J., Hulihan, M., Peuralinna, T., Dutra, A., Nussbaum, R., Lincoln, S., Crawley, A., Hanson, M., Maraganore, D., Adler, C., Cookson, M. R., Muenter, M., Baptista, M., Miller, D., et al. (2003). Alpha-Synuclein locus triplication causes Parkinson's disease. *Science (New York, N.Y.), 302*(5646), 841. https://doi.org/10.1126/science.1090278

Stacy, M., & Jankovic, J. (1992). Differential diagnosis of Parkinson's disease and the parkinsonism plus syndromes. *Neurologic Clinics, 10*(2), 341–359.

Sulkava, R., Haltia, M., Paetau, A., Wikström, J., & Palo, J. (1983). Accuracy of clinical diagnosis in primary degenerative dementia: Correlation with neuropathological findings. *Journal of Neurology, Neurosurgery, and Psychiatry, 46*(1), 9–13. https://doi.org/10.1136/jnnp.46.1.9

Sweeney, M. D., Kisler, K., Montagne, A., Toga, A. W., & Zlokovic, B. V. (2018). The role of brain vasculature in neurodegenerative disorders. *Nature Neuroscience, 21*(10), 1318–1331. https://doi.org/10.1038/s41593-018-0234-x

Tanner, C. M. (1989). The role of environmental toxins in the etiology of Parkinson's disease. *Trends in Neurosciences, 12*(2), 49–54. https://doi.org/10.1016/0166-2236(89)90135-5

Tanner, C. M. (1992). Epidemiology of Parkinson's disease. *Neurologic Clinics, 10*(2), 317–329.

Tanner, C. M., Ottman, R., Goldman, S. M., Ellenberg, J., Chan, P., Mayeux, R., & Langston, J. W. (1999). Parkinson disease in twins: An etiologic study. *JAMA, 281*(4), 341–346. https://doi.org/10.1001/jama.281.4.341

Tomlinson, B. E. (1977). The pathology of dementia. *Contemporary Neurology Series, 15*, 113–153.

Trapp, B. D., & Nave, K.-A. (2008). Multiple sclerosis: An immune or neurodegenerative disorder? *Annual Review of Neuroscience, 31*, 247–269. https://doi.org/10.1146/annurev.neuro.30.051606.094313

Uchihara, T., & Giasson, B. I. (2016). Propagation of alpha-synuclein pathology: Hypotheses, discoveries, and yet unresolved questions from experimental and human brain studies. *Acta Neuropathologica, 131*(1), 49–73. https://doi.org/10.1007/s00401-015-1485-1

Van Dijk, P. T., Dippel, D. W., Van Der Meulen, J. H., & Habbema, J. D. (1996). Comorbidity and its effect on mortality in nursing home patients with dementia. *The Journal of Nervous and Mental Disease, 184*(3), 180–187. https://doi.org/10.1097/00005053-199603000-00007

Vogelgsang, J., Wolff-Menzler, C., Kis, B., Abdel-Hamid, M., Wiltfang, J., & Hessmann, P. (2018). Cardiovascular and metabolic comorbidities in patients with Alzheimer's disease and vascular dementia compared to a psychiatric control cohort. *Psychogeriatrics, 18*(5), 393–401. https://doi.org/10.1111/psyg.12338

Walker, L., Stefanis, L., & Attems, J. (2019). Clinical and neuropathological differences between Parkinson's disease, Parkinson's disease dementia and dementia with Lewy bodies—Current issues and future directions. *Journal of Neurochemistry, 150*(5), 467–474. https://doi.org/10.1111/jnc.14698

Wang, P., Yao, L., Luo, M., Zhou, W., Jin, X., Xu, Z., Yan, S., Li, Y., Xu, C., Cheng, R., Huang, Y., Lin, X., Ma, K., Cao, H., Liu, H., Xue, G., Han, F., Nie, H., & Jiang, Q. (2021). Single-cell transcriptome and TCR profiling reveal activated and expanded T cell populations in Parkinson's disease. *Cell Discovery, 7*(1), 52. https://doi.org/10.1038/s41421-021-00280-3

Wenning, G. K., Colosimo, C., Geser, F., & Poewe, W. (2004). Multiple system atrophy. *The Lancet. Neurology, 3*(2), 93–103. https://doi.org/10.1016/s1474-4422(03)00662-8

West, A. B., Moore, D. J., Biskup, S., Bugayenko, A., Smith, W. W., Ross, C. A., Dawson, V. L., & Dawson, T. M. (2005). Parkinson's disease-associated mutations in leucine-rich repeat kinase 2 augment kinase activity. *Proceedings of the National Academy of Sciences of the United States of America, 102*(46), 16842–16847. https://doi.org/10.1073/pnas.0507360102

Wyss-Coray, T., & Mucke, L. (2002). Inflammation in neurodegenerative disease—A double-edged sword. *Neuron, 35*(3), 419–432. https://doi.org/10.1016/s0896-6273(02)00794-8

Zimprich, A., Biskup, S., Leitner, P., Lichtner, P., Farrer, M., Lincoln, S., Kachergus, J., Hulihan, M., Uitti, R. J., Calne, D. B., Stoessl, A. J., Pfeiffer, R. F., Patenge, N., Carbajal, I. C., Vieregge, P., Asmus, F., Müller-Myhsok, B., Dickson, D. W., Meitinger, T., et al. (2004). Mutations in LRRK2 cause autosomal-dominant parkinsonism with pleomorphic pathology. *Neuron, 44*(4), 601–607. https://doi.org/10.1016/j.neuron.2004.11.005

Zlokovic, B. V. (2010). Neurodegeneration and the neurovascular unit. *Nature Medicine, 16*(12), 1370–1371. https://doi.org/10.1038/nm1210-1370

Alzheimer's Disease

22

Tsuneya Ikezu

Abstract

Alzheimer's disease (AD) is the most frequent neurodegenerative disease and the most common cause of senile dementia in humans. While AD affects multiple systems of the central nervous system (CNS), the classic and most frequent initial symptom is the loss of short-term memory that progresses to profound memory failure. As initially observed by Alzheimer, the brains of AD patients exhibit two hallmark neuropathological features: (1) senile plaques containing depositions of amyloid-β peptide (Aβ) and (2) neurofibrillary tangles (NFT) containing hyperphosphorylated microtubule-associated protein tau. Quantitative evaluation has revealed that in the early stages of AD, there is a significant loss of neurons in brain regions specifically involved in memory. Genetic studies have identified causative genes, amyloid precursor protein (*APP*), presenilin-1 (*PSEN1*), and presenilin-2 (*PSEN2*), which established that aberrant proteolytic processing of the APP leading to the formation of the Aβ 1-40 and 1-42 is primarily responsible for the pathogenesis of familial AD. However, the mechanisms by which APP processing leads to senile plaque deposition, NFT formation, and neuronal cell death remain elusive. The objective of this chapter is to review the current understanding of these mechanisms and the role played by neuroimmune interactions, primarily innate immunity in the CNS, in the disease process

Keywords

Amyloid precursor protein (APP) · Apolipoprotein E (apoE) · β-amyloid peptide (Aβ) · Beta-amyloid precursor protein converting enzyme (BACE) · Insulin degrading enzyme (IDE) · Neurofibrillary tangle (NFT) · Oxidative damage · Presenilin-1 (PS1)

22.1 Clinical Features, Epidemiology, and Pathology

22.1.1 Symptoms (Memory Loss to Progressive Dementia)

AD is the leading cause of senile dementia without effective therapy (Sisodia, 1999). The symptoms of classic AD begin with a loss of short-term memory that slowly progresses to affect all cognitive functions, including long-term memory and executive functions (Corey-Bloom et al., 1994). Initially, impaired short-term memory often manifests itself as losing objects or forgetfulness in conversations. Subsequently, other early cognitive deficits occur; they include difficulty in word-finding, problem-solving, and spatiotemporal awareness. These symptoms are commonly accompanied by mental disorders, such as depression and apathy, and alternatively or subsequently agitation and irritability may occur. With further disease progression, memory failure becomes more profound and longer-term memory is affected. Patients with these deficits are frequently disoriented and can easily become lost; thus, a close supervision of patients become necessary. About this stage, their language and complex motor skills also deteriorate. Psychosis and hallucinations can occur along with bradykinesia and rigidity. Patients ultimately become bedridden, unable to speak, and die from infection (such as pneumonia) or other medical conditions. Average survival time of Alzheimer's patients is about 8 years after onset of initial symptoms, although the rate of decline is quite variable.

T. Ikezu (✉)
Department of Neuroscience, Mayo Clinic Florida, Jacksonville, FL, USA
e-mail: Ikezu.tsuneya@mayo.edu

H. E. Gendelman, T. Ikezu (eds.), *Neuroimmune Pharmacology and Therapeutics*,
https://doi.org/10.1007/978-3-031-68237-7_22

22.1.2 Diagnosis

The elements for the diagnosis of AD include several components: (1) a history obtained from someone who knows the patient well, (2) a physical and psychiatric exam to exclude other neurodegenerative diseases, (3) blood tests to exclude potential metabolic disorders involving the kidneys, liver, or thyroid, (4) a mental status test to determine dementia progression, and (5) brain imaging. The first cognitive assessment widely used was the Mini-Mental Sate Examination (MMSE). MMSE assesses multiple cognitive tasks including spatiotemporal orientation, immediate/delayed word recall, naming, verbal repetition, reading, writing, and spatial ability (Folstein et al., 1975). There are multiple reports of reduced cognitive performance in preclinical AD, notably impaired function in abstract reasoning, episodic memory, and new learning (for review see Fratiglioni et al., 2001). Episodic memory deficits are dominant in the early clinical and preclinical phases of AD. A diagnosis of dementia is usually determined using the DSM-5-TR (Diagnostic and Statistical Manual of Mental Disorders) of the American Psychiatric Association (Association, 2022). The final diagnosis is typically determined by autopsy using the criteria of NINDS-ADRDA (National Institute of Neurological Disorders and Stroke and Alzheimer's Disease and Related Disorders Association), which has been developed to document the neuropathology and underlying etiology (McKhann et al., 1984). Another diagnosis standard is the International Classification of Diseases (ICD)-10 guideline (Organization, 1992). Since DSM-IV and ICD-10 criteria closely follow NINDS-ADRDA guidelines, diagnostic procedures for AD are notably consistent throughout the world. Given that AD is the primary cause of dementia, it is often considered to be the likely cause of dementia while clinical tests are used to exclude other causes.

22.1.3 Epidemiology

AD affected an estimated 6.7 million in the United States in 2023, and prevalence studies demonstrate an exponential rise in dementia with age (Rajan et al., 2021). By 2060, the number of people aged 65 and older with AD may nearly double, from 6.7 million to as many as 14 million, barring the development of medical breakthroughs to prevent, slow or stop the disease (Rajan et al., 2021). AD is officially the 5th leading cause of death in the United States and the 5th leading cause of death for those aged 65 and older (US Department of Health and Human Services et al., 2022).

In addition to age, other factors are associated with an increased risk of AD. In developed countries, AD appears to be more common in women than men (Mielke et al., 2014). Lack of education is a risk factor for senile dementia in China and Europe (Schmand et al., 1997; Zhang et al., 1990). Head trauma is also a risk factor for both sporadic (Mortimer et al., 1991) and familial AD (Guo et al., 2000). Silent myocardial infarcts and coronary stenosis triple the risk for AD (Aronson et al., 1990; Sparks et al., 1990), suggesting the importance of vascular risk factors. Other potential risk factors being studied include diabetes and hypertension (Ott et al., 1999; Peila et al., 2002; Qiu et al., 2005; Reitz & Mayeux, 2014). As discussed below, many genetic mutations are now associated with either early-onset AD or with increased risk of late-onset AD (LOAD).

Identification of genetic mutations that cause early onset AD has helped to enhance our understanding of the pathogenesis of AD. Although cases of familial AD are relatively rare, they represent identified molecular mutations that cause AD. The genes identified to be responsible for early onset familial AD (EOAD) encode *APP*, *PSEN1,* and *PSEN2*; these are found on chromosomes 21, 14, and 1, respectively. In addition to these dominantly transmitted disease-causing mutations in early-onset AD, still other studies have identified over 50 genetic risk factors that influence the appearance of the more common sporadic late-onset AD. Current research is actively using genetically modified animals to determine how specific molecules are involved in the development of AD.

22.1.4 Genetic Studies of AD

AD is classified into EOAD (onset <65 years) and LOAD (onset ≥65 years). EOAD and LOAD account for 1–5% and >95% of all cases, respectively. EOAD is generally associated with a more rapid rate of progression and is mostly familial AD. Three genes (*APP*, *PSEN1,* and *PSEN2*) are identified in the EOAD pedigrees. All of them encode proteins involved in APP breakdown and Aβ generation. AD-linked mutations in these three genes exhibit high penetrance (>85%) and are mostly autosomal dominantly inherited.

In contrast, the genes involved in LOAD increase disease risk in a non-Mendelian fashion. For two decades, only one genetic risk factor, the apolipoprotein E4 isoform (*APOE*ε4) allele, located on chromosome 19q13, was an established risk gene in LOAD. APOE is a lipid-binding protein and is expressed in humans as three common isoforms coded for by three alleles, APOEε2, ε3, and ε4. A single *APOE-ε4* allele is associated with a 2- to 3-fold increased risk of LOAD; having two copies is associated with a fivefold or more increased risk (Kuusisto et al., 1994). A detailed discussion of ApoE biology can be found in Sect. 3.5.

Extensive genome-wide association studies (GWAS) have been conducted in a large cohort of LOAD subject in multiple institutions (Bellenguez et al., 2022; Hardy &

Singleton, 2009; Sherva & Farrer, 2011; Wightman et al., 2021). The first set of studies identified *CLU*, *PICALM*, *CR1* and *BIN1* as susceptibility loci (Harold et al., 2009; Lambert et al., 2009; Seshadri et al., 2010). These loci mainly cluster in four pathways, namely immune response, APP processing, lipid metabolism and endocytosis. The second set of large GWAS studies identified additional susceptibility genes (*CD33*, *MS4A4A/MS4A4E/MS4A6E cluster*, *ABCA7*, *CD2AP*, and *EPHA1*) (Hollingworth et al., 2011; Naj et al., 2011). All of these five loci are likely involved in the immune system, while the ABCA7 is in addition involved in lipid metabolism and APP processing. Finally, the largest GWAS to date, the International Genomics of Alzheimer's Project, performed a mega-meta-analysis that included 74,046 subjects (Lambert et al., 2013; Wightman et al., 2021). In addition to *APOE*, *CR1*, *BIN1*, *CD2AP*, *EPHA1*, *CLU*, *MS4A6A*, *PICALM*, *ABCA7*, and *CD33* this GWAS identified 12 additional susceptibility loci at genome-wide significance (*HLA-DRB5/HLA-DRB1*, *PTK2B*, *SORL1*, *SLC24A4/RIN3*, *DSG2*, *INPP5D*, *MEF2C*, *NME8*, *ZCWPW1*, *CELF1*, *FERMT2*, *CASS4*). Most of the 12 novel loci cluster in the specific pathways identified by the earlier GWAS, that is, immune response (*HLA-DRB5/DRB1*, *INPP5D*, *MEF2C*), APP processing (*SORL1*, *CASS4*), Tau pathology (*CASS4*, *FERMT2*), cell migration (*PTK2B*), and lipid transport and endocytosis (*SORL1*). These results strongly reinforce the importance of these pathways in the etiology of LOAD.

In the context of neuroimmune mechanism of LOAD, triggering receptor expressed by myeloid cells-2 (TREM2) is a transmembrane signaling molecule for phagocytosis and anti-inflammatory response and is genetically linked to LOAD, polycystic lipomembranous osteodysplasia with sclerosing leukoencephalopathy (PLOSL, also known as Nasu-Hakola disease), behavioral variant frontotemporal lobar degeneration (FTLD), Parkinson's disease (PD), and amyotrophic lateral sclerosis (ALS) (Cady et al., 2014; Giraldo et al., 2013; Guerreiro et al., 2013a, b; Jonsson et al., 2013; Paloneva et al., 2002; Rayaprolu et al., 2013). The detailed biology of TREM2 in microglia is discussed in Sect. 4.1.

22.1.5 Pathology

At autopsy, AD brains appear to have widespread neuronal and neuropil loss that thins cortical layers and expands ventricular and sulci spaces. However, there is not an obvious global loss of neurons in the cerebral cortex. Overall neuronal number appears relatively unchanged despite significant loss of brain tissue. Some of this apparent discrepancy is due to the preferential loss of large pyramidal neurons in the cortex and the relative sparing of smaller, densely packed neurons. Particularly vulnerable regions such as the entorhinal cortex and the CA1 of the hippocampus, show profound neuronal loss early in the disease. Wider involvement of temporal, frontal, and parietal cortices and a preferential loss of large neurons in these areas are seen as the disease progresses.

AD is characterized by an accumulation of senile plaques containing aggregated protein deposits of Aβ fibrils, numerous NFTs, reactive astrocytes, and activated microglia in the neocortex and hippocampus (Alzheimer, 1907; Selkoe, 1997). Senile plaques can contain either a diffuse amyloid deposition (presumably early senile plaques) or a dense core of insoluble Aβ with neuritic structures (mature senile plaques). Although Aβ deposition and NFT are commonly observed in nondemented elderly 80 years and older, increased levels of senile plaques are seen in AD brains (Berg et al., 1993; Schmitt et al., 2000). The degree of both Aβ deposition and NFT formation are correlated with cognitive decline (Berg et al., 1993; Braak & Braak, 1995; Naslund et al., 2000; Wilcock & Esiri, 1982).

Neuroinflammation is a well-recognized phenomenon and is intensively investigated, although the precise definition of brain inflammation has not yet been established (McGeer & McGeer, 2001). Clinically, AD patients exhibit few of the classic signs of inflammation. Histopathologically, the classical acute inflammatory response (neutrophil recruitment), immunoglobulin precipitation, and T-cell accumulation are not seen suggesting that humoral or classical cellular immune-mediated responses are not involved in AD progression (Eikelenboom et al., 1994). However, AD brains show an increase in activated microglial clusters on senile plaques surrounded by astrocytes, indicating focal glial inflammation (Eikelenboom et al., 2002; Wyss-Coray & Mucke, 2002). In particular, the volume density of $CD68^+$ mononuclear phagocytes (brain perivascular macrophages and microglia) is correlated with the volume density of Congophilic compact plaque deposits (Arends et al., 2000). $CD68^+$ cells are present from very early preclinical phases to terminal phases of AD and are involved in disease progression, which will be further described later (Akiyama et al., 2000; Fonseca et al., 2004a).

22.2 Molecular Pathogenesis, Animal Models, and Neuroinflammation

The molecular pathogenesis of AD is centered around Aβ production and its clearance. Aβ is generated by the processing of APP by β and γ-processing enzymes, and cleared from the brain by its diffusion, export to vascular system, phagocytosis, or degradation. The following molecules are essential for understanding the molecular mechanism of Aβ processing.

22.2.1 Amyloid Precursor Protein (APP)

APP is the first gene to be identified as a causative gene of familial AD (FAD) and has been a key molecule in the study of the molecular mechanism of AD. As shown in Fig. 22.1, the APP molecule can be cleaved at different specific cleavage sites by various proteases to generate fragments of different sizes. APP is primarily expressed in neurons and cleaved at the α, β, and γ sites by APP processing enzymes (called α, β, and γ−secretases), producing Aβ (β- and γ-processing product), p3 (α- and γ-processing product), and other APP fragments (Selkoe, 1994). Aβ is composed of 39 to 43 amino acids, mainly 1-40 (Aβ40) and 1-42 (Aβ42). Aβ42 is more amyloidogenic (faster to be aggregated) than Aβ40, and early Aβ deposits are usually detectable with anti-Aβ42 antibodies, but not with anti-Aβ40 antibodies (Selkoe, 1994). FAD mutations on the APP gene are all linked to abnormal APP processing (increase in Aβ42/Aβ40 ratio), Aβ aggregation, and Aβ deposition (Scheuner et al., 1996).

22.2.2 PS1 and γ-Processing Enzyme Complex

A combination of the presenilin genes (PS1 and 2), aph-1, pen-2, and nicastrin forms the γ-secretase complex (De Strooper, 2003). The γ-secretase complex cleaves not only APP but also other transmembrane molecules, such as Notch, ErbB-4, and sterol regulatory element binding proteins (SREBPs). PS1 and PS2 have been established as causative genes of early-onset FAD (Levy-Lahad et al., 1995; Sherrington et al., 1995). Transgene expression of PS1 FAD mutants enhances the Aβ42/Aβ40 ratio and accelerates Aβ aggregation and deposition in APP mice, which overexpress Swedish FAD APP mutant in brains (Borchelt et al., 1997; Duff et al., 1996; Holcomb et al., 1998). PS2 FAD mutant overexpression converts Aβ40 to Aβ42, whereas PS2 wild-type overexpression reduces both Aβ40 and Aβ42 (Mastrangelo et al., 2005). Taken together, these in vivo studies demonstrate the effect of PS1/2 FAD mutations on Aβ aggregation through the transition of Aβ40 to Aβ42.

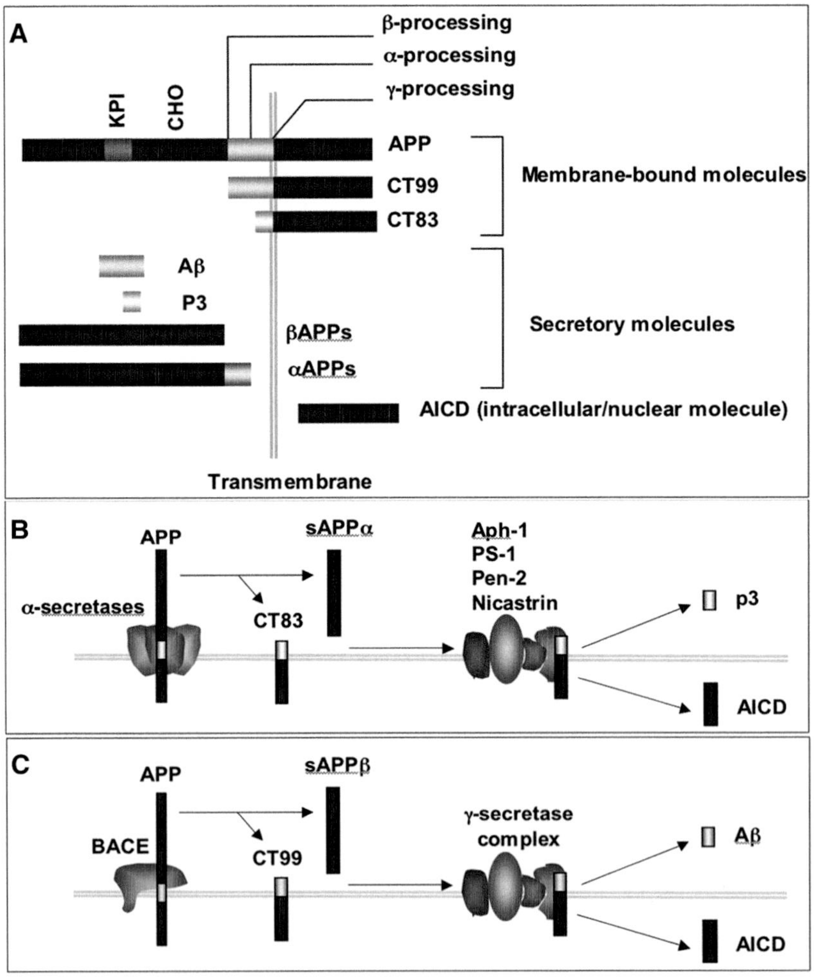

Fig. 22.1 Scheme of APP processing. (**a**) Cleavage sites and processing enzymes of APP. (**b**) P3 and sAPPα production by α and γ-processing of APP. (**c**) Aβ and sAPPβ production by β and γ-processing of APP

22.2.3 APP and PS1 Animal Models

22.2.3.1 Conventional APP and PS1 Mouse Models

Several established AD mouse models have made significant contributions to the understanding of AD pathogenesis (Price et al., 1998; Spires & Hyman, 2005). Analyses of AD brain tissue obtained from autopsies provide only limited insight into the dynamic nature of the disease process. Currently, no AD mouse model has all of the characteristics of the human disease. However, several lines of transgenic mice expressing mutated APP, such as PDAPP (Games et al., 1995), Tg2576 (Hsiao et al., 1996), APP23 (Sturchler-Pierrat et al., 1997), TgCRND8 (Chishti et al., 2001), APPswe TgC3-3 (Borchelt et al., 1997), PSAPP (Holcomb et al., 1998), and 3xTg-AD mice (Oddo et al., 2003), simulate prominent behavioral and pathological features of AD. Phenotypes displayed by these mice include age-related impairments in learning and memory, electrophysiological abnormalities, neuronal loss, microgliosis, astrogliosis, neuritic changes, amyloid deposition, oxidative stress, and abnormal tau phosphorylation (Price et al., 1998; Smith et al., 1998) (Fig. 22.2). Microglia in close proximity to fibrillar Aβ deposits are immunoreactive to interleukin (IL)-1β and TNF-α, whereas activated astrocytes are immunoreactive to IL-1β, transforming growth factor (TGF)-1β, and IL-10 in Tg2576 mice (Apelt & Schliebs, 2001; Benzing et al., 1999; Frautschy et al., 1998).

Tg2576 mice which express Swedish familial AD mutant (K670N/M671L) of APP695 under a hamster prion promoter show memory and learning impairments as determined by Morris water maze (Hsiao et al., 1996). The memory recall is impaired as early as 6 months of age, which persists without significant progression for up to 12 months of age, followed by impairment of both memory acquisition and recall after 12 months of age (Lesne et al., 2006). There is a strong correlation of such age-related memory decline to Aβ accumulation in the brain (Chen et al., 2000; Janus et al., 2000). A recent study in Tg2576 mice suggests that 9- or 12-mer Aβ oligomers are responsible for such memory loss, which can be isolated by Tg2576 and can induce transient memory loss in injected rodents (Lesne et al., 2006). Contextual fear conditioning tests, which represents hippocampal learning, show memory impairments in PDAPP mice as early as 4 months of age preceding the Aβ deposition (Jacobsen et al., 2006). These studies strongly suggest that onset of memory loss precedes the Aβ deposition, is more correlated with Aβ oligomer formation in the brain, may also be relevant to the clinical symptoms of AD patients.

22.2.3.2 APP Knocked-In Mouse Models

Transgene expression of recombinant molecules under the control of exogenous promoter derived from genetic fragment of other species are often associated with artificial phenotypes not predicted or commonly seen in human disease. Understanding the effects of elevated Aβ from those due to APP overexpression is a common challenge when interpreting the phenotypes of APP transgenic mice. To overcome this issue, Saito et al. created a new APP knocked-in mouse models (APP^{NL-F} and APP^{NL-G-F}) (Saito et al., 2014). These mouse models avoid potential artifacts introduced by APP overexpression by using a knock-in approach to express APP

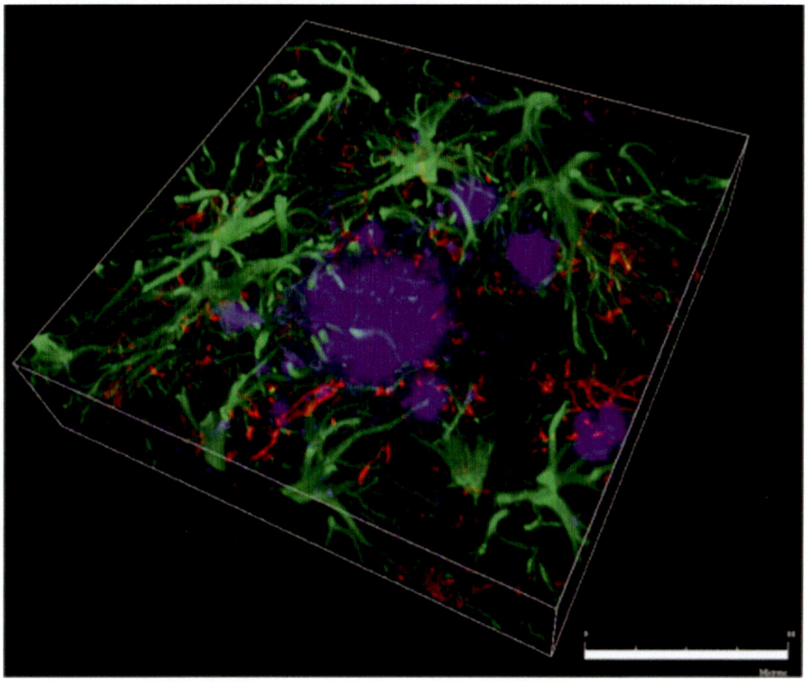

Fig. 22.2 3D reconstruction image of amyloid plaque deposition and glial accumulation in APP mouse brain. Cortical section of Tg2576 (14 months of age) was stained by thioflavin-S (compact amyloid plaque, purple), anti-IBA1 rabbit polyclonal (microglia, red), and anti-GFAP mouse monoclonal (astrocytes, green) and imaged using a laser scanning confocal microscopy (Zeiss 510 Meta). Z-stack images were deconvolved and 3D reconstructed using image processing software (AutoQuant)

at wild-type levels and with appropriate cell-type and temporal specificity. APP is not overexpressed, but levels of pathogenic Aβ are elevated due to the combined effects of two (in APP^{NL-F}) or three mutations (in APP^{NL-G-F}) linked to familial AD. Both APP^{NL-F} and APP^{NL-G-F} mice accumulate Aβ and recapitulate several AD-associated pathologies, including amyloid plaques, synaptic loss, and microgliosis and astrocytosis, especially in the vicinity of plaques. The presence of the Arctic mutation in APP^{NL-G-F} accelerates the pathology relative to APP^{NL-F} mice and leads to particularly severe phenotypes. Aβ deposition is observed starting at 2 months and is nearly saturated by 7 months. In contrast to APP^{NL-F} mice, the APP^{NL-G-F} mice develop subcortical amyloidosis in addition to cortical amyloidosis, consistent with the neuropathology observed in patients with the Arctic mutation. The mice also display age-associated cognitive impairment, specifically memory impairment as measured by the Y maze starting at 6 months. The cognitive impairment in APP^{NL-G-F} mice is more severe than the impairment in APP^{NL-F} mice. These mice will be suitable for future studies using APP mouse models.

22.2.4 β and α-Processing Enzymes

Beta-site APP-cleaving enzyme (BACE1, also called Asp2 or Memapsin2), a β-secretase, is an aspartic transmembrane proteinase (Vassar et al., 1999). BACE2 (also called Asp1 and DRAP) is a homolog of BACE1 and contains the same overall structural organization. BACE1 is essential for Aβ production as demonstrated by BACE1 gene-targeted mouse models (Cai et al., 2001). BACE overexpression enhances total Aβ production, deposition of both diffuse and compact Aβ plaques in brain parenchyma, and reduces cerebrovascular amyloid angiopathy in aged APP/BACE double transgenic mice (Mohajeri et al., 2004; Willem et al., 2004). However, this effect seems to be dose-dependent, since high BACE expression rather inhibits Aβ deposition (Lee et al., 2005). Thus, the effect of BACE upregulation in Aβ pathology in vivo is not conclusive. A Disintegrin And Metalloprotease (ADAM) 9, 10, and 17 have been identified as putative α-secretases (Asai et al., 2003). Overexpression of ADAM10 prevents Aβ deposition and hippocampal defect in APP mice (Postina et al., 2004), suggesting its therapeutic application for enhancing α-processing of APP.

22.2.5 ApoE

ApoE is the most well-characterized genetic risk factor of AD. Evidence indicates that endogenous apoE in mouse brain enhances Aβ deposition in APP animal models (Bales et al., 1999; Sadowski et al., 2004). Tg2576 mice lacking mouse apoE exhibit diminished compact plaque deposition, while diffuse plaque load remained unchanged (Bales et al., 1999). Mouse apoE is rather amyloidogenic since apoE disruption significantly reduces the compact plaque formation in APP mice (Bales et al., 1999). In contrast, human apoE3 enhances Aβ clearance in the mouse brain, while apoE4 weakly promotes Aβ clearance (Fagan et al., 2002; Holtzman et al., 2000; Holtzman et al., 1999). Although many studies concluded that astrocytes are the main apoE-producing cells in brain (Danik et al., 1999), recent reports suggest that activated microglial cells are also a source of apoE (Keren-Shaul et al., 2017; Krasemann et al., 2017; Mori et al., 2004; Saura et al., 2003). A recent study of EGFP-ApoE knock-in mice demonstrated that apoE is expressed in 75% of astrocytes, and <10% of microglia after stimulation. Thus, apoE is primarily an astrocytic protein (Xu et al., 2006). A recent study identified *Apoe* as one of the representative molecules of disease-associated/neurodegenerative microglia (DAM/MGnD) (Keren-Shaul et al., 2017; Krasemann et al., 2017) along with *Axl, Bhlhe40, Clec7a, Csf1, Cst7, Ctsb, Itgax, Gpnmb, Il1b, lagals3, Lpl,* Mir155*, Nos2, Spp1, Trem2,* and *Tyrobp* (for review, see Butovsky and Weiner (2018)).

22.2.6 Aβ Degrading Enzymes

The current knowledge of Aβ degradation biology is limited since only a few of the molecules involved have been characterized, namely, insulin degrading enzyme (IDE), neprilysin, which is also known as enkephalinase, and neutral endopeptidase (NEP) (Iwata et al., 2000; Selkoe et al., 2001; Yamin et al., 1999). IDE was originally identified as a microglia-secreted Aβ degrading enzyme (Qiu et al., 1998) and is also found in neurons (Vekrellis et al., 2000). Neprilysin protein level is downregulated during aging but is locally upregulated in activated astrocytes surrounding Aβ plaques in APP mice (Apelt et al., 2003). Thus, the amount of IDE and neprilysin can be altered by chronic inflammatory reaction. Neprilysin, a neutral zinc metallopeptidase, plays a role in degrading insoluble and/or aggregated Aβ42 (Iwata et al., 2000; Selkoe et al., 2001). Transgenic expression of both IDE and neprilysin show inhibition of Aβ accumulation in APP mice (Leissring et al., 2003). In addition, endothelin converting enzyme and matrix metalloprotease-2 and 3 are new candidates of Aβ degrading enzymes (Choi et al., 2006; White et al., 2006).

22.2.7 Neurofibrillary Tangle Formation

Presence of NFT is one of the original neuropathological hallmarks of AD. NFT are fibrous tangles composed of insoluble, conformationally abnormal, hyperphosphorylated tau protein deposited in neuronal cell bodies. Hyperphosphorylated tau can form a specific insoluble structure known as a paired helical filament (PHF) (Buee et al., 2000; Lee et al., 2001). Six tau protein isoforms are generated by alternative mRNA splicing of the *tau* gene. Based on the repeat of the intermediate microtubule-binding domain, the majority of tau is classified as three repeat (3R) or four repeat (4R) tau. The primary function of tau is to bind to and stabilize microtubules, thereby promoting microtubule polymerization. Many tau phosphorylation sites flank microtubule binding domains and are believed to play a significant role in its microtubule binding (Lee et al., 2001; Planel et al., 2002). Hyperphosphorylation of tau appears to cause tau to dissociate from microtubules and form tau protein aggregates, which becomes PHF (Alonso et al., 1996; Biernat et al., 1993; Bramblett et al., 1993). Glycogen synthase kinase 3-β (GSK3β), cyclin dependent protein kinase 5/p25 complex, microtubule affinity regulating kinase, and recently tau-tubulin kinase 1 (TTBK1) have been extensively characterized as tau kinases that seem to be involved in the hyperphosphorylation of tau (Asai et al., 2014; Drewes et al., 1997; Ikezu et al., 2020; Ishiguro et al., 1992; Patrick et al., 1999; Sato et al., 2006, 2008; Takashima et al., 1993; Xu et al., 2010). For the review of TTBK1, see Ikezu & Ikezu (2014).

Tau-related neurodegenerative disorders, or "tauopathies," include AD, argyrophilic grain dementia, corticobasal degeneration, diffuse neurofibrillary tangles with calcification, Down's syndrome, frontotemporal dementia, and Parkinsonism linked to chromosome 17 (FTDP-17), Gerstmann–Straussler–Scheinker disease, Niemann–Pick disease, non-Guamanian motor neuron disease with neurofibrillary tangles, Pick's disease, progressive supranuclear palsy, subacute sclerosing panencephalitis, and tangle only dementia (Lee et al., 2001). Genetic studies show that missense mutations in the *tau* gene (*Mtapt*) induce FTDP-17 supporting its importance in tauopathies (Hutton et al., 1998; Lee et al., 2001), although such mutations have not been found in AD.

These tau mutations develop similar pathology in transgenic animal models, such as P301L tau mouse (Gotz et al., 2001a; Lewis et al., 2000), R406W tau mouse (Sato et al., 2002), G272V tau mouse (Gotz et al., 2001b), and V337M tau mouse (Tanemura et al., 2002). However, P301L tau mice develop NFT-like tau aggregation not only in CNS but also in peripheral motor neurons, astrocytes, and oligodendrocytes, which is correlated with gait disturbances (Lewis et al., 2000).

rTg4510 mice express FTDP-17-linked tau P301L mutant under the control of tetracycline response element (TRE), to which Ca^{2+}/calmodulin-dependent kinase II (CaMKII) promoter-driven tTA binds and suppresses the gene transcription in the presence of doxycycline (Dox) (Mayford et al., 1996; Santacruz et al., 2005). The CaMKII promoter is specifically activated in and largely restricted to the hippocampus, cortex, olfactory bulb, and striatum, and Tg4510 shows age- and Dox-dependent tau aggregation and neurodegeneration in the brain subregions (Santacruz et al., 2005). This is, so far, one of the most advanced tauopathy animal models with a clear neurodegeneration phenotype, which can be controlled by Dox treatment.

In the context of neuroinflammation, P301S tau mice, another FTDP-17 mouse model, show age-dependent neurodegeneration in entorhinal cortical and hippocampal regions, which is preceded by extensive microglial activation and can be suppressed by an immunosuppressive agent, FK506 (Yoshiyama et al., 2007), indicative of nonautonomous neuronal cell loss. Similarly, another P301S tau mouse model shows extensive neurodegeneration in the brain stem and spinal cord, which is accompanied by neuroinflammation and M1-type mononuclear phagocyte accumulation (Allen et al., 2002; Bellucci et al., 2004). These studies reveal that a specific type of neuroinflammation may be one of the mechanisms in tauopathy-related neurodegeneration and a potential therapeutic target.

How tau propagates from neurons to neurons has been extensively studied in the last 10 years, and it mainly falls into trans-synaptic or exo-synaptic spread of misfolded tau protein in soluble or extracellular vesicle (EV) forms (for review, see DeLeo & Ikezu, 2018; Delpech et al., 2019; You et al., 2020; You & Ikezu, 2019). Our group pioneered in microglia and EV-mediated spread of tau pathology in animal models (Asai et al., 2015; Clayton et al., 2021) and using human AD brain-derived EVs (Ruan et al., 2021). The existence of tau as paired helical filament is shown in human AD brain-derived EVs by cryo-electron tomography (Fowler et al., 2023). Another cryo-electron tomography study demonstrates the accumulation of EVs in the amyloid plaques of aged APP^{NL-G-F} mouse brain (Leistner et al., 2023). The ultrastructural analysis of the subcellular and membrane-bound compartments surrounding the MX04-stained plaques is consistent with those of microglia and astrocytes. This is in line with our recent study, showing that plaque-associated microglia are activated and exhibit enhanced biogenesis and secretion of EVs, which we visualized in single-vesicle resolution in the APP^{NL-G-F} mouse brain (Clayton et al., 2021).

22.2.8 Neurotoxicity and Synaptic Dysfunction

22.2.8.1 Aβ-Mediated Neurotoxicity

It has been widely accepted that toxicity of Aβ requires aggregation of native Aβ monomers (Hardy & Higgins, 1992; Pike et al., 1993). Aβ (Aβ40 or Aβ42) self-assembles to form low-n oligomers (dimers-hexamers), protofibrils, and fibrils (Harper et al., 1997; Lambert et al., 1998; Roher et al., 1996; Walsh et al., 1997). Since intermediates can further associate into higher-ordered aggregates, it is difficult to determine the oligomerization state of Aβ that causes pathogenicity. Nonetheless, several groups succeeded in isolating spherical nonfibrillar assemblies of synthetic Aβ (Aβ-derived diffusible ligands; ADDLs and amylospheroid; ASPD) (Hoshi et al., 2003; Lambert et al., 1998). These Aβ oligomers show more potent toxicity than fibrillar Aβ at lower concentrations (Lambert et al., 1998) and induce reversible synaptic loss through N-methyl-D-aspartate (NMDA) receptor interaction in organotypic hippocampal culture systems (Shankar et al., 2007). Thus, Aβ oligomers are a likely therapeutic target (De Felice et al., 2004).

Aβ can also potentially mediate neurotoxicity by activating microglia and glia and hereby promoting their secretion of a variety of neurotoxic factors. Aβ and secreted β-amyloid precursor protein (sAPP)-induce microglial activation and neurotoxicity (Barger & Harmon, 1997; Giulian & Baker, 1986; Ikezu et al., 2003; Meda et al., 1995). Neurotoxicity can be mediated through the secretion of neurotoxic agents such as FasL, tumor necrosis factor (TNF)-α, reactive oxygen species (ROS), proteases, excitatory amino acids, and nitric oxides. Excitatory amino acids (glutamate, quinolinic acid, D-serine, among others) activate NMDA receptors, which are significantly involved in microglial-mediated neurotoxicity (Giulian et al., 1994; Piani et al., 1992). Indeed, both Aβ and sAPP-activated microglia increase glutamate secretion (Barger & Basile, 2001; Ikezu et al., 2003). Glutamate and ROS can exacerbate each other's neurotoxic effects due to their relationships with the cystine/glutamate exchange system (Xc^-) antiporter. Aβ/sAPP-induced microglial activation leads to ROS production and glutathione consumption, which in turn promotes cystine import to regenerate glutathione and, as a consequence, increased glutamate is exported in exchange through the Xc^- antiporter (Ikezu et al., 2003).

22.2.8.2 Aβ-Mediated Synaptic Dysfunction

There are a number of studies documenting the synaptic dysfunction of APP mice, which starts as early as 4 months of age, including enhanced paired-pulse facilitation, disturbed high-frequency stimulation burst, and rapid decay of LTP in the CA1 field of hippocampal slices (Hsia et al., 1999; Larson et al., 1999). In addition, impaired synaptic transmission and LTP in both the CA1 and DG regions are evident in Tg2576 mice (Chapman et al., 1999). Aβ oligomers block hippocampal long-term potentiation (LTP) ex vivo at nanomolar concentrations. In vivo applications of Aβ oligomers potently inhibit rat hippocampal LTP and cognitive function (Cleary et al., 2005; Walsh et al., 2002). These studies strongly suggest that Aβ oligomers induce memory impairment through inhibition of synaptic transmission and long-term potentiation in the brain. One of the potential mechanisms of Aβ-induced LTP inhibition is through the activation of a number of inflammation-related molecules, such as c-jun N-terminal kinase, CDK5, p38 mitogen-activated kinase, inducible nitric oxide synthase, and superoxide, (Wang et al., 2004a, b). These data suggest significant involvement of microglial activation in the inhibitory mechanism.

22.2.8.3 Neurotoxicity by Tau Aggregation

Although NFT is a hallmark of AD, its direct involvement in neuronal cell death is still unclear. Presence of NFT does not appear to be highly toxic since clinical studies suggest long-term survival of NFT-bearing neurons (Hof et al., 2003; Morsch et al., 1999). These data are also supported by studies using FTDP-17 tau mutant mice showing NFT formation in the absence of neuronal cell loss (Santacruz et al., 2005) and a lack of cognitive impairment as assessed by a water maze test (Arendash et al., 2004). A new doxycycline-inducible tau transgenic mouse model suggests accumulation of intracellular tau protein rather than NFT formation is responsible for neurodegeneration and cognitive impairment (Ramsden et al., 2005; Santacruz et al., 2005). This model has a caveat as the tTA transgene insertion caused disruption of Fgf14, which induces neurodegeneration regardless of tau transgene expression (Gamache et al., 2019). Tau is known to be involved in Aβ aggregate-induced neurotoxicity in conjunction with protein kinase activation in vitro (Hoshi et al., 2003; Ikezu et al., 2020; Medina et al., 2005; Rapoport et al., 2002). Thus, it is possible that acute neuronal cytoskeletal destabilization by tau phosphorylation and its end-product (globular tau aggregates) are neurotoxic, but its subsequent subcellular compartmentalization as NFT/PHF is a preventive neuroprotective process used to contain neurotoxic aggregates. Further study will be necessary to characterize its effect on neurodegeneration or dementia.

22.3 Neuroimmunological Mechanism of AD

Neuronal loss, edema, and recruited mononuclear phagocytes (moving from blood to brain), are all classical hallmarks of brain inflammation in AD (Arvin et al., 1996). Gliosis, characterized by activation, proliferation, and hypertrophy of astrocytes and microglia, is one important feature of neuroinflammation. Amyloid plaques contain several proteins, including complement proteins and pro-inflammatory cytokines and chemokines. Many of these proteins are secreted by activated microglia and reactive astrocytes. Activated microglia integrate deeply into neuritic plaques and express major histocompatibility complex type II (MHC-II), a marker for microglial activation. MHC-II expression is significantly upregulated in AD brains (Rogers et al., 1988; Styren et al., 1990).

Neurotoxicity can be mediated through secretion of various factors, ROS, proteases, excitatory amino acids, and nitric oxides. Pro-inflammatory cytokines and chemokines can inhibit fibrillar Aβ phagocytosis by a murine microglial cell line (Koenigsknecht-Talboo & Landreth, 2005). Such T-cell-mediated cytokines also inhibit post-phagocytosis intracellular Aβ degradation in human monocyte-derived macrophages. Aβ-microglia interaction results in two outputs: phagocytosis and signaling for priming and activation. Aβ binding to a complex of cell surface proteins (including CD36, integrin-associated protein (CD47), and $\alpha_6\beta_1$-integrin) leads to src-like tyrosine kinase activation and ROS production (Bamberger et al., 2003). CD36 is necessary for Aβ-induced mononuclear phagocyte activation and chemotaxis in vitro (El Khoury et al., 2003). These findings strongly suggest that Aβ aggregates activates microglia by its binding to specific cell surface molecules, which activates src-like tyrosine kinases and other signaling for Aβ clearance and neuroinflammation in AD brains.

22.3.1 Microglia-Mediated Phagocytosis

22.3.1.1 Aβ Scavenger Receptors

Our current knowledge of Aβ clearance is limited compared to the more detailed understanding of Aβ synthesis and aggregation mechanisms. It is well known that Aβ stimulation enhances microglial phagocytosis, in vitro (Kopec & Carroll, 1998). The internalized Aβ in phagosomes initially migrates to acid hydrolase-containing late endosomes and lysosomal compartments and then moves into perinuclear vesicles, which are morphologically similar to lysosomes, where it stays up to 20 days (Paresce et al., 1997). The long-term accumulation of Aβ has also been observed by using unlabeled Aβ aggregates in bone marrow-derived macrophages (Yamamoto et al., 2007b). In addition to microglia, astrocytes contribute to the uptake and removal of Aβ in a manner enhanced by a CC-type chemokine, CCL2 (Wyss-Coray et al., 2003).

Microglia constantly take up Aβ via potential Aβ receptors, including scavenger receptor type A (SR-A), CD36, and receptor for advanced glycation end product (RAGE) (El Khoury et al., 1996, 2003; Yan et al., 1996). Although SR-A was characterized as a major receptor for microglial phagocytosis of Aβ (Paresce et al., 1996), APP mice deficient in SR-A shows no difference in amyloid plaque formation or synaptic degeneration (Huang et al., 1999). This suggests that other scavenger receptors or Aβ-binding molecules, such as CD36, CD47, and $\alpha_6\beta_1$-integrin, can compensate for the deficiency of SR-A. On the other hand, RAGE seems to enhance Aβ deposition. APP mice deficient in RAGE exhibit reduced Aβ deposition (unpublished observations), which is consistent with enhanced Aβ deposition in mice overexpressing RAGE (Arancio et al., 2004). P-glycoprotein (Pgp), an ATP binding cassette transporter/multidrug resistance-1 α and β, is another molecule involved in Aβ clearance from brain parenchyma. Disruption of Pgp exhibits enhanced Aβ deposition, suggesting its role in endothelial efflux of Aβ from the brain (Cirrito et al., 2005).

22.3.1.2 TREM2/TYROBP Pathway for Phagocytic Clearance and Disease

TREM2 is mostly expressed in myeloid cells (immature dendritic cells, microglia, and osteoclasts). In the CNS, TREM2 is expressed at levels greater than 300-fold in microglia compared to astrocytes (Hickman & El Khoury, 2014). Inflammation decreases TREM2 expression in microglia and their ability to phagocytize apoptotic neurons (Takahashi et al., 2007). TREM2 requires the interaction with the adaptor protein, TYRO protein tyrosine kinase binding protein (TYROBP; formerly DAP12), which facilitates the downstream signaling of phagocytosis (Paloneva et al., 2000). Thus, healthy coupling of TREM2 and TYROBP is essential for the phagocytic function of myeloid cells. No endogenous ligand has been confirmed for TREM2, but TREM2 is shown to bind to Gram^+ and Gram^- bacteria and yeast in a charge-based method (Daws et al., 2003). Stimulation of TREM2 by a cross-linked TREM2 antibody is shown to activate TREM2. This activation causes actin polymerization and restructuring of the cytoskeleton (Takahashi et al., 2005).

TYROBP is a transmembrane adaptor protein that binds to TREM2 and signals through immunoreceptor tyrosine-based activation motifs (ITAMs). The ITAM signal is important for phagocytosis activation in microglia (Linnartz & Neumann, 2013). Tyrosine phosphorylation of ITAM by Src family proteins activates TYROBP. The phosphorylated ITAM binds to protein spleen tyrosine kinase (Syk) (Fig. 22.1). Syk activates many downstream cascades linked to phagocytosis (Linnartz & Neumann, 2013). The ITAM activation is negatively regulated by the immunoreceptor tyrosine-based inhibition motifs (ITIMs). When a regulatory ligand binds to a receptor containing an ITIM, recruitment of SHP1/SHP2 occurs. SHP1/SHP2 can dephosphorylate and inhibit ITAM activation and other downstream cascades (Linnartz & Neumann, 2013). TYROBP-deficient mice show enhanced inflammatory phenotypes (Hamerman et al., 2005).

Mutations in TREM2 or TYROBP are thought to be the main cause of Nasu-Hakola (Paloneva et al., 2003). Nasu-Hakola disease is polycystic lipomembranous osteodysplasia with sclerosing leukoencephalopathy (PLOSL), a rare autosomal recessive disease (Paloneva et al., 2003). Patients who are diagnosed with Nasu-Hakola disease suffer from bone cyst-like lesions and bone fractures often on their lower legs and also experience rapid pre-senile dementia (Paloneva et al., 2001, 2003). Around the age of 40, patients have severe neurodegeneration and suffer from severe dementia, gait disturbances, sensory agnosia, and agraphia and in some cases develop tumors in the CNS (Hakola & Puranen, 1993). The link between neurodegeneration and TREM2/TYROBP function provided by Nasu-Hakola disease indicates that the phagocytic and anti-inflammatory function of TREM2/TYROBP signaling is important for healthy brain function.

The missense R47H mutation in TREM2 is associated with LOAD, has an effect on both folding and ligand binding, and decreases the clearance of damaging debris (Abduljaleel et al., 2014). Similarly, FTDL-linked missense mutations in TREM2 (T66M and Y38C) also reduce phagocytic activity (Kleinberger et al., 2014). TYROBP is recently highlighted as one of the central genetic nodes and networks in LOAD (Paloneva et al., 2001, 2000; Zhang et al., 2013). Since both TREM2 and TYROBP are specifically expressed in myeloid cells and physically interact to transduce phagocytosis and anti-inflammatory signaling (Linnartz & Neumann, 2013; N'Diaye et al., 2009; Paradowska-Gorycka & Jurkowska, 2013), the TREM2/TYROBP pathway may play an important role in regulating peripheral or central inflammation. Indeed, TREM2 plays a central role in the metabolism and DAM/MGnD phenotype induction in mouse models (for review, see Butovsky and Weiner (2018)).

22.3.2 Glial Inflammation and Innate Immunity

Innate immune responses in AD are mediated by a large number of inflammatory proteins, such as complement factors, acute-phase proteins, pro-inflammatory cytokines, and chemokines (Akiyama et al., 2000; McGeer & McGeer, 2001). Upregulation of the complement system includes activation of both classical and alternative pathways and enhanced secretion of cytokines and chemokines including, but not limited to, IFN-γ, IL-1α/β, IL-3, IL-4, IL-10, TGF-β1/2, TNF-α, CD40L, CCL2 (MCP-1), CCL3 (MIP-1α), CCL4 (MIP-1β), S100-β (for review, see Akiyama et al., 2000). The interplay between Aβ aggregates, microglia, and astrocytes results in the secretion of pro-inflammatory cytokines that, in turn, may affect local APP expression in nearby neurons. This process results in local upregulation of Aβ and seeding of "new" Aβ deposits, which may therefore be a potential mechanism for the plaque accumulation typically seen in AD brains. A detailed review on this topic can be found in our recent publication (Varnum & Ikezu, 2012).

22.3.2.1 Cytokines

IFN-γ

IFN-γ is produced by Th1 type T cells and natural killer cells (Young & Hardy, 1995), as well as macrophages (Fultz et al., 1993), astrocytes, and microglia (DeSimone et al., 1998). IFN-γ affects Aβ production, Aβ degradation (Yamamoto et al., 2007a), and neurotoxicity (Meda et al., 1999). IFN-γ, in conjunction with TNF-α, has been shown to enhance APP mRNA transcription and Aβ production (Blasko et al., 1999) and indirectly induce neuronal loss by stimulating release of NO from microglia (Chao et al., 1995). In the Tg2576 mouse model, astrocytes significantly upregulate IFN-γ mRNA in vivo (Abbas et al., 2002). We have recently found that transgenic APP mice (Tg2576) lacking IFN-γ receptor type I show marked reduction of Aβ deposition, microgliosis, astrogliosis, and astrocytic BACE expression (Yamamoto et al., 2007a). On the other hand, IFN-γ also show beneficial role in socialization (Filiano et al., 2016) and induction of DAM/MGnD, which is beneficial for the compaction of the amyloid plaques (Yin et al., 2023).

IL-1b

IL-1b is the end-product of inflammasome activation along with IL-18 and plays a pivotal role in many neuroinflammatory disorders, including AD. Deletion core inflammasome molecule *Nlrp3* results in the resolution of neuroinflammation and amyloid plaque formation in APP/PS1 mouse models (Heneka et al., 2013). IL-1b upregulation seems to occur early in Aβ plaque formation since it is associated with diffuse plaques and also found in young children with Down's syndrome (Griffin et al., 1995, 1989). IL-1 induces S100β, an astrocyte-derived neurite-promoting cytokine (Kligman & Marshak, 1985), and α1-antichymotripsin (Das & Potter, 1995). IL-1 upregulates APP synthesis in primary astrocytes (Rogers et al., 1999), and Aβ production when co-stimulated with IFN-γ in vitro (Blasko et al., 2000).

IL-3

IL-3 is a multi-potent cytokine and is involved in autoimmune diseases (Mindur & Swirski, 2019). Although its expression is not robust in the CNS, astrocytes produce IL-3 and is reported as a growth factor of myeloid cells (Frei et al., 1986). A recent study demonstrated the role of astrocytic IL-3 for microglial activation and containment of amyloid plaque (McAlpine et al., 2021), suggesting as the therapeutic target for microglia-mediated amelioration of AD pathology development.

IL-4

IL-4 is a prototypical anti-inflammatory cytokine derived from Th2 cells. IL-4 is a particularly important cytokine for M2a skewing (Lyons et al., 2007b; Maher et al., 2005; Nolan et al., 2005). Microglia stimulated by IL-4 have a decrease in TNF-α production and an increase in secretion of insulin-like growth factor-1 (IGF-1), and they also induce neurogenesis in vitro (Butovsky et al., 2005, 2006). Stimulation of microglia by IL-4 enhanced neural differentiation (Kiyota et al., 2010). Gene delivery of IL-4 into APP+PS1 mice partially suppressed glial accumulation in the hippocampus, directly enhanced neurogenesis, restored impaired spatial learning, and also reduced Aβ deposition (Kiyota et al., 2010). Gene delivery of either IL-4 also reversed the LPS-induced deficit in LTP in the rat hippocampus (Lynch et al., 2004; Nolan et al., 2005). Thus, IL-4 is an important therapeutic target of AD.

IL-10

IL-10 is an anti-inflammatory cytokine induced by either pro- or anti-inflammatory activation of immune cells, including T cells and mononuclear phagocytes. Instead of inducing alternative activation of immune cells (such as seen by IL-4 stimulation), IL-10 suppresses intracellular signaling to restore the resting status of activated cells. Our group recently identified that resting microglia express high levels of TGF-β1, extracellular matrix-related genes, and IL-10 (Freilich et al., 2013). With gene delivery of IL-10 into APP+PS1 mice, there was suppression of astro/microgliosis and a restoration of impaired spatial learning and neurogenesis, and while there was no reduction in Aβ deposition, there was enhanced vascular transport of Aβ (Kiyota et al., 2011). Gene delivery of either IL-10 also reversed the LPS-induced deficit in LTP in the rat hippocampus (Lynch et al., 2004; Nolan et al., 2005). In accord, we recently identified that IL-10-stimulated microglia enhanced proliferation but not differentiation of neural stem cells in vitro (Kiyota et al., 2011). Thus, IL-10 has a significant potential as a therapeutic target of AD.

TGF-β1/2/3

TGF-β1/2/3 are expressed in the CNS and are involved in development, homeostasis, and repair (Finch et al., 1993). TGF-β1 has been detected in plaques (van der Wal et al., 1993) and is elevated in cerebrospinal fluid and serum in AD (Chao et al., 1994a, b). TGF-β2 has also been detected in reactive astrocytes, microglia, and tangle-bearing neurons in AD brains (Wyss-Coray et al., 2000). TGF-β1 transgenic mice show cerebrovascular amyloid deposition (Wyss-Coray et al., 1997). Overexpression of TGF-β1 gene in APP mice accelerates vascular amyloid deposition although plaque burden was reduced in brain tissue (Wyss-Coray et al., 2001, 1997), suggesting its role in clearing Aβ from brain parenchyma. Since TGF-β1 increases APP expression in astrocytes and microglia (Gray & Patel, 1993; Harris-White et al., 1998; Monning et al., 1994), it may also be involved in TGF-β1 glial Aβ production. Disruption of TGF-β1 gene induces neurodegeneration and microgliosis (Brionne et al., 2003). APP mice overexpressing kinase-deficient TGF-β receptor type II, which inhibits endogenous TGF- β signaling, show enhanced Aβ deposition, neurodegeneration, and dendritic loss (Tesseur et al., 2006).

TNF-α

TNF-α, a pleiotropic pro-inflammatory cytokine, has been extensively investigated in AD. TNF-α is predominantly expressed by activated microglia and, to a lesser extent, astrocytes and neurons (Bezzi et al., 2001; Botchkina et al., 1997; Meda et al., 1995; Williams et al., 2005). In AD, TNF-α is upregulated in cerebrospinal fluid (Tarkowski et al., 1999), serum (Fillit et al., 1991), and cortex (Tarkowski et al., 1999). The effect of TNF-α on neurons is different depending on the neuronal preparation and perhaps species; it is neurotrophic to rat cortical, hippocampal, and septal neurons (Barger & Harmon, 1997; Cheng et al., 1994), while directly neurotoxic (or enhances glutamate neurotoxicity) to human cortical neurons (Chao & Hu, 1994; D'Souza et al., 1995; Williams et al., 2005). This discrepancy could be due to the diverse signaling of two distinctly different TNF-α receptors, type I (p55 TNFR-1) and type II (p75 TNFR-2) (Botchkina et al., 1997; Dziewulska & Mossakowski, 2003). TNFR-1, but not TNFR-2, contains an intracellular death domain, common to proteins in the TNF receptor family. TNFR-1 mediates both apoptosis and NF-κB-mediated cell survival (Gupta & Gollapudi, 2005; Wallach et al., 1999). Importantly, primary cultured hippocampal neurons derived from TNFR-1 null mice were resistant to Aβ-mediated neurotoxicity (Li et al., 2004). Thus, TNF-α-mediated neurotoxicity is bidirectional and the outcomes of TNF signaling depend distinctly on cell type and species.

CD40L

CD40L belongs to the TNF family and is one of the most important T-cell-mediated cytokines. CD40, the receptor of CD40L, is upregulated by Aβ-stimulation of primary cultured microglia in vitro (Tan et al., 1999) and is found to be expressed in reactive microglia in AD patient brain tissue (Togo et al., 2000). Transgenic APP (Tg2576) mice deficient in CD40L show reduced Aβ deposition, astro/microgliosis, and hyperphosphorylation of tau (Tan et al., 2002). In addition, chronic treatment of APP/PS1 bigenic mice with neutralizing anti-CD40L antibody reduces Aβ deposition and causes a modest improvement in cognitive performance (Tan et al., 2002; Todd Roach et al., 2004), suggesting its therapeutic application for AD.

CD200

CD200 is a type I transmembrane glycoprotein with an immunoglobulin superfamily (IgSF) domain (Clark et al., 1985). It is expressed in a number of cells involved in the immune response, including T cells, B cells, and particularly in neurons in the brain (Barclay, 1981; McMaster & Williams, 1979; Webb & Barclay, 1984). The CD200R is expressed largely in cells of myeloid lineage, namely, microglia (Barclay, 1981; Koning et al., 2010). It is a cell surface glycoprotein with IgSF domains and a single transmembrane region (Wright et al., 2000). Unlike CD200, however, it has a large cytoplasmic region allowing for intracellular signaling initiated by tyrosine phosphorylation (Wright et al. 2000). CD200 expression decreases with age where it is simultaneously associated with increased microglial activation (Lyons et al., 2007a). It is likewise decreased in the brains of AD patients and in the brains of Aβ-treated mice (Lyons et al., 2009; Walker et al., 2009). Interestingly, hippocampal slices from mice lacking CD200 also showed significant impairments in synaptic plasticity, specifically in LTP, tying the age-dependent decrease in CD200 with the decreased ability to form new memories in AD patients (Costello et al., 2011).

In animal models of AD, where the age-related increase in microglial activation was accompanied by an age-related or Aβ-induced decrease in CD200 expression, IL-4 treatment was able to restore that deficit (Kiyota et al., 2010; Lyons et al., 2007a, b). These tissues and glia from AD models lack IL-4 and have enhanced inflammatory responses to LPS (Lyons et al., 2009). In accord, our laboratory recently identified that overexpression of CD200 significantly reduced Aβ deposition and inflammatory changes in Tg2576 APP mice (unpublished observation). These studies indicate that CD200 is a novel therapeutic target of AD.

22.3.2.2 Chemokines

CCL2 (MCP-1)

CCL2 (MCP-1) is a member of the β chemokine subfamily and a candidate molecule for stimulating monocyte chemotaxis into the CNS (Charo et al., 1994). Activated astrocytes and cells of monocytic origin (such as microglia and macrophages) have been shown to express CCL2 in the brain (Calvo et al., 1996; Glabinski et al., 1996). CCL2 has been detected in senile plaques, reactive microglia, and microvessels in AD brains (Grammas & Ovase, 2001; Ishizuka et al., 1997; Xia & Hyman, 1999) and is upregulated in cerebrospinal fluid and serum of AD cases and AD animal models (Galimberti et al., 2005; Janelsins et al., 2005; Sun et al., 2003). Overexpression of CCL2 resulted in increased diffuse plaque deposition, microglial accumulation, and apoE expression in APP mice (Yamamoto et al., 2005), accompanied by cognitive impairment (Kiyota et al., 2009b). Intracerebral overexpression of dominant-negative CCL2 ameliorates the spatial learning of APP/PS1 mice (Kiyota et al., 2009a). On the other hand, genetic disruption of CCR2 increases Aβ accumulation and neurodegeneration when crossed with APP mouse model, suggesting that physiological CCL2/CCR2 signaling is necessary for the Ab clearance in brain (El Khoury et al., 2007). These results show that CCL2/CCR2 signaling is critical in mononuclear phagocyte accumulation and Aβ deposition in CNS.

CXCR Family

A number of other chemokines and their receptors have been identified in AD. CXCR2 is a member of the CXC chemokines receptor family and is a receptor for multiple ligands, including CXCL1/2/3 (GROα/β/δ) (Zlotnik & Yoshie, 2000). Both CXCR2 and CXCL1 have been detected in AD brains. In particular, CXCR2 expression has been identified in dystrophic neurites of senile plaques (Xia et al., 1997). CXCR3, another CXC receptor family, and its ligand CXCL10 (IP-10) have been found in both cortical and subcortical neurons in normal and AD brains. CXCL10 was significantly upregulated in AD brain (Xia et al., 2000). In addition, expression of CC chemokine receptors, such as CCR3 and CCR5, and their ligands, CCL4 (MIP-1β) and CCL5 (RANTES), is also enhanced on reactive microglia in AD brain (Xia et al., 1998). Although the exact physiological functions of these molecules have yet to be characterized in the context of AD pathogenesis, modulation of such chemokine signaling might be beneficial to correct the detrimental imbalance of metabolic homeostasis and neuroinflammation in the brain.

22.3.2.3 Toll-Like Receptors

One of the most important recent findings in the effort to understand innate immunity has been the identification of Toll-like receptors (TLRs). TLRs define a major class of pattern-recognition receptors critical to the initiation and tailoring of both innate and subsequent adaptive immune responses (Beutler, 2004; Iwasaki & Medzhitov, 2004). For example, bacterial cell wall components are recognized by TLR2, and lipopolysaccharide and viral envelope proteins are recognized by TLR4. Double-strand RNA, single-strand RNA, and unmethylated CpGs are recognized by TLR3, TLR7/8, and TLR9, respectively (for review, see Akira et al., 2006). Primary cultured human and mouse microglia express mRNA for TLR1-9, whereas human and mouse astrocytes express high levels of TLR3, and low-level TLR 1, 2, and 4-6 (Jack et al., 2005; Olson & Miller, 2004; Carpentier et al., 2005; McKimmie & Fazakerley, 2005). A recent study on aged mouse brains demonstrated that TLR1, TLR2, TLR4, TLR5, TLR7, and CD14 expression were upregulated in correlation with age, whereas TLR9 was downregulated (Letiembre et al., 2007). Interestingly, a TLR4 polymorphism was also associated with successful aging (Candore et al., 2006), which further indicates a role of innate immune receptors in aging and potentially AD.

The role of TLRs in neuroinflammation during AD is poorly characterized. However, recent reports indicate that

TLR9 ligand (CpG-containing oligonucleotide) induced enhancement of Aβ uptake by N9 microglia cell line (Iribarren et al., 2005), and Aβ stimulation modulates TLR-specific inflammation (NO and TFN-α release) in primary cultured mouse microglia (Lotz et al., 2005). These results suggest that there is crosstalk between Aβ and TLR signaling that may regulate glial inflammation and Aβ clearance in CNS.

22.3.2.4 Complement Complex

Complement C1q, the first complement of the classical complement pathway, is upregulated in AD brains and is found in senile plaques (Afagh et al., 1996). C1q binds to soluble Aβ and modulates its uptake by mononuclear phagocytes (Webster et al., 2000). C1q also directly induces neurotoxicity through its interaction with cell surface calreticulin, a C1q receptor (Luo et al., 2003). However, APP mice lacking C1q exhibited reduced glial and astroglial activation without changes in Aβ deposition (Fonseca et al., 2004b). Thus, C1q may be involved in AD pathogenesis by influencing neurotoxicity and glial inflammation, but its effect on Aβ metabolism is modest.

Complement C3 is upregulated in APP/TGF-β1 bigenic mice and AD brains (Wyss-Coray et al., 2001). Overexpression of soluble complement receptor-related protein y (sCrry), a complement C3 inhibitor, resulted in enhanced Aβ accumulation and hippocampal degeneration (Wyss-Coray et al., 2002). Since C3 is known to promote phagocytosis by binding to specific C3 receptor on specialized cells, C3 is likely to be positively involved in Aβ clearance.

A recent study demonstrates the essential role of the complement system for the synaptic tagging of inactive synapses for microglia-mediated synaptic pruning (Bialas & Stevens, 2013; Hong et al., 2016; Schafer et al., 2012; Stephan et al., 2012). Deletion of C3 or C1q suppresses synaptic loss in APP mice and restores cognitive function, although it is not beneficial to the amyloid pathology (Hong et al., 2016).

22.3.3 Oxygen Free Radicals

Production of ROS, factors involved in the aging process (Finkel & Holbrook, 2000), is elevated in AD brain and may be an important cause of AD (Martins et al., 1986). Elevated levels of oxidized lipids (lipid peroxidation, malondialdehyde, 4-hydroxynonenal) (Markesbery & Carney, 1999), proteins (advanced glycation end product modifications, tyrosine nitration) (Good et al., 1996; Takeda et al., 1998), and nucleic acids (8-hydroxy-deoxyguanosine) have been documented in AD brains (Lyras et al., 1997). Mitochondria and nicotinamide adenine dinucleotide phosphate (NADPH) oxidase complex dysfunction can generate large amounts of ROS (Beal, 1998), and downregulation of mitochondrial NADPH 15-kD gene has been found in AD cases (Beal, 1998; Manczak et al., 2004). Markers of oxidative stress are elevated not only in the pathologic lesions in AD brain (Good et al., 1996) but also in the cerebrospinal fluid of AD patients (Lovell et al., 1997). Aβ peptide can directly activate the NADPH oxidase complex of mononuclear phagocytes (Bianca et al., 1999) and microglia, which are mediated through the interaction of fibrillar Aβ to cell surface molecules (CD36, CD47, and α6β1-integrin) and tyrosine kinase Vav signaling (Wilkinson et al., 2006). ROS is also known to mediate Aβ-induced neurotoxicity (Behl et al., 1994; Ikezu et al., 2003). In addition to NADPH oxidase, myeloperoxidase (MPO) is expressed by microglia in close proximity to Aβ plaques, and, surprisingly, in neurons of AD brains (Green et al., 2004; Reynolds et al., 1999). MPO can catalyze both apoE and Aβ, generating oxidated and fragmentated apoE (Jolivalt et al., 1996) and stable Aβ42 dimers through dityrosine bridge formation (Galeazzi et al., 1999; Jolivalt et al., 1996).

22.3.3.1 Oxidative DNA Damage

DNA damage may also have a role in AD pathology. Recent studies indicate that aging-mediated DNA damage, including alterations in neuronal network and cognitive function-related gene expression profiles, is significantly increased after 40 years of age, (Lu et al., 2004). DNA damage induces cell cycle activation (Kruman et al., 2004), and cell cycle activation has been documented in AD brains (Nagy et al., 1998; Smith et al., 1999). DNA damage also potently induces p53, a tumor suppressor and transcriptional factor involved in neuronal apoptosis (Culmsee & Mattson, 2005; Kruman et al., 2004; Morrison et al., 2003). Neuronal expression of p53 is increased in AD brains (de la Monte et al., 1997) and in APP mouse brains (LaFerla et al., 1996). These data suggest that DNA damage is involved in neurodegeneration in AD brain, or the result of neurodegeneration.

22.4 Immunotherapeutics Inventions

Epidemiological studies using nonsteroidal anti-inflammatory drugs (NSAIDs) have led to the proposal that NSAIDs reduce the risk of AD (McGeer et al., 1996). These studies have been supported by data showing that treatment of APP mice with NSAIDs (such as ibuprofen and NO-flurbiprofen) in vivo results in reduced Aβ deposition and pro-inflammatory cytokine production (Jantzen et al., 2002; Lim et al., 2000; Yan et al., 2003). Microglial activation is suppressed by ibuprofen but rather enhanced by NO-flurbiprofen. Other studies have found that NSAID also affects γ-processing of APP through inhibition of the small GTP-binding protein Rho, which converts Aβ processing from Aβ42 to Aβ40 both in vitro and in vivo (Eriksen et al., 2003; Weggen et al., 2001; Zhou et al.,

2003). Based on these observations, a new series of drugs derived from NSAIDs having a more specific effect on γ-processing has been developed (Czirr & Weggen, 2006; Kukar et al., 2005; Weggen et al., 2003).

Proliferator-activated receptor γ (PPARγ) agonists, such as the thiazolinedione family, are also NSAIDs and can efficiently clear Aβ deposition in vitro and in vivo (Camacho et al., 2004; Heneka et al., 2005). PPAR is a nuclear transcription factor, and its effect is different from ibuprofen and its derivatives. PPARγ transcription factors play important physiological roles in the regulation of lipid metabolism (Lemberger et al., 1996; Mangelsdorf et al., 1995), suppression of inflammation (Heneka et al., 2000), and clinical treatment of diabetes type II (Dormandy et al., 2005). PPARγ agonists enhance Aβ degradation (Camacho et al., 2004), as well as suppress BACE expression and microglial activation (Heneka et al., 2005).

Other anti-inflammatory/anti-oxidant dietary applications include curcumin, the yellow pigment in turmeric that targets multiple AD pathogenic cascades (Lim et al., 2001; Yang et al., 2005). The dietary omega-3 fatty acid, docosahexaenoic acid (DHA), also reduced amyloid, oxidative damage and synaptic and cognitive deficits in a transgenic mouse model (Calon et al., 2004; Lim et al., 2005). Both DHA and curcumin have favorable safety profiles, epidemiology, and efficacy and may exert general anti-aging benefits (Cole et al., 2005). Ab immunotherapy is recently approved by the Food and Drug Administration and will be discussed in the chapter in details.

22.5 Summary

AD is the most common form of elderly dementia and has no effective therapy. Clinical diagnosis is based on the evaluation of cognitive function and lab tests. Pathological diagnosis is based on post-mortem neuropathology, defined by three hallmarks: senile plaque, NFT, and neuronal cell loss. Senile plaque mainly consists of Aβ aggregates and serves as a focal point of astrocyte and microglial activation, while NFT contains hyperphosphorylated tau. Aβ is produced during the processing of APP by BACE and the γ-secretase complex that includes PS1. Both APP and PS1 are causative genes of AD, while apoE4 is a risk factor allele. The Aβ load in CNS is maintained by balancing its production and clearance, which is mediated by passive diffusion to blood stream, degradation by Aβ degrading enzymes, and microglial phagocytosis. Aβ production, deposition, and neurodegeneration are significantly regulated by astroglial and microglial activation via direct contact as well as secretion of several neurotoxicants. Specific pro-inflammatory cytokines, chemokines, and excitotoxins have been characterized as mediators of neuroinflammation and neurotoxicity. Currently, their specific roles in mediating the pathogenesis in AD are being explored. Those diverse studies are generating many potential therapeutic targets for treating AD. Beneficial effects have been reported for anti-inflammatory drugs, such as NSAIDs, to treat glial inflammation and disease progression.

22.6 Review Questions

1. Which is NOT the causative gene of familial Alzheimer's disease?
   (a) APP
   (b) Presenilin-1
   (c) Presenilin-2
   (d) Tau
   (e) None of the above
2. Presenlin-1 is a member of
   (a) α-secretase
   (b) β-secretase
   (c) γ-secretase complex
   (d) Aβ degrading enzyme
   (e) None of the above
3. Amyloid-beta peptide is a processing product of amyloid precursor protein (APP) through a combination of
   (a) α-secretase and Beta-site APP cleaving enzyme (BACE1)
   (b) α-secretase and γ-secretase complex
   (c) BACE1 and γ-secretase complex
   (d) BACE1 and insulin-degrading enzyme
   (e) None of the above
4. Microglia-mediated inflammatory reaction affects AD progression via
   (a) Cytokine production
   (b) Reactive nitrogen/oxygen production
   (c) Excitotoxin production
   (d) All of the above
   (e) None of the above
5. Aβ deposition in APP mouse brain can be reduced by vaccination with the following antigen
   (a) Aβ
   (b) apoE4
   (c) Copaxon-1
   (d) All of the above
6. Nonsteroidal anti-inflammatory drugs (NSAIDs) can reduce Aβ deposition in APP mouse brain through
   (a) Suppression of brain inflammation
   (b) Suppression of Aβ42 generation
   (c) Suppression of BACE expression
   (d) All of the above
   (e) None of the above
7. Describe the available animal models for studying AD. What is the limitation of the animal models?
8. Describe the potential immunotherapy of AD.

22.7 Answers

1. d
2. c
3. c
4. d
5. a
6. b
7. Refer to Sect. 22.3.3
8. Refer to Sect. 22.5

References

Abbas, N., Bednar, I., Mix, E., Marie, S., Paterson, D., Ljungberg, A., Morris, C., Winblad, B., Nordberg, A., & Zhu, J. (2002). Up-regulation of the inflammatory cytokines IFN-gamma and IL-12 and down-regulation of IL-4 in cerebral cortex regions of APP(SWE) transgenic mice. *Journal of Neuroimmunology, 126*, 50–57.

Abduljaleel, Z., Al-Allaf, F. A., Khan, W., Athar, M., Shahzad, N., Taher, M. M., Elrobh, M., Alanazi, M. S., & El-Huneidi, W. (2014). Evidence of trem2 variant associated with triple risk of Alzheimer's disease. *PLoS One, 9*, e92648.

Afagh, A., Cummings, B. J., Cribbs, D. H., Cotman, C. W., & Tenner, A. J. (1996). Localization and cell association of C1q in Alzheimer's disease brain. *Experimental Neurology, 138*, 22–32.

Akira, S., Uematsu, S., & Takeuchi, O. (2006). Pathogen recognition and innate immunity. *Cell, 124*, 783–801.

Akiyama, H., Barger, S., Barnum, S., Bradt, B., Bauer, J., Cole, G. M., Cooper, N. R., Eikelenboom, P., Emmerling, M., Fiebich, B. L., Finch, C. E., Frautschy, S., Griffin, W. S., Hampel, H., Hull, M., Landreth, G., Lue, L., Mrak, R., Mackenzie, I. R., McGeer, P. L., O'Banion, M. K., Pachter, J., Pasinetti, G., Plata-Salaman, C., Rogers, J., Rydel, R., Shen, Y., Streit, W., Strohmeyer, R., Tooyoma, I., Van Muiswinkel, F. L., Veerhuis, R., Walker, D., Webster, S., Wegrzyniak, B., Wenk, G., & Wyss-Coray, T. (2000). Inflammation and Alzheimer's disease. *Neurobiology of Aging, 21*, 383–421.

Allen, B., Ingram, E., Takao, M., Smith, M. J., Jakes, R., Virdee, K., Yoshida, H., Holzer, M., Craxton, M., Emson, P. C., Atzori, C., Migheli, A., Crowther, R. A., Ghetti, B., Spillantini, M. G., & Goedert, M. (2002). Abundant tau filaments and nonapoptotic neurodegeneration in transgenic mice expressing human P301S tau protein. *The Journal of Neuroscience, 22*, 9340–9351.

Alonso, A. C., Grundke-Iqbal, I., & Iqbal, K. (1996). Alzheimer's disease hyperphosphorylated tau sequesters normal tau into tangles of filaments and disassembles microtubules. *Nature Medicine, 2*, 783–787.

Alzheimer, A. (1907). Ueber eine eigenartige Erkrankung der Hirnrinde. *Allg Z Psychiat Psych-Gericht Me, 64*, 146–148.

Apelt, J., & Schliebs, R. (2001). Beta-amyloid-induced glial expression of both pro- and anti- inflammatory cytokines in cerebral cortex of aged transgenic Tg2576 mice with Alzheimer plaque pathology. *Brain Research, 894*, 21–30.

Apelt, J., Ach, K., & Schliebs, R. (2003). Aging-related down-regulation of neprilysin, a putative beta-amyloid-degrading enzyme, in transgenic Tg2576 Alzheimer-like mouse brain is accompanied by an astroglial upregulation in the vicinity of beta-amyloid plaques. *Neuroscience Letters, 339*, 183–186.

Arancio, O., Zhang, H. P., Chen, X., Lin, C., Trinchese, F., Puzzo, D., Liu, S., Hegde, A., Yan, S. F., Stern, A., Luddy, J. S., Lue, L. F., Walker, D. G., Roher, A., Buttini, M., Mucke, L., Li, W., Schmidt, A. M., Kindy, M., Hyslop, P. A., Stern, D. M., & Du Yan, S. S. (2004). RAGE potentiates Abeta-induced perturbation of neuronal function in transgenic mice. *The EMBO Journal, 23*, 4096–4105.

Arendash, G. W., Lewis, J., Leighty, R. E., McGowan, E., Cracchiolo, J. R., Hutton, M., & Garcia, M. F. (2004). Multi-metric behavioral comparison of APPsw and P301L models for Alzheimer's disease: Linkage of poorer cognitive performance to tau pathology in forebrain. *Brain Research, 1012*, 29–41.

Arends, Y. M., Duyckaerts, C., Rozemuller, J. M., Eikelenboom, P., & Hauw, J. J. (2000). Microglia, amyloid and dementia in alzheimer disease. A correlative study. *Neurobiology of Aging, 21*, 39–47.

Aronson, M. K., Ooi, W. L., Morgenstern, H., Hafner, A., Masur, D., Crystal, H., Frishman, W. H., Fisher, D., & Katzman, R. (1990). Women, myocardial infarction, and dementia in the very old. *Neurology, 40*, 1102–1106.

Arvin, B., Neville, L. F., Barone, F. C., & Fewerstein, G. Z. (1996). The role of inflammation and cytokines in brain injury. *Neuroscience and Biobehavioral Reviews, 20*, 445–452.

Asai, M., Hattori, C., Szabo, B., Sasagawa, N., Maruyama, K., Tanuma, S., & Ishiura, S. (2003). Putative function of ADAM9, ADAM10, and ADAM17 as APP alpha-secretase. *Biochemical and Biophysical Research Communications, 301*, 231–235.

Asai, H., Ikezu, S., Woodbury, M. E., Yonemoto, G. M., Cui, L., & Ikezu, T. (2014). Accelerated neurodegeneration and neuroinflammation in transgenic mice expressing P301L tau mutant and tau-tubulin kinase 1. *The American Journal of Pathology, 184*, 808–818.

Asai, H., Ikezu, S., Tsunoda, S., Medalla, M., Luebke, J., Haydar, T., Wolozin, B., Butovsky, O., Kugler, S., & Ikezu, T. (2015). Depletion of microglia and inhibition of exosome synthesis halt tau propagation. *Nature Neuroscience, 18*, 1584–1593.

Association, A.P. (2022). *Diagnostic and statistical manual of mental disorders: DSM-5-TR*. American Psychiatric Association.

Bales, K. R., Verina, T., Cummins, D. J., Du, Y., Dodel, R. C., Saura, J., Fishman, C. E., DeLong, C. A., Piccardo, P., Petegnief, V., Ghetti, B., & Paul, S. M. (1999). Apolipoprotein E is essential for amyloid deposition in the APP(V717F) transgenic mouse model of Alzheimer's disease. *Proceedings of the National Academy of Sciences of the United States of America, 96*, 15233–15238.

Bamberger, M. E., Harris, M. E., McDonald, D. R., Husemann, J., & Landreth, G. E. (2003). A cell surface receptor complex for fibrillar beta-amyloid mediates microglial activation. *The Journal of Neuroscience, 23*, 2665–2674.

Barclay, A. N. (1981). Different reticular elements in rat lymphoid tissue identified by localization of Ia, Thy-1 and MRC OX 2 antigens. *Immunology, 44*, 727–736.

Barger, S. W., & Basile, A. S. (2001). Activation of microglia by secreted amyloid precursor protein evokes release of glutamate by cystine exchange and attenuates synaptic function. *Journal of Neurochemistry, 76*, 846–854.

Barger, S., & Harmon, A. (1997). Microglial activation by Alzheimer amyloid precursor protein and modulation by apolipoprotein E. *Nature, 388*, 878–881.

Beal, M. F. (1998). Mitochondrial dysfunction in neurodegenerative diseases. *Biochimica et Biophysica Acta, 1366*, 211–223.

Behl, C., Davis, J. B., Lesley, R., & Schubert, D. (1994). Hydrogen peroxide mediates amyloid beta protein toxicity. *Cell, 77*, 817–827.

Bellenguez, C., Kucukali, F., Jansen, I. E., Kleineidam, L., Moreno-Grau, S., Amin, N., Naj, A. C., Campos-Martin, R., Grenier-Boley, B., Andrade, V., Holmans, P. A., Boland, A., Damotte, V., van der Lee, S. J., Costa, M. R., Kuulasmaa, T., Yang, Q., de Rojas, I., Bis, J. C., Yaqub, A., Prokic, I., Chapuis, J., Ahmad, S., Giedraitis, V., Aarsland, D., Garcia-Gonzalez, P., Abdelnour, C., Alarcon-Martin, E., Alcolea, D., Alegret, M., Alvarez, I., Alvarez, V., Armstrong, N. J., Tsolaki, A., Antunez, C., Appollonio, I., Arcaro, M., Archetti, S., Pastor, A. A., Arosio, B., Athanasiu, L., Bailly, H., Banaj, N.,

Baquero, M., Barral, S., Beiser, A., Pastor, A. B., Below, J. E., Benchek, P., Benussi, L., Berr, C., Besse, C., Bessi, V., Binetti, G., Bizarro, A., Blesa, R., Boada, M., Boerwinkle, E., Borroni, B., Boschi, S., Bossu, P., Brathen, G., Bressler, J., Bresner, C., Brodaty, H., Brookes, K. J., Brusco, L. I., Buiza-Rueda, D., Burger, K., Burholt, V., Bush, W. S., Calero, M., Cantwell, L. B., Chene, G., Chung, J., Cuccaro, M. L., Carracedo, A., Cecchetti, R., Cervera-Carles, L., Charbonnier, C., Chen, H. H., Chillotti, C., Ciccone, S., Claassen, J., Clark, C., Conti, E., Corma-Gomez, A., Costantini, E., Custodero, C., Daian, D., Dalmasso, M. C., Daniele, A., Dardiotis, E., Dartigues, J. F., de Deyn, P. P., de Paiva Lopes, K., de Witte, L. D., Debette, S., Deckert, J., Del Ser, T., et al. (2022). New insights into the genetic etiology of Alzheimer's disease and related dementias. *Nature Genetics, 54*, 412–436.

Bellucci, A., Westwood, A. J., Ingram, E., Casamenti, F., Goedert, M., & Spillantini, M. G. (2004). Induction of inflammatory mediators and microglial activation in mice transgenic for mutant human P301S tau protein. *The American Journal of Pathology, 165*, 1643–1652.

Benzing, W. C., Wujek, J. R., Ward, E. K., Shaffer, D., Ashe, K. H., Younkin, S. G., & Brunden, K. R. (1999). Evidence for glial-mediated inflammation in aged APP(SW) transgenic mice. *Neurobiology of Aging, 20*, 581–589.

Berg, L., McKeel, D. W., Jr., Miller, J. P., Baty, J., & Morris, J. C. (1993). Neuropathological indexes of Alzheimer's disease in demented and nondemented persons aged 80 years and older. *Archives of Neurology, 50*, 349–358.

Beutler, B. (2004). Inferences, questions and possibilities in Toll-like receptor signalling. *Nature, 430*, 257–263.

Bezzi, P., Domercq, M., Brambilla, L., Galli, R., Schols, D., De Clercq, E., Vescovi, A., Bagetta, G., Kollias, G., Meldolesi, J., & Volterra, A. (2001). CXCR4-activated astrocyte glutamate release via TNFalpha: Amplification by microglia triggers neurotoxicity. *Nature Neuroscience, 4*, 702–710.

Bialas, A. R., & Stevens, B. (2013). TGF-beta signaling regulates neuronal C1q expression and developmental synaptic refinement. *Nature Neuroscience, 16*, 1773–1782.

Bianca, V. D., Dusi, S., Bianchini, E., Dal Pra, I., & Rossi, F. (1999). beta-amyloid activates the O-2 forming NADPH oxidase in microglia, monocytes, and neutrophils. A possible inflammatory mechanism of neuronal damage in Alzheimer's disease. *The Journal of Biological Chemistry, 274*, 15493–15499.

Biernat, J., Gustke, N., Drewes, G., Mandelkow, E. M., & Mandelkow, E. (1993). Phosphorylation of Ser262 strongly reduces binding of tau to microtubules: Distinction between PHF-like immunoreactivity and microtubule binding. *Neuron, 11*, 153–163.

Blasko, I., Marx, F., Steiner, E., Hartmann, T., & Grubeck-Loebenstein, B. (1999). TNFalpha plus IFNgamma induce the production of Alzheimer beta-amyloid peptides and decrease the secretion of APPs. *The FASEB Journal, 13*, 63–68.

Blasko, I., Veerhuis, R., Stampfer-Kountchev, M., Saurwein-Teissl, M., Eikelenboom, P., & Grubeck-Loebenstein, B. (2000). Costimulatory effects of interferon-gamma and interleukin-1beta or tumor necrosis factor alpha on the synthesis of Abeta1-40 and Abeta1-42 by human astrocytes. *Neurobiology of Disease, 7*, 682–689.

Borchelt, D. R., Ratovitski, T., van Lare, J., Lee, M. K., Gonzales, V., Jenkins, N. A., Copeland, N. G., Price, D. L., & Sisodia, S. S. (1997). Accelerated amyloid deposition in the brains of transgenic mice coexpressing mutant presenilin 1 and amyloid precursor proteins. *Neuron, 19*, 939–945.

Botchkina, G. I., Meistrell, M. E., 3rd, Botchkina, I. L., & Tracey, K. J. (1997). Expression of TNF and TNF receptors (p55 and p75) in the rat brain after focal cerebral ischemia. *Molecular Medicine, 3*, 765–781.

Braak, H., & Braak, E. (1995). Staging of Alzheimer's disease-related neurofibrillary changes. *Neurobiology of Aging, 16*, 271–278; discussion 278–284.

Bramblett, G. T., Goedert, M., Jakes, R., Merrick, S. E., Trojanowski, J. Q., & Lee, V. M. (1993). Abnormal tau phosphorylation at Ser396 in Alzheimer's disease recapitulates development and contributes to reduced microtubule binding. *Neuron, 10*, 1089–1099.

Brionne, T. C., Tesseur, I., Masliah, E., & Wyss-Coray, T. (2003). Loss of TGF-beta 1 leads to increased neuronal cell death and microgliosis in mouse brain. *Neuron, 40*, 1133–1145.

Buee, L., Bussiere, T., Buee-Scherrer, V., Delacourte, A., & Hof, P. R. (2000). Tau protein isoforms, phosphorylation and role in neurodegenerative disorders. *Brain Research. Brain Research Reviews, 33*, 95–130.

Butovsky, O., & Weiner, H. L. (2018). Microglial signatures and their role in health and disease. *Nature Reviews. Neuroscience, 19*, 622–635.

Butovsky, O., Talpalar, A. E., Ben-Yaakov, K., & Schwartz, M. (2005). Activation of microglia by aggregated beta-amyloid or lipopolysaccharide impairs MHC-II expression and renders them cytotoxic whereas IFN-gamma and IL-4 render them protective. *Molecular and Cellular Neurosciences, 29*, 381–393.

Butovsky, O., Ziv, Y., Schwartz, A., Landa, G., Talpalar, A. E., Pluchino, S., Martino, G., & Schwartz, M. (2006). Microglia activated by IL-4 or IFN-gamma differentially induce neurogenesis and oligodendrogenesis from adult stem/progenitor cells. *Molecular and Cellular Neurosciences, 31*, 149–160.

Cady, J., Koval, E. D., Benitez, B. A., Zaidman, C., Jockel-Balsarotti, J., Allred, P., Baloh, R. H., Ravits, J., Simpson, E., Appel, S. H., Pestronk, A., Goate, A. M., Miller, T. M., Cruchaga, C., & Harms, M. B. (2014). TREM2 variant p.R47H as a risk factor for sporadic amyotrophic lateral sclerosis. *JAMA Neurology, 71*, 449–453.

Cai, H., Wang, Y., McCarthy, D., Wen, H., Borchelt, D. R., Price, D. L., & Wong, P. C. (2001). BACE1 is the major beta-secretase for generation of Abeta peptides by neurons. *Nature Neuroscience, 4*, 233–234.

Calon, F., Lim, G. P., Yang, F., Morihara, T., Teter, B., Ubeda, O., Rostaing, P., Triller, A., Salem, N., Jr., Ashe, K. H., Frautschy, S. A., & Cole, G. M. (2004). Docosahexaenoic acid protects from dendritic pathology in an Alzheimer's disease mouse model. *Neuron, 43*, 633–645.

Calvo, C. F., Yoshimura, T., Gelman, M., & Mallat, M. (1996). Production of monocyte chemotactic protein-1 by rat brain macrophages. *The European Journal of Neuroscience, 8*, 1725–1734.

Camacho, I. E., Serneels, L., Spittaels, K., Merchiers, P., Dominguez, D., & De Strooper, B. (2004). Peroxisome-proliferator-activated receptor gamma induces a clearance mechanism for the amyloid-beta peptide. *The Journal of Neuroscience, 24*, 10908–10917.

Candore, G., Aquino, A., Balistreri, C. R., Bulati, M., Di Carlo, D., Grimaldi, M. P., Listi, F., Orlando, V., Vasto, S., Caruso, M., Colonna-Romano, G., Lio, D., & Caruso, C. (2006). Inflammation, longevity, and cardiovascular diseases: Role of polymorphisms of TLR4. *Annals of the New York Academy of Sciences, 1067*, 282–287.

Carpentier, P. A., Begolka, W. S., Olson, J. K., Elhofy, A., Karpus, W. J., & Miller, S. D. (2005). Differential activation of astrocytes by innate and adaptive immune stimuli. *Glia, 49*, 360–374.

Chao, C. C., & Hu, S. (1994). Tumor necrosis factor-alpha potentiates glutamate neurotoxicity in human fetal brain cell cultures. *Developmental Neuroscience, 16*, 172–179.

Chao, C. C., Ala, T. A., Hu, S., Crossley, K. B., Sherman, R. E., Peterson, P. K., & Frey, W. H., 2nd. (1994a). Serum cytokine levels in patients with Alzheimer's disease. *Clinical and Diagnostic Laboratory Immunology, 1*, 433–436.

Chao, C. C., Hu, S., Frey, W. H., 2nd, Ala, T. A., Tourtellotte, W. W., & Peterson, P. K. (1994b). Transforming growth factor beta in Alzheimer's disease. *Clinical and Diagnostic Laboratory Immunology, 1*, 109–110.

Chao, C., Hu, S., Erlich, L., & Peterson, P. (1995). Interleukin-1 and tumor necrosis factor-alpha synergistically mediate neurotoxicity:

Involvement of nitric oxide and of N-methyl-D-aspartate receptors. *Brain, Behavior, and Immunity, 9*, 355–365.

Chapman, P. F., White, G. L., Jones, M. W., Cooper-Blacketer, D., Marshall, V. J., Irizarry, M., Younkin, L., Good, M. A., Bliss, T. V., Hyman, B. T., Younkin, S. G., & Hsiao, K. K. (1999). Impaired synaptic plasticity and learning in aged amyloid precursor protein transgenic mice. *Nature Neuroscience, 2*, 271–276.

Charo, I. F., Myers, S. J., Herman, A., Franci, C., Connolly, A. J., & Coughlin, S. R. (1994). Molecular cloning and functional expression of two monocyte chemoattractant protein 1 receptors reveals alternative splicing of the carboxyl-terminal tails. *Proceeding of the National Academy of Science USA, 91*, 2752–2756.

Chen, G., Chen, K. S., Knox, J., Inglis, J., Bernard, A., Martin, S. J., Justice, A., McConlogue, L., Games, D., Freedman, S. B., & Morris, R. G. (2000). A learning deficit related to age and beta-amyloid plaques in a mouse model of Alzheimer's disease. *Nature, 408*, 975–979.

Cheng, B., Christakos, S., & Mattson, M. P. (1994). Tumor necrosis factors protect neurons against metabolic-excitotoxic insults and promote maintenance of calcium homeostasis. *Neuron, 12*, 139–153.

Chishti, M. A., Yang, D. S., Janus, C., Phinney, A. L., Horne, P., Pearson, J., Strome, R., Zuker, N., Loukides, J., French, J., Turner, S., Lozza, G., Grilli, M., Kunicki, S., Morissette, C., Paquette, J., Gervais, F., Bergeron, C., Fraser, P. E., Carlson, G. A., George-Hyslop, P. S., & Westaway, D. (2001). Early-onset amyloid deposition and cognitive deficits in transgenic mice expressing a double mutant form of amyloid precursor protein 695. *The Journal of Biological Chemistry, 276*, 21562–21570.

Choi, D. S., Wang, D., Yu, G. Q., Zhu, G., Kharazia, V. N., Paredes, J. P., Chang, W. S., Deitchman, J. K., Mucke, L., & Messing, R. O. (2006). PKC{varepsilon} increases endothelin converting enzyme activity and reduces amyloid plaque pathology in transgenic mice. *Proceedings of the National Academy of Sciences of the United States of America, 103*, 8215–8220.

Cirrito, J. R., Deane, R., Fagan, A. M., Spinner, M. L., Parsadanian, M., Finn, M. B., Jiang, H., Prior, J. L., Sagare, A., Bales, K. R., Paul, S. M., Zlokovic, B. V., Piwnica-Worms, D., & Holtzman, D. M. (2005). P-glycoprotein deficiency at the blood-brain barrier increases amyloid-beta deposition in an Alzheimer disease mouse model. *The Journal of Clinical Investigation, 115*, 3285–3290.

Clark, M. J., Gagnon, J., Williams, A. F., & Barclay, A. N. (1985). MRC OX-2 antigen: A lymphoid/neuronal membrane glycoprotein with a structure like a single immunoglobulin light chain. *The EMBO Journal, 4*, 113–118.

Clayton, K., Delpech, J. C., Herron, S., Iwahara, N., Ericsson, M., Saito, T., Saido, T. C., Ikezu, S., & Ikezu, T. (2021). Plaque associated microglia hyper-secrete extracellular vesicles and accelerate tau propagation in a humanized APP mouse model. *Molecular Neurodegeneration, 16*, 18.

Cleary, J. P., Walsh, D. M., Hofmeister, J. J., Shankar, G. M., Kuskowski, M. A., Selkoe, D. J., & Ashe, K. H. (2005). Natural oligomers of the amyloid-beta protein specifically disrupt cognitive function. *Nature Neuroscience, 8*, 79–84.

Cole, G. M., Lim, G. P., Yang, F., Teter, B., Begum, A., Ma, Q., Harris-White, M. E., & Frautschy, S. A. (2005). Prevention of Alzheimer's disease: Omega-3 fatty acid and phenolic anti-oxidant interventions. *Neurobiology of Aging, 26*, 133–136.

Corey-Bloom, J., Galasko, D., & Thal, L. J. (1994). *Clinical features and natural history of Alzheimer's disease* (pp. 631–645). WB Saunders.

Costello, D. A., Lyons, A., Browne, T., Denieffe, S., Cox, F. F., & Lynch, M. A. (2011). Long-term potentiation is impaired in CD200-deficient mice: A role for Toll-like receptor activation. *The Journal of Biological Chemistry, 286*(40), 34722–34732.

Culmsee, C., & Mattson, M. P. (2005). p53 in neuronal apoptosis. *Biochemical and Biophysical Research Communications, 331*, 761–777.

Czirr, E., & Weggen, S. (2006). Gamma-secretase modulation with Abeta42-lowering nonsteroidal anti-inflammatory drugs and derived compounds. *Neurodegenerative Diseases, 3*, 298–304.

D'Souza, S., Alinauskas, K., McCrea, E., Goodyer, C., & Antel, J. P. (1995). Differential susceptibility of human CNS-derived cell populations to TNF-dependent and independent immune-mediated injury. *The Journal of Neuroscience, 15*, 7293–7300.

Danik, M., Chabot, J. G., Michel, D., & Quirion, R. (1999). Clusterin and apolipoprotein E gene expression in the adult brain. In C. E. Finch (Ed.), *Clusterin in normal brain functions and during neurodegeneration* (pp. 17–34). RG Landes Company.

Das, S., & Potter, H. (1995). Expression of the Alzheimer amyloid-promoting factor antichymotrypsin is induced in human astrocytes by IL-1. *Neuron, 14*, 447–456.

Daws, M. R., Sullam, P. M., Niemi, E. C., Chen, T. T., Tchao, N. K., & Seaman, W. E. (2003). Pattern recognition by TREM-2: Binding of anionic ligands. *Journal of Immunology, 171*, 594–599.

De Felice, F. G., Vieira, M. N., Saraiva, L. M., Figueroa-Villar, J. D., Garcia-Abreu, J., Liu, R., Chang, L., Klein, W. L., & Ferreira, S. T. (2004). Targeting the neurotoxic species in Alzheimer's disease: Inhibitors of Abeta oligomerization. *The FASEB Journal, 18*, 1366–1372.

de la Monte, S. M., Sohn, Y. K., & Wands, J. R. (1997). Correlates of p53- and Fas (CD95)-mediated apoptosis in Alzheimer's disease. *Journal of the Neurological Sciences, 152*, 73–83.

De Strooper, B. (2003). Aph-1, Pen-2, and Nicastrin with presenilin generate an active gamma-secretase complex. *Neuron, 38*, 9–12.

DeLeo, A. M., & Ikezu, T. (2018). Extracellular vesicle biology in Alzheimer's disease and related tauopathy. *Journal of Neuroimmune Pharmacology, 13*, 292–308.

Delpech, J. C., Herron, S., Botros, M. B., & Ikezu, T. (2019). Neuroimmune crosstalk through extracellular vesicles in health and disease. *Trends in Neurosciences, 42*, 361–372.

DeSimone, R., Levi, G., & Aloisi, F. (1998). Interferon-gamma gene expresion in rat central nervous system glial cells. *Cytokine, 10*, 418–422.

Dormandy, J. A., Charbonnel, B., Eckland, D. J., Erdmann, E., Massi-Benedetti, M., Moules, I. K., Skene, A. M., Tan, M. H., Lefebvre, P. J., Murray, G. D., Standl, E., Wilcox, R. G., Wilhelmsen, L., Betteridge, J., Birkeland, K., Golay, A., Heine, R. J., Koranyi, L., Laakso, M., Mokan, M., Norkus, A., Pirags, V., Podar, T., Scheen, A., Scherbaum, W., Schernthaner, G., Schmitz, O., Skrha, J., Smith, U., & Taton, J. (2005). Secondary prevention of macrovascular events in patients with type 2 diabetes in the PROactive Study (PROspective pioglitAzone Clinical Trial In macroVascular Events): A randomised controlled trial. *Lancet, 366*, 1279–1289.

Drewes, G., Ebneth, A., Preuss, U., Mandelkow, E. M., & Mandelkow, E. (1997). MARK, a novel family of protein kinases that phosphorylate microtubule-associated proteins and trigger microtubule disruption. *Cell, 89*, 297–308.

Duff, K., Eckman, C., Zehr, C., Yu, X., Prada, C. M., Perez-tur, J., Hutton, M., Buee, L., Harigaya, Y., Yager, D., Morgan, D., Gordon, M. N., Holcomb, L., Refolo, L., Zenk, B., Hardy, J., & Younkin, S. (1996). Increased amyloid-beta42(43) in brains of mice expressing mutant presenilin 1. *Nature, 383*, 710–713.

Dziewulska, D., & Mossakowski, M. J. (2003). Cellular expression of tumor necrosis factor a and its receptors in human ischemic stroke. *Clinical Neuropathology, 22*, 35–40.

Eikelenboom, P., Zhan, S. S., van Gool, W. A., & Allsop, D. (1994). Inflammatory mechanisms in Alzheimer's disease. *Trends in Pharmacological Sciences, 15*, 447–450.

Eikelenboom, P., Bate, C., Van Gool, W. A., Hoozemans, J. J., Rozemuller, J. M., Veerhuis, R., & Williams, A. (2002).

Neuroinflammation in Alzheimer's disease and prion disease. *Glia, 40*, 232–239.
El Khoury, J., Hickman, S. E., Thomas, C. A., Cao, L., Silverstein, S. C., & Loike, J. D. (1996). Scavenger receptor-mediated adhesion of microglia to beta-amyloid fibrils. *Nature, 382*, 716–719.
El Khoury, J. B., Moore, K. J., Means, T. K., Leung, J., Terada, K., Toft, M., Freeman, M. W., & Luster, A. D. (2003). CD36 mediates the innate host response to beta-amyloid. *The Journal of Experimental Medicine, 197*, 1657–1666.
El Khoury, J., Toft, M., Hickman, S. E., Means, T. K., Terada, K., Geula, C., & Luster, A. D. (2007). Ccr2 deficiency impairs microglial accumulation and accelerates progression of Alzheimer-like disease. *Nature Medicine, 13*, 432–438.
Eriksen, J. L., Sagi, S. A., Smith, T. E., Weggen, S., Das, P., McLendon, D. C., Ozols, V. V., Jessing, K. W., Zavitz, K. H., Koo, E. H., & Golde, T. E. (2003). NSAIDs and enantiomers of flurbiprofen target gamma-secretase and lower Abeta 42 in vivo. *The Journal of Clinical Investigation, 112*, 440–449.
Fagan, A. M., Watson, M., Parsadanian, M., Bales, K. R., Paul, S. M., & Holtzman, D. M. (2002). Human and murine ApoE markedly alters A beta metabolism before and after plaque formation in a mouse model of Alzheimer's disease. *Neurobiology of Disease, 9*, 305–318.
Filiano, A. J., Xu, Y., Tustison, N. J., Marsh, R. L., Baker, W., Smirnov, I., Overall, C. C., Gadani, S. P., Turner, S. D., Weng, Z., Peerzade, S. N., Chen, H., Lee, K. S., Scott, M. M., Beenhakker, M. P., Litvak, V., & Kipnis, J. (2016). Unexpected role of interferon-gamma in regulating neuronal connectivity and social behaviour. *Nature, 535*, 425–429.
Fillit, H., Ding, W. H., Buee, L., Kalman, J., Altstiel, L., Lawlor, B., & Wolf-Klein, G. (1991). Elevated circulating tumor necrosis factor levels in Alzheimer's disease. *Neuroscience Letters, 129*, 318–320.
Finch, C. E., Laping, N. J., Morgan, T. E., Nichols, N. R., & Pasinetti, G. M. (1993). TGF-beta 1 is an organizer of responses to neurodegeneration. *Journal of Cellular Biochemistry, 53*, 314–322.
Finkel, T., & Holbrook, N. J. (2000). Oxidants, oxidative stress and the biology of ageing. *Nature, 408*, 239–247.
Folstein, M. F., Folstein, S. E., & McHugh, P. R. (1975). "Mini-mental state". A practical method for grading the cognitive state of patients for the clinician. *Journal of Psychiatric Research, 12*, 189–198.
Fonseca, M. I., Kawas, C. H., Troncoso, J. C., & Tenner, A. J. (2004a). Neuronal localization of C1q in preclinical Alzheimer's disease. *Neurobiology of Disease, 15*, 40–46.
Fonseca, M. I., Zhou, J., Botto, M., & Tenner, A. J. (2004b). Absence of C1q leads to less neuropathology in transgenic mouse models of Alzheimer's disease. *The Journal of Neuroscience, 24*, 6457–6465.
Fowler, S. L., Behr, T. S., Turkes, E., Cauhy, P. M., Foiani, M. S., Schaler, A., Crowley, G., Bez, S., Ficulle, E., Tsefou, E., O'Brien, D. P., Fischer, R., Geary, B., Gaur, P., Miller, C., D'Acunzo, P., Levy, E., Duff, K. E., & Ryskeldi-Falcon, B. (2023). Tau filaments are tethered within brain extracellular vesicles in Alzheimer's disease. *bioRxiv*.
Fratiglioni, L., Small, B., Winblad, B., & Backman, L. (2001). The transition from normal functioning to dementia in the aging population. In K. Iqbal, S. S. Sisodia, & B. Winblad (Eds.), *Alzheimer's disease: Advances in etiology, pathogenesis, and therapeutics* (pp. 3–10). Wiley.
Frautschy, S. A., Yang, F., Irrizarry, M., Hyman, B., Saido, T. C., Hsaio, K., & Cole, G. M. (1998). Microglial response to amyloid plaques in APPsw transgenic mice. *The American Journal of Pathology, 152*, 307–317.
Frei, K., Bodmer, S., Schwerdel, C., & Fontana, A. (1986). Astrocyte-derived interleukin 3 as a growth factor for microglia cells and peritoneal macrophages. *Journal of Immunology, 137*, 3521–3527.
Freilich, R. W., Woodbury, M. E., & Ikezu, T. (2013). Integrated expression profiles of mRNA and miRNA in polarized primary murine microglia. *PLoS One, 8*, e79416.
Fultz, M., Barber, S., Dieffenbach, C., & Vogel, S. (1993). Induction of IFN-g in macrophages by lipopolysaccharide. *International Immunology, 5*, 1383–1392.
Galeazzi, L., Ronchi, P., Franceschi, C., & Giunta, S. (1999). In vitro peroxidase oxidation induces stable dimers of beta-amyloid (1-42) through dityrosine bridge formation. *Amyloid, 6*, 7–13.
Galimberti, D., Fenoglio, C., Lovati, C., Venturelli, E., Guidi, I., Corra, B., Scalabrini, D., Clerici, F., Mariani, C., Bresolin, N., & Scarpini, E. (2005). Serum MCP-1 levels are increased in mild cognitive impairment and mild Alzheimer's disease. *Neurobiology of Aging, 27*, 1763–1768.
Gamache, J., Benzow, K., Forster, C., Kemper, L., Hlynialuk, C., Furrow, E., Ashe, K. H., & Koob, M. D. (2019). Factors other than hTau overexpression that contribute to tauopathy-like phenotype in rTg4510 mice. *Nature Communications, 10*, 2479.
Games, D., Adams, D., Alessandrini, R., Barbour, R., Berthelette, P., Blackwell, C., Carr, T., Clemens, J., Donaldson, T., Gillespie, F., et al. (1995). Alzheimer-type neuropathology in transgenic mice overexpressing V717F beta-amyloid precursor protein. *Nature, 373*, 523–527.
Giraldo, M., Lopera, F., Siniard, A. L., Corneveaux, J. J., Schrauwen, I., Carvajal, J., Munoz, C., Ramirez-Restrepo, M., Gaiteri, C., Myers, A. J., Caselli, R. J., Kosik, K. S., Reiman, E. M., & Huentelman, M. J. (2013). Variants in triggering receptor expressed on myeloid cells 2 are associated with both behavioral variant frontotemporal lobar degeneration and Alzheimer's disease. *Neurobiology of Aging, 34*(2077), e2011–e2078.
Giulian, D., & Baker, T. J. (1986). Characterization of ameboid microglia isolated from developing mammalian brain. *The Journal of Neuroscience, 6*, 2163–2178.
Giulian, D., Li, J., Leara, B., & Keenen, C. (1994). Phagocytic microglia release cytokines and cytotoxins that regulate the survival of astrocytes and neurons in culture. *Neurochemistry International, 25*, 227–233.
Glabinski, A. R., Balasingam, V., Tani, M., Kunkel, S. L., Strieter, R. M., Yong, V. W., & Ransohoff, R. M. (1996). Chemokine monocyte chemoattractant protein-1 is expressed by astrocytes after mechanical injury to the brain. *Journal of Immunology, 156*, 4363–4368.
Good, P. F., Werner, P., Hsu, A., Olanow, C. W., & Perl, D. P. (1996). Evidence of neuronal oxidative damage in Alzheimer's disease. *The American Journal of Pathology, 149*, 21–28.
Gotz, J., Chen, F., van Dorpe, J., & Nitsch, R. M. (2001a). Formation of neurofibrillary tangles in P301l tau transgenic mice induced by Abeta 42 fibrils. *Science, 293*, 1491–1495.
Gotz, J., Tolnay, M., Barmettler, R., Chen, F., Probst, A., & Nitsch, R. M. (2001b). Oligodendroglial tau filament formation in transgenic mice expressing G272V tau. *The European Journal of Neuroscience, 13*, 2131–2140.
Grammas, P., & Ovase, R. (2001). Inflammatory factors are elevated in brain microvessels in Alzheimer's disease. *Neurobiology of Aging, 22*, 837–842.
Gray, C. W., & Patel, A. J. (1993). Regulation of beta-amyloid precursor protein isoform mRNAs by transforming growth factor-beta 1 and interleukin-1 beta in astrocytes. *Brain Research. Molecular Brain Research, 19*, 251–256.
Green, P. S., Mendez, A. J., Jacob, J. S., Crowley, J. R., Growdon, W., Hyman, B. T., & Heinecke, J. W. (2004). Neuronal expression of myeloperoxidase is increased in Alzheimer's disease. *Journal of Neurochemistry, 90*, 724–733.
Griffin, W. S., Stanley, L. C., Ling, C., White, L., MacLeod, V., Perrot, L. J., White, C. L., 3rd, & Araoz, C. (1989). Brain interleukin 1 and S-100 immunoreactivity are elevated in Down syndrome

and Alzheimer disease. *Proceedings of the National Academy of Sciences of the United States of America, 86*, 7611–7615.

Griffin, W., Sheng, J., Roberts, G., & Mrak, R. (1995). Interleukin-1 expression in different plaque types in Alzheimer's diseases: Significance in plaque evolution. *Journal of Neuropathology and Experimental Neurology, 54*, 276–281.

Guerreiro, R., Wojtas, A., Bras, J., Carrasquillo, M., Rogaeva, E., Majounie, E., Cruchaga, C., Sassi, C., Kauwe, J. S., Younkin, S., Hazrati, L., Collinge, J., Pocock, J., Lashley, T., Williams, J., Lambert, J. C., Amouyel, P., Goate, A., Rademakers, R., Morgan, K., Powell, J., St George-Hyslop, P., Singleton, A., & Hardy, J. (2013a). TREM2 variants in Alzheimer's disease. *The New England Journal of Medicine, 368*, 117–127.

Guerreiro, R. J., Lohmann, E., Bras, J. M., Gibbs, J. R., Rohrer, J. D., Gurunlian, N., Dursun, B., Bilgic, B., Hanagasi, H., Gurvit, H., Emre, M., Singleton, A., & Hardy, J. (2013b). Using exome sequencing to reveal mutations in TREM2 presenting as a frontotemporal dementia-like syndrome without bone involvement. *JAMA Neurology, 70*, 78–84.

Guo, Z., Cupples, L. A., Kurz, A., Auerbach, S. H., Volicer, L., Chui, H., Green, R. C., Sadovnick, A. D., Duara, R., DeCarli, C., Johnson, K., Go, R. C., Growdon, J. H., Haines, J. L., Kukull, W. A., & Farrer, L. A. (2000). Head injury and the risk of AD in the MIRAGE study. *Neurology, 54*, 1316–1323.

Gupta, S., & Gollapudi, S. (2005). Molecular mechanisms of TNF-alpha-induced apoptosis in aging human T cell subsets. *The International Journal of Biochemistry & Cell Biology, 37*, 1034–1042.

Hakola, H. P., & Puranen, M. (1993). Neuropsychiatric and brain CT findings in polycystic lipomembranous osteodysplasia with sclerosing leukoencephalopathy. *Acta Neurologica Scandinavica, 88*, 370–375.

Hamerman, J. A., Tchao, N. K., Lowell, C. A., & Lanier, L. L. (2005). Enhanced Toll-like receptor responses in the absence of signaling adaptor DAP12. *Nature Immunology, 6*, 579–586.

Hardy, J. A., & Higgins, G. A. (1992). Alzheimer's disease: The amyloid cascade hypothesis. *Science, 256*, 184–185.

Hardy, J., & Singleton, A. (2009). Genomewide association studies and human disease. *The New England Journal of Medicine, 360*, 1759–1768.

Harold, D., Abraham, R., Hollingworth, P., Sims, R., Gerrish, A., Hamshere, M. L., Pahwa, J. S., Moskvina, V., Dowzell, K., Williams, A., Jones, N., Thomas, C., Stretton, A., Morgan, A. R., Lovestone, S., Powell, J., Proitsi, P., Lupton, M. K., Brayne, C., Rubinsztein, D. C., Gill, M., Lawlor, B., Lynch, A., Morgan, K., Brown, K. S., Passmore, P. A., Craig, D., McGuinness, B., Todd, S., Holmes, C., Mann, D., Smith, A. D., Love, S., Kehoe, P. G., Hardy, J., Mead, S., Fox, N., Rossor, M., Collinge, J., Maier, W., Jessen, F., Schurmann, B., Heun, R., van den Bussche, H., Heuser, I., Kornhuber, J., Wiltfang, J., Dichgans, M., Frolich, L., Hampel, H., Hull, M., Rujescu, D., Goate, A. M., Kauwe, J. S., Cruchaga, C., Nowotny, P., Morris, J. C., Mayo, K., Sleegers, K., Bettens, K., Engelborghs, S., De Deyn, P. P., Van Broeckhoven, C., Livingston, G., Bass, N. J., Gurling, H., McQuillin, A., Gwilliam, R., Deloukas, P., Al-Chalabi, A., Shaw, C. E., Tsolaki, M., Singleton, A. B., Guerreiro, R., Muhleisen, T. W., Nothen, M. M., Moebus, S., Jockel, K. H., Klopp, N., Wichmann, H. E., Carrasquillo, M. M., Pankratz, V. S., Younkin, S. G., Holmans, P. A., O'Donovan, M., Owen, M. J., & Williams, J. (2009). Genome-wide association study identifies variants at CLU and PICALM associated with Alzheimer's disease. *Nature Genetics, 41*, 1088–1093.

Harper, J. D., Wong, S. S., Lieber, C. M., & Lansbury, P. T. (1997). Observation of metastable Abeta amyloid protofibrils by atomic force microscopy. *Chemistry & Biology, 4*, 119–125.

Harris-White, M. E., Chu, T., Balverde, Z., Sigel, J. J., Flanders, K. C., & Frautschy, S. A. (1998). Effects of transforming growth factor-beta (isoforms 1-3) on amyloid-beta deposition, inflammation, and cell targeting in organotypic hippocampal slice cultures. *The Journal of Neuroscience, 18*, 10366–10374.

Heneka, M. T., Klockgether, T., & Feinstein, D. L. (2000). Peroxisome proliferator-activated receptor-gamma ligands reduce neuronal inducible nitric oxide synthase expression and cell death in vivo. *The Journal of Neuroscience, 20*, 6862–6867.

Heneka, M. T., Sastre, M., Dumitrescu-Ozimek, L., Hanke, A., Dewachter, I., Kuiperi, C., O'Banion, K., Klockgether, T., Van Leuven, F., & Landreth, G. E. (2005). Acute treatment with the PPARgamma agonist pioglitazone and ibuprofen reduces glial inflammation and Abeta1-42 levels in APPV717I transgenic mice. *Brain, 128*, 1442–1453.

Heneka, M. T., Kummer, M. P., Stutz, A., Delekate, A., Schwartz, S., Vieira-Saecker, A., Griep, A., Axt, D., Remus, A., Tzeng, T. C., Gelpi, E., Halle, A., Korte, M., Latz, E., & Golenbock, D. T. (2013). NLRP3 is activated in Alzheimer's disease and contributes to pathology in APP/PS1 mice. *Nature, 493*, 674–678.

Hickman, S. E., & El Khoury, J. (2014). TREM2 and the neuroimmunology of Alzheimer's disease. *Biochemical Pharmacology, 88*, 495–498.

Hof, P. R., Bussiere, T., Gold, G., Kovari, E., Giannakopoulos, P., Bouras, C., Perl, D. P., & Morrison, J. H. (2003). Stereologic evidence for persistence of viable neurons in layer II of the entorhinal cortex and the CA1 field in Alzheimer disease. *Journal of Neuropathology and Experimental Neurology, 62*, 55–67.

Holcomb, L., Gordon, M. N., McGowan, E., Yu, X., Benkovic, S., Jantzen, P., Wright, K., Saad, I., Mueller, R., Morgan, D., Sanders, S., Zehr, C., O'Campo, K., Hardy, J., Prada, C. M., Eckman, C., Younkin, S., Hsiao, K., & Duff, K. (1998). Accelerated Alzheimer-type phenotype in transgenic mice carrying both mutant amyloid precursor protein and presenilin 1 transgenes. *Nature Medicine, 4*, 97–100.

Hollingworth, P., Harold, D., Sims, R., Gerrish, A., Lambert, J. C., Carrasquillo, M. M., Abraham, R., Hamshere, M. L., Pahwa, J. S., Moskvina, V., Dowzell, K., Jones, N., Stretton, A., Thomas, C., Richards, A., Ivanov, D., Widdowson, C., Chapman, J., Lovestone, S., Powell, J., Proitsi, P., Lupton, M. K., Brayne, C., Rubinsztein, D. C., Gill, M., Lawlor, B., Lynch, A., Brown, K. S., Passmore, P. A., Craig, D., McGuinness, B., Todd, S., Holmes, C., Mann, D., Smith, A. D., Beaumont, H., Warden, D., Wilcock, G., Love, S., Kehoe, P. G., Hooper, N. M., Vardy, E. R., Hardy, J., Mead, S., Fox, N. C., Rossor, M., Collinge, J., Maier, W., Jessen, F., Ruther, E., Schurmann, B., Heun, R., Kolsch, H., van den Bussche, H., Heuser, I., Kornhuber, J., Wiltfang, J., Dichgans, M., Frolich, L., Hampel, H., Gallacher, J., Hull, M., Rujescu, D., Giegling, I., Goate, A. M., Kauwe, J. S., Cruchaga, C., Nowotny, P., Morris, J. C., Mayo, K., Sleegers, K., Bettens, K., Engelborghs, S., De Deyn, P. P., Van Broeckhoven, C., Livingston, G., Bass, N. J., Gurling, H., McQuillin, A., Gwilliam, R., Deloukas, P., Al-Chalabi, A., Shaw, C. E., Tsolaki, M., Singleton, A. B., Guerreiro, R., Muhleisen, T. W., Nothen, M. M., Moebus, S., Jockel, K. H., Klopp, N., Wichmann, H. E., Pankratz, V. S., Sando, S. B., Aasly, J. O., Barcikowska, M., Wszolek, Z. K., Dickson, D. W., Graff-Radford, N. R., Petersen, R. C., et al. (2011). Common variants at ABCA7, MS4A6A/MS4A4E, EPHA1, CD33 and CD2AP are associated with Alzheimer's disease. *Nature Genetics, 43*, 429–435.

Holtzman, D. M., Bales, K. R., Wu, S., Bhat, P., Parsadanian, M., Fagan, A. M., Chang, L. K., Sun, Y., & Paul, S. M. (1999). Expression of human apolipoprotein E reduces amyloid-beta deposition in a mouse model of Alzheimer's disease. *The Journal of Clinical Investigation, 103*, R15–R21.

Holtzman, D. M., Bales, K. R., Tenkova, T., Fagan, A. M., Parsadanian, M., Sartorius, L. J., Mackey, B., Olney, J., McKeel, D., Wozniak, D., & Paul, S. M. (2000). Apolipoprotein E isoform-dependent amyloid deposition and neuritic degeneration in a mouse model of Alzheimer's disease. *Proceedings of the National Academy of Sciences of the United States of America, 97*, 2892–2897.

Hong, S., Beja-Glasser, V. F., Nfonoyim, B. M., Frouin, A., Li, S., Ramakrishnan, S., Merry, K. M., Shi, Q., Rosenthal, A., Barres, B. A., Lemere, C. A., Selkoe, D. J., & Stevens, B. (2016). Complement and microglia mediate early synapse loss in Alzheimer mouse models. *Science, 352*, 712–716.

Hoshi, M., Sato, M., Matsumoto, S., Noguchi, A., Yasutake, K., Yoshida, N., & Sato, K. (2003). Spherical aggregates of beta-amyloid (amylospheroid) show high neurotoxicity and activate tau protein kinase I/glycogen synthase kinase-3beta. *Proceedings of the National Academy of Sciences of the United States of America, 100*, 6370–6375.

Hsia, A. Y., Masliah, E., McConlogue, L., Yu, G. Q., Tatsuno, G., Hu, K., Kholodenko, D., Malenka, R. C., Nicoll, R. A., & Mucke, L. (1999). Plaque-independent disruption of neural circuits in Alzheimer's disease mouse models. *Proceedings of the National Academy of Sciences of the United States of America, 96*, 3228–3233.

Hsiao, K., Chapman, P., Nilsen, S., Eckman, C., Harigaya, Y., Younkin, S., Yang, F., & Cole, G. (1996). Correlative memory deficits, Abeta elevation, and amyloid plaques in transgenic mice. *Science, 274*, 99–102.

Huang, F., Buttini, M., Wyss-Coray, T., McConlogue, L., Kodama, T., Pitas, R. E., & Mucke, L. (1999). Elimination of the class A scavenger receptor does not affect amyloid plaque formation or neurodegeneration in transgenic mice expressing human amyloid protein precursors. *The American Journal of Pathology, 155*, 1741–1747.

Hutton, M., Lendon, C. L., Rizzu, P., Baker, M., Froelich, S., Houlden, H., Pickering-Brown, S., Chakraverty, S., Isaacs, A., Grover, A., Hackett, J., Adamson, J., Lincoln, S., Dickson, D., Davies, P., Petersen, R. C., Stevens, M., de Graaff, E., Wauters, E., van Baren, J., Hillebrand, M., Joosse, M., Kwon, J. M., Nowotny, P., Che, L. K., Norton, J., Morris, J. C., Reed, L. A., Trojanowski, J., Basun, H., Lannfelt, L., Neystat, M., Fahn, S., Dark, F., Tannenberg, T., Dodd, P. R., Hayward, N., Kwok, J. B., Schofield, P. R., Andreadis, A., Snowden, J., Craufurd, D., Neary, D., Owen, F., Oostra, B. A., Hardy, J., Goate, A., van Swieten, J., Mann, D., Lynch, T., & Heutink, P. (1998). Association of missense and 5′-splice-site mutations in tau with the inherited dementia FTDP-17. *Nature, 393*, 702–705.

Ikezu, S., & Ikezu, T. (2014). Tau-tubulin kinase. *Frontiers in Molecular Neuroscience, 7*, 33.

Ikezu, T., Luo, X., Weber, G. A., Zhao, J., McCabe, L., Buescher, J. L., Ghorpade, A., Zheng, J., & Xiong, H. (2003). Amyloid precursor protein-processing products affect mononuclear phagocyte activation: Pathways for sAPP- and Abeta-mediated neurotoxicity. *Journal of Neurochemistry, 85*, 925–934.

Ikezu, S., Ingraham Dixie, K. L., Koro, L., Watanabe, T., Kaibuchi, K., & Ikezu, T. (2020). Tau-tubulin kinase 1 and amyloid-beta peptide induce phosphorylation of collapsin response mediator protein-2 and enhance neurite degeneration in Alzheimer disease mouse models. *Acta Neuropathologica Communications, 8*, 12.

Iribarren, P., Chen, K., Hu, J., Gong, W., Cho, E. H., Lockett, S., Uranchimeg, B., & Wang, J. M. (2005). CpG-containing oligodeoxynucleotide promotes microglial cell uptake of amyloid beta 1-42 peptide by up-regulating the expression of the G-protein- coupled receptor mFPR2. *The FASEB Journal, 19*, 2032–2034.

Ishiguro, K., Omori, A., Takamatsu, M., Sato, K., Arioka, M., Uchida, T., & Imahori, K. (1992). Phosphorylation sites on tau by tau protein kinase I, a bovine derived kinase generating an epitope of paired helical filaments. *Neuroscience Letters, 148*, 202–206.

Ishizuka, K., Kimura, T., Igata-yi, R., Katsuragi, S., Takamatsu, J., & Miyakawa, T. (1997). Identification of monocyte chemoattractant protein-1 in senile plaques and reactive microglia of Alzheimer's disease. *Psychiatry and Clinical Neurosciences, 51*, 135–138.

Iwasaki, A., & Medzhitov, R. (2004). Toll-like receptor control of the adaptive immune responses. *Nature Immunology, 5*, 987–995.

Iwata, N., Tsubuki, S., Takaki, Y., Watanabe, K., Sekiguchi, M., Hosoki, E., Kawashima-Morishima, M., Lee, H. J., Hama, E., Sekine-Aizawa, Y., & Saido, T. C. (2000). Identification of the major Abeta1-42-degrading catabolic pathway in brain parenchyma: Suppression leads to biochemical and pathological deposition. *Nature Medicine, 6*, 143–150.

Jack, C. S., Arbour, N., Manusow, J., Montgrain, V., Blain, M., McCrea, E., Shapiro, A., & Antel, J. P. (2005). TLR signaling tailors innate immune responses in human microglia and astrocytes. *Journal of Immunology, 175*, 4320–4330.

Jacobsen, J. S., Wu, C. C., Redwine, J. M., Comery, T. A., Arias, R., Bowlby, M., Martone, R., Morrison, J. H., Pangalos, M. N., Reinhart, P. H., & Bloom, F. E. (2006). Early-onset behavioral and synaptic deficits in a mouse model of Alzheimer's disease. *Proceedings of the National Academy of Sciences of the United States of America, 103*, 5161–5166.

Janelsins, M. C., Mastrangelo, M. A., Oddo, S., LaFerla, F. M., Federoff, H. J., & Bowers, W. J. (2005). Early correlation of microglial activation with enhanced tumor necrosis factor-alpha and monocyte chemoattractant protein-1 expression specifically within the entorhinal cortex of triple transgenic Alzheimer's disease mice. *Journal of Neuroinflammation, 2*, 23.

Jantzen, P. T., Connor, K. E., DiCarlo, G., Wenk, G. L., Wallace, J. L., Rojiani, A. M., Coppola, D., Morgan, D., & Gordon, M. N. (2002). Microglial activation and beta -amyloid deposit reduction caused by a nitric oxide-releasing nonsteroidal anti-inflammatory drug in amyloid precursor protein plus presenilin-1 transgenic mice. *The Journal of Neuroscience, 22*, 2246–2254.

Janus, C., Pearson, J., McLaurin, J., Mathews, P. M., Jiang, Y., Schmidt, S. D., Chishti, M. A., Horne, P., Heslin, D., French, J., Mount, H. T., Nixon, R. A., Mercken, M., Bergeron, C., Fraser, P. E., St George-Hyslop, P., & Westaway, D. (2000). A beta peptide immunization reduces behavioural impairment and plaques in a model of Alzheimer's disease. *Nature, 408*, 979–982.

Jolivalt, C., Leininger-Muller, B., Drozdz, R., Naskalski, J. W., & Siest, G. (1996). Apolipoprotein E is highly susceptible to oxidation by myeloperoxidase, an enzyme present in the brain. *Neuroscience Letters, 210*, 61–64.

Jonsson, T., Stefansson, H., Steinberg, S., Jonsdottir, I., Jonsson, P. V., Snaedal, J., Bjornsson, S., Huttenlocher, J., Levey, A. I., Lah, J. J., Rujescu, D., Hampel, H., Giegling, I., Andreassen, O. A., Engedal, K., Ulstein, I., Djurovic, S., Ibrahim-Verbaas, C., Hofman, A., Ikram, M. A., van Duijn, C. M., Thorsteinsdottir, U., Kong, A., & Stefansson, K. (2013). Variant of TREM2 associated with the risk of Alzheimer's disease. *The New England Journal of Medicine, 368*, 107–116.

Keren-Shaul, H., Spinrad, A., Weiner, A., Matcovitch-Natan, O., Dvir-Szternfeld, R., Ulland, T. K., David, E., Baruch, K., Lara-Astaiso, D., Toth, B., Itzkovitz, S., Colonna, M., Schwartz, M., & Amit, I. (2017). A unique microglia type associated with restricting development of Alzheimer's disease. *Cell, 169*, 1276–1290 e1217.

Kiyota, T., Yamamoto, M., Schroder, B., Jacobsen, M. T., Swan, R. J., Lambert, M. P., Klein, W. L., Gendelman, H. E., Ransohoff, R. M., & Ikezu, T. (2009a). AAV1/2-mediated CNS gene delivery of dominant-negative CCL2 mutant suppresses gliosis, beta-amyloidosis, and learning impairment of APP/PS1 mice. *Molecular Therapy, 17*, 803–809.

Kiyota, T., Yamamoto, M., Xiong, H., Lambert, M. P., Klein, W. L., Gendelman, H. E., Ransohoff, R. M., & Ikezu, T. (2009b). CCL2 accelerates microglia-mediated Abeta oligomer formation and progression of neurocognitive dysfunction. *PLoS One, 4*, e6197.

Kiyota, T., Okuyama, S., Swan, R. J., Jacobsen, M. T., Gendelman, H. E., & Ikezu, T. (2010). CNS expression of anti-inflammatory cytokine interleukin-4 attenuates Alzheimer's disease-like pathogenesis in APP+PS1 bigenic mice. *The FASEB Journal, 24*, 3093–3102.

Kiyota, T., Ingraham, K. L., Swan, R. J., Jacobsen, M. T., Andrews, S. J., & Ikezu, T. (2011). AAV serotype 2/1-mediated gene delivery of anti-inflammatory interleukin-10 enhances neurogenesis and cognitive function in APP+PS1 mice. *Gene Therapy, 19*, 724–733.

Kleinberger, G., Yamanishi, Y., Suarez-Calvet, M., Czirr, E., Lohmann, E., Cuyvers, E., Struyfs, H., Pettkus, N., Wenninger-Weinzierl, A., Mazaheri, F., Tahirovic, S., Lleo, A., Alcolea, D., Fortea, J., Willem, M., Lammich, S., Molinuevo, J. L., Sanchez-Valle, R., Antonell, A., Ramirez, A., Heneka, M. T., Sleegers, K., van der Zee, J., Martin, J. J., Engelborghs, S., Demirtas-Tatlidede, A., Zetterberg, H., Van Broeckhoven, C., Gurvit, H., Wyss-Coray, T., Hardy, J., Colonna, M., & Haass, C. (2014). TREM2 mutations implicated in neurodegeneration impair cell surface transport and phagocytosis. *Science Translational Medicine, 6*, 243ra286.

Kligman, D., & Marshak, D. R. (1985). Purification and characterization of a neurite extension factor from bovine brain. *Proceedings of the National Academy of Sciences of the United States of America, 82*, 7136–7139.

Koenigsknecht-Talboo, J., & Landreth, G. E. (2005). Microglial phagocytosis induced by fibrillar beta-amyloid and IgGs are differentially regulated by proinflammatory cytokines. *The Journal of Neuroscience, 25*, 8240–8249.

Koning, N., van Eijk, M., Pouwels, W., Brouwer, M. S., Voehringer, D., Huitinga, I., Hoek, R. M., Raes, G., & Hamann, J. (2010). Expression of the inhibitory CD200 receptor is associated with alternative macrophage activation. *Journal of Innate Immunity, 2*, 195–200.

Kopec, K. K., & Carroll, R. T. (1998). Alzheimer's beta-amyloid peptide 1-42 induces a phagocytic response in murine microglia. *Journal of Neurochemistry, 71*, 2123–2131.

Krasemann, S., Madore, C., Cialic, R., Baufeld, C., Calcagno, N., El Fatimy, R., Beckers, L., O'Loughlin, E., Xu, Y., Fanek, Z., Greco, D. J., Smith, S. T., Tweet, G., Humulock, Z., Zrzavy, T., Conde-Sanroman, P., Gacias, M., Weng, Z., Chen, H., Tjon, E., Mazaheri, F., Hartmann, K., Madi, A., Ulrich, J. D., Glatzel, M., Worthmann, A., Heeren, J., Budnik, B., Lemere, C., Ikezu, T., Heppner, F. L., Litvak, V., Holtzman, D. M., Lassmann, H., Weiner, H. L., Ochando, J., Haass, C., & Butovsky, O. (2017). The TREM2-APOE pathway drives the transcriptional phenotype of dysfunctional microglia in neurodegenerative diseases. *Immunity, 47*, 566–581 e569.

Kruman, R. P., II, Wersto, F., Cardozo-Pelaez, L., Smilenov, S. L., Chan, F. J., Chrest, R. E., Jr., Gorospe, M., & Mattson, M. P. (2004). Cell cycle activation linked to neuronal cell death initiated by DNA damage. *Neuron, 41*, 549–561.

Kukar, T., Murphy, M. P., Eriksen, J. L., Sagi, S. A., Weggen, S., Smith, T. E., Ladd, T., Khan, M. A., Kache, R., Beard, J., Dodson, M., Merit, S., Ozols, V. V., Anastasiadis, P. Z., Das, P., Fauq, A., Koo, E. H., & Golde, T. E. (2005). Diverse compounds mimic Alzheimer disease-causing mutations by augmenting Abeta42 production. *Nature Medicine, 11*, 545–550.

Kuusisto, J., Koivisto, K., Kervinen, K., Mykkanen, L., Helkala, E. L., Vanhanen, M., Hanninen, T., Pyorala, K., Kesaniemi, Y. A., Riekkinen, P., et al. (1994). Association of apolipoprotein E phenotypes with late onset Alzheimer's disease: Population based study. *BMJ, 309*, 636–638.

LaFerla, F. M., Hall, C. K., Ngo, L., & Jay, G. (1996). Extracellular deposition of beta-amyloid upon p53-dependent neuronal cell death in transgenic mice. *The Journal of Clinical Investigation, 98*, 1626–1632.

Lambert, M. P., Barlow, A. K., Chromy, B. A., Edwards, C., Freed, R., Liosatos, M., Morgan, T. E., Rozovsky, I., Trommer, B., Viola, K. L., Wals, P., Zhang, C., Finch, C. E., Krafft, G. A., & Klein, W. L. (1998). Diffusible, nonfibrillar ligands derived from Abeta1-42 are potent central nervous system neurotoxins. *Proceedings of the National Academy of Sciences of the United States of America, 95*, 6448–6453.

Lambert, J. C., Heath, S., Even, G., Campion, D., Sleegers, K., Hiltunen, M., Combarros, O., Zelenika, D., Bullido, M. J., Tavernier, B., Letenneur, L., Bettens, K., Berr, C., Pasquier, F., Fievet, N., Barberger-Gateau, P., Engelborghs, S., De Deyn, P., Mateo, I., Franck, A., Helisalmi, S., Porcellini, E., Hanon, O., de Pancorbo, M. M., Lendon, C., Dufouil, C., Jaillard, C., Leveillard, T., Alvarez, V., Bosco, P., Mancuso, M., Panza, F., Nacmias, B., Bossu, P., Piccardi, P., Annoni, G., Seripa, D., Galimberti, D., Hannequin, D., Licastro, F., Soininen, H., Ritchie, K., Blanche, H., Dartigues, J. F., Tzourio, C., Gut, I., Van Broeckhoven, C., Alperovitch, A., Lathrop, M., & Amouyel, P. (2009). Genome-wide association study identifies variants at CLU and CR1 associated with Alzheimer's disease. *Nature Genetics, 41*, 1094–1099.

Lambert, J. C., Ibrahim-Verbaas, C. A., Harold, D., Naj, A. C., Sims, R., Bellenguez, C., DeStafano, A. L., Bis, J. C., Beecham, G. W., Grenier-Boley, B., Russo, G., Thorton-Wells, T. A., Jones, N., Smith, A. V., Chouraki, V., Thomas, C., Ikram, M. A., Zelenika, D., Vardarajan, B. N., Kamatani, Y., Lin, C. F., Gerrish, A., Schmidt, H., Kunkle, B., Dunstan, M. L., Ruiz, A., Bihoreau, M. T., Choi, S. H., Reitz, C., Pasquier, F., Cruchaga, C., Craig, D., Amin, N., Berr, C., Lopez, O. L., De Jager, P. L., Deramecourt, V., Johnston, J. A., Evans, D., Lovestone, S., Letenneur, L., Moron, F. J., Rubinsztein, D. C., Eiriksdottir, G., Sleegers, K., Goate, A. M., Fievet, N., Huentelman, M. W., Gill, M., Brown, K., Kamboh, M. I., Keller, L., Barberger-Gateau, P., McGuiness, B., Larson, E. B., Green, R., Myers, A. J., Dufouil, C., Todd, S., Wallon, D., Love, S., Rogaeva, E., Gallacher, J., St George-Hyslop, P., Clarimon, J., Lleo, A., Bayer, A., Tsuang, D. W., Yu, L., Tsolaki, M., Bossu, P., Spalletta, G., Proitsi, P., Collinge, J., Sorbi, S., Sanchez-Garcia, F., Fox, N. C., Hardy, J., Deniz Naranjo, M. C., Bosco, P., Clarke, R., Brayne, C., Galimberti, D., Mancuso, M., Matthews, F., Moebus, S., Mecocci, P., Del Zompo, M., Maier, W., Hampel, H., Pilotto, A., Bullido, M., Panza, F., Caffarra, P., Nacmias, B., Gilbert, J. R., Mayhaus, M., Lannefelt, L., Hakonarson, H., Pichler, S., et al. (2013). Meta-analysis of 74,046 individuals identifies 11 new susceptibility loci for Alzheimer's disease. *Nature Genetics, 45*, 1452–1458.

Larson, J., Lynch, G., Games, D., & Seubert, P. (1999). Alterations in synaptic transmission and long-term potentiation in hippocampal slices from young and aged PDAPP mice. *Brain Research, 840*, 23–35.

Lee, V. M., Goedert, M., & Trojanowski, J. Q. (2001). Neurodegenerative tauopathies. *Annual Review of Neuroscience, 24*, 1121–1159.

Lee, E. B., Zhang, B., Liu, K., Greenbaum, E. A., Doms, R. W., Trojanowski, J. Q., & Lee, V. M. (2005). BACE overexpression alters the subcellular processing of APP and inhibits Abeta deposition in vivo. *The Journal of Cell Biology, 168*, 291–302.

Leissring, M. A., Farris, W., Chang, A. Y., Walsh, D. M., Wu, X., Sun, X., Frosch, M. P., & Selkoe, D. J. (2003). Enhanced proteolysis of beta-amyloid in APP transgenic mice prevents plaque formation, secondary pathology, and premature death. *Neuron, 40*, 1087–1093.

Leistner, C., Wilkinson, M., Burgess, A., Lovatt, M., Goodbody, S., Xu, Y., Deuchars, S., Radford, S. E., Ranson, N. A., & Frank, R. A. W. (2023). The in-tissue molecular architecture of beta-amyloid pathology in the mammalian brain. *Nature Communications, 14*, 2833.

Lemberger, T., Desvergne, B., & Wahli, W. (1996). Peroxisome proliferator-activated receptors: A nuclear receptor signaling pathway in lipid physiology. *Annual Review of Cell and Developmental Biology, 12*, 335–363.

Lesne, S., Koh, M. T., Kotilinek, L., Kayed, R., Glabe, C. G., Yang, A., Gallagher, M., & Ashe, K. H. (2006). A specific amyloid-beta protein assembly in the brain impairs memory. *Nature, 440*, 352–357.

Letiembre, M., Hao, W., Liu, Y., Walter, S., Mihaljevic, I., Rivest, S., Hartmann, T., & Fassbender, K. (2007). Innate immune receptor expression in normal brain aging. *Neuroscience, 146*, 248–254.

Levy-Lahad, E., Wasco, W., Poorkaj, P., Romano, D. M., Oshima, J., Pettingell, W. H., Yu, C. E., Jondro, P. D., Schmidt, S. D., Wang, K., et al. (1995). Candidate gene for the chromosome 1 familial Alzheimer's disease locus. *Science, 269*, 973–977.

Lewis, J., McGowan, E., Rockwood, J., Melrose, H., Nacharaju, P., Van Slegtenhorst, M., Gwinn-Hardy, K., Paul Murphy, M., Baker, M., Yu, X., Duff, K., Hardy, J., Corral, A., Lin, W. L., Yen, S. H., Dickson, D. W., Davies, P., & Hutton, M. (2000). Neurofibrillary tangles, amyotrophy and progressive motor disturbance in mice expressing mutant (P301L) tau protein. *Nature Genetics, 25*, 402–405.

Li, R., Yang, L., Lindholm, K., Konishi, Y., Yue, X., Hampel, H., Zhang, D., & Shen, Y. (2004). Tumor necrosis factor death receptor signaling cascade is required for amyloid-beta protein-induced neuron death. *The Journal of Neuroscience, 24*, 1760–1771.

Lim, G. P., Yang, F., Chu, T., Chen, P., Beech, W., Teter, B., Tran, T., Ubeda, O., Ashe, K. H., Frautschy, S. A., & Cole, G. M. (2000). Ibuprofen suppresses plaque pathology and inflammation in a mouse model for Alzheimer's disease. *The Journal of Neuroscience, 20*, 5709–5714.

Lim, G. P., Chu, T., Yang, F., Beech, W., Frautschy, S. A., & Cole, G. M. (2001). The curry spice curcumin reduces oxidative damage and amyloid pathology in an Alzheimer transgenic mouse. *The Journal of Neuroscience, 21*, 8370–8377.

Lim, G. P., Calon, F., Morihara, T., Yang, F., Teter, B., Ubeda, O., Salem, N., Jr., Frautschy, S. A., & Cole, G. M. (2005). A diet enriched with the omega-3 fatty acid docosahexaenoic acid reduces amyloid burden in an aged Alzheimer mouse model. *The Journal of Neuroscience, 25*, 3032–3040.

Linnartz, B., & Neumann, H. (2013). Microglial activatory (immunoreceptor tyrosine-based activation motif)- and inhibitory (immunoreceptor tyrosine-based inhibition motif)-signaling receptors for recognition of the neuronal glycocalyx. *Glia, 61*, 37–46.

Lotz, M., Ebert, S., Esselmann, H., Iliev, A. I., Prinz, M., Wiazewicz, N., Wiltfang, J., Gerber, J., & Nau, R. (2005). Amyloid beta peptide 1-40 enhances the action of Toll-like receptor-2 and -4 agonists but antagonizes Toll-like receptor-9-induced inflammation in primary mouse microglial cell cultures. *Journal of Neurochemistry, 94*, 289–298.

Lovell, M. A., Ehmann, W. D., Mattson, M. P., & Markesbery, W. R. (1997). Elevated 4-hydroxynonenal in ventricular fluid in Alzheimer's disease. *Neurobiology of Aging, 18*, 457–461.

Lu, T., Pan, Y., Kao, S. Y., Li, C., Kohane, I., Chan, J., & Yankner, B. A. (2004). Gene regulation and DNA damage in the ageing human brain. *Nature, 429*, 883–891.

Luo, X., Weber, G. A., Zheng, J., Gendelman, H. E., & Ikezu, T. (2003). C1q-calreticulin induced oxidative neurotoxicity: Relevance for the neuropathogenesis of Alzheimer's disease. *Journal of Neuroimmunology, 135*, 62–71.

Lynch, A. M., Walsh, C., Delaney, A., Nolan, Y., Campbell, V. A., & Lynch, M. A. (2004). Lipopolysaccharide-induced increase in signalling in hippocampus is abrogated by IL-10--a role for IL-1 beta? *Journal of Neurochemistry, 88*, 635–646.

Lyons, A., Downer, E. J., Crotty, S., Nolan, Y. M., Mills, K. H., & Lynch, M. A. (2007a). CD200 ligand receptor interaction modulates microglial activation in vivo and in vitro: A role for IL-4. *The Journal of Neuroscience, 27*, 8309–8313.

Lyons, A., Griffin, R. J., Costelloe, C. E., Clarke, R. M., & Lynch, M. A. (2007b). IL-4 attenuates the neuroinflammation induced by amyloid-beta in vivo and in vitro. *Journal of Neurochemistry, 101*, 771–781.

Lyons, A., McQuillan, K., Deighan, B. F., O'Reilly, J. A., Downer, E. J., Murphy, A. C., Watson, M., Piazza, A., O'Connell, F., Griffin, R., Mills, K. H., & Lynch, M. A. (2009). Decreased neuronal CD200 expression in IL-4-deficient mice results in increased neuroinflammation in response to lipopolysaccharide. *Brain, Behavior, and Immunity, 23*, 1020–1027.

Lyras, L., Cairns, N. J., Jenner, A., Jenner, P., & Halliwell, B. (1997). An assessment of oxidative damage to proteins, lipids, and DNA in brain from patients with Alzheimer's disease. *Journal of Neurochemistry, 68*, 2061–2069.

Maher, F. O., Nolan, Y., & Lynch, M. A. (2005). Downregulation of IL-4-induced signalling in hippocampus contributes to deficits in LTP in the aged rat. *Neurobiology of Aging, 26*, 717–728.

Manczak, M., Park, B. S., Jung, Y., & Reddy, P. H. (2004). Differential expression of oxidative phosphorylation genes in patients with Alzheimer's disease: Implications for early mitochondrial dysfunction and oxidative damage. *Neuromolecular Medicine, 5*, 147–162.

Mangelsdorf, D. J., Thummel, C., Beato, M., Herrlich, P., Schutz, G., Umesono, K., Blumberg, B., Kastner, P., Mark, M., Chambon, P., & Evans, R. M. (1995). The nuclear receptor superfamily: The second decade. *Cell, 83*, 835–839.

Markesbery, W. R., & Carney, J. M. (1999). Oxidative alterations in Alzheimer's disease. *Brain Pathology, 9*, 133–146.

Martins, R. N., Harper, C. G., Stokes, G. B., & Masters, C. L. (1986). Increased cerebral glucose-6-phosphate dehydrogenase activity in Alzheimer's disease may reflect oxidative stress. *Journal of Neurochemistry, 46*, 1042–1045.

Mastrangelo, P., Mathews, P. M., Chishti, M. A., Schmidt, S. D., Gu, Y., Yang, J., Mazzella, M. J., Coomaraswamy, J., Horne, P., Strome, B., Pelly, H., Levesque, G., Ebeling, C., Jiang, Y., Nixon, R. A., Rozmahel, R., Fraser, P. E., St George-Hyslop, P., Carlson, G. A., & Westaway, D. (2005). Dissociated phenotypes in presenilin transgenic mice define functionally distinct gamma-secretases. *Proceedings of the National Academy of Sciences of the United States of America, 102*, 8972–8977.

Mayford, M., Bach, M. E., Huang, Y. Y., Wang, L., Hawkins, R. D., & Kandel, E. R. (1996). Control of memory formation through regulated expression of a CaMKII transgene. *Science, 274*, 1678–1683.

McAlpine, C. S., Park, J., Griciuc, A., Kim, E., Choi, S. H., Iwamoto, Y., Kiss, M. G., Christie, K. A., Vinegoni, C., Poller, W. C., Mindur, J. E., Chan, C. T., He, S., Janssen, H., Wong, L. P., Downey, J., Singh, S., Anzai, A., Kahles, F., Jorfi, M., Feruglio, P. F., Sadreyev, R. I., Weissleder, R., Kleinstiver, B. P., Nahrendorf, M., Tanzi, R. E., & Swirski, F. K. (2021). Astrocytic interleukin-3 programs microglia and limits Alzheimer's disease. *Nature, 595*, 701–706.

McGeer, P. L., & McGeer, E. G. (2001). Inflammation, autotoxicity and Alzheimer disease. *Neurobiology of Aging, 22*, 799–809.

McGeer, P., Schulzer, M., & McGeer, E. (1996). Arthritis and anti-inflammatory agents as possible protective factors for Alzheimer's disease: A review of 17 epidemiologic studies. *Neurology, 47*, 425–432.

McKhann, G., Drachman, D., Folstein, M., Katzman, R., Price, D., & Stadlan, E. M. (1984). Clinical diagnosis of Alzheimer's disease: Report of the NINCDS-ADRDA Work Group under the auspices of Department of Health and Human Services Task Force on Alzheimer's Disease. *Neurology, 34*, 939–944.

McKimmie, C. S., & Fazakerley, J. K. (2005). In response to pathogens, glial cells dynamically and differentially regulate Toll-like receptor gene expression. *Journal of Neuroimmunology, 169*, 116–125.

McMaster, W. R., & Williams, A. F. (1979). Identification of Ia glycoproteins in rat thymus and purification from rat spleen. *European Journal of Immunology, 9*, 426–433.

Meda, L., Cassatella, M. A., Szendrei, G. I., Otvos, L., Jr., Baron, P., Villalba, M., Ferrari, D., & Rossi, F. (1995). Activation of microglial cells by b-amyloid protein and interferon-g. *Nature, 374*, 647–650.

Meda, L., Baron, P., Prat, E., Scarpini, E., Scarlato, G., Cassatella, M. A., & Rossi, F. (1999). Proinflammatory profile of cytokine production by human monocytes and murine microglia stimulated with beta-amyloid[25-35]. *Journal of Neuroimmunology, 93*, 45–52.

Medina, M. G., Ledesma, M. D., Dominguez, J. E., Medina, M., Zafra, D., Alameda, F., Dotti, C. G., & Navarro, P. (2005). Tissue plasminogen activator mediates amyloid-induced neurotoxicity via Erk1/2 activation. *The EMBO Journal, 24*, 1706–1716.

Mielke, M. M., Vemuri, P., & Rocca, W. A. (2014). Clinical epidemiology of Alzheimer's disease: Assessing sex and gender differences. *Clinical Epidemiology, 6*, 37–48.

Mindur, J. E., & Swirski, F. K. (2019). Growth factors as immunotherapeutic targets in cardiovascular disease. *Arteriosclerosis, Thrombosis, and Vascular Biology, 39*, 1275–1287.

Mohajeri, M. H., Saini, K. D., & Nitsch, R. M. (2004). Transgenic BACE expression in mouse neurons accelerates amyloid plaque pathology. *Journal of Neural Transmission, 111*, 413–425.

Monning, U., Sandbrink, R., Banati, R. B., Masters, C. L., & Beyreuther, K. (1994). Transforming growth factor beta mediates increase of mature transmembrane amyloid precursor protein in microglial cells. *FEBS Letters, 342*, 267–272.

Mori, K., Yokoyama, A., Yang, L., Maeda, N., Mitsuda, N., & Tanaka, J. (2004). L-serine-mediated release of apolipoprotein E and lipids from microglial cells. *Experimental Neurology, 185*, 220–231.

Morrison, R. S., Kinoshita, Y., Johnson, M. D., Guo, W., & Garden, G. A. (2003). p53-dependent cell death signaling in neurons. *Neurochemical Research, 28*, 15–27.

Morsch, R., Simon, W., & Coleman, P. D. (1999). Neurons may live for decades with neurofibrillary tangles. *Journal of Neuropathology and Experimental Neurology, 58*, 188–197.

Mortimer, J. A., van Duijn, C. M., Chandra, V., Fratiglioni, L., Graves, A. B., Heyman, A., Jorm, A. F., Kokmen, E., Kondo, K., Rocca, W. A., et al. (1991). Head trauma as a risk factor for Alzheimer's disease: A collaborative re-analysis of case-control studies. EURODEM Risk Factors Research Group. *International Journal of Epidemiology, 20*(Suppl 2), S28–S35.

Nagy, Z., Esiri, M. M., & Smith, A. D. (1998). The cell division cycle and the pathophysiology of Alzheimer's disease. *Neuroscience, 87*, 731–739.

Naj, A. C., Jun, G., Beecham, G. W., Wang, L. S., Vardarajan, B. N., Buros, J., Gallins, P. J., Buxbaum, J. D., Jarvik, G. P., Crane, P. K., Larson, E. B., Bird, T. D., Boeve, B. F., Graff-Radford, N. R., De Jager, P. L., Evans, D., Schneider, J. A., Carrasquillo, M. M., Ertekin-Taner, N., Younkin, S. G., Cruchaga, C., Kauwe, J. S., Nowotny, P., Kramer, P., Hardy, J., Huentelman, M. J., Myers, A. J., Barmada, M. M., Demirci, F. Y., Baldwin, C. T., Green, R. C., Rogaeva, E., St George-Hyslop, P., Arnold, S. E., Barber, R., Beach, T., Bigio, E. H., Bowen, J. D., Boxer, A., Burke, J. R., Cairns, N. J., Carlson, C. S., Carney, R. M., Carroll, S. L., Chui, H. C., Clark, D. G., Corneveaux, J., Cotman, C. W., Cummings, J. L., DeCarli, C., DeKosky, S. T., Diaz-Arrastia, R., Dick, M., Dickson, D. W., Ellis, W. G., Faber, K. M., Fallon, K. B., Farlow, M. R., Ferris, S., Frosch, M. P., Galasko, D. R., Ganguli, M., Gearing, M., Geschwind, D. H., Ghetti, B., Gilbert, J. R., Gilman, S., Giordani, B., Glass, J. D., Growdon, J. H., Hamilton, R. L., Harrell, L. E., Head, E., Honig, L. S., Hulette, C. M., Hyman, B. T., Jicha, G. A., Jin, L. W., Johnson, N., Karlawish, J., Karydas, A., Kaye, J. A., Kim, R., Koo, E. H., Kowall, N. W., Lah, J. J., Levey, A. I., Lieberman, A. P., Lopez, O. L., Mack, W. J., Marson, D. C., Martiniuk, F., Mash, D. C., Masliah, E., McCormick, W. C., McCurry, S. M., McDavid, A. N., McKee, A. C., Mesulam, M., Miller, B. L., et al. (2011). Common variants at MS4A4/MS4A6E, CD2AP, CD33 and EPHA1 are associated with late-onset Alzheimer's disease. *Nature Genetics, 43*, 436–441.

Naslund, J., Haroutunian, V., Mohs, R., Davis, K. L., Davies, P., Greengard, P., & Buxbaum, J. D. (2000). Correlation between elevated levels of amyloid beta-peptide in the brain and cognitive decline. *Journal of the American Medical Association, 283*, 1571–1577.

N'Diaye, E. N., Branda, C. S., Branda, S. S., Nevarez, L., Colonna, M., Lowell, C., Hamerman, J. A., & Seaman, W. E. (2009). TREM-2 (triggering receptor expressed on myeloid cells 2) is a phagocytic receptor for bacteria. *The Journal of Cell Biology, 184*, 215–223.

Nolan, Y., Maher, F. O., Martin, D. S., Clarke, R. M., Brady, M. T., Bolton, A. E., Mills, K. H., & Lynch, M. A. (2005). Role of interleukin-4 in regulation of age-related inflammatory changes in the hippocampus. *The Journal of Biological Chemistry, 280*, 9354–9362.

Oddo, S., Caccamo, A., Shepherd, J. D., Murphy, M. P., Golde, T. E., Kayed, R., Metherate, R., Mattson, M. P., Akbari, Y., & LaFerla, F. M. (2003). Triple-transgenic model of Alzheimer's disease with plaques and tangles: Intracellular Abeta and synaptic dysfunction. *Neuron, 39*, 409–421.

Olson, J. K., & Miller, S. D. (2004). Microglia initiate central nervous system innate and adaptive immune responses through multiple TLRs. *Journal of Immunology, 173*, 3916–3924.

Organization, W.H. (1992). *International statistical classification of diseases and related health problems*. World Health Organization.

Ott, A., Stolk, R. P., van Harskamp, F., Pols, H. A., Hofman, A., & Breteler, M. M. (1999). Diabetes mellitus and the risk of dementia: The Rotterdam Study. *Neurology, 53*, 1937–1942.

Paloneva, J., Kestila, M., Wu, J., Salminen, A., Bohling, T., Ruotsalainen, V., Hakola, P., Bakker, A. B., Phillips, J. H., Pekkarinen, P., Lanier, L. L., Timonen, T., & Peltonen, L. (2000). Loss-of-function mutations in TYROBP (DAP12) result in a presenile dementia with bone cysts. *Nature Genetics, 25*, 357–361.

Paloneva, J., Autti, T., Raininko, R., Partanen, J., Salonen, O., Puranen, M., Hakola, P., & Haltia, M. (2001). CNS manifestations of Nasu-Hakola disease: A frontal dementia with bone cysts. *Neurology, 56*, 1552–1558.

Paloneva, J., Manninen, T., Christman, G., Hovanes, K., Mandelin, J., Adolfsson, R., Bianchin, M., Bird, T., Miranda, R., Salmaggi, A., Tranebjaerg, L., Konttinen, Y., & Peltonen, L. (2002). Mutations in two genes encoding different subunits of a receptor signaling complex result in an identical disease phenotype. *American Journal of Human Genetics, 71*, 656–662.

Paloneva, J., Mandelin, J., Kiialainen, A., Bohling, T., Prudlo, J., Hakola, P., Haltia, M., Konttinen, Y. T., & Peltonen, L. (2003). DAP12/TREM2 deficiency results in impaired osteoclast differentiation and osteoporotic features. *The Journal of Experimental Medicine, 198*, 669–675.

Paradowska-Gorycka, A., & Jurkowska, M. (2013). Structure, expression pattern and biological activity of molecular complex TREM-2/DAP12. *Human Immunology, 74*, 730–737.

Paresce, D. M., Ghosh, R. N., & Maxfield, F. R. (1996). Microglial cells internalize aggregates of the Alzheimer's disease amyloid beta-protein via a scavenger receptor. *Neuron, 17*, 553–565.

Paresce, D. M., Chung, H., & Maxfield, F. R. (1997). Slow degradation of aggregates of the Alzheimer's disease amyloid beta-protein by microglial cells. *The Journal of Biological Chemistry, 272*, 29390–29397.

Patrick, G. N., Zukerberg, L., Nikolic, M., de la Monte, S., Dikkes, P., & Tsai, L. H. (1999). Conversion of p35 to p25 deregulates Cdk5 activity and promotes neurodegeneration. *Nature, 402*, 615–622.

Peila, R., Rodriguez, B. L., & Launer, L. J. (2002). Type 2 diabetes, APOE gene, and the risk for dementia and related pathologies: The Honolulu-Asia Aging Study. *Diabetes, 51*, 1256–1262.

Piani, D., Spranger, M., Frei, K., Schaffner, A., & Fontana, A. (1992). Macrophage-induced cytotoxicity of N-methyl-D-aspartate receptor positive neurons involves excitatory amino acids rather than reactive oxygen intermediates and cytokines. *European Journal of Immunology, 22*, 2429–2436.

Pike, C. J., Burdick, D., Walencewicz, A. J., Glabe, C. G., & Cotman, C. W. (1993). Neurodegeneration induced by beta-amyloid peptides in vitro: The role of peptide assembly state. *The Journal of Neuroscience, 13*, 1676–1687.

Planel, E., Sun, X., & Takashima, A. (2002). Role of GSK-3ß in Alzheimer's disease pathology. *Drug Development Research, 56*, 491–510.

Postina, R., Schroeder, A., Dewachter, I., Bohl, J., Schmitt, U., Kojro, E., Prinzen, C., Endres, K., Hiemke, C., Blessing, M., Flamez, P., Dequenne, A., Godaux, E., van Leuven, F., & Fahrenholz, F. (2004). A disintegrin-metalloproteinase prevents amyloid plaque formation and hippocampal defects in an Alzheimer disease mouse model. *The Journal of Clinical Investigation, 113*, 1456–1464.

Price, D. L., Tanzi, R. E., Borchelt, D. R., & Sisodia, S. S. (1998). Alzheimer's disease: Genetic studies and transgenic models. *Annual Review of Genetics, 32*, 461–493.

Qiu, W. Q., Walsh, D. M., Ye, Z., Vekrellis, K., Zhang, J., Podlisny, M. B., Rosner, M. R., Safavi, A., Hersh, L. B., & Selkoe, D. J. (1998). Insulin-degrading enzyme regulates extracellular levels of amyloid beta-protein by degradation. *The Journal of Biological Chemistry, 273*, 32730–32738.

Qiu, C., Winblad, B., & Fratiglioni, L. (2005). The age-dependent relation of blood pressure to cognitive function and dementia. *Lancet Neurology, 4*, 487–499.

Rajan, K. B., Weuve, J., Barnes, L. L., McAninch, E. A., Wilson, R. S., & Evans, D. A. (2021). Population estimate of people with clinical Alzheimer's disease and mild cognitive impairment in the United States (2020-2060). *Alzheimer's & Dementia, 17*, 1966–1975.

Ramsden, M., Kotilinek, L., Forster, C., Paulson, J., McGowan, E., SantaCruz, K., Guimaraes, A., Yue, M., Lewis, J., Carlson, G., Hutton, M., & Ashe, K. H. (2005). Age-dependent neurofibrillary tangle formation, neuron loss, and memory impairment in a mouse model of human tauopathy (P301L). *The Journal of Neuroscience, 25*, 10637–10647.

Rapoport, M., Dawson, H. N., Binder, L. I., Vitek, M. P., & Ferreira, A. (2002). Tau is essential to beta -amyloid-induced neurotoxicity. *Proceedings of the National Academy of Sciences of the United States of America, 99*, 6364–6369.

Rayaprolu, S., Mullen, B., Baker, M., Lynch, T., Finger, E., Seeley, W. W., Hatanpaa, K. J., Lomen-Hoerth, C., Kertesz, A., Bigio, E. H., Lippa, C., Josephs, K. A., Knopman, D. S., White, C. L., 3rd, Caselli, R., Mackenzie, I. R., Miller, B. L., Boczarska-Jedynak, M., Opala, G., Krygowska-Wajs, A., Barcikowska, M., Younkin, S. G., Petersen, R. C., Ertekin-Taner, N., Uitti, R. J., Meschia, J. F., Boylan, K. B., Boeve, B. F., Graff-Radford, N. R., Wszolek, Z. K., Dickson, D. W., Rademakers, R., & Ross, O. A. (2013). TREM2 in neurodegeneration: Evidence for association of the p.R47H variant with frontotemporal dementia and Parkinson's disease. *Molecular Neurodegeneration, 8*, 19.

Reitz, C., & Mayeux, R. (2014). Alzheimer disease: Epidemiology, diagnostic criteria, risk factors and biomarkers. *Biochemical Pharmacology, 88*, 640–651.

Reynolds, W. F., Rhees, J., Maciejewski, D., Paladino, T., Sieburg, H., Maki, R. A., & Masliah, E. (1999). Myeloperoxidase polymorphism is associated with gender specific risk for Alzheimer's disease. *Experimental Neurology, 155*, 31–41.

Rogers, J., Luber-Narod, J., Styren, S. D., & Civin, W. H. (1988). Expression of immune system-associated antigens by cells of the human central nervous system: Relationship to the pathology of Alzheimer's disease. *Neurobiology of Aging, 9*, 339–349.

Rogers, J. T., Leiter, L. M., McPhee, J., Cahill, C. M., Zhan, S. S., Potter, H., & Nilsson, L. N. (1999). Translation of the alzheimer amyloid precursor protein mRNA is up-regulated by interleukin-1 through 5′-untranslated region sequences. *The Journal of Biological Chemistry, 274*, 6421–6431.

Roher, A. E., Chaney, M. O., Kuo, Y. M., Webster, S. D., Stine, W. B., Haverkamp, L. J., Woods, A. S., Cotter, R. J., Tuohy, J. M., Krafft, G. A., Bonnell, B. S., & Emmerling, M. R. (1996). Morphology and toxicity of Abeta-(1-42) dimer derived from neuritic and vascular amyloid deposits of Alzheimer's disease. *The Journal of Biological Chemistry, 271*, 20631–20635.

Ruan, Z., Pathak, D., Venkatesan Kalavai, S., Yoshii-Kitahara, A., Muraoka, S., Bhatt, N., Takamatsu-Yukawa, K., Hu, J., Wang, Y., Hersh, S., Ericsson, M., Gorantla, S., Gendelman, H. E., Kayed, R., Ikezu, S., Luebke, J. I., & Ikezu, T. (2021). Alzheimer's disease brain-derived extracellular vesicles spread tau pathology in interneurons. *Brain, 144*, 288–309.

Sadowski, M., Pankiewicz, J., Scholtzova, H., Ripellino, J. A., Li, Y., Schmidt, S. D., Mathews, P. M., Fryer, J. D., Holtzman, D. M., Sigurdsson, E. M., & Wisniewski, T. (2004). A synthetic peptide blocking the apolipoprotein E/beta-amyloid binding mitigates beta-amyloid toxicity and fibril formation in vitro and reduces beta-amyloid plaques in transgenic mice. *The American Journal of Pathology, 165*, 937–948.

Saito, T., Matsuba, Y., Mihira, N., Takano, J., Nilsson, P., Itohara, S., Iwata, N., & Saido, T. C. (2014). Single App knock-in mouse models of Alzheimer's disease. *Nature Neuroscience, 17*, 661–663.

Santacruz, K., Lewis, J., Spires, T., Paulson, J., Kotilinek, L., Ingelsson, M., Guimaraes, A., DeTure, M., Ramsden, M., McGowan, E., Forster, C., Yue, M., Orne, J., Janus, C., Mariash, A., Kuskowski, M., Hyman, B., Hutton, M., & Ashe, K. H. (2005). Tau suppression in a neurodegenerative mouse model improves memory function. *Science, 309*, 476–481.

Sato, S., Tatebayashi, Y., Akagi, T., Chui, D. H., Murayama, M., Miyasaka, T., Planel, E., Tanemura, K., Sun, X., Hashikawa, T., Yoshioka, K., Ishiguro, K., & Takashima, A. (2002). Aberrant tau phosphorylation by glycogen synthase kinase-3beta and JNK3 induces oligomeric tau fibrils in COS-7 cells. *The Journal of Biological Chemistry, 277*, 42060–42065.

Sato, S., Cerny, R. L., Buescher, J. L., & Ikezu, T. (2006). Tau-tubulin kinase 1 (TTBK1), a neuron-specific tau kinase candidate, is involved in tau phosphorylation and aggregation. *Journal of Neurochemistry, 98*, 1573–1584.

Sato, S., Xu, J., Okuyama, S., Martinez, L. B., Walsh, S. M., Jacobsen, M. T., Swan, R. J., Schlautman, J. D., Ciborowski, P., & Ikezu, T. (2008). Spatial learning impairment, enhanced CDK5/p35 activity, and downregulation of NMDA receptor expression in transgenic mice expressing tau-tubulin kinase 1. *The Journal of Neuroscience, 28*, 14511–14521.

Saura, J., Petegnief, V., Wu, X., Liang, Y., & Paul, S. M. (2003). Microglial apolipoprotein E and astroglial apolipoprotein J expression in vitro: Opposite effects of lipopolysaccharide. *Journal of Neurochemistry, 85*, 1455–1467.

Schafer, D. P., Lehrman, E. K., Kautzman, A. G., Koyama, R., Mardinly, A. R., Yamasaki, R., Ransohoff, R. M., Greenberg, M. E., Barres, B. A., & Stevens, B. (2012). Microglia sculpt postnatal neural circuits in an activity and complement-dependent manner. *Neuron, 74*, 691–705.

Scheuner, D., Eckman, C., Jensen, M., Song, X., Citron, M., Suzuki, N., Bird, T. D., Hardy, J., Hutton, M., Kukull, W., Larson, E., Levy-Lahad, E., Viitanen, M., Peskind, E., Poorkaj, P., Schellenberg, G., Tanzi, R., Wasco, W., Lannfelt, L., Selkoe, D., & Younkin, S. (1996). Secreted amyloid beta-protein similar to that in the senile plaques of Alzheimer's disease is increased in vivo by the presenilin 1 and 2 and APP mutations linked to familial Alzheimer's disease. *Nature Medicine, 2*, 864–870.

Schmand, B., Smit, J., Lindeboom, J., Smits, C., Hooijer, C., Jonker, C., & Deelman, B. (1997). Low education is a genuine risk factor for accelerated memory decline and dementia. *Journal of Clinical Epidemiology, 50*, 1025–1033.

Schmitt, F. A., Davis, D. G., Wekstein, D. R., Smith, C. D., Ashford, J. W., & Markesbery, W. R. (2000). "Preclinical" AD revisited: Neuropathology of cognitively normal older adults. *Neurology, 55*, 370–376.

Selkoe, D. J. (1994). Cell biology of the amyloid beta-protein precursor and the mechanism of Alzheimer's disease. *Annual Review of Cell Biology, 10*, 373–403.

Selkoe, D. J. (1997). Alzheimer's disease: Genotypes, phenotypes, and treatments. *Science, 275*, 630–631.

Selkoe, D. J., Xia, W., Kimberly, W. T., Vekrellis, K., Walsh, D., Esler, W. P., & Wolfe, M. S. (2001). Mechanism of Abeta production and Abeta degradation: Routes to the treatment of Alzheimer's disease. In K. Iqbal, S. S. Sisodia, & B. Winblad (Eds.), *Alzheimer's disease: Advances in ethology, pathogenesis and therapeutics* (pp. 421–432). Wiley.

Seshadri, S., Fitzpatrick, A. L., Ikram, M. A., DeStefano, A. L., Gudnason, V., Boada, M., Bis, J. C., Smith, A. V., Carassquillo, M. M., Lambert, J. C., Harold, D., Schrijvers, E. M., Ramirez-Lorca, R., Debette, S., Longstreth, W. T., Jr., Janssens, A. C., Pankratz, V. S., Dartigues, J. F., Hollingworth, P., Aspelund, T., Hernandez, I., Beiser, A., Kuller, L. H., Koudstaal, P. J., Dickson, D. W., Tzourio, C., Abraham, R., Antunez, C., Du, Y., Rotter, J. I., Aulchenko, Y. S., Harris, T. B., Petersen, R. C., Berr, C., Owen, M. J., Lopez-Arrieta, J., Varadarajan, B. N., Becker, J. T., Rivadeneira, F., Nalls, M. A., Graff-Radford, N. R., Campion, D., Auerbach, S., Rice, K., Hofman, A., Jonsson, P. V., Schmidt, H., Lathrop, M., Mosley, T. H., Au, R., Psaty, B. M., Uitterlinden, A. G., Farrer, L. A., Lumley, T., Ruiz, A., Williams, J., Amouyel, P., Younkin, S. G., Wolf, P. A., Launer, L. J., Lopez, O. L., van Duijn, C. M., & Breteler, M. M. (2010). Genome-wide analysis of genetic loci associated with Alzheimer disease. *Journal of the American Medical Association, 303*, 1832–1840.

Shankar, G. M., Bloodgood, B. L., Townsend, M., Walsh, D. M., Selkoe, D. J., & Sabatini, B. L. (2007). Natural oligomers of the Alzheimer amyloid-ß protein induce reversible synapse loss by modulating an NMDA-Type glutamate receptor-dependent signaling pathway. *The Journal of Neuroscience, 27*, 2866–2875.

Sherrington, R., Rogaev, E. I., Liang, Y., Rogaeva, E. A., Levesque, G., Ikeda, M., Chi, H., Lin, C., Li, G., Holman, K., et al. (1995). Cloning of a gene bearing missense mutations in early-onset familial Alzheimer's disease. *Nature, 375*, 754–760.

Sherva, R., & Farrer, L. A. (2011). Power and pitfalls of the genome-wide association study approach to identify genes for Alzheimer's disease. *Current Psychiatry Reports, 13*, 138–146.

Sisodia, S. S. (1999). Series introduction: Alzheimer's disease: Perspectives for the new millennium. *The Journal of Clinical Investigation, 104*, 1169–1170.

Smith, M. A., Hirai, K., Hsiao, K., Pappolla, M. A., Harris, P. L., Siedlak, S. L., Tabaton, M., & Perry, G. (1998). Amyloid-beta deposition in Alzheimer transgenic mice is associated with oxidative stress. *Journal of Neurochemistry, 70*, 2212–2215.

Smith, M. Z., Nagy, Z., & Esiri, M. M. (1999). Cell cycle-related protein expression in vascular dementia and Alzheimer's disease. *Neuroscience Letters, 271*, 45–48.

Sparks, D. L., Hunsaker, J. C., 3rd, Scheff, S. W., Kryscio, R. J., Henson, J. L., & Markesbery, W. R. (1990). Cortical senile plaques in coronary artery disease, aging and Alzheimer's disease. *Neurobiology of Aging, 11*, 601–607.

Spires, T. L., & Hyman, B. T. (2005). Transgenic models of Alzheimer's disease: Learning from animals. *NeuroRx, 2*, 423–437.

Stephan, A. H., Barres, B. A., & Stevens, B. (2012). The complement system: An unexpected role in synaptic pruning during development and disease. *Annual Review of Neuroscience, 35*, 369–389.

Sturchler-Pierrat, C., Abramowski, D., Duke, M., Wiederhold, K. H., Mistl, C., Rothacher, S., Ledermann, B., Burki, K., Frey, P., Paganetti, P. A., Waridel, C., Calhoun, M. E., Jucker, M., Probst, A., Staufenbiel, M., & Sommer, B. (1997). Two amyloid precursor protein transgenic mouse models with Alzheimer disease-like pathology. *Proceedings of the National Academy of Sciences of the United States of America, 94*, 13287–13292.

Styren, S. D., Civin, W. H., & Rogers, J. (1990). Molecular, cellular, and pathologic characterization of HLA-DR immunoreactivity in normal elderly and Alzheimer's disease brain. *Experimental Neurology, 110*, 93–104.

Sun, Y. X., Minthon, L., Wallmark, A., Warkentin, S., Blennow, K., & Janciauskiene, S. (2003). Inflammatory markers in matched plasma and cerebrospinal fluid from patients with Alzheimer's disease. *Dementia and Geriatric Cognitive Disorders, 16*, 136–144.

Takahashi, K., Rochford, C. D., & Neumann, H. (2005). Clearance of apoptotic neurons without inflammation by microglial triggering receptor expressed on myeloid cells-2. *The Journal of Experimental Medicine, 201*, 647–657.

Takahashi, K., Prinz, M., Stagi, M., Chechneva, O., & Neumann, H. (2007). TREM2-transduced myeloid precursors mediate nervous tissue debris clearance and facilitate recovery in an animal model of multiple sclerosis. *PLoS Medicine, 4*, e124.

Takashima, A., Noguchi, K., Sato, K., Hoshino, T., & Imahori, K. (1993). Tau protein kinase I is essential for amyloid beta-protein-induced neurotoxicity. *Proceedings of the National Academy of Sciences of the United States of America, 90*, 7789–7793.

Takeda, A., Yasuda, T., Miyata, T., Goto, Y., Wakai, M., Watanabe, M., Yasuda, Y., Horie, K., Inagaki, T., Doyu, M., Maeda, K., & Sobue, G. (1998). Advanced glycation end products co-localized with astrocytes and microglial cells in Alzheimer's disease brain. *Acta Neuropathologica (Berl), 95*, 555–558.

Tan, J., Town, T., Paris, D., Mori, T., Suo, Z., Crawford, F., Mattson, M. P., Flavell, R. A., & Mullan, M. (1999). Microglial activation resulting from CD40-CD40L interaction after b-amyloid stimulation. *Science, 286*, 2352–2355.

Tan, J., Town, T., Crawford, F., Mori, T., DelleDonne, A., Crescentini, R., Obregon, D., Flavell, R. A., & Mullan, M. J. (2002). Role of CD40 ligand in amyloidosis in transgenic Alzheimer's mice. *Nature Neuroscience, 5*, 1288–1293.

Tanemura, K., Murayama, M., Akagi, T., Hashikawa, T., Tominaga, T., Ichikawa, M., Yamaguchi, H., & Takashima, A. (2002). Neurodegeneration with tau accumulation in a transgenic mouse expressing V337M human tau. *The Journal of Neuroscience, 22*, 133–141.

Tarkowski, E., Blennow, K., Wallin, A., & Tarkowski, A. (1999). Intracerebral production of tumor necrosis factor-alpha, a local neuroprotective agent, in Alzheimer disease and vascular dementia. *Journal of Clinical Immunology, 19*, 223–230.

Tesseur, I., Zou, K., Esposito, L., Bard, F., Berber, E., Can, J. V., Lin, A. H., Crews, L., Tremblay, P., Mathews, P., Mucke, L., Masliah, E., & Wyss-Coray, T. (2006). Deficiency in neuronal TGF-beta signaling promotes neurodegeneration and Alzheimer's pathology. *The Journal of Clinical Investigation, 116*, 3060–3069.

Todd Roach, J., Volmar, C. H., Dwivedi, S., Town, T., Crescentini, R., Crawford, F., Tan, J., & Mullan, M. (2004). Behavioral effects of CD40-CD40L pathway disruption in aged PSAPP mice. *Brain Research, 1015*, 161–168.

Togo, T., Akiyama, H., Kondo, H., Ikeda, K., Kato, M., Iseki, E., & Kosaka, K. (2000). Expression of CD40 in the brain of Alzheimer's disease and other neurological diseases. *Brain Research, 885*, 117–121.

U.S. Department of Health and Human Services, C.f., N.C.f.H. Disease Control and Prevention, and Statistics. (2022). *CDC WONDER online database: About underlying cause of death, 1999–2020.*

van der Wal, E. A., Gomez-Pinilla, F., & Cotman, C. W. (1993). Transforming growth factor-beta 1 is in plaques in Alzheimer and Down pathologies. *Neuroreport, 4*, 69–72.

Varnum, M. M., & Ikezu, T. (2012). The classification of microglial activation phenotypes on neurodegeneration and regeneration in Alzheimer's disease brain. *Archivum Immunologiae et Therapiae Experimentalis (Warsz), 60*, 251–266.

Vassar, R., Bennett, B. D., Babu-Khan, S., Kahn, S., Mendiaz, E. A., Denis, P., Teplow, D. B., Ross, S., Amarante, P., Loeloff, R., Luo, Y., Fisher, S., Fuller, J., Edenson, S., Lile, J., Jarosinski, M. A., Biere, A. L., Curran, E., Burgess, T., Louis, J. C., Collins, F., Treanor, J., Rogers, G., & Citron, M. (1999). Beta-secretase cleavage of Alzheimer's amyloid precursor protein by the transmembrane aspartic protease BACE. *Science, 286*, 735–741.

Vekrellis, K., Ye, Z., Qiu, W. Q., Walsh, D., Hartley, D., Chesneau, V., Rosner, M. R., & Selkoe, D. J. (2000). Neurons regulate extracellular levels of amyloid beta-protein via proteolysis by insulin-degrading enzyme. *The Journal of Neuroscience, 20*, 1657–1665.

Walker, D. G., Dalsing-Hernandez, J. E., Campbell, N. A., & Lue, L. F. (2009). Decreased expression of CD200 and CD200 receptor in Alzheimer's disease: A potential mechanism leading to chronic inflammation. *Experimental Neurology, 215*, 5–19.

Wallach, D., Varfolomeev, E. E., Malinin, N. L., Goltsev, Y. V., Kovalenko, A. V., & Boldin, M. P. (1999). Tumor necrosis factor receptor and Fas signaling mechanisms. *Annual Review of Immunology, 17*, 331–367.

Walsh, D. M., Lomakin, A., Benedek, G. B., Condron, M. M., & Teplow, D. B. (1997). Amyloid beta-protein fibrillogenesis. Detection of a protofibrillar intermediate. *The Journal of Biological Chemistry, 272*, 22364–22372.

Walsh, D. M., Klyubin, I., Fadeeva, J. V., Cullen, W. K., Anwyl, R., Wolfe, M. S., Rowan, M. J., & Selkoe, D. J. (2002). Naturally secreted oligomers of amyloid beta protein potently inhibit hippocampal long-term potentiation in vivo. *Nature, 416*, 535–539.

Wang, Q., Rowan, M. J., & Anwyl, R. (2004a). Beta-amyloid-mediated inhibition of NMDA receptor-dependent long-term potentiation induction involves activation of microglia and stimulation of inducible nitric oxide synthase and superoxide. *The Journal of Neuroscience, 24*, 6049–6056.

Wang, Q., Walsh, D. M., Rowan, M. J., Selkoe, D. J., & Anwyl, R. (2004b). Block of long-term potentiation by naturally secreted and synthetic amyloid beta-peptide in hippocampal slices is mediated via activation of the kinases c-Jun N-terminal kinase, cyclin-dependent kinase 5, and p38 mitogen-activated protein kinase as well as metabotropic glutamate receptor type 5. *The Journal of Neuroscience, 24*, 3370–3378.

Webb, M., & Barclay, A. N. (1984). Localisation of the MRC OX-2 glycoprotein on the surfaces of neurones. *Journal of Neurochemistry, 43*, 1061–1067.

Webster, S. D., Yang, A. J., Margol, L., Garzon-Rodriguez, W., Glabe, C. G., & Tenner, A. J. (2000). Complement component C1q modulates the phagocytosis of Abeta by microglia. *Experimental Neurology, 161*, 127–138.

Weggen, S., Eriksen, J. L., Das, P., Sagi, S. A., Wang, R., Pietrzik, C. U., Findlay, K. A., Smith, T. E., Murphy, M. P., Bulter, T., Kang, D. E., Marquez-Sterling, N., Golde, T. E., & Koo, E. H. (2001). A subset of NSAIDs lower amyloidogenic Abeta42 independently of cyclooxygenase activity. *Nature, 414*, 212–216.

Weggen, S., Eriksen, J. L., Sagi, S. A., Pietrzik, C. U., Ozols, V., Fauq, A., Golde, T. E., & Koo, E. H. (2003). Evidence that nonsteroidal anti-inflammatory drugs decrease amyloid beta 42 production by direct modulation of gamma-secretase activity. *The Journal of Biological Chemistry, 278*, 31831–31837.

White, A. R., Du, T., Laughton, K. M., Volitakis, I., Sharples, R. A., Hoke, D. E., Holsinger, R. M., Evin, G., Cherny, R. A., Hill, A. F., Barnham, K. J., Li, Q. X., Bush, A. I., & Masters, C. L. (2006). Degradation of the Alzheimer's disease amyloid beta peptide by metal-dependent up-regulation of metalloprotease activity. *The Journal of Biological Chemistry, 281*, 17670–17680.

Wightman, D. P., Jansen, I. E., Savage, J. E., Shadrin, A. A., Bahrami, S., Holland, D., Rongve, A., Borte, S., Winsvold, B. S., Drange, O. K., Martinsen, A. E., Skogholt, A. H., Willer, C., Brathen, G., Bosnes, I., Nielsen, J. B., Fritsche, L. G., Thomas, L. F., Pedersen, L. M., Gabrielsen, M. E., Johnsen, M. B., Meisingset, T. W., Zhou, W., Proitsi, P., Hodges, A., Dobson, R., Velayudhan, L., Heilbron, K., Auton, A., T. and Me Research, Sealock, J. M., Davis, L. K., Pedersen, N. L., Reynolds, C. A., Karlsson, I. K., Magnusson, S., Stefansson, H., Thordardottir, S., Jonsson, P. V., Snaedal, J., Zettergren, A., Skoog, I., Kern, S., Waern, M., Zetterberg, H., Blennow, K., Stordal, E., Hveem, K., Zwart, J. A., Athanasiu, L., Selnes, P., Saltvedt, I., Sando, S. B., Ulstein, I., Djurovic, S., Fladby, T., Aarsland, D., Selbaek, G., Ripke, S., Stefansson, K., Andreassen, O. A., & Posthuma, D. (2021). A genome-wide association study with 1,126,563 individuals identifies new risk loci for Alzheimer's disease. *Nature Genetics, 53*, 1276–1282.

Wilcock, G. K., & Esiri, M. M. (1982). Plaques, tangles and dementia. A quantitative study. *Journal of the Neurological Sciences, 56*, 343–356.

Wilkinson, B., Koenigsknecht-Talboo, J., Grommes, C., Lee, C. Y., & Landreth, G. (2006). Fibrillar beta-amyloid-stimulated intracellular signaling cascades require Vav for induction of respiratory burst and phagocytosis in monocytes and microglia. *The Journal of Biological Chemistry, 281*, 20842–20850.

Willem, M., Dewachter, I., Smyth, N., Van Dooren, T., Borghgraef, P., Haass, C., & Van Leuven, F. (2004). beta-site amyloid precursor protein cleaving enzyme 1 increases amyloid deposition in brain parenchyma but reduces cerebrovascular amyloid angiopathy in aging BACE × APP[V717I] double-transgenic mice. *The American Journal of Pathology, 165*, 1621–1631.

Williams, M. A., Turchan, J., Lu, Y., Nath, A., & Drachman, D. B. (2005). Protection of human cerebral neurons from neurodegenerative insults by gene delivery of soluble tumor necrosis factor p75 receptor. *Experimental Brain Research, 165*, 383–391.

Wright, G. J., Puklavec, M. J., Willis, A. C., Hoek, R. M., Sedgwick, J. D., Brown, M. H., & Barclay, A. N. (2000). Lymphoid/neuronal cell surface OX2 glycoprotein recognizes a novel receptor on macrophages implicated in the control of their function. *Immunity, 13*, 233–242.

Wyss-Coray, T., & Mucke, L. (2002). Inflammation in neurodegenerative disease--a double-edged sword. *Neuron, 35*, 419–432.

Wyss-Coray, T., Masliah, E., Mallory, M., McConlogue, L., Johnson-Wood, K., Lin, C., & Mucke, L. (1997). Amyloidogenic role of cytokine TGF-beta1 in transgenic mice and in Alzheimer's disease. *Nature, 389*, 603–606.

Wyss-Coray, T., Lin, C., von Euw, D., Masliah, E., Mucke, L., & Lacombe, P. (2000). Alzheimer's disease-like cerebrovascular pathology in transforming growth factor-beta 1 transgenic mice and functional metabolic correlates. *Annals of the New York Academy of Sciences, 903*, 317–323.

Wyss-Coray, T., Lin, C., Yan, F., Yu, G. Q., Rohde, M., McConlogue, L., Masliah, E., & Mucke, L. (2001). TGF-beta1 promotes microglial amyloid-beta clearance and reduces plaque burden in transgenic mice. *Nature Medicine, 7*, 612–618.

Wyss-Coray, T., Yan, F., Lin, A. H., Lambris, J. D., Alexander, J. J., Quigg, R. J., & Masliah, E. (2002). Prominent neurodegeneration and increased plaque formation in complement-inhibited Alzheimer's mice. *Proceedings of the National Academy of Sciences of the United States of America, 99*, 10837–10842.

Wyss-Coray, T., Loike, J. D., Brionne, T. C., Lu, E., Anankov, R., Yan, F., Silverstein, S. C., & Husemann, J. (2003). Adult mouse astrocytes degrade amyloid-beta in vitro and in situ. *Nature Medicine, 9*, 453–457.

Xia, M. Q., & Hyman, B. T. (1999). Chemokines/chemokine receptors in the central nervous system and Alzheimer's disease. *Journal of Neurovirology, 5*, 32–41.

Xia, M., Qin, S., McNamara, M., Mackay, C., & Hyman, B. T. (1997). Interleukin-8 receptor B immunoreactivity in brain and neuritic plaques of Alzheimer's disease. *The American Journal of Pathology, 150*, 1267–1274.

Xia, M. Q., Qin, S. X., Wu, L. J., Mackay, C. R., & Hyman, B. T. (1998). Immunohistochemical study of the beta-chemokine receptors CCR3 and CCR5 and their ligands in normal and Alzheimer's disease brains. *The American Journal of Pathology, 153*, 31–37.

Xia, M. Q., Bacskai, B. J., Knowles, R. B., Qin, S. X., & Hyman, B. T. (2000). Expression of the chemokine receptor CXCR3 on neurons and the elevated expression of its ligand IP-10 in reactive astrocytes: In vitro ERK1/2 activation and role in Alzheimer's disease. *Journal of Neuroimmunology, 108*, 227–235.

Xu, Q., Bernardo, A., Walker, D., Kanegawa, T., Mahley, R. W., & Huang, Y. (2006). Profile and regulation of apolipoprotein E (ApoE) expression in the CNS in mice with targeting of green fluorescent protein gene to the ApoE locus. *The Journal of Neuroscience, 26*, 4985–4994.

Xu, J., Sato, S., Okuyama, S., Swan, R. J., Jacobsen, M. T., Strunk, E., & Ikezu, T. (2010). Tau-tubulin kinase 1 enhances prefibrillar tau aggregation and motor neuron degeneration in P301L FTDP-17 tau-mutant mice. *The FASEB Journal, 24*, 2904–2915.

Yamamoto, M., Horiba, M., Buescher, J. L., Huang, D., Gendelman, H. E., Ransohoff, R. M., & Ikezu, T. (2005). Overexpression of monocyte chemotactic protein-1/CCL2 in beta-amyloid precursor protein transgenic mice show accelerated diffuse beta-amyloid deposition. *The American Journal of Pathology, 166*, 1475–1485.

Yamamoto, M., Kiyota, T., Horiba, M., Buescher, J. L., Walsh, S. M., Gendelman, H. E., & Ikezu, T. (2007a). Interferon-{gamma} and tumor necrosis factor-{alpha} regulate amyloid-{beta} plaque deposition and {beta}-secretase expression in Swedish mutant APP transgenic mice. *The American Journal of Pathology, 170*, 680–692.

Yamamoto, M., Kiyota, T., Walsh, S. M., & Ikezu, T. (2007b). Kinetic analysis of aggregated amyloid-b peptide clearance in adult bone-marrow-derived macrophages from APP and CCL2 transgenic mice. *Journal of Neuroimmune Pharmacology, 2*, 213–221.

Yamin, R., Malgeri, E. G., Sloane, J. A., McGraw, W. T., & Abraham, C. R. (1999). Metalloendopeptidase EC 3.4.24.15 is necessary for Alzheimer's amyloid-beta peptide degradation. *The Journal of Biological Chemistry, 274*, 18777–18784.

Yan, S. D., Chen, X., Fu, J., Chen, M., Zhu, H., Roher, A., Slattery, T., Zhao, L., Nagashima, M., Morser, J., Migheli, A., Nawroth, P., Stern, D., & Schmidt, A. M. (1996). RAGE and amyloid-beta peptide neurotoxicity in Alzheimer's disease. *Nature, 382*, 685–691.

Yan, Q., Zhang, J., Liu, H., Babu-Khan, S., Vassar, R., Biere, A. L., Citron, M., & Landreth, G. (2003). Anti-inflammatory drug therapy alters beta-amyloid processing and deposition in an animal model of Alzheimer's disease. *The Journal of Neuroscience, 23*, 7504–7509.

Yang, F., Lim, G. P., Begum, A. N., Ubeda, O. J., Simmons, M. R., Ambegaokar, S. S., Chen, P. P., Kayed, R., Glabe, C. G., Frautschy, S. A., & Cole, G. M. (2005). Curcumin inhibits formation of amyloid beta oligomers and fibrils, binds plaques, and reduces amyloid in vivo. *The Journal of Biological Chemistry, 280*, 5892–5901.

Yin, Z., Herron, S., Silveira, S., Kleeman, K., Gauthier, C., Mallah, D., Cheng, Y., Margeta, M. A., Pitts, K. M., Barry, J.-L., Subramanian, A., Shorey, H., Brandao, W., Durao, A., Delpech, J. C., Madore, C., Jedrychowski, M. P., Ajay, A. K., Murugaiyan, G., Hersh, S. W., Ikezu, S., Ikezu, T., & Butovsky, O. (2023). Identification of a protective microglial state mediated by miR-155 and interferon-γ signaling in a mouse model of Alzheimer's disease. *Nature Neuroscience, 26*, 1196–1207.

Yoshiyama, Y., Higuchi, M., Zhang, B., Huang, S. M., Iwata, N., Saido, T. C., Maeda, J., Suhara, T., Trojanowski, J. Q., & Lee, V. M. (2007). Synapse loss and microglial activation precede tangles in a P301S tauopathy mouse model. *Neuron, 53*, 337–351.

You, Y., & Ikezu, T. (2019). Emerging roles of extracellular vesicles in neurodegenerative disorders. *Neurobiology of Disease, 130*, 104512.

You, Y., Borgmann, K., Edara, V. V., Stacy, S., Ghorpade, A., & Ikezu, T. (2020). Activated human astrocyte-derived extracellular vesicles modulate neuronal uptake, differentiation and firing. *Journal of Extracellular Vesicles, 9*, 1706801.

Young, H., & Hardy, K. (1995). Role of interferon-gamma in immune cell regulation. *Journal of Leukocyte Biology, 58*, 373–381.

Zhang, M. Y., Katzman, R., Salmon, D., Jin, H., Cai, G. J., Wang, Z. Y., Qu, G. Y., Grant, I., Yu, E., Levy, P., et al. (1990). The prevalence of dementia and Alzheimer's disease in Shanghai, China: Impact of age, gender, and education. *Annals of Neurology, 27*, 428–437.

Zhang, B., Gaiteri, C., Bodea, L. G., Wang, Z., McElwee, J., Podtelezhnikov, A. A., Zhang, C., Xie, T., Tran, L., Dobrin, R., Fluder, E., Clurman, B., Melquist, S., Narayanan, M., Suver, C., Shah, H., Mahajan, M., Gillis, T., Mysore, J., MacDonald, M. E., Lamb, J. R., Bennett, D. A., Molony, C., Stone, D. J., Gudnason, V., Myers, A. J., Schadt, E. E., Neumann, H., Zhu, J., & Emilsson, V. (2013). Integrated systems approach identifies genetic nodes and networks in late-onset Alzheimer's disease. *Cell, 153*, 707–720.

Zhou, Y., Su, Y., Li, B., Liu, F., Ryder, J. W., Wu, X., Gonzalez-DeWhitt, P. A., Gelfanova, V., Hale, J. E., May, P. C., Paul, S. M., & Ni, B. (2003). Nonsteroidal anti-inflammatory drugs can lower amyloidogenic Abeta42 by inhibiting Rho. *Science, 302*, 1215–1217.

Zlotnik, A., & Yoshie, O. (2000). Chemokines: A new classification system and their role in immunity. *Immunity, 12*, 121–127.

Parkinson's Disease and Lewy Body Dementias

23

Katherine E. Olson, R. Lee Mosley, and Howard E. Gendelman

Abstract

Parkinson's disease (PD) and Lewy body dementias (LBD) are synucleinopathies in which the synaptic protein, alpha-synuclein (α-syn), becomes modified, misfolded, and aggregated, resulting primarily in dopaminergic neuronal cell loss. Along with the loss of this neuronal subpopulation are various inflammatory alterations within the periphery and central nervous system (CNS). These inflammatory responses are found in both familial and sporadic forms of the disease, as well as in animal models. Whether this inflammatory response initiates disease or occurs as a result of initial neuronal damage is still under investigation. However, a multitude of research supports its contribution to disease progression and neuropathogenesis in both diseases. Therefore, researchers are currently targeting neuroinflammatory pathways as forms of immunotherapy for both PD, LBD, and other parkinsonian-like diseases.

Keywords

Adaptive immunity · Innate Immunity · Parkinson's disease · Lewy body dementia · Neuroinflammation · Inflammation · Neurodegeneration · Alpha-synuclein · Immunotherapies · Neuroprotection · Synucleinopathies

23.1 Introduction

The goal of this chapter is to review the role of inflammation in disease initiation and progression, using data collected in patients and experimental models of Parkinson's disease (PD) and Lewy body dementias (LBD). Each of these diseases falls under the disease umbrella called synucleinopathies. Synucleinopathies are a diverse group of neurodegenerative diseases in which the protein, alpha-synuclein (α-syn), accumulates abnormally to form inclusions in neuronal cell bodies termed Lewy bodies (LB). Three main types of synucleinopathies include PD, LBD, and multiple system atrophy (MSA). Additionally, there are other parkinsonian-like diseases, termed parkinsonisms, in which α-syn also becomes aggregated. However, for this chapter, we will solely focus on PD and LBD.

23.2 Overview of Parkinson's Disease

PD is the second most common neurodegenerative disorder, second only to Alzheimer's disease (AD). However, it is the most prevalent neurodegenerative movement disorder (Kalia & Lang, 2015). The first clinical description of PD was originally described as a "shaking palsy" by James Parkinson in 1817 (Donaldson, 2015). It is a neurological disease that is characterized by two disease hallmarks, the progressive loss of dopaminergic neurons that originate within the substantia nigra (SN) and innervate into the striatum (STR) and the presence of proteinaceous inclusions called LB and Lewy neurites (LN) (Fig. 23.1). LB and LN are intracellular inclusions consisting of modified and misfolded α-syn accompanied by ubiquitin (Carrithers et al., 2000; Scanlon et al., 2008). Dopaminergic neurons are the main producers of dopamine, the key neurotransmitter needed for proper movement. Thus, the loss of this neuron population causes the majority of the motor symptoms associated with PD (Maiti et al., 2017).

The classical motor symptoms of PD are recognized as prominent features for disease diagnosis but only appear after approximately 50–60% neuronal loss (Olanow et al., 2009). These motor symptoms include bradykinesia, rigidity, resting tremor, postural instability, and gait dysfunctions.

K. E. Olson · R. L. Mosley (✉) · H. E. Gendelman
Department of Pharmacology and Experimental Neuroscience, Movement Disorders Laboratory, University of Nebraska Medical Center, Omaha, NE, USA
e-mail: rlmosley@unmc.edu

H. E. Gendelman, T. Ikezu (eds.), *Neuroimmune Pharmacology and Therapeutics*,
https://doi.org/10.1007/978-3-031-68237-7_23

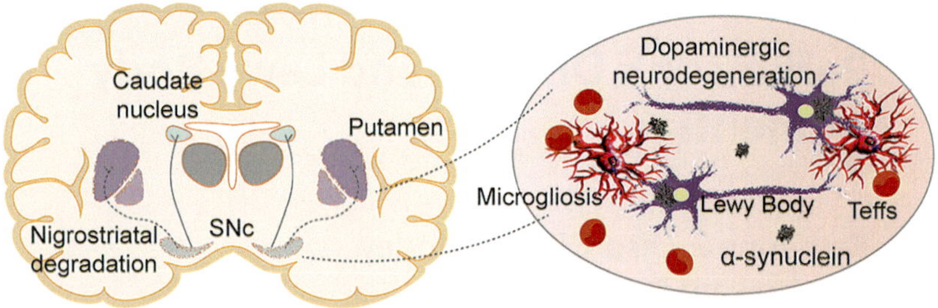

Fig. 23.1 Schematic representation of nigrostriatal neurodegeneration in Parkinson's disease. During disease progression, alpha-synuclein (α-syn) becomes posttranslationally modified, misfolded, aggregated into Lewy bodies, and released into the extracellular microenvironment following dopaminergic neuronal injury. Release of α-syn aggregates elicits microglial activation that speeds neuronal damage within the substantia nigra pars compacta (SNpc). The ensuing neuroinflammatory cascade initiates a systemic, adaptive immune response leading to increased BBB permeability and effector T cell (Teff) influx into sites of neuroinflammation resulting in increased microgliosis and neurodegeneration upon CNS entry

However, the severity and degree of these motor features remain heterogeneous within the diseased population. Apart from prevalent motor dysfunction, patients with PD display many non-motor symptoms as well, including olfactory deficits, sleep disturbances, cognitive impairments, pain, fatigue, constipation, and some psychiatric disorders (Titova et al., 2017). Frequently, these non-motor features arise well before the onset of motor dysfunction, indicating that the pathogenic cause of the disease occurs years prior to noticeable disease onset. One hypothesis for the onset of non-motor dysfunctions prior to motor dysfunction lies with Braak and colleagues (Rietdijk et al., 2017). They hypothesize that symptoms start in the periphery before entering the CNS via a nasal or gastric route and eventually initiate neuronal loss. Thus, continued evaluation of biomarkers for early diagnosis remains essential for combating the disease. To date, a variety of biomarkers are being evaluated using clinical, pathological, biochemical, genetic, and bioimaging techniques (Kalia & Lang, 2015). These include but are not limited to, detection of olfactory impairments, neuronal imaging using positron emission tomography/single photon emission computed tomography (PET/SPECT), skin biopsies, changes in biochemical makeup of the cerebrospinal fluid (CSF), saliva, or blood, and genetic sequencing.

Without easily defined and detectable diagnostics, clinicians focus on treating patients after diagnosis based on motor symptoms. Initially, motor dysfunctions are managed symptomatically. Currently, the readily available therapies for PD are palliative; however, there is no known cure or intervention to slow or halt disease progression (Olanow et al., 2009). In general, drug therapies employ dopamine agonists or the dopamine precursor, levodopa (L-DOPA), to assist with normal dopaminergic neurotransmission (Kalia & Lang, 2015). A second therapeutic avenue is through deep brain stimulation (DBS), a surgical approach aimed at stimulating either the subthalamic nucleus or globus pallidus internus to counteract motor disturbances (Follett & Torres-Russotto, 2012; Hamani et al., 2017). However, each of these therapies only targets the symptoms and shows no effect on the progressive nature of the disease. Also, as the disease progresses, it is not as easily managed and is characterized by the worsening of motor disturbances, resulting in freezing, falls, dysphagia, and slurred speech (Kalia & Lang, 2015). Thus, it remains essential that therapies are developed for combating PD progression rather than simply alleviating symptoms.

These disease hallmarks and symptoms manifest in both sporadic and familial cases of PD. Familial PD is genetically linked and accounts for approximately 10% of all PD cases. Likewise, several genes have been identified in patients with a family history of disease, including those encoding for α-syn (*SNCA*), leucine-rich repeat kinase 2 (*LRRK2*), PTEN-induced kinase 1 (*PINK1*), protein deglycase DJ-1 (*DJ-1*), and parkin (*PRKN*) (Olanow et al., 2009). Apart from genetics, other risk factors include increased age, male gender, Hispanic ethnicity, and exposure to environmental toxins (Kalia & Lang, 2015). In accordance with known genetic links and risk factors, both sporadic and familial cases are linked to dysfunctions in common cellular networks. These include dysregulation of protein homeostasis, such as abnormalities in protein aggregation, intracellular protein trafficking, ubiquitin-based protein disposal, and dysfunctional mitochondrial processes (Good et al., 1998; Moon & Paek, 2015).

The importance of mitochondrial function in parkinsonism was verified by J. William Langston in "The Case of the Frozen Addicts" (Langston et al., 1983). Men and women had injected a perceived "synthetic heroin" containing an impurity, 1-methyl-4-phenyl-1,2,3,6-tetrahydropyridine (MPTP). After injection, they suddenly became immobile, as if they had instantly acquired PD. Some continued to experience effects years after the fact (Langston et al., 1999). It was discovered that the impurity resulted in motor dysfunction and is now utilized as a model for inducing PD-like symptoms and hallmarks in rodent and primate models (Przedborski et al., 2001). MPTP, administered peripherally, rapidly crosses the blood-brain barrier (BBB), where it is metabolized into its toxic form, 1-methyl-4-phenylpyridinium (MPP+). This toxin accumulates within dopaminergic neurons via uptake by the dopamine transporter (DAT), due to the similar structure to dopamine itself. Once inside the neuron, MPP+ inhibits complex I of the electron transport chain within the mitochondria, ultimately inhibiting ATP production and generating the formation of superoxide radicals. Superoxide radicals react with nitric oxide (NO) and produce peroxynitrite, causing cellular damage. In turn, tyrosine hydroxylase (TH), the rate-limiting enzyme for dopamine production, is oxidized and nitrated causing it to become inactive. This inactivation leads to decreased dopamine production, resulting in Parkinson-like motor symptoms (Przedborski et al., 2000).

However, even with known risk factors and associations with cellular dysfunction, the underlying driving force behind the progression of PD is still unknown. The overall cause and pathogenesis of the disease are hypothesized to be multi-factorial and result in an elaborate interplay between multiple insults. For instance, genetics, environment, age, and interaction between innate and adaptive immune systems can combinatorically contribute to disease onset and progression. Recently, neuroinflammation and immune dysfunction have been shown to be a characteristic feature of PD, but whether neuroinflammation is a cause or effect of the disease is not yet known. However, while the pathobiology of PD remains in study, neuroinflammation and immunity are known to speed nigrostriatal degeneration (Anderson et al., 2014; Mosley et al., 2012).

23.2.1 Inflammation in Parkinson's Disease Patients

The neuroinflammatory cascade associated with PD begins with the aggregation of misfolded or posttranslationally modified α-syn (Marogianni et al., 2020). Such aggregation results in neuronal cell death and the presence of chronically activated microglia and astrocytes (Fig. 23.1), leading to the production of tumor necrosis factor alpha (TNFα), interleukin-1 beta (IL-1β), IL-6, reactive oxygen species (ROS), and enzymes such as nicotinamide adenine dinucleotide phosphate (NADPH) oxidase (Liu & Hong, 2003). These changes in the microglial phenotype can affect the CNS microenvironment by producing a pro-inflammatory milieu that speeds PD pathogenesis. In a healthy brain, microglia consist of 10% of all glial cells, and in their resting state, exhibit thin, elongated processes (Brawek et al., 2021). However, in postmortem analysis of PD brains, both microgliosis and astrocytosis are present in the SN and striatum (McGeer et al., 1988; Miyazaki & Asanuma, 2020). Astrocyte alterations remain relatively normal, with most exhibiting a resting-state morphology. However, the microglial reactivity is robust resulting in thick, elongated processes with amoeboid cell bodies (Badanjak et al., 2021; McGeer et al., 1988). Additionally, the number of reactive microglia is much higher in PD than in controls, and many are located in close proximity to dopaminergic neurons within the SN and striatum (Joers et al., 2017). These findings have also been confirmed in MPTP-intoxicated monkeys and mice (Bieganowska et al., 1993; Manocha et al., 2017; McGeer et al., 2003; Smeyne et al., 2016) (Fig. 23.2). Therefore, compelling evidence in both animal and human studies helps to support the idea of an overactive inflammatory state and lack of immune regulation within the PD population.

Initial studies of peripheral lymphocyte populations from PD patients show decreased frequencies and total numbers of $CD4^+$ T lymphocytes compared to controls (Baba et al., 2005; Bas et al., 2001; Saunders et al., 2012; Stevens et al., 2012). However, due to conflicting reports on T cell phenotypes, a firm consensus of other T cell subset changes in PD patients has proven difficult. For instance, the diminution of $CD4^+$ T cell numbers in PD patients was found chiefly from decreased numbers of $CD4^+CD45RA^+$ naïve T cells and, to a lesser extent, from $CD4^+CD29^+$ memory subsets (Saunders et al., 2012). On the other hand, Stevens and colleagues reported decreased levels of $CD4^+CD45R0^+$ memory T cells (Stevens et al., 2012). A second study by Saunders and colleagues showed slight, yet significant increases in frequencies of $CD4^+CD45R0^+$ memory/effector T cells with a parallel diminution of $CD4^+CD45RA^+$ resting/naive T cell levels (Saunders et al., 2012). Additionally, frequencies of peripheral $CD4^+$ T cells with effector-associated phenotypes expressing FS-7-associated surface antigen (FAS) were increased in patients, whereas those expressing α4β7 integrins and CD31 were diminished. Notably, these changes in $CD4^+$ T cell phenotypes were correlated with the severity of motor function as quantified by the Unified Parkinson's Disease Rating Scale, part III (UPDRS III) (Saunders et al., 2012). Differences in modulation of T cell immunology profiles may be the consequence of disease heterogeneity and severity but clearly require further investigation.

Postmortem studies of PD patient brain tissues show both $CD4^+$ and $CD8^+$ T cells in close proximity to dopaminergic neurons within the SN at levels exceeding tenfold those found in controls (Brochard et al., 2009). Moreover, those levels of T cells were not detected in non-lesioned brain regions. Microarray analysis of peripheral blood leukocytes and SN brain tissue showed many genes expressed were in common with those expressed by Th17-mediated immune reactions and suggests that idiopathic parkinsonism is a Th17- dominant autoimmune disease (Qian et al., 2010). However, whether T cell infiltration is primary or secondary

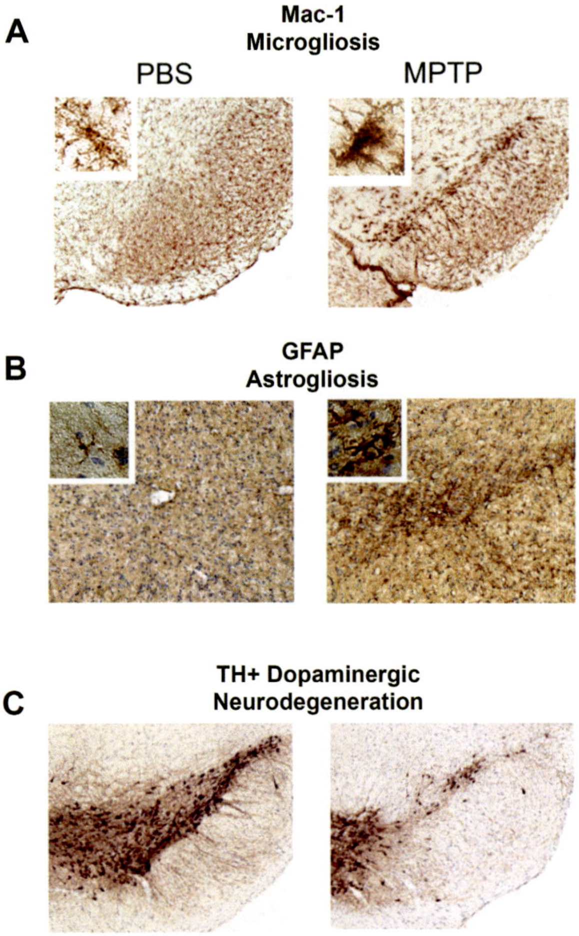

Fig. 23.2 Neuroinflammatory and neurodegenerative response in the acute 1-methyl-4-phenyl-1,2,3,6-tetrahydropyridine (MPTP) mouse model of Parkinson's disease. (**a**) Representative photomicrographs of Mac-1$^+$ activated microglia within the substantia nigra of mice treated with PBS alone or at two days post MPTP intoxication (scale bar, 25 μm). (**b**) Representative photomicrographs of GFAP immunoreactive activated astrocytes within the substantia nigra of mice treated with PBS or MPTP at seven days post-MPTP (scale bar, 50 μm). (**c**) Photomicrographs of TH$^+$Nissl$^+$ dopaminergic neurons in the substantia nigra in mice treated with PBS or MPTP at seven days post-MPTP (scale bar, 35 μm). Serial sections (30 μm) were stained with a primary antibody against Mac-1, GFAP, or TH followed by a biotinylated secondary antibody, and diaminobenzidine color development. Sections were counterstained for Nissl bodies using thionin and imaged using StereoInvestigator software

to PD progression is unknown at this time. Similarly, conflicting reports of regulatory T cells (Tregs) in PD also produced variances in detected levels ranging from increased frequencies in PD patients to little or no differences among groups (Baba et al., 2005; Fiszer et al., 1994; Saunders et al., 2012; Stevens et al., 2012). However, one study demonstrated the diminished capacity of Tregs from PD patients compared to those of controls to inhibit the proliferation of responder T cells from healthy donors (Saunders et al., 2012). This suggests that dysfunction in Tregs plays a role in hyper-activated immune state and increased disease progression. The notion that hyper-activated immune responses support increased dopaminergic loss is provided by the use of animal studies.

23.2.2 Evidence of Inflammation in Experimental Models of Parkinson's Disease

Immune involvement in PD has been extensively studied and supported by animal models. In mouse studies utilizing the neurotoxin MPTP, time course studies have defined the kinetics of dopaminergic neuronal cell loss along with the ensuing neuroinflammatory cascade following pro-neurotoxin administration (Fig. 23.2a–c). Studies indicate a robust microglial response is apparent early in the disease course with delayed astrogliosis which peaks in parallel with neuronal loss (Bieganowska et al., 1993; Kohutnicka et al., 1998; Kurkowska-Jastrzebska et al., 1999a, b; Olson et al., 2016). Additionally, multiple studies in this animal model demonstrate the involvement of the adaptive immune system in dopaminergic neurodegeneration using both active and passive transfer of immunity (Benner et al., 2008; Brochard et al., 2009; Campos-Acuna et al., 2019; Garretti et al., 2019; Gelders et al., 2018; Harms et al., 2021; Kosloski et al., 2013; Kurkowska-Jastrzebska et al., 1999b). In the MPTP mouse model, the numbers of T cells in the SN are increased after intoxication, and interestingly, numbers of CD8$^+$ T cells predominate those of CD4$^+$ T cells (Brochard et al., 2009). While in agreement with the relative proportions of T cell subsets within the SN of MPTP-treated mice, the total numbers of CD4$^+$ T cells vary widely between studies. The importance of T cells for mediating MPTP-induced neurodegeneration was found initially in adoptive transfer and reconstitution studies of functional T cells to immunodeficient mice (Benner et al., 2008; Brochard et al., 2009). While both studies confirm that immune-deficient mice were not susceptible to MPTP intoxication, reconstitution of those mice with functional naïve lymphocytes partly restored MPTP susceptibility (Benner et al., 2008) and CD4$^+$ T cells were found to be chiefly responsible for MPTP susceptibility (Brochard et al., 2009). These studies point to a deleterious role of CD4$^+$ T cells in PD. The phenotypes of CD4$^+$ populations required for increased degeneration were determined to be Th1 and Th17 cell lineages. Similarly, T cells isolated from mice immunized with a modified self-antigen, nitrated α-synuclein (N-α-syn) recognized only N-α-syn, but not unmodified α-syn in in vitro challenge assays. Moreover, N-α-syn-specific effector T cells (Teffs) exacerbated neuroinflammation and increased neuronal injury and subsequent neurodegeneration of dopaminergic neurons within the SN of MPTP mice (Benner et al., 2008). These findings indicate that N-α-syn, a modified self-protein, either evades or breaks immunological tolerance to self α-syn and induces N-α-syn-specific T cells. Likewise, another second study showed that Th17 Teffs possess a significantly greater capacity to exacerbate dopaminergic neurodegeneration than the same number of Th1 Teffs (Reynolds et al., 2010). Together, these data indicate that $CD4^+$ T cells play an important role in the neuroinflammation and subsequent neurodegeneration in models of PD (Garretti et al., 2019). It also indicates that Th17 Teffs are more potent at direct killing of neurons or alternatively, enhancing neurotoxic microglia than Th1 Teffs. These data also support the idea that increased peripherally circulating Teffs, as found in PD patients, migrate to the sites of neuroinflammation and can exacerbate and accelerate PD disease progression.

Activated microglia and Teffs are thought to be key mediators of neuroinflammatory processes in PD progression (Campos-Acuna et al., 2019; Contaldi et al., 2022; Harms et al., 2021). If left uncontrolled, these mediators support an inflammatory cascade that affects the tempo of the disease (Contaldi et al., 2022; Harms et al., 2021; Reynolds et al., 2007a; Williams et al., 2021). Initially, studies directed to harness the inflammatory cascade using an anti-inflammatory agent demonstrated that adoptive transfer of $CD4^+$ T cells from copolymer-1 (Cop-1) immunized donor mice protected dopaminergic neurons and striatal termini in MPTP-treated mice (Laurie et al., 2007). These observations supported the hypothesis that subpopulations of T cells protect from neurodegeneration in the MPTP animal model of PD. These findings are consistent with known mechanisms by which Cop-1 regulates proliferative and inflammatory responses by preferentially inducing Th2, Th3, and Tregs that secrete anti-inflammatory cytokines (Benner et al., 2004; Lalive et al., 2011; Laurie et al., 2007; Reynolds et al., 2007b). In a separate line of study, researchers found that $CD4^+CD25^+$ Tregs were capable of significantly suppressing neuroinflammation and neurodegeneration in the MPTP model with as few as 500×10^3 Tregs, and with 3.5×10^6 Tregs being sufficient to provide virtually complete neuroprotection to dopaminergic neurons along the nigrostriatal axis (Reynolds et al., 2007b). Moreover, the degree of protection afforded by Tregs seems to increase with increasing inflammatory responses as evidenced by increased neuroprotection with Treg co-transfer with N-α-syn-specific Th17 cells in the MPTP model (Reynolds et al., 2010). Also, use of vasoactive intestinal peptide (VIP), a known inducer of Treg activity (Gonzalez-Rey & Delgado, 2007), increased the neuroprotective capability of Tregs from VIP-treated donors in the MPTP model (Reynolds et al., 2010). Along those same lines, studies using granulocyte-macrophage colony-stimulating factor (GM-CSF) showed that pretreatment with GM-CSF prior to MPTP-intoxication increased Treg activity in a dose-dependent fashion and diminished the neuroinflammatory response, affording significant dopaminergic neuroprotection (Kosloski et al., 2013). Based on these findings, investigator-initiated phase I clinical trials have been carried out to target the dysregulated Treg population in PD using sargramostim, a recombinant human GM-CSF. In theory, treatment would upregulate Treg numbers and function to suppress the neuroinflammatory cascade in the PD population, with the expectation that treatment affords a neuroprotective outcome to inhibit or slow disease progression (Gendelman et al., 2017). Prolonged treatment with GM-CSF in PD patients resulted in decreased motor severity, as determined by the UPDRS III, and these changes were linked to increases in Treg function. Treatment was also associated with increased mild adverse events, such as injection site reactions, eosinophilia, increased white cell counts, bone pain, urticarial reaction, vasculitis, and stroke. However, some of these adverse events are known to be associated with the drug due to its ability to rapidly reconstitute all parts of the immune system (Vial & Descotes, 1995).

23.2.3 Protective Role of Inflammation in Parkinson's Disease

Additionally, several studies have shown the potential beneficial role of inflammation in the CNS (Marogianni et al., 2020; Vezzani et al., 2020). These benefits are usually linked to glial cell function, specifically astrocytes (Boghdadi et al., 2020). For instance, glia play an important role in homeostatic maintenance and neuronal cell survival by scavenging toxic molecules such as reactive oxygen species (ROS), which can be quite high in the substantia nigra (Boghdadi et al., 2020). Therefore, astrocytic scavenging and glutathione production aid in neuronal survival and prevent excitotoxic injury in residing neuronal populations (Drukarch et al., 1998). Additionally, glia and infiltrating T cells have the capacity to produce neurotrophic factors that are essential for neuronal survival and protection, including VIP, brain-derived neurotrophic factor (BDNF), and glial-derived neurotrophic factor (GDNF) (Poyhonen et al., 2019). For example, GDNF is released by reactive astrocytes, appears to be important in neuronal survival through the inhibition of microglial activation, and has been used as a potential therapeutic in animal models of disease (Iravani et al., 2001; Rocha et al., 2012; Schober et al., 2007) BDNF and VIP have also been shown to be effective neuroprotectants in parkinsonian models, supporting the protective effects of reactive glia in the CNS (Cao et al., 2022; Geng et al., 2023; Olson et al., 2015; Passemard et al., 2011). Moreover, adoptive transfer of Tregs has been shown to upregulate BDNF and GDNF expression by astrocytes in the MPTP model (Benner et al., 2004; Reynolds et al., 2007b).

23.3 Overview of Lewy Body Dementias

Lewy body dementia (LBD) is an umbrella term that encompasses both dementia with Lewy bodies (DLB) and Parkinson's disease with dementia (PDD) (Arnaoutoglou et al., 2019; Gomperts, 2016; Surendranathan et al., 2015). LBD is the second most common neurodegenerative form of dementia, second to AD, and accounts for up to 7% of dementia cases (Jellinger, 2018). Postmortem evaluation of the cerebral cortex confirms the presence of Lewy body–related pathologies, including LB and LN formed from α-syn concomitant with amyloid beta (Aβ) plaques and hyperphosphorylated Tau tangles (Arnaoutoglou et al., 2019; Gomperts, 2016). Additionally, the neurodegenerative relations between DLB and PDD are similar and are both neurocognitive disorders in which α-syn deposition, aggregation, and LB formation are present (Jellinger, 2018). The cognitive impairment associated with the diseases includes progressive loss of executive function, visual-spatial abnormalities, and increased memory dysfunction (Arnaoutoglou et al., 2019). Diagnosis of each depends on the onset of cognitive impairment. DLB is generally diagnosed when impairment precedes parkinsonian motor symptoms, whereas PDD occurs when cognitive impairment develops during PD (Jellinger, 2018). Almost all PD subjects will experience dementia and cognitive decline as the disease progresses, and about 75% of DLB subjects will develop parkinsonism with increased severity over time (Aarsland & Kurz, 2010). Despite the temporal differences in motor dysfunction and cognitive decline, both diseases show parallel neuropathological lesions associated with increased cortical Aβ and Tau load, as well as Lewy body deposition, making the diseases hard to distinguish from one another at postmortem evaluation (Bencze et al., 2020; Jellinger, 2018).

The clinical presentation of LBD includes cognitive impairment, behavioral and psychiatric phenomena, and physical motor symptoms (Gomperts, 2016; Jellinger, 2018). Clinically, DLB comprises four main symptoms including cognitive fluctuations, parkinsonism, visual hallucinations, and REM sleep behavior disorders along with progressive cognitive decline and memory impairment (Zupancic et al., 2011). Cognitive decline in PDD is similar to DLB, but is more variable across the population and is commonly observed in PD subjects experiencing a higher cortical LB burden (Irwin et al., 2013). The average time to dementia after PD diagnosis is about 10 years, but may take as long as 20 years post-diagnosis (Jellinger, 2018). The memory deficits associated with DLB are due to impaired neuronal signaling within cortical structures and thalamic atrophy that is often linked to AD-like pathology, whereas PDD requires cognitive decline across multiple areas of the brain leading to mood disturbances, major depression, and visual–spatial impairments (Ruffmann et al., 2016). Ultimately, the pattern and rate of cognitive decline differs among the two diseases, with DLB exhibiting a faster progression compared to PDD or AD alone (Jellinger & Korczyn, 2018). Additionally, studies in large cohorts of patients have confirmed a strong correlation between cortical Lewy body and AD-like pathologies, suggesting that phosphorylation of α-syn promotes the phosphorylation of Tau, leading to memory impairment (Shim et al., 2022). Transgenic animal models overexpressing either Aβ and α-syn or APP and presenilin also suggest this synergy (Kurata et al., 2007; Masliah et al., 2001; Moussaud et al., 2014). Additionally, α-syn aggregation is enhanced by Tau, leading to increased cellular toxicity and synaptic damage (Badiola et al., 2011; Bate et al., 2010). Taken together, postmortem human studies and animal models suggest that the synergistic relationship between AD pathology and α-syn aggregation is bidirectional. Studies also indicate that synapses are damaged within neocortical regions due to presynaptic accumulation of α-syn, due to toxic α-syn oligomer pore formation (Kramer & Schulz-Schaeffer, 2007). Phosphorylated α-syn has also been located in presynaptic terminals and associated with increased rates of cognitive decline (Bereczki et al., 2016; Colom-Cadena et al., 2017). The resulting synaptic alterations are supplemented by altered neurotransmitter signaling, impaired autophagy, and ubiquitination pathways, oxidative stress and damage, and loss of hippocampal neurons (Garcia-Esparcia et al., 2017).

23.3.1 Inflammation in Lewy Body Dementias

Along with neuronal loss and cognitive decline, neuroinflammation is increasingly considered a main contributor to the disease (Surendranathan et al., 2015, 2018). Evidence for this has been evaluated through diseases closely related to DLB, including PD and AD, as previously discussed. Current evidence suggests that neuroinflammation occurs in the early stages of disease and is linked to mild cognitive impairments (Surendranathan et al., 2018). As the disease progresses, increased microglial activation is observed and appears to be widespread (Hamelin et al., 2016; Hogl et al., 2018; Okello et al., 2009; Surendranathan et al., 2018). Postmortem evidence of inflammation in DLB and PDD also indicates microglial activation, elevated cytokine levels, and increased C-reactive protein levels after the onset of dementia (King et al., 2018; Mackenzie, 2000; Song et al., 2013; Togo et al., 2001). Quantification of microglial reactivity and LB formation is also positively correlated (Mackenzie, 2000). LBs are surrounded by ramified and reactive microglia, and increased $CD68^+$ microglial levels are found in DLB cases compared to non-demented controls (Streit & Xue, 2016). Additionally, inflammatory markers within the peripheral blood are also altered compared to controls, including higher levels of macrophage inflammatory protein-3 alpha (MIP-3α),

IL-17A, and IL-2, and lower levels of IL-8 (Surendranathan et al., 2018). Bioimaging studies also suggest an early neuroinflammatory cascade associated with decreased cognition in DLB (Donaghy et al., 2020; Dupont et al., 2017; Nedelska et al., 2015; Nicastro et al., 2020; Watson et al., 2012). However, some more recent studies have shown minimal microglial reactivity and immune cell infiltration.

Apart from microglial cells, astrocytes and T cells can contribute to inflammatory responses within the CNS and have been implicated in several neurodegenerative diseases, including LBD pathogenesis (Acioglu et al., 2021). The astrocyte mechanisms of action in LBD are likely linked to roles in α-syn degradation at the synapse with decreased activity early in the disease course that leads to increased α-syn accumulation (Myers et al., 2023). Additionally, as observed in PD, adaptive immune cell populations, including B and T cells, are closely linked to neurodegeneration in LBD. In transgenic mouse models of disease, immunohistochemical analysis indicates perivascular and parenchymal infiltration of $CD4^+$ T helper cells within the neocortex, hippocampus, and striatum (Iba et al., 2020). T cells were also found in close proximity to activated microglia and astrocytes, and increased pro-inflammatory cytokine levels such as IFN-y were also observed. Postmortem evaluation of DLB brains confirmed the presence of T cell infiltration, and infiltrating $CD4^+$ T cells were also found to contribute to the neurodegenerative process and progressive dementia (Gate et al., 2021). These cells displayed increased levels of C-X-C motif chemokine receptor 4 (CXCR4) and C-X-C motif chemokine ligand 12 (CXCL12) for homing to sites of inflammation as well as increased IL-17A expression when stimulated with a phosphorylated α-syn epitope. These increases were also associated with neuroaxonal damage.

23.4 Current Immunotherapies in Parkinson's Disease and Lewy Body Dementias

While immunosuppression is associated with neuroprotection for PD and LBD, the direct cause-and-effect relationships have not yet been realized. Thus, modulating the immune system for therapeutic gain has been openly debated. Currently, researchers are revealing how innate and adaptive immunity can affect neurodegenerative disease pathology, and similarly, how the immune response can be harnessed for successful treatments. The overarching goal is to employ immune-modulating agents as a neuroprotective strategy for disease treatment. To date, many disease-modifying treatments have been or are currently being developed as neuroprotective strategies for alpha-synucleinopathies in both experimental animal models and for human disease translation, as discussed below. Overall, the long-term goal of this research is to effectively harness the immune system to slow or prevent PD and LBD pathobiology.

Upon observational evaluation of large cohorts of PD individuals, those who use non-steroidal anti-inflammatory drugs (NSAIDs), specifically ibuprofen, exhibit a lower risk of developing PD, and treatment with NSAIDs is protective in both MPTP- and 6-hydroxydopamine (6-OHDA)-induced neuronal lesions (Rees et al., 2011). Ultimately, this suggests an association with anti-inflammatory use and diminished potential for developing PD. One of the first anti-inflammatory treatments to be investigated was minocycline. Minocycline possesses potent neuroprotective and anti-inflammatory activities in various inflammatory rodent models of PD, and it has been carried into clinical trials as well. Its mechanisms of action are mediated by blocking the activation of NADPH oxidase, decreasing microglial activation, and decreasing the production of iNOS, IL-1β, and TNFα (Du et al., 2001; He et al., 2001; Kim & Suh, 2009; Lu et al., 2010; Tikka et al., 2001). In both the MPTP and 6-OHDA models, minocycline reduced the numbers of reactive microglia and spared nigral dopaminergic neurons along with their projections into the striatum in a dose-dependent manner. However, in the clinical setting, neuroimaging analysis of PD patients following minocycline treatment suggests that microglial activation is perturbed, but motor assessments failed to show any clinical benefit (Dodel et al., 2010; NINDS, 2006, 2008). In conclusion, anti-inflammatory agents, while having been extensively studied, have failed to provide improved clinical outcomes.

Many natural and/or endogenous compounds have recently been shown to be effective in rodent models of PD as well, such as resveratrol (Lin et al., 2014), tanshinone (Ren et al., 2015), silymarin (Borah et al., 2013), resolvins (Tian et al., 2015), daidzein (Chinta et al., 2013), and apocynin derivatives (t Hart et al., 2014). The majority of these compounds act as both anti-inflammatory agents and antioxidants. Collectively, they mediate downregulating glial activation, decreasing pro-inflammatory cytokine production, suppressing M1 microglial phenotypes, and reducing NF-κB activation. Specifically, resveratrol and tanshinone treatment both show decreases in IL-1β, IL-6, and TNFα (Lofrumento et al., 2014; Wang et al., 2015b; Zhang et al., 2010). Also, resveratrol causes an upregulation of the suppressor of cytokine signaling (SOCS-1), and tanshinone treatment inhibited nitric oxide (NO) production as well as NADPH and iNOS expression. In the MPTP model, silymarin treatment stabilizes mitochondrial membrane potential, decreases levels of TNFα, iNOS, and IL-1β, and ultimately, decreases striatal levels of caspase-3 and NF-κB indicating an anti-apoptotic and anti-inflammatory function (Geed et al., 2014; Jung et al., 2014; Lee et al., 2015). Taken together, these findings help to justify the potential beneficial use of natural anti-inflammatory agents for the treatment of

PD. However, these agents have yet to be translated into clinical investigation for efficacy in PD patients.

To date, there have also been many recent investigations into the role of peroxisome proliferator-activated receptor (PPAR) agonists and their neuroprotective mechanisms within the CNS. PPAR is a receptor superfamily of transcription factors activated by small molecules such as hormones and steroids with activation playing important roles in mediating the inflammatory response. PPAR agonists, such as pioglitazone and rosiglitazone, are shown to possess neuroprotective and anti-inflammatory activity, both in vitro and in vivo by selectively targeting neurotoxic factors within reactive microglia (Carta & Pisanu, 2013; Pisanu et al., 2014). Their use in the MPTP model of PD led to decreases in microglial activation, iNOS production, NF-κB activation, and ROS production (Garrido-Gil et al., 2012; Pisanu et al., 2014). Along with decreases in pro-inflammatory mediators, their function is associated with increases in anti-inflammatory mediators, such as IL-10 and transforming growth factor-β (TGF-β) as well as restored levels of CD206, a marker for anti-inflammatory M2 microglial responses (Pisanu et al., 2014). These findings suggest a shift from a pro-inflammatory response to an anti-inflammatory role, perhaps resulting in a more protective and neurotrophic environment. To further confirm neuroprotective effects and the ability to modulate the inflammatory response, pioglitazone was tested in parkinsonian monkeys. Oral administration led to higher stereological counts of tyrosine hydroxylase-positive neurons and modulated $CD68^+$ inflammatory cells in a dose-dependent manner (Swanson et al., 2011). To this end, these findings prompted the translation of PPAR agonists, specifically pioglitazone, into a multi-center, double-blind phase 2 clinical trial (NINDS, 2015). This trial also attempted to find peripheral biomarkers for PD that included leukocyte PGC-1α, IL-6, and urine 8-hydroxydeoxyguanosine (Simon et al., 2015). No changes in peripheral biomarkers were observed. Additionally, the agonist failed to show benefit as indicated by UPDRS III scores.

Explanations for the failure of PPAR agonists in PD studies likely reflect the possibility that neurotoxin-inducing neuroinflammatory models incompletely reflect human disease. In addition, multiple inflammatory response targets are likely needed, along with secondary drug-induced neuroprotective activities. Also, UPDRS III scores may show other limitations. Indeed, the potential for using motor tests as a singular end point may not preclude other disease events that could be influenced by PD-modifying treatments. The span of time for the assessment was also relatively short. This would not allow evaluation as to whether the treatment had long-term beneficial effects. Potentially, the anti-inflammatory activities of PPAR agonists may slow the rate of progression in the long term. However, the lack of observed efficacy in treating motor dysfunction should prompt researchers to continue to explore other therapeutic alternatives.

Due to the lack of clinical efficacy using anti-inflammatory agents, potential therapies aimed at directly modulating immune aberrancies are actively being studied and investigated in the clinical setting. Currently, two main therapeutic avenues are being explored and have shown promise both in rodent models of disease and in PD patients. These two strategies are based on specific targeting of aberrantly misfolded proteins and peripheral immune modulation to elicit immune shifts into a protective phenotype.

The therapeutic strategy of targeting α-syn is aimed at ameliorating PD-associated neuroinflammation and disease progression via targeting misfolded α-syn. This would likely be beneficial for other synucleinopathies including LBD. In recent years, it has been shown that toxic versions of α-syn can accumulate within neurons and be secreted into the extracellular environment where the misfolded, oligomeric protein can cause a neurotoxic, inflammatory immune response (Benner et al., 2008). Thus, immunotherapeutic approaches are being developed to target and enhance the clearing of misfolded or posttranslationally modified α-syn. Thus, targeting both the N-terminal and/or C-terminal region of the protein may provide a potential benefit to α-syn clearance. A recent study using an adeno-associated virus (AAV)-α-syn rat model of PD indicates that anti-human α-syn N-terminal peptide antibodies can protect against dopaminergic neuron loss and decrease microglial activation to some extent (Shahaduzzaman et al., 2015). Treatment with the vaccine also led to altered IgG production, enhanced MHCII expression, and increased $CD4^+$ T cell infiltration into the CNS. Recently, it has also been shown that the use of monoclonal antibodies against the C-terminal region of α-syn or misfolded α-syn reduces levels of protein propagation, improves PD-like pathologies, ameliorates dopaminergic neuronal cell loss, and attenuates motor deficits in mouse models of the disease (Fagerqvist et al., 2013; Games et al., 2014; Lindstrom et al., 2014; Tran et al., 2014). Based on these findings, Roche and Prothena further developed and commercialized PRX002 to specifically target α-syn. A phase I clinical trial was initiated with the finding that the vaccine was safe and well-tolerated, prompting a second phase I trial to assess dose–response, tolerability, pharmacokinetics, exploratory biomarkers, and immunogenicity (NCT02095171 and NCT02157714).

With a similar directive in mind, a current study using a vaccine against α-syn demonstrated that vaccination prevented Parkinson-like inclusions as well as increased Treg recruitment, increased GDNF production, and increased antibody formation (Sanchez-Guajardo et al., 2013). Excitingly, another vaccine that has been developed to produce antibodies that are specific for α-syn while sparing neuroprotective beta-synuclein, AFFITOPE PD01A, has entered clinical trials (Mandler et al., 2014; Schneeberger et al., 2012). PD01A targets the phosphorylated form of α-syn that is responsible for the induction of a neurotoxic pro-inflammatory cascade through microglial activation. The first phase is completed and has indicated that the vaccine is safe and tolerable in two doses, 15 and 75 μg (NCT01568099). Half of the patients developed α-syn antibodies suggesting that the vaccine is effective as well. However, further investigations are underway to test the clinical benefit of PDO1A in PD as well as other parkinsonian-associated diseases (Mandler et al., 2015) (NCT02216188 and NCT02270489).

Collectively, these studies show that α-syn-targeted immunotherapies may have the potential to attenuate neuroinflammatory-associated neurotoxicity and as such, delay dopaminergic neurodegeneration associated with PD. Alternatively, targeting α-syn may have some potential pitfalls. Vaccination and antibody-mediated therapies are associated with antigen-specific T cell responses that may contribute to an inflammatory state operative during PD. This concern was realized in studies of immunotherapies for AD. Here, extensive T cell reactivity and an autoimmune meningoencephalitis followed immune-stimulating strategies for Aβ clearance (Orgogozo et al., 2003). This, coupled with the presence of intracellular misfolded proteins, makes it imperative to generate vaccination designs that are long-lasting and highly specific for α-syn. The robustness of the immune response generated needs to be evaluated together with any potential cross-reactivity of common nonpathogenic forms of α-syn or members of its protein families.

Finally, with little success of anti-inflammatory therapies translating to the clinic, therapeutic strategies aimed at modulating the aberrant immune response during disease progression may be more appropriate. The majority of immunotherapies are designed to protect the host from foreign entities; however, in neurodegenerative diseases, the target is a self-protein, thus rendering the type and duration of the peripheral immune response elicited extremely important. In the acute MPTP model of PD, nitrated α-syn enhances the neurotoxic element of microglia activation, ultimately leading to neuronal cell loss (Reynolds et al., 2008). In PD, data support that with inflammation and chronic immune stimulation in the brain, concomitant increases are evident in effector and memory T cell populations that can secrete pro-inflammatory and neurotoxic cytokines (Anderson et al., 2014; Mosley et al., 2012). This can drive microglial reactivity, producing even greater levels of pro-inflammatory mediators. Thus, inhibiting Teff subsets microglial activation may be of clinical benefit (Benner et al., 2004). One such strategy utilizes the immune-suppressing capabilities of Tregs to target neuroinflammation and neurodegeneration. In the MPTP mouse model, adoptive transfer of Tregs attenuates microgliosis, protects against dopaminergic neuronal death, and spares striatal termini (Kosloski et al., 2013). Tregs can also elicit downregulation of pro-inflammatory mediators such as iNOS, TNFα, IL-1β, and IFN-γ as well as decrease levels of ROS production and NF-κB activation (Lan et al., 2012; Lowther & Hafler, 2012). Through these mechanisms, Tregs may also mediate protective actions by switching microglial responses from neurotoxic M1 responses to neurotrophic M2 responses (Reynolds et al., 2007b). However, it should be noted that even resting microglia can exist in a spectrum of phenotypic states rather than solely existing in M1 and M2 conditions that are exclusively pro- or anti-inflammatory (Cherry et al., 2014). Rather, microglia may simply be shifted into a more or less anti-inflammatory state. Nonetheless, these findings indicate a phenotypic transition during Teff or Treg interactions with microglia and/or peripheral monocyte-macrophages. Furthermore, as discussed previously, Tregs can act directly on activated Teff populations and antigen-presenting cells by cytolysis, metabolic disruption, inhibiting maturation, and/or secreting suppressive cytokines (Dasgupta & Saxena, 2012). However, in the overall PD population, Treg deficits and increases in Teff populations are found associated with increased movement dysfunction and disease severity (Saunders et al., 2012), thus indicating a need for therapies that induce or enhance Treg functions. One appropriate therapeutic avenue would be to enhance Treg numbers or function through the use of potent immune-modulating agents such as VIP, pituitary adenylate cyclase-activating polypeptide (PACAP), or GM-CSF, or through the use of vaccine strategies that induce Treg populations. Notably, the use of PACAP, VIP, GM-CSF, and Bacillus Calmette-Guerin (BCG) vaccination has shown promising results in PD models. Such therapeutic interventions have also shown effectiveness in the treatments of other chronic inflammatory conditions and as such, support the ability to restore immune homeostasis and repair tissue injuries (Delgado & Ganea, 2013; Fraccaroli et al., 2015; Kelso et al., 2015; Kosloski et al., 2013; Lacan et al., 2013; Reynolds et al., 2010; Schabitz et al., 2008; Shivers et al., 2014; Waschek, 2013).

For instance, adoptive transfer of VIP- or GM-CSF-induced Tregs following MPTP intoxication leads to enhanced neuronal survival of nearly 100% and a decrease in microglia activation (Kosloski et al., 2013; Reynolds et al., 2010). The potential use of this therapeutic strategy, along with clearance of aberrant protein, could possibly lead to the induction of Tregs that will induce a phenotypic shift in effector subsets and microglia, ultimately ameliorating disease. Although, there remains the possibility that the neuroinflammatory response that is occurring with progressive PD will not be ameliorated simply through the induction of Tregs. Tregs may not have the ability to suppress the detrimental immune response to a point in which protection is observed. However, pre-clinical studies have already shown the potential promise of immune modulatory therapies that transform effector to regulatory T cell functions.

For LBD specifically, decreasing inflammation and targeting aberrant protein accumulation also represents a potential means to modify disease progression in dementia. As for PD, currently no disease-modifying therapies are clinically available. However, evidence indicates that the use of cholinesterase inhibitors can treat the cognitive and visual-spatial symptoms of the disease (Rolinski et al., 2012; Wild et al., 2003). This is linked to the known reduction of cholinergic markers in LBD (Klein et al., 2010). For instance, both donepezil and rivastigmine have been shown to be efficacious in LBD (Wang et al., 2015a). Additionally, strategies to decrease the abundance of α-syn aggregation by decreasing α-syn levels or promoting its degradation are currently underway. For example, preliminary results of anti-α-syn-immunotherapy in a transgenic and viral vector model of synucleinopathy have shown promise and indicate a reduction in α-syn propagation within the axon following treatment (Spencer et al., 2017).

23.5 Review Questions

1. What are synucleinopathies?
2. What are the clinical hallmarks of Parkinson's disease?
3. Describe the neuroinflammatory cascade that occurs in Parkinson's disease.
4. In what ways could inflammation be beneficial for synucleinopathies?
5. Compare and contrast the similarities and differences of Dementia with Lewy bodies (DLB) and Parkinson's disease dementia (PDD).
6. What are the current immunotherapies for synucleinopathies?

23.6 Answers

1. Synucleinopathies are a group of neurodegenerative disorders that share the abnormal accumulation of α-synuclein in neurons and glial cells.
2. Four main clinical hallmarks of Parkinson's disease that result from loss of dopaminergic neurons and dopamine neurotransmitters include at-rest tremor, bradykinesia, postural instability, and rigidity.
3. Central to the inflammatory response in Parkinson's disease is the activation of microglia by modified and misfolded α-synuclein. Activated microglia attempt to clear accumulated protein, but with increasing accumulation and microglial activation, clearance becomes less efficient with microglia becoming more activated and inducing neurotoxic and pro-inflammatory cytokines and chemokines. These toxic mediators injure or kill surrounding neurons and induce misfolding and modifications of α-syn to feed a perpetual cycle of misfolding, microglia activation, and neuronal killing. Moreover, as modified α-syn and inflammatory milieu drain into peripheral lymph nodes, effector T cells are induced which migrate across the blood-brain barrier and interact with α-syn presented by microglia to further drive inflammatory response.
4. Inflammation serves a vital function to increase the phagocytic ability of mononuclear phagocytes such as microglia and macrophages to aid in the clearance of misfolded or modified proteins such as α-syn. However, with increasing activation, the inflammatory responses transform into more neurotoxic responses that injure or kill neurons and enhance aberrant modification and misfolding of α-synuclein.
5. DLB and PDD are both characterized by Lewy bodies in the brain, brain cell atrophy, and amyloid beta protein deposits. They also share similar symptoms including cognitive decline, but the order in which they appear depends on where the Lewy bodies first form. In DLB, cognitive decline occurs before or at the same time as motor symptoms. In PDD, motor symptoms manifest at least one year before cognitive decline.
6. While no immunotherapeutic modality has yet been approved by the FDA, several experimental approaches that include pharmacological-, cellular-, and humoral-mediated strategies are being pursued. Anti-inflammatory pharmaceuticals have been assessed includes NSAIDs, minocycline, and PPAR agonists, and have shown promise in animal models, but have not attained efficacy in clinical trials. Of course, interventional studies by genetic manipulation of inflammatory pathways have also shown promise in pre-clinical studies. Strategies targeting clearance of α-syn with antibodies either by passive immunization or by active immunization approaches with α-syn mimetopes are currently being pursued. Several cellular immune approaches also target the inflammatory and neurotoxic components of PD via targeting regulatory T cells (Tregs) by their adoptive transfer or by using immune modulators to boost Treg number and/or function and include agents such as GM-CSF, VIPR agonists, and low dose IL-2.

References

Aarsland, D., & Kurz, M. W. (2010). The epidemiology of dementia associated with Parkinson's disease. *Brain Pathology, 20*(3), 633–639. https://doi.org/10.1111/j.1750-3639.2009.00369.x

Acioglu, C., Li, L., & Elkabes, S. (2021). Contribution of astrocytes to neuropathology of neurodegenerative diseases. *Brain Research, 1758*, 147291. https://doi.org/10.1016/j.brainres.2021.147291

Anderson, K. M., Olson, K. E., Estes, K. A., Flanagan, K., et al. (2014). Dual destructive and protective roles of adaptive immunity in neurodegenerative disorders. *Translational Neurodegeneration, 3*(1), 25. https://doi.org/10.1186/2047-9158-3-25

Arnaoutoglou, N. A., O'Brien, J. T., & Underwood, B. R. (2019). Dementia with Lewy bodies – From scientific knowledge to clinical insights. *Nature Reviews. Neurology, 15*(2), 103–112. https://doi.org/10.1038/s41582-018-0107-7

Baba, Y., Kuroiwa, A., Uitti, R. J., Wszolek, Z. K., et al. (2005). Alterations of T-lymphocyte populations in Parkinson disease. *Parkinsonism & Related Disorders, 11*(8), 493–498. https://doi.org/10.1016/j.parkreldis.2005.07.005

Badanjak, K., Fixemer, S., Smajic, S., Skupin, A., et al. (2021). The contribution of microglia to neuroinflammation in Parkinson's disease. *International Journal of Molecular Sciences, 22*(9). https://doi.org/10.3390/ijms22094676

Badiola, N., de Oliveira, R. M., Herrera, F., Guardia-Laguarta, C., et al. (2011). Tau enhances alpha-synuclein aggregation and toxicity in cellular models of synucleinopathy. *PLoS One, 6*(10), e26609. https://doi.org/10.1371/journal.pone.0026609

Bas, J., Calopa, M., Mestre, M., Mollevi, D. G., et al. (2001). Lymphocyte populations in Parkinson's disease and in rat models of parkinsonism. *Journal of Neuroimmunology, 113*(1), 146–152. https://doi.org/10.1016/s0165-5728(00)00422-7

Bate, C., Gentleman, S., & Williams, A. (2010). alpha-synuclein induced synapse damage is enhanced by amyloid-beta1-42. *Molecular Neurodegeneration, 5*, 55. https://doi.org/10.1186/1750-1326-5-55

Bencze, J., Seo, W., Hye, A., Aarsland, D., et al. (2020). Dementia with Lewy bodies – A clinicopathological update. *Free Neuropathology, 1*, 1–7. https://doi.org/10.17879/freeneuropathology-2020-2613

Benner, E. J., Mosley, R. L., Destache, C. J., Lewis, T. B., et al. (2004). Therapeutic immunization protects dopaminergic neurons in a mouse model of Parkinson's disease. *Proceedings of the National Academy of Sciences of the United States of America, 101*(25), 9435–9440. https://doi.org/10.1073/pnas.0400569101

Benner, E. J., Banerjee, R., Reynolds, A. D., Sherman, S., et al. (2008). Nitrated alpha-synuclein immunity accelerates degeneration of nigral dopaminergic neurons. *PLoS One, 3*(1), e1376. https://doi.org/10.1371/journal.pone.0001376

Bereczki, E., Francis, P. T., Howlett, D., Pereira, J. B., et al. (2016). Synaptic proteins predict cognitive decline in Alzheimer's disease and Lewy body dementia. *Alzheimers Dement, 12*(11), 1149–1158. https://doi.org/10.1016/j.jalz.2016.04.005

Bieganowska, K., Czlonkowska, A., Bidzinski, A., Mierzewska, H., et al. (1993). Immunological changes in the MPTP-induced Parkinson's disease mouse model. *Journal of Neuroimmunology, 42*(1), 33–37. https://doi.org/10.1016/0165-5728(93)90209-h

Boghdadi, A. G., Teo, L., & Bourne, J. A. (2020). The neuroprotective role of reactive astrocytes after central nervous system injury. *Journal of Neurotrauma, 37*(5), 681–691. https://doi.org/10.1089/neu.2019.6938

Borah, A., Paul, R., Choudhury, S., Choudhury, A., et al. (2013). Neuroprotective potential of silymarin against CNS disorders: Insight into the pathways and molecular mechanisms of action. *CNS Neuroscience & Therapeutics, 19*(11), 847–853. https://doi.org/10.1111/cns.12175

Brawek, B., Skok, M., & Garaschuk, O. (2021). Changing functional signatures of microglia along the axis of brain aging. *International Journal of Molecular Sciences, 22*(3). https://doi.org/10.3390/ijms22031091

Brochard, V., Combadiere, B., Prigent, A., Laouar, Y., et al. (2009). Infiltration of CD4+ lymphocytes into the brain contributes to neurodegeneration in a mouse model of Parkinson disease. *The Journal of Clinical Investigation, 119*(1), 182–192. https://doi.org/10.1172/JCI36470

Campos-Acuna, J., Elgueta, D., & Pacheco, R. (2019). T-cell-driven inflammation as a mediator of the gut-brain axis involved in Parkinson's disease. *Frontiers in Immunology, 10*, 239. https://doi.org/10.3389/fimmu.2019.00239

Cao, Q., Zou, Q., Zhao, X., Zhang, Y., et al. (2022). Regulation of BDNF transcription by Nrf2 and MeCP2 ameliorates MPTP-induced neurotoxicity. *Cell Death Discovery, 8*(1), 267. https://doi.org/10.1038/s41420-022-01063-9

Carrithers, M. D., Visintin, I., Kang, S. J., & Janeway, C. A., Jr. (2000). Differential adhesion molecule requirements for immune surveillance and inflammatory recruitment. *Brain, 123*(Pt 6), 1092–1101. https://doi.org/10.1093/brain/123.6.1092

Carta, A. R., & Pisanu, A. (2013). Modulating microglia activity with PPAR-gamma agonists: A promising therapy for Parkinson's disease? *Neurotoxicity Research, 23*(2), 112–123. https://doi.org/10.1007/s12640-012-9342-7

Cherry, J. D., Olschowka, J. A., & O'Banion, M. K. (2014). Are "resting" microglia more "m2"? *Frontiers in Immunology, 5*, 594. https://doi.org/10.3389/fimmu.2014.00594

Chinta, S. J., Ganesan, A., Reis-Rodrigues, P., Lithgow, G. J., et al. (2013). Anti-inflammatory role of the isoflavone diadzein in lipopolysaccharide-stimulated microglia: Implications for Parkinson's disease. *Neurotoxicity Research, 23*(2), 145–153. https://doi.org/10.1007/s12640-012-9328-5

Colom-Cadena, M., Pegueroles, J., Herrmann, A. G., Henstridge, C. M., et al. (2017). Synaptic phosphorylated alpha-synuclein in dementia with Lewy bodies. *Brain, 140*(12), 3204–3214. https://doi.org/10.1093/brain/awx275

Contaldi, E., Magistrelli, L., & Comi, C. (2022). T lymphocytes in Parkinson's disease. *Journal of Parkinson's Disease, 12*(s1), S65–S74. https://doi.org/10.3233/JPD-223152

Dasgupta, A., & Saxena, R. (2012). Regulatory T cells: A review. *National Medical Journal of India, 25*(6), 341–351.

Delgado, M., & Ganea, D. (2013). Vasoactive intestinal peptide: A neuropeptide with pleiotropic immune functions. *Amino Acids, 45*(1), 25–39. https://doi.org/10.1007/s00726-011-1184-8

Dodel, R., Spottke, A., Gerhard, A., Reuss, A., et al. (2010). Minocycline 1-year therapy in multiple-system-atrophy: Effect on clinical symptoms and [(11)C] (R)-PK11195 PET (MEMSA-trial). *Movement Disorders, 25*(1), 97–107. https://doi.org/10.1002/mds.22732

Donaghy, P. C., Firbank, M., Petrides, G., Lloyd, J., et al. (2020). Diffusion imaging in dementia with Lewy bodies: Associations with amyloid burden, atrophy, vascular factors and clinical features. *Parkinsonism & Related Disorders, 78*, 109–115. https://doi.org/10.1016/j.parkreldis.2020.07.025

Donaldson, I. M. (2015). James Parkinson's essay on the shaking palsy. *The Journal of the Royal College of Physicians of Edinburgh, 45*(1), 84–86. https://doi.org/10.4997/JRCPE.2015.118

Drukarch, B., Schepens, E., Stoof, J. C., Langeveld, C. H., et al. (1998). Astrocyte-enhanced neuronal survival is mediated by scavenging of extracellular reactive oxygen species. *Free Radical Biology & Medicine, 25*(2), 217–220. https://doi.org/10.1016/s0891-5849(98)00050-1

Du, Y., Ma, Z., Lin, S., Dodel, R. C., et al. (2001). Minocycline prevents nigrostriatal dopaminergic neurodegeneration in the MPTP model of Parkinson's disease. *Proceedings of the National Academy of Sciences of the United States of America, 98*(25), 14669–14674. https://doi.org/10.1073/pnas.251341998

Dupont, A. C., Largeau, B., Santiago Ribeiro, M. J., Guilloteau, D., et al. (2017). Translocator protein-18 kDa (TSPO) positron emission tomography (PET) imaging and its clinical impact in neurodegenerative diseases. *International Journal of Molecular Sciences, 18*(4). https://doi.org/10.3390/ijms18040785

Fagerqvist, T., Lindstrom, V., Nordstrom, E., Lord, A., et al. (2013). Monoclonal antibodies selective for alpha-synuclein oligomers/protofibrils recognize brain pathology in Lewy body disorders and alpha-synuclein transgenic mice with the disease-causing A30P mutation. *Journal of Neurochemistry, 126*(1), 131–144. https://doi.org/10.1111/jnc.12175

Fiszer, U., Mix, E., Fredrikson, S., Kostulas, V., et al. (1994). Parkinson's disease and immunological abnormalities: Increase of HLA-DR expression on monocytes in cerebrospinal fluid and of CD45RO+ T cells in peripheral blood. *Acta Neurologica Scandinavica, 90*(3), 160–166. https://doi.org/10.1111/j.1600-0404.1994.tb02699.x

Follett, K. A., & Torres-Russotto, D. (2012). Deep brain stimulation of globus pallidus interna, subthalamic nucleus, and pedunculopontine nucleus for Parkinson's disease: Which target? *Parkinsonism & Related Disorders, 18*(Suppl 1), S165–S167. https://doi.org/10.1016/S1353-8020(11)70051-7

Fraccaroli, L., Grasso, E., Hauk, V., Paparini, D., et al. (2015). VIP boosts regulatory T cell induction by trophoblast cells in an in vitro model of trophoblast-maternal leukocyte interaction. *Journal of Leukocyte Biology, 98*(1), 49–58. https://doi.org/10.1189/jlb.1A1014-492RR

Games, D., Valera, E., Spencer, B., Rockenstein, E., et al. (2014). Reducing C-terminal-truncated alpha-synuclein by immunotherapy attenuates neurodegeneration and propagation in Parkinson's disease-like models. *The Journal of Neuroscience, 34*(28), 9441–9454. https://doi.org/10.1523/JNEUROSCI.5314-13.2014

Garcia-Esparcia, P., Lopez-Gonzalez, I., Grau-Rivera, O., Garcia-Garrido, M. F., et al. (2017). Dementia with Lewy dodies: Molecular pathology in the frontal cortex in typical and rapidly progressive forms. *Frontiers in Neurology, 8*, 89. https://doi.org/10.3389/fneur.2017.00089

Garretti, F., Agalliu, D., Lindestam Arlehamn, C. S., Sette, A., et al. (2019). Autoimmunity in Parkinson's disease: The role of alpha-synuclein-specific T cells. *Frontiers in Immunology, 10*, 303. https://doi.org/10.3389/fimmu.2019.00303

Garrido-Gil, P., Joglar, B., Rodriguez-Perez, A. I., Guerra, M. J., et al. (2012). Involvement of PPAR-gamma in the neuroprotective and anti-inflammatory effects of angiotensin type 1 receptor inhibition: Effects of the receptor antagonist telmisartan and receptor deletion in a mouse MPTP model of Parkinson's disease. *Journal of Neuroinflammation, 9*, 38. https://doi.org/10.1186/1742-2094-9-38

Gate, D., Tapp, E., Leventhal, O., Shahid, M., et al. (2021). CD4(+) T cells contribute to neurodegeneration in Lewy body dementia. *Science, 374*(6569), 868–874. https://doi.org/10.1126/science.abf7266

Geed, M., Garabadu, D., Ahmad, A., & Krishnamurthy, S. (2014). Silibinin pretreatment attenuates biochemical and behavioral changes induced by intrastriatal MPP+ injection in rats. *Pharmacology, Biochemistry, and Behavior, 117*, 92–103. https://doi.org/10.1016/j.pbb.2013.12.008

Gelders, G., Baekelandt, V., & Van der Perren, A. (2018). Linking neuroinflammation and neurodegeneration in Parkinson's disease. *Journal of Immunology Research, 2018*, 4784268. https://doi.org/10.1155/2018/4784268

Gendelman, H. E., Zhang, Y., Santamaria, P., Olson, K. E., et al. (2017). Evaluation of the safety and immunomodulatory effects of sargramostim in a randomized, double-blind phase 1 clinical Parkinson's disease trial. *NPJ Parkinsons Disease, 3*, 10. https://doi.org/10.1038/s41531-017-0013-5

Geng, X., Zou, Y., Li, J., Li, S., et al. (2023). BDNF alleviates Parkinson's disease by promoting STAT3 phosphorylation and regulating neuronal autophagy. *Cell and Tissue Research, 393*(3), 455–470. https://doi.org/10.1007/s00441-023-03806-1

Gomperts, S. N. (2016). Lewy body dementias: Dementia with Lewy bodies and Parkinson disease dementia. *Continuum (Minneapolis, Minn.), 22*(2 Dementia), 435–463. https://doi.org/10.1212/CON.0000000000000309

Gonzalez-Rey, E., & Delgado, M. (2007). Vasoactive intestinal peptide and regulatory T-cell induction: A new mechanism and therapeutic potential for immune homeostasis. *Trends in Molecular Medicine, 13*(6), 241–251. https://doi.org/10.1016/j.molmed.2007.04.003

Good, P. F., Hsu, A., Werner, P., Perl, D. P., et al. (1998). Protein nitration in Parkinson's disease. *Journal of Neuropathology and Experimental Neurology, 57*(4), 338–342. https://doi.org/10.1097/00005072-199804000-00006

Hamani, C., Florence, G., Heinsen, H., Plantinga, B. R., et al. (2017). Subthalamic nucleus deep brain stimulation: Basic concepts and novel perspectives. *eNeuro, 4*(5), ENEURO.0140-17.2017. https://doi.org/10.1523/ENEURO.0140-17.2017

Hamelin, L., Lagarde, J., Dorothee, G., Leroy, C., et al. (2016). Early and protective microglial activation in Alzheimer's disease: A prospective study using 18F-DPA-714 PET imaging. *Brain, 139*(Pt 4), 1252–1264. https://doi.org/10.1093/brain/aww017

Harms, A. S., Ferreira, S. A., & Romero-Ramos, M. (2021). Periphery and brain, innate and adaptive immunity in Parkinson's disease. *Acta Neuropathologica, 141*(4), 527–545. https://doi.org/10.1007/s00401-021-02268-5

He, Y., Appel, S., & Le, W. (2001). Minocycline inhibits microglial activation and protects nigral cells after 6-hydroxydopamine injection into mouse striatum. *Brain Research, 909*(1–2), 187–193. https://doi.org/10.1016/s0006-8993(01)02681-6

Hogl, B., Stefani, A., & Videnovic, A. (2018). Idiopathic REM sleep behaviour disorder and neurodegeneration – An update. *Nature Reviews. Neurology, 14*(1), 40–55. https://doi.org/10.1038/nrneurol.2017.157

Iba, M., Kim, C., Sallin, M., Kwon, S., et al. (2020). Neuroinflammation is associated with infiltration of T cells in Lewy body disease and alpha-synuclein transgenic models. *Journal of Neuroinflammation, 17*(1), 214. https://doi.org/10.1186/s12974-020-01888-0

Iravani, M. M., Costa, S., Jackson, M. J., Tel, B. C., et al. (2001). GDNF reverses priming for dyskinesia in MPTP-treated, L-DOPA-primed common marmosets. *The European Journal of Neuroscience, 13*(3), 597–608. https://doi.org/10.1046/j.1460-9568.2001.01408.x

Irwin, D. J., Lee, V. M., & Trojanowski, J. Q. (2013). Parkinson's disease dementia: Convergence of alpha-synuclein, tau and amyloid-beta pathologies. *Nature Reviews. Neuroscience, 14*(9), 626–636. https://doi.org/10.1038/nrn3549

Jellinger, K. A. (2018). Dementia with Lewy bodies and Parkinson's disease-dementia: Current concepts and controversies. *Journal of Neural Transmission (Vienna), 125*(4), 615–650. https://doi.org/10.1007/s00702-017-1821-9

Jellinger, K. A., & Korczyn, A. D. (2018). Are dementia with Lewy bodies and Parkinson's disease dementia the same disease? *BMC Medicine, 16*(1), 34. https://doi.org/10.1186/s12916-018-1016-8

Joers, V., Tansey, M. G., Mulas, G., & Carta, A. R. (2017). Microglial phenotypes in Parkinson's disease and animal models of the disease. *Progress in Neurobiology, 155*, 57–75. https://doi.org/10.1016/j.pneurobio.2016.04.006

Jung, U. J., Jeon, M. T., Choi, M. S., & Kim, S. R. (2014). Silibinin attenuates MPP(+)-induced neurotoxicity in the substantia nigra in vivo. *Journal of Medicinal Food, 17*(5), 599–605. https://doi.org/10.1089/jmf.2013.2926

Kalia, L. V., & Lang, A. E. (2015). Parkinson's disease. *Lancet, 386*(9996), 896–912. https://doi.org/10.1016/S0140-6736(14)61393-3

Kelso, M. L., Elliott, B. R., Haverland, N. A., Mosley, R. L., et al. (2015). Granulocyte-macrophage colony stimulating factor exerts protective and immunomodulatory effects in cortical trauma. *Journal of Neuroimmunology, 278*, 162–173. https://doi.org/10.1016/j.jneuroim.2014.11.002

Kim, H. S., & Suh, Y. H. (2009). Minocycline and neurodegenerative diseases. *Behavioural Brain Research, 196*(2), 168–179. https://doi.org/10.1016/j.bbr.2008.09.040

King, E., O'Brien, J. T., Donaghy, P., Morris, C., et al. (2018). Peripheral inflammation in prodromal Alzheimer's and Lewy body dementias. *Journal of Neurology, Neurosurgery, and Psychiatry, 89*(4), 339–345. https://doi.org/10.1136/jnnp-2017-317134

Klein, J. C., Eggers, C., Kalbe, E., Weisenbach, S., et al. (2010). Neurotransmitter changes in dementia with Lewy bodies and Parkinson disease dementia in vivo. *Neurology, 74*(11), 885–892. https://doi.org/10.1212/WNL.0b013e3181d55f61

Kohutnicka, M., Lewandowska, E., Kurkowska-Jastrzebska, I., Czlonkowski, A., et al. (1998). Microglial and astrocytic involvement in a murine model of Parkinson's disease induced by 1-methyl-4-phenyl-1,2,3,6-tetrahydropyridine (MPTP). *Immunopharmacology, 39*(3), 167–180. https://doi.org/10.1016/s0162-3109(98)00022-8

Kosloski, L. M., Kosmacek, E. A., Olson, K. E., Mosley, R. L., et al. (2013). GM-CSF induces neuroprotective and anti-inflammatory responses in 1-methyl-4-phenyl-1,2,3,6-tetrahydropyridine intoxicated mice. *Journal of Neuroimmunology, 265*(1–2), 1–10. https://doi.org/10.1016/j.jneuroim.2013.10.009

Kramer, M. L., & Schulz-Schaeffer, W. J. (2007). Presynaptic alpha-synuclein aggregates, not Lewy bodies, cause neurodegeneration in dementia with Lewy bodies. *The Journal of Neuroscience, 27*(6), 1405–1410. https://doi.org/10.1523/JNEUROSCI.4564-06.2007

Kurata, T., Kawarabayashi, T., Murakami, T., Miyazaki, K., et al. (2007). Enhanced accumulation of phosphorylated alpha-synuclein in double transgenic mice expressing mutant beta-amyloid precursor protein and presenilin-1. *Journal of Neuroscience Research, 85*(10), 2246–2252. https://doi.org/10.1002/jnr.21352

Kurkowska-Jastrzebska, I., Wronska, A., Kohutnicka, M., Czlonkowski, A., et al. (1999a). The inflammatory reaction following 1-methyl-4-phenyl-1,2,3, 6-tetrahydropyridine intoxication in mouse. *Experimental Neurology, 156*(1), 50–61. https://doi.org/10.1006/exnr.1998.6993

Kurkowska-Jastrzebska, I., Wronska, A., Kohutnicka, M., Czlonkowski, A., et al. (1999b). MHC class II positive microglia and lymphocytic infiltration are present in the substantia nigra and striatum in mouse model of Parkinson's disease. *Acta Neurobiologiae Experimentalis (Wars), 59*(1), 1–8. https://doi.org/10.55782/ane-1999-1289

Lacan, G., Dang, H., Middleton, B., Horwitz, M. A., et al. (2013). Bacillus Calmette-Guerin vaccine-mediated neuroprotection is associated with regulatory T-cell induction in the 1-methyl-4-phenyl-1,2,3,6-tetrahydropyridine mouse model of Parkinson's disease. *Journal of Neuroscience Research, 91*(10), 1292–1302. https://doi.org/10.1002/jnr.23253

Lalive, P. H., Neuhaus, O., Benkhoucha, M., Burger, D., et al. (2011). Glatiramer acetate in the treatment of multiple sclerosis: Emerging concepts regarding its mechanism of action. *CNS Drugs, 25*(5), 401–414. https://doi.org/10.2165/11588120-000000000-00000

Lan, Q., Fan, H., Quesniaux, V., Ryffel, B., et al. (2012). Induced Foxp3(+) regulatory T cells: A potential new weapon to treat autoimmune and inflammatory diseases? *Journal of Molecular Cell Biology, 4*(1), 22–28. https://doi.org/10.1093/jmcb/mjr039

Langston, J. W., Ballard, P., Tetrud, J. W., & Irwin, I. (1983). Chronic Parkinsonism in humans due to a product of meperidine-analog synthesis. *Science, 219*(4587), 979–980. https://doi.org/10.1126/science.6823561

Langston, J. W., Forno, L. S., Tetrud, J., Reeves, A. G., et al. (1999). Evidence of active nerve cell degeneration in the substantia nigra of humans years after 1-methyl-4-phenyl-1,2,3,6-tetrahydropyridine exposure. *Annals of Neurology, 46*(4), 598–605. https://doi.org/10.1002/1531-8249(199910)46:4<598::aid-ana7>3.0.co;2-f

Laurie, C., Reynolds, A., Coskun, O., Bowman, E., et al. (2007). CD4+ T cells from Copolymer-1 immunized mice protect dopaminergic neurons in the 1-methyl-4-phenyl-1,2,3,6-tetrahydropyridine model of Parkinson's disease. *Journal of Neuroimmunology, 183*(1–2), 60–68. https://doi.org/10.1016/j.jneuroim.2006.11.009

Lee, Y., Park, H. R., Chun, H. J., & Lee, J. (2015). Silibinin prevents dopaminergic neuronal loss in a mouse model of Parkinson's disease via mitochondrial stabilization. *Journal of Neuroscience Research, 93*(5), 755–765. https://doi.org/10.1002/jnr.23544

Lin, T. K., Chen, S. D., Chuang, Y. C., Lin, H. Y., et al. (2014). Resveratrol partially prevents rotenone-induced neurotoxicity in dopaminergic SH-SY5Y cells through induction of heme oxygenase-1 dependent autophagy. *International Journal of Molecular Sciences, 15*(1), 1625–1646. https://doi.org/10.3390/ijms15011625

Lindstrom, V., Fagerqvist, T., Nordstrom, E., Eriksson, F., et al. (2014). Immunotherapy targeting alpha-synuclein protofibrils reduced pathology in (Thy-1)-h[A30P] alpha-synuclein mice. *Neurobiology of Disease, 69*, 134–143. https://doi.org/10.1016/j.nbd.2014.05.009

Liu, B., & Hong, J. S. (2003). Role of microglia in inflammation-mediated neurodegenerative diseases: Mechanisms and strategies for therapeutic intervention. *The Journal of Pharmacology and Experimental Therapeutics, 304*(1), 1–7. https://doi.org/10.1124/jpet.102.035048

Lofrumento, D. D., Nicolardi, G., Cianciulli, A., De Nuccio, F., et al. (2014). Neuroprotective effects of resveratrol in an MPTP mouse model of Parkinson's-like disease: Possible role of SOCS-1 in reducing pro-inflammatory responses. *Innate Immunity, 20*(3), 249–260. https://doi.org/10.1177/1753425913488429

Lowther, D. E., & Hafler, D. A. (2012). Regulatory T cells in the central nervous system. *Immunological Reviews, 248*(1), 156–169. https://doi.org/10.1111/j.1600-065X.2012.01130.x

Lu, L., Li, F., & Wang, X. (2010). Novel anti-inflammatory and neuroprotective agents for Parkinson's disease. *CNS & Neurological Disorders Drug Targets, 9*(2), 232–240. https://doi.org/10.2174/187152710791012035

Mackenzie, I. R. (2000). Activated microglia in dementia with Lewy bodies. *Neurology, 55*(1), 132–134. https://doi.org/10.1212/wnl.55.1.132

Maiti, P., Manna, J., & Dunbar, G. L. (2017). Current understanding of the molecular mechanisms in Parkinson's disease: Targets for potential treatments. *Translational Neurodegeneration, 6*, 28. https://doi.org/10.1186/s40035-017-0099-z

Mandler, M., Valera, E., Rockenstein, E., Weninger, H., et al. (2014). Next-generation active immunization approach for synucleinopathies: Implications for Parkinson's disease clinical trials. *Acta Neuropathologica, 127*(6), 861–879. https://doi.org/10.1007/s00401-014-1256-4

Mandler, M., Valera, E., Rockenstein, E., Mante, M., et al. (2015). Active immunization against alpha-synuclein ameliorates the degenerative pathology and prevents demyelination in a model of multiple system atrophy. *Molecular Neurodegeneration, 10*, 10. https://doi.org/10.1186/s13024-015-0008-9

Manocha, G. D., Floden, A. M., Puig, K. L., Nagamoto-Combs, K., et al. (2017). Defining the contribution of neuroinflammation to Parkinson's disease in humanized immune system mice. *Molecular Neurodegeneration, 12*(1), 17. https://doi.org/10.1186/s13024-017-0158-z

Marogianni, C., Sokratous, M., Dardiotis, E., Hadjigeorgiou, G. M., et al. (2020). Neurodegeneration and inflammation-an interesting interplay in Parkinson's disease. *International Journal of Molecular Sciences, 21*(22). https://doi.org/10.3390/ijms21228421

Masliah, E., Rockenstein, E., Veinbergs, I., Sagara, Y., et al. (2001). Beta-amyloid peptides enhance alpha-synuclein accumulation and neuronal deficits in a transgenic mouse model linking Alzheimer's disease and Parkinson's disease. *Proceedings of the National Academy of Sciences of the United States of America, 98*(21), 12245–12250. https://doi.org/10.1073/pnas.211412398

McGeer, P. L., Itagaki, S., Boyes, B. E., & McGeer, E. G. (1988). Reactive microglia are positive for HLA-DR in the substantia nigra of Parkinson's and Alzheimer's disease brains. *Neurology, 38*(8), 1285–1291. https://doi.org/10.1212/wnl.38.8.1285

McGeer, P. L., Schwab, C., Parent, A., & Doudet, D. (2003). Presence of reactive microglia in monkey substantia nigra years after 1-methyl-4-phenyl-1,2,3,6-tetrahydropyridine administration. *Annals of Neurology, 54*(5), 599–604. https://doi.org/10.1002/ana.10728

Miyazaki, I., & Asanuma, M. (2020). Neuron-astrocyte interactions in Parkinson's disease. *Cells, 9*(12). https://doi.org/10.3390/cells9122623

Moon, H. E., & Paek, S. H. (2015). Mitochondrial dysfunction in Parkinson's disease. *Experimental Neurobiology, 24*(2), 103–116. https://doi.org/10.5607/en.2015.24.2.103

Mosley, R. L., Hutter-Saunders, J. A., Stone, D. K., & Gendelman, H. E. (2012). Inflammation and adaptive immunity in Parkinson's disease. *Cold Spring Harbor Perspectives in Medicine, 2*(1), a009381. https://doi.org/10.1101/cshperspect.a009381

Moussaud, S., Jones, D. R., Moussaud-Lamodiere, E. L., Delenclos, M., et al. (2014). Alpha-synuclein and tau: Teammates in neurodegeneration? *Molecular Neurodegeneration, 9*, 43. https://doi.org/10.1186/1750-1326-9-43

Myers, A. J., Brahimi, A., Jenkins, I. J., & Koob, A. O. (2023). The synucleins and the astrocyte. *Biology (Basel), 12*(2). https://doi.org/10.3390/biology12020155

Nedelska, Z., Schwarz, C. G., Boeve, B. F., Lowe, V. J., et al. (2015). White matter integrity in dementia with Lewy bodies: A voxel-based analysis of diffusion tensor imaging. *Neurobiology of Aging, 36*(6), 2010–2017. https://doi.org/10.1016/j.neurobiolaging.2015.03.007

Nicastro, N., Mak, E., Williams, G. B., Surendranathan, A., et al. (2020). Correlation of microglial activation with white matter changes in dementia with Lewy bodies. *Neuroimage Clinical, 25*, 102200. https://doi.org/10.1016/j.nicl.2020.102200

NINDS. (2006). A randomized, double-blind, futility clinical trial of creatine and minocycline in early Parkinson disease. *Neurology, 66*(5), 664–671. https://doi.org/10.1212/01.wnl.0000201252.57661.e1

NINDS. (2008). A pilot clinical trial of creatine and minocycline in early Parkinson disease: 18-month results. *Clinical Neuropharmacology, 31*(3), 141–150. https://doi.org/10.1097/WNF.0b013e3181342f32

NINDS. (2015). Pioglitazone in early Parkinson's disease: A phase 2, multicentre, double-blind, randomised trial. *Lancet Neurology, 14*(8), 795–803. https://doi.org/10.1016/S1474-4422(15)00144-1

Okello, A., Edison, P., Archer, H. A., Turkheimer, F. E., et al. (2009). Microglial activation and amyloid deposition in mild cognitive impairment: A PET study. *Neurology, 72*(1), 56–62. https://doi.org/10.1212/01.wnl.0000338622.27876.0d

Olanow, C. W., Stern, M. B., & Sethi, K. (2009). The scientific and clinical basis for the treatment of Parkinson disease (2009). *Neurology, 72*(21 Suppl 4), S1–136. https://doi.org/10.1212/WNL.0b013e3181a1d44c

Olson, K. E., Kosloski-Bilek, L. M., Anderson, K. M., Diggs, B. J., et al. (2015). Selective VIP receptor agonists facilitate immune transformation for dopaminergic neuroprotection in MPTP-intoxicated mice. *The Journal of Neuroscience, 35*(50), 16463–16478. https://doi.org/10.1523/JNEUROSCI.2131-15.2015

Olson, K. E., Bade, A. N., Schutt, C. R., Dong, J., et al. (2016). Manganese-enhanced magnetic resonance imaging for detection of vasoactive intestinal peptide receptor 2 agonist therapy in a model of Parkinson's disease. *Neurotherapeutics, 13*(3), 635–646. https://doi.org/10.1007/s13311-016-0449-z

Orgogozo, J. M., Gilman, S., Dartigues, J. F., Laurent, B., et al. (2003). Subacute meningoencephalitis in a subset of patients with AD after Abeta42 immunization. *Neurology, 61*(1), 46–54. https://doi.org/10.1212/01.wnl.0000073623.84147.a8

Passemard, S., Sokolowska, P., Schwendimann, L., & Gressens, P. (2011). VIP-induced neuroprotection of the developing brain. *Current Pharmaceutical Design, 17*(10), 1036–1039. https://doi.org/10.2174/138161211795589409

Pisanu, A., Lecca, D., Mulas, G., Wardas, J., et al. (2014). Dynamic changes in pro- and anti-inflammatory cytokines in microglia after PPAR-gamma agonist neuroprotective treatment in the MPTPp mouse model of progressive Parkinson's disease. *Neurobiology of Disease, 71*, 280–291. https://doi.org/10.1016/j.nbd.2014.08.011

Poyhonen, S., Er, S., Domanskyi, A., & Airavaara, M. (2019). Effects of neurotrophic factors in glial cells in the central nervous system: Expression and properties in neurodegeneration and injury. *Frontiers in Physiology, 10*, 486. https://doi.org/10.3389/fphys.2019.00486

Przedborski, S., Jackson-Lewis, V., Djaldetti, R., Liberatore, G., et al. (2000). The parkinsonian toxin MPTP: Action and mechanism. *Restorative Neurology and Neuroscience, 16*(2), 135–142.

Przedborski, S., Jackson-Lewis, V., Naini, A. B., Jakowec, M., et al. (2001). The parkinsonian toxin 1-methyl-4-phenyl-1,2,3,6-tetrahydropyridine (MPTP): A technical review of its utility and safety. *Journal of Neurochemistry, 76*(5), 1265–1274. https://doi.org/10.1046/j.1471-4159.2001.00183.x

Qian, L., Flood, P. M., & Hong, J. S. (2010). Neuroinflammation is a key player in Parkinson's disease and a prime target for therapy. *Journal of Neural Transmission (Vienna), 117*(8), 971–979. https://doi.org/10.1007/s00702-010-0428-1

Rees, K., Stowe, R., Patel, S., Ives, N., et al. (2011). Non-steroidal anti-inflammatory drugs as disease-modifying agents for Parkinson's disease: Evidence from observational studies. *Cochrane Database of Systematic Reviews*, (11), CD008454. https://doi.org/10.1002/14651858.CD008454.pub2

Ren, B., Zhang, Y. X., Zhou, H. X., Sun, F. W., et al. (2015). Tanshinone IIA prevents the loss of nigrostriatal dopaminergic neurons by inhibiting NADPH oxidase and iNOS in the MPTP model of Parkinson's disease. *Journal of the Neurological Sciences, 348*(1–2), 142–152. https://doi.org/10.1016/j.jns.2014.11.026

Reynolds, A., Laurie, C., Mosley, R. L., & Gendelman, H. E. (2007a). Oxidative stress and the pathogenesis of neurodegenerative disorders. *International Review of Neurobiology, 82*, 297–325. https://doi.org/10.1016/S0074-7742(07)82016-2

Reynolds, A. D., Banerjee, R., Liu, J., Gendelman, H. E., et al. (2007b). Neuroprotective activities of CD4+CD25+ regulatory T cells in an animal model of Parkinson's disease. *Journal of Leukocyte Biology, 82*(5), 1083–1094. https://doi.org/10.1189/jlb.0507296

Reynolds, A. D., Kadiu, I., Garg, S. K., Glanzer, J. G., et al. (2008). Nitrated alpha-synuclein and microglial neuroregulatory activities. *Journal of Neuroimmune Pharmacology, 3*(2), 59–74. https://doi.org/10.1007/s11481-008-9100-z

Reynolds, A. D., Stone, D. K., Hutter, J. A., Benner, E. J., et al. (2010). Regulatory T cells attenuate Th17 cell-mediated nigrostriatal dopaminergic neurodegeneration in a model of Parkinson's disease. *Journal of Immunology, 184*(5), 2261–2271. https://doi.org/10.4049/jimmunol.0901852

Rietdijk, C. D., Perez-Pardo, P., Garssen, J., van Wezel, R. J., et al. (2017). Exploring Braak's hypothesis of Parkinson's disease. *Frontiers in Neurology, 8*, 37. https://doi.org/10.3389/fneur.2017.00037

Rocha, S. M., Cristovao, A. C., Campos, F. L., Fonseca, C. P., et al. (2012). Astrocyte-derived GDNF is a potent inhibitor of microglial activation. *Neurobiology of Disease, 47*(3), 407–415. https://doi.org/10.1016/j.nbd.2012.04.014

Rolinski, M., Fox, C., Maidment, I., & McShane, R. (2012). Cholinesterase inhibitors for dementia with Lewy bodies, Parkinson's disease dementia and cognitive impairment in Parkinson's disease. *Cochrane Database of Systematic Reviews, 2012*(3), CD006504. https://doi.org/10.1002/14651858.CD006504.pub2

Ruffmann, C., Calboli, F. C., Bravi, I., Gveric, D., et al. (2016). Cortical Lewy bodies and Abeta burden are associated with prevalence and timing of dementia in Lewy body diseases. *Neuropathology and Applied Neurobiology, 42*(5), 436–450. https://doi.org/10.1111/nan.12294

Sanchez-Guajardo, V., Annibali, A., Jensen, P. H., & Romero-Ramos, M. (2013). alpha-Synuclein vaccination prevents the accumulation of parkinson disease-like pathologic inclusions in striatum in association with regulatory T cell recruitment in a rat model. *Journal of Neuropathology and Experimental Neurology, 72*(7), 624–645. https://doi.org/10.1097/NEN.0b013e31829768d2

Saunders, J. A., Estes, K. A., Kosloski, L. M., Allen, H. E., et al. (2012). CD4+ regulatory and effector/memory T cell subsets profile motor dysfunction in Parkinson's disease. *Journal of Neuroimmune Pharmacology, 7*(4), 927–938. https://doi.org/10.1007/s11481-012-9402-z

Scanlon, B. R., Reedy, R. C., Jackson, W. A., & Rao, B. (2008). Mobilization of naturally occurring perchlorate related to land-use change in the southern High Plains, Texas. *Environmental Science & Technology, 42*(23), 8648–8653. https://doi.org/10.1021/es800361h

Schabitz, W. R., Kruger, C., Pitzer, C., Weber, D., et al. (2008). A neuroprotective function for the hematopoietic protein granulocyte-macrophage colony stimulating factor (GM-CSF). *Journal of Cerebral Blood Flow and Metabolism, 28*(1), 29–43. https://doi.org/10.1038/sj.jcbfm.9600496

Schneeberger, A., Mandler, M., Mattner, F., & Schmidt, W. (2012). Vaccination for Parkinson's disease. *Parkinsonism & Related Disorders, 18*(Suppl 1), S11–S13. https://doi.org/10.1016/S1353-8020(11)70006-2

Schober, A., Peterziel, H., von Bartheld, C. S., Simon, H., et al. (2007). GDNF applied to the MPTP-lesioned nigrostriatal system requires TGF-beta for its neuroprotective action. *Neurobiology of Disease, 25*(2), 378–391. https://doi.org/10.1016/j.nbd.2006.10.005

Shahaduzzaman, M., Nash, K., Hudson, C., Sharif, M., et al. (2015). Anti-human alpha-synuclein N-terminal peptide antibody protects against dopaminergic cell death and ameliorates behavioral deficits in an AAV-alpha-synuclein rat model of Parkinson's disease. *PLoS One, 10*(2), e0116841. https://doi.org/10.1371/journal.pone.0116841

Shim, K. H., Kang, M. J., Youn, Y. C., An, S. S. A., et al. (2022). Alpha-synuclein: A pathological factor with Abeta and tau and biomarker in Alzheimer's disease. *Alzheimer's Research & Therapy, 14*(1), 201. https://doi.org/10.1186/s13195-022-01150-0

Shivers, K. Y., Nikolopoulou, A., Machlovi, S. I., Vallabhajosula, S., et al. (2014). PACAP27 prevents Parkinson-like neuronal loss and motor deficits but not microglia activation induced by prostaglandin J2. *Biochimica et Biophysica Acta, 1842*(9), 1707–1719. https://doi.org/10.1016/j.bbadis.2014.06.020

Simon, D. K., Simuni, T., Elm, J., Clark-Matott, J., et al. (2015). Peripheral biomarkers of Parkinson's disease progression and pioglitazone effects. *Journal of Parkinson's Disease, 5*(4), 731–736. https://doi.org/10.3233/JPD-150666

Smeyne, R. J., Breckenridge, C. B., Beck, M., Jiao, Y., et al. (2016). Assessment of the effects of MPTP and paraquat on dopaminergic neurons and microglia in the substantia nigra pars compacta of C57BL/6 mice. *PLoS One, 11*(10), e0164094. https://doi.org/10.1371/journal.pone.0164094

Song, I. U., Kim, Y. D., Cho, H. J., & Chung, S. W. (2013). Is neuroinflammation involved in the development of dementia in patients with Parkinson's disease? *Internal Medicine, 52*(16), 1787–1792. https://doi.org/10.2169/internalmedicine.52.0474

Spencer, B., Valera, E., Rockenstein, E., Overk, C., et al. (2017). Anti-alpha-synuclein immunotherapy reduces alpha-synuclein propagation in the axon and degeneration in a combined viral vector and transgenic model of synucleinopathy. *Acta Neuropathologica Communications, 5*(1), 7. https://doi.org/10.1186/s40478-016-0410-8

Stevens, C. H., Rowe, D., Morel-Kopp, M. C., Orr, C., et al. (2012). Reduced T helper and B lymphocytes in Parkinson's disease. *Journal of Neuroimmunology, 252*(1–2), 95–99. https://doi.org/10.1016/j.jneuroim.2012.07.015

Streit, W. J., & Xue, Q. S. (2016). Microglia in dementia with Lewy bodies. *Brain, Behavior, and Immunity, 55*, 191–201. https://doi.org/10.1016/j.bbi.2015.10.012

Surendranathan, A., Rowe, J. B., & O'Brien, J. T. (2015). Neuroinflammation in Lewy body dementia. *Parkinsonism & Related Disorders, 21*(12), 1398–1406. https://doi.org/10.1016/j.parkreldis.2015.10.009

Surendranathan, A., Su, L., Mak, E., Passamonti, L., et al. (2018). Early microglial activation and peripheral inflammation in dementia with Lewy bodies. *Brain, 141*(12), 3415–3427. https://doi.org/10.1093/brain/awy265

Swanson, C. R., Joers, V., Bondarenko, V., Brunner, K., et al. (2011). The PPAR-gamma agonist pioglitazone modulates inflammation and induces neuroprotection in parkinsonian monkeys. *Journal of Neuroinflammation, 8*, 91. https://doi.org/10.1186/1742-2094-8-91

t Hart, B. A., Copray, S., & Philippens, I. (2014). Apocynin, a low molecular oral treatment for neurodegenerative disease. *BioMed Research International, 2014*, 298020. https://doi.org/10.1155/2014/298020

Tian, Y., Zhang, Y., Zhang, R., Qiao, S., et al. (2015). Resolvin D2 recovers neural injury by suppressing inflammatory mediators expression in lipopolysaccharide-induced Parkinson's disease rat model. *Biochemical and Biophysical Research Communications, 460*(3), 799–805. https://doi.org/10.1016/j.bbrc.2015.03.109

Tikka, T., Fiebich, B. L., Goldsteins, G., Keinanen, R., et al. (2001). Minocycline, a tetracycline derivative, is neuroprotective against excitotoxicity by inhibiting activation and proliferation of microglia. *The Journal of Neuroscience, 21*(8), 2580–2588. https://doi.org/10.1523/JNEUROSCI.21-08-02580.2001

Titova, N., Qamar, M. A., & Chaudhuri, K. R. (2017). The nonmotor features of Parkinson's disease. *International Review of Neurobiology, 132*, 33–54. https://doi.org/10.1016/bs.irn.2017.02.016

Togo, T., Iseki, E., Marui, W., Akiyama, H., et al. (2001). Glial involvement in the degeneration process of Lewy body-bearing neurons and the degradation process of Lewy bodies in brains of dementia with Lewy bodies. *Journal of the Neurological Sciences, 184*(1), 71–75. https://doi.org/10.1016/s0022-510x(00)00498-6

Tran, H. T., Chung, C. H., Iba, M., Zhang, B., et al. (2014). Alpha-synuclein immunotherapy blocks uptake and templated propagation of misfolded alpha-synuclein and neurodegeneration. *Cell Reports, 7*(6), 2054–2065. https://doi.org/10.1016/j.celrep.2014.05.033

Vezzani, B., Carinci, M., Patergnani, S., Pasquin, M. P., et al. (2020). The dichotomous role of inflammation in the CNS: A mitochondrial point of view. *Biomolecules, 10*(10). https://doi.org/10.3390/biom10101437

Vial, T., & Descotes, J. (1995). Clinical toxicity of cytokines used as haemopoietic growth factors. *Drug Safety, 13*(6), 371–406. https://doi.org/10.2165/00002018-199513060-00006

Wang, H. F., Yu, J. T., Tang, S. W., Jiang, T., et al. (2015a). Efficacy and safety of cholinesterase inhibitors and memantine in cognitive impairment in Parkinson's disease, Parkinson's disease dementia, and dementia with Lewy bodies: Systematic review with meta-analysis and trial sequential analysis. *Journal of Neurology, Neurosurgery, and Psychiatry, 86*(2), 135–143. https://doi.org/10.1136/jnnp-2014-307659

Wang, S., Jing, H., Yang, H., Liu, Z., et al. (2015b). Tanshinone I selectively suppresses pro-inflammatory genes expression in activated microglia and prevents nigrostriatal dopaminergic neurodegeneration in a mouse model of Parkinson's disease. *Journal of Ethnopharmacology, 164*, 247–255. https://doi.org/10.1016/j.jep.2015.01.042

Waschek, J. A. (2013). VIP and PACAP: Neuropeptide modulators of CNS inflammation, injury, and repair. *British Journal of Pharmacology, 169*(3), 512–523. https://doi.org/10.1111/bph.12181

Watson, R., Blamire, A. M., Colloby, S. J., Wood, J. S., et al. (2012). Characterizing dementia with Lewy bodies by means of diffusion tensor imaging. *Neurology, 79*(9), 906–914. https://doi.org/10.1212/WNL.0b013e318266fc51

Wild, R., Pettit, T., & Burns, A. (2003). Cholinesterase inhibitors for dementia with Lewy bodies. *Cochrane Database of Systematic Reviews, 2003*(3), CD003672. https://doi.org/10.1002/14651858.CD003672

Williams, G. P., Schonhoff, A. M., Jurkuvenaite, A., Gallups, N. J., et al. (2021). CD4 T cells mediate brain inflammation and neurodegeneration in a mouse model of Parkinson's disease. *Brain, 144*(7), 2047–2059. https://doi.org/10.1093/brain/awab103

Zhang, F., Shi, J. S., Zhou, H., Wilson, B., et al. (2010). Resveratrol protects dopamine neurons against lipopolysaccharide-induced neurotoxicity through its anti-inflammatory actions. *Molecular Pharmacology, 78*(3), 466–477. https://doi.org/10.1124/mol.110.064535

Zupancic, M., Mahajan, A., & Handa, K. (2011). Dementia with lewy bodies: Diagnosis and management for primary care providers. *The Primary Care Companion for CNS Disorders, 13*(5). https://doi.org/10.4088/PCC.11r01190

Neuropathies

24

Helmar C. Lehmann

Abstract

Peripheral neuropathies encompass a spectrum of conditions that are clinically characterized by a sensory or motor neuropathy with acute or chronic onset. In 8–10% of patients with peripheral neuropathy, the underlying cause is considered an autoimmune condition. This can be a systemic autoimmune disease, for example, vasculitis or a primary autoimmune condition. In this chapter, the underlying pathologenesis of the most common chronic autoimmune neuropathies will be reviewed. Chronic immune-mediated demyelinating polyradiculopathy (CIDP) is the most common form and research over the last decades has identified crucial steps in its pathogenesis. These include aberrant T cell responses as well as autoantibodies against glycolipids and (para)nodal proteins. Autoantibodies against myelin-associated glycoprotein (MAG) are involved in the pathogenesis of anti-MAG neuropathy. Another rarer condition is multifocal motor neuropathy in which clinical and experimental evidence suggests a crucial role of antibodies against gangliosides in its pathogenesis.

Keywords

Neuropathy · Demyelination · Autoantibody · Paranodopathy · Cytokines

H. C. Lehmann (✉)
Medical Faculty, University of Cologne, Cologne, Germany

Department of Neurology, Klinikum Leverkusen, Leverkusen, Germany

24.1 Introduction

Neuropathy refers to a spectrum of conditions resulting from damage to the peripheral nerves, which play a critical role in transmitting signals between the central nervous system (brain and spinal cord) and the rest of the body. Among the various forms of neuropathy, Chronic Inflammatory Demyelinating Polyneuropathy (CIDP) is a particularly significant subtype, characterized by immune-mediated inflammation and demyelination of peripheral nerves. Understanding CIDP within the broader context of neuropathic disorders underscores its relevance as a potentially treatable cause of peripheral nerve dysfunction, offering hope for improved outcomes when managed appropriately.

24.2 Chronic Inflammatory Demyelinating Polyradiculoneuropathy (CIDP)

Chronic inflammatory demyelinating polyradiculoneuropathy (CIDP) is the most common chronic immune-mediated peripheral neuropathy. The term CIDP was introduced by James Dyck and summarizes the main clinical and pathological features of the disease. Patients with CIDP suffer from a chronic stepwise progressive or relapsing neuropathy that manifests with weakness and sensory deficits. Pathologically, endoneural edema, mononuclear cell infiltrates, and macrophage-induced segmental demyelination can be found (Dyck et al., 1975). The disease can be separated on clinical and electrophysiological grounds into classic CIDP and several CIDP variants. It is conceivable but currently unknown whether differences in the clinical phenotype may reflect differences in their underlying pathogenesis and hence responsiveness to immunosuppressive treatment. The prevalence rates are reported to be between 0.5 and 9 per 100,000 per year (Lehmann et al., 2019). It is unknown whether these differences reflect underlying geographical or

H. E. Gendelman, T. Ikezu (eds.), *Neuroimmune Pharmacology and Therapeutics*,
https://doi.org/10.1007/978-3-031-68237-7_24

genetic differences or whether they just reflect methodological differences.

24.2.1 Clinical Presentation

The clinical presentation of CIDP varies. The most common classic variant presents with symmetric muscle weakness and sensory deficits. Muscle weakness can occur in distal as well as proximal muscles and distinguishes CIDP from other classic length-dependent neuropathies (Lehmann et al., 2020). Sensory deficits are hypoesthesia and pallhypoesthesia. Dysesthesia or hyperalgesia is less common. The muscle tendon reflexes are reduced or entirely absent often in all extremities. Usually, these neuropathic symptoms evolve in a subacute fashion over more than 8 weeks (Van den Bergh et al., 2021). The disease course is slowly progressive in most patients, about 15% present with a relapsing-remitting disease course. Cranial nerves are less frequently affected compared to Guillain-Barré syndrome (GBS, see Chap. 34). Occasionally patients may present with an acute onset and are initially misdiagnosed as having GBS. Compared to the latter, patients develop over time a chronic neuropathy with relapses (Ruts et al., 2010). Apart from this classical presentation, there are variants of CIDP (formerly denoted as atypical CIDP, Table 24.1). Distal CIDP is also denoted as distal acquired demyelinating symmetric (DADS) neuropathy (Katz et al., 2000). These patients often have an IgM paraprotein, and there is an overlap with anti-MAG neuropathy.

24.2.2 Autoimmune Pathogenesis

Like GBS, there is clinical, pathological, and experimental evidence that strongly supports the concept that CIDP is an inflammatory neuropathy caused by an aberrant immune response.

Clinical Evidence There are several clinical observations that suggest that CIDP may be an autoimmune disease although the evidence is weaker than for GBS. Several studies indicate that in about 10% of CIDP patients, the disease occurred after an infection similar to GBS (Doneddu et al., 2020; Rajabally et al., 2021). However, in contrast to GBS, definite proof or link to a particular microorganism that could cause CIDP is absent. There are numerous case reports and series that describe CIDP associated with other neurological and non-neurological autoimmune diseases, for example, myasthenia gravis (reviewed in (Rodríguez et al., 2019)). It is well known that having one autoimmune disease increases the risk of developing another (Rodríguez et al., 2019).

Lastly, the clinical experience that CIDP responds well to different immunomodulatory and suppressive treatments provides further clinical evidence for an autoimmune origin.

Extensive research over the last decades has focused on possible target antigens of the aberrant immune response. A major achievement in this regard was the discovery that approximately 10% of patients with CIDP harbor antibodies against (para)nodal proteins (Querol et al., 2017).

A brief description of these proteins might be helpful in understanding their functional consequences: These are located at the region of the node of Ranvier, which can be further distinguished into paranodal and juxtaparanodal regions (Mathey et al., 2017). The nodal cell adhesion molecules (CAMs) gliomedin, neuron glia-related CAM (NrCAM) and neurofascin 186 (NF186) are essential for the structure of the node of Ranvier, particularly for the maintenance of Na+ channel clustering, which is vital for the saltatory nerve conduction (Amor et al., 2014). In the paranodal region, there are protein complexes of contactin-1/caspr-1 complexes on the myelin loops which bind to neurofascin 155 (NF155) expressed by Schwann cells (Fig. 24.1). Patients with CIDP can have antibodies against gliomedin, neurofascin 155, contactin-1, or Caspr1. Some authors consider paranodopathies as a clinical entity distinct from CIDP.

Pathological Findings Pathological studies on nerve biopsy specimens report the presence of epi- and endoneurial mononuclear cell infiltrates consisting of T cells and macrophages (Sommer & Toyka, 2011). These findings are not specific for CIDP and are observed in numerous other neuropathic conditions. Notably, in CIDP often a characteristic clustering of macrophages can be observed (Sommer et al., 2005). T cells are predominantly CD8+, and les CD4+ cells (Matsumuro et al., 1994; Schmidt et al., 1996; Schneider-Hohendorf et al., 2012). Occasionally B cells are also present. Regarding humoral autoimmunity, deposition of immunoglobulin and

Table 24.1 Clinical characteristics of CIDP variants

	Classical CIDP	Distal CIDP	(Multi)focal CIDP	Motor CIDP	Sensory CIDP
Motor deficits	Symmetrical, proximal, and distal	Mostly in the lower extremity, distally, in more than two limbs	In more than two limbs, multifocal, or in one limb (focal)	Symmetrical in four limbs, proximal and distal	None
Sensory deficits	In more than two limbs	In more than two limbs	In supply area of the affected nerves	None	Symmetrical in four limbs

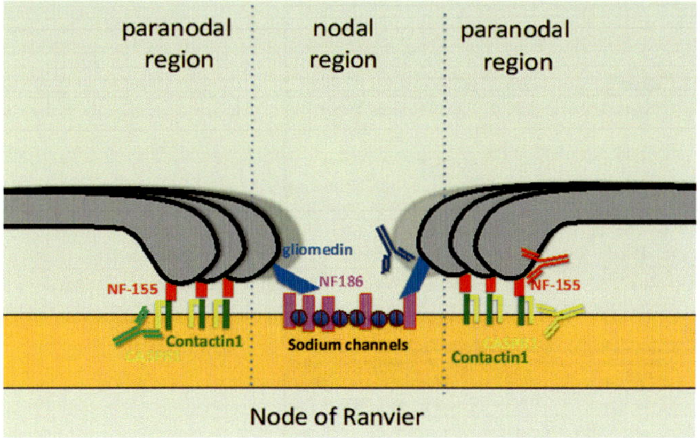

Fig. 24.1 Scheme of the node of Ranvier. Autoantibodies against nodal and paranodal proteins have been described in CIDP. NF Neurofascin, CASPR Contactin-associated protein

complement on myelinated nerve fibers suggests that autoantibodies contribute to the process of demyelination and axonal injury (Dalakas & Engel, 1980; Hays et al., 1988). A representative overview of pathological findings found in CIDP is provided in Fig. 24.2.

Experimental Evidence Experimental approaches to the pathogenesis of CIDP include animal models and passive transfer experiments. The animal model that most closely resembles CIDP is spontaneous autoimmune peripheral polyneuropathy (SAP). SAP is based on non-obese diabetic (NOD) mice that are deficient for costimulatory molecule B7-2 (B7-2−/−), ICAM-1, or mutations in the autoimmune regulator gene (AIRE) (Kim et al., 2008; Salomon et al., 2001). Female mice develop a condition that replicates many clinical and histological features of CIDP. Notably, only NOD mice but not other strains (e.g., C57/Bl6) develop SAP, which indicates that the role of B7–2 in the pathogenesis of SAP is influenced by the genetic background (Su et al., 2012).

One caveat of this model is the presumed target antigen which is myelin protein P0 (Kim et al., 2008; Louvet et al., 2009). P0 is a minor constituent of peripheral myelin. Cellular and humoral responses against P0 were described (Sanvito et al., 2009), but can only be found in a very minor proportion of patients therefore its relevance remains unclear. Nevertheless, the model shares many features that can also be observed in affected humans, including expression of Interferon-γ (IFN-γ)-secreting Th1 cells in nerves and this model has been used to explore novel immunosuppressive treatments or regenerative approaches. Gao and colleagues demonstrated that insulin-like growth factor 1 (IGF-1) can be successfully delivered to neurons by use of a recombinant adeno-associated virus serotype 9 (rAAV9) (Gao et al., 2018). Functional and morphological measures significantly improved in pre-symptomatic and symptomatic mice that were treated with IGF-1-rAAV9.

Passive transfer experiments further support the concept of an autoimmune pathogenesis. Sera or purified IgG-injected nerves can induce demyelination and functionally a conduction block (Yan et al., 2000).

The discovery that some patients with CIDP harbor antibodies against (para)nodal proteins paved the way to explore the pathogenicity of these antibodies in animal studies. In ex vivo assays, anti-contactin-1 antibodies bind to the paranodal regions of peripheral nerves and the passive transfer of these IgG into a rat model of autoimmune neuritis (experimental autoimmune neuritis, EAN) exacerbated the clinical symptoms. Notably pathological changes were restricted to the paranodal region with loss of contactin-associated protein 1/contactin-1/neurofascin-155 complex at paranodes (Manso et al., 2016). In contrast, demyelination and endoneural inflammation were not observed (Manso et al., 2016). The same group also reported that the passive transfer (intrathecal injection) of antibodies against neurofascin155 resulted in the loss of this protein and subsequent nerve conduction changes (Manso et al., 2019).

24.2.3 Treatment

CIDP is an autoimmune disorder and as such amenable to immunomodulatory and suppressive treatments. Treatment options can be classified into first-line and second-line treatments. First-line treatments are immunoglobulin, corticosteroids, and plasma exchange. The use of any one of these approaches can achieve a treatment response in 85–80% of patients (Tackenberg et al., 2007). The mechanisms by which these therapies exert their therapeutic effects are diverse and not fully understood. Immunoglobulin can be administered either intravenously or subcutaneously, the two administrations have been shown to be efficacious in randomized controlled trials (Lehmann et al., 2019).

For immunoglobulin, it is assumed that several mechanisms of action are relevant in CIDP. It includes Fc-dependent mechanisms, for example, Fc-receptor blockage or modulation of Fcγ receptor expression, and F(ab)2-dependent mechanisms such as binding of pathological autoantibodies by anti-idiotypic antibodies and neutralization of complement (Créange et al., 2003; Lehmann & Hartung, 2011; Matà et al., 2006). IVIg has also been shown to modulate inflammatory cytokines including TNF-alpha, IFN-gamma, and B cell activating factor (BAFF), which plays an important role in the survival and maturation of B cells (Ritter et al., 2014). Plasma exchange is beneficial in CIDP presumably by removing circulating autoantibodies, immune complexes, cytokines, and other inflammatory mediators. Several studies

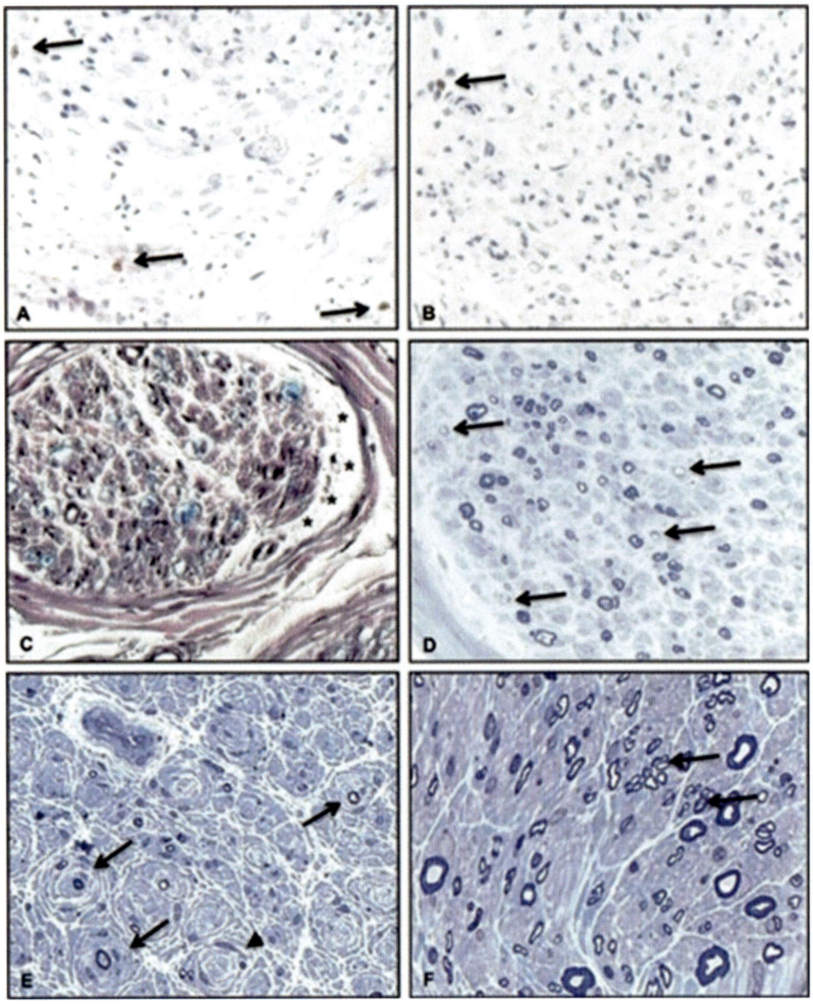

Fig. 24.2 Morphological characteristics of CIDP. Sparse inflammatory infiltrates are mostly to CD4+ T cells (A, arrows). Rarely, singular B cells can also be identified (B, arrow). A frequent indirect sign of inflammation is endoneural edema, mostly subperineural (C, asterisk). Nerve fibers that are too thinly myelinated in relation to their axon caliber (D, arrows). Several relapses of CIDP also exhibit so-called onion bulb formations, in which Schwann cell processes are arranged concentrically around an axon, as a sign of recurrent de- and remyelination (E, arrows). Axonal damage may also be evident, primarily manifested by small regenerating groups (F, arrows). (**a**, **b**) Immunohistochemistry with anti-CD4 and anti-CD20 with slight counterstaining with hemalaun; magnification ×200. (**c**): Masson aniline blue histochemical staining; magnification ×200. (**d**) Semi-thin section of glutaraldehyde-fixed, Epon-embedded nerve biopsy, stained with toluidine blue; magnification ×200. (**e**): Immunohistochemistry using anti-S100 protein (DCS) with light counterstaining with hemalaun; magnification ×400. (**f**) Semi-thin section of glutaraldehyde-fixed, Epon-embedded nerve biopsy, stained with toluidine blue; magnification ×400. (From: Balke et al., 2016)

also demonstrate that Th1/Th2 cell function is modulated by plasma exchange (De Luca et al., 1999; Goto et al., 2001).

The presence of (para)nodal antibodies does not predict a favorable or better outcome for plasma exchange. In contrast, a recent retrospective study even found that these patients tended to respond less well to plasma exchange or immunoadsorption compared to seronegative patients (Davies et al., 2020).

Second-line treatment of CIDP mainly includes immunosuppressive treatments. They are used in case a patient does not respond satisfactorily to any of the first line treatments or it is used as add on treatment to reduce corticosteroid treatment or in conjunction with plasma exchange. Commonly used drugs are cyclophosphamide, azathioprine, and cyclosporine. None of them has been evaluated in larger controlled trials, except azathioprine that was used in a small randomized controlled trial, but did not yield any additional benefit when added to prednisone (Dyck et al., 1985). Other treatments that have been explored in case studies were mycophenolate mofetil (Gorson et al., 2004), alemtuzumab (Marsh

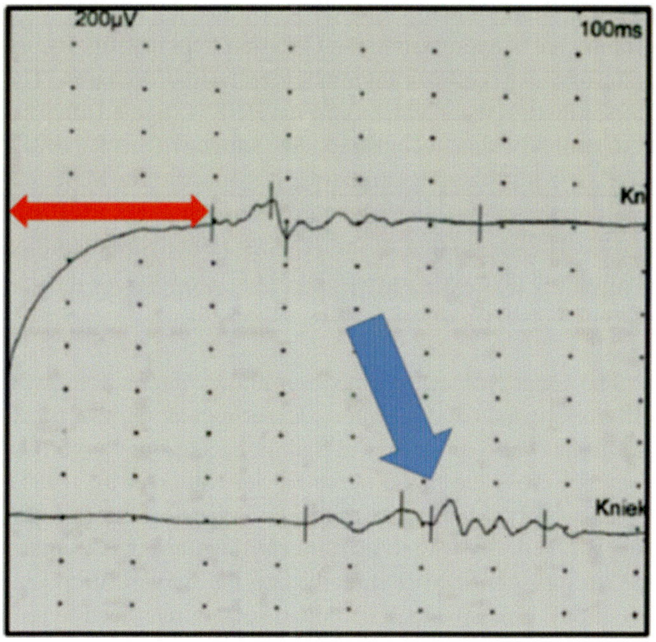

Fig. 24.3 Nerve conduction studies in a patient with MAG-associated demyelinating neuropathy. Note the enlarged distal motor latencies (red arrow) which are 29 ms (normal <5.8 ms), and the dispersed potentials (blue arrow)

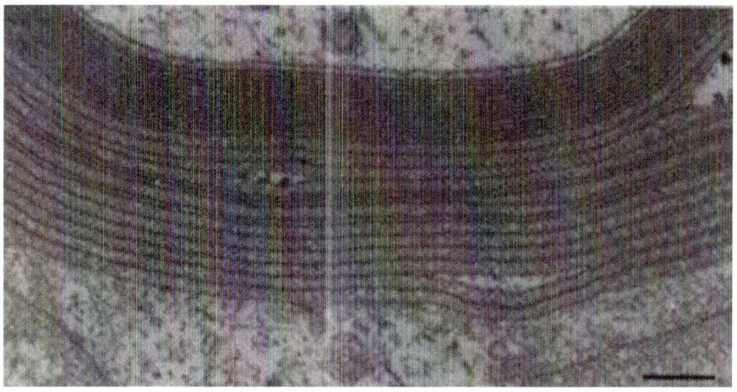

Fig. 24.4 Electron micrograph showing outer myelin lamellae with increased periodicity due to widening of the intraperiod line. Bar = 0.2 gm. (From Jacobs & Scadding, 1990)

et al., 2010), methotrexate (Fialho et al., 2006), and bortezomib (Pitarokoili et al., 2017).

24.3 Anti-MAG Neuropathy

24.3.1 Clinical Presentation and Diagnosis

In approximately 10% of patients with idiopathic polyneuropathy a monoclonal immunoglobulin can be detected in serum or urine (Kelly et al., 1981). The presence of a monoclonal gammopathy of unknown significance almost doubles the risk to develop peripheral neuropathy according to a recent study from Sweden that included more than 15.000 participants with MGUS and more than 58.000 matched controls. The prevalence of peripheral neuropathy was 6.5% in participants with MGUS compared to 2.85% in the control group (Rögnvaldsson et al., 2020). Diseases that go along with a monoclonal gammopathy are multiple myeloma, or malignant lymphoproliferative diseases such as Waldenström's macroglobulinemia, lymphoma, or chronic lymphatic leukemia.

The peripheral neuropathy that is associated with either IgG, IgA, or IgM monoclonal gammopathy is not uniform and some of these associations are controversial. For example, an axonal neuropathy in a patient who also has an IgG paraprotein is likely to be coincidental (Ramchandren & Lewis, 2009). The best-established causal association is the presence of a monoclonal IgM paraproteinemia with a demyelinating, distal predominant, sensory neuropathy. In those patients, approximately 50% have IgM antibodies against myelin-associated glycoprotein (MAG) (Nobile-Orazio et al., 1994, 2000).

Clinically, patients with anti-MAG neuropathy often show a slowly progressive, predominant sensory neuropathy. Patients often complain of gait disorder, sensory ataxia, and tremor. Nerve conduction studies often show marked prolonged distal motor latencies (Fig. 24.3), as a sign of extensive nerve demyelination. Other demyelinating features that can be found are nerve conduction blocks and reduced nerve conduction velocities.

Diagnosis is established by the detection of antibodies against MAG in patients` serum. Although antibody titers do not correlate with the severity of the neuropathy, tests that provide antibody titers are preferred. Only highly elevated anti-MAG antibody titers are of diagnostic value. Cut-off values are provided in most assays although they are not formally validated. One must consider that antibody binding to MAG is temperature and conformation dependent, so that the test conditions do not replicate the binding activity in vivo (Latov, 2021).

24.3.2 Autoimmune Pathogenesis

Clinical Evidence The finding of a specific antibody that reacts with an antigen present in the peripheral nerve myelin supports the concept that MAG-associated demyelinating peripheral neuropathy is an autoimmune disorder. Autoantibodies tend to decrease during treatment in responsive patients according to a recent meta-analysis of 50 studies of patients with MAG demyelinating neuropathy (Hänggi et al., 2022). In contrast to many other autoimmune diseases, the male gender is predominant in most cohort studies (Bardel et al., 2022; Tang et al., 2020). Patients with anti-

MAG-neuropathy respond to B cell–depleting therapies and plasma exchange (Bardel et al., 2022; Parisi et al., 2022).

Pathological Findings Histological studies in patients with MAG neuropathy demonstrate a decrease in the density of myelinated nerve fibers. The most striking pathological feature is the widened space of the myelin lamellae (Ellie et al., 1996; Jacobs & Scadding, 1990; Steck, 2021; Vital et al., 1997) (Fig. 24.4).

Usually, the outer part of the myelin sheath is affected by this phenomenon, and it can be observed in 5–10% up to 19% of myelinated axons (Kawagashira et al., 2020; Magy et al., 2015). Widened myelin space is usually found in the paranodal regions (Kawagashira et al., 2020). It is specific for anti-MAG neuropathy, but only around 60% of the patients show this finding. In the latter, features of demyelination can be observed, such as onion bulb formations and/or completely demyelinated axons (Kawagashira et al., 2020). Cellular infiltrates can be observed in some patients. They contain lymphocytes and macrophages. Nerve biopsies can demonstrate deposition of IgM and anti-MAG antibodies (Ritz et al., 1999). It is predominant in the paranodal regions and Schmidt-Lanterman incisures, but the periaxonal space, another region where MAG is expressed, is usually spared (Kawagashira et al., 2020). Teased fiber studies demonstrated that in anti-MAG neuropathies, the nodal and paranodal architecture comprising axonal proteins and ion channels is severely disrupted. In addition, deposition of complement components C1q, C3d, and C5 colocalizes with IgM deposition (Monaco et al., 1990).

These pathological findings indicate that MAG antibodies recognize epitopes expressed in the nodal and paranodal region and by fixing complement disrupt the nodal and paranodal architecture and initiate deadhesion of the terminal loops of the myelin from the adjacent axon in the paranodal region (Steck, 2021).

Deposition of IgM has also been reported in skin biopsies in patients with MAG neuropathies. However, it remains controversial whether findings of deposition of IgM on skin nerves are specific to MAG neuropathies or are just an epiphenomenon that can also occur in patients without neuropathy (Al-Zuhairy et al., 2015; Lombardi et al., 2005). In addition, in MAG neuropathy patients the axonal caliber is reduced which is associated with reduced spacing of neurofilaments in demyelinated fibers (Lunn et al., 2002). Since MAG is also considered an important signaling molecule between myelin and axons, alteration of MAG in the periaxonal space could contribute to axonal loss caused by decreased neurofilament phosphorylation.

Experimental Evidence MAG is a transmembrane glycoprotein that is expressed in the CNS and PNS myelin. It contains five extracellular immunoglobulin-like domains. The autoantibodies react with an epitope HNK-1 (named after the first antibody that recognized this antigen) (Steck, 2021). HNK-1 is a sulfated trisaccharide not specific for MAG but is also expressed in other glycoproteins such as P0, PMP, and SGPG. The expression of the HNK-1 carbohydrate epitope on multiple glycosylation sites of MAG accounts for its high immunoreactivity (Steck, 2021).

The large extracellular domain allows interactions with different ligands and receptors. Experimental passive transfer studies were unsuccessful mainly because the large IgM does not cross the blood-nerve-barrier (Steck et al., 1985). Consequently, MAG autoantibodies failed to bind to intraneural myelin. A way to circumvent this barrier is to inject the serum or antibodies directly into the nerve (intraneural injection). This approach is limited since it may lead to artificial damage to the nerve by the injection per se. In addition, this approach does not induce a primarily demyelinating pathology like that observed in humans (Willison et al., 1988). A study that explored the effects of direct injection of anti-MAG-antibodies plus complement, however, reported that this treatment resulted in similar pathological changes observed in humans including widely spaced myelin (Monaco et al., 1995). Probably, the closest resemblance to the syndrome in humans can be achieved by continuous infusion of purified MAG antibodies into chicken, as demonstrated by Tatum (1993).

One of the hurdles that prevents the development of feasible animal models is that MAG is not conserved in species. Rodents, for example, have a MAG that has much less of the HNk-1 epitope compared to human, bovine, or feline MAG, and therefore rodents are not susceptible to MAG-associated demyelination (Ilyas et al., 1986; Tatum, 1993).

24.3.3 Treatment

Treatment of anti-MAG neuropathy varies according to the underlying disease. In case of a malignant lymphoproliferative disorder, hematological treatment is necessary, which may include chemoimmunotherapy (e.g., anti-CD20 monoclonal antibody rituximab along with an alkylating agent like bendamustine or cyclophosphamide), protease inhibitors such as bortezomib, or Bruton's tyrosine kinase inhibitors, and/or hematopoietic stem cell transplantation (Zanwar & Abeykoon, 2022).

B cell–depleting therapies are long-term treatments of choice in anti-MAG peripheral neuropathy associated with MGUS. Several case series and open trials evaluated the therapeutic potential of rituximab on IgM paraproteinaemic neuropathies with anti-MAG reactivity (Benedetti et al., 2008; Renaud et al., 2003).

There are also two RCTs that have evaluated the effects of rituximab in anti-MAG-antibody positive polyneuropathy (Dalakas et al., 2009; Léger et al., 2013). They showed moderate beneficial effects but were considered of low quality because of the low number of patients. Rituximab only depletes B cells but spares the MAG antibody-producing plasma cells. This is reflected by the anti-MAG antibody titers that only show moderate reduction after rituximab treatment. In a large retrospective analysis, rituximab was effective in approximately one-third of the patients (Svahn et al., 2018).

24.3.4 Multifocal Motor Neuropathy

24.3.4.1 Clinical Presentation and Diagnosis

Multifocal motor neuropathy (MMN) is the rarest of the three conditions discussed in this chapter. A few epidemiological studies reported prevalence rates between 0.29 and 1.3 per 100,000 (Mahdi-Rogers et al., 2017; Miyashiro et al., 2014; Park et al., 2022). It manifests clinically by a slowly progressive, muscle weakness and starts often distally and at the arms without any or only minor sensory deficits. The disease course is usually slow, over time the muscles become atrophic and fasciculations may occur. Clinically it may resemble a degenerative motor neuron disease, particularly flail-arm syndrome. The electrophysiological hallmark of MMN is the occurrence of proximal nerve conduction blocks. These may occur in different limbs and are associated with signs of axonal degeneration in the distal proportion of the affected nerves. In almost 50% of the cases, IgM antibodies against GM1 and or GD1a may be present (Taylor et al., 2000).

24.3.4.2 Autoimmune Pathogenesis

Clinical Evidence About 50% of patients with MMN have antibodies against the gangliosides GM1 and GD1a. These gangliosides are not restricted to motor nerves only; on the contrary, the ganglioside content of motor and sensory nerves is similar (Gong et al., 2002). Thus, other factors such as accessibility and affinity may play a relevant role, provided that these antibodies are pathogenic and not only an epiphenomenon. These IgM-GM1 antibodies are monoclonal, in contrast to other IgM anti-ganglioside antibodies (Cats et al., 2015). In MMN, there is also evidence of high activity of the classical pathway of complement, and the high complement-activating capacity of anti-GM1 IgM antibodies correlates with clinical and electrophysiological measures of disease severity (Vlam et al., 2015).

Autoimmune diseases such as Type 1 diabetes, Hashimoto's thyroid disease, and celiac disease are slightly more common in patients with MMN (Cats et al., 2012). Furthermore, in a Dutch cohort, an increased frequency of HLA-DRB1*15 in patients with MMN and their first-degree family members were described (Sutedja et al., 2010).

Several RCTs demonstrated that patients with MMN respond to IVIG with an increase in muscle force (Keddie et al., 2022). Other immunosuppressive treatments such as cyclophosphamide, plasma exchange, or corticosteroids are not effective and may even worsen the disease course in individual patients. Nerve conduction studies showed that IVIG improves reinnervation and demyelination; even conduction blocks may be resolved by continued IVIG treatment (Vucic et al., 2004).

Pathological Findings There are only a few biopsy or autopsy studies in MMN. Early studies reported pathological findings in nerve biopsies from patients with MMN at the site of conduction block consisting of demyelinated axons with nerve edema and onion bulb formation but without any cellular infiltrates' characteristic for nerve inflammation (Auer et al., 1989; Kaji et al., 1993).

In sural nerve biopsies, only mild signs of demyelination can be observed which reflect the clinical observation that MMN is predominantly a motor neuron disease. However, sensory deficits can occur during the disease course and do not exclude the condition (Corse et al., 1996). In a fascicular nerve biopsy study, Taylor and colleagues reported that signs of demyelination such as onion bulb formation were almost completely absent, whereas most specimens showed multifocal axonal loss. Similarly, in a series of nerve biopsies, Tracey and colleagues reported that signs of demyelination were absent and axonal injury was predominant, suggesting that the conduction block in MMN is due to an antibody-mediated attack on components of the paranodal axon, rather than inflammatory demyelination (Tracy et al., 2020).

Experimental Evidence Several experimental studies demonstrated the pathogenicity of anti-GM1-IgG in models of peripheral nerve disease. However, MMN is associated with IgM antibodies and hence only studies that explored the role of IgM ganglioside-antibodies are discussed here. IgM is a relatively large molecule with a molecular weight of 900kDA (Keyt et al., 2020) and in order to exert pathological function it would have to cross the blood-nerve barrier (BNB). One study suggests that the

BNB is disrupted by autocrine secretion of VEGF from endothelial cells (Shimizu et al., 2014). In the nerve, GM1-IgM antibodies are thought to activate complement to mediate neuronal injury (Shimizu et al., 2014; Yuki et al., 2011). Harschnitz and colleagues demonstrated in an in vitro model of human-derived motoneuron cultures, that sera from MMN patients containing GM1-IgM bind specifically to GM1 and that these anti-GM1-IgM activate complement. By immunocytochemistry, it could be demonstrated that there were deposits of membrane attack complex (MAC) following incubation with these sera. Notably, neurite degeneration was also observed in the absence of complement, pointing to complement independent neuronal injury (Harschnitz et al., 2016).

Binding of IgM antibodies from patients with MMN to iPSC-derived motoneurons induces complement activation (Budding et al., 2022).

24.3.4.3 Treatment

MMN is treated with IVIG. Efficacy of IVIg has been demonstrated in several RCTs (Keddie et al., 2022). GM1 antibody titers do not correlate with treatment response (Léger et al., 2001). Maintenance therapy is usually required, and even then, the majority of patients show a slow decline in motor function, suggesting that the effectiveness of IVIg tends to decrease over time (Taylor et al., 2000; Terenghi et al., 2004).
In contrast to other chronic immune-mediated neuropathies, corticosteroids and plasma exchange are considered ineffective, and may even worsen the clinical course (Lehmann et al., 2008).

24.4 Review Questions

1. Clinical symptoms that are not compatible for CIDP are
   (a) Sensory deficits
   (b) Absent muscle tendon reflexes
   (c) *Positive Babinski reflexes*
   (d) Paresis
   (e) Walking difficulties
2. What phrase is true about the pathogenesis of CIDP
   (a) *There is strong evidence that Campylobacter jejuni is involved in the pathogenesis*
   (b) CIDP is often associated with other autoimmune diseases
   (c) CIDP responds to intravenous immunglobulin
   (d) Approximately 10% of patients with CIDP have antibodies against (para)nodal proteins
   (e) The blood-nerve-barrier remains intact
3. What is NOT considered to be standard treatment of CIDP
   (a) Immunoglobulin
   (b) Corticosteroids
   (c) Plasma exchange
   (d) *Tocilizumab*
   (e) Azathioprine
4. MAG-associated demyelinating neuropathy
   (a) Is associated with antibodies against contactin-1
   (b) responds to immunoglobulin
   (c) Is associated with antibodies against myelin-associated protein
   (d) MAG is highly conserved throughout species
   (e) *c and c*
5. Multifocal motor neuropathy.
   (a) Starts commonly at the legs
   (b) Autonomic dysfunction is a hallmark of the disease
   (c) Requires neuroimaging confirmation
   (d) *Is associated with antibodies against GM1*

References

Al-Zuhairy, A., Schrøder, H. D., Plesner, T., Abildgaard, N., & Sindrup, S. H. (2015). Immunostaining of skin biopsy adds no diagnostic value in MGUS-associated peripheral neuropathy. *Journal of the Neurological Sciences, 349*, 60–64.

Amor, V., Feinberg, K., Eshed-Eisenbach, Y., Vainshtein, A., Frechter, S., Grumet, M., Rosenbluth, J., & Peles, E. (2014). Long-term maintenance of Na+ channels at nodes of Ranvier depends on glial contact mediated by gliomedin and NrCAM. *The Journal of Neuroscience, 34*, 5089–5098.

Auer, R. N., Bell, R. B., & Lee, M. A. (1989). Neuropathy with onion bulb formations and pure motor manifestations. *The Canadian Journal of Neurological Sciences, 16*, 194–197.

Balke, M., Wunderlich, G., Brunn, A., Fink, G. R., & Lehmann, H. C. (2016). Chronic inflammatory demyelinating polyneuropathy. *Fortschritte der Neurologie-Psychiatrie, 84*, 756–769.

Bardel, B., Molinier-Frenkel, V., Le Bras, F., Ayache, S. S., Nordine, T., Lefaucheur, J.-P., & Planté-Bordeneuve, V. (2022). Revisiting the spectrum of IgM-related neuropathies in a large cohort of IgM monoclonal gammopathy. *Journal of Neurology, 269*, 4955–4960.

Benedetti, L., Briani, C., Franciotta, D., Carpo, M., Padua, L., Zara, G., Zambello, R., Sormani, M. P., Mancardi, G. L., Nobile-Orazio, E., & Schenone, A. (2008). Long-term effect of rituximab in anti-mag polyneuropathy. *Neurology, 71*, 1742–1744.

Budding, K., Johansen, L. E., Van de Walle, I., Dijkxhoorn, K., de Zeeuw, E., Bloemenkamp, L. M., Bos, J. W., Jansen, M. D., Curial, C. A. D., Silence, K., de Haard, H., Blanchetot, C., et al. (2022). Anti-C2 antibody ARGX-117 inhibits complement in a disease model for multifocal motor neuropathy. *Neurol Neuroimmunol Neuroinflamm, 9*, e1107.

Cats, E. A., Bertens, A. S., Veldink, J. H., van den Berg, L. H., & van der Pol, W. L. (2012). Associated autoimmune diseases in patients with multifocal motor neuropathy and their family members. *Journal of Neurology, 259*, 1137–1141.

Cats, E. A., van der Pol, W.-L., Tio-Gillen, A. P., Diekstra, F. P., van den Berg, L. H., & Jacobs, B. C. (2015). Clonality of anti-GM1 IgM antibodies in multifocal motor neuropathy and the Guillain-Barré syndrome. *Journal of Neurology, Neurosurgery, and Psychiatry, 86*, 502–504.

Corse, A. M., Chaudhry, V., Crawford, T. O., Cornblath, D. R., Kuncl, R. W., & Griffin, J. W. (1996). Sensory nerve pathology in multifocal motor neuropathy. *Annals of Neurology, 39*, 319–325.

Créange, A., Gregson, N. A., & Hughes, R. A. C. (2003). Intravenous immunoglobulin modulates lymphocyte CD54 and monocyte FcgammaRII expression in patients with chronic inflammatory neuropathies. *Journal of Neuroimmunology, 135*, 91–95.

Dalakas, M. C., & Engel, W. K. (1980). Immunoglobulin and complement deposits in nerves of patients with chronic relapsing polyneuropathy. *Archives of Neurology, 37*, 637–640.

Dalakas, M. C., Rakocevic, G., Salajegheh, M., Dambrosia, J. M., Hahn, A. F., Raju, R., & McElroy, B. (2009). Placebo-controlled trial of rituximab in IgM anti-myelin-associated glycoprotein antibody demyelinating neuropathy. *Annals of Neurology, 65*, 286–293.

Davies, A. J., Fehmi, J., Senel, M., Tumani, H., Dorst, J., & Rinaldi, S. (2020). Immunoadsorption and plasma exchange in seropositive and seronegative immune-mediated neuropathies. *Journal of Clinical Medicine, 9*, 2025.

De Luca, G., Lugaresi, A., Iarlori, C., Marzoli, F., Di Iorio, A., Gambi, D., & Uncini, A. (1999). Prednisone and plasma exchange improve suppressor cell function in chronic inflammatory demyelinating polyneuropathy. *Journal of Neuroimmunology, 95*, 190–194.

Doneddu, P. E., Bianchi, E., Cocito, D., Manganelli, F., Fazio, R., Filosto, M., Mazzeo, A., Cosentino, G., Cortese, A., Jann, S., Clerici, A. M., Antonini, G., et al. (2020). Risk factors for chronic inflammatory demyelinating polyradiculoneuropathy (CIDP): Antecedent events, lifestyle and dietary habits. Data from the Italian CIDP Database. *European Journal of Neurology, 27*, 136–143.

Dyck, P. J., Lais, A. C., Ohta, M., Bastron, J. A., Okazaki, H., & Groover, R. V. (1975). Chronic inflammatory polyradiculoneuropathy. *Mayo Clinic Proceedings, 50*, 621–637.

Dyck, P. J., O'Brien, P., Swanson, C., Low, P., & Daube, J. (1985). Combined azathioprine and prednisone in chronic inflammatory-demyelinating polyneuropathy. *Neurology, 35*, 1173–1176.

Ellie, E., Vital, A., Steck, A., Boiron, J. M., Vital, C., & Julien, J. (1996). Neuropathy associated with "benign" anti-myelin-associated glycoprotein IgM gammopathy: Clinical, immunological, neurophysiological pathological findings and response to treatment in 33 cases. *Journal of Neurology, 243*, 34–43.

Fialho, D., Chan, Y.-C., Allen, D. C., Reilly, M. M., & Hughes, R. A. C. (2006). Treatment of chronic inflammatory demyelinating polyradiculoneuropathy with methotrexate. *Journal of Neurology, Neurosurgery, and Psychiatry, 77*, 544–547.

Gao, T., Bogdanova, N., Ghauri, S., Zhang, G., Lin, J., & Sheikh, K. (2018). Systemic IGF-1 gene delivery by rAAV9 improves spontaneous autoimmune peripheral polyneuropathy (SAPP). *Scientific Reports, 8*, 5408.

Gong, Y., Tagawa, Y., Lunn, M. P. T., Laroy, W., Heffer-Lauc, M., Li, C. Y., Griffin, J. W., Schnaar, R. L., & Sheikh, K. A. (2002). Localization of major gangliosides in the PNS: Implications for immune neuropathies. *Brain, 125*, 2491–2506.

Gorson, K. C., Amato, A. A., & Ropper, A. H. (2004). Efficacy of mycophenolate mofetil in patients with chronic immune demyelinating polyneuropathy. *Neurology, 63*, 715–717.

Goto, H., Matsuo, H., Nakane, S., Izumoto, H., Fukudome, T., Kambara, C., & Shibuya, N. (2001). Plasmapheresis affects T helper type--1/T helper type-2 balance of circulating peripheral lymphocytes. *Therapeutic Apheresis, 5*, 494–496.

Hänggi, P., Aliu, B., Martin, K., Herrendorff, R., & Steck, A. J. (2022). Decrease in serum anti-MAG autoantibodies is associated with therapy response in patients with anti-MAG neuropathy: Retrospective study. *Neurology Neuroimmunology & Neuroinflammation, 9*, e1109.

Harschnitz, O., van den Berg, L. H., Johansen, L. E., Jansen, M. D., Kling, S., Vieira de Sá, R., Vlam, L., van Rheenen, W., Karst, H., Wierenga, C. J., Pasterkamp, R. J., & van der Pol, W. L. (2016). Autoantibody pathogenicity in a multifocal motor neuropathy induced pluripotent stem cell-derived model. *Annals of Neurology, 80*, 71–88.

Hays, A. P., Lee, S. S., & Latov, N. (1988). Immune reactive C3d on the surface of myelin sheaths in neuropathy. *Journal of Neuroimmunology, 18*, 231–244.

Ilyas, A. A., Dalakas, M. C., Brady, R. O., & Quarles, R. H. (1986). Sulfated glucuronyl glycolipids reacting with anti-myelin-associated glycoprotein monoclonal antibodies including IgM paraproteins in neuropathy: Species distribution and partial characterization of epitopes. *Brain Research, 385*, 1–9.

Jacobs, J. M., & Scadding, J. W. (1990). Morphological changes in IgM paraproteinaemic neuropathy. *Acta Neuropathologica, 80*, 77–84.

Kaji, R., Oka, N., Tsuji, T., Mezaki, T., Nishio, T., Akiguchi, I., & Kimura, J. (1993). Pathological findings at the site of conduction block in multifocal motor neuropathy. *Annals of Neurology, 33*, 152–158.

Katz, J. S., Saperstein, D. S., Gronseth, G., Amato, A. A., & Barohn, R. J. (2000). Distal acquired demyelinating symmetric neuropathy. *Neurology, 54*, 615–620.

Kawagashira, Y., Koike, H., Takahashi, M., Ohyama, K., Iijima, M., Katsuno, M., Niwa, J.-I., Doyu, M., & Sobue, G. (2020). Aberrant expression of nodal and Paranodal molecules in neuropathy associated with IgM monoclonal Gammopathy with anti-myelin-associated glycoprotein antibodies. *Journal of Neuropathology and Experimental Neurology, 79*, 1303–1312.

Keddie, S., Eftimov, F., van den Berg, L. H., Brassington, R., de Haan, R. J., & van Schaik, I. N. (2022). Immunoglobulin for multifocal motor neuropathy. *Cochrane Database of Systematic Reviews, 1*, CD004429.

Kelly, J. J., Kyle, R. A., O'Brien, P. C., & Dyck, P. J. (1981). Prevalence of monoclonal protein in peripheral neuropathy. *Neurology, 31*, 1480–1483.

Keyt, B. A., Baliga, R., Sinclair, A. M., Carroll, S. F., & Peterson, M. S. (2020). Structure, function, and therapeutic use of IgM antibodies. *Antibodies (Basel), 9*, 53.

Kim, H.-J., Jung, C.-G., Jensen, M. A., Dukala, D., & Soliven, B. (2008). Targeting of myelin protein zero in a spontaneous autoimmune polyneuropathy. *Journal of Immunology, 181*, 8753–8760.

Latov, N. (2021). Antibody testing in neuropathy associated with anti-myelin-associated glycoprotein antibodies: Where we are after 40 years. *Current Opinion in Neurology, 34*, 625–630.

Léger, J. M., Chassande, B., Musset, L., Meininger, V., Bouche, P., & Baumann, N. (2001). Intravenous immunoglobulin therapy in multifocal motor neuropathy: a double-blind, placebo-controlled study. *Brain, 124*, 145–153.

Léger, J.-M., Viala, K., Nicolas, G., Créange, A., Vallat, J.-M., Pouget, J., Clavelou, P., Vial, C., Steck, A., Musset, L., Marin, B., & RIMAG Study Group (France and Switzerland). (2013). Placebo-controlled trial of rituximab in IgM anti-myelin-associated glycoprotein neuropathy. *Neurology, 80*, 2217–2225.

Lehmann, H. C., & Hartung, H.-P. (2011). Plasma exchange and intravenous immunoglobulins: Mechanism of action in immune-mediated neuropathies. *Journal of Neuroimmunology, 231*, 61–69.

Lehmann, H. C., Hoffmann, F. R., Fusshoeller, A., Meyer zu Hörste, G., Hetzel, R., Hartung, H.-P., Schroeter, M., & Kieseier, B. C. (2008). The clinical value of therapeutic plasma exchange in multifocal motor neuropathy. *Journal of the Neurological Sciences, 271*, 34–39.

Lehmann, H. C., Burke, D., & Kuwabara, S. (2019). Chronic inflammatory demyelinating polyneuropathy: Update on diagnosis, immunopathogenesis and treatment. *Journal of Neurology, Neurosurgery, and Psychiatry, 90*, 981–987.

Lehmann, H. C., Wunderlich, G., Fink, G. R., & Sommer, C. (2020). Diagnosis of peripheral neuropathy. *Neurological Research and Practice, 2*, 20.

Lombardi, R., Erne, B., Lauria, G., Pareyson, D., Borgna, M., Morbin, M., Arnold, A., Czaplinski, A., Fuhr, P., Schaeren-Wiemers, N., & Steck, A. J. (2005). IgM deposits on skin nerves in anti-myelin-associated glycoprotein neuropathy. *Annals of Neurology, 57*, 180–187.

Louvet, C., Kabre, B. G., Davini, D. W., Martinier, N., Su, M. A., DeVoss, J. J., Rosenthal, W. L., Anderson, M. S., Bour-Jordan, H., & Bluestone, J. A. (2009). A novel myelin P0-specific T cell receptor transgenic mouse develops a fulminant autoimmune peripheral neuropathy. *The Journal of Experimental Medicine, 206*, 507–514.

Lunn, M. P. T., Crawford, T. O., Hughes, R. A. C., Griffin, J. W., & Sheikh, K. A. (2002). Anti-myelin-associated glycoprotein antibodies alter neurofilament spacing. *Brain, 125*, 904–911.

Magy, L., Kaboré, R., Mathis, S., Lebeau, P., Ghorab, K., Caudie, C., & Vallat, J.-M. (2015). Heterogeneity of polyneuropathy associated with anti-MAG antibodies. *Journal of Immunology Research, 2015*, 450391.

Mahdi-Rogers, M., Brassington, R., Gunn, A. A., van Doorn, P. A., & Hughes, R. A. (2017). Immunomodulatory treatment other than corticosteroids, immunoglobulin and plasma exchange for chronic inflammatory demyelinating polyradiculoneuropathy. *Cochrane Database of Systematic Reviews, 5*, CD003280.

Manso, C., Querol, L., Mekaouche, M., Illa, I., & Devaux, J. J. (2016). Contactin-1 IgG4 antibodies cause paranode dismantling and conduction defects. *Brain, 139*, 1700–1712.

Manso, C., Querol, L., Lleixà, C., Poncelet, M., Mekaouche, M., Vallat, J.-M., Illa, I., & Devaux, J. J. (2019). Anti-Neurofascin-155 IgG4 antibodies prevent paranodal complex formation in vivo. *The Journal of Clinical Investigation, 129*, 2222–2236.

Marsh, E. A., Hirst, C. L., Llewelyn, J. G., Cossburn, M. D., Reilly, M. M., Krishnan, A., Doran, M., Ryan, A. M., Coles, A. J., Jones, J. L., & Robertson, N. P. (2010). Alemtuzumab in the treatment of IVIG-dependent chronic inflammatory demyelinating polyneuropathy. *Journal of Neurology, 257*, 913–919.

Matà, S., Borsini, W., Caldini, A., De Scisciolo, G., Piacentini, S., & Taiuti, R. (2006). Long-term evolution of anti-ganglioside antibody levels in patient with chronic dysimmune neuropathy under IVIg therapy. *Journal of Neuroimmunology, 181*, 141–144.

Mathey, E. K., Garg, N., Park, S. B., Nguyen, T., Baker, S., Yuki, N., Yiannikas, C., Lin, C. S., Spies, J. M., Ghaoui, R., Barnett, M. H., Vucic, S., et al. (2017). Autoantibody responses to nodal and paranodal antigens in chronic inflammatory neuropathies. *Journal of Neuroimmunology, 309*, 41–46.

Matsumuro, K., Izumo, S., Umehara, F., & Osame, M. (1994). Chronic inflammatory demyelinating polyneuropathy: Histological and immunopathological studies on biopsied sural nerves. *Journal of the Neurological Sciences, 127*, 170–178.

Miyashiro, A., Matsui, N., Shimatani, Y., Nodera, H., Izumi, Y., Kuwabara, S., Imai, T., Baba, M., Komori, T., Sonoo, M., Mezaki, T., Kawamata, J., et al. (2014). Are multifocal motor neuropathy patients underdiagnosed? An epidemiological survey in Japan. *Muscle & Nerve, 49*, 357–361.

Monaco, S., Bonetti, B., Ferrari, S., Moretto, G., Nardelli, E., Tedesco, F., Mollnes, T. E., Nobile-Orazio, E., Manfredini, E., & Bonazzi, L. (1990). Complement-mediated demyelination in patients with IgM monoclonal gammopathy and polyneuropathy. *The New England Journal of Medicine, 322*, 649–652.

Monaco, S., Ferrari, S., Bonetti, B., Moretto, G., Kirshfink, M., Nardelli, E., Nobile-Orazio, E., Zanusso, G., Rizzuto, N., & Tedesco, F. (1995). Experimental induction of myelin changes by anti-MAG antibodies and terminal complement complex. *Journal of Neuropathology and Experimental Neurology, 54*, 96–104.

Nobile-Orazio, E., Manfredini, E., Carpo, M., Meucci, N., Monaco, S., Ferrari, S., Bonetti, B., Cavaletti, G., Gemignani, F., & Durelli, L. (1994). Frequency and clinical correlates of anti-neural IgM antibodies in neuropathy associated with IgM monoclonal gammopathy. *Annals of Neurology, 36*, 416–424.

Nobile-Orazio, E., Meucci, N., Baldini, L., Di Troia, A., & Scarlato, G. (2000). Long-term prognosis of neuropathy associated with anti-MAG IgM M-proteins and its relationship to immune therapies. *Brain, 123*(Pt 4), 710–717.

Parisi, M., Dogliotti, I., Clerico, M., Bertuzzo, D., Benevolo, G., Orsucci, L., Schiavetti, I., Cavallo, R., Cavallo, F., Ragaini, S., Di Liberto, A., Ferrante, M., et al. (2022). Efficacy of rituximab in anti-myelin-associated glycoprotein demyelinating polyneuropathy: Clinical, hematological and neurophysiological correlations during 2 years of follow-up. *European Journal of Neurology, 29*, 3611–3622.

Park, S. B., Li, T., Kiernan, M. C., Garg, N., Wilson, I., White, R., Boggild, M., McNabb, A., Lee-Archer, M., & Taylor, B. V. (2022). Prevalence of chronic inflammatory demyelinating polyneuropathy and multifocal motor neuropathy in two regions of Australia. *Muscle & Nerve, 66*, 576–582.

Pitarokoili, K., Yoon, M.-S., Kröger, I., Reinacher-Schick, A., Gold, R., & Schneider-Gold, C. (2017). Severe refractory CIDP: A case series of 10 patients treated with bortezomib. *Journal of Neurology, 264*, 2010–2020.

Querol, L., Devaux, J., Rojas-Garcia, R., & Illa, I. (2017). Autoantibodies in chronic inflammatory neuropathies: Diagnostic and therapeutic implications. *Nature Reviews. Neurology, 13*, 533–547.

Rajabally, Y. A., Peric, S., Bozovic, I., Loo, L. K., Kalac, A., Palibrk, A., & Basta, I. (2021). Antecedent infections and vaccinations in chronic inflammatory demyelinating polyneuropathy: A European collaborative study. *Muscle & Nerve, 64*, 657–661.

Ramchandren, S., & Lewis, R. A. (2009). Monoclonal gammopathy and neuropathy. *Current Opinion in Neurology, 22*, 480–485.

Renaud, S., Gregor, M., Fuhr, P., Lorenz, D., Deuschl, G., Gratwohl, A., & Steck, A. J. (2003). Rituximab in the treatment of polyneuropathy associated with anti-MAG antibodies. *Muscle & Nerve, 27*, 611–615.

Ritter, C., Förster, D., Albrecht, P., Hartung, H.-P., Kieseier, B. C., & Lehmann, H. C. (2014). IVIG regulates BAFF expression in patients with chronic inflammatory demyelinating polyneuropathy (CIDP). *Journal of Neuroimmunology, 274*, 225–229.

Ritz, M. F., Erne, B., Ferracin, F., Vital, A., Vital, C., & Steck, A. J. (1999). Anti-MAG IgM penetration into myelinated fibers correlates with the extent of myelin widening. *Muscle & Nerve, 22*, 1030–1037.

Rodríguez, Y., Vatti, N., Ramírez-Santana, C., Chang, C., Mancera-Páez, O., Gershwin, M. E., & Anaya, J.-M. (2019). Chronic inflammatory demyelinating polyneuropathy as an autoimmune disease. *Journal of Autoimmunity, 102*, 8–37.

Rögnvaldsson, S., Steingrímsson, V., Turesson, I., Björkholm, M., Landgren, O., & Yngvi, K. S. (2020). Peripheral neuropathy and monoclonal gammopathy of undetermined significance: a population-based study including 15,351 cases and 58,619 matched controls. *Haematologica, 105*, 2679–2681.

Ruts, L., Drenthen, J., Jacobs, B. C., van Doorn, P. A., & Dutch GBS Study Group. (2010). Distinguishing acute-onset CIDP from fluctuating Guillain-Barre syndrome: a prospective study. *Neurology, 74*, 1680–1686.

Salomon, B., Rhee, L., Bour-Jordan, H., Hsin, H., Montag, A., Soliven, B., Arcella, J., Girvin, A. M., Padilla, J., Miller, S. D., & Bluestone, J. A. (2001). Development of spontaneous autoimmune peripheral polyneuropathy in B7-2-deficient NOD mice. *The Journal of Experimental Medicine, 194*, 677–684.

Sanvito, L., Makowska, A., Mahdi-Rogers, M., Hadden, R. D. M., Peakman, M., Gregson, N., Nemni, R., & Hughes, R. A. C. (2009). Humoral and cellular immune responses to myelin protein peptides in chronic inflammatory demyelinating polyradiculoneuropathy. *Journal of Neurology, Neurosurgery, and Psychiatry, 80*, 333–338.

Schmidt, B., Toyka, K. V., Kiefer, R., Full, J., Hartung, H. P., & Pollard, J. (1996). Inflammatory infiltrates in sural nerve biopsies in Guillain-Barre syndrome and chronic inflammatory demyelinating neuropathy. *Muscle & Nerve, 19*, 474–487.

Schneider-Hohendorf, T., Schwab, N., Uçeyler, N., Göbel, K., Sommer, C., & Wiendl, H. (2012). CD8+ T-cell immunity in chronic inflammatory demyelinating polyradiculoneuropathy. *Neurology, 78*, 402–408.

Shimizu, F., Omoto, M., Sano, Y., Mastui, N., Miyashiro, A., Tasaki, A., Maeda, T., Koga, M., Kaji, R., & Kanda, T. (2014). Sera from patients with multifocal motor neuropathy disrupt the blood-nerve barrier. *Journal of Neurology, Neurosurgery, and Psychiatry, 85*, 526–537.

Sommer, C., & Toyka, K. (2011). Nerve biopsy in chronic inflammatory neuropathies: In situ biomarkers. *Journal of the Peripheral Nervous System, 16*(Suppl 1), 24–29.

Sommer, C., Koch, S., Lammens, M., Gabreels-Festen, A., Stoll, G., & Toyka, K. V. (2005). Macrophage clustering as a diagnostic marker in sural nerve biopsies of patients with CIDP. *Neurology, 65*, 1924–1929.

Steck, A. J. (2021). Anti-MAG neuropathy: From biology to clinical management. *Journal of Neuroimmunology, 361*, 577725.

Steck, A. J., Murray, N., Justafre, J. C., Meier, C., Toyka, K. V., Heininger, K., & Stoll, G. (1985). Passive transfer studies in demyelinating neuropathy with IgM monoclonal antibodies to myelin-associated glycoprotein. *Journal of Neurology, Neurosurgery, and Psychiatry, 48*, 927–929.

Su, M. A., Davini, D., Cheng, P., Giang, K., Fan, U., DeVoss, J. J., Johannes, K. P. A., Taylor, L., Shum, A. K., Valenzise, M., Meloni, A., Bour-Jordan, H., et al. (2012). Defective autoimmune regulator-dependent central tolerance to myelin protein zero is linked to autoimmune peripheral neuropathy. *Journal of Immunology, 188*, 4906–4912.

Sutedja, N. A., Otten, H. G., Cats, E. A., Piepers, S., Veldink, J. H., van der Pol, W. L., & van den Berg, L. H. (2010). Increased frequency of HLA-DRB1*15 in patients with multifocal motor neuropathy. *Neurology, 74*, 828–832.

Svahn, J., Petiot, P., Antoine, J.-C., Vial, C., Delmont, E., Viala, K., Steck, A. J., Magot, A., Cauquil, C., Zarea, A., Echaniz-Laguna, A., Iancu Ferfoglia, R., et al. (2018). Anti-MAG antibodies in 202 patients: Clinicopathological and therapeutic features. *Journal of Neurology, Neurosurgery, and Psychiatry, 89*, 499–505.

Tackenberg, B., Lünemann, J. D., Steinbrecher, A., Rothenfusser-Korber, E., Sailer, M., Brück, W., Schock, S., Zschenderlein, R., Zipp, F., & Sommer, N. (2007). Classifications and treatment responses in chronic immune-mediated demyelinating polyneuropathy. *Neurology, 68*, 1622–1629.

Tang, M.-A. H., Mathis, S., Duffau, P., Cazenave, P., Solé, G., Duval, F., Soulages, A., & Le Masson, G. (2020). Prognostic factor of poor outcome in anti-MAG neuropathy: Clinical and electrophysiological analysis of a French cohort. *Journal of Neurology, 267*, 561–571.

Tatum, A. H. (1993). Experimental paraprotein neuropathy, demyelination by passive transfer of human IgM anti-myelin-associated glycoprotein. *Annals of Neurology, 33*, 502–506.

Taylor, B. V., Wright, R. A., Harper, C. M., & Dyck, P. J. (2000). Natural history of 46 patients with multifocal motor neuropathy with conduction block. *Muscle & Nerve, 23*, 900–908.

Terenghi, F., Cappellari, A., Bersano, A., Carpo, M., Barbieri, S., & Nobile-Orazio, E. (2004). How long is IVIg effective in multifocal motor neuropathy? *Neurology, 62*, 666–668.

Tracy, J. A., Taylor, B. V., Kiernan, M., Dyck, P. J., Crum, B. A., Mauermann, M. L., Amrami, K. K., Spinner, R. J., & Dyck, P. J. B. (2020). Nerve pathology distinguishes focal motor chronic inflammatory demyelinating Polyradiculoneuropathy from multifocal motor neuropathy. *Journal of Clinical Neuromuscular Disease, 22*, 1–10.

Van den Bergh, P. Y. K., van Doorn, P. A., Hadden, R. D. M., Avau, B., Vankrunkelsven, P., Allen, J. A., Attarian, S., Blomkwist-Markens, P. H., Cornblath, D. R., Eftimov, F., Goedee, H. S., Harbo, T., et al. (2021). European Academy of Neurology/Peripheral Nerve Society guideline on diagnosis and treatment of chronic inflammatory demyelinating polyradiculoneuropathy: Report of a joint Task Force-Second revision. *Journal of the Peripheral Nervous System, 26*, 242–268.

Vital, C., Vital, A., Deminiere, C., Julien, J., Lagueny, A., & Steck, A. J. (1997). Myelin modifications in 8 cases of peripheral neuropathy with Waldenström's macroglobulinemia and anti-MAG activity. *Ultrastructural Pathology, 21*, 509–516.

Vlam, L., Cats, E. A., Harschnitz, O., Jansen, M. D., Piepers, S., Veldink, J. H., Franssen, H., Stork, A. C. J., Heezius, E., Rooijakkers, S. H. M., Herpers, B. L., van Strijp, J. A., et al. (2015). Complement activity is associated with disease severity in multifocal motor neuropathy. *Neurology Neuroimmunology & Neuroinflammation, 2*, e119.

Vucic, S., Black, K. R., Chong, P. S. T., & Cros, D. (2004). Multifocal motor neuropathy: Decrease in conduction blocks and reinnervation with long-term IVIg. *Neurology, 63*, 1264–1269.

Willison, H. J., Trapp, B. D., Bacher, J. D., Dalakas, M. C., Griffin, J. W., & Quarles, R. H. (1988). Demyelination induced by intraneural injection of human antimyelin-associated glycoprotein antibodies. *Muscle & Nerve, 11*, 1169–1176.

Yan, W. X., Taylor, J., Andrias-Kauba, S., & Pollard, J. D. (2000). Passive transfer of demyelination by serum or IgG from chronic inflammatory demyelinating polyneuropathy patients. *Annals of Neurology, 47*, 765–775.

Yuki, N., Watanabe, H., Nakajima, T., & Späth, P. J. (2011). IVIG blocks complement deposition mediated by anti-GM1 antibodies in multifocal motor neuropathy. *Journal of Neurology, Neurosurgery, and Psychiatry, 82*, 87–91.

Zanwar, S., & Abeykoon, J. P. (2022). Treatment paradigm in Waldenström macroglobulinemia: Frontline therapy and beyond. *Therapeutic Advances in Hematology, 13*, 20406207221093960.

Neuroimmune Dysfunction Facilitates COVID-19-Associated Neurological Complications

25

Samuel D. Johnson, Jeremiah J. Hudson, Atul K. Verma, and Siddappa N. Byrareddy

Abstract

The novel coronavirus SARS-CoV-2 was discovered in 2019 and spread quickly across the globe. Although most people will recover from the resultant respiratory illness COVID-19, many others develop neurological complications or manifestations. The most common nervous system symptom is protracted anosmia, the loss of smell. However, a small percentage of individuals also experience acute cerebrovascular disease caused by disrupted blood flow to the central nervous system, resulting in severe damage and often death. Additionally, despite recovery from acute disease, many people develop post-acute sequelae of COVID-19 (PASC), which can last for several months to years following initial recovery of respiratory symptoms. In this chapter, the mechanisms of each of these neurological manifestations of SARS-CoV-2 are reviewed, and emerging opportunities are highlighted for prevention and therapeutic intervention.

Keywords

SARS-CoV-2 · COVID-19 · Anosmia · Cerebrovascular disease · Blood-brain barrier · Post-acute sequelae of COVID-19 (PASC) · Long COVID

25.1 Introduction

Several infectious diseases of public health concern have emerged or re-emerged in the past decade. However, the most consequential has been the SARS-CoV-2 pandemic, which significantly disrupted the economy and continues to cause severe morbidity and mortality across the planet, even with widespread vaccination implementation. Despite considerable debate, the origins of SARS-CoV-2 remain murky. First reported in Wuhan, China, in December 2019, early descriptions documented a novel viral pneumonia of unknown etiology (Huang et al., 2020). Subsequently, the SARS-CoV-2 genome was sequenced and named after the first severe acute respiratory syndrome coronavirus (SARS-CoV), a betacoronavirus described in 2003, with which SARS-CoV-2 shares significant homology. It is now thought to have transmitted to humans from an intermediate host after originally circulating in bats, a process known as spillover. Following this early description and similarity to the 2003 outbreak, Wuhan Huanan Seafood Wholesale Market, a wet market near the Wuhan Center for Disease Control, was permanently closed on January 1, 2020, based on a suspected similar spillover event. More recently, SARS-CoV-2 has been detected in samples collected following this shutdown, indicating that this scenario remains possible. However, despite this early containment effort by Chinese authorities, SARS-CoV-2 was seen in Washington (state), USA, in January 2020 (Holshue et al., 2020) and continued its rapid spread internationally. On January 30, 2020, the World Health Organization (WHO) declared SARS-CoV-2 a Public Health Emergency of International Concern ("WHO Director-General's opening remarks at the media briefing on COVID-19," 2020). Additional efforts were made at further containment without success. SARS-CoV-2 was officially declared a pandemic on March 11, 2020, by the WHO ("WHO Director-General's opening remarks at the media briefing on COVID-19," 2020). Current estimates from the WHO place the cumulative confirmed infections at over 700 million and cumulative confirmed deaths at over six million since the pandemic's start (WHO COVID-19 Dashboard, 2023). However, the numbers are undoubtedly higher due to low testing rates.

As was predicted early in the pandemic, SARS-CoV-2 genetically diversified rapidly, as other viral infections have been well documented to do (Korber et al., 2020). During such diversification, well-adapted variants emerge and

S. D. Johnson · J. J. Hudson · A. K. Verma · S. N. Byrareddy (✉)
Department of Pharmacology and Experimental Neuroscience, University of Nebraska Medical Center, Omaha, NE, USA
e-mail: sid.byrareddy@unmc.edu

H. E. Gendelman, T. Ikezu (eds.), *Neuroimmune Pharmacology and Therapeutics*,
https://doi.org/10.1007/978-3-031-68237-7_25

spread quickly, often out-competing their predecessor variants and overcoming adaptive immune responses to previously encountered viruses (or vaccine antigens). The WHO classifies SARS-CoV-2 Variants as Variants of Concern (VOCs) and Variants of Interest (VOIs) (Table 25.1).

Mutations to the receptor-binding domain can increase the virus's ability to enter the cell and avoid detection from neutralizing antibodies. Omicron differs from the original Wuhan-Hu-1 through mutations in its receptor-binding domain. Usually, the receptor binding domain has regions that are highly conserved, and are the target of monoclonal antibodies. Since Omicron and its subvariants have changes to these usually highly conserved regions, they possess an unusual ability to avoid being targeted by monoclonal antibodies. (https://www.ncbi.nlm.nih.gov/pmc/articles/PMC9349777/) The subvariant BA.2.75 is similar to BA.5 but exhibits an increased ability to avoid neutralizing antibodies due to additional mutations in the receptor-binding domain (https://www.ncbi.nlm.nih.gov/pmc/articles/PMC9560757/). BA.1, BA.1.1, BA.2, and BA.3 all share a single point mutation at the residue N440K in the RBD, increasing their affinity for ACE2, the receptor that SARS-CoV-2 binds to in order to enter the host cell. This means that Omicron can more easily infect than prior iterations of the virus. (https://www.ncbi.nlm.nih.gov/pmc/articles/PMC9347785/).

Of these VOIs, the WHO has further acknowledged specific variants that are exceptionally infectious or pathogenic, defining VOCs as a SARS-CoV-2 variant that, when compared with other variants, has a detrimental change in clinical disease severity, a shift in epidemiology causing a substantial impact on the ability of health systems to care for patients or a significant decrease in the effectiveness of vaccines in protecting against the disease. This designation has been assigned to Alpha, Beta, Gamma, Delta, and Omicron. Further, since the emergence and subsequent predominance of the Omicron variant, a new naming convention has been developed to differentiate the emergence of new strains. The most important of these more recent subvariants have been Omicron sublineages BA.1, BA 1.1, and BA.2 (WHO). Currently, it is well known that the two subvariants of Omicron, BA.1 and BA.2 were known to evade antibodies generated during natural infection or after vaccination (Iketani et al., 2022; Pather et al., 2023). Similar phenomena were found with BA.3, BA.4, and BA.5 subvariants (Cao et al., 2022). Other sublineages, namely, BA.2.75 and BA.2.75.2, also emerged with greater virulence, transmissibility, and some extent of antibody escape (Gruell et al., 2022). Mutations in the receptor binding domain (RBD) of the spike protein of Omicron and its variants are thought to facilitate this evasion, thereby overcoming both naturally

Table 25.1 Emergence of strains of SARS-CoV-2. The WHO has classified the different strains of SARS-CoV-2 as Variants of Interest (VOIs) or Variants of Concern (VOCs). VOCs are a more severe classification of variants than VOIs

Variant	Location	Date	VOI or VOC
Zeta	Brazil[a]	April 2020	VOI
Epsilon	United States[a]	June 2020	VOI
Eta/iota	New York[a]	November 2020	VOI
Alpha	United Kingdom[a]	December 2020	VOC
Beta	South Africa[a]	December 2020	VOC
Delta	India[a]	December 2020	VOC
Gamma	Brazil[a]	January 2021	VOC
Theta	Japan[a]	February 2021	VOI
Lambda	Peru[a]	June 2021	VOI
Mu	Colombia[a]	August 2021	VOI
Omicron	South Africa[a]	November 2021	VOC
Omicron B.1.1.529	South Africa[b,c]	November 2021	VOC[d,e]
Kappa	India[a]	December 2021	VOI
Omicron BA.4	South Africa[b]	January 2022	VOC-LUM
Omicron BA.2.75	India[f]	January 2022	VOI
Omicron BA.1	Botswana[b,g]	February 2022	VOC[e]
Omicron BA.2	South Africa[b,g]	February 2022	VOC[e]
Omicron BA.5	South Africa[b]	February 2022	VOC-LUM[e]
Omicron BA.3	South Africa[b,g]	May 2022	None[e]
Omicron BA.2.75.2	India[f]	December 2022	VOI

[a]https://pubmed.ncbi.nlm.nih.gov/34033342/
[b]https://www.ncbi.nlm.nih.gov/pmc/articles/PMC9866114/
[c]https://www.ncbi.nlm.nih.gov/pmc/articles/PMC8675659/
[d]https://www.ncbi.nlm.nih.gov/pmc/articles/PMC9347785/
[e]https://www.ncbi.nlm.nih.gov/pmc/articles/PMC9254723/
[f]https://www.ncbi.nlm.nih.gov/pmc/articles/PMC9671875/
[g]https://www.ncbi.nlm.nih.gov/pmc/articles/PMC9015590/

occurring antibodies and infusions of exogenous monoclonal antibodies (Shrestha et al., 2022). This is why sotrovimab had its emergency use authorization revoked, thus limiting the effective therapeutics for said variants and highlighting their clinical impacts. Additionally, BA.1, BA.1.1, BA.2, and BA.3 share a single point mutation at the residue N440K in the RBD, increasing their affinity for angiotensin-converting enzyme-2 (ACE2), the receptor for SARS-CoV-2, thus facilitating more efficient viral entry into the host cell (Kumar et al., 2022). Subsequent omicron variants such as BA2.75 and BA.5 have further RBD mutations (Sheward et al., 2022). Together, these variations introduced by the evolution of SARS-CoV-2 contribute to ongoing changes to the infection's transmissibility, treatability, and pathogenicity.

SARS-CoV-2 primarily spreads through the respiratory droplets of an infected person. The most frequent manifestations of COVID-19, the disease caused by SARS-CoV-2 infection, include fever, fatigue, and dry cough (Baj et al., 2020; Zaim et al., 2020). However, many patients' condition deteriorates into complications like ARDS (acute respiratory distress syndrome), arrhythmia, and shock. Another study showed typical respiratory infection symptoms like fever, cough, and shortness of breath among the significant symptoms. At the same time, some had muscle aches, confusion, sore throat, headache, rhinorrhea, diarrhea, chest pain, nausea, and vomiting (Baj et al., 2020; Zaim et al., 2020). Notably, headache, confusion, and other symptoms, including dizziness, numbness, and lethargy, suggest that SARS-CoV-2 infections may impact the brain. While less severe CNS symptoms can be expected in other respiratory infections, early reports indicated that SARS-CoV-2 may affect the brain more severely. Clinical signs in Wuhan first reported initial neurological complications due to the SARS-CoV-2 (Mao et al., 2020). Neurological symptoms were acute cerebrovascular diseases, impaired consciousness, and skeletal muscle injury (Mao et al., 2020). Convulsions accompanied by unconsciousness in Japan were also described as the first case of COVID-19-associated encephalitis (Moriguchi et al., 2020). Viral RNA was detected in cerebrospinal fluid though it was negative for nasopharyngeal swabs (Moriguchi et al., 2020). Radiological analysis through MRI showed abnormalities in the medial temporal lobe and hippocampus (Moriguchi et al., 2020). Similar reports of encephalopathy emerged from China and USA, which was later diagnosed with COVID-19 (Kihira et al., 2020; Poyiadji et al., 2020). The earliest report of brain magnetic resonance imaging (MRI) scans during COVID-19 encephalopathy showed lesions in the subinsular regions and the thalamus and medial temporal lobes (Poyiadji et al., 2020). Although the neurological disorders caused by SARS-CoV-2 infection are diverse, this chapter will focus on three common examples unique to COVID-19: anosmia, acute cerebrovascular disease, and post-acute sequelae of COVID-19 (Fig. 25.1).

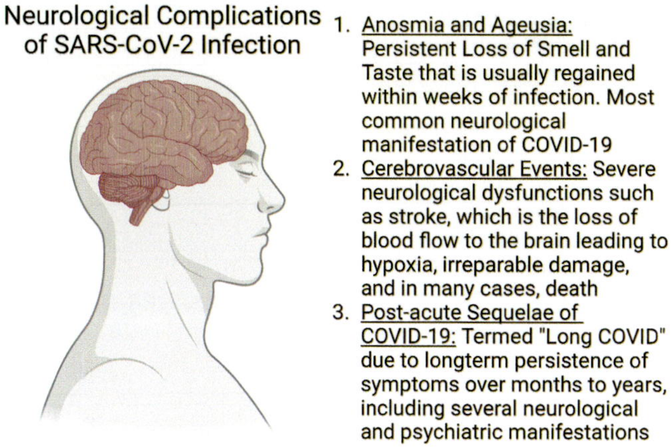

Fig. 25.1 Neurologic complications of SARS-CoV-2

25.2 Anosmia and Dysgeusia

Although endemic human coronaviruses have many similar symptomologies to SARS-CoV-2, two significant nervous system symptoms seemed to set SARS-CoV-2 apart, the severity of infection aside. These symptoms are anosmia (loss of smell) or hyposmia (impaired smell) and ageusia (loss of taste) or dysgeusia (impaired taste). Very early reports in the pandemic suggested that the incidence of olfactory disturbance could range from 5% to 98%. A later 2020 study of COVID-19 patients identified between March 22 and June 3 from 18 European hospitals determined that milder forms of COVID-19 were associated with higher rates (85.9%) of subjective olfactory dysfunction compared with moderate-to-critical conditions (4.5–6.9%) (Lechien et al., 2021). This finding was validated using objective measures of dysfunction (determined by psychophysical olfactory evaluations with smell pens) (Lechien et al., 2021).

Additionally, a meta-analysis determined a higher incidence of objective impairment across studies (77%) compared with subjective reporting (44%), suggesting that self-reported incidence of anosmia and hyposmia may underestimate the actual impact on olfactory function (Hannum et al., 2020). The high prevalence of these disorders contributes significantly to the decreased quality of life caused by these infections when subjectively reported (Khan et al., 2022a). However, more far-reaching public health consequences have been noted regardless of subjective quality of life, including the inability to smell dangerous gases or smoke, impaired food intake, increased depressive symptom incidence, changes in hygiene practices (including anxiety behaviors associated with failure to gauge hygiene), and loss of other sociocultural indicators mediated by smell (Croy et al., 2014; Khan et al., 2022a).

During upper respiratory infections such as those caused by endemic human coronaviruses, rhinorrhea, and nasal congestion can insulate olfactory sensory neurons (OSNs) in the

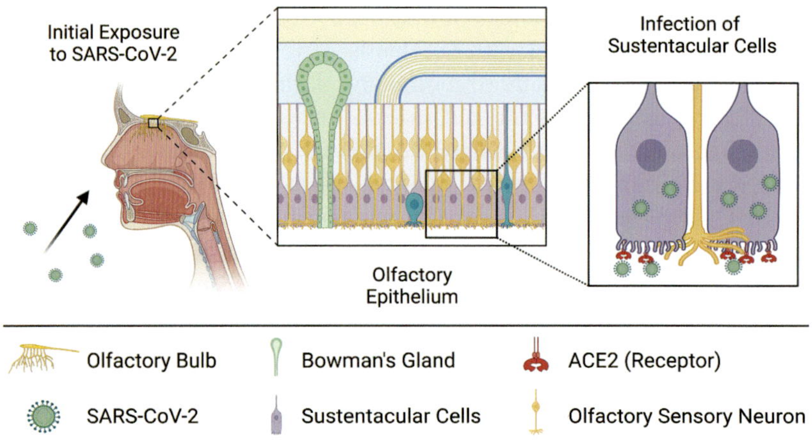

Fig. 25.2 SARS-CoV-2 infection at the respiratory epithelium. Initial SARS-CoV-2 exposure begins at the respiratory epithelium where, in addition to infecting respiratory cells, SARS-CoV-2 also infects sustentacular cells. Once infected, sustentacular cells are lost and contribute to olfactory sensory neuron dysfunction, thus leading to anosmia, one of the most common symptoms of COVID-19

olfactory epithelium (OE), causing masking of odorants from binding to their receptors, thus causing temporary anosmia or hyposmia by physical barrier interference. In contrast, SARS-CoV-2 infections are independent of such interference, suggesting that alternative mechanisms must be causing the impaired smell. This is highlighted by anosmia, which is frequently the only symptom experienced during mild infections. Adding to confusion on the role of OSNs in anosmia were early reports that OSNs, like other neurons examined across diverse anatomical niches, lacked important receptors for SARS-CoV-2 infection, including ACE2. In the elucidation process, an early report characterized the expression of ACE2 on several cell types in the olfactory mucosa, islands of specialized neuroepithelium embedded in the nasal cavity surrounded by respiratory epithelium (Mengfei et al., 2020). The apical surface of the neuroepithelium is composed primarily of sustentacular (SUS) cells which support the neuronal dendritic projections, which themselves bear nonmotile cilia on the end of their dendritic knobs that express odorant receptors (Mengfei et al., 2020). The axons of the bipolar OSNs congregate to form the olfactory nerve (cranial nerve I) that passes through the cribriform plate of the skull and terminates to form synapses with the glomeruli of the olfactory bulb of the brain. Although the PGP9.5+ OSNs did not express ACE2, the supporting keratin 18+ SUS cells expressed most of the ACE2 detected at the olfactory epithelium (Mengfei et al., 2020). Importantly, ACE2 was also not detected on doublecortin+ cells either, suggesting that SARS-CoV-2 was not directly interfering with neuron maturation by infecting immature OSNs (Mengfei et al., 2020). Single-cell RNA sequencing later determined that ACE2 and transmembrane protease, serine 2 (TMPRSS2) genes are expressed primarily in SUS cells but not in OSNs, further implicating SUS cell infection as the primary target of SARS-CoV-2 infections (Brann, 2020) (Fig. 25.2).

While ACE2 and TMPRSS2 expression would suggest SUS cells as a likely target of SARS-CoV-2, direct evidence of infection was lacking until more recently. One major obstacle preventing the acquisition of such data is the rapid degradation of samples following death. Additionally, biopsy samples from living patients are untenable due to the olfactory mucosa's proximity to the cribriform plate. To address this problem, a research team developed a rapid postmortem bedside surgical procedure to harvest endoscopic samples of (1) the respiratory mucosa, (2) the olfactory mucosa, and (3) the olfactory bulb. Unsurprisingly, the ciliated cells in the respiratory mucosa were infected with SARS-CoV-2. More pertinent, though, was the finding that consistent with previous evidence that they express entry receptors, SUS cells, but not OSNs, do get infected by SARS-CoV-2. SUS cells are nonneuronal cells with glia-like functions, including physical and physiological support. The latter roles include absorptive, detoxifying, metabolic, nourishing, phagocytic, and secretory functions, although complete elucidation of their relationship with OSNs requires further investigation (Acevedo et al., 2019; Khan et al., 2021; Suzuki et al., 1996; Villar et al., 2017). It has been posited that when these glial functions are lost during SUS cell infection, the lost support results in OSN dysfunction that is only regained after several days following SUS cell regeneration. Thus, the OSNs are significantly affected without technically being infected. This proposal remains controversial, however, because other human coronaviruses, including SARS-CoV and HCoV-NL63, use ACE2 as an entry receptor but do not induce anosmia during infection (Khan et al., 2021; Žugaj et al., 2021). Additionally, SUS cells are phagocytic, so positive staining may be more reflective of the clearance of virion debris than active infection (Khan et al., 2021; Suzuki et al., 1996). However, sense puncta in 30% of postmortem samples analyzed suggests ongoing replication at the time of death (Khan et al., 2021).

Since the establishment of SARS-CoV-2 replication in SUS cells, additional evidence has emerged for the role such infections have on olfaction. Using golden hamsters (*Mesocricetus auratus*) as a model, significant loss of SUS cells occurs on day 1 postinfection that continues through day 3. This loss was associated with a significant increase in microglia and other immune cells infiltrating the olfactory epithelium. Despite the loss of SUS cells and an increase in immune cells, the relative number of OSNs remained constant across the infection. SUS cells and microglia returned to baseline levels by day 10, when viral infection was no longer detected (Zazhytska et al., 2022). On day 1, 47% of infected cells were identified as SUS, compared with only 6% of OSNs. By day 3, microglia and other immune cells contain SARS-CoV-2 transcripts (Zazhytska et al., 2022). In addition to changes in mucosal architecture, bulk RNA-seq determined significant downregulation in olfactory receptor genes, providing further insight into the mechanism of dysfunction (Zazhytska et al., 2022). These downregulated olfactory receptor gene expression findings were later validated in autopsy samples from humans infected with SARS-CoV-2. Additionally, SARS-CoV-2 was found to infect human microglia, and infection was associated with elevated gene expression of several cytokines and antiviral genes, such as interferon-gamma (IFN-γ) (Zazhytska et al., 2022). Together, these data suggest that infection and depletion of SUS are primary determinants of anosmia during SARS-CoV-2 pathogenesis.

Although some reports have suggested insulation of the olfactory bulb from infection, SARS-CoV-2 has been detected frequently at the leptomeningeal layers (pia and arachnoid mater) protecting the parenchyma (Khan et al., 2021). The authors speculate that virions may arrive at the leptomeninges through the flow of cerebrospinal fluid through the subarachnoid space instead of migrating along the OSN intracellularly (Khan et al., 2021). Like SUS cell infection and dysfunction, resultant meningitis could, in theory, further contribute to anosmia. However, evidence for this is currently lacking. Similarly, other pro-inflammatory responses, such as cytokine release, could contribute to neuronal dysfunction at the neuroepithelium or the olfactory bulb.

Significant progress has been made on the mechanism of anosmia, but less headway has been made in understanding ageusia. Used broadly, ageusia or dysgeusia are necessarily impacted by anosmia due to the functional integration of odor and taste in the perception of flavor (Small & Prescott, 2005). However, more technically, ageusia and dysgeusia typically refer to the tongue's specific functions, including sensations of sweet, sour, salty, bitter, and umami. Few studies have indicated these as the specific impaired sensory phenomenon. However, one study reported the incidence as high as 55.9%, compared with 83.9% aroma dysfunction in the same cohort (Lechien et al., 2021). Without clear definitions, the epidemiological quantification of ageusia and hypogeusia remains poorly characterized, and mechanisms are still unclear as many sources treat these disorders as a continuum, with ageusia/dysgeusia as a subset or more severe manifestation of anosmia, likely due to the abovementioned integration of odor and taste. Thus, more work is required to fully elucidate the mechanisms by which taste is disrupted during COVID-19.

25.3 COVID-19 and Acute Cerebrovascular Disease

Due partially to its typically temporary nature, the loss of smell has generally been characterized as a less severe neurological manifestation of COVID-19. However, as previously mentioned, SARS-CoV-2 infections have also been associated with more significant neurological sequelae. Although active infection of SUS cells can occur as soon as virions are inhaled, additional infection occurs in other respiratory cells, most notably alveolar epithelial cells in the lungs. Once an individual has active replication, the virus is transported through the blood, where both the presence of the virus and the host response can significantly contribute to major neurologic events, regardless of whether the virus can enter the brain. A major meta-analysis in 2020 (including studies from March 31 to June 25, 2020) characterizing the prevalence of several neurological manifestations early during the pandemic found approximately 2.3% of COVID-19 patients experienced acute cerebrovascular disease (Favas et al., 2020). The most commonly reported were ischemic stroke in 2.1% of patients, hemorrhagic stroke in 0.4%, and cerebral venous thrombosis in 0.3% of patients. Ischemic stroke refers to disrupting blood flow to the CNS, while hemorrhagic stroke refers to brain bleeding (Favas et al., 2020). The last of these, cerebral venous thrombosis, refers specifically to the formation of a blood clot in the brain's venous sinuses that, like other causes of ischemia, blocks the blood flow of the brain. A similar meta-analysis (until August 7) found 1.3% risk of cerebrovascular event, 1.1% risk of ischemic stroke, and 0.2% hemorrhagic stroke. These statistics were compared with contemporary and historical noninfected controls. The odds ratio was 3.58 for ischemic stroke, which partially contributed to a fivefold increase in mortality in SARS-CoV-2-infected patients suffering from cerebrovascular events (Katsanos et al., 2021). It has been hypothesized that a third primary type of stroke caused by atherosclerosis is also elevated in people infected by SARS-CoV-2. Like thrombosis, atherosclerosis can block blood flow to the brain, causing ischemia. It has been proposed that the significant increase in cytokines and intraplaque inflammation may contribute to the destabilization of atherosclerotic plaques in high-risk individuals, resulting in plaque rupture at the fibrous cap and an increase in hypercoagulability further contributing to increased risk of local thrombosis (Aghayari Sheikh Neshin et al., 2021).

To better determine the mechanism by which SARS-CoV-2 increases stroke risk, a more recent study spanning the beginning of the pandemic until December 31, 2020, found a similar proportion of ischemic stroke (26 of 30 cases) to hemorrhagic stroke (4 of 30 points) as earlier meta-analyses (Mbonde et al., 2022). Importantly, though, and despite its lower number of patients, the authors provided additional information about the mechanism of ischemic stroke. They found that 35% of ischemic stroke patients suffered cardioembolism, and 35% were considered cryptogenic (unknown etiology). Among hemorrhagic stroke patients, 75% were lobar (Mbonde et al., 2022). Surprisingly, over half of the patients were asymptomatic for SARS-CoV-2. Traditional risk factors such as hypertension (80%), hyperlipidemia (60%), history of smoking (43%), and diabetes (37%) were all common among stroke patients [additional risk factors available in (Mbonde et al., 2022)]. Stroke patients in this study had elevated C-reactive protein (CRP), interleukin (IL)-6, D-dimer, fibrinogen, and ferritin, all risk or prognostic factors associated with stroke. This was consistent with the earliest reports from Wuhan that noted elevated D-dimer and fibrin degradation product levels associated with worse prognosis in COVID-19 patients (Tang et al., 2020; Wang et al., 2020). Such a hypercoagulable state was also found in another early Chinese study with COVID-19 patients specifically experiencing acute cerebrovascular disease having elevated CRP and D-dimer (Li et al., 2020; Wang et al., 2020).

After exposure to pro-inflammatory stimuli such as SARS-CoV-2 or other pathogens, T lymphocytes and myeloid cells begin secreting IL-6 (and other pro-inflammatory cytokines). IL-6 increases the liver's synthesis of CRP and fibrinogen as part of the acute phase response (Asakura & Ogawa, 2021; Pepys & Hirschfield, 2003). CRP is critical for sensitizing macrophages to pathogens by activating the complement system, facilitating increased phagocytosis and clearance of infection (Thompson et al., 1999). In tandem, the fibrinogen produced by the liver can be catalyzed by thrombin into fibrin, which allows its polymerization to form clots to plug wounds (Mosesson, 2005). This offers an adaptive response to injury by stopping bleeding. However, overactive clotting during a hypercoagulable state can contribute to thrombosis, blocking blood flow and resulting in ischemia (Adam et al., 2022). In contrast, D-dimer measures small protein fragments generated by fibrin degradation during fibrinolysis following blood clotting, such as thrombosis (Asakura & Ogawa, 2021; Hunt & Levi, 2020). D-dimer is thus a post hoc measure of clotting that has already begun and has subsequently been degraded. However, this measure does not rule out ongoing clotting; it only measures clotting that has already occurred. The upregulation of several components of the clotting response due to SARS-CoV-2 infection suggests it plays a critical role in ischemic stroke risk during infection. Compared to ischemic stroke, the elevated risk of hemorrhagic stroke may be relatively simple and SARS-CoV-2-specific. ACE2 expressed in endothelial cells is a fundamental component of the renin–angiotensin system that regulates blood pressure, electrolyte concentration, and vascular resistance (Santos et al., 2019; Wang et al., 2020). The downregulation of ACE2 during SARS-CoV-2 could increase angiotensin II, a regulator of vasoconstriction, in the serum. It has been hypothesized that the resultant increased blood pressure would thus increase the risk of blood vessels bursting (Wang et al., 2020) (Fig. 25.3).

Regardless of stroke type, the loss of blood flow means loss of oxygen and nutrients, and subsequent damage and neuronal death should flow not be adequately restored. A large North American cohort study following ischemic stroke patients (n = 216) sought to determine their long-term prognosis both objectively and in comparison to other (non-SARS-CoV-2-infected) historical ischemic stroke cohorts (Adam et al., 2022). In this study, the observed mortality rate was 39.1%, a considerably higher level than the 4.8% mortality found in the Get With the Guidelines (GWTG)-Stroke Acute Ischemic Stroke and COVID-19 Registry data for

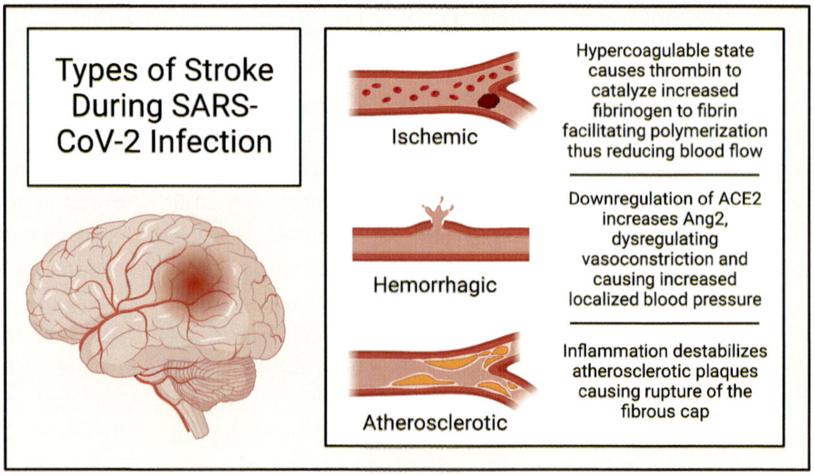

Fig. 25.3 Types of stroke during SARS-CoV-2 infection. Three main types of stroke (ischemic, hemorrhagic, and atherosclerotic) have increased incidence, contributing to a significantly elevated risk of acute cerebrovascular events during COVID-19

noninfected patients (Adam et al., 2022). Additionally, COVID-19 patients had worse outcomes at discharge, with 51.3% of patients scoring 5 or 6 on the modified Rankin Scale (mRS), a standardized measure of disability following stroke, to assess recovery. Important determinants of the outcome included age (≥60 years old), diabetes mellitus, higher NIH Stroke Scale score at presentation, large vessel occlusion, and blood neutrophil-to-lymphocyte ratio. Interestingly, it was found that D-dimer was not associated with subsequent outcomes following stroke. Instead, the blood NLR (described as a measure of inflammation) was predictive of outcome, with lower ratios associated with reduced mortality and lower mRS scores (Adam et al., 2022). Further, NLR was not associated with D-dimer levels, suggesting that following initial stroke onset, D-dimer and associated hypercoagulability were not predictive of subsequent outcomes (Adam et al., 2022). Together, these data demonstrate the long-lasting impact of elevated ischemic stroke incidence in COVID-19 patients.

Notably, all the above data were generated before the introduction and widespread implementation of large-scale vaccine administration. A large 2022 retrospective cohort study using Korean databases found a reduced risk of ischemic stroke after COVID-19 diagnosis (adjusted hazard ratio 0.40) following a two-dose vaccination compared with unvaccinated patients (Kim et al., 2022). A more recent meta-analysis, including 16 previous studies, found that following vaccination, people vaccinated against SARS-CoV-2 had no greater risk of ischemic stroke than the general population but had a significantly lower risk than SARS-CoV-2 patients (Maria-Ioanna et al., 2022). Together, these data suggest a benefit to vaccination in preventing severe neurological outcomes from SARS-CoV-2 infections.

25.4 Additional Disruptions to the Blood-Brain Barrier (BBB)

The BBB primarily comprises endothelial cells associated with pericytes and astrocyte end feet. In addition to this primary architecture, several immune cell types, such as perivascular macrophages, further protect the CNS from invasion and maintain homeostasis (Wu et al., 2021). The endothelium of the BBB, like other endothelium and the aforementioned SUS cells, express ACE2 and TMPRSS2, making them prime targets for SARS-CoV-2 infection (Wan et al., 2021). Under homeostatic conditions, tight junctions of claudin, occludin, and ZO-1 prevent crossing toxins, pathogens, and certain immune cells that could contribute to CNS damage (Wu et al., 2021). Aside from oxygen, water, and small lipid-soluble molecules, all other molecules, including glucose, require active transport (Abbott et al., 2010). Therefore, in the absence of stroke, the ability of SARS-CoV-2 to get into the brain has remained a puzzle for researchers. Importantly, in contrast with other anatomical sites, the endothelium of the BBB expresses low levels of leukocyte adhesion molecules (LAMs) (Henninger et al., 1997; Rössler et al., 1992). This suppression of LAMs requires functioning pericytes, a cell type also disrupted during SARS-CoV-2 infection (Choe et al., 2022). Thus, the ability of SARS-CoV-2 to replicate in the endothelium comprising the blood-brain barrier may contribute to both significant damage and weakening of the BBB (increasing the probability of stroke) and facilitate it crossing from the bloodstream into the CNS, where it can further interfere with brain homeostasis.

Early reports for SARS-CoV-2 suggested limited endothelial infection (Schimmel et al., 2021). However, some little evidence in both animal and in vitro BBB models (including brain microvascular endothelial cells and astrocytes), suggested that brain microvascular endothelial cells can occasionally be infected by SARS-CoV-2. However, tight junction integrity in these models, including claudin5, ZO-1, and occludin, remained unchanged. Notably, the BBB model was missing pericytes (Zhang et al., 2021). However, when other researchers added pericytes to the BBB model, similar findings were observed with little direct infection of human brain microvascular endothelial cells. In this model, human pericytes were only transiently infected, further complicating current understandings of BBB dysfunction (Constant et al., 2021). However, more recent data examining gene transcription and subsequent protein expression have found BBB dysfunction due to decreases in tight junction proteins (Yang et al., 2022).

Although current understandings of endotheliopathy during COVID-19 are conflicting, emerging evidence suggests that, at least in the lungs of COVID-19 patients, senescent endothelial cells are more susceptible to SARS-CoV-2 (Urata et al., 2022). Such a finding may indicate that models need to pay better attention to the populations most affected by SARS-CoV-2 infection since in vitro endothelial cells are, by nature, extremely young compared to the cells in elderly patients in vivo. Similarly, it has been confirmed that aged mice are significantly more susceptible to SARS-CoV-2 than young mice and are more susceptible to microthrombi and barrier dysfunction (Camell et al., 2021; Urata et al., 2022). Further, in addition to increased infectivity, endothelial senescence is associated with increased expression of LAMs such as ICAM-1 and VCAM-1 when exposed to SARS-CoV-2 spike protein (Meyer et al., 2021). Notably, VCAM-1 is the ligand of $\alpha4\beta1$ integrin, the target of natalizumab used to reduce lymphocyte trafficking to the CNS during multiple sclerosis (Polman et al., 2006). Thus, SARS-CoV-2 increases leukocyte adhesion to endothelium and likely contributes to CNS dysfunction (Meyer et al., 2021). Pro-inflammatory mediators released from macrophages following spike protein exposure further exacerbate this upregulation of ICAM-1 and VCAM-1. Although this finding was demonstrated in ex vivo monocyte-derived macrophages, perivascular macrophages closely associated with endothelial cells in vivo, unlike CNS resident microglia, are thought to be derived from monocytes suggesting increased relevance to endothelium function and LAM expression in vivo (Rotoli et al., 2021).

Despite ongoing debate over endothelium infectivity and dysfunction, organoid modeling provides evidence that supports SARS-CoV-2 replication in closely associated pericytes, the support cells of the endothelium. One model derived pericyte-like cells expressing classic pericyte markers neural/glial antigen 2 and platelet-derived growth factor receptor beta from neural crest stem cells. These cells expressed ACE2 and were permissive to productive SARS-CoV-2 infection (Wang et al., 2021). When these cells were added to cortical brain organoids, pericyte-like cells facilitated the maturation (thus improving the previous model), development of endfeet, and subsequent infection of astrocytes but not neurons (Wang et al., 2021). Together, this improved model was able to be productively infected while the cortical brain organoids without pericyte-like cells could not (Wang et al., 2021). A different model for endotheliitis found that pericyte infection led to detachment, damage, and cell death, thus disrupting the pericyte–endothelial cell crosstalk, and allowing increased microvascular permeability (Khan et al., 2022b). These data suggest that pericytes may be infected during COVID-19, thereby contributing to vascular damage, including BBB dysfunction. However, further research is still required.

25.5 SARS-CoV-2 Infiltration into the CNS

Between evidence of olfactory mucosa infection, a niche closely associated with the olfactory bulb, and evidence of extensive disruption to the BBB, the question of how SARS-CoV-2 enters the brain has remained incredibly controversial. One of the earliest reports tracking SARS-CoV-2 infiltration utilizing postmortem samples from COVID-19 patients in Germany found significant viral infiltration in most patients with exceptionally high levels in the frontal lobe and medulla oblongata, as determined by immunohistochemical staining (Matschke et al., 2020). Using qPCR, 48% of patients had positive results for at least one region analyzed, further validating their findings. Further staining found astrogliosis in 86% of patients. Microgliosis and cytotoxic T lymphocyte infiltration were most pronounced in the perivascular regions of the medulla oblongata and cerebellum (Matschke et al., 2020). Notably, the medulla oblongata is the part of the brainstem from which the cranial nerves (CNs) IX and X (glossopharyngeal and vagus nerves) project. CN X innervates the lungs, heart, and gut, locations of significant viral replication. Presumably with this in mind, the researchers also found viral proteins in the CNs. The combined findings of high virus levels in the medulla oblongata and its presence in the associated cranial nerves suggest these as a possible alternative route of CNS infiltration. A subsequent study validated this finding, further implicating CN IX and X associated with the medulla oblongata, and not CN I associated with the olfactory bulb, as a possible mechanism of SARS-CoV-2 CNS entry (Meinhardt et al., 2021). Notably, ACE-2, neuropilin-1, and TMPRSS2 are expressed in the Schwann cells, axons, vascular endothelium, and connective tissue of CNs IX and X near the medulla suggesting susceptibility to infection (Vitale-Cross et al., 2022). It has also been hypothesized that the often-overlooked nervous terminalis (alternatively called cranial nerve 0 or XIII), which, unlike the OSNs and olfactory nerve, expresses ACE2, may serve as an additional mechanism of viral delivery to the CNS near both the olfactory bulb and hypothalamus (Butowt & von Bartheld, 2022). Although still inconclusive, cranial nerves offer an alternative direct route from the periphery to the CNS.

While the mechanism(s) by which SARS-CoV-2 enters the CNS remains controversial, the impact may be long-lasting. Recently, extensive autopsy sampling was performed on the CNS of 11 patients who died of COVID-19 (Stein et al., 2022). Viral RNA was found in 10/11 patients analyzed, with 5/6 patients having CNS viral load after over 31 days of illness duration. In one patient who died after 230 days of illness, the virus was found in all the CNS samples analyzed (cervical spinal cord, cerebral cortex, brainstem, cerebellum, thalamus, hypothalamus, and corpus callosum) but not in CN I or the dura mater (Stein et al., 2022) (Fig. 25.4).

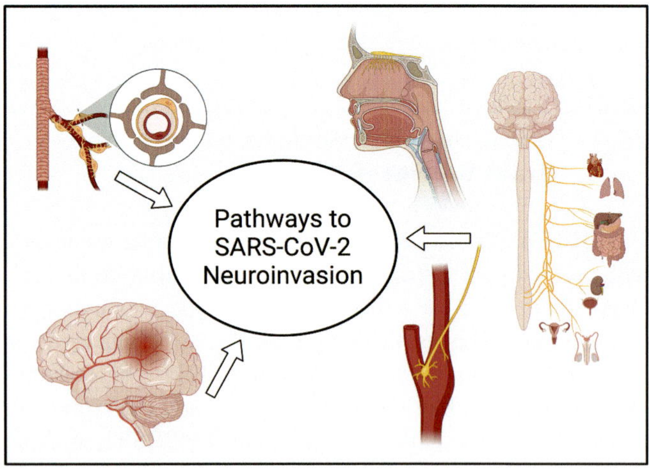

Fig. 25.4 Pathways to SARS-CoV-2 Invasion. (Clockwise from upper left) Possible mechanisms of SARS-CoV-2 entering the CNS include disruption to the BBB, limited infection of the endothelium, and dysfunction and loss of pericytes in the neurovascular unit. Olfactory, vagus, and glossopharyngeal nerves (CNs I, X, and XI) may provide nervous tissue conduits that bypass the BBB by crossing the skull and entering the CNS along the cribriform plate or jugular foramen, respectively. Finally, more drastic loss of BBB integrity such as hemorrhagic strokes may also facilitate SARS-CoV-2 entering the CNS where it persists after recovery

25.6 Long-Term Impacts of SARS-CoV-2: Post-acute Sequelae of COVID-19

In most individuals, SARS-CoV-2 is largely cleared from the body within a few weeks of infection. Despite this clearance, many individuals experience persistent symptoms months and years after their initial infection. First termed "Long COVID" by patient Elisa Perego on Twitter in May 2020, this chronic syndrome has since been more formally named post-acute sequelae of COVID-19 (PASC) (Callard & Perego, 2021). Atypical of infectious disease, over 200 symptoms of PASC have been described, and formal symptomology remains elusive. The WHO lists fatigue, shortness of breath, and cognitive dysfunction as the most common symptoms. However, it clearly states that this list is not exhaustive, specifies the presence of these symptoms after 2 months, and is not explainable by alternative diagnoses (Soriano et al., 2022). The risk factors of PASC mirror those of more severe acute COVID-19, including gender, sex, obesity, and asthma (Chen et al., 2022). Notably, PASC impacts diverse organ systems, including cardiovascular, renal, musculoskeletal, respiratory, nervous, and digestive systems. This breadth contributes to a large number of systems and likely also points to the reason the etiology of PASC remains unknown: there are, in fact, many etiologies.

There have been many comparisons between PASC and chronic fatigue syndrome/myalgic encephalomyelitis (CFS/ME). Common symptoms of CFS/ME include post-exertion malaise, difficulty completing routine tasks, sleep disturbances, and noticeable cognitive dysfunction, all similar to the most common symptoms of PASC. The pathogenesis of CFS/ME, like PASC, has remained elusive. One proposed mechanism suggests that CFS/ME may result from brainstem dysfunction, with characteristic changes in one MRI study suggesting a loss of brainstem gray matter volume. In CFS patients, brainstem gray matter volume was positively associated with seated pulse pressure (difference between systolic and diastolic pressure). The authors posit this finding suggests impaired cerebrovascular autoregulation (Barnden et al., 2011). Another study performing 3 T MRI T1 and T2 weighted spin-echo (T1wSE and T2wSE) scans, which can be used to determine levels of myelination, found that T1wSE was significantly decreased in the brainstem of CFS patients compared with healthy controls. The authors found lower T1wSE in the brainstem was inversely associated with T1wSE in sensorimotor white matter in CFS patients and healthy controls. They suggest that impaired brainstem signal conduction may cause an increase in sensorimotor myelination as compensation (Barnden et al., 2018). Thus, the hypothesis that PASC shares a common etiology with or causes CFS/ME would be consistent with the aforementioned studies suggesting that SARS-CoV-2 primarily enters and is maintained in the brainstem. Adding support to this hypothesis was a recent MRI study comparing PASC patients to CFS/ME patients and healthy control volunteers. It was determined that CFS/ME patients had significantly larger whole brainstems and pons regions compared to healthy controls. Similarly, PASC patients had larger whole brainstems and pons regions than healthy controls. There was no statistical difference between the patient groups. Additionally, larger whole brainstems were significantly positively associated with worse pain scores and breathing difficulties in CFS/ME, suggesting these differences in size may contribute to symptomology (Thapaliya et al., 2023).

Another possible etiology of PASC similar to CFS/ME is immunological. PASC is associated with elevated levels of interleukin (IL)-1β, IL-6, and tumor necrosis factor alpha (TNF-α) in the plasma relative to control participants. Plasma levels of several cytokines are also higher in CFS/ME patients than in controls, including IL-12p40, IL-12p70, IFN-γ, TNF-α, and granulocyte-macrophage colony-stimulating factor 2 (Hornig et al., 2015). IFN-γ was associated with a 104.77 odds ratio of CFS/ME, the most dramatic association with the disease ($P = 0.001$) (Hornig et al., 2015). Another study of CFS/ME patients found reductions in cytotoxic NK cells lysis activity and increases in killer cell immunoglobulin-like receptor 3DL1, IL-4, TNF-α, and IFN-γ, further implicating antiviral response dysfunction as a contributing factor to CFS/ME etiology (Brenu et al., 2013). One therapy targeting the peripheral immune system currently in clinical trials for CFS/ME recently modified its inclusion criteria to include PASC. This dsRNA called rintatolimod agonizes toll-like receptor 3 similar to polyinosinic: polycytidylic acid, thereby inducing downstream effects such as the expression of ribonuclease L, which is thought to play a role in CFS/ME (Mitchell, 2016). Thus, targeting dysfunctional antiviral dysfunction resulting from COVID-19 may provide a pathway to improving PASC patient symptoms. Complicating this view, however, is a recent report finding a relative absence of IFN-γ in the plasma of long-COVID-19 patients, the opposite of findings during CFS/ME (Williams et al., 2022). While this finding also indicates an antiviral dysfunction, more work is still required to elucidate the mechanisms of both diseases fully.

As previously mentioned, the possibility of SARS-CoV-2 reservoirs remains a possibility in contributing to chronic inflammation. However, as in other chronic viral infections, SARS-CoV-2 and associated inflammation may contribute to gut barrier dysfunction and dysbiosis (Johnson et al., 2021). Like the endothelium, intestinal epithelium expresses both ACE2 and TMPRSS2, which are upregulated during local inflammation (Hamming et al., 2004; Suárez-Fariñas et al., 2021). Further, intranasal and intragastric inoculation of rhesus macaques with SARS-CoV-2 resulted in gastrointestinal dysfunction, including impaired barrier function and loss of mucin-containing goblet cells

(Jiao et al., 2021). Barrier dysfunction facilitates microbial translocation, which is particularly problematic during COVID-19 because patients also experience dysbiosis characterized in several studies as loss of Lachnospiraceae, Ruminococcaceae, and Eubacteriaceae, especially species in *Faecalibacterium*, *Eubacterium*, *Coprococcus*, *Ruminococcus*, *Lachnospira*, and *Roseburia*, all of which are notable as short-chain fatty acid-producing bacteria (systematically reviewed in: (Zhang et al., 2023)). These losses are associated with increases in the relative abundances of Enterobacteriaceae families, many of which are pathogenic (Zhang et al., 2023). When comparing the fecal composition of deceased COVID-19 patients with survivors, four important metabolites, deoxycholic acid, lithocholic acid, isodeoxycholic acid, and desaminotyrosine were significantly lower (Stutz et al., 2022). The three primary fecal short-chain fatty acids also trended lower in the deceased patients (Stutz et al., 2022). These changes were accompanied by taxonomic changes found in previous studies, primarily an expansion of Enterobacteriaceae and loss of Bacteroidaceae and Lachnospiraceae relative abundances (Stutz et al., 2022). Notably, similar changes, namely, loss of butyrate-producing bacteria *Faecalibacterium*, *Eubacterium*, and *Roseburia*, are also found in CFS/ME patients (Guo et al., 2023; Xiong et al., 2023). Beyond the bacteriome, SARS-CoV-2 is also associated with increased fungal load, including *Aspergillus*, *Penicillium*, and *Candida* species (Zhang et al., 2023). All these microbiome changes and increased microbial translocation would synergistically exacerbate the peripheral and neuro-inflammation described above. Disruptions to the gut-brain axis would further implicate the vagus nerve (CN X) and its dysfunction in contributing to PASC beyond simply facilitating viral entry into the CNS. Pilot trials for self-administration of transcutaneous auricular vagus nerve stimulation have been performed with positive preliminary results (Badran et al., 2022). Pre-, pro-, and post-biotics and fecal microbiota transplants are all also being considered for ameliorating COVID-19 symptoms (Lau et al., 2022). However, more work is still required on how these interventions impact PASC.

Although the cause and possible therapeutics remain largely elusive, vaccination strategies offer some hope in preventing PASC. A recent systematic review (heterogeneity prevented meta-analysis) reported that of 12 studies describing long-COVID incidence, 10 found a significant reduction in long-COVID if vaccination was received before infection (Oyungerel et al., 2023). Interestingly, vaccination may also offer some therapeutic benefits. The systematic review also reported that vaccination after infection reduced PASC, with reported odds ratios ranging from 0.38 to 0.91 (Oyungerel et al., 2023). Together these data highlight the role of adaptive immune response in PASC resolution and offer meaningful prevention strategies that can be implemented (Fig. 25.5).

25.7 Conclusions

Although vaccination strategies have significantly reduced morbidity and mortality, many challenges exist in understanding SARS-CoV-2 pathogenesis. Luckily, early reports

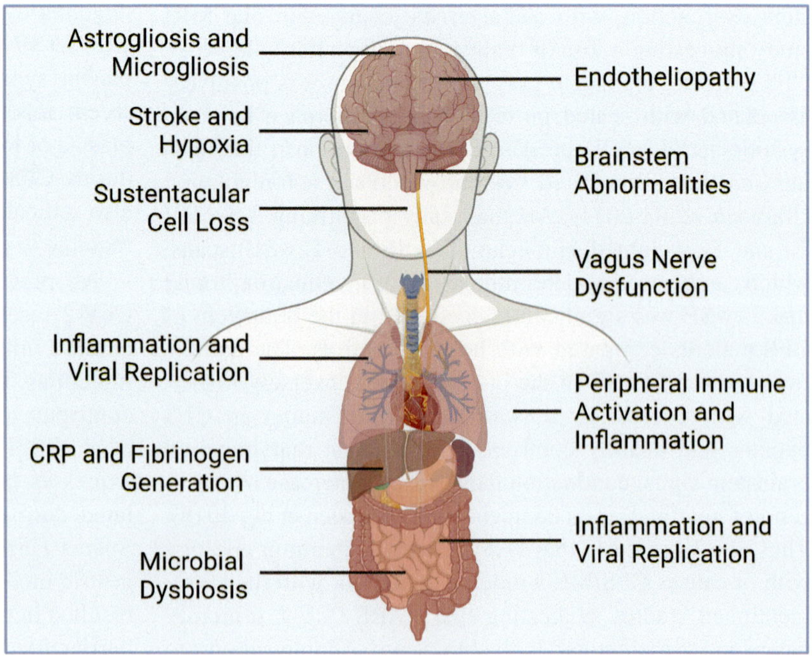

Fig. 25.5 Multiorgan neuroimmune contributions to COVID-19 neurological dysfunction. SARS-CoV-2 causes multisystem disruptions, pro-inflammatory responses, and viral replication and dissemination. These processes synergistically contribute to neurological complications mediated by diverse neuroimmune crosstalk

of anosmia have resolved mainly due to the regenerative capabilities of respiratory epithelium. However, acute cerebrovascular disease remains a problem, with many people still recovering years after the initial infection. Further, as the initial pandemic slowly transitions from acute to long-lasting impacts of SARS-CoV-2 infection with recovery from stroke and PASC being at the forefront of attention, additional research and new therapeutic interventions are still urgently needed.

25.8 Review Questions

1. What is the most common neurological symptom of COVID-19?
   (a) Acute cerebrovascular accidents (stroke)
   (b) Post-acute sequelae of COVID-19
   (c) *Anosmia*
   (d) Dysgeusia
2. Which cells in the olfactory epithelium are most likely to be infected with SARS-CoV-2
   (a) Pneumocytes of the alveoli
   (b) *Sustentacular cells (SUS)*
   (c) Olfactory sensory neurons (OSNs)
   (d) Bowman's glands
3. What cell types maintain the blood-brain barrier, preventing toxins and pathogens like SARS-CoV-2 from entering the central nervous system?
   (a) *Pericytes, endothelial cells, and astrocytes*
   (b) Claudins, occludins, and ZO-1s
   (c) Microglia, oligodendrocytes, and neurons
   (d) Olfactory sensory neurons, sustentacular cells, and respiratory epithelium cells
4. Which is a contributing factor to the incidence of SARS-CoV-2-associated acute cerebrovascular accident (stroke)?
   (a) Cribriform plate blockage
   (b) Sustentacular cell dysfunction
   (c) Decreases in D-dimer, a product of fibrinolysis
   (d) *Increases in C-reactive protein and fibrinogen production by the liver*
5. Which is not a proposed mechanism of invasion of SARS-CoV-2 to the central nervous system?
   (a) Breech of the blood-brain barrier
   (b) *Infection of cortical neurons*
   (c) Cranial nerves as viral conduits
   (d) Transport through cerebrospinal fluid

Acknowledgments Figures created with Biorender.com.

References

Abbott, N. J., Patabendige, A. A., Dolman, D. E., Yusof, S. R., & Begley, D. J. (2010). Structure and function of the blood–brain barrier. *Neurobiology of Disease, 37*(1), 13–25.

Acevedo, C., Blanchard, K., Bacigalupo, J., & Vergara, C. (2019). Possible ATP trafficking by ATP-shuttles in the olfactory cilia and glucose transfer across the olfactory mucosa. *FEBS Letters, 593*(6), 601–610. https://doi.org/10.1002/1873-3468.13346

Adam, A. D., Mahmoud, D., Kevin, P., Aslan, E., Johanna, O., Clemens, S., et al. (2022). Acute ischaemic stroke associated with SARS-CoV-2 infection in North America. *Journal of Neurology, Neurosurgery, and Psychiatry, 93*(4), 360. https://doi.org/10.1136/jnnp-2021-328354

Aghayari Sheikh Neshin, S., Shahjouei, S., Koza, E., Friedenberg, I., Khodadadi, F., Sabra, M., et al. (2021). Stroke in SARS-CoV-2 infection: A pictorial overview of the pathoetiology. *Frontiers in Cardiovascular Medicine, 8*, 649922. https://doi.org/10.3389/fcvm.2021.649922

Asakura, H., & Ogawa, H. (2021). COVID-19-associated coagulopathy and disseminated intravascular coagulation. *International Journal of Hematology, 113*(1), 45–57. https://doi.org/10.1007/s12185-020-03029-y

Badran, B. W., Huffman, S. M., Dancy, M., Austelle, C. W., Bikson, M., Kautz, S. A., & George, M. S. (2022). A pilot randomized controlled trial of supervised, at-home, self-administered transcutaneous auricular vagus nerve stimulation (taVNS) to manage long COVID symptoms. *Bioelectronic Medicine, 8*(1), 13. https://doi.org/10.1186/s42234-022-00094-y

Baj, J., Karakuła-Juchnowicz, H., Teresiński, G., Buszewicz, G., Ciesielka, M., Sitarz, R., et al. (2020). COVID-19: Specific and non-specific clinical manifestations and symptoms: The current state of knowledge. *Journal of Clinical Medicine, 9*(6). https://doi.org/10.3390/jcm9061753

Barnden, L. R., Crouch, B., Kwiatek, R., Burnet, R., Mernone, A., Chryssidis, S., et al. (2011). A brain MRI study of chronic fatigue syndrome: Evidence of brainstem dysfunction and altered homeostasis. *NMR in Biomedicine, 24*(10), 1302–1312. https://doi.org/10.1002/nbm.1692

Barnden, L. R., Shan, Z. Y., Staines, D. R., Marshall-Gradisnik, S., Finegan, K., Ireland, T., & Bhuta, S. (2018). Hyperintense sensorimotor T1 spin echo MRI is associated with brainstem abnormality in chronic fatigue syndrome. *NeuroImage: Clinical, 20*, 102–109. https://doi.org/10.1016/j.nicl.2018.07.011

Brann, D. H., Tsukahara, T., Weinreb, C., Lipovsek, M., Van den Berge, K., Gong, B., Chance, R., Macaulay, I. C., Chou, H. J., Fletcher, R. B., Das, D., Street, K., de Bezieux, H. R., Choi, Y. G., Risso, D., Dudoit, S., Purdom, E., Mill, J., Hachem, R. A., Matsunami, H., Logan, D. W., Goldstein, B. J., Grubb, M. S., Ngai, J., & Datta, S. R. (2020). Non-neuronal expression of SARS-CoV-2 entry genes in the olfactory system suggests mechanisms underlying COVID-19-associated anosmia. *Science Advances, 6*(31), eabc5801. https://doi.org/10.1126/sciadv.abc5801. PMID: 32937591; PMCID: PMC10715684.

Brenu, E. W., Hardcastle, S. L., Atkinson, G. M., van Driel, M. L., Kreijkamp-Kaspers, S., Ashton, K. J., et al. (2013). Natural killer cells in patients with severe chronic fatigue syndrome. *Autoimmunity Highlights, 4*(3), 69–80. https://doi.org/10.1007/s13317-013-0051-x

Butowt, R., & von Bartheld, C. S. (2022). The route of SARS-CoV-2 to brain infection: Have we been barking up the wrong tree? *Molecular Neurodegeneration, 17*(1), 20. https://doi.org/10.1186/s13024-022-00529-9

Callard, F., & Perego, E. (2021). How and why patients made long Covid. *Social Science & Medicine, 268*, 113426. https://doi.org/10.1016/j.socscimed.2020.113426

Camell, C. D., Yousefzadeh, M. J., Zhu, Y., Prata, L., Huggins, M. A., Pierson, M., et al. (2021). Senolytics reduce coronavirus-related mortality in old mice. *Science, 373*(6552). https://doi.org/10.1126/science.abe4832

Cao, Y., Yisimayi, A., Jian, F., Song, W., Xiao, T., Wang, L., et al. (2022). BA.2.12.1, BA.4 and BA.5 escape antibodies elicited by omicron infection. *Nature, 608*(7923), 593–602. https://doi.org/10.1038/s41586-022-04980-y

Chen, C., Haupert, S. R., Zimmermann, L., Shi, X., Fritsche, L. G., & Mukherjee, B. (2022). Global prevalence of post-coronavirus disease 2019 (COVID-19) condition or long COVID: A meta-analysis and systematic review. *The Journal of Infectious Diseases, 226*(9), 1593–1607. https://doi.org/10.1093/infdis/jiac136

Choe, Y. G., Yoon, J. H., Joo, J., Kim, B., Hong, S. P., Koh, G. Y., et al. (2022). Pericyte loss leads to capillary stalling through increased leukocyte-endothelial cell interaction in the brain. *Frontiers in Cellular Neuroscience, 16*, 848764. https://doi.org/10.3389/fncel.2022.848764

Constant, O., Barthelemy, J., Bolloré, K., Tuaillon, E., Gosselet, F., Chable-Bessia, C., et al. (2021). SARS-CoV-2 poorly replicates in cells of the human blood-brain barrier without associated deleterious effects. *Frontiers in Immunology, 12*. https://doi.org/10.3389/fimmu.2021.697329

Croy, I., Nordin, S., & Hummel, T. (2014). Olfactory disorders and quality of life—An updated review. *Chemical Senses, 39*(3), 185–194. https://doi.org/10.1093/chemse/bjt072

Favas, T. T., Dev, P., Chaurasia, R. N., Chakravarty, K., Mishra, R., Joshi, D., et al. (2020). Neurological manifestations of COVID-19: A systematic review and meta-analysis of proportions. *Neurological Sciences, 41*(12), 3437–3470. https://doi.org/10.1007/s10072-020-04801-y

Gruell, H., Vanshylla, K., Tober-Lau, P., Hillus, D., Sander, L. E., Kurth, F., & Klein, F. (2022). Neutralisation sensitivity of the SARS-CoV-2 omicron BA.2.75 sublineage. *The Lancet Infectious Diseases, 22*(10), 1422–1423. https://doi.org/10.1016/s1473-3099(22)00580-1

Guo, C., Che, X., Briese, T., Ranjan, A., Allicock, O., Yates, R. A., et al. (2023). Deficient butyrate-producing capacity in the gut microbiome is associated with bacterial network disturbances and fatigue symptoms in ME/CFS. *Cell Host & Microbe, 31*(2), 288–304.e288. https://doi.org/10.1016/j.chom.2023.01.004

Hamming, I., Timens, W., Bulthuis, M. L., Lely, A. T., Navis, G., & van Goor, H. (2004). Tissue distribution of ACE2 protein, the functional receptor for SARS coronavirus. A first step in understanding SARS pathogenesis. *The Journal of Pathology, 203*(2), 631–637. https://doi.org/10.1002/path.1570

Hannum, M. E., Ramirez, V. A., Lipson, S. J., Herriman, R. D., Toskala, A. K., Lin, C., et al. (2020). Objective sensory testing methods reveal a higher prevalence of olfactory loss in COVID-19 positive patients compared to subjective methods: A systematic review and meta-analysis. *medRxiv*. https://doi.org/10.1101/2020.07.04.20145870

Henninger, D. D., Panes, J., Eppihimer, M., Russell, J., Gerritsen, M., Anderson, D. C., & Granger, D. N. (1997). Cytokine-induced VCAM-1 and ICAM-1 expression in different organs of the mouse. *The Journal of Immunology, 158*(4), 1825–1832.

Holshue, M. L., DeBolt, C., Lindquist, S., Lofy, K. H., Wiesman, J., Bruce, H., et al. (2020). First case of 2019 novel coronavirus in the United States. *New England Journal of Medicine, 382*(10), 929–936. https://doi.org/10.1056/NEJMoa2001191

Hornig, M., Montoya, J. G., Klimas, N. G., Levine, S., Felsenstein, D., Bateman, L., et al. (2015). Distinct plasma immune signatures in ME/CFS are present early in the course of illness. *Science Advances, 1*(1), e1400121. https://doi.org/10.1126/sciadv.1400121

Huang, C., Wang, Y., Li, X., Ren, L., Zhao, J., Hu, Y., et al. (2020). Clinical features of patients infected with 2019 novel coronavirus in Wuhan, China. *Lancet, 395*(10223), 497–506. https://doi.org/10.1016/s0140-6736(20)30183-5

Hunt, B. J., & Levi, M. (2020). Re the source of elevated plasma D-dimer levels in COVID-19 infection. *British Journal of Haematology, 190*(3), e133–e134. https://doi.org/10.1111/bjh.16907

Iketani, S., Liu, L., Guo, Y., Liu, L., Chan, J. F., Huang, Y., et al. (2022). Antibody evasion properties of SARS-CoV-2 omicron sublineages. *Nature, 604*(7906), 553–556. https://doi.org/10.1038/s41586-022-04594-4

Jiao, L., Li, H., Xu, J., Yang, M., Ma, C., Li, J., et al. (2021). The gastrointestinal tract is an alternative route for SARS-CoV-2 infection in a nonhuman primate model. *Gastroenterology, 160*(5), 1647–1661. https://doi.org/10.1053/j.gastro.2020.12.001

Johnson, S. D., Olwenyi, O. A., Bhyravbhatla, N., Thurman, M., Pandey, K., Klug, E. A., et al. (2021). Therapeutic implications of SARS-CoV-2 dysregulation of the gut-brain-lung axis. *World Journal of Gastroenterology, 27*(29), 4763–4783. https://doi.org/10.3748/wjg.v27.i29.4763

Katsanos, A. H., Palaiodimou, L., Zand, R., Yaghi, S., Kamel, H., Navi, B. B., et al. (2021). The impact of SARS-CoV-2 on stroke epidemiology and care: A meta-analysis. *Annals of Neurology, 89*(2), 380–388. https://doi.org/10.1002/ana.25967

Khan, M., Yoo, S. J., Clijsters, M., Backaert, W., Vanstapel, A., Speleman, K., et al. (2021). Visualizing in deceased COVID-19 patients how SARS-CoV-2 attacks the respiratory and olfactory mucosae but spares the olfactory bulb. *Cell, 184*(24), 5932–5949. e5915. https://doi.org/10.1016/j.cell.2021.10.027

Khan, A. M., Kallogjeri, D., & Piccirillo, J. F. (2022a). Growing public health concern of COVID-19 chronic olfactory dysfunction. *JAMA Otolaryngology. Head & Neck Surgery, 148*(1), 81–82. https://doi.org/10.1001/jamaoto.2021.3379

Khan, A. O., Reyat, J. S., Hill, H., Bourne, J. H., Colicchia, M., Newby, M. L., et al. (2022b). Preferential uptake of SARS-CoV-2 by pericytes potentiates vascular damage and permeability in an organoid model of the microvasculature. *Cardiovascular Research, 118*(15), 3085–3096. https://doi.org/10.1093/cvr/cvac097

Kihira, S., Delman, B. N., Belani, P., Stein, L., Aggarwal, A., Rigney, B., et al. (2020). Imaging features of acute encephalopathy in patients with COVID-19: A case series. *AJNR. American Journal of Neuroradiology, 41*(10), 1804–1808. https://doi.org/10.3174/ajnr.A6715

Kim, Y. E., Huh, K., Park, Y. J., Peck, K. R., & Jung, J. (2022). Association between vaccination and acute myocardial infarction and ischemic stroke after COVID-19 infection. *JAMA, 328*(9), 887–889. https://doi.org/10.1001/jama.2022.12992

Korber, B., Fischer, W. M., Gnanakaran, S., Yoon, H., Theiler, J., Abfalterer, W., et al. (2020). Tracking changes in SARS-CoV-2 spike: Evidence that D614G increases infectivity of the COVID-19 virus. *Cell, 182*(4), 812–827.e819. https://doi.org/10.1016/j.cell.2020.06.043

Kumar, S., Karuppanan, K., & Subramaniam, G. (2022). Omicron (BA.1) and sub-variants (BA.1.1, BA.2, and BA.3) of SARS-CoV-2 spike infectivity and pathogenicity: A comparative sequence and structural-based computational assessment. *Journal of Medical Virology, 94*(10), 4780–4791. https://doi.org/10.1002/jmv.27927

Lau, H. C.-H., Ng, S. C., & Yu, J. (2022). Targeting the gut microbiota in coronavirus disease 2019: Hype or Hope? *Gastroenterology, 162*(1), 9–16. https://doi.org/10.1053/j.gastro.2021.09.009

Lechien, J. R., Chiesa-Estomba, C. M., Beckers, E., Mustin, V., Ducarme, M., Journe, F., et al. (2021). Prevalence and 6-month recovery of olfactory dysfunction: A multicentre study of 1363 COVID-19 patients. *Journal of Internal Medicine, 290*(2), 451–461. https://doi.org/10.1111/joim.13209

Li, Y., Li, M., Wang, M., Zhou, Y., Chang, J., Xian, Y., et al. (2020). Acute cerebrovascular disease following COVID-19: A single center, retrospective, observational study. *Stroke and Vascular Neurology, 5*(3), 279–284. https://doi.org/10.1136/svn-2020-000431

Mao, L., Jin, H., Wang, M., Hu, Y., Chen, S., He, Q., et al. (2020). Neurologic manifestations of hospitalized patients with coronavirus disease 2019 in Wuhan, China. *JAMA Neurology, 77*(6), 683–690. https://doi.org/10.1001/jamaneurol.2020.1127

Maria-Ioanna, S., Lina, P., de Diana Aguiar, S., Aikaterini, T., Eleni, B., Dimitrios Eleftherios, K., et al. (2022). Acute arterial ischemic stroke following COVID-19 vaccination. *Neurology, 99*(14), e1465. https://doi.org/10.1212/WNL.0000000000200996

Matschke, J., Lütgehetmann, M., Hagel, C., Sperhake, J. P., Schröder, A. S., Edler, C., et al. (2020). Neuropathology of patients with COVID-19 in Germany: A post-mortem case series. *The Lancet Neurology, 19*(11), 919–929. https://doi.org/10.1016/S1474-4422(20)30308-2

Mbonde, A. A., O'Carroll, C. B., Grill, M. F., Zhang, N., Butterfield, R., & Demaerschalk, B. M. (2022). Stroke features, risk factors, and pathophysiology in SARS-CoV-2-infected patients. *Mayo Clinic Proceedings: Innovations, Quality & Outcomes, 6*(2), 156–165. https://doi.org/10.1016/j.mayocpiqo.2022.01.003

Meinhardt, J., Radke, J., Dittmayer, C., Franz, J., Thomas, C., Mothes, R., et al. (2021). Olfactory transmucosal SARS-CoV-2 invasion as a port of central nervous system entry in individuals with COVID-19. *Nature Neuroscience, 24*(2), 168–175. https://doi.org/10.1038/s41593-020-00758-5

Mengfei, C., Wenjuan, S., Nicholas, R. R., Heather, K., Alexander, H., Ramanathan, M., Jr., & Andrew, P. L. (2020). Elevated ACE2 expression in the olfactory neuroepithelium: Implications for anosmia and upper respiratory SARS-CoV-2 entry and replication. *European Respiratory Journal*, 2001948. https://doi.org/10.1183/13993003.01948-2020

Meyer, K., Patra, T., Vijayamahantesh, & Ray, R. (2021). SARS-CoV-2 spike protein induces paracrine senescence and leukocyte adhesion in endothelial cells. *Journal of Virology, 95*(17), e0079421. https://doi.org/10.1128/jvi.00794-21

Mitchell, W. M. (2016). Efficacy of rintatolimod in the treatment of chronic fatigue syndrome/myalgic encephalomyelitis (CFS/ME). *Expert Review of Clinical Pharmacology, 9*(6), 755–770. https://doi.org/10.1586/17512433.2016.1172960

Moriguchi, T., Harii, N., Goto, J., Harada, D., Sugawara, H., Takamino, J., et al. (2020). A first case of meningitis/encephalitis associated with SARS-Coronavirus-2. *International Journal of Infectious Diseases, 94*, 55–58. https://doi.org/10.1016/j.ijid.2020.03.062

Mosesson, M. W. (2005). Fibrinogen and fibrin structure and functions. *Journal of Thrombosis and Haemostasis, 3*(8), 1894–1904. https://doi.org/10.1111/j.1538-7836.2005.01365.x

Oyungerel, B., Paulina, S., Justin, C., Kylie, A., & Paul, G. (2023). Effect of covid-19 vaccination on long covid: Systematic review. *BMJ Medicine, 2*(1), e000385. https://doi.org/10.1136/bmjmed-2022-000385

Pather, S., Madhi, S. A., Cowling, B. J., Moss, P., Kamil, J. P., Ciesek, S., et al. (2023). SARS-CoV-2 omicron variants: Burden of disease, impact on vaccine effectiveness and need for variant-adapted vaccines. *Frontiers in Immunology, 14*, 1130539. https://doi.org/10.3389/fimmu.2023.1130539

Pepys, M. B., & Hirschfield, G. M. (2003). C-reactive protein: a critical update. *The Journal of Clinical Investigation, 111*(12), 1805–1812. https://doi.org/10.1172/jci18921

Polman, C. H., O'Connor, P. W., Havrdova, E., Hutchinson, M., Kappos, L., Miller, D. H., et al. (2006). A randomized, placebo-controlled trial of natalizumab for relapsing multiple sclerosis. *New England Journal of Medicine, 354*(9), 899–910.

Poyiadji, N., Shahin, G., Noujaim, D., Stone, M., Patel, S., & Griffith, B. (2020). COVID-19–associated acute hemorrhagic necrotizing encephalopathy: Imaging features. *Radiology, 296*(2), E119–E120. https://doi.org/10.1148/radiol.2020201187

Rössler, K., Neuchrist, C., Kitz, K., Scheiner, O., Kraft, D., & Lassmann, H. (1992). Expression of leucocyte adhesion molecules at the human blood-brain barrier (BBB). *Journal of Neuroscience Research, 31*(2), 365–374.

Rotoli, B. M., Barilli, A., Visigalli, R., Ferrari, F., & Dall'Asta, V. (2021). Endothelial cell activation by SARS-CoV-2 spike S1 protein: A crosstalk between endothelium and innate immune cells. *Biomedicine, 9*(9). https://doi.org/10.3390/biomedicines9091220

Santos, R. A. S., Oudit, G. Y., Verano-Braga, T., Canta, G., Steckelings, U. M., & Bader, M. (2019). The renin-angiotensin system: Going beyond the classical paradigms. *American Journal of Physiology. Heart and Circulatory Physiology, 316*(5), H958–h970. https://doi.org/10.1152/ajpheart.00723.2018

Schimmel, L., Chew, K. Y., Stocks, C. J., Yordanov, T. E., Essebier, P., Kulasinghe, A., et al. (2021). Endothelial cells are not productively infected by SARS-CoV-2. *Clinical & Translational Immunology, 10*(10), e1350. https://doi.org/10.1002/cti2.1350

Sheward, D. J., Kim, C., Fischbach, J., Sato, K., Muschiol, S., Ehling, R. A., et al. (2022). Omicron sublineage BA.2.75.2 exhibits extensive escape from neutralising antibodies. *The Lancet Infectious Diseases, 22*(11), 1538–1540. https://doi.org/10.1016/s1473-3099(22)00663-6

Shrestha, L. B., Foster, C., Rawlinson, W., Tedla, N., & Bull, R. A. (2022). Evolution of the SARS-CoV-2 omicron variants BA.1 to BA.5: Implications for immune escape and transmission. *Reviews in Medical Virology, 32*(5), e2381. https://doi.org/10.1002/rmv.2381

Small, D. M., & Prescott, J. (2005). Odor/taste integration and the perception of flavor. *Experimental Brain Research, 166*(3), 345–357. https://doi.org/10.1007/s00221-005-2376-9

Soriano, J. B., Murthy, S., Marshall, J. C., Relan, P., & Diaz, J. V. (2022). A clinical case definition of post-COVID-19 condition by a Delphi consensus. *The Lancet Infectious Diseases, 22*(4), e102–e107. https://doi.org/10.1016/s1473-3099(21)00703-9

Stein, S. R., Ramelli, S. C., Grazioli, A., Chung, J. Y., Singh, M., Yinda, C. K., et al. (2022). SARS-CoV-2 infection and persistence in the human body and brain at autopsy. *Nature, 612*(7941), 758–763. https://doi.org/10.1038/s41586-022-05542-y

Stutz, M. R., Dylla, N. P., Pearson, S. D., Lecompte-Osorio, P., Nayak, R., Khalid, M., et al. (2022). Immunomodulatory fecal metabolites are associated with mortality in COVID-19 patients with respiratory failure. *Nature Communications, 13*(1), 6615. https://doi.org/10.1038/s41467-022-34260-2

Suárez-Fariñas, M., Tokuyama, M., Wei, G., Huang, R., Livanos, A., Jha, D., et al. (2021). Intestinal inflammation modulates the expression of ACE2 and TMPRSS2 and potentially overlaps with the pathogenesis of SARS-CoV-2-related disease. *Gastroenterology, 160*(1), 287–301.e220. https://doi.org/10.1053/j.gastro.2020.09.029

Suzuki, Y., Takeda, M., & Farbman, A. I. (1996). Supporting cells as phagocytes in the olfactory epithelium after bulbectomy. *Journal of Comparative Neurology, 376*(4), 509–517.

Tang, N., Li, D., Wang, X., & Sun, Z. (2020). Abnormal coagulation parameters are associated with poor prognosis in patients with novel coronavirus pneumonia. *Journal of Thrombosis and Haemostasis, 18*(4), 844–847. https://doi.org/10.1111/jth.14768

Thapaliya, K., Marshall-Gradisnik, S., Barth, M., Eaton-Fitch, N., & Barnden, L. (2023). Brainstem volume changes in myalgic encephalomyelitis/chronic fatigue syndrome and long COVID patients. *Frontiers in Neuroscience, 17*. https://doi.org/10.3389/fnins.2023.1125208

Thompson, D., Pepys, M. B., & Wood, S. P. (1999). The physiological structure of human C-reactive protein and its complex with phosphocholine. *Structure, 7*(2), 169–177. https://doi.org/10.1016/S0969-2126(99)80023-9

Urata, R., Ikeda, K., Yamazaki, E., Ueno, D., Katayama, A., Shin-Ya, M., et al. (2022). Senescent endothelial cells are predisposed to SARS-CoV-2 infection and subsequent endothelial dysfunction. *Scientific Reports, 12*(1), 11855. https://doi.org/10.1038/s41598-022-15976-z

Villar, P. S., Delgado, R., Vergara, C., Reyes, J. G., & Bacigalupo, J. (2017). Energy requirements of odor transduction in the chemosensory cilia of olfactory sensory neurons rely on oxidative phosphorylation and glycolytic processing of extracellular glucose. *The Journal of Neuroscience, 37*(23), 5736–5743. https://doi.org/10.1523/jneurosci.2640-16.2017

Vitale-Cross, L., Szalayova, I., Scoggins, A., Palkovits, M., & Mezey, E. (2022). SARS-CoV-2 entry sites are present in all structural elements of the human glossopharyngeal and vagal nerves: Clinical implications. *eBioMedicine, 78*, 103981. https://doi.org/10.1016/j.ebiom.2022.103981

Wan, D., Du, T., Hong, W., Chen, L., Que, H., Lu, S., & Peng, X. (2021). Neurological complications and infection mechanism of SARS-COV-2. *Signal Transduction and Targeted Therapy, 6*(1), 406. https://doi.org/10.1038/s41392-021-00818-7

Wang, Z., Yang, Y., Liang, X., Gao, B., Liu, M., Li, W., et al. (2020). COVID-19 associated ischemic stroke and hemorrhagic stroke: Incidence, potential pathological mechanism, and management. *Frontiers in Neurology, 11*, 571996. https://doi.org/10.3389/fneur.2020.571996

Wang, L., Sievert, D., Clark, A. E., Lee, S., Federman, H., Gastfriend, B. D., et al. (2021). A human three-dimensional neural-perivascular 'assembloid' promotes astrocytic development and enables modeling of SARS-CoV-2 neuropathology. *Nature Medicine, 27*(9), 1600–1606. https://doi.org/10.1038/s41591-021-01443-1

WHO COVID-19 Dashboard. (2023). Retrieved from https://covid19.who.int/

WHO Director-General's opening remarks at the media briefing on COVID-19. (2020). Retrieved from https://www.who.int/director--general/speeches/detail/who-director-general-s-opening-remarks-at-the-media-briefing-on-covid-19%2D%2D-11-march-2020

Williams, E. S., Martins, T. B., Shah, K. S., Hill, H. R., Coiras, M., Spivak, A. M., & Planelles, V. (2022). Cytokine deficiencies in patients with long-COVID. *Journal of Clinical and Cellular Immunology, 13*(6), 672.

Wu, Y. C., Sonninen, T. M., Peltonen, S., Koistinaho, J., & Lehtonen, Š. (2021). Blood-brain barrier and neurodegenerative diseases-modeling with iPSC-derived brain cells. *International Journal of Molecular Sciences, 22*(14). https://doi.org/10.3390/ijms22147710

Xiong, R., Gunter, C., Fleming, E., Vernon, S. D., Bateman, L., Unutmaz, D., & Oh, J. (2023). Multi-'omics of gut microbiome-host interactions in short- and long-term myalgic encephalomyelitis/chronic fatigue syndrome patients. *Cell Host & Microbe, 31*(2), 273–287.e275. https://doi.org/10.1016/j.chom.2023.01.001

Yang, R.-C., Huang, K., Zhang, H.-P., Li, L., Zhang, Y.-F., Tan, C., et al. (2022). SARS-CoV-2 productively infects human brain microvascular endothelial cells. *Journal of Neuroinflammation, 19*(1), 149. https://doi.org/10.1186/s12974-022-02514-x

Zaim, S., Chong, J. H., Sankaranarayanan, V., & Harky, A. (2020). COVID-19 and multiorgan response. *Current Problems in Cardiology, 45*(8), 100618. https://doi.org/10.1016/j.cpcardiol.2020.100618

Zazhytska, M., Kodra, A., Hoagland, D. A., Frere, J., Fullard, J. F., Shayya, H., et al. (2022). Non-cell-autonomous disruption of nuclear architecture as a potential cause of COVID-19-induced anosmia. *Cell, 185*(6), 1052–1064.e1012. https://doi.org/10.1016/j.cell.2022.01.024

Zhang, L., Zhou, L., Bao, L., Liu, J., Zhu, H., Lv, Q., et al. (2021). SARS-CoV-2 crosses the blood-brain barrier accompanied with basement membrane disruption without tight junctions alteration. *Signal Transduction and Targeted Therapy, 6*(1), 337. https://doi.org/10.1038/s41392-021-00719-9

Zhang, F., Lau, R. I., Liu, Q., Su, Q., Chan, F. K. L., & Ng, S. C. (2023). Gut microbiota in COVID-19: Key microbial changes, potential mechanisms and clinical applications. *Nature Reviews Gastroenterology & Hepatology, 20*(5), 323–337. https://doi.org/10.1038/s41575-022-00698-4

Žugaj, M., Ditzhuijzen, N. S., Golebski, K., & Fokkens, W. (2021). The effect of coronaviruses on olfaction: Systematic review. *Rhinology*. https://doi.org/10.4193/Rhin20.610

Immunopathogenesis of HAM/TSP and Radiological Features of Disease Progression

26

Emily H. Stack, Yair Mina, Govind Nair, and Steven Jacobson

Abstract

Human T cell lymphotropic virus type 1 (HTLV-1) is an oncogenic retrovirus associated with the development of several human diseases including the progressive neurological disease HTLV-1-associated myelopathy/tropical spastic paraparesis (HAM/TSP). While the complete pathogenesis of HAM/TSP is not fully understood, epidemiological, virological, immunological, and clinical studies have elucidated key components of disease development. Widely recognized characteristics of HAM/TSP include high proviral load (PVL) and hyperinflammatory cellular immune responses. Histological studies have demonstrated that proinflammatory immune infiltrates, most notably HTLV-1-specific $CD8^+$ T cells, accumulate within the spinal cord and are associated with tissue degradation and symptom severity. This review will summarize the pathophysiological features of HAM/TSP. Particular attention will be paid to recent advancements in magnetic resonance imaging (MRI) of the central nervous system (CNS) that have expanded our understanding of disease progression, especially changes to the spinal cord. These imaging advances have demonstrated that HAM/TSP is characterized by a disease-specific pattern of spinal cord atrophy that is well correlated with clinical, virological, and immunological parameters of disease development.

E. H. Stack · S. Jacobson (✉)
Viral Immunology Section, National Institute of Neurological Disorders and Stroke, National Institutes of Health, Bethesda, MD, USA
e-mail: jacobsons@ninds.nih.gov

Y. Mina
Department of Neurology, Tel Aviv Sourasky Medical Center, Tel Aviv, Israel

G. Nair
Quantitative MRI Core, National Institute of Neurological Disorders and Stroke, National Institutes of Health, Bethesda, MD, USA

Keywords

HTLV-1 · HAM/TSP · MRI · Atrophy · Central nervous system

26.1 Introduction

26.1.1 History and Epidemiology of HTLV-1

Following the discovery of retroviral reverse transcriptase in 1970, as well as the suggestion that some human cancers, most notably Burkitt's lymphoma, may have a viral etiology, the search for oncogenic retroviruses captivated the field of cancer research (Coffin, 2015). After years of false discoveries, in 1980 an oncogenic retrovirus was isolated from a T cell line derived from a patient with cutaneous T cell lymphoma (Coffin, 2015; Forlani et al., 2021; Poiesz et al., 1980). The discovery was made by the Gallo laboratory and the retrovirus was eventually termed human T cell lymphotropic virus type 1 (HTLV-1) (Poiesz et al., 1980). HTLV-1 was later identified as the etiological agent in both adult T cell leukemia (ATL) and the progressive neurological disease HTLV-1-associated myelopathy/tropical spastic paraparesis (HAM/TSP) (Gessain et al., 1985; Hinuma et al., 1981). It has been reported that HTLV-1 infects between 10 and 20 million people worldwide, although this may be an underestimate of its true prevalence. The virus is clustered in discrete foci of high endemicity in regions of southern Japan, the Caribbean, Central and South America, the Middle East, Central and West Africa, and Melanesia (Fig. 26.1) (Bomford et al., 1996; Gessain & Cassar, 2012). Seroprevalence rates are variable, but typically range from 5% to 30% of the population in endemic areas, increase gradually with age, and are higher in women than men (Carneiro-Proietti et al., 2014; Gessain & Cassar, 2012). While most infected individuals remain asymptomatic carriers (AC) of HTLV-1 throughout

H. E. Gendelman, T. Ikezu (eds.), *Neuroimmune Pharmacology and Therapeutics*,
https://doi.org/10.1007/978-3-031-68237-7_26

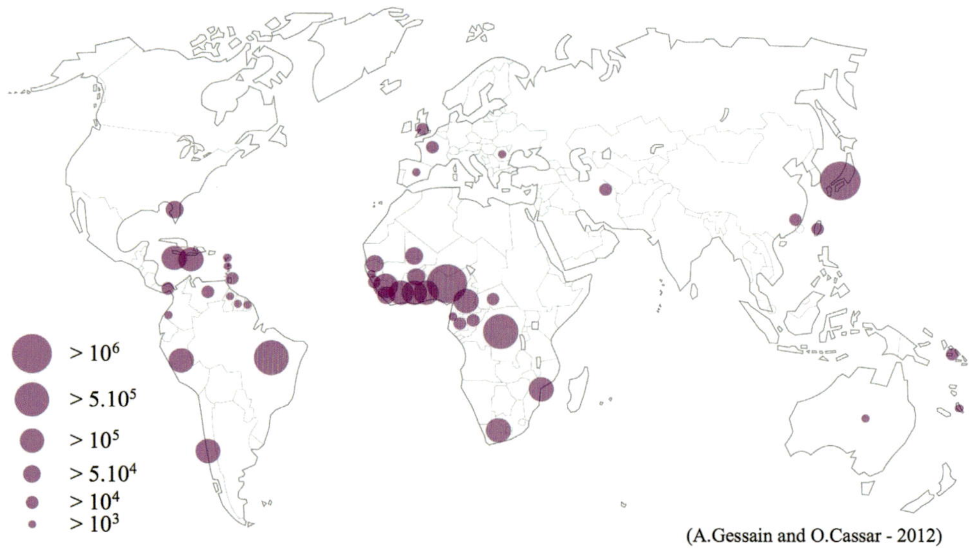

Fig. 26.1 *Global distribution of HTLV-1.* HTLV-1 is endemic to various regions of Central and South America, West Africa, Europe, Japan, and Melanesia. (Reprinted from Gessain & Cassar, 2012. Epidemiological Aspects and World Distribution of HTLV-1)

their lifetime, between 2% and 5% and 0.25% to 3.8% will develop ATL and HAM/TSP, respectively (Nagasaka et al., 2020). The latter will be the focus of this chapter.

26.1.2 HTLV-1 Transmission

There are three routes of HTLV-1 viral transmission, all of which require contact with infected cells of various biofluids. (1) Vertical transmission from mother to child is mediated by the transfer of infected T lymphocytes in breast milk (Carneiro-Proietti et al., 2014). Of breastfed children born to HTLV-1 infected mothers, between 10% and 25% will become infected (Gessain & Cassar, 2012). The risk of transmission increases with a longer duration of breast feeding and higher maternal proviral load (PVL) (Gessain & Cassar, 2012). There is limited evidence of transmission occurring during gestation or birth. (2) Sexual transmission of the virus is also possible, but this is generally not the predominant mode of transmission except in certain endemic regions (Martel & Gotuzzo, 2022; Nunes et al., 2017). Sexual transmission of HTLV-1 is reported to be more efficient from male to female (Martel & Gotuzzo, 2022; Stuver et al., 1993). The final, and most efficient, route of HTLV-1 transmission is (3) contact with infected blood or blood components (parenteral). It is most common for parenteral transmission to occur through the direct donation and transfusion of infected blood, but there are also reports of increased seroprevalence of HTLV-1 among intravenous drug users (Lee et al., 1990). The first report of HTLV-1 transmission via blood transfusion was a retrospective study conducted in Japan from 1981 to 1983. Researchers reported that 63% of patients who had received a blood donation from HTLV-1-positive donors seroconverted to HTLV-1-positive (Okochi et al., 1984). There was no seroconversion in patients who received cell-free plasma alone. Recognition of transfusion-mediated HTLV-1 transmission has caused many countries to adopt anti-HTLV-1 antibody screening for blood donations (Murphy, 2016). The high cost and low availability of these screens, however, preclude some areas of high endemicity from adopting these same protocols despite a clear need.

26.1.3 HTLV-1 Classification, Structure, and Genome

HTLV-1 is classified in the genus deltaretrovirus of the subfamily Orthoretrovirinae. Its genome comprises two identical strands of single stranded positive sense RNA that encode for several protein products: structural proteins Gag, Pro, polymerase, and Env are formed from unspliced or singly spliced mRNA, and accessory proteins p12, p13, and p30, as well as transactivating and regulatory proteins Tax and Rex, are translated from alternatively spliced mRNA transcripts (Kannian & Green, 2010). Evidenced by the routes of transmission discussed previously, viral infection is dependent on direct cell-to-cell contact and the formation of a virological synapse, which is largely driven by the activity of HTLV-1 Tax. The virological synapse is formed by binding of intercellular adhesion molecule 1 (ICAM-1), upregulated by Tax, on the HTLV-1 infected cell to its ligand integrin lymphocyte function-associated antigen (LFA-1) on the recipient cell membrane. The microtubule organizing center (MTOC) of the infected cell is then polarized and the viral RNA genome and Gag protein accumulate at the area of cell–cell contact and are transferred to the uninfected cell as budding virions (Bangham et al., 2015; Kannian & Green, 2010; Mohanty & Harhaj, 2020). Following delivery to the target cell cytoplasm, the viral RNA is reverse transcribed into double-stranded DNA and transported to the nucleus for integration into the host genome as provirus. HTLV-1 provirus can then be transcribed back into mRNA, allowing for expression of structural and regulatory viral proteins in the newly infected cell.

It has been reported that provirus integration into transcriptionally active genomic regions determines the rate of HTLV-1 provirus expression (Meekings et al., 2008). The genomic integration of HTLV-1 to certain regions may induce high HTLV-1 Tax expression and the subsequent Tax-induced proliferation that defines HTLV-1 PVL and that may drive the expansion of HTLV-1-specific CTLs (Bangham et al., 2009). More recently, Melamed et al. (2013) showed that HTLV-1 preferentially integrated within 1 kb of a host transcription start site and was strongly biased to specific transcription factor binding sites, including signal transducer and activator of transcription 1 (STAT1), p53, histone deacetylases (HDACs), and histone acetyltransferases (HATs). The relative position of the HTLV-1 integration site and the transcriptional orientation of the nearest host gene influenced both clonal abundance and spontaneous Tax expression in infected cells (Melamed et al., 2013). Furthermore, Niederer et al. (2014) demonstrated that clones isolated from asymptomatic carriers of HTLV-1 (ACs) were more frequently integrated in transcriptionally active areas than clones from HAM/TSP patients or individuals who have HLA class I alleles that were able to effectively present HBZ peptides to cytotoxic T lymphocytes (CTLs) (Niederer et al., 2014).

26.2 HAM/TSP Disease Course and Cellular Mechanism of Neurodegeneration

26.2.1 Clinical Presentation

HAM/TSP is a chronic neurodegenerative myelopathy that develops in a small subset of individuals infected with HTLV-1. The clinical presentation of HAM/TSP is heterogenous, but symptoms typically manifest as gradually progressing spasticity of the lower limbs, neurogenic bladder, bowel dysfunction, mild sensory disturbance, and lower back pain (Araujo et al., 2021; Román & Román, 1988). Since symptoms are often bilateral and symmetrical, later stages of the disease are characterized by severe motor impairments and either a complete inability to ambulate or reliance on assistive devices. Some studies have reported an occurrence of cognitive deficits in HAM/TSP, namely deviations in attention and information processing, but such deficits are often overshadowed by the more readily apparent and severe motor dysfunction (Fukushima et al., 1994; Gascón et al., 2017; Kalil et al., 2021). HAM/TSP is two to three times more common in women than men and the mean age of onset is 40–50 years old (Yamano & Sato, 2012). Like the primary progressive phenotype of multiple sclerosis (PPMS), patients with HAM/TSP gradually accrue neurological deficits. This similarity in presentation may cause HAM/TSP patients to initially be misdiagnosed with PPMS. Further complicating the clinical presentation of HAM/TSP, the rate of disease progression can vary significantly between patients. Most patients will progress slowly over the course of decades, while a small subset will progress rapidly and be unable to walk or become bedbound less than 2 years after symptom onset (Araujo et al., 2021; Matsuura et al., 2016). Management of HAM/TSP is primarily supportive, including medications used to reduce spasticity (e.g., baclofen), although corticosteroids have been reported to slow progression in some studies (Araujo et al., 2021; Croda et al., 2008). More recently, a small preliminary trial with mogamulizumab, an anti-CCR4 monoclonal antibody used for ATL, demonstrated both clinical and laboratory responses in HAM/TSP, but larger studies are warranted to evaluate its safety and clinical efficacy (Sato et al., 2018).

26.2.2 HTLV-1 Infection Precipitates a Robust Proinflammatory Immune Response in HAM/TSP

Compared to ACs, HTLV-1 infection in HAM/TSP is associated with a strong antibody response with elevated HTLV-1-specific IgM, IgA, and IgG titers (Kira et al., 1992; Manns et al., 1999). There is also a marked increase in the number of HTLV-1-specific HLA class I restricted cytotoxic T lymphocytes (CTLs) (Jacobson et al., 1990). These elevated immunological markers suggest that HTLV-1 PVL and/or HTLV-1 proteins must be persistently expressed in vivo during infection and subsequent disease progression. Accordingly, numerous studies have reported that HTLV-1 PVL is consistently higher in peripheral blood mononuclear cells (PBMCs) of HAM/TSP patients than ACs and that PVL is strongly correlated with disease pathogenesis (Jeffery et al., 1999; Kalil et al., 2021; Manns et al., 1999; Nagai et al., 1998). A study conducted in Kagoshima, Japan, an endemic region for HTLV-1, showed that the prevalence of HAM/TSP rose steeply as the PVL exceeded 1% of PBMCs (Nagai et al., 1998). The wide reproducibility of elevated PVL being associated with HAM/TSP has caused it to be recognized as the main risk factor for disease development in HTLV-1-infected individuals. PVL can vary widely between patients but is usually stable longitudinally within an individual. To determine the cellular reservoir of HTLV-1 in vivo, PBMCs were isolated from HAM/TSP patients and analyzed by PCR. It was reported that over 90% of the HTLV-1 DNA was carried by $CD4^{+}$ T cells (Richardson et al., 1990). Importantly, there was no difference in the cellular distribution of HTLV-1 DNA between HAM/TSP and ACs. The virus can also be detected in $CD8^+$ T cells, dendritic cells, and monocytes (Jones et al., 2008; Koralnik et al., 1992; Nagai al., 2001a).

26.2.3 HTLV-1 Tax Dysregulates Several Immunological Cascades

PVL is directly correlated with *tax* mRNA levels ex vivo and Tax protein expression in vitro, which are significantly higher in HAM/TSP than ACs (Yamano et al., 2002). HTLV-1 Tax protein influences several cellular cascades responsible for inflammation, including the NF-κB and CREB/ATF pathways, and has been correlated with disease severity in HAM/TSP (Matsuoka & Jeang, 2011; Yamano et al., 2002). In addition to the promotion of viral infection by facilitating VS formation discussed previously, Tax expression is associated with aberrant immunoregulatory capacity in HAM/TSP. HAM/TSP patients had significantly decreased forkhead box P3 (FOXP3) expression in $CD4^+CD25^+$ T cells compared to ACs and uninfected healthy controls (HCs), and this decrease was inversely correlated with proviral *tax* DNA load (Oh et al., 2006). Additionally, studies reported that in HAM/TSP there was a reduction in the number of Th17 cells and conversion of immunosuppressive $CD39^+CD25^+CD4^+$ cells to proinflammatory cells expressing interferon gamma (IFN-γ), tumor necrosis factor (TNF)-α, and interleukin-2 (IL-2) (Leal et al., 2013). An in vitro study using $CD4^+CD25^+CCR4^+$ regulatory T cells (Tregs) isolated from HAM/TSP patients demonstrated that the change in FOXP3 expression was mediated by Tax-induced upregulation of T box transcription activator (T-bet) (Araya et al., 2014). This was associated with a subsequent downregulation of FOXP3, loss of suppressive function, and promotion of a proinflammatory Th1 phenotype characterized by production of IFN-γ. Furthermore these proinflammatory Th1 cells were capable of inducing the proliferation of HTLV-1-specific $CD8^+$ T cells (Yamano et al., 2004). Additionally, numerous γ chain family cytokines and receptors have been shown to be upregulated by HTLV-1 Tax expression including IL-2/IL-2R, IL-9, IL-21/IL-21R, and IL-15/IL-15R (Enose-Akahata et al., 2017; Tendler et al., 1990). Previous reports have demonstrated that IL-2 and IL-15 enhanced both the proliferation and cytolytic activity of $CD8^+$ T cells in HAM/TSP by establishing an autocrine loop (Azimi et al., 1999).

26.2.4 $CD8^+$ T Cells Contribute to Disease Pathogenesis in HAM/TSP

While $CD8^+$ T cells carry only 5% of HTLV-1 PVL in vivo, they have been shown to be crucial mediators of both viral control and disease pathogenesis (Melamed et al., 2015). One of the most striking features of HAM/TSP is the elevation of HTLV-1-specific HLA class I restricted $CD8^+$ CTLs in both the periphery and the CNS (Jacobson et al., 1990; Melamed et al., 2015; Nagai et al., 2001b, c; Yamano et al., 2004; Yamano et al., 2002). $CD8^+$ CTLs are responsible for the direct killing of HTLV-1-infected cells through the induction of apoptosis. Upon recognition of viral antigen loaded on MHC I complexes of an infected antigen-presenting cell, CTLs induce apoptosis by releasing soluble perforin and granzymes or binding of the Fas receptor (Imre, 2020). HTLV-1-specific CTLs are also detectable in PBMCs of ACs, but the magnitude and frequency of these responses are significantly higher in patients with HAM/TSP (Elovaara et al., 1993; Nagai et al., 2001c; Parker et al., 1992). The immune response initiated by HTLV-1 infection can be characterized by the antigen specificity of CTLs, which is defined using tetramers. Tetramers are complexes of major histocompatibility complex (MHC) and peptide antigen that are cross-linked by streptavidin and bind stably to the TCR of antigen-specific $CD8^+$ cells (Altman et al., 1996; Greten et al., 1998). $CD8^+$ cells directed against HTLV-1 Tax 11–19 peptide (LLFGYPVYV) were found to represent an extraordinarily high proportion of the total $CD8^+$ cell population in PBMCs of HAM/TSP patients with the HLA-A*201 allele (Greten et al., 1998; Nagai et al., 2001b). Therefore, Tax is widely recognized as an immunodominant antigen in HAM/TSP and the frequency of Tax-specific CTLs is strongly correlated with *tax* mRNA load (Jacobson et al., 1990; Yamano et al., 2002). Tax-specific $CD8^+$ CTL clones have been demonstrated to secrete various inflammatory cytokines, chemokines, and matrix metalloproteinase (MMP) such as IFN-γ, TNF-α, monocyte inflammatory protein (MIP)-1α, MIP-1β, interleukin(IL)-16, and MMP-9 (Biddison et al., 1997). Interestingly, the efficacy of the CTL response, not its magnitude, is most strongly correlated with the control of PVL. Variations in class I MHC alleles have been associated with differential effects on disease development and progression, but such effects are likely population specific and mediated by other genetic and environmental factors (Catalan-Soares et al., 2009; Deschamps et al., 2010; Jeffery et al., 1999; Talledo et al., 2010; Yamano et al., 2002).

26.2.5 Inflammatory Markers Are Elevated in the CNS Compared to the Periphery

In HAM/TSP, several virological and immunological markers are elevated in the CNS compartment compared to the periphery. First, HTLV-1 PVL is significantly elevated in the cerebrospinal fluid (CSF) compared to the peripheral blood of HAM/TSP patients (Brunetto et al., 2014; Lezin et al., 2005; Nagai et al., 2001c; Nozuma et al., 2021). Studies have demonstrated that the ratio of PVL in CSF to PBMCs was significantly associated with clinically progressive disease and recent onset of HAM/TSP (Takenouchi et al., 2003). In addition, infected lymphocytes shared the same HTLV-1 integration site in the cellular DNA of both the CSF cells and PBMCs, which suggested that HTLV-1 infected lymphocytes migrated from the periphery into the CNS (Cavrois et al., 2000). Nozuma et al. (2021) investigated T cell receptor (TCR) repertoire signatures in the peripheral blood and CSF of HAM/TSP patients and HTLV-1–seronegative normal donors (NDs). They reported that compared to NDs, TCR β repertoires of HAM/TSP patients were more clonally expanded in both PBMCs and CSF. In HAM/TSP, there was considerable overlap of TCR β clonotypes between the expanded clones of PBMCs and CSF. Importantly, 9.1% of expanded CSF clonotypes were unique to the CNS compartment, suggesting some degree of intrathecal enrichment (Nozuma et al., 2021). Studies have also investigated the extent to which HTLV-1 infects resident neurological cells. They demonstrated that HTLV-1 *tax* mRNA colocalized with glial fibrillary acidic protein (GFAP), which suggested that astrocytes were susceptible to direct HTLV-1 infection (Lehky & Jacobson, 1995; Méndez et al., 1997). Additionally, astrocytic glioma transfected with HTLV-1 in vitro upregulated the expression of TNF-α and may therefore serve as immunological targets for CTLs (Méndez et al., 1997). To date, however, HTLV-1 has not been shown to productively infect neurons, oligodendrocytes, or microglia in vivo (Matsuura et al., 2015). While astrocytic infection may contribute to tissue damage, it is largely suggested that most of the pathophysiology in HAM/TSP is driven by aberrant immune responses.

26.2.6 Proinflammatory Lymphocytes Accumulate in the CNS in HAM/TSP

An interesting case report appeared in the New England Journal of Medicine in 1997 in which a spinal cord biopsy was obtained from a living patient with rapidly progressive HAM/TSP (Levin et al., 1997). They reported that the meninges, parenchyma, and perivascular tissue demonstrated hypercellularity with a high concentration of mononuclear infiltrates. The infiltrates largely consisted of activated T cells, most of which were $CD8^+$ cells with a smaller subset of $CD4^+$ cells, and macrophages. Cultures derived from the biopsy showed HTLV-1-specific CTL activity. They also reported that PVL of the CSF was more than ten times higher than that of the peripheral blood. Later studies have validated and expanded these findings. Pathological studies using PCR demonstrated that in the spinal cord of HAM/TSP patients, the majority of HTLV-1 DNA was localized to inflammatory T lymphocyte infiltrates (Matsuoka et al., 1998). The amount of HTLV-1 DNA was directly correlated with the frequency of $CD4^+$ infiltrates, but not with the frequency of infiltrating $CD8^+$ cells or macrophages (Kubota et al., 1994). Using in situ hybridization, it has been shown that HTLV-1 *tax* mRNA was also detected in $CD4^+$ infiltrates in active lesions of the spinal cord (Moritoyo et al., 1996). While early in the disease course infiltrates contain equal numbers of $CD4^+$ T cells, $CD8^+$ T cells, and foamy macrophages, $CD8^+$ T cells predominate in later stages of disease progression (Nakagawa et al., 1995). Matsuura et al. (2015) investigated the frequency of HTLV-1 Tax-specific CTL infiltrates in the spinal cord of HAM/TSP patients using confocal scanning laser microscopy and MHC/Tax tetramers. They reported that more than 20% of the $CD8^+$ infiltrates in the spinal cord were specific for HTLV-1 Tax. The infiltration of $CD8^+$ CTLs in the affected spinal cord has also been characterized as TIA-1 positive (Anderson et al., 1990; Umehara et al., 1994). TIA-1 is a monoclonal antibody that recognizes a 15-kDa granule-associated protein contained in CTLs and natural killer (NK) cells. Double staining for anti-CD8 and TIA revealed that 80% of TIA-1^+ infiltrating cells throughout the parenchyma and perivascular area were positive for CD8. Collectively, these findings suggest that there is a robust focus of inflammation in the CSF and spinal cord of patients with HAM/TSP.

26.2.7 Blood-Brain Barrier Dysregulation and Bystander Degradation of Neuronal Tissue

As depicted in Fig. 26.2, the migration of immune cells from the periphery into the CNS is dependent upon permeabilization of the blood-brain barrier (BBB). Under physiological conditions, the BBB regulates the passage of ions, molecules, and cells into the CNS to prevent perturbation and maintain homeostasis. The barrier separates peripheral blood from CSF and is maintained by endothelial cells, pericytes, and astrocytes. Studies have suggested that endothelial cells of the BBB may be susceptible to HTLV-1 infection (Ho et al., 1984). Additionally, the co-culture of endothelial cells with infected T cells increased paracellular permeability, which would allow for the migration of both infected and uninfected lymphocytes from peripheral compartments into the CNS (Afonso et al., 2008). This increased permeability was associated with decreased expression of tight junction-associated proteins ZO-1 and occludin (Afonso et al., 2008). High expression of vascular cell adhesion molecule-1 (VCAM-I) on the endothelium has been demonstrated (Umehara et al., 1996). Expression of various late antigen-4 (VLA-4) and monocyte chemoattractant protein-1 (MCP-1) was also upregulated in the infiltrating cells in active-chronic lesions of affected spinal cords. ICAM-1 and its counterpart molecule LFA-1 are also suggested to be

Immunopathogenesis of HAM/TSP

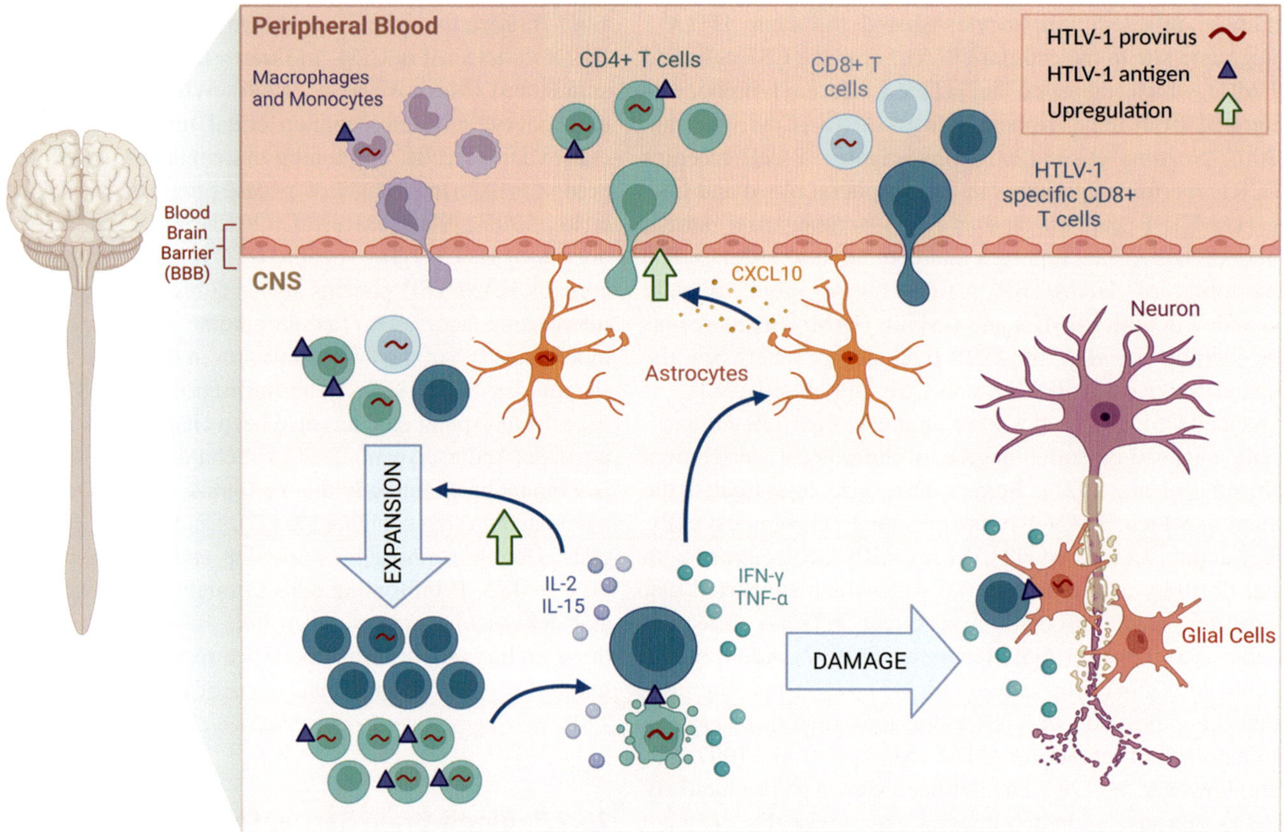

Fig. 26.2 *Proposed immunological cascade in HAM/TSP.* In HAM/TSP, damage to the CNS is suggested to be mediated by bystander degradation. HTLV-1-infected CD4+ and CD8+ T cells migrate from peripheral blood into the CNS compartment across a dysregulated blood-brain barrier (BBB). Uninfected T cells directed against HTLV-1 proteins (namely Tax) also migrate into the CNS. Immune infiltrates undergo clonal expansion within the CNS. CD8+ and CD4+ T cells may then secrete proinflammatory chemokines and cytokines (including IL-2, IL-15, IFN-γ, and TNF-α), which upregulate both the migration of immune cells into the CNS, by promoting astrocytic secretion of CXCL10, and the further expansion of immune populations. In this way, a self-perpetuating focus of inflammation is established within the CNS. Finally, cytotoxic cytokines are thought to contribute to demyelination, neuronal death, and tissue atrophy. (Adapted from "Pathogenesis of Multiple Sclerosis" by BioRender.com (2023). Retrieved from https://app.biorender.com/biorender-templates)

involved in lymphocyte infiltration into the spinal cord (Cabre et al., 1999). Additionally, it has been suggested that initial entry of HTLV-1-infected $CD4^+$ T cells can generate a positive feedback loop that results in the recruitment of more immune cells. Using a co-culture of astrocytoma and $CD4^+$ T cells, Ando et al. (2013) demonstrated that astrocytoma released chemokine interferon-γ inducible protein 10 kDa (CXCL10) in response to the IFN-γ secreted by $CD4^+$ T cells. On histology, they confirmed that astrocytes were the primary producers of CXCL10 in the human spinal cord. CXCL10 was then capable of binding the receptor CXCR3, preferentially expressed on activated $CD8^+$ T cells and Th1 cells, to induce further lymphocyte migration into the CNS (Ando et al., 2013; Groom & Luster, 2011; Hancock et al., 2000). In this way, the initial entry of immune cells into the CNS can establish a self-perpetuating focus of inflammation.

In HAM/TSP, these immune infiltrates may secrete cytotoxic products that in turn precipitate the death and damage of resident neurological cells in a mechanism referred to as bystander degradation (Fig. 26.2). Postmortem analysis has demonstrated the presence of $CD4^+$ and $CD8^+$ T cells, foamy macrophages, and fibrillary gliosis in spinal cord lesions of HAM/TSP patients (Levin et al., 1997; Takenouchi et al., 2004). Tissue damage is primarily manifested as chronic meningomyelitis and axonal degeneration and is associated with an elevation in proinflammatory cytokines including TNF-α, IFN-γ, and interleukin (IL)-1β. Lymphocyte exudation occurs at random in both gray and white matter but appears more frequently in deeper portions of the spinal cord compared to more superficial regions. Analysis of spinal cord tissue also demonstrated that $CD8^+$ infiltrates were often in contact with apoptotic oligodendrocytes, likely contributing to demyelination in lesions (Matsuura et al., 2015).

B cells are also found in the lesions but mainly located in perivascular spaces of larger vessels (Iwasaki, 1993). Several infiltrating macrophages and microglia are activated with the expression of MRP14 and MRP8 (Abe et al., 1999). Chronic inflammation is primarily seen in the mid-to-lower spinal cord, especially at the thoracic region. Lymphocytic infiltrates and foamy macrophages are seen in the thickened meninges, parenchyma, and perivascular area of affected regions.

26.3 Characterizing Neurodegeneration in HAM/TSP Using MRI

26.3.1 HAM/TSP Is Characterized by Profound Spinal Cord Atrophy and Lesion Load

It has been widely demonstrated that spinal cord atrophy is a robust and reproducible characteristic of HAM/TSP (Alcindor et al., 1992; Azodi et al., 2017; Evangelou et al., 2014; Kermode et al., 1990; Mina et al., 2021; Ogata et al., 1993; Taniguchi et al., 2017; Vilchez et al., 2014). One of the earliest reports of spinal cord atrophy in HAM/TSP using MRI came in 1990. The study was limited in sample size and included only 9 HAM/TSP patients and 9 MS patients (Kermode et al., 1990). They reported that 6 of the 9 HAM/TSP patients evaluated presented with thoracic cord atrophy, but that this was not disease specific as a similar proportion of MS patients (5 of 9 patients) also presented with thoracic cord atrophy. A similar study was also unable to differentiate between HAM/TSP patients and ACs when atrophy was evaluated in either the cervical or thoracic spine (Howard et al., 2003). Early studies did have success, however, in correlating spinal cord atrophy with pathological changes. This is evidenced by a case report of a 61-year-old male with HAM/TSP who had undergone MRI imaging shortly before his death (Ogata et al., 1993). The patient, who had a slowly progressive disease course, had significant atrophy of the thoracic cord by MRI acquired 1 year prior to death. Subsequent postmortem analysis confirmed severe atrophy of the thoracic cord and suggested that atrophy was due to a loss of myelin sheaths and axons (Ogata et al., 1993). Most reports of spinal cord atrophy in HAM/TSP have focused on the thoracic cord, but some studies have also looked exclusively at the cervical spine (Alcindor et al., 1992). It is reported that a small proportion of HAM/TSP patients had abnormal cervical spines characterized by either atrophy or lesion formation. To summarize, these incipient reports suggested that HAM/TSP was characterized by some degree of spinal cord atrophy, primarily in the thoracic cord, but that this atrophy may not have been disease specific.

Differences in spinal cord volume that are clinically significant, however, may not be visually apparent. New technological advances have allowed researchers to quantify spinal cord volume, eliminating the reliance on qualitative evaluation. Using these techniques, several studies have more precisely examined the extent and pattern of atrophy in HAM/TSP and revisited comparisons to ACs, uninfected HCs and MS (Alcindor et al., 1992). Studies from our group, Azodi et al., (2017) characterized spinal cord atrophy in HAM/TSP by implementing two complementary methods of comparison; first, we used a cross-sectional approach to compare atrophy across patient and control groups, and second, we used a longitudinal approach to track changes in cord atrophy as the disease progressed. The input images used in these analyses were T1-weighted scans of the cervical and thoracolumbar spine. Representative mid-sagittal images of the thoracolumbar spine are shown for a healthy volunteer and a patient with HAM/TSP in Fig. 26.3a. As shown in the figure, there is clear spinal cord thinning in the thoracic cord of the HAM/TSP patient compared to the healthy volunteer. Cervical and thoracolumbar scans were then stitched together to create a single scan. A novel analysis tool was used to delineate spinal cord edges on the sagittal image, which allowed for selection of the ventral edge of the spinal cord. Cross-sectional area was calculated from axial images perpendicular to the selected ventral cord edge and plotted from C1 to T10 (Fig. 26.3b). While mean SCCSA did not differ between ACs and HCs at any region of interest, it was significantly reduced in HAM/TSP at all regions evaluated (Fig. 26.3b). Spinal cord volume was also characterized longitudinally in two rapidly progressing HAM/TSP patients over the course of 32 and 38 months (Azodi et al., 2017). Both patients presented with spinal cord volumes indistinguishable from HCs at the first timepoint. As the disease progressed, however, they displayed significant atrophy that began in the thoracic region and rose to the cervical cord. Volume loss in the thoracic and cervical cord of HAM/TSP patients was as great as 38% and 32%, respectively (Evangelou et al., 2014). A general schematic of disease progression in HAM/TSP with respect to spinal cord volume is summarized in Fig. 26.4b. Patients with HAM/TSP may present with healthy spinal cord volumes early in the disease course in which atrophy progresses from the thoracic region and is associated with neurodegeneration (Fig. 26.4b). By comparison, spinal cord atrophy in MS appears to begin in the cervical region and is often less severe than in HAM/TSP (Liu et al., 2014; Mina et al., 2021). Future research will include larger cohorts of HAM/TSP patients at multiple timepoints.

MRI as well as pathological studies have also shown that atrophy of the spinal cord predominates in certain anatomical regions. Atrophy is most severe in the anteroposterior direction and preferentially occurs in the white matter of the lateral columns, and to a lesser extent in the posterior columns (Akizuki et al., 1988; Taniguchi et al., 2017; Yoshioka

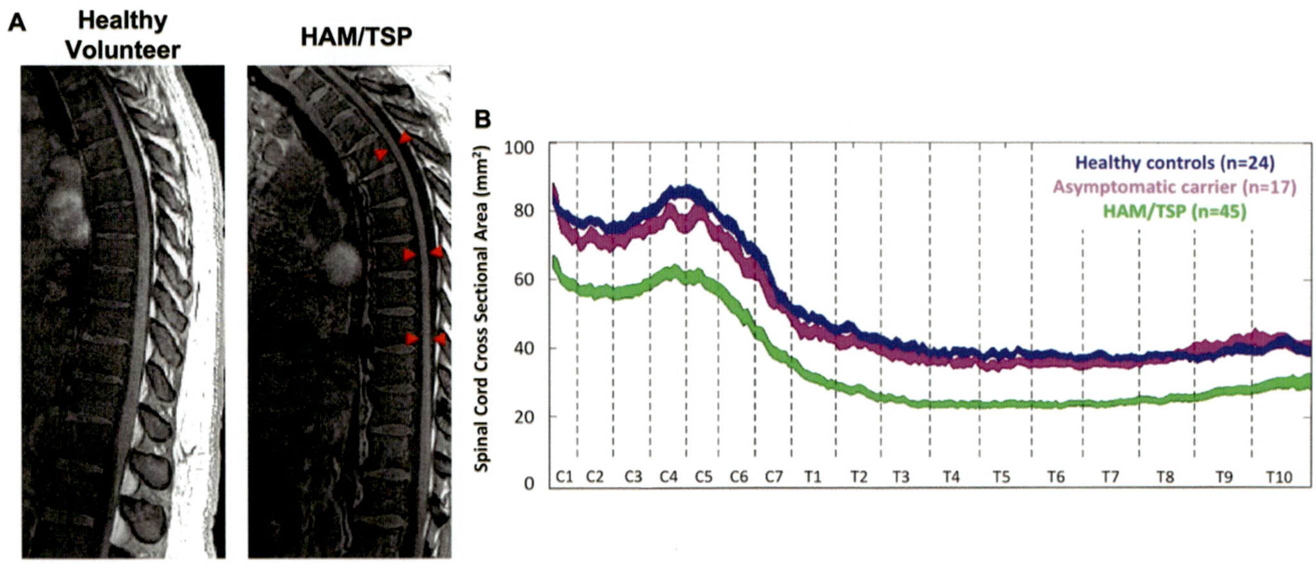

Fig. 26.3 *HAM/TSP is characterized by profound spinal cord atrophy.* (**a**) Representative T1-weighted MRI of the thoracolumbar spine from a healthy volunteer and a patient with HAM/TSP. Red arrow heads point to regions with visible cord thinning in the HAM/TSP patient compared to the healthy volunteer. (**b**) SCCSA profiles were generated by plotting cross-sectional area from C1 to T10. Average area profiles of healthy controls (blue), asymptomatic carriers of HTLV-1 (pink), and patients with HAM/TSP (green) are shown. While there is no difference between the cross-sectional area averages for healthy controls and asymptomatic carriers, the HAM/TSP averages are significantly lower than both groups at all levels of the cord. Panel b adapted from Azodi et al., 2017. (Reprinted with permission from Wiley)

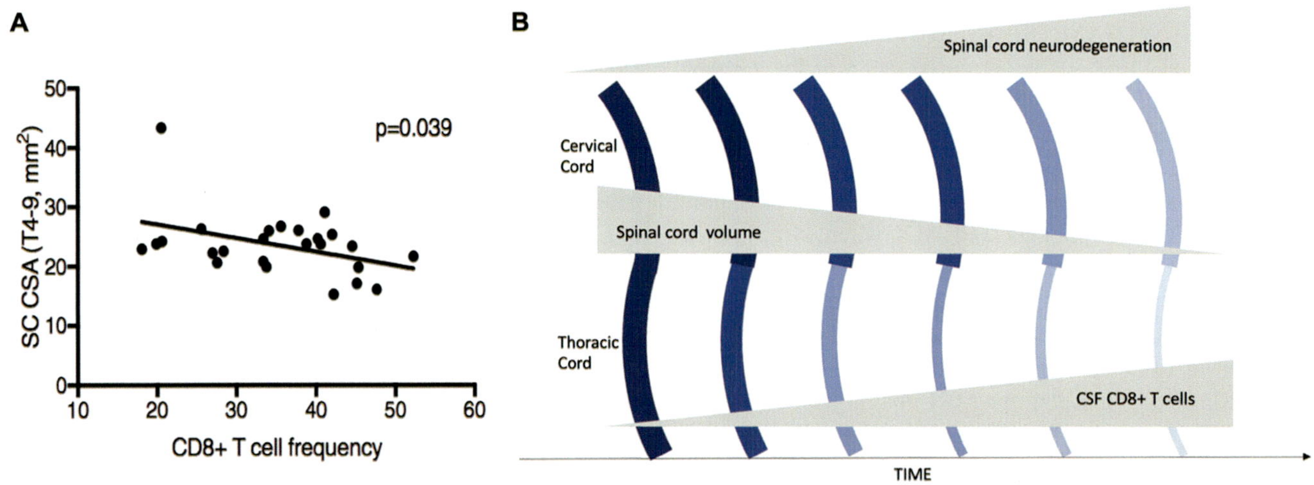

Fig. 26.4 *SCCSA in HAM/TSP is associated with clinical and immunological markers of disease progression.* (**a**) Average SCCSA from T4-9 is significantly negatively correlated with $CD8^+$ T cell frequency in the CSF. (**b**) In HAM/TSP, spinal cord neurodegeneration begins in the thoracic cord and gradually rises to the cervical cord as the disease progresses. This degeneration is associated with loss of spinal cord volume and an increase in $CD8^+$ T cells in the CSF. Panel a reproduced from Azodi et al. (2017). (Reprinted with permission from Wiley. Panel b reprinted with permission from Azodi (2018). *Updates in Myelopathy Through the Lens of Training and Research*)

et al., 1993). Most immediately relevant to the motor deficits observed in the disease is this loss of neuronal integrity in the lateral columns. The lateral columns of the spinal cord house the corticospinal tract, which gives rise to the nerves that innervate distal extremities for voluntary motor function. Pathological studies reveal that this atrophy is largely attributable to myelin pallor and loss of astroglia, neuronal cell bodies, and axons (Akizuki et al., 1988). Degradation of the lateral columns, therefore, likely underscores the gradual lower extremity weakness and spasticity that characterizes HAM/TSP. In addition to gross volume loss, lesions are also observed in the spinal cord by MRI in HAM/TSP (Vilchez et al., 2014). Lesions tend to be bilateral and appear most frequently in antero-lateral sections of the cord (Umehara et al., 2007; Vilchez et al., 2014). Less frequently, these lesions also form in the posterior columns and posterior horns (Umehara et al., 2007). Well aligned with this finding, diffuse tensor imaging (DTI) studies have demonstrated decreased fractional anisotropy scores of the HAM/TSP spinal cord, indicative of reduced myelin integrity (Vilchez et al., 2014).

26.3.2 Brain Abnormalities in HAM/TSP

Although HAM/TSP is predominately a myelopathic disease, brain abnormalities have also been described. Several studies have attempted to characterize possible abnormalities in the HAM/TSP brain compared to HCs, ACs, or other neurodegenerative diseases. The most frequently reported structural abnormality observed in HAM/TSP is elevated lesion load, but results have been contradictory and as of yet, no consensus has been reached regarding the prevalence or role of lesions in disease development (Bagnato et al., 2005; Fukushima et al., 1994; Godoy et al., 1995; Kalil et al., 2021; Morgan et al., 2007; Ogata et al., 1993; Vilchez et al., 2014). Kalil et al. (2021) reported that nearly twice the number of HAM/TSP patients had brain white matter lesions compared to the number of ACs (HAM/TSP: 11/19, 84% vs. AC: 5/17, 47%). Lesions were most commonly periventricular, but were also found in the centrum semiovale, corona radiata, frontal lobes, occipito-parietal lobe, and pons. Lesions, however, were assessed on a simple binary (present or absent) and no information on the severity of lesion load was included in the report. Additionally, lesions are observed in neurologically healthy aging populations and the average age of patients with HAM/TSP (51.5 years) was significantly higher than ACs (43.4 years; $p = 0.047$). As discussed previously, a subset of patients with HAM/TSP present with cognitive deficits, indicating a potential for some degree of brain involvement (Fukushima et al., 1994; Gascón et al., 2017; Kalil et al., 2021; Román & Román, 1988). The tendency of lesions in HAM/TSP to form more frequently in periventricular regions than other locations, however, is a reproducible finding (Godoy et al., 1995; Ogata et al., 1993; Puccioni-Sohler et al., 2012; Vilchez et al., 2014). It is hypothesized that the tendency of lesions to form around ventricles may result from the migration of proinflammatory infiltrates from circulating CSF into the brain parenchyma.

In addition to parenchymal white matter lesions, preliminary research suggests that a subset of HAM/TSP patients have enhancement of the brain surface by MRI. A recent study reported that 56% (14/25) of HAM/TSP patients had leptomeningeal enhancement (LME) on postcontrast T2 images compared to 23% (3/13) of ACs and 8% (5/66) of HVs (Absinta et al., 2017). They reported that LMEs were frequently supratentorial and located near one or more blood vessels. LMEs were not disease specific, however, and were reported in a variety of inflammatory neurological conditions including MS, immune-mediated encephalitis, and Susac syndrome. Some studies have demonstrated that LMEs were associated with T cells, B cells, and macrophages (Absinta et al., 2015). LMEs, therefore, may serve as another radiological marker of neuroinflammatory disease. Finally, studies have suggested that whole brain atrophy is apparent in a subset of patients with HAM/TSP. Puccioni-Sohler et al. (2012) evaluated the degree of brain atrophy by qualitatively assessing the prominence of cortical sulci and the relative size of the lateral ventricles. They reported that 14% of patients had some signs of brain atrophy, but no direct comparison was made between patients and any control group.

While the occurrence of brain abnormalities in patients with HAM/TSP is varied, most studies suggest that there is no correlation between these abnormalities and symptom presentation. Fukushima et al. (1994) investigated the relationship between cerebral abnormalities and cognitive impairment assessed by P300 and N200 wave latency, an indicator of cognitive event related potentials (ERPs). They observed a delay in P300 and N200 in patients with HAM/TSP using auditory tones, but this was not correlated with the presence of lesions on MRI. Subsequent studies, which included other cognitive and motor deficits, also reported no association between brain lesions and cognitive impairment in HAM/TSP (Bagnato et al., 2005; Morgan et al., 2007).

26.3.3 Radiological Findings Correlate with Clinical and Biological Markers in HAM/TSP

Radiological abnormalities in HAM/TSP are well correlated with clinical parameters of disease progression. Since HAM/TSP is largely characterized by motor deficits, most disease progression rating systems are heavily skewed toward grading these symptoms. Two common systems for rating disability in HAM/TSP are the Expanded Disability Status Scale (EDSS) and Instituto de Pesquisas de Cananeia (IPEC), both of which indicate increasing disability with higher scores. There is a significant negative correlation between SCCSA in the thoracic cord (T4-9) and disease duration, EDSS, and IPEC score. Additionally, as shown in Fig. 26.4, spinal cord atrophy is correlated with several laboratory-defined virological and immunological markers. When quantified, spinal cord atrophy in both the cervical and thoracic cord was positively correlated with HTLV-1 PVL (Charlip et al., 2015). Furthermore, we characterized cellular immunophenotypes ($CD4^+$ T cells, $CD8^+$ T cells, B cells, monocytes, and natural killer cells) in the peripheral blood and CSF of patients with HAM/TSP using flow cytometry (Azodi et al., 2017). We compared the pattern of immune markers with each patients' spinal cord volume in cervical and thoracic regions. While there was no correlation between immune markers in peripheral blood and spinal cord atrophy, there was a statistically significant negative correlation between $CD8^+$ T cell frequency in the CSF and thoracic SCCSA (Fig. 26.4a), i.e., the greater the frequency of $CD8^+$ T cell in the CSF, the more atrophic the spinal cord, consistent with an immunopathogenic role of $CD8^+$ CTLs in the HAM/TSP disease progression (Fig. 26.4b). These findings suggest that imaging of the spinal cord, specifically the thoracic cord, using MRI may be a noninvasive and robust indicator of disease severity in HAM/TSP that may reflect the level of intrathecal proinflammatory immune infiltrates.

26.4 Concluding Remarks

HAM/TSP is a progressive myelopathy characterized by profound inflammatory responses, which are elevated in the CNS compared to the periphery. The neurodegeneration that underscores disease progression is thought to be mediated by these aberrant anti-HTLV-1 immune responses. Localization of HTLV-1 in $CD4^+$ cells in the CNS may lead to the production of cytokines/chemokines thought to adversely affect resident cells in the CNS (bystander hypothesis) resulting in neurodegeneration of the long tract. Histological studies have demonstrated that although $CD4^+$ cells are the primary cellular reservoir of HTLV-1 provirus in vivo, other cells may also be infected with HTLV-1 including $CD8^+$ T cells, epithelial cells of the BBB, and astrocytes. All these cells could then express HTLV-1 antigens (processed HTLV-1 peptides in association with HLA class I molecules) and be recognized as targets by inflammatory, HTLV-1-specific CTLs. Recognition by CTLs could initiate a molecular cascade that results in the secretion of cytotoxic products within the CNS, precipitating the death or dysfunction of neuronal tissue and leading to clinical symptoms of HAM/TSP. Although it has long been recognized that HTLV-1 induces chronic central nervous system dysfunction, several mechanisms associated with infection remain unclear. New factors for the development of HAM/TSP have recently been demonstrated. Differences between HAM/TSP patients and ACs may include host HLA allelic expression, HTLV-1 PVL, HTLV-1 gene expression rate, infected $CD4^+$ cell phenotype, and CTL activating factors. Elevated proliferation of $CD4^+$ and $CD8^+$ cells in vivo may explain the continuous Tax expression, which may, in turn, serve to drive strong T cell responses. Further information regarding the frequency, distribution, activity, and function of CTLs in the CNS of HAM/TSP, as well as that of other HTLV-1-associated disorders, is needed to better understand the role that this virus plays in these diseases. Disease progression in HAM/TSP has also been radiologically characterized, demonstrating progressive loss of spinal cord volume, beginning in the thoracic cord, and rising to the cervical cord, the formation of spinal cord lesions, and some degree of brain involvement, although disease specificity has yet to be demonstrated for these brain abnormalities. Spinal cord atrophy is strongly correlated with both clinical and laboratory-defined parameters of HAM/TSP including disease duration, EDSS score, HTLV-1 PVL, and $CD8^+$ T cell frequency in the CSF, making it a robust indicator of disease progression.

26.5 Review Questions

1. Which cell population serves as the primary reservoir of HTLV-1 PVL in vivo?
   (a) Neurons
   (b) $CD8^+$ T cells
   (c) *CD 4^+ T cells*
2. Which region of the world is endemic for HTLV-1?
   (a) Europe
   (b) *The Caribbean*
   (c) North America
3. What is the most productive/infectious route of transmission of HTLV-1?
   (a) Vertical
   (b) Sexual
   (c) *Parenteral*
4. What is a major immunological difference observed between asymptomatic carriers of HTLV-1 and patients with HAM/TSP?
   (a) Asymptomatic carriers never have detectable viral DNA in peripheral blood or CSF, while HAM/TSP patients always have both
   (b) Viral DNA is primarily detected in $CD8^+$ T cells in asymptomatic carriers and in $CD4^+$ T cells in HAM/TSP
   (c) *PVL is elevated in HAM/TSP compared to asymptomatic carriers, but there is no evidence for a difference in the cellular distribution of viral DNA*
5. Which level of the spinal cord is predominantly affected in patients with HAM/TSP?
   (a) Cervical
   (b) *Thoracic*
   (c) Lumbar
6. Which resident CNS cells are suggested to be susceptible to HTLV-1 infection?
   (a) Oligodendrocytes
   (b) Neurons
   (c) *Astrocytes*
7. Approximately what percent of HTLV-1 carriers will develop HAM/TSP?
   (a) *Less than 5%*
   (b) 5–10%
   (c) Greater than 25%

8. What CNS changes are most strongly associated with HAM/TSP?
   (a) The appearance of large, confluent lesions in the brain that generally form superior to the lateral ventricles and increase in size as the disease progresses
   (b) *Gradual atrophy of the spinal cord that begins in the thoracic region and rises to the cervical cord as the disease progresses*
   (c) The CNS is grossly unremarkable in HAM/TSP
9. The frequency of which immune cell within the CSF has been shown to be positively correlated with spinal cord atrophy?
   (a) Macrophages
   (b) CD 4^+ T cells
   (c) *CD8⁺ T cells*
10. What is the suggested mechanism of neurodegeneration and tissue damage in HAM/TSP?
   (a) Direct HTLV-1 infection of oligodendrocytes results in progressive demyelination and neuronal loss
   (b) *Bystander damage of resident tissue due to the secretion of proinflammatory cytokines and chemokines by infiltrating immune cells*
   (c) $CD8^+$ T cells specific for HTLV-1 proteins are recruited to the CNS and initiate direct killing of neurons and glia through Fas-mediated apoptosis-FundingThis research was supported by the Division of Intramural Research of NINDS.

References

Abe, M., Umehara, F., Kubota, R., Moritoyo, T., Izumo, S., & Osame, M. (1999). Activation of macrophages/microglia with the calcium-binding proteins MRP14 and MRP8 is related to the lesional activities in the spinal cord of HTLV-I associated myelopathy. *Journal of Neurology, 246*(5), 358–364. https://doi.org/10.1007/s004150050363

Absinta, M., Vuolo, L., Rao, A., Nair, G., Sati, P., Cortese, I. C. M., Ohayon, J., Fenton, K., Reyes-Mantilla, M. I., Maric, D., Calabresi, P. A., Butman, J. A., Pardo, C. A., & Reich, D. S. (2015). Gadolinium-based MRI characterization of leptomeningeal inflammation in multiple sclerosis. *Neurology, 85*(1), 18–28. https://doi.org/10.1212/wnl.0000000000001587

Absinta, M., Cortese, I. C., Vuolo, L., Nair, G., de Alwis, M. P., Ohayon, J., Meani, A., Martinelli, V., Scotti, R., Falini, A., Smith, B. R., Nath, A., Jacobson, S., Filippi, M., & Reich, D. S. (2017). Leptomeningeal gadolinium enhancement across the spectrum of chronic neuroinflammatory diseases. *Neurology, 88*(15), 1439–1444. https://doi.org/10.1212/wnl.0000000000003820

Afonso, P. V., Ozden, S., Cumont, M.-C., Seilhean, D., Cartier, L., Rezaie, P., Mason, S., Lambert, S., Huerre, M., Gessain, A., Couraud, P.-O., Pique, C., Ceccaldi, P.-E., & Romero, I. A. (2008). Alteration of blood–brain barrier integrity by retroviral infection. *PLoS Pathogens, 4*(11), e1000205. https://doi.org/10.1371/journal.ppat.1000205

Akizuki, S., Setoguchi, M., Nakazato, O., Yoshida, S., Higuchi, Y., Yamamoto, S., & Okajima, T. (1988). An autopsy case of human T-lymphotropic virus type I-associated myelopathy. *Human Pathology, 19*(8), 988–990. https://doi.org/10.1016/s0046-8177(88)80017-0

Alcindor, F., Valderrama, R., Canavaggio, M., Lee, H., Katz, A., Montesinos, C., Madrid, R. E., Merino, R. R., & Pipia, P. A. (1992). Imaging of human T-lymphotropic virus type I-associated chronic progressive myeloneuropathies. *Neuroradiology, 35*(1), 69–74. https://doi.org/10.1007/bf00588283

Altman, J. D., Moss, P. A., Goulder, P. J., Barouch, D. H., McHeyzer-Williams, M. G., Bell, J. I., McMichael, A. J., & Davis, M. M. (1996). Phenotypic analysis of antigen-specific T lymphocytes. *Science, 274*(5284), 94–96. https://doi.org/10.1126/science.274.5284.94

Anderson, P., Nagler-Anderson, C., O'Brien, C., Levine, H., Watkins, S., Slayter, H. S., Blue, M. L., & Schlossman, S. F. (1990). A monoclonal antibody reactive with a 15-kDa cytoplasmic granule-associated protein defines a subpopulation of CD8+ T lymphocytes. *Journal of Immunology, 144*(2), 574–582.

Ando, H., Sato, T., Tomaru, U., Yoshida, M., Utsunomiya, A., Yamauchi, J., Araya, N., Yagishita, N., Coler-Reilly, A., Shimizu, Y., Yudoh, K., Hasegawa, Y., Nishioka, K., Nakajima, T., Jacobson, S., & Yamano, Y. (2013). Positive feedback loop via astrocytes causes chronic inflammation in virus-associated myelopathy. *Brain, 136*(Pt 9), 2876–2887. https://doi.org/10.1093/brain/awt183

Araujo, A., Bangham, C. R. M., Casseb, J., Gotuzzo, E., Jacobson, S., Martin, F., Penalva de Oliveira, A., Puccioni-Sohler, M., Taylor, G. P., & Yamano, Y. (2021). Management of HAM/TSP: Systematic review and consensus-based recommendations 2019. *Neurology Clinical Practice, 11*(1), 49–56. https://doi.org/10.1212/cpj.0000000000000832

Araya, N., Sato, T., Ando, H., Tomaru, U., Yoshida, M., Coler-Reilly, A., Yagishita, N., Yamauchi, J., Hasegawa, A., Kannagi, M., Hasegawa, Y., Takahashi, K., Kunitomo, Y., Tanaka, Y., Nakajima, T., Nishioka, K., Utsunomiya, A., Jacobson, S., & Yamano, Y. (2014). HTLV-1 induces a Th1-like state in CD4+CCR4+ T cells. *The Journal of Clinical Investigation, 124*(8), 3431–3442. https://doi.org/10.1172/jci75250

Azimi, N., Jacobson, S., Leist, T., & Waldmann, T. A. (1999). Involvement of IL-15 in the pathogenesis of human T lymphotropic virus type I-associated myelopathy/tropical spastic paraparesis: Implications for therapy with a monoclonal antibody directed to the IL-2/15R beta receptor. *Journal of Immunology, 163*(7), 4064–4072.

Azodi, S., Nair, G., Enose-Akahata, Y., Charlip, E., Vellucci, A., Cortese, I., Dwyer, J., Billioux, B. J., Thomas, C., Ohayon, J., Reich, D. S., & Jacobson, S. (2017). Imaging spinal cord atrophy in progressive myelopathies: HTLV-I-associated neurological disease (HAM/TSP) and multiple sclerosis (MS). *Annals of Neurology, 82*(5), 719–728. https://doi.org/10.1002/ana.25072

Bagnato, F., Butman, J. A., Mora, C. A., Gupta, S., Yamano, Y., Tasciyan, T. A., Solomon, J. M., Santos, W. J., Stone, R. D., McFarland, H. F., & Jacobson, S. (2005). Conventional magnetic resonance imaging features in patients with tropical spastic paraparesis. *Journal of Neurovirology, 11*(6), 525–534. https://doi.org/10.1080/13550280500385039

Bangham, C. R., Meekings, K., Toulza, F., Nejmeddine, M., Majorovits, E., Asquith, B., & Taylor, G. P. (2009). The immune control of HTLV-1 infection: Selection forces and dynamics. *Frontiers in Bioscience (Landmark Ed), 14*(8), 2889–2903. https://doi.org/10.2741/3420

Bangham, C. R., Araujo, A., Yamano, Y., & Taylor, G. P. (2015). HTLV-1-associated myelopathy/tropical spastic paraparesis. *Nature Reviews. Disease Primers, 1*, 15012. https://doi.org/10.1038/nrdp.2015.12

Biddison, W. E., Kubota, R., Kawanishi, T., Taub, D. D., Cruikshank, W. W., Center, D. M., Connor, E. W., Utz, U., & Jacobson, S. (1997). Human T cell leukemia virus type I (HTLV-I)-specific CD8+ CTL clones from patients with HTLV-I-associated neurologic disease secrete proinflammatory cytokines, chemokines, and matrix metalloproteinase. *Journal of Immunology, 159*(4), 2018–2025.

Bomford, R., Kazanji, M., & De The, G. (1996). Vaccine against human T cell leukemia-lymphoma virus type I: Progress and prospects. *AIDS Research and Human Retroviruses, 12*(5), 403–405. https://doi.org/10.1089/aid.1996.12.403

Brunetto, G. S., Massoud, R., Leibovitch, E. C., Caruso, B., Johnson, K., Ohayon, J., Fenton, K., Cortese, I., & Jacobson, S. (2014). Digital droplet PCR (ddPCR) for the precise quantification of human T-lymphotropic virus 1 proviral loads in peripheral blood and cerebrospinal fluid of HAM/TSP patients and identification of viral mutations. *Journal of Neurovirology, 20*(4), 341–351. https://doi.org/10.1007/s13365-014-0249-3

Cabre, P., al-Fahim, A., & Oger, J. (1999). Enhanced adherence of endothelial cells to blood mononuclear cells in HAM/TSP. *Revue Neurologique (Paris), 155*(4), 273–279. (Hyperadhérence aux cellules endothéliales des cellules mononucléées sanguines au cours de la HAM/TSP.).

Carneiro-Proietti, A. B., Amaranto-Damasio, M. S., Leal-Horiguchi, C. F., Bastos, R. H., Seabra-Freitas, G., Borowiak, D. R., Ribeiro, M. A., Proietti, F. A., Ferreira, A. S., & Martins, M. L. (2014). Mother-to-child transmission of human T-cell Lymphotropic Viruses-1/2: What we know, and what are the gaps in understanding and preventing this route of infection. *Journal of the Pediatric Infectious Diseases Society, 3*(Suppl 1), S24–S29. https://doi.org/10.1093/jpids/piu070

Catalan-Soares, B. C., Carneiro-Proietti, A. B., Da Fonseca, F. G., Correa-Oliveira, R., Peralva-Lima, D., Portela, R., Ribas, J. G., Gonçalves, D. U., & Proietti, F. A. (2009). HLA class I alleles in HTLV-1-associated myelopathy and asymptomatic carriers from the Brazilian cohort GIPH. *Medical Microbiology and Immunology, 198*(1), 1–3. https://doi.org/10.1007/s00430-008-0096-z

Cavrois, M., Gessain, A., Gout, O., Wain-Hobson, S., & Wattel, E. (2000). Common human T cell leukemia virus type 1 (HTLV-1) integration sites in cerebrospinal fluid and blood lymphocytes of patients with HTLV-1—Associated myelopathy/tropical spastic Paraparesis indicate that HTLV-1 crosses the blood-brain barrier via clonal HTLV-1—Infected cells. *The Journal of Infectious Diseases, 182*(4), 1044–1050. https://doi.org/10.1086/315844

Charlip, E., Liu, W., Massoud, R., Billioux, J., Caruso, B., Ohayon, J., Nair, G., & Jacobson, S. (2015). Patterns of spinal cord atrophy in HTLV-1 associated myelopathy/ tropical spastic paraparesis (HAM/TSP). *Retrovirology, 12*(1), P32. https://doi.org/10.1186/1742-4690-12-S1-P32

Coffin, J. M. (2015). The discovery of HTLV-1, the first pathogenic human retrovirus. *Proceedings of the National Academy of Sciences of the United States of America, 112*(51), 15525–15529. https://doi.org/10.1073/pnas.1521629112

Croda, M. G., de Oliveira, A. C., Vergara, M. P., Bonasser, F., Smid, J., Duarte, A. J., & Casseb, J. (2008). Corticosteroid therapy in TSP/HAM patients: The results from a 10 years open cohort. *Journal of the Neurological Sciences, 269*(1–2), 133–137. https://doi.org/10.1016/j.jns.2008.01.004

Deschamps, R., Béra, O., Belrose, G., Lezin, A., Bellance, R., Signate, A., Cabre, P., Smadja, D., Césaire, R., & Olindo, S. (2010). Absence of consistent association between human leukocyte antigen-I and -II alleles and human T-lymphotropic virus type 1 (HTLV-1)-associated myelopathy/tropical spastic paraparesis risk in an HTLV-1 French Afro-Caribbean population. *International Journal of Infectious Diseases, 14*(11), e986–e990. https://doi.org/10.1016/j.ijid.2010.05.020

Elovaara, I., Koenig, S., Brewah, A. Y., Woods, R. M., Lehky, T., & Jacobson, S. (1993). High human T cell lymphotropic virus type 1 (HTLV-1)-specific precursor cytotoxic T lymphocyte frequencies in patients with HTLV-1-associated neurological disease. *The Journal of Experimental Medicine, 177*(6), 1567–1573. https://doi.org/10.1084/jem.177.6.1567

Enose-Akahata, Y., Vellucci, A., & Jacobson, S. (2017). Role of HTLV-1 Tax and HBZ in the pathogenesis of HAM/TSP. *Frontiers in Microbiology, 8*, 2563. https://doi.org/10.3389/fmicb.2017.02563

Evangelou, I. E., Massoud, R., & Jacobson, S. (2014). HTLV-I-associated myelopathy/tropical spastic paraparesis: Semiautomatic quantification of spinal cord atrophy from 3-dimensional MR images. *Journal of Neuroimaging, 24*(1), 74–78. https://doi.org/10.1111/j.1552-6569.2011.00648.x

Forlani, G., Shallak, M., Accolla, R. S., & Romanelli, M. G. (2021). HTLV-1 infection and pathogenesis: New insights from cellular and animal models. *International Journal of Molecular Sciences, 22*(15). https://doi.org/10.3390/ijms22158001

Fukushima, T., Ikeda, T., Uyama, E., Uchino, M., Okabe, H., & Ando, M. (1994). Cognitive event-related potentials and brain magnetic resonance imaging in HTLV-1 associated myelopathy (HAM). *Journal of the Neurological Sciences, 126*(1), 30–39. https://doi.org/10.1016/0022-510x(94)90091-4

Gascón, M. R. P., Casseb, J., Smid, J., Vidal, J. E., Fonseca, L. A. M., Paiva, A., Haziot, M. J., & Penalva de Oliveira, A. C. (2017). Cognitive impairment is frequent among symptomatic carriers of human T-cell lymphotropic virus type 1 (HTLV-1), regardless of their clinical status. *Journal of the Neurological Sciences, 377*, 185–189. https://doi.org/10.1016/j.jns.2017.04.019

Gessain, A., & Cassar, O. (2012). Epidemiological aspects and world distribution of HTLV-1 infection. *Frontiers in Microbiology, 3*, 388. https://doi.org/10.3389/fmicb.2012.00388

Gessain, A., Barin, F., Vernant, J. C., Gout, O., Maurs, L., Calender, A., & de The, G. (1985). Antibodies to human T-lymphotropic virus type-I in patients with tropical spastic paraparesis. *Lancet, 2*(8452), 407–410. https://doi.org/10.1016/s0140-6736(85)92734-5

Godoy, A. J., Kira, J., Hasuo, K., & Goto, I. (1995). Characterization of cerebral white matter lesions of HTLV-I-associated myelopathy/tropical spastic paraparesis in comparison with multiple sclerosis and collagen-vasculitis: A semiquantitative MRI study. *Journal of the Neurological Sciences, 133*(1–2), 102–111. https://doi.org/10.1016/0022-510x(95)00161-t

Greten, T. F., Slansky, J. E., Kubota, R., Soldan, S. S., Jaffee, E. M., Leist, T. P., Pardoll, D. M., Jacobson, S., & Schneck, J. P. (1998). Direct visualization of antigen-specific T cells: HTLV-1 Tax11–19-specific CD8+ T cells are activated in peripheral blood and accumulate in cerebrospinal fluid from HAM/TSP patients. *Proceedings of the National Academy of Sciences, 95*(13), 7568–7573. https://doi.org/10.1073/pnas.95.13.7568

Groom, J. R., & Luster, A. D. (2011). CXCR3 in T cell function. *Experimental Cell Research, 317*(5), 620–631. https://doi.org/10.1016/j.yexcr.2010.12.017

Hancock, W. W., Lu, B., Gao, W., Csizmadia, V., Faia, K., King, J. A., Smiley, S. T., Ling, M., Gerard, N. P., & Gerard, C. (2000). Requirement of the chemokine receptor CXCR3 for acute allograft rejection. *The Journal of Experimental Medicine, 192*(10), 1515–1520. https://doi.org/10.1084/jem.192.10.1515

Hinuma, Y., Nagata, K., Hanaoka, M., Nakai, M., Matsumoto, T., Kinoshita, K. I., Shirakawa, S., & Miyoshi, I. (1981). Adult T-cell leukemia: Antigen in an ATL cell line and detection of antibodies to the antigen in human sera. *Proceedings of the National Academy of Sciences of the United States of America, 78*(10), 6476–6480. https://doi.org/10.1073/pnas.78.10.6476

Ho, D. D., Rota, T. R., & Hirsch, M. S. (1984). Infection of human endothelial cells by human T-lymphotropic virus type I. *Proceedings of the National Academy of Sciences of the United States of America, 81*(23), 7588–7590. https://doi.org/10.1073/pnas.81.23.7588

Howard, A. K., Li, D. K., & Oger, J. (2003). MRI contributes to the differentiation between MS and HTLV-I associated myelopathy in British Columbian coastal natives. *The Canadian Journal of Neurological Sciences, 30*(1), 41–48. https://doi.org/10.1017/s0317167100002420

Imre, G. (2020). Cell death signalling in virus infection. *Cellular Signalling, 76*, 109772. https://doi.org/10.1016/j.cellsig.2020.109772

Iwasaki, Y. (1993). Human T cell leukemia virus type I infection and chronic myelopathy. *Brain Pathology, 3*(1), 1–10. https://doi.org/10.1111/j.1750-3639.1993.tb00719.x

Jacobson, S., Shida, H., McFarlin, D. E., Fauci, A. S., & Koenig, S. (1990). Circulating CD8+ cytotoxic T lymphocytes specific for HTLV-I pX in patients with HTLV-I associated neurological disease. *Nature, 348*(6298), 245–248. https://doi.org/10.1038/348245a0

Jeffery, K. J., Usuku, K., Hall, S. E., Matsumoto, W., Taylor, G. P., Procter, J., Bunce, M., Ogg, G. S., Welsh, K. I., Weber, J. N., Lloyd, A. L., Nowak, M. A., Nagai, M., Kodama, D., Izumo, S., Osame, M., & Bangham, C. R. (1999). HLA alleles determine human T-lymphotropic virus-I (HTLV-I) proviral load and the risk of HTLV-I-associated myelopathy. *Proceedings of the National Academy of Sciences of the United States of America, 96*(7), 3848–3853. https://doi.org/10.1073/pnas.96.7.3848

Jones, K. S., Petrow-Sadowski, C., Huang, Y. K., Bertolette, D. C., & Ruscetti, F. W. (2008). Cell-free HTLV-1 infects dendritic cells leading to transmission and transformation of CD4(+) T cells. *Nature Medicine, 14*(4), 429–436. https://doi.org/10.1038/nm1745

Kalil, R. S., Vasconcellos, I., Rosadas, C., Cony, A., Lima, D. P., Gonçalves, C. C. A., Batista, E., Grassi, M. F., Galvão-Castro, B. G. P. T., & Puccioni-Sohler, M. (2021). Association between high proviral load, cognitive impairment, and white matter brain lesions in HTLV-1-infected individuals. *Journal of Neurovirology, 27*(6), 810–819. https://doi.org/10.1007/s13365-021-00944-6

Kannian, P., & Green, P. L. (2010). Human T Lymphotropic virus type 1 (HTLV-1): Molecular biology and Oncogenesis. *Viruses, 2*(9), 2037–2077. https://doi.org/10.3390/v2092037

Kermode, A. G., Rudge, P., Thompson, A. J., du Boulay, E. P., & McDonald, W. I. (1990). MRI of thoracic cord in tropical spastic paraparesis. *Journal of Neurology, Neurosurgery, and Psychiatry, 53*(12), 1110–1111. https://doi.org/10.1136/jnnp.53.12.1110-a

Kira, J., Nakamura, M., Sawada, T., Koyanagi, Y., Ohori, N., Itoyama, Y., Yamamoto, N., Sakaki, Y., & Goto, I. (1992). Antibody titers to HTLV-I-p40tax protein and gag-env hybrid protein in HTLV-I-associated myelopathy/tropical spastic paraparesis: Correlation with increased HTLV-I proviral DNA load. *Journal of the Neurological Sciences, 107*(1), 98–104. https://doi.org/10.1016/0022-510x(92)90215-7

Koralnik, I. J., Lemp, J. F., Jr., Gallo, R. C., & Franchini, G. (1992). In vitro infection of human macrophages by human T-cell leukemia/lymphotropic virus type I (HTLV-I). *AIDS Research and Human Retroviruses, 8*(11), 1845–1849. https://doi.org/10.1089/aid.1992.8.1845

Kubota, R., Umehara, F., Izumo, S., Ijichi, S., Matsumuro, K., Yashiki, S., Fujiyoshi, T., Sonoda, S., & Osame, M. (1994). HTLV-I proviral DNA amount correlates with infiltrating CD4+ lymphocytes in the spinal cord from patients with HTLV-I-associated myelopathy. *Journal of Neuroimmunology, 53*(1), 23–29. https://doi.org/10.1016/0165-5728(94)90060-4

Leal, F. E., Ndhlovu, L. C., Hasenkrug, A. M., Bruno, F. R., Carvalho, K. I., Wynn-Williams, H., Neto, W. K., Sanabani, S. S., Segurado, A. C., Nixon, D. F., & Kallas, E. G. (2013). Expansion in CD39+ CD4+ Immunoregulatory T cells and rarity of Th17 cells in HTLV-1 infected patients is associated with neurological complications. *PLoS Neglected Tropical Diseases, 7*(2), e2028. https://doi.org/10.1371/journal.pntd.0002028

Lee, H. H., Weiss, S. H., Brown, L. S., Mildvan, D., Shorty, V., Saravolatz, L., Chu, A., Ginzburg, H. M., Markowitz, N., Des Jarlais, D. C., et al. (1990). Patterns of HIV-1 and HTLV-I/II in intravenous drug abusers from the middle Atlantic and central regions of the USA. *The Journal of Infectious Diseases, 162*(2), 347–352. https://doi.org/10.1093/infdis/162.2.347

Lehky, T. J., & Jacobson, S. (1995). Induction of HLA class II in HTLV-I infected neuronal cell lines. *Journal of Neurovirology, 1*(2), 145–156. https://doi.org/10.3109/13550289509113961

Levin, M. C., Lehky, T. J., Flerlage, A. N., Katz, D., Kingma, D. W., Jaffe, E. S., Heiss, J. D., Patronas, N., McFarland, H. F., & Jacobson, S. (1997). Immunologic analysis of a spinal cord–biopsy specimen from a patient with human T-cell lymphotropic virus Type I–associated neurologic disease. *New England Journal of Medicine, 336*(12), 839–845. https://doi.org/10.1056/nejm199703203361205

Lezin, A., Olindo, S., Olière, S., Varrin-Doyer, M., Marlin, R., Cabre, P., Smadja, D., & Cesaire, R. (2005). Human T Lymphotropic virus type I (HTLV-I) Proviral load in cerebrospinal fluid: A new criterion for the diagnosis of HTLV-I–associated myelopathy/tropical spastic Paraparesis? *The Journal of Infectious Diseases, 191*(11), 1830–1834. https://doi.org/10.1086/429962

Liu, W., Nair, G., Vuolo, L., Bakshi, A., Massoud, R., Reich, D. S., & Jacobson, S. (2014). In vivo imaging of spinal cord atrophy in neuroinflammatory diseases. *Annals of Neurology, 76*(3), 370–378. https://doi.org/10.1002/ana.24213

Manns, A., Miley, W. J., Wilks, R. J., Morgan, O. S., Hanchard, B., Wharfe, G., Cranston, B., Maloney, E., Welles, S. L., Blattner, W. A., & Waters, D. (1999). Quantitative proviral DNA and antibody levels in the natural history of HTLV-I infection. *The Journal of Infectious Diseases, 180*(5), 1487–1493. https://doi.org/10.1086/315088

Martel, M., & Gotuzzo, E. (2022). HTLV-1 is also a sexually transmitted infection. *Frontiers in Public Health, 10*, 840295. https://doi.org/10.3389/fpubh.2022.840295

Matsuoka, M., & Jeang, K. T. (2011). Human T-cell leukemia virus type 1 (HTLV-1) and leukemic transformation: Viral infectivity, Tax, HBZ and therapy. *Oncogene, 30*(12), 1379–1389. https://doi.org/10.1038/onc.2010.537

Matsuoka, E., Takenouchi, N., Hashimoto, K., Kashio, N., Moritoyo, T., Higuchi, I., Isashiki, Y., Sato, E., Osame, M., & Izumo, S. (1998). Perivascular T cells are infected with HTLV-I in the spinal cord lesions with HTLV-I-associated myelopathy/tropical spastic paraparesis: Double staining of immunohistochemistry and polymerase chain reaction in situ hybridization. *Acta Neuropathologica, 96*(4), 340–346. https://doi.org/10.1007/s004010050903

Matsuura, E., Kubota, R., Tanaka, Y., Takashima, H., & Izumo, S. (2015). Visualization of HTLV-1-specific cytotoxic T lymphocytes in the spinal cords of patients with HTLV-1-associated myelopathy/tropical spastic paraparesis. *Journal of Neuropathology and Experimental Neurology, 74*(1), 2–14. https://doi.org/10.1097/nen.0000000000000141

Matsuura, E., Nozuma, S., Tashiro, Y., Kubota, R., Izumo, S., & Takashima, H. (2016). HTLV-1 associated myelopathy/tropical spastic paraparesis (HAM/TSP): A comparative study to identify factors that influence disease progression. *Journal of the Neurological Sciences, 371*, 112–116. https://doi.org/10.1016/j.jns.2016.10.030

Meekings, K. N., Leipzig, J., Bushman, F. D., Taylor, G. P., & Bangham, C. R. (2008). HTLV-1 integration into transcriptionally active genomic regions is associated with proviral expression and with HAM/TSP. *PLoS Pathogens, 4*(3), e1000027. https://doi.org/10.1371/journal.ppat.1000027

Melamed, A., Laydon, D. J., Gillet, N. A., Tanaka, Y., Taylor, G. P., & Bangham, C. R. (2013). Genome-wide determinants of proviral targeting, clonal abundance and expression in natural HTLV-1 infection. *PLoS Pathogens, 9*(3), e1003271. https://doi.org/10.1371/journal.ppat.1003271

Melamed, A., Laydon, D. J., Al Khatib, H., Rowan, A. G., Taylor, G. P., & Bangham, C. R. M. (2015). HTLV-1 drives vigorous clonal expansion of infected CD8+ T cells in natural infection. *Retrovirology, 12*(1), 91. https://doi.org/10.1186/s12977-015-0221-1

Méndez, E., Kawanishi, T., Clemens, K., Siomi, H., Soldan, S. S., Calabresi, P., Brady, J., & Jacobson, S. (1997). Astrocyte-specific expression of human T-cell lymphotropic virus type 1 (HTLV-1) Tax: Induction of tumor necrosis factor alpha and susceptibility to lysis by CD8+ HTLV-1-specific cytotoxic T cells. *Journal of Virology, 71*(12), 9143–9149. https://doi.org/10.1128/jvi.71.12.9143-9149.1997

Mina, Y., Azodi, S., Dubuche, T., Andrada, F., Osuorah, I., Ohayon, J., Cortese, I., Wu, T., Johnson, K. R., Reich, D. S., Nair, G., & Jacobson, S. (2021). Cervical and thoracic cord atrophy in multiple sclerosis phenotypes: Quantification and correlation with clinical disability. *NeuroImage: Clinical, 30*, 102680. https://doi.org/10.1016/j.nicl.2021.102680

Mohanty, S., & Harhaj, E. W. (2020). Mechanisms of Oncogenesis by HTLV-1 Tax. *Pathogens, 9*(7), 543. https://www.mdpi.com/2076-0817/9/7/543

Morgan, D. J., Caskey, M. F., Abbehusen, C., Oliveira-Filho, J., Araujo, C., Porto, A. F., Santos, S. B., Orge, G. O., Joia, M. J., Muniz, A. L., Siqueira, I., Glesby, M. J., & Carvalho, E. (2007). Brain magnetic resonance imaging white matter lesions are frequent in HTLV-I carriers and do not discriminate from HAM/TSP. *AIDS Research and Human Retroviruses, 23*(12), 1499–1504. https://doi.org/10.1089/aid.2007.0077

Moritoyo, T., Reinhart, T. A., Haase, A. T., Moritoyo, H., Izumo, S., Osame, M., & Sato, E. (1996). Human T-lymphotropic virus type I-associated myelopathy and tax gene expression in CD4+ T lymphocytes. *Annals of Neurology, 40*(1), 84–90. https://doi.org/10.1002/ana.410400114

Murphy, E. L. (2016). Infection with human T-lymphotropic virus types-1 and -2 (HTLV-1 and -2): Implications for blood transfusion safety. *Transfusion Clinique et Biologique, 23*(1), 13–19. https://doi.org/10.1016/j.tracli.2015.12.001

Nagai, M., Usuku, K., Matsumoto, W., Kodama, D., Takenouchi, N., Moritoyo, T., Hashiguchi, S., Ichinose, M., Bangham, C. R., Izumo, S., & Osame, M. (1998). Analysis of HTLV-I proviral load in 202 HAM/TSP patients and 243 asymptomatic HTLV-I carriers: High proviral load strongly predisposes to HAM/TSP. *Journal of Neurovirology, 4*(6), 586–593. https://doi.org/10.3109/13550289809114225

Nagai, M., Brennan, M. B., Sakai, J. A., Mora, C. A., & Jacobson, S. (2001a). CD8(+) T cells are an in vivo reservoir for human T-cell lymphotropic virus type I. *Blood, 98*(6), 1858–1861. https://doi.org/10.1182/blood.v98.6.1858

Nagai, M., Kubota, R., Greten, T. F., Schneck, J. P., Leist, T. P., & Jacobson, S. (2001b). Increased activated human T cell lymphotropic virus type I (HTLV-I) Tax11-19-specific memory and effector CD8+ cells in patients with HTLV-I-associated myelopathy/tropical spastic paraparesis: Correlation with HTLV-I provirus load. *The Journal of Infectious Diseases, 183*(2), 197–205. https://doi.org/10.1086/317932

Nagai, M., Yamano, Y., Brennan, M. B., Mora, C. A., & Jacobson, S. (2001c). Increased HTLV-I proviral load and preferential expansion of HTLV-I tax-specific CD8+ T cells in cerebrospinal fluid from patients with HAM/TSP. *Annals of Neurology, 50*(6), 807–812. https://doi.org/10.1002/ana.10065

Nagasaka, M., Yamagishi, M., Yagishita, N., Araya, N., Kobayashi, S., Makiyama, J., Kubokawa, M., Yamauchi, J., Hasegawa, D., Coler-Reilly, A. L. G., Tsutsumi, S., Uemura, Y., Arai, A., Takata, A., Inoue, E., Hasegawa, Y., Watanabe, T., Suzuki, Y., Uchimaru, K., et al. (2020). Mortality and risk of progression to adult T cell leukemia/lymphoma in HTLV-1-associated myelopathy/tropical spastic paraparesis. *Proceedings of the National Academy of Sciences of the United States of America, 117*(21), 11685–11691. https://doi.org/10.1073/pnas.1920346117

Nakagawa, M., Izumo, S., Ijichi, S., Kubota, H., Arimura, K., Kawabata, M., & Osame, M. (1995). HTLV-I-associated myelopathy: Analysis of 213 patients based on clinical features and laboratory findings. *Journal of Neurovirology, 1*(1), 50–61. https://doi.org/10.3109/13550289509111010

Niederer, H. A., Laydon, D. J., Melamed, A., Elemans, M., Asquith, B., Matsuoka, M., & Bangham, C. R. M. (2014). HTLV-1 proviral integration sites differ between asymptomatic carriers and patients with HAM/TSP. *Virology Journal, 11*(1), 172. https://doi.org/10.1186/1743-422X-11-172

Nozuma, S., Enose-Akahata, Y., Johnson, K. R., Monaco, M. C., Ngouth, N., Elkahloun, A., Ohayon, J., Zhu, J., & Jacobson, S. (2021). Immunopathogenic CSF TCR repertoire signatures in virus-associated neurologic disease. *JCI Insight, 6*(4). https://doi.org/10.1172/jci.insight.144869

Nunes, D., Boa-Sorte, N., Grassi, M. F., Taylor, G. P., Teixeira, M. G., Barreto, M. L., Dourado, I., & Galvão-Castro, B. (2017). HTLV-1 is predominantly sexually transmitted in Salvador, the city with the highest HTLV-1 prevalence in Brazil. *PLoS One, 12*(2), e0171303. https://doi.org/10.1371/journal.pone.0171303

Ogata, A., Nagashima, K., Tashiro, K., Miyakawa, A., & Mikuni, C. (1993). MRI-pathological correlate of brain lesions in a necropsy case of HTLV-I associated myelopathy. *Journal of Neurology, Neurosurgery, and Psychiatry, 56*(2), 194–196. https://doi.org/10.1136/jnnp.56.2.194

Oh, U., Grant, C., Griffith, C., Fugo, K., Takenouchi, N., & Jacobson, S. (2006). Reduced Foxp3 protein expression is associated with inflammatory disease during human t lymphotropic virus type 1 infection. *The Journal of Infectious Diseases, 193*(11), 1557–1566. https://doi.org/10.1086/503874

Okochi, K., Sato, H., & Hinuma, Y. (1984). A retrospective study on transmission of adult T cell leukemia virus by blood transfusion: Seroconversion in recipients. *Vox Sanguinis, 46*(5), 245–253. https://doi.org/10.1111/j.1423-0410.1984.tb00083.x

Parker, C. E., Daenke, S., Nightingale, S., & Bangham, C. R. (1992). Activated, HTLV-1-specific cytotoxic T-lymphocytes are found in healthy seropositives as well as in patients with tropical spastic paraparesis. *Virology, 188*(2), 628–636. https://doi.org/10.1016/0042-6822(92)90517-s

Poiesz, B. J., Ruscetti, F. W., Gazdar, A. F., Bunn, P. A., Minna, J. D., & Gallo, R. C. (1980). Detection and isolation of type C retrovirus particles from fresh and cultured lymphocytes of a patient with cutaneous T-cell lymphoma. *Proceedings of the National Academy of Sciences of the United States of America, 77*(12), 7415–7419. https://doi.org/10.1073/pnas.77.12.7415

Puccioni-Sohler, M., Gasparetto, E., Cabral-Castro, M. J., Slatter, C., Vidal, C. M., Cortes, R. D., Rosen, B. R., & Mainero, C. (2012). HAM/TSP: Association between white matter lesions on magnetic resonance imaging, clinical and cerebrospinal fluid findings. *Arquivos de Neuro-Psiquiatria, 70*(4), 246–251. https://doi.org/10.1590/s0004-282x2012000400004

Richardson, J. H., Edwards, A. J., Cruickshank, J. K., Rudge, P., & Dalgleish, A. G. (1990). In vivo cellular tropism of human T-cell leukemia virus type 1. *Journal of Virology, 64*(11), 5682–5687. https://doi.org/10.1128/jvi.64.11.5682-5687.1990

Román, G. C., & Román, L. N. (1988). Tropical spastic paraparesis. A clinical study of 50 patients from Tumaco (Colombia) and review of the worldwide features of the syndrome. *Journal of the Neurological Sciences, 87*(1), 121–138. https://doi.org/10.1016/0022-510x(88)90059-7

Sato, T., Coler-Reilly, A. L. G., Yagishita, N., Araya, N., Inoue, E., Furuta, R., Watanabe, T., Uchimaru, K., Matsuoka, M., Matsumoto, N., Hasegawa, Y., & Yamano, Y. (2018). Mogamulizumab (Anti-CCR4) in HTLV-1-associated myelopathy. *The New England Journal of Medicine, 378*(6), 529–538. https://doi.org/10.1056/NEJMoa1704827

Stuver, S. O., Tachibana, N., Okayama, A., Shioiri, S., Tsunetoshi, Y., Tsuda, K., & Mueller, N. E. (1993). Heterosexual transmission of human T cell leukemia/lymphoma virus type I among married couples in southwestern Japan: An initial report from the Miyazaki Cohort Study. *The Journal of Infectious Diseases, 167*(1), 57–65. https://doi.org/10.1093/infdis/167.1.57

Takenouchi, N., Yamano, Y., Usuku, K., Osame, M., & Izumo, S. (2003). Usefulness of proviral load measurement for monitoring of disease activity in individual patients with human T-lymphotropic virus type I-associated myelopathy/tropical spastic paraparesis. *Journal of Neurovirology, 9*(1), 29–35. https://doi.org/10.1080/13550280390173418

Takenouchi, N., Yao, K., & Jacobson, S. (2004). Immunopathogensis of HTLV-I associated neurologic disease: Molecular, histopathologic, and immunologic approaches. *Frontiers in Bioscience, 9*, 2527–2539. https://doi.org/10.2741/1414

Talledo, M., López, G., Huyghe, J. R., Verdonck, K., Adaui, V., González, E., Best, I., Clark, D., Vanham, G., Gotuzzo, E., Van Camp, G., & Van Laer, L. (2010). Evaluation of host genetic and viral factors as surrogate markers for HTLV-1-associated myelopathy/tropical spastic paraparesis in Peruvian HTLV-1-infected patients. *Journal of Medical Virology, 82*(3), 460–466. https://doi.org/10.1002/jmv.21675

Taniguchi, A., Mochizuki, H., Yamashita, A., Shiomi, K., Asada, Y., & Nakazato, M. (2017). Spinal cord anteroposterior atrophy in HAM/TSP: Magnetic resonance imaging and neuropathological analyses. *Journal of the Neurological Sciences, 381*, 135–140. https://doi.org/10.1016/j.jns.2017.08.3243

Tendler, C. L., Greenberg, S. J., Blattner, W. A., Manns, A., Murphy, E., Fleisher, T., Hanchard, B., Morgan, O., Burton, J. D., Nelson, D. L., et al. (1990). Transactivation of interleukin 2 and its receptor induces immune activation in human T-cell lymphotropic virus type I-associated myelopathy: Pathogenic implications and a rationale for immunotherapy. *Proceedings of the National Academy of Sciences of the United States of America, 87*(13), 5218–5222. https://doi.org/10.1073/pnas.87.13.5218

Umehara, F., Nakamura, A., Izumo, S., Kubota, R., Ijichi, S., Kashio, N., Hashimoto, K., Usuku, K., Sato, E., & Osame, M. (1994). Apoptosis of T lymphocytes in the spinal cord lesions in HTLV-I-associated myelopathy: A possible mechanism to control viral infection in the central nervous system. *Journal of Neuropathology and Experimental Neurology, 53*(6), 617–624. https://doi.org/10.1097/00005072-199411000-00009

Umehara, F., Izumo, S., Takeya, M., Takahashi, K., Sato, E., & Osame, M. (1996). Expression of adhesion molecules and monocyte chemoattractant protein -1 (MCP-1) in the spinal cord lesions in HTLV-I-associated myelopathy. *Acta Neuropathologica, 91*(4), 343–350. https://doi.org/10.1007/s004010050435

Umehara, F., Nose, H., Saito, M., Fukuda, M., Ogino, M., Toyota, T., Yuhi, T., Arimura, K., & Osame, M. (2007). Abnormalities of spinal magnetic resonance images implicate clinical variability in human T-cell lymphotropic virus type I-associated myelopathy. *Journal of Neurovirology, 13*(3), 260–267. https://doi.org/10.1080/13550280701258431

Vilchez, C., Gonzalez-Reinoso, M., Cubbison, C., Perez-Then, E., Roa, P., Martínez, A., Foerster, B., Oviedo, J., & Stoeter, P. (2014). Atrophy, focal spinal cord lesions and alterations of diffusion tensor imaging (DTI) parameters in asymptomatic virus carriers and patients suffering from human T-lymphotrophic virus type 1 (HTLV-1)-associated myelopathy/tropical spastic paraparesis (HAM/TSP). *Journal of Neurovirology, 20*(6), 583–590. https://doi.org/10.1007/s13365-014-0282-2

Yamano, Y., & Sato, T. (2012). Clinical pathophysiology of human T-lymphotropic virus-type 1-associated myelopathy/tropical spastic paraparesis. *Frontiers in Microbiology, 3*, 389. https://doi.org/10.3389/fmicb.2012.00389

Yamano, Y., Nagai, M., Brennan, M., Mora, C. A., Soldan, S. S., Tomaru, U., Takenouchi, N., Izumo, S., Osame, M., & Jacobson, S. (2002). Correlation of human T-cell lymphotropic virus type 1 (HTLV-1) mRNA with proviral DNA load, virus-specific CD8+ T cells, and disease severity in HTLV-1–associated myelopathy (HAM/TSP). *Blood, 99*(1), 88–94. https://doi.org/10.1182/blood.V99.1.88

Yamano, Y., Cohen, C. J., Takenouchi, N., Yao, K., Tomaru, U., Li, H.-C., Reiter, Y., & Jacobson, S. (2004). Increased Expression of Human T Lymphocyte Virus Type I (HTLV-I) Tax11-19 Peptide–Human Histocompatibility Leukocyte Antigen A*201 Complexes on CD4+ CD25+T Cells Detected by Peptide-specific, Major Histocompatibility Complex–restricted Antibodies in Patients with HTLV-I–associated Neurologic Disease. *Journal of Experimental Medicine, 199*(10), 1367–1377. https://doi.org/10.1084/jem.20032042

Yoshioka, A., Hirose, G., Ueda, Y., Nishimura, Y., & Sakai, K. (1993). Neuropathological studies of the spinal cord in early stage HTLV-I-associated myelopathy (HAM). *Journal of Neurology, Neurosurgery, and Psychiatry, 56*(9), 1004–1007. https://doi.org/10.1136/jnnp.56.9.1004

Role of Epstein-Barr Virus in Multiple Sclerosis

27

Abaigeal Donaldson, William Frazier, and Steven Jacobson

Abstract

Multiple sclerosis (MS) is a progressive neuroinflammatory disease marked by concentrated areas of demyelinated lesions that are disseminated in space and time in the central nervous system (CNS). MS is currently the most prevalent cause of atraumatic neurologic disability in the United States and is estimated to affect more than 2.8 million individuals worldwide. Neuroinflammatory diseases, including MS, often have multifactorial etiologies that make them difficult to understand and treat. We propose that the onset of MS occurs in individuals of a particular genetic background who encounter an environmental trigger. This dual-hit hypothesis leads to a dysregulation of the immune system that is characteristic of MS. Epstein-Barr virus (EBV) is of prominent interest in the research for a virus that could be one of many environmental triggers of MS. Recently, there has been a significant re-emergent interest in the role of EBV in MS pathogenesis. This spans hypotheses that suggest EBV may be a necessary but insufficient trigger of the disease to a causative role for EBV in MS. It is still unclear as to which, if any, of these assumptions may be true. In this chapter, we review MS and EBV independently before citing various forms of research that provide evidence for and against the link between MS and EBV. Ultimately, we believe that in order to claim that EBV is a primary environmental trigger for MS, researchers must conduct more clinical trials to understand and support the role played by EBV in the progression of MS.

A. Donaldson · W. Frazier · S. Jacobson (✉)
Viral Immunology Section, National Institutes of Neurological Disorders and Stroke, National Institutes of Health, Bethesda, MD, USA
e-mail: jacobsons@ninds.nih.gov

Keywords

Multiple sclerosis · Epstein-Barr virus · Neuroinflammation · Autoimmunity · Neurodegeneration

27.1 Introduction

Multiple sclerosis (MS) is a chronic, progressive neuroinflammatory disease of the central nervous system (CNS) characterized by recurrent episodes of oligodendrocyte apoptosis leading to foci of demyelinated sclerotic lesions that are disseminated in time and space ('t Hart & Massacesi, 2009). MS is the most prevalent cause of atraumatic neurologic disability in the United States with an incidence rate of 309.2 per 100,000 population, based on 2010 data, and more recent estimates suggest that greater than 900,000 Americans carry a diagnosis of MS (Wallin et al., 2019). MS is currently estimated to affect at least 2.8 million individuals worldwide (Walton et al., 2020).

We propose, along with others, that the etiology of MS is complex and multifactorial in which individuals of a specific genetic background are triggered by an environmental exposure, which culminates in a dysregulation of the immune system that is associated with MS disease progression (Olsson et al., 2017). It has been suggested that a preceding viral infection likely serves, at minimum, as one factor in the pathogenesis of MS (Steelman, 2015). There are a multitude of epidemiological studies that have revealed higher incidence rates in more northern latitudes, suggesting a potential role of Vitamin D levels (Sintzel et al., 2018) as well as genetic factors specific to populations from certain regions of the world (particularly northern Europe). In combination with a genetic predisposition secondary to geographical location or genetic lineage, multiple family studies have demonstrated a significant increase in the risk of developing MS in disease-free individuals who have first-degree relatives (4–5%) or a monozygotic twin (30%) carrying a diag-

H. E. Gendelman, T. Ikezu (eds.), *Neuroimmune Pharmacology and Therapeutics*,
https://doi.org/10.1007/978-3-031-68237-7_27

nosis of MS (Ermann & Fathman, 2001). In individuals who carry HLA-DRB*15:01, the risk of developing MS is approximately threefold to sixfold higher in heterozygous versus homozygous individuals (Jacobs et al., 2020). Further, environmental factors pose an increased risk, including a history of tobacco smoking, obesity during early childhood or adolescence, and increased salt intake (Thompson et al., 2018; Zostawa et al., 2017). In summary, the pathogenesis of MS is now thought to be related to genetic risk factors, environmental exposures (infectious and otherwise), and immune dysregulation, ultimately leading to an autoimmune process associated with the diagnosis of MS.

A recent epidemiological study from a subset of >10 million United States military personnel has again reinforced the observations that Epstein-Barr virus (EBV) exposure is associated with the development of MS (Bjornevik et al., 2022). The relative risk ratio of young adults in this study who contracted an EBV infection during early adulthood was estimated to have a 32-fold higher risk of progression to MS. All MS patients were seropositive for EBV preceding the development of MS, and had increased serum levels of a neurodegenerative marker, neurofilament light chain, compared to controls. It is hypothesized that the pathogenesis of MS is associated with a preceding viral infection, likely EBV in some cases, that may lead to a neuroinflammatory disorder culminating in the development of MS.

27.2 Pathophysiology of MS

Histologically, MS is characterized by focal regions of demyelination known as plaques. Prominent in acute lesions, infiltrates of mononuclear cells (CD8+ T cells, macrophages, microglia, B cells, and plasma cells) are commonly observed upon biopsy. In chronic lesions, inflammatory infiltrates are diminished, and reactive gliosis becomes the predominant histological feature (Popescu et al., 2013). Recent data suggest that ribbon-like subpial cortical demyelination is pathognomonic of MS (Junker et al., 2020).

The demyelination seen in these lesions results from separation of the outer lamellae of the oligodendrocyte processes surrounding the axon, which disrupts the integrity of the myelin sheath but does not necessarily indicate degeneration of the oligodendrocyte itself. Nonetheless, oligodendrocytes may still show variable levels of degeneration during disease progression. Axons, however, are relatively spared (Popescu et al., 2013). Axonal loss may become more significant in the irreversible phases of the disease (Jadidi-Niaragh & Mirshafiey, 2011).

Demyelination leads to loss of saltatory conduction, which results in slowed conduction and hence longer conduction times. Sodium channels are upregulated and redistributed along the length of the demyelinated axonal membrane to compensate for the loss of saltatory conduction and maintain electrical transmission, but this also results in an increase in metabolic demand, which may cause the patient to experience fatigue, even when electrical transmission is compensated. Conduction may be further compromised by heat, leading to conduction block at increased temperatures, a symptom known as Uhthoff's phenomenon (Humm et al., 2004). As the disease progresses, increasing conduction capacity is no longer sufficient to compensate for lost saltatory conduction, and electrical transmission is compromised (Compston & Coles, 2008).

27.3 Clinical Manifestations and Diagnostic Criteria

Clinical hallmarks of MS include MRI findings of chronic and/or active enhancing lesions disseminated in time (DIT) and space (DIS), per the 2017 revisions of the McDonald Criteria (Thompson et al., 2018). Lesions often occur in the white matter in specific locations (periventricular, juxtacortical or cortical, and infratentorial) as well as within the spinal cord (Filippi et al., 2019). Additional findings, not included in the current McDonald criteria, include central vein signs (CVSs), subpial demyelination, and lesional rims, as well as brain atrophy (Filippi et al., 2019). Cerebrospinal fluid analysis also plays a key role if oligoclonal bands (OCBs) are demonstrated, which are suggestive of a chronic inflammatory/autoimmune process (Farina et al., 2017).

Pathognomonic features of this disease include fully or partially reversible episodes of neurological dysfunction often lasting days to weeks, if reversible. The most common presenting symptoms include limb weakness and/or symptoms at a sensory level due to transverse myelitis, monocular loss of vision secondary to optic neuritis, diplopia secondary to brain stem involvement, and ataxia/disequilibrium due to cerebellar involvement (Reich et al., 2018). Other symptoms include increased fatiguability, extraocular muscle impairment, and bladder, bowel, and/or sexual dysfunction (Rodgers et al., 1999).

Following initial diagnosis, patients typically develop a progressive disease course marked by episodes of new symptoms that occasionally resolve fully or partially [hence, relapsing-remitting multiple sclerosis (RRMS)] within the next 10–20 years. These symptoms commonly manifest as impaired mobility, abnormal sensorium, disequilibrium, and, in advanced cases, impaired cognition. A smaller cohort of patients, approximately 15%, develop a progressive course from disease onset, which is classified as either primary or secondary progressive multiple sclerosis (PPMS or SPMS). Further, a small subset of patients presents with clinically isolated syndrome (CIS), with CSF often demonstrating OCBs. CIS commonly progresses to a diagnosis of MS and some patients are placed on disease-modifying therapy to reduce the risk of future exacerbations (Miller et al., 2012).

27.4 Traditional and Current FDA-Approved MS Treatments

MS therapies generally adopt the strategy of (1) blocking the action of cellular and molecular mediators of inflammation to combat the progression of inflammatory damage or (2) promoting the repair of existing damage. Examples of the former include numerous immunosuppressive therapies, such as corticosteroid treatments and antibody-mediated therapies for the depletion of T cells and B cells (Constantinescu et al., 2011). Other pharmacologic strategies include immunomodulation via the induction of regulatory T cells that tolerize immune cells to myelin-derived self-antigens with glatiramer acetate (Zozulya & Wiendl, 2008). Still, other avenues of research seek to promote oligodendrogenesis and myelin repair (Deshmukh et al., 2013; Kokaia et al., 2012). At the intersection of these preventive and reparative strategies are cell-based treatments such as stem cell transplantation, which is hypothesized to play both a regenerative role in cell replacement and an immune regulatory role in the creation of a tolerogenic environment favoring repair (Hoang et al., 2022). In this section, we will give a broad overview of the evolution of MS treatment by covering a range of therapeutic strategies, from early treatments and established mainstays to newer treatments, noting both successes and challenges.

Glucocorticoids suppress inflammation by both inhibiting the synthesis of molecular mediators of inflammation and inhibiting lymphocyte proliferation. Glucocorticoid treatments such as intravenous prednisone and methylprednisolone are favored in the treatment of acute exacerbations, for which they have been shown to reduce recovery time. However, they have not been shown to improve the ultimate degree of recovery (Goodin, 2014). These medications are also associated with significant side effects, including derangements in lipid and sugar metabolism, which may produce hyperglycemia, resulting in a condition known as "steroid diabetes." The excess corticosteroid may produce Cushing's syndrome, which is characterized by weight gain, fat redistribution, and skin changes. Other manifestations include osteoporosis, gastrointestinal disturbances, and steroid-induced psychosis. Patients are also at risk of adrenal insufficiency should they fail to recover adrenal function during steroid withdrawal (Shaikh et al., 2012).

Alemtuzumab is a human anti-CD52 monoclonal antibody used for the depletion of T cells and monocytes. It was established clinically for the treatment of chronic lymphocytic leukemia (CLL) and, in 1994, was established for the prevention of relapse in RRMS. More recently, alemtuzumab has been explored in the treatment of early multiple sclerosis and the prevention of secondary progression. A study by Coles et al. reported reduced accumulation of disability and reduced lesion burden on T2-weighted MRI in patients taking 24-mg doses of the drug. However, adverse effects associated with its use in this trial included the development of autoimmunity, with 23% of patients developing autoimmune thyroid disease and 3% developing immune thrombocytopenic purpura (Coles et al., 2008).

B cell–depleting therapies, Rituximab and ocrelizumab, are chimeric anti-CD20 monoclonal antibodies engineered for the depletion of CD19/20+ B cells. Predominantly, these medications have been used in the treatment of B cell lymphomas but more recently studied in the treatment of multiple sclerosis (Carson et al., 2009; Sorensen & Blinkenberg, 2016). A 2008 study demonstrated a significant decrease in disease relapse and reduced counts of gadolinium-enhancing lesions on T1-weighted MRI in patients treated with rituximab, and similar findings have been noted with ocrelizumab (Bourdette & Yadav, 2008; Bayas et al., 2023).

Natalizumab is a treatment which binds α4 and β1-integrins, or very late antigen (VLA)-4 integrin, a glycoprotein expressed by activated lymphocytes and monocytes that mediates their adhesion to the vascular endothelium of the blood-brain barrier and facilitates their transmigration into the CNS. This transmigration is mediated by the binding of VLA-4 to vascular cell adhesion molecule 1 (VCAM-1). Natalizumab is a human monoclonal antibody that binds the VLA-4 integrin and prevents its interaction with VCAM-1, thus blocking lymphocytic infiltration of the CNS. A placebo-controlled trial showed a reduction in both lesion load and frequency of clinical relapse over a 6-month period. Suppression of lymphocyte migration, however, may result in the reactivation of latent infections such as JC virus (Constantinescu et al., 2011; Kleinschmidt-DeMasters & Tyler, 2005; Langer-Gould et al., 2005).

Fingolimod is an oral medication derived from myriocin, a metabolite produced by Isaria sinclairii, a fungus (Schmitz et al., 2015). Once phosphorylated, fingolimod binds the sphingosine-1 receptor of lymphocytes and prevents lymphocyte migration from the lymph nodes to other compartments of the body, effectively blocking their infiltration of the CNS (Kappos et al., 2006, 2010). Fingolimod is thought to exert additional immunomodulatory effects through its action on the sphingosine-1 receptor and may also play a neuroprotective role by promoting remyelination (Kappos et al., 2006; Schmitz et al., 2015). A 2010 placebo-controlled study on fingolimod for the treatment of RRMS showed significant decreases in relapse, progression to disability, and gadolinium-enhancing lesions on MRI for patients receiving the drug (Kappos et al., 2010).

Dimethyl fumarate is a natural compound previously approved for the treatment of psoriasis that has shown moderate efficacy in the reduction of clinical scores in animal models of multiple sclerosis (Schmitz et al., 2015). In clinical trials, it has been associated with a significant reduction in active MRI lesions and relapse rate (Hutchinson et al., 2014). Glatiramer acetate (GA) is an acetate salt of a polymer consisting of l-alanine, l-lysine, l-glutamic acid, and l-tyrosine, an amino acid sequence found in myelin basic protein (MBP) (Aharoni et al., 2013). This synthetic polypeptide, formerly known as copolymer-1, has been found to induce T regulatory (Treg) cells promoting tolerance of MBP. Mechanistically, GA competes with MBP and is incorporated in its place at the MBP-binding site of the MHC-II molecule. The MHC-bound GA (MHC-GA) competes with MHC molecules presenting MBP (MHC-MBP) for binding to T cell receptors (TCRs). Binding of MHC-GA to the TCR during antigen presentation induces a tolerogenic, regulatory Th2-like phenotype. In animal models of MS, GA has been shown to promote oligodendrogenesis and is suggested to play a role in neuroprotection and repair. However, while GA carries relatively little risk of significant side effects, its clinical efficacy has yet to be clearly determined. A recent multicenter, placebo-controlled trial of GA failed to show efficacy in slowing disease progression, but the lack of treatment effect in this study may be attributable to discontinuation of medication as well as low rates of disease progression in all study participants (Wolinsky et al., 2007).

Interferon-β-1a is a cytokine that is typically administered to patients intramuscularly or subcutaneously and decreases T cell infiltration of the CNS. It also increases nerve growth and repair via increases in nerve growth factor production. While a systematic review of clinical studies found a slight decrease in the frequency of exacerbations in patients receiving interferon therapy (Rice et al., 2001), a 2012 study reported no reduction in progression to disability with interferon administration (Shirani et al., 2014). Moreover, interferon therapy is associated with several adverse effects, including flu-like symptoms and hyperthermia, likely due to cytokine release (Compston & Coles, 2008).

Finally, cell-based therapies, in limited clinical trials, have recently demonstrated that autologous hematopoietic stem cell transplantation (HSCT), following induction chemotherapy, leads to immunosuppression by the newly engrafted autologous immune system, likely due to antigen-specific tolerance (Atkins et al., 2016). Given the risk of HSCT in patients due to induction therapy prior to the graft, this clinically aggressive approach is typically limited to patients who follow an aggressive clinical course and do not respond to traditional disease-modifying therapy therapies (Freedman & Atkins, 2016). Nonetheless, clinical trials have suggested that this small cohort of patients generally follows a less aggressive disease course compared to a placebo (Soelberg Sorensen, 2016).

27.5 Role of EBV Infection and the Development of MS

Given the re-emergent interest in the role of infection agents in chronic, neurologic diseases (Levine et al., 2023), a focus of this review is the potential role of EBV infection in the pathogenesis of MS. As previously noted, recent data have demonstrated mechanistic and epidemiological evidence that prior infection with EBV is strongly associated with future development of MS. A major challenge in establishing that a preceding EBV infection may be one factor that leads to MS is that greater than 95% of the population have already been exposed to this ubiquitous virus, yet the vast majority of these individuals do not develop this disease. We will highlight the many approaches that have attempted to establish a role for EBV in MS pathogenesis. These include epidemiological, pathological, virological, immunological, animal models, and newer bioinformatical studies that have been reported to either support or refute such an association. Ultimately, we feel that well-designed clinical trials that target EBV or the host immune response to EBV will be needed to determine if such interventions may have a clinical impact.

27.5.1 Epstein-Barr Virus

EBV is one of the eight known human herpes virus infections (human herpes virus 4). With a 173-kb double-stranded DNA genome, the virus contains approximately 100 protein-coding genes in addition to many non-coding RNAs and micro-RNAs (Baer et al., 1984). There is a lytic cycle and a nonproductive (latent) cycle that is commonly shared amongst all human herpes viruses. The virus establishes lifelong latent infections within B cells in the oral mucosa as well as in the peripheral blood. As such, the virus is typically transferred through oral secretions during childhood or young adulthood (5–8 years of age in underdeveloped countries and primarily during the teenage years to early 20s in developed countries) (Dunmire et al., 2015).

EBV enters human lymphocytes via gp350/gp220 and gp42. Gp350 binds CD21, resulting in endocytosis of the of the viral particle. Fusion with the human genome is facilitated by viral proteins gB, gH, gL, gp350, and gp42 (Speck et al., 2000). Following infection, in individuals with MS, multiple longitudinal studies have now demonstrated high anti-EBV titers [anti-EBV nuclear antigen (EBNA-1) as well as EBNA complex] (Jacobs et al., 2020; Levin et al., 2010).

EBV was the first human virus identified to cause cancer, initially discovered in tumor cells of pediatric patients with Burkitt lymphoma (Rowe et al., 2014). More recently, EBV has been detected in tumor cells driving subtypes of Hodgkin and non-Hodgkin lymphomas, gastric carcinomas, pharyngeal carcinoma, natural killer cell and T cell lymphomas as well as in leiomyosarcomas (Al-Tarawneh et al., 2023; Kanda et al., 2019; Urayama et al., 2012).

27.5.2 Pathology

Studies of MS tissue samples have provided evidence for the presence of EBV infection in the brains of MS patients. An increase in the number of $CD8^{+}$, cytotoxic T cells specific to EBV was shown to be present in active MS brain lesions relative to control HLA-matched non-MS brains, concomitant with an increase in lytic EBV proteins (Angelini et al., 2013). This was also associated with severe inflammation and degeneration in the brain lesions (Serafini et al., 2019). These lesions appeared specific for EBV since no evidence for either CMV or influenza-specific CD8+ T cells were found (Serafini et al., 2019).

Additionally, researchers used in situ hybridization to detect interferon-α (IFNα) and EBV-encoded RNA (EBER) in both MS and control postmortem brain tissue, and they were shown to be overexpressed in active white matter lesions of MS patients (Tzartos et al., 2012). The co-expression of EBER and IFNα suggested that the presence of EBV in the brain could elicit a pro-inflammatory response that may contribute to the neurodegeneration associated with MS. Viral EBV RNAs were also reported to be present in gray matter lesions of MS patients, again using in situ hybridization, further supporting the hypothesis that B lineage cells infected with EBV may sustain an immunopathologic response in the MS brain (Magliozzi et al., 2013).

In another study, 85% of MS patient samples had recurrent $EBER^{+}$ cells as opposed to very few control patients (Moreno et al., 2018). Both latent and lytic EBV were found in the CNS, with lytic virus restricted to chronic MS lesions. However, while this study failed to find a significant difference in the number of samples containing viral proteins between controls and MS patients, the authors suggest an association between EBV-related viral protein and MS.

Despite histological evidence showing that EBV is frequently expressed in the MS brain, some studies have failed to demonstrate such EBV reactivity in MS lesions. In one study, brain tissue was collected from MS samples with white matter lesions. Following in situ hybridization, researchers observed that all samples were negative for EBERs (Willis et al., 2009). Even if B-lymphocytes had infiltrated the parenchyma, which would suggest a breakdown of the BBB, EBV was absent from the CNS tissue and lesions. Additionally, other reports have also failed to detect evidence of EBV in MS brain samples. One study screened a large number of MS cases ($n = 94$), in which only 16 cases were enriched for B cells and no EBERs were found in any MS sample (Peferoen et al., 2010). In another study, researchers observed that EBER-1 was the only, and rarely, detectable EBV transcript in MS plaques, and concluded that the samples they tested showed no significant evidence of EBV infection in the CNS of MS patients (Sargsyan et al., 2010). Collectively, although these observations do not clearly demonstrate the presence of EBV in MS brains compared to control material, studies using different techniques and different MS samples are difficult to compare nor do they rule out the possibility for a role of EBV in MS pathogenesis.

27.5.3 Serology

The most compelling argument for the association of EBV with MS is seroepidemiological studies in which higher anti-EBV seropositivity is observed in virtually 100% of MS patients compared with matched controls (Munch et al., 2009; Shirodaria, 1987), including higher concentrations of serum antibodies against both the EBV viral capsid antigen (VCA) and nuclear antigens (EBNA-1), and more frequent histories of infectious mononucleosis, a marker of late age at infection with EBV. (Ascherio & Munch, 2000). In a systematic review of eight published studies that included a total of 1005 cases and 1060 controls, there was a strong association between the presence of antibodies against EBV and risk of MS (Ascherio & Munch, 2000), with an estimated odds ratio of 13.5 (95% confidence interval = 6.3–31.4). This was later supported by an even more comprehensive report that identified 44 unique meta-analyses that included 416 primary studies exploring the correlation between various risk factors and MS, including risk factors such as vaccinations, comorbid diseases, surgeries, traumatic events and accidents, exposure to environmental agents, and biochemical, infectious, and musculoskeletal biomarkers (Belbasis et al., 2015). Importantly, anti-EBNA IgG antibodies, infectious mononucleosis, and smoking were the only three risk factors that showed the strongest consistent evidence of an association with MS (Belbasis et al., 2015). The researchers also observed the lack of association between MS and other viral pathogens, suggesting a specificity for EBV in MS pathogenesis (Belbasis et al., 2015). In another study of 901 MS patients that included both RRMS patients and patients with CIS, one of the earliest phases of the disease (Miller et al., 2012), all patients were seropositive for at least one EBV-specific protein (Abrahamyan et al., 2020). The presence of antibodies to EBV in CIS patients suggests that infection with EBV may even occur before the onset of MS symptoms and could therefore contribute to the earlier manifestations of symptoms in patients. This was further supported by a landmark report that used a nested case-control study from among more than 3 million US military personnel with blood samples collected between 1988 and 2000 (Levin, 2005) from which 83 individuals could be identified to develop MS compared to 166 controls matched by age, sex, race/ethnicity, and dates of blood sample collection. Importantly, serial samples were collected before the onset of symptoms with the average time between blood collection and MS onset being 4 years (range, 1–11 years). The strongest predictors of MS were serum levels of IgG antibodies to EBNA complex or EBNA-1. The risk of MS increased with antibody titers; the relative risk (RR) in persons with EBNA complex was 9.4 (95% confidence interval [CI], 2.5–35.4; P for trend .001). In longitudinal analyses, a fourfold increase in anti-EBNA complex or anti-EBNA-1 titers during the follow-up was associated with a threefold increase in MS risk (EBNA complex: RR,3.0; 95% CI, 1.3–6.5; EBNA-1: RR, 3.0; 95% CI, 1.2–7.3). No association was found between cytomegalovirus antibodies and MS (Levin, 2005).

27.5.4 Immunological Observations

EBV remains latent in host cells and predominantly targets B cells (Hatton et al., 2014; Long et al., 2019; Johnson & Tarakanova, 2020). The infection of B cells provides a compelling argument for the association of EBV with MS, particularly due to the effectiveness of B cell–depleting therapies used in MS treatments (Cencioni et al., 2021). Furthermore, the increased detection of EBV-specific antibodies in MS has supported the hypothesis that anti-EBV antibodies may recognize host self-antigens, the concept of molecular mimicry (Fujinami et al., 2006), and may therefore be associated with MS pathogenesis. This is supported by the observations that EBV epitopes have similarity with MBP (Wucherpfennig & Strominger, 1995; Lang et al., 2002), an autoantigen that has been shown to be involved in both MS immune responses and the animal model of MS, experimental autoimmune encephalomyelitis (EAE) (Martinsen & Kursula, 2022).

An increase in EBV viral load has also been reported in B cells of MS patients, as well as the chemokine CXCR3, which allows cells to cross the BBB (Langelaar et al., 2021). This is consistent with the pathological observations that EBV latent membrane protein 2A (LMP-2A) was co-expressed with the B cell–specific CD-20 in MS brain tissue samples, suggesting the infiltration of latently infected B cells in the brain (Serafini et al., 2010; Magliozzi et al., 2013). LMP-2A is one of the latent EBV proteins responsible for inducing the expression of B cell–activating factor of the tumor necrosis factor family (BAFF). Since BAFF is responsible for cell maturation and T cell–independent antibody production, these findings support a role for EBV infection in B cell activation in the MS brain and suggest that BAFF produced by EBV-infected B cells may contribute to this process resulting in viral persistence and, possibly, disruption of B cell tolerance (Serafini et al., 2010).

In addition, it has recently been reported that EBV copy numbers were higher in the peripheral blood mononuclear cells (PBMCs) and sorted B cells of active MS patients in comparison to PBMCs and B cells of stable MS patients and healthy controls (Monaco et al., 2023). Moreover, when spontaneous EBV-transformed lymphoblastoid B cell lines were generated from MS patients and controls using the endogenous EBV present from each individual, it was observed that these EBV-infected B cell lines were more readily generated from patients than controls. The higher frequency of EBV detection in $CD19^{+}$ B cells of active MS patients and the greater rate of spontaneously transformed EBV B cell lines in MS patients suggest that there may be a fundamental failure of host control of EBV latent infection in MS (Monaco et al., 2023). These findings are consistent with known abnormalities in the immune response to EBV in MS, including deficient cytotoxic T lymphocyte (CTL) control of EBV infection (Soldan & Lieberman, 2023).

T cell responses to EBV in MS patients have also been reported to be abnormal compared to controls. Activated T cells are known to cross the BBB (Goverman, 2009), and are associated with a pro-inflammatory response that is thought to be involved in the pathogenesis of MS, including disease progression, degeneration, lesion development, and the integrity of the BBB (Minagar & Alexander, 2003). In earlier studies, latent EBNA1-infected T cells have been observed at a greater frequency in MS patients compared to healthy controls and had a higher avidity for EBNA1 with a concomitant increased proliferative capacity (Lunemann, 2006). Further, these T cells were skewed toward a pro-inflammatory response with a Th1 phenotype, supporting the hypothesis that EBV-specific T cell infiltration in the CNS of MS patients could contribute to MS progression. In other reports, the presence of EBV infection was also observed in 99% of patient samples concomitant with increased CD8+ EBV-specific T cells associated with significantly higher pro-inflammatory cytokine, IFN-γ production (Jilek et al., 2008). Further, they demonstrated that the EBV-specific cytotoxic T lymphocytes (CTL) were more active in early cases of MS. In studies that found increased lytic EBV proteins in MS lesions, with an increased number of EBV-specific $CD8^{+}$ CTLs in these lesions (Angelini et al., 2013), they also demonstrated relatively greater frequencies of these CD8+ cells in peripheral blood samples of MS patients specific to lytic and latent EBV proteins (Angelini et al., 2013).

In contrast to the evidence demonstrating abnormally elevated T cell responses to EBV in MS patients, other reports have shown a decrease in EBV-specific responses in MS compared to controls. Seminal studies have shown that although MS patients had increased levels of anti-EBNA1 IgG consistent with the seroepidemiological results (Kreft et al., 2017), there was a decreased frequency of EBV-specific T cells compared with healthy subjects (Pender et al., 2009). The authors concluded that this quantitative deficiency in CD8+ T cell immunity to EBV might be responsible for the accumulation of EBV-infected B cells in the brains of patients with MS (Pender et al., 2009). These observations are supported by our recent findings of defective T cell control of EBV-infected B cells in MS patients (Ragheb et al., 2023). Using a novel nanoparticle (np) based Artificial Immune Modulation (AIM) platform, consisting of a cocktail of six EBV peptides presented on nanoparticles decorated with HLA-A2 and anti-CD28 molecules to enrich, expand, and characterize EBV-specific $CD8^+$ T cells from healthy and MS HLA-A*02:01 donors, the expanded $CD8^+$ T cells showed significant antigen specificity as shown by intracellular cytokine release and/or cytotoxic activity but exhibited functional defects in their responses to select EBV peptides (Ragheb et al., 2023).

27.5.5 Molecular Analysis

Polymerase chain reaction (PCR) techniques allow for the amplification of small amounts of DNA from patient samples and are useful for detecting the low quantities of viral DNA and RNA, particularly in samples in which viruses may be difficult to detect by other means. EBV transcripts have been demonstrated in over 90% of MS patients compared to 60% of control healthy blood samples (Santón et al., 2011). Statistical analysis indicated a correlation between the expression of EBNA2 and MS. An examination of 1055 specimens (440 DNA samples and 615 brain tissues) from 101 MS and 21 non-MS cases for the presence of EBV using PCR and EBER-in situ hybridization (EBER-ISH) was conducted (Hassani et al., 2018). They demonstrated 91/101 (90%) of MS cases compared to only 5/21 (24%) of non-MS cases were PCR EBV reactive. None of the samples were PCR positive for other common herpesviruses including HSV-1, CMV, and HHV-6. By quantitative PCR, EBV viral load in MS brain was mainly low to moderate in most cases. However, in 18/101 (18%) of MS cases, a widespread but scattered presence of EBV-infected cells was noted in the affected tissues by EBER-ISH, which demonstrated that astrocytes and microglia, in addition to B cells, can also be infected (Hassani et al., 2018). This large and comprehensive study shows that EBV is present and transcriptionally active in the brain of most cases of MS and supports a role for the virus in MS pathogenesis. Similarly, EBV genes were also found in approximately 80% of MS patient samples, with a greater incidence of EBV DNA observed in the meningeal tissue relative to white matter (Veroni et al., 2018). Multivariate analysis indicated a significant association between MS and EBV.

Conversely, others have failed to demonstrate increased EBV PCR reactivity in MS samples. While EBV DNA and RNA could be detected in peripheral blood leukocytes in one study, there was no significant difference in the prevalence of EBV positive samples between MS patients and controls (Lindsey et al., 2009), consistent with studies that have also indicated no difference in the expression of latent proteins or lytic DNA transcripts by PCR in MS patients (Hon et al., 2012). Moreover, some studies have failed to detect any EBV transcripts by PCR. One report failed to show the presence of any EBV RNA in CSF B cells and plasma cells (Sargsyan et al., 2010). While there was occasional indication of EBER1 expression in active MS plaques, no other EBV RNAs were observed, and the occurrence of EBER1 expression was rare. An investigation for the presence of both DNA and RNA in MS patient samples failed to detect EBV by quantitative real time PCR in MS brain lesions (Willis et al., 2009). Even using the same MS tissue that was previously reported to be EBV positive (Serafini et al., 2007), investigators using real-time PCR methodologies, capable of detecting very few EBV-infected cells, detected EBV at low levels in only 2 of the 12 multiple sclerosis meningeal specimens examined. The authors concluded that EBV infection was rare in multiple sclerosis brain tissue and suggested that EBV infection may be unlikely to contribute directly to multiple sclerosis brain pathology in the vast majority of cases (Willis et al., 2009).

27.5.6 Cerebrospinal Fluid (CSF)

CSF has been considered an important anatomical compartment to study due to its closer proximity to the central nervous system relative to serum or plasma (Hrishi & Sethuraman, 2019). Within the CSF, OCBs have been observed in virtually all patients with MS, and before the advent and use of clinical MRI, OCBs were considered a diagnostic criterion for the disease (Poser et al., 1983). OCBs are immunoglobulin proteins that indicate the presence of inflammation in the CNS. The IgG in MS patients' brain tissue and CSF is synthesized intrathecally and may more accurately reflect the immune response at the site of the disease (Gilden, 2001). While OCBs have been examined in MS research for decades, the specificity of these bands remained largely unknown. Importantly, the presence of OCBs is not unique to MS and has been demonstrated in other infectious and autoimmune diseases of the CNS (Virtanen et al., 2014). Indeed, in virtually all chronic inflammatory diseases of the CNS with an infectious etiology, OCBs have been shown to be, in part, directed toward the inciting agent (Burgoon et al., 2003). Indeed, it has been proposed that if EBV or any other virus causes MS, it should be possible to demonstrate that MS OCBs contain antibodies directed against the suspected agent (Gilden, 2001).

For many years, researchers have observed strong intrathecal antibody reactivity to EBV EBNA1 and BRRF2 proteins in MS patients (Rand et al., 2000; Cepok et al., 2005; Serafini et al., 2007). Even in the absence of detectable EBV DNA in CSF samples, MS patients had OCBs reactive for EBV where the patterns of CSF OCBs were stable over time (Serafini et al., 2007). Our studies have examined the relative reactivity of MS patient OCBs to two human herpesviruses, HHV-6 and EBV. We observed that 38% of MS patients had OCBs reactive to either virus (Virtanen et al., 2014), which appeared to be specific for MS as patients with other neurological diseases did not have OCBs to EBV or HHV-6. In support of these observations, it was reported that patients in the early stages of MS had detectable levels of intrathecal antibodies that were reactive specifically to EBNA1 but no other viral proteins, including VCA, Rubella, or VZV (Pfuhl et al., 2015). In this study, there was also a correlation between the presence of EBV-specific OCBs and higher levels of EBNA1-specific antibody levels in patient sera samples. Consistent with these results, more recent studies have also shown that CSF antibodies in MS patients were specific to EBNA1 and EBNA2 epitopes (Wang et al., 2021).

Despite demonstrating that OCBs in subsets of MS patients had specificity for herpesviruses, others have failed to confirm these observations. While one study did show that OCBs in MS reacted specifically to BAFF and EBV, this was not unique to MS as there were also EBV-specific OCBs in patients with other neurological diseases (Franciotta et al., 2011). This is consistent with other reports that have observed no significant difference in the production of anti-EBV intrathecal antibody production between MS patients and healthy controls (Sargsyan et al., 2010). Moreover, clonally expanded plasma cells, extracted from the CSF of MS patients, also had no reactivity to EBV proteins (Sargsyan et al., 2010). It has also been shown that although EBV-specific OCBs were present in 24% of the RRMS patients and absent in the controls, these EBV-specific intrathecal oligoclonal IgG productions were of low affinity, which the authors concluded can occur in a subset of MS patients as part of humoral polyreactivity driven by chronic brain inflammation (Castellazzi et al., 2014).

27.5.7 Animal Models

An important approach in attempts to establish a role for EBV (or any virus) in MS pathogenesis would be the use of an animal model system that might reflect the MS disease process. The most common murine model of MS is experimental autoimmune encephalomyelitis (EAE), in which stimulation with autoantigens such as myelin oligodendrocyte glycoprotein (MOG) or myelin basic protein (MBP) is associated with demyelination and clinical disease (Baxter, 2007; Hasselmann et al., 2017). While rodents lack the receptor for EBV and cannot be directly infected with this virus, reports have used a murine gamma herpesvirus, γHV-68, that can latently infect B cells and was associated with a significant CD8+ T cell response (Casiraghi et al., 2012). Subsequent induction of EAE in these latently infected mice was associated with more severe degeneration than EAE in uninfected mice. In addition, the expression of CD40, which is known to contribute to autoimmunity and EAE pathology by downregulating the number of regulatory T lymphocytes, was observed to be increased in dendritic cells and therefore skewed the T cell response to a Th1 pro-inflammatory phenotype (Casiraghi et al., 2015). In another study, mice treated with human EBV EBNA-1, which is thought to be cross-reactive with MBP, developed symptoms similar to EAE (Jog et al., 2020). Extracted splenocytes from these EBNA-1 treated mice were reactive to MBP, a result suggestive of molecular mimicry (Lang et al., 2002).

Of particular interest is a report of the spontaneous occurrence of an inflammatory demyelinating disease in a colony of Japanese macaques (JM, *Macaca fuscata*) at the Oregon National Primate Research Center (ONPRC), which was associated with a previously undescribed simian herpesvirus isolated from these animals (Axthelm et al., 2011). DNA sequence analysis revealed the virus was most similar to rhesus macaque gamma-2 herpesvirus rhadinovirus (RRV). Animals with this disease had clinical and pathologic similarities with MS (Axthelm et al., 2011).

The role of human herpesviruses has been most extensively studied in the nonhuman primate *Callithrix jacchus*, the common marmoset ('t Hart, 2016). It has been demonstrated that these nonhuman primates can be infected with a lymphocryptovirus gamma-herpesvirus termed Callitrichine Herpesvirus 3 (CalHV3) (Rivailler et al., 2002). In these infected animals, anti-CD20 treatment (a treatment commonly used in MS) significantly decreased the viral load of CalHV3 DNA in marmoset samples (Anwar Jagessar et al., 2013). The results of the anti-CD20 treatment that reduces the viral load in these latently CalHV3 infected marmosets may be extrapolated to the clinically beneficial effects of anti-CD20 in MS in which EBV has been suggested to latently infect B cells in patients (De Sèze et al., 2023). Additionally, we have shown that marmosets inoculated intranasally with the beta herpesvirus HHV-6 exhibited significantly accelerated clinical EAE compared with control uninfected animals (Leibovitch et al., 2018). Expansion of a pro-inflammatory CD8 subset correlated with post-EAE survival in latently infected animals who developed EAE, suggesting that a peripheral antigen-driven expansion may have occurred post-EAE induction. Virus antigen was markedly elevated and concentrated in brain lesions, similar to previously reported localizations of HHV-6 in MS brain lesions. Collectively, these results suggest that asymptomatic intranasal viral acquisition accelerates subsequent neuroinflammation in a nonhuman primate model of MS (Leibovitch et al., 2018).

27.5.8 Bioinformatics

There has been an increase in the use of large data sets and bioinformatic tools to explore the association between EBV and MS. Of particular importance was the recent study that again used the Department of Defense cohort of more than 10 million young adults on active duty in the US military, 955 of whom were diagnosed with MS during their period of service. For each MS case, serum samples were collected before the date of MS onset, and cases were matched to two randomly selected individuals without MS of the same age, sex, race/ethnicity, branch of military service, and dates of collection of blood samples who were on active military duty (Bjornevik et al., 2022). To interrogate the serum antibody reactivity to viruses, 30 serum samples from MS patients and matched controls without MS were used on a novel platform (VirScan) for the comprehensive serologic profiling of the entire known human virome (Xu et al., 2015). The technique is based on phage immunoprecipitation sequencing (PhIP-seq) technology that uses a bacteriophage library that displays over 200 known human pathogenic viruses in 56 amino acid peptide tiles with a 28 amino acid overlap between adjacent peptides (n = 110,000) (Xu et al., 2015). Using a virus discovery pipeline, the authors concluded that the risk of developing MS increased 32-fold compared to controls after contracting EBV and observed no significant association between MS and any other virus, such as CMV (Bjornevik et al., 2022).

In contrast, using the same platform but a different set of MS patients and controls, and using serum matched with CSF, we could also demonstrate that although patients with MS showed increased antibody responses associated with EBV peptide tiles in both serum and CSF by VirScan using a virus discovery pipeline, this may not be specific for MS since increased antibodies against EBV and CMV were also detected in patients with another chronic, progressive disease of the CNS (Enose-Akahata et al., 2023). This suggested that increased virus-specific immune responses, particularly against ubiquitous human herpesviruses, may be common in patients with neuroinflammatory diseases. In addition to using VirScan as a virus discovery analysis methodology, we used a viral signature pipeline that revealed a disease-specific antibody signature in CSF and serum of patients with MS (Enose-Akahata et al., 2023). Our findings suggest that antibody responses against multiple viral peptides, including EBV, can be identified. Rather than a single causative agent that has long been elusive to prove in a disease like MS, antibody profiling may better represent a virological exposure history that suggests multiple triggers, potentially cross-reactive with self-peptides, associated with immune-mediated neuroinflammatory disease (Libbey et al., 2007).

27.5.9 EBV Clinical Trials

Of the many approaches that have been used to support a role for EBV in the pathogenesis of MS and that have been highlighted in this chapter, ultimately, we feel that only through well-designed clinical trials that target EBV via antiviral medications or that attempt to correct an impaired host immune response to EBV can we ever truly establish a causal link between this ubiquitous virus and MS. Various antiviral agents have been explored and continue to be investigated. One hypothesis for the role of EBV in MS is the transactivation of human endogenous retroviruses (HERVs) by EBV (Sutkowski et al., 2001). This has led to the use of antiretroviral agents such as Raltegravir, an oral integrase strand transfer inhibitor approved for treating HIV infection (Pandey, 2011), in a Phase 2a clinical trial in patients with MS (Gold et al., 2018). Twenty participants were enrolled in a 3-month baseline phase followed by 3 months of treatment with 400 mg of Raltegravir twice a day. Patients had a monthly Gd-enhanced MRI, saliva collection to test for EBV shedding, blood sampling for safety monitoring, virology (including HERVs), measurement of immunological and inflammatory markers, and physical, neurological, and quality-of-life assessments (Gold et al., 2018). Raltegravir failed to demonstrate any beneficial effects in terms of relapse rate, disease progression of MS, and number of lesions as measured by MRI, nor did it reduce the amount of shedding of EBV in saliva (Gold et al., 2018). Similarly, Tenofovir has demonstrated some favorable outcomes in patients with concomitant HIV infections and the diagnosis of MS; however, this data is limited only to case reports (Skarlis et al., 2018). Valacyclovir, an anti-herpes antiviral medication, has also been used off label to suppress MS progression with the hypothesis that EBV could potentially trigger MS and/or cause the progression of the disease, yet no formal clinical trials have been conducted (Bergström, 2000). Brincidofovir (BCV) is a potent inhibitor of DNA polymerase, resulting in decreased DNA synthesis and chain termination. BCV is an orally available antiviral agent that is active against smallpox and several other DNA viruses, is currently approved for use against smallpox virus (variola) infection in humans, and has been given emergency use authorization for the treatment of monkeypox infection. Because of its broad antiviral activity against several DNA viruses, BCV is being investigated for the prevention and treatment of cytomegalovirus (CMV), BK Virus (BKV), adenoviruses (AdV), and Epstein-Barr virus (EBV). BCV has been shown, in vivo, to be effective in reducing EBV viral loads in EBV-infected B cell lines generated from MS patients and controls (Monaco et al., 2023; Steven Jacobson, personal communication). It is therefore a reasonable choice for a clinical trial in MS. However, it should be noted that, to date, antiviral agents in general have failed to demonstrate satisfactory outcomes in MS disease progression.

EBV-targeted CTL therapy has also been explored by utilizing EBV-specific cell lines, following HSCT in the setting of EBV-driven lymphomas, testing vaccines targeted to EBNA1, LMP1, and LMP2A214, and other clinical trials. One paper reported the analysis of 114 patients who had received infusions of EBV-specific CTLs at three different centers to prevent or treat EBV^+ lymphoproliferative disease arising after HSCT (Heslop et al., 2010). None of the 101 patients who received CTL prophylaxis developed EBV^+ lymphoproliferative disease, whereas 11 of 13 patients treated with CTLs for biopsy-proven or probable lymphoproliferative disease achieved sustained complete remissions (Heslop et al., 2010). This was also supported by a meta-analysis of EBV-specific cytotoxic T lymphocytes used for refractory EBV-associated post-transplant lymphoproliferative disorder in solid organ transplant recipients, which included 11 studies with 76 patients (Liu et al., 2021). Thirty-five (46.1%) patients were diagnosed with monomorphic post-transplant lymphoproliferative disorder where B lymphocyte type was the most common. All patients received primary treatment for post-transplant lymphoproliferative disorder that was ineffective. However, the response rate for EBV-CTL treatment of refractory PTLD was 66%. Of 50 patients, 36 achieved complete remission and 14 achieved partial remission (Liu et al., 2021). Based on these findings, it was therefore concluded that EBV-targeted CTL therapy was safe as an effective prophylaxis or treatment for lymphoproliferative disease in transplantation recipients (Heslop, 2015; Heslop et al., 2010).

Similar approaches are being used in progressive MS with in vitro-expanded autologous EBV-specific $CD8^+$ T cells directed against EBV viral latent proteins (Pender et al., 2014). This is based on the hypothesis that defective control of EBV infection by cytotoxic $CD8^+$ T cells might be associated with the pathogenesis of MS by allowing EBV-infected autoreactive B cells to accumulate in the central nervous system (Pender et al., 2009; Ragheb et al., 2023). One report treated an MS patient with autologous EBV-specific $CD8^+$ T cells with no adverse effects, and the patient showed clinical improvement with reduced disease activity on MRI and decreased intrathecal immunoglobulin production (Pender et al., 2014). In subsequent years, the same group analyzed the T cell response to EBV in MS by using flow cytometry and intracellular cytokine staining to measure T cell responses to EBV-infected autologous lymphoblastoid cell lines in 95 patients and 56 EBV-seropositive healthy subjects (Pender et al., 2017). They observed a decreased $CD8^+$ T cell response specific to EBV lytic antigens at the onset of MS and at all subsequent disease stages. $CD8^+$ T cells directed against EBV latent antigens were increased but had reduced cytokine polyfunctionality, indicating T cell exhaustion. Importantly, during MS exacerbations, the EBV-specific $CD4^+$ and $CD8^+$ T cell populations expanded, with increased functionality of latent-specific $CD8^+$ T cells. It was hypothesized that the lack of $CD8^+$ T cell control of EBV reactivation could lead to an expanded population of latently infected cells, including autoreactive B cells (Pender et al., 2017). Collectively, these results have led to a phase 2 EMBOLD clinical study to evaluate the safety and efficacy of an off-the-shelf, allogeneic, EBV T cell immunotherapy that targets EBV-infected B cells and plasma cells, including those within the central nervous system, in patients with progressive MS (Kresa-Reahl, 2023).

An alternative approach that is receiving considerable attention is the use of EBV-specific vaccines, either for the prevention or treatment of MS (Soldan & Lieberman, 2023). There are a number of EBV vaccines that are currently being developed (Zhong et al., 2022) that include an investigational EBV gp350-Ferritin nanoparticle vaccine (Durkee-Shock, 2023), chimeric virus-like particles engineered to present combinations of three EBV peptides derived from the receptor binding domain of EBV gp350 (Zhang et al., 2020), a gH/gL nanoparticle vaccine (Malhi et al., 2022), and a mRNA vaccine platform used for mRNA COVID vaccines that targets four EBV glycoprotein antigens on the virus particle (ModernaTX, Inc., 2023; Zhong et al., 2022). While a vaccine treatment trial would be optimal, the design of such a study is challenging. What form of MS should be targeted? Relapsing-remitting or progressive disease; What would inclusion criteria be? Increased antibody responses in blood or CSF, EBV oligoclonal bands, abnormal EBV immune responses, elevated viral loads; What would primary or exploratory outcomes of such a trial measure? Clinical or radiological changes, viral loads, immune response, etc. Importantly, some have cautioned that giving MS patients an EBV vaccine may have an adverse consequence by driving the immune system toward a heightened state of autoimmunity (Bjornevik et al., 2023). How would an EBV vaccine prevention of MS trial be performed? What risk group would be targeted (early MS, CIS, pediatric MS)? What would a control arm look like and, ultimately, how many individuals would need to be vaccinated to prevent how many cases of MS? These are all complex questions that need to be addressed.

We propose that an additional frontier in the research of MS, or other virally induced autoimmune disorders, will be dedicated to innovative treatments. These may include specific antivirals, vaccinations during childhood, and/or other small molecules that have activity against EBV (anti-EBV siRNA). Recently a small molecule compound that selectively inhibits the DNA binding activity of EBNA1 has been reported that selectively blocks EBV gene transcription (Messick et al., 2019). This compound was shown to block the proliferation and metabolic activity of autologous EBV-derived B cell lines from patients with MS and controls (Monaco et al., 2023). These studies may reveal targeted and specific therapies against EBV, a potential trigger of MS, without the multitude of side effects that accompany more traditional aggressive treatment, such as B cell–depleting agents (Sabatino et al., 2019). Finally, artificial antigen-presenting cells, nanoparticles, that serve to present protein fragments of the EBV virus in ex vivo expanded $CD3^+/CD8^+$ T cells, have shown promising results in priming the immune system against specific malignant or viral antigens (Ragheb et al., 2023).

27.6 Conclusion/Future Directions

Much additional information needs to be gathered to further support the role of EBV (or other viral infections) in MS. Rather than being the "cause" of this disease, we reiterate that EBV may be one of the environmental triggers for MS (maybe even the major trigger) in humans and that viral reactivation may be associated with disease progression. Early data, in terms of potent antivirals and/or triggering the host immune system to be more responsive against EBV, appears to be promising. Continued collaboration between clinicians and basic scientists will need to confirm this hypothesis. Genetic (e.g., HLA-DRB1*1501) and environmental risk factors certainly play a role in the development of MS following a viral infection. Ultimately, the interplay between EBV (or other viral infections), host genetic risk factors, and environmental exposures will need to be more clearly elucidated in order to determine the primary drivers of the development and progression of MS.

27.7 Review Questions

1. Which of the following was not mentioned as a risk factor for developing multiple sclerosis?
   (a) Former viral infection
   (b) *HLA-A*02:01 genotype*
   (c) Irregular levels of Vitamin D
   (d) MS diagnosis in a first-degree relative
   (e) HLA-DRB*15:01 genotype
   (f) Tobacco smoking
   (g) Obesity during childhood or adolescence
   (h) Increased salt intake
2. What do the McDonald criteria list as being hallmarks of MS?
   (a) Lesion dissemination in space
   (b) Lesion dissemination in time
   (c) Subpial demyelination
   (d) Brain atrophy
   (e) b and d
   (f) a, b, and d
   (g) *a and b*
   (h) a, b, c, and d
3. What is the primary form of transmission of Epstein-Barr virus?
   (a) Zoonotic transmission
   (b) *Oral transmission*
   (c) Airborne transmission
   (d) Pathogen ingestion
   (e) Hereditary transmission
4. What cells are latently infected by EBV?
   (a) Neurons
   (b) Microglia
   (c) Astrocytes
   (d) Macrophages
   (e) CD4+ T cells
   (f) CD8+ T cells
   (g) *B cells*
   (h) Killer T cells
   (i) Dendritic cells
5. Which is not a common histological feature of MS?
   (a) Demyelinated plaques
   (b) Infiltrates of mononuclear cells (CD8+ T cells, macrophages, microglia, B cells, plasma cells)
   (c) Reactive gliosis
   (d) *Early onset axon degeneration*

References

't Hart, B. A. (2016). Why does multiple sclerosis only affect human primates? *Multiple Sclerosis Journal, 22*(4), 559–563. https://doi.org/10.1177/1352458515591862

't Hart, B. A., & Massacesi, L. (2009). Clinical, pathological, and immunologic aspects of the multiple sclerosis model in common marmosets (*Callithrix jacchus*). *Journal of Neuropathology & Experimental Neurology, 68*(4), 341–355. https://doi.org/10.1097/NEN.0b013e31819f1d24

Abrahamyan, S., Eberspächer, B., Hoshi, M.-M., Aly, L., Luessi, F., Groppa, S., Klotz, L., Meuth, S. G., Schroeder, C., Grüter, T., Tackenberg, B., Paul, F., Then-Bergh, F., Kümpfel, T., Weber, F., Stangel, M., Bayas, A., Wildemann, B., Heesen, C., et al. (2020). Complete Epstein-Barr virus seropositivity in a large cohort of patients with early multiple sclerosis. *Journal of Neurology, Neurosurgery & Psychiatry, 91*(7), 681–686. https://doi.org/10.1136/jnnp-2020-322941

Aharoni, R., Sasson, E., Blumenfeld-Katzir, T., Eilam, R., Sela, M., Assaf, Y., & Arnon, R. (2013). Magnetic resonance imaging characterization of different experimental autoimmune encephalomyelitis models and the therapeutic effect of glatiramer acetate. *Experimental Neurology, 240*, 130–144. https://doi.org/10.1016/j.expneurol.2012.11.004

Al-Tarawneh, H., Alp, A., Gedikoglu, G., & Kosemehmetoglu, K. (2023). Epstein-Barr virus-positive Leiomyosarcoma in immunocompetent patients. *Turk Patoloji Dergisi, 40*, 56–62. https://doi.org/10.5146/tjpath.2023.01600

Angelini, D. F., Serafini, B., Piras, E., Severa, M., Coccia, E. M., Rosicarelli, B., Ruggieri, S., Gasperini, C., Buttari, F., Centonze, D., Mechelli, R., Salvetti, M., Borsellino, G., Aloisi, F., & Battistini, L. (2013). Increased CD8+ T cell response to Epstein-Barr virus lytic antigens in the active phase of multiple sclerosis. *PLoS Pathogens, 9*(4), e1003220. https://doi.org/10.1371/journal.ppat.1003220

Anwar Jagessar, S., Fagrouch, Z., Heijmans, N., Bauer, J., Laman, J. D., Oh, L., Migone, T., Verschoor, E. J., & 'T Hart, B. A. (2013). The different clinical effects of anti-BLyS, anti-APRIL and anti-CD20 antibodies point at a critical pathogenic role of γ-herpesvirus infected B cells in the marmoset EAE model. Journal of Neuroimmune Pharmacology, 8(3), 727–738. https://doi.org/10.1007/s11481-013-9448-6.

Ascherio, A., & Munch, M. (2000). Epstein-Barr virus and multiple sclerosis. *Epidemiology, 11*(2), 220–224. https://doi.org/10.1097/00001648-200003000-00023

Atkins, H. L., Bowman, M., Allan, D., Anstee, G., Arnold, D. L., Bar-Or, A., Bence-Bruckler, I., Birch, P., Bredeson, C., Chen, J., Fergusson, D., Halpenny, M., Hamelin, L., Huebsch, L., Hutton, B., Laneuville, P., Lapierre, Y., Lee, H., Martin, L., et al. (2016). Immunoablation and autologous haemopoietic stem-cell transplantation for aggressive multiple sclerosis: A multicentre single-group phase 2 trial. *The Lancet, 388*(10044), 576–585. https://doi.org/10.1016/S0140-6736(16)30169-6

Axthelm, M. K., Bourdette, D. N., Marracci, G. H., Su, W., Mullaney, E. T., Manoharan, M., Kohama, S. G., Pollaro, J., Witkowski, E., Wang, P., Rooney, W. D., Sherman, L. S., & Wong, S. W. (2011). Japanese macaque encephalomyelitis: A spontaneous multiple sclerosis-like disease in a nonhuman primate. *Annals of Neurology, 70*(3), 362–373. https://doi.org/10.1002/ana.22449

Baer, R., Bankier, A. T., Biggin, M. D., Deininger, P. L., Farrell, P. J., Gibson, T. J., Hatfull, G., Hudson, G. S., Satchwell, S. C., & Séguin, C. (1984). DNA sequence and expression of the B95-8 Epstein-Barr virus genome. *Nature, 310*(5974), 207–211. https://doi.org/10.1038/310207a0

Baxter, A. G. (2007). The origin and application of experimental autoimmune encephalomyelitis. *Nature Reviews Immunology, 7*(11), 904–912. https://doi.org/10.1038/nri2190

Bayas, A., Christ, M., Faissner, S., Klehmet, J., Pul, R., Skripuletz, T., & Meuth, S. G. (2023). Disease-modifying therapies for relapsing/active secondary progressive multiple sclerosis—A review of population-specific evidence from randomized clinical trials. *Therapeutic Advances in Neurological Disorders, 16*, 175628642211468. https://doi.org/10.1177/17562864221146836

Belbasis, L., Bellou, V., Evangelou, E., Ioannidis, J. P. A., & Tzoulaki, I. (2015). Environmental risk factors and multiple sclerosis: An umbrella review of systematic reviews and meta-analyses. *The Lancet Neurology, 14*(3), 263–273. https://doi.org/10.1016/S1474-4422(14)70267-4

Bergström, T. (2000). Several options for antiviral treatment trials in multiple sclerosis – But which targets should be selected? *Expert Opinion on Pharmacotherapy, 1*(6), 1087–1090. https://doi.org/10.1517/14656566.1.6.1087

Bjornevik, K., Cortese, M., Healy, B. C., Kuhle, J., Mina, M. J., Leng, Y., Elledge, S. J., Niebuhr, D. W., Scher, A. I., Munger, K. L., & Ascherio, A. (2022). Longitudinal analysis reveals high prevalence of Epstein-Barr virus associated with multiple sclerosis. *Science, 375*(6578), 296–301. https://doi.org/10.1126/science.abj8222

Bjornevik, K., Münz, C., Cohen, J. I., & Ascherio, A. (2023). Epstein-Barr virus as a leading cause of multiple sclerosis: Mechanisms and implications. *Nature Reviews. Neurology, 19*(3), 160–171. https://doi.org/10.1038/s41582-023-00775-5

Bourdette, D., & Yadav, V. (2008). B-cell depletion with rituximab in relapsing-remitting multiple sclerosis. *Current Neurology and Neuroscience Reports, 8*(5), 417–418. https://doi.org/10.1007/s11910-008-0064-4

Burgoon, M. P., Hammack, B. N., Owens, G. P., Maybach, A. L., Eikelenboom, M. J., & Gilden, D. H. (2003). Oligoclonal immunoglobulins in cerebrospinal fluid during varicella zoster virus (VZV) vasculopathy are directed against VZV. *Annals of Neurology, 54*(4), 459–463. https://doi.org/10.1002/ana.10685

Carson, K. R., Evens, A. M., Richey, E. A., Habermann, T. M., Focosi, D., Seymour, J. F., Laubach, J., Bawn, S. D., Gordon, L. I., Winter, J. N., Furman, R. R., Vose, J. M., Zelenetz, A. D., Mamtani, R., Raisch, D. W., Dorshimer, G. W., Rosen, S. T., Muro, K., Gottardi-Littell, N. R., et al. (2009). Progressive multifocal leukoencephalopathy after rituximab therapy in HIV-negative patients: A report of 57 cases from the research on adverse drug events and reports project. *Blood, 113*(20), 4834–4840. https://doi.org/10.1182/blood-2008-10-186999

Casiraghi, C., Shanina, I., Cho, S., Freeman, M. L., Blackman, M. A., & Horwitz, M. S. (2012). Gammaherpesvirus latency accentuates EAE pathogenesis: Relevance to Epstein-Barr virus and multiple sclerosis. *PLoS Pathogens, 8*(5), e1002715. https://doi.org/10.1371/journal.ppat.1002715

Casiraghi, C., Citlali Márquez, A., Shanina, I., & Horwitz, M. S. (2015). Latent virus infection upregulates CD40 expression facilitating enhanced autoimmunity in a model of multiple sclerosis. *Scientific Reports, 5*(1), 13995. https://doi.org/10.1038/srep13995

Castellazzi, M., Contini, C., Tamborino, C., Fasolo, F., Roversi, G., Seraceni, S., Rizzo, R., Baldi, E., Tola, M. R., Bellini, T., Granieri, E., & Fainardi, E. (2014). Epstein-Barr virus-specific intrathecal oligoclonal IgG production in relapsing-remitting multiple sclerosis is limited to a subset of patients and is composed of low-affinity antibodies. *Journal of Neuroinflammation, 11*(1), 188. https://doi.org/10.1186/s12974-014-0188-1

Cencioni, M. T., Mattoscio, M., Magliozzi, R., Bar-Or, A., & Muraro, P. A. (2021). B cells in multiple sclerosis—From targeted depletion to immune reconstitution therapies. *Nature Reviews Neurology, 17*(7), 399–414. https://doi.org/10.1038/s41582-021-00498-5

Cepok, S., Zhou, D., Srivastava, R., Nessler, S., Stei, S., Büssow, K., Sommer, N., & Hemmer, B. (2005). Identification of Epstein-Barr virus proteins as putative targets of the immune response in multiple sclerosis. *Journal of Clinical Investigation, 115*(5), 1352–1360. https://doi.org/10.1172/JCI200523661

Coles, A., Compston, A., Selmaj, K., Lake, S., Moran, S., Margolin, D., Norris, K., & Tandon, P. K. (2008). Alemtuzumab vs. interferon beta-1a in early multiple sclerosis. *New England Journal of Medicine, 359*(17), 1786–1801. https://doi.org/10.1056/NEJMoa0802670

Compston, A., & Coles, A. (2008). Multiple sclerosis. *The Lancet, 372*(9648), 1502–1517. https://doi.org/10.1016/S0140-6736(08)61620-7

Constantinescu, C. S., Farooqi, N., O'Brien, K., & Gran, B. (2011). Experimental autoimmune encephalomyelitis (EAE) as a model for multiple sclerosis (MS): EAE as model for MS. *British Journal of Pharmacology, 164*(4), 1079–1106. https://doi.org/10.1111/j.1476-5381.2011.01302.x

De Sèze, J., Maillart, E., Gueguen, A., Laplaud, D. A., Michel, L., Thouvenot, E., Zephir, H., Zimmer, L., Biotti, D., & Liblau, R. (2023). Anti-CD20 therapies in multiple sclerosis: From pathology to the clinic. *Frontiers in Immunology, 14*, 1004795. https://doi.org/10.3389/fimmu.2023.1004795

Deshmukh, V. A., Tardif, V., Lyssiotis, C. A., Green, C. C., Kerman, B., Kim, H. J., Padmanabhan, K., Swoboda, J. G., Ahmad, I., Kondo, T., Gage, F. H., Theofilopoulos, A. N., Lawson, B. R., Schultz, P. G., & Lairson, L. L. (2013). A regenerative approach to the treatment of multiple sclerosis. *Nature, 502*(7471), 327–332. https://doi.org/10.1038/nature12647

Dunmire, S. K., Hogquist, K. A., & Balfour, H. H. (2015). Infectious mononucleosis. In C. Münz (Ed.), *Epstein Barr virus volume 1* (Vol. 390, pp. 211–240). Springer. https://doi.org/10.1007/978-3-319-22822-8_9

Durkee-Shock, J. (2023). *Safety and immunogenicity of an Epstein-Barr virus (EBV) gp350-ferritin nanoparticle vaccine in healthy adults with or without EBV infection*. National Library of Medicine. Identification No. NCT04645147. Retrieved from https://clinicaltrials.gov/study/NCT04645147

Enose-Akahata, Y., Wang, L., Almsned, F., Johnson, K. R., Mina, Y., Ohayon, J., Wang, X. W., & Jacobson, S. (2023). The repertoire of CSF antiviral antibodies in patients with neuroinflammatory diseases. *Science Advances, 9*(1), eabq6978. https://doi.org/10.1126/sciadv.abq6978

Ermann, J., & Fathman, C. G. (2001). Autoimmune diseases: Genes, bugs and failed regulation. *Nature Immunology, 2*(9), 759–761. https://doi.org/10.1038/ni0901-759

Farina, G., Magliozzi, R., Pitteri, M., Reynolds, R., Rossi, S., Gajofatto, A., Benedetti, M. D., Facchiano, F., Monaco, S., & Calabrese, M. (2017). Increased cortical lesion load and intrathecal inflammation is associated with oligoclonal bands in multiple sclerosis patients: A combined CSF and MRI study. *Journal of Neuroinflammation, 14*(1), 40. https://doi.org/10.1186/s12974-017-0812-y

Filippi, M., Preziosa, P., Banwell, B. L., Barkhof, F., Ciccarelli, O., De Stefano, N., Geurts, J. J. G., Paul, F., Reich, D. S., Toosy, A. T., Traboulsee, A., Wattjes, M. P., Yousry, T. A., Gass, A., Lubetzki, C., Weinshenker, B. G., & Rocca, M. A. (2019). Assessment of lesions on magnetic resonance imaging in multiple sclerosis: Practical guidelines. *Brain, 142*(7), 1858–1875. https://doi.org/10.1093/brain/awz144

Franciotta, D., Di Stefano, A. L., Jarius, S., Zardini, E., Tavazzi, E., Ballerini, C., Marchioni, E., Bergamaschi, R., & Ceroni, M. (2011). Cerebrospinal BAFF and Epstein-Barr virus-specific oligoclonal bands in multiple sclerosis and other inflammatory demyelinating neurological diseases. *Journal of Neuroimmunology, 230*(1–2), 160–163. https://doi.org/10.1016/j.jneuroim.2010.10.027

Freedman, M., & Atkins, H. L. (2016). Haematopoietic stem cell transplants should be a second-line therapy for highly active MS – YES. *Multiple Sclerosis Journal, 22*(10), 1258–1259. https://doi.org/10.1177/1352458516654311

Fujinami, R. S., Von Herrath, M. G., Christen, U., & Whitton, J. L. (2006). Molecular mimicry, bystander activation, or viral persistence: Infections and autoimmune disease. *Clinical Microbiology Reviews, 19*(1), 80–94. https://doi.org/10.1128/CMR.19.1.80-94.2006

Gilden, D. H. (2001). Viruses and multiple Sclerosis. *JAMA, 286*(24), 3127. https://doi.org/10.1001/jama.286.24.3127

Gold, J., Marta, M., Meier, U. C., Christensen, T., Miller, D., Altmann, D., Holden, D., Bianchi, L., Adiutori, R., MacManus, D., Yousry, T., Schmierer, K., Turner, B., & Giovannoni, G. (2018). A phase II baseline versus treatment study to determine the efficacy of raltegravir (Isentress) in preventing progression of relapsing remitting multiple sclerosis as determined by gadolinium-enhanced MRI: The INSPIRE study. *Multiple Sclerosis and Related Disorders, 24*, 123–128. https://doi.org/10.1016/j.msard.2018.06.002

Goodin, D. S. (2014). Glucocorticoid treatment of multiple sclerosis. In *Handbook of clinical neurology* (Vol. 122, pp. 455–464). Elsevier. https://doi.org/10.1016/B978-0-444-52001-2.00020-0

Goverman, J. (2009). Autoimmune T cell responses in the central nervous system. *Nature Reviews Immunology, 9*(6), 393–407. https://doi.org/10.1038/nri2550

Hassani, A., Corboy, J. R., Al-Salam, S., & Khan, G. (2018). Epstein-Barr virus is present in the brain of most cases of multiple sclerosis and may engage more than just B cells. *PLoS One, 13*(2), e0192109. https://doi.org/10.1371/journal.pone.0192109

Hasselmann, J. P. C., Karim, H., Khalaj, A. J., Ghosh, S., & Tiwari-Woodruff, S. K. (2017). Consistent induction of chronic experimental autoimmune encephalomyelitis in C57BL/6 mice for the longitudinal study of pathology and repair. *Journal of Neuroscience Methods, 284*, 71–84. https://doi.org/10.1016/j.jneumeth.2017.04.003

Hatton, O. L., Harris-Arnold, A., Schaffert, S., Krams, S. M., & Martinez, O. M. (2014). The interplay between Epstein-Barr virus and B lymphocytes: Implications for infection, immunity, and disease. *Immunologic Research, 58*(2–3), 268–276. https://doi.org/10.1007/s12026-014-8496-1

Heslop, H. (2015). *Giving Epstein-Barr virus (EBV) specific killer T lymphocytes to patients who have had donor marrow grafts (ETNA)*. National Library of Medicine. Identification No. NCT00058812. Retrieved from https://clinicaltrials.gov/study/NCT00058812

Heslop, H. E., Slobod, K. S., Pule, M. A., Hale, G. A., Rousseau, A., Smith, C. A., Bollard, C. M., Liu, H., Wu, M.-F., Rochester, R. J., Amrolia, P. J., Hurwitz, J. L., Brenner, M. K., & Rooney, C. M. (2010). Long-term outcome of EBV-specific T-cell infusions to prevent or treat EBV-related lymphoproliferative disease in transplant recipients. *Blood, 115*(5), 925–935. https://doi.org/10.1182/blood-2009-08-239186

Hoang, D. M., Pham, P. T., Bach, T. Q., Ngo, A. T. L., Nguyen, Q. T., Phan, T. T. K., Nguyen, G. H., Le, P. T. T., Hoang, V. T., Forsyth, N. R., Heke, M., & Nguyen, L. T. (2022). Stem cell-based therapy for human diseases. *Signal Transduction and Targeted Therapy, 7*(1), 272. https://doi.org/10.1038/s41392-022-01134-4

Hon, G. M., Hassan, M. S., Van Rensburg, S. J., Erasmus, R. T., & Matsha, T. E. (2012). Assessment of Epstein-Barr virus in blood from patients with multiple sclerosis. *Metabolic Brain Disease, 27*(3), 311–318. https://doi.org/10.1007/s11011-012-9292-z

Hrishi, A. P., & Sethuraman, M. (2019). Cerebrospinal fluid (CSF) analysis and interpretation in neurocritical care for acute neurological conditions. *Indian Journal of Critical Care Medicine, 23*(S2), 115–119. https://doi.org/10.5005/jp-journals-10071-23187

Humm, A. M., Beer, S., Kool, J., Magistris, M. R., Kesselring, J., & Rösler, K. M. (2004). Quantification of Uhthoff's phenomenon in multiple sclerosis: A magnetic stimulation study. *Clinical Neurophysiology, 115*(11), 2493–2501. https://doi.org/10.1016/j.clinph.2004.06.010

Hutchinson, M., Fox, R. J., Havrdova, E., Kurukulasuriya, N. C., Sarda, S. P., Agarwal, S., Siddiqui, M. K., Taneja, A., & Deniz, B. (2014). Efficacy and safety of BG-12 (dimethyl fumarate) and other disease-modifying therapies for the treatment of relapsing–remitting multiple sclerosis: A systematic review and mixed treatment comparison. *Current Medical Research and Opinion, 30*(4), 613–627. https://doi.org/10.1185/03007995.2013.863755

Jacobs, B. M., Giovannoni, G., Cuzick, J., & Dobson, R. (2020). Systematic review and meta-analysis of the association between Epstein-Barr virus, multiple sclerosis and other risk factors. *Multiple Sclerosis Journal, 26*(11), 1281–1297. https://doi.org/10.1177/1352458520907901

Jadidi-Niaragh, F., & Mirshafiey, A. (2011). Th17 cell, the new player of neuroinflammatory process in multiple sclerosis: New player of neuroinflammatory process in multiple Sclerosis. *Scandinavian Journal of Immunology, 74*(1), 1–13. https://doi.org/10.1111/j.1365-3083.2011.02536.x

Jilek, S., Schluep, M., Meylan, P., Vingerhoets, F., Guignard, L., Monney, A., Kleeberg, J., Le Goff, G., Pantaleo, G., & Du Pasquier, R. A. (2008). Strong EBV-specific CD8+ T-cell response in patients with early multiple sclerosis. *Brain, 131*(7), 1712–1721. https://doi.org/10.1093/brain/awn108

Jog, N. R., McClain, M. T., Heinlen, L. D., Gross, T., Towner, R., Guthridge, J. M., Axtell, R. C., Pardo, G., Harley, J. B., & James, J. A. (2020). Epstein Barr virus nuclear antigen 1 (EBNA-1) peptides recognized by adult multiple sclerosis patient sera induce neurologic symptoms in a murine model. *Journal of Autoimmunity, 106*, 102332. https://doi.org/10.1016/j.jaut.2019.102332

Johnson, K. E., & Tarakanova, V. L. (2020). Gammaherpesviruses and B cells: A relationship that lasts a lifetime. *Viral Immunology, 33*(4), 316–326. https://doi.org/10.1089/vim.2019.0126

Junker, A., Wozniak, J., Voigt, D., Scheidt, U., Antel, J., Wegner, C., Brück, W., & Stadelmann, C. (2020). Extensive subpial cortical demyelination is specific to multiple sclerosis. *Brain Pathology, 30*(3), 641–652. https://doi.org/10.1111/bpa.12813

Kanda, T., Yajima, M., & Ikuta, K. (2019). Epstein-Barr virus strain variation and cancer. *Cancer Science, 110*(4), 1132–1139. https://doi.org/10.1111/cas.13954

Kappos, L., Antel, J., Comi, G., Montalban, X., O'Connor, P., Polman, C. H., Haas, T., Korn, A. A., Karlsson, G., & Radue, E. W. (2006). Oral Fingolimod (FTY720) for relapsing multiple sclerosis. *New England Journal of Medicine, 355*(11), 1124–1140. https://doi.org/10.1056/NEJMoa052643

Kappos, L., Radue, E.-W., O'Connor, P., Polman, C., Hohlfeld, R., Calabresi, P., Selmaj, K., Agoropoulou, C., Leyk, M., Zhang-Auberson, L., & Burtin, P. (2010). A placebo-controlled trial of oral fingolimod in relapsing multiple sclerosis. *New England Journal of Medicine, 362*(5), 387–401. https://doi.org/10.1056/NEJMoa0909494

Kleinschmidt-DeMasters, B. K., & Tyler, K. L. (2005). Progressive multifocal leukoencephalopathy complicating treatment with natalizumab and interferon beta-1a for multiple Sclerosis. *New England Journal of Medicine, 353*(4), 369–374. https://doi.org/10.1056/NEJMoa051782

Kokaia, Z., Martino, G., Schwartz, M., & Lindvall, O. (2012). Cross-talk between neural stem cells and immune cells: The key to better brain repair? *Nature Neuroscience, 15*(8), 1078–1087. https://doi.org/10.1038/nn.3163

Kreft, K. L., Van Nierop, G. P., Scherbeijn, S. M. J., Janssen, M., Verjans, G. M. G. M., & Hintzen, R. Q. (2017). Elevated EBNA-1 IgG in MS is associated with genetic MS risk variants. *Neurology Neuroimmunology & Neuroinflammation, 4*(6), e406. https://doi.org/10.1212/NXI.0000000000000406

Kresa-Reahl, K. (2023). *Phase 1/2 study to evaluate the safety and efficacy of ATA188 in subjects with progressive multiple Sclerosis (EMBOLD).* National Library of Medicine. Identification No. NCT03283826. Retrieved from https://clinicaltrials.gov/study/NCT03283826

Lang, H. L. E., Jacobsen, H., Ikemizu, S., Andersson, C., Harlos, K., Madsen, L., Hjorth, P., Sondergaard, L., Svejgaard, A., Wucherpfennig, K., Stuart, D. I., Bell, J. I., Jones, E. Y., & Fugger, L. (2002). A functional and structural basis for TCR cross-reactivity in multiple sclerosis. *Nature Immunology, 3*(10), 940–943. https://doi.org/10.1038/ni835

Langelaar, J., Wierenga-Wolf, A. F., Samijn, J. P. A., Luijks, C. J. M., Siepman, T. A., Doorn, P. A., Bell, A., Zelm, M. C., Smolders, J., & Luijn, M. M. (2021). The association of Epstein-Barr virus infection with CXCR3 + B-cell development in multiple sclerosis: Impact of immunotherapies. *European Journal of Immunology, 51*(3), 626–633. https://doi.org/10.1002/eji.202048739

Langer-Gould, A., Atlas, S. W., Green, A. J., Bollen, A. W., & Pelletier, D. (2005). Progressive multifocal leukoencephalopathy in a patient treated with natalizumab. *New England Journal of Medicine, 353*(4), 375–381. https://doi.org/10.1056/NEJMoa051847

Leibovitch, E. C., Caruso, B., Ha, S. K., Schindler, M. K., Lee, N. J., Luciano, N. J., Billioux, B. J., Guy, J. R., Yen, C., Sati, P., Silva, A. C., Reich, D. S., & Jacobson, S. (2018). Herpesvirus trigger accelerates neuroinflammation in a nonhuman primate model of multiple sclerosis. *Proceedings of the National Academy of Sciences, 115*(44), 11292–11297. https://doi.org/10.1073/pnas.1811974115

Levin, L. I. (2005). Temporal relationship between elevation of Epstein-Barr virus antibody titers and initial onset of neurological symptoms in multiple sclerosis. *JAMA, 293*(20), 2496. https://doi.org/10.1001/jama.293.20.2496

Levin, L. I., Munger, K. L., O'Reilly, E. J., Falk, K. I., & Ascherio, A. (2010). Primary infection with the Epstein-Barr virus and risk of multiple sclerosis. *Annals of Neurology, 67*(6), 824–830. https://doi.org/10.1002/ana.21978

Levine, K. S., Leonard, H. L., Blauwendraat, C., Iwaki, H., Johnson, N., Bandres-Ciga, S., Ferrucci, L., Faghri, F., Singleton, A. B., & Nalls, M. A. (2023). Virus exposure and neurodegenerative disease risk across national biobanks. *Neuron, 111*(7), 1086–1093.e2. https://doi.org/10.1016/j.neuron.2022.12.029

Libbey, J. E., McCoy, L. L., & Fujinami, R. S. (2007). Molecular mimicry in multiple sclerosis. In *International review of neurobiology* (Vol. 79, pp. 127–147). Elsevier. https://doi.org/10.1016/S0074-7742(07)79006-2

Lindsey, J., Hatfield, L., Crawford, M., & Patel, S. (2009). Quantitative PCR for Epstein-Barr virus DNA and RNA in multiple sclerosis. *Multiple Sclerosis Journal, 15*(2), 153–158. https://doi.org/10.1177/1352458508097920

Liu, J., Zhang, J., Zhan, H., Sun, L., & Wei, L. (2021). EBV-specific cytotoxic T lymphocytes for refractory EBV-associated post-transplant lymphoproliferative disorder in solid organ transplant recipients: A systematic review. *Transplant International, 34*(12), 2483–2493. https://doi.org/10.1111/tri.14107

Long, H. M., Meckiff, B. J., & Taylor, G. S. (2019). The T-cell response to Epstein-Barr virus – New tricks from an old dog. *Frontiers in Immunology, 10*, 2193. https://doi.org/10.3389/fimmu.2019.02193

Lunemann, J. D. (2006). Increased frequency and broadened specificity of latent EBV nuclear antigen-1-specific T cells in multiple sclerosis. *Brain, 129*(6), 1493–1506. https://doi.org/10.1093/brain/awl067

Magliozzi, R., Serafini, B., Rosicarelli, B., Chiappetta, G., Veroni, C., Reynolds, R., & Aloisi, F. (2013). B-cell enrichment and Epstein-Barr virus infection in inflammatory cortical lesions in secondary progressive multiple Sclerosis. *Journal of Neuropathology & Experimental Neurology, 72*(1), 29–41. https://doi.org/10.1097/NEN.0b013e31827bfc62

Malhi, H., Homad, L. J., Wan, Y.-H., Poudel, B., Fiala, B., Borst, A. J., Wang, J. Y., Walkey, C., Price, J., Wall, A., Singh, S., Moodie, Z., Carter, L., Handa, S., Correnti, C. E., Stoddard, B. L., Veesler, D., Pancera, M., Olson, J., et al. (2022). Immunization with a self-assembling nanoparticle vaccine displaying EBV gH/gL protects humanized mice against lethal viral challenge. *Cell Reports Medicine, 3*(6), 100658. https://doi.org/10.1016/j.xcrm.2022.100658

Martinsen, V., & Kursula, P. (2022). Multiple sclerosis and myelin basic protein: Insights into protein disorder and disease. *Amino Acids, 54*(1), 99–109. https://doi.org/10.1007/s00726-021-03111-7

Messick, T. E., Smith, G. R., Soldan, S. S., McDonnell, M. E., Deakyne, J. S., Malecka, K. A., Tolvinski, L., Van Den Heuvel, A. P. J., Gu, B.-W., Cassel, J. A., Tran, D. H., Wassermann, B. R., Zhang, Y., Velvadapu, V., Zartler, E. R., Busson, P., Reitz, A. B., & Lieberman, P. M. (2019). Structure-based design of small-molecule inhibitors of EBNA1 DNA binding blocks Epstein-Barr virus latent infection and tumor growth. *Science Translational Medicine, 11*(482), eaau5612. https://doi.org/10.1126/scitranslmed.aau5612

Miller, D. H., Chard, D. T., & Ciccarelli, O. (2012). Clinically isolated syndromes. *The Lancet Neurology, 11*(2), 157–169. https://doi.org/10.1016/S1474-4422(11)70274-5

Minagar, A., & Alexander, J. S. (2003). Blood-brain barrier disruption in multiple sclerosis. *Multiple Sclerosis Journal, 9*(6), 540–549. https://doi.org/10.1191/1352458503ms965oa

ModernaTX, Inc. (2023). *A study of an Epstein-Barr virus (EBV) candidate vaccine, mRNA-1189, in 12- to 30-year-old healthy adolescents and adults.* National Library of Medicine. Identification No. NCT05164094. Retrieved from https://clinicaltrials.gov/study/NCT05164094

Monaco, M. C. G., Soldan, S. S., Su, C., Clauze, A., Cooper, J. F., Patel, R. J., Lu, F., Hughes, R. J., Messick, T. E., Andrada, F. C., Ohayon, J., Lieberman, P. M., & Jacobson, S. (2023). EBNA1 inhibitors block proliferation of spontaneous lymphoblastoid cell lines from patients with multiple Sclerosis and healthy controls. *Neurology Neuroimmunology & Neuroinflammation, 10*(5), e200149. https://doi.org/10.1212/NXI.0000000000200149

Moreno, M. A., Or-Geva, N., Aftab, B. T., Khanna, R., Croze, E., Steinman, L., & Han, M. H. (2018). Molecular signature of Epstein-Barr virus infection in MS brain lesions. *Neurology Neuroimmunology & Neuroinflammation, 5*(4), e466. https://doi.org/10.1212/NXI.0000000000000466

Munch, M., Riisom, K., Christensen, T., Møller-Larsen, A., & Haahr, S. (2009). The significance of Epstein-Barr virus seropositivity in multiple sclerosis patients? *Acta Neurologica Scandinavica, 97*(3), 171–174. https://doi.org/10.1111/j.1600-0404.1998.tb00632.x

Olsson, T., Barcellos, L. F., & Alfredsson, L. (2017). Interactions between genetic, lifestyle and environmental risk factors for multiple sclerosis. *Nature Reviews Neurology, 13*(1), 25–36. https://doi.org/10.1038/nrneurol.2016.187

Pandey, K. K. (2011). Raltegravir in HIV-1 infection: Safety and efficacy in treatment-naïve patients. *Clinical Medicine Reviews in Therapeutics, 4*, 13–30. https://doi.org/10.4137/CMRT.S5022

Peferoen, L. A. N., Lamers, F., Lodder, L. N. R., Gerritsen, W. H., Huitinga, I., Melief, J., Giovannoni, G., Meier, U., Hintzen, R. Q., Verjans, G. M. G. M., Van Nierop, G. P., Vos, W., Peferoen-Baert, R. M. B., Middeldorp, J. M., Van Der Valk, P., & Amor, S. (2010). Epstein Barr virus is not a characteristic feature in the central nervous system in established multiple sclerosis. *Brain, 133*(5), e137–e137. https://doi.org/10.1093/brain/awp296

Pender, M. P., Csurhes, P. A., Lenarczyk, A., Pfluger, C. M. M., & Burrows, S. R. (2009). Decreased T cell reactivity to Epstein-Barr virus infected lymphoblastoid cell lines in multiple sclerosis. *Journal of Neurology, Neurosurgery & Psychiatry, 80*(5), 498–505. https://doi.org/10.1136/jnnp.2008.161018

Pender, M. P., Csurhes, P. A., Smith, C., Beagley, L., Hooper, K. D., Raj, M., Coulthard, A., Burrows, S. R., & Khanna, R. (2014). Epstein-Barr virus-specific adoptive immunotherapy for progressive multiple sclerosis. *Multiple Sclerosis Journal, 20*(11), 1541–1544. https://doi.org/10.1177/1352458514521888

Pender, M. P., Csurhes, P. A., Burrows, J. M., & Burrows, S. R. (2017). Defective T-cell control of Epstein-Barr virus infection in multiple sclerosis. *Clinical & Translational Immunology, 6*(1), e126. https://doi.org/10.1038/cti.2016.87

Pfuhl, C., Oechtering, J., Rasche, L., Gieß, R. M., Behrens, J. R., Wakonig, K., Freitag, E., Pache, F. C., Otto, C., Hofmann, J., Eberspächer, B., Bellmann-Strobl, J., Paul, F., & Ruprecht, K. (2015). Association of serum Epstein-Barr nuclear antigen-1 antibodies and intrathecal immunoglobulin synthesis in early multiple sclerosis. *Journal of Neuroimmunology, 285*, 156–160. https://doi.org/10.1016/j.jneuroim.2015.06.012

Popescu, B. F. G., Pirko, I., & Lucchinetti, C. F. (2013). Pathology of multiple sclerosis: Where do we stand? *CONTINUUM: Lifelong Learning in Neurology, 19*, 901–921. https://doi.org/10.1212/01.CON.0000433291.23091.65

Poser, C. M., Paty, D. W., Scheinberg, L., McDonald, W. I., Davis, F. A., Ebers, G. C., Johnson, K. P., Sibley, W. A., Silberberg, D. H., & Tourtellotte, W. W. (1983). New diagnostic criteria for multiple sclerosis: Guidelines for research protocols. *Annals of Neurology, 13*(3), 227–231. https://doi.org/10.1002/ana.410130302

Ragheb, J. A., Frazier, W., Suarez, L., Monaco-Kushner, M., Wang, R., Oelke, M., & Jacobson, S. (2023). *Expansion of EBV peptide-specific CD8 T cells from multiple sclerosis patients and healthy donors reveals dysregulation of effector responses that may be associated with disease pathogenesis.* Poster presented at: The Federation of Clinical Immunology Societies (FOCIS), Boston, MA.

Rand, K. H., Houck, H., Denslow, N. D., & Heilman, K. M. (2000). Epstein-Barr virus nuclear antigen-1 (EBNA-1) associated oligoclonal bands in patients with multiple sclerosis. *Journal of the Neurological Sciences, 173*(1), 32–39. https://doi.org/10.1016/S0022-510X(99)00298-1

Reich, D. S., Lucchinetti, C. F., & Calabresi, P. A. (2018). Multiple sclerosis. *New England Journal of Medicine, 378*(2), 169–180. https://doi.org/10.1056/nejmra1401483

Rice, G. P., Incorvaia, B., Munari, L. M., Ebers, G., Polman, C., D'Amico, R., Parmelli, E., & Filippini, G. (2001). Interferon in relapsing-remitting multiple sclerosis. *Cochrane Database of Systematic Reviews, 2001*, CD002002. https://doi.org/10.1002/14651858.CD002002

Rivailler, P., Cho, Y., & Wang, F. (2002). Complete genomic sequence of an Epstein-Barr virus-related herpesvirus naturally infecting a new world primate: A defining point in the evolution of oncogenic lymphocryptoviruses. *Journal of Virology, 76*(23), 12055–12068. https://doi.org/10.1128/JVI.76.23.12055-12068.2002

Rodgers, M. M., Mulcare, J. A., King, D. L., Mathews, T., Gupta, S. C., & Glaser, R. M. (1999). Gait characteristics of individuals with multiple sclerosis before and after a 6-month aerobic training program. *Journal of Rehabilitation Research and Development, 36*(3), 183–188.

Rowe, M., Fitzsimmons, L., & Bell, A. I. (2014). Epstein-Barr virus and Burkitt lymphoma. *Chinese Journal of Cancer., 33*, 609. https://doi.org/10.5732/cjc.014.10190

Sabatino, J. J., Zamvil, S. S., & Hauser, S. L. (2019). B-cell therapies in multiple sclerosis. *Cold Spring Harbor Perspectives in Medicine, 9*(2), a032037. https://doi.org/10.1101/cshperspect.a032037

Santón, A., Cristóbal, E., Aparicio, M., Royuela, A., Villar, L. M., & Álvarez-Cermeño, J. C. (2011). High frequency of co-infection by Epstein-Barr virus types 1 and 2 in patients with multiple sclerosis. *Multiple Sclerosis Journal, 17*(11), 1295–1300. https://doi.org/10.1177/1352458511411063

Sargsyan, S. A., Shearer, A. J., Ritchie, A. M., Burgoon, M. P., Anderson, S., Hemmer, B., Stadelmann, C., Gattenlohner, S., Owens, G. P., Gilden, D., & Bennett, J. L. (2010). Absence of Epstein-Barr virus in the brain and CSF of patients with multiple sclerosis. *Neurology, 74*(14), 1127–1135. https://doi.org/10.1212/WNL.0b013e3181d865a1

Schmitz, K., Barthelmes, J., Stolz, L., Beyer, S., Diehl, O., & Tegeder, I. (2015). "Disease modifying nutricals" for multiple sclerosis. *Pharmacology & Therapeutics, 148*, 85–113. https://doi.org/10.1016/j.pharmthera.2014.11.015

Serafini, B., Rosicarelli, B., Franciotta, D., Magliozzi, R., Reynolds, R., Cinque, P., Andreoni, L., Trivedi, P., Salvetti, M., Faggioni, A., & Aloisi, F. (2007). Dysregulated Epstein-Barr virus infection in the multiple sclerosis brain. *Journal of Experimental Medicine, 204*(12), 2899–2912. https://doi.org/10.1084/jem.20071030

Serafini, B., Severa, M., Columba-Cabezas, S., Rosicarelli, B., Veroni, C., Chiappetta, G., Magliozzi, R., Reynolds, R., Coccia, E. M., & Aloisi, F. (2010). Epstein-Barr virus latent infection and BAFF expression in B cells in the multiple Sclerosis brain: Implications for viral persistence and intrathecal B-cell activation. *Journal of Neuropathology & Experimental Neurology, 69*(7), 677–693. https://doi.org/10.1097/NEN.0b013e3181e332ec

Serafini, B., Rosicarelli, B., Veroni, C., Mazzola, G. A., & Aloisi, F. (2019). Epstein-Barr virus-specific CD8 T cells selectively infiltrate the brain in multiple sclerosis and interact locally with virus-infected cells: Clue for a virus-driven immunopathological mechanism. *Journal of Virology, 93*(24), e00980–e00919. https://doi.org/10.1128/JVI.00980-19

Shaikh, S., Verma, H., Yadav, N., Jauhari, M., & Bullangowda, J. (2012). Applications of steroid in clinical practice: A review. *ISRN Anesthesiology, 2012*, 1–11. https://doi.org/10.5402/2012/985495

Shirani, A., Zhao, Y., Karim, M. E., Petkau, J., Gustafson, P., Evans, C., Kingwell, E., Van Der Kop, M. L., Oger, J., & Tremlett, H. (2014). Investigation of heterogeneity in the association between interferon beta and disability progression in multiple sclerosis: An observational study. *European Journal of Neurology, 21*(6), 835–844. https://doi.org/10.1111/ene.12324

Shirodaria, P. V. (1987). Viral antibody titers: Comparison in patients with multiple sclerosis and rheumatoid arthritis. *Archives of Neurology, 44*(12), 1237. https://doi.org/10.1001/archneur.1987.00520240019006

Sintzel, M. B., Rametta, M., & Reder, A. T. (2018). Vitamin D and multiple sclerosis: A comprehensive review. *Neurology and Therapy, 7*(1), 59–85. https://doi.org/10.1007/s40120-017-0086-4

Skarlis, C., Gontika, M., Katsavos, S., Velonakis, G., Toulas, P., & Anagnostouli, M. (2018). Multiple sclerosis and subsequent human immunodeficiency virus infection: A case with the rare comorbidity, focus on novel treatment issues and review of the literature. *In Vivo, 31*(5), 10.21873/invivo.11167.

Soelberg Sorensen, P. (2016). Haematopoietic stem cell transplants should be a second-line therapy for highly active MS – NO. *Multiple Sclerosis Journal, 22*(10), 1260–1263. https://doi.org/10.1177/1352458516644341

Soldan, S. S., & Lieberman, P. M. (2023). Epstein-Barr virus and multiple sclerosis. *Nature Reviews Microbiology, 21*(1), 51–64. https://doi.org/10.1038/s41579-022-00770-5

Sorensen, P. S., & Blinkenberg, M. (2016). The potential role for ocrelizumab in the treatment of multiple sclerosis: Current evidence and future prospects. *Therapeutic Advances in Neurological Disorders, 9*(1), 44–52. https://doi.org/10.1177/1756285615601933

Speck, P., Haan, K. M., & Longnecker, R. (2000). Epstein-Barr virus entry into cells. *Virology, 277*(1), 1–5. https://doi.org/10.1006/viro.2000.0624

Steelman, A. J. (2015). Infection as an environmental trigger of multiple sclerosis disease exacerbation. *Frontiers in Immunology, 6*. https://doi.org/10.3389/fimmu.2015.00520

Sutkowski, N., Conrad, B., Thorley-Lawson, D. A., & Huber, B. T. (2001). Epstein-Barr virus transactivates the human endogenous retrovirus HERV-K18 that encodes a superantigen. *Immunity, 15*(4), 579–589. https://doi.org/10.1016/S1074-7613(01)00210-2

Thompson, A. J., et al. (2018). Diagnosis of multiple sclerosis: 2017 revisions of the McDonald criteria. *Lancet Neurology, 17*, 162–173. https://doi.org/10.1016/S1474-4422(17)30470-2

Tzartos, J. S., Khan, G., Vossenkamper, A., Cruz-Sadaba, M., Lonardi, S., Sefia, E., Meager, A., Elia, A., Middeldorp, J. M., Clemens, M., Farrell, P. J., Giovannoni, G., & Meier, U.-C. (2012). Association of innate immune activation with latent Epstein-Barr virus in active MS lesions. *Neurology, 78*(1), 15–23. https://doi.org/10.1212/WNL.0b013e31823ed057

Urayama, K. Y., Jarrett, R. F., Hjalgrim, H., Diepstra, A., Kamatani, Y., Chabrier, A., Gaborieau, V., Boland, A., Nieters, A., Becker, N., Foretova, L., Benavente, Y., Maynadié, M., Staines, A., Shield, L., Lake, A., Montgomery, D., Taylor, M., Smedby, K. E., Amini, R. M., et al. (2012). Genome-wide association study of classical Hodgkin lymphoma and Epstein-Barr virus status-defined subgroups. *Journal of the National Cancer Institute, 104*(3), 240–253. https://doi.org/10.1093/jnci/djr516

Veroni, C., Serafini, B., Rosicarelli, B., Fagnani, C., & Aloisi, F. (2018). Transcriptional profile and Epstein-Barr virus infection status of laser-cut immune infiltrates from the brain of patients with progressive multiple sclerosis. *Journal of Neuroinflammation, 15*(1), 18. https://doi.org/10.1186/s12974-017-1049-5

Virtanen, J., Wohler, J., Fenton, K., Reich, D., & Jacobson, S. (2014). Oligoclonal bands in multiple sclerosis reactive against two herpesviruses and association with magnetic resonance imaging findings. *Multiple Sclerosis Journal, 20*(1), 27–34. https://doi.org/10.1177/1352458513490545

Wallin, M. T., Culpepper, W. J., Campbell, J. D., Nelson, L. M., Langer-Gould, A., Marrie, R. A., Cutter, G. R., Kaye, W. E., Wagner, L., Tremlett, H., Buka, S. L., Dilokthornsakul, P., Topol, B., Chen, L. H., & LaRocca, N. G. (2019). The prevalence of MS in the United States: A population-based estimate using health claims data. *Neurology, 92*(10), e1029–e1040. https://doi.org/10.1212/WNL.0000000000007035

Walton, C., King, R., Rechtman, L., Kaye, W., Leray, E., Marrie, R. A., Robertson, N., La Rocca, N., Uitdehaag, B., Van Der Mei, I., Wallin, M., Helme, A., Angood Napier, C., Rijke, N., & Baneke, P. (2020). Rising prevalence of multiple sclerosis worldwide: Insights from the Atlas of MS, third edition. *Multiple Sclerosis Journal, 26*(14), 1816–1821. https://doi.org/10.1177/1352458520970841

Wang, Z., Kennedy, P. G., Dupree, C., Wang, M., Lee, C., Pointon, T., Langford, T. D., Graner, M. W., & Yu, X. (2021). Antibodies from multiple sclerosis brain identified Epstein-Barr virus nuclear antigen 1 & 2 epitopes which are recognized by oligoclonal bands. *Journal of Neuroimmune Pharmacology, 16*(3), 567–580. https://doi.org/10.1007/s11481-020-09948-1

Willis, S. N., Stadelmann, C., Rodig, S. J., Caron, T., Gattenloehner, S., Mallozzi, S. S., Roughan, J. E., Almendinger, S. E., Blewett, M. M., Brück, W., Hafler, D. A., & O'Connor, K. C. (2009). Epstein-Barr virus infection is not a characteristic feature of multiple sclerosis brain. *Brain, 132*(12), 3318–3328. https://doi.org/10.1093/brain/awp200

Wolinsky, J. S., Narayana, P. A., O'Connor, P., Coyle, P. K., Ford, C., Johnson, K., Miller, A., Pardo, L., Kadosh, S., Ladkani, D., & PROMiSe Trial Study Group. (2007). Glatiramer acetate in primary progressive multiple sclerosis: Results of a multinational, multicenter, double-blind, placebo-controlled trial. *Annals of Neurology, 61*(1), 14–24. https://doi.org/10.1002/ana.21079

Wucherpfennig, K. W., & Strominger, J. L. (1995). Molecular mimicry in T cell-mediated autoimmunity: Viral peptides activate human T cell clones specific for myelin basic protein. *Cell, 80*(5), 695–705. https://doi.org/10.1016/0092-8674(95)90348-8

Xu, G. J., Kula, T., Xu, Q., Li, M. Z., Vernon, S. D., Ndung'u, T., Ruxrungtham, K., Sanchez, J., Brander, C., Chung, R. T., O'Connor, K. C., Walker, B., Larman, H. B., & Elledge, S. J. (2015). Comprehensive serological profiling of human populations using a synthetic human virome. *Science, 348*(6239), aaa0698. https://doi.org/10.1126/science.aaa0698

Zhang, X., Zhao, B., Ding, M., Song, S., Kang, Y., Yu, Y., Xu, M., Xiang, T., Gao, L., Feng, Q., Zhao, Q., Zeng, M.-S., Krummenacher, C., & Zeng, Y.-X. (2020). A novel vaccine candidate based on chimeric virus-like particle displaying multiple conserved epitope peptides induced neutralizing antibodies against EBV infection. *Theranostics, 10*(13), 5704–5718. https://doi.org/10.7150/thno.42494

Zhong, L., Krummenacher, C., Zhang, W., Hong, J., Feng, Q., Chen, Y., Zhao, Q., Zeng, M.-S., Zeng, Y.-X., Xu, M., & Zhang, X. (2022). Urgency and necessity of Epstein-Barr virus prophylactic vaccines. *npj Vaccines, 7*(1), 159. https://doi.org/10.1038/s41541-022-00587-6

Zostawa, J., Adamczyk, J., Sowa, P., & Adamczyk-Sowa, M. (2017). The influence of sodium on pathophysiology of multiple sclerosis. *Neurological Sciences, 38*(3), 389–398. https://doi.org/10.1007/s10072-016-2802-8

Zozulya, A. L., & Wiendl, H. (2008). The role of regulatory T cells in multiple sclerosis. *Nature Clinical Practice Neurology, 4*(7), 384–398. https://doi.org/10.1038/ncpneuro0832

Virus-Induced Encephalitis

28

Clinton Jones

Abstract

Certain neurotropic RNA and DNA viruses can induce a life-threatening disease in the central nervous system, which is referred to as encephalitis. Viruses that cause encephalitis generally induce inflammation in brain tissue (cerebral edema) and kill neurons via a direct or indirect mechanism. These cellular insults culminate in intracerebral hemorrhage, bleeding in the brain, and brain damage. While encephalitis is a relatively rare disease, this disease costs the United States approximately $2 billion/year. RNA viruses that cause encephalitis include certain arboviruses, Astroviruses, and the severe acute respiratory syndrome coronavirus 2 (SARS-CoV-2). Herpes simplex virus type 1 (HSV-1) causes 38% of all encephalitis cases, which is more than the other viruses that cause encephalitis. Two other herpesviruses, varicella zoster virus and human cytomegalovirus, can also induce encephalitis. This chapter discusses the main virus families that cause encephalitis, their replication cycle, animal models used to examine encephalitis, clinical features of encephalitis, and therapeutic strategies. Finally, mechanisms by which these viruses cause encephalitis are discussed.

Keywords

Respiratory syndrome coronavirus 2 (SARS-CoV-2) · Herpes simplex virus type 1 (HSV-1) · Herpes simplex virus-mediated encephalitis (HSE) · West Nile virus (WNV) · Nonstructural proteins (NS1 · NS2A · NS2B · NS3 · NS4A · NS4B · and NS5) · Open reading frames (ORFs) · RNA-dependent RNA polymerase (RdRP) · Angiotensin-converting enzyme 2 (ACE2) · Toll-like receptor 2 (TLR2) · Nuclear factor-kB (NF-kB) · Tumor necrosis factor receptor (TNFR) · Tumor necrosis factor alpha (TNF-a) · Central nervous system (CNS) · Herpes simplex virus type 1 (HSV-1) · Debranching RNA lariats 1 gene (DBR1) · Magnetic resonance imaging (MRI) · Immediate early (IE) · early (E) · or late (L) · Latency-associated transcript (LAT) · Micro-RNAs (mi-RNAs) · Small non-coding RNAs (LAT sncRNA) · Locus coeruleus (LC) · Spinal trigeminal tract (Pr5) · Toll-like receptor 3 (TLR3) · Toll/interleukin-1 receptor (TIR) · Toll/interleukin-1 receptor domain-containing adaptor inducing IFN-β (TRIF) · Tumor necrosis factor receptor-associated factor 3 (TRAF3) · Interferon-alpha receptor (IFNAR1) · Receptor-interacting protein kinase 3 (RIPK3) · Receptor-interacting protein kinase 1 (RIPK1) · Varicella Zoster Virus (VZV) · Centers for Disease Control (CDC) · Glycoprotein E (gE) · In situ hybridization (ISH) · Humanized immuno-deficient mice (SCID-hu) · Human cytomegalovirus (HCMV) · Hematopoietic progenitor cells (HPCs) · Human immunodeficiency virus (HIV) · Human herpesvirus 6 (HHV-6) · Epstein-Barr virus (EBV) · Thymidine kinase (TK) · Toll-interleukin-1 receptor domain-containing adaptor protein (TIRAP) · NOD-like receptor (NLR) family pyrin domain-containing 3 (NLRP3) · Cellular inhibitor of apoptosis protein 1 (cIAP1) · TNF receptor-associated factor 2 (TRAF2) · Terminal long repeat (TR_L) · The internal (or inverted) long repeat (IR_L) · Terminal short repeat (TR_S) · The internal (or inverted) short repeat (IR_S) · Unique long (U_L) · Unique short (U_S) · Upstream of LAT transcript (UOL) · Antisense of LAT1 (AL1) · Pathogen-associated molecular patterns (PAMPs) · Danger-associated molecular patterns (DAMPs) · Pro-interleukin-1b (Pro-IL-1b) · Pro interleukin-18 (Pro-IL-18) · Gasdermin D (Gsdm D) · Mixed lineage kinase like (MLKL)

C. Jones (✉)
Oklahoma State University, College of Veterinary Medicine, Department of Veterinary Pathobiology, Stillwater, OK, USA
e-mail: clint.jones10@okstate.edu

H. E. Gendelman, T. Ikezu (eds.), *Neuroimmune Pharmacology and Therapeutics*,
https://doi.org/10.1007/978-3-031-68237-7_28

28.1 Introduction to Encephalitis

World-wide there are approximately 500,000 children and adults affected by encephalitis every year. Symptoms of encephalitis are initially nondescript and include flu-like illness, headache, odd behavior, drowsiness, trouble speaking, and/or controlling movement. The clinical hallmark of acute encephalitis is inflammation of brain parenchyma and neuronal death or dysfunction, reviewed in (Alam et al., 2022; Ellul & Solomon, 2018). It is also clear that inflammatory-mediated programmed cell death (pyroptosis and necroptosis) triggered by viruses is important during this process, as discussed below. Regardless of age, sex, or ethnicity, almost anyone can be affected by encephalitis and the mortality rate is high. Herpes simplex virus (HSV)-mediated encephalitis (HSE), the most prevalent viral encephalitis, affects 2–4000 people/year in the USA (Kumar & Mendez, 2022), and mortality rates are as high as 70% (Corey, 2005; Gilden et al., 2007; Lahat et al., 1999; McGrath et al., 1997; Skoldenberg, 1991). HSE is associated with necrotic cell death resulting from virus replication and inflammation (DeBiasi et al., 2002). Long-term complications occur despite antiviral treatment because ~2/3 of all cases are the result of reactivation from latency (Yamada et al., 2002). Acute encephalitis costs the USA approximately $2 billion/year (Kiyani et al., 2020); approximately 38% is due to HSE. Approximately 80% of people who survive acute encephalitis exhibit life-long brain injury that can lead to memory loss, personality changes, memory issues, epilepsy, and/or fatigue. Although viruses usually cause encephalitis, bacterial, fungal, or parasitic infections can also lead to encephalitis. Finally, autoimmune-related encephalitis can occur; but its frequency is low relative to infectious agents.

28.2 RNA Viruses That Can Cause Encephalitis

28.2.1 Arboviruses

Arboviruses are a group of RNA viruses transmitted by arthropod vectors. Clinical symptoms occur between 3 and 15 days after exposure and typically consist of fever, headache, and malaise, as reviewed in (Kuno & Chang, 2005). Life-threatening encephalitis and hemorrhagic fever can also occur; however, less frequently. Arboviruses are maintained in nature by cycling between a vector (mosquitoes, ticks, sandflies, as well as other arthropods) that consumes the blood of vertebrates. Vertebrates that have their blood consumed act as the host, with each vector preferring the blood of a specific species. Transmission between the vector and host occurs when the vector feeds on the blood of the vertebrate; consequently, the virus establishes infection in the salivary glands of the vector. High levels of infectious virus must be present in the blood of the host to allow transmission to the vector. If high levels of virus are not present in the host, viremia is not achieved and this infection is referred to as a "dead end host," which means it is rarely transmitted back to the vector. Below is a discussion of the major arboviruses that pose a health threat to humans.

28.2.2 West Nile Virus (WNV)-Induced Encephalitis

West Nile virus (WNV) is a member of the Flaviviridae family (Lindenback & Rice, 2001), and is a positive-strand RNA virus that is approximately 11,000 bases long. Genomic RNA serves as a messenger RNA that is translated into a single polyprotein. This polyprotein is then cleaved into at least 10 discrete proteins by cellular proteases and a virally encoded serine protease. Three structural proteins (C, prM, and E) are produced from the polyprotein (Fig. 28.1). Seven nonstructural proteins (NS1, NS2A, NS2B, NS3, NS4A, NS4B, and NS5) are generated from the polyprotein. NS5 is the most conserved protein in the Flavirviridae family because this protein is the viral encoded RNA-dependent RNA polymerase. The other nonstructural proteins are not as well conserved suggesting they have virus-specific functions.

WNV has a broad range of antigenic variation and restriction-length polymorphisms indicating WNV has a highly variable genome. Strains from Africa, Europe, and the Middle East form a distinct group relative to strains isolated in India and the Far East (Price & O'Leary, 1967). WNV is widely disseminated throughout Africa, Europe, the Middle East, and the Far East. During the summer of 1999, WNV was introduced to the east coast of the United States and has since spread westward. WNV can be readily grown in numerous mammalian cell lines and Drosophila cells (Lindenback & Rice, 2001). In the wild, WNV is vectored by female mosquitoes of the genus *Culex* and can infect a variety of small rodents and birds. In addition, WNV can infect soft and hard ticks under natural and experimental conditions. Almost all species of birds examined (wild species, chickens, and pigeons) develop viremia. During recent US outbreaks, numerous crows were killed by the virus, which allows tracking of virus spread (Anderson et al., 1999). Sporadic cases of naturally acquired infections can occur in horses, and these infections can lead to encephalitis. Bovine species do not develop viremia, but antibodies in cattle are prevalent. Although dogs are susceptible to infection, low levels of virus shedding preclude a significant role in virus transmission. Humans are a dead end host and human-to-human transmission is rare.

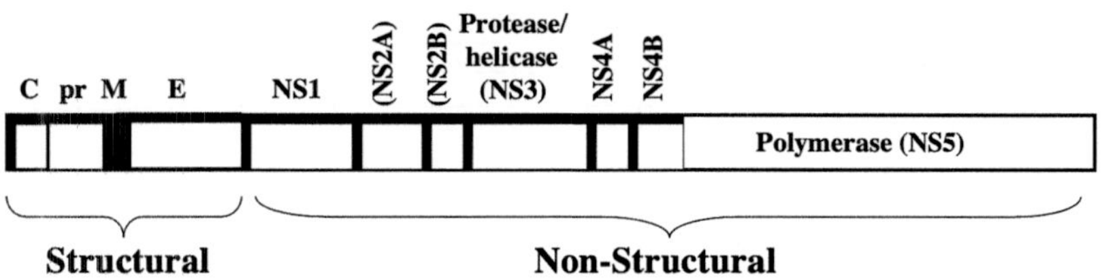

Fig. 28.1 Schematic of West Nile Virus genome. The various genes of the WNV genome are presented, and those comprising the nonstructural or structural proteins are denoted. (For details, see text)

Mouse models of infection have been used extensively to examine virus–host interactions, and virus-induced encephalitis. Adaptive immune responses are crucial for controlling WNV infections in mice, including encephalitis (Diamond et al., 2003; Shresta & Diamond, 2004; Wang et al., 2003). WNV infection leads to a toll-like receptor 3 (TLR3) dependent inflammatory response, which promotes brain penetration of the virus, neuronal injury, and enhanced encephalitis (Wang et al., 2004). Innate immune responses also protect against WNV spread and encephalitis because mice lacking the alpha/beta interferon receptor exhibit a higher incidence of encephalitis after WNV infection when compared to the parental wild-type mouse strain (Samuel & Diamond, 2005).

28.2.3 Other Arboviruses Associated with Encephalitis

In addition to WNV, there are several additional encephalitic viruses transmitted via an insect vector, reviewed in (Stahl et al., 2011). With respect to North and South America, arboviruses that cause encephalitis include Saint Louis encephalitis, Eastern encephalitic virus, Western equine encephalitic virus, La Crosse virus, and Colorado Tick Fever virus. In addition, several other RNA arboviruses can cause encephalitis in different parts of the world. These include Japanese encephalitis virus, Tick-borne encephalitis virus, and Toscana virus.

The incubation period is between 1 and 6 days, and the typical case is mild. Clinical features of patients with mild cases include fever, headache, backache, generalized myalgia, and anorexia. About 1 in 5 people who are infected develop a fever and other symptoms. A rash occurs in about 1/2 of the cases; the rash is usually roseolar or maculopapular and usually involves the chest, back, and upper extremities. The disease usually runs its course in 3–6 days and is usually milder in children. Approximately 1 out of 150 infected people can cause severe, potentially fatal neurological diseases, including encephalitis, meningitis, paralysis, and anterior myelitis. Although neurons are the primary target of WNV infection, a hallmark of WNV encephalitis is an accumulation of inflammatory infiltrates extending from the meninges in the brain parenchyma. These infiltrates contain lymphocytes and macrophages (Kelley et al., 2003). Most cases of WNV-induced encephalitis generally occur in older individuals. Currently, there is no specific treatment for arboviruses and there is no WNV vaccine currently available. Treatment consists of supportive and symptomatic care, including support of respiration, intravenous fluids, and prevention of secondary infections. In addition, insect repellents and other antivectorial measures can reduce the incidence of human transmission.

28.3 Specific Astrovirus Strains Have Encephalitic Properties

Astrovirus genomes contain a positive-sense single-stranded RNA genome that is approximately 6800 bases and contains three open reading frames (ORFs), reviewed in (Bosch et al., 2014). Although Astroviruses were previously considered to be pathogens of the gastrointestinal tract (Bosch et al., 2014), 12 cases of human encephalitis due to astrovirus infection have been identified (Lum et al., 2016; Naccache et al., 2015; Quan et al., 2010; Wunderli et al., 2011). The human astrovirus strain most frequently associated with encephalitis in humans is VA1 (Lum et al., 2016; Naccache et al., 2015; Quan et al., 2010; Wunderli et al., 2011). VA1 can productively infect primary human astrocytes and a human neuroblastoma cell line, SK-N-SH. However, primary human neurons are not productively infected (Janowski et al., 2017; Janowski et al., 2019). VA1 infects astrocytes and triggers neuronal apoptosis in immunocompromised individuals, culminating in neuronal injury and apoptosis. Two cases occurred in patients with X-linked agammaglobulinemia and seven cases in patients who underwent allogenic hematopoietic stem cell transplants. VA1 infection increases expression of CXCL10/IP-10 from astrocytes and SK-N-SH cells (Janowski et al., 2019). CXCL10 is a proinflammatory chemokine and neurons express the CXCL10 receptor, CXCR3 (Liu et al., 2011). Consequently, CXCL10 can induce neuronal hyperexcitability and apoptosis (Sui et al., 2006; Xia et al., 2000). These studies suggest VA1 infection of astrocytes indirectly triggers neuronal apoptosis in immunocompromised individuals, which can progress to encephalitis. Interestingly, neuroinvasive astrovirus infections occur in cattle (Li et al., 2013), mink (Blomstrom et al., 2010), swine (Arruda et al., 2017), sheep (Pfaff et al., 2017), Alpaca (Kuchler et al., 2020), and muskox (Boujon et al., 2019), which supports the premise that specific human Astrovirus strains can induce serious neurological disease in humans. Collectively, these studies indicate VA1 is an emerging virus that can in encephalitis in immunocompromised people and physicians should look for the presence of this virus in these individuals.

28.4 SARS-CoV-2 Infections Can Cause Inflammation and Neurological Complications, Including Encephalitis

The global COVID-19 pandemic that began in 2019 is having extraordinary effects on health care throughout the world. Severe acute respiratory syndrome coronavirus 2 (SARS-CoV-2) is the causative agent of COVID-19. At the nucleotide level, SARS-CoV-2 is closely related to certain bat coronaviruses. While the origins of SARS-CoV-2 are controversial, recent studies concluded this virus evolved via zoonotic spread through intermediate hosts, including bats (Coehn, 2022; Keusch et al., 2022).

28.4.1 Growth Properties and Clinical Disease Induced by SARS-CoV-2

SARS-CoV-2 has a linear, positive-sense, single-stranded RNA genome that is approximately 30,000 bases long (Fig. 28.2). Coronaviridae family members are the largest known RNA viruses. SARS-CoV-2 genomic RNA contains a 5′ cap structure and 3′ poly-A tail, which is typical for a positive-stranded RNA virus. Due to the large size of the genome, numerous viral proteins mediate pathogenesis. A large open reading frame (ORF1) is cleaved into ORF1a and ORF1b. ORF1b encodes the RNA-dependent RNA polymerase (RdRP), which drives viral RNA replication and transcription. Proteases further cleave ORF1a and ORF1b to generate 16 nonstructural proteins (NSPs), denoted NSP1 to NSP16. Seven NSPs (NSP1, NSP6, NSP8, NSP12, NSP13, NSP14, NSP15) and seven accessory proteins (ORF6, ORF7a, ORF7b, ORF8, ORF9b, ORF10, M) antagonize interferon induction and interferon signaling (Diamond & Kanneganti, 2022). Hence, SARS-CoV-2 is well-armed to evade innate immune responses, which promotes viral replication and virus transmission.

The SARS-CoV-2 encoded protein, spike protein (S), is required for virus entry and specifically binds to the host protein, angiotensin-converting enzyme 2 (ACE2). ACE2 converts angiotensin II into angiotensin, which has vasodilator, antiproliferative, and antifibrotic properties. These virus–host interactions can trigger cell damage to vascular endothelial cells by inhibiting mitochondrial function and inducing endothelial nitric oxide synthetase activity, which culminates in secondary cerebrovascular effects; for example, dizziness, nausea, vomiting, severe headache, confusion, disorientation, and/or memory loss. Additional symptoms include numbness, weakness in an arm, leg, or face, and these ailments can occur on only one side (Lei et al., 2021). Hence, altering ACE2 functions can contribute to neurological complications following infection. Recent studies revealed the spike protein induces inflammation by activating toll-like receptor 2 (TLR2): consequently, the nuclear factor-kB (NF-kB) signaling pathway is activated (Khan et al., 2020; Zheng et al., 2021) (Fig. 28.3). NF-kB is a latent transcription factor present in the cytoplasm until activation. Activation leads to NF-kB localization to the nucleus and specific binding to sequences in key transcriptional promoters of genes that drive expression of proinflammatory cytokines and chemokines (Liu et al., 2017). The tumor necrosis factor receptor (TNFR) is also activated because expression of tumor necrosis factor alpha (TNF-a), a soluble cytokine that binds to TNFR, is increased by the NF-kB signaling pathway. Expression of NLRP3, a potent inflammasome that drives inflammatory responses, is also stimulated by NF-kB. Increased NLRP3 levels and activation of this inflammasome activates proinflammatory-mediated cells death, pyroptosis, and several forms of programmed cell death including apoptosis, ferroptosis, and necroptosis, reviewed in (Huang et al., 2021). Collectively, SARS-CoV-2 infection induces inflammation in infected cells and can trigger bystander effects on surrounding cells, in large part due

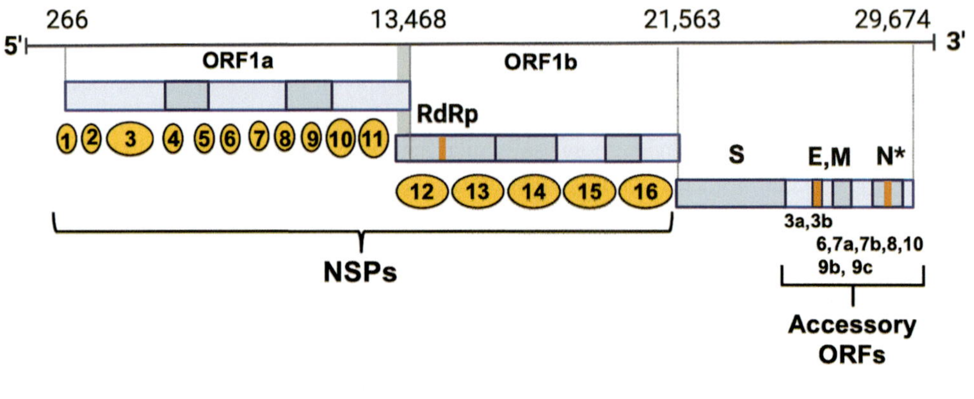

Fig. 28.2 Schematic of the SARS-CoV2 genome. The various genes of the SARS-CoV2 genome are presented, and those comprising the nonstructural proteins (NSP), or accessory open reading frames (ORF) are denoted. (For additional details, see text)

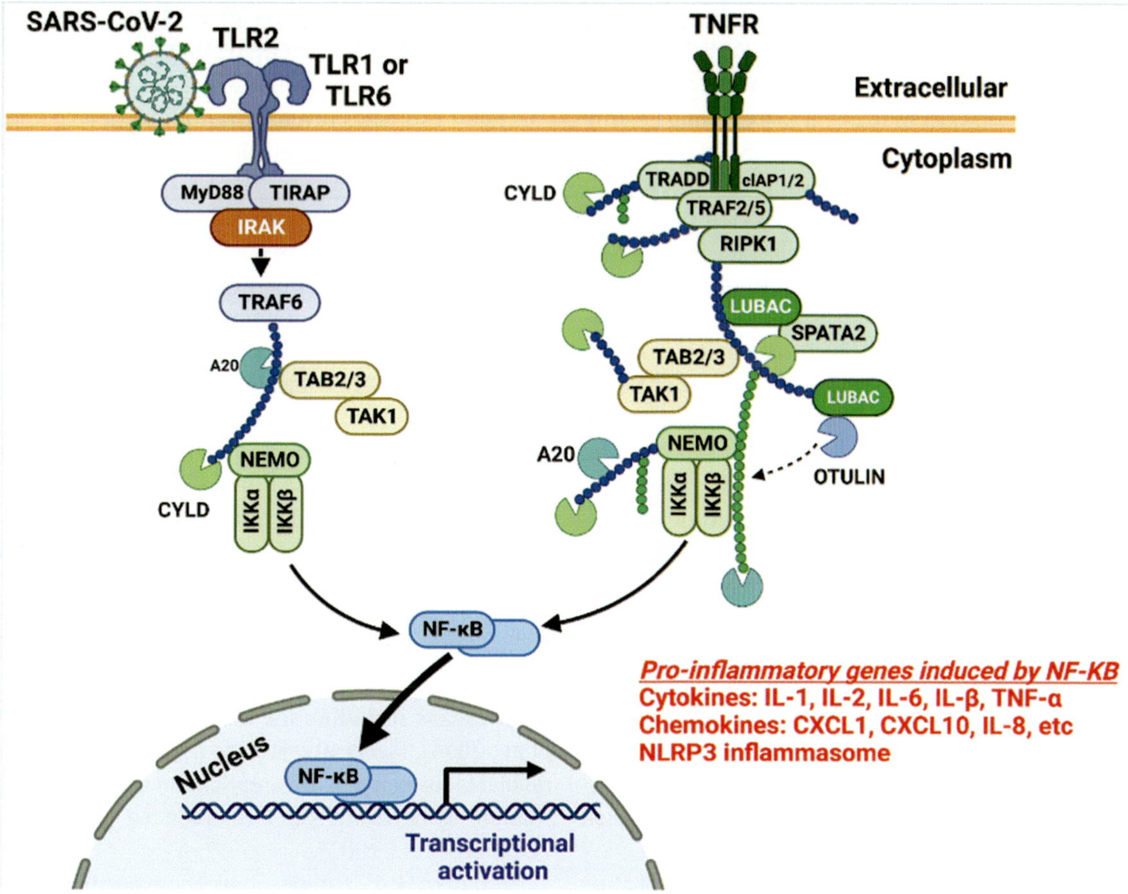

Fig. 28.3 Summary of SARS-CoV-2-mediated inflammation. Entry of a SARS-CoV-2 viral particle into a permissive cell is sensed by a TLR2/TLR1 or TLR2/TLR6 heterodimer. The viral S protein is required for virus entry, and TLR2 is the specific pattern recognition receptor (PRR). Consequently, toll-interleukin-1 Receptor (TIR) domain-containing adaptor protein (TIRAP), MyD88, and a member of the interleukin 1 (IL-1) receptor-associated kinase (IRAK) is recruited to TLR2/TLR1 or TLR2/TLR1 heterodimer bound to the viral S protein. Autophosphorylation of IRAK leads to dissociation from this complex, which triggers TRAF6 association and activation. These signaling events lead to NF-kB activation and nuclear localization. The NF-kB transcription factor is a homodimer or heterodimer of members of the Rel family of DNA binding proteins, which include RelA (p65), c-rel, RelB, p105 (precursor of p50), and p100 (precursor of p52). Nuclear NF-kB specifically binds to certain DNA motifs, 5′-GGGRNYYYCC-3′ (R is a purine, Y is a pyrimidine, and N is any nucleotide) on promoters that drive expression of proinflammatory cytokines, chemokines, and NOD-like receptor (NLR) family pyrin domain-containing 3 (NLRP3). NLP3 is a crucial inflammasome that contributes to inflammation and triggers programmed cell death (Huang et al., 2021). The genes listed are a small subset of genes activated by NF-kB: however, these genes are crucial for driving inflammation

to NF-kB-mediated activation of genes that encode proteins that have proinflammatory functions.

28.4.2 Neuropathogenesis Caused by SARS-CoV-2

Although the most common symptoms of COVID-19 are fever and dry cough, other manifestations, including runny nose, and gastrointestinal symptoms occur in a subset of patients. Furthermore, recent studies concluded that approximately 0.2% of COVID-19 patients develop encephalitis, reviewed in (Pezzini & Padovani, 2020; Siow et al., 2021). Autopsy studies revealed viral RNA can be detected in the brains of patients but at reduced RNA levels compared to the respiratory tract (Schaller et al., 2020). Since viral RNA is not always detected in brain tissues of humans, indirect effects of SARS-CoV-2 can play crucial roles during neurological complications, including encephalitis. For example, hypoxia is frequently observed in severe cases, which could facilitate encephalitis in certain individuals. Autopsies of hypoxic COVID-19 patients frequently exhibit ischemic damage; an area of bleeding within necrosing cerebral tissue (Thakur et al., 2021). Microglial activation with microglial nodules, and neuronophagia frequently is also present. In a subset of severe cases of COVID-19, patients exhibit severe systemic inflammatory responses, including microglial activation and neuronophagia in brain tissues but no evidence of viral invasion (Al-Dalahmah et al., 2020). Finally, stroke is associated with infection in 1 to 3% of hospitalized patients.

A recent study examined the effects of infecting 2–4-year-old rhesus macaques (*Macaca mulatta*) with SARS-Co-2 for 7 days, which this is operationally defined as acute infection (Beckman et al., 2022). As a comparison to the young healthy macaques, type II diabetic aged monkeys (18–24 years old) were also infected. All macaques were infected intratracheally and via the nasal cavity, which mimics how people are exposed to SARS-CoV-2. The viral nucleocapsid (N) protein, spike protein, and double-stranded RNA are readily detected in the frontal lobe of all infected macaques, regardless of age. The presence of double-stranded RNA in the brain indicates viral replication occurred in these animals. In general, neurons exhibited the highest degree of intracellular N expression regardless of age. Astrocytes and microglia were also infected in young and aged animals. Regardless of age, trans-neural spread occurred via corticocortical pathways, culminating in dissemination within the central nervous system (CNS). Notably, higher levels of inflammation were detected in aged animals relative to infected young animals, which is consistent with SARS-CoV-2 infections in humans. Aged macaques also displayed enhanced disruption of the blood-brain barrier after infection when compared to infected young macaques. This study provides compelling evidence that SARS-CoV-2 readily enters the nervous system, infects neurons, and spreads within the CNS of infected macaques.

While SARS-CoV-2 may not directly cause encephalitis in all patients, the ability of the virus to dysregulate immune responses and trigger hyper-inflammation can cause neurological complications. Finally, autoimmune encephalitis is associated with a subset of encephalitis cases following SARS-CoV-2 infection (Nabizadeh et al., 2022).

Activation of the NF-kB signaling pathway leads to increased levels of tumor necrosis factor-alpha (TNF-a), which then activates the tumor necrosis factor receptor (TNFR). Following interactions between TNF-a and TNFR, a multi-protein complex assembles. This complex contains tumor necrosis factor receptor type 1-associated DEATH domain protein (TRADD), cIAP1 (cellular inhibitor of apoptosis protein 1) or cIAP2, TNF receptor-associated factor 2 (TRAF2), TRAF5, and receptor-interacting protein kinase 1 (RIPK1). Formation of this multi-protein complex also culminates in NF-kB activation and nuclear localization. This Figure was created using BioRender.com.

28.5 Herpesvirus-Mediated Encephalitis

The Herpesviridae family is divided into three subfamilies: α-herpesvirinae, β-herpesvirinae, and γ-herpesvirinae. Each subfamily member contains a human virus that can induce encephalitis. These encephalitic human herpesviruses are discussed below.

28.5.1 Herpes Simplex Virus Type 1 (HSV-1)-Mediated Encephalitis

The latest estimates revealed 3.7 billion people under the age of 50, or 67% of the population were infected with HSV-1 (Corey, 2005). Generally, HSV-1 is acquired during childhood. In the world, approximately 491 million (13%) of people from 15–49 years old are infected with HSV-2, and infections typically are initiated in the genital tract. Interestingly, HSV-2 infections occur almost twice as frequently in women when compared to men. HSV-mediated encephalitis (HSE) is the most common causative agent of encephalitis.

HSE is the most common cause of acute, sporadic viral encephalitis in the United States accounting for 10–20% of all cases (Corey, 2005). Approximately 2000 newly diagnosed HSE cases per year occur in the United States. HSE incidence is higher between 5 and 30 years of age and after 50 years old. HSV-1 and HSV-2 can cause acute necrotizing encephalitis in infants, children, and adults. HSV-2 encephalitis in newborns is a widespread disease in the brain and other organs including the skin, eyes, and lungs (Stanberry et al., 2004). Generally, HSE is associated with necrotic cell death resulting from virus replication and widespread inflammation as a result of virus-induced immune responses (DeBiasi et al., 2002).

Initially, HSE patients have nonspecific clinical symptoms that include fever, headache, meningeal irritation, nausea, vomiting, confusion, generalized seizures, and altered levels of consciousness. The second stage of HSE is referred to as focal necrosis of the orbitofrontal and temporal cortexes and the limbic system. Symptoms associated with this phase include anosmia, memory loss, peculiar behavior, speech defects, hallucinations, and focal seizures. There is rapid progression in some cases with the appearance of reflex asymmetry, focal paralysis, hemi-paresis, and coma. Cerebral edema contributes to these symptoms and is important for the outcome (Stanberry et al., 2004).

HSE is characterized by severe destruction of temporal and frontal lobe structures. Furthermore, HSE can affect brainstem (Benbir et al., 2010; Gorospe et al., 2004; Livorsi et al., 2010; Tyler et al., 1995). Some patients with a predisposition of encephalitis in brainstem were found to contain mutations in the debranching RNA lariats 1 gene (DBR1), the only RNA lariat debranching enzyme that has been identified (Shen-Ying et al., 2018). This study also demonstrated that cells expressing the DBR1 mutation are more susceptible to HSV-1 infection. Without antiviral therapy, the mortality rate can be as high as 70%. Even after aggressive antiviral therapy, 20% of the patients die. Despite early treatment, chronic progressive tissue damage in magnetic resonance imaging (MRI) can be found up to 6 months following the onset of symptoms. Approximately 2/3 of HSE cases are due to HSV reactivation from latency (Yamada et al., 2002), which is why there is high morbidity and long-term complications despite antiviral treatment (Lahat et al., 1999; McGrath et al., 1997; Skoldenberg, 1991).

28.5.2 Acute Infection and Latency-Reactivation Cycle of HSV

HSV-1 and HSV-2 are double-stranded DNA viruses with a genome size of approximately 152,000 base pairs and they encode at least 84 proteins. Binding and entry of HSV-1 into permissive cells is mediated by viral proteins that interact with cellular receptors (Spear, 1993). After uncoating, the viral genome is present in the nucleus and viral gene expression ensues. HSV gene expression is temporally regulated in three distinct phases: immediate early (IE), early (E), or late (L) (Honess & Roizman, 1974). IE gene expression does not require protein synthesis and is stimulated by VP16 (O'Hare, 1993). In general, proteins encoded by IE genes regulate viral gene expression, and as such are important for productive infection. E gene expression is dependent on at least one IE protein, and generally, E genes encode nonstructural proteins that control viral DNA synthesis. L gene expression is maximal after viral DNA replication, requires IE protein expression, and L proteins comprise the virion particle.

Acute infection is typically initiated in the mucosal epithelium, and then HSV-1 establishes latency in sensory neurons located in the trigeminal ganglia (TG) or sacral dorsal root ganglia. HSV-1 genomic DNA is also detected in the central nervous system (CNS) of most latently infected people (Fraser et al., 1981). Despite a vigorous immune response during acute infection, latency is established. Approximately 20–30% of sensory neurons are latently infected, reviewed in Jones (1998, 2003). The latency-reactivation cycle is operationally divided into three distinct steps: establishment, maintenance, and reactivation (Perng & Jones, 2010; Phelan et al., 2017).

The only HSV-1 transcript abundantly expressed during latency is the latency-associated transcript (LAT) (Perng & Jones, 2010; Phelan et al., 2017), which is present in the large repeats (TR_L and IR_L; Fig. 28.4a). Mice, rabbits, or humans latently infected with HSV-1 express LAT. In latently infected mice, rabbits, or people an 8.3 Kb LAT transcript is expressed (Fig. 28.4b). Splicing of the 8.3 Kb transcript yields an abundant stable 2 Kb LAT and an unstable 6.3 Kb LAT (Fig. 28.4b, c). The majority of LAT is not capped, is poly A- and the stable 2 kb LAT appears to be circular and is designated as a stable intron. In small animal models, LAT is important but not required for the latency-reactivation cycle, reviewed in Jones (1998, 2003). The first 1.5 Kb of LAT coding sequences are important for reactivation from latency in rabbits and certain mouse strains. LAT functions are mediated by non-coding RNAs (Phelan et al., 2017), which includes the unstable 8.3 kb transcript, the stable 2 Kb LAT intron, six micro-RNAs (mi-RNAs; H1–H6), 2 small non-coding RNAs (LAT sncRNA1 and 2), 3 transcripts antisense to the major LAT (AL1–3), and upstream of LAT transcript referred to as UOL. In general, LAT products promote lifelong latency by inhibiting apoptosis (Inman et al., 2001; Perng et al., 2000), enhancing neuronal differentiation (Hamza et al., 2007; Li et al., 2010), and impairing productive infection (Shen et al., 2009; Umbach et al., 2008). Notably, genes that inhibit apoptosis can substitute for LAT and restore wild-type reactivation levels (Jin et al., 2008; Jin et al., 2005; Perng et al., 2002) indicating LAT's anti-apoptosis functions are necessary for the latency-reactivation cycle. Although the mechanism is not well understood, LAT clearly promotes lifelong latency by regulating establishment and maintenance of latency (Perng & Jones, 2010; Phelan et al., 2017; Thompson & Sawtell, 2011). Consequently, a pool of latently infected neurons have the potential to support reactivation from latency.

28.5.3 Viral and Cellular Factors Important for HSE and Neuropathogenesis

The frequency of HSE in experimental infections is dependent on the pathogenic potential of HSV-1, and mouse strain used for experimental infection. For HSE to occur after ocular infection, the virus must enter the TG and spread to the CNS, or the virus directly gains access into the brain via the optic nerve. HSV can also be directly inoculated into the brain. In this model, transport from the peripheral tissue→peripheral nervous system→CNS is not important. Viral genes necessary for productive infection, inhibiting apoptosis, or inhibiting immune recognition play a crucial role in initiating encephalitis. Viral genes necessary for neuronal transport and spread are not as important if the brain is inoculated. Innate immune responses play a significant role in lethal encephalitis because HSV-1 interactions with toll-like receptor 2 contribute to HSE in a mouse model of infection (Kurt-Jones et al., 2004).

Deletion of the first 1.5 Kb of LAT reduces the frequency of encephalitis in male Balb/C mice (Jones et al., 2005). Furthermore, the same HSV-1 LAT mutant induces senescence and expression of key cellular inflammatory markers in the locus coeruleus (LC) of female, but not male, mice latently infected with the LAT null mutant or parental wild-type HSV-1 strain (Sivasubramanian et al., 2022). Conversely, wild-type HSV-1 and LAT null mutants induce higher levels of senescence and inflammatory markers in the spinal trigeminal tract (Pr5) of latently infected female mice versus age-matched latently infected male mice or uninfected mice. Pr5 and LC are brainstem regions that have direct or indirect

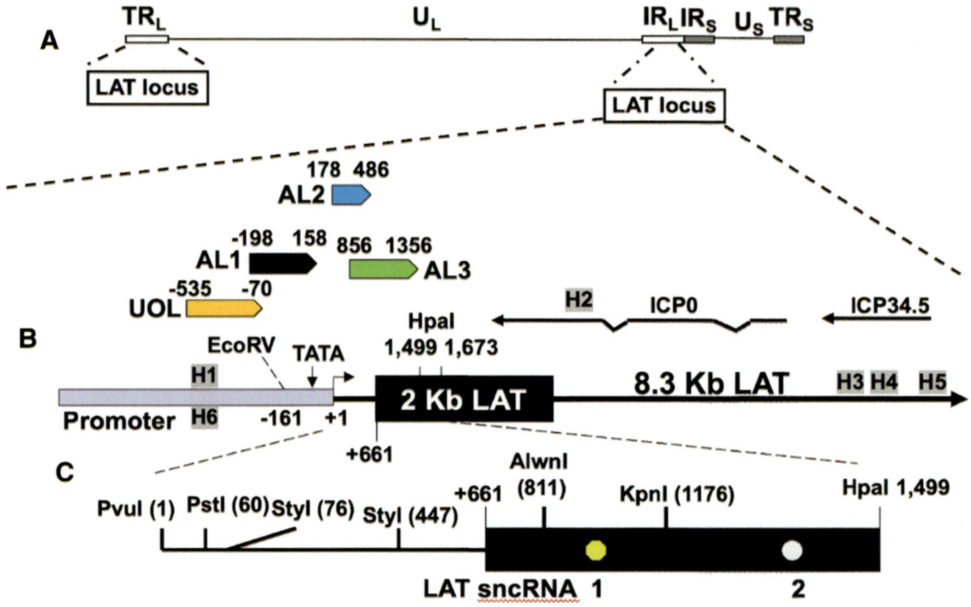

Fig. 28.4 Schematic of HSV-1 and organization of the LAT locus. (**a**) The prototypic HSV-1 genomic structure is shown at the top. The viral repeat regions are denoted as open rectangles. TR_L is the terminal long repeat, IR_L the internal (or inverted) long repeat, TR_S the terminal short repeat, and IR_S the internal (or inverted) short repeat. The unique long (U_L) and unique short (U_S) regions are denoted, and the LAT locus is within TR_L or IR_L. (**b**) The primary 8.3 kb LAT is denoted as a long arrow. The stable 2 kb LAT is shown as a black rectangle. The LAT TATA box is denoted by TATA, and the small arrow and + 1 indicate the start of LAT transcription (genomic nucleotide 118,801). The relative locations of mRNAs encoding ICP0 and ICP34.5 are shown for reference. Location of the upstream of LAT transcript (UOL), antisense of LAT1 (AL1), AL2, and AL3 are also denoted. Numbers at the beginning of UOL, AL1, AL2, and AL3 denote the beginning and end of the respective RNAs. The position of the LAT micro-RNAs (miRNAs) is denoted as H1-H6. (**c**) The position of certain restriction enzyme sites within the first 1.5 kb of LAT coding sequences and relative locations of the two LAT sncRNA1 and sncRNA2 are shown (Peng et al., 2008)

synaptic connections with TG, an important site for HSV-1 latency. This intriguing study predicts LAT products impair neuro-inflammation in a neuronal and sex-specific manner.

Predisposition to HSE in humans is linked to mutations that impair the function of six genes associated with toll-like receptor 3 (TLR3) signaling (see Fig. 28.5 for schematic summarizing TLR3-mediated signaling). These genes include UNC-93B (Casrouge et al., 2006), TRIF (Sancho-Shimizu et al., 2011), TRAF3 (de Diego et al., 2010), TBK1 (de Diego et al., 2010), and IRF3 (Andersen et al., 2015). Mutations in the cellular gene, UNC-93B, was discovered in a family having a high incidence of encephalitis. The UNC-93B gene apparently plays a role in interferon signaling by interacting with TLR3, TLR7, or TLR9. TLR3 recognizes double-stranded RNA (ds-RNA) and initiates the beta-IFN (IFN-β) response. TLR3 localizes to endosomes and two adaptors, toll/interleukin-1 receptor (TIR) domain-containing adaptor inducing IFN-β (TRIF) and tumor necrosis factor receptor-associated factor 3 (TRAF3) are essential for downstream signaling, reviewed in (Pothlicet et al., 2013). Since HSV-1 is a DNA virus, it is surprising that TLR3 is an important factor in impairing HSE. However, HSV-1 infection was reported to produce viral ds-RNAs (Weber et al., 2006) suggesting these RNAs trigger TLR3. A recent study also revealed a homozygous deletion from intron 10 to exon 11 of the interferon-alpha receptor (IFNAR1) gene of a child who died from HSE (Bastard et al., 2021). This deletion would prevent the IFN-a protein from interacting with the tyrosine kinase TYK2, which is essential for downstream signaling. Thus, the antiviral properties of interferon play an important role in HSE, presumably by impairing viral replication and spread to CNS.

Interestingly, IFN-induction can also activate necroptosis and pyroptosis, two proinflammatory forms of programmed cell death (Bergsbaken et al., 2009; Fink & Cookson, 2005; Green, 2019) (Fig. 28.5). A recent study revealed the RIPK3 scaffold functions suppress HSE-associated inflammation and subsequent neuropathology via interactions with RIPK1 and caspase 8. Interestingly, RIPK3 protein kinase activity is not important for blocking HSE. Conversely, RIPK3-mediated necroptosis, which requires kinase activity, restricts HSV-1 replication in corneal tissues. These findings identified two distinct RIPK3-dependent mechanisms that act in a tissue-specific manner to impair HSV-1 replication and HSE. HSV-1 also induces pyroptosis in cultured cells and microglia within the brains of infected mice (Hu et al., 2022)

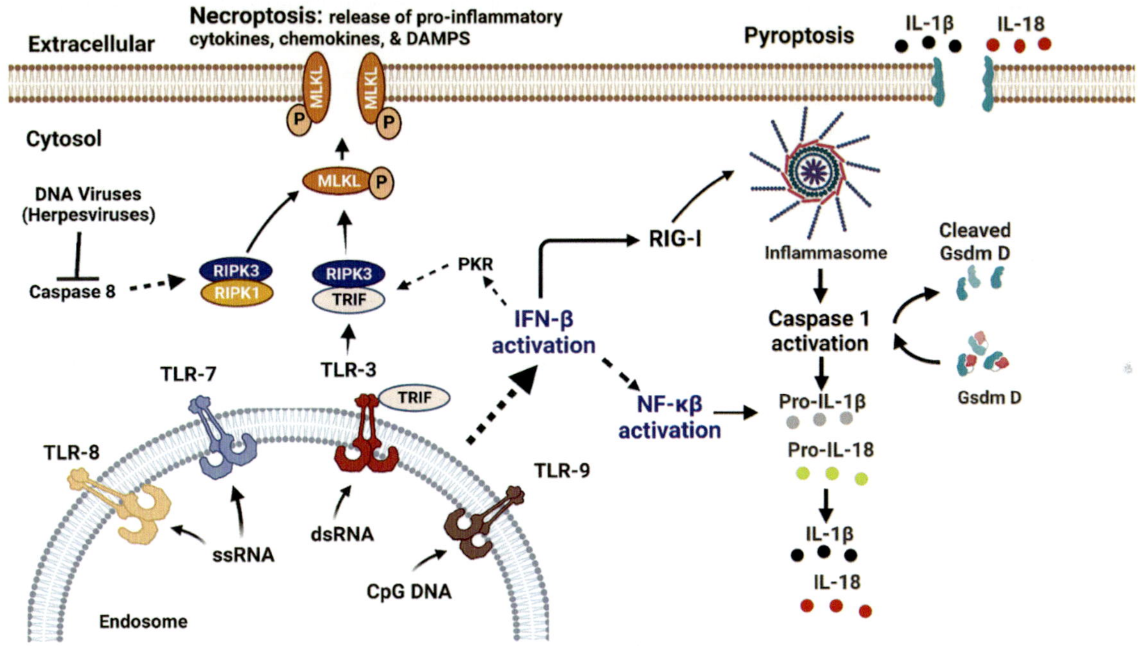

Fig. 28.5 Synopsis of TLR-3-mediated induction of interferon production and proinflammatory-mediated PCD. It is well established that viruses and other intracellular pathogens serve as pathogen associated molecular patterns (PAMPs) and danger associated molecular patterns (DAMPs): consequently, toll-like receptors (TLR) family members, including those associated with the endosome (TLR3, TLR-7, TLR-8, or TLR-9) are activated. Consequently, IFN-b and NF-kB signaling pathways are activated, which culminates in activation of NLRP3, protein kinase R (PKR), and Caspase 1. Caspase 1 cleaves Pro-interluekin-1b (Pro-IL-1b), Pro interleukin-18 (Pro-IL-18), Gasdermin D (Gsdm D), and other substrates. Cleaved Gsdm D forms pores in cells, which releases proinflammatory cytokines, including cleaved IL-b, and IL-18. Release of IL-b and IL-18 culminates in inflammation in surrounding cells and pyroptosis. TLR-3 signaling also leads to activation of RIP3K via association with TRIF family member, which phosphorylates the mixed lineage kinase like (MLKL) protein. Blocking caspase 8, which occurs after infection with large DNA viruses (including herpesviruses), leads to release of a RIPK1/RIPK3 complex and MLKL phosphorylation, reviewed in He (2016). Consequently, phosphorylated MLKL forms pores, which triggers Necroptosis and release of proinflammatory factors. (For additional details, see text. This Figure was created using BioRender.com)

suggesting pyroptosis-mediated inflammation may be important during the development of HSE. In summary, the ability of IFNs to restrict viral replication, reviewed in (Katze et al., 2002), plus activating necroptosis and pyroptosis are important for impairing HSV-1 replication in the CNS, spread in the nervous system, and HSE. These same pathways may also initiate HSE if a hyper-inflammatory environment occurs in the CNS after acute infection or during reactivation from latency. Viral load in CNS and unknown cellular factors will likely determine whether inflammation leads to viral clearance or HSE.

28.6 Varicella Zoster Virus (VZV)-Induced Encephalitis

28.6.1 Epidemiology of VZV

Humans are the only known reservoir for the neurotropic Varicella Zoster Virus (VZV), a ubiquitous human pathogen. A serologic study of 1201 US military trainees indicated 95% were exposed to the virus (Jerant et al., 1998).

28.6.2 Clinical Features of VZV

Acute VZV infection leads to chickenpox, also referred to as varicella, and results in an extensive vesicular rash (Abendorth & Arvin, 2000). Chickenpox is readily spread via coughing and sneezing is common in childhood and affects both genders equally. Prior to the rash, the incubation period varies from 10–21 days. VSV is spread by droplet or airborne transmission and is highly contagious. Other symptoms include malaise, itching, anorexia, weakness, and exhaustion, which gradually resolve as the illness improves.

Chickenpox in immunocompetent children is usually benign and associated with lethargy, and a temperature of 100–103 °F for 1–2 days. However, chickenpox can cause serious disease. For example, an epidemic of 292 chickenpox cases occurred in rural India and resulted in 3 deaths (Balraj & John, 1994). Furthermore, about 25 people die from chickenpox every year in the United Kingdom, in part because VZV vaccination is not mandatory (Rawson et al., 2001). The most frequent organ affected other than the skin is the CNS (Liesegang, 1999). The neurologic abnormalities are often seen as acute cerebellar ataxia or encephalitis.

Encephalitis is the most serious complication of chickenpox and is life-threatening in 0.1% to 0.2% of patients with chickenpox (Johnson & Milbourn, 1970).

Following primary infection, VZV becomes latent in neurons within TG, cranial nerve, dorsal root ganglia, and autonomic ganglia along the entire neuro-axis. Reactivation can occur decades later resulting in zoster (shingles). VZV vaccines effectively prevent varicella; however, breakthrough varicella and virus reactivation can still occur, as reviewed in (Arvin & Gershon, 1996).

28.6.3 Neurological Disorders Associated with VZV

VZV-induced neurological diseases generally occur because of reactivation from latency. For example, reactivation of latent VZV in dorsal root ganglia results in a localized cutaneous eruption termed "herpes zoster" (or shingles). The annual incidence of Zoster is about 1.5–3.0 cases per 1000 for a total of 500,000 cases each year in the United States. Zoster occurs during the lifetime of 10–20% of all people (Liesegang, 1999). Zoster is frequently followed by postherpetic neuralgia (pain that persists for months and often years after the rash disappears), myelitis, or unifocal or multifocal vasculopathy. VZV is the causal agent in 29% of 3231 cases of encephalitis, meningitis, and myelitis, and the most common cause of encephalitis in patients over the age 65 (Rantalaiho et al., 2001). The neurological complications of VZV reactivation can be serious and life-threatening. However, proper diagnosis is critical because antiviral treatment can be curative, even after weeks to months of chronic VZV infection.

Decreasing VZV-specific cell-mediated immune responses, which occur naturally as a result of aging or from immunosuppressive illness or medical treatments, increase the risk of shingles (zoster) (Ganann & Whitley, 2002). Most patients who develop herpes zoster have no history of exposure to other persons with VZV infection following the appearance of lesions. Over 90% of adults in the United States have serologic evidence of VZV infection and are at risk for herpes zoster. Patients with cancer, those receiving immunosuppressive agents, and organ transplant recipients have a higher risk of shingles. Furthermore, individuals who are positive for human immunodeficiency virus (HIV) also develop herpes zoster at a higher frequency than those who are negative (Liesegang, 1999). Since VZV is latent in most ganglia, herpes zoster can occur nearly anywhere on the body. The most common sites are the thoracic and cutaneous distribution of the ophthalmic branch of the trigeminal nerve. Zoster can also cause meningoencephalitis, myelitis, and Guillain-Barre syndrome (Liesegang, 1999). Extracutaneous sites of involvement include the CNS as shown by meningoencephalitis or encephalitis. Clinical symptoms are like those of other viral infections of the brain. Involvement of the CNS with cutaneous herpes zoster is probably more common than clinically recognized. Motor paralysis can occur because of the involvement of the anterior horn cells. These patients can have severe pain. Herpes zoster in patients with an intact immune system can lead to disseminated herpes zoster: however, it is rarely fatal.

Acute VZV retinal necrosis occurs occasionally in immunocompetent patients, including those infected by HIV. In HIV-infected patients, ocular lesions can rapidly increase in size and coalesce. These lesions respond poorly to antiviral therapy and almost inevitably cause blindness in the involved eye. Following a zoster outbreak, VZV can spread to blood vessels of the brain, leading to vasculopathy, particularly in immunocompromised patients.

Therapy for herpes zoster is aimed at accelerating healing, limiting the severity and duration of acute or chronic pain, and reducing complications associated with the infection. In patients who are immunocompromised, therapy is focused on reducing the risk of viral dissemination. Acyclovir, valacyclovir, and famciclovir are all used in the United States for the treatment of herpes zoster. Acyclovir is approved in the United States for the treatment of both chickenpox and herpes zoster. Oral acyclovir therapy in normal children, adolescents, and adults shortens the duration of lesion formation by about a day, reduces the number of new lesions by about 25%, and reduces symptoms in about one-third of patients (Ganann & Whitley, 2002; Snoeck et al., 1999).

Several effective vaccines have been developed against VZV. Initially, a live attenuated varicella vaccine (OKA strain) was developed in the 1970s. This vaccine was very efficacious at preventing severe or fatal varicella in children (Takahashi et al., 1974; Takahashi et al., 1975). Concerns were raised that since OKA established a latent infection, it could cause herpes zoster later in life due to reactivation in elderly people. Concerns were also raised that the OKA strain may generate "gain of function" mutations over time and cause disease in people with immune dysfunction. A new subunit vaccine, Shingrix, is now available. It is an adjuvanted subunit vaccine consisting of the VZV glycoprotein E (gE), and thus not a live vaccine (Cunningham et al., 2008). CDC recommends that adults 50 years and older get two doses of Shingrix to prevent shingles and complications resulting from the disease. Adults 19 years and older who have weakened immune systems because of disease or therapy should also get two doses of Shingrix, as they have a higher risk of getting shingles and related complications. A recent study demonstrated Shingrix is more than 90% effective at preventing shingles and postherpetic neuralgia (PHN) in adults 50 years and older who have healthy immune systems (Carryn et al., 2022). The glycoprotein E (gE) vaccine is also used for chickenpox but a different formulation is used than Shingrix. While Postherpetic neuralgia (PHN) and shingles will continue to be significant problems, the Shingrix vaccine is an effective means to reduce the effects of these serious problems in the elderly.

28.6.4 Summary of Productive Infection and Latency

VZV contains many genes like HSV-1 suggesting many of the shared genes are functional homologues. Thus, the general steps during productive infection are similar for VZV and HSV-1 and will not be discussed in detail. In contrast to HSV-1, VZV infectious virus produced during productive infection is tightly cell associated making it difficult to obtain high yields of cell-free virus.

VZV appears to be latent only in ganglia and viral genomes are primarily found in sensory neurons. Analysis of latent VZV is restricted to human ganglia obtained at autopsy. Based on in situ hybridization (ISH) studies combined with sequencing, four transcripts corresponding to VZV genes 21, 29, 62, and 63 have been identified in latently infected human ganglia (Gilden et al., 2007; Abendorth & Arvin, 2000). A monospecific polyclonal antiserum directed against VZV ORF 63 protein detected this protein in the cytoplasm of neurons, reviewed in (Mitchell et al., 2003). These VZV proteins are primarily in the nucleus during productive infection (zoster), suggesting that the cytoplasmic location might maintain VZV in a latent state. A VZV LAT (VLT) was recently identified, which is antisense with respect to the viral transactivator gene 61 (Depledge et al., 2018). The spliced VLT encodes a protein that is expressed late during productive infection. To date, it is not clear whether VLT mediates the VZV latency-reactivation cycle.

28.6.5 Potential Models to Study VZV Neuropathogenesis

Unlike HSV-1, VZV does not reactivate from ganglia after experimental infection of primates or rodents. However, after footpad inoculation of rats with VZV, the protein encoded by Immediate Early gene 63 can be detected in lumbar ganglia 1 month after infection (Debrus et al., 1995). The protein is also detected in neurons, both in the nuclei and cytoplasm of infected cells. An independent study using the same rat model detected VZV gene 63 DNA in 5% to 10% of neurons and VZV RNA in neurons and non-neuronal cells (Kennedy et al., 1999). Simian varicella virus may also be a valuable model to study the pathogenesis of varicella virus-host interactions. Finally, VZV infection of humanized immuno-deficient mice (SCID-hu) has provided new information about viral genes that regulate viral growth and pathogenesis (Ku et al., 2005).

28.7 Human Cytomegalovirus (HCMV)-Induced Encephalitis

28.7.1 HCMV Lifecycle and Prevalence

HCMV is a β-herpesvirinae subfamily member that is an important causative agent of congenital disease and significant opportunistic infection (Knipe & Howley, 2001). Like other herpesviruses, HCMV is a large DNA virus that contains more than 200 open reading frames and numerous non-coding RNAs. In contrast to HSV-1, HCMV exhibits a restricted host range in cell culture. Primary differentiated human fibroblasts show the greatest susceptibility to infection. As with other herpesvirus members, HCMV's immediate early genes regulate viral transcription and inhibit cellular responses to infection (apoptosis and interferon induction for example). HCMV is widely distributed among humans (Pass, 1995) and the prevalence of HCMV generally increases with age (Pass, 2001).

28.7.2 Clinical Features of HCMV

Although HCMV acute infection is usually asymptomatic in patients with intact immune systems, it is a common opportunistic pathogen in immunocompromised patients, including HIV-infected patients and organ transplant recipients. For example, CMV disease following transplantation of stem cells or solid organs from a CMV-seropositive individual into a CMV-seronegative individual is particularly problematic (Boeckh, 2017). HCMV infection is also a significant problem in patients who exhibit heightened inflammation and are at risk of vascular disease, including atherosclerosis and cardiovascular mortality, as reviewed in Goodrum (2016, 2022)). Notably, there is a correlation between HCMV infections and cardiovascular mortality in women older than 70 years. Retinitis is also a well-known disease associated with infection. Other issues associated with infection include gastrointestinal disease, pneumonitis, and neurologic disease.

28.7.3 Pathogenesis and Persistent/Latent Infections of HCMV

In contrast to HSV-1 and VZV, HCMV does not establish latency in the nervous system. The known HCMV latent reservoirs include CD34+ hematopoietic progenitor cells (HPCs) and CD14+ monocytes, reviewed in Goodrum

(2022), Poole et al. (2014). Low passage HCMV strains (Toledo and FIX for example) can efficiently establish latency in CD34+ cells whereas strains extensively passaged in cell culture (AD169 and Towne) do not because low passage strains contain additional genes (UL133-UL138). Compelling evidence that HCMV establishes and maintains a life-long latent infection is reactivation from latency and disease occurs in people with immune deficiency or immune suppressive therapy associated with transplantation, HIV infection, or chemotherapy regimens (Goodrum, 2022).

The viral UL138 ORF is necessary for establishing and maintaining a latent infection in CD34+ hematopoietic progenitor cells infected in vitro (Goodrum et al., 2007). The UL138 protein localizes to the Golgi apparatus and a mutant that expresses the UL138 transcripts, but not the protein, has an impaired loss of latency phenotype (Petrucelli et al., 2012). HCMV also expresses numerous viral proteins, miRNAs, and long non-coding RNAs that collectively mediate the latency-reactivation cycle (Goodrum et al., 2007). Since HCMV latently infects dividing cells and have the potential to differentiate, additional viral functions must control cellular differentiation, cell growth, preventing immune responses designed to kill latently infected cells, and ensuring the viral genome is partitioned to each daughter cell following cell division. Conversely, latently infected HSV-1 and VZV are harbored in terminally differentiated neurons, which precludes expressing certain viral factors that HCMV requires.

28.7.4 HCMV-Induced Encephalitis

Approximately, two out of three newborn children that exhibit symptomatic congenital CMV infection will have CNS involvement (Bopanna et al., 1992). CNS pathology includes microcephaly, elevation of cerebrospinal fluid protein, and neurologic abnormalities (poor feeding, lethargy, or generalized hypotonia for example). Cranial computed tomography scans are abnormal in 75% of symptomatic newborns, with the most common abnormality being periventricular calcification (Bopanna et al., 1997). Fortunately, only 5–10% of newborns with congenital HCMV infection are symptomatic.

With respect to neurologic infections in people infected with human immunodeficiency virus (HIV), HCMV is associated with inflammation of the ventricle accompanied by encephalitis (ventriculoencephalitis) and ascending polyradiculopathy (disease or injury involving multiple nerve roots). Ventriculoencephalitis usually occurs in patients with advanced HIV infection and prior HCMV diagnosis. Polyradiculopathy is the most common CNS infection caused by HCMV and is characterized by urinary retention and progressive bilateral leg weakness. Clinical symptoms generally progress over several weeks to include loss of bowel and bladder control, and paralysis. The cerebral spinal fluid contains higher levels normal number of cells, reduced glucose content, and elevated protein levels (Cheeseman & Gibson, 2004; Cheung & Teich, 1999).

28.8 Human Herpesvirus 6 (HHV-6)-Induced Encephalitis

Like HCMV, HHV-6 is a β-herpesvirinae subfamily member originally isolated from peripheral blood leukocytes of patients with lymphoproliferative disease (Salahuddin et al., 1986). It is now recognized as the causative agent of roseola, also known as exanthem subitem (Yamanishi et al., 1988). Following acute infection, HHV-6 can persist in several different cell types, including peripheral lymphocytes. The receptor for HHV-6 (complement-regulatory transmembrane protein CD46), is widely expressed in humans; hence, HHV-6 has the potential to infect many cell types (Santoro et al., 1999).

HHV-6 DNA and certain viral transcripts, but not viral antigens, have been detected in brains of healthy immunocompetent humans suggesting the virus is latent, reviewed in Reynaud and Horvat (2013). Several studies have also concluded HHV-6 can cause encephalitis (Nagasawa et al., 2007; Reynaud & Horvat, 2013; Yoshikawa & Asano, 2000). The clinical symptoms associated with HHV-6-induced encephalitis appear to be distinct relative to other herpesviruses. The features of HHV-6-induced encephalitis are summarized below (Nagasawa et al., 2007). First, primary infection occurs in a normally developing infant followed by the initial convulsion with the onset of encephalopathy occurring on the second day of febrile illness. Following the first convulsion, consciousness is disturbed. Frequent convulsions occur within one day after fever is resolved, which is associated with the onset of exanthem subitem (eruptive stage). Conversely, "typical" encephalitis that features convulsions and severe disturbance of consciousness has been described following initial stages of disease: then a slow recovery ensues.

In contrast to other human herpesviruses, HHV-6A/B strains can lead to integration into the host genome, reviewed in Collin and Flamand (2017). HHV-6 integration also occurs in germinal cells, which leads to transmission in 50% of descendants. HHV-6 contains multiple copies of telomeres in the genome and consequently integration appears to occur via homologous recombination with cellular telomeres that are at the end of chromosomes. Integration maintains the viral genome in its host and since the viral genome is generally intact reactivation may occur; thus, viral transmission is possible. The impact of viral integration on pathogenesis and encephalitis is not well understood.

28.9 γ-Herpesvirus-Induced Encephalitis

Although Epstein Barr Virus (EBV) can cause encephalitis (Bhatti et al., 1990; Fujimoto et al., 2003; Hussain & Hussain, 2013), it is a rare event relative to HSV-1 or CMV. Since EBV acute infection is primarily initiated in pharyngeal tonsils (infectious mononucleosis) and then establishes a latent infection in B cells, reviewed by Cohen (2000), Young and Rickinson (2004), encephalitis is not common.

28.10 Therapeutic Agents Available to Treat Herpesvirus Infections and Encephalitis

28.10.1 Acyclovir

Acyclovir is an acyclic guanine nucleoside analog that lacks a 3′-hydroxyl on the side chain (Fig. 28.6) and was the first drug shown to be effective against HSV infections (Wagstaff & Bryson, 1994). Acyclovir selectively the herpes virus-encoded thymidine kinase (TK) protein and consequently inhibits viral DNA synthesis. The affinity of acyclovir for HSV-1 thymidine kinase (TK) is approximately 200-fold greater than the mammalian enzyme. The herpesvirus TK adds the first phosphate to acyclovir. Cellular enzymes subsequently convert the monophosphate to acyclovir triphosphate. The acyclovir triphosphate competitively inhibits viral DNA polymerases to a much greater extent than cellular DNA polymerases. The terminated DNA template containing acyclovir binds the DNA polymerase leading to irreversible inactivation (Scholar and Noble, 2000, Wagstaff & Bryson, 1994). The oral bioavailability of acyclovir is poor and ranges from 10 to 30%. The drug is widely distributed throughout body fluids and tissues including cerebrovascular fluid. It is primarily excreted unchanged in the urine (Wagstaff & Bryson, 1994).

Acyclovir is useful for treating infections caused by HSV, VZV infections, including herpes zoster, and is the major therapy for HSE (Whitley & Roizman, 2001). It should be given intravenously at a dose of 10 mg/kg every 8 hours for 14–21 days. Although HCMV is relatively resistant to acyclovir, some HCMV infections have responded marginally to large doses of acyclovir.

Oral acyclovir is recommended for treating VZV infection (chickenpox) in patients over 13 years of age who are otherwise healthy, children over 12 months of age with a chronic cutaneous or pulmonary condition, and in children receiving short or intermittent courses of aerosolized corticosteroids. Intravenous acyclovir should be used for treatment of varicella infection in immunocompromised children, including those receiving high doses of corticosteroids (Snoeck et al., 1999).

Acyclovir is the drug of choice for treating initial and recurrent mucosal or cutaneous HSV infections in immunocompromised patients and for treatment of disseminated, neonatal, encephalitic, and severe acute episodes of genital HSV infections in immunocompetent patients (Whitley, 1997). Intravenous acyclovir should also be used for severe diseases such as encephalitis (Brady & Bernstein, 2004). Acyclovir is generally well tolerated: however, gastrointestinal disturbances, headache, and rash may occur. Renal dysfunction due to crystalline nephropathy is more likely with IV administration. CNS effects are rare but include encephalopathy, tremors, hallucinations, seizures, and coma. Due to an elevated pH, intravenous administration may also cause phlebitis and inflammation at sites of extravasation (Wagstaff & Bryson, 1994; Wagstaff et al., 1994).

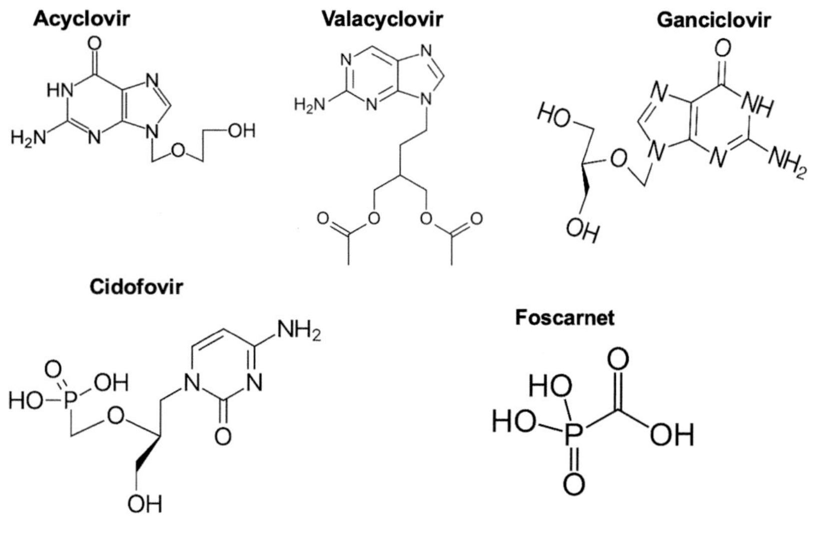

Fig. 28.6 Schematic of drugs that are available for combating herpesvirus infections and disease

28.10.2 Valacyclovir

Valacyclovir is an ester prodrug of acyclovir that provides significantly better oral bioavailability compared to acyclovir (Fig. 28.6). This advantage results in substantially higher serum concentrations than oral acyclovir. Fewer daily doses are also required with valacyclovir (Curran & Noble, 2001). Oral administration of valacyclovir results in plasma acyclovir concentrations comparable to those observed with intravenous acyclovir. Valacyclovir and famciclovir are alternative drugs used for treating HSE (Griffiths, 1995; Stanberry et al., 2004; Stanberry et al., 2002). After oral administration and absorption, valacyclovir is converted to acyclovir which is the active antiviral component of valacyclovir. The antiviral activity and mechanism of action of valacyclovir is identical to that of acyclovir (Perry & Faulds, 1996).

Valacyclovir is effective for treating herpes zoster (shingles), treating initial and recurrent episodes of genital herpes, and suppression of recurrent genital herpes in immunocompetent and HIV-infected patients. It is also used for reducing transmission of genital herpes in immunocompetent individuals and for treatment of cold sores. Valacyclovir appears to be equally effective in treating herpes zoster, recurrent genital herpes in immunocompetent adults, and HCMV infections in transplant patients (Perry & Faulds, 1996).

Like acyclovir, valacyclovir is a well-tolerated drug. The most common adverse effects of valacyclovir are headache and nausea. Other adverse effects associated with valacyclovir administration include vomiting, weakness, dizziness, and abdominal pain. Thrombotic thrombocytopenic purpura/hemolytic uremic syndrome has also been reported in a few patients after high doses of valacyclovir as has confusion, hallucinations, and nephrotoxicity (Curran & Noble, 2001; Perry & Faulds, 1996).

28.10.3 Ganciclovir

Ganciclovir is an acyclic guanine nucleotide analog like acyclovir, which inhibits all herpesviruses, including HCMV. Ganciclovir inhibits viral DNA synthesis with the same mechanism as acyclovir. Intracellular ganciclovir triphosphate concentrations are tenfold higher than those of acyclovir triphosphate and decrease much more slowly with an intracellular half-life of elimination greater than 24 hours. These differences probably explain at least in part the greater activity of ganciclovir against HCMV and provide a rationale for single daily doses in suppressing HCMV infections (Markham & Faulds, 1994; Scholar & Pratt, 2000). Ganciclovir-triphosphate acts as an inhibitor and substrate for the cytomegalovirus DNA polymerase.

The bioavailability of orally administered ganciclovir is quite low. In patients with normal renal function, the plasma half-life is about 2–4 hours. Concentrations of ganciclovir in cerebrospinal fluid are lower than those in serum after intravenous administration (Markham & Faulds, 1994). More than 90% of the drug is eliminated unchanged in the urine. The plasma half-life increases as creatinine clearance declines and may reach 28–40 hours in patients with severe renal insufficiency. Ganciclovir is an effective antiviral agent for treating life-threatening HCMV infections in immunocompromised patients. Intravenous ganciclovir, foscarnet, or a combination of both are recommended for the treatment of HCMV neurological syndromes (Markham & Faulds, 1994).

The dose-limiting and most common adverse effect of intravenous and oral ganciclovir is bone marrow suppression (anemia, leukopenia, neutropenia, and thrombocytopenia). These effects are usually reversible upon withdrawal of the drug. CNS side effects are less common and range in severity from headaches to behavioral changes to convulsions and coma. Fever, edema, phlebitis, disorientation, nausea, anorexia, rash, and myalgias have also been reported with ganciclovir therapy (Markham & Faulds, 1994). Furthermore, lengthy and inconsistent use of acyclovir and similar drugs can lead to selection of drug-resistant strains that are less responsive to this family of drugs, reviewed in (Collin & Flamand, 2017). Drug resistance is significantly more problematic in immunosuppressed individuals, including AIDS patients.

28.10.4 Cidofovir

Cidofovir is an antiviral cytidine nucleotide analog with inhibitory activity against HCMV and other herpesviruses. Cidofovir is first converted to an active diphosphate form by cellular enzymes. Antiviral effects of cidofovir are due to inhibition of viral DNA polymerase by the diphosphate metabolite (Neyts & De Clercq, 1994; Plosker & Noble, 1999). The diphosphate probably interacts with DNA polymerase either as an alternate substrate (incorporation at the 3′-end or within the interior of DNA chain) or as a competitive inhibitor (with respect to the normal substrate dCTP). Cidofovir inhibits HCMV DNA synthesis at intracellular concentrations 1000-fold lower when compared to cellular DNA synthesis (Neyts & De Clercq, 1994). For HSV-1 and HSV-2, corresponding concentrations are at least 50-fold lower.

Cidofovir has poor oral bioavailability (<5%) and is therefore administered intravenously. It has a long intracellular half-life that allows for a prolonged interval between maintenance doses. Cidofovir is excreted extensively by kidneys and is eliminated almost entirely as an unchanged drug in urine (Plosker & Noble, 1999). The major treatment-limiting toxicity of this drug is irreversible nephrotoxicity (Plosker & Noble, 1999). Intravenous prehydration with normal saline and administration of oral probenecid must be used with each cidofovir infusion to lessen the effects on the kidney. Serum creatinine and urine protein must be monitored with each infusion and adjusted accordingly. Other adverse effects associated with its use are neutropenia and peripheral neuropathy (Plosker & Noble, 1999).

Cidofovir, ganciclovir, and foscarnet are effective at treating HCMV disease and are approved for use in the United States (Cheeseman & Gibson, 2004; Cheung & Teich, 1999). Antiviral therapy for esophagitis, enterocolitis, encephalitis, peripheral neuropathy, polyradiculoneuropathy, and pneumonitis is usually treated with intravenous ganciclovir or foscarnet regimens like those used for retinitis. Patients with CNS or neurologic disease often respond best to combination therapy (Arribas et al., 1996). While acyclovir and acyclovir-like drugs are effective anti-herpetic drugs, they do not impair reactivation from latency.

28.10.5 Foscarnet

Foscarnet is an inorganic pyrophosphate analog that has antiviral activity against all herpes viruses and HIV (Fig. 28.6). Foscarnet is a pyrophosphate analog that acts as a noncompetitive inhibitor of many viral RNA and DNA polymerases as well as HIV reverse transcriptase (Chrisp & Clissold, 1991). It is approximately 100 fold more effective against the herpesvirus DNA polymerase than cellular DNA polymerase-α; however, human cell growth suppression has been observed with high concentrations in vitro. Inhibition of DNA polymerase results in inhibition of pyrophosphate exchange which prevents elongation of the DNA chain (Scholar & Pratt, 2000). Foscarnet reversibly blocks the pyrophosphate-binding site of the viral DNA polymerase in a noncompetitive manner and inhibits the cleavage of pyrophosphate from deoxynucleotide triphosphates.

The oral bioavailability of foscarnet is poor so intravenous therapy is needed to treat viral infections. It is well distributed throughout the body with cerebrospinal fluid levels averaging two-thirds of those in plasma. Foscarnet is taken up slowly by cells, and biotransformation of foscarnet does not occur. The drug is excreted mainly unchanged in the urine (Chrisp & Clissold, 1991). Foscarnet is frequently used for the treatment of HCMV retinitis and mucocutaneous acyclovir-resistant HSV infections. It may also be beneficial in other types of CMV or HSV infections (Wagstaff & Bryson, 1994). In contrast to ganciclovir, foscarnet is not associated with dose-limiting neutropenia, enabling it to be used in combination with zidovudine and other bone marrow suppressant drugs. Nephrotoxicity and hypocalcemia are the major dose-limiting toxicities. Renal toxicity can be minimized with adequate hydration. Changes in serum calcium and phosphate levels may be related to incorporation of the drug into bone. Other adverse effects of the drug include tremor, headache, fatigue, nausea, and vomiting (Isogai et al., 2020).

References

Abendorth, A., & Arvin, A. M. (2000). Host responses to primary infection. In *Varicella-Zoster virus* (pp. 142–156). Cambridge University Press.

Alam, A. M., Eastman, A., Nicholson, T. R., et al. (2022). Encephalitis: Diagnosis, management and recent advances in the field of encephalitides. *Postgraduate Medical Journal, 99*(1174), 815–825.

Al-Dalahmah, O., Thakur, K. T., Nordvig, A. S., et al. (2020). Neuronophagia and microglial nodules in a SARS-CoV-2 patient with cerebellar hemorrhage. *Acta Neuropathologica Communications, 8*, 147.

Andersen, L. L., Mørk, N., Reinert, L. S., Kofod-Olsen, E., et al. (2015). Functional IRF3 deficiency in a patient with herpes simplex encephalitis. *The Journal of Experimental Medicine, 212*, 1371–1379.

Anderson, J. F., Andreadis, T. G., & Vissbrinck, C. R. (1999). Isolation of West Nile virus from mosquitoes, crows, and a Cooper's hawk in Connecticut. *Science, 286*, 2331–2333.

Arribas, J. R., Storch, G. A., Clifford, D. B., et al. (1996). Cytomegalovirus encephalitis. *Annals of Internal Medicine, 125*, 577–587.

Arruda, B., Arruda, P., Hensch, M., et al. (2017). Porcine Astrovirus type 3 in central nervous system of swine with Polioencephalomyelitis. *Emerging Infectious Diseases, 23*, 2097–2100.

Arvin, A. M., & Gershon, A. A. (1996). Live attenuated varicella vaccine. *Annual Review of Microbiology, 50*, 59–100.

Balraj, V., & John, T. J. (1994). An epidemic of varicella in rural southern India. *The Journal of Tropical Medicine and Hygiene, 97*, 113–116.

Bastard, P., Manry, J., Chen, J., et al. (2021). Herpes simplex encephalitis in a patient with a distinctive form of inherited IFNAR1 deficiency. *Journal of Clinical Investigations, 131*, e139980.

Beckman, B., Bonillas, A. B., Diniz, G. B., et al. (2022). SARS-CoV-2 infects neurons and induces neuroinflammation in a non-human primate model of COVID-19. *Cell Reports, 41*, 111573.

Benbir, G., Goskan, B., & Kocer, N. (2010). Brainstem lesions in herpes encephalitis. *Turkish Journal of Neurology, 16*, 211–214.

Bergsbaken, T., Fink, S. L., & Cookson, B. T. (2009). Pyroptosis: Host cell death and inflammation. *Nature Reviews Microbiology, 7*, 99–109.

Bhatti, N., Larson, E., Hickey, M., & Seal, D. (1990). Encephalitis due to Epstein-Barr virus. *The Journal of Infection, 20*, 69–72.

Blomstrom, A. L., Widen, F., Hammer, A. S., Belak, S., & Berg, M. (2010). Detection of a novel astrovirus in brain tissue of mink suffering from shaking mink syndrome by use of viral metagenomics. *Journal of Clinical Microbiology, 48*, 4392–4396.

Boeckh, M. (2017). New insights into cytomegalovirus infection after allogeneic hematopoietic stem cell transplant. *Clinical Advances in Hematology & Oncology, 15*, 515–517.

Bopanna, S. B., Pass, R. F., & Britt, W. J. (1992). Symptomatic congen nital cytomegalovirus infection: Neonatal morbidity and mortality. *The Pediatric Infectious Disease Journal, 11*, 93–99.

Bopanna, S. B., Fowler, K. B., & Vaid, Y. (1997). Neuroradiographic fndings in the newborn period and long-term outcome in children with symptmatic congenital cytomegalovirus infection. *Pediatrics, 99*, 409–414.

Bosch, A., Pintó, R. M., Guix, S., et al. (2014). Human astrovirsues. *Clinical Microbiology Reviews, 27*, 1048–1074.

Boujon, C. L., Koch, M. C., Kauer, R. V., et al. (2019). Novel encephalomyelitis-associated astrovirus in a muskox (Ovibos moschatus): A surprise from the archives. *Acta Veterinaria Scandinavica, 61*, 31.

Brady, R. C., & Bernstein, D. I. (2004). Treatment of herpes simplex virus infections. *Antiviral Research, 61*, 73–81.

Carryn, S., Cheuvart, B., Povey, M., et al. (2022). No consistent evidence of decreased exposure to varicella-zoster virus among older adults in countries with universal varicella vaccination. *The Journal of Infectious Diseases, 225*, 413–421.

Casrouge, A., Zhang, S.-Y., Eidenschenk, C., et al. (2006). Herpes simplex virus encephalitis in human UNC-93B deficiency. *Science, 314*, 308–312.

Cheeseman, S. H., & Gibson, L. L. (2004). Cytomegalovirus infections. In S. L. Gorbach, J. G. Bartlett, & N. R. Blacklow (Eds.), *Infectious diseases* (pp. 1543–1551). Lippincott Williams and Wilkins.

Cheung, T. W., & Teich, S. A. (1999). Cytomegalovirus infection in patients with HIV infection. *Mount Sinai Journal of Medicine, 66*, 113–124.

Chrisp, P., & Clissold, S. P. (1991). Foscarnet: A review of its antiviral activity, pharmacokinetic properties and therapeutic use in immunocompromised patients with cytomegalovirus retinitis. *Drugs, 41*, 104–129.

Coehn, J. (2022). Evidence backs natural origin for pandemic, report asserts. *Science, 378*, 126.

Cohen, J. I. (2000). Epstein-Barr virus infection. *New England Journal of Medicine, 343*, 481–492.

Collin, V., & Flamand, L. (2017). HHV-6A/B integration and the pathogenesis associated with the reactivation of chromosomally integrated HHV-6A/B. *Viruses, 9*, 160.

Corey, L. (2005). Herpes simplex virus. In G. L. Mandell, J. E. Bennett, & R. Dolin (Eds.), *Principles and practice of infectious diseases* (pp. 1762–1780). Elsevier Churchill Livingston.

Cunningham, A. L., Heineman, T. C., Lal, H., et al. (2008). Immune responses to a recombinant glycoprotein E herpes zoster vaccine in adults aged 50 years or older. *JID, 217*, 1750–1760.

Curran, M., & Noble, S. (2001). Valganciclovir. *Drugs, 61*, 1145–1150.

de Diego, R. P., Sancho-Shimizu, V., Lorenzo, L., et al. (2010). Human TRAF3 adaptor molecule deficiency leads to impaired toll-like receptor 3 response and susceptibility to herpes simplex encephalitis. *Immunity, 33*, 400–411.

DeBiasi, R. L., Kleinschmidt-DeMasters, B. K., Richardson-Burns, S., & Tyler, K. L. (2002). Central nervous system apoptosis in human herpes simplex virus and cytomegalovirus encephalitis. *The Journal of Infectious Diseases, 186*, 1547–1557.

Debrus, S., Sadzot-Delvaux, C., Nikkels, A. F., et al. (1995). Varicella-zoster virus gene 63 encodes an immediate-early protein that is abundantly expressed during latency. *Journal of Virology, 69*, 3240–3245.

Depledge, D. P., Ouwendijk, J. D. W., Sadaoka, T., et al. (2018). A spliced latency-associated VZV transcript maps antisense to the viral transactivator gene 61. *Nature Communications, 9*, 1167.

Diamond, M. S., & Kanneganti, T.-D. (2022). Innate immunity: The first line of defense against SARS-CoV-2. *Nature Immunology, 23*, 165–176.

Diamond, M. S., Shrestha, B., Marri, A., & Engle, M. (2003). B cells and antibody play critical roles in the immediate defense of disseminated infection by West Nile encephalitis virus. *Journal of Virology, 77*, 2578–2586.

Ellul, M., & Solomon, T. (2018). Acute encephalitis—Diagnosis and management. *Clinical Medicine, 18*, 155–159.

Fink, S. L., & Cookson, B. T. (2005). Apoptosis, pyroptosis, and necrosis: Mechanistic description of dead and dying cells. *Infection and Immunity, 73*, 197–1916.

Fraser, N. W., Lawrence, W. C., Wroblewska, Z., et al. (1981). Herpes simplex virus type 1 DNA in human brain tissue. *Proceedings of the National Academy of Sciences of the United States of America*, 6461–6465.

Fujimoto, H., Asaoka, K., Imaizumi, T., et al. (2003). Epstein-Barr virus infections of the central nervous system. *Internal Medicine, 42*, 33–40.

Ganann, J. W., & Whitley, R. J. (2002). Herpes Zoster. *The New England Journal of Medicine, 347*, 340–346.

Gilden, D. H., Mahalingam, R., Cohrs, R. J., & Tyler, K. L. (2007). Herpesvirus infections of the nervous system. *Nature Clinical Practice Neurology, 3*, 82–94.

Goodrum, F. (2016). Human cytomegalovirus latency: Approaching the Gordian knot. *Annual Review of Virology, 3*, 333–357.

Goodrum, F. (2022). The complex biology of human cytomegalovirus latency. *Advances in Virus Research, 112*, 31–85.

Goodrum, F., Reeves, M., Sinclair, J., et al. (2007). Human cytomegalovirus sequences expressed in latently infected individuals promote a latent infection in vitro. *Blood, 110*, 937–945.

Gorospe, W., Rawls, R. A., & Kochs, K. A. (2004). Herpes brainstem encephalitis: A cause of intractable emesis. *Hospital Physician, 4*, 39–42.

Green, D. (2019). The coming decade of cell death research: Five riddles. *Cell, 177*, 1094–1107.

Griffiths, P. D. (1995). Progress in the clinical management of herpes virus infections. *Antiviral Chemistry and Chemotherapy, 6*, 191–209.

Hamza, M. A., Higgins, D. M., Feldman, L. T., & Ruyechan, W. T. (2007). The latency-associated transcript of herpes simplex virus type 1 promotes survival and stimulates axonal regeneration in sympathetic and trigeminal ganglia. *Journal of Neurovirology, 13*, 56–66.

He, Y., Hara, H., & Núñez, G. (2016). Mechanism and regulation of NLRP3 inflammasome activation. *Trends in Biochemical Sciences, 41*, 1012–1021.

Honess, R. W., & Roizman, R. (1974). Regulation of herpes virus macromalecular synthesis: Cascade regulation of three groups of viral proteins. *Journal of Virology, 14*, 8–19.

Hu, X., Zhen, Q., Xiao, J., et al. (2022). Herpes simplex virus 1 induces microglia gasdermin D-dependent pyroptosis through activating the NLR family pyrin domain containing 3 inflammasome. *Frontiers in Microbiology, 13*, 838808.

Huang, Y., Xhu, W., & Zhou, Z. (2021). NLRP3 inflammasome activation and cell death. *Cellular & Molecular Immunology, 18*, 2114–2127.

Hussain, R. S., & Hussain, N. A. (2013). Ataxia and encephalitis in a young adult with EBV mononucleosis: A case report. *Case Reports in Neurological Medicine, 2013*, 3.

Inman, M., Perng, G.-C., Henderson, G., et al. (2001). Region of herpes simplex virus type 1 latency-associated transcript sufficient for wild-type spontaneous reactivation promotes cell survival in tissue culture. *Journal of Virology, 75*, 3636–3646.

Isogai, Z., Ono, R. N., Ushiro, S., et al. (2020). Herpes simplex virus: Global infection prevalence and incidence estimates, 2016. *Bulletin of the World Health Organization, 983*, 315–329.

Janowski, A. B., Bauer, I. K., Holtz, L. R., et al. (2017). Propagation of astrovirus VA1, a neurotropic human astrovirus, in cell culture. *Journal of Virology*. https://doi.org/10.1128/JVI.00740-00717

Janowski, A. B., Klein, R. S., & Wang, D. (2019). Differential in vitro infection of neural cells by astroviruses. *MBio, 10*. https://doi.org/10.1128/mBio.01455-01419

Jerant, A. F., DeGaetano, J. S., Epperly, T. D., et al. (1998). Varicella susceptibility and vaccination strategies in young adults. *Journal of the American Board of Family Medicine, 11*, 296–306.

Jin, L. D., Carpenter, D., Moerdyk-Schauwecker, M., et al. (2008). Cellular FLIP can substitute for the herpes simplex virus type 1 LAT

gene to support a wild type virus reactivation phenotype in mice. *Journal of Neurovirology, 14*, 389–400.
Jin, L., Perng, G.-C., Nesburn, A. B., et al. (2005). The baculovirus inhibitor of apoptosis gene (cpIAP) can restore reactivation of latency to a herpes simplex virus type 1 that does not express the latency associated transcript (LAT). *Journal of Virology, 79*, 12286–12295.
Johnson, R., & Milbourn, P. E. (1970). CNS manifestations of chickenpox. *Canadian Medical Association Journal, 102*, 831.
Jones, C. (1998). Alphaherpesvirus latency: Its role in disease and survival of the virus in nature. *Advances in Virus Research, 51*, 81–133.
Jones, C. (2003). Herpes simplex virus type 1 and bovine herpesvirus 1 latency. *Clinical Microbiology Reviews, 16*, 79–95.
Jones, C., Inman, M., Peng, W., et al. (2005). The herpes simplex virus type 1 (HSV-1) locus that encodes the latency-associated transcript (LAT) enhances the frequency of encephalitis in male Balb/C mice. *Journal of Virology, 79*, 14465–14469.
Katze, M. G., Heng, Y., & Gale, M. (2002). Viruses and interferon: Fight for supremacy. *Nature Reviews. Immunology, 2*, 675–686.
Kelley, T. W., Prayson, R. A., Ruiz, A. I., et al. (2003). The neuropathology of West Nile virus meningoencephalitis: A report of two cases and review of the literature. *American Journal of Clinical Pathology, 119*, 749–753.
Kennedy, P. G. E., Grinfield, E., & Gow, J. W. (1999). Latent varicella-zoster virus in human dorsal root ganglia. *Vriology, 258*, 451–454.
Keusch, G. T., Amuasib, J. H., Andersone, D. E., et al. (2022). Pandemic origins and a one health approach to preparedness and prevention: Solutions based on SARS-CoV-2 and other RNA viruses. *Proceedings of the National Academy of Sciences of the United States of America, 119*, e2202871119.
Khan, K., Shafiei, M. S., Longoria, C., et al. (2020). SARS-CoV-2 spike protein induces inflammation via TLR2-dependent activation of the NF-κB pathway. *eLife, 10*, e68563.
Kiyani, M., Liu, B., Charalambous, L. T., et al. (2020). The longitudinal health economic impact of viral encephalitis in the United States. *Journal of Medical Microbiology, 69*, 270–279.
Knipe, D. M., & Howley, P. M. (2001). Cytomegaloviruses and their replication. In E. S. Mockarski & C. T. Courcelle (Eds.), *Field's virology* (Vol. 2, pp. 2629–2675). Lippincott Williams and Wilkins.
Ku, C. C., Besser, J., Abendroth, A., Grose, C., & Arvin, A. M. (2005). Varicella-zoster virus pathogenesis and immunobiology: New concepts emerging from investigations with the SCIDhu mouse model. *Journal of Virology, 79*, 2651–2658.
Kuchler, L., Rufli, I., Koch, M. C., et al. (2020). Astrovirus-associated polioencephalomyelitis in an Alpaca viruses. *Viruses, 13*. https://doi.org/10.3390/v13010050
Kumar, A. K., & Mendez, M. D. (2022). Herpes simplex encephalitis. In StatPeals. StatPearls Publishing.
Kuno, G., & Chang, G.-J. J. (2005). Biological transmissions of arboviruses: Reexamination of and new insights into components, mechanisms, and unique traits as well as their evolutionaly traits. *Clinical Microbiology Reviews, 18*, 608–637.
Kurt-Jones, E. A., Chan, M., Zhou, S., et al. (2004). Herpes simplex virus 1 interaction with toll-like receptor 2 contributes to lethal encephalitis. *Proceedings of the National Academy of Sciences of the United States of America, 101*, 1315–1320.
Lahat, E., Barr, J., Barkai, G., et al. (1999). Long term neurological outcome of herpes encephalitis. *Archives of Disease in Childhood, 80*, 69–71.
Lei, Y., Zhang, J., Schiavon, C. R., et al. (2021). SARS-CoV-2 spike protein impairs endothelial function via downregulation of ACE 2. *Circulation Research, 128*, 1323.
Li, L., Diab, S., McGraw, S., Barr, B., et al. (2013). Divergent astrovirus associated with neurologic disease in cattle. *Emerging Infectious Diseases, 19*, 1385–1392.
Li, S., Carpenter, C., Hsiang, C., Wechsler, S. L., & Jones, C. (2010). The herpes simplex virus type 1 latency-associated transcript (LAT) locus inhibits apoptosis and promotes neurite sprouting in neuroblastoma cells following serum starvation by maintaining active AKT (protein kinase B). *The Journal of General Virology, 91*, 858–866.
Liesegang, T. J. (1999). Varicella Zoster viral disease. *Mayo Clinic Proceedings, 74*, 983–998.
Lindenback, B. D., & Rice, C. M. (2001). *Flaviviridae: The viruses and their replication* (pp. 991–1125). Lippincott Williams and Wilkins.
Liu, M., Guo, S., & Hibbert, J. M. (2011). CXCL10/IP-10 in infectious diseases pathogenesis and potential therapeutic implications. *Cytokine & Growth Factor Reviews, 22*, 121–130.
Liu, T., Zhang, L., Joo, D., & Sun, S.-C. (2017). NF-κB signaling in inflammation. *Signal Transduction and Targeted Therapy, 2*, e17023.
Livorsi, D., Anderson, E., & Qureshi, S. (2010). Brainstem encephalitis: An unusual presentation of herpes simplex virus infection. *Journal of Neurology, 257*, 1432–1437.
Lum, S. H., Turner, A., & Guiver, M. (2016). An emerging opportunistic infection: Fatal astrovirus (VA1/HMO-C) encephalitis in a pediatric stem cell transplant recipient. *Transplant Infectious Disease*. Epub 2016/2009/2016.
Markham, A., & Faulds, D. (1994). Ganciclovir. An update of its therapeutic use in cytomegalovirus infection. *Drugs, 48*, 455–484.
McGrath, N., Anderson, N. E., Croxson, M. C., & Powell, K. F. (1997). Herpes simplex encephalitis treated with acyclovir: Diagnosis and long term outcome. *Journal of Neurology, Neurosurgery, and Psychiatry, 63*, 321–326.
Mitchell, B. M., Bloom, D. C., Cohrs, R. J., et al. (2003). Herpes simplex virus-1 and varicella-zoster virus latency in ganglia. *Journal of Neurovirology, 9*, 194–204.
Nabizadeh, F., Balabandian, M. B., Sodeifian, F., et al. (2022). Autoimmune encephalitis associated with COVID-19: A systematic review. *Multiple Sclerosis and Related Disorders, 62*, 103795.
Naccache, S. N., Peggs, K. S., Mattes, F. M., et al. (2015). Diagnosis of neuroinvasive astrovirus infection in an immunocompromised adult with encephalitis by unbiased next-generation sequencing. *Clinical Infectious Diseases, 60*, 919–923.
Nagasawa, T., Kimura, I., Abe, Y., & Oka, A. (2007). HHV-6 encephalopathy with cluster of convulsions during eruptive stage. *Pediatric Neurology, 36*, 61–63.
Neyts, J., & De Clercq, E. (1994). Mechanism of action of acyclic nucleoside phosphonates against herpes virus replication. *Biochemical Pharmacology, 47*, 39–41.
O'Hare, P. (1993). The virion transactivator of herpes simplex virus. *Seminars in Virology, 4*, 145–155.
Pezzini, A., & Padovani, A. (2020). Lifting the mask on neurological manifestations of COVID-19. *Nature Reviews. Neurology, 16*, 636–644.
Pass, R. F. (1995). Epidemiology and transmission of cytomegalovirus infection. *The Journal of Infectious Diseases, 152*, 243–256.
Pass, R. F. (2001). Cytomegalovirus. In D. M. Knipe & P. M. Howley (Eds.), *Field's virology* (4th ed., pp. 2675–2705). Lippincott Williams & Wilkins.
Peng, W., Vitvitskaia, O., Carpenter, D., et al. (2008). Identification of two small RNAs within the first 1.5-kb of the herpes simplex virus type 1 (HSV-1) encoded latency-associated transcript (LAT). *Journal of Neurovirology, 14*, 41–52.
Perng, G.-C., Maguen, B., & Jin, L. (2002). A gene capable of blocking apoptosis can substitute for the herpes simplex virus type 1 latency-associated transcript gene and restore wild-type reactivation levels. *Journal of Virology, 76*, 1224–1235.
Perng, G.-C., Jones, C., & Ciacci-Zanella, J. (2000). Virus-induced neuronal apoptosis blocked by the herpes simplex virus latency-associated transcript (LAT). *Science, 287*, 1500–1503.

Perng, G.-C., & Jones, C. (2010). Towards an understanding of the Herpes Simplex Virus Type 1 latency-reactivation cycle. *Interdisciplinary Perspectives on Infectious Diseases., 2010*, 1–18.

Perry, C. M., & Faulds, D. (1996). Valaciclovir. A review of its antiviral activity, pharmacokinetic properties and therapeutic efficacy in herpesvirus infections. *Drugs, 52*, 754–772.

Petrucelli, A., Rak, M., Grainger, L., & Goodrum, F. (2012). Characterization of a novel Golgi apparatus-localized latency determinant encoded by human cytomegalovirus. *Journal of Virology, 83*, 5615–5629.

Pfaff, F., Schlottau, K., Scholes, S., & Courtenay, A. (2017). A novel astrovirus associated with encephalitis and ganglionitis in domestic sheep. *Transboundary and Emerging Diseases, 64*, 677. https://doi.org/10.1111/tbed.12623

Phelan, D., Barrozo, E. R., & Bloom, D. C. (2017). HSV1 latent transcription and non-coding RNA: A critical retrospective. *Journal of Neuroimmunology, 308*, 65–101.

Plosker, G. L., & Noble, S. (1999). Cidofovir. A review of its use in cytomegalovirus retinitis in patients with AIDS. *Drugs, 58*, 325–345.

Poole, E., Wills, M., & Sinclair, J. (2014). Human cytomegalovirus latency: targeting differences in the latently infected cell with a view to clearing latent infection. In *Advances in Virology* (Vol. 2014, p. 10). Hindawi.

Pothlicet, J., Meunier, I., & Davis, B. K. (2013). Type I IFN triggers RIG-I/TLR3/NLRP3-dependedent inflammasome activation in influenza A virus infected cells. *PloS Pathogens, 9*, e1003256.

Price, W. H., & O'Leary, W. (1967). Geographical variation in the antigenic character of West Nile virus. *American Journal of Epidemiology, 85*, 84–87.

Quan, P. L., Wagner, T. A., & Briese, T. (2010). Astrovirus encephalitis in boy with X-linked agammaglobulinemia. *Emerging Infectious Diseases, 16*, 918–925.

Rantalaiho, T., Farkhila, M., Vaheri, A., & Koskiniemi, M. (2001). Acute encephalitis from 1967 to 1991. *Journal of the Neurological Sciences, 184*, 169–177.

Rawson, H., Crampin, A., & Noah, N. (2001). Deaths from chickenpox in Engalnd and Wales 1995-1997: Analysis of routine mortalitiy data. *British Medical Journal., 323*, 1091–1093.

Reynaud, J. M., & Horvat, B. H. (2013). Human herpesvirus 6 and neuroinflammation. *ISRN Virology, 2013*, 11.

Salahuddin, S. Z., Ablashi, D. V., Markham, P. D., et al. (1986). Isolation of a new virus, HBLV, in patients with lymphoproliferative disorders. *Science, 234*, 596–601.

Samuel, M. A., & Diamond, M. S. (2005). Alpha/beta interferon protects against lethal West Nile virus infection by restricting cellular tropism and enhancing neuronal survival. *Journal of Virology, 79*, 13350–13361.

Sancho-Shimizu, V., de Diego, R. P., Halwani, L. L. R., et al. (2011). Herpes simplex encephalitis in children with autosomal recessive and dominant TRIF deficiency. *Journal of Clinical Investigation, 121*, 4889–4901.

Santoro, F., Kennedy, P. E., Locatelli, G., Malnati, M. S., et al. (1999). CD46 is a cellular receptor for human herpesvirus 6. *Cell, 99*, 817–827.

Schaller, T., Hirschbühl, K., Burkhardt, K., et al. (2020). Postmortem examination of patients with COVID-19. *JAVMA, 323*, 2518–2520.

Scholar, E. M., & Pratt, W. B. (2000). Chemotherapy of viral infections. I. Drugs used to treat influenza virus infections, herpes virus infections, and drugs with broad-spectrum antiviral activity. In *The antimicrobial drugs* (pp. 491–549). Oxford University Press.

Shen, W., Sa e Silva, M., Jaber, T., et al. (2009). Two small RNAs encoded within the first 1.5 kb of the herpes simplex virus type 1 (HSV-1) latency-associated transcript (LAT) can inhibit productive infection, and cooperate to inhibit apoptosis. *Journal of Virology, 90*, 9131–9139.

Shen-Ying, Z., Clark, N. E., Freije, C. A., et al. (2018). Inborn errors of RNA lariat metabolism in humans with brainstem viral infection. *Cell, 172*, 952–965.

Shresta, B., & Diamond, M. S. (2004). Role of CD8+ T cells in control of West Nile virus infection. *Journal of Virology, 78*, 8312–8321.

Siow, I., Lee, K. S., & Zhang, J. J. Y. (2021). Encephalitis as a neurological complication of COVID-19: A systematic review and meta-analysis of incidence, outcomes, and predictors. *European Journal of Neurology, 28*, 2491–3502.

Sivasubramanian, M. K., Monteiro, R., Harrison, K. S., et al. (2022). Herpes simplex virus type 1 preferentially enhances neuro-inflammation and senescence in brainstem of female mice. *Journal of Virology, 97*, e01081–e01022.

Skoldenberg, B. (1991). Herpes simplex encephalitis. *Scandinavian Journal of Infectious Diseases. Supplementum, 80*, 40–46.

Snoeck, R., Andrei, G., & De Clercq, E. (1999). Current pharmacological approaches to the therapy of varicella zoster virus infection. A Guide to treatment. *Drugs., 57*, 187–206.

Spear, P. G. (1993). Entry of alphaherpesviruses into cells. *Seminars in Virology, 4*, 167–180.

Stahl, J.-P., Mailles, A., Dacheux, L., & Morand, P. (2011). Epidemiology of viral encephalitis in 2011. *Medecine et maladies infectieuse., 41*, 453–464.

Stanberry, L. R., Oxman, M. N., & Simmons, A. (2004). Herpes simplex viruses. In S. L. Gorbach, J. G. Bartlett, & N. R. Blacklow (Eds.), *Infectious diseases* (pp. 1905–1917). Lippincott Williams and Wilkins.

Stanberry, L. R., Spruance, S. L., Cunningham, A. L., et al. (2002). Glycoprotein-D-adjuvant vaccine to prevent genital herpes. *The New England Journal of Medicine, 347*, 1652–1661.

Sui, Y., Stehno-Bittel, L., Li, S., et al. (2006). CXCL10-induced cell death in neurons: Role of calcium dysregulation. *The European Journal of Neuroscience, 23*, 957–964.

Takahashi, M., Otsuka, T., Okuno, Y., et al. (1974). Live vaccine used to prevent the spread of varicella in children in hospital. *Lancet, 2*, 1288–1290.

Takahashi, M., Otsuka, T., et al. (1975). Development of a live attenuated varicella vaccine. *Biken Journal, 18*, 25–33.

Thakur, K. T., Miller, E. H., Glendinning, M. D., et al. (2021). COVID-19 neuropathology at Columbia University Irving Medical Center/New York Presbyterian Hospital. *Brain, 144*, 2696.

Thompson, R. L., & Sawtell, N. M. (2011). The herpes simplex virus type 1 latency associated transcript is required for the maintenace of reactivation competent latent infection. *Journal of Neurovirology, 17*, 552–558.

Tyler, K., Tedder, D. G., Yamamoto, L. J., et al. (1995). Recurrent brainstem encephalitis associated with herpes simplex virus type 1 DNA in cerebrospinal fluid. *Neurology, 45*, 2246–2250.

Umbach, J. L., Kramer, M., & Jurak, I. (2008). MicroRNAs expressed by herpes simplex virus 1 during latent infection regulate viral mRNAs. *Nature, 454*, 780–785.

Wagstaff, A. J., & Bryson, H. M. (1994). Foscarnet: A reappraisal of its antiviral activity, pharmacokinetic properties and therapeutic use in immunocompromised patients with viral infections. *Drugs, 48*, 153–205.

Wagstaff, A. J., Faulds, D., & Goa, K. L. (1994). Aciclovir. A reappraisal of its antiviral activity, pharmacokinetic properties and therapeutic efficacy. *Drugs, 47*, 153–205.

Wang, T., Town, T., Alexopoulou, L., et al. (2004). Toll-liker receptor 3 mediates West Nile virus entry into the brain causing lethal encephlaitis. *Nature Medicine, 10*, 1366–1373.

Wang, Y., Lobigs, M., Lee, E., & Mullbacher, A. (2003). CD8+ T cells mediate recovery and immunopathoglogy in West Nile virus encephalitis. *Journal of Virology, 77*, 13323–13334.

Weber, F., Wagner, V., & Rasmussen, S. B. (2006). Double-stranded RNA is produced by positive-Strand RNA viruses and DNA viruses but not in detectable amounts by negative-strand RNA viruses. *Journal of Virology, 80*, 5059–5064.

Whitley, R. (1997). *Herpes simplex virus* (pp. 73–190). Lippincott-Raven Publishers.

Whitley, R. J., & Roizman, B. (2001). Herpes simplex virus infections. *Lancet, 357*, 1513–1518.

Wunderli, W., Meerbach, A., Gungor, T., et al. (2011). Astrovirus infection in hospitalized infants with severe combined immunodeficiency after allogeneic hematopoietic stem cell transplantation. *PLoS One, 6*, e27483.

Xia, M. Q., Bacskai, B., Knowles, R. B., et al. (2000). Expression of the chemokine receptor CXCR3 on neurons and the elevated expression of its ligand IP-10 in reactive astrocytes: In vitro ERK1/2 activation and role in Alzheimer's disease. *Journal of Neuroimmunology, 108*, 227–235.

Yamada, S., Kameyama, T., Nagaya, T., et al. (2002). Relapsing herpes simplex encephalitis: Pathological confirmation of viral reactivation. *Journal of Neurology, Neurosurgery, and Psychiatry, 74*, 262–264.

Yamanishi, K., Okuno, T., Shiraki, K., et al. (1988). Identification of human herpesvirus-6 as the causal agent for exanthem subitum. *The Lancet, 1*, 1065–1067.

Yoshikawa, T., & Asano, Y. (2000). Central nervous system complications in human herpesvirus-6 infection. *Brain and Development, 22*, 307–314.

Young, L. S., & Rickinson, A. B. (2004). Epstein-Barr virus: 40 years on. *Nature Reviews Cancer, 4*, 757–768.

Zheng, M., Karki, R., Williams, E. P., et al. (2021). TLR2 senses the SARS-CoV-2 envelope protein to produce inflammatory cytokines. *Nature Immunology, 22*, 829–838.

Prion Diseases of Animals and Humans

29

Benjamin S. Steadman and Jason C. Bartz

Abstract

Prions are unique infectious agents that result from the conformational conversion of normal host protein PrP^C to an abnormal infectious self-templating form of the prion protein, PrP^{Sc}. PrP^{Sc} accumulation in the host results in a triad of neuropathological signs including vacuolation, gliosis, and neuronal degeneration that inevitably lead to the death of the host. Prion diseases have long incubation periods, are naturally found in animals and humans, and, due to the protein-only nature of the infectious agent, can have sporadic, genetic, or acquired etiologies. Some prion diseases are particularly promiscuous, such as bovine spongiform encephalopathy (BSE), which has crossed species barriers into ungulates, felines, primates, and mink. BSE has also resulted in zoonotic transmission to humans causing variant Creutzfeldt-Jakob disease (CJD). Prions from scrapie of sheep and goats and chronic wasting disease of cervids can persist in the environment for years, are horizontally and vertically transmissible, and have host ranges that are still incompletely described. All prion diseases are inevitably fatal and, to date, effective treatments have not been available. Recent work, however, has led to the development of reliable diagnostic tests that can rapidly detect PrP^{Sc}, which hold promise for practical diagnosis of animals and patients.

Keywords

Prion · PrP^{Sc} · Bovine spongiform encephalopathy · Chronic wasting disease · Interspecies transmission · Creutzfeldt-Jakob disease

B. S. Steadman · J. C. Bartz (✉)
Department of Medical Microbiology and Immunology, School of Medicine, Creighton University, Omaha, NE, USA
e-mail: jbartz@creighton.edu

29.1 An Introduction to Prion Diseases

The term, prion, was introduced by Stanley B. Prusiner in 1982 to distinguish the unique properties of the infectious agent of the transmissible spongiform encephalopathies, which are abnormally resistant to normal processes that break down nucleic acids, viruses, or plasmids (Prusiner, 1982). Infectious prions are physically, chemically, and biologically unusually stable (Hunter, 1979). Prions remain infectious after exposure to ionizing radiation, ultraviolet light, ultrasonication (Hunter & Millson, 1964), acidic and basic conditions ranging from a pH of 2 to 10 (Kimberlin, 1976), or after 500 cycles of freezing at −80 °C and dethawing (Yuan et al., 2018). Some prions are also still infectious after 8 h of boiling (Stamp et al., 1959), or after exposure to temperatures over 600 °C (Brown et al., 2000). Furthermore, prions are highly resistant to inactivation by formaldehyde, detergents, disinfectants (Brown et al., 1982), chemicals and salt solutions (Hunter, 1979). Finally, prions are resistant to digestion from nucleases, proteases, lipases, and carbohydrases (Hunter, 1979). It was this unique combined hardiness that first distinguished prions from other infectious agents, such as bacteria and viruses.

Prion formation is caused by the biochemical conversion of normal host protein PrP^C to abnormal PrP^{Sc} at the cell surface (Caughey & Raymond, 1991), or in the endosomal/lysosomal pathway (Chesebro & Caughey, 1993). As PrP^{Sc} converts PrP^C template into more PrP^{Sc}, the secondary structure of PrP^C transforms from primarily consisting of α-helices, to predominantly consisting of β-sheets in PrP^{Sc} (Safar et al., 1993a, b). After years of computationally predicted PrP^{Sc} structures, recent studies using cryo-electron microscopy suggest that PrP^{Sc} structure consists of an amyloid parallel-in register intermolecular sheet (PIRIBS) (Manka et al., 2023). The conformation of PrP^{Sc} with increased β-sheets is thought to result in the resistance of the prion agent to inactivation as explained above (Safar et al., 1993a, b). For example, the prion precursor PrP^C is suscep-

H. E. Gendelman, T. Ikezu (eds.), *Neuroimmune Pharmacology and Therapeutics*,
https://doi.org/10.1007/978-3-031-68237-7_29

tible to digestion with phospholipase C, Proteinase K (PK), or trypsin, whereas PrP^{Sc} is resistant to degradation with these enzymes (Caughey & Raymond, 1991). PrP^{Sc} resistance develops at a hydrodynamic radius of 15 nm (Cortez et al., 2021), suggesting that it is the smaller subfibrillar particles, made up of PrP^{Sc} aggregates, that are more infectious than the larger amyloid fibrils (Kim et al., 2012).

There are several requirements for successful pathogenic prion propagation to occur, although much of this process remains poorly understood. For prions to cause disease, PrP^{C} conversion to PrP^{Sc} must take place within neurons, as non-neuronal PrP^{Sc} propagation and accumulation are nonpathogenic (Mallucci et al., 2003). For example, astrocyte-restricted PrP^{Sc} propagation does not result in prion disease (Lakkaraju et al., 2022). PrP^{Sc} can accumulate on the cell membrane, within intracellular vesicles, and in some cases, as extracellular amyloid plaques (Chesebro & Caughey, 1993). Although the "protein-only" hypothesis states that PrP^{Sc} protein molecules are the sole infectious prion agents (Prusiner, 1982), experimental evidence to support this strict view is lacking and cellular cofactors appear to be involved (Supattapone, 2010). Cofactors that influence prion propagation include polyanions, such as RNA (Geoghegan et al., 2007), phospholipids (Supattapone, 2010), and phosphatidylethanolamine, some of which are required to maintain prion infectivity (Deleault et al., 2012). Additionally, post-translational modifications, such as glycosylphosphatidylinositol (GPI) anchors and/or N-linked glycans, while not required, are needed for maximum prion propagation and infectivity (Chesebro et al., 2005). PrP^{C} is normally hyperglycosylated whereas PrP^{Sc} is under glycosylated (Kang et al., 2020). However, the requirement for PrP^{C} glycosylation in prion propagation depends upon species and/or prion strain (Nishina et al., 2006). Likewise, prion strain diversity partially results from differences in prion strain affinities for various cofactor types and combinations (Deleault et al., 2012).

As PrP^{C} propagates into PrP^{Sc}, it can misfold into different conformations, referred to as prion strains (Manka et al., 2023). Prion strains are operationally defined as a heritable disease phenotype under controlled conditions of experimental transmission (Bartz, 2016). The disease phenotype can include clinical signs, incubation periods, lesion profiles, and tissue tropism (Parchi et al., 1999b; Castilla et al., 2008). Ultimately, prion strains are constantly evolving (Bian et al., 2021), and multiple strains can co-exist within one host at the same time (Cali et al., 2009). Prion strains are distinguished by their biological, chemical, and physical properties (Hill et al., 1997). Strain specificity is conferred through differences in the conformational structure of PrP^{Sc} (Manka et al., 2023).

A more thorough understanding of PrP^{C} can provide the groundwork necessary to generate targets for prion prevention and treatments. PrP^{C} is a normal cellular protein highly conserved in most organisms (Wopfner et al., 1999). PrP^{C} is encoded by the *PRNP* gene, found on chromosome 20 in humans (Forloni et al., 2015). Like many other proteins, PrP^{C} is generated in the endoplasmic reticulum and transported to the cell surface through vesicles released from the Golgi body (Caughey et al., 1989). PrP^{C} is typically located on the cell membrane as a monomeric GPI-anchored glycoprotein (Stahl et al., 1987). PrP^{C} is primarily found in the central nervous system (CNS) (Ford et al., 2002a), but is also distributed throughout the entire body, from the peripheral nervous system (PNS) to lymphoid tissues, spleen, heart, lungs, kidneys, bladder, testis, and skin (Ford et al., 2002b). Functions for PrP^{C} are not universally agreed upon but appear to have redundancy with other proteins. PrP^{C} has several key roles in vitro, including metal ion homeostasis (Brown et al., 1997), antioxidation (White et al., 1999), signal transduction (Mouillet-Richard et al., 2000), apoptosis mediation (Bounhar et al., 2001; Paitel et al., 2003), antiviral properties (Zhang et al., 2017), and cell adhesion, migration, and differentiation (Gao et al., 2016). Additionally, PrP^{C} has many physiological functions from stress protection to neuronal excitability, myelin maintenance, circadian rhythm, immunity, mitochondrial homeostasis, and interaction with several signaling pathways (Castle & Gill, 2017). Of significance is the capacity to experimentally generate PrP^{C} knockout mice that develop and behave normally (Bueler et al., 1992), yet are completely resistant to prion formation and subsequent disease development (Brandner et al., 1996). Likewise, in mice with established neuroinvasive prion infection, decreasing endogenous neuronal PrP^{C} reduced spongiosis and protected the mice from both neuronal loss and clinical disease progression (Mallucci et al., 2003). These findings suggest that stopping the continuous neuronal conversion of PrP^{C} to PrP^{Sc} prevents prion neurotoxicity during infection (Mallucci et al., 2003). Overall, PrP^{C} may provide a promising target for therapeutic intervention against prion diseases (Mallucci et al., 2003).

Developing rapid reliable diagnostic tests that are both sensitive and specific to prion diseases is challenging. Two powerful novel technologies have been developed that can detect minute levels of PrP^{Sc} protein. The first is protein misfolding cyclic amplification (PMCA) which adds PrP^{Sc} into uninfected brain homogenate that serves as a source of the template for conversion, PrP^{C}. Serial rounds of incubation and sonication result in the conversion of PrP^{C} template into PrP^{Sc} in vitro (Saborio et al., 2001). PMCA products are infectious, maintain strain properties, and are best detected by Western blotting (Castilla et al., 2005b). Since PMCA recapitulates prion formation in vitro, it accurately recapitulates species barriers (Castilla et al., 2008). Finally, compared to in vivo bioassay, PMCA is both timely and cost-effective (Saunders et al., 2012). The second method is real-time quaking-induced conversion (RT-QuIC), which utilizes purified recombinant prion protein as the template of conversion and, through a process of shaking, prion seeding activity is detected via thioflavin T fluorescent dye binding (Wilham et al., 2010). RT-QuIC works well with samples from the CNS or PNS (Dong & Satoh, 2021). However, RT-QuIC is unsuccessful at detecting prion seeding activity from blood samples unless the blood samples are first processed, which suggests that blood contains RT-QuIC inhibitors (Dong & Satoh, 2021). In contrast to PMCA, RT-QuIC products are not infectious, and cannot recapitulate the species barrier (Escobar et al., 2020). However, RT-QuIC is practical compared to bioassay because it is faster, safer, more sensitive, and cost-effective (Wilham et al., 2010; Escobar et al., 2020). Together, these two techniques have proven helpful in prion research and hold the potential for rapid practical diagnosis of patients and animals in the future.

29.2 Animal Prion Diseases

29.2.1 Scrapie

Scrapie was first described as early as 1693 in sheep of Southwestern England and is considered the prototype prion disease (Table 29.1) (Ness et al., 2023). Over time, scrapie has been called by over 20 names, such as rida (Sigurdsson, 1954), goggles, rickets, rubbers, shakings (Ness et al., 2023), cuddie trot, gnawers, trotters, and trembling disease (M'Gowan, 1914). Scrapie was later discovered naturally in goats in 1942 (Chelle, 1942), and in moufflon in 1992 (Wood et al., 1992). Scrapie is found worldwide (Hourrigan et al., 1979), however, incidence rates and prevalence are difficult to determine as surveillance for scrapie is inconsistent between countries (Bradley & Matthews, 1992). Scrapie is an economically important pathogen and eradication of scrapie in the United States alone is estimated to raise revenue by $10.8 million USD annually (Seitzinger et al., 2006).

Scrapie can transmit vertically and maternally (Hourrigan et al., 1979; Foster et al., 1992; Hoinville et al., 2010). In sheep, parents with scrapie had statistically more offspring that contracted scrapie compared to parents without scrapie (Dickinson et al., 1966; Hourrigan et al., 1979), even when controlling for genotype (Hoinville et al., 2010). Additionally, many of these offspring developed clinical signs at ages much younger than the average for animals infected with scrapie (Dickinson et al., 1966). Furthermore, scrapie transmits via embryos (Foster et al., 1992), fetal membranes, cotyledons (Alverson et al., 2006), the placenta (Andreoletti et al., 2002; Spiropoulos et al., 2014), the uterus (Tuo et al., 2001), and amniotic fluids (Garza et al., 2011). Scrapie transmission to the placenta and cotyledons is strongly influenced by fetal PrP genotype (Alverson et al., 2006). In utero transmission has been observed through the detection of scrapie in an unborn fetus (Spiropoulos et al., 2014). Finally, scrapie can transmit maternally through mammary glands (Garza et al., 2014), colostrum, and milk (Lacroux et al., 2008).

Scrapie transmits horizontally and environmentally (Greig, 1940; Brotherston et al., 1968; Hourrigan et al., 1979). Infectious PrP^{Sc} is present in the skin (Garza et al., 2014), urine, feces, blood, and saliva (Tamguney et al., 2012). Furthermore, evidence suggests that scrapie can transmit environmentally in pastures (Greig, 1940), as well as through soil (Maddison et al., 2015), and common agricultural reservoirs, such as plastic scratching posts and metal fencing (Maddison, Baker et al., 2010a). Scrapie can be transmitted to new locations by moving sheep in the preclinical phase of the disease (Hourrigan et al., 1979). Finally, iatrogenic transmission of scrapie occurred to thousands of sheep through a batch of scrapie-contaminated louping-ill vaccines (Gordon, 1959).

Clinical signs of scrapie include motor and behavioral changes (M'Gowan, 1914). The most common clinical sign of scrapie is pruritus, with animals rubbing or scraping against structures which led to the name of the disease (M'Gowan, 1914). Motor manifestations of scrapie infection can include an abnormal gait, locomotor incoordination, hyperexcitability, ataxia, tremors, asthenia, lethargy, recumbency, and weight loss (M'Gowan, 1914; Sigurdsson, 1954; Brotherston et al., 1968; Palsson, 1979). Vision abnormalities and behavioral or temperament changes are also frequently observed (M'Gowan, 1914). Death is inevitable, usually within 2 weeks to 4 months after the onset of clinical signs (M'Gowan, 1914). Scrapie incubation takes 1–5 years to develop (Gajdusek, 1979). Scrapie infection is observed in sheep and goats of 3–5 years of age on average (Hourrigan et al., 1979). In field settings, animals are rarely diagnosed with scrapie earlier than 2 years or age, but scrapie prions have been isolated from sheep as young as 4 months (Dickinson et al., 1966), and goats as young as 11 months (Hourrigan et al., 1979).

Postmortem diagnosis of scrapie is performed through the histopathological confirmation of spongiform vacuolation, astrogliosis, and neurodegeneration (Zlotnik, 1958; Brotherston et al., 1968), immunodetection of PrP^{Sc} (Miller et al., 1993), and negative-stain electron microscope detection of scrapie-associated fibrils (SAF), which are largely made up of PrP (Stack et al., 1993). Neuropathological lesions are frequently observed in the gray matter of the CNS, particularly in the cerebral cortex, hippocampus, thalamus, hypothalamus, and cerebellum (Brotherston et al., 1968; Salvesen et al., 2020). Vacuolation is also normally detected in the brainstem, including the mesencephalon, pons, medulla oblongata, and the spinal cord (Sigurdsson, 1954; Brotherston et al., 1968). Vacuoles are typically diffuse (Wood et al., 1992), severe, widespread (Brotherston

Table 29.1 Naturally occurring prion diseases of animals and humans

	Year identified	Suspected natural transmission routes	Age in years median, *mean* (range)	Incubation period in years *mean* (range)	Clinical duration in months[a] median, *mean* (range)	Species and *PRNP* codon polymorphisms or mutations[b]
Animal prion diseases						
Scrapie	1693	Environmental, horizontal, vertical, and maternal	(0–12)	*3–5* (1–5)	*2–4*	Sheep—112, 136, 137, 141, 142, 154, 171, & 176 Moufflon; goat—222 and natural knockouts
Chronic wasting disease	1967	Environmental, horizontal, vertical, and maternal	*3–5.5* (1.5–20)	(1–5)	(0.5–9)	Mule deer—20, 225; elk—132 & 226; White-tailed deer—95, 96, 185; moose; Reindeer—138; red deer—226
Bovine spongiform encephalopathy	1981	Oral and possibly vertical	*5* (2–17)	*5* (2.5–8)	*1–4* (1–14)	Cattle—12 & 23 base pair deletion Cattle—211 for H-type atypical BSE
Exotic ungulate encephalopathy	1986	Oral, and possibly maternal, horizontal or environmental	(1.5–4)	*2.5* (0.25–4)	*0.5*	Nyala; gemsbok; elands; Arabian oryx; greater kudu; scimitar-horned oryx; bison; ankole cattle
Feline spongiform encephalopathy	1990	Oral and maternal	*6–7* (1–14)	Domestic—*5* Zoo—(4.5–8)	*3* (0.5–5+)	Domestic cats—153; Cheetahs; pumas; ocelots; tigers; lions; Asiatic golden cat; Asian leopard cat
Prion disease in non-human primates	1996	Oral	(4–18)		12	Rhesus macaque; Mayotte brown lemurs; White fronted brown lemur; mongoose lemur
Transmissible mink encephalopathy	1947	Oral	(1+)	(0.33–1)	(0–2)	Mink
Camel prion disease	2018	Unknown	(8–13)		(3–8)	Dromedary camels—69? And 134?
Human prion diseases						
Sporadic Creutzfeldt-Jakob disease	1920	n.a.	66–68, *57–67* (14–97)	Unknown	5, *6–8.6* (1–132)	129 as follows: MM1/MV1—40-70%; VV2—14.8-16%; MV2—8-13.6%; VV1—1-4.1%; MM2—2-8.5%
Familial Creutzfeldt-Jakob disease	1924	n.a.	56.5, *49.5–57* (20–98)	Unknown	*8–22* (1.5–120)	105, 129, 148, 171, 178, 180, 183, 188, 196, 203, 208, 200, 210, 211, & 232 Insertions or extra octapeptide coding repeats
Iatrogenic Creutzfeldt-Jakob disease	1974	Corneal transplantations, neurosurgical equipment, dura mater grafts, and pituitary hormones	36–58, *56* (10–81)	*12–30+* (1–40+)	*14* (5–32)	129
Variant Creutzfeldt-Jakob disease	1996	Oral and blood transfusions	28, *29–30* (11–74)	(5–10)	14, (9–35)	129—Generally methionine homozygous
Kuru	1957	Oral	(4–64)	(4–50+)	*3–9* (0–24+)	127—Glycine homozygous 129—Methionine or valine homozygous

(continued)

Table 29.1 (continued)

	Year identified	Suspected natural transmission routes	Age in years median, *mean* (range)	Incubation period in years *mean* (range)	Clinical duration in months[a] median, *mean* (range)	Species and *PRNP* codon polymorphisms or mutations[b]
Gerstmann-Straussler-Scheinker syndrome	1936	n.a.	49, *46–52* (15–87)	Unknown	40–48, *48* (3–240+)	102, 105, 117, 129, 131, 145, 187, 198, 202, 212, 217, 219 & 232
Fatal familial insomnia	1986	n.a.	49–56, *49–53* (17–83)	Unknown	12.4, *11–18* (4–97)	129—At least one methionine 178—Aspartic acid
Variably protease-sensitive prionopathy	2008	n.a.	*65–70* (46–93)	Unknown	*24–36* (6–120)	129

[a]Clinical duration refers to the total time from disease onset until death
[b]PrP amino acid sequence numbering is based on the species described
n.a. not applicable

et al., 1968), symmetrical, and variable in size and number (Zlotnik, 1958). Neuronal degeneration is commonly observed, varying from animal to animal, and can be accompanied by hyperchromasia, chromatolysis, or neuronal shrinkage (Wood et al., 1992). In the CNS, scrapie PrP^{Sc} is present in the cerebral cortex, thalamus, hippocampus, granular, and molecular layers of the cerebellum, brainstem, and spinal cord (Bencsik et al., 2001; Salvesen et al., 2020). PrP^{Sc} has a widespread distribution in the host and can be found in the retina (Greenlee et al., 2006), cheeks (Maddison et al., 2010b), tongue, tonsils, enteric nervous system (ENS) (van Keulen et al., 1999), muscles, liver, kidneys, spleen (Garza et al., 2014), and lymph nodes (Andreoletti et al., 2002). PrP^{Sc} deposition can be multifocal, intracytoplasmic (Andreoletti et al., 2002), perivascular, granular, aggregated (Miller et al., 1993), linear, or perineuronal (Garza et al., 2014). Amyloid PrP plaques are rarely observed in scrapie (McBride et al., 1988).

Atypical scrapie, (a.k.a. Nor98), was formally recognized in 2003 (Benestad et al., 2003). Atypical scrapie has a higher prevalence in older animals (Benestad et al., 2003), and clinical signs are less pronounced in atypical scrapie compared to classical forms of scrapie (Gavier-Widen et al., 2004). Compared to classical forms of scrapie, atypical scrapie has different neuropathology, and PrP^{Sc} is primarily localized to the CNS (Gavier-Widen et al., 2004), leading to the suggestion that it may have a sporadic etiology (Benestad et al., 2003). PrP^{Sc} from atypical scrapie produces a distinct western blot glycotype banding pattern with a faster migration of the nonglycosylated PrP^{Sc} polypeptide compared to classical forms of scrapie (Benestad et al., 2003; Gavier-Widen et al., 2004). Selective breeding of sheep may be favoring atypical scrapie (Aguilar-Calvo et al., 2015), as *PRNP* polymorphisms associated with resistance to classical scrapie infection in sheep are associated with the occurrence of atypical scrapie (Benestad et al., 2003).

Genetics plays an important role in scrapie susceptibility (Gordon, 1959; Dickinson & Fraser, 1979; Hoinville et al., 2010). *PRNP* genotype impacts scrapie disease development and incubation period (Dickinson & Fraser, 1979) and may influence why some sheep breeds are more susceptible to scrapie than others (Gordon, 1959). A three-codon polymorphism at residues 136A/V, 154R/H, and 171Q/R/H indicates that ARR is the most resistant haplotype and VRQ is the most susceptible haplotype to prion infection (van Keulen et al., 1999). Other polymorphisms in *PRNP* that impact scrapie resistance in sheep include codons 112, 137, 141, 142, and 176 (Orge et al., 2021). In goats, the wild type is the most susceptible to scrapie with the K222 polymorphism having the least risk of developing scrapie (Imran & Mahmood, 2011). Interestingly, healthy goats that are null for the PrP^{C} allele have been discovered in nature which may provide a solution to reducing incidence rates of goat scrapie (Salvesen et al., 2020). Finally, genetics and strain type influence what animals are experimentally susceptible to scrapie.

Scrapie has been experimentally transmitted to numerous species. Under laboratory conditions, scrapie has been transmitted to mice, hamsters, rats, mink (Hanson et al., 1971; Hillier & Salmon, 2000), cattle (Cutlip et al., 1994), raccoons (Hamir et al., 2003), bank voles (Di Bari et al., 2008), and white-tailed deer (Greenlee et al., 2023). Experimentally, scrapie has been transmitted to several non-human primates (NHPs) (Gajdusek, 1977), and there are contradictory reports of scrapie transmission to transgenic mice expressing human PrP^{C} (Wilson et al., 2012; Wadsworth et al., 2013; Cassard et al., 2014). There is no direct evidence that scrapie has crossed the transmission barrier to humans, even though people have been exposed to scrapie-contaminated tissue for centuries (Palsson, 1979). While this remains a theoretical possibility, even if scrapie were to be transmitted to humans, the incidence of sporadic Creutzfeldt-Jakob disease is low enough that it would be difficult to consider scrapie as posing a significant risk.

29.2.2 Chronic Wasting Disease

Chronic wasting disease (CWD) was first observed in a Colorado mule deer in 1967 (Table 29.1) (Williams & Young, 1980). In North America, CWD has also been detected in Rocky Mountain elk (Williams & Young, 1982), white-tailed deer (Spraker et al., 1997), and moose (Baeten et al., 2007), in 35 US States and 5 Canadian provinces (Center, 2024). CWD is the only known prion disease that affects both wild and captive animals (Spraker et al., 1997; Center, 2024). Outside of North America, CWD has been detected in South Korea in cervids that were imported from North America (Kim et al., 2005). Since 2016, CWD has been identified in cervids in Norway (Benestad et al., 2016), Finland, and Sweden (Osterholm et al., 2019). Additional cervid species found naturally with CWD in Europe include reindeer (Benestad et al., 2016), and red deer (Vikoren et al., 2019). European CWD cases consist of multiple evolving prion strains, which differ from North American types in transmission efficiency, and important biological and biochemical properties of PrP^{Sc} (Bian et al., 2021; Sola et al., 2023). In CWD-endemic areas, prevalence rates can reach more than 50% (Miller & Wild, 2004). Since CWD is inevitably fatal, CWD infection can negatively affect cervid population health (Edmunds et al., 2016).

CWD transmits maternally (Nalls et al., 2013), horizontally, and environmentally (Miller et al., 2004). In cervids, the first evidence to suggest maternal CWD transmission came from the observation of mother-to-offspring CWD transmission in experimentally infected Reeves' muntjac deer (Nalls et al., 2013). Consistent with this observation, prion seeding activity is detected in reproductive and fetal tissues such as placentomes (Nalls et al., 2013), ovaries (Spraker et al., 2023), uterus, amniotic sac, cotyledon, umbilical cord, and testis (Bravo-Risi et al., 2021). Prion seeding activity is also detected in utero-derived fetal tissue from naturally infected, presymptomatic, free-ranging CWD-infected cervids providing significant support for the importance of maternal CWD transmission (Nalls et al., 2021). The distribution of PrP^{Sc} and/or CWD infectivity is widespread in CWD-infected hosts (Sigurdson et al., 2001; Fox et al., 2006). CWD prions are shed via decaying carcasses (Miller et al., 2004), blood, saliva, feces (Tamguney et al., 2009), urine (Haley et al., 2009), and antler velvet (Angers et al., 2009). The shedding of CWD prions in excreta occurs prior to the onset of clinical signs of disease, and due to the highly social nature of cervids, is thought to contribute to the horizontal transmission of CWD within a population (Tamguney et al., 2009). Additionally, CWD prions shed into the environment can bind to soil (Saunders et al., 2009; Kuznetsova et al., 2020). The interaction of prions with soil is influenced by the matrix containing the prions, prion strain, and soil type (Saunders et al., 2009). Soil-bound prions are resistant to natural weathering processes and remain highly infectious (Kuznetsova et al., 2020). Consistent with these observations, CWD naïve deer exposed to fomites from pens of CWD-infected deer subsequently developed CWD infection (Mathiason et al., 2009). Prions applied to plants can bind to the surface of the plant and remain infectious (Pritzkow et al., 2015). While maternal, horizontal, and environmental transmission of CWD prions likely exists in natural settings, the relative contribution of each to the dynamics of CWD infection in a population is unknown.

CWD incubation periods and clinical signs are variable. Incubation periods vary from 1–5 years (Bunk, 2004; Imran & Mahmood, 2011). CWD impacts animals ranging from 1 ½ to 20 years in age depending upon species (Sola et al., 2023), with an average disease onset of 3–5.5 years old (Williams & Young, 1992; Spraker et al., 2023). CWD is prevalent in almost twice as many males as females (Miller et al., 2008). Clinical signs of disease last 2 weeks to 8 months before fatality (Williams & Young, 1980). Commonly observed clinical signs include a rough, patchy pelage, posterior ataxia, weakness, lethargy, emaciation, and a reduced fear of humans (Williams & Young, 1982, 1992). Other neurological features involve fine head tremors, drooping heads, and listlessness (Imran & Mahmood, 2011). These clinical features may underlie the observed increased frequency of CWD-infected deer in vehicle collisions (Krumm et al., 2005), and predation, such as from mountain lions (Miller et al., 2008).

The neuropathology of CWD is characterized by spongiform degeneration, PrP^{Sc} deposition, and florid amyloid plaques (Haley et al., 2009). Spongiform degeneration is widely distributed bilaterally in the CNS (Williams & Young, 1992). Lesions are primarily present in the parasympathetic vagal nucleus, olfactory cortex, and hypothalamus (Williams & Young, 1980). Following CWD inoculation per os, PrP^{Sc} is first observed in the dorsal motor nucleus of the vagus, consistent with neuroinvasion via sympathetic innervation of the alimentary tract (Sigurdson et al., 2001). Immunohistochemical detection of PrP^{Sc} in the CNS is primarily located in the hippocampus, thalamus, cerebellar cortex, vagus nerve, obex, and spinal cord (Fox et al., 2006; Baeten et al., 2007). In the PNS, PrP^{Sc} is commonly detected in the adrenal medulla, tonsils, several lymph nodes, spleen, muscles, and the intestinal tract (Fox et al., 2006; Spraker et al., 2023). The morphology of PrP^{Sc} immunodetection can include, plaque-like, subpial, subependymal, granular, coarse, stellate, perivascular, linear, intraneuronal, and perineuronal deposition patterns (Sigurdson et al., 2001; Benestad et al., 2016; Orge et al., 2021). Amyloid plaques in the CNS are a common feature of CWD and, in one study, were observed in 62% of affected cervids (Bahmanyar et al., 1985). Amyloid plaques have a clustered morphology under Congo red birefringence and are often surrounded by small vacuoles (Bahmanyar et al., 1985).

Polymorphisms of cervid *PRNP* affect CWD disease pathogenesis. In mule deer, *PRNP* codons 20 and 225 influence CWD susceptibility (Jewell et al., 2005). In white-tailed deer, codons 95, 96, and 185 impact disease resistance (Wilson et al., 2009; Johnson et al., 2011). Over 14 additional *PRNP* polymorphisms have been recognized in white-tailed deer (Wilson et al., 2009; Chafin et al., 2020). Elk differ from other cervids in having a Glutamate at codon 226 instead of Glutamine (Cervenakova et al., 1997). Elk that are homozygous for leucine at codon 132 has an extended incubation period (White et al., 2010), are favored by natural selection (Monello et al., 2017), and have unique PrP^{Sc} properties (O'Rourke et al., 2007). However, these codon 132 polymorphisms appear to be specific to only some elk populations (Roh et al., 2022). There are 74 other known *PRNP* polymorphisms in elk, but none of these significantly affect CWD susceptibility (White et al., 2010). Finally, a study in gene-targeted mice suggests that S138N heterozygosity in reindeer provides resistance to the development of clinical signs for CWD but may result in animals silently spreading CWD (Arifin et al., 2023).

The host range of CWD prions is incompletely described (Spraker et al., 1997). CWD has been experimentally transmitted to many animals, including Reeves' muntjac deer, sheep, cattle, mink, ferrets, cats, hamsters, mice, meadow voles, bank voles, prairie voles, red-backed voles (Kurt & Sigurdson, 2016), goats (Williams & Young, 1992), pigs (Moore et al., 2017), and raccoons (Moore et al., 2019). Of these, several are sympatric to cervids, however, it is unclear if they are susceptible to CWD infection naturally. Of note, however, both pigs and sheep are susceptible to experimental CWD prion infection by per os or oral nasal routes (Moore et al., 2017; Cassmann et al., 2021).

The risk of transmission of CWD to humans is of growing concern. To date, there have been no known cases of human disease caused by CWD (Wang et al., 2021). Evidence conflicts as to the susceptibility of humans to CWD infection. In non-human primates, CWD was successfully transmitted to squirrel monkeys, but experimental transmission to cynomolgus macaques failed to produce clinical disease and none of the animals had evidence of a subclinical infection (Race et al., 2009). Several studies in transgenic mice overexpressing human PrP^{C} suggest there is a large species barrier for CWD transmission to humans (Race et al., 2009; Wilson et al., 2012), as CWD only rarely transmits after a prolonged incubation period (Wang et al., 2021; Tranulis & Tryland, 2023). How these results translate into assessing the relative risk of transmission to humans is unknown. While numerous studies have investigated human susceptibility to direct CWD infection, nothing is known about how the passage of CWD via an intermediate host species, which can alter the host range, affects the zoonotic potential of CWD (Bartz et al., 1998).

29.2.3 Bovine Spongiform Encephalopathy

Bovine spongiform encephalopathy (BSE), also known as mad cow disease (Marsh et al., 1991), is a deadly neurodegenerative disease in cattle (Wells et al., 1987). BSE has existed since 1981 in the UK (Table 29.1) (Wilesmith et al., 1988). By 1992, BSE had impacted 18.2% of all British cattle (Bradley & Matthews, 1992), ultimately infecting as many as three million cattle, and costing about $7.1 billion US dollars (Babelhadj et al., 2018). Twenty-nine countries have reported a total of over 190,000 confirmed cases of BSE (European Food Safety, 2023).

BSE is acquired primarily via oral transmission. BSE infection arises from the consumption of prion-infected bovine bonemeal and/or ruminant-derived meat supplements (Wilesmith et al., 1988), including the spleen, thymus, intestines, tonsils, spinal cord, and brain (Cunningham, 1991). In an attempt to control the spread of BSE, several bans against the inclusion of potentially contaminated animal products in feed have been implemented internationally, collectively resulting in a vast drop in BSE incidence rates (Oraby et al., 2016). While there is no evidence of horizontal transmission for BSE in cattle (Wilesmith et al., 1988), there is conflicting evidence suggesting maternal or vertical transmission may occur (Curnow & Hau, 1996; Castilla et al., 2005a). BSE infectivity was not detected in udders, milk (Castilla et al., 2005a), colostrum, amniotic fluid (Buschmann & Groschup, 2005), embryos (Wrathall et al., 2002), semen, or the placenta, as demonstrated by failed attempts to transmit disease from these tissues to mice (Bradley & Matthews, 1992). BSE origin remains uncertain but did not result from contact with scrapie-infected sheep (Wilesmith et al., 1988).

Clinical signs of BSE appear in cattle ranging from 2–17 years old (Bradley & Matthews, 1992), with an average age of 5 years old (Cranwell et al., 1988). The incubation period of BSE is between 2.5 and 8 years for infected cattle (Wilesmith et al., 1988). Clinical signs usually last 1–4 months long (Cranwell et al., 1988), but can occasionally last up to 14 months (Wilesmith et al., 1988). The clinical signs of the disease can include incoordination or gait abnormalities, awkward turning, stumbling, weakness, hind limb ataxia, weight loss, and tremors (Cranwell et al., 1988; Wilesmith et al., 1988). Common neurological signs and behavioral changes in BSE often involve hyperaesthesia, apprehension, nervousness, panic, and aggressive unmanageable behavior (Wells et al., 1987; Cranwell et al., 1988;

Wilesmith et al., 1988). BSE inevitably results in the death of the host (Cranwell et al., 1988).

Histopathological examination of the CNS reveals vacuolation, astrogliosis, and neurodegeneration (Wells et al., 1987). Intense vacuolation for BSE is found most frequently in the brain stem neuropil and medulla oblongata (Wells et al., 1987). Other lesions are commonly observed in the vestibular nuclei, solitary tract nucleus, red nucleus, vagus nerve dorsal motor nuclei (Wells et al., 1987), thalamus (Casalone et al., 2004), cerebellum, spinal cord, and sometimes in the hippocampus (Hope et al., 1988). Vacuolation patterns are bilaterally symmetrical, with single or multiple round vacuoles (Wells et al., 1987). Vacuolation is usually associated with PrP^{Sc} in BSE, although the latter is typically more widespread (Orge et al., 2021). PrP^{Sc} is primarily located in the brainstem, specifically the medulla oblongata, as well as the trigeminal ganglion (Orge et al., 2021), thoracic dorsal root ganglia, retina, and distal ileum (Buschmann & Groschup, 2005). PrP^{Sc} is also infrequently observed in bone marrow (Imran & Mahmood, 2011), muscles, and the cerebral cortex (Orge et al., 2021). PrP^{Sc} deposition can be fine granular or in aggregates, intraglial, or intraneuronal, stellate, linear, or perineuronal (Orge et al., 2021). Detection of scrapie-associated fibril-like structures commonly accompanies PrP^{Sc} (Hope et al., 1988). PrP^{Sc} amyloid plaques have been observed in about 5% of cases with classical BSE (Wells et al., 1991).

There are two recognized forms of atypical BSE, referred to as H-type and L-type strains (Buschmann et al., 2006). H-type BSE is characterized by a higher molecular mass of unglycosylated PrP^{Sc} than classical BSE (Biacabe et al., 2004; Buschmann et al., 2006), and an extra PrP^{Sc} band at about 14 kDa as determined by Western blot analysis after PK digestion (Biacabe et al., 2008). L-type BSE PrP^{Sc} has a lower molecular mass of unglycosylated PrP^{Sc} and less diglycosylated PrP^{Sc} compared to classical BSE (Casalone et al., 2004; Capobianco et al., 2007). Atypical BSE cases occur at the rate of about 1 in a million cattle (Biacabe et al., 2008). Atypical BSE differs from classical BSE in clinical signs (Biacabe et al., 2004), biological properties (Capobianco et al., 2007), interspecies transmission potential (Buschmann et al., 2006; Beringue et al., 2008a), and histopathological lesions (Aguilar-Calvo et al., 2015). Additionally, amyloid plaques are easily detected in L-type BSE, also called bovine amyloidotic spongiform encephalopathy (BASE) (Casalone et al., 2004). These plaques are round, dense, unicentric, or uncommonly multicentric, with a dark periphery and pale center (Orge et al., 2021). Furthermore, PrP^{Sc} from atypical BSE accumulates in different neuroanatomical regions (Casalone et al., 2004). Atypical BSE is of unknown origin but is hypothesized to be sporadic (Biacabe et al., 2004; Casalone et al., 2004). It is primarily found in aged cattle (Biacabe et al., 2004, Casalone et al., 2004), which may have a genetic predisposition (Heaton et al., 2008). Interestingly, both L-type BSE and H-type BSE produced classical BSE when passaged through mice, which collectively suggests that the origin of classical BSE was atypical BSE (Capobianco et al., 2007; Baron et al., 2011).

Genetic predisposition largely impacts BSE susceptibility as determined by theoretical models (Ferguson et al., 1997), but BSE is not determined by simple Mendelian inheritance (Wilesmith et al., 1988). Cattle with a 12-base pair deletion in the *PRNP* gene are at a significantly higher risk for contracting classical BSE (Sander et al., 2004). Likewise, cattle with a 23 base pair deletion are more susceptible to BSE (Sander et al., 2004). However, neither of these two deletions is associated with susceptibility in atypical BSE (Brunelle et al., 2007). Instead, animals with H-type BSE have the heterozygous E211K germline polymorphism, which is found at a prevalence rate of less than one in two thousand cattle (Heaton et al., 2008).

BSE directly led to four other prion diseases by oral transmission. In animals, these include exotic ungulate encephalopathy (EUE), feline spongiform encephalopathy (FSE), and a TSE in French non-human primates (NHPs). BSE has also been transmitted to humans, known formally as variant Creutzfeldt-Jakob Disease (vCJD). Please see individual sections for specific details.

29.2.4 Exotic Ungulate Encephalopathy

Exotic ungulate encephalopathy (EUE) was first identified in a nyala in 1986 (Table 29.1) (Jeffrey & Wells, 1988). Since then, EUE has been recognized in one gemsbok in 1987 (Jeffrey & Wells, 1988), six elands since 1989, two Arabian oryx since 1989, at least six greater kudu from 1989 to 1992, one scimitar-horned oryx in 1993 (Kirkwood & Cunningham, 1994), one bison and two ankole cattle (Imran & Mahmood, 2011). Ungulates ranged in age from 19 to 48 months old (Kirkwood & Cunningham, 1999).

The most likely origin for EUE is BSE. EUE has the same geographical distribution as BSE and most cases of EUE were exposed to ruminant-derived protein similar to cattle with BSE (Kirkwood & Cunningham, 1994). Likewise, EUE produced similar results to BSE when transmitted to mouse bioassay, such as very similar vacuolation patterns and incubation periods in multiple mouse types (Bruce et al., 1997). However, EUE in greater kudu has a different epidemiology compared to BSE in cattle (Kirkwood & Cunningham, 1994). Ultimately, EUE differs from BSE in clinical presentation, neuropathological distribution, and severity of vacuolation in specific nuclei (Wells & McGill, 1992), enough to classify EUE as a separate prion disease from BSE.

EUE results primarily from oral transmission, but other routes may occur. Oral transmission resulted when ungulates ate food contaminated with BSE (Wilesmith et al., 1988; Kirkwood et al., 1990). While many of these animals consumed contaminated food before the 1988 UK ban against ruminant proteins and the 1991 updated ban against specified offals, it is possible that other ungulates contracted EUE through BSE-infected food or feed supplements that slipped past this ban (Kirkwood & Cunningham, 1994, 1999). Evidence suggests that maternal transmission of EUE in greater kudu occurred (Kirkwood et al., 1992). Furthermore, lateral or environmental transmission is possible for EUE, as several cases have been reported where ungulates were not fed ruminant-derived protein (Kirkwood & Cunningham, 1994). Alternatively, uncontaminated food bought after the ban became subsequently contaminated from the environment and was consumed by ungulates leading to food-borne infection. Ultimately, despite a few outstanding cases of EUE, eliminating potentially contaminated feeds through bans and increasing awareness at zoos effectively prevented new cases. The last case of EUE was reported in 1996 (Kirkwood & Cunningham, 2006).

EUE clinical presentation is variable. The most frequently observed clinical signs are hindlimb ataxia and high stepping, hypermetria, circling, collapsing, recumbency, fine muscle tremors, abnormal posture, depression, dullness, unresponsive staring, mutilation, frequent nibbling, micturition, constipation, emaciation, appetite loss, drooling, frequent nose twitching (Kirkwood et al., 1990), nasal discharge, and hyperaesthesia (Kirkwood & Cunningham, 1994). Disease duration lasted 2 weeks on average before each ungulate was euthanized (Wells & McGill, 1992). Since the time of prion exposure is unknown, incubation periods are estimated for ungulates (Kirkwood & Cunningham, 1999). These incubation periods vary greatly between species, ranging from 3 to 48 months (Wilesmith et al., 1988, Williams et al., 2000), with a 30-month average incubation period (Kirkwood & Cunningham, 2006).

EUE is commonly diagnosed from postmortem histopathology, the presence of fibrils by electron microscopy, and PrP detection by immunoblotting (Kirkwood et al., 1992). Histopathological changes demonstrated bilaterally symmetric neuronal vacuolation, particularly in neurons, neuropil (Jeffrey & Wells, 1988), perikarya, and dendrites (Kirkwood & Cunningham, 2006). In the brains of animals infected with EUE, large vacuolation is most frequently observed in the white matter, thalamic gray matter, and brainstem, primarily in the medulla oblongata (Jeffrey & Wells, 1988). Vacuoles are empty, round, single, or many, and variable in size but are usually small (Jeffrey & Wells, 1988). Perivascular focal plaques are also sometimes observed in the cerebellum (Jeffrey & Wells, 1988).

Many tissues from ungulates with EUE contain prions. Several tissues from greater kudu infected with prions were injected into mice and had over a 50% attack rate, including the spinal cord, rostral cerebrum, brainstem, distal ileum, and retropharyngeal and visceral lymph nodes (Cunningham et al., 2004). Tissues from the skin, salivary glands, conjunctiva, and spleen or lungs can also contain infectious prions as determined by bioassay in mice (Cunningham et al., 2004).

29.2.5 Feline Spongiform Encephalopathy

Feline spongiform encephalopathy (FSE) was first reported in a Siamese cat from the UK in 1990 (Table 29.1) (Aldhous, 1990). FSE spread from the UK, mostly through exports, to Australia, Norway, France, Italy (Zanusso et al., 1998), Liechtenstein (Ryder et al., 2001), Switzerland, Ireland (2003), Portugal (Silva et al., 2006), and Germany (Eiden et al., 2010). FSE resulted in at least 96 confirmed cases in domestic cats (Zanusso et al., 1998; 2003; Silva et al., 2006). In addition to domestic cats, FSE has been recognized in several exotic captive feline species including 15 cheetahs (Imran & Mahmood, 2011), 3 pumas, 3 ocelots, 3 tigers, 5 lions, 1 Asiatic golden cat, and an Asian leopard cat (Eiden et al., 2010).

FSE shares many characteristics with BSE and is different from scrapie (Lezmi et al., 2006; Hilbe et al., 2009; Eiden et al., 2010). Cats inoculated intracerebrally with sheep scrapie did not develop neurologic disease nor spongiform encephalopathy (Hamir et al., 2002), suggesting scrapie is not the origin of FSE. Remarkably, data are not available on whether felines are experimentally susceptible to BSE transmission. FSE overlaps geographically with BSE, and both diseases result from oral transmission of prion-contaminated ruminant-derived protein (Kirkwood & Cunningham, 1994). Furthermore, when compared to BSE in cattle, FSE produces similar incubation periods (2003), and clinical signs (Leggett et al., 1990), as well as PrP^{Sc} deposition patterns (Ryder et al., 2001), distribution (Hilbe et al., 2009), proteinase cleavage site, and glycoform profile (Eiden et al., 2010). These biological and molecular properties are maintained upon transmission to mice, where FSE is similar to BSE in transmission efficiency, incubation periods, lesion profile (Bruce et al., 1997), PrP distribution (Lezmi et al., 2006), and glycoform patterns (Collinge et al., 1996). Likewise, in transgenic mice expressing ovine PrP^{C}, vacuolation patterns, PrP distribution, and florid plaques are similar between BSE and FSE (Bencsik et al., 2009). This evidence collectively suggests that FSE has a BSE origin. Other findings, however, demonstrate subtle significant differences between BSE and FSE in vacuolation (Wyatt et al., 1991), glycoform ratios, and incubation periods in transgenic mice (Eiden et al., 2010), enough to classify FSE as a novel prion disease.

Most felines contracted FSE orally through prion-infected food (Aldhous, 1990), but other forms of transmission have been implicated. As a result of BSE-related oral transmission of prion disease, in 1990 the UK banned specified bovine offals from pet food (1996). However, some felines were diagnosed with FSE after the ban (2003). Although the bans against offals took years to implement, they were successful. By the turn of the twenty-first century, the FSE epidemic had slowed down to single or no reported cases each year (Kirkwood & Cunningham, 2006). There have been findings suggesting that FSE can be transmitted maternally (Bencsik et al., 2009), but no evidence to suggest horizontal transmission (Kirkwood & Cunningham, 2006). In 1998, a unique case of prion disease was identified in a domestic cat, with a different phenotype from typical FSE, and this case correlated with its owner's death by sporadic Creutzfeldt-Jakob Disease (CJD) (Zanusso et al., 1998). Although this case is of unknown origin (Zanusso et al., 1998), CJD has been experimentally transmitted to cats (Hillier & Salmon, 2000), suggesting other possible modes of transmission.

FSE results in a clinical course lasting from several weeks to more than 5 months before death, with a 3-month average (1996). FSE impacts cats ranging from 1 to 14 years old with a mean of 6–7 years (Pearson et al., 1993; 1996). Clinical signs can include circling, incoordination, dysmetria, hypermetria, gait abnormalities, ataxia, muscle spasms, altered grooming, hypersalivation, bruxism, weight loss, or constipation (Wyatt et al., 1991). Common neurological signs and temperament changes include an abnormal head posture, head tremors, hyperaesthesia, vision impairments, anxiety, apprehension, aggression, and attacking (Wyatt et al., 1991). The average incubation period for domestic cats is 5 years (2003), whereas zoo felines have longer incubation periods ranging from 4.5 to 8 years (Kirkwood et al., 1995).

FSE diagnosis is based upon postmortem histological examination of vacuolation, gliosis, axonal degeneration (Wyatt et al., 1991), immunohistochemical detection of PrP^{Sc} (Ryder et al., 2001), and identification of fibrils via electron microscopy (Pearson et al., 1992). FSE vacuolation is most frequently observed in the gray matter neuropil and neuronal soma of the cerebral cortex, corpus striatum, thalamus, cerebellum, midbrain, medulla oblongata, and spinal cord (Wyatt et al., 1991; Eiden et al., 2010). Vacuoles are variable in size and can be plural or singular, discrete (Wyatt et al., 1991), round, cytoplasmic, and/or diffuse (Hilbe et al., 2009). FSE is accompanied by both astrogliosis and increased microglia (Wyatt et al., 1991). PrP^{Sc} from FSE is identified by immunoblotting (Pearson et al., 1992), and is typically found in gray matter neuropil and neuronal perikarya throughout the CNS, such as in the cerebral cortex, granular and molecular layers of the cerebellum, brainstem, spinal cord, and retina (Hilbe et al., 2009, Eiden et al., 2010). PrP^{Sc} is widely distributed throughout the body and is detected in the PNS, ENS, lymphoreticular system, adrenal gland, and kidneys (Hilbe et al., 2009, Eiden et al., 2010). PrP^{Sc} deposition patterns are variable and can be diffuse, intense, particulate, globular, intracytoplasmic, linear (Eiden et al., 2010), or punctate, but are almost always granular (Ryder et al., 2001; Hilbe et al., 2009).

Genetics brings further insight into FSE origin and feline susceptibility to prion disease. Although FSE cases have not revealed a specific disease-associated *PRNP* polymorphism, feline PrP^C differs in some important amino acid residues compared to other species (Stewart et al., 2012). 15 *PRNP* polymorphisms have been identified in cats, and codon 153 likely impacts FSE resistance (Kim et al., 2020). Cat prion proteins share 93.4–97.3% similarity with tigers (Wu et al., 2006a), 96.7% with lions (Wu et al., 2006b), 93.5–97.6% with dogs, 94.4–95.2% with mink (Lysek et al., 2004), 91.9% with goats and sheep, 89.5% with cattle and nyala, 85.5% with dromedary camels, and 84.7% with humans (Wopfner et al., 1999). A potential explanation for why FSE is like BSE is because felines have a polymorphism at codon 187 comparable to a bovine mutation at codon 186, which is likely involved with intermolecular interactions and transmission barrier susceptibility (Wopfner et al., 1999). Canids are resistant to prions (2003), differing from felids at multiple important codons that are involved in prion conversion and result in changes in PrP^C charge distribution (Stewart et al., 2012), amyloid propensity, and hydrogen bond number, length, and strength (Kim et al., 2020), creating a substantial transmission barrier (Lysek et al., 2004). These include canid codons 163, 181, and 189 corresponding to felid codons 166, 184, and 192, respectively (Lysek et al., 2004; Stewart et al., 2012; Kim et al., 2020). Cat PrP^C is less stable than dog PrP^C, appears to have weaker hydrogen bonds, and is more likely to convert to misfolded PrP^{Sc} (Kim et al., 2020), which may explain why felines have contracted prions, but canids have not.

29.2.6 Prion Disease in Non-Human Primates

A prion disease in French non-human primates (NHPs) was first diagnosed in 1996 (Table 29.1) (Bons et al., 1996). Prion disease in NHPs has been recognized in at least one rhesus macaque (Bons et al., 1996), two Mayotte brown lemurs (Bons et al., 1997), one white fronted brown lemur, and a mongoose lemur (Bons et al., 1999). NHPs with prion disease likely contracted the disease by eating BSE-contaminated cattle meat food products (Bons et al., 1996, 1997, 1999). Lesion profiles and PrP^{Sc} distribution of lemurs naturally infected with prions are similar to lemurs experimentally infected with BSE (Bons et al., 1999), suggesting that the lemurs with prion disease were naturally infected with BSE. While both diseases are initially characterized by behavioral changes (Bons, 1996), prion disease in naturally infected NHPs differs in clinical course from primates experimentally infected with BSE (Baker et al., 1996). Clinical signs of natural prion disease in NHPs last up to a year and include rapid hair loss, lethargy, recumbency, increased aging, appetite loss, and mood changes, including aggressiveness (Bons et al., 1997), hiding and isolation (Bons et al., 1996). Whereas the clinical course for primates experimentally infected with BSE is much shorter, and clinical signs primarily involve hindlimb ataxia, balancing issues, an abnormal gait, and the use of their hands to sit up (Baker et al., 1996).

As in other prion diseases, diagnosis of prion disease in NHPs is confirmed upon postmortem investigation of vacuolation, astrogliosis, and protein aggregates, particularly PrP^{Sc} deposits (Bons et al., 1999). Spongiform changes, accompanied by PrP^{Sc} deposition, were observed in gray matter neurons and the deeper layers of the cerebral cortex, corpus callosum, striatum, nucleus accumbens, thalamus, cerebellum (Bons, 1996), reticular formation, and spinal cord of NHPs with prion disease (Bons et al., 1996, 1999). Vacuoles are generally large, but variable in size, and enlarged nerve-cell processes are common (Bons, 1996; Bons et al., 1996). Astrogliosis was observed in the cerebral cortex, thalamus (Bons et al., 1997), pia mater, and spinal cord (Bons et al., 1999). Further immunohistochemistry (IHC) revealed finely particulate PrP^{Sc} in the spleen, lymph nodes, tonsils, and throughout the gastrointestinal tract, from the pharynx (Bons et al., 1997), to the colon (Bons et al., 1999). The neuropil of the cerebral cortex also contains β-amyloid protein deposition (Bons, 1996, Bons et al., 1996). Additionally, several neurons of the cerebral cortex, superior colliculus, thalamus, brain stem, and spinal cord contained aggregated tau in NHPs with prion disease (Bons et al., 1999).

29.2.7 Transmissible Mink Encephalopathy

Transmissible mink encephalopathy (TME) was first identified in ranch-raised mink in 1947 in Wisconsin and Minnesota (Table 29.1) (Hartsough & Burger, 1965). Additionally, TME has been identified in Idaho, Canada, Finland, Germany, and Russia (Marsh & Bessen, 1993). TME has not been recognized in wild animals (Eckroade et al., 1979). The last TME outbreak occurred in 1985 in Stetsonville, Wisconsin (Marsh et al., 1991). TME incubation periods range from 4 to 12 months, and the disease is clinically characterized by self-mutilation that includes tail-biting, somnolence, weakness, incoordination, an unusual gait, circling, and ataxia (Burger & Hartsough, 1965; Hartsough & Burger, 1965; Marsh & Hanson, 1969). TME does not transmit vertically and TME outbreaks are self-limiting, therefore, mink are dead-end hosts (Hartsough & Burger, 1965). However, horizontal transmission to other mink has occasionally occurred through cannibalism (Burger & Hartsough, 1965).

The neuropathology of TME is characterized by a spongiform encephalopathy that appears prior to the onset of clinical signs for the disease and becomes widely distributed at terminal disease (Eckroade et al., 1979). Lesions in the CNS are most frequently observed in gray matter, and can also include the cerebral cortex, hippocampus, thalamus, hypothalamus, corpus striatum, cerebellar cortex, and brainstem (Burger & Hartsough, 1965; Hartsough & Burger, 1965; Marsh & Hanson, 1969; Eckroade et al., 1979). TME neuropathology also includes astrogliosis, microgliosis, and neuronal degeneration throughout the CNS (Hartsough & Burger, 1965, Marsh & Hanson, 1969, Eckroade et al., 1979). The distribution of TME infectivity is widespread in mink and, in addition to the CNS and PNS, has been detected in lymph nodes, spleen, thymus, salivary glands, kidneys, liver, and intestines (Marsh & Hanson, 1969; Orge et al., 2021).

The etiology of TME was difficult to identify. Epidemiological evidence indicated that TME is a food-borne disease and that mink are dead-end hosts (Hartsough & Burger, 1965). It was originally hypothesized that TME originated from feed contaminated with sheep scrapie as scrapie is present in the affected countries and may have been fed to captive mink (Hanson et al., 1971). Inconsistent with this hypothesis, experimental inoculation of mink with scrapie isolates could cause disease in mink only when directly inoculated into the CNS and then, only from a subset of scrapie strains tested (Hanson et al., 1971). Additionally, the incubation periods observed in mink infected with scrapie are longer than those of mink naturally infected with TME, and the resultant pathology differs from TME in mink (Hanson et al., 1971). Finally, in the Stetsonville outbreak of TME, it was reported that these mink were not fed sheep, but their diet did include downer cattle (Marsh et al., 1991). Downer cattle are unable to rise without assistance which can be caused by metabolic, toxic, or neurological disorders. Overall, this evidence led to the hypothesis that an unrecognized prion disease of cattle is the source of TME, and that TME does not originate from scrapie (Marsh et al., 1991). Strong evidence to support this hypothesis has come from studies where transmission of cattle passaged TME into transgenic mice that express ovine PrP^{C} shares incubation period, neuropathological features, and biochemical properties of PrP^{Sc} from L-type BSE and, importantly, differs from classical or H-type BSE (Baron et al., 2007). Since L-type BSE is likely a spontaneous prion disease (Casalone et al., 2004), has been identified in North America (Dudas et al., 2010), and shares similar properties with cattle passaged TME, this suggests that the feeding of L-type BSE-infected cattle to mink resulted in the emergence of TME (Baron et al., 2007).

29.2.8 Camel Prion Disease

Camel prion disease (CPrD) was recently identified in 2018 in 3.1% of dromedary camels (*Camelus dromedarius*) from the Ouargla slaughterhouse located in southeastern Algeria (Table 29.1) (Babelhadj et al., 2018). CPrD has also spread to at least one dromedary camel in Tataouine, Tunisia (Babelhadj, 2019). Camels are becoming a sustainable livestock in parts of Asia, the Middle East, and Northern and Eastern Africa (Babelhadj et al., 2018), so understanding this disease will be important for camel husbandry and human safety. Although The World Organization for Animal Health has decided there has not yet been enough evidence to consider CPrD an emerging disease (Babelhadj, 2019), there is concern that CPrD will spread to other nations. For example, CPrD could enter the UK through imports (Horigan et al., 2020), increasing the range of CPrD. However, if CPrD surveillance testing were to be performed in the UK for camel dairy products, the risk of CPrD entering the UK would be decreased to negligible (Horigan et al., 2020).

The etiology of CPrD is unknown (Babelhadj et al., 2018). The distribution of PrP^{Sc} in camels infected with CPrD can include lymphoreticular tissue, a prerequisite for environmental prion shedding (Babelhadj, 2019). CPrD origin is unknown but may have developed from meat and bone meal imports possibly contaminated with BSE, which rapidly increased to 30,000 tons after the ban on ruminant proteins made by the UK in 1988 (Babelhadj et al., 2018). Inconsistent with this hypothesis is that dromedary camels are usually not raised on feedstuff (Babelhadj et al., 2018). A second possibility is the zoonotic transmission of scrapie from sheep or goats to camels in common pastures, but a lack of effective surveillance makes this difficult to determine (Babelhadj et al., 2018).

Clinical signs of CPrD began with irritability and appetite loss that developed over 3–8 months to include: abnormal behaviors, weight loss, hyperreactivity, aggressiveness, tremors, head nodding, uncertain walking, occasional falls, and difficulty standing up (Babelhadj et al., 2018). CPrD resulting in hind limb ataxia was the eventual cause of camel death (Babelhadj et al., 2018). CPrD camel brains demonstrated gliosis, neuronal loss, spongiform degeneration, and vacuolation in the neuropil and neuronal bodies (Babelhadj et al., 2018). CPrD neurodegeneration varied but was primarily found in gray matter including:the thalamus, striatum, midbrain, pons, and the medulla oblongata (Babelhadj et al., 2018). CPrD was proven to be statistically different from BSE and scrapie through Western blot comparison with PrP^{Sc} epitope-specific antibodies and examination of glycosylation patterns (Babelhadj et al., 2018).

Genetics plays an important role in determining zoonotic potential for prion diseases and comparing cameline genetics to other known susceptible species may provide evidence concerning how camels became susceptible to prions. While CPrD is the only known prion disease to infect camels, dromedary cameline PrP amino acid sequences share 98.5–99.5% similarity with bactrian camels, 86.0–94.5% with sheep, 83.8–93.7% with cattle, 89.2–93.3% with swine, 85.2–90.4% with goats, 87.5–92.5% with deer (Tahmoorespur & Jelokhani Niaraki, 2013), 90.0% with mink (Kaluz et al., 1997), 87.2% with cats, and have 88–88.5% homology with human PrP (Wopfner et al., 1999; Xu et al., 2012). Camel *PRNP* variance to other ruminant species includes the following genotypic codon differences: A16V, M17T, N120S, R176K, I215V, S234Y, Y237S, and Q239G (Xu et al., 2012). Most dromedaries share 100% similarity in amino acids within species (Tahmoorespur & Jelokhani Niaraki, 2013). However, some rarely exhibit two non-synonymous polymorphic variances at Gly69Ser or Gly134Glu (Zoubeyda et al., 2020), but the potential impact of these low-frequency genetic differences is unknown.

29.3 Human Prion Diseases

29.3.1 Creutzfeldt-Jakob Disease

Creutzfeldt-Jakob disease (CJD) was named after the two scientists who first reported this human disease in 1920 (Creutzfeldt, 1920; Jakob, 1920). CJD occurs at an incidence, prevalence, and death rate of 1–2 cases per million people annually, but these rates are higher in some specific localities (Malmgren et al., 1979; 2021; Unit, 2021). This makes CJD the most frequent prion disease found in humans (Holman et al., 2010). CJD cases are distributed on all continents worldwide (Masters et al., 1979b), without differences in incidence rates between rural and urban regions (Malmgren et al., 1979). The age-adjusted incidence and mortality rates are slightly higher for males than females in age groups above 60 years old (Holman et al., 2010), and the incidence rates are generally similar across ethnicities worldwide (Malmgren et al., 1979). However, in the United States, CJD incidence rates are over 2 ½ times greater in Caucasians than in other groups (Holman et al., 2010). Another major exception is Libyan Jews, who have incidence rates over 100 times greater than other ethnicities (Hsiao et al., 1991). Genetic predisposition is likely in Libyan Jews, and familial CJD accounts for some of those cases (Hsiao et al., 1991). Ultimately, CJD can be categorized into four major types depending upon origin: sporadic, familial, iatrogenic, or variant.

Sporadic CJD (sCJD) was first identified in 1920 (Table 29.1) (Creutzfeldt, 1920, Jakob, 1920). Roughly 85–90% of all CJD cases are sporadic (Piconi et al., 2019). Patients vary greatly in age for sCJD, ranging from 14 to 97 years old (Berman et al., 1988), with a median of

66–68 years old at death (Holman et al., 2010; Sequeira et al., 2022). Sporadic CJD is of an unknown etiology and may result from a somatic mutation, or the spontaneous conversion of PrP^{C} to PrP^{Sc} (Imran & Mahmood, 2011). Strains of sCJD have been identified and are classified based primarily on differences in clinical manifestations and genotype (Parchi et al., 1996, 1999a). Polymorphisms in PrP codon 129 influence sCJD disease susceptibility (Parchi et al., 1996). Based upon codon 129 polymorphisms and the PrP^{Sc} protease-resistant fragments observed by Western blot analysis, sCJD is classified into six different subtypes (Parchi et al., 1999b). The frequencies of these codon 129 polymorphisms in cases of sCJD are 40–70% for the MM1 or MV1 subtype, 14.8–16% for the VV2 subtype, 8–13.6% for the MV2 subtype, 1–4.1% for the VV1 subtype, and 2–8.5% for the MM2 subtype (Parchi et al., 1999b; Parchi & Saverioni, 2012; Belay & Bartz, 2014). The latter can be further characterized into two subtypes of equal frequencies, referred to as the MM2-thalamic and MM2-cortical subtypes, based respectively upon whether the disease primarily affects the thalamus, or if sCJD results in dementia and cortical pathology (Parchi et al., 1999b; Parchi & Saverioni, 2012).

Two glycoforms of PrP^{Sc}, referred to as types 1 and 2, result from PK digestion of PrP^{Sc} from sCJD during Western blotting (Parchi et al., 1996). Type 1 produces a 21 kDa unglycosylated PrP^{Sc} isoform from PK digestion and type 2 produces a 19 kDa unglycosylated PrP^{Sc} isoform (Cali et al., 2009). Additionally, Type 2 often has an additional 18 kDa PrP^{Sc} polypeptide as determined by Western blot analysis (Cali et al., 2009). Interestingly, up to 44% of patients harbor both type 1 and type 2 sCJD PrP^{Sc} (Parchi & Saverioni, 2012). These type 1 and 2 mixed PrP^{Sc} cases of sCJD have clinical signs and histopathology that are intermediary between type 1 and type 2 and should be considered a distinct type of CJD (Cali et al., 2009). In addition to *PRNP* genotype, the variation observed in clinical pathology for sCJD is likely due to a continuum of prion strains that differ from each other in properties such as conformational stability, conversion efficiency, and protease-sensitivity (Kim et al., 2012).

Familial CJD (fCJD) was first identified in 1924 (Masters et al., 1979a), and is an inheritable prion disease caused by mutations in the *PRNP* gene (Table 29.1) (Kovacs et al., 2005). Roughly 10–15% of CJD cases are familial (Hillier & Salmon, 2000). fCJD results from autosomal dominant transmission (Masters et al., 1979a). Codon E200K, where glutamate is substituted for lysine, is the most common mutation found in *PRNP* that impacts fCJD susceptibility (Heaton et al., 2008). Another familial form results from an asparagine, instead of aspartic acid, at polymorphic codon 178 (Brown et al., 1991). Over a dozen other *PRNP* mutations impacting fCJD phenotype have been identified, including codons 105, 129, 148, 171, 180, 183, 188, 196, 203, 208, 210, 211 (Kovacs et al., 2005), and 232 (Gambetti et al., 2003). Finally, some cases of fCJD appear to be due to insertions of additional octapeptide coding repeats in the open reading frame of *PRNP* (Brown et al., 1991). The average age for cases of fCJD is 49.5–57 years old (Unit, 2021; Chu et al., 2022), with a range of 20–98 years old (Masters et al., 1979a). Siblings have a 41% incidence rate for fCJD, whereas the rate reduces to 12% in cousins (Masters et al., 1979a). Females have a higher incidence rate than males for fCJD (Kovacs et al., 2005), but no maternal lineage has been observed (Masters et al., 1979a).

Iatrogenic CJD (iCJD) was first discovered in 1974 (Table 29.1), and less than 1% of CJD cases are iatrogenic (Blossom et al., 2007). Cases of iCJD have been associated with corneal transplantation, depth electrodes, neurosurgical equipment (Masters et al., 1979b), dura mater grafts (Blossom et al., 2007), growth hormones and other human pituitary hormones (Abrams et al., 2011). Patients with iCJD from dura mater grafts range in age from 10 to 81 years old (Sakai et al., 2020), with an average of 56 years (Ae et al., 2018). iCJD incubation periods vary greatly, dependent upon codon 129 genotype and transmission route, lasting from 1 year to over 40 years for patients with iCJD contracted from dura mater grafts (Rudge et al., 2015; Ae et al., 2018). To prevent hormonal iCJD cases, hormones began to be purified in 1977 through size exclusion chromatography, and patients have not contracted CJD from hormones since that date (Abrams et al., 2011). Similarly, a cluster of dura mater graft iCJD cases in Japan was traced to practices of a commercial supplier (Ae et al., 2018), but in 1987 product revisions were made to reduce risks of iCJD transmission (Blossom et al., 2007). Ultimately, all tissues from prion-infected humans and animals should be assumed to be potentially infectious, not just the CNS (Traub et al., 1975).

Although prions are resistant to many ordinary sterilization techniques, they can be inactivated (Hunter, 1979). Sodium hypochlorite can inactivate prions after an hour; however, it corrodes metals, tissue, and fabrics (Brown et al., 1984). Thus, steam autoclaving for an hour at 132 °C wherever possible is the preferred means for sterilizing against prions (Taylor, 1986). Surgical tools should be soaked in distilled water before and during autoclaving to prevent corrosion (Traub et al., 1975). 1 N sodium hydroxide can inactivate prions after an hour is less likely to corrode metals, plastics, or fabrics, and is a viable decontamination option (Brown et al., 1984). In summary, any instruments or equipment that have come into direct contact with blood or tissues from demented patients should either be incinerated (Traub et al., 1975), autoclaved, or washed with .5% sodium hypochlorite or 1 N sodium hydroxide for an hour to be adequately sterilized (Brown et al., 1984).

Variant CJD (vCJD) was first diagnosed in 1996 (Table 29.1) (Collinge et al., 1996; Will et al., 1996). As of

April 2021, there have been a total of 232 cases of vCJD reported worldwide, comprising the UK, Ireland, France, Spain, Portugal, the Netherlands, Italy, Saudi Arabia, Canada, the United States, Japan, and Taiwan (2021). vCJD impacts people of younger ages compared to sCJD, ranging from 11 to 74 years old (Barbot et al., 2010), with a median age of 28 years at death (Unit, 2021). Incubation periods for vCJD are 5–10 years (Will et al., 1996). vCJD differs in clinical signs and neuropathology from other forms of CJD (Will et al., 1996; Zeidler et al., 1997). Interestingly, vCJD is lymphotropic whereas sCJD is not (Beringue et al., 2008b). vCJD also produces a unique Western blot banding pattern for PK-digested PrP^{Sc} (Collinge et al., 1996). Patients with vCJD are generally methionine homozygous at codon 129 (Collinge et al., 1996).

Interspecies transmission of BSE appears to be the origin of vCJD in humans (Padilla et al., 2011). This origin was verified when molecular and biological transmission properties were similar between BSE and vCJD, and distinct from scrapie and sCJD, suggesting the same prion strain causes BSE and vCJD (Bruce et al., 1997; Hill et al., 1997). Additionally, BSE has been transmitted to several different lines of transgenic mice expressing human PrP (Hill et al., 1997; Beringue et al., 2008a; Padilla et al., 2011). BSE also converts human PrP^C into PrP^{Sc} in both human brain homogenate and humanized transgenic mouse brain homogenate using PMCA (Barria et al., 2014). These multiple converging lines of evidence suggest that vCJD is the first known zoonotic transmission of prions to humans. Importantly, vCJD can be transmitted horizontally between humans via contaminated blood-derived products (Barria et al., 2018). This is in contrast to sCJD and likely reflects the larger involvement of prion replication in spleen and lymph nodes in vCJD compared to sCJD. As there is an estimated prevalence as high as 1 in 2000 for vCJD asymptomatic carriers in the UK alone (Barria et al., 2018), it is important to inform donor recipients of the risks associated with potentially contracting CJD (Wilson & Ricketts, 2006).

CJD can be transmitted experimentally, iatrogenically, and possibly through other routes. CJD has been experimentally transmitted to mice, Guinea pigs, hamsters, cats, mink, ferrets, goats, and at least 16 different primate species (Gajdusek, 1977; Hillier & Salmon, 2000). There has been no evidence to suggest that CJD can transmit vertically (Masters et al., 1979a; Bryan, 1984). CJD also does not spread via airborne transmission (Bryan, 1984). CJD generally does not transmit horizontally through non-invasive bodily contact (Masters et al., 1979a, Bryan, 1984). However, on several independent occasions, sporadic CJD cases of unknown origin have been identified in closely associated people with no consanguinity, suggesting other possible modes for transmission are possible (Hillier & Salmon, 2000).

CJD clinical course is variable between individuals (Bernoulli et al., 1979), particularly between each CJD type. However, most patients' clinical signs are assigned into a few larger shared groupings, including motor impairments, behavioral abnormalities, involuntary movements, autonomic dysfunctions, personality changes, sensory disturbances, and higher cortical function impairments (Bernoulli et al., 1979; Ford et al., 2019). Common motor dysfunctions include incoordination, gait abnormalities, apraxia (Bernoulli et al., 1979; Parchi et al., 1996), and difficulty performing simple tasks (Ford et al., 2019). Posture is often abnormal, and muscles can have rigidity, weakness, atrophy, bradykinesia (Sequeira et al., 2022), ataxia, or even paralysis (Bernoulli et al., 1979). CJD commonly exhibits choreoathetosis, manifested as myoclonus, dyskinesia, tremors, hyperreflexia, and occasionally seizures or syncope (Bernoulli et al., 1979). Several patients develop autonomic disturbances, such as asthenia, somnolence, sleep disorders, dysphagia, respiratory problems, and/or incontinence (Bernoulli et al., 1979, Ford et al., 2019).

Neurological symptom progression is gradual in most patients with CJD, but 12–20% can develop rapidly (Bernoulli et al., 1979; Brown & Cathala, 1979). The moods of patients can shift to restlessness, irritability, anxiety, aggressiveness, nervousness, paranoia, deceitfulness, or depression (Bernoulli et al., 1979). Sensory disturbances can include pain, headaches, hyperesthesia, paresthesia, vertigo, confusion, comatose, psychosis, or hallucinations (Bernoulli et al., 1979; Parchi et al., 1999b). Visual impairments, such as nystagmus, diplopia (Parchi et al., 1996), or partial to complete blindness are occasionally identified (Bernoulli et al., 1979). Speech detriments have been observed, from dysarthria to slurred speech, dysphasia, and ultimately akinetic mutism (Bernoulli et al., 1979, Ford et al., 2019). Furthermore, there is generally an accelerated mental deterioration, as manifested through a major inability to analyze, dyscalculia, and typically severe progressive dementia (Brown & Cathala, 1979; Baldwin & Correll, 2019). Finally, pathophysiology suggests the involvement of the following brain regions: the pyramidal tracts, the extrapyramidal system, and the cerebellar pathways (Masters et al. 1979a, b). Death is inevitable for CJD, generally within a year from the onset of clinical symptoms, however, there have been some exceptions with longer durations lasting up to 11 years (Brown & Cathala, 1979; Unit, 2021). No effective treatments or prophylaxis for CJD exist (Ford et al., 2019).

Identifying biomarkers that are sensitive, specific, and non-intrusive is difficult for CJD and other prion diseases. Many cerebrospinal fluid (CSF) biomarkers have been tested at various degrees of sensitivity and specificity for CJD, including 14–3-3, Tau, S100b, neuron-specific enolase (NSE) (Sanchez-Juan et al., 2006), and neurofilament light (NfL) (Staffaroni et al., 2019). The World Health Organization has included the 14-3-3 protein immunoassay as a criteria for premortem diagnosis of patients with sCJD (Hermann et al., 2021). However, 14-3-3 protein abundance is a sign of accelerated neuronal death in many neurodegenerative diseases, and most fluid biomarkers are similarly nonspecific for CJD. Currently, neither 14-3-3, tau, S100b, nor NSE can be used to differentiate between different CJD types (Sanchez-Juan et al., 2006). Age, polymorphisms at *PRNP* codon 129, and PrP^{Sc} subtype also impact biomarker sensitivity (Sanchez-Juan et al., 2006). In general, combining biomarkers increases sensitivity and/or specificity (Sanchez-Juan et al., 2006). One longitudinal cohort study found that CSF NSE, total tau, NfL, and plasma NfL are effectively associated biomarkers for determining degree of functional impairment in sCJD, whereas CSF 14-3-3 protein, NSE, total tau, and plasma total tau were all efficient the correlational biomarkers for determining survival time in patients with sCJD (Staffaroni et al., 2019).

Electroencephalography (EEG) and magnetic resonance imaging (MRI) are the primary non-invasive brain tests performed for diagnosing CJD. EEG can identify many cases of CJD, first detecting an increase in background activity, followed by dysrhythmia, and ultimately occasional sharp discharges over a non-specific slow-wave baseline (Brown & Cathala, 1979; Malmgren et al., 1979). However, for vCJD, EEG rarely detects the periodic complexes typically observed in other forms of CJD (Zeidler et al., 1997). Instead, MRI is used to detect a "pulvinar sign" in vCJD, with the appearance of a hyperintense symmetrical hockey-stick-like pattern in the pulvinar region, which is not normally observed in classical CJD (Collie et al., 2003). MRI also detects bilateral, cortical, thalamic, and basal ganglia hyperintensity patterns in sCJD and iCJD (Meissner et al., 2009; Rudge et al., 2015). Thus, MRI is another important technique for ante mortem CJD diagnosis, and these patterns are most easily observed with fluid-attenuated inversion recovery (FLAIR), or when MRI includes both diffusion-weighted imaging (DWI) and T2-weighted imaging (Collie et al., 2003; Meissner et al., 2009). MRI can also help differentiate between the sCJD codon 129 genetic subtypes (Meissner et al., 2009).

Other highly sensitive and specific biotechnological techniques hold promise for the future in diagnosing CJD, including PMCA and RT-QuIC. PMCA can detect PrP^{Sc} from patients with vCJD at 100% specificity, with 92.9% sensitivity from urine samples (Moda et al., 2014), and 100% sensitivity from CSF (Barria et al., 2018), and blood samples (Bougard et al., 2016). However, PMCA was not an effective technique for the diagnosis of sCJD as it failed to detect PrP^{Sc} from many patients (Moda et al., 2014; Bougard et al., 2016). In contrast, RT-QuIC can detect prions from CSF of patients with sCJD ranging from 73–100% sensitivity with near 100% specificity (Dong & Satoh, 2021; Hermann et al., 2021). RT-QuIC can also work as a relatively non-intrusive diagnostic tool in quickly detecting prions from the deep nasal cavity of living patients (Bongianni et al., 2017). RT-QuIC has been helpful at increasing surveillance, confirming diagnosis, and monitoring trials (Hermann et al., 2021). However, RT-QuIC is limited in its prognostic ability to distinguish different subtypes (Piconi et al., 2019). Furthermore, RT-QuIC recombinant PrP substrate is relatively expensive, and the equipment required for RT-QuIC needs to become more widespread for practical use (Hermann et al., 2021).

CJD is diagnosed postmortem based upon histopathological identification of spongiform encephalopathy, gliosis, neuronal degeneration, and PrP^{Sc} deposition (Parchi et al., 1999b). Vacuolated lesions are highly variable within and between patients (Beck & Daniel, 1979), and are often observed in the neuropil, particularly in the gray matter, cerebral cortex, basal ganglia, striatum, thalamus, hippocampus, cerebellum, and brain stem (Malamud, 1979; Parchi et al., 1999b). Vacuoles are frequently large and confluent, but can also be fine, focal, severe (Cali et al., 2009), coarse, and/or widespread (Parchi et al., 1996). Astrogliosis and neuronal degeneration are found predominantly in the cerebral cortex, thalamus, striatum, cerebellum, and brainstem (Parchi et al., 1999b), and can be accompanied by mild microgliosis (Malamud, 1979). In CJD, PrP^{Sc} is regularly distributed throughout the cerebral cortex, basal ganglia, thalamus, striatum, cerebellum (Will et al., 1996), and brainstem (Parchi et al., 1996; Cali et al., 2009). PrP^{Sc} has also been detected in CSF, eyes, liver, urinary tract, spleen, lymph nodes (Gajdusek et al., 1977; Masters et al., 1979a), and in the blood (Bougard et al., 2016). PrP^{Sc} deposition patterns are highly variable, but commonly synaptic, laminar, focal, coarse, fine, perivacuolar, punctate, and/or widespread (Parchi et al., 1996, 1999b). Amyloid plaques are present in 10–20% of CJD cases (Masters et al., 1979a), and these plaques are most typically observed in vCJD (Will et al., 1996; Alperovitch et al., 1999), and in the sCJD codon 129 subtype MV2 (Parchi et al., 1999b). Plaques have been observed throughout the brain but most frequently in the cerebral cortex and cerebellum (Will et al., 1996; Parchi et al., 1999b).

29.3.2 Kuru

Kuru was first reported in humans in 1957 (Table 29.1) (Gajdusek & Zigas, 1957). Kuru was endemic primarily to the linguistic Fore people of the Highlands in Papua New Guinea with few other cases nearby (Gajdusek & Zigas, 1957). Overall, more than 2700 people have developed Kuru disease (Collinge et al., 2006), primarily from the South Fore, but around 500 cases were from the North Fore and almost 500 cases from nine other linguistic groups, namely, the Auyana, Awa, Gimi, Kamano, Kanite, Keiagana, Usurufa, Yagaria, and Yate (Alpers, 1979). Kuru predominantly affected adult women and children, who were as young as 4 (Gajdusek & Zigas, 1957), until the 1960s when cases in women and children vastly dropped off (Gajdusek & Alpers, 1966). Genetic resistance to Kuru disease is influenced by heterozygosity at *PRNP* codon 129, and valine at codon 127 (Mead et al., 2009).

Kuru etiology and transmission were difficult to resolve. In 1959, William J. Hadlow suggested that Kuru had similar clinical and neuropathological features to scrapie and may therefore be transmissible (Hadlow, 1995). Since then, Kuru has been experimentally transmitted to several primates including chimpanzees, spider monkeys, squirrel monkeys, capuchin monkeys, rhesus macaques, woolly monkeys, marmosets, cynomolgus macaques, pig-tailed macaques, African green monkeys, and bonnet macaques (Gajdusek, 1977). In addition to primates, Kuru has been transmitted to mink and ferrets (Gajdusek, 1977). From this developed the hypothesis that Kuru was an infectious disease resulting from cannibalistic mortuary feasts (Gajdusek, 1979), which were commonly performed in ritualistic mourning out of respect for dead relatives (Collinge et al., 2006). In particular, Kuru transmits orally from ingestion of brain tissue (Liberski et al., 2012). After abandoning endocannibalistic mourning rituals in the 1950s, Kuru cases have declined, and humans born after 1959 have not developed Kuru (Collinge et al., 2006). Kuru does not transmit from close contact with Kuru patients (Gajdusek & Alpers, 1966), nor is their evidence for environmental transmission (Gajdusek et al., 1977). Additionally, vertical transmission of Kuru from mother to child has not been observed, and consistent with this observation, infectivity in milk, the placenta, embryonal membrane (Gajdusek, 1977), or amniotic fluid has not been detected (Gajdusek et al., 1977). Kuru likely originated from a case of sCJD that was transmitted orally through mortuary feasts (Wadsworth et al., 2008).

The clinical disease course of Kuru is remarkably consistent (Gajdusek & Alpers, 1966). However, patients with Kuru have a subclinical incubation period ranging greatly from less than 4 years (Masters et al., 1979b), to over 5 decades (Collinge et al., 2006). The disease develops quicker in children with a larger variability in clinical signs compared to adults (Hornabrook, 1979). Clinical symptoms of Kuru consistently include motor dysfunctions from incoordination to an unsteady gait, stumbling, hyperactive reflexes, clonus in the legs, excessive involuntary movements, pain, tremors, ataxia, dystonia, asthenia, and the inability to walk, stand or sit without support, ultimately resulting in incapacitation (Gajdusek & Zigas, 1957; Klatzo et al., 1959; Hornabrook, 1979). Field deployable screening tests for Kuru included heel-to-toe walking and the finger-to-nose test (Hornabrook, 1979). Other common physical manifestations are decubitus ulcers, dysphagia, incontinence, and weight loss or obesity (Gajdusek & Zigas, 1957; Klatzo et al., 1959). Neurological and emotional disturbances are frequently observed, such as headaches, mood swings, euphoria, inappropriate laughing, sullenness, or late-stage dementia (Zigas & Gajdusek, 1957; Hornabrook, 1979). Visual impairments often involve strabismus and nystagmus (Hornabrook, 1979). Speech disparities digress from slurred speech to dysarthria, aphonia, or mutism (Gajdusek & Zigas, 1957; Zigas & Gajdusek, 1957). Death is inevitable, usually within a year from the onset of clinical signs (Gajdusek & Zigas, 1957), but can last over 2 years (Zigas & Gajdusek, 1957). Many drug treatments have been attempted, but none have shown an effect (Gajdusek & Zigas, 1957, Klatzo et al., 1959).

Postmortem neuropathological diagnosis of Kuru is based upon the presence of vacuolation, gliosis, neurodegeneration, and PrP^{Sc} amyloid plaques (Klatzo et al., 1959; Beck & Daniel, 1979). Vacuolation is more frequently observed in neurons of the gray matter compared to the white matter of the brain, particularly in the cerebral cortex, thalamus, striatum, cerebellum (Beck & Daniel, 1979), pons (Neumann et al., 1964), and the inferior olives of the medulla oblongata (Klatzo et al., 1959). Vacuoles are generally larger in Kuru than in CJD (Beck & Daniel, 1979). Astrogliosis is observed in Kuru, both in the white and gray matter of the cerebral cortex, striatum, thalamus, cerebellar molecular layer, pons, inferior olives of the medulla oblongata, and the spinal cord (Neumann et al., 1964; Hainfellner et al., 1997). Amyloid plaques, consisting of fibril aggregates, are a defining feature of Kuru and are found in 50–75% of patients (Hadlow, 1995). These plaques have been identified in the cerebral cortex, thalamus, striatum, and the granular and molecular layers of the cerebellum (Neumann et al., 1964, Hainfellner et al., 1997). These plaques are typically numerous, dense, compact, non-argyrophilic, round, variable in size, and usually unicentric but sometimes multicentric (Hainfellner et al., 1997). Microgliosis is most frequently observed in the cerebral cortex and molecular layer of the cerebellum (Klatzo et al., 1959). Neuronal degeneration is widespread in patients with Kuru, particularly in the cerebral cortex, striatum, thalamus, brainstem, spinal cord, and prototypically in the Purkinje cells of the cerebellum (Hainfellner et al., 1997; Liberski et al., 2012). Myelin loss is primarily seen in the medulla oblongata (Klatzo et al., 1959; Neumann et al., 1964). EEG patterns were typically normal, and any activity or slowing detected was nonspecific in patients living with Kuru (Cobb et al., 1973). Due to the historical nature of Kuru, less is known about the biochemical characteristics of PrP^{Sc} compared to other prion diseases. PrP^{Sc} has been observed throughout the entire brain, including the central cortex (Hainfellner et al., 1997), and prion infectivity is occasionally detected in the spleen and liver (Gajdusek, 1977; Gajdusek et al., 1977). Prion infectivity has not been detected from CSF, blood, leukocytes, or in urine of patients with Kuru (Gajdusek, 1977, Gajdusek et al., 1977).

29.3.3 Gerstmann-Straussler-Scheinker Syndrome

Gerstmann-Straussler-Scheinker (GSS) syndrome was named after the three scientists that formally identified this human disease in 1936 (Table 29.1) (Gerstmann et al., 1935). GSS has been recognized in Mexico (Zomosa-Signoret et al., 2011), the United States, Canada, Australia, the UK (Kovacs et al., 2005), Austria, Germany, France (Tranchant et al., 1992), Spain, Sweden (Hsiao et al., 1992), Poland, the Czech Republic (Tesar et al., 2019), Italy, Israel (Goldhammer et al., 1993), Japan, China, South Korea (Ahn et al., 2022; Chu et al., 2022), and Russia (Rudenskaya et al., 2023). GSS has an incidence rate as high as one to fourteen cases per ten million people (Hsiao et al., 1989; Tesar et al., 2019). Reported patients range in age from 20 to 87 years old (Tranchant et al., 1992), with a mean age of 46–52 (Kovacs et al., 2005; Tesar et al., 2019; Chu et al., 2022). There is a higher incidence rate of GSS in females compared to males (Kovacs et al., 2005; Tesar et al., 2019). In experimental settings, GSS has been transmitted to mice, rats, guinea pigs, squirrel monkeys, and spider monkeys (Masters et al., 1981).

GSS is a familial disease (Collinge et al., 1989), follows a pattern of autosomal dominant inheritance (Hsiao et al., 1989), and is associated with more than a dozen hereditary mutations in *PRNP* (Baldwin & Correll, 2019). The most common mutation for GSS is a single base proline to leucine substitution in codon 102 (Hsiao et al., 1989). Other important *PRNP* polymorphisms related to GSS include codons 105, 117, 129, 131, 145, 187, 198, 202, 212, 217, 219 (Wang et al., 2019), and 232 (Liberski, 2012). A proline to leucine substitution at codon 105 causes spastic paraparesis (Kitamoto et al., 1993), codon 129 influences the age of death (Kovacs et al., 2005), and mutations at codons 198 and 217 results in the concurrent development of neurofibrillary tangles in addition to the typical amyloid plaques found with GSS (Hsiao et al., 1992). Thus, genotype impacts the clinical course observed for patients with GSS.

Patients with GSS can have a long clinical disease course ranging from 3 months to over 20 years (Belay & Bartz, 2014), with a 4-year average, before inevitable death (Tesar et al., 2019; Chu et al., 2022). The most common physical manifestations include motor disturbances, gait abnormalities, incoordination, bradykinesia, apraxia, ataxia, hemiparesis, hyperreflexia, myoclonus, dysphagia, and intercurrent pneumonia (Schlote et al., 1980; Ishizawa et al., 2018; Risacher et al., 2018). Standard neurological signs include cerebellar problems, dementia, depression, irritability, visual impairments, and speech disturbances, particularly dysarthria (Masters et al., 1981; Ishizawa et al., 2018). Although up to half of some populations were 14-3-3 positive in their CSF (Kovacs et al., 2005; Wang et al., 2019), this biomarker remains too non-specific for effective clinical diagnosis, and useful biomarkers have yet to be identified for GSS (Baldwin & Correll, 2019).

Diagnosis of GSS is based upon neuropathological recognition of vacuolation, gliosis, and neurodegeneration (Ishizawa et al., 2018). Compared to other prion diseases, spongiosis for GSS is only mildly detected in the cerebral cortex and cerebellum (Ishizawa et al., 2018). Astrogliosis, microglia, and oligodendrogenesis often accompany GSS (Schlote et al., 1980), primarily noticeable in the cerebral cortex, thalamus, and cerebellum (Masters et al., 1981; Erana et al., 2022). Neuronal loss is common, particularly in the cerebral cortex, thalamus, basal ganglia, cerebellum, medulla oblongata, spinal cord, and in association with myelin loss (Masters et al., 1981, Ishizawa et al., 2018). EEG can detect nonspecific slowing in GSS (Schlote et al., 1980), or inconsistent quick discharges (Collinge et al., 1989), but appears to be ineffective at consistently diagnosing patients with GSS (Kovacs et al., 2005). Computed tomography (CT) scans, ventriculograms, and MRI (Collinge et al., 1989), particularly diffusion-weighted imaging (Ahn et al., 2022), are much more reliable techniques for antemortem diagnosis of patients with GSS based upon atrophy in the brain (Wang et al., 2019). Additionally, [^{18}F]flortaucipir PET scans can detect impairments in the cerebral cortex, basal ganglia, striatum and thalamus of living patients with GSS (Risacher et al., 2018). Finally, RT-QuIC can detect PrP^{Sc} seeding activity from the CSF of living patients with GSS (Ahn et al., 2022).

GSS diagnosis is confirmed upon postmortem immunohistochemical detection of PrP^{Sc} deposition and plaques. PrP^{Sc} deposition accumulates in the frontal cortex, thalamus, hippocampus, and cerebellum (Hallinan et al., 2022). PMCA can detect PrP^{Sc} amplification from GSS (Peden et al., 2014). Amyloid plaques, composed of PrP^{Sc} fibril aggregates, are more frequent in GSS than in other prion diseases, and are found most notably in the white matter, cerebral cortex, hippocampus, basal ganglia, putamen, thalamus, and cerebellum (Schlote et al., 1980; Masters et al., 1981; Ishizawa et al., 2018). These plaques vary in type and can be large stellate, compound multicentric, primitive, unicentric, small precipitates (Schlote et al., 1980), argyrophilic (Masters et al., 1981), or blurry confluent (Ishizawa et al., 2018). Additionally, tau and Aβ co-localize with plaques throughout the cerebral cortex in several cases of GSS (Ishizawa et al., 2018, Risacher et al., 2018). Cryogenic electron microscopy has been performed to determine the structure of GSS PrP^{Sc} amyloid protein filaments (Hallinan et al., 2022). Interestingly, these filaments differ in structure from rodent-adapted prion strains and are composed of dimeric, trimeric and tetrameric left-hand filaments with a similar PrP^{Sc} protein fold (Hallinan et al., 2022). This finding highlights that the structural diversity of PrP^{Sc} is just beginning to be understood.

29.3.4 Fatal Familial Insomnia

Fatal familial insomnia (FFI) was first diagnosed in 1986 (Table 29.1) (Lugaresi et al., 1986). FFI has been identified in Italy (Lugaresi et al., 1986), Macedonia, Austria, Spain, France, Germany, the UK (Kovacs et al., 2005), Australia, the United States, Canada, Brazil, Argentina, Morocco, Japan, China, Korea, India (Chu et al., 2022), Singapore, Taiwan, and Malaysia (Chen et al., 2023). FFI is an inherited prion disease (Lugaresi et al., 1986), with incomplete penetrance of disease (Peng et al., 2015). This disease is caused by autosomal dominant germline mutations in the prion protein gene (Little et al., 1986). FFI results from a replacement of aspartic acid by asparagine at *PRNP* codon 178, and at least one methionine at codon 129 (Parchi et al., 1995). Unlike other genetic human prion diseases, males are predominantly affected over females (Kovacs et al., 2005). Patients with FFI have a wide age range of disease onset from 17 to 83 years (Kovacs et al., 2005), with a 49-to-53-year average (Chu et al., 2022; Forloni et al., 2022).

In addition to the genetic version of FFI, a sporadic form was discovered in 1999, termed sporadic fatal familial insomnia (sFI) (Parchi et al., 1999a). sFI frequency is estimated at an incidence rate of 1–2 cases per hundred million persons (Capellari et al., 2008), accounting for around 2% of sporadic human prion disease cases (Gambetti et al., 2003). The origin for sFI is uncertain, but likely comes from either a somatic mutation in *PRNP* or the spontaneous conversion of PrP^{C} to PrP^{Sc} (Imran & Mahmood, 2011), as all cases of sFI were methionine homozygous at codon 129 and mutations in the *PRNP* gene were not identified (Parchi et al., 1999a). sFI is identical to FFI in histopathology and transmission characteristics (Parchi et al., 1999a). Likewise, Western blot analysis of PK-digested brain homogenate of both FFI and sFI revealed a similar type 2 PrP^{Sc} unglycosylated band that migrates to a molecular mass of about 19 kDa (Parchi et al., 1999a), which differs from the typical 21 kDa type 1 PrP^{Sc} unglycosylated band observed in most prion diseases. However, whereas the glycoform profile for FFI underrepresents the unglycosylated PrP^{Sc} polypeptide compared to its mono- and diglycosylated PrP^{Sc} polypeptides, the unglycosylated PrP^{Sc} in sFI has greater intensity than the diglycosylated band (Parchi et al., 1999a). Patients with sFI also have different clinical signs and a longer disease duration than patients with FFI (Chen et al., 2023). In addition to FFI and sFI, atypical cases of FFI have been reported (Takeuchi et al., 2019). Atypical cases differ from classical FFI and sFI in clinical signs, neuropathology, amplification efficiency, and transmission potential (Takeuchi et al., 2019). Despite these differences, FFI, sFI, and atypical FFI have all been experimentally transmitted to mouse models expressing prion proteins (Takeuchi et al., 2019).

Unlike most prion diseases, FFI can be diagnosed from familial history and clinical signs (Krasnianski et al., 2008; Khan & Bollu, 2022). Physical abnormalities typically include motor dysfunctions, clumsiness, an unusual gait, slowness, ataxia, dystonia, impotence, myoclonus, tremors, lacrimation, dysphagia, weight loss, sexual dysfunction, incontinence, and constipation (Little et al., 1986; Lugaresi et al., 1986; Krasnianski et al., 2008). Some physiological alterations unique to FFI are tachycardia, hyperhidrosis, and pyrexia (Krasnianski et al., 2008; Chen et al., 2023). The prototypical signs for FFI are sleep abnormalities, including progressive insomnia, somnolence, restless sleep (Little et al., 1986), sleep attacks, lucid dreams, and dream mimicking, often accompanied by tachypnea or apnea (Lugaresi et al., 1986, Krasnianski et al., 2008). Cognitive impairments, such as amnesia, dementia, comas, stupor, confusion, or inattentiveness are also common (Lugaresi et al., 1986; Khan & Bollu, 2022). Personality changes are frequently manifested as mood swings, depression, disinhibition, bizarre behavior, aggressiveness, anxiety (Krasnianski et al., 2008), and panic attacks (Parchi et al., 1999a). Psychiatric and visual problems are usually observed, from paranoia to hallucinations, upward gazing, nystagmus, and diplopia (Lugaresi et al., 1986, Krasnianski et al., 2008). Finally, patients with FFI often have speech impairments, particularly dysarthria or akinetic mutism (Lugaresi et al., 1986; Parchi et al., 1999a). Clinical course varies based upon polymorphisms at codon 129 (Krasnianski et al., 2008). The clinical disease duration lasts from 4 months to 8 years (Kovacs et al., 2005), with an 11–18-month average before inevitable death (Chu et al., 2022; Khan & Bollu, 2022). 14-3-3 protein and other biomarkers are ineffective for diagnosing FFI (Kovacs et al., 2005; Baldwin & Correll, 2019), with the exception of plasma NfL, which correlates with a shorter disease duration (Hermann et al., 2022). Several drug treatments have been attempted, but none have been successful (Peng et al., 2015; Khan & Bollu, 2022). However, there is an ongoing 10-year clinical trial on the possible preventative nature of doxycycline which should have results by 2023 (Forloni et al., 2015, 2022). Doxycycline treatment was unsuccessful at preventing fatal disease in presymptomatic transgenic mice, which normally develop FFI, but shows promise in restoring recognition memory and circadian rhythm (Lavigna et al., 2021).

Several brain tests have been examined for diagnosing FFI in living patients, each with various efficacies. Electromyogram (Lugaresi et al., 1986), EEG (Little et al., 1986), and MRI (Kovacs et al., 2005), including DWI and

FLAIR, are all unhelpful in diagnosing living patients with FFI, but can be useful in distinguishing other prion diseases from FFI (Chu et al., 2022). However, both EEG and MRI are significantly more effective at diagnosing cases of sFI (Chen et al., 2023). Polysomnography is effective at detecting a reduction in REM sleep for patients with FFI and sFI (Chen et al., 2023). Likewise, fluorodeoxyglucose positron emission tomography (FDG-PET) reveals hypometabolism in patients with FFI or sFI, particularly in the thalamus (Baldwin & Correll, 2019; Chen et al., 2023). In one study, 41 out of 49 patients with FFI were positive for either PET and/or single-photon emission computed tomography (Chu et al., 2022).

FFI neuropathology is characterized by vacuolation, astrogliosis, neurodegeneration, and hypometabolism within the thalamus (Parchi et al., 1995). Lesions are most frequently observed in the thalamus of patients with FFI (Little et al., 1986; Lugaresi et al., 1986), but spongiosis has also been seen in the white matter of the cerebral cortex, caudate nucleus, cerebellum, brainstem (Gambetti et al., 2003), and occasionally in the hippocampus (Parchi et al., 1995). Astrogliosis is observed predominantly throughout the thalamus but also is detected in the cerebral cortex, hypothalamus, cerebellum, midbrain (Parchi et al., 1995), and inferior olives (Parchi et al., 1999a). Atrophy and neuronal loss are commonly observed in the cerebrum, entorhinal cortex, cerebellum, inferior olives, and most clearly throughout the thalamus (Little et al., 1986; Parchi et al., 1999a). Amyloid plaques are rarely observed in the cerebral cortex (Little et al., 1986). PrP^{Sc} deposition levels are significantly lower compared to CJD but have been identified faintly in the cerebral cortex, hippocampus, thalamus (Parchi et al., 1999a), and cerebellum (Parchi et al., 1995). PMCA can effectively amplify PrP^{Sc} from patients with FFI (Takeuchi et al., 2019), and prion seeding activity can also be detected from the olfactory mucosa using RT-QuIC (Redaelli et al., 2017).

29.3.5 Variably Protease-Sensitive Prionopathy

In 2008, a new human prion disease was reported termed variably protease-sensitive prionopathy (VPSPr) (Table 29.1) (Gambetti et al., 2008; Zou et al., 2010). At least 43 cases of VPSPr have been formally reported (Huang et al., 2021; Baiardi et al., 2022), with an estimated total of 80 plus cases worldwide (Notari et al., 2018). VPSPr has been recognized in the United States (Gambetti et al., 2008), the UK, the Netherlands (Jansen et al., 2010), Spain, Italy (Zou et al., 2010), Austria (Assar et al., 2015), Germany (Notari et al., 2018), and Japan (Uchino et al., 2023). VPSPr has been identified in patients ranging from 46–93 years old (Ghoshal et al., 2015), with a 65–70-year mean (Notari et al., 2018; Huang et al., 2021).

VPSPr is a novel sporadic prion disease of unknown etiology (Zou et al., 2010). The most probable origins for VPSPr are either the spontaneous conversion of PrP^{C} to PrP^{Sc} or a somatic mutation of PrP (Imran & Mahmood, 2011), as germline mutations have not been identified in the *PRNP* coding sequence (Gambetti et al., 2008). Like other prion diseases, VPSPr is transmissible under experimental conditions. VPSPr has been experimentally transmitted to bank voles (Nonno et al., 2019), and multiple transgenic mouse models (Notari et al., 2014). VPSPr differs from sCJD in clinical duration, signs, pathology (Jansen et al., 2010), biochemical properties of PrP^{Sc} (Gambetti et al., 2008), and PrP codon 129 genotype frequencies (Zou et al., 2010). VPSPr was named because its disease-associated PrP^{Sc} is characteristically more susceptible to protease digestion than other prion diseases (Gambetti et al., 2008), and this susceptibility is variable based upon host genotype (Zou et al., 2010). After digestion with PK, Western blot analysis reveals 5–7 major bands of varying molecular weights, depending upon the anti-PrP primary antibody used (Zou et al., 2010; Notari et al., 2018). Each of three polymorphisms in codon 129 results in different VPSPr characteristics (Zou et al., 2010), with valine homozygosity accounting for 63–65% of all patients (Notari et al., 2018, Huang et al., 2021).

The clinical duration of VPSPr disease ranges from 6 months to 10 years (Baiardi et al., 2022), with a 2–3-year average (Notari et al., 2018; Huang et al., 2021). Clinical presentation of disease often includes motor difficulties, unsteadiness, apraxia, gait abnormalities, ataxia, paraplegia, myoclonus, tremors, seizures, and incontinence (Jansen et al., 2010; Assar et al., 2015), and behavioral changes (Gambetti et al., 2008). Mental and emotional changes can include mood swings, depression, personality changes, neuropsychiatric symptoms, mental deterioration, forgetfulness, dementia (Gambetti et al., 2008), and loss of perception (Huang et al., 2021). Additionally, language deficits in VPSPr are common, specifically aphasia, anomia (Aizpurua et al., 2019), and akinetic mutism (Assar et al., 2015). Currently, sensitive CSF biomarkers for identifying VPSPr are not available, despite uncommonly slightly positive 14-3-3 protein levels (Gambetti et al., 2008), and S100b levels in some patients (Baldwin & Correll, 2019). Likewise, effective treatments for VPSPr do not exist, although one patient with VPSPr was administered 100 mg of doxycycline daily and survived longer than average (Assar et al., 2015).

VPSPr diagnosis is confirmed upon postmortem neuropathological identification of vacuolation, astrogliosis, neurodegeneration, and PrP^{Sc} deposition, specifically in the frontal cortex (Ghoshal et al., 2015). Vacuolation in the CNS of patients with VPSPr is typically observed in the cerebral cortex, basal ganglia, putamen, thalamus, midbrain, and cerebellum (Ghoshal et al., 2015; Aizpurua et al., 2019). These vacuoles can be finer or larger than those typically observed in CJD (Gambetti et al., 2008). However, vacuolation has not been detected in the hippocampus of patients with VPSPr (Gambetti et al., 2008, Zou et al., 2010). MRI can pick up non-specific atrophy in VPSPr, but neither MRI nor EEG is generally helpful in diagnosing VPSPr (Gambetti et al., 2008), except in rare cases (Assar et al., 2015; Uchino et al., 2023). IHC generally results in the detection of PrP^{Sc} in the cerebral cortex, basal ganglia, thalamus, hippocampus, and molecular layer of the cerebellum (Gambetti et al., 2008; Peden et al., 2014). The PrP^{Sc} deposition pattern is typically widespread, intense, and spherical, with small granules surrounding a larger center of coarse granular aggregates (Gambetti et al., 2008). Amyloid PrP microdeposits can uniquely accompany VPSPr in the cerebellar molecular layer (Gambetti et al., 2008). These deposits tend to be compact, spherical, and clearly bordered (Notari et al., 2018). Both PMCA and RT-QuIC are effective techniques for detecting PrP^{Sc} seeding activity from VPSPr (Peden et al., 2014). Aβ, α synuclein, tau (Baiardi et al., 2022), FUS-positive and TDP-43 inclusions have also been identified infrequently in patients with VPSPr (Gambetti et al., 2011).

29.4 Review Questions

1. Scrapie and chronic wasting disease (CWD) are difficult animal prion diseases to control because of how they transmit. In what way(s) do these diseases naturally transmit?
   (a) Experimentally
   (b) Horizontally
   (c) Environmentally
   (d) Vertically and maternally
   (e) a and b
   (f) a and d
   (g) a, b, and d
   (h) *All the above*
2. Classical and atypical bovine spongiform encephalopathy (BSE) in cattle has transmitted to several different species, including the first known zoonotic transmission to humans. Identify the correct name for this human prion disease and additional animal prion diseases that have resulted from the BSE epidemic.
   (a) Kuru; scrapie, exotic ungulate encephalopathy, prion disease in non-human primates, transmissible mink encephalopathy, and camel prion disease
   (b) Kuru; chronic wasting disease, exotic ungulate encephalopathy, prion disease in non-human primates, camel prion disease, and feline spongiform encephalopathy
   (c) Iatrogenic Creutzfeldt-Jakob disease (iCJD); exotic ungulate encephalopathy, feline spongiform encephalopathy, prion disease in non-human primates, and transmissible mink encephalopathy
   (d) Iatrogenic Creutzfeldt-Jakob disease (iCJD); scrapie, exotic ungulate encephalopathy, prion disease in non-human primates, transmissible mink encephalopathy, and feline spongiform encephalopathy
   (e) *Variant Creutzfeldt-Jakob disease (vCJD); prion disease in non-human primates, transmissible mink encephalopathy, exotic ungulate encephalopathy, and feline spongiform encephalopathy*
   (f) Variant Creutzfeldt-Jakob disease (vCJD); exotic ungulate encephalopathy, feline spongiform encephalopathy, prion disease in non-human primates, and chronic wasting disease.
3. Select all of the following that are appropriate means for prion decontamination:
   (a) *Autoclaving for one hour at 132 °C*
   (b) Disinfecting with alcohol or ethanol for 15 minutes
   (c) Treatment with ultraviolet light for one hour
   (d) *Treatment with sodium hypochlorite or sodium hydroxide for one hour*
   (e) Spraying with Lysol detergent for one hour
   (f) Fixation with formaldehyde
   (g) Boiling for 8 hours
4. A 70-year-old female consults a physician about an unusual gait and about how she can longer do all the things she used to be able to do. Her muscles are tense, her head is shaking slightly, and she is noticeably tired. As she attempts to speak, you notice her voice is unsteady. She reports having a headache and double vision. You perform an MRI with diffusion-weighted imaging (DWI) and detect a bilateral hyperintensity pattern. A spinal tap is completed and analysis for fluid biomarkers reveals she is positive for 14-3-3 protein and tau. To confirm your suspicions of a prion disease, you decide to have her deep nasal cavity swabbed and perform RT-QuIC, which produces positive seeding activity. Brittney is a vegetarian, has no family history of neurodegenerative diseases, and has never had surgery in her life. A genetic test for *PRNP* revealed she was methionine-valine heterozygous at codon 129 but has no other outstanding mutations. What prion disease does Brittney have?
   (a) Gerstmann-Straussler-Scheinker (GSS) syndrome
   (b) *Sporadic Creutzfeldt-Jakob disease (sCJD)*
   (c) Variably protease-sensitive prionopathy (VPSPr)
   (d) Fatal familial insomnia (FFI)
   (e) Kuru
5. A 32-year-old male client walks into the clinic yawning and indicates that he has not been sleeping well the past couple of weeks. After taking vital signs you notice he has a slight fever, and his heart rate is elevated for his age group. His family history reports that when his father passed away at 60 years old, he appeared thin, had trem-

ors, complained about seeing things, had difficulty communicating, struggled with sleep attacks, had dementia and became aggressive. A genetic test reveals that the patient is methionine homozygous at *PRNP* codon 129, and has an asparagine, instead of aspartic acid, at codon 178. An MRI was performed, but nothing unusual was noted. What prion disease does the patient have?
(a) Familial Creutzfeldt-Jakob disease (fCJD)
(b) Kuru
(c) Gerstmann-Straussler-Scheinker (GSS) syndrome
(d) *Fatal familial insomnia (FFI)*
(e) Variably protease-sensitive prionopathy (VPSPr)

References

(1996). Feline spongiform encephalopathy. BSAVA Scientific Committee. *The Journal of Small Animal Practice, 37*(4), 198.

(2003). Swiss FSE case raises questions. *Animal Pharm, 525*, 4.

(2021). Creutzfeldt-Jakob Disease International Surveillance Network. *vCJD Cases Worldwide*. Retrieved June 22, 2023, from https://www.eurocjd.ed.ac.uk/data_tables

Abrams, J. Y., Schonberger, L. B., Belay, E. D., Maddox, R. A., Leschek, E. W., Mills, J. L., Wysowski, D. K., & Fradkin, J. E. (2011). Lower risk of Creutzfeldt-Jakob disease in pituitary growth hormone recipients initiating treatment after 1977. *The Journal of Clinical Endocrinology and Metabolism, 96*(10), E1666–E1669.

Ae, R., Hamaguchi, T., Nakamura, Y., Yamada, M., Tsukamoto, T., Mizusawa, H., Belay, E. D., & Schonberger, L. B. (2018). Update: Dura mater graft-associated Creutzfeldt-Jakob disease – Japan, 1975–2017. *MMWR. Morbidity and Mortality Weekly Report, 67*(9), 274–278.

Aguilar-Calvo, P., Garcia, C., Espinosa, J. C., Andreoletti, O., & Torres, J. M. (2015). Prion and prion-like diseases in animals. *Virus Research, 207*, 82–93.

Ahn, S. J., Lee, H. S., Moon, J., & Chu, K. (2022). First familial cases of P102L Gerstmann-Straussler-Scheinker syndrome in South Korea: Diffusion-weighted imaging might reflect intrafamilial phenotypic variability. *Neurological Sciences, 43*(5), 3419–3422.

Aizpurua, M., Selvackadunco, S., Yull, H., Kipps, C. M., Ironside, J. W., & Bodi, I. (2019). Variably protease-sensitive prionopathy mimicking frontotemporal dementia. *Neuropathology, 39*(2), 135–140.

Aldhous, P. (1990). BSE: Spongiform encephalopathy found in cat. *Nature, 345*(6272), 194.

Alperovitch, A., Zerr, I., Pocchiari, M., Mitrova, E., de Pedro Cuesta, J., Hegyi, I., Collins, S., Kretzschmar, H., van Duijn, C., & Will, R. G. (1999). Codon 129 prion protein genotype and sporadic Creutzfeldt-Jakob disease. *Lancet, 353*(9165), 1673–1674.

Alpers, M. P. (1979). Epidemiology and ecology of Kuru. In S. B. Prusiner & W. J. Hadlow (Eds.), *Slow transmissible diseases of the nervous system. Volume 1: Clinical, epidemiological, genetic, and pathological aspects of the spongiform encephalopathies* (pp. 67–90). Academic.

Alverson, J., O'Rourke, K. I., & Baszler, T. V. (2006). PrPSc accumulation in fetal cotyledons of scrapie-resistant lambs is influenced by fetus location in the uterus. *The Journal of General Virology, 87*(Pt 4), 1035–1041.

Andreoletti, O., Lacroux, C., Chabert, A., Monnereau, L., Tabouret, G., Lantier, F., Berthon, P., Eychenne, F., Lafond-Benestad, S., Elsen, J. M., & Schelcher, F. (2002). PrP(Sc) accumulation in placentas of ewes exposed to natural scrapie: Influence of foetal PrP genotype and effect on ewe-to-lamb transmission. *The Journal of General Virology, 83*(Pt 10), 2607–2616.

Angers, R. C., Seward, T. S., Napier, D., Green, M., Hoover, E., Spraker, T., O'Rourke, K., Balachandran, A., & Telling, G. C. (2009). Chronic wasting disease prions in elk antler velvet. *Emerging Infectious Diseases, 15*(5), 696–703.

Arifin, M. I., Kaczmarczyk, L., Zeng, D., Hannaoui, S., Lee, C., Chang, S. C., Mitchell, G., McKenzie, D., Beekes, M., Jackson, W., & Gilch, S. (2023). Heterozygosity for cervid S138N polymorphism results in subclinical CWD in gene-targeted mice and progressive inhibition of prion conversion. *Proceedings of the National Academy of Sciences of the United States of America, 120*(15), e2221060120.

Assar, H., Topakian, R., Weis, S., Rahimi, J., Trenkler, J., Hoftberger, R., Aboulenein-Djamshidian, F., Strobel, T., Budka, H., Yull, H., Head, M. W., Ironside, J. W., & Kovacs, G. G. (2015). A case of variably protease-sensitive prionopathy treated with doxycyclin. *Journal of Neurology, Neurosurgery, and Psychiatry, 86*(7), 816–818.

Babelhadj, B. (2019). *Camel prion disease: A possible emerging disease in dromedary camel populations?* Retrieved June 6, 2023, from https://oiebulletin.com/wp-content/uploads/2019/12/OIE-News-December-2019-Camel-prion-disease.pdf

Babelhadj, B., Di Bari, M. A., Pirisinu, L., Chiappini, B., Gaouar, S. B. S., Riccardi, G., Marcon, S., Agrimi, U., Nonno, R., & Vaccari, G. (2018). Prion disease in dromedary camels, Algeria. *Emerging Infectious Diseases, 24*(6), 1029–1036.

Baeten, L. A., Powers, B. E., Jewell, J. E., Spraker, T. R., & Miller, M. W. (2007). A natural case of chronic wasting disease in a free-ranging moose (Alces alces shirasi). *Journal of Wildlife Diseases, 43*(2), 309–314.

Bahmanyar, S., Williams, E. S., Johnson, F. B., Young, S., & Gajdusek, D. C. (1985). Amyloid plaques in spongiform encephalopathy of mule deer. *Journal of Comparative Pathology, 95*(1), 1–5.

Baiardi, S., Mammana, A., Rossi, M., Ladogana, A., Carla, B., Gambetti, P., Capellari, S., & Parchi, P. (2022). Phenotypic heterogeneity of variably protease-sensitive prionopathy: A report of three cases carrying different genotypes at PRNP codon 129. *Viruses, 14*(2).

Baker, H. F., Ridley, R. M., Wells, G. A., & Ironside, J. W. (1996). Spontaneous spongiform encephalopathy in a monkey. *Lancet, 348*(9032), 955–956.

Baldwin, K. J., & Correll, C. M. (2019). Prion disease. *Seminars in Neurology, 39*(4), 428–439.

Barbot, C., Castro, L., Oliveira, C., & Carpenter, S. (2010). Variant Creutzfeldt-Jakob disease: The first confirmed case from Portugal shows early onset, long duration and unusual pathology. *Journal of Neurology, Neurosurgery, and Psychiatry, 81*(1), 112–114.

Baron, T., Bencsik, A., Biacabe, A. G., Morignat, E., & Bessen, R. A. (2007). Phenotypic similarity of transmissible mink encephalopathy in cattle and L-type bovine spongiform encephalopathy in a mouse model. *Emerging Infectious Diseases, 13*(12), 1887–1894.

Baron, T., Vulin, J., Biacabe, A. G., Lakhdar, L., Verchere, J., Torres, J. M., & Bencsik, A. (2011). Emergence of classical BSE strain properties during serial passages of H-BSE in wild-type mice. *PLoS One, 6*(1), e15839.

Barria, M. A., Balachandran, A., Morita, M., Kitamoto, T., Barron, R., Manson, J., Knight, R., Ironside, J. W., & Head, M. W. (2014). Molecular barriers to zoonotic transmission of prions. *Emerging Infectious Diseases, 20*(1), 88–97.

Barria, M. A., Lee, A., Green, A. J., Knight, R., & Head, M. W. (2018). Rapid amplification of prions from variant Creutzfeldt-Jakob disease cerebrospinal fluid. *The Journal of Pathology. Clinical Research, 4*(2), 86–92.

Bartz, J. C. (2016). Prion strain diversity. *Cold Spring Harbor Perspectives in Medicine, 6*(12).

Bartz, J. C., Marsh, R. F., McKenzie, D. I., & Aiken, J. M. (1998). The host range of chronic wasting disease is altered on passage in ferrets. *Virology, 251*(2), 297–301.

Beck, E., & Daniel, P. M. (1979). Kuru and Creutzfeldt-Jakob disease; neuropathological lesions and their significance. In S. B. Prusiner & W. J. Hadlow (Eds.), *Slow transmissible diseases of the nervous system. Volume 1: Clinical, epidemiological, genetic, and pathological aspects of the spongiform encephalopathies* (pp. 253–270). Academic.

Belay, E. D., & Bartz, J. C. (2014). Prion diseases. In R. A. Kaslow, L. R. Stanberry, & J. L. Le Duc (Eds.), *Viral infections of humans epidemiology and control* (pp. 1165–1186). Springer Science+Business Media.

Bencsik, A., Lezmi, S., & Baron, T. (2001). Autonomic nervous system innervation of lymphoid territories in spleen: A possible involvement of noradrenergic neurons for prion neuroinvasion in natural scrapie. *Journal of Neurovirology, 7*(5), 447–453.

Bencsik, A., Debeer, S., Petit, T., & Baron, T. (2009). Possible case of maternal transmission of feline spongiform encephalopathy in a captive cheetah. *PLoS One, 4*(9), e6929.

Benestad, S. L., Sarradin, P., Thu, B., Schonheit, J., Tranulis, M. A., & Bratberg, B. (2003). Cases of scrapie with unusual features in Norway and designation of a new type, Nor98. *The Veterinary Record, 153*(7), 202–208.

Benestad, S. L., Mitchell, G., Simmons, M., Ytrehus, B., & Vikoren, T. (2016). First case of chronic wasting disease in Europe in a Norwegian free-ranging reindeer. *Veterinary Research, 47*(1), 88.

Beringue, V., Herzog, L., Reine, F., Le Dur, A., Casalone, C., Vilotte, J. L., & Laude, H. (2008a). Transmission of atypical bovine prions to mice transgenic for human prion protein. *Emerging Infectious Diseases, 14*(12), 1898–1901.

Beringue, V., Le Dur, A., Tixador, P., Reine, F., Lepourry, L., Perret-Liaudet, A., Haik, S., Vilotte, J. L., Fontes, M., & Laude, H. (2008b). Prominent and persistent extraneural infection in human PrP transgenic mice infected with variant CJD. *PLoS One, 3*(1), e1419.

Berman, P. H., Davidson, G. S., & Becker, L. E. (1988). Progressive neurological deterioration in a 14-year-old girl. *Pediatric Neuroscience, 14*(1), 42–49.

Bernoulli, C. C., Masters, C. L., Gajdusek, D. C., Gibbs, C. J., Jr., & Harris, J. O. (1979). Early clinical features of Creutzfeldt-Jakob disease: Subacute spongiform encephalopathy. In S. B. Prusiner & W. J. Hadlow (Eds.), *Slow transmissible diseases of the nervous system. Volume 1: Clinical, epidemiological, genetic, and pathological aspects of the spongiform encephalopathies* (pp. 229–251). Academic.

Biacabe, A. G., Laplanche, J. L., Ryder, S., & Baron, T. (2004). Distinct molecular phenotypes in bovine prion diseases. *EMBO Reports, 5*(1), 110–115.

Biacabe, A. G., Morignat, E., Vulin, J., Calavas, D., & Baron, T. G. (2008). Atypical bovine spongiform encephalopathies, France, 2001–2007. *Emerging Infectious Diseases, 14*(2), 298–300.

Bian, J., Kim, S., Kane, S. J., Crowell, J., Sun, J. L., Christiansen, J., Saijo, E., Moreno, J. A., DiLisio, J., Burnett, E., Pritzkow, S., Gorski, D., Soto, C., Kreeger, T. J., Balachandran, A., Mitchell, G., Miller, M. W., Nonno, R., Vikoren, T., Vage, J., Madslien, K., Tran, L., Vuong, T. T., Benestad, S. L., & Telling, G. C. (2021). Adaptive selection of a prion strain conformer corresponding to established North American CWD during propagation of novel emergent Norwegian strains in mice expressing elk or deer prion protein. *PLoS Pathogens, 17*(7), e1009748.

Blossom, D. B., Maddox, R. A., Beavers, S. F., Church, K. A., Thoroughman, D. A., Schonberger, L. B., & Belay, E. D. (2007). A case of Creutzfeldt-Jakob disease associated with a dura mater graft in the United States. *Infection Control and Hospital Epidemiology, 28*(12), 1396–1397.

Bongianni, M., Orru, C., Groveman, B. R., Sacchetto, L., Fiorini, M., Tonoli, G., Triva, G., Capaldi, S., Testi, S., Ferrari, S., Cagnin, A., Ladogana, A., Poleggi, A., Colaizzo, E., Tiple, D., Vaianella, L., Castriciano, S., Marchioni, D., Hughson, A. G., Imperiale, D., Cattaruzza, T., Fabrizi, G. M., Pocchiari, M., Monaco, S., Caughey, B., & Zanusso, G. (2017). Diagnosis of human prion disease using real-time quaking-induced conversion testing of olfactory mucosa and cerebrospinal fluid samples. *JAMA Neurology, 74*(2), 155–162.

Bons, N. (1996). Spontaneous spongiform encephalopathy in a monkey. *Lancet, 348*(9032), 956.

Bons, N., Mestre-Frances, N., Charnay, Y., Salmona, M., & Tagliavini, F. (1996). Spontaneous spongiform encephalopathy in a young adult rhesus monkey. *Comptes Rendus de l'Académie des Sciences. Série III, 319*(8), 733–736.

Bons, N., Mestre-Frances, N., Guiraud, I., & Charnay, Y. (1997). Prion immunoreactivity in brain, tonsil, gastrointestinal epithelial cells, and blood and lymph vessels in lemurian zoo primates with spongiform encephalopathy. *Comptes Rendus de l'Académie des Sciences. Série III, 320*(12), 971–979.

Bons, N., Mestre-Frances, N., Belli, P., Cathala, F., Gajdusek, D. C., & Brown, P. (1999). Natural and experimental oral infection of nonhuman primates by bovine spongiform encephalopathy agents. *Proceedings of the National Academy of Sciences of the United States of America, 96*(7), 4046–4051.

Bougard, D., Brandel, J. P., Belondrade, M., Beringue, V., Segarra, C., Fleury, H., Laplanche, J. L., Mayran, C., Nicot, S., Green, A., Welaratne, A., Narbey, D., Fournier-Wirth, C., Knight, R., Will, R., Tiberghien, P., Haik, S., & Coste, J. (2016). Detection of prions in the plasma of presymptomatic and symptomatic patients with variant Creutzfeldt-Jakob disease. *Science Translational Medicine, 8*(370), 370ra182.

Bounhar, Y., Zhang, Y., Goodyer, C. G., & LeBlanc, A. (2001). Prion protein protects human neurons against Bax-mediated apoptosis. *The Journal of Biological Chemistry, 276*(42), 39145–39149.

Bradley, R., & Matthews, D. (1992). Sub-acute, transmissible spongiform encephalopathies: Current concepts and future needs. *Revue Scientifique et Technique, 11*(2), 605–634.

Brandner, S., Isenmann, S., Raeber, A., Fischer, M., Sailer, A., Kobayashi, Y., Marino, S., Weissmann, C., & Aguzzi, A. (1996). Normal host prion protein necessary for scrapie-induced neurotoxicity. *Nature, 379*(6563), 339–343.

Bravo-Risi, F., Soto, P., Eckland, T., Dittmar, R., Ramirez, S., Catumbela, C. S. G., Soto, C., Lockwood, M., Nichols, T., & Morales, R. (2021). Detection of CWD prions in naturally infected white-tailed deer fetuses and gestational tissues by PMCA. *Scientific Reports, 11*(1), 18385.

Brotherston, J. G., Renwick, C. C., Stamp, J. T., Zlotnik, I., & Pattison, I. H. (1968). Spread and scrapie by contact to goats and sheep. *Journal of Comparative Pathology, 78*(1), 9–17.

Brown, P., & Cathala, F. (1979). Creutzfeldt-Jakob disease in France. In S. B. Prusiner & W. J. Hadlow (Eds.), *Slow transmissible diseases of the nervous system. Volume 1: Clinical, epidemiological, genetic, and pathological aspects of the spongiform encephalopathies* (pp. 213–227). Academic.

Brown, P., Rohwer, R. G., Green, E. M., & Gajdusek, D. C. (1982). Effect of chemicals, heat, and histopathologic processing on high-infectivity hamster-adapted scrapie virus. *The Journal of Infectious Diseases, 145*(5), 683–687.

Brown, P., Rohwer, R. G., & Gajdusek, D. C. (1984). Sodium hydroxide decontamination of Creutzfeldt-Jakob disease virus. *The New England Journal of Medicine, 310*(11), 727.

Brown, P., Goldfarb, L. G., Gibbs, C. J., Jr., & Gajdusek, D. C. (1991). The phenotypic expression of different mutations in transmissible familial Creutzfeldt-Jakob disease. *European Journal of Epidemiology, 7*(5), 469–476.

Brown, D. R., Qin, K., Herms, J. W., Madlung, A., Manson, J., Strome, R., Fraser, P. E., Kruck, T., von Bohlen, A., Schulz-Schaeffer, W., Giese, A., Westaway, D., & Kretzschmar, H. (1997). The cellular prion protein binds copper in vivo. *Nature, 390*(6661), 684–687.

Brown, P., Rau, E. H., Johnson, B. K., Bacote, A. E., Gibbs, C. J., Jr., & Gajdusek, D. C. (2000). New studies on the heat resistance of hamster-adapted scrapie agent: Threshold survival after ashing at 600 degrees C suggests an inorganic template of replication. *Proceedings of the National Academy of Sciences of the United States of America, 97*(7), 3418–3421.

Bruce, M. E., Will, R. G., Ironside, J. W., McConnell, I., Drummond, D., Suttie, A., McCardle, L., Chree, A., Hope, J., Birkett, C., Cousens, S., Fraser, H., & Bostock, C. J. (1997). Transmissions to mice indicate that 'new variant' CJD is caused by the BSE agent. *Nature, 389*(6650), 498–501.

Brunelle, B. W., Hamir, A. N., Baron, T., Biacabe, A. G., Richt, J. A., Kunkle, R. A., Cutlip, R. C., Miller, J. M., & Nicholson, E. M. (2007). Polymorphisms of the prion gene promoter region that influence classical bovine spongiform encephalopathy susceptibility are not applicable to other transmissible spongiform encephalopathies in cattle. *Journal of Animal Science, 85*(12), 3142–3147.

Bryan, J. A. (1984). Recommendations for handling specimens from patients with confirmed or suspected Creutzfeldt-Jakob disease. *Laboratory Medicine, 15*(1), 50–51.

Bueler, H., Fischer, M., Lang, Y., Bluethmann, H., Lipp, H. P., DeArmond, S. J., Prusiner, S. B., Aguet, M., & Weissmann, C. (1992). Normal development and behaviour of mice lacking the neuronal cell-surface PrP protein. *Nature, 356*(6370), 577–582.

Bunk, S. (2004). Chronic wasting disease—Prion disease in the wild. *PLoS Biology, 2*(4), E121.

Burger, D., & Hartsough, G. R. (1965). Encephalopathy of mink. II. Experimental and natural transmission. *The Journal of Infectious Diseases, 115*(4), 393–399.

Buschmann, A., & Groschup, M. H. (2005). Highly bovine spongiform encephalopathy-sensitive transgenic mice confirm the essential restriction of infectivity to the nervous system in clinically diseased cattle. *The Journal of Infectious Diseases, 192*(5), 934–942.

Buschmann, A., Gretzschel, A., Biacabe, A. G., Schiebel, K., Corona, C., Hoffmann, C., Eiden, M., Baron, T., Casalone, C., & Groschup, M. H. (2006). Atypical BSE in Germany—Proof of transmissibility and biochemical characterization. *Veterinary Microbiology, 117*(2–4), 103–116.

Cali, I., Castellani, R., Alshekhlee, A., Cohen, Y., Blevins, J., Yuan, J., Langeveld, J. P., Parchi, P., Safar, J. G., Zou, W. Q., & Gambetti, P. (2009). Co-existence of scrapie prion protein types 1 and 2 in sporadic Creutzfeldt-Jakob disease: Its effect on the phenotype and prion-type characteristics. *Brain, 132*(Pt 10), 2643–2658.

Capellari, S., Parchi, P., Cortelli, P., Avoni, P., Casadei, G. P., Bini, C., Baruzzi, A., Lugaresi, E., Pocchiari, M., Gambetti, P., & Montagna, P. (2008). Sporadic fatal insomnia in a fatal familial insomnia pedigree. *Neurology, 70*(11), 884–885.

Capobianco, R., Casalone, C., Suardi, S., Mangieri, M., Miccolo, C., Limido, L., Catania, M., Rossi, G., Di Fede, G., Giaccone, G., Bruzzone, M. G., Minati, L., Corona, C., Acutis, P., Gelmetti, D., Lombardi, G., Groschup, M. H., Buschmann, A., Zanusso, G., Monaco, S., Caramelli, M., & Tagliavini, F. (2007). Conversion of the BASE prion strain into the BSE strain: The origin of BSE? *PLoS Pathogens, 3*(3), e31.

Casalone, C., Zanusso, G., Acutis, P., Ferrari, S., Capucci, L., Tagliavini, F., Monaco, S., & Caramelli, M. (2004). Identification of a second bovine amyloidotic spongiform encephalopathy: Molecular similarities with sporadic Creutzfeldt-Jakob disease. *Proceedings of the National Academy of Sciences of the United States of America, 101*(9), 3065–3070.

Cassard, H., Torres, J. M., Lacroux, C., Douet, J. Y., Benestad, S. L., Lantier, F., Lugan, S., Lantier, I., Costes, P., Aron, N., Reine, F., Herzog, L., Espinosa, J. C., Beringue, V., & Andreoletti, O. (2014). Evidence for zoonotic potential of ovine scrapie prions. *Nature Communications, 5*, 5821.

Cassmann, E. D., Moore, S. J., & Greenlee, J. J. (2021). Experimental oronasal transmission of chronic wasting disease agent from white-tailed deer to Suffolk sheep. *Emerging Infectious Diseases, 27*(12), 3156–3158.

Castilla, J., Brun, A., Diaz-San Segundo, F., Salguero, F. J., Gutierrez-Adan, A., Pintado, B., Ramirez, M. A., del Riego, L., & Torres, J. M. (2005a). Vertical transmission of bovine spongiform encephalopathy prions evaluated in a transgenic mouse model. *Journal of Virology, 79*(13), 8665–8668.

Castilla, J., Saa, P., Hetz, C., & Soto, C. (2005b). In vitro generation of infectious scrapie prions. *Cell, 121*(2), 195–206.

Castilla, J., Gonzalez-Romero, D., Saa, P., Morales, R., De Castro, J., & Soto, C. (2008). Crossing the species barrier by PrP(Sc) replication in vitro generates unique infectious prions. *Cell, 134*(5), 757–768.

Castle, A. R., & Gill, A. C. (2017). Physiological functions of the cellular prion protein. *Frontiers in Molecular Biosciences, 4*, 19.

Caughey, B., & Raymond, G. J. (1991). The scrapie-associated form of PrP is made from a cell surface precursor that is both protease- and phospholipase-sensitive. *The Journal of Biological Chemistry, 266*(27), 18217–18223.

Caughey, B., Race, R. E., Ernst, D., Buchmeier, M. J., & Chesebro, B. (1989). Prion protein biosynthesis in scrapie-infected and uninfected neuroblastoma cells. *Journal of Virology, 63*(1), 175–181.

Center, N. W. H. (2024). *Expanding distribution of chronic wasting disease*. Retrieved January 22, 2024, from https://www.usgs.gov/centers/nwhc/science/expanding-distribution-chronic-wasting-disease

Cervenakova, L., Rohwer, R., Williams, S., Brown, P., & Gajdusek, D. C. (1997). High sequence homology of the PrP gene in mule deer and Rocky Mountain elk. *Lancet, 350*(9072), 219–220.

Chafin, T. K., Douglas, M. R., Martin, B. T., Zbinden, Z. D., Middaugh, C. R., Ballard, J. R., Gray, M. C., Don, W., Jr., & Douglas, M. E. (2020). Age structuring and spatial heterogeneity in prion protein gene (PRNP) polymorphism in white-tailed deer. *Prion, 14*(1), 238–248.

Chelle, P. L. (1942). A case of trembling in the goat. *Bulletin de l'Academie Veterinaire de France, 15*, 294–295.

Chen, Z., Chu, M., Zhang, J., Kong, Y., Xie, K., Cui, Y., Ye, H., Liu, L., Li, J., Wang, L., & Wu, L. (2023). Clinical profiles and ethnic heterogeneity of sporadic fatal insomnia. *European Journal of Neurology, 30*(4), 813–822.

Chesebro, B., & Caughey, B. (1993). Scrapie agent replication without the prion protein? *Current Biology, 3*(10), 696–698.

Chesebro, B., Trifilo, M., Race, R., Meade-White, K., Teng, C., LaCasse, R., Raymond, L., Favara, C., Baron, G., Priola, S., Caughey, B., Masliah, E., & Oldstone, M. (2005). Anchorless prion protein results in infectious amyloid disease without clinical scrapie. *Science, 308*(5727), 1435–1439.

Chu, M., Xie, K., Zhang, J., Chen, Z., Ghorayeb, I., Rupprecht, S., Reder, A. T., Garay, A., Honda, H., Nagayama, M., Shi, Q., Zhan, S., Nan, H., Zhang, J., Guan, H., Cui, L., Guo, Y., Rosa-Neto, P., Gauthier, S., Wang, J., Dong, X., & Wu, L. (2022). Proposal of new diagnostic criteria for fatal familial insomnia. *Journal of Neurology, 269*(9), 4909–4919.

Cobb, W. A., Hornabrook, R. W., & Sanders, S. (1973). The EEG of kuru. *Electroencephalography and Clinical Neurophysiology, 34*(4), 419–427.

Collie, D. A., Summers, D. M., Sellar, R. J., Ironside, J. W., Cooper, S., Zeidler, M., Knight, R., & Will, R. G. (2003). Diagnosing variant Creutzfeldt-Jakob disease with the pulvinar sign: MR imaging findings in 86 neuropathologically confirmed cases. *AJNR. American Journal of Neuroradiology, 24*(8), 1560–1569.

Collinge, J., Harding, A. E., Owen, F., Poulter, M., Lofthouse, R., Boughey, A. M., Shah, T., & Crow, T. J. (1989). Diagnosis of Gerstmann-Straussler syndrome in familial dementia with prion protein gene analysis. *Lancet, 2*(8653), 15–17.

Collinge, J., Sidle, K. C., Meads, J., Ironside, J., & Hill, A. F. (1996). Molecular analysis of prion strain variation and the aetiology of 'new variant' CJD. *Nature, 383*(6602), 685–690.

Collinge, J., Whitfield, J., McKintosh, E., Beck, J., Mead, S., Thomas, D. J., & Alpers, M. P. (2006). Kuru in the 21st century—An acquired human prion disease with very long incubation periods. *Lancet, 367*(9528), 2068–2074.

Cortez, L. M., Nemani, S. K., Duque Velasquez, C., Sriraman, A., Wang, Y., Wille, H., McKenzie, D., & Sim, V. L. (2021). Asymmetric-flow field-flow fractionation of prions reveals a strain-specific continuum of quaternary structures with protease resistance developing at a hydrodynamic radius of 15 nm. *PLoS Pathogens, 17*(6), e1009703.

Cranwell, M. P., Hancock, R. D., Hindson, J. R., Hall, S. A., Daniel, N. J., Hopkins, A. R., Wonnacott, B., Vivian, M., & Hunt, B. (1988). Bovine spongiform encephalopathy. *The Veterinary Record, 122*(8), 190.

Creutzfeldt, H. G. (1920). Uber eine eigenartige herdfurmige Erkrankung des Zentral-nervensystems. *Zeitschrift für die Gesamte Neurologie und Psychiatrie, 57*, 1–18.

Cunningham, A. A. (1991). Bovine spongiform encephalopathy and British zoos. *Journal of Zoo and Wildlife Medicine, 22*(3), 304–308.

Cunningham, A. A., Kirkwood, J. K., Dawson, M., Spencer, Y. I., Green, R. B., & Wells, G. A. (2004). Bovine spongiform encephalopathy infectivity in greater kudu (Tragelaphus strepsiceros). *Emerging Infectious Diseases, 10*(6), 1044–1049.

Curnow, R. N., & Hau, C. M. (1996). The incidence of bovine spongiform encephalopathy in the progeny of affected sires and dams. *The Veterinary Record, 138*(17), 407–408.

Cutlip, R. C., Miller, J. M., Race, R. E., Jenny, A. L., Katz, J. B., Lehmkuhl, H. D., DeBey, B. M., & Robinson, M. M. (1994). Intracerebral transmission of scrapie to cattle. *The Journal of Infectious Diseases, 169*(4), 814–820.

Deleault, N. R., Walsh, D. J., Piro, J. R., Wang, F., Wang, X., Ma, J., Rees, J. R., & Supattapone, S. (2012). Cofactor molecules maintain infectious conformation and restrict strain properties in purified prions. *Proceedings of the National Academy of Sciences of the United States of America, 109*(28), E1938–E1946.

Di Bari, M. A., Chianini, F., Vaccari, G., Esposito, E., Conte, M., Eaton, S. L., Hamilton, S., Finlayson, J., Steele, P. J., Dagleish, M. P., Reid, H. W., Bruce, M., Jeffrey, M., Agrimi, U., & Nonno, R. (2008). The bank vole (Myodes glareolus) as a sensitive bioassay for sheep scrapie. *The Journal of General Virology, 89*(Pt 12), 2975–2985.

Dickinson, A. G., & Fraser, H. (1979). An assessment of the genetics of scrapie in sheep and mice. In S. B. Prusiner & W. J. Hadlow (Eds.), *Slow transmissible diseases of the nervous system. Volume 1: Clinical, epidemiological, genetic, and pathological aspects of the spongiform encephalopathies* (pp. 367–385). Academic.

Dickinson, A. G., Young, G. B., & Renwick, C. C. (1966). Paper No. 29 Scrapie: Experiments involving maternal transmission in sheep. Report of Scrapie seminar, ARS 91-53, US Dept. of Agr., pp. 244–248.

Dong, T. T., & Satoh, K. (2021). The latest research on RT-QuIC assays – A literature review. *Pathogens, 10*(3).

Dudas, S., Yang, J., Graham, C., Czub, M., McAllister, T. A., Coulthart, M. B., & Czub, S. (2010). Molecular, biochemical and genetic characteristics of BSE in Canada. *PLoS One, 5*(5), e10638.

Eckroade, R. J., ZuRhein, G. M., & Hanson, R. P. (1979). Experimental transmissible mink encephalopathy brain lesions and their sequential development in mink. In S. B. Prusiner & W. J. Hadlow (Eds.), *Slow transmissible diseases of the nervous system. Volume 1: Clinical, epidemiological, genetic, and pathological aspects of the spongiform encephalopathies* (pp. 409–449). Academic.

Edmunds, D. R., Kauffman, M. J., Schumaker, B. A., Lindzey, F. G., Cook, W. E., Kreeger, T. J., Grogan, R. G., & Cornish, T. E. (2016). Chronic wasting disease drives population decline of White-tailed deer. *PLoS One, 11*(8), e0161127.

Eiden, M., Hoffmann, C., Balkema-Buschmann, A., Muller, M., Baumgartner, K., & Groschup, M. H. (2010). Biochemical and immunohistochemical characterization of feline spongiform encephalopathy in a German captive cheetah. *The Journal of General Virology, 91*(Pt 11), 2874–2883.

Erana, H., San Millan, B., Diaz-Dominguez, C. M., Charco, J. M., Rodriguez, R., Vieitez, I., Pereda, A., Yanez, R., Geijo, M., Navarro, C., Perez de Nanclares, G., Teijeira, S., & Castilla, J. (2022). Description of the first Spanish case of Gerstmann-Straussler-Scheinker disease with A117V variant: Clinical, histopathological and biochemical characterization. *Journal of Neurology, 269*(8), 4253–4263.

Escobar, L. E., Pritzkow, S., Winter, S. N., Grear, D. A., Kirchgessner, M. S., Dominguez-Villegas, E., Machado, G., Townsend Peterson, A., & Soto, C. (2020). The ecology of chronic wasting disease in wildlife. *Biological Reviews of the Cambridge Philosophical Society, 95*(2), 393–408.

European Food Safety, A. (2023). The European Union summary report on surveillance for the presence of transmissible spongiform encephalopathies (TSE) in 2022. *EFSA Journal, 21*(11), e08384.

Ferguson, N. M., Donnelly, C. A., Woolhouse, M. E., & Anderson, R. M. (1997). A genetic interpretation of heightened risk of BSE in offspring of affected dams. *Proceedings of the Biological Sciences, 264*(1387), 1445–1455.

Ford, M. J., Burton, L. J., Li, H., Graham, C. H., Frobert, Y., Grassi, J., Hall, S. M., & Morris, R. J. (2002a). A marked disparity between the expression of prion protein and its message by neurones of the CNS. *Neuroscience, 111*(3), 533–551.

Ford, M. J., Burton, L. J., Morris, R. J., & Hall, S. M. (2002b). Selective expression of prion protein in peripheral tissues of the adult mouse. *Neuroscience, 113*(1), 177–192.

Ford, L., Rudge, P., Robinson, K., Collinge, J., Gorham, M., & Mead, S. (2019). The most problematic symptoms of prion disease – An analysis of carer experiences. *International Psychogeriatrics, 31*(8), 1181–1190.

Forloni, G., Tettamanti, M., Lucca, U., Albanese, Y., Quaglio, E., Chiesa, R., Erbetta, A., Villani, F., Redaelli, V., Tagliavini, F., Artuso, V., & Roiter, I. (2015). Preventive study in subjects at risk of fatal familial insomnia: Innovative approach to rare diseases. *Prion, 9*(2), 75–79.

Forloni, G., Roiter, I., Artuso, V., Marcon, M., Colesso, W., Luban, E., Lucca, U., Tettamanti, M., Pupillo, E., Redaelli, V., Mariuzzo, F., Boscolo Buleghin, G., Mariuzzo, A., Tagliavini, F., Chiesa, R., & Ambrosini, A. (2022). Preventive pharmacological treatment in subjects at risk for fatal familial insomnia: Science and public engagement. *Prion, 16*(1), 66–77.

Foster, J. D., McKelvey, W. A., Mylne, M. J., Williams, A., Hunter, N., Hope, J., & Fraser, H. (1992). Studies on maternal transmission of scrapie in sheep by embryo transfer. *The Veterinary Record, 130*(16), 341–343.

Fox, K. A., Jewell, J. E., Williams, E. S., & Miller, M. W. (2006). Patterns of PrPCWD accumulation during the course of chronic wasting disease infection in orally inoculated mule deer (Odocoileus hemionus). *The Journal of General Virology, 87*(Pt 11), 3451–3461.

Gajdusek, D. C. (1977). Unconventional viruses and the origin and disappearance of kuru. *Science, 197*(4307), 943–960.

Gajdusek, D. C. (1979). Observations of the early history of Kuru investigation. In S. B. Prusiner & W. J. Hadlow (Eds.), *Slow transmissible diseases of the nervous system. Volume 1: Clinical, epidemiological, genetic, and pathological aspects of the spongiform encephalopathies* (pp. 7–35). Academic.

Gajdusek, D. C., & Alpers, M. (1966). Kuru in childhood: Disappearance of the disease in the younger age group. *The Journal of Pediatrics, 69*(5), 886–887.

Gajdusek, D. C., & Zigas, V. (1957). Degenerative disease of the central nervous system in New Guinea; the endemic occurrence of kuru in the native population. *The New England Journal of Medicine, 257*(20), 974–978.

Gajdusek, D. C., Gibbs, C. J., Jr., Asher, D. M., Brown, P., Diwan, A., Hoffman, P., Nemo, G., Rohwer, R., & White, L. (1977). Precautions in medical care of, and in handling materials from, patients with transmissible virus dementia (Creutzfeldt-Jakob disease). *The New England Journal of Medicine, 297*(23), 1253–1258.

Gambetti, P., Kong, Q., Zou, W., Parchi, P., & Chen, S. G. (2003). Sporadic and familial CJD: Classification and characterisation. *British Medical Bulletin, 66*, 213–239.

Gambetti, P., Dong, Z., Yuan, J., Xiao, X., Zheng, M., Alshekhlee, A., Castellani, R., Cohen, M., Barria, M. A., Gonzalez-Romero, D., Belay, E. D., Schonberger, L. B., Marder, K., Harris, C., Burke, J. R., Montine, T., Wisniewski, T., Dickson, D. W., Soto, C., Hulette, C. M., Mastrianni, J. A., Kong, Q., & Zou, W. Q. (2008). A novel human disease with abnormal prion protein sensitive to protease. *Annals of Neurology, 63*(6), 697–708.

Gambetti, P., Puoti, G., & Zou, W. Q. (2011). Variably protease-sensitive prionopathy: A novel disease of the prion protein. *Journal of Molecular Neuroscience, 45*(3), 422–424.

Gao, Z., Zhang, H., Hu, F., Yang, L., Yang, X., Zhu, Y., Sy, M. S., & Li, C. (2016). Glycan-deficient PrP stimulates VEGFR2 signaling via glycosaminoglycan. *Cellular Signalling, 28*(6), 652–662.

Garza, M. C., Fernandez-Borges, N., Bolea, R., Badiola, J. J., Castilla, J., & Monleon, E. (2011). Detection of PrPres in genetically susceptible fetuses from sheep with natural scrapie. *PLoS One, 6*(12), e27525.

Garza, M. C., Monzon, M., Marin, B., Badiola, J. J., & Monleon, E. (2014). Distribution of peripheral PrP(Sc) in sheep with naturally acquired scrapie. *PLoS One, 9*(5), e97768.

Gavier-Widen, D., Noremark, M., Benestad, S., Simmons, M., Renstrom, L., Bratberg, B., Elvander, M., & af Segerstad, C. H. (2004). Recognition of the Nor98 variant of scrapie in the Swedish sheep population. *Journal of Veterinary Diagnostic Investigation, 16*(6), 562–567.

Geoghegan, J. C., Valdes, P. A., Orem, N. R., Deleault, N. R., Williamson, R. A., Harris, B. T., & Supattapone, S. (2007). Selective incorporation of polyanionic molecules into hamster prions. *The Journal of Biological Chemistry, 282*(50), 36341–36353.

Gerstmann, J., Straussler, E., & Scheinker, I. (1935). Über eine eigenartige hereditär-familiäre Erkrankung des Zentralnervensystems. Zugleich ein Beitrag zur Frage des vorzeitigen lokalen Alterns. *Zeitschrift für die gesamte Neurologie und Psychiatrie, 154*, 736–762.

Ghoshal, N., Perry, A., McKeel, D., Schmidt, R. E., Carter, D., Norton, J., Zou, W. Q., Xiao, X., Puoti, G., Notari, S., Gambetti, P., Morris, J. C., & Cairns, N. J. (2015). Variably protease-sensitive prionopathy in an apparent cognitively normal 93-year-old. *Alzheimer Disease and Associated Disorders, 29*(2), 173–176.

Goldhammer, Y., Gabizon, R., Meiner, Z., & Sadeh, M. (1993). An Israeli family with Gerstmann-Straussler-Scheinker disease manifesting the codon 102 mutation in the prion protein gene. *Neurology, 43*(12), 2718–2719.

Gordon, W. S. (1959). Scrapie panel. Proceedings of the 63rd annual meeting of the United States Live Stock Sanitary Association.

Greenlee, J. J., Hamir, A. N., & West Greenlee, M. H. (2006). Abnormal prion accumulation associated with retinal pathology in experimentally inoculated scrapie-affected sheep. *Veterinary Pathology, 43*(5), 733–739.

Greenlee, J. J., Moore, S. J., Cassmann, E. D., Lambert, Z. J., Kokemuller, R. D., Smith, J. D., Kunkle, R. A., Kong, Q., & Greenlee, M. H. W. (2023). Characterization of classical sheep scrapie in white-tailed deer after experimental oronasal exposure. *The Journal of Infectious Diseases, 227*(12), 1386–1395.

Greig, J. R. (1940). Scrapie: Observations on the transmission of the disease by mediate contact. *The Veterinary Journal, 96*, 203–206.

Hadlow, W. J. (1995). Neuropathology and the scrapie-kuru connection. *Brain Pathology, 5*(1), 27–31.

Hainfellner, J. A., Liberski, P. P., Guiroy, D. C., Cervenakova, L., Brown, P., Gajdusek, D. C., & Budka, H. (1997). Pathology and immunocytochemistry of a kuru brain. *Brain Pathology, 7*(1), 547–553.

Haley, N. J., Mathiason, C. K., Zabel, M. D., Telling, G. C., & Hoover, E. A. (2009). Detection of sub-clinical CWD infection in conventional test-negative deer long after oral exposure to urine and feces from CWD+ deer. *PLoS One, 4*(11), e7990.

Hallinan, G. I., Ozcan, K. A., Hoq, M. R., Cracco, L., Vago, F. S., Bharath, S. R., Li, D., Jacobsen, M., Doud, E. H., Mosley, A. L., Fernandez, A., Garringer, H. J., Jiang, W., Ghetti, B., & Vidal, R. (2022). Cryo-EM structures of prion protein filaments from Gerstmann-Straussler-Scheinker disease. *Acta Neuropathologica, 144*(3), 509–520.

Hamir, A. N., Clark, W. W., Sutton, D. L., Miller, J. M., Stack, M. J., Chaplin, M. J., & Jenny, A. L. (2002). Resistance of domestic cats to a US sheep scrapie agent by intracerebral route. *Journal of Veterinary Diagnostic Investigation, 14*(5), 444–445.

Hamir, A. N., Miller, J. M., Cutlip, R. C., Stack, M. J., Chaplin, M. J., Jenny, A. L., & Williams, E. S. (2003). Experimental inoculation of scrapie and chronic wasting disease agents in raccoons (Procyon lotor). *The Veterinary Record, 153*(4), 121–123.

Hanson, R. P., Eckroade, R. J., Marsh, R. F., Zu Rhein, G. M., Kanitz, C. L., & Gustafson, D. P. (1971). Susceptibility of mink to sheep scrapie. *Science, 172*(3985), 859–861.

Hartsough, G. R., & Burger, D. (1965). Encephalopathy of mink. I. Epizootiologic and clinical observations. *The Journal of Infectious Diseases, 115*(4), 387–392.

Heaton, M. P., Keele, J. W., Harhay, G. P., Richt, J. A., Koohmaraie, M., Wheeler, T. L., Shackelford, S. D., Casas, E., King, D. A., Sonstegard, T. S., Van Tassell, C. P., Neibergs, H. L., Chase, C. C., Jr., Kalbfleisch, T. S., Smith, T. P., Clawson, M. L., & Laegreid, W. W. (2008). Prevalence of the prion protein gene E211K variant in U.S. cattle. *BMC Veterinary Research, 4*, 25.

Hermann, P., Appleby, B., Brandel, J. P., Caughey, B., Collins, S., Geschwind, M. D., Green, A., Haik, S., Kovacs, G. G., Ladogana, A., Llorens, F., Mead, S., Nishida, N., Pal, S., Parchi, P., Pocchiari, M., Satoh, K., Zanusso, G., & Zerr, I. (2021). Biomarkers and diagnostic guidelines for sporadic Creutzfeldt-Jakob disease. *Lancet Neurology, 20*(3), 235–246.

Hermann, P., Canaslan, S., Villar-Pique, A., Bunck, T., Goebel, S., Llorens, F., Schmitz, M., & Zerr, I. (2022). Plasma neurofilament light chain as a biomarker for fatal familial insomnia. *European Journal of Neurology, 29*(6), 1841–1846.

Hilbe, M. M., Soldati, G. G., Zlinszky, K. K., Wunderlin, S. S., & Ehrensperger, F. F. (2009). Immunohistochemical study of PrP(Sc) distribution in neural and extraneural tissues of two cats with feline spongiform encephalopathy. *BMC Veterinary Research, 5*, 11.

Hill, A. F., Desbruslais, M., Joiner, S., Sidle, K. C., Gowland, I., Collinge, J., Doey, L. J., & Lantos, P. (1997). The same prion strain causes vCJD and BSE. *Nature, 389*(6650), 448-450, 526.

Hillier, C. E., & Salmon, R. L. (2000). Is there evidence for exogenous risk factors in the aetiology and spread of Creutzfeldt-Jakob disease? *QJM, 93*(9), 617–631.

Hoinville, L. J., Tongue, S. C., & Wilesmith, J. W. (2010). Evidence for maternal transmission of scrapie in naturally affected flocks. *Preventive Veterinary Medicine, 93*(2–3), 121–128.

Holman, R. C., Belay, E. D., Christensen, K. Y., Maddox, R. A., Minino, A. M., Folkema, A. M., Haberling, D. L., Hammett, T. A., Kochanek, K. D., Sejvar, J. J., & Schonberger, L. B. (2010). Human prion diseases in the United States. *PLoS One, 5*(1), e8521.

Hope, J., Reekie, L. J., Hunter, N., Multhaup, G., Beyreuther, K., White, H., Scott, A. C., Stack, M. J., Dawson, M., & Wells, G. A. (1988).

Fibrils from brains of cows with new cattle disease contain scrapie-associated protein. *Nature, 336*(6197), 390–392.

Horigan, V., Gale, P., Adkin, A., Konold, T., Cassar, C., Spiropoulos, J., & Kelly, L. (2020). Assessing the aggregated probability of entry of a novel prion disease agent into the United Kingdom. *Microbial Risk Analysis, 16*, 100134.

Hornabrook, R. W. (1979). Kuru and clinical neurology. In S. B. Prusiner & W. J. Hadlow (Eds.), *Slow transmissible diseases of the nervous system. Volume 1: Clinical, epidemiological, genetic, and pathological aspects of the spongiform encephalopathies* (pp. 7–35). Academic.

Hourrigan, J., Klingsporn, A., Clark, W. W., & de Camp, M. (1979). Epidemiology of scrapie in the United States. In S. B. Prusiner & W. J. Hadlow (Eds.), *Slow transmissible diseases of the nervous system. Volume 1: Clinical, epidemiological, genetic, and pathological aspects of the spongiform encephalopathies* (pp. 351–356). Academic.

Hsiao, K., Baker, H. F., Crow, T. J., Poulter, M., Owen, F., Terwilliger, J. D., Westaway, D., Ott, J., & Prusiner, S. B. (1989). Linkage of a prion protein missense variant to Gerstmann-Straussler syndrome. *Nature, 338*(6213), 342–345.

Hsiao, K., Meiner, Z., Kahana, E., Cass, C., Kahana, I., Avrahami, D., Scarlato, G., Abramsky, O., Prusiner, S. B., & Gabizon, R. (1991). Mutation of the prion protein in Libyan Jews with Creutzfeldt-Jakob disease. *The New England Journal of Medicine, 324*(16), 1091–1097.

Hsiao, K., Dlouhy, S. R., Farlow, M. R., Cass, C., Da Costa, M., Conneally, P. M., Hodes, M. E., Ghetti, B., & Prusiner, S. B. (1992). Mutant prion proteins in Gerstmann-Straussler-Scheinker disease with neurofibrillary tangles. *Nature Genetics, 1*(1), 68–71.

Huang, J., Cohen, M., Safar, J., & Auchus, A. P. (2021). Variably protease-sensitive prionopathy in a middle-aged man with rapidly progressive dementia. *Cognitive and Behavioral Neurology, 34*(3), 220–225.

Hunter, G. D. (1979). The enigma of the scrapie agent: Biochemical approaches and the involvement of membranes and nucleic acids. In S. B. Prusiner & W. J. Hadlow (Eds.), *Slow transmissible diseases of the nervous system. Volume 2: Pathogenesis, immunology, virology, and molecular biology of the spongiform encephalopathies* (pp. 365–385). Academic.

Hunter, G. D., & Millson, G. C. (1964). Studies on the heat stability and chromatographic behaviour of the scrapie agent. *Journal of General Microbiology, 37*, 251–258.

Imran, M., & Mahmood, S. (2011). An overview of human prion diseases. *Virology Journal, 8*, 559.

Ishizawa, K., Mitsufuji, T., Shioda, K., Kobayashi, A., Komori, T., Nakazato, Y., Kitamoto, T., Araki, N., Yamamoto, T., & Sasaki, A. (2018). An autopsy report of three kindred in a Gerstmann-Straussler-Scheinker disease P105L family with a special reference to prion protein, tau, and beta-amyloid. *Brain and Behavior: A Cognitive Neuroscience Perspective, 8*(10), e01117.

Jakob, A. (1920). Uber eigenartige Erkrankungen des Zentralnervensystems mit bemerkenswertem anatomischen Befunde (Spastische Pseudosklerose – Encephalomyelopathie mit disseminierten Degenerationsherden). *Zeitschrift für die Gesamte Neurologie und Psychiatrie, 64*, 147–228.

Jansen, C., Head, M. W., van Gool, W. A., Baas, F., Yull, H., Ironside, J. W., & Rozemuller, A. J. (2010). The first case of protease-sensitive prionopathy (PSPr) in The Netherlands: A patient with an unusual GSS-like clinical phenotype. *Journal of Neurology, Neurosurgery, and Psychiatry, 81*(9), 1052–1055.

Jeffrey, M., & Wells, G. A. (1988). Spongiform encephalopathy in a nyala (Tragelaphus angasi). *Veterinary Pathology, 25*(5), 398–399.

Jewell, J. E., Conner, M. M., Wolfe, L. L., Miller, M. W., & Williams, E. S. (2005). Low frequency of PrP genotype 225SF among free-ranging mule deer (Odocoileus hemionus) with chronic wasting disease. *The Journal of General Virology, 86*(Pt 8), 2127–2134.

Johnson, C. J., Herbst, A., Duque-Velasquez, C., Vanderloo, J. P., Bochsler, P., Chappell, R., & McKenzie, D. (2011). Prion protein polymorphisms affect chronic wasting disease progression. *PLoS One, 6*(3), e17450.

Kaluz, S., Kaluzova, M., & Flint, A. P. (1997). Sequencing analysis of prion genes from red deer and camel. *Gene, 199*(1–2), 283–286.

Kang, H. E., Bian, J., Kane, S. J., Kim, S., Selwyn, V., Crowell, J., Bartz, J. C., & Telling, G. C. (2020). Incomplete glycosylation during prion infection unmasks a prion protein epitope that facilitates prion detection and strain discrimination. *The Journal of Biological Chemistry, 295*(30), 10420–10433.

Khan, Z., & Bollu, P. C. (2022). *Fatal familial insomnia*. StatPearls. NCBI Bookshelf, StatPearls Publishing.

Kim, T. Y., Shon, H. J., Joo, Y. S., Mun, U. K., Kang, K. S., & Lee, Y. S. (2005). Additional cases of chronic wasting disease in imported deer in Korea. *The Journal of Veterinary Medical Science, 67*(8), 753–759.

Kim, C., Haldiman, T., Surewicz, K., Cohen, Y., Chen, W., Blevins, J., Sy, M. S., Cohen, M., Kong, Q., Telling, G. C., Surewicz, W. K., & Safar, J. G. (2012). Small protease sensitive oligomers of PrPSc in distinct human prions determine conversion rate of PrP(C). *PLoS Pathogens, 8*(8), e1002835.

Kim, H. H., Kim, Y. C., Kim, K., Kim, A. D., & Jeong, B. H. (2020). Novel polymorphisms and genetic features of the prion protein gene (PRNP) in cats, hosts of feline spongiform encephalopathy. *Genes (Basel), 12*(1).

Kimberlin, R. H. (1976). Experimental scrapie in the mouse: A review of an important model disease. *Science Progress, 63*(252), 461–481.

Kirkwood, J. K., & Cunningham, A. A. (1994). Epidemiological observations on spongiform encephalopathies in captive wild animals in the British Isles. *The Veterinary Record, 135*(13), 296–303.

Kirkwood, J. K., & Cunningham, A. A. (1999). Scrapie-like spongiform encephalopathies (prion diseases) in nondomesticated species. In M. E. Fowler & R. E. Miller (Eds.), *Fowler's zoo and wild animal medicine current therapy* (pp. 662–668). Saunders.

Kirkwood, J. K., & Cunningham, A. A. (2006). Portrait of prion diseases in zoo animals. In B. Hörnlimann, D. Riesner, & H. A. Kretzschmar (Eds.), *Prions in humans and animals* (pp. 250–256). Walter de Gruyter.

Kirkwood, J. K., Wells, G. A., Wilesmith, J. W., Cunningham, A. A., & Jackson, S. I. (1990). Spongiform encephalopathy in an arabian oryx (Oryx leucoryx) and a greater kudu (Tragelaphus strepsiceros). *The Veterinary Record, 127*(17), 418–420.

Kirkwood, J. K., Wells, G. A., Cunningham, A. A., Jackson, S. I., Scott, A. C., Dawson, M., & Wilesmith, J. W. (1992). Scrapie-like encephalopathy in a greater kudu (Tragelaphus strepsiceros) which had not been fed ruminant-derived protein. *The Veterinary Record, 130*(17), 365–367.

Kirkwood, J. K., Cunningham, A. A., Flach, E. J., Thornton, S. M., & Wells, G. A. H. (1995). Spongiform encephalopathy in another captive cheetah (Acinonyx jubatus): Evidence for variation in susceptibility or incubation periods between species? *Journal of Zoo and Wildlife Medicine, 26*(4), 577–582.

Kitamoto, T., Amano, N., Terao, Y., Nakazato, Y., Isshiki, T., Mizutani, T., & Tateishi, J. (1993). A new inherited prion disease (PrP-P105L mutation) showing spastic paraparesis. *Annals of Neurology, 34*(6), 808–813.

Klatzo, I., Gajdusek, D. C., & Zigas, V. (1959). Pathology of Kuru. *Laboratory Investigation, 8*(4), 799–847.

Kovacs, G. G., Puopolo, M., Ladogana, A., Pocchiari, M., Budka, H., van Duijn, C., Collins, S. J., Boyd, A., Giulivi, A., Coulthart, M., Delasnerie-Laupretre, N., Brandel, J. P., Zerr, I., Kretzschmar, H. A., de Pedro-Cuesta, J., Calero-Lara, M., Glatzel, M., Aguzzi, A., Bishop, M., Knight, R., Belay, G., Will, R., Mitrova, E., &

EUROCJD. (2005). Genetic prion disease: The EUROCJD experience. *Human Genetics, 118*(2), 166–174.

Krasnianski, A., Bartl, M., Sanchez Juan, P. J., Heinemann, U., Meissner, B., Varges, D., Schulze-Sturm, U., Kretzschmar, H. A., Schulz-Schaeffer, W. J., & Zerr, I. (2008). Fatal familial insomnia: Clinical features and early identification. *Annals of Neurology, 63*(5), 658–661.

Krumm, C. E., Conner, M. M., & Miller, M. W. (2005). Relative vulnerability of chronic wasting disease infected mule deer to vehicle collisions. *Journal of Wildlife Diseases, 41*(3), 503–511.

Kurt, T. D., & Sigurdson, C. J. (2016). Cross-species transmission of CWD prions. *Prion, 10*(1), 83–91.

Kuznetsova, A., McKenzie, D., Cullingham, C., & Aiken, J. M. (2020). Long-term incubation PrP(CWD) with soils affects prion recovery but not infectivity. *Pathogens, 9*(4).

Lacroux, C., Simon, S., Benestad, S. L., Maillet, S., Mathey, J., Lugan, S., Corbiere, F., Cassard, H., Costes, P., Bergonier, D., Weisbecker, J. L., Moldal, T., Simmons, H., Lantier, F., Feraudet-Tarisse, C., Morel, N., Schelcher, F., Grassi, J., & Andreoletti, O. (2008). Prions in milk from ewes incubating natural scrapie. *PLoS Pathogens, 4*(12), e1000238.

Lakkaraju, A. K. K., Sorce, S., Senatore, A., Nuvolone, M., Guo, J., Schwarz, P., Moos, R., Pelczar, P., & Aguzzi, A. (2022). Glial activation in prion diseases is selectively triggered by neuronal PrP(Sc). *Brain Pathology, 32*(5), e13056.

Lavigna, G., Masone, A., Bouybayoune, I., Bertani, I., Lucchetti, J., Gobbi, M., Porcu, L., Zordan, S., Rigamonti, M., Imeri, L., Restelli, E., & Chiesa, R. (2021). Doxycycline rescues recognition memory and circadian motor rhythmicity but does not prevent terminal disease in fatal familial insomnia mice. *Neurobiology of Disease, 158*, 105455.

Leggett, M. M., Dukes, J., & Pirie, H. M. (1990). A spongiform encephalopathy in a cat. *The Veterinary Record, 127*(24), 586–588.

Lezmi, S., Bencsik, A., & Baron, T. (2006). PET-blot analysis contributes to BSE strain recognition in C57Bl/6 mice. *The Journal of Histochemistry and Cytochemistry, 54*(10), 1087–1094.

Liberski, P. P. (2012). Gerstmann-Straussler-Scheinker disease. *Advances in Experimental Medicine and Biology, 724*, 128–137.

Liberski, P. P., Sikorska, B., Lindenbaum, S., Goldfarb, L. G., McLean, C., Hainfellner, J. A., & Brown, P. (2012). Kuru: Genes, cannibals and neuropathology. *Journal of Neuropathology and Experimental Neurology, 71*(2), 92–103.

Little, B. W., Brown, P. W., Rodgers-Johnson, P., Perl, D. P., & Gajdusek, D. C. (1986). Familial myoclonic dementia masquerading as Creutzfeldt-Jakob disease. *Annals of Neurology, 20*(2), 231–239.

Lugaresi, E., Medori, R., Montagna, P., Baruzzi, A., Cortelli, P., Lugaresi, A., Tinuper, P., Zucconi, M., & Gambetti, P. (1986). Fatal familial insomnia and dysautonomia with selective degeneration of thalamic nuclei. *The New England Journal of Medicine, 315*(16), 997–1003.

Lysek, D. A., Nivon, L. G., & Wuthrich, K. (2004). Amino acid sequence of the Felis catus prion protein. *Gene, 341*, 249–253.

M'Gowan, J. P. (1914). *Investigation into the disease of sheep called "scrapie"*. William Blackwood and Sons.

Maddison, B. C., Baker, C. A., Terry, L. A., Bellworthy, S. J., Thorne, L., Rees, H. C., & Gough, K. C. (2010a). Environmental sources of scrapie prions. *Journal of Virology, 84*(21), 11560–11562.

Maddison, B. C., Rees, H. C., Baker, C. A., Taema, M., Bellworthy, S. J., Thorne, L., Terry, L. A., & Gough, K. C. (2010b). Prions are secreted into the oral cavity in sheep with preclinical scrapie. *The Journal of Infectious Diseases, 201*(11), 1672–1676.

Maddison, B. C., Spiropoulos, J., Vickery, C. M., Lockey, R., Owen, J. P., Bishop, K., Baker, C. A., & Gough, K. C. (2015). Incubation of ovine scrapie with environmental matrix results in biological and biochemical changes of PrP(Sc) over time. *Veterinary Research, 46*(1), 46.

Malamud, N. (1979). Creutzfeldt-Jakob's disease: A clinicopathologic study. In S. B. Prusiner & W. J. Hadlow (Eds.), *Slow transmissible diseases of the nervous system. Volume 1: Clinical, epidemiological, genetic, and pathological aspects of the spongiform encephalopathies* (pp. 271–285). Academic.

Mallucci, G., Dickinson, A., Linehan, J., Klohn, P. C., Brandner, S., & Collinge, J. (2003). Depleting neuronal PrP in prion infection prevents disease and reverses spongiosis. *Science, 302*(5646), 871–874.

Malmgren, R., Kurland, L., Mokri, B., & Kurtzke, J. (1979). The epidemiology of Creutzfeldt-Jakob disease. In S. B. Prusiner & W. J. Hadlow (Eds.), *Slow transmissible diseases of the nervous system. Volume 1: Clinical, epidemiological, genetic, and pathological aspects of the spongiform encephalopathies* (pp. 93–112). Academic.

Manka, S. W., Wenborn, A., Betts, J., Joiner, S., Saibil, H. R., Collinge, J., & Wadsworth, J. D. F. (2023). A structural basis for prion strain diversity. *Nature Chemical Biology, 19*(5), 607–613.

Marsh, R. F., & Bessen, R. A. (1993). Epidemiologic and experimental studies on transmissible mink encephalopathy. *Developments in Biological Standardization, 80*, 111–118.

Marsh, R. F., & Hanson, R. P. (1969). Transmissible mink encephalopathy: Behavior of the disease agent in mink. *American Journal of Veterinary Research, 30*(9), 1637–1642.

Marsh, R. F., Bessen, R. A., Lehmann, S., & Hartsough, G. R. (1991). Epidemiological and experimental studies on a new incident of transmissible mink encephalopathy. *The Journal of General Virology, 72*(Pt 3), 589–594.

Masters, C. L., Gajdusek, D. C., Gibbs, C. J. J., Bernoulli, C., & Asher, D. M. (1979a). Familial Creutzfeldt-Jakob disease and other familial dementias: An inquiry into possible modes of transmission or virus-induced familial diseases. In S. B. Prusiner & W. J. Hadlow (Eds.), *Slow transmissible diseases of the nervous system. Volume 1: Clinical, epidemiological, genetic, and pathological aspects of the spongiform encephalopathies* (pp. 143–194). Academic.

Masters, C. L., Harris, J. O., Gajdusek, D. C., Gibbs, C. J. J., Bernoulli, C., & Asher, D. M. (1979b). Creutzfeldt-Jakob disease: Patterns of worldwide occurrence. In S. B. Prusiner & W. J. Hadlow (Eds.), *Slow transmissible diseases of the nervous system. Volume 1: Clinical, epidemiological, genetic, and pathological aspects of the spongiform encephalopathies* (pp. 113–142). Academic.

Masters, C. L., Gajdusek, D. C., & Gibbs, C. J., Jr. (1981). Creutzfeldt-Jakob disease virus isolations from the Gerstmann-Straussler syndrome with an analysis of the various forms of amyloid plaque deposition in the virus-induced spongiform encephalopathies. *Brain, 104*(3), 559–588.

Mathiason, C. K., Hays, S. A., Powers, J., Hayes-Klug, J., Langenberg, J., Dahmes, S. J., Osborn, D. A., Miller, K. V., Warren, R. J., Mason, G. L., & Hoover, E. A. (2009). Infectious prions in pre-clinical deer and transmission of chronic wasting disease solely by environmental exposure. *PLoS One, 4*(6), e5916.

McBride, P. A., Bruce, M. E., & Fraser, H. (1988). Immunostaining of scrapie cerebral amyloid plaques with antisera raised to scrapie-associated fibrils (SAF). *Neuropathology and Applied Neurobiology, 14*(4), 325–336.

Mead, S., Whitfield, J., Poulter, M., Shah, P., Uphill, J., Campbell, T., Al-Dujaily, H., Hummerich, H., Beck, J., Mein, C. A., Verzilli, C., Whittaker, J., Alpers, M. P., & Collinge, J. (2009). A novel protective prion protein variant that colocalizes with kuru exposure. *The New England Journal of Medicine, 361*(21), 2056–2065.

Meissner, B., Kallenberg, K., Sanchez-Juan, P., Collie, D., Summers, D. M., Almonti, S., Collins, S. J., Smith, P., Cras, P., Jansen, G. H., Brandel, J. P., Coulthart, M. B., Roberts, H., Van Everbroeck, B., Galanaud, D., Mellina, V., Will, R. G., & Zerr, I. (2009). MRI lesion profiles in sporadic Creutzfeldt-Jakob disease. *Neurology, 72*(23), 1994–2001.

Miller, M. W., & Wild, M. A. (2004). Epidemiology of chronic wasting disease in captive white-tailed and mule deer. *Journal of Wildlife Diseases, 40*(2), 320–327.

Miller, J. M., Jenny, A. L., Taylor, W. D., Marsh, R. F., Rubenstein, R., & Race, R. E. (1993). Immunohistochemical detection of prion protein in sheep with scrapie. *Journal of Veterinary Diagnostic Investigation, 5*(3), 309–316.

Miller, M. W., Williams, E. S., Hobbs, N. T., & Wolfe, L. L. (2004). Environmental sources of prion transmission in mule deer. *Emerging Infectious Diseases, 10*(6), 1003–1006.

Miller, M. W., Swanson, H. M., Wolfe, L. L., Quartarone, F. G., Huwer, S. L., Southwick, C. H., & Lukacs, P. M. (2008). Lions and prions and deer demise. *PLoS One, 3*(12), e4019.

Moda, F., Gambetti, P., Notari, S., Concha-Marambio, L., Catania, M., Park, K. W., Maderna, E., Suardi, S., Haik, S., Brandel, J. P., Ironside, J., Knight, R., Tagliavini, F., & Soto, C. (2014). Prions in the urine of patients with variant Creutzfeldt-Jakob disease. *The New England Journal of Medicine, 371*(6), 530–539.

Monello, R. J., Galloway, N. L., Powers, J. G., Madsen-Bouterse, S. A., Edwards, W. H., Wood, M. E., O'Rourke, K. I., & Wild, M. A. (2017). Pathogen-mediated selection in free-ranging elk populations infected by chronic wasting disease. *Proceedings of the National Academy of Sciences of the United States of America, 114*(46), 12208–12212.

Moore, S. J., West Greenlee, M. H., Kondru, N., Manne, S., Smith, J. D., Kunkle, R. A., Kanthasamy, A., & Greenlee, J. J. (2017). Experimental transmission of the chronic wasting disease agent to swine after oral or intracranial inoculation. *Journal of Virology, 91*(19).

Moore, S. J., Smith, J. D., Richt, J. A., & Greenlee, J. J. (2019). Raccoons accumulate PrP(Sc) after intracranial inoculation of the agents of chronic wasting disease or transmissible mink encephalopathy but not atypical scrapie. *Journal of Veterinary Diagnostic Investigation, 31*(2), 200–209.

Mouillet-Richard, S., Ermonval, M., Chebassier, C., Laplanche, J. L., Lehmann, S., Launay, J. M., & Kellermann, O. (2000). Signal transduction through prion protein. *Science, 289*(5486), 1925–1928.

Nalls, A. V., McNulty, E., Powers, J., Seelig, D. M., Hoover, C., Haley, N. J., Hayes-Klug, J., Anderson, K., Stewart, P., Goldmann, W., Hoover, E. A., & Mathiason, C. K. (2013). Mother to offspring transmission of chronic wasting disease in reeves' muntjac deer. *PLoS One, 8*(8), e71844.

Nalls, A. V., McNulty, E. E., Mayfield, A., Crum, J. M., Keel, M. K., Hoover, E. A., Ruder, M. G., & Mathiason, C. K. (2021). Detection of chronic wasting disease prions in fetal tissues of free-ranging white-tailed deer. *Viruses, 13*(12).

Ness, A., Aiken, J., & McKenzie, D. (2023). Sheep scrapie and deer rabies in England prior to 1800. *Prion, 17*(1), 7–15.

Neumann, M. A., Gajdusek, D. C., & Zigas, V. (1964). Neuropathologic findings in exotic neurologic disorders among natives of the highlands of New Guinea. *Journal of Neuropathology and Experimental Neurology, 23*, 486–507.

Nishina, K. A., Deleault, N. R., Mahal, S. P., Baskakov, I., Luhrs, T., Riek, R., & Supattapone, S. (2006). The stoichiometry of host PrPC glycoforms modulates the efficiency of PrPSc formation in vitro. *Biochemistry, 45*(47), 14129–14139.

Nonno, R., Notari, S., Di Bari, M. A., Cali, I., Pirisinu, L., d'Agostino, C., Cracco, L., Kofskey, D., Vanni, I., Lavrich, J., Parchi, P., Agrimi, U., & Gambetti, P. (2019). Variable protease-sensitive prionopathy transmission to bank voles. *Emerging Infectious Diseases, 25*(1), 73–81.

Notari, S., Xiao, X., Espinosa, J. C., Cohen, Y., Qing, L., Aguilar-Calvo, P., Kofskey, D., Cali, I., Cracco, L., Kong, Q., Torres, J. M., Zou, W., & Gambetti, P. (2014). Transmission characteristics of variably protease-sensitive prionopathy. *Emerging Infectious Diseases, 20*(12), 2006–2014.

Notari, S., Appleby, B. S., & Gambetti, P. (2018). Variably protease-sensitive prionopathy. *Handbook of Clinical Neurology, 153*, 175–190.

O'Rourke, K. I., Spraker, T. R., Zhuang, D., Greenlee, J. J., Gidlewski, T. E., & Hamir, A. N. (2007). Elk with a long incubation prion disease phenotype have a unique PrPd profile. *Neuroreport, 18*(18), 1935–1938.

Oraby, T., Al-Zoughool, M., Elsaadany, S., & Krewski, D. (2016). A stochastic model of the bovine spongiform encephalopathy epidemic in Canada. *Journal of Toxicology and Environmental Health. Part A, 79*(16–17), 677–689.

Orge, L., Lima, C., Machado, C., Tavares, P., Mendonca, P., Carvalho, P., Silva, J., Pinto, M. L., Bastos, E., Pereira, J. C., Goncalves-Anjo, N., Gama, A., Esteves, A., Alves, A., Matos, A. C., Seixas, F., Silva, F., Pires, I., Figueira, L., Vieira-Pinto, M., Sargo, R., & Pires, M. D. A. (2021). Neuropathology of animal prion diseases. *Biomolecules, 11*(3).

Osterholm, M. T., Anderson, C. J., Zabel, M. D., Scheftel, J. M., Moore, K. A., & Appleby, B. S. (2019). Chronic wasting disease in cervids: Implications for prion transmission to humans and other animal species. *MBio, 10*(4).

Padilla, D., Beringue, V., Espinosa, J. C., Andreoletti, O., Jaumain, E., Reine, F., Herzog, L., Gutierrez-Adan, A., Pintado, B., Laude, H., & Torres, J. M. (2011). Sheep and goat BSE propagate more efficiently than cattle BSE in human PrP transgenic mice. *PLoS Pathogens, 7*(3), e1001319.

Paitel, E., Fahraeus, R., & Checler, F. (2003). Cellular prion protein sensitizes neurons to apoptotic stimuli through Mdm2-regulated and p53-dependent caspase 3-like activation. *The Journal of Biological Chemistry, 278*(12), 10061–10066.

Palsson, P. A. (1979). Rida (scrapie) in Iceland and its epidemiology. In S. B. Prusiner & W. J. Hadlow (Eds.), *Slow transmissible diseases of the nervous system, Volume 1: Clinical, epidemiological, genetic, and pathological aspects of the spongiform encephalopathies* (pp. 357–366). Academic.

Parchi, P., & Saverioni, D. (2012). Molecular pathology, classification, and diagnosis of sporadic human prion disease variants. *Folia Neuropathologica, 50*(1), 20–45.

Parchi, P., Castellani, R., Cortelli, P., Montagna, P., Chen, S. G., Petersen, R. B., Manetto, V., Vnencak-Jones, C. L., McLean, M. J., Sheller, J. R., et al. (1995). Regional distribution of protease-resistant prion protein in fatal familial insomnia. *Annals of Neurology, 38*(1), 21–29.

Parchi, P., Castellani, R., Capellari, S., Ghetti, B., Young, K., Chen, S. G., Farlow, M., Dickson, D. W., Sima, A. A., Trojanowski, J. Q., Petersen, R. B., & Gambetti, P. (1996). Molecular basis of phenotypic variability in sporadic Creutzfeldt-Jakob disease. *Annals of Neurology, 39*(6), 767–778.

Parchi, P., Capellari, S., Chin, S., Schwarz, H. B., Schecter, N. P., Butts, J. D., Hudkins, P., Burns, D. K., Powers, J. M., & Gambetti, P. (1999a). A subtype of sporadic prion disease mimicking fatal familial insomnia. *Neurology, 52*(9), 1757–1763.

Parchi, P., Giese, A., Capellari, S., Brown, P., Schulz-Schaeffer, W., Windl, O., Zerr, I., Budka, H., Kopp, N., Piccardo, P., Poser, S., Rojiani, A., Streichemberger, N., Julien, J., Vital, C., Ghetti, B., Gambetti, P., & Kretzschmar, H. (1999b). Classification of sporadic Creutzfeldt-Jakob disease based on molecular and phenotypic analysis of 300 subjects. *Annals of Neurology, 46*(2), 224–233.

Pearson, G. R., Wyatt, J. M., Gruffydd-Jones, T. J., Hope, J., Chong, A., Higgins, R. J., Scott, A. C., & Wells, G. A. (1992). Feline spongiform encephalopathy: Fibril and PrP studies. *The Veterinary Record, 131*(14), 307–310.

Pearson, G. R., Wyatt, J. M., Henderson, J. R., & Gruffydd-Jones, T. J. (1993). Feline spongiform encephalopathy: A review. *The Veterinary Annual, 33*, 1–10.

Peden, A. H., Sarode, D. P., Mulholland, C. R., Barria, M. A., Ritchie, D. L., Ironside, J. W., & Head, M. W. (2014). The prion protein protease sensitivity, stability and seeding activity in variably protease sensitive prionopathy brain tissue suggests molecular overlaps with sporadic Creutzfeldt-Jakob disease. *Acta Neuropathologica Communications, 2*, 152.

Peng, B., Zhang, S., Dong, H., & Lu, Z. (2015). Clinical, histopathological and genetic studies in a case of fatal familial insomnia with review of the literature. *International Journal of Clinical and Experimental Pathology, 8*(9), 10171–10177.

Piconi, G., Peden, A. H., Barria, M. A., & Green, A. J. E. (2019). Epitope mapping of the protease resistant products of RT-QuIC does not allow the discrimination of sCJD subtypes. *PLoS One, 14*(6), e0218509.

Pritzkow, S., Morales, R., Moda, F., Khan, U., Telling, G. C., Hoover, E., & Soto, C. (2015). Grass plants bind, retain, uptake, and transport infectious prions. *Cell Reports, 11*(8), 1168–1175.

Prusiner, S. B. (1982). Novel proteinaceous infectious particles cause scrapie. *Science, 216*(4542), 136–144.

Race, B., Meade-White, K. D., Miller, M. W., Barbian, K. D., Rubenstein, R., LaFauci, G., Cervenakova, L., Favara, C., Gardner, D., Long, D., Parnell, M., Striebel, J., Priola, S. A., Ward, A., Williams, E. S., Race, R., & Chesebro, B. (2009). Susceptibilities of nonhuman primates to chronic wasting disease. *Emerging Infectious Diseases, 15*(9), 1366–1376.

Redaelli, V., Bistaffa, E., Zanusso, G., Salzano, G., Sacchetto, L., Rossi, M., De Luca, C. M., Di Bari, M., Portaleone, S. M., Agrimi, U., Legname, G., Roiter, I., Forloni, G., Tagliavini, F., & Moda, F. (2017). Detection of prion seeding activity in the olfactory mucosa of patients with Fatal Familial Insomnia. *Scientific Reports, 7*, 46269.

Risacher, S. L., Farlow, M. R., Bateman, D. R., Epperson, F., Tallman, E. F., Richardson, R., Murrell, J. R., Unverzagt, F. W., Apostolova, L. G., Bonnin, J. M., Ghetti, B., & Saykin, A. J. (2018). Detection of tau in Gerstmann-Straussler-Scheinker disease (PRNP F198S) by [(18)F]Flortaucipir PET. *Acta Neuropathologica Communications, 6*(1), 114.

Roh, I. S., Kim, Y. C., Won, S. Y., Park, K. J., Park, H. C., Hwang, J. Y., Kang, H. E., Sohn, H. J., & Jeong, B. H. (2022). Association study of the M132L single nucleotide polymorphism with susceptibility to chronic wasting disease in Korean Elk: A meta-analysis. *Frontiers in Veterinary Science, 8*, 804325.

Rudenskaya, G. E., Konovalov, F. A., Illarioshkin, S. N., & Shchagina, O. A. (2023). Gerstmann-Straussler disease: A familial case with common PRNP mutation and atypical features. *Zhurnal Nevrologii i Psikhiatrii Imeni S.S. Korsakova, 123*(2), 138–143.

Rudge, P., Jaunmuktane, Z., Adlard, P., Bjurstrom, N., Caine, D., Lowe, J., Norsworthy, P., Hummerich, H., Druyeh, R., Wadsworth, J. D., Brandner, S., Hyare, H., Mead, S., & Collinge, J. (2015). Iatrogenic CJD due to pituitary-derived growth hormone with genetically determined incubation times of up to 40 years. *Brain, 138*(Pt 11), 3386–3399.

Ryder, S. J., Wells, G. A., Bradshaw, J. M., & Pearson, G. R. (2001). Inconsistent detection of PrP in extraneural tissues of cats with feline spongiform encephalopathy. *The Veterinary Record, 148*(14), 437–441.

Saborio, G. P., Permanne, B., & Soto, C. (2001). Sensitive detection of pathological prion protein by cyclic amplification of protein misfolding. *Nature, 411*(6839), 810–813.

Safar, J., Roller, P. P., Gajdusek, D. C., & Gibbs, C. J., Jr. (1993a). Conformational transitions, dissociation, and unfolding of scrapie amyloid (prion) protein. *The Journal of Biological Chemistry, 268*(27), 20276–20284.

Safar, J., Roller, P. P., Gajdusek, D. C., & Gibbs, C. J., Jr. (1993b). Thermal stability and conformational transitions of scrapie amyloid (prion) protein correlate with infectivity. *Protein Science, 2*(12), 2206–2216.

Sakai, K., Hamaguchi, T., Sanjo, N., Murai, H., Iwasaki, Y., Hamano, T., Honma, M., Noguchi-Shinohara, M., Nozaki, I., Nakamura, Y., Kitamoto, T., Harada, M., Mizusawa, H., & Yamada, M. (2020). Diffusion-weighted magnetic resonance imaging in dura mater graft-associated Creutzfeldt-Jakob disease. *Journal of the Neurological Sciences, 418*, 117094.

Salvesen, O., Espenes, A., Reiten, M. R., Vuong, T. T., Malachin, G., Tran, L., Andreoletti, O., Olsaker, I., Benestad, S. L., Tranulis, M. A., & Ersdal, C. (2020). Goats naturally devoid of PrP(C) are resistant to scrapie. *Veterinary Research, 51*(1), 1.

Sanchez-Juan, P., Green, A., Ladogana, A., Cuadrado-Corrales, N., Saanchez-Valle, R., Mitrovaa, E., Stoeck, K., Sklaviadis, T., Kulczycki, J., Hess, K., Bodemer, M., Slivarichova, D., Saiz, A., Calero, M., Ingrosso, L., Knight, R., Janssens, A. C., van Duijn, C. M., & Zerr, I. (2006). CSF tests in the differential diagnosis of Creutzfeldt-Jakob disease. *Neurology, 67*(4), 637–643.

Sander, P., Hamann, H., Pfeiffer, I., Wemheuer, W., Brenig, B., Groschup, M. H., Ziegler, U., Distl, O., & Leeb, T. (2004). Analysis of sequence variability of the bovine prion protein gene (PRNP) in German cattle breeds. *Neurogenetics, 5*(1), 19–25.

Saunders, S. E., Bartz, J. C., & Bartelt-Hunt, S. L. (2009). Influence of prion strain on prion protein adsorption to soil in a competitive matrix. *Environmental Science & Technology, 43*(14), 5242–5248.

Saunders, S. E., Bartz, J. C., & Shikiya, R. A. (2012). Protein misfolding cyclic amplification of prions. *Journal of Visualized Experiments*, (69), 4075.

Schlote, W., Boellaard, J. W., Schumm, F., & Stohr, M. (1980). Gerstmann-Straussler-Scheinker's disease. Electron-microscopic observations on a brain biopsy. *Acta Neuropathologica, 52*(3), 203–211.

Seitzinger, A. H., Paarlberg, P. L., & Lee, J. G. (2006). *Economic impacts of eradicating scrapie, ovine progressive pneumonia and Johne's disease on US sheep, lamb, sheep meat and lamb meat markets*. CABI Books, CABI International.

Sequeira, D., Nihat, A., Mok, T., Coysh, T., Rudge, P., Collinge, J., & Mead, S. (2022). Prevalence and treatments of movement disorders in prion diseases: A longitudinal cohort study. *Movement Disorders, 37*(9), 1893–1903.

Sigurdson, C. J., Spraker, T. R., Miller, M. W., Oesch, B., & Hoover, E. A. (2001). PrP(CWD) in the myenteric plexus, vagosympathetic trunk and endocrine glands of deer with chronic wasting disease. *The Journal of General Virology, 82*(Pt 10), 2327–2334.

Sigurdsson, B. (1954). RIDA, A chronic encephalitis of sheep. *British Veterinary Journal, 110*(9), 341–354.

Silva, J. F., Correia, J. J., Ribeiro, J., Carmo, S., & Orge, L. (2006). *Feline spongiform encephalopathy: First confirmed case in Portugal*. Proceedings of the Prion 2006. Torino, Italy. Abstract Book: 392.

Sola, D., Tran, L., Vage, J., Madslien, K., Vuong, T. T., Korpenfelt, S. L., Agren, E. O., Averhed, G., Noremark, M., Soren, K., Isaksson, M., Acin, C., Badiola, J. J., Gavier-Widen, D., & Benestad, S. L. (2023). Heterogeneity of pathological prion protein accumulation in the brain of moose (Alces alces) from Norway, Sweden and Finland with chronic wasting disease. *Veterinary Research, 54*(1), 74.

Spiropoulos, J., Hawkins, S. A., Simmons, M. M., & Bellworthy, S. J. (2014). Evidence of in utero transmission of classical scrapie in sheep. *Journal of Virology, 88*(8), 4591–4594.

Spraker, T. R., Miller, M. W., Williams, E. S., Getzy, D. M., Adrian, W. J., Schoonveld, G. G., Spowart, R. A., O'Rourke, K. I., Miller, J. M., & Merz, P. A. (1997). Spongiform encephalopathy in free-ranging mule deer (Odocoileus hemionus), white-tailed deer (Odocoileus virginianus) and Rocky Mountain elk (Cervus elaphus

nelsoni) in northcentral Colorado. *Journal of Wildlife Diseases, 33*(1), 1–6.
Spraker, T. R., Gidlewski, T., Powers, J. G., Nichols, T. A., & Wild, M. A. (2023). Distribution of the misfolded isoform of the prion protein in peripheral tissues and spinal cord of Rocky Mountain elk (Cervus elaphus nelsoni) with naturally occurring chronic wasting disease. *Veterinary Pathology, 60*(4), 420–433.
Stack, M. J., Scott, A. C., Done, S. H., & Dawson, M. (1993). Scrapie associated fibril detection on decomposed and fixed ovine brain material. *Research in Veterinary Science, 55*(2), 173–178.
Staffaroni, A. M., Kramer, A. O., Casey, M., Kang, H., Rojas, J. C., Orru, C. D., Caughey, B., Allen, I. E., Kramer, J. H., Rosen, H. J., Blennow, K., Zetterberg, H., & Geschwind, M. D. (2019). Association of blood and cerebrospinal fluid tau level and other biomarkers with survival time in sporadic Creutzfeldt-Jakob disease. *JAMA Neurology, 76*(8), 969–977.
Stahl, N., Borchelt, D. R., Hsiao, K., & Prusiner, S. B. (1987). Scrapie prion protein contains a phosphatidylinositol glycolipid. *Cell, 51*(2), 229–240.
Stamp, J. T., Brotherston, J. G., Zlotnik, I., Mackay, J. M., & Smith, W. (1959). Further studies on scrapie. *Journal of Comparative Pathology, 69*, 268–280.
Stewart, P., Campbell, L., Skogtvedt, S., Griffin, K. A., Arnemo, J. M., Tryland, M., Girling, S., Miller, M. W., Tranulis, M. A., & Goldmann, W. (2012). Genetic predictions of prion disease susceptibility in carnivore species based on variability of the prion gene coding region. *PLoS One, 7*(12), e50623.
Supattapone, S. (2010). Biochemistry. What makes a prion infectious? *Science, 327*(5969), 1091–1092.
Tahmoorespur, M., & Jelokhani Niaraki, S. (2013). Analysis of sequence variations of prion protein gene in dromedary camels in Iran. *Journal of Applied Animal Research, 42*(2), 238–243.
Takeuchi, A., Mohri, S., Kai, H., Tamaoka, A., Kobayashi, A., Mizusawa, H., Iwasaki, Y., Yoshida, M., Shimizu, H., Murayama, S., Kuroda, S., Morita, M., Parchi, P., & Kitamoto, T. (2019). Two distinct prions in fatal familial insomnia and its sporadic form. *Brain Communications, 1*(1), fcz045.
Tamguney, G., Miller, M. W., Wolfe, L. L., Sirochman, T. M., Glidden, D. V., Palmer, C., Lemus, A., DeArmond, S. J., & Prusiner, S. B. (2009). Asymptomatic deer excrete infectious prions in faeces. *Nature, 461*(7263), 529–532.
Tamguney, G., Richt, J. A., Hamir, A. N., Greenlee, J. J., Miller, M. W., Wolfe, L. L., Sirochman, T. M., Young, A. J., Glidden, D. V., Johnson, N. L., Giles, K., DeArmond, S. J., & Prusiner, S. B. (2012). Salivary prions in sheep and deer. *Prion, 6*(1), 52–61.
Taylor, D. (1986). Decontamination of Creutzfeldt-Jakob disease agent [letter]. *Annals of Neurology, 20*(6), 749–750.
Tesar, A., Matej, R., Kukal, J., Johanidesova, S., Rektorova, I., Vyhnalek, M., Keller, J., Eliasova, I., Parobkova, E., Smetakova, M., Musova, Z., & Rusina, R. (2019). Clinical variability in P102L Gerstmann-Straussler-Scheinker syndrome. *Annals of Neurology, 86*(5), 643–652.
Tranchant, C., Doh-ura, K., Warter, J. M., Steinmetz, G., Chevalier, Y., Hanauer, A., Kitamoto, T., & Tateishi, J. (1992). Gerstmann-Straussler-Scheinker disease in an Alsatian family: Clinical and genetic studies. *Journal of Neurology, Neurosurgery, and Psychiatry, 55*(3), 185–187.
Tranulis, M. A., & Tryland, M. (2023). The zoonotic potential of chronic wasting disease-A review. *Food, 12*(4).
Traub, R. D., Gajdusek, D. C., & Gibbs, C. J. (1975). Letter: Precautions in autopsies on Creutzfeldt-Jakob disease. *American Journal of Clinical Pathology, 64*(2), 287.
Tuo, W., Zhuang, D., Knowles, D. P., Cheevers, W. P., Sy, M. S., & O'Rourke, K. I. (2001). Prp-c and Prp-Sc at the fetal-maternal interface. *The Journal of Biological Chemistry, 276*(21), 18229–18234.
Uchino, A., Saito, Y., Oonuma, S., Murayama, S., Yagishita, S., Kitamoto, T., & Hasegawa, K. (2023). An autopsy case of variably protease-sensitive prionopathy with Met/Met homogeneity at codon 129. *Neuropathology, 43*(6), 486–495.
Unit, N. C. R. S. (2021). *Creutzfeldt-Jakob disease surveillance in the UK, 30th annual report 2021* (pp. 1–31). University of Edinburgh.
van Keulen, L. J., Schreuder, B. E., Vromans, M. E., Langeveld, J. P., & Smits, M. A. (1999). Scrapie-associated prion protein in the gastrointestinal tract of sheep with natural scrapie. *Journal of Comparative Pathology, 121*(1), 55–63.
Vikoren, T., Vage, J., Madslien, K. I., Roed, K. H., Rolandsen, C. M., Tran, L., Hopp, P., Veiberg, V., Heum, M., Moldal, T., Neves, C. G. D., Handeland, K., Ytrehus, B., Kolbjornsen, O., Wisloff, H., Terland, R., Saure, B., Dessen, K. M., Svendsen, S. G., Nordvik, B. S., & Benestad, S. L. (2019). First detection of chronic wasting disease in a Wild Red Deer (Cervus elaphus) in Europe. *Journal of Wildlife Diseases, 55*(4), 970–972.
Wadsworth, J. D., Joiner, S., Linehan, J. M., Desbruslais, M., Fox, K., Cooper, S., Cronier, S., Asante, E. A., Mead, S., Brandner, S., Hill, A. F., & Collinge, J. (2008). Kuru prions and sporadic Creutzfeldt-Jakob disease prions have equivalent transmission properties in transgenic and wild-type mice. *Proceedings of the National Academy of Sciences of the United States of America, 105*(10), 3885–3890.
Wadsworth, J. D., Joiner, S., Linehan, J. M., Balkema-Buschmann, A., Spiropoulos, J., Simmons, M. M., Griffiths, P. C., Groschup, M. H., Hope, J., Brandner, S., Asante, E. A., & Collinge, J. (2013). Atypical scrapie prions from sheep and lack of disease in transgenic mice overexpressing human prion protein. *Emerging Infectious Diseases, 19*(11), 1731–1739.
Wang, J., Xiao, K., Zhou, W., Shi, Q., & Dong, X. P. (2019). Analysis of 12 Chinese patients with proline-to-leucine mutation at codon 102-associated Gerstmann-Straussler-Scheinker disease. *Journal of Clinical Neurology, 15*(2), 184–190.
Wang, Z., Qin, K., Camacho, M. V., Cali, I., Yuan, J., Shen, P., Greenlee, J., Kong, Q., Mastrianni, J. A., & Zou, W. Q. (2021). Generation of human chronic wasting disease in transgenic mice. *Acta Neuropathologica Communications, 9*(1), 158.
Wells, G. A., & McGill, I. S. (1992). Recently described scrapie-like encephalopathies of animals: Case definitions. *Research in Veterinary Science, 53*(1), 1–10.
Wells, G. A., Scott, A. C., Johnson, C. T., Gunning, R. F., Hancock, R. D., Jeffrey, M., Dawson, M., & Bradley, R. (1987). A novel progressive spongiform encephalopathy in cattle. *The Veterinary Record, 121*(18), 419–420.
Wells, G. A., Wilesmith, J. W., & McGill, I. S. (1991). Bovine spongiform encephalopathy: A neuropathological perspective. *Brain Pathology, 1*(2), 69–78.
White, A. R., Collins, S. J., Maher, F., Jobling, M. F., Stewart, L. R., Thyer, J. M., Beyreuther, K., Masters, C. L., & Cappai, R. (1999). Prion protein-deficient neurons reveal lower glutathione reductase activity and increased susceptibility to hydrogen peroxide toxicity. *The American Journal of Pathology, 155*(5), 1723–1730.
White, S. N., Spraker, T. R., Reynolds, J. O., & O'Rourke, K. I. (2010). Association analysis of PRNP gene region with chronic wasting disease in Rocky Mountain elk. *BMC Research Notes, 3*, 314.
Wilesmith, J. W., Wells, G. A., Cranwell, M. P., & Ryan, J. B. (1988). Bovine spongiform encephalopathy: Epidemiological studies. *The Veterinary Record, 123*(25), 638–644.
Wilham, J. M., Orru, C. D., Bessen, R. A., Atarashi, R., Sano, K., Race, B., Meade-White, K. D., Taubner, L. M., Timmes, A., & Caughey, B. (2010). Rapid end-point quantitation of prion seeding activity with sensitivity comparable to bioassays. *PLoS Pathogens, 6*(12), e1001217.
Will, R. G., Ironside, J. W., Zeidler, M., Cousens, S. N., Estibeiro, K., Alperovitch, A., Poser, S., Pocchiari, M., Hofman, A., & Smith,

P. G. (1996). A new variant of Creutzfeldt-Jakob disease in the UK. *Lancet, 347*(9006), 921–925.
Williams, E. S., & Young, S. (1980). Chronic wasting disease of captive mule deer: A spongiform encephalopathy. *Journal of Wildlife Diseases, 16*(1), 89–98.
Williams, E. S., & Young, S. (1982). Spongiform encephalopathy of Rocky Mountain elk. *Journal of Wildlife Diseases, 18*(4), 465–471.
Williams, E. S., & Young, S. (1992). Spongiform encephalopathies in Cervidae. *Revue Scientifique et Technique, 11*(2), 551–567.
Williams, E. S., Kirkwood, J. K., & Miller, M. W. (2000). Transmissible spongiform encephalopathies. In E. S. Williams & I. K. Barker (Eds.), *Infectious diseases of wild mammals* (pp. 292–301). Wiley.
Wilson, K., & Ricketts, M. N. (2006). A third episode of transfusion-derived vCJD. *Lancet, 368*(9552), 2037–2039.
Wilson, G. A., Nakada, S. M., Bollinger, T. K., Pybus, M. J., Merrill, E. H., & Coltman, D. W. (2009). Polymorphisms at the PRNP gene influence susceptibility to chronic wasting disease in two species of deer (Odocoileus Spp.) in western Canada. *Journal of Toxicology and Environmental Health. Part A, 72*(17–18), 1025–1029.
Wilson, R., Plinston, C., Hunter, N., Casalone, C., Corona, C., Tagliavini, F., Suardi, S., Ruggerone, M., Moda, F., Graziano, S., Sbriccoli, M., Cardone, F., Pocchiari, M., Ingrosso, L., Baron, T., Richt, J., Andreoletti, O., Simmons, M., Lockey, R., Manson, J. C., & Barron, R. M. (2012). Chronic wasting disease and atypical forms of bovine spongiform encephalopathy and scrapie are not transmissible to mice expressing wild-type levels of human prion protein. *The Journal of General Virology, 93*(Pt 7), 1624–1629.
Wood, J. L., Lund, L. J., & Done, S. H. (1992). The natural occurrence of scrapie in moufflon. *The Veterinary Record, 130*(2), 25–27.
Wopfner, F., Weidenhofer, G., Schneider, R., von Brunn, A., Gilch, S., Schwarz, T. F., Werner, T., & Schatzl, H. M. (1999). Analysis of 27 mammalian and 9 avian PrPs reveals high conservation of flexible regions of the prion protein. *Journal of Molecular Biology, 289*(5), 1163–1178.
Wrathall, A. E., Brown, K. F., Sayers, A. R., Wells, G. A., Simmons, M. M., Farrelly, S. S., Bellerby, P., Squirrell, J., Spencer, Y. I., Wells, M., Stack, M. J., Bastiman, B., Pullar, D., Scatcherd, J., Heasman, L., Parker, J., Hannam, D. A., Helliwell, D. W., Chree, A., & Fraser, H. (2002). Studies of embryo transfer from cattle clinically affected by bovine spongiform encephalopathy (BSE). *The Veterinary Record, 150*(12), 365–378.
Wu, C., Pang, W., & Zhao, D. (2006a). Amino acid sequence of the Amur tiger prion protein. *Virus Research, 121*(1), 93–96.
Wu, C. D., Pang, W. Y., & Zhao, D. M. (2006b). Comparative analysis of the prion protein gene sequences in African lion. *Virus Genes, 33*(2), 213–214.
Wyatt, J. M., Pearson, G. R., Smerdon, T. N., Gruffydd-Jones, T. J., Wells, G. A., & Wilesmith, J. W. (1991). Naturally occurring scrapie-like spongiform encephalopathy in five domestic cats. *The Veterinary Record, 129*(11), 233–236.
Xu, L., Zhang, Z., Wang, L., Feng, D., Zhou, X., Xu, B., & Zhao, D. (2012). Cloning and polymorphism analysis of prion protein gene in domestic bactrian camel in China. *Gene, 491*(2), 256–259.
Yuan, Q., Telling, G., Bartelt-Hunt, S. L., & Bartz, J. C. (2018). Dehydration of prions on environmentally relevant surfaces protects them from inactivation by freezing and thawing. *Journal of Virology, 92*(8).
Zanusso, G., Nardelli, E., Rosati, A., Fabrizi, G., Ferrari, S., Carteri, A., De Simone, F., Rizzuto, N., & Monaco, S. (1998). Simultaneous occurrence of spongiform encephalopathy in a man and his cat in Italy. *Lancet, 352*(9134), 1116–1117.
Zeidler, M., Stewart, G. E., Barraclough, C. R., Bateman, D. E., Bates, D., Burn, D. J., Colchester, A. C., Durward, W., Fletcher, N. A., Hawkins, S. A., Mackenzie, J. M., & Will, R. G. (1997). New variant Creutzfeldt-Jakob disease: Neurological features and diagnostic tests. *Lancet, 350*(9082), 903–907.
Zhang, H., Gao, S., Pei, R., Chen, X., & Li, C. (2017). Hepatitis C virus-induced prion protein expression facilitates hepatitis C virus replication. *Virologica Sinica, 32*(6), 503–510.
Zigas, V., & Gajdusek, D. C. (1957). Kuru: Clinical study of a new syndrome resembling paralysis agitans in natives of the eastern highlands of Australian New Guinea. *The Medical Journal of Australia, 44*(21), 745–754.
Zlotnik, I. (1958). The histopathology of the brain stem of sheep affected with natural scrapie. *Journal of Comparative Pathology, 68*(2), 148–166.
Zomosa-Signoret, V., Mayoral, M., Limon, D., Espinosa, B., Calvillo, M., Zenteno, E., Martinez, V., & Guevara, J. (2011). Sialylated and O-glycosidically linked glycans in prion protein deposits in a case of Gerstmann-Straussler-Scheinker disease. *Neuropathology, 31*(2), 162–169.
Zou, W. Q., Puoti, G., Xiao, X., Yuan, J., Qing, L., Cali, I., Shimoji, M., Langeveld, J. P., Castellani, R., Notari, S., Crain, B., Schmidt, R. E., Geschwind, M., Dearmond, S. J., Cairns, N. J., Dickson, D., Honig, L., Torres, J. M., Mastrianni, J., Capellari, S., Giaccone, G., Belay, E. D., Schonberger, L. B., Cohen, M., Perry, G., Kong, Q., Parchi, P., Tagliavini, F., & Gambetti, P. (2010). Variably protease-sensitive prionopathy: A new sporadic disease of the prion protein. *Annals of Neurology, 68*(2), 162–172.
Zoubeyda, K., Imane, M., Youcef, C., Baaissa, B., Suheil, G. S. B., Michela, C., Antonio, C., Umberto, A., Barbara, C., & Gabriele, V. (2020). Variability of the prion protein gene (PRNP) in Algerian dromedary populations. *Animal Gene, 17–18*, 200106.

30 Guillain-Barrè Syndrome and Axonal Degeneration

Chiara Gallo, Domizia Vecchio, and Cristoforo Comi

Abstract

Guillain-Barré syndrome (GBS), the acute inflammatory neuropathy prototype and its heterogeneous variants, represents the most common cause of acute neuromuscular weakness and paralysis worldwide. An immune-mediated attack, directed against myelin, axons, or both, is presumed to be triggered by molecular mimicry, with both cell- and humoral-dependent mechanisms driving the pathogenic damage. Axonal degeneration in GBS may be primary or secondary to inflammatory demyelination in nerve trunks, and this process could be closely associated with current and future treatment approaches. Considerable progress has been made in recent years; new immunological concepts and treatments are only partly known, but some etiopathological mechanisms of inflammatory neuropathies remain poorly understood.

Keywords

Guillain-Barrè syndrome · Axonal degeneration · Bystander effect · Demyelination · Molecular mimicry

30.1 Introduction

Immune-mediated neuropathies consist of a wide range of peripheral nerve disorders, which can be classified according to current diagnostic guidelines and criteria based on clinical symptoms and signs (motor/sensory involvement), onset and time course, laboratory and electrophysiological parameters (Kieseier et al., 2018; Gwathmey & Smith, 2020).

Both in acute and chronic forms or variants, humoral and cellular immunity are likely playing a predominant role in pathogenesis: the improvement with immunomodulating and immunosuppressive treatments strongly suggest an aberrant immune response toward components of the peripheral nerve causing demyelination and axonal damage (Mathey et al., 2015; Doneddu et al., 2019). Axonal degeneration, considered initially only the "bystander" effect of a primary intense inflammatory response, represents an independent pathological process. The axon could be the first and foremost vulnerable target in those acute neuropathies with an "axonal pattern" on electroneurography (ENG) findings (Berciano, 2021; Kimura, 2013). Autoantibodies against peripheral nerve molecules, such as gangliosides or proteins of the Ranvier node, have been described, allowing the identification of subgroups of patients with specific clinical phenotypes and potential targeted treatment (Schafflick et al., 2017). Considerable progress has been made in recent years, thanks to novel animal models, and new immunological concepts and treatments are only in part known. However, some etiopathological mechanisms of inflammatory neuropathies remain poorly understood (Schafflick et al., 2017). This chapter focuses on *Guillain-Barré syndrome* (GBS), the acute inflammatory neuropathy prototype, and its heterogeneous variants, and describes immunological and functional disease mechanisms, particularly axonal degeneration. This process could be closely associated with current and future treatment approaches.

30.2 Guillain-Barrè Syndrome

GBS is an acute-onset and rapidly evolving immuno-mediated neuropathy that often occurs after an acute infection (Willison et al., 2016; Hughes & Cornblath, 2005). Its

C. Gallo (✉) · D. Vecchio
Department of Translational Medicine and Neurology, AOU Maggiore della Carità, University of Piemonte Orientale, Novara, Italy

C. Comi (✉)
Department of Translational Medicine and Neurology, AOU Maggiore della Carità, University of Piemonte Orientale, Novara, Italy

Neurology Unit, Sant'Andrea Hospital, Vercelli, Italy
e-mail: cristofo.comi@med.uniupo.it

H. E. Gendelman, T. Ikezu (eds.), *Neuroimmune Pharmacology and Therapeutics*,
https://doi.org/10.1007/978-3-031-68237-7_30

annual global incidence is approximately 1–2 per 100,000 person-years. GBS has a male predominance, and its incidence increases with age (Sejvar et al., 2011a, b). GBS is a monophasic ascending sensorimotor neuropathy characterized by progressive limb weakness and areflexia of rapid progression. Diagnosis is mainly clinical, and electrophysiological tests and CSF samples are used as supportive criteria, although specific diagnostic biomarkers are yet to be available for most syndrome variants (Martín-Aguilar et al., 2019; Shahrizaila et al., 2021). GBS is often preceded by an infectious triggering event (Yuki & Hartung, 2012), and relapses can occur in 2–5% of patients (Ruts et al., 2007). Maximum disability is reached in 2–3 weeks, followed by stabilization and recovery (Willison et al., 2016). The disease nadir could require artificial ventilation, and since its introduction, the prognosis has ameliorated: Between 50% and 70% of the patients recovered with independent walking within 6 months with only supportive therapy. Despite this, almost 20% of patients remain disabled after 6 months. However, several patients may continue to recover, often incompletely, over 2–3 years (Nobile-Orazio et al., 2017). Electrophysiological studies provide evidence of peripheral nervous system (PNS) dysfunction. They help distinguish among the subtypes of GBS, most commonly acute demyelinating inflammatory polyneuropathy (AIDP) and acute motor axonal neuropathy (AMAN) (Kimura, 2013). Other clinical syndromes, such as *Miller Fisher syndrome (MFS), acute sensory ataxic neuropathy (ASAN), acute motor–sensory axonal neuropathy (AMSAN)*, and *acute motor conduction block neuropathy (AMCBN)*, are pathophysiologically considered a part of the GBS spectrum (Sejvar et al., 2011a, b).

Patients with AMAN or AMSAN frequently have serum antibodies against *gangliosides (GM1 or GD1a)*, whereas reactivity against *GQ1b* occurs in 80–95% of patients with MFS. By contrast, no consistent anti-ganglioside reactivity has been found in AIDP patients (Asbury & Cornblath, 1990; Lopez et al., 2008; Chiba et al., 1993). Not only autoantibody-mediated damage but multiple mechanisms have been implicated in GBS pathogenesis, including humoral and cellular immune responses, complement, activated macrophages, and lymphocytes, leading to nerve infiltration, demyelination, and axonal loss, depending on the site of the target antigen (Kaida et al., 2004; Asbury et al., 1969).

Axonal degeneration in GBS may be primary or secondary to inflammatory demyelination in nerve trunks (Berciano, 2021). Many studies found that AIDP is characterized by the presence of primary inflammatory infiltrating T cells and macrophages involved in demyelination (Wanschitz et al., 2003); by contrast with AIDP, patients with AMAN show primary axonal injury without substantial T cell-mediated inflammation or demyelination (Berciano, 2021).

30.2.1 Immunopathogenesis

30.2.1.1 Molecular Mimicry Hypothesis

The autoimmune production of cross-reactive antibodies against specific nerve structures represents a crucial step in GBS pathogenesis (Fig. 30.1, Van den Berg et al., 2014). A *"molecular mimicry hypothesis"* has been demonstrated for *Campylobacter jejuni* and the motor variants of GBS. Lipo-oligosaccharides on the *C. jejuni* outer membrane elicit the production of antibodies that cross-react with gangliosides on peripheral nerves (Yuki et al., 2011). The synthesis of these ganglioside-mimicking carbohydrate structures depends on patient-related factors susceptibility, a set of polymorphic genes and enzymes that vary greatly between different *C. jejuni* strains (Gilbert et al., 2002; Van Belkum et al., 2001) (Thr51 variant of the alpha-2,3-sialyltransferase (*cstII*) gene shows GBS association, while the Asn51 variant of *cstII* is described in MFS), (Godschalk et al., 2007).

Campylobacter jejuni lipo-oligosaccharides bind to siglec-7 (sialic acid-binding immunoglobulin-like lectin 7) and activate dendritic cells via Toll-like receptor 4 and CD14. These dendritic cells produce type 1 interferon and tumor necrosis factor (TNF), which induce the proliferation of B cells. To date, genetic factors have only been studied in small cohorts of patients. Future genome-wide association studies in large, well-described, and adequately controlled cohorts are required to establish the role of host factors in the pathogenesis of GBS (Heikema et al., 2013). In AMAN, for example, the GM1 and GD1a antibodies bind to the nodal axolemma. This leads to complement activation followed by MAC formation and disappearance of voltage-gated sodium channels with detachment of paranodal myelin and nerve conduction failure. Macrophages then invade from the nodes into the periaxonal space, scavenging the injured axons (Yuki et al., 2011; Gilbert et al., 2002).

Infections like *C. Jejuni* can trigger humoral and autoimmune responses, leading to nerve dysfunction. Lipo-oligosaccharides on the *C. Jejuni* membrane may induce antibody production that crossreacts with peripheral nerve gangliosides. In AMAN, antiGM1 or anti-GD1 bind nodal axolemma, activate complement and MAC complex formation, leading to detachment of paranodal myelin; then macrophages scavenge the injury axons. AIDP target is located presumably on the myelin sheath; injury is completed by complement activation, MAC complex formation on Schwann cells, and macrophage invasion of myelin.

30.2.2 Axonal Degeneration as Secondary "Bystander Effect"

The immunopathogenesis of GBS is only partially known. Until 20 years ago, it was regarded as a homogeneous disorder driven by an autoimmune response to peripheral nerve anti-

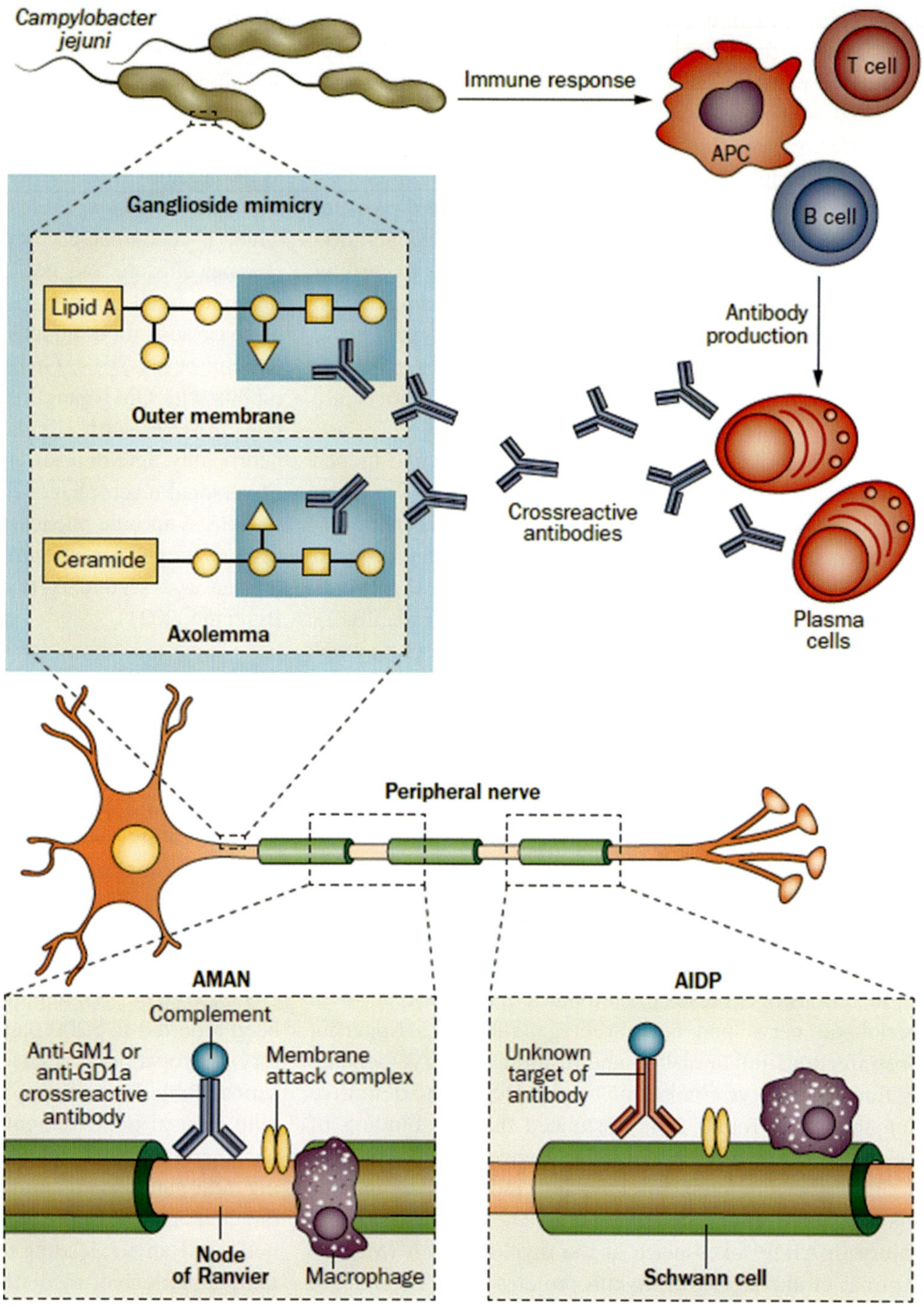

Fig. 30.1 Immunopathogenesis of GBS (Van den Berg et al., 2014)

gens causing a primary inflammatory demyelinating disease. According to severity, its outcome was believed to depend on the extension of axonal injury arising secondarily to adjacent intense inflammatory response due to demyelination as the so-called *"bystander" effect* (Kieseier et al., 2018; Berciano, 2021).

First, pathological data were generated using the *acute experimental allergic neuritis (EAN) model*. It was the first available animal model of inflammatory peripheral neuropathies, generated by immunizing a susceptible Lewis rat with PNS myelin, peripheral myelin homogenates, myelin proteins protein zero (P0), and peripheral myelin protein 22 (PMP22), or myelin-derived peptides. Immunization of animals leads to a myelin-directed T cell response characterized by T cell and macrophage infiltration, cytokine production mediated peripheral nerve damage, and attraction of secondary inflammatory cells. The monophasic disease is characterized by a flaccid tail and weakness of the hind limb (Schafflick et al., 2017). The initial pathogenic change associated with GBS was first described by Haymaker and Kernohan as the presence of inflammatory edema in the ventral and dorsal rami join of the spinal nerves (Haymaker & Kernohan, 1949). The pathological sequence is started by macrophage-mediated inflammatory infiltrates and followed by other cellular effectors that contribute to nerve damage, leading to segmental peripheral nerve demyelination and secondary axonal loss. A critical elevation of endoneural fluid pressure is a pre-requisite for ischemic axonal degeneration (Haymaker & Kernohan, 1949; Krücke, 1955). Regarding the axonal degeneration in EAN, Izumo et al. commented on the possible *"bystander damage," proposing the pathogenic role of ischemia, as axonal degeneration was observed just 1–2 days after intense endoneurial* edema (Izumo et al., 1985).

The immunological cascade that may trigger and induce demyelination in peripheral nerves and nerve roots must be completed. The direct relationship with previous infection, however, supported the role of molecular mimicry between the infectious agent's antigens and the peripheral nerve as the driver of autoimmunity in GBS (Leonhard et al., 2019). About two-thirds of patients who develop GBS report symptoms of an infection within 6 weeks from the onset (Jacobs et al., 1998). Six pathogens have been temporally associated with GBS in case-control studies: *C. jejuni*, Cytomegalovirus, Hepatitis E virus, *Mycoplasma pneumonia*, Epstein–Barr virus, and Zika virus. Other pathogens, such as Influenza A (p.H1N1), Influenza B Virus, and HTLV-III have also been linked to GBS in case or epidemiological studies. However, their role in the pathogenesis of GBS is uncertain (Vellozzi et al., 2014; Cornblath et al., 1987).

According to this model, the immune response targeting an infectious agent leads to antibodies that cross-react with peripheral nerve and nerve root gangliosides, triggering postinfectious inflammation and resulting in nerve damage or functional nerve conduction blocks. Despite no definitive autoantibodies found, it is postulated that the binding of unknown autoantibodies targeting Schwann cell epitopes could trigger complement activation, induction of inflammation, and eventual demyelination. The first pathological injury in AIDP takes place at the myelin sheath (Martín-Aguilar et al., 2019), as myelin proteins (PMP22 and P0) and the nodal protein, neurofascin, are probable targets based on the EAN model study (Berciano, 2021; Schafflick et al., 2017).

30.2.3 Axonal Degeneration as a Primary Immunopathological Event

Patients with AMAN show *primary axonal injury* without substantial T cell inflammation or demyelination. IgG and activated complement, however, are deposited on nodal and internodal axolemma (Hafer-Macko et al., 1996). Macrophages appear to invade the periaxonal space, and there is nodal lengthening following paranodal myelin detachment (Kuwabara & Yuki, 2013). This process slows conduction due to an increase in nodal membrane capacitance. In the early stage of AMAN, there is an immune attack to the node of Ranvier (as well as paranodal axoglial junctions, the nodal cytoskeleton, and Schwann cell microvilli), which is accompanied by the destruction of sodium-voltage channel clusters and nerve conduction failure. In case the immune attack progresses, calcium enters the axons, and protease activation induces axonal damage and degeneration (Uncini et al., 2017; Griffin et al., 1996). Immunization of rabbits with GM1 ganglioside can produce a disease resembling AMAN, with circulating anti-GM1 antibodies, motor neuropathy, IgG deposition on motor axons, and infiltration of personal macrophages (Yuki et al., 2001). The distinction of GBS syndrome phenotypes clarifies that axons can be the primary target for an autoimmune injury rather than being injured as a secondary phenomenon (Kieseier et al., 2018; Berciano, 2021).

30.2.4 Antigangliosides Antibodies Role

Recent serological investigations using a gangliosides panel revealed that approximately 60% of the acute-phase GBS sera had antibodies, predominantly of IgG class, against at least one ganglioside (Kusunoki, 2000; Willison & Yuki, 2002). Antibodies to a wide range of glycolipids including *GM1, GM1(NeuGc), GM1b, GalNAc-GM1b, GD1a, GalNAc-GD1a, GD1b, 9-Oacetyl GD1b, GD3, GT1a, GT1b, GQ1b, GQ1ba, LM1, galactocerebroside,* and *SGPG* have been reported in >200 papers on GBS and other inflammatory neuropathies, as case reports and in more extensive series [Table 30.1, (Willison & Yuki, 2002)]. Titers decrease with clinical improvement. Molecular mimicry mechanism has, therefore, been suggested as the mechanism of producing anti-ganglioside antibodies (Willison & Yuki, 2002), directed against sialic acid-containing glycolipids, at the nodes of Ranvier, leading to modulation of ion channel function, complement-mediated cytotoxicity and interference with nerve demyelination and axonal loss (Van den Berg et al., 2014). Regarding other subtypes, patients with AMAN often develop IgG against GM1, GD1a, and GalNAc-GD1a. Although studies suggest similar amounts of GM1 and GD1a in human sensory and motor nerves, glycolipids' fine specificity and structural orientation can contribute to the preferential involvement of motor neurons (Lopez et al., 2008). MFS patients have reactivity against GQ1b, which is expressed in the paranodal regions of extraocular motor nerves (Chiba et al., 1993). Table 30.2 (Kusunoki et al., 2008) illustrates the primary antibodies associated with clinical subtypes of acute inflammatory neuropathy.

Table 30.1 Clinical syndrome associated with specific anti-glycolipid antibodies

Clinical syndrome	Target antigen	Antibody isotype
Chronic sensory-motor demyelinating neuropathy	SGPG, SGLPG	IgM (monoclonal)
Chronic ataxic neuropathy	GD1b, GD2,GD3 GT1b, GQ1b	IgM (monoclonal)
Multifocal motor neuropathy (MMN)	GM1, GD1b, asialo-GM1	IgM (monoclonal or polyclonal)
Acute motor axonal neuropathy (AMAN)	GM1, GM1b, GD1a Gal-NAc-GD1a	IgG
Miller Fisher syndrome (MFS) Bickerstaff's brainstem encephalitis	GQ1b, GT1a	IgG
Pharyngeal-cervical-brachial weakness	GT1a, (GQ1b)	IgG

GalNAc N-acetylgalactosamine; *SGLPG* sulfated glucuronyl lactosaminyl paragloboside; *SGPG* sulfated glucuronyl paragloboside

Willison and Yuki (2002)

Table 30.2 Clinical subtypes of GBS and main associated anti-ganglioside antibodies

Clinical syndrome	Target antigen	Antigen localization	Antibody isotype
AIDP *Ataxic GBS*	GD1b (LM1) GD1b	Paranodal myelin Myelin Large neurons in DRG	IgG
AMAN *Pure motor GBS*	GM1, GM1b,GD1a Gal-NAc-GD1a	Not determinated Periaxonal membrane	IgG
Miller Fisher syndrome (MFS) *Bickerstaff's brainstem encephalitis* *GBS with ophthalmoplegia*	GQ1b (GT1a)	Paranodal myelin of III, IV, and VI nerves Some neurons in DRG	IgG

DRG dorsal root ganglia

Kusunoki et al. (2008)

Overall, a positive test result for anti-ganglioside antibody development can be helpful, especially when the diagnosis is in doubt, but a negative test result does not rule out GBS. When GBS is suspected, however, treatment should start before the antibody results (Leonhard et al., 2019). The role of anti-ganglioside antibody development in the GBS variants is more established. In fact, anti-GQ1b antibodies are found in up to 90% of patients with MFS and, therefore, have a greater diagnostic value.

30.2.5 Other Autoantibodies

In AIDP, the precise targets for the immune attack remain unknown. As found in CIDP, a subset of GBS patients with aggressive presentations and poor response to conventional therapies were found to be associated with antibodies against nodal and paranodal proteins (neurofascin 140/186 (Delmont et al., 2017), neurofascin 155 (Malotka et al., 2012), CASPR1, and contactin 1 (Appeltshauser, 2020). Other autoantibodies are reported to target myelin antigen, heat shock proteins, moesin, and tubulin, although their relevance to pathogenesis has yet to be demonstrated (Sawai et al., 2014; Yonekura et al., 2004; Helgeland et al., 2013).

30.2.6 Axonal Damage as Prognostic Criteria

The different underlying pathogenesis of AMAN and AIDP results in different patterns of clinical and neurophysiological recovery. In AIDP, recovery depends on the remyelination process and the degree of secondary axonal degeneration. In AMAN, recovery depends on the degree of axonal alterations caused by the deposition of antibodies. This process is further complicated by antibody binding at axonal sprouts that prevents axonal regeneration. In a small proportion of patients with AMAN, recovery is rapid when there is resolution of autoantibody-mediated conduction block (Willison et al., 2016).

30.3 Treatments

GBS treatments usually combine multidisciplinary supportive medical care and immunotherapy. Immunomodulatory therapy should be started if patients cannot walk independently for 10 m (GBS Disability Scale score $\geq$ 3) since the efficacy of immunomodulatory therapy on less affected patients is limited. Nonetheless, treatment should be considered for rapidly progressive weakness or the co-occurrence of autonomic dysfunction, bulbar failure, or respiratory insufficiency (Hughes et al., 2014; Chevret, 2017).

Immunomodulatory strategies include *intravenous immunoglobulin (IVIg)* and *plasma exchange (PLEx)* as equally effective treatments for GBS (Verboon et al., 2017). The dosage recommended is 0.4 g/kg body weight daily for 5 days for IVIg and 200–250 mL plasma/kg (body weight) in five sessions for PLEx (Verboon et al., 2017). IVIg and PLEx carry comparable risks of adverse events, although early studies showed that the latter is more likely than IVIg to be discontinued due to poor hemodynamic tolerance (Raphael et al., 2012). Although most patients improve significantly, prognosis can remain poor with severe disability or death in 9–17% of all GBS patients. Moreover, significant proportions of patients show long-term fatigue or pain (Leonhard et al., 2019). There is a drive to identify drugs that safely treat GBS with superior outcomes to PLEx and IVIg, especially in suboptimal responders. New targeted biological therapies for immune-mediated disorders have been identified for GBS [Fig. 30.2, (Liu et al., 2018)]. Among them, complement inhibitors and leukocyte traffic-antagonists may represent important agents in future GBS therapeutic strategies (Liu et al., 2018).

Etiopathogenetic mechanisms associated with currently therapeutic strategies for AIDP and AMAN treatment are described: antibody neutralisation by intravenous immunoglobulin (IVIg), membrane attack complex remal by plasmapheresis (PLEx) and complement inhibition by monoclonal antibodies (Eculizumab) represents the main approches.

30.3.1 Plasma Exchange (PLEx)

PLEx removes circulating neurotoxic antibodies, immune complexes, complement factors, cytokines, and other pro-inflammatory humoral mediators contributing to GBS immunopathogenesis (Van den Berg et al., 2014). In 1959, a patient with thrombotic thrombocytopenic purpura recovered after fresh frozen plasma exchange, implying that PLEx may be effective for autoimmune disorders (Rubinstein et al., 1959). In 1978, a patient with acute polyneuropathy rapidly recovered after the PLEx treatment, suggesting the potential efficacy of GBS (Brettle et al., 1978). In a randomized clinical trial of 245 GBS patients with recent onset, PLEx was established as an effective treatment compared to other conventional therapies (The Guillain-Barre syndrome Study Group, 1985): The PLEx group showed better prognosis, including as variables improvement at 4 weeks, time to one clinical grade, time to independent walking, and outcome at 6 months. PLEx was more likely to be helpful in patients who received this treatment within 7 days of onset and in

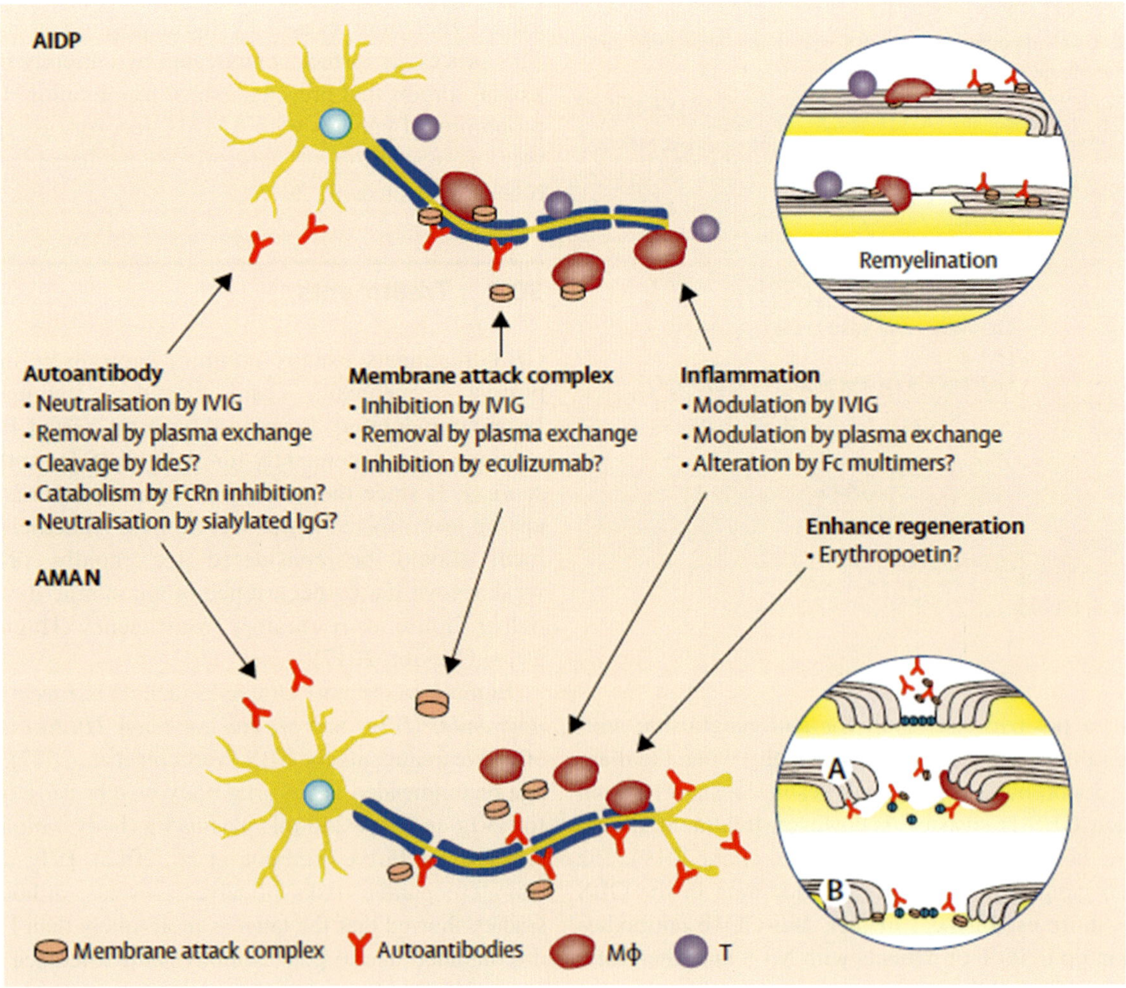

Fig. 30.2 Pathogenesis Mechanism and therapeutic targets in GBS (Liu et al., 2018)

cases requiring mechanical ventilation. A large clinical trial in 220 patients further confirmed the efficacy of PLEx in GBS patients (The French Cooperative Group on Plasma Exchange in Guillain-Barre syndrome, 1997): significant short-term benefits appeared in the group that received PLEx, as demonstrated by the reduction of assisted ventilation, and shortened time before weaning from ventilator to onset of motor recovery, and to walk with and without assistance.

According to these results, PLEx became the first validated therapy for GBS and was then considered the "gold standard": It is recommended for nonambulatory GBS patients within 4 weeks of symptom onset (Level A, Class II evidence—The Quality Standards Subcommittee of the American Academy of Neurology' guidelines), the patients who are unable to walk independently for 10 m (GBS Disability Scale score ≥ 3), and for ambulatory patients within 2 weeks of onset, Level B, limited Class II evidence (Hughes et al., 2003; Cortese et al., 2011).

Because the PLEx treatment scheme typically consists of five exchanges, the effect is temporary and administered over 1–2 weeks (Donofrio, 2003). At least four exchanges are needed in more severe cases for improved outcomes (The French Cooperative Group on Plasma Exchange in Guillain-Barre syndrome, 1997). The timing is crucial as the therapeutic response may be better within 2 weeks from disease onset, especially in nonambulatory patients. The mechanism of action of PLEx has yet to be known. However, several processes are implicated in the clearance of autoantibodies, circulating immune complexes, complement factors, cytokines (tumor necrosis factor alpha or interleukin 6) and adhesion molecules, which may result in improvement of monocyte-macrophage function, as occur in other autoimmune disorders (systemic lupus erythematosus, vasculitis), (Raphael et al., 2002; Reeves et al., 2013). The general use of PLEx is limited by the lack of access to plasma exchange (typically restricted to tertiary care hospitals), the need for close monitoring due to its potentially serious adverse effects, and the cardiovascular instability risk during autonomic deregulation (Hughes et al., 2011; Chevret, 2017).

30.3.2 Intravenous Immunoglobulin (IVIg)

Intravenous immunoglobulins (IVIg) derived from pooled human-purified immunoglobulins are an alternative to PLEx in GBS. This treatment is considered equally effective and has important pleiotropic immunomodulatory effects (Shahrizaila et al., 2021). In 1988, a study reported the use of IVIg in eight severe cases of GBS, with beneficial effects in at least some of the patients with severe GBS (Kleyweg et al., 1988). Van der Meché confirmed the result. In this study, 150 patients with GBS for less than 2 weeks, unable to walk independently, were assigned to receive either five plasma exchanges (each of 200 to 250 ml/ kg) or five doses of a preparation of intravenous immune globulin (0.4 g/kg per day). In the PLEx group, 34% of the patients improved by one or more functional grades (primary outcome) after 4 weeks, compared to 53% improvement in the IVIg (Van der Meché et al., 1992).

This evidence thus proposed IVIg to be effective in patients who are unable to walk independently for 10 m (GBS Disability Scale score ≥ 3) and when started within 2 weeks of the onset of motor symptoms (Hughes et al., 2007, 2014). Randomized controlled trials showed that IVIg at a dose of 0.4 g/kg daily for 5 consecutive days (or 1 g/kg daily for 2 days) has the same efficacy as a PLEx cycle applied over 2 weeks (Van der Meché et al., 1992; Sandoglobulin Guillain–Barré Syndrome Trial Group, 1997).

IVIg modulate the immune system in GBS in several ways aiming to reduce demyelination and axonal injury. Even though the definitive mechanism of IVIg therapy has not been established, some possible mechanisms of IVIg-mediated protection in GBS were found: neutralization of autoantibody injury, interference with pathogenic antigen recognition by excess immunoglobulin, modulation of T cell functions and Fc receptors expression on macrophages, which triggers a broad range of autoimmune response (Van der Meché et al., 1992; Zhang et al., 2004). Receptors for the Fc part of IgG constitute a family of cell-surface molecules expressed on almost every immune cells. They are also expressed on nonlymphoid cells such as Schwann cells and endothelial cells (Vedeler et al., 2001). In GBS, macrophages, recruited and activated by cytokines derived from lymphocytes, can attach to myelin-bound antibodies through FcγR (Fragment crystallizable gamma Receptor). By activating complement, antibodies bound to Schwann cells or myelin, which may contribute to demyelination by allowing the formation of a membrane attack complex. Moreover, GM1 antibodies have complement-activating capacity. C3d, a complement activation product, seems to have a role in the inhibition of complement activation and membrane attack complex formation on the outer surface of the Schwann cells and at the node of Ranvier (Vedeler et al., 2001). Furthermore, recent studies show significant association of the FcγR polymorphisms and the pathogenesis of GBS in some susceptible populations that include the Dutch and Indians (Hayat et al., 2020).

30.3.3 Complement Inhibitors

Complement deposition and activation have been described in both GBS patients and animal models (Halstead et al., 2008). In a GBS mouse model, pathological and electrophysiological alterations were prevented by the administration of

Eculizumab, a humanized monoclonal antibody that specifically binds to complement component 5 (C5) (McGonigal et al., 2010).

It specifically binds to C5 and potently inhibits the generation of pro-inflammatory C5a and C5b. Eculizumab was later tested in clinical trials to determine efficacy and safety in GBS (Davidson et al., 2017; Yamaguchi et al., 2016).

Misawa reported (Misawa et al., 2018) a placebo-controlled phase-2 study that was completed in 34 patients with severe GBS. Eculizumab was an add-on treatment to intravenous immunoglobulin at the standard dose (0,4 g/kg daily for 5 consecutive days). *Eculizumab* (900 mg) or placebo was administered before or during IVIg treatment, starting within 2–13 days from disease onset. The primary outcome, defined as the proportion of patients able to walk independently for 5 m within 4 weeks, showed no difference in the treatment arms. In particular, the proportion of patients with a functional grade of 2 or less at week 4 was 61% (90% CI 42–78; $n = 14$) in the eculizumab group compared to 45% (20–73; $n = 5$) in the placebo group. Consequently, the primary efficacy endpoint was not met, since the response rate did not exceed the predefined threshold response rate of 50%; being the lower bound of the 90% CI in the eculizumab group as 42%. A possible explanation could be that the study participants were more severely affected by the disease than expected. In the 62 historical control patients with GBS, the proportion of patients with functional grade 3–5 who regained the ability to walk independently by week 4 was 52% (90% CI 41–62). Based on these data, the response rate was set as 80%. In the present study, 88% of patients had a functional grade of 4 or 5.

Secondary endpoints also showed no difference between the treatment groups, although the proportion of patients who became able to run by week 24 (i.e., functional grade ≤ 1) was significantly greater in the *Eculizumab* group (74% vs. 18% in the placebo group).

The limitations of these studies include the small number of sample size, the absence of a direct comparison with a control group for the primary outcome, and a less severe disability of the historical control group (Nobile-Orazio et al., 2017).

30.3.4 Corticosteroids

Steroids were first successfully reported in treating a GBS patient in 1952 and used as a treatment option for many years. In the 1970s. many studies did not confirm efficacy in GBS. Cochrane systematic review including 587 GBS patients treated with different forms and dosage of corticosteroids showed no significant difference in mortality or disability grade compared with patients taking placebo (Hughes et al., 2010, 2016). The IVIg and intravenous methylprednisolone combination or methylprednisolone alone demonstrated no significant effect on long-term outcomes as compared with IVIg alone (Van Koningsveld et al., 2004).

Consequently, steroids are not recommended for the treatment of GBS patients (Level A, Class I evidence) (Hughes et al., 2003).

It is hypothesized this is due to an anti-reparative effect of corticosteroids following axonal injury (Hughes et al., 2016).

30.3.5 Other Treatments

The absence of effective additional treatments for PLEx and IVIg in recent years has brought attention to the use of monoclonal antibodies, already widely used in other neuroimmunological pathologies, such as multiple sclerosis (Rajabally, 2022; Motamed-Gorji et al., 2017). IgG monoclonal antibodies (mAbs) binding the target with their Fab antigen-binding region, result in specific function-inhibition or intracellular signaling and may also induce cytotoxicity, binding their fragment crystallizable (Fc) region. Antibody and complement-dependent lymphocyte depletion are the main mechanisms of *Alemtuzumab* and *Rituximab* (Rajabally, 2022).

Alemtuzumab is a humanized monoclonal antibody against CD52 expressed on the surface of lymphocytes, monocytes, and macrophages. This targeting of CD52 leads to complement-mediated cell lysis, antibody-dependent cytotoxicity, and apoptosis which results in a 12-month-lasting lymphopenia.

It has been seen to have prolonged remission in young patients with short disease duration in IVIG-dependent CIDP (Hirst et al., 2006). A study reported the successful treatment of a 79-year-old patient with overlapping features of Miller Fisher syndrome and Bickerstaff brainstem encephalitis with *Alemtuzumab* (Tzachanis et al., 2014).

Rituximab, a monoclonal antibody directed against CD20, expressed on the surface of B cells from pre-B cell stage to mature B cells was reported to be effective in an anecdotal case report suggesting benefit (Ostronoff et al., 2008): *Ostronoff* et al. reported a case of AMAN in a patient who had developed acute myeloid leukemia and had undergone T cell–depleted hematopoietic stem-cell transplantation as treatment. *Rituximab* improved clinical signs, but its efficacy has never been extensively investigated. Despite being effective in many autoimmune diseases, anecdotal observations indicate a paradoxical effect with GBS development or worsening of neuropathy [e.g., Anti-MAG (Broglio & Lauria, 2005)] in immunocompromised patients treated with Rituximab or chemotherapy agents (Carmona et al., 2006; Terenghi et al., 2007).

30.3.6 Potential Immunotarget and the Role of Leukocytes in Microvessels

Leukocyte trafficking [Fig. 30.3, (Liu et al., 2018)], across endoneurial microvessels, represents a complex and multistep process, pathologically relevant to AIDP and other GBS variants. According to animal models, human in situ and in vitro data, several phases of this process could be targeted as potential strategies in GBS therapy. Potential approaches could be represented by chemoattraction and haptotaxis inhibition (blocking the directional motility or outgrowth of cells, endoneurial–endothelial cells activation (e.g., chemokine receptor CCR2 antagonists), leukocyte adhesion (e.g., CD11b antagonists) or diapedesis (with platelet-endothelial cell adhesion molecule-1, CD99, CD99L2, and junctional adhesion molecules-A, -B, and –C).

Otherwise, IgG antibody transfer from blood to peripheral nerves, by Fc gamma receptor, may represent another important pathogenic mechanism that could be potentially blocked. So that, Chiang and Ubogu (2013) demonstrated an important role for chemokine receptor (CCR2) in AIDP pathogenesis of EAN murine models, observing pharmacologic efficacy of its blockade meaning the reduction of peripheral nerve demyelination, inflammation, and almost complete recovery in motor function.

Moreover, clusters of endoneurial CD11b + (alpha M-integrin) leukocytes, associated with demyelinating axons, were observed in AIDP patient sural nerves. Neutralizing monoclonal anti-mouse CD11b antibody administered in the EAN model significantly ameliorated disease severity, as well as histopathological parameters of inflammatory demyelination (Chiang & Ubogu, 2013).

Chemoattraction and haptotaxis inhibition (e.g. chemokine receptor CCR2 antagonists), leukocyte adhesion (e.g. CD11b antagonists) or diapedesis (with platelet-endothelial cell adhesion molecule-1, CD99, CD99L2 and junctional adhesion molecules-A, -B and –C), could repepresent promising targets for future treatment strategies.

30.4 Conclusions

GBS, the most common cause of acute neuromuscular weakness and paralysis worldwide, represents a group of acute immune-mediated disorders of peripheral nerves. An immune-mediated attack, directed against myelin, axons, or both, is presumed to be triggered by molecular mimicry, with both cell- and humoral-dependent mechanisms driving the pathogenic damage.

Recommended treatment is based on IVIg and PLEx, with variable efficacy dependent on subtype, clinical severity at onset, and other clinical and electrophysiologic prognostic factors. Although their mechanisms of action are not fully clarified, several monoclonal antibodies are emerging as new possibilities for immune neuropathies. Due to the pathogenic differences between demyelinating and axonal variants, future trials may require randomization based on electrophysiologic characterization and need subgroup analyses dependent to define efficacy.

30.5 Review Questions

1. Axonal degeneration:
   (a) Originally considered only the "bystander" effect of a primary intense inflammatory response, represents an independent pathological process

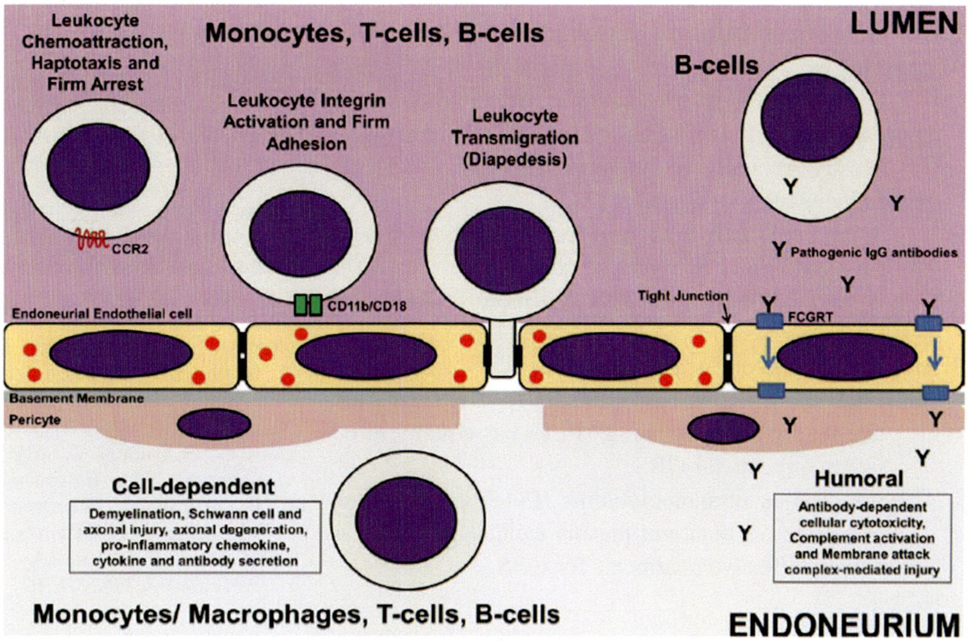

Fig. 30.3 Leukocyte trafficking as a multi-step target for future immunotherapies (Liu et al., 2018)

(b) Is based on the "bystander" effect of a primary intense inflammatory response, and is not an independent pathological process
(c) Is the primary pathological process, where the inflammatory response is considered a "bystander" effect
(d) Frequently occurs in immune-mediated neuropathies, as the most common pathological process in post-infective cases
(e) Occurs in not-immune-mediated neuropathies, with a the "bystander" effect that could trigger secondary inflammation

2. GBS:
(a) Has a female predominance, and incidence decreases with age.
(b) GBS is a recurrent ascending sensorymotor neuropathy, characterized by progressive limb paraesthesia and areflexia, of rapid progression.
(c) Is an acute-onset and rapidly evolving immunomediated neuropathy that often occurs after an acute infection.
(d) Maximum disability is reached in 6 months, followed by stabilization and recovery but disease nadir never requires artificial ventilation.
(e) Almost none remains disabled after 12 months.
3. Patients with AMAN or AMSAN:
(a) Frequently have serum antibodies against gangliosides (GQ1b)
(b) Frequently have serum antibodies against gangliosides (GM1 or GD1a)
(c) Has no consistent anti-ganglioside
(d) Frequently have serum antibodies against myelin-associated glycoprotein (MAG)
(e) As the same probability of serum antibodies as GBS
4. GBS patients report symptoms of an infection within 6 weeks from the onset:
(a) Only after COVID-19 pandemic
(b) Mostly related to *Campylobacter jejuni*, Cytomegalovirus, Hepatitis E virus, *Mycoplasma pneumonia*, Epstein–Barr virus, and Zika virus
(c) Mostly related to Influenza A (e.g., p.H1N1), Influenza B Virus, and HTLV-III
(d) Since the infective agent directly targets peripheral nerve and nerve root gangliosides
(e) Is rarely described in GBS, but more frequently in AMAN
5. Immunomodulatory strategies include:
(a) Intravenous immunoglobulins (IVIg) are more effective than plasma exchange (PLEx) as equally effective treatments for GBS.
(b) Intravenous immunoglobulins (IVIg) are definitely less effective than and plasma exchange (PLEx) as equally effective treatments for GBS.
(c) Intravenous immunoglobulin (IVIg) and plasma exchange (PLEx) as equally effective treatments for GBS.
(d) IVIg and PLEx carry comparable a great risk of adverse events because poor hemodynamic tolerance.
(e) IVIg and PLEx carry greater risks of adverse events that eculizumab and alemtuzumab.

30.6 Answers

1. a
2. c
3. b
4. b
5. c

References

Appeltshauser, L. (2020). Anti-paranodal antibodies and IgG subclasses in acute autoimmune neuropathy. *Neurology: Neuroimmunology & Neuroinflammation, 7*(5), e817.

Asbury, A. K., & Cornblath, D. R. (1990). Assessment of current diagnostic criteria for Guillain-Barré syndrome. *Annals of Neurology, 27*, S21–S24.

Asbury, A. K., Arnason, B. G., & Adams, R. D. (1969). The inflammatory lesion in idiopathic polyneuritis. Its role in pathogenesis. *Medicine, 48*, 173–215.

Berciano, J. (2021). Axonal degeneration in Guillain-Barré syndrome: A reappraisal. *Journal of Neurology, 268*(10), 3728–3743. https://doi.org/10.1007/s00415-020-10034-y. Epub 2020 Jun 30.

Brettle, R. P., Gross, M., Legg, N. J., Lockwood, M., & Pallis, C. (1978). Treatment of acute polyneuropathy by plasma exchange. *Lancet, 2*, 1100. https://doi.org/10.1016/S0140-6736(78)91837-8

Broglio, L., & Lauria, G. (2005). Worsening after rituximab treatment in anti-MAG neuropathy. *Muscle & Nerve, 32*, 378–379.

Carmona, A., Alonso, J. D., de las Heras, M., & Navarrete, A. (2006). Guillain–Barre´ syndrome in a patient with diffuse large B-cell lymphoma, and rituximab maintenance therapy. An association beyond anecdotal evidence? *Clinical & Translational Oncology, 8*, 764–766.

Chevret, S. (2017). Plasma exchange for Guillain- Barré syndrome. *Cochrane Database of Systematic Reviews, 2*, CD001798.

Chiang, S., & Ubogu, E. E. (2013). The role of chemokines in Guillain-Barre syndrome. *Muscle & Nerve, 48*, 320–330. https://doi.org/10.1002/mus.23829

Chiba, A., Kusunoki, S., Obata, H., Machinami, R., & Kanazawa, I. (1993). Serum anti-GQ1b IgG antibody is associated with ophthalmoplegia in Miller Fisher syndrome and Guillain-Barré syndrome: Clinical and immunohistochemical studies. *Neurology, 43*, 1911–1917.

Cornblath, D. R., McArthur, J. C., Kennedy, P. G., Witte, A. S., & Griffin, J. W. (1987). Inflammatorydemyelinating peripheral neuropathies associated with human T- cell lymphotropic virus type III infection. *Annals of Neurology, 21*, 32–40.

Cortese, I., Chaudhry, V., So, Y. T., Cantor, F., Cornblath, D. R., & Rae-Grant, A. (2011). Evidence-based guideline update: Plasmapheresis in neurologic disorders: Report of the Therapeutics and Technology Assessment Subcommittee of the American Academy of Neurology. *Neurology, 76*, 294–300. https://doi.org/10.1212/WNL.0b013e318207b1f6

Davidson, A. I., Halstead, S. K., Goodfellow, J. A., Chavada, G., Mallik, A., Overell, J., et al. (2017). Inhibition of complement in Guillain-Barre syndrome: The ICA-GBS study. *Journal of the Peripheral Nervous System, 22*, 4–12. https://doi.org/10.1111/jns.12194

Delmont, E., et al. (2017). Autoantibodies to nodal isoforms of neurofascin in chronic inflammatory demyelinating polyneuropathy. *Brain, 140*(7), 1851–1858. https://doi.org/10.1093/brain/awx124

Doneddu, P. E., et al. (2019). Atypical CIDP: Diagnostic criteria, progression and treatment response. Data from the Italian CIDP database. *Journal of Neurology, Neurosurgery, and Psychiatry, 90*(2), 125–132. https://doi.org/10.1136/jnnp-2018-318714

Donofrio, P. D. (2003). Immunotherapy of idiopathic inflammatory neuropathies. *Muscle & Nerve, 28*, 273–292. https://doi.org/10.1002/mus.10402

Gilbert, M., et al. (2002). The genetic bases for the variation in the lipo-oligosaccharide of the mucosal pathogen, Campylobacter jejuni. Biosynthesis of sialylated ganglioside mimics in the core oligosaccharide. *The Journal of Biological Chemistry, 277*, 327–337.

Godschalk, P. C., et al. (2007). Structural characterization of Campylobacter jejuni lipooligosaccharide outer cores associated with Guillain–Barré and Miller Fisher syndromes. *Infection and Immunity, 75*, 1245–1254.

Griffin, J. W., et al. (1996). Early nodal changes in the acute motor axonal neuropathy pattern of the Guillain-Barré syndrome. *Journal of Neurocytology, 25*(1), 33–51. https://doi.org/10.1007/BF02284784

Gwathmey, K. G., & Smith, A. G. (2020). Immune-mediated neuropathies. *Neurologic Clinics, 38*(3), 711–735. https://doi.org/10.1016/j.ncl.2020.03.008. Epub 2020 Jun 11.

Hafer-Macko, C., Hsieh, S. T., Li, C. Y., et al. (1996). Acute motor axonal neuropathy: An antibody-mediated attack on axolemma. *Annals of Neurology, 40*, 635–644.

Halstead, S. K., Zitman, F. M., Humphreys, P. D., et al. (2008). Eculizumab prevents anti-ganglioside antibody-mediated neuropathy in a murine model. *Brain, 131*, 1197–1208.

Hayat, S., Babu, G., Das, A., Howlader, Z. H., Mahmud, I., & Islam, Z. (2020). Fc-gamma IIIa-V158F receptor polymorphism contributes to the severity of Guillain-Barré syndrome. *Annals of Clinical Translational Neurology, 7*(6), 1040–1049. https://doi.org/10.1002/acn3.51072. Epub 2020 Jun 2. PMID: 32484314; PMCID: PMC7317642.

Haymaker, W. E., & Kernohan, J. W. (1949). The Landry–Guillain–Barré syndrome; a clinicopathologic report of 50 fatal cases and a critique of the literature. *Medicine (Baltimore), 28*, 59–141.

Heikema, A. P., Jacobs, B. C., Horst-Kreft, D., Huizinga, R., Kuijf, M. L., Endtz, H. P., Samsom, J. N., & van Wamel, W. J. (2013). Siglec-7 specifically recognizes Campylobacter jejuni strains associated with oculomotor weakness in Guillain-Barré syndrome and Miller Fisher syndrome. *Clinical Microbiology and Infection, 19*(2), E106–E112. https://doi.org/10.1111/1469-0691.12073. Epub 2012 Nov 23.

Helgeland, G., Petzold, A., Hoff, J. M., Gilhus, N. E., Plant, G. T., & Romi, F. R. (2013). *Anti-Heat Shock Protein 70 antibody levels are increased in myasthenia gravis and Guillain-Barré syndrome*.

Hirst, C., Raasch, S., Llewelyn, G., & Robertson, N. (2006). Remission of chronic inflammatory demyelinating polyneuropathy after alemtuzumab (Campath 1H). *Journal of Neurology, Neurosurgery, and Psychiatry, 77*, 800–802.

Hughes, R. A., & Cornblath, D. R. (2005). Guillain-Barré syndrome. *Lancet, 366*(9497), 1653–1666. https://doi.org/10.1016/S0140-6736(05)67665-9

Hughes, R. A., Wijdicks, E. F., Barohn, R., Benson, E., Cornblath, D. R., Hahn, A. F., et al. (2003). Practice parameter: Immunotherapy for Guillain-Barre syndrome: Report of the Quality Standards Subcommittee of the American Academy of Neurology. *Neurology, 61*, 736–740. https://doi.org/10.1212/WNL.61.6.736

Hughes, R. A., et al. (2007). Immunotherapy for Guillain–Barré syndrome: A systematic review. *Brain, 130*, 2245–2257.

Hughes, R. A., Swan, A. V., & van Doorn, P. A. (2010). Corticosteroids for Guillain-Barre syndrome. *Cochrane Database of Systematic Reviews, 2*, CD001446.

Hughes, R. A., Swan, A. V., & van Doorn, P. A. (2014). Intravenous immunoglobulin for Guillain- Barré syndrome. *Cochrane Database of Systematic Reviews, 9*, CD002063.

Hughes, R. A., Brassington, R., Gunn, A. A., & van Doorn, P. A. (2016). Corticosteroids for Guillain-Barre syndrome. *Cochrane Database of Systematic Reviews, 10*, CD001446.

Izumo, S., Linington, C., Wekerle, H., & Meyermann, R. (1985). Morphologic study on experimental allergic neuritis mediated by T cell line specific for bovine P2 protein in Lewis rats. *Laboratory Investigation, 53*, 209–218.

Jacobs, B. C., Rothbarth, P. H., van der Meché, F. G., et al. (1998). The spectrum of antecedent infections in Guillain-Barré syndrome: A case-control study. *Neurology, 51*, 1110–1115.

Jacobs BC, van den Berg B, Verboon C, Chavada G, Cornblath DR, Gorson KC, Harbo T, Hartung HP, Hughes RAC, Kusunoki S, van Doorn PA, Willison HJ; IGOS Consortium. 2Cite Share. International Guillain-Barre Syndrome Outcome Study: protocol of a prospective observational cohort study on clinical and biological predictors of disease course and outcome in Guillain-Barre syndrome. J Peripher Nerv Syst. 2017 Jun;22(2):68–76. https://doi.org/10.1111/jns.12209

Kaida, K., Morita, D., Kanzaki, M., et al. (2004). Ganglioside complexes as new target antigens in Guillain-Barré syndrome. *Annals of Neurology, 56*, 567–571.

Kieseier, B. C., Mathey, E. K., Sommer, C., & Hartung, H. P. (2018). Immune-mediated neuropathies. *Nature Reviews Disease Primers, 4*(1), 31. https://doi.org/10.1038/s41572-018-0027-2

Kimura, J. (2013). *Electrodiagnosis in diseases of nerve and muscle: Principles and practice* (4th ed.). Oxford University Press.

Kleyweg, R. P., van der Meche, F. G., & Meulstee, J. (1988). Treatment of Guillain- Barre syndrome with high-dose gammaglobulin. *Neurology, 38*, 1639–1641. https://doi.org/10.1212/WNL.38.10.1639

Krücke, W. (1955). Die primär-entzündliche Polyneuritis unbekannter Ursache. In O. Lubarsch et al. (Eds.), *Handbuch der speziallen pathologischen Anatomie und Histologie, Vol XIII/5, Erkrankungen des peripheren und des vegetativen Nerven* (pp. 164–182). Springer-Verlag.

Kusunoki, S. (2000). Antiglycolipid antibodies in Guillain–Barré syndrome. *The American Journal of the Medical Sciences, 319*, 234–239.

Kusunoki, S., Kaida, K., & Ueda, M. (2008). Antibodies against gangliosides and ganglioside complexes in Guillain-Barré syndrome: New aspects of research. *Biochimica et Biophysica Acta, 1780*(3), 441–444. https://doi.org/10.1016/j.bbagen.2007.10.001. Epub 2007 Oct 12.

Kuwabara, S., & Yuki, N. (2013). Axonal Guillain-Barré syndrome: Concepts and controversies. *Lancet Neurology, 12*, 1180–1188.

Leonhard, S. E., Mandarakas, M. R., Gondim, F. A. A., et al. (2019). Diagnosis and management of Guillain-Barré syndrome in ten steps. *Nature Reviews Neurology, 15*(11), 671–683. https://doi.org/10.1038/s41582-019-0250-9

Liu, S., Dong, C., & Ubogu, E. E. (2018). Immunotherapy of Guillain-Barré syndrome. *Human Vaccines & Immunotherapeutics, 14*(11), 2568–2579. https://doi.org/10.1080/21645515.2018.1493415. Epub 2018 Jul 12. PMID: 29953326; PMCID: PMC6314401.

Lopez, P. H., Zhang, G., Bianchet, M. A., Schnaar, R. L., & Sheikh, K. A. (2008). Structural requirements of anti-GD1a antibodies determine their target specificity. *Brain, 131*, 1926–1939.

Malotka, N., et al. (2012). Neurofascinas a target for autoantibodies in peripheral neuropathies. *Neurology, 79*(23), 2241–2248. https://doi.

org/10.1212/WNL.0b013e31827689ad. Epub 2012 Oct 24. PMID: 23100406; PMCID: PMC3542349.

Martín-Aguilar, L., Pascual-Goñi, E., & Querol, L. (2019). Autoantibodies in immune-mediated infammatory neuropathies. *Medicina Clínica (English Ed.), 153*(9), 360–367. https://doi.org/10.1016/j.medcle.2019.06.015

Mathey, E. K., Park, S. B., Hughes, R. A., Pollard, J. D., Armati, P. J., Barnett, M. H., Taylor, B. V., Dyck, P. J., Kiernan, M. C., & Lin, C. S. (2015). Chronic inflammatory demyelinating polyradiculoneuropathy: From pathology to phenotype. *Journal of Neurology, Neurosurgery, and Psychiatry, 86*(9), 973–985. https://doi.org/10.1136/jnnp-2014-309697. Epub 2015 Feb 12. PMID: 25677463; PMCID: PMC4552934.

McGonigal, R., et al. (2010). Anti-GD1a antibodies activate complement and calpain to injure distal motor nodes of Ranvier in mice. *Brain, 133*(7), 1944–1960. https://doi.org/10.1093/brain/awq119

Misawa, S., Kuwabara, S., Sato, Y., Yamaguchi, N., Nagashima, K., Katayama, K., Sekiguchi, Y., Iwai, Y., Amino, H., Suichi, T., Yokota, T., Nishida, Y., Kanouchi, T., Kohara, N., Kawamoto, M., Ishii, J., Kuwahara, M., Suzuki, H., Hirata, K., Kokubun, N., Masuda, R., Kaneko, J., Yabe, I., Sasaki, H., Kaida, K. I., Takazaki, H., Suzuki, N., Suzuki, S., Nodera, H., Matsui, N., Tsuji, S., Koike, H., Yamasaki, R., & Kusunoki, S. (2018). Japanese Eculizumab Trial for GBS (JET-GBS) Study Group. Safety and efficacy of eculizumab in Guillain-Barré syndrome: A multicentre, double-blind, randomised phase 2 trial. *Lancet Neurology, 17*(6), 519–529. https://doi.org/10.1016/S1474-4422(18)30114-5. Epub 2018 Apr 21.

Motamed-Gorji, N., Matin, N., Tabatabaie, O., Pavone, P., Romano, C., Falsaperla, R., & Vitaliti, G. (2017). Biological drugs in Guillain-Barré syndrome: An update. *Current Neuropharmacology, 15*(7), 938–950. https://doi.org/10.2174/1570159X14666161213114904. PMID: 27964705; PMCID: PMC5652014.

Ostronoff, F., Perales, M. A., Stubblefield, M. D., & Hsu, K. C. (2008). Rituximab-responsive Guillain-Barré syndrome following allogeneic hematopoietic SCT. *Bone Marrow Transplantation, 42*(1), 71–72.

Plasma Exchange/Sandoglobulin Guillain–Barré Syndrome Trial Group. (1997). Randomised trial of plasma exchange, intravenous immunoglobulin, and combined treatments in Guillain–Barré syndrome. *Lancet, 349*, 225–230.

Rajabally, Y. A. (2022). Immunoglobulin and monoclonal antibody therapies in Guillain-Barré syndrome. *Neurotherapeutics, 19*, 885–896. https://doi.org/10.1007/s13311-022-01253-4

Raphael, J. C., Chevret, S., Hughes, R. A., & Annane, D. (2002). Plasma exchange for Guillain-Barre syndrome. *Cochrane Database of Systematic Reviews, 2*, CD001798.

Raphael, J. C., Chevret, S., Hughes, R. A., & Annane, D. (2012). Plasma exchange for Guillain- Barré syndrome. *Cochrane Database of Systematic Reviews, 7*, CD001798.

Rubinstein, M. A., Kagan, B. M., Macgillviray, M. H., Merliss, R., & Sacks, H. (1959). Unusual remission in a case of thrombotic thrombocytopenic purpurasyndrome following fresh blood exchange transfusions. *Annals of Internal Medicine, 51*, 1409–1419. https://doi.org/10.7326/0003-4819-51-6-1409

Ruts, L., Drenthen, J., Jacobs, B. C., van Doorn, P. A., & Dutch Guillain- Barré syndrome Study Group. (2007). *Distinguishing acute- onset CIDP from fluctuating Guillain- Barré syndrome: A prospective study.*

Sawai, S., et al. (2014). Moesin is a possible target molecule for cytomegalovirus- related Guillain-Barré syndrome. *Neurology, 83*(2), 113–117. https://doi.org/10.1212/WNL.0000000000000566

Schafflick, D., Kieseier, B. C., Wiendl, H., & Meyer Zu Horste, G. (2017). Novel pathomechanisms in inflammatory neuropathies. *Journal of Neuroinflammation, 14*(1), 232. https://doi.org/10.1186/s12974-017-1001-8. PMID: 29179723; PMCID: PMC5704548.

Sejvar, J. J., Baughman, A. L., Wise, M., & Morgan, O. W. (2011a). Population incidence of Guillain- Barré syndrome: A systematic review and meta- analysis. *Neuroepidemiology, 36*, 123–133.

Sejvar, J. J., et al. (2011b). Guillain-Barré syndrome and fisher syndrome: Case definitions and guidelines for collection, analysis, and presentation of immunization safety data. *Vaccine, 29*, 599–612.

Shahrizaila, N., Lehmann, H. C., & Kuwabara, S. (2021). Guillain-Barré syndrome. *Lancet, 397*(10280), 1214–1228. https://doi.org/10.1016/S0140-6736(21)00517-1. Epub 2021 Feb 26.

Terenghi, F., Ardolino, G., & Nobile-Orazio, E. (2007). Guillain–Barre´ syndrome after combined CHOP and rituximab therapy in non-Hodgkin lymphoma. *Journal of the Peripheral Nervous System, 12*, 142–143.

The French Cooperative Group on Plasma Exchange in Guillain- Barre Syndrome. (1997). Appropriate number of plasma exchanges in Guillain-Barre syndrome. *Annals of Neurology, 41*, 298–306.

The Guillain- Barre syndrome Study Group. (1985). Plasmapheresis and acute Guillain-Barre syndrome. *Neurology, 35*, 1096–1104. https://doi.org/10.1212/WNL.35.8.1096

Tzachanis, D., Hamdan, A., Uhlmann, E. J., & Joyce, R. M. (2014). Successful treatment of refractory Guillain-Barre syndrome with alemtuzumab in a patient with chronic lymphocytic leukemia. *Acta Haematologica, 132*, 240–243.

Uncini, A., Ippoliti, L., Shahrizaila, N., Sekiguchi, Y., Kuwabara, S. (2017). Optimizing the electrodiagnostic accuracy in Guillain-Barré syndrome subtypes: Criteria sets and sparse linear discriminant analysis. *Clin Neurophysiolactions, 128*(7), 1176–1183. https://doi.org/10.1016/j.clinph.2017.03.048

Van Belkum, A., et al. (2001). A Campylobacter jejuni gene associated with immune-mediated neuropathy. *Nature Medicine, 7*, 752–753.

Van den Berg, B., Walgaard, C., Drenthen, J., Fokke, C., Jacobs, B. C., & van Doorn, P. A. (2014). Guillain-Barré syndrome: Pathogenesis, diagnosis, treatment and prognosis. *Nature Reviews Neurology, 10*(8), 469–482. https://doi.org/10.1038/nrneurol.2014.121. Epub 2014 Jul 15.

Van Koningsveld, R., Schmitz, P. I., Meche, F. G., Visser, L. H., Meulstee, J., van Doorn, P. A., et al. (2004). Effect of methylprednisolone when added to standard treatment with intravenous immunoglobulin for Guillain- Barre syndrome: Randomised trial. *Lancet, 363*, 192–196. https://doi.org/10.1016/S0140-6736(03)15324-X

Van der Meché, F. G. A., Schmitz, P. I. M., & Dutch Guillain-Barré Study Group. (1992). A randomized trial comparing intravenous immune globulin and plasma exchange in Guillain-Barré syndrome. *New England Journal of Medicine, 326*(17), 1123–1129. [PUBMED: 1552913].

Vedeler, C. A., Myhr, K. M., & Nyland, H. (2001). Fc receptors for immunoglobulin G--a role in the pathogenesis of Guillain-Barré syndrome and multiple sclerosis. *Journal of Neuroimmunology, 118*(2), 187–193. https://doi.org/10.1016/s0165-5728(01)00344-7

Vellozzi, C., Iqbal, S., & Broder, K. (2014). Guillain- Barré syndrome, influenza, and influenza vaccination: The epidemiologic evidence. *Clinical Infectious Diseases, 58*, 1149–1155.

Verboon, C., van Doorn, P. A., & Jacobs, B. C. (2017). Treatment dilemmas in Guillain- Barré syndrome. *Journal of Neurology, Neurosurgery, and Psychiatry, 88*, 346–352.

Wanschitz, J., Maier, H., Lassmann, H., Budka, H., & Berger, T. (2003). Distinct time pattern of complement activation and cytotoxic T cell response in Guillain-Barré syndrome. *Brain, 126*, 2034–2042.

Willison, H. J., & Yuki, N. (2002). Peripheral neuropathies and anti-glycolipid antibodies. *Brain, 125*, 2591–2625.

Willison, H. J., Jacobs, B. C., & van Doorn, P. A. (2016). Guillain-Barré syndrome. *Lancet, 388*, 717–727.

Yamaguchi, N., Misawa, S., Sato, Y., Nagashima, K., Katayama, K., Sekiguchi, Y., et al. (2016). A prospective, multicenter, randomized phase II study to evaluate the efficacy and safety of eculizumab in patients with Guillain-Barre syndrome (GBS): Protocol of Japanese Eculizumab trial for GBS (JET-GBS). *JMIR Research Protocols, 5*, e210. https://doi.org/10.2196/resprot.6610

Yonekura, K., et al. (2004). Prevalence of anti-heat shock protein antibodies in cerebrospinal fluids of patients with Guillain-Barré syndrome. *Journal of Neuroimmunology, 156*(1–2), 204–209. https://doi.org/10.1016/j.jneuroim.2004.07.017

Yuki, N., Shahrizaila, N. (2011). Keeping it simple: is there a need for the various subtyping of axonal forms of Guillain-Barre syndrome? *Journal of Neurology, Neurosurgery and Psychiatry, 82*(6), 592. https://doi.org/10.1136/jnnp.2011.241562

Yuki, N., & Hartung, H. P. (2012). Guillain-Barré syndrome. *The New England Journal of Medicine, 366*, 2294–2304.

Yuki, N., Yamada, M., Koga, M., et al. (2001). Animal model of axonal Guillain-Barré syndrome induced by sensitization with GM1 ganglioside. *Ann Neurol, 49*, 712–720.

Zhang, G., Lopez, P. H., Li, C. Y., Mehta, N. R., Griffin, J. W., Schnaar, R. L., et al. (2004). Anti-ganglioside antibody-mediated neuronal cytotoxicity and its protection by intravenous immunoglobulin: Implications for immune neuropathies. *Brain., 127*, 1085–1100. https://doi.org/10.1093/brain/awh127

Glaucoma

31

Meghal Gagrani, Vikas Gulati, Deepta A. Ghate, and Shane J. Havens

Abstract

Glaucoma is defined as a disturbance of the structural or functional integrity of the optic nerve that causes characteristic changes in the optic nerve leading to permanent defects in the visual field. If left untreated, it can lead to blindness. Glaucoma is the leading cause of irreversible blindness throughout the world and is second only to macular degeneration in the USA. Amongst the risk factors, elevated intraocular pressure, thin corneal thickness, increasing age, and a large cup-to-disc ratio are established. Concerning its pathogenesis, mechanical and vascular theories are fundamental. In addition, many other mechanisms contribute; such as immune factors, apoptosis, glutamate-induced excitotoxicity, free radicals, nitric acid synthase, and genetic mutations. To assess glaucoma; tonometry, gonioscopy, examination of optic disc, and visual field testing are essential. Animal models for glaucoma have been used to study the pathophysiological basis of glaucoma, aqueous humor dynamics, the trabecular meshwork, the ciliary muscle, and retinal ganglion cells. Animal experiments have provided a worthwhile means of evaluating diagnostic modalities and treatments, both medical and surgical.

Keywords

Glaucoma · Intraocular pressure · Optic nerve · Retinal ganglion cell

31.1 Introduction

The word glaucoma has its origin in ancient Greek (Hippocrates, approximately 400 BC), meaning clouded or blue-green hue, most likely describing a patient having a cloudy cornea or advanced cataract precipitated by chronic elevated pressure. Over the last 100 years, further understanding of the concept of glaucoma has accelerated.

Primary open-angle glaucoma (POAG) is a multifactorial chronic optic neuropathy with a characteristic acquired loss of optic nerve fibers. Cupping and atrophy of the optic disc occur in the absence of other known causes. Such damage develops in the presence of open anterior chamber angles and produces characteristic visual field abnormalities. Some subsets of patients can exhibit the characteristic optic nerve damage and visual field defects while having an intraocular pressure (IOP) within the statistically normal range. IOP is the only risk factor that can be currently modified in POAG. Elevated IOP in the absence of any clinically apparent optic nerve damage is referred to as ocular hypertension (OHT).

A neurodegenerative disease has loss of specific neuronal populations with axonal loss and a characteristic clinical syndrome that is progressive in nature. Glaucoma is a degenerative disorder of the retinal ganglion cells (RGCs) leading to characteristic optic nerve morphology and visual field defects. This loss of axons continues in the lateral geniculate nucleus (Chen et al., 2013; Gupta et al., 2009) and in the visual cortex (Yu et al., 2013; Yucel & Gupta, 2008).

M. Gagrani
Department of Ophthalmology, University of Pittsburgh Medical Center, Pittsburgh, PA, USA

V. Gulati (✉)
Truhlsen Eye Institute, University of Nebraska Medical Center, Omaha, NE, USA
e-mail: vgulati@unmc.edu

D. A. Ghate
Emory Eye Center, Emory University, Atlanta, GA, USA

S. J. Havens (✉)
Eye Consultants PC, Omaha, NE, USA

Omaha VA Medical Center Eye Clinic, Omaha, NE, USA

Truhlsen Eye Institute, University of Nebraska Medical Center, Omaha, NE, USA
e-mail: drhavens@eyeconsultantspc.com

H. E. Gendelman, T. Ikezu (eds.), *Neuroimmune Pharmacology and Therapeutics*, https://doi.org/10.1007/978-3-031-68237-7_31

Untreated glaucoma progresses due to continued loss of RGC and visual field constriction.

Other forms of glaucoma include angle closure glaucoma, in which the outflow of aqueous humor is physically obstructed by the iris. Secondary glaucomas have obstruction of the aqueous humor pathways secondary to other disease processes such as inflammation, trauma, bleeding, pigment, and pseudoexfoliation. The only glaucoma that fits the definition of a neurodegenerative disorder is POAG and the rest of the chapter will primarily refer to POAG.

Patients with glaucoma have preserved central vision with loss of peripheral vision. Glaucoma can be asymptomatic with 24% of patients with advanced glaucoma noting that they have vision compromise at all (Hoste, 2003). Vision loss from glaucoma does not cause either a black tunnel or big black commas in an individual's vision as commonly depicted in simulations. Vision loss from glaucoma manifests as a blur in the regions of peripheral field loss and this blur worsens as the scotoma deepens (Gagrani et al., 2022).

POAG is a major worldwide health concern because of its usually asymptomatic but progressive nature, and because it is the leading cause of irreversible blindness in the world. With appropriate screening and treatment, glaucoma usually can be identified and treated to prevent and slow significant visual loss.

31.2 Epidemiology

31.2.1 Prevalence

In the USA, multiple population studies [e.g., Framingham (Leibowitz et al., 1980), Baltimore (Tielsch et al., 1991), and Barbados (Leske et al., 1997)] have been performed to estimate the prevalence of eye diseases, including POAG and OHT.

Estimates of the prevalence of glaucoma [(Friedman et al., 2004a); Table 31.1] and blindness [(Congdon et al., 2004); Fig. 31.1] in the USA demonstrate the following: glaucoma is a leading cause of irreversible blindness, second only to macular degeneration; 56–75% of glaucoma in the USA is undiagnosed (Varma et al., 2004; Sommer et al., 1991); and more than 2.25 million Americans aged 40 years and older have POAG.

More than 1.6 million people have significant visual impairment from glaucoma, with 150,000 bilaterally blind in the USA alone (Tielsch et al., 1991). These statistics emphasize the need to identify and closely monitor those at risk for glaucomatous damage.

Approximately 6.7 million people are bilaterally blind from POAG worldwide (Quigley, 1996) and the number of people with glaucoma is projected to be approximately 118 million by 2040 (Tham et al., 2014).

Table 31.1 Prevalence of glaucoma[a] by age, gender, and race

Age, years	Prevalence/100 population (95% CI)		
	White subjects	Black subjects	Hispanic subjects
Women			
40–49	0.83 (0.65–1.06)	1.51 (0.94–2.41)	0.34 (0.15–0.72)
50–54	0.89 (0.78–1.02)	2.24 (1.59–3.14)	0.65 (0.37–1.15)
55–59	1.02 (0.89–1.16)	2.86 (2.16–3.78)	0.98 (0.61–1.58)
60–64	1.23 (1.07–1.41)	3.65 (2.83–4.69)	1.49 (0.97–2.28)
65–69	1.58 (1.37–1.82)	4.64 (3.54–6.05)	2.24 (1.43–3.49)
70–74	2.16 (1.87–2.49)	5.89 (4.28–8.05)	3.36 (2.00–5.60)
75–79	3.12 (2.68–3.63)	7.45 (5.06–10.84)	5.01 (2.68–9.15)
≥80	6.94 (5.40–8.88)	9.82 (6.08–15.48)	10.05 (4.35–21.52)
Men			
40–49	0.36 (0.27–0.47)	0.55 (0.31–0.95)	0.39 (0.18–0.85)
50–54	0.61 (0.50–0.74)	1.71 (1.25–2.32)	0.69 (0.39–1.25)
55–59	0.85 (0.72–1.00)	3.06 (2.30–4.04)	1.00 (0.61–1.64)
60–64	1.18 (1.02–1.37)	4.94 (3.69–6.59)	1.44 (0.92–2.24)
65–69	1.64 (1.40–1.91)	7.24 (5.40–9.63)	2.07 (1.32–3.23)
70–74	2.27 (1.90–2.72)	9.62 (7.29–12.59)	2.97 (1.79–4.89)
75–79	3.14 (2.53–3.90)	11.65 (8.81–15.25)	4.23 (2.32–7.60)
≥80	5.58 (4.15–7.47)	13.21 (7.85–21.38)	7.91 (3.53–16.77)

Reproduced with permission from: Friedman et al. (2004a)
CI confidence interval

[a]Glaucoma indicates primary open-angle glaucoma

31.2.2 Prognosis

Several studies (Gordon et al., 2002) have shown the incidence of new onset of glaucomatous damage in previously unaffected patients to be approximately 2.6–3% in patients with IOPs 21–25 mm Hg, 12–26% with IOPs 26–30 mm Hg, and approximately 42% with IOP higher than 30 mm Hg. The ocular hypertension treatment study (OHTS) found that approximately 10% of patients with IOPs ranging from 24 to 31 mm Hg but with no clinical signs of glaucoma at baseline develop glaucoma over 5 years and 29.5% developed glaucoma at 20 years (Kass et al., 2021), with conversion to glaucoma reduced by 50% if patients are preemptively started on IOP-lowering therapy (Kass et al., 2002). Significant subsets of higher and lower risk exist when central corneal thickness (CCT) is taken into account.

31.2.3 Risk Factors

31.2.3.1 Intraocular Pressure

In eyes suspected of being affected with glaucoma, loss of visual field develops at a rate of approximately 1–3% per year of observation (Gordon et al., 2002). High IOP is the most important risk factor (Table 31.2). The relative importance of other risk factors also is indicated in Table 31.2. In population-based surveys, between 25% and 50% of those with glaucomatous damage to the optic nerve have IOP in the normal range. However, the probability of

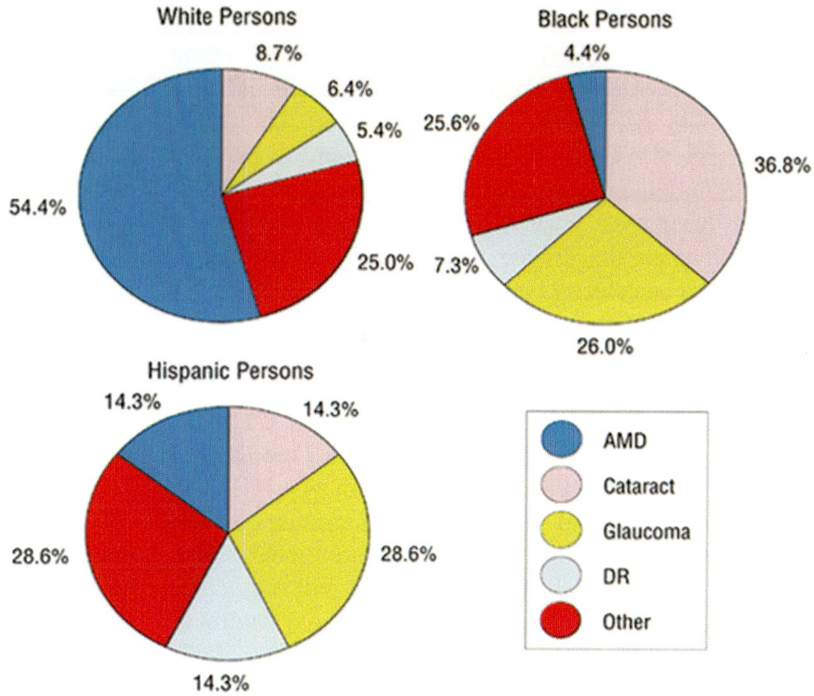

Fig. 31.1 Causes of blindness (best-corrected visual acuity <6/60 [<20/200] in the better-seeing eye) by race/ethnicity. AMD indicates age-related macular degeneration, DR diabetic retinopathy. (Reproduced with permission from Congdon et al. (2004))

Table 31.2 Risk factors for glaucoma

Established	Highly probable	Possible
Elevated IOP	Race (African descent[a])	Presence (or absence) of diabetes
Thin CCT	Positive family history	Hypertension
Increasing age	Low diastolic perfusion pressure	
Large cup to disc ratio	Migraines	
	IOP fluctuations	
	Sleep apnea	
	Immunological disorders	
	Myopia	

IOP intraocular pressure, *CCT* central corneal thickness
[a]In USA, West Africa, Caribbean

injury increases exponentially with higher IOP (Heijl et al., 2002). Large diurnal fluctuation in IOP is another independent risk factor for glaucoma (Nouri-Mahdavi et al., 2004).

31.2.3.2 Race

The prevalence of POAG is 3–4 times higher in blacks than in whites in the USA (Friedman et al., 2004b; Tielsch et al., 1991). Glaucoma is the most common cause of irreversible blindness in the USA among people of African or Hispanic descent [(Congdon et al., 2004); Fig. 31.1]. Blacks in the USA are more likely to develop glaucoma early in life, and they tend to have a more aggressive form of the disease.

The Barbados Eye Study (Leske et al., 1997) over 4 years showed a 5 times higher incidence of developing glaucoma in a group of black ocular hypertensives as compared with a predominantly white population. Some population studies have found the mean IOP in blacks to be higher than in Caucasian controls. Other studies (Tielsch et al., 1991) found no difference. Consequently, further studies are required to clarify this issue.

The OHTS (Brandt et al., 2001) demonstrated that black patients in the US overall may have reduced CCT, thereby leading to underestimation of true IOP. The reduced CCT was associated with a higher risk of developing glaucoma. Therefore, CCT measurement is particularly important in African-American patients who are glaucoma suspects. The OHTS also demonstrated a larger cup-to-disc ratio in Blacks, which was an independent risk factor for the development of glaucoma [(Feuer et al., 2002; Gordon et al., 2002); Table 31.2]. These two confounding variables discovered in the multivariate analysis appeared to account for the apparent increased risk in blacks demonstrated in the univariate analysis (Gordon et al., 2002).

Race also impacts intraocular pressure. Our understanding of normal intraocular pressure (15.5 mmHg) and a cutoff of 21 mmHg as two standard deviations above the population mean comes from studies in predominantly white populations (Armaly, 1965; Leydhecker et al., 1958). Japanese populations on the other hand have a mean IOP of 14.5 mmHg (standard deviation 2.7) and the Hispanic population in the USA has a mean IOP of 14.3 mmHg (SD of 3.3). This leads to the labeling of 92% and 82% of all glaucomas as normal tension glaucoma in the Tajimi study in Japan and the Los Angeles Latino Study in the USA, respectively (Iwase et al., 2004; Varma et al., 2004).

31.2.3.3 Age

Age older than 40 years is a risk factor for the development of POAG. Up to 15% of African-American men are affected by the ninth decade of life [(Friedman et al., 2004a); Table 31.1]. Consequently, glaucoma is found to be more prevalent in the aging population, even after compensating for the mean rise in IOP with increasing age in the USA. However, the disease itself is not limited to only middle-aged and elderly individuals.

31.2.3.4 Central Corneal Thickness

CCT is an important risk factor for the development of glaucoma [(Gordon et al., 2002); Fig. 31.2]. CCT likely influences the measurement of IOP with many tonometers, including applanation techniques (Brandt et al., 2001). Increased CCT beyond the mean of 545 μm causes overestimation of IOP; lower CCT translates into underestimation of the IOP. A thin cornea (e.g., 480 μm) may occur with glaucomatous visual field loss despite "normal" applanation IOP because the measurements are fallaciously low. Conversely, a thick cornea (e.g., 620 μm) might occur in an eye with high IOP, normal visual fields, and a normal optic nerve because a thicker cornea results in false overestimation of true IOP. Ehlers et al. (1975) extrapolated that applanation tonometry is overestimated or underestimated by approximately 5 mm Hg for every 70 μm difference in measured CCT from mean thickness. It is also possible that CCT may itself constitute an intrinsic risk (or protective) factor for glaucomatous optic nerve damage or a marker of changes in aqueous humor dynamics (Gulati et al., 2011), independent of its ability to affect the IOP measurement.

31.2.3.5 Family History

Relatives of those with glaucoma are more likely to have the disease (Drance et al., 1981). However, there is no clear Mendelian pattern of inheritance for glaucoma. The Rotterdam Eye Study (Wolfs et al., 2000) concluded that the prevalence of glaucoma was 10.4% in siblings of patients with glaucoma and the relative risk of having POAG increased 9.2-fold for individuals with a first-degree relative with POAG. High IOP may be the inherited feature of glaucoma, or inherited risk factors independent of IOP may be involved. The genetics of glaucoma will be discussed further in Sect. 31.3.

31.2.3.6 Systemic Risk Factors

The association between diabetes mellitus (DM) and glaucoma has been unclear. Epidemiology studies in different populations have given conflicting results. The Blue Mountain Eye Study in Australia (Mitchell et al., 1997), The Rotterdam Study (Dielemans et al., 1996) and the Beaver Dam Study (Klein et al., 1994) in the USA have all shown DM to be associated with glaucoma. Other studies such as the Baltimore Eye Study (Tielsch et al., 1995) in the USA or the Diabetes Audit and Research in Tayside Study in the United Kingdom (Ellis et al., 2000) have failed to show an association of POAG with DM. Since DM is a disease of small vessels of the body, it is plausible that it has an association with glaucoma.

Some reports have suggested an association between high blood pressure and glaucoma but this relation is not fully understood and is not uniformly observed. There appears to be a subset of patients with low diastolic perfusion pressures who are at higher risk for POAG (Wilson et al., 1987). Some studies indicate that patients with DM have higher IOPs and

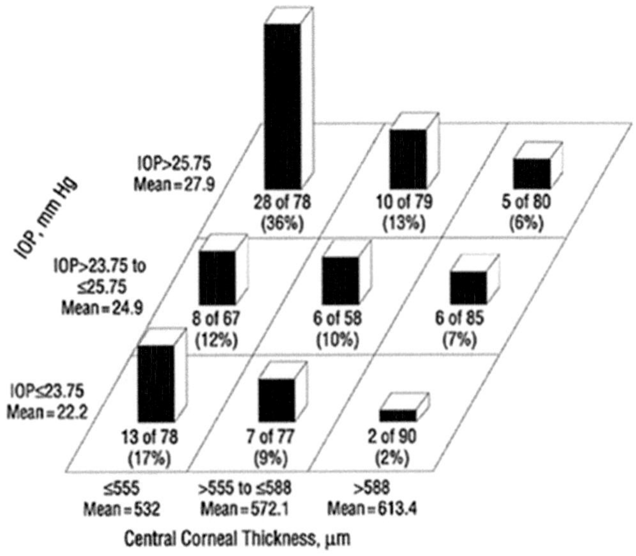

Fig. 31.2 Risk for developing primary open glaucoma varies with central corneal thickness. The numbers and percentage of participants in the observation group who developed primary open-angle glaucoma (median follow-up, 72 months) are indicated below each bar. Participants are grouped by baseline intraocular pressure of ≤23.75 mm Hg, >23.75 mm Hg to ≤25.75 mm Hg, and > 25.75 mm Hg and by central corneal thickness measurements of ≤555 μm, >555 μm to ≤588 μm, and > 588 μm. These percentages are not adjusted for the length of follow-up. (Reproduced with permission from: Gordon et al. (2002))

IOP= Intraocular pressure

a higher rate of glaucoma than people without DM. However, other studies fail to demonstrate this relationship. In a study in which the presence of DM was self-reported, diabetes protected against the development of glaucoma (Gordon et al., 2002). Other abnormalities in vascular function, including migraine headache (Cursiefen et al., 2000) and vasospasm in the extremities (Broadway & Drance, 1998), appear to be associated with glaucoma as demonstrated in some studies.

Sleep apnea has been shown to be a risk factor for glaucoma in some studies (Mojon et al., 1999). Myopia may be another risk factor for glaucoma (Daubs & Crick, 1981). The relatively thin eye wall and large globe in severely nearsighted people may account for a possible susceptibility to optic nerve damage under the influence of IOP. The association between factors such as concurrent cardiovascular disease (Tielsch et al., 1991) has not been demonstrated consistently.

In a retrospective study from the Mayo Clinic, patients with POAG were found to have lower intracranial pressure as compared to patients without POAG (Berdahl et al., 2008). This link between cerebrospinal fluid (CSF) pressure and glaucoma may be related to the mechanical effects of the intracranial pressure on the lamina cribrosa of the optic nerve.

31.3 Pathophysiology

As introduced at the beginning of this chapter, POAG is an optic neuropathy wherein IOP is the best-characterized risk factor. Therefore, the pathophysiology can be viewed at two levels: factors leading to elevated IOP and neuronal injury and death consequent to elevated IOP.

31.3.1 Elevated Intraocular Pressure

The physiologic processes related to maintenance of physiologic IOP and the probable causes of its dysregulation are largely limited to the anterior segment of the eye. IOP is maintained by a balance between aqueous inflow and outflow. Aqueous humor in the eye is produced by the ciliary body in the posterior chamber, from where it passes through the pupil to the anterior chamber and exits the eye through structures in the anterior chamber angle, a point of confluence of outer corneoscleral layer and middle uveal layer of the eye (Fig. 31.3).

The outflow pathways for the aqueous can broadly be divided into two categories: conventional outflow and uveoscleral outflow. Conventional outflow is believed to be through the trabecular meshwork to the Schlemm's canal, to collector channels and aqueous veins, to the episcleral network of vascular channels called the episcleral veins. The trabecular meshwork and Schlemm's canal complex occupy the scleral sulcus. The site of most resistance in this complex primary determinant of IOP is believed to be the region along the inner wall of the Schlemm's canal and juxtacanalicular tissue (Johnson, 2006; Rosenquist et al., 1989). The outflow through this pathway is believed to be pressure dependent and is driven by the pressure differential between the IOP and the pressure in the episcleral vessels. Outflow facility is a measure of the ease with which aqueous can flow through this system and is the inverse of the resistance in the pathways involved. Removal of the trabecular meshwork and the inner wall of the Schlemm's canal can eliminate approximately 75% of the resistance in the system with the remaining 25% believed to be contributed by more distal structures (Grant, 1963; Rosenquist et al., 1989). The second, unconventional, egress path of aqueous from the eye is less well defined and is believed to be routed through the root of the iris, to the interstitial spaces of the ciliary body and into the suprachoroidal space (Alm & Nilsson, 2009). This uveoscleral outflow is pressure independent in the physiological range of IOP.

The relationship between these inflow and outflow parameters of the eye can be summarized by the modified Goldmann equation as follows:

$$\text{IOP} = \left(\text{Aqueous inflow} - \text{uveoscleral outflow}\right) / \text{Outflow facility} + \text{Episcleral venous pressure.}$$

The daytime aqueous inflow rate has not been found to be elevated in primary and secondary glaucomas (Brown & Brubaker, 1989; Larsson et al., 1993, 1995; Levene et al., 1976; Ziai et al., 1993). A slightly higher nighttime flow rate has been reported in patients with primary open-angle glaucoma (Larsson et al., 1995). The primary pathology involved in elevated IOP has been an increased resistance in both the conventional pathway and reduced flow through the uveoscleral pathway (Grant, 1951, 1958; Johnson et al., 2008; Toris et al., 2002, 2010). Elevated episcleral venous pressure has not been found in cases of POAG or OHT (Podos et al., 1968) but plays a role in some forms of secondary glaucoma such as that found in association with Sturge-Weber syndrome (Cibis et al., 1984; Phelps, 1978; Shiau et al., 2012).

Several anatomic correlates of increased resistance in the trabecular pathways have been described. Some of these are thickening and fusion of trabecular lamellae secondary to trabecular endothelial cell loss, collapse of trabecular beams, increase in electron dense plaques and possible changes in the glycosaminoglycan composition of the basement membranes and stroma of trabecular meshwork (Tektas & Lutjen-Drecoll, 2009). The anatomic correlates of reduced uveoscleral outflow in OHT and POAG are not well defined at this time. Uveoscleral outflow has been shown to decrease with age (Toris et al., 1999) and administration of cholinergic

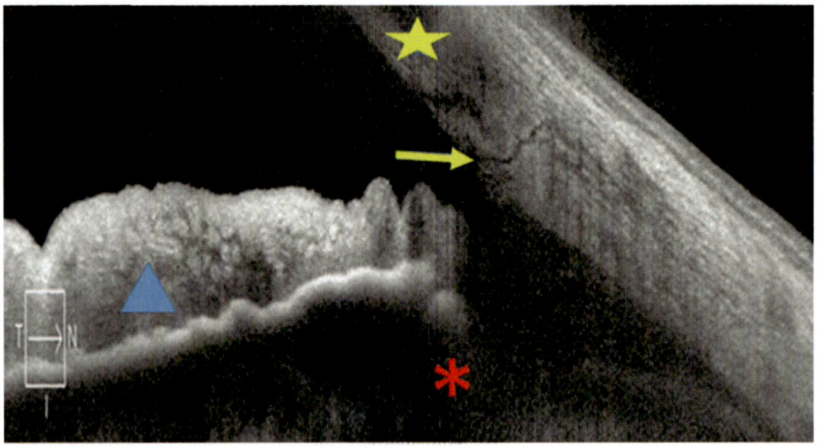

Fig. 31.3 Optical coherence tomography of a normal anterior chamber angle. *Star* = cornea, *asterisk* = ciliary body, *arrowhead* = iris, *arrow* = Schlemm's canal with collector channel

agents (Crawford & Kaufman, 1987) and increased with clinically used doses of prostaglandin analogs (Crawford et al., 1987; Gulati et al., 2012; Hayashi et al., 1987; Nilsson et al., 1989). An effect on matrix metalloproteinases and tissue inhibitors of metalloproteinases is believed to be the mechanism involved in the improvement of uveoscleral outflow with prostaglandins (Alexander et al., 1991; Brew et al., 2000; Murphy & Docherty, 1992).

31.3.2 Optic Nerve Damage

Despite significantly elevated intraocular pressure, only a small percentage of subjects develop clinically manifest glaucomatous damage. At the same time, glaucomatous optic nerve damage and consequent visual field loss have been shown to progress despite substantial IOP lowering. There is no clear explanation of this individual variability in the susceptibility of the optic nerve to varying levels of IOP. In general, the cause of glaucomatous optic neuropathy is unknown.

A unifying pathogenic mechanism, among differing hypothesis regarding the preceding mediator, appears to be the stasis of axoplasmic flow at the level of the optic nerve head (ONH). The obstruction to the flow involves both the slow and the fast phases and both retrograde and anterograde transport (Minckler et al., 1976, 1977, 1978; Quigley et al., 1979). In a nonhuman primate model of chronic IOP elevation, the flow obstruction preferentially affected the magnocellular layers of the lateral geniculate nucleus (Dandona et al., 1991).

The RGC death occurs as a consequence of triggering apoptotic pathways, or programmed cell death. This could result from the loss of essential trophic stimuli as a consequence of disruption of axoplasmic flow. Loss of axons involves early loss of axons by Wallerian degeneration and subsequent somal death after an abortive attempt at regeneration (Buckingham et al., 2008). Besides damage to the primary ganglion cells injured by the glaucomatous process, the apoptotic cells create a toxic environment leading to injury and death of surrounding cells by mechanisms that may involve glutamate excitotoxicity or free radical-mediated injury (Hare et al., 2001). Glaucomatous damage appearing as loss of discrete groups of nerve fiber bundles may be reflective of this proximity of cells being injured or a primary pathological process at the lamina cribrosa where these axons come together.

Many pathophysiological mediators common to all neurodegenerative diseases, including glaucoma, have been identified, such as increase in glutamate, conformationally altered self-proteins, increase in inflammation-associated factors [cyclooxygenase-2 (COX-2), tumor necrosis factor (TNF-alpha), nitric oxide (NO)], increase in extracellular matrix proteins and growth-associated inhibitors (myelin-related proteins), oxidative stress, and malfunctioning of local immune cells (microglia). These mediators evoke a response for which the tolerance of the neural tissue is minimal.

Multiple theories exist concerning how IOP can be one of the factors that initiates glaucomatous damage in a patient (Fig. 31.4). Two of the fundamental theories include the mechanical and vascular theory.

The mechanical theory proposes that elevated IOP causes a backward bowing of the lamina cribrosa, resulting in kinking of the axons as they exit through the lamina pores (Müller, 1858). Damaged RGC axons are affected by neurotrophin deprivation leading to subsequent apoptosis and cell death. The vascular theory proposes that cell death is triggered by ischemia, whether induced by elevated IOP or as a primary insult (Jaeger, 1858). It is possible that in any given patient one or both factors may play out to varying degrees. Intuitively, the mechanical dam-

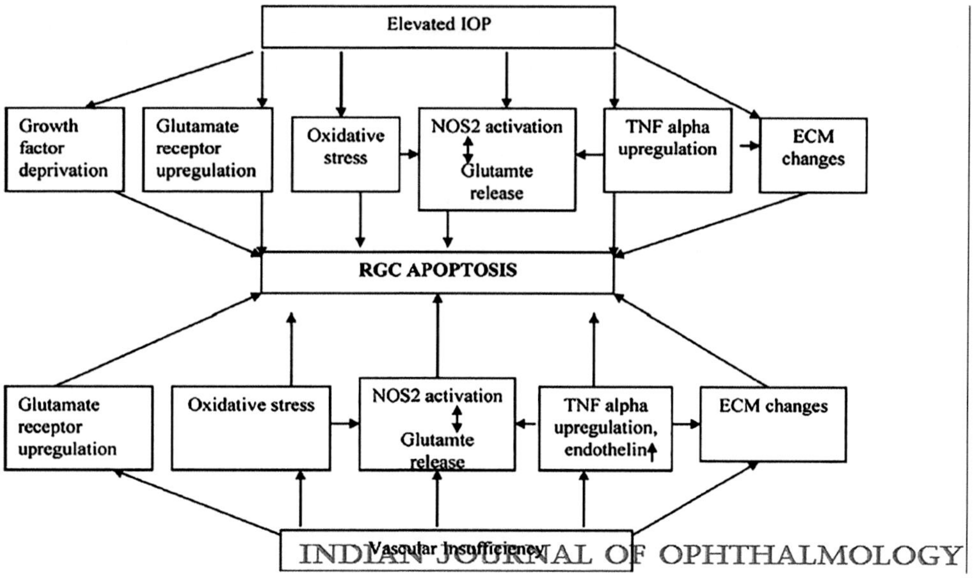

Fig. 31.4 The complex interplay of primary and secondary factors leading to RGC apoptosis in glaucoma. TNF-A - Tumor necrosis factor-alpha, ECM - Extracellular matrix, NOS-2 - Nitric oxide synthase-2 (Reproduced with permission from Agarwal et al, 2009)

age hypothesis may play a larger role when the IOP is significantly elevated above normal levels and the vascular predispositions may play a prominent role in cases where glaucoma continues to progress despite adequate IOP lowering.

31.3.3 Role of Intracranial Pressure

Recently, there has been an interest in the role that CSF pressure may have in the pathophysiology of glaucoma (Jonas et al., 2015). This is based on the theory that higher CSF pressure may provide some sort of tectonic support to lamina cribrosa faced with elevated IOP, and glaucomatous damage at the ONH may be consequent to an imbalance between IOP and CSF pressure (Ren et al., 2011). Experimental reduction of CSF pressure in monkeys has been shown to produce nerve fiber and optic nerve damage similar to glaucoma (Yang et al., 2014). Intracranial pressure has been found to be significantly lower in POAG and normal tension glaucoma as compared to control subjects with no glaucoma (Berdahl et al., 2008).

31.3.4 Glaucoma Therapy

Based on the above pathophysiology there are broadly speaking two therapeutic targets in glaucoma management. One is to lower the IOP. The other is to facilitate ganglion cell survival independent of IOP lowering, also referred to as neuroprotection (Fig. 31.5). Almost all clinical glaucoma management at this time targets IOP lowering. Numerous well-designed clinical trials have proven the benefit of IOP lowering in the preservation of visual function in glaucoma patients (AGIS Investigators, 2000; Collaborative Normal-Tension Glaucoma Study Group, 1998; Gordon et al., 2002; Leske et al., 2003; Lichter et al., 2001). Despite tremendous interest in neuroprotection, there are currently no well-proven neuroprotective strategies available for glaucoma management. Various agents have shown promise in preclinical studies. Some of these are *N*-methyl-D-aspartate (NMDA) antagonists (memantine), alpha-2 agonists, calcium channel blockers, antioxidants, and nitric oxide synthetase inhibitors. Even though a clinical trial comparing the efficacy of timolol and brimonidine did report better visual field preservation with brimonidine, the study is limited by several potential biases involved, as detailed in a prior Cochrane review (Krupin et al., 2011). The review concluded that the current evidence in support of neuroprotective strategies in the management of open-angle glaucoma is inconclusive (Sena & Lindsley, 2013). The review also comments on the current lack of availability of adequately sensitive outcome measures to be able to detect a neuroprotective effect of agents, if there was one.

31.3.5 Potential Therapeutic Targets for Neuroprotection

Figure 31.5 is a diagrammatic representation of some of the potential neuroprotective targets characterized at this time.

31.3.5.1 Immune Mechanisms

Neural injury has long been considered self-destructive and progressive primarily due to the autoimmunity to self-antigen. The anti-self-antigen response has been shown in these experiments to have a coincidental beneficial and modulating effect (Schwartz, 2005). The autoim-

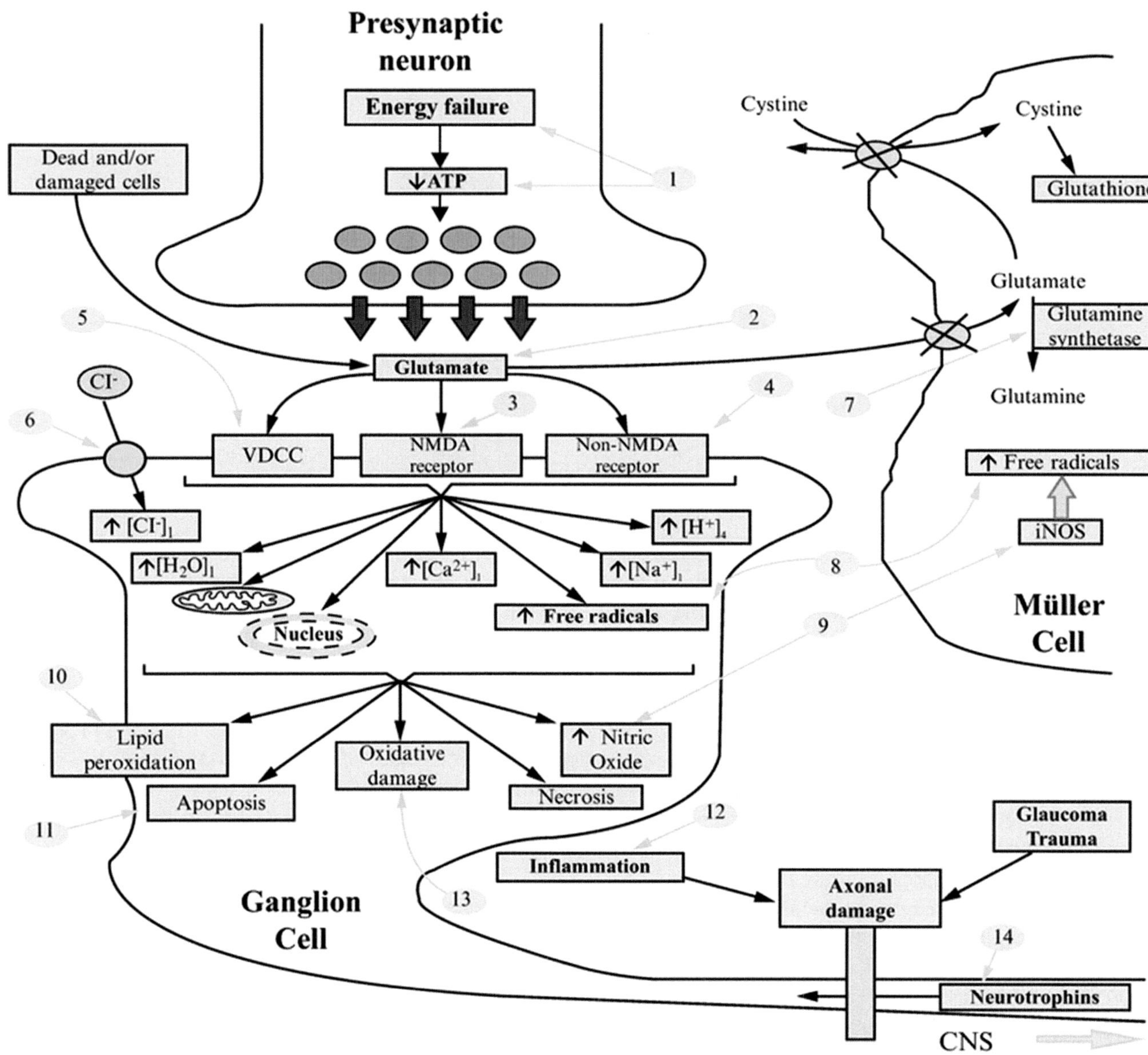

Fig. 31.5 Diagrammatic summary of potential targets for intervention in ganglion cell death. (*1*) Prevent energy loss, (*2*) reduce extracellular glutamate, (*3*) block ionotropic NMDA receptors, (*4*) block non-NMDA ionotropic glutamate receptors, (*5*) block excessive influx of Ca^{2+} through voltage-gated channels, (*6*) prevent influx of chloride and water, (*7*) enhance intracellular glutamate catabolism, (*8*) prevent excessive production of free radicals, (*9*) prevent detrimental effects of nitric oxide, (*10*) prevent lipid peroxidation, (*11*) prevent induction of apoptosis, (*12*) reduce any inflammation, (*13*) stop oxidative damage, and (*14*) maintain neurotrophin support. (Reproduced with permission from: Osborne et al. (1999))

mune T cells limit the activity of the microglial framework and blood-borne macrophages, affecting their ability to deal with a local excitatory situation. Such a controlling effect has a beneficial impact on both survival and regrowth as excessive microglial activity may be a significant factor for increased cell death (Grotegut et al., 2020). Boosting the beneficial effect by the use of a safe antigen that can cross-react with self-antigens at the site under stress might provide an effective therapeutic approach that influences both rescue and regeneration (Schwartz, 2005). Copolymer 1 (Cop-1), an approved drug for the treatment of multiple sclerosis, can be used as a treatment for autoimmune diseases and as a therapeutic vaccine for neurodegenerative diseases (Kipnis & Schwartz, 2002). Blocking microglial adenosine2A receptors (Aires et al., 2019) and P2X purinoceptor 7 (P2X7) receptor antagonists (Dong et al., 2018) have been shown to reduce microglial response to elevated IOP and decrease RGC death in rat models. These results have special implications for glaucoma therapy since glaucoma is the most common cause of optic neuropathy.

31.3.5.2 Apoptosis

Apoptosis is classically defined by an orderly pattern of internucleosomal DNA fragmentation, chromosome clumping, cell shrinkage, and membrane blebbing (Wyllie et al., 1980). This is followed by disassembly of the cell into multiple membrane-enclosed vesicles that are engulfed by neighboring cells without inciting inflammation. Apoptosis has often been labeled as "*programmed or physiological cell death.*"

Several gene families have been identified that play either positive or negative roles in determining whether a cell (in glaucoma, the RGC) will undergo apoptosis. Caspases are cysteine proteases that play a central role in both propagating apoptotic signals and carrying out disassembly of the cell (Hutchins & Barger, 1998). Many triggers activate caspases, including increased intracellular calcium, changes in cellular energy balance, and adenosine 3′5′ cyclic phosphate.

The main inhibitors of apoptosis are Bcl-2 and related proteins. They have multiple complex functions, such as inhibiting intermediate proteins that activate caspases, gauging intracellular damage, and maintaining organelle integrity in ways that are not fully understood. For example, overexpression of Bcl-2 inhibits both apoptotic and necrotic cell death by suppressing lipid peroxidation (Adams & Cory, 1998). Statins have been shown to reduce apoptosis-induced RGC cell death in rat models of ocular hypertension (Kim et al., 2021).

31.3.5.3 Glutamate-Induced Excitotoxicity

The excitatory amino acid glutamate is known to cause neurotoxicity via the binding of glutamate to the NMDA receptor and the kainate receptors. This can be secondary to many types of cellular damage, including other final common pathways of damage such as oxidative stress or free radical damage (Osborne et al., 1999). Activation of these receptors allows entry of excessive amounts of calcium. Abnormally high concentrations of calcium lead to inappropriate activation of complex cascades of proteases, nucleases, and lipases. These directly damage cellular constituents and lead to ganglion cell death.

One neuroprotective strategy, therefore, is to interrupt the resultant excitotoxic cascade by blocking the NMDA receptor with drugs such as memantine (Lipton, 2003).

31.3.5.4 Free Radicals

Free radicals are generated not only through activation of glutamate receptors but also as the inevitable by-product of normal oxidative metabolism. This is especially true in the retina and the RGCs, which have an extremely high metabolic rate. This triggers further changes including calcium release that causes cell damage (as summarized in Sect. 31.3.5.3. above). Endogenous antioxidants such as vitamins E and C, superoxide dismutase (Greenlund et al., 1995), and glutathione normally inactivate free radicals.

Coenzyme Q10 (CoQ10) is an antioxidant shown to reduce RGC apoptosis and glutamate-induced injury (Nucci et al., 2007). In a small study done in patients with POAG, an improvement in inner retinal function on pattern electroretinogram and visual cortical responses were noted after treatment with topical CoQ10 and vitamin E (Parisi et al., 2014). Further evidence, however, is needed before they can be used therapeutically.

Mitochondrial abnormalities may be an early driver of neuronal dysfunction in glaucoma (Tribble et al., 2019) and nicotinamide adenine dinucleotide (NAD) is an essential component of the electron transport chain in the mitochondria. Initial animal studies in ocular hypertension rat models showed a decrease in, retinal and optic nerve NAD as a function of IOP and neuroprotective effect with nicotinamide supplementation (Tribble et al., 2021). Encouraging results were seen in a recent phase 2 RCT which showed improvement in visual fields of patients with moderate open glaucoma taking a combination of oral nicotinamide and pyruvate (De Moraes et al., 2022).

31.3.5.5 Nitric Oxide Synthase

Nitric oxide synthase (NOS) is postulated to have a role in RGC axonal toxicity. Inducible NOS is known to be upregulated in the optic nerve in glaucoma (Liu & Neufeld, 2001). Inhibitors of NOS, such as 3-aminoguanidine, decreased the RGC loss by 70% in rat eyes with raised IOP (Neufeld et al., 1999).

31.3.5.6 Dopamine Deficiency

Dopamine has been recognized to have neuroprotective actions in glaucoma. Retinal dopaminergic cells can be detected by the immunohistochemical staining of tyrosine hydroxylase, the rate-limiting enzyme in dopamine synthesis. It has been observed that NMDA-induced RGC toxicity dramatically decreased tyrosine hydroxylase immunostaining at the junction between the inner nuclear layer and inner plexiform layer in glaucoma (Kitaoka & Kumai, 2004).

31.3.5.7 Heat-Shock Proteins

Heat-shock proteins (Hsp) are known to be an intrinsic protection mechanism associated with an early response against physiologic and environmental stress (Wax & Tezel, 2002). Certain Hsp have special effects upon neurons, including Hsp 72, 27 and 60. Hsp 27 and 60 have been demonstrated to be upregulated in RGCs of glaucoma (Tezel et al., 2000). Recent work has shown that geranylgeranylacetone can induce Hsp 72 in RGCs and protect them from glaucomatous damage in a rat glaucoma model (Ishii et al., 2003).

31.3.5.8 Calcium Channel Blockers

Calcium channel blockers may reduce the ischemic apoptotic stimulus by improving perfusion through their vasodilatory effects (Osborne et al., 2004a, b). In addition, calcium channel blockers may interfere with the calcium influx mentioned in the glutamate-induced cytotoxicity above (Choi, 1988; Schurr & Rigor, 1992; Siesjo & Bengtsson, 1989). The vascular dilation effects associate with beta 1 agonists may also be mediated through inhibition of calcium influx of the cell membrane through the voltage-gated calcium channels (Bessho et al., 1991; Hester et al., 1994; Setoguchi et al., 1995). At this time, there is some limited evidence to support the effect of calcium channel blockers in preservation of visual fields in glaucoma patients (Koseki et al., 1999; Sawada et al., 1996).

31.3.5.9 Neurotrophins

Neurotrophins encourage axonal regeneration and neuronal cell function and have had promising outcomes in other neurodegenerative disorders of the central nervous system (Mirowska-Guzel, 2009; Nasrolahi et al., 2022; Sullivan & Toulouse, 2011). Several neurotrophins have been reported to be associated with glaucoma, like nerve growth factor, brain-derived neurotrophic factor (Oddone et al., 2017), ciliary neurotrophic factor (Ji et al., 2004), glial-derived growth factor, and neurturin (Koeberle & Ball, 2002), and are appealing as therapeutic targets.

31.3.5.10 Genetic Mutations

Myocilin (MYOC) is one gene that has been associated with early- and late-onset POAG. The MYOC gene was mapped to the open-angle glaucoma locus (GLC1A) region of chromosome 1q25, and mutations were discovered in patients with adult and juvenile open-angle glaucoma. Mutations in this gene are found in 3–5% of adult-onset POAG patients and 8–20% of juvenile-onset POAG patients (Wiggs et al., 1998). Different mutations are associated with juvenile and adult glaucoma. In particular, the most common mutation associated with adult glaucoma is a nonsense mutation, whereas all the mutations associated with juvenile glaucoma are missense mutations.

Mutation studies and genetic linkage analyses have shown a causative effect of the CYP1B1 gene in the development of primary congenital glaucoma. The CYP1B1 locus has also been implicated as a modifier gene in the development of POAG, and in some cases as the causative gene in POAG and other anterior segment dysgeneses (Vasiliou & Gonzalez, 2008).

Using single large pedigrees affected by POAG, seven genetic loci have been described (GLC1A-G), and glaucoma-predisposing genes have been identified in three of these loci: GLC1A, myocilin (Wiggs et al., 1998); GLC1E, optineurin (Rezaie et al., 2002); GLC1G and WDR36 (Monemi et al., 2005). Each of these genes is only responsible for a small fraction of cases of POAG, reflecting the small percentage of POAG that is inherited as a Mendelian trait rather than as a complex trait.

The optineurin protein may function to protect the optic nerve from TNF-alpha–mediated apoptosis, and the loss of function of this protein may decrease the threshold for RGC apoptosis in patients with glaucoma. Surprisingly, mutations in this gene do not appear to contribute significantly to the optic nerve degeneration that is a component of typical high-pressure POAG (Wiggs et al., 2003). Optineurin has been associated with subjects that experience progressive glaucomatous optic nerve changes at low or normal pressures, so-called normal pressure glaucoma. Optineurin has also been implicated in sporadic amyotrophic lateral sclerosis with or without superoxide dismutase-1 mutations, confirming optineurin's role in a separate progressive degenerative neuropathy (Maruyama & Kawakami, 2013; Deng et al., 2011). Identifying these gene targets has led to research involving gene therapy utilizing these genes as therapeutic targets in experimental models of glaucoma (Jain et al., 2017; Fang et al., 2022; Visuvanathan et al., 2022); however, there are still many challenges till these can become clinically feasible.

More recent genome-wide association studies have demonstrated a strong POAG association at the SIX1/SIX6 locus (Wiggs et al., 2012). Deletions of SIX1 and SIX4 have been associated with bilateral anophthalmia and pituitary abnormalities (Bennett et al., 1991). In animal models, mutations at the SIX1/SIX6 locus result in early exit from the cell cycle among RGC progenitors. A common risk variant at the SIX6 locus, His141, is present in approximately 40% of subjects of European descent, and 99% of subjects of African descent. In POAG and non-glaucomatous eyes that are homozygous for the risk allele, there is a significantly thinner retinal nerve fiber layer at the superior and inferior poles of the optic nerve, the site of the earliest nerve fiber loss in glaucoma (Carnes et al., 2014; Cheng et al., 2014; Kuo et al., 2015).

The exact role that IOP plays in combination with these other factors and the latter's significance in the initiation and progression of subsequent glaucomatous neuronal damage and cell death requires further studies to elucidate.

31.4 Clinical Features

31.4.1 History

The initial patient interview is important in the evaluation of glaucoma. Because of the "silent" or asymptomatic nature of

POAG, patients usually will not present with any visual complaints until late in the course of the disease. A history should include the following:

1. Past ocular and medical history including:
   (a) Eye pain
   (b) Redness
   (c) Multicolored halos (indicative of intermittent IOP elevation and related corneal edema)
   (d) Headache
   (e) Previous ocular diseases such as cataracts, uveitis, diabetic retinopathy, or vascular occlusions
   (f) Previous ocular laser or surgery
   (g) Ocular or head trauma
   (h) Pertinent vasculopathic systemic illnesses, including Raynaud's phenomenon or migraine
2. Current medications, including any anti-hypertensive medications or corticosteroids used topically or systemically. Systemic and topical medications can affect IOP and ONH perfusion. Sulfa-derivative medications can result in a rare idiosyncratic reaction that causes ciliary body effusion and secondary angle closure-related IOP elevation.
3. Family history of glaucoma in first-degree relatives.

31.4.2 Physical Exam

Screening the general population for POAG is most effective if targeted at those at higher risk, such as African Americans, those with a family history of glaucoma, and elderly individuals, especially if the screening includes IOP measurements, assessment of the optic nerve, and screening visual fields.

For individuals of age 65 years or above, recommended frequency for a comprehensive adult medical eye evaluation is 6–12 months when risk factors for glaucoma are present and 1–2 years in the absence of risk factors. For glaucoma suspect individuals under 40 years, the frequency of the eye evaluation should be 2–4 years and 5–10 years in the presence and absence of risk factors for glaucoma, respectively (American Academy of Ophthalmology Glaucoma Panel, 2010a).

A standard comprehensive eye examination, including gonioscopy, is performed at the initial visit. If any visual field or optic nerve changes consistent with glaucoma are present, additional testing should be done to establish the diagnosis, establish a baseline for future follow-up, help in defining target (or safe) levels of IOP, and developing a treatment and follow-up plan.

In patients with glaucoma, factors that determine the frequency of evaluations include the severity of damage, the stage of the disease, rate of progression, extent to which the IOP exceeds the target pressure, and the number and significance of other risk factors for damage to the optic nerve. If the target IOP has been achieved and there is no progression of damage, it is recommended to follow up with an ONH exam and visual field testing at 6 to 18-month intervals in those with mild to moderate glaucoma, and as frequently as every 4 months in a patient with severe stage glaucoma (American Academy of Ophthalmology Glaucoma Panel, 2010b). On the contrary, if there is evidence of progression of damage and the target IOP has not been achieved, more frequent testing is required until stability is confirmed.

31.4.2.1 Tonometry

When checking IOP, record measurements for both eyes, the method used (Goldmann applanation is most commonly used), and the time of the measurement. IOP varies from hour to hour in any individual. Many studies indicate that the circadian rhythm of IOP peaks in the early hours of the morning just after waking. Intraocular pressure measurements should be made at different times of the day to check for diurnal variation. A diurnal variation of more than 5–6 mm Hg indicates an increased risk for glaucoma.

IOP can also change with body position, rising when changing from an upright to a supine position. The positional change in IOP is greater in people with OHT and glaucoma (Prata et al., 2010). The aqueous humor outflow facility does not change with the body position, implying other ocular parameters are causing the IOP change observed (Selvadurai et al., 2010).

A difference between contralateral eyes of 3 mm Hg or more indicates greater suspicion of glaucoma. An average of a 10% difference between individual measurements is expected. In most circumstances, the measurements should be repeated on at least 2–3 occasions before deciding on a treatment plan.

31.4.2.2 Gonioscopy

The drainage angle of the eye is examined by gonioscopy, which requires the use of special lenses. Gonioscopy is performed to rule out angle closure or secondary causes of IOP elevation, such as angle recession, pigmentary glaucoma, exfoliation syndrome, and trabeculitis. The peripheral contour of the iris is examined for plateau iris, and the trabecular meshwork for peripheral anterior synechiae, as well as for neovascular or inflammatory membranes and cell aggregates. Schlemm's canal may be seen if blood refluxes into the canal. This might indicate elevated episcleral venous pressure caused by conditions such as a carotid-cavernous fistula, Graves orbitopathy, idiopathic elevated episcleral venous pressure, or Sturge-Weber syndrome.

Ultrasound biomicroscopy (Fig. 31.6) and anterior segment optical coherence tomography (Fig. 31.3) are also helpful in assessing the angle, iris, and processes of the ciliary body to rule out anatomical pathology and secondary causes of elevated IOP.

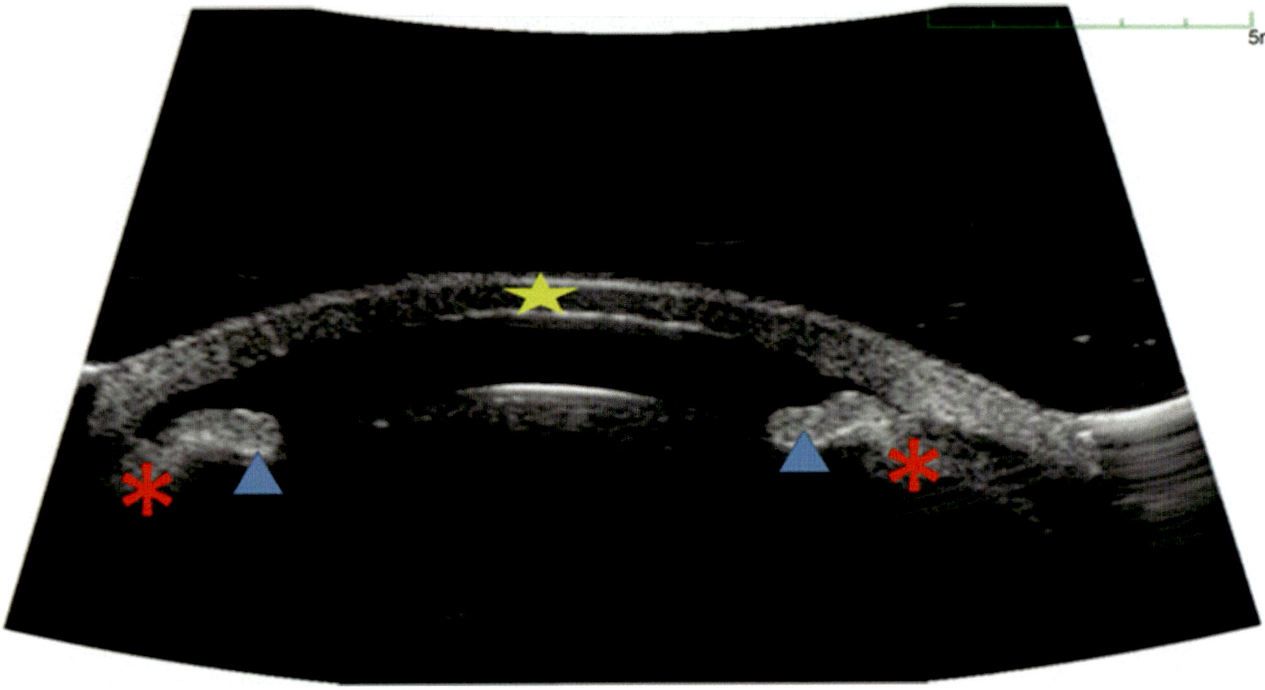

Fig. 31.6 Ultrasound biomicroscopy in an eye with narrow angle and bombé configuration of the iris. *Star* = cornea, *asterisk* = ciliary body, *arrowhead* = iris

31.4.2.3 Optic Disc and Nerve Fiber Layer

The optic disc is examined to estimate the cup-to-disc ratio and to identify cup excavation and enlargement typical of glaucoma. The optic disc is a complex, three-dimensional anatomic structure. Stereoscopic viewing during slit-lamp examination improves the accuracy of disc evaluation. Single cutoff values for the cup-to-disc ratio designed to distinguish between normal subjects and those with glaucoma are imperfect for screening since the normal ratio varies considerably with optic disc size. When the cup-to-disc ratio equals or exceeds 0.6, the probability of glaucoma increases dramatically. In screening for glaucoma, however, even combined criteria that include the level of IOP and cup-to-disc ratio identifies only two-thirds of patients with glaucoma.

The disc should be viewed stereoscopically to assess for evidence of glaucomatous damage. The following characteristics should be noted: cup-to-disc ratio in the horizontal and vertical meridians; color and slope of the cup; the appearance of the disc; progressive enlargement of the cup (comparison to prior photos); evidence of nerve fiber layer damage using a red-free filter; notching or thinning of disc rim, particularly at the superior and inferior poles; pallor; the presence of nerve fiber layer hemorrhage; asymmetry between cups; peripapillary atrophy (association with the development of glaucoma when Beta-zone atrophy is present); or congenital nerve abnormalities. Stereo fundus photographs are acquired as a baseline for future comparisons (Fig. 31.7).

Other imaging techniques that utilize different physical properties of light for optical analysis can document the status of the optic nerve and the nerve fiber layer. Furthermore, they can be used to detect changes over time (Medeiros et al., 2004). The value of these technologies for diagnosing glaucoma and for determining progression over time has been demonstrated in early to moderate glaucoma but their utility in advanced glaucoma is diminished with severe loss of the retinal nerve fiber layer making serial measurements difficult and less predictable of glaucoma progression.

Confocal scanning laser ophthalmoscopy can evaluate the optic disc and peripapillary retina in three dimensions and provide quantitative information about the cup, neuroretinal rim, and contour of the nerve fiber layer. Scanning laser polarimetry measures the change in the polarization state of an incident laser light passing through the naturally birefringent nerve fiber layer to provide indirect estimates of peripapillary nerve fiber layer thickness. Optical coherence tomography uses reflected light in a manner analogous to the use of sound waves in ultrasonography to create computerized cross-sectional images of the retina and optic disc and also gives quantitative information about the peripapillary retinal nerve fiber layer thickness that can be compared to age-matched controls (Fig. 31.8).

Fluorescein angiography, ocular blood flow analysis via laser Doppler flowmetry, color vision measurements, contrast sensitivity testing, and electrophysiological tests (e.g.,

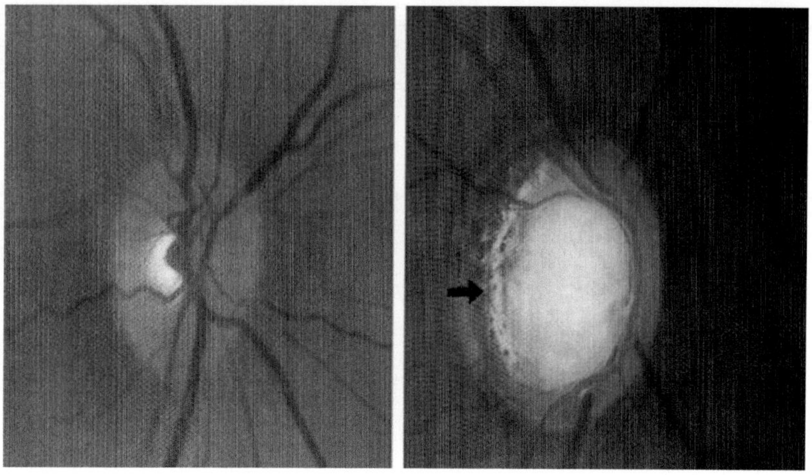

Fig. 31.7 Stereo photographs representing normal optic nerve head with a small central cup (*left*) and typical glaucomatous changes with advanced cup and visualization of the temporal lamina cribrosa pores and peripapapillary atrophy (*right*)

pattern electroretinography, visual evoked potentials) are used as research tools in the evaluation of POAG patients. Routine clinical use is not advocated at this time.

31.4.2.4 Visual Field Testing

Automated threshold testing (e.g., Humphrey 24-2) is performed to evaluate for glaucomatous visual field defects. If the patient is unable to perform automated testing, Goldmann perimetry may be substituted. New onset glaucomatous defects are found most commonly as an early nasal step, temporal wedge, or paracentral scotoma (more frequent superiorly). Generalized depression also may occur but is less specific for glaucomatous damage.

Swedish Interactive Threshold Algorithm (SITA)-based software (Fig. 31.9) decreases testing time and boosts reliability, especially in older patients (Schimiti et al., 2002). Short wavelength automated perimetry (SWAP) or blue-yellow perimetry may provide a more sensitive method of detecting visual field deficits, especially in patients with OHT without defects demonstrated by standard testing. If the Humphrey standard automated perimetry results are normal, SWAP can be utilized in detecting visual field loss earlier. Prior studies suggest SWAP may detect visual loss or progression up to 3–5 years earlier than conventional perimetry, as well as in 12–42% of patients previously diagnosed with only OHT (Racette & Sample, 2003). Because the testing time may be lengthened, SWAP testing may be strenuous for some patients. However, SITA-SWAP algorithm software may speed up the testing time and improve reliability (Bengtsson, 2003).

The pupil size should be documented at each testing session since pupillary constriction can reduce retinal sensitivity and mimic progressive visual field loss. This is especially important in patients that are using miotic therapy with pilocarpine.

To establish a reliable baseline or confirm worsening, visual fields may have to be repeated several times on successive visits, especially if initial testing shows low reliability indices. Newer glaucoma progression analysis software can help identify reliable perimetric baselines, and probability-based analyses of subsequent fields can assist in distinguishing true progression over time from fluctuation or artifact. In follow-up, if the patient is at low risk for onset of glaucomatous damage, then repeat testing may be performed once a year. If a high risk of impending glaucomatous damage is present, then testing frequency is adjusted accordingly (as frequent as every few months).

Frequency doubling perimetry (also termed frequency doubling technology or FDT) is a relatively newer technology that may detect functional visual loss at an earlier stage in the glaucomatous disease process, thereby detecting more patients who currently are misdiagnosed as having OHT instead of early POAG. Current sensitivities and specificities are continuing to improve but more baseline data are needed to determine under what circumstances these new techniques will prove to be most useful.

31.5 Differential Diagnosis

Several disorders may mimic glaucomatous ONH and/or visual field changes. Those that resemble structural glaucomatous optic neuropathy or interfere with assessment of glaucomatous cupping include optic disc anomalies, tilted disc syndrome, optic nerve drusen, optic nerve pits, optic nerve colobomas, myelinated nerve fibers, ischemic optic neuropathy, and optic atrophy.

Factors that mimic glaucomatous visual field loss include: branch retinal artery and vein occlusions; chorioretinal scars; retinal areas treated by photocoagulation or cryotherapy;

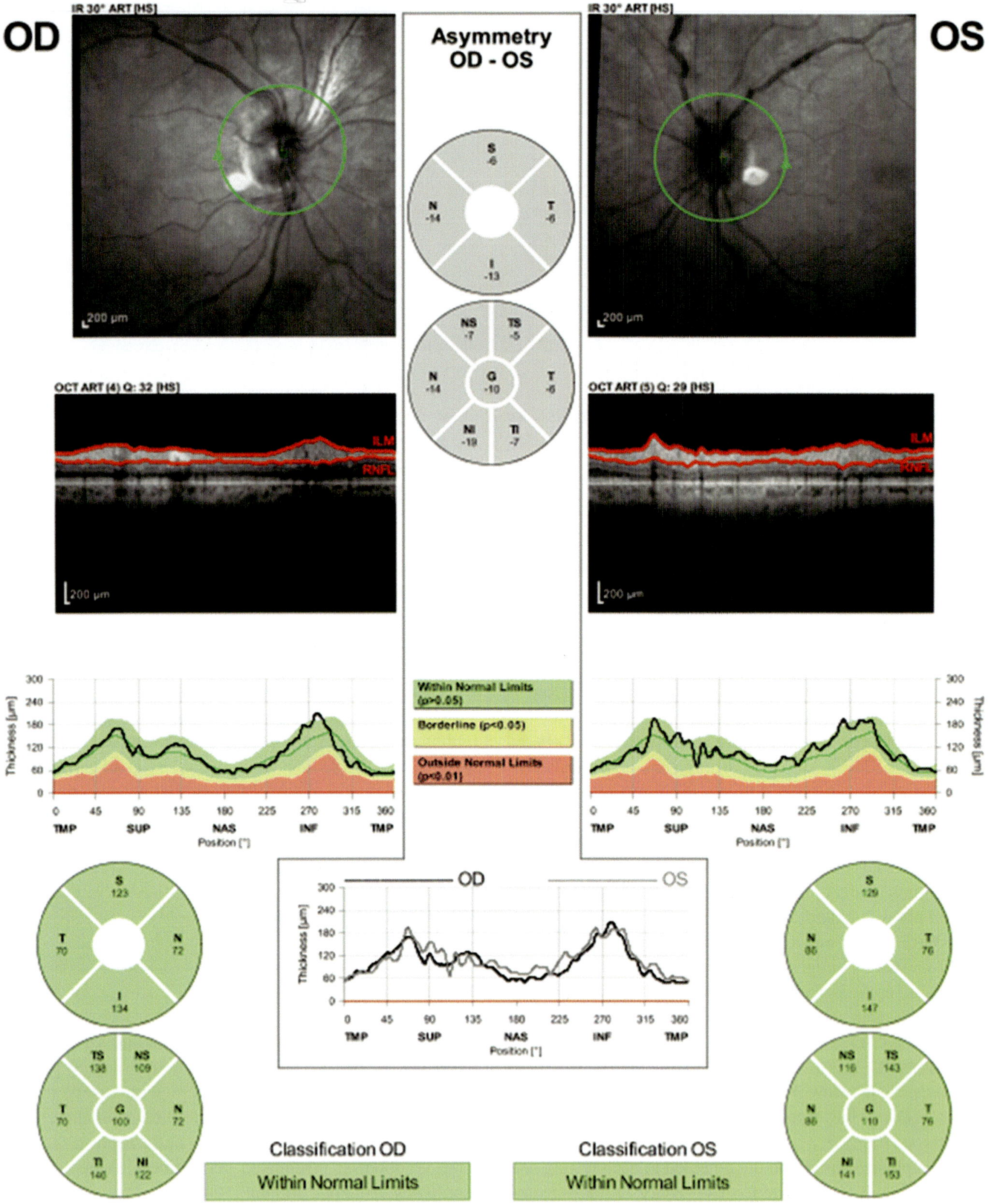

Fig. 31.8 Optic nerve optical coherence tomography in a patient with a normal optic nerve. *Top row*—red free optic nerve photo. *Second row*-cross sectional identification of the retinal nerve fiber layer (outlined in *red*). *Bottom section*-comparison with age-matched control population

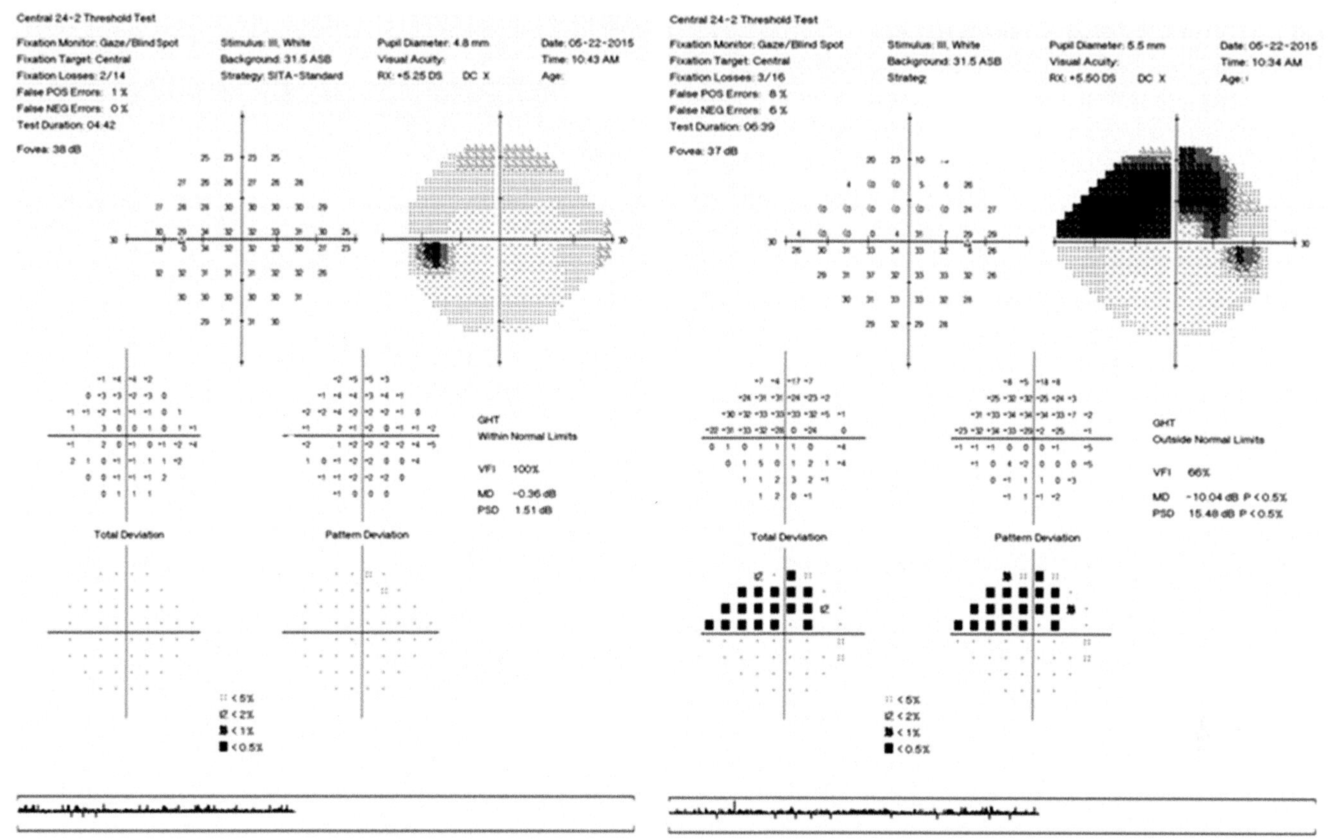

Fig. 31.9 Humphrey 24-2 SITA standard visual field test. Normal right eye visual field. Left eye demonstrating a superior arcuate scotoma of severe stage glaucoma related visual field loss in a patient with exfoliative glaucoma

demyelinating disorders; cerebrovascular accidents, tumors, or other lesions affecting the optic nerve, chiasm, optic tract, optic radiation, and/or the remaining course of nerve fibers to the occipital cortex. Other abnormalities that could account for "pseudo-glaucomatous" visual field defects or vision loss include vitreous hemorrhage, proliferative retinopathy or other retinal disorders.

31.6 Animal Models

Animal models for glaucoma have been used to study the pathophysiological basis of glaucoma. Animal-based experiments are more representative than cell culture models particularly in the study of aqueous humor dynamics, the trabecular meshwork-ciliary muscle complex, RGC, ONH axons, ONH framework, and vascular flow dynamics in glaucoma subtypes (Weinreb & Lindsey, 2005).

Conducting studies in animal models often are challenging and occasionally require anesthesia, which might affect the parameter being assessed. Conditioned reflex training may help elicit cooperation. Use of remote controlled implants and newer drug delivery options represent advances made on this front.

Animal experiments have provided excellent means to study medical and surgical treatments for glaucoma prior to human trials. Animal models are required to study de novo drugs and their structural and functional influence on ocular tissues.

Genetic modulation of animal models is another area of research. Mice are genetically engineered to suit experimental needs. New models are required and will be developed to advance our understanding of the glaucomatous disease process. Mouse mutants with glaucoma provide models for human glaucoma. From the clinical standpoint, some mouse model shares occasional similarities with various aspects of glaucoma occurring in humans. The life span and replication time (one mouse year is equal to 30 human years) allow developmental and invasive studies and make it possible to follow glaucoma progression in a reasonable span of time (Chang et al., 2005). Some of the new mouse glaucoma models used are as follows: Gpnmb [glycoprotein (transmembrane)] with mutation on chromosome 6 to study iris pigment dispersion (Anderson et al., 2002); Trypl with chromosome 4 mutation to study iris stromal atrophy (Anderson et al., 2002); Nm2702 with chromosome 11 mutation to study inner retina loss and optic nerve cupping (Chang et al., 2005); CALB/Rk and si/si model to study early enlargement of optic cup or coloboma (Chang et al., 2005); AKXD-28 for ganglion cell loss and optic nerve cupping (Anderson et al., 2001); E50K optineurin transgenic mouse for normal tension glaucoma (Tseng et al., 2015); and nm1839-model for Axenfeld's syndrome, anterior segment dysgenesis or buphthalmos (Chang et al., 2005). Due to complex regulatory gene interactions and a still incomplete understanding of the effects of SIX6, no mouse model yet exists. The effect of glaucoma on the electroretinogram response has been evaluated in mice and is currently being evaluated for its role in diagnosis and monitoring glaucoma in humans.

The pig as a model for glaucoma shares many phylogenetic similarities with that of the human and is more readily available compared to the nonhuman primate model (Ruiz-Ederra et al., 2005). The porcine retina is more similar to human than other large mammals such as dog, goat, or cow. Techniques used for glaucoma diagnosis in humans such as optical coherence tomography or multi-focal electroretinography can be reliably studied in pig eyes. Studies of the pig aqueous outflow system demonstrate it to be an acceptable model for specific glaucomas (McMenamin & Steptoe, 1991).

The nonhuman primate model of OHT with its resultant optic neuropathy closely reflects the optic neurodegeneration associated with human glaucoma. Utilization of the glaucoma model in monkeys has helped to further the understanding of aqueous humor dynamics, glaucomatous damage in the RGCs and other structures along the visual pathway, and the effects of pharmacological agents on ocular tissues (Rasmussen & Kaufman, 2005). Fundus photographs of glaucomatous monkeys show the morphologic changes consistent with human OHT with the advantage of observing progressive changes over a shorter period of time by withholding treatment despite very high IOPs (Gaasterland & Kupfer, 1974). The ability to correlate histologic changes in glaucomatous monkey eyes with functional and imaging data is a major advantage of the monkey model. Humphrey visual field defects can be performed by trained glaucomatous monkeys (Harwerth et al., 2002). The types of visual field defects are similar to those in glaucomatous humans.

Utilization of an experimental glaucoma model in nonhuman primates has led to advances in the understanding of glaucomatous changes in the visual pathways from photoreceptors to the visual cortex (Rasmussen & Kaufman, 2005). Newer investigative modalities such as contemporary perimetry techniques have been studied in primate models. Electroretinography can be performed in monkeys to evaluate changes produced by glaucomatous optic neuropathy.

The choice of animal is governed by the objectives and requirements of the study (Table 31.3). Size of the animal eye, similarity with human eyes, cost, and availability are factors that influence the selection of the animal type.

In summary, retinal ganglion cell apoptosis in glaucoma is caused by a complex interplay of raised IOP, vascular dysregulation and oxidative stress. Various molecular mechanisms have been implicated in the pathophysiology of glaucoma including glutamate excitotoxicity, reactive oxygen species, neurotrophic factors, NOS expression etc. Current treatment modalities primarily target IOP reduction but further advancement of neuroprotective pharmacological agents could help provide better therapeutic optics and improve outcomes.

Table 31.3 Advantages and limitations of several animal models of glaucoma

	Monkey	Rabbit	Rat	Mouse
General similarity to human eye	++++	++	++	++
Similarity to human aqueous dynamics	++++	+	+++	++++
Experimental IOP elevation	+++	++	+++	+++
Spontaneous IOP elevation	–	+	–	+++
Cupping, optic axon loss with IOP	++++	–	+++	+++
Genome sequenced	–[a]	–	–	+++
Many transgenic strains available	–	–	+	++++
Readily available, low cost	–	+	+++	++++
Ease of maintenance and handling	–	++	+++	+++
Ease of testing	+	+++	++	+

Reproduced with permission from: Weinreb and Lindsey (2005)
[a]The human and monkey genomes are sufficiently similar that data from the human genome project often can be applied to experiments using nonhuman primates

31.7 Review Questions

1. The estimated world population that is bilaterally blind from primary open-angle glaucoma is:
   (a) 2–3 million
   (b) 3–4 million
   (c) 4–5 million
   (d) 5–6 million
   (e) *6–7 million*
2. The primary pathology in open-angle glaucoma is believed to be
   (a) Increased aqueous production
   (b) *Increased resistance in conventional outflow*
   (c) Increased episcleral venous pressure
   (d) Decreased intra cranial pressure
3. The pathway followed by aqueous in eye sequentially involves
   (a) Episcleral veins, anterior chamber, trabecular meshwork, Schlemm's canal, ciliary body
   (b) Anterior chamber, trabecular meshwork, Schlemm's canal, ciliary body, episcleral veins
   (c) *Ciliary body, anterior chamber, trabecular meshwork, Schlemm's canal, episcleral veins*
   (d) Ciliary body, anterior chamber, Schlemm's canal, trabecular meshwork, episcleral veins

4. Mechanisms that might contribute to optic nerve damage in glaucoma include:
   (a) Apoptosis
   (b) Glutamate-induced excitotoxicity
   (c) Nitric oxide synthase
   (d) Free radicals
   (e) *All of the above*
5. Mutations in which of the following genes have been shown to be associated with the pathogenesis of glaucoma?
   (a) GLC1A
   (b) Myocilin
   (c) Optineurin
   (d) GLC1G
   (e) *All of the above*
6. Which gene product is implicated is associated with normal pressure glaucoma and amyotrophic lateral sclerosis?
   (a) GLC1A
   (b) Myocilin
   (c) *Optineurin*
   (d) GLC1G
   (e) SIX6
7. Which of the following are characteristic changes that might be observed in the optic nerve head in primary open-angle glaucoma except:
   (a) Increased cup-to-disc ratio
   (b) Notching of the neural rim
   (c) Disc hemorrhage
   (d) New vessels within the disc
   (e) *All of the above are changes associated with glaucoma*
8. Which racial background portends the greatest risk for glaucoma and glaucoma related blindness?
   (a) Caucasian
   (b) *African*
   (c) Latino
   (d) Asian
9. All of the following are examples of tests to evaluate the anterior chamber angle, EXCEPT:
   (a) Gonioscopy
   (b) Anterior segment OCT
   (c) Ultrasound biomicroscopy (UBM)
   (d) *A-scan ultrasound*
10. All of the following are methods to assess for visual field loss when glaucoma is suspected, EXCEPT:
   (a) Humphrey automated perimeter
   (b) Goldmann perimeter
   (c) *Full-field electroretinogram (ERG)*
   (d) Frequency doubling technology (FDT)

Acknowledgments Supported in part by NEI K23EY023266 (VG) and an unrestricted grant from Research to Prevent Blindness, New York, NY.

References

Adams, J. M., & Cory, S. (1998). The Bcl-2 protein family: Arbiters of cell survival. *Science, 281*(5381), 1322–1326.

Agarwal, R., Gupta, S. K., Agarwal, P., Saxena, R., Agrawal, S. S. (2009). Current concepts in the pathophysiology of glaucoma. *Indian Journal of Ophthalmology 57*(4), 257–266. https://doi.org/10.4103/0301-4738.53049

Aires, I. D., Boia, R., Rodrigues-Neves, A. C., Madeira, M. H., Marques, C., Ambrósio, A. F., & Santiago, A. R. (2019). Blockade of microglial adenosine A2A receptor suppresses elevated pressure-induced inflammation, oxidative stress, and cell death in retinal cells. *Glia, 67*(5), 896–914.

Alexander, J. P., Samples, J. R., Van Buskirk, E. M., & Acott, T. S. (1991). Expression of matrix metalloproteinases and inhibitor by human trabecular meshwork. *Investigative Ophthalmology & Visual Science, 32*(1), 172–180.

Alm, A., & Nilsson, S. F. (2009). Uveoscleral outflow--a review. *Experimental Eye Research, 88*(4), 760–768.

American Academy of Ophthalmology Glaucoma Panel. (2010a). *Preferred practice pattern®guidelines. Primary open-angle glaucoma suspect*. American Academy of Ophthalmology.

American Academy of Ophthalmology Glaucoma Panel. (2010b). *Preferred practice pattern®guidelines. Primary open-angle glaucoma*. American Academy of Ophthalmology.

Anderson, M. G., Smith, R. S., Savinova, O. V., Hawes, N. L., Chang, B., Zabaleta, A., et al. (2001). Genetic modification of glaucoma associated phenotypes between AKXD-28/Ty and DBA/2J mice. *BMC Genetics, 2*, 1.

Anderson, M. G., Smith, R. S., Hawes, N. L., Zabaleta, A., Chang, B., Wiggs, J. L., et al. (2002). Mutations in genes encoding melanosomal proteins cause pigmentary glaucoma in DBA/2J mice. *Nature Genetics, 30*(1), 81–85.

Armaly, M. F. (1965). On the distribution of applanation pressure. I. Statistical features and the effect of age, sex, and family history of glaucoma. *Archives of Ophthalmology (Chicago, Illl.: 1960), 73*, 11–18.

Bengtsson, B. (2003). A new rapid threshold algorithm for short-wavelength automated perimetry. *Investigative Ophthalmology & Visual Science, 44*(3), 1388–1394.

Bennett, C. P., Betts, D. R., & Seller, M. J. (1991). Deletion 14q (q22q23) associated with anophthalmia, absent pituitary, and other abnormalities. *Journal of Medical Genetics, 28*(4), 280–281.

Berdahl, J. P., Fautsch, M. P., Stinnett, S. S., & Allingham, R. R. (2008). Intracranial pressure in primary open angle glaucoma, normal tension glaucoma, and ocular hypertension: A case–control study. *Investigative Ophthalmology & Visual Science, 49*(12), 5412–5418.

Bessho, H., Suzuki, J., & Tobe, A. (1991). Vascular effects of betaxolol, a cardioselective beta-adrenoceptor antagonist, in isolated rat arteries. *Japanese Journal of Pharmacology, 55*(3), 351–358.

Brandt, J. D., Beiser, J. A., Kass, M. A., & Gordon, M. O. (2001). Central corneal thickness in the ocular hypertension treatment study (OHTS). *Ophthalmology, 108*(10), 1779–1788.

Brew, K., Dinakarpandian, D., & Nagase, H. (2000). Tissue inhibitors of metalloproteinases: Evolution, structure and function. *Biochimica et Biophysica Acta, 1477*(1–2), 267–283.

Broadway, D. C., & Drance, S. M. (1998). Glaucoma and vasospasm. *The British Journal of Ophthalmology, 82*(8), 862–870.

Brown, J. D., & Brubaker, R. F. (1989). A study of the relation between intraocular pressure and aqueous humor flow in the pigment dispersion syndrome. *Ophthalmology, 96*(10), 1468–1470.

Buckingham, B. P., Inman, D. M., Lambert, W., Oglesby, E., Calkins, D. J., Steele, M. R., et al. (2008). Progressive ganglion cell degeneration precedes neuronal loss in a mouse model of glaucoma. *The Journal of Neuroscience, 28*(11), 2735–2744.

Carnes, M. U., Liu, Y. P., Allingham, R. R., Whigham, B. T., Havens, S., Garrett, M. E., et al. (2014). Discovery and functional annotation of SIX6 variants in primary open-angle glaucoma. *PLoS Genetics, 10*(5), e1004372.

Chang, B., Hawes, N. L., Hurd, R. E., Wang, J., Howell, D., Davisson, M. T., et al. (2005). Mouse models of ocular diseases. *Visual Neuroscience, 22*(5), 587–593.

Chen, Z., Wang, J., Lin, F., Dai, H., Mu, K., & Zhang, H. (2013). Correlation between lateral geniculate nucleus atrophy and damage to the optic disc in glaucoma. *Journal of Neuroradiology, 40*(4), 281–287.

Cheng, C. Y., Allingham, R. R., Aung, T., Tham, Y. C., Hauser, M. A., Vithana, E. N., et al. (2014). Association of common SIX6 polymorphisms with peripapillary retinal nerve fiber layer thickness: The Singapore Chinese Eye Study. *Investigative Ophthalmology & Visual Science, 56*(1), 478–483.

Choi, D. W. (1988). Calcium-mediated neurotoxicity: Relationship to specific channel types and role in ischemic damage. *Trends in Neurosciences, 11*(10), 465–469.

Cibis, G. W., Tripathi, R. C., & Tripathi, B. J. (1984). Glaucoma in Sturge-Weber syndrome. *Ophthalmology, 91*(9), 1061–1071.

Collaborative Normal-Tension Glaucoma Study Group. (1998). Comparison of glaucomatous progression between untreated patients with normal-tension glaucoma and patients with therapeutically reduced intraocular pressures. *American Journal of Ophthalmology, 126*(4), 487–497.

Congdon, N., O'Colmain, B., Klaver, C. C., Klein, R., Munoz, B., Friedman, D. S., et al. (2004). Causes and prevalence of visual impairment among adults in the United States. *Archives of Ophthalmology, 122*(4), 477–485.

Crawford, K., & Kaufman, P. L. (1987). Pilocarpine antagonizes prostaglandin F2 alpha-induced ocular hypotension in monkeys. Evidence for enhancement of Uveoscleral outflow by prostaglandin F2 alpha. *Archives of Ophthalmology, 105*(8), 1112–1116.

Crawford, K., Kaufman, P. L., & Gabelt, B. T. (1987). Effects of topical PGF2 alpha on aqueous humor dynamics in cynomolgus monkeys. *Current Eye Research, 6*(8), 1035–1044.

Cursiefen, C., Wisse, M., Cursiefen, S., Junemann, A., Martus, P., & Korth, M. (2000). Migraine and tension headache in high-pressure and normal-pressure glaucoma. *American Journal of Ophthalmology, 129*(1), 102–104.

Dandona, L., Hendrickson, A., & Quigley, H. A. (1991). Selective effects of experimental glaucoma on axonal transport by retinal ganglion cells to the dorsal lateral geniculate nucleus. *Investigative Ophthalmology & Visual Science, 32*(5), 1593–1599.

Daubs, J. G., & Crick, R. P. (1981). Effect of refractive error on the risk of ocular hypertension and open angle glaucoma. *Transactions of the Ophthalmological Societies of the United Kingdom, 101*(1), 121–126.

De Moraes, C. G., John, S. W. M., Williams, P. A., Blumberg, D. M., Cioffi, G. A., & Liebmann, J. M. (2022). Nicotinamide and pyruvate for neuroenhancement in open-angle glaucoma: A phase 2 randomized clinical trial. *JAMA Ophthalmology, 140*(1), 11–18.

Deng, H. X., Bigio, E. H., Zhai, H., Fecto, F., Ajroud, K., Shi, Y., Yan, J., Mishra, M., Ajroud-Driss, S., Heller, S., Siddique, N., Mugnaini, E., & Siddique, T. (2011). Differential involvement of optineurin in amyotrophic lateral sclerosis with or without SOD1 mutations. *Archives of Neurology, 68*(8), 1057–1061.

Dielemans, I., de Jong, P. T., Stolk, R., Vingerling, J. R., Grobbee, D. E., & Hofman, A. (1996). Primary open-angle glaucoma, intraocular pressure, and diabetes mellitus in the general elderly population. The Rotterdam study. *Ophthalmology, 103*(8), 1271–1275.

Dong, L., Hu, Y., Zhou, L., & Cheng, X. (2018). P2X7 receptor antagonist protects retinal ganglion cells by inhibiting microglial activation in a rat chronic ocular hypertension model. *Molecular Medicine Reports, 17*(2), 2289–2296.

Drance, S. M., Schulzer, M., Thomas, B., & Douglas, G. R. (1981). Multivariate analysis in glaucoma. Use of discriminant analysis in predicting glaucomatous visual field damage. *Archives of Ophthalmology, 99*(6), 1019–1022.

Ehlers, N., Bramsen, T., & Sperling, S. (1975). Applanation tonometry and central corneal thickness. *Acta Ophthalmologica, 53*(1), 34–43.

Ellis, J. D., Evans, J. M., Ruta, D. A., Baines, P. S., Leese, G., MacDonald, T. M., et al. (2000). Glaucoma incidence in an unselected cohort of diabetic patients: Is diabetes mellitus a risk factor for glaucoma? DARTS/MEMO collaboration. Diabetes audit and research in Tayside study. Medicines monitoring unit. *The British Journal of Ophthalmology, 84*(11), 1218–1224.

Fang, F., Zhuang, P., Feng, X., Liu, P., Liu, D., Huang, H., Li, L., Chen, W., Liu, L., Sun, Y., Jiang, H., Ye, J., & Hu, Y. (2022). NMNAT2 is downregulated in glaucomatous RGCs, and RGC-specific gene therapy rescues neurodegeneration and visual function. *Molecular Therapy: The Journal of the American Society of Gene Therapy, 30*(4), 1421–1431.

Feuer, W. J., Parrish, R. K., II, Schiffman, J. C., Anderson, D. R., Budenz, D. L., Wells, M. C., et al. (2002). The ocular hypertension treatment study: Reproducibility of cup/disk ratio measurements over time at an optic disc reading center. *American Journal of Ophthalmology, 133*(1), 19–28.

Friedman, D. S., Wolfs, R. C., O'Colmain, B. J., Klein, B. E., Taylor, H. R., West, S., et al. (2004a). Prevalence of open-angle glaucoma among adults in the United States. *Archives of Ophthalmology, 122*(4), 532–538.

Friedman, D. S., West, S. K., Munoz, B., Park, W., Deremeik, J., Massof, R., et al. (2004b). Racial variations in causes of vision loss in nursing homes: The Salisbury eye evaluation in nursing home groups (SEEING) study. *Archives of Ophthalmology, 122*(7), 1019–1024.

Gaasterland, D., & Kupfer, C. (1974). Experimental glaucoma in the rhesus monkey. *Investigative Ophthalmology, 13*(6), 455–457.

Gagrani, M., Ndulue, J., Anderson, D., Kedar, S., Gulati, V., Shepherd, J., High, R., Smith, L., Fowler, Z., Khazanchi, D., Nawrot, M., & Ghate, D. (2022). What do patients with glaucoma see: A novel iPad app to improve glaucoma patient awareness of visual field loss. *The British Journal of Ophthalmology, 106*(2), 218–222.

Gordon, M. O., Beiser, J. A., Brandt, J. D., Heuer, D. K., Higginbotham, E. J., Johnson, C. A., et al. (2002). The ocular hypertension treatment study: Baseline factors that predict the onset of primary open-angle glaucoma. *Archives of Ophthalmology, 120*(6), 714–720, discussion 829–30.

Grant, W. M. (1951). Clinical measurements of aqueous outflow. *Archives of Ophthalmology, 46*(2), 113–131.

Grant, W. M. (1958). Further studies on facility of flow through the trabecular meshwork. *Archives of Ophthalmology, 60*(4), 523–533.

Grant, W. M. (1963). Experimental aqueous perfusion in enucleated human eyes. *Archives of Ophthalmology, 69*, 783–801.

Greenlund, L. J., Deckwerth, T. L., & Johnson, E. M., Jr. (1995). Superoxide dismutase delays neuronal apoptosis: A role for reactive oxygen species in programmed neuronal death. *Neuron, 14*(2), 303–315.

Grotegut, P., Perumal, N., Kuehn, S., Smit, A., Dick, H. B., Grus, F. H., & Joachim, S. C. (2020). Minocycline reduces inflammatory response and cell death in a S100B retina degeneration model. *Journal of Neuroinflammation, 17*(1), 375.

Gulati, V., Ghate, D. A., Camras, C. B., & Toris, C. B. (2011). Correlations between parameters of aqueous humor dynamics and the influence of central corneal thickness. *Investigative Ophthalmology & Visual Science, 52*(2), 920–926.

Gulati, V., Fan, S., Zhao, M., Maslonka, M. A., Gangahar, C., & Toris, C. B. (2012). Diurnal and nocturnal variations in aqueous humor dynamics of patients with ocular hypertension undergoing medical therapy. *Archives of Ophthalmology, 130*(6), 677–684.

Gupta, N., Greenberg, G., de Tilly, L. N., Gray, B., Polemidiotis, M., & Yucel, Y. H. (2009). Atrophy of the lateral geniculate nucleus in human glaucoma detected by magnetic resonance imaging. *The British Journal of Ophthalmology, 93*(1), 56–60.

Hare, W., WoldeMussie, E., Lai, R., Ton, H., Ruiz, G., Feldmann, B., et al. (2001). Efficacy and safety of memantine, an NMDA-type open-channel blocker, for reduction of retinal injury associated with experimental glaucoma in rat and monkey. *Survey of Ophthalmology, 45*(Suppl 3), S284–S289; discussion S295–S296.

Harwerth, R. S., Crawford, M. L., Frishman, L. J., Viswanathan, S., Smith, E. L., III, & Carter-Dawson, L. (2002). Visual field defects and neural losses from experimental glaucoma. *Progress in Retinal and Eye Research, 21*(1), 91–125.

Hayashi, M., Yablonski, M. E., & Bito, L. Z. (1987). Eicosanoids as a new class of ocular hypotensive agents. 2. Comparison of the apparent mechanism of the ocular hypotensive effects of A and F type prostaglandins. *Investigative Ophthalmology & Visual Science, 28*(10), 1639–1643.

Heijl, A., Leske, M. C., Bengtsson, B., Hyman, L., Bengtsson, B., Hussein, M., et al. (2002). Reduction of intraocular pressure and glaucoma progression: Results from the early manifest glaucoma trial. *Archives of Ophthalmology, 120*(10), 1268–1279.

Hester, R. K., Chen, Z., Becker, E. J., McLaughlin, M., & DeSantis, L. (1994). The direct vascular relaxing action of betaxolol, carteolol and timolol in porcine long posterior ciliary artery. *Survey of Ophthalmology, 38*(Suppl), S125–S134.

Hoste, A. M. (2003). New insights into the subjective perception of visual field defects. *Bulletin de la Société Belge d'Ophtalmologie, 287*, 65–71.

Hutchins, J. B., & Barger, S. W. (1998). Why neurons die: Cell death in the nervous system. *The Anatomical Record, 253*(3), 79–90.

Investigators, A. G. I. S. (2000). The advanced glaucoma intervention study (AGIS): 7. The relationship between control of intraocular pressure and visual field deterioration. *American Journal of Ophthalmology, 130*(4), 429–440.

Ishii, Y., Kwong, J. M., & Caprioli, J. (2003). Retinal ganglion cell protection with geranylgeranylacetone, a heat shock protein inducer, in a rat glaucoma model. *Investigative Ophthalmology & Visual Science, 44*(5), 1982–1992.

Iwase, A., Suzuki, Y., Araie, M., Yamamoto, T., Abe, H., Shirato, S., Kuwayama, Y., Mishima, H. K., Shimizu, H., Tomita, G., Inoue, Y., Kitazawa, Y., & Tajimi Study Group, Japan Glaucoma Society. (2004). The prevalence of primary open-angle glaucoma in Japanese: The Tajimi study. *Ophthalmology, 111*(9), 1641–1648.

Jaeger, E. (1858). *Ueber Glaucom Und Seine Heilung Durch Iridectomie*. Druck von C. Gerold's Sohn.

Jain, A., Zode, G., Kasetti, R. B., Ran, F. A., Yan, W., Sharma, T. P., Bugge, K., Searby, C. C., Fingert, J. H., Zhang, F., Clark, A. F., & Sheffield, V. C. (2017). CRISPR-Cas9-based treatment of myocilin-associated glaucoma. *Proceedings of the National Academy of Sciences of the United States of America, 114*(42), 11199–11204.

Ji, J. Z., Elyaman, W., Yip, H. K., Lee, V. W., Yick, L. W., Hugon, J., & So, K. F. (2004). CNTF promotes survival of retinal ganglion cells after induction of ocular hypertension in rats: The possible involvement of STAT3 pathway. *The European Journal of Neuroscience, 19*(2), 265–272.

Johnson, M. (2006). What controls aqueous humour outflow resistance? *Experimental Eye Research, 82*(4), 545–557.

Johnson, T. V., Fan, S., Camras, C. B., & Toris, C. B. (2008). Aqueous humor dynamics in exfoliation syndrome. *Archives of Ophthalmology, 126*(7), 914–920.

Jonas, J. B., Wang, N., Yang, D., Ritch, R., & Panda-Jonas, S. (2015). Facts and myths of cerebrospinal fluid pressure for the physiology of the eye. *Progress in Retinal and Eye Research, 46*, 67–83.

Kass, M. A., Heuer, D. K., Higginbotham, E. J., Johnson, C. A., Keltner, J. L., Miller, J. P., et al. (2002). The ocular hypertension treatment study: A randomized trial determines that topical ocular hypotensive medication delays or prevents the onset of primary open-angle glaucoma. *Archives of Ophthalmology, 120*(6), 701–713. discussion 829–30.

Kass, M. A., Heuer, D. K., Higginbotham, E. J., Parrish, R. K., Khanna, C. L., Brandt, J. D., Soltau, J. B., Johnson, C. A., Keltner, J. L., Huecker, J. B., Wilson, B. S., Liu, L., Miller, J. P., Quigley, H. A., Gordon, M. O., & Ocular Hypertension Study Group. (2021). Assessment of cumulative incidence and severity of primary open-angle glaucoma among participants in the ocular hypertension treatment study after 20 years of follow-up. *JAMA Ophthalmology, 139*(5), 1–9. Advance online publication.

Kim, M. L., Sung, K. R., Kwon, J., Choi, G. W., & Shin, J. A. (2021). Neuroprotective effect of statins in a rat model of chronic ocular hypertension. *International Journal of Molecular Sciences, 22*(22), 12500.

Kipnis, J., & Schwartz, M. (2002). Dual action of glatiramer acetate (Cop-1) in the treatment of CNS autoimmune and neurodegenerative disorders. *Trends in Molecular Medicine, 8*(7), 319–323.

Kitaoka Y, Kumai T (2004) Modulation of retinal dopaminergic cells by nitric oxide. A protective effect on NMDA-induced retinal injury. In Vivo (Athens, Greece) 18(3):311–315.

Klein, B. E., Klein, R., & Jensen, S. C. (1994). Open-angle glaucoma and older-onset diabetes. The Beaver Dam eye study. *Ophthalmology, 101*(7), 1173–1177.

Koeberle, P. D., & Ball, A. K. (2002). Neurturin enhances the survival of axotomized retinal ganglion cells in vivo: Combined effects with glial cell line-derived neurotrophic factor and brain-derived neurotrophic factor. *Neuroscience, 110*(3), 555–567.

Koseki, N., Araie, M., Yamagami, J., Shirato, S., & Yamamoto, S. (1999). Effects of oral brovincamine on visual field damage in patients with normal-tension glaucoma with low-normal intraocular pressure. *Journal of Glaucoma, 8*(2), 117–123.

Krupin, T., Liebmann, J. M., Greenfield, D. S., Ritch, R., Gardiner, S., & Low-Pressure Glaucoma Study Group. (2011). A randomized trial of brimonidine versus timolol in preserving visual function: Results from the low-pressure glaucoma treatment study. *American Journal of Ophthalmology, 151*(4), 671–681.

Kuo, J. Z., Zangwill, L. M., Medeiros, F. A., Liebmann, J. M., Girkin, C. A., Hammel, N., et al. (2015). Quantitative trait locus analysis of six1-six6 with retinal nerve fiber layer thickness in individuals of European descent. *American Journal of Ophthalmology, 160*(1), 123–130.

Larsson, L. I., Rettig, E. S., Sheridan, P. T., & Brubaker, R. F. (1993). Aqueous humor dynamics in low-tension glaucoma. *American Journal of Ophthalmology, 116*(5), 590–593.

Larsson, L. I., Rettig, E. S., & Brubaker, R. F. (1995). Aqueous flow in open-angle glaucoma. *Archives of Ophthalmology, 113*(3), 283–286.

Leibowitz, H. M., Krueger, D. E., Maunder, L. R., Milton, R. C., Kini, M. M., Kahn, H. A., et al. (1980). The Framingham eye study monograph: An ophthalmological and epidemiological study of cataract, glaucoma, diabetic retinopathy, macular degeneration, and visual acuity in a general population of 2631 adults 1973–1975. *Survey of Ophthalmology, 24*(Suppl), 335–610.

Leske, M. C., Connell, A. M., Wu, S. Y., Hyman, L., & Schachat, A. P. (1997). Distribution of intraocular pressure. The Barbados Eye Study. *Archives of Ophthalmology, 115*(8), 1051–1057.

Leske, M. C., Heijl, A., Hussein, M., Bengtsson, B., Hyman, L., & Komaroff, E. (2003). Factors for glaucoma progression and the effect of treatment: The early manifest glaucoma trial. *Archives of Ophthalmology, 121*(1), 48–56.

Levene, R. Z., Bloom, J. N., & Kimura, R. (1976). Fluorophotometry and the rate of aqueous flow in man. II primary open angle glaucoma. *Archives of Ophthalmology, 94*(3), 444–447.

Leydhecker, W., Akiyama, K., & Neumann, H. G. (1958). Der intraokular druck gesunder menschlicher augen [intraocular pressure in normal human eyes]. *Klinische monatsblatter fur augenheilkunde und fur augenarztliche fortbildung, 133*(5), 662–670.

Lichter, P. R., Musch, D. C., Gillespie, B. W., Guire, K. E., Janz, N. K., Wren, P. A., et al. (2001). Interim clinical outcomes in the collaborative initial glaucoma treatment study comparing initial treatment randomized to medications or surgery. *Ophthalmology, 108*(11), 1943–1953.

Lipton, S. A. (2003). Possible role for memantine in protecting retinal ganglion cells from glaucomatous damage. *Survey of Ophthalmology, 48*(Suppl 1), S38–S46.

Liu, B., & Neufeld, A. H. (2001). Expression of nitric oxide synthase-2 in specific cells of human glaucomatous optic nerve head. *[Zhonghua Yan Ke Za Zhi] Chinese Journal of Ophthalmology, 37*(5), 381–383.

Maruyama, H., & Kawakami, H. (2013). Optineurin and amyotrophic lateral sclerosis. *Geriatrics & Gerontology International, 13*(3), 528–532.

McMenamin, P. G., & Steptoe, R. J. (1991). Normal anatomy of the aqueous humour outflow system in the domestic pig eye. *Journal of Anatomy, 178*, 65–77.

Medeiros, F. A., Zangwill, L. M., Bowd, C., & Weinreb, R. N. (2004). Comparison of the GDx VCC scanning laser polarimeter, HRT II confocal scanning laser ophthalmoscope, and stratus OCT optical coherence tomograph for the detection of glaucoma. *Archives of Ophthalmology, 122*(6), 827–837.

Minckler, D. S., Tso, M. O., & Zimmerman, L. E. (1976). A light microscopic, autoradiographic study of axoplasmic transport in the optic nerve head during ocular hypotony, increased intraocular pressure, and papilledema. *American Journal of Ophthalmology, 82*(5), 741–757.

Minckler, D. S., Bunt, A. H., & Johanson, G. W. (1977). Orthograde and retrograde axoplasmic transport during acute ocular hypertension in the monkey. *Investigative Ophthalmology & Visual Science, 16*(5), 426–441.

Minckler, D. S., Bunt, A. H., & Klock, I. B. (1978). Radioautographic and cytochemical ultrastructural studies of axoplasmic transport in the monkey optic nerve head. *Investigative Ophthalmology & Visual Science, 17*(1), 33–50.

Mirowska-Guzel, D. (2009). The role of neurotrophic factors in the pathology and treatment of multiple sclerosis. *Immunopharmacology and Immunotoxicology, 31*(1), 32–38.

Mitchell, P., Smith, W., Chey, T., & Healey, P. R. (1997). Open-angle glaucoma and diabetes: The Blue Mountains eye study, Australia. *Ophthalmology, 104*(4), 712–718.

Mojon, D. S., Hess, C. W., Goldblum, D., Fleischhauer, J., Koerner, F., Bassetti, C., & Mathic, J. (1999). High prevalence of glaucoma in patients with sleep apnea syndrome. *Ophthalmology, 106*(5), 1009–1012.

Monemi, S., Spaeth, G., DaSilva, A., Popinchalk, S., Ilitchev, E., Liebmann, J., et al. (2005). Identification of a novel adult-onset primary open-angle glaucoma (POAG) gene on 5q22.1. *Human Molecular Genetics, 14*(6), 725–733.

Müller, H. (1858). Anatomische. *Beitr├ñge zur Ophthalmologie, 4*(2), 1–54.

Murphy, G., & Docherty, A. J. (1992). The matrix metalloproteinases and their inhibitors. *American Journal of Respiratory Cell and Molecular Biology, 7*(2), 120–125.

Nasrolahi, A., Javaherforooshzadeh, F., Jafarzadeh-Gharehziaaddin, M., Mahmoudi, J., Asl, K. D., & Shabani, Z. (2022). Therapeutic potential of neurotrophic factors in Alzheimer's disease. *Molecular Biology Reports, 49*(3), 2345–2357.

Neufeld, A. H., Sawada, A., & Becker, B. (1999). Inhibition of nitric-oxide synthase 2 by aminoguanidine provides neuroprotection of retinal ganglion cells in a rat model of chronic glaucoma. *Proceedings of the National Academy of Sciences of the United States of America, 96*(17), 9944–9948.

Nilsson, S. F., Samuelsson, M., Bill, A., & Stjernschantz, J. (1989). Increased uveoscleral outflow as a possible mechanism of ocular hypotension caused by prostaglandin F2 alpha-1-isopropylester in the cynomolgus monkey. *Experimental Eye Research, 48*(5), 707–716.

Nouri-Mahdavi, K., Hoffman, D., Coleman, A. L., Liu, G., Li, G., Gaasterland, D., et al. (2004). Predictive factors for glaucomatous visual field progression in the advanced glaucoma intervention study. *Ophthalmology, 111*(9), 1627–1635.

Nucci, C., Tartaglione, R., Cerulli, A., Mancino, R., Spanò, A., Cavaliere, F., Rombolà, L., Bagetta, G., Corasaniti, M. T., & Morrone, L. A. (2007). Retinal damage caused by high intraocular pressure-induced transient ischemia is prevented by coenzyme Q10 in rat. *International Review of Neurobiology, 82*, 397–406.

Oddone, F., Roberti, G., Micera, A., Busanello, A., Bonini, S., Quaranta, L., Agnifili, L., & Manni, G. (2017). Exploring serum levels of brain derived neurotrophic factor and nerve growth factor across glaucoma stages. *PLoS One, 12*(1), e0168565.

Osborne, N. N., Ugarte, M., Chao, M., Chidlow, G., Bae, J. H., Wood, J. P., et al. (1999). Neuroprotection in relation to retinal ischemia and relevance to glaucoma. *Survey of Ophthalmology, 43*(Suppl 1), S102–S128.

Osborne, N. N., Wood, J. P., Chidlow, G., Casson, R., DeSantis, L., & Schmidt, K. G. (2004a). Effectiveness of levobetaxolol and timolol at blunting retinal ischaemia is related to their calcium and sodium blocking activities: Relevance to glaucoma. *Brain Research Bulletin, 62*(6), 525–528.

Osborne, N. N., Casson, R. J., Wood, J. P., Chidlow, G., Graham, M., & Melena, J. (2004b). Retinal ischemia: Mechanisms of damage and potential therapeutic strategies. *Progress in Retinal and Eye Research, 23*(1), 91–147.

Parisi, V., Centofanti, M., Gandolfi, S., Marangoni, D., Rossetti, L., Tanga, L., Tardini, M., Traina, S., Ungaro, N., Vetrugno, M., & Falsini, B. (2014). Effects of coenzyme Q10 in conjunction with vitamin E on retinal-evoked and cortical-evoked responses in patients with open-angle glaucoma. *Journal of Glaucoma, 23*(6), 391–404.

Phelps, C. D. (1978). The pathogenesis of glaucoma in Sturge-Weber syndrome. *Ophthalmology, 85*(3), 276–286.

Podos, S. M., Minas, T. F., & Macri, F. J. (1968). A new instrument to measure episcleral venous pressure. Comparison of normal eyes and eyes with primary open-angle glaucoma. *Archives of Ophthalmology, 80*(2), 209–213.

Prata, T. S., De Moraes, C. G., Kanadani, F. N., Ritch, R., & Paranhos, A., Jr. (2010). Posture-induced intraocular pressure changes: Considerations regarding body position in glaucoma patients. *Survey of Ophthalmology, 55*(5), 445–453.

Quigley, H. A. (1996). Number of people with glaucoma worldwide. *The British Journal of Ophthalmology, 80*(5), 389–393.

Quigley, H. A., Guy, J., & Anderson, D. R. (1979). Blockade of rapid axonal transport. Effect of intraocular pressure elevation in primate optic nerve. *Archives of Ophthalmology, 97*(3), 525–531.

Racette, L., & Sample, P. A. (2003). Short-wavelength automated perimetry. *Ophthalmology Clinics of North America, 16*(2), 227–236, vi-vii.

Rasmussen, C. A., & Kaufman, P. L. (2005). Primate glaucoma models. *Journal of Glaucoma, 14*(4), 311–314.

Ren, R., Zhang, X., Wang, N., Li, B., Tian, G., & Jonas, J. B. (2011). Cerebrospinal fluid pressure in ocular hypertension. *Acta Ophthalmologica, 89*(2), e142–e148.

Rezaie, T., Child, A., Hitchings, R., Brice, G., Miller, L., Coca-Prados, M., et al. (2002). Adult-onset primary open-angle glaucoma caused by mutations in optineurin. *Science, 295*(5557), 1077–1079.

Rosenquist, R., Epstein, D., Melamed, S., Johnson, M., & Grant, W. M. (1989). Outflow resistance of enucleated human eyes at two different perfusion pressures and different extents of trabeculotomy. *Current Eye Research, 8*(12), 1233–1240.

Ruiz-Ederra, J., Garcia, M., Hernandez, M., Urcola, H., Hernandez-Barbachano, E., Araiz, J., et al. (2005). The pig eye as a novel model of glaucoma. *Experimental Eye Research, 81*(5), 561–569.

Sawada, A., Kitazawa, Y., Yamamoto, T., Okabe, I., & Ichien, K. (1996). Prevention of visual field defect progression with brovincamine in eyes with normal-tension glaucoma. *Ophthalmology, 103*(2), 283–288.

Schimiti, R. B., Avelino, R. R., Kara-Jose, N., & Costa, V. P. (2002). Full-threshold versus Swedish interactive threshold algorithm (SITA) in normal individuals undergoing automated perimetry for the first time. *Ophthalmology, 109*(11), 2084–2092. discussion 2092.

Schurr, A., & Rigor, B. M. (1992). The mechanism of cerebral hypoxic-ischemic damage. *Hippocampus, 2*(3), 221–228.

Schwartz, M. (2005). Lessons for glaucoma from other neurodegenerative diseases: Can one treatment suit them all? *Journal of Glaucoma, 14*(4), 321–323.

Selvadurai, D., Hodge, D., & Sit, A. J. (2010). Aqueous humor outflow facility by tonography does not change with body position. *Investigative Ophthalmology & Visual Science, 51*(3), 1453–1457.

Sena, D. F., & Lindsley, K. (2013). Neuroprotection for treatment of glaucoma in adults. *Cochrane Database of Systematic Reviews, 2*, CD006539.

Setoguchi, M., Ohya, Y., Abe, I., & Fujishima, M. (1995). Inhibitory action of betaxolol, a beta 1-selective adrenoceptor antagonist, on voltage-dependent calcium channels in Guinea-pig artery and vein. *British Journal of Pharmacology, 115*(1), 198–202.

Shiau, T., Armogan, N., Yan, D. B., Thomson, H. G., & Levin, A. V. (2012). The role of episcleral venous pressure in glaucoma associated with Sturge-Weber syndrome. *Journal of AAPOS, 16*(1), 61–64.

Siesjo, B. K., & Bengtsson, F. (1989). Calcium fluxes, calcium antagonists, and calcium-related pathology in brain ischemia, hypoglycemia, and spreading depression: A unifying hypothesis. *Journal of Cerebral Blood Flow & Metabolism, 9*(2), 127–140.

Sommer, A., Tielsch, J. M., Katz, J., Quigley, H. A., Gottsch, J. D., Javitt, J., & Singh, K. (1991). Relationship between intraocular pressure and primary open angle glaucoma among white and black Americans. The Baltimore Eye Survey. *Archives of Ophthalmology (Chicago, Ill.: 1960), 109*(8), 1090–1095.

Sullivan, A. M., & Toulouse, A. (2011). Neurotrophic factors for the treatment of Parkinson's disease. *Cytokine & Growth Factor Reviews, 22*(3), 157–165.

Tektas, O. Y., & Lutjen-Drecoll, E. (2009). Structural changes of the trabecular meshwork in different kinds of glaucoma. *Experimental Eye Research, 88*(4), 769–775.

Tezel, G., Hernandez, R., & Wax, M. B. (2000). Immunostaining of heat shock proteins in the retina and optic nerve head of normal and glaucomatous eyes. *Archives of Ophthalmology, 118*(4), 511–518.

Tham, Y. C., Li, X., Wong, T. Y., Quigley, H. A., Aung, T., & Cheng, C. Y. (2014). Global prevalence of glaucoma and projections of glaucoma burden through 2040: A systematic review and meta-analysis. *Ophthalmology, 121*(11), 2081–2090.

Tielsch, J. M., Katz, J., Singh, K., Quigley, H. A., Gottsch, J. D., Javitt, J., et al. (1991). A population-based evaluation of glaucoma screening: The Baltimore eye survey. *American Journal of Epidemiology, 134*(10), 1102–1110.

Tielsch, J. M., Katz, J., Quigley, H. A., Javitt, J. C., & Sommer, A. (1995). Diabetes, intraocular pressure, and primary open-angle glaucoma in the Baltimore eye survey. *Ophthalmology, 102*(1), 48–53.

Toris, C. B., Yablonski, M. E., Wang, Y. L., & Camras, C. B. (1999). Aqueous humor dynamics in the aging human eye. *American Journal of Ophthalmology, 127*(4), 407–412.

Toris, C. B., Koepsell, S. A., Yablonski, M. E., & Camras, C. B. (2002). Aqueous humor dynamics in ocular hypertensive patients. *Journal of Glaucoma, 11*(3), 253–258.

Toris, C. B., Haecker, N. R., Teasley, L. A., Zhan, G., Gulati, V., & Camras, C. B. (2010). Aqueous humor dynamics in pigment dispersion syndrome. *Archives of Ophthalmology, 128*(9), 1115–1118.

Tribble, J. R., Vasalauskaite, A., Redmond, T., Young, R. D., Hassan, S., Fautsch, M. P., Sengpiel, F., Williams, P. A., & Morgan, J. E. (2019). Midget retinal ganglion cell dendritic and mitochondrial degeneration is an early feature of human glaucoma. *Brain Communications, 1*(1), fcz035.

Tribble, J. R., Otmani, A., Sun, S., Ellis, S. A., Cimaglia, G., Vohra, R., Jöe, M., Lardner, E., Venkataraman, A. P., Domínguez-Vicent, A., Kokkali, E., Rho, S., Jóhannesson, G., Burgess, R. W., Fuerst, P. G., Brautaset, R., Kolko, M., Morgan, J. E., Crowston, J. G., Votruba, M., et al. (2021). Nicotinamide provides neuroprotection in glaucoma by protecting against mitochondrial and metabolic dysfunction. *Redox Biology, 43*, 101988.

Tseng, H. C., Riday, T. T., McKee, C., Braine, C. E., Bomze, H., Barak, I., et al. (2015). Visual impairment in an optineurin mouse model of primary open-angle glaucoma. *Neurobiology of Aging, 36*(6), 2201–2212.

Varma, R., Ying-Lai, M., Francis, B. A., Nguyen, B. B., Deneen, J., Wilson, M. R., Azen, S. P., & Los Angeles Latino Eye Study Group. (2004). Prevalence of open-angle glaucoma and ocular hypertension in Latinos: The Los Angeles Latino eye study. *Ophthalmology, 111*(8), 1439–1448.

Vasiliou, V., & Gonzalez, F. J. (2008). Role of CYP1B1 in glaucoma. *Annual Review of Pharmacology and Toxicology, 48*, 333–358.

Visuvanathan, S., Baker, A. N., Lagali, P. S., Coupland, S. G., Miller, G., Hauswirth, W. W., & Tsilfidis, C. (2022). XIAP gene therapy effects on retinal ganglion cell structure and function in a mouse model of glaucoma. *Gene Therapy, 29*(3–4), 147–156.

Wax, M., & Tezel, G. (2002). Neurobiology of glaucomatous optic neuropathy: Diverse cellular events in neurodegeneration and neuroprotection. *Molecular Neurobiology, 26*(1), 45–55.

Weinreb, R. N., & Levin, L. A. (1999). Is neuroprotection a viable therapy for glaucoma? *Arch Ophthalmol, 117*, 1540–1544.

Weinreb, R. N., & Lindsey, J. D. (2005). The importance of models in glaucoma research. *Journal of Glaucoma, 14*(4), 302–304.

Wiggs, J. L., Allingham, R. R., Vollrath, D., Jones, K. H., De La Paz, M., Kern, J., et al. (1998). Prevalence of mutations in TIGR/Myocilin in patients with adult and juvenile primary open-angle glaucoma. *American Journal of Human Genetics, 63*(5), 1549–1552.

Wiggs, J. L., Auguste, J., Allingham, R. R., Flor, J. D., Pericak-Vance, M. A., Rogers, K., et al. (2003). Lack of association of mutations in optineurin with disease in patients with adult-onset primary open-angle glaucoma. *Archives of Ophthalmology, 121*(8), 1181–1183.

Wiggs, J. L., Yaspan, B. L., Hauser, M. A., Kang, J. H., Allingham, R. R., Olson, L. M., et al. (2012). Common variants at 9p21 and 8q22 are associated with increased susceptibility to optic nerve degeneration in glaucoma. *PLoS Genetics, 8*(4), e1002654.

Wilson, M. R., Hertzmark, E., Walker, A. M., Childs-Shaw, K., & Epstein, D. L. (1987). A case–control study of risk factors in open angle glaucoma. *Archives of Ophthalmology, 105*(8), 1066–1071.

Wolfs, R. C., Borger, P. H., Ramrattan, R. S., Klaver, C. C., Hulsman, C. A., Hofman, A., et al. (2000). Changing views on open-angle glaucoma: Definitions and prevalences--The Rotterdam Study. *Investigative Ophthalmology & Visual Science, 41*(11), 3309–3321.

Wyllie, A. H., Kerr, J. F., & Currie, A. R. (1980). Cell death: The significance of apoptosis. *International Review of Cytology, 68*, 251–306.

Yang, D., Fu, J., Hou, R., Liu, K., Jonas, J. B., Wang, H., et al. (2014). Optic neuropathy induced by experimentally reduced cerebrospinal fluid pressure in monkeys. *Investigative Ophthalmology & Visual Science, 55*(5), 3067–3073.

Yu, L., Xie, B., Yin, X., Liang, M., Evans, A. C., Wang, J., et al. (2013). Reduced cortical thickness in primary open-angle glaucoma and its relationship to the retinal nerve fiber layer thickness. *PLoS One, 8*(9), e73208.

Yucel, Y., & Gupta, N. (2008). Glaucoma of the brain: A disease model for the study of transsynaptic neural degeneration. *Progress in Brain Research, 173*, 465–478.

Ziai, N., Dolan, J. W., Kacere, R. D., & Brubaker, R. F. (1993). The effects on aqueous dynamics of PhXA41, a new prostaglandin F2 alpha analogue, after topical application in normal and ocular hypertensive human eyes. *Archives of Ophthalmology, 111*(10), 1351–1358.

Adult Neurogenesis

32

Larisa Y. Poluektova

Abstract

There is scientifically proven evidence that new neurons are generated and incorporated into circuits of the adult human brain. However, their amount is limited and significantly declines with age and pathologic conditions. The contribution of adult neurogenesis in cognitive function in humans remains unknown, compared to a wide variety of animal models that support the significant role of neurogenesis in cognition. This chapter describes the history of the discovery, isolation, and differentiation of neural stem cells. The regulation of hippocampal neurogenesis known from rodent models and how neurogenesis is affected by pathologic conditions are also addressed.

Keywords

Neural stem cells · Neural progenitors · Neurons · Astrocytes · Hippocampus

32.1 Introduction

The production of new neurons—neurogenesis—is active during embryonic development. What about the adult brain? The central dogma of neuroscience for a long time was formulated by Santiago Ramón y Cajal, the father of modern neuroscience, “In adult centers the nerve paths are something fixed, ended, immutable. Everything may die, nothing may be regenerated” (Cajal, 1894). In the 1980s, studies on male canary brains showed seasonal changes in size related to breeding seasons and singing serenades to females, learning new songs (Goldman & Nottebohm, 1983). However, the first observation of new neurons production by adult rodent brains was done in the 1960s (Altman & Das, 1966, 1967). Even 60 years later, we have limited knowledge of how neurons born in adulthood are incorporated into existing circuits and contribute to adult brain regeneration during pathologic conditions in humans. On the contrary, thousands of studies have now been published, and it is widely accepted that the adult mouse brain continues to produce new neurons and these cells contribute to brain recovery (Kempermann & Gage, 1999; Kempermann et al., 1997; Reynolds & Weiss, 1992; Shingo et al., 2003; van Praag et al., 1999). The place where new neurons are generated in the adult brain is related to two regions: lateral ventricles that produce immature neurons that migrate in chains within the rostral migratory stream (RMS) to the olfactory bulb; and the dentate gyrus of the hippocampus. Newborn neurons from these sites contribute to local circuits and participate in the processing of smell information and play important roles in learning and memory, respectively. The commonly accepted fate of these cells differentiates into mature neurons that integrate into existing circuits to provide additional information-processing capabilities. Neurogenesis also was documented Old World primate’s prefrontal, inferior temporal, and posterior parietal cortex. It was important for cognitive function. The new neurons were originated in the subventricular zone and migrated through the white matter to the neocortex, where they extended axons (Gould et al., 1999). The explanation of new neurons generation is based on the presence of stem cells in the adult brain (Gage & Temple, 2013). These stem cells undergo asymmetric division and generate self-like cells for their renewal and multilineage progenitors, which differentiate into neuronal, astroglia, and oligodendroglia committed precursors. Unfortunately, the adult human central and peripheral nervous systems have a minimal capacity for neural stem cells (NSC), and their contribution to neural tissue regeneration remains low. The production of multilineage progenitors and committed precursors is highly regulated by the local tissue environment, hormonal and metabolic homeostasis. Responses to infections and inflammatory con-

L. Y. Poluektova (✉)
Department of Pharmacology and Experimental Neuroscience, University of Nebraska Medical Center, Omaha, NE, USA
e-mail: lpoluekt@unmc.edu

H. E. Gendelman, T. Ikezu (eds.), *Neuroimmune Pharmacology and Therapeutics*,
https://doi.org/10.1007/978-3-031-68237-7_32

ditions affect neurogenesis, and treatment reduces its harmful effects.

32.2 Neural Stem Cells and Their Niches in the Adult Human Brain

The NSCs were identified as cells that do not have lineage-specific markers—neuronal, astrocytes, or oligodendrocytes. However, these cells can be isolated from the brain based on neutrosphere formation. This was shown by enzymatically dissociated and plated in serum-free media mouse striatum. The critical component of media was the epidermal growth factor (EGF). After two days in culture majority of cells died, but by day 6, proliferated clusters as spheres were observed. Continued culturing reveals specific properties of neuroepithelial stem cells. Under appropriate conditions, these cells were induced to differentiate into astrocytes and neurons with phenotypes characteristics of the adult striatum in vivo (Reynolds & Weiss, 1992).

The isolation of NSC from the adult human brain is rare, but cells isolated from the fetal human brain are well-characterized (Uchida et al., 2000). These cells are positive for cluster of differentiation 133 (CD133) or prominin-1, a pentaspan transmembrane glycoprotein, first known as a surface marker for hematopoietic stem cells (Yin et al., 1997). These cells are negative for common hematopoietic lineage markers CD45 and CD34. When isolated by flow cytometry sorting and placed in a culture medium containing N2 supplement and fibroblast growth factor-2, epidermal growth factor, lymphocyte inhibitory factor, and neural survival factor-1, these cells make spheres, which can be dissociated, re-plated, and expanded (Fig. 32.1). These cells differentiated in vitro in neurons and astrocytes (Fig. 32.1).

Transplantation of these cells into the brain of newborn immunodeficient mice gives rise to human neurons, astrocytes, and oligodendrocytes. The differentiation and migration were observed by labeling NSC with the green fluorescent protein, as shown in Fig. 32.2.

The detection of proliferation and new neuron formation was done in live animals and humans by using thymidine analog bromodeoxyuridine (BrdU) (Eriksson et al., 1998). This nucleoside is incorporated into the cells during the S phase of the cell cycle and can be detected with specific antibodies. Immunohistochemistry to detect the BrdU in newborn cells and co-labeling with specific neuronal markers allows identify these cells (Miller & Nowakowski, 1988). The main markers used to identify NSC in the human hippocampus are shown in Fig. 32.3.

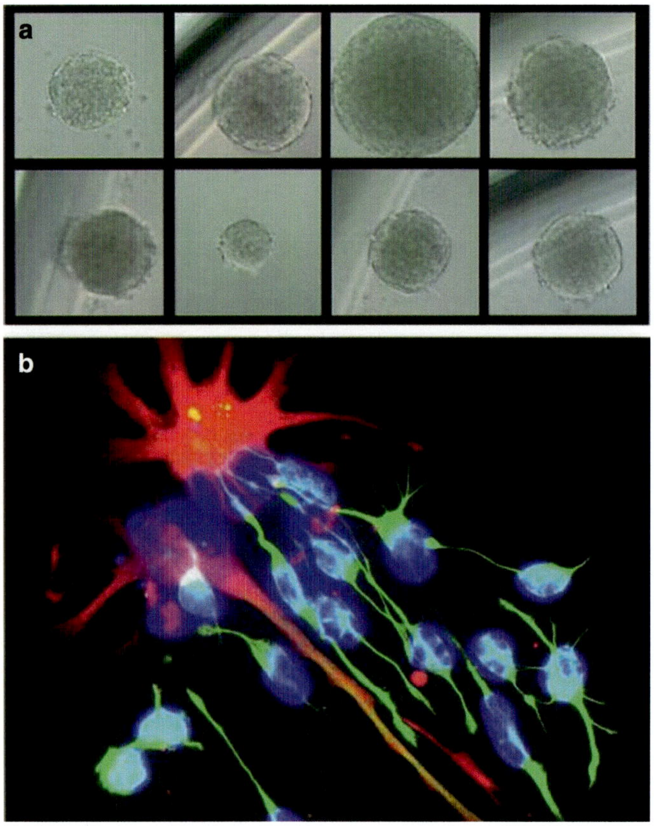

Fig. 32.1 Clonal expansion from single CD133+ cells. (**a**) Clonal expansion of neural stem/progenitor cells. Neurospheres can be derived from a single sorted CD133+ cell directly isolated from the fetal brain or from a single CD133+ cell from sorted/expanded cultured neurospheres. (**b**) Differentiation capacity of clonally derived neurosphere cells. The progeny of single cell-derived neurospheres can be differentiated into neurons (β-tubulin III, green) and astrocytes (GFAP, red). Nuclei were counterstained with Hoechst 33342 (blue). (From Uchida et al. (2000))

32.3 Contribution of Adult Neurogenesis to Cognitive Function

Humans benefit enormously from animal experimentation, and without it, a large portion of medical and pharmaceutical advances would not be possible. So, animal experimentation is both necessary and valuable for the advancement of science. However, it is imperative to consider that there are limitations inherent to the different species, which may dictate different biological behaviors under the same conditions. Thus, after >10,000 papers were published on adult neurogenesis in rodents, the data on its time course, rate, or functional characteristics are contradictory (La Rosa et al., 2020). The suggestion that adult-generated neurons make essential functional contributions to neural plasticity and cognition

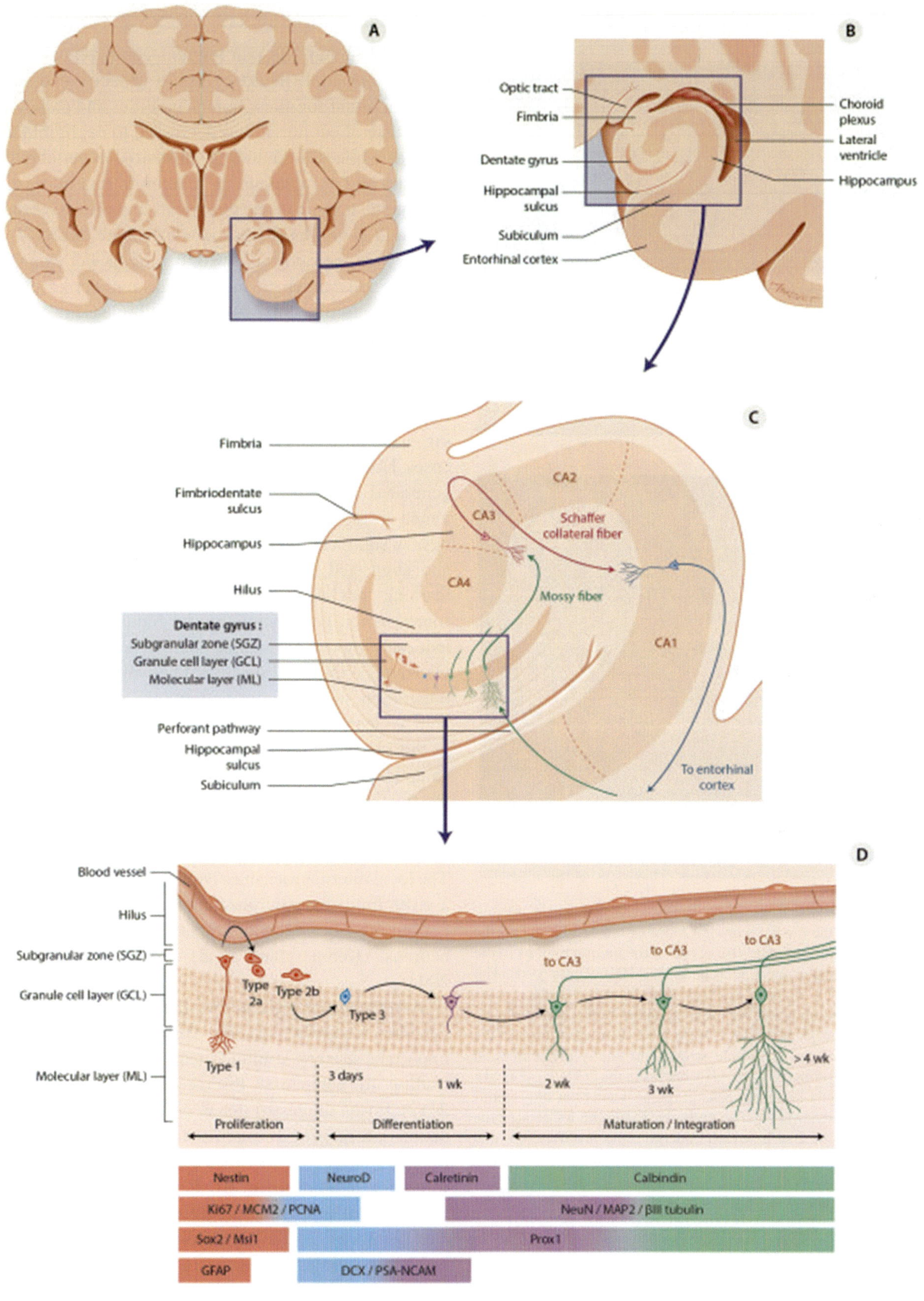

Fig. 32.2 Detection of GFP+ human cells after transplantation into neonatal NOD-SCID mice. GFP+ human neurosphere cells were transplanted into the lateral ventricles of neonatal NOD-SCID mice. Engraftment of human cells based on GFP fluorescence was analyzed 6–30 weeks after transplantation. Migration of human GFP+ cells in the olfactory bulb. (**a**) GFP+ cells detected at the end of RMS/entry of the olfactory bulb 6 weeks posttransplantation. (**b**) GFP+ cells detected in the granular cell layer of the olfactory bulb 23 weeks posttransplantation. Higher magnification from the same area shows that GFP+ cells (green) colocalize with β-tubulin III (red) staining (inset). Scale bars = 50 μm. (From Tamaki et al. (2002))

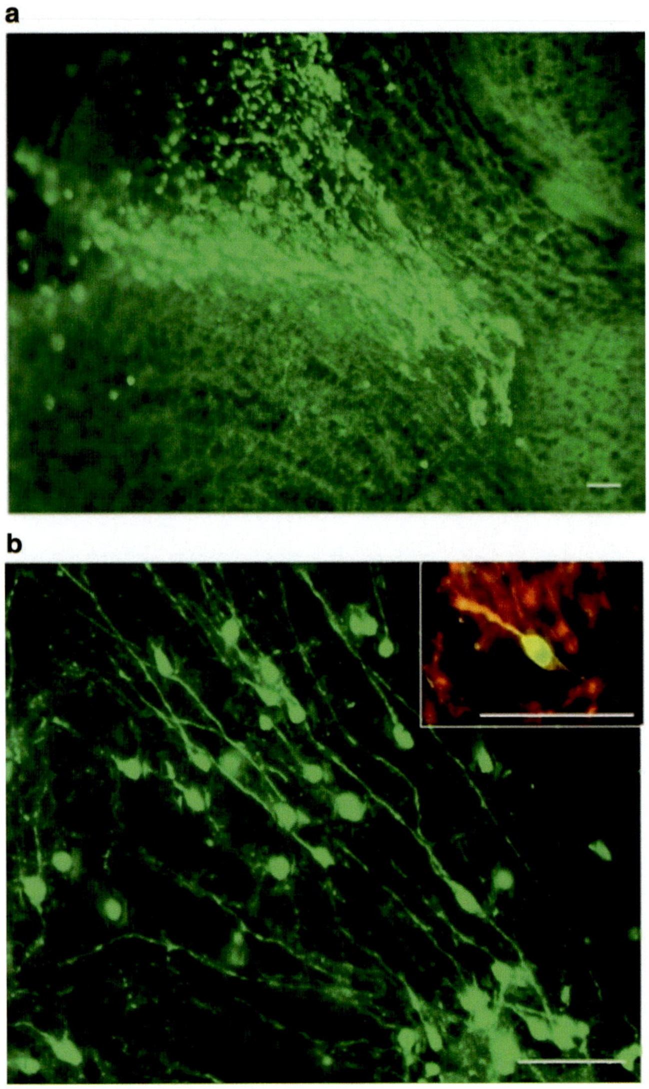

Fig. 32.3 Stages and markers of hippocampal neurogenesis in an adult human. Located in the medial temporal lobes of the cerebrum (**a**), the hippocampus, a limbic structure (**b**), is one of the unique regions in the brain where adult hippocampal neurogenesis (AHN) occurs (**c–d**). The "trisynaptic loop" (**c**), includes the projection from the dentate gyrus (**DG**) to the Cornu Ammonis (**CA**) 3 region via the mossy fiber pathway (green arrow), the projection from CA3 to the CA1 region through the Schaffer Collateral pathway (purple arrow) and finally the projection from CA1 to the entorhinal cortex. AHN occurs in the dentate gyrus (**d**). The pool of radial glia-like stem cells (Type 1; red) is not only maintained through self-renewal but also gives rise to progenitor cells (Type 2 A and B; red). Type 2 cells, which undergo rapid proliferation, differentiate and generate neuroblasts (Type 3; blue). The surviving neuroblasts (green cells) develop a stable dendritic arborization that extends in a specific way in time toward the molecular layer and from the thousands of neuroblasts generated. In the days following the exit from the cell cycle, new cells emanate their axons to their target region CA3 to form appropriate synapses. The gradual transition in marker expression for each AHN stage: proliferation (GFAP: glial fibrillary acidic protein, KI67, MCM22: minichromosome maintenance protein 2, Nestin, Sox2: Sex determining region Y-box2), differentiation/maturation/integration (βIII tubulin, Calbindin, DCX: doublecortin, MAP2a/b: microtubule associated protein 2a/b, NeuN: neuron-specific nuclear protein, Msi1: Musashi-1, NeuroD: neuronal differentiation, PCNA: proliferating cell nuclear antigen, PROX1: prospero homeobox protein 1, PSA-NCAM: polysialylated neural cell adhesion molecule. (From Mendez-David et al. (2022))

across the human lifespan remains controversial. It is paradoxical that adult neurogenesis, a capacity claimed to be very important for many human qualities and pathologies, in the course of evolution, has decreased to negligible levels in humans while still remaining substantial in vertebrates, such as fishes and amphibians. In relation to this issue, it should be noted that some of the treatments and processes that regulate adult neurogenesis in rodents also are expected to have an effect on their physiology or cognition. For example, running is well known to increase the number of capillaries, improve cardiovascular function, and release trophic factors and endorphins. Those changes are naturally expected to produce an improvement in motor and cognitive output both in rodents and in humans, irrespective of neurogenesis. Improvement can be addressed to the enhancement of existing neuronal circuits' function and generation of new synapses. However, it does not exclude the contribution of newly generated neurons. Therefore, research might be better aimed at studying those shared features of hippocampal biology. A significant amount of data was collected on rodents and suggested that adult neurogenesis is not needed for learning per se but rather for an advanced level of functionality. The spatiotemporal contextualization of information is important and new neurons help avoid catastrophic interference in the hippocampal network. They facilitate the integration of new information into pre-existing contexts help to clear the dentate gyrus at the circuit level and, at least in this sense, support forgetting. In addition, as the hippocampus is part of the limbic system, new neurons are involved in affective behaviors. These new neurons contribute synaptic plasticity to the dentate gyrus, measured as an increased long-term potentiation (Marin-Burgin & Schinder, 2012). The local interneurons massively inhibit all other neurons. At a given time, synaptic plasticity in the dentate gyrus is thus concentrated in a defined, functionally naive sub-set of (new) neurons. (Kempermann et al., 2018; Kempermann et al., 2002).

Adult neurogenesis allows the dentate gyrus to rapidly generate a context-specific, distributed representation of important sensory stimuli, such as spatial cues, ultimately leading to behavioral discrimination (Tuncdemir et al., 2019). It was suggested that neurogenesis leads to the degradation or forgetting of established memories. Increasing neurogenesis after memory formation was sufficient to induce forgetting in adult mice. By contrast, during infancy, when hippocampal neurogenesis levels are high and freshly generated memories tend to be rapidly forgotten (infantile amnesia), decreasing neurogenesis after memory formation mitigates the forgetting (Welberg, 2014). Apart from the lack of scientifically proven evidence that new neurons are generated and incorporated into circuits of the adult human brain, even if a few were added, there are many logical problems with interpreting its possible significance (Duque et al.,

2022). Duque's critical review asks a reasonable question: if adult neurogenesis is essential for human memory and cognition, which are evolutionarily new functions, why do humans not have higher rates of neurogenesis than lower vertebrates, such as fishes and frogs? And why does adult neurogenesis, even in rodents, occur only in the phylogenetically older olfactory bulb and hippocampus (archicortex) and not in the neocortex? To answer this question we need to manipulate a live human brain and efficiently detect newborn neurons and their incorporation into existing circuits to confirm their functionality. As of today, it is not possible, and we will continue to review information obtained primarily on mice brains.

32.4 Regulation and Neurogenesis

The process of adult neurogenesis in the hippocampus is marked with Nestin-, GFAP- and SRY-box 2 (Sox2)-expressing quiescent NSCs, also referred to as radial glia-like cells because of their morphology and ontogeny (Type 1, Fig. 32.3). The direct cell-cell contacts within niche cells and secreted molecules regulate whether NSCs remain quiescent or enter an active state (Matsubara et al., 2021). Neurogenic locus notch homolog protein 1 (Notch 1) is a representative factor regulating NSC behavior via Notch ligand Jagged1 (Jag1). Jag1 is expressed by NPCs and granule neurons in the dentate gyrus and upon binding of this ligand, the cleaved and activated Notch intracellular domain is released from the cell membrane into the cytoplasm and translocate to the nucleus. It interacts there with the DNA-binding CSL proteins. CSL (CBF1, Suppressor of Hairless, Lag-1) is a transcription factor that is responsible for activating the genes downstream of the Notch signaling pathway, a pathway that is essential for the development of the nervous system and the differentiation of the hematopoietic system, among others (Pursglove & Mackay, 2005). This interaction induces the expression of target genes associated with NSC quiescence.

Human Wnt signaling pathway is another regulatory factor for stem cells. It consists of 19 identified proteins secreted from local niches by NSC and astrocytes and induces the activation of quiescent cells. Wnt7a regulates multiple steps of neurogenesis, and in rodents, loss of Wnt7a expression led to reduced neural stem cell self-renewal, an increased rate of cell cycle exit in neural progenitors, and decreased neuronal differentiation and maturation (Qu et al., 2013). Another example is bone morphogenetic proteins (BMPs) and milk-fat globule-epidermal growth factor EGF factor 8 (Mfge8) has an important role in maintaining NSC quiescence. Mfge8 secreted by SGZ NSCs and astrocytes maintains NSC quiescence through integrinβ1-mediated mechanistic target of rapamycin (mTOR) signaling in autocrine and paracrine manners (Zhou et al., 2018).

Serotonin, dopamine, glutamate, acetylcholine, noradrenaline, and gamma-aminobutyric acid (GABA) secreted by neurons, have been shown to regulate the quiescent to active state transition of NSCs as well as subsequent neurogenesis in the adult dentate gyrus. Recent advances in optogenetics have enabled researchers to further uncover neuronal circuitry mechanisms controlling NSC behavior spatiotemporally through these neurotransmitters. Using the combination of optogenetics and lineage tracing technology, recent studies have identified DG parvalbumin (PV)-positive interneurons acting as a cellular niche component that signals to quiescent NSCs through GABA type A receptors ($GABA_ARs$) in an activity-dependent fashion in the adult DG. Moreover, optogenetic control of DG PV-positive neuron activity, but not somatostatin- and vasoactive intestinal polypeptide-positive interneurons, dictates the NSC's decision of whether they remain quiescent or enter activation states (Song et al., 2012).

In the brain, the extracellular matrix (ECM) in the basement membrane lining tissues and the intercellular spaces provides a functional scaffold for maintaining signaling gradients and stiffness. The ECM plays critical roles in neuronal migration, axon outgrowth, myelination, synaptogenesis, synaptic plasticity, and NSC performance. Members of the transmembrane protein integrin family form a complex composed of an α and β subunit that allows cells to recognize and communicate with the ECM. The largest integrin subfamily is composed of complexes containing integrin β1 (Itgb1), which is highly expressed in NSCs in the adult dentate gyrus. Specific deletion of Itgb1 in adult NSCs resulted in extensive cellular disorganization of the SGZ as well as the non-neurogenic region in the DG. Itgb1-deficient NSCs rapidly lost their NSC properties and differentiated into astrocytes, indicating that Itgb1 maintains the NSC population and prevents the commitment of NSCs into the astrocytic lineage (Brooker et al., 2016). The direction NSC to determine quiescence or activated state is conducted by the interaction between the ECM and the proteins recognizing it on the cell surface.

32.5 Aging and Microglial Activation

Many environmental and cell-intrinsic factors regulate adult neurogenesis in the dentate gyrus of the hippocampus. In rodents, an enriched environment, including a larger cage area, novel objects, and running wheels, has been shown to significantly increase the number of adult-born neurons and the volume of the granule cell layer of the hippocampus and to improve the speed of spatial learning (Kempermann et al., 1997). The role of a complex environment and exercise was different (Perez-Dominguez et al., 2018). The voluntary running alone selectively increased the proliferation of neuroblasts, whereas environmental enrichment promoted the survival of adult-born dentate gyrus cells through the increased integration of immature neurons (van Praag et al., 1999). In humans, the decline of neurogenesis was attributed to aging. The generation of new neurons in the hippocampus of humans is difficult to study. The immunohistological evaluation of NSC and newborn neurons, their maturation, and migration path on human samples obtained from individuals at 22 gestational weeks, birth, and 7, 18, and 48 years of age, which express progenitor/stem cell markers, are mostly depleted from the hilus of dentate gyrus by 7 years of age (Sorrells et al., 2018). In more recent studies of several human samples (aged 52, 67, 78, and 85 years), some cells were expressing NSC markers and proliferating markers in the hippocampus, which implied that NSCs still exist and were proliferating in aged humans (Wang et al., 2022). To further investigate adult neurogenesis and the regulatory continuum in aged human hippocampi, these four human samples were collected and analyzed using single-cell/single-nucleus RNA sequencing (scRNA-seq/snRNA-seq). Most NSCs were in quiescent state in aged humans. However, a small group of NSC highly and specifically expressed the active NSC markers, such as ASCL1 (Achaete-Scute Family BHLH Transcription Factor 1), and HES6 (Hes Family BHLH Transcription Factor 6), suggesting they might be active NSCs. Moreover, some clusters expressed immature neuron markers, including *DCX (*doublecortin), *STMN1* (stathmin 1), *STMN2* (Stathmin 2), *PROX1* (Prospero Homeobox 1), *NNAT* (Neuronatin), and *SOX4* (SRY-Box Transcription Factor 4). There was significant variability in the number and state of activation between the analyzed four samples.

Wang et al., analyzed the properties of the neurogenic niche in the aged hippocampus and compared with aged macaques. They observed a close negative correlation between neurogenesis and inflammation in macaques. Similarly, the astrocytes in humans were enriched in genes related to the inflammatory response (e.g., the JAK/STAT pathway, NFKB1, and TNFSF4), confirming that astrocytes were activated during brain aging. Also, some astrocytes expressed genes related to regulating NSC quiescence (e.g., platelet-derived growth factors-A, -B, -C, milk-fat globule-epidermal growth factor 8, and the Ephrin family). The analysis of microglia revealed a cluster of active microglia that expressed CD74, Interferon Gamma Receptor 1, Toll-Like Receptor 2, and TNF Receptor Superfamily Member 1B as higher is levels of neuroinflammatory signals, such as Nuclear Factor Kappa B Subunit 1, Signal transducer and activator of transcription 3 and 5B, as well as the inflammation score, as more hinder hippocampal neurogenesis. Since inflammation is a common pathological finding in many human disorders, that might explain why many different syndromes can be associated with impaired neurogenesis.

Microglial cells are closely associated with NSC in the neurogenic niche and engaged in bidirectional communication with neurons. Microglial and neuronal crosstalk is partly mediated by fractalkine (CX3CL1) and its receptor CX3CR1 signaling, and disruption in this pathway has been associated with impaired neurogenesis (de Miranda et al., 2017). Neurons and astrocytes constitutively express CX3CL1 as a transmembrane molecule that is proteolytically cleavable into a soluble chemoattractant form. Brain CX3CR1 expression is highest in microglia, although it is also expressed by some infiltrating peripheral immune cells. CX3CL1 signaling modulates immune activity, demonstrating broadly neuroprotective effects in the spinal cord and brain while promoting microglial phagocytosis of apoptotic cells. In other words, brain CX3CL1 signaling promotes a nonactivated housekeeping microglial state (Parkitny & Maletic-Savatic, 2021). In addition, CX3CL1 sets forth a survival advantage to in vitro cultured human NPCs in the absence of growth factors (Krathwohl & Kaiser, 2004).

The classic pathway of microglial activation is accompanied by increased production of proinflammatory cytokines IL-1, IL-6, and TNF-α, which have been shown to decrease the survival of young neurons (Cacci et al., 2005). On the contrary, an alternative way of microglial activation realized in some cases results in the synthesis and secretion of Il-10 and prostaglandin E2, as well as BDNF and glial-derived neurotrophic factor (GDNF), which contribute to the survival of new neurons (Ekdahl et al., 2009). Interestingly, a regional heterogeneity of microglial populations exists, depending on their ability to support the adult neurogenesis (Walton et al., 2006). In addition to a prominent effect on neurogenesis, activated microglia may influence the properties of developing neurons in the dentate gyrus by changing the structure of dendritic spines and enhancing the inhibitory inputs on newly generated synapses (Jakubs et al., 2008). Microglial phagocytosis is critical for normal adult neurogenesis, where most newborn cells undergo apoptosis and clearance, and a reduced capacity increases proinflammatory response.

32.6 Neurogenesis and Brain Repair

32.6.1 Stroke

Clusters of dividing NPCs lie adjacent to blood vessels, mainly capillaries, and arteries and veins are within the neural stem cell niche. The vasculature provides nutrients to NPCs and regulates their proliferation and migration. An abrupt reduction of cerebral blood flow, known as a stroke, results in hypoxia that triggers excitotoxicity, oxidative stress, and neuroinflammation. After the ischemic process, neural precursor cells present in the subventricular zone of the lateral ventricle and subgranular zone of the dentate gyrus proliferate and migrate toward the lesion, contributing to the brain repair (Chrostek et al., 2019; Marques et al., 2019). However, the neural differentiation of newborn cells is limited, and mostly astrocytes that form glial scars will fill damaged areas. To stimulate the repair and restore function, transplantation of NSC was proposed and tested in rodent models. Recently, similar studies were performed in the clinic. The human neural stem cell line CTX0E03, clonally derived from human fetal cortical neuroepithelial cells, was transplanted in stroke-affected patients over 40 years of age. In a phase I safety trial, 11 patients with stable, moderate-to-severe functional neurological impairments after ischemic stroke underwent intracerebral implantation of CTX0E03 cells an average of 30 months after stroke in doses of up to 20 million cells. No cell-related safety issues were observed up to 24 months post implantation, and modest neurological and functional improvements were observed in some patients. Continuing follow-up has identified no cell-related safety issues up to 8 years post implantation. Functionally relevant improvement in arm movement was observed in patients with residual upper limb motor function at baseline (Muir et al., 2020).

32.6.2 Traumatic Brain Injury (TBI)

In experimental settings in rodent models, TBI increases hippocampal neurogenesis, and newborn neurons contribute to functional recovery. However, this process is not normal, and overall, there is reduced proliferation capacity of neuroblasts, aberrant migration, and sprouting. As newborn neurons mature into excitatory neurons, aberrant migration and formation of inappropriate synaptic connections may be a part of the mechanisms leading to seizure susceptibility in animals following the TBI (Redell et al., 2020). With advances in medical care, more children and adolescents are living with TBI-related cognitive and emotional impairments that negatively affect the quality of their lives (Rizk et al., 2021). Neuronal loss after TBI is a consequence of acute inflammatory responses which persist from weeks to months. Apoptosis and necrosis of neuronal cells, along with long-term neuroinflammation, increase the risk of neuronal dysfunction, including cognitive impairment and motor dysfunction after TBI. Intrinsic neurogenesis does not compensate for lost function, and a proinflammatory state has a negative impact on neurogenesis during recovery. Exogenous human neural stem cells could have the potential for therapy after acute TBI and chronic post-TBI conditions (Ahmed et al., 2016). However, their survival and engagement in the neuronal network are not efficient. There are some other experimental approaches to compensate for neuronal loss already tested on mice. For example, transplantation of human embryonic stem cell-derived cerebral organoids in a mild TBI model in mice (Kim et al., 2022) showed reduced neuronal cell death, recovered microvessel density, and promoted neurogenesis in the ipsilateral subventricular zone and dentate gyrus of the hippocampus. In addition, increased angiogenesis into the engrafted organoid was observed, and higher expression of early brain development proteins was associated with neurogenesis, angiogenesis, and extracellular matrix remodeling. Ultimately, organoid transplantation resulted in the reconstruction of the damaged cortex and improved cognitive function after TBI (Bao et al., 2021; Lin et al., 2018).

32.6.3 Pathologic Conditions

A wide variety of pathological conditions are associated with altered neurogenesis. Current evidence indicates that adult hippocampal neurogenesis may not be a significant contributor to the development of depression but may be required for some of the behavioral effects of antidepressants (Eisch & Petrik, 2012). Synaptogenesis and re-organization or re-integration of new neurons rather than a simple addition of new neurons may underlie the role of antidepressant drugs in reversing some but not all symptoms of depression. The importance of the neurogenesis hypothesis of depression and antidepressant action lies in stimulating further research into the possible roles played by the new neurons and synapses generated (Tang et al., 2012). Moreover, neurogenesis in the ventral dentate gyrus may be preferentially involved in emotion regulation (Sahay & Hen, 2007; Tanti & Belzung, 2013).

Due to the clinical symptoms that often accompany Alzheimer's disease (AD, dementia, memory loss, etc.), and since the hippocampus is one of the most significant neurogenic niches, many studies have examined the hippocampus to investigate how alterations in neurogenesis are implicated in AD. Indeed, decreased neurogenesis correlates with cognitive decline and compromised long-term memory. Interestingly, neuron loss preferentially in the hippocampus is one of the most important pathological findings in AD (Essa et al., 2022; Jin et al., 2004). Recently, using stereological cell counts, Moreno-Jimenez et al., demonstrated a progressive decline in the number of doublecortin-positive newborn neurons in the dentate gyrus of persons with Alzheimer's disease as compared to age-matched healthy controls. Interestingly, this reduction appeared prior to the appearance of neurofibrillary tangles and Aß deposition in the hippocampus (Moreno-Jiménez et al., 2019).

The accumulation of aberrant alpha-synuclein is one of the features of Parkinson's disease (PD) (Salmina et al., 2021). Findings on dopamine-stimulated hippocampal and striatal neurogenesis suggest that PD-associated impairment of neurogenesis might have links to insufficiency of dopaminergic mechanisms of neurogenesis regulation (L'Episcopo et al., 2018; Le Grand et al., 2015; Marchetti et al., 2020).

Viral invasion of NSCs can negatively affect neurogenesis and synaptogenesis, with consequent cell death, impairment of cell cycle progression, and early differentiation, which cause neural progenitors depletion in the cortical layer of the brain. A group of congenital infections that can be transmitted from the mother to the fetus directly involved in the development of the congenital disease: Toxoplasma, Rubella, Cytomegalovirus, Herpes simplex 1 and 2, and Others (Chlamydia, HIV, Coxsackievirus, Syphilis, Hepatitis B, Chickenpox, and ZIKV). Although viral transmission during the third trimester of pregnancy has a reduced impact on the developing fetus, infection during the first trimester is extremely disruptive, with severe congenital neurological defects in the developing fetus, which include microcephaly, cognitive and intellectual disabilities, sensorineural hearing loss, and blindness. Evidence suggests that NSCs are directly affected by viral infections, which lead to developmental defects in the cerebral cortex mainly by interfering with their differentiation into mature neural cells (Ihunwo et al., 2022).

Exposure to air pollution—and particularly to particulate matter (PM)—is strongly associated with a higher risk of neurodevelopmental disorders, poor mental health, and cognitive defects. In animal models, disruption of CNS development and disturbances of adult neurogenesis contribute to PM neurotoxicity. Recent studies show that gestational PM exposure not only affects embryonic neurodevelopment but also disturbs postnatal brain growth and maturation by interfering with neurogenic/gliogenic events, myelination, and synaptogenesis (Boda et al., 2020).

32.7 Pharmacological Corrections

In depression, it has been postulated that antidepressants, in addition to directly increasing monoamine levels in synapses, may also improve depression indirectly through increasing hippocampal neurogenesis (the neurogenic hypothesis) (Eisch & Petrik, 2012). Supporting evidence includes data showing that antidepressant efficacy can be blocked by radiation-induced ablation of hippocampal progenitor cell proliferation and triggering the apoptosis of newborn neurons (Seo et al., 2015). Much research is needed to support such a hypothesis.

Several pieces of evidence have shown that metformin, widely used for treating type 2 diabetes, can improve cognition by promoting adult hippocampal neurogenesis in mice with metabolic disorders. Metformin might show a positive effect on the maintenance of neural progenitors. BDNF, as a critical neurotrophic factor, is reported to promote neurogenesis. Stimulation of AMPK-CREB-BDNF pathway is a possible mechanism by which metformin correlated with adult neurogenesis. In vitro evidence showed that metformin promotes neuronal differentiation. It was found that metformin can promote adult NPC proliferation and self-renewal by regulating p53 family members and transcription factor TAp73. A recent study demonstrated that Gadd45g, a novel antidepressant target, mediates the promotion of metformin-induced neuronal differentiation by regulating the DNA demethylation (Cao et al., 2022).

Allopregnanolone (Allo), a neurosteroid, has emerged as a promising promoter of endogenous regeneration in the brain. In a mouse model of Alzheimer's disease, Allo induced neurogenesis, oligodendrogenesis, white matter generation, and cholesterol homeostasis while simultaneously reducing beta-amyloid and neuroinflammatory burden. Allo activates signaling pathways and gene expression required for the regeneration of neural stem cells and their differentiation into neurons (Irwin et al., 2014).

32.8 Conclusion

In conclusion, a significant amount of research has focused on elucidating adult neurogenesis. The presence of neuronal stem cells in the adult brain is widely accepted. However, the role of neurogenesis in pathologic conditions and involvement in mechanistic causes of neurodegeneration remains minimal in humans. Impaired neurogenesis needs further investigation to understand whether targeting neurogenesis-related pathways could serve as a viable therapeutic approach. The major obstacles that need addressing are localized neurotrophic factor expression and the formation of a microenvironment conducive to generation, maturation, and axonal regeneration. Today, it is difficult to consider the healing ability of neurons as a potential therapy method. Moreover, as the decline of neurogenesis cannot currently be reversed, researchers must focus on identifying a panel of biomarkers to facilitate early intervention and thereby allow more time to protect neurogenesis.

32.9 Review Questions

1. Where are the major areas of new neurons generation in the adult human brain? Select all that applies.
   (a) Subventricular zone
   (b) The dentate gyrus of the hippocampus
   (c) Neocortex
   (d) All the above

2. How newborn neurons and their progenies could be identified?
   (a) By the detection of dividing cells with the use of thymidine analog BrdU
   (b) By staining for proliferation markers like Ki-67
   (c) By staining for newborn neuron-specific proteins
   (d) By a combination of A & C
   (e) By a combination of B & C
3. How can neural stem cells be isolated from the adult brain and their stemness confirmed?
   (a) By dissociation of the brain tissue sample, staining for stem cell markers with fluorescent antibodies, and sorting of positive cells by flow cytometry
   (b) By survival in culture, dividing and forming spheres, which can be dissociated and generate new spheres
   (c) By transplantation of selected cells into the brain of newborn immune-deficient mice and confirmation of the generation of new neurons, astrocytes, and oligodendrocytes of human origin
   (d) All the above
4. What are stage-specific molecular markers for stem cells and newborn neurons? What is the order in which they appear during maturation? Put a number in appropriate order.
   (a) NeuN, Map-2, post-mitotic neuronal markers
   (b) Nestin and GFAP, radial glia, Type 1 cells
   (c) CD133, prominin-1, not-committed precursors
   (d) Doublecortin (DCX) or polysialylated neural cell adhesion molecule (PSA-NCAM), intermediate progenitor cells, Type 2 cells
   (e) DCX, PSA-NCAM, Tuj-1b, and TUC-4 or NeuroD, neuronal lineage committed neuroblasts, Type 3 cells
   (f) All the above
5. How are microglia involved in the regulation of adult neurogenesis?
   (a) Secretion of cytokines
   (b) Elimination of nonintegrated newborn neurons
   (c) Secretion of growth factors
   (d) All the above

32.10 Answers

1. a and b
2. d
3. d
4. c, b, d, e, a
5. d

References

Ahmed, A. I., Gajavelli, S., Spurlock, M. S., et al. (2016). Stem cells for therapy in TBI. *Journal of the Royal Army Medical Corps, 162*(2), 98–102. https://doi.org/10.1136/jramc-2015-000475

Altman, J., & Das, G. D. (1966). Autoradiographic and histological studies of postnatal neurogenesis. I. A longitudinal investigation of the kinetics, migration and transformation of cells incorporating tritiated thymidine in neonate rats, with special reference to postnatal neurogenesis in some brain regions. *The Journal of Comparative Neurology, 126*(3), 337–389. https://doi.org/10.1002/cne.901260302

Altman, J., & Das, G. D. (1967). Postnatal neurogenesis in the Guinea-pig. *Nature, 214*(5093), 1098–1101. https://doi.org/10.1038/2141098a0

Bao, Z., Fang, K., Miao, Z., Li, C., et al. (2021). Human cerebral organoid implantation alleviated the neurological deficits of traumatic brain injury in mice. *Oxidative Medicine and Cellular Longevity, 2021*, 6338722. https://doi.org/10.1155/2021/6338722

Boda, E., Rigamonti, A. E., & Bollati, V. (2020). Understanding the effects of air pollution on neurogenesis and gliogenesis in the growing and adult brain. *Current Opinion in Pharmacology, 50*, 61–66. https://doi.org/10.1016/j.coph.2019.12.003

Brooker, S. M., Bond, A. M., Peng, C. Y., et al. (2016). β1-integrin restricts astrocytic differentiation of adult hippocampal neural stem cells. *Glia, 64*(7), 1235–1251.

Cacci, E., Claasen, J. H., & Kokaia, Z. (2005). Microglia-derived tumor necrosis factor-alpha exaggerates death of newborn hippocampal progenitor cells in vitro. *Journal of Neuroscience Research, 80*(6), 789–797. https://doi.org/10.1002/jnr.20531

Cajal, S. (1894). The Croonian lecture: The fine structure of the nerve centres. *Proceedings of the Royal Society of London, 55*, 444–468.

Cao, G., Gong, T., Du, Y., et al. (2022). Mechanism of metformin regulation in central nervous system: Progression and future perspectives. *Biomedicine & Pharmacotherapy, 156*, 113686. https://doi.org/10.1016/j.biopha.2022.113686

Chrostek, M. R., Fellows, E. G., Crane, A. T., et al. (2019). Efficacy of stem cell-based therapies for stroke. *Brain Research, 1722*, 146362. https://doi.org/10.1016/j.brainres.2019.146362

de Miranda, A. S., Zhang, C. J., Katsumoto, A., et al. (2017). Hippocampal adult neurogenesis: Does the immune system matter? *Journal of the Neurological Sciences, 372*, 482–495. https://doi.org/10.1016/j.jns.2016.10.052

Duque, A., Arellano, J. I., & Rakic, P. (2022). An assessment of the existence of adult neurogenesis in humans and value of its rodent models for neuropsychiatric diseases. *Molecular Psychiatry, 27*(1), 377–382. https://doi.org/10.1038/s41380-021-01314-8

Eisch, A. J., & Petrik, D. (2012). Depression and hippocampal neurogenesis: A road to remission? *Science, 338*(6103), 72–75. https://doi.org/10.1126/science.1222941

Ekdahl, C. T., Kokaia, Z., & Lindvall, O. (2009). Brain inflammation and adult neurogenesis: The dual role of microglia. *Neuroscience, 158*(3), 1021–1029. https://doi.org/10.1016/j.neuroscience.2008.06.052

Eriksson, P. S., Perfilieva, E., Björk-Eriksson, T., Alborn, A. M., Nordborg, C., Peterson, D. A., & Gage, F. H. (1998). Neurogenesis in the adult human hippocampus. *Nature Medicine, 4*(11), 1313–1317. https://doi.org/10.1038/3305

Essa, H., Peyton, L., Hasan, W., et al. (2022). Implication of adult hippocampal neurogenesis in Alzheimer's disease and potential therapeutic approaches. *Cells, 11*(2). https://doi.org/10.3390/cells11020286

Gage, F. H., & Temple, S. (2013). Neural stem cells: Generating and regenerating the brain. *Neuron, 80*(3), 588–601. https://doi.org/10.1016/j.neuron.2013.10.037

Goldman, S. A., & Nottebohm, F. (1983). Neuronal production, migration, and differentiation in a vocal control nucleus of the adult female canary brain. *Proceedings of the National Academy of Sciences of the United States of America, 80*(8), 2390–2394. https://doi.org/10.1073/pnas.80.8.2390

Gould, E., Reeves, A. J., Graziano, M. S., et al. (1999). Neurogenesis in the neocortex of adult primates. *Science, 286*(5439), 548–552. https://doi.org/10.1126/science.286.5439.548

Ihunwo, A. O., Perego, J., Martino, G., et al. (2022). Neurogenesis and viral infection. *Frontiers in Immunology, 13*, 826091. https://doi.org/10.3389/fimmu.2022.826091

Irwin, R. W., Solinsky, C. M., & Brinton, R. D. (2014). Frontiers in therapeutic development of allopregnanolone for Alzheimer's disease and other neurological disorders. *Frontiers in Cellular Neuroscience, 8*, 203. https://doi.org/10.3389/fncel.2014.00203

Jakubs, K., Bonde, S., Iosif, R. E., et al. (2008). Inflammation regulates functional integration of neurons born in adult brain. *The Journal of Neuroscience, 28*(47), 12477–12488. https://doi.org/10.1523/jneurosci.3240-08.2008

Jin, K., Peel, A. L., Mao, X. O., et al. (2004). Increased hippocampal neurogenesis in Alzheimer's disease. *Proceedings of the National Academy of Sciences of the United States of America, 101*(1), 343–347. https://doi.org/10.1073/pnas.2634794100

Kempermann, G., & Gage, F. H. (1999). Experience-dependent regulation of adult hippocampal neurogenesis: Effects of long-term stimulation and stimulus withdrawal. *Hippocampus, 9*(3), 321–332. https://doi.org/10.1002/(SICI)1098-1063(1999)9:3<321::AID-HIPO11>3.0.CO;2-C

Kempermann, G., Kuhn, H. G., & Gage, F. H. (1997). More hippocampal neurons in adult mice living in an enriched environment. *Nature, 386*(6624), 493–495. https://doi.org/10.1038/386493a0

Kempermann, G., Gast, D., & Gage, F. H. (2002). Neuroplasticity in old age: Sustained fivefold induction of hippocampal neurogenesis by long-term environmental enrichment. *Annals of Neurology, 52*(2), 135–143. https://doi.org/10.1002/ana.10262

Kempermann, G., Gage, F. H., Aigner, L., et al. (2018). Human adult neurogenesis: Evidence and remaining questions. *Cell Stem Cell, 23*(1), 25–30. https://doi.org/10.1016/j.stem.2018.04.004

Kim, J. T., Kim, T. Y., Youn, D. H., et al. (2022). Human embryonic stem cell-derived cerebral organoids for treatment of mild traumatic brain injury in a mouse model. *Biochemical and Biophysical Research Communications, 635*, 169–178. https://doi.org/10.1016/j.bbrc.2022.10.045

Krathwohl, M. D., & Kaiser, J. L. (2004). Chemokines promote quiescence and survival of human neural progenitor cells. *Stem Cells, 22*(1), 109–118. https://doi.org/10.1634/stemcells.22-1-109

La Rosa, C., Parolisi, R., & Bonfanti, L. (2020). Brain structural plasticity: From adult neurogenesis to immature neurons. *Frontiers in Neuroscience, 14*, 75. https://doi.org/10.3389/fnins.2020.00075

Le Grand, J. N., Gonzalez-Cano, L., Pavlou, M. A., et al. (2015). Neural stem cells in Parkinson's disease: A role for neurogenesis defects in onset and progression. *Cellular and Molecular Life Sciences, 72*(4), 773–797. https://doi.org/10.1007/s00018-014-1774-1

L'Episcopo, F., Tirolo, C., Serapide, M. F., et al. (2018). Microglia polarization, gene-environment interactions and Wnt/beta-catenin signaling: Emerging roles of glia-neuron and glia-stem/neuroprogenitor crosstalk for dopaminergic neurorestoration in aged parkinsonian brain. *Frontiers in Aging Neuroscience, 10*, 12. https://doi.org/10.3389/fnagi.2018.00012

Lin, G. Q., He, X. F., Liang, F. Y., et al. (2018). Transplanted human neural precursor cells integrate into the host neural circuit and ameliorate neurological deficits in a mouse model of traumatic brain injury. *Neuroscience Letters, 674*, 11–17. https://doi.org/10.1016/j.neulet.2018.02.064

Marchetti, B., Tirolo, C., L'Episcopo, F., et al. (2020). Parkinson's disease, aging and adult neurogenesis: Wnt/beta-catenin signalling as the key to unlock the mystery of endogenous brain repair. *Aging Cell, 19*(3), e13101. https://doi.org/10.1111/acel.13101

Marin-Burgin, A., & Schinder, A. F. (2012). Requirement of adult-born neurons for hippocampus-dependent learning. *Behavioural Brain Research, 227*(2), 391–399. https://doi.org/10.1016/j.bbr.2011.07.001

Marques, B. L., Carvalho, G. A., Freitas, E. M. M., et al. (2019). The role of neurogenesis in neurorepair after ischemic stroke. *Seminars in Cell & Developmental Biology, 95*, 98–110. https://doi.org/10.1016/j.semcdb.2018.12.003

Matsubara, S., Matsuda, T., & Nakashima, K. (2021). Regulation of adult mammalian neural stem cells and neurogenesis by cell extrinsic and intrinsic factors. *Cells, 10*(5), 1145. Retrieved from https://www.mdpi.com/2073-4409/10/5/1145

Mendez-David, I., Schofield, R., Tritschler, L., et al. (2022). Reviving through human hippocampal newborn neurons. *L'Encéphale, 48*(2), 179–187. https://doi.org/10.1016/j.encep.2021.09.001

Miller, M. W., & Nowakowski, R. S. (1988). Use of bromodeoxyuridine-immunohistochemistry to examine the proliferation, migration and time of origin of cells in the central nervous system. *Brain Research, 457*(1), 44–52. https://doi.org/10.1016/0006-8993(88)90055-8

Moreno-Jiménez, E. P., Flor-García, M., Terreros-Roncal, J., et al. (2019). Adult hippocampal neurogenesis is abundant in neurologically healthy subjects and drops sharply in patients with Alzheimer's disease. *Nature Medicine, 25*(4), 554–560. https://doi.org/10.1038/s41591-019-0375-9

Muir, K. W., Bulters, D., Willmot, M., et al. (2020). Intracerebral implantation of human neural stem cells and motor recovery after stroke: Multicentre prospective single-arm study (PISCES-2). *Journal of Neurology, Neurosurgery & Psychiatry, 91*(4), 396. https://doi.org/10.1136/jnnp-2019-322515

Parkitny, L., & Maletic-Savatic, M. (2021). Glial PAMPering and DAMPening of adult hippocampal neurogenesis. *Brain Sciences, 11*(10). https://doi.org/10.3390/brainsci11101299

Perez-Dominguez, M., Tovar, Y. R. L. B., & Zepeda, A. (2018). Neuroinflammation and physical exercise as modulators of adult hippocampal neural precursor cell behavior. *Reviews in the Neurosciences, 29*(1), 1–20. https://doi.org/10.1515/revneuro-2017-0024

Pursglove, S. E., & Mackay, J. P. (2005). CSL: A notch above the rest. *The International Journal of Biochemistry & Cell Biology, 37*(12), 2472–2477. https://doi.org/10.1016/j.biocel.2005.06.013

Qu, Q., Sun, G., Murai, K., et al. (2013). Wnt7a regulates multiple steps of neurogenesis. *Molecular and Cellular Biology, 33*(13), 2551–2559. https://doi.org/10.1128/MCB.00325-13

Redell, J. B., Maynard, M. E., Underwood, E. L., et al. (2020). Traumatic brain injury and hippocampal neurogenesis: Functional implications. *Experimental Neurology, 331*, 113372. https://doi.org/10.1016/j.expneurol.2020.113372

Reynolds, B. A., & Weiss, S. (1992). Generation of neurons and astrocytes from isolated cells of the adult mammalian central nervous system. *Science, 255*(5052), 1707–1710. https://doi.org/10.1126/science.1553558

Rizk, M., Vu, J., & Zhang, Z. (2021). Impact of pediatric traumatic brain injury on hippocampal neurogenesis. *Neural Regeneration Research, 16*(5), 926–933. https://doi.org/10.4103/1673-5374.297057

Sahay, A., & Hen, R. (2007). Adult hippocampal neurogenesis in depression. *Nature Neuroscience, 10*(9), 1110–1115. https://doi.org/10.1038/nn1969

Salmina, A. B., Kapkaeva, M. R., Vetchinova, A. S., et al. (2021). Novel approaches used to examine and control neurogenesis in Parkinson's disease. *International Journal of Molecular Sciences, 22*(17). https://doi.org/10.3390/ijms22179608

Seo, D. O., Carillo, M. A., Chih-Hsiung Lim, S., et al. (2015). Adult hippocampal neurogenesis modulates fear learning through associative and nonassociative mechanisms. *The Journal of Neuroscience, 35*(32), 11330–11345. https://doi.org/10.1523/jneurosci.0483-15.2015

Shingo, T., Gregg, C., Enwere, E., et al. (2003). Pregnancy-stimulated neurogenesis in the adult female forebrain mediated by prolactin. *Science, 299*(5603), 117–120. https://doi.org/10.1126/science.1076647

Song, J., Zhong, C., Bonaguidi, M. A., et al. (2012). Neuronal circuitry mechanism regulating adult quiescent neural stem-cell fate decision. *Nature, 489*(7414), 150–154.

Sorrells, S. F., Paredes, M. F., Cebrian-Silla, A., et al. (2018). Human hippocampal neurogenesis drops sharply in children to undetectable levels in adults. *Nature, 555*(7696), 377–381. https://doi.org/10.1038/nature25975

Tamaki, S., Eckert, K., He, D., et al. (2002). Engraftment of sorted/expanded human central nervous system stem cells from fetal brain. *Journal of Neuroscience Research, 69*(6), 976–986. https://doi.org/10.1002/jnr.10412

Tang, S. W., Helmeste, D., & Leonard, B. (2012). Is neurogenesis relevant in depression and in the mechanism of antidepressant drug action? A critical review. *The World Journal of Biological Psychiatry, 13*(6), 402–412. https://doi.org/10.3109/15622975.2011.639800

Tanti, A., & Belzung, C. (2013). Neurogenesis along the septo-temporal axis of the hippocampus: Are depression and the action of antidepressants region-specific? *Neuroscience, 252*, 234–252. https://doi.org/10.1016/j.neuroscience.2013.08.017

Tuncdemir, S. N., Lacefield, C. O., & Hen, R. (2019). Contributions of adult neurogenesis to dentate gyrus network activity and computations. *Behavioural Brain Research, 374*, 112112. https://doi.org/10.1016/j.bbr.2019.112112

Uchida, N., Buck, D. W., He, D., et al. (2000). Direct isolation of human central nervous system stem cells. *Proceedings of the National Academy of Sciences of the United States of America, 97*(26), 14720–14725. https://doi.org/10.1073/pnas.97.26.14720

van Praag, H., Kempermann, G., & Gage, F. H. (1999). Running increases cell proliferation and neurogenesis in the adult mouse dentate gyrus. *Nature Neuroscience, 2*(3), 266–270. https://doi.org/10.1038/6368

Walton, N. M., Sutter, B. M., Laywell, E. D., et al. (2006). Microglia instruct subventricular zone neurogenesis. *Glia, 54*(8), 815–825. https://doi.org/10.1002/glia.20419

Wang, W., Wang, M., Yang, M., et al. (2022). Transcriptome dynamics of hippocampal neurogenesis in macaques across the lifespan and aged humans. *Cell Research, 32*(8), 729–743. https://doi.org/10.1038/s41422-022-00678-y

Welberg, L. (2014). Learning and memory: Neurogenesis erases existing memories. *Nature Reviews. Neuroscience, 15*(7), 428. https://doi.org/10.1038/nrn3768

Yin, A. H., Miraglia, S., Zanjani, E. D., et al. (1997). AC133, a novel marker for human hematopoietic stem and progenitor cells. *Blood, 90*(12), 5002–5012.

Zhou, Y., Bond, A. M., Shade, J. E., et al. (2018). Autocrine Mfge8 signaling prevents developmental exhaustion of the adult neural stem cell pool. *Cell Stem Cell, 23*(3), 444–452.e444. https://doi.org/10.1016/j.stem.2018.08.005

33 Drugs of Abuse

Toby K. Eisenstein and Thomas J. Rogers

Abstract

Opioids and cannabinoids have been shown to modulate many different parameters of immune responses. In general, opioids and cannabinoids have been found to be immunosuppressive, both in vitro and in vivo in a variety of assays, including the response of lymphocytes to mitogens, antibody formation, activation of T cells, natural killer cell activity, and phagocytosis by macrophages and neutrophils. Drugs in these two classes have also been shown to inhibit the migration of cells from the circulatory system to areas of inflammation, including passage across the blood-brain barrier. Opioids and cannabinoids have both been shown to sensitize mice to a variety of infectious agents, which is commensurate with an immunosuppressed state. In specialized rodent models and monkeys, opioids sensitized to HIV and SIV, possibly related to their ability to cause heterologous desensitization of chemokine co-receptors for these viruses. The literature covered in this chapter supports the existence of a robust neural-immune interaction.

Keywords

Cannabinoids · Chemokines · Cocaine · Cytokines · Immunosuppression · Interleukins · Mitogen · Morphine · Natural killer cell · Opioids · Th1/Th2 · Δ^9-THC · Tolerance · Withdrawal

T. K. Eisenstein (deceased)
Center for Substance Abuse Research, Department of Microbiology, Immunology and Inflammation, Department of Neural Sciences, Lewis Katz School of Medicine at Temple University, Philadelphia, PA, USA
e-mail: tke@temple.edu

T. J. Rogers (✉)
Center for Inflammation and Lung Research, Center for Substance Abuse Research, Lewis Katz School of Medicine at Temple University, Philadelphia, PA, USA
e-mail: rogerst@temple.edu

33.1 Introduction

33.1.1 Opioid and Cannabinoid Receptors Are on Cells of the Immune System

33.1.1.1 Opioid Receptors

There is now indisputable evidence of natural physiological connections between the neural and immune systems. Much of the evidence supporting this link has come from studies of opioids and cannabinoids and their effects on immune responses. Opioid receptors were first demonstrated biochemically in the brain in the early 1970s. Three distinct receptors were discovered in neural tissue and were designated mu, kappa, and delta (Simon, 1991). Their respective natural ligands were found to be the neuropeptides, β-endorphin (predominantly mu), dynorphin (kappa), and methionine-enkephalin (delta) (Akil et al., 1984), although there is considerable cross-reactivity. Alkaloid opioids, including morphine and heroin, bind primarily, but not exclusively, to the mu receptor. Morphine is extracted from opium produced by the poppy plant, and heroin is synthetically diacetylated morphine. In vivo, the major metabolite of heroin is morphine. Most laboratory studies employ morphine because it is less lipophilic than heroin, is used therapeutically, and is easier to dissolve. Since the mu receptor is the major target of both compounds, results obtained with morphine are believed to be directly translatable to heroin abuse, although there is some evidence for unique biologically active heroin metabolites (Rossi et al., 1996).

Credit for recognition of the potential existence of receptors for opioids on cells of the immune system is given to Joseph Wybran in the laboratory of Govaerts. In 1979 he reported modulation of the function of human T cells, purified from normal peripheral blood, by exogenous and endogenous opioids (Wybran et al., 1979). He observed that if morphine was added to human T cells in culture, they lost the ability to rosette with sheep red blood cells (SRBCs), a fortuitous property that was used before flow cytometry to enumerate T cells. In contrast, when the opioid neuropeptide,

H. E. Gendelman, T. Ikezu (eds.), *Neuroimmune Pharmacology and Therapeutics*,
https://doi.org/10.1007/978-3-031-68237-7_33

met-enkephalin, was added to the T cells, rosetting increased. The actions of both opioids were blocked by naloxone, an antagonist for all types of opioid receptors, showing the pharmacological specificity of the effects. Wybran proposed that "normal human blood T lymphocytes bear surface receptor-like structures for morphine and methionine-enkephalin." He concluded that "Such findings may provide a link between the central nervous system and the immune system." This paper was seminal because it presented strong pharmacologic data to support the existence of opioid receptors on cells of the immune system. However, it has been difficult to demonstrate opioid receptors routinely by binding studies using cells of the immune system (Sibinga & Goldstein, 1988; Sharp et al., 1998). The difficulties may be due to low receptor numbers on cells of the immune system, to inability to obtain enough purified subsets of immune cells to do adequate binding studies, or the need to activate immune cells before the receptors are expressed. Nevertheless, a number of studies have shown opioid receptor expression on cells of the immune system at both the mRNA and protein levels (see recent review (Machelska & Celik, 2020)). With the advent of cloning of the opioid receptors from brain tissue, mRNA transcripts (Machelska & Celik, 2020) have been detected in primary cells of the immune system or lymphoid or monocytic cell lines for all three classes of opioid receptors (Loh & Smith, 1990; Sharp et al., 1998; McCarthy et al., 2001).

33.1.1.2 Cannabinoid Receptors

Cannabinoid receptors were identified and cloned several years after the opioid receptors. Cannabinoids, like the psychotropic active ingredient extracted from marijuana, Δ^9-tetrahydrocannabinol (Δ^9-THC), were initially believed by some to simply interdigitate into cell membranes. Subsequently, a cannabinoid receptor was cloned from neural tissue (Zhu et al., 1993). When lymphoid tissue was probed for cannabinoid receptor transcripts, a second receptor (designated CB2) was discovered that is primarily expressed in the immune system and has only 44% homology to the first receptor (designated CB1) (Munro et al., 1993). Cells of the immune system express predominantly CB2 receptors, although mRNA for CB1 has also been detected. The brain expresses predominantly CB1 receptors, although there are a few reports of CB2 expression and functional activity. mRNA for CB2 is expressed by B-cells > natural killer cells > monocytes > neutrophils > T cells in rat, mouse, and human tissues and cells (Bouaboula et al., 1993; Galiegue et al., 1995). Endogenous cannabinoid ligands have also been discovered, but they are not peptides like endogenous opioids. Anandamide, or arachidonoylethanolamide, is a polyunsaturated N-acylethanolamine (Devane et al., 1992), as is 2-arachidonoyl-glycerol (Mechoulam et al., 1995; Di Marzo & Deutsch, 1998). Radiolabeled cannabinoids have been shown to bind to mouse spleen cells and human T-cell lines (Kaminski et al., 1994; Schatz et al., 1997). Δ^9-THC and endogenous cannabinoids all bind to both CB1 and CB2 receptors. However, it is becoming increasingly apparent, using selective CB1 and CB2 antagonists, as well as cannabinoid receptor knock-out mice, that most of the immunomodulating effects of these compounds on the immune system are mediated via the CB2 receptor.

33.1.1.3 Summary

It is now widely accepted that immune cells express receptors for opioids and cannabinoids. This conclusion is based on an impressive body of literature demonstrating immune modulation by exogenous and endogenous opioids and cannabinoids. Evidence from in vitro experiments suggests that the effects of these drugs on the cells of the immune system occur, at least partially, by binding opioid and cannabinoid receptors on these cells. Cocaine has also been found to affect immune function, but the mechanisms have not been defined. Other commonly used and abused drugs include nicotine and alcohol, both of which have marked effects on the immune system, and are frequently part of the panoply of artificial substances used by addicts. In addition, methamphetamines can alter immune status. Due to space limitations, the reader is referred to key papers on these subjects (Sopori, 2002; Molina et al., 2010).

33.1.2 Mechanisms by Which Drugs of Abuse Can Affect Cells of the Immune System

Testing of whether drugs of abuse can alter immune function has been carried out by in vivo administration of the drug, with the assessment of in vivo or ex vivo immune function, or by harvesting immune cells from a drug naïve (normal) host and exposing them to the drug in vitro. If a drug is given in vivo, there is the possibility that effects on the immune system are not due to a direct interaction of the drug with cells of the immune system, but that immunomodulatory effects are instead mediated by additional systems in the body. For example, opioids, cannabinoids, and other drugs of abuse can modulate the Hypothalamic-Pituitary-Adrenal (HPA) axis, which can lead to the release of immunosuppressive corticosteroids or activation of the sympathetic nervous (SNS) system, resulting in downstream alterations in immune function. Effects on immune function could thus be a combination of indirect and direct effects if the drug is given in vivo. An additional and important consideration is that both the neural system and cells of the immune system produce opioid peptides, which could lead to complicated interacting pathways, especially in vivo (Carr, 1991; Sharp & Yaksh, 1997). In evaluating the studies discussed in this chapter, it should be recognized that most immune responses were assessed 2–3 h after a single injection of a drug. For studies testing longer exposure to morphine, the drug is frequently delivered by subcutaneous implantation of a pellet that slowly releases the drug over a period of a week. Alternatively, mini-pumps dispensing drugs can be implanted. In the case of morphine or other addictive opioids, these methods are used to avoid episodes of withdrawal, which have immunomodulatory effects on immune function. In fact, a complication of studies with opioids is that repeated administration can result in tolerance to their analgesic effects, and in many, but not all, cases tolerance to the effects of the drugs on the immune system (Eisenstein et al., 2006). There are a limited number of studies that have addressed the effects of tolerance and withdrawal from morphine on immune responses (Eisenstein et al., 2006).

33.2 Functional Consequences of Drugs of Abuse on the Immune System

33.2.1 Effects on Innate Immunity and the Inflammatory Response

33.2.1.1 Cells and Molecules of the Innate Immune Response

It is well established that drugs of abuse are involved in a wide range of immunomodulatory activities. Among the areas in immune system function in which these alterations have been observed is innate immunity. This arm of the immune system does not exhibit specificity in recognizing foreign pathogens. It encompasses natural killer (NK) cells and the "professional" phagocytic cells that are the first responders to inflammation or infection. Natural Killer cells are lymphocytes that are involved in the recognition and killing of mammalian cells that express foreign antigens, such as tumor antigens or viral antigens, or which fail to express histocompatibility antigens (MHC Class I) of the host. They also produce cytokines (small proteins produced by immune cells that act in an autocrine, paracrine, or endocrine fashion) that can activate the mononuclear phagocytic cells (monocytes and macrophages). The phagocytic cells engulf microbes invading the host and include neutrophils, monocytes, and macrophages. Monocytes are formed in the bone marrow and released into the peripheral blood. They are mature, but unactivated cells. When they are stimulated in various ways they can develop into macrophages in vivo or in vitro. Macrophages have several states of activation that are characterized by their production of cytokines and by their levels of microbicidal compounds, including reactive oxygen intermediates (ROIs) and reactive nitrogen intermediates (RNIs). Macrophages also play an important role in the acquisition and presentation of foreign antigens to lymphocytes, which triggers the adaptive immune response.

33.2.1.2 Effects of Opioids, Cannabinoids, and Cocaine on NK Cells

In the cellular repertoire of the innate immune system, NK cells have the unique ability to kill infected target cells in a manner independent of antigen recognition. Alteration in NK cell function was among the earliest and has been one of the most consistently observed, effects of drugs on the immune system. Many studies have shown that the administration of opioids to rodents results in the suppression of NK cell activity. Leibeskind's laboratory (Shavit et al., 1986b) was the first to show that morphine given to rats subcutaneously for 4 days suppressed NK cell activity. They later concluded that this effect was not directly on the NK cells in the periphery, because N-methylmorphine, which does not pass the blood-brain barrier, did not depress NK function (Shavit et al., 1986a). Subsequently, Weber and Pert (1989) reported that injection of morphine into the periaqueductal gray region of the brains of rats resulted in suppression of NK cell activity in the spleens of these animals 3 h later. Morphine administered into five other brain sites was without effect. Moreover, the opioid-specific antagonist, naltrexone, blocked the effects in the studies by both groups, implicating the involvement of classic opioid receptors in modulating NK cell activity. Studies have also been carried out in which morphine has been given to normal human subjects under controlled conditions. The results show that peripheral blood NK cell activity is depressed 24 h after morphine administration (Yeager et al., 1995). The exact mechanism by which the engagement of opioid receptors in certain regions of the brain alters the peripheral functional capacity of NK cells is not known. However, more recent work implicates a neural pathway involving neuropeptide Y in modulating NK activity (Saurer et al., 2006). Nonetheless, the mu-opioid receptor (MOR) is certainly involved in the effects on NK cell activity, as mice with a genetic deletion of MOR do not respond to morphine with depressed NK cell activity (Gavériaux-Ruff et al., 1998). Thus, morphine has been shown to depress NK cell activity in rats, humans, and mice.

The cannabinoids represent another class of abused drugs affecting immune cell function. Although fewer studies have explored the effects of cannabinoids on NK cells, the literature to date shows that the activity of splenic NK cells is suppressed following in vivo injection of Δ^9-THC into rats and mice (Patel et al., 1985; Klein et al., 1987), and also after in vitro treatment of human peripheral blood NK cells (Specter et al., 1986). Δ^9-THC has also been shown to block the proliferation of NK cells in response to the cytokine, IL-2, by downregulating the IL-2 receptor (Zhu et al., 1993).

There are a limited number of reports of immunomodulation by cocaine. In vivo, administration of cocaine i.v. to humans elevated NK cell activity in peripheral blood, but exposure of leukocytes to the drug in vitro had no effect (Van Dyke et al., 1986). It was hypothesized that the in vivo effects occurred via release of epinephrine from the SNS (Van Dyke et al., 1986). Contradictory results have been reported for the effect of cocaine on NK activity in rats, with no effect in vivo (Bayer et al., 1996) or suppression of activity after acute or chronic in vivo administration (Pacifici et al., 2003).

33.2.1.3 Drugs of Abuse Alter Functions of Neutrophils and Monocytes/Macrophages

Opioids given in vivo have been demonstrated to inhibit ex vivo phagocytic and microbicidal activity against the fungus, *Candida albicans* (Tubaro et al., 1983). Opioids selective for all three classes of opioid receptors (mu, kappa, and delta), when added to mouse peritoneal macrophages in vitro, inhibited phagocytosis of the yeast *Candida albicans*, and the effects were blocked by the respective receptor-selective antagonists (Szabo et al., 1993). Impairment of ex vivo phagocytosis of the bacterium *Streptococcus pneumoniae* by alveolar macrophages harvested from mice treated with morphine pellets, has also been reported (Wang et al., 2005). Others have shown that mu and delta agonists inhibited Fc-mediated phagocytosis of antibody-coated sheep red blood cells by mouse peritoneal cells exposed in vitro to morphine (Tomassini et al., 2004). These results showed that the activation of each of the opioid receptors resulted in reduced phagocytic activity. More recent studies have provided a mechanism by which phagocytosis is inhibited by morphine, namely through disruption of actin polymerization resulting from an increase in intracellular cAMP, activation of phosphokinase A, and inhibition of Rac1-GTPase, and p38 mitogen-activated protein kinase (MAPK) (Ninković & Roy, 2012). In addition, morphine was given in vivo by slow-release pellet, or when added to murine bone marrow cells in vitro, blocked the maturation of stem cells into more mature macrophages (Roy et al., 1991). Morphine pellets also inhibited the adhesion of leukocytes to vascular endothelium, thus reducing inflammation (Ni et al., 2000). Further work has shown that morphine also inhibited the ability of phagocytic cells to produce superoxide and peroxide (Peterson et al., 1987a), two metabolites that have strong antimicrobial activity. Peterson and colleagues (Chao et al., 1992) have shown that the inhibitory effects of morphine on phagocytic cell activity are dependent on the production of Transforming Growth Factor (TGF)-β. Reactive oxygen intermediates (ROI), such as superoxide, contribute to the microbicidal activity of phagocytes, so impairment of their production would be consonant with reduced capacity to kill ingested yeast or other microbes. Macrophages taken from mice exposed to morphine can undergo apoptosis (Nair et al., 1997; Yin et al., 1999), and evidence supports the role of ROI and nitric oxide (NO) in this process (Singhal et al., 1998; Bhat et al., 2004). Usually, depression of NO is associated with decreased macrophage function, so these findings are dichotomous with the other evidence on the effects of opioids on these cells.

Morphine treatment has been reported to result in reduced recruitment of both neutrophils and macrophages to growing tumors, using a murine Lewis Lung Carcinoma model system (Koodie et al., 2014). Chronic morphine administration has also been shown to inhibit the migration of neutrophils and monocytes which are a part of the process of wound healing (Clark et al., 2007; Martin et al., 2010a, b). The attenuated recruitment of neutrophils was attributed to morphine-induced inhibition of the expression of the chemokines CXCL3 and CCL2 (Martin et al., 2010a).

Δ^9-THC is also reported to decrease the phagocytic activity of human peripheral blood monocytes and mouse peritoneal macrophages (Lopez-Cepero et al., 1986; Specter et al., 1991a). Long-term use of Δ^9-THC in nonhuman primates has also been associated with depressed macrophage function (Cabral et al., 1991). Alveolar macrophages of marijuana smokers exhibited decreased microbicidal activity against *Staphylococcus aureus*, which was associated with decreased production of NO (Roth et al., 2004). Δ^9-THC also affects macrophage processing of exogenous protein antigens through the CB2 receptor, as shown pharmacologically and using CB2-deficient mice (McCoy et al., 1999; Buckley et al., 2000).

Cocaine, like Δ^9-THC and morphine, has been reported to decrease the antimicrobial activity of alveolar macrophages obtained from chronic crack cocaine smokers (Roth et al., 2004) and to decrease the parameters of mouse macrophage activation (Ou et al., 1989; Pacifici et al., 1993). However, in contrast to the effects of opioids, cocaine has been shown to increase the activation of neutrophils in human subjects, as evidenced by the increased killing of *Staphylococcus aureus* (Baldwin et al., 1997).

33.2.1.4 Effects of Drugs of Abuse on the Production of Cytokines and Chemokines

One mechanism by which drugs of abuse modulate immune function is through effects on expression of cytokines and chemokines, or on the functional capacity of their receptors. These proteins act as messengers to direct appropriate cellular activation and differentiation (cytokines) and coordinate the migration of cells to particular sites of inflammation or infection (chemokines). Cytokines have a number of functions. They may alter the innate immune response, act as a bridge to influence the type of adaptive immune response, and can also amplify adaptive immune responses. Chemokines can participate in innate immunity and also influence adaptive immunity by orchestrating the type of cells that migrate to a pathological site. In addition, some of the chemokines can act as co-activators of immune cell function. Pro-inflammatory and anti-inflammatory cytokines, and also chemokines, which are part of the innate immune response, mediate generalized symptoms of inflammation, including fever (IL-1), organ damage (tumor necrosis factor (TNF)-α), and migration of phagocytic cells (IL-8, CCL2), or dampen these responses (IL-10).

Morphine and other ligands for the mu receptor have been shown to modulate the production of both cytokines and chemokines, although the results have been complex. For example, treatment of stimulated peripheral blood mononuclear cells with morphine decreased IFN-γ production (Peterson et al., 1987b), but increased TGF-β expression (Chao et al., 1992). Since TGF-β can be immunosuppressive, it is possible that the upregulation of TGF-β may be responsible for the depressed IFN-γ expression. A biphasic dose-response relationship was shown for the release of pro-inflammatory cytokines from stimulated mouse peritoneal macrophages treated with morphine in vitro, with increased release of IL-6 and TNF-α and reduced production at high doses (Roy et al., 1998a). Bussiere et al. reported that morphine pellets reduced ex vivo production of the pro-inflammatory cytokines IL-1b, IL-6, and IFN-g by murine spleen cells (Bussiere et al., 1993), a conclusion that was supported by Sacredote, who found that murine peritoneal macrophages harvested after in vivo morphine had reduced production of IL-1b, TNF-a, and IL-12 (Martucci et al., 2007). The transcription factor, NF-kB, is a major controlling factor for the production of these pro-inflammatory cytokines. It has been shown that morphine-mediated suppression of these cytokines occurs via another transcription factor, RelB, which interacts with NF-kB. Cells taken from relB knock-out mice do not respond to morphine in the presence of bacterial lipopolysaccharide with downregulation of pro-inflammatory cytokine production (Martucci et al., 2007). Very recent studies with human monocytes have shown that morphine induces the expression of the microRNA miR-21 (Chang et al., 2021). The miRNA is upregulated in numerous inflammatory diseases and has been referred to as a "rheostat" to control inflammatory responses (Sen & Roy, 2012). Furthermore, miR-21 indirectly downregulates TNF-α and IL-12 but upregulates IL-10 expression, and this may partially explain the immunosuppressive activity of morphine (Kuroda et al., 2008).

The pattern of pro-inflammatory cytokines produced has implications for the character of the adaptive immune response that will occur if the host is exposed to a foreign antigen. A T-helper type 1 (Th1) cytokine profile polarizes the immune response toward a more pro-inflammatory cellular immune response, whereas a T-helper type 2 (Th2) profile favors certain types of antibody responses (Mosmann & Coffman, 1987). IFN-γ is often considered a marker of a Th1 response, and IL-4 is indicative of a Th2 response. There is evidence that morphine given in vivo biases the immune response toward a Th2 response (Roy et al., 2004). The evidence supporting this shift toward Th2-type immunity is mixed since other studies have shown that antibody production, which would be boosted by a shift to a Th2 response is suppressed in animals treated similarly with morphine (Bussiere et al., 1993). Further, Khabbazi found that morphine blocked the IL-4-mediated transition of macrophages to the M2 state, which is associated with less antimicrobial activity (Khabbazi et al., 2015).

In addition, recent studies of the murine immune response to *Acinetobacter baumannii* and pulmonary *Streptococcus pneumonia* infection show that chronic morphine administration inhibits the development of T-helper type 17 (Th17) immunity, and decreases the production of the Th17-associated cytokines IL-17 and IL-23 (Ma et al., 2010; Breslow et al., 2011b; Wang et al., 2011). These studies are in contrast to experiments with rhesus macaques receiving 3 months of daily morphine which show that both Th17 cell function (IL-17 production) and circulating Treg numbers are significantly increased (Cornwell et al., 2013). An explanation for the disparity in the Th17 results is currently unclear but may be due in part to differences in the animal models being studied (mouse vs rhesus macaque).

Regarding chemokines, exposure of peripheral blood mononuclear cells (PBMCs) to DAMGO, a synthetic peptide selective for the μ-opioid receptor, has been shown to increase the expression of pro-inflammatory chemokines CCL2, CCL5, and CXCL10 (Wetzel et al., 2000), and the chemokine receptors CXCR4 and CCR5, the major HIV-1 co-receptors (Choi et al., 1999; Steele et al., 2003), the latter leading to increased replication of HIV. Interestingly, the activation of MOR induces TGF-β expression, and this cytokine is required for the upregulation of both CCL5 and CXCR4 (Happel et al., 2008). In addition, work has shown that morphine and other μ-opioid agonists induce CCL2 expression by inducing NF-kB activation in human monocytes in vitro (Happel et al., 2011). It is difficult to reconcile some of these in vitro observations with the in vivo studies cited above showing that morphine inhibits the migration of neutrophils and monocytes that are a part of the process of wound healing through inhibition of chemokine expression (Clark et al., 2007; Martin et al., 2010a, b). The in vitro studies have the advantage of measuring the direct effects of opioids on immune cells. The in vivo studies evaluate the effect of opioids on complex outcomes that may be the product of several input mechanisms.

A major and important finding concerning opioids and chemokines is that there is heterologous desensitization between certain opioid and chemokine receptors (Adler & Rogers, 2005). Heterologous desensitization of G-Protein Coupled Receptors (GPCRs) is a process in which the engagement of one receptor leads to the inactivation of a second receptor (in the absence of the agonist for the second receptor) to function. This process is mediated by the initiation of a signaling pathway which leads to the cross-phosphorylation of the second receptor (Szabo et al., 2002, 2003; Chen et al., 2004), reviewed in (Steele et al., 2002). These studies show, for example, that activation of MOR leads to the cross-desensitization of CCR1, CCR2, and CCR5, and these cross-talk interactions are bi-directional (Fig. 33.1). Heterologous desensitization of chemokine receptors by opioids blocks chemotaxis of human peripheral blood mononuclear cells (Grimm et al., 1998; Szabo et al., 2002) and also blocks HIV infection of primary human monocytes in vitro (Szabo et al., 2003). Moreover, studies conducted in vivo show that pretreatment of rats with CCL5 or CXCL12 in the periaqueductal gray (PAG) region of the brain, blocks the antinociceptive effects of morphine similarly administered (Szabo et al., 2002). These results are consistent with cross-talk between the chemokine receptors CCR5 or CXCR4, respectively, and MOR. More recent work has shown that at least 90% of neurons in the rodent PAG co-express CXCR4 and MOR (Patel et al., 2006; Heinisch et al., 2011). These results are very important since the process of heterologous desensitization requires that the receptors must be co-expressed in order for the cross-phosphorylation to take place (Steele et al., 2002). More recent work has demonstrated cross-talk between a number of chemokine receptors and MOR in the brain, and these results have potential clinical relevance since MOR plays an important role in pain sensitivity (Parsadaniantz et al., 2015). Chemokine receptor antagonists (CRAs) which target CXCR4 (AMD3100) and CCR5 (Maraviroc) given to rats using a paw incisional pain model shift the morphine dose-response curve 3.3-fold to the left (Inan et al., 2018), showing that blocking chemokines released by the incisional insult increases the efficacy of morphine. CRAs were also effective in increasing the potency of morphine in the cold water tail-flick test in rats and the formalin test in the paw of mice (Inan et al., 2021). Even more intriguing are the reports that a CRA can inhibit opioid reinforcement and physical dependence in rats (Bongiovanni et al., 2022), as well as attenuate oxycodone-seeking in abstinent rats (Iriah et al., 2021). This result suggests that chemokines may play an intrinsic role in opioid addiction.

There are striking parallels between the effects of opioids and cannabinoids on cytokine production. Δ^9-THC has also been shown to suppress pro-inflammatory cytokines crucial in host defense, such as TNF-α (Fischer-Stenger et al., 1993). Further studies showed a biphasic dose response to cytokine production by Δ^9-THC on human PBMCs, but it was the opposite of that observed with morphine, with low doses inhibiting TNF-α, IL-6, and IL-, and high doses increasing these cytokines (Berdyshev et al., 1997). In a mouse model of Legionnaires' Disease, Δ^9-THC blocked IFN-γ, IL-12, and IL-12 receptor expression and increased IL-4 levels, reflecting a polarization of the immune response toward a Th2 phenotype (Perez-Castrillon et al., 1992; Klein & Cabral, 2006). In an immune assay of graft rejection, Δ^9-THC added into the cultures depressed IFN-γ production by human peripheral blood T cells (Yuan et al., 2002). When lymph node cells of mice were stimulated with Staphylococcal enterotoxin B, histone methylation and acetylation patterns

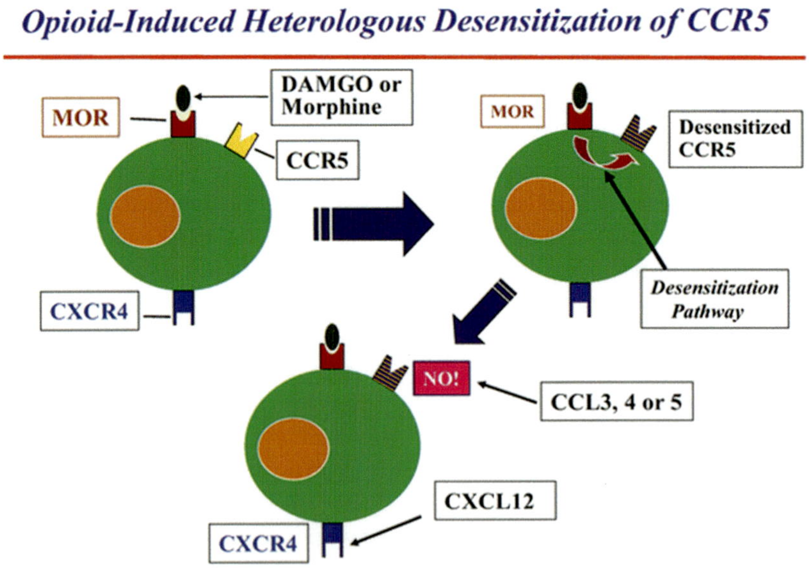

Fig. 33.1 Heterologous desensitization of chemokine receptors following activation of the MOR. The activation of MOR by either morphine or a MOR-selective agonist such as DAMGO leads to the induction of a cross-desensitization pathway which rapidly leads to the inactivation of some of the chemokine receptors, including CCR5. At this point, the cell is unresponsive to CCR5-selective chemokines such as CCL3, 4, or 5. However, the MOR-induced cross-desensitization does not affect all chemokine receptors. For example, CXCR4 remains fully responsive to activation by the CXCR4-specific chemokine CXCL12. The cell is also protected from infection with an R5 strain, but not an X4 strain, of HIV-1

were consistent with upregulation of Th2 cytokine genes and parallel downregulation of Th1 genes (Yang et al., 2014). These results are interesting given the ability of Δ^9-THC to induce elevated expression of the anti-inflammatory cytokines IL-10 and TGFβ in tumor-bearing mice (Zhu et al., 2000). Work with the endocannabinoid, anandamide, has demonstrated an inhibitory effect on the production of the pro-inflammatory cytokines IL-12 and IL-23, with a modest upregulation of the anti-inflammatory cytokine IL-10 in microglial cells (Correa et al., 2009, 2010).

Like morphine, cannabinoids have also been shown to increase the mRNA for the chemokines CCL2, and also for IL-8, in a monocytic cell line transfected with the CB2 receptor (Jbilo et al., 1999). Further, cannabinoids are reported to be chemotactic for macrophages (and microglial cells), which suggests that the CB2 receptor mediates activation of the cell migration machinery (Derocq et al., 2000; Walter et al., 2003). In contrast and somewhat anomalously, several reports have shown that treatment with Δ^9-THC or CB2 agonists depressed the capacity of macrophages or microglial cells to migrate in response to certain chemokines, including CCL2, CCL3, and CCL5 (Montecucco et al., 2008; Raborn et al., 2008; Romero-Sandoval et al., 2009; Fraga et al., 2011). Treatment with high doses of CB2 agonists has been reported to attenuate the response of human T cells to CXCL12 (Ghosh et al., 2006), and taken together these results suggest that CB2 cannabinoids can inhibit leukocyte traffic. This would be consistent with studies that have shown that CB2 agonists inhibit leukocyte adhesion to brain vascular endothelium and reduce the migration of monocytes across the blood-brain barrier (Ramirez et al., 2013; Rom et al., 2013; Persidsky et al., 2015).

Cocaine is also reported to modulate the production of cytokines and chemokines. Analysis of in vitro cocaine treatment on cytokine and chemokine expression by mouse and human cells has revealed a decreased ability of stimulated immune cells to produce INF-γ, TNF-α, IL-1, and IL-8 (Watzl et al., 1992; Shen et al., 1994). Other studies did not agree and reported an increase in TNF-α, IL-8, IL-2, and other chemokines (Wang et al., 1994; Zhang et al., 1998). Calipari reported that cocaine given intraperitoneally to mice, or self-administered, reduced IL-1a but increased granulocyte colony-stimulating factor (G-CSF) in serum. Furthermore, the drug also induced G-CSF in the nucleus accumbens (Calipari et al., 2018). Chen and Watson found that chronic cocaine in vivo decreased mouse splenic TNF-a ex vivo (Chen & Watson, 1991). In contrast, in another study, in vivo infusion of cocaine resulted in an increase in IFN-γ and a decrease in the anti-inflammatory cytokine, IL-10 (Gan et al., 1998). Fox reported that human chronic cocaine users had decreased baseline IL-10, but when exposed to stress images displayed increased TNF-a (Fox et al., 2012). The Wigdahl laboratory found that cocaine users had elevated IL-4 and IL-10 (Parikh et al., 2014) The lack of concordance between in vitro and in vivo studies again suggests that drugs may have complex effects in vivo, partly because they can engage mediators from other systems that might affect immune function. Interestingly, the interaction between chemokines and cocaine has also been reported, with the chemokine CXCL12 increasing cocaine-induced locomotion (Trecki & Unterwald, 2009). The chemokine CXCL1 was found to be a key regulator of cocaine-induced reward in mice (Saika et al., 2018). Further, the CRAs, AMD3100 and maraviroc, attenuated cocaine place preference and locomotor stimulation in rats (Kim et al., 2017; Nayak et al., 2020). In a new development, G-CSF, which is involved in upregulating the production of neutrophils from the bone marrow, has been shown to enhance cocaine self-administration and conditioned place preference in mice when administered systemically (Calipari et al., 2018).

33.2.2 Effects on Adaptive Immunity

33.2.2.1 Humoral Immunity (Antibody Responses)

Morphine given by injection (Lefkowitz & Chiang, 1975) or slow-release pellet has been shown to block the capacity of mouse spleen cells to mount an ex vivo antibody response to the artificial antigen, sheep red blood cells, and naltrexone-releasing pellet blocks the effect of the opioid (Bussiere et al., 1992). Serum antibody responses to tetanus toxoid (Eisenstein et al., 1990) and to trinitrophenylated bovine serum albumin (Weber et al., 1987) were also suppressed in mice by morphine given in vivo. Morphine, U50,488H (kappa agonist), and deltorphin II (delta$_2$ agonist) have all been shown to be immunosuppressive when administered by an osmotic mini-pump. All three agonists gave inverted U-shaped dose-response curves over a dose range of 0.1–10 mg/kg/day, with maximal immunosuppression between 0.5 and 2 mg/kg/day (Rahim et al., 2001). It has been suggested that part of the suppressive effect of morphine pellets on humoral immune responses is due to the activation of the HPA axis (Pruett et al., 1992). However, all of the effects of opioids on the capacity to mount an antibody response cannot be mediated, since spleen cells taken from normal animals that were never exposed to opioids in vivo, have a depressed capacity to make antibodies to sheep red blood cells when treated with various opioid agonists in vitro (Taub et al., 1991; Eisenstein et al., 1995). Also, quaternary naloxone, which does not pass the blood-brain barrier, blocks the immunosuppressive effect of a kappa agonist given in vivo (Radulovic et al., 1995), suggesting that the drug is acting in the periphery. The suppressive effects of morphine and U50,488H in vivo and in vitro appear to result from individual effects on T cells and macrophages (Bussiere et al.,

1993; Guan et al., 1994) which are needed to cooperate with B-cells to generate antibody production. Morphine has also been shown to have an effect on the mucosal antibody response. Cultured ileal segments from mice receiving morphine and given cholera toxin orally were suppressed in their ability to produce IgA specific for cholera toxin compared with mice receiving placebo (Peng et al., 2001). The effects of chronic administration of morphine and withdrawal from the drug have been examined for effects on responses to a B-cell mitogen (LPS) (Bryant et al., 1987) and on antibody responses (Rahim et al., 2002). Exposure to morphine from slow-release pellets for 3 to 4 days led to tolerance to the immunosuppressive effects in both assays, and withdrawal led to renewed immunosuppression. Extensive investigation into the mechanisms of withdrawal-induced immunosuppression showed that there was a deficit of macrophage function (Rahim et al., 2003, 2005). In contrast, other data suggest that withdrawal biases the immune response to a Th2 cytokine phenotype (Kelschenbach et al., 2005). At present it is not known how to reconcile these observations.

Δ^9-THC, like morphine, has been shown to suppress in vitro antibody formation by mouse spleen cells (Kaminski et al., 1992, 1994; Eisenstein et al., 2007). Interestingly, Δ^9-THC is reported to increase IgG1 responses to Legionella in drug-treated animals, presumably by polarizing the immune response toward a Th2 phenotype (see below under T-cell responses) (Newton et al., 1994).

Cocaine has been investigated for its capacity to alter antibody responses and has been found to be immunosuppressive for some antigens, but not others (Watson et al., 1983; Havas et al., 1987; Ou et al., 1989).

33.2.2.2 T-Cell Responses

Morphine given by slow-release pellet also inhibits the ability of splenic T cells to respond to mitogens (plant products that cause nonspecific proliferation of immune cells). Thus, at 72 h after pellet implantation splenic murine T cells had a marked inhibition of replication when exposed to Concanavalin A (Con A) (Bryant et al., 1987). Reduction in response to Con A has also been observed using peripheral blood T cells of rats 2 h after a single subcutaneous injection of morphine (Bayer et al., 1990). Morphine has also been reported to inhibit thymocyte and lymph node–derived T-cell proliferation when added to the cells in vitro (Roy et al., 1997; Wang et al., 2001). Other measures of T-cell function relate to adaptive immunity in the cellular arm of the immune system. Several studies have shown that chronic morphine exposure inhibits the development of delayed-type hypersensitivity (DTH) in rats, mice, and pigs (Pellis et al., 1986; Bryant & Roudebush, 1990; Molitor et al., 1992), and cytotoxic T cells in mice (Carpenter & Carr, 1995).

Δ^9-THC has also been shown to inhibit a panel of immune responses that are T-cell mediated (Eisenstein & Meissler, 2015). Among these is the inhibition of the response to Con A, a mitogen that specifically causes the proliferation of T cells in a protocol that exposed mouse spleen cells in vitro to Δ^9-THC (Klein et al., 1985). Further, Δ^9-THC and CB2 selective agonists have been shown to inhibit T-cell responses to foreign tissues in an in vitro assay of graft rejection called the Mixed Lymphocyte Reaction (MLR) (Robinson et al., 2013). A major mechanism of this immunosuppression was found to be by inhibition of IL-2, a critical autocrine growth factor for T cells (Robinson et al., 2013). The CB2 receptor has also been studied in murine models of a number of autoimmune diseases including Experimental Autoimmune Encephalitis (EAE), colitis, inflammatory bowel disease, and autoimmune uveoretinitis (Maresz et al., 2007; Xu et al., 2007; Storr et al., 2009; Zhang et al., 2009; Singh et al., 2012). The use of synthetic cannabinoids with selectivity for this receptor (as opposed to CB1) and the use of CB2 receptor k/o mice demonstrated that CB2 activation has marked activity in depressing the T-cell responses in these autoimmune models. Other effects of the CB2 system were to depress T-cell trafficking by depressing adhesion molecule expression (Xu et al., 2007; Zhang et al., 2009) and by upregulating the immunosuppressive cytokine, IL-10 (Storr et al., 2009). Cannabinoids have been studied for their ability to polarize T cells toward Th1 or Th2 responses. It has been reported that Δ^9-THC given to mice that are subsequently infected with *Legionella* suppressed IL-12Rβ2 expression by a CB1-mediated mechanism and increased GATA3 mRNA via CB2 signaling (Klein et al., 2004). The IL-12 receptor participates in Th1 responses and GATA3 biases toward a Th2 response, leading to the conclusion that cannabinoids can polarize T-cell responses toward a Th2 phenotype. Evidence from exposure of activated human peripheral blood T cells to Δ^9-THC also supports a Th2 polarization (Yuan et al., 2002). Δ^9-THC has also been shown to suppress cytotoxic T-cell activity against both allogeneic cells (Klein et al., 1991) and virus-infected cells (Fischer-Stenger et al., 1992).

An acute i.v. dose of cocaine has been reported to have an effect similar to that of morphine in suppressing the proliferation of T cells in the peripheral blood of rats exposed to Con A 2 h after injection of the drug (Bayer et al., 1996). Cocaine has been shown to inhibit the production of Th2 cytokines which would be expected to reduce antibody responses (Wang et al., 1994). Cocaine given orally also suppressed DTH responses in mice (Watson et al., 1983).

33.3 Effects of Drugs of Abuse on Infection

33.3.1 Overview

The majority of publications cited above concluded that the various drugs of abuse were immunosuppressive. A logical conclusion from these observations would be that exposure to the drugs would sensitize them to infection. Clinical observations of intravenous drug-using (IVDU) populations have long noted an increased incidence of bacterial infections (Risdahl et al., 1996; Honarvar et al., 2013). However, it has not been possible to determine whether more infection occurs in this group because of unsanitary use of needles or because the drugs render the person more susceptible. For example, there have been recent outbreaks of injection-associated anthrax and botulism in IVDUs in Europe, which have been attributed to needle contamination (MacDonald et al., 2013; Berger et al., 2014). This question was important regarding Human Immunodeficiency Virus (HIV) infection, since in the United States, at the beginning of the epidemic, one-third of all cases of HIV occurred in IVDUs, and estimates suggest that approximately 20% of global IVDUs are infected with HIV (Mathers et al., 2010; Vlahov et al., 2010). The current estimate in the United States is that 7% of HIV cases occur in IVDUs (Spiller et al., 2015; CDC, 2020). Epidemiologic studies have been compromised because drug abusers usually abuse more than drugs, and also smoke and abuse alcohol. Studies have not adequately teased apart these factors, nor controlled for drug dosage or frequency of drug use. Several other reviews explore these issues in depth (Kapadia et al., 2005; Rogers et al., 2005). Laboratory studies using animal models of infections can more definitively clarify whether drugs of abuse can be cofactors in susceptibility to infection. Yet, surprisingly there are only a limited number of studies on drugs of abuse and experimental infections of any type.

33.3.2 Effects of Drugs of Abuse on Infections Other Than HIV

33.3.2.1 Effects of Morphine on Infection

There is clinical literature documenting the increased incidence of infection in opioid abusers (Hussey & Katz, 1950; Louria et al., 1967; Scheidegger & Zimmerli, 1989; Haverkos & Lange, 1990), and in a hospital setting where controlled doses of opioids were administered to burn patients (Schwacha et al., 2006). As mentioned in the introduction to this chapter, these correlations do not prove causality. In laboratory studies, morphine has been shown to sensitize mice to fungal, bacterial, parasitic, and viral infections. In a seminal paper, morphine was shown to increase replication of the fungus, *Candida albicans*, and the bacterium, *Klebsiella pneumoniae*, apparently at least partly due to decreased phagocytosis and killing of the organisms by neutrophils (Tubaro et al., 1983). Other early laboratory studies showed that morphine pretreatment led to a shock-like syndrome when mice were injected with the parasite, *Toxoplasma gondii* (Chao et al., 1990). In a murine model of *Leishmania donovani* infection, morphine was found to have a biphasic dose-response effect, with low doses affording increased protection, and high doses sensitizing to infection (Singh & Singal, 2007). Further, morphine and withdrawal from morphine were reported to induce sepsis through increased permeability of the intestinal epithelial barrier, loosening of tight junctions in the ileum, and enhanced sensitivity to bacterial LPS (Hilburger et al., 1997; Roy et al., 1998b). Morphine was also shown to markedly sensitize to oral *Salmonella* infection (MacFarlane et al., 2000; Feng et al., 2005). Morphine has been reported to increase the virulence of *Pseudomonas aeruginosa* in the mouse intestinal tract with a direct effect on the microbe (Babrowski et al., 2012). Epidemiology studies link the use of opioid analgesics to an increased risk of infection with *Clostridium difficile* (Feng et al., 2006; Mora et al., 2012). Thus, morphine seems to have a significant effect on intestinal microbes, possibly through its well-know action as an inhibitor of gastrointestinal transit (Feng et al., 2006). Sensitization to infection could also occur by decreasing innate immune responses. As mentioned previously, morphine inhibits phagocytosis of Candida, which is one mechanism for potentiating infection (Tubaro et al., 1983; Szabo et al., 1993). Inhibition of neutrophil mobilization by morphine was shown to be a factor in sensitizing to infection with *Streptococcus pneumoniae* and *Acinetobacter baumannii* infections (Ma et al., 2010; Breslow et al., 2011a). In the case of the pneumococcal infection, depression of the IL-17/IL-23 pathway by the opioid was found to be a mechanism for impaired neutrophil responses (Ma et al., 2010), but IL-17 was not found to be important in sensitization to *Acinetobacter* (Breslow et al., 2011b). For the pneumococcus, morphine was also shown to decrease macrophage signaling through Toll-like receptor-9 (TLR-9), which would dampen the protective inflammatory response (Wang et al., 2008, 2011).

Several papers also support the effect of morphine on adaptive immune responses related to infection. Meningitis is a devastating manifestation of tuberculosis. Morphine has been shown to depress microglial functions that are necessary for initiating T-cell responses to *Mycobacterium bovis* (Olin et al., 2011). This opioid has also been shown to inhibit secretory antibody formation to cholera toxin in the gastrointestinal tract (Peng et al., 2001), which could be a mechanism of sensitization to gastrointestinal infections. Further, morphine was found to potentiate the development of Herpes simplex virus, type 1 encephalitis in a murine model of Herpes simplex virus infection, (Lioy et al., 2006), and to potentiate systemic infection with this virus by decreasing IFN-γ and IL-12 levels, and inhibiting cytotoxic T-lymphocyte responses (Mojadadi et al., 2009).

33.3.2.2 Effects of Morphine on the Microbiome

As described above, morphine is well known to inhibit gastrointestinal transit and result in constipation. It is also established that morphine can result in sepsis in mice, which is due to leakage of the organisms that make up the microbiome into the circulation and scavenging of them by organs like the liver and spleen (Hilburger et al., 1997; Roy et al., 1998b). Two laboratories have also reported that tolerance to morphine is mediated by the microbiome. Antibiotic-treated mice with a resultant reduction in microbiota load, and also germ-free mice, do not display tolerance to morphine (Kang et al., 2017; Zhang et al., 2019). Repopulation of the mouse intestine with fecal transplants restored the ability to develop morphine tolerance. The mechanism by which the microbiome causes this effect has been suggested to be by reduction in short-chain fatty acids normally supplied to the gut by the microbiota (Hofford et al., 2021).

33.3.2.3 Effects of Δ^9-THC on Infection

Δ^9-THC has been shown to sensitize to bacterial, viral, and parasitic infections. The first observations were in guinea pigs showing that the cannabinoid augmented infection with Herpes simplex virus (Cabral et al., 1986). In addition, Δ^9-THC administration has been shown to lead to greater viral replication and disease severity in murine models of Friend Leukemia Virus (Specter et al., 1991b), influenza infection (Buchweitz et al., 2007), vaccinia infection (Huemer et al., 2011) and Kaposi's sarcoma virus (KSV) (Zhang et al., 2007). At least one mechanism of viral potentiation was a decreased influx of T cells and macrophages into the lungs of influenza-infected animals (Buchweitz et al., 2007). In contrast, treatment with Δ^9-THC has been shown to depress the reactivation of KSV or Epstein Barr virus in an immortalized B-lymphocyte line (Medveczky et al., 2004). In regard to bacterial infections, Δ^9-THC sensitized to *Listeria monocytogenes* (Morahan et al., 1979), and inhibited the adaptive immune response to *Legionella pneumophila*. It was found that if the cannabinoid was given 18 h prior to a primary sublethal infection, it sensitized to a secondary challenge with Legionella (Newton et al., 1994). Mechanistic studies showed that Th1 responses, which are necessary for host defense against Legionella, were depressed by Δ^9-THC (see above section on cannabinoids and T-cell responses), and the antibody response was shifted from an IgG2a to an IgG1 isotype, indicative of a switch from a Th1 to a Th2 response (Klein et al., 2000; Newton et al., 2009). Δ^9-THC has also been reported to sensitize mice to encephalitis with the parasite, *Acanthamoeba castellanii*, by inhibiting the antimicrobial functions of microglia (Cabral & Marciano-Cabral, 2004).

33.3.2.4 Effects of Drugs of Abuse on HIV or Related Infections

There are several papers showing that morphine upregulates HIV replication when added to infected human peripheral blood cells or microglia in vitro (Peterson et al., 1990, 1994; Li et al., 2002). Heroin was shown to enhance HIV replication in human macrophages by downregulating an inhibitory miRNA (Wang et al., 2015). As mentioned above in Sect. 33.2.1.4. on cytokines and chemokines, the effects of opioids on HIV infection may be related to drug-induced alterations in cytokine or chemokine production or on the expression of chemokine receptors (Peterson et al., 1994; Li et al., 2002; Steele et al., 2003). Pro-inflammatory cytokines, particularly TNF-α, can activate T cells and lead to increased HIV replication, and chemokine receptors are co-receptors for the virus. It is to be expected that modulation of chemokine and/or chemokine receptor levels can alter HIV infectivity by increasing the pool of infectable cells or by altering the availability of the number of co-receptors.

Similar immunodeficiency syndromes may result in monkeys and cats infected with SIV (simian immunodeficiency virus) and FIV (feline immunodeficiency virus), but studies with these viruses have not provided a clear indication of the influence of opioids on the development of these immunodeficiency diseases. In one study monkeys maintained on morphine had a slower progression of the SIV infection, although temporary withdrawal augmented viral load (Donahoe et al., 1993), but in two other studies, viral progression was more rapid (Chuang et al., 1997; Kumar et al., 2006). Neither acute or chronic morphine, nor morphine withdrawal, altered FIV infection in cats, but the number of animals was small, and the variance in FIV load was large (Barr et al., 2003).

Morphine administration has been reported to exacerbate the recruitment of leukocytes to the CNS during an inflammatory response (Olin et al., 2012). A number of studies have shown that opiate administration promotes the neuroinflammatory and neurodegenerative influences associated with HIV infection. Morphine promotes the cytotoxicity of HIV viral products through a variety of mechanisms including the upregulation of pro-inflammatory cytokines, oxidative stress, and other processes (Ellis et al., 2007; Mahajan et al., 2008; Turchan-Cholewo et al., 2009). Furthermore, morphine exacerbates the response of astrocytes to HIV viral proteins, leading to an increase in the production of pro-inflammatory cytokines and chemokines, which lead, in turn, to the recruitment of additional inflammatory cells to the CNS (El-Hage et al., 2005, 2006; Hauser et al., 2012). A comprehensive review of the effects of opioids on the neuropathogenesis of HIV has recently been published (Murphy et al., 2019).

Cannabinoids have also been investigated for their effects on HIV and related retroviruses. The humanized severe combined immunodeficient mouse treated with Δ^9-THC was found to have a 50-fold increase in viral load of an engineered HIV reporter virus (Roth et al., 2005). Studies using the SIV model have shown that chronic Δ^9-THC depressed the progression of infection in male macaques (Molina et al., 2011). However, more recent studies from the same laboratory group showed that chronic Δ^9-THC does not substantially alter the progression of SIV-associated disease or viral replication in female macaques (Amedee et al., 2014). The basis for the disagreement in the results with the male and female animals remains uncertain at this time. Several studies have tested the effect of CB2 selective cannabinoids on HIV replication in vitro and have shown that these agonists reduced the replication of HIV in both human T cells and macrophages (Costantino et al., 2012; Ramirez et al., 2012). Evidence was presented to support a mechanism in the T cells whereby the CB2 agonist reduced F-actin and blocked viral integration (Costantino et al., 2012). It is interesting that a cannabinoid should be found to interfere with actin polymerization pathways, because, as noted previously in this review, a similar observation has been made for morphine (Ninković & Roy, 2012). In an in vivo study, a CB2 selective agonist was found to be protective against HIV encephalitis using a humanized SCID mouse model of intra-cranial infection (Gorantla et al., 2010). The CB2 selective cannabinoid blocked T-cell infiltration into the brain by downregulating adhesion molecules and altering the blood-brain barrier.

Cocaine has also been investigated regarding its effects on HIV infection and was found to potentiate replication in in vitro systems (Peterson et al., 1992; Bagasra & Pomerantz, 1993). Cocaine has also been shown to upregulate DC-SIGN, a molecule on the surface of human dendritic cells that can act as a receptor for HIV (Nair et al., 2005). In addition, it has been reported that mice with severe combined immunodeficiency, that were reconstituted with human leukocytes and infected with HIV, demonstrated enhanced replication of the virus after cocaine treatment (Roth et al., 2002). However, a report using the SIV model shows that chronic cocaine administration resulted in minimal effects on viral replication and neuroinflammation, including inflammatory cytokine mediator expression, or neurodegeneration (Weed et al., 2012). In contrast, in a comprehensive study, the Wigdahl laboratory reported that cocaine potentiated viral load and decreased T-cell numbers in patients infected with HIV (Parikh et al., 2014). A more complete accounting of the impact of cocaine abuse on HIV pathogenesis is clearly warranted.

33.3.2.5 Effects of Drugs of Abuse in Combination with Tobacco Smoke and HIV Infection

Tobacco smoke use is common among intravenous opiate abusers (Duaso & Duncan, 2012), and people who use illicit drugs have a higher rate of smoking (up to 23-fold) than non-drug abusers (Gingo et al., 2013; Kohut, 2017). Recent statistical analysis shows that 70–95% of people with opioid use disorder smoke tobacco (Chun et al., 2009). Current work suggests that co-incident use of tobacco and opiates creates an additive effect with increased inflammation, leading to elevated morbidity and mortality (Morris & Garver-Apgar, 2020). The consequences of opioid abuse and tobacco smoking are particularly significant for patients who are infected with HIV. For example, HIV-infected smokers have a much higher incidence of emphysema and related airway dysfunction (Diaz et al., 2003), exacerbating the capacity of HIV infection to promote emphysema in the absence of either opioid or tobacco use (Crothers et al., 2011).

The combination of tobacco use and opioid abuse, in the context of HIV infection, is an area of research that has not been extensively studied. However, recent studies using HIV-infected humanized mice have shown that chronic morphine administration, in combination with daily tobacco exposure, results in a significant increase in the activation of microglial cells (Cornwell et al., 2020). The results from this study indicate that HIV infection, tobacco smoke exposure, or chronic morphine administration individually elevate microglial cell activation, but the combination of the infection and both morphine and tobacco results in an additive induction of microglial cell activity. The activation of microglial cells is an essential element of the neuroinflammation that is associated with HIV infection and contributes directly to the neurodegeneration in these patients.

Recent reports using conventional mice have addressed the effects of the combination of chronic morphine and tobacco smoke exposure on the immune system. The results show that this combination leads to a downregulation in the functional activity of Treg cells (Shao et al., 2022a, b; Saaoud et al., 2023). Based on a combination of multiparameter flow cytometry and RNAseq analysis, it is apparent from these reports that the morphine and tobacco combination promotes the differentiation of Treg cells to a Th17 phenotype. These cells continue to express several of the Treg cell markers, express a portion of the Th17 cell markers, and lose Treg cell functional activity (Shao et al., 2022a, b). These studies are consistent with recent reports which show that morphine treatment alone of mice induces the differentiation of Treg cells to exhibit the "fragile-like" Treg phenotype (Zhu et al., 2023). The fragile-like Treg cells continue to express the Treg transcription factor FoxP3, fail to manifest immunosuppressive activity, and produce much higher levels of IFNγ

(Sharma et al., 2013; Overacre-Delgoffe et al., 2017). The opioid treatment resulted in an induction of fragile-like Treg cells in the circulation and in the nucleus accumbens, a brain region that is pivotal for the development and maintenance of drug addiction (Zhu et al., 2023). Moreover, it is conceivable that these Treg cells may participate in the upregulation of microglial cell activation described above. The opioid-induced fragile-like Treg cells are a new discovery and should receive much greater attention in order to more fully understand their role in the neuroinflammation associated with chronic opioid abuse, tobacco use, and HIV infection.

The fact that opioids, cannabinoids, and cocaine can alter the replication of HIV in cell culture in vitro certainly suggests that the drugs should be able to alter HIV progression in vivo. Isolation of variables in the laboratory permits the elimination of confounding factors in epidemiologic studies in humans, such as poly-drug abuse, and also standardizes drug dosage. Thus, there is considerable evidence from laboratory studies supporting the role of drugs of abuse in potentiating HIV infection. There are, however, some inexplicable contradictions in the literature showing that opioids and cannabinoids, in general, inhibit immune cell migration, yet seem to potentiate HIV infection in the brain.

33.4 Is Opioid Addiction Due to Inflammation?

Papers published in the last decade mostly by a single laboratory, have posited the hypothesis that morphine induces inflammation by binding to the Toll-like Receptor 4 (TLR4). Engagement of TLR4 by bacterial lipopolysaccharide induces a panel of pro-inflammatory cytokines including IL-1β, TNF-α, and IL-6. Various papers showed that morphine binds to TLR4 by the accessory molecule, MD-2 and that both the active and inactive enantiomers of morphine and naloxone are active in these and other assays. It has been concluded that morphine induces inflammation and as a corollary, that addiction is a consequence of the inflammation (Hutchinson et al., 2010, 2012; Lewis et al., 2012; Hutchinson & Watkins, 2014). Clearly, these findings do not fit with most of the rest of the literature cited above in this chapter (see Sect. 33.2.1.4) which concludes that morphine is generally immunosuppressive. Direct evidence against the hypothesis that morphine engages TLR4 comes from a number of directions. Early in the research on the effects of drugs of abuse on immune responses, it was reported that C3H/HeJ mice, which have a defect in signaling by TLR4, still produce immunosuppression when given morphine in vivo (Bussiere et al., 1992, 1993). Other investigators have reported that mice with defects in TRL4 that do not signal, or TLR4 knock-out mice, still become tolerant to morphine and also develop hyperalgesia and physical dependence, showing that the TLR4 gene is not necessary for the pharmacological actions of the opioid (Fukagawa et al., 2013; Mattioli et al., 2014). Skolnick et al. transfected HEK293 cells with the TLR4 receptor and found that neither the plus nor the minus enantiomers of naloxone or naltrexone inhibited TLR4 signaling in response to its natural ligand, bacterial lipopolysaccharide (Skolnick et al., 2014). Eisenstein has proposed two possible explanations for the findings that morphine interacts with TLR and the resulting seeming correlation between inflammation and addiction (Eisenstein, 2019). As noted above (Sect. 33.3.2.1), morphine can cause sepsis. It is possible that in vivo studies with morphine-induced sepsis which is accompanied by the production of pro-inflammatory cytokines. For the in vitro studies, it is possible that morphine, or the diluent used in the experiments, was contaminated with lipopolysaccharide, which is active in picogram amounts and is difficult to eliminate, and that the occult LPS triggered the TLR. This explanation would reconcile with the need to use very high doses of morphine, in the nonphysiologic range, to demonstrate the involvement of TLR4.

33.5 Synopsis

It has been clearly demonstrated that opioids, cannabinoids, and cocaine alter a variety of assays of immune function when added to cells of the immune system in vitro. Receptors and/or mRNA for the receptors for opioids and cannabinoids have been demonstrated in a variety of cells of the immune system. In many cases, especially for opioids, the pharmacological specificity of the action of the drugs has been verified using appropriate antagonists. For the cannabinoids, mice with genetic defects in CB1 or CB2 receptors have been used. These observations provide a biological basis for concluding that drugs of abuse have direct effects on immune cells. In addition, all three drug classes alter the immune responses of rodents, and in some cases, humans, when treated in vivo. In vivo, mechanisms of action may be harder to determine as the drugs may activate other physiologic systems in the body that may also impinge on immune responsiveness, such as the hypothalamic-pituitary axis and the sympathetic nervous system. Nonetheless, drug abusers suffer the cumulative effects of direct and indirect effects of the drugs on the immune system. Assays of immune function that have been measured include effects on the innate immune system, which encompass effects on neutrophils, macrophages, and natural killer cells, and effects on adaptive immunity, which include the humoral and cellular arms of the immune system. Considerable attention has been focused on cytokine and chemokine mediators of immune function that are modulated by the drugs and on their capacity to modulate activation of the various arms of the immune system by

polarization of the cytokine profile. In general, opioids and cannabinoids have been found to be suppressive in a wide variety of assays of immune responsiveness.

The literature on cocaine is smaller and less consistent. An extension of these observations is the assessment of the effects of the drugs on resistance to infection, a consequence of alterations in immune function. Both opioids and cannabinoids have been shown to sensitize animals to a variety of experimental infections. Interest in the effects of opioids on HIV progression is sparked by the epidemiological intersection of intravenous drug abuse and the incidence of HIV. In vitro studies give robust results showing inhibition of HIV replication in the presence of opioids. Studies with Δ^9-THC and cocaine, using an immunodeficient mouse strain repopulated with human lymphoid cells, showed enhanced replication of HIV in the presence of the drugs. The research implies that drug abuse increases vulnerability to infection. Together, the vast majority of the literature suggests substantial effects of drugs of abuse on immune competence in general, resulting in immunosuppression. However, several studies are showing that opioids increase chemokine production, which would increase inflammation. At present, these aspects of the literature have not been reconciled. Taken as a large overview, the studies reviewed in this chapter reinforce the existence of physiological pathways between the neural and immune systems mediated by opioids and cannabinoids.

33.6 Review Questions

1. In regard to cannabinoids from plants:
   (a) They are proteins.
   (b) They exert their effects primarily by interdigitating into cell membranes.
   (c) There is evidence that they polarize the immune response toward a Th1-type phenotype.
   (d) There is evidence that they polarize the immune response toward a Th2-type phenotype.
   (e) They have no direct effects on cells of the immune system.
2. In regard to the immune system, morphine has been shown to:
   (a) Elevate antibody responses to various antigens
   (b) Elevate responses to the B-cell mitogen, lipopolysaccharide (LPS)
   (c) Elevate delayed-type hypersensitivity (DTH) responses
   (d) Increase phagocytosis by macrophages
   (e) Depress natural killer (NK) cell activity
3. Evaluation of the effects of drugs of abuse can be complicated by a number of factors, including:
   (a) The effects of the sympathetic nervous system on immune cell function
   (b) The effects of the HPA axis on immune cell function
   (c) The capacity of cells of the immune system to produce endogenous opioids
   (d) The production of endogenous cannabinoids
   (e) All of the above
4. The function of G-protein-coupled receptors, such as the opioid and cannabinoid receptors, can be regulated by the process of heterologous desensitization. Which of the following statements is true about this regulatory mechanism?
   (a) This process only occurs between receptors expressed on the surfaces of adjacent cells.
   (b) This process cannot occur between receptors expressed on the same cell.
   (c) This is a process that appears to involve transphosphorylation of G-protein-coupled receptors.
   (d) All of the above.
5. Both opioids and cannabinoids have been shown to sensitize animals to a variety of experimental infections. The impact of drugs of abuse on resistance of humans to infectious agents can be difficult to evaluate because:
   (a) Drug abuse rarely involves the administration of a single drug, and the effects of poly-drug abuse are poorly understood.
   (b) Drug abusers are exposed more frequently to pathogenic agents than nonabusers.
   (c) A number of additional factors, which are hard to control and impact ~~on~~ measurement of the immune competence of drug abusers, including the dose of the drug and the time since it was last taken.
   (d) The contributions of legal drug use, including nicotine and alcohol, can complicate the effects of illegal drug abuse.
   (e) All of the above.

33.7 Answers to the Questions

1. d
2. e
3. e
4. d
5. e

Acknowledgments We thank Mr. Joseph Meissler for his assistance with this manuscript.

This work was supported by NIDA grants DA013429, DA014230, DA023860, DA025532, DA040619, and DA049745 and a grant from the Pennsylvania Department of Health.

References

Adler, M. W., & Rogers, T. J. (2005). Are chemokines the third major system in the brain? *Journal of Leukocyte Biology, 78*, 1204–1209.

Akil, H., Watson, S. J., Young, E., Lewis, M. E., Khachaturian, H., & Walker, J. M. (1984). Endogenous opioids: Biology and function. *Annual Review of Neuroscience, 7*, 223–255.

Amedee, A. M., Nichols, W. A., LeCapitaine, N. J., Vande Stouwe, C., Birke, L. L., Lacour, N., Winsauer, P. J., & Molina, P. E. (2014). Chronic Δ9-tetrahydrocannabinol administration may not attenuate Simian Immunodeficiency Virus disease progression in female rhesus macaques. *AIDS Research and Human Retroviruses, 30*, 1216–1225.

Babrowski, T., Holbrook, C., Moss, J., Gottlieb, L., Valuckaite, V., Zaborin, A., Poroyko, V., Liu, D. C., Zaborina, O., & Alverdy, J. C. (2012). Pseudomonas aeruginosa virulence expression is directly activated by morphine and is capable of causing lethal gut-derives sepsis in mice during morphine administration. *Annals of Surgery, 255*, 386–393.

Bagasra, O., & Pomerantz, R. J. (1993). Human immunodeficiency virus type 1 replication in peripheral blood mononuclear cells in the presence of cocaine. *Journal of Infectious Diseases, 168*, 1157–1164.

Baldwin, G. C., Buckley, D. M., Roth, M. D., Kleerup, E. C., & Tashkin, D. P. (1997). Acute activation of circulating PMNs following in vivo administration of cocaine: A potential etiology for pulmonary injury. *Chest, 111*, 698–705.

Barr, M. C., Huitron-Resendiz, S., Sanchez-Alavez, M., Hendriksen, S. J., & Phillips, T. R. (2003). Escalating morphine exposures followed by withdrawal in feline immunodeficincy virus-infected cats: A model for HIV infection in chronic opiate abusers. *Drug and Alcohol Dependence, 72*, 141–149.

Bayer, B. M., Daussin, S., Hernandez, M., & Irvin, L. (1990). Morphine inhibition of lymphocyte activity is mediated by an opioid dependent mechanism. *Neuropharmacology, 29*, 369–374.

Bayer, B. M., Hernandez, M. C., & Ding, X. Z. (1996). Tolerance and cross-tolerance to the suppressive effects of cocaine and morphine on lymphocyte proliferation. *Pharmacology, Biochemistry, and Behavior, 53*, 227–234.

Berdyshev, E. V., Boichot, E., Germain, N., Allain, N., Anger, J. P., & Lagente, V. (1997). Influence of fatty acid ethanolamides and Δ9-tetrahydrocannabinol on cytokine and arachidonate release by mononuclear cells. *European Journal of Pharmacology, 330*, 231–240.

Berger, T., Kassirer, M., & Aran, A. A. (2014). Injectional anthrax - new presentation of an old disease. *Euro Surveillance, 19*, 20877.

Bhat, R. S., Bhaskaran, M., Mongia, A., Hitosugi, N., & Singhal, P. C. (2004). Morphine-induced macrophage apoptosis: Oxidative stress and strategies for modulation. *Journal of Leukocyte Biology, 75*, 1131–1138.

Bongiovanni, A. R., Zhao, P., Inan, S., Wiah, S., Shekarabi, A., Farkas, D. J., Watson, M. N., Wimmer, M. E., Ruff, M. R., & Rawls, S. M. (2022). Multi-chemokine receptor antagonist RAP-103 inhibits opioid-derived respiratory depression, reduces opioid reinforcement and physical dependence, and normalizes opioid-induced dysregulation of mesolimbic chemokine receptors in rats. *Drug and Alcohol Dependence, 238*, 109556.

Bouaboula, M., Rinaldi, M., Carayon, P., Carillon, C., Delpech, B., Shire, D., Le Fur, G., & Casellas, P. (1993). Cannabinoid-receptor expression in human leukocytes. *European Journal of Biochemistry/FEBS, 214*, 173–180.

Breslow, J. M., Meissler, J. J., Hartzell, R. H., Spence, P. B., Truant, A., Gaughan, J., & Eisenstein, T. K. (2011a). Innate immune responses to systemic Acinetobacter baumannii infection in mice: Neutrophils, but not IL-17, mediate host resistance. *Infection and Immunity, 79*, 3317–3327.

Breslow, J. M., Monroy, M. A., Daly, J. M., Meissler, J. J., Gaughan, J., Adler, M. W., & Eisenstein, T. K. (2011b). Morphine, but not trauma, sensitizes to systemic Acinetobacter baumannii infection. *Journal of Neuroimmune Pharmacology, 6*, 551–565.

Bryant, H. U., & Roudebush, R. E. (1990). Suppressive effects of morphine pellet implants on in vivo parameters of immune function. *The Journal of Pharmacology and Experimental Therapeutics, 255*, 410–414.

Bryant, H. U., Bernton, E. W., & Holaday, J. W. (1987). Immunosuppressive effects of chronic morphine treatment in mice. *Life Sciences, 41*, 1731–1738.

Buchweitz, J. P., Karmaus, P. W. F., Harkema, J. R., Williams, K. J., & Kaminski, N. (2007). Modulation of airway responses to influenza A/PR/8/34 by Δ9-tetrahydrocannabinol in C57BL/6 mice. *Journal of Pharmacology and Experimental Therapeutics, 323*, 675–683.

Buckley, N. E., McCoy, K. L., Mezey, E., Bonner, T., Zimmer, A., Felder, C. C., & Glass, M. (2000). Immunomodulation by cannabinoids is absent in mice deficient for the cannabinoid CB(2) receptor. *European Journal of Pharmacology, 396*, 141–149.

Bussiere, J. L., Adler, M. W., Rogers, T. J., & Eisenstein, T. K. (1992). Differential effects of morphine and naltrexone on the antibody response in various mouse strains. *Immunopharmacology and Immunotoxicology, 14*, 657–673.

Bussiere, J. L., Adler, M. W., Rogers, T. J., & Eisenstein, T. K. (1993). Cytokine reversal of morphine-induced suppression of the antibody response. *The Journal of Pharmacology and Experimental Therapeutics, 264*, 591–597.

Cabral, G. A., & Marciano-Cabral, F. (2004). Cannabinoid-mediated exacerbation of brain infection by opportunistic amebae. *Journal of Neuroimmunology, 147*, 127–130.

Cabral, G. A., Mishkin, E. M., Marciano-Cabral, F., Coleman, P., Harris, L., & Munson, A. E. (1986). Effect of delta-9-tetrahydrocannabinol on herpes simplex virus type 2 vaginal infections in the Guinea pig. *Proceedings of the Society for Experimental Biology and Medicine Society for Experimental Biology and Medicine (New York, N Y), 182*, 181–186.

Cabral, G. A., Stinnett, A. L., Bailey, J., Ali, S. F., Paule, M. G., Scallet, A. C., & Slikker, W. (1991). Chronic marijuana smoke alters alveolar macrophage morphology and protein expression. *Pharmcology Biochemistry and Behavior, 40*, 643–649.

Calipari, E. S., Godino, A., Peck, E. G., Salery, M., Mervosh, N. L., Landry, J. A., Russo, S. J., Hurd, Y. L., Nestler, E. J., & Kiraly, D. D. (2018). Granulocyte-colony stimulating factor controls neural and behavioral plasticity in response to cocaine. *Nature Communications, 9*, 9.

Carpenter, G. W., & Carr, D. J. J. (1995). Pretreatment with β-funaltrexamine blocks morphine-mediated suppression of CTL activity in alloimmunized mice. *Immunopharmacology, 29*, 129–140.

Carr, D. J. J. (1991). The role of endogenous opioids and their receptors in the immune system. *Proceedings of the Society for Experimental Biology and Medicine Society for Experimental Biology and Medicine (New York, N Y), 198*, 710–720.

CDC. (2020). *HIV surveillance report, 2018 (Updated)*, in http://www.cdc.gov/hiv/library/reports/hiv-surveillance.html

Chang, J. K., Cornwell, W. D., & Rogers, T. J. (2021). The mu-opioid receptor induces miR-21 expression and is ERK/PKCmu-dependent. *Journal of Neuroimmunology, 356*, 577585.

Chao, C. C., Sharp, B. M., Pomeroy, C., Filice, G. A., & Peterson, P. K. (1990). Lethality of morphine in mice infected with Toxoplasma gondii. *The Journal of Pharmacology and Experimental Therapeutics, 252*, 605–609.

Chao, C. C., Hu, S., Molitor, T. W., Zhou, Y., Murtaugh, M. P., Tsang, M., & Peterson, P. K. (1992). Morphine potentiates transforming growth factor-β release from human peripheral blood mononuclear cell cultures. *The Journal of Pharmacology and Experimental Therapeutics, 262*, 19–24.

Chen, G.-J., & Watson, R. R. (1991). Modulation of tumor necrosis factor and gamma interferon production by cocaine and morphine in aging mice infected with LP-BM5, a murine retrovirus. *Journal of Leukocyte Biology, 50*, 349–355.

Chen, C., Li, J., Bot, G., Szabo, I., Rogers, T. J., & Liu-Chen, L. Y. (2004). Heterodimerization and cross-desensitization between the μ opioid receptor and the chemokine CCR5 receptor. *European Journal of Pharmacology, 483*, 175–186.

Choi, Y., Chuang, L. F., Lam, K. M., Kung, H. F., Wang, J. M., Osburn, B. I., & Chuang, R. Y. (1999). Inhibition of chemokine-induced chemotaxis of monkey leukocytes by mu-opioid receptor agonists. *in vivo, 13*, 389–396.

Chuang, L. F., Killam, K. F., Jr., & Chuang, R. Y. (1997). SIV infection of macaques: A model for studying AIDS and drug abuse. *Addiction Biology, 2*, 421–430.

Chun, J., Haug, N. A., Guydish, J. R., Sorensen, J. L., & Delucchi, K. (2009). Cigarette smoking among opioid-dependent clients in a therapeutic community. *The American Journal on Addictions, 18*, 316–320.

Clark, J. D., Shi, X., Li, X., Qiao, Y., Liang, D. Y., Angst, M. S., & Yeomans, D. C. (2007). Morphine reduces local cytokine expresssion and neutrophil infiltration after incision. *Molecular Pain, 3*, 28–40.

Cornwell, W. D., Lewis, M. G., Fan, X., Rappaport, J., & Rogers, T. J. (2013). Effect of chronic morphine administration on circulating T cell population dynamics in rhesus macaques. *Journal of Neuroimmunology, 265*, 43–50.

Cornwell, W. D., Sriram, U., Seliga, A., Zuluaga-Ramirez, V., Gajghate, S., Rom, S., Winfield, M., Heldt, N. A., Ambrose, D., Rogers, T. J., & Persidsky, Y. (2020). Tobacco smoke and morphine alter peripheral and CNS inflammation following HIV infection in a humanized mouse model. *Scientific Reports, 10*, 13977.

Correa, F., Docagne, F., Mestre, L., Clemente, D., Hernangomez, M., Loria, F., & Guaza, C. (2009). A role for CB2 receptors in anandamide signalling pathways involved in the regulation of IL-12 and IL-23 in microglial cells. *Biochemical Pharmacology, 77*, 86–100.

Correa, F., Hernangomez, M., Mestre, L., Loria, F., Spagnolo, A., Docagne, F., Di Marzo, V., & Guaza, C. (2010). Anandamide enhances IL-10 production in activated microglia by targeting CB(2) receptors: Roles of ERK1/2, JNK, and NF-kappaB. *Glia, 58*, 135–147.

Costantino, C. M., Gupta, A., Yewdall, A. W., Dale, B. M., & Devi, L. A. (2012). Cannabinoid receptor 2-mediated attenuation of CXCR4-tropic HIV infection in primary CD4+ T cells. *PLoS ONE, 7*, e33961.

Crothers, K., Huang, L., Goulet, J. L., Goetz, M. B., Brown, S. T., Rodriguez-Barradas, M. C., Oursler, K. K., Rimland, D., Gibert, C. L., Butt, A. A., & Justice, A. C. (2011). HIV infection and risk for incident pulmonary diseases in the combination antiretroviral therapy era. *American Journal of Respiratory and Critical Care Medicine, 183*, 388–395.

Derocq, J. M., Jbilo, O., Bouaboula, M., Ségui, M., Clère, C., & Casellas, P. (2000). Genomic and functional changes induced by the activation of the peripheral cannabinoid receptor CB2 in the promyelocytic cells HL-60. *Journal of Biological Chemistry, 275*, 15621–15628.

Devane, W. A., Hanus, L., Breuer, A., Pertwee, R. G., Stevenson, L. A., Griffin, G., Gibson, D., Mandelbaum, A., Etinger, A., & Mechoulam, R. (1992). Isolation and structure of a brain constituent that binds to the cannabinoid receptor. *Science, 258*, 1946–1949.

Di Marzo, V., & Deutsch, D. G. (1998). Biochemistry of the endogenous ligands of cannabinoid receptors. *Neurobiology of Disease, 5*, 386–404.

Diaz, P. T., Wewers, M. D., Pacht, E., Drake, J., Nagaraja, H. N., & Clanton, T. L. (2003). Respiratory symptoms among HIV-seropositive individuals. *Chest, 123*, 1977–1982.

Donahoe, R. M., Byrd, L. D., McClure, H. M., Fultz, P., Brantley, M., Marsteller, F., Ansari, A. A., Wenzel, D., & Aceto, M. (1993). Consequences of opiate-dependency in a monkey model of AIDS. *Advances in Experimental Medicine and Biology, 335*, 21–28.

Duaso, M., & Duncan, D. (2012). Health impact of smoking and smoking cessation strategies: Current evidence. *British Journal of Community Nursing, 17*, 356–363.

Eisenstein, T. K. (2019). The role of opioid receptors in immune system function. *Frontiers in Immunology, 10*, 2904.

Eisenstein, T. K., & Meissler, J. J. (2015). Effects of cannabinoids on T-cell function and resistance to infection. *Journal of Neuroimmune Pharmacology, 10*, 204–216.

Eisenstein, T. K., Meissler, J. J., Jr., Geller, E. B., & Adler, M. W. (1990). Immunosuppression to tetanus toxoid induced by implanted morphine pellets. *Annals of the New York Academy of Sciences, 594*, 377–379.

Eisenstein, T. K., Meissler, J. J., Jr., Rogers, T. J., Geller, E. B., & Adler, M. W. (1995). Mouse strain differences in immunosuppression by opioids in vitro. *The Journal of Pharmacology and Experimental Therapeutics, 275*, 1484–1489.

Eisenstein, T. K., Rahim, R. T., Feng, P., Thingalaya, N. K., & Meissler, J. J. (2006). Effects of opioid tolerance and withdrawal on the immune system. *Journal of Neuroimmune Pharmacology, 1*, 237–249.

Eisenstein, T. K., Meissler, J. J., Wilson, Q., Gaughan, J. P., & Adler, M. W. (2007). Anandamide and Δ9-tetrahydrocannabinol directly inhibit cells of the immune system via CB2 receptors. *Journal of Neuroimmunology, 189*, 17–22.

El-Hage, N., Gurwell, J. A., Singh, I. N., Knapp, P. E., Nath, A., & Hauser, K. F. (2005). Synergisitc increases in intracellular Ca2+, and the release of MCP-1, RANTES, and IL-6 by astrocytes treated with opiates and HIV-1 Tat. *Glia, 50*, 91–106.

El-Hage, N., Wu, G., Wang, J., Ambati, J., Knapp, P. E., Reed, J. L., Bruce-Keller, A. J., & Hauser, K. F. (2006). HIV-1 Tat and opiate-induced changes in astrocytes promote chemotaxis of microglia through the expression of MCP-1 and alternative chemokines. *Glia, 53*, 132–146.

Ellis, R., Langford, D., & Masliah, E. (2007). HIV and antiretroviral therapy in the brain: Neuronal injury and repair. *Nature Reviews Neurosciences, 8*, 33–44.

Feng, P., Wilson, Q. M., Meissler, J. J., Jr., Adler, M. W., & Eisenstein, T. K. (2005). Increased sensitivity to Salmonella enterica serovar Typhimurium infection in mice undergoing withdrawal from morphine is associated with suppression of interleukin-12. *Infection and Immunity, 73*, 7953–7959.

Feng, P., Rahim, R. T., Cowan, A., Liu-Chen, L. Y., Peng, X., Gaughan, J., Meissler, J. J., Jr., Adler, M. W., & Eisenstein, T. K. (2006). Effects of mu, kappa or delta opioids administered by pellet or pump on oral Salmonella infection and gastrointestinal transit. *European Journal of Pharmacology, 534*, 250–257.

Fischer-Stenger, K., Updegrove, A. W., & Cabral, G. A. (1992). Δ9-tetrahydrocannabinol decreases cytotoxic T lymphocyte activity to herpes simplex virus type1-infected cells. *Proceedings of the Society for Experimental Biology and Medicine Society for Experimental Biology and Medicine (New York, N Y), 200*, 422–430.

Fischer-Stenger, K., Dove Pettit, D. A., & Cabral, G. A. (1993). Δ9-tetrahydrocannabinol inhibition of tumor necrosis factor-α: Suppression of post-translational events. *The Journal of Pharmacology and Experimental Therapeutics, 267*, 1558–1565.

Fox, H. C., D'Sa, C., Kimmerling, A., Siedlarz, K. M., Tuit, K. L., Stowe, R., & Sinha, R. (2012). Immune system inflammation in cocaine dependent individuals: Implications for medications development. *Human Psychopharmacology, 27*, 156–166.

Fraga, D., Raborn, E., Ferreira, G., & Cabral, G. (2011). Cannabinoids inhibit migration of microglial-like cells to the HIV protein Tat. *Journal of Neuroimmune Pharmacology, 6*, 566–577.

Fukagawa, H., Koyama, T., Kakuyama, M., & Fukuda, K. (2013). Microglial activation involved in morphine tolerance is not mediated by toll-like receptor 4. *Journal of Anesthesia, 27*, 93–97.

Galiegue, S., Mary, S., Marchand, J., Dussossoy, D., Carriere, D., Carayon, P., Bouaboula, M., Shire, D., Le Fur, G., & Casellas, P. (1995). Expression of central and peripheral cannabinoid receptors in human immune tissues and leukocyte subpopulations. *European Journal of Biochemistry/FEBS, 232*, 54–61.

Gan, X., Zhang, L., Newton, T., Chang, S. L., Ling, W., Kermani, V., Berger, O., Graves, M. C., & Fiala, M. (1998). Cocaine infusion increases interferon-γ and decreases interleukin-10 in cocaine-dependent subjects. *Clinical Immunology and Immunopathology, 89*, 181–190.

Gavériaux-Ruff, C., Matthes, H. W. D., Peluso, J., & Kieffer, B. L. (1998). Abolition of morphine-immunosuppression in mice lacking the μ-opioid receptor gene. *Proceedings of the National Academy of Sciences, USA, 95*, 6326–6330.

Ghosh, S., Preet, A., Groopman, J. E., & Ganju, R. K. (2006). Cannabinoid receptor CB2 modulates the CXCL12/CXCR4-mediated chemotaxis of T lymphocytes. *Molecular Immunology, 43*, 2169–2179.

Gingo, M. R., Morris, A., & Crothers, K. (2013). Human immunodeficiency virus-associated obstructive lung diseases. *Clinics in Chest Medicine, 34*, 273–282.

Gorantla, S., Makarov, E., Roy, D., Finke-Dwyer, J., Murrin, L. C., Gendelman, H. E., & Poluektova, L. (2010). Immunoregulation of a CB2 receptor agonist in a murine model of neuroAIDS. *Journal of Neuroimmune Pharmacology, 5*, 456–468.

Grimm, M. C., Ben-Baruch, A., Taub, D. D., Howard, O. M. Z., Resau, J. H., Wang, J. M., Ali, H., Richardson, R., Snyderman, R., & Oppenheim, J. J. (1998). Opiates trans-deactivate chemokine receptors: δ and μ opiate receptor-mediated heterologous desensitization. *The Journal of Experimental Medicine, 188*, 317–325.

Guan, L., Townsend, R., Eisenstein, T. K., Adler, M. W., & Rogers, T. J. (1994). Both T cells and macrophages are targets of κ-opioid-induced immunosuppression. *Brain, Behavior, and Immunity, 8*, 229–240.

Happel, C., Steele, A. D., Finley, M. J., Kutzler, M. A., & Rogers, T. J. (2008). DAMGO-induced expression of chemokines and chemokine receptors: The role of TGF-{beta}1. *Journal of Leukocyte Biology, 83*, 956–963.

Happel, C., Kutzler, M., & Rogers, T. J. (2011). Opioid-induced chemokine expression requires NF-kappa-B activity: the role of PKCzeta. *Journal of Leukocyte Biology, 89*, 301–309.

Hauser, K. F., Fitting, S., Dever, S. M., Podhaizer, E. M., & Knapp, P. E. (2012). Opiate drug use and the pathophysiology of neuroAIDS. *Current HIV Research, 10*, 435–452.

Havas, H. F., Dellaria, M., Schiffman, G., Geller, E. B., & Adler, M. W. (1987). Effect of cocaine on the immune response and host resistance in BALB/c mice. *International Archives of Allergy & Immunology, 83*, 377–383.

Haverkos, H. W., & Lange, R. W. (1990). Serious infections other than human immunodeficiency virus among intravenous drug users. *The Journal of Infectious Diseases, 161*, 894–902.

Heinisch, S., Palma, J., & Kirby, L. G. (2011). Interactions between chemokine and mu-opioid receptors: Anatomical findings and electrophysiological studies in the rat periacqueductal gray. *Brain, Behavior, and Immunity, 25*, 360–372.

Hilburger, M. E., Adler, M. W., Truant, A. L., Meissler, J. J., Jr., Satishchandran, V., Rogers, T. J., & Eisenstein, T. K. (1997). Morphine induces sepsis in mice. *The Journal of Infectious Diseases, 176*, 183–188.

Hofford, R. S., Mervosh, N. L., Euston, T. J., Meckel, K. R., Orr, A. T., & Kirary, D. D. (2021). Alterations in microbiome composition and metabolic byproducts drive behavioral and transcriptional responses to morphine. *Neuropsychopharmacology, 46*, 2062–2072.

Honarvar, B., Lankarani, K. B., Odoomi, N., Roudgari, A., Moghadami, M., Kazerooni, P. A., & Abadi, A. H. (2013). Pulmonary and latent tuberculosis screening in opiate drug users. An essential and neglected approach for harm-reduction facilities. *Journal of Addiction Medicine, 7*, 230–235.

Huemer, H. P., Lassnig, C., Bernhard, D., Sturm, S., Nowotny, N., Kitchen, M., & Pavlic, M. (2011). Cannabinoids lead to enhanced virulence of the smallpox vaccine (vaccinia) virus. *Immunobiology, 216*, 670–677.

Hussey, H. H., & Katz, S. (1950). Infections resulting from narcotic addiction. *The American Journal of Medicine, 9*, 186–193.

Hutchinson, M. R., & Watkins, L. R. (2014). Why is neuroimmunophamacolgy crucial for the future of addiction research? *Neuropharmacology, 76*, 218–227.

Hutchinson, M. R., Zhang, Y., Shridhar, M., Evans, J. H., Buchanan, M. M., Zhao, T. X., Slivka, P. F., Coats, B. D., Rezvani, N., Wieseler, J., Hughes, T. S., Landgraf, K. E., Chan, S., Fong, S., Phipps, S., Falke, J. J., Leinwand, L. A., Maier, S. F., Yin, H., & Rice, K. C. W. L. R. (2010). Evidence that opioids may have toll-like receptor 4 and MD-2 effects. *Brain Behavior and Immunity, 24*, 83–95.

Hutchinson, M. R., Northcutt, A. L., Hiranita, T., Wang, X., Lewis, S. S., Thomas, J., van Steeg, K., Kopajtic, T. A., Loram, L. C., Sfregola, C., Galer, E., Miles, N. E., Bland, S. T., Amat, J., Rozeske, R. R., Maslanik, T., Chapman, T. R., Strand, K. A., Fleshner, M., Bachtell, R. K., Somogyi, A. A., Yin, H., Katz, J. L., Rice, K. C., Maier, S. F., & Watkins, L. R. (2012). Opioid activation of toll-like receptor 4 contributes to drug reinforcement. *Journal of Neuroscience, 32*, 11187–11200.

Inan, S., Eisenstein, T. K., Watson, M. N., Doura, M., Meissler, J. J., Tallarida, C. S., Chen, X., Geller, E. B., Rawls, S. M., Cowan, A., & Adler, M. W. (2018). Coadministration of chemokine receptor antagonists with morphine potentiates morphine's analgesic effect on incisional pain in rats. *Journal of Pharmacology and Experimental Therapeutics, 367*, 433–441.

Inan, S., Chen, X., Eisenstein, E. M., Meissler, J. J., Geller, E. B., Tallarida, C., Watson, M., Doura, M., Barrett, J. E., Cowan, A., Rawls, S. M., Adler, M. W., & Eisenstein, T. K. (2021). Chemokine receptor antagonists enhance morphine's antinociceptive effect but not respiratory depression. *Life Sciences, 285*, 120014.

Iriah, S. C., Borges, C., Shalev, U., Cai, X., Madularu, D., Kulkarni, P. P., & Ferris, C. F. (2021). The utility of maraviroc, an antiretroviral agent used to treat HIV, as treatment for opioid abuse? Data from MRI and behavioural testing in rats. *Journal of Psychiatry and Neuroscience, 46*, E548–E558.

Jbilo, O., Derocq, J. M., Segui, M., Le Fur, G., & Casellas, P. (1999). Stimulation of peripheral cannabinoid receptor CB2 induces MCP-1 and IL-8 gene expression in human promyelocytic cell line HL60. *FEBS Letters, 448*, 273–277.

Kaminski, N. E., Abood, M. E., Kessler, F. K., Martin, B. R., & Schatz, A. R. (1992). Identification of a functionally relevant cannabinoid receptor on mouse spleen cells that is involved in cannabinoid-mediated immune modulation. *Molecular Pharmacology, 42*, 736–742.

Kaminski, N. E., Koh, W. S., Yang, K. H., Lee, M., & Kessler, F. K. (1994). Suppression of the humoral immune response by cannabinoids is partially mediated through inhibition of adenylate cyclase by a pertussis toxin-sensitive G-protein coupled mechanism. *Biochemical Pharmacology, 48*, 1899–1908.

Kang, M., Mischel, R. A., Bhave, S., Komia, E., Cho, A., Huang, C., Dewey, W. L., & Akbarali, H. I. (2017). The effect of gut microbiome on tolerance to morphine mediated antinociception in mice. *Scientific Reports, 7*, 42658.

Kapadia, F., Vlahov, D., Donahoe, R. M., & Friedland, G. (2005). The role of substance abuse in HIV disease progression: Reconciling differences from laboratory and epidemiological investigations. *Clinical Infectious Diseases, 41*, 1027–1034.

Kelschenbach, J., Barke, R. A., & Roy, S. (2005). Morphine withdrawal contributes to Th cell differentiation by biasing cells toward the Th2 lineage. *Journal of immunology (Baltimore, Md : 1950), 175*, 2655–2665.

Khabbazi, S., Goumon, Y., & Parat, M. O. (2015). Morphine modulates interleukin-4 or breast cancer cell-induced pro-metastatic activation of macrophages. *Scientific Reports, 5*, 11389.

Kim, J., Conneley, K. L., Unterwald, E. M., & Rawls, S. M. (2017). Chemokines and cocaine: CXCR4 receptor antagonist AMD3100 attenuates cocaine place preference and locomotor stimulation in rats. *Brain, Behavior, and Immunity, 62*, 30–34.

Klein, T. W., & Cabral, G. A. (2006). Cannabinoid-induced immune suppression and modulation of antigen-presenting cells. *Journal of Neuroimmune Pharmacology, 1*, 50–64.

Klein, T. W., Newton, C. A., Widen, R., & Friedman, H. (1985). The effect of delta-9-tetrahydrocannabinol and 11-hydroxy-delta-9-tetrahydrocannabinol on T-lymphocyte and B-lymphocyte mitogen responses. *Journal of Immunopharmacology, 7*, 451–466.

Klein, T. W., Newton, C., & Friedman, H. (1987). Inhibition of natural killer cell function by marijuana components. *Journal of Toxicology and Environmental Health, 20*, 321–332.

Klein, T. W., Kawakami, Y., Newton, C., & Friedman, H. (1991). Marijuana components suppress inducion and cytolytic function of murine cytotoxic T-cells in vitro and in vivo. *Journal of Toxicology and Environmental Health, 32*, 465–477.

Klein, T. W., Newton, C. A., Nakachi, N., & Friedman, H. (2000). Δ9-tetrahydrocannabinol treatment suppresses immunity and early IFN-γ, IL-12 and IL-12 receptor β2 responses to Legionella pneumophila infection. *Journal of Immunology (Baltimore, Md : 1950), 164*, 6461–6466.

Klein, T. W., Newton, C., Larsen, K., Chou, J., Perkins, I., Lu, L., Nong, L., & Friedman, H. (2004). Cannabinoid receptors and T helper cells. *Journal of Neuroimmunology, 147*, 91–94.

Kohut, S. J. (2017). Interactions between nicotine and drugs of abuse: A review of preclinical findings. *The American Journal of Drug and Alcohol Abuse, 43*, 155–170.

Koodie, L., Yuan, H., Pumper, J. A., Yu, H., Charboneau, R., Ramkrishnan, S., & Roy, S. (2014). Morphine inhibits migration of tumor-infiltrating leukocytes and suppresses angiogenesis associated with tumor growth in mice. *The American Journal of Pathology, 184*, 1073–1084.

Kumar, R., Orsoni, S., Norman, L., Verma, A. S., Tirado, G., Giavedoni, L. D., Staprans, S., Miller, G. M., Buch, S. J., & Kumar, A. (2006). Chronic morphine exposure causes pronounced virus replication in cerebral compartment and accelerated onset of AIDS in SIV/SHIV-infected Indian rhesus macaques. *Virology, 354*, 192–206.

Kuroda, S., Nishio, M., Sasaki, T., Horie, Y., Kawahara, K., Sasaki, M., Natsui, M., Matozaki, T., Tezuka, H., Ohteki, T., Forster, I., Mak, T. W., Nakano, T., & Suzuki, A. (2008). Effective clearance of intracellular Leishmania major in vivo requires Pten in macrophages. *European Journal of Immunology, 38*, 1331–1340.

Lefkowitz, S. S., & Chiang, C. Y. (1975). Effects of certain abused drugs on hemolysin forming cells. *Life Sciences, 17*, 1763–1768.

Lewis, S. S., Loram, L. C., Hutchinson, M. R., Li, C. M., Zhang, Y., Maier, S. F., Huang, Y., Rice, K. C., & Watkins, L. R. (2012). (+)-naloxone, an opioid-inactive toll-like receptor 4 signaling inhibitor, reverses multiple models of chronic neuropathic pain in rats. *The Journal of Pain, 13*, 498–506.

Li, Y., Wang, X., Tian, S., Guo, C. J., Douglas, S. D., & Ho, W. Z. (2002). Methadone enhances human immunodeficiency virus infection of human immune cells. *The Journal of Infectious Diseases, 185*, 118–122.

Lioy, D. T., Sheridan, P. A., Hurley, S. D., Walton, J. R., Martin, A. M., Olschowka, J. A., & Moynihan, J. A. (2006). Acute morphine exposure potentiates the development of HSV-1-induced encephalitis. *Journal of Neuroimmunology, 172*, 9–17.

Loh, H. H., & Smith, A. P. (1990). Molecular characterization of opioid receptors. *Annual Review of Pharmacology and Toxicology, 30*, 123–147.

Lopez-Cepero, M., Friedman, M., Klein, T., & Friedman, H. (1986). Tetrahydrocannabinol-induced suppression of macrophage spreading and phagocytic activity in vitro. *Journal of Leukocyte Biology, 39*, 679–686.

Louria, D. B., Hensle, T., & Rose, J. (1967). The major medical complications of heroin addiction. *Annals of Internal Medicine, 67*, 1–22.

Ma, J., Wang, J., Wan, J., Charboneau, R., Chang, Y., Barke, R. A., & Roy, S. (2010). Morphine disrupts interleukin-23 (IL-23)/IL-17-mediated pulmonary mucosal host defense against Streptococcus pneumoniae infection. *Infection and Immunity, 78*, 830–837.

MacDonald, E., Arnesen, T. M., Brantsaeter, A. B., Gerlyng, P., Grepp, M., Hansen, B. Å., Jensrud, K., Lundgren, P., Mellegård, H., Møller-Stray, J., Rønning, K., Vestrhelm, D. F., & Vold, L. (2013). Outbreak of wound botulism in people who inject drugs, Norway, October to November 2013. *Euro Surveillance, 18*, 20630.

MacFarlane, A. S., Peng, X., Meissler, J. J., Jr., Rogers, T. J., Geller, E. B., Adler, M. W., & Eisenstein, T. K. (2000). Morphine increases susceptibility to oral Salmonella typhimurium infection. *The Journal of Infectious Diseases, 181*, 1350–1358.

Machelska, H., & Celik, M. O. (2020). Opioid receptors in immune and glial cells-implications for pain control. *Frontiers in Immunology, 11*, 300.

Mahajan, S. D., Aalinkeel, R., Sykes, D. E., Reynolds, J. L., Bindukumar, B., Fernandez, S. F., Chawda, R., Shanahan, T. C., & Schwartz, S. A. (2008). Tight junction regulation by morphine and HIV-1 tat modulates blood-brain barrier permeability. *Journal of Clinical Immunology, 28*, 528–541.

Maresz, K., Pryce, G., Ponomarev, E. D., Marsicano, G., Croxford, J. L., Shriver, L. P., Ledent, C., Cheng, X., Carrier, E. J., Mann, M. K., Giovannoni, G., Pertwee, R. G., Yamamura, T., Buckley, N. E., Hillard, C. J., Lutz, B., Baker, D., & Dittel, B. N. (2007). Direct suppression of CNS autoimmune inflammation via the cannabinoid receptor CB1 on neurons and CB2 on autoreactive T cells. *Nature Medicine, 13*, 492–497.

Martin, J. L., Charboneau, R., Barke, R. A., & Roy, S. (2010a). Chronic morphine treatment inhibits LPS-induced angiogenesis: Implications in wound healing. *Cellular Immunology, 265*, 139–145.

Martin, J. L., Koodie, L., Krishnan, A. G., Charboneau, R., Barke, R. A., & Roy, S. (2010b). Chronic morphine administration delays wound healing by inhibiting immune cell recruitment to the wound site. *American Journal of Pathology, 176*, 786–799.

Martucci, C., Franchi, S., Lattauda, L., Panerai, A. E., & Sacerdote, P. (2007). Differential involvement of RelB in morphine-induced modulation of chemotaxis, NO, and cytokine production in murine macrophages and lymphocytes. *Journal of Leukocyte Biology, 81*, 344–354.

Mathers, B. M., Degenhardt, L., Ali, H., Wiessing, L., Hickman, M., Mattick, R. P., Myers, B., Ambekar, A., & Strathdee, S. A. (2010). HIV prevention, treatment, and care services for people who inject drugs: A systematic review of global, regional, and national coverage. *Lancet, 375*, 1014–1028.

Mattioli, T. A., Leduc-Pessah, H., Skelhorne-Gross, G., Nicol, C. J. B., Milne, B., Trang, T., & Cahill, C. M. (2014). Toll-like receptor 4 mutant and null mice retain morphine-induced tolerance, hyperalgesia, and physical dependence. *PLoS ONE [Electronic Resource], 9*, e97361.

McCarthy, L., Wetzel, M., Sliker, J., Eisenstein, T. K., & Rogers, T. J. (2001). Opioids, opioid receptors, and the immune response. *Drug and Alcohol Dependence, 62*, 111–123.

McCoy, K. L., Matveyeva, M., Carlisle, S. J., & Cabral, G. A. (1999). Cannabinoid inhibition of the processing of intact lysozyme by macrophages: Evidence for CB2 receptor participation. *The Journal of Pharmacology and Experimental Therapeutics, 289*, 1620–1625.

Mechoulam, R., Ben-Shabat, S., Hanus, L., Ligumsky, M., Kaminski, N. E., Schatz, A. R., Gopher, A., Almog, S., Martin, B. R., Compton, D. R., Pertwee, R. G., Griffin, G., Bayewitch, M., Barg, J., & Vogel, Z. (1995). Identification of an endogenous 2-monoglyceride, present in canine gut, that binds to cannabinoid receptors. *Biochemical Pharmacology, 50*, 83–90.

Medveczky, M. M., Sherwood, T. A., Klein, T. W., Friedman, H., & Medveczky, P. G. (2004). Delta-9-tetrahydrocannabinol (THC) inhibits lytic replication of gamma oncogenic herpesviruses in vitro. *BMC Medicine, 2*, 34.

Mojadadi, S., Jamali, A., Khansarinejad, B., Soleimanjahi, H., & Bamdad, T. (2009). Acute morphine adminstration reduces cell-mediated immunity and induces reactivation of latent herpes simplex virus type 1 in BALB/c mice. *Cellular and molecular immunology, 6*, 111–116.

Molina, P. E., Happel, K. I., Zhang, P., Kolis, J. K., & Nelson, S. (2010). Focus on: Alcohol and the immune system. *Alcohol Research & Health, 33*, 97–108.

Molina, P. E., Winsauer, P., Zhang, P., Walker, E., Birke, L., Amedee, A., Stouwe, C. V., Troxclair, D., McGoey, R., Varner, K., Byerley, L., & LaMotte, L. (2011). Cannabinoid administration attenuates the progression of simian immunodeficiency virus. *AIDS Research and Human Retroviruses, 27*, 585–592.

Molitor, T. W., Morilla, A., Risdahl, J. M., Murtaugh, M. P., Chao, C. C., & Peterson, P. K. (1992). Chronic morphine administration impairs cell-mediated immune responses in swine. *The Journal of Pharmacology and Experimental Therapeutics, 260*, 581–586.

Montecucco, F., Burger, F., Mach, F., & Steffens, S. (2008). CB2 cannabinoid receptor agonist JWH-015 modulates human monocyte migration through defined intracellular signaling pathways. *American Journal of Physiology - Heart & Circulatory Physiology, 294*, H1145–H1155.

Mora, A. L., Salazar, M., Pablo-Caeiro, J., Frost, C. P., Yadav, Y., DuPont, H. L., & Garey, K. W. (2012). Moderate to high use of opioid analgesics are associated with an increased risk of Clostridium difficile infection. *American Journal of the Medical Sciences, 343*, 277–280.

Morahan, P. S., Klykken, P. C., Smith, S. H., Harris, L. S., & Munson, A. E. (1979). Effects of cannabinoids of host resistance to Listeria monocytogenes and herpes simples virus. *Infection and Immunity, 23*, 670–674.

Morris, C. D., & Garver-Apgar, C. E. (2020). Nicotine and opioids: A call for Co-treatment as the standard of care. *The Journal of Behavioral Health Services & Research, 47*, 601–613.

Mosmann, T. R., & Coffman, R. L. (1987). Two types of mouse helper T-cell clone: Implications for immune regulation. *Immunology Today, 8*, 223–227.

Munro, S., Thomas, K. K., & Abu-Shaar, M. (1993). Molecular characterization of a peripheral receptor for cannabinoids. *Nature, 365*, 61–65.

Murphy, A., Barbaro, J., Martinez-Aguado, P., Chilunda, V., Jaureguiberry-Bravo, M., & Berman, J. W. (2019). The effects of opioids on HIV neuropathogenesis. *Frontiers in Immunology, 10*, 2445.

Nair, M. P. N., Schwartz, S. A., Polasani, R., Hou, J., Sweet, A., & Chadha, K. C. (1997). Immunoregulatory effects of morphine on human lymphocytes. *Clinical and Diagnostic Laboratory Immunology, 4*, 127–132.

Nair, M. P. N., Mahajan, S. D., Schwartz, S. A., Reynolds, J., Whitney, R., Bernstein, Z., Chawda, R. P., Sykes, D., Hewitt, R., & Hsiao, C. B. (2005). Cocaine modulates dendritic cell-specific C type intercellular adhesion molecule-3-grabbing nonintegrin expression by dendritic cells in HIV-1 patients. *Journal of Immunology (Baltimore, Md : 1950), 174*, 6617–6626.

Nayak, S. U., Cicalese, S., Tallarida, C., Oliver, C. F., & Rawls, S. M. (2020). Chemokine CCR5 and cocaine interactions in the brain: Cocaine enhances mesolimbic CCR5 mRNA levels and produces place preference and locomotor activation that are reduced by a CCR5 antagonist. *Brain, Behavior, and Immunity, 83*, 288–292.

Newton, C., Klein, T. W., & Friedman, H. (1994). Secondary immunity to Legionella pneumophila and Th1 activity are suppressed by delta-9-tetrahydrocannabinol injection. *Infection and Immunity, 62*, 4015–4020.

Newton, C. A., Chou, P. J., Perkins, I., & Klein, T. W. (2009). CB1 and CB2 cannabinoid receptors mediate different aspects of Delta-9-tetrahydrocannabinol (THC)-induced T helper cell shift following immune activation by Legionella pneumophila infection. *Journal of Neuroimmune Pharmacology, 4*, 92–102.

Ni, X., Gritman, K. R., Eisenstein, T. K., Adler, M. W., Arfors, K. E., & Tuma, R. F. (2000). Morphine attenuates leukocyte/endothelial interactions. *Microvascular Research, 60*, 121–130.

Ninković, J., & Roy, S. (2012). Morphine decreases bacterial phagocytosis by inhibiting actin polymerization through cAMP-. Rac-1-, and p38 MAPK-dependent mechanisms. *American Journal of Pathology, 180*, 1068–1079.

Olin, M., Choi, K., & Molitor, T. W. (2011). Morphine alters M. bovis infected microglia's ability to activate γδ T lymphocytes. *Journal of Neuroimmune Pharmacology, 6*, 578–584.

Olin, M., Oh, S., Roy, S., Peterson, P., & Molitor, T. (2012). Morphine induces splenocyte trafficking into the CNS. *Journal of Neuroimmune Pharmacology, 7*, 436–443.

Ou, D. W., Shen, M. L., & Luo, Y. D. (1989). Effects of cocaine on the immune system of Balb/c mice. *Clinical Immunology and Immunopathology, 52*, 305–312.

Overacre-Delgoffe, A. E., Chikina, M., Dadey, R. E., Yano, H., Brunazzi, E. A., Shayan, G., Horne, W., Moskovitz, J. M., Kolls, J. K., Sander, C., Shuai, Y., Normolle, D. P., Kirkwood, J. M., Ferris, R. L., Delgoffe, G. M., Bruno, T. C., Workman, C. J., & Vignali, D. A. A. (2017). Interferon-gamma drives T(reg) fragility to promote anti-tumor immunity. *Cell, 169*, 1130–1141 e1111.

Pacifici, R., di Carlo, S., Bacosi, A., & Zuccaro, P. (1993). Macrophage functions in drugs of abuse-treated mice. *International Journal of Immunopharmacology, 15*, 711–716.

Pacifici, R., Fiaschi, A. I., Micheli, L., Centini, F., Giorgi, G., Zuccaro, P., Pichini, S., di Carlo, S., Bacosi, A., & Cerretani, D. (2003). Immunosuppression and oxidative stress induced by acute and chronic exposure to cocaine in rat. *International Immunopharmacology, 3*, 581–592.

Parikh, N., Dampier, W., Feng, R., Passic, S. R., Zhong, W., Frantz, B., Blakey, B., Aiamkitsumrit, B., Pirrone, V., Nonnemacher, M. R., Jacobson, J. M., & Wigdahl, B. (2014). Cocaine alters cytokine profiles in HIV-1-infected African American individuals in the DrexelMed HIV/AIDS genetic analysis cohort. *Journal of Acquired Immune Deficiency Syndrome, 66*, 256–264.

Parsadaniantz, S. M., Rivat, C., Rostene, W., & Goazigo, A. (2015). Opioid and chemokine receptor crosstalk: A promising target for pain therapy? *Nature Reviews Neuroscience, 16*, 69–78.

Patel, V., Borysenko, M., Kumar, M. S., & Millard, W. J. (1985). Effects of acute and subchronic delta 9-tetrahydrocannabinol administration on the plasma catecholamine, beta-endorphin, and corticosterone levels and splenic natural killer cell activity in rats. *Proceedings of the Society for Experimental Biology and Medicine Society for Experimental Biology and Medicine (New York, N Y), 180*, 400–404.

Patel, J. P., Sengupta, R., Bardi, G., Khan, M. Z., Mullen-Przeworski, A., & Meucci, O. (2006). Modulation of neuronal CXCR4 by the μ-opioid agonist DAMGO. *Journal of Neurovirology, 12*, 492–500.

Pellis, N. R., Harper, C., & Dafny, N. (1986). Suppression of the induction of delayed hypersensitivity in rats by repetitive morphine treatments. *Experimental Neurology, 93*, 92–97.

Peng, X., Cebra, J. J., Adler, M. W., Meissler, J. J., Jr., Cowan, A., Feng, P., & Eisenstein, T. K. (2001). Morphine inhibits mucosal antibody responses and TGF-β mRNA in gut-associated lymphoid tissue (GALT) following oral cholera toxin in mice. *Journal of Immunology (Baltimore, Md : 1950), 167*, 3677–3681.

Perez-Castrillon, J. L., Perez-Arellanos, J. L., Carcia-Palomo, J. D., Jimeniz-Lopez, A., & de Castro, S. (1992). Opioids depress in vitro human monocyte chemotaxis. *Immunopharmacology, 23*, 57–61.

Persidsky, Y., Fan, S., Dykstra, H., Reichenbach, N. L., Rom, S., & Ramirez, S. H. (2015). Activation of cannabinoid type two receptors (CB2) diminish inflammatory responses in macrophages and brain endothelium. *Journal of Neuroimmune Pharmacology, 10*, 302–308.

Peterson, P. K., Sharp, B., Gekker, G., Brummitt, C., & Keane, W. F. (1987a). Opioid-mediated suppression of cultured peripheral blood mononuclear cell respiratory burst activity. *Journal of immunology (Baltimore, Md : 1950), 138*, 3907–3912.

Peterson, P. K., Sharp, B., Gekker, G., Brummitt, C., & Keane, W. F. (1987b). Opioid-mediated suppression of interferon-γ production by cultured peripheral blood mononuclear cells. *The Journal of Clinical Investigation, 80*, 824–831.

Peterson, P. K., Sharp, B. M., Gekker, G., Portoghese, P. S., Sannerud, K., & Balfour, H. H., Jr. (1990). Morphine promotes the growth of HIV-1 in human peripheral blood mononuclear cell cocultures. *AIDS, 4*, 869–873.

Peterson, P. K., Gekker, G., Chao, C. C., Schut, R., Verhoef, J., Edelman, C. K., Erice, A., & Balfour, H. H., Jr. (1992). Cocaine amplifies HIV-1 replication in cytomegalovirus-stimulated peripheral blood mononuclear cell cocultures. *Journal of immunology (Baltimore, Md : 1950), 149*, 676–680.

Peterson, P. K., Gekker, G., Hu, S., Anderson, W. R., Kravitz, F., Portoghese, P. S., Balfour, H. H., Jr., & Chao, C. C. (1994). Morphine amplifies HIV-1 expression in chronically infected promonocytes cocultured with human brain cells. *Journal of Neuroimmunology, 50*, 167–175.

Pruett, S. B., Han, Y. C., & Fuchs, B. A. (1992). Morphine suppresses primary humoral immune responses by a predominantly indirect mechanism. *The Journal of Pharmacology and Experimental Therapeutics, 262*, 923–928.

Raborn, E. S., Marciano-Cabral, F., Buckley, N. E., Martin, B. R., & Cabral, G. A. (2008). The cannabinoid delta-9-tetrahydrocannabinol mediates inhibition of macrophage chemotaxis to RANTES/CCL5: Linkage to the CB2 receptor. *Journal of Neuroimmune Pharmacology, 3*, 117–129.

Radulovic, J., Miljevic, C., Djergovic, D., Vujic, V., Antic, J., von Horsten, S., & Jankovic, B. D. (1995). Opioid receptor-mediated suppression of humoral immune response in vivo and in vitro: Involvement of κ opioid receptors. *Journal of Neuroimmunology, 57*, 55–62.

Rahim, R. T., Meissler, J. J., Jr., Cowan, A., Rogers, T. J., Geller, E. B., Gaughan, J., Adler, M. W., & Eisenstein, T. K. (2001). Administration of mu-, kappa- or delta2-receptor agonists via osmotic minipumps suppresses murine splenic antibody responses. *International Immunopharmacology, 1*, 2001–2009.

Rahim, R. T., Adler, M. W., Meissler, J. J., Jr., Cowan, A., Rogers, T. J., Geller, E. B., & Eisenstein, T. K. (2002). Abrupt or precipitated withdrawal from morphine induces immunosuppression. *Journal of Neuroimmunology, 127*, 88–95.

Rahim, R. T., Meissler, J. J., Jr., Zhang, L., Adler, M. W., Rogers, T. J., & Eisenstein, T. K. (2003). Withdrawal from morphine in mice suppresses splenic macrophage function, cytokine production, and costimulatory molecules. *Journal of Neuroimmunology, 144*, 16–27.

Rahim, R. T., Meissler, J. J., Jr., Adler, M. W., & Eisenstein, T. K. (2005). Splenic macrophages and B-cells mediate immunosuppression following abrupt withdrawal from morphine. *Journal of Leukocyte Biology, 78*, 1185–1191.

Ramirez, S. H., Hasko, J., Skuba, A., Fan, S., Dykstra, H., McCormick, R., Reichenbach, N., Krizbai, I., Mahedevan, A., Zhang, M., Tuma, R., Son, Y. J., & Persidsky, Y. (2012). Activation of cannabinoid receptor 2 attenuates leukocyte-endothelial cell interactions and blood-brain barrier dysfunction under inflammatory conditions. *Journal of Neuroscience, 32*, 4004–4016.

Ramirez, S. H., Reichenbach, N. L., Fan, S., Rom, S., Merkel, S. F., Wang, X., Ho, W. Z., & Persidsky, Y. (2013). Attenuation of HIV-1 replication in macrophages by cannibinoid receptor 2 agonists. *Journal of Leukocyte Biology, 93*, 801–810.

Risdahl, J. M., Peterson, P. K., & Molitor, T. W. (1996). Opiates, infection and immunity. In H. Friedman, T. W. Klein, & S. Specter (Eds.), *Drugs of abuse, immunity, and infections* (pp. 1–42). CRC Press, Inc.

Robinson, R. H., Meissler, J. J., Breslow-Deckman, J. M., Gaughan, J., Adler, M. W., & Eisenstein, T. K. (2013). Cannabinoids inhibit T-cells via cannabinoid receptor 2 in an in vitro assay for graft rejection, the mixed lymphocyte reaction. *Journal of Neuroimmuno Pharmacology, 8*, 1239–1250.

Rogers, T. J., Bednar, F., Kaminsky, D. E., Davey, P. C., Meissler, J. J., Jr., & Eisenstein, T. K. (2005). Laboratory model systems of drug abuse and their relevance to HIV infection and dementia. In H. E. Gendelman, I. Grant, I. P. Everall, S. A. Lipton, & S. Swindells (Eds.), *The neurology of AIDS* (pp. 310–320). Oxford University Press.

Rom, S., Zuluaga-Ramirez, V., Dykstra, H., Reichenbach, N. L., Pacher, P., & Persidsky, Y. (2013). Selective activation of cannabinoid receptor 2 in leukocytes suppresses their engagement of the brain endothelium and protects the blood-brain barrier. *American Journal of Pathology, 183*, 1548–1558.

Romero-Sandoval, E. A., Horvath, R., Landry, R. P., & DeLeo, J. A. (2009). Cannabinoid receptor type 2 activation induces a microglial anti-inflammatory phenotype and reduces migration via MKP induction and ERK dephosphorylation. *Molecular Pain, 5*, 25–39.

Rossi, G. C., Brown, G. P., Leventhal, L., Yang, K., & Pasternak, G. W. (1996). Novel receptor mechanisms for heroin and morphine-6β-glucoronide analgesia. *Neuroscience Letters, 216*, 1–4.

Roth, M. D., Tashkin, D. P., Choi, R., Jamieson, B. D., Zack, J. A., & Baldwin, G. C. (2002). Cocaine enhances human immunodeficiency virus replication in a model of severe combined immunodeficient mice implanted with human peripheral blood leukocytes. *The Journal of Infectious Diseases, 185*, 701–705.

Roth, M. D., Whittaker, K., Salehi, K., Tashkin, D. P., & Baldwin, G. C. (2004). Mechanisms for impaired effector function in alveolar macrophages from marijuana and cocaine smokers. *Journal of Neuroimmunology, 147*, 82–86.

Roth, M. D., Tashkin, D. P., Whittaker, K. M., Choi, R., & Baldwin, G. C. (2005). Tetrahydrocannabinol suppresses immune function and enhances HIV replication in the huPBL-SCID mouse. *Life Sciences, 77*, 1711–1722.

Roy, S., Ramakrishnan, S., Loh, H. H., & Lee, N. M. (1991). Chronic morphine treatment selectively suppresses macrophage colony formation in bone marrow. *European Journal of Pharmacology, 195*, 359–363.

Roy, S., Chapin, R. B., Cain, K. J., Charboneau, R. G., Ramakrishnan, S., & Barke, R. A. (1997). Morphine inhibits transcriptional activation of IL-2 in mouse thymocytes. *Cellular Immunology, 179*, 1–9.

Roy, S., Cain, K. J., Chapin, R. B., Charboneau, R. G., & Barke, R. A. (1998a). Morphine modulates NF-κB activation in macrophages. *Biochemical and Biophysical Research Communications, 245*, 392–396.

Roy, S., Cain, K. J., Charboneau, R. G., & Barke, R. A. (1998b). Morphine accelerates the progression of sepsis in an experimental sepsis model. *Advances in Experimental Medicine and Biology, 437*, 21–31.

Roy, S., Wang, J., Gupta, S., Charboneau, R., Loh, H. H., & Barke, R. A. (2004). Chronic morphine treatment differentiates T helper cells to Th2 effector cells by modulating transcription factors GATA 3 and T-bet. *Journal of Neuroimmunology, 147*, 78–81.

Saaoud, F., Shao, Y., Cornwell, W., Wang, H., Rogers, T. J., & Yang, X. (2023). Cigarette smoke modulates inflammation and immunity via reactive oxygen species-regulated trained immunity and trained tolerance mechanisms. *Antioxidants & Redox Signaling, 38*, 1041–1069.

Saika, F., Matsuzaki, S., Kobayashi, D., & Kigushi, N. (2018). Chemokine CXCL12 is responsible for cocaine-induced reward in mice. *Neuropsychopharmacology Reports, 38*, 145–148.

Saurer, T. B., Carrigan, K. A., Ijames, S. G., & Lysle, D. T. (2006). Suppression of natural killer cell activity by morphine is mediated by the nucleus accumbens shell. *Journal of Neuroimmunology, 173*, 3–11.

Schatz, A. R., Lee, M., Condie, R. B., Pulaski, J. T., & Kaminski, N. E. (1997). Cannabinoid receptors CB1 and CB2: A characterization of expression and adenylate cyclase modulation within the immune system. *Toxicology and Applied Pharmacology, 142*, 278–287.

Scheidegger, C., & Zimmerli, W. (1989). Infectious complications in drug addicts: Seven year review of 269 hospitalized narcotic abusers in Switzerland. *Reviews of Infectious Diseases, 11*, 486–493.

Schwacha, M. G., McGwin, G., Hutchinson, C. B., Cross, J. M., MacLennan, P. A., & Rue, L. W. (2006). The contribution of opiate analgesics to the development of infectious complication in burn patients. *American Journal of Surgery, 192*, 82–86.

Sen, C. K., & Roy, S. (2012). MicroRNA 21 in tissue injury and inflammation. *Cardiovascular Research, 96*, 230–233.

Shao, Y., Cornwell, W., Xu, K., Kirchhoff, A., Saasoud, F., Lu, Y., Jiang, X., Criner, G. J., Wang, H., Rogers, T. J., & Yang, X. (2022a). Chronic exposure to the combination of cigarette smoke and morphine decreases CD4(+) regulatory T cell numbers by reprogramming the treg cell transcriptome. *Frontiers in Immunology, 13*, 887681.

Shao, Y., Saaoud, F., Cornwell, W., Xu, K., Kirchhoff, A., Lu, Y., Jiang, X., Wang, H., Rogers, T. J., & Yang, X. (2022b). Cigarette smoke and morphine promote treg plasticity to Th17 via enhancing trained immunity. *Cells, 11*, 2810.

Sharma, M. D., Huang, L., Choi, J. H., Lee, E. J., Wilson, J. M., Lemos, H., Pan, F., Blazar, B. R., Pardoll, D. M., Mellor, A. L., Shi, H., & Munn, D. H. (2013). An inherently bifunctional subset of Foxp3+ T helper cells is controlled by the transcription factor eos. *Immunity, 38*, 998–1012.

Sharp, B., & Yaksh, T. (1997). Pain killers of the immune system. *Nature Medicine, 3*, 831–832.

Sharp, B. M., Roy, S., & Bidlack, J. M. (1998). Evidence for opioid receptors on cells involved in host defense and the immune system. *Journal of Neuroimmunology, 83*, 45–56.

Shavit, Y., Depaulis, A., Martin, F. C., Terman, G. W., Pechnick, R. N., Zane, C. J., Gale, R. P., & Liebeskind, J. C. (1986a). Involvement of brain opiate receptors in the immune-suppressive effect of morphine. *Proceedings of the National Academy of Sciences, USA, 83*, 7114–7117.

Shavit, Y., Terman, G. W., Lewis, J. W., Zane, C. J., Gale, R. P., & Liebeskind, J. C. (1986b). Effects of foot-shock stress and morphine on natural killer lymphocytes in rats: studies of tolerance and cross-tolerance. *Brain Research, 372*, 382–385.

Shen, H. M., Kennedy, J. L., & Ou, D. W. (1994). Inhibition of cytokine release by cocaine. *International Journal of Immunopharmacology, 16*, 295–300.

Sibinga, N. E. S., & Goldstein, A. (1988). Opioid peptides and opioid receptors in cells of the immune system. *Annual Review of Immunology, 6*, 219–249.

Simon, E. (1991). Opioid receptors and endogenous opioid peptides. *Medicinal Research Reviews, 11*, 357–374.

Singh, P. P., & Singal, P. (2007). Morphine-induced neuroimmunomodulation in murine visceral leishmaniasis: The role(s) of cytokines and nitric oxide. *Journal of Neuroimmune Pharmacology, 2*, 338–351.

Singh, U. P., Singh, N. P., Singh, B., Price, R. L., Nagarkatti, M., & Nagarkatti, P. S. (2012). Cannabinoid receptor-2 (CB2) agonist ameliorates colitis in IL-10−/− mice by attenuating the activation of T cells and promoting their apoptosis. *Toxicology and Applied Pharmacology, 258*, 256–267.

Singhal, P. C., Sharma, P., Kapasi, A. A., Reddy, K., Franki, N., & Gibbons, N. (1998). Morphine enhances macrophage apoptosis. *Journal of Immunology (Baltimore, Md : 1950), 160*, 1886–1893.

Skolnick, P., Davis, H., Arnelle, D., & Deaver, D. (2014). Translational potential of naloxone and naltrexone as TLR4 antagonists. *Trends in Pharmacological Sciences, 35*, 431–433.

Sopori, M. (2002). Effects of cigarette smoke on the immune system. *Nature Reviews Immunology, 2*, 372–377.

Specter, S. C., Klein, T. W., Newton, C., Mondragon, M., Widen, R., & Friedman, H. (1986). Marijuana effects on immunity: Suppression of human natural killer cell activity of delta-9-tetrahydrocannabinol. *International Journal of Immunopharmacology, 8*, 741–745.

Specter, S., Lancz, G., & Goodfellow, D. (1991a). Suppression of human macrophage function in vitro by Δ9-tetrahydrocannabinol. *Journal of Leukocyte Biology, 50*, 423–426.

Specter, S., Lancz, G., Westrich, G., & Friedman, H. (1991b). Delta-9-tetrahydrocannabinol augments murine retroviral induced immunosuppression and infection. *International Journal of Immunopharmacology, 13*, 411–417.

Spiller, M. W., Broz, D., Wejnert, C., Nerlander, L., & Paz-Bailey, G. (2015). HIV infection and HIV-associated behaviors among persons who inject drugs—20 cities, United States, 2012. *Morbidity and Mortality Weekly Report, 64*, 270–275.

Steele, A. D., Szabo, I., Bednar, F., & Rogers, T. J. (2002). Interactions between opioid and chemokine receptors: Heterologous desensitization. *Cytokine & Growth Factor Reviews, 13*, 209–222.

Steele, A. D., Henderson, E. E., & Rogers, T. J. (2003). μ-Opioid modulation of HIV-1 coreceptor expression and HIV-1 replication. *Virology, 309*, 99–107.

Storr, M. A., Keenan, C. M., Zhang, H., Patel, K. D., Makriyannis, A., & Sharkey, K. A. (2009). Activation of the cannabinoid 2 receptor (CB2) protects against experimental colitis. *Inflammatory Bowel Disease, 15*, 1678–1685.

Szabo, I., Rojavin, M., Bussiere, J. L., Eisenstein, T. K., Adler, M. W., & Rogers, T. J. (1993). Suppression of peritoneal macrophage phagocytosis of Candida albicans by opioids. *The Journal of Pharmacology and Experimental Therapeutics, 267*, 703–706.

Szabo, I., Chen, X. H., Xin, L., Adler, M. W., Howard, O. M. Z., Oppenheim, J. J., & Rogers, T. J. (2002). Heterologous desensitization of opioid receptors by chemokines inhibits chemotaxis and enhances the perception of pain. *Proceedings of the National Academy of Sciences, USA, 99*, 10276–10281.

Szabo, I., Wetzel, M. A., Zhang, N., Steele, A. D., Kaminsky, D. E., Chen, C., Liu-Chen, L. Y., Bednar, F., Henderson, E. E., Howard, O. M. Z., Oppenheim, J. J., & Rogers, T. J. (2003). Selective inactivation of CCR5 and decreased infectivity of R5 HIV-1 strains mediated by opioid-induced heterologous desensitization. *Journal of Leukocyte Biology, 74*, 1074–1082.

Taub, D. D., Eisenstein, T. K., Geller, E. B., Adler, M. W., & Rogers, T. J. (1991). Immunomodulatory activity of μ- and κ-selective opioid agonists. *Proceedings of the National Academy of Sciences, USA, 88*, 360–364.

Tomassini, N., Renaud, F., Roy, S., & Loh, H. H. (2004). Morphine inhibits Fc-mediated phagocytosis through mu and delta opioid receptors. *Journal of Neuroimmunology, 147*, 131–133.

Trecki, J., & Unterwald, E. M. (2009). Modulation of cocaine-induced activity by intracerebral administration of CXCL12. *Neuroscience, 161*, 13–22.

Tubaro, E., Borelli, G., Croce, C., Cavallo, G., & Santiangeli, C. (1983). Effect of morphine on resistance to infection. *Journal of Infectious Diseases, 148*, 656–666.

Turchan-Cholewo, J., Dimayuga, F. O., Gupta, S., Keller, J. N., Knapp, P. E., Hauser, K. F., & Bruce-Keller, A. J. (2009). Morphine and HIV-Tat increase microglial-free radical production and oxidative stress: Possible role in cytokine regulation. *Journal of Neurochemistry, 108*, 202–215.

Van Dyke, C., Stesin, A., Jones, R., Chuntharapai, A., & Seaman, W. (1986). Cocaine increases natural killer cell activity. *The Journal of Clinical Investigation, 77*, 1387–1390.

Vlahov, D., Robertson, A. M., & Strathdee, S. A. (2010). Prevention of HIV infection among injection drug users in resource-limited settings. *Clinical Infectious Diseases, 50*(S3), S114–S121.

Walter, L., Franklin, A., Witting, A., Wade, C., Xie, Y., Kunos, G., Mackie, K., & Stella, N. (2003). Nonpsychotropic cannabinoid receptors regulate microglial cell migration. *Journal of Neuroscience, 23*, 1398–1405.

Wang, Y., Huang, D. S., & Watson, R. R. (1994). In vivo and in vitro cocaine modulation on production of cytokines in C57BL/6 mice. *Life Sciences, 54*, 401–411.

Wang, J., Charboneau, R., Balasubramanian, S., Barke, R. A., Loh, H. H., & Roy, S. (2001). Morphine modulates lymph node derived T lymphocyte function: Role of caspase-3, −8, and nitric oxide. *Journal of Leukocyte Biology, 70*, 527–536.

Wang, J., Barke, R. A., Charboneau, R., & Roy, S. (2005). Morphine impairs host innate immune response and increases susceptibility to Streptococcus pneumoniae lung infection. *Journal of Immunology (Baltimore, Md : 1950), 174*, 426–434.

Wang, J., Barke, R. A., Charboneau, R., Schwendener, R., & Roy, S. (2008). Morphine induces defects in early response of alveolar macrophages to Streptococcus pneumoniae by modulating TLR9-NFkB signaling. *Journal of Immunology (Baltimore, Md : 1950), 180*, 3594–3600.

Wang, J., Ma, J., Charboneau, R., Barke, R., & Roy, S. (2011). Morphine inhibits dendritic cell IL-23 production by modulating toll-like receptor 2 and Nod2 signaling. *Journal of Biological Chemistry, 286*, 10225–10232.

Wang, X., Ma, T. C., Li, J. L., Zhou, Y., Geller, E. B., Adler, M. W., Peng, J. S., Zhou, W., Zhou, D. J., & Ho, W. Z. (2015). Heroin inhibits HIV-restriction miRNAs and enhances HIV infection of macrophages. *Frontiers in Immunology, 6*, 1230.

Watson, E. S., Murphy, J. C., El Sohly, H. N., El Sohly, M. A., & Turner, C. E. (1983). Effects of the administration of coca alkaloids on the primary immune responses of mice: Interaction with Δ9-tetrahydrocannabinol and ethanol. *Toxicology and Applied Pharmacology, 71*, 1–13.

Watzl, B., Chen, G., Scuderi, P., Pirozhkov, S., & Watson, R. R. (1992). Cocaine-induced suppression of interferon-gamma secretion in leukocytes from young and old C57BL/6 mice. *International Journal of Immunopharmacology, 14*, 1125–1131.

Weber, R. J., & Pert, A. (1989). The periaqueductal gray matter mediates opiate-induced immunosuppression. *Science, 245*, 188–190.

Weber, R. J., Ikejiri, B., Rice, K. C., Pert, A., & Hagan, A. A. (1987). Opiate receptor mediated regulation of the immune response in vivo. *NIDA Research Monograph Series, 48*, 341–348.

Weed, M., Adams, R. J., Hienz, R. D., Meulendyke, K. A., Linde, M. E., Clements, J. E., Mankowski, J. L., & Zink, M. C. (2012). SIV/Macaque model of HIV infection in cocaine users: minimal effects of cocaine on behavior, virus replication, and CNS inflammation. *Journal of Neuroimmune Pharmacology, 7*, 401–411.

Wetzel, M. A., Steele, A. D., Eisenstein, T. K., Adler, M. W., Henderson, E. E., & Rogers, T. J. (2000). μ-opioid induction of monocyte chemoattractant protein-1, RANTES, and IFN-γ-inducible protein-10 expression in human peripheral blood mononuclear cells. *Journal of Immunology (Baltimore, Md : 1950), 165*, 6519–6524.

Wybran, J., Appelboom, T., Famaey, J. P., & Govaerts, A. (1979). Suggestive evidence for receptors for morphine and methionine-enkephalin on normal human blood T lympohcytes. *Journal of Immunology (Baltimore, Md : 1950), 123*, 1068–1070.

Xu, H., Cheng, C. L., Chen, M., Manivannan, A., Cabay, L., Pertwee, R. G., Coutts, A., & Forrester, J. V. (2007). Anti-inflammatory property of the cannabinoid receptor-2-selective agonist JWH-133 in a rodent model of autoimmune uveoretinitis. *Journal of Leukocyte Biology, 82*, 532–541.

Yang, X., Hegde, V. L., Rao, R., Zhang, J., Nagarkatti, P. S., & Nagarkatti, M. (2014). Histone modifications are associated with Δ9-tetrahydrocannabinol-mediated alterations in antigen-specific T cell responses. *Journal of Biological Chemistry, 289*, 18707–18718.

Yeager, M. P., Colacchio, T. A., Yu, C. T., Hildebrandt, L., Howell, A. L., Weiss, J., & Guyre, P. M. (1995). Morphine inhibits spontaneous and cytokine-enhanced natural killer cell cytotoxicity in volunteers. *Anesthesiology, 83*, 500–508.

Yin, D., Mufson, A., Wang, R., & Shi, Y. (1999). Fas-mediated cell death promoted by opioids. *Nature, 397*, 218.

Yuan, M., Kiertscher, S. M., Cheng, Q., Zoumalan, R., Tashkin, D. P., & Roth, M. D. (2002). Δ9-tetrahydrocannabinol regulates Th1/Th2 cytokine balance in activated human T cells. *Journal of Neuroimmunology, 133*, 124–131.

Zhang, L., Looney, D., Taub, D., Chang, S. L., Way, D., Witte, M. H., Graves, M. C., & Fiala, M. (1998). Cocaine opens the blood-brain barrier to HIV-1 invasion. *Journal of NeuroVirology, 4*, 619–626.

Zhang, X., Wang, J. F., Kunos, G., & Groopman, J. E. (2007). Cannabinoid modulation of Kaposi's sarcoma-associated herpesvirus infection and transformation. *Cancer Research, 67*, 7230–7237.

Zhang, M., Martin, B. R., Adler, M. W., Razdan, R. J., Kong, W., Ganea, D., & Tuma, R. F. (2009). Modulation of cannabinoid receptor activation as a neuroprotective strategy for EAE and stroke. *Journal of Neuroimmune Pharmacology, 4*, 249–259.

Zhang, L., Meng, J., Ban, Y., Jalodia, R., Chupikova, I., Fernandez, I., Brito, N., Sharma, U., Abreu, M. T., Ramakrishnan, S., & Roy, S. (2019). Morphine tolerance is attenuated in germfree mice and reversed by probiotics, implicating the role of gut microbiome. *Proceedings of the National Academy of Sciences of the USA, 116*, 13523–13532.

Zhu, W., Igarashi, T., Qi, Z. T., Newton, C., Widen, R. E., Friedman, H., & Klein, T. W. (1993). delta-9-tetrahydrocannabinol (THC) decreases the number of high and intermediate affinity IL-2 receptors of the IL-2 dependent cell line NKB61A2. *International Journal of Immunopharmacology, 15*, 401–408.

Zhu, L. X., Sharma, S., Stolina, M., Gardner, B., Roth, M. D., Tashkin, D. P., & Dubinett, S. M. (2000). Δ9-tetrahydrocannabinol inhibits antitumor immunity by a CB2 receptor-mediated, cytokine-dependent pathway. *Journal of immunology (Baltimore, Md : 1950), 165*, 373–380.

Zhu, Y., Yan, P., Wang, R., Lai, J., Tang, H., Xiao, X., Yu, R., Bao, X., Zhu, F., Wang, K., Lu, Y., Dang, J., Zhu, C., Zhang, R., Dang, W., Zhang, B., Fu, Q., Zhang, Q., Kang, C., Chen, Y., Chen, X., Liang, Q., & Wang, K. (2023). Opioid-induced fragile-like regulatory T cells contribute to withdrawal. *Cell, 186*, 591–606 e523.

Part III

Therapies and Diagnostics

Pharmacogenomics of Neurodegenerative Diseases: Roles in Personalized Medicines

34

Aditya Gupta, Ruby E. Evande, Rinku Dutta, Chalet Tan, and Ram I. Mahato

Abstract

Pharmacogenomics is the study of the role of genetic variations in drug response. In the study of pharmacogenomics, one can relate variations in drug exposure and/or drug response (pharmacology) to the variations in genes (genomics). An administered drug may fail to work for one patient, but show the expected result for another. In some cases, a drug may work well for a larger group of patients while causing a subset of patients to experience unexpected or severe side effects. These adverse drug reactions can be harmful and sometimes deadly. Factors such as the environment, diet, age, lifestyle, and comorbid conditions all may influence the individual's response to drugs. Conventional drug treatments involve a one-size-fits-all approach which can be beneficial for some patients but not for others. Personalized medicine takes into account differences in patient's genetic makeup and environment to customize patient-specific treatment. Each is covered in this chapter with an eye toward how best to administer drugs to individual patients.

Keywords

Gene variants · Metabolizers · Neurodegenerative · Pharmacogenomics

A. Gupta · R. Dutta · R. I. Mahato (✉)
Department of Pharmaceutical Sciences, University of Nebraska Medical Center, Omaha, NE, USA
e-mail: ram.mahato@unmc.edu

R. E. Evande
Department of Medicine, University of Nebraska Medical Center, Omaha, NE, USA

C. Tan
Department of Pharmaceutical Sciences, University of Tennessee Health Science Center, Memphis, TN, USA

34.1 Introduction

Since the beginning of medical practice, different individuals have responded differently to the same drug (Roden & George Jr, 2002). In line with this idea, several studies have shown that genetic variability clearly has implications for both drug safety and efficacy (Li et al., 2011). Neurodegenerative diseases are complex disorders of neurons, that include genetic, epigenetic, and environmental factors, and are characterized by the deterioration of higher activities of the central nervous system. Over 60% of Alzheimer's disease (AD) risk is estimated to be a result of genetic factors; several genes have been associated with AD pathogenesis and neurodegeneration (Gatz et al., 2006; Eisenstein, 2011). Studies have shown that patients with AD respond to drug treatment depending on which of the three genetic variants of the apolipoprotein E gene a person carries (Guerreiro et al., 2012; Cacabelos, 2008; Choi et al., 2008). Genes involved in the absorption, distribution, metabolism, and excretion of drugs and various drug target genes make up the most important genetic determinants (Ma et al., 2002). The pharmacologic treatment of neurodegenerative diseases has seen some progress, but more effort needs to be made to understand the genetic components and the genome-drug interactions. In this chapter, we will review the roles of several defective genes involved in neurodegenerative diseases and correlate them with clinical outcomes.

Pharmacogenomics is still in its infancy when it comes to treating neurodegenerative illnesses. In fact, the etiopathogenetic pathways underlying the majority of these diseases are still poorly known, which makes it challenging to identify the genes whose variation may explain the varied responses to medications seen in various people. Determining the most pertinent molecules we believe are likely to play a crucial role in the disease development, we will attempt to summarize current knowledge of the pathogenetic mechanisms of the three main neurodegenerative diseases amyotrophic lat-

H. E. Gendelman, T. Ikezu (eds.), *Neuroimmune Pharmacology and Therapeutics*,
https://doi.org/10.1007/978-3-031-68237-7_34

eral sclerosis, Parkinson's disease, and Alzheimer's disease in this chapter.

34.2 Classifications of Anti-neurodegenerative Drugs in Various Disease Phenotypes

Neurodegenerative diseases include AD, multiple sclerosis (MS), Parkinson's disease (PD), Huntington's disease (HD), and amyotrophic lateral sclerosis (ALS).

34.2.1 Alzheimer's Disease

Alzheimer's disease (AD) is one of the most common causes of dementia among the aging population and is characterized by declining cognitive failure and loss of communication and thinking capabilities. AD has become a major health challenge in both developed and developing countries (Gatz et al., 2006). Memory impairment is an integral feature of AD and is often the initial symptom. Deficits in other cognitive domains may occur with or after the onset of memory impairment. Language function and visual-spatial skills tend to be affected relatively early on, while loss of executive function and changes in personality and behavior generally occur later in the disease course. The pattern of memory impairment in AD is distinctive; deterioration in the ability to remember facts and events is a cardinal feature. Neuropathologic findings include extracellular amyloid plaques, intracellular neurofibrillary tangles, synaptic deterioration, and neuronal death (Mathisen, 2003). Amyloid plaques result from the accumulation of β-amyloid peptides (Aβ). Aß is an amino acid peptide formed by proteolytic cleavage of amyloid precursor protein (APP) by beta- and gamma-secretase (Mathisen, 2003). Tau, a protein involved in microtubule assembly, is critical for normal axonal growth and neuronal development. Hyper-phosphorylated tau protein aggregates into helical filamentous neurofibrillary tangles that are deposited within neurons. Death of neurons in critical locations in the brain leads to a deficiency of acetylcholine, a neurotransmitter important for memory (Hellstrom-Lindahl, 2000). Familial AD occurs earlier in life and involves mutations in genes involved in the processing of APP including *presenilin 1* and *presenilin 2*, whereas the risk for late-onset AD is mostly influenced by the apolipoprotein E (*APOE*) gene (Guerreiro et al., 2012). According to the Alzgene database, among hundreds of genes associated with AD, the top ten are *APOE* (19q13.2), *BIN1* (2q14), *CLU* (8p21–p12), *ABCA7* (19p13.3), *CR1* (1q32), *PICALM* (11q14), *MS4A6A* (11q12.1), *CD33* (19q13.3), *MS4A4E* (11q12.2), and *CD2AP* (6p12) (Cacabelos et al., 2014). Other genes associated with AD are included in Table 34.1.

Table 34.1 Some Alzheimer's disease–associated genes and their genomic locations

Genes	Genomic location
Familial genes	
APP	21q21.3
PSEN1	14q24.3
PSEN2	1q31–q42
Sporadic genes	
APOE	19q13.2
TREM2	6p21.1
PLD3	19q13.2
AKAP9	7q21–q22
GRN	17q21.32
MAPT	17q21.1
PRNP	20p13
ADAM10	15q22
TOMM40	19q13
BIN1	2q14
CRI	1q32
ABCA7	19p13.3
FERMT2	14q22.1
HLA-DRB5-DRB1	6p21.3
CD2AP	6p12
PTK2B	8p21.1
INPP5D	2q37.1
CUGBP-CELF1	11P11
S100B	21q22.3
CD33	19q13.3
SLC24A4-RIN3	14q32.12
ZCWPW1	7q22.1
MEF2C	5q14.3
NME8	7p14.1
MS4A cluster	11q12.2
EPHA1	7q34
PICALM	11q14
CLU	8p21–p12
CASS4	20q13.31
SORLI	11q23.2–q24.2
DSG2	18q12.1

Table adapted from Calero et al. (2015) and Cacabelos et al. (2014)

The US Food and Drug Administration (FDA) has approved four drugs for the treatment of AD: donepezil, rivastigmine, galantamine, and memantine. Donepezil, rivastigmine, and galantamine are cholinesterase inhibitors that are therapeutically active in ameliorating the clinical pathologies of mild-to-moderate AD (Loveman et al., 2006). Inhibition of acetylcholinesterase, the enzyme responsible for the destruction of acetylcholine, increases the availability of acetylcholine. Memantine is a noncompetitive N-methyl-D-aspartate receptor antagonist that is approved for the treatment of moderate to severe AD. Glutamate activation at the NMDA receptor is needed for learning and memory processes in the brain. Glutamate is the dominant excitatory amino acid in the brain and stimulates NMDA receptors. This in turn leads to increased intraneuronal concentrations of calcium. Excessive glutamate, and therefore calcium, in the CNS, may cause neuronal damage. The blockade of NMDA receptors by memantine slows intracellular calcium accumulation and helps to prevent further nerve damage (Danysz & Parsons, 2003). A higher dose of donepezil, with or without memantine was approved to treat moderate to severe AD (Cummings et al., 2013). However, these drugs have different pharmacokinetic and pharmacodynamics profiles with donepezil having a relatively longer half-life ($t_{1/2}$) of 70 h than galantamine ($t_{1/2}$ = 6–8 h) and rivastigmine ($t_{1/2}$ = 1–2 h), whereas memantine has a ($t_{1/2}$ = 60–70 h) (Noetzli & Eap, 2013). Studies have shown that the combination therapy of memantine and acetylcholinesterase inhibitors in patients with AD may be associated with the benefits of slowing cognitive impairment and preventing the onset of agitation and aggression (Shao, 2015; Gareri et al., 2014). Hepatic CYP2D6 and CYP3A4 are mainly responsible for the metabolism of donepezil and galantamine, whereas rivastigmine, a potent, selective inhibitor of brain acetylcholinesterase (AChE) and butylcholinesterase (BChE), is metabolized in the liver and intestine independent of CYP450 system. Reports show that polymorphisms in genes involved in the cholinergic system, including acetylcholinesterase, butylcholinesterase, choline acetyltransferase, and paraoxonase, are involved in the clinical response to AChE inhibitor (Noetzli & Eap, 2013).

It has been an arduous task for researchers to develop new drugs for the treatment of AD. Since 2003, there has been no new drug for AD that was approved by the FDA. Consequently, approximately 200 research projects in the last decade have failed. However, there are still plenty of drugs in the pipeline for the treatment of AD that either target disease symptoms or modification. Currently, elenbecestat (E2609) and umibecestat (CNP520) are two β-secretase inhibitors (BACE) that are still in phase 2 and phase 3 of clinical trials. An effective personalized medicine concept, including the elimination of risk factors and comorbid disease treatment, could be executed in every patient if all the current research efforts come to success (Cacabelos, 2020).

34.2.2 Multiple Sclerosis

Multiple sclerosis (MS) is a common neurodegenerative disease that typically affects young adults to mid-age groups. It is a *demyelinating disease* in which the the *insulating covers* of the *nerve cells* of the *brain* and *spinal cord* are damaged. It is mediated by T-cell autoimmunity whereby these cells migrate across the blood–brain barrier and subsequently induce inflammatory responses resulting in myelin destruction, oligodendrocyte destruction, axonal breakdown, and gliosis in the central nervous system (Tsareva et al., 2016). MS is characterized by clinical features including fatigue, paralysis, visual field impairments, chronic pain, and cognitive changes. Most MS patients experience a relapsing-remitting disease form characterized by varying periods of disease intensity and with time progress to a steady progression of symptoms known as secondary progressive MS. When the disease is characterized by a rapid increase in disability rate, it is known as primary progressive MS. Over 100 genetic risk factors have been associated to MS, but the class II region of the human leukocyte antigen (HLA)-DRB1*15:01, encoded within the major histocompatibility complex has been shown to be the largest contributing genetic risk component (Hollenbach & Oksenberg, 2015). Both HLA and non-HLA genes have been associated with the genetic susceptibility of MS. Treatment options for MS are very wide because the FDA has approved several disease-modifying therapies (DMTs) with proven efficacies compared with placebo in preventing relapses associated with MS (English & Aloi, 2015). The list of DMTs includes interferon beta (IFNβ), glatiramer acetate, tumor necrosis factor inhibitors etanercept and adalimumab, fingolimod, mitoxantrone, natalizumab, teriflunomide, dimethyl fumarate, peginterferon-β-1a, and alemtuzumab. Interferon injectables and glatiramer acetate were the first FDA-approved DMTs that specifically decrease disease activity and are still considered first-line treatments for relapsing-remitting forms of MS due to their desired long term safety and efficacy profiles (English & Aloi, 2015). Studies have identified some gene variants that are associated with therapy response. However, HLA class II genes which are the major MS susceptibility genes, showed associations with glatiramer acetate but not with IFNβ. Some of those results for IFNβ and glatiramer acetate are highlighted in Table 34.2.

There is yet no meaningful pharmacogenomic biomarker that would distinguish between MS therapy responders and nonresponders despite the expanding range of treatment choices available to individuals with MS and the wide variation in response to these treatments. Given that MS is a chronic, progressive illness that necessitates lifelong treatment, choosing the best therapy early on may have significant therapeutic benefits for MS patients (Grossman et al., 2017).

34.2.3 Parkinson's Disease

Parkinson's disease (PD) is a neurodegenerative disorder that results in selective loss of dopamine neurons in the substantia nigra. Dopamine is a catecholamine formed in the body by the decarboxylation of dopa; it is an intermediate product in the synthesis of norepinephrine and acts as a neurotransmitter in the central nervous system. PD is characterized by several clinical extrapyramidal features: resting tremor, rigidity, bradykinesia, postural instability, and freezing. PD has a low incidence among Asians and Africans as

Table 34.2 Genes and their polymorphisms associated with multiple sclerosis therapy

Drug	Gene	Polymorphism
Interferon β	*IFNAR1*	rs1012334, rs55884088
	LMP7	rs2071543
	CTSS	rs1136774
	MXA	rs2071430, s17000900
	IRF5	rs2004640
	IRF8	rs17445836
	USP18	rs2542109
	IFNG	Polymorphic microsatellites in the first intron
	IL10	rs1800896/ rs1800871/rs1800872
	CCR5	rs333
	TGFB1	rs1800469
	TRAILR1	rs20576
	CD58	rs12044852
	CD46	rs2724385
	GPC5	rs10492503, rs1411751
Glatiramer acetate	*HLA-DRB1*	rs3135388
	CCR5	rs333
	TCRB	rs71878
	IL12RB2	rs946685
	MBP	rs470929
	IL1R1	rs956730
	CD86	rs1129055
	CTSS	rs2275235, rs1415148
	FAS	rs982764

Table adapted from Tsareva et al. (2016)

Table 34.3 Genes and their polymorphisms associated with Parkinson's disease

Gene		Polymorphism
Levodopa	*DRD2*	rs1800497(*Taq*IA)
	DRD3	rs6280(*Bal*I), rs4646996(*Msp*I)
	DRD2	(CA)n STR in intron 2
	DRD3	rs6280(Ser9Gly)
	GRIN2B	rs1806201, rs7301328, rs1019385
	SLC6A4	VNTR
	DRD2	various (9)
	DRD3	various (2)
	DRD4	various (3)
	DAT	40-bp VNTR
	OPRM1	rs1799971 (118A > G, Asn40Asp)
	COMT	rs4680(Val158Met)
	MAOB	rs1799836 intron 13
	BDNF	rs6265 (Val66Met)
	GBA	various mutations
	ACE	rs4646994 (I/D-ins/del)
	APOE	rs429358 rs7412 (ε2,ε3,ε4)
	CCK	rs1799923 (−45 C > T)
	CCKAR	rs1800857 (779 T > C)
	CCKBR	rs1805002 (rs1550G > A)
	SLC22A1 (*OCT1*)	rs622342
COMT inhibitors	*UGT1A*	A528G
	COMT	rs4680(Val158Met)
Pyridoxine	*COMT*	rs4680(Val158Met)

Table from Drozdzik et al. (2013)

opposed to Caucasians. The various targets of antiparkinson drugs are dopamine receptors (DR), dopamine transporter (DAT), catechol-O-methyltransferase (COMT), and monoamine oxidase B (Table 34.3).

Levodopa (*l*-dopa) is the gold standard for Parkinson's disease therapy and is highly effective when given in combination with a dopa-decarboxylase inhibitor (carbidopa). This combination reduces the decarboxylation of levodopa to dopamine outside of the blood–brain barrier, thereby allowing more efficient dosing of levodopa. However, mutations in the COMT gene drastically reduce the efficacy of this therapy. L-dopa medication in PD is known to have a side effect called hyper-homocysteinemia (Redenšek & Dolžan, 2020). Numerous investigations have constantly shown a relationship between hyperhomocysteinemia brought on L-dopa medication and the MTHFR gene SNP C667T(rs1801133). This mutation causes the MTHFR enzyme to be temperature-labile, which ultimately results in hyperhomocysteinemia. The addition of COMT inhibitors to the therapy gives a clear opportunity for the application of this knowledge to patient care by offering a potential option for homocysteine level decrease and alleviation of systemic oxidative stress (Vuletić et al., 2021).

Dopamine receptors are proteins on the surfaces of certain cells that bind specifically to the neurotransmitter dopamine. Stimulation of these receptors on vascular epithelial cells in the brain by dopamine causes the cerebral arteries to dilate, resulting in increased flow of blood.

The dopaminergic system consists of the vital dopamine receptors (DR), which are classified into five categories, namely, D1–D5, which are co-expressed at various levels in the CNS and peripheral tissues and are located at chromosome 5, 11, 3, 11, and 4, respectively. The following polymorphisms may affect antiparkinson drug therapy. *DRD1* genetic polymorphism includes −48 A > G (D1.1; B1/B2) and 1403 T > C (D1.7; C1/C2), *DRD2* with *TaqIA* located in kinase *ANKK1* (ankyrin repeat and kinase domain containing 1), *DRD3* associated with Ser9Gly (or *MscI* or *BalI*), which in turn is responsible for the insertion of DRD3 receptor membrane and subsequent intracellular signal transduction. The different susceptible loci are the familial atypical forms of PD including *SNCA* (*PARK1*), *LRRK2*, *PRKN* (*PARK2*), *PINK1* (*PARK6*), *DJ-1* (*PARK7*), and recently *ATP13A2* (*PARK9*), *PLA2G6* (*PARK14*) and *FBX07* (*PARK15*). Sporadic PD is associated with common genetic variants such as N-acetyltransferase 2 (*NAT2*), monoamine oxidase B (*MAOB*), glutathione transferase (*GST*), mitochondrial tRNA, S18Y variant of ubiquitin carboxy-terminal hydrolase L1 (*UCHL1*), Rep variant of alpha-synuclein (*SNCA*) and tau (*MAPT*)H1 haplotype and leucine-rich repeat kinase-2 (*LRRK2*). Other newly recognized loci have been identified using genome-wide association studies (GWAS), namely, *BST1, GAK, HLA-DR, ACMSD, STK39, MCCC1lLAMP3, SYT11, PARK16, FGF20, and GPNMB*. However, the anti-PD drugs targeting the motor systems lose their efficacy over time due to the emergence of side effects such as dyskinesias and psychosis. The genes and their SNPs that induced dyskinesias are *DRD2* with nine polymorphisms (*TaqIA*, *TaqIB*, *TaqID*, Val96Ala, Leu141Leu, Pro310Ser, Ser311Cys, A > G231, −141C ins/del), *DRD3* with 2 (Ser9Gly, MspI), *DRD4* with 3 (48-bp VNTR, 13-bp repeat, 13-bp deletion) and 40-bp VNTR in *DET* gene. Among the COMT gene polymorphism Val158Met, (rs4680) is the most elaborately studied one and has a correlation with dyskinesia (Payami & Factor, 2014; Drozdzik et al., 2013).

34.2.4 Amyotrophic Lateral Sclerosis

Amyotrophic lateral sclerosis (ALS), also known as Lou Gehrig's disease, is a rapidly progressive, invariably fatal neurological disease that attacks the nerve cells responsible for controlling voluntary muscles and belongs to a group of disorders known as motor neuron diseases, which are characterized by the gradual degeneration and death of motor neurons. In the vast majority of all ALS cases (more than 90%), the disease occurs apparently at random with no clearly associated risk factors. The remainder is hereditary. The familial form of ALS usually results from a pattern of inheritance that requires only one parent to carry the gene responsible for the disease. Mutations in superoxide dismutase 1(SOD1) have been identified as the highest genetic risk factor to cause familial ALS. Other genes that have been linked to familial cases of ALS include *TARDBP*, *VCP*, *C9ORF72*, and *FUS*. It is important to note that these genes are associated with familial ALS and also to a smaller degree sporadic ALS (Renton et al., 2014). Other genes such as *IMPG2*, *TBK1*, *ANO9*, *LGALSL*, *CRLF3*, *DNMT3A*, *GPR158*, and *GREB1L* with their coding variants are among the top ALS-associated genes (Cirulli et al., 2015; Cady et al., 2015). The prevalence of variants in other ALS-associated genes is listed in Table 34.4.

Table 34.4 Lists of some genes and their genomic locations associated with ALS

Gene name	Genomic location
Familial genes	
SOD1	21q22
VCP	9p13
TARDBP	1p36
C9ORF72	9p21
FUS	16p11
Sporadic genes	
ATXN2	12q24
FIG4	6q21
OPTN	10p13
SETX	9q34
ANG	14q11
DCTN1	2p13
SQSTM1	5q35
VAPB	20q13

Table adapted from Cady et al. (2015) and Renton et al. (2014)

Although more than 20 genes have been identified to be associated with ALS, effective treatment is still under investigation. Therefore, most treatments are prescribed according to the specific genotype variation and their corresponding therapy is decided. For instance, variations in *SOD-1* are treated with Cu^{2+} chelators, and neurotrophic factors are treated with exogenous neurotrophic factors. The only licensed drug for ALS is riluzole. Riluzole has several pharmacological properties, but its precise mode of action is unclear. Riluzole seems to interfere with the damaging effects of excitatory amino acids, mainly glutamate, caused by excessive stimulation of motor neurons and nerve cells. Postulated mechanisms for this effect include the inhibition of glutamate release, blockade or inactivation of voltage-dependent sodium channels, and/or inactivation of a G-protein mediated pathway (Cifra et al., 2013).

34.2.5 Huntington's Disease

Huntington's disease (HD) is a dominantly inherited condition in which every affected person carries the exact same mutation, which lengthens the variable polyglutamine tract in the Huntington protein by expanding the typically polymorphism CAG trinucleoside repeat in the HD gene. Although thousands of interactions with other proteins have been reported, the precise function of huntingtin is still unknown.

The degeneration causes uncontrolled movements, loss of intellectual faculties, and emotional disturbance. HD is a familial disease, passed from parents to children through mutation in the huntingtin gene (*Htt*). Each child of an HD parent has a 50–50 chance of inheriting the HD gene. The involuntary movements associated with this disorder have

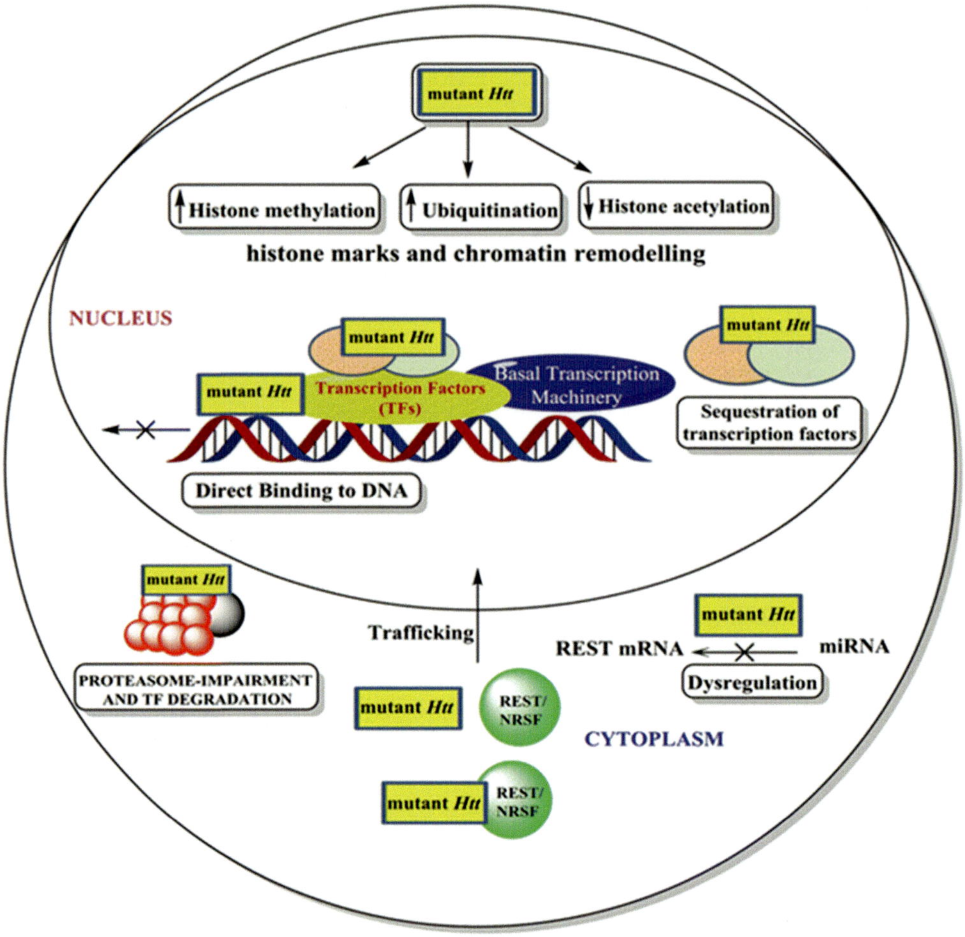

Fig. 34.1 Schematic representation of genetic dysregulation in Huntington's disease (Seredenina & Luthi-Carter, 2012). *Htt* Huntingtin gene, *TFs* transcription factors, *REST/NRSF* RE1-silencing transcription factor/neural-restrictive silencer factor

been described as "chorea". A 2006 systematic review of clinical studies evaluating pharmacologic treatments for HD concluded that the typical neuroleptics, haloperidol, and fluphenazine, were possibly useful for treating chorea in patients with HD (Bonelli & Wenning, 2006). Neuroleptics (antipsychotic medications) act by blocking dopamine transmission. Tetrabenazine is a selective, reversible, centrally-acting dopamine-depleting drug that works by inhibiting vesicular monoamine transporter 2 (VMAT2). Tetrabenazine depletes presynaptic dopamine, norepinephrine, and serotonin storage and antagonizes post-synaptic dopamine receptors. In vitro data indicate that tetrabenazine exhibits a weak binding affinity at the dopamine-2 receptor. Clinically, tetrabenazine improves the hyperkinetic movement disorder symptoms associated with HD (Chen et al., 2012).

One of the earlier theories of the pathology of the disease is the formation of intracellular inclusion bodies due to the aggregation of mutant Huntington protein, which in turn binds and sequesters proteins and transcription factors that dysregulate the normal transcriptional activities (Fig. 34.1). Different pathways and the concerned genes such as neurotransmitter receptors, G-protein coupled receptor signaling, calcium signaling and homeostasis, neurotrophin receptor signaling, ubiquitin-proteasome system that are dysregulated in the HD has been listed in Table 34.5. Other systems that influence the genetic pathways of the disease are retinoid signaling, lipid metabolism, transcription and chromatin remodeling, and synaptic transmission. Genes of these pathways have altered expression levels that lead to disease pathologies. Researchers have also found that histone deacetylase (HDAC) inhibitors might have a positive effect on the reversal of the Huntington-related gene (Chen et al., 2012; Seredenina & Luthi-Carter, 2012).

34.3 Role of Genetic Variations in Cytochrome P450 Drug Metabolizing Enzymes

Most of the CNS drugs are metabolized by enzymes of the cytochrome P450 (CYP) family. The genes encoding *CYP2D6* and *CYP3A4* isoenzymes have a high degree of polymorphism leading to marked variability in the activity of the encoded enzymes. There is a high level of allelic variation in different ethnic groups that have been linked to inter-individual differences in drug response (Seripa et al., 2010).

Table 34.5 Lists of genes dysregulated in HD and their associated genetic pathways

Genetic pathways	Genes dysregulated in HD
Neurotransmitter receptors	*DRD1A, DRD2, TH, CNR1,ADORA2A, GABRA5, GABRA4, GABRG2, GABRD, GAD1, GLRB, GRIA1, GRIA3, GRM1, GRM2, GRM3, GRIN2A, GRIN2B, SLC1A4, HEH3, OPRM1, OPRK1, OPRD1, HRH3*
G-protein receptor signaling	*PPP1R1B, ARPP21, GNAL, ADCY5, GNG3, GPR88, GPR6, PDE10A, PDE2A, PDE1B, RGS2, RGS4, RGS9, RASD2, RAS, RAPGEF3, PTPN5, GUCY1B3, PRKCB1, NGEF, PAK1*
Calcium signaling and homeostasis	*RYR1, CALB1, PPP3R1, PCP4/PEP-19, HPCA, RASGRP2, CPNE5, STRN, KCNIP2, KCNA1, SLC24A3, PLCB1, ATP2A2, ITPR1, CAMKK2, CABP1, SYNJ1, TESC*
Neurotrophin receptor signaling	*BDNF, NTRKB, NTRKA*
Neuropeptides	*PENK, NPY, PDYN, OXT, AVP, CART, POMC, SST, NTS, CRH*

Table adapted from Seredenina and Luthi-Carter (2012)

34.3.1 CYP2D6

The *CYP2D6* gene (cytochrome P450, family 2, subfamily D, polypeptide 6) lies at locus 22q13.1 (Gough et al., 1993) and is involved in the metabolism of an estimated 25% of all drugs (Wang et al., 2009). The CYP2D6 enzyme catalyzes the oxidative metabolism of over 100 clinically important and commonly prescribed drugs, such as cholinesterase inhibitors (tacrine, donepezil, galantamine), antidepressants, neuroleptics, opioids, class I anti-arrhythmics, analgesics and many other drug categories, acting as substrates, inhibitors or inducers with which cholinesterase inhibitors may potentially interact (Cacabelos, 2008). Cholinesterase inhibitors are the first-line symptomatic therapy for AD which increase active levels of acetylcholine in the synaptic cleft by inhibiting the enzymes that degrade acetylcholine (Mayeux, 2010). Tetrabenazine, the first FDA-approved drug for the management of HD-associated chorea, is a monoamine storage inhibitor extensively metabolized by the hepatic CYP2D6 to its primary active metabolite, alpha-dihydrotetrabenazine (Mehanna et al., 2013). More than 100 allelic variants of CYP2D6 have been identified with the enzyme activities ranging from deficient (poor metabolizers, PM), intermediate (intermediate metabolizers, IM), normal (extensive metabolizers, EM) to increased enzymatic activity (ultrarapid metabolizers, UM) with frequencies ranging from 5 to 10%, 10 to 17%, 70 to 80%, 3 to 5% respectively in Caucasian populations. PM have little or no CYP2D6 function because they have two null alleles and therefore unable to use the CYP2D6-dependent metabolic pathway for drug elimination. The most common variants associated with PM phenotype are *CYP2D6*3*, *CYP2D6*4*, *CYP2D6*5*, and *CYP2D6*6* in Caucasians, *CYP2D6*10* in Asian, and *CYP2D6*17* in Africans (Bradford, 2002; Zhou, 2009). The IMs metabolize drugs at a rate somewhere between the poor and extensive metabolizers whereas those classified as EMs have one or two alleles with normal CYP2D6 function. Expression of UMs genes leads to greater than normal CYP2D6 function. Consequently, drugs are metabolized rapidly in UMs, causing decreased efficacy, whereas slow metabolism of drugs in PMs causes toxicities owing to drug accumulation (Noetzli & Eap, 2013). The CYP2D6 enzyme is the major enzyme involved in the hepatic metabolism of donepezil to its active metabolite 6-O-desmethyldonepezil (Varsaldi et al., 2006). Donepezil is an inhibitor of acetyl cholinesterase, and is currently used for the symptomatic treatment of mild-to-moderate AD. Studies show that when administered with or without food, donepezil is well absorbed with a bioavailability of almost 100% (Cacabelos, 2008; Seripa et al., 2011; Benjamin & Burns, 2007; Varsaldi et al., 2006). Studies have also shown that functional polymorphisms in the CYP2D6 gene may affect enzyme activity and thus, the metabolism of donepezil, indicating that the plasma concentration of donepezil is dependent on *CYP2D6* polymorphism (Pilotto et al., 2009; Varsaldi et al., 2006). In the general population, EMs account for 55.71%, IMs for 34.7%, PMs 2.28%, and UMs 7.31% whereas in AD, EMs, IMs, PMs, and UMs account for 56.38%, 27.66%, 7.45%, and 8.51%, respectively. The therapeutic response to conventional drugs in patients with AD is genotype-specific, with *CYP2D6*-PMs, and *CYP2D6*-UMs having the worst drug response (Cacabelos et al., 2012). Studies investigating the relation between clinical response to donepezil and a common variant (rs1080985) of CYP2D6, confirmed that rs1080985 is associated with poor response to the drug (Albani et al., 2012). Recently, studies have shown that CYP2D6 is present in neurons in numerous human brain areas, including the thalamus, hypothalamus, hippocampus, substantia nigra, cerebellum, and in several layers of the frontal neocortex (Dutheil et al., 2009). The enzyme was implicated in metabolism of the endogenous compounds 5-methoxytryptamine, anandamide, progesterone, and tyramine and in generation of serotonin and dopamine from trace amines; these findings raise questions about its potential role in the function of these neurons (Miksys & Tyndale, 2013). The hypothesis that CYP2D6 may have a role in the brain was strengthened by a transgenic mouse model study: *CYP2D6* expression in the brain was associated with higher levels of serotonin measured in several brain regions, including the cerebellum and hippocampus (Cheng et al., 2013) (Table 34.6).

Table 34.6 Summary of pharmacogenetic studies on anti-dementia drugs

Drug	Gene, variant	Number of patients	Results
Donepezil	CYP2D6 Predicted phenotype	108	Significantly decreased clearance in PMs, increased clearance in UM compared with EMs
	CYP2D6 Predicted phenotype	94	EMs and IMs best responders in a combination therapy of donepezil with citicoline paracetamol and nicergoline
	CYP2D6 rs1080985 C > G	127	Significant higher frequency of the G-allele, associated with a higher enzyme activity, in nonresponders than in responders
	CYP2D6 16 functional SNPs	57	Significant higher frequency of alleles conferring decreased or absent enzyme activity in responders than in nonresponders
	CYP3A CYP3A4/CYP3A5 alleles	54	Genetic variants have no influence on plasma concentration and treatment outcome
	APOE ε 2, ε 3, ε 4	286	No significant influence of the APOE genotypes on treatment response
	APOE ε 2, ε 3, ε 4	40	APOE ε 4 carriers demonstrated a poorer response on the AD Assessment Scale-Jcog score after 3 years of therapy
	APOE ε 2, ε 3, ε 4	81	Better response in APOE ε4 carriers in specific cognitive domains and on mini-mental state examination score
Galantamine	CYP2D6 Predicted phenotype	356	The population pharmacokinetic model indicates a reduced clearance in PMs compared with EMs
	CYP2D6 Predicted phenotype	27	Significantly increased plasma concentrations in PMs compared with EMs
	APOE ε 2, ε 3, ε 4	202	No significant influence of the APOE genotypes on treatment response
Rivastigmine	APOE ε 2, ε 3, ε 4	214	No significant influence of the APOE genotypes on treatment response
Donepezil, rivastigmine	BCHE rs1803274	114	Wild-type carriers showed better response to rivastigmine than to donepezil, whereas the treatment response was similar within carriers of the variant allele
Menantine	SNPs in SLC22A1/2/5, SLC47A1, ABCB1, NR1I2, NR1I3, PPAR	108	NR1I2 rs1523130 significantly affected memantine clearance, with carriers of the CT/TT genotypes presenting a slower elimination than carriers of the CC genotype

Table adapted from Noetzli and Eap (2013)

34.3.2 The Cytochrome P450 (CYP) 3A4 Gene

The *CYP 3A4* gene (cytochrome P450, family 3, subfamily A, polypeptide 4) lies at locus 7q21.1. The *CYP3A* gene includes a cluster of *CYP3A4*, *CYP3A5*, *CYP3A7*, and *CYP3A43* and two pseudogenes named *CYP3A5P1* and *CYP3A5P2* (Seripa et al., 2010). These genes encode monooxygenases that catalyze many reactions involved in drug metabolism and synthesis of cholesterol, steroids and other lipids. The *CYP3A4* is the predominant hepatic form that demonstrates a large inter-individual variability in expression and activity and is involved in the metabolism of acetylcholinesterase inhibitors donepezil, and galantamine (Lamba et al., 2002; Noetzli et al., 2013). In a small study that evaluated the impact of CYP3A4 and CYP3A5 allele variants on donepezil plasma concentration and clinical outcome, there was no association.

34.4 Role of Defective Genes in Apolipoprotein E

Apolipoprotein E (*APOE*) gene mediates cholesterol metabolism in an isoform-dependent manner and is involved in lipid transport and injury repair in the brain involving beta-amyloid proteins (Eisenstein, 2011). The human *APOE* gene is polymorphic, and has three allelic variants namely ε2, ε3 and ε4 at a frequency of 11%, 72% and 17% respectively, and five common genotypes 2/3, 3/3, 2/4, 3/4, and 4/4 (Farrer et al., 1997; Guerreiro et al., 2012). The *APOE* gene has been shown to be the main genetic determinant of late-onset AD risk (Saunders et al., 1993; Bertram et al., 2007). Genome-wide association studies (GWAS) have repeatedly shown that the ε4 allele is the strongest known genetic risk factor for AD. Patients were found to respond to drug treatment depending on which of the ε2, ε3 or ε4 genetic variants of the

APOE gene a person carried. Studies showed that individuals with the *APOE* ε4 allele relative to those without the *APOE* ε4 allele have at least a twofold risk for developing AD (Tang et al., 1998). Generally, those carrying the ε4 allele are at increased risk of AD compared with those carrying the more common ε3 allele, whereas the ε2 allele has been shown to decrease the risk (Liu et al., 2013; Eisenstein, 2011). The *APOE* ε4 allele frequency was found to be significantly higher in the AD groups compared with the control group (Tsuang et al., 2013). Studies further indicate that *APOE* ε4 carriers preferentially benefit from receiving acetylcholinesterase inhibitors (tacrine, donepezil, galantamine, and rivastigmine) as AD treatment that improve their cholinergic deficit (Hanson et al., 2015). Other studies indicate that the presence of the APOE ε4 allele differentially affects the beneficial outcome and level of drug sensitivity in AD patients treated with neuroprotective compounds, nootropics ("smart drugs" or "memory enhancers"), endogenous nucleotides (CDP-choline), immunotrophins (such as Polypodium leucotomos: anapsos), neurotrophic factors (cerebrolysin), rosiglitazone, or combination therapies (Cacabelos et al., 2012; Cacabelos, 2009). The presence or absence of APOE ε4 allele modifies treatment response to a wide variety of therapeutics therefore its role is very important and should be carefully considered when clinical studies are being conducted for future drug optimization.

Apolipoprotein E (ApoE) mimetic peptides may be helpful in the treatment of Alzheimer's disease because they reduce the negative effects of oxidative stress on ApoE secretion and decrease the buildup of amyloid plaque. Endogenous apoE generated within the brain affects the neuroinflammatory response of the acutely damaged brain, according to strong preclinical and clinical data. Therefore, administering exogenous apoE-mimetic medicines designed to breach the blood–brain barrier is a cutting-edge strategy to enhance outcomes after acute brain damage (Minami et al., 2010; James et al., 2020). AD is a complex disorder involving many different gene clusters. However, several studies have shown that the *APOE* gene seems to be a major risk factor for both degenerative and vascular dementia. *APOE-4/4* carriers show a faster disease progression and a poorer therapeutic response to all available treatments than any other polymorphic variant. Studies analyzing the therapeutic response of AD patients with both the *APOE* and *CYP2D6* genes showed that it is genotype-specific, with *CYP2D6*-PMs, *CYP2D6*-UMs, and *APOE-4/4* carriers categorized as the worst responders. This indicates that *APOE* and *CYP2D6* may influence each other, as pleiotropic genes, in the metabolism of drugs and hepatic function. It has been observed that liver metabolism and transaminase activity were affected differently depending on whether patients were *APOE-4/4* carriers or not. Therefore, *APOE* may influence liver function and drug metabolism by modifying hepatic steatosis and transaminase activity (Cacabelos, 2008, 2009).

34.5 Role of Acetylcholinesterase Inhibitors

Acetylcholinesterase (AChE) is an enzyme found mainly in neuromuscular junctions and cholinergic brain synapses that is involved in hydrolyzing the neurotransmitter acetylcholine into acetate and choline, thereby terminating neurotransmission. In humans, AChE and butyrylcholinesterase (BChE) are the two major forms of cholinesterase enzymes. Donepezil, galantamine, and rivastigmine are AChE inhibitors that prevent hydrolysis of acetylcholine, and are commonly used for the treatment of AD (Noetzli & Eap, 2013). Donepezil is a potent noncompetitive and reversible AChE inhibitor and functions by increasing the acetylcholine levels that are available for synaptic transmission in the central nervous system. Donepezil is the most widely used drug for mild-to-moderate to severe AD treatment. When donepezil is administered orally, over 90% undergoes first-pass metabolism by the CYP2D6 enzyme leading to one active metabolite, 6-O-desmethyldonepezil and several inactive metabolites. The active metabolite of donepezil 6-O-desmethyldonepezil has a half-life of about 70 h, easily crosses the blood–brain barrier and is slowly excreted from the body (Noetzli & Eap, 2013; Benjamin & Burns, 2007; Seripa et al., 2011; Cummings et al., 2013). Studies analyzing an effect of CYP2D6 gene variants on treatment response show that it is genotype-specific, and patients with *CYP2D6*-PMs and *CYP2D6*-UMs were categorized as the worst responders. The pharmacokinetic profiles of these patients showed significantly decreased drug clearance in PMs and increased clearance in UMs when compared with EMs (Noetzli & Eap, 2013; Benjamin & Burns, 2007; Seripa et al., 2011; Varsaldi et al., 2006; Pilotto et al., 2009). A study investigating the effect of CYP2D6*10 and APOE polymorphisms on both steady state plasma concentrations and clinical response of donepezil in patients with mild-to-moderate AD found that patients with mutant allele (*10) in CYP2D6 gene may respond better to donepezil than those with wild allele (*1) (Zhong et al., 2013). However they did not find any relationship between *APOE* ε4 status and the efficacy of donepezil. Recently, studies analyzing the influence of *CYP2D6* and *APOE* polymorphisms on the therapeutic effect of donepezil in patients with mild-to-moderate AD showed no significant differences between responders and nonresponders (Klimkowicz-Mrowiec et al., 2013; Liu et al., 2014). Other studies looking at the *APOE* gene variants showed that *APOE* ε4 carriers displayed worst treatment response (Cacabelos et al., 2012), whereas the *CYP3A* genetic variants showed no influence on plasma concentration and treatment outcome (Noetzli & Eap, 2013). These results encourage a continued evaluation of the role of various polymorphisms in response to donepezil treatment in different ethnic groups, and to assess other genetic variants of CYP2D6.

Galantamineis a competitive, rapidly-reversible AChE inhibitor that functions by increasing the acetylcholine levels at cholinergic synapses. It also interacts with the nicotinic acetylcholine receptor at binding sites separate from those for AChE, and may increase the levels of glutamate and serotonin. Galantamine is metabolized primarily by O-demethylation by CYP2D6, then O-oxidation by CYP3A4 and glucuronidation. Studies investigating the clinical effectiveness and tolerability of galantamine with respect to plasma concentrations and *CYP2D6* polymorphisms showed that CYP2D6 PMs displayed 45% and 61% higher dose-adjusted plasma concentrations than CYP2D6 hetEMs and homEMs, respectively (Noetzli et al., 2013). However, other studies have failed to find any significant influence of the *CYP2D6* or *APOE* genotypes on galantamine treatment response (Clarke et al., 2011; Noetzli & Eap, 2013).

Rivastigmine is a noncompetitive and very slowly reversible AChE inhibitor. The metabolism of rivastigmine is primarily by esterases in both the liver and the intestinal tract to an inactive and nontoxic metabolite NAP226-90. It is the only AChEI that inhibits both BuChE and AChE. This is an important aspect for selectivity since the activity of AChE is higher in the CNS, and the BChE activity is higher in the periphery; additional inhibition of BChE might be beneficial (Noetzli & Eap, 2013). It has a bioavailability of 40% and a half-life of about 2 h. In order to investigate the clinical efficacy and safety of rivastigmine for patients with AD, a database of double-blinded, randomized, and controlled trials was analyzed, and results point that rivastigmine improved the rate of decline of cognitive function, activities of daily living, and severity of dementia for people with mild-to-moderate AD in comparison with placebo (Birks & Grimley Evans, 2015).

34.6 Personalized Medicine in Neurodegenerative Diseases

34.6.1 Potential Therapeutic Benefits

Personalized medicine in pharmacogenomics focuses on the ability to predict how a patient would respond when a drug is administered. In recent years, the US Food and Drug Administration (FDA) has aggressively pursued drug-label modification when studies have shown that excess risk can be convincingly linked to a genetic marker. Several examples are listed in the FDA's Table of Pharmacogenomic Biomarkers in Drug Labels, http://www.fda.gov/Drugs/ScienceResearch/ResearchAreas/Pharmacogenetics/ucm083378.htm. The US FDA granted market approval for the first pharmacogenetic test known as The AmpliChip CYP450 test that uses DNA microarray technology to identify genotypes in cytochrome P450 *CYP2D6* and *CYP2C19*. The test uses software to predict phenotypes, and tests for 27 *CYP2D6* alleles, including the deletions and duplications, and three *CYP2C19* alleles. The AmpliChip CYP450 test classifies individuals into the EM and PM CYP2C19 phenotypes by testing three alleles, and into four CYP2D6 phenotypes namely UMs, EMs, IMs, and PMs (de Leon et al., 2006a, b). Most of the treatments so far have been symptomatic but with the understanding of genetic components to the disease, personalized therapeutics based on individual genomic profile will improve treatment and reduce adverse side effects. Examples of potential therapeutic benefits are shown in a study with phamacogenomic testing whereby a patient who was highly sensitive to medications showed slow action of the Cytochrome P450 enzyme encoded by the *CYP2C19* gene, meaning the patient was a PM of most antidepressants. With these results, doctors were able to improve the therapeutic outcome by reducing her medication dose of venlafaxine (25 mg BID) and giving her a low-dose short-acting atypical antipsychotic (quetiapine IR, 25 mg/day). Another patient's pharmacogenomic profile showed increased function of enzymes encoded by *CYP2D6* and *CYP2C19* genes indicating an UM phenotype that called for adjusting treatment with escitalopram at 30 mg/day, higher than the recommended 20 mg/day (Dinama et al., 2014). As more GWAS are performed, scientists will be able to assign specific genomic patterns to EM, IM, PM or UM leading to a more studies geared toward optimized drug therapy. This is very important for the future of personalized medicine as targeted therapies are designed for patients with specific gene variations (Pacanowski et al., 2014).

34.6.2 Current Clinical Trials of Neurodegenerative Diseases

In order to understand the preclinical and clinical course of AD, a multinational collaborative project also known as The Dominantly Inherited Alzheimer Network (DIAN) was initiated where cohorts who are at risk for autosomal dominant Alzheimer disease (ADAD) with mutations in either of the *APP*, *PSEN1* or *PSEN2* genes are studied. The outcome of the study concluded there was a difference in overall pattern among younger DIAN cohorts compared to late-onset AD population (Storandt et al., 2014). A study comparing changes in brain imaging and fluid biomarkers for *PSEN1* E280A mutation carriers versus non-carriers showed that mutation carriers had lower precuneus cerebral metabolic rates for glucose, smaller hippocampal volume, lower cerebrospinal fluid (CSF) Aβ1-42, higher CSF total tau and phosphorylated tau, and significantly higher plasma Aβ1-42 that was not affected by age (Fleisher et al., 2015). Other clinical trials investigated Bapineuzumab, a humanized anti-amyloid-beta monoclonal antibody against *Aβ* (1–6) to treat mild-to-moderate AD patients categorized as *APOE* ε4 allele carriers and noncarriers. Results showed reduction in brain amyloid plaques and phosphorylated tau in CSF, but no improvement in cognitive or functional outcome between the *APOE* ε4 allele carriers compared with noncarriers was observed (Folch et al., 2016). These results are in agreement to the view that AD pathology begins before symptoms of dementia develop and largely influenced be amyloid plaques. However, more work needs to be done to incorporate the genetic aspect of neurodegenerative diseases.

Moreover, Center for Biologics Evaluation and Research (CBER) approved eight new Biological Licence Application (BLAs) in 2020. The FDA has approved novel therapies in neurology for patients with multiple sclerosis, tuberous sclerosis complex, spinal muscular atrophy, Dravet syndrome, Parkinson's disease and neuromyelitis optica spectrum disorder. Soliris (eculizumab) is the first approved drug for a rare autoimmune disease which affects brain and optic nerve known as neuromyelitis optica spectrum disorder. Subsequent approvals for Uplinza (inebilizumab-cdon) and Enspryng (satralizumab-mwge) in 2020 brought the total number of the FDA-approved therapies for this disorder to three (FDA).

34.6.3 Challenges to Overcome

Clinical studies of neurodegenerative disease presents major hurdles because of lack of molecular defects that can be targeted through drugs and translational animal models. In disease such as amyotrophic lateral sclerosis (ALS), there are no easily quantifiable biomarker that indicates the rate of disease progression. Such tools could enhance the accuracy with which drug response can be measured. There are diligent researchers, who are working towards understanding the pathogenesis of neurodegenerative diseases that could lead to several better biomarkers. A novel strategy based on induced pluripotent stem cells (iPSCs) has been developed to get around these restrictions. A new road map for clinical translational research and regeneration therapy could be made with iPSCs (Pandey et al., 2022).

Several challenges still need to be addressed for personalized medicine to become more effective. One is the use of combination therapy, which makes it difficult to study a large enough sample of patients treated with a single agent. In addition, the dosage of the drug may vary by regimen, further complicating efforts to study the pharmacogenomics of a specific drug of interest. Furthermore, replication of findings made in genome-wide association studies from large randomized clinical trial is often difficult, because high costs and ethical considerations may mean that a second identical trial is not feasible. Other challenges to overcome involve environmental factors, the interactions between different genetic loci and between different genes. When data from multiple studies are combined, there is increased potential for confounding variables that may affect how results are interpreted. Influence of sex or race or inadequate sample size may lead to the lack of inclusion of a causal genetic variation leading to an erroneous interpretation.

34.7 Review Questions

1. What is pharmacogenomics?
2. Identify the four CYP2D6 phenotypes.
3. Explain the role of acetylcholine in the AD pathology.
4. What is the goal of personalized medicine?

34.8 Answers

1. Pharmacogenomics is the study of how specific genetic variations influence the therapeutic outcome.
2. Normal activity (extensive metabolizers, EM); low or no activity (poor metabolizers, PM); intermediate activity (intermediate metabolizers, IM); and high enzymatic activity (ultrarapid metabolizers, UM).
3. Neuronal death in the brain at key locations leads to the deficiency of the neurotransmitter, acetylcholine thus causing memory impairment, hence leading to AD.
4. To combine clinical observations with information obtained from GWAS and candidate gene studies to predict how an individual will respond to a particular drug so that the problem of disease etiology is approached based on individual genomic profile. With this, clinicians can optimize drug dosage at the beginning of treatment and find effective ways to diagnose, treat, and prevent complex disorders with personalized therapeutics.

References

Albani, D., Martinelli Boneschi, F., Biella, G., et al. (2012). Replication study to confirm the role of CYP2D6 polymorphism rs1080985 on donepezil efficacy in Alzheimer's disease patients. *Journal of Alzheimer's Disease, 30*(4), 745–749. https://doi.org/10.3233/JAD-2012-112123

Benjamin, B., & Burns, A. (2007). Donepezil for Alzheimer's disease. *Expert Review of Neurotherapeutics, 7*(10), 1243–1249. https://doi.org/10.1586/14737175.7.10.1243

Bertram, L., McQueen, M. B., Mullin, K., et al. (2007). Systematic meta-analyses of Alzheimer disease genetic association studies: The AlzGene database. *Nature Genetics, 39*(1), 17–23.

Birks, J. S., & Grimley Evans, J. (2015). Rivastigmine for Alzheimer's disease. *Cochrane Database of Systematic Reviews, 4*, CD001191. https://doi.org/10.1002/14651858.CD001191.pub3

Bonelli, R. M., & Wenning, G. K. (2006). Pharmacological management of Huntington's disease: An evidence-based review. *Current Pharmaceutical Design, 12*(21), 2701–2720.

Bradford, L. D. (2002). CYP2D6 allele frequency in European Caucasians, Asians, Africans and their descendants. *Pharmacogenomics, 3*(2), 229–243.

Cacabelos, R. (2008). Influence of pharmacogenetic factors on Alzheimer's disease therapeutics. *Neurodegenerative Diseases, 5*(3–4), 176–178. https://doi.org/10.1159/000113695

Cacabelos, R. (2009). Pharmacogenomics and therapeutic strategies for dementia. *Expert Review of Molecular Diagnostics, 9*(6), 567–611. https://doi.org/10.1586/erm.09.42

Cacabelos, R. (2020). Pharmacogenomics of Alzheimer's and Parkinson's diseases. *Neuroscience Letters, 726*, 133807.

Cacabelos, R., Martinez, R., Fernandez-Novoa, L., et al. (2012). Genomics of dementia: APOE- and CYP2D6-related pharmacogenetics. *International Journal of Alzheimer's Disease, 2012*, 518901. https://doi.org/10.1155/2012/518901

Cacabelos, R., Cacabelos, P., Torrellas, C., et al. (2014). Pharmacogenomics of Alzheimer's disease: Novel therapeutic strategies for drug development. *Methods in Molecular Biology, 1175*, 323–556. https://doi.org/10.1007/978-1-4939-0956-8_13

Cady, J., Allred, P., Bali, T., et al. (2015). Amyotrophic lateral sclerosis onset is influenced by the burden of rare variants in known amyotrophic lateral sclerosis genes. *Annals of Neurology, 77*(1), 100–113. https://doi.org/10.1002/ana.24306

Calero, M., Gomez-Ramos, A., Calero, O., et al. (2015). Additional mechanisms conferring genetic susceptibility to Alzheimer's disease. *Frontiers in Cellular Neuroscience, 9*, 138. https://doi.org/10.3389/fncel.2015.00138

Chen, J. J., Ondo, W. G., Dashtipour, K., et al. (2012). Tetrabenazine for the treatment of hyperkinetic movement disorders: A review of the literature. *Clinical Therapeutics, 34*(7), 1487–1504. https://doi.org/10.1016/j.clinthera.2012.06.010

Cheng, J., Zhen, Y., Miksys, S., et al. (2013). Potential role of CYP2D6 in the central nervous system. *Xenobiotica, 43*(11), 973–984. https://doi.org/10.3109/00498254.2013.791410

Choi, S. H., Kim, S. Y., Na, H. R., et al. (2008). Effect of ApoE genotype on response to donepezil in patients with Alzheimer's disease. *Dementia and Geriatric Cognitive Disorders, 25*(5), 445–450. https://doi.org/10.1159/000124752

Cifra, A., Mazzone, G. L., & Nistri, A. (2013). Riluzole: what it does to spinal and brainstem neurons and how it does it. *The Neuroscientist, 19*(2), 137–144. https://doi.org/10.1177/1073858412444932

Cirulli, E. T., Lasseigne, B. N., Petrovski, S., et al. (2015). Exome sequencing in amyotrophic lateral sclerosis identifies risk genes and pathways. *Science, 347*(6229), 1436–1441. https://doi.org/10.1126/science.aaa3650

Clarke, J. A., Cutler, M., Gong, I., et al. (2011). Cytochrome P450 2D6 phenotyping in an elderly population with dementia and response to galantamine in dementia: A pilot study. *The American Journal of Geriatric Pharmacotherapy, 9*(4), 224–233. https://doi.org/10.1016/j.amjopharm.2011.07.003

Cummings, J. L., Geldmacher, D., Farlow, M., et al. (2013). High-dose donepezil (23 mg/day) for the treatment of moderate and severe Alzheimer's disease: Drug profile and clinical guidelines. *CNS Neuroscience & Therapeutics, 19*(5), 294–301. https://doi.org/10.1111/cns.12076

Danysz, W., & Parsons, C. G. (2003). The NMDA receptor antagonist memantine as a symptomatological and neuroprotective treatment for Alzheimer's disease: Preclinical evidence. *International Journal of Geriatric Psychiatry, 18*(Suppl 1), S23–S32. https://doi.org/10.1002/gps.938

de Leon, J., Armstrong, S. C., & Cozza, K. L. (2006a). Clinical guidelines for psychiatrists for the use of pharmacogenetic testing for CYP450 2D6 and CYP450 2C19. *Psychosomatics, 47*(1), 75–85.

de Leon, J., Susce, M. T., & Murray-Carmichael, E. (2006b). The AmpliChip CYP450 genotyping test: Integrating a new clinical tool. *Molecular Diagnosis & Therapy, 10*(3), 135–151.

Dinama, O., Warren, A. M., & Kulkarni, J. (2014). The role of pharmacogenomic testing in psychiatry: Real world examples. *The Australian and New Zealand Journal of Psychiatry, 48*(8), 778.

Drozdzik, M., Bialecka, M., & Kurzawski, M. (2013). Pharmacogenetics of Parkinson's disease—through mechanisms of drug actions. *Current Genomics, 14*(8), 568–577. https://doi.org/10.2174/1389202914666131210212521

Dutheil, F., Dauchy, S., Diry, M., et al. (2009). Xenobiotic-metabolizing enzymes and transporters in the normal human brain: Regional and cellular mapping as a basis for putative roles in cerebral function. *Drug Metabolism and Disposition, 37*(7), 1528–1538. https://doi.org/10.1124/dmd.109.027011

Eisenstein, M. (2011). Genetics: Finding risk factors. *Nature, 475*(7355), S20–S22. https://doi.org/10.1038/475S20a

English, C., & Aloi, J. J. (2015). New FDA-approved disease-modifying therapies for multiple sclerosis. *Clinical Therapeutics, 37*(4), 691–715. https://doi.org/10.1016/j.clinthera.2015.03.001

Farrer, L. A., Cupples, L. A., Haines, J. L., et al. (1997). Effects of age, sex, and ethnicity on the association between apolipoprotein E genotype and Alzheimer disease. A meta-analysis. APOE and Alzheimer Disease Meta Analysis Consortium. *JAMA, 278*(16), 1349–1356.

Fleisher, A. S., Chen, K., Quiroz, Y. T., et al. (2015). Associations between biomarkers and age in the presenilin 1 E280A autosomal dominant Alzheimer disease kindred: A cross-sectional study. *JAMA Neurology, 72*(3), 316–324. https://doi.org/10.1001/jamaneurol.2014.3314

Folch, J., Petrov, D., Ettcheto, M., et al. (2016). Current research therapeutic strategies for Alzheimer's disease treatment. *Neural Plasticity, 2016*, 8501693. https://doi.org/10.1155/2016/8501693

Gareri, P., Putignano, D., Castagna, A., et al. (2014). Retrospective study on the benefits of combined Memantine and cholinEsterase inhibitor treatMent in AGEd patients affected with Alzheimer's Disease: The MEMAGE study. *Journal of Alzheimer's Disease, 41*(2), 633–640. https://doi.org/10.3233/JAD-132735

Gatz, M., Reynolds, C. A., Fratiglioni, L., et al. (2006). Role of genes and environments for explaining Alzheimer disease. *Archives of General Psychiatry, 63*(2), 168–174.

Gough, A. C., Smith, C. A., Howell, S. M., et al. (1993). Localization of the CYP2D gene locus to human chromosome 22q13.1 by polymerase chain reaction, in situ hybridization, and linkage analysis. *Genomics, 15*(2), 430–432.

Grossman, I., et al. (2017). Pharmacogenomics strategies to optimize treatments for multiple sclerosis: Insights from clinical research. *Progress in Neurobiology, 152*, 114–130.

Guerreiro, R. J., Gustafson, D. R., & Hardy, J. (2012). The genetic architecture of Alzheimer's disease: Beyond APP, PSENs and APOE. *Neurobiol Aging, 33*(3), 437–456. https://doi.org/10.1016/j.neurobiolaging.2010.03.025

Hanson, A. J., Craft, S., & Banks, W. A. (2015). The APOE genotype: Modification of therapeutic responses in Alzheimer's disease. *Current Pharmaceutical Design, 21*(1), 114–120.

Hellstrom-Lindahl, E. (2000). Modulation of beta-amyloid precursor protein processing and tau phosphorylation by acetylcholine receptors. *European Journal of Pharmacology, 393*(1–3), 255–263.

Hollenbach, J. A., & Oksenberg, J. R. (2015). The immunogenetics of multiple sclerosis: A comprehensive review. *Journal of Autoimmunity, 64*, 13–25. https://doi.org/10.1016/j.jaut.2015.06.010

James, M. L., et al. (2020). Therapeutic development of apolipoprotein E mimetics for acute brain injury: Augmenting endogenous responses to reduce secondary injury. *Neurotherapeutics, 17*(2), 475–483.

Klimkowicz-Mrowiec, A., Wolkow, P., Sado, M., et al. (2013). Influence of rs1080985 single nucleotide polymorphism of the CYP2D6 gene on response to treatment with donepezil in patients with alzheimer's disease. *Neuropsychiatric Disease and Treatment, 9*, 1029–1033. https://doi.org/10.2147/NDT.S46689

Lamba, J. K., Lin, Y. S., Thummel, K., et al. (2002). Common allelic variants of cytochrome P4503A4 and their prevalence in different populations. *Pharmacogenetics, 12*(2), 121–132.

Li, J., Zhang, L., Zhou, H., et al. (2011). Global patterns of genetic diversity and signals of natural selection for human ADME genes. *Human Molecular Genetics, 20*(3), 528–540. https://doi.org/10.1093/hmg/ddq498

Liu, C. C., Kanekiyo, T., Xu, H., et al. (2013). Apolipoprotein E and Alzheimer disease: Risk, mechanisms and therapy. *Nature Reviews. Neurology, 9*(2), 106–118. https://doi.org/10.1038/nrneurol.2012.263

Liu, M., Zhang, Y., Huo, Y. R., et al. (2014). Influence of the rs1080985 single nucleotide polymorphism of the CYP2D6 Gene and APOE polymorphism on the response to donepezil treatment in patients with Alzheimer's disease in China. *Dementia and Geriatric Cognitive Disorders Extra, 4*(3), 450–456. https://doi.org/10.1159/000367596

Loveman, E., Green, C., Kirby, J., et al. (2006). The clinical and cost-effectiveness of donepezil, rivastigmine, galantamine and memantine for Alzheimer's disease. *Health Technology Assessment, 10*(1), iii–iv, ix–xi, 1–160.

Ma, M. K., Woo, M. H., & McLeod, H. L. (2002). Genetic basis of drug metabolism. *American Journal of Health-System Pharmacy, 59*(21), 2061–2069.

Mathisen, P. M. (2003). Gene discovery and validation for neurodegenerative diseases. *Drug Discovery Today, 8*(1), 39–46.

Mayeux, R. (2010). Clinical practice. Early Alzheimer's disease. *The New England Journal of Medicine, 362*(23), 2194–2201. https://doi.org/10.1056/NEJMcp0910236

Mehanna, R., Hunter, C., Davidson, A., et al. (2013). Analysis of CYP2D6 genotype and response to tetrabenazine. *Movement Disorders, 28*(2), 210–215. https://doi.org/10.1002/mds.25278

Miksys, S., & Tyndale, R. F. (2013). Cytochrome P450-mediated drug metabolism in the brain. *Journal of Psychiatry & Neuroscience, 38*(3), 152–163. https://doi.org/10.1503/jpn.120133

Minami, S. S., et al. (2010). ApoE mimetic peptide decreases Aβ production in vitro and in vivo. *Molecular Neurodegeneration, 5*(1), 1–12.

Noetzli, M., & Eap, C. B. (2013). Pharmacodynamic, pharmacokinetic and pharmacogenetic aspects of drugs used in the treatment of Alzheimer's disease. *Clinical Pharmacokinetics, 52*(4), 225–241. https://doi.org/10.1007/s40262-013-0038-9

Noetzli, M., Guidi, M., Ebbing, K., et al. (2013). Relationship of CYP2D6, CYP3A, POR, and ABCB1 genotypes with galantamine plasma concentrations. *Therapeutic Drug Monitoring, 35*(2), 270–275. https://doi.org/10.1097/FTD.0b013e318282ff02

Pacanowski, M. A., Leptak, C., & Zineh, I. (2014). Next-generation medicines: Past regulatory experience and considerations for the future. *Clinical Pharmacology and Therapeutics, 95*(3), 247–249. https://doi.org/10.1038/clpt.2013.222

Pandey, S., et al. (2022). iPSCs in neurodegenerative disorders: A unique platform for clinical research and personalized medicine. *Journal of Personalized Medicine, 12*(9), 1485.

Payami, H., & Factor, S. A. (2014). Promise of pharmacogenomics for drug discovery, treatment and prevention of Parkinson's disease. A perspective. *Neurotherapeutics, 11*(1), 111–116. https://doi.org/10.1007/s13311-013-0237-y

Pilotto, A., Franceschi, M., D'Onofrio, G., et al. (2009). Effect of a CYP2D6 polymorphism on the efficacy of donepezil in patients with Alzheimer disease. *Neurology, 73*(10), 761–767. https://doi.org/10.1212/WNL.0b013e3181b6bbe3

Redenšek, S., & Dolžan, V. (2020). The role of pharmacogenomics in the personalization of Parkinson's disease treatment. *Pharmacogenomics, 21*(14), 1033–1043.

Renton, A. E., Chio, A., & Traynor, B. J. (2014). State of play in amyotrophic lateral sclerosis genetics. *Nature Neuroscience, 17*(1), 17–23. https://doi.org/10.1038/nn.3584

Roden, D. M., & George, A. L., Jr. (2002). The genetic basis of variability in drug responses. *Nature Reviews. Drug Discovery, 1*(1), 37–44. https://doi.org/10.1038/nrd705

Saunders, A. M., Strittmatter, W. J., Schmechel, D., et al. (1993). Association of apolipoprotein E allele epsilon 4 with late-onset familial and sporadic Alzheimer's disease. *Neurology, 43*(8), 1467–1472.

Seredenina, T., & Luthi-Carter, R. (2012). What have we learned from gene expression profiles in Huntington's disease? *Neurobiology of Disease, 45*(1), 83–98. https://doi.org/10.1016/j.nbd.2011.07.001

Seripa, D., Pilotto, A., Panza, F., et al. (2010). Pharmacogenetics of cytochrome P450 (CYP) in the elderly. *Ageing Research Reviews, 9*(4), 457–474. https://doi.org/10.1016/j.arr.2010.06.001

Seripa, D., Bizzarro, A., Pilotto, A., et al. (2011). Role of cytochrome P4502D6 functional polymorphisms in the efficacy of donepezil in patients with Alzheimer's disease. *Pharmacogenetics and Genomics, 21*(4), 225–230. https://doi.org/10.1097/FPC.0b013e32833f984c

Shao, Z. Q. (2015). Comparison of the efficacy of four cholinesterase inhibitors in combination with memantine for the treatment of Alzheimer's disease. *International Journal of Clinical and Experimental Medicine, 8*(2), 2944–2948.

Storandt, M., Balota, D. A., Aschenbrenner, A. J., et al. (2014). Clinical and psychological characteristics of the initial cohort of the Dominantly Inherited Alzheimer Network (DIAN). *Neuropsychology, 28*(1), 19–29. https://doi.org/10.1037/neu0000030

Tang, M. X., Stern, Y., Marder, K., et al. (1998). The APOE-epsilon4 allele and the risk of Alzheimer disease among African Americans, whites, and Hispanics. *JAMA, 279*(10), 751–755.

Tsareva, E., Kulakova, O., Boyko, A., et al. (2016). Pharmacogenetics of multiple sclerosis: Personalized therapy with immunomodulatory drugs. *Pharmacogenetics and Genomics, 26*(3), 103–115. https://doi.org/10.1097/FPC.0000000000000194

Tsuang, D., Leverenz, J. B., Lopez, O. L., et al. (2013). APOE epsilon4 increases risk for dementia in pure synucleinopathies. *JAMA Neurology, 70*(2), 223–228. https://doi.org/10.1001/jamaneurol.2013.600

Varsaldi, F., Miglio, G., Scordo, M. G., et al. (2006). Impact of the CYP2D6 polymorphism on steady-state plasma concentrations and clinical outcome of donepezil in Alzheimer's disease patients. *European Journal of Clinical Pharmacology, 62*(9), 721–726. https://doi.org/10.1007/s00228-006-0168-1

Vuletić, V., et al. (2021). A systematic review of Parkinson's disease pharmacogenomics: Is there time for translation into the clinics? *International Journal of Molecular Sciences, 22*(13), 7213.

Wang, B., Yang, L. P., Zhang, X. Z., et al. (2009). New insights into the structural characteristics and functional relevance of the human cytochrome P450 2D6 enzyme. *Drug Metabolism Reviews, 41*(4), 573–643. https://doi.org/10.1080/03602530903118729

Zhong, Y., Zheng, X., Miao, Y., et al. (2013). Effect of CYP2D6*10 and APOE polymorphisms on the efficacy of donepezil in patients with Alzheimer's disease. *The American Journal of the Medical Sciences, 345*(3), 222–226. https://doi.org/10.1097/MAJ.0b013e318255a8f9

Zhou, S. F. (2009). Polymorphism of human cytochrome P450 2D6 and its clinical significance: Part I. *Clinical Pharmacokinetics, 48*(11), 689–723. https://doi.org/10.2165/11318030-000000000-00000

Immunomodulatory Therapy for Multiple Sclerosis

35

Guang-Xian Zhang

Abstract

As an understanding of the mechanisms underlying the pathogenesis of multiple sclerosis (MS) grows, in particular in those immune-related, novel therapeutic approaches have significantly improved the outcome for MS in the last decades. Rapid progress in the discovery of disease-modifying therapies (DMTs) has been made in more recent years: among the total number of 21 DMTs approved since 1993, only nine were present in the market in 2012; eight new ones have been approved since 2017. More novel approaches have been tested in preclinical and clinical trials with the goal of significantly enhancing the therapeutic efficacy and safety profile for patients with MS.

Keywords

Multiple sclerosis · Disease-modifying therapies · Immunomodulation · Central nervous system · Inflammation · Demyelination

35.1 Introduction

Multiple sclerosis (MS) is an inflammatory demyelinating disease of the central nervous system (CNS), and about three million people worldwide have been diagnosed with MS (Charabati et al., 2023). MS typically affects young adults, and the female-to-male ratio is about 3:1. The damage to myelin disrupts the normal flow of electrical impulses along the nerves, leading to a wide range of neurological deficits. MS is characterized by an overall reduction in CNS volume, an accumulation of immune cells mainly in the parenchyma of various CNS areas, and a substantial number of plaques. Inflammatory processes in these foci cause myelin damage, oligodendrocyte death, axon injury, and neuron loss (Charabati et al., 2023). The most common form of MS is relapsing-remitting MS (RRMS), diagnosed in 85–90% of patients between the ages of 20–40 and characterized by episodes of worsening neurological symptoms followed by periods of partial or complete recovery. RRMS often transitions into secondary progressive MS (SPMS), which involves a gradual worsening of symptoms over time with or without relapses. Primary-progressive MS (PPMS), the rarest form, is characterized by a steadily worsening progression of symptoms from the onset without distinct relapses or remissions (Charabati et al., 2023).

The exact cause of MS is still unclear, but it is believed to involve a combination of genetic and environmental factors. Black, Asian, and Hispanic individuals may be at risk for more severe MS-related disability. Environmental factors such as infections, vitamin D deficiency, smoking, and certain dietary factors have also been implicated in the development of MS (Ward & Goldman, 2022). The pathogenesis of MS remains largely unknown; however, immune mechanisms, especially those linked to the aberrant lymphocyte activity, have been considered mainly responsible for myelin and axonal damage in the CNS. It is believed that $CD4^+$ T cells specific for CNS antigens, most likely myelin components, play a pivotal role in the initiation and perpetuation of MS. The prevailing view of MS pathogenesis has been largely formed by analogy to experimental autoimmune encephalomyelitis (EAE), an animal model of MS (Gharagozloo et al., 2022; Glatigny & Bettelli, 2018). In this model, myelin-specific $CD4^+$ T cells that have developed in peripheral lymphoid organs crossed through the blood–brain barrier (BBB) to infiltrate the CNS, where they encounter their cognate antigens presented by infiltrating antigen-presenting cells (APC), e.g., DCs and macrophages, as well as local APCs, such as microglia and probably astrocytes (Charabati et al., 2023). This interaction with APCs leads to

G.-X. Zhang (✉)
Department of Neurology, Thomas Jefferson University, Philadelphia, PA, USA
e-mail: guang-xian.zhang@jefferson.edu

H. E. Gendelman, T. Ikezu (eds.), *Neuroimmune Pharmacology and Therapeutics*,
https://doi.org/10.1007/978-3-031-68237-7_35

the reactivation of myelin-specific CD4+ T cells, which in turn activate APCs by cell-cell contact and by secreted proinflammatory products, such as cytokines and chemokines. Secretion of proinflammatory mediators attracts various immune cells into the CNS, where they are activated and start secreting mediators that damage surrounding CNS tissue, leading to the formation of lesions and eventually to neurologic deficits (Charabati et al., 2023; Ward & Goldman, 2022). Th1 and Th17 populations of lymphocytes, which produce IFN-γ, IL-17, TNF, and GM-CSF, were primarily associated with MS pathogenesis (El-behi et al., 2010; Krishnarajah & Becher, 2022). In contrast, Th2 and regulatory T cells (Tregs), by producing anti-inflammatory/immunomodulatory mediators, e.g., IL-4, IL-9, and IL-10, are protective against MS (Krishnarajah & Becher, 2022). In addition to their antigen-presentation function, APCs may have two major phenotypes: e.g., proinflammatory, which expresses iNOS and produce IL-1, IL-12, IL-23, and TNF, or anti-inflammatory, which express Arg-1 and produce IL-10, IL-27, or IL-35. These cells/molecules play important roles in the pathogenesis and immunoregulation of MS. While the functional role of B cells is less understood, it is generally believed that B cells contribute to MS pathogenesis by producing autoantibodies, expressing inflammatory cytokines, and presenting antigens to T cells (Kumar & Axtell, 2023; von Budingen et al., 2011). On the other hand, regulatory B cells, which have been found in PBMCs of both healthy and MS patients (Geladaris et al., 2023), negatively regulate immune responses by the production of anti-inflammatory cytokines such as IL-10 and IL-35 (Matsushita, 2019). Thus, restricting these inflammatory cells/mediators from the CNS and/or making them anti-inflammatory is optimal for MS therapy.

35.2 FDA-Approved Disease Modifying Therapies (DMTs)

DMTs, with the goal of reducing the frequency and severity of relapses, slowing disease progression, and improving quality of life, are the mainstay of treatment for MS. There are >20 distinct DMT options currently approved by the U.S. Food and Drug Administration (FDA) for the treatment of relapsing MS (RMS) (Table 35.1), while in 2012, only nine were present in the market. This rapid development of new DMTs largely improved the efficacy and long-term safety of MS treatments (Stamatellos & Papazisis, 2023). DMTs can be administered via intramuscular (i.m.) and subcutaneous (s.c.) injections, oral, or intravenous (i.v.) infusion.

Table 35.1 FDA-approved medications for MS

Injectable therapies
IFN-β
IFN-β-1b, Betaseron (July1993)
IFN-β-1b, Extravia (August 2009)
IFN-β-1a, Avonex (May 1996)
IFN-β-1a, Rebif (March 2002)
PegIFN-β-1a, Plegridy (August 2014)
Others
Glatiramer acetate, Copaxone (December 1996)
Ofatumumab, Kesimpta (August 2020)
Oral therapies
S1P receptor modulators
Fingolimod, Gilenya (September 2010)
Siponimod, Mayzent (March 2019)
Ozanimod, Zeposia (March 2020)
Ponesimod, Ponvory (March 2021)
Fumarates
Dimethyl fumarate, Tecfidera (March 2013)
Diroximel fumarate, Vumerity (October 2019)
Monomethyl fumarate, Bafiertam (April 2020)
Others
Cladribine, Mavenclad (March 2019)
Teriflunomide, Aubagio (September 2012)
Infusion therapies
Mitoxantrone, Novantrone (October 2000)
Natalizumab, Tysabri (November 1997; off-label use)
Alemtuzumab, Lemtrada (November 2014)
Ocrelizumab, Ocrevus (March 2017)
Ublituximab, Briumvi (January 2023)

35.2.1 Injectable DMTs

35.2.1.1 IFN-β

IFN-β is produced in response to viral infections by a range of cell types, including dendritic cells, monocytes, and macrophages (Ali et al., 2019). In 1993, injectable recombinant IFN-β became the first FDA-approved disease-modifying therapy for MS (Lublin, 2005). The logic of using IFN-β for MS was based on its inhibitory action of IFN-γ, an inflammatory cytokine (Jacobs et al., 1996), and later studies identified its immunomodulatory activities (Clerico et al., 2008). IFN-β induced long-term, potentially beneficial, molecular effects on immune and possibly neuroprotective pathways in MS (Raftopoulou et al., 2022). Indeed, IFN-β downregulates expression of HLA Class II, thereby inhibiting Th1/Th17 cell activation (van Holten et al., 2002), and induces Foxp3+ regulatory T cells (Treg) (Biernacki et al., 2005; Raftopoulou et al., 2022). Our studies showed that IFN-β drives IL-27 production in activated monocytes and IL-10 in T-cell-activated PBMC cultures (Fitzgerald et al., 2013) and significantly suppressed GM-CSF production by T cells (Rasouli et al., 2015). IFN-β treatment in patients with MS also drives B cells toward a regulatory phenotype (Hedegaard et al., 2011).

Five recombinant IFN-β preparations have been approved worldwide for the treatment of RMS: IFN-β-1a (Avonex for i.m. and Rebif for s.c. injections, IFN-β-1b, under the brand names Betaseron and Extavia, is administered s.c., and the longer half-life, polyethylene glycol-conjugated IFN-β-1a (PEG-IFN-β-1a, Plegridy), injectable by i.m. and s.c. (Raftopoulou et al., 2022). The major differences between these medications are their forms of injections and frequency. IFN-β slows the progression of neurologic and cognitive disability, reduces the frequency of relapses and inflammatory lesion burden, and preserves cognitive function. Treatment with s.c. IFN-β-1a for 24 months reduced the number of new lesions evolving into black holes in patients with a first clinical demyelinating event (Vrenken et al., 2023). While these IFN-β preparations have similar effects through the same mechanism of action, PEGylated-IFN-β-1a has an eight-fold longer half-life of non-PEGylated-IFN-β-1a (Hu et al., 2012) and increases the duration of gene and protein induction after injection (Nguyen et al., 2023). In an ongoing, 5-year, phase 4 real-world study, the effectiveness of Plegridy in treating RMS has been indicated in all subgroups enrolled and is consistent with the established clinical safety profile (Salvetti et al., 2022).

IFN-β remains a valid first-line option due to its good long-term safety profile and cost-efficacy. Adverse effects of IFN-β treatment included injection site reactions, flu-like symptoms, abnormal laboratory tests, and the production of neutralizing antibodies (NAbs) (Raftopoulou et al., 2022). NAbs appear after 9–18 months of therapy in up to 40% of patients treated with IFN-β (Goldschmidt & Hua, 2020), and persistently high titers of NAbs result in an absence of therapeutic efficacy (Sorensen, 2022). Further, minimal concentrations of Plegridy were detected in the breast milk samples of lactating patients with MS receiving Plegridy at days 1–14 post-injection (Houtchens et al., 2022). Nevertheless, IFN-β therapy is currently less commonly prescribed as first-line therapy, because the newer therapies are more effective and better tolerated as treatment options in the field have evolved (Goldschmidt & Hua, 2020).

35.2.1.2 Glatiramer Acetate (GA)

Glatiramer acetate is an immunomodulating drug consisting of a complex polypeptide mixture (nonbiological complex drug) comprised of four amino acids in random order, resembling myelin basic protein (Messina & Patti, 2013). GA was approved in 1996 in the United States for the treatment of RMS and in 2014 for clinically isolated syndrome (CIS) under the brand name Copaxone. The efficacy of GA has been consistently shown in the treatment of patients with RMS or CIS, with an approximately 30% reduction in relapse and benefits on MRI measures of disease activity (Prod'homme & Zamvil, 2019). GA treatment switched T cells to an anti-inflammatory Th2 phenotype (Duda et al., 2000), induced FoxP3+ Tregs (Haas et al., 2009), and inhibited Th17 T-cell differentiation (Melnikov et al., 2020). CD8+ T cells generated by antigen presentation of GA contribute to inhibiting myelin degradation (Prod'homme & Zamvil, 2019). GA can also promote the differentiation of regulatory B cells, with increased production of IL-10 (Kala et al., 2011), and switch APC activation to an anti-inflammatory phenotype differentiation (Molnarfi et al., Molnarfi, 2015). GA-specific Th2 cells can cross the blood–brain barrier (BBB) and release anti-inflammatory cytokines, which can terminate an immune response and increase the secretion of protective neurotrophic factors including insulin-like growth factor-1 (IGF-1), IGF-2, and brain-derived neurotrophic factor (BDNF) (Kala et al., 2011; Kasindi et al., 2022), thus mediating tissue repair and regeneration.

GA has an overall favorable risk-benefit profile and is not associated with immunosuppression, autoimmune disease, infections, or the development of neutralizing antibodies (Boster et al., 2015). A database analysis of all patients with MS who have ever been exposed to GA 20 mg/mL daily in clinical trials, including patients with up to 20 years of continuous treatment, also showed GA to be safe and generally well tolerated (Ziemssen et al., 2017). Due to its excellent safety record, GA may be appropriate in the disease course in response to evolving patient conditions (e.g., aging, accumulating comorbidities, chronic treatment with corticosteroids and other immunosuppressants), or as a bridging therapy during conception, pregnancy, and breastfeeding. GA may also be used together with vaccinations, e.g., the use of live or attenuated vaccines as well as vaccines against hepatitis B virus or SARS-CoV-2 (Mirabella et al., 2022). Although one study showed a trend for reduced clinical efficacy in 36% of MS patients with high-level of GA-binding antibodies (Salama et al., 2003), the majority of patients treated with GA have been shown to develop GA-reactive antibodies, which do not seem to interfere with the effect of GA (Sellebjerg et al., 2012). Moreover, relapse-free patients develop higher IgG titer (Brenner et al., 2001), and GA-specific antibodies in an animal model of CNS demyelinating disease were shown to promote myelin repair (Ure & Rodriguez, 2002). These results suggest a possible beneficial effect rather than a neutralizing one for GA-reactive antibodies (Sorensen, 2022).

35.2.1.3 Ofatumumab

Ofatumumab, a fully human anti-CD20 monoclonal antibody that selectively depletes B cells, was approved by the FDA in 2020 for the treatment of adults with RMS with the brand name Kesimpta. Unlike other anti-CD20 therapies, ofatumumab can be self-administered (s.c. injection) by patients. Ofatumumab treatment also reduced the level of peripheral CD20 + T cells and CNS-migratory capacity of T cells, which may contribute to the treatment effect of ofatumumab in RMS (von Essen et al., 2022). Ofatumumab has been shown to be highly effective for the treatment of RMS, with significantly reduced relapse rate, disability worsening and MRI lesion activity with a favorable safety and tolerability profile. Ofatumumab treatment was associated with lower annualized relapse rates than teriflunomide, an oral treatment of RMS that served as a treatment control (Hauser et al., 2020). A network meta-analysis has shown that ofatumumab was as effective as other highly efficacious monoclonal antibody DMTs (i.e., alemtuzumab, natalizumab, and ocrelizumab) (Samjoo et al., 2020). The most common adverse events included nasopharyngitis, headache, upper respiratory tract infections, and urinary tract infections, and these events were generally manageable (Hauser et al., 2020). The efficacy and safety of ofatumumab vs. teriflunomide have been further accessed in recently diagnosed, treatment-naive participants, which showed that ofatumumab had a superior benefit-risk profile compared with teriflunomide, with an almost complete abrogation of inflammatory disease activity and no unexpected safety signals. Thus, teriflunomide could be used as a first-line treatment in early MS (Gartner et al., 2022). In a multicenter survey (PERITIA) European neurologists considered ofatumumab to be a very efficient and safe treatment option in RMS, with the significant advantage of convenient monthly self-administered s.c. injections for patients (Rau et al., 2023).

35.2.2 Oral DMTs

35.2.2.1 Sphingosine 1-Phosphate (S1P) Receptor Modulators

Signaling pathways controlled by S1P, with five already known subtypes (S1P1-5), are important for regulating a wide range of cellular processes, especially immune responses, and inflammation (Rosen et al., 2013). S1P receptor modulators, including fingolimod, siponimod, ozanimod, and ponesimod, are superagonists (functional antagonists) of S1P1, leading to S1P1 receptor internalization and degradation. Given that the S1P1 receptor is mainly expressed in lymphocytes, neural cells, and vasculature, these functional S1P1 receptor antagonists prevent lymphocyte egress from secondary lymphoid organs such as the lymph nodes and spleen to enter the circulatory fluids, thus reducing CNS infiltration and alleviating MS severity (Kandjani et al., 2023). These compounds can cross the BBB and directly exert cytoprotection and anti-inflammatory effects in the CNS (Prager et al., 2015).

Fingolimod

Fingolimod (FTY720), marketed under the brand name Gilenya, was the first approved oral MS medication in 2010. Fingolimod phosphate binds to four of the five already known S1P subtypes (S1P1-5), namely S1P1, S1P3, S1P4, and S1P5. Among them, subtype S1P1 is mainly expressed in lymphocytes, neural cells, and vasculature. After the binding of fingolimod phosphate in the S1P1 in lymphocytes, their egress from lymphatic tissues and migration into the CNS is prevented (Kandjani et al., 2023). Oral fingolimod in patients with RMS showed a significant reduction in the annualized relapse rate and risk of disability progression, with significantly fewer new or enlarged lesions on T2-weighted images (Singer et al., 2011). While the increased level of neurofilament light chain in the plasma of MS patients was associated with relapses and active/new T2 lesions, fingolimod treatment significantly decreased its plasma level. These results demonstrated a statistically significant effect of neurofilament light chain levels as a prognostic marker for the treatment effect of fingolimod in MS (Ashkar et al., 2023). The anti-inflammatory effects of fingolimod did not slow disease progression in primary progressive MS (PPMS) (Lublin et al., 2016).

The lipophilic nature of fingolimod allows it to easily cross the BBB and reach physiologically meaningful concentrations in CNS tissue (Miron et al., 2010). Fingolimod, after crossing the BBB, can bind to receptors of S1P3, S1P4, and S1P5, which are expressed on all cell types within the CNS (Miron et al., 2010), thus having the potential to exert effects directly on CNS cells to provide a neuroprotective effect (Pitteri et al., 2018). We showed that fingolimod, at nanomolar concentrations, effectively protected the survival of neural stem cells enhanced their development into mature oligodendrocytes in vitro, and enhanced the neural recovery effect of neural stem cells in the secondary progressive stage of remitting-relapsing EAE in vivo (Zhang et al., 2017).

Potential adverse events such as bradycardia, infections, macular edema, teratogenicity, and progressive multifocal leukoencephalopathy (Ayzenberg et al., 2016). The pooled prevalence of cancer was 2% in MS patients who received fingolimod, and the risk is higher in MS patients treated with a higher dose of fingolimod (Askari et al., 2022). Further, disease reactivation may occur upon fingolimod cessation, and patients should be closely monitored for MS activity for several months after stopping fingolimod (Ayzenberg et al., 2016). Overall, fingolimod 0.5 mg was well tolerated by patients (Singer et al., 2011). Nevertheless, the adverse events associated with fingolimod, the ancestor of the S1P receptor modulator, have stimulated a search for an S1P modulator with greater selectivity for $S1P_1$, with the most important immune mechanism to prevent MS-related neuroinflammation (Ruggieri et al., 2022).

Siponimod

Siponimod is a newer generation of S1P receptor modulators that selectively internalize S1P1 and S1P5 receptors to inhibit efflux of lymphocytes from lymph nodes into the CNS (Goodman et al., 2019). Siponimod may also have a direct neuroprotective property, through modulating S1P5 receptor independent of peripheral lymphocytopenia. Siponimod, under the brand name Mayzant, was approved in March 2019 by the FDA for the treatment of adults with relapsing forms of MS, including CIS, RMS, and active secondary progressive MS (SPMS), and became the first oral DMT specifically approved for active forms of SPMS (Bravo et al., 2022). Siponimod significantly reduced the risk of confirmed disability progression, worsening in cognitive processing speed, relapses, and MRI measures of brain atrophy and inflammation vs. placebo in SPMS patients in the Phase 3 EXPAND study (Kappos et al., 2018b). The consistent efficacy and long-term safety profile of siponimod support its clinical utility for long-term treatment of SPMS. The better treatment effects in the continuous siponimod compared to the placebo-siponimod group indicate the importance of earlier treatment initiation (Cree et al., 2022). Siponimod reduced the risk of disability progression in SPMS and improved cognition and MRI outcomes (Gold et al., 2022). Compared to fingolimod, which modulates a broader range of S1P receptors, the selective effect of siponimod on the S1P1 and S1P5 receptors offers a favorable side-effect profile and transient bradycardia can be avoided by dose titration (Goodman et al., 2019).

Ozanimod

Ozanimod, with the brand name Zeposia, was approved in March 2020 by the FDA for the treatment of adults with relapsing forms of MS, including CIS, RMS, and active SPMS. The efficacy and safety of ozanimod was superior to that of DMF (Cohan et al., 2021). Compared with ponesimod, ozanimod is comparable in reducing relapse rates, more effective in preserving brain volume, and has a favorable safety profile (Swallow et al., 2023). Up to 5 years of the study showed that 71% of patients treated with ozanimod remained relapse-free over 48 months, and confirmed a favorable safety/tolerability profile and sustained benefit on clinical and magnetic resonance imaging measures of disease activity (Cree et al., 2022). Ozanimod treatment reduced plasma neurofilament light chain concentration, which was associated with improved MRI and, a lower annualized relapse rate (Harris et al., 2021). Adverse events of ozanimod include increased risk of infections, liver injury, increased blood pressure, respiratory effects, macular edema, and posterior reversible encephalopathy syndrome (Lassiter et al., 2020). In a meta-analysis of the efficacy and acceptability of the S1P receptor modulators in the treatment of MS, ozanimod was considered the best and ponesimod the worst for acceptability (Tong et al., 2021). More head-to-head comparisons should be carried out to evaluate the efficacy and reverse effects of ozanimod vs. other S1P1 receptor modulators.

Ponesimod

Ponesimod is an orally active, revisable, highly selective modulator of the sphingosine-1-phosphate receptor 1 (S1P1). Similar to other S1P1 receptor modulators, ponesimod blocks the egress of lymphocytes from lymphoid organs into the systemic circulation, while rapid elimination of ponesimod and the reversibility of its effects on lymphocyte levels allows the rapid return of normal immune system function, which may be beneficial in terms of safety for pregnancy planning, serious infections, or vaccinations (Alnaif et al., 2023; Krause et al., 2014). Ponesimod was approved by the FDA in March 2021 under the brand name Ponvory on the basis of a Phase III, double-blind, double-dummy, randomized clinical trial that demonstrated the superiority of ponesimod to teriflunomide on clinical and MRI improvement, without unexpected safety concerns (Kappos et al., 2021). In addition to its effect on the immune system, ponesimod exhibited a synergistic anti-apoptotic effect on oligodendrocytes with amniotic epithelial cells as a neuroprotective and neurogenic agent (Safaeinejad et al., 2021). Further, a gene ontology relationship of reduced neuroinflammation and reduction in known astrocyte disease-related genes, including those of immediate early astrocytes that have been strongly associated with disease progression in MS animal models, have been identified scRNA-seq in primary human astrocytes cultured with ponesimod (Kihara et al., 2022). These results indicate that ponesimod may have a direct effect on these CNS-resident cells, which express certain locus of S1P1 (Kihara et al., 2022).

The safety of a flat 20 mg maintenance dose for ponesimod in adult patients with MS has been identified (Valenzuela et al., 2022). Given its favorable risk-benefit and convenience profile, ponesimod might become a leading option among S1P receptor modulators used for RMS (Baldin & Lugaresi, 2020). When the efficacy and safety of S1PR modulators (fingolimod, ozanimod, and siponimod) vs. IFN-β in the treatment of RMS was compared, it was found that the annualized relapse rate in the S1PR modulator group was 20% less than that in the IFN-β group, with a significant reduction in MRI lesions (Yang et al., 2022). However, it has also been found that compared with ponesimod was less effective than ozanimod (Swallow et al., 2023). This observation needs to be further followed up in future studies.

35.2.2.2 Fumarates

Dimethyl fumarate (DMF) is an oral agent approved by the FDA for the treatment of MS in 2013 under the brand name Tecfidera, and its efficacy has been well demonstrated. DMF reduced clinical relapse and lesion frequency while improving health-related quality of life in adults with RMS (Havrdova et al., 2017). Twice-daily DMF reduced clinical relapse and MRI measures of disease activity and improved some aspects of health-related quality of life. DMF had an acceptable tolerability profile. The most common adverse events were flushing and gastrointestinal events, which were of mild or moderate severity and appeared to be largely manageable (Blair, 2019). The risk of a first clinical demyelinating event during the 96-week study period was highly reduced in DMF-treated patients with radiologically isolated syndrome (RIS), the earliest detectable preclinical phase of MS (Okuda et al., 2023).

The mechanism of DMF action is believed through activating the nuclear factor erythroid 2-related factor 2 (Nrf2) pathway, which has antioxidant and anti-inflammatory effects (Majkutewicz, 2022). DMF decreases absolute lymphocyte counts, with the most profound reduction of $CD8^+$ T cells. DMF treatment increased the proportion of Th2 cells among memory T cells but with a decrease in the proportion of Th1 and with no change in Th17 cell proportion among memory T cells (Gross et al., 2016). Another study, however, showed a decrease in $CD4^+$ T cells producing IFN-γ, IL-17, and GM-CSF, with no significant change in IL-10 and IL-4 after 12 months (Montes Diaz et al., 2018). We also found that show that DMF specifically suppresses Th1 and GM-CSF–producing Th1 cells in PBMCs from healthy donors (Safavi et al., 2020). In vitro and animal models demonstrate a role for DMF within the CNS in promoting neuronal survival in an Nrf2 pathway-dependent or pathway-independent manner (Yadav et al., 2019). Gut microbiota may be involved in therapeutic action and adverse events of DMF, representing a potential target for improving disease course and DT tolerability (Ferri et al., 2023). Further, DMF may have neuroprotective effects, including reducing spinal cord inflammation and protecting myelin and neurons (Linker et al., 2011).

Both DF and *diroximel fumarate* (DRF), which was approved in the USA for RMS in 2019 under the brand name Vumerity, are prodrugs that are rapidly metabolized to monomethyl fumarate, the active metabolite after oral administration. *Monomethyl fumarate* was approved for use in adults with RMS in the United States in 2020 under the brand name Bafiertam (Berger et al., 2021). These three fumarates showed similar treatment effects, anti-inflammatory and immunomodulatory mechanisms, as well as adverse effects. Common adverse events of DMF are flushing, gastrointestinal symptoms of nausea, diarrhea or abdominal pain, dizziness, erythema, and skin rash. Rare but potentially severe adverse events include anaphylaxis, angioedema, severe lymphopenia, herpes zoster, liver injury, and progressive multifocal leukoencephalopathy. The adverse event profile of DRF and monomethyl fumarate is similar to that of DMF, with less frequency of gastrointestinal adverse reactions and better overall tolerance (Liseno et al., 2021). DRF-treated patients also showed high overall persistence, low discontinuation rate due to gastrointestinal adverse events, and high adherence to therapy compared to DMF treatment (Liseno et al., 2021).

35.2.2.3 Cladribine

Cladribine, a deoxyadenosine analog prodrug that selectively depletes lymphocytes, was approved by the FDA in March 2019 under the brand name Mavenclad. Cladribine is the first short-course oral therapy for the treatment of adults with RMS and active SPMS, in particular those who have had an inadequate response to or cannot tolerate an alternate MS treatment (Deeks, 2018). Cladribine may act as an immune reconstitution therapy, exhibiting prolonged effects after a short treatment dose through transient reduction of T cells, B cells, and to a lesser extent innate immune cells, with a gradual recovery over years (Patti et al., 2020). Unlike most other DMTs for MS that require continual treatment, cladribine is administered in short courses at the beginning of years 1 and 2, and the effect of the drug could last for many months or years (Giovannoni et al., 2019). Importantly, Cladribine can cross the BBB and thus may have the potential to deplete CNS-resident immune cells (Leist & Weissert, 2011).

The most common serious adverse events of cladribine include lymphopenia and uterine leiomyoma, and other adverse events include leukopenia, alopecia, and viral upper respiratory tract infection (Giovannoni et al., 2019). Overall, cladribine tablets were well tolerated with a favorable safety profile (Cook et al., 2011).

35.2.2.4 Teriflunomide

Teriflunomide is a reversible, selective inhibitor of dihydroorotate dehydrogenase, which is required for the expansion of antigen-activated lymphocytes (Miller, 2021). Teriflunomide was approved by the FDA in 2012 as an oral DMT for RMS treatment under the brand name Aubagio. Teriflunomide treatment has demonstrated significant improvements in clinical and MRI outcomes in relapsing MS, with long-term safety and tolerability (Miller, 2021). It was also found that 79.0% of RMS patients achieved NEDA 3 at 12-month consecutive treatment, with a significantly reduced mean annualized relapse rate, while the mean EDSS remained stable (Zhang et al., 2022).

As a mechanism of its action, teriflunomide-mediated blockade of pyrimidine synthesis interrupts the cell cycle and exerts a cytostatic effect on proliferating T and B cells, limiting their involvement in the inflammatory processes involved in the MS pathogenesis (Miller, 2021). Serum CXCL13 and TNF-α levels were significantly decreased after treatment with teriflunomide, while the change in

CXCL13 levels during the first year of treatment can be used as a biomarker to identify optimal teriflunomide responders (Fissolo et al., 2023).

Interestingly, in vitro evidence showed that low concentrations of teriflunomide have a direct effect on promoting oligodendroglial cell differentiation, thus enhancing remyelination (Gottle et al., 2018). In the mouse demyelination models, teriflunomide promoted the differentiation of newly generated oligodendrocytes and significantly increased remyelination (Martin et al., 2021). Moreover, teriflunomide restored mitochondrial integrity within oligodendroglial cells (Gottle et al., 2023). Teriflunomide attenuates neuroinflammation-related neural damage in mice carrying human PLP1 mutations, suggesting teriflunomide may also be a medication for inflammation-related genetic diseases of the CNS, for which causal treatment options are presently lacking (Groh et al., 2018).

The most frequent adverse events of teriflunomide were hair thinning, alanine aminotransferase elevation, and leukopenia (Stamatellos & Papazisis, 2023). Teriflunomide treatment may be specifically beneficial for female patients, patients with fewer relapses and less disability before treatment (Zhang et al., 2022). Overall, teriflunomide has a good record for efficacy and tolerability for treatment-naive RMS patients (Miller, 2021).

35.2.3 Infusion DMTs

35.2.3.1 Natalizumab

Natalizumab is a recombinant humanized IgG4κ mAb that selectively blocks the α4 chain (CD49d) of the VLA4 antigen of lymphocytes, thus inhibiting their migration through the BBB into the CNS (Khoy et al., 2020). Natalizumab was approved by the FDA in 2004 for RMS under the brand name Tysabri. Several long-term studies have confirmed that natalizumab is a highly effective treatment option for MS (Morrow et al., 2022). There is no evidence for natalizumab to prevent or slow the sustained progressive disability experienced by persons with SPMS (Morrow et al., 2022), although a clinical trial showed a positive effect on the progression of upper limb disability (Kapoor et al., 2018).

Natalizumab treatment has excellent long-term safety (Morrow et al., 2022). The most common serious adverse events were infections and infestations (4.1%); PML, pneumonia, urinary tract infection, and herpes zoster were the most common infections reported, with very low rates of development of malignancies and opportunistic infections (Butzkueven et al., 2020). Neutralizing antibodies against natalizumab may occur in certain patients (up to 9%), and patients aged 45–60 years had a higher rate of neutralizing antibodies (Bachelet et al., 2016; Link et al., 2017). In conclusion, these long-term studies further support the effectiveness and well-established safety profile of natalizumab in RMS (Butzkueven et al., 2020; Morrow et al., 2022).

35.2.3.2 Alemtuzumab

Alemtuzumab is a humanized mAb that targets CD52, a protein expressed on the surface of lymphocytes, to deplete circulating T and B lymphocytes. Lymphocyte depletion is followed by a distinctive pattern of T- and B-cell repopulation, changing the balance of the immune system (Hartung et al., 2015). After an almost complete depletion of all lymphocytes, the cell specificities showed different reconstitution patterns, resulting in different cell fractions, e.g., a significant reduction of the number of IFN-γ-secreting cells (Hilger et al., 2020). B cells from treated patients secreted higher levels of IL-10 and BDNF and inhibited the proliferation of autologous $CD4^{+}CD25^{-}$ T-cell targets (Kashani et al., 2021). The long-lasting therapeutic benefit of alemtuzumab in RMS may also involve a shift in the cytokine balance towards inhibition of inflammation associated with a reconstitution of the peripheral blood CD4+ T-cell subsets, e.g., expansion of Treg cells with increased suppressive function. Indeed, a significant increase in Treg cell percentages was observed at month 24 of treatment, accompanied by an increase in Treg cell suppressive activity against MBP-specific Th1 and Th17 cells (De Mercanti et al., 2016).

Alemtuzumab was approved by the FDA in 2014 for RMS under the brand name Lemtrada, generally reserved for patients who have had an inadequate response to 2 or more drugs indicated for the treatment of MS [Genzyme Corporation, 2014]. Alemtuzumab significantly reduced the rate of sustained accumulation of disability, mean disability score, and lesion burden of MRI examination compared with that in the IFN-β-1a group (Coles et al., 2011). Efficacy with alemtuzumab was maintained over 7 years in subsequent extension studies with a consistent safety profile (NCT00930553; NCT02255656).

There is an increased incidence of autoimmune events in patients treated with alemtuzumab, mainly thyroid events, but also immune thrombocytopenia and nephropathy, indicating a need for mandatory safety monitoring following the last treatment (Barclay et al., 2019). Although alemtuzumab depletes lymphocytes in most individuals, some people deplete poorly, probably due to biological-response variation and neutralizing Abs, and this may lead to treatment failure (Dubuisson et al., 2018). Consistent with product labeling post-marketing data, low absolute risk with alemtuzumab treatment was also confirmed for peri-infusion, non-autoimmune adverse events, including cardiovascular and pulmonary events (Coles et al., 2011). Together, alemtuzumab represents a new option for the treatment of active RMS (Hartung et al., 2015).

35.2.3.3 Anti-CD20 Antibodies

The involvement of B cells in the pathogenesis of MS has been suggested in the last decades, through several antibody-dependent (e.g., secreting intrathecal IgG) and -independent (e.g. as antigen presenting cells) mechanisms (Wanleenuwat & Iwanowski, 2019). CD20 is a transmembrane, nonglycosylated phosphoprotein expressed on the surface of human B cells and is involved in the generation of T-cell-independent antibody responses (Cragg et al., 2005). Based on these findings, several B-cell-targeted therapies through anti-CD20 mechanism have been developed, and anti-CD20 mAbs currently in use for the treatment of MS include rituximab, ocrelizumab, ofatumumab and ublituximab. Among them, ofatumumab can be administrated via s.c. injection (see Sect. 1.4), while others need infusion.

Rituximab

Rituximab is a chimeric monoclonal B-cell-depleting anti-CD20 antibody that has been extensively used off label for the treatment of MS. Rituximab selectively binds to the CD20 antigen expressed on the surface of pre-B cells, mature and memory B cells, but not B-cell progenitors (pro-B cells) and differentiated plasma cells. Therefore, rituximab treatment selectively induces cytotoxicity of circulating and tissue-based B cells that are responsible for antigen presentation and cytokine production, without affecting pre-existing humoral immunity (Greenfield & Hauser, 2018). Rituximab significantly reduced the annualized relapse rates and the number of MRI lesions in RMS studies and delayed time to confirmed disease progression in PPMS, with lower T2 lesion volume at week 96 of treatment (Castillo-Trivino et al., 2013). In a retrospective observational study on safety and efficacy, infusion-related adverse events occurred during 7.8% of infusions and most were mild, and total of 89 non–infusion-related adverse events including infections ($n = 76$), grade 2 malignancies ($n = 3$), and death ($n = 4$) with cardiac arrhythmia, respiratory failure, vascular surgery, and suicide by intoxication (Salzer et al., 2016). Rituximab treatment is superior to dimethyl fumarate in preventing relapses in patients with early RMS (Svenningsson et al., 2022). Thus, rituximab is considered a valid treatment option for MS patients considering its high efficacy and the safety profile despite the status of being an off-label drug, and therefore being subjected to variable regulations (Chisari et al., 2022).

Ocrelizumab

Ocrelizumab, another mAb that targets CD20, thus depleting B cells, is approved in March 2017 under the brand name Ocrevus for treatment of RMS and PPMS, and is the first FDA-approved treatment for PPMS. Significantly lower disability progression and MRI-evidenced progression had been identified in ocrelizumab-treated patients with RMS when compared to those treated with IFN-β-1a or placebo. The percentages of PPMS patients with disability progression and MRI progression after ocrelizumab treatment were also significantly lower than in patients receiving placebo (Mulero et al., 2018). The clinical and MRI efficacy in RMS of seems dose-dependent and, importantly, this effect was not associated with a higher rate of adverse events (Hauser et al., 2023). The significant reductions of the rate of global and regional brain atrophy and of serum neurofilament light chain levels suggest that this treatment may also limit neuro-axonal damage (Margoni et al., 2022). In addition to depleting $CD20^+$ B cells, ocrelizumab also reduced levels of CD8 cytotoxic T cells and serum IgG (Capasso et al., 2023). Ocrelizumab treatment also modulates B-cell regulating factors in survived B cells and $CD20^-$ plasma cells, and promotes regulatory IgA^+ plasma cells and IL-10 production in astrocytes of the CNS (Ho et al., 2023).

The common adverse events include herpetic infections, respiratory tract infections, nasopharyngitis, and infusion-related reactions, while some malignancy cases have also been reported (Hauser et al., 2021). Ocrelizumab concentration was negatively associated with $CD19^+$ B-cell count, and ocrelizumab antidrug antibodies were detectable in 5.7% of patients with corresponding low ocrelizumab concentrations (Toorop et al., 2023). Nevertheless, continuous administration of ocrelizumab for up to 7 years in clinical trials showed with a favorable and manageable safety profile, without emerging safety concerns, in a heterogeneous MS population (Hauser et al., 2021).

Ublituximab

Ublituximab is the newest CD20-targeting monoclonal antibody that is approved by the FDA (Jan. 2023) for RMS. Ublituximab was designed to have an enhanced affinity for FcgRIIIa and exert greater antibody-dependent cellular cytotoxicity activity to CD20+ B cells than other anti-CD20 medications (Fox et al., 2021). Ublituximab treatment resulted in lower annualized relapse rates and fewer brain lesions on MRI than teriflunomide/placebo over a period of 96 weeks. In the ublituximab group, 47.7% of the participants underwent infusion-related reactions vs. 12.2% in placebo, with the incidence of serious infections of about 5.0% vs. 2.9% in the teriflunomide group vs. placebo (Steinman et al., 2022). While an acceptable safety profile for the drug was suggested, long-term studies of the adverse effects of ublituximab are required.

Among the four anti-CD20 therapeutics, only ocrelizumab is approved for PPMS. Infusion/injection related reactions are the most common adverse events associated with all four therapeutics, while lower incidence of infusion/injection related reactions was observed in ofatumumab treatment than in ocrelizumab, probably due to the availability as a s.c. injection of the former (Cotchett et al., 2021). Despite the relatively modest data supporting the long-term efficacy of anti-CD20 antibodies, they represent an important therapeutic option for MS.

35.3 DMT Treatment Strategy

There are several factors that go into the selection of which DMT to use and, generally, one of two main strategies will be chosen to treat a person with MS for the first time: escalation or induction. The escalation strategy has been more common, in which patients are first treated with a moderately effective DMT, with a well-established safety profile. This so-called "first-line" therapy includes IFN-βs, GA, teriflunomide, and DMF. Then, if the person does not respond adequately to treatment, or is unable to tolerate its side effects, that patient may be switched to a more potent therapy, so-called "second-line" treatments, e.g., natalizumab, alemtuzumab, are mitoxantrone (Gajofatto & Benedetti, 2015). Fingolimod is approved as a second-line treatment in the EU and as first-line in the United States, Canada, and other countries (Sorensen, 2014). In the induction strategy, high-efficacy therapies, including ofatumumab, ocrelizumab, natalizumab, alemtuzumab, and cladribine will be used in the early stages of MS. By minimizing the accumulation of neurological damage early in the disease course, early treatment with high-efficacy therapies may enhance long-term clinical outcomes over the lifetime of the patient (Freeman et al., 2022). After identified 23 suitable trials encompassing 14,096 participants during the 2 years of follow-up, it was found that all DMTs were superior to placebo in reducing the relapse rate during the 2 years of follow-up, and alemtuzumab, ocrelizumab, natalizumab and fingolimod had a relatively higher response and lower dropout rates than did the other DMTs (Li et al., 2020).

The field of MS treatment is rapidly changing; there are a variety of FDA-approved DMTs available for the treatment of MS, each with its own mechanism of action and route of administration, and treatment decisions should be made on a case-by-case basis. Further, up to 30 recommendations for the use of DMTs have been recently provided by the U.S. Academy of Neurology, which reflects the complexity of decision-making for starting, switching, or stopping DMTs (Rae-Grant et al., 2018), and would provide a practical guideline for the treatment of adults with MS.

35.4 Emerging Experimental Therapeutic Strategies

35.4.1 Bruton's Tyrosine Kinase (BTK) Inhibitors

Almost all DMTs that control relapse activity in MS were primarily designed to target the peripheral immune system and/or to prevent CNS infiltration of immune cells. While acute relapses develop as a consequence of CNS infiltration of immune cells, MS progression may be driven by infiltrated inflammatory peripheral immune cells and CNS-resident cells (Charabati et al., 2023). Thus, medications that target both peripheral immune cells and CNS-resident cells with penetrating capacity through the BBB are ideal. Targeting BTK may achieve this goal, given that BTK, an intracellular signaling molecule involved in the regulation of maturation, survival, migration, and activation of B cells and microglia, plays an important role in the immunopathogenesis of progressive MS (Kramer et al., 2023).

The safety and efficacy of five BTK inhibitors are currently being evaluated in patients with relapsing MS or non-relapsing progressive MS whose pathological features include widespread activation of microglia and macrophages, astrogliosis, persistent lymphocyte activity in the CNS, increased cortical demyelination, deficient remyelination, chronic active lesions, and diffuse neuroaxonal damage and atrophy (Yong & Yong, 2022). As small-molecule agents, these inhibitors can be taken orally and have good intracellular access (Garcia-Merino, 2021), which are two key therapeutic advantages compared to large molecules. Among the inhibitors tested, treatment with evobrutinib, which binds BTK in a covalent manner, resulted in significantly fewer enhancing lesions in relapsing MS patients during weeks 12 through 24 treatment than placebo (Montalban et al., 2019). A Phase III trial to evaluate the efficacy and safety of evobrutinib vs. teriflunomide is ongoing (NCT04338022/NCT04338061). One 16-week, phase 2b clinical trial exhibited that tolebrutinib, a CNS-penetrant, irreversible BTK inhibitor, led to a dose-dependent reduction in new MRI lesions. The 60 mg dose of tolebrutinib was the most efficacious, and the drug was well tolerated (Reich et al., 2021). The higher CNS penetrance and potency and faster reaction rates of tolebrutinib compared to other BTK inhibitors tested, e.g., evobrutinib and fenebrutinib, indicate an important advantage of this inhibitor (Kramer et al., 2023). A Phase III, randomized, double-blind efficacy and safety study (*NCT04410978*) comparing tolebrutinib to teriflunomide is expected to be completed in September 2023. More BTK inhibitors in MS are under testing, including fenebrutinib and remibrutinib in Phase III trials and orelabrutinib in a Phase II trial. Reduction of peripheral immune response combined with the potential to cross through the BBB to directly modulate the inflammatory reaction within the CNS indicates a promising potential of BTK inhibitors in the treatment of CNS inflammatory demyelination and neurodegeneration.

35.4.2 Masitinib: A Selective Tyrosine Kinase Inhibitor

The selective tyrosine kinase inhibitor masitinib targets innate immune cells (mast cells and microglia) that are involved in the pathophysiology of progressive MS (Charabati et al., 2023). The beneficial effect of masitinib was first reported in a randomized pilot study for PPMS or relapse-free secondary progressive MS (rfSPMS) (Vermersch et al., 2012). In a randomized, double-blind, 2 parallel-group, placebo-controlled trial, oral masitinib at 4.5 mg/kg/d showed a beneficial outcome in rfSPMS patients by slowing EDSS-based disability worsening with side effects like diarrhea, nausea, rash, and hematologic events but no elevated risk of infection. In contrast, efficacy results from the independent uptitrated masitinib 6.0 mg/kg/d parallel-group were inconclusive. These data will be substantiated in an upcoming confirmatory Phase III study (Vermersch et al., 2022). Mechanistically, masitinib controls the survival, migration, and degranulation of mast cells, leading to the inhibition of proinflammatory and vasoactive molecules (Arsenault et al., 2022). Masitinib inhibits microphage infiltration into the CNS, as well as microglia and astrocyte proliferation and migration, thus potentially also being effective in several neurodegenerative disorders in addition to MS (Ketabforoush et al., 2023).

35.4.3 Amiselimod: A Novel S1PR Modulator

Given the adverse effect profile of currently used S1PR modulators, efforts have been made to identify structurally diverse lead molecules with clinical efficacy similar to or better than these currently used modulators but without the documented adverse effects. Amiselimod, as a successor of fingolimod with more specificity for S1PR1, was developed (Gusman & Shoemake, 2017). A two-year, Phase II extension study of oral amiselimod in RMS showed promising clinical and MRI results and was well tolerated, with common adverse events including headache, lymphocyte count decrease, nasopharyngitis, and MS relapse. No serious opportunistic infection, macular oedema, or malignancy was reported (Kappos et al., 2018a). As higher efficacy was observed at the 0.4 mg dose than at the 0.2 mg dose, with a similar safety profile for both doses, amiselimod at 0.4 mg may be the optimal dose for future consideration (Kappos et al., 2018a). One meta-analysis of the efficacy and acceptability of the S1PR modulators in the treatment of MS found that all S1PR modulators tested (Fingolimod, Laquinimod, Siponimod, Ozanimod, Amiselimod, and Ponesimod) were superior to placebo in terms of reducing the annualized relapse rate of MS patients, and, among these medications analyzed, amiselimod (0.4 mg) was the most effective treatment strategy as a S1PR modulator for MS patients (Tong et al., 2021).

35.4.4 Vidofludimus Calcium: A Novel DHODH Inhibitor

The mitochondrial enzyme DHODH is expressed at high levels in proliferating T cells and B cells, and its inhibition leads to metabolic stress in metabolically activated lymphocytes, thus reducing production of proinflammatory cytokines and inducing apoptosis of activated immune cells (Miller, 2021). One of the DHODH inhibitors, teriflunomide, was approved as a DMT medication of RMS in 2012. However, teriflunomide may cause off-target inhibition of a range of protein kinases, resulting in higher rates of general antiproliferative effects with side effects such as neutropenia, alopecia, and diarrhea (Stamatellos & Papazisis, 2023). Second-generation DHODH inhibitors, e.g., vidofludimus calcium, have thus been developed to separate the desired immunomodulatory effects from an undesirable side-effect profile caused by such off-target effects (Muehler et al., 2020). Vidofludimus calcium had a better safety profile in humans (Muehler et al., 2019) and was more efficacious in the inhibition of T-cell proliferation and the secretion of IL–17 and IFN–γ in vitro compared to teriflunomide (Muehler et al., 2020). A double-blind, 24-week, placebo-controlled, Phase II trial indicated that treatment with vidofludimus calcium reduced new MRI lesions in RMS and was well tolerated with a favorable safety profile (Fox et al., 2022). A longer, larger trial is required for further study of its efficacy and safety profile.

35.4.5 Attacking Epstein-Barr Virus (EBV)-Infected Cells

Evidence has suggested the involvement of EBV infection in the development of MS (Bar-Or et al., 2020; Pender et al., 2020); therefore, targeting EBV-infected immune cells in the CNS has been considered a possible approach for interrupting MS pathogenesis. Indeed, transfer of in vitro–expanded autologous EBV-specific $CD8^+$ T cells targeting EBV nuclear antigen 1, latent membrane protein 1 (LMP1), and LMP2A resulted in clinical improvement in patients with PPMS or with SPMS, and this treatment was safe and well tolerated (Pender et al., 2020). An open, single-center, Phase I proof of concept trial is ongoing, with the goal to assess the feasibility and safety of autologous transfer of several numbers of $CD8^+$ T cells directed against autologous EBV transformed B-cell lines to finally restore an efficient control of EBV in MS patients (NCT02912897).

35.4.6 Antigen-Specific Tolerance

While the current DMTs are effective, they broadly target the inflammatory immune response, often compromise protection against infections, and may even lead to secondary autoimmunity (Charabati et al., 2023; Ward & Goldman, 2022). Approaches that target autoantigen-specific immune responses without interrupting the global immune system are thus ideal. Significant progress has been made in recent decades, and more than 30 clinical trials for antigen-specific therapies for RMS, SPMS, or PPMS have been completed or are ongoing (Lutterotti et al., 2021). The approaches include DNA vaccination, peptide- and protein-based approaches, and cell-based approaches, through mechanisms of induction of tolerogenic APCs, antigen-specific Tregs, clonal anergy, or apoptosis of autoantigen-reactive T cells (Lutterotti et al., 2021). Some of these trials have shown improvement in clinical and/or MRI outcome (Bar-Or et al., 2007; Chataway et al., 2018; Garren et al., 2008; Karussis et al., 2012; Lanz et al., 2023; Loftus et al., 2009; Robinson et al., 2003; Walczak et al., 2013).

The prerequisite for antigen-specific therapy is identification of the target antigen(s); however, the daunting complexity of relevant autoantigens in MS has been an obstacle to developing antigen-specific therapy. MS patients may simultaneously develop autoimmunity against multiple myelin antigens (Kaushansky et al., 2010; 2019). This could occur through epitope spreading, whereby T-cell responses directed toward different antigens or different epitopes of the same antigen sequentially dominate overall autoimmune response and drive disease relapses/progression (McMahon et al., 2005). In preclinical studies, inhibition of EAE has been achieved by targeting multiple myelin antigens using mixtures of several myelin antigen peptides (Kaushansky et al., 2019; Liu et al., 1998; Smith & Miller, 2006) or mixtures of plasmids encoding multiple myelin proteins (Robinson et al., 2003). However, these tolerance induction approaches can cover only a few known myelin antigen peptides. Developing novel methods for inducing immune tolerance to most if not all myelin antigens in MS without the need to identify them is of the utmost importance. Such a method has recently been tested using oligodendrocyte (Ol)-derived extracellular vesicles (Ol-EVs), which naturally contain all or most myelin antigens, thus making it unnecessary to identify the relevant myelin antigen(s) in each patient. I.v. injection of Ol-EVs reduced disease severity both prophylactically and therapeutically in MBP, PLP, or MOG peptide-induced EAE models. Further, Ol-EVs induced abortive activation and proliferation only of myelin antigen-specific but not OVA-specific $CD4^+$ T cells (Casella et al., 2020). These findings provide an approach for suppressing CNS autoimmunity in a myelin antigen-specific manner without the need to identify the target antigens. A clinical trial for this strategy in MS patients is warranted.

35.5 Conclusion

While immunomodulatory therapies have revolutionized the treatment of MS by modulating the dysregulated immune responses underlying the disease, challenges remain. Treatment decisions should be individualized to each patient, considering factors such as age, disease subtype, severity, and other disease conditions. Further, long-term safety and monitoring protocols are crucial to mitigate potential risks associated with these therapies. Ongoing research and clinical trials continue to explore novel targets and therapeutic strategies, aiming to improve treatment efficacy and safety profiles. With the collective efforts of clinicians, researchers, and patients, innovative immunomodulatory therapies hold promise for a better future in the management of MS.

35.6 Review Questions

1. What is the primary goal of current DMT in MS?
   (a) Symptomatic relief
   (b) Complete cure
   (c) Immune system suppression
   (d) *Immune system modulation*
   (e) Neural repair
2. What is the role of dimethyl fumarate in the treatment of multiple sclerosis?
   (a) Suppression of interleukin-6
   (b) *Activation of the nuclear factor erythroid 2-related factor 2 (Nrf2) pathway*
   (c) Inhibition of DNA synthesis
   (d) Antiviral effects
   (e) Platelet aggregation
3. What is the mechanism of action of fingolimod in multiple sclerosis treatment?
   (a) Inhibition of DNA synthesis
   (b) Modulation of T-cell function
   (c) Suppression of interleukin-6
   (d) Blockade of TNF-alpha
   (e) *Immune cell trafficking modulation*
4. Which immunomodulatory therapy is the first FDA-approved treatment for primary-progressive multiple sclerosis (PPMS)?
   (a) *Ocrelizumab*
   (b) Rituximab
   (c) Teriflunomide
   (d) Fingolimod
   (e) Glatiramer acetate

5. Which immunomodulatory therapy requires regular monitoring for potential adverse effects such as progressive multifocal leukoencephalopathy (PML)?
   (a) Teriflunomide
   (b) Rituximab
   (c) *Natalizumab*
   (d) Alemtuzumab
   (e) Interferon beta-1b

Acknowledgments This study was supported by the NIH, USA (Grants R01-NS099594). I thank Ms. Liz Declan for editorial assistance.

Conflict of Interest The author has declared that no conflict of interest exists.

References

Ali, S., Mann-Nuttel, R., Schulze, A., Richter, L., Alferink, J., & Scheu, S. (2019). Sources of type I interferons in infectious immunity: Plasmacytoid dendritic cells not always in the driver's seat. *Frontiers in Immunology, 10*, 778.

Alnaif, A., Oiler, I., & D'Souza, M. S. (2023). Ponesimod: An oral second-generation selective sphingosine 1-phosphate receptor modulator for the treatment of multiple sclerosis. *The Annals of Pharmacotherapy, 57*(8), 956–965.

Arsenault, S., Benoit, R. Y., Clift, F., & Moore, C. S. (2022). Does the use of the Bruton Tyrosine Kinase inhibitors and the c-kit inhibitor masitinib result in clinically significant outcomes among patients with various forms of multiple sclerosis? *Multiple Sclerosis and Related Disorders, 67*, 104164.

Ashkar, A., Baig, M. M. A., Arif, A., Ali, M. M., Yousuf, F., & Ashkar, R. (2023). Prognostic significance of neurofilament light in Fingolimod therapy for Multiple Sclerosis: A systemic review and meta-analysis based on randomized control trials. *Multiple Sclerosis and Related Disorders, 69*, 104416.

Askari, M., Mirmosayyeb, O., Ghaffary, E. M., Ghoshouni, H., Shaygannejad, V., & Ghajarzadeh, M. (2022). Incidence of cancer in patients with multiple sclerosis (MS) who were treated with fingolimod: A systematic review and meta-analysis. *Multiple Sclerosis and Related Disorders, 59*, 103680.

Ayzenberg, I., Hoepner, R., & Kleiter, I. (2016). Fingolimod for multiple sclerosis and emerging indications: Appropriate patient selection, safety precautions, and special considerations. *Therapeutics and Clinical Risk Management, 12*, 261–272.

Bachelet, D., Hassler, S., Mbogning, C., Link, J., Ryner, M., Ramanujam, R., Auer, M., Jensen, P. E. H., Koch-Henriksen, N., Warnke, C., Ingenhoven, K., Buck, D., Grummel, V., Lawton, A., Donnellan, N., Hincelin-Mery, A., Sikkema, D., Pallardy, M., Kieseier, B., Hemmer, B., Hartung, H. P., Sorensen, P. S., Deisenhammer, F., Donnes, P., Davidson, J., Fogdell-Hahn, A., Broet, P., & Abirisk Consortium. (2016). Occurrence of anti-drug antibodies against interferon-beta and natalizumab in multiple sclerosis: A collaborative cohort analysis. *PLoS One, 11*, e0162752.

Baldin, E., & Lugaresi, A. (2020). Ponesimod for the treatment of relapsing multiple sclerosis. *Expert Opinion on Pharmacotherapy, 21*, 1955–1964.

Barclay, K., Carruthers, R., Traboulsee, A., Bass, A. D., LaGanke, C., Bertolotto, A., Boster, A., Celius, E. G., de Seze, J., Cruz, D. D., Habek, M., Lee, J. M., Limmroth, V., Meuth, S. G., Oreja-Guevara, C., Pagnotta, P., Vos, C., Ziemssen, T., Baker, D. P., & Wijmeersch, B. V. (2019). Best practices for long-term monitoring and follow-up of alemtuzumab-treated MS patients in real-world clinical settings. *Frontiers in Neurology, 10*, 253.

Bar-Or, A., Vollmer, T., Antel, J., Arnold, D. L., Bodner, C. A., Campagnolo, D., Gianettoni, J., Jalili, F., Kachuck, N., Lapierre, Y., Niino, M., Oger, J., Price, M., Rhodes, S., Robinson, W. H., Shi, F. D., Utz, P. J., Valone, F., Weiner, L., Steinman, L., & Garren, H. (2007). Induction of antigen-specific tolerance in multiple sclerosis after immunization with DNA encoding myelin basic protein in a randomized, placebo-controlled phase 1/2 trial. *Archives of Neurology, 64*, 1407–1415.

Bar-Or, A., Pender, M. P., Khanna, R., Steinman, L., Hartung, H. P., Maniar, T., Croze, E., Aftab, B. T., Giovannoni, G., & Joshi, M. A. (2020). Epstein-Barr virus in multiple sclerosis: Theory and emerging immunotherapies. *Trends in Molecular Medicine, 26*, 296–310.

Berger, A. A., Sottosanti, E. R., Winnick, A., Izygon, J., Berardino, K., Cornett, E. M., Kaye, A. D., Varrassi, G., Viswanath, O., & Urits, I. (2021). Monomethyl fumarate (MMF, Bafiertam) for the treatment of relapsing forms of multiple sclerosis (MS). *Neurology International, 13*, 207–223.

Biernacki, K., Antel, J. P., Blain, M., Narayanan, S., Arnold, D. L., & Prat, A. (2005). Interferon beta promotes nerve growth factor secretion early in the course of multiple sclerosis. *Archives of Neurology, 62*, 563–568.

Blair, H. A. (2019). Dimethyl fumarate: A review in relapsing-remitting MS. *Drugs, 79*, 1965–1976.

Boster, A. L., Ford, C. C., Neudorfer, O., & Gilgun-Sherki, Y. (2015). Glatiramer acetate: Long-term safety and efficacy in relapsing-remitting multiple sclerosis. *Expert Review of Neurotherapeutics, 15*, 575–586.

Bravo, G. A., Cedeno, R. R., Casadevall, M. P., & Ramio-Torrenta, L. (2022). Sphingosine-1-phosphate (S1P) and S1P signaling pathway modulators, from current insights to future perspectives. *Cells, 11*.

Brenner, T., Arnon, R., Sela, M., Abramsky, O., Meiner, Z., Riven-Kreitman, R., Tarcik, N., & Teitelbaum, D. (2001). Humoral and cellular immune responses to Copolymer 1 in multiple sclerosis patients treated with Copaxone. *Journal of Neuroimmunology, 115*, 152–160.

Butzkueven, H., Kappos, L., Wiendl, H., Trojano, M., Spelman, T., Chang, I., Kasliwal, R., Jaitly, S., Campbell, N., Ho, P. R., Licata, S., & Program, I. T. O. (2020). Long-term safety and effectiveness of natalizumab treatment in clinical practice: 10 years of real-world data from the Tysabri Observational Program (TOP). *Journal of Neurology, Neurosurgery, and Psychiatry, 91*, 660–668.

Capasso, N., Palladino, R., Cerbone, V., Spiezia, A. L., Covelli, B., Fiore, A., Lanzillo, R., Carotenuto, A., Petracca, M., Stanziola, L., Scalia, G., Brescia Morra, V., & Moccia, M. (2023). Ocrelizumab effect on humoral and cellular immunity in multiple sclerosis and its clinical correlates: A 3-year observational study. *Journal of Neurology, 270*, 272–282.

Casella, G., Rasouli, J., Boehm, A., Zhang, W., Xiao, D., Ishikawa, L. L. W., Thome, R., Li, X., Hwang, D., Porazzi, P., Molugu, S., Tang, H. Y., Zhang, G. X., Ciric, B., & Rostami, A. (2020). Oligodendrocyte-derived extracellular vesicles as antigen-specific therapy for autoimmune neuroinflammation in mice. *Science Translational Medicine, 12*.

Castillo-Trivino, T., Braithwaite, D., Bacchetti, P., & Waubant, E. (2013). Rituximab in relapsing and progressive forms of multiple sclerosis: A systematic review. *PLoS One, 8*, e66308.

Charabati, M., Wheeler, M. A., Weiner, H. L., & Quintana, F. J. (2023). Multiple sclerosis: Neuroimmune crosstalk and therapeutic targeting. *Cell, 186*, 1309–1327.

Chataway, J., Martin, K., Barrell, K., Sharrack, B., Stolt, P., Wraith, D. C., & Atx-Ms Study Group. (2018). Effects of ATX-MS-1467 immunotherapy over 16 weeks in relapsing multiple sclerosis. *Neurology, 90*, e955–ee62.

Chisari, C. G., Sgarlata, E., Arena, S., Toscano, S., Luca, M., & Patti, F. (2022). Rituximab for the treatment of multiple sclerosis: A review. *Journal of Neurology, 269*, 159–183.

Clerico, M., Rivoiro, C., Contessa, G., Viglietti, D., & Durelli, L. (2008). The therapy of multiple sclerosis with immune-modulating or immunosuppressive drug. A critical evaluation based upon evidence based parameters and published systematic reviews. *Clinical Neurology and Neurosurgery, 110*, 878–885.

Cohan, S., Kumar, J., Arndorfer, S., Zhu, X., Zivkovic, M., & Tencer, T. (2021). Comparative efficacy and safety of ozanimod and dimethyl fumarate for relapsing-remitting multiple sclerosis using matching-adjusted indirect comparison. *CNS Drugs, 35*, 795–804.

Coles, A. J., Fox, E., Vladic, A., Gazda, S. K., Brinar, V., Selmaj, K. W., Bass, A. D., Wynn, D. R., Margolin, D. H., Lake, S. L., Moran, S., Palmer, J., Smith, M. S., & Compston, D. A. (2011). Alemtuzumab versus interferon beta-1a in early relapsing-remitting multiple sclerosis: Post-hoc and subset analyses of clinical efficacy outcomes. *Lancet Neurology, 10*, 338–348.

Cook, S., Vermersch, P., Comi, G., Giovannoni, G., Rammohan, K., Rieckmann, P., Sorensen, P. S., Hamlett, A., Miret, M., Weiner, J., Viglietta, V., Musch, B., Greenberg, S. J., & Clarity Study Group. (2011). Safety and tolerability of cladribine tablets in multiple sclerosis: The CLARITY (CLAdRIbine Tablets treating multiple sclerosis orallY) study. *Multiple Sclerosis, 17*, 578–593.

Cotchett, K. R., Dittel, B. N., & Obeidat, A. Z. (2021). Comparison of the efficacy and safety of anti-CD20 B cells depleting drugs in multiple sclerosis. *Multiple Sclerosis and Related Disorders, 49*, 102787.

Cragg, M. S., Walshe, C. A., Ivanov, A. O., & Glennie, M. J. (2005). The biology of CD20 and its potential as a target for mAb therapy. *Current Directions in Autoimmunity, 8*, 140–174.

Cree, B. A., Arnold, D. L., Fox, R. J., Gold, R., Vermersch, P., Benedict, R. H., Bar-Or, A., Piani-Meier, D., Rouyrre, N., Ritter, S., Kilaru, A., Karlsson, G., Giovannoni, G., & Kappos, L. (2022). Long-term efficacy and safety of siponimod in patients with secondary progressive multiple sclerosis: Analysis of EXPAND core and extension data up to >5 years. *Multiple Sclerosis, 28*, 1591–1605.

De Mercanti, S., Rolla, S., Cucci, A., Bardina, V., Cocco, E., Vladic, A., Soldo-Butkovic, S., Habek, M., Adamec, I., Horakova, D., Annovazzi, P., Novelli, F., Durelli, L., & Clerico, M. (2016). Alemtuzumab long-term immunologic effect: Treg suppressor function increases up to 24 months. *Neurology Neuroimmunology & Neuroinflammation, 3*, e194.

Deeks, E. D. (2018). Cladribine tablets: A review in relapsing MS. *CNS Drugs, 32*, 785–796.

Dubuisson, N., Baker, D., Kang, A. S., Pryce, G., Marta, M., Visser, L. H., Hofmann, W. E., Gnanapavan, S., Giovannoni, G., & Schmierer, K. (2018). Alemtuzumab depletion failure can occur in multiple sclerosis. *Immunology, 154*, 253–260.

Duda, P. W., Schmied, M. C., Cook, S. L., Krieger, J. I., & Hafler, D. A. (2000). Glatiramer acetate (Copaxone) induces degenerate, Th2-polarized immune responses in patients with multiple sclerosis. *The Journal of Clinical Investigation, 105*, 967–976.

El-behi, M., Rostami, A., & Ciric, B. (2010). Current views on the roles of Th1 and Th17 cells in experimental autoimmune encephalomyelitis. *Journal of Neuroimmune Pharmacology, 5*, 189–197.

Ferri, C., Castellazzi, M., Merli, N., Laudisi, M., Baldin, E., Baldi, E., Mancabelli, L., Ventura, M., & Pugliatti, M. (2023). Gut microbiota changes during dimethyl fumarate treatment in patients with multiple sclerosis. *International Journal of Molecular Sciences, 24*.

Fissolo, N., Pappolla, A., Rio, J., Villar, L. M., Perez-Hoyos, S., Sanchez, A., Gutierrez, L., Montalban, X., & Comabella, M. (2023). Serum levels of CXCL13 are associated with teriflunomide response in patients with multiple sclerosis. *Neurology Neuroimmunology & Neuroinflammation, 10*, e200050.

Fitzgerald, D. C., Fonseca-Kelly, Z., Cullimore, M. L., Safabakhsh, P., Saris, C. J., Zhang, G. X., & Rostami, A. (2013). Independent and interdependent immunoregulatory effects of IL-27, IFN-beta, and IL-10 in the suppression of human Th17 cells and murine experimental autoimmune encephalomyelitis. *Journal of Immunology, 190*, 3225–3234.

Fox, E., Lovett-Racke, A. E., Gormley, M., Liu, Y., Petracca, M., Cocozza, S., Shubin, R., Wray, S., Weiss, M. S., Bosco, J. A., Power, S. A., Mok, K., & Inglese, M. (2021). A phase 2 multicenter study of ublituximab, a novel glycoengineered anti-CD20 monoclonal antibody, in patients with relapsing forms of multiple sclerosis. *Multiple Sclerosis, 27*, 420–429.

Fox, R. J., Wiendl, H., Wolf, C., De Stefano, N., Sellner, J., Gryb, V., Rejdak, K., Bozhinov, P. S., Tomakh, N., Skrypchenko, I., & Muehler, A. R. (2022). A double-blind, randomized, placebo-controlled phase 2 trial evaluating the selective dihydroorotate dehydrogenase inhibitor vidofludimus calcium in relapsing-remitting multiple sclerosis. *Annals of Clinical Translational Neurology, 9*, 977–987.

Freeman, L., Longbrake, E. E., Coyle, P. K., Hendin, B., & Vollmer, T. (2022). High-efficacy therapies for treatment-naive individuals with relapsing-remitting multiple sclerosis. *CNS Drugs, 36*, 1285–1299.

Gajofatto, A., & Benedetti, M. D. (2015). Treatment strategies for multiple sclerosis: When to start, when to change, when to stop? *World Journal of Clinical Cases, 3*, 545–555.

Garcia-Merino, A. (2021). Bruton's tyrosine kinase inhibitors: A new generation of promising agents for multiple sclerosis therapy. *Cells, 10*.

Garren, H., Robinson, W. H., Krasulova, E., Havrdova, E., Nadj, C., Selmaj, K., Losy, J., Nadj, I., Radue, E. W., Kidd, B. A., Gianettoni, J., Tersini, K., Utz, P. J., Valone, F., Steinman, L., & B. H. T. Study Group. (2008). Phase 2 trial of a DNA vaccine encoding myelin basic protein for multiple sclerosis. *Annals of Neurology, 63*, 611–620.

Gartner, J., Hauser, S. L., Bar-Or, A., Montalban, X., Cohen, J. A., Cross, A. H., Deiva, K., Ganjgahi, H., Haring, D. A., Li, B., Pingili, R., Ramanathan, K., Su, W., Willi, R., Kieseier, B., & Kappos, L. (2022). Efficacy and safety of ofatumumab in recently diagnosed, treatment-naive patients with multiple sclerosis: Results from ASCLEPIOS I and II. *Multiple Sclerosis, 28*, 1562–1575.

Geladaris, A., Hausser-Kinzel, S., Pretzsch, R., Nissimov, N., Lehmann-Horn, K., Hausler, D., & Weber, M. S. (2023). IL-10-providing B cells govern pro-inflammatory activity of macrophages and microglia in CNS autoimmunity. *Acta Neuropathologica, 145*, 461–477.

Gharagozloo, M., Mace, J. W., & Calabresi, P. A. (2022). Animal models to investigate the effects of inflammation on remyelination in multiple sclerosis. *Frontiers in Molecular Neuroscience, 15*, 995477.

Giovannoni, G., Soelberg Sorensen, P., Cook, S., Rammohan, K. W., Rieckmann, P., Comi, G., Dangond, F., Hicking, C., & Vermersch, P. (2019). Efficacy of Cladribine Tablets in high disease activity subgroups of patients with relapsing multiple sclerosis: A post hoc analysis of the CLARITY study. *Multiple Sclerosis, 25*, 819–827.

Glatigny, S., & Bettelli, E. (2018). Experimental autoimmune encephalomyelitis (EAE) as animal models of multiple sclerosis (MS). *Cold Spring Harbor Perspectives in Medicine, 8*, a028977.

Gold, R., Piani-Meier, D., Kappos, L., Bar-Or, A., Vermersch, P., Giovannoni, G., Fox, R. J., Arnold, D. L., Benedict, R. H. B., Penner, I. K., Rouyrre, N., Kilaru, A., Karlsson, G., Ritter, S., Dahlke, F., Hach, T., & Cree, B. A. C. (2022). Siponimod vs placebo in active secondary progressive multiple sclerosis: A post hoc analysis from the phase 3 EXPAND study. *Journal of Neurology, 269*, 5093–5104.

Goldschmidt, C. H., & Hua, L. H. (2020). Re-evaluating the use of IFN-beta and relapsing multiple sclerosis: Safety, efficacy and place in therapy. *Degenerative Neurological and Neuromuscular Disease, 10*, 29–38.

Goodman, A. D., Anadani, N., & Gerwitz, L. (2019). Siponimod in the treatment of multiple sclerosis. *Expert Opinion on Investigational Drugs, 28*, 1051–1057.

Gottle, P., Manousi, A., Kremer, D., Reiche, L., Hartung, H. P., & Kury, P. (2018). Teriflunomide promotes oligodendroglial differentiation and myelination. *Journal of Neuroinflammation, 15*, 76.

Gottle, P., Groh, J., Reiche, L., Gruchot, J., Rychlik, N., Werner, L., Samper Agrelo, I., Akkermann, R., Zink, A., Prigione, A., Hartung, H. P., Martini, R., & Kury, P. (2023). Teriflunomide as a therapeutic means for myelin repair. *Journal of Neuroinflammation, 20*, 7.

Greenfield, A. L., & Hauser, S. L. (2018). B-cell therapy for multiple sclerosis: Entering an era. *Annals of Neurology, 83*, 13–26.

Groh, J., Horner, M., & Martini, R. (2018). Teriflunomide attenuates neuroinflammation-related neural damage in mice carrying human PLP1 mutations. *Journal of Neuroinflammation, 15*, 194.

Gross, C. C., Schulte-Mecklenbeck, A., Klinsing, S., Posevitz-Fejfar, A., Wiendl, H., & Klotz, L. (2016). Dimethyl fumarate treatment alters circulating T helper cell subsets in multiple sclerosis. *Neurology Neuroimmunology & Neuroinflammation, 3*, e183.

Gusman, D. H., & Shoemake, C. (2017). Evaluation and optimization of in silico designed sphingosine-1-phosphate (S1P) receptor subtype 1 modulators for the management of multiple sclerosis. *The Yale Journal of Biology and Medicine, 90*, 15–23.

Haas, J., Korporal, M., Balint, B., Fritzsching, B., Schwarz, A., & Wildemann, B. (2009). Glatiramer acetate improves regulatory T-cell function by expansion of naive CD4(+)CD25(+)FOXP3(+) CD31(+) T-cells in patients with multiple sclerosis. *Journal of Neuroimmunology, 216*, 113–117.

Harris, S., Comi, G., Cree, B. A. C., Arnold, D. L., Steinman, L., Sheffield, J. K., Southworth, H., Kappos, L., Cohen, J. A., & Investigators Ozanimod Study. (2021). Plasma neurofilament light chain concentrations as a biomarker of clinical and radiologic outcomes in relapsing multiple sclerosis: Post hoc analysis of Phase 3 ozanimod trials. *European Journal of Neurology, 28*, 3722–3730.

Hartung, H. P., Aktas, O., & Boyko, A. N. (2015). Alemtuzumab: A new therapy for active relapsing-remitting multiple sclerosis. *Multiple Sclerosis, 21*, 22–34.

Hauser, S. L., Bar-Or, A., Cohen, J. A., Comi, G., Correale, J., Coyle, P. K., Cross, A. H., de Seze, J., Leppert, D., Montalban, X., Selmaj, K., Wiendl, H., Kerloeguen, C., Willi, R., Li, B., Kakarieka, A., Tomic, D., Goodyear, A., Pingili, R., Haring, D. A., Ramanathan, K., Merschhemke, M., Kappos, L., Asclepios, I., & Asclepios Ii Trial Groups. (2020). Ofatumumab versus teriflunomide in multiple sclerosis. *The New England Journal of Medicine, 383*, 546–557.

Hauser, S. L., Kappos, L., Montalban, X., Craveiro, L., Chognot, C., Hughes, R., Koendgen, H., Pasquarelli, N., Pradhan, A., Prajapati, K., & Wolinsky, J. S. (2021). Safety of ocrelizumab in patients with relapsing and primary progressive multiple sclerosis. *Neurology, 97*, e1546–e1e59.

Hauser, S. L., Bar-Or, A., Weber, M. S., Kletzl, H., Gunther, A., Manfrini, M., Model, F., Mercier, F., Petry, C., Wing, Q., Koendgen, H., Smith, T., & Kappos, L. (2023). Association of higher ocrelizumab exposure with reduced disability progression in multiple sclerosis. *Neurology Neuroimmunology & Neuroinflammation, 10*.

Havrdova, E., Giovannoni, G., Gold, R., Fox, R. J., Kappos, L., Phillips, J. T., Okwuokenye, M., & Marantz, J. L. (2017). Effect of delayed-release dimethyl fumarate on no evidence of disease activity in relapsing-remitting multiple sclerosis: Integrated analysis of the phase III DEFINE and CONFIRM studies. *European Journal of Neurology, 24*, 726–733.

Hedegaard, C. J., Sellebjerg, F., Krakauer, M., Hesse, D., Bendtzen, K., & Nielsen, C. H. (2011). Interferon-beta increases systemic BAFF levels in multiple sclerosis without increasing autoantibody production. *Multiple Sclerosis, 17*, 567–577.

Hilger, C., Riedhammer, C., Orso, E., & Weissert, R. (2020). Effects of alemtuzumab on (auto)antigen-specific immune responses. *Frontiers in Immunology, 11*, 563645.

Ho, S., Oswald, E., Wong, H. K., Vural, A., Yilmaz, V., Tuzun, E., Turkoglu, R., Straub, T., Meinl, I., Thaler, F., Kumpfel, T., Meinl, E., & Mader, S. (2023). Ocrelizumab treatment modulates B-cell regulating factors in multiple sclerosis. *Neurology Neuroimmunology & Neuroinflammation, 10*.

Houtchens, M., Mahlanza, T., Ciplea, A. I., Manieri, M. C., Ramia, N. F., Zhao, Y., England, S., Avila, R., Altincatal, A., Vignos, M., & Hellwig, K. (2022). Peginterferon beta-1a concentrations in breast milk of lactating multiple sclerosis patients. *Multiple Sclerosis and Related Disorders, 60*, 103700.

Hu, X., Miller, L., Richman, S., Hitchman, S., Glick, G., Liu, S., Zhu, Y., Crossman, M., Nestorov, I., Gronke, R. S., Baker, D. P., Rogge, M., Subramanyam, M., & Davar, G. (2012). A novel PEGylated interferon beta-1a for multiple sclerosis: Safety, pharmacology, and biology. *Journal of Clinical Pharmacology, 52*, 798–808.

Jacobs, L. D., Cookfair, D. L., Rudick, R. A., Herndon, R. M., Richert, J. R., Salazar, A. M., Fischer, J. S., Goodkin, D. E., Granger, C. V., Simon, J. H., Alam, J. J., Bartoszak, D. M., Bourdette, D. N., Braiman, J., Brownscheidle, C. M., Coats, M. E., Cohan, S. L., Dougherty, D. S., Kinkel, R. P., Mass, M. K., Munschauer, F. E., 3rd, Priore, R. L., Pullicino, P. M., Scherokman, B. J., Whitham, R. H., et al. (1996). Intramuscular interferon beta-1a for disease progression in relapsing multiple sclerosis. The Multiple Sclerosis Collaborative Research Group (MSCRG). *Annals of Neurology, 39*, 285–294.

Kala, M., Miravalle, A., & Vollmer, T. (2011). Recent insights into the mechanism of action of glatiramer acetate. *Journal of Neuroimmunology, 235*, 9–17.

Kandjani, O. J., Yaqoubi, S., Vahdati, S. S., Borhannejad, B., Dastmalchi, S., & Alizadeh, A. A. (2023). S1PR1 modulators in multiple sclerosis: Efficacy, safety, comparison, and chemical structure insights. *European Journal of Medicinal Chemistry, 250*, 115182.

Kapoor, R., Ho, P. R., Campbell, N., Chang, I., Deykin, A., Forrestal, F., Lucas, N., Yu, B., Arnold, D. L., Freedman, M. S., Goldman, M. D., Hartung, H. P., Havrdova, E. K., Jeffery, D., Miller, A., Sellebjerg, F., Cadavid, D., Mikol, D., Steiner, D., & Ascend investigators. (2018). Effect of natalizumab on disease progression in secondary progressive multiple sclerosis (ASCEND): A phase 3, randomised, double-blind, placebo-controlled trial with an open-label extension. *Lancet Neurology, 17*, 405–415.

Kappos, L., Arnold, D. L., Bar-Or, A., Camm, A. J., Derfuss, T., Sprenger, T., Davies, M., Piotrowska, A., Ni, P., & Harada, T. (2018a). Two-year results from a phase 2 extension study of oral amiselimod in relapsing multiple sclerosis. *Multiple Sclerosis, 24*, 1605–1616.

Kappos, L., Bar-Or, A., Cree, B. A. C., Fox, R. J., Giovannoni, G., Gold, R., Vermersch, P., Arnold, D. L., Arnould, S., Scherz, T., Wolf, C., Wallstrom, E., Dahlke, F., & Expand Clinical Investigators. (2018b). Siponimod versus placebo in secondary progressive multiple sclerosis (EXPAND): A double-blind, randomised, phase 3 study. *Lancet, 391*, 1263–1273.

Kappos, L., Fox, R. J., Burcklen, M., Freedman, M. S., Havrdova, E. K., Hennessy, B., Hohlfeld, R., Lublin, F., Montalban, X., Pozzilli, C., Scherz, T., D'Ambrosio, D., Linscheid, P., Vaclavkova, A., Pirozek-Lawniczek, M., Kracker, H., & Sprenger, T. (2021). Ponesimod compared with teriflunomide in patients with relapsing multiple sclerosis in the active-comparator phase 3 OPTIMUM study: A randomized clinical trial. *JAMA Neurology, 78*, 558–567.

Karussis, D., Shor, H., Yachnin, J., Lanxner, N., Amiel, M., Baruch, K., Keren-Zur, Y., Haviv, O., Filippi, M., Petrou, P., Hajag, S., Vourka-Karussis, U., Vaknin-Dembinsky, A., Khoury, S., Abramsky, O., Atlan, H., Cohen, I. R., & Abulafia-Lapid, R. (2012). T cell vaccination benefits relapsing progressive multiple sclerosis patients: A randomized, double-blind clinical trial. *PLoS One, 7*, e50478.

Kashani, N., Kelland, E. E., Vajdi, B., Anderson, L. M., Gilmore, W., & Lund, B. T. (2021). Immune regulatory cell bias following alemtuzumab treatment in relapsing-remitting multiple sclerosis. *Frontiers in Immunology, 12*, 706278.

Kasindi, A., Fuchs, D. T., Koronyo, Y., Rentsendorj, A., Black, K. L., & Koronyo-Hamaoui, M. (2022). Glatiramer acetate immunomodulation: Evidence of neuroprotection and cognitive preservation. *Cells, 11*.

Kaushansky, N., Eisenstein, M., Zilkha-Falb, R., & Ben-Nun, A. (2010). The myelin-associated oligodendrocytic basic protein (MOBP) as a relevant primary target autoantigen in multiple sclerosis. *Autoimmunity Reviews, 9*, 233–236.

Kaushansky, N., Kaminitz, A., Allouche-Arnon, H., & Ben-Nun, A. (2019). Modulation of MS-like disease by a multi epitope protein is mediated by induction of CD11c(+)CD11b(+)Gr1(+) myeloid-derived dendritic cells. *Journal of Neuroimmunology, 333*, 476953.

Ketabforoush, A., Chegini, R., Barati, S., Tahmasebi, F., Moghisseh, B., Joghataei, M. T., Faghihi, F., & Azedi, F. (2023). Masitinib: The promising actor in the next season of the amyotrophic lateral sclerosis treatment series. *Biomedicine & Pharmacotherapy, 160*, 114378.

Khoy, K., Mariotte, D., Defer, G., Petit, G., Toutirais, O., & Le Mauff, B. (2020). Natalizumab in multiple sclerosis treatment: From biological effects to immune monitoring. *Frontiers in Immunology, 11*, 549842.

Kihara, Y., Jonnalagadda, D., Zhu, Y., Ray, M., Ngo, T., Palmer, C., Rivera, R., & Chun, J. (2022). Ponesimod inhibits astrocyte-mediated neuroinflammation and protects against cingulum demyelination via S1P(1) -selective modulation. *The FASEB Journal, 36*, e22132.

Kramer, J., Bar-Or, A., Turner, T. J., & Wiendl, H. (2023). Bruton tyrosine kinase inhibitors for multiple sclerosis. *Nature Reviews. Neurology, 19*, 289–304.

Krause, A., Brossard, P., D'Ambrosio, D., & Dingemanse, J. (2014). Population pharmacokinetics and pharmacodynamics of ponesimod, a selective S1P1 receptor modulator. *Journal of Pharmacokinetics and Pharmacodynamics, 41*, 261–278.

Krishnarajah, S., & Becher, B. (2022). T(H) cells and cytokines in encephalitogenic disorders. *Frontiers in Immunology, 13*, 822919.

Kumar, G., & Axtell, R. C. (2023). Dual role of B cells in multiple sclerosis. *International Journal of Molecular Sciences, 24*.

Lanz, T. V., Robinson, W. H., Ho, P. P., & Steinman, L. (2023). Roadmap for understanding mechanisms on how Epstein-Barr virus triggers multiple sclerosis and for translating these discoveries in clinical trials. *Clinical & Translational Immunology, 12*, e1438.

Lassiter, G., Melancon, C., Rooney, T., Murat, A. M., Kaye, J. S., Kaye, A. M., Kaye, R. J., Cornett, E. M., Kaye, A. D., Shah, R. J., Viswanath, O., & Urits, I. (2020). Ozanimod to treat relapsing forms of multiple sclerosis: A comprehensive review of disease, drug efficacy and side effects. *Neurology International, 12*, 89–108.

Leist, T. P., & Weissert, R. (2011). Cladribine: Mode of action and implications for treatment of multiple sclerosis. *Clinical Neuropharmacology, 34*, 28–35.

Li, H., Hu, F., Zhang, Y., & Li, K. (2020). Comparative efficacy and acceptability of disease-modifying therapies in patients with relapsing-remitting multiple sclerosis: A systematic review and network meta-analysis. *Journal of Neurology, 267*, 3489–3498.

Link, J., Ramanujam, R., Auer, M., Ryner, M., Hassler, S., Bachelet, D., Mbogning, C., Warnke, C., Buck, D., Jensen, P. E. H., Sievers, C., Ingenhoven, K., Fissolo, N., Lindberg, R., Grummel, V., Donnellan, N., Comabella, M., Montalban, X., Kieseier, B., Sorensen, P. S., Hartung, H. P., Derfuss, T., Lawton, A., Sikkema, D., Pallardy, M., Hemmer, B., Deisenhammer, F., Broet, P., Donnes, P., Davidson, J., Fogdell-Hahn, A., & Abirisk Consortium. (2017). Clinical practice of analysis of anti-drug antibodies against interferon beta and natalizumab in multiple sclerosis patients in Europe: A descriptive study of test results. *PLoS One, 12*, e0170395.

Linker, R. A., Lee, D. H., Ryan, S., van Dam, A. M., Conrad, R., Bista, P., Zeng, W., Hronowsky, X., Buko, A., Chollate, S., Ellrichmann, G., Bruck, W., Dawson, K., Goelz, S., Wiese, S., Scannevin, R. H., Lukashev, M., & Gold, R. (2011). Fumaric acid esters exert neuroprotective effects in neuroinflammation via activation of the Nrf2 antioxidant pathway. *Brain, 134*, 678–692.

Liseno, J., Lager, B., Miller, C., Shankar, S. L., Mendoza, J. P., & Lewin, J. B. (2021). Multiple sclerosis patients treated with diroximel fumarate in the real-world setting have high rates of persistence and adherence. *Neurology and Therapy, 10*, 349–360.

Liu, J. Q., Bai, X. F., Shi, F. D., Xiao, B. G., Li, H. L., Levi, M., Mustafa, M., Wahren, B., & Link, H. (1998). Inhibition of experimental autoimmune encephalomyelitis in Lewis rats by nasal administration of encephalitogenic MBP peptides: Synergistic effects of MBP 68-86 and 87-99. *International Immunology, 10*, 1139–1148.

Loftus, B., Newsom, B., Montgomery, M., Von Gynz-Rekowski, K., Riser, M., Inman, S., Garces, P., Rill, D., Zhang, J., & Williams, J. C. (2009). Autologous attenuated T-cell vaccine (Tovaxin) dose escalation in multiple sclerosis relapsing-remitting and secondary progressive patients nonresponsive to approved immunomodulatory therapies. *Clinical Immunology, 131*, 202–215.

Lublin, F. (2005). History of modern multiple sclerosis therapy. *Journal of Neurology, 252*(Suppl 3), iii3–iii9.

Lublin, F., Miller, D. H., Freedman, M. S., Cree, B. A. C., Wolinsky, J. S., Weiner, H., Lubetzki, C., Hartung, H. P., Montalban, X., Uitdehaag, B. M. J., Merschhemke, M., Li, B., Putzki, N., Liu, F. C., Haring, D. A., Kappos, L., & Informs Study Investigators. (2016). Oral fingolimod in primary progressive multiple sclerosis (INFORMS): A phase 3, randomised, double-blind, placebo-controlled trial. *Lancet, 387*, 1075–1084.

Lutterotti, A., Hayward-Koennecke, H., Sospedra, M., & Martin, R. (2021). Antigen-specific immune tolerance in multiple sclerosis-promising approaches and how to bring them to patients. *Frontiers in Immunology, 12*, 640935.

Majkutewicz, I. (2022). Dimethyl fumarate: A review of preclinical efficacy in models of neurodegenerative diseases. *European Journal of Pharmacology, 926*, 175025.

Margoni, M., Preziosa, P., Tortorella, P., Filippi, M., & Rocca, M. A. (2022). Does ocrelizumab limit multiple sclerosis progression? Current evidence from clinical, mri, and fluid biomarkers. *Neurotherapeutics, 19*, 1216–1228.

Martin, E., Aigrot, M. S., Lamari, F., Bachelin, C., Lubetzki, C., Nait Oumesmar, B., Zalc, B., & Stankoff, B. (2021). Teriflunomide promotes oligodendroglial 8,9-unsaturated sterol accumulation and CNS Remyelination. *Neurology Neuroimmunology & Neuroinflammation, 8*.

Matsushita, T. (2019). Regulatory and effector B cells: Friends or foes? *Journal of Dermatological Science, 93*, 2–7.

McMahon, E. J., Bailey, S. L., Castenada, C. V., Waldner, H., & Miller, S. D. (2005). Epitope spreading initiates in the CNS in two mouse models of multiple sclerosis. *Nature Medicine, 11*, 335–339.

Melnikov, M., Sharanova, S., Sviridova, A., Rogovskii, V., Murugina, N., Nikolaeva, A., Dagil, Y., Murugin, V., Ospelnikova, T., Boyko, A., & Pashenkov, M. (2020). The influence of glatiramer acetate on Th17-immune response in multiple sclerosis. *PLoS One, 15*, e0240305.

Messina, S., & Patti, F. (2013). The pharmacokinetics of glatiramer acetate for multiple sclerosis treatment. *Expert Opinion on Drug Metabolism & Toxicology, 9*, 1349–1359.

Miller, A. E. (2021). An updated review of teriflunomide's use in multiple sclerosis. *Neurodegenerative Disease Management, 11*, 387–409.

Mirabella, M., Annovazzi, P., Brownlee, W., Cohen, J. A., Kleinschnitz, C., & Wolf, C. (2022). Treatment challenges in multiple sclerosis –

A continued role for glatiramer acetate? *Frontiers in Neurology, 13*, 844873.
Miron, V. E., Ludwin, S. K., Darlington, P. J., Jarjour, A. A., Soliven, B., Kennedy, T. E., & Antel, J. P. (2010). Fingolimod (FTY720) enhances remyelination following demyelination of organotypic cerebellar slices. *The American Journal of Pathology, 176*, 2682–2694.
Molnarfi, N., Prod'homme, T., Schulze-Topphoff, U., Spencer, C. M., Weber, M. S., Patarroyo, J. C., Lalive, P. H., & Zamvil, S. S. (2015). Glatiramer acetate treatment negatively regulates type I interferon signaling. *Neurology Neuroimmunology & Neuroinflammation, 2*, e179.
Montalban, X., Arnold, D. L., Weber, M. S., Staikov, I., Piasecka-Stryczynska, K., Willmer, J., Martin, E. C., Dangond, F., Syed, S., Wolinsky, J. S., & Group Evobrutinib Phase 2 Study. (2019). Placebo-controlled trial of an oral BTK inhibitor in multiple sclerosis. *The New England Journal of Medicine, 380*, 2406–2417.
Montes Diaz, G., Fraussen, J., Van Wijmeersch, B., Hupperts, R., & Somers, V. (2018). Dimethyl fumarate induces a persistent change in the composition of the innate and adaptive immune system in multiple sclerosis patients. *Scientific Reports, 8*, 8194.
Morrow, S. A., Clift, F., Devonshire, V., Lapointe, E., Schneider, R., Stefanelli, M., & Vosoughi, R. (2022). Use of natalizumab in persons with multiple sclerosis: 2022 update. *Multiple Sclerosis and Related Disorders, 65*, 103995.
Muehler, A., Kohlhof, H., Groeppel, M., & Vitt, D. (2019). The selective Oral Immunomodulator Vidofludimus in patients with active rheumatoid arthritis: Safety results from the COMPONENT study. *Drugs in R&D, 19*, 351–366.
Muehler, A., Peelen, E., Kohlhof, H., Groppel, M., & Vitt, D. (2020). Vidofludimus calcium, a next generation DHODH inhibitor for the treatment of relapsing-remitting multiple sclerosis. *Multiple Sclerosis and Related Disorders, 43*, 102129.
Mulero, P., Midaglia, L., & Montalban, X. (2018). Ocrelizumab: A new milestone in multiple sclerosis therapy. *Therapeutic Advances in Neurological Disorders, 11*, 1756286418773025.
Nguyen, K., Olcer, M., Howlett-Prieto, Q., Feng, X., & Reder, A. T. (2023). Prolonged interferon-stimulated gene and Protein signatures in multiple sclerosis induced by PEGylated IFN-beta-1a compared to non-PEGylated IFN-beta-1a. *Journal of Interferon & Cytokine Research, 43*, 108–120.
Okuda, D. T., Kantarci, O., Lebrun-Frenay, C., Sormani, M. P., Azevedo, C. J., Bovis, F., Hua, L. H., Amezcua, L., Mowry, E. M., Hotermans, C., Mendoza, J., Walsh, J. S., von Hehn, C., Vargas, W. S., Donlon, S., Naismith, R. T., Okai, A., Pardo, G., Repovic, P., Stuve, O., Siva, A., & Pelletier, D. (2023). Dimethyl fumarate delays multiple sclerosis in radiologically isolated syndrome. *Annals of Neurology, 93*, 604–614.
Patti, F., Visconti, A., Capacchione, A., Roy, S., Trojano, M., & Clarinet-Ms Study Group. (2020). Long-term effectiveness in patients previously treated with cladribine tablets: A real-world analysis of the Italian multiple sclerosis registry (CLARINET-MS). *Therapeutic Advances in Neurological Disorders, 13*, 1756286420922685.
Pender, M. P., Csurhes, P. A., Smith, C., Douglas, N. L., Neller, M. A., Matthews, K. K., Beagley, L., Rehan, S., Crooks, P., Hopkins, T. J., Blum, S., Green, K. A., Ioannides, Z. A., Swayne, A., Aftab, B. T., Hooper, K. D., Burrows, S. R., Thompson, K. M., Coulthard, A., & Khanna, R. (2020). Epstein-Barr virus-specific T cell therapy for progressive multiple sclerosis. *JCI Insight, 5*.
Pitteri, M., Magliozzi, R., Bajrami, A., Camera, V., & Calabrese, M. (2018). Potential neuroprotective effect of Fingolimod in multiple sclerosis and its association with clinical variables. *Expert Opinion on Pharmacotherapy, 19*, 387–395.
Prager, B., Spampinato, S. F., & Ransohoff, R. M. (2015). Sphingosine 1-phosphate signaling at the blood-brain barrier. *Trends in Molecular Medicine, 21*, 354–363.
Prod'homme, T., & Zamvil, S. S. (2019). The evolving mechanisms of action of glatiramer acetate. *Cold Spring Harbor Perspectives in Medicine, 9*, a029249.
Rae-Grant, A., Day, G. S., Marrie, R. A., Rabinstein, A., Cree, B. A. C., Gronseth, G. S., Haboubi, M., Halper, J., Hosey, J. P., Jones, D. E., Lisak, R., Pelletier, D., Potrebic, S., Sitcov, C., Sommers, R., Stachowiak, J., Getchius, T. S. D., Merillat, S. A., & Pringsheim, T. (2018). Practice guideline recommendations summary: Disease-modifying therapies for adults with multiple sclerosis: Report of the guideline development, dissemination, and implementation subcommittee of the American Academy of Neurology. *Neurology, 90*, 777–788.
Raftopoulou, S., Rapti, A., Karathanasis, D., Evangelopoulos, M. E., & Mavragani, C. P. (2022). The role of type I IFN in autoimmune and autoinflammatory diseases with CNS involvement. *Frontiers in Neurology, 13*, 1026449.
Rasouli, J., Ciric, B., Imitola, J., Gonnella, P., Hwang, D., Mahajan, K., Mari, E. R., Safavi, F., Leist, T. P., Zhang, G. X., & Rostami, A. (2015). Expression of GM-CSF in T cells is increased in multiple sclerosis and suppressed by IFN-beta therapy. *Journal of Immunology, 194*, 5085–5093.
Rau, D., Eichau, S., Borriello, G., Cerqueira, J., & Wagner, C. (2023). Assessment of the treating physicians' first-hand experience with handling and satisfaction of ofatumumab therapy: Findings from the PERITIA survey conducted in Europe. *BMC Neurology, 23*, 147.
Reich, D. S., Arnold, D. L., Vermersch, P., Bar-Or, A., Fox, R. J., Matta, A., Turner, T., Wallstrom, E., Zhang, X., Mares, M., Khabirov, F. A., Traboulsee, A., & Group Tolebrutinib Phase 2b Study. (2021). Safety and efficacy of tolebrutinib, an oral brain-penetrant BTK inhibitor, in relapsing multiple sclerosis: A phase 2b, randomised, double-blind, placebo-controlled trial. *Lancet Neurology, 20*, 729–738.
Robinson, W. H., Fontoura, P., Lee, B. J., de Vegvar, H. E., Tom, J., Pedotti, R., DiGennaro, C. D., Mitchell, D. J., Fong, D., Ho, P. P., Ruiz, P. J., Maverakis, E., Stevens, D. B., Bernard, C. C., Martin, R., Kuchroo, V. K., van Noort, J. M., Genain, C. P., Amor, S., Olsson, T., Utz, P. J., Garren, H., & Steinman, L. (2003). Protein microarrays guide tolerizing DNA vaccine treatment of autoimmune encephalomyelitis. *Nature Biotechnology, 21*, 1033–1039.
Rosen, H., Stevens, R. C., Hanson, M., Roberts, E., & Oldstone, M. B. (2013). Sphingosine-1-phosphate and its receptors: Structure, signaling, and influence. *Annual Review of Biochemistry, 82*, 637–662.
Ruggieri, S., Quartuccio, M. E., & Prosperini, L. (2022). Ponesimod in the treatment of relapsing forms of multiple sclerosis: An update on the emerging clinical data. *Degenerative Neurological and Neuromuscular Disease, 12*, 61–73.
Safaeinejad, F., Asadi, S., Ghafghazi, S., & Niknejad, H. (2021). The synergistic anti-apoptosis effects of amniotic epithelial stem cell conditioned medium and ponesimod on the oligodendrocyte cells. *Frontiers in Pharmacology, 12*, 691099.
Safavi, F., Thome, R., Li, Z., Zhang, G. X., & Rostami, A. (2020). Dimethyl fumarate suppresses granulocyte macrophage colony-stimulating factor-producing Th1 cells in CNS neuroinflammation. *Neurology Neuroimmunology & Neuroinflammation, 7*.
Salama, H. H., Hong, J., Zang, Y. C., El-Mongui, A., & Zhang, J. (2003). Blocking effects of serum reactive antibodies induced by glatiramer acetate treatment in multiple sclerosis. *Brain, 126*, 2638–2647.
Salvetti, M., Wray, S., Nelles, G., Altincatal, A., Kumar, A., Koster, T., Naylor, M. L., & Investigators Plegridy Observational Program. (2022). Safety and clinical effectiveness of peginterferon beta-1a for relapsing multiple sclerosis in the real-world setting: Interim results from the Plegridy Observational Program. *Multiple Sclerosis and Related Disorders, 57*, 103350.
Salzer, J., Svenningsson, R., Alping, P., Novakova, L., Bjorck, A., Fink, K., Islam-Jakobsson, P., Malmestrom, C., Axelsson, M., Vagberg, M., Sundstrom, P., Lycke, J., Piehl, F., & Svenningsson, A. (2016).

Rituximab in multiple sclerosis: A retrospective observational study on safety and efficacy. *Neurology, 87*, 2074–2081.

Samjoo, I. A., Worthington, E., Drudge, C., Zhao, M., Cameron, C., Haring, D. A., Stoneman, D., Klotz, L., & Adlard, N. (2020). Comparison of ofatumumab and other disease-modifying therapies for relapsing multiple sclerosis: A network meta-analysis. *Journal of Comparative Effectiveness Research, 9*, 1255–1274.

Sellebjerg, F., Hedegaard, C. J., Krakauer, M., Hesse, D., Lund, H., Nielsen, C. H., Sondergaard, H. B., & Sorensen, P. S. (2012). Glatiramer acetate antibodies, gene expression and disease activity in multiple sclerosis. *Multiple Sclerosis, 18*, 305–313.

Singer, B., Ross, A. P., & Tobias, K. (2011). Oral fingolimod for the treatment of patients with relapsing forms of multiple sclerosis. *International Journal of Clinical Practice, 65*, 887–895.

Smith, C. E., & Miller, S. D. (2006). Multi-peptide coupled-cell tolerance ameliorates ongoing relapsing EAE associated with multiple pathogenic autoreactivities. *Journal of Autoimmunity, 27*, 218–231.

Sorensen, P. S. (2014). New management algorithms in multiple sclerosis. *Current Opinion in Neurology, 27*, 246–259.

Sorensen, P. S. (2022). Antidrug antibodies against biological treatments for multiple sclerosis. *CNS Drugs, 36*, 569–589.

Stamatellos, V. P., & Papazisis, G. (2023). Safety and monitoring of the treatment with disease-modifying therapies (DMTs) for multiple sclerosis (MS). *Current Reviews in Clinical and Experimental Pharmacology, 18*, 39–50.

Steinman, L., Fox, E., Hartung, H. P., Alvarez, E., Qian, P., Wray, S., Robertson, D., Huang, D., Selmaj, K., Wynn, D., Cutter, G., Mok, K., Hsu, Y., Xu, Y., Weiss, M. S., Bosco, J. A., Power, S. A., Lee, L., Miskin, H. P., Cree, B. A. C., Ultimate, I., & Investigators, U. I. (2022). Ublituximab versus teriflunomide in relapsing multiple sclerosis. *The New England Journal of Medicine, 387*, 704–714.

Svenningsson, A., Frisell, T., Burman, J., Salzer, J., Fink, K., Hallberg, S., Hambraeus, J., Axelsson, M., Nimer, F. A., Sundstrom, P., Gunnarsson, M., Johansson, R., Mellergard, J., Rosenstein, I., Ayad, A., Sjoblom, I., Risedal, A., de Flon, P., Gilland, E., Lindeberg, J., Shawket, F., Piehl, F., & Lycke, J. (2022). Safety and efficacy of rituximab versus dimethyl fumarate in patients with relapsing-remitting multiple sclerosis or clinically isolated syndrome in Sweden: A rater-blinded, phase 3, randomised controlled trial. *Lancet Neurology, 21*, 693–703.

Swallow, E., Pham, T., Patterson-Lomba, O., Yin, L., Gomez-Lievano, A., Liu, J., Tencer, T., & Gupte-Singh, K. (2023). Comparative efficacy and safety of ozanimod and ponesimod for relapsing multiple sclerosis: A matching-adjusted indirect comparison. *Multiple Sclerosis and Related Disorders, 71*, 104551.

Tong, J., Zou, Q., Chen, Y., Liao, X., Chen, R., Ma, L., Zhang, D., & Li, Q. (2021). Efficacy and acceptability of the S1P receptor in the treatment of multiple sclerosis: A meta-analysis. *Neurological Sciences, 42*, 1687–1695.

Toorop, A. A., Hogenboom, L., Bloem, K., Kocyigit, M., Commandeur, N. W. M., Wijnants, A., Lissenberg-Witte, B. I., Strijbis, E. M. M., Uitdehaag, B. M. J., Rispens, T., Killestein, J., & van Kempen, Z. L. E. (2023). Ocrelizumab concentration and antidrug antibodies are associated with B-cell count in multiple sclerosis. *Journal of Neurology, Neurosurgery, and Psychiatry*.

Ure, D. R., & Rodriguez, M. (2002). Polyreactive antibodies to glatiramer acetate promote myelin repair in murine model of demyelinating disease. *The FASEB Journal, 16*, 1260–1262.

Valenzuela, B., Olsson Gisleskog, P., Poggesi, I., Sidorenko, T., Burcklen, M., Kracker, H., & Perez-Ruixo, J. J. (2022). An exposure-response analysis of ponesimod clinical efficacy in a randomized phase III study in patients with relapsing multiple sclerosis. *CPT: Pharmacometrics & Systems Pharmacology, 11*, 1294–1304.

van Holten, J., Plater-Zyberk, C., & Tak, P. P. (2002). Interferon-beta for treatment of rheumatoid arthritis? *Arthritis Research, 4*, 346–352.

Vermersch, P., Benrabah, R., Schmidt, N., Zephir, H., Clavelou, P., Vongsouthi, C., Dubreuil, P., Moussy, A., & Hermine, O. (2012). Masitinib treatment in patients with progressive multiple sclerosis: A randomized pilot study. *BMC Neurology, 12*, 36.

Vermersch, P., Brieva-Ruiz, L., Fox, R. J., Paul, F., Ramio-Torrenta, L., Schwab, M., Moussy, A., Mansfield, C., Hermine, O., Maciejowski, M., & A. B. Study Group. (2022). Efficacy and safety of masitinib in progressive forms of multiple sclerosis: A randomized, phase 3, clinical trial. *Neurology Neuroimmunology & Neuroinflammation, 9*, e1148.

von Budingen, H. C., Bar-Or, A., & Zamvil, S. S. (2011). B cells in multiple sclerosis: Connecting the dots. *Current Opinion in Immunology, 23*, 713–720.

von Essen, M. R., Hansen, R. H., Hojgaard, C., Ammitzboll, C., Wiendl, H., & Sellebjerg, F. (2022). Ofatumumab modulates inflammatory T cell responses and migratory potential in patients with multiple sclerosis. *Neurology Neuroimmunology & Neuroinflammation, 9*, e200004.

Vrenken, H., Battaglini, M., de Vos, M. L., Nagtegaal, G. J., Teixeira, B. C. A., Seitzinger, A., Jack, D., Sormani, M. P., Uitdehaag, B. M. J., Versteeg, A., Comi, G., Kappos, L., De Stefano, N., & Barkhof, F. (2023). Temporal evolution of new T1-weighted hypo-intense lesions and central brain atrophy in patients with a first clinical demyelinating event treated with subcutaneous interferon beta-1a. *Journal of Neurology, 270*, 2271–2282.

Walczak, A., Siger, M., Ciach, A., Szczepanik, M., & Selmaj, K. (2013). Transdermal application of myelin peptides in multiple sclerosis treatment. *JAMA Neurology, 70*, 1105–1109.

Wanleenuwat, P., & Iwanowski, P. (2019). Role of B cells and antibodies in multiple sclerosis. *Multiple Sclerosis and Related Disorders, 36*, 101416.

Ward, M., & Goldman, M. D. (2022). Epidemiology and pathophysiology of multiple sclerosis. *Continuum (Minneapolis, Minn.), 28*, 988–1005.

Yadav, S. K., Soin, D., Ito, K., & Dhib-Jalbut, S. (2019). Insight into the mechanism of action of dimethyl fumarate in multiple sclerosis. *Journal of Molecular Medicine (Berlin, Germany), 97*, 463–472.

Yang, S., Li, X., Wang, J., Wang, T., Xu, Z., Gao, H., & Chen, G. (2022). Sphingosine-1-phosphate receptor modulators versus interferon beta for the treatment of relapsing-remitting multiple sclerosis: Findings from randomized controlled trials. *Neurological Sciences, 43*, 3565–3581.

Yong, H. Y. F., & Yong, V. W. (2022). Mechanism-based criteria to improve therapeutic outcomes in progressive multiple sclerosis. *Nature Reviews. Neurology, 18*, 40–55.

Zhang, Y., Li, X., Ciric, B., Ma, C. G., Gran, B., Rostami, A., & Zhang, G. X. (2017). Effect of Fingolimod on neural stem cells: A novel mechanism and broadened application for neural repair. *Molecular Therapy, 25*, 401–415.

Zhang, Y., Yin, H., Zhang, D., Xu, Y., Peng, B., & Cui, L. (2022). Real-world outcomes of teriflunomide in relapsing-remitting multiple sclerosis: A prospective cohort study. *Journal of Neurology, 269*, 4808–4816.

Ziemssen, T., Ashtamker, N., Rubinchick, S., Knappertz, V., & Comi, G. (2017). Long-term safety and tolerability of glatiramer acetate 20 mg/ml in the treatment of relapsing forms of multiple sclerosis. *Expert Opinion on Drug Safety, 16*, 247–255.

Theranostics for Viral Infections

36

Bharat N. Chaudhary, Mohammad Uzair Ali, and Howard E. Gendelman

Abstract

The combination of diagnostics and therapeutics defines theranostics as a new field of translational biomedical research that seeks to improve disease outcomes. This chapter focuses on infectious disease therapeutics, demonstrating how they have successfully developed next-generation therapies. Our focus is on viral infections. Virus-based malfunctioning of physiological processes due to their aggressive progression and multiplicity is concerning. Presently, clinical resolutions for such diseases lack targeted interventions due to considerable constraints in drug tracking. Herein, novel cell-based immunomodulatory therapeutics are discussed in the context of prevention and cure. These remain a major research focus, as new diagnostic and therapeutic discoveries can help combat global health concerns. Notably, viral infections represent both diagnostic and therapeutic challenges. Newer drug delivery vehicles can overcome therapeutic limitations that include low on-target bioavailability, sub-therapeutic drug levels, and limited patient regimen adherence. Nanocarriers are being used for vaccine formulation. The current burden of infectious diseases and the immediate need for improved clinical outcomes are driving the development of novel therapeutic systems.

Keywords

Disease targeted therapeutics · Drug delivery · Nanotheranostics · Viral theranostics · Blood–brain barrier · Immunotherapy · Combinatorial therapeutics · Clinical managements

36.1 Introduction

The human immunodeficiency virus (HIV) whose inception spread from nonhuman primate to human infection began in the 1900s (Faria et al., 2014; Keele et al., 2006). It was only until the 1980s that the virus appeared as a spread of infection to mankind. This presented at the onset of an unexplained immunodeficiency syndrome (Gottlieb et al., 1981). HIV has affected more than 85.6 million people globally, being a major cause of morbidity and mortality worldwide with almost 39 million people currently living with the virus reflective of the present global statistics (HIV, 2023). Keeping aside the tremendous health consequences of this disease, there is a major loss of economic productivity in infected individuals due to the prime productive age group encompassed. The aggressive antiretroviral therapy (ART) is based on the use of three antiretroviral drugs (ARVs). Some HIV regulation is attained with chronic and adherent ART regimens and can become a problem for viral resistance, mutation, and rebound after discontinuation (Smith, 2001). The consistent adherence and the toxic secondary drug effects form the core basis of incompliance and subsequent therapeutic failure (Bonfanti et al., 1999). The ART composition was previously adult-approved due to its solid physicochemical nature and had no relevant corresponding pediatric dose which led to an unlicensed administration of these medications in liquid oral portions adjusted to neonatal and child dose. This affects nearly 40 million children every year (Conroy, 2003). However, it was only in June 2020 that the

Authors Bharat N. Chaudhary and Mohammad Uzair Ali have equally contributed to this chapter.

B. N. Chaudhary · M. U. Ali · H. E. Gendelman (✉)
Department of Pharmacology and Experimental Neuroscience, College of Medicine, University of Nebraska Medical Center, Omaha, NE, USA
e-mail: hegendel@unmc.edu

H. E. Gendelman, T. Ikezu (eds.), *Neuroimmune Pharmacology and Therapeutics*,
https://doi.org/10.1007/978-3-031-68237-7_36

US Food and Drug Administration (FDA) approved dolutegravir (DTG) suspension-based HIV-1 therapy for at least a 4-week-old pediatric population (Lee et al., 2021). HIV infection is presently a major international concern taking into consideration the epidemiological burden and increased mortality rates (Smith & Daniel, 2006). From a pathophysiological standpoint, HIV targets CD4$^+$ T cells. HIV is confined to lymphoid tissues and within a brief period is seen in brain, gut, and genitourinary tissues (Haase, 2005). The virus peaks at day 30 post-infection, this is the most infectious window. The immune system can attain a level of regulation over the virus replication making it stable for prolonged durations (Jw, 1996). The multifaceted mechanistic models of HIV are still ambiguous; however, the potent loss of CD4$^+$ T cells and multiple immunological abnormalities (McCune, 2001). As time progresses, immense immunodeficiency develops leading to oncological mutations of the disease, the average life expectancy is about 10 years with modulations in disease growth (Deeks & Walker, 2007).

The lack of compliance arising from ART physicochemical properties is often neglected causing reduced adherence and a substantial drop in the drug's bioavailability. The presence of ART ameliorates the pathophysiological consequences of the infection to a great extent, yet even complete ART adherence does not lead to viral elimination. ART rebound can occur as the major intracellular and anatomical viral reservoirs can drive replication recurrence (Geeraert et al., 2008; Park & Sinko, 2005; Romanelli & Hoven, 2006). The viral reservoirs are principally in resting CD4$^+$ T cells causing reinstitution of viral replication following ART interruption. The dynamics of the viral reservoir commonly in lymphoid, gut, and brain tissues are complex, and present limitations exist, as there are currently no agents able to eliminate it. To date, its exact location, presence throughout tissues, and cellular crosstalk remain an impediment to viral elimination (Sengupta & Siliciano, 2018). Approximately 1.5 million new cases of HIV are reported every year and 650,000 deaths from the virus-associated disease complications. Even though significant global attention is given to this disease, one-third of the population is still either undiagnosed or is not receiving effective care (HIV, 2023). A large number of those living with HIV, are unable to access ART (Cohen et al., 2019; Cohen et al., 2014). Advancements in delivery systems are essential to extend access to long-acting options. Agencies remain unable to provide the resources necessary to prevent and treat HIV infection.

The need for improved healthcare delivery is certainly timely. Keeping this in mind, a theranostic approach to HIV/acquired immunodeficiency syndrome (AIDS) diagnosis, treatment, and prevention is necessary. Newer and longer-acting medicines can improve the means to not only treat but also to track the effectiveness of ARVs in blood, spleen, lymph nodes, gut, and liver where the majority of CD4$^+$ cells are present and susceptible to viral infection. This provides a platform to precisely target and track ARV effectiveness in anatomical sites with the ability to visualize those sites and suppress viral growth in each of them (Ottemann, 2019). Nanotechnology is one means as this delivery system is capable of combining diagnostic and therapeutic applications and as such facilitates diagnosis, drug delivery, and treatment effectiveness.

36.2 The Human Immunodeficiency Virus (HIV)

HIV is a retrovirus, broadly classified under a subgroup known as the lentiviruses, often addressed as "slow viruses" (Chiu et al., 1985). The "slow" attribute was conferred due to the time lag between the initial infection and the expression of severe symptoms. As portrayed by all viruses, HIV presents intracellular replication, sabotaging the host cell's machinery driving viral protein multiplication. The initial viral exposure to the host cell involves viral attachment to the expressed CD4$^+$ receptor along with a coreceptor (CXCR4 or CCR5). This fusion is mediated through HIV-1gp120 and gp41 viral envelope proteins contributing to the membrane fusion and viral protein release in the intracellular domain (Fig. 36.1). Upon entering, the virus reverse transcriptase enzyme responsibly converts the viral RNA to DNA into a full-length copy which is subsequently dislodged into smaller functional fragments (Smith & Daniel, 2006; Zheng et al., 2005). This reverse-transcribed viral DNA translocates inside the cell's nucleus and proceeds HIV integrase-mediated splicing inside the host's DNA also known as the "HIV provirus." The host's body responds to this invasion by immune activation recruiting and producing a burst of CD4$^+$ only to increase the targetable cells by newly formed HIV protein, ultimately exhausting the host's defenses. This immune exhaustion is termed as the development of AIDS. Furthermore, the activated cells make even more viral proteins by transcribing viral DNA to mRNA that excludes the nucleus. Cytoplasmic protein-making machinery is compromised in infected cells which then use the viral mRNA as the template for protein translation. The translated protein molecules end up being large genomic products that are spliced by HIV proteases into smaller units that can be assembled within the viral envelope proteins and the membrane together with the enzymes. This molecular viral machinery then pinches off as "Viral Buds" from the cell membrane manifesting other host cells (Fig. 36.1).

ART has changed the course of modern medicine and the perception of HIV in the context of survival and fatality. People who are most susceptible to acquiring the disease can take a daily pill or monthly injectables preventing them from acquiring HIV (Landovitz et al., 2021). In line with treat-

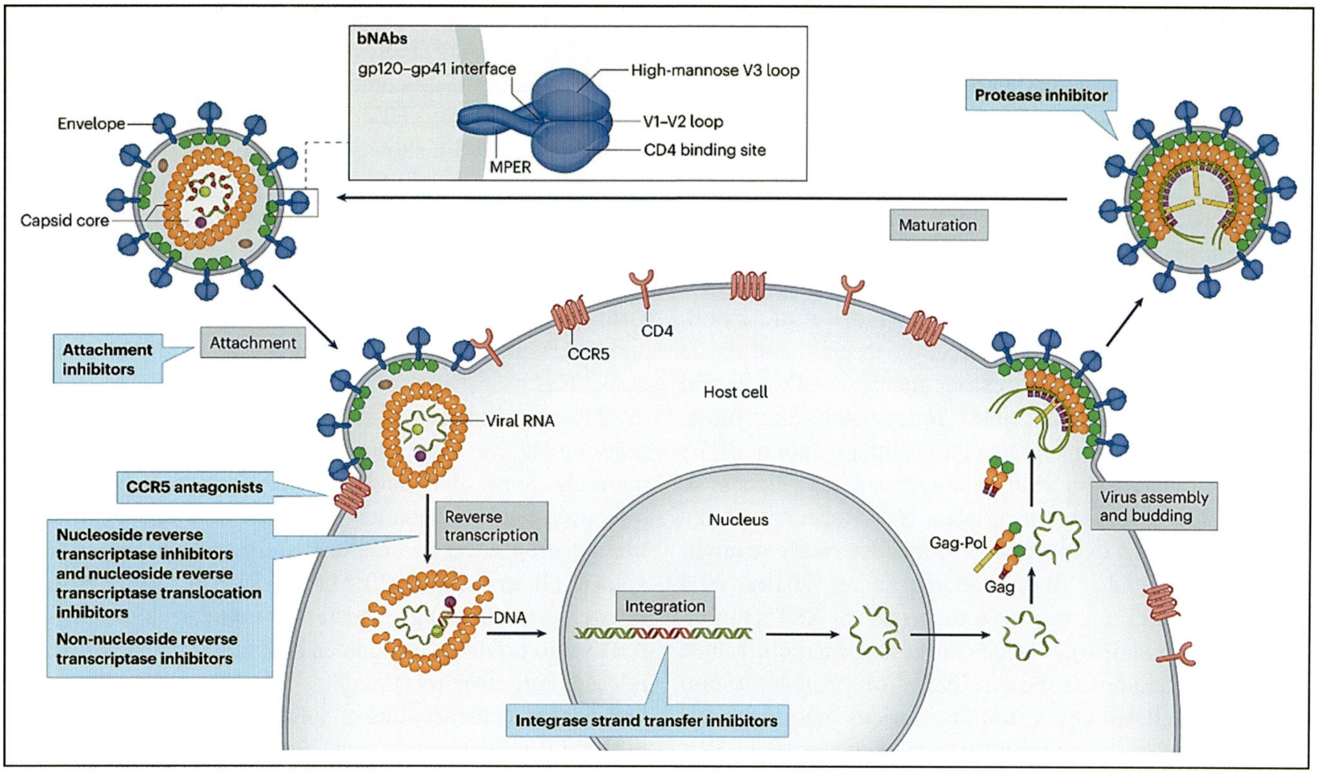

Fig. 36.1 The HIV replication cycle provides multiple opportunities for therapeutic disruption of virus-specific replicative activities. Currently, available approaches include entry inhibitors, reverse transcriptase inhibitors, integrase inhibitors, and protease inhibitors. Emerging therapies including broadly neutralizing antibodies (bNAbs) that target the envelope V1–V2 loop (PDGM1400, CAP256), the V3 loop (10-1074, PGT121), the CD4-binding site (VRC07-523LS, 3BNC117, N6), the gp120–gp41 interface and the membrane-proximal external region (MPER; 10E8V), all obstructing viral entry. Lenacapavir is a first-in-class capsid inhibitor, believed to inhibit multiple critical life cycle processes including cytoplasmic capsid dissolution and formation of functional new capsid for viral progeny. Islatravir is a first-in-class nucleoside reverse transcriptase translocation inhibitor. CCR5, CC-chemokine receptor 5. (Image license obtained from Springer Nature to reproduce image under “reuse in a springer nature imprint” (Landovitz et al., 2023))

ment, people living with HIV (PLWH) who have access to treatment services can take a daily pill or a once-monthly injection acquiring indefinite control of the infection. The optimization of these approaches is ongoing and the prospect of having a long-acting medication acting for more than 6 months per administration is very much possible. The United States government has launched a program called Ending the HIV Epidemic (EHE) that aims to reduce the number of new HIV infections by 90% by 2030.

It is theranostics, which focuses on patient-centered care, that is emerging as a targeted, safe, and efficient pharmacotherapy. Implying the diagnostic and therapeutic aspects of treatment, theranostic is a new field of disease-personalized medicine. As a modern medical approach, theranostics bridges the gap between conventional medicine and individualized precision medicine. In theranostics, nanotechnology is used to combine diagnostic and therapeutic applications into a single formulation, aiding in diagnosis, drug delivery, and treatment response monitoring. In theranostics medicine, the drug and diagnostic agents are delivered at the targeted site to diagnose and treat the disease at cellular and molecular levels by achieving enhanced systemic circulation, evading host defenses, and passing through the host's immune system.

36.3 Antiretroviral Therapeutics

36.3.1 Antiretrovirals

ART is presently the most effective treatment available for nearly two decades. However, the ART's efficiency is based on patient compliance, in complete adherence, ART can perform robust viral suppression enhancing immune function, and limiting the progression of AIDS (Davey Jr. et al., 1999). ART is used in disease prevention and treatment, but the constant efforts to develop a vaccine for HIV are failing consistently. The mechanism of action for current and emerging ART is reviewed in Fig. 36.1. The research domain is gradually transitioning towards a cure approach as the limitations of ARVs are more evident. The lack of patient adherence and associated toxicities of antiretrovirals can cause aggressive disease progression and affiliated severities. The treatment regimen includes synergistic effects of antiretrovirals when taken combinatorically making adherence complex and difficult to follow. Despite advances in ART for those people who have been infected with HIV, the medicines must be taken for the entire life span. Adherence is fraught with personal, social, economic, behavioral, drug resistance, and pharmacologic limitations (Chun et al., 1999; Davey Jr. et al., 1999). While not a cure, ART can prevent viral transmission and suppress viral replication (Finzi et al., 1999). The virus persists as a latent infection in cells and tissue compartments that include resting memory $CD4^+$ T and myeloid cells (Chun et al., 1995). Immediately after infection, the viral genome integrates into cellular genomic DNA with limited changes in viral gene expression (Siliciano & Siliciano, 2015). A reservoir of latent HIV-1 occurs within a few days after viral exposure before it is detectable in circulation (Cohen et al., 2011; Vanhamel et al., 2019). Viral rebound post-ART interruption remains one of ART's limitations. Elimination of HIV latency necessitates a multifaceted approach that conjuncts the disciplines of virology, immunology, and cell biology. Currently, various laboratory and animal models are being used for target cell studies of HIV latent reservoirs. This includes cell-based systems, humanized mice, and nonhuman primates (NHP) (Zhang et al., 2023).

Currently, ARTs include nucleoside or nucleotide reverse transcriptase inhibitors (NRTIs), nonnucleoside reverse transcriptase inhibitors (NNRTIs), protease inhibitors (PIs), integrase strand transfer inhibitors (INSTIs), CCR5 receptor antagonists, and fusion inhibitors. The number of HIV-associated deaths has reduced substantially due to ART use, however, select disease co-morbidities including specific cancer types continue (Bressan et al., 2021). Immunotherapy and chemotherapy remain a common treatment option (Ostios-Garcia et al., 2018). Combinations of chemotherapy and ART are in more widespread use but secondary co-morbidities that include hepatotoxicity and nephrotoxicity remain commonplace (Makinson et al., 2010). The former are notably prominent in patients co-infected with both HIV-1 and hepatitis B (HBV), hepatitis C virus (HCV), and or develop nonalcoholic fatty liver disease (NAFLD). Up to 25% of HIV-1 infected patients are coinfected with HBV, 30% with HCV, and 30–40% develop NAFLD (Bressan et al., 2021; Kaplan et al., 2009).

36.3.1.1 ART Challenges: Development of Long-Acting Medicines

Currently, available oral ART regimens have several drawbacks including challenges in patient adherence due to pill fatigue or side effects, the requirement of life-long daily drug intake, and limited penetration and retention in cells within lymph nodes. Progression and appropriately designed nanotherapeutics for targeted and long-acting, combination HIV chemotherapy may address these challenges (Gao et al., 2019). Unique drug-combination nanoparticles (DcNPs) composed of three HIV drugs-lopinavir, ritonavir, and tenofovir-have been shown to provide enhanced drug levels in lymph nodes; and elevated drug-combination levels in HIV-host cells in the blood and plasma for 2 weeks (Kraft et al., 2017). Additionally, co-formulated rilpivirine (RPV) and cabotegravir (CAB) (polymer NP demonstrated that the virus-mimicking NPs retain the antiviral efficacy of RPV and CAB in $CD169^+$ macrophages for 35 days (Eshaghi et al., 2022).

However, limitations of CAB and RPV LA include the requirement for large injection volumes, and short dosing intervals, none of them are active against major HIV co-infections such as hepatitis B and are also characterized by limited drug access to cellular and tissue infectious reservoirs (Kulkarni et al., 2020; Markowitz et al., 2017). Attempts to overcome these have led to the transformation of native ARVs into prodrugs designated as long-acting slow effective release antiretroviral therapy (LASER ART). These are homogeneous dispersions of solid prodrug nanocrystals stabilized by aqueous surfactant solutions (Banoub et al., 2020; Sillman et al., 2018; Smith et al., 2019). The LASER ART prodrug nanocrystals provide slow drug dissolution and rapid transport across biological membranes. This can achieve sustained plasma and tissue drug concentrations above the 90% effective concentration (EC_{90}) for periods of up to one year (Edagwa et al., 2017).

Alternatively, Bictegravir (BIC) long-acting nanoformulation exhibiting sustained drug release, limited cytotoxicity, and enhanced HIV-1 suppressive characteristics has recently been reported (Mandal et al., 2019b). HIV treatment has progressed significantly affirming that macrophages serve as viral hosts and drug targets (Herskovitz & Gendelman, 2019). Future developments in the field of macrophage-HIV eradication should apply CRISPR-Cas9 gene editing to splice out proviral sequences from host genomes. It was also reported for the first time that gold nanoparticles (AuNPs) as a novel next-generation platform to treat HIV/AIDS and a potential therapeutic approach to overcome several limitations of the current ART (Fotooh Abadi et al., 2023). Nanoencapsulation of antivirals demonstrated prolonged drug retention and HIV-1 protection over time in humanized mice after a single subcutaneous dose (Mandal et al., 2019a). Indeed, stable ProTide nanocrystals with long-acting pharmacokinetic profiles provide extended drug dosing intervals (Cobb et al., 2021). It was also demonstrated that drug-combination-nanoparticle (DcNP) technology is useful in the transformation of stable long-acting formation of 3-HIV drugs with disparate physical-chemical properties (Perazzolo et al., 2023). Through the creation of rod-shaped particles of defined sizes reflective of native LA drugs, theranostic probes can be deployed to measure particle-cell and tissue biodistribution, antiretroviral activities, and drug retention (Liu et al., 2015). Indeed, Rod-shaped theranostic nanoparticles such as radiolabelled ^{177}Lutetium-bismuth sulfide nanorods (^{177}LuBSNRs) as multimodal imaging particles can be effective imaging probes for rapid assessment tools for pharmacokinetic profiling of LA ARVs (Kevadiya et al., 2020).

A principal challenge of ART rests in adherence to complex drug regimens and in reaching viral tissue reservoirs. The recent development of long-acting nanoformulations shows advancement in antiretroviral drug delivery with higher potency, half-life, sustained plasma, and tissue-specific depot. Indeed, combination antiretrovirals delivery for the long term with the creation of cell and tissue-specific secondary depots could be possible via various nanocarriers such as solid drug nanocarriers, rod-shaped nanocarriers, polymeric nanoparticles, solid drug nanoparticles, and nanosuspension (Surve & Jindal, 2020; Wang et al., 2022). Hence, LA nanoformulations hold a future area of research and development for ARV therapy and will replace the currently available complex oral ART with a simplified regimen.

36.4 The Human Brain and HIV

There is evidence that brain cells harbor genome-integrated HIV (Kramer-Hämmerle et al., 2005). It is known that the virus invades the brain very soon following infection. Virus infection was shown in astrocytes, perivascular macrophages, and microglial cells (Churchill et al., 2006; Thompson et al., 2011). All three cell types are long-lived cells with perivascular macrophages and astrocytes with a half-life ranging from months and microglial cells with a half-life of years (Koppensteiner et al., 2012; Sofroniew & Vinters, 2010; Soulet & Rivest, 2008). All these cells are infected at high frequency in the brain. Astrocytes, the most abundant cell type in the brain, are infected in up to 19% of the cell population (Churchill et al., 2009). A similar ratio of infected cells has been found among the perivascular macrophages and the microglial cells (Gray et al., 2014; Thompson et al., 2011). In addition, several mechanisms, including epigenetic regulation, have been evoked to induce latency in these cells notably in astrocytes and microglial cells (Narasipura et al., 2014; Redel et al., 2010). Due to ethical and technical problems, it is not possible to evaluate the human brain-infected cells for their capacity to produce replication-competent viruses. However, there are several indirect evidences showing that CNS is a reservoir for HIV. Indeed, HIV DNA has been detected in brain tissues isolated from autopsies of HIV patients whose infection has been controlled by ART (Desplats et al., 2013; Thompson et al., 2011). Moreover, there is a strong correlation between the amount of HIV DNA found in astrocytes and the pathobiology of HIV-associated dementia (HAD) (Churchill et al., 2009).

There is a need to attenuate viral replication in tissue reservoirs including the CNS (Reynolds et al., 2007). The inherent toxicities associated with ART may have neurotoxic effects and are associated with cognitive complications in ART-treated patients (Rubin et al., 2021). Despite the recent advances in disease management of persons with HIV infection tissue co-morbidities continue. These include neurocognitive disorders estimated to occur in between 30% and 60% of treated infected individuals. Before the introduction of ART, patients developed end-organ diseases that included renal, hepatic, pulmonary, and neurological impairments. Most were associated with end-stage disease with severe immune suppression and infectious and cancerous opportunistic diseases. Most were seen in the final months of their illness and for the brain comprising cognitive, behavioral, and motor functions (Rosca et al., 2021). Some studies have reported that ART is associated with improved quality and longevity of life with reduced morbidities that include end-organ disease and neurocognitive function (Cysique et al., 2011). ART results in reduced HIV RNA levels in the plasma and tissue compartments. Indeed, viral reductions in the brain and cerebrospinal fluid (CSF) are associated with improvements in cognitive functions (Smit et al., 2004).

36.4.1 Targeting the Brain Reservoir

A significant constraint for ART is its inability to penetrate the CNS. The brain continues to be a major HIV reservoir even in the ART-based systemically viral-suppressed system (Fig. 36.2) (McArthur et al., 2010; Persidsky & Gendelman, 2003). Evidence reveals that ART nanomedicine formulations can cause HIV elimination from viral sanctuaries like the CNS. Literature analysis suggests that circulating monocytes-macrophages can act as circulating nanoformulated ART (nanoART) carriers and can relay the drug to human brain microvascular endothelial cells (HBMECs), which majorly constitute the blood–brain barrier (BBB) by cell-cell communication and can prospectively increase the drug's bioavailability in the brain (Kanmogne et al., 2012). Several structural modification-based approaches have been suggested in the NRTIs to facilitate BBB permeability. Zidovudine (AZT) being the first antiviral approved for clinical use in AIDS, when administered displays suppressed brain bioavailability in contrast to the periphery (Wu et al., 1998). The mechanistic phenomenon behind such low brain biodistribution lies in the active efflux from the BBB and the drug's passive diffusion route through to the brain (née Williams & Segal, 1996; Thomas & Segal, 1997). AZT has undergone multiple structural experimental alterations with

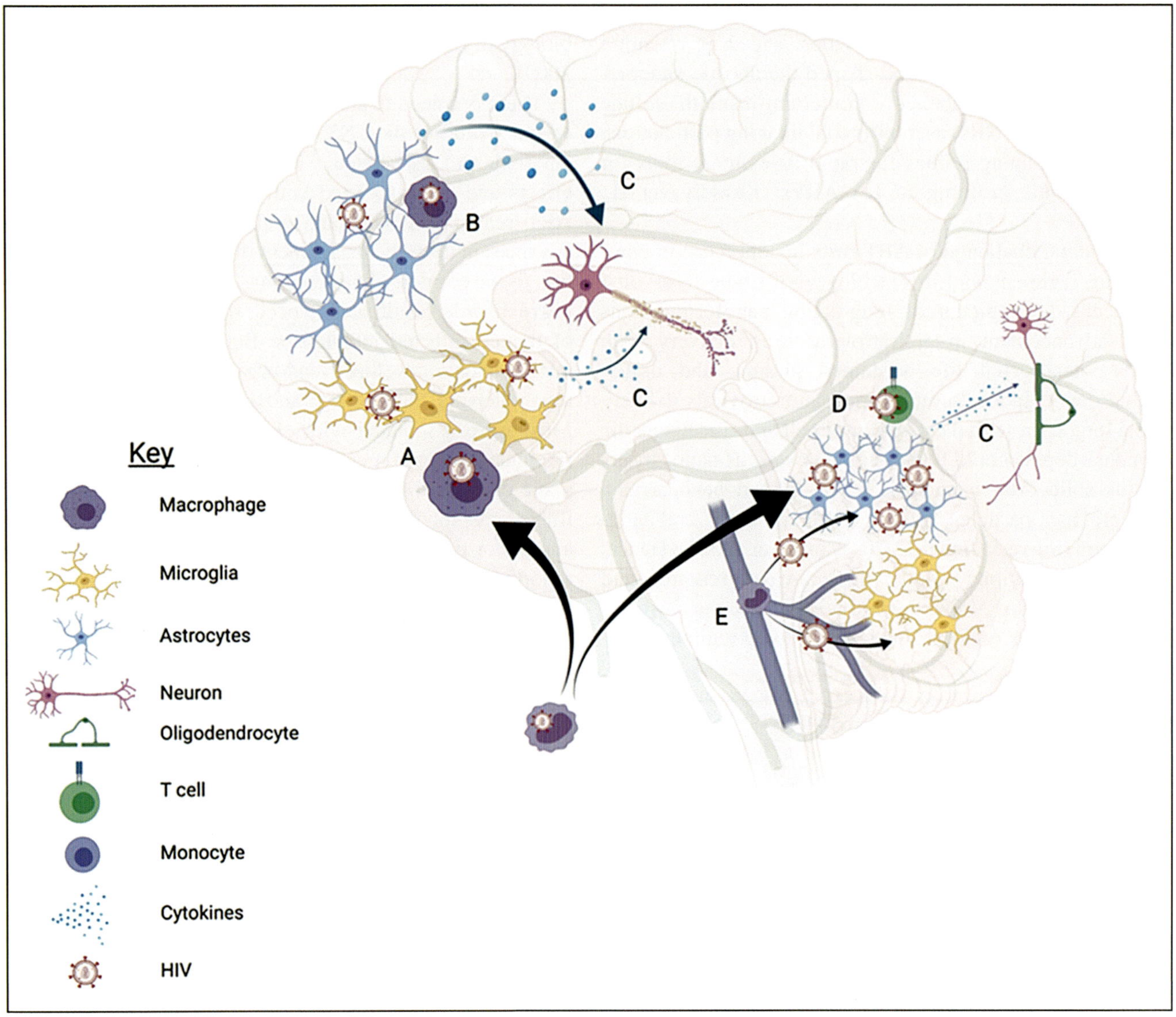

Fig. 36.2 Brain cell HIV reservoirs. (A) HIV-1 infected macrophages (purple) circulate through the blood–brain barrier and pass through into the CNS, where they can release viral particles and infect microglia (yellow). (B) direct contact with infected macrophages infects astrocytes (orange), which can pass viral infection to other astrocytes. Anti-HIV responses by astrocytes are much more potent in the cortex than in the cerebellum, therefore leading to a greater number of HIV⁺ cells in the cerebellum than in the cortex. (C) HIV⁺ macrophages and microglia, as well as astrocytes stimulate a powerful inflammatory response against infection (red dots), leading to damage and death of neurons (pink) and oligodendrocytes (green). (D) Circulating infected CD4⁺ T cells are trafficked into the brain and can infect glial cells via direct contact. (E) theorized route of infection from the meningeal lymphatic system, where HIV⁺ monocytes (blue) stimulate inflammation to permeabilize the vessels and carry infection into the CNS independent of the BBB. (Image taken from Ash et al. (2021) under CC BY open access license)

the insertion of various pro-moieties including 5′-hydroxyl connected to a labile ester linkage (McGuigan et al., 1993; Namane et al., 1992). Certain specific promoities like retinoic acid have the potential to solely suppress viral replication (Aggarwal et al., 1990). Multi-drug combinations capable of effectively crossing the blood–brain barrier (BBB) have emerged as a therapeutic possibility to improve HIV-1 treatments (Iyidogan & Anderson, 2014). Currently, a nanoformulation composed of antiretroviral drug crystals of ritonavir (RTV) and atazanavir (ATV) termed nanoART is being developed (Nowacek et al., 2009). Conventionally, micron-sized drug particles are prepared in water dispersion using different methods (e.g., high-pressure homogenization) and toxic organic solvents (Chaowanachan et al., 2013).

The most challenging aspect in brain disease treatment remains targeted drug delivery bypassing the BBB. Theranostic has shown great potential with nanoformulations to evade the BBB and facilitate effective treatment. Nanotechnological advancements have resulted in the assembly of various organic/inorganic moieties that can be modified at the surface by incorporating ligands or cell penetrating peptides (CPP) as an effective theranostic intercession. Specifically, the formulation attributes of targeted and multifunctional nanoparticles that can permeate through the BBB correlate to their optical, thermal, and magnetic properties. Some examples of real-time contrast molecules include photoacoustic Raman imaging probes that can demarcate between cancerous cells and healthy cells, assisting in precise and targeted tumor extraction-resection. These nanotheranostic systems can cause the induction of a suppressed hypoxia environment sensitizing the malignant tissue towards radiation and potentiating the effects of chemo and radiotherapeutics. Nanotechnology facilitates multiple therapeutic strategies for brain targeting. Exploiting these distinct advantages of nanoparticles to deliver photosensitizing agents, toxins, and genetic factors at the disease site.

BBB disruption-enhanced transport is an essential strategy to increase the permeation of the compounds across the BBB. A combinatorial delivery approach through focused ultrasound (FUS) in conjunction with microbubbles is a prospective strategy to facilitate the BBB opening for large pharmaceutical agent delivery as depicted in Fig. 36.4 (Fan et al., 2016).

36.4.1.1 Bypassing the BBB (Carrier-, Receptor- and Adsorptive-Mediated Transcytosis)

Carrier-mediated transcytosis (CMT) is a chief passageway to carry cellular nutrition and hormones like glucose are transported through the cell membrane via the assistance of protein moieties that act as transporters as rendered in Fig. 36.3. Glucose upon contact with the protein brings about a conformational change resulting in the formation of a concentration gradient (high-low). Several such nutrient transporters exist such as GLUT1, choline, and LAT1 (Du et al., 2014; Li et al., 2013; Lindqvist et al., 2016). Notably, GLUT1 is a prominent transporter that is present in elevated concentrations in contrast to varying receptors and transporters (Uchida et al., 2011). Glucose-mediated liposomal delivery has been shown to target GLUT1 to attain higher brain bioavailability; however, the hypothesis has not yet been proven. Glycaemic control, as a regulator of glucose-based nanocarriers, has also been implied, the nanocarrier assembly is based on oppositely charged PEG-ionomers being self-assembled. Negatively charged PEG–poly (a b-aspartic acid) (PEG-PAsp) and positively charged PEG–poly([5-aminopentyl]- a b-aspartamide) (PEG–P(Asp-AP)) block copolymers. Ester linkage progressed to the conjugation of PEG-PAsp to C6-OH (Qin et al., 2010; Xie et al., 2012).

Large endogenous molecules face substantial resistance through the CMT transport mechanism, hence receptor-mediated transcytosis (RMT) mediates a pathway for the specific transportation of such molecules as illustrated in Fig. 36.3. BBB permeability is achieved through the expression of multiple receptors onto the luminal side, ligand binding carriers, or relatively larger molecules bind to these receptors (LDLRs, TfR1, Insulin receptor, Lactoferrins) (Chen & Liu, 2012; Zhang et al., 2015). Among these receptors, TfR-1 is prominently researched as it shows potential for clinical interventions (Choudhury et al., 2018). Transportation across the BBB is attained through iron permeation through the intracellular motility of iron-binding glycoproteins (Moos & Morgan, 2000). Several Tf-operative nanocarriers have been discussed, in one such study cyclo-(Arg-Gly-Asp) (cRGD)–paclitaxel (PTX) loading in Tf-based micelles prolonged the mice survival manifested with orthotopic U87MG tumor (Zhang et al., 2012). Elevated concentrations of endogenous Tf are often observed to com-

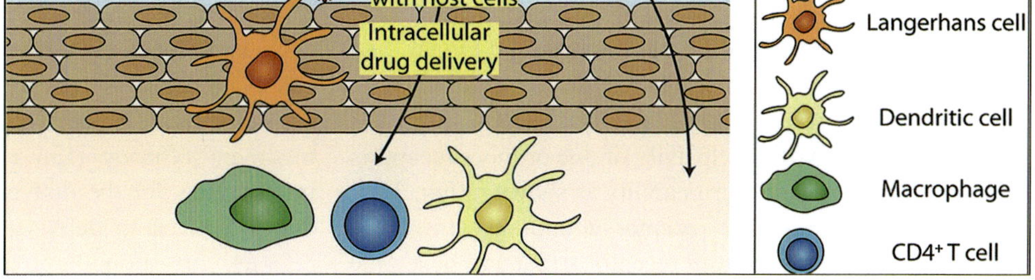

Fig. 36.4 Microbicidal nanomedicine formulations (microbicidal nanoparticles). (Image license taken from Lisziewicz and Tőke (2013) under the conditions of "Reuse in a book/textbook")

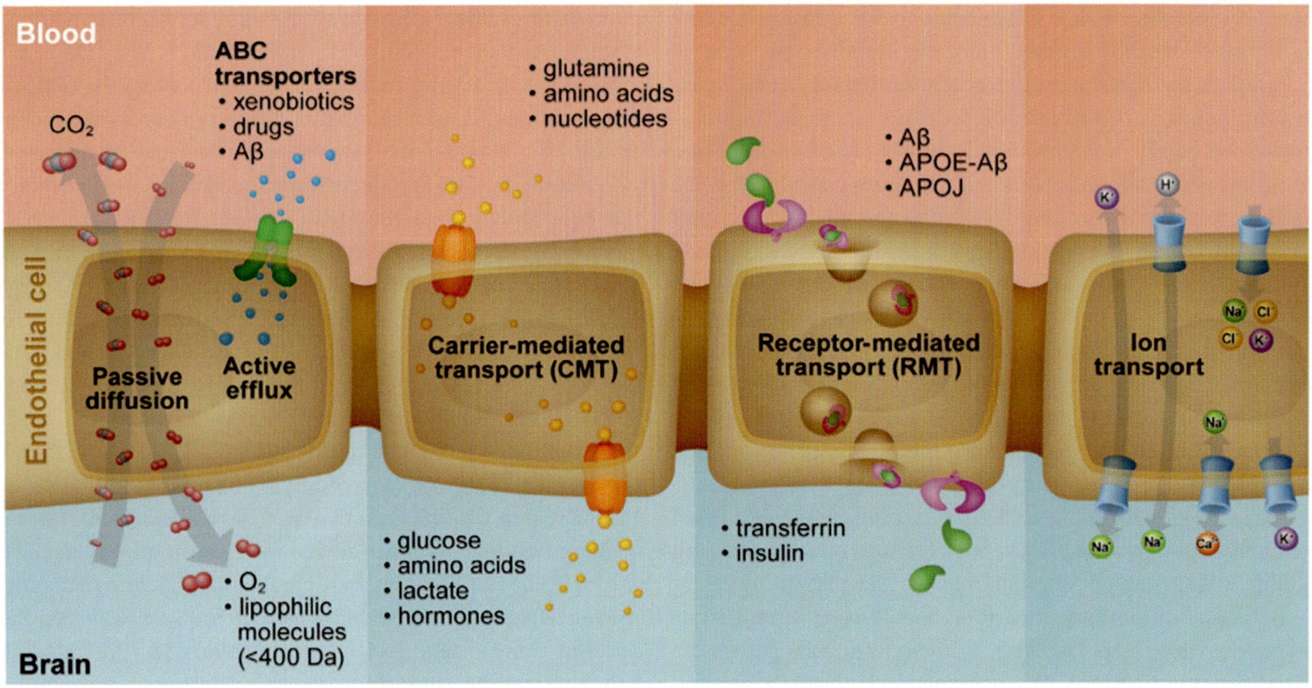

Fig. 36.3 Transport mechanism across the blood–brain barrier. O_2, CO_2, and lipophilic molecules can cross the BBB by diffusion, while active efflux by ATP-binding cassette (ABC) transporters restricts the entry of drugs and xenobiotics. Protein carriers on the luminal endothelium facilitate the uptake of specific substrates such as glucose, certain amino acids, and monocarboxylates into the CNS, while those on the abluminal side pump out glutamine and nucleotides (carrier-mediated transport, CMT). Receptor-mediated transcytosis (RMT) transfers a variety of macromolecules, such as insulin and transferrin, across the BBB by binding to a receptor, followed by endocytosis of the receptor-macromolecule complex within a vesicle, dissociation of the complex and exocytosis of the ligand. In addition, there are ion channels, which are responsible for the exchange of ions across the BBB to maintain cerebral homeostasis. (Image reused with License obtained from Elsevier publishers (Pandit et al., 2020))

pete with Tf peptide mimetic integrated nanocarriers for receptor binding. This effect compromises the pathway's efficiency.

The necessity of distinct receptors in CMT and RMT pathways aimed at BBB permeability limits their efficacy (Jiang et al., 2018). It is a significant transport system that is mediated through electrostatic interactions amidst positive substrates and negatively charged plasma membranes. Adsorptive-mediated transcytosis (AMT) transportation is not specific, and the binding affinity is also significantly decreased but the capacity of transportation is higher due to its nonspecific and non-receptor-dependent attributes. Several peptides and proteins can be carried across the BBB by AMT inclusive of the cationic proteins (Oller-Salvia et al., 2016; Tang et al., 2019). Cation proteins are usually prepared by reacting the carboxylic groups with cationic amine reagents. The simultaneous increase in kidney and liver drug absorption is witnessed in correlation with enhanced brain uptake (Bickel et al., 2001).

CMT, RMT, or AMT exclusivity of one or more receptors is required to attain BBB permeability as shown in Fig. 36.3. Specifically, the necessary receptor community for these pathways is not a part of brain receptor strata and hence the peptides or antibodies face unstable or immunogenic outcomes. Cell-mediated transcytosis has recently gained, Neural stem cells, Mesenchymal Stem cells, Macrophages, and Exosomes (Cheng et al., 2013; Gutova et al., 2013). Mesenchymal stem cell populations possess the ability to auto-renew and precisely locate the tumor cells. Multifarious intracellular cargo like HSV-tk (Herpes simplex virus thymidine kinase), CRAd (replication-competent oncolytic adenovirus), IL-2, IL-18, TNF-α, and TNF-β are delivered through MSC, they are furthermore applicable in chemotherapy, oncolytic and immunotherapy (Kosztowski et al., 2009).

36.5 Theranostic Nanomedicines

The advent of nanomedicine has laid out a platform to strategize and greatly enhance the detection and diagnosis methods. Nanoparticles constituting demarcate optical, magnetic, and electrical properties can substantially enhance the treatment efficacy (Liu et al., 2021). Nanotechnological implications for the delivery of ARVs have great prospects due to their drug delivery potential in virus-bearing tissue regions, corresponding to enhanced half-lives. The lymphoid concentration of the "latent viral reservoir" requires drug targeting to the lymphoid tissues, testes, gut, and the CNS. The nanoemedicine based HIV approaches can target and activate the $CD4^{+}$ T cells and alter the inactive dynamics of the latent reservoir to a pathologically active state, one such nanoparticle was bryostatin-2 (LNP-Bry), which was well studied in the mouse model (Kovochich et al., 2011). Nanoparticle constructs targeting the lymphoid organs through a *pH*-mediated approach were studied in macaques, the study was based on the hypothesis of drug entrapment in the lymph nodes based on their 50–80 nm size attributes. Limited bioavailability forms a major barrier in drug development, nanomedicine formulations offer flexibility in functionalization wherein, physiological aspects of the human body can be manipulated to facilitate drug delivery (Chiappetta et al., 2010). Long-acting nanomedicine formulations can help to overcome ART nonadherence based incompatibilities by long dosing intervals. These formulations can be designed to give a sustained and dose-proportional release over a longer period (Baert et al., 2009).

36.5.1 Preventing HIV Transmission

Prevention of HIV transmission can be achieved by topical vaginal and rectal microbicides. The initial clinical evidence for this has been obtained recently with a TAF-containing vaginal gel. The study highlighted the importance of maintaining sustained concentrations of ARV at the genital mucosal site, making it a critical strategy for intervention (Abdool Karim et al., 2010). Such clinical data strengthens the possibility of developing a nanoformulation to enhance the mucosal permeation and ARV half-life. Mucoadhesive materials were scrutinized to attain greater epithelial permeation (das Neves et al., 2011; Lai et al., 2009). The formulation characteristics can alter the drug-epithelium interaction patterns by either docking them or creating a diffusion barrier. Attraction-repulsion is determined by the surface chemistry attributes, ionic interactions between negatively charged mucin chains and positively charged polymers, such as chitosan, could promote mucoadhesion (Fig. 36.4).

The nanomedicine became more mucoadhesive after being modified with thiol compared to chitosan without thiol modification (Grabovac & Bernkop-Schnürch, 2007). In the presence of these multifarious therapeutic targets, the use of mucoadhesive microbicidal nanoparticles is still concerning. They tend to stimulate the mucosal translocation (Lai et al., 2009) and can potentially disrupt the mucosal barrier, furthering the entrance of foreign particles, pathogens, and toxic nanomaterials (Wang et al., 2011). These mucosal nanomedicines have an intrinsic activity to inhibit viral entry, as evident from a macaque model, either drug-loaded or empty liposomes (2-RANTES) partially prevented infection after vaginal administration (Caron et al., 2010; Kish-Catalone et al., 2007). Relative to these mucoadhesive formulations, liposomes are lipid vesicles that contain an aqueous core and are used to encapsulate hydrophilic drugs. Liposomal phospholipid bilayers are used to dissolve hydrophobic and amphiphilic drugs. Its fate includes either macrophage mediated phagocytosis or cell membrane permeation via membrane fusion. These circulating liposomes are rapidly cleared by the reticuloendothelial system (RES) (Pornpattananangkul et al., 2010).

Significantly, inhibition of HIV entry into the host cell was evident in both silver nanoparticles coated with polyvinyl pyrrolidone and gold nanoparticles coated with mannose. Topical application of nanomedicine containing L-lysine dendrimers formulated as carbomer gel (VivaGel®) showed dose-dependent resistance to viral challenge in macaques (Mamo et al., 2010). Contrasting to liposomes, whose mechanism of action is based on its lipid-aqueous attributes, dendrimer-based delivery corresponds to its surface chemistry: naphthalene disulfonate groups are attached to the polylysine branches of the dendrimer, resulting in a direct interaction with HIV envelope glycoproteins. The vivagel's safety profile has been exhibited in human subjects (Phase I) (Rupp et al., 2007). However, there is a need for a vaccine that produces a strong humoral and cellular immune response against HIV. In an attempt to formulate such a vaccine, intramuscular injection of gag and env pDNA adsorbents to the surface of cationic poly(lactide-coglycolide) microparticles has been shown to induce immune responses in mice, guinea pigs, and rhesus macaques substantially more effectively than naked pDNA vaccines (O'Hagan et al., 2001). The observed tolerance of PLG microparticles was higher in the HIV-negative population and env-targeted CD4+ T-cell responses were observed after protein boosting (Lisziewicz & Tőke, 2013). Potent neutralizing antibody response was detected countering the homologous HIV, however, neutralization breadth was limited against heterologous HIV (Spearman et al., 2011).

36.5.2 Immunotherapeutics and Nanomedicines

Immunotherapeutic nanomedicines are novel, intricate, and multitargeted vaccine systems that offer enhanced therapeutics relative to all other approaches. The physical attributes consist of a size specification of around 50 nm, which is the optimum range for immune recognition (Xiang et al., 2006). The size becomes a basis for immunogenicity, smaller than 50 nm particles are often derived from immune checkpoints. The immunotherapeutic nanomedicine correlates with virus size ranges. The human body natively has a demarcated immune response against viral particles, their manifestation derives from immune activation, the similar paradigm is followed in these nanomedicine systems that are perceived as viruses by the body (Lisziewicz & Szebeni, 2010). Apex immune response is observed in human papillomavirus (HPV) vaccines, Cervarix, and Gardasil. These vaccine compositions include the surface functionalization of viral surface protein (L1) that assembles itself in a viral-like particle (VLP). The phenotypic similarities of these vaccines proportional to wild-type vaccines, trigger a robust immune response exclusive of the adjuvants. Contrastingly, discrete purified L1 bacterial protein has soluble characteristics and cannot induce any immune response (Christensen et al., 1994). The only barrier to treatment via these VLPs is the absence of specific T-cell responses required in diseases like HPV-associated cancers (Hildesheim et al., 2007).

Also, the concept of assembling "Particulate Vaccines" has recently been employed in HIV to enhance the immune recognition of soluble antigens. This technique implies the enlargement of the molecular size of the antigen relative to the pathogen. This led to the formation of natural "particulate vaccines" that lie in the size range of 40 nm capable of inducing simultaneous humoral and cell-mediated immune responses (Martin et al., 1993; Peek et al., 2008). HIV viral-like particles developmentally are noninfective viruses competent enough to self-assemble viral envelope proteins exclusive of the viral genetic material. One distinctive approach is to increase adjuvant-assisted antigen size. In the United States, aluminum salts were approved as appropriate adjuvants to increase the antigen size, however, concerns regarding the safety profile are raised keeping in account of their particulate nature. Alternately, plasmid DNA incorporation to present one or more antigens in the body is discussed attributing to its excellent safety features, antigens expressed intracellularly are presented upon host MHC molecules in addition to their significant manufacturing scale-up potential neutralizing the economic aggravation. Unfortunately, despite all of these novel attributes, the robust immune triggering pDNA hypothesis didn't justify well in human subjects when inculcated intramuscularly or intradermally. Similar to L1 naked nature, naked pDNA is restricted to the extracellular space and is stealthed from the immune mechanisms. Ultimately, polymeric nanoparticles (bio-, and nonbiodegradable) that can be utilized to offer synthetic attributes to HIV-antigens corresponding to enhanced immunogenicity and resistance to physiological degradation were considered (Fairman et al., 2009; Lamalle-Bernard et al., 2006). It was construed that it is critical to target the dendritic cells to elicit an immune response that can be triggered by multifold nanomedicine size ranges, more than 100 nm would target the immature peripheral dendritic cells while the smaller domains 20–50 nm would drain in the lymphoid system. Surface functionalization of the nanoparticle with dendritic receptor ligands can enhance permeation via increased recognition and uptake by receptor-mediated endocytosis (Diebold et al., 1999).

Consequently, achieving cell targeting enhances the efficiency greatly, yet issues like nontargeted cell adhesion and significant resistance faced to permeate nuclear membranes are still to be addressed in these synthetic systems. The synthetic formulation development is based on distinct objectives, based on increasing the size, preventing degradation, and dendritic targeting. Achieving the former two objectives has resulted in enhanced desired immunogenicity. Ligand incorporation upon the nanomedicine surface is achieved by sophisticated techniques like gene guns, electroporation, or autologous subjecting of the formulation to dendritic cells. The specific "antigen-presenting compartment" is also crucial while targeting the dendritic cell population, formulations like DermaVir sought to differ from other formulations as they possess multiple targeting moieties to ensure precise compartmental targeting (Nátz & Lisziewicz, 2012).

36.5.2.1 CAR-T Cell Therapy

Research into chimeric antigen receptor T cells has grown exponentially over the past two decades following the tremendous clinical success these cells have had in patients with leukemia and lymphoma. CAR-T cell therapy was approved by the FDA in 2017, since then multiple attributes of genetically modified cells have been extensively examined including the clinical therapeutic potential of natural killer (NK) T cells in the treatment of lethal cancers. Furthermore, these engineered cells may exhibit superior safety profiles or offer off-the-shelf applications (e.g., gamma-delta (G-D) T cells expressing CARs) (Krause et al., 1998; Vandenberghe et al., 1992). Additionally, previous results and literature exhibit depression in alloreactivity and decreased cytokine release upon the implication of lymphocytes in contrast to CAR-T cells. Due to their ability to localize and remain in the tumor microenvironment (TME), CAR-macrophage cells can circumvent the limitations in targeting solid tumors previously faced by CAR-T cells. Presently, bioreactor technologies are revolutionizing the automatic production of CAR-T cells. These systems have enabled the possibility of decentralized and closed-system manufacturing of CAR-T cells.

Although the dramatic successes of CAR-T cell therapy in oncology have spurred enormous recent interest and investment, the first CAR-T treatments date back decades and were focused on targeting infectious diseases; specifically, HIV. These first-gen CAR constructs consisted of an extracellular (CD4 domain) in conjunction with an intracellular signaling domain (CD3). This CD4ζCAR was able to redirect T cells to HIV infected $CD4^+$ cells in vitro and in vivo, resulting in numerous clinical trials beginning in the late 90 s (Mitsuyasu et al., 2000). Such initial studies were able to rationalize and validate the safety aspects of the study but failed to suppress the infection (Scholler et al., 2012). Since that time, there have been many generations of CAR design, such as the inclusion of a costimulatory domain like CD28 or 4-1BB, to improve function and persistence (Tokarew et al., 2019). Recently there have been significant advances in the treatment of HIV with CAR-T cells, including various antigen antigen-targeting (HIV envelope protein moieties, broadly neutralizing antibodies, bispecific targeting, and others) and methods to overcome challenges such as virus escape and host antigen down-regulation (Kuhlmann et al., 2018).

Moreover, recent developments in CAR-T cell therapy exert their therapeutic potential beyond onco-therapeutics in autoimmune diseases. These diseases are a consequence of the immune cells attacking themselves unexpectedly. In certain conditions, autoantibody production is witnessed in response to a nonpathogenic protein by the B cells, out of multiple clinical scenarios only a few of the autoimmune conditions present autoantibody-induced inflammation as rendered in Fig. 36.5, majorly antigenic targets of these autoantibodies remain veiled. Apart from autoantibodies, other cases of autoimmune diseases are demonstrated because of abnormal T-cell response, that might be passively dependent on B cells (antigenic presentation), or it may be due to other immune factors (Aghajanian et al., 2022). A common treatment strategy is immunosuppression, but this is not curative and can cause side effects and safety concerns.

36.5.2.2 Immunomodulation

Significant research advances in HIV therapeutics have made one thing clear, the presence of a latent HIV reservoir that forms the basis of viral rebound in untreated and medication-incompliant populations. The cure strategies are limited and are pinned down to two of the foremost strategies one being the gene editing elimination of the latent proviral DNA and the other being the "shock and kill" strategy. The basis of this technique lies in the reactivation of latent provirus using latency reversal agents (LRAs) and their sequential elimination via reactivated immune responses or virus-induced cytopathic reaction termed as the latter "Kill" phase of this strategy. Immunomodulatory agents reported to induce latency reversal include CD8+ depleting antibodies, immune checkpoint blockers, and toll-like receptor agonists. Their alterations can target the HIV transcription associatively. However, precise molecular mechanisms are still partially perceived. This incomplete perception of the molecular mechanisms along with the heterogeneous and complex nature of the latent HIV proviruses became a prime pitfall to tread ahead towards a complete cure. Unraveling those molecular mechanisms is prime to design methodologies aiming at eliminating latent HIV reservoirs (Hui et al., 2017).

Moreover, immune checkpoint molecules like PD-1 and CTLA-4 and their mechanistic insights exhibit T-cell receptor (TCR) pathway block via CD28 inactivation. TCR activation is linked to the induction of several pathways like ZAP70M PI3K and calcineurin that can downstream activate the nuclear transcription factors including NF-κB, AP-1, and NFAT known to regulate HIV transcription. Nicholas Chomont's research group has found that when TCR-stimulated PD-1^+ T cells are treated with PD-1 present a significant reduction in active positive transcription elongation factor (P-TEFb), a master regulator of HIV-1 transcription.

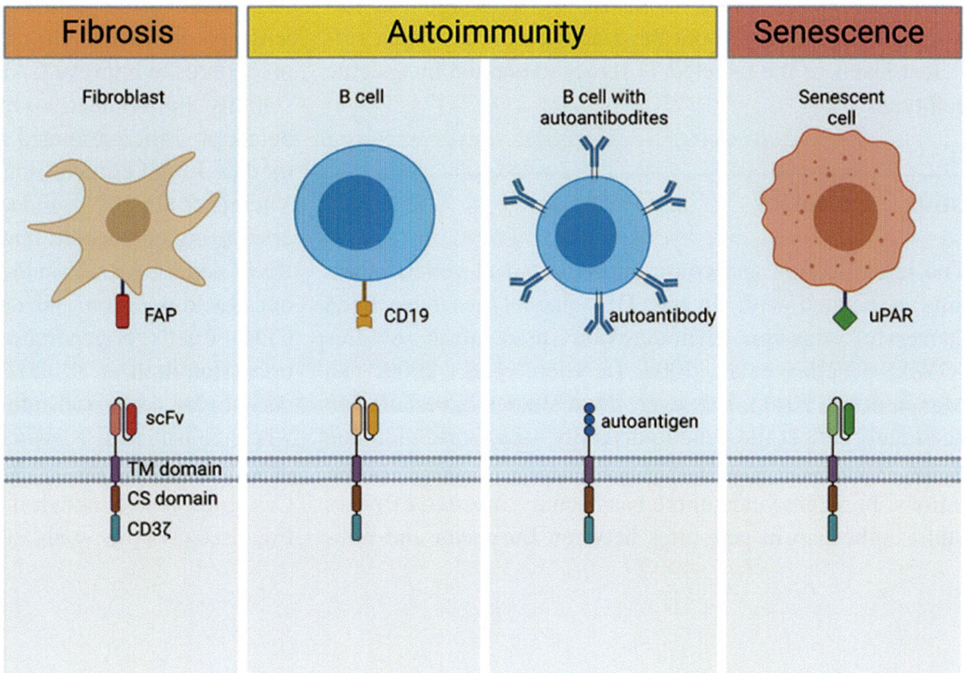

Fig. 36.5 Chimeric antigen receptor (CAR) and chimeric autoantigen receptor (CAAR) T-cell therapies to treat diseases beyond cancer. Second-generation CARs consist of a single-chain variable fragment (scFv) fused to a costimulatory domain (CS, e.g. CD28 or 4-1BB) and a CD3ζ signaling domain, while CAARs substitute a recombinant autoantigen in place of the scFv. TM—transmembrane. (Image license obtained from Springer Nature (Aghajanian et al., 2022))

Hence, they rationalized that viral reactivation through PD-1 engagement is due to P-TFEb. This unveils some mechanistic functionality of these IC molecules (Fromentin et al., 2019; Hui et al., 2017; Patsoukis et al., 2020).

Additionally, the TLR family of proteins, which belongs to the pattern recognition receptor (PRR) family, is involved in activating early innate immune responses by recognizing pathogen-associated molecular patterns (PAMP) and boosting immunity through downstream pathways. Various immune system cells express TLRs including NK cells, macrophages, B cells, dendritic cells, and T cells. Human cells have so far been discovered to possess ten functional TLRs, which can be divided into two groups based on the location of their receptors within the cell. Research is ongoing to discover small molecules that are capable of "mimicking" a TLR response for the treatment of bacterial and viral infections, inflammation, and cancer immunotherapy, and to improve vaccine efficacy. Numerous studies have shown that TLR agonists can reactivate latent HIV-1 infection as well as boost anti-HIV immunity. Nevertheless, HIV reactivation is significantly attributed to the TLR's cell type-specificity. It was shown that TLR1/2 agonists reactivated HIV-1 from latency in central memory CD4+ T-cell models of latency, while agonists of TLR5, TLR7, TLR8 and TLR9 (which are not expressed on T cells) did not (Novis et al., 2013). In vitro findings on myeloid and monocytic cells expressing TLR8 upon stimulation with TLR agonist can progress to viral reactivation in U1 and OM-10 cell lines (Schlaepfer & Speck, 2011). Studies on plasmacytoid dendritic cells (pDCs) expressing TLRN upon stimulation triggered the HIV reactivation from the latently infected T cells in both ex vivo cultures and latently infected T cells (Offersen et al., 2016; Tsai et al., 2017). Conclusively, TLR agonists can reactivate latent HIV-1, but these activities require that sufficient levels of the targeted TLRs are expressed by specific cell types.

36.6 Genetics

The identification and confirmation of novel genetic variants associated with diverse HIV phenotypes have been successful through genome-wide association studies (GWAS) (Fellay et al., 2007; Le Clerc et al., 2009; van Manen et al., 2011). However, these studies have failed to accurately reflect the genetic diversity seen worldwide due to the extensive research in predominantly European populations. In recent years, there has been a concerted effort to address the gap in genomics between European and non-European cohorts, however, the discrepancy is still pronounced. Underrepresented populations have a greater risk for poorer health outcomes due to subpar prevention, diagnosis, and treatment options for their population. Through inclusivity, genomics has the opportunity to address these health disparities and improve health globally (Gingras et al., 2020).

Moreover, the effect of HIV infection on DNA methylation has been characterized in HIV-positive individuals (Rickabaugh et al., 2015). Several studies have shown that viral infections can induce aberrant methylation patterns within the host genome (Cicchini et al., 2017; Matsusaka et al., 2014). Furthermore, methylation of HIV provirus by the host's methylation machinery can control HIV-1 transcription, replication, and persistence (Chávez et al., 2011; Trejbalová et al., 2016). For instance, DNA methylation is a potentially effective prognostic biomarker for predicting risk and type of HIV-associated lymphomas and HIV-associated cognitive impairment; however, these results are yet to be translated to a clinical setting (Corley et al., 2016; Matsunaga et al., 2014). There are still a lot of gaps in knowledge regarding the relationship between methylation and HIV. Hence, understanding the potential role of DNA methylation will provide a novel gene therapeutic approach for HIV treatment (Arumugam et al., 2021).

Furthermore, CAR-T cells engineered with gammaretroviral or lentiviral vectors (RVs/LVs) targeting B-cell lymphomas and leukemia have shown excellent clinical efficacy and no malignant transformation due to insertional mutagenesis (Larson & Maus, 2021; Leick et al., 2021). The materials used for virus-free CRISPR-Cas editing are and will be easier to produce, store, and distribute for clinical use than large-scale manufacturing of RVs/LVs. The current challenge is to further optimize and standardize the gene editing procedures to improve CAR-T cell yields and manufacturing stability. Subsequently, virus-free knock-in methods should be adopted for automated manufacturing systems to accommodate future clinical scaling (Abou-El-Enein et al., 2021; Vucinic et al., 2021). Indeed, a one-step strategy was developed using engineered lentiviral particles to introduce Cas9 RNPs and a CAR transgene into primary human T cells without electroporation and demonstrated specific targeting of CD4+ T cells for genome editing within a mixed cell population (Hamilton et al., 2021). Moreover, several challenges need to be addressed until their broad clinical application when considering a switch towards more innovative gene delivery approaches, i.e. from RV/LV systems to CRISPR-Cas gene-edited nonviral CAR-T cells as represented in Fig. 36.6 (Wagner et al., 2022).

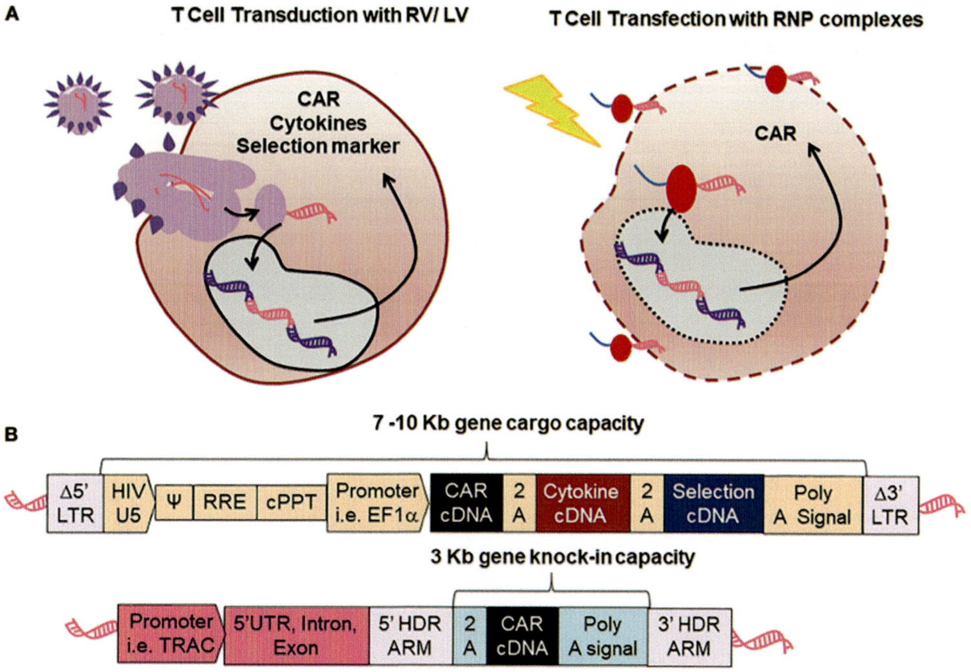

Fig. 36.6 Comparison between retroviral vector and lentiviral vector (RV/LV) gene delivery systems with CRISPR-Cas gene editing for production of CAR-T cells. LTR: Long terminal repeats; HIV: Human immune deficient virus U5: Untranslated region in the 5′ side; Y: encapsidation signal; RRE, Rev. responsive element; cPPT, polypurine tract; EF1a, Elongation factor 1 a. TRAC, Locus of T-cell receptor alpha chain; HDR, Homology-directed recombination. (Image reused under Common CC BY license from Wagner et al. (2022))

36.7 Contemporary Theranostics: Future HIV Prospects

36.7.1 Precision Medicine and Nuclear Theranostics

The application of theranostics in the medicinal field contributes to early diagnosis besides providing a holistic transition from a "trial and error" concept to "precise, targeted, and personalized" concept therapeutics. The application of nanoparticles such as quantum dots, radioisotopes, liposomes, and plasmonic nanobubbles independently or in combination, as imaging agents and therapeutic agents, enables instantaneous diagnosis and therapy. For instance, the use of nanoparticles as theranostic agents has shown promising results for the betterment of cancer diagnosis, treatments, and management (Jeyamogan et al., 2021; Jha et al., 2020). Additionally, theranostics will become an ever-increasing part of clinical nuclear medicine and will lead to the adaptation of personalized molecular targeted radionuclide therapy as a mainstream oncology practice throughout the world (Okamoto et al., 2021). For instance, radiolabelled iodine is accumulated in thyroid cancer (Luster et al., 2008; Mazzaferri & Jhiang, 1994), radiolabelled MIBG goes into cancer cells in malignant pheochromocytoma and paraganglioma (Kinuya et al., 2015; Neumann et al., 2019), and radiolabelled somatostatin receptor is trapped in neuroendocrine tumors (Strosberg et al., 2017). In the recent review, various theranostic radiopharmaceuticals and their clinical use against a variety of cancers were summarized, and theranostics for cancer imaging and treatment are rapidly evolving. Various emerging molecular targets and radiopharmaceuticals with different forms of radiation emission, such as α-emitting therapies, can emerge as next-generation theranostic. Furthermore, efforts to better estimate the dosimetry of therapeutic radiopharmaceuticals, their quantification, and treatment response monitoring are necessary steps in the direction of individualized precision medicine for theranostics (Burkett et al., 2023). This strategy may provide precise targeting for localized delivery via nanotechnology as well as improve personalized medicine in the diagnosis and treatment of viral diseases.

Moreover, theranostics can be a highly important tool in the development of precision, highly efficient, and safe therapy for personalized treatment using molecular information from patients. The pretargeting approach as portrayed in Fig. 36.7 can leverage the inherent strengths of the theranostic strategy; however, it will require orthogonal conjugation in physiological conditions. This strategy significantly reduces the circulatory time and off-target toxicity of the drug-carrier component and provides objective criteria for optimization of the treatment protocol based on the results of noninvasive imaging over single-component theranostics (Hapuarachchige & Artemov, 2020). Additionally, the nuclear theranostic approach has sparked increasing interest and gained importance in parallel to the growth in molecular imaging and personalized medicine (Gomes Marin et al., 2020). Personalized medicine, unlike traditional "one-size-fits-all" approaches, involves innovative drug therapy designs tailored to individual patient characteristics and responses.

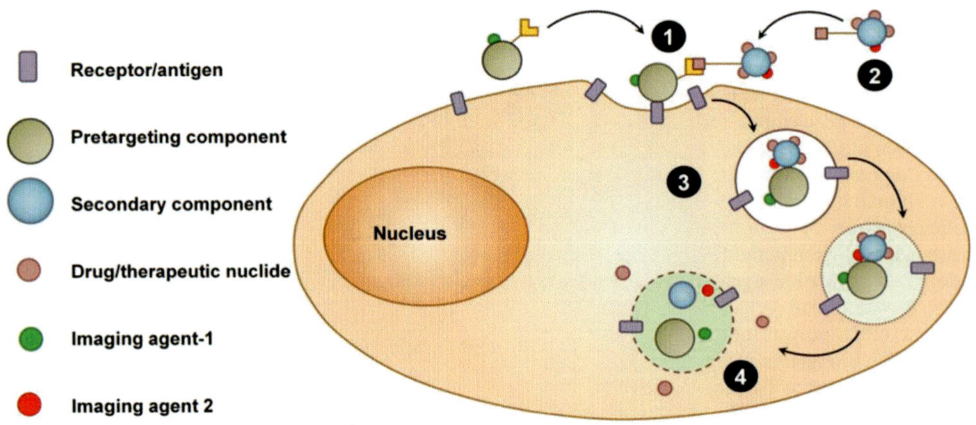

Fig. 36.7 Schematic view of the concept of pretargeting theranostic strategy. (1) Image-guided pretargeting. (2) Therapeutic/radionuclide delivery step. (3) Internalization of the complex of pretargeting-receptor-delivery components. (4) Intracellular release of therapeutics or radionuclides. (Image reused under Common CC BY license from Hapuarachchige and Artemov (2020))

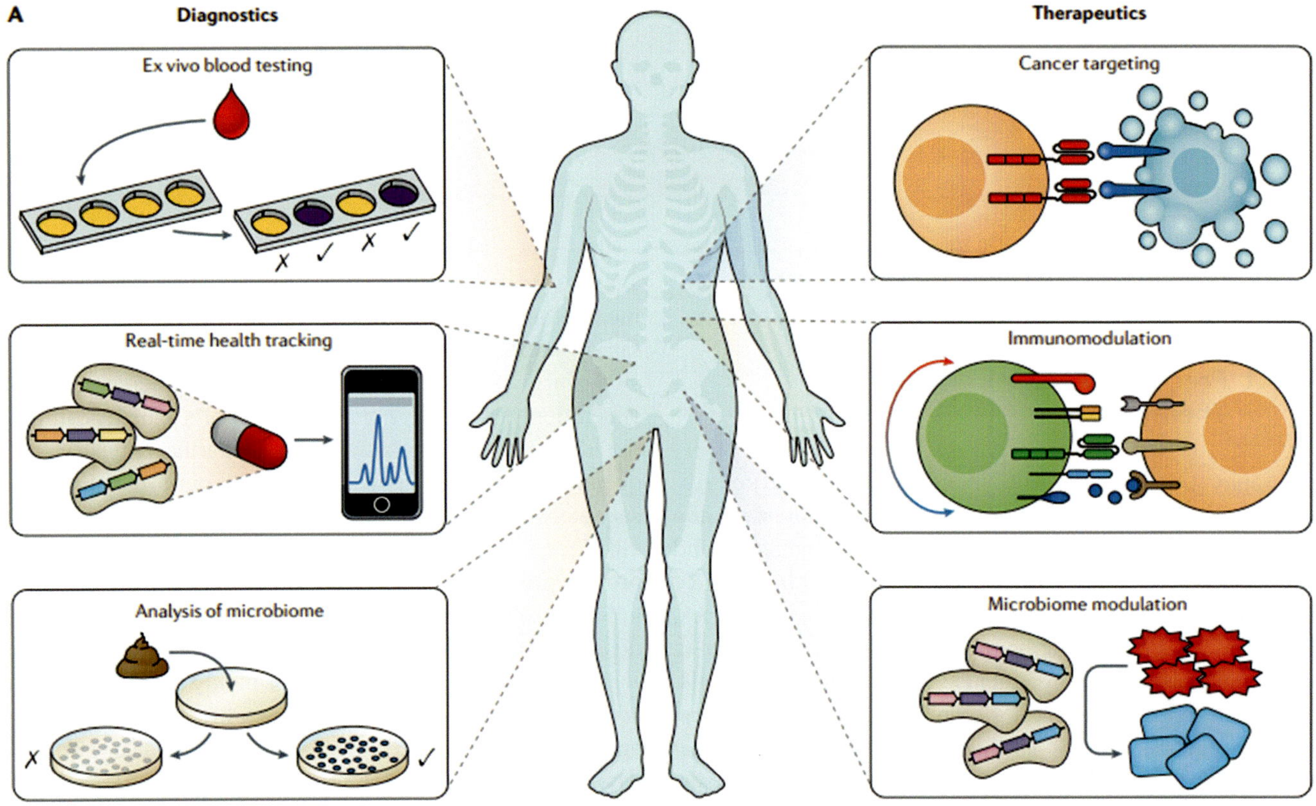

Fig. 36.8 Clinical (diagnostics and therapeutics) applications of synthetic biology. (Image license obtained from Springer Nature (McNerney et al., 2021))

36.7.2 Regenerative Medicine

The recently developed cellular diagnostics and therapeutics (theranostic cells) have shown that synthetic biology has real potential to transform healthcare paradigms as highlighted in Fig. 36.8. The expanding cellular engineering toolbox will lead to more sensitive diagnostics and novel therapeutics for previously intractable diseases (McNerney et al., 2021). Despite being a relatively young field, regenerative theranostics boasts a significant amount of work showing the promising potential of these systems in various applications, including bone defects, liver disease, cardiac dysfunction, and pancreatic islet regeneration. In the long term, regenerative theranostics biomaterials for tissue engineering hold immense potential as personalized medicine platforms, enabling patient-specific treatment plans based on diagnostic output and patient outcomes (Cuylear et al., 2021). Moreover, theranostic aims to promote safer, more effective pharmacotherapies, requiring multidisciplinary efforts to understand host responses and formulate individualized therapies (Hu, 2014). Hence, further exploring this novel field may provide a promising strategy for the diagnosis and treatment of infectious diseases such as HIV.

36.8 Summary

This chapter presents a relative description of theranostic approaches to HIV manifestation and its therapeutics. The segment elaborates upon the present constraints faced in diagnosis and therapy while discussing the prospective therapeutic improvements with the implementation of theranostic approaches. The uncertainty of treatment efficiency can cause severe outcomes including drug overdose, immunogenicity, and idiosyncratic effects, hence personalized medicine that can precisely track the physiological location as well as the potency of treatment holds great prospect. The latter portions of this segment expand upon the present understanding of immune modulation in contexts of theranostic facilitations. The extrapolation of immune activation and its manipulation to treat infectious diseases is prominently highlighted. Eventually, readers will also get to understand that though this technique has prominent clinical advantages it also holds some challenges in terms of off-target effects, therapeutic failure, and costs.

36.9 Review Questions

1. Which of the techniques can be implied for blood–brain barrier permeability?
   (a) Carrier-mediated transcytosis (CMT)
   (b) Receptor-mediated transcytosis (RMT)
   (c) Adsorptive-mediated transcytosis (AMT)
   (d) Cell-mediated transport
   (e) *All of the above*
2. Which of the following does not stand true about CAR-T cell therapy?
   (a) Engineered cells may exhibit superior safety profiles or offer off-the-shelf applications
   (b) Ability to localize and remain in the tumor microenvironment (TME)
   (c) CAR-T Cell therapy was approved by the US Food and Drug Administration in 2017
   (d) *It is being developed to load therapeutic drugs into DNA boxes with oligonucleotide locks*
3. What is the currently accepted treatment for HIV?
   (a) Theranostic nanoparticles
   (b) CAR-T cell Therapy [CAR-T]
   (c) DNA Nanotechnology
   (d) *Combinatorial Antiretroviral Therapy [cART]*
4. CAR-T cell & CRISPR-CAS9 gene editing technologies are completely efficient and have no further challenges to resolve.
   (a) True
   (b) *False*
5. HIV rebound upon ART interruption occurs majorly because of which of the following?
   (a) Latent proviral HIV reservoirs
   (b) ART-induced HIV mutations
   (c) HIV resistance to ART
   (d) *All of the Above*

References

Abdool Karim, Q., Abdool Karim, S. S., Frohlich, J. A., Grobler, A. C., Baxter, C., Mansoor, L. E., Kharsany, A. B., Sibeko, S., Mlisana, K. P., & Omar, Z. (2010). Effectiveness and safety of tenofovir gel, an antiretroviral microbicide, for the prevention of HIV infection in women. *Science, 329*(5996), 1168–1174.

Abou-El-Enein, M., Elsallab, M., Feldman, S. A., Fesnak, A. D., Heslop, H. E., Marks, P., Till, B. G., Bauer, G., & Savoldo, B. (2021). Scalable manufacturing of CAR T cells for cancer immunotherapy. *Blood Cancer Discovery, 2*(5), 408–422. https://doi.org/10.1158/2643-3230.bcd-21-0084

Aggarwal, S. K., Gogu, S. R., Rangan, S., & Agrawal, K. C. (1990). Synthesis and biological evaluation of prodrugs of zidovudine. *Journal of Medicinal Chemistry, 33*(5), 1505–1510.

Aghajanian, H., Rurik, J. G., & Epstein, J. A. (2022). CAR-based therapies: Opportunities for immuno-medicine beyond cancer. *Nature Metabolism, 4*(2), 163–169.

Arumugam, T., Ramphal, U., Adimulam, T., Chinniah, R., & Ramsuran, V. (2021). Deciphering DNA methylation in HIV infection. *Frontiers in Immunology, 12*, 795121. https://doi.org/10.3389/fimmu.2021.795121

Ash, M. K., Al-Harthi, L., & Schneider, J. R. (2021). HIV in the brain: Identifying viral reservoirs and addressing the challenges of an HIV cure. *Vaccine, 9*(8), 867.

Baert, L., van't Klooster, G., Dries, W., François, M., Wouters, A., Basstanie, E., Iterbeke, K., Stappers, F., Stevens, P., & Schueller, L. (2009). Development of a long-acting injectable formulation with nanoparticles of rilpivirine (TMC278) for HIV treatment. *European Journal of Pharmaceutics and Biopharmaceutics, 72*(3), 502–508.

Banoub, M. G., Bade, A. N., Lin, Z., Cobb, D., Gautam, N., Dyavar Shetty, B. L., Wojtkiewicz, M., Alnouti, Y., McMillan, J., Gendelman, H. E., & Edagwa, B. (2020). Synthesis and characterization of long-acting darunavir prodrugs. *Molecular Pharmaceutics, 17*(1), 155–166. https://doi.org/10.1021/acs.molpharmaceut.9b00871

Bickel, U., Yoshikawa, T., & Pardridge, W. M. (2001). Delivery of peptides and proteins through the blood–brain barrier. *Advanced Drug Delivery Reviews, 46*(1–3), 247–279.

Bonfanti, P., Capetti, A., & Rizzardini, G. (1999). HIV disease treatment in the era of HAART. *Biomedicine & Pharmacotherapy, 53*(2), 93–105.

Bressan, S., Pierantoni, A., Sharifi, S., Facchini, S., Quagliarello, V., Berretta, M., & Montopoli, M. (2021). Hepatotoxicity chemotherapy-induced in HIV patients. *Cells, 10*(11), 1–16.

Burkett, B. J., Bartlett, D. J., McGarrah, P. W., Lewis, A. R., Johnson, D. R., Berberoğlu, K., Pandey, M. K., Packard, A. T., Halfdanarson, T. R., & Hruska, C. B. (2023). A review of theranostics: Perspectives on emerging approaches and clinical advancements. *Radiology: Imaging Cancer, 5*(4), e220157.

Caron, M., Besson, G., Etenna, S. L.-D., Mintsa-Ndong, A., Mourtas, S., Radaelli, A., Morghen, C. D. G., Loddo, R., La Colla, P., & Antimisiaris, S. G. (2010). Protective properties of non-nucleoside reverse transcriptase inhibitor (MC1220) incorporated into liposome against intravaginal challenge of Rhesus macaques with RT-SHIV. *Virology, 405*(1), 225–233.

Chaowanachan, T., Krogstad, E., Ball, C., & Woodrow, K. A. (2013). Drug synergy of tenofovir and nanoparticle-based antiretrovirals for HIV prophylaxis. *PLoS One, 8*(4), e61416. https://doi.org/10.1371/journal.pone.0061416

Chávez, L., Kauder, S., & Verdin, E. (2011). In vivo, in vitro, and in silico analysis of methylation of the HIV-1 provirus. *Methods, 53*(1), 47–53. https://doi.org/10.1016/j.ymeth.2010.05.009

Chen, Y., & Liu, L. (2012). Modern methods for delivery of drugs across the blood–brain barrier. *Advanced Drug Delivery Reviews, 64*(7), 640–665.

Cheng, Y., Morshed, R., Cheng, S. H., Tobias, A., Auffinger, B., Wainwright, D. A., Zhang, L., Yunis, C., Han, Y., & Chen, C. T. (2013). Nanoparticle-programmed self-destructive neural stem cells for glioblastoma targeting and therapy. *Small, 9*(24), 4123–4129.

Chiappetta, D. A., Hocht, C., Taira, C., & Sosnik, A. (2010). Efavirenz-loaded polymeric micelles for pediatric anti-HIV pharmacotherapy with significantly higher oral bioavailability. *Nanomedicine, 5*(1), 11–23.

Chiu, I.-M., Yaniv, A., Dahlberg, J. E., Gazit, A., Skuntz, S. F., Tronick, S. R., & Aaronson, S. A. (1985). Nucleotide sequence evidence for relationship of AIDS retrovirus to lentiviruses. *Nature, 317*(6035), 366–368.

Choudhury, H., Pandey, M., Chin, P. X., Phang, Y. L., Cheah, J. Y., Ooi, S. C., Mak, K.-K., Pichika, M. R., Kesharwani, P., & Hussain, Z. (2018). Transferrin receptors-targeting nanocarriers for efficient targeted delivery and transcytosis of drugs into the brain tumors: A review of recent advancements and emerging trends. *Drug Delivery and Translational Research, 8*, 1545–1563.

Christensen, N. D., Höpfl, R., DiAngelo, S. L., Cladel, N. M., Patrick, S. D., Welsh, P. A., Budgeon, L. R., Reed, C. A., & Kreider, J. W. (1994). Assembled baculovirus-expressed human papillomavirus type 11 L1 capsid protein virus-like particles are recognized by neutralizing monoclonal antibodies and induce high titres of neutralizing antibodies. *Journal of General Virology, 75*(9), 2271–2276.

Chun, T. W., Finzi, D., Margolick, J., Chadwick, K., Schwartz, D., & Siliciano, R. F. (1995). In vivo fate of HIV-1-infected T cells: Quantitative analysis of the transition to stable latency. *Nature Medicine, 1*(12), 1284–1290. https://doi.org/10.1038/nm1295-1284

Chun, T. W., Davey, R. T., Jr., Engel, D., Lane, H. C., & Fauci, A. S. (1999). Re-emergence of HIV after stopping therapy. *Nature, 401*(6756), 874–875. https://doi.org/10.1038/44755

Churchill, M. J., Gorry, P. R., Cowley, D., Lal, L., Sonza, S., Purcell, D. F., Thompson, K. A., Gabuzda, D., McArthur, J. C., & Pardo, C. A. (2006). Use of laser capture microdissection to detect integrated HIV-1 DNA in macrophages and astrocytes from autopsy brain tissues. *Journal of Neurovirology, 12*(2), 146–152.

Churchill, M. J., Wesselingh, S. L., Cowley, D., Pardo, C. A., McArthur, J. C., Brew, B. J., & Gorry, P. R. (2009). Extensive astrocyte infection is prominent in human immunodeficiency virus–associated dementia. *Annals of Neurology: Official Journal of the American Neurological Association and the Child Neurology Society, 66*(2), 253–258.

Cicchini, L., Blumhagen, R. Z., Westrich, J. A., Myers, M. E., Warren, C. J., Siska, C., Raben, D., Kechris, K. J., & Pyeon, D. (2017). High-risk human papillomavirus E7 alters host DNA methylome and represses HLA-E expression in human keratinocytes. *Scientific Reports, 7*(1), 3633. https://doi.org/10.1038/s41598-017-03295-7

Cobb, D. A., Smith, N., Deodhar, S., Bade, A. N., Gautam, N., Shetty, B. L. D., McMillan, J., Alnouti, Y., Cohen, S. M., & Gendelman, H. E. (2021). Transformation of tenofovir into stable ProTide nanocrystals with long-acting pharmacokinetic profiles. *Nature Communications, 12*(1), 5458.

Cohen, M. S., Shaw, G. M., McMichael, A. J., & Haynes, B. F. (2011). Acute HIV-1 infection. *The New England Journal of Medicine, 364*(20), 1943–1954. https://doi.org/10.1056/NEJMra1011874

Cohen, S. M., Hu, X., Sweeney, P., Johnson, A. S., & Hall, H. I. (2014). HIV viral suppression among persons with varying levels of engagement in HIV medical care, 19 US jurisdictions. *JAIDS Journal of Acquired Immune Deficiency Syndromes, 67*(5), 519–527.

Cohen, S. E., Sachdev, D., Lee, S. A., Scheer, S., Bacon, O., Chen, M.-J., Okochi, H., Anderson, P. L., Kearney, M. F., & Coffey, S. (2019). Acquisition of tenofovir-susceptible, emtricitabine-resistant HIV despite high adherence to daily pre-exposure prophylaxis: A case report. *The Lancet HIV, 6*(1), e43–e50.

Conroy, S. (2003). Extemporaneous (magistral) preparation of oral medicines for children in European hospitals. *Acta Paediatrica, 92*(4), 408–410.

Corley, M. J., Dye, C., D'Antoni, M. L., Byron, M. M., Yo, K. L., Lum-Jones, A., Nakamoto, B., Valcour, V., SahBandar, I., Shikuma, C. M., Ndhlovu, L. C., & Maunakea, A. K. (2016). Comparative DNA methylation profiling reveals an Immunoepigenetic signature of HIV-related cognitive impairment. *Scientific Reports, 6*, 33310. https://doi.org/10.1038/srep33310

Cuylear, D., Perez, C., Desai, T., Kharbikar, B., & Zhong, J. (2021). Theranostic biomaterials for tissue engineering. *Current Opinion in Biomedical Engineering, 19*, 100299.

Cysique, L. A., Waters, E. K., & Brew, B. J. (2011). Central nervous system antiretroviral efficacy in HIV infection: A qualitative and quantitative review and implications for future research. *BMC Neurology, 11*, 1–10.

das Neves, J., Amiji, M., & Sarmento, B. (2011). Mucoadhesive nanosystems for vaginal microbicide development: Friend or foe? *Wiley Interdisciplinary Reviews: Nanomedicine and Nanobiotechnology, 3*(4), 389–399.

Davey, R. T., Jr., Bhat, N., Yoder, C., Chun, T. W., Metcalf, J. A., Dewar, R., Natarajan, V., Lempicki, R. A., Adelsberger, J. W., Miller, K. D., Kovacs, J. A., Polis, M. A., Walker, R. E., Falloon, J., Masur, H., Gee, D., Baseler, M., Dimitrov, D. S., Fauci, A. S., & Lane, H. C. (1999). HIV-1 and T cell dynamics after interruption of highly active antiretroviral therapy (HAART) in patients with a history of sustained viral suppression. *Proceedings of the National Academy of Sciences of the United States of America, 96*(26), 15109–15114. https://doi.org/10.1073/pnas.96.26.15109

Deeks, S. G., & Walker, B. D. (2007). Human immunodeficiency virus controllers: Mechanisms of durable virus control in the absence of antiretroviral therapy. *Immunity, 27*(3), 406–416.

Desplats, P., Dumaop, W., Smith, D., Adame, A., Everall, I., Letendre, S., Ellis, R., Cherner, M., Grant, I., & Masliah, E. (2013). Molecular and pathologic insights from latent HIV-1 infection in the human brain. *Neurology, 80*(15), 1415–1423.

Diebold, S. S., Kursa, M., Wagner, E., Cotten, M., & Zenke, M. (1999). Mannose polyethylenimine conjugates for targeted DNA delivery into dendritic cells. *Journal of Biological Chemistry, 274*(27), 19087–19094.

Du, D., Chang, N., Sun, S., Li, M., Yu, H., Liu, M., Liu, X., Wang, G., Li, H., & Liu, X. (2014). The role of glucose transporters in the distribution of p-aminophenyl-α-d-mannopyranoside modified liposomes within mice brain. *Journal of Controlled Release, 182*, 99–110.

Edagwa, B., McMillan, J., Sillman, B., & Gendelman, H. E. (2017). Long-acting slow effective release antiretroviral therapy. *Expert Opinion on Drug Delivery, 14*(11), 1281–1291. https://doi.org/10.1080/17425247.2017.1288212

Eshaghi, B., Fofana, J., Nodder, S. B., Gummuluru, S., Reinhard, B., & r. M. (2022). Virus-mimicking polymer nanoparticles targeting CD169+ macrophages as long-acting nanocarriers for combination antiretrovirals. *ACS Applied Materials & Interfaces, 14*(2), 2488–2500.

Fairman, J., Moore, J., Lemieux, M., Van Rompay, K., Geng, Y., Warner, J., & Abel, K. (2009). Enhanced in vivo immunogenicity of SIV vaccine candidates with cationic liposome-DNA complexes in a rhesus macaque pilot study. *Human Vaccines, 5*(3), 141–150.

Fan, C.-H., Cheng, Y.-H., Ting, C.-Y., Ho, Y.-J., Hsu, P.-H., Liu, H.-L., & Yeh, C.-K. (2016). Ultrasound/magnetic targeting with SPIO-DOX-microbubble complex for image-guided drug delivery in brain tumors. *Theranostics, 6*(10), 1542.

Faria, N. R., Rambaut, A., Suchard, M. A., Baele, G., Bedford, T., Ward, M. J., Tatem, A. J., Sousa, J. D., Arinaminpathy, N., & Pépin, J. (2014). The early spread and epidemic ignition of HIV-1 in human populations. *Science, 346*(6205), 56–61.

Fellay, J., Shianna, K. V., Ge, D., Colombo, S., Ledergerber, B., Weale, M., Zhang, K., Gumbs, C., Castagna, A., Cossarizza, A., Cozzi-Lepri, A., De Luca, A., Easterbrook, P., Francioli, P., Mallal, S., Martinez-Picado, J., Miro, J. M., Obel, N., Smith, J. P., et al. (2007). A whole-genome association study of major determinants for host control of HIV-1. *Science, 317*(5840), 944–947. https://doi.org/10.1126/science.1143767

Finzi, D., Blankson, J., Siliciano, J. D., Margolick, J. B., Chadwick, K., Pierson, T., Smith, K., Lisziewicz, J., Lori, F., Flexner, C., Quinn, T. C., Chaisson, R. E., Rosenberg, E., Walker, B., Gange, S., Gallant, J., & Siliciano, R. F. (1999). Latent infection of CD4+ T cells provides a mechanism for lifelong persistence of HIV-1, even in patients on effective combination therapy. *Nature Medicine, 5*(5), 512–517. https://doi.org/10.1038/8394

Fotooh Abadi, L., Kumar, P., Paknikar, K., Gajbhiye, V., & Kulkarni, S. (2023). Tenofovir-tethered gold nanoparticles as a novel multifunctional long-acting anti-HIV therapy to overcome deficient drug delivery-: An in vivo proof of concept. *Journal of Nanobiotechnology, 21*(1), 1–24.

Fromentin, R., DaFonseca, S., Costiniuk, C. T., El-Far, M., Procopio, F. A., Hecht, F. M., Hoh, R., Deeks, S. G., Hazuda, D. J., & Lewin, S. R. (2019). PD-1 blockade potentiates HIV latency reversal ex vivo in CD4+ T cells from ART-suppressed individuals. *Nature Communications, 10*(1), 814.

Gao, Y., Kraft, J. C., Yu, D., & Ho, R. J. (2019). Recent developments of nanotherapeutics for targeted and long-acting, combination HIV chemotherapy. *European Journal of Pharmaceutics and Biopharmaceutics, 138*, 75–91.

Geeraert, L., Kraus, G., & Pomerantz, R. J. (2008). Hide-and-seek: The challenge of viral persistence in HIV-1 infection. *Annual Review of Medicine, 59*, 487–501.

Gingras, S. N., Tang, D., Tuff, J., & McLaren, P. J. (2020). Minding the gap in HIV host genetics: Opportunities and challenges. *Human Genetics, 139*, 865–875.

Gomes Marin, J. F., Nunes, R. F., Coutinho, A. M., Zaniboni, E. C., Costa, L. B., Barbosa, F. G., Queiroz, M. A., Cerri, G. G., & Buchpiguel, C. A. (2020). Theranostics in nuclear medicine: Emerging and re-emerging integrated imaging and therapies in the era of precision oncology. *Radiographics, 40*(6), 1715–1740.

Gottlieb, M. S., Schroff, R., Schanker, H. M., Weisman, J. D., Fan, P. T., Wolf, R. A., & Saxon, A. (1981). Pneumocystis carinii pneumonia and mucosal candidiasis in previously healthy homosexual men: Evidence of a new acquired cellular immunodeficiency. *New England Journal of Medicine, 305*(24), 1425–1431.

Grabovac, V., & Bernkop-Schnürch, A. (2007). Development and in vitro evaluation of surface modified poly (lactide-co-glycolide) nanoparticles with chitosan-4-thiobutylamidine. *Drug Development and Industrial Pharmacy, 33*(7), 767–774.

Gray, L. R., Turville, S. G., Hitchen, T. L., Cheng, W.-J., Ellett, A. M., Salimi, H., Roche, M. J., Wesselingh, S. L., Gorry, P. R., & Churchill, M. J. (2014). HIV-1 entry and trans-infection of astrocytes involves CD81 vesicles. *PLoS One, 9*(2), e90620.

Gutova, M., Frank, J. A., D'Apuzzo, M., Khankaldyyan, V., Gilchrist, M. M., Annala, A. J., Metz, M. Z., Abramyants, Y., Herrmann, K. A., & Ghoda, L. Y. (2013). Magnetic resonance imaging tracking of ferumoxytol-labeled human neural stem cells: Studies leading to clinical use. *Stem Cells Translational Medicine, 2*(10), 766–775.

Haase, A. T. (2005). Perils at mucosal front lines for HIV and SIV and their hosts. *Nature Reviews Immunology, 5*(10), 783–792.

Hamilton, J. R., Tsuchida, C. A., Nguyen, D. N., Shy, B. R., McGarrigle, E. R., Espinoza, C. R. S., Carr, D., Blaeschke, F., Marson, A., & Doudna, J. A. (2021). Targeted delivery of CRISPR-Cas9 and transgenes enables complex immune cell engineering. *Cell Reports, 35*(9).

Hapuarachchige, S., & Artemov, D. (2020). Theranostic pretargeting drug delivery and imaging platforms in cancer precision medicine. *Frontiers in Oncology, 10*, 1131.

Herskovitz, J., & Gendelman, H. E. (2019). HIV and the macrophage: From cell reservoirs to drug delivery to viral eradication. *Journal of Neuroimmune Pharmacology, 14*, 52–67.

Hildesheim, A., Herrero, R., Wacholder, S., Rodriguez, A. C., Solomon, D., Bratti, M. C., Schiller, J. T., Gonzalez, P., Dubin, G., & Porras, C. (2007). Effect of human papillomavirus 16/18 L1 viruslike particle vaccine among young women with preexisting infection: A randomized trial. *JAMA, 298*(7), 743–753.

HIV, U. G. (2023). *AIDS statistics–2022 fact sheet*. Accessed 26 Apr.

Hu, T. Y. (2014). Multidisciplinary efforts driving translational theranostics. *Theranostics, 4*(12), 1209.

Hui, E., Cheung, J., Zhu, J., Su, X., Taylor, M. J., Wallweber, H. A., Sasmal, D. K., Huang, J., Kim, J. M., & Mellman, I. (2017). T cell costimulatory receptor CD28 is a primary target for PD-1–mediated inhibition. *Science, 355*(6332), 1428–1433.

Iyidogan, P., & Anderson, K. S. (2014). Current perspectives on HIV-1 antiretroviral drug resistance. *Viruses, 6*(10), 4095–4139. https://doi.org/10.3390/v6104095

Jeyamogan, S., Khan, N. A., & Siddiqui, R. (2021). Application and importance of theranostics in the diagnosis and treatment of cancer. *Archives of Medical Research, 52*(2), 131–142.

Jha, A., Viswanadh, M. K., Burande, A. S., Mehata, A. K., Poddar, S., Yadav, K., Mahto, S. K., Parmar, A. S., & Muthu, M. S. (2020). DNA biodots based targeted theranostic nanomedicine for the imaging and treatment of non-small cell lung cancer. *International Journal of Biological Macromolecules, 150*, 413–425. https://doi.org/10.1016/j.ijbiomac.2020.02.075

Jiang, Y., Yang, W., Zhang, J., Meng, F., & Zhong, Z. (2018). Protein toxin chaperoned by LRP-1-targeted virus-mimicking vesicles induces high-efficiency glioblastoma therapy in vivo. *Advanced Materials, 30*(30), 1800316.

Jw, M. (1996). Prognosis in HIV-1 infection predicted by the quantity of virus in plasma. *Science, 272*(5265), 1167–1170.

Kanmogne, G. D., Singh, S., Roy, U., Liu, X., McMillan, J., Gorantla, S., Balkundi, S., Smith, N., Alnouti, Y., & Gautam, N. (2012). Mononuclear phagocyte intercellular crosstalk facilitates transmission of cell-targeted nanoformulated antiretroviral drugs to human brain endothelial cells. *International Journal of Nanomedicine*, 2373–2388.

Kaplan, J. E., Benson, C., Holmes, K. K., Brooks, J. T., Pau, A., & Masur, H. (2009). Guidelines for prevention and treatment of opportunistic infections in HIV-infected adults and adolescents: Recommendations from CDC, the National Institutes of Health, and the HIV medicine Association of the Infectious Diseases Society of America. *MMWR Recommendations and Reports, 58*(RR-4), 1–207; quiz CE201–204.

Keele, B. F., Van Heuverswyn, F., Li, Y., Bailes, E., Takehisa, J., Santiago, M. L., Bibollet-Ruche, F., Chen, Y., Wain, L. V., & Liegeois, F. (2006). Chimpanzee reservoirs of pandemic and nonpandemic HIV-1. *Science, 313*(5786), 523–526.

Kevadiya, B. D., Ottemann, B., Mukadam, I. Z., Castellanos, L., Sikora, K., Hilaire, J. R., Machhi, J., Herskovitz, J., Soni, D., & Hasan, M. (2020). Rod-shape theranostic nanoparticles facilitate antiretroviral

drug biodistribution and activity in human immunodeficiency virus susceptible cells and tissues. *Theranostics, 10*(2), 630.
Kinuya, S., Yoshinaga, K., Higuchi, T., Jinguji, M., Kurihara, H., & Kawamoto, H. (2015). Draft guidelines regarding appropriate use of (131)I-MIBG radiotherapy for neuroendocrine tumors : Guideline drafting Committee for Radiotherapy with (131)I-MIBG, Committee for Nuclear Oncology and Immunology, The Japanese Society of Nuclear Medicine. *Annals of Nuclear Medicine, 29*(6), 543–552. https://doi.org/10.1007/s12149-015-0960-z
Kish-Catalone, T., Pal, R., Parrish, J., Rose, N., Hocker, L., Hudacik, L., Reitz, M., Gallo, R., & Devico, A. (2007). Evaluation of-2 RANTES vaginal microbicide formulations in a nonhuman primate simian/human immunodeficiency virus (SHIV) challenge model. *AIDS Research and Human Retroviruses, 23*(1), 33–42.
Koppensteiner, H., Brack-Werner, R., & Schindler, M. (2012). Macrophages and their relevance in human immunodeficiency virus type I infection. *Retrovirology, 9*, 1–11.
Kosztowski, T., Zaidi, H. A., & Quiñones-Hinojosa, A. (2009). Applications of neural and mesenchymal stem cells in the treatment of gliomas. *Expert Review of Anticancer Therapy, 9*(5), 597–612.
Kovochich, M., Marsden, M. D., & Zack, J. A. (2011). Activation of latent HIV using drug-loaded nanoparticles. *PLoS One, 6*(4), e18270.
Kraft, J. C., McConnachie, L. A., Koehn, J., Kinman, L., Collins, C., Shen, D. D., Collier, A. C., & Ho, R. J. (2017). Long-acting combination anti-HIV drug suspension enhances and sustains higher drug levels in lymph node cells than in blood cells and plasma. *AIDS, 31*(6), 765–770. https://doi.org/10.1097/qad.0000000000001405
Kramer-Hämmerle, S., Rothenaigner, I., Wolff, H., Bell, J. E., & Brack-Werner, R. (2005). Cells of the central nervous system as targets and reservoirs of the human immunodeficiency virus. *Virus Research, 111*(2), 194–213.
Krause, A., Guo, H.-F., Latouche, J.-B., Tan, C., Cheung, N.-K. V., & Sadelain, M. (1998). Antigen-dependent CD28 signaling selectively enhances survival and proliferation in genetically modified activated human primary T lymphocytes. *The Journal of Experimental Medicine, 188*(4), 619–626.
Kuhlmann, A.-S., Peterson, C. W., & Kiem, H.-P. (2018). Chimeric antigen receptor T-cell approaches to HIV cure. *Current Opinion in HIV and AIDS, 13*(5), 446–453.
Kulkarni, T. A., Bade, A. N., Sillman, B., Shetty, B. L. D., Wojtkiewicz, M. S., Gautam, N., Hilaire, J. R., Sravanam, S., Szlachetka, A., Lamberty, B. G., Morsey, B. M., Fox, H. S., Alnouti, Y., McMillan, J. M., Mosley, R. L., Meza, J., Domanico, P. L., Yue, T. Y., Moore, G., et al. (2020). A year-long extended release nanoformulated cabotegravir prodrug. *Nature Materials, 19*(8), 910–920. https://doi.org/10.1038/s41563-020-0674-z
Lai, S. K., Wang, Y.-Y., & Hanes, J. (2009). Mucus-penetrating nanoparticles for drug and gene delivery to mucosal tissues. *Advanced Drug Delivery Reviews, 61*(2), 158–171.
Lamalle-Bernard, D., Munier, S., Compagnon, C., Charles, M.-H., Kalyanaraman, V. S., Delair, T., Verrier, B., & Ataman-Önal, Y. (2006). Coadsorption of HIV-1 p24 and gp120 proteins to surfactant-free anionic PLA nanoparticles preserves antigenicity and immunogenicity. *Journal of Controlled Release, 115*(1), 57–67.
Landovitz, R. J., Donnell, D., Clement, M. E., Hanscom, B., Cottle, L., Coelho, L., Cabello, R., Chariyalertsak, S., Dunne, E. F., & Frank, I. (2021). Cabotegravir for HIV prevention in cisgender men and transgender women. *New England Journal of Medicine, 385*(7), 595–608.
Landovitz, R. J., Scott, H., & Deeks, S. G. (2023). Prevention, treatment and cure of HIV infection. zNature Reviews Microbiology, 21(10), 657–670.
Larson, R. C., & Maus, M. V. (2021). Recent advances and discoveries in the mechanisms and functions of CAR T cells. *Nature Reviews. Cancer, 21*(3), 145–161. https://doi.org/10.1038/s41568-020-00323-z
Le Clerc, S., Limou, S., Coulonges, C., Carpentier, W., Dina, C., Taing, L., Delaneau, O., Labib, T., Sladek, R., Deveau, C., Guillemain, H., Ratsimandresy, R., Montes, M., Spadoni, J. L., Therwath, A., Schächter, F., Matsuda, F., Gut, I., Lelièvre, J. D., et al. (2009). Genomewide association study of a rapid progression cohort identifies new susceptibility alleles for AIDS (ANRS Genomewide Association Study 03). *The Journal of Infectious Diseases, 200*(8), 1194–1201. https://doi.org/10.1086/605892
Lee, C., Sapasap, J., LaRochelle, J., Smith, R. O., & Badowski, M. E. (2021). Antiretroviral therapy in children and adolescents: A look into modern single tablet regimens. *The Journal of Pediatric Pharmacology and Therapeutics, 26*(8), 783–794.
Leick, M. B., Maus, M. V., & Frigault, M. J. (2021). Clinical perspective: Treatment of aggressive B cell lymphomas with FDA-approved CAR-T cell therapies. *Molecular Therapy, 29*(2), 433–441. https://doi.org/10.1016/j.ymthe.2020.10.022
Li, J., Guo, Y., Kuang, Y., An, S., Ma, H., & Jiang, C. (2013). Choline transporter-targeting and co-delivery system for glioma therapy. *Biomaterials, 34*(36), 9142–9148.
Lindqvist, A., Rip, J., van Kregten, J., Gaillard, P. J., & Hammarlund-Udenaes, M. (2016). In vivo functional evaluation of increased brain delivery of the opioid peptide DAMGO by glutathione-PEGylated liposomes. *Pharmaceutical Research, 33*, 177–185.
Lisziewicz, J., & Szebeni, J. (2010). Nanophysiology: The Janus face of immune stimulation by nanomedicines: Examples for the good and the bad. *European Journal of Nanomedicine, 3*(1), 13–18.
Lisziewicz, J., & Tőke, E. R. (2013). Nanomedicine applications towards the cure of HIV. *Nanomedicine: Nanotechnology, Biology and Medicine, 9*(1), 28–38.
Liu, J., Zheng, X., Yan, L., Zhou, L., Tian, G., Yin, W., Wang, L., Liu, Y., Hu, Z., Gu, Z., Chen, C., & Zhao, Y. (2015). Bismuth sulfide nanorods as a precision nanomedicine for in vivo multimodal imaging-guided photothermal therapy of tumor. *ACS Nano, 9*(1), 696–707. https://doi.org/10.1021/nn506137n
Liu, H., Zhong, W., Zhang, X., Lin, D., & Wu, J. (2021). Nanomedicine as a promising strategy for the theranostics of infectious diseases. *Journal of Materials Chemistry B, 9*(38), 7878–7908.
Luster, M., Clarke, S. E., Dietlein, M., Lassmann, M., Lind, P., Oyen, W. J., Tennvall, J., & Bombardieri, E. (2008). Guidelines for radioiodine therapy of differentiated thyroid cancer. *European Journal of Nuclear Medicine and Molecular Imaging, 35*(10), 1941–1959. https://doi.org/10.1007/s00259-008-0883-1
Makinson, A., Pujol, J. L., Le Moing, V., Peyriere, H., & Reynes, J. (2010). Interactions between cytotoxic chemotherapy and antiretroviral treatment in human immunodeficiency virus-infected patients with lung cancer. *Journal of Thoracic Oncology, 5*(4), 562–571. https://doi.org/10.1097/JTO.0b013e3181d3ccf2
Mamo, T., Moseman, E. A., Kolishetti, N., Salvador-Morales, C., Shi, J., Kuritzkes, D. R., Langer, R., Andrian, U. V., & Farokhzad, O. C. (2010). Emerging nanotechnology approaches for HIV/AIDS treatment and prevention. *Nanomedicine, 5*(2), 269–285.
Mandal, S., Kang, G., Prathipati, P. K., Zhou, Y., Fan, W., Li, Q., & Destache, C. J. (2019a). Nanoencapsulation introduces long-acting phenomenon to tenofovir alafenamide and emtricitabine drug combination: A comparative pre-exposure prophylaxis efficacy study against HIV-1 vaginal transmission. *Journal of Controlled Release, 294*, 216–225.
Mandal, S., Prathipati, P. K., Belshan, M., & Destache, C. J. (2019b). A potential long-acting bictegravir loaded nano-drug delivery system for HIV-1 infection: A proof-of-concept study. *Antiviral Research, 167*, 83–88.
Markowitz, M., Frank, I., Grant, R. M., Mayer, K. H., Elion, R., Goldstein, D., Fisher, C., Sobieszczyk, M. E., Gallant, J. E., Van Tieu, H., Weinberg, W., Margolis, D. A., Hudson, K. J., Stancil,

B. S., Ford, S. L., Patel, P., Gould, E., Rinehart, A. R., Smith, K. Y., & Spreen, W. R. (2017). Safety and tolerability of long-acting cabotegravir injections in HIV-uninfected men (ECLAIR): A multicentre, double-blind, randomised, placebo-controlled, phase 2a trial. *Lancet HIV, 4*(8), e331–e340. https://doi.org/10.1016/s2352-3018(17)30068-1

Martin, S. J., Vyakarnam, A., Cheingsong-Popov, R., Callow, D., Jones, K. L., Senior, J. M., Adams, S. E., Kingsman, A. J., Matear, P., & Gotch, F. M. (1993). Immunization of human HIV-seronegative volunteers with recombinant p17/p24: Ty virus-like particles elicits HIV-1 p24-specific cellular and humoral immune responses. *AIDS, 7*(10), 1315–1324.

Matsunaga, A., Hishima, T., Tanaka, N., Yamasaki, M., Yoshida, L., Mochizuki, M., Tanuma, J., Oka, S., Ishizaka, Y., Shimura, M., & Hagiwara, S. (2014). DNA methylation profiling can classify HIV-associated lymphomas. *AIDS, 28*(4), 503–510. https://doi.org/10.1097/qad.0000000000000120

Matsusaka, K., Funata, S., Fukayama, M., & Kaneda, A. (2014). DNA methylation in gastric cancer, related to Helicobacter pylori and Epstein-Barr virus. *World Journal of Gastroenterology, 20*(14), 3916–3926. https://doi.org/10.3748/wjg.v20.i14.3916

Mazzaferri, E. L., & Jhiang, S. M. (1994). Long-term impact of initial surgical and medical therapy on papillary and follicular thyroid cancer. *The American Journal of Medicine, 97*(5), 418–428. https://doi.org/10.1016/0002-9343(94)90321-2

McArthur, J. C., Steiner, J., Sacktor, N., & Nath, A. (2010). Human immunodeficiency virus-associated neurocognitive disorders: Mind the gap. *Annals of Neurology, 67*(6), 699–714.

McCune, J. M. (2001). The dynamics of CD4+ T-cell depletion in HIV disease. *Nature, 410*(6831), 974–979.

McGuigan, C., Pathirana, R. N., Balzarini, J., & De Clercq, E. (1993). Intracellular delivery of bioactive AZT nucleotides by aryl phosphate derivatives of AZT. *Journal of Medicinal Chemistry, 36*(8), 1048–1052.

McNerney, M. P., Doiron, K. E., Ng, T. L., Chang, T. Z., & Silver, P. A. (2021). Theranostic cells: Emerging clinical applications of synthetic biology. *Nature Reviews Genetics, 22*(11), 730–746.

Mitsuyasu, R. T., Anton, P. A., Deeks, S. G., Scadden, D. T., Connick, E., Downs, M. T., Bakker, A., Roberts, M. R., June, C. H., & Jalali, S. (2000). Prolonged survival and tissue trafficking following adoptive transfer of CD4ζ gene-modified autologous CD4+ and CD8+ T cells in human immunodeficiency virus–infected subjects. *Blood, The Journal of the American Society of Hematology, 96*(3), 785–793.

Moos, T., & Morgan, E. H. (2000). Transferrin and transferrin receptor function in brain barrier systems. *Cellular and Molecular Neurobiology, 20*, 77–95.

Namane, A., Gouyette, C., Fillion, M. P., Fillion, G., & Tam, H. D. (1992). Improved brain delivery of AZT using a glycosyl phosphotriester prodrug. *Journal of Medicinal Chemistry, 35*(16), 3039–3044.

Narasipura, S. D., Kim, S., & Al-Harthi, L. (2014). Epigenetic regulation of HIV-1 latency in astrocytes. *Journal of Virology, 88*(5), 3031–3038.

Nátz, E., & Lisziewicz, J. (2012). *Rational design of formulated DNA vaccines: The DermaVir approach*. Springer.

née Williams, S. A. T., & Segal, M. B. (1996). Identification of a saturable uptake system for deoxyribonucleosides at the blood-brain and blood-cerebrospinal fluid barriers. *Brain Research, 741*(1–2), 230–239.

Neumann, H. P. H., Young, W. F., Jr., & Eng, C. (2019). Pheochromocytoma and Paraganglioma. *The New England Journal of Medicine, 381*(6), 552–565. https://doi.org/10.1056/NEJMra1806651

Novis, C. L., Archin, N. M., Buzon, M. J., Verdin, E., Round, J. L., Lichterfeld, M., Margolis, D. M., Planelles, V., & Bosque, A. (2013). Reactivation of latent HIV-1 in central memory CD4+ T cells through TLR-1/2 stimulation. *Retrovirology, 10*, 1–15.

Nowacek, A. S., Miller, R. L., McMillan, J., Kanmogne, G., Kanmogne, M., Mosley, R. L., Ma, Z., Graham, S., Chaubal, M., Werling, J., Rabinow, B., Dou, H., & Gendelman, H. E. (2009). NanoART synthesis, characterization, uptake, release and toxicology for human monocyte-macrophage drug delivery. *Nanomedicine (London, England), 4*(8), 903–917. https://doi.org/10.2217/nnm.09.71

O'Hagan, D., Singh, M., Ugozzoli, M., Wild, C., Barnett, S., Chen, M., Schaefer, M., Doe, B., Otten, G. R., & Ulmer, J. B. (2001). Induction of potent immune responses by cationic microparticles with adsorbed human immunodeficiency virus DNA vaccines. *Journal of Virology, 75*(19), 9037–9043.

Offersen, R., Nissen, S. K., Rasmussen, T. A., Østergaard, L., Denton, P. W., Søgaard, O. S., & Tolstrup, M. (2016). A novel toll-like receptor 9 agonist, MGN1703, enhances HIV-1 transcription and NK cell-mediated inhibition of HIV-1-infected autologous CD4+ T cells. *Journal of Virology, 90*(9), 4441–4453.

Okamoto, S., Shiga, T., & Tamaki, N. (2021). Clinical perspectives of theranostics. *Molecules, 26*(8), 2232.

Oller-Salvia, B., Sánchez-Navarro, M., Giralt, E., & Teixidó, M. (2016). Blood–brain barrier shuttle peptides: An emerging paradigm for brain delivery. *Chemical Society Reviews, 45*(17), 4690–4707.

Ostios-Garcia, L., Faig, J., Leonardi, G. C., Adeni, A. E., Subegdjo, S. J., Lydon, C. A., Rangachari, D., Huberman, M. S., Sehgal, K., Shea, M., VanderLaan, P. A., Cheng, M. P., Marty, F. M., Hammond, S. P., Costa, D. B., & Awad, M. M. (2018). Safety and efficacy of PD-1 inhibitors among HIV-positive patients with non-small cell lung cancer. *Journal of Thoracic Oncology, 13*(7), 1037–1042. https://doi.org/10.1016/j.jtho.2018.03.031

Ottemann, B. M. (2019). Theranostics for antiretroviral biodistribution and pharmacokinetics.

Pandit, R., Chen, L., & Götz, J. (2020). The blood-brain barrier: Physiology and strategies for drug delivery. *Advanced Drug Delivery Reviews, 165*, 1–14.

Park, S., & Sinko, P. J. (2005). P-glycoprotein and mutlidrug resistance-associated proteins limit the brain uptake of saquinavir in mice. *Journal of Pharmacology and Experimental Therapeutics, 312*(3), 1249–1256.

Patsoukis, N., Duke-Cohan, J. S., Chaudhri, A., Aksoylar, H.-I., Wang, Q., Council, A, Berg, A., Freeman, G. J., & Boussiotis, V. A. (2020). Interaction of SHP-2 SH2 domains with PD-1 ITSM induces PD-1 dimerization and SHP-2 activation. *Communications Biology, 3*(1), 128.

Peek, L. J., Middaugh, C. R., & Berkland, C. (2008). Nanotechnology in vaccine delivery. *Advanced Drug Delivery Reviews, 60*(8), 915–928.

Perazzolo, S., Stephen, Z. R., Eguchi, M., Xu, X., Delle Fratte, R., Collier, A. C., Melvin, A. J., & Ho, R. J. (2023). A novel formulation enabled transformation of 3-HIV drugs tenofovir–lamivudine–dolutegravir from short-acting to long-acting all-in-one injectable. *AIDS, 37*(14), 2131–2136.

Persidsky, Y., & Gendelman, H. E. (2003). Mononuclear phagocyte immunity and the neuropathogenesis of HIV-1 infection. *Journal of Leucocyte Biology, 74*(5), 691–701.

Pornpattananangkul, D., Olson, S., Aryal, S., Sartor, M., Huang, C.-M., Vecchio, K., & Zhang, L. (2010). Stimuli-responsive liposome fusion mediated by gold nanoparticles. *ACS Nano, 4*(4), 1935–1942.

Qin, Y., Fan, W., Chen, H., Yao, N., Tang, W., Tang, J., Yuan, W., Kuai, R., Zhang, Z., & Wu, Y. (2010). In vitro and in vivo investigation of glucose-mediated brain-targeting liposomes. *Journal of Drug Targeting, 18*(7), 536–549.

Redel, L., Le Douce, V., Cherrier, T., Marban, C., Janossy, A., Aunis, D., Van Lint, C., Rohr, O., & Schwartz, C. (2010). HIV-1 regulation of latency in the monocyte-macrophage lineage and in CD4+ T lymphocytes. *Journal of Leukocyte Biology, 87*(4), 575–588.

Reynolds, A., Laurie, C., Mosley, R. L., & Gendelman, H. E. (2007). Oxidative stress and the pathogenesis of neurodegenerative disorders. *International Review of Neurobiology, 82*, 297–325. https://doi.org/10.1016/s0074-7742(07)82016-2

Rickabaugh, T. M., Baxter, R. M., Sehl, M., Sinsheimer, J. S., Hultin, P. M., Hultin, L. E., Quach, A., Martínez-Maza, O., Horvath, S., Vilain, E., & Jamieson, B. D. (2015). Acceleration of age-associated methylation patterns in HIV-1-infected adults. *PLoS One, 10*(3), e0119201. https://doi.org/10.1371/journal.pone.0119201

Romanelli, F., & Hoven, A. D. (2006). Use of virostatics as a means of targeting human immunodeficiency virus infection. *Current Pharmaceutical Design, 12*(9), 1121–1127.

Rosca, E. C., Tadger, P., Cornea, A., Tudor, R., Oancea, C., & Simu, M. (2021). International HIV dementia scale for HIV-associated neurocognitive disorders: A systematic review and meta-analysis. *Diagnostics, 11*(6), 1124.

Rubin, L. H., Li, Y., Fitzgerald, K. C., Dastgheyb, R., Spence, A. B., Maki, P. M., Sharma, A., Gustafson, D. R., Milam, J., & Weber, K. M. (2021). Associations between antiretrovirals and cognitive function in women with HIV. *Journal of Neuroimmune Pharmacology, 16*, 195–206.

Rupp, R., Rosenthal, S. L., & Stanberry, L. R. (2007). VivaGel™(SPL7013 Gel): A candidate dendrimer–microbicide for the prevention of HIV and HSV infection. *International Journal of Nanomedicine, 2*(4), 561–566.

Schlaepfer, E., & Speck, R. F. (2011). TLR8 activates HIV from latently infected cells of myeloid-monocytic origin directly via the MAPK pathway and from latently infected CD4+ T cells indirectly via TNF-α. *The Journal of Immunology, 186*(7), 4314–4324.

Scholler, J., Brady, T. L., Binder-Scholl, G., Hwang, W.-T., Plesa, G., Hege, K. M., Vogel, A. N., Kalos, M., Riley, J. L., & Deeks, S. G. (2012). Decade-long safety and function of retroviral-modified chimeric antigen receptor T cells. *Science Translational Medicine, 4*(132), 132ra153–132ra153.

Sengupta, S., & Siliciano, R. F. (2018). Targeting the latent reservoir for HIV-1. *Immunity, 48*(5), 872–895.

Siliciano, J. M., & Siliciano, R. F. (2015). The remarkable stability of the latent reservoir for HIV-1 in resting memory CD4+ T cells. *The Journal of Infectious Diseases, 212*(9), 1345–1347. https://doi.org/10.1093/infdis/jiv219

Sillman, B., Bade, A. N., Dash, P. K., Bhargavan, B., Kocher, T., Mathews, S., Su, H., Kanmogne, G. D., Poluektova, L. Y., Gorantla, S., McMillan, J., Gautam, N., Alnouti, Y., Edagwa, B., & Gendelman, H. E. (2018). Creation of a long-acting nanoformulated dolutegravir. *Nature Communications, 9*(1), 443. https://doi.org/10.1038/s41467-018-02885-x

Smit, T. K., Brew, B. J., Tourtellotte, W., Morgello, S., Gelman, B. B., & Saksena, N. K. (2004). Independent evolution of human immunodeficiency virus (HIV) drug resistance mutations in diverse areas of the brain in HIV-infected patients, with and without dementia, on antiretroviral treatment. *Journal of Virology, 78*(18), 10133–10148.

Smith, K. A. (2001). To cure chronic HIV infection, a new therapeutic strategy is needed. *Current Opinion in Immunology, 13*(5), 617–624.

Smith, J. A., & Daniel, R. (2006). Following the path of the virus: The exploitation of host DNA repair mechanisms by retroviruses. *ACS Chemical Biology, 1*(4), 217–226.

Smith, N., Bade, A. N., Soni, D., Gautam, N., Alnouti, Y., Herskovitz, J., Ibrahim, I. M., Wojtkiewicz, M. S., Dyavar Shetty, B. L., McMillan, J., Gendelman, H. E., & Edagwa, B. (2019). A long acting nanoformulated lamivudine ProTide. *Biomaterials, 223*, 119476. https://doi.org/10.1016/j.biomaterials.2019.119476

Sofroniew, M. V., & Vinters, H. V. (2010). Astrocytes: Biology and pathology. *Acta Neuropathologica, 119*, 7–35.

Soulet, D., & Rivest, S. (2008). Bone-marrow-derived microglia: Myth or reality? *Current Opinion in Pharmacology, 8*(4), 508–518.

Spearman, P., Lally, M. A., Elizaga, M., Montefiori, D., Tomaras, G. D., McElrath, M. J., Hural, J., De Rosa, S. C., Sato, A., & Huang, Y. (2011). A trimeric, V2-deleted HIV-1 envelope glycoprotein vaccine elicits potent neutralizing antibodies but limited breadth of neutralization in human volunteers. *Journal of Infectious Diseases, 203*(8), 1165–1173.

Strosberg, J., El-Haddad, G., Wolin, E., Hendifar, A., Yao, J., Chasen, B., Mittra, E., Kunz, P. L., Kulke, M. H., Jacene, H., Bushnell, D., O'Dorisio, T. M., Baum, R. P., Kulkarni, H. R., Caplin, M., Lebtahi, R., Hobday, T., Delpassand, E., Van Cutsem, E., et al. (2017). Phase 3 trial of (177)Lu-Dotatate for midgut neuroendocrine tumors. *The New England Journal of Medicine, 376*(2), 125–135. https://doi.org/10.1056/NEJMoa1607427

Surve, D. H., & Jindal, A. B. (2020). Recent advances in long-acting nanoformulations for delivery of antiretroviral drugs. *Journal of Controlled Release, 324*, 379–404.

Tang, W., Fan, W., Lau, J., Deng, L., Shen, Z., & Chen, X. (2019). Emerging blood–brain-barrier-crossing nanotechnology for brain cancer theranostics. *Chemical Society Reviews, 48*(11), 2967–3014.

Thomas, S. A., & Segal, M. B. (1997). The passage of azidodeoxythymidine into and within the central nervous system: Does it follow the parent compound, thymidine? *Journal of Pharmacology and Experimental Therapeutics, 281*(3), 1211–1218.

Thompson, K. A., Cherry, C. L., Bell, J. E., & McLean, C. A. (2011). Brain cell reservoirs of latent virus in presymptomatic HIV-infected individuals. *The American Journal of Pathology, 179*(4), 1623–1629.

Tokarew, N., Ogonek, J., Endres, S., von Bergwelt-Baildon, M., & Kobold, S. (2019). Teaching an old dog new tricks: Next-generation CAR T cells. *British Journal of Cancer, 120*(1), 26–37.

Trejbalová, K., Kovářová, D., Blažková, J., Machala, L., Jilich, D., Weber, J., Kučerová, D., Vencálek, O., Hirsch, I., & Hejnar, J. (2016). Development of 5′ LTR DNA methylation of latent HIV-1 provirus in cell line models and in long-term-infected individuals. *Clinical Epigenetics, 8*, 19. https://doi.org/10.1186/s13148-016-0185-6

Tsai, A., Irrinki, A., Kaur, J., Cihlar, T., Kukolj, G., Sloan, D. D., & Murry, J. P. (2017). Toll-like receptor 7 agonist GS-9620 induces HIV expression and HIV-specific immunity in cells from HIV-infected individuals on suppressive antiretroviral therapy. *Journal of Virology, 91*(8), 02166-02116. https://doi.org/10.1128/jvi

Uchida, Y., Ohtsuki, S., Katsukura, Y., Ikeda, C., Suzuki, T., Kamiie, J., & Terasaki, T. (2011). Quantitative targeted absolute proteomics of human blood–brain barrier transporters and receptors. *Journal of Neurochemistry, 117*(2), 333–345.

van Manen, D., Delaneau, O., Kootstra, N. A., Boeser-Nunnink, B. D., Limou, S., Bol, S. M., Burger, J. A., Zwinderman, A. H., Moerland, P. D., van 't Slot, R., Zagury, J. F., van 't Wout, A. B., & Schuitemaker, H. (2011). Genome-wide association scan in HIV-1-infected individuals identifying variants influencing disease course. *PLoS One, 6*(7), e22208. https://doi.org/10.1371/journal.pone.0022208

Vandenberghe, P., Freeman, G., Nadler, L., Fletcher, M., Kamoun, M., Turka, L., Ledbetter, J., Thompson, C., & June, C. (1992). Antibody and B7/BB1-mediated ligation of the CD28 receptor induces tyrosine phosphorylation in human T cells. *The Journal of Experimental Medicine, 175*(4), 951–960.

Vanhamel, J., Bruggemans, A., & Debyser, Z. (2019). Establishment of latent HIV-1 reservoirs: What do we really know? *Journal of Virus Eradication, 5*(1), 3–9.

Vucinic, V., Quaiser, A., Lückemeier, P., Fricke, S., Platzbecker, U., & Koehl, U. (2021). Production and application of CAR T cells: Current and future role of Europe. *Frontiers in Medicine (Lausanne), 8*, 713401. https://doi.org/10.3389/fmed.2021.713401

Wagner, D. L., Koehl, U., Chmielewski, M., Scheid, C., & Stripecke, R. (2022). Sustainable clinical development of CAR-T cells–switching from viral transduction towards CRISPR-Cas gene editing. *Frontiers in Immunology, 13*, 865424.

Wang, Y.-Y., Lai, S. K., So, C., Schneider, C., Cone, R., & Hanes, J. (2011). Mucoadhesive nanoparticles may disrupt the protective human mucus barrier by altering its microstructure. *PLoS One, 6*(6), e21547.

Wang, X., Chen, L., Wang, X., Zhang, M., Yang, F., Wu, F., Liu, J., Lu, L., & Pang, Y. (2022). Long-acting protective ocular surface by instilling adhesive dual-antiviral nanoparticles. *Advanced Healthcare Materials, 11*(14), 2200283.

Wu, D., Clement, J. G., & Pardridge, W. M. (1998). Low blood–brain barrier permeability to azidothymidine (AZT), 3TC™, and thymidine in the rat. *Brain Research, 791*(1–2), 313–316.

Xiang, S. D., Scholzen, A., Minigo, G., David, C., Apostolopoulos, V., Mottram, P. L., & Plebanski, M. (2006). Pathogen recognition and development of particulate vaccines: Does size matter? *Methods, 40*(1), 1–9.

Xie, F., Yao, N., Qin, Y., Zhang, Q., Chen, H., Yuan, M., Tang, J., Li, X., Fan, W., & Zhang, Q. (2012). Investigation of glucose-modified liposomes using polyethylene glycols with different chain lengths as the linkers for brain targeting. *International Journal of Nanomedicine, 7*, 163–175.

Zhang, P., Hu, L., Yin, Q., Feng, L., & Li, Y. (2012). Transferrin-modified c [RGDfK]-paclitaxel loaded hybrid micelle for sequential blood-brain barrier penetration and glioma targeting therapy. *Molecular Pharmaceutics, 9*(6), 1590–1598.

Zhang, F., Xu, C.-L., & Liu, C.-M. (2015). Drug delivery strategies to enhance the permeability of the blood–brain barrier for treatment of glioma. *Drug Design, Development and Therapy, 9*, 2089–2100.

Zhang, C., Zaman, L. A., Poluektova, L. Y., Gorantla, S., Gendelman, H. E., & Dash, P. K. (2023). Humanized mice for studies of HIV-1 persistence and elimination. *Pathogens, 12*(7), 879.

Zheng, Y.-H., Lovsin, N., & Peterlin, B. M. (2005). Newly identified host factors modulate HIV replication. *Immunology Letters, 97*(2), 225–234.

Advancing Long-Acting Formulations for Treating Chronic Diseases

37

Mohammad Ullah Nayan, Ashrafi Sultana, Nam Thai Hoang Le, Larisa Y. Poluektuva, Benson Edagwa, and Howard E. Gendelman

Abstract

The challenge of poor medication adherence in the management of chronic diseases remains a pivotal issue, contributing significantly to the disparity between real-world effectiveness and clinical trial outcomes. Conventional strategies aimed at improving adherence through patient education and empowerment have often proven intricate and largely inefficacious. In contrast, long-acting (LA) medicines represent a transformative approach to overcoming the adherence barrier. LA formulations offer simplified dosing regimens, reduced treatment burdens, heightened patient acceptability, and broader public health benefits. This chapter focuses on currently approved LA medicines and their global utilization in the management of chronic conditions such as psychosis, contraception, chronic ocular diseases, and HIV-1 infection. Notably, LA medicines have demonstrated a positive impact on patient acceptability and adherence rates across diverse disease categories. Furthermore, this chapter sheds light on the underexplored potential of LA medicines in addressing other chronic ailments such as osteoporosis, viral hepatitis, and various chronic viral infections, as well as cardiovascular diseases and diabetes. Emphasizing the importance of cost-effective, self-administrable, discreet, and reversible LA medicines with improved pharmacokinetics featuring a shorter PK tail, we highlight the potential for these innovations to significantly enhance the global adoption and application of LA medicines.

M. U. Nayan · A. Sultana · N. T. H. Le · L. Y. Poluektuva · B. Edagwa (✉) · H. E. Gendelman (✉)
Department of Pharmacology and Experimental Neuroscience, University of Nebraska Medical Center, Omaha, NE, USA
e-mail: benson.edagwa@unmc.edu; hegendel@unmc.edu

Keywords

Prodrugs · LASER ART · Ultra-long-acting therapies · Chronic infections

37.1 Introduction

As our knowledge of the pathobiology of chronic disease grows, we are constantly uncovering new therapeutic targets. The process of discovering and developing new drugs remains time-consuming and resource-intensive. A significant number of potential therapeutics fail during preclinical development due to suboptimal physicochemical properties and toxicities of drug candidates (Shih et al., 2018). Even though a small fraction of new therapeutics transitioned from preclinical settings to clinical trials, recent data showed that only 10% of the new therapeutics that reached the phase I clinical trials received final US Food and Drug Administration (FDA) approval, with most of the declined approvals linked to lack of optimal PK, efficacy, and safety profiles (Harrison, 2016). Safety risks were responsible for up to 50% of approval rejections (Abdelkader et al., 2021; Harrison, 2016). Furthermore, the proportion of phase II and III clinical trials that were discontinued due to insufficient efficacy and safety is nearing 75% (Harrison, 2016).

Transformation of existing medicines with established safety and efficacy profiles into long acting formulations is a de-risked drug development process that could potentially improve treatment outcomes (Harrison, 2016; Kola & Landis, 2004; Mullard, 2016). This approach presents a compelling value proposition and increases the likelihood of regulatory approvals. Additionally, the strategy can circumvent lengthy timelines associated with new drug developments. Examples of LA applications serving to improve PK and efficacy profiles of existing medicines used for treatment of chronic diseases include poly (lactic-co-glycolic) acid (PLGA) microspheres and liposomal formulations of doxo-

H. E. Gendelman, T. Ikezu (eds.), *Neuroimmune Pharmacology and Therapeutics*,
https://doi.org/10.1007/978-3-031-68237-7_37

rubicin for cancer treatment. Microparticle depot-forming risperidone, nanocrystals of cabenuva), and microparticles of naltrexone are used for treatment of psychosis, HIV-1 infections, and substance use disorders, respectively (Nieuwlaat et al., 2014; Swift et al., 2011).

There are four major ways in which LA medicines can benefit patients with chronic diseases. First, LA drugs reduce dosing frequency. Second, the drugs can mitigate concentration-dependent adverse effects by controlling plasma concentrations. LA drugs provide steady and consistent drug levels, thus reducing the risk of immediate adverse effects associated with higher drug levels. Third, LA drugs lower the barriers to continued medication use and help destigmatize medication use. Patients with conditions such as neuropsychiatric disorders, epilepsy, attention deficit hyperactivity disorder (ADHD), and HIV/AIDS may face stigma as a consequence of frequent pill-taking (DiIorio et al., 2003; Sirey et al., 2001). LA drugs can alleviate pill-taking embarrassments by reducing the number of doses required or offering discreet administration methods. Lastly, LA drugs have the potential to enhance cost-effectiveness, as they reduce the overall drug quantity required. This will be particularly beneficial for expensive biologics.

The aforementioned advantages of LA drugs have been recognized and utilized in clinical practices for several decades now. Indeed, the initial development of an LA-implanted pellet containing a testosterone derivative was used. The first depot-forming IM injectable contraceptive, DMPA, was developed in 1959 and then approved by the FDA as a contraceptive (Label, n.d.). The development of LA medications for psychosis can be traced back to the late 1960s. In 1966, the first LAI antipsychotic, fluphenazine enanthate, was approved. Eighteen months later, a newer version of fluphenazine LAI, fluphenazine deaconate LAI also became available (Johnson, 2009). LAIs were initially met with skepticism by both users and physicians due to associated stigmas and suspected side effects. Nevertheless, with the introduction of second-generation antipsychotic LAIs and the positive results from several surveys and clinical trials, they have become popular again among physicians. Success in improving adherence and treatment outcomes in contraception and psychosis also influenced the LA medicines development for chronic infectious diseases, such as HIV, hepatitis B virus (HBV), tuberculosis, and malaria. Clinical trials of LA injectable antiretroviral therapy (LA ART) for HIV prevention and treatment have been ongoing since 2010 (Li et al., 2020). Combinations of LA CAB and rilpivirine (LA RPV) were approved by the FDA in January 2021 for the treatment of HIV-1 infection (Healthcare, 2021). LA CAB was later approved for HIV pre-exposure prophylaxis (PrEP). In parallel works, encouraging preclinical research has been conducted on LA formulations for the treatment of hypertension, diabetes, cancer, and osteoporosis. Furthermore, patients, in the largest measure, have shown greater interest in LA therapeutics.

LA drugs use several approaches to deliver the drug over a long period of time, such as dissolution-based formulations, biodegradable and nonbiodegradable implants, microarray patches, and hydrogels. The selection of LA medicines depends on the physiochemical characteristics of the drugs and the nature of the diseases. For instance, drugs with low aqueous solubility are solubilized in an oil-based vehicle or converted to aqueous micro or nanosuspensions. Owing to their low aqueous solubility, they undergo slow dissolution to sustain drug release in vivo. For drugs with high aqueous solubility, prodrug or hydrophobic salt approaches can be used to alter their physicochemical properties to allow for development of sustained release formulations. Other approaches involve entrapment of drug molecules into biodegradable polymers designed to control drug release, through diffusion and polymer degradation events. These biodegradable formulations can be preformed or formed in situ after injection. Nondegradable formulations are typically macroscopic implants that facilitate prolonged release through diffusion and can be removed after use. Careful considerations of the physiochemical properties of drugs, the therapeutic drug level requirement and safety, disease nature, end-user preference, and conveniences are required to select the optimal LA DDSs for efficient clinical translation. This chapter provides a comprehensive overview of LA therapies with a focus on implants, LA injectables, and microarray patches in managing unintended pregnancies, psychosis, chronic ocular disease, and HIV-1 infection. We have also highlighted LA medicine's underexplored potential in managing other chronic illnesses such as viral hepatitis, diabetes, and cardiovascular disease (Fig. **37.1**).

37.2 Approved LA Formulations

The impact of LA DDSs on improving adherence in patients with chronic conditions is influenced by various factors, including the nature of the disease, existing treatment regimen, and the specific administration parameters of the LA DDS. Given the constant persistence and high rate of relapsing, LA DDS for psychosis, contraception, ocular diseases, and chronic viral diseases, have been widely utilized. There are several LA DDSs that have converted conventional daily oral drugs to months or sometimes year-long-acting therapeutics. However, the global uptake and perspectives on these LA drugs are not the same due to people's lack of knowledge, high cost, and requirement of frequent clinic visits. We provide an overview of clinically approved LA medicines for chronic diseases and global perspectives on their utilization.

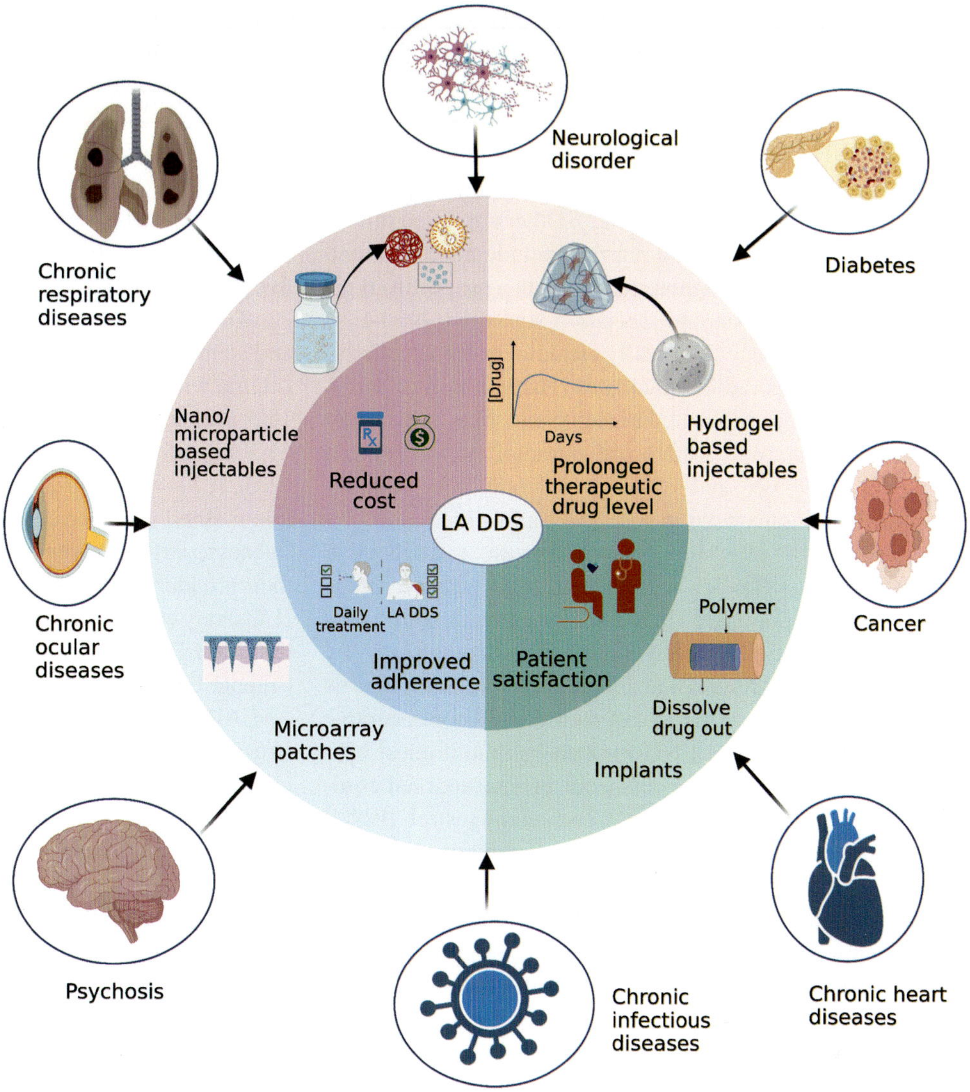

Fig. 37.1 LA DDSs for the management of chronic diseases: LA DDSs such as nano–/microparticle-based injectables, hydrogels, implants, and microarray patches can significantly improve patients' treatment adherence and satisfaction, and also potentially reduce treatment associated cost

37.2.1 LA Contraceptives

According to the UNFPA's State of World Population 2022 report, approximately 121 million unintended pregnancies occur annually, which is almost half of the total pregnancies worldwide (UNFPO, 2022). For affected women and girls, pregnancy decision is not always a choice. Almost 60% of all unintended pregnancies end in abortion, the majority of which are markedly unsafe and cause 5–13% of all maternal deaths (UNFPO, 2022). The widespread distribution of LA contraceptives is outcome-based.

The use of LA reversible contraceptives (LARC), such as intrauterine devices (IUDs), vaginal rings (VRs), and implants, varies from region to region, with higher usage rates in Latin America and Asia, but in sub-Saharan Africa (SSA), only 3% of women use LARC (Adedini et al., 2019). Compared to reversible contraceptives, LA injectable contraceptives are the dominant method of contraception in sub-Saharan Africa, with a prevalence of 9.6% among reproductive-age women. User preference is also a driving force for contraceptive use. None of the currently available contraceptive methods serve the needs of every sub-group in a population. Several surveys indicated that the number of contraceptive use sharply increases whenever a new method of contraception, specifically, LA contraceptive, has been introduced (Ross & Stover, 2013). DMPA is the most widely used injectable contraceptive worldwide (Dorflinger & Achilles, 2019). DMPA, an aqueous suspension of 17-acetoxy 6-methyl progestin, is available as a 3-month contraceptive in both IM and SC forms, with SC DMPA offering a 30% lower dose and reduced peak blood levels compared to DMPA-IM. It releases the progestin slowly from the depot, maintaining serum concentrations around 1.0 ng/mL for approximately 3 months (Mishell Jr., 1996). Metabolic effects such as weight gain, menstrual bleeding, decrease in bone mineral density, and impaired glucose tolerance are associated with DMPA use (Xiang et al., 2006). DMPA-SC has emerged as the preferred injectable method in the region due to its user-friendly all-in-one system, and ease of preparation, administration, and disposal compared to DMPA-IM. Health workers and end-users favor DMPA-SC. However, recent data showed stagnation or decline in DMPA-SC usage and low self-injection rates (Stout et al., 2018; Wood et al., 2022). Other approved injectable contraceptives in other countries include norethindrone enanthate (NET-EN) and those that contain a progestin (typically medroxyprogesterone acetate (MPA) or NET-EN) combined with estrogen (typically estradiol cypionate, estradiol valerate, or estradiol enanthate) (Halpern et al., 2015). Combination injectable contraceptives like Cyclofem (medroxyprogesterone acetate and estradiol cypionate) and Mesigyna (NET-EN and estradiol valerate) are monthly injections commonly used in Asia and Latin America.

Compared to injectable contraceptives, contraceptive implants provide longer-lasting protection against pregnancy. These implantable LARCs deliver low doses of progestin, such as ENG or LNG, to prevent pregnancy by thickening cervical mucus and suppressing ovulation (Curtis & Peipert, 2017). Among all hormonal contraceptive implants, the single rod ENG implant has the highest efficacy, as demonstrated by results from 11 international clinical trials (Darney et al., 2009). Two etonogestrel (ENG) implants are available for contraception: implanon and nexplanon. Both implants feature a single rod design, each holding a 68 mg dosage of ENG (Hohmann & Creinin, 2007). They are intended to provide up to 3 years of pregnancy prevention and are generally inserted under the skin of the upper nondominant arm (Kim et al., 2019). Despite the high efficacy of both implants, several side effects have been reported for both implants, including irregular menstrual cycles, weight gain, acne, headaches, breast tenderness, mood swings, and abdominal discomfort (Mansour et al., 2019).

Norplant, a subdermal rod-type implant containing 36 mg of LNG, provides a reliable, effective method of contraception for five years after inserting under the skin within the first week following menstruation (Coukell & Balfour, 1998). Once implanted, Norplant initiates a steady release of LNG, starting with a daily dosage of 50–80 mcg in the first year, which then decreases to a daily dosage of 30–35 mcg for the remaining four years (Lynn & Holdcroft, 1992). However, the efficacy of Norplant inversely correlates with the user's body weight—those with a higher BMI may have a higher risk of pregnancy while using the implant. A key advantage of Norplant, beyond its contraceptive effectiveness, is the swift return of fertility upon its removal. Post-removal, fertility tends to return promptly—within 3 months for half of the users and within a year for 80% of the users (International Planned Parenthood Federation IPPF, 1993). The VR is another LARC contraceptive method that forms a reservoir depot in a polymeric ring and slowly releases hormones to prevent ovulation. NuvaRing and Annovera are two hormonal contraceptive VRs approved for use in the United States.

IUDs are highly effective forms of contraception with minimal risk of failure (M. L. Kavanaugh & Jerman, 2018). There are two types of IUDs available: nonhormonal copper-containing IUDs and hormonal levonorgestrel-releasing (LNG) IUDs. In the United States, the FDA has approved five IUDs as of 2023. The copper-containing IUD, known as ParaGard, has 380 mm^2 of copper wrapped around its arms and stem. The four LNG-IUDs available include Mirena and Liletta, which contain 52 mg of levonorgestrel, Kyleena with 19.5 mg, and Skyla, which is slightly smaller and contains 13.5 mg of levonorgestrel. Compared to pills, patches, or rings, IUDs are about 20 times more effective as a contraceptive method (Winner et al., 2012). Despite their high effectiveness, adverse events associated with IUD usage limit their broader acceptance. Women who use copper-containing IUDs may experience increased menstrual bleeding, whereas the LNG-IUD generally reduces heavy menstrual bleeding (Tepper et al., 2014).

LARCs demonstrate exceptional efficacy for most women irrespective of age, parity, or BMI. LARC's "fit and forget" characteristic eliminates the need for user intervention, making them the epitome of "forgettable contraceptives" (Grimes, 2009; Higgins et al., 2016). Furthermore, research examining the sexual experiences of women using LARC methods revealed that majority reported either no changes or positive changes in their sexual experiences after 3 months of use (Higgins et al., 2016). These favorable features of LARCs significantly increase their utilization over the year. In the United States, the use of LARC methods has shown steady growth, starting from around 1% in 1995, then reaching 6% in 2008, and further rising to 16% between 2017 and 2019. Among the contraceptive methods, IUDs accounted for 12% of usage in 2014, positioning them as the fourth most popular method after pills, female sterilization, and the male condom. Additionally, there has been a substantial increase in the use of implants, representing 3% of contraceptive use, compared to just 0.5% in 2008 (Dehlendorf et al., 2018; Hubacher & Kavanaugh, 2018; Kavanaugh et al., 2011).

Low utilization of modern contraceptives in resource-limited settings is a concern due to the increased risk of unintended pregnancies and higher mortality rates. Women's individual decisions regarding contraception are influenced by a range of factors, including misperceptions and concerns about hormonal contraception, fear of pain during IUD insertion, and concerns about the invasiveness of the insertion procedure compared to other methods (Diana Mansour, 2014). However, the usage of contraception increases significantly when women are provided with comprehensive counseling and no-cost access to contraception (Birgisson et al.,

2015). The contraceptive CHOICE Project demonstrated that when women were counseled about all contraceptive options in order of effectiveness, many women, including teenagers, chose LARC methods. LARC methods were chosen by 75% of women, with 58% opting for an IUD and 17% choosing an implant (Birgisson et al., 2015; Secura et al., 2014).

37.2.2 LA Antipsychotics Agents

Psychosis disorders are often debilitating, clinically heterogeneous, and chronic conditions that require a treatment approach aimed at preventing relapses and hospitalizations. One of the major challenges in treating psychosis with medication is ensuring medication adherence. The symptoms persist over time due to their chronic or intermittent nature and frequent relapses, often caused by the discontinuation of antipsychotic treatment (de Filippis et al., 2021). Treatment nonadherence can be influenced by a lack of insight into the illness, negative beliefs about medication, and the presence of comorbid substance abuse. Over longer periods of treatment, treatment discontinuation rates range from 40% to 90% (Correll et al., 2016).

Long acting injectables (LAIs) are beneficial in improving treatment adherence in people with psychosis. However, the use of first-generation antipsychotic LAIs has been limited due to their poor tolerability and extrapyramidal side effects, despite their ability to reduce symptomatologic relapses, rehospitalization rates, and improve functional outcomes. In contrast, second-generation antipsychotic LAIs have demonstrated better tolerability and fewer extrapyramidal side effects (Correll et al., 2016).

As of now, the FDA has approved three first-generation and seven second-generation long-acting injectable antipsychotics. The first-generation antipsychotics, including flupentixol, zuclopenthixol, haloperidol, and fluphenazine, are formulated as decanoate ester prodrugs dispersed in oil-based vehicles. These vehicles commonly use either coconut-derived oil (Viscoleo) or sesame oil (Beresford & Ward, 1987; Gelenberg, 1987). Administration of these formulations is primarily done via IM injections into the gluteal muscle, resulting in the formation of a drug depot at the injection site. Following injection, the ester undergoes rapid hydrolysis, converting into active drugs within the plasma (Gelenberg, 1987). The release of these drugs from the depot relies on the slow release of the ester from the oil vehicle. The decanoate ester formulations of these antipsychotics prolong the half-life of the drugs from approximately 7 days to 21 days. However, the initial slow release from the depot leads to a delay in reaching steady-state drug concentrations, necessitating an initial oral lead-in for the first few days (Brissos et al., 2014). Individualized dosing intervals and doses should be determined based on the patient's tolerability and efficacy assessment.

Second-generation antipsychotics (SGAs) are formulated as aqueous suspensions of water-insoluble salts, ester prodrugs, or prodrug nanocrystals (de Filippis et al., 2021). Aripiprazole, for instance, comes in two forms: an aqueous suspension of aripiprazole monohydrate and aripiprazole lauroxil, which is an ester prodrug. The apparent half-life ranges from 29.9 to 46.5 days for aripiprazole monohydrate and from 53.9 to 57.2 days for aripiprazole lauroxil. Risperidone is also available in two different formulations: risperidone microspheres and risperidone 2-syringe mixing system suspension polymer (Gomeni et al., 2013; Laffont et al., 2014). The risperidone 2-syringe mixing system suspension polymer utilizes the ATRIGEL® delivery system, where risperidone is dissolved and suspended in a polymer that solidifies upon contact with tissue fluids (Gomeni et al., 2013). It exhibits two absorption peaks of plasma risperidone: one within 4–6 h after injection during depot formation and another 10–14 days after injection from slow depot release. The apparent half-life of risperidone 2-Syringe Mixing System Suspension Polymer is 9–11 days. The initial peak plasma concentration eliminates the need for an oral loading dose.

On the other hand, the delivery system for paliperidone utilizes nanocrystal technology, where it is formulated as paliperidone palmitate. It is available as a once-monthly formulation (PP1M) and a once every 3 months formulation (PP3M) (Ravenstijn et al., 2016). The PP3M formulation utilizes the same technology as PP1M but with an increased particle size. The median time to peak plasma concentration for PP1M is 13 days, and the apparent half-life ranges from 25 to 49 days (Schoretsanitis et al., 2018). Olanzapine is available as olanzapine pamoate salt. After intramuscular injection, peak plasma concentrations of olanzapine pamoate salt are achieved within 2–6 days, and the apparent half-life is 30 days (Heres et al., 2014; Mitchell et al., 2013).

Studies show that the efficacy of different LAIs within the same generation in reducing relapse rates and improving symptoms are comparable. Additionally, LAIs are preferred by a significant number of patients over oral therapy due to their less complicated regimen. A single-arm, interventional study conducted on 593 patients with schizophrenia found that the main reasons for transitioning from oral therapy to paliperidone palmitate were patients' preference (43.7%), low efficacy (24.3%), suboptimal compliance (23.3%), and tolerability (8.8%) with their prior oral antipsychotic medication(Schöttle et al., 2018; Schreiner et al., 2014).

Many clinicians worldwide believe LAIs are inappropriate for patients experiencing their first episode of psychosis, with rates of LAI prescriptions being particularly low in early intervention services (Heres et al., 2011). However, several regulatory guidelines recommend the use of LAIs when nonadherence is a priority or when patients prefer LAIs over oral formulations. Additionally, several randomized clinical trial data reflect the LAIs superiority over oral therapies (Moore et al., 2007; Taylor & Perera, 2015). Data from clinical trials showed that risperidone LAI was significantly more effective than the corresponding oral formulation on primary (psychotic exacerbations and/or relapse) and secondary (psychiatric rehospitalizations) outcome measures in a sample of patients with recent onset of a major psychotic episode (Subotnik et al., 2015). LAIs are also very effective in improving treatment completion rates compared to oral therapies. Data from a study on 1081 schizophrenic patients showed that the maintenance phase completion was 79.4% at week 52 when the patient was treated with aripiprazole once-monthly 400 mg (AOM 400)(Peters-Strickland et al., 2015).

Hospitalization, which frequently happens during a psychotic event relapse, significantly contributes to the decline in cognitive function and overall well-being in individuals with psychosis disorder. It also substantially increases the cost of managing the disease. A meta-analysis of 25 studies involving LAIs with a total of 5940 participants has demonstrated that LAIs can significantly reduce hospital readmission compared to oral antipsychotic medications. (Kishimoto et al., 2013).

Apart from LAIs' efficacy in improving clinical outcomes and cognitive function, second-generation LAIs can improve patients' psychosocial functioning. Second-generation antipsychotics LAIs have been found superior to placebo in improving psychosocial functions, demonstrating small, although significant, effects on the functional outcome compared to oral antipsychotics in long-term studies (Olagunju et al., 2019). While there is little evidence to support the superior efficacy of one LAI over another in improving quality of life and functioning, there are slight differences in head-to-head comparisons. For example, the QUALIFY study reported favorable tolerability and clinical improvements in health-related quality of life for aripiprazole once-monthly (AOM) compared to paliperidone palmitate once-monthly (PP1M) (Naber et al., 2015).

The advantages of LAIs over conventional oral antipsychotics range from better clinical outcomes to psychosocial activity improvement. However, proper administration of LAIs and follow-up management requires adequate nurse and medical training and expertise, which can increase the cost of LAIs. Despite this, LAIs are recommended by most evidence-based clinical guidelines for the maintenance treatment of psychosis. They are also explicitly preferred to oral antipsychotics for nonadherent and/or relapsing patients (Llorca et al., 2013; Taylor & Perera, 2015).

37.2.3 LA ARVs

The advent of highly effective antiretroviral therapy (ART) has transformed HIV into a manageable chronic illness. The high efficacy of these ART in viral suppression led to the development of chemoprophylaxis of HIV. PrEP with dual therapy, such as tenofovir disoproxil fumarate (TDF)/tenofovir alafenamide fumarate (TAF) and emtricitabine (FTC), has shown high efficacy when used consistently (Amico & Stirratt, 2014). The treatment-limiting factor for successful viral suppression is adherence to daily oral ART intake. Survey and clinical trial data are concerning, with about one-third of patients failing to achieve the recommended 90% adherence level for daily oral treatment (Krovi et al., 2018). Low adherence to daily oral PrEP regimen was identified as one of the main reasons for failure in the FEM-PrEP clinical trial of daily, oral TDF/FTC as PrEP for HIV prevention among women in sub-Saharan Africa, as indicated by surveys after the trial (Clement et al., 2020). Success in improving treatment adherence in psychosis and contraception has inspired the development of LAIs for treatment and prevention of HIV and other chronic viral infections. Surveys on the preference of treatment options also revealed that patients would prefer LA ART over daily oral therapy. Data from a cross-sectional study conducted on young men who have sex with men and are at risk of HIV infection showed that over 80% of people would prefer LAIs administered at least every 3 months (Meyers et al., 2014). Patient preference and demand for LAIs has accelerated the development of LAIs to manage HIV infection.

The development of LA CAB and RPV for the treatment of HIV infection is one of the most significant advancements in HIV therapeutics research. Several clinical trials, including LATTE-2, FLAIR, ATLAS, and ATLAS-2 M, have demonstrated the noninferiority efficacy of LA CAB and RPV to the standard daily oral three drug regimens. The data from these trials demonstrated that LA CAB + RPV achieved similar viral suppression to undetectable levels and was well-tolerated (Mills et al., 2022; Orkin et al., 2020; Overton et al., 2021). Furthermore, patients who received LA CAB and RPV treatment reported a higher level of satisfaction compared to those on the oral therapy (Flexner et al., 2021). In December 2021, LA CAB became the first and only long-acting PrEP agent approved by the FDA for HIV prevention in adults and adolescents weighing at least 35 kg who are at risk of sexually acquiring HIV. The approval was based on data from two phase IIb/III clinical trials: HIV Prevention Trials Network (HPTN) 083 and 084. In the HPTN-083 study, a total of 52 incidents of HIV infection were reported among 4566 participants, with 39 incidents of HIV infection in the oral PrEP group and 13 infections in the LA CAB group, indicating a 66% risk reduction with LA CAB. Similarly, in the HPTN-084 study, there were 40 incidents of HIV infection, with 36 infections in the oral PrEP group and four infections in the LA CAB group. These data demonstrate the significant superiority of LA CAB over oral PrEP agents (Delany-Moretlwe et al., 2022; Landovitz et al., 2021).

Clinical trials on LA CAB and RPV also reported significant satisfaction and treatment adherence among patients who received LA therapy. In the ATLAS-2 M clinical trial, an interim analysis at week 96 showed that all participants who responded to the treatment satisfaction questionnaire preferred LA CAB and RPV over the daily oral treatment (Swindells et al., 2022). Similarly, in the HPTN-083 study, where participants were divided into two groups, one receiving TDF/FTC and the other receiving LA CAB as PrEP, the percentage of participants with good long-term adherence was significantly higher in the LA CAB group (Landovitz et al., 2021). LA CAB and RPV reduces annual dosing schedules from 365 to only six. This significant reduction in dosage scheduling has several clinical and economic benefits, including but not limited to improved viral suppression and chances of diminishing the viral spread, cost, and resource reduction for both the patient and manufacturer. It introduces a new era in chronic viral disease management. However, further research must address challenges such as prolonged PK tail, injection volume, ease of administration, and patient comfort.

37.2.4 Intraocular Medicine and Devices

Chronic eye diseases have impacted a significant number of patients globally. Approximately 188 million individuals experienced minor vision impairment, while moderate-to-severe vision impairment affected around 216 million people (Bourne et al., 2017). Glaucoma, anterior uveitis, cataracts, and dry eye diseases are common conditions impacting the front of the eye. On the other hand, conditions such as age-related macular degeneration (AMD), diabetic retinopathy (DR), cytomegalovirus (CMV) infection, vitreoretinopathy, and posterior uveitis are frequently associated with issues affecting the back of the eye (Patel et al., 2013; Suri et al., 2020). Treatment for these conditions usually involves topical drops and intravitreal injections. However, treatment adherence plays a crucial role in the success of these therapies. Specific to asymptomatic diseases, such as glaucoma, compliance is a major challenge to efficacious therapy. Patients with glaucoma often exhibit an adherence rate of around 50%, and those who struggle with adherence are prone to higher intraocular pressure, more significant visual field loss, and an increased risk of blindness (Nordmann et al., 2010).

LA implantable drug delivery systems have been extensively utilized for adherence improvement of eye diseases, including noninfectious uveitis, macular edema for different posterior segment eye diseases, and age-related macular degeneration. Surgically implanted reservoir-based devices were the initial commercial examples of LA implantable intraocular devices. These devices were introduced through a pars plana incision and sutured to the sclera to reside within the eye for an extended period. Examples include Vitrasert® for cytomegalovirus and Retisert® for noninfectious uveitis, which slowly release therapeutic small molecules over months to years. Retisert®, an implant composed of nonbiodegradable silicone and PVA, contains fluocinolone acetonide and has shown good tolerability and acceptability in reducing noninfectious uveitis. This implant design incorporates a nonpermeable silicone-coated drug tablet with a strategically placed orifice for controlled drug release. To regulate the diffusion rate through the orifice, a permeable PVA membrane is positioned between the silicone orifice and the tablet (Jaffe et al., 2000). Similarly, Vitrasert® follows a comparable approach, involving a tablet initially coated with PVA, followed by a nonpermeable, intermittent EVA coating, and finally covered by a second coating of PVA.

The next-generation devices sought to minimize the procedure's invasiveness by introducing a custom injector for implantation, such as a biodegradable PLGA containing dexamethasone implant, Ozurdex® implant, and nonbiodegradable Iluvien®. These rod-shaped implants are deployed using a piston or fluid-based mechanism. Ozurdex® adopted Novadur® solid dosage form technology to deliver dexamethasone for 6 months (Kleiner et al., 2014; Malclès et al., 2017). The Iluvien® implant consists of a nonbiodegradable polyimide tube, PVA, and silicone adhesive. The implant releases active drugs through a permeable PVA membrane at one end, while the other end is sealed with an impermeable silicone cap (Meira et al., 2019). Both Ozurdex® and Iluvien® implants offer the advantage of being administered using syringe needles, resulting in a less invasive procedure with minimal damage to ocular tissues compared to the nonbiodegradable Retisert® implant.

Clinical surveys have provided compelling evidence of the substantial benefits of LA ocular injectables and implants in enhancing patient adherence and acceptability. In a recent survey conducted among cataract surgery patients, it was found that more than 90% of the participants expressed a clear preference for LA injectables, as opposed to the frequent administration of eye drops (Fisher & Potvin, 2016) (Fig. 37.2 and Table 37.1).

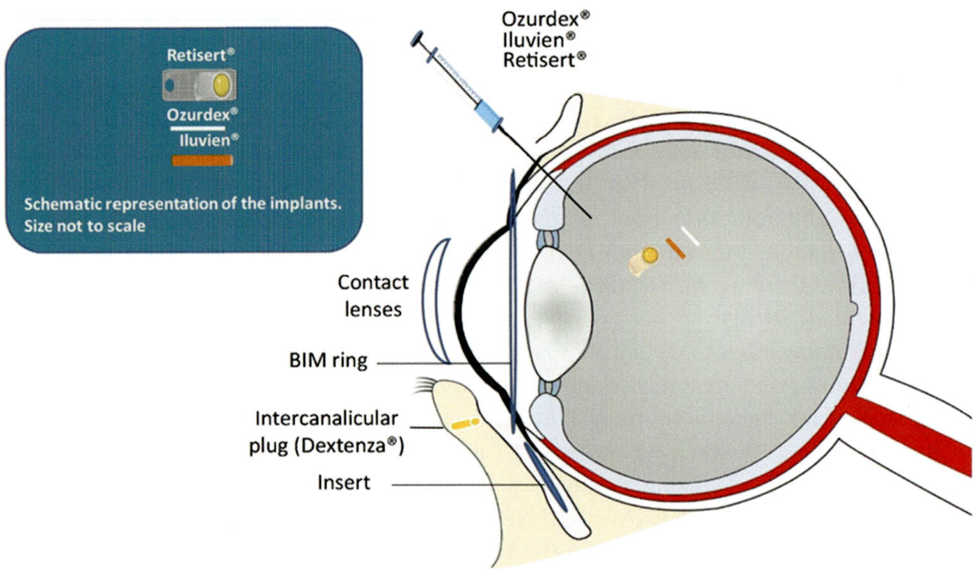

Fig. 37.2 Schematic of commercially available long-acting intraocular implant devices administration in human eyes. (Figure is adopted from Abdelkader et al., 2021 and used under the Creative Commons Attribution 4.0 International license. Copyright 2022)

37.3 Expanding the role of LA Formulations

The successful application of LA DDSs in managing diseases such as HIV-1, psychosis, and contraception has a profound influence on both clinical and preclinical developmental efforts of LA medicines for other chronic conditions. Diseases such as cancer, hypertension, osteoporosis, substance use disorder, and chronic infectious diseases like viral hepatitis, malaria, and tuberculosis can be debilitating and require consistent and effective treatment strategies. While adherence rates vary across diseases, they are often impacted by the frequency of dosing regimens. LA medicines offer a potential breakthrough in the treatment of these diseases and improving adherence. Considering the persistent nature of diseases like osteoporosis, and hypertension, the utilization of LA drugs, such as injectables or implants that deliver drugs at minimum effective levels over an extended period, can greatly improve treatment outcomes. Additionally, for conditions like substance use disorder, where medication nonadherence and early withdrawal are common, a single administration of LA DDSs that provide sustained drug release for months can significantly enhance treatment efficacy. This section highlights the current treatment options for other chronic diseases and emphasizes the need for and beneficial aspects of employing LA DDSs in managing these conditions.

37.3.1 Hepatitis

Chronic viral hepatitis is the leading cause of liver cirrhosis, hepatocellular carcinoma, and liver transplantation worldwide (European Association for the Study of the Liver. Electronic address and European Association for the Study of the L, 2017). It is defined as the persistence of serum HBsAg+ for longer than 6 months (Cox et al., 2020) and is caused by hepatitis B, C, and delta viruses.

Around 316 million individuals worldwide are impacted by chronic hepatitis B (HBV) infection (Degasperi et al., 2022). Despite the existence of a vaccine for 30 years, universal vaccine coverage has not yet been achieved. Either entecavir or TFV are the first line treatment therapies, and both are administered as once-daily oral pills. While these antivirals can control HBV replication in individuals with chronic infection, they are unable to completely eradicate the virus from infected patients (European Association for the Study of the Liver. Electronic address and European Association for the Study of the L, 2017). Therefore, lifelong treatment is required. Once the treatment is initiated, adherence and regular access to daily medicine pose significant challenges to treatment success. The use of LA medications could potentially overcome some of the challenges. This is particularly advantageous for patients who are co-infected with HIV and HBV, as medications such as TDF and TAF are commonly used as first-line treatments for both viruses, reducing the overall drug burden (Funk et al., 2021; Thompson et al., 2021).

Approximately 58 million people worldwide are infected with chronic hepatitis C (Blach et al., 2017). Newer combinatorial drug therapy administered for 3 months is now available. The regimen can effectively cure most patients. However, its success relies on strict adherence and completion of the regimen (Blach et al., 2017). A potential solution to improve adherence may be a LA medication that sustains therapeutic drug concentrations for 3 months. However, since there is no

Table 37.1 Clinically approved LA medicines for chronic diseases

Indication	Type	Brand name	API	Route	Dosing	Refs.
Psychosis	Aqueous nanocrystals	Invega hafyeratm	Paliperidone palmitate	IM	3–6 month	Morris and Tarpada (2017)
	Aqueous suspension	Aristada	Aripiprazole lauroxil	IM	1.5–3 months	Raedler (2016)
	ISFI	Perseris kit	Risperidone	SC	Monthly	Chue (2007)
	PLGA based microsphere	Risperdal Consta IM Depot®	Risperidone	IM	Every 2 weeks	
Esophageal cancer	ISFI	OncoGel™	Paclitaxel	IM	6 weeks	Karavasili and Fatouros (2016)
Prostate cancer	PLGA based microsphere	Leupron intramuscular (IM) Depot®	Leuprolide acetate	IM	Monthly	Wang and Burgess (2012)
		Trelstar IM Depot®	Triptorelin pamoate	IM	6 months	
Chemotherapy associated nausea and vomiting	Polymer-based liquid formulation	SUSTOL	Granisetron	SC	Weekly	Ngo et al. (2019)
Severe chronic pain	Transdermal patch	Durogesic® patch	Fentanyl	IM	72 h	Wright and Burgess (2012)
	ISFI	Posidur™	Bupivacaine	IM	48–72 h	Karavasili and Fatouros (2016)
Osteoarthritis pain	Microsphere based suspension	ZILRETTA®	Triamcinolone acetonide	IA injection	Montly	Paik et al. (2019)
Substance or opioid use disorder	ISFI	SUBLOCADE	Buprenorphine	SC	Monthly	Gray et al. (2023)
	Degradable microparticle depot	Vivitrol	Naltrexone	IM	Monthly	Tanum et al. (2017)
Acromegaly and Cushing's disease	Microsphere based formulation	SIGNIFOR LAR	Pasireotide pamoate	IM	Monthly	Fleseriu et al. (2017)
Growth hormone deficiency	PLGA based microsphere	Neutropin SC Depot®	Somatropin (rhGH)	SC	Monthly	Wang and Burgess (2012)
Glaucoma	Implant	Ocusert®	Pilocarpine		Weekly	
		BIM ring	Bimatoprost		6 months	Brandt et al. (2017)
Macular edema	Implant	Ozurdex®	Dexamethasone		6 months	Gaballa et al. (2021)
		Iluvien® i	Fluocinolone acetonide		36 months	Gaballa et al. (2021)
Contraceptive	Implant	Progstasert®	Progestogen		1 year	Shen and Burgess (2011)
	Implant	Norplant®	Levonorgestrel		6 months	Shen and Burgess (2011)
	IUD	Mirena/Kyleena/ Liletta/ Skyla	Levonorgestrel		3–7 years	Creinin et al. (2022)
	Intravaginal ring	NuvaRing	Etonogestrel/Ethinyl oestradiol		1 month	Wieder and Pattimakiel (2010)
HIV	Aqueous nanocrystals	Apretude	Cabotegravir	IM	1–2 month	Fonner et al. (2023)
		Cabenuva	Cabotegravir and Rilpivirine	IM	1–2 month	Smith et al. (2021)

IM Intramuscular injection, *SC* Subcutaneous injection, *IA* Intra-articular injection, *ISFI* in-situ forming implant, *IUD* Intrauterine device

protective immunity against HCV, reinfection is possible, particularly among those engaging in high-risk behaviors such as injection drug use or high-risk sexual practices. Furthermore, a considerable number of individuals infected with HCV remain undiagnosed, and many asymptomatic carriers do not have access to curative treatment. In resource-limited settings, LA medication could offer treatment benefits.

Compared to HBV or HCV, hepatitis delta virus (HDV) infection is the most severe of all chronic viral hepatitis leading to liver cirrhosis and hepatocellular carcinoma. Approximately 15–25 million people are infected with chronic hepatitis D. The disease severity and progression are worse in HIV coinfections. This coinfection group is ideal for LA therapies as they both share the same treatment modalities and follow strict lifelong adherence.

37.3.2 Tuberculosis

Tuberculosis (TB) along with HIV and malaria are three of the leading infectious diseases. Prior to the COVID-19 pandemic, tuberculosis was the single infectious agent which caused the highest number of deaths. According to WHO TB report-2022, the number of TB patients was 10.1 million, with 1.4 million deaths among HIV-negative patient (organization, 2022). The treatment of TB typically involves a combination of four first-line therapeutic drugs: rifampicin, isoniazid, pyrazinamide, and ethambutol, with a dosage regimen of 6–9 months. Long duration of treatment along with patient treatment nonadherence, diagnostic inefficiency, delay in treatment initiation, and emergence of treatment-resistant TB strains contribute to treatment failure.

In 2021, at least 450,000 cases of rifampicin-resistance TB (RR-TB) were reported (organization, 2022). The advent of increased number of RR-TB and multi-drug resistance TB (MDR-TB, RR-TB with confirmation of isoniazid resistance) has left very limited treatment options available. The emergence of these drug resistant species is increasing the morbidity and mortality due to significant challenges in implementing effective therapeutic strategies. In recent years, several anti-tuberculosis drugs with novel mechanisms of action have been reported and are either in advanced preclinical or clinical stages of development. In addition to chemotherapeutic drugs, numerous host-directed antivirals and immune-based biologics are under development. These strategies are focused on minimizing permanent lung damage, shortening treatment duration, and overcoming the challenge of drug resistance.

For drug-susceptible TB, the WHO recommends a treatment regimen that consists of an initial intensive phase lasting 2 months. During this phase, patients are typically administered a daily combination of isoniazid, rifampicin, pyrazinamide, and ethambutol. Following the intensive phase, patients transition to a 4-month continuation therapy phase, during which they receive a daily combination of isoniazid and rifampicin. Lengthy treatment duration, associated side effects, and less tolerance to a significant proportion of the population demand an effective treatment strategy. The effective treatment strategy for drug resistance TB is patient-specific and consists of at least four classes of drugs (Falzon et al., 2017). A comprehensive treatment strategy should include the use of an injectable aminoglycoside, a third or fourth generation of fluoroquinolone, and two other core drugs such as ethionamide or prothionamide, cycloserine or terizidone, linezolid, and clofazimine (Tiberi et al., 2018). The treatment duration can vary from 9 months to 18 months based on patient's response, tolerance and susceptibility to the regimen.

Based on the TB-PRACTECAL and ZeNix clinical trials results, in 2022, WHO recommends a shorter treatment regimen of 6 months for the MDR-TB or RR-TB. The recommended 6-month BPaLM regimen includes four drugs (bedaquiline, pretomanid, LEN, and moxifloxacin) (WHO, 2022). The recommended treatment strategy for TB is rapidly changing based on several ongoing clinical trial results and resistance analyses of the drugs. New therapeutic agents with increased efficiency are essential to meet the WHO's global TB commitment. Based on "Stop TB Partnership" data, at least 14 new candidates of drugs are in clinical development and have the potential to be included in the treatment regimen of MDR-TB.

Despite the chemotherapeutics drug potency and efficacy, treatment outcomes largely depend on early diagnosis and initiation of effective treatment, patient adherence, and tolerance. Improving early detection of TB and treatment initiation and adherence to therapy can help limit the emergence of drug resistance Mtb. LA medication could potentially be the game changer for TB treatment. Sustained release treatments could potentially improve patient adherence, stop the spread of drug-resistant TB, and minimize treatment cost and pill burden. Such long-acting medications can potentially improve drug availability in resource-limited settings, reduce national TB programs and healthcare systems costs, and enable the global scale-up of MDR-TB treatment.

37.3.3 Malaria

Significant successes were made in reducing malaria prevalence and deaths through the widespread dissemination of chemoprevention drugs and public health measures over the past few decades. Unfortunately, progress has stagnated since 2019. The 2022 WHO Malaria Report indicates that 247 million cases of malaria were reported in 84 malaria-endemic countries, resulting in an estimated 619,000 deaths in 2021 (WHO, 2022).

Malaria-related deaths are most common among children under 5 years old and pregnant women. Compared to non-pregnant women, pregnant women are more susceptible to experiencing more frequent and severe episodes of malaria (Moehrle, 2022). Individuals who are not previously exposed to malaria and travel to regions where the disease is prevalent, particularly in sub-Saharan Africa, are at high risk of developing severe forms of the disease (Moehrle, 2022). The fourth at-risk population are those living in areas where malaria re-emerges every year. A chemoprevention strategy can effectively protect those highly susceptible groups from malaria.

Multiple options exist for malaria chemoprophylaxis. The preferred method for preventing malaria is to administer medications that have not shown any evidence of plasmodium resistance (Plowe, 2022). Drugs that target plasmodium's asexual hepatic stages are typically used for causal prophylaxis. Atovaquone, pyrimethamine, cycloguanil, proguanil, sulfadoxine, primaquine, and tafenoquine have all been clinically demonstrated to be effective in malaria treatment (Macintyre et al., 2018). While they were originally developed for this purpose, mefloquine, doxycycline, and atovaquone-proguanil are the primary oral prophylactics currently used by travelers (Macintyre et al., 2018).

Chemoprevention of malaria requires long-term drug intake. LA medication can improve the efficacy of the chemopreventive drug by reducing pill burden and dosage nonadherence. LA malarial chemoprophylaxis could potentially benefit different vulnerable groups. First, it can effectively be used to minimize transmission of the disease among populations, such as soldiers, migrant workers, and tourists who come from malaria-free areas and travel to endemic regions. Second, nonimmune populations can combine LA malaria chemoprophylaxis therapies with other interventions to manage seasonal outbreaks. Lastly, LA chemoprophylaxis agents could potentially provide protection to vulnerable populations living in areas with a high incidence of malaria, which is somewhat distinct from the other uses.

37.3.4 Cancer

Cancer, a widespread disease affecting millions annually, is hindered by limited treatment effectiveness and side effects of available systemic anticancer drugs. Optimal dose selection for these drugs poses significant challenges and is sometimes unattainable. Administering small doses often results in ineffective concentrations within tumors, while high doses that exhibit notable tumor inhibition effects may harm healthy cells, risking patient well-being. However, cancer patients, faced with the complexities of their condition, are reluctant to compromise drug efficacy for convenience (Marshall, 2018). Recent advancements in drug delivery and sustained-release formulations offer both conveniences and enhanced safety for cancer patients. By prolonging the drug's half-life and improving drug uptake by cancerous cells, LA anticancer drugs can mitigate the side effects associated with frequent conventional dosing. Moreover, cancer patients strongly prefer LA anticancer drugs with improved efficacy. A recent survey conducted among breast cancer patients revealed that 75% of women would opt for monthly or bimonthly injectable therapeutics if these treatments demonstrated higher effectiveness (Liu et al., 1997).

Preclinical development of LA anticancer drugs has primarily focused on targeted delivery and combination therapy, as the combination of targeted therapies has shown greater potential for improving effectiveness compared to monotherapy. For instance, Arroyo-Crespo et al. achieved improved survival rates in orthotopic 4 T1 breast tumor mice by conjugating polyglutamic acid with a chemotherapeutic agent, doxorubicin (DOX), and an estrogen-modulating agent, aminoglutethimide. This conjugation increased the half-life of DOX by approximately nine-fold compared to the monotherapy (Ekladious et al., 2019). Targeted delivery of anticancer drugs is based on the unique characteristics of the tumor microenvironment and the enhanced permeability and retention (EPR) effect. Various target moieties are conjugated to the active drugs, or the surfaces of drug delivery vehicles based on the distinct molecular biology of the tumor microenvironment. Polymeric nanoparticles, nanocrystals, lipid nanoparticles, and liposomal drug delivery systems have been employed to deliver combined targeted therapies. The first FDA-approved nanosystem for cancer treatment was Doxil®, a liposome that encapsulates DOX with PEG. Since then, numerous PEGylated products have emerged in the market for various biomedical applications, including cancer treatment, significantly extending circulation time and enhancing tumor tissue accumulation (Zalba et al., 2022).

In certain cases where the tumor location is known, in situ cross-linkable hydrogels and nondegradable polymeric implants can be injected or implanted at the specific site of interest. This approach creates a depot that gradually releases the drug in a localized manner over time. It proves particularly beneficial when the site of interest is a cavity accessible after a recent surgical intervention.

37.3.5 Opiate Withdrawal and Tolerance

Opioid misuse is a significant problem in the United States, Asia, Europe, and Oceania. Worldwide consumption of opioids is on the rise, with an estimated 15.6 million illicit opioid users globally (Zoorob et al., 2018). In 2016, around 11.5 million Americans aged 12 or above misused opioid pain medications, with 1.8 million people having substance use disorder as a result of prescribed pain medications (Mansi, 2023). Opioid overdose-related death consist of almost 75% of total drug overdose-related death (CDC, 2022).

While opioid withdrawal is not typically considered a life-threatening condition, it can have significant negative effects on patients with opioid use disorder and chronic noncancer pain (CNCP). Patients taking long-term prescription opioids for CNCP may face difficulties decreasing their opioid dose or discontinuing opioid therapy altogether because of withdrawal symptoms. Pathophysiological conditions of opioid withdrawal include severe fluid loss, electrolyte abnormalities, and hemodynamic instability (Srivastava et al., 2020). Careful clinical management is necessary to manage opioid withdrawal and stabilize the patient. Further, clinical intervention is more important for treating opioid use disorder.

Treating opioid dependence with opioid agonist treatment (OAT) is a highly effective approach that has been shown to significantly reduce illicit opioid use. The standard of care for OAT includes treatment with either buprenorphine, methadone, or extended-release naltrexone (Srivastava et al., 2020). Buprenorphine (BUP) is a partial agonist of the μ-opioid receptor. It has been approved for treatment of opioid use disorder as well as medically supervised withdrawal. However, it is necessary to wait for withdrawal symptoms to emerge before starting buprenorphine treatment, as it can cause withdrawal if given too soon after ingesting a full μ-opioid receptor agonist (Koehl et al., 2019).

In contrast to BUP, methadone can be initiated without waiting for the emergence of withdrawal symptoms. However, patients transitioning from methadone to buprenorphine usually require a taper, and emergent withdrawal symptom can complicate this transition (Srivastava et al., 2020). For patients who suffer from medication nonadherence, LA naltrexone and BUP can be a very effective treatment option. In the United States and United Kingdom, clonidine, an α2-adrenergic receptor agonist, and lofexidine, an α2-receptor agonist, are approved for treating opioid withdrawal symptoms.

Although OAT is effective in managing opioid withdrawal, long-term access to this standard care can be challenging, and many treatment centers lack the capability to prescribe medications for opioid use disorder (MOUD) (Farrell et al., 2022). Moreover, this standard of care also presents additional challenges, including adverse events, overdose risk, and complex administration procedures. These medications require daily dosing, and patients in most countries must take them at a specialized clinic or pharmacy, which can be cumbersome and limiting for both patients and caregivers (Farrell et al., 2022). This process also increases medication costs. Medication nonadherence is a common issue among patients who suffers from opioid withdrawal symptoms. However, extended-release medications with weekly or monthly dosing intervals can improve retention, provide flexibility and more patient options, and reduce the burden of regular clinical visits.

37.3.6 Cardiovascular Disease

Cardiovascular diseases, including hypertension, affect a vast number of individuals worldwide, with more than two-thirds of adults in the United States taking antihypertensive medications (Snider et al., 2008). However, adherence to these medications poses significant challenges as most antihypertensive drugs require once or twice daily dosing. The adherence rates for cardiovascular medications can range from 21% to 71%, leading to adverse health outcomes and increased healthcare costs (Baroletti & Dell'Orfano, 2010). Nonadherence to antihypertensive medications has been associated with a higher rate of deaths. In fact, a survey conducted in 2014 estimated that medication nonadherence resulted in approximately 5.6 billion dollars in avoidable medical expenditures in the United States (Kumar et al., 2019). The utilization of LA DDSs holds great promise in improving adherence to antihypertensive medications and thereby reducing the health burden and related costs. Patients with cardiovascular diseases often prefer less frequent medication dosing. Even a modest improvement in medication adherence for cardiovascular diseases can significantly enhance disease outcomes (Peacock & Krousel-Wood, 2017). LA DDSs have the potential to improve medication adherence, as demonstrated by the use of LA clonidine transdermal patches, such as catapres-TTS. Surveys among users of catapres-TTS revealed that 65% of patients were able to adhere to the treatment (Hollifield, 1986).

37.3.7 Osteoporosis

As the global population continues to age, osteoporosis becomes a prevalent and concerning health condition (Asafo-Adjei et al., 2016). It is characterized by bone resorption exceeding bone formation. It poses a significant burden on aged individuals, leading to an increased risk of fractures, morbidity, mortality, and rehospitalization. Moreover, it creates a substantial financial burden (Villanueva-Martínez et al., 2022). First-line therapies for osteoporosis include bone antiresorptive agents like bisphosphonates (BPs) and estrogen. However, poor adherence to these medications is a significant concern, with studies estimating noncompliance rates ranging from one-third to one-half of patients with osteoporosis (Ettinger et al., 2006; Kendler et al., 2004; Kennel & Drake, 2009; Simon et al., 2002). Nonadherence significantly increases the risk of fractures and is associated with inferior health outcomes and increased healthcare costs. The introduction of LA osteoporosis agent may significantly improve adherence to medications. Studies showed that women with osteoporosis prefer once-weekly BPs over once-daily regimens, believing it improves long-term treatment prospects.

To address the patient preference in osteoporosis treatment, LA DDSs are being explored in clinical and preclinical development. The approval of triamcinolone acetonide ER by the FDA serves as an illustration of the ongoing research and development efforts in the field of long-acting medications for osteoporosis or osteoarthritis pain management. Triamcinolone acetonide has been widely used for the management of osteoarthritis or osteoporosis pain management for decades. However, its efficacy is limited by corticosteroid-related systemic toxicities (Hochberg et al., 1995). Triamcinolone acetonide ER, formulated as PLGA microspheres, releases triamcinolone acetonide for up to 6 months within the synovium, allowing for a prolonged presence of the medication in the joint. This localized drug delivery approach reduces systemic exposure and minimizes the occurrence of corticosteroid-related systemic adverse reactions, including elevated blood glucose levels (Paik et al., 2019). Preclinical development of LA osteoporosis medicines includes parenteral hydroxyapatite-containing scaffolds and injectable or implantable depots. These innovative formulations have the potential to minimize the need for frequent patient intervention, mitigate dose fluctuations, and bypass the gastrointestinal tract altogether.

37.4 Conclusion and Future Directions

Medication adherence remains a pivotal determinant in bridging the gap between clinical trial efficacy and real-world outcomes in the management of chronic diseases. LA medicines have emerged as a promising avenue to mitigate these disparities, with widespread recognition of their benefits by clinicians, patients, and regulatory authorities. Market research affirming patient preferences for LA medicines has spurred both the pharmaceutical industry and academic research to harness LA drug delivery systems, thereby enhancing drug efficacy and extending half-life. Looking ahead, the future of LA medicines is poised for significant evolution, driven by advancements in technology and the development of compatible biomaterials. These innovations will enable precise control over drug release rates and durations. As our understanding of drug-releasing mechanisms in LA drug delivery systems continues to grow, we anticipate a surge in the development of LA therapeutics for a broad spectrum of chronic conditions. However, it is imperative to acknowledge and address specific challenges during the design and development of LA therapeutics, as they could impede successful clinical outcomes. A primary concern is the implementation challenge, as LA medicines tend to be more expensive than their oral counterparts, posing a substantial barrier to adoption in resource-limited settings. The creation of cost-effective LA medicines holds the potential to significantly enhance disease management in such contexts. Infrastructure and logistical hurdles, including transportation and storage requirements such as cold storage and short product shelf-life, can also exert a significant influence on the implementation of LA medicines. Furthermore, the mode of administration and discretion in LA medicines play a pivotal role, particularly in diseases where social stigma is a prominent factor, such as chronic infectious diseases and contraception. LA injectable products or implants necessitating surgical procedures and regular clinic visits, coupled with the need for trained healthcare professionals, can not only elevate medication costs but also diminish patient convenience, potentially hindering product adoption. The development of self-administrable and discreet LA medicines represents a critical step forward in enhancing patient convenience, reducing social anxiety related to medication, and mitigating implementation challenges. Future endeavors in the realm of LA medicines should prioritize solutions that promote patient comfort, cost-effectiveness, and the reduction of logistical and infrastructure obstacles, thereby facilitating easier implementation and improving the overall landscape of chronic disease management.

The management of LA medicines post-administration is a crucial aspect to consider during the development of these therapeutic approaches. The reversibility of LA medicines, when necessary, emerges as a critical criterion, especially in conditions such as contraception, diabetes, and cardiovascular complications. While implantable LA medicines offer promising opportunities, they often require surgical procedures and can be susceptible to displacement from the implant site. Therefore, rigorous fitness testing before initiating LA medicines is imperative, particularly in cases where reversibility is not straightforward. Efficient management of missed doses is another paramount concern, especially in the context of infectious microbial diseases where suboptimal drug concentrations may lead to drug-resistant mutant strains. Additionally, the persistence of drug concentrations at subtherapeutic levels over extended periods, known as the pharmacokinetics tail, presents a significant challenge in LA medicine. This pharmacokinetics tail can contribute to drug-resistant microbial species and prolonged drug-drug interactions, emphasizing the importance of reducing it. Many chronic diseases necessitate the use of multiple partner drugs for effective management, particularly in cases such as HIV, tuberculosis, malaria, diabetes, and cardiovascular diseases. The development of single LA drug delivery systems capable of delivering multiple partner drugs holds substantial promise for improving uptake and disease outcomes in LA medicine approaches.

Despite these challenges, the potential of LA medicines to address obstacles in chronic disease management remains substantial. Future prospects for LA medicines lie in strategies that prioritize self-administrable discreet LA medicines with short pharmacokinetics tails and the capacity for therapeutic reversibility. These properties not only ease the implementation of LA medicines but also enhance their acceptance among a broad spectrum of patients, ultimately advancing the landscape of chronic disease management.

37.5 Review Questions

1. How do long-acting (LA) drugs benefit patients with chronic diseases?
   (a) By increasing the need for daily medication intake
   (b) By controlling plasma concentrations to mitigate concentration-dependent adverse effects and reducing the frequency of medication intake

   (c) By requiring more frequent doctor visits for administration
   (d) By increasing the overall drug quantity required for treatment
2. Given the high efficacy of long-acting formulations for chronic disease management, how might patient lifestyles and healthcare practices evolve if such treatments become the standard of care?
   (a) Patients' quality of life will improve but will require more rigorous monitoring for delayed adverse effects
   (b) There could be a shift towards self-medication practices, minimizing the role of healthcare professionals in chronic disease management
   (c) The focus of healthcare may shift from treatment to prevention, given the increased efficacy and adherence to medication regimens
   (d) Lifestyle changes may become less significant in managing chronic diseases due to the reliability of medication adherence
3. Consider the benefits and challenges of long-acting drug delivery systems. What might be a potential impact of the widespread adoption of long-acting injectable antipsychotics on the public health system?
   (a) Increased demand for specialist healthcare providers for administration
   (b) Reduction in the variety of medications available to patients
   (c) Decreased need for emergency services due to better medication adherence
   (d) Higher costs due to more frequent medication changes
4. In the context of long-acting antiretroviral therapies (LA ARTs), which of the following statements is true regarding LA cabotegravir (CAB) for HIV prevention?
   (a) It increased the number of dosing schedules required annually.
   (b) It has not shown any significant reduction in HIV infection risk compared to oral PrEP agents.
   (c) It became the first long-acting pre-exposure prophylaxis agent approved by the FDA for HIV prevention.
   (d) It is administered daily for optimal effectiveness.
5. What is a significant barrier to the widespread acceptance of intrauterine devices (IUDs) despite their high effectiveness as contraceptives?
   (a) The complexity of the device's mechanism
   (b) The adverse events associated with their usage
   (c) The cost of the device itself
   (d) The difficulty in manufacturing the devices

37.6 Answers

1. (b)
2. (a)
3. (a)
4. (c)
5. (b)

Author Contributions M.U.N., A.S, N.T.H.L, L.P., B.E., and H.E.G. contributing substantially to the writing, illustrations, editing, and or supervision of this chapter.

Funding The work is supported by National Institutes of Health grants R01 NS034239; R01 NS036126; R01 MH115860; R01 NS126089; R01 AI145542; and R01 AI158160.

References

Abdelkader, H., Fathalla, Z., Seyfoddin, A., Farahani, M., et al. (2021). Polymeric long-acting drug delivery systems (LADDS) for treatment of chronic diseases: Inserts, patches, wafers, and implants. *Advanced Drug Delivery Reviews, 177*, 113957. https://doi.org/10.1016/j.addr.2021.113957

Adedini, S. A., Omisakin, O. A., & Somefun, O. D. (2019). Trends, patterns and determinants of long-acting reversible methods of contraception among women in sub-Saharan Africa. *PLoS One, 14*(6), e0217574.

Amico, K. R., & Stirratt, M. J. (2014). Adherence to preexposure prophylaxis: Current, emerging, and anticipated bases of evidence. *Clinical Infectious Diseases, 59*(Suppl 1), S55–S60. https://doi.org/10.1093/cid/ciu266

Asafo-Adjei, T. A., Chen, A. J., Najarzadeh, A., & Puleo, D. A. (2016). Advances in controlled drug delivery for treatment of osteoporosis. *Current Osteoporosis Reports, 14*(5), 226–238. https://doi.org/10.1007/s11914-016-0321-4

Baroletti, S., & Dell'Orfano, H. (2010). Medication adherence in cardiovascular disease. *Circulation, 121*(12), 1455–1458.

Beresford, R., & Ward, A. (1987). Haloperidol decanoate: A preliminary review of its pharmacodynamic and pharmacokinetic properties and therapeutic use in psychosis. *Drugs, 33*, 31–49.

Birgisson, N. E., Zhao, Q., Secura, G. M., Madden, T., et al. (2015). Preventing unintended pregnancy: The contraceptive CHOICE project in review. *Journal of Women's Health, 24*(5), 349–353.

Blach, S., Zeuzem, S., Manns, M., Altraif, I., et al. (2017). Global prevalence and genotype distribution of hepatitis C virus infection in 2015: A modelling study. *The Lancet Gastroenterology & Hepatology, 2*(3), 161–176. https://doi.org/10.1016/S2468-1253(16)30181-9

Bourne, R. R., Flaxman, S. R., Braithwaite, T., Cicinelli, M. V., et al. (2017). Magnitude, temporal trends, and projections of the global prevalence of blindness and distance and near vision impairment: A systematic review and meta-analysis. *The Lancet Global Health, 5*(9), e888–e897.

Brandt, J. D., DuBiner, H. B., Benza, R., Sall, K. N., et al. (2017). Long-term safety and efficacy of a sustained-release bimatoprost ocular ring. *Ophthalmology, 124*(10), 1565–1566.

Brissos, S., Veguilla, M. R., Taylor, D., & Balanzá-Martinez, V. (2014). The role of long-acting injectable antipsychotics in schizophrenia: A critical appraisal. *Therapeutic Advances in Psychopharmacology, 4*(5), 198–219.

Centers for Disease Control and Prevention. (2022). State Unintentional Drug Overdose Reporting System (SUDORS). Atlanta, GA: US Department of Health and Human Services, CDC. Access at: https://www.cdc.gov/drugoverdose/fatal/dashboard

Chue, P. (2007). Long-acting risperidone injection: Efficacy, safety, and cost-effectiveness of the first long-acting atypical antipsychotic. *Neuropsychiatric Disease and Treatment, 3*(1), 13–39. https://doi.org/10.2147/nedt.2007.3.1.13

Clement, M. E., Kofron, R., & Landovitz, R. J. (2020). Long-acting injectable cabotegravir for the prevention of HIV infection. *Current Opinion in HIV and AIDS, 15*(1), 19–26. https://doi.org/10.1097/coh.0000000000000597

Correll, C. U., Citrome, L., Haddad, P. M., Lauriello, J., et al. (2016). The use of long-acting injectable antipsychotics in schizophrenia: Evaluating the evidence. *The Journal of Clinical Psychiatry, 77*(suppl 3), 1–24. https://doi.org/10.4088/JCP.15032su1

Coukell, A. J., & Balfour, J. A. (1998). Levonorgestrel subdermal implants. A review of contraceptive efficacy and acceptability. *Drugs, 55*(6), 861–887. https://doi.org/10.2165/00003495-199855060-00019

Cox, A. L., El-Sayed, M. H., Kao, J. H., Lazarus, J. V., et al. (2020). Progress towards elimination goals for viral hepatitis. *Nature Reviews. Gastroenterology & Hepatology, 17*(9), 533–542. https://doi.org/10.1038/s41575-020-0332-6

Creinin, M. D., Schreiber, C. A., Turok, D. K., Cwiak, C., et al. (2022). Levonorgestrel 52 mg intrauterine system efficacy and safety through 8 years of use. *American Journal of Obstetrics and Gynecology, 227*(6), 871.e871–871.e877. https://doi.org/10.1016/j.ajog.2022.05.022

Curtis, K. M., & Peipert, J. F. (2017). Long-acting reversible contraception. *The New England Journal of Medicine, 376*(5), 461–468. https://doi.org/10.1056/NEJMcp1608736

Darney, P., Patel, A., Rosen, K., Shapiro, L. S., et al. (2009). Safety and efficacy of a single-rod etonogestrel implant (Implanon): Results from 11 international clinical trials. *Fertility and Sterility, 91*(5), 1646–1653. https://doi.org/10.1016/j.fertnstert.2008.02.140

de Filippis, R., De Fazio, P., Gaetano, R., Steardo, L., et al. (2021). Current and emerging long-acting antipsychotics for the treatment of schizophrenia. *Expert Opinion on Drug Safety, 20*(7), 771–790. https://doi.org/10.1080/14740338.2021.1910674

Degasperi, E., Anolli, M. P., & Lampertico, P. (2022). Towards a functional cure for hepatitis B virus: A 2022 update on new antiviral strategies. *Viruses, 14*(11). https://doi.org/10.3390/v14112404

Dehlendorf, C., Reed, R., Fox, E., Seidman, D., et al. (2018). Ensuring our research reflects our values: The role of family planning research in advancing reproductive autonomy. *Contraception, 98*(1), 4–7.

Delany-Moretlwe, S., Hughes, J. P., Bock, P., Ouma, S. G., et al. (2022). Cabotegravir for the prevention of HIV-1 in women: Results from HPTN 084, a phase 3, randomised clinical trial. *Lancet, 399*(10337), 1779–1789. https://doi.org/10.1016/s0140-6736(22)00538-4

DiIorio, C., Shafer, P. O., Letz, R., Henry, T., et al. (2003). The association of stigma with self-management and perceptions of health care among adults with epilepsy. *Epilepsy & Behavior, 4*(3), 259–267.

Dorflinger, L., & Achilles, S. L. (2019). *Injectable contraception*. Speroff & Darney's Clinical Guide to Contraception.

Ekladious, I., Colson, Y. L., & Grinstaff, M. W. (2019). Polymer–drug conjugate therapeutics: Advances, insights and prospects. *Nature Reviews Drug Discovery, 18*(4), 273–294.

Ettinger, M. P., Gallagher, R., & MacCosbe, P. E. (2006). Medication persistence with weekly versus daily doses of orally administered bisphosphonates. *Endocrine Practice, 12*(5), 522–528.

European Association for the Study of the Liver. Electronic address eee and European Association for the Study of the L. (2017). EASL 2017 clinical practice guidelines on the management of hepatitis B virus infection. *Journal of Hepatology, 67*(2), 370–398. https://doi.org/10.1016/j.jhep.2017.03.021

Falzon, D., Schünemann, H. J., Harausz, E., González-Angulo, L., et al. (2017). World Health Organization treatment guidelines for drug-resistant tuberculosis, 2016 update. *The European Respiratory Journal, 49*(3), 1602308. https://doi.org/10.1183/13993003.02308-2016

Farrell, M., Shahbazi, J., Byrne, M., Grebely, J., et al. (2022). Outcomes of a single-arm implementation trial of extended-release subcutaneous buprenorphine depot injections in people with opioid dependence. *The International Journal on Drug Policy, 100*, 103492. https://doi.org/10.1016/j.drugpo.2021.103492

Fisher, B. L., & Potvin, R. (2016). Transzonular vitreous injection vs a single drop compounded topical pharmaceutical regimen after cataract surgery. *Clinical Ophthalmology, 10*, 1297–1303. https://doi.org/10.2147/opth.S112080

Fleseriu, M., Rusch, E., & Geer, E. B. (2017). Safety and tolerability of pasireotide long-acting release in acromegaly-results from the acromegaly, open-label, multicenter, safety monitoring program for treating patients who have a need to receive medical therapy (ACCESS) study. *Endocrine, 55*(1), 247–255. https://doi.org/10.1007/s12020-016-1182-4

Flexner, C., Owen, A., Siccardi, M., & Swindells, S. (2021). Long-acting drugs and formulations for the treatment and prevention of HIV infection. *International Journal of Antimicrobial Agents, 57*(1), 106220. https://doi.org/10.1016/j.ijantimicag.2020.106220

Fonner, V. A., Ridgeway, K., van der Straten, A., Lorenzetti, L., et al. (2023). Safety and efficacy of long-acting injectable cabotegravir as preexposure prophylaxis to prevent HIV acquisition. *AIDS, 37*(6), 957–966. https://doi.org/10.1097/qad.0000000000003494

Funk, A. L., Lu, Y., Yoshida, K., Zhao, T., et al. (2021). Efficacy and safety of antiviral prophylaxis during pregnancy to prevent mother-to-child transmission of hepatitis B virus: A systematic review and meta-analysis. *The Lancet Infectious Diseases, 21*(1), 70–84. https://doi.org/10.1016/S1473-3099(20)30586-7

Gaballa, S. A., Kompella, U. B., Elgarhy, O., Alqahtani, A. M., et al. (2021). Corticosteroids in ophthalmology: Drug delivery innovations, pharmacology, clinical applications, and future perspectives. *Drug Delivery and Translational Research, 11*, 866–893.

Gelenberg, A. (1987). Significant greater loss of medication from the injection site when fluphenazine (Prolixin) decanoate was injected intramuscularly as contrasted with subcutaneously. *The Journal of Clinical Psychiatry, 48*(10), 425–425.

Gomeni, R., Heidbreder, C., Fudala, P., & Nasser, A. (2013). A model-based approach to characterize the population pharmacokinetics and the relationship between the pharmacokinetic and safety profiles of RBP-7000, a new, long-acting, sustained-released formulation of risperidone. *The Journal of Clinical Pharmacology, 53*(10), 1010–1019.

Gray, J. R., Kehoe, L. G., Peckham, A. M., & Wakeman, S. E. (2023). Subcutaneous extended-release buprenorphine depot misdiagnosed as an abscess resulting in incision and drainage and disruption of opioid use disorder treatment. *Journal of Addiction Medicine, 17*(2), 227–229. https://doi.org/10.1097/adm.0000000000001067

Grimes, D. A. (2009). Forgettable contraception. *Contraception, 80*(6), 497–499.

Halpern, V., Stalter, R. M., Owen, D. H., Dorflinger, L. J., et al. (2015). Towards the development of a longer-acting injectable contraceptive: Past research and current trends. *Contraception, 92*(1), 3–9. https://doi.org/10.1016/j.contraception.2015.02.014

Harrison, R. K. (2016). Phase II and phase III failures: 2013–2015. *Nature Reviews. Drug Discovery, 15*(12), 817–818.

Healthcare, V. (2021). *Cabenuva® prescribing information*. Viiv Healthcare.

Heres, S., Reichhart, T., Hamann, J., Mendel, R., et al. (2011). Psychiatrists' attitude to antipsychotic depot treatment in patients with first-episode schizophrenia. *European Psychiatry, 26*(5), 297–301. https://doi.org/10.1016/j.eurpsy.2009.12.020

Heres, S., Kraemer, S., Bergstrom, R. F., & Detke, H. C. (2014). Pharmacokinetics of olanzapine long-acting injection: The clinical perspective. *International Clinical Psychopharmacology, 29*(6), 299.

Higgins, J. A., Sanders, J. N., Palta, M., & Turok, D. K. (2016). Women's sexual function, satisfaction, and perceptions after starting long-acting reversible contraceptives. *Obstetrics and Gynecology, 128*(5), 1143–1151. https://doi.org/10.1097/aog.0000000000001655

Hochberg, M. C., Altman, R. D., Brandt, K. D., Clark, B. M., et al. (1995). Guidelines for the medical management of osteoarthritis. Part II. Osteoarthritis of the knee. American College of Rheumatology. *Arthritis and Rheumatism, 38*(11), 1541–1546. https://doi.org/10.1002/art.1780381104

Hohmann, H., & Creinin, M. D. (2007). The contraceptive implant. *Clinical Obstetrics and Gynecology, 50*(4), 907–917. https://doi.org/10.1097/GRF.0b013e318159c2f6

Hollifield, J. (1986). Clinical acceptability of transdermal clonidine: A large-scale evaluation by practitioners. *American Heart Journal, 112*(4), 900–906.

Hubacher, D., & Kavanaugh, M. (2018). Historical record-setting trends in IUD use in the United States. *Contraception, 98*(6), 467–470.

International Planned Parenthood Federation IPPF. (1993). IMAP statement on Norplant subdermal contraceptive implant system. *IPPF Medical Bulletin, 27*(2), 1–3.

Jaffe, G. J., Ben-nun, J., Guo, H., Dunn, J. P., et al. (2000). Fluocinolone acetonide sustained drug delivery device to treat severe uveitis. *Ophthalmology, 107*(11), 2024–2033.

Johnson, D. A. (2009). Historical perspective on antipsychotic long-acting injections. *The British Journal of Psychiatry. Supplement, 52*, S7–S12. https://doi.org/10.1192/bjp.195.52.s7

Karavasili, C., & Fatouros, D. G. (2016). Smart materials: in situ gel-forming systems for nasal delivery. *Drug Discovery Today, 21*(1), 157–166.

Kavanaugh, M. L., & Jerman, J. (2018). Contraceptive method use in the United States: Trends and characteristics between 2008, 2012 and 2014. *Contraception, 97*(1), 14–21. https://doi.org/10.1016/j.contraception.2017.10.003

Kavanaugh, M. L., Jerman, J., Hubacher, D., Kost, K., et al. (2011). Characteristics of women in the United States who use long-acting reversible contraceptive methods. *Obstetrics and Gynecology, 117*(6), 1349–1357.

Kendler, D., Kung, A. W. C., Fuleihan, G. E.-H., González, J. G. G., et al. (2004). Patients with osteoporosis prefer once weekly to once daily dosing with alendronate. *Maturitas, 48*(3), 243–251.

Kennel, K. A., & Drake, M. T. (2009). *Adverse effects of bisphosphonates: Implications for osteoporosis management.* Paper presented at the Mayo Clinic Proceedings.

Kim, S., Choi, Y. S., Kim, J. S., Kim, S., et al. (2019). Experiences of localization and removal of non-palpable subdermal contraceptive implants with ultrasound. *Obstetrics & Gynecology Science, 62*(3), 166–172. https://doi.org/10.5468/ogs.2019.62.3.166

Kishimoto, T., Nitta, M., Borenstein, M., Kane, J. M., et al. (2013). Long-acting injectable versus oral antipsychotics in schizophrenia: A systematic review and meta-analysis of mirror-image studies. *The Journal of Clinical Psychiatry, 74*(10), 957–965. https://doi.org/10.4088/JCP.13r08440

Kleiner, L. W., Wright, J. C., & Wang, Y. (2014). Evolution of implantable and insertable drug delivery systems. *Journal of Controlled Release, 181*, 1–10.

Koehl, J. L., Zimmerman, D. E., & Bridgeman, P. J. (2019). Medications for management of opioid use disorder. *American Journal of Health-System Pharmacy, 76*(15), 1097–1103. https://doi.org/10.1093/ajhp/zxz105

Kola, I., & Landis, J. (2004). Can the pharmaceutical industry reduce attrition rates? *Nature Reviews Drug Discovery, 3*(8), 711–716.

Krovi, S. A., Gallovic, M. D., Keller, A. M., Bhat, M., et al. (2018). Injectable long-acting human immunodeficiency virus antiretroviral prodrugs with improved pharmacokinetic profiles. *International Journal of Pharmaceutics, 552*(1–2), 371–377. https://doi.org/10.1016/j.ijpharm.2018.10.017

Kumar, N., Simek, S., Garg, N., Vaduganathan, M., et al. (2019). Thirty-day readmissions after hospitalization for hypertensive emergency. *Hypertension, 73*(1), 60–67.

Label, F. (n.d.). Retrieved from https://www.accessdata.fda.gov/drugsatfda_docs/label/2010/020246s036lbl.pdf

Laffont, C. M., Gomeni, R., Zheng, B., Heidbreder, C., et al. (2014). Population pharmacokinetics and prediction of dopamine D 2 receptor occupancy after multiple doses of RBP-7000, a new sustained-release formulation of risperidone, in schizophrenia patients on stable Oral risperidone treatment. *Clinical Pharmacokinetics, 53*, 533–543.

Landovitz, R. J., Donnell, D., Clement, M. E., Hanscom, B., et al. (2021). Cabotegravir for HIV prevention in cisgender men and transgender women. *The New England Journal of Medicine, 385*(7), 595–608. https://doi.org/10.1056/NEJMoa2101016

Li, L., Johnson, L. M., Krovi, S. A., Demkovich, Z. R., et al. (2020). Performance and stability of Tenofovir Alafenamide formulations within subcutaneous biodegradable implants for HIV pre-exposure prophylaxis (PrEP). *Pharmaceutics, 12*(11). https://doi.org/10.3390/pharmaceutics12111057

Liu, G., Franssen, E., Fitch, M. I., & Warner, E. (1997). Patient preferences for oral versus intravenous palliative chemotherapy. *Journal of Clinical Oncology, 15*(1), 110–115.

Llorca, P. M., Abbar, M., Courtet, P., Guillaume, S., et al. (2013). Guidelines for the use and management of long-acting injectable antipsychotics in serious mental illness. *BMC Psychiatry, 13*, 340. https://doi.org/10.1186/1471-244x-13-340

Lynn, M. M., & Holdcroft, C. (1992). New concepts in contraception: Norplant subdermal implant. *The Nurse Practitioner, 17*(3), 85–89. https://doi.org/10.1097/00006205-199203000-00016

Macintyre, F., Ramachandruni, H., Burrows, J. N., Holm, R., et al. (2018). Injectable anti-malarials revisited: Discovery and development of new agents to protect against malaria. *Malaria Journal, 17*(1), 402. https://doi.org/10.1186/s12936-018-2549-1

Malclès, A., Dot, C., Voirin, N., Vié, A.-L., et al. (2017). Safety of intravitreal dexamethasone implant (Ozurdex): The SAFODEX study. Incidence and risk factors of ocular hypertension. *Retina, 37*(7), 1352–1359.

Mansi, S., Martin, R., Huecker, M. R. (2023). Opioid Withdrawal In: StatPearls. Treasure Island (FL): StatPearls Publishing; 2024 Jan. 2023 Jul 21.

Mansour, D. (2014). International survey to assess women's attitudes regarding choice of daily versus nondaily female hormonal contraception. *International Journal Of Women's Health, 6*, 367–375.

Mansour, D., Fraser, I. S., Edelman, A., Vieira, C. S., et al. (2019). Can initial vaginal bleeding patterns in etonogestrel implant users predict subsequent bleeding in the first 2 years of use? *Contraception, 100*(4), 264–268. https://doi.org/10.1016/j.contraception.2019.05.017

Marshall, V. K. (2018). Factors associated with medication beliefs in patients with cancer: An integrative review. *Oncology Nursing Forum, 45*(4), 508–526.

Meira, J., Madeira, C., Falcão-Reis, F., & Figueira, L. (2019). Sustained control from recurring non-infectious uveitic macular edema with 0.19 mg fluocinolone acetonide intravitreal implant–a case report. *Ophthalmology and therapy, 8*, 635–641.

Meyers, K., Rodriguez, K., Moeller, R. W., Gratch, I., et al. (2014). High interest in a long-acting injectable formulation of pre-exposure prophylaxis for HIV in young men who have sex with men in NYC: A P18 cohort substudy. *PLoS One, 9*(12), e114700. https://doi.org/10.1371/journal.pone.0114700

Mills, A., Richmond, G. J., Newman, C., Osiyemi, O., et al. (2022). Long-acting cabotegravir and rilpivirine for HIV-1 suppression: Switch to 2-monthly dosing after 5 years of daily oral therapy. *AIDS, 36*(2), 195–203. https://doi.org/10.1097/qad.0000000000003085

Mishell, D. R., Jr. (1996). Pharmacokinetics of depot medroxyprogesterone acetate contraception. *The Journal of Reproductive Medicine, 41*(5 Suppl), 381–390.

Mitchell, M., Kothare, P., Bergstrom, R., Zhao, F., et al. (2013). Single- and multiple-dose pharmacokinetic, safety, and tolerability profiles of olanzapine long-acting injection: An open-label, multicenter, nonrandomized study in patients with schizophrenia. *Clinical Therapeutics, 35*(12), 1890–1908.

Moehrle, J. J. (2022). Development of new strategies for malaria chemoprophylaxis: From monoclonal antibodies to long-acting injectable drugs. *Tropical Medicine and Infectious Disease, 7*(4). https://doi.org/10.3390/tropicalmed7040058

Moore, T. A., Buchanan, R. W., Buckley, P. F., Chiles, J. A., et al. (2007). The Texas medication algorithm project antipsychotic algorithm for schizophrenia: 2006 update. *The Journal of Clinical Psychiatry, 68*(11), 1751–1762. https://doi.org/10.4088/jcp.v68n1115

Morris, M. T., & Tarpada, S. P. (2017). Long-acting injectable Paliperidone palmitate: A review of efficacy and safety. *Psychopharmacology Bulletin, 47*(2), 42–52.

Mullard, A. (2016). Parsing clinical success rates. *Nature Reviews Drug Discovery, 15*(7), 447–448.

Naber, D., Hansen, K., Forray, C., Baker, R. A., et al. (2015). Qualify: A randomized head-to-head study of aripiprazole once-monthly and paliperidone palmitate in the treatment of schizophrenia. *Schizophrenia Research, 168*(1–2), 498–504. https://doi.org/10.1016/j.schres.2015.07.007

Ngo, A. L., Orhurhu, V., Urits, I., Delfin, E. O., et al. (2019). Extended release granisetron: Review of pharmacologic considerations and clinical role in the perioperative setting. *Saudi Journal of Anaesthesia, 13*(3), 231–236. https://doi.org/10.4103/sja.SJA_817_18

Nieuwlaat, R., Wilczynski, N., Navarro, T., Hobson, N., et al. (2014). Interventions for enhancing medication adherence. *Cochrane Database of Systematic Reviews, 2014*(11), CD000011.

Nordmann, J. P., Baudouin, C., Renard, J. P., Denis, P., et al. (2010). Measurement of treatment compliance using a medical device for glaucoma patients associated with intraocular pressure control: A survey. *Clinical Ophthalmology, 4*, 731–739. https://doi.org/10.2147/opth.s11799

Olagunju, A. T., Clark, S. R., & Baune, B. T. (2019). Long-acting atypical antipsychotics in schizophrenia: A systematic review and meta-analyses of effects on functional outcome. *The Australian and New Zealand Journal of Psychiatry, 53*(6), 509–527. https://doi.org/10.1177/0004867419837358

organization wh. (2022). *Global tuberculosis program.* Retrieved from https://www.who.int/teams/global-tuberculosis-programme/tb-reports/global-tuberculosis-report-2022

Orkin, C., Arasteh, K., Górgolas Hernández-Mora, M., Pokrovsky, V., et al. (2020). Long-acting Cabotegravir and Rilpivirine after Oral induction for HIV-1 infection. *The New England Journal of Medicine, 382*(12), 1124–1135. https://doi.org/10.1056/NEJMoa1909512

Overton, E. T., Richmond, G., Rizzardini, G., Jaeger, H., et al. (2021). Long-acting cabotegravir and rilpivirine dosed every 2 months in adults with HIV-1 infection (ATLAS-2M), 48-week results: A randomised, multicentre, open-label, phase 3b, non-inferiority study. *Lancet, 396*(10267), 1994–2005. https://doi.org/10.1016/s0140-6736(20)32666-0

Paik, J., Duggan, S. T., & Keam, S. J. (2019). Triamcinolone Acetonide extended-release: A review in osteoarthritis pain of the knee. *Drugs, 79*(4), 455–462. https://doi.org/10.1007/s40265-019-01083-3

Patel, A., Cholkar, K., Agrahari, V., & Mitra, A. K. (2013). Ocular drug delivery systems: An overview. *World Journal of Pharmacology, 2*(2), 47.

Peacock, E., & Krousel-Wood, M. (2017). Adherence to antihypertensive therapy. *Medical Clinics, 101*(1), 229–245.

Peters-Strickland, T., Baker, R. A., McQuade, R. D., Jin, N., et al. (2015). Aripiprazole once-monthly 400 mg for long-term maintenance treatment of schizophrenia: A 52-week open-label study. *NPJ Schizophrenia, 1*, 15039. https://doi.org/10.1038/npjschz.2015.39

Plowe, C. V. (2022). Malaria chemoprevention and drug resistance: A review of the literature and policy implications. *Malaria Journal, 21*(1), 104. https://doi.org/10.1186/s12936-022-04115-8

Raedler, L. A. (2016). Aripiprazole Lauroxil (Aristada): Long-acting atypical antipsychotic injection approved for the treatment of patients with schizophrenia. *American Health & Drug Benefits, 9*(Spec Feature), 40–43.

Ravenstijn, P., Remmerie, B., Savitz, A., Samtani, M. N., et al. (2016). Pharmacokinetics, safety, and tolerability of paliperidone palmitate 3-month formulation in patients with schizophrenia: A phase-1, single-dose, randomized, open-label study. *The Journal of Clinical Pharmacology, 56*(3), 330–339.

Ross, J., & Stover, J. (2013). Use of modern contraception increases when more methods become available: Analysis of evidence from 1982–2009. *Global Health: Science and Practice, 1*(2), 203–212. https://doi.org/10.9745/ghsp-d-13-00010

Schoretsanitis, G., Spina, E., Hiemke, C., & de Leon, J. (2018). A systematic review and combined analysis of therapeutic drug monitoring studies for long-acting paliperidone. *Expert Review of Clinical Pharmacology, 11*(12), 1237–1253.

Schöttle, D., Janetzky, W., Luedecke, D., Beck, E., et al. (2018). Effectiveness of aripiprazole once-monthly in schizophrenia patients pretreated with oral aripiprazole: A 6-month, real-life non-interventional study. *BMC Psychiatry, 18*(1), 365. https://doi.org/10.1186/s12888-018-1946-x

Schreiner, A., Bergmans, P., Cherubin, P., Keim, S., et al. (2014). A prospective flexible-dose study of paliperidone palmitate in non-acute but symptomatic patients with schizophrenia previously unsuccessfully treated with oral antipsychotic agents. *Clinical Therapeutics, 36*(10), 1372–1388.e1371. https://doi.org/10.1016/j.clinthera.2014.08.014

Secura, G. M., Madden, T., McNicholas, C., Mullersman, J., et al. (2014). Provision of no-cost, long-acting contraception and teenage pregnancy. *The New England Journal of Medicine, 371*(14), 1316–1323.

Shen, J., & Burgess, D. J. (2011). Drugs for long acting injections and implants. In *Long acting injections and implants* (pp. 73–91). Springer.

Shih, H.-P., Zhang, X., & Aronov, A. M. (2018). Drug discovery effectiveness from the standpoint of therapeutic mechanisms and indications. *Nature Reviews Drug Discovery, 17*(1), 19–33.

Simon, J. A., Lewiecki, E. M., Smith, M. E., Petruschke, R. A., et al. (2002). Patient preference for once-weekly alendronate 70 mg versus once-daily alendronate 10 mg: A multicenter, randomized, open-label, crossover study. *Clinical Therapeutics, 24*(11), 1871–1886.

Sirey, J. A., Bruce, M. L., Alexopoulos, G. S., Perlick, D. A., et al. (2001). Stigma as a barrier to recovery: Perceived stigma and patient-rated severity of illness as predictors of antidepressant drug adherence. *Psychiatric Services, 52*(12), 1615–1620.

Smith, G. H. R., Henry, W. K., Podzamczer, D., Masiá, M. D. M., et al. (2021). Efficacy, safety, and durability of long-acting Cabotegravir and Rilpivirine in adults with human immunodeficiency virus type 1 infection: 5-year results from the LATTE-2 study. *Open Forum*

Infectious Diseases, 8(9), ofab439. https://doi.org/10.1093/ofid/ofab439

Snider, M. E., Nuzum, D. S., & Veverka, A. (2008). Long-acting nifedipine in the management of the hypertensive patient. *Vascular Health and Risk Management, 4*(6), 1249–1257. https://doi.org/10.2147/vhrm.s3661

Srivastava, A. B., Mariani, J. J., & Levin, F. R. (2020). New directions in the treatment of opioid withdrawal. *Lancet, 395*(10241), 1938–1948. https://doi.org/10.1016/s0140-6736(20)30852-7

Stout, A., Wood, S., Barigye, G., Kaboré, A., et al. (2018). Expanding access to injectable contraception: Results from pilot introduction of subcutaneous depot medroxyprogesterone acetate (DMPA-SC) in 4 African countries. *Global Health: Science and Practice, 6*(1), 55–72.

Subotnik, K. L., Casaus, L. R., Ventura, J., Luo, J. S., et al. (2015). Long-acting injectable risperidone for relapse prevention and control of breakthrough symptoms after a recent first episode of schizophrenia. A randomized clinical trial. *JAMA Psychiatry, 72*(8), 822–829. https://doi.org/10.1001/jamapsychiatry.2015.0270

Suri, R., Beg, S., & Kohli, K. (2020). Target strategies for drug delivery bypassing ocular barriers. *Journal of Drug Delivery Science and Technology, 55*, 101389.

Swift, R., Oslin, D. W., Alexander, M., & Forman, R. (2011). Adherence monitoring in naltrexone pharmacotherapy trials: A systematic review. *Journal of Studies on Alcohol and Drugs, 72*(6), 1012–1018.

Swindells, S., Lutz, T., Van Zyl, L., Porteiro, N., et al. (2022). Week 96 extension results of a phase 3 study evaluating long-acting cabotegravir with rilpivirine for HIV-1 treatment. *AIDS, 36*(2), 185–194. https://doi.org/10.1097/qad.0000000000003025

Tanum, L., Solli, K. K., Z-e-H, L., Benth, J. Š., et al. (2017). Effectiveness of injectable extended-release naltrexone vs daily buprenorphine-naloxone for opioid dependence: A randomized clinical noninferiority trial. *JAMA Psychiatry, 74*(12), 1197–1205. https://doi.org/10.1001/jamapsychiatry.2017.3206

Taylor, M., & Perera, U. (2015). NICE CG178 psychosis and schizophrenia in adults: Treatment and management - an evidence-based guideline? *The British Journal of Psychiatry, 206*(5), 357–359. https://doi.org/10.1192/bjp.bp.114.155945

Tepper, N. K., Marchbanks, P. A., & Curtis, K. M. (2014). US selected practice recommendations for contraceptive use, 2013. *Journal of Women's Health, 23*(2), 108–111.

Thompson, P., Morgan, C. E., Ngimbi, P., Mwandagalirwa, K., et al. (2021). Arresting vertical transmission of hepatitis B virus (AVERT-HBV) in pregnant women and their neonates in The Democratic Republic of the Congo: A feasibility study. *The Lancet Global Health, 9*(11), e1600–e1609. https://doi.org/10.1016/S2214-109X(21)00304-1

Tiberi, S., du Plessis, N., Walzl, G., Vjecha, M. J., et al. (2018). Tuberculosis: Progress and advances in development of new drugs, treatment regimens, and host-directed therapies. *The Lancet Infectious Diseases, 18*(7), e183–e198. https://doi.org/10.1016/s1473-3099(18)30110-5

UNFPO. (2022). *Seeing the unseen: The case for action in the neglected crisis of unintended pregnancy*. https://www.unfpa.org/swp2022

Villanueva-Martínez, A., Merino, V., & Ganem-Rondero, A. (2022). Transdermal formulations and strategies for the treatment of osteoporosis. *Journal of Drug Delivery Science and Technology, 69*, 103111. https://doi.org/10.1016/j.jddst.2022.103111

Wang, Y., & Burgess, D. J. (2012). Microsphere technologies. In J. C. Wright & D. J. Burgess (Eds.), *Long acting injections and implants* (pp. 167–194). Springer US.

WHO. (2022). Retrieved from https://www.newtbdrugs.org/news/who--updates-treatment-guidance-shorter-novel-regimens-drug-resistant-tuberculosis

Wieder, D. R., & Pattimakiel, L. (2010). Examining the efficacy, safety, and patient acceptability of the combined contraceptive vaginal ring (NuvaRing). *International Journal of Women's Health, 2*, 401–409. https://doi.org/10.2147/ijwh.S6162

Winner, B., Peipert, J. F., Zhao, Q., Buckel, C., et al. (2012). Effectiveness of long-acting reversible contraception. *The New England Journal of Medicine, 366*(21), 1998–2007.

Wood, S. N., Magalona, S., Zimmerman, L. A., OlaOlorun, F., et al. (2022). Self-injected contraceptives: Does the investment reflect women's preferences? *BMJ Global Health, 7*(7). https://doi.org/10.1136/bmjgh-2022-008862

Wright, J. C., & Burgess, D. J. (2012). *Long acting injections and implants*. Springer.

Xiang, A. H., Kawakubo, M., Kjos, S. L., & Buchanan, T. A. (2006). Long-acting injectable progestin contraception and risk of type 2 diabetes in Latino women with prior gestational diabetes mellitus. *Diabetes Care, 29*(3), 613–617. https://doi.org/10.2337/diacare.29.03.06.dc05-1940

Zalba, S., Ten Hagen, T. L., Burgui, C., & Garrido, M. J. (2022). Stealth nanoparticles in oncology: Facing the PEG dilemma. *Journal of Controlled Release, 351*, 22–36.

Zoorob, R., Kowalchuk, A., & Mejia de Grubb, M. (2018). Buprenorphine therapy for opioid use disorder. *American Family Physician, 97*(5), 313–320.

Ultra-Long-Acting Drug Delivery Formulations

38

Mohammad Ullah Nayan, Ashrafi Sultana, Nam Thai Hoang Le, Larisa Y. Poluektuva, Howard E. Gendelman, and Benson Edagwa

Abstract

The emergence of novel infectious pathogens and the growing burden of drug-resistant infections underscore the need for new treatments. However, the discovery and development of new safely administered drugs with acceptable pharmacokinetics and pharmacodynamic profiles are both costly and time-dependent. Knowing inherent drug development risks underlies the protracted time required in drug development. Years between discovery to broad usage is common prior to any new therapy comes available to humans at risk for acquisition or disease. Therefore, extended-action medicines combined with novel drug delivery strategies are timely. Both are increasingly being researched to speed up drug discovery and development challenges. This chapter focuses on one component of drug development which is prodrugs. Such drug development strategies can be used to improve drug-like properties of potent parent compounds with poor Absorption, Distribution, Metabolism, Elimination, and Toxicity (ADMET) profiles, thereby reducing costs and drug development timelines.

Keywords

Long-acting drug delivery · Psychosis · Contraception · Implants · Injectables · Microarray needles

38.1 Introduction

Prodrugs have successfully enabled improvements in the physicochemical and pharmacological properties in up to 10% of the clinically available therapeutic agents (Rautio et al., 2008). Prodrugs are produced through chemical derivatization of parent drug molecules to produce inactive forms that require chemical and/or enzymatic processing to produce metabolites or original forms of drugs that are pharmacologically active (Rautio et al., 2018). The rate and site of prodrug activation depend on the nature of cleavable chemical bonds and the type of derivatizing promoiety. An ideal prodrug must be pharmacologically inert and be designed to undergo breakdown to release the active form of the drug and non-toxic byproducts upon target site delivery. The nature of functional groups in the parent molecule determines the type of cleavable prodrug bonds created and these include esters, carbonates, carbamates, amides, phosphates, and oximes. The presence of receptive handles on the parent drugs to be modified facilitate covalent linkage of cleavable promoieties to create prodrugs (Rautio et al., 2008). Other factors to be considered when designing ideal prodrugs include defining the purpose of derivatization and intended route of administration where orally administered drugs seek to optimize enhanced water solubility while prodrugs intended to simplify dosing schedules for intramuscularly or subcutaneously administered compounds focus on enhancing lipophilicity and hydrophobicity to slow dissolution. Understanding ADMET profiles of both the parent drug and prodrug is critical during prodrug design since not all modifications produce the desired improvements in the parent molecules. For instance, if a prodrug strategy is introduced to improve poor bioavailability caused by high affinity for efflux transporters in the gastrointestinal tract by the parent drug, then the prodrug should remain stable in the intestines to allow for absorption. For parent compounds that are known to have high first-pass metabolism in the liver, optimal prodrug strategies would be aimed at facilitating drug biodistribution to other tissue compartments. A comprehensive understanding of the mechanisms involved in the prodrug activation process helps in defining to what degree the prodrug enables the desired improvements. Importantly, bioactivation of the prodrug should occur at sufficient rates to produce therapeutic concentrations of the parent drug.

M. U. Nayan · A. Sultana · N. T. H. Le · L. Y. Poluektuva ·
H. E. Gendelman (✉) · B. Edagwa (✉)
Department of Pharmacology and Experimental Neuroscience,
University of Nebraska Medical Center, Omaha, NE, USA
e-mail: hegendel@unmc.edu; benson.edagwa@unmc.edu

H. E. Gendelman, T. Ikezu (eds.), *Neuroimmune Pharmacology and Therapeutics*,
https://doi.org/10.1007/978-3-031-68237-7_38

The prodrug strategy offers opportunities for overcoming parent drug challenges, and this has been thoroughly reviewed elsewhere (Rautio et al., 2008). This chapter will focus on recent advancements for a broad range of long-acting (LA) drugs administered by implants, injectables, and microarray needles with relevant examples where various prodrug strategies have been explored. Chronic diseases aided by these new treatments include psychosis, viral infections, contraception, and ocular diseases. Recent basic and translational studies serve to outline the potential for improved treatment, convenience, and accessibility.

Over the past decade, increases in the global person, family, and community burden of chronic degenerative, infectious, psychiatric, cardiovascular, and inflammatory diseases have been realized. These account for three out of five deaths on a global scale (Marengoni et al., 2011; Wang et al., 2016). While effective medications are available, frequent dosing leads to poor adherence. Therapeutic efficacy is limited by forgetfulness, toxicity, motivation, health literacy, and social stigmas (Iuga & McGuire, 2014). It is estimated that more than 50% of patients stricken with chronic diseases struggle with taking the prescribed medicines regularly, which poses substantial challenges for disease management (Alosaimi et al., 2022). The consequences are significant in health and economic outcomes. In the United States alone, medications not taken are associated with repeated hospitalization, rises in healthcare costs, and up to 125,000 annual deaths (Osterberg & Blaschke, 2005). Stoppage of antipsychotic medications alone results in up to a fivefold increase in relapse rate and hospitalizations for psychosis compared to continued treatments (Robinson et al., 1999). Moreover, medication non-adherence is the leading cause of effectiveness gaps in clinical trial assessments (Breckenridge et al., 2017; Carls et al., 2017).

Developing an LA medicine is a promising strategy to diminish barriers to regimen adherence. Such barriers include frequent dosing, side effects, cognitive impairment and complexity of treatment. LA medicines serve to improve pharmacokinetic (PK) and pharmacodynamic (PD) profiles by extending the apparent half-life of drugs to days, weeks, or even months. For example, once-a-week medication for women with osteoporosis showed improved outcomes over daily oral medicines (Kendler et al., 2004; Simon et al., 2002). Similar trends were reported for diabetes, cardiac, psychiatric, and viral disease management (Blackwood et al., 2020; Flexner et al., 2021; Hollifield, 1986). Nevertheless, LA drug discovery is time-consuming and resource-intensive, with a high likelihood of failure during the preclinical or clinical development stages (Garber, 2011; Ma et al., 2020). The use of LA drug delivery systems to prolong the half-life of already approved medicines provides alternatives that can ensure safety, efficacy and regulatory approvals.

LA formulations use drug-eluting "depot" platforms to permit sustained drug release. Advances in medicinal and polymer chemistry have been harnessed to improve PK and PD profiles seen through implants, polymeric nano or microparticles, microarray needles, nanocrystals, and liposomal drug formulations, with each carrying unique characteristics and challenges. For instance, biodegradable implant or injectable formulations, while desirable, may exhibit high burst release, posing challenges for drugs with narrow therapeutic windows and potentially impacting formulation longevity (Fu & Kao, 2010). Additionally, injectable LA micro or nanoformulations are suitable for hydrophobic drugs. Hydrophilic drugs and biologics may be formulated into LA micro- or nanoformulation by hydrophobic compound conversions (Buwalda et al., 2017).

Key formulation design considerations for implants and injectables are developed. Implants are utilized for contraceptive and ocular usage. This is based on their ability to slowly release drugs at low concentrations over long time periods and the opportunity for removal. Recent advances in drug-eluting polymer and manufacturing techniques widen implant applications. Numerous technologies have been employed to formulate LA injectable (LAI) drugs. These include the use of oil-based formulations for hydrophobic and lipophilic drugs, PLGA-based microspheres for hydrophilic drugs, and in situ forming gel-based depots. Additionally, advancements in surface tailoring have led to the development of highly effective aqueous nanocrystal formulations. Each shows unique developmental challenges.

Patients' preferences directly influence treatment adherence and, as such, disease outcomes. To maximize therapeutic impact, LA drugs must be designed in consideration of the disease being treated as well as patient preferences. For example, patients facing life-threatening diseases may prioritize efficacy. Patients may tolerate more adverse effects of cancer treatment if improvements in efficacy can be realized. On the other hand, preventative diseases that require self-administration demand ease of administration with limited toxicities. One notable example is the self-administrable subcutaneous (SC) depot of medroxyprogesterone acetate (DMPA) used in the prevention of pregnancy. The SC injections are preferred over intramuscular (IM) DMPA injections that require healthcare professionals' (HCPs') assistance (Stout et al., 2018; Wood et al., 2022). This chapter offers an overview of the latest advancements in LA drug delivery devices for the management of chronic diseases. Furthermore, it provides valuable insights into the critical design considerations of these LA drug delivery devices to facilitate their seamless and widespread adoption and utilization (Fig. 38.1).

38.2 LA Medicines for Chronic Infections

38.2.1 Implants

The development of the first SC hormonal implant in the 1950s marked the beginning of extensive research. Subsequent advancements in polymer chemistry have propelled the evolution of implantable drug delivery systems, resulting in increasingly refined and effective approaches (Reinstatler et al., 2015; Santos et al., 2014). These implants can be made from biodegradable polymers like PLGA, or PCL or non-biodegradable polymers like polyurethane, and silicones. Both biodegradable and non-biodegradable polymers have their unique advantages. Degradable polymers offer enhanced safety by breaking down into non-toxic byproducts. These degraded products can be easily eliminated from the body, eliminating the need for secondary surgery to retrieve the implant after treatment completion. On the other hand, non-degradable polymers exhibit excellent biocompatibility and utilize simpler drug release mechanisms, such as polymer swelling or diffusion, that allow tuning drug releases from implants (Kumar & Pillai, 2018).

There are several types of implants that are available or are in development. These include monolithic matrix or reservoir-type, swelling control, injectable in situ forming implants (ISFIs) or hydrogels. Monolithic systems involve a matrix where the drug is uniformly dispersed. Reservoir systems consist of non-degradable membranes, surrounding a

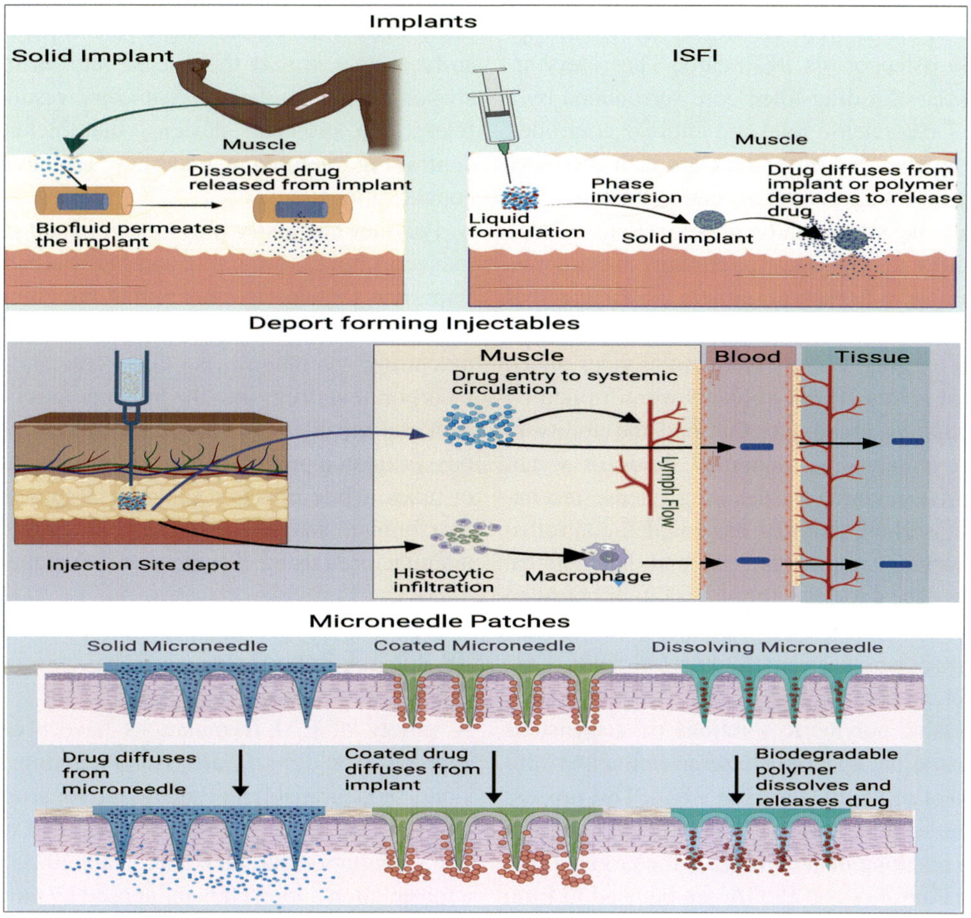

Fig. 38.1 Drug release mechanism of different LA formulations. (**a**) Solid non-biodegradable implants: Biofluid permeates the semipermeable polymer membrane and dissolves the drugs. Then, the dissolved drugs are released from the implants. The drug release rate depends on the nature of the polymer and the dissolution properties of drugs. In-situ forming implants: Drugs and polymers are dissolved in a biocompatible organic solvent and injected as a liquid formulation. After injection, the liquid formulation undergoes a phase inversion to form a solid implant. Drugs are released from the solid implant either by diffusion or polymer degradation. (**b**) Depot-forming injectables: After IM or SC injection, the formulation forms a depot at the injection site, and the drug release rate from the depot depends on the dissolution of drugs in biological fluids. Mononuclear phagocytes can phagocytize the drug formulations and form a secondary cellular depot. (**c**) Microneedle patches: Solid microneedles made with non-biodegradable polymer: Drugs diffuse from the semipermeable membrane to biological fluid. Coated microneedles: Drugs are coated in the outer layer of the microneedle, and drugs diffuse from the outer layers after administration. Dissolving microneedles: They contain biodegradable polymers that dissolve upon application to release drugs

drug-filled core reservoir. Swelling-controlled systems are constructed using water-soluble cross-linked polymers.

Drug-dispersed polymer matrices in monolithic systems can be prepared by a physical mixture of the drug and polymer, compression of polymer and drug, or physicochemical activation without solvents. These matrix-type implants can increase the dosage regimen of drugs from days to months. For instance, several bisphosphonate alendronate sodium (ALN) loaded monolithic matrix-type implants were reported. These are made of bioerodible poly(lactic acid) (PLA) and PLGA. All implants exhibit sustained ALN release between 25 to 130 days; whereas implants of greater glycolide composition and higher ALN loadings release ALN rapidly (Dharmayanti et al., 2020). The antiretroviral drug islatravir was produced with an eluting matrix EVA implant with the potential to extend dosing intervals for six months (Ullah Nayan et al., 2023).

Reservoir-type implants were developed for contraception and HIV and osteoporosis treatments. The reservoir-type implants consist of a drug-filled core surrounded by a polymer coating. Drug release rates are entirely controlled by the polymer properties. By varying factors such as molecular weight, chemical configuration, or coating thickness, the release profile can be significantly altered (Pons-Faudoa et al., 2019). Jadelle®, a levonorgestrel-eluting contraceptive implant, is an example of a reservoir system with a five-year effective use period (Cullins, 1992). Swelling-controlled hydrogel implants for drug delivery were initially developed in 1960 for contact lenses. These implants were formed by crosslinking hydrophilic monomers that had the ability to absorb significant amounts of biological fluids or water. Since then, extensive research on the drug release mechanism from water-absorbing polymer has enabled controlled drug release, by varying temperature, pH, and the chemical polymer properties. The development of swelling-controlled implants was reported for chronic conditions like cardiovascular diseases, contact lenses, and antibiotics (Almeida et al., 2012; Gustafson et al., 2016).

ISFIs are injectable polymeric solutions or suspensions that, upon administration, undergo phase inversion and convert into semisolid or solid implants (Fig. 38.2). This process allows the drug molecules/particles to be effectively encapsulated within the resulting in situ matrix at the targeted site. Polymers such as Pluronics and PLGA can be used to form ISFI, and various factors such as temperature, ionic strength, pH, solvent exchange, and light can trigger the phase transition of these systems. ISFIs have been extensively used for contraception, ocular, and infectious diseases (Abdelkader et al., 2020; Dai et al., 2020; Ge et al., 2020). A novel PLGA and N-methyl-2-pyrrolidone (NMP) based LA ISFI system was developed for delivering antiretroviral drugs, such as dolutegravir (DTG) and CAB. In addition, hormonal contraceptives such as medroxyprogesterone acetate (MPA) were also developed using this approach (Benhabbour et al., 2019; Young et al., 2023).

The design processes for chronic disease implants require a consideration of the disease, required drug levels, potential adverse reactions, reliability, affordability, and ease of use. These considerations often influence decisions in implant manufacturing, such as choosing between matrix and reservoir systems, drug release kinetics, polymers, drug loading, manufacturing techniques, drug release, and in vivo validation. Whether it is a monolithic or reservoir system determines the drug release kinetics (Abdelkader et al., 2021). For instance, a drug reservoir implant coated with a rate-controlling polymer ensures a constant release rate without an initial burst. In contrast, a monolithic matrix implant design demonstrates an initial burst release. Drug loading within the implant influences release rates, with higher drug loading leading to a nonlinear increase in initial burst release. Additionally, the dimensions of the implant, particularly the surface area, impact the release rate. In matrix-based rod-shaped implants, larger diameters result in faster drug release. In reservoir designs, the thickness of the rate-controlling membrane underlies accelerated drug release considerations.

Polymer chemistry plays a significant role in the performance of both bio-durable and bioresorbable implants. It is important to select polymers that are compatible with specific drugs or degrade at the desired rate. The choice of manufacturing technique is also crucial, particularly for incorporating drugs into the polymer matrix. Hot melt extrusion and injection molding are commonly employed methods. Extrusion processes enable the formation of rods, films, or tubes, while injection molding is suitable for more complex implant shapes. Core-sheath implant designs are often manufactured using co-extrusion techniques.

38.2.2 LA Parentals

A variety of LAI formulations have been developed and approved for clinical use, encompassing aqueous suspensions, biodegradable polymeric microparticles or nanoparticles, oily solutions and suspensions, and micro/nanocrystal line formulations (Abdelkader et al., 2021). The selection of a formulation approach is influenced by the specific physicochemical characteristics of the drugs, with emphasis on its aqueous solubility, and the intended therapeutic application. For example, hydrophobic drugs may be appropriate for aqueous nanosuspensions or oil-based solutions, whereas highly water-soluble drugs like peptides may be better suited for microsphere-based formulations. PLGA and PLA-based polymeric microspheres are extensively studied for delivering hydrophilic drugs. The abundance of PLGA and PLA-based microspheres is attributed to their biodegradability and

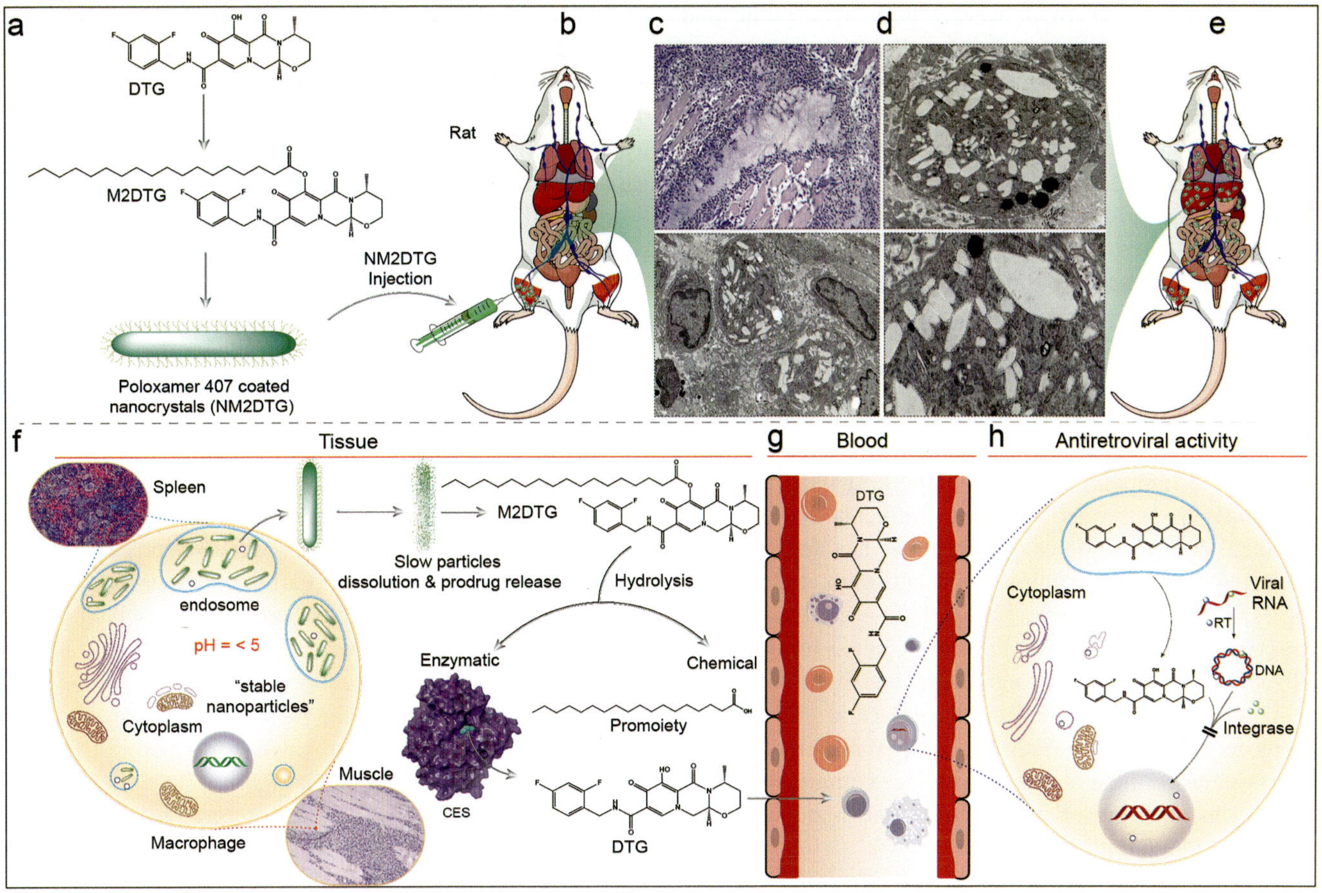

Fig. 38.2 Illustration of LASER ART mechanism of action: (**a**) ARVs are modified to lipophilic prodrug by conjugating cleavable lipophiles and then converted to aqueous nanocrystal formulations. (**b**) IM injection of LASER ART causes the formation of the primary depot at the site of injection and (**c**) and (**d**) demonstrating histiocytic infiltration of the site of injection and nanoformulation uptake by macrophages. (**e**) Biodistribution of the LASER ART to the HIV reservoir sites (**f**) Extended stability of LASER ART prodrugs in macrophages and bioconversion of LASER ART prodrugs in plasma through enzymatic and chemical hydrolysis. (The figure is adopted from (Deodhar et al., 2022) and used under the Creative Commons Attribution 4.0 International license. Copyright 2022 Nature Publishers)

the safety of the degraded products. However, such microsphere formulations are produced then stored as dry powders to minimize hydrolytic byproducts (Su et al., 2021). Examples of polymeric microsphere-based formulations in clinical practice include Risperdal consta® for risperidone delivery in the treatment of psychosis, Bydureon bcise (exenatide) consisting of PLGA-based microspheres suspended in medium-chain triglycerides (MCT), and Signifor® for LA pasireotide delivery.

Oil-based solutions and suspensions were also extensively explored in the past for the sustained release delivery of hydrophobic drugs, and many had been approved for clinical use. However, few are used in clinical settings now (Vemuri, 2016). The first-generation antipsychotic LA drugs were oil-based solutions. Oily drug solutions were commonly chosen for parenteral delivery due to the convenience of sterilization through filtration. However, due to the local injection site allergic reactions, and increased injection pain associated with oily vehicles, there has been a shift towards the use of an aqueous-based nanocrystal formulations (Murdan & Florence, 2000). These nanocrystals are prepared using top-down techniques such as high-pressure homogenization or wet milling and are stabilized with FDA-approved surfactants in an aqueous medium. Upon SC or IM injections, the nanocrystal formulations form a drug depot at the injection site, enabling slow release. The aqueous solubility of the nanocrystals plays a crucial role in their release kinetics from the depot. Prodrug strategies, such as conversion of FDA-approved drugs to hydrophobic prodrugs, have also been employed to exert control over drug release, as demonstrated by paliperidone palmitate. In preclinical development, several other anti-HIV drugs are being explored for conversion to prodrugs to regulate their release from the depot (Fig. 38.2) (Cobb et al., 2021; Deodhar et al., 2022; Kulkarni et al., 2020).

Lipid-based and polymeric nanoparticles have also been extensively explored for LA drug delivery for chronic disease treatments. Lipid nanoparticles (LNPs) are versatile in that they are capable of delivering both hydrophilic and hydrophobic drugs. For instance, Jennifer P. Freeling et al. developed LNPs loaded with three antiretroviral drugs of different physicochemical properties, including hydrophilic and hydrophobic drugs, demonstrating their prolonged release (Freeling et al., 2014). LNPs can maintain a hydrophobic and lipophilic balance when used to deliver hydrophilic drugs. Moreover, the biocompatibility, safety, and high drug loading of LNPs make them more attractive than polymeric nanoparticle delivery systems. Additionally, LNPs have the ability to form secondary drug depots within PBMCs, thereby extending the duration of therapeutic effects (Kim et al., 2018; Surve & Jindal, 2020). Like LNPs, liposomes have also been widely employed to extend the half-lives of both lipophilic and hydrophilic drugs and proteins, particularly in cancer and infection treatments. For instance, liposomal encapsulation of oxymorphone has demonstrated prolonged pain relief in animal models (Smith et al., 2003). The PLGA-based polymeric nanoparticle is another form of nanoparticle technology that has been studied for sustained delivery of antiretrovirals, anticancer, and hormones (Cao et al., 2021; Ho et al., 2021; Mandal et al., 2017). For instance, encapsulating native L-asparaginase into PLGA nanoparticles resulted in decreased tumor volume, improved survival, and drug presence at the tumor site than recorded in free drug formulations (Singh et al., 2020).

LA parenteral formulations have the potential to greatly enhance patients' compliance. However, the need for frequent clinic visits for the administration of LA parenteral products can be a barrier to patients' interest. Therefore, self-injectable LAIs with a dosage regimen of at least one week are generally preferred by patients. Additionally, designing parental LA formulations presents several challenges that need to be addressed. First, the non-removability of parental LA drugs poses additional concerns and requires prior assessments of drug tolerability. This needs to be considered before administering LA parentals. The second challenge lies in controlling the initial burst release of the drugs from formulations. Lastly, LA parental formulations may result in the persistence of drugs in the body at subtherapeutic levels after treatment cessation, leading to terminal phase PK tail. The prolonged PK tail can potentially induce side effects and drug-drug interactions when co-administering other drugs. Additionally, antimicrobial LA agents with prolonged PK tails can contribute to the development of drug-resistant mutants. A thorough investigation of PK tail for LA parentals is essential for a comprehensive understanding and management of LA regiments.

38.2.2.1 LA Prodrug Nanoformulations

Prodrugs are inert chemical entities that undergo transformation into their active form through either chemical reactions or enzymatic processes within the body (Rautio et al., 2008). Traditionally, prodrugs feature a linkage between the active drug and a promoiety through a cleavable bond. The rate and location of prodrug activation depend significantly on the design of the prodrug and the strength of the cleavable chemical bond (Stella, 2006). These factors, along with potential variations in absorption, distribution, metabolism, and excretion (ADME) properties compared to the native drugs, necessitate careful consideration during drug development (Stella, 2004). Moreover, prodrug design should aim for a scalable, cost-effective synthesis process and stable chemical properties. Although this cleavable bond allows for the eventual release of the active moiety through enzymatic or chemical hydrolysis upon administration, prodrugs must maintain sufficient chemical stability during the manufacturing and storage processes (Testa, 2004). It's important to note that the native drugs to be converted into prodrugs should possess functional groups amenable to easy conversion. These adaptable functional groups include carboxylic, hydroxyl, amine, phosphate/phosphonate, and carbonyl groups, which can be transformed into esters, carbonates, carbamates, amides, phosphates, and oximes (Rautio et al., 2008). Among these options, ester prodrugs are the most commonly used due to their ease of conversion into active drugs within the body after absorption (Ettmayer et al., 2004).

Conventional prodrug strategies primarily focus on enhancing the ADME properties of active molecules. A prime example of conventional prodrug development aimed at improving ADME properties is the transformation of tenofovir, an anti-HIV drug, into its prodrugs, tenofovir disoproxil fumarate (TDF) and tenofovir alafenamide (TAF) (Ullah Nayan et al., 2023). Tenofovir needs to convert intracellularly into tenofovir-diphosphate (TFV-DP). However, due to high hydrophilicity, the intracellular concentration of TFV-DP was minimal after native TFV treatment. To improve the oral absorption of TFV, TFV was first converted to TDF. However, the high and fast conversion rate of TDF to TFV was associated with renal and bone toxicities (Hill et al., 2018). To further improve upon, TFV was converted to TAF, a phosphoramidate prodrug of TFV. TAF improves oral absorption of TFV and also enhances intracellular concentration of TFV and TFV-diphosphate (Hill et al., 2018). Conventional prodrug design may also improve the duration of action of therapeutic compounds by altering ADME properties. However, optimal therapeutic compound PK and biodistribution profiles are dependent of prodrug design and drug delivery strategy.

Typically, drugs approved for oral administration exhibit an optimal balance between hydrophilicity and lipophilicity, ensuring efficient dissolution in the gastrointestinal tract and subsequent absorption into the bloodstream. However, these drugs may not always be suitable for use in LA DDSs due to their inherent physiochemical properties. One widely embraced approach in the development of LA DDSs involves converting established clinical drugs into prodrugs. Prodrugs designed for LA DDSs concentrate on improving their physiochemical characteristics and their interactions with the polymers and excipients employed in LA DDSs (Rautio et al., 2008). An illustrative example is the conversion of paliperidone, an oral antipsychotic drug, into a prodrug aqueous nanocrystal suspension known as paliperidone palmitate, enabling dosing at two-month intervals for psychosis treatment (Alphs et al., 2015).

LA prodrugs have found widespread application in two key areas: enhancing drug loading within LA drug carriers and improving drug dissolution properties from oil or aqueous-based injectable depots. To achieve prolonged drug release, various drug carriers like polymeric microspheres and micelle liposomal carriers have been employed. However, these hydrophobic drug carriers face challenges, particularly when formulating hydrophilic drug molecules (Kita & Dittrich, 2011). Weak interaction between the hydrophobic drug carriers and hydrophilic small molecules causes poor encapsulation during manufacturing and also enhances rapid partitioning and dissociation of drug molecules in biological fluids (Shah et al., 2012). These challenges can ultimately reduce the effectiveness of the drug's extended-release properties. LA prodrugs, specifically designed to enhance hydrophobic interactions with the carriers have the potential to significantly improve drug loading and encapsulation efficiency. For instance, *Creighton* et al. developed acyl ester prodrugs of Raltegravir (RAL), that significantly improved encapsulation of RAL in poly(lactic-glycolic acid) (PLGA) nanoparticles by as much as 20–25 fold when compared to free RAL (Creighton et al., 2019). The improved encapsulation efficiency was linked to increased hydrophobic interactions between RAL prodrugs and the polymer/solvent oil phase of the emulsion. A similar strategy yielded improved encapsulation of paclitaxel prodrugs in poly(ethylene glycol)–co-poly(caprolactone) (PEG-b-PCL), particularly when paclitaxel was transformed into 7′hexanoate ester prodrugs (Forrest et al., 2008). This approach can also be applied to enhance the encapsulation of drugs within liposomal carriers for sustained delivery. For instance, Pignatello et al. (2016) reported that converting luteinizing hormone-release hormone (LHRH) peptide into C-12 alkyl chain conjugated amide prodrug improves its encapsulation within liposomes (Pignatello et al., 2016). Another excellent example of higher encapsulation efficiency of lipophilic prodrugs into polymeric carriers for sustained release is gemcitabine. Gemcitabine is a deoxycytidine analog anticancer drug used in pancreatic cancer, requiring multiple daily doses due to its short half-life. Converision of gemcitabine into a stearoyl-gemcitabine prodrug improved encapsulation of the compound within poly(ethylene glycol)-b-poly(D,L-lactic acid) (PEG-PLA) micelles significantly extending the drug half-life (Immordino et al., 2003). This improved encapsulation was attributed to stronger interactions between the lipophilic stearoyl chain and the hydrophobic core of PLA (Daman et al., 2014).

LA prodrugs have been extensively used in depot-forming oily solutions and depot-forming solid drug dispersions (SDD). These two approaches have perhaps harnessed the prodrug strategy more widely than any other LA DDSs, extending the duration of drug dosing significantly. In these formulations, drug release occurs either through gradual partitioning of the drugs from an oily vehicle into the surrounding aqueous biological fluids or via slow dissolution of the drugs from solid dispersions into the biological medium (Wilkinson et al., 2022). Notably, the first-generation LA antipsychotic drugs broadly utilized the oily dispersion formulation approach (Surve & Jindal, 2020). The rate at which drugs are released from these dispersions depends on their ability to move from the oily vehicle into the body fluids (Wilkinson et al., 2022). However, the duration of drug release from these oily vehicles is typically limited to 2–3 weeks (Kalicharan et al., 2017). To extend the duration of drug release from these oily vehicles, native drugs are often transformed into acyl fatty acid ester prodrugs. This transformation enhances their lipophilicity, slowing down their partitioning from the oily vehicle (François et al., 2000). Common fatty acid acyl promoiety choices for this purpose include decanoate, enanthate, and caproate (Rahnfeld & Luciani, 2020). By altering the length of the fatty acid promoiety and the lipophilicity of the prodrugs, it becomes possible to adjust the rate of drug release in these oily formulations. These formulations are not only cost-effective but also scalable. However, it's important to note that many oily formulations can lead to skin irritation, inflammation, and adverse reactions at the injection site.

The limitations of oily formulations have prompted the development of aqueous-based SDD formulations. Most micro/nanosized SDD formulations are created using top-down techniques like wet milling or homogenization. Drug release from these SDD formulations relies on gradual dissolution of the drugs in bodily fluids from the SDD (Brough & Williams Iii, 2013). SDDs are particularly suitable for hydrophobic drugs, as they allow for the creation of LA dosages. Drug release rates for SDD can be tuned by optimizing particle size and surface morphology (Brough & Williams Iii, 2013). However, this formulation approach is not compatible with water soluble molecules. To facilitate formulation of hydrophilic drugs by SDD approach, lipophilic prodrugs are synthesized and transformed into aqueous solid drug suspensions (Surve & Jindal, 2020). The lipophilic solid prodrug dispersions within the aqueous buffers are stabilized by water soluble surfactants and other formulation additives. Drug release from the nanosuspension dependends on particle dissolution rates and nature of prodrug linkage (Overton et al., 2020). An example of a marketed product where this technology has been used successfully is a nanocrystalline formulation of paliperidone palmitate, an antipsychotic drug, where dosing intervals have been extended to every six months (Schreiner et al., 2015). The application of a prodrug approach to significantly extend the duration of action of therapeutic agents represents a paradigm shift in Long-Acting Drug Delivery Systems (LA DDSs) for treatment of chronic infections.

The success of aqueous nanocrystal formulation technology in antipsychotic treatment has inspired preclinical development of lipophilic prodrug nanocrystals for other chronic diseases, such as HIV-1 treatment. Our laboratory has developed a library of ARV prodrugs and transformed them into LA aqueous nanocrystal formulations, coining the term "Long-Acting Slow Effective Release Antiretroviral Therapy" (LASER ART) (Edagwa et al., 2017). LASER ART offers the capability to convert both hydrophobic and hydrophilic ARV drugs into LA prodrug nanocrystal formulations, potentially extending the HIV treatment regimen to six months. In this technology, ARVs are initially converted into hydrophobic and lipophilic prodrugs by conjugating lipophiles to ARV molecules through cleavable bonds (Edagwa et al., 2017). The prodrug approach is typically scalable, high-yielding and the produced prodrugs are stable during formulation manufacturing, terminal sterilization and storage. Notably, the LASER ART formulations exhibit stability at room temperature for extended periods and can be administered using a 22G needle. Throughout the preclinical evaluation of these formulations, no injection site reactions or adverse events were observed during the testing period (Ullah Nayan et al., 2023).

The creation of an ARV prodrug library with a diverse range of lipophiles enables screening of compounds with variable physiochemical characteristics. Intriguingly, the prolonged and sustained release of LASER ART nanocrystals is the outcome of combined effects from optimal physiochemical properties of LASER ART prodrugs and their ideal interactions with biological systems (Deodhar et al., 2022). Such physiochemical properties and interactions include but not limited to: (i) hydrophobicity and slow prodrug and particle dissolution rates in body fluids, (ii) controlled prodrug hydrolysis rates to release the pharmacologically active parent drugs and non-toxic prodrug breakdown products, (iii) interaction of the prodrugs with the mononuclear phagocytic cells and their extended stability with in the phagocytic systems, (iv) distribution of the prodrugs to the lymphatic systems and establishment of secondary drug depots within the lymphatic systems (Fig. 38.2.2.1).

LASER ART has successfully been applied to abroad range of ARV classes and other small molecules. Notable progress has been made applying this technology to integrase strand transfer inhibitors (INSTIs). An esterification process involving phenolic alcohols within INSTIs, coupled with varying chain-length fattyacids, has paved the way for the development of ultra-long-acting (ULA) nanoformulations. The fatty acids used in this process are naturally occurring and commonly found in nutritional and pharmaceutical products. This derisks the formulation strategy through application of materials and ARVs with established safety profiles. The first generation of LASER ART INSTI prodrugs, known as MCAB and MDTG, were produced by esterifying CAB and DTG with a 14-carbon chain fatty acid known as myristic acid (Sillman et al., 2018; Zhou et al., 2018). Pharmacokinetic studies conducted in Balb/cJ mice demonstrated that these nanoformulated prodrugs maintained plasma levels of CAB and DTG, exceeding the pharmacologically active IC90 (PA-IC90), for a minimum of two months and one month, respectively.

Further investigations exploring different carbon chain lengths in the fatty acid modifications revealed that extending the hydrocarbon chain length from 14 to 18 leads to a 21-fold and 23-fold increase in the plasma half-life of CAB and DTG, respectively (Deodhar et al., 2022; Kulkarni et al., 2020). A single intramuscular (IM) injection of NM2CAB, featuring the C-18 modified CAB, in NSG mice maintained plasma CAB levels above the PA-IC90 for a remarkable 364 days (Kulkarni et al., 2020). Similar pharmacokinetic properties were replicated in rhesus macaques, affirming NM2CAB's ability to extend CAB's half-life. Similarly, a single IM injection of NM2DTG in Sprague Dawley rats produced sustained plasma DTG levels above the PA-IC90 for at least 10 months (Deodhar et al., 2022). However, extending beyond an 18-carbon chain length led to suboptimal DTG PK profile, suggesting poor prodrug absorption and activation. For instance, M3DTG, a DTG prodrug with a 22-carbon chain fatty acid, exhibited plasma DTG levels above the PA-IC90 for only one week, falling short of the unmodified native DTG formulation. These findings underscored the remarkable impact of subtle structural changess in prodrug linkages on the pharmacokinetic profile of nanoformulated small molecules, with the C-18 fatty acid promoiety exhibiting optimal properties for an ultra-long acting DTG prodrug nanosuspension (Deodhar et al., 2022). Additionally, to assess the impact of solubility, a C18 dimer DTG prodrug, consisting of two DTG molecules linked via a C18 diester bond, was synthesized. Interestingly, the C18 dimer DTG prodrug displayed lower hydrophobicity and a shorter duration of DTG exposure, with plasma concentrations dropping below the pharmacologically active IC90 (PA-IC90) threshold in under 140 days (Deodhar et al., 2022). This outcome suggests that alterations in drug release rate were primarily driven by solubility rather than solely by the length of the alkyl carbon chain.

Beyond INSTIs, LASER ART has also proven effective in extending the apparent half-lives of polymerase inhibitors such as TFV where a lipophilic TFV ProTide named

M1TFV was produced and evaluated for HBV efficacy (Cobb et al., 2021). This formulation maintained high levels of the active metabolite TFV-diphosphate (TFV-DP) for over two months in peripheral blood mononuclear cells (PBMCs) and other crucial HIV and HBV target tissues and cells following a single IM injection in rats. Similarly, long-acting FTC formulations, including carboxylate ester prodrugs of FTC (MFTC), exhibited improved pharmacokinetic profiles, resulting in significantly higher FTC triphosphate levels in PBMCs compared to unmodified FTC (Ibrahim et al., 2019; Soni et al., 2019).

In conclusion, LASER ART has emerged as a promising approach for extending the half-life of antiretroviral drugs and enhancing their therapeutic efficacy, with potential benefits for the treatment of HIV and HBV infections (Cobb et al., 2021; Das et al., 2022).

38.2.3 LA Microneedle Patches

Microneedles (MNs) have emerged as a promising alternative to conventional injections (Harvey et al., 2011; Permana et al., 2019a). MNs are typically composed of tiny projections that are less than 2000 μm in height and arranged on a single side of a supportive base or patch. Initially, the development of MN formulations primarily concentrated on immediate drug release into the dermis and ocular tissues. However, recent advancements in the manufacturing techniques of MNs, particularly with the utilization of tunable biodegradable polymers, have enabled the use of MNs for LA drug delivery (Chen et al., 2013, 2020b). Combining MNs with other drug delivery technologies, such as PLGA-based nano/microparticles, LNPs, metallic nanoparticles, and nanocrystals, can significantly improve the permeability of the drugs through the skin or eye and improve the PK profile. For instance, it was reported that a combination of MNs and PLGA-loaded polymeric nanoparticles increased the half-life of vitamin D3 (VD3). PLGA nanoparticles of VD3 loaded in MNs showed sustained drug release over several days (Vora et al., 2017). Lipid-based nanocarriers, including liposomes, solid lipid nanoparticles (SLNs), and nanostructured lipid carriers (NLCs), have also been incorporated into MNs (Chen et al., 2015; Permana et al., 2019b; Qiu et al., 2008). Chen et al. developed an MN system loaded with liposomes containing triptolide, an anti-arthritis drug. This combination approach resulted in an improved plasma half-life of the triptolide, leading to enhanced effectiveness when compared to oral treatment (Chen et al., 2015). Magnetic nanoparticles, such as iron oxide nanoparticles (IONPs), when combined with MN technology, can also be used for site-specific targeting and monitoring drug distribution using magnetic resonance imaging (MRI) (Chen et al., 2009; Haj-Ahmad et al., 2015). Nanocrystals offer increased dissolution rate and mucoadhesion, enhancing drug absorption and efficacy. When loaded into MNs, these drug crystals enable enhanced drug permeation across the skin, improving therapeutic outcomes (He et al., 2018).

Based on the drug release mechanisms and MN administrations, MNs can be divided into several categories (Fig. 38.2). Solid MNs commonly use a "poke with patch" approach to create temporary microchannels in the skin (Chen et al., 2020a). Coated MNs, on the other hand, are modified versions of solid MNs, where the needle arrays are coated with drug formulations. Once inserted into the skin, the coating dissolves, releasing the drug into the subcutaneous tissue. Dissolving MNs, made of biodegradable polymers, have gained attention for their ability to completely dissolve in the skin, eliminating the need for disposal concerns (Lee et al., 2008). When designing optimal MN delivery systems, certain factors should be considered, such as the mechanical strength of MNs, drug loading capacity, and release rate. MNs, being smaller than hypodermic needles, have limitations in terms of drug loading capacity. Utilizing hydrophobic forms or drug nanocrystals can maximize drug loading. Biocompatible and biodegradable polymers are used below 60,000 Da to preclude toxicities. The drug release is influenced by MN geometry, the number of MNs in the patch, the drug's physicochemical properties, and the polymer used. MNs geometry, including microneedles radius, height, width, interspacing, and shape, also influences their mechanical properties, insertion capability, and drug loading capacity (Wang et al., 2017). For controlled drug release, MNs need to deliver the drugs inside the skin, requiring optimum mechanical strength. Therefore, when designing MNs, they must possess sufficient mechanical and insertion capabilities (Fig. 38.3 and Table 38.1).

38.3 Future Directions

LA DDSs have emerged as a highly effective method for chronic disease management, requiring only one or a few doses annually in some cases. The current landscapes of LA evolved from the treatment of chronic diseases to prophylactic interventions. LA antipsychotics were approved decades ago, and several worldwide surveys have demonstrated that long-term use of these LA medicines can significantly improve the patient's quality of life and psychosocial behavior (Kishimoto et al., 2013). Recent approvals of LA agents such as LA CAB have revolutionized treatment of HIV infection and demonstrated the potential future role of LA DDSs in dynamic chronic infectious diseases. LA medicines offer significant benefits and value propositions for patients, clinicians, and pharmaceutical manufacturers. These medications provide favorable PK and enhanced efficacy while minimizing side effects, ultimately benefiting patients the most.

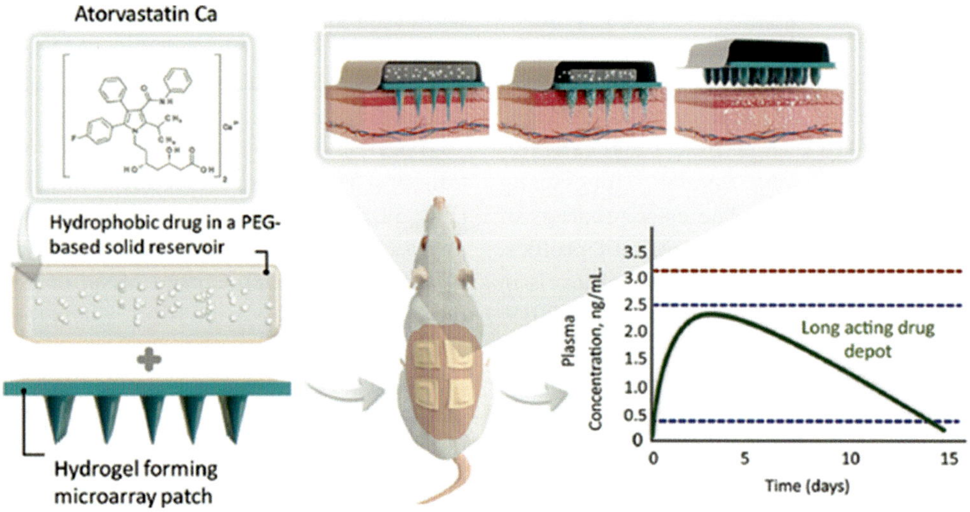

Fig. 38.3 Example of long acting micro-array patches (HF-MAP): Hydrogel forming MAP of Atrovastaton, a lipid lowering agent was developed using Gantrex S-97 (a coplymer of methyvinylether and maleic acid), PEG. MAP was able to complely dissolved within 90s in-vitro. In-vivo preclinical studies demonstrated that MAP can dissolved quickly and formed a drug depot at the adminitration site to deliver atrovastatin for 14 days in sprague dawley rats. (Figure was adopted from (Naser et al., 2023) and used under Creative Commons CC-BY license. Copyright 2022, Elsevier)

Additionally, the utilization of clinically approved medicines in drug development allows pharmaceutical companies to invest less in the development phase and greater chances of regulatory approvals.

The promising improved therapeutic outcomes of LA medicines has sparked significant research on different drug delivery modalities to meet the needs of the patients and healthcare providers. However, there are crucial developmental factors that must be considered in drug design. Such factors include the nature of disease and target population. In certain cases, the reversibility of drug delivery devices holds great importance in the decision-making process for end-users. This is particularly relevant in the context of contraception, where the ability for a smooth and rapid reversal of treatment upon termination is highly desirable. Similarly, in conditions such as diabetes, cardiovascular diseases, and asthma, very short terminal PK tail and fast termination of treatment are crucial. Sudden hypoglycemic episodes, for example, pose a significant risk to diabetic patients taking blood glucose-lowering medications, with approximately 2–4% of diabetic patient deaths attributed to sudden hypoglycemia (Agrawal et al., 2022). In such events, removability of glucose-lowering and rapid reversal of pharmacotherapy becomes essential. Conversely, for lifelong infectious diseases, the reversibility of LA DDSs may hold less importance compared to their ability to sustain therapeutic drug levels and forgiveness for missed doses.

Moreover, the assurance of safety and simplicity in administration is vital to facilitate greater acceptance and uptake of LA medicines. Patients who have been taking conventional oral medications for years may have become accustomed to the associated side effects and the routine of daily medication intake. Therefore, for LA medicines to be widely accepted, they need to offer substantial convenience and safety advantages over traditional oral treatments. In low-resource settings, self-administrable LA medicines have greater potential for wide acceptance as frequent clinic visits for LA medicines administration may increase treatment cost and stymie uptake. Additionally, for diseases where social stigma plays a crucial role, discreet LA medicines are most desirable. Logistic hurdles also pose significant challenges in clinical translation. For large-scale distribution, it is crucial to ensure sufficient stability and ease of transportation and storage at ambient conditions. LA drug formulations offer a promising alternative for those who struggle with daily medication adherence or need sustained medication release over time. Considering the impact of LA contraceptive and ocular disease management formulations, the use of LA therapies for cancer, osteoporosis, cardiovascular diseases, diabetes, and infectious diseases, will potentially bridge the unmet medical needs for many notably those in resource-limited settings.

38.4 Review Questions

1. *What is the primary purpose of long-acting formulations for chronic diseases?*
   (a) To provide a rapid onset of action
   (b) To minimize the frequency of dosing
   (c) To reduce the side effects for better monitoring
   (d) To enhance the taste and appearance of medication

Table 38.1 Comparison of a spectrum of LA formulations

Criteria	Implants	Injectables	Microneedle Patches
Drug selection criteria	Highly potent drugs, molecular weight ≤500 da	Mostly hydrophobic drugs; hydrophilic drugs convert to hydrophobic prodrugs	Both hydrophilic and hydrophobic drugs
Release mechanism	Biodegradable implant: The polymer coating of the implant degrades, and the drugs come out from the reservoir. Non-biodegradable implant: Biofluid permeates the semipermeable polymer coating, and dissolved drugs are released from the implant. ISFI: injectable liquid formulation goes through a phase inversion process to solid the liquid the formulation.	Depot-forming IM or SC injectables form a depot at the site of the injection site. The drug is released from the depot slowly owing to its low aqueous solubility. The released drug can be sequestered into a mononuclear phagocytosed system, forming a secondary depot.	Solid MNs: Biofluid permeates the MNs after insertion and dissolved drugs released from MNs Dissolving MNs: Biodegradable polymers dissolved after skin insertion and release the drug from the matrix Coated MNs: Coated drug on the outer surface of MNs dissolved in the biofluid and released after skin insertion
Advantage	Implants can release low doses of drugs for prolonged periods. Implants can be removed in settings of adverse events. Biodegradable implants do not require removal after the treatment. Consistent and predictable drug release kinetics	Compatible with a wide range of drugs Less injection site reaction Discreet Do not require to remove after treatment termination. Can be self-injected.	Discreet User controlled. Can be combined with other drug delivery technologies for improved PK profile. Predicted drug release profile. Less administration site reaction Discreet
Disadvantage	Non-biodegradable implants necessities surgical removal after the treatment termination. Local site reactions can happen. Requires HCP assistance for implantation. Implants can migrate from the original insertion site to a place where palpation is difficult. The amount of drug loaded into the implant is limited.	Non- removability in the event of an adverse event Less predicted drug release Prolong terminal phase PK tail. PK profile depends on the injection site. Based on the injection site, HCP's assistance may be required for injection.	Subject to user error. It may be affected by daily activities such as bathing and swimming. Limited drug loading capacity May induce an initial burst release, limiting the duration of microneedles.
Duration of action	From days to years	From days to months	From days to weeks
Drug loading capacity	Limited	High	Limited
Design consideration	Required therapeutic drug level and release rate of the drug from implant. Drug loading capacity of the implant Dimension of implant and ease of administration Ease of manufacturing process	Drug loading and injection volume Opportunity of self-administration Controlling of initial burst release of drugs Duration of terminal phase PK tails	Design and geometry of microneedles Manufacturing ease Drug loading capacity Self-administration capacity Ease of removal if non-biodegradable polymer used

PK Pharmacokinetic, *da* Dalton, *HCPs* healthcare professionals

2. *What is a key advantage of using implants in drug delivery?*
   (a) They can be easily self-administered
   (b) They provide an immediate release of drugs
   (c) They offer reversibility of the therapeutics in case of adverse events
   (d) They primarily utilize non-biodegradable polymers
3. *Which of the following is a characteristic of microarray needle patches?*
   (a) They are mainly used for immediate drug release
   (b) They require non-biodegradable polymers to formulate
   (c) They can be self-administered
   (d) They are less effective than traditional injections
4. *What is a critical consideration in the design of prodrugs for long-acting formulations?*
   (a) Ensuring rapid drug activation
   (b) Balancing the hydrophobicity and activation rate to achieve therapeutic drug levels for an extended period of time

   (c) Balancing the drug's stability and activation rate
   (d) Focusing on immediate therapeutic effects
5. *In the context of implants for drug delivery, what does the term "ISFI" stand for?*
   (a) Immediate Soluble Formulated Implant
   (b) Injectable Solid Formulation Implant
   (c) In Situ Forming Implant
   (d) Integrated Systemic Formulation Implant

38.5 Answers

1. b
2. c
3. c
4. b
5. c

Author Contributions M.U.N., A.S, N.T.H.L, L.P., B.E., and H.E.G. contributing substantially to the writing, illustrations, editing, and or supervision of this manuscript.

Funding The work is supported by National Institutes of Health grants R01 NS034239; R01 NS036126; R01 MH115860; R01 NS126089; R01 AI145542; and R01 AI158160.

References

Abdelkader, H., Wertheim, D., Pierscionek, B., & Alany, R. G. (2020). Curcumin in situ gelling polymeric insert with enhanced ocular performance. *Pharmaceutics, 12*(12), 1158.

Abdelkader, H., Fathalla, Z., Seyfoddin, A., Farahani, M., et al. (2021). Polymeric long-acting drug delivery systems (LADDS) for treatment of chronic diseases: Inserts, patches, wafers, and implants. *Advanced Drug Delivery Reviews, 177*, 113957.

Agrawal, S., Makuch, S., Dróżdż, M., Dudzik, T., et al. (2022). The impact of hypoglycemia on patients with diabetes mellitus: A cross-sectional analysis. *Journal of Clinical Medicine, 11*(3). https://doi.org/10.3390/jcm11030626

Almeida, A., Brabant, L., Siepmann, F., De Beer, T., et al. (2012). Sustained release from hot-melt extruded matrices based on ethylene vinyl acetate and polyethylene oxide. *European Journal of Pharmaceutics and Biopharmaceutics, 82*(3), 526–533.

Alosaimi, K., Alwafi, H., Alhindi, Y., Falemban, A., et al. (2022). Medication adherence among patients with chronic diseases in Saudi Arabia. *International Journal of Environmental Research and Public Health, 19*(16). https://doi.org/10.3390/ijerph191610053

Alphs, L., Benson, C., Cheshire-Kinney, K., Lindenmayer, J. P., et al. (2015). Real-world outcomes of paliperidone palmitate compared to daily oral antipsychotic therapy in schizophrenia: A randomized, open-label, review board-blinded 15-month study. *The Journal of Clinical Psychiatry, 76*(5), 554–561. https://doi.org/10.4088/JCP.14m09584

Benhabbour, S. R., Kovarova, M., Jones, C., Copeland, D. J., et al. (2019). Ultra-long-acting tunable biodegradable and removable controlled release implants for drug delivery. *Nature Communications, 10*(1), 4324. https://doi.org/10.1038/s41467-019-12141-5

Blackwood, C., Sanga, P., Nuamah, I., Keenan, A., et al. (2020). Patients' preference for long-acting injectable versus oral antipsychotics in Schizophrenia: Results from the patient-reported medication preference questionnaire. *Patient Preference and Adherence, 14*, 1093–1102. https://doi.org/10.2147/ppa.S251812

Breckenridge, A., Aronson, J. K., Blaschke, T. F., Hartman, D., et al. (2017). Poor medication adherence in clinical trials: Consequences and solutions. *Nature Reviews Drug Discovery, 16*(3), 149–150.

Brough, C., & Williams Iii, R. (2013). Amorphous solid dispersions and nano-crystal technologies for poorly water-soluble drug delivery. *International Journal of Pharmaceutics, 453*(1), 157–166.

Buwalda, S. J., Vermonden, T., & Hennink, W. E. (2017). Hydrogels for therapeutic delivery: Current developments and future directions. *Biomacromolecules, 18*(2), 316–330.

Cao, Z., Tang, X., Zhang, Y., Yin, T., et al. (2021). Novel injectable progesterone-loaded nanoparticles embedded in SAIB-PLGA in situ depot system for sustained drug release. *International Journal of Pharmaceutics, 607*, 121021.

Carls, G. S., Tuttle, E., Tan, R.-D., Huynh, J., et al. (2017). Understanding the gap between efficacy in randomized controlled trials and effectiveness in real-world use of GLP-1 RA and DPP-4 therapies in patients with type 2 diabetes. *Diabetes Care, 40*(11), 1469–1478.

Chen, X., Prow, T. W., Crichton, M. L., Jenkins, D. W., et al. (2009). Dry-coated microprojection array patches for targeted delivery of immunotherapeutics to the skin. *Journal of Controlled Release, 139*(3), 212–220.

Chen, M.-C., Huang, S.-F., Lai, K.-Y., & Ling, M.-H. (2013). Fully embeddable chitosan microneedles as a sustained release depot for intradermal vaccination. *Biomaterials, 34*(12), 3077–3086.

Chen, G., Hao, B., Ju, D., Liu, M., et al. (2015). Pharmacokinetic and pharmacodynamic study of triptolide-loaded liposome hydrogel patch under microneedles on rats with collagen-induced arthritis. *Acta Pharmaceutica Sinica B, 5*(6), 569–576.

Chen, M., Quan, G., Sun, Y., Yang, D., et al. (2020a). Nanoparticles-encapsulated polymeric microneedles for transdermal drug delivery. *Journal of Controlled Release, 325*, 163–175.

Chen, Z., He, J., Qi, J., Zhu, Q., et al. (2020b). Long-acting microneedles: A progress report of the state-of-the-art techniques. *Drug Discovery Today, 25*(8), 1462–1468.

Cobb, D. A., Smith, N., Deodhar, S., Bade, A. N., et al. (2021). Transformation of tenofovir into stable ProTide nanocrystals with long-acting pharmacokinetic profiles. *Nature Communications, 12*(1), 5458.

Creighton, R. L., Suydam, I. T., Ebner, M. E., Afunugo, W. E., et al. (2019). Sustained intracellular raltegravir depots generated with prodrugs designed for nanoparticle delivery. *ACS Biomaterials Science & Engineering, 5*(8), 4013–4022.

Cullins, V. E. (1992). Injectable and implantable contraceptives. *Current Opinion in Obstetrics & Gynecology, 4*(4), 536–543.

Dai, M., Bai, L., Zhang, H., Ma, Q., et al. (2020). A novel flunarizine hydrochloride-loaded organogel for intraocular drug delivery in situ: Design, physicochemical characteristics and inspection. *International Journal of Pharmaceutics, 576*, 119027.

Daman, Z., Ostad, S., Amini, M., & Gilani, K. (2014). Preparation, optimization and in vitro characterization of stearoyl-gemcitabine polymeric micelles: A comparison with its self-assembled nanoparticles. *International Journal of Pharmaceutics, 468*(1–2), 142–151.

Das, S., Wang, W., Ganesan, M., Fonseca-Lanza, F., et al. (2022). An ultralong-acting tenofovir ProTide nanoformulation achieves monthslong HBV suppression. *Science Advances, 8*(51), eade9582.

Deodhar, S., Sillman, B., Bade, A. N., Avedissian, S. N., et al. (2022). Transformation of dolutegravir into an ultra-long-acting parenteral prodrug formulation. *Nature Communications, 13*(1), 3226.

Dharmayanti, C., Gillam, T. A., Williams, D. B., & Blencowe, A. (2020). Drug-eluting biodegradable implants for the sustained release of bisphosphonates. *Polymers (Basel), 12*(12). https://doi.org/10.3390/polym12122930

Edagwa, B., McMillan, J., Sillman, B., & Gendelman, H. E. (2017). Long-acting slow effective release antiretroviral therapy. *Expert Opinion on Drug Delivery, 14*(11), 1281–1291. https://doi.org/10.1080/17425247.2017.1288212

Ettmayer, P., Amidon, G. L., Clement, B., & Testa, B. (2004). Lessons learned from marketed and investigational prodrugs. *Journal of Medicinal Chemistry, 47*(10), 2393–2404.

Flexner, C., Owen, A., Siccardi, M., & Swindells, S. (2021). Long-acting drugs and formulations for the treatment and prevention of HIV infection. *International Journal of Antimicrobial Agents, 57*(1), 106220. https://doi.org/10.1016/j.ijantimicag.2020.106220

Forrest, M. L., Yáñez, J. A., Remsberg, C. M., Ohgami, Y., et al. (2008). Paclitaxel prodrugs with sustained release and high solubility in poly (ethylene glycol)-b-poly (ε-caprolactone) micelle nanocarriers: Pharmacokinetic disposition, tolerability, and cytotoxicity. *Pharmaceutical Research, 25*, 194–206.

François, M. K. J., Embrechts, R. C. A., Borghijs, H. K., & Monbaliu, J. (2000). *Aqueous suspensions of 9-hydroxyrisperidone fatty acid esters*. Google Patents.

Freeling, J. P., Koehn, J., Shu, C., Sun, J., et al. (2014). Long-acting three-drug combination anti-HIV nanoparticles enhance drug exposure in primate plasma and cells within lymph nodes and blood. *AIDS, 28*(17), 2625–2627. https://doi.org/10.1097/qad.0000000000000421

Fu, Y., & Kao, W. J. (2010). Drug release kinetics and transport mechanisms of non-degradable and degradable polymeric delivery systems. *Expert Opinion on Drug Delivery, 7*(4), 429–444.

Garber, A. J. (2011). Long-acting glucagon-like peptide 1 receptor agonists: A review of their efficacy and tolerability. *Diabetes Care, 34*(Suppl 2), S279–S284. https://doi.org/10.2337/dc11-s231

Ge, Y., Zhang, A., Sun, R., Xu, J., et al. (2020). Penetratin-modified lutein nanoemulsion in-situ gel for the treatment of age-related macular degeneration. *Expert Opinion on Drug Delivery, 17*(4), 603–619.

Gustafson, C. T., Boakye-Agyeman, F., Brinkman, C. L., Reid, J. M., et al. (2016). Controlled delivery of vancomycin via charged hydrogels. *PLoS One, 11*(1), e0146401.

Haj-Ahmad, R., Rasekh, M., Nazari, K., Li, Y., et al. (2015). EHDA spraying: A multi-material nano-engineering route. *Current Pharmaceutical Design, 21*(22), 3239–3247.

Harvey, A. J., Kaestner, S. A., Sutter, D. E., Harvey, N. G., et al. (2011). Microneedle-based intradermal delivery enables rapid lymphatic uptake and distribution of protein drugs. *Pharmaceutical Research, 28*, 107–116.

He, M., Yang, G., Zhang, S., Zhao, X., et al. (2018). Dissolving microneedles loaded with etonogestrel microcrystal particles for intradermal sustained delivery. *Journal of Pharmaceutical Sciences, 107*(4), 1037–1045.

Hill, A., Hughes, S. L., Gotham, D., & Pozniak, A. L. (2018). *Tenofovir alafenamide versus tenofovir disoproxil fumarate: Is there a true difference in efficacy and safety?* (Vol. 4, pp. 72–79). Elsevier.

Ho, D.-K., LeGuyader, C., Srinivasan, S., Roy, D., et al. (2021). Fully synthetic injectable depots with high drug content and tunable pharmacokinetics for long-acting drug delivery. *Journal of Controlled Release, 329*, 257–269.

Hollifield, J. (1986). Clinical acceptability of transdermal clonidine: A large-scale evaluation by practitioners. *American Heart Journal, 112*(4), 900–906.

Ibrahim, I. M., Bade, A. N., Lin, Z., Soni, D., et al. (2019). Synthesis and characterization of a long-acting emtricitabine prodrug nanoformulation. *International Journal of Nanomedicine, 14*, 6231–6247.

Immordino, M. L., Brusa, P., Arpicco, S., Stella, B., et al. (2003). Preparation, characterization, cytotoxicity and pharmacokinetics of liposomes containing docetaxel. *Journal of Controlled Release, 91*(3), 417–429.

Iuga, A. O., & McGuire, M. J. (2014). Adherence and health care costs. *Risk Management and Healthcare Policy, 7*, 35–44.

Kalicharan, R., Oussoren, C., Schot, P., de Rijk, E., et al. (2017). The contribution of the in-vivo fate of an oil depot to drug absorption. *International Journal of Pharmaceutics, 528*(1–2), 595–601.

Kendler, D., Kung, A. W. C., Fuleihan, G. E.-H., González, J. G. G., et al. (2004). Patients with osteoporosis prefer once weekly to once daily dosing with alendronate. *Maturitas, 48*(3), 243–251.

Kim, M. W., Kwon, S.-H., Choi, J. H., & Lee, A. (2018). A promising biocompatible platform: Lipid-based and bio-inspired smart drug delivery systems for cancer therapy. *International Journal of Molecular Sciences, 19*(12), 3859.

Kishimoto, T., Nitta, M., Borenstein, M., Kane, J. M., et al. (2013). Long-acting injectable versus oral antipsychotics in schizophrenia: A systematic review and meta-analysis of mirror-image studies. *The Journal of Clinical Psychiatry, 74*(10), 957–965. https://doi.org/10.4088/JCP.13r08440

Kita, K., & Dittrich, C. (2011). Drug delivery vehicles with improved encapsulation efficiency: Taking advantage of specific drug–carrier interactions. *Expert Opinion on Drug Delivery, 8*(3), 329–342.

Kulkarni, T. A., Bade, A. N., Sillman, B., Shetty, B. L. D., et al. (2020). A year-long extended release nanoformulated cabotegravir prodrug. *Nature Materials, 19*(8), 910–920.

Kumar, A., & Pillai, J. (2018). Implantable drug delivery systems: An overview. In *Nanostructures for the engineering of cells, tissues and organs* (pp. 473–511).

Lee, J. W., Park, J.-H., & Prausnitz, M. R. (2008). Dissolving microneedles for transdermal drug delivery. *Biomaterials, 29*(13), 2113–2124.

Ma, T., Huo, S., Xu, B., Li, F., et al. (2020). A novel long-acting oxyntomodulin analogue eliminates diabetes and obesity in mice. *European Journal of Medicinal Chemistry, 203*, 112496. https://doi.org/10.1016/j.ejmech.2020.112496

Mandal, S., Belshan, M., Holec, A., Zhou, Y., et al. (2017). An enhanced emtricitabine-loaded long-acting nanoformulation for prevention or treatment of HIV infection. *Antimicrobial Agents and Chemotherapy, 61*(1), e01475–e01416.

Marengoni, A., Angleman, S., Melis, R., Mangialasche, F., et al. (2011). Aging with multimorbidity: A systematic review of the literature. *Ageing Research Reviews, 10*(4), 430–439.

Murdan, S., & Florence, A. T. (2000). *Non-aqueous solutions and suspensions as sustained-release injectable formulations. Sustained-release injectable products*. InterPharm.

Naser, Y. A., Tekko, I. A., Vora, L. K., Peng, K., et al. (2023). Hydrogel-forming microarray patches with solid dispersion reservoirs for transdermal long-acting microdepot delivery of a hydrophobic drug. *Journal of Controlled Release, 356*, 416–433. https://doi.org/10.1016/j.jconrel.2023.03.003

Osterberg, L., & Blaschke, T. (2005). Adherence to medication. *The New England Journal of Medicine, 353*(5), 487–497.

Overton, E. T., Richmond, G., Rizzardini, G., Jaeger, H., et al. (2020). Long-acting cabotegravir and rilpivirine dosed every 2 months in adults with HIV-1 infection (ATLAS-2M), 48-week results: A randomised, multicentre, open-label, phase 3b, non-inferiority study. *The Lancet, 396*(10267), 1994–2005.

Permana, A. D., McCrudden, M. T., & Donnelly, R. F. (2019a). Enhanced intradermal delivery of nanosuspensions of antifilariasis drugs using dissolving microneedles: A proof of concept study. *Pharmaceutics, 11*(7), 346.

Permana, A. D., Tekko, I. A., McCrudden, M. T., Anjani, Q. K., et al. (2019b). Solid lipid nanoparticle-based dissolving microneedles: A promising intradermal lymph targeting drug delivery system with potential for enhanced treatment of lymphatic filariasis. *Journal of Controlled Release, 316*, 34–52.

Pignatello, R., Musumeci, T., Graziano, A. C., Lo Furno, D., et al. (2016). A study on liposomal encapsulation of a lipophilic prodrug of LHRH. *Pharmaceutical Development and Technology, 21*(6), 664–671.

Pons-Faudoa, F. P., Ballerini, A., Sakamoto, J., & Grattoni, A. (2019). Advanced implantable drug delivery technologies: Transforming the clinical landscape of therapeutics for chronic diseases. *Biomedical Microdevices, 21*, 1–22.

Qiu, Y., Gao, Y., Hu, K., & Li, F. (2008). Enhancement of skin permeation of docetaxel: A novel approach combining microneedle and elastic liposomes. *Journal of Controlled Release, 129*(2), 144–150.

Rahnfeld, L., & Luciani, P. (2020). Injectable lipid-based depot formulations: Where do we stand? *Pharmaceutics, 12*(6), 567.

Rautio, J., Kumpulainen, H., Heimbach, T., Oliyai, R., et al. (2008). Prodrugs: Design and clinical applications. *Nature Reviews Drug Discovery, 7*(3), 255–270. https://doi.org/10.1038/nrd2468

Rautio, J., Meanwell, N. A., Di, L., & Hageman, M. J. (2018). The expanding role of prodrugs in contemporary drug design and development. *Nature Reviews Drug Discovery, 17*(8), 559–587. https://doi.org/10.1038/nrd.2018.46

Reinstatler, L., Klaassen, Z., McCraw, C. O., Chen, R., et al. (2015). Robert Benjamin Greenblatt and his many pursuits: An unlikely founder of the testosterone pellet. *Urology, 85*(2), 279–282.

Robinson, D., Woerner, M. G., Alvir, J. M., Bilder, R., et al. (1999). Predictors of relapse following response from a first episode of schizophrenia or schizoaffective disorder. *Archives of General Psychiatry, 56*(3), 241–247. https://doi.org/10.1001/archpsyc.56.3.241

Santos, A., Aw, M. S., Bariana, M., Kumeria, T., et al. (2014). Drug-releasing implants: Current progress, challenges and perspectives. *Journal of Materials Chemistry B, 2*(37), 6157–6182.

Schreiner, A., Aadamsoo, K., Altamura, A. C., Franco, M., et al. (2015). Paliperidone palmitate versus oral antipsychotics in recently diagnosed schizophrenia. *Schizophrenia Research, 169*(1), 393–399. https://doi.org/10.1016/j.schres.2015.08.015

Shah, M., Agrawal, Y., Garala, K., & Ramkishan, A. (2012). Solid lipid nanoparticles of a water soluble drug, ciprofloxacin hydrochloride. *Indian Journal of Pharmaceutical Sciences, 74*(5), 434.

Sillman, B., Bade, A. N., Dash, P. K., Bhargavan, B., et al. (2018). Creation of a long-acting nanoformulated dolutegravir. *Nature Communications, 9*(1), 443.

Simon, J. A., Lewiecki, E. M., Smith, M. E., Petruschke, R. A., et al. (2002). Patient preference for once-weekly alendronate 70 mg versus once-daily alendronate 10 mg: A multicenter, randomized, open-label, crossover study. *Clinical Therapeutics, 24*(11), 1871–1886.

Singh, M., Hassan, N., Verma, D., Thakur, P., et al. (2020). Design of expert guided investigation of native L-asparaginase encapsulated long-acting cross-linker-free poly (lactic-co-glycolic) acid nanoformulation in an Ehrlich ascites tumor model. *Saudi Pharmaceutical Journal, 28*(6), 719–728. https://doi.org/10.1016/j.jsps.2020.04.014

Smith, L. J., Krugner-Higby, L., Clark, M., Wendland, A., et al. (2003). A single dose of liposome-encapsulated oxymorphone or morphine provides long-term analgesia in an animal model of neuropathic pain. *Comparative Medicine, 53*(3), 280–287.

Soni, D., Bade, A. N., Gautam, N., Herskovitz, J., et al. (2019). Synthesis of a long acting nanoformulated emtricitabine ProTide. *Biomaterials, 222*, 119441. https://doi.org/10.1016/j.biomaterials.2019.119441

Stella, V. J. (2004). *Prodrugs as therapeutics* (Vol. 14, pp. 277–280). Taylor & Francis.

Stella, V. (2006). *Optimizing the "drug-like" properties of leads in drug discovery* (Biotechnol: Pharm Aspects). Springer.

Stout, A., Wood, S., Barigye, G., Kaboré, A., et al. (2018). Expanding access to injectable contraception: Results from pilot introduction of subcutaneous depot medroxyprogesterone acetate (DMPA-SC) in 4 African countries. *Global Health: Science and Practice, 6*(1), 55–72.

Su, Y., Zhang, B., Sun, R., Liu, W., et al. (2021). PLGA-based biodegradable microspheres in drug delivery: Recent advances in research and application. *Drug Delivery, 28*(1), 1397–1418. https://doi.org/10.1080/10717544.2021.1938756

Surve, D. H., & Jindal, A. B. (2020). Recent advances in long-acting nanoformulations for delivery of antiretroviral drugs. *Journal of Controlled Release, 324*, 379–404.

Testa, B. (2004). Prodrug research: Futile or fertile? *Biochemical Pharmacology, 68*(11), 2097–2106.

Ullah Nayan, M., Sillman, B., Hasan, M., Deodhar, S., et al. (2023). Advances in long-acting slow effective release antiretroviral therapies for treatment and prevention of HIV infection. *Advanced Drug Delivery Reviews, 200*, 115009. https://doi.org/10.1016/j.addr.2023.115009

Vemuri, N. M. (2016). Preformulation. In *Pharmaceutical dosage forms-parenteral medications* (pp. 71–89). CRC Press.

Vora, L. K., Donnelly, R. F., Larrañeta, E., González-Vázquez, P., et al. (2017). Novel bilayer dissolving microneedle arrays with concentrated PLGA nano-microparticles for targeted intradermal delivery: Proof of concept. *Journal of Controlled Release, 265*, 93–101.

Wang, H., Naghavi, M., Allen, C., Barber, R. M., et al. (2016). Global, regional, and national life expectancy, all-cause mortality, and cause-specific mortality for 249 causes of death, 1980–2015: A systematic analysis for the Global Burden of Disease Study 2015. *The Lancet, 388*(10053), 1459–1544.

Wang, M., Hu, L., & Xu, C. (2017). Recent advances in the design of polymeric microneedles for transdermal drug delivery and biosensing. *Lab on a Chip, 17*(8), 1373–1387.

Wilkinson, J., Ajulo, D., Tamburrini, V., Le Gall, G., et al. (2022). Lipid based intramuscular long-acting injectables: Current state of the art. *European Journal of Pharmaceutical Sciences, 178*, 106253.

Wood, S. N., Magalona, S., Zimmerman, L. A., OlaOlorun, F., et al. (2022). Self-injected contraceptives: Does the investment reflect women's preferences? *BMJ Global Health, 7*(7). https://doi.org/10.1136/bmjgh-2022-008862

Young, I. C., Massud, I., Cottrell, M. L., Shrivastava, R., et al. (2023). Ultra-long-acting in-situ forming implants with cabotegravir protect female macaques against rectal SHIV infection. *Nature Communications, 14*(1), 708. https://doi.org/10.1038/s41467-023-36330-5

Zhou, T., Su, H., Dash, P., Lin, Z., et al. (2018). Creation of a nanoformulated cabotegravir prodrug with improved antiretroviral profiles. *Biomaterials, 151*, 53–65.

Therapeutic Strategies in Neurodegenerative Diseases

39

Abdulmunaim M. Eid, Alireza Faridar, and Stanley H. Appel

Abstract

Neurodegenerative diseases are characterized by progressive loss of specific populations of neurons giving rise to clinical syndromes of neurological dysfunction. The World Health Organization declared neurodegenerative diseases the biggest health issue in the twenty-first century and predicted the number of people with neurodegenerative diseases to triple over the next three decades adding to the mortality and morbidity risk and the associated economic cost to society. Most of the clinically available therapeutics are symptomatic and supportive leaving a tremendous need for therapeutics that lead to slowing of disease progression rate (disease-modification). In this chapter, we discuss key therapeutic strategies in neurodegenerative diseases including gene therapy, cell therapy, and immunotherapy. These strategies address various components of the pathogenic mechanisms thought to underlie neurodegeneration including proteinopathy, neurotransmitter and key molecule deficiencies, neuroinflammation and systemic inflammation, and neuronal degeneration. We also summarize symptomatic therapies that are currently available in the clinic.

Keywords

Neurodegenerative diseases · Gene therapy · Cell therapy · Immunotherapy

39.1 Introduction

Neurodegenerative diseases are characterized by the progressive loss of specific populations of neurons giving rise to clinical syndromes of neurological dysfunction. These diseases are common, and their prevalence is postulated to grow in the next few decades with the aging population. The two most common neurodegenerative diseases are Alzheimer's disease and Parkinson's disease; about 25 million patients are living with Alzheimer's disease worldwide, and a similar number living with Parkinson's disease (Erkkinen et al., 2018). Frontotemporal dementia, amyotrophic lateral sclerosis, corticobasal degeneration, progressive supranuclear palsy, multiple system atrophy, Huntington's disease, and spinal muscular atrophy are less frequent, but their incidence and prevalence are increasing owing to increased recognition and detection. In addition to the mortality risk, neurodegenerative diseases reduce the quality of life of those affected and their families, and they burden society with a huge economic cost, e.g., $203 billion is estimated to be the annual cost of Alzheimer's disease in the United States (Seckan & June, 2013).

The clinical syndrome of neurodegenerative disorders typically correlates with the anatomical pattern and temporal evolution of neuronal loss rather than the underlying pathology. For example, parkinsonism which is defined as two or more cardinal motor manifestations (resting tremor, bradykinesia, rigidity, and postural instability), is often caused by loss of the majority of dopaminergic neurons in the substan-

A. M. Eid
Stanley H Appel Department of Neurology, Houston Methodist Neurological Institute, Houston Methodist Hospital, Houston, TX, USA

A. Faridar
Stanley H Appel Department of Neurology, Houston Methodist Neurological Institute, Houston Methodist Hospital, Houston, TX, USA

Weill Cornell Medicine, New York, NY, USA

S. H. Appel (✉)
Stanley H Appel Department of Neurology, Houston Methodist Neurological Institute, Houston Methodist Hospital, Houston, TX, USA

Weill Cornell Medicine, New York, NY, USA

Ann Kimball & John W. Johnson Center for Cellular Therapeutics, Houston Methodist Hospital, Houston, TX, USA
e-mail: sappel@houstonmethodist.org

H. E. Gendelman, T. Ikezu (eds.), *Neuroimmune Pharmacology and Therapeutics*,
https://doi.org/10.1007/978-3-031-68237-7_39

tia nigra of the midbrain due to accumulation of misfolded alpha-synuclein protein. However, parkinsonism could also be caused by loss of the same group of neurons due to a different type of pathology, e.g., tauopathy (accumulation of abnormal tau protein) such as in progressive supranuclear palsy, or even due to a variety of striatal insults. This discordance between anatomy/clinical presentation and pathology is a major hurdle in the process of developing and testing therapeutics, especially at the clinical trial level. Recently, there has been a large effort to develop and optimize biomarkers for detecting pathology in vivo to anchor the diagnostic process on pathological rather than the typical clinical or clinicopathologic entities. In 2018, the US National Institute on Aging and the Alzheimer's Association proposed new biological criteria for Alzheimer's disease diagnosis (Jack Jr. et al., 2018). The new criteria are based on three biomarker groups: amyloid pathology, tau pathology, and neurodegeneration, and it is agnostic to the clinical status of the person when it comes to diagnosing Alzheimer's disease (Jack Jr. et al., 2018). This approach has not yet been extended to most neurodegenerative diseases because of the lack of optimization of relevant biomarkers.

In this chapter, we will highlight three categories of therapeutic strategies in neurodegenerative diseases: gene therapy, cell therapy, and immunotherapy. These are three of the most extensively studied approaches, and they seem to represent those therapeutics closer to the end of the pipeline at this time. Gene therapy and cell therapy are tools that could be applied to tackle different molecular mechanisms underlying neurodegeneration. Similar to gene and cell therapies, immunotherapy represents a therapeutic tool with multiple potential applications, albeit most commonly used to target proteinopathy with less than encouraging outcomes thus far. Because the present central paradigm in neurodegenerative pathogenesis is that neuronal dysfunction and death result from the accumulation of abnormal proteins (proteinopathy), targeting abnormal proteins using immunotherapy is a straightforward way to study such governing hypothesis and, thus, we opted to discuss it. At the end of the chapter, we summarize currently available symptomatic treatments for neurodegenerative diseases.

39.2 Gene Therapy

Gene therapy is a therapeutic approach that aims to modify the person's genetic makeup. There are two major uses of gene therapy in neurodegenerative diseases: (1) Directly altering the genetic basis of a neurodegenerative disorder, such as modifying the gene that causes Huntington's disease or its messenger RNA (mRNA); and (2) Altering specific nervous system-related protein expression in order to deliver a high local concentration of a target molecule while avoiding the systemic toxicity that comes from peripherally administering high doses of that molecule. To achieve either goal, there are generally two gene editing approaches: (1) DNA editing, e.g., using CRISPR-based systems; and (2) RNA editing, e.g., using antisense oligonucleotides (ASOs). Modified viral vectors, specifically adeno-associated viruses (AAVs), are often used along with interventional MRI-guided convection-enhanced delivery (iMRI-CED) to deliver molecules to specific central nervous system (CNS) locations. Other delivery approaches are also used.

39.2.1 Challenges to Gene Therapy in Neurodegenerative Diseases

There are several challenges facing gene therapies in neurodegenerative diseases. Some strategies have a relatively high risk associated with the method of administration, e.g., ASOs and AAVs need to be delivered intrathecally or through an invasive local approach such as iMRI-CED. Toxicity may also accompany some therapies, e.g., hepatotoxicity with AAVs. Another form of toxicity is related to altered splicing as with high-dose risdiplam (splicing modulator) leading to off-target FOXM1 splicing which produces a protein that blocks cell division (Sun & Roy, 2021). Generally, off-target binding is a risk with nearly all gene therapeutic approaches and can cause a host of issues. Immunogenicity induced by some therapies, e.g., viral proteins in AAVs and bacterial nucleases in Cas-proteins can decrease the efficacy of the intervention and/or cause a harmful inflammatory response. The permanency of DNA editing is a concern for this therapeutic modality, though it is also considered a major advantage as a single long-lasting intervention is more convenient and less costly than a recurring treatment. In fact, because neurons are post-mitotic (i.e., not mitotically active), AAV expression can last decades, even though it doesn't get integrated into the host genome. Despite the challenges and owing to the promise of gene therapy in neurodegenerative disease and the availability of tools to mitigate some of those challenges, the field is more active than ever. The United States Food and Drug Administration (FDA) has issued a draft titled "Human Gene Therapy for Neurodegenerative Diseases: Draft Guidance for Industry" that will provide recommendations to sponsors developing human gene therapy once finalized (Research, 2022). The guidance focuses on product development, preclinical testing, clinical trial design, and marketing approval pathways for investigational gene therapy products. This highlights the exceptional status of such therapies and the need for special considerations for development and use. In the remainder of this section, we will highlight the most important approaches of gene therapy for a variety of neurodegenerative disorders. It is important to realize that despite our attempt to be comprehensive, there is a vast array of studies in this area that can't be covered in an overview; some of which are elaborated upon in chapters specific to each disease.

39.2.2 Huntington's Disease: A Monogenic Disease

Huntington's disease (HD) is caused by CAG trinucleotide repeat expansion in exon 1 of the huntingtin (HTT) gene on chromosome 4, resulting in a polyglutamine-mutated huntingtin (mtHTT) protein. The monogenic nature of the disease makes it an attractive target for gene therapies. Reducing mtHTT protein is the focus of almost all gene therapies in the HD field.

Reducing the levels of mtHTT protein using RNA interference (RNAi) is one of the most promising strategies employed to develop therapeutics for HD. The general mechanism of RNAi therapies, a.k.a. post-transcriptional gene silencing (PTGS), involves small non-coding RNAi molecules, e.g., short interfering RNA (siRNA), which target specific mRNA molecules directing them for destruction through the RNA-induced silencing complex (Hutvagner & Simard, 2008). An engineered microRNA targeting human huntingtin, delivered via AAV5 (AAV5-miHTT), was investigated in both rat and minipig HD models (Evers et al., 2018; Miniarikova et al., 2017). Delivery and widespread distribution of AAV5-miHTT throughout animal brains were evident (Evers et al., 2018; Miniarikova et al., 2017). Both mutant huntingtin mRNA and protein were significantly reduced in all brain regions transduced by AAV5-miHTT (Evers et al., 2018; Miniarikova et al., 2017). Perhaps, more importantly, this approach allows for an allele-specific mechanism in lowering huntingtin, which rescues the wild-type and prevents loss of its physiologic function. In June 2020, a phase 1/2 clinical trial evaluating AMT-130, an AAV5-miHTT microRNA administered through iMRI-CED into the caudate and putamen, was launched (Post, 2022). In addition to immune responses to the viral vector, and the irreversibility of the approach as discussed above, another major challenge of this method is variable tissue distribution. In addition, delivery to areas outside of the striatum, e.g., the cortex may be limited which is essential for substantial benefit.

Another approach for HD gene therapy that reached phase 3 clinical trial is the non-allele-selective ASO tominersen. This nucleic acid targets exon 36 of the HTT mRNA resulting in the degradation of both the mutant and the wild-type transcripts by RNase H. Although the first-in-human trial was a cautious success demonstrating that tominersen lowered CSF mtHTT protein concentration in a dose-dependent manner while increasing ventricular volume and temporarily increasing neurofilament light, phase 3 (GENERATION-HD1) was halted early as one of the treatment arms demonstrated significant worsening compared to the placebo group, while the other treatment arm was not significantly different from the placebo group on various outcome measures (Ltd, 2021; Tabrizi et al., 2019). It's still not clear why the trial failed. Explanations include off-target binding or proinflammatory events (evidenced by increased CSF protein and white blood cells) leading to accelerated neurodegeneration, as well as reduction of normal HTT owing to the non-allele-specificity of the ASO. A posthoc analysis showed that younger patients with lower CAG repeats did better when compared to placebo, providing hope for a sub-population benefiting from the drug, which will be tested in a new clinical trial. Another program leveraging an allele-selective approach used WVE-120101 and WVE-120102 which are ASOs directed against single nucleotide polymorphisms (SNPs) that are unique to the mutant HTT mRNA. The phase 1/2 trials were disappointing as no statistically significant lowering of mtHTT was detected in the CSF at any of the doses studied (Wave Life Sciences, 2022b, c). Other ASOs are being developed, including VO659, which not only targets mtHTT but also, theoretically, targets all polyglutamine repeat expansion diseases with preclinical studies in spinocerebellar ataxia type 1 and 3 as well as dentatorubropallidoluysian atrophy. This drug was granted Orphan-Drug Designation by the US FDA and European Commission in 2021 (Therapeutics, 2021).

Perhaps the most ambitious and direct approach to treat HD is to directly target the HTT gene. Zinc finger proteins (ZFP) and CRISPR-based systems are two modalities being investigated to achieve this goal. Zinc finger proteins consist of two elements: (1) an active effector element, such as a transcriptional repressor protein or nuclease, that acts directly on the gene; and (2) a DNA-binding element that enables selective targeting of specific DNA sequences (Tabrizi et al., 2022). AAV-ZFP selectively repressed >99% of HD-causing alleles over a wide dose range while preserving the expression of >86% of normal alleles. Infusions led to improvements in molecular, histopathological, electrophysiological, and functional endpoints in three HD mouse models (Zeitler et al., 2019). CRISPR-based systems can selectively and permanently inactivate mutant HTT genes and have been demonstrated to do so in animal and in vitro models including human fibroblasts (Shin et al., 2016; Monteys et al., 2017).

Splicing modulators are also being investigated in HD. One advantage of this approach is that these drugs could be administered orally. However, the non-specificity of such agents could be a limitation. Branaplam (LMI070) is a medication that was originally developed to target the expression of the survival motor neuron 2 (SMN2) gene as a treatment for spinal muscular atrophy but was found to non-selectively decrease the expression of HTT gene through induction of mis-splicing between exon 49 and exon 50 leading to decay of HTT mRNA and subsequent decrease in HTT protein (Keller et al., 2022; Krach et al., 2022). Other agents are being identified to act through similar mechanisms, including PTC518.

39.2.3 Parkinson's Disease

Using gene therapy for symptomatic relief or disease modification in Parkinson's disease (PD) is an active and promising endeavor. Alpha-Synuclein (α-Syn) is a neuronal protein, though also expressed in a variety of nonneuronal cells, strongly linked to PD pathogenesis. α-Syn is also aggregated in neurons in dementia with Lewy bodies (DLB) and in oligodendroglia in multiple system atrophy (MSA). Point mutations and gene multiplications of the α-Syn gene SNCA lead to aggregation of α-Syn and cause PD in a dose-dependent manner (Chartier-Harlin et al., 2004). The most important hypotheses explaining the link between α-Syn and PD are: (1) α-Syn aggregation in Lewy pathology disrupts normal cellular functions, (2) α-Syn trans-neuronal spread leads to propagation through the brain, and (3) α-Syn depletion causes loss of its normal and potentially essential role in synaptic transmission (Sandoval et al., 2021). Gene therapy for α-Syn pathology focuses on reducing α-Syn aggregation or targeting its consequences. One approach is targeted gene silencing of α-Syn or modulating its expression which could be attempted using ASOs, RNAi, zinc finger nucleases, and CRISPR-based technologies. However, due to the potentially detrimental effects of excessive reductions of monomeric α-Syn with this strategy, other strategies have emerged to achieve reduction of aggregated α-Syn without affecting the physiological levels of its monomeric form. One such strategy is using AAV5 to deliver nanobodies, which are single-domain antibody fragments expressed intracellularly as intrabodies, against the non-amyloid component (NAC) or the C-terminal component of α-Syn (Chatterjee et al., 2018). Similarly, AAV delivery of chaperones could prevent fibril formation. However, it's also possible that it's the depletion of "normal" α-Syn that promotes the pathogenesis of PD, and therapeutic efforts to reduce α-Syn may be counterproductive (Espay & Okun, 2022). Another important strategy to mitigate α-Syn pathology in PD is restoring balance in the autophagy-lysosomal pathway. Gaucher disease genetic variants, the most common genetic risk factors for PD, lead to the loss of lysosomal glucocerebrosidase enzyme and the accumulation of its substrate glycosylceramide (Sidransky et al., 2009). Interestingly, glycosylceramide acts as a scaffold for α-Syn, while α-Syn inhibits glucocerebrosidase activity leading to the accumulation of glycosylceramide creating a vicious cycle (Mazzulli et al., 2011). Addressing this pathway through gene therapy by overexpressing glucocerebrosidase or regulators of the autophagy-lysosomal pathway that work in the same direction is an ongoing effort (Xilouri et al., 2013; Decressac et al., 2013; Sardi et al., 2011). A Phase 1/2 clinical trial of intracisternal injection of AAV9-GBA1 to correct the autophagy-lysosomal imbalance in patients with moderate to severe Parkinson's disease with at least 1 pathogenic GBA1 mutation is ongoing (Prevail, 2022).

α-Syn is not the only pathogenic aspect of PD, and it is possible to slow down PD progression without directly or at all addressing proteinopathy. For example, glial-derived neurotrophic factor (GDNF) is a member of the transforming growth factor-β (TGF-β) superfamily that is thought to have promise in ameliorating PD pathology in animal models. Specifically, it's thought that this and other neurotrophic factors could prevent dopaminergic neuronal degeneration (Dass et al., 2006). A phase 1 trial of AAV2-GDNF showed tolerability, stability of the motor scores during the study period, and increased F-DOPA uptake in the infused areas on positron emission tomography (PET) imaging in a preliminary report (McFarthing et al., 2019; NINDS, 2022). Neurturin (NTN) is a functional and structural homolog to GDNF that binds to the GDNF receptor (GFRα2) and to the GFRα1 receptor. Despite reproducibly positive data in preclinical models (McFarthing et al., 2019) and a phase 1 clinical trial (Marks Jr. et al., 2008), the results of phase 2 clinical trials were disappointing (Marks Jr. et al., 2010; Bartus et al., 2013) failing to meet primary endpoints. It is worth noting, however, that low infusion volumes and lack of real-time MRI may have led to insufficient or inaccurate delivery of the vector (Sudhakar & Richardson, 2019).

Another gene therapy approach in PD focuses on delivering genes to local basal ganglia structures to improve symptoms. Delivering glutamic acid decarboxylase (GAD) gene via AAV2-GAD to the subthalamic nucleus (STN) has been tried in phase 1 and 2 clinical trials (Kaplitt et al., 2007; LeWitt et al., 2011). This enzyme catalyzes the synthesis of γ-aminobutyric acid (GABA) from glutamate achieving elevation of local GABA levels which is postulated to correct the pathological overactivity in the circuit. Although the trial showed improvements in the Unified Parkinson's Disease Rating Scale (UPDRS) scores, the same results could be achieved by deep brain stimulation (DBS) targeting STN. Delivering aromatic l-amino acid decarboxylase (AADC) gene, encoding a rate-limiting enzyme for dopamine synthesis that converts levodopa to dopamine, through AAV2-AADC to the post-commissural sensorimotor striatum had positive effects in non-human primates (Bankiewicz et al., 2000, 2006; Hadaczek et al., 2010) and was tried in two phase 1 clinical trials. The first trial demonstrated safety and modest efficacy (Christine et al., 2009; Mittermeyer et al., 2012; Eberling et al., 2008), while the second demonstrated safety and dose-dependent decrease in antiparkinsonian medication utilization along with increases in patient-reported ON-time without troublesome dyskinesia (Christine et al., 2019, 2022). ProSavin, a lentiviral vector delivering AADC, tyrosine hydroxylase, and cyclohydrolase 1 gene was studied in a phase 1/2 open-label clinical trial with dose-independent improvements in dopaminergic medications requirement, off-dose UPDRS scores, and off-dose dyskinesia (Palfi et al., 2014). However, biomarker verification of reliable and predictable delivery is a challenge.

39.2.4 Alzheimer's Disease

To address the etiology of Alzheimer's disease (AD) through gene therapy, three main genes are considered: amyloid precursor protein (APP), apolipoprotein E (ApoE), and microtubule-associated protein tau (MAPT) genes. The APP gene is responsible for a portion of familial AD when mutated and is strongly implicated in sporadic AD. Since beta-amyloid (Aβ) accumulation is thought to represent the initial key event in a cascade that leads to AD pathogenesis and the development of dementia (Hardy & Higgins, 1992), modulating the APP gene through gene therapy approaches is an obvious target. Due to its physiologic roles (Müller et al., 2017), including normal synaptic functioning, targeting APP is a highly intricate endeavor. For example, using ASOs to skip exon 17 that encodes the γ-secretase cleavage site (to prevent generation Aβ) impairs membrane-anchoring of APP as this exon encodes, in addition to the γ-secretase cleavage site, a large portion of the transmembrane protein (Chang et al., 2018). Other strategies attempt to target the Aβ generating properties of APP without negatively impairing the physiological function of APP. One such strategy is using CRISPR/Cas9 to edit APP C-terminus attenuating β-cleavage and in turn, Aβ production while upregulating the neuroprotective APP α-cleavage (Sun et al., 2019). Furthermore, this approach leaves the N-terminus and transmembrane domains of APP, which are important for its physiologic function, intact (Sun et al., 2019). The ApoE gene has three polymorphic alleles: ε2 (protective), ε3 (neutral), and ε4 (the greatest risk factor for developing sporadic AD). ApoE ε4, unlike other genetic risk factors for AD, is common in the general population (14–25%), and extremely common in the sporadic AD population (50–75%) (Alzforum). The mechanism of ApoE alleles in modulating the risk of AD has been linked to both amyloid and tau pathologies. One of the most accepted models is that these alleles differentially modulate APP transcription and Aβ secretion (Huang et al., 2017). It was reported that ApoE and TREM2 cooperate to activate microglia into a "neurodegenerative phenotype" that is closely associated with Aβ plaques (Krasemann et al., 2017). ApoE ε4 allele, like ApoE KO, traps microglia in the homeostatic state rendering them unable to respond to neuronal damage or Aβ accumulation (Alzforum). Another AD-related crime ApoE ε4 is accused of is dysregulation of lipid metabolism. This stems, in principle, from the fact that ApoE is a lipid transporter protein. Specifically, ApoE ε4, but not ApoE ε3, disrupted the cellular lipidomes of human induced pluripotent stem cell (iPSC)-derived astrocytes generated from fibroblasts of ApoE ε4 or ApoE ε3 carriers, and of yeast expressing human ApoE isoforms (Sienski et al., 2021). Furthermore, using CRISPR-edited iPSCs, a study examined the transcriptomic signatures of ApoE ε4 versus ApoE ε3 microglia in response to neuronal conditioned media, revealing that neuronal cues differentially induce a lipogenic program in ApoE ε4 microglia that exacerbates proinflammatory signal (Victor et al., 2022). In addition, ApoE ε4 microglia disrupted the coordinated activity of neurons through decreased uptake of extracellular fatty acids and lipoproteins (Victor et al., 2022). The number of possible mechanisms of how ApoE alleles modify the risk for AD is incredible (see citations for key reviews) (Steele et al., 2022; Martens et al., 2022; Koutsodendris et al., 2022; Serrano-Pozo et al., 2021; Belloy et al., 2019; Yamazaki et al., 2019). Increasing the protective allele or decreasing the risk allele are, thus, contemplated as gene therapy approaches to treat AD. For example, AAV-APOE ε2 as a method to increase the amount of the protective allele has been studied in animal models (Hudry et al., 2013; Rosenberg et al., 2018) with demonstrated efficacy and has been advanced in early clinical trial phases (Lexeo, 2022). MAPT gene encodes tau protein which is thought to be a major contributor to neuronal death in AD when phosphorylated and aggregated. To target the MAPT gene, an approach using ASOs has been assessed in animal models (Sud et al., 2014; DeVos et al., 2017), and is currently in a phase 1/2 clinical trial (Ionis Pharmaceuticals, 2022b). This approach aims to reduce tau phosphorylation and aggregation by reducing MAPT mRNA and, in turn, tau protein. BIIB080 is another ASO that acts similarly and is being tested in a phase 1 trial (Biogen, 2022c). Similar to non-allele-specific ASOs in HD, lowering MAPT mRNA may lead to deficits related to impaired tau physiological function.

The loss of cholinergic neurons in the basal forebrain is thought to be the main neurotransmitter alteration that contributes to symptoms of AD. Therefore, there are attempts to induce the expression of nerve growth factor (NGF), which regulates the growth and survival of cholinergic neurons (Mobley et al., 1986; Nagahara et al., 2009). AAV2-NGF administration to the nucleus basalis of Meynert was tried in phase 1 and 2 clinical trials without demonstration of clinical efficacy (Rafii et al., 2018). Although these results were disappointing, there are design shortcomings that still need to be addressed before declaring the failure of this target including the failure of AAV2-NGF to reach the targeted cholinergic neurons (Castle et al., 2020).

39.2.5 Amyotrophic Lateral Sclerosis

Amyotrophic lateral sclerosis (ALS) is characterized by progressive degeneration of upper and lower motor neurons. Although most cases of ALS are sporadic, about 10–15% are familial. More than 30 ALS-causing and related genes have been identified since the early 1990s (Suzuki et al., 2022). Mutations in chromosome 9 open reading frame 72 (C9orf72), superoxide dismutase 1 (SOD1), and fused in sarcoma (FUS) genes are the most common genetic etiologies responsible for 40–50% and 3–7% of familial and sporadic ALS respectively (Kim et al., 2020; Suzuki et al., 2022). Most gene therapy efforts in the ALS field are, thus, directed toward the manipulation of these three genes. GGGGCC hexanucleotide repeat expansion in intron 1 of C9orf72 is responsible for about a third of familial and 5% of sporadic ALS cases in the United States and Europe (Kim et al., 2020; Suzuki et al., 2022). The pathogenic mechanism is thought to be related to neuronal toxicity from repeat-associated non-AUG (RAN) RNA foci, or from translated non-AUG RAN dipeptide repeat proteins (Kim et al., 2020; Haeusler et al., 2016). Other mechanisms such as haploinsufficiency and loss of function (LOF) have been implicated. The ASOs WVE-004 and BIIB078 target toxic RAN RNA foci for degradation and have been successful in animal models (Jiang et al., 2016). These medications are being evaluated in phase 1/2 clinical trials (Biogen, 2022b; Wave Life Sciences, 2022a). In March 2022, the BIIB078 trial results were announced in a press release stating that the medication was well-tolerated but showed no difference from placebo on clinical efficacy endpoints at lower doses and trended toward more decline across multiple endpoints at high doses (Biogen and Ionis Pharmaceuticals), which led the companies to discontinue developing this particular agent for ALS.

Mutations in the SOD1 gene are responsible for 15% of familial ALS cases in the United States and Europe and 30% of familial ALS cases in Japan (Suzuki et al., 2022). SOD1 protein is important in protecting cells from toxic reactive oxygen species (ROS). LOF mutation leading to decreased enzymatic activity of SOD1, and subsequently decreased protection against ROS was initially proposed as a mechanism of action, but the fact that ALS-causing SOD1 mutations retain normal levels of enzymatic activity made a LOF explanation far less likely (Borchelt et al., 1994; Hayward et al., 2002). Furthermore, a systematic review of ALS-causing SOD1 mutants found no correlation between mutant SOD1 enzymatic activity and ALS onset or progression (Cleveland et al., 1995). There is in fact evidence of a gain of function (GOF) pathogenic mechanism in ALS-causing SOD1 mutations leading to SOD1 toxicity (Kim et al., 2020). This prompted designing ASOs to target SOD1 mRNA, which prolonged survival in ALS transgenic mouse models (Smith et al., 2006). A study of ASOs for mSOD1 mutations in ALS patients is now underway. Tofersen (BIIB067) is an SOD1 ASO that has been studied in phases 1, 2, and 3 clinical trials demonstrating decreased CSF SOD1 protein levels, but failing to significantly slow progression (Miller et al., 2013, 2020, 2022), although a signal in the revised ALS functional rating scale (ALSFRS-R) and other endpoints suggested decreased disease progression in the early-start cohort as compared to the delayed-start cohort in the phase 3 trial (Miller et al., 2022). The investigators will further assess the potential effects of earlier versus delayed initiation of tofersen in the open-label extension phase of the trial (Miller et al., 2022). Tofersen is also being tested in a phase 3 trial (ATLAS) in asymptomatic carriers to assess whether it could delay the onset of the disease (Biogen, 2022a). Other modalities, such as RNAi and CRISPR-based systems, are also candidates for SOD1 targeting and have been tested in animal models (Gaj et al., 2017; Iannitti et al., 2018; Foust et al., 2013). SOD1-targeting miRNA carried on an AAV has been tried in two ALS patients in a feasibility study the results of which could help design further trials to assess safety and efficacy (Mueller et al., 2020).

FUS is an RNA-binding protein whose genetic mutations are rare causes of aggressive forms of ALS in the United States and Europe albeit relatively not infrequent in Japan (10% of familial cases) (Suzuki et al., 2022). ION363, a non-allele-specific FUS ASO, efficiently silenced FUS and reduced postnatal levels of FUS protein in the brain and spinal cord in an ALS-FUS mouse model, which resulted in delaying motor neuron degeneration (Korobeynikov et al., 2022). The same experimental medication, ION363, was administered through repeated intrathecal infusions in one ALS patient with a FUSP525L mutation, resulting in lowered wild-type and mutant FUS levels, which in turn resulted in a reduction of FUS aggregates (Korobeynikov et al., 2022). A trial to evaluate the clinical efficacy of ION363 on clinical function and survival in carriers of FUS mutations with ALS has been launched (Ionis Pharmaceuticals, 2022a).

39.2.6 Frontotemporal Dementia

Frontotemporal dementia (FTD) is a spectrum of clinical syndromes that stem from frontotemporal lobar degeneration (FTLD) with various underlying pathologic features. The most common syndromes include the behavioral variant (bvFTD), non-fluent primary progressive aphasia (nfPPA), and semantic dementia (SD). Other separate disease entities often manifest with a clinical FTD syndrome and share pathologic features with the FTLD; those include progressive supranuclear palsy (PSP) and corticobasal degeneration (CBD). One of the main pathologic features of a subset of FTD cases is abnormal tau protein, often hyperphosphorylated and with four repeated microtubule-binding domains (4R tau). BIIB080 is an ASO that reduces MAPT mRNA and subsequently tau protein and is being tested in AD (see before) but is also being considered as an option for FTD. Another ASO-based strategy is targeting MAPT exon 10, which is included in 4R tau isoforms, in order to induce alternative splicing and restore the balance between 3R tau and 4R tau (Rodriguez-Martin et al., 2009). Similar to familial ALS, C9orf72 hexanucleotide repeat expansion is the most common cause for familial FTD; patients could present with the ALS-FTD combination, and families with the inherited mutation could have members with ALS and others with FTD. Modalities that address this genetic abnormality are potential future candidates for FTD treatment (see Sect. 2.5 for details). In fact, the FOCUS-C9 trial of WVE-004 recruits patients with FTD in addition to those with ALS with C9orf72 mutation (Wave Life Sciences, 2022a).

39.2.7 Spinal Muscular Atrophy: The First Success Story

Spinal muscular atrophy (SMA) is a monogenic motor neuron disease affecting lower motor neurons in infants, children, and, rarely, adults. Genetic therapies for SMA are the first successful neurological disease gene therapies. A homozygous deletion of the survival motor neuron 1 (SMN1) gene leads to decreased levels of SMN1 protein and, subsequently, lower motor neuron degeneration. SMN2 gene is a paralog to SMN1 with a single nucleotide substitution leading to altered splicing and exclusion of exon 7. SMN2 normally encodes the minority of SMN protein because most of the protein produced is truncated and degraded. Altering the splicing of the SMN2 pre-mRNA was the basis for the first successful treatment for SMA, nusinersen (Passini et al., 2011; Finkel et al., 2017). Nusinersen is an ASO that pairs with a specific target sequence on the SMN2 pre-mRNA to displace heterogeneous ribonucleoproteins (hnRNPs) at the intronic splice silencing site-1 (ISS-1) between exons 7 and 8, leading to the inclusion of exon 7, and thus producing full-length SMN transcript and restoring SMN protein levels in the central nervous system. After a phase 3 study in infants demonstrated clear efficacy and suggested better outcomes with earlier administration of the ASO, the FDA issued its approval for the first gene therapy in neurodegenerative disease (Finkel et al., 2017). Another medication is the oral small molecule, risdiplam, which shifts the balance of SMN2 splicing toward the production of full-length SMN2 messenger RNA and, subsequently, SMN functional protein (Naryshkin et al., 2014). The FDA-approved risdiplam in 2020 (Dhillon, 2020). Another approach to treating SMA through gene therapy is targeting the SMN1 deletion itself. An AAV9 vector carrying the SMN1 gene (onasemnogene abeparvovec) was approved in 2019 by the FDA as a single-dose gene-modulating therapy after remarkable improvement in survival and morbidity, including the need for respiratory support in infants with SMA (Al-Zaidy et al., 2019; Mendell et al., 2017). The work on improving the current modalities and developing others in gene therapy for SMA has not been concluded, as long-term follow-up in treated populations is ongoing. In addition, the assessment of these gene therapies in adults, and the development of other therapies that are less invasive and with more favorable side effect profiles are other goals. In conclusion, the success of gene therapy in SMA is a testament to the potential value this approach can bring to the field of neurodegenerative disease therapeutics.

39.3 Cell Therapy

Cell therapy is the transfer of intact live cells into an individual to help lessen or cure a disease. The infused cells can facilitate the repair or renewal of damaged tissue, and protect against further damage and disease by immunomodulation, releasing growth factors, protective cytokines, or direct cell contact. The cells could be autologous (from the patient) or allogeneic (from a donor). Autologous cells may be expanded or modified ex-vivo, e.g., chimeric antigen receptor (CAR)-T-cell therapy and ex-vivo expanded regulatory T-cell (Treg) therapy before being transferred back to the patient. Cell therapy has been in practice since the seventeenth century, around the time of the discovery of the cell itself, and independent of this knowledge, in the form of blood transfusions (Zimmerman & Howell, 1932). More recently, hematopoietic stem cell transplantation was developed and became a standard therapeutic tool in modern medicine (Gratwohl & Niederwieser, 2012). Cell therapy is a promising approach to treat neurodegenerative diseases in two ways: (1) transfer or implantation of cells that will assume the function of degenerated neurons, and (2) transfer or implantation of cells that support existing vulnerable neurons either directly, e.g., by releasing trophic molecules, or indirectly, e.g., by dampening inflammation. In this section, we will discuss cell replacement therapy endeavors in Parkinson's disease and Alzheimer's disease, as well as immune cell therapeutics in neurodegenerative diseases. One type of cell therapy being investigated in neurodegenerative diseases is stem cell therapy. Given the variability of stem cell types, it may be suitable to review the nomenclature used to describe their spectrum here. Totipotent cells originate from blastomeres and could give rise to all embryonic and extraembryonic tissues. Pluripotent stem cells give rise to all three germ layers, the endoderm, ectoderm, and mesoderm. There are several types of pluripotent stem cells including embryonic stem cells (ESCs) and induced pluripotent stem cells (iPSCs). Adult stem cells (ASCs) are undifferentiated multipotent cells that could give rise to specific differentiated cells, e.g., mesenchymal stem cells (MSCs), and neural stem cells (NSCs).

39.3.1 Cell Replacement Therapy

39.3.1.1 Parkinson's Disease: A Model for Cell Therapy in Neurodegeneration

Parkinson's disease (PD) is characterized by the loss of dopaminergic neurons in the substantia nigra. These dopaminergic neurons project to the striatum. The loss of this striatal input is thought to be responsible for most of the motor manifestations of PD. Although dopamine replacement using L-dopa and the use of dopamine agonists are helpful, especially in early disease stages, this approach typically loses efficacy with the progression of the disease, and the medications "produce" unwanted side effects such as dyskinesia. Dyskinesia is thought to be proportional to the duration of the disease and the pulsatile nature of levodopa administration rather than the molecule of levodopa itself. Cell therapy to replace the degenerated dopaminergic neurons is, thus, an attractive idea to achieve a uniform supply of dopamine that is sustained for longer periods of time. Cell therapy in PD is perhaps the most promising venue for using restorative cell therapy in neurodegenerative disease at the current time owing to the clarity of the pathogenic effect of loss of dopaminergic neurons in the substantia nigra and the experience with using other dopamine-enhancing approaches to improve the symptoms.

Although the chief location of neuronal loss in PD is the midbrain substantia nigra, the most common approach is to implant dopaminergic neurons directly into the striatum (heterotopic graft) to bypass the need to establish longer axonal connections between the new cell and the striatum. Fetal neuronal cells from the ventral mesencephalon were the first source of dopaminergic neurons used in PD animal models and humans in the last quarter of the twentieth century. Fetal neuronal cells are harvested from aborted fetuses after dissecting their brains and processed into cell preparations that are then transplanted. Stem cells, on the other hand, are produced by harvesting pluripotent embryonic stem cells from the center of a blastocyst, then inducing their differentiation into a specific cell line, in this case, dopaminergic neurons. Fetal neuronal cells have the advantage of being already neuronally differentiated and region-specific, while their limited availability, high variability, and ethical considerations associated with their use make them a less sought-after approach for neuronal cell therapy. On the contrary, pluripotent stem cells are readily available and less variable, but they are not neuronally differentiated or region-specific. In addition, pluripotent stem cells are less safe to use as they have the potential for tumorigenesis, unlike fetal neuronal cells.

The first attempt to study cell transplantation in PD in animal models started in 1979 using fetal ventral mesencephalic cell implants (Chmielarz et al., 2020). Fetal mesencephalic tissue implantations into the striatum have been tried in patients with variable degrees of success; younger patients had better results (Barker et al., 2015; Lindvall et al., 1989). Despite some patients benefiting greatly from this approach, e.g., not needing medications for over 20 years (Kefalopoulou et al., 2014), some had troublesome dyskinesia or little benefit. Furthermore, large sham-controlled studies failed to show clear benefits (Freed et al., 2001; Olanow et al., 2003). It is suggested that improvements in cell survival and standardization of the transplantation protocols may improve outcomes. TransEuro research consortium attempts to standardize and optimize fetal tissue transplantation procedures ('TRANSEURO | Home'). Human embryonic stem cells showed promise in animal models of PD as they could be induced to differentiate into functional dopaminergic neurons in culture (Cooper et al., 2010; Fasano et al., 2010; Kirkeby et al., 2012), expanded, and sorted based on cell markers to ensure the exclusion of non-dopaminergic cells (Bye et al., 2015; Doi et al., 2014; Samata et al., 2016), before being transplanted into animal brains with efficient engrafting and functional recovery (Grealish et al., 2014; Kriks et al., 2011). The ethical dilemmas associated with using fetal cells or tissues inspired consideration of other approaches. In 1985 Backlund and colleagues conducted the first-in-human trial using autologous adrenal medullary tissue implantation to the striatum (Backlund et al., 1985). This was followed by several other studies in hundreds of patients (Freed et al., 1990). Adrenal chromaffin cells are capable of producing low doses of dopamine as well as other catecholamines which makes the case for this choice. These trials, however, demonstrated variable degrees of improvement in patients, and some had serious side effects (Freed et al., 1990).

Human induced pluripotent stem cells (hiPSCs) are currently the most promising cell therapeutic avenue in PD as they solve the problem of availability, and their use is unburdened by ethical questions. Current methods involve the differentiation of hiPSCs from rodent or human fibroblasts to dopaminergic neurons either through the use of small molecules to activate or inhibit specific developmental pathways or through the use of viral vectors to overexpress specific developmental transcription factors (Chambers et al., 2009; Theka et al., 2013). These cells were safely implanted into PD rodent models (Kikuchi et al., 2017; Doi et al., 2014). Furthermore, hiPSC-derived dopaminergic neurons were shown to function in a primate PD model (Kikuchi et al., 2017; Morizane et al., 2017). For clinical use, two approaches could be employed regarding the source of hiPSCs: (1) autologous hiPSCs, which have perfect immunological matching and, thus, a low probability of rejection, but the high cost and time associated with this process are disadvantages, and (2) allogenic hiPSCs which could be obtained, differentiated, and biobanked on a large scale followed by further differentiation and transplantation in individual patients based on human leukocyte antigen-matching (Chmielarz et al., 2020). In 2018, a Japanese neurosurgeon implanted 2.4 million dopamine precursor cells differentiated from hiPSC into the brain of a patient with PD for the first time (Cyranoski). This trial has planned to enroll seven patients and has completed recruitment (UMIN-CTR). The cells used in this Japanese trial were allogenic from a stem cell bank with variable degrees of HLA matching to the hosts (Barbuti et al., 2021). In addition to the technical difficulties, a major issue emerges as we embark on this era of stem cell transplantation in PD and other neurodegenerative diseases; that is the vulnerability of transplanted cells to the underlying host brain pathology. In clinical trials of fetal tissue transplants, Lewy body pathology as well as reduced dopamine transporter and tyrosine hydroxylase expression were observed in some patients (Chu & Kordower, 2010; Hallett et al., 2014; Kordower et al., 2008; Kurowska et al., 2011; Li et al., 2008; Barbuti et al., 2021). This may be due to the pathogenic environment into which these cells are implanted (e.g., propagation of alpha-synuclein pathology into the implanted neurons by a prion-like mechanism). It is yet to be determined if this will represent a difficulty in trials using hiPSC-derived dopaminergic progenitors. Furthermore, autologous hiPSC-derived cells will have the same genetic and potentially epigenetic features that degenerated neurons had. This latter difficulty may be addressed using genetic manipulations to the cells prior to transplantation in order to render them less vulnerable to degeneration once implanted (Chmielarz et al., 2020). Other challenges include the possibility of rejection and the complications that may accompany immunosuppression as well as the presence of pluripotent stem cells in the grafts or injections which carries a risk of tumorigenesis. Cell therapy may also need to control the neuroinflammatory process in addition to or prior to "stem cell" transfer.

39.3.1.2 Stem Cell Therapy in Alzheimer's Disease

Numerous preclinical studies have focused on understanding and optimizing stem cell therapy in AD. Hippocampal neural stem cell transplantation was shown to rescue spatial learning and memory deficits in an AD mouse model (triple transgenic mice; 3xTg-AD) that expresses pathogenic forms of amyloid precursor protein, presenilin, and tau (Blurton-Jones et al., 2009). In this particular study, the effect was not mediated by altering amyloid or tau pathology but was rather dependent on the robust enhancement of hippocampal synaptic density, mediated by brain-derived neurotrophic factor (BDNF) (Blurton-Jones et al., 2009). As reviewed by Qin and colleagues, the body of preclinical research suggests that several mechanisms mediate the improvement of cognitive deficits with stem cell therapy in AD animal models, including: (1) stimulation of neurogenesis and inhibition of apoptosis, (2) mitigation of beta-amyloid (Aβ) plaques and neurofibrillary tau tangles, (3) inhibition of neuroinflammation which is potentially mediated by restoring the balance between microglial phenotypes, (4) immunoregulation through modulation of immune-related genes, e.g., TREM2, and (5) synaptogenesis (Qin et al., 2021).

The use of ESCs in AD faces similar challenges to those discussed with its use in PD, most importantly the ethical challenges. Nevertheless, ESCs-derived basal forebrain cholinergic neurons were shown to improve learning and memory when transplanted into 5xFAD and APP/PS1 transgenic mice (two commonly used AD mouse models) (Bissonnette et al., 2011; Yue et al., 2015; Liu et al., 2013). Fetal and adult NSCs have been studied preclinically in AD but faced challenges, including failure to differentiate into mature cells in one study, and the lack of improvement in cognitive performance in another (Marsh et al., 2017). Despite these challenges, other studies demonstrated improvement in cognition, increased synaptic connectivity, and neurotrophic factors in different AD mouse models (Lee et al., 2015; Ager et al., 2015; McGinley et al., 2018). Furthermore, early-phase clinical trials have been launched to investigate NSCs in patients with AD dementia in China (Karvelas et al., 2022).

ASCs therapy, particularly using mesenchymal and hematopoietic stem cells, is the most studied form of stem cell therapy in AD to date. The relative ease of access, availability, and the low immunogenicity and carcinogenicity are important factors encouraging the study of ASC-derived MSCs as a potential therapy in AD. The general mechanisms underlying the potential benefits of stem cells in AD apply, more or less, to MSCs, e.g., promoting neurogenesis, modification of immune cell responses through upregulating anti-inflammatory cytokines such as IL-10 and downregulating proinflammatory ones such as TNF-α and IL-1β, decreasing oxidative stress, and mitigating Aβ proteinopathy (Cui et al., 2017; Zhang et al., 2016; Lee et al., 2010, 2012; Colpo et al., 2015; Naaldijk et al., 2017). In several AD animal models such as APP/PS1, 3xTg-AD, 5xFAD, APdE9, and Tg2576 mice, MSCs were shown to improve cognitive performance (Xie et al., 2016; Cui et al., 2017; Boutajangout et al., 2017; Ma et al., 2013; Kim et al., 2012; Nasiri et al., 2019; Ruzicka et al., 2016; Harach et al., 2017; Yokokawa et al., 2019; Naaldijk et al., 2017; Matchynski-Franks et al., 2016; Wei et al., 2018; Zhang et al., 2016). It is worth noting that the source of MSCs in these studies was variable including umbilical cord, adipose tissue, and bone marrow. The mode of cell transfer was also variable including implantation into parenchymal structures, intranasal administration, and intravenous infusion (Xie et al., 2016; Cui et al., 2017; Boutajangout et al., 2017; Ma et al., 2013; Kim et al., 2012; Nasiri et al., 2019; Ruzicka et al., 2016; Harach et al., 2017; Yokokawa et al., 2019; Naaldijk et al., 2017; Matchynski-Franks et al., 2016; Wei et al., 2018; Zhang et al., 2016). There is a myriad of phase 1 and 2 clinical trials investigating MSCs in AD. In one study, human umbilical cord blood-derived MSCs (hUCB-MSCs) a.k.a. NEUROSTEM®-AD were administered via open brain surgery to the bilateral hippocampi and right precuneus of patients with AD which was feasible and safe, but because of lack of controls and small sample size, a confident interpretation of efficacy analysis is not possible (Kim et al., 2015). Another study investigated the administration of hUCB-MSCs through an Ommaya reservoir to patients with mild-moderate AD dementia (Kim et al., 2021). Although 2/9 patients had severe adverse events, the overall safety assessment was encouraging and the intervention moved into phase 2a which is expected to yield more information on the clinical efficacy (Kim et al., 2021). A promising phase 1 trial investigated the injection of autologous adipose-derived stromal vascular fraction (which contains MSCs) through an implanted Ommaya reservoir or a ventriculoperitoneal (VP) shunt (Duma et al., 2019). The trial revealed that 8/10 AD patients remained stable or improved in cognitive tests over several months (Duma et al., 2019).

Unlike MSCs, the investigation of iPSCs in AD is still in the preclinical stage. Transplantation of cholinergic neuron precursors derived from human iPSCs into the bilateral hippocampi of APP transgenic AD mouse model resulted in the thriving of choline acetyltransferase-positive cholinergic human neurons and alpha7 nicotinic acetylcholine receptor-positive neurons throughout the cortex, and vesicle GABA transporter-positive GABAergic human neurons in the hippocampus (Fujiwara et al., 2013, 2015). Furthermore, significant improvement in spatial memory dysfunction compared to controls was demonstrated (Fujiwara et al., 2013, 2015). Another group injected iPSC-derived neural precursors (iPSC-NPCs) into the hippocampus of aged 3xTg-AD mice, which resulted in improved memory, synaptic plasticity, and reduction in amyloid and tau tangle deposits (Armijo et al., 2021). A third model (5xFAD transgenic AD mouse model) received iPSCs differentiated into glial cells (exhibiting oligodendroglial markers in vivo), leading to decreased plaque depositions and mitigation of the cognitive dysfunction (Cha et al., 2017). Induced pluripotent stem cell-derived macrophage-like myeloid lineage cells (iPS-ML) reduced the level of Aβ added into the culture medium and alleviated its neurotoxicity (Takamatsu et al., 2014). These effects were enhanced by iPS-ML expressing Neprilysin-2 (iPS-ML/NEP2), a protease that is able to degrade Aβ, in vitro (Takamatsu et al., 2014). Furthermore, intracerebral injections of iPS-ML/NEP2 decreased Aβ concentration in 5xFAD mice brain interstitial fluid (Takamatsu et al., 2014). For a comprehensive review of clinical trials of stem cell therapy in AD, please see cited review (Karvelas et al., 2022).

39.3.2 Immune Cell Therapy

39.3.2.1 Regulatory T-Cell Therapy in Neurodegenerative Diseases

Regulatory T cells (Tregs) are a group of CD4+ (and to a lesser extent CD8+) T cells that are involved in suppressing inflammation and maintaining self-tolerance throughout the body including the central nervous system (Okeke et al., 2014; Xie et al., 2015; He & Balling, 2013). There is robust evidence of inflammatory processes contributing to the pathogenesis of neurodegeneration (Calsolaro & Edison, 2016; Colonna & Butovsky, 2017; Ransohoff, 2016). A review of inflammation in neurodegenerative diseases is beyond the scope of this chapter but we will highlight major aspects of inflammation in AD, PD, ALS, and FTD in what follows. In these neurodegenerative diseases especially AD and PD, the association between risk genes, and inflammation have provided the basic fundamental evidence for the association between neurodegeneration and inflammation (Hamza et al., 2010; Sliter et al., 2018; Wallings et al., 2020; Cabezudo et al., 2020; Wallings & Tansey, 2019; Cook et al., 2017; Rohn, 2013; Jonsson et al., 2013; Tai et al., 2015; Leyns et al., 2017; Zhang et al., 2013; Efthymiou & Goate, 2017; McCauley & Baloh, 2019; Wei et al., 2014). In AD, PD, ALS, and FTD, increased brain microglial activation and peripheral immune infiltration are noted using neuroimaging and in postmortem brains as well as in animal models which also demonstrate amelioration of neurodegeneration upon exposure to immunomodulatory approaches (Appel et al., 2010; Brochard et al., 2009; Gerhard et al., 2006; McGeer et al., 1988; Zhou et al., 2021; Allen Reish & Standaert, 2015; Harms et al., 2013, 2018; Thome et al., 2016; Volpicelli-Daley et al., 2016; Williams et al., 2018; Eriksen et al., 2003; Yan et al., 2003; Chakrabarty et al., 2015; Cho et al., 2007; Sy et al., 2011; Simard et al., 2006; Butovsky et al., 2007; Varvel et al., 2015; Prokop et al., 2015; Giubilei et al., 2003; Zhang et al., 2003; Ferretti et al., 2016a; Merlini et al., 2018; Henkel et al., 2004; Graves et al., 2004; McCauley & Baloh, 2019; Corcia et al., 2012; Bevan-Jones et al., 2020). Peripheral monocytic phenotype is shifted to a proinflammatory state and shows overexpression of the chemokine receptor CCR2 in patients with PD suggesting potential migration to the CNS, which has been confirmed in animal models (Funk et al., 2013; Gliem et al., 2016; Harms et al., 2017; Grozdanov et al., 2014; Harms et al., 2018; Henkel et al., 2004; McCauley & Baloh, 2019; Murdock et al., 2017; Zhang et al., 2005). Lymphocytic populations including CD4+ T cells and cytotoxic T lymphocytes were also shown to be altered in PD, AD and ALS and their animal models (Baba et al., 2005; Bas et al., 2001; Gruden et al., 2011; Hisanaga et al., 2001; Stevens et al., 2012; Kustrimovic et al., 2018; Sulzer et al., 2017; Lindestam Arlehamn et al., 2020; Marsh et al., 2016; Yokoyama et al., 2016; Togo et al., 2002; Rogers et al., 1988; Itagaki et al., 1988; Browne et al., 2013; Monsonego et al., 2001; Ferretti et al., 2016a, b; Giubilei et al., 2003; Zhang et al., 2003; Merlini et al., 2018; Graves et al., 2004). In summary, both CNS and systemic compartments show altered immune states in neurodegenerative diseases; specifically, myeloid, and lymphocytic cells show altered phenotype, function, and localization. Since Tregs normally induce the myeloid cells shift towards a protective state and suppress the chemokinesis and proinflammatory function of T lymphocytes, studying Treg phenotype and function in neurodegenerative diseases emerged as a promising approach.

In patients with ALS and PD, the expression of the transcription factor facilitating the immunosuppressive function of Tregs, FOXP3, is reduced in rapidly progressing patients (Henkel et al., 2013; Beers et al., 2017; Thome et al., 2021). In AD and PD, the expression of the alpha subunit of interleukin 2 (IL-2) receptor (CD25) is reduced (Faridar et al., 2022; Thome et al., 2021). Tregs isolated from the blood of patients with AD, PD, ALS, and FTD fail to suppress responder T-cell proliferation in vitro (Faridar et al., 2022; Thome et al., 2021; Alsuliman et al., 2016; Eid et al., 2022). IL-2 is a primary trophic factor for T lymphocytes including Tregs. When expanded ex-vivo using interleukin-2, rapamycin, and a physiologic activator, Tregs from patients with neurodegenerative diseases restore their typical phenotypic characteristics and suppressive function (Faridar et al., 2022; Thome et al., 2021; Alsuliman et al., 2016; Eid et al., 2022). Preclinically, human Tregs were expanded ex-vivo and administered to 5xFAD-Rag2KO AD mice which, at baseline, lacked mature lymphocytes, upregulated 95 inflammation genes, and showed an amplified number of reactive microglia (Faridar et al., 2022). Treg administration in these AD mice reduced brain amyloid burden and reactive glial cells and led to the down-regulation of proinflammatory cytokines, complement cascade, toll-like receptors, and microglial activation markers (Faridar et al., 2022). In ALS, infusions of Tregs in mice models demonstrated suppression of responder T lymphocytes (CD+CD25- T cells) proliferation and of microglial activation and slowed disease progression (Zhao et al., 2012; Beers et al., 2011). A phase 1 clinical trial enrolled three patients with sporadic ALS and tested autologous ex-vivo expanded Treg administration at early and later stages with concomitant subcutaneous IL-2 administrations, demonstrating safety and slowed progression during the periods of infusions compared to the intervals (Thonhoff et al., 2018). A randomized controlled phase 1 clinical trial in seven rapidly progressing ALS patients (3 in the treatment arm and four in the placebo arm) used a similar approach (ex-vivo expanded Treg infusions and IL-2 infusions) followed by an open-label phase with escalating doses (Thonhoff et al., 2022). The treatment was safe and tolerable, but efficacy end-point analysis was not possible due to the low number of participants as the study was hindered by the early COVID-19 pandemic circumstances (Thonhoff et al., 2022). Nevertheless, there were some indications of promising efficacy, e.g., two of the three participants in the treatment arm showed relative stabilization throughout the trial and the open-label phases, while the third participant showed rapid progression throughout the two phases (Thonhoff et al., 2022). In the open-label 6 months, six of the eight ALS patients had slowed progression of −0.45 points/month on the ALSFRS clinical scale, while two of the eight patients were unresponsive (progressing at a rate of greater than 1.7 points/month) despite the autologous infusions of expanded Tregs (Thonhoff et al., 2022). Larger studies are clearly needed to test efficacy.

39.3.2.2 In Vivo Enhancement of Immunomodulatory Function

Although this section is dedicated to cell therapy, we opted to highlight two small molecules that act on the immune system to expand the regulatory T cells (Tregs) population in vivo in neurodegenerative diseases for a therapeutic purpose. As discussed before, interleukin 2 (IL-2) has been used in two trials in ALS along with autologous Tregs. Aldesleukin, a recombinant IL-2 cytokine that is essential for T-cell immunity, was the first effective FDA-approved cancer immunotherapy. It is indicated for both metastatic melanoma and metastatic renal cell carcinoma (NCI, 2008). IL-2 activates IL-2 receptor (IL-2R) on Tregs, CD4+ helper T cells (Th cells), and CD8+ memory T cells. IL-2R consists of three subunits, α chain (CD25), β chain (CD122), and γ chain (CD132). Because of the variety of T cells that express IL-2R, and the fact that different IL-2R subunit combinations have different effects on the activation of these cells, IL-2 can have an anti-inflammatory, proinflammatory, or other specific immunomodulatory functions depending on the context (Li et al., 2022). IL-2Rα has a much higher affinity to IL-2 than IL-2Rβ and IL2-Rγ (Li et al., 2022). Because Tregs express an abundant amount of IL-2Rα chain in addition to the β and γ chains, low doses of IL-2 preferentially activate Treg IL-2Ra, which induces full expression of FOXP3 (a transcription factor essential for Treg immunomodulatory function) (Li et al., 2022). This is why low-dose recombinant IL-2 is being assessed clinically in many autoimmune and rheumatologic diseases (and in ALS as mentioned) (Graßhoff et al., 2021), while high-dose IL-2 is used as cancer immunotherapy as it induces an effector proinflammatory T-cell response through its activity on other more prevalent T cells (specifically Th and memory T cells) (Cheng et al., 2011). Another medication that could enhance Treg immunomodulatory function is sargramostim. Sargramostim is a granulocyte macrophage-colony stimulating factor (GM-CSF) that decreases microglial activation and protects dopaminergic neurons in 1-methyl-4-phenyl-1,2,3,6-tetrahydropyridine (MPTP) intoxicated mice (a mouse model for PD) (Olson et al., 2020; Kosloski et al., 2013). Specifically, daily peripheral rGM-CSF administration induces Tregs in a dose-dependent manner, decreasing microglial activation responses and associated neuronal destruction in the substantia nigra (Kosloski et al., 2013). Because of the short half-life of and limited bioavailability of GM-CSF, a long-acting GM-CSF (mPDM608) has been studied in the same mouse model resulting in a dose-dependent increase in Treg numbers with enhancement of their immunomodulatory function that could be sustained up to 10 days after a single injection (Olson et al., 2020).

Furthermore, a single injection scheme of mPDM608 achieved a nearly identical level of neuroprotection in the MPTP mouse model (Olson et al., 2020). A phase 1 randomized controlled trial of high-dose sargramostim (6 µg/kg/day) in PD reported modest improvements in UPDRS part III (the motor examination portion of the UPDRS scale) scores after 6–8 weeks of treatment when compared to placebo which paralleled magnetoencephalography-recorded cortical motor activities and Treg numbers and function (Gendelman et al., 2017). Although the medication was well-tolerated, side effects including injection site reactions, bone pain, urticaria, and vasculitis inspired another trial to assess a lower dose for an extended trial time i.e., 3 µg/kg/day (5 days on, 2 days off) for 12 months (Gendelman et al., 2017). The low-dose regimen resulted in no serious or severe adverse events (which was significantly different from the high-dose regimen in the previous study), and no participants withdrew from the study. Furthermore, UPDRS part III scores reductions or stabilization were evident and similar improvements in Treg immunomodulatory function were recorded (Gendelman et al., 2017).

39.4 Immunotherapy

Immunotherapy is a therapeutic approach that uses either active or passive immunization to target molecules that are key players in disease pathogenesis. Active immunotherapies (vaccines) are very successful in the prevention or mitigation of severity/complications of some infectious diseases and are actively being studied in cancer and neurodegenerative disease research. The idea is to trigger the body's own immune response to generate long-term endogenous antibodies against a specific target (or memory immune cells that are capable of producing those antibodies when needed). Passive immunotherapies, on the other hand, are simply antibodies or their alternatives, e.g., aptamers, that are typically more specific to the target and carry the advantage of being temporary as they don't generate long-term modification to the immune system (could be a disadvantage due to the need for repeated administration).

39.4.1 Immunotherapy in Alzheimer's Disease

Immunotherapy in AD aims to target those molecules thought to be key pathogenic factors leading to neurodegeneration, including beta-amyloid (Aβ), abnormal tau, and certain other immunological molecules such as triggering receptor expressed on myeloid cells 2 (TREM2). The amyloid cascade hypothesis is still the dominating hypothesis attempting to explain the pathogenesis of AD. The hypothesis posits that increased Aβ production or failure of clearance leads to accumulation and oligomerization of the protein in the limbic and association cortices (Selkoe & Hardy, 2016). These oligomers have subtle effects on synaptic efficacy but are, more importantly, gradually deposited as diffuse plaques triggering inflammatory responses, altered ionic homeostasis, oxidative stress, and altered kinase/phosphatase activities, which subsequently trigger abnormal (mostly hyperphosphorylated) tau tangles (Selkoe & Hardy, 2016). The tau tangles, in turn, lead to widespread synaptic dysfunction and neuronal loss (Selkoe & Hardy, 2016). This hypothesis provides the rationale for using immunotherapy against the early key molecule (Aβ), the intermediate key inflammatory molecules, and the later-stage key molecule (abnormal tau).

39.4.1.1 Anti-Amyloid Immunotherapies

To understand immunotherapies directed against Aβ, it's essential to have an overview of the APP and its processing. APP is a transmembrane protein that has an amino-terminal (N-terminus) in the extracellular space, and a carboxyl-terminal (C-terminus) in the intracellular space. There are three enzymes that cleave APP: α, β, and γ secretases. If APP is first cleaved by β secretase, a β C-terminus fragment stays on the membrane, and a soluble APP Aβ is produced in the extracellular space. The β C-terminus fragment subsequently gets cleaved by γ secretase, producing an APP intracellular domain (AICD) and an Aβ peptide that gets released into the extracellular space, and can vary in length; typically, Aβ1-40 is more ubiquitous than Aβ1-42 which is more toxic especially when it forms oligomers and polymeric insoluble fibrils. This is called the amyloidogenic pathway. If the protein is cleaved first by α secretase, the cleavage site is in the middle of the β fragment preventing the production of Aβ peptides from further cleavage by γ secretase (the nonamyloidogenic pathway). The idea behind anti-amyloid immunotherapies is to target the process of production and aggregation of Aβ at any of its key turns to halt the formation of Aβ monomers, oligomers, or plaques or to clear them.

Since the 1990s, the field has known many active immunotherapeutic agents that proved effective in clearing amyloid plaques but failed to be good clinical candidates either because of lack of clinical efficacy, presence of considerable side effects, or both. AN1792 vaccine, a synthetic full-length Aβ peptide with the vaccine adjuvant QS-21, was the first to be tested clinically and was found to lower Aβ deposition and slow functional and cognitive decline in a select group of patients (antibody responders) (Gilman et al., 2005; Ferrer et al., 2004; Vellas et al., 2009). However, it led to T-cell-mediated meningoencephalitis in 6% of patients, which resulted in the termination of the vaccine's further development (Gilman et al., 2005; Ferrer et al., 2004; Vellas et al., 2009). CAD106 (amilomotide) combines the N-terminus Aβ1-6 peptide which is a B-cell epitope to a QB-virus-like particle, and was developed to lack a T-cell epitope in order to prevent the meningoencephalitis experienced with AN1792. Amilomotide had a good safety profile and showed a favorable antibody "response" in patients with mild-moderate AD dementia in multiple trials (Winblad et al., 2012). However, the developing company retired the program in 2019 as the vaccine caused unpredicted changes in cognitive function and brain volume (Song et al., 2022). UB-311 is another vaccine that has Aβ1–14 peptide along with a helper T-cell epitope designed to stimulate T-helper type 2 regulatory immune response over T-helper type 1 pro-inflammatory response ('UB-311', 2020; Wang et al., 2007, 2017). A phase 2 placebo-controlled clinical trial testing UB-311 was conducted, and although the secondary endpoints (clinical efficacy measures) were not statistically significant, people who received four boosters declined only half as much on the CDR-SB, ADCS-ADL, ADADS-Cog (cognitive scores commonly used in AD trials) as those who received two boosters or placebo ('UB-311', 2020; Neuroscience, 2019). UB-311 was safe and tolerable and, surprisingly, did not cause amyloid-related imaging abnormalities despite resulting in a modest reduction in brain amyloid (Neuroscience, 2019; 'UB-311', 2020). This vaccine is planned to enter phase 3 in patients with mild cognitive impairment (MCI) or mild AD dementia.

There is a myriad of Aβ monoclonal antibodies being developed and tested in AD. Bapineuzumab is a humanized monoclonal IgG1 antibody that targets the Aβ N-terminus (Aβ1-7), binding soluble and fibrillar Aβ, and was one of the first to enter phase 3 clinical trials, but clinical efficacy was not established (Vandenberghe et al., 2016; Salloway et al., 2014). Furthermore, about a third of ApoE ε4-noncarriers screened for the amyloid PET substudy had negative PIB-PET suggesting a large number of patients with no amyloid pathology potentially diluting any modest clinical effect (Vandenberghe et al., 2016; Salloway et al., 2014). Solanezumab is humanized monoclonal IgG1 antibody directed against the mid-domain of Aβ (Aβ16-26). Solanezumab only binds soluble monomeric Aβ, and it doesn't penetrate into the CNS as it depends on what is called 'the peripheral sink hypothesis'. The peripheral sink hypothesis posits that reducing the free (unbound) fraction of Aβ in the plasma would lead to the increased clearance of Aβ from the brain. This is the same idea of plasma exchange (plasmapheresis), but with specificity to Aβ. The idea is to reduce the availability of soluble Aβ to aggregate and deposit in plaques which would reduce plaque burden over time. Despite some cautious optimism stemming from some benefit in a subgroup with mild AD in the EXPEDITION 1 and 2 trials pooled data, EXPEDITION 3 designed to explicitly test the medication in mild AD was a negative study (Doody et al., 2014; Honig et al., 2018). Crenezumab is a humanized IgG4 monoclonal antibody directed against the mid-domain of Aβ (Aβ16-26) similar to solanezumab, but it targets oligomers and fibrils with high affinity, in addition to the monomeric form with low affinity. Using an IgG4 backbone was intended to induce phagocytic activity (to clear Aβ) while limiting subsequent effector cell inflammatory responses, as IgG4 may be a weak activator of effector cells compared to IgG1 (Nirula et al., 2011). Crenezumab underwent two phase 3 clinical trials in patients with MCI or mild AD, but the trial was terminated with only 13% of participants completing the study and no significant changes in amyloid PET or CSF tau measures were found; a preventive study is now underway ('Crenezumab', 2022). Gantenerumab is a fully human IgG1 monoclonal antibody that is designed to specifically target fibrillar Aβ conformational epitope (both N-terminus and mid-domain Aβ (Aβ1-11)), and was tested in two phase 3 trials in MCI associated with amyloid pathology, but the trial was terminated due to futility ('Gantenerumab', 2022). It's important to note that initially no significant difference in amyloid PET at the endpoint was detected ('Gantenerumab', 2022), but a substudy of patients enrolled in these phase 3 trials suggested robust reduction in amyloid plaque burden as evidenced by amyloid PET (Klein et al., 2019). Gantenerumab has also failed to meet the primary endpoint in a phase 2 trial in patients with inherited AD (WUSTL, 2022; Washington University School of Medicine, 2022). The drug is still being actively studied in multiple trials. Aducanumab is a human IgG1 monoclonal antibody that binds to the N-terminus of Aβ (Aβ3-6) targeting soluble oligomers and insoluble fibrils in aggregates. The drug showed promising benefits in PRIME (phase 1b trial) and went into phase 3 (EMERGE and ENGAGE) (Budd Haeberlein et al., 2022). These trials were designed to avoid some of the inadequacies that were learned from the first-generation antibodies (e.g., bapineuzumab) and the phase 1b aducanumab trial. For example, amyloid pathology had to be confirmed in all participants by amyloid PET to increase the accuracy of the diagnosis. In addition, a titration schedule was designed and implemented as it was thought with the phase 1b trial that patients who started on lower doses had fewer amyloid-related imaging abnormalities (ARIA) than those who received higher doses. Also, patients with ApoE ε4 alleles received an overall lower dose to mitigate the development of ARIA. Not all these measures, however, were in place from the study onset, and two major protocol addendums were invoked, differentially affecting the trials because enrollment was in different stages. This complexity led, at least in one analysis, to one trial, EMERGE, to be positive exhibiting clinical benefits in the high-dose group, and the other, ENGAGE, to be negative (Budd Haeberlein et al., 2022). Despite this, and due to the favorable effect on the biomarker profile (reduction in amyloid plaques), the FDA has approved the medication in patients with MCI or mild dementia due to AD resulting in a wave of controversy.

Because of the accelerated approval provisions, a new phase 3 clinical trial is planned to evaluate the efficacy and safety of aducanumab within 9 years of the approval. Lecanemab is a humanized IgG1 monoclonal antibody that is similar to aducanumab in targeting both soluble and insoluble forms of Aβ within aggregates, but it specifically has higher affinity to protofibrils i.e., aducanumab binds more highly aggregated forms compared to lecanemab (ALZFORUM). CLARITY-AD phase 3 trial met primary and all key secondary endpoints showing reduced clinical decline at 18 months by 27% in the intention to treat population ($p = 0.00005$) and reduced amyloid burden measured by PET, as well as, several other clinical and functional scales (van Dyck et al., 2022). Lecanemab is expected to be evaluated by the FDA and perhaps approved by the time of publishing this book. Donanemab is a humanized IgG1 monoclonal antibody that binds Aβp3-42 (a pyroglutamate form of Aβ that is aggregated in plaques); it preferentially binds to amyloid plaques and strongly reacts with the plaque core (Irizarry et al., 2016). Donanemab was investigated in phase 1 trials showing promising biological efficacy with significantly lowering amyloid plaque burden as measured by PET (Lowe et al., 2021a, b). This was followed by a phase 2 trial (TRAILBLAZER-ALZ) which met its primary efficacy endpoint (Shcherbinin et al., 2022; Pontecorvo et al., 2022). TRAILBLAZER-ALZ 2 was launched in 2020 to investigate the efficacy and safety of donanemab in a large sample of patients with early AD and is set to run until early 2024 (Eli and Company, 2022a).

Some of the anti-Aβ monoclonal antibodies are being studied in secondary prevention trials in cognitively normal people with evidence of amyloid pathology. For example, a phase 3 trial, AHEAD 3–45, is evaluating lecanemab in people with amyloid pathology and normal cognition (Eisai, 2022). Donanemab is also being evaluated in TRAILBLAZER-ALZ 3 trial in cognitively normal people with a phosphorylated tau result consistent with the presence of amyloid and early-tau pathology (Eli and Company, 2022b). The idea behind these trials emerged from the amyloid cascade hypothesis itself, where amyloid accumulation is viewed as an early initiator of the disease with a less clear role in fueling the disease progression at more advanced stages. In animal models, anti-Aβ interventions are less likely to succeed once downstream neurodegeneration has begun (Hyman, 2011). Later stages of the AD pathophysiology may become increasingly independent of the toxic effects of Aβ; this happens when other mechanisms predominate (for example, calcium dyshomeostasis, tau-mediated neurodegeneration, and mitochondrial dysfunction) (Hyman, 2011). In APP transgenic mice, short-term inhibition of Aβ generation in amyloid-laden mice reduced Aβ deposition and associated glial changes, but failed to reduce Aβ seeding activity; CSF neurofilament light chain (Nfl), which is a marker for neurodegeneration, continued to increase albeit at a slower pace (Rother et al., 2022). In the same study, when short-term or long-term inhibition of Aβ generation was started at pre-amyloid stages, CSF Nfl did not increase despite some Aβ deposition, microglial activation, and robust brain Aβ seeding activity (Rother et al., 2022). This strongly suggests that anti-Aβ interventions have differential effects (likely related to the pathogenic stage) on pathogenesis depending on the timing of their administration in relationship to amyloid accumulation. Furthermore, in an exploratory analysis of the donanemab TRAILBLAZER-ALZ trial, patients with lower baseline tau did cognitively and functionally better than those with middle baseline or upper baseline tau pathology (Mintun et al., 2021).

ARIA are adverse events reported in trials of amyloid immunotherapies. They can be observed on MRI as two distinct presentations: ARIA-E (vasogenic edema and/or sulcal effusion), and ARIA-H (microhemorrhages and/or localized superficial siderosis). The exact pathogenic mechanism of these abnormalities is not known, but some suggest blood-brain barrier disruption and extravasation of fluid/iron to the parenchyma or subarachnoid space due to increased amyloid trafficking as it gets cleared through a perivascular pathway as likely mechanism (Barakos et al., 2022). The incidence of ARIA varies considerably, 0.9–40.6% for ARIA-E, and 0.5–28.4% for ARIA-H (Barakos et al., 2022). Risk factors include ApoE ε4 carrier state, higher anti-Aβ mAb treatment dose, and baseline microhemorrhages. Surprisingly, baseline amyloid burden by PET hasn't been evaluated as a risk factor in most of the monoclonal antibody treatment trials to date. Because most of these events are asymptomatic and almost always transient, the general consensus in the most recent trials is to continue the protocol except if they are accompanied by moderate to severe symptoms or there is radiologically severe ARIA-H i.e., >/= 10 new incident microhemorrhages or >2 focal areas of superficial siderosis. This is also the same recommendation by the FDA for aducanumab clinical use (Aduhelm Label: Reference ID: 4822820). For mitigation, both the FDA Aduhelm label and the designs of the most recent trials implement a titration regimen, and a monitoring protocol using MRI prior to starting the treatment and at certain intervals throughout.

It is not clear why some trials of monoclonal antibodies directed against Aβ have failed while others (specifically in phase 3, aducanumab EMERGE and lecanemab CLARITY-AD) have showed positive results. Certainly, the complexity of the study design is a factor as earlier trials relied more on the AD clinical syndrome for patient inclusion, while later trials had specific and strict biomarker criteria. Furthermore, the properties of the different antibodies may have a lot to do with both their efficacy and side effect profiles. For example, lecanemab's tenfold stronger binding to protofibrils compared to fibrils may be responsible for the lower rate of ARIA compared to aducanumab, and for the clinical efficacy results (Söderberg et al., 2022). Also, both aducanumab and lecanemab have the lowest binding affinity to monomers compared to the other antibodies (Söderberg et al., 2022), perhaps supporting the idea that soluble Aβ has an important role in maintaining normal cognition, and that AD toxicity may be, in part, secondary to decreased levels of functional Aβ rather than the toxic effects of the aggregates (Espay et al., 2021). Finally, the current and planned trials to test these medications in the pre-symptomatic/pre-tau stage will further clarify the merit of the existence of amyloid-dependent and amyloid-independent stages in the pathogenesis of AD and whether anti-amyloid immunotherapy is going to be more effective in secondary prevention.

39.4.1.2 Anti-Tau Immunotherapies

Tau is a cytoplasmic protein composed of 352–441 amino acid residues that is mainly found in neurons but is also expressed at low levels in glia (LoPresti et al., 1995). The N-terminal domain, which does not bind microtubules, projects outward from the microtubules (projection domain). The C-terminal domain contains repeats responsible for microtubule binding and assembly (assembly domain). Normally, tau is highly hydrophilic and unfolded with no secondary structure. The normal tau protein's main role is to stabilize and support the function of microtubules, but it also interacts with various cell components serving other processes, including neuronal activity, neurogenesis, iron export and long-term depression (Wang & Mandelkow, 2016). Several neurodegenerative diseases, including AD, are characterized by the presence of misfolded, aggregated, overexpressed, and mislocalized tau. Abnormal tau protein has a reduced ability to bind microtubules and is considered a toxic molecule as it affects synaptic function and other molecular events likely through its soluble oligomeric form (Ait-Bouziad et al., 2020; Chen & Mobley, 2019). There are six major tau isoforms generated by alternative splicing of exons E2, E3 and E10 of its pre-mRNA. Isoforms are commonly categorized depending on whether they contain three or four carboxyl-terminal repeat domains (3R or 4R). The presence of a fourth repeat depends on whether E10 is expressed or not. Tauopathies can be classified into three groups based on the tau isoforms found in aggregates: 4R tauopathies (e.g., progressive supranuclear palsy), 3R tauopathies (e.g., Pick's disease) and 3R+4R tauopathies (e.g., AD) (Wang & Mandelkow, 2016).

Tau immunotherapies are effective in animal models, and several of them are in clinical trials (Vaz & Silvestre, 2020). Active immunization (vaccination) targets single or multiple phosphorylated epitopes, N-terminus, full-length normal and mutant tau, or aggregated tau (Congdon & Sigurdsson, 2018). Two vaccines are active in clinical trials: AADvac1 and ACI-35 (Kontsekova et al., 2014a; Axon Neuroscience, 2019). Both vaccines target the assembly domain. Specifically, AADvac1 targets a six-amino-acid sequence, HXPGGG, that is found in each of the microtubule-binding repeats (Kontsekova et al., 2014b), while ACI-35 targets the pSer396/404 epitope (Theunis et al., 2013). AADvac1 resulted in reduced brain trophy, CSF p-tau1818 and p-tau217, and cognitive decline in patients with mild-to-moderate AD in a phase 1 trial (Axon Neuroscience, 2017; Novak et al., 2018). Another trial showed no significant change in cognition as compared to the placebo group at 24 months (Novak et al., 2021). Nevertheless, the vaccine is being investigated with plans for larger trials. ACI-35 is being investigated in a phase 1/2 clinical trial (SA, 2022).

Several monoclonal antibodies against tau protein are being investigated in AD. The regions mostly targeted are pSer396/404, pSer202, pThr231, pSer409, pSer413, pSer422, N-terminus, repeat domain, oligomeric tau, and misfolded tau (Congdon & Sigurdsson, 2018). For example, semorinemab is an anti-tau IgG4 monoclonal antibody that binds the N-terminus of all tau isoforms. Although safe, no efficacy has been demonstrated in two trials (Ayalon et al., 2021; Mullard, 2021); another trial showed mixed results (Genentech, 2022). Gosuranemab, tilavonemab, bepranemab, and several others are in development with somewhat disappointing early-phase 1 and 2 trials from a clinical efficacy perspective (Song et al., 2022). Large phase 3 trials investigating tau monoclonal antibodies are needed to study efficacy with more concrete conclusions.

Tau immunotherapies have several challenges including the intracellular location of the protein limiting the monoclonal antibodies' access to it, the potential for negatively affecting the levels and function of the normal unfolded protein, and the inflammatory response that may accompany some of these therapeutics. Nevertheless, pathologic tau is thought to be a major component of AD and other neurodegenerative disease pathogenesis and efforts to mitigate its effect through immunotherapy are worth pursuing.

39.4.1.3 Anti-Inflammation Immunotherapies

Another target for immunotherapy in AD is inflammation. TREM2 is a receptor selectively expressed by microglia; loss of function of TREM2 is associated with increased risk for AD. AL002 is a monoclonal antibody that binds microglial TREM2 receptor activating its downstream signaling and leading to microglial proliferation and activation. In mouse models, this led to increased expression of repair and proinflammatory genes, reduced neurotoxicity, rapid reduction in amyloid plaques (specifically with AL002a, not with AL002c), and improved performance in behavioral tests (Wang et al., 2020; 'AL002 | ALZFORUM'). A placebo-controlled phase 2 trial is currently underway evaluating AL002 in patients with early AD (Alector, 2022). Another drug that targets inflammation is daratumumab. It is a monoclonal antibody that binds CD38 which is a glycoprotein found on the surface of multiple immune cells including cytotoxic T lymphocytes i.e., CD8+ T cells. The medication is approved by the FDA for multiple myeloma (MM); MM cells overexpress CD38. In early AD, cytotoxic T lymphocytes are thought to invade CSF and brain locations and contribute to toxicity (Gate et al., 2020; Schindowski et al., 2007), thus providing a rationale for a phase 2 trial currently investigating daratumumab (Md, 2021).

39.4.2 Alpha-Synuclein: A Target for Immunotherapy in Synucleinopathies

Alpha-synuclein (α-syn) is a 140-amino acid intraneuronal protein that is natively unfolded. A highly pathologic α-syn motif was first described in amyloid plaques in Alzheimer's disease and thus termed the non-Aβ component (NAC) (Uéda et al., 1993). NAC is highly hydrophobic and is believed to promote aggregation of α-syn and leads to amyloidogenic fibril maturation (Du et al., 2003; Giasson et al., 2001). Abnormal misfolded α-syn monomers form oligomers that subsequently turn into protofibrils and then insoluble fibrils which are the main components of the inclusions in PD and other synucleinopathies. Whether oligomers or fibrils are cytotoxic is debatable, but most of the current literature favors the former, especially when it comes to immunotherapy targets. As mentioned before, it is also possible that the depletion of "normal" α-Syn promotes the pathogenesis of synucleinopathies, and therapeutic efforts to reduce α-Syn may be counterproductive (Espay & Okun, 2022).

Immunotherapies that target α-syn are divided into active immunotherapies (vaccines) and passive immunotherapies (monoclonal antibodies). Affitope PD01A/PD03A, and UB-312 are vaccines targeting α-syn; the former two are currently being investigated in PD and MSA and the latter is being investigated in PD (Schneeberger et al., 2012). These were designed to avoid activation of T-cell responses responsible for the development of inflammatory reactions such as meningoencephalitis reported with early Aβ vaccinations as previously discussed. Affitope PD01A/PD03A were tested in a phase 1 trial which demonstrated sustained IgG titers and specific antibody response with clear reactivity to the α-synuclein target epitope in the PD01A treatment group (Meissner et al., 2020). The vaccines were safe, but no clinical benefit in this one-year trial was observed; longer studies designed to test efficacy are planned. UB-312 is similar to the amyloid-specific UB-311 discussed previously, but designed to target oligomeric and fibrillar α-syn (Yu et al., 2022). UB-312 was tested in a first-in-human trial which suggested an excellent safety profile and demonstrated an antibody response (antibodies detectable in the serum and CSF) (Yu et al., 2022). Efficacy is yet to be tested in other trials. Cinpanemab (BII054) is a monoclonal antibody that binds to aggregated α-syn and was tested in a phase 2 trial revealing no changes in clinical measures of disease progression or DaT-SPECT measures over one year (Lang et al., 2022). Prasinezumab (PRX002) is another monoclonal antibody that showed a dose-dependent reduction in free serum α-syn in a phase 1 trial (Jankovic et al., 2018). This encouraged a phase 2 trial which reported that prasinezumab did not affect global or imaging measures of PD progression as compared with placebo and was associated with infusion reactions (Pagano et al., 2022). Other drugs in the same category such as MEDI1341, ABBV-0805, and Lu AF82422 are currently being investigated as well. No active or passive immunization strategy has reached phase 3 in the synucleinopathies to the time of writing this chapter.

39.5 Symptomatic Therapeutics

There is no clear dichotomy between what is considered symptomatic and what is considered disease-modifying treatment in neurodegeneration. For example, cholinesterase inhibitors, which are used in Alzheimer's disease and considered symptomatic treatments, interact with the pathogenic process i.e., they improve the availability of acetylcholine in the synapse. These medications lead to better cognitive performance, but the rate of disease progression beyond the initial period of administration remains similar to the rate of progression in patients who are not taking the medication. Thus, cholinesterase inhibitors are considered symptomatic treatments. On the other hand, riluzole and edaravone are ALS medications that increase survival. This effect could certainly be considered disease modification, but because of the modest degree of disease modification, some experts label them as symptomatic treatments. Therefore, the term symptomatic treatment is often loosely used to denote those medications that don't significantly slow the rate of progression. It may be useful to think about any particular treatment on a spectrum from 'symptomatic' to 'disease-modification' rather than under one category. In Table 39.1, we summarize the currently available medications, interventions, and approaches used to treat neurodegenerative diseases. Apart from medications for spinal muscular atrophy and perhaps aducanumab for AD (previously discussed), most, if not all, medications currently used to treat neurodegenerative diseases are on the 'symptomatic' end of the treatment spectrum, highlighting the need for developing and testing disease-modifying treatments for neurodegeneration.

Table 39.1 Symptomatic therapeutics for neurodegenerative diseases

Therapy	Mechanism of action	Clinical use	Side effects	Comments
Alzheimer's disease and other dementias				
Donepezil, rivastigmine, galantamine	Acetylcholinesterase inhibitors à increase levels of acetylcholine	Mild-to-moderate AD dementia. Typically, donepezil is prescribed throughout the course of the disease unless not tolerated or no initial benefit is noticed. Rivastigmine is FDA-approved for PD dementia, and donepezil is approved in Japan and used frequently in the United States for DLB; either medication could be useful for dementia associated with degenerative parkinsonism.	Vivid dreams, diarrhea, nausea, bradycardia & heart block (rare)	GI side effects usually subside with time; rivastigmine may have more GI side effects than donepezil. Caution is recommended with galantamine in MCI because of its cardiac-related deaths in the trials.
Memantine	NMDA receptor antagonist à decreases neuronal hyperexcitation	Moderate to severe AD dementia. Also, could be useful in DLB.	Infrequently dizziness, constipation, and flu-like symptoms	
Parkinson's disease and degenerative parkinsonism				
Levodopa (+carbidopa)	Levodopa is a precursor of dopamine. Carbidopa is a peripheral aromatic L-amino acid decarboxylase inhibitor that inhibits the conversion of levodopa to dopamine in the periphery allowing sufficient levels of levodopa to enter the striatum.	Very effective in improving certain motor manifestations (bradykinesia and rigidity), but with no to modest efficacy in improving others (tremor, freezing of gait, and postural instability).	Nausea is the most common side effect; slowly titrating the dose is helpful in the prevention of nausea. Peripheral neuropathy due to vitamin B complex deficiency with intestinal levodopa gel pump is another.	Dyskinesia and motor fluctuations are the main limiting factors for levodopa efficacy in advanced disease stages.
Pramipexole, ropinirole, rotigotine, apomorphine	Dopamine (D2) receptor agonists that mimic the function of dopamine in the striatum.	Similar to levodopa in their clinical use, but less efficacious and have more side effects. Apomorphine is used for sudden off periods or in a continuous release pump (not approved in the United States).	Orthostatic hypotension, nausea, impulse control disorders, and sleep attacks.	Upon discontinuation or significant dose reduction, withdrawal symptoms, e.g., agitation, anxiety, fatigue, and autonomic symptoms could be problematic.
Rasagiline, selegiline, safinamide	MAO-B inhibitors	Less effective in improving motor symptoms compared to levodopa and dopamine agonists. Safinamide is FDA-approved for OFF periods with levodopa.	There is a risk of serotonin syndrome when combined with serotonergic medications, but this is extremely rare.	
Entacapone, opicapone	COMT inhibitors that enhance levodopa's duration of action.	Reduce motor fluctuations.		
Amantadine	NMDA receptor antagonist	Less commonly used to treat degenerative parkinsonism nowadays, but useful in treating levodopa-induced dyskinesia.	Livedo reticularis, ankle edema, constipation, confusion, and hallucinations	

(continued)

Table 39.1 (continued)

Therapy	Mechanism of action	Clinical use	Side effects	Comments
Trihexyphenidyl, benzotropine mesylate	Anticholinergics that act centrally to restore the balance between striatal dopamine and acetylcholine activities in PD.	They improve motor symptoms, especially tremor, in PD.	Cognitive impairment, dry mouth, constipation, and urinary retention.	Should be avoided in patients with cognitive impairment.
Deep brain stimulation	A neurosurgical intervention (implantable system) that targets the basal ganglia with frequency stimulation.	It is mainly used for bradykinesia, and rigidity in patients with advanced PD that had responded to levodopa, as well as those with troublesome dyskinesia. PD tremor is also a major indication for DBS even if it didn't respond to levodopa.	Infection, hemorrhage, cognitive decline, worsening gait impairment, and worsening depression.	Cognitive impairment or major disability from motor symptoms other than the ones that could respond to DBS, e.g., postural instability are relative contraindications.
MR-guided focused ultrasound	A non-invasive intervention that targets the basal ganglia with thermal coagulation lesioning. The ventral intermediate nucleus of the thalamus or the globus pallidus internus are common targets.	It has similar indications to those of DBS but used in patients who can't or won't do surgery. However, it's only FDA-approved for unilateral tremor.	Permanent neurologic deficits, e.g., weakness, ataxia, dysarthria.	Contraindicated in patients who have an MRI contraindication.
Droxidopa, midodrine, fludrocortisone	Droxidopa is a norepinephrine precursor, midodrine is an α1 receptor agonist, and fludrocortisone is a steroid.	Used to treat orthostatic hypotension. Droxidopa and midodrine are especially useful for short-term management.	Supine hypertension.	Droxidopa and midodrine should be taken 20 minutes before upright activity and fludrocortisone is taken daily.
Solifenacin, mirabegron	Solifenacin is a competitive muscarinic receptor antagonist that dampens detrusor overactivity. Mirabegron is a sympathomimetic beta-3 adrenergic receptor agonist with a similar effect.	Used to treat urinary incontinence.		Should only be used after consideration of the cause for urinary incontinence which could be motor, sensory, and/or autonomic.
Huntington's disease				
Tetrabenazine, deutetrabenazine, valbenazine	VMAT2 inhibitors that deplete dopamine presynaptically by preventing its packaging into vesicles.	Tetrabenazine and deutetrabenazine are approved for HD-related chorea while valbenazine is used off-label and is being studied for this particular indication.	Depression, suicidal ideation, sedation, and parkinsonism are more prominent with tetrabenazine than the newer medications.	Deutetrabenazine and velbanazine have longer half-lives and are more stable compared to tetrabenazine.
Amantadine, risperidone, olanzapine	These medications could be used for HD chorea. The desired mechanism of action of the antipsychotics (i.e., risperidone and olanzapine) for this indication is their D2 receptor antagonism; their antipsychotic effect could be an added value in HD.			
Amyotrophic lateral sclerosis and motor neuron diseases				
Riluzole	Sodium channel blocker that inhibits neuronal glutamatergic transmission decreasing neuronal hyperexcitability.	Modest efficacy in ALS- Slows loss of respiratory function.	Can cause hepatic transaminitis (liver function monitoring is warranted), GI side effects, and dizziness.	Although these three medications have a disease-modifying effect, it is typically modest at best.
Edaravone	Free radical scavenger that decreases neuronal oxidative stress.	Approved for all patients with ALS, but efficacy is more established in patients without respiratory insufficiency.	Most side effects are related to the infusion mode of administration; a pill form has been developed and is FDA-approved.	

(continued)

Table 39.1 (continued)

Therapy	Mechanism of action	Clinical use	Side effects	Comments
Dextromethorphan and quinidine	This is a combined medication. Dextromethorphan is a sigma-1 receptor agonist and an uncompetitive NMDA receptor antagonist. Quinidine increases plasma levels of dextromethorphan by competitively inhibiting cytochrome P4502D6.	Used for pseudobulbar affect in patients with ALS.	Diarrhea is the most common side effect. Dizziness, peripheral edema, and cough are other possible side effects.	
Pharmacologic supportive therapy for airway and secretion management	Botulinum toxin injections to the salivary glands could also be used to decrease sialorrhea if supportive measures (mentioned below) are ineffective. Guaifenesin and nebulized medications are used to decrease the thickness of secretions.			
Pharmacologic supportive therapy for spasticity and cramps	Centrally acting muscle relaxants, e.g., baclofen and tizanidine are first-line therapy for spasticity in patients with motor neuron disease along with stretching regimen. Mexiletine, levetiracetam, and phenytoin could be used for cramps.			
Psychiatric and behavioral symptoms in neurodegenerative diseases				
Supportive therapy	Because a large percentage of patients with behavioral and psychiatric symptoms associated with dementia improve within 3 months without intervention (Ballard et al., 1997), it's important to first consider secondary causes for these symptoms, e.g., pain, medication side effect, and change in environment, and attempt conservative management with treating/minimizing the potential secondary causes, discussing the matter thoroughly with the caregivers, and trying non-pharmacologic methods such as avoiding triggers, limiting changes to the environment, regular safe exercise, shifting attention, and music therapy.			
Acetylcholinesterase inhibitors and memantine	They are considered the first-line therapy for behavioral symptoms in AD (typically a modest benefit; please see before for information on mechanisms of action and side effects)			
SSRIs, SNRIs, and related medications	Serotonin +/− norepinephrine reuptake inhibitors	SSRIs, especially citalopram, are the preferred pharmacologic therapy for depression and a second line therapy for behavioral symptoms in AD. SNRIs especially paroxetine and venlafaxine are particularly useful to treat depression in PD. Trazodone is also a good option for patients with dementia especially that it could also improve insomnia. Tricyclics have anticholinergic effects rendering them less favorable in AD and patients with degenerative parkinsonism associated with cognitive impairment.	QT interval prolongation, seizure threshold reduction, decreased libido, weight gait, and sleep disturbance.	

(continued)

Table 39.1 (continued)

Therapy	Mechanism of action	Clinical use	Side effects	Comments
Atypical antipsychotics (risperidone, olanzapine, quetiapine, clozapine)	Dopamine receptor antagonism. The specific antipsychotic action is thought to be related to blocking dopamine receptors in the mesolimbic pathway.	Used for agitation, psychosis, and aggressive behavior as a last resort in AD. In parkinsonian syndromes, most antipsychotics exacerbate parkinsonism except for quetiapine and clozapine as they have low D2 receptor antagonism; the latter's efficacy is established by clinical trials unlike the former (Ballard et al., 1997; Pollak et al., 2004).	Parkinsonism, tardive dyskinesia upon withdrawal, and other movement disorders. Also, sedation, falls, and increased risk of hospitalization and frequent side effects. Agranulocytosis is an infrequent side effect of clozapine; monitoring blood count is warranted.	In a meta-analysis of randomized placebo-controlled trials, the use of atypical antipsychotics in dementia was associated with an odds ratio of 1.54 for increased mortality (Kales et al., 2012). The FDA has issued a black box warning for the use of these medications in dementia.
Pimavanserin	5-HT2A receptor inverse agonist/antagonist.	FDA-approved for treatment of PD psychosis. Being studied in AD	Typically, well tolerated, but infrequently associated with peripheral edema and confusion.	35% less mortality risk compared to atypical antipsychotics in the first 6 months of treatment (Mosholder et al., 2022).
Melatonin, trazodone, clonazepam	Melatonin is a natural hormone that regulates sleep-wake cycle by binding to melatonin receptor 1 and 2. Trazodone is a serotonin receptor antagonist and reuptake inhibitor, but it also acts on histamine and other receptors. Clonazepam is a benzodiazepine that enhances GABA receptor responses.	Melatonin and trazodone could be used to treat insomnia, while melatonin and clonazepam could be tried for REM sleep behavioral disorder if non-pharmacologic measures are inadequate.	Sedation. Clonazepam is associated with the potential for drug abuse.	
Supportive therapy (non-pharmacological and non-procedural therapy)				
Supporting independence in dementia	Environmental modifications can support and supplement for the patient's deficit without direct confrontation which could be trigger anxiety especially early in the disease. These include phone alarms for different reminders, e.g., for medications and appointments, large-sized calendars throughout the house with the date and the day's activities placed on it to support orientation, and written instructions for tasks. In addition, if a patient does something unexpected or risky, it's typically not useful to confront the patient or point out the problem after the situation has passed; instead, moving on and trying to mitigate any damage is recommended, e.g., when a patient leaves the stove on after they have finished using it, it's best to just turn it off without confrontation.			
Mitigating alterations of circadian rhythm	Sundowning/alteration of circadian rhythm could be improved by maximizing the patient's activity and exposing the patient to light during the day, discouraging daytime napping, and limiting caffeine. Decreasing noise and interruptions during the night is also helpful.			
Driving considerations	Patients with dementia are at increased risk for unsafe driving if they have one or more of the following: CDR >/= 0.5 (Level B evidence), caregiver rating driving safety as marginal or unsafe (Level B evidence), MMSE score $\leq$24 (Level C evidence), recent history of crashes or traffic violation (Level C evidence), aggressive or impulsive personalities (Level C evidence), and reduced driving mileage or self-reported situational avoidance (Level C) (Iverson et al., 2010).			
Other safety and planning considerations	Designation of healthcare and financial power of attorney especially in patients with cognitive impairment is important. A discussion of whether extraordinary measures such as invasive ventilation is something that the patient desires should occur early during the disease course and involve the patient's family. Firearms should be removed from the homes of patients with dementia especially bvFTD.			
Physical therapy	Physical therapy could be used in patients with motor symptoms associated with neurodegeneration such as PD and ALS (with different strategies) to maintain and improve mobility. In diseases such as AD where there is typically no direct effect on the motor system, physical exercise is still useful in keeping the patient active, creating routine, and improving durability. There's also some evidence suggesting enhanced cognition as a direct effect of physical exercise, but application in neurodegenerative disease is most promising as a preventive measure rather than a treatment (Liu et al., 2019).			
Occupational therapy	Occupational therapy assessments can help identify safety issues in the home and suggest modifications, such as to assist someone with significant visual impairment, e.g., posterior cortical atrophy.			

(continued)

Table 39.1 (continued)

Therapy	Mechanism of action	Clinical use	Side effects	Comments
Speech and feeding therapy	Identifying communication strategies, early voice banking, and augmented communication devices, e.g., in amyotrophic lateral sclerosis and logopenic variant of PPA. Attention to food consistencies and feeding technique are helpful in patients with dysphagia; feeding tube placement is commonly utilized in advanced dysphagia associated with ALS and other neurodegenerative diseases.			
Supporting respiration and airway	Early non-invasive ventilation has a disease-modifying benefit in patients with ALS (Bourke et al., 2006; Kleopa et al., 1999; Jacobs et al., 2016). Invasive ventilation, on the other hand, is problematic as it is associated with tracheal injury and recommendations regarding its use should be individualized. Aggressive secretion management to improve/increase cough strength, decrease sialorrhea, and mitigate aspiration are important in motor neuron disease.			
Discontinuing harmful medications	Systemic anticholinergics and sedative may contribute to memory deficits or somnolence in patients with dementia, so it's recommended to discontinue such medications unless benefit deemed to outweigh risk when another condition is in play.			
Non-pharmacologic treatment of orthostatic hypotension	Adequate hydration, slow transitions from recumbency to standing, specific leg-strengthening exercises, and supportive stockings are important initial measures to improve orthostatic hypotension especially in degenerative parkinsonism.			
Support services	The social worker is an indispensable resource in any clinic for patients with neurodegenerative diseases. The social worker makes recommendations, helps patients and families navigate social problems, and provides access for financial benefits. In addition, patients and their families are encouraged to get involved with the local, state, and sometimes national organizations such as Alzheimer's Association, the Association for Frontotemporal Degeneration, and the Parkinson's Foundation. These organizations not only provide resources and education but are also facilitators for the patient and caregivers to get involved with a community of other patients and families dealing with the same or similar conditions.			

Symptomatic therapeutics in neurodegenerative diseases including pharmacologic, procedural, and supportive measures with comments on mechanism of action, clinical use, side effects, and other notes when applicable

5-HT2A 5-hydryoxytryptamine 2A, *AD* Alzheimer's disease, *ALS* amyotrophic lateral sclerosis, *bvFTD* the behavioral variant of frontotemporal dementia, *COMT* Catechol-O-methyltransferase, *D2* dopamine receptor 2, *DBS* Deep brain stimulation, *DLB* diffuse Lewy body disease, *FDA* United States Food and Drug Administration, *GABA* gamma-aminobutyric acid, *GI* gastrointestinal, *HD* Huntington's disease, *MAO-B* Monoamine oxidase type B, *NMDA* N-methyl-D-aspartic acid, *PD* Parkinson's disease, *PPA* primary progressive aphasia, *REM* rapid eye movement, *SNRIs* serotonin and norepinephrine reuptake inhibitors, *SSRIs* selective serotonin reuptake inhibitors, *VMAT2* vesicular monoamine transporter 2

39.6 Review Questions

1. Which of the following is not an ALS-causing gene?
   (a) C9orf72
   (b) SOD1
   (c) *ApoE*
   (d) FUS
2. Nusinersen, the first FDA-approved gene therapy for SMA, acts through which of the following mechanisms?
   (a) It is an AAV9 vector carrying the SMN1 gene
   (b) *It leads to the inclusion of exon 7 in the SMN2 pre-mRNA producing a full-length transcript*
   (c) It is based on RNA interference causing post-transcriptional gene silencing of an mRNA for a toxic protein
   (d) It's a zinc finger protein that leads to higher expression levels of SMN protein
3. Which of the following is the most common location to implant dopaminergic neurons as cell replacement therapy in PD?
   (a) Substantia nigra
   (b) *Striatum*
   (c) Cortex
   (d) Hippocampus
4. Blood regulatory T cells are dysfunctional in which of the following diseases?
   (a) Alzheimer's disease
   (b) Parkinson's disease
   (c) Amyotrophic lateral sclerosis
   (d) Frontotemporal dementia
   (e) *All of the above*
5. Which of the following is not established as a method to mitigate or treat anti-amyloid immunotherapy-induced amyloid-related imaging abnormalities?
   (a) Titration regimen
   (b) MRI monitoring program
   (c) *Corticosteroids*
   (d) Stopping the medication when there are severe symptoms associated with ARIA
   (e) Stopping the medication when there is evidence for radiologically severe ARIA-H
6. Which of the following is abundant in tau aggregates in Alzheimer's disease?
   (a) 3R tau, but not 4R tau
   (b) 4R tau, but not 3R tau
   (c) *Both 3R tau and 4R tau*
   (d) Beta-amyloid

7. Which one of the following medications for Alzheimer's disease is not a cholinesterase inhibitor?
   (a) Galantamine
   (b) Rivastigmine
   (c) Donepezil
   (d) *Memantine*
8. Which of the following is FDA-approved for the treatment of ALS?
   (a) Amantadine, rasagiline, entacapone
   (b) *Riluzole, edaravone*
   (c) Tetrabenazine, deutetrabenazine, valbenazine
   (d) donepezil, rivastigmine, and memantine
9. What is the first-line treatment for Parkinson's disease motor symptoms?
   (a) Pramipexole
   (b) Deep brain stimulation
   (c) *Levodopa*
   (d) Droxidopa

References

Ager, R. R., Davis, J. L., Agazaryan, A., Benavente, F., Poon, W. W., LaFerla, F. M., & Blurton-Jones, M. (2015). Human neural stem cells improve cognition and promote synaptic growth in two complementary transgenic models of Alzheimer's disease and neuronal loss. *Hippocampus, 25*, 813–826.

Ait-Bouziad, N., Chiki, A., Limorenko, G., Xiao, S., Eliezer, D., & Lashuel, H. A. (2020). Phosphorylation of the overlooked tyrosine 310 regulates the structure, aggregation, and microtubule- and lipid-binding properties of tau. *The Journal of Biological Chemistry, 295*, 7905–7922.

AL002 | ALZFORUM.

Alector, Inc. (2022). A phase 2 randomized, double-blind, placebo-controlled, multicenter study to evaluate the efficacy and safety of AL002 in participants with early Alzheimer's disease. In: clinicaltrials.gov

Allen Reish, H. E., & Standaert, D. G. (2015). Role of α-synuclein in inducing innate and adaptive immunity in Parkinson disease. *Journal of Parkinson's Disease, 5*, 1–19.

Alsuliman, A., Appel, S. H., Beers, D. R., Basar, R., Shaim, H., Kaur, I., Zulovich, J., Yvon, E., Muftuoglu, M., Imahashi, N., Kondo, K., Liu, E., Shpall, E. J., & Rezvani, K. (2016). A robust, good manufacturing practice–compliant, clinical-scale procedure to generate regulatory T cells from patients with amyotrophic lateral sclerosis for adoptive cell therapy. *Cytotherapy (Oxford, England), 18*, 1312–1324.

Al-Zaidy, S., Pickard, A. S., Kotha, K., Alfano, L. N., Lowes, L., Paul, G., Church, K., Lehman, K., Sproule, D. M., Dabbous, O., Maru, B., Berry, K., Arnold, W. D., Kissel, J. T., Mendell, J. R., & Shell, R. (2019). Health outcomes in spinal muscular atrophy type 1 following AVXS-101 gene replacement therapy. *Pediatric Pulmonology, 54*, 179–185.

Alzforum. APOE C130R (ApoE4). In: ALZFORUM.

Alzforum. Lecanemab Sweeps Up Toxic Aβ Protofibrils, Catches Eyes of Trialists | ALZFORUM.

Alzforum. On The Docket at AD/PD: The Many Crimes of ApoE4. In.

Appel, S. H., Beers, D. R., & Henkel, J. S. (2010). T cell-microglial dialogue in Parkinson's disease and amyotrophic lateral sclerosis: Are we listening? *Trends in Immunology, 31*, 7–17.

Armijo, E., Edwards, G., Flores, A., Vera, J., Shahnawaz, M., Moda, F., Gonzalez, C., Sanhueza, M., & Soto, C. (2021). Induced Pluripotent stem cell-derived neural precursors improve memory, synaptic and pathological abnormalities in a Mouse Model of Alzheimer's disease. *Cells, 10*, 10.

Axon Neuroscience, S. E. (2017). An 18-months open label phase i follow-up study on patients with Alzheimer's disease who have completed the AADvac1 phase i study "AXON CO 18700. In: clinicaltrials.gov

Axon Neuroscience, S. E. (2019). A 24 months randomised, placebo-controlled, parallel group, double blinded, multi centre, phase 2 study to assess safety and efficacy of AADvac1 applied to patients with Mild Alzheimer's disease. In: clinicaltrials.gov

Ayalon, G., Lee, S. H., Adolfsson, O., Foo-Atkins, C., Atwal, J. K., Blendstrup, M., Booler, H., Bravo, J., Brendza, R., Brunstein, F., Chan, R., Chandra, P., Couch, J. A., Datwani, A., Demeule, B., DiCara, D., Erickson, R., Ernst, J. A., Foreman, O., He, D., Hötzel, I., Keeley, M., Kwok, M. C. M., Lafrance-Vanasse, J., Lin, H., Lu, Y., Luk, W., Manser, P., Muhs, A., Ngu, H., Pfeifer, A., Pihlgren, M., Rao, G. K., Scearce-Levie, K., Schauer, S. P., Smith, W. B., Solanoy, H., Teng, E., Wildsmith, K. R., Bumbaca Yadav, D., Ying, Y., Fuji, R. N., & Kerchner, G. A. (2021). Antibody semorinemab reduces tau pathology in a transgenic mouse model and engages tau in patients with Alzheimer's disease. *Science Translational Medicine, 13*, eabb2639.

Baba, Y., Kuroiwa, A., Uitti, R. J., Wszolek, Z. K., & Yamada, T. (2005). Alterations of T-lymphocyte populations in Parkinson disease. *Parkinsonism & Related Disorders, 11*, 493–498.

Backlund, E. O., Granberg, P. O., Hamberger, B., Knutsson, E., Mårtensson, A., Sedvall, G., Seiger, A., & Olson, L. (1985). Transplantation of adrenal medullary tissue to striatum in parkinsonism. First clinical trials. *Journal of Neurosurgery, 62*, 169–173.

Ballard, C., O'Brien, J., Coope, B., Fairbairn, A., Abid, F., & Wilcock, G. (1997). A prospective study of psychotic symptoms in dementia sufferers: Psychosis in dementia. *International Psychogeriatrics, 9*, 57–64.

Bankiewicz, K. S., Eberling, J. L., Kohutnicka, M., Jagust, W., Pivirotto, P., Bringas, J., Cunningham, J., Budinger, T. F., & Harvey-White, J. (2000). Convection-enhanced delivery of AAV vector in parkinsonian monkeys; in vivo detection of gene expression and restoration of dopaminergic function using pro-drug approach. *Experimental Neurology, 164*, 2–14.

Bankiewicz, K. S., Forsayeth, J., Eberling, J. L., Sanchez-Pernaute, R., Pivirotto, P., Bringas, J., Herscovitch, P., Carson, R. E., Eckelman, W., Reutter, B., & Cunningham, J. (2006). Long-term clinical improvement in MPTP-lesioned primates after gene therapy with AAV-hAADC. *Molecular Therapy, 14*, 564–570.

Barakos, J., Purcell, D., Suhy, J., Chalkias, S., Burkett, P., Marsica Grassi, C., Castrillo-Viguera, C., Rubino, I., & Vijverberg, E. (2022). Detection and management of amyloid-related imaging abnormalities in patients with Alzheimer's disease treated with anti-amyloid beta therapy. *The Journal of Prevention of Alzheimer's Disease, 9*, 211–220.

Barbuti, P. A., Barker, R. A., Brundin, P., Przedborski, S., Papa, S. M., Kalia, L. V., Mochizuki, H., & the MDS Scientific Issues Committee. (2021). Recent advances in the development of stem-cell-derived dopaminergic neuronal transplant therapies for Parkinson's disease. *Movement Disorders, 36*, 1772–1780.

Barker, R. A., Drouin-Ouellet, J., & Parmar, M. (2015). Cell-based therapies for Parkinson disease—Past insights and future potential. *Nature Reviews Neurology, 11*, 492–503.

Bartus, R. T., Baumann, T. L., Siffert, J., Herzog, C. D., Alterman, R., Boulis, N., Turner, D. A., Stacy, M., Lang, A. E., Lozano, A. M., & Olanow, C. W. (2013). Safety/feasibility of targeting the substantia nigra with AAV2-neurturin in Parkinson patients. *Neurology, 80*, 1698–1701.

Bas, J., Calopa, M., Mestre, M., Molleví, D. G., Cutillas, B., Ambrosio, S., & Buendia, E. (2001). Lymphocyte populations in Parkinson's disease and in rat models of parkinsonism. *Journal of Neuroimmunology, 113*, 146–152.

Beers, D. R., Henkel, J. S., Zhao, W., Wang, J., Huang, A., Wen, S., Liao, B., & Appel, S. H. (2011). Endogenous regulatory T lymphocytes ameliorate amyotrophic lateral sclerosis in mice and correlate with disease progression in patients with amyotrophic lateral sclerosis. *Brain, 134*, 1293–1314.

Beers, D. R., Zhao, W., Wang, J., Zhang, X., Wen, S., Neal, D., Thonhoff, J. R., Alsuliman, A. S., Shpall, E. J., Rezvani, K., & Appel, S. H. (2017). ALS patients' regulatory T lymphocytes are dysfunctional, and correlate with disease progression rate and severity. *JCI Insight, 2*, e89530.

Belloy, M. E., Napolioni, V., & Greicius, M. D. (2019). A quarter century of APOE and Alzheimer's disease: Progress to date and the path forward. *Neuron, 101*, 820–838.

Bevan-Jones, W. R., Cope, T. E., Jones, P. S., Kaalund, S. S., Passamonti, L., Allinson, K., Green, O., Hong, Y. T., Fryer, T. D., Arnold, R., Coles, J. P., Aigbirhio, F. I., Larner, A. J., Patterson, K., O'Brien, J. T., & Rowe, J. B. (2020). Neuroinflammation and protein aggregation co-localize across the frontotemporal dementia spectrum. *Brain, 143*, 1010–1026.

Biogen. (2022a). Global early access program to provide Tofersen to patients with amyotrophic lateral sclerosis (ALS) associated With a mutation in the Superoxide Dismutase 1 (SOD1) gene. In: clinicaltrials.gov

Biogen. (2022b). A phase 1 multiple-ascending-dose study to assess the safety, tolerability, and pharmacokinetics of BIIB078 administered intrathecally to adults with C9ORF72-associated amyotrophic lateral sclerosis. In: clinicaltrials.gov

Biogen. (2022c). A randomized, double-blind, placebo-controlled, parallel-group study to assess the efficacy, safety, and tolerability of BIIB080 in subjects with mild cognitive impairment due to Alzheimer's disease or Mild Alzheimer's disease dementia. In: clinicaltrials.gov

Biogen, and Inc Ionis Pharmaceuticals. Biogen and ionis announce Topline phase 1 study results of investigational drug in C9orf72 Amyotrophic Lateral Sclerosis | Biogen.

Bissonnette, C. J., Lyass, L., Bhattacharyya, B. J., Belmadani, A., Miller, R. J., & Kessler, J. A. (2011). The controlled generation of functional basal forebrain cholinergic neurons from human embryonic stem cells. *Stem Cells, 29*, 802–811.

Blurton-Jones, M., Kitazawa, M., Martinez-Coria, H., Castello, N. A., Müller, F. J., Loring, J. F., Yamasaki, T. R., Poon, W. W., Green, K. N., & LaFerla, F. M. (2009). Neural stem cells improve cognition via BDNF in a transgenic model of Alzheimer disease. *Proceedings of the National Academy of Sciences of the United States of America, 106*, 13594–13599.

Borchelt, D. R., Lee, M. K., Slunt, H. S., Guarnieri, M., Xu, Z. S., Wong, P. C., Brown, R. H., Jr., Price, D. L., Sisodia, S. S., & Cleveland, D. W. (1994). Superoxide dismutase 1 with mutations linked to familial amyotrophic lateral sclerosis possesses significant activity. *Proceedings of the National Academy of Sciences of the United States of America, 91*, 8292–8296.

Bourke, S. C., Tomlinson, M., Williams, T. L., Bullock, R. E., Shaw, P. J., & Gibson, G. J. (2006). Effects of non-invasive ventilation on survival and quality of life in patients with amyotrophic lateral sclerosis: A randomised controlled trial. *Lancet Neurology, 5*, 140–147.

Boutajangout, A., Noorwali, A., Atta, H., & Wisniewski, T. (2017). Human umbilical cord stem cell xenografts improve cognitive decline and reduce the amyloid burden in a Mouse Model of Alzheimer's disease. *Current Alzheimer Research, 14*, 104–111.

Brochard, V., Combadière, B., Prigent, A., Laouar, Y., Perrin, A., Beray-Berthat, V., Bonduelle, O., Alvarez-Fischer, D., Callebert, J., Launay, J. M., Duyckaerts, C., Flavell, R. A., Hirsch, E. C., & Hunot, S. (2009). Infiltration of CD4+ lymphocytes into the brain contributes to neurodegeneration in a mouse model of Parkinson disease. *The Journal of Clinical Investigation, 119*, 182–192.

Browne, T. C., McQuillan, K., McManus, R. M., O'Reilly, J. A., Mills, K. H., & Lynch, M. A. (2013). IFN-γ production by amyloid β-specific Th1 cells promotes microglial activation and increases plaque burden in a mouse model of Alzheimer's disease. *Journal of Immunology, 190*, 2241–2251.

Budd Haeberlein, S., Aisen, P. S., Barkhof, F., Chalkias, S., Chen, T., Cohen, S., Dent, G., Hansson, O., Harrison, K., von Hehn, C., Iwatsubo, T., Mallinckrodt, C., Mummery, C. J., Muralidharan, K. K., Nestorov, I., Nisenbaum, L., Rajagovindan, R., Skordos, L., Tian, Y., van Dyck, C. H., Vellas, B., Wu, S., Zhu, Y., & Sandrock, A. (2022). Two randomized phase 3 studies of Aducanumab in early Alzheimer's disease. *The Journal of Prevention of Alzheimer's Disease, 9*, 197–210.

Butovsky, O., Kunis, G., Koronyo-Hamaoui, M., & Schwartz, M. (2007). Selective ablation of bone marrow-derived dendritic cells increases amyloid plaques in a mouse Alzheimer's disease model. *The European Journal of Neuroscience, 26*, 413–416.

Bye, C. R., Jönsson, M. E., Björklund, A., Parish, C. L., & Thompson, L. H. (2015). Transcriptome analysis reveals transmembrane targets on transplantable midbrain dopamine progenitors. *Proceedings of the National Academy of Sciences of the United States of America, 112*, E1946–E1955.

Cabezudo, D., Baekelandt, V., & Lobbestael, E. (2020). Multiple-hit hypothesis in Parkinson's disease: LRRK2 and inflammation. *Frontiers in Neuroscience, 14*, 376.

Calsolaro, V., & Edison, P. (2016). Neuroinflammation in Alzheimer's disease: Current evidence and future directions. *Alzheimers & Dementia, 12*, 719–732.

Castle, M. J., Baltanás, F. C., Kovacs, I., Nagahara, A. H., Barba, D., & Tuszynski, M. H. (2020). Postmortem analysis in a clinical trial of AAV2-NGF gene therapy for Alzheimer's disease identifies a need for improved vector delivery. *Human Gene Therapy, 31*, 415–422.

Cha, M. Y., Kwon, Y. W., Ahn, H. S., Jeong, H., Lee, Y. Y., Moon, M., Baik, S. H., Kim, D. K., Song, H., Yi, E. C., Hwang, D., Kim, H. S., & Mook-Jung, I. (2017). Protein-induced pluripotent stem cells ameliorate cognitive dysfunction and reduce Aβ deposition in a Mouse Model of Alzheimer's disease. *Stem Cells Translational Medicine, 6*, 293–305.

Chakrabarty, P., Li, A., Ceballos-Diaz, C., Eddy, J. A., Funk, C. C., Moore, B., DiNunno, N., Rosario, A. M., Cruz, P. E., Verbeeck, C., Sacino, A., Nix, S., Janus, C., Price, N. D., Das, P., & Golde, T. E. (2015). IL-10 alters immunoproteostasis in APP mice, increasing plaque burden and worsening cognitive behavior. *Neuron, 85*, 519–533.

Chambers, S. M., Fasano, C. A., Papapetrou, E. P., Tomishima, M., Sadelain, M., & Studer, L. (2009). Highly efficient neural conversion of human ES and iPS cells by dual inhibition of SMAD signaling. *Nature Biotechnology, 27*, 275–280.

Chang, J. L., Hinrich, A. J., Roman, B., Norrbom, M., Rigo, F., Marr, R. A., Norstrom, E. M., & Hastings, M. L. (2018). Targeting amyloid-β precursor protein, APP, splicing with antisense oligonucleotides reduces toxic amyloid-β production. *Molecular Therapy, 26*, 1539–1551.

Chartier-Harlin, M. C., Kachergus, J., Roumier, C., Mouroux, V., Douay, X., Lincoln, S., Levecque, C., Larvor, L., Andrieux, J., Hulihan, M., Waucquier, N., Defebvre, L., Amouyel, P., Farrer, M., & Destée, A. (2004). Alpha-synuclein locus duplication as a cause of familial Parkinson's disease. *Lancet, 364*, 1167–1169.

Chatterjee, D., Bhatt, M., Butler, D., De Genst, E., Dobson, C. M., Messer, A., & Kordower, J. H. (2018). Proteasome-targeted nanobodies alleviate pathology and functional decline in an α-synuclein-based Parkinson's disease model. *NPJ Parkinsons Disease, 4*, 25.

Chen, X. Q., & Mobley, W. C. (2019). Alzheimer disease pathogenesis: Insights from molecular and cellular biology studies of oligomeric a beta and tau species. *Frontiers in Neuroscience, 13*, 21.

Cheng, G., Aixin, Y., & Malek, T. R. (2011). T-cell tolerance and the multi-functional role of IL-2R signaling in T-regulatory cells. *Immunological Reviews, 241*, 63–76.

Chmielarz, P., Konovalova, J., & Domanskyi, A. (2020). Chapter 36 - Cell transplantation therapy for Parkinson's disease. In C. R. Martin & V. R. Preedy (Eds.), *Diagnosis and Management in Parkinson's disease*. Academic Press.

Cho, H. J., Kim, S. K., Jin, S. M., Hwang, E. M., Kim, Y. S., Huh, K., & Mook-Jung, I. (2007). IFN-gamma-induced BACE1 expression is mediated by activation of JAK2 and ERK1/2 signaling pathways and direct binding of STAT1 to BACE1 promoter in astrocytes. *Glia, 55*, 253–262.

Christine, C. W., Starr, P. A., Larson, P. S., Eberling, J. L., Jagust, W. J., Hawkins, R. A., VanBrocklin, H. F., Wright, J. F., Bankiewicz, K. S., & Aminoff, M. J. (2009). Safety and tolerability of putaminal AADC gene therapy for Parkinson disease. *Neurology, 73*, 1662–1669.

Christine, C. W., Bankiewicz, K. S., Van Laar, A. D., Richardson, R. M., Ravina, B., Kells, A. P., Boot, B., Martin, A. J., Nutt, J., Thompson, M. E., & Larson, P. S. (2019). Magnetic resonance imaging-guided phase 1 trial of putaminal AADC gene therapy for Parkinson's disease. *Annals of Neurology, 85*, 704–714.

Christine, C. W., Richardson, R. M., Van Laar, A. D., Thompson, M. E., Fine, E. M., Khwaja, O. S., Li, C., Liang, G. S., Meier, A., Roberts, E. W., Pfau, M. L., Rodman, J. R., Bankiewicz, K. S., & Larson, P. S. (2022). Safety of AADC gene therapy for moderately advanced Parkinson disease: Three-year outcomes from the PD-1101 trial. *Neurology, 98*, e40–e50.

Chu, Y., & Kordower, J. H. (2010). Lewy body pathology in fetal grafts. *Annals of the New York Academy of Sciences, 1184*, 55–67.

Cleveland, D. W., Laing, N., Hurse, P. V., & Brown, R. H., Jr. (1995). Toxic mutants in Charcot's sclerosis. *Nature, 378*, 342–343.

Colonna, M., & Butovsky, O. (2017). Microglia function in the central nervous system during health and neurodegeneration. In D. R. Littman & W. M. Yokoyama (Eds.), *Annual review of immunology* (Vol. 35, p. 441). Annual Reviews.

Colpo, G. D., Ascoli, B. M., Wollenhaupt-Aguiar, B., Pfaffenseller, B., Silva, E. G., Cirne-Lima, E. O., Quevedo, J., Kapczinski, F., & Rosa, A. R. (2015). Mesenchymal stem cells for the treatment of neurodegenerative and psychiatric disorders. *Anais da Academia Brasileira de Ciências, 87*, 1435–1449.

Congdon, E. E., & Sigurdsson, E. M. (2018). Tau-targeting therapies for Alzheimer disease. *Nature Reviews. Neurology, 14*, 399–415.

Cook, D. A., Kannarkat, G. T., Cintron, A. F., Butkovich, L. M., Fraser, K. B., Chang, J., Grigoryan, N., Factor, S. A., West, A. B., Boss, J. M., & Tansey, M. G. (2017). LRRK2 levels in immune cells are increased in Parkinson's disease. *NPJ Parkinsons Disease, 3*, 11.

Cooper, O., Hargus, G., Deleidi, M., Blak, A., Osborn, T., Marlow, E., Lee, K., Levy, A., Perez-Torres, E., Yow, A., & Isacson, O. (2010). Differentiation of human ES and Parkinson's disease iPS cells into ventral midbrain dopaminergic neurons requires a high activity form of SHH, FGF8a and specific regionalization by retinoic acid. *Molecular and Cellular Neurosciences, 45*, 258–266.

Corcia, P., Tauber, C., Vercoullie, J., Arlicot, N., Prunier, C., Praline, J., Nicolas, G., Venel, Y., Hommet, C., Baulieu, J. L., Cottier, J. P., Roussel, C., Kassiou, M., Guilloteau, D., & Ribeiro, M. J. (2012). Molecular imaging of microglial activation in amyotrophic lateral sclerosis. *PLoS One, 7*, e52941.

'Crenezumab'. 2022. ALZFORUM. https://www.alzforum.org/therapeutics/crenezumab

Cui, Y., Ma, S., Zhang, C., Cao, W., Liu, M., Li, D., Lv, P., Xing, Q., Qu, R., Yao, N., Yang, B., & Guan, F. (2017). Human umbilical cord mesenchymal stem cells transplantation improves cognitive function in Alzheimer's disease mice by decreasing oxidative stress and promoting hippocampal neurogenesis. *Behavioural Brain Research, 320*, 291–301.

Cyranoski, D. 'Reprogrammed' stem cells implanted into patient with Parkinson's disease.

Dass, B., Olanow, C. W., & Kordower, J. H. (2006). Gene transfer of trophic factors and stem cell grafting as treatments for Parkinson's disease. *Neurology, 66*, S89–103.

Decressac, M., Mattsson, B., Weikop, P., Lundblad, M., Jakobsson, J., & Björklund, A. (2013). TFEB-mediated autophagy rescues midbrain dopamine neurons from α-synuclein toxicity. *Proceedings of the National Academy of Sciences of the United States of America, 110*, E1817–E1826.

DeVos, S. L., Miller, R. L., Schoch, K. M., Holmes, B. B., Kebodeaux, C. S., Wegener, A. J., Chen, G., Shen, T., Tran, H., Nichols, B., Zanardi, T. A., Kordasiewicz, H. B., Swayze, E. E., Bennett, C. F., Diamond, M. I., & Miller, T. M. (2017). Tau reduction prevents neuronal loss and reverses pathological tau deposition and seeding in mice with tauopathy. *Science Translational Medicine, 9*, eaag0481.

Dhillon, S. (2020). Risdiplam: First approval. *Drugs, 80*, 1853–1858.

Doi, D., Samata, B., Katsukawa, M., Kikuchi, T., Morizane, A., Ono, Y., Sekiguchi, K., Nakagawa, M., Parmar, M., & Takahashi, J. (2014). Isolation of human induced pluripotent stem cell-derived dopaminergic progenitors by cell sorting for successful transplantation. *Stem Cell Reports, 2*, 337–350.

Doody, R. S., Thomas, R. G., Farlow, M., Iwatsubo, T., Vellas, B., Joffe, S., Kieburtz, K., Raman, R., Sun, X., Aisen, P. S., Siemers, E., Liu-Seifert, H., & Mohs, R. (2014). Phase 3 trials of solanezumab for mild-to-moderate Alzheimer's disease. *The New England Journal of Medicine, 370*, 311–321.

DrugBank. 'Nusinersen'. https://go.drugbank.com/drugs/DB13161

Du, H. N., Tang, L., Luo, X. Y., Li, H. T., Hu, J., Zhou, J. W., & Hu, H. Y. (2003). A peptide motif consisting of glycine, alanine, and valine is required for the fibrillization and cytotoxicity of human alpha-synuclein. *Biochemistry, 42*, 8870–8878.

Duma, C., Kopyov, O., Kopyov, A., Berman, M., Lander, E., Elam, M., Arata, M., Weiland, D., Cannell, R., Caraway, C., Berman, S., Scord, K., Stemler, L., Chung, K., Khoudari, S., McRory, R., Duma, C., Farmer, S., Bravo, A., Yassa, C., Sanathara, A., Singh, E., & Rapaport, B. (2019). Human intracerebroventricular (ICV) injection of autologous, non-engineered, adipose-derived stromal vascular fraction (ADSVF) for neurodegenerative disorders: Results of a 3-year phase 1 study of 113 injections in 31 patients. *Molecular Biology Reports, 46*, 5257–5272.

Eberling, J. L., Jagust, W. J., Christine, C. W., Starr, P., Larson, P., Bankiewicz, K. S., & Aminoff, M. J. (2008). Results from a phase I safety trial of hAADC gene therapy for Parkinson disease. *Neurology, 70*, 1980–1983.

Efthymiou, A. G., & Goate, A. M. (2017). Late onset Alzheimer's disease genetics implicates microglial pathways in disease risk. *Molecular Neurodegeneration, 12*, 43.

Eid, A. M., Pascual, M. B., Masdeu, J. C., Appel, S. H., & Faridar, A. (2022). The potential contribution of the adaptive immune system in frontotemporal dementia. *Alzheimer's & Dementia, 18*, e063793.

Eisai, Inc. (2022). AHEAD 3–45 study: A placebo-controlled, double-blind, parallel-treatment arm, 216 week study to evaluate efficacy and safety of treatment with BAN2401 in subjects with preclinical Alzheimer's disease and elevated amyloid (A45 Trial) and in subjects with early preclinical Alzheimer's disease and intermediate Amyloid (A3 Trial). In: clinicaltrials.gov

Eli, Lilly, and Company. (2022a). Assessment of safety, tolerability, and efficacy of Donanemab in early symptomatic Alzheimer's disease. In: clinicaltrials.gov

Eli, Lilly, and Company. (2022b). A study of Donanemab versus placebo in participants at risk for cognitive and functional decline of Alzheimer's disease. In: clinicaltrials.gov

Eriksen, J. L., Sagi, S. A., Smith, T. E., Weggen, S., Das, P., McLendon, D. C., Ozols, V. V., Jessing, K. W., Zavitz, K. H., Koo, E. H., & Golde, T. E. (2003). NSAIDs and enantiomers of flurbiprofen target gamma-secretase and lower Abeta 42 in vivo. *The Journal of Clinical Investigation, 112*, 440–449.

Erkkinen, M. G., Kim, M. O., & Geschwind, M. D. (2018). Clinical neurology and epidemiology of the major neurodegenerative diseases. *Cold Spring Harbor Perspectives in Biology, 10*, a033118.

Espay, A. J., & Okun, M. S. (2022). Abandoning the proteinopathy paradigm in Parkinson disease. *JAMA Neurology, 80*, 123.

Espay, A. J., Sturchio, A., Schneider, L. S., & Ezzat, K. (2021). Soluble Amyloid-β consumption in Alzheimer's disease. *Journal of Alzheimer's Disease, 82*, 1403–1415.

Evers, M. M., Miniarikova, J., Juhas, S., Vallès, A., Bohuslavova, B., Juhasova, J., Skalnikova, H. K., Vodicka, P., Valekova, I., Brouwers, C., Blits, B., Lubelski, J., Kovarova, H., Ellederova, Z., van Deventer, S. J., Petry, H., Motlik, J., & Konstantinova, P. (2018). AAV5-miHTT gene therapy demonstrates broad distribution and strong human mutant huntingtin lowering in a Huntington's disease Minipig Model. *Molecular Therapy, 26*, 2163–2177.

Faridar, A., Vasquez, M., Thome, A. D., Yin, Z., Xuan, H., Wang, J. H., Wen, S., Li, X., Thonhoff, J. R., Zhao, W., Zhao, H., Beers, D. R., Wong, S. T. C., Masdeu, J. C., & Appel, S. H. (2022). Ex vivo expanded human regulatory T cells modify neuroinflammation in a preclinical model of Alzheimer's disease. *Acta Neuropathologica Communications, 10*, 144.

Fasano, C. A., Chambers, S. M., Lee, G., Tomishima, M. J., & Studer, L. (2010). Efficient derivation of functional floor plate tissue from human embryonic stem cells. *Cell Stem Cell, 6*, 336–347.

Ferrer, I., Boada Rovira, M., Sánchez Guerra, M. L., Rey, M. J., & Costa-Jussá, F. (2004). Neuropathology and pathogenesis of encephalitis following amyloid-beta immunization in Alzheimer's disease. *Brain Pathology, 14*, 11–20.

Ferretti, M. T., Merlini, M., Späni, C., Gericke, C., Schweizer, N., Enzmann, G., Engelhardt, B., Kulic, L., Suter, T., & Nitsch, R. M. (2016a). T-cell brain infiltration and immature antigen-presenting cells in transgenic models of Alzheimer's disease-like cerebral amyloidosis. *Brain, Behavior, and Immunity, 54*, 211–225.

Ferretti, M. T., Merlini, M., Späni, C., Gericke, C., Schweizer, N., Enzmann, G., Britta Engelhardt, L., Kulic, T. S., & Nitsch, R. M. (2016b). T-cell brain infiltration and immature antigen-presenting cells in transgenic models of Alzheimer's disease-like cerebral amyloidosis. *Brain, Behavior, and Immunity, 54*, 211–225.

Finkel, R. S., Mercuri, E., Darras, B. T., Connolly, A. M., Kuntz, N. L., Kirschner, J., Chiriboga, C. A., Saito, K., Servais, L., Tizzano, E., Topaloglu, H., Tulinius, M., Montes, J., Glanzman, A. M., Bishop, K., Zhong, Z. J., Gheuens, S., Bennett, C. F., Schneider, E., Farwell, W., & De Vivo, D. C. (2017). Nusinersen versus sham control in infantile-onset spinal muscular atrophy. *The New England Journal of Medicine, 377*, 1723–1732.

Foust, K. D., Salazar, D. L., Likhite, S., Ferraiuolo, L., Ditsworth, D., Ilieva, H., Meyer, K., Schmelzer, L., Braun, L., Cleveland, D. W., & Kaspar, B. K. (2013). Therapeutic AAV9-mediated suppression of mutant SOD1 slows disease progression and extends survival in models of inherited ALS. *Molecular Therapy, 21*, 2148–2159.

Freed, W. J., Poltorak, M., & Becker, J. B. (1990). Intracerebral adrenal medulla grafts: a review. *Experimental Neurology, 110*, 139–166.

Freed, C. R., Greene, P. E., Breeze, R. E., Tsai, W. Y., DuMouchel, W., Kao, R., Dillon, S., Winfield, H., Culver, S., Trojanowski, J. Q., Eidelberg, D., & Fahn, S. (2001). Transplantation of embryonic dopamine neurons for severe Parkinson's disease. *The New England Journal of Medicine, 344*, 710–719.

Fujiwara, N., Shimizu, J., Takai, K., Arimitsu, N., Saito, A., Kono, T., Umehara, T., Ueda, Y., Wakisaka, S., Suzuki, T., & Suzuki, N. (2013). Restoration of spatial memory dysfunction of human APP transgenic mice by transplantation of neuronal precursors derived from human iPS cells. *Neuroscience Letters, 557*(Pt B), 129–134.

Fujiwara, N., Shimizu, J., Takai, K., Arimitsu, N., Ueda, Y., Wakisaka, S., Suzuki, T., & Suzuki, N. (2015). Cellular and molecular mechanisms of the restoration of human APP transgenic mouse cognitive dysfunction after transplant of human iPS cell-derived neural cells. *Experimental Neurology, 271*, 423–431.

Funk, N., Wieghofer, P., Grimm, S., Schaefer, R., Bühring, H. J., Gasser, T., & Biskup, S. (2013). Characterization of peripheral hematopoietic stem cells and monocytes in Parkinson's disease. *Movement Disorders, 28*, 392–395.

Gaj, T., Ojala, D. S., Ekman, F. K., Byrne, L. C., Limsirichai, P., & Schaffer, D. V. (2017). In vivo genome editing improves motor function and extends survival in a mouse model of ALS. *Science Advances, 3*, eaar3952.

Gantenerumab. (2022). ALZFORUM. https://www.alzforum.org/therapeutics/gantenerumab

Gate, D., Saligrama, N., Leventhal, O., Yang, A. C., Unger, M. S., Middeldorp, J., Chen, K., Lehallier, B., Channappa, D., De Los Santos, M. B., McBride, A., Pluvinage, J., Elahi, F., Tam, G. K., Kim, Y., Greicius, M., Wagner, A. D., Aigner, L., Galasko, D. R., Davis, M. M., & Wyss-Coray, T. (2020). Clonally expanded CD8 T cells patrol the cerebrospinal fluid in Alzheimer's disease. *Nature, 577*, 399–404.

Gendelman, H. E., Zhang, Y., Santamaria, P., Olson, K. E., Schutt, C. R., Bhatti, D., Shetty, B. L. D., Lu, Y., Estes, K. A., Standaert, D. G., Heinrichs-Graham, E., Larson, L., Meza, J. L., Follett, M., Forsberg, E., Siuzdak, G., Wilson, T. W., Peterson, C., & Mosley, R. L. (2017). Evaluation of the safety and immunomodulatory effects of sargramostim in a randomized, double-blind phase 1 clinical Parkinson's disease trial. *NPJ Parkinsons Disease, 3*, 10.

Genentech, Inc. (2022). A phase II, multicenter, randomized, double-blind, placebo-controlled, parallel-group, efficacy, and safety study of MTAU9937A in patients with moderate Alzheimer's disease. In: clinicaltrials.gov

Gerhard, A., Pavese, N., Hotton, G., Turkheimer, F., Es, M., Hammers, A., Eggert, K., Oertel, W., Banati, R. B., & Brooks, D. J. (2006). In vivo imaging of microglial activation with [11C](R)-PK11195 PET in idiopathic Parkinson's disease. *Neurobiology of Disease, 21*, 404–412.

Giasson, B. I., Murray, I. V., Trojanowski, J. Q., & Lee, V. M. (2001). A hydrophobic stretch of 12 amino acid residues in the middle of alpha-synuclein is essential for filament assembly. *The Journal of Biological Chemistry, 276*, 2380–2386.

Gilman, S., Koller, M., Black, R. S., Jenkins, L., Griffith, S. G., Fox, N. C., Eisner, L., Kirby, L., Rovira, M. B., Forette, F., & Orgogozo, J. M. (2005). Clinical effects of Abeta immunization (AN1792) in patients with AD in an interrupted trial. *Neurology, 64*, 1553–1562.

Giubilei, F., Antonini, G., Montesperelli, C., Sepe-Monti, M., Cannoni, S., Pichi, A., Tisei, P., Casini, A. R., Buttinelli, C., Prencipe, M., Salvetti, M., & Ristori, G. (2003). T cell response to amyloid-beta and to mitochondrial antigens in Alzheimer's disease. *Dementia and Geriatric Cognitive Disorders, 16*, 35–38.

Gliem, M., Schwaninger, M., & Jander, S. (2016). Protective features of peripheral monocytes/macrophages in stroke. *Biochimica et Biophysica Acta, 1862*, 329–338.

Graßhoff, H., Comdühr, S., Monne, L. R., Müller, A., Lamprecht, P., Riemekasten, G., & Humrich, J. Y. (2021). Low-dose IL-2 therapy in autoimmune and rheumatic diseases. *Frontiers in Immunology, 12*, 648408.

Gratwohl, A., & Niederwieser, D. (2012). History of hematopoietic stem cell transplantation: Evolution and perspectives. *Current Problems in Dermatology, 43*, 81–90.

Graves, M. C., Fiala, M., Dinglasan, L. A., Liu, N. Q., Sayre, J., Chiappelli, F., van Kooten, C., & Vinters, H. V. (2004). Inflammation in amyotrophic lateral sclerosis spinal cord and brain is mediated by activated macrophages, mast cells and T cells. *Amyotrophic Lateral Sclerosis and Other Motor Neuron Disorders, 5*, 213–219.

Grealish, S., Diguet, E., Kirkeby, A., Mattsson, B., Heuer, A., Bramoulle, Y., Van Camp, N., Perrier, A. L., Hantraye, P., Björklund, A., & Parmar, M. (2014). Human ESC-derived dopamine neurons show similar preclinical efficacy and potency to fetal neurons when grafted in a rat model of Parkinson's disease. *Cell Stem Cell, 15*, 653–665.

Grozdanov, V., Bliederhaeuser, C., Ruf, W. P., Roth, V., Fundel-Clemens, K., Zondler, L., Brenner, D., Martin-Villalba, A., Hengerer, B., Kassubek, J., Ludolph, A. C., Weishaupt, J. H., & Danzer, K. M. (2014). Inflammatory dysregulation of blood monocytes in Parkinson's disease patients. *Acta Neuropathologica, 128*, 651–663.

Gruden, M. A., Sewell, R. D., Yanamandra, K., Davidova, T. V., Kucheryanu, V. G., Bocharov, E. V., Bocharova, O. A., Polyschuk, V. V., Sherstnev, V. V., & Morozova-Roche, L. A. (2011). Immunoprotection against toxic biomarkers is retained during Parkinson's disease progression. *Journal of Neuroimmunology, 233*, 221–227.

Hadaczek, P., Eberling, J. L., Pivirotto, P., Bringas, J., Forsayeth, J., & Bankiewicz, K. S. (2010). Eight years of clinical improvement in MPTP-lesioned primates after gene therapy with AAV2-hAADC. *Molecular Therapy, 18*, 1458–1461.

Haeusler, A. R., Donnelly, C. J., & Rothstein, J. D. (2016). The expanding biology of the C9orf72 nucleotide repeat expansion in neurodegenerative disease. *Nature Reviews. Neuroscience, 17*, 383–395.

Hallett, P. J., Cooper, O., Sadi, D., Robertson, H., Mendez, I., & Isacson, O. (2014). Long-term health of dopaminergic neuron transplants in Parkinson's disease patients. *Cell Reports, 7*, 1755–1761.

Hamza, T. H., Zabetian, C. P., Tenesa, A., Laederach, A., Montimurro, J., Yearout, D., Kay, D. M., Doheny, K. F., Paschall, J., Pugh, E., Kusel, V. I., Collura, R., Roberts, J., Griffith, A., Samii, A., Scott, W. K., Nutt, J., Factor, S. A., & Payami, H. (2010). Common genetic variation in the HLA region is associated with late-onset sporadic Parkinson's disease. *Nature Genetics, 42*, 781–785.

Harach, T., Jammes, F., Muller, C., Duthilleul, N., Cheatham, V., Zufferey, V., Cheatham, D., Lukasheva, Y. A., Lasser, T., & Bolmont, T. (2017). Administrations of human adult ischemia-tolerant mesenchymal stem cells and factors reduce amyloid beta pathology in a mouse model of Alzheimer's disease. *Neurobiology of Aging, 51*, 83–96.

Hardy, J. A., & Higgins, G. A. (1992). Alzheimer's disease: The amyloid cascade hypothesis. *Science, 256*, 184–185.

Harms, A. S., Cao, S., Rowse, A. L., Thome, A. D., Li, X., Mangieri, L. R., Cron, R. Q., Shacka, J. J., Raman, C., & Standaert, D. G. (2013). MHCII is required for α-synuclein-induced activation of microglia, CD4 T cell proliferation, and dopaminergic neurodegeneration. *The Journal of Neuroscience, 33*, 9592–9600.

Harms, A. S., Delic, V., Thome, A. D., Bryant, N., Liu, Z., Chandra, S., Jurkuvenaite, A., & West, A. B. (2017). α-Synuclein fibrils recruit peripheral immune cells in the rat brain prior to neurodegeneration. *Acta Neuropathologica Communications, 5*, 85.

Harms, A. S., Thome, A. D., Yan, Z., Schonhoff, A. M., Williams, G. P., Li, X., Liu, Y., Qin, H., Benveniste, E. N., & Standaert, D. G. (2018). Peripheral monocyte entry is required for alpha-Synuclein induced inflammation and neurodegeneration in a model of Parkinson disease. *Experimental Neurology, 300*, 179–187.

Hayward, L. J., Rodriguez, J. A., Kim, J. W., Tiwari, A., Goto, J. J., Cabelli, D. E., Valentine, J. S., & Brown, R. H., Jr. (2002). Decreased metallation and activity in subsets of mutant superoxide dismutases associated with familial amyotrophic lateral sclerosis. *The Journal of Biological Chemistry, 277*, 15923–15931.

He, F., & Balling, R. (2013). The role of regulatory T cells in neurodegenerative diseases. *Wiley Interdisciplinary Reviews. Systems Biology and Medicine, 5*, 153–180.

Henkel, J. S., Engelhardt, J. I., Siklós, L., Simpson, E. P., Kim, S. H., Pan, T., Goodman, J. C., Siddique, T., Beers, D. R., & Appel, S. H. (2004). Presence of dendritic cells, MCP-1, and activated microglia/macrophages in amyotrophic lateral sclerosis spinal cord tissue. *Annals of Neurology, 55*, 221–235.

Henkel, J. S., Beers, D. R., Wen, S., Rivera, A. L., Toennis, K. M., Appel, J. E., Zhao, W., Moore, D. H., Powell, S. Z., & Appel, S. H. (2013). Regulatory T-lymphocytes mediate amyotrophic lateral sclerosis progression and survival. *EMBO Molecular Medicine, 5*, 64–79.

Hisanaga, K., Asagi, M., Itoyama, Y., & Iwasaki, Y. (2001). Increase in peripheral CD4 bright+ CD8 dull+ T cells in Parkinson disease. *Archives of Neurology, 58*, 1580–1583.

Honig, L. S., Vellas, B., Woodward, M., Boada, M., Bullock, R., Borrie, M., Hager, K., Andreasen, N., Scarpini, E., Liu-Seifert, H., Case, M., Dean, R. A., Hake, A., Sundell, K., Hoffmann, V. P., Carlson, C., Khanna, R., Mintun, M., DeMattos, R., Selzler, K. J., & Siemers, E. (2018). Trial of Solanezumab for mild dementia due to Alzheimer's disease. *New England Journal of Medicine, 378*, 321–330.

Huang, Y. A., Zhou, B., Wernig, M., & Südhof, T. C. (2017). ApoE2, ApoE3, and ApoE4 differentially stimulate APP transcription and Aβ secretion. *Cell, 168*, 427–41.e21.

Hudry, E., Dashkoff, J., Roe, A. D., Takeda, S., Koffie, R. M., Hashimoto, T., Scheel, M., Spires-Jones, T., Arbel-Ornath, M., Betensky, R., Davidson, B. L., & Hyman, B. T. (2013). Gene transfer of human Apoe isoforms results in differential modulation of amyloid deposition and neurotoxicity in mouse brain. *Science Translational Medicine, 5*, 212ra161.

Hutvagner, G., & Simard, M. J. (2008). Argonaute proteins: Key players in RNA silencing. *Nature Reviews. Molecular Cell Biology, 9*, 22–32.

Hyman, B. T. (2011). Amyloid-dependent and amyloid-independent stages of Alzheimer disease. *Archives of Neurology, 68*, 1062–1064.

Iannitti, T., Scarrott, J. M., Likhite, S., Coldicott, I. R. P., Lewis, K. E., Heath, P. R., Higginbottom, A., Myszczynska, M. A., Milo, M., Hautbergue, G. M., Meyer, K., Kaspar, B. K., Ferraiuolo, L., Shaw, P. J., & Azzouz, M. (2018). Translating SOD1 gene silencing toward the clinic: A highly efficacious, off-target-free, and biomarker-supported strategy for fALS. *Molecular Therapy - Nucleic Acids, 12*, 75–88.

Ionis Pharmaceuticals, Inc. (2022a). A phase 1–3 study to evaluate the efficacy, safety, pharmacokinetics and pharmacodynamics of intrathecally administered ION363 in amyotrophic lateral sclerosis patients with fused in sarcoma mutations (FUS-ALS). In.: clinicaltrials.gov

Ionis Pharmaceuticals, Inc. (2022b). A randomized, double-blind, placebo-controlled study, followed by an open-label extension, to evaluate the safety, tolerability, pharmacokinetics and pharmacodynamics of multiple ascending doses of intrathecally administered ISIS 814907 in patients with mild Alzheimer's disease. In.: clinicaltrials.gov

Irizarry, M. C., Sims, J. R., Lowe, S. L., Nakano, M., Hawdon, A., Willis, B. A., Gonzales, C., Liu, P., Fujimoto, S., Dean, R. A., Konrad, R. J., Shcherbinin, S., Schwarz, A. J., Mintun, M. A., Devous Sr, M. D., DeMattos, R. B., & Investigators, A. A. C. C. (2016). O4-08-06: Safety, Pharmacokinetics (PK), and Florbetapir F-18 positron emission tomography (PET) after multiple dose administration of LY3002813, a β-amyloid plaque-specific antibody, in Alzheimer's disease (AD). *Alzheimer's & Dementia, 12*, P352–PP53.

Itagaki, S., McGeer, P. L., & Akiyama, H. (1988). Presence of T-cytotoxic suppressor and leucocyte common antigen positive cells in Alzheimer's disease brain tissue. *Neuroscience Letters, 91*, 259–264.

Iverson, D. J., Gronseth, G. S., Réger, M., Classen, S., Dubinsky, R., & Rizzo, M. (2010). Practice parameter update: Evaluation and management of driving risk in dementia. *Neurology, 74*, 1316–1324.

Jack, C. R., Jr., Bennett, D. A., Blennow, K., Carrillo, M. C., Dunn, B., Haeberlein, S. B., Holtzman, D. M., Jagust, W., Jessen, F., Karlawish, J., Liu, E., Molinuevo, J. L., Montine, T., Phelps, C., Rankin, K. P., Rowe, C. C., Scheltens, P., Siemers, E., Snyder, H. M., & Sperling, R. (2018). NIA-AA Research Framework: Toward a biological definition of Alzheimer's disease. *Alzheimers Dement, 14*, 535–562.

Jacobs, T. L., Brown, D. L., Baek, J., Migda, E. M., Funckes, T., & Gruis, K. L. (2016). Trial of early noninvasive ventilation for ALS: A pilot placebo-controlled study. *Neurology, 87*, 1878–1883.

Jankovic, J., Goodman, I., Safirstein, B., Marmon, T. K., Schenk, D. B., Koller, M., Zago, W., Ness, D. K., Griffith, S. G., Grundman, M., Soto, J., Ostrowitzki, S., Boess, F. G., Martin-Facklam, M., Quinn, J. F., Isaacson, S. H., Omidvar, O., Ellenbogen, A., & Kinney, G. G. (2018). Safety and tolerability of multiple ascending doses of PRX002/RG7935, an anti-α-Synuclein monoclonal antibody, in patients with Parkinson disease: A randomized clinical trial. *JAMA Neurology, 75*, 1206–1214.

Jiang, J., Zhu, Q., Gendron, T. F., Saberi, S., McAlonis-Downes, M., Seelman, A., Stauffer, J. E., Jafar-Nejad, P., Drenner, K., Schulte, D., Chun, S., Sun, S., Ling, S. C., Myers, B., Engelhardt, J., Katz, M., Baughn, M., Platoshyn, O., Marsala, M., Watt, A., Heyser, C. J., Ard, M. C., De Muynck, L., Daughrity, L. M., Swing, D. A., Tessarollo, L., Jung, C. J., Delpoux, A., Utzschneider, D. T., Hedrick, S. M., de Jong, P. J., Edbauer, D., Van Damme, P., Petrucelli, L., Shaw, C. E., Bennett, C. F., Da Cruz, S., Ravits, J., Rigo, F., Cleveland, D. W., & Lagier-Tourenne, C. (2016). Gain of toxicity from ALS/FTD-linked repeat expansions in C9ORF72 is alleviated by antisense oligonucleotides targeting GGGGCC-containing RNAs. *Neuron, 90*, 535–550.

Jonsson, T., Stefansson, H., Steinberg, S., Jonsdottir, I., Jonsson, P. V., Snaedal, J., Bjornsson, S., Huttenlocher, J., Levey, A. I., Lah, J. J., Rujescu, D., Hampel, H., Giegling, I., Andreassen, O. A., Engedal, K., Ulstein, I., Djurovic, S., Ibrahim-Verbaas, C., Hofman, A., Ikram, M. A., van Duijn, C. M., Thorsteinsdottir, U., Kong, A., & Stefansson, K. (2013). Variant of TREM2 associated with the risk of Alzheimer's disease. *The New England Journal of Medicine, 368*, 107–116.

Kales, H. C., Kim, H. M., Zivin, K., Valenstein, M., Seyfried, L. S., Chiang, C., Cunningham, F., Schneider, L. S., & Blow, F. C. (2012). Risk of mortality among individual antipsychotics in patients with dementia. *The American Journal of Psychiatry, 169*, 71–79.

Kaplitt, M. G., Feigin, A., Tang, C., Fitzsimons, H. L., Mattis, P., Lawlor, P. A., Bland, R. J., Young, D., Strybing, K., Eidelberg, D., & During, M. J. (2007). Safety and tolerability of gene therapy with an adeno-associated virus (AAV) borne GAD gene for Parkinson's disease: An open label, phase I trial. *Lancet, 369*, 2097–2105.

Karvelas, N., Bennett, S., Politis, G., Kouris, N. I., & Kole, C. (2022). Advances in stem cell therapy in Alzheimer's disease: A comprehensive clinical trial review. *Stem Cell Investigation, 9*, 2.

Kefalopoulou, Z., Politis, M., Piccini, P., Mencacci, N., Bhatia, K., Jahanshahi, M., Widner, H., Rehncrona, S., Brundin, P., Björklund, A., Lindvall, O., Limousin, P., Quinn, N., & Foltynie, T. (2014). Long-term clinical outcome of fetal cell transplantation for Parkinson disease: Two case reports. *JAMA Neurology, 71*, 83–87.

Keller, C. G., Shin, Y., Monteys, A. M., Renaud, N., Beibel, M., Teider, N., Peters, T., Faller, T., St-Cyr, S., Knehr, J., Roma, G., Reyes, A., Hild, M., Lukashev, D., Theil, D., Dales, N., Cha, J. H., Borowsky, B., Dolmetsch, R., Davidson, B. L., & Sivasankaran, R. (2022). An orally available, brain penetrant, small molecule lowers huntingtin levels by enhancing pseudoexon inclusion. *Nature Communications, 13*, 1150.

Kikuchi, T., Morizane, A., Doi, D., Magotani, H., Onoe, H., Hayashi, T., Mizuma, H., Takara, S., Takahashi, R., Inoue, H., Morita, S., Yamamoto, M., Okita, K., Nakagawa, M., Parmar, M., & Takahashi, J. (2017). Human iPS cell-derived dopaminergic neurons function in a primate Parkinson's disease model. *Nature, 548*, 592–596.

Kim, S., Chang, K. A., Ja Kim, H. G., Park, J. C., Ra, H. S., Kim, & Suh, Y. H. (2012). The preventive and therapeutic effects of intravenous human adipose-derived stem cells in Alzheimer's disease mice. *PLoS One, 7*, e45757.

Kim, H. J., Seo, S. W., Chang, J. W., Lee, J. I., Kim, C. H., Chin, J., Choi, S. J., Kwon, H., Yun, H. J., Lee, J. M., Kim, S. T., Choe, Y. S., Lee, K. H., & Na, D. L. (2015). Stereotactic brain injection of human umbilical cord blood mesenchymal stem cells in patients with Alzheimer's disease dementia: A phase 1 clinical trial. *Alzheimers Dement (N Y), 1*, 95–102.

Kim, G., Gautier, O., Tassoni-Tsuchida, E., Ma, X. R., & Gitler, A. D. (2020). ALS genetics: Gains, losses, and implications for future therapies. *Neuron, 108*, 822–842.

Kim, H. J., Cho, K. R., Jang, H., Lee, N. K., Jung, Y. H., Kim, J. P., Lee, J. I., Chang, J. W., Park, S., Kim, S. T., Moon, S. W., Seo, S. W., Choi, S. J., & Na, D. L. (2021). Intracerebroventricular injection of human umbilical cord blood mesenchymal stem cells in patients with Alzheimer's disease dementia: A phase I clinical trial. *Alzheimer's Research & Therapy, 13*, 154.

Kirkeby, A., Grealish, S., Wolf, D. A., Nelander, J., Wood, J., Lundblad, M., Lindvall, O., & Parmar, M. (2012). Generation of regionally specified neural progenitors and functional neurons from human embryonic stem cells under defined conditions. *Cell Reports, 1*, 703–714.

Klein, G., Delmar, P., Voyle, N., Rehal, S., Hofmann, C., Abi-Saab, D., Andjelkovic, M., Ristic, S., Wang, G., Bateman, R., Kerchner, G. A., Baudler, M., Fontoura, P., & Doody, R. (2019). Gantenerumab reduces amyloid-β plaques in patients with prodromal to moderate Alzheimer's disease: A PET substudy interim analysis. *Alzheimer's Research & Therapy, 11*, 101.

Kleopa, K. A., Sherman, M., Neal, B., Romano, G. J., & Heiman-Patterson, T. (1999). Bipap improves survival and rate of pulmonary function decline in patients with ALS. *Journal of the Neurological Sciences, 164*, 82–88.

Kontsekova, E., Zilka, N., Kovacech, B., Novak, P., & Novak, M. (2014a). First-in-man tau vaccine targeting structural determinants essential for pathological tau-tau interaction reduces tau oligomerisation and neurofibrillary degeneration in an Alzheimer's disease model. *Alzheimer's Research & Therapy, 6*, 44.

Kontsekova, E., Zilka, N., Kovacech, B., Skrabana, R., & Novak, M. (2014b). Identification of structural determinants on tau protein essential for its pathological function: Novel therapeutic target for tau immunotherapy in Alzheimer's disease. *Alzheimer's Research & Therapy, 6*, 45.

Kordower, J. H., Chu, Y., Hauser, R. A., Freeman, T. B., & Olanow, C. W. (2008). Lewy body-like pathology in long-term embryonic nigral transplants in Parkinson's disease. *Nature Medicine, 14*, 504–506.

Korobeynikov, V. A., Lyashchenko, A. K., Blanco-Redondo, B., Jafar-Nejad, P., & Shneider, N. A. (2022). Antisense oligonucleotide silencing of FUS expression as a therapeutic approach in amyotrophic lateral sclerosis. *Nature Medicine, 28*, 104–116.

Kosloski, L. M., Kosmacek, E. A., Olson, K. E., Lee Mosley, R., & Gendelman, H. E. (2013). GM-CSF induces neuroprotective and anti-inflammatory responses in 1-methyl-4-phenyl-1,2,3,6-tetrahydropyridine intoxicated mice. *Journal of Neuroimmunology, 265*, 1–10.

Koutsodendris, N., Nelson, M. R., Rao, A., & Huang, Y. (2022). Apolipoprotein E and Alzheimer's disease: Findings, hypotheses, and potential mechanisms. *Annual Review of Pathology, 17*, 73–99.

Krach, F., Stemick, J., Boerstler, T., Weiss, A., Lingos, I., Reischl, S., Meixner, H., Ploetz, S., Farrell, M., Hehr, U., Kohl, Z., Winner, B., & Winkler, J. (2022). An alternative splicing modulator decreases mutant HTT and improves the molecular fingerprint in Huntington's disease patient neurons. *Nature Communications, 13*, 6797.

Krasemann, S., Madore, C., Cialic, R., Baufeld, C., Calcagno, N., El Fatimy, R., Beckers, L., O'Loughlin, E., Yang, X., Fanek, Z., Greco, D. J., Smith, S. T., Tweet, G., Humulock, Z., Zrzavy, T., Conde-Sanroman, P., Gacias, M., Weng, Z., Chen, H., Tjon, E., Mazaheri, F., Hartmann, K., Madi, A., Ulrich, J. D., Glatzel, M., Worthmann, A., Heeren, J., Budnik, B., Lemere, C., Ikezu, T., Heppner, F. L., Litvak, V., Holtzman, D. M., Lassmann, H., Weiner, H. L., Ochando, J., Haass, C., & Butovsky, O. (2017). The TREM2-APOE pathway drives the transcriptional phenotype of dysfunctional microglia in neurodegenerative diseases. *Immunity, 47*, 566–81.e9.

Kriks, S., Shim, J. W., Piao, J., Ganat, Y. M., Wakeman, D. R., Xie, Z., Carrillo-Reid, L., Auyeung, G., Antonacci, C., Buch, A., Yang, L., Beal, M. F., Surmeier, D. J., Kordower, J. H., Tabar, V., & Studer, L. (2011). Dopamine neurons derived from human ES cells efficiently engraft in animal models of Parkinson's disease. *Nature, 480*, 547–551.

Kurowska, Z., Englund, E., Widner, H., Lindvall, O., Li, J. Y., & Brundin, P. (2011). Signs of degeneration in 12-22-year old grafts of mesencephalic dopamine neurons in patients with Parkinson's disease. *Journal of Parkinson's Disease, 1*, 83–92.

Kustrimovic, N., Comi, C., Magistrelli, L., Rasini, E., Legnaro, M., Bombelli, R., Aleksic, I., Blandini, F., Minafra, B., Riboldazzi, G., Sturchio, A., Mauri, M., Bono, G., Marino, F., & Cosentino, M. (2018). Parkinson's disease patients have a complex phenotypic and functional Th1 bias: Cross-sectional studies of CD4+ Th1/Th2/T17 and Treg in drug-naïve and drug-treated patients. *Journal of Neuroinflammation, 15*, 205.

Lang, A. E., Siderowf, A. D., Macklin, E. A., Poewe, W., Brooks, D. J., Fernandez, H. H., Rascol, O., Giladi, N., Stocchi, F., Tanner, C. M., Postuma, R. B., Simon, D. K., Tolosa, E., Mollenhauer, B., Cedarbaum, J. M., Fraser, K., Xiao, J., Evans, K. C., Graham, D. L., Sapir, I., Inra, J., Hutchison, R. M., Yang, M., Fox, T., Budd Haeberlein, S., & Dam, T. (2022). Trial of Cinpanemab in early Parkinson's disease. *The New England Journal of Medicine, 387*, 408–420.

Lee, J. K., Jin, H. K., Endo, S., Schuchman, E. H., Carter, J. E., & Bae, J. S. (2010). Intracerebral transplantation of bone marrow-derived mesenchymal stem cells reduces amyloid-beta deposition and rescues memory deficits in Alzheimer's disease mice by modulation of immune responses. *Stem Cells, 28*, 329–343.

Lee, H. J., Lee, J. K., Lee, H., Carter, J. E., Chang, J. W., Oh, W., Yang, Y. S., Suh, J. G., Lee, B. H., Jin, H. K., & Bae, J. S. (2012). Human umbilical cord blood-derived mesenchymal stem cells improve neuropathology and cognitive impairment in an Alzheimer's disease mouse model through modulation of neuroinflammation. *Neurobiology of Aging, 33*, 588–602.

Lee, I. S., Jung, K., Kim, I. S., Lee, H., Kim, M., Yun, S., Hwang, K., Shin, J. E., & Park, K. I. (2015). Human neural stem cells alleviate Alzheimer-like pathology in a mouse model. *Molecular Neurodegeneration, 10*, 38.

LeWitt, P. A., Rezai, A. R., Leehey, M. A., Ojemann, S. G., Flaherty, A. W., Eskandar, E. N., Kostyk, S. K., Thomas, K., Sarkar, A., Siddiqui, M. S., Tatter, S. B., Schwalb, J. M., Poston, K. L., Henderson, J. M., Kurlan, R. M., Richard, I. H., Van Meter, L., Sapan, C. V., During, M. J., Kaplitt, M. G., & Feigin, A. (2011). AAV2-GAD gene therapy for advanced Parkinson's disease: A double-blind, sham-surgery controlled, randomised trial. *Lancet Neurology, 10*, 309–319.

Lexeo, Therapeutics.(2022). A 52-week, multicenter, phase 1/2 open-label study to evaluate the safety of LX1001 in participants with APOE4 Homozygote Alzheimer's disease. In: clinicaltrials.gov

Leyns, C. E. G., Ulrich, J. D., Finn, M. B., Stewart, F. R., Koscal, L. J., Remolina Serrano, J., Robinson, G. O., Anderson, E., Colonna, M., & Holtzman, D. M. (2017). TREM2 deficiency attenuates neuroinflammation and protects against neurodegeneration in a mouse model of tauopathy. *Proceedings of the National Academy of Sciences of the United States of America, 114*, 11524–11529.

Li, J. Y., Englund, E., Holton, J. L., Soulet, D., Hagell, P., Lees, A. J., Lashley, T., Quinn, N. P., Rehncrona, S., Björklund, A., Widner, H., Revesz, T., Lindvall, O., & Brundin, P. (2008). Lewy bodies in grafted neurons in subjects with Parkinson's disease suggest host-to-graft disease propagation. *Nature Medicine, 14*, 501–503.

Li, Y., Li, X., Geng, X., & Zhao, H. (2022). The IL-2A receptor pathway and its role in lymphocyte differentiation and function. *Cytokine & Growth Factor Reviews, 67*, 66–79.

Lindestam Arlehamn, C. S., Dhanwani, R., Pham, J., Kuan, R., Frazier, A., Rezende Dutra, J., Phillips, E., Mallal, S., Roederer, M., Marder, K. S., Amara, A. W., Standaert, D. G., Goldman, J. G., Litvan, I., Peters, B., Sulzer, D., & Sette, A. (2020). α-Synuclein-specific T cell reactivity is associated with preclinical and early Parkinson's disease. *Nature Communications, 11*, 1875.

Lindvall, O., Rehncrona, S., Brundin, P., Gustavii, B., Astedt, B., Widner, H., Lindholm, T., Björklund, A., Leenders, K. L., Rothwell, J. C., Frackowiak, R., Marsden, D., Johnels, B., Steg, G., Freedman, R., Hoffer, B. J., Seiger, A., Bygdeman, M., Strömberg, I., & Olson, L. (1989). Human fetal dopamine neurons grafted into the striatum in two patients with severe Parkinson's disease. A detailed account of methodology and a 6-month follow-up. *Archives of Neurology, 46*, 615–631.

Liu, Y., Weick, J. P., Liu, H., Krencik, R., Zhang, X., Ma, L., Zhou, G. M., Ayala, M., & Zhang, S. C. (2013). Medial ganglionic eminence-like cells derived from human embryonic stem cells correct learning and memory deficits. *Nature Biotechnology, 31*, 440–447.

Liu, Y., Yan, T., Chu, J. M.-T., Chen, Y., Dunnett, S., Ho, Y.-S., Wong, G. T.-C., & Chang, R. C.-C. (2019). The beneficial effects of physical exercise in the brain and related pathophysiological mechanisms in neurodegenerative diseases. *Laboratory Investigation, 99*, 943–957.

LoPresti, P., Szuchet, S., Papasozomenos, S. C., Zinkowski, R. P., & Binder, L. I. (1995). Functional implications for the microtubule-associated protein tau: Localization in oligodendrocytes. *Proceedings of the National Academy of Sciences of the United States of America, 92*, 10369–10373.

Lowe, S. L., Duggan Evans, C., Shcherbinin, S., Cheng, Y. J., Willis, B. A., Gueorguieva, I., Lo, A. C., Fleisher, A. S., Dage, J. L., Ardayfio, P., Aguiar, G., Ishibai, M., Takaichi, G., Chua, L., Mullins, G., & Sims, J. R. (2021a). Donanemab (LY3002813) phase 1b study in Alzheimer's disease: Rapid and sustained reduction of brain amyloid measured by Florbetapir F18 imaging. *The Journal of Prevention of Alzheimer's Disease, 8*, 414–424.

Lowe, S. L., Willis, B. A., Hawdon, A., Natanegara, F., Chua, L., Foster, J., Shcherbinin, S., Ardayfio, P., & Sims, J. R. (2021b). Donanemab (LY3002813) dose-escalation study in Alzheimer's disease. *Alzheimers Dement (N Y), 7*, e12112.

Ltd, F. Hoffmann-La Roche. (2021). Roche provides update on tominersen programme in manifest Huntington's disease. In https://www.roche.com/media/releases/med-cor-2021-03-22b

Ma, T., Gong, K., Ao, Q., Yan, Y., Song, B., Huang, H., Zhang, X., & Gong, Y. (2013). Intracerebral transplantation of adipose-derived mesenchymal stem cells alternatively activates microglia and ameliorates neuropathological deficits in Alzheimer's disease mice. *Cell Transplantation, 22*(Suppl 1), S113–S126.

Marks, W. J., Jr., Ostrem, J. L., Verhagen, L., Starr, P. A., Larson, P. S., Bakay, R. A., Taylor, R., Cahn-Weiner, D. A., Stoessl, A. J.,

Olanow, C. W., & Bartus, R. T. (2008). Safety and tolerability of intraputaminal delivery of CERE-120 (adeno-associated virus serotype 2-neurturin) to patients with idiopathic Parkinson's disease: An open-label, phase I trial. *Lancet Neurology, 7*, 400–408.

Marks, W. J., Jr., Bartus, R. T., Siffert, J., Davis, C. S., Lozano, A., Boulis, N., Vitek, J., Stacy, M., Turner, D., Verhagen, L., Bakay, R., Watts, R., Guthrie, B., Jankovic, J., Simpson, R., Tagliati, M., Alterman, R., Stern, M., Baltuch, G., Starr, P. A., Larson, P. S., Ostrem, J. L., Nutt, J., Kieburtz, K., Kordower, J. H., & Olanow, C. W. (2010). Gene delivery of AAV2-neurturin for Parkinson's disease: A double-blind, randomised, controlled trial. *Lancet Neurology, 9*, 1164–1172.

Marsh, S. E., Abud, E. M., Lakatos, A., Karimzadeh, A., Yeung, S. T., Davtyan, H., Fote, G. M., Lau, L., Weinger, J. G., Lane, T. E., Inlay, M. A., Poon, W. W., & Blurton-Jones, M. (2016). The adaptive immune system restrains Alzheimer's disease pathogenesis by modulating microglial function. *Proceedings of the National Academy of Sciences of the United States of America, 113*, E1316–E1325.

Marsh, S. E., Yeung, S. T., Torres, M., Lau, L., Davis, J. L., Monuki, E. S., Poon, W. W., & Blurton-Jones, M. (2017). HuCNS-SC human NSCs fail to differentiate, form ectopic clusters, and provide no cognitive benefits in a transgenic model of Alzheimer's disease. *Stem Cell Reports, 8*, 235–248.

Martens, Y. A., Zhao, N., Liu, C. C., Kanekiyo, T., Yang, A. J., Goate, A. M., Holtzman, D. M., & Bu, G. (2022). ApoE Cascade hypothesis in the pathogenesis of Alzheimer's disease and related dementias. *Neuron, 110*, 1304–1317.

Matchynski-Franks, J. J., Pappas, C., Rossignol, J., Reinke, T., Fink, K., Crane, A., Twite, A., Lowrance, S. A., Song, C., & Dunbar, G. L. (2016). Mesenchymal stem cells as treatment for behavioral deficits and neuropathology in the 5xFAD Mouse Model of Alzheimer's disease. *Cell Transplantation, 25*, 687–703.

Mazzulli, J. R., Xu, Y. H., Sun, Y., Knight, A. L., McLean, P. J., Caldwell, G. A., Sidransky, E., Grabowski, G. A., & Krainc, D. (2011). Gaucher disease glucocerebrosidase and α-synuclein form a bidirectional pathogenic loop in synucleinopathies. *Cell, 146*, 37–52.

McCauley, M. E., & Baloh, R. H. (2019). Inflammation in ALS/FTD pathogenesis. *Acta Neuropathologica, 137*, 715–730.

McFarthing, K., Prakash, N., & Simuni, T. (2019). Clinical trial highlights: 1. Gene therapy for Parkinson's, 2. Phase 3 study in focus - INTEC Pharma's accordion pill, 3. Clinical trials resources. *Journal of Parkinson's Disease, 9*, 251–264.

McGeer, P. L., Itagaki, S., Akiyama, H., & McGeer, E. G. (1988). Rate of cell death in parkinsonism indicates active neuropathological process. *Annals of Neurology, 24*, 574–576.

McGinley, L. M., Kashlan, O. N., Bruno, E. S., Chen, K. S., Hayes, J. M., Kashlan, S. R., Raykin, J., Johe, K., Murphy, G. G., & Feldman, E. L. (2018). Human neural stem cell transplantation improves cognition in a murine model of Alzheimer's disease. *Scientific Reports, 8*, 14776.

Md, Marc L. Gordon. (2021). An open-label, pilot study of Daratumumab SC in patients with mild to moderate Alzheimer's disease. In: clinicaltrials.gov

Meissner, W. G., Traon, A. P., Foubert-Samier, A., Galabova, G., Galitzky, M., Kutzelnigg, A., Laurens, B., Lührs, P., Medori, R., Péran, P., Sabatini, U., Vergnet, S., Volc, D., Poewe, W., Schneeberger, A., Staffler, G., & Rascol, O. (2020). A phase 1 randomized trial of specific active α-Synuclein immunotherapies PD01A and PD03A in multiple system atrophy. *Movement Disorders, 35*, 1957–1965.

Mendell, J. R., Al-Zaidy, S., Shell, R., Arnold, W. D., Rodino-Klapac, L. R., Prior, T. W., Lowes, L., Alfano, L., Berry, K., Church, K., Kissel, J. T., Nagendran, S., L'Italien, J., Sproule, D. M., Wells, C., Cardenas, J. A., Heitzer, M. D., Kaspar, A., Corcoran, S., Braun, L., Likhite, S., Miranda, C., Meyer, K., Foust, K. D., Burghes, A. H. M., & Kaspar, B. K. (2017). Single-dose gene-replacement therapy for spinal muscular atrophy. *The New England Journal of Medicine, 377*, 1713–1722.

Merlini, M., Kirabali, T., Kulic, L., Nitsch, R. M., & Ferretti, M. T. (2018). Extravascular CD3+ T cells in brains of Alzheimer disease patients correlate with tau but not with amyloid pathology: An immunohistochemical study. *Neurodegenerative Diseases, 18*, 49–56.

Miller, T. M., Pestronk, A., David, W., Rothstein, J., Simpson, E., Appel, S. H., Andres, P. L., Mahoney, K., Allred, P., Alexander, K., Ostrow, L. W., Schoenfeld, D., Macklin, E. A., Norris, D. A., Manousakis, G., Crisp, M., Smith, R., Bennett, C. F., Bishop, K. M., & Cudkowicz, M. E. (2013). An antisense oligonucleotide against SOD1 delivered intrathecally for patients with SOD1 familial amyotrophic lateral sclerosis: A phase 1, randomised, first-in-man study. *Lancet Neurology, 12*, 435–442.

Miller, T., Cudkowicz, M., Shaw, P. J., Andersen, P. M., Atassi, N., Bucelli, R. C., Genge, A., Glass, J., Ladha, S., Ludolph, A. L., Maragakis, N. J., McDermott, C. J., Pestronk, A., Ravits, J., Salachas, F., Trudell, R., Van Damme, P., Zinman, L., Bennett, C. F., Lane, R., Sandrock, A., Runz, H., Graham, D., Houshyar, H., McCampbell, A., Nestorov, I., Chang, I., McNeill, M., Fanning, L., Fradette, S., & Ferguson, T. A. (2020). Phase 1-2 trial of antisense oligonucleotide Tofersen for SOD1 ALS. *The New England Journal of Medicine, 383*, 109–119.

Miller, T. M., Cudkowicz, M. E., Genge, A., Shaw, P. J., Sobue, G., Bucelli, R. C., Chiò, A., Van Damme, P., Ludolph, A. C., Glass, J. D., Andrews, J. A., Babu, S., Benatar, M., McDermott, C. J., Cochrane, T., Chary, S., Chew, S., Zhu, H., Wu, F., Nestorov, I., Graham, D., Sun, P., McNeill, M., Fanning, L., Ferguson, T. A., & Fradette, S. (2022). Trial of antisense oligonucleotide Tofersen for SOD1 ALS. *The New England Journal of Medicine, 387*, 1099–1110.

Miniarikova, J., Zimmer, V., Martier, R., Brouwers, C. C., Pythoud, C., Richetin, K., Rey, M., Lubelski, J., Evers, M. M., van Deventer, S. J., Petry, H., Déglon, N., & Konstantinova, P. (2017). AAV5-miHTT gene therapy demonstrates suppression of mutant huntingtin aggregation and neuronal dysfunction in a rat model of Huntington's disease. *Gene Therapy, 24*, 630–639.

Mintun, M. A., Lo, A. C., Evans, C. D., Wessels, A. M., Ardayfio, P. A., Andersen, S. W., Shcherbinin, S., Sparks, J. D., Sims, J. R., Brys, M., Apostolova, L. G., Salloway, S. P., & Skovronsky, D. M. (2021). Donanemab in early Alzheimer's disease. *New England Journal of Medicine, 384*, 1691–1704.

Mittermeyer, G., Christine, C. W., Rosenbluth, K. H., Baker, S. L., Starr, P., Larson, P., Kaplan, P. L., Forsayeth, J., Aminoff, M. J., & Bankiewicz, K. S. (2012). Long-term evaluation of a phase 1 study of AADC gene therapy for Parkinson's disease. *Human Gene Therapy, 23*, 377–381.

Mobley, W. C., Rutkowski, J. L., Tennekoon, G. I., Gemski, J., Buchanan, K., & Johnston, M. V. (1986). Nerve growth factor increases choline acetyltransferase activity in developing basal forebrain neurons. *Brain Research, 387*, 53–62.

Monsonego, A., Maron, R., Zota, V., Selkoe, D. J., & Weiner, H. L. (2001). Immune hyporesponsiveness to amyloid beta-peptide in amyloid precursor protein transgenic mice: Implications for the pathogenesis and treatment of Alzheimer's disease. *Proceedings of the National Academy of Sciences of the United States of America, 98*, 10273–10278.

Monteys, A. M., Ebanks, S. A., Keiser, M. S., & Davidson, B. L. (2017). CRISPR/Cas9 editing of the mutant huntingtin allele in vitro and in vivo. *Molecular Therapy, 25*, 12–23.

Morizane, A., Kikuchi, T., Hayashi, T., Mizuma, H., Takara, S., Doi, H., Mawatari, A., Glasser, M. F., Shiina, T., Ishigaki, H., Itoh, Y., Okita, K., Yamasaki, E., Doi, D., Onoe, H., Ogasawara, K., Yamanaka, S., & Takahashi, J. (2017). MHC matching improves engraftment of iPSC-derived neurons in non-human primates. *Nature Communications, 8*, 385.

Mosholder, A. D., Ma, Y., Akhtar, S., Podskalny, G. D., Feng, Y., Lyu, H., Liao, J., Wei, Y., Wernecke, M., Leishear, K., Nelson, L. M., MaCurdy, T. E., Kelman, J. A., & Graham, D. J. (2022). Mortality among Parkinson's disease patients treated with Pimavanserin or atypical antipsychotics: An observational study in medicare beneficiaries. *American Journal of Psychiatry, 179*, 553–561.

Mueller, C., Berry, J. D., McKenna-Yasek, D. M., Gernoux, G., Owegi, M. A., Pothier, L. M., Douthwright, C. L., Gelevski, D., Luppino, S. D., Blackwood, M., Wightman, N. S., Oakley, D. H., Frosch, M. P., Flotte, T. R., Cudkowicz, M. E., & Brown, R. H., Jr. (2020). SOD1 suppression with Adeno-associated virus and MicroRNA in familial ALS. *The New England Journal of Medicine, 383*, 151–158.

Mullard, A. (2021). Failure of first anti-tau antibody in Alzheimer disease highlights risks of history repeating. *Nature Reviews. Drug Discovery, 20*, 3–5.

Müller, U. C., Deller, T., & Korte, M. (2017). Not just amyloid: Physiological functions of the amyloid precursor protein family. *Nature Reviews. Neuroscience, 18*, 281–298.

Murdock, B. J., Zhou, T., Kashlan, S. R., Little, R. J., Goutman, S. A., & Feldman, E. L. (2017). Correlation of peripheral immunity with rapid amyotrophic lateral sclerosis progression. *JAMA Neurology, 74*, 1446–1454.

Naaldijk, Y., Jäger, C., Fabian, C., Leovsky, C., Blüher, A., Rudolph, L., Hinze, A., & Stolzing, A. (2017). Effect of systemic transplantation of bone marrow-derived mesenchymal stem cells on neuropathology markers in APP/PS1 Alzheimer mice. *Neuropathology and Applied Neurobiology, 43*, 299–314.

Nagahara, A. H., Bernot, T., Moseanko, R., Brignolo, L., Blesch, A., Conner, J. M., Ramirez, A., Gasmi, M., & Tuszynski, M. H. (2009). Long-term reversal of cholinergic neuronal decline in aged non-human primates by lentiviral NGF gene delivery. *Experimental Neurology, 215*, 153–159.

Naryshkin, N. A., Weetall, M., Dakka, A., Narasimhan, J., Zhao, X., Feng, Z., Ling, K. K., Karp, G. M., Qi, H., Woll, M. G., Chen, G., Zhang, N., Gabbeta, V., Vazirani, P., Bhattacharyya, A., Furia, B., Risher, N., Sheedy, J., Kong, R., Ma, J., Turpoff, A., Lee, C. S., Zhang, X., Moon, Y. C., Trifillis, P., Welch, E. M., Colacino, J. M., Babiak, J., Almstead, N. G., Peltz, S. W., Eng, L. A., Chen, K. S., Mull, J. L., Lynes, M. S., Rubin, L. L., Fontoura, P., Santarelli, L., Haehnke, D., McCarthy, K. D., Schmucki, R., Ebeling, M., Sivaramakrishnan, M., Ko, C. P., Paushkin, S. V., Ratni, H., Gerlach, I., Ghosh, A., & Metzger, F. (2014). Motor neuron disease. SMN2 splicing modifiers improve motor function and longevity in mice with spinal muscular atrophy. *Science, 345*, 688–693.

Nasiri, E., Alizadeh, A., Roushandeh, A. M., Gazor, R., Hashemi-Firouzi, N., & Golipoor, Z. (2019). Melatonin-pretreated adipose-derived mesenchymal stem cells efficeintly improved learning, memory, and cognition in an animal model of Alzheimer's disease. *Metabolic Brain Disease, 34*, 1131–1143.

NCI. (2008). 'Aldesleukin'.

Neuroscience, United. (2019). 'United Neuroscience Announces Positive Top-Line Results from Phase 2a Clinical Study of UB-311 Vaccine in Patients with Alzheimer's Disease', PRNewswire. https://www.prnewswire.com/news-releases/united-neuroscience-announces-positive-top-line-results-from-phase-2a-clinical-study-of-ub-311-vaccine-in-patients-with-alzheimers-disease-300779315.html.

NINDS. (2022). A phase 1 open-label dose escalation safety study of convection enhanced delivery (CED) of Adeno-associated virus encoding Glial cell line-derived neurotrophic factor (AAV2-GDNF) in subjects with advanced Parkinson's Disease. In: clinicaltrials.gov

Nirula, A., Glaser, S. M., Kalled, S. L., & Taylor, F. R. (2011). 'What is IgG4? A review of the biology of a unique immunoglobulin subtype. *Current Opinion in Rheumatology, 23*, 119–124.

Novak, P., Schmidt, R., Kontsekova, E., Kovacech, B., Smolek, T., Katina, S., Fialova, L., Prcina, M., Parrak, V., Dal-Bianco, P., Brunner, M., Staffen, W., Rainer, M., Ondrus, M., Ropele, S., Smisek, M., Sivak, R., Zilka, N., Winblad, B., & Novak, M. (2018). FUNDAMANT: An interventional 72-week phase 1 follow-up study of AADvac1, an active immunotherapy against tau protein pathology in Alzheimer's disease. *Alzheimer's Research & Therapy, 10*, 108.

Novak, P., Kovacech, B., Katina, S., Schmidt, R., Scheltens, P., Kontsekova, E., Ropele, S., Fialova, L., Kramberger, M., & Paulenka-Ivanovova, N. (2021). ADAMANT: A placebo-controlled randomized phase 2 study of AADvac1, an active immunotherapy against pathological tau in Alzheimer's disease. *Nature Aging, 1*, 521–534.

Okeke, E. B., Okwor, I., & Uzonna, J. E. (2014). Regulatory T cells restrain CD4+ T cells from causing unregulated immune activation and hypersensitivity to lipopolysaccharide challenge. *Journal of Immunology, 193*, 655–662.

Olanow, C. W., Goetz, C. G., Kordower, J. H., Jon Stoessl, A., Sossi, V., Brin, M. F., Shannon, K. M., Michael Nauert, G., Perl, D. P., Godbold, J., & Freeman, T. B. (2003). A double-blind controlled trial of bilateral fetal nigral transplantation in Parkinson's disease. *Annals of Neurology, 54*, 403–414.

Olson, K. E., Namminga, K. L., Schwab, A. D., Thurston, M. J., Yaman, L., Woods, A., Lei, L., Shen, W., Wang, F., Joseph, S. B., Gendelman, H. E., & Lee Mosley, R. (2020). Neuroprotective activities of long-acting granulocyte–macrophage Colony-stimulating Factor (mPDM608) in 1-Methyl-4-Phenyl-1,2,3,6-Tetrahydropyridine-intoxicated mice. *Neurotherapeutics, 17*, 1861–1877.

Pagano, G., Taylor, K. I., Anzures-Cabrera, J., Marchesi, M., Simuni, T., Marek, K., Postuma, R. B., Pavese, N., Stocchi, F., Azulay, J. P., Mollenhauer, B., López-Manzanares, L., Russell, D. S., Boyd, J. T., Nicholas, A. P., Luquin, M. R., Hauser, R. A., Gasser, T., Poewe, W., Ricci, B., Boulay, A., Vogt, A., Boess, F. G., Dukart, J., D'Urso, G., Finch, R., Zanigni, S., Monnet, A., Pross, N., Hahn, A., Svoboda, H., Britschgi, M., Lipsmeier, F., Volkova-Volkmar, E., Lindemann, M., Dziadek, S., Holiga, Š., Rukina, D., Kustermann, T., Kerchner, G. A., Fontoura, P., Umbricht, D., Doody, R., Nikolcheva, T., & Bonni, A. (2022). Trial of Prasinezumab in early-stage Parkinson's disease. *The New England Journal of Medicine, 387*, 421–432.

Palfi, S., Gurruchaga, J. M., Ralph, G. S., Lepetit, H., Lavisse, S., Buttery, P. C., Watts, C., Miskin, J., Kelleher, M., Deeley, S., Iwamuro, H., Lefaucheur, J. P., Thiriez, C., Fenelon, G., Lucas, C., Brugières, P., Gabriel, I., Abhay, K., Drouot, X., Tani, N., Kas, A., Ghaleh, B., Le Corvoisier, P., Dolphin, P., Breen, D. P., Mason, S., Guzman, N. V., Mazarakis, N. D., Radcliffe, P. A., Harrop, R., Kingsman, S. M., Rascol, O., Naylor, S., Barker, R. A., Hantraye, P., Remy, P., Cesaro, P., & Mitrophanous, K. A. (2014). Long-term safety and tolerability of ProSavin, a lentiviral vector-based gene therapy for Parkinson's disease: A dose escalation, open-label, phase 1/2 trial. *Lancet, 383*, 1138–1146.

Passini, M. A., Bu, J., Richards, A. M., Kinnecom, C., Sardi, S. P., Stanek, L. M., Hua, Y., Rigo, F., Matson, J., Hung, G., Kaye, E. M., Shihabuddin, L. S., Krainer, A. R., Bennett, C. F., & Cheng, S. H. (2011). Antisense oligonucleotides delivered to the mouse CNS ameliorate symptoms of severe spinal muscular atrophy. *Science Translational Medicine, 3*, 72ra18.

Pollak, P., Tison, F., Rascol, O., Destée, A., Péré, J. J., Senard, J. M., Durif, F., & Bourdeix, I. (2004). Clozapine in drug induced psychosis in Parkinson's disease: A randomised, placebo controlled study with open follow up. *Journal of Neurology, Neurosurgery, and Psychiatry, 75*, 689–695.

Pontecorvo, M. J., Lu, M., Burnham, S. C., Schade, A. E., Dage, J. L., Shcherbinin, S., Collins, E. C., Sims, J. R., & Mintun, M. A. (2022). Association of Donanemab treatment with exploratory plasma biomarkers in early symptomatic Alzheimer disease: A secondary analysis of the TRAILBLAZER-ALZ randomized clinical trial. *JAMA Neurology, 79*, 1250–1259.

Post, Anke. (2022). Safety and efficacy of AMT-130 in European adults with early manifest Huntington's disease, ClinicalTrials.gov Identifier: NCT05243017, clinicaltrials.gov. Accessed 3 Oct.

Prevail, Therapeutics. (2022). A phase 1/2a open-label ascending dose study to evaluate the safety and effects of LY3884961 in patients with Parkinson's disease with at least one GBA1 mutation (PROPEL). In: clinicaltrials.gov

Prokop, S., Miller, K. R., Drost, N., Handrick, S., Mathur, V., Luo, J., Wegner, A., Wyss-Coray, T., & Heppner, F. L. (2015). Impact of peripheral myeloid cells on amyloid-β pathology in Alzheimer's disease-like mice. *The Journal of Experimental Medicine, 212*, 1811–1818.

Qin, C., Li, Y., & Wang, K. (2021). Novel balance mechanism participates in stem cell therapy to alleviate neuropathology and cognitive impairment in animal models with Alzheimer's disease. *Cells, 10*, 2757.

Rafii, M. S., Tuszynski, M. H., Thomas, R. G., Barba, D., Brewer, J. B., Rissman, R. A., Siffert, J., & Aisen, P. S. (2018). Adeno-associated viral vector (Serotype 2)-nerve growth factor for patients with Alzheimer disease: A randomized clinical trial. *JAMA Neurology, 75*, 834–841.

Ransohoff, R. M. (2016). How neuroinflammation contributes to neurodegeneration. *Science, 353*, 777–783.

Research, Center for Biologics Evaluation and. (2022). *Human gene therapy for neurodegenerative diseases*. U.S. Food and Drug Administration.

Rodriguez-Martin, T., Anthony, K., Garcia-Blanco, M. A., Mansfield, S. G., Anderton, B. H., & Gallo, J. M. (2009). Correction of tau mis-splicing caused by FTDP-17 MAPT mutations by spliceosome-mediated RNA trans-splicing. *Human Molecular Genetics, 18*, 3266–3273.

Rogers, J., Luber-Narod, J., Styren, S. D., & Civin, W. H. (1988). Expression of immune system-associated antigens by cells of the human central nervous system: Relationship to the pathology of Alzheimer's disease. *Neurobiology of Aging, 9*, 339–349.

Rohn, T. T. (2013). The triggering receptor expressed on myeloid cells 2: "TREM-ming" the inflammatory component associated with Alzheimer's disease. *Oxidative Medicine and Cellular Longevity, 2013*, 860959.

Rosenberg, J. B., Kaplitt, M. G., De, B. P., Chen, A., Flagiello, T., Salami, C., Pey, E., Zhao, L., Ricart Arbona, R. J., Monette, S., Dyke, J. P., Ballon, D. J., Kaminsky, S. M., Sondhi, D., Petsko, G. A., Paul, S. M., & Crystal, R. G. (2018). AAVrh.10-mediated APOE2 central nervous system gene therapy for APOE4-associated Alzheimer's disease. *Human Gene Therapy. Clinical Development, 29*, 24–47.

Rother, C., Uhlmann, R. E., Müller, S. A., Schelle, J., Skodras, A., Obermüller, U., Häsler, L. M., Lambert, M., Baumann, F., Ying, X., Bergmann, C., Salvadori, G., Loos, M., Brzak, I., Shimshek, D., Neumann, U., Walker, L. C., Schultz, S. A., Chhatwal, J. P., Kaeser, S. A., Lichtenthaler, S. F., Staufenbiel, M., Jucker, M., & Network Dominantly Inherited Alzheimer. (2022). Experimental evidence for temporal uncoupling of brain Aβ deposition and neurodegenerative sequelae. *Nature Communications, 13*, 7333.

Ruzicka, J., Kulijewicz-Nawrot, M., Rodrigez-Arellano, J. J., Jendelova, P., & Sykova, E. (2016). Mesenchymal stem cells preserve working memory in the 3xTg-AD Mouse Model of Alzheimer's disease. *International Journal of Molecular Sciences, 17*, 152.

SA, A. C. Immune. (2022). A phase Ib/IIa multicenter, double-blind, randomized, placebo-controlled study to evaluate the safety, tolerability and immunogenicity of different doses, regimens and combinations of tau targeted vaccines in subjects with early Alzheimer's disease. In: clinicaltrials.gov

Salloway, S., Sperling, R., Fox, N. C., Blennow, K., Klunk, W., Raskind, M., Sabbagh, M., Honig, L. S., Porsteinsson, A. P., Ferris, S., Reichert, M., Ketter, N., Nejadnik, B., Guenzler, V., Miloslavsky, M., Wang, D., Yuan, L., Lull, J., Tudor, I. C., Liu, E., Grundman, M., Yuen, E., Black, R., & Robert Brashear, H. (2014). Two phase 3 trials of Bapineuzumab in mild-to-moderate Alzheimer's disease. *New England Journal of Medicine, 370*, 322–333.

Samata, B., Doi, D., Nishimura, K., Kikuchi, T., Watanabe, A., Sakamoto, Y., Kakuta, J., Ono, Y., & Takahashi, J. (2016). Purification of functional human ES and iPSC-derived midbrain dopaminergic progenitors using LRTM1. *Nature Communications, 7*, 13097.

Sandoval, I. M., Marmion, D. J., Meyers, K. T., & Manfredsson, F. P. (2021). Gene therapy to modulate alpha-Synuclein in Synucleinopathies. *Journal of Parkinson's Disease, 11*, S189–Ss97.

Sardi, S. P., Clarke, J., Kinnecom, C., Tamsett, T. J., Li, L., Stanek, L. M., Passini, M. A., Grabowski, G. A., Schlossmacher, M. G., Sidman, R. L., Cheng, S. H., & Shihabuddin, L. S. (2011). CNS expression of glucocerebrosidase corrects alpha-synuclein pathology and memory in a mouse model of Gaucher-related synucleinopathy. *Proceedings of the National Academy of Sciences of the United States of America, 108*, 12101–12106.

Schindowski, K., Eckert, A., Peters, J., Gorriz, C., Schramm, U., Weinandi, T., Maurer, K., Frölich, L., & Müller, W. E. (2007). Increased T-cell reactivity and elevated levels of CD8+ memory T-cells in Alzheimer's disease-patients and T-cell hyporeactivity in an Alzheimer's disease-mouse model: Implications for immunotherapy. *Neuromolecular Medicine, 9*, 340–354.

Schneeberger, A., Mandler, M., Mattner, F., & Schmidt, W. (2012). Vaccination for Parkinson's disease. *Parkinsonism & Related Disorders, 18*, S11–S13.

Seckan, B., & The Journalist's Resource June. (2013). Monetary costs of dementia in the United States. In *The Journalist's resource*.

Selkoe, D. J., & Hardy, J. (2016). The amyloid hypothesis of Alzheimer's disease at 25 years. *EMBO Molecular Medicine, 8*, 595–608.

Serrano-Pozo, A., Das, S., & Hyman, B. T. (2021). APOE and Alzheimer's disease: Advances in genetics, pathophysiology, and therapeutic approaches. *Lancet Neurology, 20*, 68–80.

Shcherbinin, S., Evans, C. D., Lu, M., Andersen, S. W., Pontecorvo, M. J., Willis, B. A., Gueorguieva, I., Hauck, P. M., Brooks, D. A., Mintun, M. A., & Sims, J. R. (2022). Association of Amyloid reduction after Donanemab treatment with tau pathology and clinical outcomes: The TRAILBLAZER-ALZ randomized clinical trial. *JAMA Neurology, 79*, 1015–1024.

Shin, J. W., Kim, K. H., Chao, M. J., Atwal, R. S., Gillis, T., MacDonald, M. E., Gusella, J. F., & Lee, J. M. (2016). Permanent inactivation of Huntington's disease mutation by personalized allele-specific CRISPR/Cas9. *Human Molecular Genetics, 25*, 4566–4576.

Sidransky, E., Nalls, M. A., Aasly, J. O., Aharon-Peretz, J., Annesi, G., Barbosa, E. R., Bar-Shira, A., Berg, D., Bras, J., Brice, A., Chen, C. M., Clark, L. N., Condroyer, C., De Marco, E. V., Dürr, A., Eblan, M. J., Fahn, S., Farrer, M. J., Fung, H. C., Gan-Or, Z., Gasser, T., Gershoni-Baruch, R., Giladi, N., Griffith, A., Gurevich, T., Januario, C., Kropp, P., Lang, A. E., Lee-Chen, G. J., Lesage, S., Marder, K., Mata, I. F., Mirelman, A., Mitsui, J., Mizuta, I., Nicoletti, G., Oliveira, C., Ottman, R., Orr-Urtreger, A., Pereira, L. V., Quattrone, A., Rogaeva, E., Rolfs, A., Rosenbaum, H., Rozenberg, R., Samii, A., Samaddar, T., Schulte, C., Sharma, M., Singleton, A., Spitz, M., Tan, E. K., Tayebi, N., Toda, T., Troiano, A. R., Tsuji, S., Wittstock, M., Wolfsberg, T. G., Wu, Y. R., Zabetian, C. P., Zhao, Y., & Ziegler, S. G. (2009). Multicenter analysis of glucocerebrosidase mutations in Parkinson's disease. *The New England Journal of Medicine, 361*, 1651–1661.

Sienski, G., Narayan, P., Bonner, J. M., Kory, N., Boland, S., Arczewska, A. A., Ralvenius, W. T., Akay, L., Lockshin, E., He, L., Milo, B., Graziosi, A., Baru, V., Lewis, C. A., Kellis, M., Sabatini, D. M., Tsai, L. H., & Lindquist, S. (2021). APOE4 disrupts intracellular lipid homeostasis in human iPSC-derived glia. *Science Translational Medicine, 13*, eaaz4564.

Simard, A. R., Soulet, D., Gowing, G., Julien, J. P., & Rivest, S. (2006). Bone marrow-derived microglia play a critical role in restricting senile plaque formation in Alzheimer's disease. *Neuron, 49*, 489–502.
Sliter, D. A., Martinez, J., Hao, L., Chen, X., Sun, N., Fischer, T. D., Burman, J. L., Li, Y., Zhang, Z., Narendra, D. P., Cai, H., Borsche, M., Klein, C., & Youle, R. J. (2018). Parkin and PINK1 mitigate STING-induced inflammation. *Nature, 561*, 258–262.
Smith, R. A., Miller, T. M., Yamanaka, K., Monia, B. P., Condon, T. P., Hung, G., Lobsiger, C. S., Ward, C. M., McAlonis-Downes, M., Wei, H., Wancewicz, E. V., Bennett, C. F., & Cleveland, D. W. (2006). Antisense oligonucleotide therapy for neurodegenerative disease. *The Journal of Clinical Investigation, 116*, 2290–2296.
Söderberg, L., Johannesson, M., Nygren, P., Laudon, H., Eriksson, F., Osswald, G., Möller, C., & Lannfelt, L. (2022). Lecanemab, Aducanumab, and Gantenerumab — Binding profiles to different forms of amyloid-Beta might explain efficacy and side effects in clinical trials for Alzheimer's disease. *Neurotherapeutics, 20*, 195.
Song, C., Shi, J., Zhang, P., Zhang, Y., Jianrong, X., Zhao, L., Zhang, R., Wang, H., & Chen, H. (2022). Immunotherapy for Alzheimer's disease: Targeting β-amyloid and beyond. *Translational Neurodegeneration, 11*, 18.
Steele, O. G., Stuart, A. C., Minkley, L., Shaw, K., Bonnar, O., Anderle, S., Penn, A. C., Rusted, J., Serpell, L., Hall, C., & King, S. (2022). A multi-hit hypothesis for an APOE4-dependent pathophysiological state. *The European Journal of Neuroscience, 56*, 5476–5515.
Stevens, C. H., Rowe, D., Morel-Kopp, M. C., Orr, C., Russell, T., Ranola, M., Ward, C., & Halliday, G. M. (2012). Reduced T helper and B lymphocytes in Parkinson's disease. *Journal of Neuroimmunology, 252*, 95–99.
Sud, R., Geller, E. T., & Schellenberg, G. D. (2014). Antisense-mediated Exon skipping decreases tau protein expression: A potential therapy for tauopathies. *Molecular Therapy - Nucleic Acids, 3*, e180.
Sudhakar, V., & Richardson, R. M. (2019). Gene therapy for neurodegenerative diseases. *Neurotherapeutics, 16*, 166–175.
Sulzer, D., Alcalay, R. N., Garretti, F., Cote, L., Kanter, E., Agin-Liebes, J., Liong, C., McMurtrey, C., Hildebrand, W. H., Mao, X., Dawson, V. L., Dawson, T. M., Oseroff, C., Pham, J., Sidney, J., Dillon, M. B., Carpenter, C., Weiskopf, D., Phillips, E., Mallal, S., Peters, B., Frazier, A., Lindestam Arlehamn, C. S., & Sette, A. (2017). T cells from patients with Parkinson's disease recognize α-synuclein peptides. *Nature, 546*, 656–661.
Sun, J., & Roy, S. (2021). Gene-based therapies for neurodegenerative diseases. *Nature Neuroscience, 24*, 297–311.
Sun, J., Carlson-Stevermer, J., Das, U., Shen, M., Delenclos, M., Snead, A. M., Koo, S. Y., Wang, L., Qiao, D., Loi, J., Petersen, A. J., Stockton, M., Bhattacharyya, A., Jones, M. V., Zhao, X., McLean, P. J., Sproul, A. A., Saha, K., & Roy, S. (2019). CRISPR/Cas9 editing of APP C-terminus attenuates β-cleavage and promotes α-cleavage. *Nature Communications, 10*, 53.
Suzuki, N., Nishiyama, A., Warita, H., & Aoki, M. (2022). Genetics of amyotrophic lateral sclerosis: Seeking therapeutic targets in the era of gene therapy. *Journal of Human Genetics, 68*, 131.
Sy, M., Kitazawa, M., Medeiros, R., Whitman, L., Cheng, D., Lane, T. E., & Laferla, F. M. (2011). Inflammation induced by infection potentiates tau pathological features in transgenic mice. *The American Journal of Pathology, 178*, 2811–2822.
Tabrizi, S. J., Leavitt, B. R., Landwehrmeyer, G. B., Wild, E. J., Saft, C., Barker, R. A., Blair, N. F., Craufurd, D., Priller, J., Rickards, H., Rosser, A., Kordasiewicz, H. B., Czech, C., Swayze, E. E., Norris, D. A., Baumann, T., Gerlach, I., Schobel, S. A., Paz, E., Smith, A. V., Bennett, C. F., & Lane, R. M. (2019). Targeting huntingtin expression in patients with Huntington's disease. *The New England Journal of Medicine, 380*, 2307–2316.
Tabrizi, S. J., Estevez-Fraga, C., van Roon-Mom, W. M. C., Flower, M. D., Scahill, R. I., Wild, E. J., Muñoz-Sanjuan, I., Sampaio, C., Rosser, A. E., & Leavitt, B. R. (2022). Potential disease-modifying therapies for Huntington's disease: Lessons learned and future opportunities. *Lancet Neurology, 21*, 645–658.
Tai, L. M., Ghura, S., Koster, K. P., Liakaite, V., Maienschein-Cline, M., Kanabar, P., Collins, N., Ben-Aissa, M., Lei, A. Z., Bahroos, N., Green, S. J., Hendrickson, B., Van Eldik, L. J., & LaDu, M. J. (2015). APOE-modulated Aβ-induced neuroinflammation in Alzheimer's disease: Current landscape, novel data, and future perspective. *Journal of Neurochemistry, 133*, 465–488.
Takamatsu, K., Ikeda, T., Haruta, M., Matsumura, K., Ogi, Y., Nakagata, N., Uchino, M., Ando, Y., Nishimura, Y., & Senju, S. (2014). Degradation of amyloid beta by human induced pluripotent stem cell-derived macrophages expressing Neprilysin-2. *Stem Cell Research, 13*, 442–453.
Theka, I., Caiazzo, M., Dvoretskova, E., Leo, D., Ungaro, F., Curreli, S., Managò, F., Dell'Anno, M. T., Pezzoli, G., Gainetdinov, R. R., Dityatev, A., & Broccoli, V. (2013). Rapid generation of functional dopaminergic neurons from human induced pluripotent stem cells through a single-step procedure using cell lineage transcription factors. *Stem Cells Translational Medicine, 2*, 473–479.
Therapeutics, Vico. (2021). US FDA grants VICO therapeutics orphan-drug designation for VO659, an investigational therapy for Huntington disease. In *VICO therapeutics*.
Theunis, C., Crespo-Biel, N., Gafner, V., Pihlgren, M., López-Deber, M. P., Reis, P., Hickman, D. T., Adolfsson, O., Chuard, N., Ndao, D. M., Borghgraef, P., Devijver, H., Van Leuven, F., Pfeifer, A., & Muhs, A. (2013). Efficacy and safety of a liposome-based vaccine against protein tau, assessed in tau.P301L mice that model tauopathy. *PLoS One, 8*, e72301.
Thome, A. D., Harms, A. S., Volpicelli-Daley, L. A., & Standaert, D. G. (2016). microRNA-155 regulates alpha-Synuclein-induced inflammatory responses in models of Parkinson disease. *The Journal of Neuroscience, 36*, 2383–2390.
Thome, A. D., Atassi, F., Wang, J., Faridar, A., Zhao, W., Thonhoff, J. R., Beers, D. R., Lai, E. C., & Appel, S. H. (2021). 'Ex vivo expansion of dysfunctional regulatory T lymphocytes restores suppressive function in Parkinson's disease. *npj Parkinson's Disease, 7*, 41.
Thonhoff, J. R., Beers, D. R., Zhao, W., Pleitez, M., Simpson, E. P., Berry, J. D., Cudkowicz, M. E., & Appel, S. H. (2018). Expanded autologous regulatory T-lymphocyte infusions in ALS: A phase I, first-in-human study. *Neurology Neuroimmunology & Neuroinflammation, 5*, e465.
Thonhoff, J. R., Berry, J. D., Macklin, E. A., Beers, D. R., Mendoza, P. A., Zhao, W., Thome, A. D., Triolo, F., Moon, J. J., Paganoni, S., Cudkowicz, M., & Appel, S. H. (2022). Combined regulatory T-lymphocyte and IL-2 treatment is safe, tolerable, and biologically active for 1 year in persons with amyotrophic lateral sclerosis. *Neurology Neuroimmunology & Neuroinflammation, 9*, e200019.
Togo, T., Akiyama, H., Iseki, E., Kondo, H., Ikeda, K., Kato, M., Oda, T., Tsuchiya, K., & Kosaka, K. (2002). Occurrence of T cells in the brain of Alzheimer's disease and other neurological diseases. *Journal of Neuroimmunology, 124*, 83–92.
'TRANSEURO | Home'.
'UB-311. (2020). Alzforum, Accessed 21 Nov 2022. https://www.alzforum.org/therapeutics/ub-311
Uéda, K., Fukushima, H., Masliah, E., Xia, Y., Iwai, A., Yoshimoto, M., Otero, D. A., Kondo, J., Ihara, Y., & Saitoh, T. (1993). Molecular cloning of cDNA encoding an unrecognized component of amyloid in Alzheimer disease. *Proceedings of the National Academy of Sciences of the United States of America, 90*, 11282–11286.
UMIN-CTR. Kyoto trial to evaluate the safety and efficacy of iPSC-derived dopaminergic progenitors in the treatment of Parkinson's disease.
van Dyck, C. H., Swanson, C. J., Aisen, P., Bateman, R. J., Chen, C., Gee, M., Kanekiyo, M., Li, D., Reyderman, L., Cohen, S., Froelich, L., Katayama, S., Sabbagh, M., Vellas, B., Watson, D., Dhadda, S.,

Irizarry, M., Kramer, L. D., & Iwatsubo, T. (2022). Lecanemab in early Alzheimer's disease. *New England Journal of Medicine, 388*, 9.

Vandenberghe, R., Rinne, J. O., Boada, M., Katayama, S., Scheltens, P., Vellas, B., Tuchman, M., Gass, A., Fiebach, J. B., Hill, D., Lobello, K., Li, D., McRae, T., Lucas, P., Evans, I., Booth, K., Luscan, G., Wyman, B. T., Hua, L., Yang, L., Brashear, H. R., & Black, R. S. (2016). Bapineuzumab for mild to moderate Alzheimer's disease in two global, randomized, phase 3 trials. *Alzheimer's Research & Therapy, 8*, 18.

Varvel, N. H., Grathwohl, S. A., Degenhardt, K., Resch, C., Bosch, A., Jucker, M., & Neher, J. J. (2015). Replacement of brain-resident myeloid cells does not alter cerebral amyloid-β deposition in mouse models of Alzheimer's disease. *The Journal of Experimental Medicine, 212*, 1803–1809.

Vaz, M., & Silvestre, S. (2020). Alzheimer's disease: Recent treatment strategies. *European Journal of Pharmacology, 887*, 173554.

Vellas, B., Black, R., Thal, L. J., Fox, N. C., Daniels, M., McLennan, G., Tompkins, C., Leibman, C., Pomfret, M., & Grundman, M. (2009). Long-term follow-up of patients immunized with AN1792: Reduced functional decline in antibody responders. *Current Alzheimer Research, 6*, 144–151.

Victor, M. B., Leary, N., Luna, X., Meharena, H. S., Scannail, A. N., Bozzelli, P. L., Samaan, G., Murdock, M. H., von Maydell, D., Effenberger, A. H., Cerit, O., Wen, H. L., Liu, L., Welch, G., Bonner, M., & Tsai, L. H. (2022). Lipid accumulation induced by APOE4 impairs microglial surveillance of neuronal-network activity. *Cell Stem Cell, 29*, 1197–212.e8.

Volpicelli-Daley, L. A., Kirik, D., Stoyka, L. E., Standaert, D. G., & Harms, A. S. (2016). How can rAAV-α-synuclein and the fibril α-synuclein models advance our understanding of Parkinson's disease? *Journal of Neurochemistry, 139*(Suppl 1), 131–155.

Wallings, R. L., & Tansey, M. G. (2019). LRRK2 regulation of immune-pathways and inflammatory disease. *Biochemical Society Transactions, 47*, 1581–1595.

Wallings, R. L., Herrick, M. K., & Tansey, M. G. (2020). LRRK2 at the interface between peripheral and central immune function in Parkinson's. *Frontiers in Neuroscience, 14*, 443.

Wang, Y., & Mandelkow, E. (2016). Tau in physiology and pathology. *Nature Reviews Neuroscience, 17*, 22–35.

Wang, C. Y., Finstad, C. L., Walfield, A. M., Sia, C., Sokoll, K. K., Chang, T. Y., Fang, X. D., Hung, C. H., Hutter-Paier, B., & Windisch, M. (2007). Site-specific UBITh amyloid-beta vaccine for immunotherapy of Alzheimer's disease. *Vaccine, 25*, 3041–3052.

Wang, C. Y., Wang, P. N., Chiu, M. J., Finstad, C. L., Lin, F., Lynn, S., Tai, Y. H., De Fang, X., Zhao, K., Hung, C. H., Tseng, Y., Peng, W. J., Wang, J., Yu, C. C., Kuo, B. S., & Frohna, P. A. (2017). UB-311, a novel UBITh(®) amyloid β peptide vaccine for mild Alzheimer's disease. *Alzheimers Dement (N Y), 3*, 262–272.

Wang, S., Mustafa, M., Yuede, C. M., Salazar, S. V., Kong, P., Long, H., Ward, M., Siddiqui, O., Paul, R., Gilfillan, S., Ibrahim, A., Rhinn, H., Tassi, I., Rosenthal, A., Schwabe, T., & Colonna, M. (2020). Anti-human TREM2 induces microglia proliferation and reduces pathology in an Alzheimer's disease model. *The Journal of Experimental Medicine, 217*.

Washington University School of, Medicine. (2022). A phase II/III randomized, double-blind, placebo-controlled, cognitive endpoint, multi-center study of potential disease modifying therapies in individuals at risk for and with dominantly inherited Alzheimer's disease. In: clinicaltrials.gov

Wave Life Sciences, Ltd. (2022a). A multicenter, randomized, double-blind, placebo-controlled, phase 1b/2a study of WVE-004 administered intrathecally to patients with C9orf72-associated Amyotrophic Lateral Sclerosis (ALS) or Frontotemporal Dementia (FTD). In: clinicaltrials.gov

Wave Life Sciences, Ltd. (2022b). A multicenter, randomized, double-blind, placebo-controlled, Phase 1b/2a Study of WVE-120101 administered intrathecally in patients with Huntington's disease. In: clinicaltrials.gov

Wave Life Sciences, Ltd. (2022c). A multicenter, randomized, double-blind, placebo-controlled, phase 1b/2a study of WVE-120102 administered intrathecally in patients with Huntington's disease. In: clinicaltrials.gov

Wei, F., Zhang, Y., Zhao, W., Yu, X., & Liu, C. J. (2014). Progranulin facilitates conversion and function of regulatory T cells under inflammatory conditions. *PLoS One, 9*, e112110.

Wei, Y., Xie, Z., Bi, J., & Zhu, Z. (2018). Anti-inflammatory effects of bone marrow mesenchymal stem cells on mice with Alzheimer's disease. *Experimental and Therapeutic Medicine, 16*, 5015–5020.

Williams, G. P., Schonhoff, A. M., Jurkuvenaite, A., Thome, A. D., Standaert, D. G., & Harms, A. S. (2018). Targeting of the class II transactivator attenuates inflammation and neurodegeneration in an alpha-synuclein model of Parkinson's disease. *Journal of Neuroinflammation, 15*, 244.

Winblad, B., Andreasen, N., Minthon, L., Floesser, A., Imbert, G., Dumortier, T., Maguire, R. P., Blennow, K., Lundmark, J., Staufenbiel, M., Orgogozo, J. M., & Graf, A. (2012). Safety, tolerability, and antibody response of active Aβ immunotherapy with CAD106 in patients with Alzheimer's disease: Randomised, double-blind, placebo-controlled, first-in-human study. *Lancet Neurology, 11*, 597–604.

WUSTL, School of Medicine. (2022). A phase II/III randomized, double-blind, placebo-controlled, cognitive endpoint, multi-center study of potential disease modifying therapies in individuals at risk for and with dominantly inherited Alzheimer's disease. In.: clinicaltrials.gov

Xie, L., Choudhury, G. R., Winters, A., Yang, S. H., & Jin, K. (2015). Cerebral regulatory T cells restrain microglia/macrophage-mediated inflammatory responses via IL-10. *European Journal of Immunology, 45*, 180–191.

Xie, Z. H., Liu, Z., Zhang, X. R., Yang, H., Wei, L. F., Wang, Y., Xu, S. L., Sun, L., Lai, C., Bi, J. Z., & Wang, X. Y. (2016). Wharton's Jelly-derived mesenchymal stem cells alleviate memory deficits and reduce amyloid-β deposition in an APP/PS1 transgenic mouse model. *Clinical and Experimental Medicine, 16*, 89–98.

Xilouri, M., Brekk, O. R., Landeck, N., Pitychoutis, P. M., Papasilekas, T., Papadopoulou-Daifoti, Z., Kirik, D., & Stefanis, L. (2013). Boosting chaperone-mediated autophagy in vivo mitigates α-synuclein-induced neurodegeneration. *Brain, 136*, 2130–2146.

Yamazaki, Y., Zhao, N., Caulfield, T. R., Liu, C. C., & Bu, G. (2019). Apolipoprotein E and Alzheimer disease: Pathobiology and targeting strategies. *Nature Reviews. Neurology, 15*, 501–518.

Yan, Q., Zhang, J., Liu, H., Babu-Khan, S., Vassar, R., Biere, A. L., Citron, M., & Landreth, G. (2003). Anti-inflammatory drug therapy alters beta-amyloid processing and deposition in an animal model of Alzheimer's disease. *The Journal of Neuroscience, 23*, 7504–7509.

Yokokawa, K., Iwahara, N., Hisahara, S., Emoto, M. C., Saito, T., Suzuki, H., Manabe, T., Matsumura, A., Matsushita, T., Suzuki, S., Kawamata, J., Sato-Akaba, H., Fujii, H. G., & Shimohama, S. (2019). Transplantation of mesenchymal stem cells improves amyloid-β pathology by modifying microglial function and suppressing oxidative stress. *Journal of Alzheimer's Disease, 72*, 867–884.

Yokoyama, J. S., Wang, Y., Schork, A. J., Thompson, W. K., Karch, C. M., Cruchaga, C., McEvoy, L. K., Witoelar, A., Chen, C. H., Holland, D., Brewer, J. B., Franke, A., Dillon, W. P., Wilson, D. M., Mukherjee, P., Hess, C. P., Miller, Z., Bonham, L. W., Shen, J., Rabinovici, G. D., Rosen, H. J., Miller, B. L., Hyman, B. T., Schellenberg, G. D., Karlsen, T. H., Andreassen, O. A., Dale, A. M., & Desikan, R. S. (2016). Association between genetic traits for immune-mediated diseases and Alzheimer disease. *JAMA Neurology, 73*, 691–697.

Yu, H. J., Thijssen, E., van Brummelen, E., van der Plas, J. L., Radanovic, I., Moerland, M., Hsieh, E., Groeneveld, G. J., & Dodart, J.-C. (2022). A randomized first-in-human study with UB-312, a UBITh® α-Synuclein peptide vaccine. *Movement Disorders, 37*, 1416–1424.

Yue, W., Li, Y., Zhang, T., Jiang, M., Qian, Y., Zhang, M., Sheng, N., Feng, S., Tang, K., Yu, X., Shu, Y., Yue, C., & Jing, N. (2015). ESC-derived basal forebrain cholinergic neurons ameliorate the cognitive symptoms associated with Alzheimer's disease in Mouse Models. *Stem Cell Reports, 5*, 776–790.

Zeitler, B., Froelich, S., Marlen, K., Shivak, D. A., Qi, Y., Li, D., Pearl, J. R., Miller, J. C., Zhang, L., Paschon, D. E., Hinkley, S. J., Ankoudinova, I., Lam, S., Guschin, D., Kopan, L., Cherone, J. M., Nguyen, H.-O. B., Qiao, G., Ataei, Y., Mendel, M. C., Amora, R., Surosky, R., Josee Laganiere, B., Joseph, V., Narayanan, A., Sedaghat, Y., Tillack, K., Thiede, C., Gärtner, A., Kwak, S., Bard, J., Mrzljak, L., Park, L., Heikkinen, T., Lehtimäki, K. K., Svedberg, M. M., Häggkvist, J., Tari, L., Tóth, M., Varrone, A., Halldin, C., Kudwa, A. E., Ramboz, S., Day, M., Jyothisri Kondapalli, D., Surmeier, J., Urnov, F. D., Gregory, P. D., Rebar, E. J., Muñoz-Sanjuán, I., & Steve Zhang, H. (2019). Allele-selective transcriptional repression of mutant HTT for the treatment of Huntington's disease. *Nature Medicine, 25*, 1131–1142.

Zhang, J., Kong, Q., Zhang, Z., Ge, P., Ba, D., & He, W. (2003). Telomere dysfunction of lymphocytes in patients with Alzheimer disease. *Cognitive and Behavioral Neurology, 16*, 170–176.

Zhang, R., Gascon, R., Miller, R. G., Gelinas, D. F., Mass, J., Hadlock, K., Jin, X., Reis, J., Narvaez, A., & McGrath, M. S. (2005). Evidence for systemic immune system alterations in sporadic amyotrophic lateral sclerosis (sALS). *Journal of Neuroimmunology, 159*, 215–224.

Zhang, B., Gaiteri, C., Bodea, L. G., Wang, Z., McElwee, J., Podtelezhnikov, A. A., Zhang, C., Xie, T., Tran, L., Dobrin, R., Fluder, E., Clurman, B., Melquist, S., Narayanan, M., Suver, C., Shah, H., Mahajan, M., Gillis, T., Mysore, J., MacDonald, M. E., Lamb, J. R., Bennett, D. A., Molony, C., Stone, D. J., Gudnason, V., Myers, A. J., Schadt, E. E., Neumann, H., Zhu, J., & Emilsson, V. (2013). Integrated systems approach identifies genetic nodes and networks in late-onset Alzheimer's disease. *Cell, 153*, 707–720.

Zhang, Q., Wu, H. H., Wang, Y., Gu, G. J., Zhang, W., & Xia, R. (2016). Neural stem cell transplantation decreases neuroinflammation in a transgenic mouse model of Alzheimer's disease. *Journal of Neurochemistry, 136*, 815–825.

Zhao, W., Beers, D. R., Liao, B., Henkel, J. S., & Appel, S. H. (2012). Regulatory T lymphocytes from ALS mice suppress microglia and effector T lymphocytes through different cytokine-mediated mechanisms. *Neurobiology of Disease, 48*, 418–428.

Zhou, R., Ji, B., Kong, Y., Qin, L., Ren, W., Guan, Y., & Ni, R. (2021). PET imaging of neuroinflammation in Alzheimer's disease. *Frontiers in Immunology, 12*, 739130.

Zimmerman, L. M., & Howell, K. M. (1932). History of blood transfusion. *Annals of Medical History, 4*, 415–433.

Abdulmunaim M Eid is a Burroughs Wellcome Fund Fellow supported by a Burroughs Wellcome Fund Physician Scientist Institutional Award to the Texas A&M University Academy of Physician Scientists, Houston, TX, USA

Amyotrophic Lateral Sclerosis and Frontotemporal Dementia

40

Jason R. Thonhoff, Ericka P. Greene, and Stanley H. Appel

Abstract

Amyotrophic lateral sclerosis (ALS) is a fatal motor neuron disease characterized by progressive degeneration of upper and lower motor neurons, culminating in paralysis and death usually due to respiratory failure. Executive function and problem-solving are compromised in 50% of patients with ALS, and 15–20% of ALS patients develop significant frontotemporal dementia (FTD) with behavioral, language, and executive dysfunction. ALS and FTD exist on a neurodegenerative continuum with substantial clinical overlap that is influenced by genetic factors. Over the last two decades, with the advances in gene discovery technologies, more than 40 disease-causing mutations have been identified representing about 10% of cases of ALS. In most cases of ALS, however, a single genetic cause has not been discovered and the disease progresses through a complex cascade involving dysfunction of a multitude of intracellular processes that result in motor neuron degeneration. Accumulation of these intracellular insults causes the initiation of symptomatic disease. Immune and inflammatory factors then contribute to the burden of disease, rates of disease progression, and survival. These immune/inflammatory factors are increased in the blood and characterize ALS as a systemic disorder, and one not just confined to the central nervous system. Despite intensive efforts, no known pharmacotherapy substantially influences either disease progression or survival. Current therapeutic efforts are aimed at a variety of targets including modifying the expression of mutant genes, ameliorating defective intracellular pathways, and modulating neuroinflammation in attempts to slow disease progression, prolong survival, and improve quality of life. The major objectives of this chapter are to recognize the diverse clinical presentations of ALS/FTD and the many mechanisms of disease pathogenesis. ALS is a heterogeneous disorder with no single cause, but a complex cascade that compromises motor and non-motor neurons. The discovery of successful therapeutics will rely on a better understanding of these mechanisms and our ability to stratify ALS populations based on their predicted responses to specific interventions.

Keywords

Amyotrophic lateral sclerosis · Frontotemporal dementia · Neuroinflammation · Therapeutics

40.1 Clinical Characteristics

ALS, also known as Lou Gehrig's disease, is a relentlessly progressive disorder caused by the degeneration of motor neurons in the brain and spinal cord leading to widespread paralysis and death typically due to respiratory failure. Following the onset of symptoms, the mean survival time is between just 3 and 5 years (Brown & Al-Chalabi, 2017). Disease onset is insidious occurring over months with the mean age of onset in the middle-to-late 50s. However, there is wide phenotypic heterogeneity in ALS. The first symptoms can present during the late teenage years or in the nineties. Further, the disease duration can be as short as 3 months or as long as 50 years. The mean survival of younger individuals is higher than older individuals. Symptom onset usually occurs as focal weakness in one of the limbs, but approximately one-third of the cases start as bulbar weakness manifesting as slurred speech or trouble swallowing. Less commonly, the initial symptom is difficulty breathing or neck weakness. Motor neurons that innervate the eye and sphincter muscles are typically spared until very late in the disease course. Currently, there are only three FDA-approved medications for ALS, each showing only a modest slowing of disease progression.

J. R. Thonhoff · E. P. Greene · S. H. Appel (✉)
Houston Methodist Neurological Institute, Houston, TX, USA
e-mail: sappel@houstonmethodist.org

H. E. Gendelman, T. Ikezu (eds.), *Neuroimmune Pharmacology and Therapeutics*,
https://doi.org/10.1007/978-3-031-68237-7_40

Upper motor neurons are located in the motor cortex and make up the corticospinal tract. Lower motor neurons arise from the brainstem and spinal cord and innervate muscles. Degeneration of upper motor neurons leads to increased tone and hyperreflexia. The earliest sign of lower motor neuron dysfunction is spontaneous electrical activity caused by denervation/reinnervation of the motor neuron with the muscle fiber leading to muscle twitching or fasciculations. Atrophy and weakness ensue as lower motor neurons degenerate and retract from their muscle targets. The clinical hallmark of ALS is the presence of upper motor neuron and lower motor neuron disease in the brainstem and spinal regions (Rowland & Shneider, 2001). Primary lateral sclerosis (PLS) results from selective upper motor neuron degeneration whereas progressive muscular atrophy (PMA) results when there is only involvement of lower motor neurons (Fig. 40.1). The prognosis for patients with PLS or PMA is much better compared to ALS as these conditions progress much more slowly and have improved survival times (Gordon et al., 2006; Brown & Al-Chalabi, 2017).

The diagnosis of ALS relies on the history and physical examination in combination with electromyography (EMG) to confirm lower motor neuron dysfunction, and neuroimaging and laboratory testing to rule out conditions that may mimic ALS (Al-Chalabi et al., 2016). A diagnosis of definite ALS is made when there is evidence of lower motor neuron and upper motor neuron compromise in at least three areas, including the bulbar, cervical, thoracic, and lumbar regions. Needle EMG and nerve conduction studies of an upper and lower limb, thoracic paraspinous muscles, and tongue are performed to confirm the extent of lower motor neuron dysfunction. In addition, nerve conduction studies should show no evidence of conduction block indicating an ALS mimic called multifocal motor neuropathy, which is a reversible autoimmune condition that responds to treatment with intravenous immune globulin (Arcila-Londono & Lewis, 2013). Imaging of the brain and cervical spinal cord is obtained to rule out lesions of the corticospinal tract. The differential diagnoses for ALS also include other inflammatory neuropathies such as CIDP, thyroid dysfunction, infections such as West Nile Virus and Lyme disease, vitamin B_{12} deficiency, and metal toxicity, all of which can be evaluated and diagnosed through history, physical examinations, and laboratory testing (Rezania & Roos, 2013).

The clinical presentations of ALS are remarkably diverse. As such, it is not yet possible to predict the clinical course of a single patient. Symptoms and signs may start with complaints of arm or hand weakness, wasting, or functional impairment such as difficulty writing, buttoning, or holding onto objects, or leg weakness with foot drop and stumbling, tripping, and falling. The upper motor neuron compromise is often accompanied by loss of balance and increased falling. Onset is usually asymmetrical, progressing from one leg to the ipsilateral arm or the contralateral leg, and then proceeding rostrally. In patients with bulbar involvement, symptoms of hoarseness, slurred speech, and drooling often precede choking or swallowing problems. Generally, the initial onset of bulbar symptoms, in association with signs of tongue atrophy and fibrillations, carries a poorer prognosis compared to those who initially develop symptoms in the limbs. Degeneration of the frontopontine motor neurons that provide bulbar functions may also result in symptoms of emotional lability or pseudobulbar affect (Floeter et al., 2014). Pseudobulbar affect is characterized by spontaneous epi-

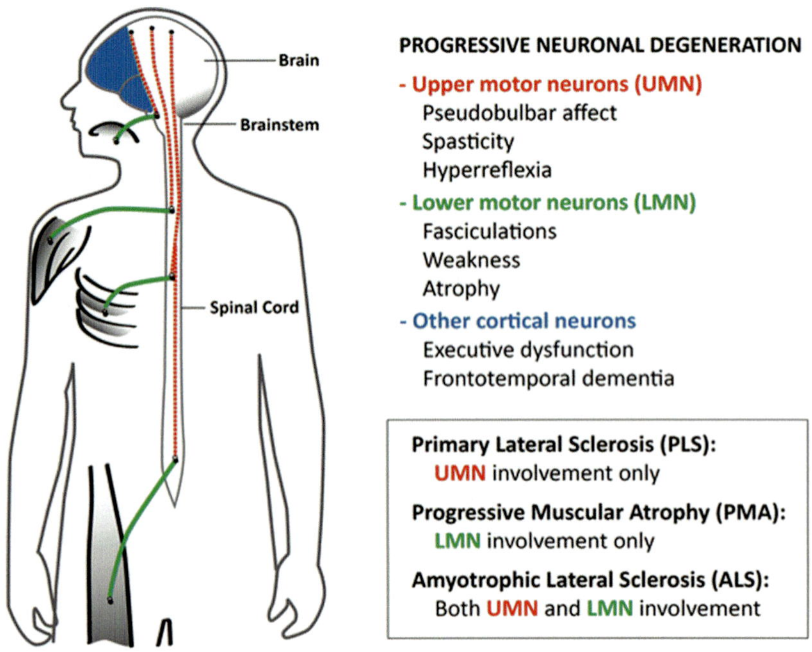

Fig. 40.1 Neuroanatomy of amyotrophic lateral sclerosis

sodes of excessive and uncontrollable laughing or crying out of proportion to the emotional stimuli encountered. This symptom can be effectively treated with Nuedexta®, which decreases the number and severity of these outbursts (Pioro, 2014).

Clinically, ALS may be misdiagnosed as painless cervical radiculopathy and myelopathy when initial signs and symptoms are restricted to a single limb or adjacent root accompanied by hyperreflexia in the legs and before EMG studies are performed. Ruling out significant cervical spinal cord compression with neuroimaging is mandatory. However, many patients may have the presence of both ALS and suggestive cervical cord compromise. By recognizing the common asymmetrical presentation of ALS as well as the slow but relentless progression of atrophy and weakness, in addition to EMG evidence of lower motor neuron involvement, these patients could avoid surgery that could aggravate the clinical syndrome. An additional dilemma in the diagnosis of ALS is the patient who presents with progressive atrophy and weakness and exhibits EMG findings characteristic of lower motor neuron involvement, yet with reflexes that are slightly decreased. The absence of hyperreflexia suggests a lower motor neuron syndrome such as progressive muscular atrophy. However, in this clinical context, corticospinal tract involvement cannot be simply ruled out with a reflex hammer. Neuroimaging modalities using PET imaging to identify signs of neuroinflammation from microglial activation within the motor cortex is currently under development and could prove to be a meaningful surrogate marker of upper motor neuron degeneration to help resolve this diagnostic dilemma (Alshikho et al., 2018; Ratai et al., 2018).

40.2 Frontotemporal Dementia

We will discuss frontotemporal dementia (FTD) not only as a clinical syndrome but also as genetically or pathologically confirmed, although frontotemporal lobar degeneration (FTLD) is a term often used for patients with genetically or pathologically confirmed diagnoses. Clinically, FTD patients exhibit progressive changes in behavior, language, and executive control. In FTD, the TAR DNA-binding protein of 43 kDa (TDP-43), fused in sarcoma/translated in liposarcoma (FUS/TLS) and tau are three major causative proteins. TDP-43 and FUS are RNA-binding proteins that are associated with similar clinicopathological phenotypes. Approximately 15 to 20% of people with ALS develop behavior and language alterations that are diagnostic of FTD. However, milder forms of frontotemporal lobe involvement with compromised executive function may be present in up to 50% of ALS patients (Ringholz et al., 2005). From the perspective of FTD, up to 40% of FTD cases have measurable motor neuron dysfunction with 15% of cases developing outright ALS, supporting the notion that ALS and FTD represent an overlapping spectrum of disease (Bennion Callister & Pickering-Brown, 2014; Boeve et al., 2022). Patients with ALS can have normal executive function (50%), frontal executive dysfunction (50%), or severe FTD (15–20%). The presence and severity of frontal lobe dysfunction also contributes to the clinical heterogeneity of ALS. Patients who develop ALS-FTD typically have onset of symptoms in their sixth decade. Similarly to ALS, it is slightly more common in men than women. Symptoms of ALS may precede, co-occur or follow the symptoms of FTD, although commonly the cognitive changes occur first followed by weakness. The interval between the cognitive symptoms and motor weakness may be a few months or up to 7 years, with a mean of 2 years (Achi & Rudnicki, 2012).

FTD is the most common cause of dementia under the age of 60, and the third most common cause of dementia over the age of 65 behind Alzheimer's disease and Lewy body dementia (Knopman & Roberts, 2011; Lambert et al., 2014). It has three clinical variants: behavioral variant FTD (bvFTD), language dysfunction of semantic dementia (SD) and progressive nonfluent aphasia (PNFA), and executive dysfunction (Bang et al., 2015; Boeve et al., 2022). In patients with bvFTD, apathy, disinhibition and changes in personality are often the first symptoms reported. These patients exhibit a lack of empathy and are unaware of the emotions of others. Some patients may start to exhibit stereotyped and/or compulsive behaviors. They may be more withdrawn and self-centered, and lack concern for their personal appearance. They may develop impulsive or socially inappropriate behavior. Their reasoning, judgment and organizational skills become markedly diminished, and their personal hygiene and overall care represent significant problems for their caregivers. These patients typically have decreased insight into their own behavioral changes. FTD patients with SD present with impairment of naming and word comprehension. Speech output remains fluent and effortless. Patients with PNFA present with effortful and hesitant speech production as well as phonologic and grammatical errors and difficulties in word retrieval, reading and writing. Patients with executive function deficits have compromised planning, reasoning and problem solving. As the degeneration spreads to include vast areas in the frontal and temporal lobes, symptoms of the three clinical variants may converge. The median survival time after symptom onset is about 6 years and death usually occurs due to complications from secondary infections (Hodges et al., 2003; Bang et al., 2015). During the course of progression, patients develop global cognitive impairment as well as the loss of motor skills. In addition to overlapping clinical syndromes with ALS, there is overlap with parkinsonism, and patients will develop profound difficulty moving and swallowing over time.

40.2.1 Genetics of FTD and Overlap with ALS

Familial FTD is present in 10–20% of FTD cases. Mutations in three genes represent 60% of familial FTD cases; *C9orf72*, *MAPT* (encoding the tau protein) and *GRN* (encoding progranulin). TDP-43, FUS and tau are three major pathogenetic proteins. *C9orf72* is the most common cause of both familial FTD and familial ALS, and provides the most compelling evidence for overlap of ALS and FTD (Abramzon et al., 2020). *C9orf72* is present in 40% of patients with familial ALS and 25% of patients with familial FTD (Van Mossevelde et al., 2018). Whether an asymptomatic person possessing the *C9orf72* mutation will develop FTD or ALS or a combination cannot be predicted. Three specific mechanisms have been implicated in the pathophysiology of neurological disease in *C9orf72* cases; haploinsufficiency, RNA toxicity and dipeptide repeat protein toxicity (Taylor et al., 2016). There is substantial evidence supporting all three mechanisms, but no compelling evidence for any one of these alone, perhaps suggesting that several of these mechanisms together with other mechanisms may promote the pathophysiology of ALS/FTD.

In both FTD and ALS, pathological protein accumulation is noted in neurons and glia. The microtubule-associated protein tau is hyperphosphorylated in about 40–50% of all patients with FTD. Tau pathology can be detected in both FTD and ALS regardless of the presence or absence of *MAPT* mutations. However, *MAPT* mutations are not present in ALS. Nevertheless, the presence of tau pathology in both FTD and ALS strengthens the concept of pathological overlap. TDP-43 cytoplasmic and nuclear inclusions are characteristic of both ALS and FTD. In ALS, TDP-43 accumulates in the cytoplasm in neurons and glia of the primary motor cortex, brainstem nuceli and spinal cord (Abramzon et al., 2020). Among FTD cases, 90% of the different syndromes show either TDP-43 proteinopathy (50%) or tauopathy (40%). In ALS, the presence of TDP-43 proteinopathy is a significant pathological hallmark that is present in over 90% of patients. Thus, TDP-43 pathology seems to be consistently associated with 50% of FTD cases and with the overwhelming majority of sporadic amyotrophic lateral sclerosis (sALS) and familiar amyotrophic lateral sclerosis (fALS) cases. However, TDP-43 pathology is not present in ALS patients with *SOD1* mutations. The lack of a specific correlation between *TARDBP* genetic variants (*TARDBP*) and disease suggests that TDP-43 cytoplasmic accumulation may be a consequence rather than a cause of FTD and ALS, and the loss of TDP-43 in the nucleus may play a significant pathophysiological role.

MAPT mutations usually cause bvFTD, but can also cause other neurodegenerative diseases including progressive supranuclear palsy or corticobasal degeneration. *MAPT* mutations do not cause ALS. *GRN* mutations usually cause bvFTD, but can also cause primary progressive aphasia (PPA) and rarely cause ALS. Up to 20% of familial FTD cases are caused by progranulin (*GRN*) haploinsufficiency (FTD-*GRN*). Mutant *FUS* also can result in both ALS and FTD by compromising neuronal splicing mechanisms. Rare genes (<5%) associated with FTD and ALS include the following; *TARDBP* mutations (encoding TDP-43) cause FTD, ALS or a combination of both; *SQSTM1* mutations (encoding a protein called p62) cause FTD, ALS or a combination of both. *UBQLN1* mutations (encoding ubiquilin 1) cause FTD, ALS or a combination of both. *TBK1* mutations (encoding TANK binding kinase 1) can cause FTD, ALS or a combination of both. Rare genes associated only with FTD include *VCP* (encoding Valosin-containing protein) and *CHMP2B* (encoding Charged Multivesicular Body Protein 2B) where a mutation may present with bvFTD (Liscic et al., 2020). The genes associated with ALS, FTD or both are shown in Table 40.1.

Table 40.1 Genes associated with ALS, FTD or both. The most common gene mutations are in bold

Disease	#	Mutant Genes
ALS	44	*ALS2, ALS7, ANG, ANXA11, ATXN2,* ***C9orf72* (40%)**, *C21orf2, CCNF, CHCHD10, CHMP2B, CSF1R, DAO, DCTN1, ERBB4, FIG4,* ***FUS* (5%)**, *HNRNPA1, HNRNPA2B1, KIF5A, MATR3, NEFH, NEK1, OPTN, PFN1, PON1, PON2, PON3, PRPH, SETX, SIGMAR1,* ***SOD1* (20%)**, *SPG11, SQSTM1, TAF15,* ***TARDBP* (5%)**, *TBK1, TIA1, TREM2, TRPM7, TUBA4A, TYROBP, UBQLN2, VAPB, VCP*
FTD	14	***C9orf72* (25%)**, *CHCHD10, CHMP2B, FUS,* ***GRN* (5–20%)**, ***MAPT* (5–20%)**, *OPTN, SQSTM1, TARDBP, TBK1, TMEM106B, TREM2, UBQLN2, VCP*
ALS/FTD	11	***C9orf72***, *CHCHD10, CHMP2B, FUS, OPTN, SQSTM1, TARDBP, TBK1, TREM2, UBQLN2, VCP*

Modified from Liscic et al. (2020)

40.3 Epidemiology

The incidence of ALS is 1–2 cases/100,000 people and the prevalence is approximately 3–5/100,000 (Chio et al., 2013; Robberecht & Philips, 2013). Similar to other neurodegenerative diseases, the incidence and prevalence of ALS increases with age. The cumulative life time risk of developing ALS in the US and Europe is about 1 in 400 (Johnston et al., 2006). Approximately 90% of ALS cases are sporadic (sALS) meaning that there is no family history, while about 10% are familial (fALS) and usually inherited in a dominant manner (Brown & Al-Chalabi, 2017). Overall, males are more likely to be affected by ALS than females in sALS by a ratio approaching 2:1 whereas males and females are equally affected in fALS. The younger the onset of ALS, the higher the propensity to affect males as the ratio may exceed 3:1

males to females in the third and fourth decades. In the population over the age of 65, the ratio approaches 1:1.

40.4 Pathology

The pathological hallmarks of ALS include the loss of upper and lower motor neurons, surrounded by astrogliosis and microgliosis. In ALS-FTD, neuronal degeneration is also observed in the frontotemporal cortical neurons. Sclerosis or scarring of the lateral spinal cord is observed due to degeneration of the corticospinal tracts. The loss of lower motor neurons leads to thinning of the ventral roots and amyotrophy of the tongue, oropharynx, and limb muscles due to denervation (Taylor et al., 2016; Brown & Al-Chalabi, 2017). Intracellular inclusions are also present in degenerating neurons and glia. Bunina bodies are eosinophilic Cystatin C+ intraneuronal inclusions found in ALS (Okamoto et al., 2008). Inclusions containing a variety of aggregated cytoplasmic proteins are also found in motor neurons. As ALS has numerous underlying causes, the pathology of inclusions in sALS and fALS is quite diverse. The hallmark inclusion is ubiquitinated TDP-43, which mislocalizes and aggregates in the cytoplasm (Neumann et al., 2006). Aggregated deposits containing ubiquilin-2 or wild-type superoxide dismutase 1 (SOD1) are also commonly found in ALS (Bosco et al., 2010; Deng et al., 2011). Inclusions with neurofilament are seen in fALS due to mutations in *SOD1* (Rouleau et al., 1996). Fused in sarcoma (FUS)-positive inclusions have been identified in sALS and fALS, but not fALS due to mutations in *SOD1* (Deng et al., 2010). Dipeptide aggregates and intranuclear RNA deposits are uniquely found in fALS due to mutations in *C9orf72* (Freibaum & Taylor, 2017).

40.5 Genetics

Advances in next generation gene sequencing techniques and genome wide association studies over the last two decades have led to the discovery of numerous disease-causing genes. More than 40 mutant genes have now been implicated in fALS. In 1993, a mutation in the gene encoding copper-zinc superoxide dismutase 1 (SOD1) was the first ALS gene discovered and represents approximately 20% of fALS cases (Rosen et al., 1993). The most common gene mutation in fALS is a hexanucleotide repeat (GGGGCC) in *C9orf72* representing about 40% of fALS (Dejesus-Hernandez et al., 2011; Renton et al., 2011). Expansions in *C9orf72* have also been reported in 5–10% of sALS patients without a family history of ALS. These genetic discoveries have dramatically accelerated the elucidation of pathogenic pathways relevant in ALS and enhanced approaches to therapeutic development through the generation of transgenic mouse models of ALS based on these genetic mutations. Several genetic mutations have been shown to initiate changes in critical motor neuron pathways that lead to an ALS clinical phenotype. Transgenic expression of mutant *SOD1* recapitulates many aspects of human ALS including the neurodenerative and paralytic processes (Gurney et al., 1994). SOD1 is a ubiquitously expressed protein that catalyses the detoxification of superoxide. More than 160 mutations in *SOD1* have been described, all of which result in SOD1 protein misfolding and a toxic gain-of-function that leads to impaired mitochondrial function, impaired axonal transport and degeneration, impaired proteasomal function, and endoplasmic reticulum stress. The cytotoxic effects of aggregated SOD1 are not confined to mutant *SOD1*, but are also noted with misfolded wild-type SOD1, which has been identified in motor neurons of sALS patients and could contribute to disease pathogenesis in sporadic cases.

Transgenic models based on mutations in *TARDBP* and *FUS* have contributed to new insights regarding the tight regulation of protein levels in ALS (Mitchell et al., 2015). TDP-43 inclusions are found in the brain and spinal cord in human ALS autopsies. TDP-43 is ubiquitously expressed as a DNA//RNA-binding protein involved in transcription, and RNA splicing and transport. In fALS, mutant TDP-43 is cleaved and aggregated and is mislocalized from the nucleus to the cytoplasm impairing nuclear function (Kabashi et al., 2008; Sreedharan et al., 2008). Mutations in FUS-TLS result in cytoplasmic aggregates, and impaired nuclear function and transport, similar to the functional changes noted with mutations in TDP-43 (Kwiatkowski Jr. et al., 2009; Vance et al., 2009). Mouse models based on mutant *C9orf72* have also been developed and recapitulate many of the molecular and pathological features found in human ALS, but the mice do not develop a prominent paralytic process (O'rourke et al., 2015; Peters et al., 2015). The C9orf72 protein is important in nuclear and endosomal membrane trafficking and autophagy. Hexanucleotide repeat expansions to the 100s to 1000s in *C9orf72*, which normally contains up to about 30 repeats, have been shown to contribute to the pathogenesis of ALS (familial and sporadic) as well as FTD through three distinct mechanisms; haploinsufficiency due to impaired *C9orf72* transcription; the abnormal formation of RNA foci in the nucleus that sequester nuclear proteins; and the generation of toxic repeat dipeptides through noncanonical translation (Taylor et al., 2016). A common denominator in many of the pathologies caused by these diverse genetic mutations is the presence of misfolded aggregated proteins.

In cases involving FTD, there is an autosomal dominant inheritance pattern in approximately 10% of cases. As noted previously, mutations in *C9orf72* make up about 25% of all familial cases of FTD. Mutations in genes encoding microtubule-associated protein tau (*MAPT*) and progranulin (*GRN*) each represent about 5–20% of FTD cases. A mutation in *C9orf72* is the most common cause of the ALS-FTD clinical syndrome. *TARDBP* and *FUS* mutations are also associated with a low percentage of ALS-FTD cases. Mutations in *MAPT* cause microtubule destabilization and an increased propensity of tau to self-aggregate. A haploinsufficiency mechanism is suspected in *GRN* mutations leading to a loss of functional progranulin. Progranulin is involved in regulating the cell cycle, modifying inflammation by competitively binding the tumor necrosis factor (TNF) receptor and axonal growth (Bang et al., 2015).

The diverse genetic mutations alone do not fully explain the phenotypic heterogeneity observed in ALS. Although each distinctive gene mutation may result in a specific defect in a critical motor neuron pathway, it is not the case that each specific gene mutation results in a specific clinical phenotype. The same genetic mutations can give rise to different clinical phenotypes, and conversely, different gene mutations can give rise to the same clinical phenotype. Phenotypic heterogeneity is noted even within affected members of a family with the same mutation. Further, in different families, the same genetic mutation can give rise to multiple clinical phenotypes. Thus, the clinical picture alone cannot classify the patients as having either fALS or sALS. Familial ALS and sALS can only be differentiated by taking a careful family history and performing genetic testing.

One longstanding perspective to explain the phenotypic heterogeneity in ALS has been that the disease develops through a complex interplay between genetic predisposition and environment factors. The strongest environmental exposure that has been associated with ALS is service in the military (Weisskopf et al., 2005; Beard et al., 2016). The activation of the human endogenous retrovirus K may also play a role in some cases of ALS (Li et al., 2015). As many attempts to identify environmental exposures that may increase the risk of ALS have been inconclusive, an alternative viewpoint could be that all ALS cases arise from a complex interplay of genetic factors, both risk genes and protective genes. The phenotypic heterogeneity could be explained by epistasis, the interactions of multiple genes, and epigenetic modifications including DNA methylation, histone acetylation, chromatin remodeling, and non-coding RNA regulation. Sporadic ALS may have an oligogenic inheritance that determines the timing of disease development and progression. Different combinations of gene variants posing a risk for ALS may be the cause of sporadic ALS. Nonetheless, the interactions between several genes could certainly influence many disease parameters such as penetrance, onset, progression, and survival in ALS.

40.6 Pathogenesis

The pathogenesis of ALS proceeds through multifactorial processes. In fALS, different genetic mutations initiate the disease by interfering with different vital motor neuron pathways. For example, mutations in *SOD1* may initially compromise mitochondria function, increase the generation of free radicals, alter calcium homeostasis, compromise axonal function, alter proteasomal function, autophagy, and RNA metabolic pathways, and lead to endoplasmic reticulum stress. On the other hand, mutations in *TARDBP* or *FUS* may initially compromise transcription, RNA splicing, and transport, and subsequently impair multiple diverse pathways within the motor neuron. Sporadic ALS is presumed to be caused by a complex interplay between several genetic and environmental factors. However, fALS and sALS have striking similarities in their pathological and clinical characteristics indicating that there may be a convergence of dysfunctional metabolic and cellular events in all types of ALS that ultimately result in motor neuron degeneration. This convergence of destructive metabolic pathways within the motor neuron may provide one reason why diverse mutations do not solely explain phenotypic heterogeneity and can give rise to similar clinical syndromes.

Based on the known ALS genes, their gene functions can be divided into three groups of genes that affect protein homeostasis, RNA homeostasis, and trafficking, or cytoskeletal proteins and axonal transport (Brown & Al-Chalabi, 2017). However, the destructive downstream mechanisms from these gene mutations are not mutually exclusive to each group and eventually lead to the impairment of many intracellular processes. In both fALS and sALS, the intracellular alterations that cause motor neuron injury are extremely diverse. It is unlikely that any one particular alteration results in motor neuron injury, but rather, an accumulation of insults to numerous intracellular pathways. Collectively, downstream alterations in diverse cellular pathways that contribute to disease pathogenesis have been documented in ALS including altered neuronal excitability, excitotoxicity, mitochondrial dysfunction, oxidative stress, diminished energy supply, defective axonal transport, altered RNA metabolism, splicing and transport, overload of the proteasome system, reduced autophagy, and endoplasmic reticulum stress (Taylor et al., 2016).

One inciting destructive mechanism may include the aggregation of misfolded proteins, both mutated proteins and wild-type proteins such as SOD1 and TDP-43. These aggregated proteins may sequester and decrease the functional levels of other proteins critical to RNA homeostasis and trafficking (Taylor et al., 2016). The accumulation of SOD1 in the mitochondria can disrupt the normal functioning of mitochondria and consequently lead to an increased generation of reactive oxygen species (ROS). Glutamate excitotoxicity and increased intracellular calcium have also been implicated in ALS pathogenesis. Increased glutamate in the synaptic cleft leads to increased firing of motor neurons and resultant increased levels of intracellular calcium. In addition, levels of calbindin and parvalbumin are normally decreased to absent in susceptible motor neurons, which further contributes to increased intracellular calcium levels and excitotoxicity (Appel et al., 2001). Mitochondrial dysfunction leading to increased free radical generation has been well-documented in ALS. Altered calcium homeostasis, free radical stress, and glutamate excitotoxicity may all participate in a cell injury cascade leading to motor neuron death. In fact, alterations in one of these parameters can lead to alterations in the other parameters, and each can enhance and propagate the injury cascade. Increased intracellular calcium can enhance free radical production and glutamate release, which in turn can further increase intracellular calcium. Increased free radicals can impair the glial uptake of glutamate by decreasing the expression of the excitatory amino acid transporter 2 (EAAT2) in astroglia, thus increasing the synaptic glutamate available to interact with neuronal AMPA/kainate receptors, which in turn causes excessive calcium influx and ROS production. Free radicals in turn can enhance lipid peroxidation leading to increased intracellular calcium. These changes of increased intracellular calcium, increased generation of free radicals, and enhanced glutamate excitotoxicity could critically impair intracellular structures such as mitochondria or neurofilaments, and compromise energy production and axoplasmic flow. Once initiated, the defective changes could become self-propagating.

The initiation of the disease involves motor neuron injury due to the accumulation of insults to multiple metabolic pathways within the motor neuron. However, transgenic ALS animal models based on mutations in *SOD1* have shown that ALS is not a cell-autonomous disease. In order for the disease to progress, dysfunction of surrounding nonneuronal cells is required (Ilieva et al., 2009). When mutant *SOD1* is expressed primarily in motor neurons, the transgenic mice do not develop motor neuron disease, or do so only very late in life . In these models, neurons do not die alone as the disease process is non-cell autonomous. Motor neuron injury caused by an accumulation of intracellular insults is an early event that may lead to the retraction of the distal axon from the neuromuscular junction, but it is not sufficient to cause severe motor neuron degeneration and death. These destructive intracellular events activate and recruit nonneuronal cells within the CNS including astroglia, oligodendroglia, innate immune microglia, and adaptive immune T lymphocytes through as yet unidentified mechanisms. An injured motor neuron could possibly signal surrounding microglia through the release of misfolded aggregrated proteins such as mutant SOD1 or TDP-43 (Zhao et al., 2010, 2015). Dysfunction of surrounding nonneuronal cells results in neuroinflammation, which promotes progression of disease and its acceleration, but does not initiate disease.

40.7 Neuroinflammation

The presence of activated innate (microglia) and adaptive immune cells (T lymphocytes) within the brain and spinal cord is characteristic of neuroinflammation. Neuroinflammation was originally thought to be a secondary effect or consequence of neuronal injury, but is now recognized to contribute significantly to motor neuron pathophysiology and disease propagation. The most cogent evidence for this was derived from transgenic mouse models of ALS in which central inflammatory responses involving microglia and astroglia as well as peripheral inflammatory responses involving monocytes and lymphocytes contribute to rates of disease progression. During the slow phase of the disease, shortly after the initial injuries to the motor neuron, neuroprotective anti-inflammatory mediators are present that attempt to stabilize the disease. However, the slow phase of the disease eventually converts to a fast phase, which is characterized by a reduction in the neuroprotective mediators and an increase in cytotoxic proinflammatory mediators (Beers et al., 2011a, b). The early stages of the disease are marked by anti-inflammatory microglia and suppressive regulatory T cells (Tregs) whereas proinflammatory microglia and Th1 cells are upregulated during the rapid phase. In the animal models, the conversion from slow phase to the fast phase occurs at a predictable time, but in humans, the conversion point varies from months to many years in individual patients, thus contributing to the phenotypic heterogeneity. The conversion point is likely determined by the volume of intracellular injuries to the motor neuron at which point the crosstalk signaling between motor neurons and nonneuronal inflammatory cells changes from those of providing support to signals that accelerate the demise and removal of the injured motor neuron.

Microglia isolated from the early and late states of the transgenic ALS mouse spinal cords exhibit anti-inflammatory and proinflammatory properties, respectively. When cultured with motor neurons in vitro, the early stage microglia are protective whereas the late stage microglia contribute to motor neuron injury (Liao et al., 2012). Additionally, transgenic mice were produced by crossing mutant *SOD1* mice with PU.1 knockout mice and transplanted with either mutant *SOD1* or wild-type bone marrows. The mice transplanted with mutant *SOD1* bone marrows were populated with CNS microglia that were proinflammatory, and they showed reduced motor neuron viability and decreased survival compared to the mice that received wild-type bone marrows (Beers et al., 2006). These results provided evidence for the neuroprotective nature of early anti-inflammatory microglia and the neurotoxic properties of later stage proinflammatory microglia.

Cumulative evidence now also suggests that ALS is a systemic disease, given the presence of peripheral proinflammatory macrophages, dysfunctional regulatory T lymphocytes, and increased proinflammatory cytokines, chemokines, and markers of oxidative stress as well as increased acute phase proteins (APPs) (Keizman et al., 2009; Lu et al., 2016; Won et al., 2016; Lunetta et al., 2017; Thonhoff et al., 2018b; Beers et al., 2020, 2021, 2022; Appel et al., 2021). Monocytes isolated from the peripheral blood show a proinflammatory phenotype compared to healthy controls. Additionally, the proinflammatory nature of the monocytes intensifies as the disease progresses (Zhao et al., 2017). These monocytes/macrophages can readily enter the CNS, and the crosstalk signaling between peripheral monocytes and central microglia could enhance the neurotoxic inflammation in ALS. Later in the disease, in both ALS transgenic models and patients, the blood-brain-barrier (BBB) may also become compromised, which could permit further migration of proinflammatory cells from the periphery into areas of motor neuron degeneration and exacerbate motor neuron demise (Garbuzova-Davis & Sanberg, 2014). Additional evidence for ALS as a systemic disease is the elevated levels of APPs, which are synthesized in the liver following signaling from the serum proinflammatory cytokines released by activated macrophages. The APPs are increased compared to healthy controls and are not increased in other neurodegenerative diseases (Beers et al., 2020).

The motor neuron projects outside the BBB to the muscle, and the neuromuscular junction may be one of the early sites of pathology, initiating a "dying back" of the motor neuron from the neuromuscular junction. Biopsies of the neuromuscular junction in ALS documented the presence in the motor axon terminal of calcium-containing swollen mitochondria, increased density of synaptic vesicles, increased active-zone vesicle density, and increased intraluminal calcium precipitates within membranous organelles. These changes were not present in either denervating or non-neuropathic controls (Siklos et al., 1996). The dramatically altered mitochondria together with decreased glutathione and increased iron, which are present in ALS patients, suggest the process of ferroptosis, a form of programmed cell death driven by lipid peroxidation (Wang et al., 2022). Lipid peroxides such as 4-hydroxynonenal (4-HNE) are elevated in ALS, and could further attract circulating innate immune cells to the neuromuscular junction (Smith et al., 1998; Simpson et al., 2004). ALS is unique compared to other neurodegenerative diseases in that the initial injuries and deleterious events resulting in clinical symptoms occur outside of the BBB at the neuromuscular junction. Thus, monocytes and macrophages can surround the degenerating distal motor neuron axons. These infiltrating cells were documented in the mutant *SOD1* transgenic mouse model after the act of denervation had occurred indicating that their presence is a consequence of denervation (Kano et al., 2012). The major purpose of these infiltrating cells into the injured peripheral axon is to remove the axonal debris to permit regrowth, but in this process, infiltration of these cells could enhance systemic inflammation that could then further exacerbate neurotoxic inflammation in the CNS through crosstalk signaling mechanisms.

A critical contribution for T lymphocytes in the pathogenesis of ALS was discovered when transgenic mutant *SOD1* mice were crossed with *RAG2* knockout mice, which lacked mature T and B cells, or with *CD4* knockout mice that lacked $CD4^+$ T cells. The resultant transgenic mice from both crossing experiments showed an accelerated disease progression and reduced lifespan, which suggested that an important protective factor had been lost, namely the $CD4^+$ Treg population. This was proven in subsequent studies by the passive transfer of regulatory T lymphocytes into the tail veins of transgenic mutant *SOD1* and *RAG2* knockout mice, which slowed progression and enhanced lifespan (Beers et al., 2008, 2011a). The normal function of Tregs is to suppress proinflammatory responses and Treg signals are reduced in the later stages of the disease. Subsequent studies in transgenic mutant *SOD1* mice further confirmed that enhancing the endogenous Treg population also exerts benefits in slowing disease progression and prolonging lifespan possibly through the reduction in microgliosis and upregulation of anti-inflammatory factors (Sheean et al., 2018). Collectively, these studies provide strong evidence for the protective role of Tregs in ALS pathogenesis, especially during the early stage of the disease. Tregs likely provide beneficial effects through the suppression of both the proliferation of responder T lymphocytes and monocytes/macrophages in the periphery, thus, reducing proinflammatory signals that could potentially cross the BBB and activate CNS microglia (Zhao et al., 2012). Tregs are also found in the spinal cord in areas of motor neuron degeneration and thus, may enter the CNS and potentially directly suppress proinflammatory microglia.

Diminished suppression of both innate and adaptive proinflammatory immune responses contributes to the acceleration of disease progression. Tregs express high levels of Forkhead box protein P3 (FOXP3) (a member of the forkhead transcription factor family), which is the transcription factor that confers function and survival to Tregs, suppresses neurotoxic inflammation, and protects motor neurons in ALS. However, the numbers and function of Tregs in the peripheral blood, and the expression levels of FOXP3, are all reduced in ALS patients. The reductions in these Treg parameters are correlated with increased progression rates and disease burden and decreased survival (Henkel et al., 2013; Beers et al., 2017). The loss of Treg numbers and function promotes and enhances the proinflammatory toxic milleiu marked by activated myeloid cells in the periphery that can transmit proinflammatory signals to the CNS resulting in microglia activation. These deleterious changes in Tregs are possibly mediated through adverse interactions with proinflammatory innate and adaptive immune cells. As the disease progresses, the neuroprotective effects of Tregs cannot overcome the progressively worsening proinflammatory environment. Thus, a vicious cycle ensues resulting in deteriorating Treg function and enhanced neurotoxic inflammation that further contributes to the acceleration of motor neuron injury and death.

40.8 Therapeutic Efforts

To date, there is no markedly effective pharmacotherapy to treat patients with ALS. Three drugs have been approved by the Food and Drug Administration: riluzole, edaravone, and Relyvrio™ (a combination of two agents, sodium phenylbutyrate and taurursodiol). All three drugs provide modest clinical benefits in slowing disease progression and prolonging survival (Bensimon et al., 1994; Writing & Edaravone, 2017; Paganoni et al., 2020). Riluzole acts on sodium channels to reduce excitability and thereby limit glutamate excitotoxicty; edaravone scavenges reactive oxygen species; and Relyvrio™ exerts its protective effects by restoring mitochondria function and reducing stress to the endoplasmic reticulum. There are also currently no disease-modifying therapeutics for FTD. Treatment is centered on the management of behavioral symptoms with the use of SSRIs and atypical antipsychotics. Memantine shows no benefit in FTD (Bang et al., 2015). The current mainstay of treatment consists of a multidisciplinary approach to care and symptomatic management to prevent falls and trauma, choking episodes, and aspiration, all of which can exacerbate the progression of the disease. Timely interventions of a percutaneous endoscopic gastrostomy (PEG) placement for nutritional support when patients are having trouble swallowing and losing weight, and ventilatory support with non-invasive ventilation usually with bilevel positive airway pressure when diaphragm muscle function and forced vital capacity deteriorate are critical.

The advances in the elucidations of the complex genetic, molecular, and cellular components of ALS pathogenesis have led to an increasing number of potential therapeutic targets. A pipeline of drugs addressing some of these mechanisms has been developed and is currently being tested systematically in an ALS platform trial (Paganoni et al., 2022). Prior to this platform trial, numerous potential ALS drugs have gone through late-phase clinical trials and were shown to be ineffective (Petrov et al., 2017). The phenotypic heterogeneity in ALS has historically posed a major challenge in designing clinical trials and stratifying the ideal ALS patient populations to enter into these clinical trials. In addition, biological or imaging markers that could indicate the extent of disease and/or responses to therapies have been lacking in the ALS field in general. There have recently been profound developments in biological markers such as neurofilament light chain that indicate axonal degeneration, and peripheral makers of inflammation that include proinflammatory cytokines and markers of oxidative stress and lipid peroxides (Benatar et al., 2018; Poesen & Van Damme, 2018; Beers et al., 2020, 2022). Most therapies will likely require early intervention to obtain the maximal clinical benefit. The time from symptom onset until diagnosis is still too delayed in this modern age, being approximately 12 months, which also limits the potential success of therapeutic interventions (Brown & Al-Chalabi, 2017). Finally, it is also likely that a successful therapy for ALS will require simultaneous interventions against multiple pathways in order to provide a substantial clinical benefit. When therapeutics are tested against a single pathogenic mechanism, there are still numerous mechanisms through which the disease can progress.

Genetic therapies utilizing antisense oligonucleotide (ASO) technologies to reduce the expression of mutant genes are currently being evaluated in clinical trials in many types of fALS including *SOD1*, *C9orf72,* and *FUS* (Ly & Miller, 2018; Korobeynikov et al., 2022; Miller et al., 2022; Tran et al., 2022). In patients afflicted with these specific mutations, these therapies have the potential to inhibit the initiating event in the pathogenic cascade. There is also considerable interest in the effects of ASOs targeting *C9orf72* in the ALS-FTD syndrome, especially in the responses of the behavioral symptoms to the therapy. Other therapeutic approaches to FTD target the autosomal dominant forms of FTD as well as sporadic FTD with tau pathology. Along with ASOs and adeno-associated virus-based gene therapies, other drug candidates include inhibitors of tau aggregation and trans-neuronal spreading. Several monoclonal antibodies, tau vaccines, and microtubule-stabilizing drugs are currently under investigation (Bang et al., 2015; Boeve et al., 2022).

In regards to neuroinflammation, broad immunosuppression has been ineffective in halting the progression of ALS. This failure of immunosuppressant therapy has been cited as evidence against an immune-mediated mechanism of motor neuron injury, but other reasons may exist for the failure of immunomodulatory therapy. It is claimed that immunotherapies may not adequately access the CNS and suppress the selective inflammatory reactions of activated microglia, T cells, and/or dendritic cells. However, given the widespread systemic immune alterations in ALS, a more plausible explanation is the lack of a therapeutic effect on the dysfunctional peripheral immune system. Conventional immunotherapies target all populations of T lymphocytes, including Treg/Th2 and Th1/Th17 lymphocytes, and simultaneously suppress both neuroprotective and neurotoxic populations. Therefore, the ratio of proinflammatory Th1/Th17 lymphocytes to anti-inflammatory Treg/Th2 lymphocytes is unchanged, and the proinflammatory advantage persists. Potentially beneficial therapeutic efforts should focus on tipping the balance toward anti-inflammatory mediators (Treg/Th2), thus reversing the ratio to slow disease progression. Peripheral infusions of mesenchymal stem cells or suppressive Tregs may alter the balance toward an anti-inflammatory milieu and are currently being evaluated in clinical trials (Thonhoff et al., 2018a, 2022; Cudkowicz et al., 2022). However, one drawback to these cellular therapies is that these cells are not end-stage differentiated and as a result, an anti-inflammatory cellular therapy could be readily converted to a proinflammatory state when introduced into a proinflammatory milieu such as in ALS. Thus, the repeated dosing and timing of these cellular immune therapies will be critical to their success. Like most therapies, the earlier they are initiated, the more likely they are to be successful. The cellular doses may need to be repeated at regular intervals and potentially escalate as the disease progresses to overcome a progressively worsening proinflammatory state. Similar therapeutic strategies targeting microglia or monocyte/macrophage populations would require changing the ratio of neurotoxic (M1) to neuroprotective (M2) activity by enhancing M2 and suppressing M1 macrophages/microglia.

40.9 Summary

ALS is a complex disorder characterized by broad phenotypic heterogeneity and a wide clinical spectrum that overlaps with FTD. A vast array of genetic, molecular, and cellular mechanisms contribute to the pathogenesis of this disorder, which could contribute to the clinical diversity. Advances in next-generation sequencing and genomewide mapping have led to a rapid rate in the discoveries of gene-causing mutations, with currently more than 40 known causative genes. Some of these genes have been implicated in both fALS and sALS, and several of the genes are associated with the ALS-FTD clinical syndrome, most prominently mutant *C9orf72*. Individual ALS patients may have unique underlying genetic initiators of the disease. These inciting factors lead to an accumulation of insults to critical intracellular pathways within the motor neuron. These pathogenic pathways may converge into common pathways implicating many dysfunctional processes that cause injury to the motor neuron and the resultant initial retraction of the motor neuron from the neuromuscular junction. The aggregation of misfolded proteins is encountered in most ALS cases and leads to the disruption of many intracellular processes. In this manner, diverse gene mutations can lead to the same clinical phenotype. Although genetic factors and the accumulation of intracellular insults are responsible for the initiation of the disease, the progression of the disease requires the dysfunction of surrounding nonneuronal cells. Neuroinflammation is critical in propagating the disease.

ALS is a systemic disease resulting in a major contribution to the pathogenesis of ALS from alterations at the motor axon terminal of the neuromuscular junction outside the BBB. Dysfunction of mitochondria in the motor axon terminal, associated with calcium and iron dyshomeostasis and reduced neuroprotective glutathione, promotes programmed cell death of ferroptosis and lipid peroxide formation. This motor neuron terminal pathology can activate circulating macrophages and T lymphocytes, thereby enhancing the proinflammatory milieu. The initial response to the motor neuron injuries is neuroprotective. However, as the intracellular injuries accumulate from the "dying back" pathology, neuroprotective signals are replaced by neurotoxic signals triggering monocyte/macrophage activation in the periphery and microglial activation in the CNS. The peripheral systemic inflammatory milieu is communicated to the CNS and is toxic to motor neurons and results in self-propagating neurotoxicity. Thus, activated innate macrophages and adaptive immune cells in the periphery, and activated microglia and astroglia within the CNS, the pathological hallmarks of neuroinflammation, do not simply represent a reactive response to motor neuron degeneration, but rather, influence the rate of disease progression through their active maintenance of the balance between neuroprotection and neurotoxicity.

As our understanding of the multitude of genetic factors, defective intracellular pathways, and extracellular neuroinflammatory processes improves, additional potential therapeutic targets come into focus. Genetic therapies utilizing ASOs have the potential to drastically reduce the expression of mutant genes in fALS as well as some forms of sALS, thus curtailing the initiation of motor neuron injury by averting the production and aggregation of misfolded proteins. Cellular and pharmacologic therapeutic efforts aimed to enhance the neuroprotective inflammatory and immune mediators could maintain the slow phase of the disease for a prolonged period of time, thus preventing the conversion to a rapidly progressive state and improving survival time. Once the neurotoxic milieu is successfully converted into an environment that promotes motor neuron survival, therapeutic strategies aimed at motor neuron regeneration and/or replacement and reinnervation of muscle targets could be explored. A combination therapy addressing multiple factors and pathways that contribute to disease pathogenesis will be needed to halt this relentlessly progressive and devastating disease and to provide hope for improvement for all people with ALS.

40.10 Review Questions

1. Clinical heterogeneity in ALS is readily explained by different specific gene disease-causing mutations.
   (a) True
   (b) *False*
2. A single mutant gene can cause multiple clinical phenotypes, and many gene mutations can cause the same clinical phenotype.
   (a) *True*
   (b) False
3. The most common genetic cause of the ALS-FTD clinical syndrome is
   (a) mutant SOD1
   (b) *mutant C9orf72*
   (c) mutant FUS
   (d) mutant TDP-43
4. Neuroinflammation in ALS is characterized by
   (a) Activated microglia
   (b) Activated peripheral monocytes and macrophages
   (c) Altered circulating regulatory T cells
   (d) *All of the above*
   (e) None of the above
5. The most common genetic causes of FTD are mutations in
   (a) *C9orf72, MAPT and GRN*
   (b) SOD1 and ATXN2
   (c) TDP-43 and FUS
   (d) KIF5A and ANG
6. ALS and/or FTD caused by mutations in the *C9orf72* have the following proposed mechanisms of action
   (a) Haploinsufficiency
   (b) RNA toxicity
   (c) Dipeptide repeat toxicity
   (d) *All of the above*
   (e) None of the above

Acknowledgments We are grateful to the Muscular Dystrophy Association, the ALS Association, and Peggy and Gary Edwards for their support of our ALS clinical and research programs.

References

Abramzon, Y. A., Fratta, P., Traynor, B. J., & Chia, R. (2020). The overlapping genetics of amyotrophic lateral sclerosis and frontotemporal dementia. *Frontiers in Neuroscience, 14*, 42.

Achi, E. Y., & Rudnicki, S. A. (2012). ALS and frontotemporal dysfunction: A review. *Neurology Research International, 2012*, 806306.

Al-Chalabi, A., Hardiman, O., Kiernan, M. C., Chio, A., Rix-Brooks, B., & Van Den Berg, L. H. (2016). Amyotrophic lateral sclerosis: Moving towards a new classification system. *Lancet Neurology, 15*, 1182–1194.

Alshikho, M. J., Zurcher, N. R., Loggia, M. L., Cernasov, P., Reynolds, B., Pijanowski, O., Chonde, D. B., Izquierdo Garcia, D., Mainero, C., Catana, C., Chan, J., Babu, S., Paganoni, S., Hooker, J. M., & Atassi, N. (2018). Integrated magnetic resonance imaging and [(11) C]-PBR28 positron emission tomographic imaging in amyotrophic lateral sclerosis. *Annals of Neurology, 83*, 1186–1197.

Appel, S. H., Beers, D., Siklos, L., Engelhardt, J. I., & Mosier, D. R. (2001). Calcium: The Darth Vader of ALS. *Amyotrophic Lateral Sclerosis and Other Motor Neuron Disorders, 2*(Suppl 1), S47–S54.

Appel, S. H., Beers, D. R., & Zhao, W. (2021). Amyotrophic lateral sclerosis is a systemic disease: Peripheral contributions to inflammation-mediated neurodegeneration. *Current Opinion in Neurology, 34*, 765–772.

Arcila-Londono, X., & Lewis, R. A. (2013). Multifocal motor neuropathy. *Handbook of Clinical Neurology, 115*, 429–442.

Bang, J., Spina, S., & Miller, B. L. (2015). Frontotemporal dementia. *Lancet, 386*, 1672–1682.

Beard, J. D., Engel, L. S., Richardson, D. B., Gammon, M. D., Baird, C., Umbach, D. M., Allen, K. D., Stanwyck, C. L., Keller, J., Sandler, D. P., Schmidt, S., & Kamel, F. (2016). Military service, deployments, and exposures in relation to amyotrophic lateral sclerosis etiology. *Environment International, 91*, 104–115.

Beers, D. R., Henkel, J. S., Xiao, Q., Zhao, W., Wang, J., Yen, A. A., Siklos, L., Mckercher, S. R., & Appel, S. H. (2006). Wild-type microglia extend survival in PU.1 knockout mice with familial amyotrophic lateral sclerosis. *Proceedings of the National Academy of Sciences of the United States of America, 103*, 16021–16026.

Beers, D. R., Henkel, J. S., Zhao, W., Wang, J., & Appel, S. H. (2008). CD4+ T cells support glial neuroprotection, slow disease progression, and modify glial morphology in an animal model of inherited ALS. *Proceedings of the National Academy of Sciences of the United States of America, 105*, 15558–15563.

Beers, D. R., Henkel, J. S., Zhao, W., Wang, J., Huang, A., Wen, S., Liao, B., & Appel, S. H. (2011a). Endogenous regulatory T lymphocytes ameliorate amyotrophic lateral sclerosis in mice and correlate with disease progression in patients with amyotrophic lateral sclerosis. *Brain, 134*, 1293–1314.

Beers, D. R., Zhao, W., Liao, B., Kano, O., Wang, J., Huang, A., Appel, S. H., & Henkel, J. S. (2011b). Neuroinflammation modulates distinct regional and temporal clinical responses in ALS mice. *Brain, Behavior, and Immunity, 25*, 1025–1035.

Beers, D. R., Zhao, W., Wang, J., Zhang, X., Wen, S., Neal, D., Thonhoff, J. R., Alsuliman, A. S., Shpall, E. J., Rezvani, K., & Appel, S. H. (2017). ALS patients' regulatory T lymphocytes are dysfunctional, and correlate with disease progression rate and severity. *JCI Insight, 2*, e89530.

Beers, D. R., Zhao, W., Neal, D. W., Thonhoff, J. R., Thome, A. D., Faridar, A., Wen, S., Wang, J., & Appel, S. H. (2020). Elevated acute phase proteins reflect peripheral inflammation and disease severity in patients with amyotrophic lateral sclerosis. *Scientific Reports, 10*, 15295.

Beers, D. R., Zhao, W., Thonhoff, J. R., Faridar, A., Thome, A. D., Wen, S., Wang, J., & Appel, S. H. (2021). Serum programmed cell death proteins in amyotrophic lateral sclerosis. *Brain, Behavior, and Immunity - Health, 12*, 100209.

Beers, D. R., Thonhoff, J. R., Faridar, A., Thome, A. D., Zhao, W., Wen, S., & Appel, S. H. (2022). Tregs attenuate peripheral oxidative stress and acute phase proteins in ALS. *Annals of Neurology, 92*, 195–200.

Benatar, M., Wuu, J., Andersen, P. M., Lombardi, V., & Malaspina, A. (2018). Neurofilament light: A candidate biomarker of presymptomatic amyotrophic lateral sclerosis and phenoconversion. *Annals of Neurology, 84*, 130–139.

Bennion Callister, J., & Pickering-Brown, S. M. (2014). Pathogenesis/genetics of frontotemporal dementia and how it relates to ALS. *Experimental Neurology, 262*(Pt B), 84–90.

Bensimon, G., Lacomblez, L., & Meininger, V. (1994). A controlled trial of riluzole in amyotrophic lateral sclerosis. ALS/Riluzole Study Group. *The New England Journal of Medicine, 330*, 585–591.

Boeve, B. F., Boxer, A. L., Kumfor, F., Pijnenburg, Y., & Rohrer, J. D. (2022). Advances and controversies in frontotemporal dementia: Diagnosis, biomarkers, and therapeutic considerations. *Lancet Neurology, 21*, 258–272.

Bosco, D. A., Morfini, G., Karabacak, N. M., Song, Y., Gros-Louis, F., Pasinelli, P., Goolsby, H., Fontaine, B. A., Lemay, N., Mckenna-Yasek, D., Frosch, M. P., Agar, J. N., Julien, J. P., Brady, S. T., & Brown, R. H., Jr. (2010). Wild-type and mutant SOD1 share an aberrant conformation and a common pathogenic pathway in ALS. *Nature Neuroscience, 13*, 1396–1403.

Brown, R. H., & Al-Chalabi, A. (2017). Amyotrophic lateral sclerosis. *The New England Journal of Medicine, 377*, 162–172.

Chio, A., Logroscino, G., Traynor, B. J., Collins, J., Simeone, J. C., Goldstein, L. A., & White, L. A. (2013). Global epidemiology of amyotrophic lateral sclerosis: A systematic review of the published literature. *Neuroepidemiology, 41*, 118–130.

Cudkowicz, M. E., Lindborg, S. R., Goyal, N. A., Miller, R. G., Burford, M. J., Berry, J. D., Nicholson, K. A., Mozaffar, T., Katz, J. S., Jenkins, L. J., Baloh, R. H., Lewis, R. A., Staff NP, Owegi, M. A., Berry, D. A., Gothelf, Y., Levy, Y. S., Aricha, R., Kern, R. Z., Windebank, A. J., & Brown, R. H., Jr. (2022). A randomized placebo-controlled phase 3 study of mesenchymal stem cells induced to secrete high levels of neurotrophic factors in amyotrophic lateral sclerosis. *Muscle & Nerve, 65*, 291–302.

Dejesus-Hernandez, M., Mackenzie, I. R., Boeve, B. F., Boxer, A. L., Baker, M., Rutherford, N. J., Nicholson, A. M., Finch, N. A., Flynn, H., Adamson, J., Kouri, N., Wojtas, A., Sengdy, P., Hsiung, G. Y., Karydas, A., Seeley, W. W., Josephs, K. A., Coppola, G., Geschwind, D. H., Wszolek, Z. K., Feldman, H., Knopman, D. S., Petersen, R. C., Miller, B. L., Dickson, D. W., Boylan, K. B., Graff-Radford, N. R., & Rademakers, R. (2011). Expanded GGGGCC hexanucleotide repeat in noncoding region of C9ORF72 causes chromosome 9p-linked FTD and ALS. *Neuron, 72*, 245–256.

Deng, H. X., Zhai, H., Bigio, E. H., Yan, J., Fecto, F., Ajroud, K., Mishra, M., Ajroud-Driss, S., Heller, S., Sufit, R., Siddique, N., Mugnaini, E., & Siddique, T. (2010). FUS-immunoreactive inclusions are a common feature in sporadic and non-SOD1 familial amyotrophic lateral sclerosis. *Annals of Neurology, 67*, 739–748.

Deng, H. X., Chen, W., Hong, S. T., Boycott, K. M., Gorrie, G. H., Siddique, N., Yang, Y., Fecto, F., Shi, Y., Zhai, H., Jiang, H., Hirano, M., Rampersaud, E., Jansen, G. H., Donkervoort, S., Bigio, E. H., Brooks, B. R., Ajroud, K., Sufit, R. L., Haines, J. L., Mugnaini, E., Pericak-Vance, M. A., & Siddique, T. (2011). Mutations in UBQLN2 cause dominant X-linked juvenile and adult-onset ALS and ALS/dementia. *Nature, 477*, 211–215.

Floeter, M. K., Katipally, R., Kim, M. P., Schanz, O., Stephen, M., Danielian, L., Wu, T., Huey, E. D., & Meoded, A. (2014). Impaired corticopontocerebellar tracts underlie pseudobulbar affect in motor neuron disorders. *Neurology, 83*, 620–627.

Freibaum, B. D., & Taylor, J. P. (2017). The role of dipeptide repeats in C9ORF72-related ALS-FTD. *Frontiers in Molecular Neuroscience, 10*, 35.

Garbuzova-Davis, S., & Sanberg, P. R. (2014). Blood-CNS barrier impairment in ALS patients versus an animal model. *Frontiers in Cellular Neuroscience, 8*, 21.

Gordon, P. H., Cheng, B., Katz, I. B., Pinto, M., Hays, A. P., Mitsumoto, H., & Rowland, L. P. (2006). The natural history of primary lateral sclerosis. *Neurology, 66*, 647–653.

Gurney, M. E., Pu, H., Chiu, A. Y., Dal Canto, M. C., Polchow, C. Y., Alexander, D. D., Caliendo, J., Hentati, A., Kwon, Y. W., Deng, H. X., et al. (1994). Motor neuron degeneration in mice that express a human Cu,Zn superoxide dismutase mutation. *Science, 264*, 1772–1775.

Henkel, J. S., Beers, D. R., Wen, S., Rivera, A. L., Toennis, K. M., Appel, J. E., Zhao, W., Moore, D. H., Powell, S. Z., & Appel, S. H. (2013). Regulatory T-lymphocytes mediate amyotrophic lateral sclerosis progression and survival. *EMBO Molecular Medicine, 5*, 64–79.

Hodges, J. R., Davies, R., Xuereb, J., Kril, J., & Halliday, G. (2003). Survival in frontotemporal dementia. *Neurology, 61*, 349–354.

Ilieva, H., Polymenidou, M., & Cleveland, D. W. (2009). Non-cell autonomous toxicity in neurodegenerative disorders: ALS and beyond. *The Journal of Cell Biology, 187*, 761–772.

Johnston, C. A., Stanton, B. R., Turner, M. R., Gray, R., Blunt, A. H., Butt, D., Ampong, M. A., Shaw, C. E., Leigh, P. N., & Al-Chalabi, A. (2006). Amyotrophic lateral sclerosis in an urban setting: A population based study of inner city London. *Journal of Neurology, 253*, 1642–1643.

Kabashi, E., Valdmanis, P. N., Dion, P., Spiegelman, D., Mcconkey, B. J., Vande Velde, C., Bouchard, J. P., Lacomblez, L., Pochigaeva, K., Salachas, F., Pradat, P. F., Camu, W., Meininger, V., Dupre, N., & Rouleau, G. A. (2008). TARDBP mutations in individuals with sporadic and familial amyotrophic lateral sclerosis. *Nature Genetics, 40*, 572–574.

Kano, O., Beers, D. R., Henkel, J. S., & Appel, S. H. (2012). Peripheral nerve inflammation in ALS mice: Cause or consequence. *Neurology, 78*, 833–835.

Keizman, D., Rogowski, O., Berliner, S., Ish-Shalom, M., Maimon, N., Nefussy, B., Artamonov, I., & Drory, V. E. (2009). Low-grade systemic inflammation in patients with amyotrophic lateral sclerosis. *Acta Neurologica Scandinavica, 119*, 383–389.

Knopman, D. S., & Roberts, R. O. (2011). Estimating the number of persons with frontotemporal lobar degeneration in the US population. *Journal of Molecular Neuroscience, 45*, 330–335.

Korobeynikov, V. A., Lyashchenko, A. K., Blanco-Redondo, B., Jafar-Nejad, P., & Shneider, N. A. (2022). Antisense oligonucleotide silencing of FUS expression as a therapeutic approach in amyotrophic lateral sclerosis. *Nature Medicine, 28*, 104–116.

Kwiatkowski, T. J., Jr., Bosco, D. A., Leclerc, A. L., Tamrazian, E., Vanderburg, C. R., Russ, C., Davis, A., Gilchrist, J., Kasarskis, E. J., Munsat, T., Valdmanis, P., Rouleau, G. A., Hosler, B. A., Cortelli, P., De Jong, P. J., Yoshinaga, Y., Haines, J. L., Pericak-Vance, M. A., Yan, J., Ticozzi, N., Siddique, T., Mckenna-Yasek, D., Sapp, P. C., Horvitz, H. R., Landers, J. E., & Brown, R. H., Jr. (2009). Mutations in the FUS/TLS gene on chromosome 16 cause familial amyotrophic lateral sclerosis. *Science, 323*, 1205–1208.

Lambert, M. A., Bickel, H., Prince, M., Fratiglioni, L., Von Strauss, E., Frydecka, D., Kiejna, A., Georges, J., & Reynish, E. L. (2014). Estimating the burden of early onset dementia; systematic review of disease prevalence. *European Journal of Neurology, 21*, 563–569.

Li, W., Lee, M. H., Henderson, L., Tyagi, R., Bachani, M., Steiner, J., Campanac, E., Hoffman, D. A., Von Geldern, G., Johnson, K., Maric, D., Morris, H. D., Lentz, M., Pak, K., Mammen, A., Ostrow, L., Rothstein, J., & Nath, A. (2015). Human endogenous retrovirus-K contributes to motor neuron disease. *Science Translational Medicine, 7*, 307ra153.

Liao, B., Zhao, W., Beers, D. R., Henkel, J. S., & Appel, S. H. (2012). Transformation from a neuroprotective to a neurotoxic microglial phenotype in a mouse model of ALS. *Experimental Neurology, 237*, 147–152.

Liscic, R. M., Alberici, A., Cairns, N. J., Romano, M., & Buratti, E. (2020). From basic research to the clinic: Innovative therapies for ALS and FTD in the pipeline. *Molecular Neurodegeneration, 15*, 31.

Lu, C. H., Allen, K., Oei, F., Leoni, E., Kuhle, J., Tree, T., Fratta, P., Sharma, N., Sidle, K., Howard, R., Orrell, R., Fish, M., Greensmith,

L., Pearce, N., Gallo, V., & Malaspina, A. (2016). Systemic inflammatory response and neuromuscular involvement in amyotrophic lateral sclerosis. *Neurology Neuroimmunology & Neuroinflammation, 3*, e244.

Lunetta, C., Lizio, A., Maestri, E., Sansone, V. A., Mora, G., Miller, R. G., Appel, S. H., & Chio, A. (2017). Serum C-reactive protein as a prognostic biomarker in amyotrophic lateral sclerosis. *JAMA Neurology, 74*, 660–667.

Ly, C. V., & Miller, T. M. (2018). Emerging antisense oligonucleotide and viral therapies for amyotrophic lateral sclerosis. *Current Opinion in Neurology, 31*, 648–654.

Miller, T. M., Cudkowicz, M. E., Genge, A., Shaw, P. J., Sobue, G., Bucelli, R. C., Chio, A., Van Damme, P., Ludolph, A. C., Glass, J. D., Andrews, J. A., Babu, S., Benatar, M., Mcdermott, C. J., Cochrane, T., Chary, S., Chew, S., Zhu, H., Wu, F., Nestorov, I., Graham, D., Sun, P., Mcneill, M., Fanning, L., Ferguson, T. A., Fradette, S., & Valor, Group OLEW. (2022). Trial of antisense Oligonucleotide Tofersen for SOD1 ALS. *The New England Journal of Medicine, 387*, 1099–1110.

Mitchell, J. C., Constable, R., So, E., Vance, C., Scotter, E., Glover, L., Hortobagyi, T., Arnold, E. S., Ling, S. C., Mcalonis, M., Da Cruz, S., Polymenidou, M., Tessarolo, L., Cleveland, D. W., & Shaw, C. E. (2015). Wild type human TDP-43 potentiates ALS-linked mutant TDP-43 driven progressive motor and cortical neuron degeneration with pathological features of ALS. *Acta Neuropathologica Communications, 3*, 36.

Neumann, M., Sampathu, D. M., Kwong, L. K., Truax, A. C., Micsenyi, M. C., Chou, T. T., Bruce, J., Schuck, T., Grossman, M., Clark, C. M., Mccluskey, L. F., Miller, B. L., Masliah, E., Mackenzie, I. R., Feldman, H., Feiden, W., Kretzschmar, H. A., Trojanowski, J. Q., & Lee, V. M. (2006). Ubiquitinated TDP-43 in frontotemporal lobar degeneration and amyotrophic lateral sclerosis. *Science, 314*, 130–133.

Okamoto, K., Mizuno, Y., & Fujita, Y. (2008). Bunina bodies in amyotrophic lateral sclerosis. *Neuropathology, 28*, 109–115.

O'rourke, J. G., Bogdanik, L., Muhammad, A., Gendron, T. F., Kim, K. J., Austin, A., Cady, J., Liu, E. Y., Zarrow, J., Grant, S., Ho, R., Bell, S., Carmona, S., Simpkinson, M., Lall, D., Wu, K., Daughrity, L., Dickson, D. W., Harms, M. B., Petrucelli, L., Lee, E. B., Lutz, C. M., & Baloh, R. H. (2015). C9orf72 BAC transgenic mice display typical pathologic features of ALS/FTD. *Neuron, 88*, 892–901.

Paganoni, S., Macklin, E. A., Hendrix, S., Berry, J. D., Elliott, M. A., Maiser, S., Karam, C., Caress, J. B., Owegi, M. A., Quick, A., Wymer, J., Goutman, S. A., Heitzman, D., Heiman-Patterson, T., Jackson, C. E., Quinn, C., Rothstein, J. D., Kasarskis, E. J., Katz, J., Jenkins, L., Ladha, S., Miller, T. M., Scelsa, S. N., Vu, T. H., Fournier, C. N., Glass, J. D., Johnson, K. M., Swenson, A., Goyal, N. A., Pattee, G. L., Andres, P. L., Babu, S., Chase, M., Dagostino, D., Dickson, S. P., Ellison, N., Hall, M., Hendrix, K., Kittle, G., Mcgovern, M., Ostrow, J., Pothier, L., Randall, R., Shefner, J. M., Sherman, A. V., Tustison, E., Vigneswaran, P., Walker, J., Yu, H., Chan, J., Wittes, J., Cohen, J., Klee, J., Leslie, K., Tanzi, R. E., Gilbert, W., Yeramian, P. D., Schoenfeld, D., & Cudkowicz, M. E. (2020). Trial of sodium Phenylbutyrate-Taurursodiol for amyotrophic lateral sclerosis. *The New England Journal of Medicine, 383*, 919–930.

Paganoni, S., Berry, J. D., Quintana, M., Macklin, E., Saville, B. R., Detry, M. A., Chase, M., Sherman, A. V., Yu, H., Drake, K., Andrews, J., Shefner, J., Chibnik, L. B., Vestrucci, M., Cudkowicz, M. E., & Healey, A. L. S. P. T. S. G. (2022). Adaptive platform trials to transform amyotrophic lateral sclerosis therapy development. *Annals of Neurology, 91*, 165–175.

Peters, O. M., Cabrera, G. T., Tran, H., Gendron, T. F., Mckeon, J. E., Metterville, J., Weiss, A., Wightman, N., Salameh, J., Kim, J., Sun, H., Boylan, K. B., Dickson, D., Kennedy, Z., Lin, Z., Zhang, Y. J., Daughrity, L., Jung, C., Gao, F. B., Sapp, P. C., Horvitz, H. R., Bosco, D. A., Brown, S. P., De Jong, P., Petrucelli, L., Mueller, C., & Brown, R. H., Jr. (2015). Human C9ORF72 Hexanucleotide expansion reproduces RNA foci and dipeptide repeat proteins but not neurodegeneration in BAC transgenic mice. *Neuron, 88*, 902–909.

Petrov, D., Mansfield, C., Moussy, A., & Hermine, O. (2017). ALS clinical trials review: 20 years of failure. Are we any closer to registering a new treatment? *Frontiers in Aging Neuroscience, 9*, 68.

Pioro, E. P. (2014). Review of Dextromethorphan 20 mg/Quinidine 10 mg (NUEDEXTA((R))) for Pseudobulbar affect. *Neurology and Therapy, 3*, 15–28.

Poesen, K., & Van Damme, P. (2018). Diagnostic and prognostic performance of neurofilaments in ALS. *Frontiers in Neurology, 9*, 1167.

Ratai, E. M., Alshikho, M. J., Zurcher, N. R., Loggia, M. L., Cebulla, C. L., Cernasov, P., Reynolds, B., Fish, J., Seth, R., Babu, S., Paganoni, S., Hooker, J. M., & Atassi, N. (2018). Integrated imaging of [(11)C]-PBR28 PET, MR diffusion and magnetic resonance spectroscopy (1)H-MRS in amyotrophic lateral sclerosis. *Neuroimage Clinical, 20*, 357–364.

Renton, A. E., Majounie, E., Waite, A., Simon-Sanchez, J., Rollinson, S., Gibbs, J. R., Schymick, J. C., Laaksovirta, H., Van Swieten, J. C., Myllykangas, L., Kalimo, H., Paetau, A., Abramzon, Y., Remes, A. M., Kaganovich, A., Scholz, S. W., Duckworth, J., Ding, J., Harmer, D. W., Hernandez, D. G., Johnson, J. O., Mok, K., Ryten, M., Trabzuni, D., Guerreiro, R. J., Orrell, R. W., Neal, J., Murray, A., Pearson, J., Jansen, I. E., Sondervan, D., Seelaar, H., Blake, D., Young, K., Halliwell, N., Callister, J. B., Toulson, G., Richardson, A., Gerhard, A., Snowden, J., Mann, D., Neary, D., Nalls, M. A., Peuralinna, T., Jansson, L., Isoviita, V. M., Kaivorinne, A. L., Holtta-Vuori, M., Ikonen, E., Sulkava, R., Benatar, M., Wuu, J., Chio, A., Restagno, G., Borghero, G., Sabatelli, M., Consortium, I., Heckerman, D., Rogaeva, E., Zinman, L., Rothstein, J. D., Sendtner, M., Drepper, C., Eichler, E. E., Alkan, C., Abdullaev, Z., Pack, S. D., Dutra, A., Pak, E., Hardy, J., Singleton, A., Williams, N. M., Heutink, P., Pickering-Brown, S., Morris, H. R., Tienari, P. J., & Traynor, B. J. (2011). A hexanucleotide repeat expansion in C9ORF72 is the cause of chromosome 9p21-linked ALS-FTD. *Neuron, 72*, 257–268.

Rezania, K., & Roos, R. P. (2013). Spinal cord: Motor neuron diseases. *Neurologic Clinics, 31*, 219–239.

Ringholz, G. M., Appel, S. H., Bradshaw, M., Cooke, N. A., Mosnik, D. M., & Schulz, P. E. (2005). Prevalence and patterns of cognitive impairment in sporadic ALS. *Neurology, 65*, 586–590.

Robberecht, W., & Philips, T. (2013). The changing scene of amyotrophic lateral sclerosis. *Nature Reviews. Neuroscience, 14*, 248–264.

Rosen, D. R., Siddique, T., Patterson, D., Figlewicz, D. A., Sapp, P., Hentati, A., Donaldson, D., Goto, J., O'regan, J. P., Deng, H. X., et al. (1993). Mutations in Cu/Zn superoxide dismutase gene are associated with familial amyotrophic lateral sclerosis. *Nature, 362*, 59–62.

Rouleau, G. A., Clark, A. W., Rooke, K., Pramatarova, A., Krizus, A., Suchowersky, O., Julien, J. P., & Figlewicz, D. (1996). SOD1 mutation is associated with accumulation of neurofilaments in amyotrophic lateral sclerosis. *Annals of Neurology, 39*, 128–131.

Rowland, L. P., & Shneider, N. A. (2001). Amyotrophic lateral sclerosis. *The New England Journal of Medicine, 344*, 1688–1700.

Sheean, R. K., Mckay, F. C., Cretney, E., Bye, C. R., Perera, N. D., Tomas, D., Weston, R. A., Scheller, K. J., Djouma, E., Menon, P., Schibeci, S. D., Marmash, N., Yerbury, J. J., Nutt, S. L., Booth, D. R., Stewart, G. J., Kiernan, M. C., Vucic, S., & Turner, B. J. (2018). Association of Regulatory T-cell expansion with progression of amyotrophic lateral sclerosis: A study of humans and a Transgenic Mouse Model. *JAMA Neurology, 75*, 681–689.

Siklos, L., Engelhardt, J., Harati, Y., Smith, R. G., Joo, F., & Appel, S. H. (1996). Ultrastructural evidence for altered calcium in motor nerve terminals in amyotropic lateral sclerosis. *Annals of Neurology, 39*, 203–216.

Simpson, E. P., Henry, Y. K., Henkel, J. S., Smith, R. G., & Appel, S. H. (2004). Increased lipid peroxidation in sera of ALS patients: A potential biomarker of disease burden. *Neurology, 62*, 1758–1765.

Smith, R. G., Henry, Y. K., Mattson, M. P., & Appel, S. H. (1998). Presence of 4-hydroxynonenal in cerebrospinal fluid of patients with sporadic amyotrophic lateral sclerosis. *Annals of Neurology, 44*, 696–699.

Sreedharan, J., Blair, I. P., Tripathi, V. B., Hu, X., Vance, C., Rogelj, B., Ackerley, S., Durnall, J. C., Williams, K. L., Buratti, E., Baralle, F., De Belleroche, J., Mitchell, J. D., Leigh, P. N., Al-Chalabi, A., Miller, C. C., Nicholson, G., & Shaw, C. E. (2008). TDP-43 mutations in familial and sporadic amyotrophic lateral sclerosis. *Science, 319*, 1668–1672.

Taylor, J. P., Brown, R. H., Jr., & Cleveland, D. W. (2016). Decoding ALS: From genes to mechanism. *Nature, 539*, 197–206.

Thonhoff, J. R., Beers, D. R., Zhao, W., Pleitez, M., Simpson, E. P., Berry, J. D., Cudkowicz, M. E., & Appel, S. H. (2018a). Expanded autologous regulatory T-lymphocyte infusions in ALS: A phase I, first-in-human study. *Neurology Neuroimmunology & Neuroinflammation, 5*, e465.

Thonhoff, J. R., Simpson, E. P., & Appel, S. H. (2018b). Neuroinflammatory mechanisms in amyotrophic lateral sclerosis pathogenesis. *Current Opinion in Neurology, 31*, 635–639.

Thonhoff, J. R., Berry, J. D., Macklin, E. A., Beers, D. R., Mendoza, P. A., Zhao, W., Thome, A. D., Triolo, F., Moon, J. J., Paganoni, S., Cudkowicz, M., & Appel, S. H. (2022). Combined regulatory T-lymphocyte and IL-2 treatment is safe, tolerable, and biologically active for 1 year in persons with amyotrophic lateral sclerosis. *Neurology Neuroimmunology & Neuroinflammation, 9*, e200019.

Tran, H., Moazami, M. P., Yang, H., Mckenna-Yasek, D., Douthwright, C. L., Pinto, C., Metterville, J., Shin, M., Sanil, N., Dooley, C., Puri, A., Weiss, A., Wightman, N., Gray-Edwards, H., Marosfoi, M., King, R. M., Kenderdine, T., Fabris, D., Bowser, R., Watts, J. K., & Brown, R. H., Jr. (2022). Suppression of mutant C9orf72 expression by a potent mixed backbone antisense oligonucleotide. *Nature Medicine, 28*, 117–124.

Van Mossevelde, S., Engelborghs, S., Van Der Zee, J., & Van Broeckhoven, C. (2018). Genotype-phenotype links in frontotemporal lobar degeneration. *Nature Reviews. Neurology, 14*, 363–378.

Vance, C., Rogelj, B., Hortobagyi, T., De Vos, K. J., Nishimura, A. L., Sreedharan, J., Hu, X., Smith, B., Ruddy, D., Wright, P., Ganesalingam, J., Williams, K. L., Tripathi, V., Al-Saraj, S., Al-Chalabi, A., Leigh, P. N., Blair, I. P., Nicholson, G., De Belleroche, J., Gallo, J. M., Miller, C. C., & Shaw, C. E. (2009). Mutations in FUS, an RNA processing protein, cause familial amyotrophic lateral sclerosis type 6. *Science, 323*, 1208–1211.

Wang, T., Tomas, D., Perera, N. D., Cuic, B., Luikinga, S., Viden, A., Barton, S. K., Mclean, C. A., Samson, A. L., Southon, A., Bush, A. I., Murphy, J. M., & Turner, B. J. (2022). Ferroptosis mediates selective motor neuron death in amyotrophic lateral sclerosis. *Cell Death and Differentiation, 29*, 1187–1198.

Weisskopf, M. G., O'reilly, E. J., Mccullough, M. L., Calle, E. E., Thun, M. J., Cudkowicz, M., & Ascherio, A. (2005). Prospective study of military service and mortality from ALS. *Neurology, 64*, 32–37.

Won, Y. H., Lee, M. Y., Choi, Y. C., Ha, Y., Kim, H., Kim, D. Y., Kim, M. S., Yu, J. H., Seo, J. H., Kim, M., Cho, S. R., & Kang, S. W. (2016). Elucidation of relevant neuroinflammation mechanisms using gene expression profiling in patients with amyotrophic lateral sclerosis. *PLoS One, 11*, e0165290.

Writing, G., & Edaravone, A. L. S. S. G. (2017). Safety and efficacy of edaravone in well defined patients with amyotrophic lateral sclerosis: A randomised, double-blind, placebo-controlled trial. *Lancet Neurology, 16*, 505–512.

Zhao, W., Beers, D. R., Henkel, J. S., Zhang, W., Urushitani, M., Julien, J. P., & Appel, S. H. (2010). Extracellular mutant SOD1 induces microglial-mediated motoneuron injury. *Glia, 58*, 231–243.

Zhao, W., Beers, D. R., Liao, B., Henkel, J. S., & Appel, S. H. (2012). Regulatory T lymphocytes from ALS mice suppress microglia and effector T lymphocytes through different cytokine-mediated mechanisms. *Neurobiology of Disease, 48*, 418–428.

Zhao, W., Beers, D. R., Bell, S., Wang, J., Wen, S., Baloh, R. H., & Appel, S. H. (2015). TDP-43 activates microglia through NF-kappaB and NLRP3 inflammasome. *Experimental Neurology, 273*, 24–35.

Zhao, W., Beers, D. R., Hooten, K. G., Sieglaff, D. H., Zhang, A., Kalyana-Sundaram, S., Traini, C. M., Halsey, W. S., Hughes, A. M., Sathe, G. M., Livi, G. P., Fan, G. H., & Appel, S. H. (2017). Characterization of gene expression phenotype in amyotrophic lateral sclerosis monocytes. *JAMA Neurology, 74*, 677–685.

Therapeutic Considerations in HIV-Associated Neurocognitive Disorders (HAND)

41

Adam Adelsberg and Dennis L. Kolson

Abstract

HIV-associated neurocognitive disorders (HAND) persist globally despite the effectiveness of suppressive combination antiretroviral therapy (cART) in dramatically reducing the prevalence of HIV-associated dementia, the severest sub-type of HAND. Nonetheless, less severe HAND can be disabling for employment and other activities of daily living, and it remains a large health burden. Success in preventing and treating HAND requires maximizing viral suppression in the central nervous system and other body compartments, modulating the potentially toxic and metabolic effects of early blood–brain barrier injury and virus persistence even in the absence of replication, and enhancing the potential for nervous system recovery. Recent pilot study successes in improving cART regimens for CNS efficacy (improving penetrance, targeting the HIV chemokine co-receptor CCR5 and delivery (nanoformulations) suggest that more complete and sustained suppression of HIV replication is achievable, and this may indeed even further reduce the prevalence of symptomatic HAND. Targeting associated pathogenic pathways with currently available FDA-approved drugs used for other disorders can also serve as adjunctive treatments to cART by adding broad suppression of neuroinflammation and oxidative stress, modulating neurotransmitter metabolism, and perhaps supporting neuroregeneration. Interest in neurotrophic factors is rapidly growing, with limited study thus far. Adjunctive cognitive behavioral therapy and aerobic exercise combined with each of these may further improve outcomes.

A. Adelsberg · D. L. Kolson (✉)
Department of Neurology, Perelman School of Medicine, University of Pennsylvania, Philadelphia, PA, USA
e-mail: kolsond@pennmedicine.upenn.edu

Keywords

HIV-associated neurocognitive disorders · HAND · Neuroprotection · HIV · Treatment · Brain

41.1 Introduction

Neurocognitive impairment (NCI) in persons living with HIV (PWH), known by the Frascati criteria as HIV-associated cognitive disorders (HAND), or more generally as HIV-NCI, is a persisting and evolving global health burden (Antinori et al., 2007; Wang et al., 2020). As defined, these disorders encompass a spectrum of cognitive impairment with sub-syndromes of varying severity, including asymptomatic neurocognitive impairment (ANI), mild neurocognitive disorder (MND), and HIV-associated dementia (HAD), which are characterized by cognitive, motor, and behavioral abnormalities (Kaul et al., 2005). Classification depends upon patient neuropsychological test performance and neurological and behavioral functioning in activities of daily living (Antinori et al., 2007). Since the inception of the Frascati criteria, more refined HIV-NCI criteria have been proposed (described as neurocognitive 'phenotypes') to attempt to account for multiple different and distinguishing features and clinical trajectories of HIV-NCI, with and without other co-morbid conditions that contribute to cognitive impairment among PWH now living longer lives in a state of viral suppression (Dastgheyb et al., 2019, 2021; Paul et al., 2021, 2022; Rubin et al., 2019; Sacktor & Robertson, 2014). Herein the term HIV-NCI will be used to acknowledge and cite the many published studies of patient cohorts often defined by overlapping criteria.

H. E. Gendelman, T. Ikezu (eds.), *Neuroimmune Pharmacology and Therapeutics*,
https://doi.org/10.1007/978-3-031-68237-7_41

41.1.1 HIV-NCI in the Era of Combination Antiretroviral Therapy (cART)

Numerous published studies have described HIV-NCI in the era of combination antiretroviral therapy (cART), with a consensus conclusion that cART has markedly reduced the risk of developing severe, dementing cognitive dysfunction worldwide (McIntosh et al., 2023; Winston & Spudich, 2020). Among the seminal studies of HIV-NCI is the CNS HIV Antiretroviral Therapy Effects Research (CHARTER) study (Heaton et al., 2010). In this longitudinal multicenter study of HIV-NCI in the United States, cognitive impairment was found in 814 (52%) of 1555 enrolled PWH. Since the introduction of cART, the prevalence of severe HIV-NCI (HIV-associated dementia) has been reduced from approximately 20% to approximately 2–5% (Heaton et al., 2010). This reduction of HIV-NCI severity by cART represents the only consistent treatment success to date (reviewed in Nightingale et al. (2014)). Nonetheless, milder HIV-NCI associated with impairment in activities of daily living persists in 10–12% of virally suppressed PWH (Heaton et al., 2010; Wang et al., 2020). This morbidity burden will likely grow as the population of PWH ages, emphasizing a need for additional treatment strategies to improve brain protection in PWH (Wilson et al., 2021; McIntosh et al., 2023). A major contributing factor for NCI risk in aging PWH is cerebrovascular disease and associated risk factors (McIntosh et al., 2021). Surprisingly, a recent meta-analysis of HIV-NCI prevalence suggested a higher burden of HIV-NCI among older (>50 years of age) PWH in higher-income countries, where overall stroke risk remains the highest (Avan et al., 2019; Deng et al., 2021).

41.1.2 Role of HIV Replication in HIV-NCI

The role of uncontrolled HIV replication within the CNS in driving risk for HIV-NCI is clear, and suppression of HIV replication is the foundation for HIV-NCI risk reduction that has achieved significant success, albeit incomplete (reviewed in Sillman et al. (2018)). Uncontrolled HIV replication within the CNS is associated with a high risk (~25%) of severe HIV-NCI associated with the pathological finding (in most cases) of encephalitis, whereas under conditions of HIV suppression with cART, the risk of severe HIV-NCI is ~2% (Saylor et al., 2016). Uncontrolled HIV replication associated with low peripheral blood CD4+ T-lymphocyte counts (CD4 nadir <200 cells/μL) is the most consistently identified risk factor for the development of HIV-NCI (Ellis et al., 2011). The persistent risk of developing less severe (non-dementing) clinical manifestations of HIV-NCI in the setting of HIV suppression, which is achieved in most PWH receiving cART, has shifted attention towards the therapeutic development of newer neuroprotective strategies that complement the protective effects of HIV suppression. As discussed, such strategies may target persistent or episodic neuroinflammation, oxidative stress, viral blipping, or other potential pathogenic factors (Eden et al., 2007; Gisslen et al., 2021; Spudich, 2016).

HIV enters the CNS compartment early (days) in the course of infection and yet not all PWH develop HIV-NCI (Zayyad & Spudich, 2015; Hagberg & Gisslen, 2023). A growing number of studies confirm early entry (within days of systemic infection) of HIV into the CNS (Davis et al., 1992; Valcour et al., 2012), and these studies have also confirmed a predominance of T-lymphocyte-tropic, non-macrophage-tropic R5 (CCR5 receptor-utilizing) HIV strains in the CSF of newly infected individuals within the first 1–2 years after infection (Sturdevant et al., 2015). During this time CNS reservoirs are thought to be established (Chaillon et al., 2020; Estes et al., 2017; Leon-Rivera et al., 2021; Rose et al., 2018; Veenhuis et al., 2019; Valcour et al., 2012). The emergence of R5 macrophage-tropic strains, is a common observation, with a predominance of such strains later in the course of infection (Sturdevant et al., 2015; Dunfee et al., 2006; Martin-Garcia et al., 2006; Ohagen et al., 2003). Such emergence of macrophage-tropic strains reflects an adaptation of the virus to replication in brain macrophages and microglia, which express markedly lower levels of CD4 in comparison to T lymphocytes (Joseph & Swanstrom, 2018; Martin-Garcia et al., 2005, 2006; Shieh et al., 1998).

The process of macrophage/microglial HIV adaptation remains enigmatic, and predicting risk for developing HIV-NCI based upon early entry events remains uncertain, despite intensive study (Chan & Spudich, 2022). As eradication of the reservoir(s) of replication-competent HIV proviruses is critical for HIV cure strategies, the controversial role of the CNS as a potential HIV reservoir for re-seeding of peripheral reservoirs, perhaps during periods of CSF escape, is receiving attention (Board et al., 2022; Cohn et al., 2020; Joseph et al., 2019; Kreider & Bar, 2022). Interrupting early events in CNS entry and reservoir establishment is a major target of newer therapeutic investigations (Hellmuth et al., 2019; Kessing et al., 2017; Kore et al., 2015; Peluso et al., 2013; Peluso & Spudich, 2014).

A seminal study reported the prevalence of neurocognitive deficits in patients with undetectable serum HIV RNA over a median time of 48 months. Within this cohort, 27% had cognitive complaints and within this group, HAND was diagnosed in 84% (ANI 24%, MND 52%, and HAD 8%). Of the 73% of patients without cognitive complaints, an additional 64% were found to have HAND on formal neuropsychological testing (ANI 60%, MND 4%, and HAD 0%) (Simioni et al., 2010). Furthermore, data suggest that neurological injury can progress even while patients are on virally suppressive ART. Several studies have demonstrated significantly elevated CSF neurofilament light-chain (NFL, a specific biomarker of neuronal injury) not only in ART-naïve individuals but also in individuals on suppressive ART, in comparison with age-matched HIV-negative controls (Jessen Krut et al., 2014). Additional evidence indicates that neuronal damage occurring prior to initiation of ART is not fully reversible (Tozzi et al., 2007). In a cohort of individuals with AIDS, 21% of those who had undetectable blood and CSF HIV RNA progressed to HAD (Sevigny et al., 2007). And initiation of neuronal damage and clinically evident HAND can occur before AIDS criteria are met, as is evident by a prospective study that identified HAND in 8–34% of aviremic patients who had no comorbidities and who had a CD4 nadir of greater than 200 cells/μL (Cysique et al., 2006b). As such, HAND cannot be mediated solely by uncontrolled HIV replication in the CNS compartment, and the neuropathogenesis of this syndrome must account for accumulating neuronal damage even in the absence of clinically detectable HIV replication. These new insights underscore the need for adjunctive neuroprotective therapies, as there are no current therapies for improving neurologic outcomes other than ART, which targets only HIV replication (Antinori et al., 2007; Brew, 2004; Sacktor et al., 2002).

41.2 General Themes for Reducing the Risk of HIV-NCI

Although suppression of HIV replication reduces the risk for severe HIV-NCI, the effects of co-morbid conditions, CNS viral 'blipping' or escape, chronic neuroinflammation and oxidative stress, and persistent effects of early injury in acute HIV infection (before cART achieves full viral suppression) probably contribute to long-term HIV-NCI risk (Chan et al., 2021; Heaton et al., 2015). Complete neuroprotection and reduction of HIV-NCI risk have not yet been achieved (Bougea et al., 2019). Because HIV replication is the initiating factor for HIV-NCI, neuroprotection approaches must include complete suppression of HIV replication and HIV gene expression with modified cART regimens, and also modulation of cell/virus trafficking and spread, disordered metabolism, and inappropriate activation of oxidative stress and inflammation pathways that impact neuronal cell function (Fig. 41.1) (Killingsworth & Spudich, 2022).

This chapter presents previously tested and newly emerging neuroprotective strategies to improve neurocognitive outcomes in PWH, for which the foundation is suppressive cART. More effective neuroprotection strategies will require the continued evolution of cART regimens and the development of adjunctive therapies. Preliminary successes in reducing HIV-NCI have been observed in pilot clinical trials, which have often later been recognized as treatment failures when examined in larger well-controlled validation studies. Nonetheless, as knowledge of the functional, biochemical, and structural integrity of the CNS in PWH continues to grow, and as new neuroprotective/neuromodulating medications and drug delivery systems are developed for other CNS disorders, future success seems likely. This review will target selected areas of cART strategies, therapies for neuroinflammation and oxidative stress, neurodegenerative disease therapies, neuropsychiatric medications, trophic factors, and exercise therapy, with the recognition that some areas of potential interest are outside of the selected scope of this discussion.

41.3 Improving Consistency of Viral Suppression by Modifying cART Regimens

It is clear that suppression of HIV replication in the peripheral blood compartment and the cerebrospinal fluid (considered a surrogate for the CNS compartment) dramatically reduces the risk for severe cognitive impairment in PWH. Improving the consistency of cART suppression through enhancing the accessibility of ART drugs to the CNS compartment has been intensively pursued as a primary approach for reducing HIV-NCI risk. Improving ART penetration and/or retention of ART drugs should not only decrease CNS HIV replication but also reduce the resulting production of directly- and indirectly acting neurotoxins and immune-activating factors from macrophages/microglia and astrocytes (Ellis et al., 2007; Lin et al., 2021). Reduction of HIV replication in the productive CNS reservoir (macrophages/microglia) would likely reduce the amplification of pathways of toxin production in uninfected bystander cells. In addition to choosing ART drugs that more readily enter and remain in the CNS compartment, some ART regimens may reduce HIV replication in monocyte and macrophage lineages more efficiently than others, and thus be more effective in reducing HIV-NCI pathogenesis (Shikuma et al., 2012).

41.3.1 The Goal of Reducing CNS Viral 'Blipping'

HIV 'blipping' (or CSF 'escape') in the CSF is defined as the detection of HIV genomic RNA copies in the CSF during systemic HIV suppression (undetectable plasma HIV RNA) (Grennan et al., 2012). Such blipping occurs at a relatively high rate (~10–20%) (Hagberg & Gisslen, 2023), but whether HIV blipping is a risk factor for NCI in virally suppressed PWH remains uncertain (Chen et al., 2014; Filippidis et al., 2023; Killingsworth & Spudich, 2022). Notably, the potential for the evolution of ART resistance mutations in contrib-

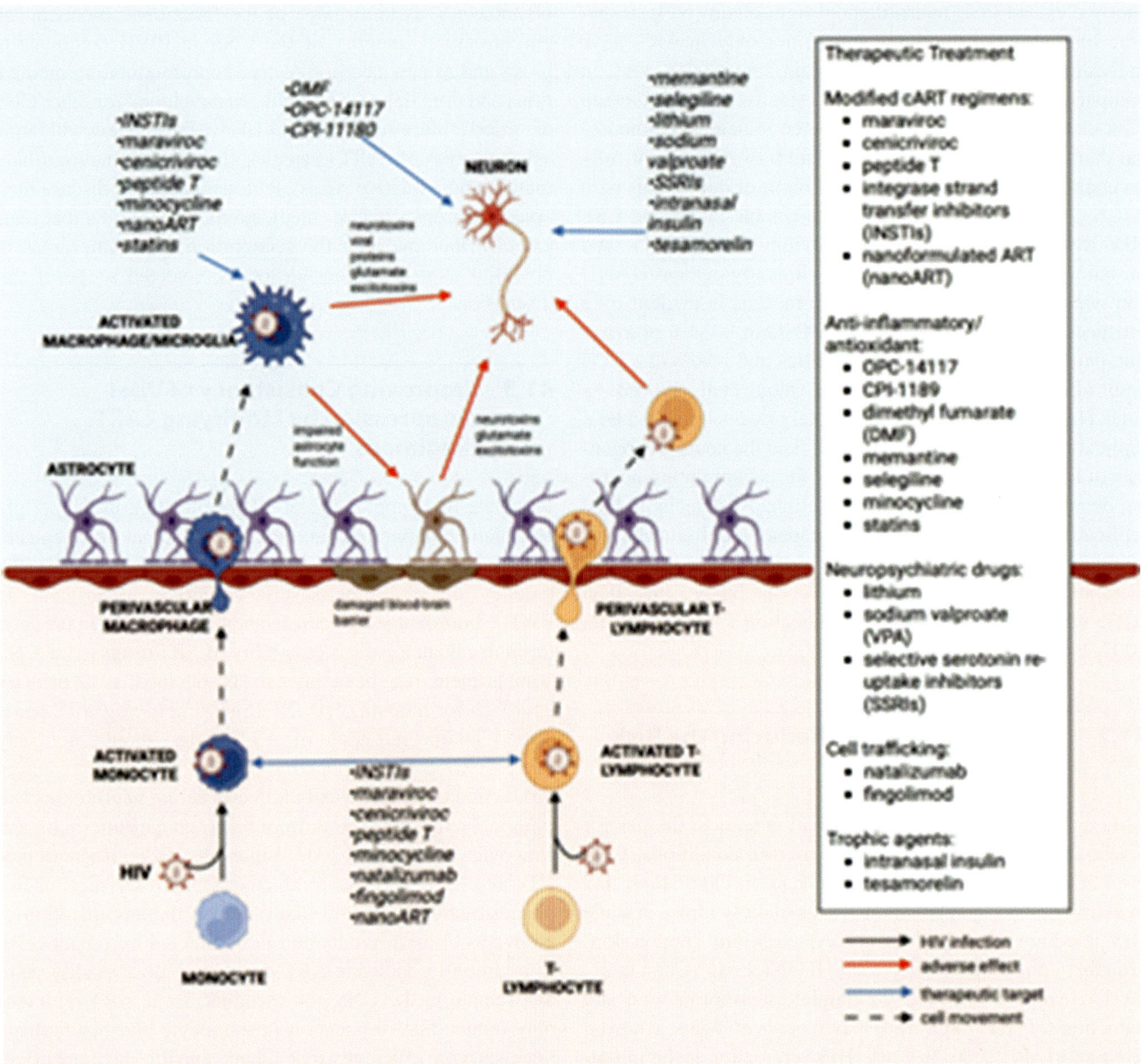

Fig. 41.1 Following systemic HIV infection, HIV-infected immune cells migrate from the circulation across the blood–brain barrier into the CNS. Virus replication continues within these cells and proinflammatory and oxidative processes are amplified in both infected and noninfected cells. These processes lead to the release of neurotoxins, viral proteins, glutamate, and excitotoxins that damage neurons and glial cells. Therapeutic treatments for HIV-NCI attempt to target HIV-infected cells in the systemic circulation and CNS. (Created with BioRedner.com)

uting to HIV blipping is a concern (Trunfio et al., 2023). The development of more consistently suppressive cART regimens may lead to reduced risk for the emergence of ART resistance mutations (or vice versa) in the CSF, which may be required for more effective neuroprotection (Eden et al., 2010). The suppression of HIV blipping is one goal for improving cART efficacy, but whether this would result in reducing the risk for HIV-NCI is not known.

41.3.1.1 Enhancing ART CNS Penetrance

Enhancing the penetration of ART drugs into the CNS is a long-standing theme in achieving the goal of more consistent CNS viral suppression. An ART ranking system known as CNS penetrance efficacy (CPE) ranking has been developed (Cusini et al., 2013; Letendre et al., 2008). The use of favorably-ranked cART regimens may reduce CSF neuroinflammatory markers and CSF HIV load, but there has not been a consistently observed effect on HIV-NCI progression (Dahl et al., 2011; Force et al., 2021; Llibre et al., 2012) (Arentoft et al., 2022; Webb et al., 2022). It seems likely that continued surveillance of ART CPE ranking associations with HIV-NCI outcomes in different PWH cohorts will continue. One feasible means of monitoring CNS responses to different ART regimens is the application of brain magnetic resonance spectroscopy (MRS) in longitudinal follow-up of treated PWH, although several MRS studies have failed to demonstrate an association between CPE ranking and preservation of neuronal integrity (Brkic et al., 2023; Harezlak et al., 2014).

In addition to altering the choice of sub-classes of ART drugs to improve the cART regimen CPE, a complementary approach for improving CNS ART efficacy is to enhance cART delivery vehicles. One such approach is the development of nanoparticulate-ART (nanoART). Such an approach has produced promising results, at least in humanized HIV-infected mice, wherein effective sustained delivery and suppression of HIV replication in tissue macrophages have been observed. Moreover, the delivery of several different classes of ART drugs, including INSTI and NNRTI, into the humanized mouse brain through nanoformulations has been successful (Gong et al., 2020; Herskovitz et al., 2021). Recently, delivery into muscle tissue macrophages in the rhesus macaque has been demonstrated (Deodhar et al., 2022). has which suggests that enhancing CNS ART delivery can have neuroprotective effects (Dou et al., 2009; Gomes et al., 2014; Nowacek & Gendelman, 2009) These and other re-formulated ART preparations that produce longer-lasting effective drug release may soon be ready for new human clinical trials for HIV-NCI neuroprotection (Edagwa et al., 2017; McMillan et al., 2018).

41.3.1.2 ART Intensification with CCR5 Chemokine Receptor Blockade

The chemokine receptor CCR5, serves as a co-receptor to CD4 to medicate HIV-1 entry into the major cellular targets for infection, CD4 T lymphocytes and macrophages/microglia (Riviere-Cazaux et al., 2022). In such cells, maraviroc, a CCR5 chemokine receptor antagonist is effective at inhibiting HIV-1 entry and also CCR5-mediated chemotaxis in response to endogenous CCR5 ligands, each of which is central to HIV neuropathogenesis (Nickoloff-Bybel et al., 2021). Maraviroc has been investigated as an ART intensification drug for the reduction of HIV-NCI risk, with some promising results (Gates et al., 2016; Ndhlovu et al., 2014). Furthermore, maraviroc has been examined as a therapeutic in other neuroinflammation-associated disorders, including traumatic brain injury and multiple sclerosis (Liu et al., 2023; Martin-Blondel et al., 2016). A single-arm, open-label intensification study of maraviroc in 12 PWH on suppressive cART for 6 months showed suppressive effects on monocyte inflammation (lower sCD163 levels, lower levels of CD16+ circulating monocytes) and improvement in neuropsychological (NP) performance (in 6 of 12 subjects with mild to moderate HIV-NCI) (Ndhlovu et al., 2014). The results of this study suggest that antiretroviral therapy with potency against monocytes may have efficacy against HAND. To date, a large-scale maraviroc intensification study for HIV-NCI risk reduction has not been developed, but further investigations of maraviroc as a neuroprotectant in HIV-NCI could be justified (Martin-Blondel et al., 2016).

A broader spectrum CCR5 blocker, cenicriviroc, has also shown promise as a neuroprotectant in HIV-NCI. Cenicriviroc blocks both CCR5 and CCR2 chemokine receptors; notably, CCL2 (the cognate ligand for CCR2) is a chemotactic chemokine that recruits CD14+CD16+ monocytes across the blood–brain barrier. CCL2 is elevated in the CNS of virally suppressed PWH with HIV-NCI (Williams et al., 2014). Cenicriviroc thus can suppress HIV entry into cells and also inflammation and monocyte migration(Leon-Rivera et al., 2021). A pilot 24-week trial of cenicriviroc in 17 cART-suppressed PWH with mild to moderate HIV-NCI showed modest beneficial effects (D'Antoni et al., 2018). Based on this preliminary study, the potential neuroprotective effects of combination blockade of chemokine receptors (CCR5, CCR2) with cenicriviroc should be further tested in larger controlled studies in PWH with HIV-NCI. Furthermore, the ability of cenicriviroc to suppress the replication of SARS CoV-2 in vitro has raised interest in its potential use in the treatment of COVID-19 (Files et al., 2022; Okamoto et al., 2020). This additionally suggests a rationale for the evaluation of cenicriviroc in PWH with COVID-19.

A unique CCR5-targeting HIV-1 entry inhibitor, Peptide T (D-ala-peptide-T-amide/DAPTA) was examined in a phase II randomized double-blind placebo-controlled study (6 months duration) of PWH with HIV-NCI ($n = 215$) of varying levels of severity (Heseltine et al., 1998; Ruff et al., 2003). This was the first trial of an intranasally-delivered therapeutic drug for the treatment of HIV-NCI. Although several neurocognitive outcome parameters suggested some improvement (without meeting statistical significance) no significant difference in neurocognitive performance (neuropsychometric testing in seven domains of function, global z score change from baseline) between DAPTA and placebo arms was observed. A follow-up sub-group analysis of study patients demonstrated a modest reduction of plasma viral load and no effect on CSF viral load (Goodkin et al., 2006). Thus, evidence for a significant beneficial effect of DAPTA for neuroprotection against HIV is lacking, and future therapeutic studies do not appear warranted.

41.3.1.3 Integrase Strand Transfer Inhibitors (INSTI)

Integrase strand transfer inhibitors (INSTI) are the foundation of initial ART regimens (Gandhi et al., 2023). They act by blocking the integration of the HIV provirus into cellular DNA, thereby preventing productive HIV replication in the target cell. Generally, INSTI are widely used because of their effectiveness and relatively limited side effect profile. Their widespread use has generated interest in defining their potential benefits and risks/adverse effects for the CNS. Two categories of adverse effects may limit their use however: weight gain and neuropsychiatric symptoms (Christensen & Tan, 2022; Diggins et al., 2022) (Kolakowska et al., 2019). Neuropsychiatric symptoms of currently used second-generation INSTI, dolutegravir (DTG), and bictegravir (BIC) include sleep disturbances, anxiety, and depression which may lead to discontinuation rates up to ~12% in some studies (Perez-Valero et al., 2023).

The efficacy of INSTI in modulating HIV-NCI has been assessed in several recent studies. The inclusion of INSTI, including dolutegravir (DTG), in cART regimens in early HIV treatment did not associate with improved neurocognitive performance or neuroimaging markers of brain integrity (O'Halloran et al., 2019; Prats et al., 2021; Chan et al., 2020). Additionally, a recent 96-week, randomized, placebo-controlled ART intensification study (ACTG 5234; 25 national and international sites; ClinicalTrials.gov Identifier: NCT02519777) examined the addition of DTG with or without maraviroc to suppressive cART. This study (n = 191 PWH) stands in distinction from prior studies with the inclusion of a DTG arm alone. The investigators found no significant differences from placebo effects on neurocognitive performance (preliminary results: https://www.natap.org/2022/CROI/croi_187.htm). An important additional observation was the lack of worsening of symptoms of depression in individuals in the DTG alone arm. This is of considerable interest in view of reports of adverse neuropsychiatric effects of DTG in some studies (Chan et al., 2020; Lanman et al., 2019; Perez-Valero et al., 2023). Thus, although current INSTI implementation strategies consistently reduce HIV-NCI risk further study of INSTI is warranted (Table 41.1).

41.4 ART Toxicity

There are inescapable risks of toxicity of long-term ART use, including hepatic steatosis, peripheral neuropathy, cardiomyopathy, pancreatitis, ototoxicity, retinopathy, and lipodystrophy (Gandhi et al., 2023; Lanman et al., 2019). Substantial toxicity of various nucleoside reverse transcriptase inhibitors (NRTIs) in different tissue types has been observed, and ART toxicity within the CNS compartment is a concern. In vitro studies demonstrate that numerous ART drugs are directly neurotoxic at concentrations detected in the CSF of PWH on cART, and certain cART regimens, including those with INSTI, may be associated with poorer neurocognitive outcomes (Lanman et al., 2019; Marra et al., 2009; Robertson et al., 2012). Notably, one major issue about the possible neurotoxicity of DTG during pregnancy has fortunately been conclusively addressed. Concerns over the potential risk for fetal neural tube defects with dolutegravir (DTG) use during pregnancy have proved to be unfounded, and DTG use during pregnancy is recommended (Abrams & Myer, 2021; Gandhi et al., 2023; Pereira et al., 2021). As new ART regimens are implemented, continued surveillance of cognitive functioning in PWH is essential for defining and ultimately mitigating toxicity risks, which are currently clearly superseded by the beneficial effects of cART.

41.5 Inflammation and Oxidative Stress in HIV-NCI

41.5.1 Neuroinflammation

Neuroinflammation is a persistent feature of HIV infection, which, in ART-naïve subjects produces a robust expression of proinflammatory cytokines and chemokines, reactive oxygen species, glutamate, and other excitotoxins, lipid peroxidation products and other non-proteinaceous neurotoxins, and possibly some HIV proteins, resulting in neuronal dysfunction and varying degrees of injury (Killingsworth & Spudich, 2022; McArthur & Johnson, 2020; Uzasci et al., 2013). Other anatomic sites of tissue injury that contribute to brain parenchyma injury include the blood–brain barrier and the gut, which may promote the ingress of activated immune cells and toxic blood products, and, in the gut, bacterial translocation into the systemic circulation; each of these processes has been linked to HIV-NCI (Ancuta et al., 2008; Brenchley et al., 2006). Thus, reducing systemic and CNS inflammation may be expected to reduce the risk for HIV-NCI risk in PWH, although how effective such a strategy might be has been challenged (Price et al., 2014). The association of CSF HIV blipping with 'blips' of neuroinflammation is another consideration, and each should be further investigated both in PWH and in non-human primate models to inform therapeutic neuroprotection strategies (Stam et al., 2013).

41.5.2 Contribution of Systemic Inflammation: The Gut

HIV infection is associated with the translocation of microbial products across a damaged gastrointestinal tract and dysbiosis of bacterial populations, and the systemic and CNS sequelae of gut dysbiosis remain intensively studied PWH

Table 41.1 Therapeutic clinical trials for HIV-associated neurocognitive disorders

Treatment	Molecular target/Action	Proposed effects	Clinical data	References
Modified cART regimens				
Maraviroc	CCR5 receptor blocker	Block HIV entry; block CCR5 receptor signaling	Open-label 15-subject maraviroc intensification of cART showed neurocognitive improvement over 24 weeks; RCT of 191 patients showed no significant effect on neurocognitive improvement over 96 weeks.	Ndhlovu et al. (2014) (https://www.natap.org/2022/CROI/croi_187.htm)
Cenicriviroc	CCR5 and CCR2 receptor blocker	Block HIV entry and CCL2-mediated monocyte chemotaxis	Open-label 17-paitent study of cenicriviroc intensification of cART showed modest neurocognitive improvement over 24 weeks	D'Antoni et al. (2018)
Peptide T	CCR5 receptor blocker	Block HIV entry; block CCR5 receptor signaling	Phase II randomized double-blind placebo-controlled study (6 months duration) of PWH with HIV-NCI ($n = 215$) of varying levels of severity. No significant difference in neurocognitive performance from the placebo arm	Heseltine et al. (1998) Ruff et al. (2003) Goodkin et al. (2006)
Dolutegravir (DTG)	Integrase strand transfer inhibitor (INSTI)	Block final ligation (integration) of HIV provirus into cellular genome	Cross-sectional 202-patient study of DTG inclusion in cART associated with worse neurocognitive improvement and lower brain volume; Longitudinal 254-patient study up to 37 weeks showed a switch to DTG associated with worsened depression; practice effects obscured NP test results; 30-patient study initiating INSTI <3 months vs >6 months after HIV acquisition showed no difference in neurocognitive performance after 48 weeks; DTG arm alone, randomized, double-blinded, placebo-controlled in 191 PWH. No significant difference from placebo effects on neurocognitive performance	Perez-Valero et al. (2023) O'Halloran et al. (2019) Chan et al. (2020) Prats et al. (2021) Lanman et al. (2009)
Anti-inflammatory/antioxidant				
OPC-14117	Free radical scavenger	Decrease oxidative injury by free radicals	Phase I/II study in 30 patients with mild HIV-NCI for 12 weeks, neurocognitive improvement did not reach significance.	Disorders (1997)
CPI-1189	TNFa	Inhibit p38 MAPK and inhibit TNFa proinflammatory signaling; antioxidant properties	RCT of 64 patients with mild-moderate HIV-NCI treated for 10 weeks: no neurocognitive improvement	Clifford et al. (2002)
Memantine	NMDA receptor antagonist	Block excessive calcium influx through NMDA receptor channel; increase BDNF levels	RCT of 98 patients with mild HIV-CI treated for 20 weeks, with open-label extension up to 60 weeks; modest preservation of NAA levels (neuronal integrity) by MR spectroscopy; no neurocognitive improvement	Schifitto et al. (2007a), Zhao et al. (2010)
Selegiline	MAO-B inhibitor	May reduce the antioxidant burden of cell	RCT of 124 patients with mild-moderate HIV-CI treated for 24 weeks, with open-label extension up to 60 weeks; No neurocognitive improvement overall; some improvement in NP subtests	Schifitto et al. (2007b) Schifitto et al. (2009a)
Minocycline	5-lipoxygenase and others	Decrease CCL2 levels in CSF, improve SIV encephalitis, suppress HIV replication, and inhibit secretion of TNF-α, IFN-γ, and IL-2 by lymphocytes	RCT of 107 patients with progressive mild-moderate HIV-NCI treated for 24 weeks: no neurocognitive improvement	Sacktor et al. (2011)

(continued)

Table 41.1 (continued)

Treatment	Molecular target/Action	Proposed effects	Clinical data	References
Neuropsychiatric drugs				
Lithium	Inhibits dopamine, glutamate neurotransmission (NMDA receptor), promotes GABA-ergic transmission, induce BDNF, Bcl2 PKC, GSK-3β inhibitor	Reduces excitotoxicity, trophic effects	Open-label 21-patient study for 12 weeks showed improved neurocognitive performance; Open-label 15-patient study for 10 weeks showed no neurocognitive improvement. RCT of 66 patients with moderate-severe HIV-NCI treated for 24 weeks; no neurocognitive performance difference compared to placebo	Letendre et al. (2006) Schifitto et al. (2009b) Decloedt et al. (2016)
Sodium valproate (VPA)	GSK-3β inhibitor, HDAC inhibitor	Provides neuroprotection against HIV-induced toxicity	40-patient trial (8 VPA, 32 no VPA in advanced HIV) showed worse neurocognitive performance with VPA;	Cysique et al. (2006a) Schifitto et al. (2006)
Selective serotonin reuptake inhibitors (SSRIs): Citalopram Sertraline Trazadone Paroxetine	Selective serotonin reuptake inhibitors	Some (citalopram, sertraline, trazadone) associated with lower HIV viral levels in CSF	RCT of 45 patients subdivided to paroxetine and/or fluconazole for 24 weeks; in those receiving paroxetine, neurocognitive improvement in 4/6 tests, worsening in 2/6; Citalopram, sertraline, and trazadone use linked historically to better neurocognitive performance in PLWH	Sacktor et al. (2018) Letendre et al. (2007) Stewart et al. (2020)
Galantamine (GAL)	Acetylcholinesterase (AChE) inhibitor, nicotinic receptor agonist	Increase available acetylcholine at the synaptic junction through AChE inhibition	12-week, double-blind, placebo-controlled crossover study of GAL treatment in PWH, results pending	(ClinicalTrials.gov identifier: NCT03384784)
Trophic agents targeting metabolism				
Intranasal insulin	Binding to insulin receptor A (IR-A) on neurons	Activation of p_AKT and p-GSK3B expression	In PWH, intranasal insulin has been associated with improved cognitive performance in patients with diabetes mellitus, while a randomized clinical trial in patients with Alzheimer's disease showed no benefit; A preliminary report of a pilot study (of intranasal insulin administration for the treatment of HIV-NCI in 21 PWH indicated improvement in neuropsychological tests of memory and attention.	Craft et al. (2020) Gaddam et al (2021) (ClinicalTrials.gov Identifier: NCT03081117) (https://www.natap.org/2021/CROI/croi_91.htm)
Tesamorelin	GHRH activator	Induces release of insulin-like growth factor-1 (IGF-1)	randomized, double-blind, placebo-controlled trial (20 weeks of treatment) of 152 adults, tesamorelin was shown to improve cognitive functioning in both healthy adults and in those with mild cognitive impairment; (Baker et al., 2012) a phase II trial of tesamorelin for the treatment (6 months of treatment) of HIV-NCI in PWH is now underway	(128) (ClinicalTrials.gov identifier: NCT02572323)
Aerobic exercise		Increasing transcription of genes related to trophic factors, neuronal survival, and regeneration	Longitudinal study of 291 PWH showed that consistent physical activity (≥ 50% of study visits) was associated with better performance in verbal fluency, working memory, speed of information processing, executive function, and motor function over a mean period of 35 months	Dufour et al. (2018)

(Vujkovic-Cvijin et al., 2020) (Burgener et al., 2015; Mak et al., 2021; Marchetti et al., 2013; Zevin et al., 2016). Disordered gut integrity is associated with decreased microbial diversity and alterations of abundance of different bacterial genera. Increased abundance of the genus *Prevotella* and decreased prevalence of the genus *Bacteroides* are characteristic of HIV infection (Zevin et al., 2016). Generally, HIV infections drive changes towards decreased abundance of commensal, protective bacteria and an increased abundance of proinflammatory bacteria. Increased systemic inflammation has been linked to the translocated microbial product LPS, which associates with monocyte activation as indicated by elevated soluble CD14 (sCD14) and sCD163 in plasma, and elevated expression of monocyte-associated CD14 and CD16 (Jiang et al., 2021; Klatt et al., 2013). Each of these markers has been correlated with an increased risk for HIV-NCI. Therefore, the gut microbiome and LPS are considered targets for reducing the neuropathogenic effects of microbial translocation (Valcour et al., 2013).

Preliminary studies targeted plasma LPS by attempting sequestration with the drug sevelamer (Kristoff et al., 2014; Sandler et al., 2014; Zevin et al., 2016). Evidence from the non-human primate model suggests a suppressive effect on systemic inflammation in SIV infection and a clinical study in 36 PWH not receiving ART revealed lower levels of D-dimer and oxidized LDL, but no reduction of markers of inflammation or T-lymphocyte activation(Kristoff et al., 2014; Sandler et al., 2014). Targeting LPS has not yet been explored in HIV-NCI risk reduction. Also, therapies targeted towards directly reducing microbial translocation (e.g., preserving gut mucosal integrity) in PWH may have a role in neuroprotection, although no such specific therapies currently exist (Rich et al., 2020). Focusing on correcting the gut microbiome dysbiosis as a means of promoting gut integrity and additionally reducing systemic inflammation is a strategy for attempting to reduce HIV-NCI risk and preliminary evidence suggests that probiotic dietary supplementation may reduce neuroinflammation and promote neurocognitive recovery in PWH (Ceccarelli et al. 2017a, b; Rich et al., 2020).

41.5.3 Cannabinoids: Targeting Neuroinflammation

Emerging evidence suggests a potential neuroprotective effect of cannabinoids against HIV- and SIV-induced neuroinflammation, neurotoxicity, and neurocognitive dysfunction, although clinical use of cannabinoids remains controversial (Chu et al., 2023; Ellis et al., 2020; Kaddour et al., 2022; McDew-White et al., 2023; Rosario-Rodriguez et al., 2022; Watson et al., 2021). In a study of SIV-infected rhesus macaques Kaddour et al. (Kaddour et al., 2022) treated chronically with low-dose delta-9-tetrohydrocannabinol (THC) there was decreased expression of miRNAs associated with proinflammatory signaling expressed in basal ganglia-derived extracellular vesicles. This suggests a potential neuroprotective response against the proinflammatory signaling induced by SIV (and, presumably HIV) infection. Similarly, McDew-White et al. (McDew-White et al., 2023) examined the effects of low-dose THC in SIV-infected rhesus macaques and showed that treatment associated with reduced basal ganglia expression of genes associated with type I interferon responses, excitotoxicity, and oxidative stress. Additionally, these investigators showed that THC treatment altered the gut microbiota in a pattern that may modulate tryptophan metabolism in a beneficial manner, thus further potentially linking gut microbiota alterations to clinical benefits in PWH. Studies of PWH have shown that cannabis use may be associated with lower proinflammatory profiles in both peripheral blood and cerebrospinal fluid in PWH (Ellis et al., 2020; Watson et al., 2021).

Nonetheless, published studies of cannabis use in PWH have provided conflicting evidence for the effects on cognitive functioning (Chu et al., 2023; Skalski et al., 2016, 2018). Skalski et al. (2016, 2018) provided a literature review of cannabis use in PWH and concluded that memory dysfunction of mild to moderate effect sizes was associated with marijuana use, perhaps by stressing brain regions also impaired by HIV infection. Memory impairment with marijuana use was associated with a younger age of marijuana use. Watson et al. (2021) observed a statistically non-significant trend in improved cognitive test performance with certain patterns of marijuana use associated with lower inflammatory profiles in PWH. The potential therapeutic benefits and risks of cannabinoids in cognitive performance in PWH will require comprehensive longitudinal, stratified cohort studies in PWH to define the long-term effects on HIV-NCI.

41.5.4 Oxidative Stress

Oxidative stress is typically linked to neuroinflammation. Oxidative stress is an imbalance between oxidant production and antioxidant responses, and when such responses are inadequate, oxidative damage to proteins, nucleic acids, and other biomolecules results in cellular injury (Golden & Patel, 2009; Morris et al., 2019; Negre-Salvayre et al., 2008). Damaging molecules include free radicals (superoxide anion, nitric oxide, hydroxyl radical), reactive oxygen species (including free radicals and non-radicals such as H_2O_2 and peroxynitrite), and reactive nitrogen species (Aguilera et al., 2018; Biswas, 2016). Oxidative stress and inflammation are closely related processes that often co-exist in chronic diseases, and neuroprotection strategies should target both processes concurrently (Biswas, 2016). HIV infection robustly induces oxidative stress, marked by low plasma levels of reduced glutathione and elevated levels of biomarkers of lipid peroxidation products such as malondialdehyde and hydroperoxide, which correlate with disease progression and mortality (Malvy et al., 1994; Mollace et al., 2001; Suresh et al., 2009; Valcour & Shiramizu, 2004; Wanchu et al., 2009). HIV-associated oxidative stress can drive inflammation through enhancing NF-κB-driven HIV replication and release of proinflammatory cytokines (Driscoll, 2000; Fuchs et al., 1998; Israel et al., 1992; Palamara et al., 1996).

HIV suppression with cART significantly reduces immune activation and oxidative stress but not completely (Bandaru et al., 2007; Eden et al., 2007; Roc et al., 2007). This suggests that these pathways associated with the production of neurotoxins can drive accumulating neuronal injury, even during HIV suppression. Despite some previous therapeutic failures in treating HIV-NCI with free radical scavenging drugs, newer drugs with broader targeting of inflammation and oxidative stress pathways may offer new therapeutic options (Lin et al., 2021).

41.6 Targeting Oxidative Stress and Associated Inflammation

Therapeutic trials involving drugs that target inflammation with or without oxidative stress in PWH have thus far failed to confirm a clinical benefit, despite some positive effects on expression of markers of inflammation or neuronal injury. Among these agents are [OPC-14117 (free radical scavenger), CPI-1189 (TNFα blocker), selegiline (MAO-B inhibitor), minocycline (antibiotic, antioxidant), paroxetine (SSRI), lexipafant (platelet-activating factor receptor blocker), and maraviroc (CCR5 blocker)] (Angelovich et al., 2021; Lin et al., 2021; Bougea et al., 2019). However, consistent evidence for chronic inflammation and oxidative stress in multiple body compartments in PWH suggests a need for investigating more robust therapeutics (Ambrosius et al., 2019). These previous treatment failures involved agents that are either highly target-specific or relatively weakly effective with their targets. Broader targeting of inflammatory and oxidative pathways with wide-spectrum agents currently in use to treat other neuroinflammatory diseases (Ambrosius et al., 2019; Thaney & Kaul, 2019).

41.6.1 Broadening the Target for Better Efficacy: The Host Antioxidant Response

The host protective antioxidant response is mediated in part through activation of the transcription factor NF-E2-related factor 2 (Nrf2), which drives the antioxidant response element (ARE), a cis-acting regulatory sequence in the promoter region of numerous antioxidant and anti-inflammatory genes (Joshi & Johnson, 2012). Targeting Nrf2 is considered a broadly applicable therapeutic approach for neuroinflammatory and neurodegenerative disorders (Dinkova-Kostova et al., 2018; Joshi & Johnson, 2012; Loboda et al., 2016; Neilson et al., 2020; Robledinos-Anton et al., 2019). Among Nrf2 targets are several antioxidant enzymes, including heme oxygenase-1 (HO-1), an inducible, detoxifying enzyme that is critical for limiting oxidative stress, inflammation, and cellular injury within the CNS and other tissues. Transcriptional regulation of HO-1 depends partly upon *HMOX1* promoter region genetic variations, including a $(GT)_n$ dinucleotide repeat, which is associated with higher basal *HMOX1* transcriptional activity and inducibility (Fig. 41.2), and better outcomes in inflammatory and oxidative stress-associated diseases (Chen et al., 2012; Joshi & Johnson, 2012; Seu et al., 2012). Associations between HO-1 expression, variation in the *HMOX1* promoter, and risk for HIV neuroinflammation and HIV-NCI suggest a strong rationale for Nrf2 activation as a therapeutic option for the prevention and treatment of HIV-NCI (Garza et al., 2020; Gill et al., 2014, 2018; Gill & Kolson, 2014; Kovacsics et al., 2017).

41.6.2 Nrf2 Activators: Targeting the ARE Antioxidant Response

Success in therapeutic targeting of ARE-(Nrf2) driven gene expression to enhance antioxidant responses and associated inflammation cascades has proved successful in the treatment of multiple sclerosis (Dubey et al., 2015). Target ARE gene expression for the treatment and/or prevention of HIV-NCI has been proposed (Ambrosius et al., 2019; Gill et al., 2018; Gill & Kolson, 2013). The Nrf2 transcriptional activator, dimethyl fumarate (DMF), is FDA-approved for the treatment of multiple sclerosis, and it has excellent CNS penetration(Deeks, 2016). DMF, and its major in vivo metabolite monomethyl fumarate (MMF), effectively induce HO-1

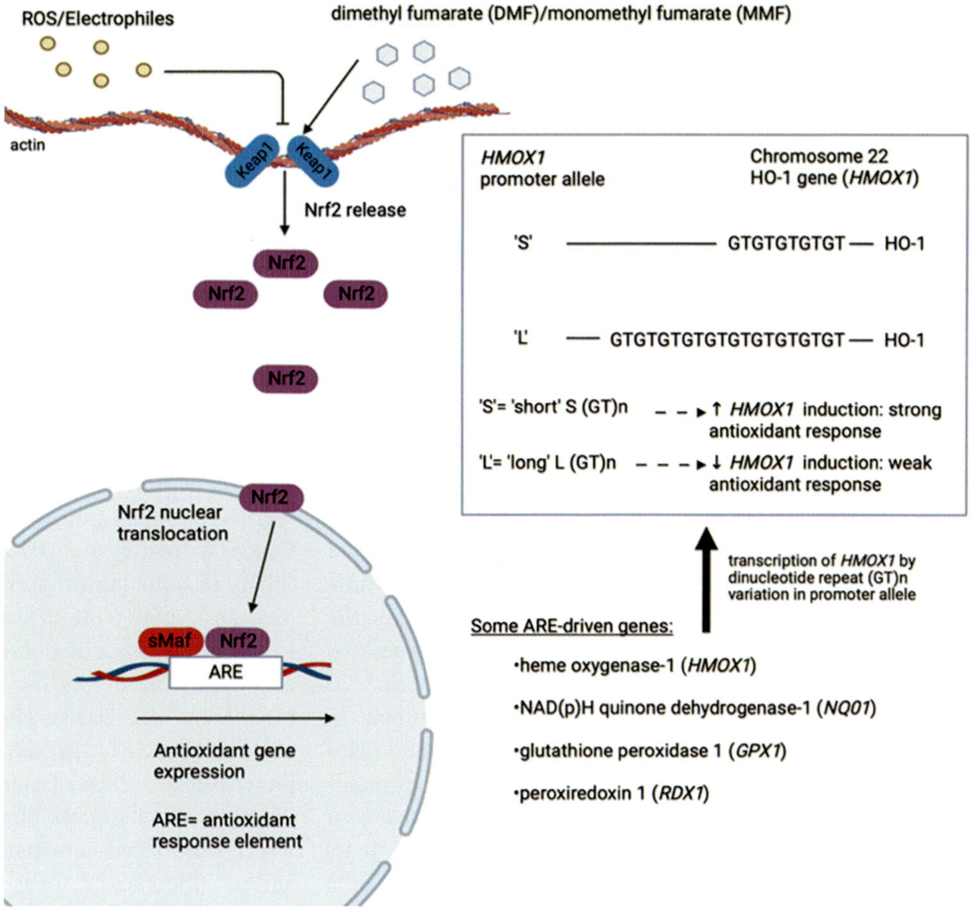

Fig. 41.2 Upon oral ingestion, DMF undergoes demethylation to produce its in vivo active metabolite MMF. Nrf-2 release into the cytoplasm is induced by MMF interacting with KEAP1. Nrf-2 nuclear translocation follows, followed by binding to the antioxidant response element (ARE) in the promoter region of multiple antioxidant enzymes. Heme oxygenase-1 (HO-1, HMOX1 gene) is a phase II detoxifying enzyme with antioxidant functions. HMOX1 promoter expression is modulated by the length of an internal dinucleotide (GT) repeat sequence (GT)n. A short dinucleotide repeat (GT) n promotes increased HMOX1 induction, resulting in increased HO-1 expression and antioxidant response. A long (GT)n repeat (GT)n results in decreased HMOX1 induction and a weaker HO-1 antioxidant response. Short HMOX1 (GT)n repeat lengths are associated with reduced risk for HIV-NCI. (Created with BioRender.com)

expression, inhibit NFκB nuclear translocation, and attenuate macrophage-mediated neurotoxicity in in vitro models of HIV neurotoxicity (Ambrosius et al., 2017; Cross et al., 2011). Furthermore, DMF suppresses the Warburg effect (proinflammatory shift of metabolism to aerobic glycolysis) in the mammalian brain, which has been proposed as a mechanism underlying its neuroprotective effect in multiple sclerosis patients (Kornberg et al., 2018). In SIV-infected rhesus macaques, daily oral delivery (>100 days) of DMF increased the expression of ARE-driven antioxidant enzymes (including HO-1) and reduced oxidative DNA and protein damage in multiple brain regions without apparent toxicity in immunodeficient animals (Garcia-Mesa et al., 2021). Thus, DMF is an attractive candidate for HIV-NCI neuroprotection studies. Additionally, several other Nrf2-activating drugs (Bardoxolone, Resveratrol, others) are also being examined in clinical trials for various inflammatory/oxidative disease states, further supporting Nrf2-directed therapeutics as a promising target for HIV-NCI neuroprotection (Robledinos-Anton et al., 2019).

41.6.3 Statins: Targeting Inflammation and Oxidative Stress

Statins have been extensively examined for their potential to protect the brain against a variety of insults, with mixed results. They are widely used as lipid-lowering agents that block the enzymatic activity of hydroxymethlyglutaryl-CoA (HMG-CoA) reductase. They also demonstrate anti-inflammatory, antioxidant, and immunomodulatory effects, which make them particularly attractive as cytoprotectants (Mansouri et al., 2022). High-dose (80 mg daily) Atorvastatin has been shown to reduce monocyte activation in cART-treated PWH, but beneficial effects on reducing HIV-NCI risk have not been confirmed (Ganesan et al., 2011; Funderburg et al., 2014). In clinical studies not involving PWH randomized controlled trials and observational studies indicate that initiation of statin therapy late in life does not prevent cognitive decline or overt dementia (Power et al., 2015). In traumatic brain injury, some studies have revealed protective effects (Susanto et al., 2023). In studies of multiple sclerosis, which is characterized by neuroinflammation and oxidative stress that contribute to neurodegeneration, statins have also failed to demonstrate protective effects (Stefanou et al., 2022; Williams et al., 2022). Thus, lacking any additional evidence for statin neuroprotective effects in humans, the likelihood of beneficial effects of statins on HIV-NCI risk appears low.

41.7 Other Therapies Applied to Neurodegenerative Diseases and Neuropsychiatric Disorders

41.7.1 Neurodegenerative Disease Therapies

Drug therapies used in neurodegenerative diseases (Alzheimer's disease, Parkinson's disease) are directed towards neuroprotection while others are directed towards treating symptoms. Memantine, which is approved for the treatment of Alzheimer's disease, is a non-competitive NMDA glutamate receptor antagonist. A short-term (20 weeks) placebo-controlled trial of memantine treatment in HIV-NCI patients provided some evidence (magnetic resonance spectroscopy) for the preservation of neuronal integrity in the absence of significant neurocognitive improvement (Schifitto et al. 2007a). A longer follow-up study (1 year) showed no clinically demonstrable neurocognitive benefits (Zhao et al., 2010). Selegiline, a monoamine oxidase B (MAO-B) inhibitor used in the treatment of early-stage Parkinson's disease, is proposed to act as a neuroprotectant by reducing oxidative stress (Magyar & Szende, 2004). However, clinical studies (24 weeks) of transdermal selegiline have shown neither a reduction of oxidative stress nor improvement in HIV-NCI (Schifitto et al. 2007b, 2009a). In each of these clinical trials, PWH with an established diagnosis of HIV-NCI were enrolled, and treatment duration was 1 year or less, which constrains projections about potential long-term efficacy.

41.7.2 Neuropsychiatric Drug Therapies

Neuropsychiatric drug therapies have been studied for effects on HIV-NCI for a number of years, with variable results. Among such drugs are sodium valproate (VPA) and lithium (approved for the treatment of bipolar disorder and related mood disorders). Each has shown to have neuroprotective capabilities against HIV-induced toxicity in vitro and in mouse models, perhaps by their ability to inhibit glycogen synthase kinase-3 (Dou et al., 2003; Everall et al., 2002; Tong et al., 2001). Several open-label pilot studies of lithium therapy in PWH demonstrated improved neurocognitive performance, a more recent randomized, placebo-controlled study of lithium showed no beneficial effect on HIV-NCI (Letendre et al., 2006; Schifitto et al. 2007b; Decloedt et al., 2016). In an observational study of VPA use in PWH over an average of 18 months, VPA users vs. non-users demonstrated poorer neurocognitive performance (Cysique et al., 2006a). Subsequently, a placebo-controlled pilot study of VPA also failed to demonstrate cognitive improvement in PLWH and with HIV-NCI (Schifitto et al., 2006). A possible confounder in these studies is how to account for VPA activity as a histone deacetylase inhibitor (HDAC), which may reactivate HIV from latency. Nonetheless, available clinical trial data to date do not convincingly support further trials for either lithium or VPA for HIV-NCI reduction risk.

A more promising class of neuropsychiatric drugs is the selective serotonin reuptake inhibitors (SSRIs) class. These are widely used for beneficial effects on depression and their relatively limited side effects in PWH (Medeiros et al., 2020; Eshun-Wilson et al., 2018). Additionally, some benefits of SSRIs in reducing immune activation were observed in the Veterans Aging Cohort Study (1546 HIV-positive veterans). In that study, investigators showed a significant, albeit modest, reduction in plasma CD14 and IL-6 in SSRI users (Stewart et al., 2020). However, there is limited research evaluating SSRIs as adjunctive therapies for HIV-NCI (Ances et al., 2008; Letendre et al., 2007; Sacktor et al., 2018). A recent double-blind, placebo-controlled study of 45 PWH subdivided into patient groups receiving paroxetine and/or fluconazole over 24 weeks showed significant improvement in four neuropsychological tests (including the NPZ8 summary measure), with worse performance in two neuropsychological tests, in those receiving paroxetine (Sacktor et al., 2018). Considering the results of this first-of-its kind, controlled study of SSRI and HIV-NCI outcomes in PLWH, additional SSRI studies should be considered. At least one such study has been proposed (Ojagbemi, 2021).

41.7.3 Other Neurotransmitter Modulators: Galantamine (GAL)

Galantamine (GAL) is an acetylcholinesterase (AChE) inhibitor and nicotinic receptor agonist that is FDA-approved for the symptomatic treatment of Alzheimer's disease (Jamali, 2021; Thakral et al., 2023). The AChE inhibitor class of drugs presumably increase available acetylcholine levels at the synaptic junction; these levels are thought to be reduced as a result of loss of cholinergic neurons, and the use of AChE inhibitors associates with a temporary improvement in memory and cognitive function through increasing acetylcholine levels. The dual-actions of GAL as an AChE inhibitor and selective nicotinic receptor ($\alpha4/\beta2$ and $\alpha7n$) agonist also underlie its use in the treatment of nicotine withdrawal, and it has also been studied in combination with memantine (NMDA receptor open channel antagonist) for the treatment of cognitive dysfunction in schizophrenia and other neuropsychiatric disorders (Bai et al., 2021; Koola, 2018). A 12-week, double-blind, placebo-controlled crossover study of GAL treatment in PWH for effects on neurocognition and quality of life outcomes has recently been completed, with results currently pending (ClinicalTrials.gov identifier: NCT03384784).

41.8 Therapies Targeting Cell Trafficking and Trophic Factors

41.8.1 Cell Trafficking

The rationale for limiting trafficking of HIV-infected cells into the CNS as a treatment strategy for HIV-NCI is controversial, but potentially useful for limiting the size of the CNS HIV reservoir and preventing continual re-introduction of HIV into the CNS even after infection is established (Campbell et al., 2014; Lakritz et al., 2016; Pino et al., 2019, 2022; Resop et al., 2020). Fingolimod (FTY720), an FDA-approved treatment for multiple sclerosis, is a lysophospholipid sphingosine-1 phosphate receptor modulator that limits egress of CD4+ and CD8+ T lymphocytes from lymph nodes (McGinley & Cohen, 2021). Although no studies of fingolimod in PWH or studies of effects in the CNS of SIV-infected macaques have been published, a pilot study in SIV-infected rhesus macaques showed reduction in SIV DNA-containing lymphocytes in lymph nodes in fingolimod-treated animals, presumably reflecting the action of retained nodal cytotoxic T lymphocytes (Pino et al., 2019). A follow up to that SIV study by the same group showed that early fingolimod treatment (during the first few weeks of ART) in SIV-infected macaques failed to significantly limit SIV persistence (Pino et al., 2022). Although fingolimod is currently used for treating multiple sclerosis, recent concerns about the possibility of reduced response to SARS CoV-2 vaccination may temper enthusiasm for expanding neuroprotection studies to HIV-NCI (Kister et al., 2022).

Natalizumab, an FDA-approved treatment for multiple sclerosis, is an anti-α4 integrin-targeting monoclonal antibody that inhibits monocyte and T-lymphocyte trafficking to the brain and other tissues (De Kleijn & Martens, 2020; Haanstra et al., 2013). In the rhesus macaque model of multiple sclerosis (experimental autoimmune encephalomyelitis) natalizumab blocked inflammation, demyelination, and ingress of monocytes and T lymphocytes into the CNS (Haanstra et al., 2013). In a pilot study in rhesus macaques, natalizumab treatment prior to SIV inoculation prevented CNS infection (Campbell et al., 2014). Furthermore, natalizumab treatment post-SIV infection significantly reduced progressive neuronal injury, which suggests that natalizumab could have some neuroprotective effect in PWH by reducing continual trafficking of HIV-infected cells into the CNS. However, the relatively high risk of reactivation of JC virus and associated risk of progressive multifocal leukoencephalopathy (PML) with prolonged treatment with natalizumab (observed in treated multiple sclerosis patients), will limit enthusiasm for neuroprotection studies in HIV-NCI.

41.8.2 Trophic Factors

41.8.2.1 Intranasal Insulin

To date, published studies of clinical investigations of trophic factors for the treatment of HIV-NCI are limited. In patients without HIV, administration of insulin intranasally has been associated with improved cognitive performance in patients with diabetes mellitus, while a randomized clinical trial in patients with Alzheimer's disease showed no benefit (Craft et al., 2020; Gaddam et al., 2021). A preliminary report of a pilot study (ClinicalTrials.gov Identifier: NCT03081117) of intranasal insulin administration for the treatment of HIV-NCI in 21 PWH indicated improvement in neuropsychological tests of memory and attention (https://www.natap.org/2021/CROI/croi_91.htm). A full report of clinical and biomarker outcomes in this pilot study is anticipated.

41.8.2.2 Tesamorelin

Tesamorelin is an injectable synthetic 44 amino acid polypeptide analogue of growth hormone releasing hormone (GHRH) that is FDA-approved for the treatment of HIV lipodystrophy, and it also under intense investigation as a treatment for HIV-associated non-alcoholic-fatty liver disease (NAFLD) (Yen & Sherman, 2022). Its activation of GHRH receptors in the pituitary leads to release of growth hormone and subsequent release of insulin-like growth factor-1 (IGF-1), which has potent neuroprotective properties in models of peripheral nerve injury (Slavin et al., 2021; Tuffaha et al., 2016). In a randomized, double-blind, placebo-controlled trial (20 weeks of treatment) of 152 adults, tesamorelin was shown to improve cognitive functioning in both healthy adults and in those with mild cognitive impairment (Baker et al., 2012). A phase II trial of tesamorelin for the treatment (6 months of treatment) of HIV-NCI in PWH is now underway (ClinicalTrials.gov identifier: NCT02572323).

41.9 Alternative Adjunctive Therapies: Cognitive Therapy and Exercise

Non-pharmacologic interventions such as cognitive therapy and exercise may have a role in improving HIV-NCI through direct or indirect effects (Dufour et al., 2018; Monroe et al., 2017; Vance et al., 2021). A recent meta-analysis of randomized controlled trials (n = 22 studies; 643 patients treated, 633 control patients; age range 63.8–80.1 years) concluded that working memory training has been shown to improve processing speed and cognitive reasoning in older, healthy adults (Hou et al., 2020). Numerous clinical trials of cognitive behavioral therapy (CBT), which is focused on mood and anxiety outcomes, have been initiated in PWH. Studies of cognitive therapy focused on cognitive dysfunction (memory, attention) are limited, however. Nonetheless, cognitive training in PWH, has demonstrated beneficial effects in specific cognitive domains that are targeted by the training, and an interventional clinical trial with individualized targeting of specific domains of deficit in 109 PWH has been completed (Vance et al., 2019, 2021) (ClinicalTrials.gov identifier: NCT03122288).

Neuroprotective effects of aerobic exercise on brain integrity have been well-documented (Tarumi et al., 2021). In PWH, physical exercise indeed associates with lower risk for HIV-NCI (Dickens et al., 2015; Dufour et al., 2018). Both interventional and non-interventional studies of physical activity in PWH have shown preservation or improvement in cognitive functioning in PWH (Quigley et al., 2019). A longitudinal study of 291 PWH demonstrated that individuals with consistent physical activity (physical activity and >50% of study visits) maintained better neurocognitive functioning in domains of verbal fluency, working memory, speed of information processing executive function, and motor function over a mean period of 35 months, even when adjusting for confounding factors (Dufour et al., 2018). Initiating cognitive training with regular aerobic exercise is a low-cost, low risk strategy that clearly should be implemented in PWH for reducing risk for HIV-NCI.

41.10 Conclusions

Reducing the persistent risk of HIV-NCI in PWH despite suppressive cART requires improvement of cART CNS effectiveness and the implementation of adjunctive therapies for risk reduction based upon preservation of brain integrity. Optimum treatment of HIV-NCI has not yet been achieved, but there is promise for the future. Overall, the strategy of increasing cART CNS effectiveness and reducing CNS ART toxicity risks are worthy goals to develop as newer ART drugs of all classes are brought forth. Lessons from past clinical trials also suggest that early intervention with agents that more broadly target relevant pathogenic pathways effects should be prioritized, and among such attractive targets are pathways of inflammation, oxidative stress, neurotransmitter metabolism, and metabolism-modulating trophic factors. Each of these is targetable in patients receiving ART, which will always be the mainstay of neuroprotection against HIV. Early intervention with neuroprotective strategies at the time of diagnosis and initiation of cART is likely the soundest approach for limiting risk of HIV-NCI. For drug therapies, it is likely that concurrent initiation of an adjunctive therapeutic drug in combination with cART, and even short-duration treatment with a robust neuroprotectant, may have long-term benefits. Agents that robustly activate endogenous antioxidant response genes while suppressing associated inflammatory responses should be pursued, and one example is dimethyl fumarate, widely used in the treatment of multiple sclerosis. Additionally, therapies initiated later in the course of HIV infection may also have an impact. Minimizing effects of impaired gut mucosal integrity and altered gut microflora with microbiome modifications also hold future promise. Other currently approved neuromodulating drugs that limit immune cell trafficking into the CNS in multiple sclerosis patients are probably too risky for use in PWH. Re-purposing of other available drugs, including SSRIs such as paroxetine, which also directly or indirectly suppress inflammatory and oxidative stress processes, also deserves immediate attention. Development of new pharmacological agents targeting these and other newly identified pathologic pathways should also be pursued as a longer-term strategy. Each approach should be combined with aerobic exercising and cognitive therapy, which can be implemented regardless of other treatment paradigms. In total, these approaches will likely also mitigate effects of comorbidities that contribute to HIV-NCI and enhance the quality of life for PWH.

41.11 Review Questions

1. Of the following statements, which *incorrectly* describes the effects of ART on HIV-NCI?
   (a) Since the introduction of cART, the overall prevalence of severe HIV-NCI (HIV-associated dementia) has decreased from approximately 20% to between 2 and 5%.
   (b) Milder HIV-NCI that is associated with impaired activities of daily living is found in 10–12% of virally suppressed PWH.
   (c) CSF HIV 'blipping' or 'escape' occurs in approximately 10–20% of virally suppressed PWH.
   (d) Chemokine receptor antagonists, such as maraviroc and cenicriviroc, act exclusively by blocking HIV binding to chemokine receptors.
2. Which of the following statements *incorrectly* describes the use of INSTIs in the treatment of PWH?
   (a) INSTIs block the integration of HIV provirus into cellular DNA, preventing productive HIV replication in the target cell.
   (b) Integrase strand transfer inhibitors (INSTI) are not recommended for initial ART regimens.
   (c) Dolutegravir (DTG) and bictegravir (BIC) have, in some studies, been noted to cause neuropsychiatric symptoms such as sleep disturbances, anxiety, and depression.
   (d) Concerns over the potential risk for fetal neural tube defects with the use of dolutegravir (DTG) during

pregnancy have been proven to be unfounded, and DTG use during pregnancy is recommended.

3. The enhancement of ART CNS penetrance is a goal in achieving CNS viral suppression. Which of the following statements relating to this goal is *incorrect*?
   (a) An ART ranking system known as CNS penetrance efficacy (CPE score) was developed for ART drugs to assess their relative activity in the CNS.
   (b) The use of favorably-ranked (CPE score) cART regimens consistently reduces CSF neuroinflammatory markers and CSF HIV load.
   (c) Nanoformulations of ART drugs have demonstrated improved tissue delivery and viral suppression in animal studies
   (d) Brain magnetic resonance spectroscopy (MRS) has been used to assess neuronal integrity to attempt to monitor the CNS response to different ART regimens.
4. Which statement regarding dimethyl fumarate (DMF) is *true*?
   (a) DMF is a Nrf2 transcriptional inhibitor and is an FDA-approved treatment commonly used for patients with Parkinson's disease.
   (b) DMF's primary in vivo metabolite methyl fumarate (MMF) has no efficacy against macrophage-mediated neurotoxicity in in vitro models of HIV neurotoxicity.
   (c) DMF is not an FDA-approved treatment for any neurological disorder but has shown promise in in vivo models of demyelinating diseases.
   (d) DMF is a Nrf2 transcriptional activator and is an FDA-approved treatment for patients with multiple sclerosis.
5. Which pharmacologic/adjunctive therapy is correctly described below?
   (a) Statins block hydroxymethylglutaryl-CoA (HMG-CoA) reductase enzymatic activity and increase blood lipid levels.
   (b) Cenicriviroc blocks both CCR5 and CCR2 chemokine receptors to block HIV entry into the cell, inflammation, and monocyte migration.
   (c) Memantine is a non-competitive NMDA glutamate receptor agonist approved for the use in treatment of Alzheimer's disease.
   (d) Aerobic exercise in PWH does not associate with lower risk of HIV-NCI, despite the documented neuroprotective effects found in other neurodegenerative diseases.
6. Many therapies for other neurodegenerative or neuropsychiatric diseases are being investigated for use in HIV-NCI. Which of the drug therapies below is incorrectly paired with its molecular target?
   (a) Sodium Valproate (VPA) is a glycogen synthase kinase-3β inhibitor.
   (b) Selegiline is a monoamine oxidase B (MAO-B) inhibitor.
   (c) Galantamine (GAL) is a selective serotonin reuptake inhibitor (SSRI).
   (d) Lithium is a glycogen synthase kinase-3β inhibitor.

41.12 Answers

1. (d). CCR5 chemokine receptor blockers act by inhibiting HIV-1 entry and by inhibiting chemokine-mediated chemotaxis of immune cells.
2. (b). INSTIs are recommended for initial ART regimens
3. (b). CNS penetrance efficacy (CPE) is a ranking system developed for ART drugs that penetrate the CNS. Answer B is incorrect. Although some studies have shown that some favorably ranked cART regimens may reduce CSF neuroinflammatory markers and CSF HIV load, the effect is not consistent among many studies.
4. (d). DMF is a Nrf2 transcriptional activator and is an FDA-approved treatment for patients with multiple sclerosis. Answer A is incorrect because DMF is not a Nrf2 inhibitor and not an FDA-approved treatment for Parkinson's disease. Answer B is incorrect because the primary in vivo metabolite of DMF, monomethyl fumarate (MMF) has efficacy against macrophage-mediated neurotoxicity in in vitro models of HIV neurotoxicity. Answer C is incorrect because DMF is approved by the FDA to treat multiple sclerosis,
5. (b). Cenicriviroc is a CCR5 and CCR2 inhibitor that functions to block HIV entry into the cell, inflammation, and monocyte migration. Answer A is incorrect because although statins do block HMG-CoA reductase enzyme activity, they do so to decrease lipid levels, not increase them. Answer C in incorrect because memantine is a non-competitive NMDA glutamate receptor antagonist. It is used to treat Alzheimer's diseases, but its action is as an inhibitor. Answer D is incorrect because in PWH physical exercise does indeed associate with lower risk of HIV-NCI.
6. (c). Galantamine (GAL) is not a selective serotonin reuptake inhibitor, it is an acetylcholinesterase (AChE) inhibitor and nicotinic receptor agonist.

Compliance with Ethical Standards

Conflict of Interest

Dennis Kolson declares that he has no conflict of interest.
Adam Adelsberg declares that he has no conflict of interest.

Human and Animal Rights

All reported studies/experiments with human or animal subjects performed by the authors have been previously published and complied with all applicable ethical standards (including the Helsinki declaration and its amendments, institutional/national research committee standards, and international/national/institutional guidelines).

References

Abrams, E., & Myer, L. (2021). Lessons from dolutegravir and neural tube defects. *Lancet HIV, 8*(1), e3–e4. https://doi.org/10.1016/S2352-3018(20)30280-0

Aguilera, G., Colin-Gonzalez, A. L., Rangel-Lopez, E., Chavarria, A., & Santamaria, A. (2018). Redox signaling, neuroinflammation, and neurodegeneration. *Antioxidants & Redox Signaling, 28*(18), 1626–1651. https://doi.org/10.1089/ars.2017.7099

Ambrosius, B., Faissner, S., Guse, K., von Lehe, M., Grunwald, T., Gold, R., et al. (2017). Teriflunomide and monomethylfumarate target HIV-induced neuroinflammation and neurotoxicity. *Journal of Neuroinflammation, 14*(1), 51. https://doi.org/10.1186/s12974-017-0829-2

Ambrosius, B., Gold, R., Chan, A., & Faissner, S. (2019). Antineuroinflammatory drugs in HIV-associated neurocognitive disorders as potential therapy. *Neurology, Neuroimmunology & Neuroinflammation, 6*(3), e551. https://doi.org/10.1212/NXI.0000000000000551

Ances, B. M., Letendre, S. L., Alexander, T., & Ellis, R. J. (2008). Role of psychiatric medications as adjunct therapy in the treatment of HIV associated neurocognitive disorders. *International Review of Psychiatry, 20*(1), 89–93. https://doi.org/10.1080/09540260701877670

Ancuta, P., Kamat, A., Kunstman, K. J., Kim, E. Y., Autissier, P., Wurcel, A., et al. (2008). Microbial translocation is associated with increased monocyte activation and dementia in AIDS patients. *PLoS One, 3*(6), e2516. https://doi.org/10.1371/journal.pone.0002516

Angelovich, T. A., Churchill, M. J., Wright, E. J., & Brew, B. J. (2021). New potential axes of HIV neuropathogenesis with relevance to biomarkers and treatment. In L. A. Cysique & S. B. Rourke (Eds.), *Neurocognitive complications of HIV-infection: Neuropathogenesis to implications for clinical practice* (pp. 3–39). Springer International Publishing. https://doi.org/10.1007/7854_2019_126

Antinori, A., Arendt, G., Becker, J. T., Brew, B. J., Byrd, D. A., Cherner, M., et al. (2007). Updated research nosology for HIV-associated neurocognitive disorders. *Neurology, 69*(18), 1789–1799. https://doi.org/10.1212/01.WNL.0000287431.88658.8b

Arentoft, A., Troxell, K., Alvarez, K., Aghvinian, M., Rivera Mindt, M., Cherner, M., et al. (2022). HIV antiretroviral medication neuropenetrance and neurocognitive outcomes in HIV+ adults: A review of the literature examining the central nervous system penetration effectiveness score. *Viruses, 14*(6). https://doi.org/10.3390/v14061151

Avan, A., Digaleh, H., Di Napoli, M., Stranges, S., Behrouz, R., Shojaeianbabaei, G., et al. (2019). Socioeconomic status and stroke incidence, prevalence, mortality, and worldwide burden: An ecological analysis from the Global Burden of Disease Study 2017. *BMC Medicine, 17*(1), 191. https://doi.org/10.1186/s12916-019-1397-3

Bai, M. Y., Lovejoy, D. B., Guillemin, G. J., Kozak, R., Stone, T. W., & Koola, M. M. (2021). Galantamine-memantine combination and kynurenine pathway enzyme inhibitors in the treatment of neuropsychiatric disorders. *Complex Psychiatry, 7*(1–2), 19–33. https://doi.org/10.1159/000515066

Baker, L. D., Barsness, S. M., Borson, S., Merriam, G. R., Friedman, S. D., Craft, S., & Vitiello, M. V. (2012). Effects of growth hormone-releasing hormone on cognitive function in adults with mild cognitive impairment and healthy older adults: Results of a controlled trial. *Archives of Neurology, 69*(11), 1420–1429. https://doi.org/10.1001/archneurol.2012.1970

Bandaru, V. V., McArthur, J. C., Sacktor, N., Cutler, R. G., Knapp, E. L., Mattson, M. P., & Haughey, N. J. (2007). Associative and predictive biomarkers of dementia in HIV-1-infected patients. *Neurology, 68*(18), 1481–1487. https://doi.org/10.1212/01.wnl.0000260610.79853.47

Biswas, S. K. (2016). Does the interdependence between oxidative stress and inflammation explain the antioxidant paradox? *Oxidative Medicine and Cellular Longevity, 2016*, 5698931. https://doi.org/10.1155/2016/5698931

Board, N. L., Moskovljevic, M., Wu, F., Siliciano, R. F., & Siliciano, J. D. (2022). Engaging innate immunity in HIV-1 cure strategies. *Nature Reviews. Immunology, 22*(8), 499–512. https://doi.org/10.1038/s41577-021-00649-1

Bougea, A., Spantideas, N., Galanis, P., Gkekas, G., & Thomaides, T. (2019). Optimal treatment of HIV-associated neurocognitive disorders: Myths and reality. A critical review. *Therapeutic Advances in Infectious Disease, 6*, 2049936119838228. https://doi.org/10.1177/2049936119838228

Brenchley, J. M., Price, D. A., Schacker, T. W., Asher, T. E., Silvestri, G., Rao, S., et al. (2006). Microbial translocation is a cause of systemic immune activation in chronic HIV infection. *Nature Medicine, 12*(12), 1365–1371. https://doi.org/10.1038/nm1511

Brew, B. J. (2004). Evidence for a change in AIDS dementia complex in the era of highly active antiretroviral therapy and the possibility of new forms of AIDS dementia complex. *AIDS, 18 Suppl 1*, S75–S78.

Brkic, S., Veres, B., Thurnher, M. M., Boban, J., Radovanovic, B., Tomic, S., & Kozic, D. (2023). CNS efficacy parameters of combination antiretroviral therapy in chronic HIV infection: A multi-voxel magnetic resonance spectroscopy study. *Frontiers in Neurology, 14*, 943183. https://doi.org/10.3389/fneur.2023.943183

Burgener, A., McGowan, I., & Klatt, N. R. (2015). HIV and mucosal barrier interactions: Consequences for transmission and pathogenesis. *Current Opinion in Immunology, 36*, 22–30. https://doi.org/10.1016/j.coi.2015.06.004

Campbell, J. H., Ratai, E. M., Autissier, P., Nolan, D. J., Tse, S., Miller, A. D., et al. (2014). Anti-alpha4 antibody treatment blocks virus traffic to the brain and gut early, and stabilizes CNS injury late in infection. *PLoS Pathogens, 10*(12), e1004533. https://doi.org/10.1371/journal.ppat.1004533

Ceccarelli, G., Brenchley, J. M., Cavallari, E. N., Scheri, G. C., Fratino, M., Pinacchio, C., et al. (2017a). Impact of high-dose multi-Strain probiotic supplementation on neurocognitive performance and central nervous system immune activation of HIV-1 infected individuals. *Nutrients, 9*(11). https://doi.org/10.3390/nu9111269

Ceccarelli, G., Fratino, M., Selvaggi, C., Giustini, N., Serafino, S., Schietroma, I., et al. (2017b). A pilot study on the effects of probiotic supplementation on neuropsychological performance and microRNA-29a-c levels in antiretroviral-treated HIV-1-infected patients. *Brain and Behavior: A Cognitive Neuroscience Perspective, 7*(8), e00756. https://doi.org/10.1002/brb3.756

Chaillon, A., Gianella, S., Dellicour, S., Rawlings, S. A., Schlub, T. E., De Oliveira, M. F., et al. (2020). HIV persists throughout deep tissues with repopulation from multiple anatomical sources. *The Journal of Clinical Investigation, 130*(4), 1699–1712. https://doi.org/10.1172/JCI134815

Chan, P., & Spudich, S. (2022). HIV compartmentalization in the CNS and its impact in treatment outcomes and cure strategies. *Current HIV/AIDS Reports, 19*(3), 207–216. https://doi.org/10.1007/s11904-022-00605-1

Chan, P., Goh, O., Kroon, E., Colby, D., Sacdalan, C., Pinyakorn, S., et al. (2020). Neuropsychiatric outcomes before and after switching to dolutegravir-based therapy in an acute HIV cohort. *AIDS Research and Therapy, 17*(1), 1. https://doi.org/10.1186/s12981-019-0257-8

Chan, P., Kerr, S. J., Kroon, E., Colby, D., Sacdalan, C., Hellmuth, J., et al. (2021). Cognitive trajectories after treatment in acute HIV infection. *AIDS*. https://doi.org/10.1097/QAD.0000000000002831

Chen, M., Zhou, L., Ding, H., Huang, S., He, M., Zhang, X., et al. (2012). Short (GT) (n) repeats in heme oxygenase-1 gene promoter are associated with lower risk of coronary heart disease in subjects with high levels of oxidative stress. *Cell Stress & Chaperones, 17*(3), 329–338. https://doi.org/10.1007/s12192-011-0309-z

Chen, M. F., Gill, A. J., & Kolson, D. L. (2014). Neuropathogenesis of HIV-associated neurocognitive disorders: Roles for immune activation, HIV blipping and viral tropism. *Current Opinion in HIV and AIDS, 9*(6), 559–564. https://doi.org/10.1097/COH.0000000000000105

Christensen, B. L., & Tan, D. H. (2022). An up-to-date evaluation of dolutegravir/abacavir/lamivudine for the treatment of HIV. *Expert Opinion on Pharmacotherapy, 23*(4), 439–446. https://doi.org/10.1080/14656566.2022.2029409

Chu, L., Shu, Z., Gu, X., Wu, Y., Yang, J., & Deng, H. (2023). The endocannabinoid system as a potential therapeutic target for HIV-1-associated neurocognitive disorder. *Cannabis and Cannabinoid Research, 8*(3), 445–463. https://doi.org/10.1089/can.2022.0267

Clifford, D. B., McArthur, J. C., Schifitto, G., Kieburtz, K., McDermott, M. P., Letendre, S., et al. (2002). A randomized clinical trial of CPI-1189 for HIV-associated cognitive-motor impairment. *Neurology, 59*(10), 1568–1573. http://www.ncbi.nlm.nih.gov/entrez/query.fcgi?cmd=Retrieve&db=PubMed&dopt=Citation&list_uids=12451199

Cohn, L. B., Chomont, N., & Deeks, S. G. (2020). The biology of the HIV-1 latent reservoir and implications for cure strategies. *Cell Host & Microbe, 27*(4), 519–530. https://doi.org/10.1016/j.chom.2020.03.014

Craft, S., Raman, R., Chow, T. W., Rafii, M. S., Sun, C. K., Rissman, R. A., et al. (2020). Safety, efficacy, and feasibility of intranasal insulin for the treatment of mild cognitive impairment and Alzheimer disease dementia: A randomized clinical trial. *JAMA Neurology, 77*(9), 1099–1109. https://doi.org/10.1001/jamaneurol.2020.1840

Cross, S. A., Cook, D. R., Chi, A. W., Vance, P. J., Kolson, L. L., Wong, B. J., et al. (2011). Dimethyl fumarate, an immune modulator and inducer of the antioxidant response, suppresses HIV replication and macrophage-mediated neurotoxicity: A novel candidate for HIV neuroprotection [Research Support, N.I.H., Extramural]. *Journal of Immunology, 187*(10), 5015–5025. https://doi.org/10.4049/jimmunol.1101868

Cusini, A., Vernazza, P. L., Yerly, S., Decosterd, L. A., Ledergerber, B., Fux, C. A., et al. (2013). Higher CNS penetration-effectiveness of long-term combination antiretroviral therapy is associated with better HIV-1 viral suppression in cerebrospinal fluid. *Journal of Acquired Immune Deficiency Syndromes, 62*(1), 28–35. https://doi.org/10.1097/QAI.0b013e318274e2b0

Cysique, L. A., Maruff, P., & Brew, B. J. (2006a). Valproic acid is associated with cognitive decline in HIV-infected individuals: A clinical observational study. *BMC Neurology, 6*, 42. https://doi.org/10.1186/1471-2377-6-42

Cysique, L. A., Maruff, P., & Brew, B. J. (2006b). Variable benefit in neuropsychological function in HIV-infected HAART-treated patients. *Neurology, 66*(9), 1447–1450. https://doi.org/10.1212/01.wnl.0000210477.63851.d3

Dahl, V., Lee, E., Peterson, J., Spudich, S. S., Leppla, I., Sinclair, E., et al. (2011). Raltegravir treatment intensification does not alter cerebrospinal fluid HIV-1 infection or immunoactivation in subjects on suppressive therapy. *The Journal of Infectious Diseases, 204*(12), 1936–1945. https://doi.org/10.1093/infdis/jir667

D'Antoni, M. L., Paul, R. H., Mitchell, B. I., Kohorn, L., Fischer, L., Lefebvre, E., et al. (2018). Improved cognitive performance and reduced monocyte activation in virally suppressed chronic HIV after dual CCR2 and CCR5 antagonism. *Journal of Acquired Immune Deficiency Syndromes, 79*(1), 108–116. https://doi.org/10.1097/QAI.0000000000001752

Dastgheyb, R. M., Sacktor, N., Franklin, D., Letendre, S., Marcotte, T., Heaton, R., et al. (2019). Cognitive trajectory phenotypes in human immunodeficiency virus-infected patients. *Journal of Acquired Immune Deficiency Syndromes, 82*(1), 61–70. https://doi.org/10.1097/QAI.0000000000002093

Dastgheyb, R. M., Buchholz, A. S., Fitzgerald, K. C., Xu, Y., Williams, D. W., Springer, G., et al. (2021). Patterns and predictors of cognitive function among virally suppressed women with HIV. *Frontiers in Neurology, 12*, 604984. https://doi.org/10.3389/fneur.2021.604984

Davis, L. E., Hjelle, B. L., Miller, V. E., Palmer, D. L., Llewellyn, A. L., Merlin, T. L., et al. (1992). Early viral brain invasion in iatrogenic human immunodeficiency virus infection. *Neurology, 42*(9), 1736–1739. http://www.ncbi.nlm.nih.gov/entrez/query.fcgi?cmd=Retrieve&db=PubMed&dopt=Citation&list_uids=1513462

De Kleijn, K. M. A., & Martens, G. J. M. (2020). Molecular effects of FDA-approved multiple sclerosis drugs on glial cells and neurons of the central nervous system. *International Journal of Molecular Sciences, 21*(12). https://doi.org/10.3390/ijms21124229

Decloedt, E. H., Freeman, C., Howells, F., Casson-Crook, M., Lesosky, M., Koutsilieri, E., et al. (2016). Moderate to severe HIV-associated neurocognitive impairment: A randomized placebo-controlled trial of lithium. *Medicine (Baltimore), 95*(46), e5401. https://doi.org/10.1097/MD.0000000000005401

Deeks, E. D. (2016). Dimethyl fumarate: A review in relapsing-remitting MS. *Drugs, 76*(2), 243–254. https://doi.org/10.1007/s40265-015-0528-1

Deng, L., Zhang, X., Gao, Y., Turner, D., Qian, F., Lu, H., et al. (2021). Association of HIV infection and cognitive impairment in older adults: A meta-analysis. *Ageing Research Reviews, 68*, 101310. https://doi.org/10.1016/j.arr.2021.101310

Deodhar, S., Sillman, B., Bade, A. N., Avedissian, S. N., Podany, A. T., McMillan, J. M., et al. (2022). Transformation of dolutegravir into an ultra-long-acting parenteral prodrug formulation. *Nature Communications, 13*(1), 3226. https://doi.org/10.1038/s41467-022-30902-7

Dickens, A. M., Anthony, D. C., Deutsch, R., Mielke, M. M., Claridge, T. D., Grant, I., et al. (2015). Cerebrospinal fluid metabolomics implicate bioenergetic adaptation as a neural mechanism regulating shifts in cognitive states of HIV-infected patients. *AIDS, 29*(5), 559–569. https://doi.org/10.1097/QAD.0000000000000580

Diggins, C. E., Russo, S. C., & Lo, J. (2022). Metabolic consequences of antiretroviral therapy. *Current HIV/AIDS Reports*. https://doi.org/10.1007/s11904-022-00600-6

Dinkova-Kostova, A. T., Kostov, R. V., & Kazantsev, A. G. (2018). The role of Nrf2 signaling in counteracting neurodegenerative diseases. *The FEBS Journal, 285*(19), 3576–3590. https://doi.org/10.1111/febs.14379

Dou, H., Birusingh, K., Faraci, J., Gorantla, S., Poluektova, L. Y., Maggirwar, S. B., et al. (2003). Neuroprotective activities of sodium valproate in a murine model of human immunodeficiency virus-1 encephalitis. *The Journal of Neuroscience, 23*(27), 9162–9170.

Dou, H., Grotepas, C. B., McMillan, J. M., Destache, C. J., Chaubal, M., Werling, J., et al. (2009). Macrophage delivery of nanoformulated antiretroviral drug to the brain in a murine model of neuroAIDS. *Journal of Immunology, 183*(1), 661–669. https://doi.org/10.4049/jimmunol.0900274

Driscoll, K. E. (2000). TNFalpha and MIP-2: Role in particle-induced inflammation and regulation by oxidative stress. *Toxicology Letters, 112-113*, 177–183. https://doi.org/10.1016/s0378-4274(99)00282-9

Dubey, D., Kieseier, B. C., Hartung, H. P., Hemmer, B., Warnke, C., Menge, T., et al. (2015). Dimethyl fumarate in relapsing-remitting multiple sclerosis: Rationale, mechanisms of action, pharmacokinetics, efficacy and safety. *Expert Review of Neurotherapeutics, 15*(4), 339–346. https://doi.org/10.1586/14737175.2015.1025755

Dufour, C. A., Marquine, M. J., Fazeli, P. L., Umlauf, A., Henry, B. L., Zlatar, Z., et al. (2018). A longitudinal analysis of the impact of physical activity on neurocognitive functioning among HIV-infected adults. *AIDS and Behavior, 22*(5), 1562–1572. https://doi.org/10.1007/s10461-016-1643-z

Dunfee, R. L., Thomas, E. R., Gorry, P. R., Wang, J., Taylor, J., Kunstman, K., et al. (2006). The HIV Env variant N283 enhances macrophage tropism and is associated with brain infection and dementia. *Proceedings of the National Academy of Sciences of*

the United States of America, 103(41), 15160–15165. https://doi.org/10.1073/pnas.0605513103
Edagwa, B., McMillan, J., Sillman, B., & Gendelman, H. E. (2017). Long-acting slow effective release antiretroviral therapy. *Expert Opinion on Drug Delivery, 14*(11), 1281–1291. https://doi.org/10.1080/17425247.2017.1288212
Eden, A., Price, R. W., Spudich, S., Fuchs, D., Hagberg, L., & Gisslen, M. (2007). Immune activation of the central nervous system is still present after >4 years of effective highly active antiretroviral therapy. *The Journal of Infectious Diseases, 196*(12), 1779–1783. https://doi.org/10.1086/523648
Eden, A., Fuchs, D., Hagberg, L., Nilsson, S., Spudich, S., Svennerholm, B., et al. (2010). HIV-1 viral escape in cerebrospinal fluid of subjects on suppressive antiretroviral treatment. *The Journal of Infectious Diseases, 202*(12), 1819–1825. https://doi.org/10.1086/657342
Ellis, R., Langford, D., & Masliah, E. (2007). HIV and antiretroviral therapy in the brain: Neuronal injury and repair. *Nature Reviews. Neuroscience, 8*(1), 33–44. http://www.ncbi.nlm.nih.gov/entrez/query.fcgi?cmd=Retrieve&db=PubMed&dopt=Citation&list_uids=17180161
Ellis, R. J., Badiee, J., Vaida, F., Letendre, S., Heaton, R. K., Clifford, D., et al. (2011). CD4 nadir is a predictor of HIV neurocognitive impairment in the era of combination antiretroviral therapy. *AIDS, 25*(14), 1747–1751. https://doi.org/10.1097/QAD.0b013e32834a40cd
Ellis, R. J., Peterson, S. N., Li, Y., Schrier, R., Iudicello, J., Letendre, S., et al. (2020). Recent cannabis use in HIV is associated with reduced inflammatory markers in CSF and blood. *Neurology, Neuroimmunology & Neuroinflammation, 7*(5). https://doi.org/10.1212/NXI.0000000000000809
Eshun-Wilson, I., Siegfried, N., Akena, D. H., Stein, D. J., Obuku, E. A., & Joska, J. A. (2018). Antidepressants for depression in adults with HIV infection. *Cochrane Database of Systematic Reviews, 1*, CD008525. https://doi.org/10.1002/14651858.CD008525.pub3
Estes, J. D., Kityo, C., Ssali, F., Swainson, L., Makamdop, K. N., Del Prete, G. Q., et al. (2017). Defining total-body AIDS-virus burden with implications for curative strategies. *Nature Medicine, 23*(11), 1271–1276. https://doi.org/10.1038/nm.4411
Everall, I. P., Bell, C., Mallory, M., Langford, D., Adame, A., Rockestein, E., & Masliah, E. (2002). Lithium ameliorates HIV-gp120-mediated neurotoxicity. *Molecular and Cellular Neurosciences, 21*(3), 493–501.
Files, D. C., Tacke, F., O'Sullivan, A., Dorr, P., Ferguson, W. G., & Powderly, W. G. (2022). Rationale of using the dual chemokine receptor CCR2/CCR5 inhibitor cenicriviroc for the treatment of COVID-19. *PLoS Pathogens, 18*(6), e1010547. https://doi.org/10.1371/journal.ppat.1010547
Filippidis, P., Damas, J., Viala, B., Assal, F., Nawej Tshikung, O., Tarr, P., et al. (2023). Cerebrospinal fluid HIV-1 escape in patients with neurocognitive symptoms: Pooled data from a neuro-HIV platform and the NAMACO study. *Journal of Acquired Immune Deficiency Syndromes, 93*(3), 219–228. https://doi.org/10.1097/QAI.0000000000003189
Force, G., Ghout, I., Ropers, J., Carcelain, G., Marigot-Outtandy, D., Hahn, V., et al. (2021). Improvement of HIV-associated neurocognitive disorders after antiretroviral therapy intensification: The Neuro+3 study. *The Journal of Antimicrobial Chemotherapy, 76*(3), 743–752. https://doi.org/10.1093/jac/dkaa473
Fuchs, J., Oelke, N., Imhof, M., Ochsendorf, F., Schofer, H., Oromek, G., et al. (1998). Multiparameter analysis of clastogenic factors, pro-oxidant cytokines, and inflammatory markers in HIV-1-infected patients with asymptomatic disease, opportunistic infections, and malignancies. *Molecular Medicine, 4*(5), 333–343. https://www.ncbi.nlm.nih.gov/pubmed/9642683
Funderburg, N. T., Jiang, Y., Debanne, S. M., Storer, N., Labbato, D., Clagett, B., et al. (2014). Rosuvastatin treatment reduces markers of monocyte activation in HIV-infected subjects on antiretroviral therapy. *Clinical Infectious Diseases, 58*(4), 588–595. https://doi.org/10.1093/cid/cit748
Gaddam, M., Singh, A., Jain, N., Avanthika, C., Jhaveri, S., De la Hoz, I., et al. (2021). A comprehensive review of intranasal insulin and its effect on the cognitive function of diabetics. *Cureus, 13*(8), e17219. https://doi.org/10.7759/cureus.17219
Gandhi, R. T., Bedimo, R., Hoy, J. F., Landovitz, R. J., Smith, D. M., Eaton, E. F., et al. (2023). Antiretroviral drugs for treatment and prevention of HIV infection in adults: 2022 recommendations of the international antiviral society-USA panel. *JAMA, 329*(1), 63–84. https://doi.org/10.1001/jama.2022.22246
Ganesan, A., Crum-Cianflone, N., Higgins, J., Qin, J., Rehm, C., Metcalf, J., et al. (2011). High dose atorvastatin decreases cellular markers of immune activation without affecting HIV-1 RNA levels: Results of a double-blind randomized placebo controlled clinical trial. *The Journal of Infectious Diseases, 203*(6), 756–764. https://doi.org/10.1093/infdis/jiq115
Garcia-Mesa, Y., Xu, H. E., Vance, P., Gruenewald, A. L., Garza, R., Midkiff, C., et al. (2021). Dimethyl fumarate, an approved multiple sclerosis treatment, reduces brain oxidative stress in SIV-infected rhesus macaques: Potential therapeutic repurposing for HIV neuroprotection. *Antioxidants, 10*. https://doi.org/10.3390/antiox10030416
Garza, R., Gill, A. J., Bastien, B. L., Garcia-Mesa, Y., Gruenewald, A. L., Gelman, B. B., et al. (2020). Heme oxygenase-1 promoter (GT) n polymorphism associates with HIV neurocognitive impairment. *Neurology, Neuroimmunology & Neuroinflammation, 7*(3). https://doi.org/10.1212/NXI.0000000000000710
Gates, T. M., Cysique, L. A., Siefried, K. J., Chaganti, J., Moffat, K. J., & Brew, B. J. (2016). Maraviroc-intensified combined antiretroviral therapy improves cognition in virally suppressed HIV-associated neurocognitive disorder. *AIDS, 30*(4), 591–600. https://doi.org/10.1097/QAD.0000000000000951
Gill, A. J., & Kolson, D. L. (2013). Dimethyl fumarate modulation of immune and antioxidant responses: Application to HIV therapy. *Critical Reviews in Immunology, 33*(4), 307–359. http://www.ncbi.nlm.nih.gov/pubmed/23971529
Gill, A. J., & Kolson, D. L. (2014). Chronic inflammation and the role for cofactors (hepatitis C, drug abuse, antiretroviral drug toxicity, aging) in HAND persistence. *Current HIV/AIDS Reports, 11*(3), 325–335. https://doi.org/10.1007/s11904-014-0210-3
Gill, A. J., Kovacsics, C. E., Cross, S. A., Vance, P. J., Kolson, L. L., Jordan-Sciutto, K. L., et al. (2014). Heme oxygenase-1 deficiency accompanies neuropathogenesis of HIV-associated neurocognitive disorders. *The Journal of Clinical Investigation, 124*(10), 4459–4472. https://doi.org/10.1172/JCI72279
Gill, A. J., Garza, R., Ambegaokar, S. S., Gelman, B. B., & Kolson, D. L. (2018). Heme oxygenase-1 promoter region (GT)n polymorphism associates with increased neuroimmune activation and risk for encephalitis in HIV infection. *Journal of Neuroinflammation, 15*(1), 70. https://doi.org/10.1186/s12974-018-1102-z
Gisslen, M., Keating, S. M., Spudich, S., Arechiga, V., Stephenson, S., Zetterberg, H., et al. (2021). Compartmentalization of cerebrospinal fluid inflammation across the spectrum of untreated HIV-1 infection, central nervous system injury and viral suppression. *PLoS One, 16*(5), e0250987. https://doi.org/10.1371/journal.pone.0250987
Golden, T. R., & Patel, M. (2009). Catalytic antioxidants and neurodegeneration. *Antioxidants & Redox Signaling, 11*(3), 555–570. https://doi.org/10.1089/ARS.2008.2256
Gomes, M. J., Neves, J., & Sarmento, B. (2014). Nanoparticle-based drug delivery to improve the efficacy of antiretroviral therapy in the central nervous system. *International Journal of Nanomedicine, 9*, 1757–1769. https://doi.org/10.2147/IJN.S45886
Gong, Y., Zhi, K., Nagesh, P. K. B., Sinha, N., Chowdhury, P., Chen, H., et al. (2020). An Elvitegravir Nanoformulation crosses the blood-

brain barrier and suppresses HIV-1 replication in microglia. *Viruses, 12*(5). https://doi.org/10.3390/v12050564

Goodkin, K., Vitiello, B., Lyman, W. D., Asthana, D., Atkinson, J. H., Heseltine, P. N., et al. (2006). Cerebrospinal and peripheral human immunodeficiency virus type 1 load in a multisite, randomized, double-blind, placebo-controlled trial of D-Ala1-peptide T-amide for HIV-1-associated cognitive-motor impairment. *Journal of Neurovirology, 12*(3), 178–189. https://doi.org/10.1080/13550280600827344

Grennan, J. T., Loutfy, M. R., Su, D., Harrigan, P. R., Cooper, C., Klein, M., et al. (2012). Magnitude of virologic blips is associated with a higher risk for virologic rebound in HIV-infected individuals: A recurrent events analysis. *The Journal of Infectious Diseases, 205*(8), 1230–1238. https://doi.org/10.1093/infdis/jis104

Haanstra, K. G., Hofman, S. O., Lopes Estevao, D. M., Blezer, E. L., Bauer, J., Yang, L. L., et al. (2013). Antagonizing the alpha4beta1 integrin, but not alpha4beta7, inhibits leukocytic infiltration of the central nervous system in rhesus monkey experimental autoimmune encephalomyelitis. *Journal of Immunology, 190*(5), 1961–1973. https://doi.org/10.4049/jimmunol.1202490

Hagberg, L., & Gisslen, M. (2023). Cohort profile: A longitudinal study of HIV infection in the central nervous system with focus on cerebrospinal fluid - the Gothenburg HIV CSF Study Cohort. *BMJ Open, 13*(5), e070693. https://doi.org/10.1136/bmjopen-2022-070693

Harezlak, J., Cohen, R., Gongvatana, A., Taylor, M., Buchthal, S., Schifitto, G., et al. (2014). Predictors of CNS injury as measured by proton magnetic resonance spectroscopy in the setting of chronic HIV infection and CART. *Journal of Neurovirology*. https://doi.org/10.1007/s13365-014-0246-6

Heaton, R. K., Clifford, D. B., Franklin, D. R., Jr., Woods, S. P., Ake, C., Vaida, F., et al. (2010). HIV-associated neurocognitive disorders persist in the era of potent antiretroviral therapy: CHARTER Study [Research Support, N.I.H., Extramural Research Support, U.S. Gov't, Non-P.H.S.]. *Neurology, 75*(23), 2087–2096. https://doi.org/10.1212/WNL.0b013e318200d727

Heaton, R. K., Franklin, D. R., Jr., Deutsch, R., Letendre, S., Ellis, R. J., Casaletto, K., et al. (2015). Neurocognitive change in the era of HIV combination antiretroviral therapy: The longitudinal CHARTER study. *Clinical Infectious Diseases, 60*(3), 473–480. https://doi.org/10.1093/cid/ciu862

Hellmuth, J., Slike, B. M., Sacdalan, C., Best, J., Kroon, E., Phanuphak, N., et al. (2019). Very early ART initiation during acute HIV infection is associated with normalization of cerebrospinal fluid but not plasma markers of immune activation. *The Journal of Infectious Diseases*. https://doi.org/10.1093/infdis/jiz030

Herskovitz, J., Hasan, M., Machhi, J., Mukadam, I., Ottemann, B. M., Hilaire, J. R., et al. (2021). Europium sulfide nanoprobes predict antiretroviral drug delivery into HIV-1 cell and tissue reservoirs. *Nano, 5*(4), 417–430. https://doi.org/10.7150/ntno.59568

Heseltine, P. N., Goodkin, K., Atkinson, J. H., Vitiello, B., Rochon, J., Heaton, R. K., et al. (1998). Randomized double-blind placebo-controlled trial of peptide T for HIV-associated cognitive impairment. *Archives of Neurology, 55*(1), 41–51. https://doi.org/10.1001/archneur.55.1.41

Hou, J., Jiang, T., Fu, J., Su, B., Wu, H., Sun, R., & Zhang, T. (2020). The long-term efficacy of working memory training in healthy older adults: A systematic review and meta-analysis of 22 randomized controlled trials. *The Journals of Gerontology. Series B, Psychological Sciences and Social Sciences, 75*(8), e174–e188. https://doi.org/10.1093/geronb/gbaa077

Israel, N., Gougerot-Pocidalo, M. A., Aillet, F., & Virelizier, J. L. (1992). Redox status of cells influences constitutive or induced NF-kappa B translocation and HIV long terminal repeat activity in human T and monocytic cell lines. *Journal of Immunology, 149*(10), 3386–3393. http://www.ncbi.nlm.nih.gov/entrez/query.fcgi?cmd=Retrieve&db=PubMed&dopt=Citation&list_uids=1431113

Jamali, Q. (2021). Galantamine as a treatment option for nicotine addiction. *Journal of Smoking Cessation, 2021*, 9975811. https://doi.org/10.1155/2021/9975811

Jessen Krut, J., Mellberg, T., Price, R. W., Hagberg, L., Fuchs, D., Rosengren, L., et al. (2014). Biomarker evidence of axonal injury in neuroasymptomatic HIV-1 patients. *PLoS One, 9*(2), e88591. https://doi.org/10.1371/journal.pone.0088591

Jiang, W., Luo, Z., Stephenson, S., Li, H., Di Germanio, C., Norris, P. J., et al. (2021). Cerebrospinal fluid and plasma lipopolysaccharide levels in human immunodeficiency virus type 1 infection and associations with inflammation, blood-brain barrier permeability, and neuronal injury. *The Journal of Infectious Diseases, 223*(9), 1612–1620. https://doi.org/10.1093/infdis/jiaa765

Joseph, S. B., & Swanstrom, R. (2018). The evolution of HIV-1 entry phenotypes as a guide to changing target cells. *Journal of Leukocyte Biology, 103*(3), 421–431. https://doi.org/10.1002/JLB.2RI0517-200R

Joseph, S. B., Kincer, L. P., Bowman, N. M., Evans, C., Vinikoor, M. J., Lippincott, C. K., et al. (2019). Human immunodeficiency virus type 1 RNA detected in the central nervous system (CNS) after years of suppressive antiretroviral therapy can originate from a replicating CNS reservoir or clonally expanded cells. *Clinical Infectious Diseases, 69*(8), 1345–1352. https://doi.org/10.1093/cid/ciy1066

Joshi, G., & Johnson, J. A. (2012). The Nrf2-ARE pathway: A valuable therapeutic target for the treatment of neurodegenerative diseases. *Recent Patents on CNS Drug Discovery, 7*(3), 218–229. http://www.ncbi.nlm.nih.gov/pubmed/22742419

Kaddour, H., McDew-White, M., Madeira, M. M., Tranquille, M. A., Tsirka, S. E., Mohan, M., & Okeoma, C. M. (2022). Chronic delta-9-tetrahydrocannabinol (THC) treatment counteracts SIV-induced modulation of proinflammatory microRNA cargo in basal ganglia-derived extracellular vesicles. *Journal of Neuroinflammation, 19*(1), 225. https://doi.org/10.1186/s12974-022-02586-9

Kaul, M., Zheng, J., Okamoto, S., Gendelman, H. E., & Lipton, S. A. (2005). HIV-1 infection and AIDS: Consequences for the central nervous system [Research Support, N.I.H., Extramural Research Support, Non-U.S. Gov't Research Support, U.S. Gov't, P.H.S. Review]. *Cell Death and Differentiation, 12 Suppl 1*, 878–892. https://doi.org/10.1038/sj.cdd.4401623

Kessing, C. F., Spudich, S., Valcour, V., Cartwright, P., Chalermchai, T., Fletcher, J. L., et al. (2017). High number of activated CD8+ T cells targeting HIV antigens are present in cerebrospinal fluid in acute HIV infection. *Journal of Acquired Immune Deficiency Syndromes, 75*(1), 108–117. https://doi.org/10.1097/QAI.0000000000001301

Killingsworth, L., & Spudich, S. (2022). Neuropathogenesis of HIV-1: Insights from across the spectrum of acute through long-term treated infection. *Seminars in Immunopathology, 44*(5), 709–724. https://doi.org/10.1007/s00281-022-00953-5

Kister, I., Patskovsky, Y., Curtin, R., Pei, J., Perdomo, K., Rimler, Z., et al. (2022). Cellular and humoral immunity to SARS-CoV-2 infection in multiple sclerosis patients on ocrelizumab and other disease-modifying therapies: A multi-ethnic observational study. *Annals of Neurology*. https://doi.org/10.1002/ana.26346

Klatt, N. R., Funderburg, N. T., & Brenchley, J. M. (2013). Microbial translocation, immune activation, and HIV disease. *Trends in Microbiology, 21*(1), 6–13. https://doi.org/10.1016/j.tim.2012.09.001

Kolakowska, A., Maresca, A. F., Collins, I. J., & Cailhol, J. (2019). Update on adverse effects of HIV integrase inhibitors. *Current Treatment Options in Infectious Diseases, 11*(4), 372–387. https://doi.org/10.1007/s40506-019-00203-7

Koola, M. M. (2018). Potential role of antipsychotic-galantamine-memantine combination in the treatment of positive, cognitive, and negative symptoms of schizophrenia. *Molecular Neuropsychiatry, 4*(3), 134–148. https://doi.org/10.1159/000494495

Kore, I., Ananworanich, J., Valcour, V., Fletcher, J. L., Chalermchai, T., Paul, R., et al. (2015). Neuropsychological impairment in acute HIV and the effect of immediate antiretroviral therapy. *Journal of Acquired Immune Deficiency Syndromes, 70*(4), 393–399. https://doi.org/10.1097/QAI.0000000000000746

Kornberg, M. D., Bhargava, P., Kim, P. M., Putluri, V., Snowman, A. M., Putluri, N., et al. (2018). Dimethyl fumarate targets GAPDH and aerobic glycolysis to modulate immunity. *Science*. https://doi.org/10.1126/science.aan4665

Kovacsics, C. E., Gill, A. J., Ambegaokar, S. S., Gelman, B. B., & Kolson, D. L. (2017). Degradation of heme oxygenase-1 by the immunoproteasome in astrocytes: A potential interferon-gamma-dependent mechanism contributing to HIV neuropathogenesis. *Glia, 65*(8), 1264–1277. https://doi.org/10.1002/glia.23160

Kreider, E. F., & Bar, K. J. (2022). HIV-1 reservoir persistence and decay: Implications for cure strategies. *Current HIV/AIDS Reports, 19*(3), 194–206. https://doi.org/10.1007/s11904-022-00604-2

Kristoff, J., Haret-Richter, G., Ma, D., Ribeiro, R. M., Xu, C., Cornell, E., et al. (2014). Early microbial translocation blockade reduces SIV-mediated inflammation and viral replication. *The Journal of Clinical Investigation, 124*(6), 2802–2806. https://doi.org/10.1172/JCI75090

Lakritz, J. R., Thibault, D. M., Robinson, J. A., Campbell, J. H., Miller, A. D., Williams, K. C., & Burdo, T. H. (2016). alpha4-integrin antibody treatment blocks monocyte/macrophage traffic to, vascular cell adhesion molecule-1 expression in, and pathology of the dorsal root ganglia in an SIV macaque model of HIV-peripheral neuropathy. *The American Journal of Pathology, 186*(7), 1754–1761. https://doi.org/10.1016/j.ajpath.2016.03.007

Lanman, T., Letendre, S., Ma, Q., Bang, A., & Ellis, R. (2019). CNS neurotoxicity of antiretrovirals. *Journal of Neuroimmune Pharmacology*. https://doi.org/10.1007/s11481-019-09886-7

Leon-Rivera, R., Veenstra, M., Donoso, M., Tell, E., Eugenin, E. A., Morgello, S., & Berman, J. W. (2021). Central nervous system (CNS) viral seeding by mature monocytes and potential therapies to reduce CNS viral reservoirs in the cART era. *MBio, 12*(2). https://doi.org/10.1128/mBio.03633-20

Letendre, S. L., Woods, S. P., Ellis, R. J., Atkinson, J. H., Masliah, E., van den Brande, G., et al. (2006). Lithium improves HIV-associated neurocognitive impairment. *AIDS, 20*(14), 1885–1888. https://doi.org/10.1097/01.aids.0000244208.49123.1b

Letendre, S. L., Marquie-Beck, J., Ellis, R. J., Woods, S. P., Best, B., Clifford, D. B., et al. (2007). The role of cohort studies in drug development: Clinical evidence of antiviral activity of serotonin reuptake inhibitors and HMG-CoA reductase inhibitors in the central nervous system. *Journal of Neuroimmune Pharmacology, 2*(1), 120–127. https://doi.org/10.1007/s11481-006-9054-y

Letendre, S., Marquie-Beck, J., Capparelli, E., Best, B., Clifford, D., Collier, A. C.,... Ellis, R. J. (2008). Validation of the CNS penetration-effectiveness rank for quantifying antiretroviral penetration into the central nervous system. Archives of Neurology, 65(1), 65–70. https://doi.org/10.1001/archneurol.2007.31.

Lin, S. P., Calcagno, A., Letendre, S. L., & Ma, Q. (2021). Clinical treatment options and randomized clinical trials for neurocognitive complications of HIV infection: Combination antiretroviral therapy, central nervous system penetration effectiveness, and adjuvants. *Current Topics in Behavioral Neurosciences, 50*, 517–545. https://doi.org/10.1007/7854_2020_186

Liu, X. L., Sun, D. D., Zheng, M. T., Li, X. T., Niu, H. H., Zhang, L., et al. (2023). Maraviroc promotes recovery from traumatic brain injury in mice by suppression of neuroinflammation and activation of neurotoxic reactive astrocytes. *Neural Regeneration Research, 18*(1), 141–149. https://doi.org/10.4103/1673-5374.344829

Llibre, J. M., Buzon, M. J., Massanella, M., Esteve, A., Dahl, V., Puertas, M. C., et al. (2012). Treatment intensification with raltegravir in subjects with sustained HIV-1 viraemia suppression: A randomized 48-week study. *Antiviral Therapy, 17*(2), 355–364. https://doi.org/10.3851/IMP1917

Loboda, A., Damulewicz, M., Pyza, E., Jozkowicz, A., & Dulak, J. (2016). Role of Nrf2/HO-1 system in development, oxidative stress response and diseases: An evolutionarily conserved mechanism. *Cellular and Molecular Life Sciences, 73*(17), 3221–3247. https://doi.org/10.1007/s00018-016-2223-0

Magyar, K., & Szende, B. (2004). (-)-Deprenyl, a selective MAO-B inhibitor, with apoptotic and anti-apoptotic properties. *Neurotoxicology, 25*(1–2), 233–242. https://doi.org/10.1016/S0161-813X(03)00102-5

Mak, G., Zaunders, J. J., Bailey, M., Seddiki, N., Rogers, G., Leong, L., et al. (2021). Preservation of gastrointestinal mucosal barrier function and microbiome in patients with controlled HIV infection. *Frontiers in Immunology, 12*, 688886. https://doi.org/10.3389/fimmu.2021.688886

Malvy, D. J., Richard, M. J., Arnaud, J., Favier, A., & Amedee-Manesme, O. (1994). Relationship of plasma malondialdehyde, vitamin E and antioxidant micronutrients to human immunodeficiency virus-1 seropositivity. *Clinica Chimica Acta, 224*(1), 89–94. https://www.ncbi.nlm.nih.gov/pubmed/8174282

Mansouri, A., Reiner, Z., Ruscica, M., Tedeschi-Reiner, E., Radbakhsh, S., Bagheri Ekta, M., & Sahebkar, A. (2022). Antioxidant effects of statins by modulating Nrf2 and Nrf2/HO-1 signaling in different diseases. *Journal of Clinical Medicine, 11*(5). https://doi.org/10.3390/jcm11051313

Marchetti, G., Tincati, C., & Silvestri, G. (2013). Microbial translocation in the pathogenesis of HIV infection and AIDS. *Clinical Microbiology Reviews, 26*(1), 2–18. https://doi.org/10.1128/CMR.00050-12

Marra, C. M., Zhao, Y., Clifford, D. B., Letendre, S., Evans, S., Henry, K., et al. (2009). Impact of combination antiretroviral therapy on cerebrospinal fluid HIV RNA and neurocognitive performance. *AIDS, 23*(11), 1359–1366. https://doi.org/10.1097/QAD.0b013e32832c4152

Martin-Blondel, G., Brassat, D., Bauer, J., Lassmann, H., & Liblau, R. S. (2016). CCR5 blockade for neuroinflammatory diseases--beyond control of HIV. Nature Reviews. Neurology, 12(2), 95–105. https://doi.org/10.1038/nrneurol.2015.248.

Martin-Garcia, J., Cocklin, S., Chaiken, I. M., & Gonzalez-Scarano, F. (2005). Interaction with CD4 and antibodies to CD4-induced epitopes of the envelope gp120 from a microglial cell-adapted human immunodeficiency virus type 1 isolate. *Journal of Virology, 79*(11), 6703–6713. https://doi.org/10.1128/JVI.79.11.6703-6713.2005

Martin-Garcia, J., Cao, W., Varela-Rohena, A., Plassmeyer, M. L., & Gonzalez-Scarano, F. (2006). HIV-1 tropism for the central nervous system: Brain-derived envelope glycoproteins with lower CD4 dependence and reduced sensitivity to a fusion inhibitor. *Virology, 346*(1), 169–179. https://doi.org/10.1016/j.virol.2005.10.031

McArthur, J. C., & Johnson, T. P. (2020). Chronic inflammation mediates brain injury in HIV infection: Relevance for cure strategies. *Current Opinion in Neurology, 33*(3), 397–404. https://doi.org/10.1097/WCO.0000000000000807

McDew-White, M., Lee, E., Premadasa, L. S., Alvarez, X., Okeoma, C. M., & Mohan, M. (2023). Cannabinoids modulate the microbiota-gut-brain axis in HIV/SIV infection by reducing neuroinflammation and dysbiosis while concurrently elevating endocannabinoid and indole-3-propionate levels. *Journal of Neuroinflammation, 20*(1), 62. https://doi.org/10.1186/s12974-023-02729-6

McGinley, M. P., & Cohen, J. A. (2021). Sphingosine 1-phosphate receptor modulators in multiple sclerosis and other conditions. *Lancet, 398*(10306), 1184–1194. https://doi.org/10.1016/S0140-6736(21)00244-0

McIntosh, E. C., Tureson, K., Rotblatt, L. J., Singer, E. J., & Thames, A. D. (2021). HIV, vascular risk factors, and cognition in the combination antiretroviral therapy era: A systematic review and meta-analysis. *Journal of the International Neuropsychological Society, 27*(4), 365–381. https://doi.org/10.1017/S1355617720001022

McIntosh, R. C., Clark, U. S., Cherner, M., Cysique, L. A., Heaton, R. K., Levin, J., et al. (2023). The evolution of assessing central nervous system complications in human immunodeficiency virus: Where do we go from here? *The Journal of Infectious Diseases, 227*(Suppl 1), S30–S37. https://doi.org/10.1093/infdis/jiac316

McMillan, J., Szlachetka, A., Slack, L., Sillman, B., Lamberty, B., Morsey, B., et al. (2018). Pharmacokinetics of a long-acting nanoformulated dolutegravir prodrug in rhesus macaques. *Antimicrobial Agents and Chemotherapy, 62*(1). https://doi.org/10.1128/AAC.01316-17

Medeiros, G. C., Smith, F. A., Trivedi, M. H., & Beach, S. R. (2020). Depressive disorders in HIV/AIDS: A clinically focused narrative review. *Harvard Review of Psychiatry, 28*(3), 146–158. https://doi.org/10.1097/HRP.0000000000000252

Mollace, V., Nottet, H. S., Clayette, P., Turco, M. C., Muscoli, C., Salvemini, D., & Perno, C. F. (2001). Oxidative stress and neuroAIDS: Triggers, modulators and novel antioxidants. *Trends in Neurosciences, 24*(7), 411–416.

Monroe, A. K., Zhang, L., Jacobson, L. P., Plankey, M. W., Brown, T. T., Miller, E. N., et al. (2017). The association between physical activity and cognition in men with and without HIV infection. *HIV Medicine, 18*(8), 555–563. https://doi.org/10.1111/hiv.12490

Morris, G., Puri, B. K., Walker, A. J., Berk, M., Walder, K., Bortolasci, C. C., et al. (2019). The compensatory antioxidant response system with a focus on neuroprogressive disorders. *Progress in Neuro-Psychopharmacology & Biological Psychiatry, 95*, 109708. https://doi.org/10.1016/j.pnpbp.2019.109708

Ndhlovu, L. C., Umaki, T., Chew, G. M., Chow, D. C., Agsalda, M., Kallianpur, K. J., et al. (2014). Treatment intensification with maraviroc (CCR5 antagonist) leads to declines in CD16-expressing monocytes in cART-suppressed chronic HIV-infected subjects and is associated with improvements in neurocognitive test performance: Implications for HIV-associated neurocognitive disease (HAND). *Journal of Neurovirology, 20*(6), 571–582. https://doi.org/10.1007/s13365-014-0279-x

Negre-Salvayre, A., Coatrieux, C., Ingueneau, C., & Salvayre, R. (2008). Advanced lipid peroxidation end products in oxidative damage to proteins. Potential role in diseases and therapeutic prospects for the inhibitors. *British Journal of Pharmacology, 153*(1), 6–20. https://doi.org/10.1038/sj.bjp.0707395

Neilson, L. E., Quinn, J. F., & Gray, N. E. (2020). Peripheral blood NRF2 expression as a biomarker in human health and disease. *Antioxidants (Basel), 10*(1). https://doi.org/10.3390/antiox10010028

Nickoloff-Bybel, E. A., Festa, L., Meucci, O., & Gaskill, P. J. (2021). Co-receptor signaling in the pathogenesis of neuroHIV. *Retrovirology, 18*(1), 24. https://doi.org/10.1186/s12977-021-00569-x

Nightingale, S., Winston, A., Letendre, S., Michael, B. D., McArthur, J. C., Khoo, S., & Solomon, T. (2014). Controversies in HIV-associated neurocognitive disorders. *Lancet Neurology, 13*(11), 1139–1151. https://doi.org/10.1016/S1474-4422(14)70137-1

Nowacek, A., & Gendelman, H. E. (2009). NanoART, neuroAIDS and CNS drug delivery. *Nanomedicine (London, England), 4*(5), 557–574. https://doi.org/10.2217/nnm.09.38

Ohagen, A., Devitt, A., Kunstman, K. J., Gorry, P. R., Rose, P. P., Korber, B., et al. (2003). Genetic and functional analysis of full-length human immunodeficiency virus type 1 env genes derived from brain and blood of patients with AIDS. *Journal of Virology, 77*(22), 12336–12345. http://www.ncbi.nlm.nih.gov/pubmed/14581570

O'Halloran, J. A., Cooley, S. A., Strain, J. F., Boerwinkle, A., Paul, R., Presti, R. M., & Ances, B. M. (2019). Altered neuropsychological performance and reduced brain volumetrics in people living with HIV on integrase strand transfer inhibitors. *AIDS, 33*(9), 1477–1483. https://doi.org/10.1097/QAD.0000000000002236

Ojagbemi, A. (2021). HIV associated neurocognitive disorders subsidence through citalopram addition in anti-retroviral therapy (HANDS-CARE): A concept note. *Frontiers in Neurology, 12*, 658705. https://doi.org/10.3389/fneur.2021.658705

Okamoto, M., Toyama, M., & Baba, M. (2020). The chemokine receptor antagonist cenicriviroc inhibits the replication of SARS-CoV-2 in vitro. *Antiviral Research, 182*, 104902. https://doi.org/10.1016/j.antiviral.2020.104902

Palamara, A. T., Perno, C. F., Aquaro, S., Bue, M. C., Dini, L., & Garaci, E. (1996). Glutathione inhibits HIV replication by acting at late stages of the virus life cycle. *AIDS Research and Human Retroviruses, 12*(16), 1537–1541. https://doi.org/10.1089/aid.1996.12.1537

Paul, R., Garcia-Egan, P., Bolzenius, J., & Mannarino, J. (2021). Deep phenotyping of HIV neurocognitive complications among individuals residing in high-income countries. *Current Topics in Behavioral Neurosciences, 50*, 245–269. https://doi.org/10.1007/7854_2020_185

Paul, R. H., Cho, K., Belden, A., Carrico, A. W., Martin, E., Bolzenius, J., et al. (2022). Cognitive phenotypes of HIV defined using a novel data-driven approach. *Journal of Neuroimmune Pharmacology*. https://doi.org/10.1007/s11481-021-10045-0

Peluso, M. J., & Spudich, S. (2014). Treatment of HIV in the CNS: Effects of antiretroviral therapy and the promise of non-antiretroviral therapeutics. *Current HIV/AIDS Reports, 11*(3), 353–362. https://doi.org/10.1007/s11904-014-0223-y

Peluso, M. J., Meyerhoff, D. J., Price, R. W., Peterson, J., Lee, E., Young, A. C., et al. (2013). Cerebrospinal fluid and neuroimaging biomarker abnormalities suggest early neurological injury in a subset of individuals during primary HIV infection. *The Journal of Infectious Diseases, 207*(11), 1703–1712. https://doi.org/10.1093/infdis/jit088

Pereira, G. F. M., Kim, A., Jalil, E. M., Fernandes Fonseca, F., Shepherd, B. E., Veloso, V. G., et al. (2021). Dolutegravir and pregnancy outcomes in women on antiretroviral therapy in Brazil: A retrospective national cohort study. *Lancet HIV, 8*(1), e33–e41. https://doi.org/10.1016/S2352-3018(20)30268-X

Perez-Valero, I., Corona, D., Martinez, N., Lopez-Cavanillas, M., Lluis, C., & Luque, I. (2023). Real-world discontinuations due to neuropsychiatric symptoms in people living with HIV treated with second-generation integrase inhibitors: A systematic review. *Expert Review of Anti-Infective Therapy, 21*(6), 655–665. https://doi.org/10.1080/14787210.2023.2203914

Pino, M., Paganini, S., Deleage, C., Padhan, K., Harper, J. L., King, C. T., et al. (2019). Fingolimod retains cytolytic T cells and limits T follicular helper cell infection in lymphoid sites of SIV persistence. *PLoS Pathogens, 15*(10), e1008081. https://doi.org/10.1371/journal.ppat.1008081

Pino, M., Pagliuzza, A., Pampena, M. B., Deleage, C., Viox, E. G., Nguyen, K., et al. (2022). Limited impact of fingolimod treatment during the initial weeks of ART in SIV-infected rhesus macaques. *Nature Communications, 13*(1), 5055. https://doi.org/10.1038/s41467-022-32698-y

Power, M. C., Weuve, J., Sharrett, A. R., Blacker, D., & Gottesman, R. F. (2015). Statins, cognition, and dementia-systematic review and methodological commentary. *Nature Reviews. Neurology, 11*(4), 220–229. https://doi.org/10.1038/nrneurol.2015.35

Prats, A., Martinez-Zalacain, I., Mothe, B., Negredo, E., Perez-Alvarez, N., Garolera, M., et al. (2021). Effects of integrase inhibitor-based antiretroviral therapy on brain outcomes according to time since acquisition of HIV-1 infection. *Scientific Reports, 11*(1), 11289. https://doi.org/10.1038/s41598-021-90678-6

Price, R. W., Spudich, S. S., Peterson, J., Joseph, S., Fuchs, D., Zetterberg, H., et al. (2014). Evolving character of chronic central nervous system HIV infection. *Seminars in Neurology, 34*(1), 7–13. https://doi.org/10.1055/s-0034-1372337

Quigley, A., O'Brien, K., Parker, R., & MacKay-Lyons, M. (2019). Exercise and cognitive function in people living with HIV: A scoping review. *Disability and Rehabilitation, 41*(12), 1384–1395. https://doi.org/10.1080/09638288.2018.1432079

Resop, R. S., Fromentin, R., Newman, D., Rigsby, H., Dubrovsky, L., Bukrinsky, M., et al. (2020). Fingolimod inhibits multiple stages of the HIV-1 life cycle. *PLoS Pathogens, 16*(8), e1008679. https://doi.org/10.1371/journal.ppat.1008679

Rich, S., Klann, E., Bryant, V., Richards, V., Wijayabahu, A., Bryant, K., et al. (2020). A review of potential microbiome-gut-brain axis mediated neurocognitive conditions in persons living with HIV. *Brain, Behavior, & Immunity - Health, 9*, 100168. https://doi.org/10.1016/j.bbih.2020.100168

Riviere-Cazaux, C., Cornell, J., Shen, Y., & Zhou, M. (2022). The role of CCR5 in HIV-associated neurocognitive disorders. *Heliyon, 8*(7), e09950. https://doi.org/10.1016/j.heliyon.2022.e09950

Robertson, K., Liner, J., & Meeker, R. B. (2012). Antiretroviral neurotoxicity. *Journal of Neurovirology, 18*(5), 388–399. https://doi.org/10.1007/s13365-012-0120-3

Robledinos-Anton, N., Fernandez-Gines, R., Manda, G., & Cuadrado, A. (2019). Activators and inhibitors of NRF2: A review of their potential for clinical development. *Oxidative Medicine and Cellular Longevity, 2019*, 9372182. https://doi.org/10.1155/2019/9372182

Roc, A. C., Ances, B. M., Chawla, S., Korczykowski, M., Wolf, R. L., Kolson, D. L., et al. (2007). Detection of human immunodeficiency virus induced inflammation and oxidative stress in lenticular nuclei with magnetic resonance spectroscopy despite antiretroviral therapy. *Archives of Neurology, 64*(9), 1249–1257. https://doi.org/10.1001/archneur.64.9.noc60125

Rosario-Rodriguez, L. J., Gerena, Y., Garcia-Requena, L. A., Cartagena-Isern, L. J., Cuadrado-Ruiz, J. C., Borges-Velez, G., & Melendez, L. M. (2022). Cannabinoid receptor type 2 agonist JWH-133 decreases cathepsin B secretion and neurotoxicity from HIV-infected macrophages. *Scientific Reports, 12*(1), 233. https://doi.org/10.1038/s41598-021-03896-3

Rose, R., Nolan, D. J., Maidji, E., Stoddart, C. A., Singer, E. J., Lamers, S. L., & McGrath, M. S. (2018). Eradication of HIV from tissue reservoirs: Challenges for the cure. *AIDS Research and Human Retroviruses, 34*(1), 3–8. https://doi.org/10.1089/AID.2017.0072

Rubin, L. H., Saylor, D., Nakigozi, G., Nakasujja, N., Robertson, K., Kisakye, A., et al. (2019). Heterogeneity in neurocognitive change trajectories among people with HIV starting antiretroviral therapy in Rakai, Uganda. *Journal of Neurovirology*. https://doi.org/10.1007/s13365-019-00768-5

Ruff, M. R., Polianova, M., Yang, Q. E., Leoung, G. S., Ruscetti, F. W., & Pert, C. B. (2003). Update on D-ala-peptide T-amide (DAPTA): A viral entry inhibitor that blocks CCR5 chemokine receptors. *Current HIV Research, 1*(1), 51–67. https://doi.org/10.2174/1570162033352066

Sacktor, N., & Robertson, K. (2014). Evolving clinical phenotypes in HIV-associated neurocognitive disorders. *Current Opinion in HIV and AIDS, 9*(6), 517–520. https://doi.org/10.1097/COH.0000000000000102

Sacktor, N., McDermott, M. P., Marder, K., Schifitto, G., Selnes, O. A., McArthur, J. C., et al. (2002). HIV-associated cognitive impairment before and after the advent of combination therapy. *Journal of Neurovirology, 8*, 136–142.

Sacktor, N., Miyahara, S., Deng, L., Evans, S., Schifitto, G., Cohen, B. A., et al. (2011). Minocycline treatment for HIV-associated cognitive impairment: Results from a randomized trial. *Neurology, 77*(12), 1135–1142. https://doi.org/10.1212/WNL.0b013e31822f0412

Sacktor, N., Skolasky, R. L., Moxley, R., Wang, S., Mielke, M. M., Munro, C., et al. (2018). Paroxetine and fluconazole therapy for HIV-associated neurocognitive impairment: Results from a double-blind, placebo-controlled trial. *Journal of Neurovirology, 24*(1), 16–27. https://doi.org/10.1007/s13365-017-0587-z

Sandler, N. G., Zhang, X., Bosch, R. J., Funderburg, N. T., Choi, A. I., Robinson, J. K., et al. (2014). Sevelamer does not decrease lipopolysaccharide or soluble CD14 levels but decreases soluble tissue factor, low-density lipoprotein (LDL) cholesterol, and oxidized LDL cholesterol levels in individuals with untreated HIV infection. *The Journal of Infectious Diseases, 210*(10), 1549–1554. https://doi.org/10.1093/infdis/jiu305

Saylor, D., Dickens, A. M., Sacktor, N., Haughey, N., Slusher, B., Pletnikov, M., et al. (2016). HIV-associated neurocognitive disorder – Pathogenesis and prospects for treatment. *Nature Reviews. Neurology, 12*(5), 309. https://doi.org/10.1038/nrneurol.2016.53

Schifitto, G., Peterson, D. R., Zhong, J., Ni, H., Cruttenden, K., Gaugh, M., et al. (2006). Valproic acid adjunctive therapy for HIV-associated cognitive impairment: A first report. *Neurology, 66*(6), 919–921. https://doi.org/10.1212/01.wnl.0000204294.28189.03

Schifitto, G., Navia, B. A., Yiannoutsos, C. T., Marra, C. M., Chang, L., Ernst, T., et al. (2007a). Memantine and HIV-associated cognitive impairment: A neuropsychological and proton magnetic resonance spectroscopy study. *AIDS, 21*(14), 1877–1886. https://doi.org/10.1097/QAD.0b013e32813384e8

Schifitto, G., Zhang, J., Evans, S. R., Sacktor, N., Simpson, D., Millar, L. L., et al. (2007b). A multicenter trial of selegiline transdermal system for HIV-associated cognitive impairment. *Neurology, 69*(13), 1314–1321. https://doi.org/10.1212/01.wnl.0000268487.78753.0f

Schifitto, G., Yiannoutsos, C. T., Ernst, T., Navia, B. A., Nath, A., Sacktor, N., et al. (2009a). Selegiline and oxidative stress in HIV-associated cognitive impairment. *Neurology, 73*(23), 1975–1981. https://doi.org/10.1212/WNL.0b013e3181c51a48

Schifitto, G., Zhong, J., Gill, D., Peterson, D. R., Gaugh, M. D., Zhu, T., et al. (2009b). Lithium therapy for human immunodeficiency virus type 1-associated neurocognitive impairment. *Journal of Neurovirology, 15*(2), 176–186. https://doi.org/10.1080/13550280902758973

Seu, L., Burt, T. D., Witte, J. S., Martin, J. N., Deeks, S. G., & McCune, J. M. (2012). Variations in the heme oxygenase-1 microsatellite polymorphism are associated with plasma CD14 and viral load in HIV-infected African-Americans. *Genes and Immunity, 13*(3), 258–267. https://doi.org/10.1038/gene.2011.76

Sevigny, J. J., Albert, S. M., McDermott, M. P., Schifitto, G., McArthur, J. C., Sacktor, N., et al. (2007). An evaluation of neurocognitive status and markers of immune activation as predictors of time to death in advanced HIV infection. *Archives of Neurology, 64*(1), 97–102. http://www.ncbi.nlm.nih.gov/entrez/query.fcgi?cmd=Retrieve&db=PubMed&dopt=Citation&list_uids=17210815

Shieh, J. T., Albright, A. V., Sharron, M., Gartner, S., Strizki, J., Doms, R. W., & Gonzalez-Scarano, F. (1998). Chemokine receptor utilization by human immunodeficiency virus type 1 isolates that replicate in microglia. *Journal of Virology, 72*(5), 4243–4249. https://doi.org/10.1128/JVI.72.5.4243-4249.1998

Shikuma, C. M., Nakamoto, B., Shiramizu, B., Liang, C. Y., DeGruttola, V., Bennett, K., et al. (2012). Antiretroviral monocyte efficacy score linked to cognitive impairment in HIV. *Antiviral Therapy, 17*(7), 1233–1242. https://doi.org/10.3851/IMP2411

Sillman, B., Woldstad, C., McMillan, J., & Gendelman, H. E. (2018). Neuropathogenesis of human immunodeficiency virus infection. *Handbook of Clinical Neurology, 152*, 21–40. https://doi.org/10.1016/B978-0-444-63849-6.00003-7

Simioni, S., Cavassini, M., Annoni, J. M., Rimbault Abraham, A., Bourquin, I., Schiffer, V., et al. (2010). Cognitive dysfunction in HIV patients despite long-standing suppression of viremia. *AIDS, 24*(9), 1243–1250. https://doi.org/10.1097/QAD.0b013e3283354a7b

Skalski, L. M., Towe, S. L., Sikkema, K. J., & Meade, C. S. (2016). The impact of marijuana use on memory in HIV-infected patients: A comprehensive review of the HIV and marijuana literatures. *Current Drug Abuse Reviews, 9*(2), 126–141. https://doi.org/10.2174/1874473709666160502124503

Skalski, L. M., Towe, S. L., Sikkema, K. J., & Meade, C. S. (2018). Memory impairment in HIV-infected individuals with early and late initiation of regular marijuana use. *AIDS and Behavior, 22*(5), 1596–1605. https://doi.org/10.1007/s10461-017-1898-z

Slavin, B. R., Sarhane, K. A., von Guionneau, N., Hanwright, P. J., Qiu, C., Mao, H. Q., et al. (2021). Insulin-like growth factor-1: A promising therapeutic target for peripheral nerve injury. *Frontiers in Bioengineering and Biotechnology, 9*, 695850. https://doi.org/10.3389/fbioe.2021.695850

Spudich, S. S. (2016). Immune activation in the central nervous system throughout the course of HIV infection. *Current Opinion in HIV and AIDS, 11*(2), 226–233. https://doi.org/10.1097/COH.0000000000000243

Stam, A. J., Nijhuis, M., van den Bergh, W. M., & Wensing, A. M. (2013). Differential genotypic evolution of HIV-1 quasispecies in cerebrospinal fluid and plasma: A systematic review. *AIDS Reviews, 15*(3), 152–161. http://www.ncbi.nlm.nih.gov/pubmed/24002199

Stefanou, M. I., Palaiodimou, L., Katsanos, A. H., Milionis, H., Kosmidou, M., Lambadiari, V., et al. (2022). The effects of HMG-CoA reductase inhibitors on disease activity in multiple sclerosis: A systematic review and meta-analysis. *Multiple Sclerosis and Related Disorders, 58*, 103395. https://doi.org/10.1016/j.msard.2021.103395

Stewart, J. C., Polanka, B. M., So-Armah, K. A., White, J. R., Gupta, S. K., Kundu, S., et al. (2020). Associations of total, cognitive/affective, and somatic depressive symptoms and antidepressant use with cardiovascular disease-relevant biomarkers in HIV: Veterans Aging Cohort Study. *Psychosomatic Medicine, 82*(5), 461–470. https://doi.org/10.1097/PSY.0000000000000808

Sturdevant, C. B., Joseph, S. B., Schnell, G., Price, R. W., Swanstrom, R., & Spudich, S. (2015). Compartmentalized replication of R5 T cell-tropic HIV-1 in the central nervous system early in the course of infection. *PLoS Pathogens, 11*(3), e1004720. https://doi.org/10.1371/journal.ppat.1004720

Suresh, D. R., Annam, V., Pratibha, K., & Prasad, B. V. (2009). Total antioxidant capacity – A novel early bio-chemical marker of oxidative stress in HIV infected individuals. *Journal of Biomedical Science, 16*, 61. https://doi.org/10.1186/1423-0127-16-61

Susanto, M., Pangihutan Siahaan, A. M., Wirjomartani, B. A., Setiawan, H., Aryanti, C., & Michael. (2023). The neuroprotective effect of statin in traumatic brain injury: A systematic review. *World Neurosurg X, 19*, 100211. https://doi.org/10.1016/j.wnsx.2023.100211

Tarumi, T., Tomoto, T., Repshas, J., Wang, C., Hynan, L. S., Cullum, C. M., et al. (2021). Midlife aerobic exercise and brain structural integrity: Associations with age and cardiorespiratory fitness. *NeuroImage, 225*, 117512. https://doi.org/10.1016/j.neuroimage.2020.117512

Thakral, S., Yadav, A., Singh, V., Kumar, M., Kumar, P., Narang, R., et al. (2023). Alzheimer's disease: Molecular aspects and treatment opportunities using herbal drugs. *Ageing Research Reviews, 88*, 101960. https://doi.org/10.1016/j.arr.2023.101960

Thaney, V. E., & Kaul, M. (2019). Type I interferons in NeuroHIV. *Viral Immunology, 32*(1), 7–14. https://doi.org/10.1089/vim.2018.0085

The Dana Consortium on the Therapy of HIV Dementia and Related Cognitive Disorders. (1997). Safety and tolerability of the antioxidant OPC-14117 in HIV-associated cognitive impairment. The Dana Consortium on the Therapy of HIV Dementia and Related Cognitive Disorders. *Neurology, 49*(1), 142–146. https://doi.org/10.1212/wnl.49.1.142

Tong, N., Sanchez, J. F., Maggirwar, S. B., Ramirez, S. H., Guo, H., Dewhurst, S., & Gelbard, H. A. (2001). Activation of glycogen synthase kinase 3 beta (GSK-3beta) by platelet activating factor mediates migration and cell death in cerebellar granule neurons. *The European Journal of Neuroscience, 13*(10), 1913–1922. ejn1572 [pii]

Tozzi, V., Balestra, P., Bellagamba, R., Corpolongo, A., Salvatori, M. F., Visco-Comandini, U., et al. (2007). Persistence of neuropsychologic deficits despite long-term highly active antiretroviral therapy in patients with HIV-related neurocognitive impairment: Prevalence and risk factors. *Journal of Acquired Immune Deficiency Syndromes, 45*(2), 174–182. https://doi.org/10.1097/QAI.0b013e318042e1ee

Trunfio, M., Pinnetti, C., Arsuffi, S., Bai, F., Celani, L., D'Ettorre, G., et al. (2023). The presence of resistance-associated mutations in reverse transcriptase gene is associated with cerebrospinal fluid HIV-1 escape: A multicentric retrospective analysis. *Journal of Medical Virology, 95*(4), e28704. https://doi.org/10.1002/jmv.28704

Tuffaha, S. H., Singh, P., Budihardjo, J. D., Means, K. R., Higgins, J. P., Shores, J. T., et al. (2016). Therapeutic augmentation of the growth hormone axis to improve outcomes following peripheral nerve injury. *Expert Opinion on Therapeutic Targets, 20*(10), 1259–1265. https://doi.org/10.1080/14728222.2016.1188079

Uzasci, L., Nath, A., & Cotter, R. (2013). Oxidative stress and the HIV-infected brain proteome. *Journal of Neuroimmune Pharmacology, 8*(5), 1167–1180. https://doi.org/10.1007/s11481-013-9444-x

Valcour, V., & Shiramizu, B. (2004). HIV-associated dementia, mitochondrial dysfunction, and oxidative stress. *Mitochondrion, 4*(2–3), 119–129. https://doi.org/10.1016/j.mito.2004.05.009

Valcour, V., Chalermchai, T., Sailasuta, N., Marovich, M., Lerdlum, S., Suttichom, D., et al. (2012). Central nervous system viral invasion and inflammation during acute HIV infection. *The Journal of Infectious Diseases, 206*(2), 275–282. https://doi.org/10.1093/infdis/jis326

Valcour, V. G., Ananworanich, J., Agsalda, M., Sailasuta, N., Chalermchai, T., Schuetz, A., et al. (2013). HIV DNA reservoir increases risk for cognitive disorders in cART-naive patients. *PLoS One, 8*(7), e70164. https://doi.org/10.1371/journal.pone.0070164

Vance, D. E., Fazeli, P. L., Cheatwood, J., Nicholson, W. C., Morrison, S. A., & Moneyham, L. D. (2019). Computerized cognitive training for the neurocognitive complications of HIV infection: A systematic review. *The Journal of the Association of Nurses in AIDS Care, 30*(1), 51–72. https://doi.org/10.1097/JNC.0000000000000030

Vance, D. E., Fazeli, P. L., Cheatwood, J., Nicholson, C., Morrison, S., & Moneyham, L. D. (2021). Targeting HIV-related neurocognitive impairments with cognitive training strategies: Insights from the cognitive aging literature. *Current Topics in Behavioral Neurosciences, 50*, 503–515. https://doi.org/10.1007/7854_2018_80

Veenhuis, R. T., Clements, J. E., & Gama, L. (2019). HIV eradication strategies: Implications for the central nervous system. *Current HIV/AIDS Reports, 16*(1), 96–104. https://doi.org/10.1007/s11904-019-00428-7

Vujkovic-Cvijin, I., Sortino, O., Verheij, E., Sklar, J., Wit, F. W., Kootstra, N. A., et al. (2020). HIV-associated gut dysbiosis is independent of sexual practice and correlates with noncommunicable diseases. *Nature Communications, 11*(1), 2448. https://doi.org/10.1038/s41467-020-16222-8

Wanchu, A., Rana, S. V., Pallikkuth, S., & Sachdeva, R. K. (2009). Short communication: Oxidative stress in HIV-infected individuals: A cross-sectional study. *AIDS Research and Human Retroviruses, 25*(12), 1307–1311. https://doi.org/10.1089/aid.2009.0062

Wang, Y., Liu, M., Lu, Q., Farrell, M., Lappin, J. M., Shi, J., et al. (2020). Global prevalence and burden of HIV-associated neurocognitive disorder: A meta-analysis. *Neurology, 95*(19), e2610–e2621. https://doi.org/10.1212/WNL.0000000000010752

Watson, C. W., Campbell, L. M., Sun-Suslow, N., Hong, S., Umlauf, A., Ellis, R. J., et al. (2021). Daily cannabis use is associated with lower CNS inflammation in people with HIV. *Journal of the International Neuropsychological Society, 27*(6), 661–672. https://doi.org/10.1017/S1355617720001447

Webb, A. J., Borrelli, E. P., Vyas, A., Taylor, L. E., & Buchanan, A. L. (2022). The effect of antiretroviral therapy with high central nervous system penetration on HIV-related cognitive impairment: A

systematic review and meta-analysis. *AIDS Care, 1-12*. https://doi.org/10.1080/09540121.2022.2098231

Williams, D. W., Veenstra, M., Gaskill, P. J., Morgello, S., Calderon, T. M., & Berman, J. W. (2014). Monocytes mediate HIV neuropathogenesis: Mechanisms that contribute to HIV associated neurocognitive disorders. *Current HIV Research, 12*(2), 85–96. http://www.ncbi.nlm.nih.gov/pubmed/24862333

Williams, T. E., Holdsworth, K. P., Nicholas, J. M., Eshaghi, A., Katsanouli, T., Wellington, H., et al. (2022). Assessing neurofilaments as biomarkers of neuroprotection in progressive multiple sclerosis: From the MS-STAT randomized controlled trial. *Neurology, Neuroimmunology & Neuroinflammation, 9*(2). https://doi.org/10.1212/NXI.0000000000001130

Wilson, S., Milanini, B., Javandel, S., Nyamayaro, P., & Valcour, V. (2021). Validity of digital assessments in screening for HIV-related cognitive impairment: A review. *Current HIV/AIDS Reports*. https://doi.org/10.1007/s11904-021-00585-8

Winston, A., & Spudich, S. (2020). Cognitive disorders in people living with HIV. *Lancet HIV, 7*(7), e504–e513. https://doi.org/10.1016/S2352-3018(20)30107-7

Yen, D. W., & Sherman, K. E. (2022). Causes and outcomes of hepatic fibrosis in persons living with HIV. *Current Opinion in HIV and AIDS, 17*(6), 359–367. https://doi.org/10.1097/COH.0000000000000760

Zayyad, Z., & Spudich, S. (2015). Neuropathogenesis of HIV: From initial neuroinvasion to HIV-associated neurocognitive disorder (HAND). *Current HIV/AIDS Reports*. https://doi.org/10.1007/s11904-014-0255-3

Zevin, A. S., McKinnon, L., Burgener, A., & Klatt, N. R. (2016). Microbial translocation and microbiome dysbiosis in HIV-associated immune activation. *Current Opinion in HIV and AIDS, 11*(2), 182–190. https://doi.org/10.1097/COH.0000000000000234

Zhao, Y., Navia, B. A., Marra, C. M., Singer, E. J., Chang, L., Berger, J., et al. (2010). Memantine for AIDS dementia complex: Open-label report of ACTG 301. *HIV Clinical Trials, 11*(1), 59–67. https://doi.org/10.1310/hct1101-59

Control of Neuroinflammation and Protecting the Integrity of the Blood-Brain Barrier

42

Slava Rom and Yuri Persidsky

Abstract

The blood-brain barrier (BBB) consists of specialized endothelial cells (ECs) surrounded by pericytes, astrocytes, and microglia. These ECs perform tight control on the bidirectional traffic of molecules from the blood to the brain parenchyma, thus guarding the brain's microenvironment. If the BBB components fail to perform their function, BBB's dysfunction can lead to a chain of neuroinflammatory events resulting in neuronal dysfunction and/or degeneration. This chapter aims to provide the readers with an overview of the evolving field of BBB and neuroinflammation, signaling pathways involved in BBB dysfunction, and the role of non-coding microRNAs in these signaling events. We conclude that further advances are needed in finding precise mechanisms regulating the BBB function to find BBB-protective reagents with neuroprotective action and their potential therapeutic application in treating neuroinflammatory and/or neurodegenerative diseases in the future.

Keywords

BBB · NVU · Neuroinflammation · microRNAs

42.1 Introduction

42.1.1 Neurovascular Unit

Microvasculature plays a significant role in physiologic and pathologic processes within the CNS. To describe the interconnection in functions between different cell types, the term neurovascular unit has been proposed. The neurovascular unit (NVU) is composed of brain microvascular endothelial cells, astrocytes, pericytes, and perivascular macrophages interacting as a functional unit that regulates blood supply to the CNS. In addition to astrocytes, microglia also regulates the NVU, occasionally being in contact with microvessels (Baeten & Akassoglou, 2011). Perivascular macrophages and microglia are antigen-presenting cells that reside within the brain parenchyma and respond to antigens crossing the BBB by initiating neuroinflammation.

Brain endothelial cells possess unique functional properties that include very low pinocytotic activity, high content of mitochondria, and high expression of junctional complexes (ensuring BBB tightness). Solute carrier (SLC), glucose transporter (Glut1), and ATP binding cassette (ABC) transporters (protecting the brain from toxins in blood and transporting by-products of CNS metabolism) are expressed in a highly polarized fashion in brain microvascular endothelial cells. The brain endothelial monolayers are connected by junctional complexes consisting of adherent, gap, and tight junctions (Abbott, 2000; Persidsky et al., 2006b). Tight junctions (TJ) are considered the main structures responsible for the establishment of the BBB. TJ integrity is a key factor to the very low paracellular permeability of hydrophilic molecules (Persidsky & Poluektova, 2006). They are composed of transmembrane proteins including occludin, claudins, junctional adhesion molecules, and endothelial cell-selective adhesion molecules (Persidsky et al., 2006b; Suzuki et al., 2002; Wegmann et al., 2006). Transmembrane TJ proteins are connected to the actin-based cytoskeleton by association with zonula occludens (ZO)-1, ZO-2, ZO-3, cingulin, AF6, and 7H6, which also may have a role in regulating TJ tightness (Chen et al., 2015; Fanning et al., 2007; Persidsky et al., 2006a).

The most expressed ABC transporters at the human BBB are P-glycoprotein (PGP, ABCB1), multidrug resistance protein 4 (MRP4, ABCC4), and breast cancer resistance protein (BCRP, ABCG2) (Taggi et al., 2022). Determination of ultrastructural localization within endothelial cells of the human

S. Rom · Y. Persidsky (✉)
Department of Pathology and Laboratory Medicine, Center for Substance Abuse Research, Lewis Katz School of Medicine, Temple University, Philadelphia, PA, USA
e-mail: yuri.persidsky@tuhs.temple.edu

H. E. Gendelman, T. Ikezu (eds.), *Neuroimmune Pharmacology and Therapeutics*,
https://doi.org/10.1007/978-3-031-68237-7_42

BBB allows prediction of their contribution towards the net transcellular transport of their substrates. Indeed, confocal and electron microscopy analysis has confirmed luminal localization of BCRP and P-gp (Bendayan et al., 2006) while suggesting luminal and abluminal expression of the monocarboxylate transporter 1 (MCT1) (Lauritzen et al., 2011) and glucose transporter 1(GLUT1) (Cornford et al., 1994).

Changes in expression for certain transporters can not only be considered a consequence of a pathological condition but may also be involved in the pathogenesis or even progression of certain diseases. This is especially true for Alzheimer's Disease (AD) in the brain. In neurological diseases, P-gp is the most-studied BBB transporter. Since P-gp is assumed to be involved in the clearance of Aβ peptides, multiple studies suggest the role of this efflux transporter in the development of AD. Various molecular mechanisms and regulatory pathways are associated with downregulation of P-gp, such as NF-κB activation, P-gp ubiquitination, and suppression of the Wnt/β-catenin pathway (Hashimoto et al., 2023).

Another important component of NVU is pericytes located on the abluminal side of brain microvascular endothelial cells. Pericytes wrapped around brain microvascular endothelial cells (BMVEC) on the "brain side" provide support for endothelial barrier function (Armulik et al., 2010; Bell et al., 2010). Pericytes regulate CNS blood flow at the level of capillaries providing neuronal coupling with the NVU (Hall et al., 2014). Pericyte number around endothelial cells has been determined to inversely correlate with BBB permeability and pericytes themselves regulate TJ formation and transendothelial vesicular transport, although the mechanism by which this occurs is unknown (Daneman et al., 2010). Reduction in brain pericyte density and detachment from brain endothelial cells lead to BBB dysfunction (Aguilera & Brekken, 2013; Zlokovic, 2010). Decreases have been reported in AD, multiple sclerosis (MS), amyotrophic lateral sclerosis (ALS), diabetic microangiopathy, HIV infection and stroke (Murakami, 2012; Persidsky et al., 2016; Sengillo et al., 2013; Vates et al., 2010; Winkler et al., 2013). It has been suggested that the extent of HIV-1-related endothelial activation may be associated with altered levels of the vascular growth factors angiopoietin (Ang) -1 and −2, recently identified as important regulators of endothelial quiescence and activation (Graham et al., 2013). Ang-1 secreted by pericytes plays a key role in maintaining vascular stability by binding to endothelial Tie-2, a receptor tyrosine kinase (Fiedler & Augustin, 2006). Diminished pericyte density, as well as decreased secretion in Ang-1, TGFβ1 and PDGF-Rβ, can reduce barrier stability. Of note, the deregulation of PDGF production in HIV-1-infected macrophages has been extensively studied before, indicating its proinflammatory and neuroprotective properties (Yao et al., 2013).

While being critical in BBB maintenance under physiologic conditions, pericytes shift to proinflammatory phenotype during neuroinflammation. Their proinflammatory reaction further amplifies the potential for leukocyte recruitment and chemotactic responses (Persidsky et al., 2016). Pericyte dysfunction can be a result of diminished interactions with brain endothelial cells via gap junctions (like connexin 43) and decreased secretion of BBB supporting factors and components of basement membrane. Further, pericytes may play a role in leukocyte migration since inflammatory cells need to cross both endothelial and pericyte barriers during their migration across BBB.

A decrease in pericyte-specific markers has been suggested to reflect a "less differentiated" phenotype of these cells (Thanabalasundaram et al., 2011). Diversity of pericyte function along the vascular hierarchy has been reported (Hall et al., 2014; Hill et al., 2014). Additional research into more thoroughly characterizing the heterogeneity of pericyte function along the vascular network is essential for the ultimate understanding of these clearly important cells in the NVU. Our results indicate that exposure to prototypic inflammatory factors, TNFα, or IL-1β, that are upregulated during neuroinflammatory responses (like HIV-1 CNS infection) (Letendre et al., 2011), leads to diminished expression of αSMA or integrin, arguing for changes in the phenotype of pericytes during inflammation (Persidsky et al., 2016). The intensity of neuroinflammation was associated with the decline of pericyte coverage even in patients treated with antiretroviral therapy. Furthermore, we found downregulation of PDGF-Rβ expression under the same conditions. This receptor possesses important functions in attracting pericytes to brain endothelium, thereby assuring BBB maturation. Pericyte loss seen in PDGF-Rβ ret/ret mice resulted in severe disruption of hemodynamics leading to impaired cerebral perfusion (Stobart et al., 2022).

Pericytes have only recently been studied for their pivotal role in BBB function during AD. During neurodegeneration, there is an increase in proinflammatory cytokines, especially IL-1β, IL-6, and TNFα (Zlokovic, 2010), and systemic inflammation correlating with impairment of low-density lipoprotein receptor-related protein-1 (LRP-1), which has been characterized as an efflux transporter of Aβ (Jaeger et al., 2009). In AD, Aβ accumulates around pericytes, especially in capillary beds causing pericyte death and detachment from endothelial cells in vitro and is seen in post-mortem tissue from patients with AD (Ozen et al., 2018). Pericyte number and coverage are decreased in AD patients; this reduction correlates with increases in both fibrin (a sign of increased BBB permeability) and Aβ accumulation (Sengillo et al., 2013).

There is a possible sequence of vascular pathology in Parkinson's Disease (PD) ranging from early pericyte activation and BBB leakage to an attempt for compensatory angio-

genesis and finally vascular rarefication (Paul & Elabi, 2022). Several different pathogenetic mechanisms play a role in these vascular alterations including perivascular inflammation and concomitant metabolic disease. BBB impairment, diminished expression of TJ proteins, increased permeability, and neuroinflammation seen in type 1 and type 2 diabetes are accompanied by pericyte loss and result in cognitive impairment (Rom et al., 2019).

Astrocytes, the most abundant glial cells, play an important role in BBB regulation. Their end feet extend to the basolateral side of brain microvessels enveloping them (Daneman & Prat, 2015). These cells express an important array of proteins including aquaporin 4 and dystrophin (Alvarez et al., 2013). The first regulates water balance and the second provides linkage to basement membranes of the BBB. A network of astrocytes serves as a signaling pathway between neurons and microvasculature. The ability of astrocytes to influence BBB tightness has been confirmed by either co-culturing brain endothelial cells and astrocytes or using astrocyte-conditioned media. Both increased BBB tightness in vitro and enhanced expression of TJ proteins in BBB. Astrocytes secrete factors from the Hedgehog (Hh) signaling cascade, including Sonic Hh (SHh) (Wang et al., 2008), and brain microvascular cells express the Hh receptor Patched-1, the signal transducer Smoothened (Smo), as well as transcription factors of the Gli family. Astrocytes and pericytes produce VEGF, controlling angiogenesis under physiologic (promoting new vessel formation) and pathologic conditions (destabilizing BBB and increasing barrier permeability). Another astrocyte-secreted factor is Ang-1 which stabilizes the BBB (Wang et al., 2008).

It is believed that astrocytes can amplify the inflammatory response of microglia/macrophages (Persidsky et al., 1999). Under inflammatory conditions (like HIV CNS infection), perivascular macrophages and microglia activated by virus and immune responses produce proinflammatory cytokines like IL-1β, which is used to treat primary human astrocytes in in vitro models of HIV infection (Banerjee et al., 2008; Fields et al., 2011; Mamik et al., 2011; Persidsky et al., 1999). IL-1β stimulates astrocytes, which then over-express multiple inflammatory factors, including MCP-1, IL-6, IL-8, TNF-α, TGF-β, NO, and Fas ligand (which initiates apoptosis in neurons) (Kou et al., 2009). Expression of these factors amplifies microglial/macrophage activation. Microarray analysis further confirms that the differential gene expression of IL-1β-treated astrocytes is associated with innate immunity and inflammation, interferon-inducible genes, and pro-apoptotic networks (Kou et al., 2009).

Microglia play an important role in neuroinflammation and BBB regulation contributing to its dysfunction and increased permeability (Nishioku et al., 2009). Perivascular microglia enhance the disintegration of the BBB in the periischemic region after experimental stroke (Jolivel et al., 2015). Inhibition of microglia activation after stroke promotes BBB integrity and viability (Yenari et al., 2006). Microglia generate a significant number of inflammatory mediators, including MMPs, proinflammatory cytokines (such as IL-1β and TNFα), and chemokines, which are known to compromise the BBB (Jolivel et al., 2015). A recent study found that abluminal (brain) stimulation of endothelial cells induces a more robust response compared to luminal activation of endothelial cells by IL-1β (Krasnow et al., 2017), suggesting that brain-derived (possibly microglia-derived) IL-1 is the main inducer of BBB dysfunction.

At late stages of injury, microglia transform into a protective phenotype (Jin et al., 2014) and can contribute to repair mechanisms including postinjury angiogenesis (Kabba et al., 2018). Release of microglia-derived soluble factors (Rymo et al., 2011) like release HMGB-1 (Chen et al., 2014) and galectin-3 (Walther et al., 2000) is a part of such a response. These factors act on endothelial progenitor cells to trigger angiogenesis after experimental stroke in rodents. TNFα secreted by microglia has been reported to induce endothelial cells to produce ephrin-A3 and ephrin-A4 that act in an autocrine manner to promote angiogenesis (Li et al., 2014). Microglia also release insulin-like growth factor-binding protein 1 (IGF-1) to promote angiogenesis (Nijaguna et al., 2015). IL-1β produced by microglia regulated the acute phase pentraxin-3 to promote BBB repair (Rodriguez-Grande et al., 2014), angiogenesis and neurogenesis (Rodriguez-Grande et al., 2015). IL-1α expressed in microglial cells (Luheshi et al., 2011) has been recently found to promote endothelial cell proliferation toward a proangiogenic phenotype in vitro.

42.1.2 Signaling Mechanisms Regulating BBB Permeability During Inflammation

The BBB is not an inactive structure, but one of great complexity responding to cellular insults coming from the basal (CNS) and luminal (blood) side of the barrier. Brain diseases, including neurological disorders and cerebrovascular diseases, are often associated with disrupted BBB integrity and various proinflammatory insults can trigger signaling mechanisms at the brain endothelial cell that can impact the permeability of the BBB. During neuroinflammation, the BBB can be under attack from either side of the barrier. While compensatory mechanisms are in place, it is not completely known at what point or what signaling thresholds need to be reached for the BBB to lose integrity during disease situations. A sizeable number of studies was dedicated to understanding how BBB integrity is regulated (Chakraborty et al., 2020). These works have led to the discovery of several signaling pathways and/or key molecules, in both brain endothelial cells and supporting cells, which play a critical role in maintaining BBB integrity: (a) endothelial junction molecules (e.g., VE-cadherin, claudin-5 and other TJ proteins), (b) platelet-derived growth factor (PDGF) signaling, and Wnt/b-catenin signaling, as well as (c) microRNAs (miRNAs) and other non-coding RNA molecules (Bernstein et al., 2021; Chakraborty et al., 2020; Li et al., 2020; Nation et al., 2019; Rom et al., 2015; Wang et al., 2021). miRNAs are a type of endogenous small non-coding RNAs (20–25 nucleotides) that regulate target genes at the post-transcriptional stage either via cleavage of mRNA or inhibition of translation (Bernstein et al., 2020a, 2021; Chakraborty et al., 2020). The role of miRNAs in regulation of the BBB integrity attracted the attention of many investigators and some of their findings will be reviewed here along with other signaling molecules. The most observable dysfunction of the BBB is that of increased permeability and enhanced surface and secreted inflammatory mediators that lead to immune infiltration of the CNS. On both counts, the TJs are greatly involved and are endpoint targets of complex intracellular signaling mechanisms. Multiple pathways have been associated with the disruption of member proteins of the TJ complex with crosstalk signaling that differs temporally (in time of initiation and range of duration) and spatially (i.e., compartmentalized signaling). The pathways that affect TJ proteins during inflammation can result in any of the following outcomes: subcellular redistribution, change in expression, and post-translational modification that affect protein half-life and protein-protein interactions. The following section examines signaling pathways known to act on the tight junction complex.

Both, chemokine MCP-1/CCL2 as well as its receptor (CCR2) have been shown to be involved in many neuroinflammatory conditions, including the neuropathogenesis of HIV-1, MS, stroke, AD, and PD (Covino et al., 2016; Georgakis et al., 2022; Rom et al., 2010; Singh et al., 2021; Zhang et al., 2019). During neuroinflammation, MCP-1 plays a dual role: (1) primary—a proinflammatory chemoattractant important for the recruitment of monocytes and microglia into the CNS (Singh et al., 2021) and (2) secondary—MCP-1's role, involved in the stimulation of the opening of the barrier (Stamatovic et al., 2016). MCP-1 enhances permeability of the barrier by internalization of both occludin and claudin-5 (subcellular re-distribution) into recycling endosomes (Stamatovic et al., 2016) as well as activation of the actin cytoskeleton machinery, which has been mapped to act on the cytoskeleton-modifying Rho GTPases and Rho kinase pathway (Persidsky et al., 2006; Rom et al., 2015; Stamatovic et al., 2016). The contribution of signaling that acts on the Rho family of GTPases is of central importance since TJ proteins are coupled to the cytoskeleton of the endothelial cell (García-Ponce et al., 2015). It is thought that one possible mechanism of inter-endothelial BBB opening comes from cytoskeletal contractile forces that retract the TJ complex and generate a "leakier" barrier. Different studies have confirmed that a variety of signaling pathways can affect TJ by regulating TJ protein expression, including Toll-like receptor 2 (TLR2)-protein kinase C (PKC), Ras-mitogen-activated protein kinase (MAPK), Wnt-β-catenin NF-κB, and other signaling pathways (Zhao et al., 2021). Recently, several miRNAs were shown to be involved in the modulation of BBB function, as well as the expression of TJ or MCP-1 proteins. miR-501-3p, miR-132, and miR-212 were shown to down-regulate ZO-1, and miR-181a affected both ZO-1 and claudin-5 proteins, consequently increasing BBB permeability (Chakraborty et al., 2020; Li et al., 2020; Wang et al., 2021; Zhao et al., 2021). Glycogen synthase kinase 3β (GSK3β) was shown to protect BBB under neuroinflammatory conditions. let-7g* and miR-98, which belong to the highly conserved let-7 family, facilitated the BBB-protective effect of GSK3β by targeting inflammatory molecules MCP-1/CCL2 and RANTES/CCL5 (Rom et al., 2015). Overexpression of let-7g* and miR-98 reduced neuroinflammation-induced BBB leakiness, as well as exacerbated monocyte attraction into the CNS (Bernstein & Rom, 2020; Bernstein et al., 2020b, 2021; Rom et al., 2015). Several miRNAs, miR-18a, miR-762, miR-19a, miR34a, miR-509, miR-200 were shown to regulate activity of the small Rho GTPases by mainly affecting their regulators, guanine nucleotide exchange factors (RhoGEFs), GTPase-activating proteins (RhoGAPs), GDP dissociation inhibitors (RhoGDIs) and guanine nucleotide exchange modifiers (GEMs) (Humphries et al., 2020).

In addition to soluble proinflammatory factors, endothelial-immune cell interaction initiates signaling to the TJ complex as a priming event for TJ disassembly and immune cell migration. During brain endothelial activation, the surface expression of adhesion molecules such as ICAM-1 and VCAM-1 increases, leading to engagement of these adhesion molecules by complementary integrin clusters on inflammatory/immune cells and initiation of signaling (Videm & Albrigtsen, 2008). Clustering of ICAM-1 and VCAM-1 triggers Rac-1 and RhoA-Rho kinase activation that leads to BBB opening by phosphorylation of TJ proteins (Persidsky et al., 2006; Yamamoto et al., 2008). miR-146a/b, miR-181b, miR-155, miR132, and miR-125a/b have been shown to modulate NF-κB signaling and thus VCAM-1 and ICAM-1 protein expression (Zhong et al., 2018). Whether in cytokine activation, as in the case with MCP-1, or immune engagement, the Rho pathway in the brain endothelial funnel signaling mechanisms modulates the TJ.

Metalloproteinases (MMPs) are also involved in BBB permeability. The MMPs at the BBB have been studied in various settings of neuroinflammation, including hemorrhage/stroke, MS, AD, and alcohol-mediated BBB compromise (Haorah et al., 2007b, 2008; Muri et al., 2019). MMPs are zinc-dependent endopeptidases, that once cleaved to the active form, degrade nearly all components of the basal lamina extracellular matrix, like collagen type IV, fibronectin, and laminins (Javadian et al., 2019; Muri et al., 2019). The result is a focal weakening in the anchoring of the vessel wall, which disrupts the BBB. It is known that depending on the inflammatory condition the source of MMPs can be immune cells (such as monocytes and neutrophils) or endothelial cells. In cerebral ischemia and in both acute and chronic brain inflammation, the upregulated and active forms of MMP-2 (activated earlier) and MMP-9 (later) are major contributors of BBB breakdown and neuronal damage (Li et al., 2018). MMP activity has also been implicated in TJ proteins occludin and claudin 5 degradation as well as of the basement membrane, and thus in facilitating leukocyte trafficking to the CNS (Haorah et al., 2007b; Smigiel & Parks, 2017). miR-29b and miR-200 were shown to regulate MMP-2, while miR-182 and miR-340 are modulating MMP-9 expression (Javadian et al., 2019). The role of MMPs in BBB integrity will continue to be defined, although it appears that MMPs may be an endpoint part of various pathways that signal to regulate the BBB (Behl et al., 2021).

The Wnt signaling pathway is known to play important roles in development and in many biological processes associated with cell proliferation and differentiation(Laksitorini et al., 2019) recently has been recognized as central to BBB regulation. The canonical Wnt/β-catenin signaling pathway is prompted by WNT ligand (i.e. WNT1 and WNT3a) binding to Frizzled (FZD) receptors and lipoprotein receptor-related protein 5 (LRP5) or LRP6 co-receptors (Laksitorini et al., 2019), thus stabilizing the β-catenin and prompting it to translocate to the nucleus where it binds to TCF/LEF and mediates transcription of BBB-specific genes (Tran et al., 2016). Axin and GSK3β in the absence of WNT activation induce the degradation of β-catenin; however, binding of WNT ligand to FZD/LRP recruits Axin and GSK3β away from β-catenin freeing β-catenin to translocate to the nucleus. WNT signaling has been placed at the center of BBB formation during development (Tran et al., 2016). It has been shown that β-catenin associates with the TJ complex due to its transcriptional role rather than the TJ-associated role that was important for increased claudin-3 expression (Ramirez et al., 2013). β-catenin stabilization is dependent on the inhibition of GSK3β; it is still not entirely clear whether the role of β-catenin on TJ protein expression via WNT could be bypassed by this inhibition. Previous studies give some validity to the notion of tightening of the barrier by inhibition of GSK3β without the involvement of WNT (Liu et al., 2002). Studies with primary human brain endothelial cells demonstrated that inhibition of GSK3β, both pharmacologically or genetically, resulted in higher transendothelial electrical resistance (TEER) on endothelial cells due to increased half-life of TJ proteins, claudin-5 and occludin, which are interrelated with the augmented association of β-catenin with ZO-1 protein (Ramirez et al., 2013). Besides the effects of GSK3β on barrier integrity, a consequence of GSK3β inhibition on human brain endothelial cells is the potent attenuation of endothelial inflammatory responses. Therefore, WNT ligands and GSK3β pharmacological inhibitors hold promise in protecting the barrier in two ways: by (1) improving barrier integrity and (2) inhibiting endothelial activation. A number of miRNAs were described recently to target the Wnt/β-catenin signaling pathway. The miR-200 family, containing 5 miRNAs (miR-200a, miR-200b, miR-200c, miR-141, and miR-429), is believed to play a significant role in Wnt/β-catenin signaling. miR-200 was found to affect β-catenin translation (Su et al., 2012). miR-122 was reported to suppress the Wnt/β-catenin signaling by targeting Snail1 and Snail2 proteins (Jin et al., 2017). miR-708 and miR-34a were shown to directly target LEF1, a pivotal transcription factor in the Wnt signaling pathway (Lei et al., 2020; Liang et al., 2015). miR-26a has been shown to target GSK3β (Lucci et al., 2020), thus indirectly it can influence the Wnt/β-catenin signaling and potentially barrier integrity.

CD40L (CD154) is a 33 kDa type II membrane glycoprotein from the TNFα family. CD40L is expressed predominantly by activated leukocytes and platelets (Li & Nord, 2009). In addition to the membrane-bound form of the protein, 31 kDa and/or 18 kDa versions of CD40L can be secreted or shed from activated cells (Mancino et al., 2008). Either form of CD40L retains the ability to activate CD40, a 45- to 50-kDa type I membrane glycoprotein expressed at a low level in resting cells of myeloid and vascular origin (Mancino et al., 2008; Pluvinet et al., 2008; Sui et al., 2007). CD40 expression is rapidly upregulated in these cells following exposure to proinflammatory mediators (Pluvinet et al., 2008; Sui et al., 2007). Elevated levels of sCD40L are found in a variety of diseases in which sCD40L is thought to initiate or potentiate inflammation (Heeschen et al., 2003; Tsakiris et al., 2000). Inflammatory conditions increase the expression of the CD40 receptor on the surface of endothelial cells and the shedding of the ligand (Chai et al., 2006). Our prior study explored the pathways that are activated by CD40/CD40L and the potential role in neuroinflammation via enhanced expression of ICAM-1 and VCAM-1. Although CD40 lacks intrinsic enzymatic activity, its trimerization with CD40L promotes interaction mainly with the TNF receptor-associated factors (TRAFs). Directly and via the TRAFs, CD40 initiates multiple and diverse signaling cascades involved in inflammation including JAK/STAT, IKK/NFκB, p38 MAPK, ERK1/2, and JNK1/AP1 (Elgueta et al., 2009). JNK-1 inhibitors have been shown to inhibit the known pathways that participate in CD40-mediated inflammatory response in BMVECs activated by sCD40L and lead to a nearly complete inhibition of the two adhesion molecules, ICAM and VCAM (Ramirez et al., 2010a). Analysis of phosphorylation status showed that not only JNK1 but also ERK1/2 was rapidly activated by sCD40L. Interestingly, the p38 MAPK appeared active in the BMVECs under resting conditions; however, it was not further activated/phosphorylated by sCD40L. Since ERK1/2 inhibition did not prevent adhesion molecule upregulation by CD40, Persidsky and colleagues investigated JNK1 activation of its downstream effectors and found that JNK1 contributes to the inflammatory response via AP-1 transcription factors (Ramirez et al., 2010a). The AP1 transcription factor c-JUN had increased phosphorylation, nuclear translocation, and DNA binding activity after CD40 activation, which suggested that signal integration from CD40 in the BMVECs may occur via MLK3. Thus, sCD40L engagement of CD40 generates signaling that activates MLK3, JNK1, and c-JUN, leading to the induction of ICAM-1 and VCAM-1 in the BMVECs. If JNK1 inhibition blocks adhesion molecule expression by sCD40L, then monocyte adhesion and transendothelial migration would also be affected. The results demonstrate that inhibition of JNK drastically reduces monocyte adhesion to and migration of monocytes across CD40-activated brain endothelial cells. Results of this study support the key concept that sCD40L is an important regulator of brain endothelial activation that could potentiate the progression of HIV-1-driven neuroinflammation and HAND (Ramirez et al., 2010a). Uncovering the signaling mechanism has significant implications beyond HIV-1 CNS infection, given the observed upregulation of CD40L in Alzheimer's disease, multiple sclerosis, stroke, and numerous inflammatory conditions outside of the brain (Desideri et al., 2008; Elgueta et al., 2009; Ishikawa et al., 2005; Rezai-Zadeh et al., 2009).

Modifications of the cytoskeleton play a crucial role in BBB integrity. In addition to RhoA and Rho kinase myosin light chain kinase (MLCK) is a potent regulator of myosin light chain phosphorylation leading to the contraction of endothelial cells. Using primary human BMVEC, it was previously demonstrated that oxidative stress induced by ethanol metabolism in BMVEC activated MLCK, resulting in the enhanced phosphorylation of either cytoskeletal or TJ proteins, and in BBB impairment (Haorah et al., 2005). Haorah and colleagues also found that alcohol or its metabolites stimulated inositol 1,4,5-triphosphate receptor (IP(3) R)-operated intracellular calcium (Ca^{2+}) release, thereby causing the activation of MLCK in BMVEC. Exposure of primary human BMVEC to alcohol or acetaldehyde resulted in the increased expression of IP(3)R protein and IP(3) R-gated intracellular Ca^{2+} release (Haorah et al., 2007a). These functional changes paralleled MLCK activation, phosphorylation of cytoskeletal/TJ proteins, loss of BBB integrity, and enhanced leukocyte migration across BMVEC monolayers. Inhibition of either alcohol metabolism or IP(3) R activation prevented BBB impairment.

Pericyte coverage of brain endothelial cells is a vital factor of BBB function. Diminished coverage results in an augmented permeability (Hill et al., 2014; Persidsky et al., 2016; Rom et al., 2020). miRNAs can modulate the BBB integrity via pericyte coverage-associated molecule regulation. miR-27 enhances BBB by targeting Sematophorin 6A/D, thus promoting the interaction of pericytes with endothelium (Wang et al., 2021). Tie-2/angiopoetin signaling is very important for pericyte recruitment to endothelial cells. miR-150 was shown to target Tie-2 as well as claudin-5 in the ischemia-reperfusion mouse stroke model, resulting in BBB dysfunction (Fang et al., 2016). miR-532 was found to regulate angiopoietin-1 and thus affect pericyte function. miRNAs are also involved in cell-to-cell communication. NF-kB-regulated miR-503 was secreted by endothelial cells and transferred to vascular pericytes via extracellular vesicles shedding during ischemia and diabetic conditions. This endothelial miRNA reduced pericyte coverage and induced permeability (Caporali et al., 2015).

Endothelial barrier integrity is affected by several circulating factors such as ATP, ADP, and adenosine (Aslam et al., 2021). Purinergic (P2) receptors are comprised of ionotropic P2X and metabotropic P2Y receptors (Burnstock, 2018). mRNA and protein of all P2X receptors have been detected in the endothelium of various types of blood vessels (Menzies et al., 2013). Activation of P2X7 receptors in the endothelium is primarily linked to a pro-inflammatory and hyperpermeability response. In an in vitro model of the BBB, ATP induced an increased production of matrix metallopeptidase 9 in an IL-1β-dependent manner leading to the degradation of TJ proteins (Yang et al., 2016). These ATP effects were abrogated by P2X7 receptor antagonists. In an in vivo model of intracranial hemorrhage, an upregulation of P2X7 receptor expression was accompanied by the development of brain edema. Inhibition of P2X7 receptors attenuated the disruption of the BBB and the resultant edema (Zhao et al., 2016). Using an in vitro coculture BBB model consisting of a human endothelial cell line with human astrocytes, Yang et al. showed that antagonist of P2X7 receptor attenuated hyperpermeability, prevented disruption of TJ proteins and IL-1β secretion after ATP exposure (Yang et al., 2016). P2X7 inhibition prevented BBB injury in an in vivo model of MDMA (ecstasy) administration (Rubio-Araiz et al., 2014).

BBB dysfunction is a common feature of many pathological CNS conditions, which can further aggravate initial CNS damage. The involvement of different regulators, including miRNAs, in BBB function has been explored in recent years. The effects of miRNAs on BBB integrity make them promising targets for both protecting integrity and transiently opening (disrupting) the BBB to improve CNS-targeted drug delivery.

42.1.3 BBB Protection

A myriad of pathologic conditions compromise the BBB which is a significant contributor to their pathogenesis. Realization of this fact in the last 10–15 years resulted in a continuous focus on the development of protective strategies. Several factors influence the potential utility of different approaches. These include the ability to diminish the inflammatory phenotype of brain endothelium, prevent changes in the cytoskeleton in these cells, and strengthen the function of TJ.

One of such multifunctional molecules is glycogen synthase kinase 3β (GSK3β), a ubiquitous serine/threonine protein kinase. It is involved in diverse biological functions including glycogen metabolism, regulation of cell division, differentiation, and apoptosis (Jope et al., 2007). Unlike most kinases, GSK3β is constitutively active in cells, and a wide range of extracellular stimuli exert their effects by inhibiting GSK3β activity (Cross et al., 1995). Suppression of GSK3 activity results in a diminution of inflammation caused by various pathogenic microorganisms (Cortés-Vieyra et al., 2021). Using primary human BMVEC, Persidsky, and colleagues studied the effects of GSK3β inhibition under inflammatory conditions (Ramirez et al., 2010b). Suppression of GSK3β activity in inflamed brain endothelium prevented the enhanced adhesion of leukocytes in vitro and in vivo, down-regulated expression of adhesion molecules, protected barrier function, and diminished migration of monocytes across BMVEC monolayers in response to relevant proinflammatory factors. Suppression of GSK3β decreased secretion of inflammatory factors in TNFα-stimulated BMVEC. Relevance of these observations was further confirmed by the presence of active GSK3β in brain endothelium in human brain tissue affected by neuroinflammation. Furthermore, Rom and colleagues showed that GSK3β-regulated let-7g* and miR-98, target several proinflammatory cytokines (e.g., MCP-1, RANTES, IP-10, etc.). Thus, the overexpression of these miRs resulted in diminished leukocyte migration into the brain, improved BBB function, and superior recovery from stroke-induced neuroinflammation (Bernstein et al., 2020a; b, 2021; Bernstein & Rom, 2020).

Additionally, GSK3β inhibition also promotes barrier tightness by affecting tight junction (TJ) protein stability (Ramirez et al., 2013). Inhibition of GSK3β produced a gradual and sustained increase in TEER with both pharmacological inhibitors and mutants of GSK3β (as much as 22% over baseline). Analysis of subcellular membrane fractions revealed an increase in the amount of essential tight junction proteins, occludin and claudin-5, but not claudin-3. This phenomenon was attributed to a decrease in TJ protein turnover and not transcriptional regulation (Ramirez et al., 2013). Using a novel cell-based assay, inactivation of GSK3β significantly increased the half-life of occludin and claudin-5 by 32% and 43%, respectively. A correlation was also established between the enhanced association of β-catenin with ZO-1 as a function of GSK3β inhibition (Ramirez et al., 2013).

Peroxisome proliferator-activated receptors (PPARs) are transcriptional regulators and members of a nuclear receptor family that include the closely related PPARα, PPARβ/δ, and PPARγ. PPARs play important roles in the regulation of adipocyte differentiation, insulin sensitivity, anti-inflammatory responses, and antiproliferative effects in certain tumors (Lehrke & Lazar, 2005). The lipid-lowering fibrates, which activate PPARα, and the insulin-sensitizing thiazolidinediones (TZDs), which activate PPARγ, have proven useful as potent anti-inflammatory agents (Zingarelli & Cook, 2005). Ligand binding of PPAR promotes the heterodimerization with the retinoid x receptor inducing PPAR transactivation of target genes. In addition, activation of PPARs can also lead to trans-repression, a process that has been shown to negatively interfere with key transcription factors (e.g., NF-κB, AP-1, STAT1, and sp1) involved in inflammatory responses. PPAR agonists diminished the expression of proinflammatory genes such as IL-1β, IL-6, inducible nitric oxide synthase, cyclooxygenase-2, and various chemokines. Additionally, PPARs were found to decrease endothelial-leukocyte interactions in models of atherosclerosis(Kurebayashi et al., 2005; Zandbergen & Plutzky, 2007).

PPARγ activation in BMVEC resulted in a significant reduction of monocyte adhesion and migration in the primary BBB culture model. Analysis of the Rho family GTPases in BMVEC stimulated with cytokine and thiazolidinediones (TZD), a synthetic ligand of PPARγ, resulted in a dose-dependent inhibition of Rac1 and RhoA that paralleled diminished monocyte adhesion and migration across BMVEC monolayers. Similarly, TZD stimulation of PPARγ in activated BMVEC resulted in a significant decrease in the adhesion and transmigration of HIV-1 infected monocytes across the brain endothelium (Ramirez et al., 2008). Together these observations provided insight into the inhibitory mechanism of monocyte adhesion/migration by PPARγ ligands acting in endothelial cells forming the BBB.

There are two well-characterized cannabinoid receptors with distinct physiological properties. The psychoactive effects of cannabinoids are associated with the cannabinoid type-1 receptor (CB1R), while the CB2R mainly mediates anti-inflammatory actions (Miller & Stella, 2008). Ramirez et al. addressed the effects of CB2R agonists on the brain endothelium and BBB function (Ramirez et al., 2012) and discovered enhanced CB2R receptor expression in cerebral vessels of human brain tissue affected by neuroinflammation, HIV-1 encephalitis (HIV-E). When tested, in an LPS-induced encephalitis model, to evaluate the effects of CB2R agonists on leukocyte–endothelial cell adhesion in both surface and ascending cortical vessels, they showed that a significant level of leukocyte adhesion was prevented in the encephalitic brain following administration of CB2R agonist. CB2R activation reduced expression of adhesion molecules needed for leukocyte engagement in TNFα- or LPS-activated primary human BMVECs and in brain endothelium of animals after LPS administration. Thus, the BBB-protective effects of CB2R agonists are due to the upregulated expression of TJ proteins (Ramirez et al., 2012).

During the last decade, inhibitors of poly(ADP-ribose) polymerase (PARP)-1 emerged as potent anti-inflammatory and immunomodulatory compounds. PARP-1 is a member of a family of NAD-dependent enzymes that are responsible for ~80% of cellular poly(ADP-ribose) formation (Rouleau et al., 2010). Poly(ADP-ribosyl)ation is a post-translational modification of proteins with widespread effects on diverse cellular functions including gene expression, differentiation and cell death (Krishnakumar & Kraus, 2010). As a scaffold protein and as an enzyme, PARP-1 (further referred to in the text as PARP) is a key component of the transcriptional machinery that controls chromatin architecture, histone shuttling, and spatial organization of transcription-regulating supramolecular complexes, thereby regulating inflammation-associated genes.

Inactivation of PARP resulted in a significant decrease in leukocyte adhesion to and migration across the CNS endothelium in vivo. In addition, Rom et al. showed that PARP inhibition caused a reduction in leukocyte adhesion and migration in an in vitro BBB model, thereby preserving barrier functions (Rom et al., 2016). These functional changes were accompanied by a decline in the expression of proinflammatory cytokines and adhesion molecules. For the first time, Rom et al. identified the mechanisms by which PARP inhibition attenuates BBB injury via effects on the activity of RhoA/Rac1 and augmentation of transcriptional expression of TJ proteins in brain endothelium, key elements controlling BBB integrity and monocyte migration across the BBB. PARP inhibition resulted in inactivation of repressor activity of nuclear factor kappa-light-chain-enhancer of activated B cells (NF-κB) and Snail-1 proteins, which in turn caused augmented expression of TJ proteins. Together, these results highlight the possible use of PARP inhibitors in BBB repair.

Besides systemic administration of BBB-protective agents, targeted delivery of protective agents to brain microvasculature has significant advantages. Such approaches include coupling drugs with affinity ligands to molecules expressed on CNS microvessels. Among these ligands are ICAM1, VCAM1, transferrin receptor (TfR), P-selectin, and others (Marcos-Contreras et al., 2020). Using this approach investigators successfully treated experimental traumatic brain injury by intravenous injection of anti-oxidative enzymes linked to ICAM (Lutton et al., 2019, 2017). In a recent study using mouse model with intracranial injection of TNFα inducing acute inflammation, there was 10-fold uptake of intravenously injected antibody to anti-VCAM in the inflamed brain when compared to antibodies to transferrin receptor-1 and ICAM-1 successfully used in prior studies (Marcos-Contreras et al., 2019). Furthermore, uptake of anti-VCAM/liposomes exceeded that of anti-TfR and anti-ICAM counterparts by ~27- and ~8-fold, respectively, achieving a brain/blood ratio >300-fold higher than that of immunoglobulin G/liposomes. Computed tomography imaging confirmed specific anti-VCAM/liposome targeting to inflamed brain. Using cranial windows, intravital microscopy, and flow cytometry investigators showed that in the inflamed brain anti-VCAM/liposomes bind to the endothelium. Anti-VCAM/LNP selectively accumulated in the inflamed brain, providing de novo expression of proteins encoded by cargo mRNA. Anti-VCAM/LNP-mRNA mediated expression of thrombomodulin (endothelial inhibitor of thrombosis, inflammation, and vascular leakage) and alleviated TNFα-induced brain edema.

42.2 Conclusions

In this chapter, we focused on BBB as the principal barrier contributing to CNS protection and sustaining brain homeostasis. Contrasting other vascular endothelial cells, that cover peripheral blood vessels, brain microvascular endothelial cells present unique structural, morphological, and functional characteristics that distinguish them apart from other vascular endothelia, including (a) the expression of tight junctions (TJs), which seal the paracellular pathways between adjacent endothelial cells, and block the unregulated passage of water-soluble molecules between the blood and the brain; (b) the lack of fenestrations; (c) the absence of pinocytic activity and (d) the existence of active transport mechanisms to control the passageway of essential molecules (e.g. essential amino acids and nutrients) whereas impeding the passage of potentially undesired substances. BBB dysfunction has been reported in stroke, TBI, AD, MS, ALS, diabetic microangiopathy, HIV infection, and other pathological conditions (Murakami, 2012; Persidsky et al., 2016; Sengillo et al., 2013; Vates et al., 2010; Winkler et al., 2013). The BBB dysfunction can range from acute mild and transient tight junction opening altering the BBB permeability, or severe changes in transport systems and enzymatic activity (e.g., MMPs, elastases, and granzymes) resulting in chronic barrier breakdown.

As reviewed above, in normal conditions, the BBB is relatively impermeable, while in pathologic conditions, numerous vasoactive agents, cytokines, and chemical mediators are released that increase BBB permeability. The BBB is an essential part of the CNS. Thus, BBB functional and structural integrity is fundamental in sustaining the homeostasis of the brain microenvironment. Different signaling pathways were shown to be involved in BBB dysfunction, and targeting these pathways, whether pharmacologically or via microRNA machinery might provide potential future treatment to preserve the BBB function in a variety of pathological conditions.

42.3 Review Questions

1. What is the major cell type comprising the BBB?
   (a) Pericyte
   (b) Endothelial
   (c) Epithelial
   (d) Adipose
   (e) b and a
2. The neurovascular unit (NVU):
   (a) Would include circulating immune cells and their products
   (b) Would not include circulating immune cells and their products
   (c) Would include pericytes and astrocytes
3. Major characteristics of the BBB:
   (a) Tight junctions
   (b) Presence of efflux transporters
   (c) Increased pinocytotic activity
   (d) All of the above
   (e) b and c
   (f) a and b

42.4 Answers

1. The blood-brain barrier (BBB) comprises endothelial cells (ECs), pericytes (PCs), the capillary basement membrane, and astrocyte end feet.
2. The neurovascular unit (NVU) would include pericytes and astrocytes.
3. A major characteristic of the BBB is the presence of tight junctions and efflux proteins.

References

Abbott, N. J. (2000). Inflammatory mediators and modulation of blood-brain barrier permeability. *Cellular and Molecular Neurobiology, 20*(2), 131–147.

Aguilera, K. Y., & Brekken, R. A. (2013). Recruitment and retention: Factors that affect pericyte migration. *Cellular and Molecular Life Sciences, 71*, 299. https://doi.org/10.1007/s00018-013-1432-z

Alvarez, J. I., Katayama, T., & Prat, A. (2013). Glial influence on the blood brain barrier. *Glia, 61*(12), 1939–1958. https://doi.org/10.1002/glia.22575

Armulik, A., Genove, G., Mae, M., Nisancioglu, M. H., Wallgard, E., Niaudet, C., He, L., Norlin, J., Lindblom, P., Strittmatter, K., Johansson, B. R., & Betsholtz, C. (2010). Pericytes regulate the blood-brain barrier. *Nature, 468*(7323), 557–561. https://doi.org/10.1038/nature09522

Aslam, M., Gunduz, D., Troidl, C., Heger, J., Hamm, C. W., & Schulz, R. (2021). Purinergic regulation of endothelial barrier function. *International Journal of Molecular Sciences, 22*(3). https://doi.org/10.3390/ijms22031207

Baeten, K. M., & Akassoglou, K. (2011). Extracellular matrix and matrix receptors in blood-brain barrier formation and stroke. *Developmental Neurobiology, 71*(11), 1018–1039. https://doi.org/10.1002/dneu.20954

Banerjee, S., Walseth, T. F., Borgmann, K., Wu, L., Bidasee, K. R., Kannan, M. S., & Ghorpade, A. (2008). CD38/cyclic ADP-ribose regulates astrocyte calcium signaling: Implications for neuroinflammation and HIV-1-associated dementia. *Journal of Neuroimmune Pharmacology, 3*(3), 154–164. https://doi.org/10.1007/s11481-008-9105-7

Behl, T., Kaur, G., Sehgal, A., Bhardwaj, S., Singh, S., Buhas, C., Judea-Pusta, C., Uivarosan, D., Munteanu, M. A., & Bungau, S. (2021). Multifaceted role of matrix metalloproteinases in neurodegenerative diseases: Pathophysiological and therapeutic perspectives. *International Journal of Molecular Sciences, 22*(3). https://doi.org/10.3390/ijms22031413

Bell, R. D., Winkler, E. A., Sagare, A. P., Singh, I., LaRue, B., Deane, R., & Zlokovic, B. V. (2010). Pericytes control key neurovascular functions and neuronal phenotypes in the adult brain and during brain aging. *Neuron, 68*(3), 409–427. https://doi.org/10.1016/j.neuron.2010.09.043

Bendayan, R., Ronaldson, P. T., Gingras, D., & Bendayan, M. (2006). In situ localization of P-glycoprotein (ABCB1) in human and rat brain. *The Journal of Histochemistry and Cytochemistry, 54*(10), 1159–1167. https://doi.org/10.1369/jhc.5A6870.2006

Bernstein, D. L., & Rom, S. (2020). Let-7g* and miR-98 reduce stroke-induced production of proinflammatory cytokines in mouse brain. *Frontiers in Cell and Development Biology, 8*, 632. https://doi.org/10.3389/fcell.2020.00632

Bernstein, D. L., Gajghate, S., Reichenbach, N. L., Winfield, M., Persidsky, Y., Heldt, N. A., & Rom, S. (2020a). let-7g counteracts endothelial dysfunction and ameliorating neurological functions in mouse ischemia/reperfusion stroke model. *Brain, Behavior, and Immunity, 87*, 543. https://doi.org/10.1016/j.bbi.2020.01.026

Bernstein, D. L., Zuluaga-Ramirez, V., Gajghate, S., Reichenbach, N. L., Polyak, B., Persidsky, Y., & Rom, S. (2020b). miR-98 reduces endothelial dysfunction by protecting blood-brain barrier (BBB) and improves neurological outcomes in mouse ischemia/reperfusion stroke model. *Journal of Cerebral Blood Flow and Metabolism, 40*(10), 1953–1965. https://doi.org/10.1177/0271678X19882264

Bernstein, D. L., Jiang, X., & Rom, S. (2021). let-7 microRNAs: Their role in cerebral and cardiovascular diseases, inflammation, cancer, and their regulation. *Biomedicine, 9*(6). https://doi.org/10.3390/biomedicines9060606

Burnstock, G. (2018). Purine and purinergic receptors. *Brain and Neuroscience Advances, 2*, 2398212818817494. https://doi.org/10.1177/2398212818817494

Caporali, A., Meloni, M., Nailor, A., Mitić, T., Shantikumar, S., Riu, F., Sala-Newby, G. B., Rose, L., Besnier, M., Katare, R., Voellenkle, C., Verkade, P., Martelli, F., Madeddu, P., & Emanueli, C. (2015). p75NTR-dependent activation of NF-κB regulates microRNA-503 transcription and pericyte–endothelial crosstalk in diabetes after limb ischaemia. *Nature Communications, 6*(1), 8024. https://doi.org/10.1038/ncomms9024

Chai, H., Yan, S., Wang, H., Zhang, R., Lin, P. H., Yao, Q., & Chen, C. (2006). CD40 ligand increases expression of its receptor CD40 in human coronary artery endothelial cells. *Surgery, 140*(2), 236–242. https://doi.org/10.1016/j.surg.2006.03.016

Chakraborty, C., Sharma, A. R., Sharma, G., Bhattacharya, M., & Lee, S. S. (2020). MicroRNAs: Possible regulatory molecular switch controlling the BBB microenvironment. *Molecular Therapy - Nucleic Acids, 19*, 933–936. https://doi.org/10.1016/j.omtn.2019.12.024

Chen, W., Sharma, R., Rizzo, A. N., Siegler, J. H., Garcia, J. G., & Jacobson, J. R. (2014). Role of claudin-5 in the attenuation of murine acute lung injury by simvastatin. *American Journal of*

Respiratory Cell and Molecular Biology, 50(2), 328–336. https://doi.org/10.1165/rcmb.2013-0058OC

Chen, C. H., Mayo, J. N., Gourdie, R. G., Johnstone, S. R., Isakson, B. E., & Bearden, S. E. (2015). The connexin 43/ZO-1 complex regulates cerebral endothelial F-actin architecture and migration. *American Journal of Physiology. Cell Physiology, 309*(9), C600–C607. https://doi.org/10.1152/ajpcell.00155.2015

Cornford, E. M., Hyman, S., & Swartz, B. E. (1994). The human brain GLUT1 glucose transporter: Ultrastructural localization to the blood-brain barrier endothelia. *Journal of Cerebral Blood Flow and Metabolism, 14*(1), 106–112. https://doi.org/10.1038/jcbfm.1994.15

Cortés-Vieyra, R., Silva-García, O., Gómez-García, A., Gutiérrez-Castellanos, S., Álvarez-Aguilar, C., & Baizabal-Aguirre, V. M. (2021). Glycogen synthase kinase 3β modulates the inflammatory response activated by bacteria, viruses, and parasites. *Frontiers in Immunology, 12*, 675751. https://doi.org/10.3389/fimmu.2021.675751

Covino, D. A., Sabbatucci, M., & Fantuzzi, L. (2016). The CCL2/CCR2 axis in the pathogenesis of HIV-1 infection: A new cellular target for therapy? *Current Drug Targets, 17*(1), 76–110. https://doi.org/10.2174/138945011701151217110917

Cross, D. A., Alessi, D. R., Cohen, P., Andjelkovich, M., & Hemmings, B. A. (1995). Inhibition of glycogen synthase kinase-3 by insulin mediated by protein kinase B. *Nature, 378*(6559), 785–789.

Daneman, R., & Prat, A. (2015). The blood-brain barrier. *Cold Spring Harbor Perspectives in Biology, 7*(1), a020412. https://doi.org/10.1101/cshperspect.a020412

Daneman, R., Zhou, L., Kebede, A. A., & Barres, B. A. (2010). Pericytes are required for blood-brain barrier integrity during embryogenesis. *Nature, 468*(7323), 562–566. https://doi.org/10.1038/nature09513

Desideri, G., Cipollone, F., Necozione, S., Marini, C., Lechiara, M. C., Taglieri, G., Zuliani, G., Fellin, R., Mezzetti, A., di Orio, F., & Ferri, C. (2008). Enhanced soluble CD40 ligand and Alzheimer's disease: Evidence of a possible pathogenetic role. *Neurobiology of Aging, 29*(3), 348–356. https://doi.org/10.1016/j.neurobiolaging.2006.10.019

Elgueta, R., Benson, M. J., de Vries, V. C., Wasiuk, A., Guo, Y., & Noelle, R. J. (2009). Molecular mechanism and function of CD40/CD40L engagement in the immune system. *Immunological Reviews, 229*(1), 152–172. https://doi.org/10.1111/j.1600-065X.2009.00782.x

Fang, Z., He, Q. W., Li, Q., Chen, X. L., Baral, S., Jin, H. J., Zhu, Y. Y., Li, M., Xia, Y. P., Mao, L., & Hu, B. (2016). MicroRNA-150 regulates blood-brain barrier permeability via Tie-2 after permanent middle cerebral artery occlusion in rats. *The FASEB Journal, 30*(6), 2097–2107. https://doi.org/10.1096/fj.201500126

Fanning, A. S., Little, B. P., Rahner, C., Utepbergenov, D., Walther, Z., & Anderson, J. M. (2007). The unique-5 and -6 motifs of ZO-1 regulate tight junction strand localization and scaffolding properties. *Molecular Biology of the Cell, 18*(3), 721–731. https://doi.org/10.1091/mbc.E06-08-0764

Fiedler, U., & Augustin, H. G. (2006). Angiopoietins: A link between angiogenesis and inflammation. *Trends in Immunology, 27*(12), 552–558. https://doi.org/10.1016/j.it.2006.10.004

Fields, J., Gardner-Mercer, J., Borgmann, K., Clark, I., & Ghorpade, A. (2011). CCAAT/enhancer binding protein beta expression is increased in the brain during HIV-1-infection and contributes to regulation of astrocyte tissue inhibitor of metalloproteinase-1. *Journal of Neurochemistry, 118*(1), 93–104. https://doi.org/10.1111/j.1471-4159.2011.07203.x

García-Ponce, A., Citalán-Madrid, A. F., Velázquez-Avila, M., Vargas-Robles, H., & Schnoor, M. (2015). The role of actin-binding proteins in the control of endothelial barrier integrity. *Thrombosis and Haemostasis, 113*(1), 20–36. https://doi.org/10.1160/th14-04-0298

Georgakis, M. K., Bernhagen, J., Heitman, L. H., Weber, C., & Dichgans, M. (2022). Targeting the CCL2-CCR2 axis for atheroprotection. *European Heart Journal, 43*(19), 1799–1808. https://doi.org/10.1093/eurheartj/ehac094

Graham, S. M., Rajwans, N., Jaoko, W., Estambale, B. B., McClelland, R. S., Overbaugh, J., & Liles, W. C. (2013). Endothelial activation biomarkers increase after HIV-1 acquisition: Plasma vascular cell adhesion molecule-1 predicts disease progression. *AIDS, 27*(11), 1803–1813. https://doi.org/10.1097/00002030-201307170-00014

Hall, C. N., Reynell, C., Gesslein, B., Hamilton, N. B., Mishra, A., Sutherland, B. A., O'Farrell, F. M., Buchan, A. M., Lauritzen, M., & Attwell, D. (2014). Capillary pericytes regulate cerebral blood flow in health and disease [In Vitro Research Support, Non-U.S. Gov't]. *Nature, 508*(7494), 55–60. https://doi.org/10.1038/nature13165

Haorah, J., Knipe, B., Leibhart, J., Ghorpade, A., & Persidsky, Y. (2005). Alcohol-induced oxidative stress in brain endothelial cells causes blood-brain barrier dysfunction. *Journal of Leukocyte Biology, 78*(6), 1223–1232.

Haorah, J., Knipe, B., Gorantla, S., Zheng, J., & Persidsky, Y. (2007a). Alcohol-induced blood-brain barrier dysfunction is mediated via inositol 1,4,5-triphosphate receptor (IP3R)-gated intracellular calcium release. *Journal of Neurochemistry, 100*(2), 324–336.

Haorah, J., Ramirez, S. H., Schall, K., Smith, D., Pandya, R., & Persidsky, Y. (2007b). Oxidative stress activates protein tyrosine kinase and matrix metalloproteinases leading to blood-brain barrier dysfunction. *Journal of Neurochemistry, 101*(2), 566–576.

Haorah, J., Schall, K., Ramirez, S. H., & Persidsky, Y. (2008). Activation of protein tyrosine kinases and matrix metalloproteinases causes blood-brain barrier injury: Novel mechanism for neurodegeneration associated with alcohol abuse. *Glia, 56*(1), 78–88.

Hashimoto, Y., Greene, C., Munnich, A., & Campbell, M. (2023). The CLDN5 gene at the blood-brain barrier in health and disease. *Fluids and Barriers of the CNS, 20*(1), 22. https://doi.org/10.1186/s12987-023-00424-5

Heeschen, C., Dimmeler, S., Hamm, C. W., van den Brand, M. J., Boersma, E., Zeiher, A. M., & Simoons, M. L. (2003). Soluble CD40 ligand in acute coronary syndromes. *The New England Journal of Medicine, 348*(12), 1104–1111. https://doi.org/10.1056/NEJMoa022600

Hill, J., Rom, S., Ramirez, S. H., & Persidsky, Y. (2014). Emerging roles of pericytes in the regulation of the neurovascular unit in health and disease. *Journal of Neuroimmune Pharmacology, 9*(5), 591–605. https://doi.org/10.1007/s11481-014-9557-x

Humphries, B. A., Wang, Z., & Yang, C. (2020). MicroRNA regulation of the small rho GTPase regulators-complexities and opportunities in targeting cancer metastasis. *Cancers (Basel), 12*(5). https://doi.org/10.3390/cancers12051092

Ishikawa, M., Vowinkel, T., Stokes, K. Y., Arumugam, T. V., Yilmaz, G., Nanda, A., & Granger, D. N. (2005). CD40/CD40 ligand signaling in mouse cerebral microvasculature after focal ischemia/reperfusion. *Circulation, 111*(13), 1690–1696. https://doi.org/10.1161/01.CIR.0000160349.42665.0C

Jaeger, L. B., Dohgu, S., Hwang, M. C., Farr, S. A., Murphy, M. P., Fleegal-DeMotta, M. A., Lynch, J. L., Robinson, S. M., Niehoff, M. L., Johnson, S. N., Kumar, V. B., & Banks, W. A. (2009). Testing the neurovascular hypothesis of Alzheimer's disease: LRP-1 antisense reduces blood-brain barrier clearance, increases brain levels of amyloid-beta protein, and impairs cognition. *Journal of Alzheimer's Disease, 17*(3), 553–570. https://doi.org/10.3233/jad-2009-1074

Javadian, M., Gharibi, T., Shekari, N., Abdollahpour-Alitappeh, M., Mohammadi, A., Hossieni, A., Mohammadi, H., & Kazemi, T. (2019). The role of microRNAs regulating the expression of matrix metalloproteinases (MMPs) in breast cancer development, progression, and metastasis. *Journal of Cellular Physiology, 234*(5), 5399–5412. https://doi.org/10.1002/jcp.27445

Jin, Q., Cheng, J., Liu, Y., Wu, J., Wang, X., Wei, S., Zhou, X., Qin, Z., Jia, J., & Zhen, X. (2014). Improvement of functional recovery by chronic metformin treatment is associated with enhanced alternative activation of microglia/macrophages and increased angiogenesis and neurogenesis following experimental stroke. *Brain, Behavior, and Immunity, 40*, 131–142. https://doi.org/10.1016/j.bbi.2014.03.003

Jin, Y., Wang, J., Han, J., Luo, D., & Sun, Z. (2017). MiR-122 inhibits its epithelial-mesenchymal transition in hepatocellular carcinoma by targeting Snail1 and Snail2 and suppressing WNT/β-cadherin signaling pathway. *Experimental Cell Research, 360*(2), 210–217. https://doi.org/10.1016/j.yexcr.2017.09.010

Jolivel, V., Bicker, F., Biname, F., Ploen, R., Keller, S., Gollan, R., Jurek, B., Birkenstock, J., Poisa-Beiro, L., Bruttger, J., Opitz, V., Thal, S. C., Waisman, A., Bäuerle, T., Schäfer, M. K., Zipp, F., & Schmidt, M. H. H. (2015). Perivascular microglia promote blood vessel disintegration in the ischemic penumbra. *Acta Neuropathologica, 129*(2), 279–295. https://doi.org/10.1007/s00401-014-1372-1

Jope, R. S., Yuskaitis, C. J., & Beurel, E. (2007). Glycogen synthase kinase-3 (GSK3): Inflammation, diseases, and therapeutics. *Neurochemical Research, 32*(4–5), 577–595. https://doi.org/10.1007/s11064-006-9128-5

Kabba, J. A., Xu, Y., Christian, H., Ruan, W., Chenai, K., Xiang, Y., Zhang, L., Saavedra, J. M., & Pang, T. (2018). Microglia: Housekeeper of the central nervous system. *Cellular and Molecular Neurobiology, 38*(1), 53–71. https://doi.org/10.1007/s10571-017-0504-2

Kou, W., Banerjee, S., Eudy, J., Smith, L. M., Persidsky, R., Borgmann, K., Wu, L., Sakhuja, N., Deshpande, M. S., Walseth, T. F., & Ghorpade, A. (2009). CD38 regulation in activated astrocytes: Implications for neuroinflammation and HIV-1 brain infection. *Journal of Neuroscience Research, 87*(10), 2326–2339. https://doi.org/10.1002/jnr.22060

Krasnow, S. M., Knoll, J. G., Verghese, S. C., Levasseur, P. R., & Marks, D. L. (2017). Amplification and propagation of interleukin-1β signaling by murine brain endothelial and glial cells. *Journal of Neuroinflammation, 14*(1), 133. https://doi.org/10.1186/s12974-017-0908-4

Krishnakumar, R., & Kraus, W. L. (2010). The PARP side of the nucleus: Molecular actions, physiological outcomes, and clinical targets. *Molecular Cell, 39*(1), 8–24. https://doi.org/10.1016/j.molcel.2010.06.017

Kurebayashi, S., Xu, X., Ishii, S., Shiraishi, M., Kouhara, H., & Kasayama, S. (2005). A novel thiazolidinedione MCC-555 downregulates tumor necrosis factor-alpha-induced expression of vascular cell adhesion molecule-1 in vascular endothelial cells. *Atherosclerosis, 182*(1), 71–77.

Laksitorini, M. D., Yathindranath, V., Xiong, W., Hombach-Klonisch, S., & Miller, D. W. (2019). Modulation of Wnt/β-catenin signaling promotes blood-brain barrier phenotype in cultured brain endothelial cells. *Scientific Reports, 9*(1), 19718. https://doi.org/10.1038/s41598-019-56075-w

Lauritzen, F., de Lanerolle, N. C., Lee, T. S., Spencer, D. D., Kim, J. H., Bergersen, L. H., & Eid, T. (2011). Monocarboxylate transporter 1 is deficient on microvessels in the human epileptogenic hippocampus. *Neurobiology of Disease, 41*(2), 577–584. https://doi.org/10.1016/j.nbd.2010.11.005

Lehrke, M., & Lazar, M. A. (2005). The many faces of PPARgamma. *Cell, 123*(6), 993–999.

Lei, Y., Chen, L., Zhang, G., Shan, A., Ye, C., Liang, B., Sun, J., Liao, X., Zhu, C., Chen, Y., Wang, J., Zhang, E., & Deng, L. (2020). MicroRNAs target the Wnt/β-catenin signaling pathway to regulate epithelial-mesenchymal transition in cancer (review). *Oncology Reports, 44*(4), 1299–1313. https://doi.org/10.3892/or.2020.7703

Letendre, S. L., Zheng, J. C., Kaul, M., Yiannoutsos, C. T., Ellis, R. J., Taylor, M. J., Marquie-Beck, J., & Navia, B. (2011). Chemokines in cerebrospinal fluid correlate with cerebral metabolite patterns in HIV-infected individuals. *Journal of Neurovirology, 17*(1), 63–69. https://doi.org/10.1007/s13365-010-0013-2

Li, H., & Nord, E. P. (2009). IL-8 amplifies CD40/CD154-mediated ICAM-1 production via the CXCR-1 receptor and p38-MAPK pathway in human renal proximal tubule cells. *American Journal of Physiology. Renal Physiology, 296*(2), F438–F445. https://doi.org/10.1152/ajprenal.90214.2008

Li, Y., Liu, D. X., Li, M. Y., Qin, X. X., Fang, W. G., Zhao, W. D., & Chen, Y. H. (2014). Ephrin-A3 and ephrin-A4 contribute to microglia-induced angiogenesis in brain endothelial cells. *Anatomical Record (Hoboken), 297*(10), 1908–1918. https://doi.org/10.1002/ar.22998

Li, W., Pan, R., Qi, Z., & Liu, K. J. (2018). Current progress in searching for clinically useful biomarkers of blood-brain barrier damage following cerebral ischemia. *Brain Circulation, 4*(4), 145–152. https://doi.org/10.4103/bc.bc_11_18

Li, J., Zhao, Y., Choi, J., Ting, K. K., Coleman, P., Chen, J., Cogger, V. C., Wan, L., Shi, Z., Moller, T., Zheng, X., Vadas, M. A., & Gamble, J. R. (2020). Targeting miR-27a/VE-cadherin interactions rescues cerebral cavernous malformations in mice. *PLoS Biology, 18*(6), e3000734. https://doi.org/10.1371/journal.pbio.3000734

Liang, J., Li, Y., Daniels, G., Sfanos, K., De Marzo, A., Wei, J., Li, X., Chen, W., Wang, J., Zhong, X., Melamed, J., Zhao, J., & Lee, P. (2015). LEF1 targeting EMT in prostate cancer invasion is regulated by miR-34a. *Molecular Cancer Research, 13*(4), 681–688. https://doi.org/10.1158/1541-7786.Mcr-14-0503

Liu, F., Schaphorst, K. L., Verin, A. D., Jacobs, K., Birukova, A., Day, R. M., Bogatcheva, N., Bottaro, D. P., & Garcia, J. G. (2002). Hepatocyte growth factor enhances endothelial cell barrier function and cortical cytoskeletal rearrangement: Potential role of glycogen synthase kinase-3beta. *The FASEB Journal, 16*(9), 950–962.

Lucci, C., Mesquita-Ribeiro, R., Rathbone, A., & Dajas-Bailador, F. (2020). Spatiotemporal regulation of GSK3beta levels by miRNA-26a controls axon development in cortical neurons. *Development, 147*(3). https://doi.org/10.1242/dev.180232

Luheshi, N. M., Kovacs, K. J., Lopez-Castejon, G., Brough, D., & Denes, A. (2011). Interleukin-1alpha expression precedes IL-1beta after ischemic brain injury and is localised to areas of focal neuronal loss and penumbral tissues. *Journal of Neuroinflammation, 8*, 186. https://doi.org/10.1186/1742-2094-8-186

Lutton, E. M., Razmpour, R., Andrews, A. M., Cannella, L. A., Son, Y. J., Shuvaev, V. V., Muzykantov, V. R., & Ramirez, S. H. (2017). Acute administration of catalase targeted to ICAM-1 attenuates neuropathology in experimental traumatic brain injury. *Scientific Reports, 7*(1), 3846. https://doi.org/10.1038/s41598-017-03309-4

Lutton, E. M., Farney, S. K., Andrews, A. M., Shuvaev, V. V., Chuang, G. Y., Muzykantov, V. R., & Ramirez, S. H. (2019). Endothelial targeted strategies to combat oxidative stress: Improving outcomes in traumatic brain injury. *Frontiers in Neurology, 10*, 582. https://doi.org/10.3389/fneur.2019.00582

Mamik, M. K., Banerjee, S., Walseth, T. F., Hirte, R., Tang, L., Borgmann, K., & Ghorpade, A. (2011). HIV-1 and IL-1beta regulate astrocytic CD38 through mitogen-activated protein kinases and nuclear factor-kappaB signaling mechanisms. *Journal of Neuroinflammation, 8*, 145. https://doi.org/10.1186/1742-2094-8-145

Mancino, A., Schioppa, T., Larghi, P., Pasqualini, F., Nebuloni, M., Chen, I. H., Sozzani, S., Austyn, J. M., Mantovani, A., & Sica, A. (2008). Divergent effects of hypoxia on dendritic cell functions. *Blood, 112*(9), 3723–3734. https://doi.org/10.1182/blood-2008-02-142091

Marcos-Contreras, O. A., Brenner, J. S., Kiseleva, R. Y., Zuluaga-Ramirez, V., Greineder, C. F., Villa, C. H., Hood, E. D., Myerson, J. W., Muro, S., Persidsky, Y., & Muzykantov, V. R. (2019). Combining vascular targeting and the local first pass provides 100-fold higher uptake of ICAM-1-targeted vs untargeted nanocarriers in the inflamed brain. *Journal of Controlled Release, 301*, 54–61. https://doi.org/10.1016/j.jconrel.2019.03.008

Marcos-Contreras, O. A., Greineder, C. F., Kiseleva, R. Y., Parhiz, H., Walsh, L. R., Zuluaga-Ramirez, V., Myerson, J. W., Hood, E. D., Villa, C. H., Tombacz, I., Pardi, N., Seliga, A., Mui, B. L., Tam, Y. K., Glassman, P. M., Shuvaev, V. V., Nong, J., Brenner, J. S., Khoshnejad, M., et al. (2020). Selective targeting of nanomedicine to inflamed cerebral vasculature to enhance the blood-brain barrier. *Proceedings of the National Academy of Sciences of the United States of America, 117*(7), 3405–3414. https://doi.org/10.1073/pnas.1912012117

Menzies, R. I., Unwin, R. J., Dash, R. K., Beard, D. A., Cowley, A. W., Jr., Carlson, B. E., Mullins, J. J., & Bailey, M. A. (2013). Effect of P2X4 and P2X7 receptor antagonism on the pressure diuresis relationship in rats. *Frontiers in Physiology, 4*, 305. https://doi.org/10.3389/fphys.2013.00305

Miller, A. M., & Stella, N. (2008). CB2 receptor-mediated migration of immune cells: It can go either way. *British Journal of Pharmacology, 153*(2), 299–308. https://doi.org/10.1038/sj.bjp.0707523

Murakami, M. (2012). Signaling required for blood vessel maintenance: Molecular basis and pathological manifestations. *International Journal of Vascular Medicine, 2012*, 293641. https://doi.org/10.1155/2012/293641

Muri, L., Leppert, D., Grandgirard, D., & Leib, S. L. (2019). MMPs and ADAMs in neurological infectious diseases and multiple sclerosis. *Cellular and Molecular Life Sciences, 76*(16), 3097–3116. https://doi.org/10.1007/s00018-019-03174-6

Nation, D. A., Sweeney, M. D., Montagne, A., Sagare, A. P., D'Orazio, L. M., Pachicano, M., Sepehrband, F., Nelson, A. R., Buennagel, D. P., Harrington, M. G., Benzinger, T. L. S., Fagan, A. M., Ringman, J. M., Schneider, L. S., Morris, J. C., Chui, H. C., Law, M., Toga, A. W., & Zlokovic, B. V. (2019). Blood-brain barrier breakdown is an early biomarker of human cognitive dysfunction. *Nature Medicine, 25*(2), 270–276. https://doi.org/10.1038/s41591-018-0297-y

Nijaguna, M. B., Patil, V., Urbach, S., Shwetha, S. D., Sravani, K., Hegde, A. S., Chandramouli, B. A., Arivazhagan, A., Marin, P., Santosh, V., & Somasundaram, K. (2015). Glioblastoma-derived macrophage colony-stimulating factor (MCSF) induces microglial release of insulin-like growth factor-binding protein 1 (IGFBP1) to promote angiogenesis. *The Journal of Biological Chemistry, 290*(38), 23401–23415. https://doi.org/10.1074/jbc.M115.664037

Nishioku, T., Dohgu, S., Takata, F., Eto, T., Ishikawa, N., Kodama, K. B., Nakagawa, S., Yamauchi, A., & Kataoka, Y. (2009). Detachment of brain pericytes from the basal lamina is involved in disruption of the blood-brain barrier caused by lipopolysaccharide-induced sepsis in mice. *Cellular and Molecular Neurobiology, 29*(3), 309–316. https://doi.org/10.1007/s10571-008-9322-x

Ozen, I., Roth, M., Barbariga, M., Gaceb, A., Deierborg, T., Genove, G., & Paul, G. (2018). Loss of regulator of G-protein signaling 5 leads to neurovascular protection in stroke. *Stroke, 49*(9), 2182–2190. https://doi.org/10.1161/strokeaha.118.020124

Paul, G., & Elabi, O. F. (2022). Microvascular changes in Parkinson's disease – Focus on the neurovascular unit. *Frontiers in Aging Neuroscience, 14*, 853372. https://doi.org/10.3389/fnagi.2022.853372

Persidsky, Y., & Poluektova, L. (2006). Immune privilege and HIV-1 persistence in the CNS. *Immunological Reviews, 213*, 180–194.

Persidsky, Y., Ghorpade, A., Rasmussen, J., Limoges, J., Liu, X. J., Stins, M., Fiala, M., Way, D., Kim, K. S., Witte, M. H., Weinand, M., Carhart, L., & Gendelman, H. E. (1999). Microglial and astrocyte chemokines regulate monocyte migration through the blood-brain barrier in human immunodeficiency virus-1 encephalitis. *The American Journal of Pathology, 155*(5), 1599–1611.

Persidsky, Y., Heilman, D., Haorah, J., Zelivyanskaya, M., Persidsky, R., Weber, G. A., Shimokawa, H., Kaibuchi, K., & Ikezu, T. (2006). Rho-mediated regulation of tight junctions during monocyte migration across the blood-brain barrier in HIV-1 encephalitis (HIVE). *Blood, 107*(12), 4770–4780.

Persidsky, Y., Ramirez, S. H., Haorah, J., & Kanmogne, G. D. (2006a). Blood brain barrier: Structural components and function under physiologic and pathologic conditions. *Journal of Neuroimmune Pharmacology, 1*, 223–236.

Persidsky, Y., Ramirez, S. H., Haorah, J., & Kanmogne, G. D. (2006b). Blood-brain barrier: Structural components and function under physiologic and pathologic conditions. *Journal of Neuroimmune Pharmacology, 1*(3), 223–236. https://doi.org/10.1007/s11481-006-9025-3

Persidsky, Y., Hill, J., Zhang, M., Dykstra, H., Winfield, M., Reichenbach, N. L., Potula, R., Mukherjee, A., Ramirez, S. H., & Rom, S. (2016). Dysfunction of brain pericytes in chronic neuroinflammation. *Journal of Cerebral Blood Flow and Metabolism, 36*(4), 794–807. https://doi.org/10.1177/0271678X15606149

Pluvinet, R., Olivar, R., Krupinski, J., Herrero-Fresneda, I., Luque, A., Torras, J., Cruzado, J. M., Grinyo, J. M., Sumoy, L., & Aran, J. M. (2008). CD40: An upstream master switch for endothelial cell activation uncovered by RNAi-coupled transcriptional profiling. *Blood, 112*(9), 3624–3637. https://doi.org/10.1182/blood-2008-03-143305

Ramirez, S. H., Heilman, D., Morsey, B., Potula, R., Haorah, J., & Persidsky, Y. (2008). Activation of peroxisome proliferator-activated receptor gamma (PPARgamma) suppresses Rho GTPases in human brain microvascular endothelial cells and inhibits adhesion and transendothelial migration of HIV-1 infected monocytes. *Journal of Immunology, 180*(3), 1854–1865.

Ramirez, S. H., Fan, S., Dykstra, H., Reichenbach, N., Del Valle, L., Potula, R., Phipps, R. P., Maggirwar, S. B., & Persidsky, Y. (2010a). Dyad of CD40/CD40 ligand fosters neuroinflammation at the blood-brain barrier and is regulated via JNK signaling: Implications for HIV-1 encephalitis. *The Journal of Neuroscience, 30*(28), 9454–9464. https://doi.org/10.1523/JNEUROSCI.5796-09.2010

Ramirez, S. H., Fan, S., Zhang, M., Papugani, A., Reichenbach, N., Dykstra, H., Mercer, A. J., Tuma, R. F., & Persidsky, Y. (2010b). Inhibition of glycogen synthase kinase 3beta (GSK3beta) decreases inflammatory responses in brain endothelial cells. *The American Journal of Pathology, 176*(2), 881–892. https://doi.org/10.2353/ajpath.2010.090671

Ramirez, S. H., Hasko, J., Skuba, A., Fan, S., Dykstra, H., McCormick, R., Reichenbach, N., Krizbai, I., Mahadevan, A., Zhang, M., Tuma, R., Son, Y. J., & Persidsky, Y. (2012). Activation of cannabinoid receptor 2 attenuates leukocyte-endothelial cell interactions and blood-brain barrier dysfunction under inflammatory conditions. *Journal of Neuroscience, 32*(12), 4004–4016. https://doi.org/10.1523/JNEUROSCI.4628-11.2012

Ramirez, S. H., Fan, S., Dykstra, H., Rom, S., Mercer, A., Reichenbach, N. L., Gofman, L., & Persidsky, Y. (2013). Inhibition of glycogen synthase kinase 3beta promotes tight junction stability in brain endothelial cells by half-life extension of Occludin and Claudin-5. *PLoS One, 8*(2), e55972. https://doi.org/10.1371/journal.pone.0055972

Rezai-Zadeh, K., Gate, D., Szekely, C. A., & Town, T. (2009). Can peripheral leukocytes be used as Alzheimer's disease biomarkers? *Expert Review of Neurotherapeutics, 9*(11), 1623–1633. https://doi.org/10.1586/ern.09.118

Rodriguez-Grande, B., Swana, M., Nguyen, L., Englezou, P., Maysami, S., Allan, S. M., Rothwell, N. J., Garlanda, C., Denes, A., & Pinteaux, E. (2014). The acute-phase protein PTX3 is an essential mediator of glial scar formation and resolution of brain edema after ischemic injury. *Journal of Cerebral Blood Flow and Metabolism, 34*(3), 480–488. https://doi.org/10.1038/jcbfm.2013.224

Rodriguez-Grande, B., Varghese, L., Molina-Holgado, F., Rajkovic, O., Garlanda, C., Denes, A., & Pinteaux, E. (2015). Pentraxin 3 mediates neurogenesis and angiogenesis after cerebral ischaemia. *Journal of Neuroinflammation, 12*, 15. https://doi.org/10.1186/s12974-014-0227-y

Rom, S., Rom, I., Passiatore, G., Pacifici, M., Radhakrishnan, S., Del Valle, L., Pina-Oviedo, S., Khalili, K., Eletto, D., & Peruzzi, F. (2010). CCL8/MCP-2 is a target for mir-146a in HIV-1-infected human microglial cells. *The FASEB Journal, 24*(7), 2292–2300. https://doi.org/10.1096/fj.09-143503

Rom, S., Dykstra, H., Zuluaga-Ramirez, V., Reichenbach, N. L., & Persidsky, Y. (2015). miR-98 and let-7g* protect the blood-brain barrier under neuroinflammatory conditions. *Journal of Cerebral Blood Flow and Metabolism, 35*(12), 1957–1965. https://doi.org/10.1038/jcbfm.2015.154

Rom, S., Zuluaga-Ramirez, V., Reichenbach, N. L., Dykstra, H., Gajghate, S., Pacher, P., & Persidsky, Y. (2016). PARP inhibition in leukocytes diminishes inflammation via effects on integrins/cytoskeleton and protects the blood-brain barrier. *Journal of Neuroinflammation, 13*(1), 254. https://doi.org/10.1186/s12974-016-0729-x

Rom, S., Zuluaga-Ramirez, V., Gajghate, S., Seliga, A., Winfield, M., Heldt, N. A., Kolpakov, M. A., Bashkirova, Y. V., Sabri, A. K., & Persidsky, Y. (2019). Hyperglycemia-driven neuroinflammation compromises BBB leading to memory loss in both diabetes mellitus (DM) type 1 and type 2 mouse models. *Molecular Neurobiology, 56*(3), 1883–1896. https://doi.org/10.1007/s12035-018-1195-5

Rom, S., Gajghate, S., Winfield, M., Reichenbach, N. L., & Persidsky, Y. (2020). Combination of HIV-1 and diabetes enhances blood brain barrier injury via effects on brain endothelium and pericytes. *International Journal of Molecular Sciences, 21*(13). https://doi.org/10.3390/ijms21134663

Rouleau, M., Patel, A., Hendzel, M. J., Kaufmann, S. H., & Poirier, G. G. (2010). PARP inhibition: PARP1 and beyond. *Nature Reviews. Cancer, 10*(4), 293–301. https://doi.org/10.1038/nrc2812

Rubio-Araiz, A., Perez-Hernandez, M., Urrutia, A., Porcu, F., Borcel, E., Gutierrez-Lopez, M. D., O'Shea, E., & Colado, M. I. (2014). 3,4-Methylenedioxymethamphetamine (MDMA, ecstasy) disrupts blood-brain barrier integrity through a mechanism involving P2X7 receptors. *The International Journal of Neuropsychopharmacology, 17*(8), 1243–1255. https://doi.org/10.1017/S1461145714000145

Rymo, S. F., Gerhardt, H., Wolfhagen Sand, F., Lang, R., Uv, A., & Betsholtz, C. (2011). A two-way communication between microglial cells and angiogenic sprouts regulates angiogenesis in aortic ring cultures. *PLoS One, 6*(1), e15846. https://doi.org/10.1371/journal.pone.0015846

Sengillo, J. D., Winkler, E. A., Walker, C. T., Sullivan, J. S., Johnson, M., & Zlokovic, B. V. (2013). Deficiency in mural vascular cells coincides with blood-brain barrier disruption in Alzheimer's disease [Research Support, N.I.H., Extramural]. *Brain Pathology, 23*(3), 303–310. https://doi.org/10.1111/bpa.12004

Singh, S., Anshita, D., & Ravichandiran, V. (2021). MCP-1: Function, regulation, and involvement in disease. *International Immunopharmacology, 101*(Pt B), 107598. https://doi.org/10.1016/j.intimp.2021.107598

Smigiel, K. S., & Parks, W. C. (2017). Matrix metalloproteinases and leukocyte activation. *Progress in Molecular Biology and Translational Science, 147*, 167–195. https://doi.org/10.1016/bs.pmbts.2017.01.003

Stamatovic, S. M., Johnson, A. M., Keep, R. F., & Andjelkovic, A. V. (2016). Junctional proteins of the blood-brain barrier: New insights into function and dysfunction. *Tissue Barriers, 4*(1), e1154641. https://doi.org/10.1080/21688370.2016.1154641

Stobart, J. L., Erlebach, E., Gluck, C., Huang, S. F., Barrett, M. J., Li, M., Vinogradov, S. A., Klohs, J., Zarb, Y., Keller, A., & Weber, B. (2022). Altered hemodynamics and vascular reactivity in a mouse model with severe pericyte deficiency. *Journal of Cerebral Blood Flow and Metabolism, 271678X221147366*, 763. https://doi.org/10.1177/0271678X221147366

Su, J., Zhang, A., Shi, Z., Ma, F., Pu, P., Wang, T., Zhang, J., Kang, C., & Zhang, Q. (2012). MicroRNA-200a suppresses the Wnt/β-catenin signaling pathway by interacting with β-catenin. *International Journal of Oncology, 40*(4), 1162–1170. https://doi.org/10.3892/ijo.2011.1322

Sui, Z., Sniderhan, L. F., Schifitto, G., Phipps, R. P., Gelbard, H. A., Dewhurst, S., & Maggirwar, S. B. (2007). Functional synergy between CD40 ligand and HIV-1 Tat contributes to inflammation: Implications in HIV type 1 dementia. *Journal of Immunology, 178*(5), 3226–3236. https://doi.org/10.4049/jimmunol.178.5.3226

Suzuki, A., Ishiyama, C., Hashiba, K., Shimizu, M., Ebnet, K., & Ohno, S. (2002). aPKC kinase activity is required for the asymmetric differentiation of the premature junctional complex during epithelial cell polarization. *Journal of Cell Science, 115*(Pt 18), 3565–3573.

Taggi, V., Riera Romo, M., Piquette-Miller, M., Meyer Zu Schwabedissen, H. E., & Neuhoff, S. (2022). Transporter regulation in critical protective barriers: Focus on brain and placenta. *Pharmaceutics, 14*(7). https://doi.org/10.3390/pharmaceutics14071376

Thanabalasundaram, G., Schneidewind, J., Pieper, C., & Galla, H. J. (2011). The impact of pericytes on the blood-brain barrier integrity depends critically on the pericyte differentiation stage. *The International Journal of Biochemistry & Cell Biology, 43*(9), 1284–1293. https://doi.org/10.1016/j.biocel.2011.05.002

Tran, K. A., Zhang, X., Predescu, D., Huang, X., Machado, R. F., Gothert, J. R., Malik, A. B., Valyi-Nagy, T., & Zhao, Y. Y. (2016). Endothelial beta-catenin signaling is required for maintaining adult blood-brain barrier integrity and central nervous system homeostasis. *Circulation, 133*(2), 177–186. https://doi.org/10.1161/CIRCULATIONAHA.115.015982

Tsakiris, D. A., Tschopl, M., Wolf, F., Labs, K. H., Jager, K. A., & Marbet, G. A. (2000). Platelets and cytokines in concert with endothelial activation in patients with peripheral arterial occlusive disease. *Blood Coagulation & Fibrinolysis, 11*(2), 165–173.

Vates, G. E., Takano, T., Zlokovic, B., & Nedergaard, M. (2010). Pericyte constriction after stroke: The jury is still out. *Nature Medicine, 16*(9), 959; author reply 960. https://doi.org/10.1038/nm0910-959

Videm, V., & Albrigtsen, M. (2008). Soluble ICAM-1 and VCAM-1 as markers of endothelial activation. *Scandinavian Journal of Immunology, 67*(5), 523–531. https://doi.org/10.1111/j.1365-3083.2008.02029.x

Walther, M., Kuklinski, S., Pesheva, P., Guntinas-Lichius, O., Angelov, D. N., Neiss, W. F., Asou, H., & Probstmeier, R. (2000). Galectin-3 is upregulated in microglial cells in response to ischemic brain lesions, but not to facial nerve axotomy. *Journal of Neuroscience Research, 61*(4), 430–435. https://doi.org/10.1002/1097-4547

Wang, Y., Imitola, J., Rasmussen, S., O'Connor, K. C., & Khoury, S. J. (2008). Paradoxical dysregulation of the neural stem cell pathway sonic hedgehog-Gli1 in autoimmune encephalomyelitis and multiple sclerosis. *Annals of Neurology, 64*(4), 417–427. https://doi.org/10.1002/ana.21457

Wang, J., Xu, F., Zhu, X., Li, X., Li, Y., & Li, J. (2021). Targeting microRNAs to regulate the integrity of the blood-brain barrier. *Frontiers in Bioengineering and Biotechnology, 9*, 673415. https://doi.org/10.3389/fbioe.2021.673415

Wegmann, F., Petri, B., Khandoga, A. G., Moser, C., Khandoga, A., Volkery, S., Li, H., Nasdala, I., Brandau, O., Fassler, R., Butz, S., Krombach, F., & Vestweber, D. (2006). ESAM supports neutrophil extravasation, activation of Rho, and VEGF-induced vascular permeability. *The Journal of Experimental Medicine, 203*(7), 1671–1677. https://doi.org/10.1084/jem.20060565

Winkler, E. A., Sengillo, J. D., Sullivan, J. S., Henkel, J. S., Appel, S. H., & Zlokovic, B. V. (2013). Blood-spinal cord barrier breakdown and pericyte reductions in amyotrophic lateral sclerosis. *Acta Neuropathologica, 125*(1), 111–120. https://doi.org/10.1007/s00401-012-1039-8

Yamamoto, M., Ramirez, S. H., Sato, S., Kiyota, T., Cerny, R. L., Kaibuchi, K., Persidsky, Y., & Ikezu, T. (2008). Phosphorylation of claudin-5 and occludin by rho kinase in brain endothelial cells. *The American Journal of Pathology, 172*(2), 521–533.

Yang, H., Zhou, J., & Lehmann, C. (2016). GPR55 – A putative "type 3" cannabinoid receptor in inflammation. *Journal of Basic and Clinical Physiology and Pharmacology, 27*(3), 297–302. https://doi.org/10.1515/jbcpp-2015-0080

Yao, H., Bethel-Brown, C., Niu, F., Yang, L., Peng, F., & Buch, S. (2013). Yin and Yang of PDGF-mediated signaling pathway in the context of HIV infection and drug abuse. *Journal of Neuroimmune Pharmacology, 9*, 161. https://doi.org/10.1007/s11481-013-9481-5

Yenari, M. A., Xu, L., Tang, X. N., Qiao, Y., & Giffard, R. G. (2006). Microglia potentiate damage to blood-brain barrier constituents: Improvement by minocycline in vivo and in vitro. *Stroke, 37*(4), 1087–1093. https://doi.org/10.1161/01.STR.0000206281.77178.ac

Zandbergen, F., & Plutzky, J. (2007). PPARalpha in atherosclerosis and inflammation. *Biochimica et Biophysica Acta, 1771*(8), 972–982. https://doi.org/10.1016/j.bbalip.2007.04.021

Zhang, A., Qian, Y., & Qian, J. (2019). MicroRNA-152-3p protects neurons from oxygen-glucose-deprivation/reoxygenation-induced injury through upregulation of Nrf2/ARE antioxidant signaling by targeting PSD-93. *Biochemical and Biophysical Research Communications, 517*(1), 69–76. https://doi.org/10.1016/j.bbrc.2019.07.012

Zhao, H. L., Zhang, X., Dai, Z. Q., Feng, Y., Li, Q., Zhang, J. H., Liu, X., Chen, Y. J., & Feng, H. (2016). P2X7 receptor suppression preserves blood-brain barrier through inhibiting RhoA activation after experimental intracerebral hemorrhage in rats. *Scientific Reports, 6*. https://doi.org/10.1038/srep23286

Zhao, X., Zeng, H., Lei, L., Tong, X., Yang, L., Yang, Y., Li, S., Zhou, Y., Luo, L., Huang, J., Xiao, R., Chen, J., & Zeng, Q. (2021). Tight junctions and their regulation by non-coding RNAs. *International Journal of Biological Sciences, 17*(3), 712–727. https://doi.org/10.7150/ijbs.45885

Zhong, L., Simard, M. J., & Huot, J. (2018). Endothelial microRNAs regulating the NF-κB pathway and cell adhesion molecules during inflammation. *The FASEB Journal, 32*(8), 4070–4084. https://doi.org/10.1096/fj.201701536R

Zingarelli, B., & Cook, J. A. (2005). Peroxisome proliferator-activated receptor-gamma is a new therapeutic target in sepsis and inflammation. *Shock, 23*(5), 393–399.

Zlokovic, B. V. (2010). Neurodegeneration and the neurovascular unit. *Nature Medicine, 16*(12), 1370–1371. https://doi.org/10.1038/nm1210-1370

43 Proteomics and Genomics in Neuroimmunological Disorders

Wojciech Rozek and Pawel S. Ciborowski

Abstract

Proteomic and genomic technologies, especially when integrated with functional genomics, offer powerful experimental strategies for generating expression profiles of proteins and mapping their links to disease-specific changes, spanning from transcription to posttranslational modifications. It seems that molecular mechanisms underlying many neurodegenerative disorders may have common features. In a short amount of time, proteomics underwent unprecedented development from its early stages of collecting high numbers of protein identifications in a single experiment. Quantitative mass spectrometry of proteins is now a standard approach in many laboratories. However, analytical techniques based on the principles of liquid chromatography are dominating and advancing the field. Further technological developments such as single-cell analysis, next-generation proteomics, or proteogenomics will further facilitate a more precise view of changes occurring in proteomes resulting from alterations in biological systems. Another big step will be to go beyond slowing down or stopping neurodegenerative processes and push the immune system to repair damage and promote regeneration. To what extent this is possible and how soon it can be accomplished remains an open question. It is certain, however, that understanding functions of complex biological systems, such as duality of neurotoxic/neurotrophic function of mononuclear phagocytes in the brain during inflammation, will require coordinated monitoring of a large number of parameters simultaneously and well-designed proteomic, genomic, and neuro-imaging experiments.

Keywords

2-dimensional electrophoresis · Bioinformatics · Biomarker · Bottom-up analysis · Cerebrospinal fluid · Gene arrays · Genomics · HIV-1 associated dementia · Hyphenated techniques · Liquid chromatography · Mass spectrometry · Neurodegeneration · Neuroproteomics · Protein fingerprinting · Proteomics · SELDI-TOF · SWATH-MS · Systems biology · Tissue profiling · Top-down analysis

43.1 Introduction

Since the last edition of this textbook, proteomics has significantly developed in two directions: technological and methodological. Designs for proteomic experiments left the pursuit of a number of protein identifications as the main driving force. They centered on targeted analytics focusing on identification of proteoforms and specific posttranslational identifications.

Upon successfully sequencing the human genome, it was shown that it consists of a considerably smaller number of genes (20,000–25,000 protein-coding genes) (International Human Genome Sequencing Consortium, 2004) than expected, with the number of proteins greatly exceeding the number of genes. This had a big impact on how we envision the human genome, its translation to the proteome, and eventually the function of living organisms. Isoforms and proteoforms, which are products of one gene modified by one or more of 354 posttranslational modifications (http://www.abrf.org/index.cfm/dm.home?AvgMass=all) link or alternative splicing events, are only one part of the complexity. Function and biological roles can also depend on cyto-

W. Rozek
Department of Pharmacology and Experimental Neuroscience, University of Nebraska Medical Center, Omaha, NE, USA

Department of Virology, National Veterinary Research Institute, Pulawy, Poland

P. S. Ciborowski (✉)
Department of Pharmacology and Experimental Neuroscience, University of Nebraska Medical Center, Omaha, NE, USA
e-mail: pciborowski@unmc.edu

H. E. Gendelman, T. Ikezu (eds.), *Neuroimmune Pharmacology and Therapeutics*,
https://doi.org/10.1007/978-3-031-68237-7_43

plasmic, sub-cellular compartments, and/or extracellular localization. Interactions with other proteins and non-proteinaceous molecules/factors add yet another layer of complexity. At the same time, we observed unprecedented development of mass analyzers with increased sensitivity, resolution, mass accuracy, and dynamic range. Considering this, it is clear that the past 25 years have had a tremendous impact on shaping the proteomic field.

43.2 Proteomics and Genomics: Omics Technologies for Global Oversight of Complex Biological Systems

While the word *genomics* was first coined by Thomas H. Roderick, Ph.D., a geneticist at the Jackson Laboratory in Bar Harbor, Maine, in 1986 (Yadav, 2007), proteomics as a scientific field emerged in mid-1990 along with technological advances in analysis of proteins and peptides. Two-dimensional polyacrylamide gel electrophoresis (2D SDS-PAGE), originally developed in the late 1970s (Barritault et al., 1976), rapidly advanced to the next level with immobilized pH gradient (IPG) strips (Corbett et al., 1994). This improvement allowed for the analysis of gene products containing over 2000 proteins in a mixture (Gianazza, 1995) with increasing reproducibility.

Mass spectrometry of proteins and peptides has developed quickly, providing new methods and reagents for quantitative analysis such as Isotope Coded Affinity Tags (ICAT) isobaric Tags for Relative and Absolute Quantitation (iTRAQ) (Ross et al., 2004), Stable Isotope Labeling with Amino acids in Culture (SILAC) (Oda et al., 1999), and Sequential Window Acquisition of all Theoretical Mass Spectra (SWATH) mass spectrometry, which was first introduced in the ABSciex qTOF (TripleTOF) system (Sturm et al., 2008). This was followed by development of TMT (tandem mass tag) platform that is currently widely used for quantitative mass spectrometry based proteomics, allowing the analysis of up to 18 samples simultaneously (Rauniyar & Yates, 2014; Li et al., 2021).

Development of user-friendly kits for protein sample preparation and extraction has enabled nonprotein chemists to isolate proteins, fractionate complex samples, and develop their own protocols in a similar way to the development of expression vectors and systems (Feuerstein et al. 2005). Growth of various databases followed by their integration, along with new and improved algorithms for database searches, has opened the door for high throughput experiments with substantially increased quality of protein identification and quantification (Ku and Yona 2005). Currently, it is standard to deposit data into publicly available databases. The PRoteomics IDEntifications (PRIDE) database (https://www.ebi.ac.uk/pride/) is the world's largest data repository of mass spectrometry-based proteomics data (Perez-Riverol et al., 2022). The deposited data can be used by other scientists, verified, and bioinformatically processed using new tools, grouped with data from various experiments, and analyzed in other aspects than originally outlined, enabling meta-analysis. PRIDE is a part of the ProteomeXchange (PX) consortium of proteomics resources (http://www.proteomexchange.org) that, since 2012, has standardized data submission and dissemination of mass spectrometry proteomics data worldwide. Since 2017, in addition to the four PX members at the time (PRIDE, PeptideAtlas including the PASSEL resource, MassIVE, and jPOST), two new resources have joined PX: iProX (China) and Panorama Public (USA) (Deutsch et al., 2020).

Thus, it became a reality to be able to identify fingerprints of change induced by infection, malignant transformation, exposure to toxic agents, or differentiation (Pang et al., 2006). As a result of this advancement, substantial progress in prognosis and monitoring of disease progression and the discovery of new biomarkers and drug targets was and is widely expected. Concurrently, gene sequencing and gene arrays, which can be used to measure changes in expression levels to detect single nucleotide polymorphisms (SNPs) or to genotype or target re-sequencing, are rapidly advancing technologies that provide fast and low-cost mapping of whole genomes. Gene array and gene sequencing techniques, which have been used to study biological systems for many years, provided enormous valuable information and led to many discoveries and advancements, including completion of the Human Genome Project (International Human Genome Sequencing Consortium, 2004). Due to the development of these technologies, unprecedented amounts of data are generated containing information about the function of biological systems under investigation. This created a bottleneck in data analysis and a demand for better statistical and bioinformatics tools for data analysis.

Reproducibility and orthogonal validation, as well as data integration, remain a substantial challenge to be tackled in years to come. Figure 43.1 shows a general overview of top–down and bottom–up experimental approaches. The top–down approach deals with system modeling leading to hypothesis, prediction, and simulation of the behavior of an organism under various physiological conditions. The bottom–up approach studies individual pieces on a discovery basis and combines all the information in an effort to understand how the system in question works as a whole (Fig. 43.1). In proteomics, a third approach known as middle–down is based on generating peptides that are longer than in bottom–up proteomics but shorter than intact proteins in top–down

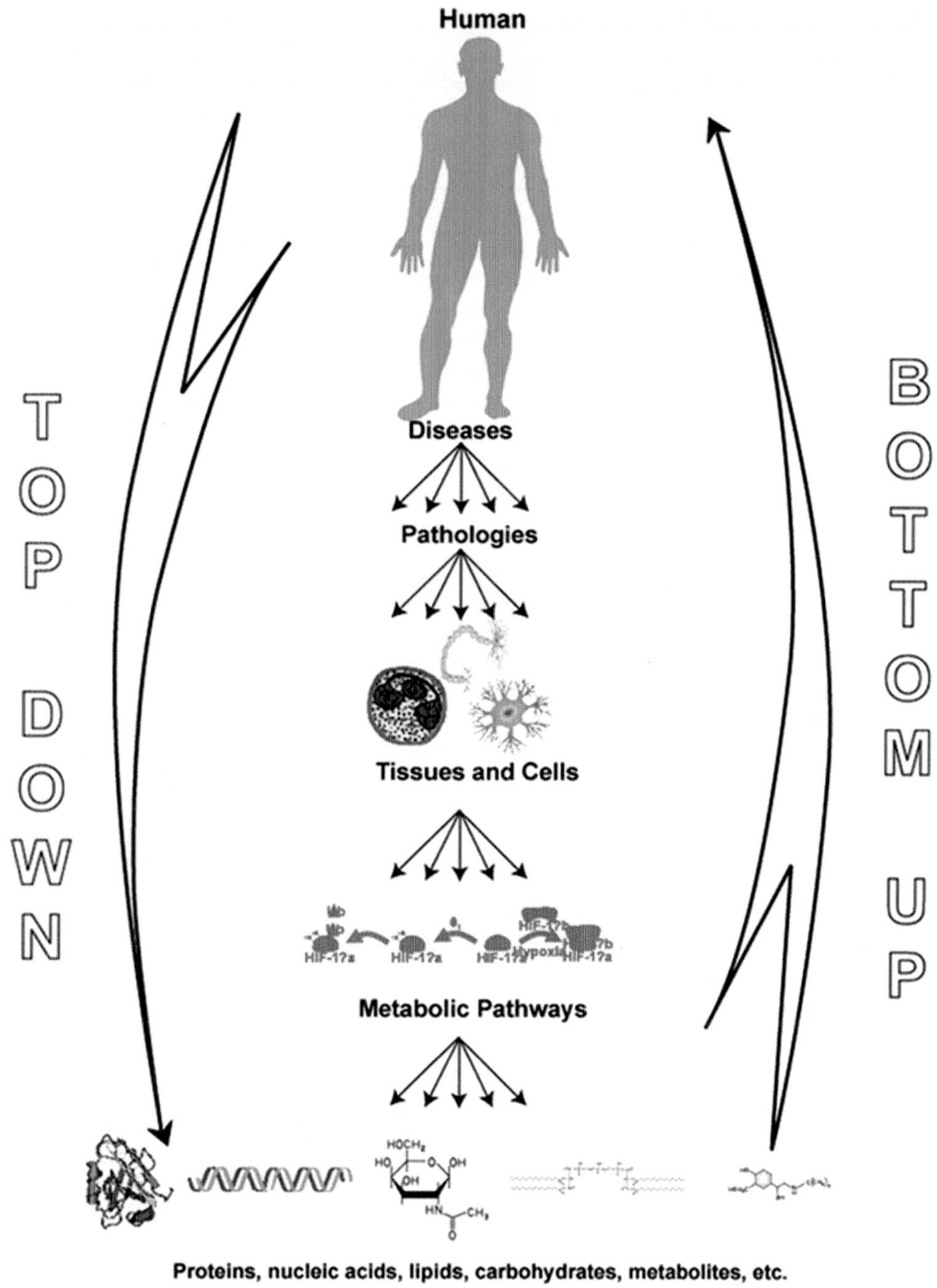

Fig. 43.1 Top–down and bottom–up approach in studying systems biology. There are three components contributing equally to study of biology systems: theoretical, conceptual, and experimental approaches. These approaches heavily depend on implementation of effective bioinformatics tools. Hypothesis-driven and discovery-driven projects equally participate in studying systems biology

proteomics that are analyzed by mass spectrometry (Forbes et al., 2001; Duong et al., 2022). Proteomics offers a number of additional and unique benefits. For example, proteins that are the functional products of an expressed gene can be post-translationally modified; and their sub-cellular localization can be altered in a different physiological state. This important information cannot be obtained using gene arrays but is available through proteomic studies. Results from studies of biological systems led to the conclusion that individualized diagnosis, clinical evaluation, and treatment are inevitable. In many instances, the sample amount available for diagnostic purposes is limited, resulting in the demand for more sensitive techniques and instrumentation and the development of single-cell analysis. Automated and simplified steps of sample processing, combined with highly sensitive and specific measurements, will improve what currently contributes to the loss of important markers in a limited sample.

In summary, it is necessary to understand the origin of changes at the level of proteoforms rather than the proteome.

43.3 Proteomics Technologies

A variety of technologies are being developed to support qualitative and quantitative measurements of protein changes in proteomes and to address important issues in proteomics as a global overview of protein makeup (Rohlff & Southan, 2002; Petricoin & Liotta, 2003; Patel, 2014; Woods et al., 2014). Therefore, the selection of the proteomic platform to be used in addressing any particular question is of utmost importance. A summary of proteomic workflows is presented in Fig. 43.2, which is divided into three major groups: top–down, middle–down, and bottom–up. Top–down analysis is based on isolation, analysis, and characterization of intact protein to reveal its function. Fourier-transformed ion cyclotron resonance mass spectrometry (FT-ICR) (Marshall et al., 1998) facilitates such an approach in protein identification as a result of the random fragmentation of an intact molecule. Conversely, the bottom–up approach is based on up–front fragmentation of the protein in question using various proteolytic enzymes with known specificity (Chalmers et al., 2005; Millea et al., 2006). In these experiments, trypsin is most commonly used; and the resulting peptides are identified by mass spectrometry. An important question remains as to whether more information will be obtained from the bottom–up approach, which is protein identification on a cost of protein characterization, or from the top–down approach, which distinguishes differences between two similar proteins/proteoforms. Of course, the ideal situation is to obtain both answers from one type of analysis. Handling mixtures of peptides generated from complex protein mixture (bottom–up approach) will be very different from sample processing at the protein level prior to analysis (top–down approach). Therefore, the integration of these two will be a technological challenge in the coming years because to make significant contributions, both approaches will require the collaboration of scientists with diverse expertise and with cross-disciplinary attributes.

Methods representing top–down approaches are used for separation of intact proteins using either gel electrophoresis or liquid chromatography. Chemical or metabolic labeling, as well as label-free techniques, are widely used for bottom–up studies. With the middle–down approach, larger size protein fragments are generated and analyzed by tandem mass spectrometry with front-end liquid chromatography separation. In these experiments, label-free techniques are used. A summary of technologies most commonly used in proteomics is presented in Table 43.1. Our intention here is that this summary not only presents past experience but also serves as a guide for the design of proteomic workflow in future studies.

The most commonly used proteomic platform is the combination of liquid chromatography and tandem mass spectrometry. In the first dimension, tryptic-digested peptides are fractionated on a strong cation exchange (SCX) column, and collected fractions are further fractionated on reverse-phase high performance liquid chromatography (RP-HPLC) with direct infusion to the mass spectrometer. Alternatively, to SCX, an OFFGEL fractionation based on isoelectric point is quite often used. The OFFGEL platform was introduced by Agilent, Inc., in 2006.

The majority of analytical techniques currently used for proteomic profiling have a liquid chromatography-based step of fractionation. This will be RP-HPLC directly interfaced with mass spectrometer and offline fractionations such

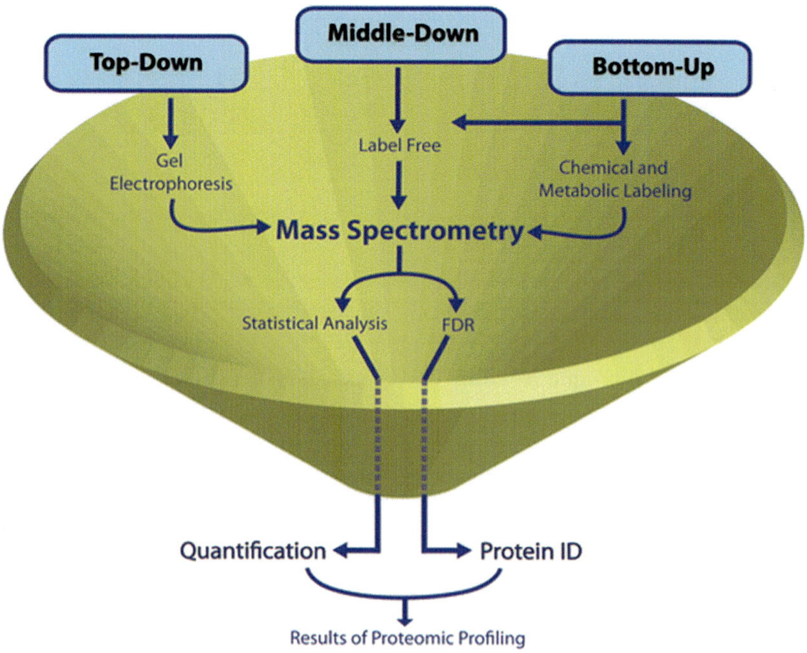

Fig. 43.2 A summary of proteomic platforms. The three primary methods of proteomic analysis involving mass spectrometry are top–down, middle–down, and bottom–up. While each requires different methods of sample preparation, they all have the same outcome—identification and quantification of samples that leads to complete proteomic profiling

Table 43.1 Summary of analytical and preparative methods most commonly used in proteomic studies

Method	Description	Applications
1D SDS PAGE	Fractionation based on approximate molecular weight (m.w.)	Preparative/ analytical
IEF	Fractionation based on isoelectric point (pI). In addition to IPG strips used for first-dimension separation as part of 2DE, includes OFFGEL technology	Preparative/ analytical
2 D electrophoresis (IEF/SDS PAGE)	High-resolution separation/fractionation based on pI and m.w., long and complex procedure. IPG strips with variety of ranges of pH are available. 2DE DIGE using differential labeling of proteins with fluorescent Cy dyes	Preparative/ analytical
LC-MS/MS	Chromatographic one- or two-dimensional fractionation followed by mass spectrometric peptide sequencing and protein identification	Analytical
Capillary Zone Electrophoresis (CZE)	Fractionation based on m.w. or pI, directly connected to tandem mass spectrometry	Analytical/ preparative
HPLC/FPLC	Fractionation based on physico-chemical features of protein (m.w., pI, hydrophobic profile)	Preparative/ analytical
SELDI-TOF	Fractionation based on selective binding to activated surfaces followed by MS determination of m.w., automatic fast sample processing combined with mass spectrometry	Analytical
MALDI-TOF	Requires high-resolution mass spectrometry for protein identification based on peptide fingerprinting. High-density plate spotters combined with micro-capillary RP-HPLC fractionation are available.	Analytical
Protein Arrays	Identification/separation techniques based on protein–protein interactions. Requires high-quality and well characterized antibodies.	Analytical

as SCX chromatography, IEF (OFFGEL technology), immunodepletion, or all configurations of pull-down and enrichment steps. Multidimensional fractionation is an essential step, but we have to keep in mind that extending pre-profiling fractionation may not always work towards our advantage, since every analytical step will introduce some degree of variability.

Enrichment of specific proteins or groups of proteins is certainly the main goal of pre-fractionation. For example, immunodepletion of the most abundant proteins from serum or cerebrospinal fluid (CSF) samples will facilitate detection of low abundant proteins in complex samples (Chromy et al., 2004; Fountoulakis et al., 2004; Ramstrom et al., 2005). However, the remaining proteins could be quite different between samples. A published report of the Human Proteome Organization (HUPO) Human Plasma Protein Project (HPPP) showed that almost all laboratories participating in this international effort used some means of sample pre-fractionation. Eventually, 3020 proteins were included in a database set from 15,677 proteins reported (Omenn et al., 2005). Enrichment using TiO_2 columns for phosphoproteins and phosphopeptides or affinity columns to selectively pull-down ubiquitination is to facilitate identification and quantification of low abundant modified proteins. If enrichment is at the protein level, samples can be used for top-down profiling. In many instances, proteins are proteolytically digested, and enrichment is performed at the level of peptides. Such samples are used for bottom-up analyses.

2D SDS-PAGE, which has been used for about 40 years, combines two modes of separation: isoelectric focusing and SDS polyacrylamide gel electrophoresis (O'Farrell, 1975; Barritault et al., 1976). Development of immobilized pH gradient (IPG) strips made this method more user-friendly, and most importantly, much more reproducible. This allowed construction of computerized protein databases based on two-dimensional separation (Celis et al., 1990). Although the resolution, thus reproducibility, of 2D SDS-PAGE remains a limitation, especially in high molecular mass ranges or for hydrophobic proteins, advancements in protein sequencing by electrospray ionization (ESI-MS/MS) or Matrix Assisted Lased Desorption Ionization Time-of-Flight/Time-of-Flight (MALDI-TOF/TOF) mass spectrometry made it possible to identify proteins directly from the protein "spots" in 2D SDS-PAGE. This method is used for top-down proteomics. *DI*fference *G*el *E*lectrophoresis (DIGE) compares two samples: one labelled with Cy 3 and the other labeled with Cy 5 fluorescent dye. These two samples are mixed and analyzed in one 2D SDS-PAGE experiment. Common proteins are represented by yellow spots while different colored spots, which are detected with a laser-based scanner, represent differentially expressed proteins. This is another example of further technological advancement.

Another tool for protein profiling is Surface Enhanced Laser Desorption Ionization Time-of-Flight (SELDI-TOF) (Hutchens TWaY, 1993). This technology uses a MALDI-TOF mass spectrometer for detection, along with one-step chromatography to reduce complexity of the analyzed sample. The latter involves the use of proprietary Protein Chips initially commercialized by Ciphergen, Inc. and now offered by BioRad Laboratories. This is an analytical tool with relatively low resolution and mass accuracy in addition to high variability of intensities. Nevertheless, the big advantage of this method is ease of use for fast screening and collection of information regarding properties and conditions associated with binding of the proteins of interest to ion-exchange, hydrophobic, Immobilized Metal Affinity Chromatography

(IMAC), and/or normal phases that can be used with success in translation to preparative methods using the corresponding HPLC column (Enose et al., 2005). This technology is also susceptible to relatively high variability in the analyses of multiple SELDI-TOF spectra. Jeffries et al. developed an algorithm that greatly improved spectra alignment and thus improved reproducibility (Jeffries, 2005). MALDI-TOF instruments with higher resolution than SELDI-TOF, combined with various chromatographic techniques, have also been used but did not produce convincingly better results than SELDI-TOF (Coombes et al., 2005).

A separate field of so-called "hyphenated methods" has been developed and deals with combining two or more methods together in succession. The first component of the combined methods will deal with separation/fractionation of an analyzed sample, followed by an in-line connected detection system. Hyphenated techniques were developed and became popular in the 1990s and 2000s (Wolfender et al., 1995; Patel et al., 2010) and remain a vital part of the current technological advancements (Murray et al., 1975; Lee et al., 1976). Examples of the most commonly used hyphenated methods in proteomics are LC-MS (liquid chromatography combined with mass spectrometry), GC-MS (gas chromatography combined with mass spectrometry) (Kazakevich et al., 2005), and LC-UV-SPE-NMR (liquid chromatography, UV detection, solid phase extraction, and nuclear magnetic resonance) (Exarchou et al., 2005). Sample fractionation prior to detection and/or analysis usually leads to enhanced analytical outcomes. Hyphenated techniques are also meeting a big challenge in proteomic analyses dealing with an extremely large and dynamic range of protein levels (Julka & Regnier, 2005; Kusnezow et al., 2006).

Protein arrays attempt to occupy another space in proteomic technologies. While this approach is similar in concept to gene arrays, protein arrays are based on antigen-antibody interaction rather than hybridization of complementary strands of nucleic acid. As much as antigen–antibody interaction is very specific, it also depends on the quality of well-standardized antibodies. This is much harder to accomplish. Therefore, accuracy and reproducibility of protein array data highly depends on proper experimental design, normalization procedures, eliminating systematic bias, and appropriate statistical analysis. Eckel-Passow and co-authors made several excellent suggestions about how experiments should be designed and analyzed to avoid false results (Eckel-Passow et al., 2005). For example, the authors propose to use several different clustering tools instead of a single cluster analysis for verification of results in discovering differences in a class of molecules, e.g., cytokines. Another obstacle in working with protein microarrays is maintaining the stability of protein structures, which is critical for proper folding and, thus, interaction with the binding partner. Another side of this problem is exposure of the interacting surface/epitope while immobilized on the chip's surface. This technology, although not as popular now, is slowly gaining ground in biomedical research with increasing potential of applications in clinical diagnosis. Similar to DNA microarrays, protein arrays are commercially available as user-friendly kits, made specifically for various microorganisms and diseases, with improved and sophisticated fluorescent tags. For those working in the area of drug development, protein microarrays are vehicles for accelerated discoveries. Protein arrays also make it possible to screen functions such as protein–antibody, protein–protein, protein–ligand, protein–drug, and enzyme–substrate interactions. They can also be used to perform multi-analyte diagnostic assays. For others, protein microarrays are envisioned as tools for the exploration and understanding of pathological processes, kinetics of changes at the molecular level, drug candidates, and interactions between molecules such as lipids, nucleic acids, and carbohydrates. Protein microarrays are a powerful tool when the proteomic analysis is narrowed to a group of proteins such as growth factors, chemokines, and cytokines. These classes of proteins already have a great number of pre-required, well-characterized, and highly specific antibodies reacting with high affinity. Such protein arrays (chips) were developed and are successfully used in monitoring markers of immune responses (Kastenbauer et al., 2005) complementing DNA arrays (Ho et al., 2005). Peptide microarrays are another promising format for proteomics research. Thousands of peptides can be synthesized directly on the chip and reacted to test their immunoreactivity (overlapping epitopes, usually linear) or enzymatic profiling of substrates (e.g., phosphorylation). Peptide microarrays have been used in the field of neurological disorders to detect autoantibodies or immune responses to infectious agents (Hecker et al., 2016; Zandian et al., 2017; Garcia-Montojo et al., 2022).

In summary, one should decide which platform to use based on questions that are asked. The experimental design and sample processing should align with the type of biological material being investigated. Investigation of proteomes of body fluids and culture supernatants or lysates of whole cells, tissues, or cellular compartments/organelles would all require a different platform, and this can be determined by reviewing already published studies. Along with improvements in sensitivity, resolution, mass accuracy, dynamic range, and speed of data acquisition of new generations of mass spectrometers, proteomic data are becoming massive, and reduction of sample complexity is inevitable. Otherwise, analysis of such massive data and interpretation becomes very complicated.

43.4 Biomarkers

Biomarkers in disease can be defined as quantifiable analytes or factors reflecting changes in an individual's physiological state. These changes, or biological responses, can be found at the molecular, cellular, and/or whole organism level. They play a critical role in assessing disease outcome, particularly in patient treatment, how these treatments will be individualized and/or modified in future clinical trials, and how they will affect directions in drug development. Thus, biomarkers must provide objective measures of normal versus pathogenic processes and responses to treatment whether this leads to cures, improvement, or only maintenance of a patient's health. Undoubtfully, biomarkers are essential for early diagnosis of any disease, whether it be a slowly progressing pathological process such as neurodegenerative disorders or acutely developing viral and/or bacterial infection(s).

Human immunodeficiency virus (HIV) associated neurocognitive disorders (HAND) and in its extreme form, HIV-associated dementia (HAD), are good examples on the importance and qualities of reliable biomarkers. HAND is a brain pathology that develops over a long period of HIV-1 infection, ranging in progression from subclinical to end stage. Currently, HAND is diagnosed by a set of neuropsychiatric and psychological tests, and it is very difficult to predict and measure the time of onset of this slowly progressing disease. The main reason for this is a lack of appropriate molecular biomarkers, which inform us about mechanisms of brain injury triggered by HIV-1 in the brain.

Given this information, what would constitute a good and reliable biomarker of HAND? We expect that regardless of whether a biomarker is found in the CSF or plasma of diseased people, it will give us an insight into the cellular mechanisms of an ongoing infection. This insight is important for those who are designing new clinical trials or regimes of existing treatments. Whether or not a biomarker enters a clinical trial is based on vague measures, which might be influenced by the subjective impact of the examiner, causing an additional risk of inaccurate results. Thus, an ideal biomarker should inform us as precisely and as objectively as possible about the mechanism, status, and prediction of therapy and disease. The ideal biomarker should also help in the design of "individualized" therapy for the most effective clinical result. In our example of HAND, an activation of mononuclear phagocytes recruited in high numbers due to ongoing inflammation led to conclusions that up- or down-regulated proteins originating from these cells due to HIV-1 infection might be good candidates for biomarkers. Such an approach, however, should be considered with caution. Our studies showed that although HIV-1 down-regulates MMP-9 expression in monocyte-derived macrophages in in vitro studies, the net level of this protein in the CSF of HAD individuals as well as its net proteolytic activity is increased due to recruitment of high numbers of these cells to the inflamed brain (Ciborowski et al., 2004).

Recent developments of nanoscale in analytics make it possible to use lower amounts of the sample for analysis while sustaining the same level of sensitivity and specificity. Single-cell analysis and tissue laser dissections are becoming a reality, although we have to be aware that not all proteins or their modified forms can be effectively detected at this level. Unlike in genomics, the proteomics approach does not have the luxury of in vitro amplification of proteins. Besides improving the yield of proteins of interest during sample preparation, limitations of detection in laboratory tests still exist, creating an analytical challenge. Nevertheless, there are numerous examples, in which proteomics was successfully used to discover new biomarkers. On the other hand, commercial tests are almost always based on assays using different principles, e.g., ELISA (Sandhu et al., 1991; Hampel et al., 2004).

A biomarker can be a specific fragment of the protein that is present in the diseased state but not under normal conditions. Currently, many biomarkers of this nature have been identified to be measures of neurodegenerative disease. In several cases, the biomarker is a misfolded protein, showing a slightly different molecular weight from its corresponding wild type. In Alzheimer's disease (AD), an existing biomarker of this type is the $A\beta_{42}$ fragment of α-amyloid protein (Olsson et al., 2003). The accumulation of internalized misfolded Tau protein has also been found to correlate with the progression of AD, as well as frontotemporal dementia (FTD) (Rademakers et al. 2012a, b; Wu et al., 2013). Likewise, the aggregation of α-synuclein into Lewy Bodies has emerged as a biomarker for Parkinson's disease (PD) (Luk et al., 2012).

43.5 Proteomics in Biomarker Discovery

There are four necessary key components for performing proteomic (global) experiments: (1) identification of differences generated within a system under various physiological conditions, (2) fractionation of complex mixtures of proteins and other molecules, (3) identification of differentially expressed proteins and their interacting partners, and (4) data analysis, which includes algorithms for database searches, tools for statistical analysis, and tools for bioinformatics analysis. The choice of method, instrumentation, and experimental approach should be selected based on the questions being asked and answers being sought.

As we collectively accumulate knowledge and experience on how proteomic experiments should be conducted in order to yield the most information, a shift from fully unbiased profiling to targeted proteomics has been observed. Therefore, the majority of studies are now focused on specific groups of proteins or even single proteins and their interacting partners whose function(s) and/or interaction(s) are significant in biological processes. The expectation is that these studies will generate results that will be translated into the development of commercial diagnostic tests. The major bottleneck in this process is the processing of increasingly complex data generated by proteomic studies. Therefore, targeted proteomics may find broader applications in the field when pathological processes are far away from explanations as they are in neurodegeneration. Limiting mass spectrometry identifications to the specific set of proteins (instead of profiling across whole proteome) enables analysis of many probes with concurrent increases in sensitivity and specificity.

Proteomics cannot exist without interdisciplinary research. Experts from a wide variety of specialties who understand and are able to communicate issues related to protein chemistry will bring proteomics to the level of functional studies. Currently, research teams combining a relatively narrow group of scientists including protein chemists, mass spectrometrists, and bioinformaticians are pushing the levels of sensitivity and accuracy of analytical instrumentation to unprecedented levels; but experiments still remain at the stage of comparing two conditions only. Therefore, to benefit more from proteomics, we need to organize international groups/consortia to tackle and prioritize problems. One example is global proteomic analysis such as the Human Plasma Proteome Project (Omenn et al., 2005), which can be accomplished only through international collaborations of laboratories equipped with various types of mass spectrometers and are capable of performing complex analyses. Undoubtedly, efforts of this and other currently ongoing international initiatives, e.g., Human Brain Proteome Project (http://www.hbpp.org/), will build a foundation for new discoveries and functional studies. Everyone in the broad research community will benefit from this work, particularly the rapidly growing number of smaller proteomics laboratories and programs with research goals focused on specific question(s) of biological importance. The most exciting potential lies in functional studies combining proteomic and genomic approaches.

43.6 Proteomics of Cerebrospinal Fluid

The total volume of cerebrospinal fluid (CSF) in an adult human is about 130–150 mL, and production is relatively constant under normal physiological conditions at the rate of approximately 0.4 mL/min (more than 500 mL daily). This means that turnover of CSF is at the rate of 4 volumes daily. A significant decrease in CSF secretion and turnover is associated with neurological diseases such as AD and chronic hydrocephalus (Tsunoda et al., 2002; Silverberg et al., 2004). Thus, CSF levels are highly reflective of the physiological status of the central nervous system (CNS).

CSF, which surrounds the brain and spinal cord, acts as an intermediate between blood and nervous tissues. Functions of the CSF include buoyancy, acid–base buffering, and delivery of electrolytes, signaling molecules, transport molecules, and micronutrients to the brain parenchyma (Silverberg et al., 2004). CSF is produced predominantly by choroid plexus, specialized vascular tissue located in the brain ventricles (Serot et al., 2003). Choroid plexus tissues are intraventricular structures composed of villi covered by a single layer of ciliated, cuboid epithelium. The plexuses secrete CSF, synthesize numerous molecules, carry nutrients from the blood to the CSF, reabsorb brain metabolism by-products, and participate in brain immunosurveillance (Serot et al., 2003). Active transport across the choroid plexus is bidirectional, which allows for macromolecular transport out of the CSF. Blood flow to the choroid plexus is approximately six times greater than that of the equal volume of brain tissue, suggesting that the choroid plexus is very metabolically active (Silverberg et al., 2004).

Changes in the biochemical composition of CSF could serve as a useful tool for investigations of pathological processes in the CNS. CSF is also in contact with blood plasma through the blood–brain barrier and, thus, resembles an ultra-filtrate of plasma in its protein constituents. CSF contains sugars, lipids, electrolytes, and proteins. The concentration of proteins in CSF ranges from 0.2 to 0.8 mg/mL (0.3–1% of serum protein concentration); but more than 70% of the proteins in CSF are isoforms of albumin, transferrin, and immunoglobulins (Ogata et al., 2005; Wittke et al., 2005). The proteome of CSF contains proteins that are expressed in and secreted from circumventricular CNS structures. Therefore, the CSF proteome could provide unique biomarkers for early-stage diagnosis or the staging of a neuronal disease. This could offer potential insight into the biochemical characterization of affected neuronal populations and clarify the molecular basis of CNS pathologies (Yuan & Desiderio, 2003, 2005).

Previous studies have used proteome comparisons to study neurological disorders such as AD, PD, FTD, schizophrenia, and Creutzfeldt-Jacob disease. Some examples of using proteomics approaches for discovery of biomarkers for neurological disorders in human CSF are given in Table 43.2.

Although insightful, the investigation of biomarkers in CSF has important limitations. Availability of large volumes of CSF samples is limited by the need for invasive lumbar punctures. The large dynamic range of protein concentrations similar to this in plasma, which can be up to 12 orders of magnitude between the highest and the lowest expressed proteins, makes the analysis difficult (Maccarrone et al.,

Table 43.2 Summary of protein biomarkers of neurodegenerative disorders discovered using a proteomics approach

Disease	Biomarker (protein)	Applied method	Reference
Alzheimer's disease	Cystatin C, B-2 microglobulin isoforms, 4.8 kDa VGF polypeptide, unknown 7.7 kDa polypeptide	SELDI TOF and LC/MS	Carrette et al. (2003)
	Apolipoprotein A1, apolipoprotein E, apolipoprotein J, B-trace, retinol-binding protein, kininogen, alpha-1 antitrypsin, cell cycle progression 8 protein, alpha 1B	2D electrophoresis and MALDI-TOF	Puchades et al. (2003)
	Glycoprotein, alpha-2—HS glycoprotein Tau, hyperphosphorylated tau (p-tau), $A\beta_{42}$	MALDI-TOF	Wildsmith et al. (2014)
Frontotemporal Dementia	Transthyretin, fragment of VGF S-cysteinylated transthyretin, truncated cystatin C, fragment of chromogranin B	SELDI-TOF and MS/MS	Ruetschi et al. (2005)
Creutzfeldt-Jacob Disease	14–3-3 protein	2 D electrophoresis	Harrington et al. (1986)
	Cystatin C	SELDI-TOF, and LC-MS/MS	Sanchez et al. (2004)
	Ubiquitin	SELDI-TOF	Steinacker et al. (2010)
Schizophrenia	Apolipoprotein A-IV	2D electrophoresis and LC/MS	Jiang et al. (2003)
	Fibrinopeptide A	MALDI-TOF	Zhou et al. (2014)
	A-defensins	SELDI-TOF	Craddock et al. (2008)
Multiple Sclerosis	Cartilage acidic protein, tetranectin, SPARC-like protein, autotaxin t	2 D electrophoresis and LCMS/MS	Hammack et al. (2004)
	Atg16L2	MALDI-TOF	Yin et al. (2014)
	Alpha-1-antichymotrypsin, contactin-1, apolipoprotein D, clusterin, kallikrein-6	ITRAQ	Kroksveen et al. (2013)

2004). Various schemes of sample pre-fractionation have been used to circumvent these problems. The widely used albumin removal method is affinity chromatography on various resins. However, this method is not specific only to albumin; and it is known to bind numerous other proteins (Gianazza & Arnaud, 1982; Maccarrone et al., 2004). A newer approach is to use immunoaffinity methods using monoclonal or polyclonal antibodies to deplete albumin and other highly abundant proteins from biological samples. Resins coupled with antibodies could be packed in HPLC columns for depletion of the most abundant proteins for simultaneous depletion of 6, 12, 14, or more proteins (Laspiur et al., 2007; Rozek et al., 2007). The application of HPLC columns for CSF preparation before 2D electrophoresis and shotgun mass spectrometry was compared with Cibacron Blue and protein G resins by Maccarrone et al. (2004). Immunoaffinity columns showed less nonspecific binding of CSF proteins. Another example of pretreating CSF samples before proteomic analysis is the use of immunoglobulin Y (IgY) antibody microbeads and spin filters for depletion of the most abundant proteins (Seppro Mixed 6, GenWay, San Diego, California). Comparison of IgY microbeads and immunoaffinity columns for pretreatment of CSF samples was performed by Ogata et al. (2005). The column method removed major proteins more effectively and approximately 50% more spots were visualized when compared to the 2D gel of CSF without protein depletion. Two other commercially available kits for removal of human serum albumin and HSA/IgG were used by Ramstrom et al. (2005) for CSF preparation prior to liquid chromatography Fourier transform ion cyclotron resonance mass spectrometry (LC-FTICR MS). Both depletion methods provided a significant reduction of albumin, and the identification of lower abundant components was clearly facilitated.

Immunodepletion of the most abundant proteins, also called albuminome because albumin is the most abundant protein in plasma, is based on interaction of antigen with immunoglobulin. Although it is a highly specific interaction, it still remains to be protein–protein interaction that might be nonspecific to some extent. On the other hand, some proteins are partially removed from plasma samples because of their interactions with proteins that are being removed. Partial removal of other proteins may affect quantification and should be taken under consideration. More information about proteins removed along with albumin can be found in (Ciborowski, 2013).

Every method of removing highly abundant proteins from CSF samples before proteomic analysis aids identification of discrete proteins, which could be important as a biomarker for classifying physiological stages of the CNS. However, the possibility of nonspecific removal of important peptides (through nonspecific binding to resins or forming complexes with target proteins) should be taken into account.

Currently, in the study of CSF, exosomes are emerging as potential sources of biomarkers in the context of neurological diseases. The role of extracellular vesicles (EV) in intercellular communication can be modified in pathological states. EVs not only play a neuroprotective role by mediating the removal of toxic proteins but also may facilitate misfolded protein propagation (Vinaiphat & Sze, 2022). An interesting example was the proteomic characteristics of EVs isolated from CSF and tears of multiple sclerosis (MS) patients (Vinaiphat & Sze, 2022). In addition to protein biomarkers of neurological diseases, mRNA, noncoding RNA, and micro RNAs in EVs are being investigated (Yagi et al., 2017; Otake et al., 2019; Yu et al., 2021).

43.7 Neuroproteomics

Proteomic studies of the brain (central nervous system) pose significant challenges. Brain tissue collection, dissection, preservation, and maintenance of biochemical and molecular integrity are all critical aspects of neuroproteomics and neurogenomics. Despite the fact that many proteins and nucleic acids are stable postmortem (Yates et al., 1990), rapid degradation of other proteins may result in changes in overall protein composition (Abbott, 2003; Choudhary & Grant, 2004). Protein profiles can change rapidly within minutes after death, making it very difficult to sort out changes related to disease from those related to death itself. In many studies done so far, protein extraction of the entire brain was used without distinguishing the specific region of origin (Wang et al., 2005). To address these issues, a variety of brain dissection techniques have been developed (Sanna et al., 2005). Manual dissection methods are still very useful and yield enough material for microarray analysis with somewhat limited "contamination" of the sample of interest with neighboring regions, which has a chance of affecting reproducibility of genomic and proteomic analyses. Ultimately, the laser dissection of brain tissues technique, a quickly emerging technique (Ly et al., 2011; He et al., 2013), is further facilitated by increasing sensitivity of mass spectrometers, will be the method of choice for retrieving small regions, or even single cells of interest, from brain tissue to improve anatomical accuracy (Evans et al., 2003). Postmortem collection of CSF has been shown to be a valuable source of information; however, it has to be analyzed with caution because of brain necrosis following death (Lescuyer et al., 2004). Lescuyer and colleagues found in postmortem collected CSF samples three proteins have been reported as potential markers of various neurodegenerative disorders: prostaglandin D synthase, glial fibrillary acidic protein, and cathepsin D. Other proteins include ubiquitin C-terminal hydrolase L1, peroxiredoxin 5 (Lescuyer et al., 2004).

Some neurological disorders such as AD, PD, HAND, MS, schizophrenia, or amyotrophic lateral sclerosis (ALS) are slowly developing diseases and will require extended longitudinal studies to identify biomarkers. In contrary, brain injury due to stroke, trauma, exposure to toxic substances, hypoxia, or microbial infections are fast-acting and require immediate intervention. Nevertheless, in all these cases, the CNS is affected and damaged irreversibly to various extents. Thus, quick diagnosis and medical intervention are absolutely critical for the outcome of CNS diseases. Slowly developing neurodegenerative disorders are diagnosed by the observation of clinical symptoms after an already existing disease has been established. Thus, as much as biomarkers to measure disease progression are important, we should not stop exploring prognostic biomarkers, which will inform us who might be highly susceptible to the development of neurodegenerative diseases. Although genetic tests help us in such prognosis, they are only applicable to a portion of cases. For example, in PD about 15% of cases are correlated with a genetic predisposition due to having a first-degree relative with this disease. Therefore, when genetic tests cannot be used as predictors, proteomics appears to be an indispensable approach in the quest for predictive biomarkers.

Aggregated and modified proteins (α-synuclein in PD and tau in AD), along with deregulation of mitochondrial functions leading to overproduction of toxic reactive oxygen species (ROS) and glutamate, are factors responsible for neuronal death that are observed in many neurodegenerative disorders (Sultana et al., 2005). Nevertheless, specific triggers of neurotoxic mechanisms are vastly unknown and various proteomic techniques are employed to address these questions. High-throughput shotgun analysis used by Soreghan et al. to identify targets of protein carbonylation in a PS1 + AbPP transgenic mouse model pointed to iNOS-integrin signaling, CRE/CBP transcription regulation, and rab-lyst vesicular trafficking pathways as future research directions on neurodegeneration (Soreghan et al., 2005).

Significant progress has been made in CSF proteomic profiling in search of biomarkers for MS, which was reviewed by Kroksveen et al. (2013). Fourteen studies reviewed in this paper showed ten verified proteins, out of which six can be linked to an immune response. Although neuroinflammation is a hallmark of MS, it is not specific for MS and its signatures have been shown in the CSF and plasma of individuals suffering from other neurodegenerative disorders. Haptoglobin, alpha-1-antichymotrypsin, Ig μ-chain C region, alpha-2-HS-glycoprotein/fetuin A, kallikrein-6, apolipoprotein D (ApoD), and chitinase 3-like protein 1 (CHI3L1) are almost uniformly detected in the CSF and plasma of individuals with all kinds of diseases, including cancer. CHI3L1 is secreted by activated macrophages from viral infections (Ciborowski et al., 2007).

Another neurodegenerative disorder is HIV-1 Associated Neurocognitive Disorder (HAND). Despite the dramatic effect of antiretroviral treatment (ART) on slowing disease progression, HAND continues to be a major cause of HIV-associated pathologies. This suggests that ART does not provide complete protection against neurological damage in HIV-infected brains. Laboratory measures of HIV-1-associated inflammation of the brain to determine HAND progression have been developed. Although valuable, these measures are not specific enough to be diagnostic at this time. Studies performed in nonhuman primate models with simian immunodeficiency virus (SIV) infection did not lead to the discovery of much sought after biomarkers and only changes in relatively abundant proteins representing processes associated with many diseases were observed (Wiederin et al., 2012). Taken together, results of this study provide an explanation as to why, despite convincing quantitative measures seen during biomarker discovery, protein candidates often fail during the subsequent validation phase. The authors propose that in chronic conditions biological but not technical variance is a major obstacle in biomarker discoveries. Therefore, significant changes of proteins with medium to high concentrations can be measured if such differences are high enough to exceed levels of natural fluctuations. The authors also envision that to overcome such a hurdle better correlates need to be made in improving experimental models, in system biologic statistical and computational approaches, and in using better defined and more relevant clinical endpoints. It was encouraging to see that the same group showed changes in some of the same proteins such as gelsolin and ceruloplasmin between humans and nonhuman primates (Pottiez et al., 2012).

Considering all obstacles related to neuroproteomics studies, questions arise as to which models are preferred. While there is no doubt that studies involving the intact nervous system are more relevant for elucidating mechanisms underlying neurodegenerative disorders, the use of proteomic profiling with in vitro culture models has many advantages. Ciborowski et al. proposed that proteomic profiling of the secretome in in vitro HIV-infected human macrophages may provide biomarker candidates to be measured directly in the CSF or plasma, omitting full unbiased profiling (Ciborowski et al., 2007).

In summary, the integration of proteomic profiling studies with in vitro, in vivo, and ex vivo models can provide key insights into mechanisms of underlying pathologies.

43.8 Gene Arrays

Due to the established principle of making proteins based on transcription of genomic DNA using mRNA as a messenger between chromosomes and ribosomes, gene arrays appear to be an indispensable link between genetic information and the functional product of genes–proteins. It was a surprise to discover that the human genome codes for only ~25,000 genes, that only approximately 75% of the genome is transcribed into RNA, and that the majority of these RNA transcripts lack protein-coding potential (Djebali et al., 2012), making them noncoding or ncRNAs. The development of next-generation sequencing (Margulies et al., 2005; Shendure et al., 2005) revolutionized and advanced our knowledge about many diseases and allowed for the identification of new genes and genetic factors associated with diseases of the CNS. Data presented by Guerreiro et al. (2014) in Table 43.3 shows examples of the impact next-generation sequencing has on the identification of novel genes associated with neurodegenerative diseases.

Regardless of the fundamental importance of genomic experiments, this technology platform is only initial results and is far from complete. Proteomics, on the other hand, is a follow-up approach, and when results acquired from both technology platforms are combined, a more comprehensive picture of functional significance arises. Integration of these two experimental platforms is another challenge because of the complexity of systems biology. *First*, a bottom–up or a top–down strategy has to be chosen as an experimental strategy. *Second*, which part will be a driving force, genomics or proteomics? *Third*, an integrated bioinformatics platform has to be secured for complementary and coherent transformation of data into information. The ability to produce custom-made DNA microarrays designed and manufactured by commercial entities and/or individual investigators to address specific biological question(s) would provide enormous support to such integrated strategies. Extensive literature can be found reviewing technical issues related to gene arrays including hybridization, post-hybridization quality control, normalization, and data processing (Boes & Neuhauser, 2005; Hartmann, 2005; Reimers, 2005). These specific tools allow for the performance of genomic experiments based on results generated in initial proteomic studies (Soumelis et al., 2015), and our experience shows that experimental design and depth of data mining are critical.

Many more comprehensive or focused studies have been performed to establish links between gene expression and neurodegenerative diseases in search of biomarkers. A thorough review of all these efforts would require a separate book; and therefore, we will highlight here those of which we found to have an impact on our collective "-omics" knowledge. Tian et al. performed one such interesting study (Tian et al., 2014). The authors made interesting comparisons between microarray data from patients with Sporadic Creutzfeldt-Jakob Disease (sCJD), Fatal Familial Insomnia (FFI), and AD. Based on the principal component analysis, they identified commonalities between global gene expressions, thus supporting the idea that all neurodegenerative diseases have, to some extent, similar underlying mechanisms. Further bioinformatics analysis showed that signal transduc-

Table 43.3 Gene mutations associated with neurodegenerative diseases discovered using next-generation sequencing approach

Disease	Gene	Mutation(s)	Effect on disease	References
Parkinson's disease	*VPS35*	Heterozygous p.Asp620Asn	Causative	Zimprich et al. (2011) and Vilarino-Guell et al. (2011)
Alzheimer's disease	*TREM2*	Heterozygous p.R47H	Increased risk (OR > 3)	Guerreiro et al. (2013) and Jonsson et al. (2013)
	SORL1	Heterozygous missense and nonsense	Potentially causative	Pottier et al. (2012)
Hereditary diffuse leukoencephalopathy with spheroids	*CSF1R*	Heterozygous missense, insertions, and deletions all affecting the protein tyrosine kinase domain	Causative	Rademakers et al. (2012a, b)
Amyotrophic lateral sclerosis	*PFN1*	Heterozygous missense	Causative	Wu et al. (2012)
Autosomal-recessive cerebellar ataxia with spasticity	*GBA2*	Homozygous missense and nonsense	Causative	Hammer et al. (2013)

Source: Guerreiro et al. (2014)

tion, synaptic transmission, and neuropeptide signaling pathways representing essential functions of the CNS are commonly affected in all three aforementioned diseases. A major weakness of this study was the limited number of samples; nevertheless, these data show the proof of principle that—omics technologies are indispensable in understanding the complexity of pathogenesis mechanisms and in searching for diagnostic and therapeutic biomarkers.

An interesting study in 2015 of gene expression by microarray technology was published by Durrenberger et al. (2014). The authors used 113 samples of postmortem brains to create six study groups and controls of the following neurodegenerative disorders: AD, Huntington's disease, PD, MS, ALS, and schizophrenia. The minimum number of samples in the diseased group was nine in the case of ALS. Bioinformatics data analysis showed that antigen processing and presentation, cell adhesion, inflammatory response, regulation of cell proliferation, pattern specification process (referring to a developmental process), response to drugs and nutrients, and skeletal development were common for all six neurodegenerative disorders. The remaining genes, which were not included in these categories, were further grouped as signal transduction, angiogenesis, nervous system development, oxidation-reduction, apoptosis, and synaptic transmission ranging from 2 to 5 genes per category. Interestingly, the authors came to a conclusion that the molecular basis of the shared features in neurodegenerative diseases may not be as common as initially thought, which reinforced their view that each disease's microenvironment is unique. Although the authors' assumption is experimentally substantiated, such a generalized conclusion should be taken with caution due to a lack of complementary proteomic studies.

In summary, it is now quite evident that all aspects of functional systems biology studies, including gene expression, mRNA, and proteins with all posttranslational modifications (PTMs), are essential for understanding molecular mechanisms underlying not only pathological processes but developmental as well.

43.9 Proteomics and Tissue Profiling

With the current state of proteomic technology, tissue can be examined for global protein expression directly without time-consuming sample preparation. However, time required for data acquisition by scanning tissue sections remains significant. Can such profiling be used effectively in rapid clinical diagnostics to aid pathologists in their work? Can sections of tissue be placed on target, covered with matrix, and profiled directly in mass spectrometers? Recent advancements indicate that regardless of limitations and challenges, this idea could lead to a breakthrough in modern clinical diagnostics in the near future (M'Koma et al., 2011; Jimenez & Verheul, 2014; Xu & Li, 2019; Piehowski et al., 2020).

Early attempts to develop patterns of specific compounds and proteins were based on tissue homogenates (Burns, 1982; Jimenez et al., 1997; Mitchell et al., 1997). Tissue profiling, or more accurately called tissue imaging, based on direct mass spectrometry measurements is a younger area of research that began its development in the late 1990s when MALDI-MS measurements were performed on intact tissue sections (Chaurand et al., 1999). Techniques already well-established in proteomic workflows, such as microwave-assisted and multi-enzyme digestion, have been adapted to tissue proteomics including formalin-fixed paraffin-embedded (FFPE) and fresh tissue samples (Turiak et al., 2014).

MALDI-TOF mass spectrometers with high resolution and sensitivity reaching the attomolar range are the choice class of instrumentation for direct proteomic tissue analyses. In addition to proteins and peptides, this method allows for the detection of lipids, carbohydrates or glycoconjugates, and small molecules. FFPE tissue can also be investigated, making this approach an ideal tool for disease diagnostics as well as biomarker discovery. Since spectra are generated very fast, a wide range of masses are acquired. Hundreds and even thousands of spots can be analyzed using one tissue section, which is further processed by computers to create an image. Besides the high throughput of this type of analysis, a major advantage is that collected spectra can be linked to a location on the tissue section. Thus, molecules in question can be localized to tissue structures supplementing diagnosis based on histology, with immunohistology adding a molecular pathology component to histopathological diagnostics (Kriegsmann et al., 2014). This, in turn, is important for monitoring processes such as distribution of therapeutics, which can be small molecules as well as peptides and/or proteins. The remaining question to be addressed in tissue imaging is how important is structural analysis (peptide sequence, drug structure) of identified species since downstream processing of spectra is based on the presence, intensity, and signal-to-noise ratio data. Scattered presence of other molecule(s) with close molecular m/z (mass-to-charge ratio) can lead to misleading results. If the analyte's m/z is within 2000, identification based on sequencing can be performed in instruments equipped with post-source decay mode (Stoeckli et al., 2002). Another question deals with absolute quantitation, which is the weakness of MALDI-TOF compared to other types of mass spectrometers. Nevertheless, meaningful information can be acquired if the system is properly validated. Despite the challenges discussed above, application of this advancing technology has resulted in reports on subjects such as classification of tumor samples using this technology platform (Chaurand et al., 1999; Moginger et al., 2020; Foll et al., 2022) and single cell analysis (Li et al., 1999; Taylor et al., 2021).

More thorough reviews of this field that go beyond the scope of this chapter can be found on PubMed. Although certainly not limited to these, the following references give examples of such review papers published to date (Espina et al., 2009; Lagarrigue et al., 2012; Rodrigo et al., 2014; Dilillo et al., 2018; Michno et al., 2019; Xu & Li, 2019; Scupakova et al., 2020; Ajith et al., 2021; Ma & Fernandez, 2022; Schnackenberg et al., 2022).

43.10 Databases and Search Engines

Expression of proteins and genes on a global scale requires new and efficient tools for database searches and data analysis. Advancements in proteomics include the development of multiple search engines to determine how likely (or unlikely) a peptide sequence is represented by a particular mass list containing both precursor and fragment masses (Mann & Wilm, 1994; Yates et al., 1995). Each search engine has inherent strengths and weaknesses, and these must be considered when choosing which search engine(s) to use in processing data. We highly recommend using more than one search engine, particularly when data are searched for PTMs. MASCOT, SEQUEST, Paragon used by ProteinPilot™, and the more recently produced Andromeda, as part of MaxQuant, are, to date, the most widely used for database searches. Other search engines used less often include Protein Prospector,!XTandem or Spectrum Mill.

After an initial effort by many to create a universal protein database, an international consortium created the Universal Protein Resource (UniProt) provides a comprehensive and freely accessible central resource of protein sequences and functional annotation (www.ebi.ac.uk/uniprot/). UniProt consists of two sections: UniProtKB/Swiss-Prot, which is manually annotated and reviewed, and UniProtKB/TrEMBL, which is automatically annotated and not reviewed. Despite this coordinated international effort, information included in protein databases is not complete; and it is impossible to even estimate when, if ever, we will be able to claim victory of completeness.

PTMs occur in different biological contexts; and therefore, complex protein mixtures extracted from biological samples invariably contain proteins carrying PTMs. The Association of Biomolecular Resource Facilities lists 355 PTMs (https://www.abrf.org/delta-mass). An additional level of complication is that PTMs can be heterogeneous and therefore consist of various modifications present on the same peptide core. A good example of this are histones that have a mosaic of PTMs dynamically changing in time as a result of their regulatory role (Burlingame et al., 2005). Variable modifications may or may not be present, e.g., phosphorylation of serine, threonine, and tyrosine occurring in a percentage of the population of protein molecules. While searching for variable modifications is a powerful tool for identifying the PTMs, one needs to be cautious when specifying the number of variable modifications since adding even a single variable modification will generate more possible peptides to be searched against. If there are multiple modifiable residues within a single peptide, the workload for searching all the possible modification permutations could be exponentially increased.

Cells, their compartments, and organelles (nucleus, mitochondria, endosomes, etc.), as well as body fluids or exosomes, are complex and dynamic systems comprising multiple networks of interacting molecules: proteins, glycoproteins, glycolipids, nucleic acids, metabolites, etc. Forming a better understanding of these systems and their mutual interactions and communications requires specific bioinformatics tools. This issue is further amplified by the amount of data/information accumulated in every experiment of high throughput analyses. A full understanding of interactions between different types of molecules such as DNA, RNA, and proteins in any biological system will require coordination of multiple experimental techniques at the same time, which will either require more samples or reduction of sample size for each individual test. Developing an application of micro fluids in new analytical instrumentation will definitely facilitate miniaturization efforts in proteomics.

43.11 Review Questions

1. Mathematical models of the molecular pathways in cells will facilitate:
   (a) Prediction of previously unknown interactions
   (b) Combining genomic and proteomic data
   (c) Discover novel proteins, cellular functions, and pathways
   (d) *All of the above*
2. Successful translation of results from laboratory to clinic depends on:
   (a) Exploratory research studies
   (b) *Validation*
   (c) Prospective screening studies on large-scale population
   (d) All of the above
3. Weakness of two-dimensional SDS-PAGE is:
   (a) *Lack of superior reproducibility*
   (b) Lack of protection of proteins from structural de-stabilization
   (c) Inability of obtaining protein sequencing data
   (d) Lack of superior methods of protein labeling
4. A major difficulty in proteomic studies of certain types of samples such as cerebrospinal fluid is:
   (a) Contamination with drugs penetrating through blood–brain barrier
   (b) *Size and availability of the sample*
   (c) Lack of correlation with clinical diagnosis
   (d) Contamination with red blood cells during spinal tap
5. Top–down approach in studies of systems biology is preferential over bottom–up approach because:
   (a) It provides more global view
   (b) It allows for faster conclusions which will lead to direct development of clinical tests
   (c) Requires less sophisticated technology and instrumentation
   (d) *Both approaches are equal*
6. A major question to be addressed when considering application of genomics and proteomics to screening and testing include:
   (a) Selection of the mouse species
   (b) Selection of the nonradioactive reagents
   (c) Selection of the reputable provider of proteomics and genomics technologies
   (d) *Selection of conditions which are true reflections of* in vivo *biological function*
7. Commercially available antibody array technology has several weaknesses. Major weakness is:
   (a) Problems associated with data analysis, interpretation, storage, and retrieval
   (b) Problems associated with high background created by not well-developed reagents
   (c) Problems associated with necessity of using radioactive materials
   (d) *Problems associated with quality and standardization of antibodies*
8. What is the best approach to determine the sensitivity and specificity of the discussed methodologies, and to validate and standardize these technologies?
   (a) To reduce "biological noise" in the samples and reduce number of replicates
   (b) To link data obtained from database searches to biological effects via fast computers
   (c) *To develop a standard approach to understanding and communicating variability in experimental data*
   (d) None of the above
9. While planning proteomics experiments, a researcher should:
   (a) Have available all types of mass spectrometers
   (b) Mandatory access to protein microarrays
   (c) *Be aware of strengths and weaknesses of each technology*
   (d) Always be an expert in statistics and computer programming for solving problems associated with bioinformatics
10. Neuroproteomics analysis is most conclusive when
   (a) Experiments are done using entire brain so that all cell types are used in one assay
   (b) Only in vitro models are conclusive
   (c) *Selected types of cells from the brain are obtained and manipulated*
   (d) Two or more cell types are mixed together to facilitate interaction mimicking in vivo situation

References

Abbott, A. (2003). Brain protein project enlists mice in 'dry run'. *Nature, 425*(6954), 110.

Ajith, A., Sthanikam, Y., & Banerjee, S. (2021). Chemical analysis of the human brain by imaging mass spectrometry. *Analyst, 146*(18), 5451–5473.

Barritault, D., Expert-Bezancon, A., Milet, M., & Hayes, D. H. (1976). Inexpensive and easily built small scale 2D electrophoresis equipment. *Analytical Biochemistry, 70*(2), 600–611.

Boes, T., & Neuhauser, M. (2005). Normalization for Affymetrix GeneChips. *Methods of Information in Medicine, 44*(3), 414–417.

Burlingame, A. L., Zhang, X., & Chalkley, R. J. (2005). Mass spectrometric analysis of histone posttranslational modifications. *Methods, 36*(4), 383–394.

Burns, M. S. (1982). Applications of secondary ion mass spectrometry (SIMS) in biological research: A review. *Journal of Microscopy, 127*(Pt 3), 237–258.

Carrette, O., Demalte, I., Scherl, A., Yalkinoglu, O., Corthals, G., Burkhard, P., Hochstrasser, D. F., & Sanchez, J. C. (2003). A panel of cerebrospinal fluid potential biomarkers for the diagnosis of Alzheimer's disease. *Proteomics, 3*(8), 1486–1494.

Celis, J. E., Honore, B., Bauw, G., & Vandekerckhove, J. (1990). Comprehensive computerized 2D gel protein databases offer a global approach to the study of the mammalian cell. *BioEssays, 12*(2), 93–97.

Chalmers, M. J., Mackay, C. L., Hendrickson, C. L., Wittke, S., Walden, M., Mischak, H., Fliser, D., Just, I., & Marshall, A. G. (2005). Combined top-down and bottom-up mass spectrometric approach to characterization of biomarkers for renal disease. *Analytical Chemistry, 77*(22), 7163–7171.

Chaurand, P., Stoeckli, M., & Caprioli, R. M. (1999). Direct profiling of proteins in biological tissue sections by MALDI mass spectrometry. *Analytical Chemistry, 71*(23), 5263–5270.

Choudhary, J., & Grant, S. G. (2004). Proteomics in postgenomic neuroscience: The end of the beginning. *Nature Neuroscience, 7*(5), 440–445.

Chromy, B. A., Gonzales, A. D., Perkins, J., Choi, M. W., Corzett, M. H., Chang, B. C., Corzett, C. H., & McCutchen-Maloney, S. L. (2004). Proteomic analysis of human serum by two-dimensional differential gel electrophoresis after depletion of high-abundant proteins. *Journal of Proteome Research, 3*(6), 1120–1127.

Ciborowski, P. (2013). *Proteomic profiling and analytical chemistry*. Elsevier.

Ciborowski, P., Enose, Y., Mack, A., Fladseth, M., & Gendelman, H. E. (2004). Diminished matrix metalloproteinase 9 secretion in human immunodeficiency virus-infected mononuclear phagocytes: Modulation of innate immunity and implications for neurological disease. *Journal of Neuroimmunology, 157*(1–2), 11–16.

Ciborowski, P., Kadiu, I., Rozek, W., Smith, L., Bernhardt, K., Fladseth, M., Ricardo-Dukelow, M., & Gendelman, H. E. (2007). Investigating the human immunodeficiency virus type 1-infected monocyte-derived macrophage secretome. *Virology, 363*(1), 198–209.

Coombes, K. R., Morris, J. S., Hu, J., Edmonson, S. R., & Baggerly, K. A. (2005). Serum proteomics profiling – a young technology begins to mature. *Nature Biotechnology, 23*(3), 291–292.

Corbett, J. M., Dunn, M. J., Posch, A., & Gorg, A. (1994). Positional reproducibility of protein spots in two-dimensional polyacrylamide gel electrophoresis using immobilised pH gradient isoelectric focusing in the first dimension: An interlaboratory comparison. *Electrophoresis, 15*(8–9), 1205–1211.

Craddock, R. M., Huang, J. T., Jackson, E., Harris, N., Torrey, E. F., Herberth, M., & Bahn, S. (2008). Increased alpha-defensins as a blood marker for schizophrenia susceptibility. *Molecular & Cellular Proteomics : MCP, 7*(7), 1204–1213.

Deutsch, E. W., Bandeira, N., Sharma, V., Perez-Riverol, Y., Carver, J. J., Kundu, D. J., Garcia-Seisdedos, D., Jarnuczak, A. F., Hewapathirana, S., Pullman, B. S., Wertz, J., Sun, Z., Kawano, S., Okuda, S., Watanabe, Y., Hermjakob, H., MacLean, B., MacCoss, M. J., Zhu, Y., et al. (2020). The ProteomeXchange consortium in 2020: Enabling 'big data' approaches in proteomics. *Nucleic Acids Research, 48*(D1), D1145–D1152.

Dilillo, M., Heijs, B., & McDonnell, L. A. (2018). Mass spectrometry imaging: How will it affect clinical research in the future? *Expert Review of Proteomics, 15*(9), 709–716.

Djebali, S., Davis, C. A., Merkel, A., Dobin, A., Lassmann, T., Mortazavi, A., Tanzer, A., Lagarde, J., Lin, W., Schlesinger, F., Xue, C., Marinov, G. K., Khatun, J., Williams, B. A., Zaleski, C., Rozowsky, J., Roder, M., Kokocinski, F., Abdelhamid, R. F., et al. (2012). Landscape of transcription in human cells. *Nature, 489*(7414), 101–108.

Duong, V. A., Park, J. M., & Lee, H. (2022). A review of suspension trapping digestion method in bottom-up proteomics. *Journal of Separation Science, 45*(16), 3150–3168.

Durrenberger, P. F., Fernando, F. S., Kashefi, S. N., Bonnert, T. P., Seilhean, D., Nait-Oumesmar, B., Schmitt, A., Gebicke-Haerter, P. J., Falkai, P., Grunblatt, E., Palkovits, M., Arzberger, T., Kretzschmar, H., Dexter, D. T., & Reynolds, R. (2014). Common mechanisms in neurodegeneration and neuroinflammation: A BrainNet Europe gene expression microarray study. *Journal of Neural Transmission*.

Eckel-Passow, J. E., Hoering, A., Therneau, T. M., & Ghobrial, I. (2005). Experimental design and analysis of antibody microarrays: Applying methods from cDNA arrays. *Cancer Research, 65*(8), 2985–2989.

Enose, Y., Destache, C. J., Mack, A. L., Anderson, J. R., Ullrich, F., Ciborowski, P. S., & Gendelman, H. E. (2005). Proteomic fingerprints distinguish microglia, bone marrow, and spleen macrophage populations. *Glia, 51*(3), 161–172.

Espina, V., Mueller, C., Edmiston, K., Sciro, M., Petricoin, E. F., & Liotta, L. A. (2009). Tissue is alive: New technologies are needed to address the problems of protein biomarker pre-analytical variability. *Proteomics. Clinical Applications, 3*(8), 874–882.

Evans, S. J., Choudary, P. V., Vawter, M. P., Li, J., Meador-Woodruff, J. H., Lopez, J. F., Burke, S. M., Thompson, R. C., Myers, R. M., Jones, E. G., Bunney, W. E., Watson, S. J., & Akil, H. (2003). DNA microarray analysis of functionally discrete human brain regions reveals divergent transcriptional profiles. *Neurobiology of Disease, 14*(2), 240–250.

Exarchou, V., Fiamegos, Y. C., van Beek, T. A., Nanos, C., & Vervoort, J. (2005). Hyphenated chromatographic techniques for the rapid screening and identification of antioxidants in methanolic extracts of pharmaceutically used plants. *Journal of Chromatography. A*.

Feuerstein, I., Rainer, M., Bernardo, K., Stecher, G., Huck, C.W., Kofler, K., Pelzer, A., Horninger, W., Klocker, H., Bartsch, G., Bonn, G. K. (2005). Derivatized cellulose combined with MALDI-TOF MS: a new tool for serum protein profiling. *Journal of Proteome Research, 4*(6), 2320–2326. https://doi.org/10.1021/pr050227z

Foll, M. C., Volkmann, V., Enderle-Ammour, K., Timme, S., Wilhelm, K., Guo, D., Vitek, O., Bronsert, P., & Schilling, O. (2022). Moving translational mass spectrometry imaging towards transparent and reproducible data analyses: A case study of an urothelial cancer cohort analyzed in the Galaxy framework. *Clinical Proteomics, 19*(1), 8.

Forbes, A. J., Mazur, M. T., Patel, H. M., Walsh, C. T., & Kelleher, N. L. (2001). Toward efficient analysis of >70 kDa proteins with 100% sequence coverage. *Proteomics, 1*(8), 927–933.

Fountoulakis, M., Juranville, J. F., Jiang, L., Avila, D., Roder, D., Jakob, P., Berndt, P., Evers, S., & Langen, H. (2004). Depletion of the high-abundance plasma proteins. *Amino Acids, 27*(3–4), 249–259.

Garcia-Montojo, M., Simula, E. R., Fathi, S., McMahan, C., Ghosal, A., Berry, J. D., Cudkowicz, M., Elkahloun, A., Johnson, K., Norato, G., Jensen, P., James, T., Sechi, L. A., & Nath, A. (2022). Antibody response to HML-2 may be protective in amyotrophic lateral sclerosis. *Annals of Neurology, 92*(5), 782–792.

Gianazza, E. (1995). Isoelectric focusing as a tool for the investigation of post-translational processing and chemical modifications of proteins. *Journal of Chromatography. A, 705*(1), 67–87.

Gianazza, E., & Arnaud, P. (1982). Chromatography of plasma proteins on immobilized Cibacron Blue F3-GA. Mechanism of the molecular interaction. *The Biochemical Journal, 203*(3), 637–641.

Guerreiro, R., Wojtas, A., Bras, J., Carrasquillo, M., Rogaeva, E., Majounie, E., Cruchaga, C., Sassi, C., Kauwe, J. S., Younkin, S., Hazrati, L., Collinge, J., Pocock, J., Lashley, T., Williams, J., Lambert, J. C., Amouyel, P., Goate, A., Rademakers, R., et al. (2013). TREM2 variants in Alzheimer's disease. *The New England Journal of Medicine, 368*(2), 117–127.

Guerreiro, R., Bras, J., Hardy, J., & Singleton, A. (2014). Next generation sequencing techniques in neurological diseases: Redefining clinical and molecular associations. *Human Molecular Genetics, 23*(R1), R47–R53.

Hammack, B. N., Fung, K. Y., Hunsucker, S. W., Duncan, M. W., Burgoon, M. P., Owens, G. P., & Gilden, D. H. (2004). Proteomic analysis of multiple sclerosis cerebrospinal fluid. *Multiple Sclerosis, 10*(3), 245–260.

Hammer, M. B., Eleuch-Fayache, G., Schottlaender, L. V., Nehdi, H., Gibbs, J. R., Arepalli, S. K., Chong, S. B., Hernandez, D. G., Sailer, A., Liu, G., Mistry, P. K., Cai, H., Shrader, G., Sassi, C., Bouhlal, Y., Houlden, H., Hentati, F., Amouri, R., & Singleton, A. B. (2013). Mutations in GBA2 cause autosomal-recessive cerebellar ataxia with spasticity. *American Journal of Human Genetics, 92*(2), 245–251.

Hampel, H., Buerger, K., Zinkowski, R., Teipel, S. J., Goernitz, A., Andreasen, N., Sjoegren, M., DeBernardis, J., Kerkman, D., Ishiguro, K., Ohno, H., Vanmechelen, E., Vanderstichele, H., McCulloch, C., Moller, H. J., Davies, P., & Blennow, K. (2004). Measurement of phosphorylated tau epitopes in the differential diagnosis of Alzheimer disease: A comparative cerebrospinal fluid study. *Archives of General Psychiatry, 61*(1), 95–102.

Harrington, M. G., Merril, C. R., Asher, D. M., & Gajdusek, D. C. (1986). Abnormal proteins in the cerebrospinal fluid of patients with Creutzfeldt-Jakob disease. *The New England Journal of Medicine, 315*(5), 279–283.

Hartmann, O. (2005). Quality control for microarray experiments. *Methods of Information in Medicine, 44*(3), 408–413.

He, J., Zhu, J., Liu, Y., Wu, J., Nie, S., Heth, J. A., Muraszko, K. M., Fan, X., & Lubman, D. M. (2013). Immunohistochemical staining, laser capture microdissection, and filter-aided sample preparation-assisted proteomic analysis of target cell populations within tissue samples. *Electrophoresis, 34*(11), 1627–1636.

Hecker, M., Fitzner, B., Wendt, M., Lorenz, P., Flechtner, K., Steinbeck, F., Schroder, I., Thiesen, H. J., & Zettl, U. K. (2016). High-density peptide microarray analysis of IgG autoantibody reactivities in serum and cerebrospinal fluid of multiple sclerosis patients. *Molecular & Cellular Proteomics : MCP, 15*(4), 1360–1380.

Ho, L., Sharma, N., Blackman, L., Festa, E., Reddy, G., & Pasinetti, G. M. (2005). From proteomics to biomarker discovery in Alzheimer's disease. *Brain Research. Brain Research Reviews, 48*(2), 360–369.

Hutchens TWaY, T.-T. (1993). New desorption strategies for the mass spectrometric analysis of macromolecules. *Rapid Communications in Mass Spectrometry, 7*(7), 576–580.

International Human Genome Sequencing Consortium. (2004). Finishing the euchromatic sequence of the human genome. *Nature, 431*(7011), 931–945.

Jeffries, N. (2005). Algorithms for alignment of mass spectrometry proteomic data. *Bioinformatics*.

Jiang, L., Lindpaintner, K., Li, H. F., Gu, N. F., Langen, H., He, L., & Fountoulakis, M. (2003). Proteomic analysis of the cerebrospinal fluid of patients with schizophrenia. *Amino Acids, 25*(1), 49–57.

Jimenez, C.R., & Verheul, H.M. (2014). Mass spectrometry-based proteomics: From cancer biology to protein biomarkers, drug targets, and clinical applications. American Society of Clinical Oncology educational book / ASCO. American Society of Clinical Oncology. Meeting: e504–e510.

Jimenez, C. R., Li, K. W., Dreisewerd, K., Mansvelder, H. D., Brussaard, A. B., Reinhold, B. B., Van der Schors, R. C., Karas, M., Hillenkamp, F., Burbach, J. P., Costello, C. E., & Geraerts, W. P. (1997). Pattern changes of pituitary peptides in rat after salt-loading as detected by means of direct, semiquantitative mass spectrometric profiling. *Proceedings of the National Academy of Sciences of the United States of America, 94*(17), 9481–9486.

Jonsson, T., Stefansson, H., Steinberg, S., Jonsdottir, I., Jonsson, P. V., Snaedal, J., Bjornsson, S., Huttenlocher, J., Levey, A. I., Lah, J. J., Rujescu, D., Hampel, H., Giegling, I., Andreassen, O. A., Engedal, K., Ulstein, I., Djurovic, S., Ibrahim-Verbaas, C., Hofman, A., et al. (2013). Variant of TREM2 associated with the risk of Alzheimer's disease. *The New England Journal of Medicine, 368*(2), 107–116.

Julka, S., & Regnier, F. E. (2005). Recent advancements in differential proteomics based on stable isotope coding. *Briefings in Functional Genomics & Proteomics, 4*(2), 158–177.

Kastenbauer, S., Angele, B., Sporer, B., Pfister, H. W., & Koedel, U. (2005). Patterns of protein expression in infectious meningitis: A cerebrospinal fluid protein array analysis. *Journal of Neuroimmunology, 164*(1-2), 134–139.

Kazakevich, Y. V., LoBrutto, R., & Vivilecchia, R. (2005). Reversed-phase high-performance liquid chromatography behavior of chaotropic counteranions. *Journal of Chromatography. A, 1064*(1), 9–18.

Kriegsmann, J., Kriegsmann, M., & Casadonte, R. (2014). MALDI TOF imaging mass spectrometry in clinical pathology: A valuable tool for cancer diagnostics (review). *International Journal of Oncology, 46*(3), 893–906.

Kroksveen, A. C., Aasebo, E., Vethe, H., Van Pesch, V., Franciotta, D., Teunissen, C. E., Ulvik, R. J., Vedeler, C., Myhr, K. M., Barsnes, H., & Berven, F. S. (2013). Discovery and initial verification of differentially abundant proteins between multiple sclerosis patients and controls using iTRAQ and SID-SRM. *Journal of Proteomics, 78*, 312–325.

Ku, C. J., Yona, G. (2005). The distance-profile representation and its application to detection of distantly related protein families. *BMC Bioinformatics, 29*(6), 282. https://doi.org/10.1186/1471-2105-6-282

Kusnezow, W., Syagailo, Y. V., Goychuk, I., Hoheisel, J. D., & Wild, D. G. (2006). Antibody microarrays: The crucial impact of mass transport on assay kinetics and sensitivity. *Expert Review of Molecular Diagnostics, 6*(1), 111–124.

Lagarrigue, M., Lavigne, R., Guevel, B., Com, E., Chaurand, P., & Pineau, C. (2012). Matrix-assisted laser desorption/ionization imaging mass spectrometry: A promising technique for reproductive research. *Biology of Reproduction, 86*(3), 74.

Laspiur, J. P., Anderson, E. R., Ciborowski, P., Wojna, V., Rozek, W., Duan, F., Mayo, R., Rodriguez, E., Plaud-Valentin, M., Rodriguez-Orengo, J., Gendelman, H. E., & Melendez, L. M. (2007). CSF proteomic fingerprints for HIV-associated cognitive impairment. *Journal of Neuroimmunology, 192*(1–2), 157–170.

Lee, M. L., Novotny, M., & Bartle, K. D. (1976). Gas chromatography/mass spectrometric and nuclear magnetic resonance determination of polynuclear aromatic hydrocarbons in airborne particulates. *Analytical Chemistry, 48*(11), 1566–1572.

Lescuyer, P., Allard, L., Zimmermann-Ivol, C. G., Burgess, J. A., Hughes-Frutiger, S., Burkhard, P. R., Sanchez, J. C., & Hochstrasser, D. F. (2004). Identification of post-mortem cerebrospinal fluid proteins as potential biomarkers of ischemia and neurodegeneration. *Proteomics, 4*(8), 2234–2241.

Li, L., Garden, R. W., Romanova, E. V., & Sweedler, J. V. (1999). In situ sequencing of peptides from biological tissues and single cells using MALDI-PSD/CID analysis. *Analytical Chemistry, 71*(24), 5451–5458.

Li, J., Cai, Z., Bomgarden, R. D., Pike, I., Kuhn, K., Rogers, J. C., Roberts, T. M., Gygi, S. P., & Paulo, J. A. (2021). TMTpro-18plex: The expanded and complete set of TMTpro reagents for sample multiplexing. *Journal of Proteome Research, 20*(5), 2964–2972.

Luk, K. C., Kehm, V., Carroll, J., Zhang, B., O'Brien, P., Trojanowski, J. Q., & Lee, V. M. (2012). Pathological alpha-synuclein transmission initiates Parkinson-like neurodegeneration in nontransgenic mice. *Science, 338*(6109), 949–953.

Ly, L., Barnett, M. H., Zheng, Y. Z., Gulati, T., Prineas, J. W., & Crossett, B. (2011). Comprehensive tissue processing strategy for quantitative proteomics of formalin-fixed multiple sclerosis lesions. *Journal of Proteome Research, 10*(10), 4855–4868.

M'Koma, A. E., Seeley, E. H., Washington, M. K., Schwartz, D. A., Muldoon, R. L., Herline, A. J., Wise, P. E., & Caprioli, R. M.

(2011). Proteomic profiling of mucosal and submucosal colonic tissues yields protein signatures that differentiate the inflammatory colitides. *Inflammatory Bowel Diseases, 17*(4), 875–883.

Ma, X., & Fernandez, F. M. (2022). Advances in mass spectrometry imaging for spatial cancer metabolomics. *Mass Spectrometry Reviews, e21804*.

Maccarrone, G., Milfay, D., Birg, I., Rosenhagen, M., Holsboer, F., Grimm, R., Bailey, J., Zolotarjova, N., & Turck, C. W. (2004). Mining the human cerebrospinal fluid proteome by immunodepletion and shotgun mass spectrometry. *Electrophoresis, 25*(14), 2402–2412.

Mann, M., & Wilm, M. (1994). Error-tolerant identification of peptides in sequence databases by peptide sequence tags. *Analytical Chemistry, 66*(24), 4390–4399.

Margulies, M., Egholm, M., Altman, W. E., Attiya, S., Bader, J. S., Bemben, L. A., Berka, J., Braverman, M. S., Chen, Y. J., Chen, Z., Dewell, S. B., Du, L., Fierro, J. M., Gomes, X. V., Godwin, B. C., He, W., Helgesen, S., Ho, C. H., Irzyk, G. P., et al. (2005). Genome sequencing in microfabricated high-density picolitre reactors. *Nature, 437*(7057), 376–380.

Marshall, A. G., Hendrickson, C. L., & Jackson, G. S. (1998). Fourier transform ion cyclotron resonance mass spectrometry: A primer. *Mass Spectrometry Reviews, 17*(1), 1–35.

Michno, W., Wehrli, P. M., Blennow, K., Zetterberg, H., & Hanrieder, J. (2019). Molecular imaging mass spectrometry for probing protein dynamics in neurodegenerative disease pathology. *Journal of Neurochemistry, 151*(4), 488–506.

Millea, K. M., Krull, I. S., Cohen, S. A., Gebler, J. C., & Berger, S. J. (2006). Integration of multidimensional chromatographic protein separations with a combined "top-down" and "bottom-up" proteomic strategy. *Journal of Proteome Research, 5*(1), 135–146.

Mitchell, A. E., Morin, D., Lakritz, J., & Jones, A. D. (1997). Quantitative profiling of tissue- and gender-related expression of glutathione S-transferase isoenzymes in the mouse. *The Biochemical Journal, 325*(Pt 1), 207–216.

Moginger, U., Marcussen, N., & Jensen, O. N. (2020). Histo-molecular differentiation of renal cancer subtypes by mass spectrometry imaging and rapid proteome profiling of formalin-fixed paraffin-embedded tumor tissue sections. *Oncotarget, 11*(44), 3998–4015.

Murray, J. F., Jr., Gordon, G. R., Gulledge, C. C., & Peters, J. H. (1975). Chromatographic-fluorometric analysis of antileprotic sulfones. *Journal of Chromatography, 107*(1), 67–72.

Oda, Y., Huang, K., Cross, F. R., Cowburn, D., & Chait, B. T. (1999). Accurate quantitation of protein expression and site-specific phosphorylation. *Proceedings of the National Academy of Sciences of the United States of America, 96*(12), 6591–6596.

O'Farrell, P. H. (1975). High resolution two-dimensional electrophoresis of proteins. *The Journal of Biological Chemistry, 250*(10), 4007–4021.

Ogata, Y., Charlesworth, M. C., & Muddiman, D. C. (2005). Evaluation of protein depletion methods for the analysis of total-, phospho- and glycoproteins in lumbar cerebrospinal fluid. *Journal of Proteome Research, 4*(3), 837–845.

Olsson, A., Hoglund, K., Sjogren, M., Andreasen, N., Minthon, L., Lannfelt, L., Buerger, K., Moller, H. J., Hampel, H., Davidsson, P., & Blennow, K. (2003). Measurement of alpha- and beta-secretase cleaved amyloid precursor protein in cerebrospinal fluid from Alzheimer patients. *Experimental Neurology, 183*(1), 74–80.

Omenn, G. S., States, D. J., Adamski, M., Blackwell, T. W., Menon, R., Hermjakob, H., Apweiler, R., Haab, B. B., Simpson, R. J., Eddes, J. S., Kapp, E. A., Moritz, R. L., Chan, D. W., Rai, A. J., Admon, A., Aebersold, R., Eng, J., Hancock, W. S., Hefta, S. A., et al. (2005). Overview of the HUPO Plasma Proteome Project: Results from the pilot phase with 35 collaborating laboratories and multiple analytical groups, generating a core dataset of 3020 proteins and a publicly-available database. *Proteomics, 5*(13), 3226–3245.

Otake, K., Kamiguchi, H., & Hirozane, Y. (2019). Identification of biomarkers for amyotrophic lateral sclerosis by comprehensive analysis of exosomal mRNAs in human cerebrospinal fluid. *BMC Medical Genomics, 12*(1), 7.

Pang, R. T., Poon, T. C., Chan, K. C., Lee, N. L., Chiu, R. W., Tong, Y. K., Wong, R. M., Chim, S. S., Ngai, S. M., Sung, J. J., & Lo, Y. M. (2006). Serum proteomic fingerprints of adult patients with severe acute respiratory syndrome. *Clinical Chemistry*.

Patel, S. (2014). Role of proteomics in biomarker discovery: Prognosis and diagnosis of neuropsychiatric disorders. *Advances in Protein Chemistry and Structural Biology, 94*, 39–75.

Patel, K. N., Patel, J. K., Patel, M. P., Rajput, G. C., & Patel, H. A. (2010). Introduction to hyphenated techniques and their applications in pharmacy. *Pharmaceutical Methods, 1*(1), 2–13.

Perez-Riverol, Y., Bai, J., Bandla, C., Garcia-Seisdedos, D., Hewapathirana, S., Kamatchinathan, S., Kundu, D. J., Prakash, A., Frericks-Zipper, A., Eisenacher, M., Walzer, M., Wang, S., Brazma, A., & Vizcaino, J. A. (2022). The PRIDE database resources in 2022: A hub for mass spectrometry-based proteomics evidences. *Nucleic Acids Research, 50*(D1), D543–D552.

Petricoin, E. F., & Liotta, L. A. (2003). Clinical applications of proteomics. *Journal of Nutrition, 133*(7 suppl), 2476S–2484S.

Piehowski, P. D., Zhu, Y., Bramer, L. M., Stratton, K. G., Zhao, R., Orton, D. J., Moore, R. J., Yuan, J., Mitchell, H. D., Gao, Y., Webb-Robertson, B. M., Dey, S. K., Kelly, R. T., & Burnum-Johnson, K. E. (2020). Automated mass spectrometry imaging of over 2000 proteins from tissue sections at 100-mum spatial resolution. *Nature Communications, 11*(1), 8.

Pottier, C., Hannequin, D., Coutant, S., Rovelet-Lecrux, A., Wallon, D., Rousseau, S., Legallic, S., Paquet, C., Bombois, S., Pariente, J., Thomas-Anterion, C., Michon, A., Croisile, B., Etcharry-Bouyx, F., Berr, C., Dartigues, J. F., Amouyel, P., Dauchel, H., Boutoleau-Bretonniere, C., et al. (2012). High frequency of potentially pathogenic SORL1 mutations in autosomal dominant early-onset Alzheimer disease. *Molecular Psychiatry, 17*(9), 875–879.

Pottiez, G., Jagadish, T., Yu, F., Letendre, S., Ellis, R., Duarte, N. A., Grant, I., Gendelman, H. E., Fox, H. S., & Ciborowski, P. (2012). Plasma proteomic profiling in HIV-1 infected methamphetamine abusers. *PLoS One, 7*(2), e31031.

Puchades, M., Hansson, S. F., Nilsson, C. L., Andreasen, N., Blennow, K., & Davidsson, P. (2003). Proteomic studies of potential cerebrospinal fluid protein markers for Alzheimer's disease. *Brain Research. Molecular Brain Research, 118*(1–2), 140–146.

Rademakers, R., Baker, M., Nicholson, A. M., Rutherford, N. J., Finch, N., Soto-Ortolaza, A., Lash, J., Wider, C., Wojtas, A., DeJesus-Hernandez, M., Adamson, J., Kouri, N., Sundal, C., Shuster, E. A., Aasly, J., MacKenzie, J., Roeber, S., Kretzschmar, H. A., Boeve, B. F., et al. (2012a). Mutations in the colony stimulating factor 1 receptor (CSF1R) gene cause hereditary diffuse leukoencephalopathy with spheroids. *Nature Genetics, 44*(2), 200–205.

Rademakers, R., Neumann, M., & Mackenzie, I. R. (2012b). Advances in understanding the molecular basis of frontotemporal dementia. *Nature Reviews. Neurology, 8*(8), 423–434.

Ramstrom, M., Hagman, C., Mitchell, J. K., Derrick, P. J., Hakansson, P., & Bergquist, J. (2005). Depletion of high-abundant proteins in body fluids prior to liquid chromatography fourier transform ion cyclotron resonance mass spectrometry. *Journal of Proteome Research, 4*(2), 410–416.

Rauniyar, N., & Yates, J. R., 3rd. (2014). Isobaric labeling-based relative quantification in shotgun proteomics. *Journal of Proteome Research, 13*(12), 5293–5309.

Reimers, M. (2005). Statistical analysis of microarray data. *Addiction Biology, 10*(1), 23–35.

Rodrigo, M. A., Zitka, O., Krizkova, S., Moulick, A., Adam, V., & Kizek, R. (2014). MALDI-TOF MS as evolving cancer diagnostic tool: A review. *Journal of Pharmaceutical and Biomedical Analysis, 95*, 245–255.

Rohlff, C., & Southan, C. (2002). Proteomic approaches to central nervous system disorders. *Current Opinion in Molecular Therapeutics, 4*(3), 251–258.

Ross, P. L., Huang, Y. N., Marchese, J. N., Williamson, B., Parker, K., Hattan, S., Khainovski, N., Pillai, S., Dey, S., Daniels, S., Purkayastha, S., Juhasz, P., Martin, S., Bartlet-Jones, M., He, F., Jacobson, A., & Pappin, D. J. (2004). Multiplexed protein quantitation in Saccharomyces cerevisiae using amine-reactive isobaric tagging reagents. *Molecular & Cellular Proteomics : MCP, 3*(12), 1154–1169.

Rozek, W., Ricardo-Dukelow, M., Holloway, S., Gendelman, H. E., Wojna, V., Melendez, L. M., & Ciborowski, P. (2007). Cerebrospinal fluid proteomic profiling of HIV-1-infected patients with cognitive impairment. *Journal of Proteome Research, 6*(11), 4189–4199.

Ruetschi, U., Zetterberg, H., Podust, V. N., Gottfries, J., Li, S., Hviid Simonsen, A., McGuire, J., Karlsson, M., Rymo, L., Davies, H., Minthon, L., & Blennow, K. (2005). Identification of CSF biomarkers for frontotemporal dementia using SELDI-TOF. *Experimental Neurology, 196*(2), 273–281.

Sanchez, J. C., Guillaume, E., Lescuyer, P., Allard, L., Carrette, O., Scherl, A., Burgess, J., Corthals, G. L., Burkhard, P. R., & Hochstrasser, D. F. (2004). Cystatin C as a potential cerebrospinal fluid marker for the diagnosis of Creutzfeldt-Jakob disease. *Proteomics, 4*(8), 2229–2233.

Sandhu, F. A., Salim, M., & Zain, S. B. (1991). Expression of the human beta-amyloid protein of Alzheimer's disease specifically in the brains of transgenic mice. *The Journal of Biological Chemistry, 266*(32), 21331–21334.

Sanna, P. P., King, A. R., van der Stap, L. D., & Repunte-Canonigo, V. (2005). Gene profiling of laser-microdissected brain regions and sub-regions. *Brain Research. Brain Research Protocols, 15*(2), 66–74.

Schnackenberg, L. K., Thorn, D. A., Barnette, D., & Jones, E. E. (2022). MALDI imaging mass spectrometry: An emerging tool in neurology. *Metabolic Brain Disease, 37*(1), 105–121.

Scupakova, K., Balluff, B., Tressler, C., Adelaja, T., Heeren, R. M. A., Glunde, K., & Ertaylan, G. (2020). Cellular resolution in clinical MALDI mass spectrometry imaging: The latest advancements and current challenges. *Clinical Chemistry and Laboratory Medicine, 58*(6), 914–929.

Serot, J. M., Bene, M. C., & Faure, G. C. (2003). Choroid plexus, aging of the brain, and Alzheimer's disease. *Frontiers in Bioscience, 8*, s515–s521.

Shendure, J., Porreca, G. J., Reppas, N. B., Lin, X., McCutcheon, J. P., Rosenbaum, A. M., Wang, M. D., Zhang, K., Mitra, R. D., & Church, G. M. (2005). Accurate multiplex polony sequencing of an evolved bacterial genome. *Science, 309*(5741), 1728–1732.

Silverberg, G. D., Mayo, M., Saul, T., Carvalho, J., & McGuire, D. (2004). Novel ventriculo-peritoneal shunt in Alzheimer's disease cerebrospinal fluid biomarkers. *Expert Review of Neurotherapeutics, 4*(1), 97–107.

Soreghan, B. A., Lu, B. W., Thomas, S. N., Duff, K., Rakhmatulin, E. A., Nikolskaya, T., Chen, T., & Yang, A. J. (2005). Using proteomics and network analysis to elucidate the consequences of synaptic protein oxidation in a PS1+AbetaPP mouse model of Alzheimer's disease. *Journal of Alzheimer's Disease, 8*(3), 227–241.

Soumelis, V., Pattarini, L., Michea, P., & Cappuccio, A. (2015). Systems approaches to unravel innate immune cell diversity, environmental plasticity and functional specialization. *Current Opinion in Immunology, 32C*, 42–47.

Steinacker, P., Rist, W., Swiatek-de-Lange, M., Lehnert, S., Jesse, S., Pabst, A., Tumani, H., von Arnim, C. A., Mitrova, E., Kretzschmar, H. A., Lenter, M., Wiltfang, J., & Otto, M. (2010). Ubiquitin as potential cerebrospinal fluid marker of Creutzfeldt-Jakob disease. *Proteomics, 10*(1), 81–89.

Stoeckli, M., Staab, D., Staufenbiel, M., Wiederhold, K. H., & Signor, L. (2002). Molecular imaging of amyloid beta peptides in mouse brain sections using mass spectrometry. *Analytical Biochemistry, 311*(1), 33–39.

Sturm, M., Bertsch, A., Gropl, C., Hildebrandt, A., Hussong, R., Lange, E., Pfeifer, N., Schulz-Trieglaff, O., Zerck, A., Reinert, K., & Kohlbacher, O. (2008). OpenMS – An open-source software framework for mass spectrometry. *BMC Bioinformatics, 9*, 163.

Sultana, R., Poon, H. F., Cai, J., Pierce, W. M., Merchant, M., Klein, J. B., Markesbery, W. R., & Butterfield, D. A. (2005). Identification of nitrated proteins in Alzheimer's disease brain using a redox proteomics approach. *Neurobiology of Disease*.

Taylor, M. J., Lukowski, J. K., & Anderton, C. R. (2021). Spatially resolved mass spectrometry at the single cell: Recent innovations in proteomics and metabolomics. *Journal of the American Society for Mass Spectrometry, 32*(4), 872–894.

Tian, C., Liu, D., Xiang, W., Kretzschmar, H. A., Sun, Q. L., Gao, C., Xu, Y., Wang, H., Fan, X. Y., Meng, G., Li, W., & Dong, X. P. (2014). Analyses of the similarity and difference of global gene expression profiles in cortex regions of three neurodegenerative diseases: Sporadic Creutzfeldt-Jakob disease (sCJD), fatal familial insomnia (FFI), and Alzheimer's disease (AD). *Molecular Neurobiology, 50*(2), 473–481.

Tsunoda, A., Mitsuoka, H., Bandai, H., Endo, T., Arai, H., & Sato, K. (2002). Intracranial cerebrospinal fluid measurement studies in suspected idiopathic normal pressure hydrocephalus, secondary normal pressure hydrocephalus, and brain atrophy. *Journal of Neurology, Neurosurgery, and Psychiatry, 73*(5), 552–555.

Turiak, L., Shao, C., Meng, L., Khatri, K., Leymarie, N., Wang, Q., Pantazopoulos, H., Leon, D. R., & Zaia, J. (2014). Workflow for combined proteomics and glycomics profiling from histological tissues. *Analytical Chemistry, 86*(19), 9670–9678.

Vilarino-Guell, C., Wider, C., Ross, O. A., Dachsel, J. C., Kachergus, J. M., Lincoln, S. J., Soto-Ortolaza, A. I., Cobb, S. A., Wilhoite, G. J., Bacon, J. A., Behrouz, B., Melrose, H. L., Hentati, E., Puschmann, A., Evans, D. M., Conibear, E., Wasserman, W. W., Aasly, J. O., Burkhard, P. R., et al. (2011). VPS35 mutations in Parkinson disease. *American Journal of Human Genetics, 89*(1), 162–167.

Vinaiphat, A., & Sze, S. K. (2022). Proteomics for comprehensive characterization of extracellular vesicles in neurodegenerative disease. *Experimental Neurology, 355*, 114149.

Wang, H., Qian, W. J., Mottaz, H. M., Clauss, T. R., Anderson, D. J., Moore, R. J., Camp, D. G., 2nd, Khan, A. H., Sforza, D. M., Pallavicini, M., Smith, D. J., & Smith, R. D. (2005). Development and evaluation of a micro- and nanoscale proteomic sample preparation method. *Journal of Proteome Research, 4*(6), 2397–2403.

Wiederin, J. L., Yu, F., Donahoe, R. M., Fox, H. S., Ciborowski, P., & Gendelman, H. E. (2012). Changes in the plasma proteome follows chronic opiate administration in simian immunodeficiency virus infected rhesus macaques. *Drug and Alcohol Dependence, 120*(1–3), 105–112.

Wildsmith, K. R., Schauer, S. P., Smith, A. M., Arnott, D., Zhu, Y., Haznedar, J., Kaur, S., Mathews, W. R., & Honigberg, L. A. (2014). Identification of longitudinally dynamic biomarkers in Alzheimer's disease cerebrospinal fluid by targeted proteomics. *Molecular Neurodegeneration, 9*, 22.

Wittke, S., Mischak, H., Walden, M., Kolch, W., Radler, T., & Wiedemann, K. (2005). Discovery of biomarkers in human urine and cerebrospinal fluid by capillary electrophoresis coupled to mass spectrometry: Towards new diagnostic and therapeutic approaches. *Electrophoresis, 26*(7–8), 1476–1487.

Wolfender, J. L., Hostettmann, K., Abe, F., Nagao, T., Okabe, H., & Yamauchi, T. (1995). Liquid chromatography combined with thermospray and continuous-flow fast atom bombardment mass spectrometry of glycosides in crude plant extracts. *Journal of Chromatography. A, 712*(1), 155–168.

Woods, A. G., Sokolowska, I., Ngounou Wetie, A. G., Wormwood, K., Aslebagh, R., Patel, S., & Darie, C. C. (2014). Mass spectrometry for proteomics-based investigation. *Advances in Experimental Medicine and Biology, 806*, 1–32.

Wu, C. H., Fallini, C., Ticozzi, N., Keagle, P. J., Sapp, P. C., Piotrowska, K., Lowe, P., Koppers, M., McKenna-Yasek, D., Baron, D. M., Kost, J. E., Gonzalez-Perez, P., Fox, A. D., Adams, J., Taroni, F., Tiloca, C., Leclerc, A. L., Chafe, S. C., Mangroo, D., et al. (2012). Mutations in the profilin 1 gene cause familial amyotrophic lateral sclerosis. *Nature, 488*(7412), 499–503.

Wu, J. W., Herman, M., Liu, L., Simoes, S., Acker, C. M., Figueroa, H., Steinberg, J. I., Margittai, M., Kayed, R., Zurzolo, C., Di Paolo, G., & Duff, K. E. (2013). Small misfolded Tau species are internalized via bulk endocytosis and anterogradely and retrogradely transported in neurons. *The Journal of Biological Chemistry, 288*(3), 1856–1870.

Xu, G., & Li, J. (2019). Recent advances in mass spectrometry imaging for multiomics application in neurology. *The Journal of Comparative Neurology, 527*(13), 2158–2169.

Yadav, S. P. (2007). The wholeness in suffix -omics, -omes, and the word om. *Journal of Biomolecular Techniques : JBT, 18*(5), 277.

Yagi, Y., Ohkubo, T., Kawaji, H., Machida, A., Miyata, H., Goda, S., Roy, S., Hayashizaki, Y., Suzuki, H., & Yokota, T. (2017). Next-generation sequencing-based small RNA profiling of cerebrospinal fluid exosomes. *Neuroscience Letters, 636*, 48–57.

Yates, J. R., 3rd, Eng, J. K., McCormack, A. L., & Schieltz, D. (1995). Method to correlate tandem mass spectra of modified peptides to amino acid sequences in the protein database. *Analytical Chemistry, 67*(8), 1426–1436.

Yates, C. M., Butterworth, J., Tennant, M. C., & Gordon, A. (1990). Enzyme activities in relation to pH and lactate in postmortem brain in Alzheimer-type and other dementias. *Journal of Neurochemistry, 55*(5), 1624–1630.

Yin, L., Liu, J., Dong, H., Xu, E., Qiao, Y., Wang, L., Zhang, L., Jia, J., Li, L., & Geng, X. (2014). Autophagy-related gene16L2, a potential serum biomarker of multiple sclerosis evaluated by bead-based proteomic technology. *Neuroscience Letters, 562*, 34–38.

Yu, Y., Hou, K., Ji, T., Wang, X., Liu, Y., Zheng, Y., Xu, J., Hou, Y., & Chi, G. (2021). The role of exosomal microRNAs in central nervous system diseases. *Molecular and Cellular Biochemistry, 476*(5), 2111–2124.

Yuan, X., & Desiderio, D. M. (2003). Proteomics analysis of phosphotyrosyl-proteins in human lumbar cerebrospinal fluid. *Journal of Proteome Research, 2*(5), 476–487.

Yuan, X., & Desiderio, D. M. (2005). Proteomics analysis of human cerebrospinal fluid. *Journal of Chromatography. B, Analytical Technologies in the Biomedical and Life Sciences, 815*(1–2), 179–189.

Zandian, A., Forsstrom, B., Haggmark-Manberg, A., Schwenk, J. M., Uhlen, M., Nilsson, P., & Ayoglu, B. (2017). Whole-proteome peptide microarrays for profiling autoantibody repertoires within multiple sclerosis and narcolepsy. *Journal of Proteome Research, 16*(3), 1300–1314.

Zhou, N., Wang, J., Yu, Y., Shi, J., Li, X., Xu, B., & Yu, Q. (2014). Mass spectrum analysis of serum biomarker proteins from patients with schizophrenia. *Biomedical Chromatography : BMC, 28*(5), 654–659.

Zimprich, A., Benet-Pages, A., Struhal, W., Graf, E., Eck, S. H., Offman, M. N., Haubenberger, D., Spielberger, S., Schulte, E. C., Lichtner, P., Rossle, S. C., Klopp, N., Wolf, E., Seppi, K., Pirker, W., Presslauer, S., Mollenhauer, B., Katzenschlager, R., Foki, T., et al. (2011). A mutation in VPS35, encoding a subunit of the retromer complex, causes late-onset Parkinson disease. *American Journal of Human Genetics, 89*(1), 168–175.

Biomarkers for Neurodegenerative Diseases

44

Mai M. Abdelmoaty, Eugene Lu, and Emma G. Foster

Abstract

Biomarkers for neurodegenerative diseases are needed to improve the differential diagnosis in the clinics as well as the drug discovery and development of disease-modifying therapies (DMTs) for these devastating disorders. Neurodegenerative disorders share several signs and symptoms and pathological features, which represents a challenge for the accurate diagnosis. Thus far, there is no specific method, test, or biomarker available for diagnosis, prognosis, or therapeutic evaluation. Diagnosis of neurodegenerative disorders primarily requires expensive structural and functional brain imaging techniques (such as Positron emission tomography [PET], single-photon emission computed tomography [SPECT], or magnetic resonance imaging [MRI]) or invasive tests (such as lumbar puncture) to assess the cerebrospinal fluid (CSF). Currently, FDA-approved therapies for neurodegenerative diseases mainly relieve the symptoms, they don't halt or reverse the motor and cognitive dysfunctions seen commonly in different neurodegenerative disorders. Therefore, there is an urgent need to uncover biomarkers that increase the accuracy of diagnosis, are easily accessible (such as blood-based biomarkers), and cost effective as well as track the therapeutic responses to promising DMTs in clinical trials. Revealing biomarkers with these criteria will enhance the diagnosis and treatment of neurodegenerative disorders.

Keywords

Biomarkers · Neurodegenerative diseases · Alzheimer's disease · Parkinson's disease · CSF biomarkers · Blood-based biomarkers

M. M. Abdelmoaty (✉) · E. Lu · E. G. Foster
Department of Pharmacology and Experimental Neuroscience, College of Medicine, University of Nebraska Medical Center, Omaha, NE, USA

44.1 Introduction

Neurodegenerative diseases are characterized by dysfunction and progressive loss of neurons as a result of the accumulation of pathologically altered proteins that deposit mainly in the brain and spinal cord (Kovacs, 2019). Accumulation of amyloid-beta (Aβ) and tau results in the formation of intracellular neurofibrillary tangles (NFTs) and amyloid plaques, both key pathologic features of Alzheimer's disease (AD) (Konickova et al., 2022). While synucleinopathies are pathologically and clinically heterogeneous disorders that share common pathologic features, which include the accumulation of misfolded α-synuclein (α-syn) in neurons and glia, and the formation of Lewy bodies (LBs), Lewy neurites (LNs), neuronal cytoplasmic inclusions, and glial cytoplasmic inclusions (Koga et al., 2021). Synucleinopathies include Parkinson's disease (PD), PD dementia (PDD), dementia with Lewy bodies (DLB), multiple system atrophy (MSA), and other rare disorders such as different neuroaxonal dystrophies (Goedert et al., 2017). PD, the most common synucleinopathy, is the second most common neurodegenerative disease after AD. Population prevalence of PD increases from about 1% at age 60 to 4% by age 80 ("https://www.cdc.gov/genomics/hugenet/casestudy/parkinson/parkcoffee_view.htm," Accessed September 5, 2023). This chapter will focus on the most common neurodegenerative diseases (AD and PD).

AD is the most common neurodegenerative disorder and the most common cause of dementia. It accounts for 60–70% of all dementia cases (Seshadri & Wolf, 2007). According to the 2020 Alzheimer's report, Alzheimer's patients in America will rise from 5.8 million to 13 million by 2050 ("2020 Alzheimer's disease facts and figures," 2020). This is caused by increasing the aging population and a more nuanced understanding of AD (Li et al., 2021). Regarding PD, around 90,000 people in the United States are diagnosed with this disease every year along with ten million people living with it worldwide (Hess & Okun, 2016; https://www.parkinson.

H. E. Gendelman, T. Ikezu (eds.), *Neuroimmune Pharmacology and Therapeutics*,
https://doi.org/10.1007/978-3-031-68237-7_44

org/understanding-parkinsons/statistics, Accessed September 5, 2023). The prevalence of PD has doubled in the past 25 years. According to several epidemiological studies, 20–40% of individuals living with PD also have dementia. Incidence rate of PDD increases 4–6 times as compared to non-PD controls (Aarsland & Kurz, 2010a; Aarsland et al., 1996; Astrom et al., 2022; Hanagasi et al., 2017). It has been reported that at least 75% of patients who live with PD for more than 10 years will develop dementia as part of disease progression (Aarsland et al., 2003; Aarsland & Kurz, 2010b). Globally, disability and death due to PD increase faster than for other neurological disorders ("https://www.who.int/news/item/14-06-2022-launch-of-who-s-parkinson-disease-technical-brief#:~:text=Disability%20and%20death%20due%20to,of%20over%20100%25%20since%202000," Accessed September 5, 2023).

Accurate diagnosis of neurodegenerative diseases as well as timely and cost-effective follow-up of disease progression and response to therapy are great needs. These needs extend from providing patients with diagnostic and prognostic information about their disease to optimizing treatment regimens and strategies and giving patients opportunities to join clinical studies. Misdiagnosis has disadvantages including, but not limited to, suboptimal treatment, unnecessary care-seeking, and costly investigations to confirm diagnosis (Hansson, 2021; Hunter et al., 2015). In the absence of pathology-specific biomarkers, low accuracy of clinical diagnosis in different neurodegenerative disorders increases the risk of including patients without the pathology of interest in clinical trials. At specialized dementia clinics, a comparison of clinical diagnosis to autopsy diagnosis indicated that about 25–30% of patients with a clinical diagnosis of AD were misdiagnosed. Moreover, clinical trials in mild AD dementia showed that about 25% of patients fulfilling clinical criteria for AD do not exhibit Aβ pathology (Salloway et al., 2014; Sevigny et al., 2016). Therefore, many AD trials today require diagnostic biomarkers for Aβ pathology for inclusion. In PD, the accuracy of clinical diagnosis is around 80%, but correct diagnosis of other parkinsonian disorders including MSA, progressive supranuclear palsy (PSP), and corticobasal degeneration (CBD) is more challenging (Jellinger et al., 2016; Respondek et al., 2020).

Accurate, easily accessible (such as blood-based biomarkers), and cost-effective diagnostic biomarkers have the potential for widespread clinical use to follow-up the disease progression and therapeutic response to disease-modifying therapies (DMTs). There are optimal time windows, during the disease progression, when DMTs are most effective. When targeting key pathological hallmarks, such as tau and Aβ in AD and α-syn in PD, therapies will likely be more effective during early pre-symptomatic or prodromal disease stages preceding irreversible occurrence of neurodegeneration (Masters et al., 2015; Smedinga et al., 2021). Therefore, biomarkers identifying disease pathology before the onset of disabling symptoms are needed for better selection of subjects for inclusion in clinical trials as well as follow-up of the therapeutic responses in these studies. In this chapter, we summarize the pathogenesis and clinical evaluation of the two most common neurodegenerative diseases (AD and PD). We also discuss the diagnostic/prognostic biomarkers available for both disorders. In addition, we present promising therapeutic biomarkers that can be used to track disease progression and response to therapy in DMT clinical studies.

44.2 Pathogenesis of Neurodegenerative Diseases

44.2.1 Prion Pathology

Prion theory has emerged as a molecular principle for the pathogenesis of many neurodegenerative diseases. This theory presents a fundamental cause of many neurodegenerative disorders that is the misfolding and aggregation of certain proteins. The theory came from the concept that devastating brain diseases called spongiform encephalopathies are transmissible to new hosts by agents consisting of a misfolded protein, now known as the prion protein, and consequently, prion was defined as a proteinaceous infectious particle (Walker & Jucker, 2015). This provides the hallmark feature of many neurodegenerative diseases is the overproduction of aberrant protein, which aggregates intra- and extracellularly leading to neuronal damage and death in the affected area of the brain.

44.2.1.1 Alzheimer's Disease and Prion Pathology

AD is characterized by the accumulation of Aβ and tau proteins in the form of amyloid plaques and NFTs in brain tissue (Konickova et al., 2022). Amyloid precursor protein (APP) is a ubiquitous transmembrane protein that contains an extracellular hydrophobic transmembrane domain, and an intracellular domain (Kang et al., 1987). Under physiological conditions, soluble APP (sAPP) is involved in neurite outgrowth, synaptogenesis, synaptic plasticity, and cell survival through gamma-aminobutyric acid and N-methyl-D-aspartate receptors, which maintain intracellular calcium homeostasis (Muller et al., 2017; Rice et al., 2019). There are two pathways of APP cleavage by different secretase complexes. In the first pathway, the non-amyloidogenic APP processing, both α- and γ-secretases release sAPPα, P3 peptide (which corresponds to the $A\beta_{17-40/42}$ sequence), and the APP intracellular domain (De Strooper et al., 1998). The second pathway, the amyloidogenic processing of APP, is initiated by β-secretase or β-site APP-cleaving enzyme 1 (BACE1), instead of α-secretase. Thus, sAPPα is released then γ-secretase generates Aβ peptides with different C-terminal residues and APP intracellular domain. This cleavage at the β-site can be also performed by BACE2, although BACE2 tends to cleavage at the β`-site generating an N-truncated, non-amyloidogenic β-peptide (Kimura et al., 2016; Wang et al., 2019). Therefore, cleavage by γ-secretase is the one generating heterogeneity at the C-terminus end of Aβ peptide, producing mainly $A\beta_{1-40}$ and, to a lesser extent, the most amyloidogenic $A\beta_{1-42}$ form (Roda et al., 2022). $A\beta_{1-40}$ tends to accumulate in the vasculature, while $A\beta_{1-42}$ is the predominant form in amyloid plaques. The Aβ aggregates in cortical extracellular plaques about 10–30 years before the onset of dementia (Gordon et al., 2018; Villemagne et al., 2013). Another key pathological characteristic of AD is the neuropil threads, which contain hyperphosphorylated tau (Laliberte et al., 1992). Tau-containing tangles are found in healthy people older than 60 years, but they are restricted to the medial temporal lobe. While, in AD, tau aggregates spread from the medial temporal lobe to the neocortex (Braak & Del Tredici, 2016).

44.2.1.2 Parkinson's Disease and Prion Pathology

Similarly, PD is characterized by the aggregation of α-syn into Lewy bodies (LBs) in the central and peripheral nervous systems (Spillantini & Goedert, 2018). Encoded by the *SNCA* gene, α-syn is a presynaptic protein in the central nervous system (CNS) neurons. α-syn is a small acidic protein (Molecular weight of 14 KDa) expressed in neurons of the central nervous system (CNS), peripheral nervous system (PNS), blood cells, and other tissues (Villemagne et al., 2013). α -syn is a natively unfolded monomer, however, it is found naturally in large part as a folded tetramer of 58 KDa with little or no amyloid-like aggregation potential (DeTure & Dickson, 2019). Monomer and tetramer forms exist together. When there is a tetramer: monomer ratio imbalance, this leads to the predominance of proaggregating forms. α-syn was found to have three main regions, each region has different molecular and biological properties (Peng et al., 2020). The N-terminus, the amino acid residues 1–60, is characterized by amphipathic repetitions, which form an α-helix structure. This region of the protein controls the interaction of α-syn to the membranes (Braak & Del Tredici, 2016). The non-amyloid-ß component (NAC region), the amino acid residues 61–95, is the most aggregation-prone region. The C-terminus, the amino acid residues 96–140, is involved in Ca^{2+} binding and chaperone-like activity (Hardy & Selkoe, 2002). It was found that the binding of Ca^{2+} to the C terminus of α-syn also regulates its binding to synaptic membranes (De Strooper & Karran, 2016). The exact physiological functions of α-syn remain enigmatic, but it has been found to play a role in the synaptic vesicle release and a variety of other functions. α-syn aggregates localize more in the brain stem and substantia nigra (Bousiges & Blanc, 2022; Desplats et al., 2009). In PD, α-syn has a pathological β-sheet conformation that makes additional monomers form oligomers and amyloid fibrils. α-syn aggregates into LBs, which localize in the neuron soma, or LNs, which localize to axons (De Strooper & Karran, 2016; Peng et al., 2020). These aggregates inhibit axonal transport and reduce neuronal function and survival (Nury et al., 2020).

44.2.2 Innate and Adaptive Immunity

In the early 1980s, McGeer initially reported that neuroinflammation in neurodegenerative disorders has been confirmed by several subsequent studies showing the link between neuroinflammation and pathogenesis of different neurodegenerative diseases (Bjelobaba et al., 2017; Heneka et al., 2015; Wang et al., 2015). This includes increased pro-inflammatory molecules in the blood and central nervous system (CNS) (Heneka et al., 2015; Hofmann et al., 2009; Nagatsu et al., 2000). Acute neuroinflammation is meant to protect the CNS from injury, however, prolonged or uncontrolled inflammation in the brain can be detrimental (Kempuraj et al., 2016). The resident immune cells of the CNS, microglia, contribute to neuroinflammation in response to misfolded, and aggregated proteins involved in the pathogenesis of neurodegenerative diseases. It has also been shown that innate and adaptive immune cells from the periphery can travel past the blood–brain barrier (BBB) and play a role in neuroinflammation as well (Hansen et al., 2018; Zang et al., 2022). In the acute phase, these cells may be beneficial, protecting the brain, but in the chronic phase of disease, they exert detrimental effects. Influences of the innate and adaptive immune systems on the pathogenesis of neurodegenerative disorders became a major area of research, but they remain not fully understood.

44.2.2.1 Alzheimer's Disease and Immune System

The pathophysiological hallmark of AD is the progressive degeneration of cholinergic neurons, contributing to memory and attention deficits (Ferreira-Vieira et al., 2016). The innate immune system is the first line of defense in the immune response. Cells involved in innate immunity include monocytes/macrophages, neutrophils, dendritic cells (DCs), and natural killer (NK) cells. The resident innate immune cells of the CNS are microglia. Microglia act as the macrophages of the CNS and constantly survey the environment, giving them a critical role in tissue maintenance, pathogen defense, and response to neuronal injury (Colonna & Butovsky, 2017). Furthermore, microglia play a significant role in neuroinflammation. Alterations in the functionality of microglia are implicated in neurodegeneration. Microglia are frequently classified into either a proinflammatory and neurotoxic M1 phenotype or an anti-inflammatory and neuroprotective M2 phenotype, with the M1 phenotype associated with neurodegeneration (Guo et al., 2022). Acute microglial activation is neuroprotective. In AD, activated microglia can create a barrier around Ab plaques, making these plaques less toxic and preventing further accumulation of Ab and reducing axonal dystrophy (Condello et al., 2015). Further, activated M2 microglia can release proteases to help in the phagocytosis and clearance of Ab (Jimenez et al., 2008). Although microglia seem to play a neuroprotective role in the early stages of AD, a switch to an M1 phenotype in the chronic phase of the disease appears to be detrimental. Uncontrolled microglial activation can result in the overproduction of proinflammatory cytokines like TNF-a and INF-g and push microglia toward an M1 phenotype. Because of compromised BBB integrity in AD, peripheral immune cells can infiltrate the CNS and contribute to the inflammatory responses. Not only can these immune cells traverse the BBB to enter the CNS but they are also able to return to the periphery to elicit further immune responses, establishing a crosstalk between the CNS and peripheral immune system (Huang et al., 2022a). Blood-derived monocytes can be recruited from circulation to the AD brain. Microglia activated by Ab produce monocyte chemoattractant protein 1 (MCP-1), which binds to the MCP-1 receptor—C-C chemokine receptor type 2 (CCR2)—triggering their mobilization to the inflamed brain (Theriault et al., 2015). After reaching the site of inflammation, these proinflammatory monocytes can differentiate into macrophages, associate with Ab plaques, and may be more efficient in Ab phagocytosis than resident microglia through the production of proinflammatory cytokines including IL-6 and TNF-a (Malm et al., 2010).

Supporting their key role in AD, many risk genes for AD are expressed selectively or preferentially in microglia (Srinivasan et al., 2016). Triggering receptor expressed in myeloid cells 2 (TREM2) is a cell surface protein expressed on microglia that acts to stimulate phagocytosis and suppress inflammatory cytokine production (Gratuze et al., 2018). Indeed, it was found that TREM2 facilitates Ab phagocytosis and inhibits Ab-related proinflammatory responses in vitro (Jiang et al., 2014). Similarly, overexpression of TREM2 in a transgenic mouse model of AD ameliorated AD pathology, suggesting that TREM2 modulates microglial function and is protective against AD progression. Several variants in TREM2 have been found to increase the risk of sporadic AD by 2- to four-fold, with R47H, a single nucleotide polymorphism, being the most common and well-studied (Gratuze et al., 2018). Although results vary, patients with this variant have been found to have an earlier onset of symptoms and a quicker cognitive decline (Del-Aguila et al., 2018; Slattery et al., 2014). Further, carriers of this mutation were also found to have significantly higher levels of total tau in CSF compared to controls, suggesting that TREM2 may also be involved in tau pathology (Lill et al., 2015). Similarly, TREM1 was found to facilitate microglial Aβ phagocytosis, while TREM1 variant rs6910730^{G} was found to impair this function and exacerbate AD pathogenesis (Saadipour, 2017). TREM1 was also found to contribute to AD pathogenesis via amplifying TLR2-induced inflammation and increasing the production of proinflammatory cytokines (Arts et al., 2013). A recent study showed that genetic expression of *TLR2* increased with age in a transgenic APP/PS1 mouse model of AD, suggesting the contribution of *TREM1* and *TLR2* in the development of neuroinflammatory responses in AD (Abdelmoaty et al., 2023b). In addition, myeloid differentiation primary response 88 (*MYD88*) deficiency has been found to decrease Aβ load and microglial activation in the brain of an AD mouse model (Lim et al., 2011). Genome-wide association studies (GWAS) have also revealed multiple variants of cluster of differentiation 33 (CD33) that increase the risk of AD (Dos Santos et al., 2017). Expression of CD33 is increased in the AD brain and associated with the modulation of microglial activation, leading to the inhibition of Ab clearance (Griciuc et al., 2013). These risk genes, along with a number of other risk genes associated with microglial function, like INPP5D, MS4A6A, and PLCG2, solidify the critical role microglia play in AD (Hansen et al., 2018).

Polymorphonuclear neutrophils (PMN) have also been investigated for their role in AD. Neutrophils are known for their role as a key contributor to early innate immunity with multiple strategies to defend against and kill pathogens. These include phagocytosis, production of reactive oxygen species (ROS), the release of neutrophil extracellular traps (NETs), and production of inflammatory cytokines (Nauseef & Borregaard, 2014; Tecchio et al., 2014). While these responses are beneficial for host defense, improper activation of neutrophils can result in the overproduction of ROS and inflammatory molecules (Dong et al., 2018). Indeed, blood samples collected from AD patients revealed that neutrophil hyperactivation was correlated with AD progression. Hyperactivation was associated with higher levels of ROS and NET production, as well as an increase in the ratio of harmful hyperactive CXCR4^{high}/CD62L^{low} senescent and CD16^{bright}/CD62L^{dim} immunosuppressive neutrophil subsets. Moreover, a meta-analysis of AD patients and healthy controls showed higher levels of neutrophils in AD patients (Huang et al., 2022b). Similarly, a study in AD mice showed the association of neutrophils with Ab plaques, suggesting that neutrophils could be a target for therapeutic intervention (Zenaro et al., 2015).

Both myeloid-derived DCs (mDCs) and monocyte-derived DCs (MDDCs) have been implicated in AD pathogenesis. Analysis of blood mDCs from AD patients revealed a significant depletion of mDC in AD patients compared to healthy controls, but this was only evident after clinical symptoms were observed (Ciaramella et al., 2016). This was the first evidence that dysregulation of blood mDC levels is associated with the severity of AD symptoms. In a study evaluating MDDCs derived from AD patients and healthy controls, it was revealed that MDDCs derived from AD patients release more proinflammatory IL-6 than those derived from healthy controls (Ciaramella et al., 2010). Additionally, these MDDCs had decreased expression of CD40, a co-stimulatory molecule, with reduced ability to induce T cell proliferation.

Another subtype of innate immune cells is NK cells, which possess important roles as cytotoxic effectors against infected cells and tumor cells as well as in communication between the innate and adaptive immune systems (Huang et al., 2022a). While NK cell populations have been found to be similar between AD patients and healthy controls, the activity of these NK cells was increased in patients with amnesic mild cognitive impairment (aMCI) (Le Page et al., 2015). Supporting the detrimental effects of NK cells early in the disease, a study in 3xTg-AD mice found that the depletion of NK cells can improve cognitive function (Zhang et al., 2020). This study suggested that NK cells play a proinflammatory role in AD and could be a target for therapeutic intervention. Overall, this supported AD-related innate immune dysregulation.

The adaptive immune response is much slower than the innate immune response but is far more specific and able to recognize certain antigens. The adaptive immune system is comprised of B cells and T cells. In AD patients, CD8+ cytotoxic T cells are clonally expanded in the CSF and their numbers increase in the peripheral blood (Gate et al., 2020). These CD8+ T cells have the ability to infiltrate the brain and regulate the expression of genes associated with neuronal and synaptic functions (Unger et al., 2020). Effector T cell (Teff) subsets include CD4+ T helper (Th) cells such as proinflammatory Th1 and Th17 cells and anti-inflammatory Th2 cells. Foxp3+ regulatory T cells (Tregs) are immunosuppressive T cells contributing to self-tolerance and controlling autoreactive Teffs. The balance between Teffs function "defense against infectious or neoplastic diseases" and Tregs function "maintenance of immunological tolerance with control of overactive immune responses" is crucial to maintain homeostatic conditions. In neurodegenerative diseases, Teffs recognize disease-specific modified self-proteins. These are presented by α-syn protein in PD, and Aβ and tau proteins in AD, which can break immune tolerance with the expansion of self-reactive T cells (Anderson et al., 2014; Mosley & Gendelman, 2017). In neurodegenerative disorders, Teffs were also found to have uncontrolled neurotoxic effects, while Tregs have impaired immunosuppressive functions (Chao et al., 2014; Kustrimovic et al., 2018; Lindestam Arlehamn et al., 2020). Studies in AD patients have revealed increased CD4+ Teffs, Th17, and Th9 cells, and reduced Tregs (Larbi et al., 2009; Saresella et al., 2011). Supporting the detrimental effects of proinflammatory Teffs in AD, in APP/PS1 mice, Aβ reactive Teffs accelerated memory impairment, increased amyloid burden, increased microglia activation, and exacerbated neuroinflammation. Moreover, both Th1 and Th17 Aβ-reactive Teffs progressed AD pathology by downregulating anti-inflammatory and immunosuppressive Tregs in the periphery and the CNS (Machhi et al., 2021). Besides, Treg number and function are reduced in AD patients (Ciccocioppo et al., 2019; Faridar et al., 2020). This suggests an imbalance between Teffs and Tregs in the AD brain, likely contributing to a proinflammatory and neurotoxic environment, and consequently, neurodegeneration. Supporting the neuroprotective role of Tregs in AD, adoptive therapy with Ab-specific Tregs was successful in modulating the immune response, ameliorating cognitive deficits, reducing Ab deposition, and limiting tau hyperphosphorylation in a mouse model of AD (Yang et al., 2022). This suggests that AD-associated Teff deleterious immune responses can be controlled by Aβ reactive Tregs.

44.2.3 Parkinson's Disease and Immune System

Several reports documented the activation of innate and adaptive immune systems in PD, and this was associated with neuroinflammation and neurodegeneration (Allen Reish & Standaert, 2015; Schonhoff et al., 2020). Misfolded α-syn acts as a main trigger of immune responses and a potent inducer of inflammation, which leads to neurodegeneration. Like AD, PD pathogenesis is characterized by the production of proinflammatory cytokines, which contribute to the inflammatory environment in PD. Interleukin-1beta (IL-1β) is a part of the larger IL-1 family and plays an important role in controlling many innate immune responses (Kaneko et al., 2019). IL-1β was found to increase dopaminergic neuron damage within substantia nigra in an adenoviral IL-1β expression vector model (Ferrari et al., 2004). In PD patients, serum levels of IL-6, IL-1β, and IL-1 receptor antagonist (IL-1RA) were reported to be significantly elevated (Karpenko et al., 2018). In another report, IL-1β was found to be predictive of disease progression in PD (Williams-Gray et al., 2016). Serum IL-6 has also been found to be negatively correlated with the Activities of Daily Living (ADL) scale, which reflects the severity of the disease (Hofmann et al., 2009). Chemokine ligand 5 (CCL5) has also been found to be associated with the severity and length of PD (Tang et al., 2014). Another proinflammatory cytokine of interest in PD pathogenesis is tumor necrosis factor-alpha (TNF-α), which is a crucial component of microglia-derived inflammatory responses. Serum TNF-α increases in PD patients compared to controls and it is positively correlated with Unified Parkinson's Disease Rating Scale (UPDRS) part III scores (Xiromerisiou et al., 2022). Another area of interest is the association between gut inflammation and PD as many PD patients present intestinal maladies either prior to or during motor dysfunction. Supporting that association, observations of increased expression of *TNF-α, interferon-gamma (IFN-γ), IL-6,* and *IL-1*β genes were found in ascending colons of PD patients compared to age-matched healthy controls (Devos et al., 2013).

Although microglia are morphologically and functionally similar to circulating monocytes and tissue macrophages, they originate from a different lineage in the yolk sac and migrate into the brain during early development (Ginhoux et al., 2010; Schulz et al., 2012). A mixture of resident brain microglia and infiltrating peripheral monocytes exert different effects on PD pathogenesis. Activated microglia surrounding degenerative dopaminergic neurons in the substantia nigra pars compacta (SNpc) were found in PD brains (McGeer et al., 1988), and CD68 and MHC class II (MHC-II) staining showed the degree of microglial activation in PD (Croisier et al., 2005). Additionally, complement C1q was found to be associated with activated microglia surrounding degenerating neurons (Depboylu et al., 2011), with C3d, C4d, C7, and C9 components found co-localized with α-syn aggregations and degenerating neurons in PD autopsies (Yamada et al., 1992). In addition, microglia can phagocytose extracellular aggregated α-syn and direct it to light chain 3B immunoreactive ($LC3B^+$) autophagosomes for degradation, which induces NF-κB-dependent signaling cascades (Cao et al., 2012). Oxidative stress and upregulation of ROS were also observed in response to phagocytosis of aggregated α-syn in rat primary microglia (Zhang et al., 2005). Although microglia treated with aggregated α-syn increased MHC-II expression and antigen processing, a robust MHC-II-dependent cytokine response to aggregated α-syn was only induced in primary microglia by co-culturing them with T cells (Harms et al., 2013). Moreover, two different GWAS studies showed significant associations between PD diagnosis and SNPs in pathways encoding cytokine signaling and regulation of leukocyte/lymphocyte activity (Holmans et al., 2013), and indicated that modified α-syn led to microglial or monocytic activation with the production of proinflammatory cytokines. Similarly, the expression of toll-like receptor (TLR) is affected by α-syn. Murine microglia pretreated with oligomeric α-syn showed increased TLR-2, while TLR-3 and TLR-7 were downregulated (Roodveldt et al., 2013).

The involvement of adaptive immunity in PD pathogenesis is explicit with evidence of the ability of modified α-syn forms to modulate adaptive responses in PD animal models. Several studies showed increased numbers of T-cells in both α-syn and toxin animal models of PD with their involvement in neurodegeneration (Kurkowska-Jastrzebska et al., 1999; Theodore et al., 2008; Watson et al., 2012). With the polarization of T cells responsive to the nitrated C-terminus of α-syn to the Th1, Th2, and Th17 subtypes before adoptive transfer, both proinflammatory Th1 and Th17 cells were found to increase neurodegeneration in response to 1-methyl-4-phenyl-1,2,3,6-tetrahydropyridine (MPTP) in animals. Th17 cells showed a greater effect than Th1 cells, while Th2 cells had no effect (Reynolds et al., 2010). Similarly, in passive transfer studies into Rag1−/− mice, CD4+ T cells mediated MPTP neurotoxicity. In mice treated with MPTP, dopaminergic neurodegeneration was attenuated in CD4−/− animals, while neurodegeneration was unaffected in CD8a−/− animals(Brochard et al., 2009). In addition, overexpression of α-syn in AAV constructs in rat brains leads to dopaminergic neuronal loss with increasing CD4+ and CD8+ T cells within the substantia nigra 8 weeks post-injection (Sanchez-Guajardo et al., 2010). Studies from our group showed that increased Teff phenotypes are associated with worsened UPDRS Part III scores in PD patients (Saunders et al., 2012). In addition, our research group, along with others, demonstrated that Tregs attenuate neuroinflammation and protect dopaminergic neurons from injury and loss in the MPTP animal model (Huang et al., 2020; Kosloski et al., 2013; Olson et al., 2015). Clinically, we have shown that sargramostim—recombinant human granulocyte-macrophage colony-stimulating factor—can increase Treg numbers or functions as well as improve motor functions in PD patients (Olson et al., 2023, 2021).

44.2.4 Genetics

44.2.4.1 Alzheimer's Disease and Genetics

AD, as a complex multifactorial neurodegenerative disease, is highly heritable. Genetically, AD can be divided into two forms: A rare familial form (with autosomal dominant inheritance) accounting for 5–10% of all patients, and a multifactorial sporadic form in which specific environmental factors with genetic susceptibility contribute to the disease exacerbation. The familial form of the disease usually develops before the age of 65 years and is referred to as early-onset AD (EOAD), while the sporadic form usually appears in individuals older than 65 years and is referred to as late-onset AD (LOAD) (Blennow et al., 2006). Three genes have been identified for their responsibility for AD familial form: *APP*, presenilin-1 (*PSEN1*), and presenilin-2 (*PSEN2*). Altogether, they account for 5–10% of cases of EOAD. They map on three different chromosomes but share a common biological pathway (Lill & Bertram, 2011). Familial AD cases explained by mutations in these genes vary widely from 12 to 77% (Jarmolowicz et al., 2015; Wallon et al., 2012), this suggests that there are additional genes that contribute to the EOAD pathogenesis. Other studies involving EOAD found an association between triggering receptor expressed on the myeloid cells 2 (*TREM2*) gene, and the risk of developing EOAD (Gratuze et al., 2018; Pottier et al., 2013). TREM2 acts as a phagocytic receptor expressed in brain microglia, it modulates microglial phagocytosis and inflammatory pathways (Jiang et al., 2013). Other genes, related to EOAD, were also identified such as the neurogenic locus notch homolog protein 3 (*NOTCH3) gene*. A missense mutation in *NOTCH3* (R1231C variant) was previously linked to cerebral autosomal dominant arteriopathy with subcortical infarcts and leukoencephalopathy (Guerreiro et al., 2012). Another gene is the sortilin-related receptor 1 (*SORL1*) gene whose mutations were identified in the EOAD (Pottier et al., 2012). This gene encodes a neuronal sorting protein, which binds to APP driving it toward the endosomal recycling pathways (Mishra et al., 2022).

The majority of AD cases (90–95%) are represented by LOAD, with multiple genes involved with the disease. Several genome-wide association studies (GWAS) studies have succeeded in expanding our understanding of AD genetics. Recent GWAS studies have identified more than 74 AD risk loci, including the Apolipoprotein E glycoprotein (*APOE*) ε4 locus, and suggested different biological processes such as innate immunity and amyloid processing in the development of AD (Andrews et al., 2020; Bellenguez et al., 2022; Schwartzentruber et al., 2021). *APOE* plays a crucial role in lipid metabolism, it maintains cholesterol hemostasis by facilitating the transfer of phospholipids and cholesterol between cells. In CNS, *APOE* is mainly produced by astrocytes and to a lesser extent by microglia. This might explain the upregulation of *APOE* in neurodegenerative diseases and associated neuroinflammation because of astrocytes and microglia activation. *APOE* lipidation mediates Ab clearance via ATP-binding cassette A1 (ABCA1) cascade in astrocytes and inhibits Ab plaque formation. A potential explanation of Ab plaques formation is that APOE protein is overexpressed, undergoes mutations, and acids and triglycerides accumulate forming a complex, which binds with the Ab aggregates to form Ab plaques. *APOE* has different isoforms (1,2,3,4); studies showed that APOE4 is associated with DLB and PDD pathology (Zhao et al., 2020). In a study on 100 PD patients, APOE4 allele carriers were found to be associated with a higher risk of early dementia (Tropea et al., 2018). In AD, despite different studies identifying several risk variants of the *APOE* gene, many variants showed to have a very low risk of the disease, except the ε4 allele of the *APOE* gene. The risk associated with *APOE* ε4 allele was proved in different studies considering many ethnic groups (Farrer et al., 1997). Nevertheless, known common AD variants explain only about 30% of the AD genetic variance (Ridge et al., 2016) and a high percentage of the underlying heritability remains unexplained. Recently, Ali and colleagues (Ali et al., 2023) utilized amyloid PET levels as a quantitative feature to conduct the race-specific GWAS across eight different cohorts (n = 7557) as well as six additional cohorts to perform a multiethnic meta-analysis to identify novel genetic associations with brain amyloidosis in AD (n = 13,409). The results uncovered a very strong significant genome-wide locus on the *APOE* gene region, which is in agreement with previous studies (Li et al., 2015; Raghavan et al., 2020; Yan et al., 2021). In accordance with the presence of at least two independent signals of the *APOE* gene (*ε4* and *ε2*) contributing to AD pathogenesis (Raghavan et al., 2020; Yan et al., 2021), Ali and colleagues found a strong and positive association of *APOE ε4* across populations. They also found significant heterogeneity in the effect of *APOE ε2*, with a strong protective effect of this variant in non-Hispanic white population.

In a study of 12,162 LOAD cases and 13,614 healthy controls from 31 epidemiological studies of LOAD, other genes including tetraspanin 14 (*TSPAN14*), tRNA isopentenyltransferase 1 (*TRIT1*), and metadherin (*MTDH*) have been identified for their contribution to the disease (Chen et al., 2021). *TSPAN14* is a tetraspanin. Tetraspanins are compact and glycosylated transmembrane proteins that were reported previously to be associated with the APP and their potential role in the development of LOAD (Seipold & Saftig, 2016). *TRIT1* is targeted to mitochondria, it modifies tRNA. It has been known as a tumor suppressor for different forms of cancer (Spinola et al., 2005). Previously, *TRIT1* pleiotropy between LOAD and major depressive disorder (Lutz et al., 2020) or cardiovascular risk factors (Broce et al., 2019) have been reported. *MTDH* is also known as astrocyte elevated gene-1 protein (AEG-1). Its contribution to neurodegeneration, especially HIV-induced dementia, has been previously reported (Noch & Khalili, 2013). Research uncovering genes involved in AD pathogenesis is an interesting area of research that aims to understand the complex underlying disease mechanisms.

44.2.4.2 Parkinson's Disease and Genetics

In the last two decades, several studies have identified multiple autosomal recessive and dominant genes that cause familial PD. Mutations in or duplications of the α-syn gene were found to be the cause of some dominantly inherited forms of PD (Singleton et al., 2013). In addition, GWAS have revealed that variations in two of the identified familial PD genes (Synuclein Alpha; *SNCA* and leucine-rich repeat kinase 2; *LRRK2*) are significant risk factors for the development of sporadic PD (Edwards et al., 2010; Gandhi & Wood, 2010; Satake et al., 2009; Simon-Sanchez et al., 2009). Furthermore, several studies have identified potential associations between PD and common genes, single nucleotide polymorphisms (SNPs), and both *PARK*-designated and non-

PARK-designated genes including *BST1, CCDCC2/HIP1R, DGKQ/GAK, HLA-DRA, APOE, GBA, LRRK2, MAPT, MCCC1/LAMP3, SNCA, STK39, SYT11/ RAB25SNCA, UCHL1, PARK16, NAT2, INOS2A,* and *GAK* (Kouli et al., 2018).

Intensive research implies the impaired protein and mitochondrial homeostasis as key contributors to the development and progression of PD, with oxidative stress as an important link between the two pathogenic processes. Besides overproduction and aggregation of abnormal protein, impaired degradation pathways such as lysosomal and autophagy dysfunctions are known contributors to PD pathogenesis (Tanji et al., 2011; Zhu et al., 2003). Multiple studies suggest that aggregation of α-syn is a consequence of dysfunction of the autophagic-lysosomal degradative system. In a row, α-syn has also been found to affect mitochondrial, autophagic, and lysosomal functions in PD (Caballero et al., 2018; Ryan et al., 2015; Xilouri et al., 2016). Dopaminergic neurons are metabolically active and need a high mitochondrial energy source; therefore, they can have insufficient clearance of damaged mitochondria (Haddad & Nakamura, 2015). Accumulation of defective mitochondria will increase the levels of ROS, which damage surrounding healthy mitochondria to further accelerate disease progression. Several studies uncovered genes related to these biological processes and their contribution to PD pathogenesis and progression. GWAS and colocalization analysis identified two loci risks, *SNCA*, and *SNCA* antisense RNA 1 (*SNCA-AS1*)—a noncoding RNA playing a role in regulating the expression of α-syn protein. Recent studies characterizing overexpression of both genes showed their effect on synaptogenesis as well as PD-related pathology (Chia et al., 2021; Rey et al., 2021). In addition, *GBA* gene variants were the first genetic risk factor identified for PD (Pradas & Martinez-Vicente, 2023). *GBA* is encoding for lysosomal hydrolase enzyme b-glucocerebrosidase, which hydrolyzes glucocerebrosidase (GCase) to ceremide and glucose, leading to an increase in GCase misfolding, trapping in endoplasmic reticulum (ER), inducing ER stress and ubiquitin-proteasome system (UPS), which followed by triggering ER-associated degradation (ERAD) and/or unfolded protein response (UPR). The prolonged activation of ERAD and UPR will increase apoptosis, and the presence of misfolded GCase in lysosomes will cause accumulation and aggregation of α-syn via decreasing its degradation through chaperon-mediated autophagy (CMA). *GBA* mutations cause also mitochondrial dysfunction, increasing ROS, and decreasing ATP production and oxygen consumption (Migdalska-Richards & Schapira, 2016).

For optimal lysosomal activity, acidic PH is essential. In several neurodegenerative disorders, abnormal PH was found with impaired lysosomal degradation and aggregation of neurotoxic proteins (Bonam et al., 2019). The optimization of PH is determined by influx and efflux proton pump dynamics. Vacuolar-type H+ ATPase (V-ATPase) is a proton influx (Mindell, 2012). While transmembrane protein 175 (TMEM175)—an endolysosomal potassium channel—is a selective preamble efflux to H^+ protons especially when its luminal face is facing acidic PH, which balances the V-ATPase effect. It is controversial about TMEM175's role in PD pathology, whether decrease or increase is pathological in PD. Studies showed that TMEM175 deficiency in PD animal models was associated with an increased dopaminergic neuron loss and deposition of α-syn aggregates (Jinn et al., 2017; Wie et al., 2021). While other studies showed the TMEM175 activity correlation with the Bcl-2 apoptosis regulator factor, which plays a role in mitophagy. Upon binding to Bcl-2, TMEM175 is activated and induces ROS, and exacerbates dopaminergic neuron loss and motor dysfunction in PD animal models (Qu et al., 2022). Moreover, other studies investigated which alleles of TMEM175 are associated with synucleinopathies' risk. TMEM175 variant rs2290402 allele was the risk allele for PD pathology (Palomba et al., 2023), and TMEM175 rs6599388-T was the risk allele for DLB (Chia et al., 2021). More studies are still required to illustrate how the TMEM175 variants, mutations, and PH changes affect PD pathology.

It is challenging to determine if *APOE4* contributes directly, or if it depends on Ab pathology to affect α-syn pathology. It was found that in the transgenic *APOE4* AAV-α-syn mouse model, increased neurodegeneration, behavior deficits, and neuroinflammation were independent of Ab pathology in the *APOE4* variant more than other APOE variants (Zhao et al., 2020). These results were accordant when it was compared with human postmortem brain sections from DLB patients with minimal AD pathology, this showed higher α-syn pathology by staining in *APOE4* carriers more than noncarriers (Zhao et al., 2020). More research studies are still required to study and confirm the role of *APOE* in different neurodegenerative diseases.

44.3 Clinical Evaluation of Neurodegenerative Diseases

Neurodegenerative diseases are characterized by abnormal aggregation and accumulation of misfolded proteins that are specific to given groups of diseases and are reflected in the designation of proteinopathies. Simultaneously, they involve dysfunctions of a range of cellular mechanisms, including mitochondrial and/or lysosomal dysfunction, oxidative stress, and inflammation, that ultimately lead to neuronal damage. Neurodegenerative diseases employ different categories of diagnostic and/or therapeutic biomarkers such as clinical signs and symptoms, neuroimaging methods, biological, and biochemical testing in CSF, blood, or certain tissues, and genetics. Omics—including transcriptomics, proteomics, and metabolomics potentially involved in neurodegenerative processes—are widely used for diagnostic and/or prognostic purposes (Cova & Priori, 2018; He et al., 2018). Diagnostic, prognostic, and/or therapeutic biomarkers became an interesting area of research for neurodegenerative diseases. The main aim is to create a signature profile—a combination of biomarkers that are specific to a specific disease.

44.3.1 Signs and Symptoms

44.3.1.1 Alzheimer's Disease Signs and Symptoms

Dementia is characterized in general terms as the impaired ability to remember or make decisions that impact daily activities. Dementia can be subdivided into different types with Lewy body dementia affecting 5% of dementia patients, frontotemporal dementia affecting another 5%, vascular dementia affecting 20%, and AD affecting the majority with 70% (Cunningham et al., 2015). AD can be divided into early-onset AD (EOAD), which affects individuals younger than 65, and late-onset AD (LOAD), which affects individuals older than 65. EOAD only consists of 5–10% of all AD cases (Sirkis et al., 2022). Symptoms of AD can manifest in many ways. The most common and distinct symptom is cognitive degeneration, which includes forgetfulness, poor decision-making, and long-term memory loss (Bature et al., 2017; Porsteinsson et al., 2021). With new diagnostic techniques, connections are being made with newer noncognitive symptoms, which include neuropsychiatric symptoms (NPS) that are prevalent within the population of AD patients within all stages and with variation between sexes (Eikelboom et al., 2021, 2022). These symptoms include depression, delusion, some psychosis, and most commonly apathy (Devanand et al., 2022). These NPS symptoms have been linked to the APOE4 gene (Burke et al., 2016). Gut dysbiosis is also being studied as a more significant symptom of AD with its relationship to the gut–brain axis (Chandra et al., 2023). Alteration in the gut microbiome has also been seen in AD patients with higher levels of proinflammatory bacteria (Kesika et al., 2021). There are also a lot of comorbidities relating to symptoms involved in AD, which include diabetes, cardiovascular diseases, and gastrointestinal diseases (Santiago & Potashkin, 2021). One of the most common ways to classify the AD progression is the Braak stages (Braak & Braak, 1991). The first stage involves NFTs that are sequestered in the transentorhinal cortex region of the brain. Then in stages 2 and 3, the NFTs will spread to the hippocampus and temporal cortex, respectively. This eventually leads to spread to the cerebral cortex in stage 4, visual association cortex in stage 5, and lastly, primary visual cortex in stage 6. While these observations were made in postmortem studies, these stages have been re-validated with PET tau imaging in AD Patients (Therriault et al., 2022).

44.3.1.2 Parkinson's Disease Signs and Symptoms

In PD, α-syn accumulation into LBs with associated loss of SNpc dopaminergic neurons heralds the motor and non-motor symptoms. Motor symptoms begin insidiously and progress gradually over time (Greenland & Barker, 2018), they include bradykinesia, tremors, rigidity, and postural instability. Bradykinesia is a slowness of movement interrupted by halts in movement (Berardelli et al., 2001). A unilateral tremor also occurs while the afflicted limb is at rest and can manifest in the legs, chin, and lips (Jankovic, 2008). Rigidity in the flexion and extension of limbs is another motor symptom of PD. Posture instability is due to the inability to maintain equilibrium in a static state, such as sitting or standing, or instability in the transition from a static state to a moving state (Palakurthi & Burugupally, 2019). Non-motor symptoms of PD often precede the onset of motor symptoms by 5–10 years (Kalia & Lang, 2015). The non-motor symptoms can range from mood changes to alterations in sleep habits (Poewe, 2008). Rapid-eye-movement (REM) sleep behavior disorder (RBD) has been seen as a preclinical indicator of PD (Iranzo et al., 2013). Olfactory dysfunction has also been noted in PD patients (Fullard et al., 2017). Higher UPDRS score PD patients showed higher levels of olfactory dysfunction compared to PD patients with lower UPDRS scores (He et al., 2020). Constipation is also associated with PD, which shows the importance of the gut–brain axis due to the hypothesis that PD etiology may originate in the gut and then migrate to the brain (Stocchi & Torti, 2017).

44.3.2 Diagnosis

44.3.2.1 Alzheimer's Disease Diagnosis

Diagnosis of AD is mainly done in the clinic with cognitive tests such as the ascertain dementia 8 (AD8), mini-mental state exam (MMSE), or Montreal Cognitive Assessment (MoCA) (Arevalo-Rodriguez et al., 2021). Once cognition is tested, a neurological examination is then also tested such as reflexes, coordination, speech, and sensation. For further confirmation, noninvasive brain imaging can be done. The timeline of cognitive impairments and diagnosable physical symptoms of AD is very different and has been seen to be delayed compared to the pathological hallmarks of AD such as Ab plaques and NFTs (Hampel et al., 2023). This means when AD is diagnosed via manifested symptoms, the patient is farther along with Ab formations and NTFs already present. This creates a challenge and a necessity to find alternative methods of detection even before AD symptoms have occurred. One such noninvasive method is neuroimaging (Young et al., 2020). One of the most common imaging methods for internal structures within the body is magnetic resonance imaging (MRI), which utilizes a powerful magnet to generate a magnetic field that forces protons to align in a certain direction. A radio frequency is then pulsed in and pushes the protons out of alignment and when the radio frequency is turned off, the protons realign to the magnetic field. As the proton realigns back to its original orientation, it will release electromagnetic energy, which is detected by the MRI machine. Faster proton alignments lead to a brighter image (Berger, 2002). This contrast of alignment within tissue types provides distinct MRI images. Different imaging methods can also be applied such as T1 or T2 weighted imaging with T1 having shorter echo times and repetition times compared to T2 weighted imaging. In the context of AD, there are structural changes within the brain between normal healthy individuals and AD patients (Vemuri & Jack Jr., 2010). These differences have allowed MRI to be leveraged as a useful tool to help diagnose AD. A Longitudinal MRI study showed that an AD diagnosis and progression could be tracked 24 months out using structural T1-weighted MRI along with neuropsychological measures as indicators with 85% accuracy, 83% sensitivity, and 87% specificity (Salvatore et al., 2018). A retrospective longitudinal study using a large MRI data set found that regions within the medial temporal such as the hippocampus and amygdala atrophied first; followed by the middle temporal gyrus in stage 2; then entorhinal cortex, parahippocampal cortex, and other temporal ages in stage 3; striatum and thalamus in stage 4; and lastly middle frontal, cingular, parietal, and insular cortices (Planche et al., 2022). Brain atrophy, and more specifically the medial temporal lobe, has been seen to be strongly correlated with other pathological AD hallmarks such as tau deposition and amyloid plaque formation. Brain atrophy is also a strong indicator and key biomarker for early neurodegenerative changes in AD (Chincarini et al., 2011; Lombardi et al., 2020; Rajmohan & Reddy, 2017; Whitwell et al., 2008). Functional MRI (fMRI) has also been utilized to be used as a possible diagnostic tool for AD. One specific fMRI technique called blood oxygenated level dependent (BOLD) imaging shows the difference in magnetic properties of oxygenated atrial blood and deoxygenated venous blood and is used as an indirect way of measuring activity in different regions of the brain with results showing variability between brain regions (Berger, 2002; Chen & Glover, 2015; Lv et al., 2018; Scarapicchia et al., 2018). A newer direction of machine learning and deep learning are starting to be explored to be used in conjunction with MRI imaging for better sensitivity and earlier detection of AD in life (Frizzell et al., 2022; Li et al., 2020). Because of the similar atrophy patterns, some limitations still include struggles to differentiate between specific types of dementia along with mild cognitive impairment (MCI).

Positron emission tomography (PET) is another tool within the physician's repertoire to diagnose AD. PET imaging works by using a radioactive isotope that decays and produces positron particles. When these positron particles collide with an electron, it will create an annihilation event that leads to the emission of gamma rays in directly opposite directions that are detected by the photomultiplier rings within the machine. The machine can cross-reference when the two-photon signals are detected and back-reference when the annihilation even happened. These signals can be reconstructed, and a 3D image of the activity can be generated. One way that PET imaging is being utilized is by looking at cerebral glucose metabolism via fluorine −18 fluorodeoxyglucose (^{18}F-FDG), which is a glucose analog that contains an F-18 isotope (Marcus et al., 2014). Studies have shown that ^{18}F-FDG can mimic the functions of glucose in vivo and has been correlated with neuronal activity (Mosconi, 2013; Quigg & Kundu, 2022; Sarikaya et al., 2020). With the brain accounting for only 2% of the body weight and consuming around 20% of energy from glucose, any irregularities in this metabolic process can provide useful insight (Erbsloh et al., 1958; Mergenthaler et al., 2013). This was first utilized in 1979 and has become commonplace for providing helpful insight for physicians diagnosing AD (Phelps et al., 1979). ^{18}F-FDG-PET has been used for diagnostic workup of several neurodegenerative diseases due, in part, to certain patterns of regional hypometabolism that reflect particular neurodegenerative diseases. For example, frontal and anterior temporal hypometabolism is typically observed in the behavioral variant of frontotemporal dementia (FTD) (Chetelat et al., 2020). It has been seen that AD patients have hypometabolism, which can help with predicting disease progression (Arlt et al., 2009). The hypometabolic regions include bilateral temporal lobes, limbic system, and parietal lobes (Mosconi, 2005; Sanabria-Diaz et al., 2013). A large cohort study found that there was a correlation between phosphorylated tau deposition and hypometabolism within the temporal, parietal, and frontal regions of the brain and other CSF biomarkers for AD such as Ab 1–12 (Chiaravalloti et al., 2015, 2018; Klunk et al., 2004; Rubí et al., 2018; Schilling et al., 2016). The ^{18}F-FDG imaging was also able to differentiate between AD and other types of dementia and MCI (Bao et al., 2021; Shivamurthy et al., 2014). Another way PET imaging is being utilized is through radiolabeled tracers for specific pathological hallmarks of AD such as NFTs and amyloid plaques. This allows for the potential observation of disease progression and can be used as an alternative for ^{18}F-FDG (Boccalini et al., 2022). Certain radiolabeled formulations with different isotopes have already been used. One such example is Pittsburgh compound B (11C-PiB), which is an analog to thioflavin T bound to a radiolabeled C-11. It was developed by William E. Klunk and Chester A. Mathis from the University of Pittsburgh in 2004 (Klunk et al., 2004). It was the only compound to date that was able to cross the BBB and was the first foray into the bioimaging for Ab plaques in humans (Suppiah et al., 2019). F-18-linked isotopes have become very popular and have been used to target Ab (Mantel & Williams, 2019). In 2012, F-18 Florbetapir was the first FDA-approved F-18-radiolabeled tracer for amyloid plaque. F-18 Flutemetamol was next in 2013 (Martínez et al., 2017; Okamura & Yanai, 2010) and the latest one was in 2014 and called F-18 Florbetaben (Librizzi et al., 2021). In a head-to-head comparison, it was seen that F-18 Flutemetamol was slightly better at detecting amyloid burden in the striatum compared to F-18 Florbetaben but had overall similar properties of sensitivity (Cho et al., 2020). Newer targets, such as tau tangles have also been explored. F-18 Flortaucipir (TAUVID®) is used for imaging of tau and was approved by the FDA in 2020 (Fleisher et al., 2020; Jie et al., 2021) and was able to differentiate between AD and other neurodegenerative diseases in a multi-site study (Ossenkoppele et al., 2018). Another set of tau-specific radiolabeled ligands includes F-18-labeled aryl quinoline derivatives, they have been studied but still need to be clinically validated (Bischof et al., 2021; Harada et al., 2016; Okamura et al., 2013). While the isotopes that are being used (F-18 and C-11) have half-lives of 110 and 20 minutes, respectively, and have been FDA approved, some of the disadvantages of PET imaging is the radioactive burden associated with nuclear medicine. Other challenges associated with radiolabeled tracers are their ability to cross the BBB, and for tau tracers, to be able to cross the plasma membrane of neurons. With Ab and tau formation being a hallmark for AD, these hallmarks are able to help not only in detecting AD earlier on but can also help in determining the efficacy of therapeutics in the near future.

44.3.2.2 Parkinson's Disease Diagnosis

Physicians usually diagnose PD, similar to AD, by taking the medical history and then performing a neurological examination. If PD cardinal motor symptoms improve after starting to take the medication (such as levodopa), this is considered another indication that the person has PD. People having PD-like symptoms that result from other disorders, such as DLB and MSA, are sometimes misdiagnosed as PD. Therefore, certain medical tests, such as seeding aggregation assays of α-syn, and response to drug treatment may help to better diagnose the disease. Neuroimaging techniques allow the assessment of structural and functional alterations of neurodegenerative patients and enhance the criteria for differential diagnoses for different neurodegenerative diseases (Prell, 2018). In PD, different imaging techniques can verify the validity of a PD diagnosis from the motor and non-motor symptoms. Degradation of the dopamine transporter in the substantia nigra has gained much attention as a possible biomarker for PD. Synaptic connections between the dopaminergic neurons for PD diagnosis and progression have been assessed (Stoessl et al., 2011). PET imaging can be used to determine the density of dopaminergic nerve terminals in the basal ganglia, which is typically reduced in PD and other neurodegenerative diseases such as PSP, CBD, and MSA (Brooks, 2010). In addition, the binding of PET ligands to the synaptic vesicle protein 2A (SV2A) was found to detect regionally decreased synaptic density in PD as well as AD (Matuskey et al., 2020; Mecca et al., 2020). Imaging disease-specific patterns of regional glucose metabolism with PET and ^{18}F-FDG allows for a highly accurate distinction between PD and atypical parkinsonian syndromes, including MSA, PSP, and CBD (Meyer et al., 2017). ^{11}C-dihydrotetrabenazine (^{11}C-DTBZ) is another PET ligand for vesicular monoamine type 2 transporter (VMAT2) of presynaptic dopaminergic neurons. In PD patients, ^{11}C-DTBZ PET imaging along with correlational tractography showed diminished nigrostriatal tract integrity with lower ^{11}C-DTBZ distribution, which suggested nigrostriatal axonal dysfunction (Sanchez-Catasus et al., 2021). Another popular imaging probe is ^{123}I-ioflupane with single-photon emission computed tomography (SPECT) is a combination also known as dopamine transporter scan (DaTSCAN) (Pagano et al., 2016). This type of imaging allows better differentiation between PD and other types of parkinsonism. Other nuclear medicine techniques have also been applied along with subsequent radiolabeled tracers. One such example is the ^{18}F-FE-PE2I PET imaging, which showed comparable sensitivity to mainstream ^{123}I-ioflupane SPECT imaging (Delva et al., 2020; Jakobson Mo et al., 2018).

Despite MRI scans may not provide a full picture, deeper insight into the progression of PD can be obtained and differentiate PD from other neurodegenerative diseases. For instance, thin section T2 and proton density spin echo images have been able to discriminate between PD and MSA with 88% sensitivity and 89% specificity (Righini et al., 2002). The PD onset leads to neuronal changes in the brain as well as large-scale physical changes such as atrophy in many areas of the cortex and subcortex with decreases in brain volume (Prakash et al., 2016) and increased frontal lobe volume (Cerasa et al., 2011). Structural MRI, including T1- and T2-weighted imaging, can show brain atrophy patterns as an estimate of regional neurodegeneration. Measuring structural characteristics associated with PD makes MRI an instrument capable of capturing small spatial differences within different areas of the brain. Differences in atrophy rates can be detected in longitudinal clinical trials because MRI measures of gray matter volumes and cortical thickness often exhibit low intra-individual variability over time (Cullen et al., 2020). Another MIR application for PD diagnostics is iron mapping. Iron content increases in the substantia nigra of PD patients compared to healthy controls (Ghassaban et al., 2019). This provides a reference for diseased brains compared to normal brains. A meta-analysis study showed that MRI measures of iron content in PD patients across multiple studies provided a reliable marker for PD, which correlated with the severity of motor symptoms (Cho et al., 2021). Another application of MIR for several neurodegenerative diseases depends on that BBB integrity is disrupted and this can be evaluated by certain MRI techniques such as dynamic contrast-enhanced and dynamic susceptibility contrast MRI. This technique depends on a more permeable BBB and leakage of gadolinium-based contrast agents into the brain (Elschot et al., 2021). Another promising MRI method is diffusion tensor imaging (DTI), which is used for measuring fractional anisotropy and microstructural indices of brain white matter in different neurodegenerative diseases (Alexander et al., 2007). Lower than normal fractional anisotropy and higher than normal diffusivity are associated with loss of microstructural integrity and neurodegeneration. Previously, DTI studies in PD demonstrated abnormal fractional anisotropy in multiple regions of white matter, especially in the dopaminergic nuclei pathways (Zhang & Burock, 2020). DTI has also shown promise in differentiating PD patients from healthy subjects. In a meta-analysis of 39 studies, mean fractional anisotropy and diffusivity found differences in the substantia nigra, corpus callosum, and cingulate and temporal cortices of PD patients (Atkinson-Clement et al., 2017). Additionally, DTI has been used to differentiate PD patients from PDD patients by assessing microstructural changes in the nucleus basalis of Meynert (NBM). The NBM of PDD patients showed increased mean diffusivity and decreased gray matter volume compared to PD patients without cognitive impairment. This microstructural change in the NBM was shown to precede gray matter volumetric loss suggesting an early biomarker for PDD (Schulz et al., 2018). Other imaging studies have found that cortical thinning of the frontal, anterior cingulate cortex, and right precentral in combination with gray matter atrophy are predictive of cognitive decline in PD (Kandiah et al., 2014; Lee et al., 2014). Another imaging tool; ^{123}iodine-meta-iodobenzylguanidine (^{123}I-MIBG) myocardial scintigraphy; is used to assess sympathetic denervation and function in organs highly innervated by sympathetic nervous system such as the postganglionic presynaptic of cardiac sympathetic nerve endings. The radiolabeled ^{123}I-MIBG—a norepinephrine analog—is taken up by postganglionic postsynaptic nerve endings and this uptake was found to be correlated to adrenergic innervation and integrity of substantia nigra neurons in PD patients (Sakakibara et al., 2014).

44.4 Biomarkers

Currently, a large number of drug candidates are evaluated in clinical studies as DMTs for neurodegenerative diseases (Budd Haeberlein et al., 2022; Olson et al., 2023). Therefore, there is an urgent need for cost-effective and scalable methods for accurate diagnosis of patients with early neurodegeneration. Neurodegenerative disease diagnostics is considered not complete when biomarkers are not used to support the clinical diagnosis. Biomarkers are defined as measurable indicators that give information about normal biological processes, pathological processes, and/or pharmacological responses to therapies. Therefore, biomarkers can help in diagnosing a disease, estimating disease stage and progression, and reflecting therapeutic responses (Konickova et al., 2022). Studies uncovering biomarkers for neurodegenerative disorders aim to improve the clinical diagnosis, which leads to increasing the accuracy of differential diagnosis between different neurodegenerative diseases. Important criteria of biomarkers for early diagnosis of neurodegenerative disorders are being sensitive and specific for early disease changes that differentiate between disease changes and normal aging changes. Moreover, the discovery and development of pharmacodynamic biomarkers that can identify drug targets in vivo are crucial (Hansson, 2021). These biomarkers represent an important stone in medicine as well as the pharmaceutical industry. While therapeutic biomarkers aim to provide beneficial information in clinical studies, which is very important for the discovery and development of effective disease-modifying therapies (DMTs) (Bateman & Klunk, 2008). Multiple techniques and technologies such as transcriptomics, proteomics, metabolomics, and different bioinformatic tools are used to reveal unique candidates for biomarker analysis and validation.

In neurodegenerative diseases, CSF is considered the most useful biological fluid for biomarker research because it directly reflects biochemical processes within the CNS and can enable early preclinical diagnosis (Matuskey et al., 2020). However, monitoring biomarkers for neurodegenerative diseases would be more advantageous and painless with a more accessible and less invasive option such as blood, which communicates with the brain via the hematoencephalic barrier, lymphatic vessels, and glymphatic system (Konickova et al., 2022). Despite that, blood has some limitations as a sample source for neurodegenerative disease biomarker research. These limitations include the possibility of low levels of CNS biomarkers, which would be even lower in the blood and become undetectable in the periphery due to the substantial analyte dilution effect from the CSF: blood volume ratio. Another limitation is the possibility of interfering with blood proteins such as albumins, globulins, and transferrin (Abdelmoaty et al., 2023a; Konickova et al., 2022). A third limitation is that the biomarker could be not specific to the CNS and is expressed in peripheral tissues. Therefore, measuring biomarkers of neurodegenerative diseases in the blood requires sensitive and specific techniques that are not confounded by blood or blood components.

44.4.1 Alzheimer's Disease Diagnostic/ Prognostic Biomarkers

It has been found that the percentage of AD misdiagnosis reaches 25–30% when not using AD-specific biomarkers. Additionally, around 50% of AD cases are not routinely recognized or correctly diagnosed, resulting in suboptimal treatment and care (Hansson et al., 2023). In AD, the most used biomarkers are CSF Ab, total tau, phosphorylated tau (p-tau), and neurofilament light, which have been studied for diagnosis, prognosis, and therapeutic follow-up as well as along with genetics and neuroimaging. Aβ is derived from the amyloid precursor protein (APP), which is concentrated in neuronal synapses and expressed in many other tissues. APP is cleaved by β-APP cleaving enzyme and then by γ-secretase to form Aβ. Based on the cleavage by γ-secretase, Aβ peptide contains 37–43 amino acids. Aβ40 is the most common isoform, while the isoform Aβ42 tends to aggregate into smaller oligomers and larger insoluble fibrils (Hansson, 2021). In AD, the accumulation of Aβ starts approximately 20 years before dementia onset (Gordon et al., 2018; Villemagne et al., 2013). Imaging has shown that PET ligands—(^{18}F) florbetapir, (^{18}F) flutemetamol, and (^{18}F) florbetaben—can detect insoluble Aβ fibrils in plaques very accurately (Clark et al., 2012; Curtis et al., 2015). Therefore, a negative Aβ-PET scan in a patient with cognitive impairment will exclude AD as the underlying disease. Different Aβ species are measured in CSF with Ab42 as the dominant biomarker associated with AD disease. Its level in CSF is negatively correlated with the Ab plaques in the brain tissue. The CSF Aβ42 levels are decreased by approximately 50% in AD patients (Olsson et al., 2016), and this drop occurs before Aβ-PET starts to become abnormal (Palmqvist et al., 2016). The ratio of Ab42/Ab40 is highly concordant with Aβ-PET results. This ratio is associated with progression from presymptomatic to MCI and from MCI to AD (Gunes et al., 2022).

Plasma levels of Aβ42/Aβ40 are associated with the presence of Aβ plaques in the brain (Salvado et al., 2023). Multiple studies showed that the plasma Aβ42/Aβ40 ratio is lower in Aβ-positive subjects than in Aβ-negative subjects, regardless of the cognitive status of the cohort (Li et al., 2022; Ovod et al., 2017; West et al., 2021). There is a wide variation in the performance of different plasma Aβ42/Aβ40 assays; a recent study showed that certain mass spectrometry-based assays could detect Aβ pathology with areas under the receiver operating characteristic curve (AUC) of 0.84–0.87. While many commonly used immunoassays performed much worse (AUC, 0.64–0.69) (Janelidze et al., 2021). A combination of *APOE* genotyping with plasma Aβ42/Aβ40 prediction models has been found to increase the AUC by 10% (Janelidze et al., 2021; Palmqvist et al., 2019b). Modest correlations have been found between the Aβ42/Aβ40 levels in CSF and plasma (r_s of 0.56–0.65) (Janelidze et al., 2021), probably because much of plasma Aβ is derived from peripheral tissues (Roher et al., 2009). According to the NIA-AA research framework for AD, Aβ pathology (A) can be determined using either CSF Aβ or Aβ PET interchangeably (Jack Jr. et al., 2018), but subtle differences between these two measures have been found. First, CSF Aβ42 level, and potentially Aβ42/Aβ40 ratio, changes somewhat earlier than Aβ PET signals. This is similar to what is seen in Aβ42/Aβ40 levels in blood (Palmqvist et al., 2016, 2019a). While the Aβ PET signal increases with disease progression because it measures insoluble Aβ plaques, whereas, in CSF and blood, the Aβ42/Aβ40 ratio decreases with disease progression (Hansson et al., 2023).

Tau is a microtubule-associated protein involved in the assembly and stabilization of microtubules, it is localized mainly in axons in the CNS. There are six tau isoforms that are expressed from the microtubule-associated protein tau (*MAPT*) gene via alternative mRNA splicing. Tau isoform composition and posttranslational modifications determine which tau filament structures are formed. In AD, certain posttranslational modifications such as phosphorylation of the proline-rich region, might promote aggregation (Wesseling et al., 2020). Pathogenic tau might transmit from cell to cell leading to the transformation of the recipient cell's normal tau into misfolded tau as well as the formation of tau aggregates (Peng et al., 2020). Tau-PET ligands—such as (^{18}F) flortaucipir, (^{18}F) RO948, and (^{18}F) MK6240—bind with insoluble tau fibrils in AD brain tissues (Scholl et al., 2019), in vivo uptake of (^{18}F) flortaucipir has been found to be correlated with the density of tau pathology measured postmortem in AD patients (Pontecorvo et al., 2020; Smith et al., 2019). Furthermore, quantitative assessments of (^{18}F) flortaucipir PET were found to accurately identify neocortical AD-like tau pathology (Gatto et al., 2023; Soleimani-Meigooni et al., 2020). Total tau (T-tau) and tau phosphorylated at threonine 181 (P-tau181) have been widely studied in AD patients. The latter is selectively increased in AD but not in other neurodegenerative diseases including other tauopathies (Uribe et al., 1988). Tau can be phosphorylated at more than 40 different positions, and tau phosphorylated at threonine 217 (P-tau217) exhibits a stronger association with tau-tangle load and disease severity than P-tau181 (Janelidze et al., 2020). Furthermore, CSF P-tau217 might distinguish AD dementia from other dementias with even higher accuracy than other p-tau isoforms (Hanes et al., 2020; Leuzy et al., 2021).

High-sensitivity assays that can detect different p-tau isoforms in plasma have been developed. These assays can measure p-tau181 (Mielke et al., 2018; Thijssen et al., 2020), p-tau217 (Palmqvist et al., 2020), and p-tau231 (Ashton et al., 2021). A few comparisons of these assays using plasma from patients having cognitive impairment showed that the assays that quantify plasma p-tau217 are better at detecting AD pathogenesis and predicting the progression of dementia (Bayoumy et al., 2021; Janelidze et al., 2023; Mielke et al., 2021). A high correlation between CSF and plasma levels (correlation coefficient of 0.89) has been shown using the best-performing p-tau217 assay (Janelidze et al., 2023). On the other hand, plasma p-tau231 is associated with Aβ plaque load but not tau tangle load (Salvado et al., 2023; Therriault et al., 2023) because plasma p-tau231was found recently to increase at very low Aβ plaque levels (Ashton et al., 2021, 2022). Dissimilarly, p-tau181 and p-tau217 are associated with both Aβ plaques and tau tangles, with stronger correlations of p-tau217 (Mattsson-Carlgren et al., 2021; Salvado et al., 2023). For tau pathology (T), it has been found that using tau PET and CSF p-tau interchangeably would be more complex. For instance, with absence of cognitive impairment, more subjects are identified as T-positive when using CSF p-tau versus tau PET (Mattsson-Carlgren et al., 2020c) because CSF p-tau levels and plasma p-tau levels increase much earlier than tau PET signal reaches the threshold for detection during the preclinical stages of AD (Mattsson-Carlgren et al., 2020a). Therefore, Aβ-positive individuals who are positive for CSF p-tau but still negative for tau PET might have an early AD and are just about to start accumulating tau aggregates in the neocortex (Groot et al., 2023). Thus, it has been suggested that the NIA-AA research framework be updated to include p-tau and tau PET as separate biomarkers using "APT" instead of "AT," where P stands for p-tau (soluble hyperphosphorylated tau) and T stands for tau PET (density of insoluble tau fibrils) (Groot et al., 2023).

Currently, certified reference materials and methods are available to establish uniform thresholds for AD diagnosis (Kuhlmann et al., 2017). An international consensus protocol for the collection and handling of CSF samples has been established to avoid false results (Hansson et al., 2021a). Although CSF AD biomarkers are used clinically in many countries, the lumbar puncture is still invasive and not widely available. Thus, many efforts have been made to develop blood-based assays to detect cerebral Aβ pathology. Similarly, standardization of blood sample collection for this purpose is crucial for clinical implementation. For blood AD biomarkers, the same type of sampling tubes (EDTA plasma) should be used for all biomarkers, all the blood biomarkers can withstand up to three freeze-thaw cycles (Rozga et al., 2019; Verberk et al., 2022). Not alike CSF Aβ, plasma Aβ is not sensitive to collection tubes made of glass, and tubes with a gel separator can be used. Importantly, both Aβ40 and Aβ42 are not stable in the whole blood, and their levels decrease after 2 hours. Therefore, samples should be centrifuged early within 1 hour and plasma should be separated. After that, plasma can be stored at 4 °C for up to 6 hours before freezing (Rozga et al., 2019; Winston et al., 2023).

44.4.2 Parkinson's Disease Diagnostic/Prognostic Biomarkers

In PD, several studies for biomarker identification and validation focused on measuring total α-syn species in CSF and blood (Besong-Agbo et al., 2013; Foulds et al., 2013; Shi et al., 2014). The attractiveness of α-syn as a biomarker for PD depends mainly on its increased aggregation and accumulation in the CNS. The level of total α-syn in CSF can be used to differentiate synucleinopathies from other proteinopathies; however, it is not useful for differentiating between the different synucleinopathies, even though significantly lower concentrations have been observed in MSA than in PD and DLB (Parnetti et al., 2014). Different α-syn species, such as oligomeric α-syn, proaggregated α-syn, and phosphorylated α-syn, in CSF and blood, have been considered as potential diagnostic biomarkers for PD (Parnetti et al., 2019). Real-time quaking-induced conversion (RT-QuIC) analysis can directly detect pathogenic proteins with prion-like properties and is very sensitive even in the early stages of diseases that express those types of proteins. This method has been used to detect α-syn misfolding in PD and other synucleinopathies (Zerr et al., 2020). However, complex techniques and cost concerns have ruled out wider use of this ultrasensitive method. The level of oligomeric α-syn in CSF is generally higher in PD patients than in healthy controls. Additionally, a higher level of phosphorylated α-syn was observed in PD patients than in healthy controls (He et al., 2018; Wang et al., 2012). Moreover, RT-QuIC showed the presence of α-syn aggregates in CSF of PD patients with 92% sensitivity and 100% specificity compared to AD patients and healthy controls (Fairfoul et al., 2016). Interestingly, the properties of α-syn aggregates were different between PD and MSA patients suggesting that different conformational strains of α-syn exist as distinct species for each disease (Shahnawaz et al., 2020). Different seeding aggregation assays show high concordance for the detection of misfolded CSF α-syn when evaluated in the same cohort, indicating high reproducibility among the assays (Kang et al., 2019). Recent evidence shows that α-syn seed amplification assay (SAA) of CSF has the potential to differentiate PD patients from healthy controls. In this study, 1123 participants were enrolled from 33 participating academic neurology outpatient practices worldwide (Austria, Canada,

France, Germany, Greece, Israel, Italy, Netherlands, Norway, Spain, UK, and USA) between July 7, 2010, and July 4, 2019. Sensitivity for PD was 87.7% (95% CI 84.9–90.5%), and specificity for healthy controls was 96.3% (93.4–99.2%). In addition, the sensitivity of α-syn SAA in sporadic PD with the typical olfactory deficit was 98.6% (96.4–99.4%) (Siderowf et al., 2023). Another method that detects tandem α-syn and tau aggregates is surface-based fluorescence intensity distribution analysis (sFIDA). This technique features single-particle sensitivity through a microscopy-based readout and uses linear epitopes to detect and count all subtypes of aggregated protein irrespective of higher-ordered structures (Herrmann et al., 2017). Whereas other assays using structural epitopes can only determine subfractions of oligomers, fibrils, or other aggregates from a heterogeneous pool of structures (Blomeke et al., 2022). A recent study showed that sFIDA could discriminate between α-syn from CSF of PD patients and DLB patients with 73% sensitivity and 65% specificity (Blomeke et al., 2022).

Detection of α-syn in less invasive and from more easily accessible fluids and tissues—instead of CSF samples—would be a great step forward. Recent studies showed that skin biopsies, which include nerve terminals, are a reliable source of samples for seeding aggregation assays to detect misfolded α-syn in PD (Manne et al., 2020b; Wang et al., 2020). Other potential tissues for the detection of misfolded α-syn included olfactory mucosa and submandibular gland tissues. In addition, salivary RT-QuIC has been used to detect α-syn and showed 76% sensitivity and 94.4% specificity in differentiating PD patients from healthy controls (Luan et al., 2022). In another study (13 PD patients, 3 incidental LBD, and 16 control cases), postmortem submandibular gland tissue was used to sample α-syn with RT-QuIC in 32 cases with 100% sensitivity and 95% specificity for PD with 100% concordance for elevated levels of pathological α-syn seeding activity in both PD and incidental LBD tissues compared to control tissues (Manne et al., 2020a). Even so, more data from different tissues are required to verify the sensitivity and specificity of this ultrasensitive RT-QuIC method.

Although α-syn is expressed and measured in the blood (El-Agnaf et al., 2003), its levels in the blood are strongly influenced by levels of red blood cell (RBC) contamination and hemolysis, even more than in CSF. RBCs represent >99% of the source of blood α-syn and their abundance and fragility make it possible that even low RBC contamination could result in a false positive increase of α-syn in serum or plasma (Barbour et al., 2008). For this reason, levels of intracellular RBC α-syn are studied as an alternative measurement. Serum and plasma total α-syn levels have been reported to be either higher (Shi et al., 2014), lower (Besong-Agbo et al., 2013), or not significantly different (Foulds et al., 2013) in PD patients compared to healthy controls. This limits the utility of plasma or serum total α-syn measurements for PD diagnostic purposes. On the other hand, studies of blood oligomeric α-syn showed consistent results of increased levels in serum and RBCs of PD patients, (Daniele et al., 2018; Williams et al., 2016; Zhao et al., 2016) with remarkable diagnostic accuracy in serum (sensitivity 75% and specificity 100%) compared to controls (Williams et al., 2016). However, this requires confirmation in larger cohorts. Moreover, plasma phosphorylated α-syn levels have been found higher in PD patients compared to healthy controls (Foulds et al., 2013). While RBC measurements of posttranslational modified forms of α-syn—such as nitrated, SUMOylated, Tyr-125 phosphorylated, and glycated species—were controversial in distinguishing PD patients from healthy controls (Vicente Miranda et al., 2017). In part due to RBC contamination and overlap of α-syn forms found in CSF for different neurodegenerative disorders, α-syn measurements in CSF and/or blood can't be confirmed as a definitive biomarker for PD.

44.4.3 Alzheimer's Disease Therapeutic Biomarkers

As discussed above, extensive research identified promising diagnostic and prognostic biomarkers for different neurodegenerative diseases that can be measured in CSF or blood. However, the knowledge and identification of therapeutic biomarkers, which track therapeutic responses, remain enigmatic. For different neurodegenerative disorders, the currently approved therapies are symptomatic relief and have little or no curative effects on motor and cognitive dysfunctions seen in neurodegenerative disorders. Therefore, there is an urgent need to develop DMTs that better control the progression of neurodegenerative diseases. Identifying blood-based biomarkers that can facilitate recruiting patients for clinical studies is an unmet need for drug development. Similarly, identifying therapeutic biomarkers that can measure therapy responses and efficacy in the blood is an urgent need for clinical studies wherein DMTs are assessed.

In AD, plasma AD biomarkers have been found to facilitate recruitment of patients with MCI or dementia due to AD for clinical trials (Hansson et al., 2023). Approximately 20–30% of patients diagnosed with AD dementia and 40–60% of patients with MCI do not have brain Aβ pathology (Hansson, 2021). When recruiting patients with prodromal AD or mild AD dementia for clinical studies, prescreening the potential candidates, for example, using plasma p-tau217, would reduce the need for confirmatory investigations using Aβ PET or CSF AD biomarkers. This would substantially reduce the costs and increase the scalability of such trials (Hansson et al., 2023). In two independent cohort studies, plasma p-tau217 levels could accurately predict future cognitive decline in preclinical AD. Plasma p-tau217 showed better prediction than other CSF and plasma biomarkers (p-tau181, p-tau231, glial fibrillary acidic protein; GFAP, and NfL) and amyloid PET (Mattsson-Carlgren et al., 2023). Interestingly, power calculations showed that using plasma p-tau217 measurements to enrich cognitively normal subjects likely to show cognitive decline largely reduced the required sample size (Hansson et al., 2023). Moreover, using plasma Aβ42/Aβ40 assays for prescreening the potential candidates resulted in substantial cost and time savings (Cullen et al., 2023; Schindler et al., 2023). This was reported when the plasma test was incorporated early in the enrollment process before the screening visit (Schindler et al., 2023). Therefore, the combination of plasma Aβ42/Aβ40 with p-tau217 (Hansson et al., 2022) or p-tau231 (Ashton et al., 2022) measurements might result in even more efficient detection of preclinical AD.

In the area of identifying potential therapeutic biomarkers for AD, previous studies showed that subjects with increased CSF tau levels showed faster AD progression, suggesting this marker reflects the disease progression and therapeutic responses in AD clinical studies (Buchhave et al., 2012; Samgard et al., 2010). Similarly, several studies with new blood tests for p-tau forms showed that changes in plasma p-tau levels are associated with both cognitive decline and brain atrophy in AD patients (Hansson et al., 2021b; Karikari et al., 2022; Leuzy et al., 2022; Mattsson-Carlgren et al., 2020b; Moscoso et al., 2021; Teunissen et al., 2022). Importantly, promising anti-Aβ antibody clinical studies have shown reductions in plasma p-tau markers associated with less clinical deterioration, supporting disease modification and slowing the neurodegeneration (Budd Haeberlein et al., 2022; Pontecorvo et al., 2022). Based on this evidence, clinically, plasma p-tau forms might be used to assess the therapeutic effect of anti-Aβ antibody treatments in AD. Another potential therapeutic biomarker for AD is plasma GFAP. Plasma GFAP levels increase with AD progression (Cicognola et al., 2021), and plasma GFAP levels decrease after efficient removal of Aβ plaques by anti-Aβ immunotherapy (Pontecorvo et al., 2022). However, there is a challenge that therapeutic antibodies may change the half-life of the biomarkers, which makes data interpretation difficult (Portelius et al., 2017). This was reported for biofluid-based tau biomarkers in anti-tau antibody trials (Yanamandra et al., 2017).

44.4.4 Parkinson's Disease Therapeutic Biomarkers

For PD, there are similar efforts to identify therapeutic biomarkers that can reflect the disease modification in clinical studies. We and other research groups showed the neuroprotective effects of Tregs, which can attenuate neuroinflammation and protect dopaminergic neurons from damage (Kosloski et al., 2013; Olson et al., 2015, 2021). Over the last two decades, our studies demonstrated that granulocyte-macrophage colony-stimulating factor (GM-CSF, sargramostim, Leukine®) increases Treg numbers and function, protects dopaminergic neurons in PD animal models (Kosloski et al., 2013; Olson et al., 2020), and improves motor function in PD patients (Gendelman et al., 2017; Olson et al., 2021, 2023). *Our research group was the first to assess the transcriptomic and proteomic profiles of monocytes* (Abdelmoaty et al., 2022) *and peripheral blood lymphocytes (PBLs)* (Olson et al., 2021) *in PD patients treated with an immune modulator drug (sargramostim). In PBLs, significant* increases in IL-10 gene expression by 2 and 6 months after treatment initiation were found, supporting the immunosuppressive biomarker expression observed in Treg function. Additionally, proteomic studies showed that sargramostim treatment downregulates calcineurin and NF-κB expression significantly after 2 months of treatment and their levels remained reduced after 6 months of treatment, this suggests reducing the inflammatory environment in PD patients (Olson et al., 2021). In monocytes, *we studied* the association between monocyte profiles and clinical motor function and disease progression during immune modulatory therapy with sargramostim in PD patients (Abdelmoaty et al., 2022). The monocyte transcriptomic and proteomic profiles showed neuroprotective functions, which include antioxidant, anti-inflammatory, and autophagy-related genes and proteins (LRRK2, HMOX1, TLR2, TLR8, RELA, ATG7, and GABARAPL2). Our results showed the potential of LRRK2 gene and protein levels for UPDRS, Part III scores, and changes in scores. Moreover, *HMOX1*, *TLR2*, and *ATG7* gene levels as well as RELA, TLR2, and ATG7 protein levels, showed predictive potential for UPDRS, Part III scores and changes in scores. This suggested the possibility of using these genes/proteins as putative biomarkers for sargramostim therapy in PD. Therefore, these genes/proteins may serve as potential biomarkers to predict therapeutic responses in neurodegenerative disorders treated with sargramostim or similar immunomodulatory therapies. These biomarkers are currently being validated in a larger number of PD patients through a 12-month study in our laboratories (NCT05677633) ("US National Library of Medicine. ClinicalTrials.gov. https://classic.clinicaltrials.gov/ct2/show/NCT05677633 (2023),"). Obviously, the research area of identifying and confirming the utility of therapeutic biomarkers for neurodegenerative disorders still needs further investigation in current/future clinical studies.

44.5 Summary

The need for clinical biomarkers for different neurodegenerative disorders is extremely needed. To reveal potential clinical biomarkers, different aspects need to be taken into consideration including. First, easily accessible and less invasive sample collection such as blood-based samples. Second, the mechanisms by which peripheral biomarkers interact with

CNS compartments and contribute to disease pathology should be known and well-established (For example, the link between peripheral monocytes and CNS microglia in neurodegenerative diseases). Third, the availability of high-performance technologies/techniques, which are sensitive and specific to the target biomarker. One of the most important technologies is high-throughput "omics," which can investigate the entire genome, proteome, and metabolome for biomarker discovery. Fourth, mechanistic studies to identify potential drug targets such as Tregs (Olson et al., 2021) and monocytes (Abdelmoaty et al., 2022) for sargramostim therapy. This will facilitate uncovering potential pharmacodynamic and therapeutic biomarkers for clinical trials. Biomarkers for neurodegenerative diseases are needed in the clinic to improve the differential diagnosis, and in drug discovery to facilitate the development and monitoring of potential DMTs.

44.6 Review Questions

1. Clinical evaluation of neurodegenerative diseases depends mainly on:
   (a) Signs and symptoms
   (b) Neuroimaging methods
   (c) Biomarker measurements
   (d) *a and b*
   (e) a, b, and c
2. Which statement is correct about the definition of biomarkers?
   (a) Measurable indicators, which give information about normal biological and physiological processes
   (b) Measurable indicators that give information about pathological processes
   (c) Measurable indicators that give information about pharmacological responses to therapy
   (d) a and b
   (e) b and c
   (f) *All of the above*
   (g) None of the above
3. Therapeutic biomarkers:
   (a) Detect or confirm the presence of a disease or condition of interest
   (b) Reflect the disease pathobiology
   (c) Reflect the response to therapy
   (d) Provide information about the disease recurrence irrespective of the treatment
   (e) *Reflect the disease pathobiology and response to therapy*
   (f) a and b
   (g) a and c
   (h) All of the above
   (i) None of the above

References

Aarsland, D., & Kurz, M. W. (2010a). The epidemiology of dementia associated with Parkinson disease. *Journal of the Neurological Sciences, 289*(1–2), 18–22. https://doi.org/10.1016/j.jns.2009.08.034

Aarsland, D., & Kurz, M. W. (2010b). The epidemiology of dementia associated with Parkinson's disease. *Brain Pathology, 20*(3), 633–639. https://doi.org/10.1111/j.1750-3639.2009.00369.x

Aarsland, D., Tandberg, E., Larsen, J. P., & Cummings, J. L. (1996). Frequency of dementia in Parkinson disease. *Archives of Neurology, 53*(6), 538–542. https://doi.org/10.1001/archneur.1996.00550060082020

Aarsland, D., Andersen, K., Larsen, J. P., Lolk, A., & Kragh-Sorensen, P. (2003). Prevalence and characteristics of dementia in Parkinson disease: An 8-year prospective study. *Archives of Neurology, 60*(3), 387–392. https://doi.org/10.1001/archneur.60.3.387

Abdelmoaty, M. M., Machhi, J., Yeapuri, P., Shahjin, F., Kumar, V., Olson, K. E., et al. (2022). Monocyte biomarkers define sargramostim treatment outcomes for Parkinson's disease. *Clinical and Translational Medicine, 12*(7), e958. https://doi.org/10.1002/ctm2.958

Abdelmoaty, M. M., Lu, E., Kadry, R., Foster, E. G., Bhattarai, S., Mosley, R. L., & Gendelman, H. E. (2023a). Clinical biomarkers for Lewy body diseases. *Cell & Bioscience, 13*(1), 209. https://doi.org/10.1186/s13578-023-01152-x

Abdelmoaty, M. M., Yeapuri, P., Machhi, J., Lu, Y., Namminga, K. L., Kadry, R., et al. (2023b). Immune senescence in aged APP/PS1 mice. *NeuroImmune Pharmacology and Therapeutics, 2*(3), 317–330. https://doi.org/10.1515/nipt-2023-0015

Alexander, A. L., Lee, J. E., Lazar, M., & Field, A. S. (2007). Diffusion tensor imaging of the brain. *Neurotherapeutics, 4*(3), 316–329. https://doi.org/10.1016/j.nurt.2007.05.011

Ali, M., Archer, D. B., Gorijala, P., Western, D., Timsina, J., Fernandez, M. V., et al. (2023). Large multi-ethnic genetic analyses of amyloid imaging identify new genes for Alzheimer disease. *Acta Neuropathologica Communications, 11*(1), 68. https://doi.org/10.1186/s40478-023-01563-4

Allen Reish, H. E., & Standaert, D. G. (2015). Role of alpha-synuclein in inducing innate and adaptive immunity in Parkinson disease. *Journal of Parkinson's Disease, 5*(1), 1–19. https://doi.org/10.3233/JPD-140491

Alzheimer's disease facts and figures. (2020). *Alzheimers dement.* doi:https://doi.org/10.1002/alz.12068.

Anderson, K. M., Olson, K. E., Estes, K. A., Flanagan, K., Gendelman, H. E., & Mosley, R. L. (2014). Dual destructive and protective roles of adaptive immunity in neurodegenerative disorders. *Translational Neurodegeneration, 3*(1), 25. https://doi.org/10.1186/2047-9158-3-25

Andrews, S. J., Fulton-Howard, B., & Goate, A. (2020). Interpretation of risk loci from genome-wide association studies of Alzheimer's disease. *Lancet Neurology, 19*(4), 326–335. https://doi.org/10.1016/S1474-4422(19)30435-1

Arevalo-Rodriguez, I., Smailagic, N., Roqué-Figuls, M., Ciapponi, A., Sanchez-Perez, E., Giannakou, A., et al. (2021). Mini-Mental State Examination (MMSE) for the early detection of dementia in people with mild cognitive impairment (MCI). *Cochrane Database of Systematic Reviews, 7*(7), CD010783. https://doi.org/10.1002/14651858.CD010783.pub3

Arlt, S., Brassen, S., Jahn, H., Wilke, F., Eichenlaub, M., Apostolova, I., et al. (2009). Association between FDG uptake, CSF biomarkers and cognitive performance in patients with probable Alzheimer's disease. *European Journal of Nuclear Medicine and Molecular Imaging, 36*(7), 1090–1100. https://doi.org/10.1007/s00259-009-1063-7

Arts, R. J., Joosten, L. A., van der Meer, J. W., & Netea, M. G. (2013). TREM-1: Intracellular signaling pathways and interaction with pattern recognition receptors. *Journal of Leukocyte Biology, 93*(2), 209–215. https://doi.org/10.1189/jlb.0312145

Ashton, N. J., Pascoal, T. A., Karikari, T. K., Benedet, A. L., Lantero-Rodriguez, J., Brinkmalm, G., et al. (2021). Plasma p-tau231: A new biomarker for incipient Alzheimer's disease pathology. *Acta Neuropathologica, 141*(5), 709–724. https://doi.org/10.1007/s00401-021-02275-6

Ashton, N. J., Janelidze, S., Mattsson-Carlgren, N., Binette, A. P., Strandberg, O., Brum, W. S., et al. (2022). Differential roles of Abeta42/40, p-tau231 and p-tau217 for Alzheimer's trial selection and disease monitoring. *Nature Medicine, 28*(12), 2555–2562. https://doi.org/10.1038/s41591-022-02074-w

Astrom, D. O., Simonsen, J., Raket, L. L., Sgarbi, S., Hellsten, J., Hagell, P., et al. (2022). High risk of developing dementia in Parkinson's disease: A Swedish registry-based study. *Scientific Reports, 12*(1), 16759. https://doi.org/10.1038/s41598-022-21093-8

Atkinson-Clement, C., Pinto, S., Eusebio, A., & Coulon, O. (2017). Diffusion tensor imaging in Parkinson's disease: Review and meta-analysis. *NeuroImage: Clinical, 16*, 98–110. https://doi.org/10.1016/j.nicl.2017.07.011

Bao, W., Xie, F., Zuo, C., Guan, Y., & Huang, Y. H. (2021). PET neuroimaging of Alzheimer's disease: Radiotracers and their utility in clinical research. *Frontiers in Aging Neuroscience, 13*, 624330.

Barbour, R., Kling, K., Anderson, J. P., Banducci, K., Cole, T., Diep, L., et al. (2008). Red blood cells are the major source of alpha-synuclein in blood. *Neurodegenerative Diseases, 5*(2), 55–59. https://doi.org/10.1159/000112832

Bateman, R. J., & Klunk, W. E. (2008). Measuring target effect of proposed disease-modifying therapies in Alzheimer's disease. *Neurotherapeutics, 5*(3), 381–390. https://doi.org/10.1016/j.nurt.2008.05.009

Bature, F., Guinn, B. A., Pang, D., & Pappas, Y. (2017). Signs and symptoms preceding the diagnosis of Alzheimer's disease: A systematic scoping review of literature from 1937 to 2016. *BMJ Open, 7*(8), e015746. https://doi.org/10.1136/bmjopen-2016-015746

Bayoumy, S., Verberk, I. M. W., den Dulk, B., Hussainali, Z., Zwan, M., van der Flier, W. M., et al. (2021). Clinical and analytical comparison of six Simoa assays for plasma P-tau isoforms P-tau181, P-tau217, and P-tau231. *Alzheimer's Research & Therapy, 13*(1), 198. https://doi.org/10.1186/s13195-021-00939-9

Bellenguez, C., Kucukali, F., Jansen, I. E., Kleineidam, L., Moreno-Grau, S., Amin, N., et al. (2022). New insights into the genetic etiology of Alzheimer's disease and related dementias. *Nature Genetics, 54*(4), 412–436. https://doi.org/10.1038/s41588-022-01024-z

Berardelli, A., Rothwell, J. C., Thompson, P. D., & Hallett, M. (2001). Pathophysiology of bradykinesia in Parkinson's disease. *Brain, 124*(Pt 11), 2131–2146. https://doi.org/10.1093/brain/124.11.2131

Berger, A. (2002). Magnetic resonance imaging. *BMJ, 324*(7328), 35. https://doi.org/10.1136/bmj.324.7328.35

Besong-Agbo, D., Wolf, E., Jessen, F., Oechsner, M., Hametner, E., Poewe, W., et al. (2013). Naturally occurring alpha-synuclein autoantibody levels are lower in patients with Parkinson disease. *Neurology, 80*(2), 169–175. https://doi.org/10.1212/WNL.0b013e31827b90d1

Bischof, G. N., Dodich, A., Boccardi, M., van Eimeren, T., Festari, C., Barthel, H., et al. (2021). Clinical validity of second-generation tau PET tracers as biomarkers for Alzheimer's disease in the context of a structured 5-phase development framework. *European Journal of Nuclear Medicine and Molecular Imaging, 48*(7), 2110–2120. https://doi.org/10.1007/s00259-020-05156-4

Bjelobaba, I., Savic, D., & Lavrnja, I. (2017). Multiple sclerosis and neuroinflammation: The overview of current and prospective therapies. *Current Pharmaceutical Design, 23*(5), 693–730. https://doi.org/10.2174/1381612822666161214153108

Blennow, K., de Leon, M. J., & Zetterberg, H. (2006). Alzheimer's disease. *Lancet, 368*(9533), 387–403. https://doi.org/10.1016/S0140-6736(06)69113-7

Blomeke, L., Pils, M., Kraemer-Schulien, V., Dybala, A., Schaffrath, A., Kulawik, A., et al. (2022). Quantitative detection of alpha-synuclein and tau oligomers and other aggregates by digital single particle counting. *npj Parkinson's Disease, 8*(1), 68. https://doi.org/10.1038/s41531-022-00330-x

Boccalini, C., Peretti, D. E., Ribaldi, F., Scheffler, M., Stampacchia, S., Tomczyk, S., … Garibotto, V. (2022). Early-phase ¹⁸F-Florbetapir and ¹⁸F-Flutemetamol images as proxies of brain metabolism in a memory clinic setting. Journal of Nuclear Medicine. doi:https://doi.org/10.2967/jnumed.122.264256.

Bonam, S. R., Wang, F., & Muller, S. (2019). Lysosomes as a therapeutic target. *Nature Reviews Drug Discovery, 18*(12), 923–948. https://doi.org/10.1038/s41573-019-0036-1

Bousiges, O., & Blanc, F. (2022). Biomarkers of dementia with Lewy bodies: Differential diagnostic with Alzheimer's disease. *International Journal of Molecular Sciences, 23*(12). https://doi.org/10.3390/ijms23126371

Braak, H., & Braak, E. (1991). Neuropathological stageing of Alzheimer-related changes. *Acta Neuropathologica, 82*(4), 239–259. https://doi.org/10.1007/BF00308809

Braak, H., & Del Tredici, K. (2016). Potential pathways of abnormal tau and alpha-Synuclein dissemination in sporadic Alzheimer's and Parkinson's diseases. *Cold Spring Harbor Perspectives in Biology, 8*(11). https://doi.org/10.1101/cshperspect.a023630

Broce, I. J., Tan, C. H., Fan, C. C., Jansen, I., Savage, J. E., Witoelar, A., et al. (2019). Dissecting the genetic relationship between cardiovascular risk factors and Alzheimer's disease. *Acta Neuropathologica, 137*(2), 209–226. https://doi.org/10.1007/s00401-018-1928-6

Brochard, V., Combadiere, B., Prigent, A., Laouar, Y., Perrin, A., Beray-Berthat, V., et al. (2009). Infiltration of CD4+ lymphocytes into the brain contributes to neurodegeneration in a mouse model of Parkinson disease. *The Journal of Clinical Investigation, 119*(1), 182–192. https://doi.org/10.1172/JCI36470

Brooks, D. J. (2010). Imaging approaches to Parkinson disease. *Journal of Nuclear Medicine, 51*(4), 596–609. https://doi.org/10.2967/jnumed.108.059998

Buchhave, P., Minthon, L., Zetterberg, H., Wallin, A. K., Blennow, K., & Hansson, O. (2012). Cerebrospinal fluid levels of beta-amyloid 1-42, but not of tau, are fully changed already 5 to 10 years before the onset of Alzheimer dementia. *Archives of General Psychiatry, 69*(1), 98–106. https://doi.org/10.1001/archgenpsychiatry.2011.155

Budd Haeberlein, S., Aisen, P. S., Barkhof, F., Chalkias, S., Chen, T., Cohen, S., et al. (2022). Two randomized phase 3 studies of Aducanumab in early Alzheimer's disease. *The Journal of Prevention of Alzheimer's Disease, 9*(2), 197–210. https://doi.org/10.14283/jpad.2022.30

Burke, S. L., Maramaldi, P., Cadet, T., & Kukull, W. (2016). Neuropsychiatric symptoms and Apolipoprotein E: Associations with eventual Alzheimer's disease development. *Archives of Gerontology and Geriatrics, 65*, 231–238. https://doi.org/10.1016/j.archger.2016.04.006

Caballero, B., Wang, Y., Diaz, A., Tasset, I., Juste, Y. R., Stiller, B., et al. (2018). Interplay of pathogenic forms of human tau with different autophagic pathways. *Aging Cell, 17*(1). https://doi.org/10.1111/acel.12692

Cao, S., Standaert, D. G., & Harms, A. S. (2012). The gamma chain subunit of fc receptors is required for alpha-synuclein-induced pro-inflammatory signaling in microglia. *Journal of Neuroinflammation, 9*, 259. https://doi.org/10.1186/1742-2094-9-259

Cerasa, A., Messina, D., Pugliese, P., Morelli, M., Lanza, P., Salsone, M., et al. (2011). Increased prefrontal volume in PD with levodopa-induced dyskinesias: A voxel-based morphometry study.

Movement Disorders, 26(5), 807–812. https://doi.org/10.1002/mds.23660

Chandra, S., Sisodia, S. S., & Vassar, R. J. (2023). The gut microbiome in Alzheimer's disease: What we know and what remains to be explored. *Molecular Neurodegeneration, 18*(1), 9. https://doi.org/10.1186/s13024-023-00595-7

Chao, Y., Wong, S. C., & Tan, E. K. (2014). Evidence of inflammatory system involvement in Parkinson's disease. *BioMed Research International, 2014*, 308654. https://doi.org/10.1155/2014/308654

Chen, J. E., & Glover, G. H. (2015). Functional magnetic resonance imaging methods. *Neuropsychology Review, 25*(3), 289–313. https://doi.org/10.1007/s11065-015-9294-9

Chen, H. H., Petty, L. E., Sha, J., Zhao, Y., Kuzma, A., Valladares, O., et al. (2021). Genetically regulated expression in late-onset Alzheimer's disease implicates risk genes within known and novel loci. *Translational Psychiatry, 11*(1), 618. https://doi.org/10.1038/s41398-021-01677-0

Chetelat, G., Arbizu, J., Barthel, H., Garibotto, V., Law, I., Morbelli, S., et al. (2020). Amyloid-PET and (18)F-FDG-PET in the diagnostic investigation of Alzheimer's disease and other dementias. *Lancet Neurology, 19*(11), 951–962. https://doi.org/10.1016/S1474-4422(20)30314-8

Chia, R., Sabir, M. S., Bandres-Ciga, S., Saez-Atienzar, S., Reynolds, R. H., Gustavsson, E., et al. (2021). Genome sequencing analysis identifies new loci associated with Lewy body dementia and provides insights into its genetic architecture. *Nature Genetics, 53*(3), 294–303. https://doi.org/10.1038/s41588-021-00785-3

Chiaravalloti, A., Martorana, A., Koch, G., Toniolo, S., di Biagio, D., di Pietro, B., & Schillaci, O. (2015). Functional correlates of t-tau, p-tau and $A\beta_{1-42}$ amyloid cerebrospinal fluid levels in Alzheimer's disease: A ^{18}F-FDG PET/CT study. *Nuclear Medicine Communications, 36*(5), 461–468. https://doi.org/10.1097/mnm.0000000000000272

Chiaravalloti, A., Barbagallo, G., Ricci, M., Martorana, A., Ursini, F., Sannino, P., et al. (2018). Brain metabolic correlates of CSF tau protein in a large cohort of Alzheimer's disease patients: A CSF and FDG PET study. *Brain Research, 1678*, 116–122. https://doi.org/10.1016/j.brainres.2017.10.016

Chincarini, A., Bosco, P., Calvini, P., Gemme, G., Esposito, M., Olivieri, C., et al. (2011). Local MRI analysis approach in the diagnosis of early and prodromal Alzheimer's disease. *NeuroImage, 58*(2), 469–480. https://doi.org/10.1016/j.neuroimage.2011.05.083

Cho, S. H., Choe, Y. S., Kim, Y. J., Kim, H. J., Jang, H., Kim, Y., et al. (2020). Head-to-head comparison of 18F-florbetaben and 18F-flutemetamol in the cortical and striatal regions. *Journal of Alzheimer's Disease, 76*, 281–290. https://doi.org/10.3233/JAD-200079

Cho, S. J., Bae, Y. J., Kim, J. M., Kim, H. J., Baik, S. H., Sunwoo, L., et al. (2021). Iron-sensitive magnetic resonance imaging in Parkinson's disease: A systematic review and meta-analysis. *Journal of Neurology, 268*(12), 4721–4736. https://doi.org/10.1007/s00415-021-10582-x

Ciaramella, A., Bizzoni, F., Salani, F., Vanni, D., Spalletta, G., Sanarico, N., et al. (2010). Increased pro-inflammatory response by dendritic cells from patients with Alzheimer's disease. *Journal of Alzheimer's Disease, 19*(2), 559–572. https://doi.org/10.3233/JAD-2010-1257

Ciaramella, A., Salani, F., Bizzoni, F., Orfei, M. D., Caltagirone, C., Spalletta, G., & Bossu, P. (2016). Myeloid dendritic cells are decreased in peripheral blood of Alzheimer's disease patients in association with disease progression and severity of depressive symptoms. *Journal of Neuroinflammation, 13*, 18. https://doi.org/10.1186/s12974-016-0483-0

Ciccocioppo, F., Lanuti, P., Pierdomenico, L., Simeone, P., Bologna, G., Ercolino, E., et al. (2019). The characterization of regulatory T-cell profiles in Alzheimer's disease and multiple sclerosis. *Scientific Reports, 9*(1), 8788. https://doi.org/10.1038/s41598-019-45433-3

Cicognola, C., Janelidze, S., Hertze, J., Zetterberg, H., Blennow, K., Mattsson-Carlgren, N., & Hansson, O. (2021). Plasma glial fibrillary acidic protein detects Alzheimer pathology and predicts future conversion to Alzheimer dementia in patients with mild cognitive impairment. *Alzheimer's Research & Therapy, 13*(1), 68. https://doi.org/10.1186/s13195-021-00804-9

Clark, C. M., Pontecorvo, M. J., Beach, T. G., Bedell, B. J., Coleman, R. E., Doraiswamy, P. M., et al. (2012). Cerebral PET with florbetapir compared with neuropathology at autopsy for detection of neuritic amyloid-beta plaques: A prospective cohort study. *Lancet Neurology, 11*(8), 669–678. https://doi.org/10.1016/S1474-4422(12)70142-4

Colonna, M., & Butovsky, O. (2017). Microglia function in the central nervous system during health and neurodegeneration. *Annual Review of Immunology, 35*, 441–468. https://doi.org/10.1146/annurev-immunol-051116-052358

Condello, C., Yuan, P., Schain, A., & Grutzendler, J. (2015). Microglia constitute a barrier that prevents neurotoxic protofibrillar Abeta42 hotspots around plaques. *Nature Communications, 6*, 6176. https://doi.org/10.1038/ncomms7176

Cova, I., & Priori, A. (2018). Diagnostic biomarkers for Parkinson's disease at a glance: Where are we? *Journal of Neural Transmission (Vienna), 125*(10), 1417–1432. https://doi.org/10.1007/s00702-018-1910-4

Croisier, E., Moran, L. B., Dexter, D. T., Pearce, R. K., & Graeber, M. B. (2005). Microglial inflammation in the parkinsonian substantia nigra: Relationship to alpha-synuclein deposition. *Journal of Neuroinflammation, 2*, 14. https://doi.org/10.1186/1742-2094-2-14

Cullen, N. C., Zetterberg, H., Insel, P. S., Olsson, B., Andreasson, U., Alzheimer's Disease Neuroimaging, I., et al. (2020). Comparing progression biomarkers in clinical trials of early Alzheimer's disease. *Annals of Clinical Translational Neurology, 7*(9), 1661–1673. https://doi.org/10.1002/acn3.51158

Cullen, N. C., Janelidze, S., Stomrud, E., Bateman, R. J., Palmqvist, S., Hansson, O., & Mattsson-Carlgren, N. (2023). Plasma amyloid-beta42/40 and apolipoprotein E for amyloid PET prescreening in secondary prevention trials of Alzheimer's disease. *Brain Communications, 5*(2), fcad015. https://doi.org/10.1093/braincomms/fcad015

Cunningham, E. L., McGuinness, B., Herron, B., & Passmore, A. P. (2015). Dementia. *The Ulster Medical Journal, 84*(2), 79–87.

Curtis, C., Gamez, J. E., Singh, U., Sadowsky, C. H., Villena, T., Sabbagh, M. N., et al. (2015). Phase 3 trial of flutemetamol labeled with radioactive fluorine 18 imaging and neuritic plaque density. *JAMA Neurology, 72*(3), 287–294. https://doi.org/10.1001/jamaneurol.2014.4144

Daniele, S., Frosini, D., Pietrobono, D., Petrozzi, L., Lo Gerfo, A., Baldacci, F., et al. (2018). Alpha-synuclein heterocomplexes with beta-amyloid are increased in red blood cells of Parkinson's disease patients and correlate with disease severity. *Frontiers in Molecular Neuroscience, 11*, 53. https://doi.org/10.3389/fnmol.2018.00053

De Strooper, B., & Karran, E. (2016). The cellular phase of Alzheimer's disease. *Cell, 164*(4), 603–615. https://doi.org/10.1016/j.cell.2015.12.056

De Strooper, B., Saftig, P., Craessaerts, K., Vanderstichele, H., Guhde, G., Annaert, W., et al. (1998). Deficiency of presenilin-1 inhibits the normal cleavage of amyloid precursor protein. *Nature, 391*(6665), 387–390. https://doi.org/10.1038/34910

Del-Aguila, J. L., Fernandez, M. V., Schindler, S., Ibanez, L., Deming, Y., Ma, S., et al. (2018). Assessment of the genetic architecture of Alzheimer's disease risk in rate of memory decline. *Journal of Alzheimer's Disease, 62*(2), 745–756. https://doi.org/10.3233/JAD-170834

Delva, A., Van Weehaeghe, D., van Aalst, J., Ceccarini, J., Koole, M., Baete, K., et al. (2020). Quantification and discriminative power of (18)F-FE-PE2I PET in patients with Parkinson's disease. *European*

Journal of Nuclear Medicine and Molecular Imaging, 47(8), 1913–1926. https://doi.org/10.1007/s00259-019-04587-y
Depboylu, C., Schafer, M. K., Arias-Carrion, O., Oertel, W. H., Weihe, E., & Hoglinger, G. U. (2011). Possible involvement of complement factor C1q in the clearance of extracellular neuromelanin from the substantia nigra in Parkinson disease. *Journal of Neuropathology and Experimental Neurology, 70*(2), 125–132. https://doi.org/10.1097/NEN.0b013e31820805b9
Desplats, P., Lee, H. J., Bae, E. J., Patrick, C., Rockenstein, E., Crews, L., et al. (2009). Inclusion formation and neuronal cell death through neuron-to-neuron transmission of alpha-synuclein. *Proceedings of the National Academy of Sciences of the United States of America, 106*(31), 13010–13015. https://doi.org/10.1073/pnas.0903691106
DeTure, M. A., & Dickson, D. W. (2019). The neuropathological diagnosis of Alzheimer's disease. *Molecular Neurodegeneration, 14*(1), 32. https://doi.org/10.1186/s13024-019-0333-5
Devanand, D. P., Lee, S., Huey, E. D., & Goldberg, T. E. (2022). Associations between neuropsychiatric symptoms and neuropathological diagnoses of Alzheimer disease and related dementias. *JAMA Psychiatry, 79*(4), 359–367. https://doi.org/10.1001/jamapsychiatry.2021.4363
Devos, D., Lebouvier, T., Lardeux, B., Biraud, M., Rouaud, T., Pouclet, H., et al. (2013). Colonic inflammation in Parkinson's disease. *Neurobiology of Disease, 50*, 42–48. https://doi.org/10.1016/j.nbd.2012.09.007
Dong, Y., Lagarde, J., Xicota, L., Corne, H., Chantran, Y., Chaigneau, T., et al. (2018). Neutrophil hyperactivation correlates with Alzheimer's disease progression. *Annals of Neurology, 83*(2), 387–405. https://doi.org/10.1002/ana.25159
Dos Santos, L. R., Pimassoni, L. H. S., Sena, G. G. S., Camporez, D., Belcavello, L., Trancozo, M., et al. (2017). Validating GWAS variants from microglial genes implicated in Alzheimer's disease. *Journal of Molecular Neuroscience, 62*(2), 215–221. https://doi.org/10.1007/s12031-017-0928-7
Edwards, T. L., Scott, W. K., Almonte, C., Burt, A., Powell, E. H., Beecham, G. W., et al. (2010). Genome-wide association study confirms SNPs in SNCA and the MAPT region as common risk factors for Parkinson disease. *Annals of Human Genetics, 74*(2), 97–109. https://doi.org/10.1111/j.1469-1809.2009.00560.x
Eikelboom, W. S., van der Berg, E., Singleton, E. H., Baart, S. J., Coesmans, M., Leeuwis, A. E., et al. (2021). Neuropsychiatric and cognitive symptoms across the Alzheimer disease clinical spectrum. *Cross-sectional and Longitudinal Associations, 97*(13), e1276–e1287. https://doi.org/10.1212/wnl.0000000000012598
Eikelboom, W. S., Pan, M., Ossenkoppele, R., Coesmans, M., Gatchel, J. R., Ismail, Z., et al. (2022). Sex differences in neuropsychiatric symptoms in Alzheimer's disease dementia: A meta-analysis. *Alzheimer's Research & Therapy, 14*(1), 48. https://doi.org/10.1186/s13195-022-00991-z
El-Agnaf, O. M., Salem, S. A., Paleologou, K. E., Cooper, L. J., Fullwood, N. J., Gibson, M. J., et al. (2003). Alpha-synuclein implicated in Parkinson's disease is present in extracellular biological fluids, including human plasma. *The FASEB Journal, 17*(13), 1945–1947. https://doi.org/10.1096/fj.03-0098fje
Elschot, E. P., Backes, W. H., Postma, A. A., van Oostenbrugge, R. J., Staals, J., Rouhl, R. P. W., & Jansen, J. F. A. (2021). A comprehensive view on MRI techniques for imaging blood-brain barrier integrity. *Investigative Radiology, 56*(1), 10–19. https://doi.org/10.1097/RLI.0000000000000723
Erbsloh, F., Bernsmeier, A., & Hillesheim, H. (1958). The glucose consumption of the brain & its dependence on the liver. *Archiv fur Psychiatrie und Nervenkrankheiten, vereinigt mit Zeitschrift fur die gesamte Neurologie und Psychiatrie, 196*(6), 611–626.
Fairfoul, G., McGuire, L. I., Pal, S., Ironside, J. W., Neumann, J., Christie, S., et al. (2016). Alpha-synuclein RT-QuIC in the CSF of patients with alpha-synucleinopathies. *Annals of Clinical Translational Neurology, 3*(10), 812–818. https://doi.org/10.1002/acn3.338
Faridar, A., Thome, A. D., Zhao, W., Thonhoff, J. R., Beers, D. R., Pascual, B., et al. (2020). Restoring regulatory T-cell dysfunction in Alzheimer's disease through ex vivo expansion. *Brain Communications, 2*(2), fcaa112. https://doi.org/10.1093/braincomms/fcaa112
Farrer, L. A., Cupples, L. A., Haines, J. L., Hyman, B., Kukull, W. A., Mayeux, R., et al. (1997). Effects of age, sex, and ethnicity on the association between apolipoprotein E genotype and Alzheimer disease. A meta-analysis. APOE and Alzheimer Disease Meta Analysis Consortium. *JAMA, 278*(16), 1349–1356. Retrieved from https://www.ncbi.nlm.nih.gov/pubmed/9343467
Ferrari, C. C., Depino, A. M., Prada, F., Muraro, N., Campbell, S., Podhajcer, O., et al. (2004). Reversible demyelination, blood-brain barrier breakdown, and pronounced neutrophil recruitment induced by chronic IL-1 expression in the brain. *The American Journal of Pathology, 165*(5), 1827–1837. https://doi.org/10.1016/s0002-9440(10)63438-4
Ferreira-Vieira, T. H., Guimaraes, I. M., Silva, F. R., & Ribeiro, F. M. (2016). Alzheimer's disease: Targeting the cholinergic system. *Current Neuropharmacology, 14*(1), 101–115. https://doi.org/10.2174/1570159x13666150716165726
Fleisher, A. S., Pontecorvo, M. J., Devous, M. D., Sr., Lu, M., Arora, A. K., Truocchio, S. P., et al. (2020). Positron emission tomography imaging with [18F]flortaucipir and postmortem assessment of Alzheimer disease neuropathologic changes. *JAMA Neurology, 77*(7), 829–839. https://doi.org/10.1001/jamaneurol.2020.0528
Foulds, P. G., Diggle, P., Mitchell, J. D., Parker, A., Hasegawa, M., Masuda-Suzukake, M., et al. (2013). A longitudinal study on alpha-synuclein in blood plasma as a biomarker for Parkinson's disease. *Scientific Reports, 3*, 2540. https://doi.org/10.1038/srep02540
Frizzell, T. O., Glashutter, M., Liu, C. C., Zeng, A., Pan, D., Hajra, S. G., et al. (2022). Artificial intelligence in brain MRI analysis of Alzheimer's disease over the past 12 years: A systematic review. *Ageing Research Reviews, 77*, 101614. https://doi.org/10.1016/j.arr.2022.101614
Fullard, M. E., Morley, J. F., & Duda, J. E. (2017). Olfactory dysfunction as an early biomarker in Parkinson's disease. *Neuroscience Bulletin, 33*(5), 515–525. https://doi.org/10.1007/s12264-017-0170-x
Gandhi, S., & Wood, N. W. (2010). Genome-wide association studies: The key to unlocking neurodegeneration? *Nature Neuroscience, 13*(7), 789–794. https://doi.org/10.1038/nn.2584
Gate, D., Saligrama, N., Leventhal, O., Yang, A. C., Unger, M. S., Middeldorp, J., et al. (2020). Clonally expanded CD8 T cells patrol the cerebrospinal fluid in Alzheimer's disease. *Nature, 577*(7790), 399–404. https://doi.org/10.1038/s41586-019-1895-7
Gatto, R. G., Carlos, A. F., Reichard, R. R., Lowe, V. J., Whitwell, J. L., & Josephs, K. A. (2023). Comparative assessment of regional tau distribution by tau-PET and Post-mortem neuropathology in a representative set of Alzheimer's & frontotemporal lobar degeneration patients. *PLoS One, 18*(5), e0284182. https://doi.org/10.1371/journal.pone.0284182
Gendelman, H. E., Zhang, Y., Santamaria, P., Olson, K. E., Schutt, C. R., Bhatti, D., et al. (2017). Evaluation of the safety and immunomodulatory effects of sargramostim in a randomized, double-blind phase 1 clinical Parkinson's disease trial. *npj Parkinson's Disease, 3*, 10. https://doi.org/10.1038/s41531-017-0013-5
Ghassaban, K., He, N., Sethi, S. K., Huang, P., Chen, S., Yan, F., & Haacke, E. M. (2019). Regional high iron in the substantia nigra differentiates Parkinson's disease patients from healthy controls. *Frontiers in Aging Neuroscience, 11*, 106. https://doi.org/10.3389/fnagi.2019.00106
Ginhoux, F., Greter, M., Leboeuf, M., Nandi, S., See, P., Gokhan, S., et al. (2010). Fate mapping analysis reveals that adult microglia

derive from primitive macrophages. *Science, 330*(6005), 841–845. https://doi.org/10.1126/science.1194637

Goedert, M., Jakes, R., & Spillantini, M. G. (2017). The synucleinopathies: Twenty years on. *Journal of Parkinson's Disease, 7*(s1), S51–S69. https://doi.org/10.3233/JPD-179005

Gordon, B. A., Blazey, T. M., Su, Y., Hari-Raj, A., Dincer, A., Flores, S., et al. (2018). Spatial patterns of neuroimaging biomarker change in individuals from families with autosomal dominant Alzheimer's disease: A longitudinal study. *Lancet Neurology, 17*(3), 241–250. https://doi.org/10.1016/S1474-4422(18)30028-0

Gratuze, M., Leyns, C. E. G., & Holtzman, D. M. (2018). New insights into the role of TREM2 in Alzheimer's disease. *Molecular Neurodegeneration, 13*(1), 66. https://doi.org/10.1186/s13024-018-0298-9

Greenland, J. C., & Barker, R. A. (2018). The differential diagnosis of Parkinson's disease. In T. B. Stoker & J. C. Greenland (Eds.), *Parkinson's disease: Pathogenesis and clinical aspects*. Exon Publications. https://exonpublications.com/index.php/exon/article/view/191

Griciuc, A., Serrano-Pozo, A., Parrado, A. R., Lesinski, A. N., Asselin, C. N., Mullin, K., et al. (2013). Alzheimer's disease risk gene CD33 inhibits microglial uptake of amyloid beta. *Neuron, 78*(4), 631–643. https://doi.org/10.1016/j.neuron.2013.04.014

Groot, C., Smith, R., Stomrud, E., Binette, A. P., Leuzy, A., Wuestefeld, A., et al. (2023). Phospho-tau with subthreshold tau-PET predicts increased tau accumulation rates in amyloid-positive individuals. *Brain, 146*(4), 1580–1591. https://doi.org/10.1093/brain/awac329

Guerreiro, R. J., Lohmann, E., Kinsella, E., Bras, J. M., Luu, N., Gurunlian, N., et al. (2012). Exome sequencing reveals an unexpected genetic cause of disease: NOTCH3 mutation in a Turkish family with Alzheimer's disease. *Neurobiology of Aging, 33*(5), 1008.e1017–1008.e1023. https://doi.org/10.1016/j.neurobiolaging.2011.10.009

Gunes, S., Aizawa, Y., Sugashi, T., Sugimoto, M., & Rodrigues, P. P. (2022). Biomarkers for Alzheimer's disease in the current state: A narrative review. *International Journal of Molecular Sciences, 23*(9). https://doi.org/10.3390/ijms23094962

Guo, S., Wang, H., & Yin, Y. (2022). Microglia polarization from M1 to M2 in neurodegenerative diseases. *Frontiers in Aging Neuroscience, 14*, 815347. https://doi.org/10.3389/fnagi.2022.815347

Haddad, D., & Nakamura, K. (2015). Understanding the susceptibility of dopamine neurons to mitochondrial stressors in Parkinson's disease. *FEBS Letters, 589*(24 Pt A), 3702–3713. https://doi.org/10.1016/j.febslet.2015.10.021

Hampel, H., Hu, Y., Hardy, J., Blennow, K., Chen, C., Perry, G., et al. (2023). The amyloid-beta pathway in Alzheimer's disease: A plain language summary. *Neurodegenerative Disease Management, 13*, 141. https://doi.org/10.2217/nmt-2022-0037

Hanagasi, H. A., Tufekcioglu, Z., & Emre, M. (2017). Dementia in Parkinson's disease. *Journal of the Neurological Sciences, 374*, 26–31. https://doi.org/10.1016/j.jns.2017.01.012

Hanes, J., Kovac, A., Kvartsberg, H., Kontsekova, E., Fialova, L., Katina, S., et al. (2020). Evaluation of a novel immunoassay to detect p-tau Thr217 in the CSF to distinguish Alzheimer disease from other dementias. *Neurology, 95*(22), e3026–e3035. https://doi.org/10.1212/WNL.0000000000010814

Hansen, D. V., Hanson, J. E., & Sheng, M. (2018). Microglia in Alzheimer's disease. *The Journal of Cell Biology, 217*(2), 459–472. https://doi.org/10.1083/jcb.201709069

Hansson, O. (2021). Biomarkers for neurodegenerative diseases. *Nature Medicine, 27*(6), 954–963. https://doi.org/10.1038/s41591-021-01382-x

Hansson, O., Batrla, R., Brix, B., Carrillo, M. C., Corradini, V., Edelmayer, R. M., et al. (2021a). The Alzheimer's Association international guidelines for handling of cerebrospinal fluid for routine clinical measurements of amyloid beta and tau. *Alzheimer's & Dementia, 17*(9), 1575–1582. https://doi.org/10.1002/alz.12316

Hansson, O., Cullen, N., Zetterberg, H., & Alzheimer's Disease Neuroimaging Initiative; Blennow, K., Mattsson-Carlgren, N. (2021b). Plasma phosphorylated tau181 and neurodegeneration in Alzheimer's disease. *Annals of Clinical Translational Neurology, 8*(1), 259–265. https://doi.org/10.1002/acn3.51253

Hansson, O., Edelmayer, R. M., Boxer, A. L., Carrillo, M. C., Mielke, M. M., Rabinovici, G. D., et al. (2022). The Alzheimer's Association appropriate use recommendations for blood biomarkers in Alzheimer's disease. *Alzheimer's & Dementia, 18*(12), 2669–2686. https://doi.org/10.1002/alz.12756

Hansson, O., Blennow, K., Zetterberg, H., & Dage, J. (2023). Blood biomarkers for Alzheimer's disease in clinical practice and trials. *Nature Aging, 3*(5), 506–519. https://doi.org/10.1038/s43587-023-00403-3

Harada, R., Okamura, N., Furumoto, S., Furukawa, K., Ishiki, A., Tomita, N., et al. (2016). 18F-THK5351: A novel PET radiotracer for imaging neurofibrillary pathology in Alzheimer disease. *Journal of Nuclear Medicine, 57*(2), 208–214. https://doi.org/10.2967/jnumed.115.164848

Hardy, J., & Selkoe, D. J. (2002). The amyloid hypothesis of Alzheimer's disease: Progress and problems on the road to therapeutics. *Science, 297*(5580), 353–356. https://doi.org/10.1126/science.1072994

Harms, A. S., Cao, S., Rowse, A. L., Thome, A. D., Li, X., Mangieri, L. R., et al. (2013). MHCII is required for alpha-synuclein-induced activation of microglia, CD4 T cell proliferation, and dopaminergic neurodegeneration. *The Journal of Neuroscience, 33*(23), 9592–9600. https://doi.org/10.1523/JNEUROSCI.5610-12.2013

He, R., Yan, X., Guo, J., Xu, Q., Tang, B., & Sun, Q. (2018). Recent advances in biomarkers for Parkinson's disease. *Frontiers in Aging Neuroscience, 10*, 305. https://doi.org/10.3389/fnagi.2018.00305

He, R., Zhao, Y., He, Y., Zhou, Y., Yang, J., Zhou, X., et al. (2020). Olfactory dysfunction predicts disease progression in Parkinson's disease: A longitudinal study. *Frontiers in Neuroscience, 14*, 569777. https://doi.org/10.3389/fnins.2020.569777

Heneka, M. T., Carson, M. J., El Khoury, J., Landreth, G. E., Brosseron, F., Feinstein, D. L., et al. (2015). Neuroinflammation in Alzheimer's disease. *Lancet Neurology, 14*(4), 388–405. https://doi.org/10.1016/S1474-4422(15)70016-5

Herrmann, Y., Bujnicki, T., Zafiu, C., Kulawik, A., Kuhbach, K., Peters, L., et al. (2017). Nanoparticle standards for immuno-based quantitation of alpha-synuclein oligomers in diagnostics of Parkinson's disease and other synucleinopathies. *Clinica Chimica Acta, 466*, 152–159. https://doi.org/10.1016/j.cca.2017.01.010

Hess, C. W., & Okun, M. S. (2016). Diagnosing Parkinson disease. *Continuum, 22*(4 Movement Disorders), 1047–1063. https://doi.org/10.1212/CON.0000000000000345

Hofmann, K. W., Schuh, A. F., Saute, J., Townsend, R., Fricke, D., Leke, R., et al. (2009). Interleukin-6 serum levels in patients with Parkinson's disease. *Neurochemical Research, 34*(8), 1401–1404. https://doi.org/10.1007/s11064-009-9921-z

Holmans, P., Moskvina, V., Jones, L., Sharma, M., International Parkinson's Disease Genomics, C., Vedernikov, A., et al. (2013). A pathway-based analysis provides additional support for an immune-related genetic susceptibility to Parkinson's disease. *Human Molecular Genetics, 22*(5), 1039–1049. https://doi.org/10.1093/hmg/dds492

https://www.cdc.gov/genomics/hugenet/casestudy/parkinson/parkcoffee_view.htm. Accessed 5 Sept 2023.

https://www.parkinson.org/understanding-parkinsons/statistics. Accessed 5 Sept 2023.

https://www.who.int/news/item/14-06-2022-launch-of-who-s-parkinson-disease-technical-brief#:~:text=Disability%20and%20death%20due%20to,of%20over%20100%25%20since%202000. Accessed 5 Sept 2023.

Huang, Y., Liu, Z., Cao, B. B., Qiu, Y. H., & Peng, Y. P. (2020). Treg cells attenuate neuroinflammation and protect neurons in a mouse model of Parkinson's disease. *Journal of Neuroimmune Pharmacology, 15*(2), 224–237. https://doi.org/10.1007/s11481-019-09888-5

Huang, B., Zhenxin, Y., Chen, S., Tan, Z., Zong, Z., Zhang, H., & Xiong, X. (2022a). The innate and adaptive immune cells in Alzheimer's and Parkinson's diseases. *Oxidative Medicine and Cellular Longevity, 2022*, 1315248. https://doi.org/10.1155/2022/1315248

Huang, L. T., Zhang, C. P., Wang, Y. B., & Wang, J. H. (2022b). Association of peripheral blood cell profile with Alzheimer's disease: A meta-analysis. *Frontiers in Aging Neuroscience, 14*, 888946. https://doi.org/10.3389/fnagi.2022.888946

Hunter, C. A., Kirson, N. Y., Desai, U., Cummings, A. K., Faries, D. E., & Birnbaum, H. G. (2015). Medical costs of Alzheimer's disease misdiagnosis among US Medicare beneficiaries. *Alzheimers Dement, 11*(8), 887–895. https://doi.org/10.1016/j.jalz.2015.06.1889

Iranzo, A., Tolosa, E., Gelpi, E., Molinuevo, J. L., Valldeoriola, F., Serradell, M., et al. (2013). Neurodegenerative disease status and post-mortem pathology in idiopathic rapid-eye-movement sleep behaviour disorder: An observational cohort study. *Lancet Neurology, 12*(5), 443–453. https://doi.org/10.1016/S1474-4422(13)70056-5

Jack, C. R., Jr., Bennett, D. A., Blennow, K., Carrillo, M. C., Dunn, B., Haeberlein, S. B., et al. (2018). NIA-AA research framework: Toward a biological definition of Alzheimer's disease. *Alzheimers Dement, 14*(4), 535–562. https://doi.org/10.1016/j.jalz.2018.02.018

Jakobson Mo, S., Axelsson, J., Jonasson, L., Larsson, A., Ögren, M. J., Ögren, M., et al. (2018). Dopamine transporter imaging with [(18)F]FE-PE2I PET and [(123)I]FP-CIT SPECT-a clinical comparison. *EJNMMI Research, 8*(1), 100. https://doi.org/10.1186/s13550-018-0450-0

Janelidze, S., Stomrud, E., Smith, R., Palmqvist, S., Mattsson, N., Airey, D. C., et al. (2020). Cerebrospinal fluid p-tau217 performs better than p-tau181 as a biomarker of Alzheimer's disease. *Nature Communications, 11*(1), 1683. https://doi.org/10.1038/s41467-020-15436-0

Janelidze, S., Teunissen, C. E., Zetterberg, H., Allue, J. A., Sarasa, L., Eichenlaub, U., et al. (2021). Head-to-head comparison of 8 plasma amyloid-beta 42/40 assays in Alzheimer disease. *JAMA Neurology, 78*(11), 1375–1382. https://doi.org/10.1001/jamaneurol.2021.3180

Janelidze, S., Bali, D., Ashton, N. J., Barthelemy, N. R., Vanbrabant, J., Stoops, E., et al. (2023). Head-to-head comparison of 10 plasma phospho-tau assays in prodromal Alzheimer's disease. *Brain, 146*(4), 1592–1601. https://doi.org/10.1093/brain/awac333

Jankovic, J. (2008). Parkinson's disease: Clinical features and diagnosis. *Journal of Neurology, Neurosurgery, and Psychiatry, 79*(4), 368–376. https://doi.org/10.1136/jnnp.2007.131045

Jarmolowicz, A. I., Chen, H. Y., & Panegyres, P. K. (2015). The patterns of inheritance in early-onset dementia: Alzheimer's disease and frontotemporal dementia. *American Journal of Alzheimer's Disease and Other Dementias, 30*(3), 299–306. https://doi.org/10.1177/1533317514545825

Jellinger, K. A., Logroscino, G., Rizzo, G., Copetti, M., Arcuti, S., Martino, D., & Fontana, A. (2016). Accuracy of clinical diagnosis of Parkinson disease: A systematic review and meta-analysis. *Neurology, 87*(2), 237–238. https://doi.org/10.1212/WNL.0000000000002876

Jiang, T., Yu, J. T., Zhu, X. C., & Tan, L. (2013). TREM2 in Alzheimer's disease. *Molecular Neurobiology, 48*(1), 180–185. https://doi.org/10.1007/s12035-013-8424-8

Jiang, T., Tan, L., Zhu, X. C., Zhang, Q. Q., Cao, L., Tan, M. S., et al. (2014). Upregulation of TREM2 ameliorates neuropathology and rescues spatial cognitive impairment in a transgenic mouse model of Alzheimer's disease. *Neuropsychopharmacology, 39*(13), 2949–2962. https://doi.org/10.1038/npp.2014.164

Jie, C., Treyer, V., Schibli, R., & Mu, L. (2021). Tauvid™: The first FDA-approved PET tracer for imaging tau pathology in Alzheimer's disease. *Pharmaceuticals (Basel), 14*(2). https://doi.org/10.3390/ph14020110

Jimenez, S., Baglietto-Vargas, D., Caballero, C., Moreno-Gonzalez, I., Torres, M., Sanchez-Varo, R., et al. (2008). Inflammatory response in the hippocampus of PS1M146L/APP751SL mouse model of Alzheimer's disease: Age-dependent switch in the microglial phenotype from alternative to classic. *The Journal of Neuroscience, 28*(45), 11650–11661. https://doi.org/10.1523/JNEUROSCI.3024-08.2008

Jinn, S., Drolet, R. E., Cramer, P. E., Wong, A. H., Toolan, D. M., Gretzula, C. A., et al. (2017). TMEM175 deficiency impairs lysosomal and mitochondrial function and increases alpha-synuclein aggregation. *Proceedings of the National Academy of Sciences of the United States of America, 114*(9), 2389–2394. https://doi.org/10.1073/pnas.1616332114

Kalia, L. V., & Lang, A. E. (2015). Parkinson's disease. *Lancet, 386*(9996), 896–912. https://doi.org/10.1016/S0140-6736(14)61393-3

Kandiah, N., Zainal, N. H., Narasimhalu, K., Chander, R. J., Ng, A., Mak, E., et al. (2014). Hippocampal volume and white matter disease in the prediction of dementia in Parkinson's disease. *Parkinsonism & Related Disorders, 20*(11), 1203–1208. https://doi.org/10.1016/j.parkreldis.2014.08.024

Kaneko, N., Kurata, M., Yamamoto, T., Morikawa, S., & Masumoto, J. (2019). The role of interleukin-1 in general pathology. *Inflammation and Regeneration, 39*(1), 12. https://doi.org/10.1186/s41232-019-0101-5

Kang, J., Lemaire, H. G., Unterbeck, A., Salbaum, J. M., Masters, C. L., Grzeschik, K. H., et al. (1987). The precursor of Alzheimer's disease amyloid A4 protein resembles a cell-surface receptor. *Nature, 325*(6106), 733–736. https://doi.org/10.1038/325733a0

Kang, U. J., Boehme, A. K., Fairfoul, G., Shahnawaz, M., Ma, T. C., Hutten, S. J., et al. (2019). Comparative study of cerebrospinal fluid alpha-synuclein seeding aggregation assays for diagnosis of Parkinson's disease. *Movement Disorders, 34*(4), 536–544. https://doi.org/10.1002/mds.27646

Karikari, T. K., Ashton, N. J., Brinkmalm, G., Brum, W. S., Benedet, A. L., Montoliu-Gaya, L., et al. (2022). Blood phospho-tau in Alzheimer disease: Analysis, interpretation, and clinical utility. *Nature Reviews. Neurology, 18*(7), 400–418. https://doi.org/10.1038/s41582-022-00665-2

Karpenko, M. N., Vasilishina, A. A., Gromova, E. A., Muruzheva, Z. M., Miliukhina, I. V., & Bernadotte, A. (2018). Interleukin-1beta, interleukin-1 receptor antagonist, interleukin-6, interleukin-10, and tumor necrosis factor-alpha levels in CSF and serum in relation to the clinical diversity of Parkinson's disease. *Cellular Immunology, 327*, 77–82. https://doi.org/10.1016/j.cellimm.2018.02.011

Kempuraj, D., Thangavel, R., Natteru, P. A., Selvakumar, G. P., Saeed, D., Zahoor, H., et al. (2016). Neuroinflammation induces neurodegeneration. *Journal of Neurosurgery: Spine, 1*(1) Retrieved from https://www.ncbi.nlm.nih.gov/pubmed/28127589

Kesika, P., Suganthy, N., Sivamaruthi, B. S., & Chaiyasut, C. (2021). Role of gut-brain axis, gut microbial composition, and probiotic intervention in Alzheimer's disease. *Life Sciences, 264*, 118627. https://doi.org/10.1016/j.lfs.2020.118627

Kimura, A., Hata, S., & Suzuki, T. (2016). Alternative selection of beta-site APP-cleaving enzyme 1 (BACE1) cleavage sites in amyloid beta-protein precursor (APP) harboring protective and pathogenic mutations within the abeta sequence. *The Journal of Biological Chemistry, 291*(46), 24041–24053. https://doi.org/10.1074/jbc.M116.744722

Klunk, W. E., Engler, H., Nordberg, A., Wang, Y., Blomqvist, G., Holt, D. P., et al. (2004). Imaging brain amyloid in Alzheimer's disease with Pittsburgh compound-B. *Annals of Neurology, 55*(3), 306–319. https://doi.org/10.1002/ana.20009

Koga, S., Sekiya, H., Kondru, N., Ross, O. A., & Dickson, D. W. (2021). Neuropathology and molecular diagnosis of synucleinopathies. *Molecular Neurodegeneration, 16*(1), 83. https://doi.org/10.1186/s13024-021-00501-z

Konickova, D., Mensikova, K., Tuckova, L., Henykova, E., Strnad, M., Friedecky, D., et al. (2022). Biomarkers of neurodegenerative diseases: Biology, taxonomy, clinical relevance, and current research status. *Biomedicine, 10*(7). https://doi.org/10.3390/biomedicines10071760

Kosloski, L. M., Kosmacek, E. A., Olson, K. E., Mosley, R. L., & Gendelman, H. E. (2013). GM-CSF induces neuroprotective and anti-inflammatory responses in 1-methyl-4-phenyl-1,2,3,6-tetrahydropyridine intoxicated mice. *Journal of Neuroimmunology, 265*(1–2), 1–10. https://doi.org/10.1016/j.jneuroim.2013.10.009

Kouli, A., Torsney, K. M., & Kuan, W. L. (2018). Parkinson's disease: Etiology, neuropathology, and pathogenesis. In T. B. Stoker & J. C. Greenland (Eds.), *Parkinson's disease: Pathogenesis and clinical aspects*. Exon Publications. https://exonpublications.com/index.php/exon/article/view/191

Kovacs, G. G. (2019). Molecular pathology of neurodegenerative diseases: Principles and practice. *Journal of Clinical Pathology, 72*(11), 725–735. https://doi.org/10.1136/jclinpath-2019-205952

Kuhlmann, J., Andreasson, U., Pannee, J., Bjerke, M., Portelius, E., Leinenbach, A., et al. (2017). CSF Abeta(1-42) – An excellent but complicated Alzheimer's biomarker – A route to standardisation. *Clinica Chimica Acta, 467*, 27–33. https://doi.org/10.1016/j.cca.2016.05.014

Kurkowska-Jastrzebska, I., Wronska, A., Kohutnicka, M., Czlonkowski, A., & Czlonkowska, A. (1999). MHC class II positive microglia and lymphocytic infiltration are present in the substantia nigra and striatum in mouse model of Parkinson's disease. *Acta Neurobiologiae Experimentalis (Wars), 59*(1), 1–8. Retrieved from https://www.ncbi.nlm.nih.gov/pubmed/10230070

Kustrimovic, N., Comi, C., Magistrelli, L., Rasini, E., Legnaro, M., Bombelli, R., et al. (2018). Parkinson's disease patients have a complex phenotypic and functional Th1 bias: Cross-sectional studies of CD4+ Th1/Th2/T17 and Treg in drug-naive and drug-treated patients. *Journal of Neuroinflammation, 15*(1), 205. https://doi.org/10.1186/s12974-018-1248-8

Laliberte, J., Marquez, V. E., & Momparler, R. L. (1992). Potent inhibitors for the deamination of cytosine arabinoside and 5-aza-2′-deoxycytidine by human cytidine deaminase. *Cancer Chemotherapy and Pharmacology, 30*(1), 7–11. https://doi.org/10.1007/BF00686478

Larbi, A., Pawelec, G., Witkowski, J. M., Schipper, H. M., Derhovanessian, E., Goldeck, D., & Fulop, T. (2009). Dramatic shifts in circulating CD4 but not CD8 T cell subsets in mild Alzheimer's disease. *Journal of Alzheimer's Disease, 17*(1), 91–103. https://doi.org/10.3233/JAD-2009-1015

Le Page, A., Bourgade, K., Lamoureux, J., Frost, E., Pawelec, G., Larbi, A., et al. (2015). NK cells are activated in amnestic mild cognitive impairment but not in mild Alzheimer's disease patients. *Journal of Alzheimer's Disease, 46*(1), 93–107. https://doi.org/10.3233/JAD-143054

Lee, J. E., Cho, K. H., Song, S. K., Kim, H. J., Lee, H. S., Sohn, Y. H., & Lee, P. H. (2014). Exploratory analysis of neuropsychological and neuroanatomical correlates of progressive mild cognitive impairment in Parkinson's disease. *Journal of Neurology, Neurosurgery, and Psychiatry, 85*(1), 7–16. https://doi.org/10.1136/jnnp-2013-305062

Leuzy, A., Janelidze, S., Mattsson-Carlgren, N., Palmqvist, S., Jacobs, D., Cicognola, C., et al. (2021). Comparing the Clinical utility and diagnostic performance of CSF P-Tau181, P-Tau217, and P-Tau231 assays. *Neurology, 97*(17), e1681–e1694. https://doi.org/10.1212/WNL.0000000000012727

Leuzy, A., Mattsson-Carlgren, N., Palmqvist, S., Janelidze, S., Dage, J. L., & Hansson, O. (2022). Blood-based biomarkers for Alzheimer's disease. *EMBO Molecular Medicine, 14*(1), e14408. https://doi.org/10.15252/emmm.202114408

Li, J., Zhang, Q., Chen, F., Yan, J., Kim, S., Wang, L., et al. (2015). Genetic interactions explain variance in cingulate amyloid burden: An AV-45 PET genome-wide association and interaction study in the ADNI cohort. *BioMed Research International, 2015*, 647389. https://doi.org/10.1155/2015/647389

Li, W., Lin, X., & Chen, X. (2020). Detecting Alzheimer's disease based on 4D fMRI: An exploration under deep learning framework. *Neurocomputing, 388*, 280–287. https://doi.org/10.1016/j.neucom.2020.01.053

Li, Z., Zhang, Z., Ren, Y., Wang, Y., Fang, J., Yue, H., et al. (2021). Aging and age-related diseases: From mechanisms to therapeutic strategies. *Biogerontology, 22*(2), 165–187. https://doi.org/10.1007/s10522-021-09910-5

Li, Y., Schindler, S. E., Bollinger, J. G., Ovod, V., Mawuenyega, K. G., Weiner, M. W., et al. (2022). Validation of plasma amyloid-beta 42/40 for detecting Alzheimer disease amyloid plaques. *Neurology, 98*(7), e688–e699. https://doi.org/10.1212/WNL.0000000000013211

Librizzi, D., Cabanel, N., Zavorotnyy, M., Riehl, E., Kircher, T., Luster, M., & Hooshyar Yousefi, B. (2021). Clinical relevance of [(18)F] Florbetaben and [(18)F]FDG PET/CT imaging on the management of patients with dementia. *Molecules, 26*(5). https://doi.org/10.3390/molecules26051282

Lill, C. M., & Bertram, L. (2011). Towards unveiling the genetics of neurodegenerative diseases. *Seminars in Neurology, 31*(5), 531–541. https://doi.org/10.1055/s-0031-1299791

Lill, C. M., Rengmark, A., Pihlstrom, L., Fogh, I., Shatunov, A., Sleiman, P. M., et al. (2015). The role of TREM2 R47H as a risk factor for Alzheimer's disease, frontotemporal lobar degeneration, amyotrophic lateral sclerosis, and Parkinson's disease. *Alzheimer's & Dementia, 11*(12), 1407–1416. https://doi.org/10.1016/j.jalz.2014.12.009

Lim, J. E., Kou, J., Song, M., Pattanayak, A., Jin, J., Lalonde, R., & Fukuchi, K. (2011). MyD88 deficiency ameliorates beta-amyloidosis in an animal model of Alzheimer's disease. *The American Journal of Pathology, 179*(3), 1095–1103. https://doi.org/10.1016/j.ajpath.2011.05.045

Lindestam Arlehamn, C. S., Dhanwani, R., Pham, J., Kuan, R., Frazier, A., Rezende Dutra, J., et al. (2020). alpha-Synuclein-specific T cell reactivity is associated with preclinical and early Parkinson's disease. *Nature Communications, 11*(1), 1875. https://doi.org/10.1038/s41467-020-15626-w

Lombardi, G., Crescioli, G., Cavedo, E., Lucenteforte, E., Casazza, G., Bellatorre, A. G., et al. (2020). Structural magnetic resonance imaging for the early diagnosis of dementia due to Alzheimer's disease in people with mild cognitive impairment. *Cochrane Database of Systematic Reviews, 3*(3), CD009628. https://doi.org/10.1002/14651858.CD009628.pub2

Luan, M., Sun, Y., Chen, J., Jiang, Y., Li, F., Wei, L., et al. (2022). Diagnostic value of salivary real-time quaking-induced conversion in Parkinson's disease and multiple system atrophy. *Movement Disorders, 37*(5), 1059–1063. https://doi.org/10.1002/mds.28976

Lutz, M. W., Sprague, D., Barrera, J., & Chiba-Falek, O. (2020). Shared genetic etiology underlying Alzheimer's disease and major depressive disorder. *Translational Psychiatry, 10*(1), 88. https://doi.org/10.1038/s41398-020-0769-y

Lv, H., Wang, Z., Tong, E., Williams, L. M., Zaharchuk, G., Zeineh, M., et al. (2018). Resting-state functional MRI: Everything that nonexperts have always wanted to know. *AJNR. American Journal of Neuroradiology, 39*(8), 1390–1399. https://doi.org/10.3174/ajnr.A5527

Machhi, J., Yeapuri, P., Lu, Y., Foster, E., Chikhale, R., Herskovitz, J., et al. (2021). CD4+ effector T cells accelerate Alzheimer's disease in mice. *Journal of Neuroinflammation, 18*(1), 272. https://doi.org/10.1186/s12974-021-02308-7

Malm, T., Koistinaho, M., Muona, A., Magga, J., & Koistinaho, J. (2010). The role and therapeutic potential of monocytic cells in Alzheimer's disease. *Glia, 58*(8), 889–900. https://doi.org/10.1002/glia.20973

Manne, S., Kondru, N., Jin, H., Anantharam, V., Huang, X., Kanthasamy, A., & Kanthasamy, A. G. (2020a). α-Synuclein real-time quaking-induced conversion in the submandibular glands of Parkinson's disease patients. *Movement Disorders, 35*(2), 268–278. https://doi.org/10.1002/mds.27907

Manne, S., Kondru, N., Jin, H., Serrano, G. E., Anantharam, V., Kanthasamy, A., et al. (2020b). Blinded RT-QuIC analysis of alpha-synuclein biomarker in skin tissue from Parkinson's disease patients. *Movement Disorders, 35*(12), 2230–2239. https://doi.org/10.1002/mds.28242

Mantel, E., & Williams, J. (2019). An introduction to newer PET diagnostic agents and related therapeutic radiopharmaceuticals. *Journal of Nuclear Medicine Technology, 47*(3), 203–209. https://doi.org/10.2967/jnmt.118.224022

Marcus, C., Mena, E., & Subramaniam, R. M. (2014). Brain PET in the diagnosis of Alzheimer's disease. *Clinical Nuclear Medicine, 39*(10), e413-422; quiz e423-416. https://doi.org/10.1097/rlu.0000000000000547

Martínez, G., Vernooij, R. W., Fuentes Padilla, P., Zamora, J., Flicker, L., & Bonfill Cosp, X. (2017). 18F PET with flutemetamol for the early diagnosis of Alzheimer's disease dementia and other dementias in people with mild cognitive impairment (MCI). *Cochrane Database of Systematic Reviews, 11*(11), Cd012884. https://doi.org/10.1002/14651858.Cd012884

Masters, C. L., Bateman, R., Blennow, K., Rowe, C. C., Sperling, R. A., & Cummings, J. L. (2015). Alzheimer's disease. *Nature Reviews. Disease Primers, 1*, 15056. https://doi.org/10.1038/nrdp.2015.56

Mattsson-Carlgren, N., Andersson, E., Janelidze, S., Ossenkoppele, R., Insel, P., Strandberg, O., et al. (2020a). Abeta deposition is associated with increases in soluble and phosphorylated tau that precede a positive Tau PET in Alzheimer's disease. *Science Advances, 6*(16), eaaz2387. https://doi.org/10.1126/sciadv.aaz2387

Mattsson-Carlgren, N., Janelidze, S., Palmqvist, S., Cullen, N., Svenningsson, A. L., Strandberg, O., et al. (2020b). Longitudinal plasma p-tau217 is increased in early stages of Alzheimer's disease. *Brain, 143*(11), 3234–3241. https://doi.org/10.1093/brain/awaa286

Mattsson-Carlgren, N., Leuzy, A., Janelidze, S., Palmqvist, S., Stomrud, E., Strandberg, O., et al. (2020c). The implications of different approaches to define AT(N) in Alzheimer disease. *Neurology, 94*(21), e2233–e2244. https://doi.org/10.1212/WNL.0000000000009485

Mattsson-Carlgren, N., Janelidze, S., Bateman, R. J., Smith, R., Stomrud, E., Serrano, G. E., et al. (2021). Soluble P-tau217 reflects amyloid and tau pathology and mediates the association of amyloid with tau. *EMBO Molecular Medicine, 13*(6), e14022. https://doi.org/10.15252/emmm.202114022

Mattsson-Carlgren, N., Salvado, G., Ashton, N. J., Tideman, P., Stomrud, E., Zetterberg, H., et al. (2023). Prediction of longitudinal cognitive decline in preclinical Alzheimer disease using plasma biomarkers. *JAMA Neurology, 80*(4), 360–369. https://doi.org/10.1001/jamaneurol.2022.5272

Matuskey, D., Tinaz, S., Wilcox, K. C., Naganawa, M., Toyonaga, T., Dias, M., et al. (2020). Synaptic changes in Parkinson disease assessed with in vivo imaging. *Annals of Neurology, 87*(3), 329–338. https://doi.org/10.1002/ana.25682

McGeer, P. L., Itagaki, S., Boyes, B. E., & McGeer, E. G. (1988). Reactive microglia are positive for HLA-DR in the substantia nigra of Parkinson's and Alzheimer's disease brains. *Neurology, 38*(8), 1285–1291. https://doi.org/10.1212/wnl.38.8.1285

Mecca, A. P., Chen, M. K., O'Dell, R. S., Naganawa, M., Toyonaga, T., Godek, T. A., et al. (2020). In vivo measurement of widespread synaptic loss in Alzheimer's disease with SV2A PET. *Alzheimer's & Dementia, 16*(7), 974–982. https://doi.org/10.1002/alz.12097

Mergenthaler, P., Lindauer, U., Dienel, G. A., & Meisel, A. (2013). Sugar for the brain: The role of glucose in physiological and pathological brain function. *Trends in Neurosciences, 36*(10), 587–597. https://doi.org/10.1016/j.tins.2013.07.001

Meyer, P. T., Frings, L., Rucker, G., & Hellwig, S. (2017). (18)F-FDG PET in parkinsonism: Differential diagnosis and evaluation of cognitive impairment. *Journal of Nuclear Medicine, 58*(12), 1888–1898. https://doi.org/10.2967/jnumed.116.186403

Mielke, M. M., Hagen, C. E., Xu, J., Chai, X., Vemuri, P., Lowe, V. J., et al. (2018). Plasma phospho-tau181 increases with Alzheimer's disease clinical severity and is associated with tau- and amyloid-positron emission tomography. *Alzheimer's & Dementia, 14*(8), 989–997. https://doi.org/10.1016/j.jalz.2018.02.013

Mielke, M. M., Frank, R. D., Dage, J. L., Jeromin, A., Ashton, N. J., Blennow, K., et al. (2021). Comparison of plasma phosphorylated tau species with amyloid and tau positron emission tomography, neurodegeneration, vascular pathology, and cognitive outcomes. *JAMA Neurology, 78*(9), 1108–1117. https://doi.org/10.1001/jamaneurol.2021.2293

Migdalska-Richards, A., & Schapira, A. H. (2016). The relationship between glucocerebrosidase mutations and Parkinson disease. *Journal of Neurochemistry, 139 Suppl 1*(Suppl Suppl 1), 77–90. https://doi.org/10.1111/jnc.13385

Mindell, J. A. (2012). Lysosomal acidification mechanisms. *Annual Review of Physiology, 74*(1), 69–86. https://doi.org/10.1146/annurev-physiol-012110-142317

Mishra, S., Knupp, A., Szabo, M. P., Williams, C. A., Kinoshita, C., Hailey, D. W., et al. (2022). The Alzheimer's gene SORL1 is a regulator of endosomal traffic and recycling in human neurons. *Cellular and Molecular Life Sciences, 79*(3), 162. https://doi.org/10.1007/s00018-022-04182-9

Mosconi, L. (2005). Brain glucose metabolism in the early and specific diagnosis of Alzheimer's disease. *European Journal of Nuclear Medicine and Molecular Imaging, 32*(4), 486–510. https://doi.org/10.1007/s00259-005-1762-7

Mosconi, L. (2013). Glucose metabolism in normal aging and Alzheimer's disease: Methodological and physiological considerations for PET studies. *Clinical and Translational Imaging, 1*(4). https://doi.org/10.1007/s40336-013-0026-y

Moscoso, A., Grothe, M. J., Ashton, N. J., Karikari, T. K., Rodriguez, J. L., Snellman, A., et al. (2021). Time course of phosphorylated-tau181 in blood across the Alzheimer's disease spectrum. *Brain, 144*(1), 325–339. https://doi.org/10.1093/brain/awaa399

Mosley, R. L., & Gendelman, H. E. (2017). T cells and Parkinson's disease. *Lancet Neurology, 16*(10), 769–771. https://doi.org/10.1016/S1474-4422(17)30276-4

Muller, U. C., Deller, T., & Korte, M. (2017). Not just amyloid: Physiological functions of the amyloid precursor protein family. *Nature Reviews. Neuroscience, 18*(5), 281–298. https://doi.org/10.1038/nrn.2017.29

Nagatsu, T., Mogi, M., Ichinose, H., & Togari, A. (2000). Changes in cytokines and neurotrophins in Parkinson's disease. *Journal of Neural Transmission, Suppl*(60), 277–290. https://doi.org/10.1007/978-3-7091-6301-6_19

Nauseef, W. M., & Borregaard, N. (2014). Neutrophils at work. *Nature Immunology, 15*(7), 602–611. https://doi.org/10.1038/ni.2921

Noch, E. K., & Khalili, K. (2013). The role of AEG-1/MTDH/LYRIC in the pathogenesis of central nervous system disease. *Advances in Cancer Research, 120*, 159–192. https://doi.org/10.1016/B978-0-12-401676-7.00006-1

Nury, T., Lizard, G., & Vejux, A. (2020). Lipids nutrients in Parkinson and Alzheimer's diseases: Cell death and cytoprotection. *International Journal of Molecular Sciences, 21*(7). https://doi.org/10.3390/ijms21072501

Okamura, N., & Yanai, K. (2010). Florbetapir (18F), a PET imaging agent that binds to amyloid plaques for the potential detection of Alzheimer's disease. *IDrugs, 13*(12), 890–899.

Okamura, N., Furumoto, S., Harada, R., Tago, T., Yoshikawa, T., Fodero-Tavoletti, M., et al. (2013). Novel 18F-labeled arylquinoline derivatives for noninvasive imaging of tau pathology in Alzheimer disease. *Journal of Nuclear Medicine, 54*(8), 1420–1427.

Olson, K. E., Kosloski-Bilek, L. M., Anderson, K. M., Diggs, B. J., Clark, B. E., Gledhill, J. M., Jr., et al. (2015). Selective VIP receptor agonists facilitate immune transformation for dopaminergic neuroprotection in MPTP-intoxicated mice. *The Journal of Neuroscience, 35*(50), 16463–16478. https://doi.org/10.1523/JNEUROSCI.2131-15.2015

Olson, K. E., Namminga, K. L., Schwab, A. D., Thurston, M. J., Lu, Y., Woods, A., et al. (2020). Neuroprotective activities of long-acting granulocyte-macrophage colony-stimulating factor (mPDM608) in 1-methyl-4-phenyl-1,2,3,6-tetrahydropyridine-intoxicated mice. *Neurotherapeutics, 17*(4), 1861–1877. https://doi.org/10.1007/s13311-020-00877-8

Olson, K. E., Namminga, K. L., Lu, Y., Schwab, A. D., Thurston, M. J., Abdelmoaty, M. M., et al. (2021). Safety, tolerability, and immune-biomarker profiling for year-long sargramostim treatment of Parkinson's disease. *eBioMedicine, 67*, 103380. https://doi.org/10.1016/j.ebiom.2021.103380

Olson, K. E., Abdelmoaty, M. M., Namminga, K. L., Lu, Y., Obaro, H., Santamaria, P., et al. (2023). An open-label multiyear study of sargramostim-treated Parkinson's disease patients examining drug safety, tolerability, and immune biomarkers from limited case numbers. *Translational Neurodegeneration, 12*(1), 26. https://doi.org/10.1186/s40035-023-00361-1

Olsson, B., Lautner, R., Andreasson, U., Ohrfelt, A., Portelius, E., Bjerke, M., et al. (2016). CSF and blood biomarkers for the diagnosis of Alzheimer's disease: A systematic review and meta-analysis. *Lancet Neurology, 15*(7), 673–684. https://doi.org/10.1016/S1474-4422(16)00070-3

Ossenkoppele, R., Rabinovici, G. D., Smith, R., Cho, H., Schöll, M., Strandberg, O., et al. (2018). Discriminative accuracy of [18F] flortaucipir positron emission tomography for Alzheimer disease vs other neurodegenerative disorders. *JAMA, 320*(11), 1151–1162. https://doi.org/10.1001/jama.2018.12917

Ovod, V., Ramsey, K. N., Mawuenyega, K. G., Bollinger, J. G., Hicks, T., Schneider, T., et al. (2017). Amyloid beta concentrations and stable isotope labeling kinetics of human plasma specific to central nervous system amyloidosis. *Alzheimers Dement, 13*(8), 841–849. https://doi.org/10.1016/j.jalz.2017.06.2266

Pagano, G., Niccolini, F., & Politis, M. (2016). Imaging in Parkinson's disease. *Clinical Medicine (London, England), 16*(4), 371–375. https://doi.org/10.7861/clinmedicine.16-4-371

Palakurthi, B., & Burugupally, S. P. (2019). Postural instability in Parkinson's disease: A review. *Brain Sciences, 9*(9). https://doi.org/10.3390/brainsci9090239

Palmqvist, S., Mattsson, N., Hansson, O., & Alzheimer's Disease Neuroimaging Initiative. (2016). Cerebrospinal fluid analysis detects cerebral amyloid-beta accumulation earlier than positron emission tomography. *Brain, 139*(Pt 4), 1226–1236. https://doi.org/10.1093/brain/aww015

Palmqvist, S., Insel, P. S., Stomrud, E., Janelidze, S., Zetterberg, H., Brix, B., et al. (2019a). Cerebrospinal fluid and plasma biomarker trajectories with increasing amyloid deposition in Alzheimer's disease. *EMBO Molecular Medicine, 11*(12), e11170. https://doi.org/10.15252/emmm.201911170

Palmqvist, S., Janelidze, S., Stomrud, E., Zetterberg, H., Karl, J., Zink, K., et al. (2019b). Performance of fully automated plasma assays as screening tests for Alzheimer disease-related beta-amyloid status. *JAMA Neurology, 76*(9), 1060–1069. https://doi.org/10.1001/jamaneurol.2019.1632

Palmqvist, S., Janelidze, S., Quiroz, Y. T., Zetterberg, H., Lopera, F., Stomrud, E., et al. (2020). Discriminative accuracy of plasma Phospho-tau217 for Alzheimer disease vs other neurodegenerative disorders. *JAMA, 324*(8), 772–781. https://doi.org/10.1001/jama.2020.12134

Palomba, N. P., Fortunato, G., Pepe, G., Modugno, N., Pietracupa, S., Damiano, I., et al. (2023). Common and rare variants in TMEM175 gene concur to the pathogenesis of Parkinson's disease in Italian patients. *Molecular Neurobiology, 60*(4), 2150–2173. https://doi.org/10.1007/s12035-022-03203-9

Parnetti, L., Chiasserini, D., Persichetti, E., Eusebi, P., Varghese, S., Qureshi, M. M., et al. (2014). Cerebrospinal fluid lysosomal enzymes and alpha-synuclein in Parkinson's disease. *Movement Disorders, 29*(8), 1019–1027. https://doi.org/10.1002/mds.25772

Parnetti, L., Gaetani, L., Eusebi, P., Paciotti, S., Hansson, O., El-Agnaf, O., et al. (2019). CSF and blood biomarkers for Parkinson's disease. *Lancet Neurology, 18*(6), 573–586. https://doi.org/10.1016/S1474-4422(19)30024-9

Peng, C., Trojanowski, J. Q., & Lee, V. M. (2020). Protein transmission in neurodegenerative disease. *Nature Reviews. Neurology, 16*(4), 199–212. https://doi.org/10.1038/s41582-020-0333-7

Phelps, M. E., Huang, S. C., Hoffman, E. J., Selin, C., Sokoloff, L., & Kuhl, D. E. (1979). Tomographic measurement of local cerebral glucose metabolic rate in humans with (F-18)2-fluoro-2-deoxy-D--glucose: Validation of method. *Annals of Neurology, 6*(5), 371–388. https://doi.org/10.1002/ana.410060502

Planche, V., Manjon, J. V., Mansencal, B., Lanuza, E., Tourdias, T., Catheline, G., & Coupé, P. (2022). Structural progression of Alzheimer's disease over decades: The MRI staging scheme. *Brain Communications, 4*(3). https://doi.org/10.1093/braincomms/fcac109

Poewe, W. (2008). Non-motor symptoms in Parkinson's disease. *European Journal of Neurology, 15*(Suppl 1), 14–20. https://doi.org/10.1111/j.1468-1331.2008.02056.x

Pontecorvo, M. J., Keene, C. D., Beach, T. G., Montine, T. J., Arora, A. K., Devous, M. D., Sr., et al. (2020). Comparison of regional flortaucipir PET with quantitative tau immunohistochemistry in three subjects with Alzheimer's disease pathology: A clinicopathological study. *EJNMMI Research, 10*(1), 65. https://doi.org/10.1186/s13550-020-00653-x

Pontecorvo, M. J., Lu, M., Burnham, S. C., Schade, A. E., Dage, J. L., Shcherbinin, S., et al. (2022). Association of donanemab treatment with exploratory plasma biomarkers in early symptomatic Alzheimer disease: A secondary analysis of the TRAILBLAZER-ALZ randomized clinical trial. *JAMA Neurology, 79*(12), 1250–1259. https://doi.org/10.1001/jamaneurol.2022.3392

Porsteinsson, A. P., Isaacson, R. S., Knox, S., Sabbagh, M. N., & Rubino, I. (2021). Diagnosis of early Alzheimer's disease: Clinical practice in 2021. *The Journal of Prevention of Alzheimer's Disease, 8*(3), 371–386. https://doi.org/10.14283/jpad.2021.23

Portelius, E., Mattsson, N., Pannee, J., Zetterberg, H., Gisslen, M., Vanderstichele, H., et al. (2017). Ex vivo (18)O-labeling mass spectrometry identifies a peripheral amyloid beta clearance pathway. *Molecular Neurodegeneration, 12*(1), 18. https://doi.org/10.1186/s13024-017-0152-5

Pottier, C., Hannequin, D., Coutant, S., Rovelet-Lecrux, A., Wallon, D., Rousseau, S., et al. (2012). High frequency of potentially pathogenic SORL1 mutations in autosomal dominant early-onset Alzheimer disease. *Molecular Psychiatry, 17*(9), 875–879. https://doi.org/10.1038/mp.2012.15

Pottier, C., Wallon, D., Rousseau, S., Rovelet-Lecrux, A., Richard, A. C., Rollin-Sillaire, A., et al. (2013). TREM2 R47H variant as a risk factor for early-onset Alzheimer's disease. *Journal of Alzheimer's Disease, 35*(1), 45–49. https://doi.org/10.3233/JAD-122311

Pradas, E., & Martinez-Vicente, M. (2023). The consequences of GBA deficiency in the autophagy-lysosome system in Parkinson's disease associated with GBA. *Cells, 12*(1). https://doi.org/10.3390/cells12010191

Prakash, K. G., Bannur, B. M., Chavan, M. D., Saniya, K., Sailesh, K. S., & Rajagopalan, A. (2016). Neuroanatomical changes in Parkinson's disease in relation to cognition: An update. *Journal of Advanced Pharmaceutical Technology & Research, 7*(4), 123–126. https://doi.org/10.4103/2231-4040.191416

Prell, T. (2018). Structural and functional brain patterns of non-motor syndromes in Parkinson's disease. *Frontiers in Neurology, 9*, 138. https://doi.org/10.3389/fneur.2018.00138

Qu, L., Lin, B., Zeng, W., Fan, C., Wu, H., Ge, Y., et al. (2022). Lysosomal K(+) channel TMEM175 promotes apoptosis and aggravates symptoms of Parkinson's disease. *EMBO Reports, 23*(9), e53234. https://doi.org/10.15252/embr.202153234

Quigg, M., & Kundu, B. (2022). Dynamic FDG-PET demonstration of functional brain abnormalities. *Annals of Clinical and Translational Neurology, 9*(9), 1487–1497. https://doi.org/10.1002/acn3.51546

Raghavan, N. S., Dumitrescu, L., Mormino, E., Mahoney, E. R., Lee, A. J., Gao, Y., et al. (2020). Association between common variants in RBFOX1, an RNA-binding protein, and brain amyloidosis in early and preclinical Alzheimer disease. *JAMA Neurology, 77*(10), 1288–1298. https://doi.org/10.1001/jamaneurol.2020.1760

Rajmohan, R., & Reddy, P. H. (2017). Amyloid-Beta and Phosphorylated tau accumulations cause abnormalities at synapses of Alzheimer's disease neurons. *Journal of Alzheimer's Disease, 57*(4), 975–999. https://doi.org/10.3233/jad-160612

Respondek, G., Grimm, M. J., Piot, I., Arzberger, T., Compta, Y., Englund, E., et al. (2020). Validation of the movement disorder society criteria for the diagnosis of 4-repeat tauopathies. *Movement Disorders, 35*(1), 171–176. https://doi.org/10.1002/mds.27872

Rey, F., Pandini, C., Messa, L., Launi, R., Barzaghini, B., Zangaglia, R., et al. (2021). Alpha-synuclein antisense transcript SNCA-AS1 regulates synapses- and aging-related genes suggesting its implication in Parkinson's disease. *Aging Cell, 20*(12), e13504. https://doi.org/10.1111/acel.13504

Reynolds, A. D., Stone, D. K., Hutter, J. A., Benner, E. J., Mosley, R. L., & Gendelman, H. E. (2010). Regulatory T cells attenuate Th17 cell-mediated nigrostriatal dopaminergic neurodegeneration in a model of Parkinson's disease. *Journal of Immunology, 184*(5), 2261–2271. https://doi.org/10.4049/jimmunol.0901852

Rice, H. C., de Malmazet, D., Schreurs, A., Frere, S., Van Molle, I., Volkov, A. N., et al. (2019). Secreted amyloid-beta precursor protein functions as a GABA(B)R1a ligand to modulate synaptic transmission. *Science, 363*(6423). https://doi.org/10.1126/science.aao4827

Ridge, P. G., Hoyt, K. B., Boehme, K., Mukherjee, S., Crane, P. K., Haines, J. L., et al. (2016). Assessment of the genetic variance of late-onset Alzheimer's disease. *Neurobiology of Aging, 41*, 200 e213–200 e220. https://doi.org/10.1016/j.neurobiolaging.2016.02.024

Righini, A., Antonini, A., Ferrarini, M., de Notaris, R., Canesi, M., Triulzi, F., & Pezzoli, G. (2002). Thin section MR study of the basal ganglia in the differential diagnosis between striatonigral degeneration and Parkinson disease. *Journal of Computer Assisted Tomography, 26*(2), 266–271. https://doi.org/10.1097/00004728-200203000-00018

Roda, A. R., Serra-Mir, G., Montoliu-Gaya, L., Tiessler, L., & Villegas, S. (2022). Amyloid-beta peptide and tau protein cross-talk in Alzheimer's disease. *Neural Regeneration Research, 17*(8), 1666–1674. https://doi.org/10.4103/1673-5374.332127

Roher, A. E., Esh, C. L., Kokjohn, T. A., Castano, E. M., Van Vickle, G. D., Kalback, W. M., et al. (2009). Amyloid beta peptides in human plasma and tissues and their significance for Alzheimer's disease. *Alzheimer's & Dementia, 5*(1), 18–29. https://doi.org/10.1016/j.jalz.2008.10.004

Roodveldt, C., Labrador-Garrido, A., Gonzalez-Rey, E., Lachaud, C. C., Guilliams, T., Fernandez-Montesinos, R., et al. (2013). Preconditioning of microglia by alpha-synuclein strongly affects the response induced by toll-like receptor (TLR) stimulation. *PLoS One, 8*(11), e79160. https://doi.org/10.1371/journal.pone.0079160

Rozga, M., Bittner, T., Batrla, R., & Karl, J. (2019). Preanalytical sample handling recommendations for Alzheimer's disease plasma biomarkers. *Alzheimer's & Dementia, 11*, 291–300. https://doi.org/10.1016/j.dadm.2019.02.002

Rubí, S., Noguera, A., Tarongí, S., Oporto, M., García, A., Vico, H., et al. (2018). Concordance between brain (18)F-FDG PET and cerebrospinal fluid biomarkers in diagnosing Alzheimer's disease. *Revista Española de Medicina Nuclear e Imagen Molecular, 37*(1), 3–8. https://doi.org/10.1016/j.remn.2017.05.003

Ryan, B. J., Hoek, S., Fon, E. A., & Wade-Martins, R. (2015). Mitochondrial dysfunction and mitophagy in Parkinson's: From familial to sporadic disease. *Trends in Biochemical Sciences, 40*(4), 200–210. https://doi.org/10.1016/j.tibs.2015.02.003

Saadipour, K. (2017). TREM1: A potential therapeutic target for Alzheimer's disease. *Neurotoxicity Research, 32*(1), 14–16. https://doi.org/10.1007/s12640-017-9716-y

Sakakibara, R., Tateno, F., Kishi, M., Tsuyusaki, Y., Terada, H., & Inaoka, T. (2014). MIBG myocardial scintigraphy in pre-motor Parkinson's disease: A review. *Parkinsonism & Related Disorders, 20*(3), 267–273. https://doi.org/10.1016/j.parkreldis.2013.11.001

Salloway, S., Sperling, R., Fox, N. C., Blennow, K., Klunk, W., Raskind, M., et al. (2014). Two phase 3 trials of bapineuzumab in mild-to-moderate Alzheimer's disease. *The New England Journal of Medicine, 370*(4), 322–333. https://doi.org/10.1056/NEJMoa1304839

Salvado, G., Ossenkoppele, R., Ashton, N. J., Beach, T. G., Serrano, G. E., Reiman, E. M., et al. (2023). Specific associations between plasma biomarkers and postmortem amyloid plaque and tau tangle loads. *EMBO Molecular Medicine, 15*(5), e17123. https://doi.org/10.15252/emmm.202217123

Salvatore, C., Cerasa, A., & Castiglioni, I. (2018). MRI characterizes the progressive course of AD and predicts conversion to Alzheimer's dementia 24 months before probable diagnosis. *Frontiers in Aging Neuroscience, 10*, 135. https://doi.org/10.3389/fnagi.2018.00135

Samgard, K., Zetterberg, H., Blennow, K., Hansson, O., Minthon, L., & Londos, E. (2010). Cerebrospinal fluid total tau as a marker of Alzheimer's disease intensity. *International Journal of Geriatric Psychiatry, 25*(4), 403–410. https://doi.org/10.1002/gps.2353

Sanabria-Diaz, G., Martínez-Montes, E., & Melie-Garcia, L. (2013). Glucose metabolism during resting state reveals abnormal brain networks organization in the Alzheimer's disease and mild cognitive impairment. *PLoS One, 8*(7), e68860. https://doi.org/10.1371/journal.pone.0068860

Sanchez-Catasus, C. A., Bohnen, N. I., Yeh, F. C., D'Cruz, N., Kanel, P., & Muller, M. (2021). Dopaminergic nigrostriatal connectivity in early Parkinson disease: In vivo neuroimaging study of (11) C-DTBZ PET combined with correlational tractography. *Journal of Nuclear Medicine, 62*(4), 545–552. https://doi.org/10.2967/jnumed.120.248500

Sanchez-Guajardo, V., Febbraro, F., Kirik, D., & Romero-Ramos, M. (2010). Microglia acquire distinct activation profiles depending on the degree of alpha-synuclein neuropathology in a rAAV based model of Parkinson's disease. *PLoS One, 5*(1), e8784. https://doi.org/10.1371/journal.pone.0008784

Santiago, J. A., & Potashkin, J. A. (2021). The impact of disease comorbidities in Alzheimer's disease. *Frontiers in Aging Neuroscience, 13*. https://doi.org/10.3389/fnagi.2021.631770

Saresella, M., Calabrese, E., Marventano, I., Piancone, F., Gatti, A., Alberoni, M., et al. (2011). Increased activity of Th-17 and Th-9 lymphocytes and a skewing of the post-thymic differentiation pathway are seen in Alzheimer's disease. *Brain, Behavior, and Immunity, 25*(3), 539–547. https://doi.org/10.1016/j.bbi.2010.12.004

Sarikaya, I., Albatineh, A. N., & Sarikayaa, A. (2020). Effect of various blood glucose levels on regional FDG uptake in the brain. *Asia Oceania Journal of Nuclear Medicine and Biology, 8*(1), 46–53. https://doi.org/10.22038/aojnmb.2019.14418

Satake, W., Nakabayashi, Y., Mizuta, I., Hirota, Y., Ito, C., Kubo, M., et al. (2009). Genome-wide association study identifies common variants at four loci as genetic risk factors for Parkinson's disease. *Nature Genetics, 41*(12), 1303–1307. https://doi.org/10.1038/ng.485

Saunders, J. A., Estes, K. A., Kosloski, L. M., Allen, H. E., Dempsey, K. M., Torres-Russotto, D. R., et al. (2012). CD4+ regulatory and effector/memory T cell subsets profile motor dysfunction in Parkinson's disease. *Journal of Neuroimmune Pharmacology, 7*(4), 927–938. https://doi.org/10.1007/s11481-012-9402-z

Scarapicchia, V., Mazerolle, E. L., Fisk, J. D., Ritchie, L. J., & Gawryluk, J. R. (2018). Resting state BOLD variability in Alzheimer's disease: A marker of cognitive decline or cerebrovascular status? *Frontiers in Aging Neuroscience, 10*, 39. https://doi.org/10.3389/fnagi.2018.00039

Schilling, L. P., Zimmer, E. R., Shin, M., Leuzy, A., Pascoal, T. A., Benedet, A. L., et al. (2016). Imaging Alzheimer's disease pathophysiology with PET. *Dementia & Neuropsychologia, 10*(2), 79–90. https://doi.org/10.1590/s1980-5764-2016dn1002003

Schindler, S. E., Li, Y., Li, M., Despotis, A., Park, E., Vittert, L., et al. (2023). Using Alzheimer's disease blood tests to accelerate clinical trial enrollment. *Alzheimers Dement, 19*(4), 1175–1183. https://doi.org/10.1002/alz.12754

Scholl, M., Maass, A., Mattsson, N., Ashton, N. J., Blennow, K., Zetterberg, H., & Jagust, W. (2019). Biomarkers for tau pathology. *Molecular and Cellular Neurosciences, 97*, 18–33. https://doi.org/10.1016/j.mcn.2018.12.001

Schonhoff, A. M., Williams, G. P., Wallen, Z. D., Standaert, D. G., & Harms, A. S. (2020). Innate and adaptive immune responses in Parkinson's disease. *Progress in Brain Research, 252*, 169–216. https://doi.org/10.1016/bs.pbr.2019.10.006

Schulz, C., Gomez Perdiguero, E., Chorro, L., Szabo-Rogers, H., Cagnard, N., Kierdorf, K., et al. (2012). A lineage of myeloid cells independent of Myb and hematopoietic stem cells. *Science, 336*(6077), 86–90. https://doi.org/10.1126/science.1219179

Schulz, J., Pagano, G., Fernandez Bonfante, J. A., Wilson, H., & Politis, M. (2018). Nucleus basalis of Meynert degeneration precedes and predicts cognitive impairment in Parkinson's disease. *Brain, 141*(5), 1501–1516. https://doi.org/10.1093/brain/awy072

Schwartzentruber, J., Cooper, S., Liu, J. Z., Barrio-Hernandez, I., Bello, E., Kumasaka, N., et al. (2021). Author correction: Genome-wide meta-analysis, fine-mapping and integrative prioritization implicate new Alzheimer's disease risk genes. *Nature Genetics, 53*(4), 585–586. https://doi.org/10.1038/s41588-021-00822-1

Seipold, L., & Saftig, P. (2016). The emerging role of Tetraspanins in the proteolytic processing of the amyloid precursor protein. *Frontiers in Molecular Neuroscience, 9*, 149. https://doi.org/10.3389/fnmol.2016.00149

Seshadri, S., & Wolf, P. A. (2007). Lifetime risk of stroke and dementia: Current concepts, and estimates from the Framingham study. *Lancet Neurology, 6*(12), 1106–1114. https://doi.org/10.1016/S1474-4422(07)70291-0

Sevigny, J., Suhy, J., Chiao, P., Chen, T., Klein, G., Purcell, D., et al. (2016). Amyloid PET screening for enrichment of early-stage Alzheimer disease Clinical trials: Experience in a phase 1b clinical trial. *Alzheimer Disease and Associated Disorders, 30*(1), 1–7. https://doi.org/10.1097/WAD.0000000000000144

Shahnawaz, M., Mukherjee, A., Pritzkow, S., Mendez, N., Rabadia, P., Liu, X., et al. (2020). Discriminating alpha-synuclein strains in Parkinson's disease and multiple system atrophy. *Nature, 578*(7794), 273–277. https://doi.org/10.1038/s41586-020-1984-7

Shi, M., Liu, C., Cook, T. J., Bullock, K. M., Zhao, Y., Ginghina, C., et al. (2014). Plasma exosomal alpha-synuclein is likely CNS-derived and increased in Parkinson's disease. *Acta Neuropathologica, 128*(5), 639–650. https://doi.org/10.1007/s00401-014-1314-y

Shivamurthy, V. K. N., Tahari, A. K., Marcus, C., & Subramaniam, R. M. (2014). Brain FDG PET and the diagnosis of dementia. *American Journal of Roentgenology, 204*(1), W76–W85. https://doi.org/10.2214/AJR.13.12363

Siderowf, A., Concha-Marambio, L., Lafontant, D. E., Farris, C. M., Ma, Y., Urenia, P. A., et al. (2023). Assessment of heterogeneity among participants in the Parkinson's progression markers initiative cohort using alpha-synuclein seed amplification: A cross-sectional study. *Lancet Neurology, 22*(5), 407–417. https://doi.org/10.1016/S1474-4422(23)00109-6

Simon-Sanchez, J., Schulte, C., Bras, J. M., Sharma, M., Gibbs, J. R., Berg, D., et al. (2009). Genome-wide association study reveals genetic risk underlying Parkinson's disease. *Nature Genetics, 41*(12), 1308–1312. https://doi.org/10.1038/ng.487

Singleton, A. B., Farrer, M. J., & Bonifati, V. (2013). The genetics of Parkinson's disease: Progress and therapeutic implications. *Movement Disorders, 28*(1), 14–23. https://doi.org/10.1002/mds.25249

Sirkis, D. W., Bonham, L. W., Johnson, T. P., La Joie, R., & Yokoyama, J. S. (2022). Dissecting the clinical heterogeneity of early-onset Alzheimer's disease. *Molecular Psychiatry, 27*(6), 2674–2688. https://doi.org/10.1038/s41380-022-01531-9

Slattery, C. F., Beck, J. A., Harper, L., Adamson, G., Abdi, Z., Uphill, J., et al. (2014). R47H TREM2 variant increases risk of typical early-onset Alzheimer's disease but not of prion or frontotemporal dementia. *Alzheimers Dement, 10*(6), 602–608 e604. https://doi.org/10.1016/j.jalz.2014.05.1751

Smedinga, M., Darweesh, S. K. L., Bloem, B. R., Post, B., & Richard, E. (2021). Towards early disease modification of Parkinson's disease: A review of lessons learned in the Alzheimer field. *Journal of Neurology, 268*(2), 724–733. https://doi.org/10.1007/s00415-020-10162-5

Smith, R., Wibom, M., Pawlik, D., Englund, E., & Hansson, O. (2019). Correlation of in vivo [18F]Flortaucipir with postmortem Alzheimer disease tau pathology. *JAMA Neurology, 76*(3), 310–317. https://doi.org/10.1001/jamaneurol.2018.3692

Soleimani-Meigooni, D. N., Iaccarino, L., La Joie, R., Baker, S., Bourakova, V., Boxer, A. L., et al. (2020). 18F-flortaucipir PET to autopsy comparisons in Alzheimer's disease and other neurodegenerative diseases. *Brain, 143*(11), 3477–3494. https://doi.org/10.1093/brain/awaa276

Spillantini, M. G., & Goedert, M. (2018). Neurodegeneration and the ordered assembly of alpha-synuclein. *Cell and Tissue Research, 373*(1), 137–148. https://doi.org/10.1007/s00441-017-2706-9

Spinola, M., Galvan, A., Pignatiello, C., Conti, B., Pastorino, U., Nicander, B., et al. (2005). Identification and functional characterization of the candidate tumor suppressor gene TRIT1 in human lung cancer. *Oncogene, 24*(35), 5502–5509. https://doi.org/10.1038/sj.onc.1208687

Srinivasan, K., Friedman, B. A., Larson, J. L., Lauffer, B. E., Goldstein, L. D., Appling, L. L., et al. (2016). Untangling the brain's neuroinflammatory and neurodegenerative transcriptional responses. *Nature Communications, 7*, 11295. https://doi.org/10.1038/ncomms11295

Stocchi, F., & Torti, M. (2017). Constipation in Parkinson's disease. *International Review of Neurobiology, 134*, 811–826. https://doi.org/10.1016/bs.irn.2017.06.003

Stoessl, A. J., Martin, W. W., McKeown, M. J., & Sossi, V. (2011). Advances in imaging in Parkinson's disease. *Lancet Neurology, 10*(11), 987–1001. https://doi.org/10.1016/s1474-4422(11)70214-9

Suppiah, S., Didier, M. A., & Vinjamuri, S. (2019). The who, when, why, and how of PET amyloid imaging in Management of Alzheimer's disease-review of literature and interesting images. *Diagnostics (Basel), 9*(2). https://doi.org/10.3390/diagnostics9020065

Tang, P., Chong, L., Li, X., Liu, Y., Liu, P., Hou, C., & Li, R. (2014). Correlation between serum RANTES levels and the severity of Parkinson's disease. *Oxidative Medicine and Cellular Longevity, 2014*, 208408. https://doi.org/10.1155/2014/208408

Tanji, K., Mori, F., Kakita, A., Takahashi, H., & Wakabayashi, K. (2011). Alteration of autophagosomal proteins (LC3, GABARAP and GATE-16) in Lewy body disease. *Neurobiology of Disease, 43*(3), 690–697. https://doi.org/10.1016/j.nbd.2011.05.022

Tecchio, C., Micheletti, A., & Cassatella, M. A. (2014). Neutrophil-derived cytokines: Facts beyond expression. *Frontiers in Immunology, 5*, 508. https://doi.org/10.3389/fimmu.2014.00508

Teunissen, C. E., Verberk, I. M. W., Thijssen, E. H., Vermunt, L., Hansson, O., Zetterberg, H., et al. (2022). Blood-based biomarkers for Alzheimer's disease: Towards clinical implementation. *Lancet Neurology, 21*(1), 66–77. https://doi.org/10.1016/S1474-4422(21)00361-6

Theodore, S., Cao, S., McLean, P. J., & Standaert, D. G. (2008). Targeted overexpression of human alpha-synuclein triggers microglial activation and an adaptive immune response in a mouse model of Parkinson disease. *Journal of Neuropathology and Experimental Neurology, 67*(12), 1149–1158. https://doi.org/10.1097/NEN.0b013e31818e5e99

Theriault, P., ElAli, A., & Rivest, S. (2015). The dynamics of monocytes and microglia in Alzheimer's disease. *Alzheimer's Research & Therapy, 7*(1), 41. https://doi.org/10.1186/s13195-015-0125-2

Therriault, J., Pascoal, T. A., Lussier, F. Z., Tissot, C., Chamoun, M., Bezgin, G., et al. (2022). Biomarker modeling of Alzheimer's disease using PET-based Braak staging. *Nature Aging, 2*(6), 526–535. https://doi.org/10.1038/s43587-022-00204-0

Therriault, J., Vermeiren, M., Servaes, S., Tissot, C., Ashton, N. J., Benedet, A. L., et al. (2023). Association of phosphorylated tau biomarkers with amyloid positron emission tomography vs tau positron emission tomography. *JAMA Neurology, 80*(2), 188–199. https://doi.org/10.1001/jamaneurol.2022.4485

Thijssen, E. H., La Joie, R., Wolf, A., Strom, A., Wang, P., Iaccarino, L., et al. (2020). Diagnostic value of plasma phosphorylated tau181 in Alzheimer's disease and frontotemporal lobar degeneration. *Nature Medicine, 26*(3), 387–397. https://doi.org/10.1038/s41591-020-0762-2

Tropea, T. F., Xie, S. X., Rick, J., Chahine, L. M., Dahodwala, N., Doshi, J., et al. (2018). APOE, thought disorder, and SPARE-AD predict cognitive decline in established Parkinson's disease. *Movement Disorders, 33*(2), 289–297. https://doi.org/10.1002/mds.27204

Unger, M. S., Li, E., Scharnagl, L., Poupardin, R., Altendorfer, B., Mrowetz, H., et al. (2020). CD8(+) T-cells infiltrate Alzheimer's disease brains and regulate neuronal- and synapse-related gene expression in APP-PS1 transgenic mice. *Brain, Behavior, and Immunity, 89*, 67–86. https://doi.org/10.1016/j.bbi.2020.05.070

Uribe, M., Garcia Compean, D., Feria, G., Gabilondo, F., Mendoza, A., & de la Rosa, C. (1988). Extracorporeal lithotripsy in patients with biliary lithiasis. Study and fragmentation in 6 patients. *Revista de Gastroenterología de México, 53*(2), 89–94. Retrieved from https://www.ncbi.nlm.nih.gov/pubmed/3212338

US National Library of Medicine. (2023). ClinicalTrials.gov. https://classic.clinicaltrials.gov/ct2/show/NCT05677633.

Vemuri, P., & Jack, C. R., Jr. (2010). Role of structural MRI in Alzheimer's disease. *Alzheimer's Research & Therapy, 2*(4), 23. https://doi.org/10.1186/alzrt47

Verberk, I. M. W., Misdorp, E. O., Koelewijn, J., Ball, A. J., Blennow, K., Dage, J. L., et al. (2022). Characterization of pre-analytical sample handling effects on a panel of Alzheimer's disease-related blood-based biomarkers: Results from the Standardization of Alzheimer's Blood Biomarkers (SABB) working group. *Alzheimers Dement, 18*(8), 1484–1497. https://doi.org/10.1002/alz.12510

Vicente Miranda, H., Cassio, R., Correia-Guedes, L., Gomes, M. A., Chegao, A., Miranda, E., et al. (2017). Posttranslational modifications of blood-derived alpha-synuclein as biochemical markers for Parkinson's disease. *Scientific Reports, 7*(1), 13713. https://doi.org/10.1038/s41598-017-14175-5

Villemagne, V. L., Burnham, S., Bourgeat, P., Brown, B., Ellis, K. A., Salvado, O., et al. (2013). Amyloid beta deposition, neurodegeneration, and cognitive decline in sporadic Alzheimer's disease: A prospective cohort study. *Lancet Neurology, 12*(4), 357–367. https://doi.org/10.1016/S1474-4422(13)70044-9

Walker, L. C., & Jucker, M. (2015). Neurodegenerative diseases: Expanding the prion concept. *Annual Review of Neuroscience, 38*(1), 87–103. https://doi.org/10.1146/annurev-neuro-071714-033828

Wallon, D., Rousseau, S., Rovelet-Lecrux, A., Quillard-Muraine, M., Guyant-Marechal, L., Martinaud, O., et al. (2012). The French series of autosomal dominant early onset Alzheimer's disease cases: Mutation spectrum and cerebrospinal fluid biomarkers. *Journal of Alzheimer's Disease, 30*(4), 847–856. https://doi.org/10.3233/JAD-2012-120172

Wang, Y., Shi, M., Chung, K. A., Zabetian, C. P., Leverenz, J. B., Berg, D., et al. (2012). Phosphorylated alpha-synuclein in Parkinson's disease. *Science Translational Medicine, 4*(121), 121ra120. https://doi.org/10.1126/scitranslmed.3002566

Wang, Q., Liu, Y., & Zhou, J. (2015). Neuroinflammation in Parkinson's disease and its potential as therapeutic target. *Translational Neurodegeneration, 4*, 19. https://doi.org/10.1186/s40035-015-0042-0

Wang, Y., Zhang, Y., Zhang, X., Yang, T., Liu, C., & Wang, P. (2019). Alcohol dehydrogenase 1B suppresses beta-amyloid-induced neuron apoptosis. *Frontiers in Aging Neuroscience, 11*, 135. https://doi.org/10.3389/fnagi.2019.00135

Wang, Z., Becker, K., Donadio, V., Siedlak, S., Yuan, J., Rezaee, M., et al. (2020). Skin alpha-synuclein aggregation seeding activity as a novel biomarker for Parkinson disease. *JAMA Neurology, 78*(1), 1–11. https://doi.org/10.1001/jamaneurol.2020.3311

Watson, M. B., Richter, F., Lee, S. K., Gabby, L., Wu, J., Masliah, E., et al. (2012). Regionally-specific microglial activation in young mice over-expressing human wildtype alpha-synuclein. *Experimental Neurology, 237*(2), 318–334. https://doi.org/10.1016/j.expneurol.2012.06.025

Wesseling, H., Mair, W., Kumar, M., Schlaffner, C. N., Tang, S., Beerepoot, P., et al. (2020). Tau PTM profiles identify patient heterogeneity and stages of Alzheimer's disease. *Cell, 183*(6), 1699–1713 e1613. https://doi.org/10.1016/j.cell.2020.10.029

West, T., Kirmess, K. M., Meyer, M. R., Holubasch, M. S., Knapik, S. S., Hu, Y., et al. (2021). A blood-based diagnostic test incorporating plasma Abeta42/40 ratio, ApoE proteotype, and age accurately identifies brain amyloid status: Findings from a multi cohort validity analysis. *Molecular Neurodegeneration, 16*(1), 30. https://doi.org/10.1186/s13024-021-00451-6

Whitwell, J. L., Josephs, K. A., Murray, M. E., Kantarci, K., Przybelski, S. A., Weigand, S. D., et al. (2008). MRI correlates of neurofibrillary tangle pathology at autopsy: A voxel-based morphometry study. *Neurology, 71*(10), 743–749. https://doi.org/10.1212/01.wnl.0000324924.91351.7d

Wie, J., Liu, Z., Song, H., Tropea, T. F., Yang, L., Wang, H., et al. (2021). A growth-factor-activated lysosomal K(+) channel regulates Parkinson's pathology. *Nature, 591*(7850), 431–437. https://doi.org/10.1038/s41586-021-03185-z

Williams, S. M., Schulz, P., & Sierks, M. R. (2016). Oligomeric alpha-synuclein and beta-amyloid variants as potential biomarkers for Parkinson's and Alzheimer's diseases. *The European Journal of Neuroscience, 43*(1), 3–16. https://doi.org/10.1111/ejn.13056

Williams-Gray, C. H., Wijeyekoon, R., Yarnall, A. J., Lawson, R. A., Breen, D. P., Evans, J. R., et al. (2016). Serum immune markers and disease progression in an incident Parkinson's disease cohort (ICICLE-PD). *Movement Disorders, 31*(7), 995–1003. https://doi.org/10.1002/mds.26563

Winston, C. N., Langford, O., Levin, N., Raman, R., Yarasheski, K., West, T., et al. (2023). Evaluation of blood-based plasma biomarkers as potential markers of amyloid burden in preclinical Alzheimer's disease. *Journal of Alzheimer's Disease, 92*(1), 95–107. https://doi.org/10.3233/JAD-221118

Xilouri, M., Brekk, O. R., & Stefanis, L. (2016). Autophagy and alpha-synuclein: Relevance to Parkinson's disease and related synucleopathies. *Movement Disorders, 31*(2), 178–192. https://doi.org/10.1002/mds.26477

Xiromerisiou, G., Marogianni, C., Lampropoulos, I. C., Dardiotis, E., Speletas, M., Ntavaroukas, P., et al. (2022). Peripheral inflammatory markers TNF-alpha and CCL2 revisited: Association with Parkinson's disease severity. *International Journal of Molecular Sciences, 24*(1). https://doi.org/10.3390/ijms24010264

Yamada, T., McGeer, P. L., & McGeer, E. G. (1992). Lewy bodies in Parkinson's disease are recognized by antibodies to complement proteins. *Acta Neuropathologica, 84*(1), 100–104. https://doi.org/10.1007/BF00427222

Yan, Q., Nho, K., Del-Aguila, J. L., Wang, X., Risacher, S. L., Fan, K. H., et al. (2021). Genome-wide association study of brain amyloid deposition as measured by Pittsburgh compound-B (PiB)-PET imaging. *Molecular Psychiatry, 26*(1), 309–321. https://doi.org/10.1038/s41380-018-0246-7

Yanamandra, K., Patel, T. K., Jiang, H., Schindler, S., Ulrich, J. D., Boxer, A. L., et al. (2017). Anti-tau antibody administration increases plasma tau in transgenic mice and patients with tauopathy. *Science Translational Medicine, 9*(386). https://doi.org/10.1126/scitranslmed.aal2029

Yang, H., Park, S. Y., Baek, H., Lee, C., Chung, G., Liu, X., et al. (2022). Adoptive therapy with amyloid-beta specific regulatory T cells alleviates Alzheimer's disease. *Theranostics, 12*(18), 7668–7680. https://doi.org/10.7150/thno.75965

Young, P. N. E., Estarellas, M., Coomans, E., Srikrishna, M., Beaumont, H., Maass, A., et al. (2020). Imaging biomarkers in neurodegeneration: Current and future practices. *Alzheimer's Research & Therapy, 12*(1), 49. https://doi.org/10.1186/s13195-020-00612-7

Zang, X., Chen, S., Zhu, J., Ma, J., & Zhai, Y. (2022). The emerging role of central and peripheral immune Systems in Neurodegenerative Diseases. *Frontiers in Aging Neuroscience, 14*, 872134. https://doi.org/10.3389/fnagi.2022.872134

Zenaro, E., Pietronigro, E., Della Bianca, V., Piacentino, G., Marongiu, L., Budui, S., et al. (2015). Neutrophils promote Alzheimer's disease-like pathology and cognitive decline via LFA-1 integrin. *Nature Medicine, 21*(8), 880–886. https://doi.org/10.1038/nm.3913

Zerr, I., Cramm, M., da Silva Correia, S. M., Zafar, S., Villar-Pique, A., Llorens, F., & Schmitz, M. (2020). Optimization of the real-time quaking-induced conversion assay for prion disease diagnosis. *Frontiers in Bioengineering and Biotechnology, 8*, 586890. https://doi.org/10.3389/fbioe.2020.586890

Zhang, Y., & Burock, M. A. (2020). Diffusion tensor imaging in Parkinson's disease and parkinsonian syndrome: A systematic review. *Frontiers in Neurology, 11*, 531993. https://doi.org/10.3389/fneur.2020.531993

Zhang, W., Wang, T., Pei, Z., Miller, D. S., Wu, X., Block, M. L., et al. (2005). Aggregated alpha-synuclein activates microglia: A process leading to disease progression in Parkinson's disease. *The FASEB Journal, 19*(6), 533–542. https://doi.org/10.1096/fj.04-2751com

Zhang, Y., Fung, I. T. H., Sankar, P., Chen, X., Robison, L. S., Ye, L., et al. (2020). Depletion of NK cells improves cognitive function in the Alzheimer disease mouse model. *Journal of Immunology, 205*(2), 502–510. https://doi.org/10.4049/jimmunol.2000037

Zhao, H. Q., Li, F. F., Wang, Z., Wang, X. M., & Feng, T. (2016). A comparative study of the amount of alpha-synuclein in ischemic stroke and Parkinson's disease. *Neurological Sciences, 37*(5), 749–754. https://doi.org/10.1007/s10072-016-2485-1

Zhao, N., Attrebi, O. N., Ren, Y., Qiao, W., Sonustun, B., Martens, Y. A., et al. (2020). APOE4 exacerbates alpha-synuclein pathology and related toxicity independent of amyloid. *Science Translational Medicine, 12*(529). https://doi.org/10.1126/scitranslmed.aay1809

Zhu, J. H., Guo, F., Shelburne, J., Watkins, S., & Chu, C. T. (2003). Localization of phosphorylated ERK/MAP kinases to mitochondria and autophagosomes in Lewy body diseases. *Brain Pathology, 13*(4), 473–481. https://doi.org/10.1111/j.1750-3639.2003.tb00478.x

Bioimaging of Neuroimmune Disorders: Physical Principles of Neuroimaging Modalities and Clinical Applications

45

Mahsa Mayeli, Thomas Ernst, and Linda Chang

Abstract

A wide range of bioimaging techniques have been applied to evaluate and study the neuroimmune aspects of neurological and psychiatric disorders. The development of reconstruction and visualization techniques for magnetic resonance imaging (MRI), as well as novel radiotracers for positron emission tomography (PET), enhanced their applications in this field. In this chapter, we provide a brief introduction to the neuroimaging techniques used to evaluate neuroimmune-mediated pathology in the brain. We will first discuss the general physical principles and some limitations of each modality. We will also briefly discuss the neuroimaging findings from various modalities or techniques for several neurological and psychiatric disorders, including those from viral infections, neurodegenerative diseases, substance use disorders, and chronic pain. This chapter illustrates how various imaging techniques can assess neuropathology, provide insights regarding mechanisms of disease, monitor disease progression, and assess treatment efficacies.

Keywords

Magnetic resonance imaging · Nuclear medicine imaging · Bioimaging · Imaging biomarkers · Neuroimmunology

45.1 An Introduction to Bioimaging

Bioimaging refers to imaging technologies that utilize different physical principles, such as radiation, magnetism, or certain biological mechanisms, to provide anatomical or physiological information in certain organs or selected regions of the living organism. In the previous edition of this book, this chapter focused on the physical principles of neuroimaging (Boska & White, 2008). Here, we will again review the physical principles of various in vivo neuroimaging techniques and additionally provide examples of the clinical applications of each modality and some of the latest preclinical and clinical research using these techniques to facilitate translational research. Although bioimaging can be used to study any tissue/organ, we will focus only on imaging studies of the brain. In addition, since different postprocessing tools are increasingly used for three-dimensional visualization and quantification of neuroimaging, we will discuss some of these tools. Further, we will review some hybrid techniques that can facilitate the simultaneous measurements of physiological or molecular changes in the brain. Lastly, we will illustrate and describe how some of these techniques are used to improve diagnoses and guide neurointerventional procedures to treat specific brain disorders in clinical settings.

45.2 Principles of Imaging Techniques

We will introduce morphological or structural imaging methods and proceed to discuss several physiological and functional imaging techniques, including magnetic resonance (MR) and nuclear medicine imaging techniques that are used in clinical research and clinical settings. Lastly, we will briefly discuss some preclinical imaging techniques. Table 45.1 summarizes various neuroimaging modalities and the corresponding physical principles involved.

45.2.1 Morphological or Structural Neuroimaging Methods

45.2.1.1 X-Ray and Computed Tomography (CT)

Morphological imaging refers to the assessment of the structure or shape of the organ of interest (i.e., the brain). The earliest and most common techniques employed are X-rays,

M. Mayeli · T. Ernst · L. Chang (✉)
Department of Diagnostic Radiology and Nuclear Medicine, University of Maryland School of Medicine, Baltimore, MD, USA
e-mail: LChang@som.umaryland.edu

H. E. Gendelman, T. Ikezu (eds.), *Neuroimmune Pharmacology and Therapeutics*, https://doi.org/10.1007/978-3-031-68237-7_45

Table 45.1 Physical principles of various clinical neuroimaging techniques

Technique			Physical principles	What can it measure?
Morphological neuroimaging techniques	X-Ray and computed tomography (CT)		Rotating X-ray beams and radially placed detectors are used to create two-dimensional or three-dimensional images based on the density of the object and its attenuation of the X-ray	Gross brain structures, skull, ventricular space, any lesions
	Structural magnetic resonance imaging (MRI)		Brain tissue contrast is based on differences in proton density (based on the water content) and relaxation times among brain structures	Skull, ventricles, cortical (volume, surface area, thickness) or subcortical (volume or shape) brain structures, large blood vessels, any lesions
	Microstructural MRI	Diffusion tensor imaging (DTI)	Applies diffusion-sensitizing gradients to the MRI pulse sequence and provides information about the extent and directionality of water diffusion in tissues	The directionality of the water diffusivity in tissues, especially along white matter tracts, which may be restricted by myelin or cells
		Magnetization transfer imaging (MTI)	Compares the magnetization exchange between protons bound to water and those bound to macromolecules to determine the magnetization transfer ratio (MTR)	The MTR creates a map that shows brain regions with loss of macromolecules (e.g., demyelination)
		Chemical exchange saturation transfer (CEST)	Based on the chemical exchange of the magnetization to surrounding water protons after an RF pulse is applied to certain nuclei (detects the reduced water signal intensity)	Detects and quantifies specific molecules or metabolites in the brain
		Quantitative susceptibility mapping (QSM)	Based on different magnetic susceptibilities of different tissues that cause variable local magnetic field distortions in the presence of an external magnetic field	The magnetic susceptibility of tissues can provide a map of the spatial distribution of tissue susceptibility and provides information about the underlying tissue composition, such as iron content
Physiological MRI techniques	Perfusion MRI	Dynamic susceptibility contrast (DSC)	Based on signal loss due to the susceptibility changes during the first pass of a bolus of gadolinium (Gd)-based contrast agent through the brain tissue, with high-speed MR scans	Relative cerebral blood flow, mean transit time of the Gd bolus, cerebral blood volume
		Arterial spin labeling (ASL)	Detects signal change resulting from the inversion of spins at an input artery (e.g., carotid artery) that flow into the higher levels in the brain	Absolute quantitative cerebral blood flow in selected brain regions or the whole brain
	Cerebrovascular reactivity (CVR) imaging		Based on the hemodynamic changes in the brain vasculature in response to a physiological challenge or changes in oxygenation	Cerebral hemodynamic changes in response to a vasodilatory stimulus (e.g., inhaled CO_2 or acetazolamide) or changes in oxygenation

(continued)

Table 45.1 (continued)

Technique		Physical principles	What can it measure?
Functional MRI	Blood-oxygen level dependent (BOLD) fMRI	The BOLD contrast is based on changes in the MRI signals in response to changes in the proportions of diamagnetic oxygenated hemoglobin (Hgb) and paramagnetic deoxygenated hemoglobin (dHgb) that occur during brain activation, either during a task (task-fMRI) or during the rest state (rsfMRI).	Changes in blood flow and oxygenation in the brain that provide an indirect assessment of changes in regional brain activity within a particular brain network during task-fMRI. For rsfMRI, the low-level brain activities across the brain networks can be assessed to evaluate brain connectivity.
MR spectroscopy (MRS)	(^{1}H, ^{31}P, ^{13}C) MRS	Based on the slightly different resonance frequencies (ppm) of different nuclei of various molecules, and the local influence of the particular chemical structure of the molecule, the different MR signals can be separated by Fourier transformation into a spectrum of peaks. The area of each peak reflects the concentration of the corresponding chemical compound.	Concentrations of neurometabolites; ^{1}H-MRS can detect N-acetyl compounds, total creatine, choline compounds, glutamate+glutamine, and myoinositol. ^{31}P-MRS can measure concentrations of ATP, inorganic phosphate (Pi), and the intracellular pH, and ^{19}F can evaluate the biodistribution and pharmacokinetics of drug candidates labeled with fluorine-19.
Nuclear medicine imaging	Single-photon emission computed tomography (SPECT)	Based on the radioactive decay of SPECT tracers that emit gamma rays that are detected by radially placed detectors in the SPECT scanner; the distribution of these single photons is reconstructed into a SPECT image	Depending on the labeled tracer, changes in regional cerebral blood flow, regional cerebral metabolism, or neurotransmitter systems can be assessed
	Positron emission tomography (PET)	Injected radiotracers labeled with one of the isotopes (^{18}F, ^{11}C, ^{15}O) emit positrons, which collide with electrons and give off gamma rays in opposite directions that are detected by radially placed detectors in a PET scanner; the distributions of the positrons are reconstructed into a PET image	Detecting gamma rays emitted indirectly by a positron-emitting radionuclide, which is injected into the body

ASL arterial spin labeling, *BOLD* blood-oxygen level dependent, *CEST* chemical exchange saturation transfer, *CT* computed tomography, *CVR* cerebrovascular reactivity, *dHgb* deoxygenated hemoglobin, *DSC* dynamic susceptibility contrast, *DTI* diffusion tensor imaging, *Hgb* oxygenated hemoglobin, *Gd* gadolinium, *MRI* structural magnetic resonance imaging, *MRS* magnetic resonance spectroscopy, *MTI* magnetization transfer imaging, *MTR* magnetization transfer ratio, *PET* positron emission tomography, *QSM* quantitative susceptibility mapping, *SPECT* single-photon emission computed tomography

which can generate a planar image of the head and the skull, or head computed tomography (CT). Both X-rays of the skull and head CTs are based on the tissue's attenuation of X-ray beams to create the images. While X-ray is a 2-D projection, head CT uses a rotating X-ray tube to generate a 3-D image with axial (cross-sectional) slices (Fig. 45.1a). In combination with a contrast agent (i.e., an iodinated compound), blood vessels can also be assessed with 2-D or 3-D angiography.

During a skull X-ray or head CT procedure, a controlled amount of X-ray radiation is directed through the head from an X-ray tube toward a detector on the opposite side. Dense tissues, such as skull bone, attenuate more of the X-rays and will appear white on the resulting X-ray image, while less dense tissues, such as brain tissue, allow more X-rays to pass through and appear darker or gray. The captured X-ray intensity is converted into an image that can be viewed on a computer screen or printed on film. This tech-

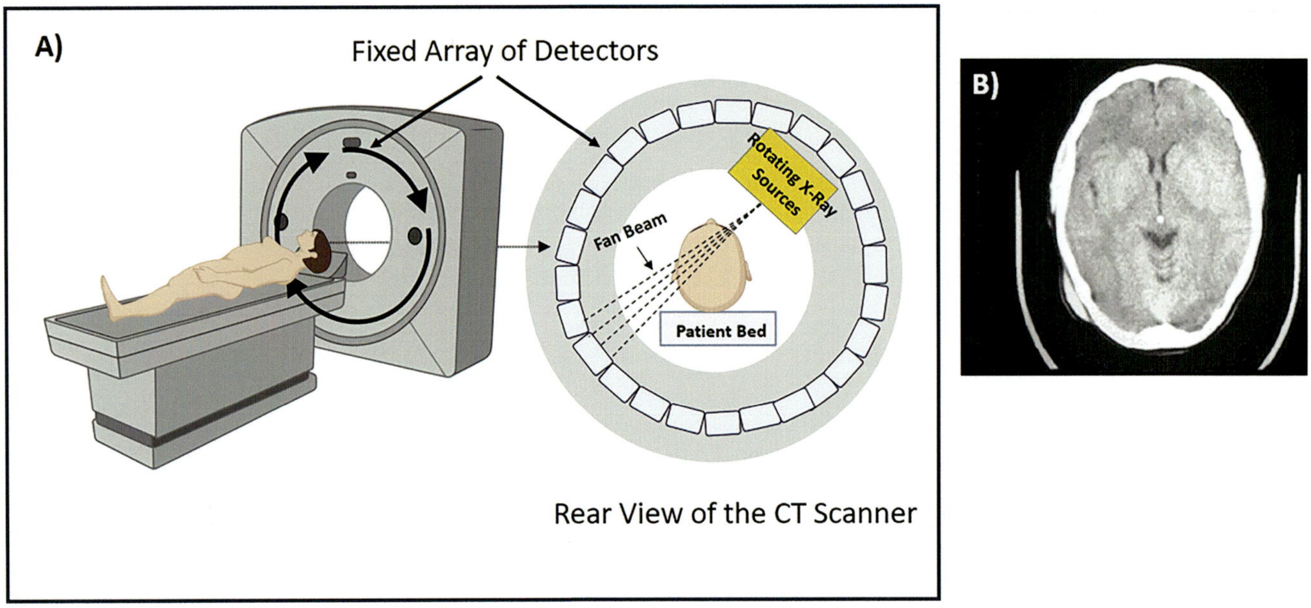

Fig. 45.1 (**a**) Computerized tomography: X-rays are attenuated after radiating from the X-ray source and passing through the tissue and are detected using detectors in the rotating plate. (**b**) A normal CT scan of the head can then be reconstructed in the computer from the X-ray detected by the detectors

nique is most useful for evaluating bone structures or abnormalities (e.g., skull or neck fractures) and density differences but has limited ability to differentiate soft tissue properties. Another limitation is that X-ray imaging only produces 2D projection images.

CT imaging combines X-ray technology with computerized reconstruction to produce detailed cross-sectional images using a CT scanner (Fig. 45.1b). CT provides more detailed tissue contrast compared to conventional X-rays. Modern CT scanners commonly feature a configuration with multiple X-ray tubes distributed in a circular or semi-circular arrangement within the CT scanner gantry. The X-ray tubes rotate with the gantry to allow for faster imaging and data acquisition from multiple angles simultaneously. As in projection X-ray images, the X-rays are attenuated variably by different tissues in the head, and the resulting data are processed by a computer using reconstruction algorithms to create multiple cross-sectional images ("slices") of the head. The 2-D slices can be further combined to create three-dimensional (3D) images. Therefore, CT imaging is a fast-imaging method that provides significantly more information about soft tissues and various pathologies than conventional X-ray (Withers et al., 2021). Head CT scans, including dual-energy, sequential, and helical acquisition methods, are commonly used in clinical settings due to their fast scanning and reconstruction times, which also make them the modality of choice in clinical emergency settings. The main limitation of CT scans is their low tissue contrast.

45.2.1.2 Structural Magnetic Resonance Imaging (MRI) Techniques

MRI was first introduced into the clinical arena in the 1980s, and dramatic advancements have led to the development of many methods to assess the brain in numerous clinical applications. MRI employs powerful magnets producing a strong magnetic field caused by the electrical currents passing through the helical MRI coil. The MRI signal originates from the interaction of hydrogen (or other) nuclei in tissues with the external magnetic field. Technically, any nucleus with an odd atomic number or an odd mass number will have a non-zero spin quantum number ($I \neq 0$) and a nuclear magnetic moment, which can produce a signal. Since the human body comprises mostly of water, the hydrogen nuclei in water are used for structural MRI. Figure 45.2a, b shows that in the MRI scanner, the magnetic moments of these hydrogen nuclei (protons) become aligned with the magnetic field; the sum of magnetic moments of hydrogen nuclei in a given volume is called the net magnetization vector. A radiofrequency (RF) field can alter the direction of the magnetization vector, and eventually cause the vector to precess around the magnetic field. The frequency of these precessions is called the Larmor frequency, which is proportional to the magnetic field: $\omega = \gamma B$, where ω is the rotational frequency of the magnetization after excitation, γ is the gyromagnetic ratio in MHz/Tesla (depends on the nucleus) and B is the strength of the static magnetic field in Tesla (T) or Gauss (G). Based on this equation, the Larmor frequency increases proportionally with the field strength. Therefore, a magnetic field that varies linearly in space (a "gradient") causes a position-dependent

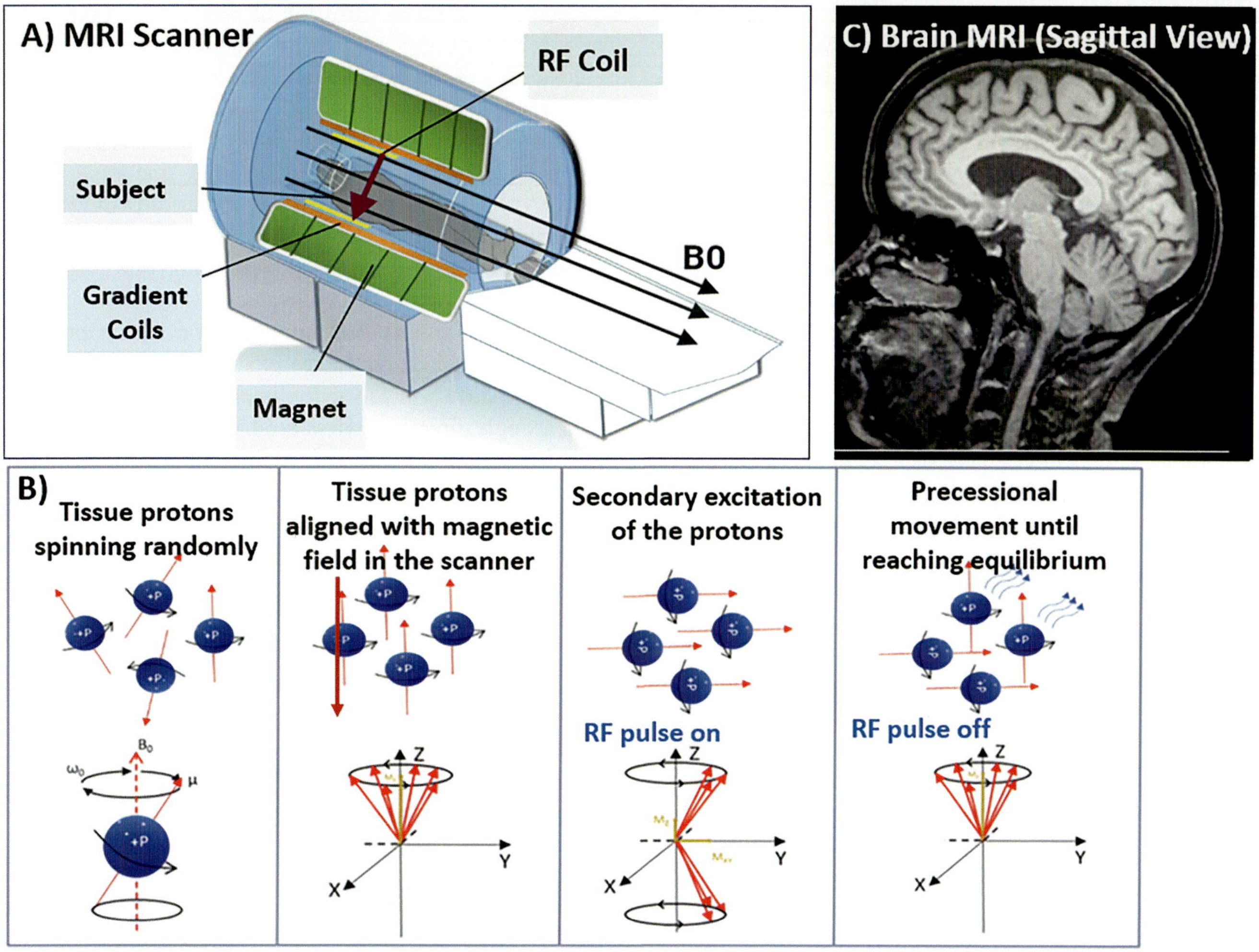

Fig. 45.2 Physical principles of MRI and examples of structural MRI. (**a**) A magnetic resonance (MR) scanner and its components. (**b**) Diagrams of protons that are randomly aligned (*bottom left*) and then become aligned with the direction of the magnetic field in the scanner (*bottom left middle*). Next, the radiofrequency (RF) coil will generate an RF pulse, which leads to the magnetization perpendicular to the direction of the magnetic field in the scanner, which will lead to precessional movements of the protons (*bottom right middle*) that will eventually return to their equilibrium state (*bottom right*). These changes in the MR signal can be detected by the MRI scanner and the computer reconstructs the MRI. (**c**) A typical brain MRI from a healthy 26-year-old female, sagittal view

shift in the magnetic field; this provides the basis for imaging. Therefore, a single species such as water can be used as a signal source for imaging (Yousaf et al., 2018).

When a brief RF pulse at Larmor frequency is applied to the subject, the magnetization is rotated out of equilibrium ("excitation") (Grover et al., 2015). However, the energy absorbed by the magnetic nuclei during the RF pulse is slowly absorbed by the nuclei's environment; this T1 relaxation (spin-lattice relaxation) eventually causes a loss of signal. Additionally, adjacent spins interact with each other; this spin-spin relaxation (T2) does not involve a loss of energy but still causes a signal decay over time. Importantly, soft tissues show a substantial variation in T1 and T2 values; this provides the basis for the excellent soft tissue contrast of MRI. By varying the timing and amplitude of repeated RF pulses (the "pulse sequence"), a given MRI scan can be optimized to reflect spatial differences (tissue contrast) in T1 or T2 values. Consequently, T1-weighted (T1W) and T2-weighted (T2W) images are two common types of contrasts created using pulse sequences in MRI. Fluid-attenuated inversion recovery (FLAIR) is another sequence created by applying an inversion pulse to null the signal from fluids with a long T2 relaxation time like cerebrospinal fluid (CSF). This inversion pulse is timed to attenuate the signal from CSF to zero during the image acquisition, thereby suppressing the CSF signal and improving contrast for detecting pathology (Fig. 45.2c).

In T1W imaging of the brain, the goal is to maximize the image contrast between tissues with different T1-values, for instance, those between brain grey matter and white matter, while minimizing the signals of spins with long T1 relaxation times, such as CSF. To create T1W images, a short time between the application of successive RF pulses is used, allowing the tissues with shorter T1 relaxation times to recover their alignment with the magnetic field and produce a strong signal. Consequently, white matter, which has a shorter T1 relaxation time than gray matter, appears brighter in T1W images relative to gray matter. Additionally, CSF has a very long T1 relaxation time and appears darker.

In T2W imaging, the goal is to maximize the signal from tissues with longer T2 relaxation times; most importantly, many brain lesions have increased T2 relaxation times relative to healthy brain tissue (due to the typically higher water content in lesions). To create T2W images, the time between the RF pulse and signal acquisition is increased; this allows for the T2 decay of magnetization. Consequently, tissues with long T2 relaxation times, such as brain lesions, show lesser signal decay and appear brighter in T2W images relative to normal brain tissue, which has shorter T2 relaxation times.

Overall, T1W and T2W images provide complementary information about the composition and characteristics of various tissues in the brain. The combination of T1W and T2W images allows for improved visualization and differentiation of tissues and abnormalities in clinical imaging.

45.2.1.3 Microstructural MRI Techniques—Diffusion-Weighted Imaging (DWI)

One of the most widely used microstructural MRI methods is DWI, which assesses the Brownian motion of water molecules to provide contrast for the images. The motion of water molecules within tissues is normally caused by thermal energy that leads to the dephasing of individual magnetic moments of nuclei under the symmetrical magnetic field gradient of a DWI sequence. The diffusion of water molecules in the brain is influenced and restricted by cellular structures, such as cell membranes and fibers. By applying diffusion-sensitizing gradients to a DWI pulse sequence, DWI can measure the diffusivity of water molecules in different directions within tissues inside and outside of these cells' barriers. Hence, DWI provides information about the extent and directionality of water diffusion in tissues, which reflects the integrity of the brain tissue (Lebel & Deoni, 2018).

Diffusion tensor imaging (DTI) is a DWI technique that quantifies the directionality of water diffusion using a mathematical model called the diffusion tensor. The diffusion tensor is a 3x3 symmetric matrix that characterizes the diffusion properties in multiple directions within each voxel of the image. The tensor can be diagonalized to obtain its eigenvalues and eigenvectors. The three eigenvectors represent the principal diffusion directions, which relate to the orientation of fibers within the voxel. The corresponding eigenvalues (e1, e2, and e3) of the diffusion tensor indicate the magnitude or intensity of diffusion in each eigenvector direction. The eigenvector associated with the largest eigenvalue (e1) represents the primary diffusion direction within the voxel. The eigenvalues also provide information about the degree of anisotropy or the extent to which water diffusion is directionally restricted.

The eigenvectors and eigenvalues obtained from DTI can be converted into other metrics, such as fractional anisotropy (FA), mean diffusivity (MD), radial diffusivity (RD), and axial diffusivity (AD). FA measures the degree of anisotropy of diffusion in a voxel, using the normalized standard deviation of the eigenvalues. FA ranges from 0 to 1, where 0 represents isotropic diffusion (diffusion in all directions is equal) and values close to 1 represent highly anisotropic diffusion (diffusion is predominantly restricted along a single direction). Hence, a higher FA value indicates more organized and coherent diffusion, reflecting more coherent axonal fibers in the white matter tract. MD is calculated by averaging the eigenvalues of the diffusion tensor and represents the average magnitude of diffusivity within a voxel, without considering directionality. Elevated MD indicates greater overall diffusion and is commonly associated with increased water content, tissue damage, and reduced cellular density. RD is calculated as the average of the two smallest eigenvalues of the diffusion tensor; it represents the diffusivity perpendicular to the principal diffusion. RD generally reflects the integrity of the myelin sheath or axonal membranes; elevated RD often indicates myelin degradation or disruption of the axonal structure. AD measures the diffusivity parallel to the principal diffusion direction (largest eigenvalue). AD reflects axonal structural integrity, including axonal density or caliber; elevated AD values indicate axonal damage or alterations. Overall, these DTI metrics provide valuable information about white matter integrity, the myelination status of axons in the tracts, and microstructural abnormalities associated with neuropathology.

45.2.1.4 Magnetization Transfer Imaging (MTI)

MTI assesses the magnetization exchange between protons bound to water and those bound to macromolecules and generates a magnetization transfer ratio (MTR) map. Hence, MTI is useful for detecting alterations in macromolecules, such as the loss of lipoproteins from myelin sheaths with demyelination. MTI is also highly sensitive for detecting early demyelination. MTI is based on the spin-spin relaxation time (T2), which is much shorter for larger molecules with slow molecular motion. Therefore, the spectral peak of large molecules is markedly broader than that of free water molecules, and the protons of large molecules and membranes can be saturated selectively without affecting the water molecules (Tambasco et al., 2015). However, the saturation of macromolecular protons affects (reduces) the signal of free water, which is quantified by MTI. Since macromolecules, such as proteins and lipids, contribute to the microstructural organization of tissues, MTI can indirectly probe the macromolecular environment and provide information about tissue integrity and composition such as myelin content.

45.2.1.5 Chemical Exchange Saturation Transfer (CEST)

CEST imaging detects and quantifies specific molecules or metabolites in the brain. CEST takes advantage of the exchange of protons between specific molecules in the soluble state, such as protons in water. When an RF pulse is applied to saturate the protons of a specific molecule, such as an endogenous metabolite or an exogenously administered contrast agent, the saturation can be transferred to the surrounding water protons through chemical exchange. This transfer reduces the water signal intensity, which can be detected and quantified using CEST MRI. Due to the higher mobility of the contributing protons in CEST imaging compared to MTI, CEST imaging has a higher signal-to-noise ratio (SNR) than MTI (van Zijl et al., 2018).

45.2.1.6 Quantitative Susceptibility Mapping (QSM)

QSM focuses on measuring the magnetic susceptibility of tissues and provides a map of the spatial distribution of tissue susceptibility. QSM provides information about the underlying tissue composition that contributes to the susceptibility effect, such as iron content and is particularly useful for studying the brain. Different tissues have different magnetic susceptibilities that cause local magnetic field distortions in the presence of the external magnetic field generated by the MRI scanner. QSM scans measure the resulting magnetic field variations, which are influenced by various factors, including iron content, myelin concentration, and cellular density. QSM utilizes complex image processing algorithms to convert MRI phase data into quantitative maps of tissue susceptibility.

In summary, microstructural MRI techniques, such as DWI/DTI, MTI, CEST, and QSM, are advanced MRI techniques that can assess tissue microarchitecture, organization, and composition. These techniques offer valuable insights into the cellular and subcellular characteristics, enabling researchers and clinicians to quantitatively and noninvasively study and evaluate brain tissues that change with age or pathological conditions at the microscopic level.

45.2.2 Physiological MRI Techniques

Advanced MRI techniques can assess various physiological parameters. Here, we will discuss perfusion MRI (pMRI) techniques, including dynamic susceptibility contrast (DSC) and arterial spin labeling (ASL).

45.2.2.1 Perfusion MRI (pMRI)

pMRI measures blood flow and provides information about cerebral blood volume (CBV), cerebral blood flow (CBF), and mean transit time (MTT) of the CBV; these parameters reflect metabolic activity and the hemodynamic function of the brain. Various mathematical models are used to analyze pMRI data and derive quantitative perfusion parameters. pMRI can provide valuable information about blood flow patterns, perfusion deficits, and vascular abnormalities in the brain and is widely used in the diagnosis, evaluation, and monitoring of various neurological conditions.

Dynamic Susceptibility Contrast (DSC) pMRI

DSC MRI is one of the most frequently used pMRI techniques and requires the injection of a bolus of a gadolinium (Gd) based contrast agent, which traverses the vessels and capillary beds. Since Gd is paramagnetic, it affects the T1 and T2* relaxation of nearby water protons. After rapid intravenous administration, the bolus of the contrast agent passes through the blood vessels and affects the surrounding water molecules. On a fast T2*-weighted image, the bolus of the Gd contrast agent transiently decreases the signal intensity, proportional to the concentration of the contrast agent, in selected brain region; the resulting changes in the MR signal in each voxel are used to generate a pMRI map (Fig. 45.3). Various modeling approaches, such as deconvolution, are applied to the signal intensity data to estimate the perfusion parameters. These models can quantify the relative CBV, CBF, and MTT values for different brain regions.

Arterial Spin Labeling (ASL)

ASL is a pMRI technique that does not require the use of a Gd contrast agent and measures CBF noninvasively using arterial blood as an endogenous tracer. To acquire an ASL scan, arterial blood water in the neck is first selectively labeled by magnetically tagging the protons in the blood by applying an RF pulse to invert the magnetization of the arterial blood, typically in the neck region. As the labeled blood reaches the brain tissue (after typically 1–2 s), the inverted blood signal causes a drop in the MRI signal in the voxels and provides a measure of blood flow. The mean-transit time (MTT) is the average time it takes for the labeled blood to reach the brain. Accurate measurement and consideration of the transit time are essential for the proper evaluation of the ASL signal since the inverted blood magnetization decays over a few seconds. A quantitative CBF map can be derived from the differences between images with the labeled blood and reference images without the labeling (Fig. 45.4). CBF can be quantified regionally, to evaluate cerebral perfusion in various brain disorders. Furthermore, another advantage compared to DSC-pMRI is that ASL can provide absolute CBF measures while DSC can only provide relative CBF measures in the brain. However, the signal changes detected

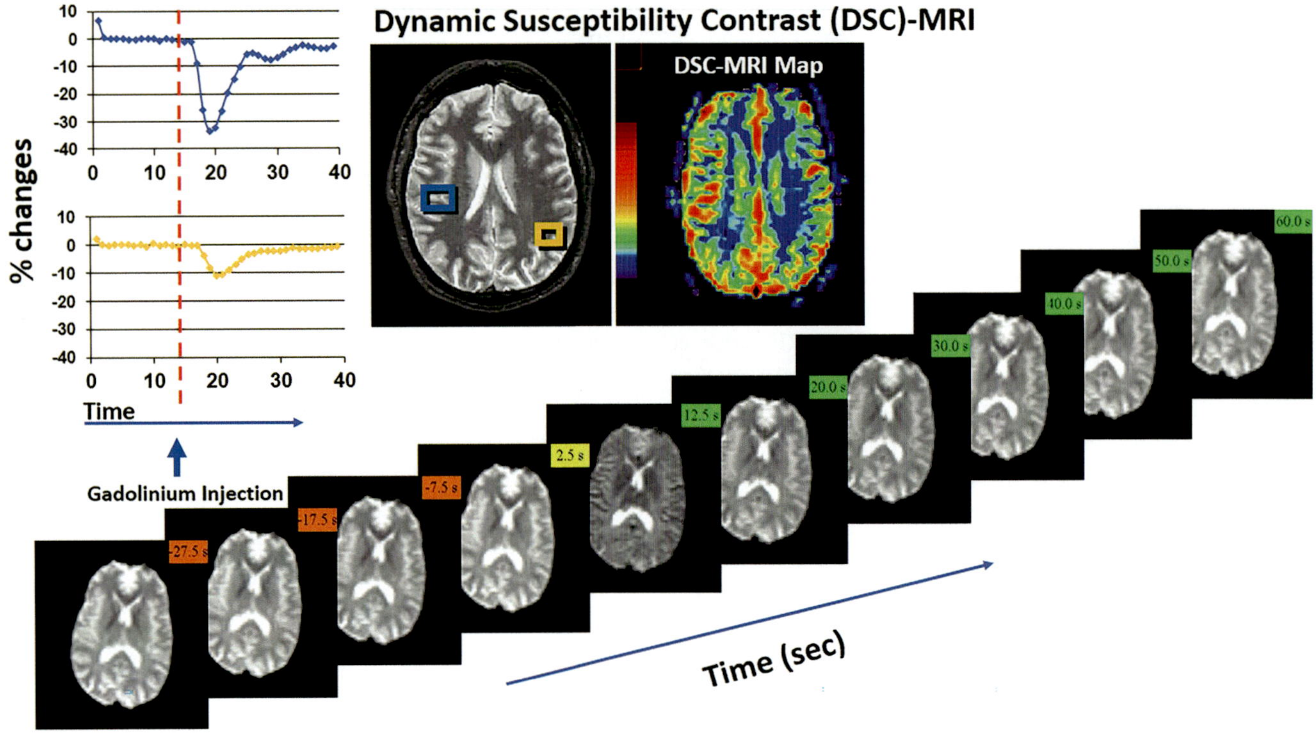

Fig. 45.3 Dynamic susceptibility contrast imaging. *Left panel*, time-dependent concentration for two regions of interest, showing changes in the MR signals associated with passing of the bolus and the resulting % changes in the contrast agent concentration, which generated the final pMRI map (*top panel*). pMRI images for 10-time points at every 10 seconds showing the changes in MR signals, especially at 2.5 seconds

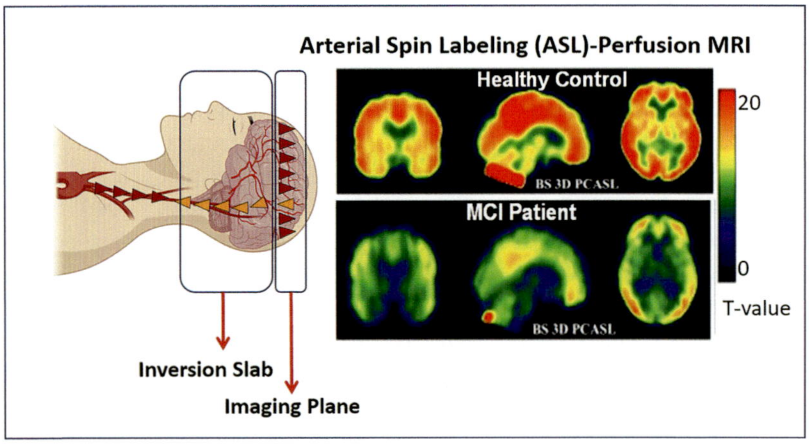

Fig. 45.4 Physical principles of arterial spin labeling (ASL). *Left*: ASL signal is calculated by subtracting the cerebral blood flow (CBF) map of the control image (not inverted) from the image labeled by an RF pulse (inverted). *Right panel*: Examples of ASL maps. Compared to the ASL map in a healthy control, the patient with MCI has markedly lower regional cerebral blood flow (rCBF). (Adapted from Camargo and Wang (2023))

with ASL (approximately 1%) are much smaller than those measured with DSC-pMRI (approximately 20%), leading to a lower SNR for ASL (Narayanan et al., 2020).

45.2.2.2 Cerebrovascular Reactivity (CVR) Mapping

MRI can probe cerebral hemodynamic changes in response to a vasodilatory stimulus using similar MR sequences as functional MRI (fMRI), which is based on the blood-oxygen-level-dependent (BOLD) signal. These cerebral hemodynamic changes are termed CVR and can be assessed and quantified using MRI techniques such as ASL or BOLD techniques. CVR mapping of the brain typically requires a controlled physiological challenge to cause vasodilation, such as inhalation of a CO_2 mixture to alter the concentration of CO_2 in the blood, while monitoring the changes in the MR signals. CO_2 is a good agent for the CVR challenge since it can be given easily as a gas and has a potent vasodilatory effect with rapid onset and washout (Liu et al., 2019). Hypercapnia leads to vasodilation in the brain to increase blood flow and ensure adequate oxygen supply. Conversely, with hypocapnia, vasoconstriction decreases the blood flow. Intravenous administration of acetazolamide, a carbonic anhydrase inhibitor that increases levels of CO_2, can also be used as a pharmacological challenge for CVR mapping. Therefore, CVR can be determined by contrasting the baseline blood flow and oxygenation levels with those observed during the stimulus with a gas mixture or with acetazolamide. A higher CVR indicates better vasoreactivity and suggests healthier cerebrovascular function (Sleight et al., 2021).

45.2.3 Blood-Oxygenation Level Dependent-Functional MRI (BOLD-fMRI)

BOLD-fMRI is a noninvasive neuroimaging technique that measures changes in blood flow and oxygenation in the brain, providing an indirect visualization of brain activity. The BOLD contrast is the basis of the fMRI signal and is generated by changes in the proportions of oxygenated hemoglobin (oxy-Hb), which is diamagnetic, and deoxygenated hemoglobin (deoxy-Hb), which is paramagnetic. When neurons in a specific brain region are activated while a subject performs a mental task, regional oxygen and glucose demands increase to support the local metabolic needs, leading initially to higher deoxy-Hb concentration in that brain region. In response, the regional CBF increases to deliver an excess amount of oxygenated blood, which greatly dilutes the deoxy-Hb and ultimately increases oxy-Hb and hence the BOLD signal. The assumption is that the regional increases in CBF and oxygenation are linked to increased neuronal activity in a particular brain region. Therefore, fMRI measures the indirect effects of neural activity, and the BOLD signal can be affected by oxidative metabolism, blood flow, blood volume, and pathological changes in the neurovascular system. By comparing the BOLD signals between task-induced activated states and baseline nonactivated states, task-fMRI can localize the specific brain regions or brain networks engaged during a given task (Khanna et al., 2015).

fMRI techniques can assess brain activity either during a cognitive or simple motor task or during the resting state (rs-fMRI). In contrast to task-fMRI, rs-fMRI examines BOLD signal variations across different brain regions at rest without a specific task. The brain's ongoing neural activity even at rest requires a continuous supply of oxygenated blood. However, the distribution of blood flow to different brain regions is not uniform and fluctuates over time. rs-fMRI assesses correlated activity with similar fluctuations between different brain regions, known as resting state networks (RSNs). Hence, RSNs reflect synchronized fluctuations in neural activity between interconnected brain regions and represent intrinsic functional connections that occur at the rest. These temporal correlations of spontaneous BOLD signal fluctuations represent functional organization, or brain networks and connectivity of the brain, which can be used to evaluate brain function in healthy individuals or in those with brain disorders. A limitation of this technique is that the BOLD signal changes at rest from the RSNs are substantially smaller than those during a task; therefore, rs-fMRI studies typically require a larger sample size for group comparisons and cannot be applied in the clinical setting with a single subject (Smitha et al., 2017).

45.2.4 Magnetic Resonance Spectroscopy (MRS)

MRS is a noninvasive imaging technique that provides quantitative information about the chemical composition and metabolic processes within tissues. In the MRI scanner, each nucleus is impacted by the magnetic field differently depending on its associated molecular structure, leading to chemical shifts of the nuclei associated with the neurochemicals and the electron density of their environment. After applying an RF pulse, the nuclei emit signals while returning to their equilibrium state. This signal is called the free induction decay (FID). Fourier transformation is used to convert the FID signals associated with the various neurochemicals into an MR spectrum. The area under each spectral peak reflects the concentration of the associated neurochemical (Fig. 45.5a). Hence, the unique resonance pattern of the nucleus associated with each molecule or metabolite, such as neurotransmitters (e.g., glutamate), energy substrates (e.g., glucose), and biochemical markers, reflects cellular metabolism, tissue composition, and disease processes. MRS is most often performed using hydrogen (^{1}H) nuclei (i.e., protons) due to their abundance in biological tissues, which generates a strong signal. Other atomic nuclei, such as carbon-13 (^{13}C), phosphorus-31 (^{31}P), and fluorine-19 (^{19}F) can also be measured using MRS. The selection of the nucleus for MRS is based on the metabolite or chemical of interest.

The concentration of various metabolites can be assessed by comparing the integrated areas of specific spectral peaks with the area of known reference compounds or internal standards. The major brain metabolites evaluated with ^{1}H-MRS in the clinical settings include N-acetylaspartate (NAA), creatine (Cr), phosphocreatine (PCr), choline (Cho), glutamate (Glu), glutamine (Gln), myoinositol (mI), as well as lactate and lipids (Fig. 45.5b). ^{31}P-MRS can assess the concentration of phosphorus compounds, such as duplex oligonucleotides, other phosphate esters, and energy metabolites [e.g., phosphoric acid (H_3PO_4) and adenosine triphosphate (ATP)] (Fig. 45.5c). Since fluorine is relatively rare in biological tissues, ^{19}F MRS can evaluate the biodistribution and pharmacokinetics of drug candidates labeled with

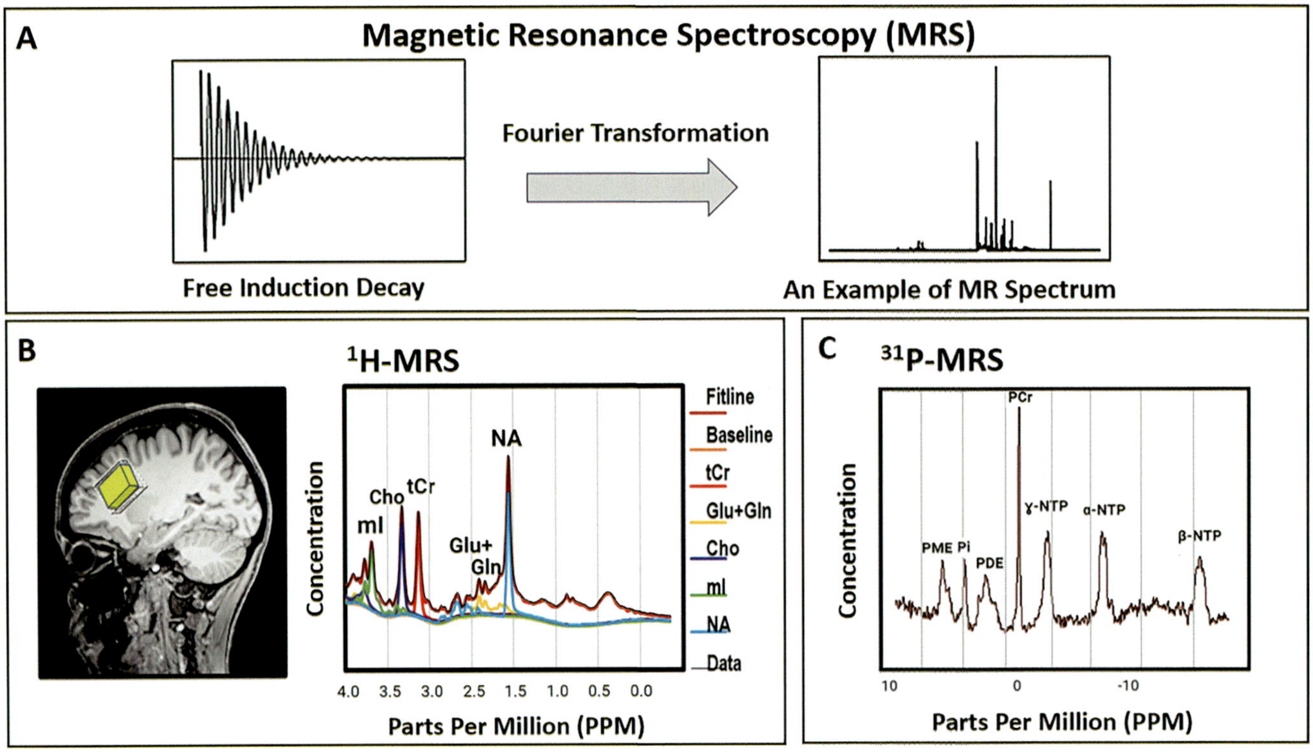

Fig. 45.5 Examples of magnetic resonance spectroscopy (MRS). (**a**) The nuclei of different molecules in a particular voxel are impacted differently in the magnetic field, leading to distinguishable chemical shifts of the protons. Fourier transformation then converts the free induction decay (FID) signals of each nucleus associated with the neurochemicals resonating at different frequencies into the MR spectrum. (**b**) An example of a proton (^{1}H) MRS spectrum from the frontal brain region. (**c**) An example of a phosphorus-31 (^{31}P) MRS spectrum of the whole brain at 3 Tesla. (Adapted from Kondo et al. (2011))

fluorine, such as hexafluorobenzene (C6F6). MRS can also characterize tissue composition by assessing the concentrations of metabolites that reflect neuronal or glial density, which is useful for assessing various neuropathological processes. ^{1}H-MRS is used in both clinical and research settings to evaluate the brain since NAA and Glu are neuronal markers, while mI is a glial marker. tCr and Cho may also reflect glial changes due to their relatively higher concentrations in glia than in neurons, and Cho may also be elevated in conditions that lead to cell membrane turnover, such as demyelination or neoplasm.

45.2.5 Nuclear Medicine Techniques for Brain Imaging

Nuclear medicine techniques for brain mapping require the administration of radiopharmaceuticals that emit gamma rays or positrons (which are eventually converted into a pair of gamma rays). Specialized detectors, such as gamma cameras or scintillation detectors, are then used to detect the gamma rays emitted from the radiopharmaceutical within the body and provide quantitative information such as blood flow, metabolic activity, or receptor binding in the brain. The two most common nuclear imaging techniques are single-photon emission computed tomography (SPECT) and positron emission tomography (PET).

45.2.5.1 SPECT

The gamma rays emitted by SPECT tracers are recorded and eventually reconstructed into a SPECT image that shows the distribution of the single photons. SPECT uses a collimator with an array of parallel holes that allow only gamma rays traversing along specific trajectories to reach the detectors while blocking gamma rays emitted at other angles. These signals are detected by a rotating gamma camera that consists of one or more detectors rotating around the subject and acquiring a series of projection images at different angles. These detectors are typically scintillation crystals coupled to photomultiplier tubes (PMTs). When a gamma ray interacts with the scintillation crystal, it produces flashes of light. The PMTs convert these light signals into electrical signals for further processing and reconstruction of a 3D image showing the radiopharmaceutical distribution in the brain (Fig. 45.6a, *left panel*) (Newberg & Alavi, 2017).

Common radioactive isotopes used to synthesize the SPECT tracers include Iodine-123, Technetium-99m, Indium-111, Xenon-133, and Thallium-201. These isotopes are combined with specific compounds to form SPECT tracers tailored to target specific receptors, enzymes, transporters, or other molecular targets, and are injected intravenously. Table 45.3 provides some examples of SPECT tracers used for brain imaging and Fig. 45.6a, right shows the cerebral blood flow measured using technetium-99m HMPAO tracer in a SPECT image from a healthy control. The selection of a specific SPECT tracer depends on factors such as its half-life, energy of emitted radiation, and imaging characteristics of the isotope, as well as the targeted organ or process of interest (Newberg & Alavi, 2017).

45.2.5.2 PET

In PET, the positron emitted from an unstable isotope collides with an electron, leading to the emission of a pair of gamma rays moving in opposite directions. These gamma rays are detected by the PET scanner and transformed into images showing precise location of the radioactivity and concentration throughout the imaged tissue. Again, collimators are used to control the direction of gamma rays detected by the PET scanner camera to provide spatial resolution. Additionally, PET scanners are designed to accept only pairs of photons (Fig. 45.6b, *left panel*), which results in improved spatial resolution compared to SPECT (Heurling et al., 2017).

PET tracers involve radiopharmaceuticals synthesized with a short-lived positron-emitting isotope, such as fluorine-18 (F-18), carbon-11 (C-11), oxygen-15 (O-15), or nitrogen-15 (N-15). These isotopes are elements of life that are incorporated into molecules involved in the physiological processes in the living system. However, the isotopes have different half-lives, 110 minutes for F-18, 20.4 minutes for C-11, 10 minutes for N-15, and only 2 minutes for O-15. Rapid measurements with precise timing are required for quantitative analyses.

PET images can be quantified using standardized uptake values (SUVs), which are important indices that measure the level of tracer uptake within the brain tissues or lesions.

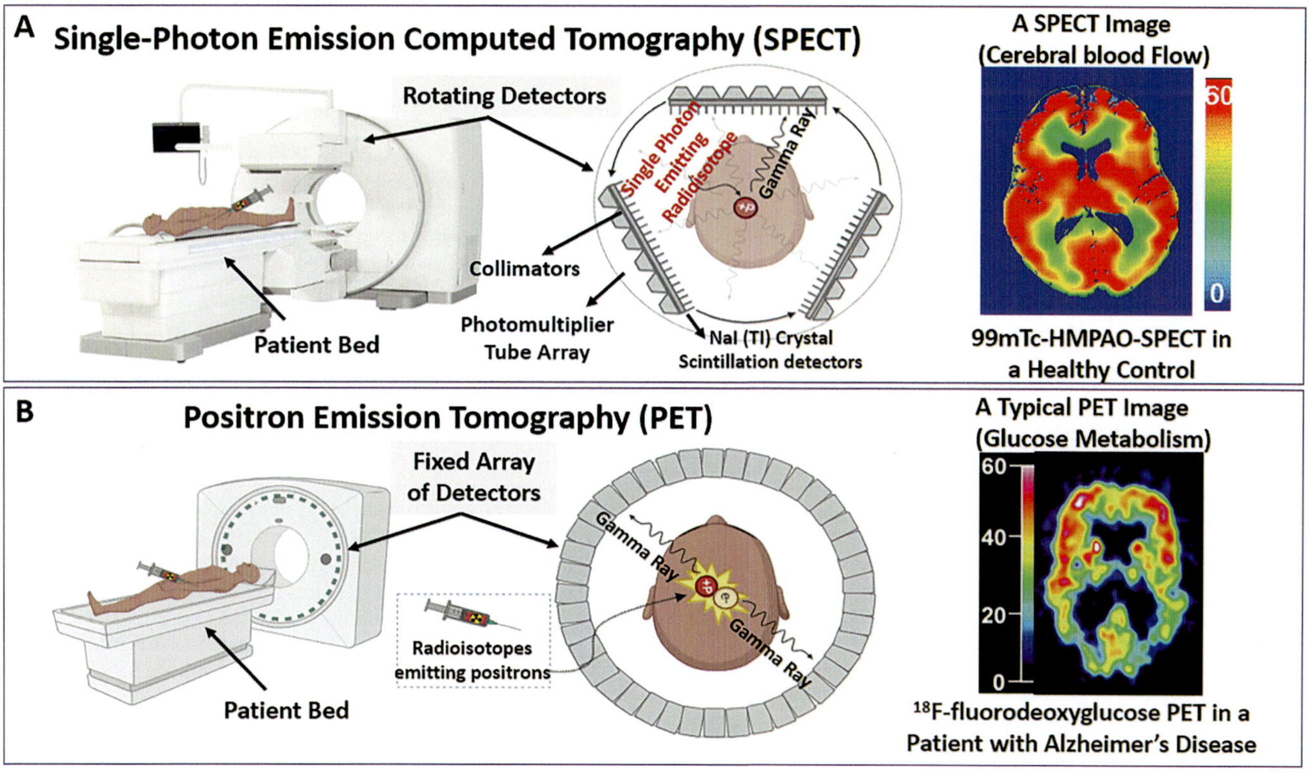

Fig. 45.6 Physical principles of nuclear imaging techniques. (**a**) *Left panel*: single photon emission computed tomography (SPECT), showing the scanner with 2–3 rotating detectors (*left*) and enlarged picture to show the structure of the rotating detectors, including collimators to absorb and direct gamma rays, sodium iodide (NaI) (TI) crystal scintillation detectors to detect the gamma rays and photomultiplier tube array to detect the visible light produced by scintillators and converting it to a measurable electrical signal (see enlarged picture). *Right panel* shows a SPECT image acquired using 99mTc-HMPAO to measure cerebral blood flow from a healthy control. (**b**) *Left panel*: Positron emission tomography (PET), depicting the scanner with a fixed array of detectors (*left*) and enlarged picture to showcase the annihilation of the radioisotope in the tissue, generating two gamma rays in opposite directions. *Right panel* shows a typical 18F-fluorodeoxyglucose (FDG) PET in a Patient with Alzheimer's Disease

Table 45.3 Clinical applications of neuroimaging techniques to assess neuroinflammatory processes

Neuroviral infections	COVID-19	CT and MRI	Enhancements of the cranial nerves, particularly optic nerves, and olfactory bulbs, thickening of the optic nerves, signal changes in the splenium of the corpus callosum, cytotoxic lesions of the corpus callosum	
		DTI	Lower than normal MD and elevated FA, suggest cytotoxic edema, enhanced myelination, or iron deposition	
		fMRI	Greater BOLD activation in the working memory network, decreased static functional connectivity (FC) between both lingual gyri and the right superior occipital gyrus, and a reduction in dynamic FC with the precentral gyrus on rs-fMRI	
		^{1}H-MRS	Reduced NAA in the frontal brain regions consistent with neuronal injury and reduced mI suggesting glial dystrophy, but not glial activation	
	HIV	MRI	Generalized or focal brain atrophy, white matter lesions (leukoencephalopathy)	
		DTI	Higher than normal MD in the frontal and parietal tracts, putamen, and genu of corpus callosum. Reduced anisotropy (FA) and elevated diffusivity (AD and RD) in those with HAND	
		MRS	Elevated glial metabolites (mI, CHO, and tCr) and reduced neuronal markers (NAA or glutamate)	
		fMRI	Increased BOLD signal on rs-fMRI in attention, working memory, and hippocampal function brain networks	
Neurodegenerative diseases	Alzheimer's disease	MRI	Widespread cortical and subcortical atrophy, particularly cortical thinning in fusiform, precentral, and paracentral regions	
		DTI	Widespread decreased FA and increased MD	
		PET	Amyloid PET with tracers such as Flortaucipir (18F), Florbetaben (18F), Flutemetamol (18F), Pittsburgh Compound B (PiB), and Tau PET with Flortaucipir (18F) (also known as AV-1451 or T807), PI-2620 (18F)	Aβ deposits or tau protein accumulation in the frontal and temporal lobes
			TSPO PET such as 11C-(R)-PK11195, 11C-DAA1106, or ^{3}H-(R)-PK11195	Higher than normal global TSPO uptake
	Parkinson's disease	MRI	Widespread cortical and subcortical atrophy	
		DTI	Lower than normal FA in multiple tracts and higher than normal MD in multiple tracts	
		PET	Lesser uptake of dopamine tracer on PET [^{11}C]-(R)-PK11195 (first generation) and several other second-generation ligands bind to the translocator protein 18 kDa, indicating neuroinflammation; increased [^{18}F]NOS uptake on NOS PET	

(continued)

Table 45.3 (continued)

Substance use disorders	MRI	General cortical and subcortical atrophy consistent with premature brain aging
	MRS	Higher than normal myoinositol levels, indicating glial activation
	PET/SPECT	Higher TSPO uptake for CFT-PET and 123I-CIT-SPECT in TSPO
Chronic pain	MRI/fMRI	Greater than normal BOLD signals in the nucleus accumbens (NAc) and medial prefrontal cortex (mPFC)
	MRS	Increased mI levels in ACC, thalamus, and insula and increased glutamate levels in ACC, thalamus, and insula.
	DTI	Widespread increased MD and decreased FA

ACC anterior cingulate cortex, *AD* Alzheimer's disease, *FA* fractional anisotropy, *MD* mean diffusivity, *mPFC* medial prefrontal cortex, *mI* myo-inositol, *NAA* N-acetylaspartate, *PD* Parkinson's disease, *SUDs* substance use disorders

An SUV is calculated by normalizing the tissue radioactivity concentration to the injected dose of radiotracers and the patient's body weight (Beattie et al., 2022). PET is particularly useful in oncology; for example, F-18 fluorodeoxyglucose (FDG) PET is routinely used in the clinical settings to evaluate tumors or to screen for metastases since neoplasms have increased glucose metabolism. Figure 45.6b shows a typical FDG-PET image in a patient with Alzheimer's disease. Since F-18 has the longest half-life amongst the isotopes used for PET tracers, FDG and other F-18 labeled tracers are also often used to evaluate neurodegenerative disorders, both for clinical evaluations and for clinical research studies. Other statistical models for different indices, such as the hypometabolism convergence index (HCI), are also useful for assessing degenerative processes at the cellular level (Ottoy et al., 2019). Furthermore, PET tracers can provide highly specific binding and localization in particular cell types or molecules, such as cellular transporters or receptors (Table 45.2). Several PET tracers can assess microglia activation, including those that bind to the translocator protein (TSPO), such as ^{11}C-PK11195, ^{11}C-PBR28, ^{11}C-DAA1106, and ^{18}F-DP714, or other receptors on microglia or astrocytes (Boche et al., 2019). PET studies can be performed both in humans and in animals, using a microPET system and the same radiotracers as those used in humans for preclinical studies.

45.2.6 In Vivo Imaging System (IVIS) for Preclinical Imaging

IVIS uses 2D optical and 3D optical tomography for noninvasive longitudinal monitoring of disease progression in living animals, primarily in small animals like mice and rats, and can track the distribution of certain cells, monitor gene expression, and to assess drug delivery.

45.2.6.1 Bioluminescent Imaging (BLI)

BLI is a noninvasive imaging technique used widely in preclinical research to investigate gene expression in animal models since the 1990s. BLI can visualize the biological processes and activities within living organisms using bioluminescent signals. Modeled after the natural phenomenon of light-emitting organisms (e.g., fireflies), bioluminescence can be produced in organisms through the chemical reaction of luciferase enzymes. In BLI, a genetically engineered luciferase enzyme is introduced into an organism, or the mRNA for luciferase can be administered to the animal, which will be translated by the animal to produce luciferase. The luciferase acts as a reporter and will emit light when it interacts with a specific substrate (i.e., luciferin), which localizes the reaction that can be visualized in the animal. Specialized imaging systems with highly sensitive detectors employ optical filters, such as charge-coupled device (CCD) cameras or photomultiplier tubes (PMTs) to detect and capture the specific wavelengths emitted by low-level bioluminescent light, while reducing background unwanted noise. To obtain BLI, the organism is placed in a dark environment to minimize background light interference. The emitted bioluminescent light is then detected by the imaging system, and the signal is captured as an image. Multiple images can be acquired over time to track dynamic biological processes. Advanced algorithms and image processing techniques are used to enhance the SNR, correct for background noise, and generate quantitative information about the intensity and localization of the bioluminescent signal. BLI can visualize biological processes such as gene expression, protein–protein interactions, tumor growth, and response to treatment in live animals (Fig. 45.7a).

45.2.6.2 Fluorescent Imaging

Fluorescent imaging is another technique for visualizing the location and activity of specific molecules or cells in live organisms. This is one of the most widely used tools in bio-

Table 45.2 Some common FDA-approved and experimental SPECT and PET tracers, compounds, and applications in brain imaging

Technique	Tracers and compounds (radiopharmaceuticals)		What can it measure?
Single-photon emission computed tomography (SPECT)	Technetium-99m	Hexamethylpropyleneamine oxime (HMPAO)	Cerebral blood flow patterns and brain perfusion
		Ethylene cysteinate dimer (ECD)	
	Ioflupane 123 (DaTSCAN)		Dopamine transporters in dopamine-containing neurons in the brain
	Indium-111, pentetreotide (octreoscan)		Somatostatin receptors, which are located primarily on the surface of neuroendocrine tumors
	Thallium-201 (Tl-201), thallous chloride Tl-201		Brain perfusion can be useful in detecting brain tumors
Positron emission tomography (PET)	^{18}F-Fluorodeoxyglucose (FDG)		Glucose metabolism and brain function, mostly used to diagnose neoplasms or neurodegenerative disorders
	Carbon-11 (^{11}C) raclopride		Dopamine receptor density can aid in the diagnosis of Parkinson's disease and schizophrenia
	Carbon-11 (^{11}C)Pittsburgh compound B (PiB)		Beta-amyloid plaques detect and quantify beta-amyloid burden in the brain in patients with Alzheimer's disease
	^{18}F-Fluorbetaben (Neuraceq) ^{18}F-Florbetapir (Amyvid) ^{18}F-Flutemetamol (Vizamyl)		Beta-amyloid aggregates in the brain allow visualization or quantification of plaque deposition in patients with Alzheimer's disease
	^{18}F-Flortaucipir (Tauvid)		Abnormal tau protein aggregates in the brain allow visualization and quantification of tau protein distribution in the brain in patients with Alzheimer's disease
	O-(2-[18F]Fluoroethyl)-L-Tyrosine (FET)		Characterize, stage, and monitor brain tumors by assessing their amino acid metabolism
	11C-(R)-PK11195		TSPO tracer is the best PET ligand that binds to activated microglia in vivo and may also bind to astrocytes (controversial reports)
	11C-DAA1106		Alternative TSPO tracer detects signal in Alzheimer's disease patients in all regions but does not correspond as well as 11C-(R)-PK11195 to cognitive scores in AD
	^{3}H-(R)-PK11195		TSPO tracer binds to CD68-phagocytic microglia in postmortem tissue

AD Alzheimer's disease, *ECD* ethylene cysteinate dimer, *FET* O-(2-[18F]Fluoroethyl)-L-Tyrosine, *FDG* ^{18}F-Fluorodeoxyglucose, *HMPAO* hexamethylpropyleneamine oxime, *PD* Parkinson's disease, *PET* positron emission tomography, *PiB* Carbon-11 (^{11}C)Pittsburgh compound B, *SPECT* single-photon emission computed tomography

logical research due to its high sensitivity and specificity in imaging the labeled molecules and structures of interest, and it can be applied to living cells and organisms. Fluorescent imaging uses photographic methods, using a light source to illuminate tissue at a wavelength appropriate for fluorochrome excitation, along with a highly sensitive digital camera (e.g., CCD camera) to detect fluorescence emitted back from tissues. Fluorescent imaging has been used to study pharmacokinetics or drug distributions in living animals (Lázaro-Ibánez et al., 2021) (Fig. 45.7b).

45.3 Clinical Applications of Neuroimaging Techniques

Brain imaging plays a crucial role in the clinical evaluation and management of neuroimmune disorders and provides valuable insights into the structural and functional abnormalities or changes in the brains of patients with these disorders. Abnormalities in neuroimaging can guide differential diagnosis and provide biomarkers for disease monitoring and response to treatments, which are critical for patient manage-

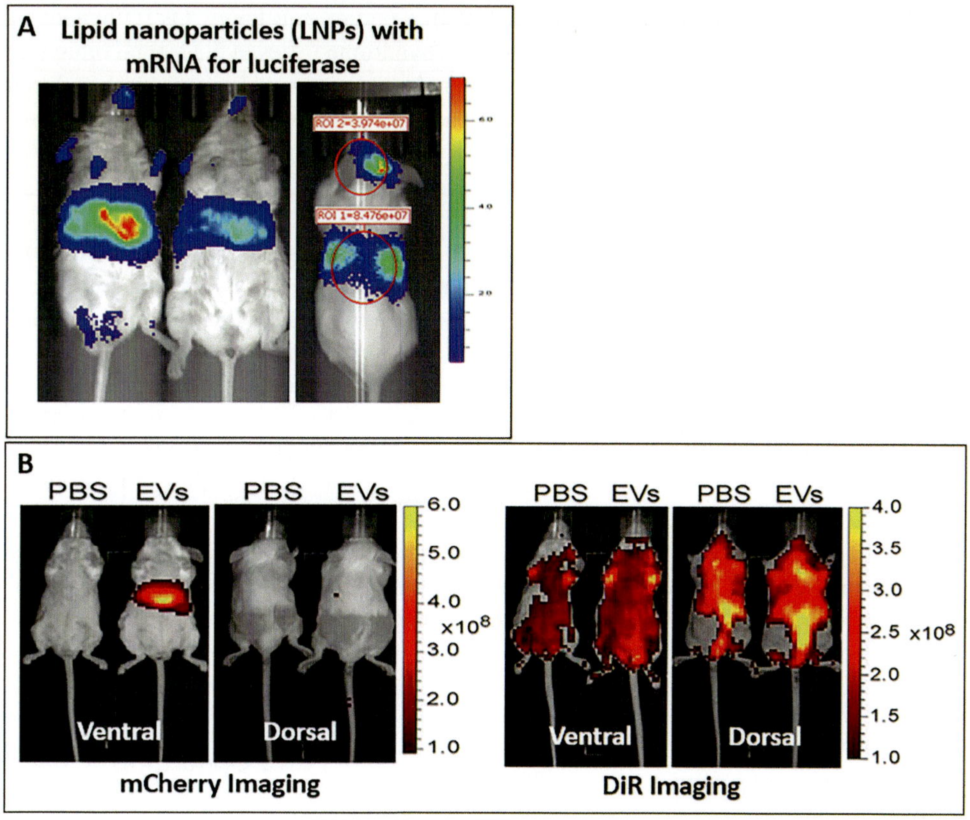

Fig. 45.7 In vivo imaging of animals can be performed with IVIS Spectrum for bioluminescent Imaging (BLI) and Fluorescence imaging. (**a**) Typical bioluminescent imaging of mice that received lipid nanoparticles (LNPs) that are labeled with mRNA for luciferase. Note high signals in the liver and spleen regions in all three mice. However, the mouse on the right also received the LNPs directly injected into the brain. (**b**) Left panels show in vivo images of phosphate-buffered saline (PBS)-treated or mCherry labeled extracellular vesicles (EVs)-treated mice, *Right panel* shows mice treated with DiR-labeled EVs or PBS. (Adapted from Lázaro-Ibánez et al. (2021))

ment. Here, we will briefly discuss how various neuroimaging techniques are used in six types of brain disorders and their associated neuroinflammation, and provide examples to illustrate the neuroimaging abnormalities in these conditions, both in clinical and research settings.

45.3.1 Neuroviral Infections

Neuroviral infections can trigger neuroimmune responses, which can be detected with specific neuroimaging techniques, based on the type of virus, the affected brain regions, and the stage of the infection. MRI is sensitive for assessing neuroinflammation in the brain parenchyma and meninges. These brain abnormalities may be seen as areas of hyperintensity on T2-weighted or FLAIR images, indicating edema or neuroinflammation. Contrast-enhanced MRI may reveal Gd-enhancement of affected regions, indicating blood–brain barrier disruption. With chronic neuroinflammation, both MRI and CT scans may reveal enlargement of the ventricles, periventricular edema, and enhancement of the ependymal lining, especially in those with severe neuroinflammation. However, these imaging findings can be nonspecific in many neuroviral infections. Moreover, hemorrhagic brain manifestations can be seen in some neuroviral infections. For instance, the Zika virus infection may lead to areas of hemorrhage or petechial hemorrhagic foci in the basal ganglia, thalamus, or cerebral cortex. Additionally, neuroviral infections can cause white matter abnormalities, such as demyelination that can be visualized or assessed using microstructural MRI techniques (Table 45.3) (Sriwastava et al., 2021).

45.3.1.1 COVID-19

Brain abnormalities in patients with both acute and long-term COVID have been studied with various neuroimaging techniques over the past 4 years. The most common neuroimaging findings using CT and structural MRI modalities were acute infarcts with large clot burden and intracranial hemorrhage, including microhemorrhages. Other MRI abnormalities in those with acute COVID-19 included evidence of global hypoxic injury, with symmetric T2 FLAIR hyperintensities and mild restricted diffusion in the bilateral globus pallidi, acute disseminated encephalomyelitis, and cytotoxic lesions of the corpus callosum. These brain imaging abnormalities largely represented the immune activation associated with a prothrombotic state and cytokine storm (Moonis et al., 2021). Other specific MRI findings up to 1 month after confirmed COVID-19 infection included greater enhancement and thickening of the optic nerves, as well as enhancements of some other cranial nerves and olfactory bulbs (Sriwastava et al., 2021). In children, specific signal changes in the splenium of the corpus callosum, including T2 hyperintensities in three children and splenial lesions in one child, were detected and were associated with a poorer prognosis, leading to multi-system inflammatory syndrome (Hacohen et al., 2020). In addition, clinical research studies found long-term microstructural changes in patients with long COVID-19, including restricted diffusivity (reduced MD) and higher FA (Fig. 45.8a), that can be associated with iron deposition or enhanced myelination, which might be related to ferroptosis or the stress response to the illness (Liang et al., 2023). Furthermore, BOLD-fMRI studies of those with persistent neuropsychiatric symptoms after acute COVID-19 showed greater BOLD activation in frontal regions of the working memory network, with lesser deactivation in the default mode region, indicating greater usage of frontal cortical reserve but lesser reallocation of other reserve brain resources to maintain cognitive performance (Chang et al., 2023). rs-fMRI was also used to study the effects of COVID-19 on brain function, showing decreased static functional connectivity (FC) between both lingual gyri and the right superior occipital gyrus and a reduction in dynamic FC with the precentral gyrus (Han et al., 2024). Lastly, ^{1}H-MRS in long COVID patients showed neurochemical abnormalities in the frontal brain regions that were consistent with neuronal injury (with reduced NAA) and glial dystrophy (with reduced myoinositol (mI)), instead of glial activation, suggesting possible mitochondrial dysfunction (Ernst et al., 2023).

45.3.1.2 Human Immunodeficiency Virus-1 (HIV-1)

Over the past four decades, various brain imaging abnormalities were reported in persons with HIV (PWH), both with and without opportunistic infections, in clinical settings and in multiple research studies (Chang & Shukla, 2018). The HIV-1 virus enters the brain early after acute infection and is not eradicated from the brain despite antiretroviral treatments. Although HIV does not infect neurons directly, when HIV-infected monocytes cross the blood-brain barrier, they spread the virus to the brain. The HIV-infected macrophages and microglia in the brain release cytokines and chemokines, along with viral particles, that can lead to neuronal apoptosis

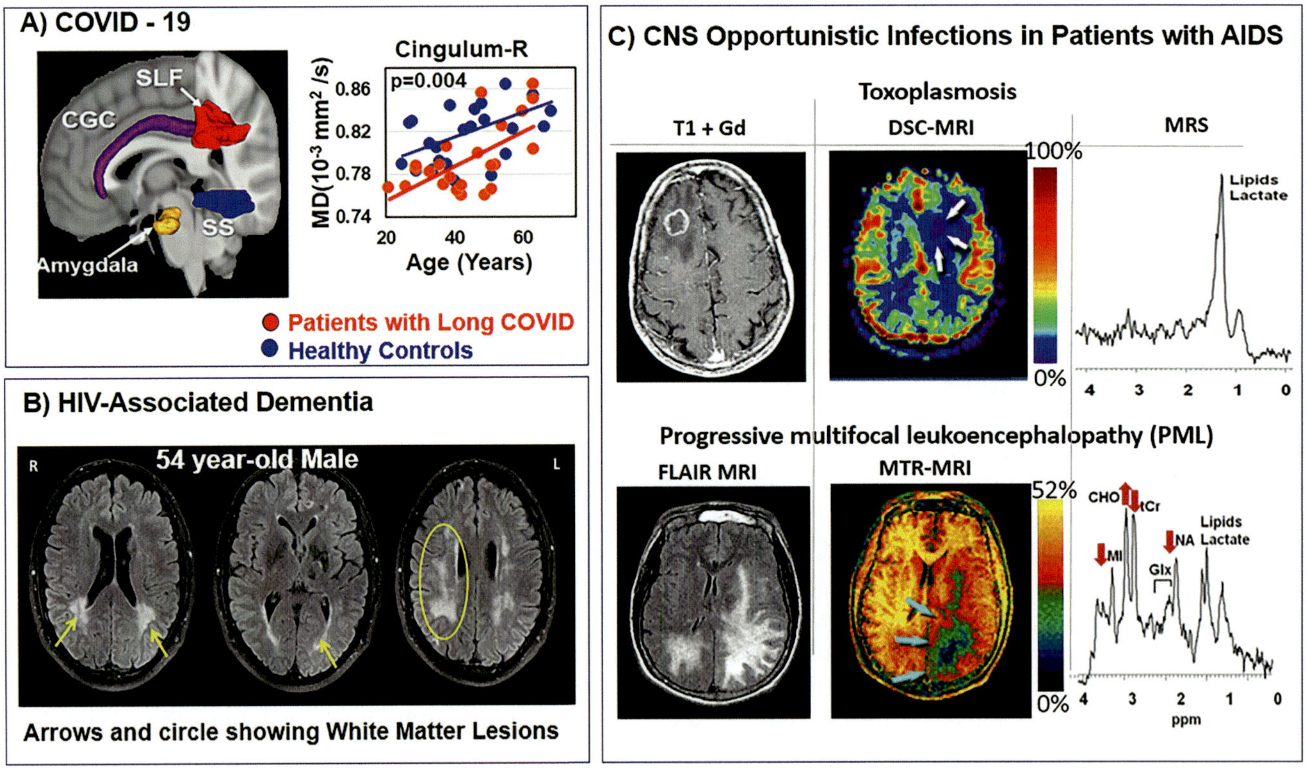

Fig. 45.8 Some clinical applications of different imaging modalities for neuroviral infections. (**a**) Cingulum showing reduced mean diffusivity (MD) in patients with long COVID compared to healthy controls. (Adapted from Liang et al. (2023).) (**b**) FLAIR scan showing white matter lesions in a 54-year-old male patient with HIV-associated dementia (HAD). (**c**) Examples of imaging findings in AIDS patients with opportunistic infections; *Top panel*: imaging findings from a patient with toxoplasmosis, a ring-enhancing lesion on post-gadolinium T1-weighted image (*left*), low or no perfusion within the lesion on dynamic susceptibility contrast MRI (DSC-MRI) (*middle*) and primarily lipids and lactate within and at the rim of the lesion on a ^{1}H-MRS spectrum (*right*). *Bottom panel*: imaging findings from a patient with progressive multifocal leukoencephalopathy (PML), showing bilateral confluent white matter lesions on an axial fluid-attenuated inversion recovery (FLAIR) scan (*left*), low magnetization transfer ratio (MTR) in the demyelinated brain regions (*middle*), and elevated choline-containing compounds (CHO) but reduced N-acetyl aspartate (NAA) and myoinositol (MI) in PML (*right*)

and ongoing aberrant neuroinflammation. The persistence of viral reservoirs in the brain leads to ongoing glial activation, and neuronal dysfunction with dystrophic microglia, which ultimately results in premature aging of the brain in PWH. Various MRI techniques can be used to detect signs of these processes. Structural MRI often showed brain atrophy and discrete or diffuse white matter lesions known as leukoencephalopathy. The ongoing brain infection may also lead to HIV-associated neurocognitive disorder (HAND), and some of these patients with HIV-associated dementia (HAD) show generalized or focal brain atrophy and white matter lesions on MRI (Fig. 45.8b).

Since neuroinflammation is associated with increased brain water, DTI, which detects the microscopic motion of water, is a sensitive technique to identify microstructural WM abnormalities. DWI shows decreased FA and increased diffusion in PWH, especially those with HAND (Holt et al., 2012; Stubbe-Drger et al., 2012). However, even PWH without cognitive deficits or impairments may have significantly higher mean diffusion (MD) in the frontal and parietal WM, corpus callosum, and the putamen, as well as lower FA in the parietal and frontal WM, which correlated with changes in cognitive deficit scores. Furthermore, the relatively higher MD and lower FA in the WM are consistent with neuroinflammation, which was evident even in younger PWH (Chang et al., 2008). In addition, PWH showed a steeper age-related decline in FA within fibers projecting to the temporal and frontal lobes, suggesting accelerated age-related brain abnormalities due to HIV (Kuhn et al., 2017, 2019).

In addition, numerous MRS studies demonstrated neuroinflammation in PWH, as shown by elevated glial metabolites (mI, tCho, and tCr) (Ernst et al., 2003). Altogether, various imaging modalities have confirmed the presence of neuroinflammation in PWH. These findings may have diagnostic utility and are particularly useful for monitoring therapeutic interventions.

Brain imaging is also useful to assess opportunistic infections in the brain of patients with AIDS when their CD4 cell counts are below 200/mm^3. For instance, toxoplasmosis or lymphoma may present with ring-enhancing lesions on post-gadolinium T1 images (Fig. 45.8c, *top panel*). Progressive multifocal leukoencephalopathy (PML), caused by the JC virus in immunocompromised hosts, typically leads to large confluent white matter lesions on T2-weighted or FLAIR images (Chang et al., 1997). Ventriculitis or encephalitis caused by cytomegalovirus may be seen as periventricular enhancement or diffuse hyperintensity in the white matter (Holt et al., 2012). In addition, MRS and perfusion MRI are particularly useful for differentiating toxoplasmosis from CNS lymphoma, as well as evaluating white matter lesions to differentiate those associated with PML from those caused by more fulminant HIV infection (Chang et al., 1995; Ernst et al., 1998; 1999) (Fig. 45.8c, *bottom panel*).

Functional modalities, such as perfusion MRI, fMRI, and nuclear imaging techniques were also used to investigate brain changes in PWH without opportunistic brain lesions (Chang & Shukla, 2018). DSC-pMRI studies demonstrated reduced rCBF in the frontal and parietal lobes but elevated rCBF in posterior inferior parietal white matter, in addition to reduced rCBV in the right temporoparietal, left frontal, and posterior cingulate cortices (Chang et al., 2001; Bladowska et al., 2014). ASL also showed that higher HIV viral load is associated with lower CBF in PWH (Petersen et al., 2021). Moreover, increased BOLD signals, indicating increased usage of brain reserve or declined cognitive efficiency, on task-fMRI were also detected in PWH compared to HIV-negative controls in task-associated regions in task fMRI (Chang et al., 2001; Ernst et al., 2009; Hakkers et al., 2017; Chang & Shukla, 2018). In addition, decreased brain connectivity on rs-fMRI was detected in HIV-positive patients in attention, working memory, and hippocampal function brain networks, although some of these abnormalities normalized with working memory training (Jia et al., 2023). PET study with the TSPO tracer was used to demonstrate neuroinflammation in PWH. Using another PET tracer for neuroinflammation, [^{11}C]PBR28 binding was not significantly different between PWH and HIV-negative controls; however, cognitive performance in PWH was inversely correlated with the [^{11}C]PBR28 binding (Boerwinkle et al., 2020).

45.3.1.3 Neurodegenerative Diseases

Brain imaging is widely used for diagnosis and prognostication in various neurodegenerative disorders. Neuroinflammation and neuronal loss are key pathologic hallmarks in many neurodegenerative diseases. The combined reactions of activated or aberrant neuroimmune cells, such as microglia, astrocytes, and infiltrating lymphocytes, along with inflammatory factors that are released, excitotoxicity from excess extracellular glutamate, and other aspects of neuronal dysfunction, can change the brain homeostasis and lead to neuronal cell death and chronic neuroinflammation (Fuzzati-Armentero et al., 2019). Here, we will briefly discuss three of the most prevalent neurodegenerative diseases and highlight the associated brain imaging findings. Table 45.3 summarizes these findings.

Alzheimer's Disease (AD)

Various imaging techniques were used to study the brain in patients with AD over the past three decades (Haage & De Jager, 2022). The most common findings are brain atrophy, including cortical thinning on structural MRI, in the medial and lateral temporal lobes of patients with AD (Fig. 45.9a). Increased diffusivity may also be found on DTI, while altered brain activation and reduced network connectivity may be detected on task-based fMRI and rs-fMRI. Furthermore,

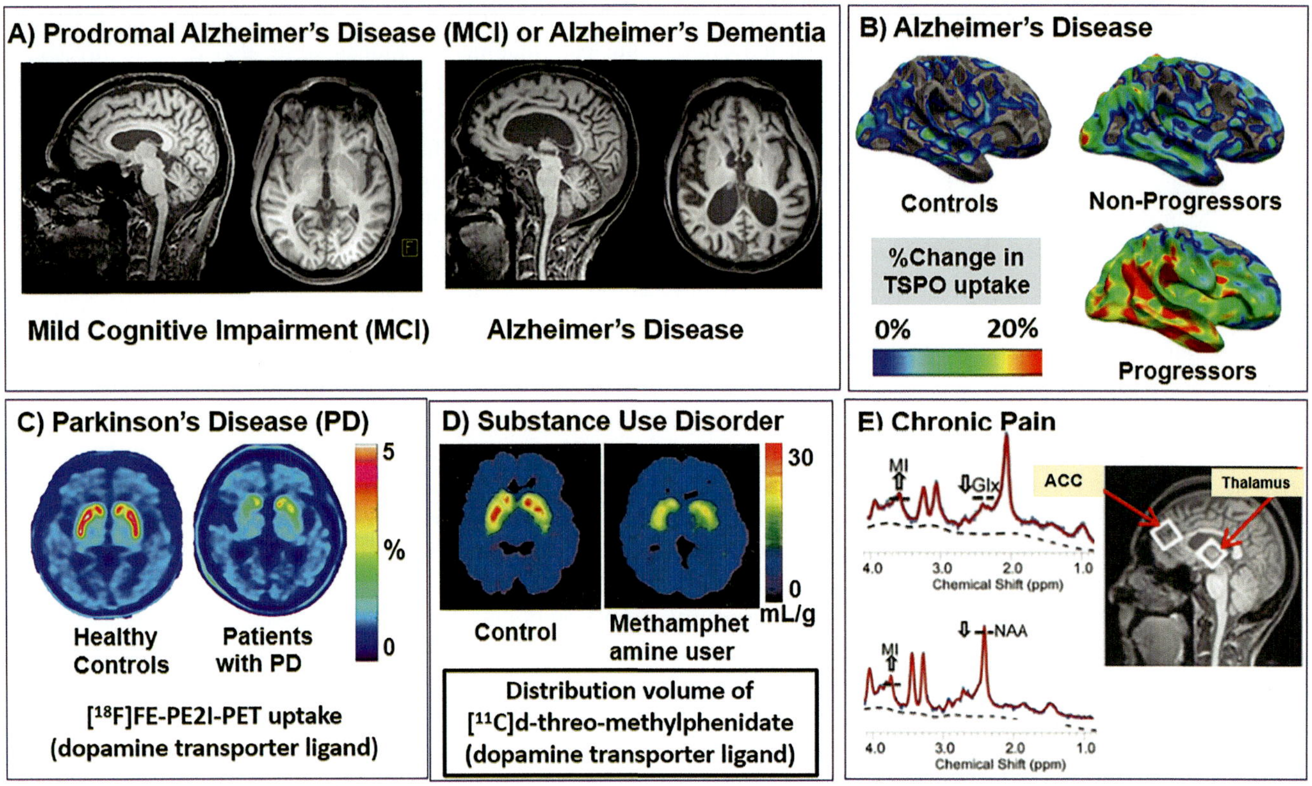

Fig. 45.9 Some clinical applications of different imaging modalities. (**a**) MPRAGE sagittal images of the brain showing moderate to severe brain atrophy and ventriculomegaly in patients with mild cognitive impairment and Alzheimer's disease. (**b**) ^{11}C-PBR28 binding to translocator protein (TSPO) increases with progression of Alzheimer's disease (AD), *top left* shows the typical TSPO uptake in healthy controls, while *top right* shows TSPO binding in patients with nonprogressive AD and *bottom right* shows the same in progressive AD, (Adapted from Kreisl et al. (2017)). (**c**) [^{18}F]FE-PE2I-PET images in a healthy control and in a patient with Parkinson's disease (PD), which shows lesser uptake indicating lower levels of dopamine transporters. (Adapted from Bidesi et al. (2021).) (**d**) Brain images of the distribution volume of [^{11}C] d-threo-methylphenidate showing significantly reduced tracer binding in the striatum in a methamphetamine user relative to the healthy control. (Adapted from Volkow et al. (2001).) (**e**) MRS findings for glutamine and GABA combined in anterior and mid cingulate cortex in a patient with chronic pain (ACC, MCC). (Adapted from Chang et al. (2013))

pathological aggregates such as abnormal levels of beta-amyloid and tau proteins can be detected by nuclear imaging techniques (Song et al., 2021), see the different tracers studied in Table 45.3. The brain regions most affected in AD include the hippocampi, the entorhinal cortices in the medial temporal lobes, the posterior cingulate cortices, the precuneus, parietal lobes, dorsolateral prefrontal cortices, anterior cingulate cortices in the frontal lobes, temporal poles, and superior temporal gyri, while the occipital cortices are often spared. DTI studies consistently reported widespread reduced FA and increased MD in these brain regions that are often affected by AD; these findings were attributed to axonal damage, myelin degradation, and loss of white matter density. Higher MD indicates increased overall water diffusion due to the breakdown of cellular structures, including axonal damage and myelin loss, as well as increased water associated with neuroinflammation. However, DTI requires post-processing to generate the quantitative diffusivity metrics and maps. ^{1}H-MRS was also applied to evaluate patients with AD. On MRS, the neuronal marker NAA is typically reduced while the glial metabolite mI is elevated indicating microglial activation, especially in the temporal parietal brain regions (Ernst et al., 1997; Miller et al., 1993). However, quantitative MRS techniques are not readily available on all clinical MR scanners. Nevertheless, structural MRI, DTI, and MRS are all useful non-invasive and non-radioactive imaging techniques to corroborate with clinical AD diagnosis, and to monitor disease progression; the quantitative measures (e.g., cortical thickness, brain diffusivity, neuro-metabolite levels) are especially useful in clinical research or to monitor treatment effects in clinical trials.

In addition, nuclear medicine techniques such as ^{18}F-FDG PET were used to assess regional metabolism in patients with AD. Hypometabolism in the temporoparietal brain regions is typically seen in these patients due to regional atrophy and dysfunctional brain tissue. However, specific tracers such as those that can assess Aβ deposits and tau protein accumulation are more sensitive and specific than ^{18}F-FDG PET for evaluating AD pathology; these tracers can provide more accurate diagnosis and monitor AD progression. Three amyloid PET tracers and one tau PET tracers are currently available for clinical use (Table 45.2). The recently FDA-approved medications for the treatment of AD include three anti-amyloid monoclonal antibody drugs (aducanumab, lecanemab, and donanemab) that relied on the use of amyloid PET imaging (with ^{18}F-florbetapir tracer) to demonstrate a reduction in brain amyloid (Hoy, 2023; Rahman et al., 2023) and to monitor treatment outcomes. Amyloid imaging can detect amyloid deposits not only in plaques but also in the cerebral blood vessels, and the levels of amyloid burden are often correlated with AD disease severity (Kolanko et al., 2020). Furthermore, tau PET correlates well with cognitive deficits and can significantly predict cognitive decline, supporting the causal role of tau accumulations as a driver of cognitive dysfunction in AD (Sintini et al., 2020) Additionally, tracers that can detect microglia activation and neuroinflammation can also predict disease progression (Valotassiou et al., 2018). Specifically, tracers that bind to the TSPO can visualize and quantify the level of neuroinflammation in the brain, since TSPO is located on the outer membranes of mitochondria mainly in glia, and the TSPO tracer binds specifically to TSPO located in activated microglia. Other tracers that bind to astrocytes have also been studied (Boche et al., 2019). Hence, PET studies can assess the degree of regional microglial or astrocytic activation associated with neuroinflammation. PET studies have consistently shown increased global and regional TSPO uptake in AD patients (Fig. 45.9b) (Li et al., 2024; Zhang et al., 2021).

Vascular Cognitive Impairment and Dementia (VCID)

VCID is the second most prevalent cause of dementia and its diagnosis relies on detecting distinct cerebrovascular pathology on brain imaging (Lang et al., 2018). CT can visualize the potential infarcts, white matter lesions, and related atrophy (Yang et al., 2023). In structural MRI, evidence of end-stage tissue damage appears as white matter hyperintensities and infarcts, such as lacunar infarcts observed as small hyperintense lesions (typically 2–15 mm) on T2 images or small hypointense lesions on T1 scans in deep white matter, basal ganglia, or pons (Ng & Singh, 2015). Additionally, cerebral microbleeds present with hypointense foci on T2 scans. MRI can also show cortical superficial siderosis potentially due to disrupted cerebral vessels. Among the microstructural MR techniques, QSM showed increased iron depositions in the putamen and caudate nucleus on brain MRI of patients with VCID. Moreover, perfusion MRI techniques such as ASL showed reduced CBF in the frontal cortex in patients with VCID (Yang et al., 2023). Lastly, PET imaging showed significant reduction of regional cerebral metabolic rates for glucose (rCMRglc) in VCID compared to cognitively normal controls in widespread cerebral regions such as middle frontal cortex, temporoparietal cortex, basal ganglia, cerebellum, and brainstem. Additionally, PET can provide complementary information to differentiate VCID from degenerative dementia and detect inflammatory changes associated with amyloid depositions in those with mixed dementias. Compared to those with AD, PET images in patients with VCID demonstrated greater hypometabolism in subcortical brain regions and primary sensorimotor cortex, with the association cortical brain regions less affected (Heiss & Zimmermann-Meinzingen, 2012).

Parkinson's Disease (PD)

Several brain imaging biomarkers are used to confirm the clinical diagnosis, monitor disease progression, and assess neuropathology in patients with PD. On structural MR, changes in cortical thickness and gyrification were used as markers to monitor PD progression, with cortical thinning in brain regions like those in AD, in addition to the fusiform, precentral, and paracentral regions. These structural brain abnormalities are often correlated with cognitive impairment, especially in PD patients with amyloid co-pathology. Since the pathological hallmark of PD involves the significant loss of dopaminergic neurons, imaging methods that can evaluate the loss of dopamine transporters (DaT), using SPECT or PET, are particularly useful to confirm the diagnosis of PD. DaT are located on the presynaptic dopaminergic neurons, and tracer binding values in the brain reflect the density of DaT, which is reduced in PD. DaT scans using SPECT or PET, such as CFT-PET and 123I-CIT-SPECT, can confirm the diagnosis, distinguish PD from other parkinsonian syndromes, and monitor disease progression (Mitchell et al., 2021). [^{18}F]fluorodopa (^{18}F-DOPA) is a highly sensitive and specific PET tracer that can also evaluate the integrity of presynaptic dopaminergic nerve terminals, while [^{11}C] raclopride can assess dopamine D2 receptor binding on postsynaptic brain cells, including those on the glia (Fig. 45.9c). These tracers primarily bind to the striatal brain regions, where the dopaminergic synapses are found at the highest concentration in the brain. Since the abnormal aggregation of α-synuclein after phagocytosis by microglia is a possible neuropatho-physiological mechanism that enhances the progression of PD, a recent new PET tracer [^{18}F]NOS was used to detect the inducible isoform of nitric oxide synthase (NOS) and provided an indirect assessment of neuroinflammation. Patients with early stages of idiopathic PD already showed increased whole-brain uptake of [^{18}F]NOS, suggesting elevated oxidative stress, a surrogate marker of neuroinflammation (Doot et al., 2022).

45.3.1.4 Substance Use Disorders (SUDs)

The role of neuroinflammation in SUDs has been investigated extensively. Imaging biomarkers of the neuroimmune mechanisms in vivo provides a vital translational bridge between preclinical and clinical research. The abused substances exert diverse and complex neuroimmune effects that may change across addiction stages (Woodcock et al., 2019). Chronic substance use, especially with alcohol, is associated with structural alterations in the brain such as reduced cortical gray matter volumes, particularly in regions involved in reward processing, decision-making, and impulse control, such as the prefrontal cortex and the striatum. Some studies also reported smaller hippocampal volumes in individuals with alcohol use disorder. The reward circuitry, including the mesolimbic dopaminergic pathway, is a key target of drugs of abuse and often shows reduced D2 receptors in individuals with SUD. These alterations include abnormalities in dopamine release and reduced dopamine D2 receptor availability, which can be assessed with PET studies (Volkow & Morales, 2015), while disrupted network connectivity between the striatum, prefrontal cortex, and other regions involved in reward processing can be studied with rs-fMRI (Xu et al., 2020).

Microglia are the resident neuroimmune cells that play a crucial role in neuroinflammation. Microglial activation also contributes to neuroimmune dysregulation in SUD. PET imaging using radiotracers targeting microglial activation markers, such as TSPO with various tracers including [^{11}C]DPA-713, [^{11}C]PBR28 or [^{18}F]FEPPA, has been used to study participants who used alcohol, opioids, tobacco, methamphetamine, cocaine, and cannabis (Notter et al., 2018). TSPO is also expressed in astrocytes and endothelial cells in the brain but is predominantly localized in microglia (Fig. 45.9d). Additionally, ^{1}H-MRS was used to study neuroinflammation and neuronal injury in persons with SUDs, including those who used opioids, cocaine, alcohol, marijuana, tobacco, and 3,4-methylenedioxymethamphetamine (MDMA) (Woodcock et al., 2019). Specifically, mI, a glial marker, is often elevated, indicating neuroinflammation, while NAA or glutamate are neuronal markers that might be reduced, reflecting the extent of neuronal injury or loss, in those with SUDs (Chang et al., 2005; Ernst and Chang, 2008; Woodcock et al., 2019) (Table 45.3).

45.3.1.5 Chronic Pain

Different neuroimaging techniques were used to study brain regions and mechanisms involved in the processing of pain (Martucci & MacKey, 2018), Table 45.3 Chronic pain conditions also involve neuroimmune mechanisms and hence may exhibit signs of neuroinflammation in the brain. ^{1}H-MRS showed elevated glial metabolite levels of mI in patients with chronic pain (Fig. 45.9e). Another neurometabolite that showed variable levels in different types of chronic pain was glutamate, which was elevated in those with migraines, but reduced in those with musculoskeletal conditions, and may be relatively normal in those with chronic pain syndromes (Peek et al., 2020). rsfMRI also demonstrated increased synchronous ultralow frequency oscillations (<0.1 Hz) between brain regions involved in pain processing, emotional regulation, and cognitive control in patients with chronic pain (Davis & Moayedi, 2012). On rs-fMRI, the nucleus accumbens (NAc) exhibited positive functional connectivity correlations with other brain regions involved in pain processing, such as the basal ganglia, amygdala, and medial prefrontal cortex (mPFc). Increased mPFc-NAc connectivity was detected both in people with chronic and acute pain. Additionally, the BOLD signal in NAc was significantly associated with the perceived magnitude of acute pain and with the intensity of ongoing chronic pain (Baliki et al., 2010). Neuroinflammation associated with chronic pain may also impact white matter integrity. On DTI, patients with chronic pain may show widespread increased MD and decreased FA, and the pain severity was predicted by the FA index in the anterior insula and the ventroposterior lateral (VPL) thalamus (Wang et al., 2017). In addition, PET imaging studies using radiotracers targeting microglial activation markers (e.g., [11C]PBR28 or [11C]DPA-713) showed increased tracer uptakes in regions associated with pain processing, such as the somatosensory cortex and insula. Furthermore, neuroinflammation associated with chronic pain can spread via axonal projections in the brain, thereby establishing neuroinflammatory tracts and secondary neuroinflammatory foci, which might be useful to monitor disease progression (Cooper & Clark, 2012). Such spreading neuroinflammation might be visualized with [^{11}C](R)-PK11195, a TSPO PET tracer that assesses glial activation in the CNS in the case of a peripheral nerve injury (Banati et al., 2001).

45.4 Summary

This chapter presented a brief overview of various brain imaging techniques and provided examples of some of their clinical applications. We discussed the major structural neuroimaging techniques, including CT and MRI, as well as different MRI methods specifically designed to detect microstructural changes in the brain, including DTI, MTI, CEST, and QSM. We also discussed several physiological imaging techniques focusing on those that evaluated cerebral perfusion or vascular reactivity, including DSC-pMRI, ASL, and CVR. We further described functional imaging techniques that can evaluate brain activation or brain network connectivity using BOLD-fMRI. In addition, we discussed how ^{1}H-MRS can measure neurochemicals and brain metabolites or neurochemicals that reflect glial or neuronal composition in particular brain regions. We also discussed and illustrated some preclinical imaging techniques. Lastly, we reviewed the physical principles and showed examples of clinical applications of PET and SPECT techniques and their associated radiotracers.

45.5 Review Questions

1. Which imaging technique is primarily used for assessing bone structures or abnormalities, such as fractures, and has the ability to generate both 2D projection and 3D images?
   (a) X-Ray and computed tomography (CT)
   (b) Structural magnetic resonance imaging (MRI)
   (c) Positron emission tomography (PET)
   (d) Ultrasound imaging
   (e) Single-photon emission computed tomography (SPECT)
2. Which microstructural MRI technique assesses the magnetization exchange between protons bound to water and macromolecules, providing valuable insights into tissue integrity and composition, particularly for detecting early demyelination?
   (a) Diffusion-weighted imaging (DWI)
   (b) Magnetization transfer imaging (MTI)
   (c) Chemical exchange saturation transfer (CEST)
   (d) Quantitative susceptibility mapping (QSM)
   (e) Functional magnetic resonance imaging (fMRI)
3. Which of the following is NOT a parameter that can be assessed using perfusion MRI (pMRI) techniques?
   (a) Cerebral blood volume (CBV)
   (b) Cerebral blood flow (CBF)
   (c) Mean transit time (MTT)
   (d) Cerebrovascular reactivity (CVR)
   (e) None of the above
4. Which of the following statements regarding arterial spin labeling (ASL) is TRUE?
   (a) ASL requires the injection of a gadolinium-based contrast agent.
   (b) ASL measures cerebral blood flow (CBF) using exogenous tracers.
   (c) ASL measures CBF by magnetically tagging the protons in the arterial blood.
   (d) ASL does not provide absolute CBF measures.
   (e) None of the above
5. What is the neuroimaging modality that is specific for assessing neuroinflammation in the brain?
   (a) DTI
   (b) MRI
   (c) ASL
   (d) CT
   (e) TSPO-PET

45.6 Correct Answers

1. (a)
2. (b)
3. (d)
4. (c)
5. (e)

References

Baliki, M. N., Geha, P. Y., Fields, H. L., & Apkarian, A. V. (2010). Predicting value of pain and analgesia: Nucleus accumbens response to noxious stimuli changes in the presence of chronic pain. *Neuron, 66*, 149–160. https://doi.org/10.1016/J.NEURON.2010.03.002

Banati, R. B., Cagnin, A., Brooks, D. J., Gunn, R. N., Myers, R., Jones, T., Birch, R., & Anand, P. (2001). Long-term trans-synaptic glial responses in the human thalamus after peripheral nerve injury. *Neuroreport, 12*, 3439–3442. https://doi.org/10.1097/00001756-200111160-00012

Beattie, B. J., Akhurst, T. J., Augensen, F., & Humm, J. L. (2022). SUVfdg: A standard-uptake-value (SUV) body habitus normalizer specific to fluorodeoxyglucose (FDG) in humans. *PLoS One, 17*, e0266704. https://doi.org/10.1371/JOURNAL.PONE.0266704

Bidesi, N. S. R., Vang Andersen, I., Windhorst, A. D., Shalgunov, V. &, Herth, M. M. (2021, November). The role of neuroimaging in Parkinson's disease. *Journal of Neurochemistry*. *159*(4):660–689. https://doi.org/10.1111/jnc.15516. Epub 2021 Oct 3. PMID: 34532856; PMCID: PMC9291628

Bladowska, J., Knysz, B., Zimny, A., Małyszczak, K., Kołtowska, A., Szewczyk, P., Gamsiorowski, J., Furdal, M., & Samsiadek, M. J. (2014). Value of perfusion-weighted MR imaging in the assessment of early cerebral alterations in neurologically asymptomatic HIV-1-positive and HCV-positive patients. *PLoS One, 9*, e102214. https://doi.org/10.1371/JOURNAL.PONE.0102214

Boche, D., Gerhard, A., Rodriguez-Vieitez, E., et al. (2019). Prospects and challenges of imaging neuroinflammation beyond TSPO in Alzheimer's disease. *European Journal of Nuclear Medicine and Molecular Imaging, 46*, 2831–2847. https://doi.org/10.1007/s00259-019-04462-w

Boerwinkle, A. H., Strain, J. F., Burdo, T., Doyle, J., Christensen, J., Su, Y., Wisch, J. K., Cooley, S. A., Vaida, F., Smith, M. D., Jafri, H., Paul, R. H., Benzinger, T. L. S., & Ances, B. M. (2020). Comparison of [11C]-PBR28 binding between persons living with HIV and HIV-uninfected individuals. *Journal of Acquired Immune Deficiency Syndromes, 1988*(85), 244–251. https://doi.org/10.1097/QAI.0000000000002435

Boska, M. D., & White, M. L. (2008). Neuroimaging. In *Neuroimmune pharmacology* (pp. 743–764). Springer. https://doi.org/10.1007/978-0-387-72573-4_50/COVER

Camargo, A., & Wang, Z. (2023, June). Alzheimer's disease neuroimaging initiative. Hypo- and hyper-perfusion in MCI and AD identified by different ASL MRI sequences. *Brain Imaging and Behavior 17*(3):306–319. https://doi.org/10.1007/s11682-023-00764-8. Epub 2023 Mar 27. PMID: 36973476; PMCID: PMC10198885.

Chang, L., & Shukla, D. K. (2018). Imaging studies of the HIV-infected brain. *Handbook of Clinical Neurology, 152*, 229–264. https://doi.org/10.1016/B978-0-444-63849-6.00018-9

Chang, L., Speck, O., Miller, E. N., Braun, J., Jovicich, J., Koch, C., Itti, L., & Ernst, T. (2001). Neural correlates of attention and working memory deficits in HIV patients. *Neurology, 57*, 1001–1007. https://doi.org/10.1212/WNL.57.6.1001/ASSET/3057F90C-AED2-49D4-98B9-8C14CE6391A5/ASSETS/GRAPHIC/G16FF4.JPEG

Chang, L., Wong, V., Nakama, H., Watters, M., Ramones, D., Miller, E. N., Cloak, C., & Ernst, T. (2008). Greater than age-related changes in brain diffusion of HIV patients after 1 year. *Journal of Neuroimmune Pharmacology, 3*, 265–274. https://doi.org/10.1007/S11481-008-9120-8/FIGURES/4

Chang, L., Ryan, M. C., Liang, H., Zhang, X., Cunningham, E., Wang, J., Wilson, E., Herskovits, E. H., Kottilil, S., & Ernst, T. M. (2023). Changes in brain activation patterns during working memory tasks in people with post-COVID condition and persistent neuropsychiatric symptoms. *Neurology, 100*, E2409–E2423. https://doi.org/10.1212/WNL.0000000000207309/ASSET/DA2260D3-9EF6-4DC4-97C5-7A78713BFF9C/ASSETS/IMAGES/LARGE/12TTU1.JPG

Chang, L., Miller, B. L., McBride, D., Cornford, M., Oropilla, G., Buchthal, S., Chiang, F., Aronow, H., Beck, K., & Ernst, T. (1995). Brain lesions in patients with AIDS: H-1 MR spectroscopy. *Radiology, 197*, 525–531.

Chang, L., Ernst, T., Tornatore, C., Aronow, H., Melchor, R., Walot, I., Singer, E., & Cornford, M. (1997). Metabolite abnormalities in progressive multifocal leukoencephalopathy by proton magnetic resonance spectroscopy. *Neurology, 48*(4), 836–845.

Chang, L., Ernst, T., Speck, O., & Grob, C. (2005, February). Additive effects of HIV and chronic methamphetamine use on brain metabolite abnormalities. *American Journal of Psychiatry, 162*(2), 1–9.

Chang, L., Munsaka, S. M., Kraft-Terry, S., et al. (2013). Magnetic resonance spectroscopy to assess neuroinflammation and neuropathic pain. *Journal of Neuroimmune Pharmacology, 8*, 576–593. https://doi.org/10.1007/s11481-013-9460-x

Cooper, M. S., & Clark, V. P. (2012). Neuroinflammation, neuroautoimmunity, and the Co-morbidities of complex regional pain syndrome. *Journal of Neuroimmune Pharmacology, 8*(3), 452–469. https://doi.org/10.1007/S11481-012-9392-X

Davis, K. D., & Moayedi, M. (2012). Central mechanisms of pain revealed through functional and structural MRI. *Journal of Neuroimmune Pharmacology, 8*(3), 518–534. https://doi.org/10.1007/S11481-012-9386-8

Doot, R. K., Young, A. J., Nasrallah, I. M., Wetherill, R. R., Siderowf, A., Mach, R. H., & Dubroff, J. G. (2022). [18F]NOS PET brain imaging suggests elevated neuroinflammation in idiopathic Parkinson's disease. *Cells, 11*, 3081. https://doi.org/10.3390/CELLS11193081/S1

Ernst, T., & Chang, L. (2008). Adaptation of brain glutamate plus glutamine during abstinence from chronic methamphetamine use. *Journal of NeuroImmune Pharmacology 3*(3), 165–172; Epub2008 Jun 3; PMID: 18521756; PMC2575014.

Ernst, T., Chang, L., Melchor, R., & Mark Mehringer, C. (1997). Frontotemporal dementia and early Alzheimer disease: Differentiation with frontal lobe H-1 MR spectroscopy. *Radiology, 203*, 829–836. https://doi.org/10.1148/RADIOLOGY.203.3.9169712

Ernst, T., Chang, L., & Arnold, S. (2003). Increased glial metabolites predict increased working memory network activation in HIV brain injury. *Neuroimage, 19*, 1686–1693. https://doi.org/10.1016/S1053-8119(03)00232-5

Ernst, T., Ryan, M. C., Liang, H.-J., Wang, J. P., Cunningham, E., Saleh, M. G., Kottilil, S., & Chang, L. (2023). Neuronal and glial metabolite abnormalities in participants with persistent neuropsychiatric symptoms after COVID-19: A brain proton MR spectroscopy study. *The Journal of Infectious Diseases, 228*, 1559. https://doi.org/10.1093/INFDIS/JIAD309

Ernst, T., Chang, L., Witt, M., Aronow, H., Cornford, M., Walot, I., McBride, D., & Goldberg, M. (1998). Perfusion MR imaging of cerebral toxoplasmosis and lymphoma in AIDS: Preliminary observations. *Radiology, 208*(3), 663–339.

Ernst, T., Chang, L., Witt, M., Aronow, H., Walot, I., Leonido-Yee, M., & Singer, E. (1999). Progressive multifocal leukoencephalopathy and HIV-associated white matter lesions in AIDS: Magnetization transfer MR imaging. *Radiology, 210*(2), 539–543.

Ernst, T., Yakupov, R., Nakama, H., Crocket, G., Cole, M., Watters, M., Ricardo-Dukelow, M., & Chang, L. (2009). Declined neural efficiency in cognitively stable HIV patients. *Annals of Neurology, 65*(3), 316–325. PMID:19334060; PMC2734503.

Fuzzati-Armentero, M. T., Cerri, S., & Blandini, F. (2019). Peripheral-central neuroimmune crosstalk in Parkinson's disease: What do patients and animal models tell us? *Frontiers in Neurology, 10*, 232. https://doi.org/10.3389/FNEUR.2019.00232/BIBTEX

Grover, V. P. B., Tognarelli, J. M., Crossey, M. M. E., Cox, I. J., Taylor-Robinson, S. D., & McPhail, M. J. W. (2015). Magnetic resonance imaging: Principles and techniques: Lessons for clinicians. *Journal of Clinical and Experimental Hepatology, 5*, 246–255. https://doi.org/10.1016/J.JCEH.2015.08.001

Haage, V., & De Jager, P. L. (2022). Neuroimmune contributions to Alzheimer's disease: A focus on human data. *Molecular Psychiatry, 27*, 3164–3181. https://doi.org/10.1038/s41380-022-01637-0

Hacohen, Y., Abdel-Mannan, O., Eyre, M., Löbel, U., Bamford, A., Eltze, C., Hameed, B., & Hemingway, C. (2020). Neurologic and radiographic findings associated with COVID-19 infection in children. *JAMA Neurology, 77*, 1440–1445. https://doi.org/10.1001/JAMANEUROL.2020.2687

Hakkers, C. S., Arends, J. E., Barth, R. E., Du Plessis, S., Hoepelman, A. I. M., & Vink, M. (2017). Review of functional MRI in HIV: Effects of aging and medication. *Journal of Neurovirology, 23*, 20–32. https://doi.org/10.1007/S13365-016-0483-Y/TABLES/5

Han, M., He, C., Li, T., Li, Q., Chu, T., Li, J., & Wang, P. (2024). Altered dynamic and static brain activity and functional connectivity in COVID-19 patients: A preliminary study. *Neuroreport, 35*, 306. https://doi.org/10.1097/WNR.0000000000002009

Heiss, W. D., & Zimmermann-Meinzingen, S. (2012). PET imaging in the differential diagnosis of vascular dementia. *Journal of the Neurological Sciences, 322*, 268–273. https://doi.org/10.1016/J.JNS.2012.09.023

Heurling, K., Leuzy, A., Jonasson, M., Frick, A., Zimmer, E. R., Nordberg, A., & Lubberink, M. (2017, September). Quantitative positron emission tomography in brain research. *Brain Research* 1;1670:220–234. https://doi.org/10.1016/j.brainres.2017.06.022. Epub 2017 Jun 23.

Holt, J. L., Kraft-Terry, S. D., & Chang, L. (2012). Neuroimaging studies of the aging HIV-1-infected brain. *Journal of Neurovirology, 18*, 291–302. https://doi.org/10.1007/S13365-012-0114-1/TABLES/2

Hoy, S. M. (2023). Lecanemab: First approval. *Drugs, 83*, 359–365. https://doi.org/10.1007/S40265-023-01851-2

Jia, C., Long, Q., Ernst, T., Shang, Y., Chang, L., & Adali, T. (2023). Independent component and graph theory analyses reveal normalized brain networks on resting-state functional MRI after working memory training in people with HIV. *Journal of Magnetic Resonance Imaging, 57*(5), 1552–1564. https://doi.org/10.1002/jmri.28439. Epub 2022 Sep 27. PMID: 36165907; PMCID: PMC10040468.

Khanna, N., Altmeyer, W., Zhuo, J., & Steven, A. (2015). Functional neuroimaging: Fundamental principles and clinical applications. *The Neuroradiology Journal, 28*, 87. https://doi.org/10.1177/1971400915576311

Kolanko, M. A., Win, Z., Loreto, F., Patel, N., Carswell, C., Gontsarova, A., Perry, R. J., & Malhotra, P. A. (2020). Amyloid PET imaging in clinical practice. *Practical Neurology, 20*, 451–462. https://doi.org/10.1136/PRACTNEUROL-2019-002468

Kondo, D. G., Hellem, T. L., Sung, Y. H., et al. (2011). Review: Magnetic resonance spectroscopy studies of pediatric major depressive disorder. *Depression Research and Treatment, 2011*, 650450. https://doi.org/10.1155/2011/650450. PMID: 21197097; PMCID: PMC3003951.

Kreisl, W. C., Lyoo, C. H., Liow, J. S., Snow, J., Page, E., Jenko, K. J., Morse, C. L., Zoghbi, S. S., Pike, V. W., Turner, R. S., & Innis, R. B. (2017, March). Distinct patterns of increased translocator protein in posterior cortical atrophy and amnestic Alzheimer's disease. *Neurobiology of Aging. 51*:132–140. https://doi.org/10.1016/j.neurobiolaging.2016.12.006. Epub 2016 Dec 16. PMID: 28068564; PMCID: PMC5381519.

Kuhn, T., Schonfeld, D., Sayegh, P., Arentoft, A., Jones, J. D., Hinkin, C. H., Bookheimer, S. Y., & Thames, A. D. (2017). The effects of HIV and aging on subcortical shape alterations: A 3D morphometric study. *Human Brain Mapping, 38*, 1025–1037. https://doi.org/10.1002/HBM.23436

Kuhn, T., Jin, Y., Huang, C., Kim, Y., Nir, T. M., Gullett, J. M., Jones, J. D., Sayegh, P., Chung, C., Dang, B. H., Singer, E. J., Shattuck, D. W., Jahanshad, N., Bookheimer, S. Y., Hinkin, C. H., Zhu, H., Thompson, P. M., & Thames, A. D. (2019). The joint effect of aging and HIV infection on microstructure of white matter bundles. *Human Brain Mapping, 40*, 4370–4380. https://doi.org/10.1002/HBM.24708

Lang, B., et al. (2018). Multi-parametric classification of vascular cognitive impairment and dementia: The impact of diverse cerebrovascular injury biomarkers. *Journal of Alzheimer's Disease, 62*(1), 39–60.

Lázaro-Ibánez, E., Al-Jamal, K. T., Dekker, N., Faruqu, F. N., Saleh, A. F., Silva, A. M., Wang, J. T. W., & Rak, J. (2021). Selection of fluorescent, bioluminescent, and radioactive tracers to accurately reflect extracellular vesicle biodistribution in vivo. *ACS Nano, 15*, 3212–3227. https://doi.org/10.1021/ACSNANO.0C09873/SUPPL_FILE/NN0C09873_SI_001.PDF

Lebel, C., & Deoni, S. (2018). The development of brain white matter microstructure. *Neuroimage, 182*, 207–218. https://doi.org/10.1016/J.NEUROIMAGE.2017.12.097

Li, Y., et al. (2024). Current status of PET tracers for the early diagnosis of Alzheimer's disease. *TrAC Trends in Analytical Chemistry*, 117546.

Liang, H., Ernst, T., Oishi, K., Ryan, M. C., Herskovits, E., Cunningham, E., Wilson, E., Kottilil, S., & Chang, L. (2023). Abnormal brain diffusivity in participants with persistent neuropsychiatric symptoms after COVID-19. *NeuroImmune Pharmacology and Therapeutics, 2*, 37–48. https://doi.org/10.1515/NIPT-2022-0016

Liu, P., de Vis, J. B., & Lu, H. (2019). Cerebrovascular reactivity (CVR) MRI with CO2 challenge: A technical review. *Neuroimage, 187*, 104–115. https://doi.org/10.1016/J.NEUROIMAGE.2018.03.047

Martucci, K. T., & MacKey, S. C. (2018). Neuroimaging of pain: Human evidence and clinical relevance of central nervous system processes and modulation. *Anesthesiology, 128*, 1241. https://doi.org/10.1097/ALN.0000000000002137

Mitchell, T., Lehéricy, S., Chiu, S. Y., Strafella, A. P., Stoessl, A. J., & Vaillancourt, D. E. (2021). Emerging neuroimaging biomarkers across disease stage in Parkinson disease: A review. *JAMA Neurology, 78*, 1262–1272. https://doi.org/10.1001/JAMANEUROL.2021.1312

Moonis, G., Filippi, C. G., Kirsch, C. F. E., Mohan, S., Stein, E. G., Hirsch, J. A., & Mahajan, A. (2021). The spectrum of neuroimaging findings on CT and MRI in adults with COVID-19. *AJR. American Journal of Roentgenology, 217*, 959–974. https://doi.org/10.2214/AJR.20.24839

Miller, B. L., Moats, R. A., Shonk, T., Ernst, T., Woolley, S., & Ross, B. D. (1993, May). Alzheimer disease: Depiction of increased cerebral myo-inositol with proton MR spectroscopy. *Radiology 187*(2):433–437. https://doi.org/10.1148/radiology.187.2.8475286. PMID: 8475286.

Narayanan, S., Schmithorst, V., & Panigrahy, A. (2020). Arterial spin labeling in pediatric neuroimaging. *Seminars in Pediatric Neurology, 33*, 100799. https://doi.org/10.1016/J.SPEN.2020.100799

Newberg, A. B., & Alavi, A. (2017). Single photon emission computed tomography. In *The curated reference collection in neuroscience and biobehavioral psychology* (pp. 861–870). Elsevier. https://doi.org/10.1016/B978-0-12-809324-5.02487-1

Notter, T., Coughlin, J. M., Sawa, A., & Meyer, U. (2018). Reconceptualization of translocator protein as a biomarker of neuroinflammation in psychiatry. *Molecular Psychiatry, 23*, 36–47. https://doi.org/10.1038/MP.2017.232

Ng, M., & Singh, P. (2015, March). Use of lacunar infarction as a marker of small vessel disease. *European Journal of Neurology 22*(3):e39. https://doi.org/10.1111/ene.12610. PMID: 25677457.

Ottoy, J., Niemantsverdriet, E., Verhaeghe, J., de Roeck, E., Struyfs, H., Somers, C., Wyffels, L., Ceyssens, S., van Mossevelde, S., van den Bossche, T., van Broeckhoven, C., Ribbens, A., Bjerke, M., Stroobants, S., Engelborghs, S., & Staelens, S. (2019). Association of short-term cognitive decline and MCI-to-AD dementia conversion with CSF, MRI, amyloid- and 18F-FDG-PET imaging. *NeuroImage. Clinical, 22*, 101771. https://doi.org/10.1016/J.NICL.2019.101771

Petersen, K. J., Metcalf, N., Cooley, S., Tomov, D., Vaida, F., Paul, R., & Ances, B. M. (2021). Accelerated brain aging and cerebral blood flow reduction in persons with human immunodeficiency virus. *Clinical Infectious Diseases, 73*, 1813. https://doi.org/10.1093/CID/CIAB169

Peek, A. L., Rebbeck, T., Puts, N. A., Watson, J., Aguila, M. R., & Leaver, A. M. (2020, April 15). Brain GABA and glutamate levels across pain conditions: A systematic literature review and meta-analysis of 1H-MRS studies using the MRS-Q quality assessment tool. *NeuroImage 210*, 116532. https://doi.org/10.1016/j.neuroimage.2020.116532. Epub 2020 Jan 18. PMID: 31958584.

Rahman, A., Hossen, M. A., Chowdhury, M. F. I., Bari, S., Tamanna, N., Sultana, S. S., Haque, S. N., Al Masud, A., & Saif-Ur-Rahman, K. M. (2023). Aducanumab for the treatment of Alzheimer's disease: A systematic review. *Psychogeriatrics, 23*, 512–522. https://doi.org/10.1111/PSYG.12944

Sintini, I., Graff-Radford, J., Senjem, M. L., Schwarz, C. G., Machulda, M. M., Martin, P. R., Jones, D. T., Boeve, B. F., Knopman, D. S., Kantarci, K., Petersen, R. C., Jack, C. R., Lowe, V. J., Josephs, K. A., & Whitwell, J. L. (2020). Longitudinal neuroimaging biomarkers differ across Alzheimer's disease phenotypes. *Brain, 143*, 2281–2294. https://doi.org/10.1093/BRAIN/AWAA155

Sleight, E., Stringer, M. S., Marshall, I., Wardlaw, J. M., & Thrippleton, M. J. (2021). Cerebrovascular reactivity measurement using magnetic resonance imaging: A systematic review. *Frontiers in Physiology, 12*, 182. https://doi.org/10.3389/FPHYS.2021.643468/BIBTEX

Smitha, K. A., Akhil Raja, K., Arun, K. M., Rajesh, P. G., Thomas, B., Kapilamoorthy, T. R., & Kesavadas, C. (2017). Resting state fMRI: A review on methods in resting state connectivity analysis and resting state networks. *The Neuroradiology Journal, 30*, 305–317. https://doi.org/10.1177/1971400917697342

Song, T., Song, X., Zhu, C., Patrick, R., Skurla, M., Santangelo, I., Green, M., Harper, D., Ren, B., Forester, B. P., Öngür, D., & Du, F. (2021). Mitochondrial dysfunction, oxidative stress, neuroinflammation, and metabolic alterations in the progression of Alzheimer's disease: A meta-analysis of in vivo magnetic resonance spectroscopy studies. *Ageing Research Reviews, 72*, 101503. https://doi.org/10.1016/J.ARR.2021.101503

Sriwastava, S., Tandon, M., Podury, S., Prasad, A., Wen, S., Guthrie, G., Kakara, M., Jaiswal, S., Subedi, R., Elkhooly, M., & Lisak, R. P. (2021). COVID-19 and neuroinflammation: A literature review of relevant neuroimaging and CSF markers in central nervous system inflammatory disorders from SARS-COV2. *Journal of Neurology, 268*, 4448–4478. https://doi.org/10.1007/S00415-021-10611-9/TABLES/6

Tambasco, N., Nigro, P., Romoli, M., Simoni, S., Parnetti, L., & Calabresi, P. (2015). Magnetization transfer MRI in dementia disorders, Huntington's disease and parkinsonism. *Journal of the Neurological Sciences, 353*, 1–8. https://doi.org/10.1016/J.JNS.2015.03.025

Valotassiou, V., Malamitsi, J., Papatriantafyllou, J., Dardiotis, E., Tsougos, I., Psimadas, D., Alexiou, S., Hadjigeorgiou, G., & Georgoulias, P. (2018, November). SPECT and PET imaging in Alzheimer's disease. *Annals of Nuclear Medicine 32*(9):583–593.

https://doi.org/10.1007/s12149-018-1292-6. Epub 2018 Aug 20. PMID: 30128693.

van Zijl, P. C. M., Lam, W. W., Xu, J., Knutsson, L., & Stanisz, G. J. (2018). Magnetization transfer contrast and chemical exchange saturation transfer MRI. Features and analysis of the field-dependent saturation spectrum. *Neuroimage, 168*, 222–241. https://doi.org/10.1016/J.NEUROIMAGE.2017.04.045

Volkow, N. D., & Morales, M. (2015). The brain on drugs: From reward to addiction. *Cell, 162*, 712–725. https://doi.org/10.1016/J.CELL.2015.07.046

Volkow, N. D., Chang, L., Wang, G. J., Fowler, J. S., Leonido-Yee, M., Franceschi, D., Sedler, M. J., Gatley, S. J., Hitzemann, R., Ding, Y. S., Logan, J., Wong, C., & Miller, E. N. (2001, March). Association of dopamine transporter reduction with psychomotor impairment in methamphetamine abusers. *The American Journal of Psychiatry, 158*(3):377–382. https://doi.org/10.1176/appi.ajp.158.3.377. PMID: 11229977.

Woodcock, E. A., Hillmer, A. T., Mason, G. F., & Cosgrove, K. P. (2019). Imaging biomarkers of the neuroimmune system among substance use disorders: A systematic review. *Molecular Neuropsychiatry, 5*, 125–146. https://doi.org/10.1159/000499621

Withers, P. J., Bouman, C., Carmignato, S., et al. (2021). X-ray computed tomography. *Nature Reviews Methods Primers, 1*, 18. https://doi.org/10.1038/s43586-021-00015-4

Wang, Y., Cao, D. Y., Remeniuk, B., Krimmel, S., Seminowicz, D. A., & Zhang, M. (2017, August). Altered brain structure and function associated with sensory and affective components of classic trigeminal neuralgia. *Pain 158*(8):1561–1570. https://doi.org/10.1097/j.pain.0000000000000951. PMID: 28520647.

Xu, X., Dai, J., Chen, Y., Liu, C., Xin, F., Zhou, X., Zhou, F., Stamatakis, E. A., Yao, S., Luo, L., Huang, Y., Wang, J., Zou, Z., Vatansever, D., Kendrick, K. M., Zhou, B., & Becker, B. (2020). Intrinsic connectivity of the prefrontal cortex and striato-limbic system respectively differentiate major depressive from generalized anxiety disorder. *Neuropsychopharmacology, 46*(4), 791–798. https://doi.org/10.1038/s41386-020-00868-5

Yang, X., Chen, C., Wang, A., Li, C., & Cheng, G. (2023). Imaging, genetic, and pathological features of vascular dementia. *European Neurology, 86*, 277–284. https://doi.org/10.1159/000531088

Yousaf, T., Dervenoulas, G., & Politis, M. (2018). Advances in MRI methodology. *International Review of Neurobiology, 141*, 31–76. https://doi.org/10.1016/BS.IRN.2018.08.008

Zhang, L., Hu, K., Shao, T., Hou, L., Zhang, S., Ye, W., Josephson, L., Meyer, J. H., Zhang, M. R., Vasdev, N., Wang, J., Xu, H., Wang, L., & Liang, S. H. (2021). Recent developments on PET radiotracers for TSPO and their applications in neuroimaging. *Acta Pharmaceutica Sinica B, 11*, 373–393. https://doi.org/10.1016/J.APSB.2020.08.006

Bioimaging Biomarkers

46

Yutong Liu and Matthew L. White

Abstract

A wide range of neuroimaging techniques have been developed which aid in the clinical diagnosis of neurological diseases and are valuable tools for research into basic physiological mechanisms and effects of new therapies to combat neurological disease. This is a rapidly evolving field as new imaging techniques with enhanced sensitivity and specificity are constantly being developed and made available to the clinician and research community. This chapter will provide an introduction to the basic mechanisms, capabilities and applications of neuroimaging methods, allowing the readers in neuroscience community to gain an appreciation of the range of available imaging techniques and the potential for use as noninvasive monitors of the progression and treatment of neurological diseases in animals and humans, especially as imaging biomarkers of neuroimmune dysfunctions.

Keywords

Magnetic resonance imaging · Magnetic resonance spectroscopy · Single photon emission computed tomography · Positron emission tomography · Neurological diseases

46.1 Introduction

A wide range of imaging techniques have been developed that aid in the clinical diagnosis of neurological diseases and are valuable tools for research in physiological and pathological mechanisms and for monitoring effects of new therapies to combat neurological diseases. Rapid advancement of imaging technologies with enhanced sensitivity and specificity in the past decades has made imaging one of the most useful tools for clinicians and research community. Imaging plays an integral role in the diagnosis of neurological diseases, monitoring of disease progression, assessment of therapeutic efficacy and treatment planning. Imaging is used not only clinically, but also in neuroscience research to understand functional and biochemical mechanisms of normal and diseased nerve systems. Preclinical imaging on laboratory animals provides insights into the nerve systems, which are usually not available in clinical settings. Preclinical imaging is especially useful to study biological processes in neurological diseases and to develop drugs and drug delivery systems.

Neuroimmune dysfunctions happen in many neurological diseases. Noninvasive imaging biomarkers of neuroinflammatory events are critical for the clinical diagnosis of neurological diseases and for the research of underlying cellular and molecular mechanisms. The purpose of this chapter is to first present state-of-the-art imaging technologies including magnetic resonance imaging (MRI), positron emission tomography (PET) and single photon emission computed tomography (SPECT), and then discuss their applications in clinical diagnosis and treatment, and in preclinical research. For each technology, we will cover the basics of signal generation, signal acquisition and analysis. a variety of methods of MRI will be introduced including diffusion tensor imaging (DTI), dynamic susceptibility contrast (DSC) and dynamic contrast enhancement (DEC), arterial spin labeling (ASL), functional MRI (fMRI), magnetization transfer (MT), chemical exchange saturation transfer (CEST) and magnetic resonance spectroscopy (MRS). The chapter will discuss the physiological information obtained using these MRI methods. Following the introduction of imaging mechanisms of these modalities, we will discuss the applications of MRI, SPECT, and PET in the diagnosis and treatment evaluations of neurodegenerative disorders, stroke, psychiatric defects, tumors, and infectious diseases. Their applications

Y. Liu (✉) · M. L. White
Department of Radiology, University of Nebraska Medical Center, Omaha, NE, USA
e-mail: yutongliu@unmc.edu

H. E. Gendelman, T. Ikezu (eds.), *Neuroimmune Pharmacology and Therapeutics*,
https://doi.org/10.1007/978-3-031-68237-7_46

in neuroinflammation in clinical and preclinical settings will also be discussed including, but not limited to, the measurements of metabolites associated with neuroinflammation such as myo-inositol and choline using MRI and PET, the imaging of immune activation, the detection of neuronal impairments induced by inflammatory events, and the characterization of disease-specific inflammatory cascade for neurological diseases. The chapter will enable readers to gain an appreciation of the range of available imaging techniques and provide insights into their potential for use as noninvasive imaging biomarkers of the progression and treatment of neurological diseases in animals and humans.

Neuroimaging is a broad and fascinating field. We will focus on MRI, SPECT, and PET as they are functional and molecular imaging technologies providing noninvasive measurement of alterations in brain physiology and biochemistry, which are early and sensitive indicators of many neurological diseases. Some other imaging techniques such as CT, ultrasound, and optical imaging are also important for clinical diagnosis and in animal research. However, they are not included in this chapter due to space limits.

46.2 Magnetic Resonance Imaging (MRI)

MRI is a versatile and informative imaging modality serving as an essential arm of clinical diagnostic imaging and a powerful tool for biomedical research. MRI can provide noninvasive three-dimensional (3D) information of organ systems with high anatomical resolution. It offers a variety of controls over image contrast, thus the flexibility and specificity to detect different physiology and pathology. The wide variety of morphological, physiological, and biochemical information available from the multitude of MR techniques allows sophisticated combinations of methods to be applied to individual neurological studies. This is one reason that MRI is often the best choice for neuroimaging studies, as the combination of multiple methods in one study can provide a wide array of information on the effects of disease and experimental therapies on brain function and biochemistry. In preclinical research, MRI has been used in the studies of a variety of neurological disease models of stroke, Parkinson's disease, Alzheimer's disease, and NeuroAIDS amongst others. Specially designed MRI systems for laboratory animals are now commercially available. Software tools for MRI acquisition and post-processing are widely accessible. The basic physics of MRI signals will be first described, followed by the principle of image generation. A variety of MRI methods and their applications in the central nervous system will be discussed.

46.2.1 Basic Principles of MRI

MRI methodology is a complex and intriguing field of study. The basic mechanisms of MRI signal production and acquisition are introduced to allow the readers in the neuroscience community to develop a base understanding.

46.2.1.1 MRI Signal Generation

Most MRI techniques rely on signals from water molecules. The percentage of water within the body changes with age and body composition. Water typically comprises about 70% of the human body and 85% of the brain. Water is also the most abundant compound within laboratory animals. The hydrogen nucleus (or proton, ^{1}H) in a water molecule possesses a quantum mechanical property termed "spin" (Fig. 46.1a, b). The spinning proton or spin (spin and proton are synonymous herein) produces a magnetic dipole similar to a tiny bar magnet. When placed in a magnetic field (we call this field B_0 thereafter, in the unit of Tesla. B_0 is also referred to as external or main magnetic field), the spins align with the direction of B_0. The protons experience a torque from B_0 that causes precession. The proton precession can be analogized to a spinning top wobbling due to the force of gravity (Fig. 46.1c). The precessional frequency is called the Larmor frequency, and is determined by the strength of the external magnetic field B_0:

$$\omega_0 = \gamma \cdot B_0 \tag{46.1}$$

where ω_0 is the precessional frequency, $\gamma = 42.58$ MHz/Tesla is the gyromagnetic ratio of proton. As a reference, the Larmor frequencies of typical clinical MRI scanner magnetic field strengths 1.5 and 3 Tesla are about 64 and 128 MHz, respectively. Preclinical scanners usually have much stronger magnetic field strength such as 7, 9.4, or even 11.7 Tesla. The Larmor frequencies are about 300, 400, and 500 MHz, respectively at these field strengths.

When a spin is exposed to an electromagnetic signal that is identical in frequency and direction to its own Larmor precession, the spin absorbs energy from the electromagnetic signal. The electromagnetic signal is applied at the Larmor frequency, which is typically in the radiofrequency (RF) range, thus the brief application of the electromagnetic signal is referred to as an RF pulse. Absorption of energy from the RF pulse results in a phase coherence among individual spins in precession. The RF pulse tips spins away from the B_0 direction (also called Z axis or longitudinal direction) towards the plane perpendicular to the B_0 direction, which is termed X–Y plane or transverse plane. After the RF pulse is turned off, the spins continue to rotate about the B_0 direction at the Larmor frequency with the same phase (phase coherence). The rotating spins generate an RF signal of Larmor frequency that can be picked up using an RF coil by magnetic induction (Fig. 46.1d, e). This RF signal is the basis of

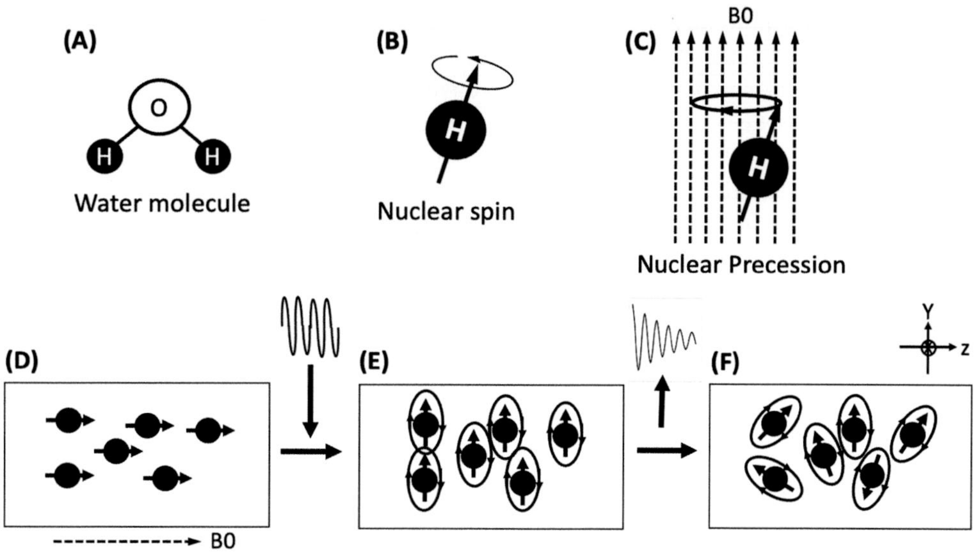

Fig. 46.1 MRI signal generation. (**a**) A water molecule has two hydrogen atoms (protons). Black circles labeled "H". The white circle with "O" is the oxygen atom. (**b**) A proton has an intrinsic property termed "nuclear spin". (**c**) When placed in a magnetic field B_0, a spin precesses about the direction of the magnetic field. The precessional frequency is called Larmor frequency, which is determined by the magnetic field strength and the gyromagnetic ratio of proton. (**d, e**) An RF signal of the Larmor frequency tips the spins away from the B_0 direction to the transversal plane. (**e**) After the RF pulse is off, the spins rotate about the B_0 direction at the Larmor frequency with the same phase (phase coherence). The rotating spins generate an RF signal of the Larmor frequency that can be picked up using an RF coil. (**e, f**) The spins lose phase coherence due to T_2 relaxation, and the RF signal decreases exponentially

magnetic resonance image generation. The resulting angle between the Z axis and the spins (typically 90°) is called *flip* angle and depends upon the amplitude and duration of the RF pulse.

Longitudinal Relaxation (T1)

A 90° RF pulse rotates spins from the B_0 direction (Z axis) to the X–Y plane. After the RF is turned off, the spins must return to their initial position over time, i.e., re-align with B_0. The time that it takes for the spins to return to the B_0 direction is called longitudinal relaxation time or T_1. T_1 relaxation results from the transfer of energy from the excited spins to the surrounding lattice. This energy transfer between spins and lattice is called spin-lattice interaction, thus T_1 relaxation is also termed as spin-lattice relaxation. The inverse of relaxation time is called the spin-lattice or longitudinal relaxation ratio ($R_1 = 1/T_1$). Different tissues may have different T_1 values, and this difference is the basis of T_1-weighted (T_1-wt) contrast in MR images. T_1 relaxation results from the transfer of energy from the excited spins to the surrounding lattice. This energy transfer between spins and lattice is called spin–lattice interaction, thus T_1 relaxation is also termed as spin-lattice relaxation. The inverse of relaxation time is called the spin-lattice or longitudinal relaxation ratio ($R_1 = 1/T_1$). Different tissues may have different T_1 values, and this difference is the basis of T_1-weighted (T_1-wt) contrast in MR images.

Transverse Relaxation (T2)

After the RF pulse is off, spins rotate about B_0 at the Larmor frequency. Due to the interaction among individual spins (that is called spin–spin interaction), each spin experiences a slightly different local magnetic field strength, whereby the individual spins process at different frequencies. Some spins travel faster and some slower, resulting in a loss of phase coherence. This process is called dephasing. The dephasing is termed transverse relaxation or T_2 relaxation, which is an exponential function of time (Fig. 46.1e, f). T_2 relaxation is also called spin-spin relaxation. The inverse of T_2 relaxation time is called spin-spin or transverse relaxation ratio ($R_2 = 1/T_2$). As in the case of T_1 relaxation, different tissues may have different T_2 values, and the difference can be used to generate T_2-weighted (T_2-wt) signal in MRI. In both humans and laboratory animals, T_2 is usually much shorter than T_1.

Time Evolution of RF Signal

As mentioned above, as spins precess at the Larmor frequency in the transverse plane, an RF signal is generated and picked up using an RF coil (also called RF probe) via electromagnetic induction. This RF signal constitutes the *MR signal*. As the spins in the transverse plane gradually lose phase coherence due to T_2 relaxation, so does the magnitude of the RF signal induced in the coil (Fig. 46.1e, f). The evolution and decay of the RF signal is termed *free induction decay (FID)*. At the same time, but independently, the spins gradually return to their initial directions and realign with B_0 due to T_1 relaxation. This process is at a rate ($1/T_1$) that is usually much slower than the rate of signal loss ($1/T_2$).

In reality, the loss in phase coherence occurs more rapidly than from the spin-spin interactions by themselves due to the presence of extrinsic magnetic inhomogeneities such as imperfect main magnetic fields or susceptibility contrast agents that are present in tissues. The overall relaxation time is designated as T_2* when both intrinsic spin-spin interactions and the extrinsic magnetic inhomogeneities are considered. T_2* is usually much shorter than T_2. Therefore, because of the T_2* effect, the FID decays faster than when only T_2 relaxation is considered.

46.2.1.2 From MR Signal to Image

The MR signal described so far is from the whole subject (a laboratory animal, a human, or any subject being examined) placed in the magnetic field B_0, and simply the sum of signals of individual spins in the subject. To generate an MR image, spatial information has to be acquired to distinguish the positions of the signal in the three-dimensional (3D) subject. The magnetic field dependence of the signal's frequency is used as a basis for spatial encoding of the magnetic resonance signal. This is done using magnetic field gradients. Magnetic field gradients cause a linear shift in magnetic field with position. Sophisticated spatial encoding schemes have been developed to allow multidimensional spatial encoding (Fig. 46.2). Magnetic field gradients are superimposed upon the homogenous main magnetic field and change the main magnetic field in a controlled and predictable pattern. As the magnetic field increases or decreases with position, the precessional frequency (Larmor frequency) increases or decreases with position as well as it is determined by the magnetic field strength. In a gradient field, the position is directly and linearly proportional to frequency. The process of image formation can be broken down into three components: slice selection, frequency encoding, and phase encoding.

Most MRI studies acquire a series of slices through the subject to be imaged. To acquire the MR signal from a slice, a slice selection gradient (G_{slice}) is applied resulting in protons precessing at different frequencies along the slice selection direction. An RF excitation pulse is simultaneously transmitted to the subject at a narrow range of frequencies so that only those protons in a slice of tissue precessing at the particular range of frequencies are excited (Fig. 46.2a, b). Slice selection gradient can be applied along any axis of the body. The slice selection direction in conjunction with the animal's position in the magnet dictates the orientation—coronal, axial, or sagittal—of the image. For an animal in the "head first prone" position, the three orientations correspond to slice selection in the *z*, the *y*, or the *x* direction, respectively.

With the application of the slice selection gradient, it is determined that the acquired signal is from a slice in the subject. The remaining job is encoding 2D spatial information in the signal. This is done using frequency encoding gradient (G_{freq}) and phase encoding gradient (G_{phase}) (Fig. 46.2c). The frequency encoding gradient (G_{freq}) is also known as the readout gradient. Using frequency encoding, the positions of protons along the G_{freq} direction are distinguished. The frequency-encoding gradient is applied in a direction perpendicular to the slice selection gradient. Position of the protons in the third dimension is determined with a phase encoding gradient (G_{phase}). The phase encoding gradient is applied along the third perpendicular axis. The result of a single measurement made with both phase and frequency encodings is that each spatial position along the phase encoding direction is represented by a unique phase, and each position along the frequency encoding direction is represented by a unique frequency. These unique combinations of frequency and phase encode the coordinates of each pixel (Fig. 46.2c).

46.2.1.3 Summary

While a subject is placed inside a strong external magnetic field, the protons in the subject, acting as small magnets,

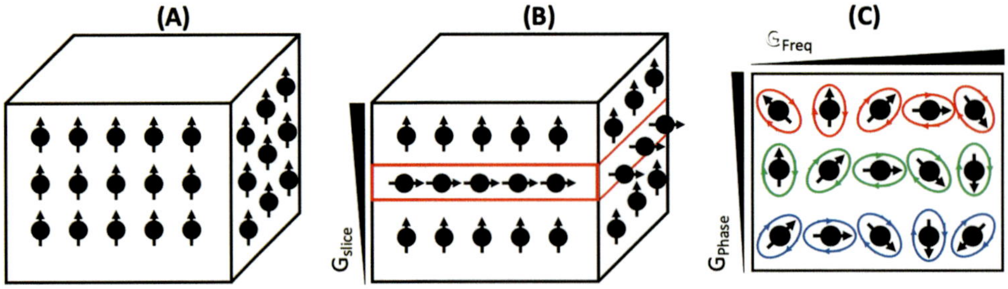

Fig. 46.2 Spatial information of spins is encoded into MR signal using magnetic field gradients. (**a**) A subject placed in a magnetic field. (**b**) A slice selection gradient (G_{slice}) is applied, resulting in only the protons within a specific slice of the subject to be flipped. (**c**) On the selected slice, the frequency encoding gradient (G_{freq}) and the phase encoding gradient (G_{phase}) are applied perpendicularly resulting in each spatial position along the phase encoding direction that is represented by a unique phase, and each position along the frequency encoding direction is represented by a unique frequency. These unique combinations of frequency and phase encode the coordinates of each pixel

align along the direction of this field. Transmitting an RF signal pulse tips the protons away from their initial directions. When the RF is turned off, the protons precess about the main magnetic field, and the precessing protons produce an RF signal. The RF signal decreases with time due to relaxation. Spatial information is encoded into the RF signal using gradient magnetic fields. These impose a linear variation in all three dimensions in the main magnetic field. As a result of the variation, the precessional frequencies and phases of the protons are affected by their spatial locations. The RF signal is detected using an RF coil (probe) and digitized. A mathematical operation is then performed to convert the signal into an image. The contents discussed in this section can be found in many books (Bushberg et al., 2002; Berry & Bulpitt, 2008; Brown et al., 2014; Webb, 2022) and online resources (Hornak, 1996).

46.2.2 MRI Methods and Applications

The power and versatility of MRI lie in the ability to modify signal intensity based on the biophysical properties of the spins of interest (primarily tissue water). Signals can be "weighted" (signal amplitude altered) based on tissue relaxation properties (T_1 and T_2), water diffusion (random molecular motion), the interaction of the hydration layers of membranes and proteins with free water (magnetization transfer), or the proton transfer between mobile biomolecules and water (chemical exchange saturation transfer). Tissue perfusion can be mapped by the introduction of contrast agents or inversion of signal from water in arterial blood supply of the tissue. In addition, signal can be suppressed from tissue based on frequency (fat or water suppression), position (suppressing motion artifacts from unwanted tissue), or preferentially viewed based on moving vs static water (angiography or suppression of motion from flowing blood). In following paragraphs, we will introduce some MRI techniques that are used as imaging biomarkers for neurological diseases.

46.2.2.1 T_1, T_2, and Proton Density Weighted MRI

As explained above, MRI signal results from water protons. Proton relaxation times including longitudinal relaxation time (T_1) and transverse relaxation time (T_2) depend on the microenvironments of the protons. Therefore, protons in different tissues may possess different relaxation times. Pathology may cause changes in proton density, and relaxation times in a tissue. Thereby tissue contrast can be generated on MR images employing the tissue properties in proton density (PD, also called spin density (SD)) and relaxation times. Pathological conditions may be visualized using MRI by emphasizing the altered PD or relaxation times. The effects of PD, T_1, and T_2 of tissues can be maximized individually, which means that an MR image can have different weightings of the three effects. MRI that maximizes the PD effect while minimizing the T_1 and T_2 effects is called a PD-weighted (PD-wt) MRI. Similarly, MRI maximizing T_1 or T_2 effects while minimizing other effects is called a T_1-weighted (T_1-wt) or T_2-weighted (T_2-wt) MRI, respectively. On a PD-wt image, the signal intensity in each voxel is proportional to the free water content. T_1 relaxation is the process of releasing the energy absorbed by the magnetic nuclei during the application of the RF pulse. This process occurs by releasing the energy to the surrounding molecules (lattice) returning the spin system to the equilibrium condition (aligned with B_0 direction). The rate at which this process occurs in tissue depends on the molecular environment and varies with tissue type, providing a means of generating contrast between tissues with otherwise similar signal intensities (water content). The natural process of energy exchange between adjacent molecules causes a loss of signal coherence by exchange of spin states with a loss of phase coherence. The exchange of spin states with no net loss of energy in the spin system is referred to as spin–spin relaxation (T_2).

Applications in Neurological Diseases

The use of T_1-, T_2- and PD-wt images comprises the bulk of clinical MRI scans due to the ability to detect morphological changes with exquisite soft tissue contrast. MRI can generate detailed anatomical images for morphological studies in neuroscience. T_1 and T_2 weighted images can be used to determine the volume and shape of multiple brain structures including grey matter, white matter, and ventricles. Altered volumes of brain structures, such as the reduced substantia nigra in Parkinson's disease, atrophy of the grey matter in dementia such as Alzheimer's disease (Risacher & Saykin, 2019; Yousaf et al., 2019), and decreased hippocampal volume in a wide range of neurological diseases (Geuze et al., 2005) have been found. The details of brain regional volume change are beyond the scope of this chapter, and interested readers can refer to the review articles listed above. Research continues into this area and as imaging and analysis software improves and the definition changes due to disease are better defined, the use of these data in clinical diagnosis and patient management will be put into practice.

Besides morphology, PD, T_1, and T_2-wt MRI can provide physicochemical information of the brain tissue. A PD-wt image directly reflects water content in a tissue. Water content also affects tissue T_1 and T_2 values. Free bulk water, such as that present in the cerebral ventricles, has high T_1 and T_2 values, whereas intracellular water, which interacts with macromolecules, has shorter T_1 and T_2 times. Many lesions including certain types of tumors, edema, hemorrhage, and damaged regions from vascular diseases contain increased water content and are hyperintense on T_2-wt images and hypointense on T_1-wt images.

Image contrasts based on relaxation times can be further accentuated using exogenous contrast agents. Gadolinium (Gd) based contrast agents for human MRI exams were first approved in 1988. In healthy brains, Gd-based contrast agents normally cannot pass through blood–brain barrier (BBB) to enter brain parenchyma. Enhancement (hyperintensity) in brain parenchyma on T_1-weighted images reflects rather simplistically BBB impairments. This type of brain parenchymal enhancement is the "delayed static" appearance of the contrast in the brain. The relative ease of imaging brain parenchymal enhancement (injecting the contrast and then obtaining a T_1-weighted image) continues to make this a powerful biomarker clinically used for monitoring BBB intactness to analyze disease processes. However, parenchymal enhancement is very non-specific and can be secondary to a number of etiologies including primary tumors, secondary tumors, active demyelination, stroke, vasculitis, granulomatous disease, and infection (viral, fungal, or bacterial).

In primary brain tumors, Gd-induced T_1 hyperintensity is useful for determining tumor grade and survival of patients with astrocytomas and oligodendrogliomas. These are the two largest categories of primary brain tumors. Hyperintensity in these tumors typically indicates higher-grade tumors and shorter survival but this relationship isn't seen for all tumors. Some tumors will enhance, have longer survival, and behave like true low-grade tumors. Some low-grade primary tumors such as pilocytic astrocytoma almost always appear hyperintensity. Secondary tumors in the brain universally show hyperintensity making Gd-enhanced T_1-wt MRI an extremely useful characteristic for disease detection. Due to the relatively high frequency of cancers that can metastasize to the brain (lung, breast, renal, ovarian, colon, testicular, etc.), and the growing elderly population, the continued analysis of "delayed static" contrast enhancement is guaranteed.

In summary, proton density-, T_1, and T_2-wt MRI are considered as conventional MRI methods in the assessment of neurologic disorders. Edema, inflammation, tumors, and cellular degeneration demonstrate increased T_1 and T_2, leading to dark appearance on T_1-wt images and bright appearance of many pathologies on T_2-wt images. The combination of T_1, T_2, and proton density are not specific to pathology. These relaxation changes are not universal and pathology can be either light, dark, or isointense with normal tissue on either T_1- or T_2-wt images, depending upon the nature of the cellular abnormality. Improved sensitivity and specificity can be obtained by using Gd-based contrast agents.

46.2.2.2 Diffusion Tensor Imaging (DTI)

DTI is an extension of diffusion-weighted imaging (DWI). Diffusion is the random (or pseudo-random) motion of water molecules due to non-zero temperature. Water diffusion depends on the microscopic structural environment in the tissue, and measurement of it can provide valuable insights into the regional functional and structural changes occurring in disease states. DWI is sensitized to the random molecular motion of water in tissues by applying diffusion weighting on acquired MR signals. The diffusion weighting is determined by a parameter called "b-value", which is in the unit of s/mm^2 (inverse of diffusion coefficient). Diffusion is a 3-dimensional process. In a uniform environment, it is isotropic in all directions and can be represented by a sphere. If water molecule movement is restricted in certain directions, the diffusion becomes anisotropic, represented by an ellipsoid (Fig. 46.3a). For example, in fiber-like cell structures, such as white matter tracts, the diffusion is relatively free along the longitudinal axis of the fiber tract but restricted in the other two dimensions along the transverse axes. DTI was developed to measure the directions and magnitudes of the three mutually perpendicular axes, in which DWI is repeated with diffusion weighting in varying non-colinear directions. A tensor is a 3×3 matrix that mathematically describes the water diffusion in cellular structures. Several metrics have been proposed to represent the diffusion anisotropy and the fractional anisotropy (FA), which is based on the ratio of the longitudinal axis to the transverse axes is a commonly used one. FA is a scalar quantity from 0 to 1. A voxel with a high FA value indicates that water diffusion is more likely anisotropic. FA provides information for identifying and quantifying anisotropic tissue structures (Fig. 46.3b). The longitudinal axis of the diffusion ellipsoid provides the directions of maximum diffusion. For white matter tracts, this is the axis parallel to fiber tract orientation. Several methods have been proposed to display the longitudinal axis, one of which is the so-called color encoding technique. In the color encoding technique, the longitudinal axis is encoded with the primary colors (red—left-right; green—up-down, and blue—in-out), and the brightness is scaled by a diffusion index such as FA. Figure 46.3b shows an example of this method on a mouse brain.

Information on diffusion anisotropy in brain tissue can be used for modeling neuronal tracts. The longitudinal axis is used to connect similar voxels and is the basis for DTI-based fiber tracking (Fig. 46.3c). Many fiber tracking algorithms (Tournier, 2019) have been developed to reconstruct fiber architecture and confirm connections among different regions of the brain of humans and animals (Fig. 46.3d). Despite certain limitations, such as relatively poor spatial resolution as compared with axon dimensions and reconstruction difficulties at fiber branching or merging sites, DTI provides a unique means to study integrity of fiber bundles in particular neurologic disorders.

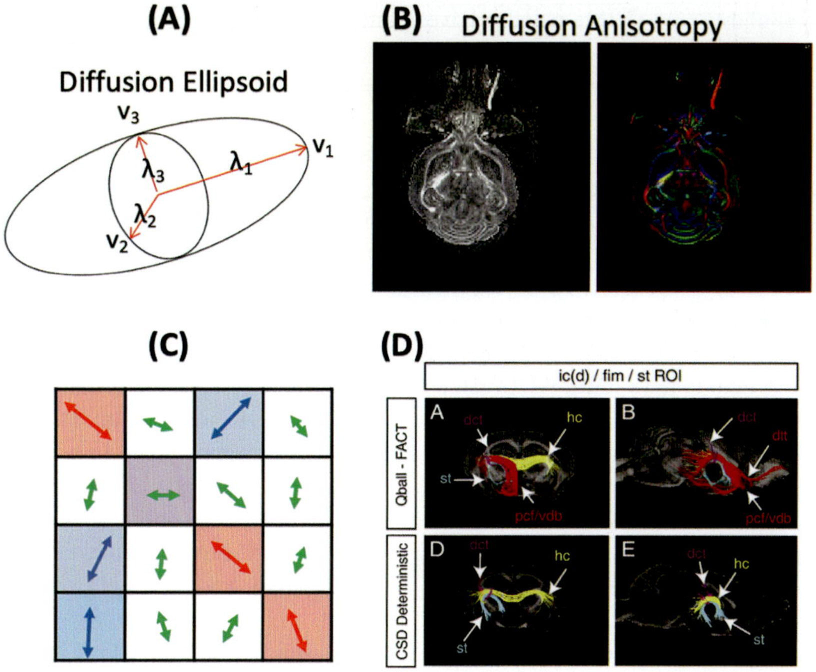

Fig. 46.3 DTI and DTI-based tractography. (**a**) Diffusion can be represented by an ellipsoid, in which the longitudinal axis v_1 provides the direction of maximum diffusion, and l_1 is the diffusion coefficient along v_1. The two transverse directions are represented using v_2 and v_3, and the diffusion coefficients are l_2 and l_3, respectively. (**b**) The FA map of a mouse brain (left), and the color map of the longitudinal diffusion direction scaled using FA (right). (**c**) Two fibers (blue and red) are generated by connecting the longitudinal axis of similar voxels. Advanced algorithms are developed to process voxels at fiber crossing. (**d**) Tractography of white matter regions of interest using Qball-FACT (top), CSD deterministic (bottom) shown in a coronal (left), and sagittal (right) view. Fibers were generated from a region-of-interest (ROI) placed at the intersection of the fimbria, stria terminalis, and dorsal internal capsule. The fibers were identified include hippocampal commissure (hc; yellow), stria terminalis (st; light blue), precommissural fornix (pcf; red), ventral diagonal band (of Broca) (vdb; red), dorsal tenia tecta (dtt; red), dorsal corticothalamic tracts (dct; dark blue in top row, purple in bottom row). Figure (**d**) is copied from a previous study with permission (Moldrich et al., 2010)

Applications

Simple DWI turns out to be a powerful imaging tool in acute stroke detection in clinical practice. Within minutes after a stroke, a low diffusion area of brain is visible in the majority of stroke cases. In a diffusion-weighted image, this appears as a bright area, as tissues with low diffusion lose less signal during the diffusion weighting. A parametric image of the apparent diffusion coefficient of water (ADC) can be calculated from DWI. Acute stroke lesions typically appear as bright on DWI and dark on ADC maps. As stroke lesions age, the cells in the lesion begin to break down after cell death and leave an area of edema. As the stroke lesion progresses toward cellular breakdown and edema, both diffusion and T_2 increase. After days to weeks, the diffusion "pseudonormalizes" to a value equivalent to normal brain tissue, progressing to high diffusion. A combination of ADC, T_1, and T_2 maps can be used to characterize the age of the stroke lesion (Campbell et al., 2019). Diffusion is altered in other types of brain lesions and is not specific for stroke. The range and types of diffusion alterations are an area of active research. Combination of DWI with other imaging modalities serves to more precisely characterize lesions from neurological diseases.

Multiple metrics can be calculated from DTI data, typically mean diffusivity (MD) and fractional anisotropy (FA). MD is equivalent to ADC, which in DTI is defined as the averaged magnitude in the three diffusion ellipsoid axes. Analysis of brain pathology using DTI metrics can be performed by lesion analysis, whole brain histograms of FA and MD, tractography, and analysis of otherwise normal appearing white matter.

DTI has been widely used in the study of neurological diseases. Restricted diffusion (lower ADC/MD values) in brain tumors has been shown to be associated with decreased survival and tumor grade (Zulfiqar et al., 2013; Kono et al., 2001). The restricted diffusion in brain tumors is thought to be primarily an effect of increased cellularity (hypercellularity) but measurement of the hypercellularity is confounded by presence of vasogenic edema, necrosis, hemorrhage, and calcifications (Gupta et al., 2013). ADC values typically in lymphoma are lower than in GBM (Toh et al., 2008). The decreased ADC in lymphoma in general is the result of extensive areas of hypercellularity and GBMs usually have limited areas of restricted diffusion secondary to hypercellularity (Hakyemez et al., 2005). DTI can help differentiate abscess, glioblastoma multiforme, and metastatic disease. Abscesses tend to have cavities of marked restricted ADC whereas the necrotic cavities of tumors have higher ADC values (Erdogan et al., 2005; Hakyemez et al., 2005). The opposite is true for the rims of these lesions due to that necrotic tumor cavities have mild to moderate restricted diffusion mainly from cellular debris and hemorrhage. The rims of abscess show elevated ADC values while the rims of GBMs and metastatic lesions tend to have lower ADC values due to cellularity (Chang et al., 2002; Hakyemez et al., 2005). The peritumoral ADC analysis helps to differentiate GBM from metastatic disease secondary to the peri-enhancing region of tumor invasion that is seen with GBM and not metastasis (Wang et al., 2009a, b).

FA has been studied in numerous diseases and detects abnormalities in otherwise normal-appearing white matter by conventional MRI. This demonstrates the increased sensitivity of DTI metrics to brain diseases. Abnormal FA in normal-appearing white matter has been found to occur in multiple sclerosis, radiation injury, aging, schizophrenia, and amyotrophic lateral sclerosis (Cosottini et al., 2005; Salat et al., 2005). FA has already been found to correlate with the cell density of brain tumors (Beppu et al., 2005), to be able to differentiate between low-grade and anaplastic astrocytomas and to help differentiate between brain metastasis and high-grade gliomas. There is also evidence that FA may help to detect tumor infiltration not visible by conventional MRI (Tsuchiya et al., 2005; Goebell et al., 2006).

DTI data can also be utilized to perform white matter tractography (Tournier, 2019). Tractography is the virtual dissection of white matter tracts within the brain and provides a novel method to analyze the white matter tracts in disease states. DTI tractography can be used to map out numerous fiber tracts including but not limited to the corticospinal tracts, superior occipitofrontal fasciculus, inferior occipitofrontal fasciculus, uncinate fasciculus, superior longitudinal fasciculus, arcuate fasciculus, and inferior longitudinal (occipitotemporal) fasciculus (Jellison et al., 2004). With advancements and refinements in acquisition of DTI data, there likely will be improvements to tractography which will provide further advances in the "virtual dissection" of the white matter tracts. A very exciting and innovative application of tractography that will have profound clinical application is analysis of brain lesions. There has already been early work defining tumor growth and metastases within major white matter tracts (Yeh et al., 2021). This application will provide measures of white matter tract displacement and tumor invasion that will be defined with great specificity. Previously estimates of white matter tract invasion could only be made by gross estimates of where the white matter tracts were expected to lie based on cross-sectional MRI or CT studies.

46.2.2.3 Dynamic Susceptibility Contrast (DSC) MRI

DSC MRI is an imaging technique to measure cerebral perfusion. Cerebral perfusion is crucial for brain functions. It is represented by the amount of blood flow (in the unit of ml g^{-1} s^{-1}) in the capillary bed to a given region of the brain (flow per unit volume of tissue). Perfusion in cerebral tissue with normal vasculature is controlled by metabolic demand, allowing use of regional quantitative perfusion measures to detect neurodegeneration in a variety of diseases. MRI has been used to assess perfusion as a noninvasive technique. Perfusion MRI has been used to demonstrate regions of tissue at risk of further damage from stroke and blood flow abnormalities and highlight regions of hyper-perfusion in tumors.

The susceptibility contrast agent used in the DSC MRI is a paramagnetic substance (e.g., gadolinium (Gd)), which has a small magnetic susceptibility. Gd, usually in a chelated form, causes fast spin dephasing (i.e., decrease in T_2*) during its passage through the tissue. In a typical DSC MRI, a series of baseline T_2*-wt MRIs is first acquired. A bolus of Gd is then injected into a vein or an artery. T_2*-wt MRI is continued for another period of time until the Gd has passed through the tissue. We here demonstrate DSC MRI using a rat model of stroke (Fig. 46.4). The signal drop and recovery can be observed in the contralateral hemisphere (left side). The signal change in the ischemic hemisphere is mild compared to the other side (Fig. 46.4a). The stroked region has a perfusion deficit with lower signal drop, delayed entering and wash-out, and prolonged pass-through. Several parameters can be derived from the signal change with time to describe the perfusion in tissue. The parameters usually include the relative cerebral blood volume (CBV), mean transit time (MTT) of the bolus through the tissue, and time to peak (TTP), which is the time it takes for the contrast agent to reach the maximum concentration. The cerebral blood flow can be calculated by dividing CBV by MTT. The perfusion parameters are usually calculated pixel by pixel in the MR images, and the maps of the parameters are generated for further analysis. For example, the maps of CBF on the rat brain in the stroke study are shown in Fig. 46.4b. The region with lower CBV is identified.

Applications

DSC MRI studies have been used primarily to study perfusion deficits in stroke lesions (Rowley, 2005). Brain tumor study is a growing area of clinical applications of DSC MRI. DSC MRI studies provide valuable information when

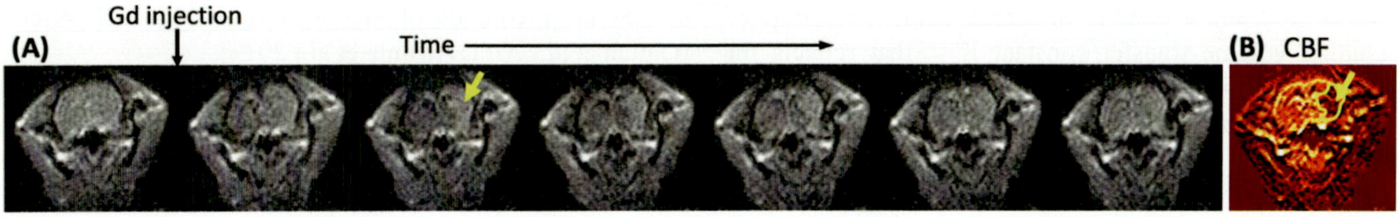

Fig. 46.4 DSC MRI of a stoked rat. (**a**) A series of baseline T_2*-wt images are first acquired followed by the injection of a bolus of Gd. T_2*-wt MRI is continued for another period of time until the Gd has passed through the tissue. Normal tissue in the contralateral hemisphere shows signal drop when Gd enters. The signal drop is not clear in the ischemic region on the ipsilateral hemisphere (yellow arrow) due to perfusion deficit. (**b**) CBF map calculated from the MRI signal change with time. The region with low CBF is identified

evaluating multiple facets of gliomas, metastatic disease, abscesses, radiation necrosis, and meningiomas. The grading of gliomas pre-operatively has been found to be enhanced by utilizing DSC perfusion (Law et al., 2003, 2004). The determination of a lesion is a high-grade glioma or a metastatic lesion can be facilitated utilizing DSC perfusion (Law et al., 2002). The differentiation of an abscess from a ring-enhancing glioma is also potentially facilitated by DSC perfusion (Holmes et al., 2004). It should be noted that in high-grade gliomas, BBB is usually impaired, and leakage of the contrast agents through the BBB may occur so that MRI signal curve is distorted and a true evaluation of the tumor perfusion is not obtained.

46.2.2.4 Arterial Spin Labeling (ASL)

The applications of DSC MRI in neurological diseases are limited because first, it requires the administration of exogenous contrast agents, and second, it cannot provide quantitative cerebral perfusion values. Arterial spin labeling (ASL) is a noninvasive technique for measuring cerebral perfusion by labeling the spins of the intravascular arterial water using one or two RF pulses. A great advantage for ASL is that no contrast agent needs to be administered so IV access isn't a problem and the exam can be safely performed in patients with poor renal function. This eliminates the risk of developing nephrogenic systemic fibrosis that can occur with GD-based contrast agents. Another advantage of ASL over the bolus tracking DSC MRI is that ASL can be repeated many times in a single study, which makes it appealing for the continuous monitoring of the perfusion. Also, ASL does not suffer from the vascular leakage phenomena that plague contrast-based perfusion techniques. In a typical ASL, RF pulses first label the spins in the arterial blood just upstream of the tissue of interest. The spins then perfuse into the tissue of interest, and thus the signal in the tissue is modified by the labeled spins. A second unlabeled acquisition serves as a reference to then calculate the perfusion images. Spin labeling can be achieved by basically two approaches, namely pulsed ASL (PASL) and continuous ASL (CASL) (Petcharunpaisan et al., 2010).

Applications

ASL research has demonstrated that DSC perfusion exams can potentially be replaced by ASL. ASL exams have been found to correlate to the histopathology of brain tumors, and others have found ASL techniques useful for grading tumors (Abdel Razek et al., 2019). ASL has been shown to be advantageous in determining glioma recurrence versus radiation necrosis (Ozsunar et al., 2010). ASL techniques have been demonstrated to be useful in stroke, Alzheimer's disease (AD), Parkinson's disease (PD), and multiple sclerosis (MS) (Joseph, 2021). Delayed bolus arrival time and reduced CBF were observed in patients with ischemic stroke (Wolf et al., 2014). The findings match those of DSC imaging (Wolf et al., 2014). CBF quantified with ASL was a good diagnostic marker for dementia (Bron et al., 2014). This was not surprising since hypoperfusion of brain tissue precedes atrophy in dementia. ASL indicated preserved perfusion in the degenerated brain regions of PD patients compared with healthy controls, and decreased perfusion in other regions, including the posterior parieto-occipital cortex, precuneus and cuneus, and middle frontal gyrus (Melzer et al., 2011). Studies have indicated that the neurovascular abnormalities observed in PD are dependent on the neurotransmitter dopamine and this tight neurovascular coupling is unique (Chen et al., 2013). In patients with multiple sclerosis (MS), ASL provides insight into the disease pathophysiology with decreased CBF found in the bilateral thalami and right frontal lobe (Ota et al., 2013). These studies demonstrate the utility and robustness of ASL in research paradigms, applicability across the pathological spectrum, and importance of correlating with other radiological modalities. An investigation of one technique compared to another radiological result can help confirm accuracy (Wolf et al., 2014).

46.2.2.5 Dynamic Contrast Enhancement (DCE)

Dynamic contrast enhancement (DCE) MRI also needs the administration of paramagnetic contrast agents like Gd. Unlike DSC in which serial T_2*/T_2-wt images are acquired before and after the contrast agent injection, T_1-wt images are acquired. The temporal T_1 enhancement characteristics are analyzed and a number of metric maps are generated including volume transfer constant K^{trans} that reflects the efflux rate of Gd from blood plasma into tissue extravascular extracellular space (EES), volume of EES per unit volume of tissue (Ve) that is a measure of fraction (%) of a tissue volume occupied by the EES (i.e., the leakage space for contrast agent), blood plasma volume per unit volume of tissue (Vp), and numerous other parameters (Bergamino et al., 2014).

Applications

The DCE technique has been widely applied to brain tumor analysis to evaluate angiogenesis and blood vessel leakiness. DCE gives insight into brain lesion etiology, brain tumor grade, survival, and tumor progression after treatment or pseudoprogression (Jain et al., 2011; Suh et al., 2013). High-grade gliomas showed higher K^{trans} and Ve than low-grade gliomas, and it has been found in patients with gliomas, survival was worse for patients with high K^{trans} and Vp (Nguyen et al., 2014). DCE analysis can be applied to looking at tumor response to chemotherapy or radiation therapy. Patients who are not responding to treatment then may have their therapy altered. DCE analysis has been found to be able to predict brain metastatic response with whole-brain radiation therapy (Farjam et al., 2014).

Analysis of DCE has been applied to other fields of work such as that related to stroke risk factors, ischemic strokes, and multiple sclerosis plaques. It has been demonstrated that DCE analysis of carotid artery plaques can provide insight into the neovascularization (Truijman et al., 2013) and this data may be predictive of future risk of recurrent stroke. Also, the analysis of permeability may help predict what ischemic stroke lesions will result in symptomatic hemorrhagic transformation and/or malignant edema (Hoffmann et al., 2011). DCE MRI has been used to evaluate treatment effects and prediction of treatment success in MS patients (Gaitan et al., 2011; Absinta et al., 2013).

BBB impairment is indicative of neuroinflammation that occurs in normal aging, Parkinson's disease, Alzheimer's disease, type II diabetes, HIV, cerebral microvascular disease, and non-enhancing multiple sclerosis lesions (Armitage et al., 2011). There is increasing interest in using DCE to identify more subtle BBB abnormalities. Using DCE MRI, subtle BBB changes have been found in patients with MS in the periventricular location, a common location for MS lesions. The abnormal permeability changed with treatment which indicated the level of permeability was correlating with the generalized MS inflammatory activity (Cramer et al., 2014).

46.2.2.6 Functional MRI (fMRI)

During the past decades, mapping of cerebral activity has become a major field in MRI research (Van Bruggen & Roberts, 2002). Functional activation of brain regions causes a flux in the oxygen level and the perfusion rate in the activated tissue. These dynamic changes generate dynamic contrast changes in the MRI signal. This was first demonstrated in 1992 (Kwong et al., 1992) and has developed into a popular method among neuroscientists for mapping task-related activity as well as brain functional connectivity. Most fMRI studies rely on blood oxygenation level-dependent (BOLD) contrast. BOLD contrast is based on the magnetic property of blood, which is dependent upon the oxygenation state of hemoglobin. Deoxygenated hemoglobin is paramagnetic, whereas oxyhemoglobin is diamagnetic. Deoxygenation results in increased local magnetic susceptibility differences between intra- and extravascular compartments and thereby causes signal loss on T_2-wt and T_2*-wt MR images (Fig. 46.5a).

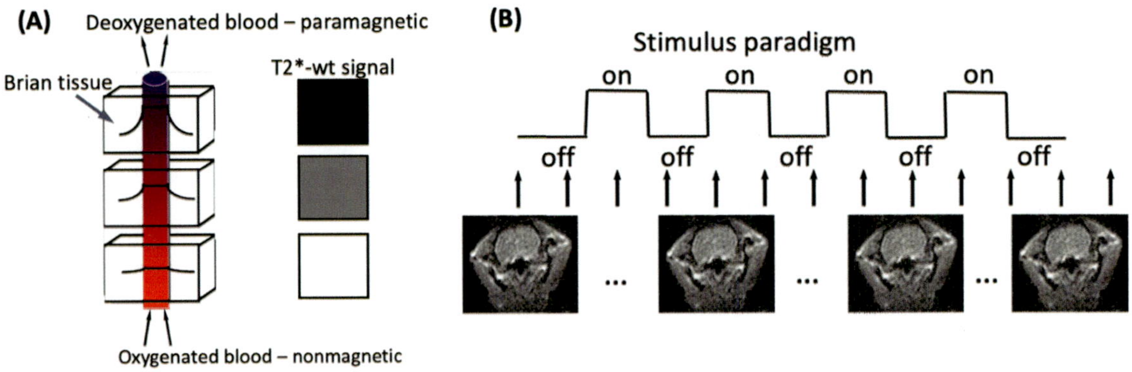

Fig. 46.5 BOLD Functional MRI. (**a**) Oxygenated blood is diamagnetic, and deoxygenated blood is paramagnetic. Deoxygenation results in signal loss on T_2-wt and T_2*-wt MR images. Neuronal activity requires increased demand for oxygen in the activated area so that the deoxygenation level in the blood drops causing regional MRI hyperintensity. (**b**) Functional MRI is used for mapping task-specific brain activation. Images are acquired continuously while a task is performed (or a stimulus is applied) with a precise timing synchronized to the image acquisitions. The paradigm of the task is usually designed as a "block-chain" with alternating "on" and "off". This design enables the statistical analysis of images with pixel-wise correlation between signal intensity changes and the timing of the task

In a typical task-related fMRI study, neural activity in the functioning brain during cognitive, perceptual, sensory, or motor processes or during a pharmacologic challenge was measured based on the BOLD contrast. The mechanism leading to MRI contrast is the reduction of deoxyhemoglobin in the blood in the activated area (Fig. 46.5a). This effect occurs because the vasculature reacts to the increased demand for oxygen in the activated area by overcompensating, a physiological process known as "luxury perfusion". During luxury perfusion, the deoxyhemoglobin levels in the blood drop causing regional MRI hyperintensity. Images are acquired continuously during an fMRI study while the task is performed with a precise timing synchronized to the image acquisitions. The paradigm of the task is usually designed as a "blockchain" with alternating "on" and "off". This design enables the statistical analysis of images with pixel-wise correlation between signal intensity changes and the timing of the task.

Functional MRI is extensively used in research for mapping task-specific brain activation and continues to find greater clinical applications. Numerous paradigms have been used to test an expansive array of brain functions. A test may map areas of function for memory, motor function (hand, face, and foot), sensory (pain, thermal, light touch, and vibratory), and multiple aspects of language and vision. Given the complexities of the brain, there are numerous potential paradigms that are utilized for any one function.

Applications
Clinically fMRI has been employed to determine areas of brain function in relationship to a tumor that is going to be resected. Imaging results in brain tumor patients are used to guide surgeons to achieve maximal tumor resection with preservation of function. However, the application of the technique for surgical guidance must be approached with caution. The tumor itself may alter the underlying physiology fMRI activation relies upon and give spurious results. The tumor may have infiltrated areas of function and the functional tissue may lie within the tumor itself and not be demonstrable by fMRI. Moreover, people with brain tumors can have difficulties with performing even simple tasks and it can be difficult to determine if subjects are successfully performing tasks such as language and memory (Chang et al., 2010). Despite these shortcomings, fMRI performs robustly and provides useful guidance. Functional MRI has been found numerous applications in studies of neurodegeneration such as in Parkinson's disease (Bardakan et al., 2022), and Alzheimer's disease (Warren & Moustafa, 2022; Bonakdarpour & Takarabe, 2023).

46.2.2.7 Resting-State fMRI
Resting-state fMRI (rsfMRI) has been broadly applied to numerous neurological conditions. Resting-state fMRI measures the baseline spontaneous low frequency (<0.1 Hz) BOLD signals at resting state. Brain regional connectivity is detected by testing the correlations of the BOLD signals on brain regions. Before the correlation calculations, a series of pre-processing needs to be performed on rsfMRI data including motion correction, baseline drifting correction, spatial smoothing, registration to standard space, and others. A number of methods are used to analyze rsfMRI data. Seed-based analysis was first developed and has been applied in numerous studies. In this method, a region-of-interest (ROI) or a seed is first selected, and the average BOLD time course of voxels within the seed is correlated with the time courses of all other voxels in the brain. A threshold is determined to identify voxels significantly correlated with the seed. Seed-based analysis is straightforward and still widely used. Independent component analysis (ICA) is data-driven approach, which attempts to find a set of statistically independent components within rsfMRI. ICA can be used to spatially identify distinct brain networks. Compared with seed-based methods, ICA does not require manually selected seeds. Graph-based analysis method provides an excellent way to present brain connectivity as a collection of nodes connected by edges. Brain regions of interest are represented as nodes, and correlation between the regions as the connectivity of the edges. The edges can be weighted with connectivity strength.

Applications
In addition to demonstrate numerous brain networks associated with motor, sensory, vision, dorsal attention, and language networks such as the executive network and default mode network (Guerra-Carrillo et al., 2014), rsfMRI can provide exquisite insight into the functional brain changes associated with neurocognitive findings in many disease processes. Resting-state fMRI has been successfully applied to analyzing brain functions in multiple sclerosis, Parkinson's disease, mild cognitive impairment, and Alzheimer's disease. Numerous correlations can be utilized in combination with rsfMRI to enrich the data analysis and deepen the understanding of neurocognitive dysfunction. Such analysis can be focused on how neurological dysfunctions correlate with rsfMRI findings, and how other studies such as nuclear medicine demonstrate physicochemical correlations to rsfMRI and rsfMRI can demonstrate changes over time. Clinical symptoms can be analyzed and correlated with cognitive changes within the brain. For instance, rsfMRI found that fatigue severity in MS patients was negatively correlated with functional connectivity of basal ganglia nuclei with medial prefrontal cortex, precuneus, and posterior cingulate cortex (Finke et al., 2014), and was positively correlated with functional connectivity between caudate nucleus and motor cortex (Finke et al., 2014). This study exemplifies how a specific symptom in a certain disease etiology can be clearly

begun to be elucidated with rsfMRI. This then allows the opportunity for treatment interventions and measuring the rsfMRI initial state and rsfMRI response after treatment. Similar analysis has been performed with aerobic exercise which was found to increase hippocampal volume and connectivity and improve memory in MS subjects (Leavitt et al., 2014).

RsfMRI has been used to analyze cognition and motor function in patients with Parkinson disease (PD) (Olde Dubbelink et al., 2014). In a rsfMRI study in PD, a progressive loss of functional connectivity was demonstrated over a period of 3 years for multiple brain regions, especially in the posterior parts of the brain. The strong correlation with decreasing cognitive performance supports the pathophysiologic role of reduced functional connectivity in cognitive decline and the development of dementia in PD (Olde Dubbelink et al., 2014). Using rsfMRI, it was found that better executive performance was associated with increased dorsal frontoparietal cortical processing and inhibited subcortical and primary sensory involvement. This profile was also characterized by a relative preservation of nigrostriatal dopaminergic function. The profile associated with better memory performance correlated with increased prefrontolimbic processing and was not associated with presynaptic striatal dopamine uptake (Lebedev et al., 2014). Freezing of gait is a paroxysmal and sporadic gait impairment in PD patients. RsfMRI studies reported cortico-striatal decoupling and disruptions in connectivity along the dorsal stream of visuomotor processing (Bardakan et al., 2022).

RsfMRI analysis of the default mode has been utilized to demonstrate in subjects with Alzheimer's disease that there is associated decreased activity in the posterior cingulate gyrus and in the hippocampus (Greicius et al., 2004). RsfMRI has further been employed to analyze behavioral effects on the hippocampus (Burdette et al., 2010). Network analysis of a group undergoing exercise training demonstrated increased connectivity of the hippocampus and the hippocampus was consistently shown to be within the same network group as the anterior cingulate gyrus in the subjects undergoing exercise training (Burdette et al., 2010).

46.2.2.8 Magnetization Transfer (MT)

Magnetization transfer (MT) effects are caused by chemical exchange or dipole-dipole interactions between protons from water and macromolecules. This phenomenon can be used as an MR contrast mechanism. The very broad MR signal from protons of macromolecules and immobile water can be suppressed by off-resonance, frequency-selective RF irradiation, which does not affect the narrow MR signal from mobile water molecules. Saturation of the immobile proton pool causes a reduction of the signal from the free water pool via chemical exchange and cross-relaxation. Absolute quantization of the signal is possible, but complex. A more common semi-quantitative description of the magnitude of the magnetization transfer effect is to provide the magnetization transfer ratio (MTR), which is the are the signal difference with and without saturation and informs on the ratio between free and bound proton pools.

Applications

MT contrast is strongest in areas of de- and re-myelination in the CNS. MT is almost absent in areas of high water content, including cerebral spinal fluid (CSF) and regions of edema (Van Bruggen & Roberts, 2002). The use of MT MRI for the detection and characterization of lesions due to multiple sclerosis has been well studied (Atalay et al., 2005). Quantitative analysis of the magnetization transfer ratio (MTR) of gray matter in the earliest stages of the disease demonstrates statistically significant reduction in the basal ganglia of these patients (Audoin et al., 2004). In general multiple advanced MR techniques, including MTR maps, ^{1}H MRSI, quantitative T_1 and T_2 mapping, diffusion tensor imaging, and studies of blood flow have led to a more sophisticated understanding of the disease from the early view of simply an immune-mediated demyelination process to the understanding of disease effects on neurons, axons, and oligodendrocytes (Minagar et al., 2005).

MT MRI has been shown to be sensitive to other minor structural abnormalities caused by neurological disease. MTR maps were analyzed with a statistical package to demonstrate that areas of epileptic activity, which were not detected visually demonstrated statistically reduced MTR in the epileptic focus (Rugg-Gunn et al., 2003). When MT MRI was combined with DWI, reductions in both MTR and ADC were found in the epileptic focus of patients (Ferini-Strambi et al., 2000).

More advanced quantitation methods for MT MRI have been shown to be sensitive to white matter tracts and axonal density (Yarnykh & Yuan, 2004) as well as being sensitive to the earliest events in stroke lesions (Jiang et al., 2001) including early detection of ruptures in the blood–brain barrier, which may be susceptible to hemorrhagic transformation (Knight et al., 2005).

46.2.2.9 Chemical Exchange Saturation Transfer (CEST) MRI

CEST is a relatively new MRI contrast mechanism based on the chemical exchange of protons (^{1}H) in mobile biomolecules and water. Saturation is a temporary state in which the magnetization of protons is saturated and not able to generate an MRI signal. This has been employed to achieve image contrast such as in ASL mentioned above, fat tissue signal suppression (termed fat saturation), and in magnetic resonance angiography. CEST contrast mechanism can be illustrated using a simple 2-pool model (Fig. 46.6) (van Zijl & Yadav, 2011; Vinogradov et al., 2013; Wu et al., 2016;

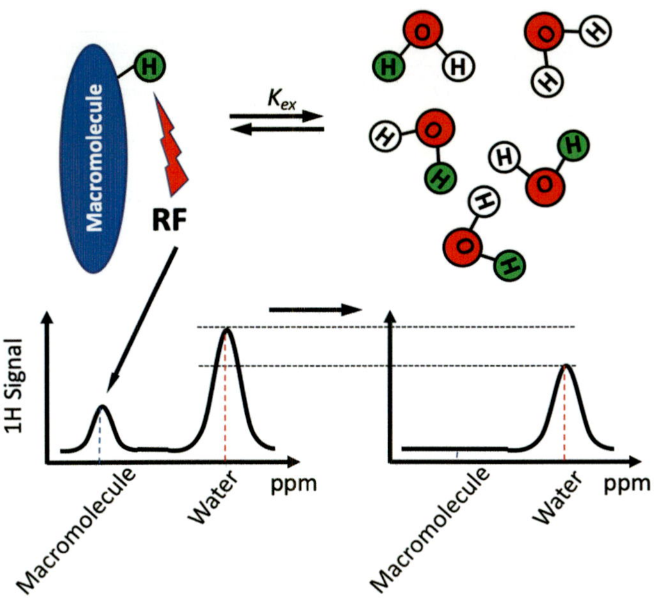

Fig. 46.6 Illustration of CEST contrast mechanism using a simple 2-pool model. One pool is free water, and the other pool is macromolecules. The macromolecules have labile/exchangeable protons that exchange with water protons. On the frequency spectrum, in addition to the signal of the free water, the signal of the macromolecule protons is at a different frequency. When an RF signal is applied to the macromolecule protons, their magnetization is saturated. The saturated protons are then exchanged with water molecules causing water signal reduction. The water signal reduction can be detected using conventional MR imaging methods and providing an indirect measure for the concentration of the macromolecules

Pankowska et al., 2019). One pool is free water, and the other pool is macromolecules. The macromolecules have labile/exchangeable protons that exchange with water protons at the exchange rate K_{ex}. On the frequency spectrum, in addition to the signal of the free water, the signal of the macromolecule protons is at a different frequency. The frequency difference between water and the macromolecule is called chemical shift (Δ). When an RF signal is applied to the macromolecule protons, their magnetization is saturated. The saturated protons are then exchanged with water molecules causing water signal reduction. The water signal reduction can be detected using conventional MR imaging methods and providing an indirect measure for the concentration of the macromolecules. When a saturated proton is transferred from a macromolecule to water, that proton will be replaced with an unsaturated proton from water, which can in turn be saturated for another transfer. The continuous exchange of protons leads to a buildup of saturation in water enabling the detection of low-concentration macromolecules and making CEST much more sensitive than directly measuring macromolecule protons such as in ^{1}H MRS. To generate CEST contrast, the exchange rate needs to be slower than the chemical shift, i.e., $K_{ex} < \Delta$ (Sherry & Wu, 2013; Wu et al., 2016).

In a typical CEST MRI, CEST contrast is measured by collecting a *z*-spectrum, which is a plot of reduced water signals at varying saturation RF frequency. A number of algorithms have been developed to analyze the *z*-spectrum including asymmetric magnetization ratio (MTRasym), Lorentzian fitting, AREX, and artificial neural network method. The goal of the analysis is to identify and quantitate CEST contrasts of biomolecules of interest from confounding factors including relayed nuclear Overhauser effect (rNOE) of macromolecules, MTC of semi-solid macromolecules, and direct bulk free water saturation.

Applications

In theory, CEST is applicable to any chemical substance with exchangeable protons. For the CNS, most CEST applications focus on brain metabolites, proteins, and peptides. Several biomolecules in the brain with exchangeable protons in amide, amine, and hydroxyl groups with optimal exchange properties under physiological conditions have been identified and imaged in vivo (Dou et al., 2019) (Table 46.1). CEST MRI shows promise as a noninvasive molecular imaging.

The most successful application of CEST MRI is a method called amide proton transfer (APT) which measures the CEST contrast from amide protons in mobile proteins and peptides (Kamimura et al., 2019; Zhou et al., 2019). APT is mainly used for brain tumors because it is believed that there is a close relationship between the proliferative activity of the tumor and mobile protein synthesis (Jones et al., 2018). APT MRI has been used to predict malignancy grade in diffuse gliomas (Zhou et al., 2008; Togao et al., 2014). It has been found that APT is superior to the DTI and perfusion MRI in distinguishing malignant gliomas from benign gliomas (Choi et al., 2017; Togao et al., 2017), and comparable with MRS (Sakata et al., 2017). The presence of isocitrate dehydrogenase (IDH) mutation is a marker of favorable prognosis in grade II and III astrocytomas and glioblastomas (Brat et al., 2015; Eckel-Passow et al., 2015). It has been shown that APT MRI is useful in noninvasively predicting the IDH mutation status in grade II gliomas (Jiang et al., 2017). There have been few studies on utility of APT imaging in the differential diagnosis of brain tumors. Jiang et al. (2016) have demonstrated the usefulness of APT imaging in differentiating high-grade gliomas from primary CNS malignant lymphomas (PCNSLs). Differentiation between recurrent tumors and treatment-related changes is crucial but often challenging with conventional MR techniques (Kanda et al., 2018). Several clinical studies have shown that recurrent tumors showed significantly higher APT contrast than treatment-related change indicating higher active proliferation in recurrent tumors (Mehrabian et al., 2017; Park et al., 2018). APT imaging also has applications in the detection of neurodegenerative diseases, e.g., Alzheimer's disease (AD)

Table 46.1 CEST methods used in neurological diseases

Proton type	Amides proton	Amine proton	Hydroxyl proton
Biomolecule (Technique)	Mobile proteins and peptides (Amide proton transfer)	Glutamate (GluCEST) Creatine (CrCEST) Glycogen (GlycoCEST)	Myoinositol (MICEST) Glucose (GlucoCEST) Glycosaminoglycans (GagCEST) Lactate (LATEST)
Applications	Cancer, stroke, pH	Neurodegenerative diseases	Neurological disorders, cancer metabolism

(Wang et al., 2015b) and Parkinson's disease (PD) (Li et al., 2017). AD patients showed increased APT intensity in the bilateral hippocampus region which correlated directly to clinical disease severity (Wang et al., 2015b). In PD patients, APT reduction was found in the substantia nigra and red nucleus relative to healthy controls, and APT intensity was consistently decreased with disease severity from early to advanced stage. In addition to amide protons, the amine protons of mobile proteins, peptides, and amino acids can generate CEST contrasts that potentially can be used for brain tumor studies (Harris et al., 2015).

Glutamate is an excitatory neurotransmitter. The anime protons of glutamate generate CEST contrast that can be used for the imaging of glutamate. The imaging of glutamate based on its CEST contrast is termed gluCEST MRI and has been successfully performed in healthy brains (Cai et al., 2012, 2013). GluCEST MRI has great potential for clinical uses. A gluCEST MRI correctly lateralized the temporal lobe seizure focus in patients with epilepsy, which was corroborated with MRS results (Davis et al., 2015). Glucose is a major energy source for the brain. The feasibility of CEST MRI to measure glucose uptake (termed glucoCEST) by tissue is being tested, and the technique has been used in some studies (Zaiss et al., 2019; Xu et al., 2020; Bender et al., 2022; Kim et al., 2022) especially in brain tumors as the tumor cells usually have active energy metabolism.

CEST MRI has become a great imaging tool for preclinical studies using animals (Dou et al., 2019; Cember et al., 2022). Among different CEST MRI techniques, preclinical studies have demonstrated the applications of gluCEST as a biomarker in a number of different neurological events and have demonstrated it has the potential for molecular imaging with exquisite spatial resolution. Studies suggested that elevated gluCEST is a biomarker of the proliferation of glial fibrillary acidic protein (GFAP), which is associated with reactive astrocytes (Bagga et al., 2013). Studies of brain abscess and neuroinflammation have also exhibited increased gluCEST contrast (Chen et al., 2016, 2018). Alzheimer's disease modeled using transgenic mice shows a sharply decreased hippocampal gluCEST signal with age in comparison with wild-type mice (Haris et al., 2013a; Crescenzi et al., 2014, 2017). In a mouse model of Huntington's disease, GluCEST signal was found decreased on corpus callosum (Crescenzi et al., 2017).

One application of preclinical APT MRI is in ischemia detection. This technique has been proposed in predicting ischemic stroke at acute stage (Wang et al., 2015a; Wu et al., 2018). In a study of AD using rats, APT signal was found decreased in brain indicating protein aggregation involved in AD (Wang et al., 2019).

Myo-inositol (MI) is a biomarker of glial activation. The CEST MRI has been tested in vivo for MI detection (Haris et al., 2011, 2013b). This technique is termed MICEST and has been proven useful for measuring microglial activation. The energy metabolism of brain tumors was tested using glucoCEST in rodents (Walker-Samuel et al., 2013; Goldenberg & Pagel, 2019; Rivlin & Navon, 2019). Reduced glucose uptake was observed in a mouse model of AD using glucoCEST (Machhi et al., 2021).

46.2.2.10 ^{1}H Magnetic Resonance Spectroscopy (MRS)

In addition to water, tissue also contains biomolecules such as metabolites and neurotransmitters. The protons on these biomolecules have Larmor frequencies that are different from the water proton. Each biomolecule has a unique "frequency fingerprint". Magnetic resonance spectroscopy acquires frequency-specific signals from the biomolecules so that MRS is unique in the ability to perform non-invasive analysis of chemical composition of tissue without biopsy and allows the serial non-invasive determination of tissue chemistry with treatment. An example of the spectrum from a region of mouse brain acquired at 7 Tesla is shown in Fig. 46.7.

Most MRS studies of brain use creatine as an internal concentration standard, as creatine has a long T_2 and has relatively constant concentration in most cell types. Advanced methods exist to allow quantitation of the metabolites measured using MRS/MRSI. These methods include the referencing of signal to the internal water signal and referencing in vivo signal amplitude to signal amplitudes from chemical samples of known concentration. Quantitation requires the correction of signal amplitude for T_1 and T_2 of the metabolites for each acquisition technique.

In a typical MRS experiment, only brain tissue in a single cuboid shape voxel can be measured. 1H MRS imaging (^{1}H MRSI) is an extension of MRS that enables the measurements of a 2D array of voxels and provides improved spatial

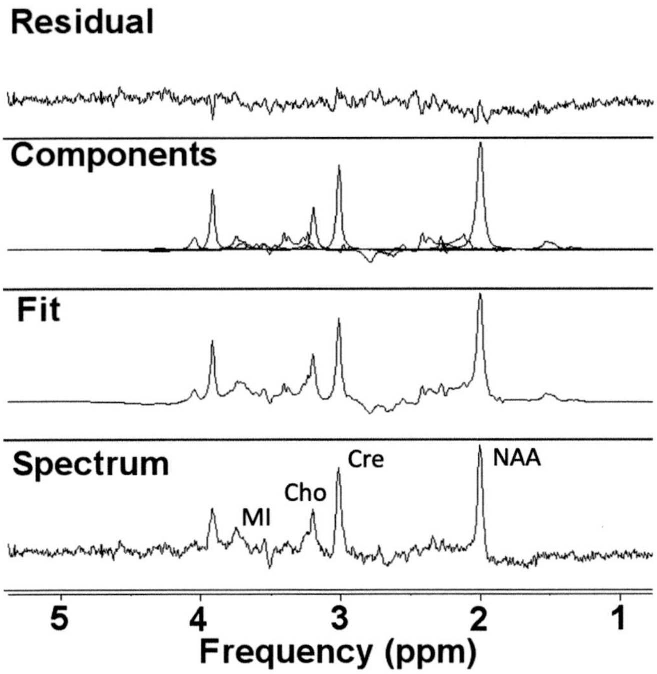

Fig. 46.7 Example of a ^{1}H MR spectrum from a ^{1}H MRS data set with a nominal spatial resolution of 2.34 μl (1.5 × 1.25 × 1.25 mm) in a mouse brain. Spectrum: Original spectrum obtained from the region. Fit: Overall fit using models of known spectra of metabolites including, but not limited to N-acetyl aspartate (NAA), glutamate (Glu), glutamine, creatine (Cre), choline (Cho), and myoinostitol (mI). Components: Individual peaks contributing to the overall spectra of metabolites and Residual: Subtraction of the fit from the original spectrum. The frequency scale is given in ppm ((frequency/Larmor frequency) × 10^6) to create a frequency scale insensitive to the magnetic field strength

resolution and the ability to reformat data post-acquisition to allow precise localization of spectra within the voxel.

Application

1H MRS detectable metabolites are primarily N-acetyl aspartate (NAA), choline (Cho), creatine (Cre), glutamate (Glu), glutamine (Gln), and myoinostitol (mI). NAA is found predominately in the neurons and the typical physiological intraneuroanal concentration is 9–12 mM, and as such is used as a marker of neuronal density. Studies also suggest multiple roles of NAA including an energy metabolism reserve, a supply of acetyl-Coenzyme-A for the tricarboxylic acid cycle in mitochondria, a source of N-acetyl-aspartylglutamate which is a neuropeptide involved in neurotransmission, a source of glutamate (a neuroexcitatory molecule), and a role as an osmotic regulator. Neuronal cell death induces an irreversible loss of NAA due to localization of NAA in the neuron, while reversible loss of NAA is likely associated with reversal of mitochondrial dysfunction or replacement of damaged mitochondria within the neurons. Cho mainly consists of glycerophosphocholine and phosphocholine, compounds involved in phospholipid metabolism in brain tissue. Choline concentrations are known to be elevated in multiple sclerosis (MS) during periods of active demyelination. Cre concentration is the amount of creatine plus phosphocreatine in brain tissue. Cre is part of the creatine kinase energy metabolism buffer system used to maintain ATP levels in times of acute mismatch between oxidative adenosine triphosphate (ATP) supply and ATP demand. The [Cre] reflects the health of systematic energy use and storage. Cre concentration tends to remain relatively constant in most cell types and is known to be elevated in multiple sclerosis and epilepsy where gliosis is significant. MI is a biomarker of glial activation and as an osmoregulator or intracellular messenger.

^{1}H MRS and MRSI have been demonstrated to be a sensitive method for early detection of neuronal dysfunction or loss, reflected by loss of NAA. The application of MRS adds specificity to the diagnosis of a lesion relative to MR imaging alone (Howe & Opstad, 2003; Nelson, 2003). The Cho/Cr and NAA/Cho ratios have been used to differentiate tumor from non-tumor lesions in the brain (Moller-Hartmann et al., 2002). Increased Cho/Cr and decreased NAA/Cho are indicative of brain tumors (Moller-Hartmann et al., 2002). Cho and the Lip/Lac ratio tend to increase as an astrocytoma progresses from low-grade to glioblastoma (GBM) multiforme (Nelson et al., 2002). Cre and NAA are decreased in low-grade GBM (Nelson et al., 2002). Myo-inositol is increased in low-grade astrocytomas and is decreased in high grade astrocytomas (Castillo et al., 2000). Lip and Lac can be used to separate low-grade from high-grade (anaplastic) oligodendrogliomas in which they have been found to be significantly elevated (Rijpkema et al., 2003). In low grade oligodendrogliomas, glutamate and glutamine have been found to be elevated and this may help to differentiate oligodendrogliomas from other brain tumors such as astrocytomas (Rijpkema et al., 2003). Gliomatosis cerebri can demonstrate subtle changes on MR spectroscopy. It has been shown that the Cho/Cr levels may appear normal and that only an elevated MI level will be detected (Mohana-Borges et al., 2004). It was also found normal Cho/Cr levels and decreased NAA and Lip/Lac in Gliomatosis cerebri (Pyhtinen, 2000). Diagnosing radiation necrosis versus tumor recurrence is also a useful and accurate application of MRS (Nelson, 2003). Radiation necrosis has high Lip and Lac without elevated Cho whereas a tumor would have elevated Cho.

It can be critical to obtain MRSI data over the entire complex heterogeneous areas of a brain lesion in order to institute appropriate therapy, plan surgeries and accurately diagnose the brain lesion. Compared to MRS, MRSI helps to define the spatial heterogeneity in brain lesions and also contributes to defining the spatial extent of brain lesions. MRSI is particularly useful for distinguishing recurrent tumor versus radiation necrosis given its ability to spatially sample multiple areas of a lesion (Chernov et al., 2005).

46.3 Single Photon Emission Computed Tomography (SPECT) and Positron Emission Tomography (PET)

Computed tomography techniques share a common mechanism for spatial encoding. A two-dimensional (2D) projection of a subject to be imaged at multiple angles can be reconstructed into two- or three-dimensional images. Data are acquired by rotation of a set of detectors around the subject that contains internal radiation sources as in SPECT and PET. Reconstruction of images from tomographic methods is performed using the reverse Radon transform, which uses a series of angular projections to reconstruct images.

SPECT detects emitted gamma rays from radioactively labeled molecules or cells (termed radioactive tracers or radiotracers). The position of the emitted gamma ray is determined by gamma-ray cameras rotating about the subject to be detected.

PET uses positron emission to increase sensitivity of detection over SPECT. Once a positron is emitted from the radioactive atom on the labeled molecule, it travels a short distance (2–5 mm, depending on the energy of the emitted positron) and undergoes an annihilation reaction with an electron. Annihilation occurs because the positron is antimatter to the electron. The reaction creates two gamma ray photons which then travel 180° away from each other. PET scanners employ a ring of detectors and coincidence detection (simultaneous detection of two gamma rays) to improve detection sensitivity and filter out background radiation. As a result of the high capture cross-section of the ring of detectors and the virtual elimination of background radiation, sensitivity is enhanced by approximately four orders of magnitude compared to SPECT. This allows detection of lower concentrations of tagged molecules and cells down to the picomolar (10^{-12} M) concentration range with lower doses of radioactivity injected into the subject (typically around 10μCi).

SPECT and PET images provide spatial distribution maps of radiotracers, but lack of anatomical details. To overcome this problem, most SPECT and PET scanners can integrate with CT or MRI systems in certain ways, so that anatomical and radiotracer studies are acquired simultaneously providing anatomical images acquired simultaneously with SPECT or PET image acquisition.

46.3.1 Single Photon Emission Computed Tomography (SPECT)

SPECT in general has a greater clinical availability compared to PET. Secondary to SPECT's greater availability, proven capabilities, and its lower cost it is an often used clinical and research imaging modality. SPECT is very useful for evaluating brain perfusion and has demonstrated utility for evaluating the brain perfusion of many disease processes including stroke, Alzheimer's disease and other dementias, Parkinson's disease, seizure patients, and brain tumors. Perfusion is just one parameter that can be measured with SPECT and there are many different radiotracers that can be used with SPECT for neurotransmitter system imaging. The Ioflupane I-123 SPECT scan (DaTscan) can help differentiate between Parkinsonian syndromes and essential tremors. An analysis of the utility of this scan demonstrated therapies were changed 63% of the time secondary to the scan results (Sadasivan & Friedman, 2015).

Thallium-201 SPECT and fluoro-deoxyglucose (FDG) PET are both used to help differentiate residual brain tumors and brain tumor recurrence from radiation necrosis. Thallium-201 SPECT has been found to be just as good or even better for tumor detection than FDG PET (Kumar et al., 2017). Thallium-201 SPECT can also be utilized to help differentiate lymphoma from toxoplasmosis which is an important question in immunocompromised patients, particularly in patients infected with human immunodeficiency virus (HIV) (Licho et al., 2002; Young et al., 2005).

SPECT has been used in numerous psychiatric studies including evaluation of major depressive disorder, obsessive compulsive disorder, and schizophrenia (Novak et al., 2005; Topcuoglu et al., 2005; Alves et al., 2006). Studies have been performed evaluating these diseases and other psychiatric disorders at their initial presentations, during disease progression, and what the associated imaging changes are with medical intervention. There are many continued opportunities for investigation to use SPECT in psychiatry.

SPECT has been found to be quite useful when evaluating patients with seizures (McNally et al., 2005; Tae et al., 2005). SPECT imaging is commonly used for the ictal study of seizure patients to define the seizure focus. It has in general been much easier to perform ictal SPECT than ictal PET. Advanced analytical techniques of SPECT imaging have helped to improve the data analysis and utilization of the SPECT data in the evaluation of seizure patients (Knowlton et al., 2004).

46.3.2 Positron Emission Tomography (PET)

A wide array of radiotracers is available for neuroscience research. Fluoro-deoxyglucose (FDG) is the most used radiotracer in neurosciences. FDG-PET has been used to study numerous diseases including tumors, strokes, dementia, addiction disorders, schizophrenia, depression, seizures, and developmental abnormalities. FDG-PET has been used extensively to investigate brain tumors including astrocytomas, oligodendrogliomas, meningiomas, lymphomas, metastatic disease, and the differentiation of tumor necrosis from tumor recurrence. In the evaluation of brain tumors, FDG-PET has been found to help determine the grade of malignancy and can provide predictive prognostic information (Borja et al., 2022; Verger et al., 2022; Zhang-Yin et al., 2022). Carbon-11-methionine has also been found useful for predicting histological grade and the prognosis of gliomas. Carbon-11-methionine seems superior to FDG in sensitivity and accuracy in delineation of tumors (Borja et al., 2022, Verger et al., 2022, Zhang-Yin et al., 2022). Thallium-201 SPECT as mentioned in the previous section also has been shown to perform as good if not better than FDG-PET in brain tumor analysis (Kumar et al., 2017).

Evaluating dementias with PET imaging is another area that has generated a lot of attention as well as clinical and research applications. The ability of PET scanning to help differentiate whether a patient has Alzheimer's disease has generated great interest due to the large number of individuals affected. FDG is the radiotracer that is routinely utilized in the evaluation of dementias. FDG-PET can help determine whether a patient has Alzheimer's disease or mild cognitive impairment, and it can also predict whether a patient with mild cognitive impairment is likely to progress to Alzheimer's disease (Anchisi et al., 2005). Frontal-temporal dementia, vascular dementia, and Lewy body dementia diagnosis can also be facilitated with FDG-PET (Kerrouche et al., 2006). Amyloid and Tau PET imaging are both adding significant new opportunities for evaluating patients who have memory dysfunction. Radiotracers that can define these abnormal proteins within the brain make it possible to detect and quantitate the disease processes leading to dementia and to determine the time of effective therapeutic intervention. Pittsburgh compound B (PIB) was the first and is the most extensively studied of such compounds for detecting AB plaques (Shokouhi et al., 2014). PIB has been demonstrated to have increased uptake in patients with Alzheimer's disease and has been confirmed to be present in AB plaques postmortem. Tau imaging has advanced rapidly and can now help to further refine the diagnostic biomarkers also being able to detect the neurofibrillary tangles appearing in the transentorhinal cortex, followed by the entorhinal cortex and the hippocampus, and subsequently the neocortex in Alzheimer's disease (Okamura et al., 2013). Other tauopathies, such as chronic traumatic encephalopathy, can also be detected and diagnosis facilitated using tau PET (Mitsis et al., 2014).

PET is a valuable tool for Parkinson's disease in multiple ways including early confirmation of the diagnosis, predicting clinical course, and to evaluate the results of therapies. When evaluating Parkinson's disease both FDG and 6-[(18) F]fluoro-L-dopa (FDOPA), a common presynaptic dopaminergic radiotracer, can be effectively employed (Kaasinen et al., 2006). FDG and FDOPA PET has been shown to be an indicator of the severity of Parkinson's disease and at patient's initial evaluations can be used to predict the clinical prognosis (Kaasinen et al., 2006). FDG-PET may help differentiate between Parkinson's disease, progressive supranuclear palsy, and multiple system atrophy due to the distinct metabolic patterns observed in these conditions (Juh et al., 2004).

Therapeutic-induced changes of brain metabolism can be evaluated for both surgical and medical interventions (Zhao et al., 2004). These types of inquiries provide knowledge that can be used to judge an intervention's success, how an intervention's success may have been produced, and potentially to predict early on in a treatment paradigm what long-term success may be obtained.

FDG-PET again has been the predominant radiotracer that has been utilized for seizure evaluation. Pre-operative evaluation of temporal lobe epilepsy is a frequently performed application of PET-FDG and FDG-PET is able to help predict the success of surgical intervention and can help guide the type of surgical resection (Kim et al., 2003). PET as already mentioned usually studies the seizure patient in the interictal state since it is difficult to perform ictal PET studies. FDG-PET can also be a predictor of verbal decline after anterior temporal lobectomy—a surgery performed to help control complex partial seizures (Griffith et al., 2000). It has also been shown that PET-FDG is beneficial for evaluating for seizure foci in a general epilepsy population (Swartz et al., 2002).

Microglial activation is a hallmark of neuroinflammation. The translocator protein-18 kDa (TSPO) is a molecular biomarker of microglial activation. PET imaging of TSPO ligands has been successfully used to detect in a wide variety of animal models of neuroinflammatory conditions, including AD, stroke, brain injury, experimental autoimmune encephalitis, and epilepsy (Werry et al., 2019). A number of TSPO ligands with 11C and 18F have been developed for PET (Best et al., 2019). Clinical trials are underway (Dupont et al., 2017). Other molecular biomarkers of microglial activation are being tested for PET imaging (Narayanaswami et al., 2018) for better specificity and sensitivity.

46.4 Summary

This chapter has presented an overview of the mechanisms and applications of neuroimaging for the readers of neuroscience with a view towards providing an understanding of the potential of neuroimaging techniques in the research of neuroscience research, especially in the neuroimmune system. As the field progresses, methods for whole brain analyses, development of new detection methods and new radiotracers combined will provide whole brain non-invasive histology for diagnosis of disease and tracking the effects of new therapies in animal models of neurological disease and, eventually, human subjects. Research techniques being developed will synergize multiple techniques combined with molecular and histological analyses to provide a sensitive and specific characterization of the gamut of neuroimaging methods for the diagnosis and monitoring of neurological diseases.

46.5 Review Questions

1. Advantages of diffusion tensor imaging include:
   (a) Demonstrates abnormalities when conventional MRI is normal.
   (b) Provides functional maps of brain activity.
   (c) It is indispensable for planning stroke therapy.
   (d) Can be used to calculate perfusion metrics.
2. Perfusion imaging data can be used to create maps that:
   (a) Allow for "virtual dissection" of white matter tracts.
   (b) Are used for creating MR angiograms.
   (c) Can be utilized to determine tissue at risk for damage in patients with stroke.
   (d) Show peaks of lactate in tissue damaged by stroke.
3. CEST MRI can be used to measure:
   (a) BBB permeability.
   (b) Brain metabolism.
   (c) Brain functional connection.
   (d) Brain atrophy.
4. MR spectroscopy does not:
   (a) Demonstrate increased myo-inositol in high-grade astrocytomas.
   (b) Potentially only demonstrates subtle findings for gliomatosis cerebri.
   (c) Demonstrate increased myo-inositol in low-grade astrocytomas.
   (d) Show glutamate and glutamine elevations in low-grade oligodendrogliomas.
5. PET imaging:
   (a) Has fewer radiotracers that can be utilized than SPECT.
   (b) Has few imaging applications for which FDG is used.
   (c) Has the ability to differentiate tumor necrosis from tumor recurrence.
   (d) Lacks clear ability to differentiate Alzheimer's disease from mild cognitive impairment.

46.5.1 Answers to Review Question

1. (a)
2. (c)
3. (b)
4. (a)
5. (c)

References

Abdel Razek, A. A. K., Talaat, M., El-Serougy, L., Gaballa, G., & Abdelsalam, M. (2019). Clinical applications of arterial spin labeling in brain tumors. *Journal of Computer Assisted Tomography, 43*(4), 525–532.

Absinta, M., Sati, P., Gaitan, M. I., Maggi, P., Cortese, I. C., Filippi, M., & Reich, D. S. (2013). Seven-tesla phase imaging of acute multiple sclerosis lesions: A new window into the inflammatory process. *Annals of Neurology, 74*(5), 669–678.

Alves, T. C., Rays, J., Fraguas, R., Jr., Wajngarten, M., Telles, R. M. S., Duran, F. L. D. S., Meneghetti, J. C., Robilotta, C. C., Prando, S., De Castro, C. C., Buchpiguel, C. A., & Busatto, G. F. (2006). Association between major depressive symptoms in heart failure and impaired regional cerebral blood flow in the medial temporal region: A study using 99m Tc-HMPAO single photon emission computerized tomography (SPECT). *Psychological Medicine, 36*, 597–608.

Anchisi, D., Borroni, B., Franceschi, M., Kerrouche, N., Kalbe, E., Beuthien-Beumann, B., Cappa, S., Lenz, O., Ludecke, S., Marcone, A., Mielke, R., Ortelli, P., Padovani, A., Pelati, O., Pupi, A., Scarpini, E., Weisenbach, S., Herholz, K., Salmon, E., Holthoff, V., Sorbi, S., Fazio, F., & Perani, D. (2005). Heterogeneity of brain glucose metabolism in mild cognitive impairment and clinical progression to Alzheimer disease. *Archives of Neurology, 62*(11), 1728–1733.

Armitage, P. A., Farrall, A. J., Carpenter, T. K., Doubal, F. N., & Wardlaw, J. M. (2011). Use of dynamic contrast-enhanced MRI to measure subtle blood-brain barrier abnormalities. *Magnetic Resonance Imaging, 29*(3), 305–314.

Atalay, K., Diren, H. B., Gelmez, S., Incesu, L., & Terzi, M. (2005). The effectiveness of magnetization transfer technique in the evaluation of acute plaques in the central nervous system of multiple sclerosis patients and its correlation with the clinical findings. *Diagnostic and Interventional Radiology, 11*(3), 137–141.

Audoin, B., Ranjeva, J. P., Au Duong, M. V., Ibarrola, D., Malikova, I., Confort-Gouny, S., Soulier, E., Viout, P., Ali-Cherif, A., Pelletier, J., & Cozzone, P. J. (2004). Voxel-based analysis of MTR images: A method to locate gray matter abnormalities in patients at the earliest stage of multiple sclerosis. *Journal of Magnetic Resonance Imaging, 20*(5), 765–771.

Bagga, P., Chugani, A. N., Varadarajan, K. S., & Patel, A. B. (2013). In vivo NMR studies of regional cerebral energetics in MPTP model of Parkinson's disease: Recovery of cerebral metabolism with acute levodopa treatment. *Journal of Neurochemistry, 127*(3), 365–377.

Bardakan, M. M., Fink, G. R., Zapparoli, L., Bottini, G., Paulesu, E., & Weiss, P. H. (2022). Imaging the neural underpinnings of freezing of gait in Parkinson's disease. *NeuroImage: Clinical, 35*, 103123.

Bender, B., Herz, K., Deshmane, A., Richter, V., Tabatabai, G., Schittenhelm, J., Skardelly, M., Scheffler, K., Ernemann, U., Kim, M., Golay, X., Zaiss, M., & Lindig, T. (2022). GLINT: GlucoCEST in neoplastic tumors at 3 T—Clinical results of GlucoCEST in gliomas. *Magma, 35*(1), 77–85.

Beppu, T., Inoue, T., Shibata, Y., Yamada, N., Kurose, A., Ogasawara, K., Ogawa, A., & Kabasawa, H. (2005). Fractional anisotropy value by diffusion tensor magnetic resonance imaging as a predictor of cell density and proliferation activity of glioblastomas. *Surgical Neurology, 63*(1), 56–61; discussion 61.

Bergamino, M., Bonzano, L., Levrero, F., Mancardi, G. L., & Roccatagliata, L. (2014). A review of technical aspects of T1-weighted dynamic contrast-enhanced magnetic resonance imaging (DCE-MRI) in human brain tumors. *Physica Medica, 30*(6), 635–643.

Berry, E., & Bulpitt, A. J. (2008). *Fundamentals of MRI: An interactive learning approach*. CRC Press.

Best, L., Ghadery, C., Pavese, N., Tai, Y. F., & Strafella, A. P. (2019). New and old TSPO PET radioligands for imaging brain microglial activation in neurodegenerative disease. *Current Neurology and Neuroscience Reports, 19*(5), 24.

Bonakdarpour, B., & Takarabe, C. (2023). Brain networks, clinical manifestations, and neuroimaging of cognitive disorders: The role of computed tomography (CT), magnetic resonance imaging (MRI), positron emission tomography (PET), and other advanced neuroimaging tests. *Clinics in Geriatric Medicine, 39*(1), 45–65.

Borja, A. J., Saini, J., Raynor, W. Y., Ayubcha, C., Werner, T. J., Alavi, A., Revheim, M. E., & Nagaraj, C. (2022). Role of molecular imaging with PET/MR imaging in the diagnosis and management of brain tumors. *PET Clinics, 17*(3), 431–451.

Brat, D. J., Verhaak, R. G., Aldape, K. D., Yung, W. K., Salama, S. R., Cooper, L. A., Rheinbay, E., Miller, C. R., Vitucci, M., Morozova, O., Robertson, A. G., Noushmehr, H., Laird, P. W., Cherniack, A. D., Akbani, R., Huse, J. T., Ciriello, G., Poisson, L. M., Barnholtz-Sloan, J. S., Berger, M. S., Brennan, C., Colen, R. R., Colman, H., Flanders, A. E., Giannini, C., Grifford, M., Iavarone, A., Jain, R., Joseph, I., Kim, J., Kasaian, K., Mikkelsen, T., Murray, B. A., O'Neill, B. P., Pachter, L., Parsons, D. W., Sougnez, C., Sulman, E. P., Vandenberg, S. R., Van Meir, E. G., von Deimling, A., Zhang, H., Crain, D., Lau, K., Mallery, D., Morris, S., Paulauskis, J., Penny, R., Shelton, T., Sherman, M., Yena, P., Black, A., Bowen, J., Dicostanzo, K., Gastier-Foster, J., Leraas, K. M., Lichtenberg, T. M., Pierson, C. R., Ramirez, N. C., Taylor, C., Weaver, S., Wise, L., Zmuda, E., Davidsen, T., Demchok, J. A., Eley, G., Ferguson, M. L., Hutter, C. M., Mills Shaw, K. R., Ozenberger, B. A., Sheth, M., Sofia, H. J., Tarnuzzer, R., Wang, Z., Yang, L., Zenklusen, J. C., Ayala, B., Baboud, J., Chudamani, S., Jensen, M. A., Liu, J., Pihl, T., Raman, R., Wan, Y., Wu, Y., Ally, A., Auman, J. T., Balasundaram, M., Balu, S., Baylin, S. B., Beroukhim, R., Bootwalla, M. S., Bowlby, R., Bristow, C. A., Brooks, D., Butterfield, Y., Carlsen, R., Carter, S., Chin, L., Chu, A., Chuah, E., Cibulskis, K., Clarke, A., Coetzee, S. G., Dhalla, N., Fennell, T., Fisher, S., Gabriel, S., Getz, G., Gibbs, R., Guin, R., Hadjipanayis, A., Hayes, D. N., Hinoue, T., Hoadley, K., Holt, R. A., Hoyle, A. P., Jefferys, S. R., Jones, S., Jones, C. D., Kucherlapati, R., Lai, P. H., Lander, E., Lee, S., Lichtenstein, L., Ma, Y., Maglinte, D. T., Mahadeshwar, H. S., Marra, M. A., Mayo, M., Meng, S., Meyerson, M. L., Mieczkowski, P. A., Moore, R. A., Mose, L. E., Mungall, A. J., Pantazi, A., Parfenov, M., Park, P. J., Parker, J. S., Perou, C. M., Protopopov, A., Ren, X., Roach, J., Sabedot, T. S., Schein, J., Schumacher, S. E., Seidman, J. G., Seth, S., Shen, H., Simons, J. V., Sipahimalani, P., Soloway, M. G., Song, X., Sun, H., Tabak, B., Tam, A., Tan, D., Tang, J., Thiessen, N., Triche, T., Jr., Van Den Berg, D. J., Veluvolu, U., Waring, S., Weisenberger, D. J., Wilkerson, M. D., Wong, T., Wu, J., Xi, L., Xu, A. W., Yang, L., Zack, T. I., Zhang, J., Aksoy, B. A., Arachchi, H., Benz, C., Bernard, B., Carlin, D., Cho, J., DiCara, D., Frazer, S., Fuller, G. N., Gao, J., Gehlenborg, N., Haussler, D., Heiman, D. I., Iype, L., Jacobsen, A., Ju, Z., Katzman, S., Kim, H., Knijnenburg, T., Kreisberg, R. B., Lawrence, M. S., Lee, W., Leinonen, K., Lin, P., Ling, S., Liu, W., Liu, Y., Liu, Y., Lu, Y., Mills, G., Ng, S., Noble, M. S., Paull, E., Rao, A., Reynolds, S., Saksena, G., Sanborn, Z., Sander, C., Schultz, N., Senbabaoglu, Y., Shen, R., Shmulevich, I., Sinha, R., Stuart, J., Sumer, S. O., Sun, Y., Tasman, N., Taylor, B. S., Voet, D., Weinhold, N., Weinstein, J. N., Yang, D., Yoshihara, K., Zheng, S., Zhang, W., Zou, L., Abel, T., Sadeghi, S., Cohen, M. L., Eschbacher, J., Hattab, E. M., Raghunathan, A., Schniederjan, M. J., Aziz, D., Barnett, G., Barrett, W., Bigner, D. D., Boice, L., Brewer, C., Calatozzolo, C., Campos, B., Carlotti, C. G., Jr., Chan, T. A., Cuppini, L., Curley, E., Cuzzubbo, S., Devine, K., DiMeco, F., Duell, R., Elder, J. B., Fehrenbach, A., Finocchiaro, G., Friedman, W., Fulop, J., Gardner, J., Hermes, B., Herold-Mende, C., Jungk, C., Kendler, A., Lehman, N. L., Lipp, E., Liu, O., Mandt, R., McGraw, M., McLendon, R., McPherson, C., Neder, L., Nguyen, P., Noss, A., Nunziata, R., Ostrom, Q. T., Palmer, C., Perin, A., Pollo, B., Potapov, A., Potapova, O., Rathmell, W. K., Rotin, D., Scarpace, L., Schilero, C., Senecal, K., Shimmel, K., Shurkhay, V., Sifri, S., Singh, R., Sloan, A. E., Smolenski, K., Staugaitis, S. M., Steele, R., Thorne, L., Tirapelli, D. P., Unterberg, A., Vallurupalli, M., Wang, Y., Warnick, R., Williams, F., Wolinsky, Y., Bell, S., Rosenberg, M., Stewart, C., Huang, F., Grimsby, J. L., Radenbaugh, A. J., & Zhang, J. (2015). Comprehensive, integrative genomic analysis of diffuse lower-grade gliomas. *The New England Journal of Medicine, 372*(26), 2481–2498.

Bron, E. E., Steketee, R. M., Houston, G. C., Oliver, R. A., Achterberg, H. C., Loog, M., van Swieten, J. C., Hammers, A., Niessen, W. J., Smits, M., Klein, S., & Alzheimer's Disease Neuroimaging Initiative. (2014). Diagnostic classification of arterial spin labeling and structural MRI in presenile early stage dementia. *Human Brain Mapping, 35*(9), 4916–4931.

Brown, R. W., Cheng, Y.-C. N., Haacke, E. M., Thompson, M. R., & Venkatesan, R. (2014). *Magnetic resonance imaging: Physical principles and sequence design*. Wiley-Blackwell.

Burdette, J. H., Laurienti, P. J., Espeland, M. A., Morgan, A., Telesford, Q., Vechlekar, C. D., Hayasaka, S., Jennings, J. M., Katula, J. A., Kraft, R. A., & Rejeski, W. J. (2010). Using network science to evaluate exercise-associated brain changes in older adults. *Frontiers in Aging Neuroscience, 2*, 23.

Bushberg, J. T., Seibert, J. A., Leidholdt, E. M. J., & Boone, J. M. (2002). *The essential physics of medical imaging*. Williams & Wilkins.

Cai, K., Haris, M., Singh, A., Kogan, F., Greenberg, J. H., Hariharan, H., Detre, J. A., & Reddy, R. (2012). Magnetic resonance imaging of glutamate. *Nature Medicine, 18*(2), 302–306.

Cai, K., Singh, A., Roalf, D. R., Nanga, R. P., Haris, M., Hariharan, H., Gur, R., & Reddy, R. (2013). Mapping glutamate in subcortical brain structures using high-resolution GluCEST MRI. *NMR in Biomedicine, 26*(10), 1278–1284.

Campbell, B. C. V., De Silva, D. A., Macleod, M. R., Coutts, S. B., Schwamm, L. H., Davis, S. M., & Donnan, G. A. (2019). Ischaemic stroke. *Nature Reviews. Disease Primers, 5*(1), 70.

Castillo, M., Smith, J. K., & Kwock, L. (2000). Correlation of myo-inositol levels and grading of cerebral astrocytomas. *AJNR. American Journal of Neuroradiology, 21*(9), 1645–1649.

Cember, A. T. J., Nanga, R. P. R., & Reddy, R. (2022). Glutamate-weighted CEST (gluCEST) imaging for mapping neurometabolism: An update on the state of the art and emerging findings from in vivo applications. *NMR in Biomedicine, 36*, e4780.

Chang, S.-C., Lai, P.-H., Chen, W.-L., Weng, H.-H., Ho, J.-T., Wang, J.-S., Chang, C.-Y., Pan, H.-B., & Yang, C.-F. (2002). Diffusion-weighted MRI features of brain abscess and cystic or necrotic brain tumors: Comparison with conventional MRI. *Clinical Imaging, 26*(4), 227–236.

Chang, C. Y., Peck, K. K., Brennan, N. M., Hou, B. L., Gutin, P. H., & Holodny, A. I. (2010). Functional MRI in the presurgical evaluation of patients with brain tumors: Characterization of the statistical threshold. *Stereotactic and Functional Neurosurgery, 88*(1), 35–41.

Chen, C. C., Shih, Y. Y., & Chang, C. (2013). Dopaminergic imaging of nonmotor manifestations in a rat model of Parkinson's disease by fMRI. *Neurobiology of Disease, 49*, 99–106.

Chen, Y. Z., Dai, Z. Z., Shen, Z. W., Lin, G. S., Zhuang, C. Y., Li, H. J., & Wu, R. H. (2016). Magnetic resonance imaging of glutamate in neuroinflammation. *Radiology of Infectious Diseases, 3*(2), 92–97.

Chen, Y., Dai, Z., Shen, Z., Guan, J., Zhuang, Z., Mao, Y., & Wu, R. (2018). Imaging of glutamate in brain abscess using GLUCEST at 7T. *Radiology of Infectious Diseases, 5*(4), 148–153.

Chernov, M., Hayashi, M., Izawa, M., Ochiai, T., Usukura, M., Abe, K., Ono, Y., Muragaki, Y., Kubo, O., Hori, T., & Takakura,

K. (2005). Differentiation of the radiation-induced necrosis and tumor recurrence after gamma knife radiosurgery for brain metastases: Importance of multi-voxel proton MRS. *Minimally Invasive Neurosurgery, 48*(4), 228–234.

Choi, Y. S., Ahn, S. S., Lee, S. K., Chang, J. H., Kang, S. G., Kim, S. H., & Zhou, J. (2017). Amide proton transfer imaging to discriminate between low- and high-grade gliomas: Added value to apparent diffusion coefficient and relative cerebral blood volume. *European Radiology, 27*(8), 3181–3189.

Cosottini, M., Giannelli, M., Siciliano, G., Lazzarotti, G., Michelassi, M. C., Del Corona, A., Bartolozzi, C., & Murri, L. (2005). Diffusion-tensor MR imaging of corticospinal tract in amyotrophic lateral sclerosis and progressive muscular atrophy. *Radiology, 237*(1), 258–264.

Cramer, S. P., Simonsen, H., Frederiksen, J. L., Rostrup, E., & Larsson, H. B. (2014). Abnormal blood-brain barrier permeability in normal appearing white matter in multiple sclerosis investigated by MRI. *NeuroImage: Clinical, 4*, 182–189.

Crescenzi, R., DeBrosse, C., Nanga, R. P., Reddy, S., Haris, M., Hariharan, H., Iba, M., Lee, V. M., Detre, J. A., Borthakur, A., & Reddy, R. (2014). In vivo measurement of glutamate loss is associated with synapse loss in a mouse model of tauopathy. *Neuroimage, 101*, 185–192.

Crescenzi, R., DeBrosse, C., Nanga, R. P., Byrne, M. D., Krishnamoorthy, G., D'Aquilla, K., Nath, H., Morales, K. H., Iba, M., Hariharan, H., Lee, V. M., Detre, J. A., & Reddy, R. (2017). Longitudinal imaging reveals subhippocampal dynamics in glutamate levels associated with histopathologic events in a mouse model of tauopathy and healthy mice. *Hippocampus, 27*(3), 285–302.

Davis, K. A., Nanga, R. P., Das, S., Chen, S. H., Hadar, P. N., Pollard, J. R., Lucas, T. H., Shinohara, R. T., Litt, B., Hariharan, H., Elliott, M. A., Detre, J. A., & Reddy, R. (2015). Glutamate imaging (GluCEST) lateralizes epileptic foci in nonlesional temporal lobe epilepsy. *Science Translational Medicine, 7*(309), 309ra161.

Dou, W., Lin, C. E., Ding, H., Shen, Y., Dou, C., Qian, L., Wen, B., & Wu, B. (2019). Chemical exchange saturation transfer magnetic resonance imaging and its main and potential applications in pre-clinical and clinical studies. *Quantitative Imaging in Medicine and Surgery, 9*(10), 1747–1766.

Dupont, A. C., Largeau, B., Santiago Ribeiro, M. J., Guilloteau, D., Tronel, C., & Arlicot, N. (2017). Translocator protein-18 kDa (TSPO) positron emission tomography (PET) imaging and its clinical impact in neurodegenerative diseases. *International Journal of Molecular Sciences, 18*(4), 785.

Eckel-Passow, J. E., Lachance, D. H., Molinaro, A. M., Walsh, K. M., Decker, P. A., Sicotte, H., Pekmezci, M., Rice, T., Kosel, M. L., Smirnov, I. V., Sarkar, G., Caron, A. A., Kollmeyer, T. M., Praska, C. E., Chada, A. R., Halder, C., Hansen, H. M., McCoy, L. S., Bracci, P. M., Marshall, R., Zheng, S., Reis, G. F., Pico, A. R., O'Neill, B. P., Buckner, J. C., Giannini, C., Huse, J. T., Perry, A., Tihan, T., Berger, M. S., Chang, S. M., Prados, M. D., Wiemels, J., Wiencke, J. K., Wrensch, M. R., & Jenkins, R. B. (2015). Glioma groups based on 1p/19q, IDH, and TERT promoter mutations in tumors. *The New England Journal of Medicine, 372*(26), 2499–2508.

Erdogan, C., Hakyemez, B., Yildirim, N., & Parlak, M. (2005). Brain abscess and cystic brain tumor: Discrimination with dynamic susceptibility contrast perfusion-weighted MRI. *Journal of Computer Assisted Tomography, 29*(5), 663–667.

Farjam, R., Tsien, C. I., Lawrence, T. S., & Cao, Y. (2014). DCE-MRI defined subvolumes of a brain metastatic lesion by principle component analysis and fuzzy-c-means clustering for response assessment of radiation therapy. *Medical Physics, 41*(1), 011708.

Ferini-Strambi, L., Bozzali, M., Cercignani, M., Oldani, A., Zucconi, M., & Filippi, M. (2000). Magnetization transfer and diffusion-weighted imaging in nocturnal frontal lobe epilepsy. *Neurology, 54*(12), 2331–2333.

Finke, C., Schlichting, J., Papazoglou, S., Scheel, M., Freing, A., Soemmer, C., Pech, L., Pajkert, A., Pfuller, C., Wuerfel, J., Ploner, C., Paul, F., & Brandt, A. (2014). Altered basal ganglia functional connectivity in multiple sclerosis patients with fatigue. *Multiple Sclerosis, 21*(7), 925–934.

Gaitan, M. I., Shea, C. D., Evangelou, I. E., Stone, R. D., Fenton, K. M., Bielekova, B., Massacesi, L., & Reich, D. S. (2011). Evolution of the blood-brain barrier in newly forming multiple sclerosis lesions. *Annals of Neurology, 70*(1), 22–29.

Geuze, E., Vermetten, E., & Bremner, J. D. (2005). MR-based in vivo hippocampal volumetrics: 2. Findings in neuropsychiatric disorders. *Molecular Psychiatry, 10*(2), 160–184.

Goebell, E., Paustenbach, S., Vaeterlein, O., Ding, X. Q., Heese, O., Fiehler, J., Kucinski, T., Hagel, C., Westphal, M., & Zeumer, H. (2006). Low-grade and anaplastic gliomas: Differences in architecture evaluated with diffusion-tensor MR imaging. *Radiology, 239*(1), 217–222.

Goldenberg, J. M., & Pagel, M. D. (2019). Assessments of tumor metabolism with CEST MRI. *NMR in Biomedicine, 32*(10), e3943.

Greicius, M. D., Srivastava, G., Reiss, A. L., & Menon, V. (2004). Default-mode network activity distinguishes Alzheimer's disease from healthy aging: Evidence from functional MRI. *Proceedings of the National Academy of Sciences of the United States of America, 101*(13), 4637–4642.

Griffith, H. R., Perlman, S. B., Woodard, A. R., Rutecki, P. A., Jones, J. C., Ramirez, L. F., DeLaPena, R., Seidenberg, M., & Hermann, B. P. (2000). Preoperative FDG-PET temporal lobe hypometabolism and verbal memory after temporal lobectomy. *Neurology, 54*(5), 1161–1165.

Guerra-Carrillo, B., Mackey, A. P., & Bunge, S. A. (2014). Resting-state fMRI: A window into human brain plasticity. *The Neuroscientist, 20*(5), 522–533.

Gupta, A., Prager, A., Young, R. J., Shi, W., Omuro, A. M., & Graber, J. J. (2013). Diffusion-weighted MR imaging and MGMT methylation status in glioblastoma: A reappraisal of the role of preoperative quantitative ADC measurements. *AJNR. American Journal of Neuroradiology, 34*(1), E10–E11.

Hakyemez, B., Erdogan, C., Yildirim, N., & Parlak, M. (2005). Glioblastoma multiforme with atypical diffusion-weighted MR findings. *The British Journal of Radiology, 78*(935), 989–992.

Haris, M., Cai, K., Singh, A., Hariharan, H., & Reddy, R. (2011). In vivo mapping of brain myo-inositol. *Neuroimage, 54*(3), 2079–2085.

Haris, M., Nath, K., Cai, K., Singh, A., Crescenzi, R., Kogan, F., Verma, G., Reddy, S., Hariharan, H., Melhem, E. R., & Reddy, R. (2013a). Imaging of glutamate neurotransmitter alterations in Alzheimer's disease. *NMR in Biomedicine, 26*(4), 386–391.

Haris, M., Singh, A., Cai, K., Nath, K., Crescenzi, R., Kogan, F., Hariharan, H., & Reddy, R. (2013b). MICEST: A potential tool for non-invasive detection of molecular changes in Alzheimer's disease. *Journal of Neuroscience Methods, 212*(1), 87–93.

Harris, R. J., Cloughesy, T. F., Liau, L. M., Prins, R. M., Antonios, J. P., Li, D., Yong, W. H., Pope, W. B., Lai, A., Nghiemphu, P. L., & Ellingson, B. M. (2015). pH-weighted molecular imaging of gliomas using amine chemical exchange saturation transfer MRI. *Neuro-Oncology, 17*(11), 1514–1524.

Hoffmann, A., Bredno, J., Wendland, M. F., Derugin, N., Hom, J., Schuster, T., Su, H., Ohara, P. T., Young, W. L., & Wintermark, M. (2011). Validation of in vivo magnetic resonance imaging blood-brain barrier permeability measurements by comparison with gold standard histology. *Stroke, 42*(7), 2054–2060.

Holmes, T. M., Petrella, J. R., & Provenzale, J. M. (2004). Distinction between cerebral abscesses and high-grade neoplasms by dynamic susceptibility contrast perfusion MRI. *AJR. American Journal of Roentgenology, 183*(5), 1247–1252.

Hornak, J. P. (1996). *The basics of MRI.* https://www.cis.rit.edu/htbooks/mri/

Howe, F. A., & Opstad, K. S. (2003). 1H MR spectroscopy of brain tumours and masses. *NMR in Biomedicine, 16*(3), 123–131.

Jain, R., Narang, J., Schultz, L., Scarpace, L., Saksena, S., Brown, S., Rock, J. P., Rosenblum, M., Gutierrez, J., & Mikkelsen, T. (2011). Permeability estimates in histopathology-proved treatment-induced necrosis using perfusion CT: Can these add to other perfusion parameters in differentiating from recurrent/progressive tumors? *AJNR. American Journal of Neuroradiology, 32*(4), 658–663.

Jellison, B. J., Field, A. S., Medow, J., Lazar, M., Salamat, M. S., & Alexander, A. L. (2004). Diffusion tensor imaging of cerebral white matter: A pictorial review of physics, fiber tract anatomy, and tumor imaging patterns. *AJNR. American Journal of Neuroradiology, 25*(3), 356–369.

Jiang, Q., Ewing, J. R., Zhang, Z. G., Zhang, R. L., Hu, J., Divine, G. W., Arniego, P., Li, Q. J., & Chopp, M. (2001). Magnetization transfer MRI: Application to treatment of middle cerebral artery occlusion in rat. *Journal of Magnetic Resonance Imaging, 13*(2), 178–184.

Jiang, S., Yu, H., Wang, X., Lu, S., Li, Y., Feng, L., Zhang, Y., Heo, H. Y., Lee, D. H., Zhou, J., & Wen, Z. (2016). Molecular MRI differentiation between primary central nervous system lymphomas and high-grade gliomas using endogenous protein-based amide proton transfer MR imaging at 3 Tesla. *European Radiology, 26*(1), 64–71.

Jiang, S., Zou, T., Eberhart, C. G., Villalobos, M. A. V., Heo, H. Y., Zhang, Y., Wang, Y., Wang, X., Yu, H., Du, Y., van Zijl, P. C. M., Wen, Z., & Zhou, J. (2017). Predicting IDH mutation status in grade II gliomas using amide proton transfer-weighted (APTw) MRI. *Magnetic Resonance in Medicine, 78*(3), 1100–1109.

Jones, K. M., Pollard, A. C., & Pagel, M. D. (2018). Clinical applications of chemical exchange saturation transfer (CEST) MRI. *Journal of Magnetic Resonance Imaging, 47*(1), 11–27.

Joseph, C. R. (2021). Utilizing 3D arterial spin labeling to identify cerebrovascular leak and glymphatic obstruction in neurodegenerative disease. *Diagnostics (Basel), 11*(10), 1888.

Juh, R., Kim, J., Moon, D., Choe, B., & Suh, T. (2004). Different metabolic patterns analysis of Parkinsonism on the 18F-FDG PET. *European Journal of Radiology, 51*(3), 223–233.

Kaasinen, V., Maguire, R. P., Hundemer, H. P., & Leenders, K. L. (2006). Corticostriatal covariance patterns of 6-[(18)F]fluoro-L-dopa and [(18)F]fluorodeoxyglucose PET in Parkinson's disease. *Journal of Neurology, 253*(3), 340–348.

Kamimura, K., Nakajo, M., Yoneyama, T., Takumi, K., Kumagae, Y., Fukukura, Y., & Yoshiura, T. (2019). Amide proton transfer imaging of tumors: Theory, clinical applications, pitfalls, and future directions. *Japanese Journal of Radiology, 37*(2), 109–116.

Kanda, T., Wakabayashi, Y., Zeng, F., Ueno, Y., Sofue, K., Maeda, T., Nogami, M., & Murakami, T. (2018). Imaging findings in radiation therapy complications of the central nervous system. *Japanese Journal of Radiology, 36*(9), 519–527.

Kerrouche, N., Herholz, K., Mielke, R., Holthoff, V., & Baron, J. C. (2006). (18)FDG PET in vascular dementia: Differentiation from Alzheimer's disease using voxel-based multivariate analysis. *Journal of Cerebral Blood Flow and Metabolism, 26*(9), 1213–1221.

Kim, Y. K., Lee, D. S., Lee, S. K., Kim, S. K., Chung, C. K., Chang, K. H., Choi, K. Y., Chung, J. K., & Lee, M. C. (2003). Differential features of metabolic abnormalities between medial and lateral temporal lobe epilepsy: Quantitative analysis of (18)F-FDG PET using SPM. *Journal of Nuclear Medicine, 44*(7), 1006–1012.

Kim, M., Eleftheriou, A., Ravotto, L., Weber, B., Rivlin, M., Navon, G., Capozza, M., Anemone, A., Longo, D. L., Aime, S., Zaiss, M., Herz, K., Deshmane, A., Lindig, T., Bender, B., & Golay, X. (2022). What do we know about dynamic glucose-enhanced (DGE) MRI and how close is it to the clinics? Horizon 2020 GLINT consortium report. *Magma, 35*(1), 87–104.

Knight, R. A., Nagesh, V., Nagaraja, T. N., Ewing, J. R., Whitton, P. A., Bershad, E., Fagan, S. C., & Fenstermacher, J. D. (2005). Acute blood-brain barrier opening in experimentally induced focal cerebral ischemia is preferentially identified by quantitative magnetization transfer imaging. *Magnetic Resonance in Medicine, 54*(4), 822–832.

Knowlton, R. C., Lawn, N. D., Mountz, J. M., & Kuzniecky, R. I. (2004). Ictal SPECT analysis in epilepsy: Subtraction and statistical parametric mapping techniques. *Neurology, 63*(1), 10–15.

Kono, K., Inoue, Y., Nakayama, K., Shakudo, M., Morino, M., Ohata, K., Wakasa, K., & Yamada, R. (2001). The role of diffusion-weighted imaging in patients with brain tumors. *AJNR. American Journal of Neuroradiology, 22*(6), 1081–1088.

Kumar, A., Kumar, P. G., Sampath, S., & Suhag, V. (2017). To study the efficacy of thallium-201 as tumor seeking agent and to study its role in therapeutic response. *Medical Journal, Armed Forces India, 73*(3), 242–249.

Kwong, K. K., Belliveau, J. W., Chesler, D. A., Goldberg, I. E., Weisskoff, R. M., Poncelet, B. P., Kennedy, D. N., Hoppel, B. E., Cohen, M. S., Turner, R., et al. (1992). Dynamic magnetic resonance imaging of human brain activity during primary sensory stimulation. *Proceedings of the National Academy of Sciences of the United States of America, 89*(12), 5675–5679.

Law, M., Cha, S., Knopp, E. A., Johnson, G., Arnett, J., & Litt, A. W. (2002). High-grade gliomas and solitary metastases: Differentiation by using perfusion and proton spectroscopic MR imaging. *Radiology, 222*(3), 715–721.

Law, M., Yang, S., Wang, H., Babb, J. S., Johnson, G., Cha, S., Knopp, E. A., & Zagzag, D. (2003). Glioma grading: Sensitivity, specificity, and predictive values of perfusion MR imaging and proton MR spectroscopic imaging compared with conventional MR imaging. *AJNR. American Journal of Neuroradiology, 24*(10), 1989–1998.

Law, M., Yang, S., Babb, J. S., Knopp, E. A., Golfinos, J. G., Zagzag, D., & Johnson, G. (2004). Comparison of cerebral blood volume and vascular permeability from dynamic susceptibility contrast-enhanced perfusion MR imaging with glioma grade. *AJNR. American Journal of Neuroradiology, 25*(5), 746–755.

Leavitt, V. M., Cirnigliaro, C., Cohen, A., Farag, A., Brooks, M., Wecht, J. M., Wylie, G. R., Chiaravalloti, N. D., DeLuca, J., & Sumowski, J. F. (2014). Aerobic exercise increases hippocampal volume and improves memory in multiple sclerosis: Preliminary findings. *Neurocase, 20*(6), 695–697.

Lebedev, A. V., Westman, E., Simmons, A., Lebedeva, A., Siepel, F. J., Pereira, J. B., & Aarsland, D. (2014). Large-scale resting state network correlates of cognitive impairment in Parkinson's disease and related dopaminergic deficits. *Frontiers in Systems Neuroscience, 8*, 45.

Li, C., Chen, M., Zhao, X., Wang, R., Chen, H., Su, W., Li, S., Lou, B., Song, G., Zhang, S., Zhang, J., & Zhou, J. (2017). Chemical exchange saturation transfer MRI signal loss of the substantia nigra as an imaging biomarker to evaluate the diagnosis and severity of Parkinson's disease. *Frontiers in Neuroscience, 11*, 489.

Licho, R., Litofsky, N. S., Senitko, M., & George, M. (2002). Inaccuracy of Tl-201 brain SPECT in distinguishing cerebral infections from lymphoma in patients with AIDS. *Clinical Nuclear Medicine, 27*(2), 81–86.

Machhi, J., Yeapuri, P., Lu, Y., Foster, E., Chikhale, R., Herskovitz, J., Namminga, K. L., Olson, K. E., Abdelmoaty, M. M., Gao, J., Quadros, R. M., Kiyota, T., Jingjing, L., Kevadiya, B. D., Wang, X., Liu, Y., Poluektova, L. Y., Gurumurthy, C. B., Mosley, R. L., & Gendelman, H. E. (2021). CD4+ effector T cells accelerate Alzheimer's disease in mice. *Journal of Neuroinflammation, 18*(1), 272.

McNally, K. A., Paige, A. L., Varghese, G., Zhang, H., Novotny, E. J., Jr., Spencer, S. S., Zubal, I. G., & Blumenfeld, H. (2005). Localizing value of ictal-interictal SPECT analyzed by SPM (ISAS). *Epilepsia, 46*(9), 1450–1464.

Mehrabian, H., Desmond, K. L., Soliman, H., Sahgal, A., & Stanisz, G. J. (2017). Differentiation between radiation necrosis and tumor progression using chemical exchange saturation transfer. *Clinical Cancer Research, 23*(14), 3667–3675.

Melzer, T. R., Watts, R., MacAskill, M. R., Pearson, J. F., Rueger, S., Pitcher, T. L., Livingston, L., Graham, C., Keenan, R., Shankaranarayanan, A., Alsop, D. C., Dalrymple-Alford, J. C., & Anderson, T. J. (2011). Arterial spin labelling reveals an abnormal cerebral perfusion pattern in Parkinson's disease. *Brain, 134*(Pt 3), 845–855.

Minagar, A., Gonzalez-Toledo, E., Pinkston, J., & Jaffe, S. L. (2005). Neuroimaging in multiple sclerosis. *International Review of Neurobiology, 67*, 165–201.

Mitsis, E. M., Riggio, S., Kostakoglu, L., Dickstein, D. L., Machac, J., Delman, B., Goldstein, M., Jennings, D., D'Antonio, E., Martin, J., Naidich, T. P., Aloysi, A., Fernandez, C., Seibyl, J., DeKosky, S. T., Elder, G. A., Marek, K., Gordon, W., Hof, P. R., Sano, M., & Gandy, S. (2014). Tauopathy PET and amyloid PET in the diagnosis of chronic traumatic encephalopathies: Studies of a retired NFL player and of a man with FTD and a severe head injury. *Translational Psychiatry, 4*, e441.

Mohana-Borges, A. V., Imbesi, S. G., Dietrich, R., Alksne, J., & Amjadi, D. K. (2004). Role of proton magnetic resonance spectroscopy in the diagnosis of gliomatosis cerebri: A unique pattern of normal choline but elevated Myo-inositol metabolite levels. *Journal of Computer Assisted Tomography, 28*(1), 103–105.

Moldrich, R. X., Pannek, K., Hoch, R., Rubenstein, J. L., Kurniawan, N. D., & Richards, L. J. (2010). Comparative mouse brain tractography of diffusion magnetic resonance imaging. *Neuroimage, 51*(3), 1027–1036.

Moller-Hartmann, W., Herminghaus, S., Krings, T., Marquardt, G., Lanfermann, H., Pilatus, U., & Zanella, F. E. (2002). Clinical application of proton magnetic resonance spectroscopy in the diagnosis of intracranial mass lesions. *Neuroradiology, 44*(5), 371–381.

Narayanaswami, V., Dahl, K., Bernard-Gauthier, V., Josephson, L., Cumming, P., & Vasdev, N. (2018). Emerging PET radiotracers and targets for imaging of neuroinflammation in neurodegenerative diseases: Outlook beyond TSPO. *Molecular Imaging, 17*, 1536012118792317.

Nelson, S. J. (2003). Multivoxel magnetic resonance spectroscopy of brain tumors. *Molecular Cancer Therapeutics, 2*(5), 497–507.

Nelson, S. J., McKnight, T. R., & Henry, R. G. (2002). Characterization of untreated gliomas by magnetic resonance spectroscopic imaging. *Neuroimaging Clinics of North America, 12*(4), 599–613.

Nguyen, T. B., Cron, G. O., Mercier, J. F., Foottit, C., Torres, C. H., Chakraborty, S., Woulfe, J., Jansen, G. H., Caudrelier, J. M., Sinclair, J., Hogan, M. J., Thornhill, R. E., & Cameron, I. G. (2014). Preoperative prognostic value of dynamic contrast-enhanced MRI-derived contrast transfer coefficient and plasma volume in patients with cerebral gliomas. *AJNR. American Journal of Neuroradiology, 36*(1), 63–69.

Novak, B., Milcinski, M., Grmek, M., & Kocmur, M. (2005). Early effects of treatment on regional cerebral blood flow in first episode schizophrenia patients evaluated with 99Tc-ECD-SPECT. *Neuro Endocrinology Letters, 26*(6), 685–689.

Okamura, N., Furumoto, S., Harada, R., Tago, T., Yoshikawa, T., Fodero-Tavoletti, M., Mulligan, R. S., Villemagne, V. L., Akatsu, H., Yamamoto, T., Arai, H., Iwata, R., Yanai, K., & Kudo, Y. (2013). Novel 18F-labeled arylquinoline derivatives for noninvasive imaging of tau pathology in Alzheimer disease. *Journal of Nuclear Medicine, 54*(8), 1420–1427.

Olde Dubbelink, K. T., Schoonheim, M. M., Deijen, J. B., Twisk, J. W., Barkhof, F., & Berendse, H. W. (2014). Functional connectivity and cognitive decline over 3 years in Parkinson disease. *Neurology, 83*(22), 2046–2053.

Ota, M., Sato, N., Nakata, Y., Ito, K., Kamiya, K., Maikusa, N., Ogawa, M., Okamoto, T., Obu, S., Noda, T., Araki, M., Yamamura, T., & Kunugi, H. (2013). Abnormalities of cerebral blood flow in multiple sclerosis: A pseudocontinuous arterial spin labeling MRI study. *Magnetic Resonance Imaging, 31*(6), 990–995.

Ozsunar, Y., Mullins, M. E., Kwong, K., Hochberg, F. H., Ament, C., Schaefer, P. W., Gonzalez, R. G., & Lev, M. H. (2010). Glioma recurrence versus radiation necrosis? A pilot comparison of arterial spin-labeled, dynamic susceptibility contrast enhanced MRI, and FDG-PET imaging. *Academic Radiology, 17*(3), 282–290.

Pankowska, A., Kochalska, K., Łazorczyk, A., Dyndor, K., Kozioł, P., Zieńczuk, B., Toborek, M., & Pietura, R. (2019). Chemical exchange saturation transfer (CEST) as a new method of signal obtainment in magnetic resonance molecular imaging in clinical and research practice. *Polish Journal of Radiology, 84*, e147–e152.

Park, J. E., Lee, J. Y., Kim, H. S., Oh, J. Y., Jung, S. C., Kim, S. J., Keupp, J., Oh, M., & Kim, J. S. (2018). Amide proton transfer imaging seems to provide higher diagnostic performance in post-treatment high-grade gliomas than methionine positron emission tomography. *European Radiology, 28*(8), 3285–3295.

Petcharunpaisan, S., Ramalho, J., & Castillo, M. (2010). Arterial spin labeling in neuroimaging. *World Journal of Radiology, 2*(10), 384–398.

Pyhtinen, J. (2000). Proton MR spectroscopy in gliomatosis cerebri. *Neuroradiology, 42*(8), 612–615.

Rijpkema, M., Schuuring, J., van der Meulen, Y., van der Graaf, M., Bernsen, H., Boerman, R., van der Kogel, A., & Heerschap, A. (2003). Characterization of oligodendrogliomas using short echo time 1H MR spectroscopic imaging. *NMR in Biomedicine, 16*(1), 12–18.

Risacher, S. L., & Saykin, A. J. (2019). Neuroimaging in aging and neurologic diseases. *Handbook of Clinical Neurology, 167*, 191–227.

Rivlin, M., & Navon, G. (2019). Molecular imaging of tumors by chemical exchange saturation transfer MRI of glucose analogs. *Quantitative Imaging in Medicine and Surgery, 9*(10), 1731–1746.

Rowley, H. A. (2005). Extending the time window for thrombolysis: Evidence from acute stroke trials. *Neuroimaging Clinics of North America, 15*(3), 575–587.

Rugg-Gunn, F. J., Eriksson, S. H., Boulby, P. A., Symms, M. R., Barker, G. J., & Duncan, J. S. (2003). Magnetization transfer imaging in focal epilepsy. *Neurology, 60*(10), 1638–1645.

Sadasivan, S., & Friedman, J. H. (2015). Experience with DaTscan at a tertiary referral center. *Parkinsonism & Related Disorders, 21*(1), 42–45.

Sakata, A., Fushimi, Y., Okada, T., Arakawa, Y., Kunieda, T., Minamiguchi, S., Kido, A., Sakashita, N., Miyamoto, S., & Togashi, K. (2017). Diagnostic performance between contrast enhancement, proton MR spectroscopy, and amide proton transfer imaging in patients with brain tumors. *Journal of Magnetic Resonance Imaging, 46*(3), 732–739.

Salat, D. H., Tuch, D. S., Hevelone, N. D., Fischl, B., Corkin, S., Rosas, H. D., & Dale, A. M. (2005). Age-related changes in prefrontal white matter measured by diffusion tensor imaging. *Annals of the New York Academy of Sciences, 1064*, 37–49.

Sherry, A. D., & Wu, Y. (2013). The importance of water exchange rates in the design of responsive agents for MRI. *Current Opinion in Chemical Biology, 17*(2), 167–174.

Shokouhi, S., Claassen, D., & Riddle, W. (2014). Imaging brain metabolism and pathology in Alzheimer's disease with positron emission tomography. *Journal of Alzheimer's Disease & Parkinsonism, 4*(2), 143.

Suh, C. H., Kim, H. S., Choi, Y. J., Kim, N., & Kim, S. J. (2013). Prediction of pseudoprogression in patients with glioblastomas using the initial and final area under the curves ratio derived from dynamic contrast-enhanced T1-weighted perfusion MR imaging. *AJNR. American Journal of Neuroradiology, 34*(12), 2278–2286.

Swartz, B. E., Brown, C., Mandelkern, M. A., Khonsari, A., Patell, A., Thomas, K., Torgersen, D., Delgado-Escueta, A. V., & Walsh, G. O. (2002). The use of 2-deoxy-2-[18F]fluoro-D-glucose (FDG-PET) positron emission tomography in the routine diagnosis of epilepsy. *Molecular Imaging and Biology, 4*(3), 245–252.

Tae, W. S., Joo, E. Y., Kim, J. H., Han, S. J., Suh, Y. L., Kim, B. T., Hong, S. C., & Hong, S. B. (2005). Cerebral perfusion changes in mesial temporal lobe epilepsy: SPM analysis of ictal and interictal SPECT. *Neuroimage, 24*(1), 101–110.

Togao, O., Yoshiura, T., Keupp, J., Hiwatashi, A., Yamashita, K., Kikuchi, K., Suzuki, Y., Suzuki, S. O., Iwaki, T., Hata, N., Mizoguchi, M., Yoshimoto, K., Sagiyama, K., Takahashi, M., & Honda, H. (2014). Amide proton transfer imaging of adult diffuse gliomas: Correlation with histopathological grades. *Neuro-Oncology, 16*(3), 441–448.

Togao, O., Hiwatashi, A., Yamashita, K., Kikuchi, K., Keupp, J., Yoshimoto, K., Kuga, D., Yoneyama, M., Suzuki, S. O., Iwaki, T., Takahashi, M., Iihara, K., & Honda, H. (2017). Grading diffuse gliomas without intense contrast enhancement by amide proton transfer MR imaging: Comparisons with diffusion- and perfusion-weighted imaging. *European Radiology, 27*(2), 578–588.

Toh, C. H., Castillo, M., Wong, A. M., Wei, K. C., Wong, H. F., Ng, S. H., & Wan, Y. L. (2008). Primary cerebral lymphoma and glioblastoma multiforme: Differences in diffusion characteristics evaluated with diffusion tensor imaging. *AJNR. American Journal of Neuroradiology, 29*(3), 471–475.

Topcuoglu, V., Comert, B., Karabekiroglu, A., Dede, F., Erdil, T. Y., & Turoglu, H. T. (2005). Right basal ganglion hypoperfusion in obsessive compulsive disorder patients demonstrated by Tc-99m-HMPAO brain perfusion spect: A controlled study. *The International Journal of Neuroscience, 115*(12), 1643–1655.

Tournier, J. D. (2019). Diffusion MRI in the brain – Theory and concepts. *Progress in Nuclear Magnetic Resonance Spectroscopy, 112–113*, 1–16.

Truijman, M. T., Kwee, R. M., van Hoof, R. H., Hermeling, E., van Oostenbrugge, R. J., Mess, W. H., Backes, W. H., Daemen, M. J., Bucerius, J., Wildberger, J. E., & Kooi, M. E. (2013). Combined 18F-FDG PET-CT and DCE-MRI to assess inflammation and microvascularization in atherosclerotic plaques. *Stroke, 44*(12), 3568–3570.

Tsuchiya, K., Fujikawa, A., Nakajima, M., & Honya, K. (2005). Differentiation between solitary brain metastasis and high-grade glioma by diffusion tensor imaging. *The British Journal of Radiology, 78*(930), 533–537.

Van Bruggen, N., & Roberts, T. P. L. (2002). *Biomedical imaging in experimental neuroscience*. CRC Press.

van Zijl, P. C., & Yadav, N. N. (2011). Chemical exchange saturation transfer (CEST): What is in a name and what isn't? *Magnetic Resonance in Medicine, 65*(4), 927–948.

Verger, A., Kas, A., Darcourt, J., & Guedj, E. (2022). PET imaging in neuro-oncology: An update and overview of a rapidly growing area. *Cancers (Basel), 14*(5), 1103.

Vinogradov, E., Sherry, A. D., & Lenkinski, R. E. (2013). CEST: From basic principles to applications, challenges and opportunities. *Journal of Magnetic Resonance, 229*, 155–172.

Walker-Samuel, S., Ramasawmy, R., Torrealdea, F., Rega, M., Rajkumar, V., Johnson, S. P., Richardson, S., Gonçalves, M., Parkes, H. G., Årstad, E., Thomas, D. L., Pedley, R. B., Lythgoe, M. F., & Golay, X. (2013). In vivo imaging of glucose uptake and metabolism in tumors. *Nature Medicine, 19*(8), 1067–1072.

Wang, S., Kim, S., Chawla, S., Wolf, R. L., Zhang, W. G., O'Rourke, D. M., Judy, K. D., Melhem, E. R., & Poptani, H. (2009a). Differentiation between glioblastomas and solitary brain metastases using diffusion tensor imaging. *Neuroimage, 44*(3), 653–660.

Wang, W., Steward, C. E., & Desmond, P. M. (2009b). Diffusion tensor imaging in glioblastoma multiforme and brain metastases: The role of p, q, L, and fractional anisotropy. *AJNR. American Journal of Neuroradiology, 30*(1), 203–208.

Wang, M., Hong, X., Chang, C. F., Li, Q., Ma, B., Zhang, H., Xiang, S., Heo, H. Y., Zhang, Y., Lee, D. H., Jiang, S., Leigh, R., Koehler, R. C., van Zijl, P. C. M., Wang, J., & Zhou, J. (2015a). Simultaneous detection and separation of hyperacute intracerebral hemorrhage and cerebral ischemia using amide proton transfer MRI. *Magnetic Resonance in Medicine, 74*(1), 42–50.

Wang, R., Li, S. Y., Chen, M., Zhou, J. Y., Peng, D. T., Zhang, C., & Dai, Y. M. (2015b). Amide proton transfer magnetic resonance imaging of Alzheimer's disease at 3.0 Tesla: A preliminary study. *Chinese Medical Journal, 128*(5), 615–619.

Wang, R., Chen, P., Shen, Z., Lin, G., Xiao, G., Dai, Z., Zhang, B., Chen, Y., Lai, L., Zong, X., Li, Y., Tang, Y., & Wu, R. (2019). Brain amide proton transfer imaging of rat with Alzheimer's disease using saturation with frequency alternating RF irradiation method. *Frontiers in Aging Neuroscience, 11*, 217.

Warren, S. L., & Moustafa, A. A. (2022). Functional magnetic resonance imaging, deep learning, and Alzheimer's disease: A systematic review. *Journal of Neuroimaging, 33*(1), 5–18.

Webb, A. (2022). *Introduction to biomedical imaging*. Wiley-IEEE Press.

Werry, E. L., Bright, F. M., Piguet, O., Ittner, L. M., Halliday, G. M., Hodges, J. R., Kiernan, M. C., Loy, C. T., Kril, J. J., & Kassiou, M. (2019). Recent developments in TSPO PET imaging as a biomarker of neuroinflammation in neurodegenerative disorders. *International Journal of Molecular Sciences, 20*(13), 3161.

Wolf, M. E., Layer, V., Gregori, J., Griebe, M., Szabo, K., Gass, A., Hennerici, M. G., Matthias, G., & Rolf, K. (2014). Assessment of perfusion deficits in ischemic stroke using 3D-GRASE arterial spin labeling magnetic resonance imaging with multiple inflow times. *Journal of Neuroimaging, 24*(5), 453–459.

Wu, B., Warnock, G., Zaiss, M., Lin, C., Chen, M., Zhou, Z., Mu, L., Nanz, D., Tuura, R., & Delso, G. (2016). An overview of CEST MRI for non-MR physicists. *EJNMMI Physics, 3*(1), 19.

Wu, Y., Zhou, I. Y., Lu, D., Manderville, E., Lo, E. H., Zheng, H., & Sun, P. Z. (2018). pH-sensitive amide proton transfer effect dominates the magnetization transfer asymmetry contrast during acute ischemia-quantification of multipool contribution to in vivo CEST MRI. *Magnetic Resonance in Medicine, 79*(3), 1602–1608.

Xu, X., Sehgal, A. A., Yadav, N. N., Laterra, J., Blair, L., Blakeley, J., Seidemo, A., Coughlin, J. M., Pomper, M. G., Knutsson, L., & van Zijl, P. C. M. (2020). GlucoCEST-based dynamic glucose enhanced (DGE) MRI at 3 T: Early experience in healthy volunteers and brain tumor patients. *Magnetic Resonance in Medicine, 84*(1), 247–262.

Yarnykh, V. L., & Yuan, C. (2004). Cross-relaxation imaging reveals detailed anatomy of white matter fiber tracts in the human brain. *Neuroimage, 23*(1), 409–424.

Yeh, F. C., Irimia, A., Bastos, D. C. A., & Golby, A. J. (2021). Tractography methods and findings in brain tumors and traumatic brain injury. *Neuroimage, 245*, 118651.

Young, R. J., Ghesani, M. V., Kagetsu, N. J., & Derogatis, A. J. (2005). Lesion size determines accuracy of thallium-201 brain single-photon emission tomography in differentiating between intracranial malignancy and infection in AIDS patients. *AJNR. American Journal of Neuroradiology, 26*(8), 1973–1979.

Yousaf, T., Dervenoulas, G., Valkimadi, P. E., & Politis, M. (2019). Neuroimaging in Lewy body dementia. *Journal of Neurology, 266*(1), 1–26.

Zaiss, M., Anemone, A., Goerke, S., Longo, D. L., Herz, K., Pohmann, R., Aime, S., Rivlin, M., Navon, G., Golay, X., & Scheffler, K. (2019). Quantification of hydroxyl exchange of D-Glucose at physiological conditions for optimization of glucoCEST MRI at 3, 7 and 9.4 Tesla. *NMR in Biomedicine, 32*(9), e4113.

Zhang-Yin, J. T., Girard, A., & Bertaux, M. (2022). What does PET imaging bring to neuro-oncology in 2022? A review. *Cancers (Basel), 14*(4), 879.

Zhao, Y. B., Sun, B. M., Li, D. Y., & Wang, Q. S. (2004). Effects of bilateral subthalamic nucleus stimulation on resting-state cerebral glucose metabolism in advanced Parkinson's disease. *Chinese Medical Journal, 117*(9), 1304–1308.

Zhou, J., Blakeley, J. O., Hua, J., Kim, M., Laterra, J., Pomper, M. G., & van Zijl, P. C. (2008). Practical data acquisition method for human brain tumor amide proton transfer (APT) imaging. *Magnetic Resonance in Medicine, 60*(4), 842–849.

Zhou, J., Heo, H. Y., Knutsson, L., van Zijl, P. C. M., & Jiang, S. (2019). APT-weighted MRI: Techniques, current neuro applications, and challenging issues. *Journal of Magnetic Resonance Imaging, 50*(2), 347–364.

Zulfiqar, M., Yousem, D. M., & Lai, H. (2013). ADC values and prognosis of malignant astrocytomas: Does lower ADC predict a worse prognosis independent of grade of tumor?--A meta-analysis. *AJR. American Journal of Roentgenology, 200*(3), 624–629.

Extracellular Vesicles: Therapeutic Delivery

47

Rong Ma, Liang Chen, and Guoku Hu

Abstract

Extracellular vesicles (EVs) are emerging and promising therapeutic delivery vehicles due to their ability to transfer bioactive molecules to recipient cells. EVs are membrane-bound particles secreted by cells and play essential roles in intercellular communication. This chapter overviews EV's origin and their potential as therapeutic delivery vehicles. The isolation and characterization of EVs, including methods of purification and quantification, are also discussed, along with approaches for loading therapeutic cargo into EVs. The chapter also examines EV modification strategies that can enhance the targeting and efficacy of EV-based therapeutics, such as surface and genetic engineering. Finally, the challenges and opportunities in developing EV-based therapeutics are discussed. Overall, this chapter highlights the potential of EVs as a versatile and promising therapeutic delivery system with broad applications in treating various human diseases.

R. Ma
The Department of Pharmacology and Experimental Neuroscience, University of Nebraska Medical Center, Omaha, NE, USA

The Department of Pharmacology, School of Basic Medicine, Tongji Medical College, Huazhong University of Science and Technology, Wuhan, Hubei, China

L. Chen
The Department of Computer Science, College of Engineering, Shantou University, Shantou, Guangdong, China

Key Laboratory of Intelligent Manufacturing Technology, Ministry of Education, Shantou University, Shantou, Guangdong, China

G. Hu (✉)
The Department of Pharmacology and Experimental Neuroscience, University of Nebraska Medical Center, Omaha, NE, USA
e-mail: guoku.hu@unmc.edu

Keywords

Extracellular vesicles · Exosomes · Drug delivery · Targeted therapy

47.1 Introduction

Extracellular vesicles (EVs) are small membrane-bound structures secreted by cells into the extracellular space. EVs were first discovered in the 1940s, but their significance in intercellular communication was not recognized until much later (Fig. 47.1) (Weibull et al., 1965; Smalley & Birss, 1987; Pan et al., 1985; Pan & Johnstone, 1983; Johnstone et al., 1987; Chargaff & West, 1946; Aaronson et al., 1971). EVs are now known to play crucial roles in various physiological and pathological processes, including immune modulation, tissue repair, and disease progression (Fujita et al., 2015; Davies, 2023; Xie et al., 2019; Chivero et al., 2021; Hu et al., 2012a; Ma et al., 2021). There are various types of EVs, including exosomes, microvesicles, and apoptotic bodies (Willms et al., 2018; Doyle & Wang, 2019; Ma et al., 2023a). Exosomes are small (30–150 nm in diameter) vesicles released by the fusion of multivesicular bodies (MVBs) with the plasma membrane (Pegtel & Gould, 2019). Exosomes are derived from the endocytic pathway and contain various proteins, lipids, and nucleic acids that reflect the origin and function of their parent cells (Kalluri & LeBleu, 2020). Microvesicles are larger (100–1000 nm in diameter) vesicles that bud off directly from the plasma membrane (Stahl & Raposo, 2019). Microvesicles contain a similar cargo to exosomes but may also include cytosolic proteins and membrane receptors (Stahl & Raposo, 2019; Lawson et al., 2016). Apoptotic bodies are larger (1000–5000 nm in diameter) vesicles released by dying cells during.

H. E. Gendelman, T. Ikezu (eds.), *Neuroimmune Pharmacology and Therapeutics*,
https://doi.org/10.1007/978-3-031-68237-7_47

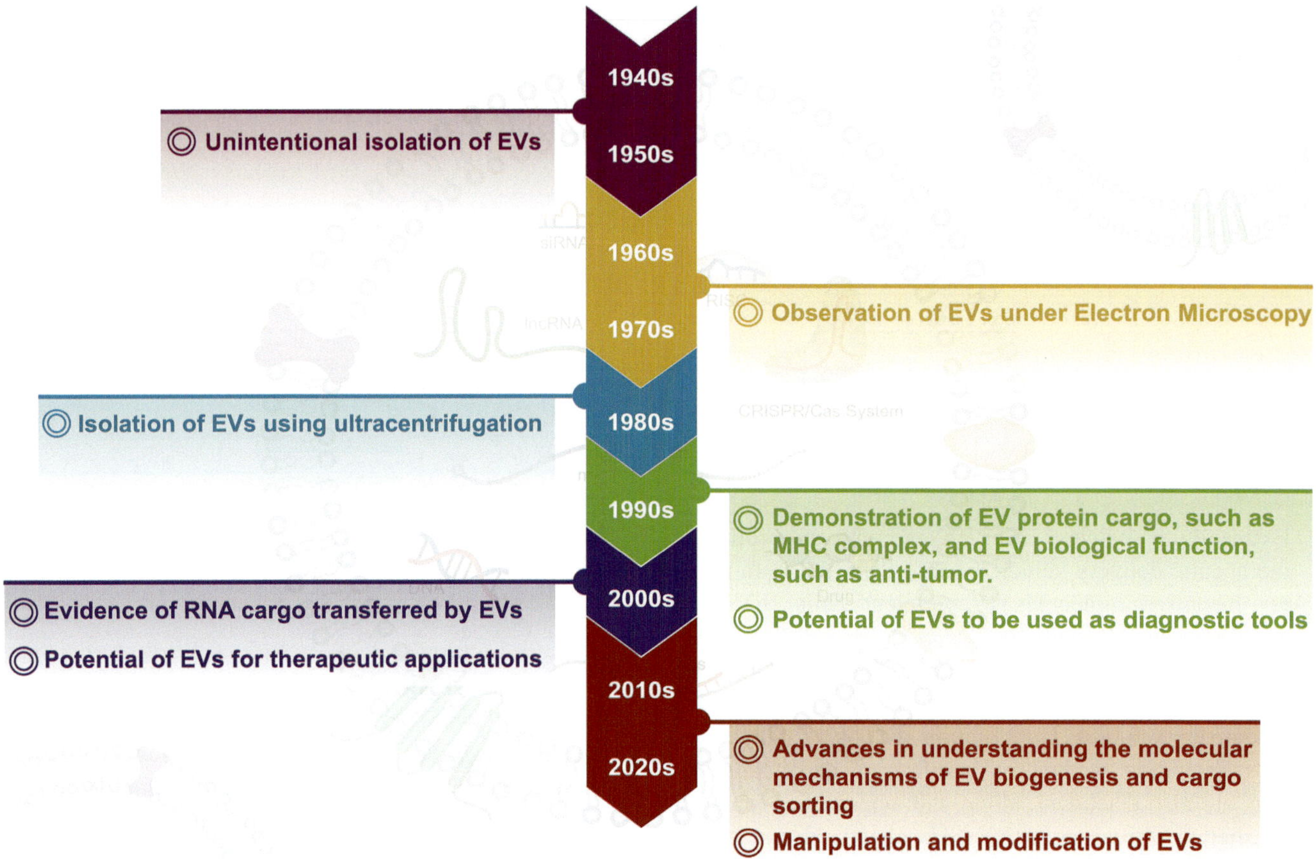

Fig. 47.1 Timeline of extracellular vesicle (EV) research. In the 1940s, the unintentional isolation of EVs was observed during the study of blood coagulation. Researchers found that platelets released membrane-bound vesicles, which were later identified as EVs. In the 1960s, the observation of EVs under electron microscopy allowed for further characterization of their size and structure. In the 1980s, the isolation of EVs using ultracentrifugation was established as a standard method for EV separation. In the 1990s, demonstration of EV protein cargo, such as MHC complex, helped to elucidate the role of EVs in immune system regulation, evidence of EV biological function, such as anti-tumor properties, was discovered, and the potential of EVs to be used as diagnostic tools were explored. In the 2000s, the discovery of RNA cargo transferred by EVs opened up new avenues of research into the role of EVs in intercellular communication, and the potential of EVs for therapeutic applications was explored. In the 2010s and 2020s, advances in understanding the molecular mechanisms of EV biogenesis and cargo sorting, as well as manipulation and modification of EVs for targeted delivery, are ongoing areas of research

Apoptosis (Battistelli & Falcieri, 2020) Apoptotic bodies contain cellular debris and fragmented DNA and are taken up by phagocytic cells for clearance (Kakarla et al., 2020; Caruso & Poon, 2018; Gregory & Dransfield, 2018; Pavlyukov et al., 2018).

The utilization of EVs as drug delivery vehicles has gained increasing attention in recent years, attributed to their capability of transporting cargo molecules to specific cells and tissues. Several unique properties of EVs make them attractive for therapeutic and diagnostic applications (Roy et al., 2018; Stremersch et al., 2016). First, EVs can be isolated from various biological fluids, such as blood, urine, and saliva, making them easily accessible and noninvasive biomarkers (Hu & Brew, 2023). Second, EVs are highly stable in circulation due to their lipid bilayer structure, which protects their cargo from degradation by nucleases and proteases (O'Brien et al., 2020). Third, EVs can cross biological barriers, such as the blood–brain barrier, allowing for targeted delivery of therapeutic cargo to specific tissues. This makes EVs particularly attractive for treating conditions impacting the central nervous system, such as Alzheimer's disease and Parkinson's disease (Elsharkasy et al., 2020; Loch-Neckel et al., 2022). Fourth, EVs can modify the function of recipient cells by transferring their cargo, making them potential agents for cell-free therapies (Vader et al., 2014; Murphy et al., 2019). Finally, EVs have low immunogenicity and toxicity, making them a safer alternative for conventional drug delivery tools (Jurgielewicz et al., 2021; Zhu et al., 2017). Therefore, using EVs as drug delivery tools holds great promise for treating various diseases. While challenges remain to be addressed, such as optimizing the loading and targeting of cargo molecules, the potential benefits of using EVs as drug delivery vehicles make them an exciting area of research for the future.

47.2 Source Cells for the Production of Drug Delivery EVs

The choice of origin cells for EV production can significantly impact the characteristics and functionality of the resulting EVs; therefore, selecting the appropriate origin cells for EV production is critical for EV-based drug delivery strategies (Sil et al., 2020; Herrmann et al., 2021). Different types of cells can produce EVs with distinct cargo profiles, membrane compositions, and physical properties (Zaborowski et al., 2015). For instance, EVs isolated from mesenchymal stem cells (MSCs) contain various growth factors and anti-inflammatory molecules, making them attractive candidates for therapeutic applications in regenerative medicine and immune disorders (Gowen et al., 2020; Martinez-Arroyo et al., 2022). On the other hand, EVs derived from immune cells, such as macrophages and dendritic cells, contain immunostimulatory molecules and antigens, making them useful for cancer immunotherapy (P. Yang et al., 2021; Marar et al., 2021).

When choosing the producing cells for EV isolation and manufacturing, an essential factor to consider is the ability to scale up and reproduce the production process (Corso et al., 2017). Generating large quantities of EVs consistent in composition and functional properties is essential for clinical translation (Silva et al., 2021). Therefore, cells that can be cultured in large quantities under good manufacturing practices (GMP) conditions are preferred (Pachler et al., 2017). MSCs, for example, are relatively easy to isolate and expand in culture, making them attractive candidates for large-scale EV production (Pachler et al., 2017).

Another factor to consider when choosing the producing cells for EV isolation and manufacturing is the safety profile of the resulting EVs. As EVs can also be derived from disease cells, including tumor cells and virus-infected cells, there are concerns regarding potential adverse effects, such as mediating the transfer of oncogenic or viral cargo to recipient cells, when using these EVs in clinical applications. Therefore, cells that are safe and non-tumorigenic, such as MSCs and adipose-derived stem cells (ADSCs), are preferred for EV production (Shukla et al., 2020; Nagelkerke et al., 2021).

The selection of EV-producing cells also depends on the therapeutic application (Fig. 47.2). EVs derived from neural stem cells (NSCs) have been proven to promote neural repair and regeneration in preclinical models simulating stroke and traumatic brain injury (Vogel et al., 2018; M.K. Sun et al., 2020; Webb et al., 2018; Rong et al., 2019). Similarly, EVs derived from cardiomyocytes have been demonstrated to enhance cardiac function in models of heart failure (Fu et al., 2020; B. Liu et al., 2018). Therefore, using NSC-derived EVs or cardiomyocyte-derived EVs may be more appropriate for these specific therapeutic applications (Table 47.1).

In summary, the choice of origin cells for EV production is a critical factor that can impact the success of EV-based drug delivery strategies. The selection of cells should consider factors such as scalability, reproducibility, safety, and therapeutic application. Furthermore, the method of EV isolation and purification should also be carefully considered to ensure the quality and functionality of the resulting EVs. With careful consideration of these factors, EV-mediated drug delivery can revolutionize the therapeutics domain by furnishing a secure and efficient mode of transporting cargo molecules to intended cells.

47.3 Isolation and Characterization of EVs

In addition to the origin cells, the method of EV isolation and purification can also impact the characteristics and functionality of the resulting EVs (Thery et al., 2018). Several methods are available for EV isolation, including ultracentrifugation, size-exclusion chromatography, and immunoaffinity capture (Thery et al., 2018; Liangsupree et al., 2021; Sidhom et al., 2020). Each approach comes with its strengths and limitations concerning yield, purity, and scalability (Table 47.2). Therefore, the choice of isolation method should also be carefully considered when developing EV-based drug delivery strategies (Konoshenko et al., 2018). For example, the isolation of EVs from cell culture media typically involves several steps to remove cells, debris, and other contaminants before purifying EVs. One commonly used method is differential ultracentrifugation, which involves multiple rounds of centrifugation at increasing speeds to pellet cells, debris, and large vesicles, followed by ultracentrifugation to pellet EVs (Szatanek et al., 2015). However, this method can be time-consuming, and EVs may co-pellet with other particles, making it difficult to obtain pure EVs (Ramirez et al., 2018; Tian et al., 2020).

Alternative methods include density gradient centrifugation, size-exclusion chromatography, and precipitation-based methods using polymers or antibodies that target specific EV surface markers (Thery et al., 2018; Liangsupree et al., 2021; Konoshenko et al., 2018; Szatanek et al., 2015). While isolation of EVs from blood or plasma is more challenging due to the presence of high levels of soluble proteins, lipids, and other particles that can interfere with EV isolation and characterization (Carnino et al., 2019; Karimi et al., 2018). Several methods have been developed for isolating EVs from blood or plasma, including gradient ultracentrifugation, size-exclusion chromatography, and immunomagnetic bead separation (Shirejini & Inci, 2022; Buschmann

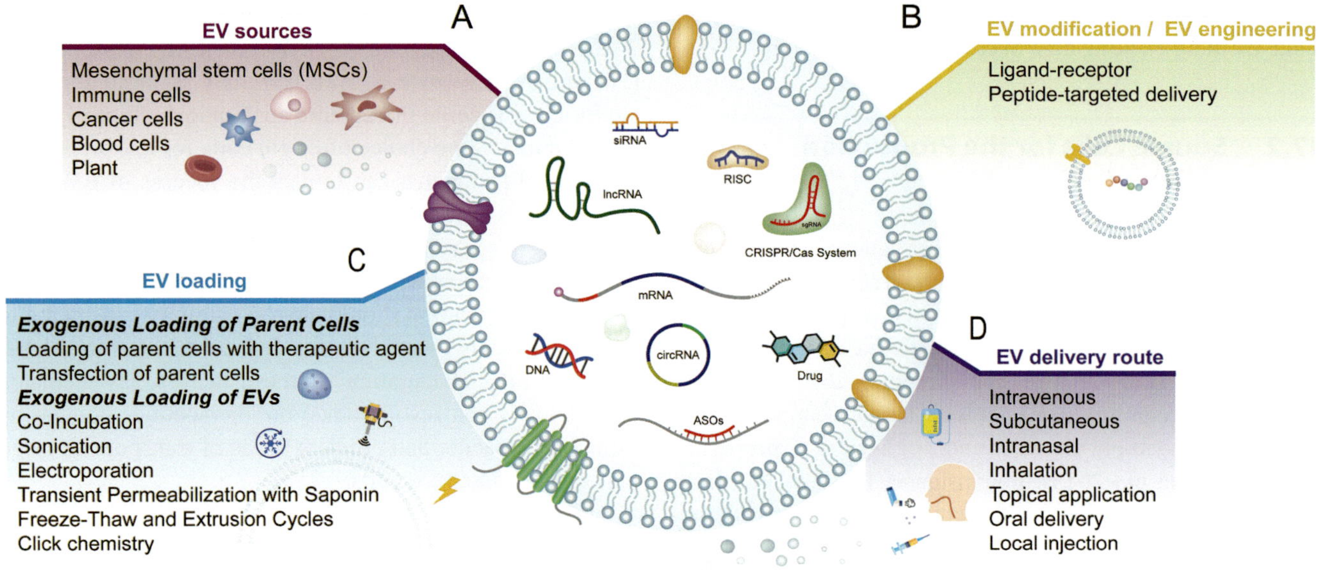

Fig. 47.2 Schematic diagram illustrating the utilization of extracellular vesicles (EVs) as a delivery system for therapeutics. (**a**) Therapeutic EVs can be prepared from various sources, including (1) mesenchymal stem cells (MSCs), including human umbilical cord mesenchymal stem cells (huc-MSCs), bone marrow mesenchymal stem cells (BM-MSCs), adipose tissue-derived mesenchymal stem cells (AD-MSCs), and human embryonic mesenchymal stem cells (ES-MSCs): MSC-derived EVs have anti-inflammatory and immunomodulatory properties and can promote tissue regeneration, making them useful for treating various inflammatory and degenerative diseases. (2) Immune cells: EVs derived from immune cells can induce or suppress immune responses, making them potential candidates for immunotherapy or autoimmune disease treatment. (3) Cancer cells: Cancer cell-derived EVs can contain biomolecules that promote tumor growth and progression, but they can also be designed to administer anti-cancer therapeutics to tumor cells specifically. (4) Blood cells: EVs derived from blood cells can carry important biomolecules such as proteins, lipids, and nucleic acids and can be used as diagnostic or therapeutic tools for various diseases. Also, red blood cells (RBCs) have surface proteins like CD47, CD59, and CR1 that prevent immune clearance mechanisms, leading to their prolonged circulation time in the bloodstream. Moreover, they have excellent targeting capabilities for delivering lipophilic drugs to lung carcinoma cells, among other advantages. (6) Plant: Plant-derived EVs can be engineered to deliver drugs, vaccines, or other therapeutics and may offer advantages over other sources such as lower immunogenicity and potential for large-scale production. (**b**) To enhance EVs' targeting and therapeutic efficacy EVs can be modified/engineered based on (1) ligand-receptor interactions to target specific cells or tissues; (2) peptide-targeted delivery, where specific peptides are incorporated into the EVs to target specific cells or tissues. (**c**) EVs can be loaded with therapeutic agents either by loading parent cells with the therapeutic agent, which is then packaged into EVs during EV biogenesis, resulting in EVs that are loaded with the therapeutic cargo, or by exogenous loading of pre-formed EVs with therapeutic agents through various mechanisms. (**d**) EVs can be administered via several routes, including the following: (1) intravenous (IV) injection: EVs can be administered directly into the bloodstream via IV injection; (2) subcutaneous injection: EVs can also be delivered via subcutaneous injection, which involves injecting the EVs under the skin; (3) intranasal delivery: EVs can be administered through the nose, which allows them to bypass the blood–brain barrier and potentially target the brain; (4) inhalation: EVs can be inhaled as a mist or aerosol, which can target the lungs; (5) topical application: EVs can be applied directly to the skin or mucous membranes; (6) oral delivery: EVs can be encapsulated in enteric-coated capsules or other formulations that protect them from degradation in the digestive system, allowing them to be delivered orally; and (7) local injection: EVs can be injected directly into a specific tissue or organ, allowing for targeted delivery

et al., 2021). Each of these techniques comes with its own set of advantages and limitations, and the selection of the appropriate method is contingent upon the particular research query and subsequent applications. For instance, gradient ultracentrifugation is considered the gold standard method, but it is time-consuming and can damage EVs, whereas size-exclusion chromatography is faster but may not be suitable for isolating small EVs. Immunomagnetic bead separation, on the other hand, allows for selective isolation of EV subpopulations based on specific surface markers, but it may introduce artifacts and require optimization for different cell types and samples (Brennan et al., 2020; Sidhom et al., 2020; Witwer et al., 2013). Recently, Sun et al. described a lipid labeling and click chemistry-based EV capture platform called "Click Beads" that enables efficient and rapid purification of EVs, making it ideal for detecting oncogenic gene alterations in cancer patients (N. Sun et al., 2022). The platform allows downstream molecular quantification using reverse transcription digital polymerase chain reaction (RT-dPCR). The effectiveness of the platform is demonstrated by detecting and quantifying EWS gene rearrangements and KRAS gene mutation status in EVs isolated from biopsy specimens of Ewing sarcoma or pancreatic cancer patients, which can also be used for monitoring treatment responses and disease progression (N. Sun et al., 2022).

After isolation, EVs can be characterized using a variety of methods to determine their size, morphology, and biochemical composition. Electron microscopy is often used to visual-

Table 47.1 The clinical trials of MSCs-EVs

Clinical trial	EV source	Disease	Intervention	Route
NCT04493242	Bone marrow mesenchymal stem cells	COVID-19 Associated acute respiratory distress syndrome (ARDS)	EVs	Intravenous
NCT03857841	Bone marrow mesenchymal stem cells	Preterm neonates at high risk for bronchopulmonary dysplasia	EVs	Intravenous
NCT05125562	Bone marrow mesenchymal stem cell	Mild-to-moderate COVID-19	EVs	Intravenous
NCT05127122	Bone marrow mesenchymal stem cell	Acute respiratory distress syndrome	EVs	Intravenous
NCT04281901	Platelet- and EV-rich plasma	Chronic postsurgical temporal bone inflammations (PvRP-ear)	Platelet- and EV-rich plasma	Ear wick soaked in platelet- and EV-rich plasma
NCT05078385	Bone marrow mesenchymal stem cell	Burn wounds	EVs	Direct application
NCT05774509	Cardiovascular progenitor cells	Non-ischemic cardiomyopathies	EVs	Intravenous
NCT04761562	Platelet- and EV-rich plasma	Chronic middle ear infections	Platelet- and EV-rich plasma	Direct application
NCT05176366	Bone marrow mesenchymal stem cell	Medically refractory ulcerative colitis	EVs	Intravenous
NCT04652531	Autologous serum	Venous trophic lesions not responsive to conventional treatments	EVs	Peri-wound injection
NCT04173650	Bone marrow mesenchymal stem cell	Lesions in subjects with epidermolysis bullosa	EVs	Direct application
NCT05116761	Bone marrow mesenchymal stem cell	Post-acute COVID-19 and chronic post-COVID-19 syndrome	EVs	Intravenous
NCT05354141	Bone marrow mesenchymal stem cell	COVID-19 Moderate-to-severe Acute respiratory distress syndrome (ARDS)	EVs	Intravenous
NCT05130983	Bone marrow mesenchymal stem cell	Medically refractory Crohn's disease	EVs	Intravenous
NCT05523011	Mesenchymal stem/stromal cell	Psoriasis in healthy volunteers	Exosome ointment	Topical application

Examples of clinical trials currently exploring EV's potential for various therapeutic applications. The source of EVs, disease targeted, and intervention vary widely among these trials, highlighting the versatility and potential of EV-based therapeutics

Table 47.2 Extracellular vesicle isolation and characterization methods

Isolation methods	*Advantages*	*Disadvantages*
Differential ultracentrifugation	Gold standard method High yield	Time-consuming EVs may co-pellet with other particles Potential damage to EVs from high centrifugal forces
Density gradient centrifugation	High yield Better separation of EVs from other particles	Time-consuming Requires specialized equipment and expertise
Size-exclusion chromatography	Good separation of EVs from other particles High yield	Time-consuming May result in loss of smaller EVs
Precipitation-based methods (e.g., ExoQuick)	Simple and quick High yield	Co-precipitation of non-EV particles
Immunomagnetic bead separation	Can target specific EV subpopulations High yield	Requires specialized antibodies and equipment May result in loss of EVs not expressing the target antigen
Characterization methods	*Advantages*	*Disadvantages*
Electron microscopy	Direct visualization of EV morphology	Labor-intensive and time-consuming Limited ability to analyze large numbers of EVs
Dynamic light scattering/nanoparticle tracking analysis	High-throughput analysis of EV size and concentration	Limited ability to distinguish EVs from other particles of similar size
Western blotting/flow cytometry	Can identify EV-specific markers	Limited ability to analyze EV cargo
Mass spectrometry	Can identify EV protein cargo	Requires specialized equipment and expertise
RNA sequencing	Can identify EV RNA cargo	Requires specialized equipment and expertise

ize the morphology of EVs, while dynamic light scattering and nanoparticle tracking analysis can be employed for the measurement of size and concentration of EVs (Chuo et al., 2018; Vestad et al., 2017; Maas et al., 2015). Western blotting and flow cytometry can be used to identify EV-specific markers such as CD63, CD81, and Alix (Geeurickx et al., 2019; Ekstrom et al., 2022). Mass spectrometry and RNA sequencing can be used for analyzing protein and RNA content of EVs, respectively (Erdbrugger et al., 2021; O'Grady et al., 2022). These methods can provide valuable information about the cargo and biological function of EVs. To be noted, the complexity of EVs in blood or plasma can make their analysis more challenging (Karimi et al., 2018). Several techniques have been widely used to address these challenges to enrich or deplete specific subpopulations of EVs. For example, immunoaffinity capture using antibodies that target specific EV surface markers can be used to enrich EVs from a particular cell type or disease state (Lo et al., 2020; Liangsupree et al., 2021). In addition, recent advances in single EV analysis techniques, such as single vesicle flow cytometry and microfluidic devices, have enabled high-throughput analysis of individual EVs, providing new insights into their heterogeneity and biological function (Hilton & White, 2021; Fortunato et al., 2021; Han et al., 2021).

In summary, the isolation and characterization of EVs from cell culture media and biological fluids, such as blood and plasma, are critical for examining EV biological functions and developing EVs as potential diagnostic and therapeutic tools. While many methods are available for EV isolation and characterization, there is still a need for standardized protocols and quality control measures to ensure reliable and reproducible results and comparability between different studies and laboratories. Without such standards, the field of EV research risks producing inconsistent and unreliable data, hindering progress in developing EV-based diagnostics and therapeutics.

47.4 Modification/Engineering of EVs for Targeted Therapeutics Delivery

EVs have emerged as a promising avenue for drug delivery, leveraging their innate capacity to transport cargo molecules across biological barriers. However, the efficiency of EVs as drug carriers is limited by their biological properties, such as their short half-life in circulation, low targeting efficiency, and low drug-loading capacity. Various strategies have been developed to modify and engineer EVs to address these limitations, thereby improving their therapeutic potential (Fig. 47.2).

Chemical conjugation of molecules onto the surface of EVs, such as click chemistry, is a viable approach for EV modification. Click chemistry is a highly efficient, specific, and biocompatible method of chemoselective ligation of molecules. This technique has been used to attach targeting ligands, such as antibodies or peptides, onto the surface of EVs for targeted drug delivery (Salunkhe et al., 2020). For example, Mardpour et al. investigated a new strategy to prolong the bioavailability of extracellular vesicles derived from MSC-EVs in the liver using clickable polyethylene glycol (PEG) macromeres to form EV-encapsulated PEG hydrogels that release EVs in a sustained manner over 1 month. This system improved the anti-fibrosis, anti-inflammation, anti-apoptosis, and regenerative effects of EVs (Mardpour et al., 2019). The study may enable the use of MSC-EVs as cell-free therapeutics for chronic liver failure, and the sustained systemic delivery strategy may extend the effects of disease-targeting EVs over time. Additionally, to control the in vivo fate of EVs, Lim et al. developed a versatile method to alter the surfaces of EVs using bioorthogonal copper-free click chemistry (BCC) and metabolically glycoengineered cells (MGE) (Lim et al., 2021). The authors introduced an azide group on donor cells using tetraacetylatedN-azidoacetyl-d-mannosamine—(Ac4ManNAz) and labeled it with dibenzocyclooctyne-terminated PEGylated hyaluronic acid (DBCO-PHA) via BCC. Intravenous injection of these EVs resulted in specific binding affinity to CD44-overexpressing tissues and prolonged circulation in the blood in animal models with rheumatoid arthritis and tumors. The study demonstrated that after systemic administration, PHA-EVs could accumulate in CD44-abundant tissues via their active targeting mechanism in vivo, making the method highly useful for surface modification of EVs to modulate their in vivo fate (Lim et al., 2021). Moreover, Lee et al. developed exosome mimetics (EMs) that were modified using bioorthogonal chemistry to target bone tissue (Lee et al., 2023). The binding efficacy of the ALD-EMs to an HA-rich environment, such as bone, increased to 85% of feeding over time. The engineered EMs were then used to create, a biocompatible scaffold with an excellent bone-binding affinity that provided a substrate for cell adhesion, proliferation, and osteogenic potential. Furthermore, the EMs played a critical role as a delivery carrier of bioactive compounds, such as smoothened agonist (SAG), which showed promising osteogenic and re-ossification potential in vitro and in vivo (Lee et al., 2023). The study concluded that these targeted EMs engineered by bioorthogonal chemistry have the potential to promote bone regeneration. Interestingly, Zheng et al. designed and synthesized five sialic acid analogs with different lengths of N-acyl side chains modified at the C-5 position for exosome labeling using bio-orthogonal click chemistry (Zheng et al., 2023). The authors found that SiaNAz-3 produced the strongest metabolic labeling efficiency of cells and exosomes. Exosomes derived from macrophages (RAW-Exo) were found to strongly co-localize with central nervous system microglia, suggesting that RAW-Exo may be a promising targeted drug delivery vector for the treatment of neurodegenerative disorders (Zheng et al., 2023). An RSV-loaded RAW-Exo formulation (RSV & Exo) was developed for the treatment of refractory experimental autoimmune encephalitis (EAE, a murine model of MS) using click chemistry, which significantly inhibited inflammatory responses in the CNS and peripherally, providing a potential therapeutic strategy for multiple sclerosis (MS) (Zheng et al., 2023). These studies suggest that the use of click chemistry to chemically conjugate molecules onto the surface of EVs is a promising approach for targeted drug delivery. Additionally, bioorthogonal click chemistry has also been used to modify the surfaces of EVs, making them highly useful for surface modification of EVs to modulate their in vivo fate and potential therapeutic applications.

Genetic modification is another approach to engineering EVs for drug delivery. This involves manipulating the genetic material of the cells that produce EVs to express specific proteins or peptides on the surface of the EVs. For example, Alvarez-Erviti et al. engineered EVs to express a fusion protein of the membrane protein Lamp2b and the neuron-specific RVG peptide, which can facilitate transport across the blood–brain barrier (Alvarez-Erviti et al., 2011). The RVG-modified EVs loaded with therapeutics have also demonstrated enhanced therapeutic efficacy in various disease models, including Alzheimer's disease (Cui et al., 2019; Y. Yu et al., 2021), morphine relapse (Y. Liu et al., 2015), brain injury (J. Yang et al., 2017), major depressive disorder, (MDD) (X. Yu et al., 2022) and neuroinflammation (Chivero et al., 2020). Additionally, the T7 peptide is a 12-amino acid peptide that binds to the transferrin receptor (TfR) overexpressed on many tumor cells. By incorporating the T7 peptide into EVs, they can be specifically targeted to tumor cells for drug delivery. Kim et al. developed a targeted carrier using T7 peptide-decorated exosomes to deliver antisense miRNA oligonucleotides against miR-21 (AMO-21) to treat glioblastoma (G. Kim et al., 2020). Incorporation of the T7 peptide into the exosome membrane was achieved through the fusion of T7 and Lamp2b proteins, resulting in proficient targeting of glioblastoma cells exhibiting elevated transferrin receptor expression. Findings from both in vitro and in vivo investigations illustrated that T7-exo-mediated delivery of AMO-21 into the brain more efficiently than unmodified exosomes and RVG-decorated exosomes, efficiently reducing miR-21 levels in the glioblastoma and tumor sizes (G. Kim et al., 2020). Overall, T7-exo is an efficient carrier for AMO-21 in treating glioblastoma. While, the GE11 peptide is a 12-amino acid sequence and has an affinity for the epidermal growth factor receptor (EGFR), which is often overexpressed on various cancer cells. EVs modified with the GE11 peptide have been shown to selectively target cancer cells and enhance the delivery of therapeutic agents. Ohno et al. investigated the potential use of exosomes as a delivery system for nucleic acid drugs to target epidermal growth factor receptor (EGFR)-expressing breast cancer cells (Ohno et al., 2013). The researchers modified the donor cells to express the transmembrane domain of the platelet-derived growth factor receptor linked with the GE11 peptide, facilitating precise targeting of cells expressing EGFR. Exosomes were found to efficiently deliver let-7a miRNA to EGFR-expressing xenograft breast cancer tissue in mice, suggesting that GE11 peptide-modified EVs could be used therapeutically to target cancerous tissues with nucleic acid drugs (Ohno et al., 2013). These findings suggest that incorporating these peptides into the EV surface makes it possible to create vesicles with enhanced targeting specificity and therapeutic efficacy. Furthermore, the extended presence of exosomes in circulation can be attributed to the protection provided by CD47, which shields them from phagocytosis by monocytes and macrophages (Kamerkar et al., 2017).

Liposome fusion and membrane permeabilization have also been used for physically modifying the membrane structure of EVs to improve their drug-loading capacity and targeting efficiency (Walker et al., 2019). Sato et al. developed hybrid exosomes as drug delivery systems by fusing the membranes of exosomes with liposomes using the freeze-thaw method to modify the performance of exosomal nanocarriers (Sato et al., 2016). The researchers found that the membrane engineering method can be combined with genetic modification techniques and that the interactions between the hybrid exosomes and cells can be modified by changing the lipid composition or properties of the exogenous lipids (Sato et al., 2016). This finding suggests that the membrane-engineering approach offers a new strategy for developing exosomes as hybrid nanocarriers for advanced drug delivery systems (Garcia-Manrique et al., 2018). Moreover, membrane permeabilization has also been utilized to physically modify the structure of EVs. The sonication method uses sound waves to permeabilize EV membranes and increases chemotherapeutic agent's loading into EVs, leading to improved anti-cancer efficacy (Deshmukh et al., 2021; Wu et al., 2021). Kim et al. developed an exosome-based drug delivery system using paclitaxel (PTX) to treat multidrug-resistant (MDR) cancer (M.S. Kim et al., 2016). Exosomes isolated from macrophages were loaded with PTX using different methods. The incorporation of PTX into exosomes increased cytotoxicity more than 50 times in drug-resistant cancer cells, and sonication was used to reform the exosomal membrane and improve loading efficiency and sustained drug release (M.S. Kim et al., 2016). This study also demonstrated nearly complete co-localization of airway-delivered exosomes with cancer cells and a potent anticancer effect in a mouse model of pulmonary metastases (M.S. Kim et al., 2016). Haney et al. conducted a study, in which they loaded exosomes with catalase (exoCAT) and used sonication to permeabilize the exosomal membrane (Haney et al., 2015). The results showed that sonication and extrusion led to high loading efficiency, sustained release, and catalase preservation against protease degradation. ExoCAT demonstrated noteworthy neuroprotective properties in both in vitro and in vivo Parkinson's disease models, indicating that exosome-based formulations hold promise as a versatile approach for addressing neurodegenerative and inflammatory disorders (Haney et al., 2015).

Overall, the modification and engineering of EVs for drug delivery is a promising area of research with potential applications in treating a wide range of diseases. While various methods are available for EV modification, each with its advantages and disadvantages, developing more efficient and specific techniques will be crucial to exploiting EVs' therapeutic potential.

47.5 EV Loading with Therapeutics

EVs have gained considerable interest as potential drug delivery vehicles due to their ability to protect their cargo from degradation. EVs can be loaded with various drugs, including small molecules, proteins, and nucleic acids. This section will cover the various techniques utilized for loading drugs into EVs. There are two main methods for loading therapeutic agents into EVs: exogenous loading of parent cells and exogenous loading of EVs. Exogenous loading of parent cells involves incubating the cells with the therapeutic agent, which is then incorporated into the EVs during biogenesis. On the other hand, exogenous loading of EVs involves incubating pre-isolated EVs with the therapeutic agent, which is then taken up by the EVs (Fig. 47.2). Both methods have advantages and limitations and are being actively researched for their potential use in treating various diseases, including cancer, neurological disorders, and cardiovascular diseases.

Loading therapeutic agents into EVs via exogenous loading of parent cells involves incubating the parent cells with the therapeutic agent, which is then incorporated into the EVs released from the cells. Exogenous loading has several advantages, including ease of use and the ability to load a wide range of therapeutic agents into EVs. Also, the loaded EVs have the same surface markers as the parent cells, which can improve targeting and uptake by specific cell types. Several studies have shown the potential of EVs as a drug delivery tool using exogenous loading of parent cells with therapeutic agents. For example, Wang and colleagues designed exosomes as a carrier for curcumin to prevent neuronal death and alleviate Alzheimer's disease symptoms (H. Wang et al., 2019a). The authors prepared exosomes from curcumin-treated cells (Exo-cur) and found that Exo-cur enhanced the solubility and bioavailability of curcumin, and facilitated drug penetration across the blood–brain barrier through specific active targeting (H. Wang et al., 2019a). The study also demonstrated that Exo-cur was more effective in preventing neuron death and inhibiting phosphorylation of the Tau protein, indicating a promising potential for targeted drug delivery and neuronal function recovery in Alzheimer's disease therapy. In another study, by culturing cells with doxorubicin (Dox), Farhat and colleagues utilized the cell's natural loading machinery to load EVs (Farhat et al., 2022). Their findings showed that delivering Dox-loaded EVs increased cytotoxicity while significantly reducing the drug dosage (Farhat et al., 2022). The study also demonstrated that Dox-loaded EVs effectively inhibited cancer cell growth, highlighting the potential of EVs as a promising drug delivery approach for treating retinoblastoma.

Transfection of parent cells is a specific method for loading therapeutic agents into EVs, which involves the transfection of parent cells with a plasmid encoding for the desired therapeutic agent. The transfected cells then produce EVs that contain the therapeutic agent. This approach has been used successfully to generate EVs that deliver siRNA and miRNA to target cells for therapy (Kanada et al., 2015; Hu et al., 2012b). For example, Li et al. engineered the exosomes by fusing the exosomal membrane protein CD9 with an RNA-binding protein called HuR (Li et al., 2019). The researchers successfully enriched miR-155 into exosomes and demonstrated that it could be efficiently delivered into recipient cells and recognize endogenous targets. The study also revealed that the CD9-HuR exosomes could enrich functional miRNA inhibitors, or CRISPR/dCas9 when the RNAs were engineered to have AU-rich elements.

EVs can also be loaded with specific cargo using exogenous loading methods, including co-incubation, sonication, electroporation, transient permeabilization with saponin, freeze-thaw and extrusion cycles, and click chemistry (Murphy et al., 2019; Haney et al., 2015; Sil et al., 2020). In the method of co-incubation, EVs are incubated with a solution containing the desired molecule or compound. Over time, the molecules are taken up by the EVs, and the loading efficiency depends on the concentration, size, and surface charge of the molecules and the EVs. Using this method, Zhuang and colleagues utilized exosomes to encapsulate curcumin (Exo-cur) or a STAT33 inhibitor (Exo-JSI124) and administered them noninvasively to microglial cells via intranasal delivery in three inflammation-mediated disease models (Zhuang et al., 2011). The rapid delivery of the exosome-encapsulated drug to the brain, selectively took up by microglial cells and subsequently induced apoptosis of microglial cells. These results suggest that intranasal delivery of Exo-cur and Exo-JSI124 offers a promising non-invasive therapeutic approach for treating brain inflammatory-related diseases, including glioblastoma and experimental autoimmune encephalomyelitis (EAE)

(Zhuang et al., 2011). While sonication involves exposing EVs to high-frequency sound waves that create small cavitation bubbles. These bubbles create temporary pores in the EV membrane, allowing the entry of cargo molecules. However, this method can also damage the EVs, leading to reduced loading efficiency and biological activity. Yerneni and colleagues developed a novel method to deliver curcumin-albumin encapsulated EVs (CA-EVs) for the treatment of inflammatory skin diseases (Yerneni et al., 2022). The researchers used mild sonication to encapsulate curcumin and albumin into EVs and integrated them into dissolvable microneedle arrays (dMNAs) for intradermal delivery. The resulting CA-EVs exhibited higher stability and anti-inflammatory bioactivity in vitro compared to naïve curcumin or curcumin-EVs without albumin. In vivo studies showed that dMNA-delivered CA-EVs effectively suppressed and reduced inflammation in rat and mouse models, demonstrating the potential of this approach as a local-delivery strategy for inflammatory skin diseases (Yerneni et al., 2022). The electroporation method applies a brief electric field to the EVs, creating transient pores in the membrane. While this approach has proven to be successful in delivering RNA molecules into EVs, it could potentially compromise the integrity and stability of the EVs. Wan et al. developed a new method called "exosomeRNP" that allows for the loading of Cas9 ribonucleoprotein (RNP) into exosomes through electroporation for precise and tissue-specific gene therapies of liver diseases (Wan et al., 2022). Targeting specific genes using exosomeRNP has demonstrated therapeutic potential in mouse models of acute liver injury, chronic liver fibrosis, and hepatocellular carcinoma. The exosomeRNP delivery system can effectively deliver RNP into cells and specifically accumulate in the liver tissue in vivo (Wan et al., 2022). This study presents a feasible platform for precise and tissue-specific gene therapies for liver diseases using exosomeRNP (Wan et al., 2022). In the method of transient permeabilization with saponin, saponin is a natural compound that can create transient pores in the EV membrane. The loading efficiency can be improved by optimizing the concentration and duration of exposure to saponin. To treat Neuronal Ceroid Lipofuscinoses 2 (CLN2), Haney and colleagues loaded TPP1 into EVs through either the permeabilization of EV membranes with saponin or sonication in the presence of TPP1 (Haney et al., 2019). Both methods effectively incorporated functional TPP1 into EVs and prevented its degradation by proteases. In vivo experiments on mice with Batten disease showed that these EV-based formulations of TPP1 significantly increased their lifespan (Haney et al., 2019). This suggests that EVs are a promising natural and biocompatible option for delivering TPP1 systemically to the brain, potentially serving as effective drug nanocarriers. Freeze-thaw and extrusion cycles involve repeatedly freezing and thawing EVs with the desired cargo. This process can disrupt the EV membrane, allowing the cargo to enter. Extrusion cycles are used to homogenize the EVs and increase their size distribution. Shi et al. assessed six distinct approaches, including the incubation, sonication, extrusion, freeze-thaw, saponin-assisted, and electroporation methods, to load the GLP-1 receptor agonist liraglutide into exosomes derived from milk (Shi et al., 2022). The authors found that the extrusion method resulted in the largest drug load, which was 2.45 times greater than the direct incubation method. Furthermore, the liraglutide-Exo obtained by the freeze-thaw cycles method had the smallest morphological change (Shi et al., 2022). Click chemistry involves covalently attaching a specific molecule to the EV surface using click chemistry. This technique can provide a stable and specific loading of the desired molecule onto the EVs. Zou et al. developed an injectable conductive hydrogel to treat myocardial injuries resulting from myocardial infarction and reperfusion (Zou et al., 2021). The hydrogel was created through a "click" reaction and designed to bind exosomes derived from human umbilical cord mesenchymal stem cells, prolonging their retention in the ischemic myocardium. The Gel@Exo composite system has several features, including controllable gelation kinetics, shear-thinning injectability, conductivity matching the native myocardium, and excellent cytocompatibility. Administration of the Gel@Exo system significantly improves cell-to-cell interactions, promotes cell proliferation and angiogenesis, and provides a remarkable therapeutic effect on myocardial infarction-ischemia/reperfusion (MI-I/R) (Zou et al., 2021). In summary, the exogenous loading of EVs has emerged as a promising method to optimize their therapeutic potential. The choice of loading technique depends on the desired cargo, the intended application, and the characteristics of the EVs. Continued development and optimization of these techniques can pave the way for new and effective EV-based therapies.

47.6 EV Administration Routes

The route of EV administration depends on the intended target and the delivery method (Fig. 47.2). The most common route is intravenous injection, which distributes EVs throughout the body and has been effective in treating cancer and cardiovascular diseases (Lu et al., 2017; Tian et al., 2014; Hergenreider et al., 2012; Jansen et al., 2013). Subcutaneous injection can be used to treat skin diseases and wound healing (Cheng et al., 2020; X. Wang et al., 2019b). Intranasal administration of EVs is a non-invasive and promising delivery method for treating neurological disorders (Kutchy et al., 2022; Liao et al., 2020; Chivero et al., 2020; Hu et al., 2018; Haney et al., 2020; Ma et al., 2023b). Preclinical research has demonstrated that administering EVs via inhalation enhances their bioavailability and improves therapeutic efficacy. This approach enables the delivery of EVs to the respiratory system, offering a potential treatment for asthma and chronic obstructive pulmonary disease (COPD) (Gomez et al., 2022). The topical application treats skin diseases and promotes wound healing (J. Liu et al., 2020; K. Zhang et al., 2020a; J. Kim et al., 2023). Oral administration is less common, but studies demonstrated that EVs can survive in harsh stomach environments and deliver cargo to the intestine (Chang et al., 2020; Shen et al., 2022; Cieslik et al., 2022). The intraperitoneal injection has been successfully used for treating liver, pancreatic, and ovarian cancer (Mardpour et al., 2019; Park et al., 2019; Patel et al., 2022). Intracerebral injection treats neurodegenerative diseases such as Parkinson's disease and Alzheimer's disease (Kojima et al., 2018; Elia et al., 2019; Yuyama et al., 2014). The therapeutic potential of EV-based therapies can be enhanced by developing new administration routes and optimizing existing methods.

47.7 Conclusions, Challenges, and Perspectives

In conclusion, EVs have emerged as a promising tool for the delivery of therapeutics due to their natural ability to transport biomolecules between cells, their biocompatibility, and low immunogenicity. EVs can be prepared and characterized from various sources, including cell culture media and blood or plasma, and can be modified to enhance their therapeutic potential. Several preclinical and clinical studies have demonstrated the efficacy of EVs utilized as a delivery mechanism for a diverse array of therapeutics—encompassing RNA, chemotherapeutic agents, and proteins—for treating a broad spectrum of illnesses, such as cancer, cardiovascular disorders, and neurological ailments. In addition, EVs have shown potential for targeted drug delivery and for overcoming drug resistance.

Despite the substantial advancements in the field, several challenges still need to be tackled. One major hurdle is the absence of universally accepted procedures for isolating, purifying, and characterizing EVs, which can impede the reproducibility and comparability of research findings. Additionally, the scalability of EV production is still a major limitation for clinical applications. The heterogeneity of EVs in size, cargo, and surface markers makes it challenging to design and engineer EVs for specific therapeutic applications. Moreover, delivering large macromolecules through the plasma membrane and endosomal barriers is a major challenge. Researchers developed a novel method, termed "Gectosomes", which employs engineered ectosomes to transport proteins and ribonucleoproteins (RNPs) to cells and mouse liver tissue (X. Zhang et al., 2020b). These Gectosomes are engineered to co-encapsulate the vesicular stomatitis virus G protein (VSV-G) with bioactive macromolecules through split GFP complementation. The technique enables efficient cargo loading, enhances the specific activity of cargo delivery, and streamlines Gectosome purification. This capability facilitates the execution of designed modifications to endogenous genes in cultured cell lines in vitro and mouse liver tissue in vivo (X. Zhang et al., 2020b). The Gectosome method thus provides a framework for delivering biologics that can regulate intracellular targets. The enhancement of this technology in the future could lead to improved therapeutics for a diverse array of human diseases.

Further research and development are necessary to optimize the therapeutic potential of EVs as a delivery system. Currently, EVs are still in their early use stage, and more studies are required to investigate their biodistribution, pharmacokinetics, and toxicity in vivo. To improve EV production and engineering, novel technologies like 3D printing and microfluidics can be explored (Casajuana Ester & Day, 2023; Zhao et al., 2019). An exciting development in the field is the creation of EXOsomal transfer into cells (EXOtic) devices, which efficiently produce designer exosomes in engineered mammalian cells (Kojima et al., 2018). These devices enhance exosome production and facilitate specific mRNA packaging and delivery of mRNA into the cytosol of target cells. In both in vitro and in vivo Parkinson's disease models, the potential therapeutic efficacy of these specially engineered exosomes was exhibited, with EXOtic therapy demonstrating a reduction in neurotoxicity and neuroinflammation (Kojima et al., 2018). In this therapy, implanted designer cells produce and secrete therapeutic molecules in situ, which could overcome the limited half-life of exosomes administered intravenously (Kojima et al., 2018). Furthermore, combining EVs with other treatment modalities like immunotherapy and gene therapy could lead to more effective treatment strategies.

Taken together, using EVs as a therapeutic delivery tool has shown great promise and has the potential to revolutionize medicine. Nonetheless, additional research and development are still required to surmount current obstacles and enhance the therapeutic capacity of EVs for treating different ailments. With continued efforts and advancements in the field, the future of EV-based therapeutics looks bright.

47.8 Review Questions

1. Which of the following characteristics can be used to separate extracellular vesicles (EVs)?
   (a) Size
   (b) Density
   (c) Surface markers
   (d) Both a and b
   (e) *All of the above*
2. Which of the following statements about extracellular vesicles (EVs) are correct?

(a) EVs are small membrane-bound nanoparticles released by cells.
(b) EVs can traverse biological barriers, including the blood–brain barrier.
(c) EVs can serve as carriers for drug delivery.
(d) Both a and b
(e) *All of the above*

3. Through which routes can extracellular vesicles (EVs) be administered?
(a) Subcutaneous injection.
(b) Intranasal administration.
(c) Intraperitoneal injection.
(d) Intravenous injection
(e) *All of the above*

References

Aaronson, S., Behrens, U., Orner, R., & Haines, T. H. (1971). Ultrastructure of intracellular and extracellular vesicles, membranes, and myelin figures produced by Ochromonas danica. *Journal of Ultrastructure Research, 35*(5), 418–430. https://doi.org/10.1016/s0022-5320(71)80003-5. https://www.ncbi.nlm.nih.gov/pubmed/4111037

Alvarez-Erviti, L., Seow, Y., Yin, H., Betts, C., Lakhal, S., & Wood, M. J. (2011). Delivery of siRNA to the mouse brain by systemic injection of targeted exosomes. *Nature Biotechnology, 29*(4), 341–345. https://doi.org/10.1038/nbt.1807. https://www.ncbi.nlm.nih.gov/pubmed/21423189

Battistelli, M., & Falcieri, E. (2020). Apoptotic bodies: Particular extracellular vesicles involved in intercellular communication. *Biology (Basel), 9*(1), 21. https://doi.org/10.3390/biology9010021. https://www.ncbi.nlm.nih.gov/pubmed/31968627

Brennan, K., Martin, K., FitzGerald, S. P., O'Sullivan, J., Wu, Y., Blanco, A., Richardson, C., & Mc Gee, M. M. (2020). A comparison of methods for the isolation and separation of extracellular vesicles from protein and lipid particles in human serum. *Scientific Reports, 10*(1), 1039. https://doi.org/10.1038/s41598-020-57497-7. https://www.ncbi.nlm.nih.gov/pubmed/31974468

Buschmann, D., Mussack, V., & Byrd, J. B. (2021). Separation, characterization, and standardization of extracellular vesicles for drug delivery applications. *Advanced Drug Delivery Reviews, 174*, 348–368. https://doi.org/10.1016/j.addr.2021.04.027. https://www.ncbi.nlm.nih.gov/pubmed/33964356

Carnino, J. M., Lee, H., & Jin, Y. (2019). Isolation and characterization of extracellular vesicles from Broncho-alveolar lavage fluid: A review and comparison of different methods. *Respiratory Research, 20*(1), 240. https://doi.org/10.1186/s12931-019-1210-z. https://www.ncbi.nlm.nih.gov/pubmed/31666080

Caruso, S., & Poon, I. K. H. (2018). Apoptotic cell-derived extracellular vesicles: More than just debris. *Frontiers in Immunology, 9*, 1486. https://doi.org/10.3389/fimmu.2018.01486. https://www.ncbi.nlm.nih.gov/pubmed/30002658

Casajuana Ester, M., & Day, R. M. (2023). Production and utility of extracellular vesicles with 3D culture methods. *Pharmaceutics, 15*(2), 663. https://doi.org/10.3390/pharmaceutics15020663. https://www.ncbi.nlm.nih.gov/pubmed/36839984

Chang, X., Wang, S. L., Zhao, S. B., Shi, Y. H., Pan, P., Gu, L., Yao, J., Li, Z. S., & Bai, Y. (2020). Extracellular vesicles with possible roles in gut intestinal tract homeostasis and IBD. *Mediators of Inflammation, 2020*, 1945832. https://doi.org/10.1155/2020/1945832. https://www.ncbi.nlm.nih.gov/pubmed/32410847

Chargaff, E., & West, R. (1946). The biological significance of the thromboplastic protein of blood. *The Journal of Biological Chemistry, 166*(1), 189–197. https://www.ncbi.nlm.nih.gov/pubmed/20273687

Cheng, S., Xi, Z., Chen, G., Liu, K., Ma, R., & Zhou, C. (2020). Extracellular vesicle-carried microRNA-27b derived from mesenchymal stem cells accelerates cutaneous wound healing via E3 ubiquitin ligase ITCH. *Journal of Cellular and Molecular Medicine, 24*(19), 11254–11271. https://doi.org/10.1111/jcmm.15692. https://www.ncbi.nlm.nih.gov/pubmed/32845084

Chivero, E. T., Liao, K., Niu, F., Tripathi, A., Tian, C., Buch, S., & Hu, G. (2020). Engineered extracellular vesicles loaded with miR-124 attenuate cocaine-mediated activation of microglia. *Frontiers in Cell and Development Biology, 8*, 573. https://doi.org/10.3389/fcell.2020.00573. https://www.ncbi.nlm.nih.gov/pubmed/32850781

Chivero, E. T., Dagur, R. S., Peeples, E. S., Sil, S., Liao, K., Ma, R., Chen, L., Gurumurthy, C. B., Buch, S., & Hu, G. (2021). Biogenesis, physiological functions and potential applications of extracellular vesicles in substance use disorders. *Cellular and Molecular Life Sciences, 78*(11), 4849–4865. https://doi.org/10.1007/s00018-021-03824-8. https://www.ncbi.nlm.nih.gov/pubmed/33821293

Chuo, S. T., Chien, J. C., & Lai, C. P. (2018). Imaging extracellular vesicles: Current and emerging methods. *Journal of Biomedical Science, 25*(1), 91. https://doi.org/10.1186/s12929-018-0494-5. https://www.ncbi.nlm.nih.gov/pubmed/30580764

Cieslik, M., Nazimek, K., & Bryniarski, K. (2022). Extracellular vesicles-oral therapeutics of the future. *International Journal of Molecular Sciences, 23*(14), 7554. https://doi.org/10.3390/ijms23147554. https://www.ncbi.nlm.nih.gov/pubmed/35886902

Corso, G., Mager, I., Lee, Y., Gorgens, A., Bultema, J., Giebel, B., Wood, M. J. A., Nordin, J. Z., & Andaloussi, S. E. (2017). Reproducible and scalable purification of extracellular vesicles using combined bind-elute and size exclusion chromatography. *Scientific Reports, 7*(1), 11561. https://doi.org/10.1038/s41598-017-10646-x. https://www.ncbi.nlm.nih.gov/pubmed/28912498

Cui, G. H., Guo, H. D., Li, H., Zhai, Y., Gong, Z. B., Wu, J., Liu, J. S., Dong, Y. R., Hou, S. X., & Liu, J. R. (2019). RVG-modified exosomes derived from mesenchymal stem cells rescue memory deficits by regulating inflammatory responses in a mouse model of Alzheimer's disease. *Immunity & Ageing, 16*, 10. https://doi.org/10.1186/s12979-019-0150-2. https://www.ncbi.nlm.nih.gov/pubmed/31114624

Davies, O. G. (2023). Extracellular vesicles: From bone development to regenerative orthopedics. *Molecular Therapy, 31*(5), 1251–1274. https://doi.org/10.1016/j.ymthe.2023.02.021. https://www.ncbi.nlm.nih.gov/pubmed/36869588

Deshmukh, S. K., Khan, M. A., Singh, S., & Singh, A. P. (2021). Extracellular nanovesicles: From intercellular messengers to efficient drug delivery systems. *ACS Omega, 6*(3), 1773–1779. https://doi.org/10.1021/acsomega.0c05539. https://www.ncbi.nlm.nih.gov/pubmed/33521418

Doyle, L. M., & Wang, M. Z. (2019). Overview of extracellular vesicles, their origin, composition, purpose, and methods for exosome isolation and analysis. *Cells, 8*(7), 727. https://doi.org/10.3390/cells8070727. https://www.ncbi.nlm.nih.gov/pubmed/31311206

Ekstrom, K., Crescitelli, R., Petursson, H. I., Johansson, J., Lasser, C., & Olofsson Bagge, R. (2022). Characterization of surface markers on extracellular vesicles isolated from lymphatic exudate from patients with breast cancer. *BMC Cancer, 22*(1), 50. https://doi.org/10.1186/s12885-021-08870-w. https://www.ncbi.nlm.nih.gov/pubmed/35012489

Elia, C. A., Tamborini, M., Rasile, M., Desiato, G., Marchetti, S., Swuec, P., Mazzitelli, S., Clemente, F., Anselmo, A., Matteoli, M., Malosio, M. L., & Coco, S. (2019). Intracerebral injection of extracellular vesicles from mesenchymal stem cells exerts reduced Abeta plaque burden in early stages of a preclinical model of Alzheimer's disease. *Cells, 8*(9), 1059. https://doi.org/10.3390/cells8091059. https://www.ncbi.nlm.nih.gov/pubmed/31510042

Elsharkasy, O. M., Nordin, J. Z., Hagey, D. W., de Jong, O. G., Schiffelers, R. M., Andaloussi, S. E., & Vader, P. (2020). Extracellular vesicles as drug delivery systems: Why and how? *Advanced Drug Delivery Reviews, 159*, 332–343. https://doi.org/10.1016/j.addr.2020.04.004. https://www.ncbi.nlm.nih.gov/pubmed/32305351

Erdbrugger, U., Blijdorp, C. J., Bijnsdorp, I. V., Borras, F. E., Burger, D., Bussolati, B., Byrd, J. B., Clayton, A., Dear, J. W., Falcon-Perez, J. M., Grange, C., Hill, A. F., Holthofer, H., Hoorn, E. J., Jenster, G., Jimenez, C. R., Junker, K., Klein, J., Knepper, M. A., Koritzinsky, E. H., Luther, J. M., Lenassi, M., Leivo, J., Mertens, I., Musante, L., Oeyen, E., Puhka, M., van Royen, M. E., Sanchez, C., Soekmadji, C., Thongboonkerd, V., van Steijn, V., Verhaegh, G., Webber, J. P., Witwer, K., Yuen, P. S. T., Zheng, L., Llorente, A., & Martens-Uzunova, E. S. (2021). Urinary extracellular vesicles: A position paper by the Urine Task Force of the International Society for Extracellular Vesicles. *Journal of Extracellular Vesicles, 10*(7), e12093. https://doi.org/10.1002/jev2.12093. https://www.ncbi.nlm.nih.gov/pubmed/34035881

Farhat, W., Yeung, V., Kahale, F., Parekh, M., Cortinas, J., Chen, L., Ross, A. E., & Ciolino, J. B. (2022). Doxorubicin-loaded extracellular vesicles enhance tumor cell death in retinoblastoma. *Bioengineering (Basel), 9*(11), 671. https://doi.org/10.3390/bioengineering9110671. https://www.ncbi.nlm.nih.gov/pubmed/36354582

Fortunato, D., Mladenovic, D., Criscuoli, M., Loria, F., Veiman, K. L., Zocco, D., Koort, K., & Zarovni, N. (2021). Opportunities and pitfalls of fluorescent labeling methodologies for extracellular vesicle profiling on high-resolution single-particle platforms. *International Journal of Molecular Sciences, 22*(19), 10510. https://doi.org/10.3390/ijms221910510. https://www.ncbi.nlm.nih.gov/pubmed/34638850

Fu, S., Zhang, Y., Li, Y., Luo, L., Zhao, Y., & Yao, Y. (2020). Extracellular vesicles in cardiovascular diseases. *Cell Death Discovery, 6*, 68. https://doi.org/10.1038/s41420-020-00305-y. https://www.ncbi.nlm.nih.gov/pubmed/32821437

Fujita, Y., Kosaka, N., Araya, J., Kuwano, K., & Ochiya, T. (2015). Extracellular vesicles in lung microenvironment and pathogenesis. *Trends in Molecular Medicine, 21*(9), 533–542. https://doi.org/10.1016/j.molmed.2015.07.004. https://www.ncbi.nlm.nih.gov/pubmed/26231094

Garcia-Manrique, P., Matos, M., Gutierrez, G., Pazos, C., & Blanco-Lopez, M. C. (2018). Therapeutic biomaterials based on extracellular vesicles: Classification of bio-engineering and mimetic preparation routes. *Journal of Extracellular Vesicles, 7*(1), 1422676. https://doi.org/10.1080/20013078.2017.1422676. https://www.ncbi.nlm.nih.gov/pubmed/29372017

Geeurickx, E., Tulkens, J., Dhondt, B., Van Deun, J., Lippens, L., Vergauwen, G., Heyrman, E., De Sutter, D., Gevaert, K., Impens, F., Miinalainen, I., Van Bockstal, P. J., De Beer, T., Wauben, M. H. M., Nolte-'t-Hoen, E. N. M., Bloch, K., Swinnen, J. V., van der Pol, E., Nieuwland, R., Braems, G., Callewaert, N., Mestdagh, P., Vandesompele, J., Denys, H., Eyckerman, S., De Wever, O., & Hendrix, A. (2019). The generation and use of recombinant extracellular vesicles as biological reference material. *Nature Communications, 10*(1), 3288. https://doi.org/10.1038/s41467-019-11182-0. https://www.ncbi.nlm.nih.gov/pubmed/31337761

Gomez, N., James, V., Onion, D., & Fairclough, L. C. (2022). Extracellular vesicles and chronic obstructive pulmonary disease (COPD): A systematic review. *Respiratory Research, 23*(1), 82. https://doi.org/10.1186/s12931-022-01984-0. https://www.ncbi.nlm.nih.gov/pubmed/35382831

Gowen, A., Shahjin, F., Chand, S., Odegaard, K. E., & Yelamanchili, S. V. (2020). Mesenchymal stem cell-derived extracellular vesicles: Challenges in clinical applications. *Frontiers in Cell and Development Biology, 8*, 149. https://doi.org/10.3389/fcell.2020.00149. https://www.ncbi.nlm.nih.gov/pubmed/32226787

Gregory, C. D., & Dransfield, I. (2018). Apoptotic tumor cell-derived extracellular vesicles as important regulators of the onco-regenerative niche. *Frontiers in Immunology, 9*, 1111. https://doi.org/10.3389/fimmu.2018.01111. https://www.ncbi.nlm.nih.gov/pubmed/29875772

Han, C., Kang, H., Yi, J., Kang, M., Lee, H., Kwon, Y., Jung, J., Lee, J., & Park, J. (2021). Single-vesicle imaging and co-localization analysis for tetraspanin profiling of individual extracellular vesicles. *Journal of Extracellular Vesicles, 10*(3), e12047. https://doi.org/10.1002/jev2.12047. https://www.ncbi.nlm.nih.gov/pubmed/33456726

Haney, M. J., Klyachko, N. L., Zhao, Y., Gupta, R., Plotnikova, E. G., He, Z., Patel, T., Piroyan, A., Sokolsky, M., Kabanov, A. V., & Batrakova, E. V. (2015). Exosomes as drug delivery vehicles for Parkinson's disease therapy. *Journal of Controlled Release, 207*, 18–30. https://doi.org/10.1016/j.jconrel.2015.03.033. https://www.ncbi.nlm.nih.gov/pubmed/25836593

Haney, M. J., Klyachko, N. L., Harrison, E. B., Zhao, Y., Kabanov, A. V., & Batrakova, E. V. (2019). TPP1 delivery to lysosomes with extracellular vesicles and their enhanced brain distribution in the animal model of batten disease. *Advanced Healthcare Materials, 8*(11), e1801271. https://doi.org/10.1002/adhm.201801271. https://www.ncbi.nlm.nih.gov/pubmed/30997751

Haney, M. J., Zhao, Y., Jin, Y. S., & Batrakova, E. V. (2020). Extracellular vesicles as drug carriers for enzyme replacement therapy to treat CLN2 batten disease: Optimization of drug administration routes. *Cells, 9*(5), 1273. https://doi.org/10.3390/cells9051273. https://www.ncbi.nlm.nih.gov/pubmed/32443895

Hergenreider, E., Heydt, S., Treguer, K., Boettger, T., Horrevoets, A. J., Zeiher, A. M., Scheffer, M. P., Frangakis, A. S., Yin, X., Mayr, M., Braun, T., Urbich, C., Boon, R. A., & Dimmeler, S. (2012). Atheroprotective communication between endothelial cells and smooth muscle cells through miRNAs. *Nature Cell Biology, 14*(3), 249–256. https://doi.org/10.1038/ncb2441. https://www.ncbi.nlm.nih.gov/pubmed/22327366

Herrmann, I. K., Wood, M. J. A., & Fuhrmann, G. (2021). Extracellular vesicles as a next-generation drug delivery platform. *Nature Nanotechnology, 16*(7), 748–759. https://doi.org/10.1038/s41565-021-00931-2. https://www.ncbi.nlm.nih.gov/pubmed/34211166

Hilton, S. H., & White, I. M. (2021). Advances in the analysis of single extracellular vesicles: A critical review. *Sensors and Actuators Reports, 3*, 100052. https://doi.org/10.1016/j.snr.2021.100052. https://www.ncbi.nlm.nih.gov/pubmed/35098157

Hu, G., & Brew, B. J. (2023). Extracellular vesicle small noncoding RNAs: A window into pathogenesis and diagnosis. *AIDS, 37*(5), 849–850. https://doi.org/10.1097/QAD.0000000000003517. https://www.ncbi.nlm.nih.gov/pubmed/36919790

Hu, G., Drescher, K. M., & Chen, X. M. (2012a). Exosomal miRNAs: Biological properties and therapeutic potential. *Frontiers in Genetics, 3*, 56. https://doi.org/10.3389/fgene.2012.00056. https://www.ncbi.nlm.nih.gov/pubmed/22529849

Hu, G., Yao, H., Chaudhuri, A. D., Duan, M., Yelamanchili, S. V., Wen, H., Cheney, P. D., Fox, H. S., & Buch, S. (2012b). Exosome-mediated shuttling of microRNA-29 regulates HIV Tat and morphine-mediated neuronal dysfunction. *Cell Death & Disease, 3*(8), e381. https://doi.org/10.1038/cddis.2012.114. https://www.ncbi.nlm.nih.gov/pubmed/22932723

Hu, G., Liao, K., Niu, F., Yang, L., Dallon, B. W., Callen, S., Tian, C., Shu, J., Cui, J., Sun, Z., Lyubchenko, Y. L., Ka, M., Chen, X. M., & Buch, S. (2018). Astrocyte EV-induced lincRNA-Cox2 regulates microglial phagocytosis: Implications for morphine-mediated neurodegeneration. *Molecular Therapy – Nucleic Acids, 13*, 450–463. https://doi.org/10.1016/j.omtn.2018.09.019. https://www.ncbi.nlm.nih.gov/pubmed/30388619

Jansen, F., Yang, X., Hoelscher, M., Cattelan, A., Schmitz, T., Proebsting, S., Wenzel, D., Vosen, S., Franklin, B. S., Fleischmann, B. K., Nickenig, G., & Werner, N. (2013). Endothelial microparticle-mediated transfer of MicroRNA-126 promotes vascular endothelial cell repair via SPRED1 and is abrogated in glucose-damaged endothelial microparticles. *Circulation, 128*(18), 2026–2038. https://doi.org/10.1161/CIRCULATIONAHA.113.001720. https://www.ncbi.nlm.nih.gov/pubmed/24014835

Johnstone, R. M., Adam, M., Hammond, J. R., Orr, L., & Turbide, C. (1987). Vesicle formation during reticulocyte maturation. Association of plasma membrane activities with released vesicles (exosomes). *The Journal of Biological Chemistry, 262*(19), 9412–9420. https://www.ncbi.nlm.nih.gov/pubmed/3597417

Jurgielewicz, B., Stice, S., & Yao, Y. (2021). Therapeutic potential of nucleic acids when combined with extracellular vesicles. *Aging and Disease, 12*(6), 1476–1493. https://doi.org/10.14336/AD.2021.0708. https://www.ncbi.nlm.nih.gov/pubmed/34527423

Kakarla, R., Hur, J., Kim, Y. J., Kim, J., & Chwae, Y. J. (2020). Apoptotic cell-derived exosomes: Messages from dying cells. *Experimental & Molecular Medicine, 52*(1), 1–6. https://doi.org/10.1038/s12276-019-0362-8. https://www.ncbi.nlm.nih.gov/pubmed/31915368

Kalluri, R., & LeBleu, V. S. (2020). The biology, function, and biomedical applications of exosomes. *Science, 367*(6478), eaau6977. https://doi.org/10.1126/science.aau6977. https://www.ncbi.nlm.nih.gov/pubmed/32029601

Kamerkar, S., LeBleu, V. S., Sugimoto, H., Yang, S., Ruivo, C. F., Melo, S. A., Lee, J. J., & Kalluri, R. (2017). Exosomes facilitate therapeutic targeting of oncogenic KRAS in pancreatic cancer. *Nature, 546*(7659), 498–503. https://doi.org/10.1038/nature22341. https://www.ncbi.nlm.nih.gov/pubmed/28607485

Kanada, M., Bachmann, M. H., Hardy, J. W., Frimannson, D. O., Bronsart, L., Wang, A., Sylvester, M. D., Schmidt, T. L., Kaspar, R. L., Butte, M. J., Matin, A. C., & Contag, C. H. (2015). Differential fates of biomolecules delivered to target cells via extracellular vesicles. *Proceedings of the National Academy of Sciences of the United States of America, 112*(12), E1433–E1442. https://doi.org/10.1073/pnas.1418401112. https://www.ncbi.nlm.nih.gov/pubmed/25713383

Karimi, N., Cvjetkovic, A., Jang, S. C., Crescitelli, R., Hosseinpour Feizi, M. A., Nieuwland, R., Lotvall, J., & Lasser, C. (2018). Detailed analysis of the plasma extracellular vesicle proteome after separation from lipoproteins. *Cellular and Molecular Life Sciences, 75*(15), 2873–2886. https://doi.org/10.1007/s00018-018-2773-4. https://www.ncbi.nlm.nih.gov/pubmed/29441425

Kim, M. S., Haney, M. J., Zhao, Y., Mahajan, V., Deygen, I., Klyachko, N. L., Inskoe, E., Piroyan, A., Sokolsky, M., Okolie, O., Hingtgen, S. D., Kabanov, A. V., & Batrakova, E. V. (2016). Development of exosome-encapsulated paclitaxel to overcome MDR in cancer cells. *Nanomedicine, 12*(3), 655–664. https://doi.org/10.1016/j.nano.2015.10.012. https://www.ncbi.nlm.nih.gov/pubmed/26586551

Kim, G., Kim, M., Lee, Y., Byun, J. W., Hwang, D. W., & Lee, M. (2020). Systemic delivery of microRNA-21 antisense oligonucleotides to the brain using T7-peptide decorated exosomes. *Journal of Controlled Release, 317*, 273–281. https://doi.org/10.1016/j.jconrel.2019.11.009. https://www.ncbi.nlm.nih.gov/pubmed/31730913

Kim, J., Kim, E. H., Lee, H., Sung, J. H., & Bang, O. Y. (2023). Clinical-scale mesenchymal stem cell-derived extracellular vesicle therapy for wound healing. *International Journal of Molecular Sciences, 24*(5), 4273. https://doi.org/10.3390/ijms24054273. https://www.ncbi.nlm.nih.gov/pubmed/36901703

Kojima, R., Bojar, D., Rizzi, G., Hamri, G. C., El-Baba, M. D., Saxena, P., Auslander, S., Tan, K. R., & Fussenegger, M. (2018). Designer exosomes produced by implanted cells intracerebrally deliver therapeutic cargo for Parkinson's disease treatment. *Nature Communications, 9*(1), 1305. https://doi.org/10.1038/s41467-018-03733-8. https://www.ncbi.nlm.nih.gov/pubmed/29610454

Konoshenko, M. Y., Lekchnov, E. A., Vlassov, A. V., & Laktionov, P. P. (2018). Isolation of extracellular vesicles: General methodologies and latest trends. *BioMed Research International, 2018*, 8545347. https://doi.org/10.1155/2018/8545347. https://www.ncbi.nlm.nih.gov/pubmed/29662902

Kutchy, N. A., Ma, R., Liu, Y., Buch, S., & Hu, G. (2022). Extracellular vesicle-mediated delivery of ultrasmall superparamagnetic iron oxide nanoparticles to mice brain. *Frontiers in Pharmacology, 13*, 819516. https://doi.org/10.3389/fphar.2022.819516. https://www.ncbi.nlm.nih.gov/pubmed/35462907

Lawson, C., Vicencio, J. M., Yellon, D. M., & Davidson, S. M. (2016). Microvesicles and exosomes: New players in metabolic and cardiovascular disease. *The Journal of Endocrinology, 228*(2), R57–R71. https://doi.org/10.1530/JOE-15-0201. https://www.ncbi.nlm.nih.gov/pubmed/26743452

Lee, C. S., Fan, J., Hwang, H. S., Kim, S., Chen, C., Kang, M., Aghaloo, T., James, A. W., & Lee, M. (2023). Bone-targeting exosome mimetics engineered by bioorthogonal surface functionalization for bone tissue engineering. *Nano Letters, 23*(4), 1202–1210. https://doi.org/10.1021/acs.nanolett.2c04159. https://www.ncbi.nlm.nih.gov/pubmed/36762874

Li, Z., Zhou, X., Wei, M., Gao, X., Zhao, L., Shi, R., Sun, W., Duan, Y., Yang, G., & Yuan, L. (2019). In vitro and in vivo RNA inhibition by CD9-HuR functionalized exosomes encapsulated with miRNA or CRISPR/dCas9. *Nano Letters, 19*(1), 19–28. https://doi.org/10.1021/acs.nanolett.8b02689. https://www.ncbi.nlm.nih.gov/pubmed/30517011

Liangsupree, T., Multia, E., & Riekkola, M. L. (2021). Modern isolation and separation techniques for extracellular vesicles. *Journal of Chromatography. A, 1636*, 461773. https://doi.org/10.1016/j.chroma.2020.461773. https://www.ncbi.nlm.nih.gov/pubmed/33316564

Liao, K., Niu, F., Dagur, R. S., He, M., Tian, C., & Hu, G. (2020). Intranasal delivery of lincRNA-Cox2 siRNA loaded extracellular vesicles decreases lipopolysaccharide-induced microglial proliferation in mice. *Journal of Neuroimmune Pharmacology, 15*(3), 390–399. https://doi.org/10.1007/s11481-019-09864-z. https://www.ncbi.nlm.nih.gov/pubmed/31325121

Lim, G. T., You, D. G., Han, H. S., Lee, H., Shin, S., Oh, B. H., Kumar, E. K. P., Um, W., Kim, C. H., Han, S., Lee, S., Lim, S., Yoon, H. Y., Kim, K., Kwon, I. C., Jo, D. G., Cho, Y. W., & Park, J. H. (2021). Bioorthogonally surface-edited extracellular vesicles based on metabolic glycoengineering for CD44-mediated targeting of inflammatory diseases. *Journal of Extracellular Vesicles, 10*(5), e12077. https://doi.org/10.1002/jev2.12077. https://www.ncbi.nlm.nih.gov/pubmed/33738083

Liu, Y., Li, D., Liu, Z., Zhou, Y., Chu, D., Li, X., Jiang, X., Hou, D., Chen, X., Chen, Y., Yang, Z., Jin, L., Jiang, W., Tian, C., Zhou, G., Zen, K., Zhang, J., Zhang, Y., Li, J., & Zhang, C. Y. (2015). Targeted exosome-mediated delivery of opioid receptor Mu siRNA for the treatment of morphine relapse. *Scientific Reports, 5*, 17543. https://doi.org/10.1038/srep17543. https://www.ncbi.nlm.nih.gov/pubmed/26633001

Liu, B., Lee, B. W., Nakanishi, K., Villasante, A., Williamson, R., Metz, J., Kim, J., Kanai, M., Bi, L., Brown, K., Di Paolo, G., Homma, S., Sims, P. A., Topkara, V. K., & Vunjak-Novakovic, G. (2018). Cardiac recovery via extended cell-free delivery of extracellular vesicles secreted by cardiomyocytes derived from induced pluripotent stem cells. *Nature Biomedical Engineering, 2*(5), 293–303. https://doi.org/10.1038/s41551-018-0229-7. https://www.ncbi.nlm.nih.gov/pubmed/30271672

Liu, J., Qiu, X., Lv, Y., Zheng, C., Dong, Y., Dou, G., Zhu, B., Liu, A., Wang, W., Zhou, J., Liu, S., Liu, S., Gao, B., & Jin, Y. (2020). Apoptotic bodies derived from mesenchymal stem cells promote cutaneous wound healing via regulating the functions of macrophages. *Stem Cell Research & Therapy, 11*(1), 507. https://doi.org/10.1186/s13287-020-02014-w. https://www.ncbi.nlm.nih.gov/pubmed/33246491

Lo, T. W., Zhu, Z., Purcell, E., Watza, D., Wang, J., Kang, Y. T., Jolly, S., Nagrath, D., & Nagrath, S. (2020). Microfluidic device for high-throughput affinity-based isolation of extracellular vesicles. *Lab on a Chip, 20*(10), 1762–1770. https://doi.org/10.1039/c9lc01190k. https://www.ncbi.nlm.nih.gov/pubmed/32338266

Loch-Neckel, G., Matos, A. T., Vaz, A. R., & Brites, D. (2022). Challenges in the development of drug delivery systems based on small extracellular vesicles for therapy of brain diseases. *Frontiers in Pharmacology, 13*, 839790. https://doi.org/10.3389/fphar.2022.839790. https://www.ncbi.nlm.nih.gov/pubmed/35422699

Lu, Z., Zuo, B., Jing, R., Gao, X., Rao, Q., Liu, Z., Qi, H., Guo, H., & Yin, H. (2017). Dendritic cell-derived exosomes elicit tumor regression in autochthonous hepatocellular carcinoma mouse models. *Journal of Hepatology, 67*(4), 739–748. https://doi.org/10.1016/j.jhep.2017.05.019. https://www.ncbi.nlm.nih.gov/pubmed/28549917

Ma, R., Kutchy, N. A., & Hu, G. (2021). Astrocyte-derived extracellular vesicle-mediated activation of primary ciliary signaling contributes to the development of morphine tolerance. *Biological Psychiatry, 90*(8), 575–585. https://doi.org/10.1016/j.biopsych.2021.06.009. https://www.ncbi.nlm.nih.gov/pubmed/34417054

Ma, R., Chen, L., Hu, N., Caplan, S., & Hu, G. (2023a). Cilia and extracellular vesicles in brain development and disease. *Biological Psychiatry, 95*(11), 1020–1029. https://doi.org/10.1016/j.biopsych.2023.11.004. https://www.ncbi.nlm.nih.gov/pubmed/37956781

Ma, R., Kutchy, N. A., Wang, Z., & Hu, G. (2023b). Extracellular vesicle-mediated delivery of anti-miR-106b inhibits morphine-induced primary ciliogenesis in the brain. *Molecular Therapy, 31*(5), 1332–1345. https://doi.org/10.1016/j.ymthe.2023.03.030. https://www.ncbi.nlm.nih.gov/pubmed/37012704

Maas, S. L., de Vrij, J., van der Vlist, E. J., Geragousian, B., van Bloois, L., Mastrobattista, E., Schiffelers, R. M., Wauben, M. H., Broekman, M. L., & Nolte-'t Hoen, E. N. (2015). Possibilities and limitations of current technologies for quantification of biological extracellular vesicles and synthetic mimics. *Journal of Controlled Release, 200*, 87–96. https://doi.org/10.1016/j.jconrel.2014.12.041. https://www.ncbi.nlm.nih.gov/pubmed/25555362

Marar, C., Starich, B., & Wirtz, D. (2021). Extracellular vesicles in immunomodulation and tumor progression. *Nature Immunology, 22*(5), 560–570. https://doi.org/10.1038/s41590-021-00899-0. https://www.ncbi.nlm.nih.gov/pubmed/33753940

Mardpour, S., Ghanian, M. H., Sadeghi-Abandansari, H., Mardpour, S., Nazari, A., Shekari, F., & Baharvand, H. (2019). Hydrogel-mediated sustained systemic delivery of mesenchymal stem cell-derived extracellular vesicles improves hepatic regeneration in chronic liver failure. *ACS Applied Materials & Interfaces, 11*(41), 37421–37433. https://doi.org/10.1021/acsami.9b10126. https://www.ncbi.nlm.nih.gov/pubmed/31525863

Martinez-Arroyo, O., Ortega, A., Forner, M. J., & Cortes, R. (2022). Mesenchymal stem cell-derived extracellular vesicles as non-coding RNA therapeutic vehicles in autoimmune diseases. *Pharmaceutics, 14*(4), 733. https://doi.org/10.3390/pharmaceutics14040733. https://www.ncbi.nlm.nih.gov/pubmed/35456567

Murphy, D. E., de Jong, O. G., Brouwer, M., Wood, M. J., Lavieu, G., Schiffelers, R. M., & Vader, P. (2019). Extracellular vesicle-based therapeutics: Natural versus engineered targeting and trafficking. *Experimental & Molecular Medicine, 51*(3), 1–12. https://doi.org/10.1038/s12276-019-0223-5. https://www.ncbi.nlm.nih.gov/pubmed/30872574

Nagelkerke, A., Ojansivu, M., van der Koog, L., Whittaker, T. E., Cunnane, E. M., Silva, A. M., Dekker, N., & Stevens, M. M. (2021). Extracellular vesicles for tissue repair and regeneration: Evidence, challenges and opportunities. *Advanced Drug Delivery Reviews, 175*, 113775. https://doi.org/10.1016/j.addr.2021.04.013. https://www.ncbi.nlm.nih.gov/pubmed/33872693

O'Brien, K., Breyne, K., Ughetto, S., Laurent, L. C., & Breakefield, X. O. (2020). RNA delivery by extracellular vesicles in mammalian cells and its applications. *Nature Reviews. Molecular Cell Biology, 21*(10), 585–606. https://doi.org/10.1038/s41580-020-0251-y. https://www.ncbi.nlm.nih.gov/pubmed/32457507

O'Grady, T., Njock, M. S., Lion, M., Bruyr, J., Mariavelle, E., Galvan, B., Boeckx, A., Struman, I., & Dequiedt, F. (2022). Sorting and packaging of RNA into extracellular vesicles shape intracellular transcript levels. *BMC Biology, 20*(1), 72. https://doi.org/10.1186/s12915--022-01277-4. https://www.ncbi.nlm.nih.gov/pubmed/35331218

Ohno, S., Takanashi, M., Sudo, K., Ueda, S., Ishikawa, A., Matsuyama, N., Fujita, K., Mizutani, T., Ohgi, T., Ochiya, T., Gotoh, N., & Kuroda, M. (2013). Systemically injected exosomes targeted to EGFR deliver antitumor microRNA to breast cancer cells. *Molecular Therapy, 21*(1), 185–191. https://doi.org/10.1038/mt.2012.180. https://www.ncbi.nlm.nih.gov/pubmed/23032975

Pachler, K., Lener, T., Streif, D., Dunai, Z. A., Desgeorges, A., Feichtner, M., Oller, M., Schallmoser, K., Rohde, E., & Gimona, M. (2017). A Good Manufacturing Practice-grade standard protocol for exclusively human mesenchymal stromal cell-derived extracellular vesicles. *Cytotherapy, 19*(4), 458–472. https://doi.org/10.1016/j.jcyt.2017.01.001. https://www.ncbi.nlm.nih.gov/pubmed/28188071

Pan, B. T., & Johnstone, R. M. (1983). Fate of the transferrin receptor during maturation of sheep reticulocytes in vitro: Selective externalization of the receptor. *Cell, 33*(3), 967–978. https://doi.org/10.1016/0092-8674(83)90040-5. https://www.ncbi.nlm.nih.gov/pubmed/6307529

Pan, B. T., Teng, K., Wu, C., Adam, M., & Johnstone, R. M. (1985). Electron microscopic evidence for externalization of the transferrin receptor in vesicular form in sheep reticulocytes. *The Journal of Cell Biology, 101*(3), 942–948. https://doi.org/10.1083/jcb.101.3.942. https://www.ncbi.nlm.nih.gov/pubmed/2993317

Park, K. S., Svennerholm, K., Shelke, G. V., Bandeira, E., Lasser, C., Jang, S. C., Chandode, R., Gribonika, I., & Lotvall, J. (2019). Mesenchymal stromal cell-derived nanovesicles ameliorate bacterial outer membrane vesicle-induced sepsis via IL-10. *Stem Cell Research & Therapy, 10*(1), 231. https://doi.org/10.1186/s13287--019-1352-4. https://www.ncbi.nlm.nih.gov/pubmed/31370884

Patel, S., Schmidt, K. F., Farhoud, M., Zi, T., Jang, S. C., Dooley, K., Kentala, D., Dobson, H., Economides, K., & Williams, D. E. (2022). In vivo tracking of [(89)Zr]Zr-labeled engineered extracellular vesicles by PET reveals organ-specific biodistribution based upon the route of administration. *Nuclear Medicine and Biology, 112–113*, 20–30. https://doi.org/10.1016/j.nucmedbio.2022.06.004. https://www.ncbi.nlm.nih.gov/pubmed/35763877

Pavlyukov, M. S., Yu, H., Bastola, S., Minata, M., Shender, V. O., Lee, Y., Zhang, S., Wang, J., Komarova, S., Wang, J., Yamaguchi, S., Alsheikh, H. A., Shi, J., Chen, D., Mohyeldin, A., Kim, S. H., Shin, Y. J., Anufrieva, K., Evtushenko, E. G., Antipova, N. V., Arapidi, G. P., Govorun, V., Pestov, N. B., Shakhparonov, M. I., Lee, L. J., Nam, D. H., & Nakano, I. (2018). Apoptotic cell-derived extracellular vesicles promote malignancy of glioblastoma via intercellular transfer of splicing factors. *Cancer Cell, 34*(1), 119–135.e10. https://doi.org/10.1016/j.ccell.2018.05.012. https://www.ncbi.nlm.nih.gov/pubmed/29937354

Pegtel, D. M., & Gould, S. J. (2019). Exosomes. *Annual Review of Biochemistry, 88*, 487–514. https://doi.org/10.1146/annurev--biochem-013118-111902. https://www.ncbi.nlm.nih.gov/pubmed/31220978

Ramirez, M. I., Amorim, M. G., Gadelha, C., Milic, I., Welsh, J. A., Freitas, V. M., Nawaz, M., Akbar, N., Couch, Y., Makin, L., Cooke, F., Vettore, A. L., Batista, P. X., Freezor, R., Pezuk, J. A., Rosa-Fernandes, L., Carreira, A. C. O., Devitt, A., Jacobs, L., Silva, I. T., Coakley, G., Nunes, D. N., Carter, D., Palmisano, G., & Dias-Neto, E. (2018). Technical challenges of working with extracellular vesicles. *Nanoscale, 10*(3), 881–906. https://doi.org/10.1039/c7nr08360b. https://www.ncbi.nlm.nih.gov/pubmed/29265147

Rong, Y., Liu, W., Wang, J., Fan, J., Luo, Y., Li, L., Kong, F., Chen, J., Tang, P., & Cai, W. (2019). Neural stem cell-derived small extracellular vesicles attenuate apoptosis and neuroinflammation after traumatic spinal cord injury by activating autophagy. *Cell Death & Disease, 10*(5), 340. https://doi.org/10.1038/s41419-019-1571-8. https://www.ncbi.nlm.nih.gov/pubmed/31000697

Roy, S., Hochberg, F. H., & Jones, P. S. (2018). Extracellular vesicles: The growth as diagnostics and therapeutics; a survey. *Journal of Extracellular Vesicles, 7*(1), 1438720. https://doi.org/10.1080/20013078.2018.1438720. https://www.ncbi.nlm.nih.gov/pubmed/29511461

Salunkhe, S., Dheeraj, Basak, M., Chitkara, D., & Mittal, A. (2020). Surface functionalization of exosomes for target-specific delivery and in vivo imaging & tracking: Strategies and significance. *Journal of Controlled Release, 326*, 599–614. https://doi.org/10.1016/j.jconrel.2020.07.042. https://www.ncbi.nlm.nih.gov/pubmed/32730952

Sato, Y. T., Umezaki, K., Sawada, S., Mukai, S. A., Sasaki, Y., Harada, N., Shiku, H., & Akiyoshi, K. (2016). Engineering hybrid exosomes by membrane fusion with liposomes. *Scientific Reports, 6*, 21933. https://doi.org/10.1038/srep21933. https://www.ncbi.nlm.nih.gov/pubmed/26911358

Shen, Q., Huang, Z., Yao, J., & Jin, Y. (2022). Extracellular vesicles-mediated interaction within intestinal microenvironment in inflammatory bowel disease. *Journal of Advanced Research, 37*, 221–233. https://doi.org/10.1016/j.jare.2021.07.002. https://www.ncbi.nlm.nih.gov/pubmed/35499059

Shi, Y., Guo, S., Liang, Y., Liu, L., Wang, A., Sun, K., & Li, Y. (2022). Construction and evaluation of liraglutide delivery system based on milk exosomes: A new idea for oral peptide delivery. *Current Pharmaceutical Biotechnology, 23*(8), 1072–1079. https://doi.org/10.2174/1389201022666210820114236. https://www.ncbi.nlm.nih.gov/pubmed/34414872

Shirejini, S. Z., & Inci, F. (2022). The Yin and Yang of exosome isolation methods: Conventional practice, microfluidics, and commercial kits. *Biotechnology Advances, 54*, 107814. https://doi.org/10.1016/j.biotechadv.2021.107814. https://www.ncbi.nlm.nih.gov/pubmed/34389465

Shukla, L., Yuan, Y., Shayan, R., Greening, D. W., & Karnezis, T. (2020). Fat therapeutics: The clinical capacity of adipose-derived stem cells and exosomes for human disease and tissue regeneration. *Frontiers in Pharmacology, 11*, 158. https://doi.org/10.3389/fphar.2020.00158. https://www.ncbi.nlm.nih.gov/pubmed/32194404

Sidhom, K., Obi, P. O., & Saleem, A. (2020). A review of exosomal isolation methods: Is size exclusion chromatography the best option? *International Journal of Molecular Sciences, 21*(18), 6466. https://doi.org/10.3390/ijms21186466. https://www.ncbi.nlm.nih.gov/pubmed/32899828

Sil, S., Dagur, R. S., Liao, K., Peeples, E. S., Hu, G., Periyasamy, P., & Buch, S. (2020). Strategies for the use of extracellular vesicles for the delivery of therapeutics. *Journal of Neuroimmune Pharmacology, 15*(3), 422–442. https://doi.org/10.1007/s11481--019-09873-y. https://www.ncbi.nlm.nih.gov/pubmed/31456107

Silva, A. K. A., Morille, M., Piffoux, M., Arumugam, S., Mauduit, P., Larghero, J., Bianchi, A., Aubertin, K., Blanc-Brude, O., Noel, D., Velot, E., Ravel, C., Elie-Caille, C., Sebbagh, A., Boulanger, C., Wilhelm, C., Rahmi, G., Raymond-Letron, I., Cherukula, K., Montier, T., Martinaud, C., Bach, J. M., Favre-Bulle, O., Spadavecchia, J., Jorgensen, C., Menasche, P., Aussel, C., Chopineau, J., Mosser, M., Ullah, M., Sailliet, N., Luciani, N., Mathieu, N., Rautou, P. E., Brouard, S., Boireau, W., Jauliac, S., Dedier, M., Trouvin, J. H., Gazeau, F., Trouillas, M., Peltzer, J., Monsel, A., & Banzet, S. (2021). Development of extracellular vesicle-based medicinal products: A position paper of the group "Extracellular Vesicle translatiOn to clinicaL perspectiVEs – EVOLVE France". *Advanced Drug Delivery Reviews, 179*, 114001. https://doi.org/10.1016/j.addr.2021.114001. https://www.ncbi.nlm.nih.gov/pubmed/34673131

Smalley, J. W., & Birss, A. J. (1987). Trypsin-like enzyme activity of the extracellular membrane vesicles of Bacteroides gingivalis W50. *Journal of General Microbiology, 133*(10), 2883–2894. https://doi.org/10.1099/00221287-133-10-2883. https://www.ncbi.nlm.nih.gov/pubmed/3449601

Stahl, P. D., & Raposo, G. (2019). Extracellular vesicles: Exosomes and microvesicles, integrators of homeostasis. *Physiology (Bethesda), 34*(3), 169–177. https://doi.org/10.1152/physiol.00045.2018. https://www.ncbi.nlm.nih.gov/pubmed/30968753

Stremersch, S., De Smedt, S. C., & Raemdonck, K. (2016). Therapeutic and diagnostic applications of extracellular vesicles. *Journal of Controlled Release, 244*(Pt B), 167–183. https://doi.org/10.1016/j.jconrel.2016.07.054. https://www.ncbi.nlm.nih.gov/pubmed/27491882

Sun, M. K., Passaro, A. P., Latchoumane, C. F., Spellicy, S. E., Bowler, M., Goeden, M., Martin, W. J., Holmes, P. V., Stice, S. L., & Karumbaiah, L. (2020). Extracellular vesicles mediate neuroprotection and functional recovery after traumatic brain injury. *Journal of Neurotrauma, 37*(11), 1358–1369. https://doi.org/10.1089/neu.2019.6443. https://www.ncbi.nlm.nih.gov/pubmed/31774030

Sun, N., Tran, B. V., Peng, Z., Wang, J., Zhang, C., Yang, P., Zhang, T. X., Widjaja, J., Zhang, R. Y., Xia, W., Keir, A., She, J. W., Yu, H. H., Shyue, J. J., Zhu, H., Agopian, V. G., Pei, R., Tomlinson, J. S., Toretsky, J. A., Jonas, S. J., Federman, N., Lu, S., Tseng, H. R., & Zhu, Y. (2022). Coupling lipid labeling and click chemistry enables isolation of extracellular vesicles for noninvasive detection of oncogenic gene alterations. *Advanced Science, 9*(14), e2105853. https://doi.org/10.1002/advs.202105853. https://www.ncbi.nlm.nih.gov/pubmed/35486030

Szatanek, R., Baran, J., Siedlar, M., & Baj-Krzyworzeka, M. (2015). Isolation of extracellular vesicles: Determining the correct approach (Review). *International Journal of Molecular Medicine, 36*(1), 11–17. https://doi.org/10.3892/ijmm.2015.2194. https://www.ncbi.nlm.nih.gov/pubmed/25902369

Thery, C., Witwer, K. W., Aikawa, E., Alcaraz, M. J., Anderson, J. D., Andriantsitohaina, R., Antoniou, A., Arab, T., Archer, F., Atkin-Smith, G. K., Ayre, D. C., Bach, J. M., Bachurski, D., Baharvand, H., Balaj, L., Baldacchino, S., Bauer, N. N., Baxter, A. A., Bebawy, M., Beckham, C., Bedina Zavec, A., Benmoussa, A., Berardi, A. C., Bergese, P., Bielska, E., Blenkiron, C., Bobis-Wozowicz, S., Boilard, E., Boireau, W., Bongiovanni, A., Borras, F. E., Bosch, S., Boulanger, C. M., Breakefield, X., Breglio, A. M., Brennan, M. A., Brigstock, D. R., Brisson, A., Broekman, M. L., Bromberg, J. F., Bryl-Gorecka, P., Buch, S., Buck, A. H., Burger, D., Busatto, S., Buschmann, D., Bussolati, B., Buzas, E. I., Byrd, J. B., Camussi, G., Carter, D. R., Caruso, S., Chamley, L. W., Chang, Y. T., Chen, C., Chen, S., Cheng, L., Chin, A. R., Clayton, A., Clerici, S. P., Cocks, A., Cocucci, E., Coffey, R. J., Cordeiro-da-Silva, A., Couch, Y., Coumans, F. A., Coyle, B., Crescitelli, R., Criado, M. F., D'Souza-Schorey, C., Das, S., Datta Chaudhuri, A., de Candia, P., De Santana, E. F., De Wever, O., Del Portillo, H. A., Demaret, T., Deville, S., Devitt, A., Dhondt, B., Di Vizio, D., Dieterich, L. C., Dolo, V., Dominguez Rubio, A. P., Dominici, M., Dourado, M. R., Driedonks, T. A., Duarte, F. V., Duncan, H. M., Eichenberger, R. M., Ekstrom, K., El Andaloussi, S., Elie-Caille, C., Erdbrugger, U., Falcon-Perez, J. M., Fatima, F., Fish, J. E., Flores-Bellver, M., Forsonits, A., Frelet-Barrand, A., Fricke, F., Fuhrmann, G., Gabrielsson, S.,

Gamez-Valero, A., Gardiner, C., Gartner, K., Gaudin, R., Gho, Y. S., Giebel, B., Gilbert, C., Gimona, M., Giusti, I., Goberdhan, D. C., Gorgens, A., Gorski, S. M., Greening, D. W., Gross, J. C., Gualerzi, A., Gupta, G. N., Gustafson, D., Handberg, A., Haraszti, R. A., Harrison, P., Hegyesi, H., Hendrix, A., Hill, A. F., Hochberg, F. H., Hoffmann, K. F., Holder, B., Holthofer, H., Hosseinkhani, B., Hu, G., Huang, Y., Huber, V., Hunt, S., Ibrahim, A. G., Ikezu, T., Inal, J. M., Isin, M., Ivanova, A., Jackson, H. K., Jacobsen, S., Jay, S. M., Jayachandran, M., Jenster, G., Jiang, L., Johnson, S. M., Jones, J. C., Jong, A., Jovanovic-Talisman, T., Jung, S., Kalluri, R., Kano, S. I., Kaur, S., Kawamura, Y., Keller, E. T., Khamari, D., Khomyakova, E., Khvorova, A., Kierulf, P., Kim, K. P., Kislinger, T., Klingeborn, M., Klinke, D. J., 2nd, Kornek, M., Kosanovic, M. M., Kovacs, A. F., Kramer-Albers, E. M., Krasemann, S., Krause, M., Kurochkin, I. V., Kusuma, G. D., Kuypers, S., Laitinen, S., Langevin, S. M., Languino, L. R., Lannigan, J., Lasser, C., Laurent, L. C., Lavieu, G., Lazaro-Ibanez, E., Le Lay, S., Lee, M. S., Lee, Y. X. F., Lemos, D. S., Lenassi, M., Leszczynska, A., Li, I. T., Liao, K., Libregts, S. F., Ligeti, E., Lim, R., Lim, S. K., Line, A., Linnemannstons, K., Llorente, A., Lombard, C. A., Lorenowicz, M. J., Lorincz, A. M., Lotvall, J., Lovett, J., Lowry, M. C., Loyer, X., Lu, Q., Lukomska, B., Lunavat, T. R., Maas, S. L., Malhi, H., Marcilla, A., Mariani, J., Mariscal, J., Martens-Uzunova, E. S., Martin-Jaular, L., Martinez, M. C., Martins, V. R., Mathieu, M., Mathivanan, S., Maugeri, M., McGinnis, L. K., McVey, M. J., Meckes, D. G., Jr., Meehan, K. L., Mertens, I., Minciacchi, V. R., Moller, A., Moller Jorgensen, M., Morales-Kastresana, A., Morhayim, J., Mullier, F., Muraca, M., Musante, L., Mussack, V., Muth, D. C., Myburgh, K. H., Najrana, T., Nawaz, M., Nazarenko, I., Nejsum, P., Neri, C., Neri, T., Nieuwland, R., Nimrichter, L., Nolan, J. P., Nolte-'t Hoen, E. N., Noren Hooten, N., O'Driscoll, L., O'Grady, T., O'Loghlen, A., Ochiya, T., Olivier, M., Ortiz, A., Ortiz, L. A., Osteikoetxea, X., Ostergaard, O., Ostrowski, M., Park, J., Pegtel, D. M., Peinado, H., Perut, F., Pfaffl, M. W., Phinney, D. G., Pieters, B. C., Pink, R. C., Pisetsky, D. S., Pogge von Strandmann, E., Polakovicova, I., Poon, I. K., Powell, B. H., Prada, I., Pulliam, L., Quesenberry, P., Radeghieri, A., Raffai, R. L., Raimondo, S., Rak, J., Ramirez, M. I., Raposo, G., Rayyan, M. S., Regev-Rudzki, N., Ricklefs, F. L., Robbins, P. D., Roberts, D. D., Rodrigues, S. C., Rohde, E., Rome, S., Rouschop, K. M., Rughetti, A., Russell, A. E., Saa, P., Sahoo, S., Salas-Huenuleo, E., Sanchez, C., Saugstad, J. A., Saul, M. J., Schiffelers, R. M., Schneider, R., Schoyen, T. H., Scott, A., Shahaj, E., Sharma, S., Shatnyeva, O., Shekari, F., Shelke, G. V., Shetty, A. K., Shiba, K., Siljander, P. R., Silva, A. M., Skowronek, A., Snyder, O. L., 2nd, Soares, R. P., Sodar, B. W., Soekmadji, C., Sotillo, J., Stahl, P. D., Stoorvogel, W., Stott, S. L., Strasser, E. F., Swift, S., Tahara, H., Tewari, M., Timms, K., Tiwari, S., Tixeira, R., Tkach, M., Toh, W. S., Tomasini, R., Torrecilhas, A. C., Tosar, J. P., Toxavidis, V., Urbanelli, L., Vader, P., van Balkom, B. W., van der Grein, S. G., Van Deun, J., van Herwijnen, M. J., Van Keuren-Jensen, K., van Niel, G., van Royen, M. E., van Wijnen, A. J., Vasconcelos, M. H., Vechetti, I. J., Jr., Veit, T. D., Vella, L. J., Velot, E., Verweij, F. J., Vestad, B., Vinas, J. L., Visnovitz, T., Vukman, K. V., Wahlgren, J., Watson, D. C., Wauben, M. H., Weaver, A., Webber, J. P., Weber, V., Wehman, A. M., Weiss, D. J., Welsh, J. A., Wendt, S., Wheelock, A. M., Wiener, Z., Witte, L., Wolfram, J., Xagorari, A., Xander, P., Xu, J., Yan, X., Yanez-Mo, M., Yin, H., Yuana, Y., Zappulli, V., Zarubova, J., Zekas, V., Zhang, J. Y., Zhao, Z., Zheng, L., Zheutlin, A. R., Zickler, A. M., Zimmermann, P., Zivkovic, A. M., Zocco, D., & Zuba-Surma, E. K. (2018). Minimal information for studies of extracellular vesicles 2018 (MISEV2018): A position statement of the International Society for Extracellular Vesicles and update of the MISEV2014 guidelines. *Journal of Extracellular Vesicles, 7*(1), 1535750. https://doi.org/10.1080/20013078.2018.1535750. https://www.ncbi.nlm.nih.gov/pubmed/30637094

Tian, Y., Li, S., Song, J., Ji, T., Zhu, M., Anderson, G. J., Wei, J., & Nie, G. (2014). A doxorubicin delivery platform using engineered natural membrane vesicle exosomes for targeted tumor therapy. *Biomaterials, 35*(7), 2383–2390. https://doi.org/10.1016/j.biomaterials.2013.11.083. https://www.ncbi.nlm.nih.gov/pubmed/24345736

Tian, Y., Gong, M., Hu, Y., Liu, H., Zhang, W., Zhang, M., Hu, X., Aubert, D., Zhu, S., Wu, L., & Yan, X. (2020). Quality and efficiency assessment of six extracellular vesicle isolation methods by nano-flow cytometry. *Journal of Extracellular Vesicles, 9*(1), 1697028. https://doi.org/10.1080/20013078.2019.1697028. https://www.ncbi.nlm.nih.gov/pubmed/31839906

Vader, P., Breakefield, X. O., & Wood, M. J. (2014). Extracellular vesicles: Emerging targets for cancer therapy. *Trends in Molecular Medicine, 20*(7), 385–393. https://doi.org/10.1016/j.molmed.2014.03.002. https://www.ncbi.nlm.nih.gov/pubmed/24703619

Vestad, B., Llorente, A., Neurauter, A., Phuyal, S., Kierulf, B., Kierulf, P., Skotland, T., Sandvig, K., Haug, K. B. F., & Ovstebo, R. (2017). Size and concentration analyses of extracellular vesicles by nanoparticle tracking analysis: A variation study. *Journal of Extracellular Vesicles, 6*(1), 1344087. https://doi.org/10.1080/20013078.2017.1344087. https://www.ncbi.nlm.nih.gov/pubmed/28804597

Vogel, A. D., Upadhya, R., & Shetty, A. K. (2018). Neural stem cell derived extracellular vesicles: Attributes and prospects for treating neurodegenerative disorders. *eBioMedicine, 38*, 273–282. https://doi.org/10.1016/j.ebiom.2018.11.026. https://www.ncbi.nlm.nih.gov/pubmed/30472088

Walker, S., Busatto, S., Pham, A., Tian, M., Suh, A., Carson, K., Quintero, A., Lafrence, M., Malik, H., Santana, M. X., & Wolfram, J. (2019). Extracellular vesicle-based drug delivery systems for cancer treatment. *Theranostics, 9*(26), 8001–8017. https://doi.org/10.7150/thno.37097. https://www.ncbi.nlm.nih.gov/pubmed/31754377

Wan, T., Zhong, J., Pan, Q., Zhou, T., Ping, Y., & Liu, X. (2022). Exosome-mediated delivery of Cas9 ribonucleoprotein complexes for tissue-specific gene therapy of liver diseases. *Science Advances, 8*(37), eabp9435. https://doi.org/10.1126/sciadv.abp9435. https://www.ncbi.nlm.nih.gov/pubmed/36103526

Wang, H., Sui, H., Zheng, Y., Jiang, Y., Shi, Y., Liang, J., & Zhao, L. (2019a). Curcumin-primed exosomes potently ameliorate cognitive function in AD mice by inhibiting hyperphosphorylation of the Tau protein through the AKT/GSK-3beta pathway. *Nanoscale, 11*(15), 7481–7496. https://doi.org/10.1039/c9nr01255a. https://www.ncbi.nlm.nih.gov/pubmed/30942233

Wang, X., Jiao, Y., Pan, Y., Zhang, L., Gong, H., Qi, Y., Wang, M., Gong, H., Shao, M., Wang, X., & Jiang, D. (2019b). Fetal dermal mesenchymal stem cell-derived exosomes accelerate cutaneous wound healing by activating notch signaling. *Stem Cells International, 2019*, 2402916. https://doi.org/10.1155/2019/2402916. https://www.ncbi.nlm.nih.gov/pubmed/31281370

Webb, R. L., Kaiser, E. E., Scoville, S. L., Thompson, T. A., Fatima, S., Pandya, C., Sriram, K., Swetenburg, R. L., Vaibhav, K., Arbab, A. S., Baban, B., Dhandapani, K. M., Hess, D. C., Hoda, M. N., & Stice, S. L. (2018). Human neural stem cell extracellular vesicles improve tissue and functional recovery in the murine thromboembolic stroke model. *Translational Stroke Research, 9*(5), 530–539. https://doi.org/10.1007/s12975-017-0599-2. https://www.ncbi.nlm.nih.gov/pubmed/29285679

Weibull, C., Mohri, T., & Afzelius, B. A. (1965). The morphology and fine structure of small particles present in cultures of a proteus L form. *Journal of Ultrastructure Research, 12*, 81–91. https://doi.org/10.1016/s0022-5320(65)80007-7. https://www.ncbi.nlm.nih.gov/pubmed/14289433

Willms, E., Cabanas, C., Mager, I., Wood, M. J. A., & Vader, P. (2018). Extracellular vesicle heterogeneity: Subpopulations, isolation techniques, and diverse functions in cancer progression. *Frontiers in Immunology, 9*, 738. https://doi.org/10.3389/fimmu.2018.00738. https://www.ncbi.nlm.nih.gov/pubmed/29760691

Witwer, K. W., Buzas, E. I., Bemis, L. T., Bora, A., Lasser, C., Lotvall, J., Nolte-'t Hoen, E. N., Piper, M. G., Sivaraman, S., Skog, J., Thery, C., Wauben, M. H., & Hochberg, F. (2013). Standardization of sample collection, isolation and analysis methods in extracellular vesicle research. *Journal of Extracellular Vesicles, 2*, 20360. https://doi.org/10.3402/jev.v2i0.20360. https://www.ncbi.nlm.nih.gov/pubmed/24009894

Wu, P., Zhang, B., Ocansey, D. K. W., Xu, W., & Qian, H. (2021). Extracellular vesicles: A bright star of nanomedicine. *Biomaterials, 269*, 120467. https://doi.org/10.1016/j.biomaterials.2020.120467. https://www.ncbi.nlm.nih.gov/pubmed/33189359

Xie, F., Zhou, X., Fang, M., Li, H., Su, P., Tu, Y., Zhang, L., & Zhou, F. (2019). Extracellular vesicles in cancer immune microenvironment and cancer immunotherapy. *Advanced Science, 6*(24), 1901779. https://doi.org/10.1002/advs.201901779. https://www.ncbi.nlm.nih.gov/pubmed/31871860

Yang, J., Zhang, X., Chen, X., Wang, L., & Yang, G. (2017). Exosome mediated delivery of miR-124 promotes neurogenesis after ischemia. *Molecular Therapy – Nucleic Acids, 7*, 278–287. https://doi.org/10.1016/j.omtn.2017.04.010. https://www.ncbi.nlm.nih.gov/pubmed/28624203

Yang, P., Peng, Y., Feng, Y., Xu, Z., Feng, P., Cao, J., Chen, Y., Chen, X., Cao, X., Yang, Y., & Jie, J. (2021). Immune cell-derived extracellular vesicles – New strategies in cancer immunotherapy. *Frontiers in Immunology, 12*, 771551. https://doi.org/10.3389/fimmu.2021.771551. https://www.ncbi.nlm.nih.gov/pubmed/34956197

Yerneni, S. S., Yalcintas, E. P., Smith, J. D., Averick, S., Campbell, P. G., & Ozdoganlar, O. B. (2022). Skin-targeted delivery of extracellular vesicle-encapsulated curcumin using dissolvable microneedle arrays. *Acta Biomaterialia, 149*, 198–212. https://doi.org/10.1016/j.actbio.2022.06.046. https://www.ncbi.nlm.nih.gov/pubmed/35809788

Yu, Y., Li, W., Mao, L., Peng, W., Long, D., Li, D., Zhou, R., & Dang, X. (2021). Genetically engineered exosomes display RVG peptide and selectively enrich a neprilysin variant: A potential formulation for the treatment of Alzheimer's disease. *Journal of Drug Targeting, 29*(10), 1128–1138. https://doi.org/10.1080/1061186X.2021.1929257. https://www.ncbi.nlm.nih.gov/pubmed/34182845

Yu, X., Bai, Y., Han, B., Ju, M., Tang, T., Shen, L., Li, M., Yang, L., Zhang, Z., Hu, G., Chao, J., Zhang, Y., & Yao, H. (2022). Extracellular vesicle-mediated delivery of circDYM alleviates CUS-induced depressive-like behaviours. *Journal of Extracellular Vesicles, 11*(1), e12185. https://doi.org/10.1002/jev2.12185. https://www.ncbi.nlm.nih.gov/pubmed/35029057

Yuyama, K., Sun, H., Sakai, S., Mitsutake, S., Okada, M., Tahara, H., Furukawa, J., Fujitani, N., Shinohara, Y., & Igarashi, Y. (2014). Decreased amyloid-beta pathologies by intracerebral loading of glycosphingolipid-enriched exosomes in Alzheimer model mice. *The Journal of Biological Chemistry, 289*(35), 24488–24498. https://doi.org/10.1074/jbc.M114.577213. https://www.ncbi.nlm.nih.gov/pubmed/25037226

Zaborowski, M. P., Balaj, L., Breakefield, X. O., & Lai, C. P. (2015). Extracellular vesicles: Composition, biological relevance, and methods of study. *Bioscience, 65*(8), 783–797. https://doi.org/10.1093/biosci/biv084. https://www.ncbi.nlm.nih.gov/pubmed/26955082

Zhang, K., Yu, L., Li, F. R., Li, X., Wang, Z., Zou, X., Zhang, C., Lv, K., Zhou, B., Mitragotri, S., & Chen, M. (2020a). Topical application of exosomes derived from human umbilical cord mesenchymal stem cells in combination with sponge spicules for treatment of photoaging. *International Journal of Nanomedicine, 15*, 2859–2872. https://doi.org/10.2147/IJN.S249751. https://www.ncbi.nlm.nih.gov/pubmed/32368058

Zhang, X., Xu, Q., Zi, Z., Liu, Z., Wan, C., Crisman, L., Shen, J., & Liu, X. (2020b). Programmable extracellular vesicles for macromolecule delivery and genome modifications. *Developmental Cell, 55*(6), 784–801.e9. https://doi.org/10.1016/j.devcel.2020.11.007. https://www.ncbi.nlm.nih.gov/pubmed/33296682

Zhao, Z., McGill, J., Gamero-Kubota, P., & He, M. (2019). Microfluidic on-demand engineering of exosomes towards cancer immunotherapy. *Lab on a Chip, 19*(10), 1877–1886. https://doi.org/10.1039/c8lc01279b. https://www.ncbi.nlm.nih.gov/pubmed/31044204

Zheng, X., Sun, K., Liu, Y., Yin, X., Zhu, H., Yu, F., & Zhao, W. (2023). Resveratrol-loaded macrophage exosomes alleviate multiple sclerosis through targeting microglia. *Journal of Controlled Release, 353*, 675–684. https://doi.org/10.1016/j.jconrel.2022.12.026. https://www.ncbi.nlm.nih.gov/pubmed/36521687

Zhu, X., Badawi, M., Pomeroy, S., Sutaria, D. S., Xie, Z., Baek, A., Jiang, J., Elgamal, O. A., Mo, X., Perle, K., Chalmers, J., Schmittgen, T. D., & Phelps, M. A. (2017). Comprehensive toxicity and immunogenicity studies reveal minimal effects in mice following sustained dosing of extracellular vesicles derived from HEK293T cells. *Journal of Extracellular Vesicles, 6*(1), 1324730. https://doi.org/10.1080/20013078.2017.1324730. https://www.ncbi.nlm.nih.gov/pubmed/28717420

Zhuang, X., Xiang, X., Grizzle, W., Sun, D., Zhang, S., Axtell, R. C., Ju, S., Mu, J., Zhang, L., Steinman, L., Miller, D., & Zhang, H. G. (2011). Treatment of brain inflammatory diseases by delivering exosome encapsulated anti-inflammatory drugs from the nasal region to the brain. *Molecular Therapy, 19*(10), 1769–1779. https://doi.org/10.1038/mt.2011.164. https://www.ncbi.nlm.nih.gov/pubmed/21915101

Zou, Y., Li, L., Li, Y., Chen, S., Xie, X., Jin, X., Wang, X., Ma, C., Fan, G., & Wang, W. (2021). Restoring cardiac functions after myocardial infarction-ischemia/reperfusion via an exosome anchoring conductive hydrogel. *ACS Applied Materials & Interfaces, 13*(48), 56892–56908. https://doi.org/10.1021/acsami.1c16481. https://www.ncbi.nlm.nih.gov/pubmed/34823355

48

Polymeric Nanomaterials for Drug Delivery to the Brain

Chinmay M. Jogdeo, Farhana Islam, Kasturi Siddhanta, and David Oupický

Abstract

The blood–brain barrier is a major hurdle in the development of systemically administered therapeutics targeting the brain. Polymers, a class of materials composed of long monomer chains, have been extensively used to tackle this problem. Polymeric systems offer an unmatched ability to be designed based on their intended use by modulating their synthesis, composition, and base monomers. This flexibility has been used not only to create various types of polymeric nanoparticles (dendrimers, polymersomes, and polyplexes), but also to create scaffolds that mimic the extracellular matrix, allowing tissue growth and regeneration. Advances in polymeric systems, particularly in the context of drug delivery to the brain, are an ever-evolving area with tremendous translational potential.

Keywords

Polymers · Polymeric nanoparticles · Blood–brain barrier · Drug delivery · Brain

48.1 Introduction

Maladies affecting the central nervous system (CNS) are among the most prevalent yet complex diseases known to humans. The global market for neurodegenerative diseases is expected to grow exponentially in the coming years (Shakeri et al., 2020). Despite this, very few drugs have been successful in the treatment of CNS diseases. The development of CNS-targeted drugs is limited by multiple factors, with ineffective transport across the blood–brain barrier (BBB) being the most significant. Unsatisfactory delivery is a significant hurdle in developing better and more effective therapeutics in the diagnosis and treatment of neurological diseases, especially in the initial stages of the disease, where the BBB is still intact. Development of novel delivery vectors, which can overcome this hurdle in drug delivery is of utmost importance for the treatment of neurological disorders. To that end, polymers and polymer-based delivery vectors have emerged as promising strategies.

Polymers are macromolecules made up of a large number of repeating units called monomers. The choice of monomers ultimately governs the properties of the final structure, enabling the design of highly tunable polymers specific to a given application. The use of polymers for controlling the release rates of therapeutics, thus affecting their pharmacokinetic properties, dates back to the late 1900s with the works of Alejandro Zaffaroni, Ph.D., and Judah Folkman, M.D., who established first-order release kinetics of drugs from polymers. This led to a tremendous boost in the research for polymers in biomedical applications leading to the development of multiple sustained-release devices for topical, ocular, and oral delivery (Hoffman, 2008). The observed benefits of polymers ultimately led to their use in the field of nanomedicine to achieve targeted delivery of therapeutics with minimal off-target effects and a desired control over their pharmacokinetic profile. Today, several polymeric drug products have been approved by the FDA, and polymers in the context of their biomedical applications are a highly researched area worldwide. Not surprisingly, polymer-based delivery systems are often considered to be better suited for delivery to the brain for multiple reasons, including ease of modification, control over mechanical properties, and a tunable drug release profile. Additionally, polymers can be designed to be biologically active, supplementing their role as a delivery vector, and thereby providing a greater therapeutic impact. Polymers can be biodegradable, improving

C. M. Jogdeo · F. Islam · K. Siddhanta · D. Oupický (✉)
Center for Drug Delivery and Nanomedicine, Department of Pharmaceutical Sciences, University of Nebraska Medical Center, Omaha, NE, USA
e-mail: david.oupicky@unmc.edu

H. E. Gendelman, T. Ikezu (eds.), *Neuroimmune Pharmacology and Therapeutics*,
https://doi.org/10.1007/978-3-031-68237-7_48

their biocompatibility. Lastly, polymers can act as scaffolds providing mechanical support to aid tissue regeneration. Polymers can be natural, found in nature with predefined chemistries, or synthetic, with precise control over monomer properties and the synthesis process, yielding polymers with predefined geometries, configurations, and physicochemical properties. Depending on polymer properties, several polymeric structures can be formed, including but not limited to polymeric micelles, polymer-drug conjugates, dendrimers, and polyplexes.

The use of polymers in the context of the brain is a vast topic, practically impossible to discuss in full within a single chapter. The present chapter aims to introduce the reader to current research in the field of polymeric nanomaterials and summarize their applications for drug delivery to the brain. Additionally, an overview of nontraditional routes of brain targeting and advances in the use of conductive polymers as scaffolds for neural tissue regeneration has been provided. The readers are directed to several excellent reviews in the literature to complement this chapter as and when required throughout the manuscript.

48.2 The Blood–Brain Barrier

The BBB is a formidable challenge for the brain-targeted delivery of systemically administered drugs. Almost all large-molecule therapeutics and more than 90% of small-molecule drugs are excluded by the BBB. The BBB is a vital semipermeable barrier between the CNS tissues and the circulating blood created by the unique structure of blood vessels in the CNS. The BBB has several essential functions, including maintenance of homeostasis, protection of the brain from blood-borne toxins, regulation of neurotransmitter levels, and separation of central and peripheral neurotransmitters to reduce crosstalk (Pandit et al., 2020). The BBB not only acts as a physical and metabolic barrier but also aids in the selective transport of molecules into the CNS via active transport and passive or facilitated diffusion. The concept of BBB originated from the works of Ehrlich, Lewandowsky, and others, all of whom demonstrated the selective exclusion of systemically injected dyes from the brain, thus establishing the concept of a barrier restricting entry to the brain (Bouffard, 1906; Ehrlich, 1885; Lewandowsky, 1909). Stern and Gautier later introduced the term "barrière hémato-encéphalique," i.e., BBB, in 1921 following thorough studies of brain penetration of several molecules from the blood (Stern & Gautier, 1921).

The BBB is created by several structural components unique to CNS blood vessels, including endothelial cells joined by tight junctions, astrocytes, pericytes, and a basement membrane made up of extracellular matrix (ECM) components such as collagens, heparin sulfate proteoglycans, and laminins (Daneman, 2012). These structures line cerebral capillaries within the brain and spinal cord to create an interface for communication and exchange between systemic blood circulation and the brain (Kadry et al., 2020). The microvascular endothelial cells forming the BBB have distinct structural, physiological, and morphological characteristics that restrict the movement of drugs across the BBB. Some of these include the presence of tight junctions blocking the paracellular pathways between adjacent endothelial cells, which inhibit the passage of polar molecules; extremely low pinocytic/transcytotic activity; and the expression of active transport mechanisms including efflux transporters such as P-glycoprotein (P-gp), breast cancer resistance protein (BCRP), and organic anion transporting polypeptide (OATP). These efflux systems actively pump substances, including drugs, out of the endothelial cells back into circulation. An additional metabolic hurdle to drug transport across the BBB is the presence of drug-metabolizing enzymes, such as the CYP 450, peptidases, and cholinesterases, within the endothelial cells of the BBB (Kadry et al., 2020; Khan et al., 2018). These features of the BBB necessitate the development of novel formulations and strategies to enable effective drug delivery to regions within the CNS. Nanoparticle-based formulations represent a promising strategy for crossing the BBB and delivering therapeutic modalities to specific areas within the brain. CNS disorders, including cancer, neurodegenerative diseases, and neuroinflammation, can affect the structural and functional integrity of the BBB, offering an opportunity for therapeutics to enter the CNS. The intact, healthy BBB can also be temporarily disrupted to improve penetration of drugs using several techniques, such as focused ultrasound, providing an opportunity to deliver formulations to the CNS. Additionally, the BBB has several receptors and transporters that are required for the uptake of essential nutrients and macromolecules. Chemical modification of drugs to resemble ligands for these receptors and decorating nanoparticles with targeting ligands utilizing these transporters is another strategy to improve BBB penetration. Several extensive reviews expounding on the physiology of the BBB and strategies to overcome it have been published (Furtado et al., 2018; Lalatsa & Butt, 2018; Pandit et al., 2020). A lot of research is also directed toward developing in vitro models that better mimic the physiological environment of the BBB, thus aiding drug development. A detailed review of advances in the development of better BBB models can be found here (Li et al., 2021; Zhang et al., 2021).

48.3 Strategies for Enhancing Transport Across the BBB

The BBB presents a significant hurdle in the development of systemically administered therapeutics targeting the brain. A molecule must fulfill several requirements to partition successfully across the BBB. The molecule needs to be small enough to pass through the intercellular space of endothelial cells to permeate via paracellular transport; it also needs to be hydrophilic enough to dissolve in the aqueous tight junction pores and partition into the aqueous interstitial fluid. Additionally, the drug should not be a substrate for the several efflux pumps or the metabolic enzymes present in the BBB (Ayub & Wettig, 2022). Several strategies have been employed for bypassing the BBB, which can be classified into invasive and noninvasive approaches discussed below.

48.3.1 Invasive Approaches

48.3.1.1 Convection-Enhanced Delivery (CED)

CED was developed in the early 1990s as a method to bypass the BBB for localized delivery to the brain (Bobo et al., 1994). CED involves stereotactic placement of one or more catheters or micro-infusion pumps in the interstitial space of the brain, which can pump drugs directly into desired regions of the brain. CED employs bulk flow instead of diffusion, which leads to drug distribution via a pressure gradient, making it concentration and size-independent. This allows the delivery of large therapeutics without needing high concentrations to achieve the desired therapeutic levels in the brain. Additionally, the pressure gradient enables tissue penetration of a few centimeters, which is greater than the few millimeters achieved by diffusion-based bolus injection delivery systems. The gradual infusion of CED can potentially avoid the rapid pressure buildup and reduce the systemic toxicity of drugs by limiting the amount of drugs in the body at a given time. Several clinical trials exploring CED for the delivery of therapeutics to the brain are currently underway (NCT04608812, NCT03927274, NCT04160494, NCT05063357, and NCT04264143). Although CED has demonstrated great potential in preclinical studies, its success in the clinic has been limited by concerns associated with its invasive nature, including the risk of infections, unpredictable distribution of drugs, inconvenience to the patient, and inability to achieve long-term treatment. Air bubbles and backflow from the catheters pose additional risks (Mehta et al., 2017). Additionally, drugs with short half-lives are typically metabolized quickly after the infusion stops. To decrease infusion time, prevent rapid metabolism, enable targeting, and achieve sustained release, thus canceling the need for repeat infusions. CED has also been used to deliver polymeric nanoparticles encapsulating therapeutic cargo in several CNS disorders (Chen et al., 2018; Choi et al., 2020; Wang et al., 2021).

48.3.1.2 Intrathecal Administration

Intrathecal delivery methods involve invasive injection of soluble therapeutics directly into the cerebrospinal fluid (CSF) that flows through the thecal sac. Intrathecal methods include the intracerebroventricular (ICV) route, which involves surgical implantation of an ICV device such as an Ommaya reservoir or a Rickham reservoir under the scalp. This is connected via catheters to ventricles within the brain. The ICV route is a globally well-established route that has been used to deliver chemotherapeutics and analgesics for pain management (Cohen-Pfeffer et al., 2017). Approved therapeutics delivered by this route include chemotherapeutics and opioid analgesics. Currently, only three drugs have been approved by the FDA for intrathecal administration–morphine (Infumorph®; Baxter Healthcare Corp., Deerfield, IL), ziconotide (Prialt®; Jazz Pharmaceuticals Inc., Palo Alto, CA), and baclofen (Lioresal®; Novartis Pharmaceuticals, New York, NY, and Gablofen®; Mallinckrodt Pharmaceuticals, St. Louis, MO). Typically, the intrathecal route is used only after the patient does not show positive effects from other approaches. The intrathecal route has been used for a long time to deliver nanoparticle-based therapeutics to the brain, with a few therapeutics approved for clinical use, including DepoCyt®—a liposomal cytarabine formulation approved for the intrathecal treatment of lymphomatous meningitis (Fowler et al., 2020). The intrathecal administration route also involves administration via the intrathecal-lumbar and intracisternal magna routes. Despite clinical progress, propensity for microbial infections, poor patient compliance, increased intracranial pressure, and the need for a skilled surgeon remain a concern for this strategy. Additionally, the slow diffusion of drugs from the ventricles into the brain parenchyma combined with the rapid CSF turnover means that the drug is cleared before reaching the target tissue leading to a need for high therapeutic doses and repeat injections. Encapsulating therapeutics in sustained-release polymeric nanoparticles offers a promising strategy to overcome these problems and enable cell-specific targeted delivery (Peviani et al., 2019).

48.3.2 Non-invasive Approaches

48.3.2.1 Focused Ultrasound

Focused ultrasound (FUS) uses ultrasound waves that are directed using devices such as a lens or a curved transducer such that the pressure is highest at a small target and significantly less elsewhere. FUS is a promising technique employed to transiently improve permeation across the BBB, using the concept of cavitations that utilizes mechanical pressure waves with frequencies above the range of human hearing. Cavitation basically refers to the oscillation of microbubbles in response to pressure waves. Focused ultrasound employs biologically inert microbubbles, which are filled with inert gases, usually perfluorocarbons. The microbubbles localize at the BBB, oscillate, and amplify the mechanical effects of the ultrasound waves on the vasculature. This transiently opens the BBB via structural and functional disruption of tight junctions, enabling transport across the BBB. The microbubbles are commercially available under trade names such as SonoVue, Definity, and Optison. Several FUS-mediated BBB opening devices are currently under development, including SonoCloud-9 (CarThera, Paris, France), NaviFUS (NaviFUS, Taipei, Taiwan), and ExAblate Neuro 4000 220 kHz (InSightec, Haifa, Israel) (Meng et al., 2021a). Ultrasound-mediated opening of the BBB has been employed to deliver traditional small molecule drugs as well as large molecules, including antibodies for treating CNS disorders with improved penetration across the BBB. Enhanced delivery of trastuzumab to Her2-positive brain metastases was seen using magnetic resonance-guided FUS in a first-in-human trial, demonstrating the promise of this technique (Meng et al., 2021c). Additionally, several trials are currently underway exploring applying FUS in the context of improving BBB penetration in neuro-oncology and neurodegenerative diseases (Meng et al., 2021b). Despite promising results, ultrasound-mediated delivery is associated with risks, including complications such as unwanted heating of the tissues leading to hemorrhages.

48.3.2.2 Nose-to-Brain Delivery

The intranasal route of delivery provides a noninvasive route for substances to bypass the BBB while potentially avoiding the adverse effects that are usually associated with systemic absorption. Substances in the nasal cavity can reach the brain via two major routes–(1) the neuronal pathway and (2) systemic circulation. The nasal cavity can be divided into three major regions for the purpose of drug absorption–the vestibule, the respiratory region, and the olfactory region. The respiratory cavity is highly vascularized, allowing for the absorption of the drugs into the systemic circulation, from which they can reach the brain. However, most of the drugs bypass the BBB and reach the brain through the olfactory neurons in the olfactory region and the trigeminal nerves of the nasal cavity. The exact mechanisms of drug transport from these regions to the brain are not yet known (Martins et al., 2019; Wang et al., 2019). The intranasal route has been used for the delivery of multiple therapeutics, including insulin, nucleotides, peptides, and traditional small-molecule drugs (Martins et al., 2019). Although the intranasal route is a promising strategy to bypass the BBB, there are limitations associated with this approach. A very limited dose can be administered through the nasal cavity making this route feasible only for potent drugs. The nasal cavity has numerous metabolic enzymes, including isoforms of CYP450, which can metabolize the drugs, rendering them ineffective. Additionally, drugs get trapped in the mucosal secretions and cilia in the nasal cavity leading to their elimination by mucociliary clearance. These hurdles contribute to the extremely low bioavailability of compounds administered by the intranasal route. Efforts have been directed to improve bioavailability by using novel nanoparticle-based delivery systems, permeation enhancers, inhibitors of mucociliary clearance, and other physical methods, including the use of devices such as ViaNasa™ (an electronic atomizer) and OptiMist™ (a breath actuator) (Agrawal et al., 2018). The nose-to-brain route is a favored route of administration for polymeric nanoparticles due to its noninvasive nature and ease of administration. Reviews discussing this route in the context of polymeric nanoparticles can be found here (Aderibigbe & Naki, 2019; Musumeci, Bonaccorso, & Puglisi, 2019; Rabiee et al., 2021).

48.3.2.3 Osmotic Disruption

Osmotic disruption employs hyperosmotic solutions made up of compounds such as mannitol, urea, and glycerol to temporarily increase permeation across the BBB. The mechanism involves dehydration of endothelial cells leading to their shrinkage and subsequent disruption of tight junctions, thus enhancing paracellular transport (Rapoport et al., 1972; Xie et al., 2019). Several clinical trials have demonstrated the utility of this strategy in the brain-targeted delivery of chemotherapeutics and antibodies (Chakraborty et al., 2016; Iorio-Morin et al., 2021). Recently, Godinho et al. employed this strategy to improve CNS delivery of hydrophobically modified siRNAs conjugated to phosphocholine-docosahexaenoic acid. Significant uptake and target gene silencing was seen in groups with BBB disruption caused by an injection of 25% mannitol compared to groups without transient osmotic BBB disruption, further proving the utility of this strategy (Godinho et al., 2018).

48.3.3 Leveraging Natural Pathways

The non-fenestrated endothelium of the BBB with tight junctions restricts the movement of molecules via traditional routes. To overcome this, the endothelial cells of the BBB employ various specialized modes of transport for the uptake of physiological cargo from the systemic circulation to maintain the normal functioning of the brain. These pathways can be leveraged for drug delivery across the BBB (Fig. 48.1). Physiological macromolecules required for the functioning of the brain, including glucose, carbohydrates, and amino

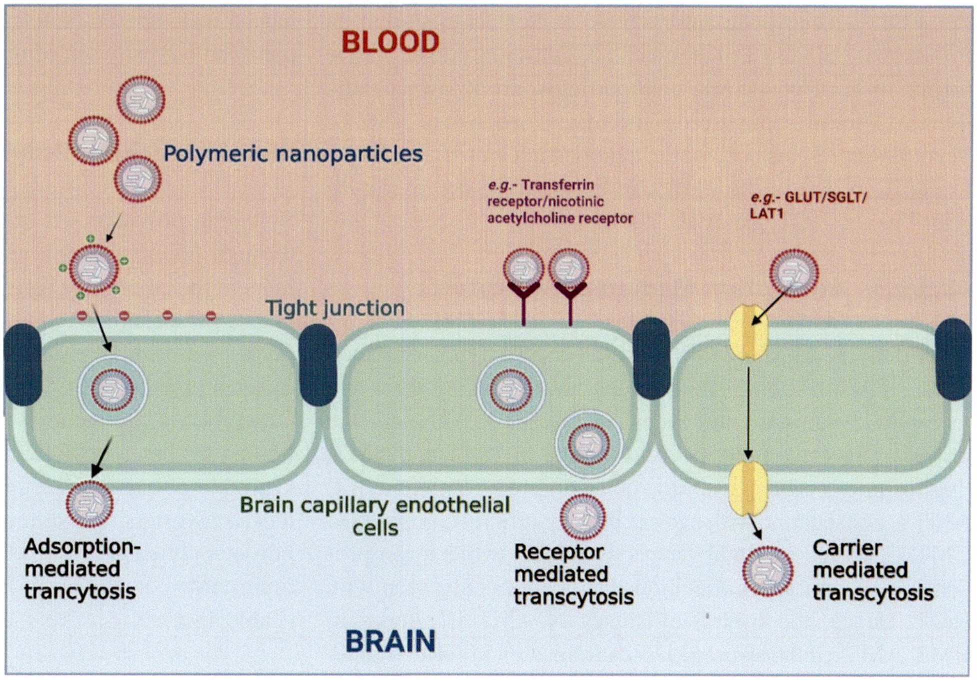

Fig. 48.1 Leveraging natural pathways to deliver polymeric nanoparticles across the blood–brain barrier

acids, are taken up by carrier-mediated endocytosis through the various solute-like carriers present at the BBB. Large proteins and peptides are taken up by either specific or nonspecific transcytosis. The former depends on specific interactions with receptors expressed on the endothelial cell membrane, such as insulin and transferrin receptors. Nonspecific transcytosis depends on adsorption of the molecule to the negatively charged endothelial cell surface and is also known as adsorptive-mediated transcytosis. This pathway is limited to cationic substances on account of the negatively charged glycocalyx barrier of the endothelial cells (Hervé et al., 2008). Drug substances and delivery systems can be modified to improve their ability to cross the BBB by leveraging these pathways.

48.3.3.1 Carrier-Mediated Transcytosis (CMT)

CMT involves leveraging the solute-like carrier (SLC) transporters to cross the BBB. SLCs are responsible for the supply of essential nutrients and energy to the brain via uptake of amino acids, glucose, and other necessary macromolecules from the blood. SLCs are specific for their substrates and are widely expressed on the BBB, enabling their use in CMT of therapeutics. Among the wide range of SLCs present at the BBB, the carbohydrate transporters (e.g., GLUTs, SGLTs), amino acid transporters (e.g., LAT1), and the peptide transporters (e.g., OATP, PepT1) are widely used for delivery of therapeutics. Drug molecules and nanoparticles can be modified to mimic the endogenous substrates for these transporters. Conjugating glucose to drugs and carrier systems is a widely used strategy. It has been used in improving CNS uptake of drugs, including ibuprofen and L-DOPA—a precursor of dopamine (Bonina et al., 2003; Chen et al., 2009). Similarly, ketoprofen was conjugated to L-lysine, leading to improved CNS delivery through the LAT1 receptor (Gynther et al., 2010; Puris et al., 2017). This strategy also translates to polymeric nanoparticles enabling targeted delivery to the CNS (Khan et al., 2019). Min et al. formulated a polyion complex micelle modified with glucose for the delivery of antisense oligonucleotides using the GLUT1 receptor and glycemic control as an external trigger. The glucose-containing formulation led to a 17-fold higher brain accumulation than the non-glucose-containing formulation indicating the promising utility of CMT in the targeted delivery of therapeutics across the BBB (Min et al., 2020). However, CMT is typically involved in the transport of small molecules, limiting its use in the delivery of large-molecule therapeutics.

48.3.3.2 Receptor-Mediated Transcytosis (RMT)

RMT is a highly specific system commonly employed for the delivery of macromolecules, mainly proteins and peptide-based drugs, across the BBB. RMT employs the vesicular trafficking system of brain endothelial cells in combination with receptor-mediated endocytosis (Ayub & Wettig, 2022). RMT is the most common pathway employed by delivery systems to cross the BBB (Lu et al., 2018). Some of the major receptors used for drug delivery across the BBB using RMT include the transferrin receptor, nicotinic acetylcholine receptor, insulin-like growth factor receptor, diphtheria toxin receptor, and the low-density lipoprotein receptor-related protein 1. Surfactants such as Polysorbate 80 and Poloxamer 188, commonly used in the formulation of polymeric nanoparticles, are known to utilize RMT for transport across the BBB (Joseph et al., 2021). Several studies on polymeric nanoparticles utilizing RMT to partition across the BBB can be found in the literature (Hartl et al., 2021; Lu et al., 2018).

48.3.3.3 Absorption-Mediated Transcytosis (AMT)

AMT delivers drugs to the brain based on endocytosis initiated by electrostatic interactions between cationic therapeutics and the anionic luminal components of the BBB. Although the process following endocytosis is similar in RMT and AMT, their affinity and capacity are quite different. Unlike RMT, AMT is nonspecific, with a lower binding affinity compared to CMT. However, its binding capacity is high, with a lower propensity for saturation inhibition that is commonly seen with RMT. Hence, the overall efficiency of AMT is similar to RMT. AMT can be leveraged by cationization of the therapeutic or delivery system to enhance electrostatic interactions with the endothelial cells of the BBB (Hervé et al., 2008; Zhu et al., 2019a). AMT has widely been used for the delivery of macromolecules, including proteins, by their cationization by various methods, including amidation of the carboxylic acid groups with positively charged amines. Similarly, drugs can be conjugated to such cationized proteins (e.g., cationized HAS, cationized BSA) to improve their delivery across the BBB. A similar strategy can be employed to achieve AMT-mediated uptake of polymeric nanoparticles (Muniswamy et al., 2019; Pang et al., 2012; Xu et al., 2009). Cell-penetrating peptides (CPPs) are a class of peptides that facilitate the uptake of attached cargo across the BBB using AMT. CPPs are short amphipathic peptides whose composition usually includes abundant cationic amino acids. Their short size and relatively low toxicity make them promising tools for enabling the delivery of therapeutics across the BBB. Commonly used CPPs conjugated to polymeric nanoparticles include TAT, penetratin, and Syn-B (Habault & Poyet, 2019; Kanazawa et al., 2019; Rao et al., 2008). Disadvantages associated with AMT include lack of selectivity, thereby leading to nonspecific biodistribution, and increased propensity of unwanted off-target effects (Hervé et al., 2008; Zhu et al., 2019a).

48.4 Polymers in Brain Targeted Drug Delivery

Polymers in nanomedicine are usually classified into natural and synthetic polymers. Natural polymers are extracted from various natural sources and are generally more biocompatible. Synthetic polymers can be designed depending on their end use, offer much better control over their properties, and usually have a well-defined architecture. Additionally, synthetic polymers can be easily scaled up with reproducible properties.

48.4.1 Synthetic Polymers

Synthetic polymers are completely man-made, synthesized through polymerization reactions using a variety of monomers in the laboratory. Synthetic polymers are highly tunable and can be designed to mimic the biological tissues, thus acting as scaffolds or tissue mimics. The choice of monomers dictates properties of the final polymer. The monomers can either be of the same type or of different types forming homo- or hetero-polymers, respectively. Commonly used synthetic polymers in drug delivery are polyethers, polyamides, polyesters, polyamines, and polyacrylamides. Select polymers of each of these classes with major applications in brain-targeted delivery systems are discussed below (Table 48.1).

48.4.1.1 Polyesters

Polyesters are a sub-class of synthetic polymers with an ester functional group in their main chain, which enables their degradation in the physiological system, thus improving biocompatibility. Polyesters such as polycaprolactone (PCL), polylactic acid (PLA), and poly(lactic-co-glycolic acid) (PLGA) have been used extensively for the development of brain-targeted nanoparticles. PLGA is an FDA-approved, biodegradable, anionic copolymer of lactic and glycolic acids. The degradation rate and, thus, the release kinetics of PLGA depend on the ratio of the two monomers, thereby offering an opportunity to tune to the release rate of the encapsulated therapeutic. PLGA has been extensively used in the development of CNS-targeted delivery systems (Del Amo et al., 2021; Li & Sabliov, 2013; Zhi et al., 2021). PCL is another FDA-approved nontoxic, biodegradable, semi-crystalline polyester synthesized by ring-opening polymerization of caprolactone that has been used to develop brain-targeted delivery systems (de Oliveira Junior et al., 2019; Kanwar et al., 2019; Lahkar & Das, 2020). PCL is hydrophobic; however, it can be combined with other materials to develop a sustained release system that degrades very slowly over months to years in the human body. PCL can be easily conjugated to a hydrophilic polymer, commonly PEG, which enables its self-assembly into micelles. Additionally, PCL is a common polymer employed for improving the biocompatibility of conductive polymers for the fabrication of cell and tissue scaffolds (Garrudo et al., 2019, 2021). PCL has been shown to maintain and support the differentiation of neural stem cells (Nisbet et al., 2008, 2009;). Similarly, PCL-based electrospun scaffolds were found to be excellent candidates as a biomimetic matrix for glioblastoma tumors (Unal et al., 2020).

Table 48.1 Common synthetic polymers employed in the development of drug delivery systems and their structures

Polymer	Structure
Poly(lactic-co-glycolic acid)(PLGA)	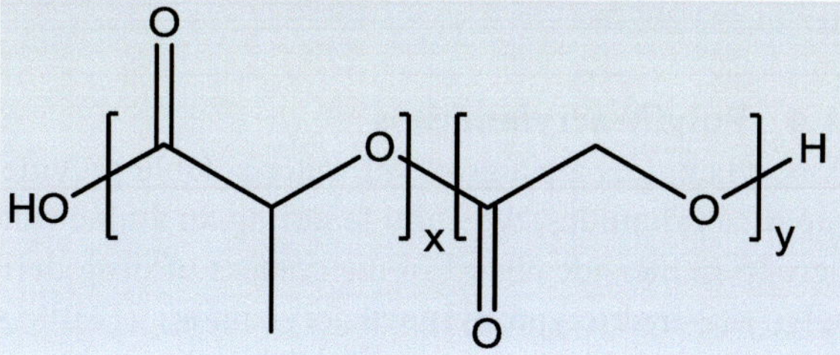
Poly(ethylenimine)(PEI)	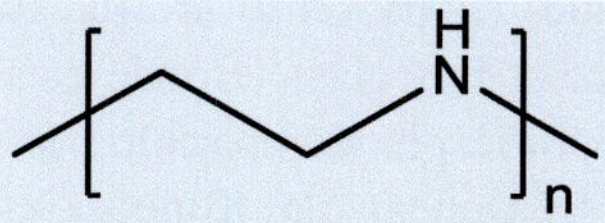
Polyethylene glycol (PEG)	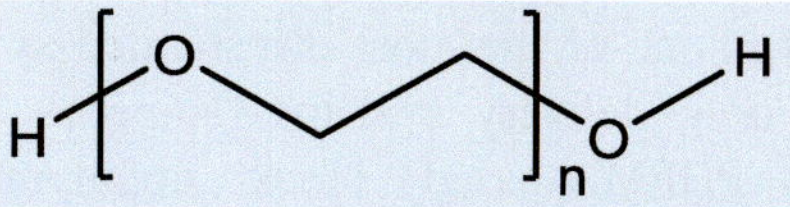
Polycaprolactone (PCL)	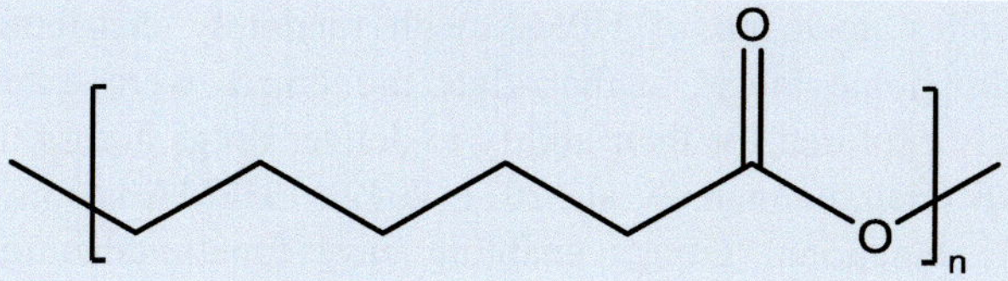
Poly(N-(2-hydroxypropyl)methacrylamide) (PHPMA)	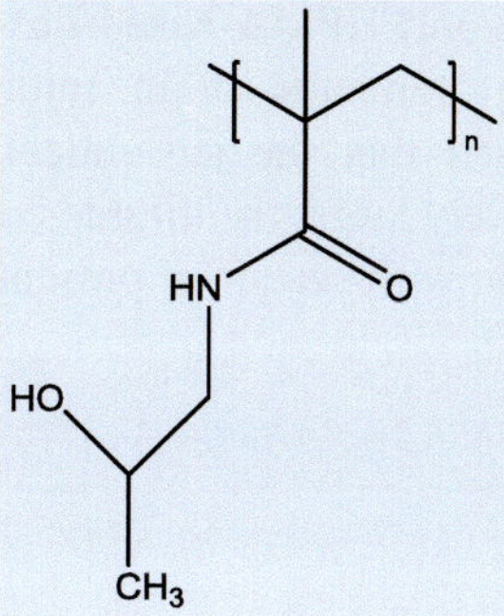

48.4.1.2 Polyethers

Polyethers are a sub-class of polymers with an ether functional group in their primary chain. The most common polyether utilized for drug delivery applications is poly(ethylene glycol) (PEG). PEG is an FDA-approved water-soluble, nonionic, and nontoxic polyether. PEG is available in a variety of molecular weights and is commonly used to impart a "stealth effect" to the nanoparticles, thus increasing their circulation time. PEGylation has been used extensively to improve nanoparticle delivery to the brain (Gajbhiye et al., 2020). One of the major drawbacks of PEG is its nonbiodegradability. Additionally, PEG has been known to elicit unwanted immune responses in the body. Similarly, PEGylation can shield the targeting ligand decorated on the nanoparticle surface and decrease its cell-specific targeting. Nevertheless, PEG remains one of the most widely explored synthetic polymers in the field of drug delivery. Poly(propylene glycol) or poly(propylene oxide) (PPG/PPO) is another commonly used polyether that, unlike PEG, is hydrophobic. In the context of drug delivery to the brain, PPG has been used in the synthesis of pluronic block copolymers along with PEG (PEG-*b*-PPG-*b*-PEG) as the central hydrophobic segment. Pluronic copolymers have been explored extensively to deliver therapeutics to the brain due to their ability to inhibit P-gp and thus limit drug efflux from the brain (Batrakova et al., 1999; Kabanov et al., 2003; Liu et al., 2014; Pitto-Barry & Barry, 2014).

48.4.1.3 Poly(ethylene imine)s

Poly(ethylene imine) (PEI) is a ubiquitous polymer in the field of drug delivery. PEI has several amine groups in its structure, imparting a highly cationic nature to the polymer. PEI has been extensively used to deliver anionic nucleic acids by forming complexes by electrostatic interactions, known as polyplexes (Boussif et al., 1995; Lungwitz et al., 2005). PEI is available in several molecular weights with both linear and branched geometries. The use of PEI is hindered by its high cytotoxicity, and these days PEI is rarely used as a stand-alone polymer. PEI is usually combined with other materials in a hybrid system to decrease its cytotoxicity (Z. Chen et al., 2020).

48.4.1.4 Poly(*N*-acrylamide)s

Poly(*N*-acrylamide)s are a group of water-soluble polymers made up of acrylamides, typically featuring an amine functional group in the side chain. In the context of drug delivery, poly(*N*-(2-hydroxypropyl)methacrylamide) (PHPMA) and poly(*N*-isopropylacrylamide) (PNIPAAm) are the two commonly employed polymers of this class (Bobde et al., 2020). PHPMA is a hydrophilic, biocompatible, nontoxic polymer, first developed by Duncan and Kopecek as a plasma expander and is one of the most investigated synthetic polymers for drug delivery systems (Kopeček & Kopečková, 2010). PHPMA-based block copolymer micelles made up of HPMA with randomly distributed hydrophobic lauryl methacrylate monomers were extensively explored for their ability to deliver drugs across the BBB (Hemmelmann et al., 2011, 2012). PHPMA has multiple functional groups enabling easy functionalization, which led to the development of an PHPMA-based dexamethasone prodrug for treating traumatic brain injury. Optical imaging results confirmed that the systemically administered prodrug could passively target the injured cortex through the ELVIS mechanism with sustained presence for over 14 days (Wei et al., 2022).

48.4.2 Natural Polymers

Natural polymers have been extensively explored to develop drug delivery systems due to their superior biocompatibility, non-immunogenicity, biodegradability, and low cost compared to synthetic polymers. Natural polymers can be broadly classified into two groups: proteins and polysaccharides (Table 48.2).

Table 48.2 Common natural polymers are employed in the development of drug delivery systems and their structures

Polymer	Structure
Chitosan	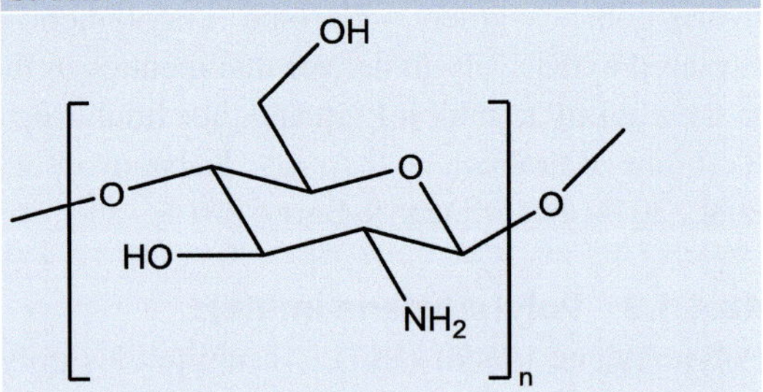
Hyaluronic Acid	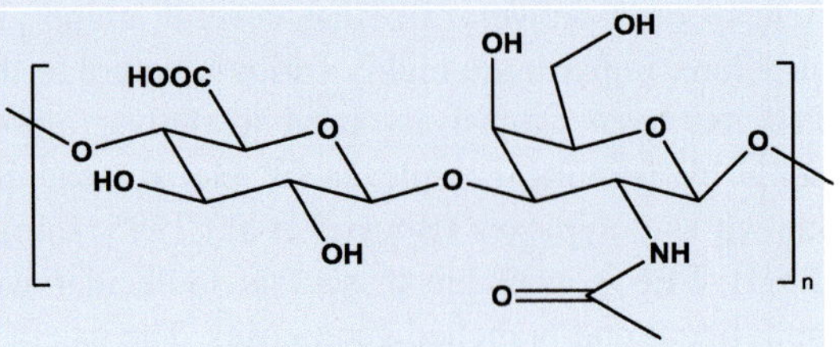

48.4.2.1 Polysaccharides

Polysaccharides occur extensively in nature and are made up of repeating monomeric units of monosaccharides. Polysaccharides have a large number of hydroxyl groups in their structure, which allows ease of modification. Polysaccharides can have linear or branched geometries and, in the simplest sense, are a string of sugar molecules attached to each other by glycosidic bonds. Chitosan, hyaluronic acid, cellulose, cyclodextrins, and alginate are some of the most common polysaccharides employed in brain-targeted drug delivery (Curcio et al., 2020).

Chitosan

Chitosan is a water-soluble cationic polysaccharide obtained from the deacetylation of chitin—a natural polymer obtained from crustacean shells and fungi. It is made up of N-glucosamine and N-acetyl glucosamine joined by β 1–4 glycosidic bonds. Chitosan breaks down inside the body by enzymatic degradation, with the degradation rate being proportional to the degree of deacetylation, offering an opportunity to tailor the release kinetics of drugs. Chitosan also has gel-forming abilities, which enable its use in the fabrication of hydrogels and scaffolds in tissue engineering. Additionally, chitosan has mucoadhesive properties allowing its use in the development of nose-to-brain delivery systems (Aderibigbe & Naki, 2019). Chitosan is a popular biomaterial for the development of delivery systems targeting the brain, and its application in this context has been extensively reviewed elsewhere (Caprifico et al., 2020; Cortés et al., 2020; Yu et al., 2019).

Hyaluronic Acid (HA)

HA is a polysaccharide made up of D-glucuronic acid and N-acetyl-D-glucosamine monomers. HA is found throughout the body as a major component of the ECM and hence has a high biocompatibility. HA is enzymatically degraded by hyaluronidases, a group of enzymes found throughout the body. Additionally, HA is a ligand for the CD44 receptor, which is often overexpressed in a number of tumors, imparting it with an innate tumor-targeting ability. Yang et al. fabricated a core-shell micelle, with HA forming the shell for intranasal siRNA delivery in the treatment of glioma. The mucoadhesive properties of HA enabling prolonged retention in the nasal cavity combined with its CD44 targeting ability led to a significantly higher uptake and better therapeutic efficacy in an intracranial GL261 mouse model (Yang et al., 2022). Similar to chitosan, HA has gel-forming ability enabling its use in hydrogel-based scaffolds in tissue engineering applications and local drug delivery to the brain (Jeong et al., 2021; Wang et al., 2022; Zhang et al., 2022a).

48.4.2.2 Proteins

Proteins are macromolecules that can be considered as polymers made up of amino acid building blocks (monomers) linked together by peptide bonds. Proteins have been used for a long time to fabricate nanoparticles for drug delivery due to their biocompatibility, biodegradability, ease of modification, and innate targeting ability. Commonly used proteins in the development of CNS-targeted delivery systems include albumin (Bilia et al., 2019; Katona et al., 2020), collagen (Rathore et al., 2020; Siddiqui et al., 2022), gelatin, silk, and elastin. Protein-based therapeutics have also been approved by the FDA, with the typical example being Abraxane, an albumin-bound chemotherapeutic that has been approved for the treatment of several cancers. Collagen is a common structural protein naturally found in humans and is a major component of the ECM. Collagen can also be easily formulated into a hydrogel and has been used extensively for the development of scaffolds to provide mechanical support and induce tissue repair and regeneration. Several collagen-based nerve conduits have already been approved by the FDA (NeuraGen®, Neuroflex™, NeuroMatrix™) (Ucar, 2021). Albumin is a well-known blood protein that has long been used to develop delivery systems. Additionally, albumin was shown to have innate neuroprotective effects, further improving its potential as a favorable biomaterial in the development of brain-targeted delivery systems (Liumin Wang et al., 2017; Yao et al., 2010).

48.5 Types of Polymeric Nanoparticles for Targeted Delivery to the Brain

Polymers can be fabricated into several types of nanoparticles based on their structure and physicochemical properties. A few select types have been discussed below (Fig. 48.2).

48.5.1 Dendrimers

Dendrimers are an emerging class of synthetic hyperbranched polymers with a structure resembling a tree or dendron. They have received widespread attention due to their multifunctional capabilities, ease of surface modification, high molecular uniformity, and monodispersibility. Dendrimers usually consist of a central core with repeating units known as generations attached to it with terminal functional groups. The terminal functional groups decide the surface properties of the dendrimer, allowing therapeutic cargo to either be covalently conjugated or electrostatically condensed within the internal cavities of the dendrimer. Several dendrimer types, including but not limited to polyamidoamine (PAMAM), polypropyleneimine, poly-l-lysine have been utilized to deliver therapeutic cargo to the brain using invasive as well as noninvasive approaches discussed above (Zhu et al., 2019b). Dendrimers have been employed for the brain-targeted delivery of traditional small molecules and biologics, including nucleic acids and proteins, both alone and in combination. Small molecules can be entrapped in the dendrimer or covalently attached to it. The encapsulation may be based on physical entrapment, electrostatic interactions, or hydrophobic and hydrogen bond interactions. Covalent conjugation of drugs to dendrimers is often through hydrolyzable, biodegradable, or stimuli-cleavable linkers, which offers enhanced control over drug release. Gene delivery using dendrimers is primarily accomplished by electrostatic condensation. Fourth-generation hydroxyl-terminated PAMAM dendrimers have been shown to cross-impaired BBB and selectively target activated microglia and astrocytes in multiple mouse models without the use of any targeting ligands. Building upon this work, Sharma et al. leveraged the CMT pathway by conjugating sugars such as mannose, glucose, and galactose to the surface of G4-OH dendrimers and assessed the impact of such modifications on the uptake and targeting in glioblastoma. The results demonstrated that the conjugation of sugars significantly increased BBB penetration and tumor specificity (A. Sharma et al., 2018, 2021). Additionally, dendrimers have been employed as carriers for imaging agents in the diagnosis of CNS disorders (Ma et al., 2018; Xu et al., 2019). Readers are pointed to

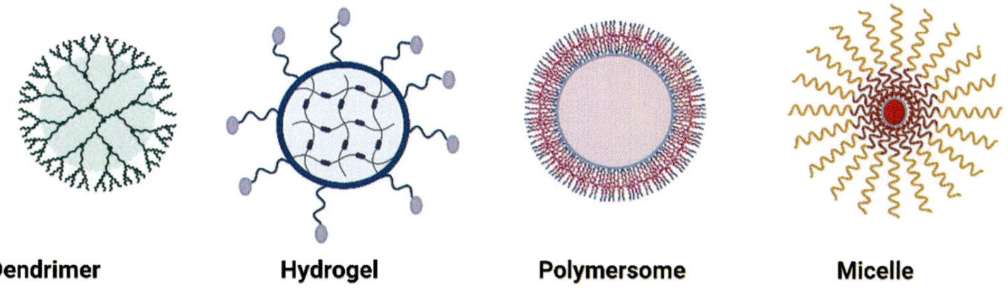

Fig. 48.2 Types of polymeric nanoparticles

an excellent review on the use of polymeric dendrimers for brain targeting by Zhu et al. (2019b).

48.5.2 Hydrogels and Nanogels

Hydrogels are crosslinked hydrophilic polymer networks that can absorb large amounts of water. Nanogels are nanometer-ranged three-dimensional hydrogel networks formed by crosslinked hydrophilic polymer chains. Nanogels offer an improved ability to cross biomembranes compared to traditional bulk hydrogels due to their small size. Their small size and corresponding large surface allow for a very high loading capacity. Additionally, the small size allows for rapid response to environmental stimuli such as changes in pH and temperature, creating opportunities for the development of stimuli-responsive delivery systems, which have tremendous potential for drug delivery in diseased states, especially cancer (Kousalová & Etrych, 2018). Nanogels have been utilized for the delivery of all types of therapeutic cargo ranging from traditional small molecule drugs to biologics, including nucleotides, proteins, and even cells. Additionally, nanogels, which can be injected in the intrathecal space, have been utilized as scaffolds to enable neural tissue regeneration. Hydrogels, and in turn nanogels have an intrinsic ability to swell in an aqueous environment, which affects their circulation time and interaction with cellular components in addition to affecting the stability and rate of drug release. Hence, a challenge with the use of polymeric nanogels is the ability to control their response to changes in the external environment (Vashist et al., 2018). Advances in nanogel formulation have led to the development of multifunctional hybrid systems. Xiao et al. reported the development of macrophage camouflaged, redox responsive poly(N-vinylcaprolactam) nanogels coloaded with manganese dioxide and cisplatin for combined magnetic resonance imaging-guided chemotherapy/chemodynamic therapy of orthotopic glioma. Coating with the macrophage membrane led to enhanced circulation time and improved BBB penetration (Xiao et al., 2021). Similarly, Zhang et al. developed a biomimetic nanogel system synthesized by crosslinking pullulan and poly(deca-4,6-diynedioic acid) (PDDA), which was further camouflaged with apolipoprotein E peptide-decorated erythrocyte membrane to enable extended blood circulation, tumor targeting, and RMT mediated BBB penetration. The nanogels, coloaded with temozolamide and ICG, were impervious to endogenous oxidative conditions and could only be selectively disintegrated by ICG generated ROS in response to NIR irradiation enabling precise spatiotemporal control on drug release. The synergistic photodynamic therapy and chemotherapy led to excellent anti-glioblastoma efficacy in both, orthotopic U87MG and GBM stem cell (CSC2) mice models, with a nearly threefold improvement in median survival time (Zhang et al., 2022b). A similar diphtheria toxin receptor ligand-mediated RMT strategy was employed by Singh et al. for brain-targeted delivery of radiopharmaceuticals (Singh et al., 2021). Nanogels have also been used for the treatment of neurodegenerative diseases and spinal cord injury (Ye et al., 2022; Yu et al., 2022). Hydrogels have also been utilized to enable CNS tissue regeneration through their capacity to deliver and control release of biomolecules through modifications in the polymer composition impacting pore size. Hydrogels have also been used to deliver several growth factors including nerve growth factor, brain-derived neurotrophic factor, and epidermal growth factor. Additionally, stem cell therapies have emerged as a promising strategy to treat neurodegenerative disorders, and the use of hydrogels as carriers for stem cells is being widely explored (Grimaudo et al., 2021). Ferreira et al. developed a hyaluronic acid-based injectable hydrogel to deliver bone marrow mesenchymal stem cells (BMSCs). The formulation led to enhanced therapeutic efficacy compared to treatment with BMSC cell suspension after intracerebroventricular injection in an experimental autoimmune encephalomyelitis rat model, demonstrating the advantage of stem cell delivery using hydrogels (Ferreira et al., 2021).

48.5.3 Polymersomes

Polymersomes are a class of nanoparticles composed of a vesicle formed by the self-assembly of synthetic amphiphilic block copolymers. Block copolymers in the context of polymersomes, are made up of two or more homopolymer blocks with alternating hydrophilic or hydrophobic characteristics. Polymersomes with varied properties can be formulated by using copolymers with different characteristics, including varied molecular weight, degradability, stimuli responsiveness, and molecular architectures. Polymersomes are like liposomes in structure and typically have a central aqueous core surrounded by a polymeric bilayer. This polymeric bilayer in polymersomes is typically thicker than liposomes allowing for a higher drug loading as well as improved stability (Sharma et al., 2020). The aqueous core can be used for encapsulation of hydrophilic therapeutic cargo, while the bilayer can integrate hydrophobic drugs. This offers a unique opportunity for the co-delivery of hydrophilic and hydrophobic therapeutics to achieve combination therapy. The long circulation times that are usually seen with polymersomes offer an opportunity for brain targeting. The circulation time can be further improved by introducing a hydrophilic PEG layer imparting so-called "stealth character" to the polymersomes (Lee & Feijen, 2012). Additionally, polymers can be chosen to enable surface modification using targeting moieties to improve BBB penetration (Lefley et al., 2020). Polymersomes have been extensively investigated for delivering therapeutics to the brain (Fan et al., 2021; Goyal et al., 2018; Ouyang et al., 2021; Shi et al., 2018). Zeng et al. synthesized a library of polymersomes with different elasticities having identical shape, size, and zeta potential to investigate the effect of elasticity on BBB penetration and tumor targeting. Angiopep-2, a commonly used glioblastoma targeting ligand was attached to the surface of these nanoparticles, and their biological performance, including BBB penetration, circulation time, and tumor penetration, was assessed. The results showed that polymersome elasticity influences brain tumor targeting. In comparison to soft polymersomes, stiff polymersomes lead to a higher cell uptake, excellent tumor penetration, and better in vitro BBB crossing ability despite comparable particle size, zeta potential, and morphology (Zheng et al., 2021).

48.5.4 Polymeric Micelles

Amphiphilic block copolymers undergo self-assembly in aqueous media at concentrations above their critical micelle concentrations, forming nanosized, colloidal particles with a hydrophobic core and a hydrophilic shell called polymeric micelles. The hydrophobic core allows for the encapsulation of a variety of non-soluble therapeutic cargo and imaging agents, thus improving solubility in the body. Block copolymers forming the micelle can be modulated to impart desired characteristics to the micelles, including attachment of targeting ligands, stimuli sensitivity, and biodegradability. Polymeric micelles have been utilized for targeted delivery to the brain due to their ability to cross the BBB, high circulation times, and ease of surface modification (Abourehab et al., 2018; Jiang et al., 2021; Kaur et al., 2022; Lu et al., 2020). Additionally, certain copolymers, especially pluronics have been widely utilized for brain-targeted delivery due to their ability to inhibit P-gp efflux transporters (Niu et al., 2020). Drug release from polymeric micelles can be controlled with internal stimuli (pH responsive or reduction-responsive) or extrinsic stimuli (magnets, light, temperature, and ultrasound) to enable targeted drug delivery. Micelles have also been employed for nose-to-brain delivery of therapeutics in addition to the traditional intravenous route (Lamptey et al., 2022; Pokharkar et al., 2020). Kanazawa et al. developed bombesin-modified PEG-PCL-Tat micelles for intranasal targeted delivery to gastrin-releasing peptide receptor (GRPR) positive gliomas. Bombesin is a peptide that is specific for GPCRs and enabled a significantly higher uptake in GPCR-positive cells in vitro, and these results were translatable to a C6-glioma model in rats (Kanazawa et al., 2020). Incorporation of stimuli-sensitive moieties in micelle forming copolymers allows for spatiotemporal control over the release of therapeutics. This is particularly useful for targeted delivery to gliomas, which, like most tumors have high levels of reactive oxygen species, reduced glutathione, and a hypoxic environment. More recently, copolymers responsive to external stimuli such as light, electricity, and magnetism are being explored for brain-targeted delivery of therapeutics. A detailed review describing the use of polymeric micelles for brain-targeted delivery of therapeutics can be found here (Kaur et al., 2022).

48.6 Effect of Physicochemical Properties of Polymeric Nanomaterials for BBB Penetration

The fate of nanoparticles on systemic administration and hence their ability to reach the desired target tissue greatly depend on their properties. Properties such as nanoparticle size, surface charge, and shape as well as properties of the polymer such as its elasticity play a major role in their ability to effectively cross the BBB and reach the intended target. A few of these properties are discussed below.

48.6.1 Effect of Size

The size of nanoparticles is a major factor affecting their biodistribution. In particular, size plays an important role in the intravenous delivery of nanoparticles to the brain. Prolonged circulation time increases the probability of nanoparticles interacting with the BBB and subsequently crossing it to deliver drugs to the CNS. In general, nanoparticles having a diameter of less than 5 nm can be filtered by the glomerulus and are easily cleared by the kidneys. On the other hand, nanoparticles with sizes greater than 200 nm tend to accumulate in the spleen, liver, and lungs. The size of polymeric nanoparticles can be easily tuned to the desired range and hence are well suited to optimize delivery to the brain. An important point to note is that the effect of size on permeability is vastly different for a healthy BBB and a compromised BBB routinely seen in CNS disorders. Several studies have tried to establish the effect of particle size on the BBB penetration of polymeric nanoparticles (Cruz et al., 2016; Kulkarni & Feng, 2013; Zhou et al., 2013). More recently, Brown et al. synthesized polystyrene nanoparticles of different shapes and sizes and investigated the effect of these parameters on penetration across the BBB in an in vitro model using human cerebral microvascular endothelial cells (hCMEC/D3). The apparent permeability was inversely proportional to the particle size in conjunction with previous studies (Brown et al., 2020). Similar results were seen in a study by Meng et al. who explored the effect of size on BBB penetration of polymeric nanoparticles carrying carbamazepine (Meng et al., 2022). However, the charge of the nanoparticle impacts this dependency as shown by Yuan et al. who found that the permeability of the bEnd3 monolayer to cationic nanoparticles of an average diameter of 80 nm is the same as that to a much smaller solute, dextran 4 k (Stokes diameter ~ 2.8 nm). Additional data from Zhang et al. showed that the permeability of neutral particles in the same size range is significantly lower than the cationic nanoparticles (L. Zhang et al., 2020). Additionally, contrasting reports disproving the dependence of BBB penetration and particle size can be found in the literature. Voigt et al. synthesized polybutylcyanoacrylate nanoparticles coated with different surfactants in a size range of 67–464 nm with varied zeta potential and explored barrier penetration using a blood–retina barrier as a substitute to BBB. The group demonstrated that the nature of surfactant on the BP surface rather than the NP size plays a role in BBB penetration (Voigt et al., 2014). These contrasting findings underscore the need for further research to establish the relationship between size and BBB penetration.

48.6.2 Effect of Surface Charge

The surface charge of nanoparticles plays an important role in their fate in vivo and influences the circulation time, protein binding, and clearance by the MPS. Neutral nanoparticles usually show higher circulation times when compared to charged nanoparticles. Positively charged nanoparticles show higher accumulation in the liver as well as higher binding to serum proteins, which in turn affects their biodistribution. As with transport across other biomembranes, the surface charge of nanoparticles plays a role in their transport across the BBB. Since the endothelial cells of the BBB are rich in negatively charged proteoglycans, cationic nanoparticles show a higher uptake via AMT. Surface charge also influences the effect of size on permeation across the BBB as seen in the section above. However, positively charged nanoparticles have a higher propensity for clearance by the MPS leading to lower circulation times. Additionally, cationic particles are usually associated with a higher toxicity than neutral or anionic particles. Hence, a careful control over surface charge is important for optimizing the transport of polymeric nanoparticles across the BBB while ensuring sufficient circulation and low cytotoxicity (Akbaba et al., 2018; Ribovski et al., 2021b).

48.6.3 Effect of Nanoparticle Shape

Spherical nanoparticles are the most common type of nanoparticles encountered in preclinical studies. However, the shape of nanoparticles can influence their behavior in blood flow as well as their interactions with cellular components, thus affecting their circulation time and biodistribution. For example, rod-shaped particles were shown to cross the GFB much easier than spherical nanoparticles. Additionally, filamentous polymeric micelles were shown to have longer circulation times than spherical micelles, which could be leveraged to improve BBB interaction. The curvature of commonly used spherical nanoparticles leads to limited binding sites with surface receptors. Accordingly, filamentous micelles were found to remain in circulation for much longer compared to their spherical counterparts. In the context of BBB permeation, Kolhar et al. reported that polystyrene nanorods-coated antibodies led to higher accumulation in the brain compared to spherical nanoparticles (Kolhar et al., 2013).

48.6.4 Polymer Stiffness

The effect of nanoparticle stiffness on its ability to cross biological membranes including the BBB is a relatively unexplored area. However, limited reports state that hydrophilic rigid nanoparticles show a higher uptake across different cell types compared to soft nanoparticles in vitro. Differences in circulation half-lives, retention, and uptake were seen in spherical PEG diacrylate nanoparticles with varied stiffness in a study by Anselmo et al. with prolonged circulation seen for softer nanoparticles at earlier timepoints. Similarly, a higher brain accumulation was seen for the softer nanoparticles at the 30 min timepoint; however, this difference was no longer significant at later timepoints (Anselmo et al., 2015). Polymeric hydrogels are nanoparticles made up of a crosslinked hydrophilic polymer network whose stiffness can be tuned by varying the extent of crosslinking. This offers an excellent opportunity to assess the impact of stiffness on the ability to cross biological membranes including the BBB. Ribovski et al. synthesized poly(*N*-isopropylmethacrylamide) (p(NIPMAM)) hydrogels of varying stiffness by changing the amount of crosslinker and evaluated its effect on BBB penetration. They showed that stiffer nanogels had a higher uptake compared to the softer nanogels in an in vitro BBB model, composed of a polarized hCMEC/D3 cell monolayer grown on a collagen gel. However, in contrast, transcytosis was favored by the softer nano gels suggesting that nanoparticle stiffness has an opposing effect on uptake and transcytosis across the BBB (Ribovski et al., 2021a). Polymer flexibility was also shown to affect FUS-mediated BBB penetration of polymer-drug conjugated as demonstrated by Sun et al. (Sun et al., 2022). Not limiting to drug delivery, polymer stiffness also affects the efficacy of polymeric scaffolds. Neural stem cells cultured in calcium-crosslinked alginate hydrogels with varying stiffness demonstrated significantly higher expression of a neuronal marker in gels with lower elasticity suggesting the correlation between neural differentiation and hydrogel stiffness (Banerjee et al., 2009; Grimaudo et al., 2021). Modulating polymer stiffness thus offers an opportunity to tune BBB penetration of polymeric nanoparticles.

48.6.5 Effect of Targeting Ligands

Surface decoration with ligands specific for a receptor is a common structural feature in almost every targeted nanoformulation over the recent years. Surface conjugation to ligands specific for receptors involved in RMT is a strategy that has been employed to improve BBB penetration of nanoparticles. However, designing a nanoparticle to achieve optimum targeting involves multiple factors in addition to the choice of a proper targeting ligand (Ribovski et al., 2021b). Factors such as conjugation strategies, ligand affinity, and extent of surface decoration play a major role in achieving successful targeting. In theory, the probability of binding should increase with an increase in ligand density. However, high ligand densities have been shown to negatively affect nanoparticle uptake causing receptor saturation and undesirable changes in the physical properties of the nanoparticles. These changes include loss of stealth properties, increase in the hydrodynamic diameter, and stearic hindrance, all of which can affect nanoparticle behavior inside the body (Alkilany et al., 2019). Anraku et al. developed polymeric micelles decorated with different densities of glucose targeted to the GLUT-1 receptor and explored the effect of ligand density on uptake and biodistribution. They showed that the ligand-conjugated micelles successfully crossed the BBB and that ligand density affects targeting at the cellular level. Although micelles with 50% ligand density had a higher overall accumulation in the brain, they were mainly found around endothelial cells. In contrast, micelles with 25% ligand density were directed to neurons and microglia with a much higher cellular uptake, confirming the importance of ligand density on specific targeting (Anraku et al., 2017). Ligand avidity is another important factor that plays a role in BBB penetration. Wiley et al. showed that nanoparticles with large amounts of the targeting ligand remained attached to the receptors on the brain endothelial cells while those with a lesser ligand density could attach to the receptor on the luminal side, cross the BBB, and successfully detach from the receptor leading to a higher uptake (Wiley et al., 2013). Similar results were seen in a study employing antibodies as targeting ligands (Yu et al., 2011). An alternative approach is to use cleavable linkers to conjugate targeting ligands to the nanoparticles, which can be cleaved off during transcytosis.

48.7 Electrically Conductive Polymers for Regeneration of Neural Tissues

The limited regenerative ability of the nervous system is a major hurdle in the development of better therapeutics. Damage to brain structures and functional tissues in neurogenerative diseases and peripheral nerve injuries is usually permanent. Treatments are limited and are typically restricted to improving lost functions. Replacing the lost cells and tissues through neural graft implantations is a promising strategy that has emerged in the recent years. However, the extensive loss of grafted cells is a major limitation to the success of this therapy (Ghasemi-Mobarakeh et al., 2011). Tissue engineering, which employs novel polymeric biomaterials that mimic the ECM of the brain to support tissue growth offers a promising alternative to traditional transplantation methods. Polymeric biomaterials could be leveraged to provide mechanical support to the grafts and enable graft growth, differentiation, and integration (Aurand et al., 2012).

Electrical stimulation is a novel strategy that has been shown to be beneficial for neural regeneration, and hence, conductive polymers are another class of biomaterials that have shown promise in nerve tissue engineering. Using electrically conductive polymers enables precise, controlled local delivery of electric currents, stimulating growth and repair (Ghasemi-Mobarakeh et al., 2011). Conductive polymers (CPs) are a special class of polymers with an ability to conduct electricity. These polymers usually have conjugated double bonds in their backbone with loosely held electrons and a dopant ion that carries charge in the form of free electrons and imparts conductivity to the polymer (Ravichandran et al., 2010). Traditionally used CPs include synthetic polymers such as polypyrrole, polyaniline, polythiophene, and polyacetylene. However, low biocompatibility, non-biodegradability, and poor mechanical properties of these polymers limit their use. To overcome these limitations, natural, biodegradable polymers including zein, chitosan, collagen, and their combinations with traditional CPs are now being widely used. The physical, electrical, and chemical properties of CPs can be easily tuned by modifying their chemistry and synthesis methods making them a promising platform in tissue engineering. Conductive polymer-based regenerative platforms are discussed below.

Hydrogels are a class of polymeric materials that have been extensively investigated in context of tissue engineering. Hydrogels, in addition to acting as growth scaffolds, have a soft, flexible nature and can be used as a local delivery system to deliver therapeutics at specific sites within the brain as discussed in section. Fabricating hydrogels from CPs offers a unique opportunity to improve their therapeutic impact (Uz & Mallapragada, 2019). Neural stem cells implanted in a PEG-based hydrogel and exposed to electrical stimuli led to the formation of differentiated cells with substantive neurite growth (Lee et al., 2018). Eftekhari et al. fabricated a PC12 cell surface morphology imprinted conductive hydrogel from chitosan and polyaniline and investigated its neural priming ability in rat adipose-derived stem cells. An elongated bipolar neuronal morphology and upregulated expression of neural-precursor-specific genes indicated successful differentiation of the stem cells (Eftekhari et al.,

2021). Zhao et al. synthesized a series of hydrogels from silk-fibroin and graphene oxide nanosheets and demonstrated that soft hydrogels with increased conductivity are more beneficial in survival, spreading, and differentiation of Schwann cells (Zhao et al., 2022). Similarly, Cho et al. demonstrated that collagen/polypyrrole hydrogel microfibers improved neurogenesis and cell growth in PC12 cells. This was attributed to the electrical conductivity of the matrix and the delivered electrical stimulation to the cells, which upregulated the expression of L-type voltage-gated ion channel leading to an increase in the intracellular calcium level, which in turn promoted neurogenesis (Wu et al., 2019). Polymeric hydrogel scaffolds have been widely used in tissue regeneration after spinal cord injury to inhibit development of a cystic cavity, enabling successful repair and regeneration (Zhang et al., 2019). Hong et al. developed an imidazole-poly(organophosphazenes) based hydrogel that could significantly eliminate cystic cavities in a clinically relevant rat spinal cord injury model. The hydrogel could enhance tissue repair by enabling ECM remodeling by interacting with macrophages in addition to providing physical support and structural stabilization (Hong et al., 2017). Electrically conductive hydrogels have also been used in the treatment of neurodegenerative diseases like Alzheimer's disease and Parkinson's disease, which show limited capacity for neural regeneration. Xu et al. fabricated conductive self-healing hydrogels with anti-inflammatory properties composed of dialdehyde polyurethane, gold nanoparticles, and O-carboxymethyl chitosan for treatment of Parkinson's disease. The hydrogels promoted neural stem cell proliferation and differentiation toward neuronal cells in vitro and led to improved motor functions in PD rats in vivo (Xu et al., 2022). Xue et al. fabricated gelatin-polyaniline hydrogels loaded with bone marrow stromal cells (BMSCs) for treatment of Parkinson's disease. The therapeutic efficacy was assessed by stereotactic injections in a MPTP-induced PD model. Histological and motor capability studies demonstrated that BMSCs-loaded hydrogels lead to improved and sustained protective effects than BMSCs alone providing support for the use of hydrogels to deliver cells in CNS disorders (Xue et al., 2019). Similarly, Feig et al. fabricated injectable, granular hydrogels from a CP–PEDOT: PSS and demonstrated its ability to form 3D stem cell niches in a neural microenvironment (Feig et al., 2021).

Nerve conduits are a special form of scaffolds with a tubular geometry, which act as nerve guidance channels and provide a path for the proper growth of regenerating nerves. Nerve conduits bridge two ends of a damaged nerve providing directional guidance, support, and protection for nerve growth. Nerve conduits are particularly useful in peripheral nerve injuries. Song et al. developed nerve conduits from polypyrrole, a conductive polymer, and explored the effect of electrical stimulation on the efficacy of implanted human neural progenitor cells in improving peripheral nerve regeneration in a sciatic nerve transaction model in rats. Electrical stimulation led to improved therapeutic efficacy through upregulation of neurotrophic factors supporting the positive impact of electrically enhanced cell-based polymeric nerve conduits in nerve regeneration (Song et al., 2021). Reviews expounding on the use of polymeric nerve conduits can be found here (Ghane et al., 2021; Jiang et al., 2020; Zarrintaj et al., 2020). Polymeric nanofibers are another class of scaffolds, which can be developed to mimic the ECM and enable nerve tissue regeneration. Electrospinning is a common technique to develop fibers using a wide range of materials including CPs. Additionally, therapeutic cargo can be loaded on these fibers to enhance regeneration. Nune et al. fabricated electroactive, melanin-incorporated silk fibroin nanofibrous scaffolds for nerve tissue engineering. Melanin incorporation imparted enhanced antioxidant and radical scavenging properties to the fibers. The neurogenic potential of these fibers was assessed by measuring the cellular viability, proliferation, adhesion, and differentiation levels of human neuroblastoma cells (SH-SY5Y) in vitro. The scaffolds stimulated cell attachment, adhesion, and promoted cell differentiation into neurons, providing evidence for the use of nanofibers as effective biomaterials for nerve regeneration and growth (Nune et al., 2019). Similar results were seen in a study employing electrospun cellulose-based conductive polymer nanofibers for growth and regeneration of nerve tissues (Zha et al., 2020).

48.8 Concluding Remarks

The use of polymeric biomaterials as delivery vectors and ECM-mimicking scaffolds has shown tremendous promise in the field of CNS diseases. Polymeric delivery systems offer significant advantages over other systems by allowing a tight control over polymer properties, which subsequently enables fine-tuning of the physicochemical properties of the delivery system and hence its fate in the body. However, clinical translation of polymeric nanoparticles for CNS diseases is still hindered by setbacks including lack of effective models to mimic the BBB and pathological complexities of the human brain that are difficult to mimic in vivo. Advances have been made in the form of dynamic "organ-on-a-chip" models, which better mimic the cellular microenvironment providing proof of concept data. However, a lot of work is still required to fully integrate these models in the industrial drug development pipeline. Similarly, regulatory guidelines specific for nanomedicines are ever-evolving. The recent approval of the nanoparticle-based COVID-19 vaccines has ushered in an era of nanomedicine and polymeric nanosystems, which have shown impressive potential in several preclinical studies and can soon be expected to be a clinical reality in the treatment of CNS disorders.

48.9 Review Questions

1. List invasive and noninvasive routes of drug delivery to the brain.
2. List common types of polymeric nanoparticles for targeted delivery to the brain.
3. Write a short note on carrier-mediated transcytosis (CMT).
4. Which physicochemical properties of polymeric nanomaterials affect blood–brain barrier penetration? (Tick all that apply)
   (a) Size
   (b) Shape
   (c) Polymer stiffness
   (d) Surface charge
   (e) All of the above
5. What are the major components of the blood–brain barrier?

48.10 Answers

1. Invasive and non-invasive routes of drug delivery to the brain are as follows -
   A. Invasive routes–convection enhanced delivery, intrathecal administration
   B. Non-invasive routes–focused ultrasound, nose-to-brain delivery, osmotic disruption
2. Polymeric nanoparticles that have been employed for targeted delivery to the brain include–dendrimers, hydrogels and nanogels, polymersomes, and micelles.
3. Carrier-mediated transcytosis (CMT) –

CMT involves leveraging the solute-like carrier (SLC) transporters to cross the BBB. SLCs are responsible for the supply of essential nutrients and energy to the brain via uptake of amino acids, glucose, and other necessary macromolecules from the blood. SLCs are specific for their substrates and are widely expressed on the BBB, enabling their use in CMT of therapeutics. Among the wide range of SLCs present at the BBB, the carbohydrate transporters (e.g., GLUTs, SGLTs), amino acid transporters (e.g., LAT1), and the peptide transporters (e.g., OATP, PepT1) are widely used for delivery of therapeutics. Drug molecules and nanoparticles can be modified to mimic the endogenous substrates for these transporters. Conjugating glucose to drugs and carrier systems is a widely used strategy. It has been used in improving CNS uptake of drugs, including ibuprofen and L-DOPA—a precursor of dopamine. This strategy also translates to polymeric nanoparticles enabling targeted delivery to the CNS. CMT is typically involved in the transport of small molecules, limiting its use in the delivery of large-molecule therapeutics.

4. (e) All the above
5. The BBB is created by several structural components unique to CNS blood vessels, including endothelial cells joined by tight junctions, astrocytes, pericytes, and a basement membrane made up of extracellular matrix components such as collagens, heparin sulfate proteoglycans, and laminins.

References

Abourehab, M. A., Ahmed, O. A., Balata, G. F., & Almalki, W. H. (2018). Self-assembled biodegradable polymeric micelles to improve dapoxetine delivery across the blood–brain barrier. *International Journal of Nanomedicine, 13*, 3679.

Aderibigbe, B. A., & Naki, T. (2019). Chitosan-based nanocarriers for nose to brain delivery. *Applied Sciences, 9*(11), 2219.

Agrawal, M., Saraf, S., Saraf, S., Antimisiaris, S. G., Chougule, M. B., Shoyele, S. A., & Alexander, A. (2018). Nose-to-brain drug delivery: An update on clinical challenges and progress towards approval of anti-Alzheimer drugs. *Journal of Controlled Release, 281*, 139–177.

Akbaba, G. E., Akbaba, H., & Kantarcı, A. G. (2018). Development and in vitro evaluation of positive-charged solid lipid nanoparticles as nucleic acid delivery system in glioblastoma treatment. *Marmara Pharmaceutical Journal, 22*(2), 299–306.

Alkilany, A. M., Zhu, L., Weller, H., Mews, A., Parak, W. J., Barz, M., & Feliu, N. (2019). Ligand density on nanoparticles: A parameter with critical impact on nanomedicine. *Advanced Drug Delivery Reviews, 143*, 22–36.

Anraku, Y., Kuwahara, H., Fukusato, Y., Mizoguchi, A., Ishii, T., Nitta, K., et al. (2017). Glycaemic control boosts glucosylated nanocarrier crossing the BBB into the brain. *Nature Communications, 8*(1), 1–9.

Anselmo, A. C., Zhang, M., Kumar, S., Vogus, D. R., Menegatti, S., Helgeson, M. E., & Mitragotri, S. (2015). Elasticity of nanoparticles influences their blood circulation, phagocytosis, endocytosis, and targeting. *ACS Nano, 9*(3), 3169–3177.

Aurand, E. R., Lampe, K. J., & Bjugstad, K. B. (2012). Defining and designing polymers and hydrogels for neural tissue engineering. *Neuroscience Research, 72*(3), 199–213.

Ayub, A., & Wettig, S. (2022). An overview of nanotechnologies for drug delivery to the brain. *Pharmaceutics, 14*(2), 224.

Banerjee, A., Arha, M., Choudhary, S., Ashton, R. S., Bhatia, S. R., Schaffer, D. V., & Kane, R. S. (2009). The influence of hydrogel modulus on the proliferation and differentiation of encapsulated neural stem cells. *Biomaterials, 30*(27), 4695–4699.

Batrakova, E. V., Li, S., Miller, D. W., & Kabanov, A. V. (1999). Pluronic P85 increases permeability of a broad spectrum of drugs in polarized BBMEC and Caco-2 cell monolayers. *Pharmaceutical Research, 16*(9), 1366–1372.

Bilia, A. R., Nardiello, P., Piazzini, V., Leri, M., Bergonzi, M. C., Bucciantini, M., & Casamenti, F. (2019). Successful brain delivery of andrographolide loaded in human albumin nanoparticles to TgCRND8 mice, an Alzheimer's disease mouse model. *Frontiers in Pharmacology, 10*, 910.

Bobde, Y., Biswas, S., & Ghosh, B. (2020). Current trends in the development of HPMA-based block copolymeric nanoparticles for their application in drug delivery. *European Polymer Journal, 139*, 110018.

Bobo, R. H., Laske, D. W., Akbasak, A., Morrison, P. F., Dedrick, R. L., & Oldfield, E. H. (1994). Convection-enhanced delivery of macromolecules in the brain. *Proceedings of the National Academy of Sciences, 91*(6), 2076–2080.

Bonina, F., Puglia, C., Rimoli, M. G., Melisi, D., Boatto, G., Nieddu, M., et al. (2003). Glycosyl derivatives of dopamine and L-dopa as anti-Parkinson prodrugs: Synthesis, pharmacological activity and in vitro stability studies. *Journal of Drug Targeting, 11*(1), 25–36.

Bouffard, G. (1906). Injection des couleurs de benzidine aux animaux normaux. *Annales de l'Institut Pasteur, 20*, 539–546.

Boussif, O., Lezoualc'h, F., Zanta, M. A., Mergny, M. D., Scherman, D., Demeneix, B., & Behr, J.-P. (1995). A versatile vector for gene and oligonucleotide transfer into cells in culture and in vivo: Polyethylenimine. *Proceedings of the National Academy of Sciences, 92*(16), 7297–7301.

Brown, T. D., Habibi, N., Wu, D., Lahann, J., & Mitragotri, S. (2020). Effect of Nanoparticle Composition, Size, Shape, and Stiffness on Penetration Across the Blood–Brain Barrier. *ACS Biomaterials Science & Engineering, 6*(9), 4916–4928.

Caprifico, A. E., Foot, P. J., Polycarpou, E., & Calabrese, G. (2020). Overcoming the blood-brain barrier: Functionalised chitosan nanocarriers. *Pharmaceutics, 12*(11), 1013.

Chakraborty, S., Filippi, C. G., Wong, T., Ray, A., Fralin, S., Tsiouris, A. J., et al. (2016). Superselective intraarterial cerebral infusion of cetuximab after osmotic blood/brain barrier disruption for recurrent malignant glioma: Phase I study. *Journal of Neuro-Oncology, 128*(3), 405–415.

Chen, Q., Gong, T., Liu, J., Wang, X., Fu, H., & Zhang, Z. (2009). Synthesis, in vitro and in vivo characterization of glycosyl derivatives of ibuprofen as novel prodrugs for brain drug delivery. *Journal of Drug Targeting, 17*(4), 318–328.

Chen, E. M., Quijano, A. R., Seo, Y.-E., Jackson, C., Josowitz, A. D., Noorbakhsh, S., et al. (2018). Biodegradable PEG-poly (ω-pentadecalactone-co-p-dioxanone) nanoparticles for enhanced and sustained drug delivery to treat brain tumors. *Biomaterials, 178*, 193–203.

Chen, Z., Lv, Z., Sun, Y., Chi, Z., & Qing, G. (2020). Recent advancements in polyethyleneimine-based materials and their biomedical, biotechnology, and biomaterial applications. *Journal of Materials Chemistry B, 8*(15), 2951–2973.

Choi, J., Rui, Y., Kim, J., Gorelick, N., Wilson, D. R., Kozielski, K., et al. (2020). Nonviral polymeric nanoparticles for gene therapy in pediatric CNS malignancies. *Nanomedicine: Nanotechnology, Biology and Medicine, 23*, 102115.

Cohen-Pfeffer, J. L., Gururangan, S., Lester, T., Lim, D. A., Shaywitz, A. J., Westphal, M., & Slavc, I. (2017). Intracerebroventricular delivery as a safe, long-term route of drug administration. *Pediatric Neurology, 67*, 23–35.

Cortés, H., Alcalá-Alcalá, S., Caballero-Florán, I. H., Bernal-Chávez, S. A., Ávalos-Fuentes, A., González-Torres, M., et al. (2020). A reevaluation of chitosan-decorated nanoparticles to cross the blood-brain barrier. *Membranes, 10*(9), 212.

Cruz, L. J., Stammes, M. A., Que, I., van Beek, E. R., Knol-Blankevoort, V. T., Snoeks, T. J., et al. (2016). Effect of PLGA NP size on efficiency to target traumatic brain injury. *Journal of Controlled Release, 223*, 31–41.

Curcio, M., Cirillo, G., Rouaen, J. R., Saletta, F., Nicoletta, F. P., Vittorio, O., & Iemma, F. (2020). Natural polysaccharide carriers in brain delivery: Challenge and perspective. *Pharmaceutics, 12*(12), 1183.

Daneman, R. (2012). The blood–brain barrier in health and disease. *Annals of Neurology, 72*(5), 648–672.

de Oliveira Junior, E. R., Nascimento, T. L., Salomão, M. A., da Silva, A. C. G., Valadares, M. C., & Lima, E. M. (2019). Increased nose-to-brain delivery of melatonin mediated by polycaprolactone nanoparticles for the treatment of glioblastoma. *Pharmaceutical Research, 36*(9), 1–10.

Del Amo, L., Cano, A., Ettcheto, M., Souto, E. B., Espina, M., Camins, A., et al. (2021). Surface functionalization of PLGA nanoparticles to increase transport across the BBB for Alzheimer's disease. *Applied Sciences, 11*(9), 4305.

Eftekhari, B. S., Eskandari, M., Janmey, P. A., Samadikuchaksaraei, A., & Gholipourmalekabadi, M. (2021). Conductive chitosan/polyaniline hydrogel with cell-imprinted topography as a potential substrate for neural priming of adipose derived stem cells. *RSC Advances, 11*(26), 15795–15807.

Ehrlich, P. (1885). Das sauerstoff-bedurfnis des organismus. *Eine farbenanalytische studie*.

Fan, Q., Liu, Y., Cui, G., Zhong, Z., & Deng, C. (2021). Brain delivery of Plk1 inhibitor via chimaeric polypeptide polymersomes for safe and superb treatment of orthotopic glioblastoma. *Journal of Controlled Release, 329*, 1139–1149.

Feig, V. R., Santhanam, S., McConnell, K. W., Liu, K., Azadian, M., Brunel, L. G., et al. (2021). Conducting polymer-based granular hydrogels for injectable 3D cell scaffolds. *Advanced Materials Technologies, 6*(6), 2100162.

Ferreira, H., Amorim, D., Lima, A. C., Pirraco, R. P., Costa-Pinto, A. R., Almeida, R., et al. (2021). A biocompatible and injectable hydrogel to boost the efficacy of stem cells in neurodegenerative diseases treatment. *Life Sciences, 287*, 120108.

Fowler, M., Cotter, J., Knight, B., Sevick-Muraca, E., Sandberg, D., & Sirianni, R. (2020). Intrathecal drug delivery in the era of nanomedicine. *Advanced Drug Delivery Reviews, 165*, 77–95.

Furtado, D., Björnmalm, M., Ayton, S., Bush, A. I., Kempe, K., & Caruso, F. (2018). Overcoming the blood–brain barrier: The role of nanomaterials in treating neurological diseases. *Advanced Materials, 30*(46), 1801362.

Gajbhiye, K., Pawar, A., Mahadik, K., & Gajbhiye, V. (2020). PEGylated nanocarriers: A promising tool for targeted delivery to the brain. *Colloids and Surfaces B: Biointerfaces, 187*, 110770.

Garrudo, F. F., Chapman, C. A., Hoffman, P. R., Udangawa, R. W., Silva, J. C., Mikael, P. E., et al. (2019). Polyaniline-polycaprolactone blended nanofibers for neural cell culture. *European Polymer Journal, 117*, 28–37.

Garrudo, F. F., Mikael, P. E., Rodrigues, C. A., Udangawa, R. W., Paradiso, P., Chapman, C. A., et al. (2021). Polyaniline-polycaprolactone fibers for neural applications: Electroconductivity enhanced by pseudo-doping. *Materials Science and Engineering: C, 120*, 111680.

Ghane, N., Khalili, S., Nouri Khorasani, S., Esmaeely Neisiany, R., Das, O., & Ramakrishna, S. (2021). Regeneration of the peripheral nerve via multifunctional electrospun scaffolds. *Journal of Biomedical Materials Research Part A, 109*(4), 437–452.

Ghasemi-Mobarakeh, L., Prabhakaran, M. P., Morshed, M., Nasr-Esfahani, M. H., Baharvand, H., Kiani, S., et al. (2011). Application of conductive polymers, scaffolds and electrical stimulation for nerve tissue engineering. *Journal of Tissue Engineering and Regenerative Medicine, 5*(4), e17–e35.

Godinho, B. M., Henninger, N., Bouley, J., Alterman, J. F., Haraszti, R. A., Gilbert, J. W., et al. (2018). Transvascular delivery of hydrophobically modified siRNAs: Gene silencing in the rat brain upon disruption of the blood-brain barrier. *Molecular Therapy, 26*(11), 2580–2591.

Goyal, K., Konar, A., Kumar, B. H., & Koul, V. (2018). Lactoferrin-conjugated pH and redox-sensitive polymersomes based on PEG-SS-PLA-PCL-OH boost delivery of bacosides to the brain. *Nanoscale, 10*(37), 17781–17798.

Grimaudo, M., Krishnakumar, G., Giusto, E., Furlani, F., Bassi, G., Rossi, A., et al. (2021). Bioactive injectable hydrogels for on demand molecule/cell delivery and for tissue regeneration in the central nervous system. *Acta Biomaterialia*.

Gynther, M., Jalkanen, A., Lehtonen, M., Forsberg, M., Laine, K., Ropponen, J., et al. (2010). Brain uptake of ketoprofen–lysine prodrug in rats. *International Journal of Pharmaceutics, 399*(1–2), 121–128.

Habault, J., & Poyet, J.-L. (2019). Recent advances in cell penetrating peptide-based anticancer therapies. *Molecules, 24*(5), 927.

Hartl, N., Adams, F., & Merkel, O. M. (2021). From adsorption to covalent bonding: Apolipoprotein E functionalization of polymeric nanoparticles for drug delivery across the blood–brain barrier. *Advanced Therapeutics, 4*(1), 2000092.

Hemmelmann, M., Knoth, C., Schmitt, U., Allmeroth, M., Moderegger, D., Barz, M., et al. (2011). HPMA based amphiphilic copolymers mediate central nervous effects of domperidone. *Macromolecular Rapid Communications, 32*(9–10), 712–717.

Hemmelmann, M., Metz, V. V., Koynov, K., Blank, K., Postina, R., & Zentel, R. (2012). Amphiphilic HPMA–LMA copolymers increase the transport of Rhodamine 123 across a BBB model without harming its barrier integrity. *Journal of Controlled Release, 163*(2), 170–177.

Hervé, F., Ghinea, N., & Scherrmann, J.-M. (2008). CNS delivery via adsorptive transcytosis. *The AAPS Journal, 10*(3), 455–472.

Hoffman, A. S. (2008). The origins and evolution of "controlled" drug delivery systems. *Journal of Controlled Release, 132*(3), 153–163.

Hong, L. T. A., Kim, Y.-M., Park, H. H., Hwang, D. H., Cui, Y., Lee, E. M., et al. (2017). An injectable hydrogel enhances tissue repair after spinal cord injury by promoting extracellular matrix remodeling. *Nature Communications, 8*(1), 1–14.

Iorio-Morin, C., Gahide, G., Morin, C., Vanderweyen, D., Roy, M.-A., St-Pierre, I., et al. (2021). Management of primary central nervous system lymphoma using intra-arterial chemotherapy with osmotic blood-brain barrier disruption: Retrospective analysis of the Sherbrooke cohort. *Frontiers in Oncology, 10*, 543648.

Jeong, D. U., Bae, S., Macks, C., Whitaker, J., Lynn, M., Webb, K., & Lee, J. S. (2021). Hydrogel-mediated local delivery of dexamethasone reduces neuroinflammation after traumatic brain injury. *Biomedical Materials, 16*(3), 035002.

Jiang, H., Qian, Y., Fan, C., & Ouyang, Y. (2020). Polymeric guide conduits for peripheral nerve tissue engineering. *Frontiers in Bioengineering and Biotechnology, 8*, 582646.

Jiang, T., Qiao, Y., Ruan, W., Zhang, D., Yang, Q., Wang, G., et al. (2021). Cation-free siRNA micelles as effective drug delivery platform and potent RNAi nanomedicines for glioblastoma therapy. *Advanced Materials, 33*(45), 2104779.

Joseph, A., Simo, G. M., Gao, T., Alhindi, N., Xu, N., Graham, D. J., et al. (2021). Surfactants influence polymer nanoparticle fate within the brain. *Biomaterials, 277*, 121086.

Kabanov, A. V., Batrakova, E. V., & Miller, D. W. (2003). Pluronic® block copolymers as modulators of drug efflux transporter activity in the blood–brain barrier. *Advanced Drug Delivery Reviews, 55*(1), 151–164.

Kadry, H., Noorani, B., & Cucullo, L. (2020). A blood–brain barrier overview on structure, function, impairment, and biomarkers of integrity. *Fluids and Barriers of the CNS, 17*(1), 1–24.

Kanazawa, T., Kurano, T., Ibaraki, H., Takashima, Y., Suzuki, T., & Seta, Y. (2019). Therapeutic effects in a transient middle cerebral artery occlusion rat model by nose-to-brain delivery of anti-TNF-alpha siRNA with cell-penetrating peptide-modified polymer micelles. *Pharmaceutics, 11*(9), 478.

Kanazawa, T., Taki, H., & Okada, H. (2020). Nose-to-brain drug delivery system with ligand/cell-penetrating peptide-modified polymeric nano-micelles for intracerebral gliomas. *European Journal of Pharmaceutics and Biopharmaceutics, 152*, 85–94.

Kanwar, N., Bhandari, R., Kuhad, A., & Sinha, V. (2019). Polycaprolactone-based neurotherapeutic delivery of rasagiline targeting behavioral and biochemical deficits in Parkinson's disease. *Drug Delivery and Translational Research, 9*(5), 891–905.

Katona, G., Balogh, G. T., Dargó, G., Gáspár, R., Márki, Á., Ducza, E., et al. (2020). Development of meloxicam-human serum albumin nanoparticles for nose-to-brain delivery via application of a quality by design approach. *Pharmaceutics, 12*(2), 97.

Kaur, J., Gulati, M., Kapoor, B., Jha, N. K., Gupta, P. K., Gupta, G., et al. (2022). Advances in designing of polymeric micelles for biomedical application in brain related diseases. *Chemico-Biological Interactions, 361*, 109960.

Khan, A. R., Yang, X., Fu, M., & Zhai, G. (2018). Recent progress of drug nanoformulations targeting to brain. *Journal of Controlled Release, 291*, 37–64.

Khan, N. U., Miao, T., Ju, X., Guo, Q., & Han, L. (2019). 6 - Carrier-mediated transportation through BBB. In H. Gao & X. Gao (Eds.), *Brain targeted drug delivery system* (pp. 129–158). Academic Press.

Kolhar, P., Anselmo, A. C., Gupta, V., Pant, K., Prabhakarpandian, B., Ruoslahti, E., & Mitragotri, S. (2013). Using shape effects to target antibody-coated nanoparticles to lung and brain endothelium. *Proceedings of the National Academy of Sciences, 110*(26), 10753–10758.

Kopeček, J., & Kopečková, P. (2010). HPMA copolymers: Origins, early developments, present, and future. *Advanced Drug Delivery Reviews, 62*(2), 122–149.

Kousalová, J., & Etrych, T. (2018). Polymeric nanogels as drug delivery systems. *Physiological Research, 67, S305.*

Kulkarni, S. A., & Feng, S.-S. (2013). Effects of particle size and surface modification on cellular uptake and biodistribution of polymeric nanoparticles for drug delivery. *Pharmaceutical Research, 30*(10), 2512–2522.

Lahkar, S., & Das, M. K. (2020). Surface-modified polycaprolactone nanoparticles for the brain-targeted delivery of nevirapine. *Journal of Nanoparticle Research, 22*(5), 1–19.

Lalatsa, A., & Butt, A. M. (2018). Physiology of the blood–brain barrier and mechanisms of transport across the BBB. In *Nanotechnology-based targeted drug delivery systems for brain tumors* (pp. 49–74). Elsevier.

Lamptey, R. N. L., Gothwal, A., Trivedi, R., Arora, S., & Singh, J. (2022). Synthesis and characterization of fatty acid grafted chitosan polymeric micelles for improved gene delivery of VGF to the brain through intranasal route. *Biomedicine, 10*(2), 493.

Lee, J. S., & Feijen, J. (2012). Polymersomes for drug delivery: Design, formation and characterization. *Journal of Controlled Release, 161*(2), 473–483.

Lee, J. M., Moon, J. Y., Kim, T. H., Lee, S. W., Ahrberg, C. D., & Chung, B. G. (2018). Conductive hydrogel/nanowire micropattern-based sensor for neural stem cell differentiation. *Sensors and Actuators B: Chemical, 258*, 1042–1050.

Lefley, J., Waldron, C., & Becer, C. R. (2020). Macromolecular design and preparation of polymersomes. *Polymer Chemistry, 11*(45), 7124–7136.

Lewandowsky, M. (1909). Zur lehre der cerebrospinalflussigkeit. *Zeitschrift für Klinische Medizin, 40*, 480–494.

Li, J., & Sabliov, C. (2013). PLA/PLGA nanoparticles for delivery of drugs across the blood-brain barrier. *Nanotechnology Reviews, 2*(3), 241–257.

Li, Q.-Y., Lee, J.-H., Kim, H.-W., & Jin, G.-Z. (2021). Research Models of the Nanoparticle-Mediated Drug Delivery across the Blood–Brain Barrier. *Tissue Engineering and Regenerative Medicine, 18*(6), 917–930.

Liu, J.-S., Wang, J.-H., Zhou, J., Tang, X.-H., Xu, L., Shen, T., et al. (2014). Enhanced brain delivery of lamotrigine with Pluronic® P123-based nanocarrier. *International Journal of Nanomedicine, 9*, 3923.

Lu, Q., Cai, X., Zhang, X., Li, S., Song, Y., Du, D., et al. (2018). Synthetic polymer nanoparticles functionalized with different ligands for receptor-mediated transcytosis across the blood–brain barrier. *ACS Applied Bio Materials, 1*(5), 1687–1694.

Lu, L., Zhao, X., Fu, T., Li, K., He, Y., Luo, Z., et al. (2020). An iRGD-conjugated prodrug micelle with blood-brain-barrier penetrability for anti-glioma therapy. *Biomaterials, 230*, 119666.

Lungwitz, U., Breunig, M., Blunk, T., & Göpferich, A. (2005). Polyethylenimine-based non-viral gene delivery systems. *European Journal of Pharmaceutics and Biopharmaceutics, 60*(2), 247–266.

Ma, W., Fu, F., Zhu, J., Huang, R., Zhu, Y., Liu, Z., et al. (2018). 64 Cu-Labeled multifunctional dendrimers for targeted tumor PET imaging. *Nanoscale, 10*(13), 6113–6124.

Martins, P. P., Smyth, H. D., & Cui, Z. (2019). Strategies to facilitate or block nose-to-brain drug delivery. *International Journal of Pharmaceutics, 570*, 118635.

Mehta, A., Sonabend, A., & Bruce, J. (2017). Convection-enhanced delivery. *Neurotherapeutics, 14*(2), 358–371.

Meng, Y., Hynynen, K., & Lipsman, N. (2021a). Applications of focused ultrasound in the brain: From thermoablation to drug delivery. *Nature Reviews Neurology, 17*(1), 7–22.

Meng, Y., Pople, C. B., Budiansky, D., Li, D., Suppiah, S., Lim-Fat, M. J., et al. (2021b). Current state of therapeutic focused ultrasound applications in neuro-oncology. *Journal of Neuro-Oncology, 156*, 1–11.

Meng, Y., Reilly, R. M., Pezo, R. C., Trudeau, M., Sahgal, A., Singnurkar, A., et al. (2021c). MR-guided focused ultrasound enhances delivery of trastuzumab to Her2-positive brain metastases. *Science Translational Medicine, 13*(615), eabj4011.

Meng, Q., Meng, H., Pan, Y., Liu, J., Li, J., Qi, Y., & Huang, Y. (2022). Influence of nanoparticle size on blood–brain barrier penetration and the accumulation of anti-seizure medicines in the brain. *Journal of Materials Chemistry B, 10*(2), 271–281.

Min, H. S., Kim, H. J., Naito, M., Ogura, S., Toh, K., Hayashi, K., et al. (2020). Systemic brain delivery of antisense oligonucleotides across the blood–brain barrier with a glucose-coated polymeric nanocarrier. *Angewandte Chemie International Edition, 59*(21), 8173–8180.

Muniswamy, V. J., Raval, N., Gondaliya, P., Tambe, V., Kalia, K., & Tekade, R. K. (2019). 'Dendrimer-Cationized-Albumin' encrusted polymeric nanoparticle improves BBB penetration and anticancer activity of doxorubicin. *International Journal of Pharmaceutics, 555*, 77–99.

Musumeci, T., Bonaccorso, A., & Puglisi, G. (2019). Epilepsy disease and nose-to-brain delivery of polymeric nanoparticles: An overview. *Pharmaceutics, 11*(3), 118.

Nisbet, D. R., Yu, L., Zahir, T., Forsythe, J. S., & Shoichet, M. (2008). Characterization of neural stem cells on electrospun poly (ε-caprolactone) submicron scaffolds: Evaluating their potential in neural tissue engineering. *Journal of Biomaterials Science, Polymer Edition, 19*(5), 623–634.

Nisbet, D. R., Rodda, A. E., Horne, M. K., Forsythe, J. S., & Finkelstein, D. I. (2009). Neurite infiltration and cellular response to electrospun polycaprolactone scaffolds implanted into the brain. *Biomaterials, 30*(27), 4573–4580.

Niu, J., Wang, L., Yuan, M., Zhang, J., Chen, H., & Zhang, Y. (2020). Dual-targeting nanocarrier based on glucose and folic acid functionalized pluronic P105 polymeric micelles for enhanced brain distribution. *Journal of Drug Delivery Science and Technology, 57*, 101343.

Nune, M., Manchineella, S., Govindaraju, T., & Narayan, K. (2019). Melanin incorporated electroactive and antioxidant silk fibroin nanofibrous scaffolds for nerve tissue engineering. *Materials Science and Engineering: C, 94*, 17–25.

Ouyang, J., Jiang, Y., Deng, C., Zhong, Z., & Lan, Q. (2021). Doxorubicin delivered via ApoE-directed reduction-sensitive polymersomes potently inhibit orthotopic human glioblastoma xenografts in nude mice. *International Journal of Nanomedicine, 16*, 4105.

Pandit, R., Chen, L., & Götz, J. (2020). The blood-brain barrier: Physiology and strategies for drug delivery. *Advanced Drug Delivery Reviews, 165*, 1–14.

Pang, Z., Gao, H., Chen, J., Shen, S., Zhang, B., Ren, J., et al. (2012). Intracellular delivery mechanism and brain delivery kinetics of biodegradable cationic bovine serum albumin-conjugated polymersomes. *International Journal of Nanomedicine, 7*, 3421.

Peviani, M., Palmiero, U. C., Cecere, F., Milazzo, R., Moscatelli, D., & Biffi, A. (2019). Biodegradable polymeric nanoparticles administered in the cerebrospinal fluid: Brain biodistribution, preferential internalization in microglia and implications for cell-selective drug release. *Biomaterials, 209*, 25–40.

Pitto-Barry, A., & Barry, N. P. (2014). Pluronic® block-copolymers in medicine: From chemical and biological versatility to rationalisation and clinical advances. *Polymer Chemistry, 5*(10), 3291–3297.

Pokharkar, V., Suryawanshi, S., & Dhapte-Pawar, V. (2020). Exploring micellar-based polymeric systems for effective nose-to-brain drug delivery as potential neurotherapeutics. *Drug Delivery and Translational Research, 10*(4), 1019–1031.

Puris, E., Gynther, M., Huttunen, J., Petsalo, A., & Huttunen, K. M. (2017). L-type amino acid transporter 1 utilizing prodrugs: How to achieve effective brain delivery and low systemic exposure of drugs. *Journal of Controlled Release, 261*, 93–104.

Rabiee, N., Ahmadi, S., Afshari, R., Khalaji, S., Rabiee, M., Bagherzadeh, M., et al. (2021). Polymeric nanoparticles for nasal drug delivery to the brain: Relevance to Alzheimer's disease. *Advanced Therapeutics, 4*(3), 2000076.

Rao, K. S., Reddy, M. K., Horning, J. L., & Labhasetwar, V. (2008). TAT-conjugated nanoparticles for the CNS delivery of anti-HIV drugs. *Biomaterials, 29*(33), 4429–4438.

Rapoport, S. I., Hori, M., & Klatzo, I. (1972). Testing of a hypothesis for osmotic opening of the blood-brain barrier. *American Journal of Physiology-Legacy Content, 223*(2), 323–331.

Rathore, P., Arora, I., Rastogi, S., Akhtar, M., Singh, S., & Samim, M. (2020). Collagen nanoparticle-mediated brain silymarin delivery: An approach for treating cerebral ischemia and reperfusion-induced brain injury. *Frontiers in Neuroscience, 14*, 538404.

Ravichandran, R., Sundarrajan, S., Venugopal, J. R., Mukherjee, S., & Ramakrishna, S. (2010). Applications of conducting polymers and their issues in biomedical engineering. *Journal of the Royal Society Interface, 7*(suppl_5), S559–S579.

Ribovski, L., de Jong, E., Mergel, O., Zu, G., Keskin, D., van Rijn, P., & Zuhorn, I. S. (2021a). Low nanogel stiffness favors nanogel transcytosis across an in vitro blood–brain barrier. *Nanomedicine: Nanotechnology, Biology and Medicine, 34*, 102377.

Ribovski, L., Hamelmann, N. M., & Paulusse, J. M. (2021b). Polymeric nanoparticles properties and brain delivery. *Pharmaceutics, 13*(12), 2045.

Shakeri, S., Ashrafizadeh, M., Zarrabi, A., Roghanian, R., Afshar, E. G., Pardakhty, A., et al. (2020). Multifunctional polymeric nanoplatforms for brain diseases diagnosis, therapy and theranostics. *Biomedicines, 8*(1), 13. Retrieved from https://www.mdpi.com/2227-9059/8/1/13

Sharma, A., Porterfield, J. E., Smith, E., Sharma, R., Kannan, S., & Kannan, R. M. (2018). Effect of mannose targeting of hydroxyl PAMAM dendrimers on cellular and organ biodistribution in a neonatal brain injury model. *Journal of Controlled Release, 283*, 175–189.

Sharma, A. K., Prasher, P., Aljabali, A. A., Mishra, V., Gandhi, H., Kumar, S., et al. (2020). Emerging era of "somes": Polymersomes as versatile drug delivery carrier for cancer diagnostics and therapy. *Drug Delivery and Translational Research, 10*(5), 1171–1190.

Sharma, R., Liaw, K., Sharma, A., Jimenez, A., Chang, M., Salazar, S., et al. (2021). Glycosylation of PAMAM dendrimers significantly improves tumor macrophage targeting and specificity in glioblastoma. *Journal of Controlled Release, 337*, 179–192.

Shi, Y., Jiang, Y., Cao, J., Yang, W., Zhang, J., Meng, F., & Zhong, Z. (2018). Boosting RNAi therapy for orthotopic glioblastoma with nontoxic brain-targeting chimaeric polymersomes. *Journal of Controlled Release, 292*, 163–171.

Siddiqui, M. A., Asad, M., Akhter, J., Hoda, U., Rastogi, S., Arora, I., et al. (2022). Resveratrol-loaded glutathione-coated collagen nanoparticles attenuate acute seizures by inhibiting HMGB1 and TLR-4 in the hippocampus of mice. *ACS Chemical Neuroscience, 13*(8), 1342–1354.

Singh, S., Drude, N., Blank, L., Desai, P. B., Königs, H., Rütten, S., et al. (2021). Protease responsive nanogels for transcytosis across the blood– brain barrier and intracellular delivery of radiopharmaceuticals to brain tumor cells. *Advanced Healthcare Materials, 10*(20), 2100812.

Song, S., McConnell, K. W., Amores, D., Levinson, A., Vogel, H., Quarta, M., et al. (2021). Electrical stimulation of human neural stem cells via conductive polymer nerve guides enhances peripheral nerve recovery. *Biomaterials, 275*, 120982.

Stern, L., & Gautier, R. (1921). Recherches sur Le liquide céphalo-rachidien: I.–Les rapports entre Le liquide céphalo-rachidien et la circulation sanguine. *Archives Internationales de Physiologie, 17*(2), 138–192.

Sun, T., Krishnan, V., Pan, D. C., Filippov, S. K., Ravid, S., Sarode, A., et al. (2022). Ultrasound-mediated delivery of flexibility-tunable polymer drug conjugates for treating glioblastoma. *Bioengineering & Translational Medicine*, e10408.

Ucar, B. (2021). Natural biomaterials in brain repair: A focus on collagen. *Neurochemistry International, 146*, 105033.

Unal, S., Arslan, S., Yilmaz, B. K., Oktar, F. N., Ficai, D., Ficai, A., & Gunduz, O. (2020). Polycaprolactone/gelatin/hyaluronic acid electrospun scaffolds to mimic glioblastoma extracellular matrix. *Materials, 13*(11), 2661.

Uz, M., & Mallapragada, S. K. (2019). Conductive polymers and hydrogels for neural tissue engineering. *Journal of the Indian Institute of Science, 99*(3), 489–510.

Vashist, A., Kaushik, A., Vashist, A., Bala, J., Nikkhah-Moshaie, R., Sagar, V., & Nair, M. (2018). Nanogels as potential drug nanocarriers for CNS drug delivery. *Drug Discovery Today, 23*(7), 1436–1443.

Voigt, N., Henrich-Noack, P., Kockentiedt, S., Hintz, W., Tomas, J., & Sabel, B. A. (2014). Surfactants, not size or zeta-potential influence blood–brain barrier passage of polymeric nanoparticles. *European Journal of Pharmaceutics and Biopharmaceutics, 87*(1), 19–29.

Wang, L., Li, M., Xie, Y., Xu, L., Ye, R., & Liu, X. (2017). Preclinical efficacy of human Albumin in subarachnoid hemorrhage. *Neuroscience, 344*, 255–264.

Wang, Z., Xiong, G., Tsang, W. C., Schätzlein, A. G., & Uchegbu, I. F. (2019). Nose-to-brain delivery. *Journal of Pharmacology and Experimental Therapeutics, 370*(3), 593–601.

Wang, Y., Jiang, Y., Wei, D., Singh, P., Yu, Y., Lee, T., et al. (2021). Nanoparticle-mediated convection-enhanced delivery of a DNA intercalator to gliomas circumvents temozolomide resistance. *Nature Biomedical Engineering, 5*(9), 1048–1058.

Wang, L., Zhang, D., Ren, Y., Guo, S., Li, J., Ma, S., et al. (2022). Injectable hyaluronic acid hydrogel loaded with BMSC and NGF for traumatic brain injury treatment. *Materials Today Bio, 13*, 100201.

Wei, X., Zhao, G., Jia, Z., Zhao, Z., Chen, N., Sun, Y., et al. (2022). Macromolecular dexamethasone prodrug ameliorates Neuroinflammation and prevents bone loss associated with traumatic brain injury. *Molecular Pharmaceutics, 19*(11), 4000–4009.

Wiley, D. T., Webster, P., Gale, A., & Davis, M. E. (2013). Transcytosis and brain uptake of transferrin-containing nanoparticles by tuning avidity to transferrin receptor. *Proceedings of the National Academy of Sciences, 110*(21), 8662–8667.

Wu, C., Liu, A., Chen, S., Zhang, X., Chen, L., Zhu, Y., et al. (2019). Cell-laden electroconductive hydrogel simulating nerve matrix to deliver electrical cues and promote neurogenesis. *ACS Applied Materials & Interfaces, 11*(25), 22152–22163.

Xiao, T., He, M., Xu, F., Fan, Y., Jia, B., Shen, M., et al. (2021). Macrophage membrane-camouflaged responsive polymer Nanogels enable magnetic resonance imaging-guided chemotherapy/chemodynamic therapy of orthotopic glioma. *ACS Nano, 15*(12), 20377–20390.

Xie, J., Shen, Z., Anraku, Y., Kataoka, K., & Chen, X. (2019). Nanomaterial-based blood-brain-barrier (BBB) crossing strategies. *Biomaterials, 224*, 119491.

Xu, F., Lu, W., Wu, H., Fan, L., Gao, X., & Jiang, X. (2009). Brain delivery and systemic effect of cationic albumin conjugated PLGA nanoparticles. *Journal of Drug Targeting, 17*(6), 423–434.

Xu, X., Liu, K., Wang, Y., Zhang, C., Shi, M., Wang, P., et al. (2019). A multifunctional low-generation dendrimer-based nanoprobe for the targeted dual mode MR/CT imaging of orthotopic brain gliomas. *Journal of Materials Chemistry B, 7*(23), 3639–3643.

Xu, J., Tai, C.-H., Chen, T.-Y., & Hsu, S.-H. (2022). An anti-inflammatory electroconductive hydrogel with self-healing property for the treatment of Parkinson's disease. *Chemical Engineering Journal*, 137180.

Xue, J., Liu, Y., Darabi, M. A., Tu, G., Huang, L., Ying, L., et al. (2019). An injectable conductive Gelatin-PANI hydrogel system serves as a promising carrier to deliver BMSCs for Parkinson's disease treatment. *Materials Science and Engineering: C, 100*, 584–597.

Yang, Y., Zhang, X., Wu, S., Zhang, R., Zhou, B., Zhang, X., et al. (2022). Enhanced nose-to-brain delivery of siRNA using hyaluronan-enveloped nanomicelles for glioma therapy. *Journal of Controlled Release, 342*, 66–80. https://doi.org/10.1016/j.jconrel.2021.12.034

Yao, X., Miao, W., Li, M., Wang, M., Ma, J., Wang, Y., et al. (2010). Protective effect of albumin on VEGF and brain edema in acute ischemia in rats. *Neuroscience Letters, 472*(3), 179–183.

Ye, C., Cheng, M., Ma, L., Zhang, T., Sun, Z., Yu, C., et al. (2022). Oxytocin Nanogels inhibit innate inflammatory response for early intervention in Alzheimer's disease. *ACS Applied Materials & Interfaces, 14*, 21822.

Yu, Y. J., Zhang, Y., Kenrick, M., Hoyte, K., Luk, W., Lu, Y., et al. (2011). Boosting brain uptake of a therapeutic antibody by reducing its affinity for a transcytosis target. *Science Translational Medicine, 3*(84), 84ra44.

Yu, S., Xu, X., Feng, J., Liu, M., & Hu, K. (2019). Chitosan and chitosan coating nanoparticles for the treatment of brain disease. *International Journal of Pharmaceutics, 560*, 282–293.

Yu, Q., Jiang, X., Liu, X., Shen, W., Mei, X., Tian, H., & Wu, C. (2022). Glutathione-modified macrophage-derived cell membranes encapsulated metformin nanogels for the treatment of spinal cord injury. *Materials Science and Engineering: C*, 112668.

Zarrintaj, P., Zangene, E., Manouchehri, S., Amirabad, L. M., Baheiraei, N., Hadjighasem, M. R., et al. (2020). Conductive biomaterials as nerve conduits: Recent advances and future challenges. *Applied Materials Today, 20*, 100784.

Zha, F., Chen, W., Hao, L., Wu, C., Lu, M., Zhang, L., & Yu, D. (2020). Electrospun cellulose-based conductive polymer nanofibrous mats: Composite scaffolds and their influence on cell behavior with electrical stimulation for nerve tissue engineering. *Soft Matter, 16*(28), 6591–6598.

Zhang, Q., Shi, B., Ding, J., Yan, L., Thawani, J. P., Fu, C., & Chen, X. (2019). Polymer scaffolds facilitate spinal cord injury repair. *Acta Biomaterialia, 88*, 57–77.

Zhang, L., Fan, J., Li, G., Yin, Z., & Fu, B. M. (2020). Transcellular model for neutral and charged nanoparticles across an in vitro blood–brain barrier. *Cardiovascular Engineering and Technology, 11*(6), 607–620.

Zhang, W., Mehta, A., Tong, Z., Esser, L., & Voelcker, N. H. (2021). Development of polymeric nanoparticles for blood–brain barrier transfer—Strategies and challenges. *Advanced Science, 8*(10), 2003937.

Zhang, D., Ren, Y., He, Y., Chang, R., Guo, S., Ma, S., et al. (2022a). In situ forming and biocompatible hyaluronic acid hydrogel with reactive oxygen species-scavenging activity to improve traumatic brain injury repair by suppressing oxidative stress and neuroinflammation. *Materials Today Bio*, 100278.

Zhang, D., Tian, S., Liu, Y., Zheng, M., Yang, X., Zou, Y., et al. (2022b). Near infrared-activatable biomimetic nanogels enabling deep tumor drug penetration inhibit orthotopic glioblastoma. *Nature Communications, 13*(1), 1–16.

Zhao, Y., Liu, J., Gao, Y., Xu, Z., Dai, C., Li, G., et al. (2022). Conductive biocomposite hydrogels with multiple biophysical cues regulate schwann cell behaviors. *Journal of Materials Chemistry B, 10*(10), 1582–1590.

Zheng, M., Du, Q., Wang, X., Zhou, Y., Li, J., Xia, X., et al. (2021). Tuning the elasticity of polymersomes for brain tumor targeting. *Advanced Science, 8*(20), 2102001.

Zhi, K., Raji, B., Nookala, A. R., Khan, M. M., Nguyen, X. H., Sakshi, S., et al. (2021). PLGA nanoparticle-based formulations to cross the blood–brain barrier for drug delivery: From R&D to cGMP. *Pharmaceutics, 13*(4), 500.

Zhou, J., Patel, T. R., Sirianni, R. W., Strohbehn, G., Zheng, M.-Q., Duong, N., et al. (2013). Highly penetrative, drug-loaded nanocarriers improve treatment of glioblastoma. *Proceedings of the National Academy of Sciences, 110*(29), 11751–11756.

Zhu, X., Jin, K., Huang, Y., & Pang, Z. (2019a). Brain drug delivery by adsorption-mediated transcytosis. In *Brain targeted drug delivery system* (pp. 159–183). Elsevier.

Zhu, Y., Liu, C., & Pang, Z. (2019b). Dendrimer-based drug delivery systems for brain targeting. *Biomolecules, 9*(12), 790.

Adjuvants for Vaccines

49

Jason P. Stewart and Joseph A. Vetro

Abstract

This chapter highlights the general characteristics of pathogen- and host-derived adjuvants, their advantages and disadvantages, and their contribution to adaptive immunity. An introduction to host-derived immunostimulants to improve immune outcomes while minimizing toxicities follows, emphasizing the generation and efficacy of conformationally restricted, response-selective analogs of complement peptide C5a. Such host-derived immunostimulants show therapeutic, manufacturing, and commercial potential as adjuvants and represent just one approach to meeting the modern-day healthcare threats posed by antibiotic-resistant strains of bacteria and the emergence of novel pathogens.

Keywords

Antigen-presenting cells · C5a · Complement Peptide · EP54 · EP67 · Adjuvant · Vaccines

49.1 Introduction

Vaccines remain one of the most significant advances in medical history, evident by the eradication of smallpox and the near eradication of polio through worldwide vaccination efforts. Vaccination is primarily a prophylactic approach to generate long-lived humoral and cellular immune responses that protect against invading pathogens. Traditionally, vaccine development relied on live-attenuated, killed, or inactivated forms of a whole pathogen as the target antigen (i.e., the immune response target). Whole-cell pathogen vaccines effectively generate protective immune responses primarily due to their close resemblance to natural infections. Unfortunately, these vaccines pose undesirable issues, including unwanted host reactions, difficulties generating, attenuating, or killing the target pathogen in a laboratory setting, the risk of pathogen reversion to virulence, and logistical challenges in maintaining a cold supply chain for vaccine distribution. These vaccines can also pose significant safety issues in immunocompromised and neonate patients (Yadav et al., 2014).

In contrast to whole-cell pathogen vaccines that use the whole pathogen as the target antigen, acellular or recombinant vaccines have been increasing in popularity. Recombinant vaccines are cellular fragments from the target pathogen, specifically proteins or mRNA, which can stimulate protective adaptive immune responses. The first of these acellular vaccines were subunit protein vaccines. Subunit protein vaccines are short, specific fragments of pathogen-associated proteins that, when administered, are safer, noninfectious, and more defined immunologically than whole-pathogen vaccine approaches. Advances in antigen discovery technologies using reverse engineering of protective adaptive immune responses from pathogen-infected subjects have led to a new generation of vaccine development for the prevention of global infectious diseases. By limiting the antigen to immunodominant proteins from the pathogen, there is better control over the induced immune response. In addition, recombinant vaccines remove many of the challenges of whole-cell vaccines as there isn't a risk of virulence, and manufacturing is significantly easier.

More recently, other vaccine technologies, such as mRNA vaccines, have been introduced to the market. In contrast to subunit vaccines, which present target protein antigens to the vaccinated host, mRNA vaccines introduce mRNA that is translated by host ribosomes to produce protein antigens as targets for adaptive immune responses by the host. This method of target antigen introduction allows for larger quantities of antigen with prolonged host exposure to the antigen with a smaller dosage, a limitation to recombinant protein vaccines. A restriction of mRNA vaccines is delivery. mRNA

J. P. Stewart · J. A. Vetro (✉)
Department of Pharmaceutical Sciences, University of Nebraska Medical Center, Omaha, NE, USA
e-mail: jvetro@unmc.edu

H. E. Gendelman, T. Ikezu (eds.), *Neuroimmune Pharmacology and Therapeutics*,
https://doi.org/10.1007/978-3-031-68237-7_49

is large (up to 10^6 Da) and negatively charged, making it difficult for the vaccine to pass through the lipid bilayer of cell membranes. The host's immune system also has nucleases that degrade non-host mRNA (Chaudhary et al., 2021). These limitations lead to poor immunogenicity and ineffective immune responses.

As mentioned above, a prominent feature that underscores the current era of modern vaccinology is the ability to generate various antigens through synthetic chemistry and genetic engineering techniques. While these subunit vaccines can be produced rapidly at large scale with well-defined structures in highly purified forms, they are ineffective as vaccines due to their inherent lack of immunogenicity (Tsoras & Champion, 2019). This lack of immunogenicity is partly due to recombinant vaccines being only a polypeptide or oligonucleotide rather than an entire pathogen, thus not mimicking a natural infection as effectively as a whole-cell vaccine. Such vaccines require the addition of adjuvants to address this reduction in immunogenicity by protecting the immune response-inducing compound (e.g., protein antigen or mRNA) and efficiently potentiating and focusing the adaptive immune response against the target antigen. As such, the discovery and development of new adjuvants are critical for designing vaccines capable of meeting the modern threats posed to human and animal populations by new or resurgent infectious and noninfectious diseases.

This chapter highlights current trends in the development of adjuvants and their current proposed mechanisms of action with an emphasis on adjuvants derived from components of the complement pathway (i.e., host-derived adjuvants) vs. more commonly used adjuvants derived from toxins and other molecules of pathogens (i.e., pathogen-derived adjuvants).

49.2 Adjuvants: Vehicles and Immunostimulants

An adjuvant is any substance or formulation that increases adaptive immune responses against a specific immunogen. Adjuvants improve the target antigen's magnitude, duration, exposure, and quality within the host. Apart from increasing vaccine effectiveness, adjuvants also reduce the amount and number of vaccine doses required to generate a protective adaptive immune response, thus contributing to improved patient compliance.

Adjuvants can be broadly classified into either antigen-delivery vehicles or immunostimulants (Mohan et al., 2013). In many cases, immunostimulants can be combined with vehicles to increase vaccine efficacy further. In the past, adjuvant use was limited by inadequate adjuvant screening, leading to suboptimal adjuvant selection. This limitation necessitates the development of new vaccine adjuvants that promote adaptive immune responses that correlate with protection against different target pathogens. For example, research has shown that the bacterium *B. pertussis* requires a strong mucosal Th1 immune response (i.e., T cell response) along with a mucosal IgA response to produce long-term protection (Fedele et al., 2015; Solans & Locht, 2018). However, the current acellular vaccine against pertussis fails to induce this effective and pathogen-specific immune response, leading to suboptimal vaccine efficacy (Zerbo et al., 2019). The development of effective adjuvants that stimulate protection correlated will optimize vaccine immune responses against the antigen and the pathogen itself.

49.2.1 Vehicles

Vaccine-delivery vehicles are adjuvants that, as the name implies, carry the target antigen within the host. This "carrying" function contains several benefits, depending on the vehicle adjuvant utilized, including protecting the antigen from degradation, producing an antigen "depot" that allows continuous antigen exposure to the host, and carrying the antigen to target regions within the host (Gao et al., 2022; Gutjahr et al., 2016). Classic examples of vehicles include liposomes (Tandrup Schmidt et al., 2016), emulsions (Shah et al., 2015), proteosomes (Stark et al., 2022), polymeric particles (Figueiredo et al., 2022), and immune-stimulating complexes (ISCOM) (Crouch et al., 2005). In these vehicle adjuvant systems, the final vaccine formulation contains just the antigen within the vehicle. Some vehicle adjuvant systems contain the antigen covalently linked to the particle's surface rather than encapsulated. Examples of such systems include virus-like particles (VLPs) (Frietze et al., 2016), biocompatible and biodegradable particles (e.g., poly(D, L-lactide-coglycolide) (PLGA)) (Diwan et al., 2004), and adsorbed onto salt carriers such as calcium and aluminum (Masson et al., 2017). As stated above, vehicle adjuvants, in addition to carrying the antigen, can be formulated with immunostimulants to increase the adaptive immune response against the target antigen.

49.2.2 Immunostimulants

Immunostimulatory adjuvants activate parts of the innate and adaptive immune systems (e.g., macrophages, dendritic cells, lymphocytes). This activation occurs when the immunostimulant binds to receptors on target cells and generates immunostimulatory responses such as cytokine secretion or phagocytic priming. Immunostimulants bind to receptors such as pathogen recognition receptors (PRRs) or other receptors associated with immune cell activation (Yang et al., 2022). Examples of immunostimulants include muramyl-dipeptide (MDP), monophosphoryl lipid A (MPL)

(Ulrich & Myers, 1995), lipopolysaccharide (LPS) (Johnson et al., 1956; Shimoyama & Fukase, 2021), bacterial cell membranes (Muhlradt et al., 1998), complement proteins (Dempsey et al., 1996; Jacquier-Sarlin et al., 1995), cytokines (Afonso et al., 1994; Kurzawa et al., 1998), and oligonucleotide mimics of bacterial DNA (Champion, 2021). Many immunostimulatory adjuvants are derivatives of pathogenic molecules or signaling molecules modified to maintain their immune system activation potency while reducing their potential toxicity (Batista-Duharte et al., 2018). Vaccine systems containing immunostimulatory adjuvants can have the antigen covalently linked to the adjuvant. In some cases, this has been more efficient than administering the adjuvant and antigen unconjugated. Some examples of this approach include the incorporation of lipids, lipopolysaccharides, and lipoamino acids into peptide antigens by synthetic methods (Defoort et al., 1992; Martinon et al., 1992; Metzger et al., 1991; Wiesmuller et al., 1992), the genetic fusion of chemokines/cytokines to protein antigens (Biragyn et al., 1999), the conjugation of TLR agonists to antigens (Zhang et al., 2022), and the conjugation of bacterial oligonucleotides to protein antigens (Weiss et al., 2021).

Some adjuvant vehicles are also immunostimulatory, which can transport and stimulate adaptive immune responses against the target antigen. For example, specific types of saponins are used to create emulsions as vehicles for the target antigen that also stimulate cellular and humoral immune responses against the emulsified antigen (Shi & Chiang, 2005).

As mentioned, combining adjuvants by conjugation or formulation to optimize their efficacy by inducing a long-term protective adaptive immune response against the target antigen. While the simplified classification helps categorize adjuvants by their broad mechanisms of action, there are numerous methods of categorizing adjuvants based on their origin, structure, or mechanisms of action. There is no specific list of characteristics regarding the ideal adjuvant for any vaccine formulation because each vaccine targets a particular pathogen, and each pathogen requires a unique immune response (McKee et al., 2010). Returning to the *B. pertussis* example, a humoral immune response (i.e., antibody response) is suboptimal, whereas a humoral response is optimal for protection against SARS-CoV-2 (Dundar et al., 2022). Developing and formulating vaccine adjuvants should consider the desired immune response specific to the target pathogen.

49.3 Adjuvant Development: General Concepts

As stated above, vaccine adjuvants potentiate adaptive immune responses against a target antigen through various means that vary from adjuvant to adjuvant. In many cases, the precise mechanism of action is unknown, although, in general, the adjuvant binds to and induces various nonspecific activities within the innate arm of the immune system. Once activated, the innate branch of immunity (e.g., complement system, phagocytes, etc.) mediates various humoral and cell-mediated adaptive responses within and between the innate and adaptive immune systems. This mediation results in generalized activation and potentiation of the immune system against the target antigen to a much greater degree than the antigen could produce alone. Origins for adjuvant design can come from sources derived from microbial pathogens and the host's immune system.

49.3.1 Pathogen-Derived Adjuvants

Due to a continual arms race between pathogenic microbes and the mammalian immune system, the human immune system has developed a high sensitivity to the presence of bacteria, viruses, fungi, and any other component that could indicate the presence of a foreign pathogen. The initial response to these microbial signals is primarily a function of the innate branch of the immune system. Therefore, many pathogen-derived adjuvants have a mechanism of action that triggers an immune response within this branch of immunity. The use of these adjuvants in vaccines provides the signals to which the innate arm of the immune system vigorously reacts, resulting in the priming/potentiation of the immune system.

The use of pathogen-derived immunostimulants does come with inherent risks. The rapid and vigorous innate response to the pathogen-derived adjuvant can result in an overly aggressive and misdirected immune response, accompanied by undesirable side effects such as anaphylaxis, fever, injection site granulomas or rashes, and local or systemic inflammation. Another concern with pathogen-derived adjuvants is the risk of inducing a generalized immune response with little specificity directed to the target antigen of interest (Petrovsky, 2015). In addition, the magnitude of the response to the adjuvant could mask the desired response to the target antigen (Petrovsky & Aguilar, 2004). Thankfully, pathogen-derived adjuvants can be modified to minimize or eliminate the adverse side effects (e.g., systemic and local inflammation, cellular toxicity, and undesired immune responses.) while retaining their adjuvanticity and inducing the desired immune responses. Notable examples of such efforts include the development of dmLT from heat-labile enterotoxin (Clements & Norton, 2018), MPL from gram-negative bacteria LPS (Wang et al., 2020), saponin-based adjuvants (SBAs) (den Brok et al., 2016), CTA1 from cholera toxin (Lycke, 2005; Schussek et al., 2020). In addition to using pathogen-derived molecules as immunostimulatory adjuvants, pathogen structures are also being used as adjuvant vehicles to deliver the antigen in a way that will improve its immunogenicity. Examples include virus-like particles (VLPs) (Cimica & Galarza, 2017), outer membrane vesicles (OMVs) (Mancini et al., 2020), and viral vectors (Travieso et al., 2022).

49.3.2 Host-Derived Adjuvants

Against the various issues associated with pathogen-derived adjuvants, another area of adjuvant discovery focuses on developing adjuvants derived from molecules within the host immune system. Such host-derived adjuvants could display a degree of immunologic control, focus, and host tolerance that pathogen-derived adjuvants may not achieve. Molecules based on cell signaling and activation, particularly within the innate and adaptive immune systems, are the basis of these immunostimulatory adjuvants. Unfortunately, one concern about host-derived immunostimulants is their function within cell signaling and activation. The mode of action of signaling molecules encompasses a wide range of cellular effects within and without the immune system. Therefore, their undesired functions, namely unwanted cell signaling and overstimulation of nonspecific effector cells, must be removed.

Nonetheless, developing an adjuvant from a molecule involved in the induction of host innate/adaptive immunity is of keen interest. As mentioned above, the primary issue will be introducing modifications to the host-derived compound to avoid undesired activity, for example, engineering a degree of specificity to the host-derived adjuvant that would limit its activity to a single (or limited) pathway of immune activation or its affinity to a single (or smaller) population of immune cells. Current ventures in developing host-derived adjuvants include cytokine-derived (Degen & Schijns, 2017) or signaling molecules such as LTB_4 (Salina et al., 2020), antimicrobial peptides (Munshi et al., 2020), and complement peptides (Villiers et al., 2003).

49.3.3 Mucosal Adjuvants

One of the challenges to developing a vaccine system (antigen and adjuvant) against a pathogen is tailoring the induced adaptive immune response to be the most optimal to counter the target pathogen. Depending on the pathogen of interest, this could require systemic adaptive immune responses or adaptive immune responses limited to specific surfaces or organs. Many currently approved vaccine adjuvants potentiate adaptive immune responses at the systemic level. Most pathogens, however, enter the host through mucosal surfaces that act as barriers between the host and the external environment.

Mucosal surfaces of interest include those of the respiratory system (e.g., nose and lungs) and the gastrointestinal tract (e.g., stomach, small intestine, and large intestine). The ability to generate humoral and cellular immune responses at mucosal surfaces can help inhibit or restrict the entry of invading pathogens (Brandtzaeg, 2009). Furthermore, adaptive immune responses generated at one mucosal site can localize to other mucosal surfaces within the host and can be exploited to induce immune responses that are not easily accessible by clinically feasible vaccine delivery systems (Brandtzaeg, 2009; Lycke, 2012). Through oral or intranasal delivery, mucosal immunization is also the most viable and non-invasive route for mass immunization.

In contrast to systemic immunization, mucosal immunization faces the challenge of generating immune responses to an antigen after being administered at a non-sterile site with constant exposure to a large quantity of harmless food and environmental antigens (Tsuji & Kosaka, 2008). To produce a successful mucosal immune response, the vaccine system, in addition to inducing a specific response to the target antigen, must also overcome the immune tolerance mechanisms at the mucosal site. These mechanisms prevent unnecessary detrimental immune responses against non-pathogens and food antigens. While necessary to prevent autoimmune responses, these immune tolerance mechanisms can be circumvented by potent immunostimulatory adjuvants (immunostimulants) or by adjuvants that present the target antigen in quantities (e.g. vehicles).

Many adjuvant systems used for systemic vaccine applications, such as oil-in-water emulsions and mineral salt adjuvants, may not be suitable for mucosal applications due to their inability to overcome the inherent tolerogenicity associated with mucosal tissues. Effective mucosal adjuvants overcome this tolerogenicity by activating antigen-presenting dendritic cells (DCs). Examples include CpG, CT/CTA1, endotoxin-derived adjuvants, dmLT, cytokine-derived adjuvants, and VLPs (Clements & Norton, 2018; Huang et al., 2008; Iho et al., 2015; Kayamuro et al., 2010; Lycke, 2012; Rosales-Mendoza et al., 2019; Schussek et al., 2020; van Ginkel et al., 2005). Of the existing mucosal adjuvants, bacterial toxin adjuvants are most potent in generating humoral and cellular immune responses but are very toxic and may not be fit for human use. CpG and other toll-like receptor (TLR) agonists have shown promise as mucosal adjuvants but are generally associated with unwanted inflammation.

49.4 Development of a Host-Derived Adjuvant

As described above, many adjuvants are typically derived from pathogen-associated molecular patterns (PAMPs) and activate pattern-recognition receptors (PRRs) on innate immunity cells. (e.g., macrophages and dendritic cells.). The complement system is one of the first biological systems to activate in response to a PAMP. Consequently, complement system components provide a rich source of host-derived adjuvant development, thus avoiding the tedious development of a PAMP derivative that reduces the risk of adverse events while maintaining its immunostimulatory activity. This section describes the development of a host-derived adjuvant developed from a human complement protein, C5a. This host-derived adjuvant embodies the attributes of an immunostimulatory adjuvant by stimulating cellular and humoral responses (adaptive immunity) and phagocytic responses (innate immunity).

49.4.1 Complement

Complement are plasma proteins that make up part of the innate immune system. They comprise interrelated proteases arranged in a cascade fashion that activate in response to pathogens and PAMPs. The role of complement is twofold: (1) to serve as an initial line of defense against invading microorganisms and (2) to enhance the systemic host immune system by activating the various humoral and cell-mediated responses within both the innate and adaptive arms of the host immune system necessary for the elimination of the microorganism.

There are three pathways to activate the complement system: The classical pathway, activated by antigen-antibody complexes; the alternative path, activated by surface molecules on cells containing carbohydrates and lipids; and the lectin-binding path, triggered by the binding of either mannose-binding lectin protein (MBL) or ficolin to bacterial or fungal carbohydrate structures (Guo & Ward, 2005). Regardless of which pathway activates the complement system. All three pathways lead to (1) activation of phagocytosis and antigen presentation, (2) chemotaxis and phagocyte activation, and (3) formation of the membrane attack complex (MAC) (Rooijakkers et al., 2005). The activation of the complement system, regardless of the pathway, leads to the production of the MAC, assembled at the end of the complement cascade from complement peptides generated by proteolytic cleavage steps along the said cascade. The MAC disrupts the bacterial outer membrane by introducing pores and causing the bacterial cell to lyse (Doorduijn et al., 2022). Complement also brings and primes phagocytes to the site of infection through Complement system-produced complement peptides.

The complement system produces more than the MAC. Some of the proteolytic by-products produced by the complement system act as complement peptides (specifically C3a, C4a, and C5a). These complement peptides interact with phagocytes and other APCs, leading to various humoral and cellular responses. Many APCs express receptors for C3a and C5a (C3aR1, C5aR1/CD88, and C5aR2). The binding of complement peptides to their respective receptors leads to a pro-inflammatory response, priming APCs to phagocytize and display foreign antigens and, eventually, an adaptive immune response. C3aR1, C5aR1, and C5aR2 are found on B-cells and T-cells, showing complement peptides' ability to stimulate adaptive immune cells directly (Li et al., 2020, 2021; Merle et al., 2015; Schanzenbacher et al., 2022). The ability of complement peptides to stimulate a more focused innate and adaptive immune response makes them attractive as immunostimulatory adjuvants.

49.4.2 Opsonic Complement Components: Host-Derived Adjuvants

Complement proteins with opsonic properties have been used as immunostimulatory adjuvants to enhance antigen-specific immune responses, specifically C3b, C4b (Arvieux et al., 1988; Jacquier-Sarlin et al., 1995; Villiers et al., 2003), and C3d (Bergmann-Leitner et al., 2006; Dempsey et al., 1996). These complement peptides promote opsonization, the ability to bind to and coat the surface of a microorganism, rendering it more susceptible to immune cell uptake via phagocytosis. The rationale with these opsonic molecules was that an antigen covalently attached to said complement protein fragments would be rendered more vulnerable to uptake by APCs and, consequently, more effectively processed and presented by APCs. In each of these examples, the immunogenicity of the target antigen in the antigen-complement vaccine complex increased significantly relative to the antigen alone, supporting the proposed mechanism of action of the opsonic complement molecule adjuvant. These studies demonstrate the immunologic potential of using complement components as adjuvants in the design of vaccines.

49.4.3 The Complement Peptides

A significant group of pharmacologically active by-products of complement activation are complement peptides C3a, C4a, and C5a. These small (74–76 residue) peptides are proteolytically cleaved from their more significant parent complement components, C3, C4, and C5. One of the principal roles of complement peptides is to recruit phagocytes, inflammatory cells, and lymphocytes to sites of tissue injury and infection and to activate these cells' various effector responses once recruited (Hugli et al., 1981; Merle et al., 2015; Schanzenbacher et al., 2022). As mentioned earlier, complement peptides play important roles in the activation and regulation of humoral and cell-mediated responses to antigens due to their ability to modulate various humoral and cell-mediated activities by their binding to their respective receptors (C3aR1, C5aR1, C5aR2) between the innate and adaptive arms of the immune system (Alshammari et al., 2020; Mastellos et al., 2005; Morgan, 1986).

While the functions of these complement peptides make them attractive as potential immunostimulatory adjuvants, they are also very potent pro-inflammatory mediators that can induce potentially toxic levels of local and systemic inflammation (Horiuchi & Tsukamoto, 2016; Sadik et al., 2018). In addition to potential inflammatory toxicity, complement peptides also appear to downregulate immune function under certain conditions (Kawamoto et al., 2004; Morgan, 1986). Due to these risks, complement peptides are not ideal for use as immunostimulatory adjuvants. As such, complement peptide derivatives that significantly decrease inflammatory and immunosuppressive functions while maintaining the desired immunostimulatory properties must be developed.

49.4.4 Inflammatory and Immunostimulatory Properties of C5a

One of the complement peptides, C5a, is a 74-residue polypeptide with pleiotropic biological functions (Fig. 49.1), including stimulation of smooth muscle contraction, vascular permeability, mast cell degranulation, and chemotaxis. C5a potentiates antibody and antigen-induced T cell proliferative responses in vitro, possibly by activating T helper cells (Kim et al., 2021; Morgan et al., 1983). Direct stimulation of mouse and human dendritic cells with C5a enhances the expression of MHC class II, co-stimulatory molecules CD80, CD86, CD40, and CD54, and induces secretion of cytokines of Th1 phenotype (Li et al., 2012; Rudilla et al., 2012; Zaal et al., 2013). C5a also induces the secretion of various immunoregulatory cytokines from APCs, including IL-1, IL-6, IL-8, IL-12, TNFα, and IFNγ (Alshammari et al., 2020; Buchner et al., 1995; Floreani et al., 2007; Gasque et al., 1995; Goodman et al., 1982; Okusawa et al., 1987; Strey et al., 2003; Vecchiarelli et al., 1998; Wetsel, 1995). C5a-mediated stimulation of DCs also decreases intracellular production of cAMP, which is a negative regulator of DC activation and function (Li et al., 2012). Furthermore, C5a interaction with C5aR1 activates PI3K, ERK1/2, and NF-kB signaling pathways, positively regulating antigen uptake, presentation, and secretion of pro-inflammatory cytokines by DCs (Li et al., 2012; Zaal et al., 2013). C5a also shows synergistic and antagonistic cross-talk with toll-like receptors (TLRs) in the functional modulation of DCs (Li et al., 2012; Zaal et al., 2013). C5a decreased TLR4-mediated IL-12 secretion by DCs (Weaver et al., 2010). C5aR1 ligation with C5a generates protective anti-viral $CD8^+$ T cells (Kim et al., 2004). In contrast, genetic ablation, or pharmacologic blocking of C5aR1 signaling, predominantly generated Th2 and regulatory T cell responses (Kim et al., 2004). Furthermore, previous studies found that blockade of the C5aR impaired the ability of the immune system to generate memory $CD4^+$ T cells (Moulton et al., 2007). Also, DCs stimulated with C5a showed increased expression of multiple co-stimulatory molecules, migration to the draining lymph nodes, and strong interaction with antigen-specific naïve T cells (Li et al., 2012, 2020, 2021).

Besides immunostimulatory properties that promote antigen presentation and functional immune responses, C5a also possesses highly pro-inflammatory properties. These include its potent chemotactic activities for the recruitment of inflammatory cells to sites of tissue injury and infection, its ability to induce smooth muscle contraction (Hugli et al., 1987; Shin et al., 1968), increase vascular permeability (Hugli et al., 1981; Hugli & Muller-Eberhard, 1978), and generate the release of a variety of secondary inflammatory mediators such as histamine, lysosomal enzymes, and vasoactive eicosanoids from responsive cells such as mast cells, neutrophils, eosinophils, and macrophages (Drapeau et al., 1993; Goldstein et al., 1974; Johnson et al., 1975; Lundberg et al., 1987; Schorlemmer & Allison, 1976). The ligation of C5a induces these biological responses to one of its specific, high-affinity receptors, C5aR1/CD88, expressed on the surface of the C5a-responsive cells. Previously, cells of myeloid origin (neutrophils, monocytes, mast cells, and eosinophils) were the only cells to express C5aR1. More recent research shows that various cells of nonmyeloid origin express C5aR1. These include hepatocytes, astrocytes, bronchial epithelial cells, epithelial cells of the gut and kidney, and osteoblasts (Ender et al., 2017; Peng et al., 2009; Wang et al., 2019). Due to its immunostimulatory properties, C5a is an ideal basis for developing a host and complement-derived adjuvant. The optimal derivative would maintain C5a's immunostimulatory effects (i.e., chemotaxis and stimulation of macrophages and dendritic cells) while removing the pro-inflammatory functions. The theorized basis for these functions lies in the structure of C5a and its binding to the C5aR1 receptor.

49.4.5 C5a: Structure-Function Considerations

Human C5a is a 74-residue glycopolypeptide comprised of two structural and functional domains. The first is the well-ordered N-terminal core domain consisting of residues 1–63 and is primarily involved in the recognition and binding of C5aR1 (Rooijakkers et al., 2005). The second is the C-terminal domain, residues 64–74, which extends from the N-terminal core as a finger-like projection. $C5a_{65-74}$

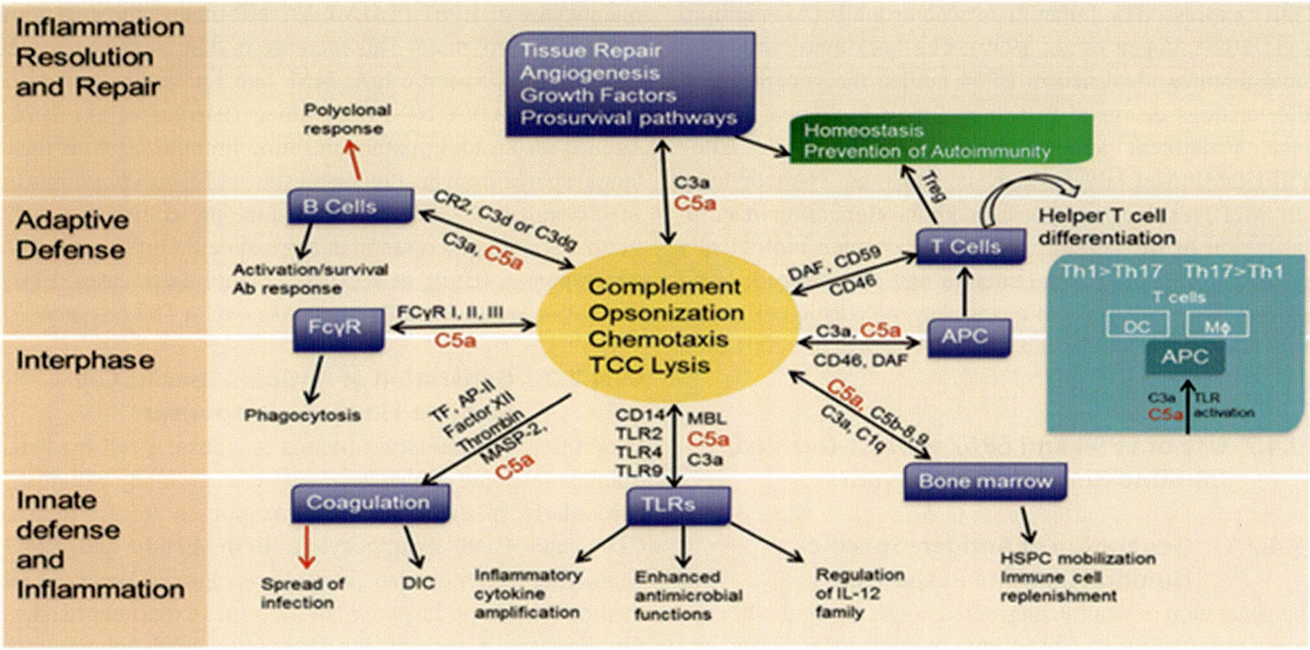

Fig. 49.1 Role of complement C5a in host defense and homeostasis. (Adapted from Ricklin et al. (2010))

(ISHKDMQLGR) is a region of considerable backbone flexibility and a poorly defined structure. The inflammatory and immunostimulatory activities characteristic of natural C5a reside in this small C-terminal domain (C5a pharmacophore) (Ember et al., 1992; Morgan et al., 1992).

Flexibility in the C-terminal region of C5a ($C5a_{65-74}$ or ISHKDMQLGR) is a dominant feature of natural C5a. This flexibility may allow C5a to form a unique conformational structure within the C5aR1 binding domain to produce the biological activity associated with C5a. Once more, this flexibility might allow the various effects (inflammation, chemotaxis, priming, etc.) associated with C5a. This flexibility-function theory suggests that restricting the flexibility of the C5a C-terminal to specific conformations will reduce the biological effects induced by C5aR1-expressing cells. For example, reducing the conformational flexibility of the C-terminus of C5a might selectively induce chemotaxis and activation but not pro-inflammatory responses in C5aR1 expressing APCs.

49.4.6 Conformationally Restricted Analogues of $C5a_{65-74}$

Over the years, researchers have produced a library of analogs of $C5a_{65-74}$ with restricted backbone flexibility by specific amino acid substitutions to bias certain conformational features that might help indicate the biologically relevant conformations for immunostimulatory activities versus those for inflammation through C5aR1.

The backbone flexibility of $C5a_{65-74}$ was restricted by introducing three principal types of residue substitutions within $C5a_{65-74}$: (1) Pro substitutions to restrict ϕ angle flexibility and to narrow the range of sterically allowed backbone conformations in the pre-proline residue (Misicka et al., 1991), (2) Ala substitutions to evaluate the biological importance of the side chains in the peptide, and (3) D-residue substitutions to assess the contribution of stereoisomeric arrangements. Such peptide modifications have an additional advantage because Pro, Ala, and D-residues occupy well-defined regions of sterically allowed Ramachandran space (Hruby et al., 1991). Thus, evaluating the changes in the biological activity of these conformationally restricted analogs of $C5a_{65-74}$ can provide information about the specific types of backbone conformation features important to specific biological activity.

Using this approach, one peptide from this library, $C5a_{65-74}$ I65Y, H67F, D69P, Q71P, G73D-Ala, or YSFKPMPLaR (EP54) exhibited about 10% of the potency of natural C5a for its ability to induce the release of spasmogenic eicosanoids from human macrophages, but only about 0.1% of C5a activity in its ability to induce the release of β-glucuronidase from human neutrophils (Finch et al., 1997). The corresponding difference in binding affinity to the C5aRs expressed on these cells explains this difference in activity (Finch et al., 1997; Vogen et al., 1999).

Structural analysis of YSFKPMPLaR indicated the presence of unique conformational features that appear to be responsible for the selective accommodation by C5aR1 expressed on antigen-presenting cells (APC) relative to

C5aR1 expressed on inflammatory neutrophils (Alshammari et al., 2020; Vogen et al., 1999). The backbone conformational features identified in EP54 guided the generation of other analogs designed to enhance their biological importance. Prominent among these new analogs is EP67 (YSFKDMP[MeL]aR), which contains an N-methylated Leu (MeL) residue to enhance backbone elongation in a crucial region of the peptide, which was deemed biologically necessary from the NMR structure of EP54. Indeed, EP67 exhibits more potency and macrophage/dendritic cell selectivity than EP54 (Taylor et al., 2001; Vogen et al., 2001).

49.4.7 Use of EP54 and EP67 as Host-Derived, Immunostimulatory Adjuvants

49.4.7.1 Generation of Antigen-Specific Humoral Immune Responses

The generation of neutralizing, antigen-specific antibodies is a necessity to treat a variety of invading pathogens. Thanks to their complement peptide-derived mechanism of action, both EP54 and EP67 can increase antibody production against co-administered antigens. EP54 induced antigen-specific antibody (Ab) responses to a B cell epitope derived from the human mucin-type 1 glycoprotein (MUC1). A vaccine was then generated by covalently attaching the MUC1 epitope (*YKQGGFLGL*) to the N-terminus of EP54 (*YKQGGFLGL*YSFKPMPLaR), and mice immunized with this vaccine generated high Ab titers specific for the MUC1 epitope. These anti-epitope Abs cross-reacted with MUC1 protein expressed on the surface of a transfected pancreatic cell line, indicating that the anti-*YKQGGFLGL* Abs recognized the epitope within intact whole MUC1 protein. Also, Ab isotypes generated by this EP54-containing vaccine were IgM, IgG2a, and IgG2b, contrasted with IgM and IgG1 Abs, generated by an analogous KLH construct (Tempero et al., 1997). These results suggest that the EP54-containing vaccine induced an Ab class switch characteristic of a Th1-like response and generated Ab with isotypes distinct from the traditional KLH construct. Over the years, we have used epitope-EP54 vaccine constructs to immunize rats, hamsters, rabbits, and cattle to a wide variety of epitopes of various lengths (8–35 residues) and small molecules such as nicotine (Sanderson et al., 2003) and methamphetamine (Duryee et al., 2009).

Similarly, EP67-containing vaccines generated a higher antibody response to target antigens (peptides and proteins). Mice immunized with these EP67-containing vaccines generated significant antigen-specific IgM, IgG2a, and IgG2b responses against LPS-free ovalbumin (OVA), a model antigen (Morgan et al., 2009). In another study, EP67 (referred to as Complement peptide-derived immunostimulant 2, or CPDI-02) was used as an adjuvant in LPS-free OVA-loaded PLGA nanoparticles (NPs) to induce significantly higher antibody-secreting cells (ASCs) against OVA in the lungs and splenocytes of EP67-PLGA-OVA NP-treated mice as compared to control mice. This increase in ASCs correlated with elevated OVA-Specific IgA, IgM, and IgG antibodies in the EP67-PLGA-OVA NP-treated mice (Parriott et al., 2022). Against an invading pathogen, mice immunized with intact fungal spores from an attenuated strain of *Coccidioides posadasii* conjugated to EP67 increased the protective efficacy by a corresponding increase in antigen-specific IgG1 and IgG2a Ab responses (Hung et al., 2012). In all these cases, EP67, like EP54, promoted a Th1 response over a Th2 response.

49.4.7.2 Generation of Antigen-Specific Cell-Mediated Immune Responses

In addition to a humoral immune response, a cell-mediated immune response is also necessary to clear many pathogens, particularly in cytotoxic T-cell responses (CTL) seen in $CD4^+$ and $CD8^+$ lymphocytes. Both EP54- and EP67-containing vaccines also promote an antigen-specific cell-mediated immune response. In the same experiment against *Coccidioides posadasii* as above, attenuated fungal spores conjugated to EP67 produced protective Th1 and Th17 CTL responses (Cole et al., 2013; Hung et al., 2012). EP67 conjugated to Hepatitis B surface antigen (HBsAg) epitopes produced increased CTLs compared to the antigen without an adjuvant (Ulrich et al., 2000). This increase in cell-mediated immune response can occur with MUC1 expressed on cancer cells (Pisarev et al., 2005) and the gp70 glycoprotein from RAW 117-H10 lymphoma cells (Kollessery et al., 2011).

EP67 can be used to target mucosal immunization, particularly through EP67-vaccine conjugates. Intranasal administration of murine cytomegalovirus (MCMV) epitopes covalently conjugated to EP67 generated functional mucosal and systemic epitope-specific $CD8^+$ T cells that increased protection against primary mucosal infection with MCMV (Karuturi et al., 2017). Moreover, many of these $CD8^+$ T cells formed a pool of long-lived memory $CD8^+$ T cells that respond strongly upon reinfection, even without $CD4^+$ Th1 help. EP67-based CTL peptides generated a higher proportion of epitope-specific mucosal (lungs) and systemic (spleen) $CD8a^+/CD44^+$ cells with cell surface phenotype ($CD127^+/KLRG1^-$) associated long-lived memory (Karuturi et al., 2017). As with the humoral response, encapsulating the antigen with EP67 surface-modified PLGA improved this response (Karuturi et al., 2017; Tallapaka et al., 2019). In addition to CD127 and KLRG1, we found that EP67-based CTL peptides generated epitope-specific $CD8^+$ T cells with an increased expression of CD27, which is a strong predictor of proliferative recall responses upon antigen encounter (Ochsenbein et al., 2004; Xiao et al., 2008).

Immunization results in mice demonstrate the unique ability of C5a/host-derived adjuvants for systemic and mucosal vaccine applications. EP67 may be acceptable as a universal adjuvant for various vaccine applications due to its

Th1 response skewed stimulation and significantly reduced pro-inflammatory properties.

49.4.7.3 Stimulation of Innate Immune Responses

Because of the mechanism of action of these complement peptide-derived adjuvants, their therapeutic uses might extend beyond modulating adaptive immunity. More recent studies have shown the ability of EP67 to protect against influenza, MRSA, and GBS infections via induction of innate immunity from its selective engagement of C5aR-bearing macrophages (Cavaco et al., 2013; Hanke et al., 2013; Sanderson et al., 2012; Sheen et al., 2011). This innate immune stimulation seems to stem from the selective binding of EP67 to C5aR1 on macrophages and other APCs over neutrophils, which also express C5aR1. These studies indicate that EP67 has significant immunostimulatory properties conducive to desirable innate and adaptive immune outcomes (Table 49.1).

49.5 Review Questions

1. True or False. The only immune response needed against all pathogens is a Th2 humoral antibody response.
2. Which of the following microbial components could serve as the basis for a vehicle adjuvant?
   (a) Cholera toxin
   (b) Virus-like particles
   (c) Lipopolysaccharides
   (d) Complement peptide C3b
3. Please explain the differences between pathogen-derived and host-derived adjuvants and the risks associated with each of them.
4. What are the differences between immunostimulatory adjuvants and vehicle adjuvants?
5. What is the primary difference between the mechanism of action between natural C5a and its C-terminal analogs, EP54 and EP67?

Table 49.1 Vaccine adjuvant and immunostimulatory properties of EP54 and EP67

Vaccine/immunostimulant construct	Species	Route	Immune response	Ref
Peptide epitope from MUC1 glycoprotein conjugated to EP54	C57BL6 and BALB/c mice	i.p.	Higher titers of IgG2b, IgG2c, and IgM Abs in sera reactive against recombinant MUC1 and MUC1 expressing cell lines	Tempero et al. (1997)
CTL peptide epitope derived from hepatitis B surface antigen (HBsAg) conjugated to EP54	BALB/c mice	s.c	Antigen-specific $CD8^+$ CTL responses against murine P815S target cells expressing an H-$2L^d$ restricted CTL epitope of the HBsAg.	Ulrich et al. (2000)
Nicotine hapten conjugated to EP54	Sprague–Dawley rats	i.p.	Nicotine-specific Abs capable of attenuating nicotine-induced behavioral effects	Sanderson et al. (2003)
Methamphetamine (meth) hapten conjugated to EP54	Rats	s.c/i.p.	Meth-specific Abs in sera capable of altering meth self-administration	Duryee et al. (2009)
Ovalbumin (OVA) conjugated to EP67	C57BL6 BALB/c mice	i.p.	OVA-specific Th1-like Ab class switch and OVA-specific proliferative responses in splenocytes	Morgan et al. (2009)
OVA conjugated to EP67	Young/aged mice	i.p.	Higher antigen-specific humoral responses compared to alum-adjuvanted OVA	Morgan et al. (2010)
rPrp1, a protein from the cell wall of *coccidioides* conjugated to EP67	Young/aged mice	i.p.	Higher antigen-specific humoral responses compared alum and CpG adjuvanted rPrp1	Hung et al. (2012)
Peptide epitope derived from gp70 glycoprotein conjugated to EP54 and EP67	BALB/c mice	s.c	Antigen-specific CTL responses and protection against metastatic RAW117-H10 lymphoma lethal challenge	Kollessery et al. (2011)
Live spores of an attenuated vaccine strain of *Coccidioides posadasii* conjugated to EP67	BALB/c mice	s.c	Increased protective efficacy of live vaccine by increasing IgG1, and IgG2a antibodies against antigens, and Th1, and Th17 immune responses	Hung et al. (2012)
EP67 was administered preventatively to the challenge, followed by a booster 24 h post-challenge	CD-1 mice	s.c.	Increased mouse KC, IL-6, and IL-1β. Reduced MRSA-induced dermal abscess surface area compared to control mice.	Sheen et al. (2011)
EP67 was administered at MRSA challenged, followed by booster 24 and 48 h post-challenge	C56BL/6	catheter	Increased IL-17, IL-1α, and IFNγ. Reduced MRSA biofilm growth. Increased macrophage infiltration of biofilm.	Hanke et al. (2013)
MCMV epitopes M84 and pp89 covalently attached to EP67	BALB/c mice	i.n.	Decrease in viral burden in lungs and salivary glands. It increased $CD8^+$ T cells.	Karuturi et al. (2015)
EP67 surface-modified nanoparticles containing OVA	C57BL/6NCrl	i.n.	Increased IgA, IgG1, and IgG2b against the antigen. Increased antibody-secreting cells compared to controls	Parriott et al. (2022)

49.6 Answers

1. False
2. (b) Virus-like particles
3. As the names imply, pathogen-derived adjuvants are immunostimulatory or vehicle molecules derived from PAMPs or other molecules found in microbes not native to the host. Host-derived adjuvants are based on DAMPs or other molecules native to the host system. The potential risk of using pathogen-derived adjuvants is the risk of overstimulating the immune system, thus masking the antigen signal (remember, most pathogen-derived adjuvants are based on toxins or virulence factors). Host-derived adjuvants risk signaling undesired pathways, thus inducing undesired side effects.
4. Immunostimulatory adjuvants strengthen the immune response to the co-administered antigen by stimulating/priming the host immune system to become more sensitive to the antigen. The binding of immunostimulants to receptors on immune cells or triggering of host immune responses produces this stimulation/priming. Vehicle adjuvants act as a depot to present the antigen to the host continuously. They can also function by presenting the antigen in such high volumes/repeated units that it's easier for the host's immune system to produce a robust immune response.
5. C5a binds to its receptor C5aR1/CD88 on whatever cell expresses it. On the other hand, the C-terminal analogs EP54 and EP67 have been developed to be selective in their binding. Thanks to their conformational limitations, they show selective binding to C5aR1 found on phagocytes and other APCs but not neutrophils or other inflammatory immune cells.

References

Afonso, L. C., Scharton, T. M., Vieira, L. Q., Wysocka, M., Trinchieri, G., & Scott, P. (1994). The adjuvant effect of interleukin-12 in a vaccine against Leishmania major. *Science, 263*, 235–237.

Alshammari, A. M., Smith, D. D., Parriott, J., Stewart, J. P., Curran, S. M., McCulloh, R. J., Barry, P. A., Iyer, S. S., Palermo, N., Phillips, J. A., Dong, Y., Ronning, D. R., Vennerstrom, J. L., Sanderson, S. D., & Vetro, J. A. (2020). Targeted amino acid substitution overcomes scale-up challenges with the human C5a-derived decapeptide immunostimulant EP67. *ACS Infectious Diseases, 6*, 1169–1181.

Arvieux, J., Yssel, H., & Colomb, M. G. (1988). Antigen-bound C3b and C4b enhance antigen-presenting cell function in activation of human T-cell clones. *Immunology, 65*, 229–235.

Batista-Duharte, A., Martinez, D. T., & Carlos, I. Z. (2018). Efficacy and safety of immunological adjuvants. Where is the cut-off? *Biomedicine & Pharmacotherapy, 105*, 616–624.

Bergmann-Leitner, E. S., Leitner, W. W., & Tsokos, G. C. (2006). Complement 3d: From molecular adjuvant to target of immune escape mechanisms. *Clinical Immunology, 121*, 177–185.

Biragyn, A., Tani, K., Grimm, M. C., Weeks, S., & Kwak, L. W. (1999). Genetic fusion of chemokines to a self tumor antigen induces protective, T-cell dependent antitumor immunity. *Nature Biotechnology, 17*, 253–258.

Brandtzaeg, P. (2009). Mucosal immunity: Induction, dissemination, and effector functions. *Scandinavian Journal of Immunology, 70*, 505–515.

Buchner, R. R., Hugli, T. E., Ember, J. A., & Morgan, E. L. (1995). Expression of functional receptors for human C5a anaphylatoxin (CD88) on the human hepatocellular carcinoma cell line HepG2. Stimulation of acute-phase protein-specific mRNA and protein synthesis by human C5a anaphylatoxin. *Journal of Immunology, 155*, 308–315.

Cavaco, C. K., Patras, K. A., Zlamal, J. E., Thoman, M. L., Morgan, E. L., Sanderson, S. D., & Doran, K. S. (2013). A novel C5a-derived immunobiotic peptide reduces Streptococcus agalactiae colonization through targeted bacterial killing. *Antimicrobial Agents and Chemotherapy, 57*, 5492–5499.

Champion, C. R. (2021). Heplisav-B: A Hepatitis B vaccine with a novel adjuvant. *The Annals of Pharmacotherapy, 55*, 783–791.

Chaudhary, N., Weissman, D., & Whitehead, K. A. (2021). mRNA vaccines for infectious diseases: Principles, delivery and clinical translation. *Nature Reviews. Drug Discovery, 20*, 817–838.

Cimica, V., & Galarza, J. M. (2017). Adjuvant formulations for virus-like particle (VLP) based vaccines. *Clinical Immunology, 183*, 99–108.

Clements, J. D., & Norton, E. B. (2018). The mucosal vaccine adjuvant LT(R192G/L211A) or dmLT. *mSphere, 3*, 10.

Cole, G. T., Hung, C. Y., Sanderson, S. D., Hurtgen, B. J., Wuthrich, M., Klein, B. S., Deepe, G. S., Ostroff, G. R., & Levitz, S. M. (2013). Novel strategies to enhance vaccine immunity against coccidioidomycosis. *PLoS Pathogens, 9*, e1003768.

Crouch, C. F., Daly, J., Henley, W., Hannant, D., Wilkins, J., & Francis, M. J. (2005). The use of a systemic prime/mucosal boost strategy with an equine influenza ISCOM vaccine to induce protective immunity in horses. *Veterinary Immunology and Immunopathology, 108*, 345–355.

Defoort, J. P., Nardelli, B., Huang, W., & Tam, J. P. (1992). A rational design of synthetic peptide vaccine with a built-in adjuvant. A modular approach for unambiguity. *International Journal of Peptide and Protein Research, 40*, 214–221.

Degen, W. G. J., & Schijns, V. E. (2017). Host derived cytokines and chemokines as vaccine adjuvants. In *Immunopotentiators in modern vaccines* (pp. 65–84). Academic Press.

Dempsey, P. W., Allison, M. E., Akkaraju, S., Goodnow, C. C., & Fearon, D. T. (1996). C3d of complement as a molecular adjuvant: Bridging innate and acquired immunity. *Science, 271*, 348–350.

den Brok, M. H., Bull, C., Wassink, M., de Graaf, A. M., Wagenaars, J. A., Minderman, M., Thakur, M., Amigorena, S., Rijke, E. O., Schrier, C. C., & Adema, G. J. (2016). Saponin-based adjuvants induce cross-presentation in dendritic cells by intracellular lipid body formation. *Nature Communications, 7*, 13324.

Diwan, M., Elamanchili, P., Cao, M., & Samuel, J. (2004). Dose sparing of CpG oligodeoxynucleotide vaccine adjuvants by nanoparticle delivery. *Current Drug Delivery, 1*, 405–412.

Doorduijn, D. J., Lukassen, M. V., van't Wout, M. F. L., Franc, V., Ruyken, M., Bardoel, B. W., Heck, A. J. R., & Rooijakkers, S. H. M. (2022). Soluble MAC is primarily released from MAC-resistant bacteria that potently convert complement component C5. *eLife, 11*, e77503.

Drapeau, G., Brochu, S., Godin, D., Levesque, L., Rioux, F., & Marceau, F. (1993). Synthetic C5a receptor agonists. Pharmacology, metabolism and in vivo cardiovascular and hematologic effects. *Biochemical Pharmacology, 45*, 1289–1299.

Dundar, B., Karahangil, K., Elgormus, C. S., & Topsakal, H. N. H. (2022). Efficacy of antibody response following the vaccination of SARS-CoV-2 infected and noninfected healthcare workers by two-dose inactive vaccine against COVID-19. *Journal of Medical Virology, 94*, 2431–2437.

Duryee, M. J., Bevins, R. A., Reichel, C. M., Murray, J. E., Dong, Y., Thiele, G. M., & Sanderson, S. D. (2009). Immune responses to methamphetamine by active immunization with peptide-based, molecular adjuvant-containing vaccines. *Vaccine, 27*, 2981–2988.

Ember, J. A., Sanderson, S. D., Taylor, S. M., Kawahara, M., & Hugli, T. E. (1992). Biologic activity of synthetic analogues of C5a anaphylatoxin. *Journal of Immunology, 148*, 3165–3173.

Ender, F., Wiese, A. V., Schmudde, I., Sun, J., Vollbrandt, T., Konig, P., Laumonnier, Y., & Kohl, J. (2017). Differential regulation of C5a receptor 1 in innate immune cells during the allergic asthma effector phase. *PLoS One, 12*, e0172446.

Fedele, G., Cassone, A., & Ausiello, C. M. (2015). T-cell immune responses to Bordetella pertussis infection and vaccination. *Pathogens and Disease, 73*, ftv051.

Figueiredo, D. B., Kaneko, K., Rodrigues, T. D. C., MacLoughlin, R., Miyaji, E. N., Saleem, I., & Goncalves, V. M. (2022). Pneumococcal surface protein A-hybrid nanoparticles protect mice from lethal challenge after mucosal immunization targeting the lungs. *Pharmaceutics, 14*, 1238.

Finch, A. M., Vogen, S. M., Sherman, S. A., Kirnarsky, L., Taylor, S. M., & Sanderson, S. D. (1997). Biologically active conformer of the effector region of human C5a and modulatory effects of N-terminal receptor binding determinants on activity. *Journal of Medicinal Chemistry, 40*, 877–884.

Floreani, A. A., Gunselman, S. J., Heires, A. J., Hauke, R. J., Tarantolo, S., & Jackson, J. D. (2007). Novel C5a agonist-based dendritic cell vaccine in a murine model of melanoma. *Cell Cycle, 6*, 2835–2839.

Frietze, K. M., Peabody, D. S., & Chackerian, B. (2016). Engineering virus-like particles as vaccine platforms. *Current Opinion in Virology, 18*, 44–49.

Gao, L., Li, J., & Song, T. (2022). Poly lactic-co-glycolic acid-based nanoparticles as delivery systems for enhanced cancer immunotherapy. *Frontiers in Chemistry, 10*, 973666.

Gasque, P., Chan, P., Fontaine, M., Ischenko, A., Lamacz, M., Gotze, O., & Morgan, B. P. (1995). Identification and characterization of the complement C5a anaphylatoxin receptor on human astrocytes. *Journal of Immunology, 155*, 4882–4889.

Goldstein, B. D., Lai, L. Y., & Cuzzi-Spada, R. (1974). Potentiation of complement-dependent membrane damage by ozone. *Archives of Environmental Health, 28*, 40–42.

Goodman, M. G., Chenoweth, D. E., & Weigle, W. O. (1982). Induction of interleukin 1 secretion and enhancement of humoral immunity by binding of human C5a to macrophage surface C5a receptors. *The Journal of Experimental Medicine, 156*, 912–917.

Guo, R. F., & Ward, P. A. (2005). Role of C5a in inflammatory responses. *Annual Review of Immunology, 23*, 821–852.

Gutjahr, A., Phelip, C., Coolen, A. L., Monge, C., Boisgard, A. S., Paul, S., & Verrier, B. (2016). Biodegradable polymeric nanoparticles-based vaccine adjuvants for lymph nodes targeting. *Vaccines (Basel), 4*, 34.

Hanke, M. L., Heim, C. E., Angle, A., Sanderson, S. D., & Kielian, T. (2013). Targeting macrophage activation for the prevention and treatment of Staphylococcus aureus biofilm infections. *Journal of Immunology, 190*, 2159–2168.

Horiuchi, T., & Tsukamoto, H. (2016). Complement-targeted therapy: Development of C5- and C5a-targeted inhibition. *Inflammation and Regeneration, 36*, 11.

Hruby, V. J., Prakash, O., Kazmierski, W., Gehrig, C., & Matsunaga, T. O. (1991). Conformational analysis of opioid receptor-selective peptides using nuclear magnetic resonance and theoretical calculations. *NIDA Research Monograph, 112*, 198–217.

Huang, C. F., Wu, T. C., Chu, Y. H., Hwang, K. S., Wang, C. C., & Peng, H. J. (2008). Effect of neonatal sublingual vaccination with native or denatured ovalbumin and adjuvant CpG or cholera toxin on systemic and mucosal immunity in mice. *Scandinavian Journal of Immunology, 68*, 502–510.

Hugli, T. E., & Muller-Eberhard, H. J. (1978). Anaphylatoxins: C3a and C5a. *Advances in Immunology, 26*, 1–53.

Hugli, T. E., Gerard, C., Kawahara, M., Scheetz, M. E., 2nd, Barton, R., Briggs, S., Koppel, G., & Russell, S. (1981). Isolation of three separate anaphylatoxins from complement-activated human serum. *Molecular and Cellular Biochemistry, 41*, 59–66.

Hugli, T. E., Marceau, F., & Lundberg, C. (1987). Effects of complement fragments on pulmonary and vascular smooth muscle. *The American Review of Respiratory Disease, 135*, S9–S13.

Hung, C. Y., Hurtgen, B. J., Bellecourt, M., Sanderson, S. D., Morgan, E. L., & Cole, G. T. (2012). An agonist of human complement fragment C5a enhances vaccine immunity against Coccidioides infection. *Vaccine, 30*, 4681–4690.

Iho, S., Maeyama, J., & Suzuki, F. (2015). CpG oligodeoxynucleotides as mucosal adjuvants. *Human Vaccines & Immunotherapeutics, 11*, 755–760.

Jacquier-Sarlin, M. R., Gabert, F. M., Villiers, M. B., & Colomb, M. G. (1995). Modulation of antigen processing and presentation by covalently linked complement C3b fragment. *Immunology, 84*, 164–170.

Johnson, A. G., Gaines, S., & Landy, M. (1956). Studies on the O antigen of Salmonella typhosa. V. Enhancement of antibody response to protein antigens by the purified lipopolysaccharide. *The Journal of Experimental Medicine, 103*, 225–246.

Johnson, A. R., Hugli, T. E., & Muller-Eberhard, H. J. (1975). Release of histamine from rat mast cells by the complement peptides C3a and C5a. *Immunology, 28*, 1067–1080.

Karuturi, B. V., Tallapaka, S. B., Phillips, J. A., Sanderson, S. D., & Vetro, J. A. (2015). Preliminary evidence that the novel host-derived immunostimulant EP67 can act as a mucosal adjuvant. *Clinical Immunology, 161*, 251–259.

Karuturi, B. V. K., Tallapaka, S. B., Yeapuri, P., Curran, S. M., Sanderson, S. D., & Vetro, J. A. (2017). Encapsulation of an EP67-conjugated CTL peptide vaccine in nanoscale biodegradable particles increases the efficacy of respiratory immunization and affects the magnitude and memory subsets of vaccine-generated mucosal and systemic CD8(+) T cells in a diameter-dependent manner. *Molecular Pharmaceutics, 14*, 1469–1481.

Kawamoto, S., Yalcindag, A., Laouini, D., Brodeur, S., Bryce, P., Lu, B., Humbles, A. A., Oettgen, H., Gerard, C., & Geha, R. S. (2004). The anaphylatoxin C3a downregulates the Th2 response to epicutaneously introduced antigen. *The Journal of Clinical Investigation, 114*, 399–407.

Kayamuro, H., Yoshioka, Y., Abe, Y., Arita, S., Katayama, K., Nomura, T., Yoshikawa, T., Kubota-Koketsu, R., Ikuta, K., Okamoto, S., Mori, Y., Kunisawa, J., Kiyono, H., Itoh, N., Nagano, K., Kamada, H., Tsutsumi, Y., & Tsunoda, S. (2010). Interleukin-1 family cytokines as mucosal vaccine adjuvants for induction of protective immunity against influenza virus. *Journal of Virology, 84*, 12703–12712.

Kim, A. H., Dimitriou, I. D., Holland, M. C., Mastellos, D., Mueller, Y. M., Altman, J. D., Lambris, J. D., & Katsikis, P. D. (2004). Complement C5a receptor is essential for the optimal generation of antiviral CD8+ T cell responses. *Journal of Immunology, 173*, 2524–2529.

Kim, S. H., Cho, B. H., Kim, K. S., & Jang, Y. S. (2021). Complement C5a promotes antigen cross-presentation by Peyer's patch monocyte-derived dendritic cells and drives a protective CD8(+) T cell response. *Cell Reports, 35*, 108995.

Kollessery, G., Nordgren, T. M., Mittal, A. K., Joshi, S. S., & Sanderson, S. D. (2011). Tumor-specific peptide-based vaccines containing the conformationally biased, response-selective C5a agonists EP54 and EP67 protect against aggressive large B cell lymphoma in a syngeneic murine model. *Vaccine, 29*, 5904–5910.

Kurzawa, H., Wysocka, M., Aruga, E., Chang, A. E., Trinchieri, G., & Lee, W. M. (1998). Recombinant interleukin 12 enhances cellular immune responses to vaccination only after a period of suppression. *Cancer Research, 58*, 491–499.

Li, K., Fazekasova, H., Wang, N., Peng, Q., Sacks, S. H., Lombardi, G., & Zhou, W. (2012). Functional modulation of human monocytes derived DCs by anaphylatoxins C3a and C5a. *Immunobiology, 217*, 65–73.

Li, X. X., Clark, R. J., & Woodruff, T. M. (2020). C5aR2 activation broadly modulates the signaling and function of primary human macrophages. *Journal of Immunology, 205*, 1102–1112.

Li, X. X., Clark, R. J., & Woodruff, T. M. (2021). Anaphylatoxin receptor promiscuity for commonly used complement C5a peptide agonists. *International Immunopharmacology, 100*, 108074.

Lundberg, C., Gardinali, M., & Hugli, T. E. (1987). Complement activation and membrane lipids in lung vascular injury. *The American Review of Respiratory Disease, 136*, 459–462.

Lycke, N. (2005). Targeted vaccine adjuvants based on modified cholera toxin. *Current Molecular Medicine, 5*, 591–597.

Lycke, N. (2012). Recent progress in mucosal vaccine development: Potential and limitations. *Nature Reviews. Immunology, 12*, 592–605.

Mancini, F., Rossi, O., Necchi, F., & Micoli, F. (2020). OMV vaccines and the role of TLR agonists in immune response. *International Journal of Molecular Sciences, 21*, 4416.

Martinon, F., Gras-Masse, H., Boutillon, C., Chirat, F., Deprez, B., Guillet, J. G., Gomard, E., Tartar, A., & Levy, J. P. (1992). Immunization of mice with lipopeptides bypasses the prerequisite for adjuvant. Immune response of BALB/c mice to human immunodeficiency virus envelope glycoprotein. *Journal of Immunology, 149*, 3416–3422.

Masson, J. D., Thibaudon, M., Belec, L., & Crepeaux, G. (2017). Calcium phosphate: A substitute for aluminum adjuvants? *Expert Review of Vaccines, 16*, 289–299.

Mastellos, D., Germenis, A. E., & Lambris, J. D. (2005). Complement: An inflammatory pathway fulfilling multiple roles at the interface of innate immunity and development. *Current Drug Targets. Inflammation and Allergy, 4*, 125–127.

McKee, A. S., MacLeod, M. K., Kappler, J. W., & Marrack, P. (2010). Immune mechanisms of protection: Can adjuvants rise to the challenge? *BMC Biology, 8*, 37.

Merle, N. S., Noe, R., Halbwachs-Mecarelli, L., Fremeaux-Bacchi, V., & Roumenina, L. T. (2015). Complement system part II: Role in immunity. *Frontiers in Immunology, 6*, 257.

Metzger, J., Wiesmuller, K. H., Schaude, R., Bessler, W. G., & Jung, G. (1991). Synthesis of novel immunologically active tripalmitoyl-S-glycerylcysteinyl lipopeptides as useful intermediates for immunogen preparations. *International Journal of Peptide and Protein Research, 37*, 46–57.

Misicka, A., Lipkowski, A. W., Fang, L., Knapp, R. J., Davis, P., Kramer, T., Burks, T. F., Yamamura, H. I., Carr, D. B., & Hruby, V. J. (1991). Topographical requirements for delta opioid ligands: Presence of a carboxyl group in position 4 is not critical for deltorphin high delta receptor affinity and analgesic activity. *Biochemical and Biophysical Research Communications, 180*, 1290–1297.

Mohan, T., Verma, P., & Rao, D. N. (2013). Novel adjuvants & delivery vehicles for vaccines development: A road ahead. *The Indian Journal of Medical Research, 138*, 779–795.

Morgan, E. L. (1986). Modulation of the immune response by anaphylatoxins. *Complement, 3*, 128–136.

Morgan, E. L., Thoman, M. L., Weigle, W. O., & Hugli, T. E. (1983). Anaphylatoxin-mediated regulation of the immune response. II. C5a-mediated enhancement of human humoral and T cell-mediated immune responses. *Journal of Immunology, 130*, 1257–1261.

Morgan, E. L., Sanderson, S., Scholz, W., Noonan, D. J., Weigle, W. O., & Hugli, T. E. (1992). Identification and characterization of the effector region within human C5a responsible for stimulation of IL-6 synthesis. *Journal of Immunology, 148*, 3937–3942.

Morgan, E. L., Morgan, B. N., Stein, E. A., Vitrs, E. L., Thoman, M. L., Sanderson, S. D., & Phillips, J. A. (2009). Enhancement of in vivo and in vitro immune functions by a conformationally biased, response-selective agonist of human C5a: Implications for a novel adjuvant in vaccine design. *Vaccine, 28*, 463–469.

Morgan, E. L., Thoman, M. L., Sanderson, S. D., & Phillips, J. A. (2010). A novel adjuvant for vaccine development in the aged. *Vaccine, 28*, 8275–8279.

Moulton, R. A., Mashruwala, M. A., Smith, A. K., Lindsey, D. R., Wetsel, R. A., Haviland, D. L., Hunter, R. L., & Jagannath, C. (2007). Complement C5a anaphylatoxin is an innate determinant of dendritic cell-induced Th1 immunity to Mycobacterium bovis BCG infection in mice. *Journal of Leukocyte Biology, 82*, 956–967.

Muhlradt, P. F., Kiess, M., Meyer, H., Sussmuth, R., & Jung, G. (1998). Structure and specific activity of macrophage-stimulating lipopeptides from mycoplasma hyorhinis. *Infection and Immunity, 66*, 4804–4810.

Munshi, T., Sparrow, A., Wren, B. W., Reljic, R., & Willcocks, S. J. (2020). The antimicrobial peptide, bactenecin 5, supports cell-mediated but not humoral immunity in the context of a mycobacterial antigen vaccine model. *Antibiotics (Basel), 9*, 926.

Ochsenbein, A. F., Riddell, S. R., Brown, M., Corey, L., Baerlocher, G. M., Lansdorp, P. M., & Greenberg, P. D. (2004). CD27 expression promotes long-term survival of functional effector-memory CD8+ cytotoxic T lymphocytes in HIV-infected patients. *The Journal of Experimental Medicine, 200*, 1407–1417.

Okusawa, S., Dinarello, C. A., Yancey, K. B., Endres, S., Lawley, T. J., Frank, M. M., Burke, J. F., & Gelfand, J. A. (1987). C5a induction of human interleukin 1. Synergistic effect with endotoxin or interferon-gamma. *Journal of Immunology, 139*, 2635–2640.

Parriott, J. E., Stewart, J. P., Smith, D. D., Curran, S. M., Bauer, C. D., Wyatt, T. A., Phillips, J. A., Lyden, E., Thiele, G. M., & Vetro, J. A. (2022). Surface modification of biodegradable microparticles with the novel host-derived immunostimulant CPDI-02 significantly increases short-term and long-term mucosal and systemic antibodies against encapsulated protein antigen in young naive mice after respiratory immunization. *Pharmaceutics, 14*, 1843.

Peng, Q., Li, K., Sacks, S. H., & Zhou, W. (2009). The role of anaphylatoxins C3a and C5a in regulating innate and adaptive immune responses. *Inflammation & Allergy Drug Targets, 8*, 236–246.

Petrovsky, N. (2015). Comparative safety of vaccine adjuvants: A summary of current evidence and future needs. *Drug Safety, 38*, 1059–1074.

Petrovsky, N., & Aguilar, J. C. (2004). Vaccine adjuvants: Current state and future trends. *Immunology and Cell Biology, 82*, 488–496.

Pisarev, V. M., Kinarsky, L., Caffrey, T., Hanisch, F. G., Sanderson, S., Hollingsworth, M. A., & Sherman, S. (2005). T cells recognize PD(N/T)R motif common in a variable number of tandem repeat and degenerate repeat sequences of MUC1. *International Immunopharmacology, 5*, 315–330.

Ricklin, D., Hajishengallis, G., Yang, K., & Lambris, J. D. (2010). Complement: a key system for immune surveillance and homeostasis. *Nature Immunology, 11*, 785–797.

Rooijakkers, S. H., van Kessel, K. P., & van Strijp, J. A. (2005). Staphylococcal innate immune evasion. *Trends in Microbiology, 13*, 596–601.

Rosales-Mendoza, S., et al. (2019). *Virus like particles based mucosal nanovacccines* (pp. 267–318). Springer.

Rudilla, F., Fayolle, C., Casares, N., Durantez, M., Arribillaga, L., Lozano, T., Villanueva, L., Pio, R., Sarobe, P., Leclerc, C., Prieto, J., & Lasarte, J. J. (2012). Combination of a TLR4 ligand and anaphylatoxin C5a for the induction of antigen-specific cytotoxic T cell responses. *Vaccine, 30*, 2848–2858.

Sadik, C. D., Miyabe, Y., Sezin, T., & Luster, A. D. (2018). The critical role of C5a as an initiator of neutrophil-mediated autoimmune inflammation of the joint and skin. *Seminars in Immunology, 37*, 21–29.

Salina, A. C. G., Brandt, S. L., Klopfenstein, N., Blackman, A., Bazzano, J. M. R., Sa-Nunes, A., Byers-Glosson, N., Brodskyn, C., Tavares, N. M., Da Silva, I. B. S., Medeiros, A. I., & Serezani, C. H. (2020). Leukotriene B(4) licenses inflammasome activation to enhance skin host defense. *Proceedings of the National Academy of Sciences of the United States of America, 117*, 30619–30627.

Sanderson, S. D., Cheruku, S. R., Padmanilayam, M. P., Vennerstrom, J. L., Thiele, G. M., Palmatier, M. I., & Bevins, R. A. (2003). Immunization to nicotine with a peptide-based vaccine composed of a conformationally biased agonist of C5a as a molecular adjuvant. *International Immunopharmacology, 3*, 137–146.

Sanderson, S. D., Thoman, M. L., Kis, K., Virts, E. L., Herrera, E. B., Widmann, S., Sepulveda, H., & Phillips, J. A. (2012). Innate immune induction and influenza protection elicited by a response-selective agonist of human C5a. *PLoS One, 7*, e40303.

Schanzenbacher, J., Kohl, J., & Karsten, C. M. (2022). Anaphylatoxins spark the flame in early autoimmunity. *Frontiers in Immunology, 13*, 958392.

Schorlemmer, H. U., & Allison, A. C. (1976). Effects of activated complement components on enzyme secretion by macrophages. *Immunology, 31*, 781–788.

Schussek, S., Bernasconi, V., Mattsson, J., Wenzel, U. A., Stromberg, A., Gribonika, I., Schon, K., & Lycke, N. Y. (2020). The CTA1-DD adjuvant strongly potentiates follicular dendritic cell function and germinal center formation, which results in improved neonatal immunization. *Mucosal Immunology, 13*, 545–557.

Shah, R. R., et al. (2015). Emulsions as Vaccines Adjuvants. In *Subunit vaccine delivery* (pp. 59–76).

Sheen, T. R., Cavaco, C. K., Ebrahimi, C. M., Thoman, M. L., Sanderson, S. D., Morgan, E. L., & Doran, K. S. (2011). Control of methicillin resistant Staphylococcus aureus infection utilizing a novel immunostimulatory peptide. *Vaccine, 30*, 9–13.

Shi, R., & Chiang, V. L. (2005). Facile means for quantifying microRNA expression by real-time PCR. *BioTechniques, 39*, 519–525.

Shimoyama, A., & Fukase, K. (2021). Lipid A-mediated bacterial-host chemical ecology: Synthetic research of bacterial lipid as and their development as adjuvants. *Molecules, 26*, 6294.

Shin, H. S., Snyderman, R., Friedman, E., Mellors, A., & Mayer, M. M. (1968). Chemotactic and anaphylatoxic fragment cleaved from the fifth component of Guinea pig complement. *Science, 162*, 361–363.

Solans, L., & Locht, C. (2018). The role of mucosal immunity in pertussis. *Frontiers in Immunology, 9*, 3068.

Stark, F. C., Akache, B., Deschatelets, L., Tran, A., Stuible, M., Durocher, Y., McCluskie, M. J., Agbayani, G., Dudani, R., Harrison, B. A., Renner, T. M., Makinen, S. R., Bavananthasivam, J., Duque, D., Gagne, M., Zimmermann, J., Zarley, C. D., Cochrane, T. R., & Handfield, M. (2022). Intranasal immunization with a proteosome-adjuvanted SARS-CoV-2 spike protein-based vaccine is immunogenic and efficacious in mice and hamsters. *Scientific Reports, 12*, 9772.

Strey, C. W., Markiewski, M., Mastellos, D., Tudoran, R., Spruce, L. A., Greenbaum, L. E., & Lambris, J. D. (2003). The proinflammatory mediators C3a and C5a are essential for liver regeneration. *The Journal of Experimental Medicine, 198*, 913–923.

Tallapaka, S. B., Karuturi, B. V. K., Yeapuri, P., Curran, S. M., Sonawane, Y. A., Phillips, J. A., David Smith, D., Sanderson, S. D., & Vetro, J. A. (2019). Surface conjugation of EP67 to biodegradable nanoparticles increases the generation of long-lived mucosal and systemic memory T-cells by encapsulated protein vaccine after respiratory immunization and subsequent T-cell-mediated protection against respiratory infection. *International Journal of Pharmaceutics, 565*, 242–257.

Tandrup Schmidt, S., Foged, C., Korsholm, K. S., Rades, T., & Christensen, D. (2016). Liposome-based adjuvants for subunit vaccines: Formulation strategies for subunit antigens and immunostimulators. *Pharmaceutics, 8*, 7.

Taylor, S. M., Sherman, S. A., Kirnarsky, L., & Sanderson, S. D. (2001). Development of response-selective agonists of human C5a anaphylatoxin: Conformational, biological, and therapeutic considerations. *Current Medicinal Chemistry, 8*, 675–684.

Tempero, R. M., Hollingsworth, M. A., Burdick, M. D., Finch, A. M., Taylor, S. M., Vogen, S. M., Morgan, E. L., & Sanderson, S. D. (1997). Molecular adjuvant effects of a conformationally biased agonist of human C5a anaphylatoxin. *Journal of Immunology, 158*, 1377–1382.

Travieso, T., Li, J., Mahesh, S., Mello, J., & Blasi, M. (2022). The use of viral vectors in vaccine development. *npj Vaccines, 7*, 75.

Tsoras, A. N., & Champion, J. A. (2019). Protein and peptide biomaterials for engineered subunit vaccines and immunotherapeutic applications. *Annual Review of Chemical and Biomolecular Engineering, 10*, 337–359.

Tsuji, N. M., & Kosaka, A. (2008). Oral tolerance: Intestinal homeostasis and antigen-specific regulatory T cells. *Trends in Immunology, 29*, 532–540.

Ulrich, J. T., & Myers, K. R. (1995). Monophosphoryl lipid A as an adjuvant. Past experiences and new directions. *Pharmaceutical Biotechnology, 6*, 495–524.

Ulrich, J. T., Cieplak, W., Paczkowski, N. J., Taylor, S. M., & Sanderson, S. D. (2000). Induction of an antigen-specific CTL response by a conformationally biased agonist of human C5a anaphylatoxin as a molecular adjuvant. *Journal of Immunology, 164*, 5492–5498.

van Ginkel, F. W., Jackson, R. J., Yoshino, N., Hagiwara, Y., Metzger, D. J., Connell, T. D., Vu, H. L., Martin, M., Fujihashi, K., & McGhee, J. R. (2005). Enterotoxin-based mucosal adjuvants alter antigen trafficking and induce inflammatory responses in the nasal tract. *Infection and Immunity, 73*, 6892–6902.

Vecchiarelli, A., Retini, C., Casadevall, A., Monari, C., Pietrella, D., & Kozel, T. R. (1998). Involvement of C3a and C5a in interleukin-8 secretion by human polymorphonuclear cells in response to capsular material of Cryptococcus neoformans. *Infection and Immunity, 66*, 4324–4330.

Villiers, M. B., Marche, P. N., & Villiers, C. L. (2003). Improvement of long-lasting response and antibody affinity by the complexation of antigen with complement C3b. *International Immunology, 15*, 91–95.

Vogen, S. M., Prakash, O., Kirnarsky, L., Sanderson, S. D., & Sherman, S. A. (1999). Determination of structural elements related to the biological activities of a potent decapeptide agonist of human C5a anaphylatoxin. *The Journal of Peptide Research, 54*, 74–84.

Vogen, S. M., Paczkowski, N. J., Kirnarsky, L., Short, A., Whitmore, J. B., Sherman, S. A., Taylor, S. M., & Sanderson, S. D. (2001). Differential activities of decapeptide agonists of human C5a: The conformational effects of backbone N-methylation. *International Immunopharmacology, 1*, 2151–2162.

Wang, Y., Zhang, H., & He, Y. W. (2019). The complement receptors C3aR and C5aR are a new class of immune checkpoint receptor in cancer immunotherapy. *Frontiers in Immunology, 10*, 1574.

Wang, Y. Q., Bazin-Lee, H., Evans, J. T., Casella, C. R., & Mitchell, T. C. (2020). MPL adjuvant contains competitive antagonists of human TLR4. *Frontiers in Immunology, 11*, 577823.

Weaver, D. J., Jr., Reis, E. S., Pandey, M. K., Kohl, G., Harris, N., Gerard, C., & Kohl, J. (2010). C5a receptor-deficient dendritic cells promote induction of Treg and Th17 cells. *European Journal of Immunology, 40*, 710–721.

Weiss, A. M., Ajit, J., Albin, T. J., Kapoor, N., Maroju, S., Berges, A., Pill, L., Fairman, J., & Esser-Kahn, A. P. (2021). Site-specific antigen-adjuvant conjugation using cell-free protein synthesis enhances antigen presentation and CD8(+) T-cell response. *Scientific Reports, 11*, 6267.

Wetsel, R. A. (1995). Expression of the complement C5a anaphylatoxin receptor (C5aR) on non-myeloid cells. *Immunology Letters, 44*, 183–187.

Wiesmuller, K. H., Bessler, W. G., & Jung, G. (1992). Solid phase peptide synthesis of lipopeptide vaccines eliciting epitope-specific B-, T-helper and T-killer cell response. *International Journal of Peptide and Protein Research, 40*, 255–260.

Xiao, Y., Peperzak, V., Keller, A. M., & Borst, J. (2008). CD27 instructs CD4+ T cells to provide help for the memory CD8+ T cell response after protein immunization. *Journal of Immunology, 181*, 1071–1082.

Yadav, D. K., Yadav, N., & Khurana, S. M. P. (2014). Chapter 26-Vaccines: Present status and applications. In *Animal biotechnology* (pp. 491–508). Academic Press.

Yang, J. X., Tseng, J. C., Yu, G. Y., Luo, Y., Huang, C. F., Hong, Y. R., & Chuang, T. H. (2022). Recent advances in the development of toll-like receptor agonist-based vaccine adjuvants for infectious diseases. *Pharmaceutics, 14*, 423.

Zaal, A., Lissenberg-Thunnissen, S. N., van Schijndel, G., Wouters, D., van Ham, S. M., & ten Brinke, A. (2013). Crosstalk between Toll like receptors and C5a receptor in human monocyte derived DCs suppress inflammatory cytokine production. *Immunobiology, 218*, 175–180.

Zerbo, O., Bartlett, J., Goddard, K., Fireman, B., Lewis, E., & Klein, N. P. (2019). Acellular pertussis vaccine effectiveness over time. *Pediatrics, 144*, e20183466.

Zhang, R. Y., Zhou, S. H., He, C. B., Wang, J., Wen, Y., Feng, R. R., Yin, X. G., Yang, G. F., & Guo, J. (2022). Adjuvant-protein conjugate vaccine with built-in TLR7 agonist on S1 induces potent immunity against SARS-CoV-2 and variants of concern. *ACS Infectious Diseases, 8*, 1367–1375.

Immune Transformations for Therapeutic Gain in Neurodegenerative Disease

50

Roshan Sapkota and R. Lee Mosley

Abstract

Neurodegenerative diseases are a significant cause of morbidity and mortality worldwide, for which effective interventional therapies are limited. Past consideration of the central nervous system (CNS) as a site of immune privilege imparted a major impediment to the interpretation of neurodegenerative pathophysiology in the context of immune processes. Current realizations are that the resident CNS immune system and those elements from the periphery that infiltrate the brain are the main drivers of inflammation, protein misfolding and aggregation, and neuronal injury and death, all of which perpetuate the degenerative cycle. Indeed, immune responses such as microglia activation and leukocytic extravasation into the CNS provide sources of pro-inflammatory and neurotoxic mediators. Thus, the biomarkers and checkpoints that define the immune processes provide candidate targets for therapeutic strategies for immune transformations that provide beneficial outcomes in neurodegenerative disease and are the topic of this chapter.

Keywords

Immunotherapy · Neurodegenerative diseases · T cells · Antibodies · Multiple sclerosis · Alzheimer's disease · Parkinson's disease · Amyotrophic lateral sclerosis · Huntington's disease.

R. Sapkota · R. L. Mosley (✉)
Department of Pharmacology and Experimental Neuroscience, Center for Neurodegenerative Disorders, University of Nebraska Medical Center, Omaha, NE, USA
e-mail: rlmosley@unmc.edu

50.1 Introduction

Neurodegenerative diseases are a group of disorders with a slow, progressive decline of the nervous system leading to disease-dependent memory loss and/or dysfunctional motor movements and ultimately resulting in the inability to perform daily tasks and ultimately death. Although recent accountings show the progressive increase and better understanding of neurodegenerative disease, reports of these diseases and their advancements have been reported throughout human history. An early account of neurodegenerative disease in the thirteenth-century Icelandic Saga, Thorlakur (Sveinbjornsdottir et al., 2014), described a woman experiencing sudden loss of vision with aphasia, presumably multiple sclerosis (MS). In an effort to ameliorate her symptoms, she prayed, walked to Skalholt (an Icelandic bishops' residency and school), and fasted until mass. On the third day of symptoms, a wick was applied to her head, which returned vision to one eye. Her speech returned later, and during mass, she recovered vision in the other eye as well. With the advent of the microscope centuries later, French neurologist Jean-Martin Charcot colloquially known as the father of neurology, characterized MS in 1868 (Zalc, 2018) and went on to classify Parkinson's disease (PD) and amyotrophic lateral sclerosis (ALS) through clinical, anatomical, and histological analysis (Nutma et al., 2019). A few years later in New York City, physician George Huntington reported distinct clinical features of a familial form of chorea, later named Huntington's disease (HD) (Lo & Hughes, 2011). At the turn of the century, advancements in histological staining techniques enabled German psychiatrist Alois Alzheimer to observe distinctive plaques and neurofibrillary tangles in the brain of a 50-year-old woman experiencing paranoia, progressive memory disturbance, and confusion—clinical characteristics that later would be collectively designated as Alzheimer's disease (AD) (Hippius & Neundorfer, 2003).

Despite the extensive history of neurodegenerative disease, these disorders were believed to result only from neu-

H. E. Gendelman, T. Ikezu (eds.), *Neuroimmune Pharmacology and Therapeutics*,
https://doi.org/10.1007/978-3-031-68237-7_50

ronal degradation without an immune component, as the CNS was thought to be a site of immune privilege (Nutma et al., 2019). Specific physiological characteristics, as well as a perceived impermeability of the blood–brain barrier (BBB), contributed to the belief that the CNS limited or even denied entry of cellular immune components (Louveau et al., 2015). However, that notion is no longer accepted, and the understanding has broadened that interactions between the central nervous and the immune systems not only play a role in the pathology of neurodegenerative disease, but also in the homeostatic regulation of the CNS (Louveau et al., 2015). Indeed, immune cells, such as monocyte-derived macrophages and specific lymphocytes have the capability to migrate through the BBB and infiltrate the CNS in response to acute or neurodegenerative injury utilizing a seemingly complex cascade of events that control immune cell extravasation into the brain (Doty et al., 2015). Cellular elements of the immune system also release inflammatory cytokines that penetrate the BBB, leading to neurotoxicity and activation of glial cells, such as microglia and astrocytes (Zang et al., 2022). Activated glial cells release more cytokines and chemokines, which recruit additional immune cells to the CNS and perpetuate a cycle of neurodegeneration and inflammation (Zang et al., 2022). In contrast, immune cells also play an essential role in controlling homeostasis by regulating neurogenesis and brain plasticity (Doty et al., 2015). Taken together, this cyclic dysregulation of the homeostatic neuroimmune system eventually resulting in neuroinflammation and subsequent neurodegeneration is characteristically shared by many neurodegenerative disorders (Zang et al., 2022). Overall, recent advancements reveal that instead of behaving as two separate entities, the immune system intimately is intertwined within the CNS and is most likely important in the etiology of several neurodegenerative diseases.

The prevalence of neurodegenerative diseases, and therefore the social and economic burdens associated with those continue to rise. In 2016, neurodegenerative disorders were attributable to nine million global deaths and 276 global disability-adjusted life years (Collaborators, 2019). Moreover, the accumulated cost in 2020 US dollars of medical, non-medical, and uncompensated caregiving hours as well as indirect lost productivity was estimated at over $650 billion (Alzheimer's Association, 2020). This increase in prevalence and consequent burden is primarily attributed to an aging population. Indeed, aging remains the most significant risk factor for most neurodegenerative diseases. Notably, one of the hallmarks of aging is altered immune function (Hou et al., 2019). Studies have found that chronic, low-grade inflammation positively correlates with age and is associated with neurodegenerative diseases (Currais, 2015). This finding was substantiated by gene expression studies demonstrating the importance of inflammation in aged human brains, especially with respect to neurodegenerative diseases (Lu et al., 2004). Given these findings and the increasingly rapid rate of societal aging (Lutz et al., 2008), the interplay between an altered immune system and neurodegenerative diseases is critical to understanding those disease processes to better strategize preventative and therapeutic approaches to relevant beneficial clinical outcomes and ultimate cures. To date, numerous therapies in clinical trials aim to modulate the immune system to ameliorate neurodegenerative diseases. For the most part, the outcomes of these interventional trials have largely been underwhelming, thus underscoring the complexity of immune function and neuroinflammation in neurodegenerative disease. This chapter will focus on the immunological basis of neurodegenerative disease and immune transformation strategies that drive current hypotheses, and the challenges and limitations of those strategies that guide future directions of immune-mediated transformations to treat neurodegenerative diseases.

50.2 Immunological Basis of Neurodegeneration

Both innate and adaptive immune systems play a role in neurodegenerative diseases. Specifically, microglia, the resident innate immune component in the brain, actively participate in neurodegenerative disease progression and possibly its etiology (Butovsky & Weiner, 2018). Microglia represent the first line of defense against potential danger in the CNS, surveying the surrounding environment for harmful stimuli and foreign pathogens. Upon recognition of pathological stimuli, microglia are activated leading to the production of pro-inflammatory cytokines, chemokines, and other immune processes that accompany immune processes such as phagocytosis and microglial proliferation (Butovsky & Weiner, 2018). Microglial activation can be protective for the brain, especially when defending against foreign pathogens. However, numerous studies have highlighted the effect of chronic activation on microglia to drive neuronal damage by creating a pro-inflammatory environment through the production of nitric oxide (NO), reactive oxygen species (ROS), prostaglandins, proteinases, and complement proteins (Dheen et al., 2007; Subhramanyam et al., 2019). Activated microglia exhibit a wide range of phenotypes with two polar extremes being M1 and M2 phenotypes. While oversimplified (Tang & Le, 2016), the M1 phenotype (classical activation) is considered to be pro-inflammatory, resulting in increased production of cytokines and neurotoxic molecules, while the M2 phenotype (alternative activation) is considered to promote tissue repair, regeneration, and synaptogenesis by promoting an immunosuppressive or anti-inflammatory state (Lee et al., 2020). Stimuli such as lipopolysaccharide (LPS), interferon (IFN)-γ, and molecules associated with neurodegenerative processes, such as β-amyloid (Aβ) and aggregated α-synuclein, have been shown to result in the activation of microglia to exhibit an M1 phenotype (Subhramanyam et al., 2019). Toll-like receptors (TLR) and other pattern recognition receptors (PRRs) are important recognition receptors used by the innate immune cells to sense pathogen- or danger-associated molecular patterns (PAMPs or DAMPs). Microglial TLR4 receptors recognize LPS from Gram-negative bacteria, which leads to a cascade of molecular events and, ultimately, the production and release of TNF-α, IL-6, NO, and neuronal injury and subsequent death (Hanisch & Kettenmann, 2007; Ransohoff & Perry, 2009). Studies found that TLR4-dependent activation is also associated with neurodegenerative diseases such as PD and AD. Alternatively, molecules such as IL-4 and IL-13 can induce M2 phenotypes. Microglia exhibiting M2 phenotypes secrete more anti-inflammatory cytokines, such as IL-10 and transforming growth factor-β (TGF-β), and other factors responsible for tissue repair, such as arginase and chitinase-3 (Subhramanyam et al., 2019).

Infiltration of peripheral immune cells into the CNS can also influence the activation of microglia and the overall pathophysiology and progression of neurodegenerative diseases. Under pathological conditions, specifically, disruption of the BBB, either physically or by inflammation, promotes infiltration of neutrophils and monocytes into the CNS. In fact, such disruptions have been shown as a putative early marker for neurodegeneration (Giannoni et al., 2020; Ishii & Iadecola, 2020). During pathological conditions, neutrophils infiltrating the CNS in response to insult or injury produce free radicals, proteolytic enzymes, and matrix metalloproteinases that can further disrupt the BBB (Zang et al., 2022). Additionally, through crosstalk with microglia and the production of neutrophil extracellular traps (NETS), neutrophils contribute to neuronal damage in AD (Zenaro et al., 2015). Also with a compromised BBB, monocyte-derived macrophages (MDMs) can also easily infiltrate the CNS and communicate with microglia to intensify their functions. Specifically, MDMs regulate acute and long-term microglia-mediated inflammation by suppressing microglia-mediated phagocytosis and inflammation through myeloid-regulatory mechanisms (Greenhalgh et al., 2018). Also, infiltrating NK cells also can modulate neuroinflammation by control of regulatory T cell (Treg) recruitment and microglial phenotypes, by secretion of IFN-γ, or by exocytosis of cytotoxic granules, the latter leading to neuronal and immune cell killing (Garofalo et al., 2020). Dendritic cells (DC) are found in the CNS under physiological conditions, but peripheral DCs can migrate to the meninges and choroid plexus in response to chemokines and adhesion molecules (Sabahi et al., 2021). Herein DCs recognize and process antigens to present to T cells by MHC-I or -II class molecules. The effect of such antigen presentation on neurodegenerative diseases has been found to be diverse and dependent upon the specific T cell phenotype activated, which in part is regulated by the co-stimulatory molecules that are concomitantly expressed at presentation. DCs can promote the activation of pro-inflammatory $CD4^+$ helper T cells (Th) or $CD8^+$ cytotoxic T cells (Tc) depending on presentation by MHC II or MHC I molecules, respectively, both of which have the potential to cause tissue damage (Colton, 2013). Alternatively, DCs can stimulate neuroprotective effects through the induction of anti-inflammatory Tregs as observed in a PD model (Schutt et al., 2018).

T lymphocytes also infiltrate the CNS and play unique, diverse roles in neurodegenerative disease depending on the specific subset of T cells that predominates. T cells bridge the innate and adaptive effector immune systems. Specifically, $CD4^+$ T cells, orchestrate immune responses by secreting cytokines and recruiting immune cells to the site of injury. $CD4^+$ T cells differentiate into effector T cells (Teffs) and regulatory T cells (Tregs). Several different Teff lineages include type 1 helper T cells (Th1) and type 17 helper T cells (Th17), which promote neuroinflammation and neurotoxicity by secreting pro-inflammatory mediators, such as IFN-γ by Th1 cells or IL-17 by Th17 cells, as well as other mediators such as IL-1, Il-6, and TNF-α) (Zang et al., 2022). Notably, effector molecules produced by Th1 cells activate and trigger M1 phenotypes (Prajeeth et al., 2014), while Th17 cells drive microglial phagocytic functions and higher levels of inflammation. Although the role of T cells in neurodegenerative disease is primarily considered one of a neurotoxic nature, recent studies illustrate that other T cell types, namely Th2 and Tregs, have greater neuroprotective and remodeling effects in the CNS by promotion of anti-inflammatory environments and enhancement of glia-mediated neuroprotection (Zang et al., 2022). Tregs regulate microglial activation and play an important role in various neurodegenerative diseases. For example, as ALS progresses, Treg frequency and function levels decrease with increasing levels of pro-inflammatory cytokines and M1 microglia phenotype (Ciccocioppo et al., 2019). Furthermore, Tregs seem to be beneficial in AD patients by slowing the disease progression of AD (Ciccocioppo et al., 2019). On the other hand, $CD8^+$ cytotoxic T cells can drive the pathophysiology to amplify axonal degeneration by T cell receptor- and granzyme B-dependent mechanisms (Zang et al., 2022). Due to the diverse effects of T cells, an equilibrium of pro-inflammatory Teffs and immunosuppressive T cells is thought to be critical for a homeostatic CNS environment. In fact, abnormal Th1/Th2 ratios are believed to exert a potential environment for the preferential influence of neurodegenerative etiology (Luckheeram et al., 2012). The other major component of the adaptive immune system is B cells, which are present in smaller levels in the CNS. However,

CNS inflammation can induce a several-fold increase in B cell levels in the affected region (Sabatino et al., 2019). B cells produce cytokines and antibodies, present antigens to T cells, and activate glial cells, which can provide either pro-inflammatory or anti-inflammatory effects depending on the neurological disease and stage of disease (Ahn et al., 2021).

Although not explicitly mentioned, clearly cytokines play an important role in neurodegenerative disease. However, the exact role of cytokines is not fully understood due to their complexity and diverse effects on many different cell types. Two main reasons for such complexity are the inherent pleiotropic nature of cytokines as well as the contextual-specific functions of cytokines. Therefore, a plethora of contradictory data exists on the influence of cytokines on the pathogenesis of neurodegeneration. Nonetheless, cytokine responses to pathogenic stimuli are well-established and are generally produced early by CNS-resident immune cells, such as microglia as previously mentioned. The early production of cytokines has proven beneficial in the clearance of proteins and repair in several neurodegenerative diseases; however, prolonged, chronic exposure to pro-inflammatory cytokines counters such benefits and can accelerate the progression of neurodegenerative disease (Becher et al., 2017). Although major cytokines implicated involved in neurodegeneration have been discussed, such discussions provide a mere snapshot into the complex interactive role of cytokines currently under study.

Recent emerging data suggests that autoimmune components exist in several neurodegenerative disorders. In general, autoimmune diseases are characterized as a breakdown of immune tolerance to self- or altered self-antigens with the consequent induction of autoreactive immune cells to those antigens (Smith & Germolec, 1999). As a result, the immune system may produce autoantibodies and autoreactive T cells, which exacerbate neurodegenerative pathology. In PD, α-synuclein, a ubiquitous unfolded CNS protein associated with neuronal vesicle trafficking, is found to be post-translationally modified, misfolded, and aggregated, which activates microglia and antigen-presenting cells to induce autoreactive T cells, and autoantibodies by B cells (Bonam & Muller, 2020). Moreover, aberrantly folded α-synuclein that escapes degradation and activates microglia has been shown to be an operative component in chronic inflammation, a common feature of the substantia nigra in PD. Furthermore, detectable levels of autoantibodies to α-synuclein fibrils in the sera and CSF of PD patients have been found (Yanamandra et al., 2011). While the origins of such autoantibodies are unknown, cross-reactivity with neuronal antigens and neuronal cells exists, and can induce toxicity to dopaminergic neurons (Bonam & Muller, 2020); however, passive immunization with low- or high-titer anti-α-synuclein antibodies do not significantly affect neurotoxin-induced dopaminergic neurodegeneration (unpublished data). On the other hand, autoreactive T cells have been reported to recognize and establish immune responses to phosphorylated Ser129-α-synuclein in PD patients (Sulzer et al., 2017). However, those T cell isolates produced primarily IL-5 with a minor population of IFN-γ-producing Th1 cells, suggesting a predominance of Th2 Teffs. Moreover, no pathogenic inducing significance has been assigned to those populations. In contrast, Th1 and Th17 Teffs that recognize nitrated α-synuclein show significant capacity to exacerbate neurodegeneration by approximately 25% and 60%, respectively, in neurotoxin models of PD (Benner et al., 2008; Reynolds et al., 2009, 2010). Taken altogether, these findings expand our understanding of the role of the immune system in PD and putative autoimmune components, not only in PD, but in other neurodegenerative diseases as well. Notably, such paradigm-shifting research has been conducted in AD and preclinical models illustrating autoimmune components associated with AD pathology (Machhi et al., 2021; Weaver, 2022).

50.3 Immune Transformation Strategies

Given the importance of microglia activation in neurodegenerative diseases, microglia have emerged as a key target in immune transformation strategies. While much debate remains about the diverse phenotypes that microglia exhibit, strategies targeting the inflammatory M1 phenotype and cell processes make attractive therapeutic candidates. Peroxisome proliferator-activated receptor gamma (PPARγ) is a nuclear hormone receptor expressed by macrophages and microglia, which, upon activation, skews monocytes to an anti-inflammatory M2 phenotype (Bouhlel et al., 2007). Multiple PPARγ agonists, such as pioglitazone, rosiglitazone, and an N-carbamylated urethane compound (SNU-BP), focus on affecting microglia phenotypes (Song & Suk, 2017). For example, pioglitazone mediates microglia transformation to M2 phenotypes, exerting neuroprotective effects in a neuro-inflammatory environment (Liu et al., 2020). Similar findings were observed with rosiglitazone, which was found to induce IL-4 expression, a key cytokine for induction of protective M2 phenotypes (Rossi et al., 2018; Song & Suk, 2017). SNU-BP, a novel PPARγ agonist, inhibits LPS-induced pro-inflammatory cytokines and nitric oxide production by M1 phenotypes (Song et al., 2016). Additionally, SNU-BP potentiates M2 phenotypes by amplifying IL-4-induced arginase-1 expression, thus providing an overall anti-neuroinflammatory effect in preclinical models (Song et al., 2016). Another substance, malibatol A (MA), a natural resveratrol oligomer from the leaves of the Chinese plant *Hopea hainanensisi* that promotes the activation of PPARγ, was shown to decrease expression of M1 markers (CD16, CD32, and CD86), increase expression of M2 markers

(CD206 and YM-1), reverse induced mitochondrial dysfunction, and provide an overall anti-inflammatory/neuroprotective effect in neurodegenerative mouse models (Pan et al., 2015; Yang et al., 2015).

Changes in metabolism also influence microglial phenotypes. For example, studies have shown that shifts of cellular metabolism from oxidative phosphorylation to aerobic glycolysis favor polarization of M1 microglial phenotypes (Song & Suk, 2017). Moreover, numerous metabolic modifying molecules, such as metformin, also induce shifts to M2 phenotypes with additional benefits of improved neurogenesis and reduced pain and inflammation. Cyclic-adenosine monophosphate (cAMP) also influences microglial activation and function by targeting cAMP-phosphodiesterase (PDE) signaling (Ghosh et al., 2012). These data indicated that cAMP plays a major role in M1/M2 polarization. Thus, several therapeutic strategies target the activation or inhibition of key enzymes in the cAMP production. Overall, these drugs show that targeting cAMP may be beneficial in inhibiting pro-inflammatory cytokine production and induction of anti-inflammatory/neuroprotective microglial phenotypes (Song & Suk, 2017). Additionally, oxidative stress plays an important factor in microglial activation. Studies showed that activation of a key producer of ROS, NADPH oxidase (NOX), induces M1 microglial phenotypes, thus inhibiting NOX with apocynin or genetic deletion promotes M2 microglial phenotype expression (Choi et al., 2012; Song & Suk, 2017). Other factors, such as inhibitors of mammalian targets of rapamycin (mTOR) or genetically severing the signal pathway influences on microglial activation, have been shown to diminish M1 microglia numbers with a concomitant increase in M2 phenotype expression and neuroprotection (Li et al., 2016). Overall, a vast amount of intellectual energy has been committed to understanding mechanisms that contribute to microglial activation and how potential molecules could transform the immunological state of microglia for therapeutic gain in neurodegenerative diseases.

Given the neuroinflammation associated with neurodegenerative diseases, anti-inflammatory agents have been proposed as therapeutic strategies. In a 1-methyl-4-phenyl-1,2,3,6-tetrahydropyridinean (MPTP) mouse model of PD, the effects of acetylsalicylic acid, a COX-1/2 inhibitor, and meloxicam, a preferential COX-2 inhibitor, showed that treatment with either inhibitor was neuroprotective for dopaminergic neurons in the substantia nigra and striatum (Teismann & Ferger, 2001). Underscoring those data was epidemiological evidence showing that nonsteroidal anti-inflammatory drugs (NSAIDs), specifically ibuprofen, diminished the risk of PD (Chen et al., 2005). However, these findings were challenged in a recent meta-analysis of 17 studies, and over 14,000 PD patients found no observed correlation between the benefits of NSAID usage and PD risk (Tahmina Nasrin Poly et al., 2019). Neuroprotective steroids are another anti-inflammatory agent explored for therapeutic management of neurodegenerative diseases. Both natural and synthetic steroids have shown improved function within the brain through reduction of oxidative stress, neuronal excitability, and inflammation associated with neurodegeneration (Bansal & Singh, 2018). Examples of these types of neuroprotective steroids include endogenous neuroactive steroids and hormonal steroids, such as dehydroepiandrosterone (DHEA), DHEA sulfate (DHEAS), testosterone, estradiol, pregnenolone, and progesterone, which are produced by steroidogenic glands in the periphery, in contrast to those directly produced in the CNS, such as allopregnanolone and tetrahydrodeoxycorticosterone (THDOC) (Melcangi et al., 2008). Limitations for endogenous neuroprotective steroids include short half-lives and fast metabolism that limit bioavailability.

Recently, immunotherapy, or more specifically, the therapeutic inhibition of immune checkpoint signaling, has been utilized in cancer to break immune tolerance allowing for the immune recognition and killing of cancer cells (Pardoll, 2012). Under immune evasion scenarios, tumor checkpoint proteins bind cognate ligand proteins on T cells, whereby an "off" signal is sent to the T cell preventing immune activation or inflammation. Thus, off-checkpoint signals are targeted to inhibit negative signal transduction in immunotherapy for tumors. For neurodegenerative diseases that are associated with chronic inflammatory components, diminution, or inhibition of those inflammatory signaling pathways may provide beneficial outcomes. Thus, blockade of such checkpoint proteins, such as programmed death-1 (PD-1), was assessed in an APP/$PS1_{M146L}$ AD mouse model and found to incite IFN-γ dependent systemic immune responses and recruit peripheral immune cells to the brain (Baruch et al., 2016). Clearance of cerebral Aβ plaques and improved cognitive performance accompanied the immune response, suggesting the utilization of immune checkpoint inhibition as a therapeutic strategy in AD. However, converse findings were reported in a 5XFAD mouse model wherein anti-PD-1 blockade also produced significant immune responses, but not immune infiltration or significant clearance of Aβ plaques (Latta-Mahieu et al., 2018). Factors leading to the divergent results are unclear but may involve model differences, background effects, or experimental design differences between the studies. In short, these findings suggest that additional factors other than the PD-1 checkpoint are involved in the clearance of Aβ plaques, thus warranting more studies to understand the mechanisms associated with PD-1 blockade therapy and AD. Other negative checkpoint signaling pathways include negative regulators such as cytotoxic T lymphocyte antigen-4 (CTLA-4) and lymphocyte activation gene-3 (LAG-3), positive co-stimulatory regulators OX40 (CD134) and 4-1BB, and T cell immunoglobulin and mucin-domain-3 (Tim-3), which has both inhibitory and stimulatory effects. Taken together, the potential of such checkpoint-based strategies is promising for immune therapeutics for neurodegenerative diseases.

Considering the important roles of pro-inflammatory cytokines in neurodegenerative disease, cytokine-based therapy manipulation of those profiles has been proposed for therapeutic gain. For example, cytokine profiles in ALS show elevated IL-6, GM-CSF, IL-10, IL-2, and IL-15 (Azodi & Jacobson, 2016). Notably, G-CSF acts as a neural ligand that has been shown to prevent neuronal apoptosis (Azodi & Jacobson, 2016). While several small studies in ALS have shown that G-CSF is well tolerated, clinical outcomes, measured by the ALS Functional Rating Scale (ALSFRS), were similar among controls (Chio et al., 2011; Duning et al., 2011). In MS, histopathological analysis of demyelinating lesions shows depositions of inflammatory cytokines such as IL-12, IL-17, IL-23, and IFN-γ (Azodi & Jacobson, 2016). Therapeutic strategies to alter this pro-inflammatory state have been proposed, namely with IFN-β treatment (Graber et al., 2007; Kurtuncu et al., 2012). IFN-β treatment increases levels of the anti-inflammatory cytokine IL-10, while decreasing levels of pro-inflammatory cytokines such as IL-18, IL-23, IL-12, and IFN-γ (Graber et al., 2007; Kurtuncu et al., 2012). Of note in MS, IL-12, and IL-23 contribute to the differentiation and proliferation of necessary pro-inflammatory Th1 and Th17 cells, respectively (Azodi & Jacobson, 2016). Thus, anti-IL-12 and anti-IL-23 therapies targeting the shared p40 subunit were assessed by monoclonal antibodies; however, while the therapies are tolerated, greater efficacies were not attained compared to the current standard of care (Segal et al., 2008; Vollmer et al., 2011). Another pro-inflammatory and neurotoxic cytokine in MS is TNF-α, which is also elevated in peripheral blood of relapsing-remitting MS patients (Larochelle et al., 2011). However, neutralization of TNF-α led to worse patient outcomes leading to the belief that TNF-α may have a dual function, pro- and anti-inflammatory effects depending on other factors (Azodi & Jacobson, 2016). Indeed, the particular TNR receptor temporally engaged may play a substantial role in outcomes. For instance, engagement of the death-domain containing TNFR1 receptor promotes pro-inflammatory and cytotoxic outcomes, whereas engagement of the TNFR2 receptor has anti-inflammatory, pro-survival, and neuroprotective effects in the CNS (Fresegna et al., 2020). Thus, combinatorial therapies using TNFR1 inhibitors and TNFR2 agonists represent promising approaches to neurodegenerative diseases wherein TNF plays a major role. IL-2 is another cytokine involved in the pathogenesis of MS as the IL-2/IL-2R pathway determines the balance between immunity and tolerance (Larochelle et al., 2011). Of note, a clinical trial in MS patients treated with daclizumab, an anti-IL-2R monoclonal antibody approved for preventing acute graft rejection in 1997 and for RRMS in 2016, reduced the number of contrast-enhanced lesions and relapse rate (Bielekova et al., 2009). However, after an extended assessment of daclizumab for up to 5 years, the study was terminated and the drug was withdrawn from the world market in 2018 due to serious adverse events, including encephalitis, despite significant efficacious outcomes in reduced number and rate of lesion formation (Kappos et al., 2021). It is important to note that approval was based on short-term results; however, these results underscore the necessity for extended assessment for immunotherapeutic modalities. Moreover, divergent and untoward findings of cytokine-based therapies in ALS and MS underscore the complex nature of cytokine networks and the need for greater elucidation of cytokine roles in neurodegenerative disease and putative disease-modifying therapies.

Another strategy for immune transformation utilizes IL-10, a key immunosuppressive cytokine, which exerts profound beneficial effects on most neurodegenerative diseases (Porro et al., 2020). It is produced by most leukocytes as well as keratinocytes and epithelial cells in response to insult or injury (Itakura et al., 2011; Saraiva et al., 2009). It primarily exerts anti-inflammatory effects by signaling through 2 receptors IL-10R1 and IL-10R2, which are members of the IFNR family (Ho et al., 1993). After receptor engagement, signal is transduced via the JAK1/TYK2/STAT3 pathway with phosphorylated STAT3 translocating to the nucleus to drive the expression of STAT3 responsive genes such as suppressor of cytokine signaling 3 (SOCS-3) and inhibiting expression of pro-inflammatory mediators such as TNF-α, IL-1, IL-6, IL-8, IL-12, and iNOS (Berlato et al., 2002). In several neurodegenerative diseases such as AD, PD, ALS, and MS, IL-10 is found to be elevated in plasma along with pro-inflammatory cytokines, which is hypothesized to indicate attempts at regulating the inflammatory state (Kwilasz et al., 2015). In animal models of PD, IL-10 overexpression proved to suppress neuroinflammation processes, reduce pro-inflammatory cytokines, and protect dopaminergic neurons (Porro et al., 2020). In models of AD, depletion of IL-10 promotes activated M1 microglial phenotypes and attenuates M2-like phenotypes with increased AD progression and exacerbated pathology, which highlights the significance of M1 microglia and the importance of IL-10 in regulating those phenotypes (Laffer et al., 2019). Furthermore, IL-10 treatment in AD models attenuates neuroinflammation, reduces amyloid deposition, enhances neurogenesis, and rescues cognitive deficits (Bagyinszky et al., 2017; Guillot-Sestier et al., 2015). However, one report showed that while IL-10 overexpression induces M2 transcriptomic signatures in microglia, Aβ deposition was increased with worsening cognitive impairment (Chakrabarty et al., 2015). Ex vivo studies indicated that at certain levels, IL-10 inhibits glial phagocytosis of Aβ with increased expression of ApoE, which exacerbates plaque burden and cognitive impairment. Together, these data suggest that a therapeutic range for

IL-10 levels necessary to provide beneficial outcomes may exist for treatment of AD patients. Similarly, in EAE models of MS, therapeutic administrations of IL10 were mixed. While IL-10 has a clear role in limiting neurodegeneration in EAE, overexpression of IL-10 transgene provides complete protection to the development of disease (Bettelli et al., 1998); however, direct administration of IL-10 in MS models has yielded conflicting results ranging from complete to partial to insignificant levels of protection depending on the model and route of delivery (Kwilasz et al., 2015). Exacerbation of EAE after intravenous delivery of IL-10 has also been reported (Cannella et al., 1996). Thus, direct administration of IL-10 appears ineffective or too variable for any beneficial effects in MS. More promising findings have been reported with IL-10 gene constructs such as adenoviral (ADV)-, adeno-associated viral (AAV)-, or lentiviral vectors that were administered directly to the CNS with most studies showing significant suppression of clinical symptoms when sufficient doses of vectors were delivered (Kwilasz et al., 2015). IL-17A is another pro-inflammatory cytokine that is a key component in neurodegenerative diseases. IL-17A is a signature cytokine of Th17 Teffs and has diverse functions such as contributing to pathogenic inflammation and also induction of innate-like immune defenses. Numerous studies have shown increased levels of IL-17A in CSF and plasma in AD, MS, and PD patients and may play a role by either direct cytotoxicity or activating glial cells (Chen et al., 2020). Deletion of IL-17A gene in mouse model of PD alleviates dopaminergic neurodegeneration and BBB disruption (Liu et al., 2019). In EAE model of MS, treatment with anti-IL-17A prevents clinical and histological manifestations of EAE (Uyttenhove et al., 2007). Phase 2 clinical trials in RRMS using secukinumab, an IL-17A monoclonal antibody, showed improved clinical outcomes, with Phase 3 results pending (McGeachy et al., 2019).

Most neurodegenerative diseases, including AD, PD, HD, and ALS, are characterized by neuronal injury with eventual loss of neurons from the CNS (Sivandzade & Cucullo, 2021). Thus, regeneration of neurons through stem cell therapy is not only an attractive therapeutic strategy, but most likely will be required to reverse the loss of functional neurons. Stem cells are differentially classed based on shared or varied characteristics and functions. Of importance to immune therapeutics is the use of mesenchymal stem cells (MSCs), noted for their immunomodulatory effects, and induced pluripotent stem cells (iPSCs) necessary to provide personalized modalities. MSCs are found in the bone marrow, umbilical cord, adipose tissue, and spleen, and can differentiate into various cell types such as bone, cartilage, fat, muscle, and neurons (Sivandzade & Cucullo, 2021). Given the relative safety of MSC transplantation, it also provides added several advantages to therapeutic intervention that include limited ethical problems, high accessibility, easy isolation methods, autologous cell generation, self-renewal capacity, and low risk of immune rejection (Sivandzade & Cucullo, 2021). Additionally, while posing a slight tumorigenic risk, tumor organization and metastasis are lower than other stem cell types such as embryonic stem cells (ESCs) (Aleynik et al., 2014). Another benefit of MSCs is the ability to cross the BBB through paracellular pathways even amid the presence of tight junctions (Matsushita et al., 2011). In the CNS, MSCs promote neuronal growth by producing neurotrophic factors, secreting angiopoietin 1, and angiogenic cytokines, which suppress neuroinflammation, oxidative stress, and apoptosis, thus promoting a neuroprotective environment for survival and growth (Kar et al., 2016; Wang et al., 2017). In contrast to MSCs, iPSCs can be converted into mature functional neural lineages using optimized differentiation protocols using a variety of transcription factors (Matsumoto et al., 2016; Sproul, 2015). iPSCs are derived from transformable non-pluripotent adult somatic cells, such as hepatocytes or circulating T cells, to express genes and transcription factors that mimic ESCs (de Lazaro et al., 2014). Much like MSCs, the advantages of iPSCs are the lack of serious ethical concerns, high accessibility, and low risk of immune rejection (Sivandzade & Cucullo, 2021). However, protocols that provided unvaried differentiation into mature neurons have yet to be perfected for clinical translation, and the risk remains for tumor formation due to unwanted viral integration as well as a low reprogramming efficiency (Sivandzade & Cucullo, 2021). Another issue with transplantation therapy is the vulnerability of the engrafted tissues. Notably, three early investigations using fetal mesencephalonic grafts into PD patients showed those cells could survive and develop into dopaminergic neurons (Kordower et al., 2008; Li et al., 2008; Mendez et al., 2008). However, upon postmortem examination, several of the grafts had developed Lewy body pathology with evidence of activated microglia around the affected neurons suggesting that susceptibility of the grafted neurons was due to non-autonomous cell processes, quite possibly associated with inflammatory response and/or transmissibility of pathogenic α-synuclein. The possibility remains with stem cell replacement therapy that suppresses neuroinflammation in the short term, regimens that target long-term attenuation of inflammation may be necessary to bring neuronal replacement therapies to fruition. While the notion of direct intervention by neuronal regeneration through stem cell therapy sounds appealing, much work defining the methods and mechanisms that provide more reliable and efficacious outcomes remains to be assessed.

50.4 Immune Transformation in Neurodegenerative Diseases

50.4.1 Alzheimer's Disease

Two histological hallmarks of AD are abnormal misfolding and aggregation of Aβ that form plaques and the hyperphosphorylation and misfolding of tau proteins that form neurofibrillary tangles (Muraoka et al., 2020). In 1999, Schenk et al. demonstrated that active immunization with Aβ42 prevented the development of amyloid plaque formation in transgenic animals (Schenk et al., 1999). Their findings were instrumental in shaping the notion that immunization may be effective in preventing and treating AD. Clinical trials for active immunizations of Aβ in humans promptly followed in the early twenty-first century. Unfortunately, due to serious adverse events such as T cell-mediated meningoencephalitis, a phase 2 trial was terminated, though analysis of postmortem tissues revealed that immunization reduced amyloid plaque aggregation, but did not provide significant improvement to quality of life indicators (Orgogozo et al., 2003). Second-generation vaccines were developed to remove dominant T cell epitopes to mitigate untoward effects, but those have yet to come to fruition. Instead, passive immunization strategies were developed using monoclonal antibodies to Aβ as disease-modifying therapies for AD. Results from clinical trials for several of anti-Aβ antibodies such as bapineuzumab, solanezumab, and crenezumab failed due to either failure to meet primary endpoints or unacceptable adverse events (van Dyck, 2018). Moreover, a finding consistent with these approaches is that despite reductions of amyloid plaques, significant clinical benefits in disease progression or quality of life were not achieved (Zhang et al., 2023). Moreover, these failures could be attributable to the inability of the vaccine to delete neurotoxic forms of Aβ, such as oligomers and fibrils, and/or to the inability of treatment to effect therapeutic benefits in late stages of disease (Tabira, 2010). Aducanumab, a monoclonal antibody against the N terminus of the Aβ conformation, binds to both soluble oligomers and insoluble fibrils (Song et al., 2022). Despite several issues with the new drug, substantial decrease in amyloid plaque accumulation determined by PET imaging provided an impetus for accelerated approval as a treatment for early AD but will require Phase 4 post-marketing monitoring to confirm any clinical benefits (Liu & Howard, 2021). Lecanemab, another monoclonal antibody, that binds to soluble Aβ protofibrils, similarly achieved accelerated approval based on reduced PET amyloid signaling and has compelling data showing clinical improvement in early AD that may provide a pathway to standard approval (Cummings, 2023; Larkin, 2023). However, aducanumab and lecanemab, both failed to meet primary Phase 2 endpoints, and have been associated with amyloid-related imaging abnormalities (ARIAs). More recently, donanemab, a monoclonal antibody that recognizes Aβ(p3-42), a pyroglutamate aggregated form in amyloid plaques, has also shown reduction of Aβ burdens and met its primary clinical endpoints. It should be noted that the clinical rating scales were different for each monoclonal antibody tested. Notably, as of January 2024, Biogen has discontinued aducanumab citing "reprioritazation" of the company's efforts in AD therapies.

Given the limited clinical success of therapies despite being able to decrease Aβ plaques, approaches that target AD biomarkers other than Aβ have been considered. Recent data suggests that misfolded tau protein more closely correlates to tissue atrophy, neuornal loss, and cognitive decline, thus motivating the development of anti-tau strategies (Schäfer et al., 2021). Active immunization against tau has been shown by numerous studies to reduce pathological tau levels and improve behavioral outcomes in mouse models (Asuni et al., 2007; Joly-Amado et al., 2020). Unfortunately, the same efficacy has not been translated into the clinic as over 30 drugs targeting tau aggregation have failed in trials to demonstrate clinical efficacy (Imbimbo et al., 2022). Another strategy that targets AD-associated neuroinflammation focuses on the protein designated as triggering receptor expressed on myeloid cells 2 (TREM2), which is involved in microglial activation (Ulrich et al., 2017). Preclinical mouse models using TREM2 agonists attenuated tau plaques and neurite dystrophy while improving behavior and reducing microglial inflammation (Wang et al., 2020). Based on these promising preclinical results, three TREM2 targeting drugs have been developed and are under clinical trial consideration (Zhang et al., 2023). In addition, numerous drugs, namely small molecule inhibitors and monoclonal antibodies that target inflammatory processes associated with AD have been developed and tested. A combination of ibuprofen and cromolyn (ALZT-OP1), a small molecule with Aβ anti-aggregation properties, in APP mutant AD mice reduced insoluble Aβ, showed increased microglial recruitment to and phagocytosis of Aβ, and is currently undergoing phase 3 clinical trials (Zhang et al., 2018). Targeting the NLRP3 inflammasome with inhibitors such as dapansutrile, desloratadine, or DW14006 in AD mouse models ameliorated innate immune response, enhanced microglial Aβ phagocytosis, reduced amyloid plaque deposition, and rescued cognitive impairment (Gao et al., 2023). Additionally, other approaches that target immune checkpoints such as PDL-1, MSC based modulation of neuroinflammation, and ERK signaling, are undergoing clinical trials. A novel approach is the use of granulocyte/macrophage colony-stimulating factor (GM-CSF or CSF2), a small protein leukocyte growth factor. The rationale is that it promotes innate cellular expansion, which facilitates phagocytosis, and promotes Treg induction that targets neuroinflammation. In AD mouse models, GM-CSF increased microglia number, reduced AD pathology, and improved cognitive behavior (Boyd et al., 2010; Kiyota et al., 2018). In a Down syndrome mouse model, GM-CSF reduced amyloid deposition, neuroinflammation, and improved memory and learning in both Down syndrome and wild type mice (Ahmed et al., 2022). In a Phase 2 trial, treatment of AD patients with sargramostim (recombinant human GM-CSF) was tolerated without amyloid-related imaging abnormalities (ARIA), improved cognition, and moved plasma levels of Aβ, tau, and markers of neurodegeneration to baseline levels (Potter et al., 2021). A Phase 2 trial is currently ongoing.

50.4.2 Parkinson's Disease/Synucleinopathies

Parkinson's disease (PD) is the second most common neurodegenerative disease and is classified as a progressive degeneration of dopaminergic neurons in the substantia nigra and termini in the striatum (Poewe et al., 2017). Lewy bodies consisting of α-synuclein aggregates are a pathological hallmark of PD. A failure to clear α-synuclein aggregates accompanies mitochondrial dysfunction, microglial activation, and spread of α-synuclein to nearby dopaminergic neurons (Calabresi et al., 2023). The activation of microglia and neuroinflammation drive α-synuclein modifications that produce toxic forms of oligomers and fibrils in PD (Calabresi et al., 2023). This not only leads to the release of neurodestructive cytokines and chemokines (Pajares et al., 2020), but also facilitates prion-like spreading of α-synuclein aggregates and enhancement of α-synuclein expression (George et al., 2019; Kasen et al., 2022). Additionally, a growing body of evidence has established the importance of the adaptive immune system in driving PD pathophysiology (Benner et al., 2008; Olson et al., 2023b; Reynolds et al., 2010; Saunders et al., 2012; Schwab et al., 2020; Sulzer et al., 2017). Importantly, α-synuclein-specific Th1 and Th17 Teffs secrete neurotoxic factors and drive microglia to hyperactive states that mediate further neurodegeneration and α-synuclein aggregation to perpetuate the neurodestructive cycle.

Understanding the role of the immune responses in PD has led to the development of numerous putative immune-based therapies for PD. Similar to experimental therapies in AD, active and passive immunization protocols have provided potential therapeutics for beneficial gain. Currently, three peptide-based vaccines have been developed that target the clearance of α-synuclein and are in clinical trials with each AFFITOPE PD01A, AFFITOPE PD03A, and UB-312 shown to be safe and tolerable, and forwarded to Phase 2 studies (Zhang et al., 2023). Additionally, seven monoclonal antibodies are currently undergoing or have undergone clinical trials for passive immunization in PD (Calabresi et al., 2023). Prasinezumab and cinpanemab are antibodies that recognize aggregated α-synuclein or the N-terminus, respectively, and have recently completed Phase 2 clinical trials. Unfortunately, neither met primary endpoints; however, the prasinezumab trial was extended. Moreover, reanalysis of data uncovered evidence of slower motor deterioration (Jankovic, 2019; Pagano et al., 2022); thus, another Phase 2B is ongoing. Another α-synuclein approach is modulation of oligomer formation. Two inhibitors of α-synuclein aggregation are anle138b and UCB0599 (aka NPT200-11), which inhibit aggregation and misfolding, respectively. In preclinical studies, the inhibitors suppressed α-synuclein aggregation and accumulation, neuronal degeneration, and disease progression (Price et al., 2018; Wagner et al., 2013), and have proven safe and tolerable in Phase 1 clinical trials with continuation into ongoing Phase 2 trials (Calabresi et al., 2023; Smit et al., 2022).

Similar to AD therapies, anti-inflammatory small molecules such as NSAIDs have been assessed in PD for potential therapeutic benefit with mixed findings (Zhang et al., 2023). Animal studies consistently show that NSAIDs and other anti-inflammatory therapies attenuate neuroinflammation and preserve dopaminergic neurons (Liu et al., 2022), which spurred an epidemiological study that showed a lower risk of PD incidence in cohorts that took ibuprofen at least twice a week (Gao et al., 2011). However, several epidemiological and meta-analytic studies failed to confirm an association between NSAIDs, ibuprofen, and risk of PD (Manthripragada et al., 2011; Poly et al., 2019; Rees et al., 2011). Some approaches have focused on anti-TLR-2 therapy to inhibit α-synuclein activation of microglia. In PD mouse models, anti-TLR-2 antibody therapy was found to decrease aggregation of α-synuclein, reduce inflammation, and improve behavioral deficits (Kim et al., 2018). Additionally, in preclinical models, a small molecule antagonist of TLR-2, NPT520-34, reduced α-synuclein pathology, attenuated neuroinflammation expression, and improved motor function (Khan et al., 2021). NPT520-34 was found to be safe and well tolerated in Phase 1 clinical trials in healthy subjects and is slated for Phase 2 consideration. Additionally, epidemiological and experimental evidence finds close associations between diabetes mellitus and PD; thus, the use of anti-diabetes therapies for PD was considered (Hassan et al., 2020). Notably, treatment with GLP-1 receptor agonists such as exenatide, NLY01, or semaglutide in mouse models of PD attenuated inflammation, reduced lipid peroxidation, inhibited mitophagic signal transduction, increased expression of neurotrophic factors, and provided neuroprotective outcomes (Gharagozloo et al., 2021; Mahapatra et al., 2022). A recent Phase 2 clinical trial of exendin-4, a long-acting GLP-1 receptor agonist, demonstrated positive beneficial outcomes (Esparza-Salazar et al., 2021). Phase 2 and 3 clinical trials are ongoing or planned for semaglutide and exenatide (Zhang et al., 2023).

Given the interplay between the adaptive immune system, neuroinflammation, and PD, the transformation of Teff profiles represents a novel therapeutic strategy. Specifically, PD progression was associated with reduced Treg function and increased Teff numbers in PD patients (Saunders et al., 2012). The importance of Tregs in PD is underscored by many animal studies showing that depletion of Tregs leads to increased neurodegeneration, while adoptive transfer of Tregs to PD mice is neuroprotective (Benner et al., 2008; Olson et al., 2023b; Reynolds et al., 2010; Schwab et al., 2020). Moreover, administration of GM-CSF to PD mice increased Treg populations, leading to nigrostriatal neuroprotection and upregulated immunosuppression (Kosloski et al., 2013; Olson et al., 2021b). These results formed the basis of a Phase 1 clinical trial to test sargramostim (recombinant human GM-CSF) in PD patients, which concluded that sargramostim was safe and well tolerated, increased Treg number and function, improved cortical motor neuronal activity as determined by magnetoencephalography, and significantly improved clinical scores (Gendelman et al., 2017). A small open-label study of sargramostim found long-term administration at a lower dosage was safe with fewer adverse event rates per patient basis, increased peripheral Treg number and function, and lowered clinical motor function scores from baseline at enrollment, which directly correlated with Treg function (Olson et al., 2021a, 2023a). These studies also demonstrated that after discontinuation of the 24-month sargramostim regimen, UPDRS part III scores returned to baseline levels prior to initiation of sargramostim. Re-start of sargramostim resulted in improved UPDRS part III scores with significant changes from baseline by 6 months. Moreover, changes in transcriptomic and proteomic biomarkers indicated that sargramostim-regulated pathways were enriched for anti-inflammatory, anti-oxidant, and pro-autophagic processes (Abdelmoaty et al., 2022). Other studies have also showcased the benefit of restoring Treg equilibrium under neuroinflammatory environments, thus further reinforcing the importance of the adaptive immunity in PD (Pacheco, 2021).

50.4.3 Amyotrophic Lateral Sclerosis

Amyotrophic lateral sclerosis (ALS) is characterized by progressive loss of motor neurons in the brain and spinal cord and is universally fatal, but the underlying etiology and disease mechanisms amid a heterogeneous presentation remain elusive (Dhasmana et al., 2022; Masrori & Van Damme, 2020). The most common hallmark identified in ALS is misfolded, hyperphosphorylated, and/or ubiquitinated proteins such as SOD1 and TDP-43, which aggregate and form intraneuronal cytoplasmic inclusions (Zhang et al., 2023). These induce cellular damage and neuronal dysfunction by invasion of mitochondria leading to mitochondrial dysfunction, promotion of neuronal inflammation through cGAS/STING pathway (Decout et al., 2021), and activation of microglia. Over 50 randomized clinical trials for ALS have been conducted with disappointing results, which may highlight the complexity underlying the disease processes. Despite the shortcomings of clinical trials for ALS drugs, the FDA in 2017 approved edaravone, a free radical scavenger of peroxyl radical and peroxynitrite. Edaravone protects neurons in the CNS by eliminating ROS produced by glia during neuroinflammation (Cho & Shukla, 2020).

Promising immunotherapies in ongoing clinical trials for ALS aim to inhibit microglial activation, influence Treg populations, and modulate mitochondrial function (Khalid et al., 2017; Masrori & Van Damme, 2020; Zhang et al., 2023). Masitinib, a c-Kit tyrosine kinase inhibitor targets pro-inflammatory myeloid cells and shows efficacy in animal models (Khalid et al., 2017). In an ALS animal model, masitinib decreased microglial activation by inhibiting colony-stimulating factor 1 receptor signaling activity (Ketabforoush et al., 2023). A Phase 2/3 clinical trial showed masitinib slowed the rate of clinical score decline, delaying disease progression, and is currently in Phase 3 clinical trials (Trias et al., 2016). Additionally, interim results of the phase 3 study show a statistically significant difference in patient outcomes with those on masitinib. Edaravone is a substituted 2-pyrazolin-5-one class of drug and is hypothesized to function as an ROS scavenger to alleviate oxidative stress (Cruz, 2018). One clinical trial showed that the drug was safe, but failed to meet primary endpoints to delay disease progression, although secondary endpoints were improved. However, in a Japanese study, a significant slowing in the rate of clinical decline was found. The reason for the contrasting results is unclear but may involve the population, study design, and/or administration of drugs. Nevertheless, the FDA approved its use in ALS in 2017. Another FDA-approved drug with antioxidant capacities is sodium phenylbutyrate-taurursodiol (SP-TS), which functions to ameliorate endoplasmic reticulum stress and reduce mitochondrial permeability, thus increasing the apoptotic trigger threshold (Paganoni et al., 2020). In a clinical trial of ALS patients, SP-TS slowed the rate of clinical decline and increased time of survival compared to placebo, and was approved for ALS in 2022 (Alqallaf et al., 2024). Mitochondrial dysfunction is hypothesized to be an important component in the pathogenesis of ALS, and sigma-1 receptor (Sig-1R) has a key role in mitochondrial calcium signaling and in ALS (Thonhoff et al., 2022). Thus, the Sig-1R agonist, pridopidine, has shown preclinical and clinical evidence to slow disease progression and is currently in Phase 3 clinical trials (Zhang et al., 2023).

Another novel immune-based strategy is rooted in several preclinical animal studies that showed the beneficial role of Tregs in disease onset and progression in ALS (Banerjee et al., 2008; Beers et al., 2011; Zhao et al., 2012). In clinical studies, Treg number and function were found to be diminished and correlated with disease progression (Beers et al., 2017; Henkel et al., 2013; Rentzos et al., 2012). A clinical trial wherein autologous Tregs from ALS patients were isolated, expanded, and adoptively transferred to the ALS donor slowed disease progression rates that were directly correlated with Treg function; however, progression rates were affected only over the time that Tregs had been administered (Alsuliman et al., 2016; Thonhoff et al., 2018), suggesting that sustainability of autologous Tregs may be an issue. As Treg numbers and function are increased with low-dose IL-2 and/or rapamycin (Yadav et al., 2013), a Phase 2 clinical in ALS is currently underway to test autologous Treg transfer for 6 months followed by a 6-month regimen of low dose IL-2 (NCT04055623). In light of the suggested sustainability of autologous Tregs, a Phase 1 trial has been initiated to test a protocol (CK080) to isolate, expand, and transfer allogeneic Tregs from umbilical cord blood to ALS patients (NCT05695521).

50.4.4 Huntington's Disease

Huntington's Disease (HD) is an autosomal dominant neurodegenerative disease caused by an abnormally elongated triplet repeat region encoding a polyQ tract in the Huntington (*HTT*) gene (Roos, 2010). The expansion of the polyQ tract leads to the formation of the mutant Htt (mHtt) protein, which can form intracellular aggregates, in part due to the expansion, which is cytotoxic to cortical and striatal neurons (Fatoba et al., 2020). Additionally, neuroinflammation is important in HD as mHTT is not only expressed in the brain but also by peripheral immune cells and serves as an inflammatory stimulus, leading to the activation of microglia and production of pro-inflammatory cytokines and chemokines (Fatoba et al., 2020).

Like other proteinopathies, the intracellular nature of mHtt aggregation presents intracellular targeting strategies. Intrabodies are intracellularly expressed single-chain variable fragments from antibodies developed to target mHtt. Transfection of mHTT-expressing cells with constructs encoding the mHtt-specific intrabody binding to the poly-Q domain reduced mHtt-induced aggregation and increased cell survival (Khoshnan et al., 2002). Furthermore, using several other highly specific anti-Htt intrabodies to various determinants in HD animal modes demonstrated that intracellular recognition of not only the poly-Q region, but also determinants in the N-terminus reduced mHtt proteins, mitigated neuropathology, and improved motor and cognitive function (Southwell et al., 2008). Despite promising preclinical findings, the safety and efficacy of such therapy in humans is uncertain (Fatoba et al., 2020; Messer & Butler, 2020). Additionally, active immunizations have been assessed, but with mixed findings that will require further investigation into the precision and mechanistic processes to determine the efficacy of active immunizations in HD.

Other immunomodulatory strategies have been evaluated for HD, such as small molecule kinase inhibitors, mesenchymal stem cell (MSC) therapy, and cytokine-modulating therapy. Laquinimod, an immunomodulatory drug that exerts anti-inflammatory and neuroprotective properties in HD mouse models decreases expression of hyper-reactive cytokines, modulates antigen-presentation-related genes, and transforms Th1 polarization to Th2 profiles (Bruck & Wegner, 2011; Dobson et al., 2016; Gurevich et al., 2010; Yang et al., 2004). A clinical trial in HD patients showed that laquinimod treatment reduced loss of caudate volume and gliosis, but no clinical improvement was discerned (Ferguson et al., 2022). In a Phase 2 trial in HD, significant differences were not detected between laquinimod and placebo groups for microgliosis as determined by PET scan nor for clinical scores (Roussakis et al., 2023). Another approach to HD therapeutics is MSC transplantation with concomitant attenuation of neuroinflammation. From several studies in HD mice, a meta-analysis noted that while MSC therapy improved muscle strength, neuromuscular electromyography activity, cortex- and striatal-associated motor functions, and cognition deficits were not changed (Liang et al., 2023). Three clinical trials investigating MSC therapy in HD are currently ongoing but have not yet reported results (Fatoba et al., 2020; Liang et al., 2023). Lastly, as increased levels of TNF-α are found in the brains and CSF of HD mice and patients, anti-TNF-α therapy has been extensively examined. In HD mice, minocycline, a second-generation tetracycline antibiotic and inhibitor of TNF-α signaling, decreases microgliosis, and TNF-α-induced activation and cell death; however, in clinical trials, minocycline had no change in disease progression or outcomes (Fatoba et al., 2020; Huntington Study Group, 2010). One anti-TNF-α approach is the inhibition of TNF-α signaling. DN-TNF-α (XPro1595) is a human-engineered inactive TNF-α variant that inactivates native homotrimeric TNF-α by exchanging complex subunits, thus blocking TNF bioactivity. In HD mice (Steed et al., 2003) intracerebroventricular (ICV) administration of DN-TNF-α reduced neuroinflammation, mHtt aggregate burden, caspase activation, and motor function deficits in R6/2 HD transgenic mice (Hsiao et al., 2014); however, clinical evaluation of DN-TNF-α has yet to be performed. Etanercept is a recombinant fusion of TNF receptors and the Fc portion of IgG1. In HD mice, systemic administration of etanercept, diminished levels of plasma TNF-α and other pro-inflammatory cytokines such as IL-1β and IL-6 (Pido-Lopez et al., 2019). Moreover, etanercept partially reduced brain atrophy, but failed to affect striatal TNF-α and IL-6 expression levels or to improve HD-related functional and cognitive deficits.

50.5 Challenges, Remaining Questions, and Future Directions

Rejection of the brain as an immune privilege site catalyzed the discovery and fundamental understanding of the interactions between the central nervous and immune systems as well as the respective role of neuroinflammation in the context of neurodegenerative disease. While the strength and impermeability of the BBB perpetuated the original belief of immune privilege, the processes associated with the breakdown of such a barrier play key roles in neurodegenerative diseases such as AD, PD, ALS, and MS (Sweeney et al., 2019). BBB breakdown is a complicated process with changes in expression of transporters such as glucose transporter 1, disruption of tight junction proteins, and increased expression of adhesion molecules allowing increased leukocyte invasion into the brain at foci of inflammation (Galea, 2021). Such disruption poses significant risk of infection by pathogens and leakage of toxins or inflammatory mediators from the periphery. Whether BBB disruption plays a causative role in the etiology of neurodegenerative diseases is unclear, but its integrity may be crucial in disease progression and the ultimate determination of neurotherapeutic efficacies (Knox et al., 2022). However, microglia and astrocyte activation with accompanying inflammatory responses are known to be important for tissue remodeling, which maintains the structural integrity of the BBB (Cabezas et al., 2014). As the predominant cell type of the CNS and a central BBB cell type, astrocytes represent an important component in the immunological crosstalk between the peripheral and the resident CNS immune systems (Linnerbauer et al., 2020). Thus, future investigation into the neuronal unit may provide better understanding of BBB disruption and the roles of astrocytes, microglia, and lymphocytes in neuroinflammation and the strategies by which interventional therapies can best target those components for BBB remodeling and restoration (Valori et al., 2021).

Given the role of neuroinflammation in neurodegenerative pathology, immunotherapy represents an important interventional tool necessary for harnessing the neurotoxic environment that perpetuates disease progression. However, immunotherapies have challenges that must be considered to develop effective therapies. Immunotherapy of neoplastic disease offers some insights into these challenges. A major challenge seen in tumor immunotherapy is off target effects and increased inflammation (Whiteside et al., 2016). For example, CAR-T cells against cancer cells have been shown to result in cytokine release syndrome, a systematic inflammatory response to immunotherapy, which can be fatal. A challenge of unique importance to immunotherapy of neurodegenerative diseases is the observed neurotoxicity in patients receiving anti-CD19 CAR-T cell therapy, resulting in delirium, confusion, and seizures (Gust et al., 2018). Although neurotoxicity has mostly been reversible, understanding the underlying mechanistic driver of neurotoxicity will be of critical importance to the development of neurodegenerative immunotherapies. Furthermore, many antibody-based treatments developed primarily to enhance clearance of misfolded and aggregated proteins can have off-target effects, which may modulate inflammation in peripheral organs leading to worsened patient outcomes (Catapano & Papadopoulos, 2013). Many immunotherapies developed for neurodegenerative diseases aim to dampen chronic immune activation and neuroinflammation. However, those may harness the activation responses so completely that phagocytosis is blocked and protein aggregates are not cleared, allowed to accumulate, and spread unchecked, thus accelerating neurodegeneration. Moreover, whether the degree of suppression by immune therapeutics at levels necessary to control neuroinflammation will inhibit immune responses to neoplasia or infection has yet to be determined (Pandya et al., 2016). Thus, future immunotherapies must consider the potential harmful effects of immune suppression on other systemic functions, and how best to appropriately target and balance immunosuppression for optimal beneficial gain.

50.6 Conclusion

In neurodegenerative diseases, certain modalities that target glial activation and chronic neuroinflammation of the immune system via a series of pro-inflammatory mediators are becoming a consistent theme in interventional therapeutics. Moreover, the adaptive immune system is now recognized as one of the major drivers of innate-mediated inflammation and neurodegeneration; thus, modalities targeting antigen-specific effector T cells are becoming more attractive. Indeed, common to most neurodegenerative diseases discussed is a perpetual cycle of protein misfolding, chronic inflammation, and neuronal injury or death. Current immune transformation approaches largely focus on targeting key biological markers through active or passive immunization, modulating microglial function, or altering immune pathways through cytokine- and stem cell-based therapies. Despite promising preclinical results in animal models, many therapies fall short of providing a clear clinical benefit to patients, suggesting that a complete immunological picture of neurodegenerative disease may not be fully realized in the short term. Challenges inherent in the development of immune therapies include consideration of the BBB, development of more accurate preclinical models, control or restricting global immunosuppression, and lack of clear clinical benefits despite clearance or blockade of pathological biomarkers. Thus, future works should consider how to alleviate these challenges to allow translation of promising preclinical results to patient populations. Even amid these challenges, immune transformation approaches offer promising novel therapeutic strategies that will best provide beneficial gains.

50.7 Questions

1. Which immunosuppressive cytokine plays a significant role in regulating microglial activation and is elevated in several neurodegenerative diseases?
   (a) IL-1
   (b) IFN-γ
   (c) TNF-α
   (d) IL-10

2. What is the potential risk associated with induced pluripotent stem cell (iPSC) therapy for neurodegenerative diseases?
   (a) Numerous ethical issues
   (b) Tumor formation
   (c) Immune rejection
   (d) Limited differentiation capacity

3. Which mechanisms and factors most contribute to the chronic activation of microglia and the potential impact on neurodegenerative diseases?
   (a) The role of pro-inflammatory cytokines and their direct cytotoxic effects.
   (b) The cross-talk between microglia and peripheral immune cells.
   (c) The impact of cytokine-based therapies on immune transformation.
   (d) The role of stem cell therapy in suppressing neuroinflammation.

4. How can the blockade of immune checkpoint signaling pathways, such as PD-1 and CTLA-4, potentially impact the progression of neurodegenerative diseases?
   (a) By inducing an autoimmune response and exacerbating neuroinflammation.
   (b) By promoting a neuroprotective environment and enhancing neuronal regeneration.
   (c) By preventing the infiltration of peripheral immune cells into the CNS.
   (d) By suppressing the production of pro-inflammatory cytokines.

5. How do cytokine-based therapies impact the complex cytokine networks involved in neurodegenerative diseases, and what challenges do they face in terms of efficacy?
   (a) They suppress production of pro-inflammatory cytokines and promote neuroprotection.
   (b) They exhibit consistent outcomes across different neurodegenerative diseases.
   (c) They face considerable challenges related to variability and the need for extended assessment.
   (d) They primarily target immune checkpoint proteins like PD-1 and CTLA-4.

6. In the context of immune transformation strategies, how can metabolic modification influence microglial activation and their phenotypic shifts?
   (a) Inhibition of nitric oxide and reactive oxygen species.
   (b) Promotion of M2 microglial phenotypes.
   (c) Shift cellular metabolism from oxidative phosphorylation to aerobic glycolysis.
   (d) Targets cytokine regulation of checkpoint proteins.

7. What has motivated the development of anti-tau strategies in the treatment of Alzheimer's disease?
   (a) The reduction of amyloid plaques
   (b) The success in clinical trials targeting tau aggregation
   (c) The correlation of misfolded tau protein with tissue atrophy and cognitive decline
   (d) The significant clinical benefit achieved with Aβ antibodies

8. What is the relationship between α-synuclein and neuroinflammation in Parkinson's disease (PD)?
   (a) α-synuclein causes neuroinflammation.
   (b) Neuroinflammation leads to α-synuclein accumulation.
   (c) α-Synuclein is a consequence of neuroinflammation.
   (d) Neuroinflammation and α-synuclein influence each other.

9. What evidence supports the role of Tregs in ALS progression, and how are Tregs being used in ALS clinical trials?
   (a) Treg depletion accelerates disease progression; clinical trials focus on their expansion and transfer.
   (b) Tregs have no impact on ALS progression; clinical trials investigate alternative treatments.
   (c) Tregs promote intraneuronal cytoplasmic inclusions; clinical trials aim to inhibit Treg function.
   (d) Treg expansion accelerates disease progression; clinical trials examine their depletion.

10. What are intrabodies, and how do they offer potential therapeutic benefits in HD?
    (a) Extracellular antibodies targeting mHtt aggregates; they prevent mHtt aggregation.

(b) Intracellularly expressed antibody fragments targeting mHtt; they inhibit mHtt-induced aggregation.
(c) Antioxidant enzymes targeting mitochondrial dysfunction; they improve motor function.
(d) Extracellular antibodies targeting pro-inflammatory cytokines; they reduce neuroinflammation.

50.8 Answers

1. d
2. b
3. b
4. a
5. c
6. c
7. c
8. d
9. a
10. b

References

Abdelmoaty, M. M., Machhi, J., Yeapuri, P., Shahjin, F., et al. (2022). Monocyte biomarkers define sargramostim treatment outcomes for Parkinson's disease. *Clinical and Translational Medicine, 12*(7), e958. https://doi.org/10.1002/ctm2.958

Ahmed, M. M., Wang, A. C., Elos, M., Chial, H. J., et al. (2022). The innate immune system stimulating cytokine GM-CSF improves learning/memory and interneuron and astrocyte brain pathology in Dp16 Down syndrome mice and improves learning/memory in wild-type mice. *Neurobiology of Disease, 168*, 105694. https://doi.org/10.1016/j.nbd.2022.105694

Ahn, J. J., Abu-Rub, M., & Miller, R. H. (2021). B cells in neuroinflammation: New perspectives and mechanistic insights. *Cells, 10*(7), 1605. https://doi.org/10.3390/cells10071605

Aleynik, A., Gernavage, K. M., Mourad, Y., Sherman, L. S., et al. (2014). Stem cell delivery of therapies for brain disorders. *Clinical and Translational Medicine, 3*, 24. https://doi.org/10.1186/2001-1326-3-24

Alqallaf, A., Cates, D. W., Render, K. P., & Patel, K. A. (2024). Sodium phenylbutyrate and taurursodiol: A new therapeutic option for the treatment of amyotrophic lateral sclerosis. *The Annals of Pharmacotherapy, 58*(2), 165–173. https://doi.org/10.1177/10600280231172802

Alsuliman, A., Appel, S. H., Beers, D. R., Basar, R., et al. (2016). A robust, good manufacturing practice-compliant, clinical-scale procedure to generate regulatory T cells from patients with amyotrophic lateral sclerosis for adoptive cell therapy. *Cytotherapy, 18*(10), 1312–1324. https://doi.org/10.1016/j.jcyt.2016.06.012

Alzheimer's Association. (2020). 2020 Alzheimer's disease facts and figures. *Alzheimers Dement, 16*(3), 391–460. https://doi.org/10.1002/alz.12068

Asuni, A. A., Boutajangout, A., Quartermain, D., & Sigurdsson, E. M. (2007). Immunotherapy targeting pathological tau conformers in a tangle mouse model reduces brain pathology with associated functional improvements. *The Journal of Neuroscience, 27*(34), 9115–9129. https://doi.org/10.1523/JNEUROSCI.2361-07.2007

Azodi, S., & Jacobson, S. (2016). Cytokine therapies in neurological disease. *Neurotherapeutics, 13*(3), 555–561. https://doi.org/10.1007/s13311-016-0455-1

Bagyinszky, E., Giau, V. V., Shim, K., Suk, K., et al. (2017). Role of inflammatory molecules in the Alzheimer's disease progression and diagnosis. *Journal of the Neurological Sciences, 376*, 242–254. https://doi.org/10.1016/j.jns.2017.03.031

Banerjee, R., Mosley, R. L., Reynolds, A. D., Dhar, A., et al. (2008). Adaptive immune neuroprotection in G93A-SOD1 amyotrophic lateral sclerosis mice. *PLoS One, 3*(7), e2740. https://doi.org/10.1371/journal.pone.0002740

Bansal, R., & Singh, R. (2018). Exploring the potential of natural and synthetic neuroprotective steroids against neurodegenerative disorders: A literature review. *Medicinal Research Reviews, 38*(4), 1126–1158. https://doi.org/10.1002/med.21458

Baruch, K., Deczkowska, A., Rosenzweig, N., Tsitsou-Kampeli, A., et al. (2016). PD-1 immune checkpoint blockade reduces pathology and improves memory in mouse models of Alzheimer's disease. *Nature Medicine, 22*(2), 135–137. https://doi.org/10.1038/nm.4022

Becher, B., Spath, S., & Goverman, J. (2017). Cytokine networks in neuroinflammation. *Nature Reviews Immunology, 17*(1), 49–59. https://doi.org/10.1038/nri.2016.123

Beers, D. R., Henkel, J. S., Zhao, W., Wang, J., et al. (2011). Endogenous regulatory T lymphocytes ameliorate amyotrophic lateral sclerosis in mice and correlate with disease progression in patients with amyotrophic lateral sclerosis. *Brain, 134*(Pt 5), 1293–1314. https://doi.org/10.1093/brain/awr074

Beers, D. R., Zhao, W., Wang, J., Zhang, X., et al. (2017). ALS patients' regulatory T lymphocytes are dysfunctional, and correlate with disease progression rate and severity. *JCI Insight, 2*(5), e89530. https://doi.org/10.1172/jci.insight.89530

Benner, E. J., Banerjee, R., Reynolds, A. D., Sherman, S., et al. (2008). Nitrated alpha-synuclein immunity accelerates degeneration of nigral dopaminergic neurons. *PLoS One, 3*(1), e1376. https://doi.org/10.1371/journal.pone.0001376

Berlato, C., Cassatella, M. A., Kinjyo, I., Gatto, L., et al. (2002). Involvement of suppressor of cytokine signaling-3 as a mediator of the inhibitory effects of IL-10 on lipopolysaccharide-induced macrophage activation. *Journal of Immunology, 168*(12), 6404–6411. https://doi.org/10.4049/jimmunol.168.12.6404

Bettelli, E., Das, M. P., Howard, E. D., Weiner, H. L., et al. (1998). IL-10 is critical in the regulation of autoimmune encephalomyelitis as demonstrated by studies of IL-10- and IL-4-deficient and transgenic mice. *Journal of Immunology, 161*(7), 3299–3306.

Bielekova, B., Howard, T., Packer, A. N., Richert, N., et al. (2009). Effect of anti-CD25 antibody daclizumab in the inhibition of inflammation and stabilization of disease progression in multiple sclerosis. *Archives of Neurology, 66*(4), 483–489. https://doi.org/10.1001/archneurol.2009.50

Bonam, S. R., & Muller, S. (2020). Parkinson's disease is an autoimmune disease: A reappraisal. *Autoimmunity Reviews, 19*(12), 102684. https://doi.org/10.1016/j.autrev.2020.102684

Bouhlel, M. A., Derudas, B., Rigamonti, E., Dievart, R., et al. (2007). PPARgamma activation primes human monocytes into alternative M2 macrophages with anti-inflammatory properties. *Cell Metabolism, 6*(2), 137–143. https://doi.org/10.1016/j.cmet.2007.06.010

Boyd, T. D., Bennett, S. P., Mori, T., Governatori, N., et al. (2010). GM-CSF upregulated in rheumatoid arthritis reverses cognitive impairment and amyloidosis in Alzheimer mice. *Journal of Alzheimer's Disease, 21*(2), 507–518. https://doi.org/10.3233/JAD-2010-091471

Bruck, W., & Wegner, C. (2011). Insight into the mechanism of laquinimod action. *Journal of the Neurological Sciences, 306*(1–2), 173–179. https://doi.org/10.1016/j.jns.2011.02.019

Butovsky, O., & Weiner, H. L. (2018). Microglial signatures and their role in health and disease. *Nature Reviews Neuroscience, 19*(10), 622–635. https://doi.org/10.1038/s41583-018-0057-5

Cabezas, R., Avila, M., Gonzalez, J., El-Bacha, R. S., et al. (2014). Astrocytic modulation of blood brain barrier: Perspectives on Parkinson's disease. *Frontiers in Cellular Neuroscience, 8*, 211. https://doi.org/10.3389/fncel.2014.00211

Calabresi, P., Mechelli, A., Natale, G., Volpicelli-Daley, L., et al. (2023). Alpha-synuclein in Parkinson's disease and other synucleinopathies: From overt neurodegeneration back to early synaptic dysfunction. *Cell Death & Disease, 14*(3), 176. https://doi.org/10.1038/s41419-023-05672-9

Cannella, B., Gao, Y. L., Brosnan, C., & Raine, C. S. (1996). IL-10 fails to abrogate experimental autoimmune encephalomyelitis. *Journal of Neuroscience Research, 45*(6), 735–746. https://doi.org/10.1002/(SICI)1097-4547(19960915)45:6<735::AID-JNR10>3.0.CO;2-V

Catapano, A. L., & Papadopoulos, N. (2013). The safety of therapeutic monoclonal antibodies: Implications for cardiovascular disease and targeting the PCSK9 pathway. *Atherosclerosis, 228*(1), 18–28. https://doi.org/10.1016/j.atherosclerosis.2013.01.044

Chakrabarty, P., Li, A., Ceballos-Diaz, C., Eddy, J. A., et al. (2015). IL-10 alters immunoproteostasis in APP mice, increasing plaque burden and worsening cognitive behavior. *Neuron, 85*(3), 519–533. https://doi.org/10.1016/j.neuron.2014.11.020

Chen, H., Jacobs, E., Schwarzschild, M. A., McCullough, M. L., et al. (2005). Nonsteroidal antiinflammatory drug use and the risk for Parkinson's disease. *Annals of Neurology, 58*(6), 963–967. https://doi.org/10.1002/ana.20682

Chen, J., Liu, X., & Zhong, Y. (2020). Interleukin-17A: The key cytokine in neurodegenerative diseases. *Frontiers in Aging Neuroscience, 12*, 566922. https://doi.org/10.3389/fnagi.2020.566922

Chio, A., Mora, G., La Bella, V., Caponnetto, C., et al. (2011). Repeated courses of granulocyte colony-stimulating factor in amyotrophic lateral sclerosis: Clinical and biological results from a prospective multicenter study. *Muscle & Nerve, 43*(2), 189–195. https://doi.org/10.1002/mus.21851

Cho, H., & Shukla, S. (2020). Role of edaravone as a treatment option for patients with amyotrophic lateral sclerosis. *Pharmaceuticals (Basel), 14*(1), 29. https://doi.org/10.3390/ph14010029

Choi, S. H., Aid, S., Kim, H. W., Jackson, S. H., et al. (2012). Inhibition of NADPH oxidase promotes alternative and anti-inflammatory microglial activation during neuroinflammation. *Journal of Neurochemistry, 120*(2), 292–301. https://doi.org/10.1111/j.1471-4159.2011.07572.x

Ciccocioppo, F., Lanuti, P., Pierdomenico, L., Simeone, P., et al. (2019). The characterization of regulatory T-cell profiles in Alzheimer's disease and multiple sclerosis. *Scientific Reports, 9*(1), 8788. https://doi.org/10.1038/s41598-019-45433-3

Collaborators, G. B. D. N. (2019). Global, regional, and national burden of neurological disorders, 1990-2016: A systematic analysis for the global burden of disease study 2016. *Lancet Neurology, 18*(5), 459–480. https://doi.org/10.1016/S1474-4422(18)30499-X

Colton, C. A. (2013). Immune heterogeneity in neuroinflammation: Dendritic cells in the brain. *Journal of Neuroimmune Pharmacology, 8*(1), 145–162. https://doi.org/10.1007/s11481-012-9414-8

Cruz, M. P. (2018). Edaravone (Radicava): A novel neuroprotective agent for the treatment of amyotrophic lateral sclerosis. *Pharmacy and Therapeutics, 43*(1), 25–28.

Cummings, J. (2023). Anti-amyloid monoclonal antibodies are transformative treatments that redefine Alzheimer's disease therapeutics. *Drugs, 83*(7), 569–576. https://doi.org/10.1007/s40265-023-01858-9

Currais, A. (2015). Ageing and inflammation – A central role for mitochondria in brain health and disease. *Ageing Research Reviews, 21*, 30–42. https://doi.org/10.1016/j.arr.2015.02.001

de Lazaro, I., Yilmazer, A., & Kostarelos, K. (2014). Induced pluripotent stem (iPS) cells: A new source for cell-based therapeutics? *Journal of Controlled Release, 185*, 37–44. https://doi.org/10.1016/j.jconrel.2014.04.011

Decout, A., Katz, J. D., Venkatraman, S., & Ablasser, A. (2021). The cGAS-STING pathway as a therapeutic target in inflammatory diseases. *Nature Reviews. Immunology, 21*(9), 548–569. https://doi.org/10.1038/s41577-021-00524-z

Dhasmana, S., Dhasmana, A., Narula, A. S., Jaggi, M., et al. (2022). The panoramic view of amyotrophic lateral sclerosis: A fatal intricate neurological disorder. *Life Sciences, 288*, 120156. https://doi.org/10.1016/j.lfs.2021.120156

Dheen, S. T., Kaur, C., & Ling, E. A. (2007). Microglial activation and its implications in the brain diseases. *Current Medicinal Chemistry, 14*(11), 1189–1197. https://doi.org/10.2174/092986707780597961

Dobson, L., Trager, U., Farmer, R., Hayardeny, L., et al. (2016). Laquinimod dampens hyperactive cytokine production in Huntington's disease patient myeloid cells. *Journal of Neurochemistry, 137*(5), 782–794. https://doi.org/10.1111/jnc.13553

Doty, K. R., Guillot-Sestier, M. V., & Town, T. (2015). The role of the immune system in neurodegenerative disorders: Adaptive or maladaptive? *Brain Research, 1617*, 155–173. https://doi.org/10.1016/j.brainres.2014.09.008

Duning, T., Schiffbauer, H., Warnecke, T., Mohammadi, S., et al. (2011). G-CSF prevents the progression of structural disintegration of white matter tracts in amyotrophic lateral sclerosis: A pilot trial. *PLoS One, 6*(3), e17770. https://doi.org/10.1371/journal.pone.0017770

Esparza-Salazar, F. J., Lezama-Toledo, A. R., Rivera-Monroy, G., & Borlongan, C. V. (2021). Exendin-4 for Parkinson's disease. *Brain Circulation, 7*(1), 41–43. https://doi.org/10.4103/bc.bc_21_21

Fatoba, O., Ohtake, Y., Itokazu, T., & Yamashita, T. (2020). Immunotherapies in Huntington's disease and alpha-synucleinopathies. *Frontiers in Immunology, 11*, 337. https://doi.org/10.3389/fimmu.2020.00337

Ferguson, M. W., Kennedy, C. J., Palpagama, T. H., Waldvogel, H. J., et al. (2022). Current and possible future therapeutic options for Huntington's disease. *Journal of Central Nervous System Disease, 14*, 11795735221092517. https://doi.org/10.1177/11795735221092517

Fresegna, D., Bullitta, S., Musella, A., Rizzo, F. R., et al. (2020). Re-examining the role of TNF in MS pathogenesis and therapy. *Cells, 9*(10). https://doi.org/10.3390/cells9102290

Galea, I. (2021). The blood-brain barrier in systemic infection and inflammation. *Cellular & Molecular Immunology, 18*(11), 2489–2501. https://doi.org/10.1038/s41423-021-00757-x

Gao, X., Chen, H., Schwarzschild, M. A., & Ascherio, A. (2011). Use of ibuprofen and risk of Parkinson disease. *Neurology, 76*(10), 863–869. https://doi.org/10.1212/WNL.0b013e31820f2d79

Gao, C., Jiang, J., Tan, Y., & Chen, S. (2023). Microglia in neurodegenerative diseases: Mechanism and potential therapeutic targets. *Signal Transduction and Targeted Therapy, 8*(1), 359. https://doi.org/10.1038/s41392-023-01588-0

Garofalo, S., Cocozza, G., Porzia, A., Inghilleri, M., et al. (2020). Natural killer cells modulate motor neuron-immune cell cross talk in models of amyotrophic lateral sclerosis. *Nature Communications, 11*(1), 1773. https://doi.org/10.1038/s41467-020-15644-8

Gendelman, H. E., Zhang, Y., Santamaria, P., Olson, K. E., et al. (2017). Evaluation of the safety and immunomodulatory effects of sargramostim in a randomized, double-blind phase 1 clinical Parkinson's disease trial. *NPJ Parkinsons Disease, 3*, 10. https://doi.org/10.1038/s41531-017-0013-5

George, S., Rey, N. L., Tyson, T., Esquibel, C., et al. (2019). Microglia affect alpha-synuclein cell-to-cell transfer in a mouse model of Parkinson's disease. *Molecular Neurodegeneration, 14*(1), 34. https://doi.org/10.1186/s13024-019-0335-3

Gharagozloo, M., Smith, M. D., Sotirchos, E. S., Jin, J., et al. (2021). Therapeutic potential of a novel glucagon-like peptide-1 receptor Agonist, NLY01, in experimental autoimmune encephalomyelitis. *Neurotherapeutics, 18*(3), 1834–1848. https://doi.org/10.1007/s13311-021-01088-5

Ghosh, M., Garcia-Castillo, D., Aguirre, V., Golshani, R., et al. (2012). Proinflammatory cytokine regulation of cyclic AMP-phosphodiesterase 4 signaling in microglia in vitro and following CNS injury. *Glia, 60*(12), 1839–1859. https://doi.org/10.1002/glia.22401

Giannoni, P., Claeysen, S., Noe, F., & Marchi, N. (2020). Peripheral routes to neurodegeneration: Passing through the blood-brain barrier. *Frontiers in Aging Neuroscience, 12*, 3. https://doi.org/10.3389/fnagi.2020.00003

Graber, J. J., Ford, D., Zhan, M., Francis, G., et al. (2007). Cytokine changes during interferon-beta therapy in multiple sclerosis: Correlations with interferon dose and MRI response. *Journal of Neuroimmunology, 185*(1–2), 168–174. https://doi.org/10.1016/j.jneuroim.2007.01.011

Greenhalgh, A. D., Zarruk, J. G., Healy, L. M., Baskar Jesudasan, S. J., et al. (2018). Peripherally derived macrophages modulate microglial function to reduce inflammation after CNS injury. *PLoS Biology, 16*(10), e2005264. https://doi.org/10.1371/journal.pbio.2005264

Guillot-Sestier, M. V., Doty, K. R., Gate, D., Rodriguez, J., Jr., et al. (2015). Il10 deficiency rebalances innate immunity to mitigate Alzheimer-like pathology. *Neuron, 85*(3), 534–548. https://doi.org/10.1016/j.neuron.2014.12.068

Gurevich, M., Gritzman, T., Orbach, R., Tuller, T., et al. (2010). Laquinimod suppress antigen presentation in relapsing-remitting multiple sclerosis: In-vitro high-throughput gene expression study. *Journal of Neuroimmunology, 221*(1–2), 87–94. https://doi.org/10.1016/j.jneuroim.2010.02.010

Gust, J., Taraseviciute, A., & Turtle, C. J. (2018). Neurotoxicity associated with CD19-targeted CAR-T cell therapies. *CNS Drugs, 32*(12), 1091–1101. https://doi.org/10.1007/s40263-018-0582-9

Hanisch, U. K., & Kettenmann, H. (2007). Microglia: Active sensor and versatile effector cells in the normal and pathologic brain. *Nature Neuroscience, 10*(11), 1387–1394. https://doi.org/10.1038/nn1997

Hassan, A., Sharma Kandel, R., Mishra, R., Gautam, J., et al. (2020). Diabetes mellitus and Parkinson's disease: Shared pathophysiological links and possible therapeutic implications. *Cureus, 12*(8), e9853. https://doi.org/10.7759/cureus.9853

Henkel, J. S., Beers, D. R., Wen, S., Rivera, A. L., et al. (2013). Regulatory T-lymphocytes mediate amyotrophic lateral sclerosis progression and survival. *EMBO Molecular Medicine, 5*(1), 64–79. https://doi.org/10.1002/emmm.201201544

Hippius, H., & Neundorfer, G. (2003). The discovery of Alzheimer's disease. *Dialogues in Clinical Neuroscience, 5*(1), 101–108. https://doi.org/10.31887/DCNS.2003.5.1/hhippius

Ho, A. S., Liu, Y., Khan, T. A., Hsu, D. H., et al. (1993). A receptor for interleukin 10 is related to interferon receptors. *Proceedings of the National Academy of Sciences of the United States of America, 90*(23), 11267–11271. https://doi.org/10.1073/pnas.90.23.11267

Hou, Y., Dan, X., Babbar, M., Wei, Y., et al. (2019). Ageing as a risk factor for neurodegenerative disease. *Nature Reviews Neurology, 15*(10), 565–581. https://doi.org/10.1038/s41582-019-0244-7

Hsiao, H. Y., Chiu, F. L., Chen, C. M., Wu, Y. R., et al. (2014). Inhibition of soluble tumor necrosis factor is therapeutic in Huntington's disease. *Human Molecular Genetics, 23*(16), 4328–4344. https://doi.org/10.1093/hmg/ddu151

Huntington Study Group DI. (2010). A futility study of minocycline in Huntington's disease. *Movement Disorders, 25*(13), 2219–2224. https://doi.org/10.1002/mds.23236

Imbimbo, B. P., Ippati, S., Watling, M., & Balducci, C. (2022). A critical appraisal of tau-targeting therapies for primary and secondary tauopathies. *Alzheimers Dement, 18*(5), 1008–1037. https://doi.org/10.1002/alz.12453

Ishii, M., & Iadecola, C. (2020). Risk factor for Alzheimer's disease breaks the blood-brain barrier. *Nature, 581*(7806), 31–32. https://doi.org/10.1038/d41586-020-01152-8

Itakura, E., Huang, R. R., Wen, D. R., Paul, E., et al. (2011). IL-10 expression by primary tumor cells correlates with melanoma progression from radial to vertical growth phase and development of metastatic competence. *Modern Pathology, 24*(6), 801–809. https://doi.org/10.1038/modpathol.2011.5

Jankovic, J. (2019). Pathogenesis-targeted therapeutic strategies in Parkinson's disease. *Movement Disorders, 34*(1), 41–44. https://doi.org/10.1002/mds.27534

Joly-Amado, A., Davtyan, H., Serraneau, K., Jules, P., et al. (2020). Active immunization with tau epitope in a mouse model of tauopathy induced strong antibody response together with improvement in short memory and pSer396-tau pathology. *Neurobiology of Disease, 134*, 104636. https://doi.org/10.1016/j.nbd.2019.104636

Kappos, L., Cohan, S., Arnold, D. L., Robinson, R. R., et al. (2021). Safety and efficacy of daclizumab beta in patients with relapsing multiple sclerosis in a 5-year open-label study (EXTEND): Final results following early termination. *Therapeutic Advances in Neurological Disorders, 14*, 1756286420987941. https://doi.org/10.1177/1756286420987941

Kar, M., Vernon Shih, Y. R., Velez, D. O., Cabrales, P., et al. (2016). Poly(ethylene glycol) hydrogels with cell cleavable groups for autonomous cell delivery. *Biomaterials, 77*, 186–197. https://doi.org/10.1016/j.biomaterials.2015.11.018

Kasen, A., Houck, C., Burmeister, A. R., Sha, Q., et al. (2022). Upregulation of alpha-synuclein following immune activation: Possible trigger of Parkinson's disease. *Neurobiology of Disease, 166*, 105654. https://doi.org/10.1016/j.nbd.2022.105654

Ketabforoush, A., Chegini, R., Barati, S., Tahmasebi, F., et al. (2023). Masitinib: The promising actor in the next season of the amyotrophic lateral sclerosis treatment series. *Biomedicine & Pharmacotherapy, 160*, 114378. https://doi.org/10.1016/j.biopha.2023.114378

Khalid, S. I., Ampie, L., Kelly, R., Ladha, S. S., et al. (2017). Immune modulation in the treatment of amyotrophic lateral sclerosis: A review of clinical trials. *Frontiers in Neurology, 8*, 486. https://doi.org/10.3389/fneur.2017.00486

Khan, A., Johnson, R., Wittmer, C., Maile, M., et al. (2021). NPT520-34 improves neuropathology and motor deficits in a transgenic mouse model of Parkinson's disease. *Brain, 144*(12), 3692–3709. https://doi.org/10.1093/brain/awab214

Khoshnan, A., Ko, J., & Patterson, P. H. (2002). Effects of intracellular expression of anti-huntingtin antibodies of various specificities on mutant huntingtin aggregation and toxicity. *Proceedings of the National Academy of Sciences of the United States of America, 99*(2), 1002–1007. https://doi.org/10.1073/pnas.022631799

Kim, C., Spencer, B., Rockenstein, E., Yamakado, H., et al. (2018). Immunotherapy targeting toll-like receptor 2 alleviates neurodegeneration in models of synucleinopathy by modulating alpha-synuclein transmission and neuroinflammation. *Molecular Neurodegeneration, 13*(1), 43. https://doi.org/10.1186/s13024-018-0276-2

Kiyota, T., Machhi, J., Lu, Y., Dyavarshetty, B., et al. (2018). Granulocyte-macrophage colony-stimulating factor neuroprotective activities in Alzheimer's disease mice. *Journal of Neuroimmunology, 319*, 80–92. https://doi.org/10.1016/j.jneuroim.2018.03.009

Knox, E. G., Aburto, M. R., Clarke, G., Cryan, J. F., et al. (2022). The blood-brain barrier in aging and neurodegeneration. *Molecular Psychiatry, 27*(6), 2659–2673. https://doi.org/10.1038/s41380-022-01511-z

Kordower, J. H., Chu, Y., Hauser, R. A., Freeman, T. B., et al. (2008). Lewy body-like pathology in long-term embryonic nigral transplants in Parkinson's disease. *Nature Medicine, 14*(5), 504–506. https://doi.org/10.1038/nm1747

Kosloski, L. M., Kosmacek, E. A., Olson, K. E., Mosley, R. L., et al. (2013). GM-CSF induces neuroprotective and anti-inflammatory responses in 1-methyl-4-phenyl-1,2,3,6-tetrahydropyridine intoxicated mice. *Journal of Neuroimmunology, 265*(1–2), 1–10. https://doi.org/10.1016/j.jneuroim.2013.10.009

Kurtuncu, M., Tuzun, E., Turkoglu, R., Petek-Balci, B., et al. (2012). Effect of short-term interferon-beta treatment on cytokines in multiple sclerosis: Significant modulation of IL-17 and IL-23. *Cytokine, 59*(2), 400–402. https://doi.org/10.1016/j.cyto.2012.05.004

Kwilasz, A. J., Grace, P. M., Serbedzija, P., Maier, S. F., et al. (2015). The therapeutic potential of interleukin-10 in neuroimmune diseases. *Neuropharmacology, 96*(Pt A), 55–69. https://doi.org/10.1016/j.neuropharm.2014.10.020

Laffer, B., Bauer, D., Wasmuth, S., Busch, M., et al. (2019). Loss of IL-10 promotes differentiation of microglia to a M1 phenotype. *Frontiers in Cellular Neuroscience, 13*, 430. https://doi.org/10.3389/fncel.2019.00430

Larkin, H. D. (2023). Lecanemab gains FDA approval for early Alzheimer disease. *JAMA, 329*(5), 363. https://doi.org/10.1001/jama.2022.24490

Larochelle, C., Alvarez, J. I., & Prat, A. (2011). How do immune cells overcome the blood-brain barrier in multiple sclerosis? *FEBS Letters, 585*(23), 3770–3780. https://doi.org/10.1016/j.febslet.2011.04.066

Latta-Mahieu, M., Elmer, B., Bretteville, A., Wang, Y., et al. (2018). Systemic immune-checkpoint blockade with anti-PD1 antibodies does not alter cerebral amyloid-beta burden in several amyloid transgenic mouse models. *Glia, 66*(3), 492–504. https://doi.org/10.1002/glia.23260

Lee, H., Fessler, M. B., Qu, P., Heymann, J., et al. (2020). Macrophage polarization in innate immune responses contributing to pathogenesis of chronic kidney disease. *BMC Nephrology, 21*(1), 270. https://doi.org/10.1186/s12882-020-01921-7

Li, J. Y., Englund, E., Holton, J. L., Soulet, D., et al. (2008). Lewy bodies in grafted neurons in subjects with Parkinson's disease suggest host-to-graft disease propagation. *Nature Medicine, 14*(5), 501–503. https://doi.org/10.1038/nm1746

Li, D., Wang, C., Yao, Y., Chen, L., et al. (2016). mTORC1 pathway disruption ameliorates brain inflammation following stroke via a shift in microglia phenotype from M1 type to M2 type. *The FASEB Journal, 30*(10), 3388–3399. https://doi.org/10.1096/fj.201600495R

Liang, X. S., Sun, Z. W., Thomas, A. M., & Li, S. (2023). Mesenchymal stem cell therapy for Huntington disease: A meta-analysis. *Stem Cells International, 2023*, 1109967. https://doi.org/10.1155/2023/1109967

Linnerbauer, M., Wheeler, M. A., & Quintana, F. J. (2020). Astrocyte crosstalk in CNS inflammation. *Neuron, 108*(4), 608–622. https://doi.org/10.1016/j.neuron.2020.08.012

Liu, K. Y., & Howard, R. (2021). Can we learn lessons from the FDA's approval of aducanumab? *Nature Reviews Neurology, 17*(11), 715–722. https://doi.org/10.1038/s41582-021-00557-x

Liu, Z., Qiu, A. W., Huang, Y., Yang, Y., et al. (2019). IL-17A exacerbates neuroinflammation and neurodegeneration by activating microglia in rodent models of Parkinson's disease. *Brain, Behavior, and Immunity, 81*, 630–645. https://doi.org/10.1016/j.bbi.2019.07.026

Liu, Y., Wang, J., Luo, S., Zhan, Y., et al. (2020). The roles of PPARgamma and its agonists in autoimmune diseases: A comprehensive review. *Journal of Autoimmunity, 113*, 102510. https://doi.org/10.1016/j.jaut.2020.102510

Liu, T. W., Chen, C. M., & Chang, K. H. (2022). Biomarker of neuroinflammation in Parkinson's disease. *International Journal of Molecular Sciences, 23*(8). https://doi.org/10.3390/ijms23084148

Lo, D. C., & Hughes, R. E. (2011). Frontiers in neuroscience. In D. C. Lo & R. E. Hughes (Eds.), *Neurobiology of Huntington's disease: Applications to drug discovery*. CRC Press/Taylor & Francis.

Louveau, A., Harris, T. H., & Kipnis, J. (2015). Revisiting the mechanisms of CNS immune privilege. *Trends in Immunology, 36*(10), 569–577. https://doi.org/10.1016/j.it.2015.08.006

Lu, T., Pan, Y., Kao, S. Y., Li, C., et al. (2004). Gene regulation and DNA damage in the ageing human brain. *Nature, 429*(6994), 883–891. https://doi.org/10.1038/nature02661

Luckheeram, R. V., Zhou, R., Verma, A. D., & Xia, B. (2012). CD4(+) T cells: Differentiation and functions. *Clinical & Developmental Immunology, 2012*, 925135. https://doi.org/10.1155/2012/925135

Lutz, W., Sanderson, W., & Scherbov, S. (2008). The coming acceleration of global population ageing. *Nature, 451*(7179), 716–719. https://doi.org/10.1038/nature06516

Machhi, J., Yeapuri, P., Lu, Y., Foster, E., et al. (2021). CD4+ effector T cells accelerate Alzheimer's disease in mice. *Journal of Neuroinflammation, 18*(1), 272. https://doi.org/10.1186/s12974-021-02308-7

Mahapatra, M. K., Karuppasamy, M., & Sahoo, B. M. (2022). Therapeutic potential of semaglutide, a newer GLP-1 receptor agonist, in abating obesity, mon-alcoholic steatohepatitis and neurodegenerative diseases: A narrative review. *Pharmaceutical Research, 39*(6), 1233–1248. https://doi.org/10.1007/s11095-022-03302-1

Manthripragada, A. D., Schernhammer, E. S., Qiu, J., Friis, S., et al. (2011). Non-steroidal anti-inflammatory drug use and the risk of Parkinson's disease. *Neuroepidemiology, 36*(3), 155–161. https://doi.org/10.1159/000325653

Masrori, P., & Van Damme, P. (2020). Amyotrophic lateral sclerosis: A clinical review. *European Journal of Neurology, 27*(10), 1918–1929. https://doi.org/10.1111/ene.14393

Matsumoto, T., Fujimori, K., Andoh-Noda, T., Ando, T., et al. (2016). Functional neurons generated from T cell-derived induced pluripotent stem cells for neurological disease modeling. *Stem Cell Reports, 6*(3), 422–435. https://doi.org/10.1016/j.stemcr.2016.01.010

Matsushita, T., Kibayashi, T., Katayama, T., Yamashita, Y., et al. (2011). Mesenchymal stem cells transmigrate across brain microvascular endothelial cell monolayers through transiently formed interendothelial gaps. *Neuroscience Letters, 502*(1), 41–45. https://doi.org/10.1016/j.neulet.2011.07.021

McGeachy, M. J., Cua, D. J., & Gaffen, S. L. (2019). The IL-17 family of cytokines in health and disease. *Immunity, 50*(4), 892–906. https://doi.org/10.1016/j.immuni.2019.03.021

Melcangi, R. C., Garcia-Segura, L. M., & Mensah-Nyagan, A. G. (2008). Neuroactive steroids: State of the art and new perspectives. *Cellular and Molecular Life Sciences, 65*(5), 777–797. https://doi.org/10.1007/s00018-007-7403-5

Mendez, I., Vinuela, A., Astradsson, A., Mukhida, K., et al. (2008). Dopamine neurons implanted into people with Parkinson's disease survive without pathology for 14 years. *Nature Medicine, 14*(5), 507–509. https://doi.org/10.1038/nm1752

Messer, A., & Butler, D. C. (2020). Optimizing intracellular antibodies (intrabodies/nanobodies) to treat neurodegenerative disorders. *Neurobiology of Disease, 134*, 104619. https://doi.org/10.1016/j.nbd.2019.104619

Muraoka, S., DeLeo, A. M., Sethi, M. K., Yukawa-Takamatsu, K., et al. (2020). Proteomic and biological profiling of extracellular vesicles from Alzheimer's disease human brain tissues. *Alzheimers Dement, 16*(6), 896–907. https://doi.org/10.1002/alz.12089

Nutma, E., Willison, H., Martino, G., & Amor, S. (2019). Neuroimmunology – The past, present and future. *Clinical and Experimental Immunology, 197*(3), 278–293. https://doi.org/10.1111/cei.13279

Olson, K. E., Namminga, K. L., Lu, Y., Schwab, A. D., et al. (2021a). Safety, tolerability, and immune-biomarker profiling for year-long sargramostim treatment of Parkinson's disease. *eBioMedicine, 67*, 103380. https://doi.org/10.1016/j.ebiom.2021.103380

Olson, K. E., Namminga, K. L., Lu, Y., Thurston, M. J., et al. (2021b). Granulocyte-macrophage colony-stimulating factor mRNA and neuroprotective immunity in Parkinson's disease. *Biomaterials, 272*, 120786. https://doi.org/10.1016/j.biomaterials.2021.120786

Olson, K. E., Abdelmoaty, M. M., Namminga, K. L., Lu, Y., et al. (2023a). An open-label multiyear study of sargramostim-treated Parkinson's disease patients examining drug safety, tolerability, and immune biomarkers from limited case numbers. *Translational Neurodegeneration, 12*(1), 26. https://doi.org/10.1186/s40035-023-00361-1

Olson, K. E., Mosley, R. L., & Gendelman, H. E. (2023b). The potential for treg-enhancing therapies in nervous system pathologies. *Clinical and Experimental Immunology, 211*(2), 108–121. https://doi.org/10.1093/cei/uxac084

Orgogozo, J. M., Gilman, S., Dartigues, J. F., Laurent, B., et al. (2003). Subacute meningoencephalitis in a subset of patients with AD after Abeta42 immunization. *Neurology, 61*(1), 46–54. https://doi.org/10.1212/01.wnl.0000073623.84147.a8

Pacheco, R. (2021). T-cell based immunotherapies for Parkinson's disease. *Exploration of Neuroprotective Therapy, 1*(2), 72–85.

Pagano, G., Taylor, K. I., Anzures-Cabrera, J., Marchesi, M., et al. (2022). Trial of prasinezumab in early-stage Parkinson's disease. *The New England Journal of Medicine, 387*(5), 421–432. https://doi.org/10.1056/NEJMoa2202867

Paganoni, S., Macklin, E. A., Hendrix, S., Berry, J. D., et al. (2020). Trial of sodium phenylbutyrate-taurursodiol for amyotrophic lateral sclerosis. *The New England Journal of Medicine, 383*(10), 919–930. https://doi.org/10.1056/NEJMoa1916945

Pajares, M., Rojo, A. I., Manda, G., Bosca, L., et al. (2020). Inflammation in Parkinson's disease: Mechanisms and therapeutic implications. *Cells, 9*(7), 1687. https://doi.org/10.3390/cells9071687

Pan, J., Jin, J. L., Ge, H. M., Yin, K. L., et al. (2015). Malibatol A regulates microglia M1/M2 polarization in experimental stroke in a PPARgamma-dependent manner. *Journal of Neuroinflammation, 12*(s1), 51. https://doi.org/10.1186/s12974-015-0270-3

Pandya, P. H., Murray, M. E., Pollok, K. E., & Renbarger, J. L. (2016). The immune system in cancer pathogenesis: Potential therapeutic approaches. *Journal of Immunology Research, 2016*, 4273943. https://doi.org/10.1155/2016/4273943

Pardoll, D. M. (2012). The blockade of immune checkpoints in cancer immunotherapy. *Nature Reviews. Cancer, 12*(4), 252–264. https://doi.org/10.1038/nrc3239

Pido-Lopez, J., Tanudjojo, B., Farag, S., Bondulich, M. K., et al. (2019). Inhibition of tumour necrosis factor alpha in the R6/2 mouse model of Huntington's disease by etanercept treatment. *Scientific Reports, 9*(1), 7202. https://doi.org/10.1038/s41598-019-43627-3

Poewe, W., Seppi, K., Tanner, C. M., Halliday, G. M., et al. (2017). Parkinson disease. *Nature Reviews Disease Primers, 3*(1), 17013. https://doi.org/10.1038/nrdp.2017.13

Poly, T. N., Islam, M. M., Yang, H.-C., & Li, Y.-C. J. (2019). Non-steroidal anti-inflammatory drugs and risk of Parkinson's disease in the elderly population: A meta-analysis. *European Journal of Clinical Pharmacology, 75*(1), 99–108. https://doi.org/10.1007/s00228-018-2561-y

Porro, C., Cianciulli, A., & Panaro, M. A. (2020). The regulatory role of IL-10 in neurodegenerative diseases. *Biomolecules, 10*(7), 1017. https://doi.org/10.3390/biom10071017

Potter, H., Woodcock, J. H., Boyd, T. D., Coughlan, C. M., et al. (2021). Safety and efficacy of sargramostim (GM-CSF) in the treatment of Alzheimer's disease. *Alzheimers Dement (N Y), 7*(1), e12158. https://doi.org/10.1002/trc2.12158

Prajeeth, C. K., Lohr, K., Floess, S., Zimmermann, J., et al. (2014). Effector molecules released by Th1 but not Th17 cells drive an M1 response in microglia. *Brain, Behavior, and Immunity, 37*, 248–259. https://doi.org/10.1016/j.bbi.2014.01.001

Price, D. L., Koike, M. A., Khan, A., Wrasidlo, W., et al. (2018). The small molecule alpha-synuclein misfolding inhibitor, NPT200-11, produces multiple benefits in an animal model of Parkinson's disease. *Scientific Reports, 8*(1), 16165. https://doi.org/10.1038/s41598-018-34490-9

Ransohoff, R. M., & Perry, V. H. (2009). Microglial physiology: Unique stimuli, specialized responses. *Annual Review of Immunology, 27*, 119–145. https://doi.org/10.1146/annurev.immunol.021908.132528

Rees, K., Stowe, R., Patel, S., Ives, N., et al. (2011). Non-steroidal anti-inflammatory drugs as disease-modifying agents for Parkinson's disease: Evidence from observational studies. *Cochrane Database of Systematic Reviews*, (11), CD008454. https://doi.org/10.1002/14651858.CD008454.pub2

Rentzos, M., Evangelopoulos, E., Sereti, E., Zouvelou, V., et al. (2012). Alterations of T cell subsets in ALS: A systemic immune activation? *Acta Neurologica Scandinavica, 125*(4), 260–264. https://doi.org/10.1111/j.1600-0404.2011.01528.x

Reynolds, A. D., Stone, D. K., Mosley, R. L., & Gendelman, H. E. (2009). Nitrated alpha-synuclein-induced alterations in microglial immunity are regulated by CD4+ T cell subsets. *Journal of Immunology, 182*(7), 4137–4149. https://doi.org/10.4049/jimmunol.0803982

Reynolds, A. D., Stone, D. K., Hutter, J. A., Benner, E. J., et al. (2010). Regulatory T cells attenuate Th17 cell-mediated nigrostriatal dopaminergic neurodegeneration in a model of Parkinson's disease. *Journal of Immunology, 184*(5), 2261–2271. https://doi.org/10.4049/jimmunol.0901852

Roos, R. A. (2010). Huntington's disease: A clinical review. *Orphanet Journal of Rare Diseases, 5*(1), 40. https://doi.org/10.1186/1750-1172-5-40

Rossi, C., Cusimano, M., Zambito, M., Finardi, A., et al. (2018). Interleukin 4 modulates microglia homeostasis and attenuates the early slowly progressive phase of amyotrophic lateral sclerosis. *Cell Death & Disease, 9*(2), 250. https://doi.org/10.1038/s41419-018-0288-4

Roussakis, A. A., Gennaro, M., Gordon, M. F., Reilmann, R., et al. (2023). A PET-CT study on neuroinflammation in Huntington's disease patients participating in a randomized trial with laquinimod. *Brain Communications, 5*(2), fcad084. https://doi.org/10.1093/braincomms/fcad084

Sabahi, M., Joshaghanian, A., Dolatshahi, M., Jabbari, P., et al. (2021). Modification of glial cell activation through dendritic cell vaccination: Promises for treatment of neurodegenerative diseases. *Journal of Molecular Neuroscience, 71*(7), 1410–1424. https://doi.org/10.1007/s12031-021-01818-6

Sabatino, J. J., Jr., Probstel, A. K., & Zamvil, S. S. (2019). B cells in autoimmune and neurodegenerative central nervous system diseases. *Nature Reviews Neuroscience, 20*(12), 728–745. https://doi.org/10.1038/s41583-019-0233-2

Saraiva, M., Christensen, J. R., Veldhoen, M., Murphy, T. L., et al. (2009). Interleukin-10 production by Th1 cells requires interleukin-12-induced STAT4 transcription factor and ERK MAP kinase activation by high antigen dose. *Immunity, 31*(2), 209–219. https://doi.org/10.1016/j.immuni.2009.05.012

Saunders, J. A., Estes, K. A., Kosloski, L. M., Allen, H. E., et al. (2012). CD4+ regulatory and effector/memory T cell subsets profile motor dysfunction in Parkinson's disease. *Journal of Neuroimmune Pharmacology, 7*(4), 927–938. https://doi.org/10.1007/s11481-012-9402-z

Schäfer, A., Chaggar, P., Thompson, T. B., Goriely, A., et al. (2021). Predicting brain atrophy from tau pathology: A summary of clinical findings and their translation into personalized models. *Brain Multiphysics, 2*, 100039. https://doi.org/10.1016/j.brain.2021.100039

Schenk, D., Barbour, R., Dunn, W., Gordon, G., et al. (1999). Immunization with amyloid-beta attenuates Alzheimer-disease-like pathology in the PDAPP mouse. *Nature, 400*(6740), 173–177. https://doi.org/10.1038/22124

Schutt, C. R., Gendelman, H. E., & Mosley, R. L. (2018). Tolerogenic bone marrow-derived dendritic cells induce neuroprotective regulatory T cells in a model of Parkinson's disease. *Molecular Neurodegeneration, 13*(1), 26. https://doi.org/10.1186/s13024-018-0255-7

Schwab, A. D., Thurston, M. J., Machhi, J., Olson, K. E., et al. (2020). Immunotherapy for Parkinson's disease. *Neurobiology of Disease, 137*, 104760. https://doi.org/10.1016/j.nbd.2020.104760

Segal, B. M., Constantinescu, C. S., Raychaudhuri, A., Kim, L., et al. (2008). Repeated subcutaneous injections of IL12/23 p40 neutralising antibody, ustekinumab, in patients with relapsing-remitting multiple sclerosis: A phase II, double-blind, placebo-controlled, randomised, dose-ranging study. *Lancet Neurology, 7*(9), 796–804. https://doi.org/10.1016/S1474-4422(08)70173-X

Sivandzade, F., & Cucullo, L. (2021). Regenerative stem cell therapy for neurodegenerative diseases: An overview. *International Journal of Molecular Sciences, 22*(4). https://doi.org/10.3390/ijms22042153

Smit, J. W., Basile, P., Prato, M. K., Detalle, L., et al. (2022). Phase 1/1b studies of UCB0599, an oral inhibitor of alpha-synuclein misfolding, including a randomized study in Parkinson's disease. *Movement Disorders, 37*(10), 2045–2056. https://doi.org/10.1002/mds.29170

Smith, D. A., & Germolec, D. R. (1999). Introduction to immunology and autoimmunity. *Environmental Health Perspectives, 107*(Suppl 5)), 661–665. https://doi.org/10.1289/ehp.99107s5661

Song, G. J., & Suk, K. (2017). Pharmacological modulation of functional phenotypes of microglia in neurodegenerative diseases. *Frontiers in Aging Neuroscience, 9*, 139. https://doi.org/10.3389/fnagi.2017.00139

Song, G. J., Nam, Y., Jo, M., Jung, M., et al. (2016). A novel small-molecule agonist of PPAR-gamma potentiates an anti-inflammatory M2 glial phenotype. *Neuropharmacology, 109*, 159–169. https://doi.org/10.1016/j.neuropharm.2016.06.009

Song, C., Shi, J., Zhang, P., Zhang, Y., et al. (2022). Immunotherapy for Alzheimer's disease: Targeting beta-amyloid and beyond. *Translational Neurodegeneration, 11*(1), 18. https://doi.org/10.1186/s40035-022-00292-3

Southwell, A. L., Khoshnan, A., Dunn, D. E., Bugg, C. W., et al. (2008). Intrabodies binding the proline-rich domains of mutant huntingtin increase its turnover and reduce neurotoxicity. *The Journal of Neuroscience, 28*(36), 9013–9020. https://doi.org/10.1523/JNEUROSCI.2747-08.2008

Sproul, A. A. (2015). Being human: The role of pluripotent stem cells in regenerative medicine and humanizing Alzheimer's disease models. *Molecular Aspects of Medicine, 43-44*, 54–65. https://doi.org/10.1016/j.mam.2015.06.007

Steed, P. M., Tansey, M. G., Zalevsky, J., Zhukovsky, E. A., et al. (2003). Inactivation of TNF signaling by rationally designed dominant-negative TNF variants. *Science, 301*(5641), 1895–1898. https://doi.org/10.1126/science.1081297

Subhramanyam, C. S., Wang, C., Hu, Q., & Dheen, S. T. (2019). Microglia-mediated neuroinflammation in neurodegenerative diseases. *Seminars in Cell & Developmental Biology, 94*, 112–120. https://doi.org/10.1016/j.semcdb.2019.05.004

Sulzer, D., Alcalay, R. N., Garretti, F., Cote, L., et al. (2017). T cells from patients with Parkinson's disease recognize alpha-synuclein peptides. *Nature, 546*(7660), 656–661. https://doi.org/10.1038/nature22815

Sveinbjornsdottir, S., Magnusson, H., & Benedikz, J. E. (2014). Multiple sclerosis in Iceland from 1900 to 2000: A total population study. *Multiple Sclerosis and Related Disorders, 3*(3), 375–383. https://doi.org/10.1016/j.msard.2013.12.002

Sweeney, M. D., Zhao, Z., Montagne, A., Nelson, A. R., et al. (2019). Blood-brain barrier: From physiology to disease and back. *Physiological Reviews, 99*(1), 21–78. https://doi.org/10.1152/physrev.00050.2017

Tabira, T. (2010). Immunization therapy for Alzheimer disease: A comprehensive review of active immunization strategies. *The Tohoku Journal of Experimental Medicine, 220*(2), 95–106. https://doi.org/10.1620/tjem.220.95

Tang, Y., & Le, W. (2016). Differential roles of M1 and M2 microglia in neurodegenerative diseases. *Molecular Neurobiology, 53*(2), 1181–1194. https://doi.org/10.1007/s12035-014-9070-5

Teismann, P., & Ferger, B. (2001). Inhibition of the cyclooxygenase isoenzymes COX-1 and COX-2 provide neuroprotection in the MPTP-mouse model of Parkinson's disease. *Synapse, 39*(2), 167–174. https://doi.org/10.1002/1098-2396(200102)39:2<167::AID-SYN8>3.0.CO;2-U

Thonhoff, J. R., Beers, D. R., Zhao, W., Pleitez, M., et al. (2018). Expanded autologous regulatory T-lymphocyte infusions in ALS: A phase I, first-in-human study. *Neurology: Neuroimmunology & Neuroinflammation, 5*(4), e465. https://doi.org/10.1212/NXI.0000000000000465

Thonhoff, J. R., Berry, J. D., Macklin, E. A., Beers, D. R., et al. (2022). Combined regulatory T-lymphocyte and IL-2 treatment is safe, tolerable, and biologically active for 1 year in persons with amyotrophic lateral sclerosis. *Neurology: Neuroimmunology & Neuroinflammation, 9*(6), e200019. https://doi.org/10.1212/NXI.0000000000200019

Trias, E., Ibarburu, S., Barreto-Nunez, R., Babdor, J., et al. (2016). Post-paralysis tyrosine kinase inhibition with masitinib abrogates neuroinflammation and slows disease progression in inherited amyotrophic lateral sclerosis. *Journal of Neuroinflammation, 13*(1), 177. https://doi.org/10.1186/s12974-016-0620-9

Ulrich, J. D., Ulland, T. K., Colonna, M., & Holtzman, D. M. (2017). Elucidating the role of TREM2 in Alzheimer's disease. *Neuron, 94*(2), 237–248. https://doi.org/10.1016/j.neuron.2017.02.042

Uyttenhove, C., Sommereyns, C., Theate, I., Michiels, T., et al. (2007). Anti-IL-17A autovaccination prevents clinical and histological manifestations of experimental autoimmune encephalomyelitis. *Annals of the New York Academy of Sciences, 1110*, 330–336. https://doi.org/10.1196/annals.1423.035

Valori, C. F., Possenti, A., Brambilla, L., & Rossi, D. (2021). Challenges and opportunities of targeting astrocytes to halt neurodegenerative disorders. *Cells, 10*(8). https://doi.org/10.3390/cells10082019

van Dyck, C. H. (2018). Anti-amyloid-beta monoclonal antibodies for Alzheimer's disease: Pitfalls and promise. *Biological Psychiatry, 83*(4), 311–319. https://doi.org/10.1016/j.biopsych.2017.08.010

Vollmer, T. L., Wynn, D. R., Alam, M. S., & Valdes, J. (2011). A phase 2, 24-week, randomized, placebo-controlled, double-blind study examining the efficacy and safety of an anti-interleukin-12 and -23 monoclonal antibody in patients with relapsing-remitting or secondary progressive multiple sclerosis. *Multiple Sclerosis, 17*(2), 181–191. https://doi.org/10.1177/1352458510384496

Wagner, J., Ryazanov, S., Leonov, A., Levin, J., et al. (2013). Anle138b: A novel oligomer modulator for disease-modifying therapy of neurodegenerative diseases such as prion and Parkinson's disease. *Acta Neuropathologica, 125*(6), 795–813. https://doi.org/10.1007/s00401-013-1114-9

Wang, Y., Ji, X., Leak, R. K., Chen, F., et al. (2017). Stem cell therapies in age-related neurodegenerative diseases and stroke. *Ageing Research Reviews, 34*, 39–50. https://doi.org/10.1016/j.arr.2016.11.002

Wang, S., Mustafa, M., Yuede, C. M., Salazar, S. V., et al. (2020). Anti-human TREM2 induces microglia proliferation and reduces pathology in an Alzheimer's disease model. *The Journal of Experimental Medicine, 217*(9). https://doi.org/10.1084/jem.20200785

Weaver, D. F. (2022). Alzheimer's disease as an innate autoimmune disease (AD(2)): A new molecular paradigm. *Alzheimers Dement, 19*, 1086–1098. https://doi.org/10.1002/alz.12789

Whiteside, T. L., Demaria, S., Rodriguez-Ruiz, M. E., Zarour, H. M., et al. (2016). Emerging opportunities and challenges in cancer

immunotherapy. *Clinical Cancer Research, 22*(8), 1845–1855. https://doi.org/10.1158/1078-0432.CCR-16-0049
Yadav, M., Stephan, S., & Bluestone, J. A. (2013). Peripherally induced tregs – role in immune homeostasis and autoimmunity. *Frontiers in Immunology, 4*, 232. https://doi.org/10.3389/fimmu.2013.00232
Yanamandra, K., Gruden, M. A., Casaite, V., Meskys, R., et al. (2011). alpha-synuclein reactive antibodies as diagnostic biomarkers in blood sera of Parkinson's disease patients. *PLoS One, 6*(4), e18513. https://doi.org/10.1371/journal.pone.0018513
Yang, J. S., Xu, L. Y., Xiao, B. G., Hedlund, G., et al. (2004). Laquinimod (ABR-215062) suppresses the development of experimental autoimmune encephalomyelitis, modulates the Th1/Th2 balance and induces the Th3 cytokine TGF-beta in Lewis rats. *Journal of Neuroimmunology, 156*(1–2), 3–9. https://doi.org/10.1016/j.jneuroim.2004.02.016
Yang, W., Chen, X., Pan, J., Ge, H., et al. (2015). Malibatol A protects against brain injury through reversing mitochondrial dysfunction in experimental stroke. *Neurochemistry International, 80*, 33–40. https://doi.org/10.1016/j.neuint.2014.11.003
Zalc, B. (2018). One hundred and fifty years ago Charcot reported multiple sclerosis as a new neurological disease. *Brain, 141*(12), 3482–3488. https://doi.org/10.1093/brain/awy287
Zang, X., Chen, S., Zhu, J., Ma, J., et al. (2022). The emerging role of central and peripheral immune systems in neurodegenerative diseases. *Frontiers in Aging Neuroscience, 14*, 872134. https://doi.org/10.3389/fnagi.2022.872134
Zenaro, E., Pietronigro, E., Della Bianca, V., Piacentino, G., et al. (2015). Neutrophils promote Alzheimer's disease-like pathology and cognitive decline via LFA-1 integrin. *Nature Medicine, 21*(8), 880–886. https://doi.org/10.1038/nm.3913
Zhang, C., Griciuc, A., Hudry, E., Wan, Y., et al. (2018). Cromolyn reduces levels of the Alzheimer's disease-associated amyloid beta-protein by promoting microglial phagocytosis. *Scientific Reports, 8*(1), 1144. https://doi.org/10.1038/s41598-018-19641-2
Zhang, W., Xiao, D., Mao, Q., & Xia, H. (2023). Role of neuroinflammation in neurodegeneration development. *Signal Transduction and Targeted Therapy, 8*(1), 267. https://doi.org/10.1038/s41392-023-01486-5
Zhao, W., Beers, D. R., Liao, B., Henkel, J. S., et al. (2012). Regulatory T lymphocytes from ALS mice suppress microglia and effector T lymphocytes through different cytokine-mediated mechanisms. *Neurobiology of Disease, 48*(3), 418–428. https://doi.org/10.1016/j.nbd.2012.07.008

Immunotherapy of Alzheimer's Disease

51

Giacomo Tondo, Fabiola De Marchi, Matteo Anselmi, and Cristoforo Comi

Abstract

Alzheimer's disease (AD) is the most common neurodegenerative dementia. Despite continuous research using biomarkers to identify pathogenic mechanisms, a complete understanding of the disease-related processes is far from being achieved. Consequently, effective disease-modifying therapies are still lacking. Several therapeutic strategies have been tested in clinical trials for AD. Misfolded protein aggregates formed by amyloid-β and tau proteins represent the pathological hallmarks of AD and are associated with synapsis alteration and neuronal loss. This evidence led to the amyloid cascade hypothesis, in which amyloid- β and tau accumulation are strictly related and associated with progressive neurodegeneration. Immunotherapeutic approaches targeting amyloid-β and tau proteins showed promising effectiveness in preclinical studies. However, clinical trial results are disappointing. The failure of clinical trials on vaccines and humanized anti-amyloid-β and anti-tau monoclonal antibodies has questioned the strength of the amyloid hypothesis. The Aducanumab, recently approved by the United States Food and Drug Administration, is an anti-amyloid-β monoclonal antibody showing a beneficial effect in AD patients. However, its use has been debated due to the negligible clinical improvement compared with the pathology resolution. Immunotherapies targeting tau, microglia, and the gut-brain axis are under development, but further research is still required to clarify the exact benefit of these approaches. The current chapter will provide an overview of the immunotherapeutic approaches studied or in use. We will focus on amyloid-β, tau, and microglia in AD, discussing future perspectives on immunotherapies in AD.

G. Tondo · F. De Marchi · M. Anselmi · C. Comi (✉)
Neurology Unit, Maggiore della Carità University Hospital, Department of Translational Medicine, University of Piemonte Orientale, Novara, Italy
e-mail: cristoforo.comi@med.uniupo.it

Keywords

Therapy · Cognition · Dementia · Immunization · Amyloid · Tau · Microglia · Clinical trials

51.1 Introduction

Alzheimer's disease (AD) is a neurological disorder that progressively leads to cognitive decline, behavioural disturbances, and functional dependency. AD is the most common neurodegenerative disease, the leading cause of cognitive impairment and dementia, and among the ten most common causes of death globally (Abubakar et al., 2022). AD prevalence is continuously increasing due to the progressive ageing of the world's population since advanced age represents the most significant risk factor. The National Institute on Ageing and Alzheimer's Association (NIA-AA) guidelines defined three AD phases: preclinical AD, characterized by pathologic brain changes without cognitive impact; mild cognitive impairment (MCI), showing mild neuropsychological impairment; dementia phase (Albert et al., 2011; McKhann et al., 2011). Considering the preclinical, prodromal (MCI), and overt dementia phases, it has been estimated that about one-fifth of people aged over 50 years lie on the AD continuum (Gustavsson et al., 2023).

Progressively worsening amnestic syndrome represents the most frequent presentation of AD, associated with neurodegeneration that usually starts in the hippocampus and spreads to cortical regions. Memory impairment is generally followed by progressive dysfunction involving multiple cognitive domains, including language, visuospatial abilities, and executive functions (Ferrari & Sorbi, 2021). A loss of functional autonomy and neuropsychiatric disturbances characterize the advanced disease stages (Scheltens et al., 2016). AD natural history can be quite variable, with a median survival after diagnosis of about 6 years, and death usually occurring due to complications like malnutrition,

H. E. Gendelman, T. Ikezu (eds.), *Neuroimmune Pharmacology and Therapeutics*,
https://doi.org/10.1007/978-3-031-68237-7_51

dehydration, infectious disease, and bedsores (Liang et al., 2021). Most AD cases involve patients aged more than 65 years, and AD prevalence increases with age, being age and genetics the main non-modifiable risk factors (Takizawa et al., 2015). Other potentially modifiable risks and protective factors have captured the researcher's attention, such as education, leisure activities, healthy lifestyle, and comorbidities, including diabetes, hypertension, and cardiovascular diseases (Frisoni et al., 2023).

The pathological hallmark of AD is the deposition of misfolded proteins, namely amyloid-β (Aβ) and hyperphosphorylated tau (p-tau). Aβ aggregates, called Aβ plaques, are found in the extracellular space, while tau deposits in intracellular neurofibrillary tangles. Studies in autosomal dominant AD forms showed that Aβ accumulates decades before symptom's onset (Frisoni et al., 2022). Tau deposition in neurofibrillary tangles, confined to the medial temporal lobe, is typical in individuals over 60 years. However, tau spreads to the cortical regions in AD, possibly driven by Aβ pathology (Peng et al., 2020). Aβ deposition has been related to the proteolytic cleavages of the amyloid precursor protein (APP), a membrane-bound protein that can be processed by two mechanisms, the non-amyloidogenic pathway and the amyloidogenic pathway. In the non-amyloidogenic pathway, the APP is cleaved by the α-secretase into a soluble N-terminal fragment and a membrane-bound C-terminal fragment, cleaved by γ-secretase in the p3 soluble peptide. In the amyloidogenic pathway, the cleavage is mediated by the β-secretase, with subsequent participation of the γ-secretase, which acts on the APP C-terminal fragment, generating a cytoplasmic polypeptide and the Aβ peptides (Wilkins & Swerdlow, 2017). The Aβ peptides are secreted into the extracellular space; among them, the Aβ42 fragments are the most self-aggregating Aβ species, generating neurotoxic oligomers, which in turn generate Aβ fibrils and Aβ plaques (Zhao et al., 2020). Aβ aggregates are toxic both in vitro and in vivo, and Aβ oligomers seem more toxic than fibrils; in addition, Aβ promotes tau hyperphosphorylation and aggregation, and microglia activation (Wilkins & Swerdlow, 2017). According to the amyloid cascade hypothesis, Aβ deposition precedes tau tangle formation and subsequent neurodegenerative changes. Further hypothesis development posed tau pathology as a primary event in the cascade, more strictly associated with neurodegenerative changes and cognitive impairment than Aβ pathology (Jack Jr et al., 2013). Tau protein is physiologically involved in microtubule stabilization and axonal trafficking. In several neurodegenerative disorders, including AD, tau protein hyperphosphorylation and deposition alter axonal transport and neuronal stability, facilitating progressive neuronal dysfunction and loss (Medeiros et al., 2011). Compared with Aβ plaque deposition, cortical tau deposition is highly correlated with synaptic dysfunction, neurodegenerative changes, and grey matter atrophy (Das et al., 2018; Ossenkoppele et al., 2016; Xia et al., 2017). These findings align with *post-mortem* results showing that neurofibrillary tangles distribution, not amyloid status, correlates with antemortem cognitive status (Nelson et al., 2012). In addition, the progressive tau deposition correlates with the progression of neurodegenerative changes and cognitive decline (Schöll et al., 2016).

The development of AD therapies has been pushed up by the identification of toxic Aβ peptides and p-tau in cerebrospinal fluid (CSF) and blood, while the brain amyloid and tau load can be in-vivo identified by positron emission tomography (PET), and thus can serve as diagnostic biomarkers. The evidence that AD-related pathological changes, including Aβ and p-tau deposition and neurodegeneration, which can occur in the brain of individuals in the prodromal and preclinical phase of dementia years before any clinical manifestations (Jack Jr et al., 2013; Janssen et al., 2022; Tondo et al., 2022), has allowed the biomarker-based stratification of subjects. The precise diagnostic and prognostic stratification based on biomarker positivity along the AD spectrum is critical to providing optimal patient management and avoiding misdiagnosis (Tondo & De Marchi, 2022). In addition, an early concern related to the development of effective therapeutic strategies relied on identifying of the right target at the optimal disease stage, meaning to treat subjects prior to irreversible neurodegenerative changes (Sperling et al., 2011).

Besides the deposition of misfolded proteins, a joint pathological event in neurodegenerative disease, especially AD, is the presence of microglia activation (Li et al., 2022). *Post-mortem*, in vitro, and animal model studies consistently reported the activation of neuroimmune cells strictly linked to the deposition of Aβ and tau (Brelstaff et al., 2018; Lee et al., 2013). Microglia and astrocytes represent the principal neuroinflammatory central nervous system (CNS) cells, capable of multiple protective or deleterious roles depending on environmental stimuli (Comi & Tondo, 2017). In vivo clinical studies suggested that activation of microglia and astrocytes represents an early phenomenon in the neurodegenerative cascade in AD (Malpetti et al., 2020; Tondo et al., 2021). In the last two decades, a large part of research in neurodegenerative diseases has focused on the mechanisms linking neuroinflammation and neurodegeneration to open new frontiers for developing successful treatments (Du et al., 2017).

51.2 Current and Future Potential Therapeutic Strategies

To date, three acetylcholinesterase enzyme (AChE) inhibitors and one N-methyl-D-aspartate (NMDA) receptor antagonist are approved for AD treatment. The cholinergic hypothesis states that the cognitive decline in AD is related to the pauperization of cholinergic neurones, which are correlated with memory function. Consequently, AChE inhibitors, including donepezil, rivastigmine, and galantamine, reducing the activity of AChE, may produce some clinical improvement in cognition (Raina et al., 2008). Memantine is an NMDA receptor antagonist blocking NMDA-mediated ion flux and inhibiting the elevated release of glutamate, which may lead to neuronal damage by excitotoxicity mechanisms (Danysz & Parsons, 2003). In several studies, memantine improved cognition, behaviour, and global functioning in AD patients (Matsunaga et al., 2015). However, the benefits of AchE inhibitors and memantine are limited to a specific period. In addition, they are ineffective in modifying the disease course or in preventing neuronal loss and cognitive decline (Vaz & Silvestre, 2020). Since the approval of memantine by the FDA dates back to 2003, it is clear that the development of disease-modifying drugs has become an unrelenting need.

The amyloid cascade hypothesis suggests that Aβ deposition represents the primum movens in the neurodegenerative cascade, leading to synaptic degeneration and neuronal loss. Thus, researchers have tried to develop strategies against Aβ pathology. The aim to promote Aβ clearance has been pursued using experimental vaccines or specific anti-Aβ antibodies. The research on disease-modifying therapies for AD has focused primarily on lowering brain Aβ levels but without reaching significant clinical benefit (Vogt et al., 2023). In addition, due to the closer relationship between neurofibrillary tangles and cognitive impairment, targeting tau pathology might with active or passive immunization be clinically more relevant (Giacobini & Gold, 2013). Similarly to anti-Aβ strategies, these treatments aim to prevent tau phosphorylation, obstacle tau aggregation, and favour tau clearance (Zheng et al., 2022).

Lastly, immunomodulation therapies targeting microglia and neuroinflammatory pathways have shown promising results. Microglia cells bind to soluble Aβ oligomers via cell surface receptors, such as Toll-like receptors (TLRs), and secrete pro-inflammatory molecules, including cytokines, chemokines, growth factors, prostaglandins, complement factors, reactive oxygen species, and proteases. All these mediators increase the permeability of the blood-brain barrier, favouring leukocyte infiltration of the brain. Local neuroinflammation, in turn, activates the expression of pro-inflammatory genes and stimulates the amyloidogenic pathway, increasing Aβ and self-stimulating the amyloid cascade (Guillot-Sestier & Town, 2018). The triggering receptor expressed on myeloid cell-2 (TREM2) is one of the microglia receptors involved in this vicious circle. TREM2 can stimulate phagocytosis of Aβ plaques and microglia activation. However, this function can be compromised in pathophysiologic conditions associated with neurodegenerative changes, such as ageing, resulting in increased Aβ production and deposition (Streit et al., 2004). Dysfunctional microglia have also been associated with tau deposition. Microglia activation can be determined by tau, aiming for tau deposit clearance. Dysregulated responses result in neuronal damage and tau spread, thus contributing to disease progression (Brelstaff et al., 2018). All these receptors can be promising targets for antibodies to modulate microglia activity (Lewcock et al., 2020).

51.3 Therapies Targeting the Amyloid Pathway

Studies on animal models showed that active and passive immunization might effectively favour amyloid clearance from the brain and improve cognitive performance. Several mechanisms have been proposed, including reducing the free concentration of Aβ by the link antibodies-Aβ, inhibiting Aβ fibril generation by blocking the Aβ conversion due to the antibody binding, activating microglia-mediated phagocytosis (Ikezu, 2017). In active immunization, the immune system produces antibodies against Aβ in response to the administration of Aβ peptides or fragments, whereas passive immunotherapy employs humanized monoclonal antibodies or polyclonal immunoglobulins that act on Aβ clearance (Song et al., 2022). In an early study in an AD transgenic mouse model, Shenk and colleagues demonstrated that active immunization against Aβ, triggered by the administration of the full-length Aβ, significantly reduced amyloid deposition (Schenk et al., 1999). Along with the evidence of Aβ clearance, a clinical improvement in cognitive performances was reported in mice (Akiyama et al., 2000; Janus et al., 2000; Morgan et al., 2000). Similarly, passive immunization effectively improved memory deficits in mice (Dodart et al., 2002; Kotilinek et al., 2002). Subsequent evidence suggested a pathogenic role for the Aβ oligomers in disrupting synaptic activity, memory, and learning (Walsh et al., 2002) and that antibodies may act on the oligomers protecting synapses (Klyubin et al., 2005). Considering passive immunization, monoclonal antibodies against Aβ increased the circulating level of Aβ. It has been suggested that antibodies can sequestrate Aβ in the periphery, favouring the brain-to-blood concentration gradient and leading to a higher passage from the brain to the periphery (DeMattos et al., 2001). Lastly, a crucial role of microglia was already evident in the early study of Shenk and colleagues, which demonstrated activated microglia surrounding the remaining Aβ plaques after active immunization in mice (Schenk et al., 1999).

51.3.1 Active Immunization

Starting from the intriguing results in animal models, active vaccination against Aβ was proposed in humans. The AN1792 was the first vaccine entering a clinical trial in humans, administering the full-length Aβ42 peptide using the saponin QS-21 adjuvant (Morgan, 2011). After a phase I trial without overt adverse effects in human volunteers, a phase II trial involved 300 patients in the vaccine arm and 60 patients in the placebo arm. A high antibody response against Aβ was reached in about 20% of the patient group and after repeated inoculations (Dickey et al., 2001), as it was demonstrated in mice (Schenk et al., 1999). However, the trial was discontinued due to the occurrence of adverse reactions in response to the vaccine, mainly T-cell-mediated aseptic meningoencephalitis (Orgogozo et al., 2003). Overall, 6% of the patients in the trial developed meningoencephalitis; most of the patients recovered after treatment with steroidal treatment (Ferrer et al., 2004; Gilman et al., 2005; Morgan, 2011). The adverse effect was likely linked to T-lymphocytes mediated responses against Aβ; thus, subsequent vaccines were developed without T-lymphocyte epitopes. In fact, the adjuvant QS-21 stimulates the Th1 reaction, which is associated with autoimmune responses, while the Th2 response suppresses autoimmunity (Cribbs et al., 2003).

Even if the vaccine administration was stopped, the patients involved in the trial continued to be followed. A subsequent study in a cohort of patients showed that the patients with the highest antibody response remained stable over 2 years after the last immunization (Hock et al., 2003). However, when the entire cohort was analysed, the results were less clear-cut (Gilman et al., 2005). Even if post-mortem data are available for only three patients, two with meningoencephalitis and one lacking adverse events (Ferrer et al., 2004; Masliah et al., 2005; Nicoll et al., 2003), all these cases found a lower amyloid burden than expected for an AD patient at each patient's stage of the disease (Ikezu, 2017).

An interesting finding in the cohort with the highest antibody level was the MRI evidence of an increased hippocampal shrinkage and associated ventricular enlargement (Fox et al., 2005). Researchers hypothesized that the loss of brain volume in the vaccine responders could be related to the clearance of Aβ, reducing the consequent inflammation and oedema associated with the glial reaction to the Aβ deposits. However, a more extended follow-up study in a small sample of patients did not find any difference in hippocampal volume between vaccine-treated patients and placebo (Vellas et al., 2009). In addition, the study showed a long-term effect on functional decline, since the antibody responders continue to show a lesser decline as determined by the disability.

51.3.1.1 Assessment for Dementia Scale

Beyond the AN17992 trial, several second-generation vaccines have been developed. Short peptide fragments have been employed to avoid non-specific immune responses observed with the full-length Aβ vaccine. On the other hand, shorter Aβ peptides are not very potent immunogens (Ikezu, 2017).

The vaccine CAD106 (Amilomotide), including the N-terminal Aβ1–6, was the first second-generation vaccine to reach the clinical phases (Wiessner et al., 2011), showing a safety profile and good antibody response (Winblad et al., 2012). The Aβ-specific antibody response, B-cell mediated and avoiding T-cell response, has been reported in 75% of treated patients, without provoking adverse inflammatory reactions (Winblad et al., 2012). However, the clinical trial was stopped due to unpredictable changes in cognitive function, brain volume, and body weight.

Vanutide cridificar (ACC-001) is a vaccine consisting of multiple Aβ1–7 fragments linked to inactivated diphtheria toxin and the QS-21 adjuvant (Arai et al., 2015). The vaccine showed an acceptable safety profile and elicited a high antibody response in patients with mild-to-moderate AD, but it was not further developed (Hull et al., 2017; Pasquier et al., 2016).

Affitope AD02 is an Aβ vaccine evaluated in early AD. It was well tolerated but failed to show significant effects on cognition (Schneeberger et al., 2015).

The UB-311 vaccine is based on two synthetic Aβ1-14 targeting peptides, and the vaccine delivery system stimulates the T-helper type 2 regulatory response rather than the type 1 response. The type 2 response stimulates immunogenicity while minimizing T-cell inflammatory reactivity. The UB-311 vaccine elicited strong immune response in early-to-mild AD patients in a phase II trial, not including a placebo arm (Wang et al., 2017a). A subsequent phase II trial including a placebo group confirmed safety, tolerability, and immunogenicity of UB-311, with a 97% antibody response rate in the treated arm (Yu et al., 2023).

The third-generation vaccines include the DNA-based vaccines. In DNA-based vaccines, the target gene is inserted in a vector, and expressing the protein of interest, Aß can trigger a humoral and cellular immune response. The DNA vaccine AV-1959D targets the N-terminal epitope of the Aβ peptide and has been shown to be immunogenic and efficacious in AD mouse models of AD (Valiukas et al., 2022).

51.3.2 Passive Immunization

Monoclonal antibodies against Aβ showed to be effective in favouring Aβ plaque clearance. The early study of Bard and colleagues reported the efficacy of passive immunization with anti-Aβ antibodies in lowering cerebral amyloid burden in the AD mouse model (Bard et al., 2000). The process of Aβ-clearance has been related to microglia activity: within 1 week, the injection of anti-Aβ antibodies into the mouse brain was associated with microglia-mediated phagocytosis of amyloid deposits (Wilcock et al., 2003). Furthermore, the hyperphosphorylation of tau can also be reduced by the injection of anti-Aβ antibodies (Oddo et al., 2004).

Several characteristics should be considered in the development of the antibody. Each antibody may have a specificity for a particular epitope domain, including the N terminal, the mild domain, or the C terminal. Several antibody subtypes exist with different possible specificities. Each antibody may have a different affinity for Aβ in general and for different Aβ conformations, including oligomers, fibrils, and plaques. Thus, each antibody treatment may result in a different ability to clear the cerebral amyloid burden.

Aducanumab is a human monoclonal antibody binding the Aβ N-terminus and targeting Aβ deposits, oligomers, and fibrils. In individuals with prodromal or mild AD, aducanumab showed to reduce brain amyloid load as assessed by amyloid-PET. At the same time, the Clinical Dementia Rating Scale Sum of Boxes (CDR SOB) and the Mini-Mental State Examination (MMSE) decline decreased with treatment (Sevigny et al., 2016). In the PRIME study, 165 patients with prodromal or mild AD and an amyloid-PET showing cerebral amyloid accumulation were treated. Treatment with aducanumab reduced brain Aβ plaques as showed by amyloid-PET in a dose- and time-dependent manner. The most common adverse effect was amyloid-related imaging abnormalities (ARIA), occurring in a variable percentage between 3 and 41% of treated patients, increasing with the aducanumab dose. The term ARIA includes alterations caused by the increased vascular permeability associated with inflammation, which can be associated with oedema (ARIA-E) or haemorrhage (ARIA-H) (Salloway et al., 2022). In the PRIME study, no patients required hospitalization for ARIA (Sevigny et al., 2016). Subsequently, two identically designed trials, EMERGE and ENGAGE, recruited a total of over 3000 subjects with MCI or early AD, with amyloid-PET or CSF evidence of amyloid pathology. Both trials were discontinued in 2019, when a preliminary analysis had deemed a significant positive effect of Aducanumab compared to placebo unlikely (Selkoe, 2019). Subsequent analyses conducted equally blinded, led to reassessment of the results. The re-analysis showed in the EMERGE trial a 22% reduction in the observed decline, measured by CDR, in patients treated with Aducanumab compared to patients in the placebo arm. The ENGAGE trial showed similar results when considering patients treated with a high dose of aducanumab, but failing to meet the primary endpoint (Haeberlein et al., 2020). Both trials demonstrated, in a subgroup of patients, the significant efficacy of Aducanumab in reducing cerebral amyloid accumulation and protecting against tau protein accumulation. Based on its biological efficacy, in 2021 the FDA approved Aducanumab for treating AD. The approval came via the 'Accelerated Approval Pathway', a faster pathway that can be used when a study obtains evidence, based on a biomarker, that reasonably predicts clinical improvement. The approval of Aducanumab created debate in the scientific community since the results of the EMERGE study were not replicated in the counter-analysis of the ENGAGE study, suggesting only questionable evidence of minimal clinical benefit from a single study (Alexander et al., 2021). Regardless of the controversies, the aducanumab approval has brought new hope in the treatment of AD, which has firmed long since the memantine approval in 2003.

Lecanemab is a monoclonal antibody binding aggregated Aβ and Aβ oligomers and protofibrils. It was well tolerated in the first phase II trial, being associated with 9.9% of patients experimenting with ARIA-E. At 12 months, follow-up showed a 64% probability of more than 25% clinical reduction in cognitive decline vs. placebo (evaluated by the clinical change on the Alzheimer's Disease Composite Score, ADCOMS), but was considered not enough to reach the primary endpoint. In addition, at 18 months, the same 10-mg/kg biweekly dose showed a lower effect on reducing cognitive decline than what was observed at 12 months. However, both amyloid-PET and CSF biomarkers were supportive of a treatment benefit (Swanson et al., 2021). By considering as the primary endpoint the change from baseline at 18 months in the CDR-SB score in patients with MCI or mild AD, a phase III trial demonstrated that lecanemab reduced markers of amyloid and was associated with moderately less cognitive decline than placebo at follow-up, but resulted in infusion-related reactions in 26.4% of cases and ARIA-E in 12.6% (Van Dyck et al., 2023).

Donanemab is a humanized monoclonal antibody that confirmed cerebral amyloid plaque reduction and was associated with less cognitive decline and decreased tau progression at 24 weeks. ARIA-E alterations were documented in 26.7% of patients in the treated group vs. 0.8% in the placebo group (Mintun et al., 2021).

Bapineuzumab, a monoclonal antibody non-selectively targeting all forms of Aβ, was tested in phase 3 trials involving patients with mild-to-moderate AD (Salloway et al., 2014). The primary outcome was the effect on the 11-item cognitive subscale of the Alzheimer's Disease Assessment Scale (ADAS-cog11) and the Disability Assessment for Dementia (DAD). The treatment was shown to be ineffective in cognitive and functional scores; no significant differences in the ADAS-cog11 and DAD scores emerged between the treated and placebo groups. In addition, despite reducing tau levels in CSF, the administration of bapineuzumab was dose-dependent to the development of amyloid-related imaging abnormalities with oedema.

Solanezumab is a humanized antibody targeting the monomeric Aβ peptides, that failed to improve cognition or functional abilities in AD patients (Honig et al., 2018). Two phase III trials (EXPEDITION 1 and EXPEDITION 2) enrolled patients with mild-to-moderate AD, considering as primary endpoints the changes from baseline to week 80 in scores on the ADAS-cog11 and the Alzheimer's Disease Cooperative Study-Activities of Daily Living scale (ADCS-ADL). Both studies showed no significant improvement in the primary outcomes (Doody et al., 2014). As a confirmation, in patients with dominantly inherited AD across asymptomatic and symptomatic disease stages, both solanezumab and gantenerumab (another monoclonal antibody tested in this study) engaged their Aβ targets but neither demonstrated a beneficial effect on cognitive measures compared to controls, and the solanezumab-treated group showed a slight aggravation of cognitive decline vs. the control group (Salloway et al., 2021). Similar results, with Solanezumab ineffective to slow cognitive decline as compared with placebo, were observed in a subgroup of subjects with preclinical AD (Sperling et al., 2023).

Crenezumab is a monoclonal antibody primarily targeting Aβ oligomers (Adolfsson et al., 2012). In two phase II trials (ABBY and BLAZE), crenezumab failed to meet the primary endpoints, based on score changes on the ADAS-cog12 and Clinical Dementia Rating-Sum of Boxes (CDR-SOB) scores (Cummings et al., 2018; Salloway et al., 2018). Nevertheless, in a post hoc analysis, a decline in the ADAS-Cog12 was observed in the group treated with the highest dosage (Cummings et al., 2018). The safety and efficacy of crenezumab were also tested in two phase III trials (CREAD and CREAD2) enrolling participants with prodromal to mild AD. The final analysis included about 800 participants in each trial. Neither study showed a significant group (treated vs. placebo) difference in mean change from baseline in CDR-SOB score (Ostrowitzki et al., 2022). Evaluating safety, amyloid-related imaging abnormalities with oedema were rare and transient.

Gantenerumab is a subcutaneously administered monoclonal antibody anti-Aβ with a high affinity for aggregated Aβ. It was tested in two phase III trials (GRADUATE I and GRADUATE II) including patients with MCI or mild AD with amyloid-PET evidence of cerebral amyloid load. The primary outcome was the change from baseline in the CDR-SOB score. The treatment led to a significant reduction of amyloid plaque burden compared with placebo, and with lower CSF levels of p-tau, but was not associated with slower clinical decline than placebo. ARIA-E occurred in about 24.9% of the participants treated with gantenerumab, with symptomatic ARIA-E in 5% of cases (Bateman et al., 2023). Ponezumab is a humanized antibody targeting the Aβ1-40 peptide, tested in patients with mild-to-moderate AD. In a phase II trial, the treatment did not affect CSF biomarkers, brain amyloid burden, or cognitive and functional abilities (Landen et al., 2017). The principal ongoing clinical trials with passive immunization for AD are summarized in Fig. 51.1.

51.4 Therapies Targeting Tau

The role of tau in the AD pathological process has been supported by consistent evidence showing the relationship between tau deposition and the topographical distribution of neurodegenerative changes, which is evident in neurodegenerative dementia (Perani et al., 2021). Consequently, the reduction of brain tau load has been associated with the attenuation of disease impact in animal models (Chai et al., 2011). For the direct toxic effect of misfolded tau, tau immunotherapy is considered more promising than amyloid immunotherapy (Bittar et al., 2020). Tau hyperphosphorylation alters microtubule stabilization and axonal trafficking and facilitates the formation of aggregates (Ait-Bouziad et al., 2020). Tau deposition is associated with neurodegeneration and correlates with cognitive decline better than Aβ, suggesting tau as a promising target for curbing disease progression (Hawkes et al., 2014). Therapies targeting tau aim to prevent tau phosphorylation, obstacle tau aggregation and deposition, and favour neurofibrillary tangle clearance (Song et al., 2022). The first effective anti-tau immunotherapy emerged in mice model in 2007. The active immunization with a p-tau epitope in a tangle mouse model, reduced tau aggregation in the brain and attenuated behavioural abnormalities (Asuni et al., 2007). After that first evidence, several active and passive immunotherapeutic strategies entered clinical trials (Vaz & Silvestre, 2020).

51.4.1 Active Immunization

Active immunization is based on administering tau protein to stimulate a patient's immune response. The use of adjuvants in tau vaccines has reduced the occurrence of severe immune reactions, compared with Aβ vaccines (Vaz & Silvestre, 2020). Two tau vaccines are currently under evaluation: the AADVac1, which is the first vaccine entering phase II trial (NCT02579252) (Kontsekova et al., 2014), and the ACI-35, targeting the pathological conformers of hyperphosphorylated tau.

AADvac1 has been the first immunotherapeutic agent targeting tau tested in humans. The vaccine is based on an immunogen peptide triggering the production of endogenous antibodies against a 12-amino-acid sequence in the microtubule-binding region of the tau protein. The peptide is coupled with a carrier providing T-cell epitopes without stimulating a T-cell response, thus reducing the develop-

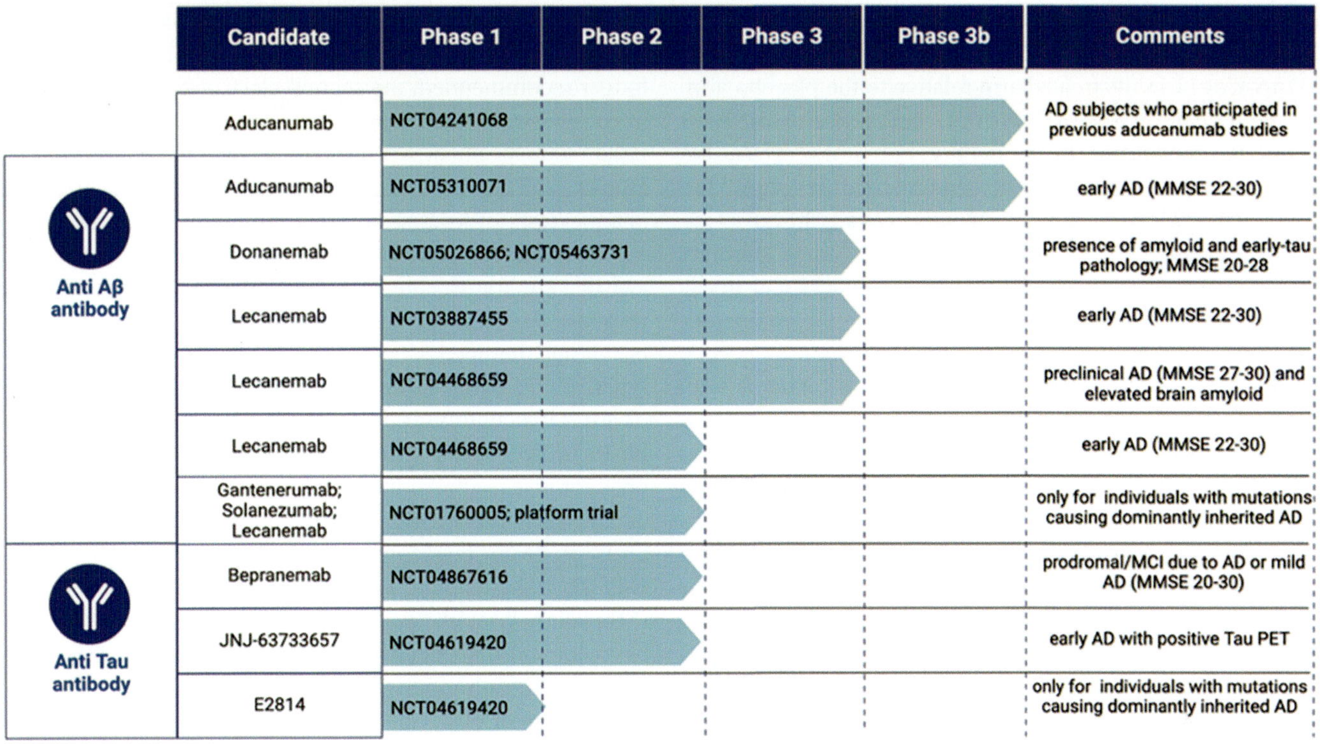

Fig. 51.1 Ongoing clinical trials using passive immunization (amyloid and tau) for Alzheimer's disease

ment of severe adverse immune reactions (Novak et al., 2018b). The safety and immunogenicity of the AADvac1 were tested by the first human study using the tau vaccine, in which 29 out of the 30 enrolled patients developed an IgG immune response and no case of meningoencephalitis was reported (Novak et al., 2018a). In the subsequent 72-week study, the same patients were followed up. The study confirmed the good safety and immunogenicity of the vaccine and revealed that the hippocampal atrophy rate was lower in patients with a high antibody level (Novak et al., 2018b). In addition, AADvac1 was shown to significantly reduce the CSF levels of two AD biomarkers, p-tau181 and p-tau217, suggesting that the vaccine affects tau pathology (Panza et al., 2016). A 24-month double-blinded placebo-controlled enrolled 196 mild AD dementia patients. The vaccine was safe and well tolerated, it induced high levels of IgG antibodies, but no significant effect was found in cognitive evaluation (Novak et al., 2021). A recent post-hoc subgroup analysis including 119 patients reported that a high antibody titer response was associated with a slowing decline in CDR-SOB and with slower development of brain atrophy (Cullen et al., 2024).

The vaccine ACI-35 targets the pathological p-tau, and its safety and efficacy are still under evaluation. In a mouse model, the vaccine demonstrated a robust immunogenic response, reducing indices of tauopathy in the mouse brain, and showing a good safety profile (Theunis et al., 2013).

51.4.2 Passive Immunotherapy

Anti-tau antibodies recognize different sites of tau protein, usually the microtubule-binding domain, and can enter neurones targeting intraneuronal tau providing higher specificity and lower adverse effects than tau vaccines (Congdon & Sigurdsson, 2018; Song et al., 2022). However, one of the major issues in anti-tau antibody development is that therapy should target both intracellular and extracellular tau, which is challenging to achieve without an adverse inflammatory reaction (Congdon & Sigurdsson, 2018; Wu et al., 2021).

Semorinemab is a humanized anti-tau monoclonal antibody binding all six human tau isoforms. In vitro and animal model tests showed that semorinemab reduced tau-mediated neuronal toxicity and tau deposition and showed a favourable safety profile in humans (Ayalon et al., 2021). However, this first anti-tau antibody early failed to improve clinical symptoms in AD (Mullard, 2021). Semorinemab was tested in a phase II trial in prodromal and mild AD, with confirmed Aβ pathology by PET or CSF, verifying as the main outcome change from baseline to week 73 on the CDR-SOB. Similar increases in the CDR-SOB were observed both in the placebo and semorinemab arms, suggesting that semorinemab could not slow clinical AD progression compared with placebo (Teng et al., 2022). A second phase II trial explored the clinical efficacy of semorinemab in patients with mild-to-moderate AD. This study tested change in the ADAS-Cog11 and in the ADCS-ADL scale as primary endpoints in 238 patients, demonstrating a 42.2% reduction in decline on the ADAS-Cog11 in the treated arm relative to the placebo arm, but no effect on the ADCS-ADL or on the MMSE or CDR-SB, considered as secondary outcomes. Thus, the study confirmed the failure of semorinemab in improving functional or global outcomes (Monteiro et al., 2023).

Gosuranemab is a humanized monoclonal antibody against N-terminal of tau, binding to monomers and fibrils with high affinity and reducing N-terminal tau from interstitial fluid and CSF and lowering tau aggregation in cells (Sopko et al., 2020). It was initially tested in patients with progressive supranuclear palsy, being well-tolerated and reducing the levels of unbound N-terminal tau but without improving clinical manifestations (Dam et al., 2021). A subsequent phase II trial in early AD patients assessed the efficacy of gosuranemab in slowing cognitive and functional impairment measuring change in the CDR-SOB after 78 weeks. The study confirmed the tolerability, showing an incidence of serious adverse events compared to the placebo arm, but failed to identify significant effects in cognitive and functional tests (Shulman et al., 2023).

Tilavonemab is a monoclonal antibody binding the tau N-terminus and targeting the extracellular spread of tau pathology (Yanamandra et al., 2013). A 96-week phase II study involving patients with early AD tested the efficacy of tilavonemab in reducing change in CDR-SOB. Again, the incidence of adverse events was comparable between the treated and the placebo arm, but tilavonemab did not demonstrate efficacy in slowing cognitive decline (Florian et al., 2023). Lastly, zagotenemab is a humanized antibody targeting extracellular aggregated tau. The phase I trials showed a good safety profile (Willis et al., 2023), but further studies are warranted.

As for drugs targeting amyloid, the principal ongoing clinical trials with passive immunization targeting tau for AD are summarized in Fig. 51.1.

51.5 Microglia

Although the exact mechanism that correlates neuroinflammation and neurodegeneration should be better investigated, the observation that inflammation may promote neuronal loss and cognitive decline supports the potential role of microglia in AD pathogenesis (Hickman et al., 2018).

Overall, medications designed to target microglia in treating neurodegenerative diseases can be categorized into five primary groups. The initial category involves the inhibition of activated microglia using immunosuppressive or anti-inflammatory drugs, thereby diminishing the generation of pro-inflammatory factors, chemokines, and cytotoxic substances and consequently alleviating neurodegeneration. Although the conventional classification of microglia into pro-inflammatory and anti-inflammatory phenotypes is no longer recommended, the second class of drugs is grounded in the property of microglia to display both pro- and anti-inflammatory phenotypes upon activation (Wendimu & Hooks, 2022). The third group of drugs emphasizes the phagocytosis of microglia rather than their phenotype, aiming to eliminate detrimental tissue or debris linked to AD swiftly. The fourth category of drugs was conceptualized in response to rapid advancements in single-cell sequencing, providing therapeutic strategies by targeting specific subtypes of microglia associated with the disease. The last class of drugs concentrates on depleting and regenerating microglia to utilize newly formed microglia instead of activated ones for disease prevention and treatment.

Drawing on various mechanisms of action, the first class of drugs tested in the neuroinflammatory pathway was the non-steroidal anti-inflammatory drugs (NSAIDs) (Stopschinski et al., 2023). NSAIDs were tested in several neurological disorders but without clear efficacy. Starting from epidemiological studies and meta-analyses suggesting a positive effect of NSAIDs on AD development and progression, randomized clinical trials in the early 2000s investigated NSAIDS such as aspirin, indomethacin, celecoxib, ibuprofen, and naproxen on cognitive status in AD (Aisen et al., 2002; De Jong et al., 2008; Scharf et al., 1999). Regrettably, these trials did not yield favourable efficacy outcomes. Nevertheless, a significant development emerged in 2020 through a comprehensive retrospective cohort study (Rivers-Auty et al., 2020) that reported that diclofenac significantly decreased the frequency of AD compared to other NSAIDs. In this study, which included more than 1500 individuals (including AD, MCI, and controls) from the ADNI database, the proportions of cognitive impairment did significantly differ based on pain-reliever use. In fact, the use of NSAIDs at baseline was associated with a reduced AD prevalence compared to the no-pain-reliever group. Also, the use of diclofenac was found to be the only NSAID associated with reduced cognitive decline as measured by the MMSE score in all groups over time.

An alternative and more sophisticated therapeutic strategy revolves around the modulation of microglial function (Kwon et al., 2022). It is widely recognized that microglia play a crucial role as immune cells within the CNS, with one of their key functions being tissue phagocytosis. A newfound optimism is linked to drugs able of regulating microglial phagocytosis. With age, there is a decline in microglial phagocytosis, accompanied by a propensity for synaptic pruning (Deczkowska et al., 2018; Rawji et al., 2020). Also, the TREM2 gene, predominantly expressed in microglia of the brain and associated with microglial phagocytosis, poses a risk factor for AD (Guerreiro et al., 2013).

Gantenerumab, a previously mentioned human monoclonal antibody with high affinity for Aβ, exemplifies a promising candidate that stimulates the phagocytosis of aggregated Aβ by microglia, yielding favourable results in extensive clinical trials. Another emerging molecule, AL002, a humanized monoclonal antibody, binds to TREM2, activating signalling pathways. This activation increases the phosphorylation of TREM2's downstream effector Syk, thereby inducing microglial proliferation. Following positive outcomes in animal models (Wang et al., 2020), the drug successfully underwent a phase I clinical trial (INVOKE), demonstrating safety and tolerability. The Phase II clinical trial (INVOKE-2) is ongoing; however, in *ad interim* analysis, notable adverse neurological events, such as AIRA, were reported in approximately 30% of patients, predominantly in E4 homozygotes.

DNL919 (ATV: TREM2), functions as a TREM2 agonist antibody, activating microglia to phagocytose amyloid by binding to the abundant transferrin receptors on endothelial cells of the BBB. A recently concluded phase I trial involving healthy participants revealed that DNL919 altered multiple microglial biomarkers, including interleukins, cytokines, and growth factors, indicating successful target engagement. As of now, however, no phase II trial has been scheduled because "safety signals of moderate, reversible hematologic effects" that were "observed at the highest dose tested, suggesting a narrow therapeutic window" for AD patients (https://www.biospace.com/article/takeda-denali-scrap-phase-i-alzheimer-s-candidate-data-shows-narrow-therapeutic-window-/, accessed in 29-Jan-2024).

Microglia also express the granulocyte-macrophage colony-stimulating factor receptor (GM-CSFR) and undergo non-inflammatory proliferation upon binding with GM-CSF (Suzumura et al., 1990). Sargramostim is a recombinant human GM-CSF, already in use for haematological use, promising for AD treatment for its ability to increase microglial proliferation and activity, and related cytokine production, with positive biological and cognitive effects (Potter et al., 2021). A phase II is currently ongoing, and in a recruiting phase, with targets on clinical scores, haematopoiesis, and immune system biomarkers.

Another intriguing target, particularly implicated in neuroinflammation, is the modulation of the regulatory T cells (Tregs). Recent findings indicate that Tregs exhibit a compromised phenotype and diminished suppressive function in AD patients, skewing the immune system towards a proinflammatory status and potentially contributing to disease progression (Faridar et al., 2020, 2022). Interleukin-2 (IL-2), a cytokine with the capacity to selectively expand and restore functional Tregs, emerges as a promising immunoregulatory drug. To date, only one phase I clinical trial published in 2023 evaluated the ability of IL-2-induced expansion of Tregs in an AD clinical setting, with positive biological results (Faridar et al., 2023). Numerous trials are currently underway and in the recruiting phase on clinicaltrials.gov, anticipating favourable outcomes.

Lastly, a promising target is the TNF-α; the chronic production of TNF-α by microglia generates a pathological neuro-inflammatory response, potentially treatable. In AD, this cytokine is elevated in the CSF and correlates directly with disease progression (Tarkowski et al., 2003). However, TNF-α is not specific for AD, and results elevated in several systemic conditions, including rheumatoid arthritis, inflammatory bowel disease, and psoriasis. In this group of patients, who were treated with TNF blocking agents resulted associated with lower risk of AD (Zhou et al., 2020).

In recent years, several clinical trials have attempted to follow the path of TNF-α, through different strategies, mainly using the TNF blocking agents (i.e. etanercept) with some interesting trends in cognition (Butchart et al., 2015). However, there are currently no active/recruiting trials with these agents. Also, relatively small human clinical trial using theracurmin, an easily absorbable form of curcumin able to low TNF-α, showed a stabilization of cognitive performance during treatment, with a minor percentage of conversion from MCI to AD (Dost et al., 2021).

The latest innovation in this context, is the development of a platform trial (Exploratory Platform Trial on Anti-Inflammatory Agents in Alzheimer's Disease, EXPLAIN-AD), a new trial design aiming to evaluate at the same time the effect of anti-inflammatory agents on cognition in early AD, paying attention both to the safety and tolerability of these drugs and the effects on central and peripheral inflammation. The first molecule in study is the Canakinumab, a human anti-IL-1beta monoclonal antibody developed by Novartis to treat MCI and AD by modulating neuroinflammation and inhibiting the inflammatory response (Dhimolea, 2010).

51.6 Discussion

AD is a severe neurodegenerative disease lacking effective treatment. The pathogenesis of AD is still far from being completely elucidated. The Aβ cascade hypothesis sustains that Aβ misfolded protein deposition in the extracellular senile plaques is the first actor in the neurodegenerative process, while tau phosphorylation and aggregation in the intracellular neurofibrillary tangles have been proposed as the main players in favouring neuronal dysfunction and synaptic loss, resulting in cognitive decline (Jack Jr et al., 2013). Microglial activation and the shift of neuroinflammatory response from a protective function to a deleterious activity have been demonstrated to contribute actively to neurodegenerative changes (Comi & Tondo, 2017).

Based on this view, the inhibition of Aβ and p-tau formation and aggregation, and the modulation of microglia activity, have been a critical strategy in the development of potential disease-modifying therapies in AD (summarized in Fig. 51.2) (Panza et al., 2019a; Wang et al., 2017b). Overall, immunotherapy relies on the development of active or passive immunization. Both active and passive immunotherapies have advantages and disadvantages. For example, active immunotherapy provokes the long-term production of endogenous antibodies by stimulating cellular and humoral responses, producing a high concentration of antibodies despite few injections and low medical cost, representing an indubitable strength point. Specific advantages of passive immunotherapies regard their effectiveness in the elderly with lower vaccine responses. In addition, antibodies can be easily stopped in case of adverse effects. Unfortunately, passive immunotherapy requires repeated administration in the hospital and higher costs than vaccines. Furthermore, both strategies may result in the over-activation of the innate and adaptive immune systems generating side effects such as encephalitis and vasculitis (Song et al., 2022).

Targeting the Aβ pathway has been promising, but developing drugs in this field has been particularly challenging (Lu et al., 2023). The Aβ vaccine therapy for AD is overloaded by the possible development of severe autoimmune reactions, including T-cell-mediated meningoencephalitis (Song et al., 2022). Passive immunization with anti-Aβ antibodies aims to bind and remove Aβ deposits. However, the activity (and efficacy) of the antibody may vary depending on the antigen epitope targeted and the form of Aβ (monomers, oligomers, protofibrils, and amyloid fibrils). Different antibodies targeting several antigenic epitopes have been tested for their ability to favour Aβ clearance and potentially slow the progression of cognitive decline in AD.

Aducanumab, lecanemab, bapineuzumab, and donanemab target the N-terminal sequence of Aβ. They act on all forms of Aβ, including monomers, oligomers, fibrils, and plaques (Van Dyck, 2018). All showed to reduce AD biomarkers, but without positively acting on cognitive decline (Lu et al., 2023). Ponezumab, solanezumab, and crenezumab target the C-terminal or central region of Aβ, binding to Aβ monomers and reducing the cytotoxic forms of Aβ (La Porte et al., 2012). Results of clinical trials are promising, suggesting that these antibodies can increase the concentration of Aβ peptides in plasma and CSF in AD patients, but fall in slowing disease progression (Doody et al., 2014; Ostrowitzki et al., 2022). These findings suggest that targeting Aβ by immunotherapy did not reduce tau pathology or influence the progression of cognitive decline (Panza et al., 2019b). The failure of Aβ immunotherapy suggests that Aβ deposition starts years before the AD symptoms onset, thus selecting patients in the symptomatic phase may miss the optimal window for anti-amyloid approaches, since effective disease-modifying therapies should start at the earliest disease stages,

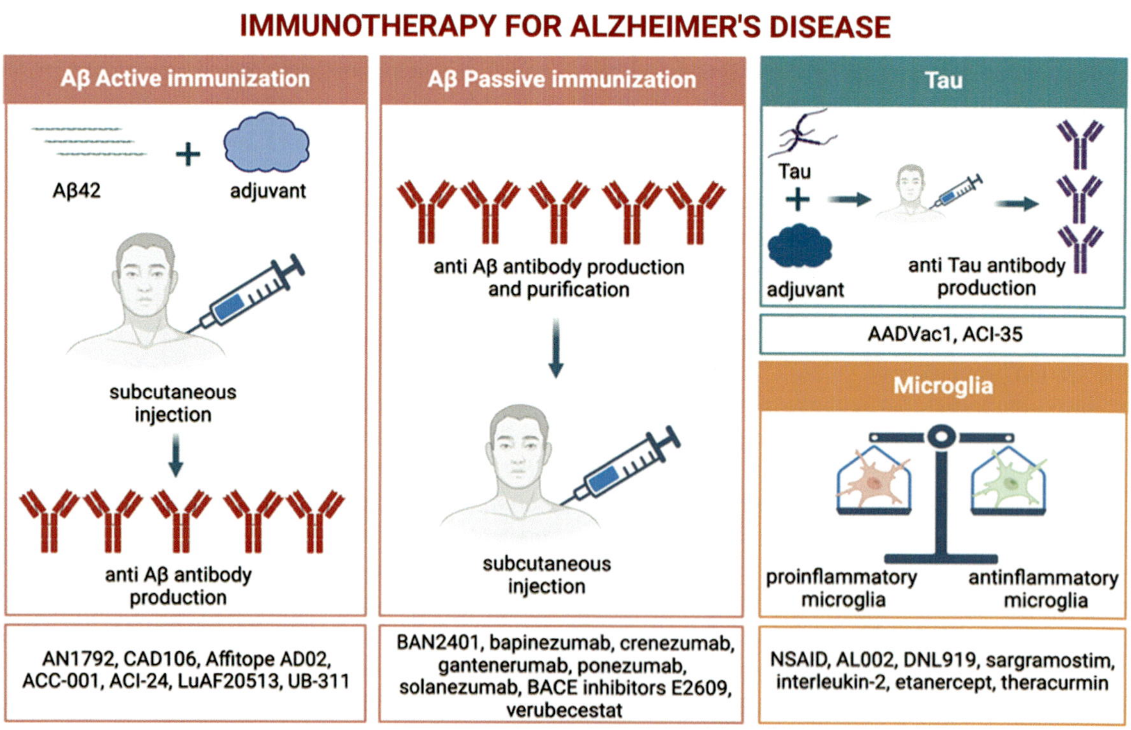

Fig. 51.2 Schematic summary of the main available immunotherapic treatment strategies for Alzheimer's disease

before the onset of neurodegenerative changes. In addition, the failure of Aβ immunotherapies may be related to the evidence that Aß deposition is irrelevant to neurodegeneration and cognitive decline.

Considering that tau pathology well correlates with neurodegeneration and dementia progression, tau immunotherapy appears hopeful (Bittar et al., 2020). However, targeting tau pathology may be extremely complex. Tau protein has multiple roles in neurones, including microtubule stabilization and axonal transport (Kadavath et al., 2015). Tau exists in 6 different isoforms in humans, depending on the number of N-terminus repeats. Tauopathies, including AD, can be classified by the tau isoform that generates pathological aggregates (Lane-Donovan & Boxer, 2024). Thus, different tauopathies are linked to different tau isoforms, and in the same tauopathy, ta can exist in different structures, implying that targeting one species of tau can be effective in one neurodegenerative disease, but not another (He et al., 2020). This aspect explains the difficulties in developing effective tau-targeted therapies in AD and why the Food and Drug Administration has approved no tau therapies for the treatment of any neurodegenerative disease. To date, two anti-tau vaccines are under evaluation in AD patients: the AADvac1 and the AVI-35, although evidence excludes an effect on cognitive function (Novak et al., 2021). In addition, several humanized tau antibodies have been developed and are currently being examined in clinical trials. The favourable safety and good tolerability of tau antibodies have been demonstrated, but still, there is poor evidence of a positive effect on cognitive decline in AD (Bittar et al., 2020).

Lastly, there are no available drugs specifically targeting microglia for the treatment of CNS disorders. However, several are undergoing clinical trials. Unfortunately, a few of these drugs have been halted in their development either due to unsuccessful outcomes or the identification of more effective alternatives. The primary challenge is that microglia-related hypotheses still need more evidence or raise lingering doubts in drug development studies. Given the pivotal role of microglia in neurodegenerative diseases, there is significant potential for developing drugs and therapies targeting these cells. Subsequent research could focus on identifying specific areas where microglia play a role in AD, leading to targeted drug design. Researchers might also investigate new classifications or functions of microglia, allowing for the targeted exploitation or blockade of specific functions.

Further experimental evidence is still needed to confirm that molecules targeting different pathological pathways may act on disease progression. Future advances in research and in diagnostic techniques may allow the adoption of multiple strategies, simultaneously impacting the Aβ pathway, the tau neurotoxic effects, and the microglia modulation. The multiple-target strategy might represent a successful intervention, along with identifying candidates at the earliest possible disease stage, offering the potential to delay the onset of AD and to modify the progression of neurodegenerative changes.

51.7 Review Questions

1. What is the role of tau proteins in Alzheimer's disease?
   (a) They help maintain synaptic plasticity by acting on synaptic pruning
   (b) They promote the formation of amyloid plaques
   (c) They act as neurotransmitters in the brain
   (d) *They stabilize microtubules within neurons*
   (e) They regulate the production of acetylcholine
2. What is the main goal of current Alzheimer's disease treatments?
   (a) Completely curving the progression of the disease
   (b) Reversing cognitive decline
   (c) *Managing symptoms and improving quality of life*
   (d) Restoring the lost memories
   (e) Preventing the formation of tau proteins
3. Which of the following mechanisms is believed to precede the formation of soluble amyloid beta (Aβ) oligomers in the pathogenesis cascade of Alzheimer's disease?
   (a) *Overexpression of amyloid precursor protein (APP)*
   (b) Impaired clearance of tau protein deposits by microglia
   (c) Mutation in the presenilin 1 or 2 genes
   (d) Disruption of synaptic transmission in the entorhinal cortex
   (e) Dysfunction of insulin-related pathways in the brain
4. Which of the following statements about active immunization against Aβ is true?
   (a) Active immunization is based on the activation of CD8 lymphocytes.
   (b) *In active immunization, administering Aβ peptides or fragments triggers the immune system to produce antibodies against Aβ.*
   (c) In active immunization, the administration of humanized monoclonal antibodies stimulates Aβ clearance.
   (d) Active immunotherapy does not induce the long-term production of endogenous antibodies.
   (e) Active immunization requires repeated administrations to obtain a satisfactory Aβ clearance.
5. What is the primary mechanism of action of Aducanumab in the treatment of Alzheimer's disease?
   (a) Inhibition of acetylcholinesterase activity
   (b) Enhancement of dopamine receptor sensitivity
   (c) Stimulation of neurogenesis in the hippocampus
   (d) Inhibition of tau protein aggregation
   (e) *Reduction of amyloid beta (Aβ) plaque formation*

References

Abubakar, M. B., Sanusi, K. O., Ugusman, A., Mohamed, W., Kamal, H., Ibrahim, N. H., Khoo, C. S., & Kumar, J. (2022). Alzheimer's disease: An update and insights into pathophysiology. *Frontiers in Aging Neuroscience, 14*, 742408.

Adolfsson, O., Pihlgren, M., Toni, N., Varisco, Y., Buccarello, A. L., Antoniello, K., Lohmann, S., Piorkowska, K., Gafner, V., & Atwal, J. K. (2012). An effector-reduced anti-β-amyloid (Aβ) antibody with unique aβ binding properties promotes neuroprotection and glial engulfment of Aβ. *Journal of Neuroscience, 32*(28), 9677–9689.

Aisen, P. S., Schmeidler, J., & Pasinetti, G. M. (2002). Randomized pilot study of nimesulide treatment in Alzheimer's disease. *Neurology, 58*(7), 1050–1054.

Ait-Bouziad, N., Chiki, A., Limorenko, G., Xiao, S., Eliezer, D., & Lashuel, H. A. (2020). Phosphorylation of the overlooked tyrosine 310 regulates the structure, aggregation, and microtubule-and lipid-binding properties of Tau. *Journal of Biological Chemistry, 295*(23), 7905–7922.

Akiyama, H., Barger, S., Barnum, S., Bradt, B., Bauer, J., Cole, G. M., Cooper, N. R., Eikelenboom, P., Emmerling, M., & Fiebich, B. L. (2000). Inflammation and Alzheimer's disease. *Neurobiology of Aging, 21*(3), 383–421.

Albert, M. S., DeKosky, S. T., Dickson, D., Dubois, B., Feldman, H. H., Fox, N. C., Gamst, A., Holtzman, D. M., Jagust, W. J., & Petersen, R. C. (2011). The diagnosis of mild cognitive impairment due to Alzheimer's disease: Recommendations from the National Institute on Aging-Alzheimer's association workgroups on diagnostic guidelines for Alzheimer's disease. *Alzheimer's & Dementia, 7*(3), 270–279.

Alexander, G. C., Emerson, S., & Kesselheim, A. S. (2021). Evaluation of aducanumab for Alzheimer disease: Scientific evidence and regulatory review involving efficacy, safety, and futility. *JAMA, 325*(17), 1717–1718.

Arai, H., Suzuki, H., & Yoshiyama, T. (2015). Vanutide cridificar and the QS-21 adjuvant in Japanese subjects with mild to moderate Alzheimer's disease: Results from two phase 2 studies. *Current Alzheimer Research, 12*(3), 242–254.

Asuni, A. A., Boutajangout, A., Quartermain, D., & Sigurdsson, E. M. (2007). Immunotherapy targeting pathological tau conformers in a tangle mouse model reduces brain pathology with associated functional improvements. *Journal of Neuroscience, 27*(34), 9115–9129.

Ayalon, G., Lee, S.-H., Adolfsson, O., Foo-Atkins, C., Atwal, J. K., Blendstrup, M., Booler, H., Bravo, J., Brendza, R., & Brunstein, F. (2021). Antibody semorinemab reduces tau pathology in a transgenic mouse model and engages tau in patients with Alzheimer's disease. *Science Translational Medicine, 13*(593), eabb2639.

Bard, F., Cannon, C., Barbour, R., Burke, R.-L., Games, D., Grajeda, H., Guido, T., Hu, K., Huang, J., & Johnson-Wood, K. (2000). Peripherally administered antibodies against amyloid β-peptide enter the central nervous system and reduce pathology in a mouse model of Alzheimer disease. *Nature Medicine, 6*(8), 916–919.

Bateman, R. J., Smith, J., Donohue, M. C., Delmar, P., Abbas, R., Salloway, S., Wojtowicz, J., Blennow, K., Bittner, T., & Black, S. E. (2023). Two phase 3 trials of gantenerumab in early Alzheimer's disease. *New England Journal of Medicine, 389*(20), 1862–1876.

Bittar, A., Bhatt, N., & Kayed, R. (2020). Advances and considerations in AD tau-targeted immunotherapy. *Neurobiology of Disease, 134*, 104707.

Brelstaff, J., Tolkovsky, A. M., Ghetti, B., Goedert, M., & Spillantini, M. G. (2018). Living neurons with tau filaments aberrantly expose phosphatidylserine and are phagocytosed by microglia. *Cell Reports, 24*(8), 1939–1948.

Butchart, J., Brook, L., Hopkins, V., Teeling, J., Püntener, U., Culliford, D., Sharples, R., Sharif, S., McFarlane, B., & Raybould, R. (2015). Etanercept in Alzheimer disease: A randomized, placebo-controlled, double-blind, phase 2 trial. *Neurology, 84*(21), 2161–2168.

Chai, X., Wu, S., Murray, T. K., Kinley, R., Cella, C. V., Sims, H., Buckner, N., Hanmer, J., Davies, P., & O'Neill, M. J. (2011). Passive immunization with anti-tau antibodies in two transgenic models: Reduction of Tau pathology and delay of disease progression. *Journal of Biological Chemistry, 286*(39), 34457–34467.

Comi, C., & Tondo, G. (2017). Insights into the protective role of immunity in neurodegenerative disease. *Neural Regeneration Research, 12*(1), 64.

Congdon, E. E., & Sigurdsson, E. M. (2018). Tau-targeting therapies for Alzheimer disease. *Nature Reviews Neurology, 14*(7), 399–415.

Cribbs, D. H., Ghochikyan, A., Vasilevko, V., Tran, M., Petrushina, I., Sadzikava, N., Babikyan, D., Kesslak, P., Kieber-Emmons, T., & Cotman, C. W. (2003). Adjuvant-dependent modulation of Th1 and Th2 responses to immunization with β-amyloid. *International Immunology, 15*(4), 505–514.

Cullen, N. C., Novak, P., Tosun, D., Kovacech, B., Hanes, J., Kontsekova, E., Fresser, M., Ropele, S., Feldman, H. H., & Schmidt, R. (2024). Efficacy assessment of an active tau immunotherapy in Alzheimer's disease patients with amyloid and tau pathology: A post hoc analysis of the "ADAMANT" randomised, placebo-controlled, double-blind, multi-centre, phase 2 clinical trial. *eBioMedicine, 99*, 104923.

Cummings, J. L., Cohen, S., van Dyck, C. H., Brody, M., Curtis, C., Cho, W., Ward, M., Friesenhahn, M., Rabe, C., & Brunstein, F. (2018). ABBY: A phase 2 randomized trial of crenezumab in mild to moderate Alzheimer disease. *Neurology, 90*(21), e1889–e1897.

Dam, T., Boxer, A. L., Golbe, L. I., Höglinger, G. U., Morris, H. R., Litvan, I., Lang, A. E., Corvol, J.-C., Aiba, I., & Grundman, M. (2021). Safety and efficacy of anti-tau monoclonal antibody gosuranemab in progressive supranuclear palsy: A phase 2, randomized, placebo-controlled trial. *Nature Medicine, 27*(8), 1451–1457.

Danysz, W., & Parsons, C. G. (2003). The NMDA receptor antagonist memantine as a symptomatological and neuroprotective treatment for Alzheimer's disease: Preclinical evidence. *International Journal of Geriatric Psychiatry, 18*(S1), S23–S32.

Das, S. R., Xie, L., Wisse, L. E. M., Ittyerah, R., Tustison, N. J., Dickerson, B. C., Yushkevich, P. A., Wolk, D. A., & Initiative, A. D. N. (2018). Longitudinal and cross-sectional structural magnetic resonance imaging correlates of AV-1451 uptake. *Neurobiology of Aging, 66*, 49–58.

De Jong, D., Jansen, R., Hoefnagels, W., Jellesma-Eggenkamp, M., Verbeek, M., Borm, G., & Kremer, B. (2008). No effect of one-year treatment with indomethacin on Alzheimer's disease progression: A randomized controlled trial. *PLoS One, 3*(1), e1475.

Deczkowska, A., Keren-Shaul, H., Weiner, A., Colonna, M., Schwartz, M., & Amit, I. (2018). Disease-associated microglia: A universal immune sensor of neurodegeneration. *Cell, 173*(5), 1073–1081.

DeMattos, R. B., Bales, K. R., Cummins, D. J., Dodart, J.-C., Paul, S. M., & Holtzman, D. M. (2001). Peripheral anti-Aβ antibody alters CNS and plasma Aβ clearance and decreases brain Aβ burden in a mouse model of Alzheimer's disease. *Proceedings of the National Academy of Sciences, 98*(15), 8850–8855.

Dhimolea, E. (2010). Canakinumab. *MAbs, 2*(1), 3–13.

Dickey, C. A., Morgan, D. G., Kudchodkar, S., Weiner, D. B., Bai, Y., Cao, C., Gordon, M. N., & Ugen, K. E. (2001). Duration and specificity of humoral immune responses in mice vaccinated with the Alzheimer's disease-associated β-amyloid 1-42 peptide. *DNA and Cell Biology, 20*(11), 723–729.

Dodart, J.-C., Bales, K. R., Gannon, K. S., Greene, S. J., DeMattos, R. B., Mathis, C., DeLong, C. A., Wu, S., Wu, X., & Holtzman, D. M. (2002). Immunization reverses memory deficits without reducing brain Aβ burden in Alzheimer's disease model. *Nature Neuroscience, 5*(5), 452–457.

Doody, R. S., Thomas, R. G., Farlow, M., Iwatsubo, T., Vellas, B., Joffe, S., Kieburtz, K., Raman, R., Sun, X., & Aisen, P. S. (2014). Phase

3 trials of solanezumab for mild-to-moderate Alzheimer's disease. *New England Journal of Medicine, 370*(4), 311–321.

Dost, F. S., Kaya, D., Ontan, M. S., Erken, N., Bulut, E. A., Aydin, A. E., & Isik, A. T. (2021). Theracurmin supplementation may be a therapeutic option for older patients with Alzheimer's disease: A 6-month retrospective follow-up study. *Current Alzheimer Research, 18*(14), 1087–1092.

Du, L., Zhang, Y., Chen, Y., Zhu, J., Yang, Y., & Zhang, H.-L. (2017). Role of microglia in neurological disorders and their potentials as a therapeutic target. *Molecular Neurobiology, 54*(10), 7567–7584.

Faridar, A., Thome, A. D., Zhao, W., Thonhoff, J. R., Beers, D. R., Pascual, B., Masdeu, J. C., & Appel, S. H. (2020). Restoring regulatory T-cell dysfunction in Alzheimer's disease through ex vivo expansion. *Brain Communications, 2*(2), fcaa112.

Faridar, A., Vasquez, M., Thome, A. D., Yin, Z., Xuan, H., Wang, J. H., Wen, S., Li, X., Thonhoff, J. R., & Zhao, W. (2022). Ex vivo expanded human regulatory T cells modify neuroinflammation in a preclinical model of Alzheimer's disease. *Acta Neuropathologica Communications, 10*(1), 144.

Faridar, A., Eid, A. M., Thome, A. D., Zhao, W., Beers, D. R., Pascual, M. B., Nakawah, M. O., Roman, G. C., Davis, C. S., & Grundman, M. (2023). A phase 1 open-label pilot study of low-dose interleukine-2 immunotherapy in patients with Alzheimer's disease. *Translational Neurodegeneration, 12*(1), 54.

Ferrari, C., & Sorbi, S. (2021). The complexity of Alzheimer's disease: An evolving puzzle. *Physiological Reviews, 101*(3), 1047–1081.

Ferrer, I., Rovira, M. B., Guerra, M. L. S., Rey, M. J., & Costa-Jussá, F. (2004). Neuropathology and pathogenesis of encephalitis following amyloid β immunization in Alzheimer's disease. *Brain Pathology, 14*(1), 11–20.

Florian, H., Wang, D., Arnold, S. E., Boada, M., Guo, Q., Jin, Z., Zheng, H., Fisseha, N., Kalluri, H. V., & Rendenbach-Mueller, B. (2023). Tilavonemab in early Alzheimer's disease: Results from a phase 2, randomized, double-blind study. *Brain, 146*(6), 2275–2284.

Fox, N. C., Black, R. S., Gilman, S., Rossor, M. N., Griffith, S. G., Jenkins, L., & Koller, M. (2005). Effects of Aβ immunization (AN1792) on MRI measures of cerebral volume in Alzheimer disease. *Neurology, 64*(9), 1563–1572.

Frisoni, G. B., Altomare, D., Thal, D. R., Ribaldi, F., van der Kant, R., Ossenkoppele, R., Blennow, K., Cummings, J., van Duijn, C., & Nilsson, P. M. (2022). The probabilistic model of Alzheimer disease: The amyloid hypothesis revised. *Nature Reviews Neuroscience, 23*(1), 53–66.

Frisoni, G. B., Altomare, D., Ribaldi, F., Villain, N., Brayne, C., Mukadam, N., Abramowicz, M., Barkhof, F., Berthier, M., & Bieler-Aeschlimann, M. (2023). Dementia prevention in memory clinics: Recommendations from the European task force for brain health services. *The Lancet Regional Health–Europe, 26*, 100576.

Giacobini, E., & Gold, G. (2013). Alzheimer disease therapy—Moving from amyloid-β to tau. *Nature Reviews Neurology, 9*(12), 677–686.

Gilman, S., Koller, M., Black, R. S., Jenkins, L., Griffith, S. G., Fox, N. C., Eisner, L., Kirby, L., Rovira, M. B., & Forette, F. (2005). Clinical effects of Aβ immunization (AN1792) in patients with AD in an interrupted trial. *Neurology, 64*(9), 1553–1562.

Guerreiro, R., Wojtas, A., Bras, J., Carrasquillo, M., Rogaeva, E., Majounie, E., Cruchaga, C., Sassi, C., Kauwe, J. S. K., Younkin, S., Hazrati, L., Collinge, J., Pocock, J., Lashley, T., Williams, J., Lambert, J.-C., Amouyel, P., Goate, A., Rademakers, R., et al. (2013). TREM2 variants in Alzheimer's disease. *The New England Journal of Medicine, 368*(2), 117–127. https://doi.org/10.1056/NEJMoa1211851

Guillot-Sestier, M.-V., & Town, T. (2018). Let's make microglia great again in neurodegenerative disorders. *Journal of Neural Transmission, 125*, 751–770.

Gustavsson, A., Norton, N., Fast, T., Frölich, L., Georges, J., Holzapfel, D., Kirabali, T., Krolak-Salmon, P., Rossini, P. M., & Ferretti, M. T. (2023). Global estimates on the number of persons across the Alzheimer's disease continuum. *Alzheimer's & Dementia, 19*(2), 658–670.

Haeberlein, S. B., von Hehn, C., Tian, Y., Chalkias, S., Muralidharan, K. K., Chen, T., Wu, S., Skordos, L., Nisenbaum, L., & Rajagovindan, R. (2020). Emerge and engage topline results: Phase 3 studies of aducanumab in early Alzheimer's disease: Developments in clinical trials and cognitive assessment. *Alzheimer's & Dementia, 16*, e047259.

Hawkes, C. A., Carare, R. O., & Weller, R. O. (2014). Amyloid and tau in the brain in sporadic Alzheimer's disease: Defining the chicken and the egg. *Acta Neuropathologica, 127*, 617–618.

He, Z., McBride, J. D., Xu, H., Changolkar, L., Kim, S., Zhang, B., Narasimhan, S., Gibbons, G. S., Guo, J. L., & Kozak, M. (2020). Transmission of tauopathy strains is independent of their isoform composition. *Nature Communications, 11*(1), 7.

Hickman, S., Izzy, S., Sen, P., Morsett, L., & El Khoury, J. (2018). Microglia in neurodegeneration. *Nature Neuroscience, 21*(10), 1359–1369.

Hock, C., Konietzko, U., Streffer, J. R., Tracy, J., Signorell, A., Müller-Tillmanns, B., Lemke, U., Henke, K., Moritz, E., & Garcia, E. (2003). Antibodies against β-amyloid slow cognitive decline in Alzheimer's disease. *Neuron, 38*(4), 547–554.

Honig, L. S., Vellas, B., Woodward, M., Boada, M., Bullock, R., Borrie, M., Hager, K., Andreasen, N., Scarpini, E., & Liu-Seifert, H. (2018). Trial of solanezumab for mild dementia due to Alzheimer's disease. *New England Journal of Medicine, 378*(4), 321–330.

Hull, M., Sadowsky, C., Arai, H., Le Prince Leterme, G., Holstein, A., Booth, K., Peng, Y., Yoshiyama, T., Suzuki, H., & Ketter, N. (2017). Long-term extensions of randomized vaccination trials of ACC-001 and QS-21 in mild to moderate alzheimer's disease. *Current Alzheimer Research, 14*(7), 696–708.

Ikezu, T. (2017). Immunotherapy for Alzheimer's disease. *Neuroimmune Pharmacology*, 753–766.

Jack, C. R., Jr., Knopman, D. S., Jagust, W. J., Petersen, R. C., Weiner, M. W., Aisen, P. S., Shaw, L. M., Vemuri, P., Wiste, H. J., & Weigand, S. D. (2013). Update on hypothetical model of Alzheimer's disease biomarkers. *Lancet Neurology, 12*(2), 207.

Janssen, O., Jansen, W. J., Vos, S. J. B., Boada, M., Parnetti, L., Gabryelewicz, T., Fladby, T., Molinuevo, J. L., Villeneuve, S., & Hort, J. (2022). Characteristics of subjective cognitive decline associated with amyloid positivity. *Alzheimer's & Dementia, 18*(10), 1832–1845.

Janus, C., Pearson, J., McLaurin, J., Mathews, P. M., Jiang, Y., Schmidt, S. D., Chishti, M. A., Horne, P., Heslin, D., & French, J. (2000). Aβ peptide immunization reduces behavioural impairment and plaques in a model of Alzheimer's disease. *Nature, 408*(6815), 979–982.

Kadavath, H., Hofele, R. V., Biernat, J., Kumar, S., Tepper, K., Urlaub, H., Mandelkow, E., & Zweckstetter, M. (2015). Tau stabilizes microtubules by binding at the interface between tubulin heterodimers. *Proceedings of the National Academy of Sciences, 112*(24), 7501–7506.

Klyubin, I., Walsh, D. M., Lemere, C. A., Cullen, W. K., Shankar, G. M., Betts, V., Spooner, E. T., Jiang, L., Anwyl, R., & Selkoe, D. J. (2005). Amyloid β protein immunotherapy neutralizes Aβ oligomers that disrupt synaptic plasticity in vivo. *Nature Medicine, 11*(5), 556–561.

Kontsekova, E., Zilka, N., Kovacech, B., Novak, P., & Novak, M. (2014). First-in-man tau vaccine targeting structural determinants essential for pathological tau–tau interaction reduces tau oligomerisation and neurofibrillary degeneration in an Alzheimer's disease model. *Alzheimer's Research & Therapy, 6*, 1–12.

Kotilinek, L. A., Bacskai, B., Westerman, M., Kawarabayashi, T., Younkin, L., Hyman, B. T., Younkin, S., & Ashe, K. H. (2002). Reversible memory loss in a mouse transgenic model of Alzheimer's disease. *Journal of Neuroscience, 22*(15), 6331–6335.

Kwon, W., Kim, D., Kim, H.-Y., Jeong, S. W., Lee, S.-G., Kim, H.-C., Lee, Y.-J., Kwon, M. K., Hwang, J.-S., & Han, J. E. (2022). Microglial phagocytosis of polystyrene microplastics results in immune alteration and apoptosis in vitro and in vivo. *Science of the Total Environment, 807*, 150817.

La Porte, S. L., Bollini, S. S., Lanz, T. A., Abdiche, Y. N., Rusnak, A. S., Ho, W.-H., Kobayashi, D., Harrabi, O., Pappas, D., & Mina, E. W. (2012). Structural basis of C-terminal β-amyloid peptide binding by the antibody ponezumab for the treatment of Alzheimer's disease. *Journal of Molecular Biology, 421*(4–5), 525–536.

Landen, J. W., Andreasen, N., Cronenberger, C. L., Schwartz, P. F., Börjesson-Hanson, A., Östlund, H., Sattler, C. A., Binneman, B., & Bednar, M. M. (2017). Ponezumab in mild-to-moderate Alzheimer's disease: Randomized phase II PET-PIB study. *Alzheimer's & Dementia: Translational Research & Clinical Interventions, 3*(3), 393–401.

Lane-Donovan, C., & Boxer, A. L. (2024). Disentangling tau: One protein, many therapeutic approaches. *Neurotherapeutics*, e00321.

Lee, D. C., Rizer, J., Hunt, J. B., Selenica, M., Gordon, M. N., & Morgan, D. (2013). Experimental manipulations of microglia in mouse models of Alzheimer's pathology: Activation reduces amyloid but hastens tau pathology. *Neuropathology and Applied Neurobiology, 39*(1), 69–85.

Lewcock, J. W., Schlepckow, K., Di Paolo, G., Tahirovic, S., Monroe, K. M., & Haass, C. (2020). Emerging microglia biology defines novel therapeutic approaches for Alzheimer's disease. *Neuron, 108*(5), 801–821.

Li, Q., Wu, Y., Chen, J., Xuan, A., & Wang, X. (2022). Microglia and immunotherapy in Alzheimer's disease. *Acta Neurologica Scandinavica, 145*(3), 273–278.

Liang, C.-S., Li, D.-J., Yang, F.-C., Tseng, P.-T., Carvalho, A. F., Stubbs, B., Thompson, T., Mueller, C., Shin, J. I., & Radua, J. (2021). Mortality rates in Alzheimer's disease and non-Alzheimer's dementias: A systematic review and meta-analysis. *The Lancet Healthy Longevity, 2*(8), e479–e488.

Lu, D., Dou, F., & Gao, J. (2023). Development of amyloid beta-directed antibodies against Alzheimer's disease: Twists and turns. *Drug Discoveries & Therapeutics, 17*(6), 440–444.

Malpetti, M., Kievit, R. A., Passamonti, L., Jones, P. S., Tsvetanov, K. A., Rittman, T., Mak, E., Nicastro, N., Bevan-Jones, W. R., & Su, L. (2020). Microglial activation and tau burden predict cognitive decline in Alzheimer's disease. *Brain, 143*(5), 1588–1602.

Masliah, E., Rockenstein, E., Adame, A., Alford, M., Crews, L., Hashimoto, M., Seubert, P., Lee, M., Goldstein, J., & Chilcote, T. (2005). Effects of α-synuclein immunization in a mouse model of Parkinson's disease. *Neuron, 46*(6), 857–868.

Matsunaga, S., Kishi, T., & Iwata, N. (2015). Memantine monotherapy for Alzheimer's disease: A systematic review and meta-analysis. *PLoS One, 10*(4), e0123289.

McKhann, G. M., Knopman, D. S., Chertkow, H., Hyman, B. T., Jack, C. R., Jr., Kawas, C. H., Klunk, W. E., Koroshetz, W. J., Manly, J. J., & Mayeux, R. (2011). The diagnosis of dementia due to Alzheimer's disease: Recommendations from the National Institute on Aging-Alzheimer's association workgroups on diagnostic guidelines for Alzheimer's disease. *Alzheimer's & Dementia, 7*(3), 263–269.

Medeiros, R., Baglietto-Vargas, D., & LaFerla, F. M. (2011). The role of tau in Alzheimer's disease and related disorders. *CNS Neuroscience & Therapeutics, 17*(5), 514–524.

Mintun, M. A., Lo, A. C., Duggan Evans, C., Wessels, A. M., Ardayfio, P. A., Andersen, S. W., Shcherbinin, S., Sparks, J., Sims, J. R., & Brys, M. (2021). Donanemab in early Alzheimer's disease. *New England Journal of Medicine, 384*(18), 1691–1704.

Monteiro, C., Toth, B., Brunstein, F., Bobbala, A., Datta, S., Ceniceros, R., Sanabria Bohorquez, S. M., Anania, V. G., Wildsmith, K. R., & Schauer, S. P. (2023). Randomized phase II study of the safety and efficacy of semorinemab in participants with mild-to-moderate Alzheimer disease: Lauriet. *Neurology, 101*(14), e1391–e1401.

Morgan, D. (2011). Immunotherapy for Alzheimer's disease. *Journal of Internal Medicine, 269*(1), 54–63.

Morgan, D., Diamond, D. M., Gottschall, P. E., Ugen, K. E., Dickey, C., Hardy, J., Duff, K., Jantzen, P., DiCarlo, G., & Wilcock, D. (2000). Aβ peptide vaccination prevents memory loss in an animal model of Alzheimer's disease. *Nature, 408*(6815), 982–985.

Mullard, A. (2021). Failure of first anti-tau antibody in Alzheimer disease highlights risks of history repeating. *Nature Reviews Drug Discovery, 20*(1), 3–6.

Nelson, P. T., Alafuzoff, I., Bigio, E. H., Bouras, C., Braak, H., Cairns, N. J., Castellani, R. J., Crain, B. J., Davies, P., & Del Tredici, K. (2012). Correlation of Alzheimer disease neuropathologic changes with cognitive status: A review of the literature. *Journal of Neuropathology & Experimental Neurology, 71*(5), 362–381.

Nicoll, J. A. R., Wilkinson, D., Holmes, C., Steart, P., Markham, H., & Weller, R. O. (2003). Neuropathology of human Alzheimer disease after immunization with amyloid-β peptide: A case report. *Nature Medicine, 9*(4), 448–452.

Novak, P., Kontsekova, E., Zilka, N., & Novak, M. (2018a). Ten years of tau-targeted immunotherapy: The path walked and the roads ahead. *Frontiers in Neuroscience, 12*, 798.

Novak, P., Schmidt, R., Kontsekova, E., Kovacech, B., Smolek, T., Katina, S., Fialova, L., Prcina, M., Parrak, V., & Dal-Bianco, P. (2018b). FUNDAMANT: An interventional 72-week phase 1 follow-up study of AADvac1, an active immunotherapy against tau protein pathology in Alzheimer's disease. *Alzheimer's Research & Therapy, 10*(1), 1–16.

Novak, P., Kovacech, B., Katina, S., Schmidt, R., Scheltens, P., Kontsekova, E., Ropele, S., Fialova, L., Kramberger, M., & Paulenka-Ivanovova, N. (2021). ADAMANT: A placebo-controlled randomized phase 2 study of AADvac1, an active immunotherapy against pathological tau in Alzheimer's disease. *Nature Aging, 1*(6), 521–534.

Oddo, S., Billings, L., Kesslak, J. P., Cribbs, D. H., & LaFerla, F. M. (2004). Aβ immunotherapy leads to clearance of early, but not late, hyperphosphorylated tau aggregates via the proteasome. *Neuron, 43*(3), 321–332.

Orgogozo, J.-M., Gilman, S., Dartigues, J.-F., Laurent, B., Puel, M., Kirby, L. C., Jouanny, P., Dubois, B., Eisner, L., & Flitman, S. (2003). Subacute meningoencephalitis in a subset of patients with AD after Aβ42 immunization. *Neurology, 61*(1), 46–54.

Ossenkoppele, R., Schonhaut, D. R., Schöll, M., Lockhart, S. N., Ayakta, N., Baker, S. L., O'Neil, J. P., Janabi, M., Lazaris, A., & Cantwell, A. (2016). Tau PET patterns mirror clinical and neuroanatomical variability in Alzheimer's disease. *Brain, 139*(5), 1551–1567.

Ostrowitzki, S., Bittner, T., Sink, K. M., Mackey, H., Rabe, C., Honig, L. S., Cassetta, E., Woodward, M., Boada, M., & van Dyck, C. H. (2022). Evaluating the safety and efficacy of crenezumab vs placebo in adults with early Alzheimer disease: Two phase 3 randomized placebo-controlled trials. *JAMA Neurology, 79*(11), 1113–1121.

Panza, F., Solfrizzi, V., Seripa, D., Imbimbo, B. P., Lozupone, M., Santamato, A., Tortelli, R., Galizia, I., Prete, C., & Daniele, A. (2016). Tau-based therapeutics for Alzheimer's disease: Active and passive immunotherapy. *Immunotherapy, 8*(9), 1119–1134.

Panza, F., Lozupone, M., Logroscino, G., & Imbimbo, B. P. (2019a). A critical appraisal of amyloid-β-targeting therapies for Alzheimer disease. *Nature Reviews Neurology, 15*(2), 73–88.

Panza, F., Lozupone, M., Seripa, D., & Imbimbo, B. P. (2019b). Amyloid-β immunotherapy for alzheimer disease: Is it now a long shot? *Annals of Neurology, 85*(3), 303–315.

Pasquier, F., Sadowsky, C., Holstein, A., Leterme, G. L. P., Peng, Y., Jackson, N., Fox, N. C., Ketter, N., Liu, E., & Ryan, J. M. (2016). Two phase 2 multiple ascending–dose studies of vanutide cridificar

(ACC-001) and QS-21 adjuvant in mild-to-moderate Alzheimer's disease. *Journal of Alzheimer's Disease, 51*(4), 1131–1143.

Peng, C., Trojanowski, J. Q., & Lee, V. M.-Y. (2020). Protein transmission in neurodegenerative disease. *Nature Reviews Neurology, 16*(4), 199–212.

Perani, D., Caminiti, S. P., Carli, G., & Tondo, G. (2021). PET neuroimaging in dementia conditions. *PET and SPECT in Neurology*, 211–282.

Potter, H., Woodcock, J. H., Boyd, T. D., Coughlan, C. M., O'Shaughnessy, J. R., Borges, M. T., Thaker, A. A., Raj, B. A., Adamszuk, K., & Scott, D. (2021). Safety and efficacy of sargramostim (GM-CSF) in the treatment of Alzheimer's disease. *Alzheimer's & Dementia: Translational Research & Clinical Interventions, 7*(1), e12158.

Raina, P., Santaguida, P., Ismaila, A., Patterson, C., Cowan, D., Levine, M., Booker, L., & Oremus, M. (2008). Effectiveness of cholinesterase inhibitors and memantine for treating dementia: Evidence review for a clinical practice guideline. *Annals of Internal Medicine, 148*(5), 379–397.

Rawji, K. S., Young, A. M. H., Ghosh, T., Michaels, N. J., Mirzaei, R., Kappen, J., Kolehmainen, K. L., Alaeiilkhchi, N., Lozinski, B., & Mishra, M. K. (2020). Niacin-mediated rejuvenation of macrophage/microglia enhances remyelination of the aging central nervous system. *Acta Neuropathologica, 139*, 893–909.

Rivers-Auty, J., Mather, A. E., Peters, R., Lawrence, C. B., & Brough, D. (2020). Anti-inflammatories in Alzheimer's disease—Potential therapy or spurious correlate? *Brain Communications, 2*(2), fcaa109.

Salloway, S., Sperling, R., Fox, N. C., Blennow, K., Klunk, W., Raskind, M., Sabbagh, M., Honig, L. S., Porsteinsson, A. P., & Ferris, S. (2014). Two phase 3 trials of bapineuzumab in mild-to-moderate Alzheimer's disease. *New England Journal of Medicine, 370*(4), 322–333.

Salloway, S., Honigberg, L. A., Cho, W., Ward, M., Friesenhahn, M., Brunstein, F., Quartino, A., Clayton, D., Mortensen, D., & Bittner, T. (2018). Amyloid positron emission tomography and cerebrospinal fluid results from a crenezumab anti-amyloid-beta antibody double-blind, placebo-controlled, randomized phase II study in mild-to-moderate Alzheimer's disease (BLAZE). *Alzheimer's Research & Therapy, 10*(1), 1–13.

Salloway, S., Farlow, M., McDade, E., Clifford, D. B., Wang, G., Llibre-Guerra, J. J., Hitchcock, J. M., Mills, S. L., Santacruz, A. M., & Aschenbrenner, A. J. (2021). A trial of gantenerumab or solanezumab in dominantly inherited Alzheimer's disease. *Nature Medicine, 27*(7), 1187–1196.

Salloway, S., Chalkias, S., Barkhof, F., Burkett, P., Barakos, J., Purcell, D., Suhy, J., Forrestal, F., Tian, Y., & Umans, K. (2022). Amyloid-related imaging abnormalities in 2 phase 3 studies evaluating aducanumab in patients with early Alzheimer disease. *JAMA Neurology, 79*(1), 13–21.

Scharf, S., Mander, A., Ugoni, A., Vajda, F., & Christophidis, N. (1999). A double-blind, placebo-controlled trial of diclofenac/misoprostol in Alzheimer's disease. *Neurology, 53*(1), 197.

Scheltens, P., Blennow, K., Breteler, M. M. B., De Strooper, B., Frisoni, G. B., Salloway, S., & Van der Flier, W. M. (2016). Alzheimer's disease. *The Lancet, 388*(10043), 505–517.

Schenk, D., Barbour, R., Dunn, W., Gordon, G., Grajeda, H., Guido, T., Hu, K., Huang, J., Johnson-Wood, K., & Khan, K. (1999). Immunization with amyloid-β attenuates Alzheimer-disease-like pathology in the PDAPP mouse. *Nature, 400*(6740), 173–177.

Schneeberger, A., Hendrix, S., Mandler, M., Ellison, N., Bürger, V., Brunner, M., Frölich, L., Mimica, N., Hort, J., & Rainer, M. (2015). For the investigators of CT AFF006 (2015): Results from a phase II study to assess the clinical and immunological activity of AFFITOPE® AD02 in patients with early Alzheimer's disease. *The Journal of Prevention of Alzheimer's Disease (JPAD)*. https://doi.org/10.14283/Jpad

Schöll, M., Lockhart, S. N., Schonhaut, D. R., O'Neil, J. P., Janabi, M., Ossenkoppele, R., Baker, S. L., Vogel, J. W., Faria, J., & Schwimmer, H. D. (2016). PET imaging of tau deposition in the aging human brain. *Neuron, 89*(5), 971–982.

Selkoe, D. J. (2019). Alzheimer disease and aducanumab: Adjusting our approach. *Nature Reviews Neurology, 15*(7), 365–366.

Sevigny, J., Chiao, P., Bussière, T., Weinreb, P. H., Williams, L., Maier, M., Dunstan, R., Salloway, S., Chen, T., & Ling, Y. (2016). The antibody aducanumab reduces Aβ plaques in Alzheimer's disease. *Nature, 537*(7618), 50–56.

Shulman, M., Kong, J., O'Gorman, J., Ratti, E., Rajagovindan, R., Viollet, L., Huang, E., Sharma, S., Racine, A. M., & Czerkowicz, J. (2023). TANGO: A placebo-controlled randomized phase 2 study of efficacy and safety of the anti-tau monoclonal antibody gosuranemab in early Alzheimer's disease. *Nature Aging, 3*, 1–11.

Song, C., Shi, J., Zhang, P., Zhang, Y., Xu, J., Zhao, L., Zhang, R., Wang, H., & Chen, H. (2022). Immunotherapy for Alzheimer's disease: Targeting β-amyloid and beyond. *Translational Neurodegeneration, 11*(1), 1–17.

Sopko, R., Golonzhka, O., Arndt, J., Quan, C., Czerkowicz, J., Cameron, A., Smith, B., Murugesan, Y., Gibbons, G., & Kim, S.-J. (2020). Characterization of tau binding by gosuranemab. *Neurobiology of Disease, 146*, 105120.

Sperling, R. A., Jack, C. R., Jr., & Aisen, P. S. (2011). Testing the right target and right drug at the right stage. *Science Translational Medicine, 3*(111), 111cm33.

Sperling, R. A., Donohue, M. C., Raman, R., Rafii, M. S., Johnson, K., Masters, C. L., van Dyck, C. H., Iwatsubo, T., Marshall, G. A., & Yaari, R. (2023). Trial of solanezumab in preclinical Alzheimer's disease. *New England Journal of Medicine, 389*(12), 1096–1107.

Stopschinski, B. E., Weideman, R. A., McMahan, D., Jacob, D. A., Little, B. B., Chiang, H.-S., Saez Calveras, N., & Stuve, O. (2023). Microglia as a cellular target of diclofenac therapy in Alzheimer's disease. *Therapeutic Advances in Neurological Disorders, 16*, 17562864231156674.

Streit, W. J., Sammons, N. W., Kuhns, A. J., & Sparks, D. L. (2004). Dystrophic microglia in the aging human brain. *Glia, 45*(2), 208–212.

Suzumura, A., Sawada, M., Yamamoto, H., & Marunouchi, T. (1990). Effects of colony stimulating factors on isolated microglia in vitro. *Journal of Neuroimmunology, 30*(2–3), 111–120.

Swanson, C. J., Zhang, Y., Dhadda, S., Wang, J., Kaplow, J., Lai, R. Y. K., Lannfelt, L., Bradley, H., Rabe, M., & Koyama, A. (2021). A randomized, double-blind, phase 2b proof-of-concept clinical trial in early Alzheimer's disease with lecanemab, an anti-Aβ protofibril antibody. *Alzheimer's Research & Therapy, 13*, 1–14.

Takizawa, C., Thompson, P. L., van Walsem, A., Faure, C., & Maier, W. C. (2015). Epidemiological and economic burden of Alzheimer's disease: A systematic literature review of data across Europe and The United States of America. *Journal of Alzheimer's Disease, 43*(4), 1271–1284.

Tarkowski, E., Andreasen, N., Tarkowski, A., & Blennow, K. (2003). Intrathecal inflammation precedes development of Alzheimer's disease. *Journal of Neurology, Neurosurgery, and Psychiatry, 74*(9), 1200.

Teng, E., Manser, P. T., Pickthorn, K., Brunstein, F., Blendstrup, M., Bohorquez, S. S., Wildsmith, K. R., Toth, B., Dolton, M., & Ramakrishnan, V. (2022). Safety and efficacy of semorinemab in individuals with prodromal to mild Alzheimer disease: A randomized clinical trial. *JAMA Neurology, 79*(8), 758–767.

Theunis, C., Crespo-Biel, N., Gafner, V., Pihlgren, M., López-Deber, M. P., Reis, P., Hickman, D. T., Adolfsson, O., Chuard, N., & Ndao, D. M. (2013). Efficacy and safety of a liposome-based vaccine against protein tau, assessed in tau. P301L mice that model tauopathy. *PLoS One, 8*(8), e72301.

Tondo, G., & De Marchi, F. (2022). From biomarkers to precision medicine in neurodegenerative diseases: Where are we? *Journal of Clinical Medicine, 11*(15), 4515. MDPI.

Tondo, G., Boccalini, C., Caminiti, S. P., Presotto, L., Filippi, M., Magnani, G., Frisoni, G. B., Iannaccone, S., & Perani, D. (2021). Brain metabolism and microglia activation in mild cognitive impairment: A combined [18F] FDG and [11C]-(R)-PK11195 PET study. *Journal of Alzheimer's Disease, 80*(1), 433–445.

Tondo, G., Boccalini, C., Vanoli, E. G., Presotto, L., Muscio, C., Ciullo, V., Banaj, N., Piras, F., Filippini, G., & Tiraboschi, P. (2022). Brain metabolism and amyloid load in individuals with subjective cognitive decline or pre–mild cognitive impairment. *Neurology, 99*(3), e258–e269.

Valiukas, Z., Ephraim, R., Tangalakis, K., Davidson, M., Apostolopoulos, V., & Feehan, J. (2022). Immunotherapies for Alzheimer's disease—A review. *Vaccine, 10*(9), 1527.

Van Dyck, C. H. (2018). Anti-amyloid-β monoclonal antibodies for Alzheimer's disease: Pitfalls and promise. *Biological Psychiatry, 83*(4), 311–319.

Van Dyck, C. H., Swanson, C. J., Aisen, P., Bateman, R. J., Chen, C., Gee, M., Kanekiyo, M., Li, D., Reyderman, L., & Cohen, S. (2023). Lecanemab in early Alzheimer's disease. *New England Journal of Medicine, 388*(1), 9–21.

Vaz, M., & Silvestre, S. (2020). Alzheimer's disease: Recent treatment strategies. *European Journal of Pharmacology, 887*, 173554.

Vellas, B., Black, R., Thal, L. J., Fox, N. C., Daniels, M., McLennan, G., Tompkins, C., Leibman, C., Pomfret, M., & Grundman, M. (2009). Long-term follow-up of patients immunized with AN1792: Reduced functional decline in antibody responders. *Current Alzheimer Research, 6*(2), 144–151.

Vogt, A.-C. S., Jennings, G. T., Mohsen, M. O., Vogel, M., & Bachmann, M. F. (2023). Alzheimer's disease: A brief history of immunotherapies targeting amyloid β. *International Journal of Molecular Sciences, 24*(4), 3895.

Walsh, D. M., Klyubin, I., Fadeeva, J. V., Cullen, W. K., Anwyl, R., Wolfe, M. S., Rowan, M. J., & Selkoe, D. J. (2002). Naturally secreted oligomers of amyloid β protein potently inhibit hippocampal long-term potentiation in vivo. *Nature, 416*(6880), 535–539.

Wang, C. Y., Wang, P.-N., Chiu, M.-J., Finstad, C. L., Lin, F., Lynn, S., Tai, Y.-H., De Fang, X., Zhao, K., & Hung, C.-H. (2017a). UB-311, a novel UBITh® amyloid β peptide vaccine for mild Alzheimer's disease. *Alzheimer's & Dementia: Translational Research & Clinical Interventions, 3*(2), 262–272.

Wang, Y., Yan, T., Lu, H., Yin, W., Lin, B., Fan, W., Zhang, X., & Fernandez-Funez, P. (2017b). Lessons from anti-amyloid-β immunotherapies in Alzheimer disease: Aiming at a moving target. *Neurodegenerative Diseases, 17*(6), 242–250.

Wang, S., Mustafa, M., Yuede, C. M., Salazar, S. V., Kong, P., Long, H., Ward, M., Siddiqui, O., Paul, R., & Gilfillan, S. (2020). Anti-human TREM2 induces microglia proliferation and reduces pathology in an Alzheimer's disease model. *Journal of Experimental Medicine, 217*(9), e20200785.

Wendimu, M. Y., & Hooks, S. B. (2022). Microglia phenotypes in aging and neurodegenerative diseases. *Cells, 11*(13), 2091.

Wiessner, C., Wiederhold, K.-H., Tissot, A. C., Frey, P., Danner, S., Jacobson, L. H., Jennings, G. T., Lüönd, R., Ortmann, R., & Reichwald, J. (2011). The second-generation active Aβ immunotherapy CAD106 reduces amyloid accumulation in APP transgenic mice while minimizing potential side effects. *Journal of Neuroscience, 31*(25), 9323–9331.

Wilcock, D. M., DiCarlo, G., Henderson, D., Jackson, J., Clarke, K., Ugen, K. E., Gordon, M. N., & Morgan, D. (2003). Intracranially administered anti-Aβ antibodies reduce β-amyloid deposition by mechanisms both independent of and associated with microglial activation. *Journal of Neuroscience, 23*(9), 3745–3751.

Wilkins, H. M., & Swerdlow, R. H. (2017). Amyloid precursor protein processing and bioenergetics. *Brain Research Bulletin, 133*, 71–79.

Willis, B. A., Lo, A. C., Dage, J. L., Shcherbinin, S., Chinchen, L., Andersen, S. W., LaBell, E. S., Perahia, D. G. S., Hauck, P. M., & Lowe, S. L. (2023). Safety, tolerability, and pharmacokinetics of zagotenemab in participants with symptomatic Alzheimer's disease: A phase I clinical trial. *Journal of Alzheimer's Disease Reports, Preprint*, 1–10.

Winblad, B., Andreasen, N., Minthon, L., Floesser, A., Imbert, G., Dumortier, T., Maguire, R. P., Blennow, K., Lundmark, J., & Staufenbiel, M. (2012). Safety, tolerability, and antibody response of active Aβ immunotherapy with CAD106 in patients with Alzheimer's disease: Randomised, double-blind, placebo-controlled, first-in-human study. *The Lancet Neurology, 11*(7), 597–604.

Wu, K.-M., Zhang, Y.-R., Huang, Y.-Y., Dong, Q., Tan, L., & Yu, J.-T. (2021). The role of the immune system in Alzheimer's disease. *Ageing Research Reviews, 70*, 101409.

Xia, C., Makaretz, S. J., Caso, C., McGinnis, S., Gomperts, S. N., Sepulcre, J., Gomez-Isla, T., Hyman, B. T., Schultz, A., & Vasdev, N. (2017). Association of in vivo [18F] AV-1451 tau PET imaging results with cortical atrophy and symptoms in typical and atypical Alzheimer disease. *JAMA Neurology, 74*(4), 427–436.

Yanamandra, K., Kfoury, N., Jiang, H., Mahan, T. E., Ma, S., Maloney, S. E., Wozniak, D. F., Diamond, M. I., & Holtzman, D. M. (2013). Anti-tau antibodies that block tau aggregate seeding in vitro markedly decrease pathology and improve cognition in vivo. *Neuron, 80*(2), 402–414.

Yu, H. J., Dickson, S. P., Wang, P.-N., Chiu, M.-J., Huang, C.-C., Chang, C.-C., Liu, H., Hendrix, S. B., Dodart, J.-C., & Verma, A. (2023). Safety, tolerability, immunogenicity, and efficacy of UB-311 in participants with mild Alzheimer's disease: A randomised, double-blind, placebo-controlled, phase 2a study. *eBioMedicine, 94*, 104665.

Zhao, J., Liu, X., Xia, W., Zhang, Y., & Wang, C. (2020). Targeting amyloidogenic processing of APP in Alzheimer's disease. *Frontiers in Molecular Neuroscience, 13*, 137.

Zheng, X., Tang, Y., Yang, Q., Wang, S., Chen, R., Tao, C., Zhang, P., Fan, B., Zhan, J., & Tang, C. (2022). Effectiveness and safety of anti-tau drugs for Alzheimer's disease: Systematic review and meta-analysis. *Journal of the American Geriatrics Society, 70*(11), 3281–3292.

Zhou, M., Xu, R., Kaelber, D. C., & Gurney, M. E. (2020). Tumor Necrosis Factor (TNF) blocking agents are associated with lower risk for Alzheimer's disease in patients with rheumatoid arthritis and psoriasis. *PLoS One, 15*(3), e0229819.

Immunotherapy for Neurodegenerative Movement Disorders

52

R. Lee Mosley, Maamoon Saleh, and Katherine E. Olson

Abstract

Neurodegenerative movement disorders represent a growing global health challenge. The most notable of these include amyotrophic lateral sclerosis (ALS), Huntington's disease (HD), and the synucleinopathies encompassing Parkinson's disease (PD), dementia with Lewy bodies (DLB), and multiple system atrophy (MSA). Recent advances in immunotherapy have opened new avenues for intervention in these devastating conditions. Protein aggregation and misfolding underlie the pathophysiology of these disorders. Immunotherapeutic approaches, such as vaccine-based strategies, monoclonal antibodies, and therapies that inhibit aggregation or enhance clearance, are being developed to target disease-specific protein aggregates, mitigating their toxic effects and offering potential disease-modifying treatments. Chronic neuroinflammation is a common hallmark of these disorders. Immunomodulatory agents, such as monoclonal antibodies against proinflammatory cytokines, hold promise in ameliorating inflammation-induced neurodegeneration. By dampening excessive immune responses, these therapies may slow disease progression and enhance neuronal survival. Finally, the role of T cells reacting to modified proteins and increased inflammation has gained prominent recognition as mechanisms for disease exacerbation and progression. Therapies that modulate T cell responses and infiltration into the central nervous system are thus under investigation. Immune checkpoint inhibitors and T cell reprogramming techniques may hold potential for halting the progression of these disorders by regulating immune-mediated damage. This chapter explores the emerging landscape of immunotherapeutic strategies, with a focus on neuroinflammation modulation, T cell targeting, and proteinopathy interventions. Immunotherapy for these disorders represents a promising frontier, offering more effective disease management and improved quality of life. Herein, we examine the research strategies aimed at immunotherapy for better clinical outcomes of these disorders. In light of the super-plethoric number of preclinical studies that focus on immune-based therapeutics for movement disorders, this chapter will limit discussions to those approaches that have initiated clinical transition or provided evidence of a transformative shift in strategic direction. Moreover, further research and clinical trials are essential to validate these innovative approaches and move closer to a future where these disorders can be successfully treated or even prevented.

Keywords

Neurodegeneration · Movement disorders · Neuroinflammation · Immunomodulatory therapy · Adaptive immunity · Innate immunity · Proteinopathy therapy · Synucleinopathy · Parkinson's disease · Dementia with Lewy bodies · Multiple system atrophy · Amyotrophic lateral sclerosis · Huntington's disease

52.1 Introduction

Movement disorders represent a diverse group of neurological conditions that affect normal movement of the body. These disorders arise from various causes, including genetic factors, brain injuries, infections, medication side effects, and neurodegenerative processes. Abnormal involuntary movements that range from insufficient to excessive with levels from mild to severe, typically characterize these disorders, yet each significantly impacts everyday performance. Common movement disorders include essential tremor, dystonia, Tourette syndrome,

R. L. Mosley (✉) · M. Saleh · K. E. Olson
Department of Pharmacology and Experimental Neuroscience, Center for Neurodegenerative Disorders, University of Nebraska Medical Center, Omaha, NE, USA
e-mail: rlmosley@unmc.edu

H. E. Gendelman, T. Ikezu (eds.), *Neuroimmune Pharmacology and Therapeutics*,
https://doi.org/10.1007/978-3-031-68237-7_52

and restless leg syndrome, with a large group of ataxias and chorea disorders. While these are certainly significant disorders, immunological associations have not been established, thus immune-based therapeutic strategies may not be appropriate targets. However, for some neurodegenerative movement disorders, immunological associations for etiology or progression are well-established, and thus, therapeutic interventions by immune-based strategies are clearly warranted. Parkinson's disease (PD), dementia with Lewy bodies (DLB), multiple system atrophy (MSA), amyotrophic lateral sclerosis (ALS), and Huntington's disease (HD) are a group of neurodegenerative movement disorders with established or hypothesized immune associations for either etiology or disease progression. Characteristically, these disorders demonstrate common hallmarks of substantial neuroinflammatory processes that are distinct for each disorder, typically accompanied with gliosis. The idiosyncratic nature of those neuroinflammatory processes is due, at least in part, to another hallmark of misfolded protein species and spatial incidence that are distinct for each disorder. PD is included in a group of disorders termed synucleinopathies such as MSA, pure autonomic failure (PAF), and DLB; the latter presenting with dementia either before or after onset of movement disorder. These disorders commonly show accumulation of misfolded α-synuclein protein principally associated with the neurons of the substantia nigrostriatal axis (PD) or oligodendroglia (MSA) (Jellinger & Wenning, 2016; Trojanowski et al., 2007) in several brain regions, and neurons of the autonomic ganglia and peripheral nerves (PAF) (Marti et al., 2003). Although PAF exhibits typical postmortem histology for synucleinopathies, Lewy body formation is predominantly confined to the autonomic pathway including the autonomic ganglia and peripheral nerves, without involvement of motor function (Kabir & Chelimsky, 2019), and thus outlies this chapter's scope. Similarly, ALS and HD have protein accumulation and spatial incidence that are also disorder-specific and account for the particular phenotype associated with each disorder. Additionally, innate- and adaptive-mediated immunity weaves a common thread through these disorders. Another important commonality with these neurodegenerative movement disorders is the paucity of approved drugs that modify disease progression, leaving modalities that only target palliative resolution. This chapter is tasked with exploring new translational strategies for immunotherapeutic intervention to provide beneficial outcome for patients afflicted with these neurodegenerative motor diseases.

52.2 Synucleinopathies

52.2.1 Parkinson's Disease

Parkinson's disease (PD) is the second most common neurodegenerative disorder and the prototypical synucleinopathy. A pivotal disease hallmark is cytoplasmic accumulations of proteinaceous aggregates or Lewy bodies (LBs) comprised primarily of misfolded α-synuclein and ubiquitin (Spillantini et al., 1997). PD presents as tremor, slowed movement (bradykinesia and hypokinesia), rigidity, mask-like facial expressions (hypomimia), rigidity of limbs, neck, and trunk, and impaired posture and balance (Dauer & Przedborski, 2003; Kalia & Lang, 2015). These signs are primarily due to the loss of dopaminergic neurons that originate in the substantia nigra (SN) pars compacta and innervate to the striatum with consequent loss of the neurotransmitter dopamine (Capriotti & Terzakis, 2016; Kikuchi et al., 2017; Marogianni et al., 2020; Parent & Parent, 2010; Rose, 2016). The contribution of LBs to disease progression is thought to be substantially due to toxic fibrils of α-synuclein oligomers and includes disruption of pre-synaptic communications, oxidative stress, mitochondrial toxicity, and intraneuronal proteomic homeostasis (Wakabayashi et al., 2007). The release of α-synuclein from neurons and interaction with microglia activate microglia resulting in neuroinflammation in those afflicter regions of the brain, which represents another hallmark of PD. Neuroinflammation and oxidative stress fuels α-synuclein modifications, misfolding, and oligomerization, which promote LB generation. Infiltration of T cells and microgliosis support the production and secretion of proinflammatory cytokines and mediators of neurotoxicity, thus these reactive T cells and microglia become proximal sources for protein modification and dopaminergic neuronal injury which perpetuate greater release of pathogenic forms of α-synuclein (Gao et al., 2011; Gordon et al., 2018; Rokad et al., 2017; Zhang et al., 2005).

52.2.1.1 Current Pharmaceutical Therapeutics

Pharmacological modalities for PD are designed to increase dopamine availability or dopaminergic signaling. The former strategy replaces dopamine with *levadopa (L-DOPA)* as a dopamine precursor and suppresses dopamine degradation with monoamine oxidase B (MAO-B) inhibitors such as *selegiline* and *rasagiline*, with catechol-O-methyl transferase (COMT) inhibitors such as *entacpone* and *tolcapone*, with or without aromatic L-amino acid decarboxylase (AAAD) inhibitors such as *carbidopa* or *benserazide* (Ferrazzoli et al., 2016). The latter strategy utilizes dopamine receptor agonists such as *pramipexole* and *ropinirole* to enhance post-synaptic dopamine receptor activation. These approaches serve to affect the dopaminergic circuits responsible for movement coordination (Armstrong & Okun, 2020; Hayes, 2019; Oertel & Schulz, 2016). Alternatively, stimulation of compromised neuronal networks can be accomplished through surgically implanted probes by deep brain stimulation (DBS) (Ferrazzoli et al., 2016; Rokad et al., 2017). In essence, current clinical approaches to treatment of PD provide symptomatic relief, but otherwise are not specifically designed to target modifications of pathological disease progression (Armstrong & Okun, 2020; Paolini Paoletti et al., 2020).

52.2.1.2 Therapeutics Targeting Neuroinflammation

Evidence of chronic neuroinflammation in the brains of PD patients is consistently observed and marked by activated microglia and infiltrating T cells in the substantia nigra and striatum (Hirsch & Hunot, 2009; Hirsch et al., 2012; Lee et al., 2019; Wang et al., 2015b). Multiple studies show that misfolded α-synuclein aggregates in PD, activate microglia and trigger a neuroinflammatory cascade that leads to dopaminergic cell death (Reynolds et al., 2008; Saleh et al., 2022; Schwab et al., 2020; Wang et al., 2023). This inflammation is characterized by increased proinflammatory cytokines like tumor necrosis factor-alpha (TNF-α), interleukin (IL)-1β, and IL-6, as well as activated transcription factors such as nuclear factor kappa B (NF-κB) (Wang et al., 2022). Thus, targeting neuroinflammation represents an attractive therapeutic strategy. This strategy was underscored in a large cohort study of healthcare workers that regularly administered nonsteroidal anti-inflammatory drugs (*NSAIDs*), specifically *ibuprofen*, and showed diminished risk of developing PD (Chen et al., 2003, 2005). However, meta-analyses of those and other studies have not confirmed those associations (Poly et al., 2019; Ren et al., 2018). A population-based case–control study of 48,295 PD patients found the use of *corticosteroids* and *inosine monophosphate dehydrogenase inhibitors* was associated with a reduced risk of PD compared to controls (Racette et al., 2018). Another study of inflammatory bowel disease (IBD) patients showed that early exposure to *anti-TNF-α* drugs also reduced the risk of developing PD in that population (Peter et al., 2018). These studies and many preclinical studies facilitated the notion that regulatory control may delay PD progression. Of primary interest are strategies that target microglia and effector T cells (Teffs) that drive inflammatory responses. Post-mortem PD patients have shown evidence of activated microglia that proximate injured or dying neurons (Kam et al., 2020; McGeer et al., 1988). With substantial evidence supporting the anti-inflammatory and neuroprotective capability of *minocycline* (Cankaya et al., 2019; Matthews et al., 1999), a Phase 2 clinical trial in PD (DATATOP) failed to show beneficial outcomes in patients (N. N.-P. Investigators, 2008).

Other possible approaches to affect inflammatory responses include glucagon-like peptide 1 receptor (GLP-1R) agonists, which have been shown to have beneficial effects in several processes associated with neurodegenerative diseases, such as increased endoplasmic reticulum stress responses and oxidative stress, diminished autophagy, and chronic inflammation. However, no single mechanism to explain those disease-modifying effects have been provided, though as most agonists seem to affect proinflammatory myeloid and lymphocytic populations (Monti et al., 2022), those agents herein will be discussed in the context of neuroinflammation targets. *Pioglitazone* and *semaglutide* that target the peroxisome proliferator-activated receptor-gamma (PPARγ), a nuclear receptor, were shown to interact with NF-κB and suppress NF-κB-mediated proinflammatory responses, attenuate microglia, and protect dopaminergic neurons in PD models (Breidert et al., 2002; Dehmer et al., 2004; Schintu et al., 2009; Swanson et al., 2011). However, treatment of PD patients with *pioglitazone* in a Phase 2 trial (NCT01280123) did not show clinical improvements (N. E. T. i. P. D. F.-Z. Investigators, 2015). Another Phase 2 clinical trial in PD for *exenatide*, another GLP-1R agonist (NCT01971242) showed improved clinical scores in treated group that were sustained after discontinuance while scores worsened in the placebo controls (Athauda et al., 2017). A possible reason for these discrepancies is postulated to relate to differential sensitivities of preclinical and clinical outcome measures, heterogeneity of microglia subtypes, and/or differences in the patient populations (Thrupp et al., 2020). A Phase 3 trial for exenatide in PD (NCT04232969) is currently ongoing. Also, a Phase 2 trial in PD is ongoing for *NLY01*, an exenatide derivative (NCT04154072). Other clinical trials for GLP-1R agonists in PD include *semaglutide* (Phase 2, NCT03659682), *liraglutide* (Phase 2, NCT02953665), and *lixisenatide* (Phase 2, NCT03439943). Other strategies utilize anti-inflammatory cells to regulate neuroinflammation and are discussed in the below section.

52.2.1.3 Therapeutics Targeting T Cells

Compelling evidence exists that demonstrate T cells, especially proinflammatory T cells, play a key role in driving neuroinflammation and neurodegeneration in PD (Olson et al., 2023b; Reynolds et al., 2010; Saleh et al., 2022). Multiple studies have found increased numbers of activated T cells in the blood and brains of PD patients, especially IL-17-producing T helper (Th17) cells that are known to promote chronic inflammation (Bolte & Lukens, 2018; Chen et al., 2018; Li et al., 2023; Prots & Winner, 2019; Saunders et al., 2012). The simplest strategy would inhibit proliferation of proinflammatory Teffs and microglia with immunosuppressive agents. *Azathioprine* inhibits synthesis of purine nucleotides and is indicated to inhibit graft rejection in transplantation (Gassmann & van Furth, 1975; Caspritz & Hadden, 1987; Maltzman & Koretzky, 2003; Moeslinger et al., 2006). Thus, a Phase 2 clinical trial with escalating doses of azathioprine in PD patients (EudraCT/CTIS 2018-003089-14;ISRCTN14616801) was initiated and is scheduled for completion in 2024 (Greenland, Cutting et al., 2020). *Sirolimus (rapamycin)* is an inhibitor of mammalian target of rapamycin (mTOR) and is an immunosuppressant initially indicated for kidney transplantation, but has been repurposed for several neurological disorders (Lui et al., 2022). Preclinical studies in PD models have shown sirolimus increases autophagy, protects dopaminergic neurons, and improves motor deficits (Pupyshev et al., 2021; Zhang et al., 2021). An ongoing Phase 1/2 clinical trial has included sirolimus for PD patients with at least one mutation in *GBA1* gene (see below) and is expected to conclude in 2029 (NCT04127578).

As regulatory T cells (Tregs) appear dysfunctional in PD patients, numbers and function of proinflammatory Teffs increase in direct correlation with disease progression (Saunders et al., 2012). Thus, another strategy would restore the functional balance between Tregs and proinflammatory Teffs as a promising approach. Indeed in preclinical PD models, adoptive transfer of Tregs attenuates neuroinflammation, provides virtually complete dopaminergic neuroprotection, and improves motor deficits (Hutter-Saunders et al., 2012; Reynolds et al., 2007, 2010; Saleh et al., 2022). In animal models of PD, various Treg inducers are able to increase Treg number and/or function with neuroprotective capacity, and include *glatiramer acetate* (Benner et al., 2004; Cui et al., 2009), *vasoactive intestinal peptide (VIP)* (Reynolds et al., 2010), *VIPR2 agonists* (Olson et al., 2015, 2016), *bee venom phospholipase A2* (Baek et al., 2018; Chung et al., 2015), *IL-2* (Markovic et al., 2022; Thome et al., 2021), and *granulocyte/macrophage colony-stimulating factor (GM-CSF)* (Kosloski et al., 2013). As bee venom phospholipase A2 directly induces Tregs (Baek et al., 2020), a Phase 2 clinical trial in PD patients using desensitizing doses of *bee venom* (NCT01341431) showed safety and tolerability, but did not change disease progression or outcomes; however, desensitizing dosage regimens may not have been sufficient to influence disease outcomes. Several of these inducers have direct Treg effects, however, for GM-CSF, Tregs express little to no receptors for GM-CSF, and thus GM-CSF appears to operate through tolerogenic myeloid dendritic cell intermediaries (Bhattacharya et al., 2011; Schutt et al., 2018). The decline of dendritic cell frequencies that correspond to motor symptom severity, may explain diminished Treg function in PD patients (Ciaramella et al., 2013; Saunders et al., 2012). Thus, a Phase 1 clinical trial in PD using *recombinant human GM-CSF (sargramostim)* (NCT01882010), a myeloid cell growth factor, showed that sargramostim was safe, increased Treg number and function, improved clinical scores of motor function, and increased electrical output in neurons of cortical motor areas (Gendelman et al., 2017). A 36-month, open-label Phase 1b trial of sargramostim with a reduced dosage regimen (NCT03790670), reduced adverse events, and affirmed that long-term sargramostim maintained increased Treg number and function as well as reduced motor function scores (Olson et al., 2021, 2023a). Moreover, transcriptomic and proteomic analysis of monocytes from sargramostim-treated patients revealed biomarkers that correlated with clinical motor function scores and yielded neuroprotective-enriched pathway signatures that were antioxidant, anti-inflammatory, and pro-autophagy (Abdelmoaty et al., 2022). Another clinical trial (NCT05677633) is ongoing to validate those findings in a larger cohort.

52.2.1.4 Therapeutics Targeting Proteinopathy

Some approaches focus on clearing pathogenic α-synuclein, using direct strategies such as vaccines, antibodies, or gene therapies to remove or limit toxic oligomer misfolding (Saleh et al., 2022; Schwab et al., 2020). One strategy is to clear oligomeric and aggregated α-synuclein by active or passive immunization with anti-α-synuclein antibodies. Active vaccination using α-synuclein or α-synuclein mimetics target α-synuclein from the brain. Two leading α-synuclein mimetic vaccines, *PD01A* and *PD03A* are composed of short peptides of eight and ten amino acids, respectively, and were selected from a library of synthetic peptides that recognize antibodies directed against α-synuclein oligomers (Mandler et al., 2015; Meissner et al., 2020). Preclinical studies established those vaccines yielded tangible neuroprotective benefits with decreased accumulation of α-synuclein (Volc et al., 2020). In clinical trials (NCT01568099 and NCT02267434), both vaccine candidates were safe and tolerable, and increased antibody titers to α-synuclein though PD01A yielded more responders and higher titers than those receiving PD03A. Another immunogen, *ACI-7014.056*, a synthetic multimeric configuration of α-synuclein epitopes, is currently in Phase 2 clinical trials (NCT06015841). Another vaccine candidate is *UB-312,* a synthetic ten amino acid peptide from the C-terminus of α-synuclein that was selected from B cell-specific α-synuclein epitopes showed increased immunoreactivity preclinical models (Nimmo et al., 2020). In an α-synuclein transgenic model of PD, UB-312 immunization-induced anti-α-synuclein antibodies, reduced α-syn oligomers, attenuated reactive microglia, and prevented development of motor impairments (Nimmo et al., 2020). From a Phase 1 clinical trial in early PD (NCT04075318), UB-312 was described to be safe and tolerable, and produced anti-α-synuclein antibodies, but results have not been posted. A Phase 1/2 trial of UB-312 in PD and MSA patients (NCT05634876) is currently ongoing.

Another approach utilizes monoclonal antibodies (mAbs) to enhance clearance of aggregated α-synuclein. *BIIB054 (cinpanemab)* is an IgG1 anti-synuclein mAb derived from memory B cells of elderly individuals with CNS diseases and recognizes aa1-10 of the N-terminus with an 800-fold higher affinity for α-synuclein (Weihofen et al., 2019). Moreover, in preclinical studies, BIIB054 inhibited α-synuclein spread and aggregation, diminished loss of striatal dopaminergic termini, and rescued motor impairments. In a Phase I trial with healthy subjects and PD patients (NCT02459886), BIIB054 displayed a half-life of 28-35 days and virtually all bioavailable plasma BIIB054 was saturated with α-synuclein (Brys et al., 2019). Two Phase 2 trials (NCT03318523 and NCT03716570) were terminated due to failure to meet endpoints to change baseline clinical scores (Lang et al., 2022). *ABVV-0805* is a mAb that preferentially recognizes aggregated α-synuclein and reduced aggregates in several mouse models of PD in a dose-dependent manner (Nordstrom et al., 2021). However, a phase 1 trial was withdrawn before subject enrollment due to "strategic considerations" (NCT04127695). *PRX002/RG7935 (PRX002, prasinezumab)* is a humanized IgG1 mAb that recognizes the C-terminus of α-synuclein and targets aggregated forms, thus reducing α-synuclein propagation, neuropathology, and behavior deficits in a PD models (Games et al., 2013, 2014). A Phase 1 trial in healthy subjects (NCT02095171) and PD patients (NCT02157714) found PX002 safe and tolerable, and showed vigorous binding of peripheral α-synuclein with dose-dependent increases of PX002 mAb levels in cerebrospinal fluid (Jankovic et al., 2018). In a Phase 2 trial of early-stage PD patients (NCT03100149), PRX002 did not lessen progression of motor and non-motor symptoms during 52 weeks of intravenous administration compared to placebo treatment, but an early start group showed diminished UPDRS, part III scores (Pagano et al., 2021). Moreover, efficacy in secondary and exploratory measures were demonstrated in the 52 week open-label extension. Based on those findings, a Phase 2b clinical trial of prasinezumab in early PD patients (NCT04777331) is currently ongoing. *MEDI1341* is a human mAb selected for high affinity to the C-terminus of α-synuclein for epitopes in the aa103-129 region and binds monomeric and aggregated α-synuclein. In preclinical models, MEDI1341 enters the brain, sequesters aggregated α-synuclein, and blocks cell-to-cell spreading of α-synuclein (Schofield et al., 2019). Two Phase 1 clinical trial of in health subjects (NCT03272165) and PD patients (NCT04449484) showed that MEDI1341 was safe and tolerable, exhibited dose-dependent serum concentrations, penetrated the CNS as indicated by concentrations in CSF, and diminished α-synuclein in CSF in a dose-dependent fashion (Shering et al., 2023). *LU AF82422* is a humanized IgG1 mAb to the C-terminus of α-synuclein that recognizes monomeric and oligomeric species (Fjord-Larsen et al., 2021). In that study, LU AF82422 treatment of rat and NHP models showed engagement of α-synuclein, but did not bind most blood cells other than a subset of monocytes, which were not activated by engagement. A Phase 1 trial is ongoing in healthy subjects and PD patients (NCT03611569).

An alternative immunotherapeutic strategy utilizes nanobodies to destabilize or inhibit intracellular α-synuclein aggregation and accumulation of toxic fibrils. Nanobodies are proteins containing only the antigen-binding portions of the light (L) and/or heavy (H) chains of the antibody variable region (V), such as the fragment antigen binding (Fab) domains, single chain antigen binding (V_L + V_H) (scFv) domain, or single-domain Ab (V_HH) domain. Therefore, these immunoglobulin-derived domains bind the antigenic epitope to which the antibody was directed. Typically, these proteins are obtained from high affinity antibodies, since efficacy depends on only the small antigen-binding portions of the antibody. Two nanobodies from camelid antibodies, *NbSyn2* and *NbSyn87*, show high degrees of affinity towards distinct epitopes of the α-synuclein C terminus, aa118-131 and aa137-140, respectively (Chatterjee et al., 2018; Guilliams et al., 2013; Iljina et al., 2017). Both nanobodies selectively and efficiently bind and inhibit α-synuclein fibrillation, but their affinities vary significantly depending on the length of time during α-synuclein fibrillation and the state of the fibril. Two scFv nanobodies (*NAC32* and *VH14*) recognize aa53-87 of the α-synuclein non-amyloid component (NAC) region and the coding sequences were incorporated into adeno-associated virus (AAV) constructs to be used as intracellularly expressed nanobodies, termed intrabodies (Lynch et al., 2008). Co-transfection of each construct with AAV-α-synuclein overexpression constructs into HEK293 and SH-SY5Y cells demonstrated that NAC32, but not VH14 intrabodies, downregulated α-synuclein protein, but had no effect on α-synuclein mRNA (Chen et al., 2020b). In α-synuclein overexpression animal models, *AAV-NAC32* reduced α-synuclein expression in the substantia nigra and striatum, spared dopaminergic termini in the striatum, and improved locomotor measurements (Chen et al., 2021). Interestingly, the efficacy of VH14 construct could be improved with the addition of constructs encoding polypeptides rich in proline, aspartate/glutamate, serine, and threonine (PEST) residues designed to enhance proteasomal degradation (Chatterjee et al., 2018). While clinical trials have yet to include anti-α-synuclein intrabody constructs, many nanobody proof-of-principle trials have been evaluated targeting other proteins, including those with specificities to IL-17A/F (bimekizumab (ALX-0761/MSB0010841)), NCT02156466) (Chiricozzi et al., 2019); IL-17A/IL-17F/human serum albumin (sonelokimab (M1095), NCT03384745) (Papp et al., 2021; Svecova et al., 2019); IL-6 receptor (vobarilizumab (ALX-0061), NCT01284569) (Dörner et al., 2017); CXCR4 (ALX-0651, NCT01374503);

and respiratory syncytial virus (gontivimab (ALX-0171), NCT01483911) (Cunningham et al., 2021).

Another broad strategy targets posttranslational modifications of α-synuclein as those can influence misfolding, aggregation, and toxicity; however, no clear consensus has been reached as to which modification(s) are necessary to transmute modified proteins to pathogenicity (Canever et al., 2023). While much attention has focused on the influence of phosphorylated S129 to promote aggregation and toxicity, substitution of serine with other amino acids suggests that pS129 is not necessary for misfolding or pathogenicity, and that phosphorylation of other moieties such as tyrosine may better serve as target candidates (Kawahata et al., 2022). Therefore, one strategic rationale is to inhibit phosphorylation of α-synuclein tyrosine residues aimed to inhibit or suppress its aggregation. One kinase, Abelson tyrosine kinase (c-Abl), serves as a sensor of cellular stresses and abnormalities in neurons, which when engaged also initiates pathways leading to neurodegeneration (Lindholm et al., 2016). c-Abl is elevated in the nigrostriatum of PD patients and promotes protein fibrilization (Fowler et al., 2019). Deletion of c-Abl prevents α-synuclein fibrilization, ROS-induced cell death, and protects dopaminergic neurons in PD models (Alvarez et al., 2004; Ko et al., 2010). Moreover, tyrosine kinase inhibitors have been shown to induce autophagy, kill proliferating cells, degrade aggregated α-synuclein and amyloid, diminish inflammation, increase dopamine levels, and improve motor outcomes in preclinical studies (Hebron et al., 2013a, b; Lonskaya et al., 2013b, 2014; Salomoni & Calabretta, 2009). Thus, inhibition of c-Abl represents a plausible target for disease-modifying therapeutics in PD (Werner & Olanow, 2022). *Nilotinib* is a c-Abl tyrosine kinase inhibitor indicated for chronic myeloid leukemia, and has been shown in animal models of PD to prevent dopaminergic neuronal death and impaired motor function (Hebron et al., 2013a, 2014; Karuppagounder et al., 2023; Lonskaya et al., 2013b, 2014; Salomoni & Calabretta, 2009; Werner & Olanow, 2022). A Phase 2 clinical trial of oral nilotinib in PD patients (NCT02954978) showed the drug to be safe and tolerable, reduced oligomerized α-synuclein, and increased dopamine metabolites in the CSF with reduced levels of hyperphosphorylated tau (Pagan et al., 2020). A separate Phase 2 trial of the same doses of nilotinib in PD (NCT03205488) also showed acceptable safety profile, however, nilotinib-treated patients exhibited significantly worse motor function scores than placebo group, and exhibited no treatment alteration of dopamine metabolites in CSF (Simuni et al., 2021). The latter study suggested that the lack of robust or reproducible effects of nilotinib on disease processes limits its further utilization in PD. *Bosutinib* is a dual inhibitor of c-Abl and c-Sarcoma (Src) tyrosine kinases and is approved to treat chronic myelogenous leukemia (CML) (Musumeci et al., 2012). In animal models of overexpressed α-human mutant A53T α-synuclein, bosutinib reduced levels of α-synuclein, facilitated dopaminergic neuron survival, and modulated immune responses to α-synuclein differentially in peripheral and central immune compartments (Hebron et al., 2014; Lonskaya et al., 2013a). Interestingly, while bosutinib has been assessed in other neurodegenerative diseases and synucleinopathies, clinical trials in PD have yet to be initiated. *IkT-148009*, is an *in silico*-designed c-Abl inhibitor, which in animal models, penetrated the brain, suppressed c-Abl activation, protected dopaminergic neurons from degeneration, and concomitantly reduced α-synuclein pathology with recovery of motor function (Karuppagounder et al., 2023). A Phase 1 clinical trial of IKT-148009 in healthy elderly subjects (NCT04350177) has been completed, but results have yet to be posted. A Phase 2 placebo-controlled trial to test the effects of IkT-148009 in untreated PD patients is ongoing (NCT05424276).

Another strategy targets the lysosomal enzyme glucocerebrosidase (GCase) encoded by the *GBA1* gene, for which loss of function mutations lead to neuropathic and non-neuropathic lysosomal storage disorders. More importantly, *GBA1* mutations pose one of the major risk factors for developing synucleinopathies such as PD and DLB by impairing autophagy with permissive accumulation of α-synuclein (Kuo et al., 2022; Lwin et al., 2004; Sidransky & Lopez, 2012; Velayati et al., 2010). Thus, agents that affect GCase and increase autophagy may prove beneficial in PD patients. *Ambroxol (ABX)* is a secretolytic agent indicated for relief of excess mucus in respiratory disorders, but also acts as an inhibitory chaperone that binds the GCase active site, mobilizes sequestered GCase from the endoplasmic reticulum, and induces conformational changes that facilitate lysosomal trafficking and degradation in the acidic lysosome (Bendikov-Bar et al., 2013; Luan et al., 2013). In animal models of synucleinopathy, ABX treatment-induced brain GCase activity, enhanced autophagy-lysosomal degradation, and decreased total and phosphorylated α-synuclein and aggregation (Migdalska-Richards et al., 2016; Velayati et al., 2010). A Phase 2 open-label trial of ABX in PD patients with or without *GBA1* mutation (NCT02941822) showed the drug was well-tolerated with detectable levels of AMX in the CSF, diminished GCase activity, improved motor function scores, but increased α-synuclein levels in CSF (Mullin et al., 2020). A Phase 2 clinical trial (NCT05287503) and a Phase 2/3 trial (NCT05830396) of ABX, both placebo-controlled in PD patients with *GBA1* mutations are currently ongoing. A Phase 3 placebo-controlled clinical trial has been initiated in PD patients with or without *GBA1* mutations (NCT05778617, CCTU/2020/365). Additionally, a Phase 2 trial of ABX in PD patients with dementia (NCT02914366) is currently ongoing (Silveira et al., 2019).

Another possible strategy targets the interaction of Rit2/Rin and b-Raf. The association between the PD risk gene

RIT2 and *BRAF* gene was discovered from an *in silico* screening of databases of meta-GWAS-derived PD risk factors and protein–protein interactions, which with a database of FDA-approved drugs included in the screen, matched *dabrafenib* as a possible candidate PD drug (Uenaka et al., 2018). *RIT2* encodes for Rit2/Rin, a GTPase that is expressed primarily in the brain, enriched in the substantia nigra (SN) pars compacta, and reduced in the SN of PD patients (Shi et al., 2005). Moreover, Rit2 modulates lysosomal and DAT functions but also is expressed by immune cells. Diminished Rit2 activity in dopaminergic neurons is sufficient to drive neuronal loss with phosphorylated α-synuclein and motor deficits (Kearney et al., 2023), while Rit2 overexpression restores autophagy-lysosomal pathways and protects against α-synuclein pathology (Obergasteiger et al., 2023). *BRAF* encodes for b-Raf, a member of the Raf serine/threonine kinase family that is involved in MEK/ERK pathways, and is structurally similar to the kinase domain of LRRK2 (Ray & Liu, 2012). The rationale for this strategy is that Rit2 activates b-Raf, which phosphorylates and activates MEK, while MEK phosphorylates and activates ERK that regulates differentiation, growth, proliferation, and survival in developing neuronal tissues. Overexpression of activated b-Raf by microglia drives proliferation with increased production of proinflammatory cytokines, aggregated α-synuclein, and killing of proximal dopaminergic neurons (Galabova-Kovacs et al., 2008; Ye et al., 2023). *Dabrafenib* is an inhibitor of b-Raf and is approved for treatment of aggressive BRAF-mutated myelomas, which also express high levels of α-synuclein. In preclinical in vitro and in vivo models of PD, dabrafenib enhanced lysosomal-autophagy pathways and protected dopaminergic neuronal loss, but had no effect on overall α-synuclein expression; however, the status of α-synuclein modifications or pathogenic species were not determined (Uenaka et al., 2018). Although no clinical trials have yet to be initiated to test this approach, its mode of discovery, utility, and potential warrants discussion.

52.2.2 Dementia with Lewy Bodies

Dementia with Lewy bodies (DLB) is the second most common form of dementia after Alzheimer's disease. Also similar to PD is the underlying hallmark of abnormal protein accumulations consisting primarily of α-synuclein that form Lewy bodies and Lewy neurites (Walker et al., 2015). In DLB, loss of dopamine-producing neurons in the substantia nigra similar to PD is operative; however, these losses accompany deficits in other neurotransmitter systems, such as acetylcholine, which contributes to cognitive and memory impairments. With progression, neuronal loss extends to other regions including the cortex and limbic system, which is thought to contribute to cognitive decline (McKeith et al., 2017). As expected, innate and adaptive immunity are implicated in the progression of DLB. Similar to other synucleinopathies, neuroinflammation is revealed by activated microglia, due in part to aberrant protein misfolding and tissue injury, and is a prominent hallmark in DLB. Patients with DLB have been shown to have higher numbers of activated microglia than control subjects, and the number of activated microglia correlates with LB burden in different gray and white matter regions (Amin et al., 2022; Mackenzie, 2000; Nicastro et al., 2020; Surendranathan et al., 2015). However, one report that failed to find significant increases in microglia in DLB (Amin et al., 2020), suggested that the role of inflammation has not definitely been established (Amin et al., 2022). Evidence of increased levels of activated microglia was found by PK11195 PET scans early in mild DLB, but levels decreased with more severe disease and cognitive decline suggesting a more important role of microglia during onset or early disease progression (Surendranathan et al., 2018). Moreover, upregulated proinflammatory cytokines in peripheral blood included IL-2, IL-17a, and MIP-3a, with diminished levels of IL-8. This is in light of overexpression of IL-1α, TNF-α, and iNOS by activated glia of the amygdala, hippocampus, entorhinal, and insular cortices of DLB brains with diminished expression of nNOS and increased Lewy body pathology (Katsuse et al., 2003). Interestingly compared to controls, the percentage of DLB patients that exhibited CD3+ T cells in parenchyma of gray and white matter were increased compared to those that showed perivascular CD3+ T cells in those tissues (Amin et al., 2020). Increased numbers of parenchymal CD3+T cells are found adjacent to α-synuclein deposits in dopaminergic neurons and neurites with increased expression of *CXCR4* and *CD69* genes in the majority of CD4+ T cells (Gate et al., 2021).

52.2.2.1 Current Pharmaceutical Therapeutic

Current pharmacological therapeutics used to treat parkinsonism are less effective in DLB and have been associated with exacerbation of psychosis, but *L-DOPA* is used at lower doses with ascending titration below psychiatric thresholds (Goldman et al., 2008). For cognitive and psychoses symptoms, drugs used for Alzheimer's disease are found to effective (Hershey & Coleman-Jackson, 2019). Cholinesterase inhibitors such as *donepezil*, *rivastigmine*, and *galantamine* effectively reduce delusional severity and hallucinations; however, the effectiveness of *memantine* is less certain and may be only beneficial in mild to moderate DLB (Wang et al., 2015a). Cautious use of antipsychotics is warranted due to acceleration of worsened disease progression and must be initiated at lower doses (Henriksen et al., 2006).

52.2.2.2 Therapeutics Targeting Neuroinflammation

Substantial evidence of inflammation in DLB implicates increased inflammatory cellularity and proinflammatory mediators in blood, CSF, vital imaging, and post-mortem brain tissues from patients compared to nonafflicted controls (Amin et al., 2022). However, meta-analysis of over 80 studies found variable results in prodromal and early DLB indicating signs of inflammation, whereas late-stage disease and postmortem evaluations suggest a more altered or varied state of inflammatory profiles (Loveland et al., 2023). Several potential therapeutics such as *candesartan*, *nilotinib*, *lenalinomide*, and *etanercept* have exhibited potential anti-inflammatory benefits in preclinical studies in DLB models, but have yet to transition to clinical testing (Pope et al., 2021). One strategy is to intervene in the proinflammatory signaling pathways of microglia. *Neflamapimod (VX-745)* is a small molecule that selectively inhibits the p38α isoform of mitogen-activated serine/threonine protein kinase (p38α MAPK), which is expressed in neurons and microglia. p38α MAPK expression in microglia induces production and release of proinflammatory cytokines such as TNF-α and IL-1β. In studies of aged animals with Rab5-dependent lysosomal pathology, VX-745 inhibited microglial produced IL-1β and TNF-α, reduced lysosomal degradation, and improved behavioral deficits (Alam, 2015; Jiang et al., 2022). Based on these and several other preclinical studies, a Phase 2 randomized controlled trial of neflamapimod in patients with DLB (NCT04001517) showed significant improvements in a custom battery of six cognitive tests over 16 weeks of treatment compared with placebo, and a greater effect was found in DLB patients without evidence of AD co-pathology (Jiang et al., 2022). A Phase 2b clinical trial to test the effects of neflamapimod on cognition compared to placebo control is ongoing (NCT05869669). Nevertheless, the overlapping pathologies exhibited in DLB that are due to extenuating effects of several protein aggregates of α-synuclein, Aβ, phosphorylated tau (ptau), and TDP-43 render many treatment strategies complicated and present a primary impediment to clinical studies.

52.2.2.3 Therapeutics Targeting T Cells

As with the involvement of T cells in neuroinflammation, studies have implicated the possible role of T cells in DLB (Amin et al., 2023; Gate et al., 2021; Iba et al., 2020). Post-mortem DLB brains show increased numbers of CD3+CD4+ T cells in proximity to blood vessels and to α-synuclein aggregates in both nigral dopaminergic neurons and hippocampal glutamatergic neurons and next to Iba1+ cells of myeloid lineage. Single-cell analyses of CSF CD4+ T cells compared to those of healthy controls showed enhanced expression of *JAK1*, *CD69*, and *CXCR4* genes, indicating those T cells were in a state of greater activation. Moreover, the role of CD4+ Th17 Teffs was implicated by increased gene expression of killer cell lectin-like receptor subfamily B, member 1 (*KLRB1*), a CD161 marker of proinflammatory Th17 memory T cells (Gate et al., 2021; Shi et al., 2022). In 2 case studies of probable DLB, immunotherapy with *methylprednisolone* using an initial intravenous high dose, then standard oral administration over 3 years, remarkably improved neuropsychiatric and motor dysfunction (Abe & Chiba, 2019), whereas in the other case, intravenous *methylprednisolone* followed by intravenous immunoglobulin *(IVIG)* therapy yielded partial recovery after 4 months (Coban et al., 2014). Despite consistent evidence for the role of T cells in DLB, therapeutic strategies that target those T cells are lacking. Many of the anti-inflammatory therapies that are now of focus in PD and AD, should be brought to bear in DLB as the triple proteinopathy from α-synuclein, amyloid, and tau aggregation, complicates individual protein targeting strategies. On the other hand, as a common hallmark regardless of proteinopathy type, T cell-driven neuroinflammation could represent a reasonable interventional therapeutic approach in DLB via Treg or Treg-like immunomodulators such as *sargramostim* or low-dose *IL-2*.

52.2.2.4 Therapeutics Targeting Proteinopathies

In DLB, α-synuclein is only one of many aberrantly folded and aggregated proteins. Thus, enhancing the clearance of α-synuclein aggregates that lead to Lewy body formation presents a plausible initial approach. In mouse models of over expression of human α-synuclein reminiscent of LB disease, active and passive immunization with human α-synuclein, decreased aggregated α-synuclein in neurons and synapses, thus reducing neurodegeneration (Masliah et al., 2005). Of particular interest to DLB would be those strategies using active immunization with α-synuclein or their mimetopes, and passive immunization using any of a number of the anti-α-synuclein mAbs under investigation for PD (see above section on PD). However, protocols using α-synuclein epitopes or anti-α-synuclein antibodies have yet to be tested in DLB clinical trials. In contrast, more numerous trials exist for protocols that strategize processes affecting pan-protein targets rather than those specific for α-synuclein. Similar to its action in PD, *ambroxol (ABX)* not only increases GCase activity but also enhances autophagy-lysosomal degradation pathways. Thus, two clinical trials, a Phase 1/2 (NCT04405596) and a Phase 2a (NCT04588285) are currently ongoing to assess the safety and efficacy of ambroxol in DLB patients (Chwiszczuk et al., 2023). This more global protein approach has favored those processes that are hypothesized to enhance protein degradation or autophagy. A related strategy targets patients with mutations of the gene *GBA1* that encodes for GCase. *LY3884961 (PR001)* is a gene replacement therapy to deliver a functional

GBA1 gene in an AAV9 construct into the CSF of the cisterna magna at the base of the brain. In mouse models of GCase deficiency, intraventricular injection of LY3884961 restored GCase activity, lowered the accumulation of insoluble α-synuclein, and improved motor function (Abeliovich et al., 2021). A Phase 1/2 clinical trial in PD patients with at least one mutation in *GBA1* will test safety and efficacy of intracisterna magna administration of LY3884961 with planned completion in 2029 (NCT04127578). Initial adverse events related to immune responses to AAV9 warranted the inclusion of the immunosuppressants prednisone and sirolimus to the protocol.

Bosutinib is a dual inhibitor of Abl/Src tyrosine kinases and approved to treat chronic myelogenous leukemia (CML) (Musumeci et al., 2012). Bosutinib has been shown in preclinical studies of DLB to clear misfolded α-synuclein, ptau, and Aβ via autophagy, attenuate inflammation, and improve motor function (Pagan et al., 2022). A Phase 2 trial in DLB patients (NCT03888222) showed that bosutinib was well-tolerated, reduced α-synuclein levels in CSF, attenuated dopamine catabolism, and improved daily living activities (Pagan et al., 2022). *Nilotinib* is an inhibitor of break point cluster (BCR)-Abl (Abelson) tyrosine kinase and also is indicated for CML. Treatment with nilotinib has reduced levels of α-synuclein aggregates in preclinical studies (Skylar-Scott & Sha, 2023). In a small Phase 1 trial in advanced PDD and DLB patients (NCT02281474), nilotinib was well tolerated with improved clinical scores throughout drug exposure which were reversed after discontinuation follow-up, thus a Phase 2 trial has been initiated in DLB patients (NCT04002674). *Vodobatinib (K0706)* is a third-generation c-Abl trysine kinase inhibitor developed as an anti-neoplastic, but has been repurposed as putative disease-modifying therapy (DMT) for PD and DBL given the rationale that inhibition of this kinase will diminish phosphorylated α-synuclein that leads toxic fibril formation and dopaminergic neuronal loss. A Phase 2 clinical trial of K0706 in DLB patients (NCT03996460) was initiated in 2019 and is currently ongoing.

52.2.3 Multiple System Atrophy

Multiple system atrophy (MSA) is a progressive neurodegenerative disorder with symptoms affecting the nervous system that controls movement and the autonomic nervous system that controls involuntary functions. Hallmarks include α-synuclein inclusions found in oligodendrocytes known as glial cytoplasmic inclusions (GCIs) and in neurons (Jellinger, 2014, 2018). These are found to accompany neuronal loss and gliosis in the inferior olives, pontine nuclei, cerebellar Purkinje cells, putamen, caudate nucleus, globus pallidus, substantia nigra, locus coeruleus, autonomic nuclei of brainstem, intermediolateral cell columns of the spinal cord, and Onuf's nucleus. Major clinical variants of MSA include one phenotype involving the nigrostriatal pathway and motor functions that manifest similar to PD (MSA-P) and another involving cerebellar regions (MSA-C) leading to olivopontocerebellar (OPC) demyelination and atrophy (Ishizawa et al., 2008). Interestingly, α-synuclein from GCIs has been shown to have greater seeding activity than α-synuclein from PD and DLB patients (Peng et al., 2018; Woerman et al., 2015), and show a heterogeneous range of seeding capacities in cell-free assays (Martinez-Valbuena et al., 2022). A prominent immune feature accompanying MSA synucleinopathy is microglial activation in degenerating regions that include white matter tracts enervating cerebellum and extrapyramidal and gray matter of the nigrostriatal axis as determined histologically (Ishizawa et al., 2008) and by PET imaging (Gerhard et al., 2006; Kubler et al., 2019). Whether greater seeding activity translates to a greater degree of microglial activation is unclear. Increased M1 phenotypes compared to M2 activation are more commonly exhibited in MSA as identified by co-localization of CD68 and translocator protein (TSPO) as well as upregulated TLR4, major histocompatibility class II (MHC-II) molecules, and myeloperoxidase (Hoffmann et al., 2016; Rydbirk et al., 2017; Stefanova et al., 2007, 2012b). Stereological analyses of MSA patient white matter compared to controls revealed that microglia were increased by 100% with a concomitant 50% diminution of interneurons, but no significant changes in astrocytes or oligodendrocytes suggesting a strong role for microglia in MSA (Stefanova et al., 2012b). Notably, with increased inflammation and permeability of the blood–brain barrier, infiltration of CD4+ and CD8+ T cells is commonly found in the MSA brain with significant increases in peripheral CD4+ T cells (Cao et al., 2020; Williams et al., 2020).

52.2.3.1 Current Pharmaceutical Therapeutics

Current therapy for MSA provides no cure or intervention of disease progression, and most treatments are palliative for primary and subtypes of disease (Goh et al., 2023). Although a poor or unsustained response to initial *L-DOPA* treatment is diagnostically supportive of MSA, it remains the first-line therapy for Parkinsonian symptoms for MSA-P, but may exacerbate orthostatic hypertension in the short-term or lead to early nonresponsiveness after only a few years (Burns & McFarland, 2020; Low et al., 2015; Wenning et al., 2013). *Dopamine receptor agonists* or *amantadine* serve as second choices or as adjuncts, but lack supporting evidence and may exacerbate other symptoms (Flabeau et al., 2010; Kollensperger et al., 2010). Orthostatic hypotension in MSA is treated with *midodrine*, an α1 adrenergic agonist, and the mineralocorticoid fludrocortisone (Schoffer et al., 2007). *Droxidopa* (L-threo-3,4-dihydroxyphenylserine or L-DOPS) is a synthetic amino acid precursor that is converted to norepinephrine and is used to increase peripheral vasoconstriction and decrease orthostatic symptoms (Elgebaly et al., 2016; Hauser et al., 2015; Kaufmann et al., 2014). Additionally, the norepinephrine transport inhibitors *atomoxetine* and *ampreloxetine* have shown efficacy for neurogenic orthostatic hypertension (Kaufmann et al., 2021; Patel et al., 2018). No efficacious therapy for MSA-C-related ataxias is currently available, but cerebellar N-methyl-D-aspartate (NMDA) receptors may have play a role in motor learning and coordination or treating ataxias (Llansola et al., 2005). Clinical trials of *rislenemda (CERC-301)*, an antagonist for the NMDA receptor 2B subunit for orthostatic hypertension, failed to meet efficacy milestones (NCT02459236). Few pharmacological therapeutics are available for dystonia and most autonomic afflictions (Meszaros et al., 2020); however, for urinary dysfunction and incontinence, the most common autonomic symptom of MSA-C, treatment includes antispasmodics and anticholinergics such as *oxybutynin*, *solifenacin*, *trospium*, or *fesoterodine* (Nygaard, 2010; Ortiz et al., 2020; Wenning et al., 2005).

52.2.3.2 Therapeutics Targeting Neuroinflammation

As a consistent component in MSA, neuroinflammation represents an attractive therapeutic target. Indeed, a register-based, case–control study of non-aspirin *NSAIDs* use showed a decreased risk of MSA compared to non-users (Starhof et al., 2020). In an open-label pilot trial, *IVIG* was tested in 7 MSA patients with good tolerability and improved clinical scores by UMSAR, parts I and II (Novak et al., 2012). Based on neuroprotective results with *minocycline* from preclinical studies in MSA models (Stefanova et al., 2007), a Phase 3 clinical trial of minocycline in 63 MSA-P patients (NCT00146809) demonstrated that while PET imaging showed diminished microglial activity in the minocycline treated group, it failed to show improvement in clinical scores of motor function (Dodel et al., 2010). *Verdiperstat (BHV3241)* is an irreversible inhibitor of myeloperoxidase and inhibits ROS production and microglial-mediated neuroinflammation in preclinical models of MSA (Stefanova et al., 2012a). However, clinical trials of verdiperstat (NCT03952806, Phase 3; NCT04616456, Phase 1; and NCT05086094), while showing reduced microglial activity and activation by TSPO PET imaging, failed to reach study endpoints of improved clinical scores (Burns & McFarland, 2020; Miglis et al., 2021). Other preclinical candidates that have been shown to affect neuroinflammation and neuroprotection in MSA models include serotonin reuptake inhibitors (*SSRIs*) (Zhang et al., 2012); *CSF1R inhibitor (PLX5622)* to inhibit microglial development (Battis et al., 2022); two inhibitory peptides that antagonize the wild-type TLR2-interaction domain of MyD88 (*wtTIDM*) and the wild-type of NEMO-biding domain (*wtNBD*) peptide to inhibit TLR2 activation and NF-κB activation (Dutta et al., 2021). A fingolimod derivative, *FTY720-Mitoxt* in MSA models reduced microglial activation and increased expression of the neurotrophic mediators, BDNF, GDNF, and NGF, by oligodendrocytes (Vidal-Martinez et al., 2020). Based on improvement of stroke patients treated with *granulocyte colony-stimulating factor (G-CSF, CSF3)*, a small clinical trial of a few MSA patients was performed with G-CSF, but results were inconclusive due to the small study size (Pezzoli et al., 2010). One strategy leveraged the anti-inflammatory capabilities of *IVIG* to also inhibit effects of putative autoreactive T cells and autoantibodies as well as interfere with proinflammatory cytokine production (Aktas & Zipp, 2003). Thus, passive immunization with IVIG in a small Phase 2 clinical trial (NCT00750867) found IVIG improved symptoms and slowed brain degeneration in MSA patients (Novak et al., 2012).

52.2.3.3 Therapeutics Targeting T Cells

As T cells are known to initiate and enhance inflammatory processes, increased levels of parenchymal CD3+, CD4+, and CD8+ T cells are found in post-mortem MSA brain tissues (Williams et al., 2020). Also increased in those tissues are levels of proinflammatory cytokines such as IL-2, IL-13, and TNF-α as well as expression of genes for stimulator of interferon genes (STING) and TRAF-associated NF-κB activator (TANK)-binding kinase 1 (TBK1) (Rydbirk et al., 2017; Williams et al., 2020), which are thought to contribute to maintaining a chronic inflammatory state (Inoue et al., 2021). These observation and preclinical studies suggest the inhibition of T cell function or blockade of T cell infiltration as a potential therapeutic target for MSA (Harms et al., 2021; Williams et al., 2020). Using a vaccine approach with *microparticle-formulated α-synuclein ± rapamycin (RAP)* or with *RAP* alone in an MSA model, increased the numbers of CD4+, CD25+, and FoxP3+ T cells with increased TGFβ1 production and diminished microglia activation; however, RAP alone was unable to rescue dopaminergic or hippocampal neurons (Rockenstein et al., 2018). While T cell-targeting approaches may have potential, these have yet to translate to clinical testing.

52.2.3.4 Therapeutics Targeting Proteinopathy

As GCIs are a major neuropathological hallmark of MSA and likely represent a major driver of the neurodegenerative progression in MSA, thus therapeutic strategies to ameliorate protein misfolding or clear protein aggregates and relieve degenerative pathology are postulated to be an effective modality. However, unlike PD and DLB, α-synuclein aggregates are found primarily in oligodendrocytes. In models of MSA and other synucleinopathies, active immunization has shown promise (Fleming et al., 2022; Mandler et al., 2015; Masliah et al., 2005; Polinsky et al., 1991). Immunization with short peptide mimotopes (AFFITOPEs, AFFiRiS AG. Vienna, Austria) that mimic α-synuclein has been shown to alleviate protein aggregation and facilitates clearance in preclinical studies of MSA (Mandler et al., 2015). In a Phase 1 clinical trial, MSA patients were immunized with either of two mimotopes, *PD01A* and *PD03A* or placebo (NCT02270489). Results showed both were safe and well tolerated with 89% and 58% of patients producing antibodies to PD01A or PD03A, respectively, but did not show beneficial outcomes in clinical scores (Meissner et al., 2020). To test antibodies with greater specificity, a clinical trial (NCT05104476) wherein MSA patients are passively immunized with an IgG1 monoclonal antibody to the C-terminal of α-synuclein, *Lu AF82422*, is currently underway (Fjord-Larsen et al., 2021). Another anti-α-synuclein antibody, *BIIB054*, has been tested in PD patients (NCT03318523), but in MSA failed to meet primary outcomes and has been discontinued (Lang et al., 2022). Using antisense oligonucleotide (ASO) to pre-mRNA of α-synuclein has reduced α-synuclein in the CNS in preclinical model studies (Cole et al., 2021), thus a Phase 1 clinical trial is ongoing in MSA patients to test *ASO BIIB101* (*ION464*) (NCT03318523). *Epigallacatechin gallate (EGCG)* is a small molecule extract from green tea that was shown to inhibit α-synuclein misfolding and provide a neuroprotective effect (Xu et al., 2017); however, a Phase 3 trial of EGCG treatment of MSA patients for 52 weeks failed to modify disease and showed evidence of hepatotoxicity (Levin et al., 2019). Another inhibitor of α-synuclein misfolding, *NPT200-11A*, was developed from structure-based drug discovery that dynamically modeled and targeted specific regions of α-synuclein that facilitate misfolded oligomer formation (Price et al., 2018). In animal models of PD, NPT200-11a improved cortical α-synuclein pathology, astrogliosis, and motor dysfunction. Results from a completed Phase 1 clinical trial in healthy subjects have yet to be posted or published (NCT02606682). As dysregulation of iron metabolism promotes α-synuclein aggregation and ROS production, a quinazolinone-based inhibitor of iron reactivity, *ATH434 (PBT434)* binds iron, reduces iron-mediated redox activity, total ROS production, and α-synuclein levels with improved motor function in PD and MSA mouse models (Finkelstein et al., 2017; Heras-Garvin et al., 2021). In a Phase 1 clinical trial in healthy subjects, ATH434 was shown safe and attained comparable therapeutic levels in blood and CSF as those observed in animal models (U1111-1211-0052) (Sidoroff et al., 2022), thus triggering two ongoing Phase 2 trials in MSA patients; one for drug efficacy (NCT05109091) and another for effects on therapeutic biomarkers (NCT05864365). Enhancement of the autophagy-lysosomal pathway with *sirolimus (rapamycin)* reduced hydroxlynonenal-protein adducts, attenuated α-synuclein-mediated synaptic injury, and improved motor function, but not clearance of α-synuclein in preclinical testing (Bai et al., 2015). A Phase 2 trial (NCT03589976) in MSA patients assessing oral sirolimus efficacy to slow disease was terminated after meeting futility criteria. Similarly, the antibiotic *rifampicin* that inhibits α-synuclein fibril formation and enhances fibril disaggregation in MSA mouse models (Ubhi et al., 2008) was tested in a Phase 3 trial (NCT01287221), but was terminated after interim analysis revealed no benefit was achieved in clinical scores and futility criteria were met (Low et al., 2014).

52.3 Amyotrophic Lateral Sclerosis

Amyotrophic lateral sclerosis (ALS) is a progressive neurodegenerative disorder that primarily affects the α-motor neurons in the brain (upper neurons) and spinal cord (lower neurons) that are responsible for transmitting signals to muscles throughout the body, thus allowing voluntary muscle movements (Goutman et al., 2022). Loss of these neurons leads to muscle weakness, atrophy, and eventually paralysis. Loss of motor function presents as progressive difficulties in speaking (dysarthria), swallowing (dysphagia), and breathing as well as various other signs and phenotypes (Goutman et al., 2022). Underlying mechanisms in ALS include protein misfolding and accumulation and mitochondrial dysfunction, which contribute to loss of subcellular trafficking, neuronal injury, and eventual cell death (Feldman et al., 2022). Several proteins have been implicated and include, but are not limited to, superoxide dismutase 1 (SOD1), TAR DNA-binding protein 43 (TDP-43), fused in sarcoma (FUS), and the protein encoded by the chromosome 9, open reading frame 72 gene (C9orf72) which is linked to both ALS and frontotemporal dementia (FTD) (Chen et al., 2022; Chia et al., 2018; McAlary et al., 2019). While aberrant protein misfolding and accumulation are just one aspect of the complex pathophysiology of ALS, interactions between genetic, environmental, and cellular factors are also operative.

52.3.1 Current Approved Therapeutics

As ALS is an incurable disease, most current therapies focus on palliative care and maximization of life quality. Four approved medications have been designed to slow disease progression and delay the necessity for mechanical ventilation and include *riluzole*, *edaravone*, *sodium phenylbutyrate–taurursodiol (SP-TS, AMX0035)*, and *tofersen* (Feldman et al., 2022); the latter three only approved since 2017, 22 years after approval of riluzole. *Riluzole* is a benzothizole that blocks tetrodotoxin (TTX)-sensitive, voltage-gated sodium channels, but has exhibited differential efficacy, which is temporally stage-dependent at the time drug is initiated (Andrews et al., 2020; Fang et al., 2018). *Edaravone* is an antioxidant designed to relieve oxidative stress and neuroinflammation by scavenging free radicals and peroxynitrite (Kawai et al., 1997). While the exact mechanism of action is unknown, in a case study analysis of ALS patients treated or not treated with intravenous edaravone, showed edaravone lowered the risk of death and increased survival by a median of 6 months (Brooks et al., 2022). Several ongoing clinical trials are further evaluating edaravone in ALS (NCT04391361, Phase 2; NCT04569084, Phase 3; NCT04577404, Phase 3; NCT04744532, Phase 1/2; NCT05151471, Phase 3; NCT05178810, Phase 3; NCT05357950, Phase 2; and NCT05568615, Phase 3). *SP-TS* is a suspension of histone deacetylase inhibitor (SP) and hydrophilic bile acid (TS) (Alqallaf et al., 2024). The mechanism of action for SP is believed to suppress cellular toxicity by relieving stress on the endoplasmic reticulum and protein misfolding with upregulation of heat shock proteins and chaperones. TS reduces mitochondrial permeability and increases the threshold for apoptosis induction by inhibiting Bcl-2-associated X-protein (BAX) translocation into the mitochondria (Paganoni et al., 2020). Treatment with SP-TS slows disease progression compared to controls with an improved median survival time of 6.5 months (Alqallaf et al., 2024). Current clinical trials are ongoing to extend effects of SP-TS inn ALS (NCT04987671, Phase 1/2; NCT05021536, Phase 3; NCT05357950, Phase 2, and NCT05619783, Phase 3). More recently, *tofersen*, an intrathecal-delivered antisense oligonucleotide, received accelerated FDA approval on April 2023, based on the diminution of plasma neurofilament light chain (NfL) levels and SOD1 in treated ALS patients, though it did not improve clinical end points (Cerillo & Parmar, 2024; Escobar et al., 2022). Tofersen is a 20mer methoxyethyl-ribose/deoxy-ribose-based antisense oligonucleotide that recognizes and binds complimentary mRNA sequences for SOD1, which facilitates mRNA degradation and reduces the expression of SOD1 protein (Miller et al., 2022b). In a Phase 3 clinical trial in ALS patients with mutated SOD1 and treated with tofersen (NCT02623699), SOD1 and NfL levels in CSF and plasma were reduced, and clinical scores were improved compared to placebo group, which triggered accessibility to an ongoing Phase 3 open-label extension (NCT03070119) (Miller et al. 2022b). Several clinical trials are underway to further assess the efficacy of toferson in ALS (NCT04856982, Phase 3; NCT05725759; and NCT05852418).

52.3.2 Therapeutics Targeting Neuroinflammation

Indeed, among the many pathogenic mechanisms in ALS, activation of microglia and release of inflammatory molecules play a prominent role. In ALS, both microglia (immune cells in the CNS) and astrocytes (support cells) can become activated and release proinflammatory mediators, thus contributing to progression of neuronal damage (Beers et al., 2011; Murdock et al., 2017; Puentes et al., 2016). Evidence in postmortem ALS tissues show microgliosis in all areas of expected disease pathology, including motor cortex, brainstem motor nuclei, along the corticospinal tract, and in the ventral horn (Cady et al., 2014; Geloso et al., 2017; Henkel et al., 2004; Philips & Robberecht, 2011; Turner et al., 2004). Interestingly, it has been proposed that microglia from early stage disease express phenotypes more associated with anti-inflammatory or M2 type, while those from end stage are more proinflammatory, neurotoxic, or M1 type (Clarke & Patani, 2020; D'Erchia et al., 2017; Deczkowska et al., 2018; Dols-Icardo et al., 2020); however, M1/M2 polarized phenotypes more probably represent a heterogeneous spectrum of phenotypes rather defined standards (Wang et al., 2023). In addition, adaptive immune responses have been implicated in significant roles in ALS disease progression. Several of the approved drugs, such as edaravone, may provide indirect anti-inflammatory or anti-ROS effects. Some of the newer therapeutics directly target regulating immune cells. *Masitinib (AB10015)* is a CSF1R-selective tyrosine kinase inhibitor that inhibits proliferation and activity of microglia and macrophages in the CNS and PNS, and attenuates neuroinflammation (Mora et al., 2020). A Phase 2/3 trial (NCT02588677) to test oral delivery of masitinib in ALS patients showed that when combined with riluzole, masitinib significantly slowed disease progression. Moreover, life extension analyses indicated that administration early in disease could prolong survival by 2 years (Mora et al., 2021). An ongoing Phase 3 trial (NCT03127267) will assess the combination of riluzole with higher doses of masitinib over 48 weeks. Another immune modulatory agent, *NP001* is a chlorite compound that is taken up by phagocytic cells and intracellularly is converted to taurine chloramine (Schempp et al., 2001). This transforms macrophages/microglia from M1 to M2 state by downregulation of NF-κB expression, thus decreasing the production of proinflammatory cytokines in preclinical studies (Miller et al., 2022a). Initial Phase 1 (NCT01091142) and Phase 2a (NCT01281631) clinical trials showed intravenous NP001 was well tolerated and slowed disease progression in patients with high C-reactive protein (CRP) levels in plasma (Miller et al., 2015), while a subsequent Phase 2 trial (NCT02794857) showed no differences in ALSFRS-R scores between NP001 and placebo groups after 6 months (Miller et al., 2022a). However, post-hoc analysis revealed that NP001 treatment significantly slowed and even halted progression in some patients that included those between 40 and 65 years of age with onset within 3 years of treatment and with higher CRP levels. Moreover, ventilation capacity in the responder group was significantly preserved. IL-6 is a proinflammatory cytokine that is upregulated in ALS patients and animal models. *Tocilizumab* is a humanized mAb that recognizes IL-6 receptor (IL-6R) blocking IL-6 from the receptor and has been shown to attenuate inflammatory activation of peripheral and CSF blood cells from ALS patients and reduce the rate of disease progression (Fiala et al., 2013; Mizwicki et al., 2012). Thus, a Phase 2 clinical trial in ALS patients (NCT02469896) showed that tocilizumab was tolerable and decreased CRP levels in plasma and CSF; however, no differences in clinical measures were discerned compared to placebo treatment (Milligan et al., 2021).

Another strategy targets the pro-oxidant enzyme myeloperoxidase (MPO) in activated myeloid cells, which are at high levels in ALS patients, and thus aims to inhibit oxidative stress and attenuate neuroinflammation (Stefanova et al., 2012a). *Verdiperstat* is a small molecule, irreversible MPO inhibitor designed to test that idea. In a Phase 2/3 clinical trial (NCT04436510) to assess oral verdiperstat in ALS, no significant differences were discerned in primary or key secondary measures between treated and placebo group, and the drug was discontinued for further testing. Blocking CSF1R function and activity of inflammatory myeloid cells presents another strategy to attenuate inflammation in ALS. *BLZ945 (sotuletinib)* is a selective inhibitor of CSF-1R that reduces the proliferation of microglial cells and infiltration of macrophages in an ALS mouse model (Martinez-Muriana et al., 2016). Oral delivery of the small molecule is under current testing in a Phase 2 clinical trial with ALS patients (NCT04066244).

52.3.3 Therapeutics Targeting T Cells

The infiltration of lymphocytes in spinal cords and areas of motor neuron injury in ALS patients, wherein T cells represent a main component, suggests a role for the adaptive immune system (Beers & Appel, 2019; Engelhardt et al., 1993; Henkel et al., 2009; Machhi et al., 2020; McCombe & Henderson, 2011; Olson et al., 2023b). Moreover, CD4+ T cells from spinal cords of patients are associated with increased levels of CCL2 expression and activated microglia. Finally, while these studies suggest that T cells are important in ALS, their role is still an area of ongoing research, and their precise contribution to disease progression remains complex and incompletely understood. Although studies of peripheral CD4+ T cells from ALS patients provided inconclusive results as to a possible role, a more recent study of CD4+ T cells from the blood and spinal cords of newly diagnosed patients revealed that in ALS patients, high frequencies of CD4+effector T cells (Teffs) with evidence of oligoclonal TCR repertoire expansions were associated with a poor prognosis, whereas high frequencies of activated Tregs were associated with better survival (Beers et al., 2017). This validates previous observations that patients with Tregs of high activity were associated with better prognosis, whereas patients with low Treg activity exhibited poorer prognosis (Henkel et al., 2013). As Tregs normally suppress proinflammatory responses, the lack of functionally suppressive Tregs increases inflammation and neurodegeneration. Thus strategies that correct or boost Treg number and/or function should provide positive clinical outcomes in ALS (Mosley & Gendelman, 2010). Indeed, adoptive transfer of Tregs in mouse models of ALS proved to delay onset of neurological symptoms and loss of motor function as well as extend survival (Banerjee et al., 2008). These studies led to a first-in-human Phase 1 clinical trial (NCT03241784) whereby *autologous Tregs* were isolated, expanded, and transferred to ALS patients with concomitant administration of low-dose IL-2 (Thonhoff et al., 2018). Transfer of expanded autologous Tregs to ALS patients was found to be safe and well-tolerated, and slowed disease progression in early stage of disease. An ongoing Phase 2a trial (NCT04055623) will test monthly infusions of expanded autologous Tregs or placebo infusions with concomitant IL-2 or placebo for 6 months and open label access with ascending Treg dosages for the last 5 months. Another Phase 2/3 clinical trial in ALS patients will test *epigenetically-reprogramed, autologous Tregs/Th2 cells (RAPA-501)* to express the transcription factors FOXP3 and GATA that control Tregs and Th2 activities, respectively, is currently ongoing (NCT04220190). A concern with this strategy is whether the inherent flaws in the Treg populations that led to diminution of number and/or function are retained in expanded for transfer and lack long-term sustainability. To counter that possibility, a Phase 1 clinical trial in ALS patients will test *in vitro expanded allogeneic umbilical cord blood-derived Tregs (CK0803)* intravenously administered in 4 weekly doses followed by 1 dose every 4 weeks for 5 doses with a 24 week follow-on (NCT05695521). However, while short-term host immune responses may be suppressed, whether allogeneic Tregs can be retained for long-term functional efficacy is uncertain.

Another strategy would simply enhance existing Tregs using various Treg inducers or modulatory agents previously discussed (Sect. 52.2.1.3). Indeed IL-2 is a major cytokine necessary for generation, activation and survival of Tregs (Moorman et al., 2021), and has been utilized to expand Treg isolates from ALS patients for Treg transfer therapy (see above section) (Alsuliman et al., 2016; Thome et al., 2021; Thonhoff et al., 2018). Low dose IL-2 regimens are becoming more promising for in vivo induction of Treg proliferation and enhancement of Treg function (Pol et al., 2020). Selective expansion of Tregs with low doses of *recombinant IL-2 (aldesleukin)* have proven safe in mice and humans (Ye et al., 2018). A Phase 2a clinical trial of low dose IL-2 therapy in ALS patients (NCT02059759) was found to be safe without changes to plasma NfL levels, but increased frequencies of peripheral blood Tregs compared to placebo treatment (Camu et al., 2020). Post-hoc analysis of gene expression from leukocyte population found dose-dependent increases Treg markers that correlated with alterations and inhibitions of inflammatory pathways that indicated a cumulative effect of repeated low-dose IL-2 doses. Two genes *TLR9* and *CD27* were correlated with magnitude of drug responsiveness and proposed as therapeutic biomarkers for low dose IL-2 responses (Giovannelli et al., 2021). An ongoing Phase 2 trial will assess efficacy and therapeutic biomarker validation for low-dose IL-2 administration to ALS patients given for 5 days every 4 weeks over an 18-month period (NCT03039673).

Another immunomodulatory agent is *dimethyl fumarate (DMF)*, an FDA-approved drug indicated for multiple sclerosis (Gold et al., 2012). DMF increases Treg number and suppresses neuroinflammation. As a potential neuroprotective agent, DMF alters the activity of the transcription factor nuclear factor erythroid 2-related factor 2 (Nrf2), transforming proinflammatory effector T cells to anti-inflammatory types, such as Tregs and Th2 (Rajabinejad et al., 2020). A Phase 2 trial (ACTRN12618000534280) showed that while DMF proved safe and tolerable, and lowered NfL blood levels, it failed to change clinical scores, respiratory function, or survival in ALS patient (Vucic et al., 2021).

As the CD40 costimulatory pathway is significantly upregulated in ALS patients (Henkel et al., 2004), CD40 ligand (CD40L) provides an attractive target to block CD40-CD40L interactions and diminish neuroinflammation. *AT-1501 (tegoprubart)* is an anti-CD40L mAb that has been shown to reduce inflammation, delay paralysis, and prolong survival in ALS mouse models (Lincecum et al., 2010). In a Phase 2a clinical trial (NCT04322149), intravenous administration of AT-1501 proved safe and well-tolerated and reduced production of proinflammatory cytokines such as TNF-α, MCP1, EN-RAGE, and CRP in a dose-dependent fashion.

One strategy would use a combined immunosuppressive regimen to inhibit T cell- and microglia-driven neuroinflammation to produce a neuroprotective response. An open-label Phase 2 trial in ALS patients (NCT01884571) tested a varied dosing regimen combining *tacrolimus*, *mycophenolate*, *prednisone*, and *basiliximab*, a monoclonal antibody that blocks the CD25 α-chain on T cells. Results indicated the regimen was generally safe, but with a high drop-out rate (Fournier et al., 2018). No changes were discerned in disease progression by clinical scores or Treg frequency and number, but levels of IL-2 and TNF-α in CSF were increased. These results argued against the use of off-label immunosuppressive therapies without immune biomarkers to monitor target engagement.

52.3.4 Therapeutics Targeting Proteinopathy

As ALS is characterized by the accumulation of several misfolded and aggregated proteins, such as SOD1, TDP-43, and C9ORF72, which lead to neurodegeneration and motor impairment, strategies that target proteins for clearance and/or de-aggregation are emerging as promising therapeutics (Poulin-Briere et al., 2021). Many monoclonal antibodies have been developed that target extracellular (antibodies) and intracellular (nanobodies) proteins and have shown therapeutic potential in animal and in vitro models of ALS (Poulin-Briere et al., 2021), yet few have merited inclusion for translational testing. *AP-101* is a human antibody that recognizes aggregated SOD1 and reported to attenuate motor symptoms and increase overall survival in mouse models. In a Phase 1 trial in ALS patients (NCT03981536) treatment with AP-101 was well tolerated and formed the basis for an ongoing Phase 2 clinical trial (NCT05039099).

A considerable amount of effort emphasizes therapeutic strategies that focus on enhancement of autophagy to clear misfolded and aggregated proteins. *AL001 (iatozinemab)* is a monoclonal antibody that recognizes sortilin, a post-Golgi trafficking receptor that regulates protein trafficking in the exocytic and endocytic pathways (Gumina et al., 2019). It was designed to increase progranulin levels which stimulate autophagic clearance of misfolded proteins (Root et al., 2021). A Phase 1 clinical trial (NCT03636204) tested intravenous administration of AL001 in patients with C9ORF72 mutant frontotemporal dementia found the drug safe and well-tolerated (Kurnellas et al., 2023). The AL001-treated group showed long-lived increase in serum and CSF progranulin concentrations at levels of healthy subjects. A Phase 2 trial (NCT05053035) is currently ongoing. *Colchicine* acts as a potent enhancer of autophagy by upregulating transcription factor EB (TFEB) and heat shock protein B8 (HSPB8), both facilitate the direction of misfolded proteins into autophagosomes (Crippa et al., 2016) and also may interfere with intracellular assembly of the inflammasome complex (Dalbeth et al., 2014). A Phase 2 clinical trial (NCT03693781) of 2 oral dosages of colchicine in ALS patients was completed, but results remain unreported. Another enhancer of HSP and autophagy is *arimoclomol*. A successful Phase 2 trial (NCT00244244) in ALS patients preceded a Phase 3 trial for efficacy, however the latter failed to attain primary and secondary goals and did not slow disease progression (NCT03491462). *Sirolimus (rapamycin)* enhances autophagy by inhibiting mTOR, an inhibitor of autophagy, which in turn, promotes the production of autophagosomes, increasing autophagy of misfolded proteins, and is a known inducer of Tregs, which regulate proinflammatory T cells and microglia/macrophages (Mandrioli et al., 2018). In a Phase 2 clinical trial (NCT03359538), oral sirolimus treatment of ALS patients showed increased Treg levels, decreased inflammasome engagement, slowed the decline in clinical scores, and improved the quality of life (Mandrioli et al., 2023). Another autophagy enhancer is *tamoxifen*, which functions through mTOR-dependent and -independent pathways to clear aberrant TDP-43 in preclinical models (Wang et al., 2013). In an initial Phase 2 trial (NCT00214110), oral delivery of tamoxifen in ALS patients failed to show any significant differences in ALSFR-R scores or vital capacity, but survival was increased in groups receiving the 3 higher doses compared to the two lower doses (Brooks et al., 2005). A second Phase 2 trial (NCT01257581) found ALS patients treated with the highest dose of tamoxifen exhibited slower decline in clinical scores and progression (Babu et al., 2020). A third Phase 1/2 trial (NCT02166944) in ALS patients without SOD1 or FUS mutations, treated subjects with oral tamoxifen or placebo for 1 year and showed that tamoxifen-treated patients exhibited slower disease progression as determined by clinical scores and forced vital capacity and were less likely to succumb or start ventilation therapy, however, those measures lacked statistical significance due to small sample size (Chen et al., 2020a). Interestingly, in a clinical study of 10,450 healthy individuals, those with known association of exposure to tamoxifen exhibited a lower risk of developing ALS (Pfeiffer et al., 2020).

Several strategies have attempted to alter cell processes that can alter autophagy. *Bosutinib* is a selective inhibitor of Src/c-Abl kinases, which promotes autophagy and reduces misfolded mutant SOD1 protein. In induced pluripotent stem cells (iPSC)-derived motor neurons from ALS patients, treatment with bosutinib prolonged survival and function (Imamura et al., 2019). An open-label Phase 1/2 trial (NCT04744532) showed the drug was well-tolerated and 5/9 patients responded as determined by reduced plasma NfL levels, less pathogenic transcriptomic biomarker signatures, and slowed or halted disease progression from clinical scores (Imamura et al., 2022). The Phase 2 portion of the trial is currently ongoing. *3K3A-Activated Protein C (3K3A-APC)* is a modified activated protein C (zymogen) with disease-modifying properties that include BBB repair and inhibition of microglia activation (Shi et al., 2019). Additionally, in iPSC-derived motor neurons from sporadic and C9ORF72 ALS patients, treatment with 3K3A-APC restored autophagosomes, reduced SOD1 synthesis, and deaccelerated disease progression in those cultures. An open-label Phase 2 trial is complete and results are submitted, but not yet published (NCT05039268).

Other approaches use antisense oligonucleotides (ASOs) to target mRNA transcripts, which promote their degradation prior to translation. An example of an approved therapeutic is *tofersen* that binds SOD1 mRNA (see current therapeutics). *ION363 (jacifusen)* is an ASO that recognizes the 5′-GCAATGTCACCTTTCATACC-3′ sequence within the sixth intron of the *FUS* gene and targets the *FUS* mRNA transcript to silence FUS protein expression. In mutant *FUS* KI-ALS mice, ION363 reduced both wild-type and mutant FUS levels in the CNS and aggregated forms of FUS (Korobeynikov et al., 2022). In a case study of one patient with FUS-ALS who obtained compassionate use permission for intrathecal ION363, the rate of disease progression was slowed from baseline rate. Post-mortem examination detected ION363 throughout brain and spinal cord tissues two months after the last infusion. Wild-type and mutant FUS levels were reduced to nearly undetectable levels, including FUS-positive neuronal cytoplasmic and nuclear inclusions (Korobeynikov et al., 2022). A Phase 3 trial (NCT04768972) with FUS-ALS patients is currently ongoing without benefit of Phase 1 or 2 trials. *WVE-004* is an ASO that targets the G4C2 repeat expansion of the C9ORF72 (C9) mutation, but preserves normal protein expression (Liu et al., 2021). Similarly, in C9ORF72-FTD/ALS mice, intracerebroventricular administration of WVE-004 produced long-term depletion of mutant protein without affecting wild-type protein expression, and protected motor neurons that express mutant mRNA from glutamate-induced toxicity. Interim results from an ongoing Phase 1/2 clinical trial of intrathecal WVE-004 in patients with C9ORF72-associated ALS or FTD (NCT04931862) showed reduced poly-glutamine-proline (GP) dipeptides, but no improvement in clinical scores. An open-label extension of that trial has been initiated (NCT05683860).

52.4 Huntington's Disease

Huntington's disease (HD) is a hereditary neurodegenerative disorder affecting more vulnerable medium spiny neurons of the striatum during early stage, but involving the thalamus and globus pallidus in latter stages (McColgan & Tabrizi, 2018). Clinical presentation is characterized by progressive motor dysfunction, cognitive decline, and psychiatric symptoms. The disease is caused by a mutation in the huntingtin gene (*HTT*), part of which encodes a trinucleotide CAG repeat that is translated to a poly-glutamine (polyQ) portion within the protein. Repeats of <36 do not generally affect the protein or host, but larger expansions are pathological causing toxic gain of function in the mHtt product. A key feature of expanded *HTT* is its propensity to fragment, misfold, and accumulate producing neurotoxic fibrillar aggregates that diminish proteostasis, sequester chaperones, impair autophagy, and disrupt organelles, particularly mitochondria (Almeida et al., 2023; Bates et al., 2015).

52.4.1 Current Approved Therapeutics

For HD, current pharmacological therapy is limited to *tetrabenazine* or a derivate, *deutetrabenazine*, both vesicular monoamine transporter 2 (VMAT-2) inhibitors that reduce neurotransmission by depleting dopamine from presynaptic vesicles with the rationale to mitigate uncontrolled movements (Wyant et al., 2017). A Phase 2/3 clinical trial is ongoing to assess the effects of deutetrabenazine on gait and speech in HD (NCT04713982). To manage psychiatric manifestations, antipsychotic and antidepressive medications have shown limited success, but evidence-based guidance on therapeutics are limited in HD. Nevertheless, a Phase 1 clinical trial is ongoing in HD patients to test the monotherapeutic effects on symptoms by the anti-psychotics, haloperidol and risperidone, and the anti-depressive, sertraline (NCT04071639). For cognitive symptoms, *cholinesterase inhibitors* or *memantine* used in Alzheimer's disease has also shown limited success.

52.4.2 Therapeutics Targeting Neuroinflammation

The concomitant neuronal destruction and aberrant protein aggregation present an environment wherein inflammation becomes operative. Indeed, substantial evidence exists to suggest that neuroinflammation contributes to the neurodegenerative processes of HD. Post-mortem evidence shows reactive microglia in the striatum and cortices of HD brains that proximate areas of neuronal loss (Sapp et al., 2001). Additionally, PET imaging using first and second-generation probes for TSPO expression by activated microglia shows increased levels in striatum that correlate with disease severity (Saba et al., 2022). Moreover, higher TSPO activities are found in striatal and cortical areas of presymptomatic gene carriers. Accompanying microgliosis in post-mortem striata of HD brains are GFAP+ astrocytes that exhibit increased staining intensities, which correlate with striatal neurodegeneration and disease progression. These intensities are found in areas of atrophy, but are not readily detected in other areas of the brain, and importantly are also found in pre-symptomatic carriers compared to controls (Saba et al., 2022; Vonsattel et al., 2008). Interestingly, despite increased GFAP reactivity with disease progression, transcriptomic studies have not determined whether HD astrocytes reflect more neurotoxic (A1) or neuroprotective (A2) phenotypes at different disease stages (Benraiss et al., 2021). Some studies suggest modulating the immune system and reducing inflammation may alleviate much of HD pathology and that early treatment may be more beneficial as inflammation is more prominent in early stages of HD (Colpo et al., 2017). A multimodal immunotherapeutic approach has been suggested that targets multiple factors (Fatoba et al., 2020); however, those approaches pose confounding risks and results have been mixed (Colpo et al., 2017). Several clinical trials (NCT00029874, NCT00277355) assessed *minocycline* in HD, which was safe and well-tolerated, but a Phase 3 trial for efficacy showed worsened disease progression and functional capacity, thus the trial was terminated after meeting futility criteria (Huntington Study Group, 2010). *Laquinimod* is an orally active immunomodulatory agent that has shown efficacy in inflammation and neurodegeneration in several disorders (Fatoba et al., 2020). In a recent Phase 2 clinical trial in HD patients (NCT02215616), treatment with laquinimod diminished caudate atrophy, but did not significantly affect disease progression or changes in ^{14}C-PBR28 PET imaging for TSPO expression by activated microglia (Roussakis et al., 2023). One strategy targets immune modification of the complement pathway. *ANX-005* is a humanized IgG4 monoclonal antibody to the complement C1q protein, which inhibits initiation of the classical complement cascade, and attenuates inflammatory responses mediated by microglia. As microglia and C1q also play roles in synaptic pruning, C1q neutralization aims to limit aberrant microglia activation preventing synaptic over-pruning and limiting neurodegeneration. An open-label Phase 2 clinical trial (NCT04514367) was initiated to assess intravenous delivery of ANX-005 in patients with early or high risk for HD. Results showed the therapy was well-tolerated, reduced levels of CSF C1q and YKL-40, a marker of neuroinflammation, and improved clinical outcomes in patients with high plasma C4a/C4 levels (Kumar et al., 2023).

52.4.3 Therapeutics Targeting T Cells

Unlike other neurodegenerative movement disorders, lymphocytic infiltration has not yet been reported in the post-mortem HD brain (Silvestroni et al., 2009); however, increased levels of soluble CD27 (sCD27), a biomarker that reflects T cell-mediated neuroinflammation, are detectable in the CSF of *HTT* gene expansion carriers, both pre-symptomatic and symptomatic (Niemela et al., 2018). Further suggesting that T cells are operative in HD and *HTT* gene expansion carriers, levels of CD4+ Th17.1 (CXCR3+CCR6+) T cells from CSF are increased compared to controls and Treg levels are correspondingly diminished (Abbaoui et al., 2023). Moreover, higher frequencies of Th17.1 T cells are found in presymptomatic HD patients compared to HD-manifested patients. Taken together, levels of Th17.1 T cells indirectly correlated with the severity of disease as determined by NfL levels and disease progression determined from CAG-age product (CAP) scores. Overall, these data support the notion that cellular immune-sourced neuroinflammation plays a role in HD, at least in early HD. This is underscored by increased expression of mRNAs and proteins for proinflammatory mediators such as IL-1β, IL-6, IL8, CCL2 (MCP-1), CCL5 (RANTES), and TNF-α in the CSF, striata, and/or cortices of HD patients that correlate with disease severity (Saba et al., 2022; von Essen et al., 2020). Moreover, upregulated mRNAs for IL-8, IL-15, and CCL26 are found in presymptomatic expanded *HTT* carriers. Semaphorin 4D (SEMA4D) (CD100) serves as a signaling molecule with interactions that govern axonal guidance and other CNS functions but also immunoregulatory interactions that activate B cells and dendritic cells, the latter of which elicit T cell activation and immune responses to presented antigens (Kuklina, 2021; Kumanogoh & Kikutani, 2001). *Pepinemab (VX15/2503)* is a monoclonal antibody that blocks SEM4D interaction with CD72 and in preclinical HD models, ameliorates neuropathological deficits with partial rescue of behavior deficits (Southwell et al., 2015). A Phase 2 trial of pepinemab in in early HD (NCT02481674) failed to meet primary outcomes, but provided improvements in secondary outcomes in cognitive, volumetric MRI, and ^{18}F-FDG PET assessments (Feigin et al., 2022).

52.4.4 Therapeutics Targeting Proteinopathy

Multiple studies suggest that reducing levels of mutant huntingtin protein (mHtt), the toxic product of the *HTT* gene, can help alleviate disease symptoms. Antibodies that specifically target mHtt have been produced, and several have been shown to improve HD phenotypes in cell cultures, fruit flies, mice, and other HD models (Butler et al., 2012; Denis et al., 2019). Extending those findings, intrabodies, engineered antibody fragments expressed intracellularly, bind mHtt within cells to reduce mHtt aggregation and toxicity in HD models (Amaro & Henderson, 2016; Butler & Messer, 2011; Southwell et al., 2008; Wang et al., 2008). This suggested a critical role not only for intracellular mHtt but also extracellular protein that spreads through the brain and is critical to HD progression. Thus, intervention of protein translation and spreading could be of clinical benefit. Initially, this idea was tested at the RNA level using antisense oligonucleotide (ASOs). *Tominersen (IONIS-HTT$_{RX}$, ISIS 44313, RG6042)* is an intrathecal administered gapmer antisense oligonucleotide which targets the mRNA for mHtt and facilitates its degradation. Based on initial Phase 1/2 clinical trial results (NCT02519036), which showed ascending doses of tominersen were safe and tolerable, and reduced levels of mHtt in CSF in a dose-dependent fashion (Tabrizi et al., 2019). Indeed, the trial was amended to extend open-label access to the maximum dose and is ongoing (NCT03342053). A Phase 3 trial (NCT03761849) was initiated using the maximum dose every 8 or 16 weeks. Although interim analysis showed tominersen lowered mHtt levels in the CSF, the study was terminated based on worsening clinical scores, increased levels of CSF NfL, and adverse events at 8 weeks, and no statistical differences in treated and placebo groups at 16 weeks of the trial (Tabrizi et al., 2022). Moreover, increased NfL levels in the CSF suggested an off-target inflammatory or neuronal injurious response by tominersen. A new Phase 3 trial is ongoing to assess lower doses of tominersen (NCT05686551). A second mHtt-lowering strategy with gapamer ASOs was to use an allele-specific approach (Aslesh & Yokota, 2020; Chan & Yokota, 2020; Svrzikapa et al., 2020). Two Phase 1/2 clinical trials were initiated to test two ASOs, *WVE-120101* (NCT03225833) and *WVE-120102* (NCT03225846), that target unique SNPs of the mHtt transcript. After monthly intrathecal administration, interim analysis of both studies revealed lack of target engagement and no significant reduction of CSF mHtt levels (34012000). Unlike tominersen, which showed mHtt lowering potential, exposure to the latter two ASOs was thought insufficient to produce an effect (Kwon, 2021). Another Phase 1/2 clinical trial has been initiated to test *VO659*, an allele-preferential, CAG repeat-targeted ASO that acts by inhibiting translation and inducing exon-skipping (NCT05822908). Another genetic strategy uses a transgene encoding miRNA that binds huntingtin mRNA (miHTT) preventing translation and decreasing production of both wt and mutant Htt. Two ongoing Phase 1/2 clinical trials will assess the safety and proof-of-concept for intrastriatal delivery of the *AMT-130* transgene construct (AAV5-miHTT) in early HD patients from the United States (NCT04120493) and Europe (NCT05243017). *WVE-003* is another is an allele-specific ASO that targets SNP3 of mHtt pre-mRNA to prevent its translation leading to the subsequent reduction of mHtt, but not wtHtt. A Phase 1/2 clinical trial (NCT05032196) is ongoing to assess the intrathecal delivery of WVE-003. Interim analysis showed the drug was well-tolerated and reduced mHTT in CSF by 22% at 85 days after one dose (Estevez-Fraga et al., 2022). An alternative strategy targets alternative splicing of mHtt to normalize splicing events under homeostatic conditions. *Branaplam (NVS-SM1, LMI070)* is a pyridazine derivative that binds to the nGA sequence and inhibits mHtt splicing by facilitating the inclusion of a pseudoexon in Htt transcripts, which reduces mHtt protein levels (Childs-Disney et al., 2022; Krach et al., 2022). To test that principle, oral delivery of branaplam was tested in a Phase 2 clinical trial (NCT05111249), but was halted due to evidence of peripheral neuropathy that included clinical manifestations, neurophysiological abnormalities, and increases in plasma NfL levels (Estevez-Fraga et al., 2022). Another protein-targeting strategy is based on the notion that HD pathology is associated with increased levels of cholesterol in neuronal membranes and influences mHttt expression and/or neuronal survival (Karasinska & Hayden, 2011). Thus, delivery of the *CYP46A1* gene that encodes cholesterol 24-hydroxylase enzyme, levels that are diminished in HD patients, can restore cholesterol homeostasis, decrease levels of mHtt aggregates and NfL, and prevent neurodegeneration. Preclinical studies in HD models support this approach (Estevez-Fraga et al., 2022; Kacher et al., 2019). A Phase 1/2 clinical trial will test the intrastriatal delivery of *AB-1001*, an AAV construct containing the *CYP46A1* gene (AAVrh10. CAG.hCYP46A1) in early HD patients with a planned 5 year follow on (NCT05541627).

52.5 Conclusions

Neurodegenerative movement disorders present an increasing complex and often idiosyncratic pathobiology. However, common hallmarks of misfolded and aggregated proteins, disorder-specific neuronal inclusions, and chronic neuroinflammation are drivers of neurodegeneration. Moreover, modified self-proteins and proinflammatory innate immune environments are conducive to presentation of modified self-antigens and elicitation of T cell-mediated immunity promoting proinflammatory effector functions that recognize modified CNS antigens. Proinflammatory Teffs provide a mediator for exacerbation of neuroinflammation, more protein modifications and misfolding, and increased neurodegeneration leading to accelerated or intensified disease progression. Thus, this proinflammatory triad that maintains perpetual cascade of toxic processes should advance strategies steeped in immune-based interventional therapies. Thus, a super-plethora of preclinical studies have targeted individual components of this triad, and from those studies designed interventional protocols that target microglial-mediated neuroinflammation, proinflammatory T cells and cytokines, and protein mis-folding and aggregation that have proceeded to Phases 1, 2, and/or 3 clinical trials (Fig. 52.1). Interestingly, a number of these interventions have been or are ongoing in Phase 2 trials. An abundance of Phase 3 trials for several immunotherapeutics are ongoing for ALS, while virtually none are in the pipeline for DLB, which in part, may reflect the complexity of the disease with presentation of multiple proteinopathies. Indeed, several disorder-specific strategies have emerged that target global processes controlling neuroinflammation or proinflammatory T cell function with limited success. A promising example of countering both innate- and adaptive-driven inflammation is the utilization of antigen-specific Tregs, which upon encountering cognate antigen initiate their anti-inflammatory program to attenuate, inactivate, or kill proinflammatory microglia and Teff cells, and to induce astrocytic neurotrophic factors. Many therapeutic approaches target the offending proteinopathies, which exhibit spatial and disorder specificity, and thus attempt to limit the proteins that drive both neuronal injury and inflammatory processes, but also to restrict the fodder of modified-self epitopes that drive proinflammatory Teffs. For synucleinopathies such as PD, DLB, and MSA, misfolded α-synuclein is found as neuronal or oligodendritic inclusions and fibrils that drive neuroinflammation. Current treatments aim to either clear or inhibit misfolded and aggregated proteins. In ALS, protein aggregates like SOD1 and TDP-43 accumulate and damage motor neurons. Approved drugs like riluzole and edaravone provide modest benefit. Experimental therapies aim to modulate neuroinflammation, boost regulatory T cells, or clear protein aggregates. In HD, mutant huntingtin protein misfolds with accumulation of aggregates that injure neuron processes, thus may be amenable to targeted intracellular therapeutic strategies to inhibit misfolding and enhance intracellular clearance through lysosomal-autophagic processes. Evidence from preclinical studies and clinical trials suggests that modulation of immune responses provides treatment options for potential disease-modifying therapies that extend beyond current palliative management and can tangibly alter the processes that underlie neurodegenerative mechanisms. However, utilization of those options in a therapeutic fashion requires a greater understanding of molecular mechanisms associated with the inflammatory troika that drive neurodegeneration. Indeed, the lack of translational potentials for many agents, particularly targeted anti-inflammatory interventions that do not meet primary or secondary endpoint measures in clinical trials underscores the complexity of these neurodegenerative processes, thus suggest that combinatorial therapeutic approaches may be required to counter these complex processes. Nonetheless, immunotherapeutic strategies that target proteinopathies and detrimental innate inflammatory responses that are initiated or potentiated by autoreactive T cells will generate new pharmacologic avenues to provide beneficial outcomes for patients afflicted with neurodegenerative motor diseases.

52.6 Review Questions

1. What is the common hallmark misfolded protein involved in pathogenesis of Parkinson's disease, dementia with Lewy bodies, and multiple system atrophy?
   (a) Aβ
   (b) Tau
   (c) α-Synuclein
   (d) Prion protein
2. What is the primary function of regulatory T cells (Tregs) in the context of Parkinson's disease?
   (a) To increase inflammation
   (b) To induce autophagy
   (c) To reduce inflammation
   (d) To reduce GCase activity
3. What is the primary impediment to successful clinical studies of therapeutics targeting proteinopathies in dementia with Lewy body disease?
   (a) Experimental protocols interfere with curative therapeutic regiments.

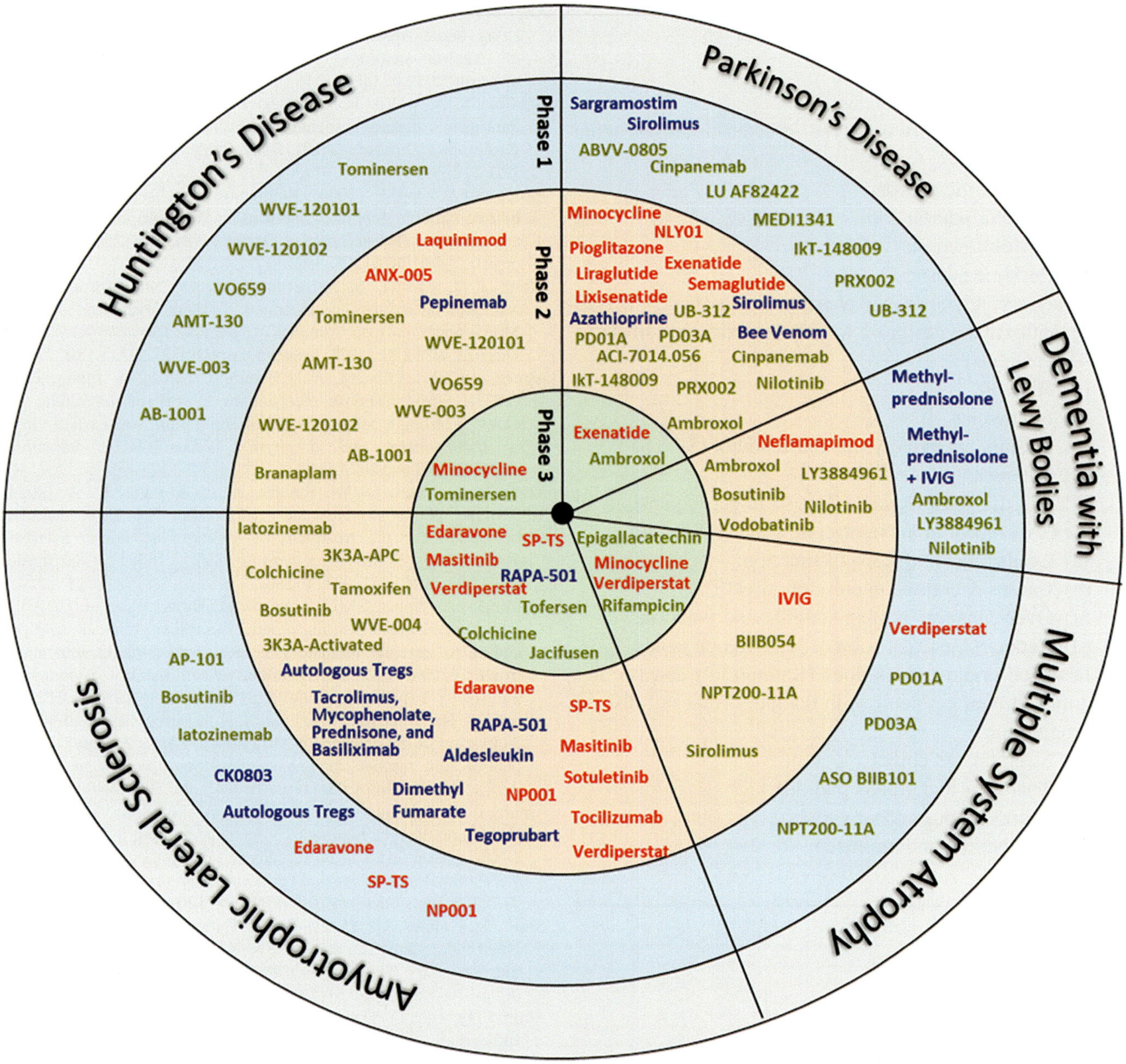

Fig. 52.1 Pipeline of immunotherapeutic strategies tested in clinical trials. Various interventional approaches are proposed for the neurodegenerative movement disorders of Parkinson's disease, dementia with Lewy bodies, multiple system atrophy, amyotrophic lateral sclerosis, and Huntington's disease. These disorders have similar common pathobiological hallmarks that include protein misfolding and aggregation, microglial-mediated chronic neuroinflammation, and infiltration of T cells that have pro-inflammatory phenotypes that further drive microglial-mediated inflammation and protein mis-folding/aggregation. Thus, intervention therapeutic approaches and modalities have incorporated strategies that target neuroinflammation (red), pro-inflammatory T cells (blue), and disorder-specific proteinopathies (green). Sufficient pre-clinical evidence of potential efficacy had spurred specific modalities on to Phase 1, Phase 2, and/or Phase 3 clinical trials

   (b) Aggregations of multiple protein species.
   (c) Ethical concerns on a terminal patient population.
   (d) Lack of patients willing to enroll in clinical trials.
4. Which brain regions are most affected by neuronal loss in dementia with Lewy bodies as the disease progresses?
   (a) Substantia nigra and hippocampus
   (b) Cortex and limbic system
   (c) Amygdala and entorhinal cortex
   (d) Frontal lobe and basal ganglia
5. What is the hallmark feature of multiple system atrophy?
   (a) Tau protein inclusions in oligodendrocytes
   (b) Amyloid plaques in hippocampal neurons
   (c) α-Synuclein inclusions in dopaminergic neurons
   (d) α-Synuclein inclusions in oligodendrocytes
6. Which part of the nervous system does MSA primarily affect?

(a) Autonomic nervous system
(b) Peripheral nervous system
(c) Central nervous system
(d) Sensory nervous system

7. What is the primary cell type affected in amyotrophic lateral sclerosis?
(a) Cholinergic neuron
(b) γ-Motor neuron
(c) α-Motor neuron
(d) Purkinje neuron

8. Which recently approved pharmaceutical therapy acts as an antioxidant designed to relieve oxidative stress in ALS patients?
(a) Riluzole
(b) Edaravone
(c) Sodium phenylbutyrate–taurursodiol (SP-TS)
(d) Tofersen

9. What is the pathological effect of larger expansions of the CAG repeat in the huntingtin gene (HTT)?
(a) Leads to a toxic gains of function.
(b) Causes decreased expression of HTT.
(c) Over-enhances protein stability and function.
(d) Reduces the expression of HTT run on genes.

10. In which brain regions does Huntington's disease predominantly affect neurons in the latter stages of disease progression?
(a) Striatum and cortex
(b) Thalamus and globus pallidus
(c) Hippocampus and amygdala
(d) Cerebellum and brainstem

52.7 Answers

1. c
2. c
3. b
4. b
5. d
6. a
7. c
8. b
9. a
10. b

References

Abbaoui, A., Fatoba, O., & Yamashita, T. (2023). Meningeal T cells function in the central nervous system homeostasis and neurodegenerative diseases. *Frontiers in Cellular Neuroscience, 17*, 1181071. https://doi.org/10.3389/fncel.2023.1181071

Abdelmoaty, M. M., Machhi, J., Yeapuri, P., Shahjin, F., et al. (2022). Monocyte biomarkers define sargramostim treatment outcomes for Parkinson's disease. *Clinical and Translational Medicine, 12*(7), e958. https://doi.org/10.1002/ctm2.958

Abe, K., & Chiba, Y. (2019). A case of treatable dementia with Lewy bodies remarkably improved by immunotherapy. *Journal of Neuroimmunology, 330*, 35–37. https://doi.org/10.1016/j.jneuroim.2019.02.003

Abeliovich, A., Hefti, F., & Sevigny, J. (2021). Gene therapy for Parkinson's disease associated with GBA1 mutations. *Journal of Parkinson's Disease, 11*(s2), S183–S188. https://doi.org/10.3233/JPD-212739

Aktas, O., & Zipp, F. (2003). Regulation of self-reactive T cells by human immunoglobulins--implications for multiple sclerosis therapy. *Current Pharmaceutical Design, 9*(3), 245–256. https://doi.org/10.2174/1381612033392152

Alam, J. J. (2015). Selective brain-targeted antagonism of p38 MAPKalpha reduces hippocampal IL-1beta levels and improves Morris water maze performance in aged rats. *Journal of Alzheimer's Disease, 48*(1), 219–227. https://doi.org/10.3233/JAD-150277

Almeida, L. M., Oliveira, A., Oliveira, J. M. A., & Pinho, B. R. (2023). Stress response mechanisms in protein misfolding diseases: Profiling a cellular model of Huntington's disease. *Archives of Biochemistry and Biophysics, 745*, 109711. https://doi.org/10.1016/j.abb.2023.109711

Alqallaf, A., Cates, D. W., Render, K. P., & Patel, K. A. (2024). Sodium phenylbutyrate and taurursodiol: A new therapeutic option for the treatment of amyotrophic lateral sclerosis. *The Annals of Pharmacotherapy, 58*(2), 165–173. https://doi.org/10.1177/10600280231172802

Alsuliman, A., Appel, S. H., Beers, D. R., Basar, R., et al. (2016). A robust, good manufacturing practice-compliant, clinical-scale procedure to generate regulatory T cells from patients with amyotrophic lateral sclerosis for adoptive cell therapy. *Cytotherapy, 18*(10), 1312–1324. https://doi.org/10.1016/j.jcyt.2016.06.012

Alvarez, A. R., Sandoval, P. C., Leal, N. R., Castro, P. U., et al. (2004). Activation of the neuronal c-Abl tyrosine kinase by amyloid-beta-peptide and reactive oxygen species. *Neurobiology of Disease, 17*(2), 326–336. https://doi.org/10.1016/j.nbd.2004.06.007

Amaro, I. A., & Henderson, L. A. (2016). An intrabody drug (rAAV6-INT41) reduces the binding of N-terminal huntingtin fragment(s) to DNA to basal levels in PC12 cells and delays cognitive loss in the R6/2 animal model. *Journal of Neurodegenerative Diseases, 2016*, 7120753. https://doi.org/10.1155/2016/7120753

Amin, J., Holmes, C., Dorey, R. B., Tommasino, E., et al. (2020). Neuroinflammation in dementia with Lewy bodies: A human post-mortem study. *Translational Psychiatry, 10*(1), 267. https://doi.org/10.1038/s41398-020-00954-8

Amin, J., Erskine, D., Donaghy, P. C., Surendranathan, A., et al. (2022). Inflammation in dementia with Lewy bodies. *Neurobiology of Disease, 168*, 105698. https://doi.org/10.1016/j.nbd.2022.105698

Amin, J., Gee, C., Stowell, K., Coulthard, D., et al. (2023). T lymphocytes and their potential role in dementia with Lewy bodies. *Cells, 12*(18), 2283. https://doi.org/10.3390/cells12182283

Andrews, J. A., Jackson, C. E., Heiman-Patterson, T. D., Bettica, P., et al. (2020). Real-world evidence of riluzole effectiveness in treating amyotrophic lateral sclerosis. *Amyotrophic Lateral Sclerosis and Frontotemporal Degeneration, 21*(7-8), 509–518. https://doi.org/10.1080/21678421.2020.1771734

Armstrong, M. J., & Okun, M. S. (2020). Diagnosis and treatment of Parkinson disease: A review. *JAMA, 323*(6), 548–560. https://doi.org/10.1001/jama.2019.22360

Aslesh, T., & Yokota, T. (2020). Development of antisense oligonucleotide gapmers for the treatment of Huntington's disease. *Methods in Molecular Biology, 2176*, 57–67. https://doi.org/10.1007/978-1-0716-0771-8_4

Athauda, D., Maclagan, K., Skene, S. S., Bajwa-Joseph, M., et al. (2017). Exenatide once weekly versus placebo in Parkinson's disease: A randomised, double-blind, placebo-controlled trial. *Lancet, 390*(10103), 1664–1675. https://doi.org/10.1016/S0140-6736(17)31585-4

Babu, S., Macklin, E. A., Jackson, K. E., Simpson, E., et al. (2020). Selection design phase II trial of high dosages of tamoxifen and creatine in amyotrophic lateral sclerosis. *Amyotrophic Lateral Sclerosis and Frontotemporal Degeneration, 21*(1-2), 15–23. https://doi.org/10.1080/21678421.2019.1672750

Baek, H., Jang, H. I., Jeon, H. N., & Bae, H. (2018). Comparison of administration routes on the protective effects of bee venom phospholipase A2 in a mouse model of Parkinson's disease. *Frontiers in Aging Neuroscience, 10*, 179. https://doi.org/10.3389/fnagi.2018.00179

Baek, H., Park, S. Y., Ku, S. J., Ryu, K., et al. (2020). Bee venom phospholipase A2 induces regulatory T cell populations by suppressing apoptotic signaling pathway. *Toxins (Basel), 12*(3), 198. https://doi.org/10.3390/toxins12030198

Bai, X., Wey, M. C., Fernandez, E., Hart, M. J., et al. (2015). Rapamycin improves motor function, reduces 4-hydroxynonenal adducted protein in brain, and attenuates synaptic injury in a mouse model of synucleinopathy. *Pathobiology of Aging & Age-Related Diseases, 5*, 28743. https://doi.org/10.3402/pba.v5.28743

Banerjee, R., Mosley, R. L., Reynolds, A. D., Dhar, A., et al. (2008). Adaptive immune neuroprotection in G93A-SOD1 amyotrophic lateral sclerosis mice. *PLoS One, 3*(7), e2740. https://doi.org/10.1371/journal.pone.0002740

Bates, G. P., Dorsey, R., Gusella, J. F., Hayden, M. R., et al. (2015). Huntington disease. *Nature Reviews: Disease Primers, 1*, 15005. https://doi.org/10.1038/nrdp.2015.5

Battis, K., Florio, J. B., Mante, M., Lana, A., et al. (2022). CSF1R-mediated myeloid cell eepletion prolongs lifespan but aggravates distinct motor symptoms in a model of multiple system atrophy. *The Journal of Neuroscience, 42*(40), 7673–7688. https://doi.org/10.1523/JNEUROSCI.0417-22.2022

Beers, D. R., & Appel, S. H. (2019). Immune dysregulation in amyotrophic lateral sclerosis: Mechanisms and emerging therapies. *Lancet Neurology, 18*(2), 211–220. https://doi.org/10.1016/S1474-4422(18)30394-6

Beers, D. R., Henkel, J. S., Zhao, W., Wang, J., et al. (2011). Endogenous regulatory T lymphocytes ameliorate amyotrophic lateral sclerosis in mice and correlate with disease progression in patients with amyotrophic lateral sclerosis. *Brain, 134*(Pt 5), 1293–1314. https://doi.org/10.1093/brain/awr074

Beers, D. R., Zhao, W., Wang, J., Zhang, X., et al. (2017). ALS patients' regulatory T lymphocytes are dysfunctional, and correlate with disease progression rate and severity. *JCI Insight, 2*(5), e89530. https://doi.org/10.1172/jci.insight.89530

Bendikov-Bar, I., Maor, G., Filocamo, M., & Horowitz, M. (2013). Ambroxol as a pharmacological chaperone for mutant glucocerebrosidase. *Blood Cells, Molecules & Diseases, 50*(2), 141–145. https://doi.org/10.1016/j.bcmd.2012.10.007

Benner, E. J., Mosley, R. L., Destache, C. J., Lewis, T. B., et al. (2004). Therapeutic immunization protects dopaminergic neurons in a mouse model of Parkinson's disease. *Proceedings of the National Academy of Sciences of the United States of America, 101*(25), 9435–9440. https://doi.org/10.1073/pnas.0400569101

Benraiss, A., Mariani, J. N., Osipovitch, M., Cornwell, A., et al. (2021). Cell-intrinsic glial pathology is conserved across human and murine models of Huntington's disease. *Cell Reports, 36*(1), 109308. https://doi.org/10.1016/j.celrep.2021.109308

Bhattacharya, P., Gopisetty, A., Ganesh, B. B., Sheng, J. R., et al. (2011). GM-CSF-induced, bone-marrow-derived dendritic cells can expand natural Tregs and induce adaptive Tregs by different mechanisms. *Journal of Leukocyte Biology, 89*(2), 235–249. https://doi.org/10.1189/jlb.0310154

Bolte, A. C., & Lukens, J. R. (2018). Th17 cells in Parkinson's disease: The bane of the midbrain. *Cell Stem Cell, 23*(1), 5–6. https://doi.org/10.1016/j.stem.2018.06.016

Breidert, T., Callebert, J., Heneka, M. T., Landreth, G., et al. (2002). Protective action of the peroxisome proliferator-activated receptor-gamma agonist pioglitazone in a mouse model of Parkinson's disease. *Journal of Neurochemistry, 82*(3), 615–624. https://doi.org/10.1046/j.1471-4159.2002.00990.x

Brooks, B., Sanajak, M., & Roelke, K. (2005). Phase 2B randomized dose-ranging clinical trial of tamoxifen, a selective estrogen receptor modulator (SERM), in amyotrophic lateral sclerosis (ALS): Sensitivity analyses of discordance between survival and functional outcomes with long-term follow-up. *Amyotrophic Lateral Sclerosis, 6*, 118.

Brooks, B. R., Berry, J. D., Ciepielewska, M., Liu, Y., et al. (2022). Intravenous edaravone treatment in ALS and survival: An exploratory, retrospective, administrative claims analysis. *EClinicalMedicine, 52*, 101590. https://doi.org/10.1016/j.eclinm.2022.101590

Brys, M., Fanning, L., Hung, S., Ellenbogen, A., et al. (2019). Randomized phase I clinical trial of anti-alpha-synuclein antibody BIIB054. *Movement Disorders, 34*(8), 1154–1163. https://doi.org/10.1002/mds.27738

Burns, M. R., & McFarland, N. R. (2020). Current management and emerging therapies in multiple system atrophy. *Neurotherapeutics, 17*(4), 1582–1602. https://doi.org/10.1007/s13311-020-00890-x

Butler, D. C., & Messer, A. (2011). Bifunctional anti-huntingtin proteasome-directed intrabodies mediate efficient degradation of mutant huntingtin exon 1 protein fragments. *PLoS One, 6*(12), e29199. https://doi.org/10.1371/journal.pone.0029199

Butler, D. C., McLear, J. A., & Messer, A. (2012). Engineered antibody therapies to counteract mutant huntingtin and related toxic intracellular proteins. *Progress in Neurobiology, 97*(2), 190–204. https://doi.org/10.1016/j.pneurobio.2011.11.004

Cady, J., Koval, E. D., Benitez, B. A., Zaidman, C., et al. (2014). TREM2 variant p.R47H as a risk factor for sporadic amyotrophic lateral sclerosis. *JAMA Neurology, 71*(4), 449–453. https://doi.org/10.1001/jamaneurol.2013.6237

Camu, W., Mickunas, M., Veyrune, J. L., Payan, C., et al. (2020). Repeated 5-day cycles of low dose aldesleukin in amyotrophic lateral sclerosis (IMODALS): A phase 2a randomised, double-blind, placebo-controlled trial. *eBioMedicine, 59*, 102844. https://doi.org/10.1016/j.ebiom.2020.102844

Canever, J. B., Soares, E. S., de Avelar, N. C. P., & Cimarosti, H. I. (2023). Targeting alpha-synuclein post-translational modifications in Parkinson's disease. *Behavioural Brain Research, 439*, 114204. https://doi.org/10.1016/j.bbr.2022.114204

Cankaya, S., Cankaya, B., Kilic, U., Kilic, E., et al. (2019). The therapeutic role of minocycline in Parkinson's disease. *Drugs Context, 8*, 212553. https://doi.org/10.7573/dic.212553

Cao, B., Chen, X., Zhang, L., Wei, Q., et al. (2020). Elevated percentage of CD3(+) T-cells and CD4(+)/CD8(+) ratios in multiple system atrophy patients. *Frontiers in Neurology, 11*, 658. https://doi.org/10.3389/fneur.2020.00658

Capriotti, T., & Terzakis, K. (2016). Parkinson disease. *Home Healthcare Now, 34*(6), 300–307. https://doi.org/10.1097/NHH.0000000000000398

Caspritz, G., & Hadden, J. (1987). The immunopharmacology of immunotoxicology, and immunorestoration. *Toxicologic Pathology, 15*(3), 320–332. https://doi.org/10.1177/019262338701500310

Cerillo, J. L., & Parmar, M. (2024). Tofersen. In *StatPearls*. StatPearls Publishing; 2023 Jan-, PMID: 37603661.

Chan, L., & Yokota, T. (2020). Development and Ccinical applications of antisense oligonucleotide gapmers. *Methods in Molecular Biology, 2176*, 21–47. https://doi.org/10.1007/978-1-0716-0771-8_2

Chatterjee, D., Bhatt, M., Butler, D., De Genst, E., et al. (2018). Proteasome-targeted nanobodies alleviate pathology and functional decline in an alpha-synuclein-based Parkinson's disease model. *NPJ Parkinsons Disease, 4*(1), 25. https://doi.org/10.1038/s41531-018-0062-4

Chen, H., Zhang, S. M., Hernan, M. A., Schwarzschild, M. A., et al. (2003). Nonsteroidal anti-inflammatory drugs and the risk of Parkinson disease. *Archives of Neurology, 60*(8), 1059–1064. https://doi.org/10.1001/archneur.60.8.1059

Chen, H., Jacobs, E., Schwarzschild, M. A., McCullough, M. L., et al. (2005). Nonsteroidal antiinflammatory drug use and the risk for Parkinson's disease. *Annals of Neurology, 58*(6), 963–967. https://doi.org/10.1002/ana.20682

Chen, Z., Chen, S., & Liu, J. (2018). The role of T cells in the pathogenesis of Parkinson's disease. *Progress in Neurobiology, 169*, 1–23. https://doi.org/10.1016/j.pneurobio.2018.08.002

Chen, P. C., Hsieh, Y. C., Huang, C. C., & Hu, C. J. (2020a). Tamoxifen for amyotrophic lateral sclerosis: A randomized double-blind clinical trial. *Medicine (Baltimore), 99*(22), e20423. https://doi.org/10.1097/MD.0000000000020423

Chen, Y. H., Yu, S. J., Wu, K. J., Wang, Y. S., et al. (2020b). Downregulation of alpha-Synuclein Protein Levels by an Intracellular Single-Chain Antibody. *Journal of Parkinson's Disease, 10*(2), 573–590. https://doi.org/10.3233/JPD-191787

Chen, Y. H., Wu, K. J., Hsieh, W., Harvey, B. K., et al. (2021). Administration of AAV-alpha synuclein NAC antibody improves locomotor behavior in rats overexpressing alpha synuclein. *Genes (Basel), 12*(6), 948. https://doi.org/10.3390/genes12060948

Chen, W., Guo, L., Li, M., Wei, C., et al. (2022). The pathogenesis of amyotrophic lateral sclerosis: Mitochondrial dysfunction, protein misfolding and epigenetics. *Brain Research, 1786*, 147904. https://doi.org/10.1016/j.brainres.2022.147904

Chia, R., Chio, A., & Traynor, B. J. (2018). Novel genes associated with amyotrophic lateral sclerosis: Diagnostic and clinical implications. *Lancet Neurology, 17*(1), 94–102. https://doi.org/10.1016/S1474-4422(17)30401-5

Childs-Disney, J. L., Yang, X., Gibaut, Q. M. R., Tong, Y., et al. (2022). Targeting RNA structures with small molecules. *Nature Reviews: Drug Discovery, 21*(10), 736–762. https://doi.org/10.1038/s41573-022-00521-4

Chiricozzi, A., De Simone, C., Fossati, B., & Peris, K. (2019). Emerging treatment options for the treatment of moderate to severe plaque psoriasis and psoriatic arthritis: Evaluating bimekizumab and its therapeutic potential. *Psoriasis (Auckl), 9*, 29–35. https://doi.org/10.2147/PTT.S179283

Chung, E. S., Lee, G., Lee, C., Ye, M., et al. (2015). Bee venom phospholipase A2, a novel Foxp3+ regulatory T cell inducer, protects dopaminergic neurons by modulating neuroinflammatory responses in a mouse model of Parkinson's disease. *Journal of Immunology, 195*(10), 4853–4860. https://doi.org/10.4049/jimmunol.1500386

Chwiszczuk, L. J., Breitve, M. H., Kirsebom, B. B., Selnes, P., et al. (2023). The ANeED study - ambroxol in new and early dementia with Lewy bodies (DLB): Protocol for a phase IIa multicentre, randomised, double-blinded and placebo-controlled trial. *Frontiers in Aging Neuroscience, 15*, 1163184. https://doi.org/10.3389/fnagi.2023.1163184

Ciaramella, A., Salani, F., Bizzoni, F., Pontieri, F. E., et al. (2013). Blood dendritic cell frequency declines in idiopathic Parkinson's disease and is associated with motor symptom severity. *PLoS One, 8*(6), e65352. https://doi.org/10.1371/journal.pone.0065352

Clarke, B. E., & Patani, R. (2020). The microglial component of amyotrophic lateral sclerosis. *Brain, 143*(12), 3526–3539. https://doi.org/10.1093/brain/awaa309

Coban, A., Ismail Kucukali, C., Bilgic, B., Yalcinkaya, N., et al. (2014). Evaluation of incidence and clinical features of antibody-associated autoimmune encephalitis mimicking dementia. *Behavioural Neurology, 2014*, 935379. https://doi.org/10.1155/2014/935379

Cole, T. A., Zhao, H., Collier, T. J., Sandoval, I., et al. (2021). alpha-Synuclein antisense oligonucleotides as a disease-modifying therapy for Parkinson's disease. *JCI Insight, 6*(5), e135633. https://doi.org/10.1172/jci.insight.135633

Colpo, G. D., Rocha, N. P., Stimming, E. F., & Teixeira, A. L. (2017). Immunomodulatory strategies for Huntington's disease treatment. *CNS & Neurological Disorders Drug Targets, 16*(8), 936–944. https://doi.org/10.2174/1871527316666170613084801

Crippa, V., D'Agostino, V. G., Cristofani, R., Rusmini, P., et al. (2016). Transcriptional induction of the heat shock protein B8 mediates the clearance of misfolded proteins responsible for motor neuron diseases. *Scientific Reports, 6*, 22827. https://doi.org/10.1038/srep22827

Cui, G., Zhang, Y., Gong, Z., Zhang, J. Z., et al. (2009). Induction of CD4+CD25+Foxp3+ regulatory T cell response by glatiramer acetate in type 1 diabetes. *Cell Research, 19*(5), 574–583. https://doi.org/10.1038/cr.2009.12

Cunningham, S., Piedra, P. A., Martinon-Torres, F., Szymanski, H., et al. (2021). Nebulised ALX-0171 for respiratory syncytial virus lower respiratory tract infection in hospitalised children: A double-blind, randomised, placebo-controlled, phase 2b trial. *The Lancet Respiratory Medicine, 9*(1), 21–32. https://doi.org/10.1016/S2213-2600(20)30320-9

Dalbeth, N., Lauterio, T. J., & Wolfe, H. R. (2014). Mechanism of action of colchicine in the treatment of gout. *Clinical Therapeutics, 36*(10), 1465–1479. https://doi.org/10.1016/j.clinthera.2014.07.017

Dauer, W., & Przedborski, S. (2003). Parkinson's disease: Mechanisms and models. *Neuron, 39*(6), 889–909. https://doi.org/10.1016/s0896-6273(03)00568-3

Deczkowska, A., Keren-Shaul, H., Weiner, A., Colonna, M., et al. (2018). Disease-associated microglia: A universal immune sensor of neurodegeneration. *Cell, 173*(5), 1073–1081. https://doi.org/10.1016/j.cell.2018.05.003

Dehmer, T., Heneka, M. T., Sastre, M., Dichgans, J., et al. (2004). Protection by pioglitazone in the MPTP model of Parkinson's disease correlates with I kappa B alpha induction and block of NF kappa B and iNOS activation. *Journal of Neurochemistry, 88*(2), 494–501. https://doi.org/10.1046/j.1471-4159.2003.02210.x

Denis, H. L., David, L. S., & Cicchetti, F. (2019). Antibody-based therapies for Huntington's disease: Current status and future directions. *Neurobiology of Disease, 132*, 104569. https://doi.org/10.1016/j.nbd.2019.104569

D'Erchia, A. M., Gallo, A., Manzari, C., Raho, S., et al. (2017). Massive transcriptome sequencing of human spinal cord tissues provides new insights into motor neuron degeneration in ALS. *Scientific Reports, 7*(1), 10046. https://doi.org/10.1038/s41598-017-10488-7

Dodel, R., Spottke, A., Gerhard, A., Reuss, A., et al. (2010). Minocycline 1-year therapy in multiple-system-atrophy: Effect on clinical symptoms and [(11)C] (R)-PK11195 PET (MEMSA-trial). *Movement Disorders, 25*(1), 97–107. https://doi.org/10.1002/mds.22732

Dols-Icardo, O., Montal, V., Sirisi, S., Lopez-Pernas, G., et al. (2020). Motor cortex transcriptome reveals microglial key events in amyotrophic lateral sclerosis. *Neurology: Neuroimmunology & Neuroinflammation, 7*(5), e829. https://doi.org/10.1212/NXI.0000000000000829

Dörner, T., Weinblatt, M., Van Beneden, K., Dombrecht, E. J., et al. (2017). FRI0239 results of a phase 2B study of vobarilizumab, an anti-interleukin-6 receptor nanobody, as monotherapy in patients with moderate to severe rheumatoid arthritis. *Annals of the Rheumatic Diseases, 76*, 575.

Dutta, D., Jana, M., Majumder, M., Mondal, S., et al. (2021). Selective targeting of the TLR2/MyD88/NF-kappaB pathway reduces alpha-synuclein spreading in vitro and in vivo. *Nature Communications, 12*(1), 5382. https://doi.org/10.1038/s41467-021-25767-1

Elgebaly, A., Abdelazeim, B., Mattar, O., Gadelkarim, M., et al. (2016). Meta-analysis of the safety and efficacy of droxidopa for neurogenic orthostatic hypotension. *Clinical Autonomic Research, 26*(3), 171–180. https://doi.org/10.1007/s10286-016-0349-7

Engelhardt, J. I., Tajti, J., & Appel, S. H. (1993). Lymphocytic infiltrates in the spinal cord in amyotrophic lateral sclerosis.

Archives of Neurology, 50(1), 30–36. https://doi.org/10.1001/archneur.1993.00540010026013

Escobar, M. F., Nassar, A. H., Theron, G., Barnea, E. R., et al. (2022). FIGO recommendations on the management of postpartum hemorrhage 2022. *International Journal of Gynaecology and Obstetrics, 157 Suppl 1(Suppl 1)*, 3–50. https://doi.org/10.1002/ijgo.14116

Estevez-Fraga, C., Tabrizi, S. J., & Wild, E. J. (2022). Huntington's Disease Ccinical trials corner: November 2022. *Journal of Huntington's Disease, 11*(4), 351–367. https://doi.org/10.3233/JHD-229006

Fang, T., Al Khleifat, A., Meurgey, J. H., Jones, A., et al. (2018). Stage at which riluzole treatment prolongs survival in patients with amyotrophic lateral sclerosis: A retrospective analysis of data from a dose-ranging study. *Lancet Neurology, 17*(5), 416–422. https://doi.org/10.1016/S1474-4422(18)30054-1

Fatoba, O., Ohtake, Y., Itokazu, T., & Yamashita, T. (2020). Immunotherapies in Huntington's disease and alpha-Synucleinopathies. *Frontiers in Immunology, 11*, 337. https://doi.org/10.3389/fimmu.2020.00337

Feigin, A., Evans, E. E., Fisher, T. L., Leonard, J. E., et al. (2022). Pepinemab antibody blockade of SEMA4D in early Huntington's disease: A randomized, placebo-controlled, phase 2 trial. *Nature Medicine, 28*(10), 2183–2193. https://doi.org/10.1038/s41591-022-01919-8

Feldman, E. L., Goutman, S. A., Petri, S., Mazzini, L., et al. (2022). Amyotrophic lateral sclerosis. *Lancet, 400*(10360), 1363–1380. https://doi.org/10.1016/S0140-6736(22)01272-7

Ferrazzoli, D., Carter, A., Ustun, F. S., Palamara, G., et al. (2016). Dopamine replacement therapy, learning and reward prediction in Parkinson's disease: Implications for rehabilitation. *Frontiers in Behavioral Neuroscience, 10*, 121. https://doi.org/10.3389/fnbeh.2016.00121

Fiala, M., Mizwicki, M. T., Weitzman, R., Magpantay, L., et al. (2013). Tocilizumab infusion therapy normalizes inflammation in sporadic ALS patients. *American Journal of Neurodegenerative Disease, 2*(2), 129–139.

Finkelstein, D. I., Billings, J. L., Adlard, P. A., Ayton, S., et al. (2017). The novel compound PBT434 prevents iron mediated neurodegeneration and alpha-synuclein toxicity in multiple models of Parkinson's disease. *Acta Neuropathologica Communications, 5*(1), 53. https://doi.org/10.1186/s40478-017-0456-2

Fjord-Larsen, L., Thougaard, A., Wegener, K. M., Christiansen, J., et al. (2021). Nonclinical safety evaluation, pharmacokinetics, and target engagement of Lu AF82422, a monoclonal IgG1 antibody against alpha-synuclein in development for treatment of synucleinopathies. *MAbs, 13*(1), 1994690. https://doi.org/10.1080/19420862.2021.1994690

Flabeau, O., Meissner, W. G., & Tison, F. (2010). Multiple system atrophy: Current and future approaches to management. *Therapeutic Advances in Neurological Disorders, 3*(4), 249–263. https://doi.org/10.1177/1756285610375328

Fleming, S. M., Davis, A., & Simons, E. (2022). Targeting alpha-synuclein via the immune system in Parkinson's disease: Current vaccine therapies. *Neuropharmacology, 202*, 108870. https://doi.org/10.1016/j.neuropharm.2021.108870

Fournier, C. N., Schoenfeld, D., Berry, J. D., Cudkowicz, M. E., et al. (2018). An open label study of a novel immunosuppression intervention for the treatment of amyotrophic lateral sclerosis. *Amyotrophic Lateral Sclerosis and Frontotemporal Degeneration, 19*(3-4), 242–249. https://doi.org/10.1080/21678421.2017.1421666

Fowler, A. J., Hebron, M., Missner, A. A., Wang, R., et al. (2019). Multikinase Abl/DDR/Src unhibition produces optimal effects for tyrosine kinase inhibition in neurodegeneration. *Drugs in R&D, 19*(2), 149–166. https://doi.org/10.1007/s40268-019-0266-z

Galabova-Kovacs, G., Catalanotti, F., Matzen, D., Reyes, G. X., et al. (2008). Essential role of B-Raf in oligodendrocyte maturation and myelination during postnatal central nervous system development. *The Journal of Cell Biology, 180*(5), 947–955. https://doi.org/10.1083/jcb.200709069

Games, D., Seubert, P., Rockenstein, E., Patrick, C., et al. (2013). Axonopathy in an alpha-synuclein transgenic model of Lewy body disease is associated with extensive accumulation of C-terminal-truncated alpha-synuclein. *The American Journal of Pathology, 182*(3), 940–953. https://doi.org/10.1016/j.ajpath.2012.11.018

Games, D., Valera, E., Spencer, B., Rockenstein, E., et al. (2014). Reducing C-terminal-truncated alpha-synuclein by immunotherapy attenuates neurodegeneration and propagation in Parkinson's disease-like models. *The Journal of Neuroscience, 34*(28), 9441–9454. https://doi.org/10.1523/JNEUROSCI.5314-13.2014

Gao, H. M., Zhang, F., Zhou, H., Kam, W., et al. (2011). Neuroinflammation and alpha-synuclein dysfunction potentiate each other, driving chronic progression of neurodegeneration in a mouse model of Parkinson's disease. *Environmental Health Perspectives, 119*(6), 807–814. https://doi.org/10.1289/ehp.1003013

Gate, D., Tapp, E., Leventhal, O., Shahid, M., et al. (2021). CD4(+) T cells contribute to neurodegeneration in Lewy body dementia. *Science, 374*(6569), 868–874. https://doi.org/10.1126/science.abf7266

Geloso, M. C., Corvino, V., Marchese, E., Serrano, A., et al. (2017). The dual role of microglia in ALS: Mechanisms and therapeutic approaches. *Frontiers in Aging Neuroscience, 9*, 242. https://doi.org/10.3389/fnagi.2017.00242

Gendelman, H. E., Zhang, Y., Santamaria, P., Olson, K. E., et al. (2017). Evaluation of the safety and immunomodulatory effects of sargramostim in a randomized, double-blind phase 1 clinical Parkinson's disease trial. *NPJ Parkinsons Disease, 3*, 10. https://doi.org/10.1038/s41531-017-0013-5

Gerhard, A., Pavese, N., Hotton, G., Turkheimer, F., et al. (2006). In vivo imaging of microglial activation with [11C](R)-PK11195 PET in idiopathic Parkinson's disease. *Neurobiology of Disease, 21*(2), 404–412. https://doi.org/10.1016/j.nbd.2005.08.002

Giovannelli, I., Bayatti, N., Brown, A., Wang, D., et al. (2021). Amyotrophic lateral sclerosis transcriptomics reveals immunological effects of low-dose interleukin-2. *Brain Communications, 3*(3), fcab141. https://doi.org/10.1093/braincomms/fcab141

Goh, Y. Y., Saunders, E., Pavey, S., Rushton, E., et al. (2023). Multiple system atrophy. *Practical Neurology, 23*(3), 208–221. https://doi.org/10.1136/pn-2020-002797

Gold, R., Kappos, L., Arnold, D. L., Bar-Or, A., et al. (2012). Placebo-controlled phase 3 study of oral BG-12 for relapsing multiple sclerosis. *The New England Journal of Medicine, 367*(12), 1098–1107. https://doi.org/10.1056/NEJMoa1114287

Goldman, J. G., Goetz, C. G., Brandabur, M., Sanfilippo, M., et al. (2008). Effects of dopaminergic medications on psychosis and motor function in dementia with Lewy bodies. *Movement Disorders, 23*(15), 2248–2250. https://doi.org/10.1002/mds.22322

Gordon, R., Albornoz, E. A., Christie, D. C., Langley, M. R., et al. (2018). Inflammasome inhibition prevents alpha-synuclein pathology and dopaminergic neurodegeneration in mice. *Science Translational Medicine, 10*(465), eaah4066. https://doi.org/10.1126/scitranslmed.aah4066

Goutman, S. A., Hardiman, O., Al-Chalabi, A., Chio, A., et al. (2022). Recent advances in the diagnosis and prognosis of amyotrophic lateral sclerosis. *Lancet Neurology, 21*(5), 480–493. https://doi.org/10.1016/S1474-4422(21)00465-8

Gassmann, A. E., & van Furth, R. (1975). The effect of azathioprine (Imuran) on the kinetics of monocytes and macrophages during the normal steady state and an acute inflammatory reaction. *Blood, 46*(1), 51–64.

Guilliams, T., El-Turk, F., Buell, A. K., O'Day, E. M., et al. (2013). Nanobodies raised against monomeric alpha-synuclein distinguish between fibrils at different maturation stages. *Journal of*

Molecular Biology, 425(14), 2397–2411. https://doi.org/10.1016/j.jmb.2013.01.040

Gumina, V., Onesto, E., Colombrita, C., Maraschi, A., et al. (2019). Inter-species differences in regulation of the progranulin-sortilin axis in TDP-43 cell models of neurodegeneration. *International Journal of Molecular Sciences, 20*(23), 5866. https://doi.org/10.3390/ijms20235866

Harms, A. S., Kordower, J. H., Sette, A., Lindestam Arlehamn, C. S., et al. (2021). Inflammation in experimental models of alpha-aynucleinopathies. *Movement Disorders, 36*(1), 37–49. https://doi.org/10.1002/mds.28264

Hauser, R. A., Isaacson, S., Lisk, J. P., Hewitt, L. A., et al. (2015). Droxidopa for the short-term treatment of symptomatic neurogenic orthostatic hypotension in Parkinson's disease (nOH306B). *Movement Disorders, 30*(5), 646–654. https://doi.org/10.1002/mds.26086

Hayes, M. T. (2019). Parkinson's disease and Parkinsonism. *The American Journal of Medicine, 132*(7), 802–807. https://doi.org/10.1016/j.amjmed.2019.03.001

Hebron, M. L., Lonskaya, I., & Moussa, C. E. (2013a). Nilotinib reverses loss of dopamine neurons and improves motor behavior via autophagic degradation of alpha-synuclein in Parkinson's disease models. *Human Molecular Genetics, 22*(16), 3315–3328. https://doi.org/10.1093/hmg/ddt192

Hebron, M. L., Lonskaya, I., Sharpe, K., Weerasinghe, P. P., et al. (2013b). Parkin ubiquitinates Tar-DNA binding protein-43 (TDP-43) and promotes its cytosolic accumulation via interaction with histone deacetylase 6 (HDAC6). *The Journal of Biological Chemistry, 288*(6), 4103–4115. https://doi.org/10.1074/jbc.M112.419945

Hebron, M. L., Lonskaya, I., Olopade, P., Selby, S. T., et al. (2014). Tyrosine kinase inhibition regulates early systemic immune changes and modulates the neuroimmune response in alpha-synucleinopathy. *Journal of Clinical & Cellular Immunology, 5*, 259. https://doi.org/10.4172/2155-9899.1000259

Henkel, J. S., Engelhardt, J. I., Siklos, L., Simpson, E. P., et al. (2004). Presence of dendritic cells, MCP-1, and activated microglia/macrophages in amyotrophic lateral sclerosis spinal cord tissue. *Annals of Neurology, 55*(2), 221–235. https://doi.org/10.1002/ana.10805

Henkel, J. S., Beers, D. R., Zhao, W., & Appel, S. H. (2009). Microglia in ALS: The good, the bad, and the resting. *Journal of Neuroimmune Pharmacology, 4*(4), 389–398. https://doi.org/10.1007/s11481-009-9171-5

Henkel, J. S., Beers, D. R., Wen, S., Rivera, A. L., et al. (2013). Regulatory T-lymphocytes mediate amyotrophic lateral sclerosis progression and survival. *EMBO Molecular Medicine, 5*(1), 64–79. https://doi.org/10.1002/emmm.201201544

Henriksen, A. L., St Dennis, C., Setter, S. M., & Tran, J. T. (2006). Dementia with lewy bodies: Therapeutic opportunities and pitfalls. *The Consultant Pharmacist, 21*(7), 563–575. https://doi.org/10.4140/tcp.n.2006.563

Heras-Garvin, A., Refolo, V., Schmidt, C., Malfertheiner, K., et al. (2021). ATH434 reduces alpha-Synuclein-related neurodegeneration in a Murine model of multiple system atrophy. *Movement Disorders, 36*(11), 2605–2614. https://doi.org/10.1002/mds.28714

Hershey, L. A., & Coleman-Jackson, R. (2019). Pharmacological management of dementia with Lewy bodies. *Drugs & Aging, 36*(4), 309–319. https://doi.org/10.1007/s40266-018-00636-7

Hirsch, E. C., & Hunot, S. (2009). Neuroinflammation in Parkinson's disease: A target for neuroprotection? *Lancet Neurology, 8*(4), 382–397. https://doi.org/10.1016/S1474-4422(09)70062-6

Hirsch, E. C., Vyas, S., & Hunot, S. (2012). Neuroinflammation in Parkinson's disease. *Parkinsonism & Related Disorders, 18*(Suppl 1), S210–S212. https://doi.org/10.1016/S1353-8020(11)70065-7

Hoffmann, A., Ettle, B., Bruno, A., Kulinich, A., et al. (2016). Alpha-synuclein activates BV2 microglia dependent on its aggregation state. *Biochemical and Biophysical Research Communications, 479*(4), 881–886. https://doi.org/10.1016/j.bbrc.2016.09.109

Huntington Study Group DI. (2010). A futility study of minocycline in Huntington's disease. *Movement Disorders, 25*(13), 2219–2224. https://doi.org/10.1002/mds.23236

Hutter-Saunders, J. A., Gendelman, H. E., & Mosley, R. L. (2012). Murine motor and behavior functional evaluations for acute 1-methyl-4-phenyl-1,2,3,6-tetrahydropyridine (MPTP) intoxication. *Journal of Neuroimmune Pharmacology, 7*(1), 279–288. https://doi.org/10.1007/s11481-011-9269-4

Iba, M., Kim, C., Sallin, M., Kwon, S., et al. (2020). Neuroinflammation is associated with infiltration of T cells in Lewy body disease and alpha-synuclein transgenic models. *Journal of Neuroinflammation, 17*(1), 214. https://doi.org/10.1186/s12974-020-01888-0

Iljina, M., Hong, L., Horrocks, M. H., Ludtmann, M. H., et al. (2017). Nanobodies raised against monomeric a-synuclein inhibit fibril formation and destabilize toxic oligomeric species. *BMC Biology, 15*(1), 57. https://doi.org/10.1186/s12915-017-0390-6

Imamura, K., Izumi, Y., Banno, H., Uozumi, R., et al. (2019). Induced pluripotent stem cell-based drug repurposing for amyotrophic lateral sclerosis medicine (iDReAM) study: Protocol for a phase I dose escalation study of bosutinib for amyotrophic lateral sclerosis patients. *BMJ Open, 9*(12), e033131. https://doi.org/10.1136/bmjopen-2019-033131

Imamura, K., Izumi, Y., Nagai, M., Nishiyama, K., et al. (2022). Safety and tolerability of bosutinib in patients with amyotrophic lateral sclerosis (iDReAM study): A multicentre, open-label, dose-escalation phase 1 trial. *EClinicalMedicine, 53*, 101707. https://doi.org/10.1016/j.eclinm.2022.101707

Inoue, Y., Ayaki, T., Ishimoto, T., Yamakado, H., et al. (2021). The stimulator of interferon genes (STING) pathway is upregulated in striatal astrocytes of patients with multiple system atrophy. *Neuroscience Letters, 757*, 135972. https://doi.org/10.1016/j.neulet.2021.135972

Investigators, N. N.-P. (2008). A pilot clinical trial of creatine and minocycline in early Parkinson disease: 18-month results. *Clinical Neuropharmacology, 31*(3), 141–150. https://doi.org/10.1097/WNF.0b013e3181342f32

Investigators NETiPDF-Z. (2015). Pioglitazone in early Parkinson's disease: A phase 2, multicentre, double-blind, randomised trial. *Lancet Neurology, 14*(8), 795–803. https://doi.org/10.1016/S1474-4422(15)00144-1

Ishizawa, K., Komori, T., Arai, N., Mizutani, T., et al. (2008). Glial cytoplasmic inclusions and tissue injury in multiple system atrophy: A quantitative study in white matter (olivopontocerebellar system) and gray matter (nigrostriatal system). *Neuropathology, 28*(3), 249–257. https://doi.org/10.1111/j.1440-1789.2007.00855.x

Jankovic, J., Goodman, I., Safirstein, B., Marmon, T. K., et al. (2018). Safety and tolerability of multiple ascending doses of PRX002/RG7935, an anti-alpha-synuclein monoclonal antibody, in patients with Parkinson disease: A randomized clinical trial. *JAMA Neurology, 75*(10), 1206–1214. https://doi.org/10.1001/jamaneurol.2018.1487

Jellinger, K. A. (2014). Neuropathology of multiple system atrophy: New thoughts about pathogenesis. *Movement Disorders, 29*(14), 1720–1741. https://doi.org/10.1002/mds.26052

Jellinger, K. A. (2018). Multiple system atrophy: An oligodendroglioneural synucleinopathy1. *Journal of Alzheimer's Disease, 62*(3), 1141–1179. https://doi.org/10.3233/JAD-170397

Jellinger, K. A., & Wenning, G. K. (2016). Multiple system atrophy: Pathogenic mechanisms and biomarkers. *Journal of Neural Transmission (Vienna), 123*(6), 555–572. https://doi.org/10.1007/s00702-016-1545-2

Jiang, Y., Alam, J. J., Gomperts, S. N., Maruff, P., et al. (2022). Preclinical and randomized clinical evaluation of the p38alpha kinase inhibitor neflamapimod for basal forebrain cholinergic degeneration. *Nature Communications, 13*(1), 5308. https://doi.org/10.1038/s41467-022-32944-3

Kabir, M. A., & Chelimsky, T. C. (2019). Pure autonomic failure. *Handbook of Clinical Neurology, 161*, 413–422. https://doi.org/10.1016/B978-0-444-64142-7.00064-3

Kacher, R., Lamaziere, A., Heck, N., Kappes, V., et al. (2019). CYP46A1 gene therapy deciphers the role of brain cholesterol metabolism in Huntington's disease. *Brain, 142*(8), 2432–2450. https://doi.org/10.1093/brain/awz174

Kalia, L. V., & Lang, A. E. (2015). Parkinson's disease. *Lancet, 386*(9996), 896–912. https://doi.org/10.1016/S0140-6736(14)61393-3

Kam, T. I., Hinkle, J. T., Dawson, T. M., & Dawson, V. L. (2020). Microglia and astrocyte dysfunction in parkinson's disease. *Neurobiology of Disease, 144*, 105028. https://doi.org/10.1016/j.nbd.2020.105028

Karasinska, J. M., & Hayden, M. R. (2011). Cholesterol metabolism in Huntington disease. *Nature Reviews. Neurology, 7*(10), 561–572. https://doi.org/10.1038/nrneurol.2011.132

Karuppagounder, S. S., Wang, H., Kelly, T., Rush, R., et al. (2023). The c-Abl inhibitor IkT-148009 suppresses neurodegeneration in mouse models of heritable and sporadic Parkinson's disease. *Science Translational Medicine, 15*(679), eabp9352. https://doi.org/10.1126/scitranslmed.abp9352

Katsuse, O., Iseki, E., & Kosaka, K. (2003). Immunohistochemical study of the expression of cytokines and nitric oxide synthases in brains of patients with dementia with Lewy bodies. *Neuropathology, 23*(1), 9–15. https://doi.org/10.1046/j.1440-1789.2003.00483.x

Kaufmann, H., Freeman, R., Biaggioni, I., Low, P., et al. (2014). Droxidopa for neurogenic orthostatic hypotension: A randomized, placebo-controlled, phase 3 trial. *Neurology, 83*(4), 328–335. https://doi.org/10.1212/WNL.0000000000000615

Kaufmann, H., Vickery, R., Wang, W., Kanodia, J., et al. (2021). Safety and efficacy of ampreloxetine in symptomatic neurogenic orthostatic hypotension: A phase 2 trial. *Clinical Autonomic Research, 31*(6), 699–711. https://doi.org/10.1007/s10286-021-00827-0

Kawahata, I., Finkelstein, D. I., & Fukunaga, K. (2022). Pathogenic impact of alpha-synuclein phosphorylation and its kinases in alpha-synucleinopathies. *International Journal of Molecular Sciences, 23*(11), 6216. https://doi.org/10.3390/ijms23116216

Kawai, H., Nakai, H., Suga, M., Yuki, S., et al. (1997). Effects of a novel free radical scavenger, MCl-186, on ischemic brain damage in the rat distal middle cerebral artery occlusion model. *The Journal of Pharmacology and Experimental Therapeutics, 281*(2), 921–927.

Kearney, P. J., Zhang, Y., Tan, Y., Kahuno, E., et al. (2023). Rit2 silencing in dopamine neurons drives a progressive Parkinsonian phenotype. *Res Sq In Press*. https://doi.org/10.21203/rs.3.rs-2944614/v1

Kikuchi, T., Morizane, A., Doi, D., Magotani, H., et al. (2017). Human iPS cell-derived dopaminergic neurons function in a primate Parkinson's disease model. *Nature, 548*(7669), 592–596. https://doi.org/10.1038/nature23664

Ko, H. S., Lee, Y., Shin, J. H., Karuppagounder, S. S., et al. (2010). Phosphorylation by the c-Abl protein tyrosine kinase inhibits parkin's ubiquitination and protective function. *Proceedings of the National Academy of Sciences of the United States of America, 107*(38), 16691–16696. https://doi.org/10.1073/pnas.1006083107

Kollensperger, M., Geser, F., Ndayisaba, J. P., Boesch, S., et al. (2010). Presentation, diagnosis, and management of multiple system atrophy in Europe: Final analysis of the European multiple system atrophy registry. *Movement Disorders, 25*(15), 2604–2612. https://doi.org/10.1002/mds.23192

Korobeynikov, V. A., Lyashchenko, A. K., Blanco-Redondo, B., Jafar-Nejad, P., et al. (2022). Antisense oligonucleotide silencing of FUS expression as a therapeutic approach in amyotrophic lateral sclerosis. *Nature Medicine, 28*(1), 104–116. https://doi.org/10.1038/s41591-021-01615-z

Kosloski, L. M., Kosmacek, E. A., Olson, K. E., Mosley, R. L., et al. (2013). GM-CSF induces neuroprotective and anti-inflammatory responses in 1-methyl-4-phenyl-1,2,3,6-tetrahydropyridine intoxicated mice. *Journal of Neuroimmunology, 265*(1-2), 1–10. https://doi.org/10.1016/j.jneuroim.2013.10.009

Krach, F., Stemick, J., Boerstler, T., Weiss, A., et al. (2022). An alternative splicing modulator decreases mutant HTT and improves the molecular fingerprint in Huntington's disease patient neurons. *Nature Communications, 13*(1), 6797. https://doi.org/10.1038/s41467-022-34419-x

Kubler, D., Wachter, T., Cabanel, N., Su, Z., et al. (2019). Widespread microglial activation in multiple system atrophy. *Movement Disorders, 34*(4), 564–568. https://doi.org/10.1002/mds.27620

Kuklina, E. (2021). Semaphorin 4D as a guidance molecule in the immune system. *International Reviews of Immunology, 40*(4), 268–273. https://doi.org/10.1080/08830185.2021.1905807

Kumanogoh, A., & Kikutani, H. (2001). The CD100-CD72 interaction: A novel mechanism of immune regulation. *Trends in Immunology, 22*(12), 670–676. https://doi.org/10.1016/s1471-4906(01)02087-7

Kumar, R., Claassen, D., Mongan, A., Hoehn, B., et al. (2023). A phase 2 open-label study to assess the safety, tolerability, pharmacokinetics, and pharmacodynamics of intravenous ANX005 in patients with, or at risk of, manifest Huntington's disease (HD) (S32.009). *Neurology, 100*, 3366.

Kuo, S. H., Tasset, I., Cheng, M. M., Diaz, A., et al. (2022). Mutant glucocerebrosidase impairs alpha-synuclein degradation by blockade of chaperone-mediated autophagy. *Science Advances, 8*(6), eabm6393. https://doi.org/10.1126/sciadv.abm6393

Kurnellas, M., Mitra, A., Schwabe, T., Paul, R., et al. (2023). Latozinemab, a novel progranulin-elevating therapy for frontotemporal dementia. *Journal of Translational Medicine, 21*(1), 387. https://doi.org/10.1186/s12967-023-04251-y

Kwon, D. (2021). Failure of genetic therapies for Huntington's devastates community. *Nature, 593*(7858), 180. https://doi.org/10.1038/d41586-021-01177-7

Lang, A. E., Siderowf, A. D., Macklin, E. A., Poewe, W., et al. (2022). Trial of cinpanemab in early Parkinson's disease. *The New England Journal of Medicine, 387*(5), 408–420. https://doi.org/10.1056/NEJMoa2203395

Lee, Y., Lee, S., Chang, S. C., & Lee, J. (2019). Significant roles of neuroinflammation in Parkinson's disease: Therapeutic targets for PD prevention. *Archives of Pharmacal Research, 42*(5), 416–425. https://doi.org/10.1007/s12272-019-01133-0

Levin, J., Maass, S., Schuberth, M., Giese, A., et al. (2019). Safety and efficacy of epigallocatechin gallate in multiple system atrophy (PROMESA): A randomised, double-blind, placebo-controlled trial. *Lancet Neurology, 18*(8), 724 735. https://doi.org/10.1016/S1474-4422(19)30141-3

Li, J., Zhao, J., Chen, L., Gao, H., et al. (2023). alpha-Synuclein induces Th17 differentiation and impairs the function and stability of Tregs by promoting RORC transcription in Parkinson's disease. *Brain, Behavior, and Immunity, 108*, 32–44. https://doi.org/10.1016/j.bbi.2022.10.023

Lincecum, J. M., Vieira, F. G., Wang, M. Z., Thompson, K., et al. (2010). From transcriptome analysis to therapeutic anti-CD40L treatment in the SOD1 model of amyotrophic lateral sclerosis. *Nature Genetics, 42*(5), 392–399. https://doi.org/10.1038/ng.557

Lindholm, D., Pham, D. D., Cascone, A., Eriksson, O., et al. (2016). c-Abl inhibitors enable insights into the pathophysiology and neuroprotection in Parkinson's disease. *Frontiers in Aging Neuroscience, 8*, 254. https://doi.org/10.3389/fnagi.2016.00254

Liu, Y., Dodart, J. C., Tran, H., Berkovitch, S., et al. (2021). Variant-selective stereopure oligonucleotides protect against pathologies associated with C9orf72-repeat expansion in preclinical models. *Nature Communications, 12*(1), 847. https://doi.org/10.1038/s41467-021-21112-8

Llansola, M., Sanchez-Perez, A., Cauli, O., & Felipo, V. (2005). Modulation of NMDA receptors in the cerebellum. 1. Properties of the NMDA receptor that modulate its function. *Cerebellum, 4*(3), 154–161. https://doi.org/10.1080/14734220510007996

Lonskaya, I., Desforges, N. M., Hebron, M. L., & Moussa, C. E. (2013a). Ubiquitination increases parkin activity to promote autophagic alpha-synuclein clearance. *PLoS One, 8*(12), e83914. https://doi.org/10.1371/journal.pone.0083914

Lonskaya, I., Hebron, M. L., Desforges, N. M., Franjie, A., et al. (2013b). Tyrosine kinase inhibition increases functional parkin-Beclin-1 interaction and enhances amyloid clearance and cognitive performance. *EMBO Molecular Medicine, 5*(8), 1247–1262. https://doi.org/10.1002/emmm.201302771

Lonskaya, I., Hebron, M. L., Desforges, N. M., Schachter, J. B., et al. (2014). Nilotinib-induced autophagic changes increase endogenous parkin level and ubiquitination, leading to amyloid clearance. *Journal of Molecular Medicine (Berlin, Germany), 92*(4), 373–386. https://doi.org/10.1007/s00109-013-1112-3

Loveland, P. M., Yu, J. J., Churilov, L., Yassi, N., et al. (2023). Investigation of inflammation in Lewy body dementia: A systematic scoping review. *International Journal of Molecular Sciences, 24*(15). https://doi.org/10.3390/ijms241512116

Low, P. A., Robertson, D., Gilman, S., Kaufmann, H., et al. (2014). Efficacy and safety of rifampicin for multiple system atrophy: A randomised, double-blind, placebo-controlled trial. *Lancet Neurology, 13*(3), 268–275. https://doi.org/10.1016/S1474-4422(13)70301-6

Low, P. A., Reich, S. G., Jankovic, J., Shults, C. W., et al. (2015). Natural history of multiple system atrophy in the USA: A prospective cohort study. *Lancet Neurology, 14*(7), 710–719. https://doi.org/10.1016/S1474-4422(15)00058-7

Luan, Z., Li, L., Higaki, K., Nanba, E., et al. (2013). The chaperone activity and toxicity of ambroxol on Gaucher cells and normal mice. *Brain Dev, 35*(4), 317–322. https://doi.org/10.1016/j.braindev.2012.05.008

Lui, A., Vanleuven, J., Perekopskiy, D., Liu, D., et al. (2022). FDA-approved kinase inhibitors in preclinical and clinical trials for neurological sisorders. *Pharmaceuticals (Basel), 15*(12). https://doi.org/10.3390/ph15121546

Lwin, A., Orvisky, E., Goker-Alpan, O., LaMarca, M. E., et al. (2004). Glucocerebrosidase mutations in subjects with parkinsonism. *Molecular Genetics and Metabolism, 81*(1), 70–73. https://doi.org/10.1016/j.ymgme.2003.11.004

Lynch, S. M., Zhou, C., & Messer, A. (2008). An scFv intrabody against the nonamyloid component of alpha-synuclein reduces intracellular aggregation and toxicity. *Journal of Molecular Biology, 377*(1), 136–147. https://doi.org/10.1016/j.jmb.2007.11.096

Machhi, J., Kevadiya, B. D., Muhammad, I. K., Herskovitz, J., et al. (2020). Harnessing regulatory T cell neuroprotective activities for treatment of neurodegenerative disorders. *Molecular Neurodegeneration, 15*(1), 32. https://doi.org/10.1186/s13024-020-00375-7

Mackenzie, I. R. (2000). Activated microglia in dementia with Lewy bodies. *Neurology, 55*(1), 132–134. https://doi.org/10.1212/wnl.55.1.132

Maltzman, J. S., & Koretzky, G. A. (2003). Azathioprine: old drug, new actions. *The Journal of clinical investigation, 111*(8), 1122-1124. https://doi.org/10.1172/jci18384

Mandler, M., Valera, E., Rockenstein, E., Mante, M., et al. (2015). Active immunization against alpha-synuclein ameliorates the degenerative pathology and prevents demyelination in a model of multiple system atrophy. *Molecular Neurodegeneration, 10*, 10. https://doi.org/10.1186/s13024-015-0008-9

Mandrioli, J., D'Amico, R., Zucchi, E., Gessani, A., et al. (2018). Rapamycin treatment for amyotrophic lateral sclerosis: Protocol for a phase II randomized, double-blind, placebo-controlled, multicenter, clinical trial (RAP-ALS trial). *Medicine (Baltimore), 97*(24), e11119. https://doi.org/10.1097/MD.0000000000011119

Mandrioli, J., D'Amico, R., Zucchi, E., De Biasi, S., et al. (2023). Randomized, double-blind, placebo-controlled trial of rapamycin in amyotrophic lateral sclerosis. *Nature Communications, 14*(1), 4970. https://doi.org/10.1038/s41467-023-40734-8

Markovic, M., Yeapuri, P., Namminga, K. L., Lu, Y., et al. (2022). Interleukin-2 expands neuroprotective regulatory T cells in Parkinson's disease. *Neuroimmune Pharmacology and Therapeutics, 1*(1), 43–50. https://doi.org/10.1515/nipt-2022-0001

Marogianni, C., Sokratous, M., Dardiotis, E., Hadjigeorgiou, G. M., et al. (2020). Neurodegeneration and inflammation-an interesting interplay in Parkinson's disease. *International Journal of Molecular Sciences, 21*(22), 8421. https://doi.org/10.3390/ijms21228421

Marti, M. J., Tolosa, E., & Campdelacreu, J. (2003). Clinical overview of the synucleinopathies. *Movement Disorders, 18*(Suppl 6), S21–S27. https://doi.org/10.1002/mds.10559

Martinez-Muriana, A., Mancuso, R., Francos-Quijorna, I., Olmos-Alonso, A., et al. (2016). CSF1R blockade slows the progression of amyotrophic lateral sclerosis by reducing microgliosis and invasion of macrophages into peripheral nerves. *Scientific Reports, 6*, 25663. https://doi.org/10.1038/srep25663

Martinez-Valbuena, I., Visanji, N. P., Kim, A., Lau, H. H. C., et al. (2022). Alpha-synuclein seeding shows a wide heterogeneity in multiple system atrophy. *Translational Neurodegeneration, 11*(1), 7. https://doi.org/10.1186/s40035-022-00283-4

Masliah, E., Rockenstein, E., Adame, A., Alford, M., et al. (2005). Effects of alpha-synuclein immunization in a mouse model of Parkinson's disease. *Neuron, 46*(6), 857–868. https://doi.org/10.1016/j.neuron.2005.05.010

Matthews, R. T., Ferrante, R. J., Klivenyi, P., Yang, L., et al. (1999). Creatine and cyclocreatine attenuate MPTP neurotoxicity. *Experimental Neurology, 157*(1), 142–149. https://doi.org/10.1006/exnr.1999.7049

McAlary, L., Plotkin, S. S., Yerbury, J. J., & Cashman, N. R. (2019). Prion-like propagation of protein misfolding and aggregation in amyotrophic lateral sclerosis. *Frontiers in Molecular Neuroscience, 12*, 262. https://doi.org/10.3389/fnmol.2019.00262

McColgan, P., & Tabrizi, S. J. (2018). Huntington's disease: A clinical review. *European Journal of Neurology, 25*(1), 24–34. https://doi.org/10.1111/ene.13413

McCombe, P. A., & Henderson, R. D. (2011). The Role of immune and inflammatory mechanisms in ALS. *Current Molecular Medicine, 11*(3), 246–254. https://doi.org/10.2174/156652411795243450

McGeer, P. L., Itagaki, S., Boyes, B. E., & McGeer, E. G. (1988). Reactive microglia are positive for HLA-DR in the substantia nigra of Parkinson's and Alzheimer's disease brains. *Neurology, 38*(8), 1285–1291. https://doi.org/10.1212/wnl.38.8.1285

McKeith, I. G., Boeve, B. F., Dickson, D. W., Halliday, G., et al. (2017). Diagnosis and management of dementia with Lewy bodies: Fourth consensus report of the DLB Consortium. *Neurology, 89*(1), 88–100. https://doi.org/10.1212/WNL.0000000000004058

Meissner, W. G., Traon, A. P., Foubert-Samier, A., Galabova, G., et al. (2020). A phase 1 randomized trial of specific active alpha-synuclein immunotherapies PD01A and PD03A in multiple system atrophy. *Movement Disorders, 35*(11), 1957–1965. https://doi.org/10.1002/mds.28218

Meszaros, L., Hoffmann, A., Wihan, J., & Winkler, J. (2020). Current symptomatic and disease-modifying treatments in multiple system atrophy. *International Journal of Molecular Sciences, 21*(8), 2775. https://doi.org/10.3390/ijms21082775

Migdalska-Richards, A., Daly, L., Bezard, E., & Schapira, A. H. (2016). Ambroxol effects in glucocerebrosidase and alpha-synuclein transgenic mice. *Annals of Neurology, 80*(5), 766–775. https://doi.org/10.1002/ana.24790

Miglis, M. G., Larsen, N., & Muppidi, S. (2021). The 2021 Nobel Prize in Medicine and its relevance to autonomic medicine-and other updates on recent autonomic research. *Clinical Autonomic Research, 31*(6), 655–658. https://doi.org/10.1007/s10286-021-00838-x

Miller, R. G., Block, G., Katz, J. S., Barohn, R. J., et al. (2015). Randomized phase 2 trial of NP001-a novel immune regulator: Safety and early efficacy in ALS. *Neurology ® Neuroimmunology & Neuroinflammation, 2*(3), e100. https://doi.org/10.1212/NXI.0000000000000100

Miller, R. G., Zhang, R., Bracci, P. M., Azhir, A., et al. (2022a). Phase 2B randomized controlled trial of NP001 in amyotrophic lateral sclerosis: Pre-specified and post hoc analyses. *Muscle & Nerve, 66*(1), 39–49. https://doi.org/10.1002/mus.27511

Miller, T. M., Cudkowicz, M. E., Genge, A., Shaw, P. J., et al. (2022b). Trial of antisense oligonucleotide tofersen for SOD1 ALS. *The New England Journal of Medicine, 387*(12), 1099–1110. https://doi.org/10.1056/NEJMoa2204705

Milligan, C., Atassi, N., Babu, S., Barohn, R. J., et al. (2021). Tocilizumab is safe and tolerable and reduces C-reactive protein concentrations in the plasma and cerebrospinal fluid of ALS patients. *Muscle & Nerve, 64*(3), 309–320. https://doi.org/10.1002/mus.27339

Mizwicki, M. T., Fiala, M., Magpantay, L., Aziz, N., et al. (2012). Tocilizumab attenuates inflammation in ALS patients through inhibition of IL6 receptor signaling. *American Journal of Neurodegenerative Disease, 1*(3), 305–315.

Moeslinger, T., Friedl, R,. & Spieckermann, P. G. (2006). Inhibition of inducible nitric oxide synthesis by azathioprine in a macrophage cell line. *Life Sciences, 79*(4), 374–381. https://doi.org/10.1016/j.lfs.2006.01.015

Monti, G., Gomes Moreira, D., Richner, M., Mutsaers, H. A. M., et al. (2022). GLP-1 receptor agonists in neurodegeneration: Neurovascular Unit in the spotlight. *Cells, 11*(13). https://doi.org/10.3390/cells11132023

Moorman, C. D., Sohn, S. J., & Phee, H. (2021). Emerging therapeutics for immune tolerance: Tolerogenic vaccines, T cell therapy, and IL-2 therapy. *Frontiers in Immunology, 12*, 657768. https://doi.org/10.3389/fimmu.2021.657768

Mora, J. S., Genge, A., Chio, A., Estol, C. J., et al. (2020). Masitinib as an add-on therapy to riluzole in patients with amyotrophic lateral sclerosis: A randomized clinical trial. *Amyotrophic Lateral Sclerosis and Frontotemporal Degeneration, 21*(1-2), 5–14. https://doi.org/10.1080/21678421.2019.1632346

Mora, J. S., Bradley, W. G., Chaverri, D., Hernandez-Barral, M., et al. (2021). Long-term survival analysis of masitinib in amyotrophic lateral sclerosis. *Therapeutic Advances in Neurological Disorders, 14*, 17562864211030365. https://doi.org/10.1177/17562864211030365

Mosley, R. L., & Gendelman, H. E. (2010). Control of neuroinflammation as a therapeutic strategy for amyotrophic lateral sclerosis and other neurodegenerative disorders. *Experimental Neurology, 222*(1), 1–5. https://doi.org/10.1016/j.expneurol.2009.12.018

Mullin, S., Smith, L., Lee, K., D'souza, G., et al. (2020). Ambroxol for the treatment of patients with Parkinson disease with and without glucocerebrosidase gene mutations: A nonrandomized, noncontrolled trial. *JAMA Neurology, 77*(4), 427–434. https://doi.org/10.1001/jamaneurol.2019.4611

Murdock, B. J., Zhou, T., Kashlan, S. R., Little, R. J., et al. (2017). Correlation of peripheral immunity with rapid amyotrophic lateral sclerosis progression. *JAMA Neurology, 74*(12), 1446–1454. https://doi.org/10.1001/jamaneurol.2017.2255

Musumeci, F., Schenone, S., Brullo, C., & Botta, M. (2012). An update on dual Src/Abl inhibitors. *Future Medicinal Chemistry, 4*(6), 799–822. https://doi.org/10.4155/fmc.12.29

Nicastro, N., Mak, E., Williams, G. B., Surendranathan, A., et al. (2020). Correlation of microglial activation with white matter changes in dementia with Lewy bodies. *Neuroimage Clinical, 25*, 102200. https://doi.org/10.1016/j.nicl.2020.102200

Niemela, V., Burman, J., Blennow, K., Zetterberg, H., et al. (2018). Cerebrospinal fluid sCD27 levels indicate active T cell-mediated inflammation in premanifest Huntington's disease. *PLoS One, 13*(2), e0193492. https://doi.org/10.1371/journal.pone.0193492

Nimmo, J. T., Verma, A., Dodart, J. C., Wang, C. Y., et al. (2020). Novel antibodies detect additional alpha-synuclein pathology in synucleinopathies: Potential development for immunotherapy. *Alzheimer's Research & Therapy, 12*(1), 159. https://doi.org/10.1186/s13195-020-00727-x

Nordstrom, E., Eriksson, F., Sigvardson, J., Johannesson, M., et al. (2021). ABBV-0805, a novel antibody selective for soluble aggregated alpha-synuclein, prolongs lifespan and prevents buildup of alpha-synuclein pathology in mouse models of Parkinson's disease. *Neurobiology of Disease, 161*, 105543. https://doi.org/10.1016/j.nbd.2021.105543

Novak, P., Williams, A., Ravin, P., Zurkiya, O., et al. (2012). Treatment of multiple system atrophy using intravenous immunoglobulin. *BMC Neurology, 12*, 131. https://doi.org/10.1186/1471-2377-12-131

Nygaard, I. (2010). Clinical practice. Idiopathic urgency urinary incontinence. *The New England Journal of Medicine, 363*(12), 1156–1162. https://doi.org/10.1056/NEJMcp1003849

Obergasteiger, J., Castonguay, A. M., Pizzi, S., Magnabosco, S., et al. (2023). The small GTPase Rit2 modulates LRRK2 kinase activity, is required for lysosomal function and protects against alpha-synuclein neuropathology. *NPJ Parkinsons Disease, 9*(1), 44. https://doi.org/10.1038/s41531-023-00484-2

Oertel, W., & Schulz, J. B. (2016). Current and experimental treatments of Parkinson disease: A guide for neuroscientists. *Journal of Neurochemistry, 139*(Suppl 1), 325–337. https://doi.org/10.1111/jnc.13750

Olson, K. E., Kosloski-Bilek, L. M., Anderson, K. M., Diggs, B. J., et al. (2015). Selective VIP receptor agonists facilitate immune transformation for dopaminergic neuroprotection in MPTP-intoxicated mice. *The Journal of Neuroscience, 35*(50), 16463–16478. https://doi.org/10.1523/JNEUROSCI.2131-15.2015

Olson, K. E., Bade, A. N., Schutt, C. R., Dong, J., et al. (2016). Manganese-enhanced magnetic resonance imaging for detection of vasoactive intestinal peptide receptor 2 agonist therapy in a model of Parkinson's disease. *Neurotherapeutics, 13*(3), 635–646. https://doi.org/10.1007/s13311-016-0449-z

Olson, K. E., Namminga, K. L., Lu, Y., Schwab, A. D., et al. (2021). Safety, tolerability, and immune-biomarker profiling for year-long sargramostim treatment of Parkinson's disease. *eBioMedicine, 67*, 103380. https://doi.org/10.1016/j.ebiom.2021.103380

Olson, K. E., Abdelmoaty, M. M., Namminga, K. L., Lu, Y., et al. (2023a). An open-label multiyear study of sargramostim-treated Parkinson's disease patients examining drug safety, tolerability, and immune biomarkers from limited case numbers. *Translational Neurodegeneration, 12*(1), 26. https://doi.org/10.1186/s40035-023-00361-1

Olson, K. E., Mosley, R. L., & Gendelman, H. E. (2023b). The potential for treg-enhancing therapies in nervous system pathologies. *Clinical and Experimental Immunology, 211*(2), 108–121. https://doi.org/10.1093/cei/uxac084

Ortiz, J. F., Bette, S., Tambo, W., Tao, F., et al. (2020). Multiple system atrophy - cerebellar type: Clinical picture and treatment of an often-overlooked disorder. *Cureus, 12*(9), e10741. https://doi.org/10.7759/cureus.10741

Pagan, F. L., Hebron, M. L., Wilmarth, B., Torres-Yaghi, Y., et al. (2020). Nilotinib effects on safety, tolerability, and potential biomarkers in Parkinson disease: A phase 2 randomized clinical trial. *JAMA Neurology, 77*(3), 309–317. https://doi.org/10.1001/jamaneurol.2019.4200

Pagan, F. L., Torres-Yaghi, Y., Hebron, M. L., Wilmarth, B., et al. (2022). Safety, target engagement, and biomarker effects of bosutinib in dementia with Lewy bodies. *Alzheimers Dement (N Y), 8*(1), e12296. https://doi.org/10.1002/trc2.12296

Pagano, G., Boess, F. G., Taylor, K. I., Ricci, B., et al. (2021). A Phase II study to evaluate the safety and efficacy of prasinezumab in early Parkinson's disease (PASADENA): Rationale, design, and baseline data. *Frontiers in Neurology, 12*, 705407. https://doi.org/10.3389/fneur.2021.705407

Paganoni, S., Macklin, E. A., Hendrix, S., Berry, J. D., et al. (2020). Trial of sodium phenylbutyrate-taurursodiol for amyotrophic lateral sclerosis. *The New England Journal of Medicine, 383*(10), 919–930. https://doi.org/10.1056/NEJMoa1916945

Paolini Paoletti, F., Gaetani, L., & Parnetti, L. (2020). The challenge of disease-modifying therapies in Parkinson's disease: Role of CSF biomarkers. *Biomolecules, 10*(2), 335. https://doi.org/10.3390/biom10020335

Papp, K. A., Weinberg, M. A., Morris, A., & Reich, K. (2021). IL17A/F nanobody sonelokimab in patients with plaque psoriasis: A multicentre, randomised, placebo-controlled, phase 2b study. *Lancet, 397*(10284), 1564–1575. https://doi.org/10.1016/S0140-6736(21)00440-2

Parent, M., & Parent, A. (2010). Substantia nigra and Parkinson's disease: A brief history of their long and intimate relationship. *The Canadian Journal of Neurological Sciences, 37*(3), 313–319. https://doi.org/10.1017/s0317167100010209

Patel, H., Simpson, A., Palevoda, G., & Hale, G. M. (2018). Evaluating the effectiveness of atomoxetine for the treatment of primary orthostatic hypotension in adults. *Journal of Clinical Hypertension (Greenwich, Conn.), 20*(4), 794–797. https://doi.org/10.1111/jch.13260

Peng, C., Gathagan, R. J., Covell, D. J., Medellin, C., et al. (2018). Cellular milieu imparts distinct pathological alpha-synuclein strains in alpha-synucleinopathies. *Nature, 557*(7706), 558–563. https://doi.org/10.1038/s41586-018-0104-4

Peter, I., Dubinsky, M., Bressman, S., Park, A., et al. (2018). Anti-tumor necrosis factor therapy and incidence of Parkinson disease among patients with inflammatory bowel disease. *JAMA Neurology, 75*(8), 939–946. https://doi.org/10.1001/jamaneurol.2018.0605

Pezzoli, G., Tesei, S., Canesi, M., Sacilotto, G., et al. (2010). The effect of repeated administrations of granulocyte colony stimulating factor for blood stem cells mobilization in patients with progressive supranuclear palsy, corticobasal degeneration and multiple system atrophy. *Clinical Neurology and Neurosurgery, 112*(1), 65–67. https://doi.org/10.1016/j.clineuro.2009.08.023

Pfeiffer, R. M., Mayer, B., Kuncl, R. W., Check, D. P., et al. (2020). Identifying potential targets for prevention and treatment of amyotrophic lateral sclerosis based on a screen of medicare prescription drugs. *Amyotrophic Lateral Sclerosis and Frontotemporal Degeneration, 21*(3-4), 235–245. https://doi.org/10.1080/21678421.2019.1682613

Philips, T., & Robberecht, W. (2011). Neuroinflammation in amyotrophic lateral sclerosis: Role of glial activation in motor neuron disease. *Lancet Neurology, 10*(3), 253–263. https://doi.org/10.1016/S1474-4422(11)70015-1

Pol, J. G., Caudana, P., Paillet, J., Piaggio, E., et al. (2020). Effects of interleukin-2 in immunostimulation and immunosuppression. *The Journal of Experimental Medicine, 217*(1), e20191247. https://doi.org/10.1084/jem.20191247

Polinsky, R. J., McRae, A., Baser, S. M., & Dahlstrom, A. (1991). Antibody in the CSF of patients with multiple system atrophy reacts specifically with rat locus ceruleus. *Journal of the Neurological Sciences, 106*(1), 96–104. https://doi.org/10.1016/0022-510x(91)90201-h

Poly, T. N., Islam, M. M. R., Yang, H. C., & Li, Y. J. (2019). Nonsteroidal anti-inflammatory drugs and risk of Parkinson's disease in the elderly population: A meta-analysis. *European Journal of Clinical Pharmacology, 75*(1), 99–108. https://doi.org/10.1007/s00228-018-2561-y

Pope, E. D., Cordes, L., Shi, J., Mari, Z., et al. (2021). Dementia with Lewy bodies: Emerging drug targets and therapeutics. *Expert Opinion on Investigational Drugs, 30*(6), 603–609. https://doi.org/10.1080/13543784.2021.1916913

Poulin-Briere, A., Rezaei, E., & Pozzi, S. (2021). Antibody-based therapeutic interventions for amyotrophic lateral Sclerosis: A systematic literature review. *Frontiers in Neuroscience, 15*, 790114. https://doi.org/10.3389/fnins.2021.790114

Price, D. L., Koike, M. A., Khan, A., Wrasidlo, W., et al. (2018). The small molecule alpha-synuclein misfolding inhibitor, NPT200-11, produces multiple benefits in an animal model of Parkinson's disease. *Scientific Reports, 8*(1), 16165. https://doi.org/10.1038/s41598-018-34490-9

Prots, I., & Winner, B. (2019). Th17 cells: A promising therapeutic target for Parkinson's disease? *Expert Opinion on Therapeutic Targets, 23*(4), 309–314. https://doi.org/10.1080/14728222.2019.1590336

Puentes, F., Malaspina, A., van Noort, J. M., & Amor, S. (2016). Non-neuronal cells in ALS: Role of glial, immune cells and blood-CNS barriers. *Brain Pathology, 26*(2), 248–257. https://doi.org/10.1111/bpa.12352

Pupyshev, A. B., Tenditnik, M. V., Ovsyukova, M. V., Akopyan, A. A., et al. (2021). Restoration of Parkinson's disease-like seficits by activating autophagy through mTOR-dependent and mTOR-independent mechanisms in pharmacological and transgenic models of Parkinson's disease in mice. *Bulletin of Experimental Biology and Medicine, 171*(4), 425–430. https://doi.org/10.1007/s10517-021-05242-z

Racette, B. A., Gross, A., Vouri, S. M., Camacho-Soto, A., et al. (2018). Immunosuppressants and risk of Parkinson disease. *Annals of Clinical Translational Neurology, 5*(7), 870–875. https://doi.org/10.1002/acn3.580

Rajabinejad, M., Ranjbar, S., Afshar Hezarkhani, L., Salari, F., et al. (2020). Regulatory T cells for amyotrophic lateral sclerosis/motor neuron disease: A clinical and preclinical systematic review. *Journal of Cellular Physiology, 235*(6), 5030–5040. https://doi.org/10.1002/jcp.29401

Ray, S., & Liu, M. (2012). Current understanding of LRRK2 in Parkinson's disease: Biochemical and structural features and inhibitor design. *Future Medicinal Chemistry, 4*(13), 1701–1713. https://doi.org/10.4155/fmc.12.110

Ren, L., Yi, J., Yang, J., Li, P., et al. (2018). Nonsteroidal anti-inflammatory drugs use and risk of Parkinson disease: A dose-response meta-analysis. *Medicine (Baltimore), 97*(37), e12172. https://doi.org/10.1097/MD.0000000000012172

Reynolds, A. D., Banerjee, R., Liu, J., Gendelman, H. E., et al. (2007). Neuroprotective activities of CD4+CD25+ regulatory T cells in an animal model of Parkinson's disease. *Journal of Leukocyte Biology, 82*(5), 1083–1094. https://doi.org/10.1189/jlb.0507296

Reynolds, A. D., Kadiu, I., Garg, S. K., Glanzer, J. G., et al. (2008). Nitrated alpha-synuclein and microglial neuroregulatory activities. *Journal of Neuroimmune Pharmacology, 3*(2), 59–74. https://doi.org/10.1007/s11481-008-9100-z

Reynolds, A. D., Stone, D. K., Hutter, J. A., Benner, E. J., et al. (2010). Regulatory T cells attenuate Th17 cell-mediated nigrostriatal dopaminergic neurodegeneration in a model of Parkinson's disease. *Journal of Immunology, 184*(5), 2261–2271. https://doi.org/10.4049/jimmunol.0901852

Rockenstein, E., Ostroff, G., Dikengil, F., Rus, F., et al. (2018). Combined active humoral and cellular immunization approaches for the treatment of synucleinopathies. *The Journal of Neuroscience, 38*(4), 1000–1014. https://doi.org/10.1523/JNEUROSCI.1170-17.2017

Rokad, D., Ghaisas, S., Harischandra, D. S., Jin, H., et al. (2017). Role of neurotoxicants and traumatic brain injury in alpha-synuclein protein misfolding and aggregation. *Brain Research Bulletin, 133*, 60–70. https://doi.org/10.1016/j.brainresbull.2016.12.003

Root, J., Merino, P., Nuckols, A., Johnson, M., et al. (2021). Lysosome dysfunction as a cause of neurodegenerative diseases: Lessons from frontotemporal dementia and amyotrophic lateral sclerosis. *Neurobiology of Disease, 154*, 105360. https://doi.org/10.1016/j.nbd.2021.105360

Rose, O. (2016). Parkinson's disease: Basic knowledge. *Medizinische Monatsschrift für Pharmazeuten, 39*(7), 277–281.

Roussakis, A. A., Gennaro, M., Gordon, M. F., Reilmann, R., et al. (2023). A PET-CT study on neuroinflammation in Huntington's disease patients participating in a randomized trial with laquinimod. *Brain Communications, 5*(2), fcad084. https://doi.org/10.1093/braincomms/fcad084

Rydbirk, R., Elfving, B., Andersen, M. D., Langbol, M. A., et al. (2017). Cytokine profiling in the prefrontal cortex of Parkinson's Disease and Multiple System Atrophy patients. *Neurobiology of Disease, 106*, 269–278. https://doi.org/10.1016/j.nbd.2017.07.014

Saba, J., Couselo, F. L., Bruno, J., Carniglia, L., et al. (2022). Neuroinflammation in Huntington's disease: A starring role for astrocyte and microglia. *Current Neuropharmacology, 20*(6), 1116–1143. https://doi.org/10.2174/1570159X19666211201094608

Saleh, M., Markovic, M., Olson, K. E., Gendelman, H. E., et al. (2022). Therapeutic strategies for immune transformation in Parkinson's disease. *Journal of Parkinson's Disease, 12*(s1), S201–S222. https://doi.org/10.3233/JPD-223278

Salomoni, P., & Calabretta, B. (2009). Targeted therapies and autophagy: New insights from chronic myeloid leukemia. *Autophagy, 5*(7), 1050–1051. https://doi.org/10.4161/auto.5.7.9509

Sapp, E., Kegel, K. B., Aronin, N., Hashikawa, T., et al. (2001). Early and progressive accumulation of reactive microglia in the Huntington disease brain. *Journal of Neuropathology and Experimental Neurology, 60*(2), 161–172. https://doi.org/10.1093/jnen/60.2.161

Saunders, J. A., Estes, K. A., Kosloski, L. M., Allen, H. E., et al. (2012). CD4+ regulatory and effector/memory T cell subsets profile motor dysfunction in Parkinson's disease. *Journal of Neuroimmune Pharmacology, 7*(4), 927–938. https://doi.org/10.1007/s11481-012-9402-z

Schempp, H., Reim, M., Dornisch, K., & Elstner, E. F. (2001). Chlorite-hemoprotein interaction as key role for the pharmacological activity of the chlorite-based drug WF10. *Arzneimittel-Forschung, 51*(7), 554–562. https://doi.org/10.1055/s-0031-1300079

Schintu, N., Frau, L., Ibba, M., Caboni, P., et al. (2009). PPAR-gamma-mediated neuroprotection in a chronic mouse model of Parkinson's disease. *The European Journal of Neuroscience, 29*(5), 954–963. https://doi.org/10.1111/j.1460-9568.2009.06657.x

Schoffer, K. L., Henderson, R. D., O'Maley, K., & O'sullivan, J. D. (2007). Nonpharmacological treatment, fludrocortisone, and domperidone for orthostatic hypotension in Parkinson's disease. *Movement Disorders, 22*(11), 1543–1549. https://doi.org/10.1002/mds.21428

Schofield, D. J., Irving, L., Calo, L., Bogstedt, A., et al. (2019). Preclinical development of a high affinity alpha-synuclein antibody, MEDI1341, that can enter the brain, sequester extracellular alpha-synuclein and attenuate alpha-synuclein spreading in vivo. *Neurobiology of Disease, 132*, 104582. https://doi.org/10.1016/j.nbd.2019.104582

Schutt, C. R., Gendelman, H. E., & Mosley, R. L. (2018). Tolerogenic bone marrow-derived dendritic cells induce neuroprotective regulatory T cells in a model of Parkinson's disease. *Molecular Neurodegeneration, 13*(1), 26. https://doi.org/10.1186/s13024-018-0255-7

Schwab, A. D., Thurston, M. J., Machhi, J., Olson, K. E., et al. (2020). Immunotherapy for Parkinson's disease. *Neurobiology of Disease, 137*, 104760. https://doi.org/10.1016/j.nbd.2020.104760

Shering, C., Pomfret, M., Kubiak, R., Pouliquen, I., et al. (2023). Reducing α-synuclein in human CSF; an evaluation of safety, tolerability, pharmacokinetics and pharmacodynamics of MEDI1341, an α-synuclein-specific antibody, in healthy volunteers and Parkinson's disease patients (P1-11.007). *Neurology, 100*(17_supplement_2), 2469. https://doi.org/10.1212/wnl.0000000000202579

Shi, G. X., Han, J., & Andres, D. A. (2005). Rin GTPase couples nerve growth factor signaling to p38 and b-Raf/ERK pathways to promote neuronal differentiation. *The Journal of Biological Chemistry, 280*(45), 37599–37609. https://doi.org/10.1074/jbc.M507364200

Shi, Y., Hung, S. T., Rocha, G., Lin, S., et al. (2019). Identification and therapeutic rescue of autophagosome and glutamate receptor defects in C9ORF72 and sporadic ALS neurons. *JCI Insight, 5*(15), e127736. https://doi.org/10.1172/jci.insight.127736

Shi, Y., Wei, B., Li, L., Wang, B., et al. (2022). Th17 cells and inflammation in neurological disorders: Possible mechanisms of action. *Frontiers in Immunology, 13*, 932152. https://doi.org/10.3389/fimmu.2022.932152

Sidoroff, V., Bower, P., Stefanova, N., Fanciulli, A., et al. (2022). Disease-modifying therapies for multiple system atrophy: Where are we in 2022? *Journal of Parkinson's Disease, 12*(5), 1369–1387. https://doi.org/10.3233/JPD-223183

Sidransky, E., & Lopez, G. (2012). The link between the GBA gene and parkinsonism. *Lancet Neurology, 11*(11), 986–998. https://doi.org/10.1016/S1474-4422(12)70190-4

Silveira, C. R. A., MacKinley, J., Coleman, K., Li, Z., et al. (2019). Ambroxol as a novel disease-modifying treatment for Parkinson's disease dementia: Protocol for a single-centre, randomized, double-blind, placebo-controlled trial. *BMC Neurology, 19*(1), 20. https://doi.org/10.1186/s12883-019-1252-3

Silvestroni, A., Faull, R. L., Strand, A. D., & Moller, T. (2009). Distinct neuroinflammatory profile in post-mortem human Huntington's disease. *Neuroreport, 20*(12), 1098–1103. https://doi.org/10.1097/WNR.0b013e32832e34ee

Simuni, T., Fiske, B., Merchant, K., Coffey, C. S., et al. (2021). Efficacy of nilotinib in Patients with moderately advanced Parkinson disease: A randomized clinical trial. *JAMA Neurology, 78*(3), 312–320. https://doi.org/10.1001/jamaneurol.2020.4725

Skylar-Scott, I. A., & Sha, S. J. (2023). Lewy body dementia: An overview of promising therapeutics. *Current Neurology and Neuroscience Reports, 23*(10), 581–592. https://doi.org/10.1007/s11910-023-01292-0

Southwell, A. L., Khoshnan, A., Dunn, D. E., Bugg, C. W., et al. (2008). Intrabodies binding the proline-rich domains of mutant huntingtin increase its turnover and reduce neurotoxicity. *The Journal of Neuroscience, 28*(36), 9013–9020. https://doi.org/10.1523/JNEUROSCI.2747 08.2008

Southwell, A. L., Franciosi, S., Villanueva, E. B., Xie, Y., et al. (2015). Anti-semaphorin 4D immunotherapy ameliorates neuropathology and some cognitive impairment in the YAC128 mouse model of Huntington disease. *Neurobiology of Disease, 76*, 46–56. https://doi.org/10.1016/j.nbd.2015.01.002

Spillantini, M. G., Schmidt, M. L., Lee, V. M., Trojanowski, J. Q., et al. (1997). Alpha-synuclein in Lewy bodies. *Nature, 388*(6645), 839–840. https://doi.org/10.1038/42166

Starhof, C., Hejl, A. M., Korbo, L., Winge, K., et al. (2020). Risk of multiple system atrophy and the use of anti-inflammatory drugs: A Danish register-based case-control study. *Neuroepidemiology, 54*(1), 58–63. https://doi.org/10.1159/000503003

Stefanova, N., Reindl, M., Neumann, M., Kahle, P. J., et al. (2007). Microglial activation mediates neurodegeneration related to oligodendroglial alpha-synucleinopathy: Implications for multiple system atrophy. *Movement Disorders, 22*(15), 2196–2203. https://doi.org/10.1002/mds.21671

Stefanova, N., Georgievska, B., Eriksson, H., Poewe, W., et al. (2012a). Myeloperoxidase inhibition ameliorates multiple system atrophy-like degeneration in a transgenic mouse model. *Neurotoxicity Research, 21*(4), 393–404. https://doi.org/10.1007/s12640-011-9294-3

Stefanova, N., Kaufmann, W. A., Humpel, C., Poewe, W., et al. (2012b). Systemic proteasome inhibition triggers neurodegeneration in a transgenic mouse model expressing human alpha-synuclein under oligodendrocyte promoter: Implications for multiple system atrophy. *Acta Neuropathologica, 124*(1), 51–65. https://doi.org/10.1007/s00401-012-0977-5

Surendranathan, A., Rowe, J. B., & O'Brien, J. T. (2015). Neuroinflammation in Lewy body dementia. *Parkinsonism & Related Disorders, 21*(12), 1398–1406. https://doi.org/10.1016/j.parkreldis.2015.10.009

Surendranathan, A., Su, L., Mak, E., Passamonti, L., et al. (2018). Early microglial activation and peripheral inflammation in dementia with Lewy bodies. *Brain, 141*(12), 3415–3427. https://doi.org/10.1093/brain/awy265

Svecova, D., Lubell, M. W., Casset-Semanaz, F., Mackenzie, H., et al. (2019). A randomized, double-blind, placebo-controlled phase 1 study of multiple ascending doses of subcutaneous M1095, an anti-interleukin 17A/F nanobody, in moderate-to-severe psoriasis. *Journal of the American Academy of Dermatology, 81*(1), 196–203. https://doi.org/10.1016/j.jaad.2019.03.056

Svrzikapa, N., Longo, K. A., Prasad, N., Boyanapalli, R., et al. (2020). Investigational assay for haplotype phasing of the huntingtin gene. *Molecular Therapy - Methods & Clinical Development, 19*, 162–173. https://doi.org/10.1016/j.omtm.2020.09.003

Swanson, C. R., Joers, V., Bondarenko, V., Brunner, K., et al. (2011). The PPAR-gamma agonist pioglitazone modulates inflammation and induces neuroprotection in parkinsonian monkeys. *Journal of Neuroinflammation, 8*, 91. https://doi.org/10.1186/1742-2094-8-91

Tabrizi, S. J., Leavitt, B. R., Landwehrmeyer, G. B., Wild, E. J., et al. (2019). Targeting huntingtin expression in patients with Huntington's disease. *The New England Journal of Medicine, 380*(24), 2307–2316. https://doi.org/10.1056/NEJMoa1900907

Tabrizi, S. J., Estevez-Fraga, C., van Roon-Mom, W. M. C., Flower, M. D., et al. (2022). Potential disease-modifying therapies for Huntington's disease: Lessons learned and future opportunities. *Lancet Neurology, 21*(7), 645–658. https://doi.org/10.1016/S1474-4422(22)00121-1

Thome, A. D., Atassi, F., Wang, J., Faridar, A., et al. (2021). Ex vivo expansion of dysfunctional regulatory T lymphocytes restores suppressive function in Parkinson's disease. *NPJ Parkinsons Dis, 7*(1), 41. https://doi.org/10.1038/s41531-021-00188-5

Thonhoff, J. R., Beers, D. R., Zhao, W., Pleitez, M., et al. (2018). Expanded autologous regulatory T-lymphocyte infusions in ALS: A phase I, first-in-human study. *Neurol Neuroimmunol Neuroinflamm, 5*(4), e465. https://doi.org/10.1212/NXI.0000000000000465

Thrupp, N., Sala Frigerio, C., Wolfs, L., Skene, N. G., et al. (2020). Single-nucleus RNA-seq Is not suitable for detection of microglial activation genes in humans. *Cell Reports, 32*(13), 108189. https://doi.org/10.1016/j.celrep.2020.108189

Trojanowski JQ, Revesz T, Neuropathology Working Group on MSA. (2007). Proposed neuropathological criteria for the post mortem diagnosis of multiple system atrophy. *Neuropathology and Applied Neurobiology, 33*(6), 615–620. https://doi.org/10.1111/j.1365-2990.2007.00907.x

Turner, M. R., Cagnin, A., Turkheimer, F. E., Miller, C. C., et al. (2004). Evidence of widespread cerebral microglial activation in amyotrophic lateral sclerosis: An [11C](R)-PK11195 positron emission tomography study. *Neurobiology of Disease, 15*(3), 601–609. https://doi.org/10.1016/j.nbd.2003.12.012

Ubhi, K., Rockenstein, E., Mante, M., Patrick, C., et al. (2008). Rifampicin reduces alpha-synuclein in a transgenic mouse model of multiple system atrophy. *Neuroreport, 19*(13), 1271–1276. https://doi.org/10.1097/WNR.0b013e32830b3661

Uenaka, T., Satake, W., Cha, P. C., Hayakawa, H., et al. (2018). In silico drug screening by using genome-wide association study data repurposed dabrafenib, an anti-melanoma drug, for Parkinson's disease. *Human Molecular Genetics, 27*(22), 3974–3985. https://doi.org/10.1093/hmg/ddy279

Velayati, A., Yu, W. H., & Sidransky, E. (2010). The role of glucocerebrosidase mutations in Parkinson disease and Lewy body disorders. *Current Neurology and Neuroscience Reports, 10*(3), 190–198. https://doi.org/10.1007/s11910-010-0102-x

Vidal-Martinez, G., Segura-Ulate, I., Yang, B., Diaz-Pacheco, V., et al. (2020). FTY720-Mitoxy reduces synucleinopathy and neuroinflammation, restores behavior and mitochondria function, and increases GDNF expression in Multiple System Atrophy mouse models. *Experimental Neurology, 325*, 113120. https://doi.org/10.1016/j.expneurol.2019.113120

Volc, D., Poewe, W., Kutzelnigg, A., Luhrs, P., et al. (2020). Safety and immunogenicity of the alpha-synuclein active immunotherapeutic PD01A in patients with Parkinson's disease: A randomised, single-blinded, phase 1 trial. *Lancet Neurology, 19*(7), 591–600. https://doi.org/10.1016/S1474-4422(20)30136-8

von Essen, M. R., Hellem, M. N. N., Vinther-Jensen, T., Ammitzboll, C., et al. (2020). Early intrathecal T helper 17.1 cell activity in Huntington disease. *Annals of Neurology, 87*(2), 246–255. https://doi.org/10.1002/ana.25647

Vonsattel, J. P., Keller, C., & Del Pilar, A. M. (2008). Neuropathology of Huntington's disease. *Handbook of Clinical Neurology, 89*, 599–618. https://doi.org/10.1016/S0072-9752(07)01256-0

Vucic, S., Henderson, R. D., Mathers, S., Needham, M., et al. (2021). Safety and efficacy of dimethyl fumarate in ALS: Randomised controlled study. *Annals of Clinical Translational Neurology, 8*(10), 1991–1999. https://doi.org/10.1002/acn3.51446

Wakabayashi, K., Tanji, K., Mori, F., & Takahashi, H. (2007). The Lewy body in Parkinson's disease: Molecules implicated in the formation and degradation of alpha-synuclein aggregates. *Neuropathology, 27*(5), 494–506. https://doi.org/10.1111/j.1440-1789.2007.00803.x

Walker, Z., Possin, K. L., Boeve, B. F., & Aarsland, D. (2015). Lewy body dementias. *Lancet, 386*(10004), 1683–1697. https://doi.org/10.1016/S0140-6736(15)00462-6

Wang, C. E., Tydlacka, S., Orr, A. L., Yang, S. H., et al. (2008). Accumulation of N-terminal mutant huntingtin in mouse and monkey models implicated as a pathogenic mechanism in Huntington's disease. *Human Molecular Genetics, 17*(17), 2738–2751. https://doi.org/10.1093/hmg/ddn175

Wang, I. F., Tsai, K. J., & Shen, C. K. (2013). Autophagy activation ameliorates neuronal pathogenesis of FTLD-U mice: A new light for treatment of TARDBP/TDP-43 proteinopathies. *Autophagy, 9*(2), 239–240. https://doi.org/10.4161/auto.22526

Wang, H. F., Yu, J. T., Tang, S. W., Jiang, T., et al. (2015a). Efficacy and safety of cholinesterase inhibitors and memantine in cognitive impairment in Parkinson's disease, Parkinson's disease dementia, and dementia with Lewy bodies: Systematic review with meta-analysis and trial sequential analysis. *Journal of Neurology, Neurosurgery, and Psychiatry, 86*(2), 135–143. https://doi.org/10.1136/jnnp-2014-307659

Wang, Q., Liu, Y., & Zhou, J. (2015b). Neuroinflammation in Parkinson's disease and its potential as therapeutic target. *Translational Neurodegeneration, 4*, 19. https://doi.org/10.1186/s40035-015-0042-0

Wang, T., Shi, C., Luo, H., Zheng, H., et al. (2022). Neuroinflammation in Parkinson's disease: Triggers, mechanisms, and immunotherapies. *The Neuroscientist, 28*(4), 364–381. https://doi.org/10.1177/1073858421991066

Wang, J., He, W., & Zhang, J. (2023). A richer and more diverse future for microglia phenotypes. *Heliyon, 9*(4), e14713. https://doi.org/10.1016/j.heliyon.2023.e14713

Weihofen, A., Liu, Y., Arndt, J. W., Huy, C., et al. (2019). Development of an aggregate-selective, human-derived alpha-synuclein antibody BIIB054 that ameliorates disease phenotypes in Parkinson's disease models. *Neurobiology of Disease, 124*, 276–288. https://doi.org/10.1016/j.nbd.2018.10.016

Wenning, G. K., Geser, F., & Poewe, W. (2005). Therapeutic strategies in multiple system atrophy. *Movement Disorders, 20*(Suppl 12), S67–S76. https://doi.org/10.1002/mds.20543

Wenning, G. K., Geser, F., Krismer, F., Seppi, K., et al. (2013). The natural history of multiple system atrophy: A prospective European cohort study. *Lancet Neurology, 12*(3), 264–274. https://doi.org/10.1016/S1474-4422(12)70327-7

Werner, M. H., & Olanow, C. W. (2022). Parkinson's Disease modification through Abl kinase inhibition: An opportunity. *Movement Disorders, 37*(1), 6–15. https://doi.org/10.1002/mds.28858

Williams, G. P., Marmion, D. J., Schonhoff, A. M., Jurkuvenaite, A., et al. (2020). T cell infiltration in both human multiple system atrophy and a novel mouse model of the disease. *Acta Neuropathologica, 139*(5), 855–874. https://doi.org/10.1007/s00401-020-02126-w

Woerman, A. L., Stohr, J., Aoyagi, A., Rampersaud, R., et al. (2015). Propagation of prions causing synucleinopathies in cultured cells. *Proceedings of the National Academy of Sciences of the United States of America, 112*(35), E4949–E4958. https://doi.org/10.1073/pnas.1513426112

Wyant, K. J., Ridder, A. J., & Dayalu, P. (2017). Huntington's disease-opdate on treatments. *Current Neurology and Neuroscience Reports, 17*(4), 33. https://doi.org/10.1007/s11910-017-0739-9

Xu, Q., Langley, M., Kanthasamy, A. G., & Reddy, M. B. (2017). Epigallocatechin gallate has a neurorescue effect in a mouse model of Parkinson disease. *The Journal of Nutrition, 147*(10), 1926–1931. https://doi.org/10.3945/jn.117.255034

Ye, C., Brand, D., & Zheng, S. G. (2018). Targeting IL-2: An unexpected effect in treating immunological diseases. *Signal Transduction and Targeted Therapy, 3*, 2. https://doi.org/10.1038/s41392-017-0002-5

Ye, Q., Srivastava, P., Al-Kuwari, N., & Chen, X. (2023). Oncogenic BRAF(V600E) induces microglial proliferation through extracellular signal-regulated kinase and neuronal death through c-Jun N-terminal kinase. *Neural Regeneration Research, 18*(7), 1613–1622. https://doi.org/10.4103/1673-5374.361516

Zhang, W., Wang, T., Pei, Z., Miller, D. S., et al. (2005). Aggregated alpha-synuclein activates microglia: A process leading to disease progression in Parkinson's disease. *The FASEB Journal, 19*(6), 533–542. https://doi.org/10.1096/fj.04-2751com

Zhang, F., Zhou, H., Wilson, B. C., Shi, J. S., et al. (2012). Fluoxetine protects neurons against microglial activation-mediated neurotoxicity. *Parkinsonism & Related Disorders, 18*(Suppl 1), S213–S217. https://doi.org/10.1016/S1353-8020(11)70066-9

Zhang, G., Yin, L., Luo, Z., Chen, X., et al. (2021). Effects and potential mechanisms of rapamycin on MPTP-induced acute Parkinson's disease in mice. *Ann Palliat Med, 10*(3), 2889–2897. https://doi.org/10.21037/apm-20-1096

Regenerative and Repair Strategies for the Central Nervous System: Progress in Basic and Clinical Research on Spinal Cord Injury

53

Alfredo Sandoval and Bo Chen

Abstract

The brain and spinal cord comprise the central nervous system (CNS), which generates motor commands that produce all bodily movements. Damage to these structures (e.g., from spinal cord injury) causes major neurological impairments, devastating patients quality of life. To date, no randomized clinical trial has shown definitive evidence of a repair strategy that improves functional recovery after spinal cord injury (SCI). Therefore, developing strategies to recover function after severe neurotrauma remains a major challenge for basic and clinical neuroscientists. In this chapter, we review fundamental principles and recent advances in our understanding of SCI biology through experimental readouts from animal models and observations from human SCI patients. Furthermore, we describe the current state of regenerative and repair strategies for restoring function after CNS trauma with an emphasis on spinal cord injuries. In summary, only by tailoring combinations of different strategies can we appropriately target multiple facets of spinal cord repair and mediate clinically meaningful functional improvements. These concepts may also apply to other CNS disorders.

Keywords

Spinal cord injury · Axonal growth · Relay circuits · Engineering strategies · Stem cell therapy · Epidural stimulation · Neuroinflammation · Neuronal connectivity · Neuroprosthetics · Rehabilitation training

A. Sandoval · Bo Chen (✉)
Department of Neurobiology, The University of Texas Medical Branch, Galveston, TX, USA
e-mail: alfsando@utmb.edu; bochen1@utmb.edu

53.1 Introduction

For thousands of years, physicians have recognized that the spinal cord cannot functionally recover (Wilkins, 1964). In the early twentieth century, neuroanatomists identified this lack of regeneration as a major culprit for the functional deficits observed below the injury site. To address CNS trauma, researchers have turned their focus towards strategies that promote CNS axon regeneration aiming to restore function after spinal cord injury (SCI) (David & Aguayo, 1981; He & Jin, 2016; Kim et al., 2003; Park et al., 2008; Raineteau & Schwab, 2001; Sofroniew, 2018). In the past century, various discoveries have greatly advanced our fundamental understanding of spinal cord injury and its associated injury and healing mechanisms. Most recently, it was discovered that activation of spared intraspinal neurons through epidrual stimulation in chronic SCI patients has been shown to promote functional recovery across various organ systems (Angeli et al., 2014; Gill et al., 2018; Wagner et al., 2018). These studies highlighted the "relay" function of spared intraspinal neurons, which have become a prime target for next generation of SCI therapies. Likewise, preclinical studies exploring stem cell transplantation, spontaneous relay circuit formation, and modulation of intrinsic spinal neuron excitability have further defined the factors that can be leveraged to activate spared intraspinal neurons and restore function after SCI (Chen et al., 2018; Courtine et al., 2008; Laliberte et al., 2019; Murray et al., 2010). In this chapter, we will review fundamental principles and recent advances in our understanding of SCI biology, via experimental readouts from animal models and observations from human SCI patients. Furthermore, we will describe the current state of regenerative and repair strategies for restoring function after central nervous system trauma with an emphasis on spinal cord injuries.

H. E. Gendelman, T. Ikezu (eds.), *Neuroimmune Pharmacology and Therapeutics*,
https://doi.org/10.1007/978-3-031-68237-7_53

53.2 Restoring Supraspinal Axonal Connectivity After SCI

Spinal cord injuries severely damage axonal connections between the brain and spinal cord. Interrupting these descending pathways suppresses the source of modulation and excitation required for lumbar circuit function. One approach to restore spinal cord function involves regenerating damaged axons to reestablish brain-derived signals to the distal spinal cord. Over time, the concepts and terminology surrounding axon regeneration has changed dramatically, now distinguishing between different types of axon regeneration and regrowth (Fig. 53.1).

53.2.1 Strategy 1: Axon Regeneration Across Lesion Sites

In cases of anatomically complete SCI, all motor and sensory connections between the brain and spinal cord are destroyed. Therefore, restoring connectivity to the brain requires promoting axon regeneration across the lesion site to the distal spinal cord. However, injured CNS axons often fail to spontaneously regrow across a lesion. Two alternative hypotheses have emerged to explain the regeneration failure after spinal cord injury (SCI), one focusing on intrinsic neuronal limitations and the other on extrinsic inhibitory factors. The first hypothesis attributes this failure to the diminished growth capacity of mature spinal neurons, which are less capable of initiating regrowth compared to developing neurons (Liu et al., 2010; Ramón Y Cajal et al., 2012). During CNS development, axon guidance is mediated by diffusible and contact-dependent molecules (Dodd & Jessell, 1988; Tessier-Lavigne & Goodman, 1996). Early studies explored the ability of chemoattractant agents to direct axon growth after SCI, but were hindered by their technical difficulty and inconclusive readouts. These studies were further complicated by the discovery that adult CNS neurons downregulate their axon growth capacity upon reaching maturity (Goldberg et al., 2002; Li et al., 2020; Poplawski et al., 2020) and demonstrate limited reactivation of intrinsic growth programs following injury (He & Jin, 2016). As axon growth capacity is critical for successful regeneration, several groups have begun developing strategies to genetically restore the intrinsic growth potential of injured neurons, aiming to stim ulate long-distance axon regrowth and facilitate preinjury target reinnervation. Seminal work from Zhigang He's lab at Harvard Medical School demonstrated that mTOR, a key regulator of cellular growth, is significantly downregulated in adult CNS neurons (Liu et al., 2010). The He Lab found that reactivating mTOR activity in corticospinal tract (CST) neurons following a T8 complete crush injury was sufficient to promote axon regeneration beyond the lesion site (Liu et al., 2010; Zukor et al., 2013). Other molecular studies have uncovered additional regulators of axon growth, including PTEN, STAT3, SOCS3, MAPK, ATF3, Jun, SOX11, GAP43, Arginase-1, the Klf family, MYC, RhoA, and various epigenetic regulators (Anderson et al., 2018; Belin et al., 2015; Benowitz et al., 2017; Geoffroy et al., 2015).

The second prominent hypothesis focuses on the presence of extrinsic inhibitory factors, both contact- and chemorepulsive molecules, around the injury site and lesion scar (Harel & Strittmatter, 2006; Klapka & Müller, 2006; Silver & Miller, 2004). While the exact role of the post-lesion scar remains debated (Anderson et al., 2016; Silver & Miller, 2004), it is generally accepted that injured spinal cord parenchyma releases growth-inhibiting molecules. Notably, chondroitin sulfate proteoglycans (CSPGs) produced by scar-forming stromal cells, oligodendrocyte precursor cells, and other cell types, inhibit axon growth (Bradbury et al., 2002; Tom et al., 2004). Several strategies targeting CSPGs have been tested. For instance, directly dissolving CSPGs

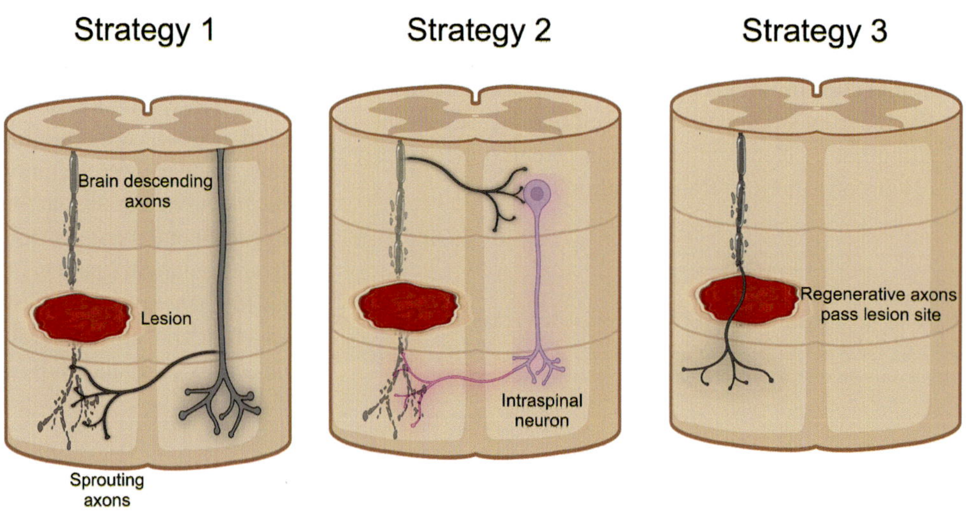

Fig. 53.1 Strategies to promote axon regrowth after SCI. Strategy 1: Sprouting of brain descending axons to bypass the lesion center. Strategy 2: Formation of intraspinal relay circuits. Strategy 3: Regeneration of long descending axons across lesion sites

with intrathecally delivered chondroitinase ABC (ChABC) in SCI rats, promoted axon regeneration in both ascending and descending tracts, and enhanced functional recovery (Bradbury et al., 2002). In another study, Lang et al. demonstrated that inhibiting the CSPG receptor protein tyrosine phosphatase σ (PTPσ) with a function blocking peptide, enhanced descending serotonergic axon regeneration, leading to improved locomotor and urinary system recovery (Lang et al., 2015).

53.2.2 Strategy 2: Long-Distance Supraspinal Axon Sprouting

Although CNS axons are unable to directly traverse the lesion site, recent studies have explored the circuit-level adaptations that occur following spinal cord injury. These studies revealed that descending motor and brainstem projections—including the corticospinal and reticulospinal tracts—form new collaterals that reinnervate their original targets in response to the loss of other synaptic input. These circuit reorganization events represent a remarkably conserved repair strategy, with most major descending axon tracts showing signficant sprouting and collateral formation after spinal cord injury (Asboth et al., 2018; Isa, 2019). For instance, Filli et al. demonstrated that reticulospinal neurons undergo spontaneous sprouting after a cervical hemisection injury (Filli et al., 2014). Following a C4-level hemisection injury, severed axons from the gigantocellular reticular nucleus (NRG) regenerated, forming new collateral sprouts that synapsed with descending cervical propriospinal neurons, leading to spontaneous functional recovery. In another landmark study, Takeoka et al., discovered that axons from brainstem nuclei, including the magnocellular, pontine, gigantocellular, and spinal vestibular nuclei, were capable of regenerating long distances following injury. Following a lateral hemisection, retrograde viral tracing in the lumbar spine revealed that these brainstem-derived axons sprouted across the cervical and thoracic spinal cords to reinnervate the lumbar cord distal to the lesion (Takeoka et al., 2014). These findings highlight the feasibility of therapeutic strategies requiring long-distance axon regrowth, suggesting that strategies to enhance long-distance axon regeneration and sprouting may further improve functional recovery following spinal cord injury. Corticospinal neurons (CSNs) provide direct cortical outputs to the spinal cord and play crucial roles in motor control across species (Wang et al., 2017). However, the corticospinal tract, which carries CSN signals, lacks the ability to sprout over long distances. Since the degree of functional recovery is linked to the extent of CST collateral formation, enhancing CST sprouting may represent a promising strategy for treating SCI. To this end, Liu et al. developed a strategy, using AAV-mediated expression of insulin-like growth factor 1 (IGF1) and osteopontin (OPN) to induce robust CST regrowth and functional recovery following a low thoracic hemisection injury. Furthermore, combining this treatment with the voltage-gated potassium channel blocker, 4-AP-MeOH (which enhances axon conduction velocity) led to further improvements in functional recovery (Liu et al., 2017).

In summary, spontaneously reorganizing supraspinal circuits are key contributors to functional recovery after spinal cord injury and represent viable targets for therapeutic intervention. While the mechanisms regulating the degree and location of collateral sprout formation require further study, the ability to experimentally enhance both axon regrowth and collateral formation represents a significant advance, likely inspiring future therapies that will leverage these processes. Moreover, the use of combinatorial therapies targeting independent mechanisms, such as 4-AP-MeOH, seems to be a practical strategy for enhancing recovery after injury and likely represents future treatment paradigms that will need to target multiple repair mechanisms for maximum benefit.

53.2.3 Strategy 3: Relay Circuits via Intraspinal Neurons

Propriospinal neurons comprise a diverse group of intraspinal projection neurons involved with several locomotor functions including forelimb–hindlimb coupling and fine motor coordination (Arber, 2017; Baek et al., 2017; Ruder et al., 2016). These neurons are integrated with other descending and intraspinal motor circuits, ideally positioned to synchronize, modulate, and, in the case of SCI, provide redundancy to existing circuits (Isa, 2019; Isa et al., 2013; Laliberte et al., 2019; Zholudeva et al., 2018b, 2021). Following injury, various spinal circuits spontaneously reorganize to reconnect supraspinal structures with the distal spinal cord. Bareyre et al. were the first to discover that transected hindlimb CST axons spontaneously form new collaterals that synapse with propriospinal neurons (Bareyre et al., 2004). Following a mid-thoracic dorsal hemisection injury, lesioned CST axons sprouted into the cervical gray matter, contacting PSNs that project to the lumbar spinal cord. This newly formed CST-PSN detour circuit relays cortical signals to their original spinal targets, restoring partial locomotor function. While newly formed relay circuits integrate with existing motor tracts, they can be silenced, rendering distal lumbar circuits nonfunctional. Since these de novo circuits may hold therapeutic relevance for SCI patients, several groups have sought a deeper understanding of their functional capacity post-injury. To reproducibly isolate these circuits, researchers have employed a double lateral hemisection injury paradigm that severs all descending connections through two staggered injuries (Chen et al., 2018; Courtine et al., 2008; Murray et al., 2010) (contralateral hemisections at spinal levels T7 and T10/12, Fig. 53.2). Following this staggered injury, propriospinal neurons and descending axons from various brain nuclei undergo sprouting, reconnecting the brain and spinal cord through a spared relay zone between the two lesions. Despite the formation of numerous detour circuits within this relay zone, many animals fail to achieve significant locomotor recovery. As a result, these findings have driven efforts to better understand the mechanisms regulating spinal circuit function post-injury. For instance, Chen et al. discovered that injury-related reductions in KCC2 (a Cl^- extruder) expression rendered relay neurons silent (Chen et al., 2018). KCC2 downregulation causes intracellular chloride to accumulate, reversing GABA—and glycinergic currents from inhibitory to excitatory. Restoring chloride balance in inhibitory pro-

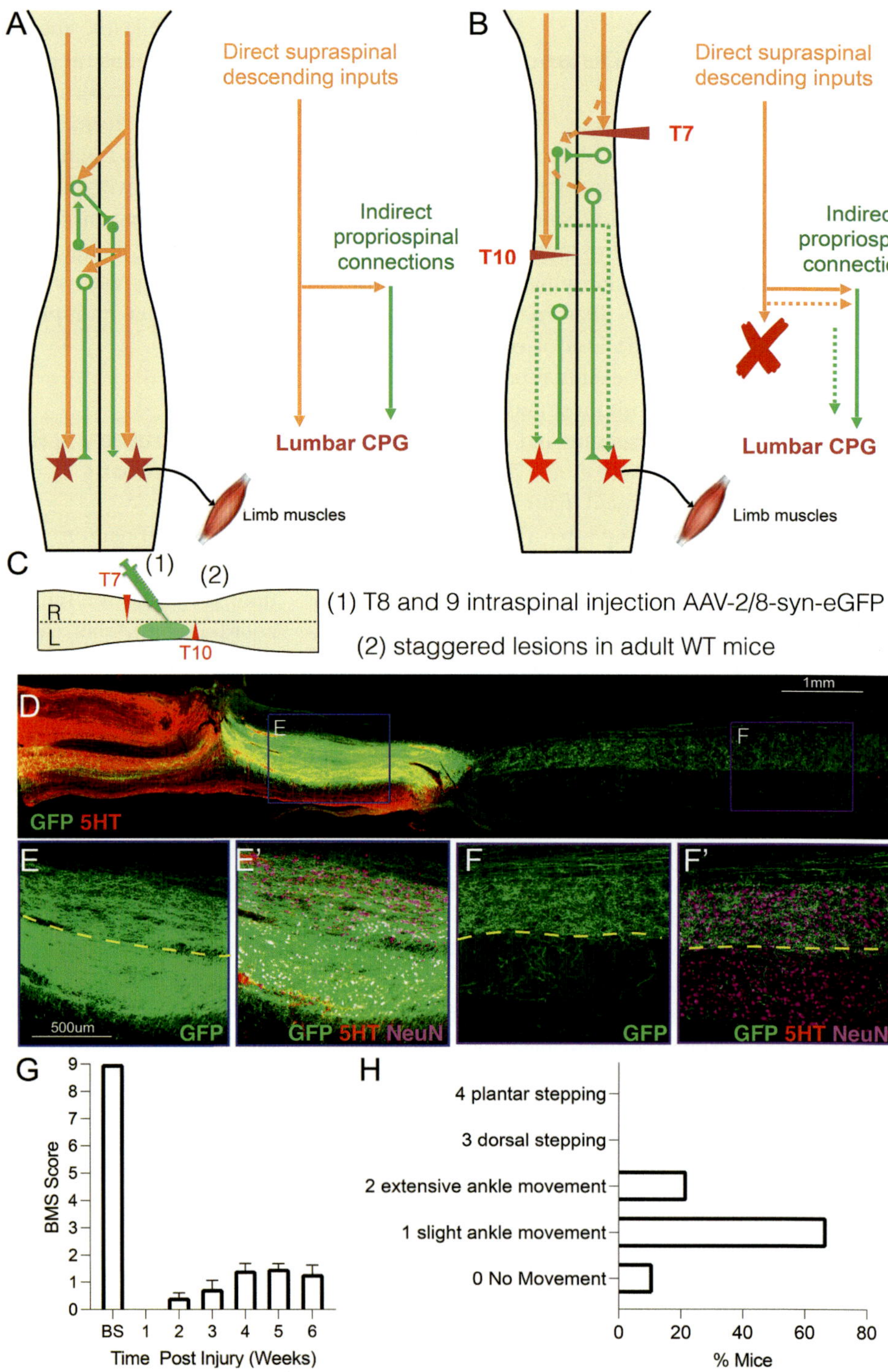

Fig. 53.2 Anatomy and detour circuits of double lateral hemisection SCI in mice model (**a**). Schematic diagram depicting propriospinal circuits innervating lumbar central pattern generators (CPGs). (**b**). Schematic representation of staggered lateral hemisections at T7 and T10, with arrowheads indicating the lesion sites. Descending inputs from detour circuits to the lumbar spinal cord can project to lumbar propriospinal neurons and/or lumbar CPGs. (**c**). Experimental scheme (above) and (**d**) representative image (below) showing intraspinal tracing in the staggered SCI model. (**e**, **e**', **f**, and **f**'). Magnified view of the relay zone (**e**, **e**') and the lumbar region (**f**, **f**') in the injured spinal cord. (**g**). BMS scores for staggered SCI mice. $n = 10$. (**h**). Breakdown of BMS scores in mice six weeks post-staggered SCI ($n = 10$)

priospinal neurons with a KCC2 agonist or by KCC2 overexpression was sufficient to restore dorsal stepping ability in severe SCI mice. While further research is needed to better understand relay circuits and discover new compounds capable of promoting beneficial circuit reorganization, these findings underscore the therapeutic value of enhancing relay circuit activity following SCI.

53.3 Restoring Lost Intraspinal Neurons After SCI

Spinal cord injury is a biphasic process, with the primary injury caused by direct trauma or compression, followed by a secondary injury response. The secondary phase involves a cascade of events that lead to spinal neuron death, including hemorrhage, edema, excitotoxicity, apoptosis, and necrosis (Alizadeh et al., 2019; Beattie et al., 2002; Cho et al., 2017; Fehlings et al., 2017). As a result, current strategies to restore lost neurons focus on neuroprotective (e.g., antiedema) and regenerative (e.g., stem cell) approaches For current and future therapies to be successful, a deep understanding of the pathophysiology of SCI will be required to target specific features of the injury cascade to maximize functional recovery.

53.3.1 Strategy 1: Prevent Intraspinal Neuron Death After SCI

Cell death is a signficant consequence of secondary injury following SCI, triggered by various mechanisms in response to injury-induced mediators. Necrosis and apoptosis are widely recognized as the two primary mechanisms of cell death after SCI (Crowe et al., 1997). However, due to the complexity and variability of spinal cord injuries, the extent and timing of cell death after CNS trauma remains unclear. This uncertainty regarding cell death after injury is evident in both clinical and preclinical research. Most neuroprotective strategies have not yet been implemented in clinical practice, and debate continues in the preclinical studies over the extent of cell death after SCI (Ahuja et al., 2017a, b). In preclinical studies, spinal edema is well recognized as pathological process contributing to cell death. Klatzo first characterized the pathological hallmarks of CNS edema, distinguishing it as either vasogenic or cytotoxic (Klatzo, 1967). Spinal vasogenic edema (tissue swelling) typically resolves within 7 days in rodents (Demediuk et al., 1987; Kitchen et al., 2020) and 14 days in primates (Yashon et al., 1973), whereas cytotoxic edema (oncotic cell swelling) occurs seconds after injury in neurons (Rungta et al., 2015) and astrocytes (Kitchen et al., 2020). Spinal edema after injury reduces blood flow and contributes to spinal cord ischemia, cell death, and functional decline (Leonard et al., 2015). Thus, preventing cell death induced by cytotoxic edema is a logical strategy for treating SCI. Building on this idea, cation chloride cotransporters (CCCs) have emerged as essential regulators of ion homeostasis and neuron volume (Ahmed et al., 2014; Jha et al., 2019), differing from the traditional mechanism to regulate cell volume via osmotically driven water influx, since CNS neurons lack the aquaporins needed for this mechanism (Liang et al., 2007). Clinically, dysregulated ion and water flux may contribute to pathologic cytotoxic edema following CNS trauma. Moreover, SCI downregulates KCC2 (a Cl^- extruder) and upregulates NKCC1 (Na-K-Cl cotransporter 1, Cl^- importer) in the injured spinal cord (Beverungen et al., 2020; Côté et al., 2014). Collectively, these findings suggest that Cl^- cotransporters are key targets for treating neuronal edema after SCI. In line with this, Punam et al. recently employed in vivo imaging and electrophysiological recordings to study the ionic mechanisms driving neuronal edema in a mouse traumatic brain injury model (Sawant-Pokam et al., 2020). They demonstrated that the CCC blocker bumetanide prevented neuronal swelling, supporting the idea that altered chloride flux is a major contributor to neuronal edema (Rungta et al., 2015). Interestingly, their electrophysiology results revealed that preventing neuronal edema increased network excitability, emphasizing that neuronal edema also disrupts injured neuronal circuits. Despite these promising new discoveries about the molecular mechanisms of neuronal edema, no therapies have yet entered clinical trials for treating spinal edema (Cho et al., 2017). Therefore, further research is needed to validate methods for reducing spinal cord edema and improve neurological function after SCI.

53.3.2 Strategy 2: Restore Lost Intraspinal Neurons to Relay Supraspinal Signals

SCI triggers complex and prolonged pathophysiological processes, including neuroinflammation and demyelination, which together lead to neuronal death and spinal circuit dysfunction. As a result, cell transplantation therapies hold promise as they offer benefits at the molecular, cellular, and circuit levels (Assinck et al., 2017; Fischer et al., 2020). Among current transplantation approaches, neuronal progenitor cell (NPC) transplantation—precursors to glial cells (astrocytes and oligodendrocytes) and neurons - has shown significant progress in the past decade. NPC-derived neurons form synaptic connections with host circuits and establish neuronal relays across the injury site (Bonner et al., 2011; Ceto et al., 2020; Dulin et al., 2018; Lu et al., 2012; Rosenzweig et al., 2018; Zholudeva et al., 2018a). NPC-derived glial cells support grafted neurons and provide additional therapeutic benefits, such as supporting injured host axons, protecting neurons, and reducing scar size (Cao et al., 2005; Cusimano et al., 2012; Kadoya et al., 2016; Yasuda et al., 2011). For example, Lu et al. demonstrated promising anatomical and behavioral results using spinal cord progenitor cells from E14 embryos transplanted into injured adult spinal cords. Anatomically, these transplanted NPCs exhibited robust long-term survival, expressing markers of mature neurons, astrocytes, oligodendrocytes, and their appropriate synaptic markers. These grafted neurons extended a large numbers of axons over significant distances, forming synapses with host cells. Behaviorally, eight weeks after NPC transplantation, grafted SCI rats demonstrated substantial locomotor recovery (Lu et al., 2012). This finding demonstrated that grafted early-stage neurons can form new relay circuits that significantly improve function. So far, human NPC transplants have been tested in various species including primates. The consistent improvement between rodents and primates suggests that human NPC transplantation is a feasible strategy for treating SCI patients. Given the potential of NPC transplants to improve connectivity through neuronal relays, several efforts have evaluated NSC transplantation in human SCI patients (Assinck et al., 2017; Fischer et al., 2020). However, due to the relatively small population of SCI patients, the cost–benefit ratio for developing and testing cell-based treatments has been unfavorable until recently. As of this writing, neural stem cell transplantation strategies have not yet received approval.

To advance the clinical translational of stem cell-based therapies, several challenges must be overcome. A critical gap in our understanding of SCI is identifying the specific cell types and circuits that are preferentially lost or become dysfunctional following the injury. This knowledge is essential for developing targeted therapies and interventions. Understanding cell-type and circuit-specific dysfunction after SCI will guide the selection of optimal cell grafts to maximize sensorimotor recovery. For example, more than 66% of human SCI patients experience spasticity and pain (Siddall & Loeser, 2001), potentially linked to reduced inhibitory tone in the lumbar spinal cord (Boulenguez et al., 2010) after SCI. Transplanting inhibitory (GABAergic) interneuron precursors has been shown to reduce pain associated behaviors after SCI, even when neurons are grafted into the lumbar region - far from the lesion site - with no evidence of neuronal loss (Fandel et al., 2016). Other significant challenges such as stem cell precursor manufacturing and cell line selection, are also under intense study, however these are outside of the scope of this chapter.

53.4 Technological Approaches to Revive Spared Circuits

Although our understanding of CNS injury pathophysiology has advanced, progress in reactivating spared circuits after SCI has lagged. Biological therapies are in development, but require further optimization before they can be used clinically. In contrast, engineering approaches for treating SCI have gained traction, with early reports showing significant clinical benefit. Seminal work by Harkema et al. demonstrated that lumbosacral epidural stimulation combined with rehabilitative therapy could restore brain-mediated movement in paraplegic patients (Harkema et al., 2011). This study highlights the potential of engineering approaches to enhance spared circuit function and promote functional recovery, establishing the clinical viability of these solutions for treating spinal cord injury.

53.4.1 Strategy 1: Combining Engineering Strategies and Rehabilitation Training to Reactivate Spared Circuits in SCI Patients

For decades, clinicians have reported moderate functional recovery using rehabilitative methods after CNS trauma (Lovely et al., 1986; Torres-Espín et al., 2018). For instance, over-ground walking, body-weight-supported treadmill training, and robot-assisted gait training can promote recovery in spinal cord injury patients (Gaspar et al., 2019; Harkema et al., 2011; Nam et al., 2017). While rehabilitative motor training is a common approach for recovery after CNS injury, improvement is often inconsistent and unpredictable, particularly in SCI patients with varying injury severity and functional deficits. Recent case reports have shown that combining rehabilitative motor training with dorsally applied electrical stimulation to the lumbosacral spine can consistently promote functional recovery (Angeli et al., 2014; Herrity et al., 2021). These findings have prompted further exploration of the therapeutic mechanisms behind spinal stimulation to maximize its benefits.

Recent evidence suggests that electrical currents spread through the dorsal roots, where large diameter proprioceptive fibers are the first to be recruited (Formento et al., 2018). This supports earlier preclinical studies showing that proprioceptive input is essential for locomotor recovery (Takeoka, 2020; Takeoka & Arber, 2019). In theory, recapitulating proprioceptive feedback during normal gait cycles could restore function in completely paralyzed SCI patients following sufficient rehabilitative training. To test this, Courtine's group analyzed the hindlimb muscle activation patterns in uninjured rats and humans. Using a closed-loop protocol, they delivered short bursts of lumbosacral stimulation that matched the temporal and spatial activity patterns observed during normal walking. When combined with extended physical training, targeted epidural stimulation enabled refined lower limb movement in both animal models and humans. Compared to continuous stimulation, spatiotemporal stimulation increased the amplitude and robustness of leg movements, enabling sustained, full weight-bearing locomotion (Wenger et al., 2016, 2018).

53.4.2 Strategy 2: Engineering Strategies to Bypass Lesions

Reconnecting the brain to the spinal cord is essential for restoring voluntary function. Both biological and engineering strategies have been employed to reconnect the CNS across injury sites, with recent neuroengineering approaches showing signficant success. The successful use of targeted epidural stimulation to recapitulate proprioceptive sensory inputs, during the gait cycle has opened new possibilities for SCI treatment. For example, can brain signals be decoded and their final targets directly stimulated to restore top-down control? This has become increasingly feasible with the rapid development of neuroprosthetic technologies such as brain-computer interfaces (BCI).

Brain–computer interfaces (BCIs) decode brain signals from neural sensing arrays and convert them into commands for output devices to perform desired actions. In practice, BCIs have been used to re-establish bidirectional communication between the brain and denervated body parts—even in patients with severe SCI (Ajiboye et al., 2017; Bouton et al., 2016; Capogrosso et al., 2016; Ethier et al., 2012; Flesher et al., n.d.). For example, Bouton et al. used previously decoded intracortical signals to extract locomotion-related signals. They demonstrated that intracortical signals could be linked to real-time muscle activation and used to restore movement in a paralyzed patient. This was the first successful control of muscle-activation using decoded intracortical signals, allowing a paralyzed SCI patient to regain hand function. Similar approaches have successfully improved other sensorimotor functions (Capogrosso et al., 2016). Collectively, these findings demonstrate the feasibility of using direct neural bypass via BCIs to circumvent an injured spinal cord. However, despite the promising results from BCIs and other neuroprostheses, these technologies have remained in the proof-of-concept stage due to practical, financial, and technological challenges. While these strategies provide modest improvements, they are currently overshadowed by the significant technical challenges and financial costs of their implementation (Borton et al., 2013). Therefore, substantial innovation is required before these tools can become widely accessible to spinal cord injury patients.

53.5 Conclusions and Closing Remarks

Current strategies have demonstrated therapeutic potential in animal models, but their ability to improve clinical outcomes after SCI remains uncertain. One limiting factor for their clinical translation is the modest effectiveness of these treatments, even in controlled laboratory environments. The injury cascade after severe SCI is multifaceted, and since current treatments are only modestly effective, combination therapy may be necessary to achieve clinically significant improvement. Based on the studies summarized in this chapter, a combination strategy could enhance axon regeneration genetically, limit cell death with neuroprotective small molecules, replace lost cells with stem cell grafts or reactivate relay circuits through neuroprosthetics and rehabilitation. However, combining treatments has proven more challenging and less fruitful than expected. Several studies have revealed unexpected interactions between treatments, so future studies must closely examine the impact of combined treatments on functional recovery. Additionally, the biphasic injury response after SCI involves distinct cellular mechanisms each with its own onset and time scale. Therefore, treatments targeting these mechanisms must adhere to strict temporal windows, as some may only be beneficial within specific time frames after injury. Unfortunately, most CNS trauma studies neglect the timing of treatments, with many beginning at the same time after injury, potentially missing the optimal window for initiation. In summary, many promising therapies for SCI on the horizon, but continued effort and optimization are needed to overcome the debilitating paralysis caused by SCI.

53.6 Review Questions

1. What is spinal cord injury and why can't it self-repair?
2. Why do researchers use different preclinical models to study SCI?
3. What are the challenges or gaps in knowledge that need to be addressed to improve the clinical translational of stem cell-based therapies.
4. What are the possible approaches to restore function after SCI? (Open question)

53.7 Answers

1. Spinal cord injuries disrupt axonal connections between the brain and spinal cord, severing the ascending and descending pathways essential for brain–spinal circuit function. Due to the limited regenerative capacity of CNS axons, these injuries often lead to permanent functional deficits, significantly impacting the quality of life for patients with SCI.
2. Animal models in research are tailored to specific hypotheses, focusing on factors that directly test these hypotheses while limiting irrelevant variables. For instance, a double lateral hemisection model, though clinically irrelevant as no patient undergoes such an injury, is ideal for studying the potential of intraspinal neurons to relay circuits. This model lacks direct connections, meaning any functional recovery must occur through relay neurons, providing clear insights into their roles.
3. We still lack a thorough understanding of the cell types and circuits that are preferentially lost or dysfunctional after spinal cord injury. Knowledge of cell-type and circuit-specific dysfunction after SCI will allow the selection of ideal cell grafts to maximize sensorimotor recovery.
4. There is no definitive answer to this question, which is precisely the challenge all researchers in this field are tackling. The complexity and severity of spinal cord injury (SCI), a condition with no current treatment, are daunting yet underscore the importance of this research. To address SCI effectively, it is critical to consider a combination of treatments, including axon regeneration, reactivation of spared circuits, and the use of advanced neuroprosthetic technologies. A comprehensive understanding of the pathological progression following SCI is essential to design targeted treatments that address each aspect of the condition. Ideally, a synergistic approach combining these strategies may eventually enable patients to regain lost functions.

References

Ahmed, M. M., Lee, H. K., Clark, Z., Miranpuri, G. S., Nacht, C., Patel, K., Liu, L., Joslin, J., Kintner, D., & Resnick, D. K. (2014). Pathogenesis of spinal cord injury induced edema and neuropathic pain: Expression of multiple isoforms of WNK1. *Annals of Neurosciences, 21*(3), 97–103. https://doi.org/10.5214/ans.0972.7531.210305

Ahuja, C. S., Nori, S., Tetreault, L., Wilson, J., Kwon, B., Harrop, J., Choi, D., & Fehlings, M. G. (2017a). Traumatic spinal cord injury – repair and regeneration. *Clinical Neurosurgery, 80*(3), S22–S90. https://doi.org/10.1093/neuros/nyw080

Ahuja, C. S., Wilson, J. R., Nori, S., Kotter, M. R. N., Druschel, C., Curt, A., & Fehlings, M. G. (2017b). Traumatic spinal cord injury. *Nature Reviews Disease Primers, 3*. https://doi.org/10.1038/nrdp.2017.18

Ajiboye, A. B., Willett, F. R., Young, D. R., Memberg, W. D., Murphy, B. A., Miller, J. P., Walter, B. L., Sweet, J. A., Hoyen, H. A., Keith, M. W., Hunter Peckham, P., Simeral, J. D., Donoghue, J. P., Hochberg, L. R., & Kirsch, R. F. (2017). Restoration of reaching and grasping movements through brain-controlled muscle stimulation in a person with tetraplegia: A proof-of-concept demonstration. *The Lancet, 389*(10081), 1821–1830. https://doi.org/10.1016/S0140-6736(17)30601-3

Alizadeh, A., Dyck, S. M., & Karimi-Abdolrezaee, S. (2019). Traumatic spinal cord injury: An overview of pathophysiology, models and acute injury mechanisms. *Frontiers in Neurology, 10*(March), 1–25. https://doi.org/10.3389/fneur.2019.00282

Anderson, M. A., Burda, J. E., Ren, Y., Ao, Y., O'Shea, T. M., Kawaguchi, R., Coppola, G., Khakh, B. S., Deming, T. J., & Sofroniew, M. V. (2016). Astrocyte scar formation AIDS central nervous system axon regeneration. *Nature, 532*(7598), 195–200. https://doi.org/10.1038/nature17623

Anderson, M. A., O'Shea, T. M., Burda, J. E., Ao, Y., Barlatey, S. L., Bernstein, A. M., Kim, J. H., James, N. D., Rogers, A., Kato, B., Wollenberg, A. L., Kawaguchi, R., Coppola, G., Wang, C., Deming, T. J., He, Z., Courtine, G., & Sofroniew, M. V. (2018). Required growth facilitators propel axon regeneration across complete spinal cord injury. *Nature, 561*(7723), 396–400. https://doi.org/10.1038/s41586-018-0467-6

Angeli, C. A., Reggie Edgerton, V., Gerasimenko, Y. P., & Harkema, S. J. (2014). Altering spinal cord excitability enables voluntary movements after chronic complete paralysis in humans. *Brain, 137*(5), 1394–1409. https://doi.org/10.1093/brain/awu038

Arber, S. (2017). Organization and function of neuronal circuits controlling movement. *EMBO Molecular Medicine, 9*(3), 281–284. https://doi.org/10.15252/emmm.201607226

Asboth, L., Friedli, L., Beauparlant, J., Martinez-Gonzalez, C., Anil, S., Rey, E., Baud, L., Pidpruzhnykova, G., Anderson, M. A., Shkorbatova, P., Batti, L., Pagès, S., Kreider, J., Schneider, B. L., Barraud, Q., & Courtine, G. (2018). Cortico-reticulo-spinal circuit reorganization enables functional recovery after severe spinal cord contusion. *Nature Neuroscience, 21*(4), 576–588. https://doi.org/10.1038/s41593-018-0093-5

Assinck, P., Duncan, G. J., Hilton, B. J., Plemel, J. R., & Tetzlaff, W. (2017). Cell transplantation therapy for spinal cord injury. *Nature Neuroscience, 20*(5), 637–647.

Baek, M., Pivetta, C., Liu, J. P., Arber, S., & Dasen, J. S. (2017). Columnar-intrinsic cues shape premotor input specificity in locomotor circuits. *Cell Reports, 21*(4), 867–877. https://doi.org/10.1016/j.celrep.2017.10.004

Bareyre, F. M., Kerschensteiner, M., Raineteau, O., Mettenleiter, T. C., Weinmann, O., & Schwab, M. E. (2004). The injured spinal cord spontaneously forms a new intraspinal circuit in adult rats. *Nature Neuroscience, 7*(3), 269–277. https://doi.org/10.1038/nn1195

Beattie, M. S., Hermann, G. E., Rogers, R. C., & Bresnahan, J. C. (2002). Cell death in models of spinal cord. *Injury, 137*, 37–47.

Belin, S., Nawabi, H., Wang, C., Tang, S., Latremoliere, A., Warren, P., Schorle, H., Uncu, C., Woolf, C. J., He, Z., & Steen, J. A. (2015). Injury-induced decline of intrinsic regenerative ability revealed by quantitative proteomics. *Neuron, 86*(4). https://doi.org/10.1016/j.neuron.2015.03.060

Benowitz, L. I., He, Z., & Goldberg, J. L. (2017). Reaching the brain: Advances in optic nerve regeneration. *Experimental Neurology, 287*, 365–373.

Beverungen, H., Klaszky, S. C., Klaszky, M., & Côté, M. P. (2020). Rehabilitation decreases spasticity by restoring chloride homeostasis through the brain-derived neurotrophic factor-KCC2 pathway after spinal cord injury. *Journal of Neurotrauma, 37*(6), 846–859. https://doi.org/10.1089/neu.2019.6526

Bonner, J. F., Connors, T. M., Silverman, W. F., Kowalski, D. P., Lemay, M. A., & Fischer, I. (2011). Grafted neural progenitors integrate and restore synaptic connectivity across the injured spinal cord. *Journal of Neuroscience, 31*(12), 4675–4686. https://doi.org/10.1523/JNEUROSCI.4130-10.2011

Borton, D., Micera, S., José, D. R., & Millán, and Grégoire Courtine. (2013). Personalized Neuroprosthetics. *Science Translational Medicine, 5*, 210rv2.

Boulenguez, P., Liabeuf, S., Bos, R., Bras, H., Jean-Xavier, C., Brocard, C., Stil, A., Darbon, P., Cattaert, D., Delpire, E., Marsala, M., & Vinay, L. (2010). Down-regulation of the potassium-chloride cotransporter KCC2 contributes to spasticity after spinal cord injury. *Nature Medicine, 16*(3), 302–307. https://doi.org/10.1038/nm.2107

Bouton, C. E., Shaikhouni, A., Annetta, N. V., Bockbrader, M. A., Friedenberg, D. A., Nielson, D. M., Sharma, G., Sederberg, P. B., Glenn, B. C., Jerry Mysiw, W., Morgan, A. G., Deogaonkar, M., & Rezai, A. R. (2016). Restoring cortical control of functional movement in a human with quadriplegia. *Nature, 533*, 247–250. https://doi.org/10.1038/nature17435

Bradbury, E. J., Moon, L. D. F., Popat, R. J., King, V. R., Bennett, G. S., Patel, P. N., Fawcett, J. W., & McMahon, S. B. (2002). Chondroitinase ABC promotes functional recovery after spinal cord injury. *Nature, 416*(6881), 636–640. https://doi.org/10.1038/416636a

Cao, Q., Xiao Ming, X., DeVries, W. H., Enzmann, G. U., Ping, P., Tsoulfas, P., Wood, P. M., Bunge, M. B., & Whittemore, S. R. (2005). Functional recovery in traumatic spinal cord injury after transplantation of multineurotrophin-expressing glial-restricted precursor cells. *Journal of Neuroscience, 25*(30), 6947–6957. https://doi.org/10.1523/JNEUROSCI.1065-05.2005

Capogrosso, M., Milekovic, T., Borton, D., Wagner, F., Moraud, E. M., Mignardot, J. B., Buse, N., Gandar, J., Barraud, Q., Xing, D., Rey, E., Duis, S., Jianzhong, Y., Wai, K. D., Ko, Q. L., Detemple, P., Denison, T., Micera, S., Bezard, E., Bloch, J., & Courtine, G. (2016). A brain-spine interface alleviating gait deficits after spinal cord injury in primates. *Nature, 539*(7628), 284–288. https://doi.org/10.1038/nature20118

Ceto, S., Sekiguchi, K. J., Takashima, Y., Nimmerjahn, A., & Tuszynski, M. H. (2020). Neural stem cell grafts form extensive synaptic networks that integrate with host circuits after spinal cord injury. *Cell Stem Cell, 27*(3), 430–440.e5. https://doi.org/10.1016/j.stem.2020.07.007

Chen, B., Li, Y., Bin, Y., Zhang, Z., Brommer, B., Williams, P. R., Liu, Y., Hegarty, S. V., Zhou, S., Zhu, J., Guo, H., Yi, L., Zhang, Y., Xiaosong, G., & He, Z. (2018). Reactivation of dormant relay pathways in injured spinal cord by KCC2 manipulations. *Cell, 174*(3), 521–535.e13. https://doi.org/10.1016/j.cell.2018.06.005

Cho, N., Hachem, L. D., & Fehlings, M. G. (2017). *Spinal cord edema after spinal cord injury: From pathogenesis to management.* Elsevier.

Côté, M. P., Gandhi, S., Zambrotta, M., & Houlé, J. D. (2014). Exercise modulates chloride homeostasis after spinal cord injury. *Journal of Neuroscience, 34*(27), 8976–8987. https://doi.org/10.1523/JNEUROSCI.0678-14.2014

Courtine, G., Song, B., Roy, R. R., Zhong, H., Herrmann, J. E., Ao, Y., Jingwei Qi, V., Edgerton, R., & Sofroniew, M. V. (2008). Recovery of supraspinal control of stepping via indirect propriospinal relay connections after spinal cord injury. *Nature Medicine, 14*(1), 69–74. https://doi.org/10.1038/nm1682

Crowe, M., Bresnahan, J. C., Shuman, S. L., Masters, J. N., & Beattie, M. (1997). Apoptosis and delayed degeneration after spinal cord injury in rats and monkeys. *Nature Medicine, 3*(1), 73–76.

Cusimano, M., Biziato, D., Brambilla, E., Doneg, M., Alfaro-Cervello, C., Snider, S., Salani, G., Pucci, F., Comi, G., Garcia-Verdugo, J. M., De Palma, M., Martino, G., & Pluchino, S. (2012). Transplanted neural stem/precursor cells instruct phagocytes and reduce secondary tissue damage in the injured spinal cord. *Brain, 135*(2), 447–460. https://doi.org/10.1093/brain/awr339

David, S., & Aguayo, A. J. (1981). Axonal elongation into peripheral nervous system 'bridges' after central nervous system injury in adult rats. *Science, 214*(4523), 931–933. https://doi.org/10.1126/science.6171034

Demediuk, P., Saunders, R. D., Anderson, D. K., Means, E. D., & Horrocks, L. A. (1987). Early membrane lipid changes in laminectomized and traumatized cat spinal cord. *Neurochemical Pathology, 7*(1), 79–89. https://doi.org/10.1007/BF02834293

Dodd, J., & Jessell, T. M. (1988). Axon guidance and the patterning of neuronal projections in vertebrates. *Science, 242*(4879). https://doi.org/10.1126/science.3055291

Dulin, J. N., Adler, A. F., Kumamaru, H., Poplawski, G. H. D., Lee-Kubli, C., Strobl, H., Gibbs, D., Kadoya, K., Fawcett, J. W., Paul, L., & Tuszynski, M. H. (2018). Injured adult motor and sensory axons regenerate into appropriate organotypic domains of neural progenitor grafts. *Nature Communications, 9*(1). https://doi.org/10.1038/s41467-017-02613-x

Ethier, C., Oby, E. R., Bauman, M. J., & Miller, L. E. (2012). Restoration of grasp following paralysis through brain-controlled stimulation of muscles. *Nature, 485*(7398), 368–371. https://doi.org/10.1038/nature10987

Fandel, T. M., Trivedi, A., Nicholas, C. R., Zhang, H., Chen, J., Martinez, A. F., Noble-Haeusslein, L. J., & Kriegstein, A. R. (2016). Transplanted human stem cell-derived interneuron precursors mitigate mouse bladder dysfunction and central neuropathic pain after spinal cord injury. *Cell Stem Cell, 19*(4), 544–557. https://doi.org/10.1016/j.stem.2016.08.020

Fehlings, M. G., Tetreault, L. A., Wilson, J. R., Kwon, B. K., Burns, A. S., Martin, A. R., Hawryluk, G., & Harrop, J. S. (2017). A clinical practice guideline for the management of acute spinal cord injury: Introduction, rationale, and scope. *Global Spine Journal, 7*(3_supplement), 84S–94S.

Filli, L., Engmann, A. K., Zörner, B., Weinmann, O., Moraitis, T., Gullo, M., Kasper, H., Schneider, R., & Schwab, M. E. (2014). Bridging the gap: A reticulo-propriospinal detour bypassing an incomplete spinal cord injury. *Journal of Neuroscience, 34*(40), 13399–13410. https://doi.org/10.1523/JNEUROSCI.0701-14.2014

Fischer, I., Dulin, J. N., & Lane, M. A. (2020). Transplanting neural progenitor cells to restore connectivity after spinal cord injury. *Nature Reviews Neuroscience, 21*(7), 366–383.

Flesher, S. N., Collinger, J. L., Foldes, S. T., Weiss, J. M., Downey, J. E., Tyler-Kabara, E. C., Bensmaia, S. J., Schwartz, A. B., Boninger, M. L., & Gaunt, R. A. (n.d.). Intracortical microstimulation of human somatosensory cortex. *Science Translational Medicine, 8*(361), 361ra14.

Formento, E., Minassian, K., Wagner, F., Mignardot, J. B., Le Goff-Mignardot, C. G., Rowald, A., Bloch, J., Micera, S., Capogrosso, M., & Courtine, G. (2018). Electrical spinal cord stimulation must preserve proprioception to enable locomotion in humans with spinal cord injury. *Nature Neuroscience, 21*(12), 1728–1741. https://doi.org/10.1038/s41593-018-0262-6

Gaspar, R., Padula, N., Freitas, T. B., de Oliveira, J. P. J., & Torriani-Pasin, C. (2019). Physical exercise for individuals with spinal cord injury: Systematic review based on the international classification of functioning, disability, and health. *Journal of Sport Rehabilitation, 28*(5), 505–516. https://doi.org/10.1123/jsr.2017-0185

Geoffroy, C. G., Lorenzana, A. O., Kwan, J. P., Lin, K., Ghassemi, O., Ma, A., Nuo, X., Creger, D., Liu, K., He, Z., & Zheng, B. (2015). Effects of PTEN and Nogo Codeletion on corticospinal axon sprouting and regeneration in mice. *Journal of Neuroscience, 35*(16), 6413–6428. https://doi.org/10.1523/JNEUROSCI.4013-14.2015

Gill, M. L., Grahn, P. J., Calvert, J. S., Linde, M. B., Lavrov, I. A., Strommen, J. A., Beck, L. A., Sayenko, D. G., Van Straaten,

M. G., Drubach, D. I., Veith, D. D., Thoreson, A. R., Lopez, C., Gerasimenko, Y. P., Reggie Edgerton, V., Lee, K. H., & Zhao, K. D. (2018). Neuromodulation of lumbosacral spinal networks enables independent stepping after complete paraplegia. *Nature Medicine, 24*(11), 1677–1682. https://doi.org/10.1038/s41591-018-0175-7

Goldberg, J. L., Klassen, M. P., Hua, Y., & Barres, B. A. (2002). Amacrine-signaled loss of intrinsic axon growth ability by retinal ganglion cells. *Science, 296*(5574), 1860–1864. https://doi.org/10.1126/science.1068428

Harel, N. Y., & Strittmatter, S. M. (2006). Can regenerating axons recapitulate developmental guidance during recovery from spinal cord injury? *Nature Reviews Neuroscience, 7*(8), 603–616.

Harkema, S., Gerasimenko, Y., Hodes, J., Burdick, J., Angeli, C., Chen, Y., Ferreira, C., Willhite, A., Rejc, E., Grossman, R. G., & Reggie Edgerton, V. (2011). Effect of epidural stimulation of the lumbosacral spinal cord on voluntary movement, standing, and assisted stepping after motor complete paraplegia: A case study. *The Lancet, 377*(9781), 1938–1947. https://doi.org/10.1016/S0140-6736(11)60547-3

He, Z., & Jin, Y. (2016). Intrinsic control of axon regeneration. *Neuron, 90*(3), 437–451.

Herrity, A. N., Aslan, S. C., Ugiliweneza, B., Mohamed, A. Z., Hubscher, C. H., & Harkema, S. J. (2021). Improvements in bladder function following activity-based recovery training with epidural stimulation after chronic spinal cord injury. *Frontiers in Systems Neuroscience, 14*, 614691. https://doi.org/10.3389/fnsys.2020.614691

Isa, T. (2019). Dexterous hand movements and their recovery after central nervous system injury. *Annual Review of Neuroscience, 42*(1), 315–335. https://doi.org/10.1146/annurev-neuro-070918-050436

Isa, T., Kinoshita, M., & Nishimura, Y. (2013). Role of direct vs. indirect pathways from the motor cortex to spinal Motoneurons in the control of hand dexterity. *Frontiers in Neurology, 4*(November), 1–9. https://doi.org/10.3389/fneur.2013.00191

Jha, R. M., Kochanek, P. M., & Marc Simard, J. (2019). Pathophysiology and treatment of cerebral edema in traumatic brain injury. *Neuropharmacology, 145*(August 2018), 230–246. https://doi.org/10.1016/j.neuropharm.2018.08.004

Kadoya, K., Paul, L., Nguyen, K., Lee-Kubli, C., Kumamaru, H., Yao, L., Knackert, J., Poplawski, G., Dulin, J. N., Strobl, H., Takashima, Y., Biane, J., James Conner, S., Zhang, C., & Tuszynski, M. H. (2016). Spinal cord reconstitution with homologous neural grafts enables robust corticospinal regeneration. *Nature Medicine, 22*(5), 479–487. https://doi.org/10.1038/nm.4066

Kim, J.-E., Li, S., GrandPré, T., Qiu, D., & Strittmatter, S. M. (2003). Axon regeneration in Young adult mice lacking Nogo-a/B. *Neuron, 38*(2), 187–199. https://doi.org/10.1016/S0896-6273(03)00147-8

Kitchen, P., Salman, M. M., Halsey, A. M., Clarke-Bland, C., MacDonald, J. A., Ishida, H., Vogel, H. J., Almutiri, S., Logan, A., Kreida, S., Al-Jubair, T., Missel, J. W., Gourdon, P., Törnroth-Horsefield, S., Conner, M. T., Ahmed, Z., Conner, A. C., & Bill, R. M. (2020). Targeting Aquaporin-4 subcellular localization to treat central nervous system edema. *Cell, 181*(4), 784–799.e19. https://doi.org/10.1016/j.cell.2020.03.037

Klapka, N., & Müller, H. W. (2006). Collagen matrix in spinal cord injury. *Journal of Neurotrauma, 23*(3–4), 422–436. https://doi.org/10.1089/neu.2006.23.422

Klatzo, I. (1967). Presidential address: Neuropathological aspects of Braix edema. *Journal of Neuropathology and Experimental Neurology, 26*(1), 1–14. https://doi.org/10.1097/00005072-196701000-00001

Laliberte, A. M., Goltash, S., Lalonde, N. R., & Bui, T. V. (2019). Propriospinal neurons: Essential elements of locomotor control in the intact and possibly the injured spinal cord. *Frontiers in Cellular Neuroscience, 13*(November), 1–16. https://doi.org/10.3389/fncel.2019.00512

Lang, B. T., Cregg, J. M., Depaul, M. A., Tran, A. P., Kui, X., Dyck, S. M., Madalena, K. M., Brown, B. P., Weng, Y. L., Li, S., Karimi-Abdolrezaee, S., Busch, S. A., Shen, Y., & Silver, J. (2015). Modulation of the proteoglycan receptor PTPσ promotes recovery after spinal cord injury. *Nature, 518*(7539), 404–408. https://doi.org/10.1038/nature13974

Leonard, A. V., Thornton, E., & Vink, R. (2015). The relative contribution of edema and hemorrhage to raised intrathecal pressure after traumatic spinal cord injury. *Journal of Neurotrauma, 32*(6), 397–402. https://doi.org/10.1089/neu.2014.3543

Li, Y., He, X., Riki Kawaguchi, Y., Zhang, Q. W., Monavarfeshani, A., Yang, Z., Chen, B., Shi, Z., Meng, H., Zhou, S., Zhu, J., Jacobi, A., Swarup, V., Popovich, P. G., Geschwind, D. H., & He, Z. (2020). Microglia-organized scar-free spinal cord repair in neonatal mice. *Nature, 587*(7835), 613–618. https://doi.org/10.1038/s41586-020-2795-6

Liang, D., Bhatta, S., Gerzanich, V., & Marc Simard, J. (2007). Cytotoxic edema: Mechanisms of pathological cell swelling. *Neurosurgical Focus, 22*(5), 1–9. https://doi.org/10.3171/foc.2007.22.5.3

Liu, K., Yi, L., Lee, J. K., Samara, R., Willenberg, R., Sears-Kraxberger, I., Tedeschi, A., Park, K. K., Jin, D., Cai, B., Bengang, X., Connolly, L., Steward, O., Zheng, B., & He, Z. (2010). PTEN deletion enhances the regenerative ability of adult corticospinal neurons. *Nature Neuroscience, 13*(9), 1075–1081. https://doi.org/10.1038/nn.2603

Liu, Y., Wang, X., Li, W., Zhang, Q., Li, Y., Zhang, Z., Zhu, J., Chen, B., Williams, P. R., Zhang, Y., Bin, Y., Xiaosong, G., & He, Z. (2017). A sensitized IGF1 treatment restores corticospinal axon-dependent functions. *Neuron, 95*(4), 817–833.e4. https://doi.org/10.1016/j.neuron.2017.07.037

Lovely, R. G., Gregor, R. J., Roy, R. R., & Edgerton, V. R. (1986). Effects of training on the recovery of full-weight-bearing stepping in the adult spinal Cat. *Experimental Neurology, 92*(2), 421–435.

Lu, P., Wang, Y., Graham, L., McHale, K., Gao, M., Di, W., Brock, J., Blesch, A., Rosenzweig, E. S., Havton, L. A., Zheng, B., Conner, J. M., Marsala, M., & Tuszynski, M. H. (2012). Long-distance growth and connectivity of neural stem cells after severe spinal cord injury. *Cell, 150*(6), 1264–1273. https://doi.org/10.1016/j.cell.2012.08.020

Murray, K. C., Nakae, A., Stephens, M. J., Rank, M., D'Amico, J., Harvey, P. J., Xiaole Li, R., Harris, L. W., Ballou, E. W., Anelli, R., Heckman, C. J., Mashimo, T., Vavrek, R., Sanelli, L., Gorassini, M. A., Bennett, D. J., & Fouad, K. (2010). Recovery of Motoneuron and locomotor function after spinal cord injury depends on constitutive activity in 5-HT2C receptors. *Nature Medicine, 16*(6), 694–700. https://doi.org/10.1038/nm.2160

Nam, K. Y., Kim, H. J., Kwon, B. S., Park, J. W., Lee, H. J., & Yoo, A. (2017). Robot-assisted gait training (Lokomat) improves walking function and activity in people with spinal cord injury: A systematic review. *Journal of Neuroengineering and Rehabilitation, 14*(1). https://doi.org/10.1186/s12984-017-0232-3

Park, K. K., Liu, K., Yang, H., Smith, P. D., Wang, C., Cai, B., Bengang, X., Connolly, L., Kramvis, I., Sahin, M., & He, Z. (2008). Promoting axon regeneration in the adult CNS by modulation of the PTEN/MTOR pathway. *Science, 322*(5903), 963–966. https://doi.org/10.1126/science.1161566

Poplawski, G. H. D., Kawaguchi, R., Van Niekerk, E., Lu, P., Mehta, N., Canete, P., Lie, R., Dragatsis, I., Meves, J. M., Zheng, B., Coppola, G., & Tuszynski, M. H. (2020). Injured adult neurons regress to an embryonic transcriptional growth state. *Nature, 581*(April 2019). https://doi.org/10.1038/s41586-020-2200-5

Raineteau, O., & Schwab, M. E. (2001). Plasticity of motor systems after incomplete spinal cord injury. *Nature Reviews Neuroscience, 2*(4), 263–273. https://doi.org/10.1038/35067570

Ramón Y Cajal, S., DeFelipe, J., Jones, E. G., & May, R. M. (2012). *Cajal's degeneration and regeneration of the nervous system.*

Rosenzweig, E. S., Brock, J. H., Paul, L., Kumamaru, H., Salegio, E. A., Kadoya, K., Weber, J. L., Liang, J. J., Moseanko, R., Stephanie Hawbecker, J., Huie, R., Havton, L. A., Nout-Lomas, Y. S., Ferguson, A. R., Beattie, M. S., Bresnahan, J. C., & Tuszynski, M. H. (2018). Restorative effects of human neural stem cell grafts on the primate spinal cord. *Nature Medicine, 24*(4), 484–490. https://doi.org/10.1038/nm.4502

Ruder, L., Takeoka, A., & Arber, S. (2016). Long-distance descending spinal neurons ensure quadrupedal locomotor stability. *Neuron, 92*(5), 1063–1078. https://doi.org/10.1016/j.neuron.2016.10.032

Rungta, R. L., Choi, H. B., Tyson, J. R., Malik, A., Dissing-Olesen, L., Lin, P. J. C., Cain, S. M., Cullis, P. R., Snutch, T. P., & Macvicar, B. A. (2015). The cellular mechanisms of neuronal swelling underlying cytotoxic edema. *Cell, 161*(3), 610–621. https://doi.org/10.1016/j.cell.2015.03.029

Sawant-Pokam, P. A., Vail, T. J., Metcalf, C. S., Maguire, J. L., McKean, T. O., McKean, N. O., & Brennan, K. C. (2020). Preventing neuronal edema increases network excitability after traumatic brain injury. *Journal of Clinical Investigation, 130*(11), 6005–6020. https://doi.org/10.1172/JCI134793

Siddall, P. J., & Loeser, J. D. (2001). Pain following spinal cord. *Injury, 39*, 63–73.

Silver, J., & Miller, J. H. (2004). Regeneration beyond the glial scar. *Nature Reviews Neuroscience, 5*(2), 146–156.

Sofroniew, M. V. (2018). Dissecting spinal cord regeneration perspective. *Nature, 557*(7705), 343–350.

Takeoka, A. (2020). Proprioception: Bottom-up directive for motor recovery after spinal cord injury. *Neuroscience Research, 154*, 1–8.

Takeoka, A., & Arber, S. (2019). Functional local proprioceptive feedback circuits initiate and maintain locomotor recovery after spinal cord injury. *Cell Reports, 27*(1), 71–85.e3. https://doi.org/10.1016/j.celrep.2019.03.010

Takeoka, A., Vollenweider, I., Courtine, G., & Arber, S. (2014). Muscle spindle feedback directs locomotor recovery and circuit reorganization after spinal cord injury. *Cell, 159*(7), 1626–1639. https://doi.org/10.1016/j.cell.2014.11.019

Tessier-Lavigne, M., & Goodman, C. S. (1996). The molecular biology of axon guidance. *Science, 274*(5290), 1123–1133.

Tom, V. J., Steinmetz, M. P., Miller, J. H., Doller, C. M., & Silver, J. (2004). Studies on the development and behavior of the dystrophic growth cone, the Hallmark of regeneration failure, in an in vitro model of the glial scar and after spinal cord injury. *Journal of Neuroscience, 24*(29), 6531–6539. https://doi.org/10.1523/JNEUROSCI.0994-04.2004

Torres-Espín, A., Beaudry, E., Fenrich, K., & Fouad, K. (2018). Rehabilitative training in animal models of spinal cord injury. *Journal of Neurotrauma, 35*(16), 1970–1985. https://doi.org/10.1089/neu.2018.5906

Wagner, F. B., Mignardot, J. B., Le Goff-Mignardot, C. G., Demesmaeker, R., Komi, S., Capogrosso, M., Rowald, A., Seáñez, I., Caban, M., Pirondini, E., Vat, M., McCracken, L. A., Heimgartner, R., Fodor, I., Watrin, A., Seguin, P., Paoles, E., Van Den Keybus, K., Eberle, G., Schurch, B., Pralong, E., Becce, F., Prior, J., Buse, N., Buschman, R., Neufeld, E., Kuster, N., Carda, S., von Zitzewitz, J., Delattre, V., Denison, T., Lambert, H., Minassian, K., Bloch, J., & Courtine, G. (2018). Targeted Neurotechnology restores walking in humans with spinal cord injury. *Nature, 563*(7729), 65–93. https://doi.org/10.1038/s41586-018-0649-2

Wang, X., Liu, Y., Li, X., Zhang, Z., Hengfu Yang, Y., Zhang, P. R., Williams, N. S. A., Alwahab, K. K., Bin, Y., Zhang, Y., Chen, M., Ding, H., Gerfen, C. R., Wang, K. H., & He, Z. (2017). Deconstruction of corticospinal circuits for goal-directed motor skills. *Cell, 171*(2), 440–455.e14. https://doi.org/10.1016/j.cell.2017.08.014

Wenger, N., Moraud, E. M., Gandar, J., Musienko, P., Capogrosso, M., Baud, L., Le Goff, C. G., Barraud, Q., Pavlova, N., Dominici, N., Minev, I. R., Asboth, L., Hirsch, A., Duis, S., Kreider, J., Mortera, A., Haverbeck, O., Kraus, S., Schmitz, F., et al. (2016). Spatiotemporal neuromodulation therapies engaging muscle synergies improve motor control after spinal cord injury. *Nature Medicine, 22*(2), 138–145. https://doi.org/10.1038/nm.4025

Wilkins, R. H. (1964). Neurosurgical Classic—XVII. *Journal of Neurosurgery, 21*(3), 240–244. https://doi.org/10.3171/jns.1964.21.3.0240

Yashon, D., George Bingham, W., Faddoul, E. M., & Hunt, W. E. (1973). Edema of the spinal cord following experimental impact trauma. *Journal of Neurosurgery, 38*(6), 693–697. https://doi.org/10.3171/jns.1973.38.6.0693

Yasuda, A., Tsuji, O., Shibata, S., Nori, S., Takano, M., Kobayashi, Y., Takahashi, Y., Fujiyoshi, K., Hara, C. M., Miyawaki, A., Okano, H. J., Toyama, Y., Nakamura, M., & Okano, H. (2011). Significance of Remyelination by neural stem/progenitor cells transplanted into the injured spinal cord. *Stem Cells, 29*(12), 1983–1994. https://doi.org/10.1002/stem.767

Zholudeva, L. V., Iyer, N., Qiang, L., Spruance, V. M., Randelman, M. L., White, N. W., Bezdudnaya, T., Fischer, I., Sakiyama-Elbert, S. E., & Lane, M. A. (2018a). Transplantation of neural progenitors and V2a interneurons after spinal cord injury. *Journal of Neurotrauma, 35*(24), 2883–2903. https://doi.org/10.1089/neu.2017.5439

Zholudeva, L. V., Qiang, L., Marchenko, V., Dougherty, K. J., Sakiyama-Elbert, S. E., & Lane, M. A. (2018b). The Neuroplastic and therapeutic potential of spinal interneurons in the injured spinal cord. *Trends in Neurosciences, 41*(9), 625–639. https://doi.org/10.1016/j.tins.2018.06.004

Zholudeva, L. V., Abraira, V. E., Satkunendrarajah, K., McDevitt, T. C., Goulding, M. D., Magnuson, D. S. K., & Lane, M. A. (2021). Spinal interneurons as gatekeepers to neuroplasticity after injury or disease. *Journal of Neuroscience, 41*(5), 845–854. https://doi.org/10.1523/JNEUROSCI.1654-20.2020

Zukor, K., Belin, S., Wang, C., Keelan, N., Wang, X., & He, Z. (2013). Short hairpin RNA against PTEN enhances regenerative growth of corticospinal tract axons after spinal cord injury. *Journal of Neuroscience, 33*(39), 15350–15361. https://doi.org/10.1523/JNEUROSCI.2510-13.2013

Macromolecular Therapeutics: Development and Delivery Engineering

54

Xin Wei, Gang Zhao, and Dong Wang

Abstract

Macromolecular therapeutics have been developed to improve the efficacy and safety profile of varied therapeutic agents for different diseases more than 50 years ago. *N*-(2-hydroxypropyl) methacrylamide (HPMA) copolymers as a drug delivery platform have been extensively studied and applied in the clinical trials. Recently, the application of HPMA copolymers in the treatment of noncancerous diseases such as infectious diseases, musculoskeletal diseases, and traumatic brain injury has also been explored widely. In this chapter, different formulations of macromolecular therapeutics will be discussed. The design principles illustrated in the development of HPMA copolymer drug conjugates are also valid for other hydrophilic polymers, and they will be the principal focus of our discussion. The linker chemistry of the HPMA copolymer drug conjugates and its fine tuning will be discussed. Active targeting of HPMA copolymers based on targeting moieties and passively targeting HPMA copolymers based on the enhanced permeability and retention effect for cancer disease and extravasation through leaky vasculature and subsequent inflammatory cell-mediated sequestration mechanism for inflammation disease will be discussed. The most recent macromolecular prodrug-based thermoresponsive hydrogels designed for local injection to provide extended release with minimized systemic distribution will also be discussed in this chapter.

X. Wei · D. Wang (✉)
Department of Pharmaceutical Sciences, College of Pharmacy, University of Nebraska Medical Center, Omaha, NE, USA
e-mail: dwang@unmc.edu

G. Zhao
Ensign Pharmaceutical, Inc, Omaha, NE, USA

Keywords

HPMA Copolymers · Targeting · Thermoresponsive Hydrogel · Cancer Therapy · Central Nervous System (CNS)

54.1 Introduction

Water-soluble polymers have been widely used as carriers of drugs, genes, oligonucleotides and for modification of proteins, liposomes, and surface of biomaterials. Drug conjugated to synthetic macromolecules was initiated by Jatzkewitz 60 years ago, when researchers first used the water-soluble polymer polyvinyl pyrrolidone as a drug depot for the biologically active primary amine mescaline (Jatzkewitz, 1955). Since then, the field was expanded and developed dramatically beyond depot effects. Numerous water-soluble polymer–drug conjugates, PEG–protein conjugates, lysosomotropic polymer–drug conjugates, and block copolymer micelles were synthesized in 1960s and 1970s (De Duve et al., 1974; Abuchowski et al., 1977). "Polymer therapeutics" was coined by Ruth Duncan to describe polymeric drugs, polymer protein conjugates, block copolymer polymeric micelles to which a drug is covalently bound and polyplexes developed as nonviral vectors for gene interfering ribonucleic acid (siRNA) delivery (Duncan, 2003). The rationales for polymer–drug conjugates include prolong the blood circulation time, reduce the side effects of the drug, improve drug targeting, and overcome drug resistance. Conjugation of low molecular weight drugs to polymers would prolong their resident time in blood circulation resulting in enriched bioavailability. The delivery system based on synthetic water-soluble polymer would render hydrophobic drugs better solubility and limit the cellular uptake to the endocytic route. Based on enhanced permeability and reten-

H. E. Gendelman, T. Ikezu (eds.), *Neuroimmune Pharmacology and Therapeutics*,
https://doi.org/10.1007/978-3-031-68237-7_54

tion (EPR) effect (Matsumura & Maeda, 1986), polymer conjugation would facilitate the passive targeting of drugs to the solid tumor (Maeda, 2010) as observed in preclinical studies and some clinical studies although the clinical result is controversial (Nichols & Bae, 2014; Danhier, 2016). The newly discovered extravasation through leaky vasculature and inflammatory cell-mediated sequestration (ELVIS) mechanism (Yuan et al., 2012; Wang & Goldring, 2011; Ren et al., 2011, 2014) explains macromolecular therapeutics' preferentially accumulate in inflammatory tissues (Liu et al., 2010) in general and inflammatory arthritis in particular (Wang et al., 2004; Yuan et al., 2012; Liu et al., 2008). In a recent study (Zhao et al., 2021), with the increase of the hydrophobic drug payload (e.g., dexamethasone) in HPMA copolymer, the macromolecular prodrug forms a thermoresponsive injectable hydrogel that allows the prodrug being retained at the pathophysiological injection site for extended period time to provide sustained release.

In this chapter, we will focus mainly on the development of *N*-(2-hydroxypropyl) methacrylamide (HPMA) copolymer drug conjugates as the design principles illustrated are also valid for other hydrophilic polymers. HPMA copolymer-based macromolecular therapeutics are active against many cancer models and have been evaluated in several clinical trials (Kopeček et al., 2000). In 1994, HPMA copolymer-doxorubicin (FCE28068) was the first HPMA copolymer drug conjugate to enter clinical evaluation (Vasey et al., 1999). To achieve the controlled release and better solubility of paclitaxel, Pharmacia developed the HPMA copolymer-paclitaxel (PNU166945) (Meerum Terwogt et al., 2001). Beyond oncology, HPMA copolymer is also applied to treat rheumatoid arthritis when conjugated with glucocorticoids (Yuan et al., 2012). A recent study (Wei et al., 2022) utilized HPMA copolymer dexamethasone conjugates in the treatment of brain injury showed amelioration of neuroinflammation and associated complications. With this efficacy study as proof of concept, there is great potential for these technologies to be extrapolated into neurological applications.

54.2 Structure of Macromolecular Therapeutics

The structure of macromolecular therapeutic agents can vary due to different types of the payload (Duncan, 2003). While micelles, dendrimers, liposomes, nanoparticles, and polymer–drug conjugates all share the concept of macromolecular therapeutics, they can be very different in their structures (Vikas et al., 2015). Different delivery systems are also applied to deliver different candidates because of their own unique properties. For example, macromolecular payloads prefer the liposomes, nanoparticles, and micelle delivery systems, while polymeric delivery systems (e.g., HPMA copolymers) favor small molecular payloads (Fig. 54.1).

Liposomes

Liposomes have been defined as a vesicle for drug delivery since 1960s (Bangham et al., 1965). Liposomes are composed of a phospholipid membrane surrounded by the aqueous environment. The structure of the liposome membrane is a double layer with fluidity. The shape of liposomes is basically spherical. Liposomes can be categorized in to multilamellar vesicle (MLV) (Gentile et al., 2014), small unilamellar liposome vesicle (SUV) (Lokappa & Ulmer, 2011), and large unilamellar vesicle (LUV) (Moyano et al., 2008) according to the number of lipid layers and the size of liposome. Most of the clinically used liposome formulations have to be decorated with inert water-soluble polymers (e.g., PEG, HPMA copolymer, etc.) to ensure their stealth against the mononuclear phagocyte system (MPS). While this modification may enlarge the size of the liposome, its half-life in circulation will be improved. The surface decorating polymers can also be used to introduce various targeting moieties or imaging probes. Liposome size can vary from 100 to 400 nm in diameter and plays a critical role in complement activation and MPS clearance of the delivery system (Drummond et al., 1999; Devine et al., 1994). Vesicles larger than 100 nm require additional strategies to prevent surface opsonization (Koo et al., 2005).

Nanoparticles

Nanoparticles can be categorized as drug nanoparticles, solid nanoparticles, polymer-based nanoparticles, lipid-based nanoparticles, and nanocapsules. Drug nanoparticles are the dispersion of water-insoluble drug particles of drug formulations in the nanosize range in an aqueous environment. Drug nanoparticles can be achieved by breaking down the bigger particles by high-pressure homogenization method or by special crystallization techniques. Polymer-based nanoparticles are commonly composed of poly(lactic acid) (PLA), poly(glycolicacid) (PGA), poly (lactic acid-co-glycolic acid) (PLGA), poly(e-caprolactone) (PCL), and poly(methylmethacrylate) (PMMA). Chemical synthesis, salting-out, emulsification-diffusion, nanoprecipitation, and freeze-drying methods are used to obtain polymer-based nanoparticles. Lipid-based nanoparticles mostly composed of fats or waxes and can be obtained by homogenization. For certain applications, nanoparticles may be tailored into different shape: namely sphere nanoparticles (Kocbek et al., 2007), cube nanoparticles (Verma & Ahuja, 2015), rod-like nanoparticles (Wang et al., 2014), hollow spherical nanoparticles (Chen et al., 2014), and random shape nanoparticles.

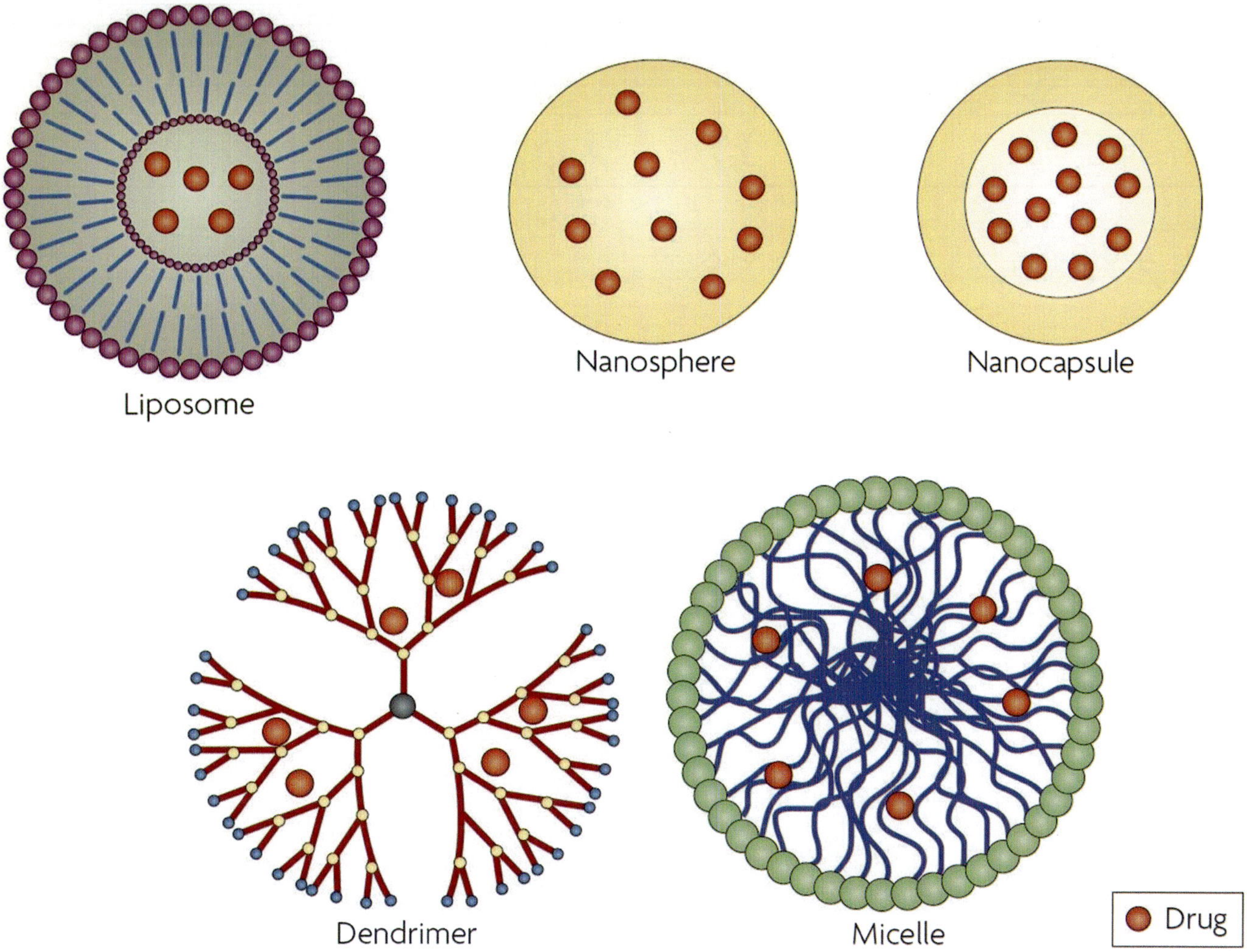

Fig. 54.1 A schematic representation of some macromolecular drug delivery systems. (**a**) Liposomes (~100–400 nm) are small spherical artificial vesicles typically made with lipid bilayers, (**b**) Nanosphere (~20–200 nm) are typically made with biodegradable polymers for sustained drug release, (**c**) nanocapsules (~10–1000 nm) can encapsulate relatively large amounts of drugs and nucleic acids such as DNA, microRNA, small interfering RNA (siRNA) and small hairpin RNA (shRNA), (**d**) Dendrimers (~3–20 nm) are monodisperse macromolecules that can be used to encapsulate or covalently conjugate drugs, targeting moieties and imaging agents, (**e**) Micelles (~10–100 nm) are self-assembled amphiphilic particles that can encapsulate both lipophilic or lipophobic drugs stabilized by surfactants. (Adapted from Orive et al., 2009, with permission)

Dendrimers

Dendrimers (Fig. 54.2) are highly uniformed, branched, or star-shaped macromolecules. Dendrimers can be synthesized by divergent (Li et al., 2014b) or convergent approaches (Kushwaha & Tiwari, 2013). They are often of uniform molecular weight and very low polydispersities. Dendrimers also have modifiable surface functional group as well as internal cavities (Jansen et al., 1994). For example, Newkome et al. (1991) have synthesized a dendrimer containing hydrophobic interior and hydrophilic surface functionality. During or after synthesis of dendrimers, the drug molecules can be physically entrapped or chemically conjugated to the dendrimer (Thanki et al., 2013).

Micelles

A formulation of micelle can be achieved by the self-assembly (aggregation) of amphipathic macromolecules when they are added to aqueous solution. Critical micelle concentration (CMC), is defined as the concentration above which the micelle will form. The amphipathic macromolecules used for micelle formulations can be classified into three types: (1) The molecule is composed of the one polar end (hydrophilic group) and one nonpolar end (hydrophobic group) (Koizumi et al., 2006). (2) Two polar ends with the non-polar center (Chen et al., 2013). (3) Two nonpolar ends with the polar center (Nederberg et al., 2011). Shapes of micelles can be sphere (Chen et al., 2013), ellipsoid cylinder, and monolayer micelle (Wiedwald et al., 2010). To improve

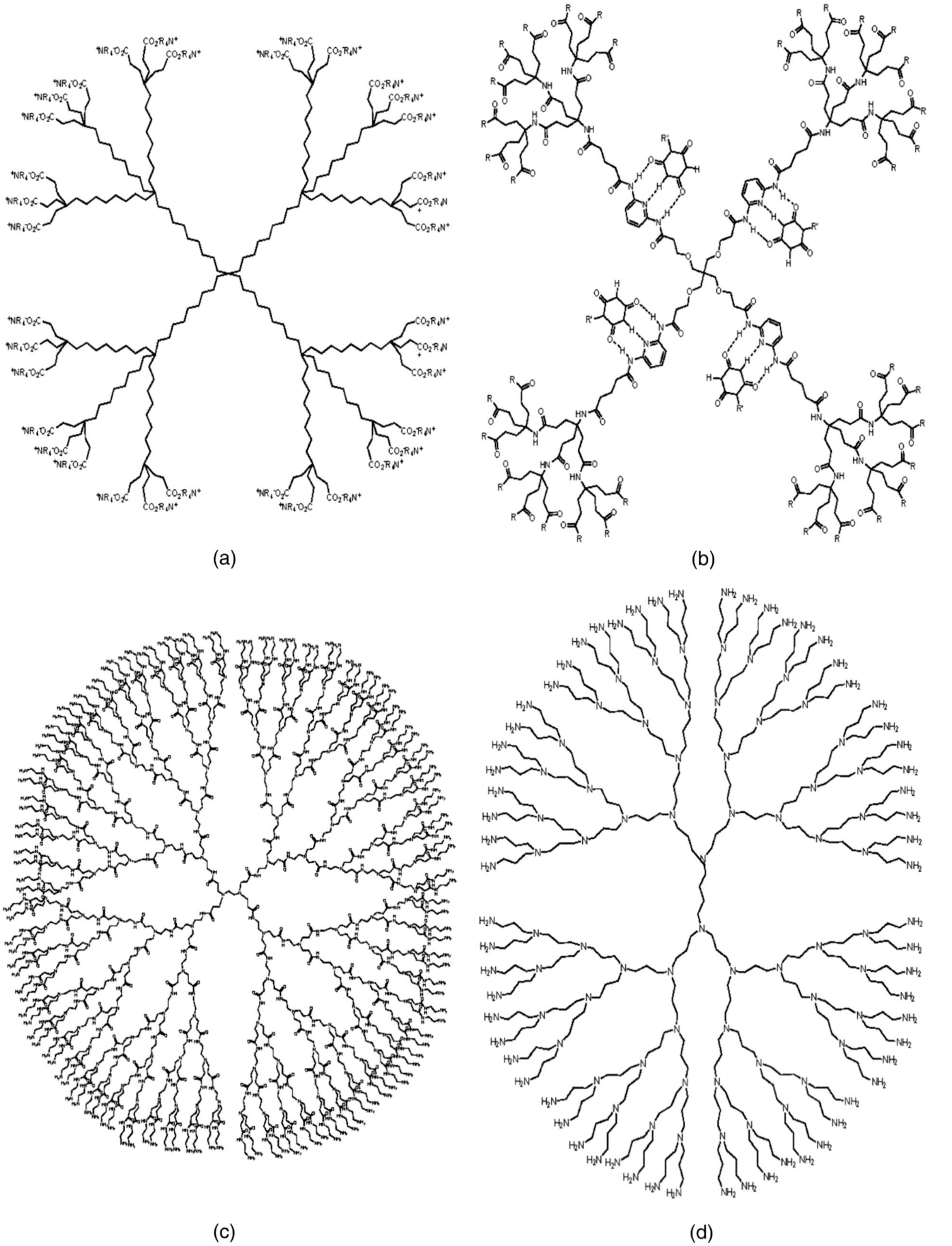

Fig. 54.2 Different families of dendrimers. (**a**) Unimolecular micelle. (**b**) Newkome's dendrimer with internal binding units. (**c**) PAMAM dendrimer. (**d**) POPAM dendrimer. (Adapted from (Patri et al., 2002), with permission)

micelle stability, Liu et al. prepared poly (2-cinnamoylethyl methacrylate-b-acrylic acid) which forms polymeric micelles in water (Henselwood & Liu, 1997; Liu, 1997) and cross-linked the core segment by photochemically induced polymerization of the cinnamoyl side groups, which decreases the free volume of the core. The core density may increase and the size of the micelles might decrease because of the cross-linking.

Polymer–Drug Conjugates

Polymer–drug conjugates are generally consisted of a polymeric backbone, active drugs, targeting moiety and signaling moiety, as illustrated in the Fig. 54.3. The primary structure of the conjugates and the chemical bonds by which the components are linked to the backbone can be designed base on the conjugates applications (Miller et al., 2008), drug properties (Liu et al., 2010; Wang et al., 2007a), and biological rationales (Kopeček, 1984; Kopeček et al., 1991; Putnam & Kopeček, 1995). The shape of HPMA–copolymer conjugates (Nakamura et al., 2015), the length of the backbone (Nakamura et al., 2015), the number of the side chain, and the length of the side chain (Kopeček et al., 2000) can be managed to some degree by different synthetic strategies. Reversible addition-fragmentation chain transfer (RAFT) polymerization is commonly used to manage the molecular weight and polydispersity of the polymer conjugates, which will affect the distribution, accumulation, and the elimination of the conjugates on living organism (Moad et al., 2008; Vikas et al., 2015).

Thermoresponsive Macromolecular Prodrug-Based Hydrogel

In a recent study (Zhao et al., 2021), it was discovered that when the hydrophobic drug payload (e.g., dexamethasone) loading ratio in of HPMA copolymer conjugates was increased to an unusually high level (over 20 wt%), its aqueous solution demonstrated a thermoresponsive phase transition behavior with the prodrug concentration maintained at 20 w/v%. This aqueous solution of this prodrug, which was designated as "ProGel-Dex," is a free-flowing liquid at 4 °C but forms a hydrogel when the temperature is elevated to ~30 °C (Fig. 54.4). After in vivo injection, ProGel-Dex would form a hydrogel, resulting in its physical entrapment at the injection site. Upon interaction with the body fluids, the hydrogel would gradually dissolve, providing sustained prodrug release, and activation at the site of pathology.

54.3 Targetability of HPMA Copolymer Conjugates

The targeting of the polymeric conjugates to pathology is essential for a drug delivery system. The drug targeting mechanisms are generally categorized into active targeting and passive targeting (Zhou & Kopeček, 2012).

54.3.1 Active Targeting

Active targeting is the method using the target moiety, such as antibody and specific low molecular weight ligands, to direct the payload to preferentially accumulate to the site of pathology. Active targeting always takes advantages of the different pathological features of the lesion when compared to the normal tissue. The preferential interaction of the antigen expressed on the cell surface of a particular disease lesion and the targeting moiety of the delivery system will facilitate the active targeting process. As one may speculate, this strategy can lower the side effect of the drug by reducing unnecessary drug exposure to the normal tissues.

Antibody-Mediated Targeting

Conjugate bound with specific antibodies is called antibody-targeted polymer conjugates. While such conjugation would significantly improve the polymer–drug conjugates' diseases targeting ability, the original antigen binding affinity of the antibodies conjugated to polymeric drug carriers can be compromised due to the chemical modification process (Tappertzhofen et al., 2014). Recently, a new method has been developed, attempting to address this problem (Pechar et al., 2011). A heterodimeric coiled coil structure can be formed with a highly specific interaction between peptide K ($(VAALKEK)_4$) and peptide E ($(VAALEKE)_4$). Benefiting from such molecular recognition, a cytostatic drug-bearing HPMA copolymer peptide E conjugate was synthesized and self-assembled with peptide K tagged BCL1 leukemia cells specific scFv of B1 mAb, to obtain an antibody–polymer drug conjugate with largely intact antibody targeting ability.

Folate-Mediated Targeting

Folate shows high affinity to its receptors, folate-binding proteins. This unique property has established folate as a popular active targeting moiety to assist the intracellular transportation and drug/drug delivery systems (Campbell et al., 1991; Kranz et al., 1995). The folate receptors are highly expressed on the cancer cells (Yoo & Park, 2004), which would facilitate the active targeting of folate bearing polymer conjugates to the cancer lesion. The conjugates can be transported intracellularly via receptor-mediated endocytosis. Unlike liposomes, micelles, and nanoparticles, hydrophobicity of folate affect polymer–drug conjugate more because it can be hidden within the polymer random coil but not on the surface like in the other nanomedicine formulations. To overcome this limitation, researchers introduced the positive charges into the HPMA copolymer conjugates (Li et al., 2014a). Folate receptor (FR) positive Hela cells were used to test the drug-cell binding affinity. In the presence of folate-modified cationic HPMA copolymers, FR antibody bounded cells decreased from 71.2% to only 34.0%. This indicates that the positive charge could probably amplify the binding efficiency of folate to its receptor due to close proximity of conjugates to the cell surface by the elec-

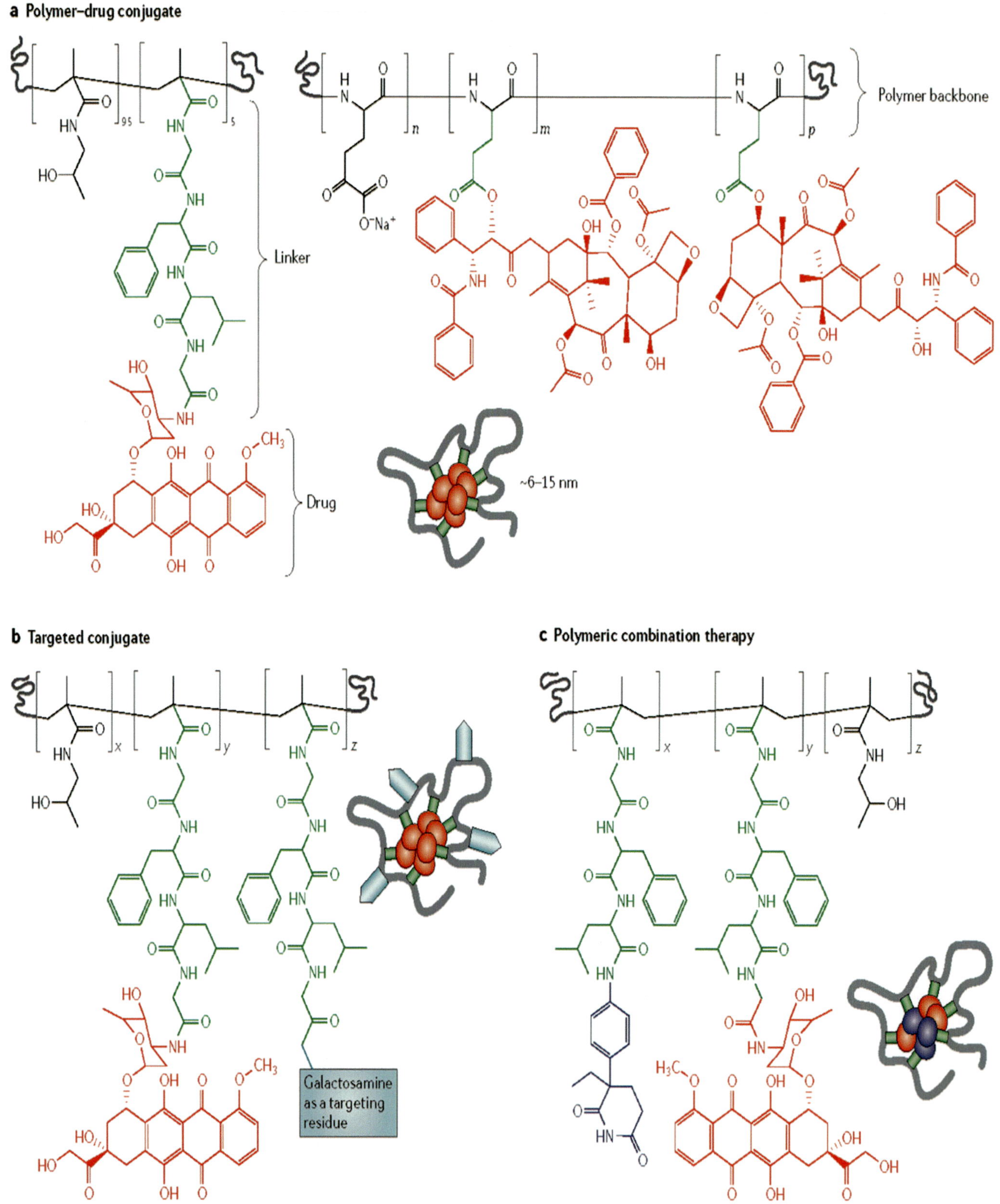

Fig. 54.3 Each panel shows both the detailed chemical structure and a cartoon of the general structure. The polymer backbone is shown in black, linker region in green, drug in red, and additional components (for example, a targeting residue) in blue. (**a**) Two examples of more "simple" polymer–drug conjugates containing doxorubicin (left) and paclitaxel (right) that have progressed to clinical trial. (**b**) A multivalent receptor-targeted conjugate containing galactosamine (light blue) to promote liver targeting. (**c**) Polymer combination therapy containing the aromatase inhibitor aminogluthethimide (red) and doxorubicin (blue). (Adapted from (Duncan, 2006), with permission)

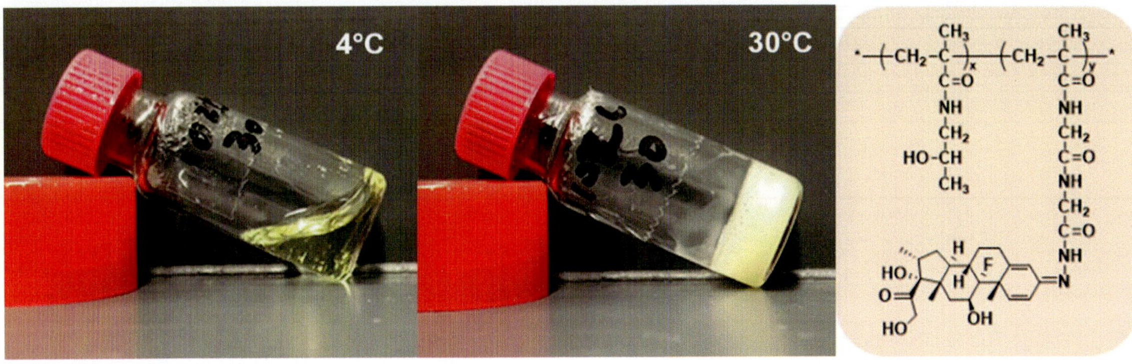

Fig. 54.4 The thermoresponsive phase transition of ProGel-Dex. Left panel: aqueous ProGel-Dex solution was free flowing at 4 °C. Middle panel: aqueous ProGel-Dex solution formed a hydrogel at 30 °C. Right panel: ProGel-Dex's chemical structure. (Adapted from (Zhao et al., 2021), with permission)

tronic adhesion. To avoid nonspecific binding by positive charge in the circulation, the researchers applied charge shielding/deshielding approach. Charge shielding groups 2,3-dimethylmaleic anhydride (DMA) can be specifically hydrolyzed at tumor extracellular environment (pH 6.8). Then the conjugates show the charge-reversible ability, leading to the faster endocytosis into the cells of cancerous lesion than the normal tissue (pH 7.4). This technique may improve the folate-mediate targeting of HPMA copolymer conjugates to cancer.

Transferrin-Mediated Targeting

Similar to folate, transferrin is another well-studied targeting moiety because of the overexpression of its receptor on the cancer cells surface (Tomita et al., 1991) and brain microvascular endothelial cells (Roberts et al., 1993). The binding of transferrin with its receptor would lead to its endocytosis. Iron in plasma will bind with transferrin and dissociate from it in the acidic compartment in the cell. Copolymers consist of the hydrophobic monomer lauryl methacrylate (LMA), a reactive anhydride functional methacrylate (TMA), and a large polyethylene glycol methacrylate monomer was copolymerized and conjugated to transferrin. This docetaxel containing transferrin receptor targeting conjugate shows the enhanced cancer cell killing properties in in vitro cytotoxicity studies (Roy et al., 2014).

54.3.2 Passive Targeting

In 1979, the antitumor protein drug neocarzinostatin (NCS) was conjugated with a synthetic copolymer of styrene maleic acid copolymer (SMA), which was called SMANCS (molecular weight 16,000 g/mol) by Maeda et al. (1979). This antitumor protein–drug conjugate exhibited many unique properties, including prolonged half-life in circulation (20-fold), improved tumor-targeting capacity (2000 fold intratumor concentration than plasma), no immunogenicity, and higher lipophilicity (Fang et al., 2011; Maeda et al., 1984; Maeda & Konno, 1997). These properties led to the conceptualization of the enhanced permeability and retention effect (EPR effect) (Matsumura & Maeda, 1986) showed in Fig. 54.5. With the advantage of EPR effect, SMANCS became the first macromolecular anticancer drug approved in 1993 in Japan (Fang et al., 2011; Maeda & Konno, 1997). The EPR effect has been observed in the applications of many nanomedicine formulations, such as polymer conjugates, polymeric micelles, and liposomes (Seymour et al., 1987; Dand et al., 2013). Most HPMA copolymers utilize the EPR effect to achieve better tumor targeting and delivery of drugs. EPR effect is a tumor vasculature-dependent phenomenon. Most tumors are well vascularized with high density, and the rapid growth of blood vessels leads to irregular vascular alignment and defects of junction between endothelial cells. The increased local fenestration of macromolecules paired with the ill-developed lymphatic drainage at the tumor lesion lead to the local accumulation of macromolecules over time (Folkman, 1995; Fang et al., 2011; Hashizume et al., 2000; Jain, 1999; Jain et al., 2002; Roberts & Palade, 1997).

Wang et al. first described macromolecules' extravasation through leaky vasculature and their subsequent inflammatory cell-mediated sequestration (ELVIS) (Wang & Goldring, 2011). It explains the passive targeting of the HPMA copolymer–dexamethasone conjugate (P-Dex) to inflammations (Yuan et al., 2012; Quan et al., 2010). Unlike the traditional understanding of EPR effect that macromolecules only can passively target to solid tumors (not inflammation) due to the leaky vasculature and impaired lymphatic drainage, they found although the extravasated macromolecules are cleared from the inflammatory tissue quickly, rapid internalization by inflammatory infiltrates and local activated cells provide the mechanism for the sustained retention of the macromolecular drug conjugate at the site of inflammation.

The macromolecular prodrug-based hydrogel mentioned above is also passive targeting. Unlike conventional water-

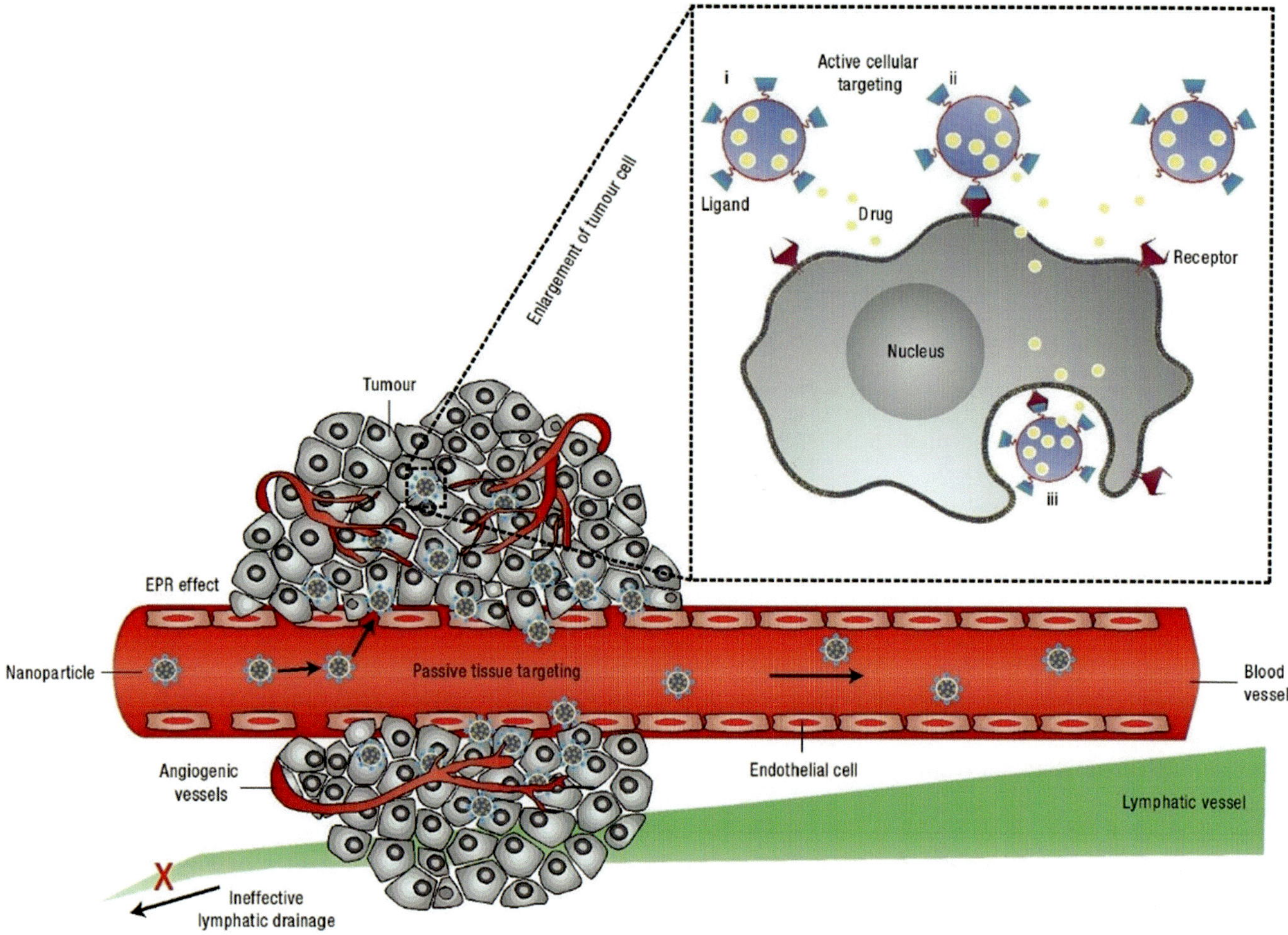

Fig. 54.5 Schematic representation of different mechanisms by which nanocarriers can deliver drugs to tumors. Polymeric nanoparticles are shown as representative nanocarriers (circles). Passive tissue targeting is achieved by extravasation of nanoparticles through increased permeability of the tumor vasculature and ineffective lymphatic drainage (EPR effect). Active cellular targeting (inset) can be achieved by functionalizing the surface of nanoparticles with ligands that promote cell-specific recognition and binding. The nanoparticles can (i) release their contents in close proximity to the target cells; (ii) attach to the membrane of the cell and act as an extracellular sustained-release drug depot; or (iii) internalize into the cell. (Adapted from (Peer et al., 2007), with permission)

soluble polymeric prodrug formulations that have been developed for systemic delivery, the thermoresponsive macromolecular prodrug-based hydrogel is designed to be administered locally at sites of pathology. After injection, it rapidly transitions into a stable hydrogel, which through gradual dissolution slowly releases the prodrug. The released prodrug will be sequestered by the activated phagocytes at inflammation site. The macromolecular prodrug offers a unique cellular pharmacology that targets the highly phagocytic cells present at sites of inflammation. Through physical entrapment and slow dissolution, the prodrug hydrogel provides sustained retention at the pathophysiological site. From this research report (Zhao et al., 2021), the thermoresponsive property of this prodrug based hydrogel is insensitive to its *Mw*. The hydrogel can be designed to a small molecular weight (< 10 kDa) without hampering the targetability. The prodrug released from the locally administered hydrogel, which is not internalized by phagocytic cells will be excreted through lymphatic drainage quickly and reduce the systemic distribution.

54.4 Clinical Tests on HPMA Copolymer–Anticancer Conjugates

Cancer is the second leading cause of death in the United States. One of every four deaths in the United States is from cancer. In the past 60 years, synthetic water-soluble polymers, including HPMA copolymers, have been used extensively as anticancer drug carriers due to their high biocompatibility and chemical functionality. All polymer conjugates used as anticancer agents in clinical trials are in effect macromolecular prodrugs. HPMA copolymers as an anticancer drug carrier are nontoxic, nonimmunogenic and have multiple side chains to carry drug and targeting moiety payload (Duncan & Kopeček, 1984; Duncan, 2009; Kopeček et al., 2000). The rationale for HPMA copolymer anticancer drug conjugation, from biological point, is to ensure that the drug would reach the pharmacological target and has limited access to sites of nonspecific toxicity. Moreover, the drug must be released at a concentration and rate, which should be able to maximize its therapeutic action. It is important to control polydispersity as the molecule weight variation has high impact on general and cellular pharmacokinetics with consequent implications for efficacy and toxicity. The antitumor HPMA copolymers target to the tumor through EPR effect, which shows a significant increased tumor drug concentration.

Many achievements have been made in the clinical evaluation of HPMA copolymer anticancer drug conjugate. For example, a Phase I clinical trial using HPMA copolymer–doxorubicin to treat breast cancer showed signs of activity coupled with fivefold decreased anthracycline toxicity in chemotherapy-refractory patients (Vasey et al., 1999). Phase II studies utilized 17 breast cancer patients, 29 nonsmall cell lung cancer patients, and 16 colorectal cancer patients (Seymour et al., 2009). The study has expanded and confirmed the early indications that HPMA copolymer doxorubicin has therapeutic activity against some advanced solid neoplasms. Of the 14 evaluable patients with breast cancer 21% had partial responses, and these were all anthracycline-naive patients, and it is possible that free doxorubicin could also have been effective in these patients. Three chemotherapy-naïve patients of 26 nonsmall cell lung cancer patients had partial responses. In contrast, none of the 16 evaluable patients with colorectal cancer responded (Seymour et al., 2009). HPMA copolymer doxorubicin has the extended plasma circulation of the polymer conjugate, an important prerequisite for passive accumulation within tumor tissues comparing with free doxorubicin and its delivery is determined largely by its kinetics and biodistribution (Duncan, 2009). Failure of traditional chemotherapy results in recurrence and development of virulent multi drug resistance (Persidis, 1999). To deal with this, the majority of cancers are being treated by combination of drugs. HPMA copolymer–doxorubicin conjugate and HPMA copolymer–photosensitizer conjugate were the first combination therapy using polymer bound to drugs (Krinick et al., 1994). Experiments results show that combination therapy achieved the better cure than either therapy alone (Kopeček et al., 2000; Zuluaga & Lange, 2008; Matthew Peterson et al., 1996).

Paclitaxel is a well-known antitumor agent with proven efficacy in ovarian, breast, and non-small cell lung cancer (Huizing et al., 1995). Intravenous administration of paclitaxel is hindered by poor water solubility of the drug and associated with unpredictable side effects. PNU166945 is a novel polymer–conjugate derivative of paclitaxel, which uses HPMA polymer as the polymeric carrier. This compound is water soluble, no toxicity, and provides a wide platform for specific tumor targeting (Meerum Terwogt et al., 2001). HPMA copolymer paclitaxel was administered as a 1 h intravenous infusion every 3 weeks to a total of 12 patients in the phase I clinical trial (Meerum Terwogt et al., 2001). However, this trail was discontinued before reaching dose-limiting toxicity because of there is potential neurotoxicity at higher doses found in preclinical animal study (Meerum Terwogt et al., 2001). Plasma pharmacokinetic was measured over 48 h using HPLC and showed a linear relation for both HPMA copolymer–paclitaxel and released paclitaxel. The results suggest an improved pharmacokinetic behavior with potential controlled release of paclitaxel.

54.5 Noncancerous Diseases

Beyond oncology, people are afflicted by many chronic conditions such as autoimmune diseases, musculoskeletal disorders, and infectious diseases. Rheumatoid arthritis (RA) is a chronic, systemic, and inflammatory autoimmune disease, which manifests in multiple joints of the body. One in five adults in the United States report having doctor diagnosed arthritis (CDC, 2013). Currently, there is no cure for rheumatoid arthritis. Commonly used medications for treatment of the disease include nonsteroidal anti-inflammatory drug (NSAIDs), glucocorticoids (GCs), and disease-modifying antirheumatic drugs (DMARDs) (Yuan et al., 2012). For oral administration of NSAIDs such as ibuprofen, only a small amount of drug can be absorbed and reach at the arthritic joints (Gallo et al., 1986; Mäkelä et al., 1981). The NSAIDs conjugated to the macromolecular carrier would target to the inflammation site after parenteral administration (Larsen et al., 1989). The therapeutic effects of NSAIDs are mediated primarily through nonselective inhibition of cyclooxygenase (COX), inhibiting both the cyclooxygenase-1 (COX-1), and cyclooxygenase-2 (COX-2). However, they will not prevent joint destruction and only show a small impact on inflammation and also the application of NSAIDs are limited due to the side effects, especially GI and renal toxicity (El Desoky, 2001). Glucocorticoids (GCs), which are immunosuppressive and anti-inflammatory, have been widely used for controlling the symptoms of inflammatory diseases (Kirwan et al., 1995). However, the side effects associated with long-term GC treatment (e.g., immunodeficiency, weight gain, growth impairment, fluid retention, and secondary osteoporosis) has hampered their clinical application (Canalis et al., 2007; Barber et al., 1993). To address this challenge, polymeric glucocorticoid have been proposed (Timofeevski et al., 1996). As a major breakthrough, HPMA copolymer–dexamethasone conjugate was designed and synthesized. Due to the high anti-inflammation potency and high toxicity, dexamethasone (Dex) was selected as the prototype drug candidate to be conjugated to HPMA copolymer via a hydrazone bond, which is acid cleavable (Kratz et al., 1999). The general idea is that after systemic administration, HPMA copolymer-Dex (P-Dex) conjugate will passively target to the inflammatory joints and be quickly internalized by the synoviocytes via endocytosis and gradually release free drug to exert its anti-inflammation effect locally without triggering systemic GC side effect of. While the in vivo data proved that this general design works very well, some chemistry problem remains. Due to the initial polymer analogous reaction approach that was taken, the drug content in P-Dex varies from batch to batch. To solve this problem, a new acid-cleavable, Dex containing monomer (MA-Gly-Gly-NHN=Dex) was synthesized to achieve precise control of Dex loading in the conjugate (Wang et al., 2007a). In the inflammatory joint, P-Dex treatment has a long-lasting inflammation resolution effect, which is critical and indirectly supports the long-term presence of Dex in the arthritic joints (Wang et al., 2007a; Liu et al., 2008). The mechanism underlining the P-Dex' targeting to arthritis is different from the classical EPR effect as the intimal lymphatic is intact in most of the arthritic joints and the passive targeting is initiated by their selective extravasation through the leaky vasculature of the inflamed synovial tissues (Yuan et al., 2012). As discussed in the passive targeting section, this novel passive targeting mechanism was called "ELVIS," which represent the macromolecular prodrugs' Extravasation through leaky vasculature and the subsequent inflammatory cell-mediated sequestration. The fast sequestration of P-Dex by the activated synoviocytes and potentially inflammatory infiltrates explains the passive targeting of P-Dex to inflammations (Quan et al., 2010; Wang & Goldring, 2011; Yuan et al., 2012).

HPMA copolymer–drug conjugates have also been used to treat musculoskeletal diseases. Bone-targeting HPMA copolymer conjugates were designed for the treatment of osteoporosis and other musculoskeletal conditions (Wang et al., 2003, 2004, 2006, 2007b; Pan et al., 2006, 2008a, b; Miller et al., 2008). HPMA copolymer-D-Asp_8 conjugates show extensive targeting potential to the skeleton. D-Asp_8 (D-aspartic acid octapeptide) is an anionic peptide being used as bone-targeting moieties for HPMA polymer conjugates (Wang et al., 2003). It was showed that FITC-labeled HPMA copolymer D-Asp_8 conjugate at 0.5 g/kg dose has a very strong binding at the high bone turnover sites such as primary spongiosa and metaphyseal areas in mice (Wang et al., 2003). The conjugate was also bound to the bones of osteopenic animal model in rat with the same dose in mice (Miller et al., 2008). Besides D-Asp_8 as bone targeting moieties, bisphosphonates are also commonly used as bone targeting moieties for HPMA copolymer conjugates (Russell, 2011; Fleisch, 1981). Instead of recognizing only resorption sites in skeletal tissues of D-Asp_8, alendronate (ALN) directs the delivery system to both formation and resorption sites (Wang et al., 2007b). Based on a previous study (Wang et al., 2003, 2006), alendronate-containing conjugates with relatively low molecular weight show higher binding to bone than D-Asp_8 containing conjugates. Prostaglandin E_1 was selected as the first anabolic compound to be conjugated to the HPMA copolymer bone targeting polymers (Pan et al., 2006, 2008a). The high molecular weight, multiblock, backbone degradable bone targeted HPMA copolymer PGE_1 conjugate showed higher accumulation in bone of OVX rat model and a higher efficacy in promoting bone formation in an osteoporosis rat model (Pan et al., 2008b, 2013).

54.6 Linker Chemistry

To achieve on-site drug release and reduce the drug toxicity to other organs, linker design between drug and HPMA copolymer is very important. There are several criteria for the linker design, such as the polymer–drug linker must be stable in the circulation until it reaches the targeted site. The polymer–drug linker should also be able to release the drug at optimum rate at the target (Duncan, 2006). The use of unsuitable polymer–drug linkers which are either too stable to release drug on target site or not stable enough that degraded before the arriving the target site would lead to undesirable treatment outcome. pH-sensitive or peptidyl polymer–drug linkers are most frequently used in drug delivery (Duncan, 2003, 2006). pH-sensitive linker was widely used to deliver antitumor agents as the tumor tissues are often associated with acidosis. Their eventual subcellular trafficking into the acidic endosomal/lysosomal compartments would provide another mechanism for the drug activation. One of the most widely used low pH sensitive linker is the hydrazone bond (Kratz et al., 1999). It is formed by a reaction between hydrazine and ketone in the presence of acidic catalyst. The hydrazone bond has been used to link doxorubicin onto HPMA copolymer to treat solid tumor. The drug was released within 48 h in vitro at pH 5, and it correlated well with the antitumor activity in vivo (Etrych et al., 2001). Another pH-sensitive linker being widely used is *cis*-aconityl linkage, which is being reported in the use of anticancer drug immunoconjugates (Choi et al., 1999). Many pH-sensitive chemical structures have been synthesized for drug delivery system. The release rate could be tuned to meet the therapeutic requirements. Besides tumor site release, pH-sensitive linker has also been used as cleavable structure in bone targeting and inflammation targeting (Wang et al., 2005; Yuan et al., 2012).

Speaking of bone targeting and on site release, cathepsin K-specific peptide linkage is also widely used in the bone-

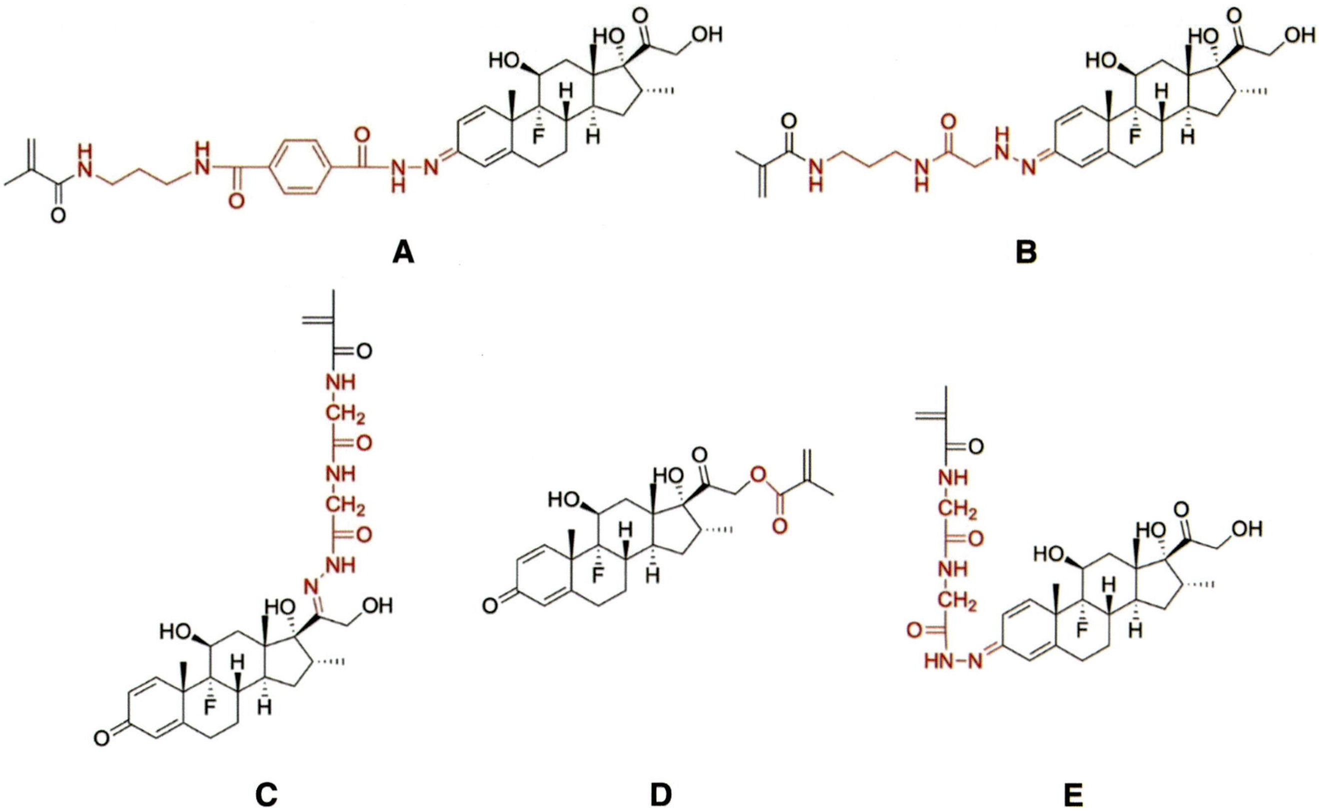

Fig. 54.6 Detailed chemistry of the Dex-containing monomers design (A, B, C, D, E) for the synthesis of HPMA copolymer-based Dex prodrugs with different releasing rates. Red colored structures represent the different linker chemistries. Releasing speed in acidic condition: B > C > E > A > D. Releasing speed in neutral condition: B > D > C > E > A. (Adapted from (Jia et al., 2020) with permission)

targeting copolymer conjugates. Cathepsin K is predominantly expressed in osteoclasts which makes the cathepsin K-specific peptide the best linker candidate to release therapeutic agents from bone-targeting copolymer conjugates (Wang et al., 2005; Miller et al., 2008). Oligopeptide Gly-Phe-Leu-Gly is designed for degradation by the lysosomal cysteine proteases, particularly cathepsin B (Duncan et al., 1983). This linker is stable in the body circulation and can release the drug over a 24–48 h period in vitro and in vivo (Rejmanová et al., 1985).

In a recent study of optimization of HPMA copolymer-based dexamethasone prodrug for improved treatment of inflammatory arthritis (Jia et al., 2020), the linker of N-(2-hydroxypropyl) methacrylamide (HPMA) copolymer-based dexamethasone (Dex) prodrugs was fine-tuned by systematically adjusting the linker chemistry between Dex and the methacryloyl group. The five polymeric prodrugs (Fig. 54.6) demonstrated a wide spectrum of activation kinetics under different conditions. When evaluated in an adjuvant-induced arthritis (AA) rat model, these polymeric Dex prodrugs exhibited different therapeutic efficacy profiles, which directly correlated with their activation kinetics. The results from this systematic structure–activity relationship study suggest that the rational design of prodrug activation mechanisms is a critical element in successful development of macromolecular prodrugs in treating different conditions.

Recently, new linkers that can be cleaved by tumor-associated protease legumain have been developed (Stern et al., 2009). Self-immolative linkers, which form stable bonds between a macromolecular scaffold and its reporter groups, are able to facilitate enhanced conjugate disassembly upon a specific activation event (Blencowe et al., 2011). These self-immolative linkers showed high loading capabilities and excellent solubility and stability characteristics. Bioresponsive coiled-coil peptide linkers that are noncovalent and biologically inspired are formed by heterodimerization of two complementary peptide sequences (Apostolovic et al., 2010). This linker offered several potential advantages including facile access to combination therapeutics and rapid production of compound libraries to screen for structure–activity relationships. Not only act as a linker, coiled coil peptides may also direct intracellular transport and trafficking (Apostolovic et al., 2010).

54.7 HPMA Copolymer for the Treatment of CNS Diseases

As described above, numerous macromolecular therapeutics, especially HPMA copolymer drug conjugates, have been developed for the treatment of cancer and noncancerous diseases. We believe some of the benefits of these delivery systems can be further extrapolated to diseases of the central nervous system (CNS). There are a wide range of pathological conditions associated with human brain, including infections, stroke, hypoxia, poisoning by teratogens (including alcohol), chronic degenerative disease, and traumatic injury. Currently, no therapies are yet available to fully restore lost function or slow ongoing neurodegeneration in the brain. The poor brain permeability limited by blood–brain barrier prevents the macromolecular therapeutics transport into the brain (Shlosberg et al., 2010; Pardridge et al., 1992). The blood–brain barrier restricts the entry of macromolecules and low molecular weight compounds from the periphery. Approximately 98% small molecular weight drugs and almost 100% larger molecular weight peptides and proteins do not cross the blood–brain barrier (Pardridge, 2004). To achieve a brain targeting drug delivery and sufficient therapeutic drug concentration in the brain, the presence of blood brain barrier seems to be the main limiting factor. In the past few years, some notable developments in systemic and local macromolecular drug delivery system for CNS diseases have been reported (Neuwelt et al., 2008). The advantage of systemic drug delivery is their noninvasive nature and use of traditional routes of administration. The drawback of the systemic drug delivery is the large dosages, which may be needed to achieve therapeutic drug concentration in the brain. To overcome this limitation, modification of a delivery vehicle with site-specific ligands that increase its penetration through blood–brain barrier is frequently used (Weinstein et al., 1978; Béduneau et al., 2007). The pharmacokinetic properties of these drug delivery systems must increase the bioavailability of the drug and the time it remains in the circulation as well as reducing the drug clearance by the MPS (Siwak et al., 2002). To fulfill these requirements, polymeric drug delivery system seems to be one of the most promising solutions to shuttle drugs into the brain effectively (Liu et al., 2010).

Amphiphilic copolymers poly(*N*-(2-hydroxypropyl methacrylamide-co-laurylmethacrylate) or P(HPMA-co-LMA) designed by Zentel group are able to mediate drug transportation to the brain (Hemmelmann et al., 2012). They show the P(HPMA-co-LMA) copolymers enhances the permeability of Rhodamine 123 across the blood–brain barrier without affecting the integrity and barrier function. With the in vitro and in vivo experiments results, they showed that the delivery system is suitable for a host of potential drugs. The random P(HPMA-co-LMA) copolymer having 10 mol% of lauryl side chains is the most efficient delivery system (Hemmelmann et al., 2011, 2012).

HPMA copolymer-based hydrogels have been invented to treat physical injury of the central nervous system specifically traumatic spinal cord injury (Fawcett & Asher, 1999; Zhong & Bellamkonda, 2008). The hydrogel is a biomaterial implantation at the injury site to support the injured neuron tissue and act as a platform for axon to attach and grow. The hydrogel could also carry drugs for the local delivery (Geller & Fawcett, 2002; Nisbet et al., 2008; Zhong & Bellamkonda, 2008). Arginine–glycine–aspartic acid (RGD) peptide sequences and aminosugar sequences, which could promote tissue organization and regeneration were conjugated to the HPMA copolymer hydrogel to treat CNS trauma (Woerly et al., 1993; Begovac & Shur, 1990; Ruoslahti & Pierschbacher, 1987; Plant et al., 1997). In vitro studies suggested that RGD containing HPMA copolymer hydrogels have the potential of facilitating tissues regeneration in the injured CNS such as spinal cord injury (SCI). In vivo studies confirmed the in vitro findings and showed that the hydrogels were able to bridge tissue defects in the acutely injured spinal cord, support cellular ingrowth, angiogenesis, and axonogenesis within the structure of the hydrogel network, which were similar to those of the developing spinal cord (Woerly et al., 1999). To further investigate whether the hydrogel could restore the functions to the chronically injured spinal cord, HPMA copolymer hydrogel was implanted into the cavity that formed in the rat spinal cord 3 months after a severe injury when the functional deficit had reached a plateau (Woerly et al., 2001). The results showed that long-term implantation not only reduce the volume of the cavity but also induced tissue repair and axonal regrowth with extended long distance throughout the graft, accompanied by gradual functional recovery. This hydrogel offers two advantages: it can be combined with either genetically engineered cells (Loh et al., 2001) or human stem cells, and these cells develop successfully within the polymer before implantation.

All the findings above show that HPMA copolymer has the potential to be a CNS drug delivery carrier candidate. HPMA copolymer hydrogel system shows the potential to repair CNS lesions caused by trauma. It is a simple and safe scaffold for neuronal regeneration. It could also incorporate different growth factors and cells to further potentiate their repair capability. Amphiphilic copolymer P(HPMA-co-LMA) has the potential to transport drugs through blood brain barrier while maintaining the barrier integrity. It is the most efficient drug delivery system to delivery various drug cross the blood brain barrier.

Traumatic brain injury (TBI) is a major cause of physical or cognitive morbidity or even mortality all over the world. The pathophysiology of TBI is complicated and consists of primary injury (mechanical damage to the head) and secondary injury mechanisms including cerebral edema, inflammation, and oxidative stress. Currently, there are no treatments. The synthetic glucocorticoid dexamethasone (Dex) is widely used for its anti-inflammatory and immunosuppressive effects. As we mentioned in the previous sections, Dex has a significant positive effect on inflammation. As a glucocorticoid, it could insert into the cell membrane and inhibit lipid peroxidation propagation reactions as part of the secondary injury of TBI. However, there are also significant challenges in using steroids as a therapeutic option for TBI, such as their poor water solubility and adverse side effects associated with the treatment. Recently, HPMA copolymer dexamethasone conjugate (P-Dex) was reported (Wei et al., 2022) as an effective treatment for severe traumatic brain injury in mice. A single dose of systemic P-Dex passively targeted to and retained at the traumatic brain tissue, ameliorating the neuroinflammation while preventing the neuronal death. Signs of functional improvements including the improvement of balance post TBI were also evident. P-Dex also protected the mice from post-TBI systemic osteopenia. Given that the compromised balance and osteopenia are often associated with the TBI patients, putting them at a higher risk of fall, secondary TBI and fracture, the promising results from this study seems to support the potential utility of P-Dex in clinical management of TBI patients.

54.8 Summary

In the past few decades, polymer therapeutics have been developed for not only antitumor treatment but also used to treat numerous other diseases. The advantages of macromolecular therapeutics have been recognized for its diverse functionality, targetability, and potential to improve the treatment of a variety of diseases (Li & Wallace, 2008; Liu et al., 2010; Yuan et al., 2012; Hemmelmann et al., 2012; Rihova & Kubackova, 2003) As being discussed in this chapter, targeted drug delivery to the skeleton, inflammation, tumor, and CNS system using HPMA copolymer has great potential for further therapeutic development.

54.9 Review Questions

1. What is the ELVIS mechanism?
2. What is the structural difference between liposomes and micelles?
3. Describe three targeting moieties used to modify the copolymers to achieve active targeting property and what they are used for.
4. Describe two linkers used for the HPMA copolymer drug conjugates and what are they designed for.

54.10 Answers

1. HPMA copolymers' passively targeting and retention at the site of inflammation can be attributed to their extravasation through leaky vasculature and subsequent inflammatory cell-mediated sequestration. This process is termed as ELVIS mechanism.
2. Liposomes are spherical vesicles composed of phospholipid bilayer membrane encapsulating an aqueous core. Both the aqueous core and the lipid bilayer can be used for drug loading purpose.

   Micelles are formed by aggregation of surfactant molecules (e.g., lipids and amphiphilic polymers) with a hydrophobic core and a hydrophilic corona. The hydrophobic core can be used for drug loading purpose.
3. Refer to 54.3.1.
4. Refer to 54.6.

References

Abuchowski, A., McCoy, J. R., Palczuk, N. C., van Es, T., & Davis, F. F. (1977). Effect of covalent attachment of polyethylene glycol on immunogenicity and circulating life of bovine liver catalase. *Journal of Biological Chemistry, 252*(11), 3582–3586.

Apostolovic, B., Deacon, S. P. E., Duncan, R., & Klok, H.-A. (2010). Hybrid polymer therapeutics incorporating bioresponsive, coiled coil peptide linkers. *Biomacromolecules, 11*(5), 1187–1195. https://doi.org/10.1021/bm901313c

Bangham, A. D., Standish, M. M., & Watkins, J. C. (1965). Diffusion of univalent ions across the lamellae of swollen phospholipids. *Journal of Molecular Biology, 13*(1), 238–252. https://doi.org/10.1016/S0022-2836(65)80093-6

Barber, A. E., Coyle, S. M., Marano, M. A., Fischer, E., Calvano, S. E., Fong, Y., Moldawer, L. L., & Lowry, S. F. (1993). Glucocorticoid therapy alters hormonal and cytokine responses to endotoxin in man. *The Journal of Immunology, 150*(5), 1999–2006.

Béduneau, A., Saulnier, P., & Benoit, J.-P. (2007). Active targeting of brain tumors using nanocarriers. *Biomaterials, 28*(33), 4947–4967. https://doi.org/10.1016/j.biomaterials.2007.06.011

Begovac, P. C., & Shur, B. D. (1990). Cell surface galactosyltransferase mediates the initiation of neurite outgrowth from PC12 cells on laminin. *The Journal of Cell Biology, 110*(2), 461–470. https://doi.org/10.1083/jcb.110.2.461

Blencowe, C. A., Russell, A. T., Greco, F., Hayes, W., & Thornthwaite, D. W. (2011). Self-immolative linkers in polymeric delivery systems. *Polymer Chemistry, 2*(4), 773–790. https://doi.org/10.1039/C0PY00324G

Campbell, I. G., Jones, T. A., Foulkes, W. D., & Trowsdale, J. (1991). Folate-binding protein is a marker for ovarian cancer. *Cancer Research, 51*(19), 5329–5338.

Canalis, E., Mazziotti, G., Giustina, A., & Bilezikian, J. P. (2007). Glucocorticoid-induced osteoporosis: Pathophysiology and therapy. *Osteoporosis International, 18*(10), 1319–1328. https://doi.org/10.1007/s00198-007-0394-0

CDC. (2013). Prevalence of doctor-diagnosed arthritis and arthritis-attributable activity limitation--United States, 2010–2012. *MMWR Morbidity and Mortality Weekly Report, 62*(44), 869–873.

Chen, F., Jia, Z., Rice, K. C., Reinhardt, R. A., Bayles, K. W., & Wang, D. (2013). The development of dentotropic micelles with biodegradable tooth-binding moieties. *Pharmaceutical Research, 30*(11), 2808–2817. https://doi.org/10.1007/s11095-013-1105-5

Chen, J., Wu, X., Hou, X., Su, X., Chu, Q., Fahruddin, N., & Zhao, J. X. (2014). Shape-tunable hollow silica nanomaterials based on a soft-templating method and their application as a drug carrier. *ACS Applied Materials & Interfaces, 6*(24), 21921–21930. https://doi.org/10.1021/am507642t

Choi, W.-M., Kopečková, P., Minko, T., & Kopeček, J. (1999). Synthesis of HPMA copolymer containing adriamycin bound via an acid-labile spacer and its activity toward human ovarian carcinoma cells. *Journal of Bioactive and Compatible Polymers, 14*(6), 447–456. https://doi.org/10.1177/088391159901400601

Dand, M., Patel, P., Ayre, A., & Vilasrau, K. (2013). Polymeric micelles as a drug carrier for tumor targeting. *Chronicles of Young Scientists, 4*(2), 94. https://doi.org/10.4103/2229-5186.115544

Danhier, F. (2016). To exploit the tumor microenvironment: Since the EPR effect fails in the clinic, what is the future of nanomedicine? *Journal of Controlled Release: Official Journal of the Controlled Release Society, 244*(Pt A), 108–121. https://doi.org/10.1016/j.jconrel.2016.11.015

De Duve, C., De Barsy, T., Poole, B., Trouet, A., Tulkens, P., & Van Hoof, F. (1974). Lysosomotropic agents. *Biochemical Pharmacology, 23*(18), 2495–2531. https://doi.org/10.1016/0006-2952(74)90174-9

Devine, D. V., Wong, K., Serrano, K., Chonn, A., & Cullis, P. R. (1994). Liposome—Complement interactions in rat serum: Implications for liposome survival studies. *Biochimica et Biophysica Acta (BBA) - Biomembranes, 1191*(1), 43–51. https://doi.org/10.1016/0005-2736(94)90231-3

Drummond, D. C., Meyer, O., Hong, K., Kirpotin, D. B., & Papahadjopoulos, D. (1999). Optimizing liposomes for delivery of chemotherapeutic agents to solid tumors. *Pharmacological Reviews, 51*(4), 691–744.

Duncan, R. (2003). The dawning era of polymer therapeutics. *Nature Reviews Drug Discovery, 2*(5), 347–360. https://doi.org/10.1038/nrd1088

Duncan, R. (2006). Polymer conjugates as anticancer nanomedicines. *Nature Reviews Cancer, 6*(9), 688–701. https://doi.org/10.1038/nrc1958

Duncan, R. (2009). Development of HPMA copolymer-anticancer conjugates: Clinical experience and lessons learnt. *Advanced Drug Delivery Reviews, 61*(13), 1131–1148. https://doi.org/10.1016/j.addr.2009.05.007

Duncan, R., & Kopeček, J. (1984). Soluble synthetic polymers as potential drug carriers. In *Polymers in medicine* (pp. 51–101). Springer Berlin Heidelberg. https://doi.org/10.1007/3-540-12796-8_10

Duncan, R., Cable, H. C., Lloyd, J. B., Rejmanová, P., & Kopeček, J. (1983). Polymers containing enzymatically degradable bonds, 7. Design of oligopeptide side-chains in poly[N-(2-hydroxypropyl) methacrylamide] copolymers to promote efficient degradation by lysosomal enzymes. *Die Makromolekulare Chemie, 184*(10), 1997–2008. https://doi.org/10.1002/macp.1983.021841005

El Desoky, E. S. (2001). Pharmacotherapy of rheumatoid arthritis: An overview. *Current Therapeutic Research, 62*(2), 92–112. https://doi.org/10.1016/S0011-393X(01)80020-5

Etrych, T., Jelínková, M., Říhová, B., & Ulbrich, K. (2001). New HPMA copolymers containing doxorubicin bound via pH-sensitive linkage: Synthesis and preliminary in vitro and in vivo biological properties. *Journal of Controlled Release, 73*(1), 89–102. https://doi.org/10.1016/S0168-3659(01)00281-4

Fang, J., Nakamura, H., & Maeda, H. (2011). The EPR effect: Unique features of tumor blood vessels for drug delivery, factors involved, and limitations and augmentation of the effect. *Advanced Drug Delivery Reviews, 63*(3), 136–151. https://doi.org/10.1016/j.addr.2010.04.009

Fawcett, J. W., & Asher, R. A. (1999). The glial scar and central nervous system repair. *Brain Research Bulletin, 49*(6), 377–391. https://doi.org/10.1016/S0361-9230(99)00072-6

Fleisch, H. (1981). Diphosphonates: History and mechanisms of action. *Metabolic Bone Disease and Related Research, 3*(4–5), 279–287. https://doi.org/10.1016/0221-8747(81)90044-8

Folkman, J. (1995). Angiogenesis in cancer, vascular, rheumatoid and other disease. *Nature Medicine, 1*(1), 27–30.

Gallo, J. M., Gall, E. P., Gillespie, W. R., Albert, K. S., & Perrier, D. (1986). Ibuprofen kinetics in plasma and synovial fluid of arthritic patients. *The Journal of Clinical Pharmacology, 26*(1), 65–70. https://doi.org/10.1002/j.1552-4604.1986.tb02905.x

Geller, H. M., & Fawcett, J. W. (2002). Building a bridge: Engineering spinal cord repair. *Experimental Neurology, 174*(2), 125–136. https://doi.org/10.1006/exnr.2002.7865

Gentile, L., Behrens, M. A., Porcar, L., Butler, P., Wagner, N. J., & Olsson, U. (2014). Multilamellar vesicle formation from a planar lamellar phase under shear flow. *Langmuir: The ACS Journal of Surfaces and Colloids, 30*(28), 8316–8325. https://doi.org/10.1021/la501071s

Hashizume, H., Baluk, P., Morikawa, S., McLean, J. W., Thurston, G., Roberge, S., Jain, R. K., & McDonald, D. M. (2000). Openings between defective endothelial cells explain tumor vessel leakiness. *The American Journal of Pathology, 156*(4), 1363–1380. https://doi.org/10.1016/S0002-9440(10)65006-7

Hemmelmann, M., Knoth, C., Schmitt, U., Allmeroth, M., Moderegger, D., Barz, M., Koynov, K., Hiemke, C., Rösch, F., & Zentel, R. (2011). HPMA based amphiphilic copolymers mediate central nervous effects of domperidone. *Macromolecular Rapid Communications, 32*(9–10), 712–717. https://doi.org/10.1002/marc.201000810

Hemmelmann, M., Metz, V. V., Koynov, K., Blank, K., Postina, R., & Zentel, R. (2012). Amphiphilic HPMA–LMA copolymers increase the transport of Rhodamine 123 across a BBB model without harming its barrier integrity. *Journal of Controlled Release, 163*(2), 170–177. https://doi.org/10.1016/j.jconrel.2012.08.034

Henselwood, F., & Liu, G. (1997). Water-soluble nanospheres of poly(2-cinnamoylethyl methacrylate)-block-poly(acrylic acid). *Macromolecules, 30*(3), 488–493. https://doi.org/10.1021/ma961401v

Huizing, M. T., Misser, V. H., Pieters, R. C., ten Bokkel Huinink, W. W., Veenhof, C. H., Vermorken, J. B., Pinedo, H. M., & Beijnen, J. H. (1995). Taxanes: A new class of antitumor agents. *Cancer Investigation, 13*(4), 381–404.

Jain, R. K. (1999). Transport of molecules, particles, and cells in solid tumors. *Annual Review of Biomedical Engineering, 1*(1), 241–263. https://doi.org/10.1146/annurev.bioeng.1.1.241

Jain, R. K., Munn, L. L., & Fukumura, D. (2002). Dissecting tumour pathophysiology using intravital microscopy. *Nature Reviews Cancer, 2*(4), 266–276. https://doi.org/10.1038/nrc778

Jansen, J. F., de Brabander-van den Berg, E. M., & Meijer, E. W. (1994). Encapsulation of guest molecules into a dendritic box. *Science, 266*(5188), 1226–1229. https://doi.org/10.1126/science.266.5188.1226

Jatzkewitz, H. (1955). Peptamin (glycyl-L-leucyl-mescalin) bound to blood plasma expander (polyvinylpyrrolidone) as a new depot form of a biologically active primary amine (mescaline). *Zeitschrift für Naturforschung, 10b*, 27–31.

Jia, Z., Zhao, G., Wei, X., Kong, D., Sun, Y., Zhou, Y., Lele, S. M., Fehringer, E. V., Garvin, K. L., Goldring, S. R., & Wang, D. (2020). Structural optimization of HPMA copolymer-based dexamethasone prodrug for improved treatment of inflammatory arthritis. *Journal of Controlled Release, 324*, 560–573. https://doi.org/10.1016/j.jconrel.2020.05.028

Kirwan, J. R., & Arthritis t, Group RCL-DGS. (1995). The effect of glucocorticoids on joint destruction in rheumatoid arthritis. *New England Journal of Medicine, 333*(3), 142–147. https://doi.org/10.1056/NEJM199507203330302

Kocbek, P., Obermajer, N., Cegnar, M., Kos, J., & Kristl, J. (2007). Targeting cancer cells using PLGA nanoparticles surface modified with monoclonal antibody. *Journal of Controlled Release: Official*

Journal of the Controlled Release Society, 120(1–2), 18–26. https://doi.org/10.1016/j.jconrel.2007.03.012

Koizumi, F., Kitagawa, M., Negishi, T., Onda, T., Matsumoto, S.-i., Hamaguchi, T., & Matsumura, Y. (2006). Novel SN-38-incorporating polymeric micelles, NK012, eradicate vascular endothelial growth factor-secreting bulky tumors. *Cancer Research, 66*(20), 10048–10056. https://doi.org/10.1158/0008-5472.CAN-06-1605

Koo, O. M., Rubinstein, I., & Onyuksel, H. (2005). Role of nanotechnology in targeted drug delivery and imaging: A concise review. *Nanomedicine, 1*(3), 193–212. https://doi.org/10.1016/j.nano.2005.06.004

Kopeček, J. (1984). Controlled biodegradability of polymers — A key to drug delivery systems. *Biomaterials, 5*(1), 19–25. https://doi.org/10.1016/0142-9612(84)90062-0

Kopeček, J., Rejmanova, P., Strohalm, J., Ulbrich, K., Rihova, B., Chytry, V., Lloyd, J. B., & Duncan, R. (1991). *Synthetic polymeric drugs*. United States Patent.

Kopeček, J., Kopečková, P., Minko, T., & Lu, Z.-R. (2000). HPMA copolymer–anticancer drug conjugates: Design, activity, and mechanism of action. *European Journal of Pharmaceutics and Biopharmaceutics, 50*(1), 61–81. https://doi.org/10.1016/S0939-6411(00)00075-8

Kranz, D. M., Patrick, T. A., Brigle, K. E., Spinella, M. J., & Roy, E. J. (1995). Conjugates of folate and anti-T-cell-receptor antibodies specifically target folate-receptor-positive tumor cells for lysis. *Proceedings of the National Academy of Sciences of the United States of America, 92*(20), 9057–9061.

Kratz, F., Beyer, U., & Schutte, M. T. (1999). Drug-polymer conjugates containing acid-cleavable bonds. *Critical Reviews™ in Therapeutic Drug Carrier Systems, 16*(3), 245–288. https://doi.org/10.1615/CritRevTherDrugCarrierSyst.v16.i3.10

Krinick, N. L., Sun, Y., Joyner, D., Spikes, J. D., Straight, R. C., & Kopeček, J. (1994). A polymeric drug delivery system for the simultaneous delivery of drugs activatable by enzymes and/or light. *Journal of Biomaterials Science, Polymer Edition, 5*(4), 303–324. https://doi.org/10.1163/156856294X00040

Kushwaha, D., & Tiwari, V. K. (2013). Click chemistry inspired synthesis of glycoporphyrin dendrimers. *The Journal of Organic Chemistry, 78*(16), 8184–8190. https://doi.org/10.1021/jo4012392

Larsen, C. S., Johansen, M., Harboe, E., Kurtzhals, P., & Olesen, H. P. (1989). *High molecular weight prodrug derivatives of antiinflammatory drugs*. European Patent.

Li, C., & Wallace, S. (2008). Polymer-drug conjugates: Recent development in clinical oncology. *Advanced Drug Delivery Reviews, 60*(8), 886–898. https://doi.org/10.1016/j.addr.2007.11.009

Li, L., Yang, Q., Zhou, Z., Zhong, J., & Huang, Y. (2014a). Doxorubicin-loaded, charge reversible, folate modified HPMA copolymer conjugates for active cancer cell targeting. *Biomaterials, 35*(19), 5171–5187. https://doi.org/10.1016/j.biomaterials.2014.03.027

Li, Y., Yang, J., Jin, J., Sun, X., Wang, L., & Chen, J. (2014b). New reversed-phase/anion-exchange/hydrophilic interaction mixed-mode stationary phase based on dendritic polymer-modified porous silica. *Journal of Chromatography A, 1337*, 133–139. https://doi.org/10.1016/j.chroma.2014.02.044

Liu, G. (1997). Diblock copolymer nanostructures. *Macromolecular Symposia, 113*(1), 233–248. https://doi.org/10.1002/masy.19971130120

Liu, X.-M., Quan, L.-D., Tian, J., Alnouti, Y., Fu, K., Thiele, G. M., & Wang, D. (2008). Synthesis and evaluation of a well-defined HPMA copolymer–dexamethasone conjugate for effective treatment of rheumatoid arthritis. *Pharmaceutical Research, 25*(12), 2910–2919. https://doi.org/10.1007/s11095-008-9683-3

Liu, X.-M., Miller, S. C., & Wang, D. (2010). Beyond oncology--application of HPMA copolymers in non-cancerous diseases. *Advanced Drug Delivery Reviews, 62*(2), 258–271. https://doi.org/10.1016/j.addr.2009.10.006

Loh, N. K., Woerly, S., Bunt, S. M., Wilton, S. D., & Harvey, A. R. (2001). The regrowth of axons within tissue defects in the CNS is promoted by implanted hydrogel matrices that contain BDNF and CNTF producing fibroblasts. *Experimental Neurology, 170*(1), 72–84. https://doi.org/10.1006/exnr.2001.7692

Lokappa, S. B., & Ulmer, T. S. (2011). Alpha-synuclein populates both elongated and broken helix states on small unilamellar vesicles. *The Journal of Biological Chemistry, 286*(24), 21450–21457. https://doi.org/10.1074/jbc.M111.224055

Maeda, H. (2010). Tumor-selective delivery of macromolecular drugs via the EPR effect: Background and future prospects. *Bioconjugate Chemistry, 21*(5), 797–802. https://doi.org/10.1021/bc100070g

Maeda, H., & Konno, T. (1997). Metamorphosis of neocarzinostatin to SMANCS: Chemistry, biology, pharmacology, and clinical effect of the first prototype anticancer polymer therapeutic. In K. Edo & N. Ishida (Eds.), *Maeda H* (pp. 227–267). Neocarzinostatin. https://doi.org/10.1007/978-4-431-66914-2_12

Maeda, H., Takeshita, J., & Kanamaru, R. (1979). A lipophilic derivative of neocarzinostatin a polymer conjugation of an antitumor protein antibiotic*. *International Journal of Peptide and Protein Research, 14*(2), 81–87. https://doi.org/10.1111/j.1399-3011.1979.tb01730.x

Maeda, H., Matsumoto, T., Konno, T., Iwai, K., & Ueda, M. (1984). Tailor-making of protein drugs by polymer conjugation for tumor targeting: A brief review on smancs. *Journal of Protein Chemistry, 3*(2), 181–193. https://doi.org/10.1007/bf01040499

Mäkelä, A.-L., Lempiäinen, M., & Ylijoki, H. (1981). Ibuprofen levels in serum and synovial fluid. *Scandinavian Journal of Rheumatology, 10*(sup39), 15–17. https://doi.org/10.3109/03009748109095329

Matsumura, Y., & Maeda, H. (1986). A new concept for macromolecular therapeutics in cancer chemotherapy: Mechanism of tumoritropic accumulation of proteins and the antitumor agent smancs. *Cancer Research, 46*(12 Part 1), 6387–6392.

Matthew Peterson, C., Lu, J. M., Sun, Y., Anthony Peterson, C., Shiah, J.-G., Straight, R. C., & Kopeček, J. (1996). Combination chemotherapy and photodynamic therapy with N-(2-Hydroxypropyl) methacrylamide copolymer-bound anticancer drugs inhibit human ovarian carcinoma heterotransplanted in nude mice. *Cancer Research, 56*(17), 3980–3985.

Meerum Terwogt, J. M., ten Bokkel Huinink, W. W., Schellens, J. H., Schot, M., Mandjes, I. A., Zurlo, M. G., Rocchetti, M., Rosing, H., Koopman, F. J., & Beijnen, J. H. (2001). Phase I clinical and pharmacokinetic study of PNU166945, a novel water-soluble polymer-conjugated prodrug of paclitaxel. *Anti-Cancer Drugs, 12*(4), 315–323.

Miller, S. C., Pan, H., Wang, D., Bowman, B. M., Kopečková, P., & Kopeček, J. (2008). Feasibility of using a bone-targeted, macromo lecular delivery system coupled with prostaglandin E1 to promote bone formation in aged, estrogen-deficient rats. *Pharmaceutical Research, 25*(12), 2889–2895. https://doi.org/10.1007/s11095-008-9706-0

Moad, G., Rizzardo, E., & Thang, S. H. (2008). Radical addition–fragmentation chemistry in polymer synthesis. *Polymer, 49*(5), 1079–1131. https://doi.org/10.1016/j.polymer.2007.11.020

Moyano, F., Silber, J. J., & Correa, N. M. (2008). On the investigation of the bilayer functionalities of 1,2-di-oleoyl-sn-glycero-3-phosphatidylcholine (DOPC) large unilamellar vesicles using cationic hemicyanines as optical probes: A wavelength-selective fluorescence approach. *Journal of Colloid and Interface Science, 317*(1), 332–345. https://doi.org/10.1016/j.jcis.2007.09.051

Nakamura, H., Koziolová, E., Etrych, T., Chytil, P., Fang, J., Ulbrich, K., & Maeda, H. (2015). Comparison between linear and star-like HPMA conjugated pirarubicin (THP) in pharmacokinetics and antitumor activity in tumor bearing mice. *European Journal of Pharmaceutics and Biopharmaceutics, 90*, 90–96. https://doi.org/10.1016/j.ejpb.2014.10.007

Nederberg, F., Zhang, Y., Tan, J. P., Xu, K., Wang, H., Yang, C., Gao, S., Guo, X. D., Fukushima, K., Li, L., Hedrick, J. L., & Yang, Y.-Y. Y. (2011). Biodegradable nanostructures with selective lysis of microbial membranes. *Nature Chemistry, 3*(5), 409–414. https://doi.org/10.1038/nchem.1012

Neuwelt, E., Abbott, N. J., Abrey, L., Banks, W. A., & Blakley, B. (2008). Strategies to advance translational research into brain barriers. *The Lancet*.

Newkome, G. R., Moorefield, C. N., Baker, G. R., Saunders, M. J., & Grossman, S. H. (1991). Unimolecular Micelles. *Angewandte Chemie International Edition in English, 30*(9), 1178–1180. https://doi.org/10.1002/anie.199111781

Nichols, J. W., & Bae, Y. H. (2014). EPR: Evidence and fallacy. *Journal of Controlled Release: Official Journal of the Controlled Release Society, 190*, 451–464. https://doi.org/10.1016/j.jconrel.2014.03.057

Nisbet, D. R., Crompton, K. E., Horne, M. K., Finkelstein, D. I., & Forsythe, J. S. (2008). Neural tissue engineering of the CNS using hydrogels. *Journal of Biomedical Materials Research Part B: Applied Biomaterials, 87B*(1), 251–263. https://doi.org/10.1002/jbm.b.31000

Orive, G., Anitua, E., Pedraz, J. L., & Emerich, D. F. (2009). Biomaterials for promoting brain protection, repair and regeneration. *Nature Reviews. Neuroscience, 10*(9), 682–692.

Pan, H., Kopečková, P., Wang, D., Yang, J., Miller, S., & Kopeček, J. (2006). Water-soluble HPMA copolymer—Prostaglandin E1 conjugates containing a cathepsin K sensitive spacer. *Journal of Drug Targeting, 14*(6), 425–435. https://doi.org/10.1080/10611860600834219

Pan, H., Liu, J., Dong, Y., Sima, M., Kopečková, P., Brandi, M. L., & Kopeček, J. (2008a). Release of prostaglandin E1 from N-(2-Hydroxypropyl)methacrylamide copolymer conjugates by bone cells. *Macromolecular Bioscience, 8*(7), 599–605. https://doi.org/10.1002/mabi.200700338

Pan, H., Sima, M., Kopečková, P., Wu, K., Gao, S., Liu, J., Wang, D., Miller, S. C., & Kopeček, J. (2008b). Biodistribution and pharmacokinetic studies of bone-targeting N-(2-Hydroxypropyl) methacrylamide copolymer–alendronate conjugates. *Molecular Pharmaceutics, 5*(4), 548–558. https://doi.org/10.1021/mp800003u

Pan, H., Sima, M., Miller, S. C., Kopečková, P., Yang, J., & Kopeček, J. (2013). Efficiency of high molecular weight backbone degradable HPMA copolymer-prostaglandin E1 conjugate in promotion of bone formation in ovariectomized rats. *Biomaterials, 34*(27), 6528–6538. https://doi.org/10.1016/j.biomaterials.2013.05.003

Pardridge, W. M. (2004). The blood-brain barrier: Bottleneck in brain drug development. *NeuroRx: The Journal of the American Society for Experimental NeuroTherapeutics, 2*(1), 3–14. https://doi.org/10.1602/neurorx.2.1.3

Pardridge, W. M., Boado, R. J., Black, K. L., & Cancilla, P. A. (1992). Blood-brain barrier and new approaches to brain drug delivery. *The Western Journal of Medicine, 156*(3), 281–286.

Patri, A. K., Majoros, I. J., & Baker, J. R. (2002). Dendritic polymer macromolecular carriers for drug delivery. *Current Opinion in Chemical Biology, 6*(4), 466–471.

Pechar, M., Pola, R., Laga, R., Ulbrich, K., Bednárová, L., Maloň, P., Sieglová, I., Král, V., Fábry, M., & Vaněk, O. (2011). Coiled coil peptides as universal linkers for the attachment of recombinant proteins to polymer therapeutics. *Biomacromolecules, 12*(10), 3645–3655. https://doi.org/10.1021/bm200897b

Peer, D., Karp, J. M., Hong, S., Farokhzad, O. C., Margalit, R., & Langer, R. (2007). Nanocarriers as an emerging platform for cancer therapy. *Nature Nanotechnology, 2*(12), 751–760. https://doi.org/10.1038/nnano.2007.387

Persidis, A. (1999). Cancer multidrug resistance. *Nature Biotechnology, 17*(1), 94–95. https://doi.org/10.1038/5289

Plant, G. W., Woerly, S., & Harvey, A. R. (1997). Hydrogels containing peptide or aminosugar sequences implanted into the rat brain: Influence on cellular migration and axonal growth. *Experimental Neurology, 143*(2), 287–299. https://doi.org/10.1006/exnr.1997.6407

Putnam, D., & Kopeček, J. (1995). Polymer conjugates with anticancer activity. In N. A. Peppas & R. S. Langer (Eds.), *Biopolymers II* (pp. 55–123). Springer Berlin Heidelberg. https://doi.org/10.1007/3540587888_14

Quan, L.-d., Purdue, P. E., Liu, X.-m., Boska, M. D., Lele, S. M., Thiele, G. M., Mikuls, T. R., Dou, H., Goldring, S. R., & Wang, D. (2010). Development of a macromolecular prodrug for the treatment of inflammatory arthritis: Mechanisms involved in arthrotropism and sustained therapeutic efficacy. *Arthritis Research & Therapy, 12*(5), 1–10. https://doi.org/10.1186/ar3130

Rejmanová, P., Kopeček, J., Duncan, R., & Lloyd, J. B. (1985). Stability in rat plasma and serum of lysosomally degradable oligopeptide sequences in N-(2-hydroxypropyl) methacrylamide copolymers. *Biomaterials, 6*(1), 45–48. https://doi.org/10.1016/0142-9612(85)90037-7

Ren, K., Purdue, P. E., Burton, L., Quan, L.-d., Fehringer, E. V., Thiele, G. M., Goldring, S. R., & Wang, D. (2011). Early detection and treatment of Wear particle-induced inflammation and bone loss in a mouse calvarial osteolysis model using HPMA copolymer conjugates. *Molecular Pharmaceutics, 8*(4), 1043–1051. https://doi.org/10.1021/mp2000555

Ren, K., Dusad, A., Yuan, F., Yuan, H., Purdue, P. E., Fehringer, E. V., Garvin, K. L., Goldring, S. R., & Wang, D. (2014). Macromolecular prodrug of dexamethasone prevents particle-induced peri-implant osteolysis with reduced systemic side effects. *Journal of Controlled Release: Official Journal of the Controlled Release Society, 175*, 1–9. https://doi.org/10.1016/j.jconrel.2013.11.024

Rihova, B., & Kubackova, K. (2003). Clinical implications of N-(2-hydroxypropyl) methacrylamide copolymers. *Current Pharmaceutical Biotechnology, 4*(5), 311–322.

Roberts, W. G., & Palade, G. E. (1997). Neovasculature induced by vascular endothelial growth factor is fenestrated. *Cancer Research, 57*(4), 765–772.

Roberts, R. L., Fine, R. E., & Sandra, A. (1993). Receptor-mediated endocytosis of transferrin at the blood-brain barrier. *Journal of Cell Science, 104*(2), 521–532.

Roy, D., Berguig, G. Y., Ghosn, B., Lane, D., Braswell, S., Stayton, P. S., & Convertine, A. J. (2014). Synthesis and characterization of transferrin-targeted chemotherapeutic delivery systems prepared via RAFT copolymerization of high molecular weight PEG macromonomers. *Polymer Chemistry, 5*(5), 1791–1799. https://doi.org/10.1039/C3PY01404E

Ruoslahti, E., & Pierschbacher, M. D. (1987). New perspectives in cell adhesion: RGD and integrins. *Science, 238*(4826), 491–497. https://doi.org/10.1126/science.2821619

Russell, R. G. G. (2011). Bisphosphonates: The first 40 years. *Bone, 49*(1), 2–19. https://doi.org/10.1016/j.bone.2011.04.022

Seymour, L. W., Duncan, R., Strohalm, J., & Kopeček, J. (1987). Effect of molecular weight (Mw) of N-(2-hydroxypropyl)methacrylamide copolymers on body distribution and rate of excretion after subcutaneous, intraperitoneal, and intravenous administration to rats. *Journal of Biomedical Materials Research, 21*(11), 1341–1358. https://doi.org/10.1002/jbm.820211106

Seymour, L. W. F. D., Kerr, D. J., Rea, D., Whitlock, M., Poyner, R., Boivin, C., Hesslewood, S., Twelves, C., Blackie, R., Schatzlein, A., Jodrell, D., Bissett, D., Calvert, H., Lind, M., Robbins, A., Burtles, S., Duncan, R., & Cassidy, J. (2009). Phase II studies of polymer-doxorubicin (PK1, FCE28068) in the treatment of breast, lung and colorectal cancer. *International Journal of Oncology, 34*(6), 1629–1636. https://doi.org/10.3892/ijo_00000293

Shlosberg, D., Benifla, M., Kaufer, D., & Friedman, A. (2010). Blood-brain barrier breakdown as a therapeutic target in traumatic brain injury. *Nature Reviews Neurology, 6*(7), 393–403. https://doi.org/10.1038/nrneurol.2010.74

Siwak, D. R., Tari, A. M., & Lopez-Berestein, G. (2002). The potential of drug-carrying immunoliposomes as anticancer agents. Commentary re: J. W. Park et al., Anti-HER2 immunoliposomes: enhanced efficacy due to targeted delivery. *Clinical Cancer Research, 8*(4), 955–956.

Stern, L., Perry, R., Ofek, P., Many, A., Shabat, D., & Satchi-Fainaro, R. (2009). A novel antitumor prodrug platform designed to be cleaved by the endoprotease legumain. *Bioconjugate Chemistry, 20*(3), 500–510. https://doi.org/10.1021/bc800448u

Tappertzhofen, K., Bednarczyk, M., Koynov, K., Bros, M., Grabbe, S., & Zentel, R. (2014). Toward anticancer immunotherapeutics: Well-defined polymer-antibody conjugates for selective dendritic cell targeting. *Macromolecular Bioscience, 14*(10), 1444–1457. https://doi.org/10.1002/mabi.201400190

Thanki, K., Gangwal, R. P., Sangamwar, A. T., & Jain, S. (2013). Oral delivery of anticancer drugs: Challenges and opportunities. *Journal of Controlled Release, 170*(1), 15–40. https://doi.org/10.1016/j.jconrel.2013.04.020

Timofeevski, S. L., Panarin, E. F., Vinogradov, O. L., & Nezhentsev, M. V. (1996). Anti-inflammatory and antishock water-soluble polyesters of glucocorticoids with low level systemic toxicity. *Pharmaceutical Research, 13*(3), 476–480. https://doi.org/10.1023/a:1016069315423

Tomita, Y., Nishiyama, T., Sato, S., & Fujiwara, M. (1991). Expression of transferrin receptor on transitional cell cancer. *Hinyokika kiyo Acta urologica Japonica, 37*(1), 11–16.

Vasey, P. A., Kaye, S. B., Morrison, R., Twelves, C., Wilson, P., Duncan, R., Thomson, A. H., Murray, L. S., Hilditch, T. E., Murray, T., & Burtles, S. (1999). Phase I clinical and pharmacokinetic study of PK1 [N-(2-hydroxypropyl)methacrylamide copolymer doxorubicin]: First member of a new class of chemotherapeutic agents-drug-polymer conjugates. Cancer research campaign phase I/II committee. *Clinical Cancer Research: An Official Journal of the American Association for Cancer Research, 5*(1), 83–94.

Verma, P., & Ahuja, M. (2015). Optimization, characterization and evaluation of Chitosan- tailored cubic nanoparticles of clotrimazole. *International Journal of Biological Macromolecules, 73*, 138–145. https://doi.org/10.1016/j.ijbiomac.2014.10.065

Vikas, J., Shikha, J., & Mahajan, S. C. (2015). Nanomedicines based drug delivery systems for anti-cancer targeting and treatment. *Current Drug Delivery, 12*(2), 177–191. https://doi.org/10.2174/1567201811666140822112516

Wang, D., & Goldring, S. R. (2011). The bone, the joints and the balm of gilead. *Molecular Pharmaceutics, 8*(4), 991–993. https://doi.org/10.1021/mp200328t

Wang, D., Miller, S., Sima, M., Kopečková, P., & Kopeček, J. (2003). Synthesis and evaluation of water-soluble polymeric bone-targeted drug delivery systems. *Bioconjugate Chemistry, 14*(5), 853–859. https://doi.org/10.1021/bc034090j

Wang, D., Miller, S. C., Sima, M., Parker, D., Buswell, H., Goodrich, K. C., Kopečková, P., & Kopeček, J. (2004). The Arthrotropism of macromolecules in adjuvant-induced arthritis rat model: A preliminary study. *Pharmaceutical Research, 21*(10), 1741–1749. https://doi.org/10.1023/B:PHAM.0000045232.18134.e9

Wang, D., Miller, S. C., Kopečková, P., & Kopeček, J. (2005). Bone-targeting macromolecular therapeutics. *Advanced Drug Delivery Reviews, 57*(7), 1049–1076. https://doi.org/10.1016/j.addr.2004.12.011

Wang, D., Sima, M., Mosley, R. L., Davda, J. P., Tietze, N., Miller, S. C., Gwilt, P. R., Kopečková, P., & Kopeček, J. (2006). Pharmacokinetic and biodistribution studies of a bone-targeting drug delivery system based on N-(2-Hydroxypropyl)methacrylamide copolymers. *Molecular Pharmaceutics, 3*(6), 717–725. https://doi.org/10.1021/mp0600539

Wang, D., Miller, S. C., Liu, X.-M., Anderson, B., Wang, X. S., & Goldring, S. R. (2007a). Novel dexamethasone-HPMA copolymer conjugate and its potential application in treatment of rheumatoid arthritis. *Arthritis Research & Therapy, 9*(1), 1–9. https://doi.org/10.1186/ar2106

Wang, D., Miller, S. C., Shlyakhtenko, L. S., Portillo, A. M., Liu, X.-M., Papangkorn, K., Kopečková, P., Lyubchenko, Y., Higuchi, W. I., & Kopeček, J. (2007b). Osteotropic peptide that differentiates functional domains of the skeleton. *Bioconjugate Chemistry, 18*(5), 1375–1378. https://doi.org/10.1021/bc7002132

Wang, Y., Wang, D., Fu, Q., Liu, D., Ma, Y., Racette, K., He, Z., & Liu, F. (2014). Shape-controlled paclitaxel nanoparticles with multiple morphologies: Rod-shaped, worm-like, spherical, and fingerprint-like. *Molecular Pharmaceutics, 11*(10), 3766–3771. https://doi.org/10.1021/mp500436p

Wei, X., Zhao, G., Jia, Z., Zhao, Z., Chen, N., Sun, Y., Kelso, M., Rathore, G., & Wang, D. (2022). Macromolecular dexamethasone prodrug ameliorates neuroinflammation and prevents bone loss associated with traumatic brain injury. *Molecular Pharmaceutics, 19*(11), 4000–4009. https://doi.org/10.1021/acs.molpharmaceut.2c00482

Weinstein, J. N., Blumenthal, R., Sharrow, S. O., & Henkart, P. A. (1978). Antibody-mediated targeting of liposomes. Binding to lymphocytes does not ensure incorporation of vesicle contents into the cells. *Biochimica et Biophysica Acta, 509*(2), 272–288.

Wiedwald, U., Han, L., Biskupek, J., Kaiser, U., & Ziemann, P. (2010). Preparation and characterization of supported magnetic nanoparticles prepared by reverse micelles. *Beilstein Journal of Nanotechnology, 1*, 24–47. https://doi.org/10.3762/bjnano.1.5

Woerly, S., Maghami, G., Duncan, R., Subr, V., & Ulbrich, K. (1993). Synthetic polymer derivatives as substrata for neuronal adhesion and growth. *Brain Research Bulletin, 30*(3), 423–432. https://doi.org/10.1016/0361-9230(93)90274-F

Woerly, S., Petrov, P., Syková, E., Roitbak, T., Simonová, Z., & Harvey, A. R. (1999). Neural tissue formation within porous hydrogels implanted in brain and spinal cord lesions: Ultrastructural, immunohistochemical, and diffusion studies. *Tissue Engineering, 5*(5), 467–488. https://doi.org/10.1089/ten.1999.5.467

Woerly, S., Pinet, E., de Robertis, L., Van Diep, D., & Bousmina, M. (2001). Spinal cord repair with PHPMA hydrogel containing RGD peptides (NeuroGel™). *Biomaterials, 22*(10), 1095–1111. https://doi.org/10.1016/S0142-9612(00)00354-9

Yoo, H. S., & Park, T. G. (2004). Folate-receptor-targeted delivery of doxorubicin nano-aggregates stabilized by doxorubicin–PEG–folate conjugate. *Journal of Controlled Release, 100*(2), 247–256. https://doi.org/10.1016/j.jconrel.2004.08.017

Yuan, F., Quan, L.-d., Cui, L., Goldring, S. R., & Wang, D. (2012). Development of macromolecular prodrug for rheumatoid arthritis. *Advanced Drug Delivery Reviews, 64*(12), 1205–1219. https://doi.org/10.1016/j.addr.2012.03.006

Zhao, G., Ren, R., Wei, X., Jia, Z., Chen, N., Sun, Y., Zhao, Z., Lele, S. M., Zhong, H. A., Goldring, M. B., Goldring, S. R., & Wang, D. (2021). Thermoresponsive polymeric dexamethasone prodrug for arthritis pain. *Journal of Controlled Release, 339*, 484–497. https://doi.org/10.1016/j.jconrel.2021.10.007

Zhong, Y., & Bellamkonda, R. V. (2008). Biomaterials for the central nervous system. *Journal of the Royal Society Interface, 5*(26), 957–975. https://doi.org/10.1098/rsif.2008.0071

Zhou, Y., & Kopeček, J. (2012). Biological rationale for the design of polymeric anti-cancer nanomedicines. *Journal of Drug Targeting, 21*(1), 1–26. https://doi.org/10.3109/1061186X.2012.723213

Zuluaga, M. F., & Lange, N. (2008). Combination of photodynamic therapy with anti-cancer agents. *Current Medicinal Chemistry, 15*(17), 1655–1673. https://doi.org/10.2174/092986708784872401

Therapeutic Gene Editing

55

Chen Zhang, Haotian Li, Keshore R. Bidasee,
Howard E. Gendelman, and Prasanta K. Dash

Abstract

Genome editing, alternatively known as gene editing, allows manipulation of genetic material in a living organism in order to develop means to treat genetic, infectious, or other acquired diseases. Here, customized DNA cutting follows natural DNA repair resulting from either precise cuts, gene insertions, deletions, partial, or full sequence replacements in specific target areas of the host genome. Low recombination rates and random integration into undesired genomic loci restricted broader application. To overcome such limitations in preclinical and clinical studies, researchers first identified the highly targeted genome engineering systems, the zinc finger nucleases (ZFNs) and transcription activator-like effector nucleases (TALENs), which failed in broader use, as each required protein reengineering for site-specific targeting. Ultimately, the cloning and protein engineering difficulties of ZFNs and TALENs sped the development of clustered regularly interspaced short palindromic repeats (CRISPR). CRISPR-Cas-nuclease-based technologies remain the most revolutionized gene-editing system available and have shown early success in genome editing. CRISPR technology has cured patients of certain genetic diseases, but not all patients can receive it due to cost and accessibility. CRISPR technology has been successful in treating a pediatric patient with T-cell acute lymphoblastic leukemia, showing feasibility of its use for cancer immunotherapy. For the human immunodeficiency virus type one (HIV-1), CRISPR technologies were successfully applied in animal model systems because of its unique RNA-based reprogramming system. The translation from animals to humans for HIV-1 infection remains in process due to a need for advanced delivery to virus target cells. The delivery of CRISPR-Cas9 cargos influences therapeutic efficacy but remains a limitation. Safe and effective delivery methods for gene editing include viral and non-viral-based vectors. Use of non-viral vectors requires that the gene-editing payload specifically reach its desired cellular designation, be protected from large-scale enzymatic degradation, and show no off-target toxicities. These challenges remain active areas of research activities.

Keywords

Gene Editing · TALENs · ZFNs · CRISPR-Cas9 · Neurodegenerative and Infectious Viral Diseases · HIV-1

55.1 Introduction

Gene or genome editing is a powerful technology that enables scientists to manipulate an organism's genetic material. This allows genome editing for targeted therapy of human diseases to uncover mechanisms and to make advances. Genes can now be altered or modified either by introducing insertions or deletions with partial or full replacements of specific sequences. The history of the gene editing began in late 1970s with the discovery of restriction enzymes that for the first time enabled scientists to manipulate DNA in the research laboratory (Szybalski & Skalka, 1978). During the late 1980s, significant advances in gene editing were made through the use of restriction enzyme-based techniques and homologous recombination in yeast and mice, as demonstrated by a number of what now is considered as groundbreaking studies (Rothstein, 1983; Smithies et al., 1985; Thomas et al., 1986). These early investigations laid the foundation for the feasibility of achieving "site-

C. Zhang · K. R. Bidasee · H. E. Gendelman · P. K. Dash (✉)
Department of Pharmacology and Experimental Neuroscience, College of Medicine, University of Nebraska Medical Center, Omaha, NE, USA
e-mail: pdash@unmc.edu

H. Li
School of Biological Sciences, University of Nebraska-Lincoln, Lincoln, NE, USA

H. E. Gendelman, T. Ikezu (eds.), *Neuroimmune Pharmacology and Therapeutics*,
https://doi.org/10.1007/978-3-031-68237-7_55

specific-editing" of the target genome. However, low recombination rate and random integration of the exogenous copy into undesired genomic loci has restricted broader application. This was realized in both preclinical and clinical studies. To overcome such limitations, a highly targeted genome engineering system was developed with the zinc finger nucleases (ZFNs) in 1985 (Miller et al., 1985). ZFNs are chimeric nucleases created by fusing a zinc finger DNA binding domain to the nuclease domain from the FokI type IIS restriction endonuclease (Kim & Chandrasegaran, 1994). Each ZFN binds specifically to a set of three base pairs and can be combined to recognize longer DNA sequences. This leads to improvements in the specificity and efficacy of gene targeting (Kim et al., 1996). Another key domain of ZFNs is the restriction endonuclease domain which mediates DNA breakage (Kim et al., 1996). The advantage of ZFNs lies in that it can be reengineered to recognize specific sites within the human genome (Porteus & Carroll, 2005). This makes it possible to stimulate genome editing in human somatic cells at high frequencies. It was shown successful in modifying human IL2RG (Urnov et al., 2005). Improved efficacies and specificities prompt ZFNs in the use for a spectrum of genomic targets throughout many different cell types and organisms. After ZFNs, other powerful nucleases represented by transcription activator-like effector nucleases (TALENs) and clustered regularly interspaced short palindromic repeats associated (CRISPR-Cas)-nucleases were identified and developed (Adli, 2018).

TALEN is a type of engineered endonuclease developed as an alternative to ZFNs for genome editing (Christian et al., 2010; Cermak et al., 2011). Like ZFNs, TALENs have a pair of engineered TALE DNA-binding domains fused to the Fok I endonuclease. TAL effectors, which are natural proteins produced by the plant bacterial pathogen Xanthomonas, play a pivotal role in interacting with the host transcriptional machinery. These proteins have the remarkable ability to bind to precise DNA sequences at single-base resolution. Their interactions with the host's genome can result in the activation of genes that can either confer resistance or susceptibility to Xanthomonad infections. Due to their high specificity, they are modified to create customized TAL DNA-binding domains that can recognize specific genome sequences (Bogdanove & Voytas, 2011). TALENs recognition sequences are typically longer than those recognized by ZFNs. This reduces the likelihood of off-target effects. In addition, TALENs modularity makes it a standout when compared to ZFNs. Their DNA-binding domains are made up of a series of repeated units, each of which can recognize a specific base pair. The modularity allows for easy customization of the TALEN to target a specific DNA sequence of interest. Moreover, the modular design of TALENs has allowed for the development of libraries of TALENs that can target multiple gene sites (Miller et al., 2011). Other than DNA binding domain, TALENs have two additional domains, a nuclear localization signal and a transcriptional activation domain, which can be designed to recognize sequences in non-coding genomic regions that might be important for regulating gene expression. Taken together, TALENs sparked additional enthusiasm by offering several advantages over ZFNs, including higher specificity, greater modularity, and less likelihood of off-target effects (Joung & Sander, 2013).

While ZFNs and TALENs achieved success in earlier genome editing studies, they always require the redesign or reengineering of proteins for targeting different sites. Thus, the cloning and protein engineering difficulties of ZFNs and TALENs have hindered their widespread implementation (Porteus, 2016). The emergence of CRISPR, a novel gene-editing technique with its distinctive RNA-based reprogramming system, has revolutionized the field of gene editing by overcoming several of the sensitivity-specificity drawbacks associated with prior genome editing systems. CRISPR was adapted from the natural antiviral (prokaryotic anti-phage) defense systems (Barrangou, 2015), then developed as the "molecular scissors" in eukaryotes (Hsu et al., 2014). In the CRISPR system, one guide RNA (gRNA) is designed to bind to the target sequence that directs the Cas9 enzyme or scissor protein to locate where the excision is made. Upon the precise excision of the targeted DNA sequence, the host cell machinery repairs the cut generated because of cleavage through cellular DNA repair mechanisms, specifically non-homologous end-joining (NHEJ) or microhomology-mediated end-joining (MMEJ) methods (Allen et al., 2018; Shen et al., 2018). This approach allows gene editing and to change precise genetic information contained in a cell. As CRISPR offers simplicity, flexibility, and robustness in facilitating editing efficiency, it is more attractive for use in high-throughput screening and for genome-wide targeting (Adli, 2018).

Most gene-editing methods rely on the deliberate creation of a DNA double strand break at designated sites using nucleases followed by involvement of the host DNA repair machinery to achieve the intended result (Xue & Greene, 2021). The three major pathways involved in the repair of double strand breaks permitting restoration of the genetic integrity are (1) homology-dependent repair (HDR), where a donor homologous sequence matching the target gets copied leading to the repair, (2) non-homologous end joining (NHEJ) in which the broken ends after removal of the targeted mutated sequence get stitched together at the break site without the need of a homologous template, and (3) microhomology-mediated end joining (MMEJ), which is less effective and more error-prone but used by cells when NHEJ pathway is not active (Xu & Xu, 2020). The NHEJ and HDR are frequently used pathways involved in repairing gene editing induced DNA breaks (Miyaoka et al., 2016). However, studies of higher eukaryotic cells have shown a preference for utilizing the NHEJ pathway, active in both non-dividing and proliferating cells. Additionally, it has also been shown that the process of knocking out desired sequences is more straightforward than incorporating sequences of interest into the genome through knock-in procedures. (Xue & Greene, 2021; Miyaoka et al., 2016; Scully et al., 2019).

Though the progress is limited in its abilities to provide safe and effective translation for human diseases, nonetheless, gene editing has been applied to a wide variety of genetic disorders that includes Rett syndrome, cystic fibrosis, hemophilia, and Krabbe disease (Jensen et al., 2021). Promising developments were also reported for treatment of cancer, cardiovascular disease, and neurodegenerative disorders (White et al., 2016; Zhang et al., 2021; Sun & Roy, 2021). In addition, antiviral gene therapy now offer more hope for chronic infectious diseases. The advantages of antiviral gene therapy include lack of development of drug resistance over the daily intake of antiviral drugs for life and reduced toxicity (White et al., 2016). Over the past decades, in vivo gene editing has received considerable attention. Researchers have applied a variety of techniques to treat chronic genetic and infectious diseases (Cornu et al., 2017; Trevisan et al., 2017). This chapter provides a summary of the advantages and limitations of currently available delivery schemes of genome editing. The challenges in using each of the technologies for translational therapeutic purposes are discussed.

55.2 Gene Editing for the Treatment of Cardiovascular Disease

For cardiovascular disease, gene editing has emerged as a therapeutic tool (Strong & Musunuru, 2017; Nishiga et al., 2021). While there are currently no approved gene-editing therapeutic interventions for cardiovascular disease, promising preclinical studies hold future therapeutic potential. One successful application is the use of CRISPR-Cas9 gene editing to target the protein convertase subtilisin/kexin 9 (PCSK9) gene, which regulates the levels of low-density lipoprotein (LDL) cholesterol (Ding et al., 2014). PCSK9 gene provides a pathway for a protein that can regulate blood cholesterol. PCSK9 mutations are linked to familial hypercholesterolemia, which can lead to early onset cardiovascular disease. Using CRISPR-Cas9 to disrupt PCSK9, reduced level of LDL was observed in animals of human disease and led to finding a novel approach for treating hypercholesterolemia. The successful application may be helpful in reducing the risk of cardiovascular disease (Ding et al., 2014; Wang et al., 2016a; Wang et al., 2018b). Another application using gene editing tools centers in the development of atherosclerosis (Jaiswal et al., 2017). By targeting genes either involved in inflammation or plaque formation, it might be applicable to slow or reverse the progression of atherosclerosis, and thus reduce the risk of myocardial infarction (MI) (Sano et al., 2018; Nishiga et al., 2021). Recently, our group found that AAV2/9-Endo-Glo1 treatment was beneficial in improving left ventricular function in models of chronic viral disease (Dash et al., 2021).

While these and other gene editing approaches to cardiovascular disease are still in their infancy, they hold promise for the development of novel targeted therapies. However, several limitations must be overcome before they can be generally used. One of the main limitations is the difficulty in delivering the gene-editing tools to the affected cells in heart or blood vessels with the current viral vectors (Strong & Musunuru, 2017). Another limitation is the heterogeneity of cardiovascular disease. Cardiovascular disease encompasses a range of conditions with varying genetic factors influenced by the environment (Nishiga et al., 2021). Gene-editing approaches that target specific mutations may not be sufficient for controlling the cardiovascular disease and may only be effective for few patients. A further limitation is the potential off-target effects of gene editing, which is a general concern both for cardiovascular disease and for all of gene editing therapeutics. In summary, improvements in gene delivery, specificity, safety, and a better understanding of the complexity of cardiovascular disease are necessary to realize the potential of gene editing.

55.3 Gene Editing-Based Therapies for the Treatment of Neurological Disorders

As the two most prevalent neurodegenerative disorders, Alzheimer's, and Parkinson's diseases (AD and PD) gained lots of research focus, and recently several gene editing-based strategies are being tested to find novel therapeutic outcomes. AD is the most common form of dementia in elder population that is characterized by the loss of cholinergic neurons, synaptic dysfunction, accumulation of amyloid beta protein (forms the senile plaques), and neurofibrillary tangles (Gandy & Dekosky, 2013; Lane et al., 2018). Recent efforts have been targeting genes/pathways that contribute to these disease hallmarks in animal models, aiming to find a potential therapeutic solution for AD (Park et al., 2019; Sun et al., 2019). For example, beta-site amyloid precursor protein cleaving enzyme 1 (BACE-1) is one of the therapeutic candidates for AD. This is because it is a protease required to produce amyloid beta protein, which plays a key role in AD progression. A recent work investigated CRISPR-Cas9 nanocomplexes targeting the key enzyme BACE1. BACE1 plays a role in APP cleavage pathway and investigated early in physiology studies of the mouse' hippocampus and midbrain. The research suggested that amyloid beta (Aβ)-associated pathologies and cognitive deficits in these APP knock-in AD mice were significantly alleviated. The study broadens the use of in vivo delivered CRISPR-Cas9 in the field of neurodegenerative disease; however, limitations such as insufficient delivery to the whole brain neural circuits, off-target effects, and maintenance of long-term efficacy of CRISPR-Cas9 therapy for AD were also posed by the authors (Park et al., 2019). Another study utilized the Adeno-associated virus (AAV)-mediated CRISPR-Cas9 to edit amyloid precursor protein (APP) in human iPSC-derived neurons in the brain targeting amyloid cleavage pathways and found efficient editing of APP in mouse brains with no detectable off-target effects. However, due to the complexity of the APP cleavage pathway, downregulation of certain downstream enzymes resulted in mild deleterious effects on the overall homeostasis of the brain, necessitating further studies with long-term behavioral and brain anatomical examinations (Sun et al., 2019). Another work utilized AAV-mediated CRISPR-Cas9 targeting of the APP in AD-transgenic mice. Following a single dose of intrahippocampal administration, those mice experienced a reduction of amyloid-beta-associated pathologies for up to six months. Interestingly, when the authors delivered an engineered brain-penetrating CRISPR-Cas9 construct through an intravenous (IV) route, the hallmark of AD pathologies including deposition of amyloid-beta, neurite dystrophy, microgliosis, and the impairment of cognitive performance were all found to be ameliorated (Duan et al., 2022). This IV route of delivery of gene-editing therapy that crossed the blood-brain barrier (BBB) is of great interest for clinical translation as peripheral routes of administration have fewer injection site reactions, and thus, it could serve as a viable option in clinical settings. As the development of improved delivery systems continues, the utilization of CRISPR-Cas9 not only holds the potential to alleviate the complex neurodegenerative diseases like AD but also brings hope for improved treatment and cure of other rare but equally important monogenetic brain diseases.

The second most prevalent progressive neurodegenerative disease, for which there's currently no curative therapy available, is PD. PD is characterized by the loss of dopaminergic neurons in striatum and substantia nigra, Lewy bodies formation by aggregated α-synuclein, and neuroinflammation (Walker et al., 2015). Researchers have discovered dozens of genes that are associated with the onset and development of PD-linked pathologies. Although there were many gene targets studied for PD such as alpha-synuclein (SNCA), leucine-rich repeat kinase 2 (LRRK2), and several parkin (PRKN) genes, which were comprehensively reviewed by Selvaraj and Safari (Safari et al., 2020; Selvaraj & Piramanayagam, 2019), most of them are biomarkers for diagnosis but not for therapeutic purposes because targeting one or a couple of them were not sufficient in slowing the disease progression. And it is also extremely challenging to target multiple genes simultaneously utilizing the current gene-editing techniques without causing much off-target effects. A review article by Bartus et al. summarized the limitations of gene therapy, describing several failure trials in PD patients (Bartus et al., 2014). Although, most of these reported CNS targeted gene therapies have proven to be safe, but have not shown clinical efficacy. One recent preclinical work reported improvement of movement disorder and reduction in neuronal death when CRISPR-Cas9 edited mesenchymal stem cells (MSC) were transplanted in a PD mice study (Lee et al., 2019). The authors first transfected the Cas9 protein, sgRNA, and the donor plasmid encoding the soluble receptor for AGEs (sRAGE) gene (which has anti-inflammatory effects in PD) into MSCs followed by integration of sRAGE gene into AAVS1 site to generate the sRAGE-secreting MSCs. The sRAGE-secreting MSCs were then implanted into corpus striatum of the PD mice, and they observed that most of the PD-related phenotypes were reversed at study end. Their findings highlighted the importance of a future potential treatment strategy for PD through combinatorial immunomodulation (via gene editing) and stem cell therapy; however, the translation of this approach into the clinic needs much more investigation as it involves the direct transplantation into the brain, and that can lead to higher risk of adverse events.

Early studies though provided the proof-of-concept that gene editing tools can be used with success in the treatment of some neurodegenerative disorders; however, many hurdles still need to be addressed before applying these approaches to treat infected patients. One major challenge is effective and safe delivery of the CRISPR-Cas9 containing gene-editing cargo, since most of them do not effectively penetrate the BBB. Meanwhile, as the progression of neurodegenerative diseases typically spans months to years, preserving the sustained efficacy of the gene-editing therapy while avoiding chronic toxicities necessitates further investigation through long-term evaluations in animal models. In addition, the complexity of the neurodegenerative disorders, which involves multiple genetic and environmental factors, also poses a significant challenge for current gene-editing techniques. Overall, addressing these challenges will be crucial for the successful translation of gene-editing technique into clinical practice for neurogenerative diseases.

55.4 Gene Editing for Viral Diseases

Other than its application in the treatment of neurodegenerative and cardiovascular diseases, gene editing-based approach have been tested for various infectious viral pathogens existing in either integrative or episomal form, which include hepatitis B virus (HBV), human papilloma viruses (HPV), herpes viruses, and Human Immunodeficiency Virus Type 1 (HIV-1) (White et al., 2016; Trevisan et al., 2017; Martinez et al., 2022; Aghamiri et al., 2020; Dash et al., 2020). While the currently available anti-viral drug regimens have positive effects in controlling the ongoing viral infection, they require good patient compliance and can result in the development of viral resistant mutants when taken for longer duration. Moreover, viruses like HIV undergo latency in the presence of long-term antiviral drug (ARV) treatments, and the virus gets reactivated once the antiviral drugs are stopped. Thus, gene-editing approaches that can directly target the viral genomes are the best candidates to potentially eliminate infectious virus from the cells and achieve disease elimination.

HBV is the pathogen responsible for causing severe liver infection in humans, and the research on preclinical therapeutics for HBV involves extensive application using CRISPR-Cas9. HBV contains a small, partially double stranded DNA (dsDNA) genome that gets converted to covalently closed circular DNA (cccDNA), which serves as the template for transcription of the viral pregenomic RNA (pgRNA) and mRNA that encode the viral proteins. PgRNA then gets reverse transcribed to mature viral DNA and gets packaged inside the viral particle (Tsukuda & Watashi, 2020). Several researchers utilized gene-editing tools to disrupt the different stages of its life cycle and have made progress in treating HBV infection. Evidences from laboratory studies suggest that CRISPR-Cas9 had significant inhibitory effect on the viral gene expression and viral production in primary cells and HepG2 cell lines without any cytotoxicity effects; however, off-target effects were not investigated in these studies (Kennedy et al., 2015; Seeger & Sohn, 2014; Lin et al., 2014). Cleavage of cccDNA appeared to block early viral infection from the initial stage in most cases; however, more concrete evidence is needed to support the findings. The investigators observed that the cccDNA that escaped the targeted gene editing was found to carry mutations at cleavage site and severely impaired steps in the virus life cycle, leading to reduction in viral protein production and pgRNA expression. Zheng et al. (Zheng et al., 2017) have shown that by using the combination of CRISPR-Cas9 and HBV reverse transcription inhibitors, there was a synergistic effect leading to reduction of the total HBV DNA and cccDNA by 86 and 85.34%, respectively, in primary hepatocytes..

The first in vivo study was reported describing the efficiency of CRISPR/Cas9 in a mouse model of HBV persistent disease. The researchers designed eight different guide RNAs targeting HBV genome and first tested in HBV transfected Huh7 cells and observed that the viral protein expression was inhibited up to 96% using the combination of multiplex sgRNA—and Cas9—expression plasmids. Further investigations in an HBV persistence mouse model detected a reduction in HBsAg expression and cleavage of approximately 5% of the intrahepatic HBV DNA (Lin et al., 2014). Although the preliminary gene editing based findings indicated the potential of CRISPR-Cas9 system in disrupting HBV infection, the mouse model they used for the study was not an ideal model system as it lacked most steps of the viral life cycle, and therefore does not reflect the real infectious disease pattern observed in HBV-infected patients (Lin et al., 2014). Later, in vivo studies by other groups achieved higher cleavage efficacy using viral vector-mediated CRISPR-Cas9 administration in combination with anti-HBV drug treatments, which includes lamivudine, tenofovir, entecavir, or the capsid assembly inhibitor heteroaryldihydropyrimidine-12. The efficacy of these combinations were tested in several pharmacodynamic studies using either HBV-mouse models or transgenic-HBV-models, without much evaluation on the off-target effects of the tested gene-editing methods (Dong et al., 2015; Zhen et al., 2015). More recent works examined the off-target effects of CRISPR-Cas9 in different HBV mouse models. Li et al. and Sakuma et al. screened for the predicted on-target and off-target sites by Sanger sequencing and validated using T7E1 assay and found no off-target effects in their CRISPR-treated mice (Sakuma et al.) However, other studies detected minimal off-target effects through deep sequencing, questioning the safety of CRISPR and concluding that higher sensitivity assays are needed to preclude off-target effects (Jiang et al., 2017; Li et al., 2017) as an alternative to plasmid transfection or AAV or lentiviral transduction for lipid nanoparticles (LNP) guide RNA (gRNA) delivery of CRISPR-Cas9. This showed higher efficacy of up to 50% in tested cell lines. The HBV DNA isolated from LNP-CRISPR treated-mouse liver demonstrated multiple indels at the cleavage site with minimal off-target effects (Jiang et al., 2017). Though the data from these initial gene-editing studies for the treatment of HBV infection seem promising; however, it lacked a proper animal model system of HBV infections that can accurately reflect human infections. HBV life cycle and CRISPR-based targeting of the viral genome is summarized in Fig. 55.1. In summary, the success made in HBV treatment has laid the foundation for the application of gene-editing technology in other infectious human diseases.

Another DNA virus that can be potentially disrupted using gene-editing tools is human papilloma Virus (HPV). HPV is a group of more than 200 related viruses, but only a small portion of high-risk strains is known to cause severe cancer diseases such as cervical cancer, oropharyngeal cancer, anal cancer, vaginal cancer, and vulvar cancer (Doorbar et al., 2012). As a small double-stranded circular DNA virus, HPV virion infect epithelial tissues through micro wounds (Doorbar et al., 2012). HPV viral genome is maintained as low copy number episomes in the infected lamina cells. During cell

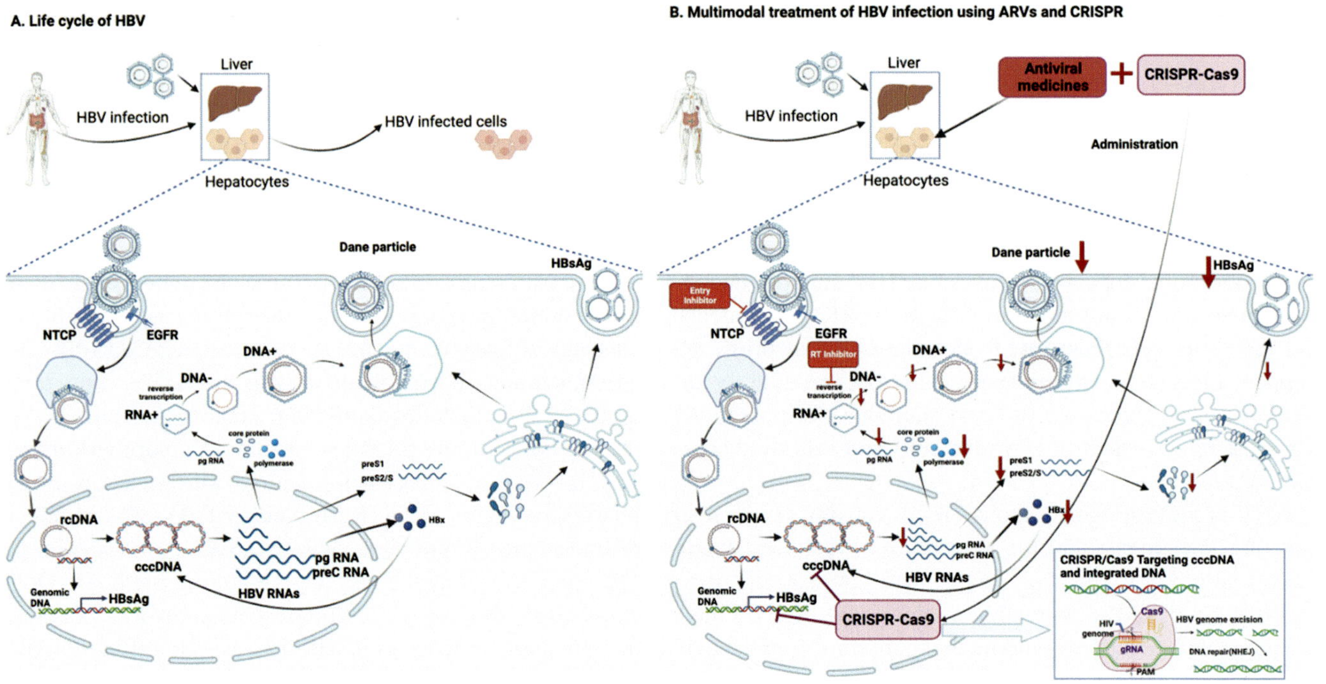

Fig. 55.1 Hepatitis B virus (HBV) life cycle and CRISPR targeting of the viral genome, (**a**) HBV is a partially double-stranded DNA virus that exclusively infects human hepatocytes. As shown in the figure, its envelope proteins interact with multiple host surface receptors to enter the host cell. Once it enters the cell through endocytosis, its nucleocapsid is subsequently released into the cytoplasm. The nucleocapsid is uncoated at the nuclear membrane and releases circular HBV DNA (rcDNA) into the nucleus. In the nucleus, the rcDNA is either converted into a covalently closed circular double-stranded DNA (cccDNA) or integrated into the host genome. The cccDNA serves as a template for transcription of multiple viral RNAs. Assembly of the HBV nucleocapsid occurs within the cytoplasm utilizing the pgRNA and the key viral proteins (core protein, and polymerase). Further assembly of hepatitis B virions occurs at the endoplasmic reticulum where the envelope protein gets packaged and matured. Mature virions are ultimately excreted through a process of budding and vesicular transport into the extracellular space and be able to infect neighboring cells. Viral proteins HBsAg without viral genetic material are also released through exocytosis to the extracellular space. Noted that cccDNA and integrated viral genome can exist for the lifetime of hepatocytes. (**b**) Multimodal treatment of HBV infection using ARVs and CRISPR. Current evidence has shown that the combinatorial administration of CRISPR/Cas9 and ARVs treatment has a synergistic effect leading to reduction of the total HBV DNA and cccDNA in in vitro system

division, virus genome gets amplified at upper epithelial layers and after the completion of the cell cycle event, viral proteins are made, allowing assembly, packaging, and release of the new virus particles (Doorbar et al., 2012). Viral E6 and E7 proteins known to inhibit tumor suppressor genes are the major factors influencing the development of HPV-induced malignancy (Mclaughlin-Drubin & Münger, 2009). When the tumor suppressor genes are switched off, the associated cells divide uncontrollably and resulting in malignancy (Mclaughlin-Drubin & Münger, 2009). CRISPR/Cas-mediated gene editing of the E6 and/or E7 gene in HPV-infected cells has shown potential to function as a highly effective and specific approach for the elimination of transformed malignant cells. Some of the early in vitro studies (Kennedy et al., 2014; Hu et al., 2014b; Zhen et al., 2014) found that HPV-specific gRNAs along with Cas9 could be able to disrupt the targeted E6 or E7 gene, making them inactivate in HPV-induced transformed cells. 90% of these cleaved and inactivated transformed cells without functional HPV-oncogenes underwent cell death, leaving HPV-negative cells unaffected (Kennedy et al., 2014).

CRISPR-mediated gene editing of E6, E7 gene, or their promoter region of HPV shows high cleavage efficacy and significant reduction of tumor growth in a BALB/c nude mice model transplanted with human cervical cancer cells as compared to the gRNA empty vector control (Zhen et al., 2014). Although this work confirmed the efficacy of CRISPR in targeting HPV E6/E7, the pre-treated transplantation animal model did not reflect the actual patient scenario where the patients will always have pre-existing tumors, thus requiring better HPV-induced cancer-bearing mouse model studies before translation to clinic. Lao et al. utilized micelle-packaged CRISPR/Cas9 system and achieved satisfactory HPV suppression in a HeLa xenograft mouse model with minimal off-target effects; however, the toxicity of the micelle was not addressed (Lao et al., 2018). A later study by Gao et al. evaluated the efficacy as well as the cellular and tissue toxicity of hyperbranched polymer-based nanoparticles carrying CRISPR-Cas9 targeting HPV E7 oncogene in a xenograft tumor-bearing mice model and observed strong tumor growth inhibition with low toxicity (Gao et al., 2020). While all these studies utilizing different delivery methods of

CRISPR gene editing for HPV have shown strong inhibitory effects on tumor growth, none of them achieved complete tumor elimination in any models, until recently complete clearance of established tumors in a cervical cancer xenograft mouse model had not been reported (Jubair et al., 2019). However, there are many unanswered queries remain on this study including the tissue toxicity introduced by the large payloads of CRISPR-containing liposomes and possible generation of immune tolerance with seven sequential administrations of payload within a month. The authors found that three doses of CRISPR injection could only prolong the animal survival by 12 days, strongly suggesting that substantial improvements are needed to improve the overall delivery, dosing, and toxicity assessment before being translated to humans. Overall, though the use of viral/nonviral vector mediated (unmodified or chemically modified nanoparticles, micelles, and liposomes) CRISPR-Cas9 payload provided early promising results for the treatment in HPV-induced cancers, the data is too preliminary in terms of efficacy and safety for clinical translation.

Another large family of DNA viruses known Herpesviruses, also known as Herpes simplex viruses (HSV), are responsible for human diseases like oral herpes, cold sores and keratitis typically caused by HSV-1, and genital herpes caused by HSV-2 (Chayavichitsilp et al., 2009). Globally, the herpes virus is widespread in humans, and either it can lyse or cause latent infections in host cells (Chayavichitsilp et al., 2009). The latter form of infection enables the virus to hide in the host cells (mostly in types 1 and 2 sensory neurons) for many years, rendering clearance of the viruses from the hosts extremely challenging. The lytic infection part of the life cycle of Herpesvirus is similar to HIV infection of the host but still has some unique signature steps (Kukhanova et al., 2014). Once the virus enters the host cell, it produces viral DNA copies using the host cell machinery leading to production of viral proteins that gets assembled into complete infectious virions and gets released from the infected cells and ultimately attack the nearby cells making them infected. Unlike HIV latency, where the HIV-DNA gets integrated into the host genome and becomes an integral part, the latent infection pathway of herpesviruses evades the host cell nucleus utilizing episomes without disturbing the host cell genome and establishes latency (Kukhanova et al., 2014). Currently, the nucleoside analogs such as Ganciclovir, Valganciclovir, and Foscarnet are preferred drugs used to block the active virus replication, but they can't block the latent viruses from reactivation (Kukhanova et al., 2014). Therefore, genome editing tools that can directly target these viruses within infected cells are more likely to provide a cure for persistent herpesvirus infections. So far, very few studies have reported the use of CRISPR-Cas9 mediated gene editing to eliminate herpesviruses as compared to other infectious viruses, mainly because of lack of an ideal animal model system to test for efficacy (Bi et al., 2014; van Diemen et al., 2016; Xu et al., 2016; Yin et al., 2021). The first report came from Bi et al. who showed the introduction of high-efficacy of indels in cell-based studies using CRISPR-Cas9 in the viral target sites (Bi et al., 2014). Upon treatment with specific gRNA and Cas9 vectors, virus replication was significantly suppressed in 7 out of 8 replicates. Later, in vitro studies targeted other key viral genes using anti-HSV guide RNAs in non-human and human cell lines and demonstrated better potency in reducing viral DNA expression and viral replication (Roehm et al., 2016; van Diemen et al., 2016). Additionally, others showed that single gRNA-Cas9 system have a strong inhibitory effect on viral replication but can render the generation of escape mutants. However, upon co-delivery of two gRNAs targeting two different genes, complete abrogation of HSV-1 replication was achieved in human lung fibroblast cells (van Diemen et al., 2016). No detectable off-target effects were reported in these dual-gRNA/Cas9-treated cells, suggesting that the dual gRNA approach can be a promising strategy in combating HSV-1 infection. In a recent HSV-inoculated mice study, the potential value of lentiviral vector-mediated CRISPR/Cas9 in treating herpetic stromal keratitis (HSK) was reported. Introduction of dual guide RNAs targeting essential HSV genes *UL8* and *UL29* effectively prevented ocular herpes progression for three months in the treated mice without recurrence of herpetic stromal keratitis (Yin et al., 2021). The mice were initially inoculated with HSV-1 followed by CRISPR-based gene-editing construct administration leading to significantly reduced virus production, and the CRISPR treatment outperformed the conventional acyclovir therapy (Yin et al., 2021). Despite the relatively small sample size in the treatment group ($n = 6$), the results reported by the researchers is a good first step to show the potential clinical utility of lentiviral vector mediated CRISPR/Cas9 in treating refractory HSK.

Other than successful interruption of the HSV genome, CRISPR-engineered HSV-1 strain inserted with either EGFP gene or other desired phenotypic gene was also developed by the co-delivery of a donor-template sharing sequence homologous to a specific HSV-1 region, greatly facilitated HSV-1 studies in vitro and in vivo (Bi et al., 2014; Russell et al., 2015). Another added potential of CRISPR-Cas9 system to generate mutant HSV-1 as reported recently could help researchers uncover some of the unanswered mechanisms of HSV–host interactions (Xu et al., 2016; Suenaga et al., 2014). In summary, more mechanistic efforts are needed to elucidate whether CRISPR can target and break HSV-1 latency in vivo before being tested as a potential therapeutic approach.

55.5 Gene Editing-Based Approaches for HIV-1 Elimination

Since the discovery of HIV in 1983, more than 38 million people worldwide are infected by the virus as per the CDC report (Control and Prevention, 2021). With significant ongoing efforts designed toward HIV-1 prevention and therapy, there is no cure for this viral disease (Ndung'u et al., 2019). While antiretroviral therapy (ART) can markedly suppress active viral replication (Althoff et al., 2010; Wittkop et al., 2009), levels of viral DNA remain unaffected following its integration into the host genome. The integrated HIV DNA termed as proviral DNA remains latent for extended time periods and is the principal source for viral rebound after ART cessation. This precludes any possibility for HIV-1 elimination by ART (Pitman et al., 2018; Dash et al., 2020). To this end, gene therapy targeting latent HIV first emerged in the early 2000s. Intensive investigation has emerged on gene editing toward achieving a functional HIV cure (Xiao et al., 2019b; Das et al., 2019; Atkins et al., 2021; Dash et al., 2019; Herskovitz et al., 2022; Soriano, 2017).

Over the past decade, several gene-editing tools were being tested that includes nucleases capable of targeting viral DNA sequences for site cleavage required to eliminate HIV-1. This includes deployment of ZFNs, TALENs, and other nucleases designed to disrupt HIV proviral DNA or inactivate host genes. The latter includes HIV-1 co-receptors (Urnov et al., 2010; Manjunath et al., 2013; Stone et al., 2013; Owens, 2014; Benjamin et al., 2016). The mechanism of action for these nucleases is associated with DNA repair after targeted excision and is governed through the non-homologous end joining (NHEJ) pathway. Any small errors made during the NHEJ repair can introduce small insertions, deletions, or frame-shift mutation and are sufficient to impair targeted gene functions and can lead to permanent viral gene inactivation.

55.6 Zinc-Finger Nucleases (ZFNs) and HIV-1 Infection

Engineered ZFNs are first-generation gene-editing tools containing zinc finger proteins that bind to DNA through the zinc finger motif linked to the Fok 1 restriction endonuclease serving as a scissor motif (Bitinaite et al., 1998). ZFNs engineered for recognizing specific DNA sequences are widely tested in cells and organisms (Urnov et al., 2010; Meng et al., 2008; Townsend et al., 2009). An early report on CCR5 disruption by using ZFN technology showed the specific binding of a zinc finger domain to CCR5, led to its cleavage (Mani et al., 2005). Since then, others have reported the success of CCR5 disruption in primary and transformed cell lines and in humanized mouse model (Perez et al., 2008; Wilen et al., 2011; Yuan et al., 2012; Kim et al., 2009) as described in Fig. 55.2. Use of a zinc-finger nuclease targeting CCR5 protects against HIV-1 infection in primary CD4+ T cells and maintain viral resistance in NOG humanized mice (Perez et al., 2008). Infusion of gene edited CD4 + T cells safely into humanized mice laid the foundation for clinical translation. To further improve upon these results, ZFN-engineered hematopoietic stem cells were developed (Yao et al., 2012; Holt et al., 2010; Li et al., 2013) resulting in multilineage reconstitution of human immune cells in mice demonstrating cell-based resistance to HIV infection. Like CD4+ T cells engraftment, the stem cell-derived CCR5-edited populations were enriched over the course of the infection (Yao et al., 2012). However, off-target events were observed at CCR2 locus in these ZFNs-treated humanized mice, still requiring higher ZFN editing specificity (Li et al., 2013; Holt et al., 2010).

HIV-1 uses two coreceptors for entry into the cell for infection (Moore et al., 1997). Generally, founding viruses are primarily CCR5(R5) tropic. Over time, there is a switch to CXCR4(X4) tropic or dual tropic strains (Simonetti & Kearney, 2015). With the emergence of CXCR4 viruses, single CCR5 (R5) disruption by ZFNs may no longer protect against HIV-1 spreading infection. Thus, editing CXCR4 by ZFN approaches is considered a therapeutic strategy. To date, three groups have demonstrated that targeting CXCR4 can promote HIV-1 resistance. The first report showed robust proliferation of the edited T cells and were resistant to CXCR4 HIV-1 infection (Wilen et al., 2011). When tested in humanized mice, CXCR4 disruption led to protection against infection but failed to protect the cells in the later stage. This was due to the emergence of R5-tropic viral mutants. Another study demonstrated that CXCR4 ZFN disruption in CD4+ T cells conferred resistance to HIV-1 challenge (Yuan et al., 2012). The results collectively show that CXCR4 disruption alone may not be sufficient to provide long-term viral protection. Another group developed a HIV resistance T cell line utilizing ZFNs to inactivate CXCR4 and CCR5 (Voit et al., 2013). The modified T cell lines were found to be resistant to X4 and R5 tropic HIV-1. However, off-target effect, long-term protection, and toxicities were not assessed.

Targeting proviral DNA editing using ZFNs remains in its infancy. Though few groups tested the effect in cells, no animal studies were reported. This failure could be due to the barriers posed to deliver the ZFNs to organs carrying latent virus (Qu et al., 2013; Ji et al., 2018). Clinical trials using zinc finger nuclease technology were reported (Jo et al., 2015). Two clinical trials performed on HIV patients used either ZFN-modified T cells or hematopoietic stem cells (Tebas et al., 2014; DiGiusto et al., 2016). ZFN-modified T cells targeting CCR5 in one of twelve patients showed

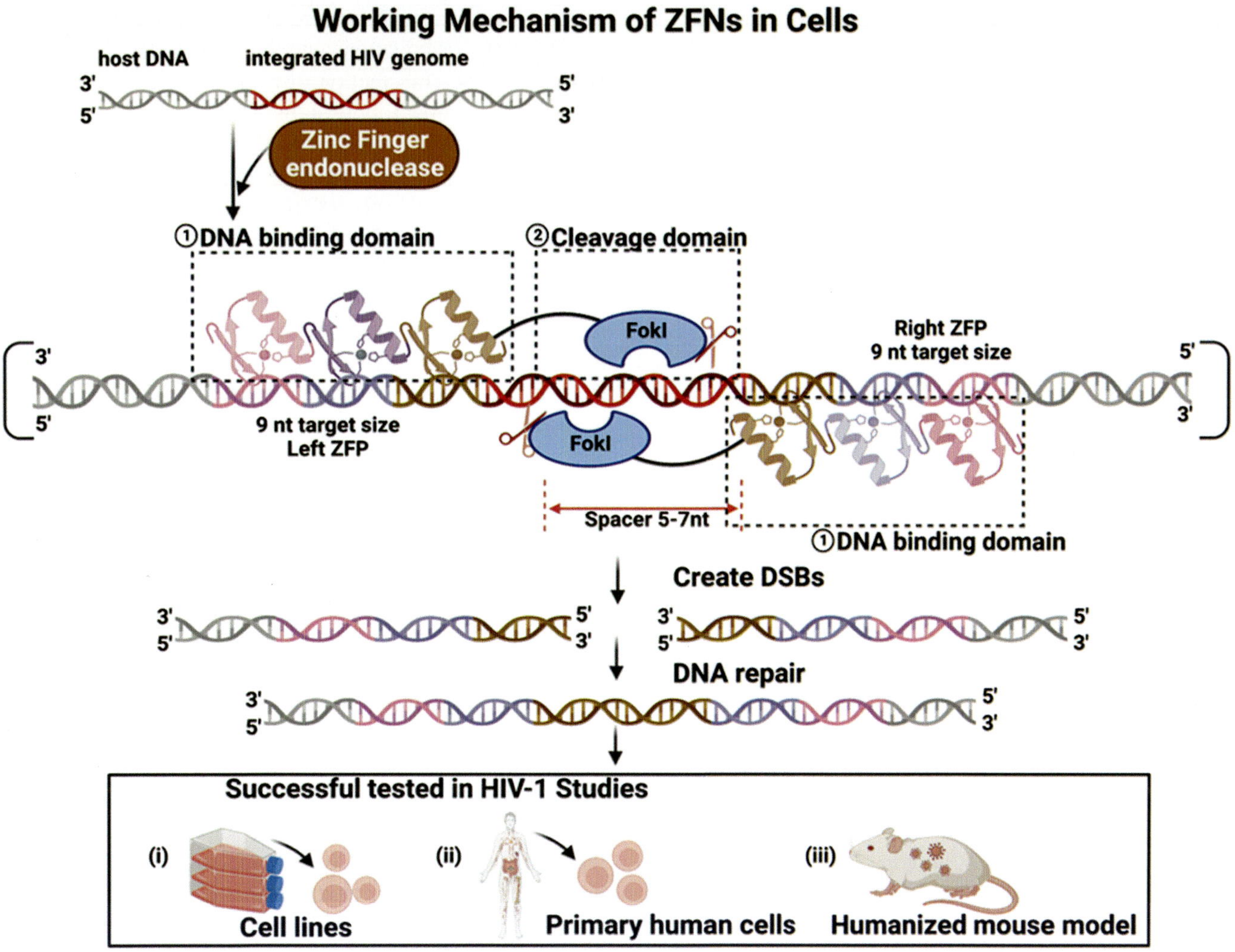

Fig. 55.2 Zinc Finger Nucleases (ZFNs). ZFNs work in pair and each ZFN is composed of two domains: zinc finger domain which is the binding domain and Fok I domain which is the excision domain. Zinc finger domain binds specifically to a set of three base pairs in the target DNA resulting in DNA double strand break at specific sites by nucleases. Host DNA repair machinery then gets involved in restoring the integrity of the newly edited double-stranded DNA. For HIV-1 studies, ZFNs have been successfully tested in cell lines, patient-derived primary cells, and in small animal models of HIV infection

adverse events (Tebas et al., 2014). ZFN-modified T cells were protected against HIV-1 infections for 48 weeks. Blood HIV DNA levels were found to be decreased in all patients. After the treatment interruption, HIV RNA was undetectable in one of four patients. Though the first clinical trial supported the safety of utilizing gene modified T cells, further investigation with additional patients is required despite encouraging results in mouse models (DiGiusto et al., 2016). ZFN-*CCR5* locus-modified human stem cells were found well tolerated with successful engraftment at twenty weeks. No evidence of tumorigenesis was observed. Minimal off-target effects with fewer biological consequences were detected at the CCR2 locus. Gene-edited human stem cell transplanted humanized mice were not yet tested for protection against HIV infection (DiGiusto et al., 2016).

55.7 TALENs for Gene Editing and HIV-1 Infection

TALEN is a naturally occurring gene-editing tool, which has shown efficacy in editing host genes including CCR5 and CXCR4, proviral integration related protein LEDGF/p75, and in the long terminal repeats (LTR) region of proviral HIV-DNA (Benjamin et al., 2016; Ho et al., 2018). Prior works have compared the gene-editing efficacy of ZFNs and TALENs for targeting human CCR5 gene. Data acquired found comparable levels of gene disruption (45%) in both the approaches. However, CCR2 off-target editing produced less cytotoxicity in TALEN-treated cells (Mussolino et al., 2011). These data provided proof-of-concept for this approach. Later studies affirmed these finding in primary and transformed T cells through mRNA transfection and peptide-mediated delivery (Ru et al., 2013; Liu et al., 2014; Mock et al., 2015).

However, due to the complexity and size of TALENs, packaging and delivery has become one obstacle for clinical translation. This is because cargo delivery that includes adeno-associated viral and lentiviral (AAV and LV) vectors have limited loading capacity. Moreover, the delivery methods such as mRNA and peptide-based systems can induce immune responses in patients. With the development of less immunogenic nonviral delivery methods that include liposomes, exosomes, and nanoparticles that can carry increased TALEN loaded cargos, it will be difficult to take AAV and LV gene editing to the next step of translation. The major challenges remain to be answered for these nuclease vectors to treat HIV, are their editing efficacy, off-target toxicities, and delivery efficacy for defined cargos.

55.8 CRISPR-Cas9 for Direct HIV-1 Excision

CRISPR-Cas9 has emerged as a revolutionary technology for the gene-editing field. HIV therapeutics have utilized it as a tool to excise integrated proviral HIV-1 DNA in in vitro and in vivo models and alter the susceptibility of host cells for viral elimination (Allers & Schneider, 2015; Ebina et al., 2013; Cornu et al., 2015; Xiao et al., 2019a; Li et al., 2022b; Zhang et al., 2023). Early attempts using CRISPR-based gene editing demonstrated promising results for HIV-1 provirus excision through optimal viral suppression. CRISPR-mediated disruption was reported that targeted the HIV transcriptional promotor utilizing lentiviral vector-treated cells (Ebina et al., 2013). Single guide RNAs targeting key LTR elements NF-κB and transactivation response element (TAR) region was tested separately demonstrating that TAR-targeting guide RNA was a potent suppressor of HIV-1 replication. TAR-targeting guide RNA CRISPR-Cas9 in HIV-1 latently infected cells showed reduction in latent proviral reactivation from 97.8% (vector control) to 35.5% (Cas9/guide RNA vector-treated). This was seen after 3 rounds of treatment. Gel electrophoretic analysis showing proviral DNA excision with one LTR footprint showed high specificity of the CRISPR-Cas9 system (Ebina et al., 2013). The data showed that CRISPR was able to target both the actively replicating and latent infected cells. However, the study achieved only limited excision even after multiple rounds of CRISPR treatment and without full testing of off-target effects. This highlighted the existing challenge in targeting all latently infected cells. Later studies affirmed the needs to target all infected cells by CRISPR-Cas9 (Hu et al., 2014a; Liao et al., 2015; Zhu et al., 2015). Studies of both single and dual gRNAs/Cas9 targeting the HIV-LTR regions in latently infected monocytic cells (Hu et al., 2014a) found that dual gRNAs/Cas9 system was more efficient in suppressing HIV transcription, reactivation, and in eliciting HIV-1 proviral DNA excision. Transfection of a dual gRNAs/Cas9 system into TZM-bI cells followed by HIV-1 challenge showed excision of HIV-DNA against HIV-1 infectious challenge. These data showed that CRISPR is capable of not only eliminating HIV from already infected cells but could also serve to prevent new infections. The same study examined predicted off-target toxicities and could not find any indel mutations in the human genome by CRISPR/Cas9 system. These results were touted as important for the field in affirming treatment safety for potential future clinical applications (Hu et al., 2014a). Yet another study screened 10 different gRNAs to look for excision utilizing HIV-1-infected Jurkat cell line and found that one gRNA targeting the Rev. region effectively reduced HIV gene expression. The study also emphasized that CRISPR was able to target both actively replicating and latently infected HIV cells at similar levels (Zhu et al., 2015), highlighting the importance of CRISPR in targeting the viral reservoir, as the latently infected cells can be eliminated without reactivation but still be efficiently targeted by the CRISPR system (Fig. 55.3).

Other investigations showed that CRISPR-Cas9 treatment can directly target viral genomic DNA [at the (preintegration stage)] and the integrated provirus in primary T cells. The findings supported excisions of the targeted HIV-1 sequence (Liao et al., 2015). The authors hypothesized that the virus could produce CRISPR-mediated escape mutants, as they found significant higher excision efficiency between a multiplex CRISPR-Cas9 than a single-gRNA CRISPR-Cas9 system. This idea was then validated by others (Wang et al., 2016b; Yoder & Bundschuh, 2016; Lebbink et al., 2017). The work also demonstrated that CRISPR/Cas9 can act as a double-edged sword in both inactivating HIV-1 but also facilitating escape mutants. The resulting escape mutants often carry indels or nucleotide substitutions in the Cas9 cleavage site within the target viral DNA. This most likely originated from the error-prone host DNA repair machinery post-cleavage then contributing to viral escape (Wang et al., 2016b, 2018d; Yoder & Bundschuh, 2016). Yet, other investigations (Yin et al., 2016; Kaminski et al., 2016a, b; Lebbink et al., 2017) showed CRISPR used against multiple gRNAs can prevent HIV-1 escape mutant generation and provide protection from subsequent infection by HIV. The HIV genome possesses two LTRs at both the ends and both have identical sequence in the 5′ and 3′ LTR region. By targeting the LTR-region and the nearby gag-region with dual gRNAs, the entire LTR region gets eliminated, making the virus replication-incompetent (Kaminski et al., 2016b). This is likely important as when CRISPR-mediated gRNA targets only a single locus of HIV, the virus can escape. If multiple gRNAs are used targeting different regions, then there are less chance for viral escape

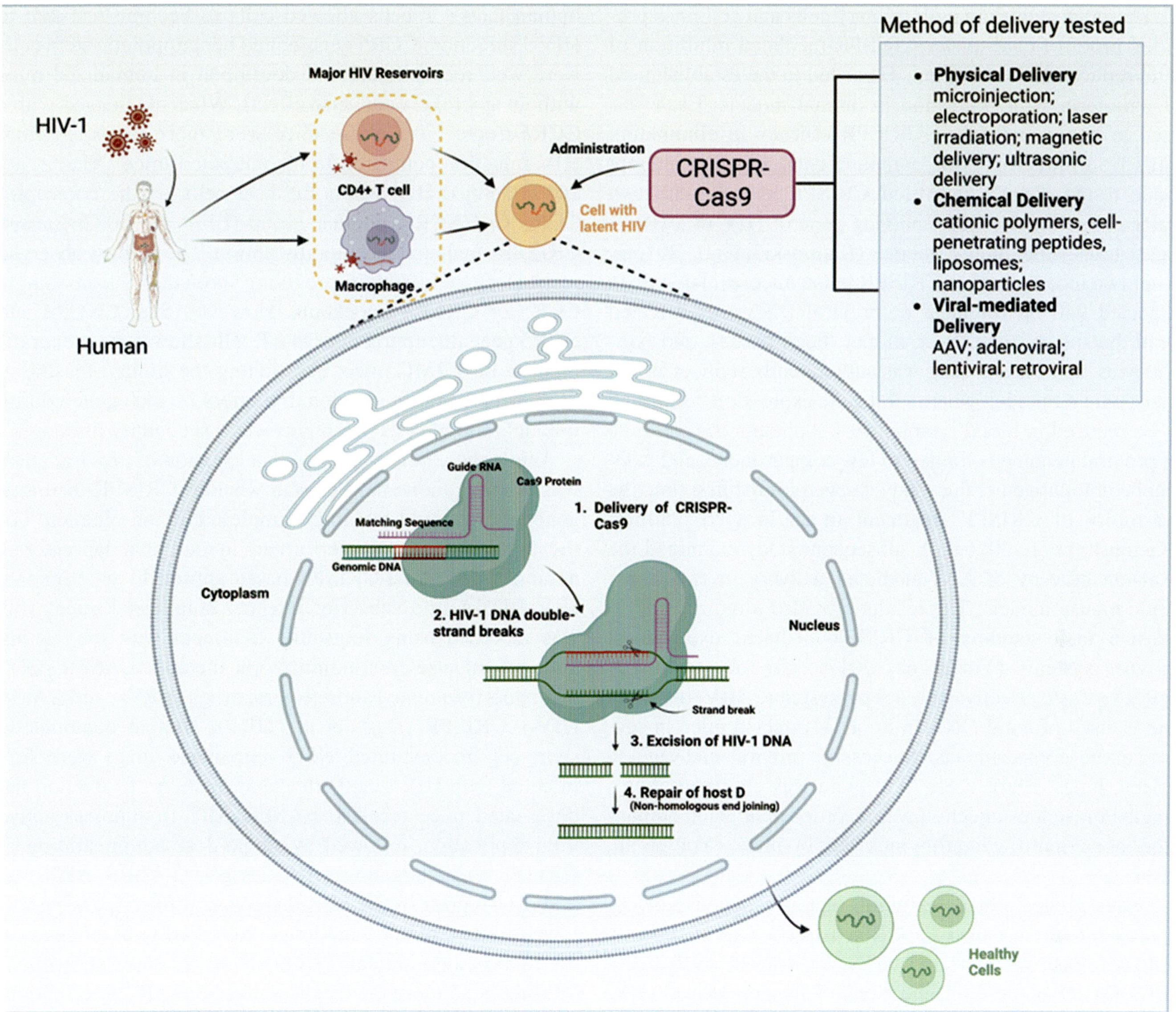

Fig. 55.3 Mechanism of CRISPR/Cas9-Mediated Excision. Clustered regulatory interspaced short palindromic repeat (CRISPR)-associated nuclease 9 (Cas9) system has emerged as a revolutionary tool for HIV-1 provirus excision and tested successfully for viral suppression studies in vitro/ex vivo. As shown in this figure, guide RNA (gRNA) is designed to recognize and specifically bind to the target sequence that directs the Cas9 enzyme (scissor protein) to locate where the excision will be made. Upon excision of the targeted DNA sequence, the host cell machinery takes care to repair the cut through cellular repair mechanisms, specifically using nonhomologous end-joining (NHEJ) pathway. Recent efforts made in HIV therapeutics have achieved success utilizing CRISPR-Cas9 tool to excise the integrated proviral HIV DNA from the primary infected cells derived from patients and cell lines without causing much off-target effects

(Lebbink et al., 2017). Rather than targeting the commonly targeted LTR region, more recent studies tested multiplex gRNAs targeting tat/rev region and achieved success (Ophinni et al., 2020; Herskovitz et al., 2021). Ophinni et al. (2020) first treated a T cell line by multiplexing three tat-targeting CRISPR gRNAs, followed by challenging the cells with HIV-1 and found protection effect for up to 45 days and then followed with a second virus infection challenge, showing for the first time the potential of CRISPR/Cas9 therapy for long-term HIV suppression (Ophinni et al., 2020). Works performed in our own laboratories designed highly conserved guide RNAs targeting five coding genes of HIV genome (TAT 1 and 2, REV 1 and 2, and gp41) and tested multiple delivery methods. This included lentiviral transduction, lipid nanoparticle delivery, and electroporation each used to deliver the guides to HIV-1 target cells. The newly designed lipid nanoparticle delivery system achieved close to complete viral excision efficacy in cell-based systems (Herskovitz et al., 2021). These studies highlight the importance of designing CRISPR gRNAs that target then inactivate the functional coding regions of the virus precluding viral escape without off-target toxicities.

Thorough studies in both primary cells and cell lines provided proof-of-concept for CRISPR-mediated inhibition of viral replication-inactivation. These led to the establishment of long-term viral protection in animal models. Each was used to better understand CRISPR efficacy in eliminating virus from infected tissue compartments. The first attempt made tested a AAV_9-delivered CRISPR system with two guides targeting the LTR and Gag gene of HIV in a HIV-1 Tg26 transgenic mouse model (Kaminski et al., 2016a). After two inoculations of CRISPR in the mice, excision of an expected 940 bp fragment of proviral DNA was detected from the spleen, lung, liver, kidney, heart tissues, and lymphocytes, and their finding was subsequently replicated in a transgenic rat model, where viral gene expression was found to be reduced in blood lymphocytes. Although the cleavage of proviral genome is limited to few compartments and AAV can be immunogenic, this study showed for the first time the feasibility of CRISPR treatment in an in vivo platform (Kaminski et al., 2016a). A subsequent study examined the excision efficacy of AAV-mediated delivery in two transgenic mouse models. The results provided a more comprehensive understanding of CRISPR-mediated excision in in vivo systems (Yin et al., 2017). Use of quadruplex sgRNAs/Cas9 effectively reduced systemic HIV infection and excised proviral DNA in EcoHIV-infected mice in several tissue compartments. Successful proviral cleavage in spleen, lungs, heart, colon, and brain was observed after a single intravenous injection was recorded in infected humanized bone marrow/liver/thymus (BLT) mice (Yin et al., 2017).

Other groups tested lentiviral vector-mediated delivery approach to deliver the CRISPR constructs. One such study utilized this method in NOD.CB17-Prkdcscid/NCrCrl (NOD/SCID) mice with injections of patient-derived HIV-infected peripheral blood mononuclear cells (PBMCs) followed by administration of multiple CRISPR guides targeting the LTR and Gag-region of HIV-1 (Bella et al., 2018). CRISPR-mediated excision was detected in PBMCs, spleen, lung, and liver, but not in the brains of these mice. Viral rescue assays performed in this study showed reduced viral growth only from peripheral blood PBMCs in these treated mice. Significant reduction in viral replication was not observed in spleen suggesting that challenges remain in CRISPR therapy to achieve complete HIV elimination from major viral sanctuaries and solid organ compartments (Bella et al., 2018).

Other than directly targeting viral genome, two recent works revealed an alternative HIV therapeutic approach utilizing CRISPR/Cas9 system by targeting HIV co-receptors in virus susceptible cells (Xiao et al., 2019a; Li et al., 2022b). Here a dual-gRNAs CRISPR system was used by targeting the HIV-1 co-receptor CCR5 in target CD4+ T cells and in humanized mice. Efficient editing of CCR5 in primary human CD4+ T cells allowed cells to become resistant to HIV-1 infection. CCR5 gene-edited hematopoietic stem cells were well reconstituted and developed in humanized mice without any toxic or adverse effects. When challenged with a CCR5-tropic virus, these mice were more protected from HIV infection compared to the untreated mice (Xiao et al., 2019a). Since HIV enters the host cells by the coreceptor CCR5 or CXCR4, another group (Li et al., 2022b) tested CRISPR-mediated receptor disruption. The authors observed resistance to infection providing protection from single CCR5 or CXCR4 knockouts. However, dual CXCR4 and CCR5 gene disruption in CD4+ T cells showed poor engraftment in hu-PBMC mice, questioning the limited feasibility of this approach. In addition, the safety of such gene-editing implantation of cells in humans was a secondary limitation.

While these studies proved that excision of proviral DNA was possible, none have proven whether CRISPR-mediated gene editing could facilitate complete HIV elimination. For the first time, studies performed in our own laboratories demonstrated that a CRISPR-based approach can eliminate HIV-1 DNA in a subset of infected humanized mice. This was recorded using sequential treatment with long-acting slow-effective release antiretroviral therapy (LASER ART) of 4 nano-formulated antiretroviral drugs (ARVs) and AAV9-HIV-1 CRISPR (Dash et al., 2019). Human hemopoietic stem cell reconstituted NSG humanized mice were first infected with HIV-1 and after confirmation of infection the designated mice received LASER ART to suppress active viral replication, followed by single-dose administration of CRISPR-Cas9 targeting HIV-LTR-gag. LASER ART was withdrawn prior to CRISPR therapy, and the mice were followed for two months and longer to record viral rebound. In the dual-treated animals, 8 out of the 23 mice showed no viral rebound more than eight weeks after ART interruption. No viral RNA or DNA was detected in any of the tissue compartments analyzed in those aviremic mice using sensitive assays with a detection limit of detecting one viral copy such as nested qPCR, digital droplet PCR, and RNAscope. Viral recovery assays confirmed viral sterilization. In vivo viral outgrowth assays employing adoptive transfer of cells isolated from spleen and bone marrow were performed from the aviremic donor mice and injecting the cells into new healthy recipient humanized mice. No HIV infection was detected in the recipient humanized mice, validating the clearance of viral reservoir in those cured animals (Dash et al., 2019). To extend these observations, we targeted both host CCR5 and HIV-LTR-Gag, using dual CRISPR systems by sequentially employing AAV6 and AAV9 delivery in ARV-suppressed HIV-infected humanized mice and showed complete HIV-1 elimination in 58% of the dual-treated animals (Dash et al., 2023). To date, these are the only two studies that have achieved a sterilizing HIV-1 cure using CRISPR-Cas9 and LASER ART (Fig. 55.4) and highlighting its potential for

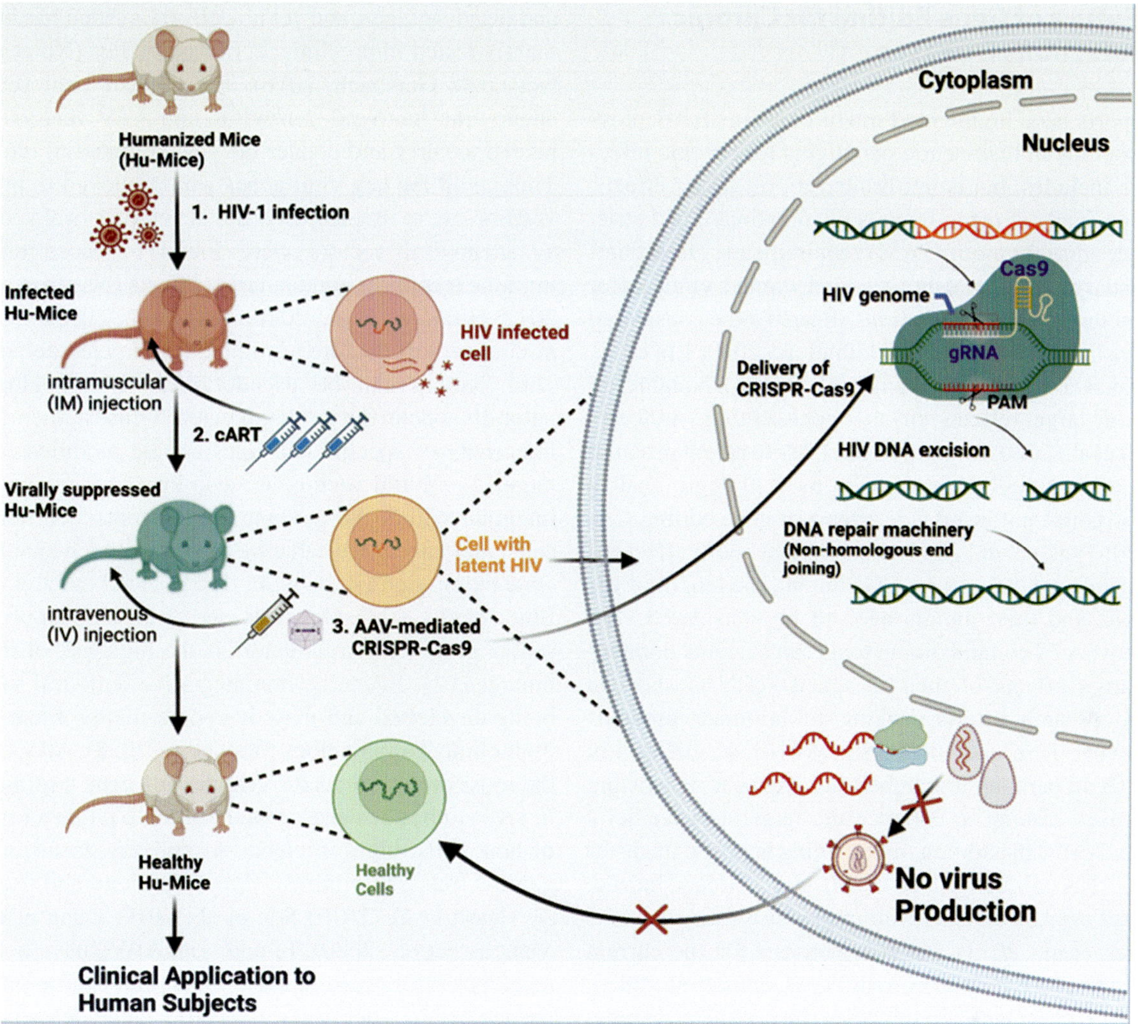

Fig. 55.4 CRISPR/Cas9-based gene therapy for HIV elimination. CRISPR/Cas9 gene editing targeting the HIV genome under ART suppression shows great potential for HIV cure. NSG-humanized mice infected with HIV-1 first received a potent combinatorial ART (cART) treatment to maintain a virally suppressed state. This was followed by administration of adeno associated virus (AAV)9-mediated CRISPR-Cas9 targeting HIV-LTR-Gag region in these virally suppressed Hu-mice to excise the integrated proviral DNA. HIV-specific guide RNAs recognized and bind to the target sites, followed by the cleavage at double strand by Cas9 which excised the integrated HIV-1 DNA from the host genome. Host DNA then got repaired by nonhomologous end joining at the cleavage site. This new sequential combination of cART and CRISPR-Cas9 has so far achieved successful viral elimination from a subset of HIV-infected humanized mice and has a great clinical translational potential in treating HIV-infected patients in near future

clinical translation. More work in the delivery of the CRISPR-loaded cargo is needed to increase HIV-elimination efficiency by improved cargo delivery.

Evidence of excision from CRISPR/Cas9 system shown in mouse models further moved its testing to nonhuman primates (NHPs). To this end, guide RNAs were designed to target specific regions of the SIV genome. The strategy first tested in PBMCs isolated from SIV-infected macaques showed up to 60% excision efficiency (Mancuso et al., 2020). When tested in three macaques by AAV-mediated delivery, the excision efficiency reached 37, 65, and 92% effectiveness. Excision was detected in blood, spleen, bone marrow, lungs, and lymph nodes of the AAV-CRISPR injected and infected NHPs. Decrease in viral SIV DNA was shown in multiple tissues, the highest recorded was a 95% reduction in the lymph nodes of treated animals. The data suggested the power of CRISPR in HIV-1 proviral cleavage. Although the sample size was small, the results remained encouraging. In all, the next step needs to be toward improving the delivery efficiency of CRISPR-loaded cargoes in order for the therapy to reach the major viral compartments as a requirement for clinical translation.

55.9 Future of Gene Editing for Chronic Infectious Diseases

Recent reports have highlighted major concerns for deploying gene therapy in therapeutic paradigms for chronic infections. This includes, but is not limited to, off-target effects, viral escape (genetic recombinants or mutations), and strict protospacer adjacent motif (PAM) requirements, DNA damage, secondary immune responses, and limited viral vector delivery methods. These limitations all need to be considered as real translational obstacles (Uddin et al., 2020; Lin et al., 2021; Hirakawa et al., 2020; Wang et al., 2016b). Notably, all point to off-target effects of the gene-editing strategies (Hirakawa et al., 2020; Lin et al., 2021). More recent research served to achieve sgRNA specificity as a strategic goal to effectively reduce untoward side effects of gene editing (Liu et al., 2020; Tycko et al., 2016). Factors that could affect the sgRNA specificity include the GC content, the length of the guide RNA, and these points need to be considered. The usage of advanced computational tools can evaluate both on- and off-target effects of their designed sgRNAs and can assist the research efforts seeking to optimize products before moving them to clinical testing (Li et al., 2022a; Liu et al., 2020). In parallel, researchers are engineering existing Cas nucleases, aiming to enhance the targeting specificity (Ran et al., 2013). In addition, maintaining appropriate doses of Cas9-sgRNA can also reduce the off-target effects. Indeed, excess Cas9 expression can result in higher off-target events (Pattanayak et al., 2013). Another concern for the current gene-editing approach is irreversibility (Uddin et al., 2020; Lin et al., 2021). Once administered in a clinical setting, there is no way to switch it off, so any off-target events generated would introduce permanent changes to the host genome, raising practical challenges for the researchers. Fortunately, the advances made in the last decade in the next-generation and deep sequencing technologies have enabled the discovery of potential off-target sites at higher resolution, allowing for better assessment of guide RNA specificity and CRISPR-Cas9 safety in quick time (Jiang et al., 2017; Li et al., 2017). Beyond these limitations, ethical issues regarding the application of germline gene editing have also been comprehensively discussed because of the potential for serious injury, disability, and misapplications if utilized without proper surveillance (Brokowski & Adli, 2019; Brokowski, 2018). In the case of the first report of gene-edited babies, the scientist who used CRISPR-Cas9 to edit the embryo was sentenced for illegal medical practice (Cyranoski, 2019).

Optimizing the delivery method of the CRISPR-loaded cargo is another way to reduce off-target effects and increase CRISPR editing efficiency. Delivery modalities used till date for gene editing generally consist of viral and nonviral-mediated strategies, each manifesting its own advantages and disadvantages, and yet no delivery system has been dominantly tested in preclinical/clinical studies (Xu et al., 2019; Nelson & Gersbach, 2016). The current viral vectors like adenoviral, lentiviral, retroviral, and AAV vectors are engineered to carry and deliver the genetic material into the cell. Though all the key viral genes were removed to make these vectors safe to use, some earlier developed viral vectors such as adenoviral vectors were found to cause undesirable immune responses and off-target effects (Nayerossadat et al., 2012; Hirakawa et al., 2020; Tong et al., 2019). Unlike adenoviral vectors that are not suited for targeted delivery, lentiviral vectors can be pseudotyped with envelopes from naturally occurring or engineered viruses, therefore their infectivity of specific cell lines can be modified to suit the target. Lentiviral vectors are also known for their ability to integrate within the genome, stably transduce nondividing cells, and have large carrying capacity, which enables multigene expression (Bella et al., 2018; Herskovitz et al., 2021; Ebina et al., 2013). However, one of the biggest problems of utilizing the lentiviral vectors is the high risk of insertional mutagenesis. Recently, nonintegrative lentiviral vectors are being developed and may overcome this problem to widen their clinical applications (Yew et al., 2022). AAVs are so far the most preferred vector vehicles for gene editing in many in vivo studies because of their ability to target a broad range of host cells, high efficiency of delivery, insertional mutagenesis free, and minimal immunogenicity and tissue toxicity (Dash et al., 2019; Sun et al., 2019; Duan et al., 2022; Mancuso et al., 2020). In addition, AAVs have a variety of serotypes with broad tissue tropism that can be selected for tissue-specific or species-specific gene-delivery studies (Zincarelli et al., 2008; Srivastava, 2016). Though it is the most common method used for preclinical models of in vivo CRISPR delivery, one of the major drawback of AAV-mediated vectors is their limited loading capacity (less than 4.7 kb) (Grieger & Samulski, 2005), which is a major challenge for its use in delivery of sgRNA and Cas9 simultaneously into the cells. In addition, the detection of preexisting neutralizing antibodies against AAVs in the nonhuman primates and humans can weaken the therapeutic effects of AAV-mediated gene therapy (Selot et al., 2014). Despite these limitations, AAVs remain the most important gene delivery vector candidate in clinical trials for its expression stability and biosafety in in vivo system (Mendell et al., 2021).

Nonviral delivery strategies tested so far falls into two distinct categories: *first* the physical delivery (which includes microinjection, electroporation, laser irradiation, magnetic delivery, ultrasonic delivery, and high-pressure tail intravenous injections, and more) and *second* the chemical delivery (use of lipids and cationic polymers, cell-penetrating peptides, liposomes-mediated delivery and nanoparticle-mediated delivery and more) (Mellott et al., 2013; Nelson & Gersbach, 2016). Physical delivery has its unique advantages in allowing the entry of gene-editing cargo into any cell regardless of the difficulty of transfection. As an example, during electroporation it opens the pores in the membrane of cells transiently via high-voltage electrical currents, sending the gene-editing cargo easily into the cells. The cons of this delivery strategy lie in the disruption of a cell's normal function; thus, it is not suitable for most of the mammalian cells that are current sensitive. Like electroporation, most of the physical delivery methods requires better tolerance by cells/tissues, thus they have been used very less in in vivo systems and mostly tested for tissue engineering purpose in in vitro or ex vivo studies.

Chemical delivery mediators are designed by natural or synthetic materials. Chemical vectors include liposomes (Iida et al., 2022; Wang et al., 2018c), modified-polymers (Gao et al., 2020; Wang et al., 2018a; Liu et al., 2019), and nanoparticles combat a broad number of diseases. Compared to viral vectors, these nonviral vectors have proven to cause lower level of immunogenicity and cytotoxicity. Moreover, it is also easier to incorporate flexible changes to the chemical composition and structure of nonviral agents for different cells/tissue targeting scheme. However, the main hurdle for nonviral vector delivery is the low delivery efficiency (Ramamoorth & Narvekar, 2015). Unlike the viral vectors possessing the natural ability to overcome systemic and local gene delivery barrier, nonviral delivery cargos need to overcome multiple barriers which involve shedding and degradation steps (from circulatory degradation, intercellular trafficking to intracellular trafficking barrier for example lysosomal degradation pathway) (Nelson & Gersbach, 2016). In addition, many nanoparticles can aggregate and can be toxic at high concentrations such as gold nanoparticles which raises safety concern and the long-term toxicity (Arvizo et al., 2010). In the future, nonviral delivery cargos need to be developed in such a way that could maximally overcome the limitations mentioned above especially having better ability to escape from the cell lysosomal degradation pathway. In general, chemical delivery mediators have minimal immunogenicity, low cost, high loading capacity, and high biocompatibility. This makes them the ideal next-generation delivery vehicles from in vitro to in vivo studies. Meantime, potential toxicity issues such as particle aggregation-mediated tissue damage needs to be addressed when higher and multiple doses of nonviral vectors are considered (Foldvari et al., 2016; Durymanov & Reineke, 2018).

Despite the success of gene editing for treating diseases and using the approaches to modify genomic organization, cargo delivery remains a challenge. Future nonviral vectors need to be developed in such a way that the total gene-editing payload (DNA/RNA) contents should be protected from large-scale degradation by plasma endonucleases, cellular endosomes, and lysosomes. Meanwhile, the overall structure carrying the gene-editing cargo also needs to be more stable to avoid shedding when crossing various cellular and intracellular barriers. Although the development of non-viral vectors for gene-editing cargo delivery is still at an infancy but growing stage in terms of tissue specificity and efficacy, the future of nonviral vector delivery is bright due to its lower cost and unique convenience. Taken together, the delivery modalities of CRISPR-Cas9 and other gene-editing engineered nucleases can greatly influence the efficacy and safety to treat many infectious diseases for which cure is still distantly achievable. Viral vectors, especially AAVs, are still being utilized as the go-to carriers of gene-editing tools for preclinical research and clinical trials, but novel nonviral vectors like lipid nano particles which can target most of the viral reservoirs including brain will be the future delivery vehicles for clinal translation and for the treatment and elimination of chronic viral infections.

Acknowledgments This work was supported by the National Institute of Health (NIH) grant R01MH121402, R56 HL1516002 and R01 HL164306. The funders have no role in the preparation of this book chapter. The figures for the book chapter were created using Biorender.com software purchased through the support provided by the Center for Heart and Vascular Research of the University of Nebraska Medical Center.

References

Adli, M. (2018). The CRISPR tool kit for genome editing and beyond. *Nature Communications, 9*, 1911.

Aghamiri, S., Talaei, S., Roshanzamiri, S., Zandsalimi, F., Fazeli, E., Aliyu, M., Kheiry Avarvand, O., Ebrahimi, Z., Keshavarz-Fathi, M., & Ghanbarian, H. (2020). Delivery of genome editing tools: A promising strategy for HPV-related cervical malignancy therapy. *Expert Opinion on Drug Delivery, 17*, 753–766.

Allen, F., Crepaldi, L., Alsinet, C., Strong, A. J., Kleshchevnikov, V., De Angeli, P., Páleníková, P., Khodak, A., Kiselev, V., Kosicki, M., Bassett, A. R., Harding, H., Galanty, Y., Muñoz-Martínez, F., Metzakopian, E., Jackson, S. P., & Parts, L. (2018). Predicting the mutations generated by repair of Cas9-induced double-strand breaks. *Nature Biotechnology*.

Allers, K., & Schneider, T. (2015). CCR5Δ32 mutation and HIV infection: Basis for curative HIV therapy. *Current Opinion in Virology, 14*, 24–29.

Althoff, K. N., Justice, A. C., Gange, S. J., Deeks, S. G., Saag, M. S., Silverberg, M. J., Gill, M. J., Lau, B., Napravnik, S., Tedaldi, E., Klein, M. B., & Gebo, K. A. (2010). Virologic and immunologic response to HAART, by age and regimen class. *AIDS, 24*, 2469–2479.

Arvizo, R., Bhattacharya, R., & Mukherjee, P. (2010). Gold nanoparticles: Opportunities and challenges in nanomedicine. *Expert Opinion on Drug Delivery, 7*, 753–763.

Atkins, A. J., Allen, A. G., Dampier, W., Haddad, E. K., Nonnemacher, M. R., & Wigdahl, B. (2021). HIV-1 cure strategies: Why CRISPR? *Expert Opinion on Biological Therapy, 21*, 781–793.

Barrangou, R. (2015). The roles of CRISPR-Cas systems in adaptive immunity and beyond. *Current Opinion in Immunology, 32*, 36–41.

Bartus, R. T., Weinberg, M. S., & Samulski, R. J. (2014). Parkinson's disease gene therapy: Success by design meets failure by efficacy. *Molecular Therapy, 22*, 487–497.

Bella, R., Kaminski, R., Mancuso, P., Young, W. B., Chen, C., Sariyer, R., Fischer, T., Amini, S., Ferrante, P., Jacobson, J. M., Kashanchi, F., & Khalili, K. (2018). Removal of HIV DNA by CRISPR from patient blood engrafts in humanized mice. *Molecular Therapy - Nucleic Acids, 12*, 275–282.

Benjamin, R., Berges, B. K., Solis-Leal, A., Igbinedion, O., Strong, C. L., & Schiller, M. R. (2016). TALEN gene editing takes aim on HIV. *Human Genetics, 135*, 1059–1070.

Bi, Y., Sun, L., Gao, D., Ding, C., Li, Z., Li, Y., Cun, W., & Li, Q. (2014). High-efficiency targeted editing of large viral genomes by RNA-guided nucleases. *PLoS Pathogens, 10*, e1004090.

Bitinaite, J., Wah, D. A., Aggarwal, A. K., & Schildkraut, I. (1998). FokI dimerization is required for DNA cleavage. *Proceedings of the National Academy of Sciences of the United States of America, 95*, 10570–10575.

Bogdanove, A. J., & Voytas, D. F. (2011). TAL effectors: Customizable proteins for DNA targeting. *Science, 333*, 1843–1846.

Brokowski, C. (2018). Do CRISPR germline ethics statements cut it? *CRISPR Journal, 1*, 115–125.

Brokowski, C., & Adli, M. (2019). CRISPR ethics: Moral considerations for applications of a powerful tool. *Journal of Molecular Biology, 431*, 88–101.

Cermak, T., Doyle, E. L., Christian, M., Wang, L., Zhang, Y., Schmidt, C., Baller, J. A., Somia, N. V., Bogdanove, A. J., & Voytas, D. F. (2011). Efficient design and assembly of custom TALEN and other TAL effector-based constructs for DNA targeting. *Nucleic Acids Research, 39*, e82.

Chayavichitsilp, P., Buckwalter, J. V., Krakowski, A. C., & Friedlander, S. F. (2009). Herpes simplex. *Pediatrics in Review, 30*, 119–129. quiz 130.

Christian, M., Cermak, T., Doyle, E. L., Schmidt, C., Zhang, F., Hummel, A., Bogdanove, A. J., & Voytas, D. F. (2010). Targeting DNA double-strand breaks with TAL effector nucleases. *Genetics, 186*, 757–761.

Control, C. F. D., & Prevention. (2021). *HIV surveillance report, 2019*, Vol. 32.

Cornu, T. I., Mussolino, C., Bloom, K., & Cathomen, T. (2015). Editing CCR5: A novel approach to HIV gene therapy. *Advances in Experimental Medicine and Biology, 848*, 117–130.

Cornu, T. I., Mussolino, C., & Cathomen, T. (2017). Refining strategies to translate genome editing to the clinic. *Nature Medicine, 23*, 415–423.

Cyranoski, D. (2019). The CRISPR-baby scandal: what's next for human gene-editing. *Nature, 566*, 440–442.

Das, A. T., Binda, C. S., & Berkhout, B. (2019). Elimination of infectious HIV DNA by CRISPR-Cas9. *Current Opinion in Virology, 38*, 81–88.

Dash, P. K., Kaminski, R., Bella, R., Su, H., Mathews, S., Ahooyi, T. M., Chen, C., Mancuso, P., Sariyer, R., Ferrante, P., Donadoni, M., Robinson, J. A., Sillman, B., Lin, Z., Hilaire, J. R., Banoub, M., Elango, M., Gautam, N., Mosley, R. L., Poluektova, L. Y., Mcmillan, J., Bade, A. N., Gorantla, S., Sariyer, I. K., Burdo, T. H., Young, W. B., Amini, S., Gordon, J., Jacobson, J. M., Edagwa, B., Khalili, K., & Gendelman, H. E. (2019). Sequential LASER ART and CRISPR treatments eliminate HIV-1 in a subset of infected humanized mice. *Nature Communications, 10*, 2753.

Dash, P. K., Kevadiya, B. D., Su, H., Banoub, M. G., & Gendelman, H. E. (2020). Pathways towards human immunodeficiency virus elimination. *eBioMedicine, 53*, 102667.

Dash, P. K., Alomar, F. A., Cox, J. L., Mcmillan, J., Hackfort, B. T., Makarov, E., Morsey, B., Fox, H. S., Gendelman, H. E., Gorantla, S., & Bidasee, K. R. (2021). A link between methylglyoxal and heart failure during HIV-1 infection. *Frontiers in Cardiovascular Medicine, 8*, 792180.

Dash, P. K., Chen, C., Kaminski, R., Su, H., Mancuso, P., Sillman, B., Zhang, C., Liao, S., Sravanam, S., Liu, H., Waight, E., Guo, L., Mathews, S., Sariyer, R., Mosley, R. L., Poluektova, L. Y., Caocci, M., Amini, S., Gorantla, S., Burdo, T. H., Edagwa, B., Gendelman, H. E., & Khalili, K. (2023). CRISPR editing of CCR5 and HIV-1 facilitates viral elimination in antiretroviral drug-suppressed virus-infected humanized mice. *Proceedings of the National Academy of Sciences of the United States of America, 120*, e2217887120.

Digiusto, D. L., Cannon, P. M., Holmes, M. C., Li, L., Rao, A., Wang, J., Lee, G., Gregory, P. D., Kim, K. A., Hayward, S. B., Meyer, K., Exline, C., Lopez, E., Henley, J., Gonzalez, N., Bedell, V., Stan, R., & Zaia, J. A. (2016). Preclinical development and qualification of ZFN-mediated CCR5 disruption in human hematopoietic stem/progenitor cells. *Molecular Therapy - Methods & Clinical Development, 3*, 16067.

Ding, Q., Strong, A., Patel, K. M., Ng, S. L., Gosis, B. S., Regan, S. N., Cowan, C. A., Rader, D. J., & Musunuru, K. (2014). Permanent alteration of PCSK9 with in vivo CRISPR-Cas9 genome editing. *Circulation Research, 115*, 488–492.

Dong, C., Qu, L., Wang, H., Wei, L., Dong, Y., & Xiong, S. (2015). Targeting hepatitis B virus cccDNA by CRISPR/Cas9 nuclease efficiently inhibits viral replication. *Antiviral Research, 118*, 110–117.

Doorbar, J., Quint, W., Banks, L., Bravo, I. G., Stoler, M., Broker, T. R., & Stanley, M. A. (2012). The biology and life-cycle of human papillomaviruses. *Vaccine, 30*(Suppl 5), F55–F70.

Duan, Y., Ye, T., Qu, Z., Chen, Y., Miranda, A., Zhou, X., Lok, K. C., Chen, Y., Fu, A. K. Y., Gradinaru, V., & Ip, N. Y. (2022). Brain-wide Cas9-mediated cleavage of a gene causing familial Alzheimer's disease alleviates amyloid-related pathologies in mice. *Nature Biomedical Engineering, 6*, 168–180.

Durymanov, M., & Reineke, J. (2018). Non-viral delivery of nucleic acids: Insight into mechanisms of overcoming intracellular barriers. *Frontiers in Pharmacology, 9*, 971.

Ebina, H., Misawa, N., Kanemura, Y., & Koyanagi, Y. (2013). Harnessing the CRISPR/Cas9 system to disrupt latent HIV-1 provirus. *Scientific Reports, 3*, 2510.

Foldvari, M., Chen, D. W., Nafissi, N., Calderon, D., Narsineni, L., & Rafiee, A. (2016). Non-viral gene therapy: Gains and challenges of non-invasive administration methods. *Journal of Controlled Release, 240*, 165–190.

Gandy, S., & Dekosky, S. T. (2013). Toward the treatment and prevention of Alzheimer's disease: Rational strategies and recent progress. *Annual Review of Medicine, 64*, 367–383.

Gao, X., Jin, Z., Tan, X., Zhang, C., Zou, C., Zhang, W., Ding, J., Das, B. C., Severinov, K., Hitzeroth, I., Debata, P. R., He, D., Ma, X., Tian, X., Gao, Q., Wu, J., Tian, R., Cui, Z., Fan, W., Huang, Z., Cao, C., Bao, Y., Tan, S., & Hu, Z. (2020). Hyperbranched poly(β-amino ester) based polyplex nanopaticles for delivery of CRISPR/Cas9 system and treatment of HPV infection associated cervical cancer. *Journal of Controlled Release, 321*, 654–668.

Grieger, J. C., & Samulski, R. J. (2005). Packaging capacity of adeno-associated virus serotypes: Impact of larger genomes on infectivity and postentry steps. *Journal of Virology, 79*, 9933–9944.

Herskovitz, J., Hasan, M., Patel, M., Blomberg, W. R., Cohen, J. D., Machhi, J., Shahjin, F., Mosley, R. L., Mcmillan, J., Kevadiya, B. D., & Gendelman, H. E. (2021). CRISPR-Cas9 mediated Exonic disruption for HIV-1 elimination. *eBioMedicine, 73*, 103678.

Herskovitz, J., Hasan, M., Patel, M., Kevadiya, B. D., & Gendelman, H. E. (2022). Pathways toward a functional HIV-1 cure: Balancing promise and perils of CRISPR therapy. *Methods in Molecular Biology, 2407*, 429–445.

Hirakawa, M. P., Krishnakumar, R., Timlin, J. A., Carney, J. P., & Butler, K. S. (2020). Gene editing and CRISPR in the clinic: Current and future perspectives. *Bioscience Reports, 40*.

Ho, B. X., Loh, S. J. H., Chan, W. K., & Soh, B. S. (2018). In vivo genome editing as a therapeutic approach. *International Journal of Molecular Sciences, 19*.

Holt, N., Wang, J., Kim, K., Friedman, G., Wang, X., Taupin, V., Crooks, G. M., Kohn, D. B., Gregory, P. D., Holmes, M. C., & Cannon, P. M. (2010). Human hematopoietic stem/progenitor cells modified by zinc-finger nucleases targeted to CCR5 control HIV-1 in vivo. *Nature Biotechnology, 28*, 839–847.

Hsu, P. D., Lander, E. S., & Zhang, F. (2014). Development and applications of CRISPR-Cas9 for genome engineering. *Cell, 157*, 1262–1278.

Hu, W., Kaminski, R., Yang, F., Zhang, Y., Cosentino, L., Li, F., Luo, B., Alvarez-Carbonell, D., Garcia-Mesa, Y., Karn, J., Mo, X., & Khalili, K. (2014a). RNA-directed gene editing specifically eradicates latent and prevents new HIV-1 infection. *Proceedings of the National Academy of Sciences of the United States of America, 111*, 11461–11466.

Hu, Z., Yu, L., Zhu, D., Ding, W., Wang, X., Zhang, C., Wang, L., Jiang, X., Shen, H., He, D., Li, K., Xi, L., Ma, D., & Wang, H. (2014b). Disruption of HPV16-E7 by CRISPR/Cas system induces apoptosis and growth inhibition in HPV16 positive human cervical cancer cells. *BioMed Research International, 2014*, 612823.

Iida, K., Tsuchiya, A., Tamura, M., Yamamoto, K., Kawata, S., Ishihara-Sugano, M., Kato, M., Kitamura, T., & Goyama, S. (2022). Runx1 inhibition using lipid nanoparticle-mediated silencing RNA delivery as an effective treatment for acute Leukemias. *Experimental Hematology, 112–113*, 1–8.

Jaiswal, S., Natarajan, P., Silver, A. J., Gibson, C. J., Bick, A. G., Shvartz, E., Mcconkey, M., Gupta, N., Gabriel, S., Ardissino, D., Baber, U., Mehran, R., Fuster, V., Danesh, J., Frossard, P., Saleheen, D., Melander, O., Sukhova, G. K., Neuberg, D., Libby, P., Kathiresan, S., & Ebert, B. L. (2017). Clonal hematopoiesis and risk of atherosclerotic cardiovascular disease. *The New England Journal of Medicine, 377*, 111–121.

Jensen, T. L., Gøtzsche, C. R., & Woldbye, D. P. D. (2021). Current and future prospects for gene therapy for rare genetic diseases affecting the brain and spinal cord. *Frontiers in Molecular Neuroscience, 14*, 695937.

Ji, H., Lu, P., Liu, B., Qu, X., Wang, Y., Jiang, Z., Yang, X., Zhong, Y., Yang, H., Pan, H., Zhao, L., Xu, J., Lu, H., & Zhu, H. (2018). Zinc-finger nucleases induced by HIV-1 tat excise HIV-1 from the host genome in infected and latently infected cells. *Molecular Therapy - Nucleic Acids, 12*, 67–74.

Jiang, C., Mei, M., Li, B., Zhu, X., Zu, W., Tian, Y., Wang, Q., Guo, Y., Dong, Y., & Tan, X. (2017). A non-viral CRISPR/Cas9 delivery system for therapeutically targeting HBV DNA and pcsk9 in vivo. *Cell Research, 27*, 440–443.

Jo, Y. I., Kim, H., & Ramakrishna, S. (2015). Recent developments and clinical studies utilizing engineered zinc finger nuclease technology. *Cellular and Molecular Life Sciences, 72*, 3819–3830.

Joung, J. K., & Sander, J. D. (2013). Talens: A widely applicable technology for targeted genome editing. *Nature Reviews. Molecular Cell Biology, 14*, 49–55.

Jubair, L., Fallaha, S., & Mcmillan, N. A. J. (2019). Systemic delivery of CRISPR/Cas9 targeting HPV oncogenes is effective at eliminating established Tumors. *Molecular Therapy, 27*, 2091–2099.

Kaminski, R., Bella, R., Yin, C., Otte, J., Ferrante, P., Gendelman, H. E., Li, H., Booze, R., Gordon, J., Hu, W., & Khalili, K. (2016a). Excision of HIV-1 DNA by gene editing: A proof-of-concept in vivo study. *Gene Therapy, 23*, 690–695.

Kaminski, R., Chen, Y., Fischer, T., Tedaldi, E., Napoli, A., Zhang, Y., Karn, J., Hu, W., & Khalili, K. (2016b). Elimination of HIV-1 genomes from human T-lymphoid cells by CRISPR/Cas9 gene editing. *Scientific Reports, 6*, 22555.

Kennedy, E. M., Kornepati, A. V., Goldstein, M., Bogerd, H. P., Poling, B. C., Whisnant, A. W., Kastan, M. B., & Cullen, B. R. (2014). Inactivation of the human papillomavirus E6 or E7 gene in cervical carcinoma cells by using a bacterial CRISPR/Cas RNA-guided endonuclease. *Journal of Virology, 88*, 11965–11972.

Kennedy, E. M., Bassit, L. C., Mueller, H., Kornepati, A. V. R., Bogerd, H. P., Nie, T., Chatterjee, P., Javanbakht, H., Schinazi, R. F., & Cullen, B. R. (2015). Suppression of hepatitis B virus DNA accumulation in chronically infected cells using a bacterial CRISPR/Cas RNA-guided DNA endonuclease. *Virology, 476*, 196–205.

Kim, Y. G., & Chandrasegaran, S. (1994). Chimeric restriction endonuclease. *Proceedings of the National Academy of Sciences of the United States of America, 91*, 883–887.

Kim, Y. G., Cha, J., & Chandrasegaran, S. (1996). Hybrid restriction enzymes: Zinc finger fusions to Fok I cleavage domain. *Proceedings of the National Academy of Sciences of the United States of America, 93*, 1156–1160.

Kim, H. J., Lee, H. J., Kim, H., Cho, S. W., & Kim, J. S. (2009). Targeted genome editing in human cells with zinc finger nucleases constructed via modular assembly. *Genome Research, 19*, 1279–1288.

Kukhanova, M. K., Korovina, A. N., & Kochetkov, S. N. (2014). Human herpes simplex virus: Life cycle and development of inhibitors. *Biochemistry (Mosc), 79*, 1635–1652.

Lane, C. A., Hardy, J., & Schott, J. M. (2018). Alzheimer's disease. *European Journal of Neurology, 25*, 59–70.

Lao, Y. H., Li, M., Gao, M. A., Shao, D., Chi, C. W., Huang, D., Chakraborty, S., Ho, T. C., Jiang, W., Wang, H. X., Wang, S., & Leong, K. W. (2018). HPV oncogene manipulation using nonvirally delivered CRISPR/Cas9 or Natronobacterium gregoryi Argonaute. *Advanced Science (Weinh), 5*, 1700540.

Lebbink, R. J., De Jong, D. C., Wolters, F., Kruse, E. M., Van Ham, P. M., Wiertz, E. J., & Nijhuis, M. (2017). A combinational CRISPR/Cas9 gene-editing approach can halt HIV replication and prevent viral escape. *Scientific Reports, 7*, 41968.

Lee, J., Bayarsaikhan, D., Arivazhagan, R., Park, H., Lim, B., Gwak, P., Jeong, G. B., Lee, J., Byun, K., & Lee, B. (2019). CRISPR/Cas9 edited sRAGE-MSCs protect neuronal death in Parkinson's disease model. *International Journal of Stem Cells, 12*, 114–124.

Li, L., Krymskaya, L., Wang, J., Henley, J., Rao, A., Cao, L. F., Tran, C. A., Torres-Coronado, M., Gardner, A., Gonzalez, N., Kim, K., Liu, P. Q., Hofer, U., Lopez, E., Gregory, P. D., Liu, Q., Holmes, M. C., Cannon, P. M., Zaia, J. A., & Digiusto, D. L. (2013). Genomic editing of the HIV-1 coreceptor CCR5 in adult hematopoietic stem and progenitor cells using zinc finger nucleases. *Molecular Therapy, 21*, 1259–1269.

Li, H., Sheng, C., Wang, S., Yang, L., Liang, Y., Huang, Y., Liu, H., Li, P., Yang, C., Yang, X., Jia, L., Xie, J., Wang, L., Hao, R., Du, X., Xu, D., Zhou, J., Li, M., Sun, Y., Tong, Y., Li, Q., Qiu, S., & Song, H. (2017). Removal of integrated hepatitis B virus DNA using CRISPR-Cas9. *Frontiers in Cellular and Infection Microbiology, 7*, 91.

Li, C., Chu, W., Gill, R. A., Sang, S., Shi, Y., Hu, X., Yang, Y., Zaman, Q. U., & Zhang, B. (2022a). Computational tools and resources for CRISPR/Cas genome editing. *Genomics, Proteomics & Bioinformatics, 21*, 108–126.

Li, S., Holguin, L., & Burnett, J. C. (2022b). CRISPR-Cas9-mediated gene disruption of HIV-1 co-receptors confers broad resistance to infection in human T cells and humanized mice. *Molecular Therapy - Methods & Clinical Development, 24*, 321–331.

Liao, H. K., Gu, Y., Diaz, A., Marlett, J., Takahashi, Y., Li, M., Suzuki, K., Xu, R., Hishida, T., Chang, C. J., Esteban, C. R., Young, J., & Izpisua Belmonte, J. C. (2015). Use of the CRISPR/Cas9 system as an intracellular defense against HIV-1 infection in human cells. *Nature Communications, 6*, 6413.

Lin, S. R., Yang, H. C., Kuo, Y. T., Liu, C. J., Yang, T. Y., Sung, K. C., Lin, Y. Y., Wang, H. Y., Wang, C. C., Shen, Y. C., Wu, F. Y., Kao, J. H., Chen, D. S., & Chen, P. J. (2014). The CRISPR/Cas9 system facilitates clearance of the intrahepatic HBV templates in vivo. *Molecular Therapy - Nucleic Acids, 3*, e186.

Lin, H., Li, G., Peng, X., Deng, A., Ye, L., Shi, L., Wang, T., & He, J. (2021). The use of CRISPR/Cas9 as a tool to study human infectious viruses. *Frontiers in Cellular and Infection Microbiology, 11*, 590989.

Liu, J., Gaj, T., Patterson, J. T., Sirk, S. J., & Barbas, C. F., 3rd. (2014). Cell-penetrating peptide-mediated delivery of TALEN proteins via bioconjugation for genome engineering. *PLoS One, 9*, e85755.

Liu, C., Wan, T., Wang, H., Zhang, S., Ping, Y., & Cheng, Y. (2019). A boronic acid-rich dendrimer with robust and unprecedented efficiency for cytosolic protein delivery and CRISPR-Cas9 gene editing. *Science Advances, 5*, eaaw8922.

Liu, G., Zhang, Y., & Zhang, T. (2020). Computational approaches for effective CRISPR guide RNA design and evaluation. *Computational and Structural Biotechnology Journal, 18*, 35–44.

Mancuso, P., Chen, C., Kaminski, R., Gordon, J., Liao, S., Robinson, J. A., Smith, M. D., Liu, H., Sariyer, I. K., Sariyer, R., Peterson, T. A., Donadoni, M., Williams, J. B., Siddiqui, S., Bunnell, B. A., Ling, B., Maclean, A. G., Burdo, T. H., & Khalili, K. (2020). CRISPR based editing of SIV proviral DNA in ART treated non-human primates. *Nature Communications, 11*, 6065.

Mani, M., Kandavelou, K., Dy, F. J., Durai, S., & Chandrasegaran, S. (2005). Design, engineering, and characterization of zinc finger nucleases. *Biochemical and Biophysical Research Communications, 335*, 447–457.

Manjunath, N., Yi, G., Dang, Y., & Shankar, P. (2013). Newer gene editing technologies toward HIV gene therapy. *Viruses, 5*, 2748–2766.

Martinez, M. G., Smekalova, E., Combe, E., Gregoire, F., Zoulim, F., & Testoni, B. (2022). Gene editing technologies to target HBV cccDNA. *Viruses, 14*.

Mclaughlin-Drubin, M. E., & Münger, K. (2009). Oncogenic activities of human papillomaviruses. *Virus Research, 143*, 195–208.

Mellott, A. J., Forrest, M. L., & Detamore, M. S. (2013). Physical non-viral gene delivery methods for tissue engineering. *Annals of Biomedical Engineering, 41*, 446–468.

Mendell, J. R., Al-Zaidy, S. A., Rodino-Klapac, L. R., Goodspeed, K., Gray, S. J., Kay, C. N., Boye, S. L., Boye, S. E., George, L. A., Salabarria, S., Corti, M., Byrne, B. J., & Tremblay, J. P. (2021). Current clinical applications of in vivo gene therapy with AAVs. *Molecular Therapy, 29*, 464–488.

Meng, X., Noyes, M. B., Zhu, L. J., Lawson, N. D., & Wolfe, S. A. (2008). Targeted gene inactivation in zebrafish using engineered zinc-finger nucleases. *Nature Biotechnology, 26*, 695–701.

Miller, J., Mclachlan, A. D., & Klug, A. (1985). Repetitive zinc-binding domains in the protein transcription factor IIIA from Xenopus oocytes. *The EMBO Journal, 4*, 1609–1614.

Miller, J. C., Tan, S., Qiao, G., Barlow, K. A., Wang, J., Xia, D. F., Meng, X., Paschon, D. E., Leung, E., Hinkley, S. J., Dulay, G. P., Hua, K. L., Ankoudinova, I., Cost, G. J., Urnov, F. D., Zhang, H. S., Holmes, M. C., Zhang, L., Gregory, P. D., & Rebar, E. J. (2011). A TALE nuclease architecture for efficient genome editing. *Nature Biotechnology, 29*, 143–148.

Miyaoka, Y., Berman, J. R., Cooper, S. B., Mayerl, S. J., Chan, A. H., Zhang, B., Karlin-Neumann, G. A., & Conklin, B. R. (2016). Systematic quantification of HDR and NHEJ reveals effects of locus, nuclease, and cell type on genome-editing. *Scientific Reports, 6*, 23549.

Mock, U., Machowicz, R., Hauber, I., Horn, S., Abramowski, P., Berdien, B., Hauber, J., & Fehse, B. (2015). Mrna transfection of a novel TAL effector nuclease (TALEN) facilitates efficient knockout of HIV co-receptor CCR5. *Nucleic Acids Research, 43*, 5560–5571.

Moore, J. P., Trkola, A., & Dragic, T. (1997). Co-receptors for HIV-1 entry. *Current Opinion in Immunology, 9*, 551–562.

Mussolino, C., Morbitzer, R., Lütge, F., Dannemann, N., Lahaye, T., & Cathomen, T. (2011). A novel TALE nuclease scaffold enables high genome editing activity in combination with low toxicity. *Nucleic Acids Research, 39*, 9283–9293.

Nayerossadat, N., Maedeh, T., & Ali, P. A. (2012). Viral and nonviral delivery systems for gene delivery. *Advanced Biomedical Research, 1*, 27.

Ndung'U, T., Mccune, J. M., & Deeks, S. G. (2019). Why and where an HIV cure is needed and how it might be achieved. *Nature, 576*, 397–405.

Nelson, C. E., & Gersbach, C. A. (2016). Engineering delivery vehicles for genome editing. *Annual Review of Chemical and Biomolecular Engineering, 7*, 637–662.

Nishiga, M., Qi, L. S., & Wu, J. C. (2021). Therapeutic genome editing in cardiovascular diseases. *Advanced Drug Delivery Reviews, 168*, 147–157.

Ophinni, Y., Miki, S., Hayashi, Y., & Kameoka, M. (2020). Multiplexed tat-targeting CRISPR-Cas9 protects T cells from acute HIV-1 infection with inhibition of viral escape. *Viruses, 12*.

Owens, B. (2014). Zinc-finger nucleases make the cut in HIV. *Nature Reviews. Drug Discovery, 13*, 321–322.

Park, H., Oh, J., Shim, G., Cho, B., Chang, Y., Kim, S., Baek, S., Kim, H., Shin, J., Choi, H., Yoo, J., Kim, J., Jun, W., Lee, M., Lengner, C. J., Oh, Y. K., & Kim, J. (2019). In vivo neuronal gene editing via CRISPR-Cas9 amphiphilic nanocomplexes alleviates deficits in mouse models of Alzheimer's disease. *Nature Neuroscience, 22*, 524–528.

Pattanayak, V., Lin, S., Guilinger, J. P., Ma, E., DouDNA, J. A., & Liu, D. R. (2013). High-throughput profiling of off-target DNA cleavage reveals RNA-programmed Cas9 nuclease specificity. *Nature Biotechnology, 31*, 839–843.

Perez, E. E., Wang, J., Miller, J. C., Jouvenot, Y., Kim, K. A., Liu, O., Wang, N., Lee, G., Bartsevich, V. V., Lee, Y. L., Guschin, D. Y., Rupniewski, I., Waite, A. J., Carpenito, C., Carroll, R. G., Orange, J. S., Urnov, F. D., Rebar, E. J., Ando, D., Gregory, P. D., Riley, J. L., Holmes, M. C., & June, C. H. (2008). Establishment of HIV-1 resistance in CD4+ T cells by genome editing using zinc-finger nucleases. *Nature Biotechnology, 26*, 808–816.

Pitman, M. C., Lau, J. S. Y., Mcmahon, J. H., & Lewin, S. R. (2018). Barriers and strategies to achieve a cure for HIV. *Lancet HIV, 5*, e317–e328.

Porteus, M. (2016). Genome editing: A new approach to human therapeutics. *Annual Review of Pharmacology and Toxicology, 56*, 163–190.

Porteus, M. H., & Carroll, D. (2005). Gene targeting using zinc finger nucleases. *Nature Biotechnology, 23*, 967–973.

Qu, X., Wang, P., Ding, D., Li, L., Wang, H., Ma, L., Zhou, X., Liu, S., Lin, S., Wang, X., Zhang, G., Liu, S., Liu, L., Wang, J., Zhang, F., Lu, D., & Zhu, H. (2013). Zinc-finger-nucleases mediate specific and efficient excision of HIV-1 proviral DNA from infected and latently infected human T cells. *Nucleic Acids Research, 41*, 7771–7782.

Ramamoorth, M., & Narvekar, A. (2015). Non viral vectors in gene therapy- An overview. *Journal of Clinical and Diagnostic Research, 9*, Ge01-6.

Ran, F. A., Hsu, P. D., Lin, C. Y., Gootenberg, J. S., Konermann, S., Trevino, A. E., Scott, D. A., Inoue, A., Matoba, S., Zhang, Y., & Zhang, F. (2013). Double nicking by RNA-guided CRISPR Cas9 for enhanced genome editing specificity. *Cell, 154*, 1380–1389.

Roehm, P. C., Shekarabi, M., Wollebo, H. S., Bellizzi, A., He, L., Salkind, J., & Khalili, K. (2016). Inhibition of HSV-1 replication by gene editing strategy. *Scientific Reports, 6*, 23146.

Rothstein, R. J. (1983). One-step gene disruption in yeast. *Methods in Enzymology, 101*, 202–211.

Ru, R., Yao, Y., Yu, S., Yin, B., Xu, W., Zhao, S., Qin, L., & Chen, X. (2013). Targeted genome engineering in human induced pluripotent stem cells by penetrating TALENs. *Cell Regeneration, 2*, 5.

Russell, T. A., Stefanovic, T., & Tscharke, D. C. (2015). Engineering herpes simplex viruses by infection-transfection methods including recombination site targeting by CRISPR/Cas9 nucleases. *Journal of Virological Methods, 213*, 18–25.

Safari, F., Hatam, G., Behbahani, A. B., Rezaei, V., Barekati-Mowahed, M., Petramfar, P., & Khademi, F. (2020). CRISPR system: A high-throughput toolbox for research and treatment of Parkinson's disease. *Cellular and Molecular Neurobiology, 40*, 477–493.

Sano, S., Oshima, K., Wang, Y., Katanasaka, Y., Sano, M., & Walsh, K. (2018). CRISPR-mediated gene editing to assess the roles of Tet2 and Dnmt3a in clonal hematopoiesis and cardiovascular disease. *Circulation Research, 123*, 335–341.

Scully, R., Panday, A., Elango, R., & Willis, N. A. (2019). DNA double-strand break repair-pathway choice in somatic mammalian cells. *Nature Reviews. Molecular Cell Biology, 20*, 698–714.

Seeger, C., & Sohn, J. A. (2014). Targeting hepatitis B virus with CRISPR/Cas9. *Molecular Therapy - Nucleic Acids, 3*, e216.

Selot, R. S., Hareendran, S., & Jayandharan, G. R. (2014). Developing immunologically inert adeno-associated virus (AAV) vectors for gene therapy: Possibilities and limitations. *Current Pharmaceutical Biotechnology, 14*, 1072–1082.

Selvaraj, S., & Piramanayagam, S. (2019). Impact of gene mutation in the development of Parkinson's disease. *Genes & Diseases, 6*, 120–128.

Shen, M. W., Arbab, M., Hsu, J. Y., Worstell, D., Culbertson, S. J., Krabbe, O., Cassa, C. A., Liu, D. R., Gifford, D. K., & Sherwood, R. I. (2018). Predictable and precise template-free CRISPR editing of pathogenic variants. *Nature, 563*, 646–651.

Simonetti, F. R., & Kearney, M. F. (2015). Review: Influence of ART on HIV genetics. *Current Opinion in HIV and AIDS, 10*, 49–54.

Smithies, O., Gregg, R. G., Boggs, S. S., Koralewski, M. A., & Kucherlapati, R. S. (1985). Insertion of DNA sequences into the human chromosomal beta-globin locus by homologous recombination. *Nature, 317*, 230–234.

Soriano, V. (2017). Gene therapy with CRISPR/Cas9 coming to age for HIV cure. *AIDS Reviews, 19*, 167–172.

Srivastava, A. (2016). In vivo tissue-tropism of adeno-associated viral vectors. *Current Opinion in Virology, 21*, 75–80.

Stone, D., Kiem, H. P., & Jerome, K. R. (2013). Targeted gene disruption to cure HIV. *Current Opinion in HIV and AIDS, 8*, 217–223.

Strong, A., & Musunuru, K. (2017). Genome editing in cardiovascular diseases. *Nature Reviews. Cardiology, 14*, 11–20.

Suenaga, T., Kohyama, M., Hirayasu, K., & Arase, H. (2014). Engineering large viral DNA genomes using the CRISPR-Cas9 system. *Microbiology and Immunology, 58*, 513–522.

Sun, J., & Roy, S. (2021). Gene-based therapies for neurodegenerative diseases. *Nature Neuroscience, 24*, 297–311.

Sun, J., Carlson-Stevermer, J., Das, U., Shen, M., Delenclos, M., Snead, A. M., Koo, S. Y., Wang, L., Qiao, D., Loi, J., Petersen, A. J., Stockton, M., Bhattacharyya, A., Jones, M. V., Zhao, X., Mclean, P. J., Sproul, A. A., Saha, K., & Roy, S. (2019). CRISPR/Cas9 editing of APP C-terminus attenuates β-cleavage and promotes α-cleavage. *Nature Communications, 10*, 53.

Szybalski, W., & Skalka, A. (1978). Nobel prizes and restriction enzymes. *Gene, 4*, 181–182.

Tebas, P., Stein, D., Tang, W. W., Frank, I., Wang, S. Q., Lee, G., Spratt, S. K., Surosky, R. T., Giedlin, M. A., Nichol, G., Holmes, M. C., Gregory, P. D., Ando, D. G., Kalos, M., Collman, R. G., Binder-Scholl, G., Plesa, G., Hwang, W. T., Levine, B. L., & June, C. H. (2014). Gene editing of CCR5 in autologous CD4 T cells of persons infected with HIV. *The New England Journal of Medicine, 370*, 901–910.

Thomas, K. R., Folger, K. R., & Capecchi, M. R. (1986). High frequency targeting of genes to specific sites in the mammalian genome. *Cell, 44*, 419–428.

Tong, S., Moyo, B., Lee, C. M., Leong, K., & Bao, G. (2019). Engineered materials for in vivo delivery of genome-editing machinery. *Nature Reviews Materials, 4*, 726–737.

Townsend, J. A., Wright, D. A., Winfrey, R. J., Fu, F., Maeder, M. L., Joung, J. K., & Voytas, D. F. (2009). High-frequency modification of plant genes using engineered zinc-finger nucleases. *Nature, 459*, 442–445.

Trevisan, M., Palù, G., & Barzon, L. (2017). Genome editing technologies to fight infectious diseases. *Expert Review of Anti-Infective Therapy, 15*, 1001–1013.

Tsukuda, S., & Watashi, K. (2020). Hepatitis B virus biology and life cycle. *Antiviral Research, 182*, 104925.

Tycko, J., Myer, V. E., & Hsu, P. D. (2016). Methods for optimizing CRISPR-Cas9 genome editing specificity. *Molecular Cell, 63*, 355–370.

Uddin, F., Rudin, C. M., & Sen, T. (2020). CRISPR gene therapy: Applications, limitations, and implications for the future. *Frontiers in Oncology, 10*, 1387.

Urnov, F. D., Miller, J. C., Lee, Y. L., Beausejour, C. M., Rock, J. M., Augustus, S., Jamieson, A. C., Porteus, M. H., Gregory, P. D., & Holmes, M. C. (2005). Highly efficient endogenous human gene correction using designed zinc-finger nucleases. *Nature, 435*, 646–651.

Urnov, F. D., Rebar, E. J., Holmes, M. C., Zhang, H. S., & Gregory, P. D. (2010). Genome editing with engineered zinc finger nucleases. *Nature Reviews. Genetics, 11*, 636–646.

Van Diemen, F. R., Kruse, E. M., Hooykaas, M. J., Bruggeling, C. E., Schürch, A. C., Van Ham, P. M., Imhof, S. M., Nijhuis, M., Wiertz, E. J., & Lebbink, R. J. (2016). CRISPR/Cas9-mediated genome editing of herpesviruses limits productive and latent infections. *PLoS Pathogens, 12*, e1005701.

Voit, R. A., Mcmahon, M. A., Sawyer, S. L., & Porteus, M. H. (2013). Generation of an HIV resistant T-cell line by targeted "stacking" of restriction factors. *Molecular Therapy, 21*, 786–795.

Walker, D. G., Lue, L. F., Serrano, G., Adler, C. H., Caviness, J. N., Sue, L. I., & Beach, T. G. (2015). Altered expression patterns of inflammation-associated and trophic molecules in substantia Nigra and striatum brain samples from Parkinson's disease, incidental Lewy body disease and normal control cases. *Frontiers in Neuroscience, 9*, 507.

Wang, X., Raghavan, A., Chen, T., Qiao, L., Zhang, Y., Ding, Q., & Musunuru, K. (2016a). CRISPR-Cas9 targeting of PCSK9 in human hepatocytes in vivo-brief report. *Arteriosclerosis, Thrombosis, and Vascular Biology, 36*, 783–786.

Wang, Z., Pan, Q., Gendron, P., Zhu, W., Guo, F., Cen, S., Wainberg, M. A., & Liang, C. (2016b). CRISPR/Cas9-derived mutations both inhibit HIV-1 replication and accelerate viral escape. *Cell Reports, 15*, 481–489.

Wang, H. X., Song, Z., Lao, Y. H., Xu, X., Gong, J., Cheng, D., Chakraborty, S., Park, J. S., Li, M., Huang, D., Yin, L., Cheng, J., & Leong, K. W. (2018a). Nonviral gene editing via CRISPR/Cas9 delivery by membrane-disruptive and endosomolytic helical polypeptide. *Proceedings of the National Academy of Sciences of the United States of America, 115*, 4903–4908.

Wang, L., Smith, J., Breton, C., Clark, P., Zhang, J., Ying, L., Che, Y., Lape, J., Bell, P., Calcedo, R., Buza, E. L., Saveliev, A., Bartsevich, V. V., He, Z., White, J., Li, M., Jantz, D., & Wilson, J. M. (2018b). Meganuclease targeting of PCSK9 in macaque liver leads to stable reduction in serum cholesterol. *Nature Biotechnology, 36*, 717–725.

Wang, P., Zhang, L., Zheng, W., Cong, L., Guo, Z., Xie, Y., Wang, L., Tang, R., Feng, Q., Hamada, Y., Gonda, K., Hu, Z., Wu, X., & Jiang, X. (2018c). Thermo-triggered release of CRISPR-Cas9 system by lipid-encapsulated gold nanoparticles for tumor therapy. *Angewandte Chemie (International Ed. in English), 57*, 1491–1496.

Wang, Z., Wang, W., Cui, Y. C., Pan, Q., Zhu, W., Gendron, P., Guo, F., Cen, S., Witcher, M., & Liang, C. (2018d). HIV-1 employs multiple

mechanisms to resist Cas9/single guide RNA targeting the viral primer binding site. *Journal of Virology, 92*.
White, M. K., Hu, W., & Khalili, K. (2016). Gene editing approaches against viral infections and strategy to prevent occurrence of viral escape. *PLoS Pathogens, 12*, e1005953.
Wilen, C. B., Wang, J., Tilton, J. C., Miller, J. C., Kim, K. A., Rebar, E. J., Sherrill-Mix, S. A., Patro, S. C., Secreto, A. J., Jordan, A. P., Lee, G., Kahn, J., Aye, P. P., Bunnell, B. A., Lackner, A. A., Hoxie, J. A., Danet-Desnoyers, G. A., Bushman, F. D., Riley, J. L., Gregory, P. D., June, C. H., Holmes, M. C., & Doms, R. W. (2011). Engineering HIV-resistant human CD4+ T cells with CXCR4-specific zinc-finger nucleases. *PLoS Pathogens, 7*, e1002020.
Wittkop, L., Breilh, D., Da Silva, D., Duffau, P., Mercié, P., Raymond, I., Anies, G., Fleury, H., Saux, M. C., Dabis, F., Fagard, C., Thiébaut, R., Masquelier, B., & Pellegrin, I. (2009). Virological and immunological response in HIV-1-infected patients with multiple treatment failures receiving raltegravir and optimized background therapy, ANRS CO3 Aquitaine Cohort. *The Journal of Antimicrobial Chemotherapy, 63*, 1251–1255.
Xiao, Q., Chen, S., Wang, Q., Liu, Z., Liu, S., Deng, H., Hou, W., Wu, D., Xiong, Y., Li, J., & Guo, D. (2019a). Ccr5 editing by Staphylococcus aureus Cas9 in human primary Cd4(+) T cells and hematopoietic stem/progenitor cells promotes HIV-1 resistance and Cd4(+) T cell enrichment in humanized mice. *Retrovirology, 16*, 15.
Xiao, Q., Guo, D., & Chen, S. (2019b). Application of CRISPR/Cas9-based gene editing in HIV-1/AIDS therapy. *Frontiers in Cellular and Infection Microbiology, 9*, 69.
Xu, Y., & Xu, D. (2020). Repair pathway choice for double-strand breaks. *Essays in Biochemistry, 64*, 765–777.
Xu, X., Fan, S., Zhou, J., Zhang, Y., Che, Y., Cai, H., Wang, L., Guo, L., Liu, L., & Li, Q. (2016). The mutated tegument protein UL7 attenuates the virulence of herpes simplex virus 1 by reducing the modulation of α-4 gene transcription. *Virology Journal, 13*, 152.
Xu, X., Wan, T., Xin, H., Li, D., Pan, H., Wu, J., & Ping, Y. (2019). Delivery of CRISPR/Cas9 for therapeutic genome editing. *The Journal of Gene Medicine, 21*, e3107.
Xue, C., & Greene, E. C. (2021). DNA repair pathway choices in CRISPR-Cas9-mediated genome editing. *Trends in Genetics, 37*, 639–656.
Yao, Y., Nashun, B., Zhou, T., Qin, L., Qin, L., Zhao, S., Xu, J., Esteban, M. A., & Chen, X. (2012). Generation of CD34+ cells from CCR5-disrupted human embryonic and induced pluripotent stem cells. *Human Gene Therapy, 23*, 238–242.
Yew, C. T., Gurumoorthy, N., Nordin, F., Tye, G. J., Wan Kamarul Zaman, W. S., Tan, J. J., & Ng, M. H. (2022). Integrase deficient lentiviral vector: Prospects for safe clinical applications. *PeerJ, 10*, e13704.
Yin, C., Zhang, T., Li, F., Yang, F., Putatunda, R., Young, W. B., Khalili, K., Hu, W., & Zhang, Y. (2016). Functional screening of guide RNAs targeting the regulatory and structural HIV-1 viral genome for a cure of AIDS. *AIDS, 30*, 1163–1174.
Yin, C., Zhang, T., Qu, X., Zhang, Y., Putatunda, R., Xiao, X., Li, F., Xiao, W., Zhao, H., Dai, S., Qin, X., Mo, X., Young, W. B., Khalili, K., & Hu, W. (2017). In vivo excision of HIV-1 provirus by saCas9 and multiplex single-guide RNAs in animal models. *Molecular Therapy, 25*, 1168–1186.
Yin, D., Ling, S., Wang, D., Dai, Y., Jiang, H., Zhou, X., Paludan, S. R., Hong, J., & Cai, Y. (2021). Targeting herpes simplex virus with CRISPR-Cas9 cures herpetic stromal keratitis in mice. *Nature Biotechnology, 39*, 567–577.
Yoder, K. E., & Bundschuh, R. (2016). Host double Strand break repair generates HIV-1 strains resistant to CRISPR/Cas9. *Scientific Reports, 6*, 29530.
Yuan, J., Wang, J., Crain, K., Fearns, C., Kim, K. A., Hua, K. L., Gregory, P. D., Holmes, M. C., & Torbett, B. E. (2012). Zinc-finger nuclease editing of human cxcr4 promotes HIV-1 CD4(+) T cell resistance and enrichment. *Molecular Therapy, 20*, 849–859.
Zhang, H., Qin, C., An, C., Zheng, X., Wen, S., Chen, W., Liu, X., Lv, Z., Yang, P., Xu, W., Gao, W., & Wu, Y. (2021). Application of the CRISPR/Cas9-based gene editing technique in basic research, diagnosis, and therapy of cancer. *Molecular Cancer, 20*, 126.
Zhang, C., Zaman, L., Poluektova, L. Y., Gorantla, S., Gendelman, H. E., & Dash, P. K. (2023). Humanized mice for studies of HIV-1 persistence and elimination. *Pathogens, 12*, 879.
Zhen, S., Hua, L., Takahashi, Y., Narita, S., Liu, Y. H., & Li, Y. (2014). In vitro and in vivo growth suppression of human papillomavirus 16-positive cervical cancer cells by CRISPR/Cas9. *Biochemical and Biophysical Research Communications, 450*, 1422–1426.
Zhen, S., Hua, L., Liu, Y. H., Gao, L. C., Fu, J., Wan, D. Y., Dong, L. H., Song, H. F., & Gao, X. (2015). Harnessing the clustered regularly interspaced short palindromic repeat (CRISPR)/CRISPR-associated Cas9 system to disrupt the hepatitis B virus. *Gene Therapy, 22*, 404–412.
Zheng, Q., Bai, L., Zheng, S., Liu, M., Zhang, J., Wang, T., Xu, Z., Chen, Y., Li, J., & Duan, Z. (2017). Efficient inhibition of duck hepatitis B virus DNA by the CRISPR/Cas9 system. *Molecular Medicine Reports, 16*, 7199–7204.
Zhu, W., Lei, R., Le Duff, Y., Li, J., Guo, F., Wainberg, M. A., & Liang, C. (2015). The CRISPR/Cas9 system inactivates latent HIV-1 proviral DNA. *Retrovirology, 12*, 22.
Zincarelli, C., Soltys, S., Rengo, G., & Rabinowitz, J. E. (2008). Analysis of AAV serotypes 1-9 mediated gene expression and tropism in mice after systemic injection. *Molecular Therapy, 16*, 1073–1080.

The Neuroimmune System in Psychiatric Disorders

56

Yezan Hassan, Sara C. Esteves, Jonna M. Leyrer-Jackson, and Mark P. Thomas

Abstract

The term "neuroimmune system" emphasizes the fact that nervous and lymphoid tissues constitute a unified system that functions in the maintenance of homeostasis. The conventional division between the two systems has blurred, as well as the distinction between neuropeptides on the one hand, and immune cytokines on the other. Two lines of research have altered our perspective on neuroimmune interactions: (a) the identification of conventional neuropeptides and their receptors, especially those related to the hypothalamic–pituitary–adrenal (HPA) axis, in most lymphoid tissues and (b) the large body of evidence that cytokines, historically associated with immune system communication, play vital roles in nervous function. The role of the HPA axis in coordinating responses to stressors and the role of the immune system in mediating the "sickness" response underscore this reciprocal relationship between the central nervous system and the immune system. Despite this nascent state of knowledge, the evidence for hyper-reactivity of the HPA axis and the alterations in cytokine levels observed in several psychiatric disorders including major depressive disorder, post-traumatic stress disorder, stress and anxiety disorders, schizophrenia, and addiction now serves as a foundation for further research establishing a causative relationship between alterations in immune system function and these debilitating psychiatric disorders. Lastly, newer studies showing the ability to target cytokines and neuropeptides as a therapeutic for these disorders have yielded promising results and provides a promising avenue of effective treatment for individuals with such disorders.

Y. Hassan · S. C. Esteves · J. M. Leyrer-Jackson
Department of Medical Education, School of Medicine, Creighton University, Phoenix, AZ, USA

M. P. Thomas (✉)
Department of Biological Sciences, University of Northern Colorado, Greeley, CO, USA
e-mail: mark.thomas@unco.edu

Keywords

α-MSH · β-Endorphin · β-Lipotropin · β-MSH · γ-MSH · ACTH · Anxiety · Autoimmune disease · Autoimmune T cells · B cells · Beta adrenergic receptors · Corticotropin Releasing Hormone (CRH) · Cytokines · Depression · Hypothalamic–Pituitary–Adrenal axis (HPA) · Interleukin-6 (IL-6) · Inflammation · Leukocytes · Leucine-Enkephalin · Lymphocytes · Melanocortin-1 receptors · Methionine-Enkephalin · Monocytes · Neuropeptides · Preproenkephalin · Proenkephalin · Proopiomelanocortin (POMC) · Regulatory T cells · Schizophrenia · Stress · Thymocytes · Tumor Necrosis Factor Alpha (TNF-α) · Urocortins · Stress · Post-Traumatic Stress Disorder (PTSD) · Major Depressive Disorder (MDD) · Addiction

56.1 Introduction

The earliest evidence for neuroimmune interaction and regulation is the classical immune conditioning experiment of Ader and Cohen in 1975 (Ader, 1987). Since then, it is has been well established that the central nervous system (CNS) and the immune system interact and regulate one another's function (Carr & Blalock, 1989; Madden & Felten, 1995), which is particularly important during times of extraordinary stress, including psychological and physical stressors (Reed & Raison, 2016). Utilizing both classical neurotransmitters as well as neuropeptides, the central nervous system (CNS) regulates immune system function (Chen et al., 2021). The immune system, however, regulates CNS function particularly through the use of cytokines (Salvador et al., 2021). While this reciprocal regulation between the CNS and the immune system is designed to maintain homeostasis, dysfunction in one can lead to aberrant function in the other, which can lead to the development of psychological disorders.

H. E. Gendelman, T. Ikezu (eds.), *Neuroimmune Pharmacology and Therapeutics*,
https://doi.org/10.1007/978-3-031-68237-7_56

Interactions between the immune and nervous system are complex and attempts to decipher these are still incomplete. One of the first empirical findings in support of an interaction between these systems was the discovery of receptors for neurohormones and neuropeptides on cells of the immune system (Hadden et al., 1970; Hazum et al., 1979). This implied that immune cells can and do respond to the endogenous ligands for these receptors. Conversely, the demonstration of many cytokine receptors and the cytokines themselves in the CNS suggests not only that these small proteins have a functional role in the CNS, but that cytokines released by the immune system may be able to regulate CNS function (Adler & Rogers, 2005; Cartier et al., 2005). Subsequent studies demonstrated that immune cells also synthesize and release many neurohormones and neuropeptides, which provides a mechanism for local control of the physiological effects of these transmitters, in addition to distant control mediated through the peripheral and central nervous systems. This sharing of ligands and receptors in the nervous and immune systems led to the suggestion that this constituted a complete biochemical information circuit and that this interactive communication system provided the basis for the immune system acting as a sensory system and the brain playing an immunoregulatory role (Blalock, 1994).

In this chapter, we will provide an overview of how the hypothalamic–pituitary–adrenal (HPA) axis, the CNS, and the immune system are integrated and regulate each other's activity. First, we will highlight classic neurotransmitters and neurohormones utilized by the CNS to regulate the immune system, including serotonin, norepinephrine, and gamma-aminobutyric acid (GABA). We will then discuss a variety of neuropeptides and cytokines involved in neuroimmune function and found to play a major role in regulating the immune system. In particular, we will examine HPA axis regulation of CNS function by release of peptides and glucocorticoids in response to stress and inflammation. Most importantly, we will provide up to date knowledge regarding the role of the immune system in a variety of psychiatric disorders including major depression, anxiety/stress disorders, schizophrenia, addiction, bipolar disorder, and posttraumatic stress disorder. Specifically, we will highlight known abnormalities in HPA axis activity, neuropeptides and cytokines associated with each disorder and how targeting these aspects of the disorder can be utilized as treatment options. Together, we believe the research summarized here clearly indicates that the dysfunction of the immune system can contribute to the development of several psychiatric disorders, however continued research is necessary to establish a cohesive foundation of the immune system's role in each disorder, a theme that will be covered in many chapters within this textbook.

56.2 The Function of the HPA Axis and Its Associated Circuitry

The occurrence of physical or psychological events, or "stressors," that threaten the well-being of an organism can lead to activation of a generalized stress response. The ability to respond to stress allows an organism to maintain homeostasis following psychological and physiological stressors, which is vital to normal physiological function. The ability to respond to stress relies on a coordinated response that is mediated by components of the CNS, the peripheral nervous system (PNS), and the endocrine system. The HPA axis defines the classical core of the neuro-endocrine stress response system. The HPA axis coordinates the component processes of the stress response by serving as the major communication route between the CNS, the PNS, and the endocrine system. The HPA axis consists of three major components: the hypothalamic paraventricular nucleus (PVN), corticotrophs in the anterior pituitary gland, and the adrenal cortex. The hypothalamus is situated at the anterior-most end of the brainstem and serves as the final common pathway for regulation of visceromotor and endocrine functions by the CNS. The PVN, situated bilaterally along the midline of the anterior hypothalamus, coordinates endocrine functions via the pituitary gland, and visceromotor functions via pathways to the brainstem. The pituitary gland contains corticotrophs, which synthesize and release the peptide, adrenocorticotrophic hormone (ACTH) in response to signals from the PVN. Lastly, cells within the outer layer of the adrenal cortex release the steroid hormone, cortisol, in response to ACTH. Cortisol released from the adrenal cortex has been shown to play a major role in coordinating the response to stress as detailed below.

Prior to the involvement of the HPA axis, the activation of a stress response begins with the ability of an individual to evaluate and recognize a stressful stimulus, which relies on regions of the CNS commonly known as the limbic system. The concept of the limbic system, which was originally defined anatomically, has taken on a functional connotation and now typically refers to CNS systems which directly modulate the hypothalamus. These brain areas include the hippocampus, the amygdala, and regions of the medial prefrontal cortex. The components of the limbic system all influence hypothalamic output and constitute a distributed system that regulates affective state and is critically involved in the appraisal and response of the organism to stressors. In general, the hippocampus and the anterior cingulate/prelimbic regions of prefrontal cortex inhibit HPA axis output, while the amygdala and the infralimbic region of prefrontal cortex potentiate HPA axis activity. Thus, different regions of the limbic system modulate the output of the HPA axis in a reciprocal fashion to coordinate the temporal sequence of the stress response, controlling the initiation, propagation, and termination of distinct processes. While the circuitry and crosstalk between these regions can be complex, the hormones and neuropeptides associated with the HPA axis are well characterized and are highlighted below.

56.3 Hormones Released by the HPA Axis

Two neuropeptides, corticotrophin-releasing hormone (CRH) and arginine vasopressin (AVP), are released from parvocellular neurons in the hypothalamic PVN to initiate a stress response. The terminal endings of these neurons, located in the median eminence of the hypothalamus, release CRH and AVP into the hypothalamic-hypophysial portal vessel system, where they travel to the anterior pituitary. The two neuropeptides act synergistically on pituitary corticotrophs to activate the release of other peptides (Bilezikjian et al., 1987). Specifically, CRH induces the release of ACTH from anterior pituitary corticotrophs, which travels via the plasma and binds to receptors in the adrenal cortex to stimulate glucocorticoid release. ACTH is cleaved from the prohormone pro-opiomelanocortin (POMC), which is highly expressed in the hypothalamus and pituitary. Aside from ACTH, POMC can also be cleaved into a variety of other bioactive peptides including β-endorphin, α-melanocyte-stimulating hormone, and β-lipotropin. Of these peptides, ACTH is of most importance given its ability to travel via the bloodstream and act on cells in the zona fasciculata layer of the adrenal cortex, where it stimulates the synthesis and release of glucocorticoids, including cortisol.

In addition to regulating the synthesis and release of ACTH from the pituitary, neurons in the PVN that produce AVP also project to areas in the brainstem that control the output of the sympathetic branch of the PNS. Activation of PVN neurons at the onset of a stress response thus leads to increased sympathetic outflow, resulting in release of norepinephrine from sympathetic nerve terminals and epinephrine from the adrenal medulla. The effects of these catecholamines are responsible for the “fight-or-flight” response.

The effects of the neuropeptides and glucocorticoids associated with the HPA axis are mediated through distinct receptor subclasses. Two classes of CRH receptors have been identified, type 1 and type 2, with distinct but overlapping distributions in the CNS (Reyes et al., 2001). It is currently thought that CRH acts through type 1 receptors (CRHR1), while novel ligands have been identified for the related type 2 receptors (CRHR2). These ligands, urocortin II (Reyes et al., 2001) and urocortin III (Lewis et al., 2001), play a role in the regulation of appetitive behaviors and are thought to play a role in the adaptive (slow) phase of the stress response (de Kloet et al., 2005) as discussed below. Expression of mRNA for the urocortins has been demonstrated in the hypothalamus, brainstem, and amygdala (Hsu & Hsueh, 2001). Interestingly, the CNS distributions of the CRHR1 and CRHR2 receptors correspond to the CRH and urocortin terminal fields, respectively.

The effects of the glucocorticoids are also mediated through two distinct subclasses of cytosolic steroid receptors. Glucocorticoid receptors (GRs) are found throughout the brain, with the highest density found within the hypothalamic PVN, ascending aminergic pathways, and in limbic brain regions (Herman et al., 2003). Mineralocorticoid receptors (MR) are named for their primary ligand, the steroid aldosterone, which is produced by the outer zona glomerulosa layer of the adrenal cortex and regulates sodium homeostasis through its actions at the level of the nephron. However, MRs in the brain have a ten-fold higher affinity for glucocorticoids than GRs and thus respond to lower levels of these hormones during the initial phase of a stress response. MRs are also expressed at high levels in limbic brain regions; thus, there is considerable overlap between these two receptors in key brain areas that regulate the HPA axis.

56.4 The Role of HPA Axis in Response to Stress

As mentioned previously, the human ability to respond to stress utilizing the HPA axis and its associated hormones and neuropeptides allows for homeostatic maintenance during both real and perceived stress. Interestingly, under basal conditions, both CRH and AVP are released from the hypothalamus in a pulsatile fashion, with a frequency of 2–3 episodes per hour (Engler et al., 1989). These pulsatile episodes are closely associated with central and peripheral clocks, resulting in circadian fluctuations in both ACTH and cortisol levels throughout the day, with cortisol levels rising following observed elevations in ACTH. Under normal resting conditions, pulsatile release frequency and the concentration of ACTH and cortisol release is highest in the early morning and decreases throughout the day (Kazakou et al., 2023). However, stress increases the magnitude of the secretory pulses, resulting in elevated hormone/neuropeptide levels throughout the day. Thus, both acute and long-term stress can have profound impacts on the HPA axis and levels of neuropeptides, ultimately resulting in fluctuations associated with neuropsychiatric disorders, which will be discussed in detail below. However, it is important to summarize and differentiate the acute and chronic stress response on the HPA axis.

56.5 An Overview of Acute and Chronic Stress Responses

Stress can be classified as either acute or chronic, depending on the duration of the event. Acute stress is classified as an experience with an immediate physical, emotional, or psychological perceived threat. Chronic stress is characterized by the exposure of an acute stressor that persists over a long period of time. The underlying mechanisms of acute and chronic stress are similar in nature but can have profound longitudinal effects that can render an individual more susceptible to psychological disorders.

Acute stress can be characterized by two phases, a fast phase that serves to mobilize and activate the fight or flight response, and the slow phase, which promotes recovery and adaptation of the organism. The initial fast phase is manifested behaviorally by increases in vigilance and arousal, and alterations in attention. This phase is associated with the release of CRH and AVP from the hypothalamic PVN, activation of sympathetic outflow and release of catecholamines from the adrenal medulla, and rising blood levels of cortisol. The effects of CRH in the fast phase are mediated through CRHR1, which are responsible for activation of the HPA axis and the sympathetic nervous system. Levels of cortisol rise rapidly due to release of ACTH from the anterior pituitary and serve in the periphery to mobilize metabolic resources and alter immune system responses, which will be further discussed below. The effects of cortisol on brain structures, and thus behavior, in the fast phase are mediated primarily by the higher-affinity MRs, which respond to the rising levels of cortisol at the initiation of a stress response. An important aspect of MR-induced changes involves their role in the appraisal of environmental stimuli in this phase of the response. The alterations in neuronal function mediated by the interaction of cortisol with MRs lead to increases in arousal level, heightened vigilance and alertness, and enhancement of attention.

The slow phase of the stress response is characterized by processes that promote recovery from, and adaptation to, the stressful conditions that prompted the initial fast phase response. At the level of the hypothalamus, this phase is hypothesized to be mediated by the urocortins acting through CRHR2 receptors (de Kloet et al., 2005; Reul & Holsboer, 2002). In a mouse model of stress, urocortins were found to play an essential role in stress recovery, and knockout of urocortin rendered animals the inability to recover properly from acute stress exposure (Neufeld-Cohen et al., 2010). In contrast to the fast phase, the slow phase is associated with activation of the parasympathetic nervous system, which promotes the appetitive and metabolic functions which help to restore homeostasis. As cortisol levels peak and then slowly subside during the slow phase, the lower-affinity GRs are activated and lead to transcriptional changes in target organs that promote recovery. In the brain, GR activation-induced transcriptional changes may be crucial for the consolidation of memory traces involving the association of stressors with sensory stimuli (in the amygdala) (Roozendaal et al., 2009) and environmental context (in the hippocampus) (Oitzl & de Kloet, 1992).

While acute stress is common in humans, chronic stress, and prolonged activation of the HPA axis can have profound effects leading to maladaptive conditions. Most importantly, chronic stress leads to an enhanced CNS drive over the HPA axis that leads to impaired negative feedback control (Checkley, 1996). A prominent feature of chronic stress is elevated cortisol levels, which is thought to induce increased expression of both CRF and AVP mRNA within the parvocellular neurons of the hypothalamus and a decrease in expression of both GR and MR within the hippocampus. In tandem, these effects lead to a hyperactive and loosely controlled HPA axis and the hormones and neuropeptides associated with it. Such changes often lead to maladaptive conditions, especially in "vulnerable" individuals with genetic predispositions, which may facilitate disease processes. We will discuss commonly observed findings in HPA axis dysfunction as it relates to multiple psychiatric disorders later in this chapter, which are often due to prolonged chronic stress. However, it is important to note that at all levels of the HPA axis, feedback mechanisms exist that control the output of neuropeptides and circulating hormones. These mechanisms regulate the levels of these factors during normal diurnal cycles but also act to limit the magnitude of the stress response and terminate its effects. Alterations to this process predispose individuals to psychiatric diseases that could otherwise be prevented and targeting these dysregulated peptides and pathways have gained traction as therapeutic targets for treating such disorders, as discussed below.

56.6 Role of HPA Axis in Inflammation

Stress and the activation of the systems discussed above can affect all tissues of the organism, where the peripheral immune system is no exception. In fact, excessive HPA axis activation has been linked to increased inflammation in the periphery. There is also evidence that injury to the CNS can make an individual more susceptible to infections (Harrison-Felix et al., 2009). It is hypothesized that these effects are mediated through at least two pathways: via the HPA axis and by virtue of the innervation of lymphatic tissues by autonomic nerve fibers, especially from the sympathetic nervous system. Interestingly, damage to the CNS increases SNS activation and promotes norepinephrine release which then activates adrenergic receptors expressed by immune cells. As well, all lymphoid tissues, including primary (bone marrow and thymus) and secondary (spleen, lymph nodes, and gut-associated lymphoid tissue) lymphatic tissue are innervated by sympathetic nerve fibers, and most lymphoid cells express catecholamine receptors rendering them responsive to SNS neurotransmitters. Specifically, B-lymphocytes, $CD4^+$ and $CD8^+$ T cells, dendritic cells, monocytes, and macrophages express catecholamine receptors, and when activated are known to release cytokines (Vazquez et al., 2015). Below we highlight multiple neuropeptides and cytokines and their relationship to the CNS, HPA-axis as well as immune function; however, it is important to note that we do not discuss all neuropeptides and cytokines utilized by these systems but a sampling of important modulators known to be dysregulated across psychiatric disorders.

56.7 Neuropeptides as Modulators of the CNS and Immune Function

There is differential expression of neurotransmitter and neurohormonal receptors among the various cell types of the immune system. This suggests that particular signaling molecules produce specific and differential effects on immune cells, and therefore immune function. It also suggests that the pharmacology of immune cells will vary from cell type to cell type and that these pharmacological differences represent potential drug treatment strategies in coping with various pathological states. While a great deal of work needs to be done in this field due to some conflicting evidence on receptor expression and the functions of particular signaling molecules, it is widely accepted that neuropeptides, tachykinins, and cytokines each play an important role in modulating immune function. Further, dysregulation of these signaling molecules can be linked to many psychiatric disorders, which will be further discussed below. Here we highlight various signaling molecules utilized by the CNS and immune system and provide an overview of their functions. However, more detailed information regarding these molecules can be found across multiple chapters of this textbook, and therefore, we direct readers to other chapters for additional background information. The endogenous opioid system consists of three peptide precursors that encode for enkephalins, dynorphins, β-endorphins, and proopiomelanocortin (POMC) (Kieffer & Gavériaux-Ruff, 2002), which are discussed here.

56.8 Peptides Derived from Proenkephalin

Two of the most abundant endogenous opioid peptides, methionine-enkephalin (Met-enk; Tyr-Gly-Gly-Phe-Met) and leucine-enkephalin (Leu-enk; Tyr-Gly-Gly-Phe-Leu) are derived from the precursor protein, proenkephalin (PENK). PENK is a 267 amino acid precursor that contains 6 copies of Met-enk, two of which are extended forms (Tyr-Gly-Gly-Phe-Met-Arg-Phe and Tyr-Gly-Gly-Phe-Met-Arg-Gly-Leu) and one copy of Leu-enk. The fact that there are two distinct peptides closely related in structure and function proved to be a major obstacle that slowed their discovery (Kosterlitz & Hughes, 1977).

PENK contains not only both pentapeptide enkephalins but also a hepta- and octapeptide enkephalin. This structure is maintained across species (Comb et al., 1982; Gubler et al., 1982; Noda et al., 1982). The processing of PENK into smaller peptides, even for an extended period of time, maintains this assortment of four distinct peptides (Fleminger et al., 1983), indicating that all are active products of PENK and that the longer peptides are not simply alternate precursors for Met-enk. Interestingly, while a great deal of attention has been paid to the role of the pentapeptides, Met-enk and Leu-enk, knowledge of the physiological role of the hepta- and octapeptides is still very limited.

There is evidence that PENK is not processed in the same way in all tissues and the suggestion has been made that alternative processing may occur in immune system cells containing PENK, compared to processing in neurons (Fricker et al., 2020). This would open the possibility that differential processing may lead to release of unique enkephalin-like peptides from the immune system that would have a different spectrum of activities, at least to some extent, than those derived from neural tissues. This, in turn, would provide the opportunity for differential regulation of components of the immune system, depending on the source of the enkephalin-like peptide release. In addition, opioid peptides generated from proenkephalin A act as cytokines that are capable of regulating granulocytes and mononuclear cell function (Sharp et al., 1998). Lastly, the role of these enkephalin-like peptides in regulating synaptic plasticity in distinct brain regions lends insight that dysregulation could lead to inappropriate long-term synaptic plasticity, leading to altered behaviors.

56.8.1 Expression and Function of Enkephalins in the Immune System: An Overview

PENK is distributed widely throughout the brain (Wamsley et al., 1980), indicating involvement of the enkephalins in a variety of physiological functions. The first suggestion that peptides derived from PENK were involved in immune system function came from the work of Wybran et al. (1979), which demonstrated specific binding of both Met-enk and morphine to lymphocytes. Early studies also demonstrated that the enkephalins are concentrated in leukocytes compared to other blood components, supporting the idea that immune cells can synthesize opioid peptides. In addition, enkephalins have been found to play a role in analgesia, angiogenesis, embryonic development, neuroprotection, as well as peristalsis (Duque-Díaz et al., 2019).

Determination of which PENK products are formed by different immune cells has been difficult due to the lack of antibodies specific to the various peptides. For instance, many studies use readily available antibodies to Met-enk, but the possibility is strong that these antibodies can also react with larger peptides containing Met-enk. However, it has been established that enkephalins have an impact on cytokine release. Due to the expression of opioid receptors on immune cells, enkephalins have the ability to bind and modulate cytokine release. For example, enkephalin activation of immune cells induces the release of proinflammatory cytokines such as TNF-α (Duque-Díaz et al., 2019). However, some studies have reported conflicting results, likely due to differences in cell types as well as the signal transduction pathway. For instance, Th2 cytokines, such as IL-4, IL-6, and TGF-β, produce many-fold increases in PENK mRNA in human thymocytes. On the other hand, Th1 cytokines have variable (IL-1β) or no (IFN-γ) effect (Kavelaars & Heijnen, 2000). Further, Th2 cytokines lead to a 4- to 5-fold increase in Met-enk-like protein in thymocytes, whereas Th1 cytokines produce no such effect. It appears, moreover, that the induced Met-enk itself is not released by thymocytes; rather, larger peptides containing Met-enk.

Similar results have been found in T cells, where PENK is synthesized and processed into smaller peptides by macrophages and activated monocytes. However, PENK is not processed in inactivated monocytes. Interestingly, T cells do not appear to secrete detectable levels of PENK-derived peptides whereas monocytes do. On the other hand, under some conditions T cells have been shown to secrete intermediate products of preproenkephalin, the precursor to PENK. These intermediate products appear to inhibit proliferative capacity via unidentified receptors (i.e., nonopioid binding sites). Interestingly, cytokine release increases the expression of μ-opioid receptors, which is thought to play a protective role in intestinal inflammation (Saccani et al., 2012). Taken together, these results suggest that enkephalins not only promote cytokine release by binding to μ-opioid receptors on immune cells, but that cytokines themselves can regulate μ-opioid receptor expression and therefore the overarching effects of enkephalins. It is of note that some classify met-enkephalin as a cytokine itself (Plotnikoff et al., 1997).

In addition to their effects on the immune response, enkephalins have profound effects on synaptic plasticity. For example, vasoactive intestinal peptide (VIP)-expressing interneurons of the hippocampus, which play a substantial role in memory storage and modulating information transfer between CA2 and CA3 pyramidal neurons, release enkephalin which is hypothesized to modulate the memory storage associated plasticity (Leroy et al., 2022). However, met-enke has been shown to induce long-term synaptic depression of glutamate-MSN connections within the nucleus accumbens (Trieu et al., 2022) and depress synaptic transmission in the striatum (Bonci & Malenka, 1999). Interestingly, PENK mRNA expression across various brain structures including the striatum (Beckstead, 1995) and hippocampus (Morris & Johnston, 1995) is dependent on N-methyl-D-aspartate (NMDA) receptor activation. Given that enkephalin has the ability to not only modulate synaptic connections but also PENK expression across multiple brain regions, it is not surprising that dysregulation of enkephalin can lead to altered connectivity between brain regions that gives rise to abnormal behaviors associated with psychiatric disorders.

56.9 Peptides Derived from Prodynorphin

A second major source of endogenous opioid peptides are those derived from prodynorphin (pDYN) the product of a single gene that contains four exons (Gein, 2014) with seven splice variants in humans. The processing of the pDYN peptide appears to be differentially regulated in CNS regions (reviewed in Schwarzer (2009)). Of several peptide fragments generated, the major components appear to be dynorphin A and dynorphin B, separately derived from pDYN. Dynorphins make up a family of opioid peptides that appear to act primarily at kappa opioid receptors in the central and peripheral nervous system, with dynorphin A (amino acids 1–17) having the highest affinity for this opioid receptor (James et al., 1984) and is produced and contained within immune cells. Lastly, evidence suggests that opioids modulate both innate and acquired immune responses (McCarthy et al., 2010).

56.9.1 Expression and Function of Dynorphins in the CNS and Immune System: An Overview

Within the central nervous system, dynorphins are primarily localized to the hippocampus, amygdala, striatum, and the spinal cord. Through the nervous system, dynorphins control memory, learning, emotions, and reactions to stress. However, dynorphins, like other opioids, are locally synthesized in various tissues including tissues of the immune system (Gein, 2014). Dynorphins are known to be present within immune cells and nerve fibers in inflamed tissues (Hassan et al., 1992). Aside from dynorphin expression, opioid receptor expression has been shown on a variety of immune cells, which allow for direct and dynamic effects on immune function (Bidlack, 2000). Specifically, k-opioid and δ-opioid receptors are found to be present on lymphocytes, and there is further evidence for both neural and non-neural opioid receptors on immune cells (Bidlack, 2000). Depending on the site of their production, dynorphins have different physiological actions, indicating that dynorphins have a wide variety of functions regarding the human immunological response.

The immunomodulatory function of dynorphins is determined by the expression levels of kappa (k-opioid) receptors. The dynamic effects of dynorphins are thought to be due to the complicated pro-opioid system and its multiple receptors. Through k-opioid receptors, dynorphins regulate proliferation as well as the production of antibodies. Dynorphins also regulate anti-inflammatory activity as well as proinflammatory activity. In addition, dynorphin peptides are capable of directly modulating lymphocytes and other cells involved in immunity (Bidlack et al., 2006) and are thought to contribute to the pathophysiology of depression, schizophrenia, epilepsy, and addiction (Schwarzer, 2009) by modulating not only immune function but also synaptic plasticity, altering network connectivity.

Dynorphins are known to suppress synaptic plasticity of direct-pathway medium spiny neurons within the striatum through activation of the k-opioid receptor, which leads to enhanced performance in re-learning a reversed reward contingency (Yang et al., 2023). Such results suggest that dynorphins play an important role in reversal learning. Furthermore, dynorphin has been shown to inhibit dopamine release through activation of k-opioid receptors within the striatum, which prevents long-term potentiation (LTP) of corticostriatal synapses (Hawes et al., 2017). This pathway is particularly important in mediating goal-directed actions and habitual learning, whereby inhibition of LTP induces deficits in such functions. Dynorphin modulation of striatal circuits has also been implicated in Parkinson's disease as well as dyskinesia (Sgroi & Tonini, 2018). Interestingly, stress, which is known to be associated with aversive mental states, is known to induce the release of dynorphins leading to prolonged k-opioid receptor activation, a mechanism characteristic of depressive disorders (Knoll & Carlezon, 2010). However, the use of dynorphin antagonists has been shown to alleviate depressive behaviors (Shirayama et al., 2004), providing evidence that therapeutic effects may be induced by preventing the over-activation of k-opioid receptors by dynorphins. Targeting dynorphins likely results in therapeutic effects by preventing inappropriate synaptic modulation across different brain regions in combination with altering immune system activation.

56.10 Peptides Derived from Proopiomelanocortin (POMC)

Another major source of endogenous opioid peptides as well as nonopioid peptides is proopiomelanocortin (POMC). This large peptide contains the sequences of, and is processed into, a number of smaller peptides, including ACTH, β-endorphin, β-lipotropin, α-melanocyte-stimulating hormone (MSH), β-MSH, and γ-MSH. Which products are derived from POMC depends upon the cell in which the processing takes place. In the anterior lobe of the pituitary POMC is processed primarily into ACTH and β-lipotropin. In the neurointermediate lobe the major products are β-endorphin, α-MSH and γ-MSH. Interestingly, POMC was the first of the peptide precursors in which cell-dependent processing was demonstrated.

56.10.1 Expression of POMC in the Immune System: An Overview

The discovery that POMC was produced not only by neuronal tissue but also by immune cells was a major event in the recognition of interactions between the immune and nervous systems [reviewed in (Blalock, 1999)]. These studies indicated that not only do neurohormones derived from the nervous system alter immune system activity, but that similar or identical hormones can be released from immune cells to autoregulate itself as well as regulate the nervous system. Like PENK-derived enkephalins, POMC-derived peptides frequently have biphasic dose–response curves, i.e., low doses stimulate immune system function whereas high doses usually suppress function, each designed to maintain homeostasis in variable conditions. As discussed above, immune cells express receptors for both opioids and ACTH, rendering them responsive to POMC derived peptides, which allow for the interaction between the immune and central nervous systems. POMC-derived peptides have a dynamic role in this interaction, and the effects of each are summarized below.

56.10.2 Function of POMC Derivatives in the CNS and Immune System: An Overview of β-Endorphin and α-MSH

56.10.2.1 β-Endorphin

β-endorphin acts at μ- and δ-opioid receptors, which are expressed throughout the CNS as well as the immune system. Various brain regions such as the bed nucleus of the stria terminalis, ventral tegmental area, and amygdala exhibit immunoreactivity for β-endorphin in fiber-like processes, suggesting that it likely modulates synaptic connectivity in these regions. A major function of β-endorphins is to inhibit communication of pain signals and produce feelings of euphoria, similar to exogenous opioid therapeutics. However, the effects of β-endorphin are seemingly more complex where they not only regulate pain through immune system regulation but also activate reward associated brain circuitry, including modulation of synaptic connectivity across and within brain regions such as the striatum, ventral tegmental area, and prefrontal cortex. More recently, immune cells including T-lymphocytes, B-lymphocytes, monocytes, and macrophages have been found to express POMC mRNA and thus have the ability to secrete POMC derived peptides during bouts of inflammation (Jain et al., 2019). Furthermore, such secretion has the ability to inhibit hypothalamic CRH production, impacting HPA-axis function (Jain et al., 2019). Stressful conditions also increase endorphin-associated receptor expression within the CNS as well as immune cells, further supporting the role of endorphins in regulating immune function as well as CNS circuitry. This is further supported by studies demonstrating that as β-endorphin levels rise innate immune cell activity is enhanced (i.e. natural killer cells and macrophages), which is likely due to increased expression of opioid receptors. Other experiments have also demonstrated significantly increased levels of both β-endorphin and Met-enk localized to the area of inflammation (Stein et al., 1990). Dysregulation of β-endorphin is associated with an abundance of psychiatric disorders, which will be discussed in more detail below.

56.10.2.2 α-MSH

The tridecapeptide, α-MSH, is also derived from POMC and was originally discovered through its effects on the skin. However, more recently, α-MSH has been shown to alter immune system function (Luger et al., 2003), yet the exact role of α-MSH in neuroimmune modulation is still predominantly unclear. α-MSH acts through melanocortin-1 receptors (MC-1R), which are expressed by immune cells, including macrophages and monocytes. Interestingly, other peptides of POMC, including β-MSH and γ-MSH, also act through other melanocortin receptors; however, these subtypes have not been found on immune cells to date. Such findings indicate that α-MSH may be the only of these three to be involved in regulating immune function through melanocortin receptors expressed by immune cells. Most studies to date suggest that α-MSH modulates or reduces inflammation, downregulating proinflammatory cytokines, and other proinflammatory molecules (Luger et al., 2003; Shahraki et al., 2022). These effects are produced, at least in part, by inhibition of nuclear factor kappa B (NFkB) activity.

In addition to its effects on immune cells, α-MSH has been shown to modulate synaptic plasticity within the hippocampus through postsynaptic activation of the melanocortin-4 receptor (Fu & Ip, 2017). Specifically, activation of the melanocortin-4 receptor promotes α-amino-3-hydroxy-5--methyl-4-isoxazolepropionic acid (AMPA) receptor insertion into synapses and increases the number of synaptic spines observed in cultured hippocampal neurons (Shen et al., 2013). Activation of the POMC/melanocortin-4 circuit alleviates synaptic plasticity impairment observed in animal models of Alzheimer's disease (Shen et al., 2013; Shen et al., 2016), suggesting that α-MSH plays an essential role in mediating normal cognitive function. The dysregulation and downstream behavioral effects will be discussed in detail below in the context of disease states.

56.10.2.3 Corticotropin-Releasing Hormone

Corticotropin-releasing hormone (CRH; corticotropin-releasing factor, CRF) plays a major role in CNS regulation of the immune system. In particular, CRH released from the hypothalamus is the first point in HPA axis regulation of immune function. As a critical player in the response to stress, the HPA axis is a primary point of coordination in the neural-immune interaction. CRH also plays an important role in many CNS functions, including regulation of the autonomic nervous system and endocrine function. CRH is involved in the central response to anxiety and stress, and hence affects immune system function via that route. There is also convincing evidence that CRH can play a role in depression. The central role of CRH has been reviewed (Dunn & Berridge, 1990; Heim & Nemeroff, 1999). Beyond these well-known roles for CRH, this neuropeptide hormone is also synthesized in immune cells and so appears to play an even more widespread role than originally thought.

Early studies demonstrated a role for CRH in augmenting the function of immune cells, including natural killer cells (Carr & Blalock, 1989). The localization and the synthesis of CRH by immune cells were initially established by the identification of mRNA for CRH in leukocytes (Stephanou et al., 1990) and significant concentrations of CRH in areas of inflammation (Karalis et al., 1991). This suggested CRH was produced locally, i.e., in the periphery in areas of inflammation, probably in part by immune cells. Further studies supported these ideas. In Lew/N rats, which are deficient in hypothalamic CRH responses to inflammatory stimuli, there are high levels of CRH in inflamed joints (Crofford et al., 1992), demonstrating the release of CRH at localized sites by immune cells. The obvious paradox is that CRH plays a powerful anti-inflammatory role as the primary CNS activator of the HPA axis, yet in local peripheral sites it has just the opposite effect. Thus, CRH expression and function with respect to inflammation appears to be site specific.

CRH has been shown to have important effects in areas of inflammation, such as augmentation of analgesia by endogenous opioid peptides. It has been further demonstrated that these effects could be blocked by local administration of a CRH antagonist, a CRH antibody, or by antisense oligonucleotides directed against CRH. This implies that local production and release of CRH by immune cells do, in fact, occur and are physiologically important (Schafer et al., 1996). Eventually, it became clear that regulation of CRH synthesis and release in immune cells differs from that in hypothalamic neurons. Specifically, immune cells synthesize and release much smaller concentrations of CRH and other neuroimmune peptides, and although their release may require de novo synthesis, an inherently slow process, the fact that immune cells release these hormones locally in the target area compensates for both of these factors to some extent. These data indicate site and tissue-specific effects of CRH, which are sometimes contradictory, point to the complex interrelationship between the nervous, endocrine, and immune systems, an interaction that has yet to be deciphered completely.

Unlike other discussed peptides thus far, CRH is not known to directly impact synaptic plasticity. While there are CRH-expressing neurons within the hypothalamus, and their connections with other circuits can be modified, there is limited evidence suggesting that CRH directly impacts plasticity across brain regions. However, the lack of findings does not rule out the idea that changes in synaptic strength induced by CRH-expressing neurons could lead to downstream changes in immune function or hypothalamic excitability.

56.11 Tachykinins

Tachykinins are a family of neuropeptides that are intensively studied. In fact, tachykinins are the largest family of neuropeptides present within mammals and amphibians. They make up a group of structurally related peptides all derived from tac genes, in alternating sequences (Steinhoff et al., 2014). Tac genes encode preprotachykinins, precursor proteins, which get chopped into smaller peptides, forming tachykinins. Tachykinins interact with three types of G protein-coupled receptors known as neurokins. In mammals, substance P, neurokinin A, and neurokinin B are the most well-known and studied tachykinins (Pinto et al., 2004). However, new tachykinins are still being discovered. Hemokinin 1, for example, is preferentially expressed in peripheral tissues and may have functions outside of the well-defined functions of tachykinin peptides as neurotransmitters. It has been suggested that aside from neurotransmitter-like actions, tachykinins can have profound influences on inflammatory responses, effecting immune cell function in many different ways (Zhang et al., 2006).

56.11.1 Expression and Function of Tachykinins in the CNS and Immune System

Tachykinins show broadly distributed expression within human tissues. It has been observed that tachykinin peptide expression is mainly found within neuronal tissues (Otsuka & Yoshioka, 1993); however recent findings have also identified tachykinin expression in a variety of peripheral tissues. Substance P and neurokinin A, for example, are found to be expressed throughout the CNS but also within primary afferent neurons that innervate peripheral tissues. Substance P and neurokinin A have also been found to be expressed in non-neural tissues, like endothelial cells, immune cells, and inflammatory cells (Lecci & Maggi, 2003; Linnik & Moskowitz, 1989). Neurokinin B has been discovered to be expressed solely in the CNS, however, its function remains largely unknown (Moussaoui et al., 1992). Tachykinin expression has also been reported in pulmonary arteries and veins, suggesting its potential role in lung physiopathology (Pinto et al., 2004). Understanding the interactions of tachykinins and immune cells may provide advancements in new therapies for inflammatory and immune-related diseases, and targeting these signaling molecules may prove beneficial for psychiatric disorders.

Tachykinins participate in physiological regulation of the nervous, immune, gastrointestinal, urogenital, respiratory, and dermal systems. Due to the variety of expression levels throughout the mammalian body, it has been found that different tachykinin expression levels, in different areas, can lead to different physiological regulations. For example, activation of substance P receptors in the spinal cord regulates synaptic transmission in response to noxious stimuli (from nociceptors), while activation of central receptors for substance P regulates cardiovascular and respiratory functions, and is responsible for emetic reflex activation (Marriott, 2004). Tachykinins also play a role in modulating synaptic plasticity across several brain regions. Substance P has been shown to induce synaptic plasticity, through an NMDA receptor and protein synthesis-dependent mechanism, of CA2 synapses of the hippocampus (Dasgupta et al., 2017). It has also been shown to induce heterosynaptic potentiation of glutamatergic synapses within the thalamus (Masterson et al., 2010). Thus, like other neuropeptides, tachykinins are released by both the CNS and immune cells, can induce inflammatory responses, and modulate synaptic plasticity. Interestingly, altered tachykinin levels, especially substance P, are especially apparent in psychiatric disorders and are thought to contribute to the underpinning physiology of the associated symptoms.

56.12 Cytokines

Cytokines are small signaling molecules (<40 kDa) utilized by the immune system in response to infections, inflammation, and trauma. Cytokines can be classified into two major categories: proinflammatory, which aide in promoting inflammation and anti-inflammatory, which serve to reduce inflammation. Cytokines are released not only by peripheral immune cells, but also microglia within the CNS. The study of cytokines has gained traction due to their known involvement in inflammatory and psychiatric disorders. In fact, new therapeutic targeting of cytokines, particularly proinflammatory cytokines, has shown considerable promise for treating multiple disorders including septic shock, autoimmune disorders, osteoarthritis, and substance use disorders.

56.12.1 Expression and Function of Cytokines in the CNS and Immune System

Cytokines are expressed by an abundance of tissues within the body including neuronal, epithelial, and immune cells. However, cytokine levels can also be detected in high concentrations within blood serum as well as cerebrospinal fluid (CSF). Two major subcategories of cytokines include interleukins (IL) and tumor necrosis factor (TNF). Interestingly, ILs) and TNF have different expression across immune cells. ILs are classified as a class of glycoproteins produced predominantly by leukocytes, whereas TNF is expressed by all white blood cells, but not dendritic cells. Two proinflammatory cytokines within these groups, IL-6 and TNFα, are expressed by most immune cells including neutrophils, macrophages, T-cells, B-cells, and monocytes (Kany et al., 2019). IL-6 is perhaps the most well-characterized cytokine in regard to autoimmune diseases, bacterial infections, and psychiatric disorders, where serum levels are often highly elevated. TNFα is also well characterized for its dysregulation across a variety of psychiatric disorders, where high serum levels are associated with autoimmune disorders, major depressive disorder, and addiction. Most often, CSF levels of TNFα are elevated in major depression and addiction, suggesting a neuroimmune response as previously described by Namba et al., 2021.

Cytokines have been repeatedly shown to regulate synaptic plasticity, learning, and memory. In fact, glial cells that release cytokines are known to play a dynamic role in modulating synaptic plasticity beyond their original suspected role of a "support cell" for tissue homeostasis. Through the release of cytokines, both microglia and astrocytes impact synaptic plasticity through associated cytokine surface receptors present on neuronal plasma membranes. Interestingly, glial released TNFα facilitates AMPA receptor plasma membrane insertion, and endocytosis/internalization of $GABA_A$ receptors within the hippocampus, facilitating excitability and enhanced synaptic transmission (Beattie et al., 2002; Stellwagen et al., 2005). However, this effect appears to be opposite within the ventral striatum, where TNFα promotes internalization of AMPA receptors (Lewitus et al., 2016). Thus, inappropriate regulation of TNFα can have profound effects, promoting excitability within some brain regions and inhibition in others. Furthermore, targeting of cytokines for therapeutic treatments has shown promising results for a variety of disorders, which are discussed below. Particularly interesting is the effect demonstrating that inhibition of cytokines can prevent inappropriate synaptic plasticity and restore synaptic transmission across animal models of addiction and depression. We have highlighted some of these studies in detail in our recent review (Namba et al., 2021) and direct the readers there for a more thorough coverage of this topic.

56.13 The Role of the Neuroimmune System in Psychiatric Disorders

56.13.1 Major Depression

56.13.1.1 Overview of the Immune System in Major Depression

Understanding the relationship between depression and the immune system is an ever-evolving area of research. It should be noted that depression is an umbrella term that includes a variety of subtypes which are themselves difficult to specifically define. These different subtypes vary in presentation, intensity, pathophysiology, chronicity, and treatment, all lending support to the idea that depression may be better represented as a complex of disease states. Thus, it stands to reason that the influence on the immune system can be distinct, and even opposite, depending on the subtype of depression in consideration (Antonijevic, 2006). For example, a recent systematic review found that activity of HPA axis is normal or depressed in the atypical depressive subtypes, while it is increased in the melancholic subtype (Juruena et al., 2018). This section will focus on major depressive disorder (MDD) as its impact on the HPA axis is better characterized and more firmly established (Keller et al., 2017; Nandam et al., 2019).

It is important to acknowledge that correlating altered immune function and depression can be confounded by several variables that exist with both. Along with age and gender, other factors that are associated with immune function and depression include sleep status, the shared set of physical or psychiatric comorbidities, and that depression may be better characterized as a complex of disease states which impact the immune system in variable ways (Irwin, 1999, 2001). In fact, research has supported that psychiatric and stress disorders, including MDD, may share overlapping characteristics which describe the immune changes found with each (Irwin, 2001; Raison & Miller, 2001). Schizophrenia, bipolar disorder, posttraumatic stress disorder, stress and anxiety, and addiction will be discussed in more detail in the later sections.

Several mechanisms have been described as contributing to the pathology of depression. One longstanding hypothesis for depression is monoamine deficiency in which there is a depletion of mood stabilizing neurotransmitters, such as dopamine, serotonin, and norepinephrine (Delgado, 2000). Many antidepressants have targeted the monoaminergic pathways to increase the levels of these neurotransmitters in the brain, which have shown promising therapeutic effects in many individuals and populations. It has become clearer, however, that monoamine depletion is not a sufficient mechanism alone to represent depression, which explains why antidepressants are not always effective (Catena-Dell'Osso et al., 2013; Xu et al., 2017). Thus, in recent years, there has been increased investigation of potential links between immune function and depression.

Many components of the immune system have been correlated to depression, including alterations in circulating inflammatory cytokines, oxidative stressors, brain-derived neurotrophic factor (BDNF) as well as altered HPA axis regulation. Evidence has continued to accumulate in support of the hypothesis that dysfunctionality of the immune system can trigger or contribute to the presentation of MDD. Abnormal immune system components, including cytokines and inflammatory markers, are generally associated with MDD, regardless of the health status of the patient, an idea supported by many studies (Leonard, 2010; Raison et al., 2006; Ruiz et al., 2022). In fact, two large meta-analyses found that patients with major depression show reliable changes in immune function (Herbert & Cohen, 1993; Irwin, 1999), including lowered proliferative response of lymphocytes to mitogens, lower NK cell activity and changes in white blood cell populations, where decreased number of lymphocytes and increased number of neutrophils and monocytes have been observed (Herbert & Cohen, 1993).

Several studies have demonstrated the relationship between proinflammatory cytokine expression and depression. For example, a meta-analysis by Dowlati and colleagues found that higher concentrations of circulating TNF-α and IL-6 was observed in patients with depression compared to healthy individuals (Dowlati et al., 2010). Furthermore, a systematic review by Orsolini et al. showed support for c-reactive protein (CRP) as an inflammatory biomarker in depression (Orsolini et al., 2022). A noteworthy influence of these proinflammatory cytokines is on the neurogenesis of the hippocampus, where both TNF-α and IL-6 have antiproliferative and antineurogenic effects of the progenitor cells that support important functions such as behavior, mood, cognition, and memory (Cacci et al., 2005; Ekdahl et al., 2003; Iosif et al., 2006; Kino, 2015; Monje et al., 2003). Thus, alterations in TNF-α and IL-6 could have profound impacts on disease progression and associated symptomology.

The hippocampus is an important modulator of the HPA axis where it expresses both MRs and GRs responsible for detecting altered levels of corticosteroids, and when activated can exert a negative feedback mechanism to shut down HPA axis activity (Furay et al., 2008; Jacobson & Sapolsky, 1991). Furthermore, the negative feedback of glucocorticoids at the hippocampus is important because corticosteroids also have anti-neurogenic properties, much like proinflammatory cytokines (Cameron & Gould, 1994). Brain-derived neurotrophic factor (BDNF), a protein that promotes neuroresilience in stress-induced synapses (Allen & Dawbarn, 2006; Brunoni et al., 2008), is reduced in both blood serum and the hippocampus in depressed patients (Duman & Monteggia, 2006; Dwivedi, 2009) and likely plays a role in perpetuating HPA axis dysregulation. BDNF has gained a spotlight as an immunological biomarker of MDD due to its impact on the regeneration of monoamine neurotransmitters and signaling pathways such as MAPK (Jin et al., 2019), which is described in further detail in the cytokine section below.

Glucocorticoids and proinflammatory cytokines have known impacts on the metabolism of tryptophan—a substrate of one of the major monoamines that is often depleted in depression, serotonin. Tryptophan 2,3-dioxygenase (TDO) and indoleamine 2,3-dioxygenase (IDO) are two enzymes that are responsible for converting tryptophan into kynurenine, ultimately reducing the tryptophan concentrations available for serotonin synthesis. Glucocorticoids have been found to induce TDO activation and proinflammatory cytokines, such as TNF-α and IL-6, which in turn induce IDO activation (Campbell et al., 2014; Danesch et al., 1983; Maes et al., 1990; Young, 1981). A study executed by Sordrager and colleagues elucidated how an imbalance between the HPA axis and the metabolism of tryptophan may influence the presentation of depression (Sorgdrager et al., 2017). The role of these enzymes and serotonin dysregulation is further discussed in the section below on bipolar disorder.

Lastly, altered circadian rhythm patterns have been described in depression. One study showed that MDD patients demonstrated an elevation of IL-6 levels, which fluctuated across the circadian cycle (Alesci et al., 2005). While serum cortisol was not significantly different between MDD patients and controls, there was a shift in serum IL-6. Furthermore, this circadian alteration of IL-6 levels was found to be significantly correlated with the mood of the patient. In addition to these findings, Khairuddin and colleagues have recently described the dysregulation of orexins, neuropeptides involved in the circadian rhythm, neural reward systems, and neuroendocrine stress response, in patients with MDD and may play a role in symptomology of patients with MDD (Khairuddin et al., 2020).

As this overview highlights, there is an appreciable and growing body of evidence supporting an intimate interplay between the immune system and MDD. While it is still not completely clear which of the two precedes the other, there is data demonstrating differences in the levels of TNF-α and IL-6 in MDD patients compared with healthy controls.

Elucidating this relationship becomes clinically relevant as it may give rise to new therapeutic developments and optimize treatment for individuals with MDD.

56.13.1.2 Impact of the HPA Axis on Major Depression

A commonality found between some presentations of chronic stress and MDD is HPA axis hyperactivity (Holsboer, 2000; Menke, 2019). This manifests with elevated levels of the HPA axis hormones CRH, AVP, ACTH, and cortisol (Holsboer, 2000; Mikulska et al., 2021; Owens & Nemeroff, 1991). The increase in glucocorticoids in depression, however, is not regulated by normal negative feedback mechanisms (Wyrwoll et al., 2011). For example, the affinity of these glucocorticoids for their receptors throughout the hippocampus, HPA axis, and other brain structures has been found to be reduced or impaired in patients diagnosed with depression (Modell et al., 1997; Slavich & Irwin, 2014). Interestingly, Boyle and colleagues found that forebrain-specific GR knock-out mice demonstrated increased depression-related behavior (Boyle et al., 2005) which has profound implications for the human phenotype. Furthermore, in humans GR antagonists, such as Mifepristone, have been found to be effective in the treatment of depression in that it takes advantage of the compensatory measures of the HPA axis to increase the expression of GR (Mikulska et al., 2021). In addition, animal studies exploring other HPA axis receptors have found that while depression and anxiety phenotypes may be elicited with CRH injection into the brain, antagonism of the CRHR1 decreases the symptoms of depression (Heinrichs & Koob, 2004). Each of these studies lends support to the idea that both the over-activity and dysregulation of the HPA axis can be implicated in the presentation of depression.

Another important feedback system worth mentioning involves FK506-binding protein 51 (FKBP51), a protein that associates with the 90 kDa heat shock protein (Hsp90) to regulate GR (Criado-Marrero et al., 2018). FKBP51 is both activated by and an inhibitor of GR (Denny et al., 2000; Hubler & Scammell, 2004; Wochnik et al., 2005). This relationship allows for an intracellular negative feedback modulation of GR (Rein, 2016), ultimately impacting downstream regulation of the HPA axis. Newer antidepressant formulations have taken advantage of this process by selectively inhibiting FKBP51 to restore appropriate functioning of GR (Gaali et al., 2015). This feedback loop provides an added layer of complexity in the regulation of the HPA axis and may influence the efficacy of traditional antidepressants.

As we have mentioned above, the stressed-related elevation of cortisol levels (Bertollo et al., 2020) can disturb the monoaminergic pathways involved in MDD. The therapeutic effectiveness of antidepressants varies between depressed patients which may be caused by individualized and unique HPA axis alterations among patients. In one neuroendocrine cohort study, depressed patients responded poorly to antidepressant treatments when HPA axis disruption continued; moreover, the relapse rate of patients was increased when dysfunction of the HPA axis returned (Zobel et al., 2001). Another study that evaluated the effect of amitriptyline on rats found that changes in expression of MRs and GRs in the limbic system were associated with altered homeostasis of the HPA axis (Reul et al., 1993). These findings together support an HPA axis-based etiology for depression.

56.13.1.3 Neuropeptides Involved in Major Depression

Neuropeptides involved in MDD have been determined based on the mechanisms discovered utilizing MDD disease models. While there are over one hundred neuropeptides in the mammalian system that have been described to date, several have been more closely associated with psychiatric disorders such as depression and anxiety (Rana et al., 2022). Some of these neuropeptides play a primary role in the HPA axis, including CRH and AVP, which are released by the paraventricular nucleus of the hypothalamus to promote downstream release of corticosteroids such as cortisol (Bao & Swaab, 2010). Several other neuropeptides secondarily affect the HPA axis, generally by means of CRH, AVP, ACTH, and/or their receptors. For example, circulation of neuropeptides, such as apelin-13, galanin (GAL), and neuropeptide Y (NPY), has been shown to ameliorate hyperactivity of the HPA axis (Dai et al., 2018; Kash & Winder, 2006; Unschuld et al., 2010), although studies on the latter have been controversial (Held et al., 2006; Hou et al., 2006). Moreover, apelin-13 has been shown to support hippocampal BDNF expression as well as GR nuclear translocation in animal models of stress-induced depression (Dai et al., 2018). Neuropeptides are also known to directly impact monoamines and play a role in their dysregulation observed in individuals with depression. For example, GAL has been shown to act as a modulator of 5-hydroxytryptamine serotonin receptors (Ogren et al., 2007), supporting the monoamine mechanism hypothesis of depression. However, it should be mentioned that the body of literature describing the relationship of neuropeptides does not solely focus on the molecules alone but their receptors as well. Depending on the specific receptor a neuropeptide binds, the outcome may be variable and even opposite, particularly in regard to the CRH and GAL receptor subtypes (Demsie et al., 2020; Zhu et al., 2011).

It is important to mention that there is growing interest in the role that neuropeptides have on circadian patterns. Given that circadian rhythms are altered in individuals with depression (Crouse et al., 2021), these findings can have profound implications. Several studies and meta-analyses have suggested a relationship between psychiatric disorders, the HPA axis, circulating levels of neuropeptides, and the time of day (Coulon et al., 2016; Engeland & Arnhold, 2005; Mikulska et al., 2021). Whereas it is normal for morning time cortisol levels to be elevated, individuals experiencing depression can show an elevation in cortisol throughout the day compared to healthy controls. In fact, recent reviews have highlighted the dysfunction of orexin neuropeptides on circadian patterns and depressive symptomology, although it is still not clear whether this is due to a hyper- or hypo-activity of the orexinergic pathways (James et al., 2017; Khairuddin et al., 2020). It is noteworthy, regardless, that orexin neurons have been found to innervate the paraventricular nucleus of the hypothalamus, suggesting some effect on the HPA axis (Smith & Vale, 2006).

Other neuropeptides involved in the pathophysiology of depression, including oxytocin (OXT), pituitary adenylate cyclase-activating polypeptide (PACAP), tyrosine melanocyte-stimulating hormone-release inhibiting factor-1 (Tyr-MIF-1), cholecystokinin (CCK), MCH, thyrotropin-releasing hormone (TRH), angiotensin II, and substance P, have been recently described in detail elsewhere and thus will not be discussed here (Rana et al., 2022; Werner & Coveñas, 2010).

56.13.1.4 The Role of the Neuroimmune System and Cytokines in Major Depression

The relationship between inflammatory cytokines in the neuroimmune system with MDD followed observations that those with MDD displayed many of the symptoms of sickness caused by infection, termed "sickness behavior." This includes lethargy, sleep disturbances, lack of appetite and hyperthermia, among others (Hart, 1988). Since then, many studies throughout the years have elucidated the role of proinflammatory cytokines in the depression phenotype including IL-6, TNF-α, IL-1β, and CRP (Brebner et al., 2000; Dowlati et al., 2010; Goldsmith et al., 2016; Kim et al., 2007b; Valkanova et al., 2013). This work ultimately supported a neuroinflammatory hypothesis in the pathophysiology of depressive illness (Maes et al., 2009; Miller et al., 2009).

It is important to understand that proinflammatory cytokines play a normal role in neurogenesis and neuroprotection. Animal studies looking at IL-6 knockout mice and TNF-α/TNF-R1 knockout mice (Camara et al., 2013) showed that hippocampal neurogenesis was significantly lower than for controls that had physiological levels of circulating cytokines. These proinflammatory cytokines work to defend neurons from potential harm and help sustain neural integrity. On the other hand, it is also the chronic dysregulation of these cytokines or their sustained release that can cause the damage seen in psychiatric disorders such as MDD (Kim et al., 2016; Leonard, 2007). The neuroinflammation seen in MDD has been shown in a large meta-analysis by Palta and colleagues to potentiate an imbalance between oxidants and antioxidants downstream (Palta et al., 2014). Neurons depend on oxidative phosphorylation for energy, and failure to produce high levels of antioxidants (Mahadik & Mukherjee, 1996) can lead to cellular damage caused by reactive oxidative species, which can be detrimental.

It is of interest that several studies have highlighted how other inflammatory diseases commonly exist as comorbidities of depression including coronary artery disease (Zellweger et al., 2004), rheumatoid arthritis (Margaretten et al., 2011), as well as other chronic inflammatory or autoimmune conditions (Pryce & Fontana, 2017). Results of such work linking inflammatory conditions with MDD have inspired interest in using anti-inflammatory medications as adjuvant treatments with antidepressants. Thus, there is a growing body of literature demonstrating a benefit to using nonsteroidal anti-inflammatory drugs or cytokine blockers to augment antidepressants and improve the efficacy of treatment in MDD (Köhler et al., 2014; Misiak et al., 2018; Shariq et al., 2018).

In the CNS, the driver of the inflammatory process relies on the brain's primary immune cell, the microglia (Fenn et al., 2014). A number of studies have supported mechanisms that describe how homeostasic disruptions of the neural environment and/or injury to the brain stimulates microglial activation and the subsequent release of proinflammatory cytokines (Lehnardt, 2010; Streit et al., 2004). For example, CRH has been thought to activate neurosupport cells, namely microglia and astrocytes, following chronic stress exposure (Rana et al., 2022). Even more interesting, a study by Bortolato and colleagues showed that TNF-α release by microglia is activated when peripheral TNF-α binds to type 4 toll-like receptors in the brain (Bortolato et al., 2015). Taken together, these studies all substantiate an argument for microglial activation in MDD and provide further insight into the cellular underpinnings of the disorder.

While we have touched on many proinflammatory cytokines throughout this chapter, there are a couple of additional factors worth mentioning. Primarily, (NF-κB), mitogen-activated kinase (MAPK), and signal transducer and activator of transcription 5 (STAT5) are considered to the major modulatory pathways of the proinflammatory cytokines discussed (Munhoz et al., 2010; Pace et al., 2007). For example, NF-κB has been implicated as a driver of IL-6 and is known to interfere with hippocampal neurogenesis (Mouihate, 2014), which are both associated with the depressive phenotype. Additionally, these cytokines have been shown to inhibit GR function (Pace et al., 2007), ultimately disrupting the negative feedback mechanism of the HPA axis, thus, creating a de facto positive feedback loop, whereby continued corticosteroid release can promote greater cytokine secretion. Lastly, while we have discussed many cytokines and peptides associated with MDD, a number of other cytokines and chemokines have been described in a review paper elsewhere (Misiak et al., 2018), and therefore will not be discussed here.

56.13.2 Schizophrenia

56.13.2.1 Overview of the Neuroimmune System in Schizophrenia

The involvement of the immune system in schizophrenia (SZ) is an even more complicated issue compared to involvement with depression. There have been suggestions for many decades that schizophrenia is associated with immune dysfunction (Vaughan et al., 1949). Particularly, the discovery of genetics and immune marker correlations to the disorder have been pertinent in the "immune hypothesis" of SZ. To date, studies highlighting the altered HPA axis function as well as neuropeptide and cytokine levels in patients with SZ compared to normal healthy individuals has solidified the idea that an altered immune system likely plays a substantial role in the etiology of SZ and the degree of symptoms associated. However, given the wide degree of symptoms associated with SZ, the spectrum of the disorder, and the potential genetic and environmental components, it is difficult to control for the numerous factors frequently associated with SZ, and that is no exception for the degree of immune system dysregulation. Simplistically, factors contributing to SZ often include age, gender, sleep status, genetic predisposition, and environmental experiences (such as stress and trauma), which likely each effect immune system function to different degrees and severities, making this field of study extremely complex.

Nevertheless, despite the difficulties, there are intriguing findings that have strengthened the hypothesis that neuroimmune interactions are altered in schizophrenia patients. Most interestingly, epidemiological data has shown a linkage between prenatal viral infection and subsequent appearance of SZ, particularly when the viral infection occurs within the third trimester (Kneeland & Fatemi, 2013). Countless studies have now demonstrated a link between HPA axis dysfunction, particularly hypofunctionality, within individuals with SZ and numerous neuropeptides and cytokines have been found to be dysregulated in these patients (Buckley, 2019; Keller et al., 2017; Momtazmanesh et al., 2019). Further, SZ has been found to be associated with autoimmune diseases and SZ, in certain instances, is likely associated with abnormal autoimmune function (Al-Diwani et al., 2017). In some cases, it is thought that dysfunctional autoimmunity is due to prenatal loss of a specific subset of T cells (Kipnis et al., 2006), based on several studies from the Schwartz lab that have shown that autoimmune T cells exert a neuroprotective role within the CNS, especially in response to stress. Such studies have yielded the hypothesis that the prenatal loss of relevant autoimmune T cells renders the human brain more prone to neurodegeneration upon stressful stimuli, which often occurs later in life, leading to the onset of SZ and its associated symptomology.

Aside from the autoimmune hypothesis, another hypothesis for the onset of SZ is the dopamine hypothesis, whereby abnormally high levels of dopamine induce psychotic symptoms associated with SZ. This hypothesis speculates that in normal individuals, dopamine release increases centrally in response to stress and this normally suppresses regulatory T cells, leading to increased activity in protective autoimmune T cells. However, in schizophrenia, there is a loss of autoimmune protection, consistent with a lack of protective autoimmune T cells, whereby such elevations in dopamine would lead to greater neurodegeneration without the protective mechanism of autoimmune T cells. Thus, in summary, the role of the immune system in the development of SZ is extremely complex. However, below we summarize major consistent findings in regard to the immune system and correlations to the disorder and have directed readers to comprehensive reviews where appropriate.

56.13.2.2 Impact of the HPA Axis on Schizophrenia

Hyperactivity of the HPA axis has been implicated in the pathology of schizophrenia, where elevated basal levels of cortisol are commonly observed (Cherian et al., 2019; Walker et al., 2008). In fact, elevated cortisol levels have been shown to augment and exaggerate psychotic symptoms in patients with SZ (Walker et al., 2008). Newer studies have found that individuals with schizophrenia exhibit altered genetic and epigenetic variations in genes related to the HPA axis, which are associated with cognitive performance in the clinical setting (Ferrer et al., 2020). In fact, individuals with schizophrenia have altered mRNA for genes encoding GRs, which has been confirmed utilizing post-mortem tissue from patients with SZ. In addition, psychosis exhibited in these patients has been linked to FKBP5, a gene that encodes a molecular co-chaperone of the GR complex important for intracellular glucocorticoid signaling and regulation of the stress response (Ferrer et al., 2020). Such alterations give rise to an inability to respond appropriately to stress and induce psychotic symptoms characteristic of the disease. Interestingly, antipsychotics and drugs targeting hormones related to the HPA axis have yielded promising results for treatment of SZ, where they have been shown to alleviate cognitive symptoms associated with the disorder (Soria et al., 2018). Although it is well known that the HPA axis is hyperactive in individuals with schizophrenia, newer research suggests that the involvement of the HPA axis in the pathogenesis depends mainly on the severity of the disease (Mikulska et al., 2021) and therefore further research is needed to identify how the HPA axis dysregulation gives rise to symptoms commonly seen within the disorder.

56.13.2.3 Neuropeptides Associated with Schizophrenia

Similar to other disorders, neuropeptides have found to be dysregulated in SZ. Until recently, neuropeptide Y (NPY) has been perhaps the most commonly studied neuropeptide in regard to SZ, which has been repeatedly shown to be reduced in individuals with the disorder (Frederiksen et al., 1991; Gabriel et al., 1996; Itokawa et al., 2003). In addition, peptide YY (PYY), a peptide exclusively expressed by endocrine cells of the digestive system, has also been shown to be reduced in SZ (Widerlöv et al., 1988). Together, NPY and PYY have been suggested to be useful trait markers in identifying patients with SZ (Widerlöv et al., 1988), and post-mortem studies have found decreases in these peptides within the temporal cortex of patients with SZ, among others including arginine vasopressin and galanin (Frederiksen et al., 1991). More recently, focus has shifted to the somatostatin and neurotensin systems and their involvement with SZ. Neurotensin has been found to primarily modulate dopaminergic neurotransmission, making it particularly relevant to SZ, where dopamine levels are often dysregulated and fluctuate greatly. Previously, it was reported that patients with SZ have substantially reduced CSF levels of neurotensin relative to control/healthy individuals (Binder et al., 2001; Breslin et al., 1994). However, it has also been shown that while some patients have low levels of neurotensin within the CSF, others have normal or slightly elevated levels (Ni et al., 2019; Widerlöv et al., 1988). From animal studies, it is suggested that those with high levels of neurotensin may have ErbB4 gene-related schizophrenia, which was correlated in a mouse model of SZ (Ni et al., 2019). Interestingly, elevations in ErbB4 mRNA have been shown to be correlated with deficits in prefrontal interneuron activity (Joshi et al., 2014) and may play a role in somatostatin dysregulation, decreased somatostatin interneurons within the prefrontal cortex and symptoms observed in SZ. Thus, while neurotensin and its role in SZ has been studied for a long period of time (Binder et al., 2001; Breslin et al., 1994), newer findings suggest that targeting this peptide can be used as a therapeutic mechanism for individuals suffering from the disorder.

In addition to those discussed above, the somatostatin system has also been implicated in SZ, where older studies have reported decreased levels of somatostatin in temporal and frontal lobes (Gabriel et al., 1996) and mRNA levels were decreased in the dorsolateral prefrontal cortex (Morris et al., 2008) in patients with SZ. Similarly, decreased levels of somatostatin-like immunoreactivity in the CSF have been observed (Bissette et al., 1986). Newer research has focused on somatostatin positive interneurons within the neocortex, where dysregulation of these neurons is thought to contribute to oscillatory and information deficits in individuals with SZ (Van Derveer et al., 2021). To further this idea, others have shown that potentiation of metabotropic glutamatergic receptors (mGluR1) can increase somatostatin interneuron activity within the prelimbic prefrontal cortex and is capable of rescuing schizophrenia-like cognitive deficits (Maksymetz et al., 2021). Also, prefrontal-specific knockdown of somatostatin is capable of producing schizophrenia-like behavioral deficits (Perez et al., 2019). It is of note that antipsychotic treatment has been shown to have beneficial effects on cortical levels of somatostatin, as well as NPY, increasing levels back to normal (Sakai et al. 1995), suggesting that antipsychotics may have therapeutic effects by targeting neuropeptides. Much work remains to be done in order to identify all neuropeptides dysregulated within the disorder; however therapeutic effects may prove beneficial by targeting dysfunctional neuropeptides of the disorder. Furthermore, neuropeptides may even modulate an individual's response to antipsychotics and may prove beneficial for guiding treatment options in certain individuals (Rodríguez et al., 2020).

56.13.2.4 Impact of Immune System Cytokines in Schizophrenia

The dysregulation of cytokines in SZ has gained attraction over the past few years. A number of cytokines have been found to be altered in individuals with the disorder, including TNF-α, IL-6, and TGF-β (Momtazmanesh et al., 2019). While some studies have shown no alterations in IL-6 in patients with SZ (Hartwig et al., 2017), a significant number of studies have reported significant increases in IL-6 in patients treated with and without antipsychotics after their first episode (for review see Momtazmanesh et al., 2019). However, these discrepancies may be due to the patients who were enrolled in the studies and their disease duration and prognosis. For example, it has been found that patients with elevated levels of IL-6 tend to have longer hospitalizations upon episode onset and longer disease durations (Dahan et al., 2018). In addition, IL-6 levels are positively correlated with more severe positive symptoms, such as hallucinations (Zhang et al., 2002). Interestingly, prolonged labor results in abnormally elevated serum IL-6 levels (De Jongh et al., 1997), which suggests a role of perinatal complications and the development of SZ.

The proinflammatory cytokine TNF-α has been shown to be elevated in first-episode and drug naïve patients with SZ as well as patients treated with antipsychotic medications but still having psychotic episodes (Falcone et al., 2015; Miller et al., 2011). Interestingly, individuals with a disease duration of under two years (Zhu et al., 2018), and patients taking antipsychotics for a duration longer than 5 years (Lv et al., 2015) were more likely to show a decrease in TNFα levels. Again, these confounding results are likely due to the complexity of the disorder including various treatment regimens and timeline of the disorder.

In addition to IL-6 and TNF-α, other proinflammatory cytokines such as TGF-α and IL1-β have also shown to be elevated in patients with SZ (Goldsmith et al., 2016; Upthegrove et al., 2014). While normal proinflammatory cytokines aide in repair of tissue damage and infection, it is interesting that abnormal elevations of TGF-α and IL1-β can suppress the normal expression of BDNF, a neurotrophin that regulates neuronal and synaptic development in the hippocampus (Lapchak et al., 1993). Conversely, IL1-β causes upregulation of nerve growth factor (NGF), another trophic factor that regulates growth and development of central neurons, and microglial cells (Heese et al., 1998). These findings compel further research into the interactions between cytokines and the brain-derived trophic factors that may lead to abnormal development of brain regions that ultimately result in the manifestation of schizophrenia in the mature brain.

Antipsychotic treatment of SZ with both typical and atypical antipsychotics has been shown to decrease levels of IL-1β and IL-6 (Miller et al., 2013a), which may be part of their therapeutic effects. While targeting of cytokines has not been well established in the field of SZ treatment, some have utilized levels of cytokines to predict the response to treatment in patients with SZ (Şimşek et al., 2016). Interestingly, high levels of IL-6 are associated with treatment resistance (Lin et al., 1998), whereas TGF-β levels had no relation to treatment response (Lee & Kim, 2010). Overall, these studies suggest that cytokines are dysregulated in patients with SZ, levels of dysregulation may predict treatment outcome and that treatment with antipsychotics may restore levels of cytokines back to basal levels. However, further studies are necessary to establish time-line changes, treatment changes as well as a causative relationship between cytokine levels in patients and the precipitation of psychotic episodes. Such studies will greatly enhance our understanding of how the entire neuroimmune system plays a role in the development of SZ and the associated symptomology.

56.13.3 Bipolar Disorder

56.13.3.1 Impact of HPA Axis on Bipolar Disorder

The abnormalities of HPA axis activity remain associated with bipolar disorder (BD); however, its association as a mechanism of the pathophysiology, consequence of BD, endophenotype, or etiological factor remains unclear (Hasler et al., 2006). In comparing several indices of HPA axis activity, BD is associated with significant changes in basal CORT and ACTH, measured via plasma, saliva, CSF and other biological fluids (van den Berg et al., 2020; Vieta et al., 1997). Hypercortisolemia has long been reported in depressed patients with mood disorders involving HPA axis activity disruption (Gillespie & Nemeroff, 2005; Umeoka et al., 2021). Interestingly, chronic hypercortisolemia can lead to downregulation in GR transcription, translation, and expression which acts as a negative feedback loop of the HPA axis, propagating hypercortisolemia (Rosenblat & McIntyre, 2017). Hypercortisolism is thought to play a central role in the pathogenesis of cognitive deficits and depressive symptoms of BD, which may result from the neurocytotoxic effects of heightened cortisol levels (Wartchow et al., 2023). Prolonged hypercortisolemia can regulate the onset of BD symptoms and elevate the risk of relapse after significant psychosocial stress (Marques et al., 2009). Although CRH regulates ACTH release from the pituitary gland, there was no statistically significant difference in CSF CRH levels between patients with BD and controls (Belvederi Murri et al., 2016).

As mentioned above, the pituitary gland is involved in HPA axis activity via the anterior pituitary gland secretion of ACTH which promotes the production of cortisol (Farrow et al., 2020). Because cortisol secretion is increased in BD, pituitary gland volume (PGV) has been associated with hyperactivity of the HPA axis (Farrow et al., 2020) leading to greater production of ACTH (Clark et al., 2014; Pariante et al., 2005). Because of this, there has been a suspected possibility of patients with BD having an increase in PGV; however, a meta-analysis of studies done to confirm this correlation has yielded conflicting results and show an insignificantly small increase in PGV (Clark et al., 2014). Yet, there have been studies that have found an increase in PGV in BD to a significant degree (MacMaster et al., 2008; Pariante et al., 2005).

The role of HPA axis activity has also been studied as a possible etiological or endophenotypic factor of BD. Surprisingly, there is more consistent data revealing HPA axis hyperactivity more often in the offspring of patients with BD than in studies on first-degree relatives (Belvederi Murri et al., 2016). This suggests environmental stressors affecting offspring rather than a heritable trait, which would have also been seen in first-degree relatives. Furthermore, the level of childhood trauma, dysregulated parenting style experiences, and subsequent transition to disease phenomena were correlated with HPA axis activity (Belvederi Murri et al., 2016; Wartchow et al., 2023). Childhood trauma is associated with epigenetic modifications of the GR gene in patients suffering from BD (Belvederi Murri et al., 2016; Fish et al., 2004; Perroud et al., 2014). In an attempt to understand the association between HPA axis activity and early life stress, Farrow et al. (2020) examined a structural component of the HPA axis, revealing greater baseline anterior PGV in those with early life stress. These findings are consistent with altered HPA axis activity as a pathogenetic mechanism mediating the interactions between genes and the environment in the presentation of BD (Belvederi Murri et al., 2016).

Increased ACTH and CORT, known otherwise as hyperactivity of the HPA axis, are also more prominent among patients assessed during the manic phase of BD; however, it is also evident in euthymia (Belvederi Murri et al., 2016). Because there was a difference in expression during the manic phase, this suggests a state-like property in which HPA axis activity abnormalities vary according to the illness state of BD (Wartchow et al., 2023). On the other hand, because hyperactivity of the HPA axis was still present in euthymia, it suggests a trait-like role characterizing BD (Girshkin et al., 2014). Hence, this may indicate pathophysiological HPA axis activity within BD. This relationship suggests that targeting the HPA axis pharmacologically could improve the outcome of BD (Belvederi Murri et al., 2016; Watson et al., 2012).

56.13.3.2 Neuropeptides Associated with Bipolar Disorder

Because of evidence that suggests circadian rhythm disruption is associated with BD, neuropeptides involved in circadian rhythm regulation were examined among patients with BD (Aydin et al., 2013; Pagani et al., 2016; Yildiz et al., 2022). Significantly decreased levels of serum somatostatin (SST), NPY, VIP, and cortisol were observed in the morning in patients with BD compared to healthy controls (Yildiz et al., 2022). These lower levels suggest a disruption in circadian rhythms and may correlate with increased morning anxiety and depression often observed in patients with BD (Murray, 2007; Yildiz et al., 2022). Serum melatonin secretion has also been determined to be significantly lower among BD patients than healthy controls, regardless of illness state (Kennedy et al., 1996; Takaesu, 2018). Because of its role in circadian rhythm regulation, it can be a culprit of the imbalance of the circadian rhythm in BD (Takaesu, 2018).

Orexin-A (hypocretin-1), a hypothalamic neuropeptide that regulates sleep, appetite, and arousal levels, has been suggested to be involved in the pathophysiology of psychiatric disorders such as BD, major depressive disorder (MDD), and schizophrenia (Saper et al., 2005; Tsuchimine et al., 2019). Studies revealed that plasma orexin-A levels are significantly lower in patients with BD when compared to healthy controls (Tsuchimine et al., 2019). However, there was no statistically significant difference in orexin-A between healthy controls and patients with MDD or patients with schizophrenia (Tsuchimine et al., 2019). An attempt to correlate orexin-A levels and clinical variables including the symptoms, the severity of depression, and the dose of medications also yielded no association (Tsuchimine et al., 2019). When another study was conducted in 2021, there were significantly higher plasma levels of orexin-A in those with BD compared to healthy controls (Li et al., 2021). Although more research is needed to understand this discrepancy, studies utilized dual hypocretin antagonists and hypocretin-2 (orexin-B) receptor antagonists in those with mood disorders and MDD, and it resulted in an overall improvement in sleep quality and a decrease in anxiety and depression (Brooks et al., 2019; Li et al., 2021; Nakamura & Nagamine, 2017), suggesting that higher levels of hypocretin may be involved in the pathophysiology of sleep disturbances, anxiety, and depression.

56.13.3.3 Impact of Immune System Cytokines in Bipolar Disorder

Evidence suggests a bidirectional link between inflammation and mood disorders (Brietzke et al., 2012; Kivimäki et al., 2014; Rosenblat et al., 2014). This is supported by studies showing that emotional, psychological, or physical stress could induce stress responses (Miller et al., 2013b). Indeed, elevated IL-1, IL-6, TNF-α, and CRP have been observed following the application of psychological stress alone. In patients with bipolar disorder, numerous studies have correlated the elevation of inflammatory markers [prostaglandin E2 (PGE2), CRP, TNF-α, IL-1-β, IL-2, IL-6] with increased incidence of mood symptoms and mood episodes (reviewed in Rosenblat et al. 2014). Recently, elevated IL-6, IL-10, IL-15, IL-27, and CXC ligand chemokine have also been correlated with BD (Martinuzzi et al., 2021). Tumor necrosis factor-α receptor 1 (TNFR1) levels were elevated significantly in patients with BD, even when compared to those with MDD (Chen et al., 2020). Interestingly, those suffering from anxiety disorders like atypical BD appear to progress to BD as a continuous disease process, and the distinctive phase at which the anxiety disorders develop into BD is characterized by an elevation of cytokines as well as other symptoms (Du et al., 2017).

Evidence also suggests that changes in gray matter (GM) volume can be used as a biomarker in mood disorders like BD and MDD (Chen et al., 2020; Han et al., 2014; Keramatian et al., 2018; Peng et al., 2016; Singh et al., 2012). Specifically, GM volume decreased in the left middle frontal cortex, right occipital fusiform cortex, and left postcentral cortex in those with BD compared with controls (Chen et al., 2020). Interestingly, GM volume differences have also been negatively correlated with IL-6 receptor levels among patients with BD and MDD. However, Chen et al. 2020 reported a positive correlation between the GM volume of the middle frontal cortex and TNFR1 levels. The Wisconsin card sorting test (WCST) was used to evaluate executive function and revealed a positive relationship between executive function and the volume of the left middle frontal cortex (Chen et al., 2020). Because the middle frontal cortex is involved in the control of attention, executive function, and mood regulation, and GM volume is decreased within this region, this suggests a cytokine-related mechanism by which GM volume changes can affect executive function in those with BD (Alesci et al., 2005; Japee et al., 2015; Niu et al., 2017).

One postulated mechanism to explain how proinflammatory cytokines can induce depressive symptoms is through the regulation of tryptophan metabolism. Tryptophan is a precursor of serotonin, melatonin, and other bioactive metabolites. Impairment of serotonin neurotransmission underlies one theory explaining the etiology of mood disorders including major depression and bipolar disorder (Cervenka et al., 2017). Specifically, depletion of brain tryptophan levels can induce depressive symptoms in individuals who are sensitive to major depression (Booij et al., 2020). Moreover, inhibition of tryptophan hydroxylase, the first step in the pathway of serotonin synthesis, can induce a relapse of depression in remitted depressive patients (Delgado et al., 1990; Smith et al., 1997), which may also limit the capacity to produce serotonin. Tryptophan is metabolized by indolamine 2,3-dioxygenase (IDO) to kynurenine (Correia & Vale, 2022). IDO activity is upregulated, at least in part, by proinflammatory cytokines such as IL-1β (Correia & Vale, 2022), interferon-α, IFN-γ (the major inducer), IL-2, TNF-α, and IL-6 (Anderson & Maes, 2015; Correia & Vale, 2022; Raison et al., 2006). Kynurenine (Kyn), kynurenic acid, and quinolinic acid are all tryptophan metabolites, and each has been shown on its own to induce depressive symptoms (Maes, 2011). Proinflammatory cytokines, such as IL-1, IL-6, and TNF-α, have also been shown to activate the HPA axis leading to increased circulating cortisol levels (Correia & Vale, 2022). Cortisol, in turn, drives higher IDO activity and tryptophan depletion (Maes, 2011). Finally, the elimination of serotonin from the brain is initiated by its conversion to 5-hydroxyindoleacetic acid, a process that is enhanced by IL-6 and TNF-α (Lotrich, 2015). Additionally, Kyn is degraded into quinolinic acid (Quin) by microglial cells and kynurenic acid (Kyna) by astrocytes. An additional pathway by which Kyn has been associated with MDD and schizophrenia is through dysregulation in the levels of these two metabolites causing neuronal excitotoxicity (Quin) or protection (Kyna) (Cervenka et al., 2017; Correia & Vale, 2022). Further, there is an overall decrease in the 5-HT signaling pathway due to the shunting of the Kyn pathway, which is implicated largely in depression (Correia & Vale, 2022; Cowen & Browning, 2015). Thus, one of the most effective and commonly prescribed antidepressants is selective 5-HT reuptake inhibitors (Correia & Vale, 2022), which leads to elevated 5-HT within the synaptic cleft and prolonged activity at its receptors.

Many studies have examined how cytokines vary in BD compared to healthy individuals. A meta-analysis, consisting of 30 studies evaluating levels of cytokines in patients with BD, has revealed consistent elevations in proinflammatory, anti-inflammatory, and regulatory cytokines (Modabbernia et al., 2013). Increases in IL-6 and TNF-α serum levels have been consistently observed in patients with BD (Brietzke et al., 2012; Ortiz-Domínguez et al., 2007), relative to healthy controls. In contrast, other cytokines have been revealed to have different levels during the manic and depressive phases of BD. For example, during manic episodes, elevated serum levels of IL-6, TNF-α and IL-4 (Kim et al., 2007a; Ortiz-Domínguez et al., 2007) have been observed; whereas during depressive phases, serum levels of IL-6 and TNF-α are elevated, while serum IL-2 levels are decreased (Ortiz-Domínguez et al., 2007). These studies suggest that serum levels of cytokines may vary depending on the phase of the illness. More recently, a systematic review was also conducted by Rowland et al. (2018) evaluating the altered levels of cytokines across the different phases of BD, including mania, euthymia, and depression, with an attempt to determine whether altered levels could discriminate between each mood phase. The analysis showed that hsCRP, IL1-RA, IL-6, sIL-2R, sIL-6R, TNF-α, and sTNFR1 were significantly elevated in the manic phase. IL-6, sIL-2R, and sIL-6R were significantly elevated in the euthymia phase (Rowland et al., 2018). Finally, TNF-α was found to be elevated during the depressive phase (Rowland et al., 2018). While the analysis revealed promising results, no biomarkers were found that could definitively discriminate each mood phase due to limited data. However, hsCRP/IL-6, sTNFR1, and BDNF/TNF-α could potentially discriminate each phase of BD (Rowland et al., 2018), and warrant further evaluation. Another study suggested that IL-10, IL-15, and IL-27 levels may be beneficial for determining when patients are in the depressive phase of BD (Martinuzzi et al., 2021). Further studies characterizing changes in cytokine levels across the different phases of BD are essential for optimizing treatments and fully understanding the role they play in mediating symptomatology associated with each phase.

56.13.4 Posttraumatic Stress Disorder (PTSD)

56.13.4.1 Impact of the HPA Axis on PTSD

PTSD is often experienced after traumatic experiences (i.e., war, sexual violation, or life-threatening experience) and is accompanied by stress-related symptoms (i.e. negative cognition and mood or reexperiencing the traumatic event (Schumacher et al., 2019)). The pathophysiology of PTSD is generally accompanied by alterations in neurobiological systems (Schumacher et al., 2019). Further, it is known that 11.0% of women and 5.4% of men exposed to traumatic events will develop PTSD at some point in their lifetime (Kilpatrick et al., 2013; Schumacher et al., 2019). Individuals who develop PTSD show symptoms such as hyperarousal and the symptoms are thought to be accompanied by, or lead to, alterations in biological stress responses (de Kloet et al., 2006).

One of the major stress response systems that has been studied links HPA function and PTSD (Morris et al., 2012; Schumacher et al., 2019). The HPA axis is one of the three systems activated in response to stress (Teicher et al., 2002), whereby exposure to stress triggers emotional responses, including activation of the limbic system that initiates HPA activity (Dunlop & Wong, 2019). This activation is mediated via the release of CRH, a hormone and neurotransmitter associated with the onset of PTSD-associated symptoms (Laryea et al., 2012) which include enhanced fear responses by activation of the CRH receptor within the amygdala (Dunlop & Wong, 2019; Kolber et al., 2008). Interestingly, some studies have reported that PTSD patients have high levels of CRH in their CSF (Baker & Shalhoub-Kevorkian, 1999; Bremner et al., 1997; Laryea et al., 2012; Schumacher et al., 2018), while diurnal plasma cortisol levels are decreased (Schumacher et al., 2018; Yehuda et al., 1994). The combination of low diurnal plasma cortisol levels and high levels of CRH in CSF has been attributed to *enhanced* HPA-axis feedback regulation.

In healthy controls, cortisol produces a negative feedback signal via binding to GRs in the pituitary and hypothalamus leading to a subsequent decrease in CRH and ACTH release (Dunlop & Wong, 2019). However, in patients with PTSD, the reason for the rise in CRH but low levels of cortisol in PTSD is due to a rise of cortisol secretion upon exposure to initial trauma and a subsequent attenuation upon chronic exposure to stressors following the trauma (Schumacher et al., 2019; Steudte-Schmiedgen et al., 2016). Two mechanisms have been postulated to explain this phenomenon. The first suggests that CRH hypersecretion leads to an adaptive decrease of GRs within the pituitary, hindering its response to altered CRH levels (Dunlop & Wong, 2019; Schumacher et al., 2018). The second mechanism suggests that the HPA axis becomes increasingly sensitive to negative feedback, resulting in cortisol suppression (Heim et al., 2000; Schumacher et al., 2018) in patients with PTSD. However, regardless of the underlying mechanism, a failure to achieve homeostasis after trauma exposure in patients with PTSD is a large contributor to observed symptomology (Schumacher et al., 2019; Yehuda & LeDoux, 2007).

Since the early 2000s, researchers have posited an "attenuation hypothesis" of HPA axis functioning in PTSD (Jones & Moller, 2011; Susman, 2006; Trickett et al., 2010). This hypothesis states that low cortisol levels seen in individuals with PTSD may be explained by allostasis, the body's adaptive efforts to maintain homeostasis (Jones & Moller, 2011). In normal individuals, stress leads to HPA axis activation and the release of cortisol to allow for adaptive coping that may have protective effects on neuroendocrine processes and blunt the impact of physiological and psychological stressors (Jones & Moller, 2011). In the case of PTSD, it is hypothesized that stress and trauma are prolonged, unpredictable, or overwhelming and the demand of the HPA axis may be excessive, leading to hyperactivity and allostatic overload (Jones & Moller, 2011). Over long periods of time, this allostatic overload may decrease the body's ability to cope with stress and increase the vulnerability to PTSD pathology (Jones & Moller, 2011).

The most significant observation concerning PTSD is the low cortisol levels consistently found in patients (Pan et al., 2018; Pervanidou et al., 2020; Zimmerman et al., 2020). Due to these findings, novel therapeutic approaches to treat PTSD are being developed to target low levels of cortisol (Yehuda & LeDoux, 2007). It is important to note, however, that increased levels of cortisol are seen in children with PTSD (Carrion et al., 2002). Additionally, given HPA axis dysregulation, preventative therapeutic approaches to stabilize the HPA axis before psychopathology development has become of great interest (Schumacher et al., 2018). To date, high-dose hydrocortisone administered within a few hours of a traumatic experience has been associated with a significant favorable alteration of the trajectory of exposure to trauma (Zohar et al., 2011) as well as preventing PTSD symptoms (Dunlop & Wong, 2019; Kearns et al., 2012; Schelling et al., 2006; Schumacher et al., 2018).

56.13.4.2 Neuropeptides Associated with PTSD

Patients with PTSD experience an imbalance between stress-related neuropeptides, including corticotropin-releasing factor (CRF or CRH) and orexin, as well as antistress neuropeptides, such as oxytocin and neuropeptide Y (NPY), which are known to induce changes in the functions of stress-responsive systems (Bangasser & Wiersielis, 2018; Lehner et al., 2021; Nahvi & Sabban, 2020). Further understanding of the roles of these neuropeptides could lead to advancements in the treatment of PTSD that currently only has complete remission rates of less than 30% (Berger et al., 2009; Kim et al., 2020). Interestingly, antidepressants used to treat PTSD have been shown to increase levels of hypothalamic neuropeptides like NPY (Lehner et al., 2021; Yehuda & LeDoux, 2007), which has been shown to help alleviate some symptoms seen within the disorder and increase the efficacy of psychotherapies (Lehner et al., 2021).

CRF is a neuropeptide that has been identified as a modulator of brain regions and that regulates stress-related behavior responses and cognitive function (Hupalo et al., 2019; Lehner et al., 2021) by activating the HPA axis (Lehner et al., 2021). In PTSD, increased CRF concentrations within the CSF have been observed (Bremner et al., 1997; Lehner et al., 2021), suggesting that increases in CRF could be an associated mechanism of the pathophysiology of the disease caused by changes in neuronal microenvironments. A suggested mechanism of neuropsychiatric disease development is the excess activation of the type 1 CRF receptor via stress-induced increases in CRF levels (Contarino et al., 1999; Howerton et al., 2014; Lehner et al., 2021). Interestingly, the pharmacological intervention of CRF1 antagonism suppresses the stress response, leading to decreases in anxiety and depression symptoms (Lehner et al., 2021), a pharmacotherapy that has yielded promising results in patients with PTSD. CRF also regulates the activity of other neuropeptides like orexin, another neuropeptide that has been identified as a modulator of stress, arousal, emotional memory, and cognitive flexibility (Grafe & Bhatnagar, 2018; Lehner et al., 2021; Porkka-Heiskanen et al., 2004). Orexin neurons are upregulated by stressful activity and CRF (Winsky-Sommerer et al., 2005), and the orexin system is known to be dysfunctional in those with PTSD (Grafe & Bhatnagar, 2018; Lehner et al., 2021). Interestingly, increased arousal, sleep disturbances, and hypervigilance are key phenotypes of PTSD, and each is mediated by orexin (Grafe & Bhatnagar, 2018). Taken together, these findings suggest that altered CRF and orexin levels may promote PTSD in patients exposed to trauma.

While CRF and orexin are elevated in individuals with PTSD, oxytocin, and NPY are reduced, causing decreased antistress behaviors. Oxytocin is an antistress neuropeptide that regulates stress-coping behaviors (Carter et al., 2020) and is thought to do so by inhibiting the stress-induced hyperactivity of the HPA axis and instead promoting a decrease of cortisol levels to baseline levels following the stressful event (Dabrowska et al., 2011; Lehner et al., 2021). Oxytocin has also been shown to attenuate the fear response by decreasing the activity of limbic regions, especially the amygdala (Lehner et al., 2021; Viviani et al., 2011). However, in those with PTSD, reduced plasma oxytocin levels (Carmassi et al., 2021) may lead to reductions in resiliency to stress (Donadon et al., 2018; Lehner et al., 2021). Interestingly, Frijling 2017 discovered that intranasal administration of oxytocin early following a traumatic experience in patients experiencing acute PTSD symptoms decreased subsequence PTSD symptom development (Frijling, 2017). Similarly, NPY regulates energy homeostasis, cognition, and has anti-anxiety properties and promotes stress-protective effects in limbic regions and the cerebral cortex (Lehner et al., 2021). PTSD, however, has been associated with lower levels of NPY in the plasma and CSF (Tural & Iosifescu, 2020). Because of this, antianxiety effects and fear extinction were associated with intracerebroventricular administration of NPY (Lach & de Lima, 2013). In fact, RNA sequencing of human postmortem brain tissue also revealed that NPY and NPY receptor type 2 was associated with differentially expressed genes known as key drivers precipitating transcriptional changes in patients with PTSD, and these could be used as genetic biomarkers of PTSD and elevated stress (Stone et al., 2020).

Several of these neuropeptides play a role in the observed sex difference of PTSD prevalence. Women are two times more likely to develop PTSD and to suffer a more chronic form of the disorder, and there are varying mechanisms suggested mediating this difference (Bangasser & Wiersielis, 2018; Galovski et al., 2013). Decreased estrogen levels have been associated with a higher risk of developing PTSD and exacerbating the symptoms of PTSD when explaining the sexual bias of PTSD (Lehner et al., 2021; Maeng & Milad, 2015). This also indicates that women with PTSD could have more severe symptoms as hormone levels fluctuate throughout their menstrual cycle. In addition, estrogen increases oxytocin receptor expression and promotes oxytocin release (Young et al., 1997), and sex differences in estrogen levels, the number of oxytocin receptors, or binding affinity to oxytocin receptors could lead to altered PTSD phenotypes in women compared to men (Lehner et al., 2021). Furthermore, lower levels of NPY peptide and its receptor in the hypothalamus of females compared to males contribute to enhanced HPA axis activity and thus greater susceptibility to stress-related disorders in females (Lehner et al., 2021; Nahvi & Sabban, 2020). While evaluation of sex differences is a newer avenue of research, the establishment of underpinning mechanisms and peptides responsible for these differences require continued research for the development of effective treatments for PTSD in both male and female patients.

56.13.4.3 The Role of the Neuroimmune System and Cytokines in PTSD

Within patients with PTSD, increased activity of the HPA axis and increased sympathetic nervous system (SNS) activity compared to parasympathetic nervous system (PNS) activity have been observed and may contribute to increased fear and anxiety often exhibited by these patients (Katrinli et al., 2022). This imbalance may lead to the increased production of proinflammatory cytokines that are often observed in patients with PTSD (Katrinli et al., 2022; Kim et al., 2019). Specifically, the imbalance within the autonomic nervous system (ANS) is likely due to chronic and recurrent stress triggering the SNS to release epinephrine and norepinephrine (Katrinli et al., 2022; Kim et al., 2019). As such, monocytes are then mobilized to danger-associated molecular patterns (DAMPs), which then activate NF-kB ultimately leading to an increase in proinflammatory cytokines (Katrinli et al., 2022; Miller & Raison, 2016). In those with PTSD, an increase in peripheral NF-kB activity may account for the enhanced inflammatory response observed in patients (Katrinli et al., 2022; Pace et al., 2012). In addition, due to long-term exposure to stress, cortisol cannot inhibit the upregulated NF-kB activity due to glucocorticoid resistance, which can lead to a decrease in negative feedback on the

HPA axis induced by glucocorticoids (Michopoulos et al., 2017).

In patients with PTSD, increased levels of proinflammatory cytokines like CRP, IL-6, IL-1β, Il-2, TNF-α, and INF-γ have been observed (Passos et al., 2015; Yang & Jiang, 2020). Others have shown increases in plasma IL-1β (Spivak et al., 1997; Tucker et al., 2014), IL-6 (Ogłodek et al., 2015), and TNF-α (Gola et al., 2013). In addition, elevated levels of IL-6 have been detected in the CSF of patients with PTSD (Baker et al., 2001). There have also been observed decreased serum levels of anti-inflammatory markers like IL-10 (Katrinli et al., 2022; Ogłodek et al., 2015). However, there has been conflicting data regarding increased versus decreased levels of both proinflammatory and anti-inflammatory cytokines, and this may be due to the difficulty in determining whether changes are due to a pre- or posttraumatic experience and the origin of the inflammatory response (Katrinli et al., 2022). For example, the acute stress response in patients with PTSD showed significantly higher serum levels of IL-10, but no difference in serum levels of IL-6 compared to healthy controls (Renner & Petrowski, 2021). These inconsistencies may be caused by individual phenotypes, where many but not all individuals with PTSD exhibit alterations in either pro- or anti-inflammatory cytokines (Hori & Kim, 2019). This inflammation state may be propagated by the increased levels of IL-6, IL-1β, and TNF-α, which may activate the complement system via CRP elevation in PTSD (Katrinli et al., 2022). Hence, anti-inflammatory therapies could help those with PTSD. In fact, cannabinoids are known for their anti-inflammatory effects and a synthetic cannabinoid, nabilone, improved PTSD symptoms including but not limited to nightmares and insomnia (Cameron et al., 2014; Katrinli et al., 2022). Although anti-inflammatory therapies remain a promising possibility for treatment, further evaluation is needed to determine their efficacy.

56.13.5 Addiction

56.13.5.1 Overview of the Immune System in Addiction

Substance addiction is a complex disorder characterized by the inability to control substance seeking or use, which includes legal or illegal drugs such as alcohol, cocaine, heroin, methamphetamine, cannabis, and others. While addictive behaviors are perpetuated by behavioral disinhibition, the disorder is extremely complex and relies on both environmental and genetic components. Environmental components, especially exposure to stressful stimuli and traumatic experiences, perpetuate substance seeking and abuse. It is thought that stressful stimuli induce alterations in brain circuitry and structure ultimately promoting impulsivity, risky behavior, and altered decision-making abilities—each of which are characteristics of those suffering from substance abuse and addiction.

Addictive substances have been shown to activate the HPA axis, inducing elevations in ACTH and glucocorticoids. Specifically, nicotine has been shown to increase corticosterone levels (Chen et al., 2008), an effect thought to rely on the direct activation of catecholaminergic and cholinergic stimulation via the autonomic nervous system (Wemm & Sinha, 2019). Interestingly, epinephrine release by the ANS is dose dependent in response to nicotine, where high consumption leads to higher levels of epinephrine release, which can alter cardiovascular output in humans (Perkins et al., 2009). Lastly, chronic nicotine consumers have elevated basal levels of cortisol, which can perpetuate consumption and neuroimmune dysregulation (al'Absi, 2006). Similar effects have been observed following cannabis (THC) consumption, where cortisol (Cuttler et al., 2017; Ranganathan et al., 2009) and ACTH levels are enhanced relative to non-THC consuming individuals. However, the effects of THC on the HPA axis are complex, exerting both direct and indirect effects on the paraventricular nuclei of the hypothalamus as well as areas such as the amygdala (Armario, 2010).

Stimulants such as methamphetamine and cocaine have also been shown to elevate cortisol levels following use (Söderpalm et al., 2003). Interestingly, chronic cocaine use results in elevated levels of both cortisol and ACTH (for review see Manetti et al. (2014)), where high levels of basal cortisol are observed and unchanged with periods of abstinence (McDougle et al., 1994). Furthermore, catecholamine levels are elevated in individuals utilizing stimulants such as cocaine, resulting in hypertension and increased heart rates maintained over a long period of time (Esel et al., 2001; Foltin et al., 1995). While these results have been convincing and many agree, the degree of substance use can alter these effects and periods of abstinence from various compounds can further complicate the implications of these results.

Alcohol exposure (prenatally) as well as usage during adolescence and adulthood can have profound impacts on the HPA axis. For example, prenatal alcohol exposure (PAE) has been shown to increase the tone of the HPA axis, predisposing these individuals to depression and anxiety-like symptoms in adulthood (Weinberg et al., 2008). Furthermore, PAE is known to deplete β-endorphin producing neurons as well as hypothalamic POMC levels, which can have profound effects on the reward circuitry (for review see Ciafrè et al. (2020)). In adults with alcohol use disorder (AUD), it has been shown that alcohol directly stimulates the HPA axis and alters GRs in areas outside of the hypothalamus including the limbic forebrain and medial prefrontal cortex, which may give rise to the propensity to seek alcohol and relapse risk (Blaine & Sinha, 2017). In fact, the reward value of alcohol is thought to be enhanced by alcohol's effects on the autonomic nervous system and the HPA axis, which in turn increases striatal synaptic plasticity through activation of GRs located within this region (Lee & Rivier, 1997). Further amplifying this idea, withdrawal from alcohol in individuals with AUD has been shown to increase CRF within the CSF and increases ACTH, cortisol, norepinephrine, and epinephrine plasma levels, each thought to contribute to relapse propensity and continued drug seeking (Blaine & Sinha, 2017; Breese et al., 2011).

Taken together, these studies highlight the profound effects of substance abuse and usage on the HPA axis and dysregulation that may significantly contribute to relapse following a state of abstinence. While these effects are extremely complex and likely have variable effects due to usage amount, duration of usage and time of exposure, it is undisputed that these compounds can each significantly alter the immune system that likely contributes to the neurological deficits and altered behaviors observed in individuals with substance abuse disorders.

56.13.5.2 Neuropeptides Associated with Addiction

Accumulating evidence has demonstrated the role in various neuropeptides in substance abuse disorders. Most commonly, altered protein levels and expression of NPY have been observed across all substances of abuse. However, more recently oxytocin and vasopressin systems have been of interest, where targeting each system has yielded therapeutic effects for treating individuals with substance abuse disorders. Here we highlight major findings regarding each of these systems as they relate to abused substances.

The alterations in NPY protein level and expression varies based on substance. Interestingly, amphetamines and nicotine have been shown to consistently increase both protein levels and mRNA levels of NYP across multiple brain regions, including the prefrontal cortex, hippocampus, striatum, and dorsomedial nuclei of the hypothalamus (Gonçalves et al., 2016). This increase in expression can be observed for upward of two days following drug exposure, demonstrating that changes to the NPY system may be long lasting, even when the drug has been metabolized and cleared from the body. Of interest, Westwood and Hanson found that antagonism of the D1 dopamine receptor prevents drug induced increases in NPY expression, suggesting an interaction between NPY and D1 receptors that may play a role in the rewarding effects of stimulatory drugs of abuse.

Alcohol and opioids have commonly been found to decrease NPY mRNA and protein expression (Gonçalves et al., 2016). Following alcohol exposure, NPY mRNA and expression has been found to be significantly reduced within both the hippocampus and cerebral cortex of rats consuming alcohol chronically (Bison & Crews, 2003), which is thought to be related to levels of intoxication. Specifically, Bison and Crews found that low levels of intoxication yielded minimal alterations in NPY across brain regions, whereas high consumption and intoxication substantially decreased NPY mRNA and expression. Interestingly, injections of NPY into extra-hypothalamic regions significantly inhibits alcohol consumption in rodent models of AUD (Sparrow et al., 2012). Withdrawal from alcohol has been shown to increase NPY levels within the hippocampus, yet further decrease NPY levels within the amygdala and frontal cortex (Barkley-Levenson et al., 2016). However, the conflicting results are likely due to degree of initial alcohol exposure and the duration of withdrawal (Robinson & Thiele, 2017). For a detailed review on the effects of alcohol on NPY, we direct the readers to a recent review by Robinson and Thiele (2017).

Aside from NPY, altered levels of oxytocin and vasopressin have also been associated with substance abuse. Many studies have shown that drugs of abuse, including opioids, cocaine, and alcohol lead to hypofunction of the OXT system, reducing levels of oxytocin within the CNS and PNS (Che et al., 2021). Additionally, the degree of hypofunction can be correlated with relapse propensity in individuals abstaining from drug use. Interestingly, oxytocin administration can reduce drug seeking and consumption in animals and humans utilizing abused substances (Che et al., 2021), which has led to oxytocin being utilized in treatment of those suffering from substance abuse. For a thorough review on substance abuse and oxytocin, we direct readers to the recent review by Che et al. (2021).

Lastly, vasopressin levels have been correlated with substance abuse, where psychostimulants lead to an increase in vasopressin in both animal and human studies (Godino & Renard, 2018). Given that vasopressin produces synergistic effects on ACTH, it is not surprising that individuals with substance abuse disorders often exhibit altered levels of ACTH. Interestingly, post-mortem brains of those with alcohol use disorders show decreases in AVP mRNA in the paraventricular nuclei, indicating that alcohol can also alter the morphology and physiology of parvocellular and magnocellular cells of the nuclei (Sivukhina et al., 2006). Further evidence suggests that vasopressin V_{1b} receptors increase drinking patterns of ethanol-dependent rats, promoting excessive drinking, whereby inhibition of vasopressin V_{1b} receptors reduced excessive levels of alcohol self-administration (Edwards et al., 2012). These results highlight the importance of vasopressin dysregulation within the brain and likely contribute to altered peripheral activity in individuals using alcohol as well as other substances. While effects of substance abuse on vasopressin have been found, they have rarely been reviewed until recently. For additional information on the specific results and correlation in both animals and humans, see (Godino & Renard, 2018).

56.13.5.3 The Role of the Neuroimmune System and Cytokines in Addiction

The neuroimmune system and its role in addiction has recently become an important area of study. In fact, these studies have shown that neuroimmune signaling provides an important mechanism by which drugs of abuse induce alterations in synaptic plasticity across multiple brain regions that may give rise to behaviors exhibited by individuals with substance abuse disorders. Below we highlight the major findings regarding cytokine dysregulation caused by psychostimulants and alcohol. For additional information on cytokines variability and dysregulation following exposure of different abused substances as well as regional specificity of these changes, we direct readers to our recent review (Namba et al., 2021) as well as the recent review by Ahearn and colleagues (Ahearn et al., 2021).

A number of cytokines have found to be dysregulated following cocaine exposure, including TNF-α, IL-6, and MCP1, and of these, the most well studied being TNF-α. It has been widely accepted that cocaine use decreases levels of TNF-α as well as IL-6 (Irwin et al., 2007). Interestingly, the decrease in TNF-α has been found to play a substantial role cocaine-induced neuroadaptations and inappropriate synaptic plasticity within the nucleus accumbens which may promote continued drug seeking (Lewitus et al., 2016). TNF-α and IL-6 release are enhanced following acute methamphetamine exposure, specifically within the nucleus accumbens, caudate-putamen, and hippocampus, which was inhibited utilizing dopamine D1 antagonists (Wang et al., 2019). Together, these results suggest that D1 receptors and TNFα may interact and allow for synaptic modulation of reward circuitry in areas like the striatum.

Like cocaine and methamphetamine, alcohol also has been shown to increase TNF and MCP1 expression. Increased expression was found within the cerebellum and was reliant on the activation of miR-155, an miRNA induced by NF-kB activation (Lippai et al., 2013). Interestingly, NF-kB expression and immune gene mRNA related to NF-kB were found in individuals with alcohol use disorders and are now thought to influence alcoholism susceptibility. Preclinically, it has been found that alcohol increases NF-kB-DNA binding both in vivo and in vitro in brains and slice cultures of rat tissue, respectively (Ahearn et al., 2021).

Given the new abundance of literature highlighting the role of dysregulated cytokines and the ability of drugs of abuse to alter the functioning of astrocytes and microglia, it is not surprising that cytokines are now being targeted and utilized as potential treatment options for those suffering from substance and alcohol use disorders. In fact, such targeted therapies have showed substantial promise in preclinical studies of conditioned place preference, self-administration, and drug seeking, where microglial inhibitors and specific cytokine antagonists decrease opioid, psychostimulant, and alcohol seeking and usage. An overview and summary of such results are highlighted within a recent review by Bachtell and colleagues (Bachtell et al., 2017). These recent findings and avenues in research will likely yield large contributions to the treatment options for those suffering from substance abuse and will continue to shed light on new findings related to substance abuse circuitry.

56.14 Conclusion

In this chapter, we have provided an overview of the circuitry involved in modulating immune responses as well as signaling molecules utilized by each aspect of this circuitry. An overview of neuropeptides and cytokines was discussed, with a particular emphasis on enkephalins, endorphins, dynorphins, interleukins, and tumor necrosis factor, each of which have been implicated to play a role in a variety of psychiatric disorders. Most importantly, this chapter covers known dysfunctions within the immune system circuitry and dysregulations of neuropeptide and cytokine levels known to give rise to symptomology observed across many psychiatric disorders. However, while additional work must be done to shed further insight into these mechanisms, the targeting of cytokines and neuropeptides as a therapeutic has shown promising results in treating not only autoimmune disorders but each of the psychiatric disorders discussed above. We believe the information discussed across this chapter provides a promising avenue of treatments and warrants additional studies to elucidate a complete mechanism and understanding of how the immune system contributes to a variety of disorders and determining best avenues of treatment for these individuals.

56.15 Review Questions

1. Name three of the four peptides for which proenkephalin is the precursor.
   (a) methionine-enkephalin
   (b) leucine-enkephalin
   (c) methionine-enkephalin-Arg-Phe
   (d) methionine-enkephalin-Arg-Gly-Leu
2. Name the three products derived from proopiomelanocortin (POMC) for which evidence is the strongest supporting involvement in regulating immune system function.
3. The site of action for α-MSH is
   (a) mu opioid receptors
   (b) melanocortin-1 receptors
   (c) glucocorticoid receptors
   (d) melatonin receptors

   (e) ACTH receptors
4. Which neuropeptide is anti-inflammatory when activating the HPA axis but is proinflammatory when released locally by immune cells?
   (a) ACTH
   (b) CRH
   (c) β-endorphin
   (d) nociceptin
   (e) γ-MSH
5. The neuropeptide synthesized and released from the CNS, particularly the hypothalamus, as well as from immune cells is
   (a) ACTH
   (b) urocortin-1
   (c) CRH
   (d) TNF-α
   (e) insulin
6. The proinflammatory cytokine found to have increased levels in plasma and CNS in major depressive illness and whose levels correlate significantly with mood rating is
   (a) IL-6
   (b) TNF-α
   (c) CRH
   (d) IL-2
   (e) β-endorphin
7. The most commonly associated neuropeptide with schizophrenia is which of the following?
   (a) Neuropeptide Y
   (b) IL-6
   (c) IL-2
   (d) ACTH
   (e) IL-10

56.16 Answers

1. Any combination of the 3 listed
2. β-endorphin; α-MSH and γ-MSH
3. (b) melanocortin-1 receptors
4. (b) CRH
5. (c) CRH
6. (a) IL-6
7. (a) Neuropeptide Y

References

Ader, R. (1987). Conditioned immune responses: adrenocortical influences. *Progress in Brain Research, 72*, 79–90. https://doi.org/10.1016/s0079-6123(08)60198-3

Adler, M. W., & Rogers, T. J. (2005). Are chemokines the third major system in the brain? *Journal of Leukocyte Biology, 78*(6), 1204–1209. https://doi.org/10.1189/jlb.0405222

Ahearn, O. C., Watson, M. N., & Rawls, S. M. (2021). Chemokines, cytokines and substance use disorders. *Drug and Alcohol Dependence, 220*, 108511. https://doi.org/10.1016/j.drugalcdep.2021.108511

al'Absi, M. (2006). Hypothalamic-pituitary-adrenocortical responses to psychological stress and risk for smoking relapse. *International Journal of Psychophysiology, 59*(3), 218–227. https://doi.org/10.1016/j.ijpsycho.2005.10.010

Al-Diwani, A. A. J., Pollak, T. A., Irani, S. R., & Lennox, B. R. (2017). Psychosis: an autoimmune disease? *Immunology, 152*(3), 388–401. https://doi.org/10.1111/imm.12795

Alesci, S., Martinez, P. E., Kelkar, S., Ilias, I., Ronsaville, D. S., Listwak, S. J., et al. (2005). Major depression is associated with significant diurnal elevations in plasma interleukin-6 levels, a shift of its circadian rhythm, and loss of physiological complexity in its secretion: clinical implications. *The Journal of Clinical Endocrinology and Metabolism, 90*(5), 2522–2530. https://doi.org/10.1210/jc.2004-1667

Allen, S. J., & Dawbarn, D. (2006). Clinical relevance of the neurotrophins and their receptors. *Clinical Science (London, England), 110*(2), 175–191. https://doi.org/10.1042/CS20050161

Anderson, G., & Maes, M. (2015). Bipolar disorder: role of immune-inflammatory cytokines, oxidative and nitrosative stress and tryptophan catabolites. *Current Psychiatry Reports, 17*(2), 8. https://doi.org/10.1007/s11920-014-0541-1

Antonijevic, I. A. (2006). Depressive disorders—Is it time to endorse different pathophysiologies? *Psychoneuroendocrinology, 31*(1), 1–15. https://doi.org/10.1016/j.psyneuen.2005.04.004

Armario, A. (2010). Activation of the hypothalamic-pituitary-adrenal axis by addictive drugs: different pathways, common outcome. *Trends in Pharmacological Sciences, 31*(7), 318–325. https://doi.org/10.1016/j.tips.2010.04.005

Aydin, A., Selvi, Y., Besiroglu, L., Boysan, M., Atli, A., Ozdemir, O., et al. (2013). Mood and metabolic consequences of sleep deprivation as a potential endophenotype' in bipolar disorder. *Journal of Affective Disorders, 150*(2), 284–294. https://doi.org/10.1016/j.jad.2013.04.007

Bachtell, R. K., Jones, J. D., Heinzerling, K. G., Beardsley, P. M., & Comer, S. D. (2017). Glial and neuroinflammatory targets for treating substance use disorders. *Drug and Alcohol Dependence, 180*, 156–170. https://doi.org/10.1016/j.drugalcdep.2017.08.003

Baker, A., & Shalhoub-Kevorkian, N. (1999). Effects of political and military traumas on children: the Palestinian case. *Clinical Psychology Review, 19*(8), 935–950. https://doi.org/10.1016/s0272-7358(99)00004-5

Baker, D. G., Ekhator, N. N., Kasckow, J. W., Hill, K. K., Zoumakis, E., Dashevsky, B. A., et al. (2001). Plasma and cerebrospinal fluid interleukin-6 concentrations in posttraumatic stress disorder. *Neuroimmunomodulation, 9*(4), 209–217. https://doi.org/10.1159/000049028

Bangasser, D. A., & Wiersielis, K. R. (2018). Sex differences in stress responses: a critical role for corticotropin-releasing factor. *Hormones (Athens, Greece), 17*(1), 5–13. https://doi.org/10.1007/s42000-018-0002-z

Bao, A. M., & Swaab, D. F. (2010). Corticotropin-releasing hormone and arginine vasopressin in depression focus on the human postmortem hypothalamus. *Vitamins and Hormones, 82*, 339–365. https://doi.org/10.1016/S0083-6729(10)82018-7

Barkley-Levenson, A. M., Ryabinin, A. E., & Crabbe, J. C. (2016). Neuropeptide Y response to alcohol is altered in nucleus accumbens of mice selectively bred for drinking to intoxication. *Behavioural Brain Research, 302*, 160–170. https://doi.org/10.1016/j.bbr.2016.01.015

Beattie, E. C., Stellwagen, D., Morishita, W., Bresnahan, J. C., Ha, B. K., Von Zastrow, M., et al. (2002). Control of synaptic strength by glial TNFalpha. *Science, 295*(5563), 2282–2285. https://doi.org/10.1126/science.1067859

Beckstead, R. M. (1995). N-methyl-D-aspartate acutely increases pro-enkephalin mRNA in the rat striatum. *Synapse, 21*(4), 342–347. https://doi.org/10.1002/syn.890210409

Belvederi Murri, M., Prestia, D., Mondelli, V., Pariante, C., Patti, S., Olivieri, B., et al. (2016). The HPA axis in bipolar disorder: Systematic review and meta-analysis. *Psychoneuroendocrinology, 63*, 327–342. https://doi.org/10.1016/j.psyneuen.2015.10.014

Berger, W., Mendlowicz, M. V., Marques-Portella, C., Kinrys, G., Fontenelle, L. F., Marmar, C. R., & Figueira, I. (2009). Pharmacologic alternatives to antidepressants in posttraumatic stress disorder: a systematic review. *Progress in Neuro-Psychopharmacology & Biological Psychiatry, 33*(2), 169–180. https://doi.org/10.1016/j.pnpbp.2008.12.004

Bertollo, A. G., Grolli, R. E., Plissari, M. E., Gasparin, V. A., Quevedo, J., Réus, G. Z., et al. (2020). Stress and serum cortisol levels in major depressive disorder: a cross-sectional study. *AIMS Neuroscience, 7*(4), 459–469. https://doi.org/10.3934/Neuroscience.2020028

Bidlack, J. M. (2000). Detection and function of opioid receptors on cells from the immune system. *Clinical and Diagnostic Laboratory Immunology, 7*(5), 719–723. https://doi.org/10.1128/CDLI.7.5.719-723.2000

Bidlack, J. M., Khimich, M., Parkhill, A. L., Sumagin, S., Sun, B., & Tipton, C. M. (2006). Opioid receptors and signaling on cells from the immune system. *Journal of Neuroimmune Pharmacology, 1*(3), 260–269. https://doi.org/10.1007/s11481-006-9026-2

Bilezikjian, L. M., Blount, A. L., & Vale, W. W. (1987). The cellular actions of vasopressin on corticotrophs of the anterior pituitary: Resistance to glucocorticoid action. *Molecular Endocrinology, 1*(7), 451–458. https://doi.org/10.1210/mend-1-7-451

Binder, E. B., Kinkead, B., Owens, M. J., & Nemeroff, C. B. (2001). The role of neurotensin in the pathophysiology of schizophrenia and the mechanism of action of antipsychotic drugs. *Biological Psychiatry, 50*(11), 856–872. https://doi.org/10.1016/s0006-3223(01)01211-2

Bison, S., & Crews, F. (2003). Alcohol withdrawal increases neuropeptide Y immunoreactivity in rat brain. *Alcoholism, Clinical and Experimental Research, 27*(7), 1173–1183. https://doi.org/10.1097/01.ALC.0000075827.74538.FE

Bissette, G., Widerlöv, E., Walléus, H., Karlsson, I., Eklund, K., Forsman, A., & Nemeroff, C. B. (1986). Alterations in cerebrospinal fluid concentrations of somatostatinlike immunoreactivity in neuropsychiatric disorders. *Archives of General Psychiatry, 43*(12), 1148–1151. https://doi.org/10.1001/archpsyc.1986.01800120034008

Blaine, S. K., & Sinha, R. (2017). Alcohol, stress, and glucocorticoids: From risk to dependence and relapse in alcohol use disorders. *Neuropharmacology, 122*, 136–147. https://doi.org/10.1016/j.neuropharm.2017.01.037

Blalock, J. E. (1994). The syntax of immune-neuroendocrine communication. *Immunology Today, 15*(11), 504–511. https://doi.org/10.1016/0167-5699(94)90205-4

Blalock, J. E. (1999). Proopiomelanocortin and the immune-neuroendocrine connection. *Annals of the New York Academy of Sciences, 885*, 161–172. https://doi.org/10.1111/j.1749-6632.1999.tb08673.x

Bonci, A., & Malenka, R. C. (1999). Properties and plasticity of excitatory synapses on dopaminergic and GABAergic cells in the ventral tegmental area. *The Journal of Neuroscience, 19*(10), 3723–3730. https://doi.org/10.1523/JNEUROSCI.19-10-03723.1999

Booij, S. H., Wigman, J. T. W., Jacobs, N., Thiery, E., Derom, C., Wichers, M., & Oravecz, Z. (2020). Cortisol dynamics in depression: Application of a continuous-time process model. *Psychoneuroendocrinology, 115*, 104598. https://doi.org/10.1016/j.psyneuen.2020.104598

Bortolato, B., Carvalho, A. F., Soczynska, J. K., Perini, G. I., & McIntyre, R. S. (2015). The involvement of TNF-α in cognitive dysfunction associated with major depressive disorder: An opportunity for domain specific treatments. *Current Neuropharmacology, 13*(5), 558–576. https://doi.org/10.2174/1570159x13666150630171433

Boyle, M. P., Brewer, J. A., Funatsu, M., Wozniak, D. F., Tsien, J. Z., Izumi, Y., & Muglia, L. J. (2005). Acquired deficit of forebrain glucocorticoid receptor produces depression-like changes in adrenal axis regulation and behavior. *Proceedings of the National Academy of Sciences of the United States of America, 102*(2), 473–478. https://doi.org/10.1073/pnas.0406458102

Brebner, K., Hayley, S., Zacharko, R., Merali, Z., & Anisman, H. (2000). Synergistic effects of interleukin-1beta, interleukin-6, and tumor necrosis factor-alpha: central monoamine, corticosterone, and behavioral variations. *Neuropsychopharmacology, 22*(6), 566–580. https://doi.org/10.1016/S0893-133X(99)00166-9

Breese, G. R., Sinha, R., & Heilig, M. (2011). Chronic alcohol neuroadaptation and stress contribute to susceptibility for alcohol craving and relapse. *Pharmacology & Therapeutics, 129*(2), 149–171. https://doi.org/10.1016/j.pharmthera.2010.09.007

Bremner, J. D., Licinio, J., Darnell, A., Krystal, J. H., Owens, M. J., Southwick, S. M., et al. (1997). Elevated CSF corticotropin-releasing factor concentrations in posttraumatic stress disorder. *The American Journal of Psychiatry, 154*(5), 624–629. https://doi.org/10.1176/ajp.154.5.624

Breslin, N. A., Suddath, R. L., Bissette, G., Nemeroff, C. B., Lowrimore, P., & Weinberger, D. R. (1994). CSF concentrations of neurotensin in schizophrenia: An investigation of clinical and biochemical correlates. *Schizophrenia Research, 12*(1), 35–41. https://doi.org/10.1016/0920-9964(94)90082-5

Brietzke, E., Mansur, R. B., Soczynska, J. K., Kapczinski, F., Bressan, R. A., & McIntyre, R. S. (2012). Towards a multifactorial approach for prediction of bipolar disorder in at risk populations. *Journal of Affective Disorders, 140*(1), 82–91. https://doi.org/10.1016/j.jad.2012.02.016

Brooks, S., Jacobs, G. E., de Boer, P., Kent, J. M., Van Nueten, L., van Amerongen, G., et al. (2019). The selective orexin-2 receptor antagonist seltorexant improves sleep: An exploratory double-blind, placebo controlled, crossover study in antidepressant-treated major depressive disorder patients with persistent insomnia. *Journal of Psychopharmacology, 33*(2), 202–209. https://doi.org/10.1177/0269881118822258

Brunoni, A. R., Lopes, M., & Fregni, F. (2008). A systematic review and meta-analysis of clinical studies on major depression and BDNF levels: Implications for the role of neuroplasticity in depression. *The International Journal of Neuropsychopharmacology, 11*(8), 1169–1180. https://doi.org/10.1017/S1461145708009309

Buckley, P. F. (2019). Neuroinflammation and Schizophrenia. *Current Psychiatry Reports, 21*(8), 72. https://doi.org/10.1007/s11920-019-1050-z

Cacci, E., Claasen, J. H., & Kokaia, Z. (2005). Microglia-derived tumor necrosis factor-alpha exaggerates death of newborn hippocampal progenitor cells in vitro. *Journal of Neuroscience Research, 80*(6), 789–797. https://doi.org/10.1002/jnr.20531

Camara, M. L., Corrigan, F., Jaehne, E. J., Jawahar, M. C., Anscomb, H., Koerner, H., & Baune, B. T. (2013). TNF-α and its receptors modulate complex behaviours and neurotrophins in transgenic mice. *Psychoneuroendocrinology, 38*(12), 3102–3114. https://doi.org/10.1016/j.psyneuen.2013.09.010

Cameron, H. A., & Gould, E. (1994). Adult neurogenesis is regulated by adrenal steroids in the dentate gyrus. *Neuroscience, 61*(2), 203–209. https://doi.org/10.1016/0306-4522(94)90224-0

Cameron, C., Watson, D., & Robinson, J. (2014). Use of a synthetic cannabinoid in a correctional population for posttraumatic stress disorder-related insomnia and nightmares, chronic pain, harm reduction, and other indications: A retrospective evaluation. *Journal of Clinical Psychopharmacology, 34*(5), 559–564. https://doi.org/10.1097/JCP.0000000000000180

Campbell, B. M., Charych, E., Lee, A. W., & Möller, T. (2014). Kynurenines in CNS disease: Regulation by inflammatory cytokines. *Frontiers in Neuroscience, 8*, 12. https://doi.org/10.3389/fnins.2014.00012

Carmassi, C., Marazziti, D., Mucci, F., Della Vecchia, A., Barberi, F. M., Baroni, S., et al. (2021). Decreased plasma oxytocin levels in patients with PTSD. *Frontiers in Psychology, 12*, 612338. https://doi.org/10.3389/fpsyg.2021.612338

Carr, D. J., & Blalock, J. E. (1989). A molecular basis for intersystem communication between the immune and neuroendocrine systems. *International Reviews of Immunology, 4*(3), 213–228. https://doi.org/10.3109/08830188909054419

Carrion, V. G., Weems, C. F., Ray, R. D., Glaser, B., Hessl, D., & Reiss, A. L. (2002). Diurnal salivary cortisol in pediatric posttraumatic stress disorder. *Biological Psychiatry, 51*(7), 575–582. https://doi.org/10.1016/s0006-3223(01)01310-5

Carter, C. S., Kenkel, W. M., MacLean, E. L., Wilson, S. R., Perkeybile, A. M., Yee, J. R., et al. (2020). Is oxytocin "nature's medicine"? *Pharmacological Reviews, 72*(4), 829–861. https://doi.org/10.1124/pr.120.019398

Cartier, L., Hartley, O., Dubois-Dauphin, M., & Krause, K. H. (2005). Chemokine receptors in the central nervous system: Role in brain inflammation and neurodegenerative diseases. *Brain Research. Brain Research Reviews, 48*(1), 16–42. https://doi.org/10.1016/j.brainresrev.2004.07.021

Catena-Dell'Osso, M., Fagiolini, A., Marazziti, D., Baroni, S., & Bellantuono, C. (2013). Non-monoaminergic targets for the development of antidepressants: Focus on neuropeptides. *Mini Reviews in Medicinal Chemistry, 13*(1), 2–10.

Cervenka, I., Agudelo, L. Z., & Ruas, J. L. (2017). Kynurenines: Tryptophan's metabolites in exercise, inflammation, and mental health. *Science, 357*(6349). https://doi.org/10.1126/science.aaf9794

Che, X., Cai, J., Liu, Y., Xu, T., Yang, J., & Wu, C. (2021). Oxytocin signaling in the treatment of drug addiction: Therapeutic opportunities and challenges. *Pharmacology & Therapeutics, 223*, 107820. https://doi.org/10.1016/j.pharmthera.2021.107820

Checkley, S. (1996). The neuroendocrinology of depression and chronic stress. *British Medical Bulletin, 52*(3), 597–617. https://doi.org/10.1093/oxfordjournals.bmb.a011570

Chen, H., Fu, Y., & Sharp, B. M. (2008). Chronic nicotine self-administration augments hypothalamic-pituitary-adrenal responses to mild acute stress. *Neuropsychopharmacology, 33*(4), 721–730. https://doi.org/10.1038/sj.npp.1301466

Chen, M. H., Kao, Z. K., Chang, W. C., Tu, P. C., Hsu, J. W., Huang, K. L., et al. (2020). Increased proinflammatory cytokines, executive dysfunction, and reduced gray matter volumes in first-episode bipolar disorder and major depressive disorder. *Journal of Affective Disorders, 274*, 825–831. https://doi.org/10.1016/j.jad.2020.05.158

Chen, C.-S., Coline, B., & Scheiermann, C. (2021). Peripheral neurotransmitters in the immune system. *Current Opinion in Physiology, 19*, 73–79. https://doi.org/10.1016/j.cophys.2020.09.009

Cherian, K., Schatzberg, A. F., & Keller, J. (2019). HPA axis in psychotic major depression and schizophrenia spectrum disorders: Cortisol, clinical symptomatology, and cognition. *Schizophrenia Research, 213*, 72–79. https://doi.org/10.1016/j.schres.2019.07.003

Ciafrè, S., Ferraguti, G., Greco, A., Polimeni, A., Ralli, M., Ceci, F. M., et al. (2020). Alcohol as an early life stressor: Epigenetics, metabolic, neuroendocrine and neurobehavioral implications. *Neuroscience and Biobehavioral Reviews, 118*, 654–668. https://doi.org/10.1016/j.neubiorev.2020.08.018

Clark, I. A., Mackay, C. E., & Goodwin, G. M. (2014). Pituitary gland volumes in bipolar disorder. *Journal of Affective Disorders, 169*, 197–202. https://doi.org/10.1016/j.jad.2014.08.022

Comb, M., Seeburg, P. H., Adelman, J., Eiden, L., & Herbert, E. (1982). Primary structure of the human Met- and Leu-enkephalin precursor and its mRNA. *Nature, 295*(5851), 663–666. https://doi.org/10.1038/295663a0

Contarino, A., Dellu, F., Koob, G. F., Smith, G. W., Lee, K. F., Vale, W., & Gold, L. H. (1999). Reduced anxiety-like and cognitive performance in mice lacking the corticotropin-releasing factor receptor 1. *Brain Research, 835*(1), 1–9. https://doi.org/10.1016/s0006-8993(98)01158-5

Correia, A. S., & Vale, N. (2022). Tryptophan metabolism in depression: A narrative review with a focus on serotonin and kynurenine pathways. *International Journal of Molecular Sciences, 23*(15). https://doi.org/10.3390/ijms23158493

Coulon, N., Brailly-Tabard, S., Walter, M., & Tordjman, S. (2016). Altered circadian patterns of salivary cortisol in individuals with schizophrenia: A critical literature review. *Journal of Physiology, Paris, 110*(4 Pt B), 439–447. https://doi.org/10.1016/j.jphysparis.2017.05.002

Cowen, P. J., & Browning, M. (2015). What has serotonin to do with depression? *World Psychiatry, 14*(2), 158–160. https://doi.org/10.1002/wps.20229

Criado-Marrero, M., Rein, T., Binder, E. B., Porter, J. T., Koren, J., & Blair, L. J. (2018). Hsp90 and FKBP51: Complex regulators of psychiatric diseases. *Philosophical Transactions of the Royal Society of London. Series B, Biological Sciences, 373*(1738). https://doi.org/10.1098/rstb.2016.0532

Crofford, L. J., Sano, H., Karalis, K., Webster, E. L., Goldmuntz, E. A., Chrousos, G. P., & Wilder, R. L. (1992). Local secretion of corticotropin-releasing hormone in the joints of Lewis rats with inflammatory arthritis. *The Journal of Clinical Investigation, 90*(6), 2555–2564. https://doi.org/10.1172/JCI116150

Crouse, J. J., Carpenter, J. S., Song, Y. J. C., Hockey, S. J., Naismith, S. L., Grunstein, R. R., et al. (2021). Circadian rhythm sleep-wake disturbances and depression in young people: Implications for prevention and early intervention. *Lancet Psychiatry, 8*(9), 813–823. https://doi.org/10.1016/S2215-0366(21)00034-1

Cuttler, C., Spradlin, A., Nusbaum, A. T., Whitney, P., Hinson, J. M., & McLaughlin, R. J. (2017). Blunted stress reactivity in chronic cannabis users. *Psychopharmacology, 234*(15), 2299–2309. https://doi.org/10.1007/s00213-017-4648-z

Dabrowska, J., Hazra, R., Ahern, T. H., Guo, J. D., McDonald, A. J., Mascagni, F., et al. (2011). Neuroanatomical evidence for reciprocal regulation of the corticotrophin-releasing factor and oxytocin systems in the hypothalamus and the bed nucleus of the stria terminalis of the rat: Implications for balancing stress and affect. *Psychoneuroendocrinology, 36*(9), 1312–1326. https://doi.org/10.1016/j.psyneuen.2011.03.003

Dahan, S., Bragazzi, N. L., Yogev, A., Bar-Gad, M., Barak, V., Amital, H., & Amital, D. (2018). The relationship between serum cytokine levels and degree of psychosis in patients with schizophrenia. *Psychiatry Research, 268*, 467–472. https://doi.org/10.1016/j.psychres.2018.07.041

Dai, T. T., Wang, B., Xiao, Z. Y., You, Y., & Tian, S. W. (2018). Apelin-13 upregulates BDNF against chronic stress-induced depression-like phenotypes by ameliorating HPA axis and hippocampal glucocorticoid receptor dysfunctions. *Neuroscience, 390*, 151–159. https://doi.org/10.1016/j.neuroscience.2018.08.018

Danesch, U., Hashimoto, S., Renkawitz, R., & Schütz, G. (1983). Transcriptional regulation of the tryptophan oxygenase gene in rat liver by glucocorticoids. *The Journal of Biological Chemistry, 258*(8), 4750–4753.

Dasgupta, A., Baby, N., Krishna, K., Hakim, M., Wong, Y. P., Behnisch, T., et al. (2017). Substance P induces plasticity and synaptic tagging/capture in rat hippocampal area CA2. *Proceedings of the National Academy of Sciences of the United States of America, 114*(41), E8741–E8749. https://doi.org/10.1073/pnas.1711267114

De Jongh, R., Vranken, J., Vundelinckx, G., Bosmans, E., Maes, M., & Heylen, R. (1997). The effects of anticoagulation and processing on assays of IL-6, sIL-6R, sIL-2R and soluble transferrin receptor. *Cytokine, 9*(9), 696–701. https://doi.org/10.1006/cyto.1997.0217

de Kloet, E. R., Joëls, M., & Holsboer, F. (2005). Stress and the brain: from adaptation to disease. *Nature Reviews. Neuroscience, 6*(6), 463–475. https://doi.org/10.1038/nrn1683

de Kloet, C. S., Vermetten, E., Geuze, E., Kavelaars, A., Heijnen, C. J., & Westenberg, H. G. (2006). Assessment of HPA-axis function in posttraumatic stress disorder: pharmacological and non-pharmacological challenge tests, a review. *Journal of Psychiatric Research, 40*(6), 550–567. https://doi.org/10.1016/j.jpsychires.2005.08.002

Delgado, P. L. (2000). Depression: The case for a monoamine deficiency. *The Journal of Clinical Psychiatry, 61*(Suppl 6), 7–11.

Delgado, P. L., Charney, D. S., Price, L. H., Aghajanian, G. K., Landis, H., & Heninger, G. R. (1990). Serotonin function and the mechanism of antidepressant action. Reversal of antidepressant-induced remission by rapid depletion of plasma tryptophan. *Archives of General Psychiatry, 47*(5), 411–418. https://doi.org/10.1001/archpsyc.1990.01810170011002

Demsie, D. G., Altaye, B. M., Weldekidan, E., Gebremedhin, H., Alema, N. M., Tefera, M. M., & Bantie, A. T. (2020). Galanin receptors as drug target for novel antidepressants: Review. *Biologics Targets and Therapy, 14*, 37–45. https://doi.org/10.2147/BTT.S240715

Denny, W. B., Valentine, D. L., Reynolds, P. D., Smith, D. F., & Scammell, J. G. (2000). Squirrel monkey immunophilin FKBP51 is a potent inhibitor of glucocorticoid receptor binding. *Endocrinology, 141*(11), 4107–4113. https://doi.org/10.1210/endo.141.11.7785

Donadon, M. F., Martin-Santos, R., & Osório, F. L. (2018). The associations between oxytocin and trauma in humans: A systematic review. *Frontiers in Pharmacology, 9*, 154. https://doi.org/10.3389/fphar.2018.00154

Dowlati, Y., Herrmann, N., Swardfager, W., Liu, H., Sham, L., Reim, E. K., & Lanctôt, K. L. (2010). A meta-analysis of cytokines in major depression. *Biological Psychiatry, 67*(5), 446–457. https://doi.org/10.1016/j.biopsych.2009.09.033

Du, N., Zhou, Y. L., Zhang, X., Guo, J., & Sun, X. L. (2017). Do some anxiety disorders belong to the prodrome of bipolar disorder? A clinical study combining retrospective and prospective methods to analyse the relationship between anxiety disorder and bipolar disorder from the perspective of biorhythms. *BMC Psychiatry, 17*(1), 351. https://doi.org/10.1186/s12888-017-1509-6

Duman, R. S., & Monteggia, L. M. (2006). A neurotrophic model for stress-related mood disorders. *Biological Psychiatry, 59*(12), 1116–1127. https://doi.org/10.1016/j.biopsych.2006.02.013

Dunlop, B. W., & Wong, A. (2019). The hypothalamic-pituitary-adrenal axis in PTSD: Pathophysiology and treatment interventions. *Progress in Neuro-Psychopharmacology & Biological Psychiatry, 89*, 361–379. https://doi.org/10.1016/j.pnpbp.2018.10.010

Dunn, A. J., & Berridge, C. W. (1990). Physiological and behavioral responses to corticotropin-releasing factor administration: Is CRF a mediator of anxiety or stress responses? *Brain Research. Brain Research Reviews, 15*(2), 71–100. https://doi.org/10.1016/0165-0173(90)90012-d

Duque-Díaz, E., Alvarez-Ojeda, O., & Coveñas, R. (2019). Enkephalins and ACTH in the mammalian nervous system. *Vitamins and Hormones, 111*, 147–193. https://doi.org/10.1016/bs.vh.2019.05.001

Dwivedi, Y. (2009). Brain-derived neurotrophic factor: Role in depression and suicide. *Neuropsychiatric Disease and Treatment, 5*, 433–449. https://doi.org/10.2147/ndt.s5700

Edwards, S., Guerrero, M., Ghoneim, O. M., Roberts, E., & Koob, G. F. (2012). Evidence that vasopressin V1b receptors mediate the transition to excessive drinking in ethanol-dependent rats. *Addiction Biology, 17*(1), 76–85. https://doi.org/10.1111/j.1369-1600.2010.00291.x

Ekdahl, C. T., Claasen, J. H., Bonde, S., Kokaia, Z., & Lindvall, O. (2003). Inflammation is detrimental for neurogenesis in adult brain. *Proceedings of the National Academy of Sciences of the United States of America, 100*(23), 13632–13637. https://doi.org/10.1073/pnas.2234031100

Engeland, W. C., & Arnhold, M. M. (2005). Neural circuitry in the regulation of adrenal corticosterone rhythmicity. *Endocrine, 28*(3), 325–332. https://doi.org/10.1385/ENDO:28:3:325

Engler, D., Pham, T., Fullerton, M. J., Ooi, G., Funder, J. W., & Clarke, I. J. (1989). Studies of the secretion of corticotropin-releasing factor and arginine vasopressin into the hypophysial-portal circulation of the conscious sheep. I. Effect of an audiovisual stimulus and insulin-induced hypoglycemia. *Neuroendocrinology, 49*(4), 367–381. https://doi.org/10.1159/000125141

Esel, E., Sofuoglu, S., Aslan, S. S., Kula, M., Yabanoglu, I., & Turan, M. T. (2001). Plasma levels of beta-endorphin, adrenocorticotropic hormone and cortisol during early and late alcohol withdrawal. *Alcohol and Alcoholism, 36*(6), 572–576. https://doi.org/10.1093/alcalc/36.6.572

Falcone, T., Carlton, E., Lee, C., Janigro, M., Fazio, V., Forcen, F. E., et al. (2015). Does systemic inflammation play a role in pediatric psychosis? *Clinical Schizophrenia & Related Psychoses, 9*(2), 65–78B. https://doi.org/10.3371/CSRP.FACA.030813

Farrow, P., Simmons, J. G., Pozzi, E., Díaz-Arteche, C., Richmond, S., Bray, K., et al. (2020). Associations between early life stress and anterior pituitary gland volume development during late childhood. *Psychoneuroendocrinology, 122*, 104868. https://doi.org/10.1016/j.psyneuen.2020.104868

Fenn, A. M., Gensel, J. C., Huang, Y., Popovich, P. G., Lifshitz, J., & Godbout, J. P. (2014). Immune activation promotes depression 1 month after diffuse brain injury: A role for primed microglia. *Biological Psychiatry, 76*(7), 575–584. https://doi.org/10.1016/j.biopsych.2013.10.014

Ferrer, A., Labad, J., Salvat-Pujol, N., Monreal, J. A., Urretavizcaya, M., Crespo, J. M., et al. (2020). Hypothalamic-pituitary-adrenal axis-related genes and cognition in major mood disorders and schizophrenia: A systematic review. *Progress in Neuro-Psychopharmacology & Biological Psychiatry, 101*, 109929. https://doi.org/10.1016/j.pnpbp.2020.109929

Fish, E. W., Shahrokh, D., Bagot, R., Caldji, C., Bredy, T., Szyf, M., & Meaney, M. J. (2004). Epigenetic programming of stress responses through variations in maternal care. *Annals of the New York Academy of Sciences, 1036*, 167–180. https://doi.org/10.1196/annals.1330.011

Fleminger, G., Ezra, E., Kilpatrick, D. L., & Udenfriend, S. (1983). Processing of enkephalin-containing peptides in isolated bovine adrenal chromaffin granules. *Proceedings of the National Academy of Sciences of the United States of America, 80*(20), 6418–6421. https://doi.org/10.1073/pnas.80.20.6418

Foltin, R. W., Fischman, M. W., & Levin, F. R. (1995). Cardiovascular effects of cocaine in humans: Laboratory studies. *Drug and Alcohol Dependence, 37*(3), 193–210. https://doi.org/10.1016/0376-8716(94)01085-y

Frederiksen, S. O., Ekman, R., Gottfries, C. G., Widerlöv, E., & Jonsson, S. (1991). Reduced concentrations of galanin, arginine vasopressin, neuropeptide Y and peptide YY in the temporal cortex but not in the hypothalamus of brains from schizophrenics. *Acta Psychiatrica Scandinavica, 83*(4), 273–277. https://doi.org/10.1111/j.1600-0447.1991.tb05539.x

Fricker, L. D., Margolis, E. B., Gomes, I., & Devi, L. A. (2020). Five decades of research on opioid peptides: Current knowledge and unanswered questions. *Molecular Pharmacology, 98*(2), 96–108. https://doi.org/10.1124/mol.120.119388

Frijling, J. L. (2017). Preventing PTSD with oxytocin: Effects of oxytocin administration on fear neurocircuitry and PTSD symptom development in recently trauma-exposed individuals. *European Journal of Psychotraumatology, 8*(1), 1302652. https://doi.org/10.1080/20008198.2017.1302652

Fu, A. K., & Ip, N. Y. (2017). Regulation of postsynaptic signaling in structural synaptic plasticity. *Current Opinion in Neurobiology, 45*, 148–155. https://doi.org/10.1016/j.conb.2017.05.016

Furay, A. R., Bruestle, A. E., & Herman, J. P. (2008). The role of the forebrain glucocorticoid receptor in acute and chronic stress. *Endocrinology, 149*(11), 5482–5490. https://doi.org/10.1210/en.2008-0642

Gaali, S., Kirschner, A., Cuboni, S., Hartmann, J., Kozany, C., Balsevich, G., et al. (2015). Selective inhibitors of the FK506-binding protein 51 by induced fit. *Nature Chemical Biology, 11*(1), 33–37. https://doi.org/10.1038/nchembio.1699

Gabriel, S. M., Davidson, M., Haroutunian, V., Powchik, P., Bierer, L. M., Purohit, D. P., et al. (1996). Neuropeptide deficits in schizophrenia vs. Alzheimer's disease cerebral cortex. *Biological Psychiatry, 39*(2), 82–91. https://doi.org/10.1016/0006-3223(95)00066-6

Galovski, T. E., Blain, L. M., Chappuis, C., & Fletcher, T. (2013). Sex differences in recovery from PTSD in male and female interpersonal assault survivors. *Behaviour Research and Therapy, 51*(6), 247–255. https://doi.org/10.1016/j.brat.2013.02.002

Gein, S. V. (2014). Dynorphins in regulation of immune system functions. *Biochemistry (Mosc), 79*(5), 397–405. https://doi.org/10.1134/S0006297914050034

Gillespie, C. F., & Nemeroff, C. B. (2005). Hypercortisolemia and depression. *Psychosomatic Medicine, 67*(Suppl 1), S26–S28. https://doi.org/10.1097/01.psy.0000163456.22154.d2

Girshkin, L., Matheson, S. L., Shepherd, A. M., & Green, M. J. (2014). Morning cortisol levels in schizophrenia and bipolar disorder: A meta-analysis. *Psychoneuroendocrinology, 49*, 187–206. https://doi.org/10.1016/j.psyneuen.2014.07.013

Godino, A., & Renard, G. M. (2018). Effects of alcohol and psychostimulants on the vasopressin system: Behavioral implications. *Journal of Neuroendocrinology*, e12611. https://doi.org/10.1111/jne.12611

Gola, H., Engler, H., Sommershof, A., Adenauer, H., Kolassa, S., Schedlowski, M., et al. (2013). Posttraumatic stress disorder is associated with an enhanced spontaneous production of pro-inflammatory cytokines by peripheral blood mononuclear cells. *BMC Psychiatry, 13*, 40. https://doi.org/10.1186/1471-244X-13-40

Goldsmith, D. R., Rapaport, M. H., & Miller, B. J. (2016). A meta-analysis of blood cytokine network alterations in psychiatric patients: Comparisons between schizophrenia, bipolar disorder and depression. *Molecular Psychiatry, 21*(12), 1696–1709. https://doi.org/10.1038/mp.2016.3

Gonçalves, J., Martins, J., Baptista, S., Ambrósio, A. F., & Silva, A. P. (2016). Effects of drugs of abuse on the central neuropeptide Y system. *Addiction Biology, 21*(4), 755–765. https://doi.org/10.1111/adb.12250

Grafe, L. A., & Bhatnagar, S. (2018). Orexins and stress. *Frontiers in Neuroendocrinology, 51*, 132–145. https://doi.org/10.1016/j.yfrne.2018.06.003

Gubler, U., Seeburg, P., Hoffman, B. J., Gage, L. P., & Udenfriend, S. (1982). Molecular cloning establishes proenkephalin as precursor of enkephalin-containing peptides. *Nature, 295*(5846), 206–208. https://doi.org/10.1038/295206a0

Hadden, J. W., Hadden, E. M., & Middleton, E. (1970). Lymphocyte blast transformation. I. Demonstration of adrenergic receptors in human peripheral lymphocytes. *Cellular Immunology, 1*(6), 583–595. https://doi.org/10.1016/0008-8749(70)90024-9

Han, K. M., Choi, S., Jung, J., Na, K. S., Yoon, H. K., Lee, M. S., & Ham, B. J. (2014). Cortical thickness, cortical and subcortical volume, and white matter integrity in patients with their first episode of major depression. *Journal of Affective Disorders, 155*, 42–48. https://doi.org/10.1016/j.jad.2013.10.021

Harrison-Felix, C. L., Whiteneck, G. G., Jha, A., DeVivo, M. J., Hammond, F. M., & Hart, D. M. (2009). Mortality over four decades after traumatic brain injury rehabilitation: A retrospective cohort study. *Archives of Physical Medicine and Rehabilitation, 90*(9), 1506–1513. https://doi.org/10.1016/j.apmr.2009.03.015

Hart, B. L. (1988). Biological basis of the behavior of sick animals. *Neuroscience and Biobehavioral Reviews, 12*(2), 123–137. https://doi.org/10.1016/s0149-7634(88)80004-6

Hartwig, F. P., Borges, M. C., Horta, B. L., Bowden, J., & Davey Smith, G. (2017). Inflammatory biomarkers and risk of schizophrenia: A 2-sample mendelian randomization study. *JAMA Psychiatry, 74*(12), 1226–1233. https://doi.org/10.1001/jamapsychiatry.2017.3191

Hasler, G., Drevets, W. C., Gould, T. D., Gottesman, I. I., & Manji, H. K. (2006). Toward constructing an endophenotype strategy for bipolar disorders. *Biological Psychiatry, 60*(2), 93–105. https://doi.org/10.1016/j.biopsych.2005.11.006

Hassan, A. H., Pzewłocki, R., Herz, A., & Stein, C. (1992). Dynorphin, a preferential ligand for kappa-opioid receptors, is present in nerve fibers and immune cells within inflamed tissue of the rat. *Neuroscience Letters, 140*(1), 85–88. https://doi.org/10.1016/0304-3940(92)90688-4

Hawes, S. L., Salinas, A. G., Lovinger, D. M., & Blackwell, K. T. (2017). Long-term plasticity of corticostriatal synapses is modulated by pathway-specific co-release of opioids through κ-opioid receptors. *The Journal of Physiology, 595*(16), 5637–5652. https://doi.org/10.1113/JP274190

Hazum, E., Chang, K. J., & Cuatrecasas, P. (1979). Specific nonopiate receptors for beta-endorphin. *Science, 205*(4410), 1033–1035. https://doi.org/10.1126/science.224457

Heese, K., Fiebich, B. L., Bauer, J., & Otten, U. (1998). NF-kappaB modulates lipopolysaccharide-induced microglial nerve growth factor expression. *Glia, 22*(4), 401–407. https://doi.org/10.1002/(sici)1098-1136(199804)22:4<401::aid-glia9>3.0.co;2-5

Heim, C., & Nemeroff, C. B. (1999). The impact of early adverse experiences on brain systems involved in the pathophysiology of anxiety and affective disorders. *Biological Psychiatry, 46*(11), 1509–1522. https://doi.org/10.1016/s0006-3223(99)00224-3

Heim, C., Ehlert, U., & Hellhammer, D. H. (2000). The potential role of hypocortisolism in the pathophysiology of stress-related bodily disorders. *Psychoneuroendocrinology, 25*(1), 1–35. https://doi.org/10.1016/s0306-4530(99)00035-9

Heinrichs, S. C., & Koob, G. F. (2004). Corticotropin-releasing factor in brain: A role in activation, arousal, and affect regulation. *The Journal of Pharmacology and Experimental Therapeutics, 311*(2), 427–440. https://doi.org/10.1124/jpet.103.052092

Held, K., Antonijevic, I., Murck, H., Kuenzel, H., & Steiger, A. (2006). Neuropeptide Y (NPY) shortens sleep latency but does not suppress ACTH and cortisol in depressed patients and normal controls. *Psychoneuroendocrinology, 31*(1), 100–107. https://doi.org/10.1016/j.psyneuen.2005.05.015

Herbert, T. B., & Cohen, S. (1993). Depression and immunity: A meta-analytic review. *Psychological Bulletin, 113*(3), 472–486. https://doi.org/10.1037/0033-2909.113.3.472

Herman, J. P., Figueiredo, H., Mueller, N. K., Ulrich-Lai, Y., Ostrander, M. M., Choi, D. C., & Cullinan, W. E. (2003). Central mechanisms of stress integration: Hierarchical circuitry controlling hypothalamo-pituitary-adrenocortical responsiveness. *Frontiers in Neuroendocrinology, 24*(3), 151–180. https://doi.org/10.1016/j.yfrne.2003.07.001

Holsboer, F. (2000). The corticosteroid receptor hypothesis of depression. *Neuropsychopharmacology, 23*(5), 477–501. https://doi.org/10.1016/S0893-133X(00)00159-7

Hori, H., & Kim, Y. (2019). Inflammation and post-traumatic stress disorder. *Psychiatry and Clinical Neurosciences, 73*(4), 143–153. https://doi.org/10.1111/pcn.12820

Hou, C., Jia, F., Liu, Y., & Li, L. (2006). CSF serotonin, 5-hydroxyindolacetic acid and neuropeptide Y levels in severe major depressive disorder. *Brain Research, 1095*(1), 154–158. https://doi.org/10.1016/j.brainres.2006.04.026

Howerton, A. R., Roland, A. V., Fluharty, J. M., Marshall, A., Chen, A., Daniels, D., et al. (2014). Sex differences in corticotropin-releasing factor receptor-1 action within the dorsal raphe nucleus in stress responsivity. *Biological Psychiatry, 75*(11), 873–883. https://doi.org/10.1016/j.biopsych.2013.10.013

Hsu, S. Y., & Hsueh, A. J. (2001). Human stresscopin and stresscopin-related peptide are selective ligands for the type 2 corticotropin-releasing hormone receptor. *Nature Medicine, 7*(5), 605–611. https://doi.org/10.1038/87936

Hubler, T. R., & Scammell, J. G. (2004). Intronic hormone response elements mediate regulation of FKBP5 by progestins and glucocorticoids. *Cell Stress & Chaperones, 9*(3), 243–252. https://doi.org/10.1379/csc-32r.1

Hupalo, S., Bryce, C. A., Bangasser, D. A., Berridge, C. W., Valentino, R. J., & Floresco, S. B. (2019). Corticotropin-Releasing Factor (CRF) circuit modulation of cognition and motivation. *Neuroscience and Biobehavioral Reviews, 103*, 50–59. https://doi.org/10.1016/j.neubiorev.2019.06.010

Iosif, R. E., Ekdahl, C. T., Ahlenius, H., Pronk, C. J., Bonde, S., Kokaia, Z., et al. (2006). Tumor necrosis factor receptor 1 is a negative regulator of progenitor proliferation in adult hippocampal neurogenesis. *The Journal of Neuroscience, 26*(38), 9703–9712. https://doi.org/10.1523/JNEUROSCI.2723-06.2006

Irwin, M. (1999). Immune correlates of depression. *Advances in Experimental Medicine and Biology, 461*, 1–24. https://doi.org/10.1007/978-0-585-37970-8_1

Irwin, M. (2001). Neuroimmunology of disordered sleep in depression and alcoholism. *Neuropsychopharmacology, 25*(5 Suppl), S45–S49. https://doi.org/10.1016/S0893-133X(01)00338-4

Irwin, M. R., Olmos, L., Wang, M., Valladares, E. M., Motivala, S. J., Fong, T., et al. (2007). Cocaine dependence and acute cocaine induce decreases of monocyte proinflammatory cytokine expression across the diurnal period: Autonomic mechanisms. *The Journal of Pharmacology and Experimental Therapeutics, 320*(2), 507–515. https://doi.org/10.1124/jpet.106.112797

Itokawa, M., Arai, M., Kato, S., Ogata, Y., Furukawa, A., Haga, S., et al. (2003). Association between a novel polymorphism in the promoter region of the neuropeptide Y gene and schizophrenia in humans. *Neuroscience Letters, 347*(3), 202–204. https://doi.org/10.1016/s0304-3940(03)00718-3

Jacobson, L., & Sapolsky, R. (1991). The role of the hippocampus in feedback regulation of the hypothalamic-pituitary-adrenocortical axis. *Endocrine Reviews, 12*(2), 118–134. https://doi.org/10.1210/edrv-12-2-118

Jain, A., Mishra, A., Shakkarpude, J., & Lakhani, P. (2019). Beta endorphins: The natural opioids. *International Journal of Chemical Studies, 7*, 323–332.

James, I. F., Fischli, W., & Goldstein, A. (1984). Opioid receptor selectivity of dynorphin gene products. *The Journal of Pharmacology and Experimental Therapeutics, 228*(1), 88–93.

James, M. H., Campbell, E. J., & Dayas, C. V. (2017). Role of the orexin/hypocretin system in stress-related psychiatric disorders. *Current Topics in Behavioral Neurosciences, 33*, 197–219. https://doi.org/10.1007/7854_2016_56

Japee, S., Holiday, K., Satyshur, M. D., Mukai, I., & Ungerleider, L. G. (2015). A role of right middle frontal gyrus in reorienting of attention: A case study. *Frontiers in Systems Neuroscience, 9*, 23. https://doi.org/10.3389/fnsys.2015.00023

Jin, Y., Sun, L. H., Yang, W., Cui, R. J., & Xu, S. B. (2019). The role of BDNF in the neuroimmune axis regulation of mood disorders. *Frontiers in Neurology, 10*, 515. https://doi.org/10.3389/fneur.2019.00515

Jones, T., & Moller, M. D. (2011). Implications of hypothalamic-pituitary-adrenal axis functioning in posttraumatic stress disorder. *Journal of the American Psychiatric Nurses Association, 17*(6), 393–403. https://doi.org/10.1177/1078390311420564

Joshi, D., Fullerton, J. M., & Weickert, C. S. (2014). Elevated ErbB4 mRNA is related to interneuron deficit in prefrontal cortex in schizophrenia. *Journal of Psychiatric Research, 53*, 125–132. https://doi.org/10.1016/j.jpsychires.2014.02.014

Juruena, M. F., Bocharova, M., Agustini, B., & Young, A. H. (2018). Atypical depression and non-atypical depression: Is HPA axis function a biomarker? A systematic review. *Journal of Affective Disorders, 233*, 45–67. https://doi.org/10.1016/j.jad.2017.09.052

Kany, S., Vollrath, J. T., & Relja, B. (2019). Cytokines in inflammatory disease. *International Journal of Molecular Sciences, 20*(23). https://doi.org/10.3390/ijms20236008

Karalis, K., Sano, H., Redwine, J., Listwak, S., Wilder, R. L., & Chrousos, G. P. (1991). Autocrine or paracrine inflammatory actions of corticotropin-releasing hormone in vivo. *Science, 254*(5030), 421–423. https://doi.org/10.1126/science.1925600

Kash, T. L., & Winder, D. G. (2006). Neuropeptide Y and corticotropin-releasing factor bi-directionally modulate inhibitory synaptic transmission in the bed nucleus of the stria terminalis. *Neuropharmacology, 51*(5), 1013–1022. https://doi.org/10.1016/j.neuropharm.2006.06.011

Katrinli, S., Oliveira, N. C. S., Felger, J. C., Michopoulos, V., & Smith, A. K. (2022). The role of the immune system in posttraumatic stress disorder. *Translational Psychiatry, 12*(1), 313. https://doi.org/10.1038/s41398-022-02094-7

Kavelaars, A., & Heijnen, C. J. (2000). Expression of preproenkephalin mRNA and production and secretion of enkephalins by human thymocytes. *Annals of the New York Academy of Sciences, 917*, 778–783. https://doi.org/10.1111/j.1749-6632.2000.tb05443.x

Kazakou, P., Nicolaides, N. C., & Chrousos, G. P. (2023). Basic concepts and hormonal regulators of the stress system. *Hormone Research in Pædiatrics, 96*(1), 8–16. https://doi.org/10.1159/000523975

Kearns, M. C., Ressler, K. J., Zatzick, D., & Rothbaum, B. O. (2012). Early interventions for PTSD: A review. *Depression and Anxiety, 29*(10), 833–842. https://doi.org/10.1002/da.21997

Keller, J., Gomez, R., Williams, G., Lembke, A., Lazzeroni, L., Murphy, G. M., & Schatzberg, A. F. (2017). HPA axis in major depression: Cortisol, clinical symptomatology and genetic variation predict cognition. *Molecular Psychiatry, 22*(4), 527–536. https://doi.org/10.1038/mp.2016.120

Kennedy, S. H., Kutcher, S. P., Ralevski, E., & Brown, G. M. (1996). Nocturnal melatonin and 24-hour 6-sulphatoxymelatonin levels in various phases of bipolar affective disorder. *Psychiatry Research, 63*(2-3), 219–222. https://doi.org/10.1016/0165-1781(96)02910-1

Keramatian, K., Dhanoa, T., McGirr, A., Lang, D. J., Honer, W. G., Lam, R. W., & Yatham, L. N. (2018). Structural brain changes in first episode mania with and without psychosis: Data from the Systematic Treatment Optimization Program for Early Mania (STOP-EM). *The World Journal of Biological Psychiatry, 19*(sup2), S30–S40. https://doi.org/10.1080/15622975.2016.1249950

Khairuddin, S., Aquili, L., Heng, B. C., Hoo, T. L. C., Wong, K. H., & Lim, L. W. (2020). Dysregulation of the orexinergic system: A potential neuropeptide target in depression. *Neuroscience and Biobehavioral Reviews, 118*, 384–396. https://doi.org/10.1016/j.neubiorev.2020.07.040

Kieffer, B. L., & Gavériaux-Ruff, C. (2002). Exploring the opioid system by gene knockout. *Progress in Neurobiology, 66*(5), 285–306. https://doi.org/10.1016/s0301-0082(02)00008-4

Kilpatrick, D. G., Resnick, H. S., Milanak, M. E., Miller, M. W., Keyes, K. M., & Friedman, M. J. (2013). National estimates of exposure to traumatic events and PTSD prevalence using DSM-IV and DSM-5 criteria. *Journal of Traumatic Stress, 26*(5), 537–547. https://doi.org/10.1002/jts.21848

Kim, Y. K., Jung, H. G., Myint, A. M., Kim, H., & Park, S. H. (2007a). Imbalance between pro-inflammatory and anti-inflammatory cytokines in bipolar disorder. *Journal of Affective Disorders, 104*(1-3), 91–95. https://doi.org/10.1016/j.jad.2007.02.018

Kim, Y. K., Na, K. S., Shin, K. H., Jung, H. Y., Choi, S. H., & Kim, J. B. (2007b). Cytokine imbalance in the pathophysiology of major depressive disorder. *Progress in Neuro-Psychopharmacology & Biological Psychiatry, 31*(5), 1044–1053. https://doi.org/10.1016/j.pnpbp.2007.03.004

Kim, Y. K., Na, K. S., Myint, A. M., & Leonard, B. E. (2016). The role of pro-inflammatory cytokines in neuroinflammation, neurogenesis and the neuroendocrine system in major depression. *Progress in Neuro-Psychopharmacology & Biological Psychiatry, 64*, 277–284. https://doi.org/10.1016/j.pnpbp.2015.06.008

Kim, Y. K., Amidfar, M., & Won, E. (2019). A review on inflammatory cytokine-induced alterations of the brain as potential neural biomarkers in post-traumatic stress disorder. *Progress in Neuro-Psychopharmacology & Biological Psychiatry, 91*, 103–112. https://doi.org/10.1016/j.pnpbp.2018.06.008

Kim, T. D., Lee, S., & Yoon, S. (2020). Inflammation in Post-Traumatic Stress Disorder (PTSD): A review of potential correlates of PTSD with a neurological perspective. *Antioxidants (Basel), 9*(2). https://doi.org/10.3390/antiox9020107

Kino, T. (2015). Stress, glucocorticoid hormones, and hippocampal neural progenitor cells: Implications to mood disorders. *Frontiers in Physiology, 6*, 230. https://doi.org/10.3389/fphys.2015.00230

Kipnis, J., Cardon, M., Strous, R. D., & Schwartz, M. (2006). Loss of autoimmune T cells correlates with brain diseases: Possible implications for schizophrenia? *Trends in Molecular Medicine, 12*(3), 107–112. https://doi.org/10.1016/j.molmed.2006.01.003

Kivimäki, M., Shipley, M. J., Batty, G. D., Hamer, M., Akbaraly, T. N., Kumari, M., et al. (2014). Long-term inflammation increases risk of common mental disorder: A cohort study. *Molecular Psychiatry, 19*(2), 149–150. https://doi.org/10.1038/mp.2013.35

Kneeland, R. E., & Fatemi, S. H. (2013). Viral infection, inflammation and schizophrenia. *Progress in Neuro-Psychopharmacology & Biological Psychiatry, 42*, 35–48. https://doi.org/10.1016/j.pnpbp.2012.02.001

Knoll, A. T., & Carlezon, W. A. (2010). Dynorphin, stress, and depression. *Brain Research, 1314*, 56–73. https://doi.org/10.1016/j.brainres.2009.09.074

Köhler, O., Benros, M. E., Nordentoft, M., Farkouh, M. E., Iyengar, R. L., Mors, O., & Krogh, J. (2014). Effect of anti-inflammatory treatment on depression, depressive symptoms, and adverse effects: A systematic review and meta-analysis of randomized clinical trials. *JAMA Psychiatry, 71*(12), 1381–1391. https://doi.org/10.1001/jamapsychiatry.2014.1611

Kolber, B. J., Roberts, M. S., Howell, M. P., Wozniak, D. F., Sands, M. S., & Muglia, L. J. (2008). Central amygdala glucocorticoid receptor action promotes fear-associated CRH activation and conditioning. *Proceedings of the National Academy of Sciences of the United States of America, 105*(33), 12004–12009. https://doi.org/10.1073/pnas.0803216105

Kosterlitz, H. W., & Hughes, J. (1977). Opiate receptors and endogenous opioid peptides in tolerance and dependence. *Advances in Experimental Medicine and Biology, 85B*, 141–154. https://doi.org/10.1007/978-1-4615-9038-5_10

Lach, G., & de Lima, T. C. (2013). Role of NPY Y1 receptor on acquisition, consolidation and extinction on contextual fear conditioning: Dissociation between anxiety, locomotion and non-emotional memory behavior. *Neurobiology of Learning and Memory, 103*, 26–33. https://doi.org/10.1016/j.nlm.2013.04.005

Lapchak, P. A., Araujo, D. M., & Hefti, F. (1993). Systemic interleukin-1 beta decreases brain-derived neurotrophic factor messenger RNA expression in the rat hippocampal formation. *Neuroscience, 53*(2), 297–301. https://doi.org/10.1016/0306-4522(93)90196-m

Laryea, G., Arnett, M. G., & Muglia, L. J. (2012). Behavioral studies and genetic alterations in Corticotropin-Releasing Hormone (CRH) neurocircuitry: Insights into human psychiatric disorders. *Behavioral Science (Basel), 2*(2), 135–171. https://doi.org/10.3390/bs2020135

Lecci, A., & Maggi, C. A. (2003). Peripheral tachykinin receptors as potential therapeutic targets in visceral diseases. *Expert Opinion on Therapeutic Targets, 7*(3), 343–362. https://doi.org/10.1517/14728222.7.3.343

Lee, H. Y., & Kim, Y. K. (2010). Effect of TGF-β1 polymorphism on the susceptibility to schizophrenia and treatment response to atypical antipsychotic agent. *Acta Neuropsychiatrica, 22*(4), 174–179. https://doi.org/10.1111/j.1601-5215.2009.00435.x

Lee, S., & Rivier, C. (1997). Alcohol increases the expression of type 1, but not type 2 alpha corticotropin-releasing factor (CRF) receptor messenger ribonucleic acid in the rat hypothalamus. *Brain Research. Molecular Brain Research, 52*(1), 78–89. https://doi.org/10.1016/s0169-328x(97)00226-x

Lehnardt, S. (2010). Innate immunity and neuroinflammation in the CNS: The role of microglia in Toll-like receptor-mediated neuronal injury. *Glia, 58*(3), 253–263. https://doi.org/10.1002/glia.20928

Lehner, M., Skórzewska, A., & Wisłowska-Stanek, A. (2021). Sex-related predisposition to post-traumatic stress disorder development-the role of neuropeptides. *International Journal of Environmental Research and Public Health, 19*(1). https://doi.org/10.3390/ijerph19010314

Leonard, B. E. (2007). Psychopathology of depression. *Drugs of today (Barcelona, Spain: 1998), 43*(10), 705–716. https://doi.org/10.1358/dot.2007.43.10.1122223

Leonard, B. E. (2010). The concept of depression as a dysfunction of the immune system. *Current Immunology Reviews, 6*(3), 205–212. https://doi.org/10.2174/157339510791823835

Leroy, F., de Solis, C. A., Boyle, L. M., Bock, T., Lofaro, O. M., Buss, E. W., et al. (2022). Enkephalin release from VIP interneurons in the hippocampal CA2/3a region mediates heterosynaptic plasticity and social memory. *Molecular Psychiatry, 27*(6), 2879–2900. https://doi.org/10.1038/s41380-021-01124-y

Lewis, K., Li, C., Perrin, M. H., Blount, A., Kunitake, K., Donaldson, C., et al. (2001). Identification of urocortin III, an additional member of the corticotropin-releasing factor (CRF) family with high affinity for the CRF2 receptor. *Proceedings of the National Academy of Sciences of the United States of America, 98*(13), 7570–7575. https://doi.org/10.1073/pnas.121165198

Lewitus, G. M., Konefal, S. C., Greenhalgh, A. D., Pribiag, H., Augereau, K., & Stellwagen, D. (2016). Microglial TNF-α suppresses cocaine-induced plasticity and behavioral sensitization. *Neuron, 90*(3), 483–491. https://doi.org/10.1016/j.neuron.2016.03.030

Li, H., Lu, J., Li, S., Huang, B., Shi, G., Mou, T., & Xu, Y. (2021). Increased hypocretin (orexin) plasma level in depression, bipolar disorder patients. *Frontiers in Psychiatry, 12*, 676336. https://doi.org/10.3389/fpsyt.2021.676336

Lin, A., Kenis, G., Bignotti, S., Tura, G. J., De Jong, R., Bosmans, E., et al. (1998). The inflammatory response system in treatment-resistant schizophrenia: Increased serum interleukin-6. *Schizophrenia Research, 32*(1), 9–15. https://doi.org/10.1016/s0920-9964(98)00034-6

Linnik, M. D., & Moskowitz, M. A. (1989). Identification of immunoreactive substance P in human and other mammalian endothelial cells. *Peptides, 10*(5), 957–962. https://doi.org/10.1016/0196-9781(89)90175-7

Lippai, D., Bala, S., Csak, T., Kurt-Jones, E. A., & Szabo, G. (2013). Chronic alcohol-induced microRNA-155 contributes to neuroinflammation in a TLR4-dependent manner in mice. *PLoS One, 8*(8), e70945. https://doi.org/10.1371/journal.pone.0070945

Lotrich, F. E. (2015). Inflammatory cytokine-associated depression. *Brain Research, 1617*, 113–125. https://doi.org/10.1016/j.brainres.2014.06.032

Luger, T. A., Scholzen, T. E., Brzoska, T., & Böhm, M. (2003). New insights into the functions of alpha-MSH and related peptides in the immune system. *Annals of the New York Academy of Sciences, 994*, 133–140. https://doi.org/10.1111/j.1749-6632.2003.tb03172.x

Lv, M. H., Tan, Y. L., Yan, S. X., Tian, L., Chen, D. C., Tan, S. P., et al. (2015). Decreased serum TNF-alpha levels in chronic schizophrenia patients on long-term antipsychotics: Correlation with psychopathology and cognition. *Psychopharmacology, 232*(1), 165–172. https://doi.org/10.1007/s00213-014-3650-y

MacMaster, F. P., Leslie, R., Rosenberg, D. R., & Kusumakar, V. (2008). Pituitary gland volume in adolescent and young adult bipolar and unipolar depression. *Bipolar Disorders, 10*(1), 101–104. https://doi.org/10.1111/j.1399-5618.2008.00476.x

Madden, K. S., & Felten, D. L. (1995). Experimental basis for neural-immune interactions. *Physiological Reviews, 75*(1), 77–106. https://doi.org/10.1152/physrev.1995.75.1.77

Maeng, L. Y., & Milad, M. R. (2015). Sex differences in anxiety disorders: Interactions between fear, stress, and gonadal hormones. *Hormones and Behavior, 76*, 106–117. https://doi.org/10.1016/j.yhbeh.2015.04.002

Maes, M. (2011). Depression is an inflammatory disease, but cell-mediated immune activation is the key component of depression. *Progress in Neuro-Psychopharmacology & Biological Psychiatry, 35*(3), 664–675. https://doi.org/10.1016/j.pnpbp.2010.06.014

Maes, M., Jacobs, M. P., Suy, E., Vandewoude, M., Minner, B., & Raus, J. (1990). Effects of dexamethasone on the availability of L-tryptophan and on the insulin and FFA concentrations in unipolar depressed patients. *Biological Psychiatry, 27*(8), 854–862. https://doi.org/10.1016/0006-3223(90)90466-f

Maes, M., Yirmyia, R., Noraberg, J., Brene, S., Hibbeln, J., Perini, G., et al. (2009). The inflammatory & neurodegenerative (I&ND) hypothesis of depression: Leads for future research and new drug developments in depression. *Metabolic Brain Disease, 24*(1), 27–53. https://doi.org/10.1007/s11011-008-9118-1

Mahadik, S. P., & Mukherjee, S. (1996). Free radical pathology and antioxidant defense in schizophrenia: A review. *Schizophrenia Research, 19*(1), 1–17. https://doi.org/10.1016/0920-9964(95)00049-6

Maksymetz, J., Byun, N. E., Luessen, D. J., Li, B., Barry, R. L., Gore, J. C., et al. (2021). $mGlu_1$ potentiation enhances prelimbic somatostatin interneuron activity to rescue schizophrenia-like physiological and cognitive deficits. *Cell Reports, 37*(5), 109950. https://doi.org/10.1016/j.celrep.2021.109950

Manetti, L., Cavagnini, F., Martino, E., & Ambrogio, A. (2014). Effects of cocaine on the hypothalamic-pituitary-adrenal axis. *Journal of Endocrinological Investigation, 37*(8), 701–708. https://doi.org/10.1007/s40618-014-0091-8

Margaretten, M., Julian, L., Katz, P., & Yelin, E. (2011). Depression in patients with rheumatoid arthritis: Description, causes and mechanisms. *International Journal of Clinical Rheumatology, 6*(6), 617–623. https://doi.org/10.2217/IJR.11.6

Marques, A. H., Silverman, M. N., & Sternberg, E. M. (2009). Glucocorticoid dysregulations and their clinical correlates. From receptors to therapeutics. *Annals of the New York Academy of Sciences, 1179*, 1–18. https://doi.org/10.1111/j.1749-6632.2009.04987.x

Marriott, I. (2004). The role of tachykinins in central nervous system inflammatory responses. *Frontiers in Bioscience, 9*, 2153–2165. https://doi.org/10.2741/1377

Martinuzzi, E., Barbosa, S., Courtet, P., Olié, E., Guillaume, S., Ibrahim, E. C., et al. (2021). Blood cytokines differentiate bipolar disorder and major depressive disorder during a major depressive episode: Initial discovery and independent sample replication. *Brain, Behavior, & Immunity – Health, 13*, 100232. https://doi.org/10.1016/j.bbih.2021.100232

Masterson, S. P., Li, J., & Bickford, M. E. (2010). Frequency-dependent release of substance P mediates heterosynaptic potentiation of glutamatergic synaptic responses in the rat visual thalamus. *Journal of Neurophysiology, 104*(3), 1758–1767. https://doi.org/10.1152/jn.00010.2010

McCarthy, M. J., Zhang, H., Neff, N. H., & Hadjiconstantinou, M. (2010). Nicotine withdrawal and kappa-opioid receptors. *Psychopharmacology (Berl), 210*(2), 221–229. https://doi.org/10.1007/s00213-009-1674-5

McDougle, C. J., Black, J. E., Malison, R. T., Zimmermann, R. C., Kosten, T. R., Heninger, G. R., & Price, L. H. (1994). Noradrenergic dysregulation during discontinuation of cocaine use in addicts. *Archives of General Psychiatry, 51*(9), 713–719. https://doi.org/10.1001/archpsyc.1994.03950090045007

Menke, A. (2019). Is the HPA axis as target for depression outdated, or is there a new hope? *Frontiers in Psychiatry, 10*, 101. https://doi.org/10.3389/fpsyt.2019.00101

Michopoulos, V., Powers, A., Gillespie, C. F., Ressler, K. J., & Jovanovic, T. (2017). Inflammation in fear- and anxiety-based disorders: PTSD, GAD, and beyond. *Neuropsychopharmacology, 42*(1), 254–270. https://doi.org/10.1038/npp.2016.146

Mikulska, J., Juszczyk, G., Gawrońska-Grzywacz, M., & Herbet, M. (2021). HPA axis in the pathomechanism of depression and schizophrenia: New therapeutic strategies based on its participation. *Brain Sciences, 11*(10). https://doi.org/10.3390/brainsci11101298

Miller, A. H., & Raison, C. L. (2016). The role of inflammation in depression: From evolutionary imperative to modern treatment target. *Nature Reviews. Immunology, 16*(1), 22–34. https://doi.org/10.1038/nri.2015.5

Miller, A. H., Maletic, V., & Raison, C. L. (2009). Inflammation and its discontents: The role of cytokines in the pathophysiology of major depression. *Biological Psychiatry, 65*(9), 732–741. https://doi.org/10.1016/j.biopsych.2008.11.029

Miller, B. J., Buckley, P., Seabolt, W., Mellor, A., & Kirkpatrick, B. (2011). Meta-analysis of cytokine alterations in schizophrenia: Clinical status and antipsychotic effects. *Biological Psychiatry, 70*(7), 663–671. https://doi.org/10.1016/j.biopsych.2011.04.013

Miller, A. H., Haroon, E., Raison, C. L., & Felger, J. C. (2013a). Cytokine targets in the brain: Impact on neurotransmitters and neurocircuits. *Depression and Anxiety, 30*(4), 297–306. https://doi.org/10.1002/da.22084

Miller, S., Hallmayer, J., Wang, P. W., Hill, S. J., Johnson, S. L., & Ketter, T. A. (2013b). Brain-derived neurotrophic factor val66met genotype and early life stress effects upon bipolar course. *Journal of Psychiatric Research, 47*(2), 252–258. https://doi.org/10.1016/j.jpsychires.2012.10.015

Misiak, B., Beszłej, J. A., Kotowicz, K., Szewczuk-Bogusławska, M., Samochowiec, J., Kucharska-Mazur, J., & Frydecka, D. (2018). Cytokine alterations and cognitive impairment in major depressive disorder: From putative mechanisms to novel treatment targets. *Progress in Neuro-Psychopharmacology & Biological Psychiatry, 80*(Pt C), 177–188. https://doi.org/10.1016/j.pnpbp.2017.04.021

Modabbernia, A., Taslimi, S., Brietzke, E., & Ashrafi, M. (2013). Cytokine alterations in bipolar disorder: A meta-analysis of 30 studies. *Biological Psychiatry, 74*(1), 15–25. https://doi.org/10.1016/j.biopsych.2013.01.007

Modell, S., Yassouridis, A., Huber, J., & Holsboer, F. (1997). Corticosteroid receptor function is decreased in depressed patients. *Neuroendocrinology, 65*(3), 216–222. https://doi.org/10.1159/000127275

Momtazmanesh, S., Zare-Shahabadi, A., & Rezaei, N. (2019). Cytokine alterations in schizophrenia: An updated review. *Frontiers in Psychiatry, 10*, 892. https://doi.org/10.3389/fpsyt.2019.00892

Monje, M. L., Toda, H., & Palmer, T. D. (2003). Inflammatory blockade restores adult hippocampal neurogenesis. *Science, 302*(5651), 1760–1765. https://doi.org/10.1126/science.1088417

Morris, B. J., & Johnston, H. M. (1995). A role for hippocampal opioids in long-term functional plasticity. *Trends in Neurosciences, 18*(8), 350–355. https://doi.org/10.1016/0166-2236(95)93927-p

Morris, H. M., Hashimoto, T., & Lewis, D. A. (2008). Alterations in somatostatin mRNA expression in the dorsolateral prefrontal cortex of subjects with schizophrenia or schizoaffective disorder. *Cerebral Cortex, 18*(7), 1575–1587. https://doi.org/10.1093/cercor/bhm186

Morris, M. C., Compas, B. E., & Garber, J. (2012). Relations among posttraumatic stress disorder, comorbid major depression, and HPA function: A systematic review and meta-analysis. *Clinical Psychology Review, 32*(4), 301–315. https://doi.org/10.1016/j.cpr.2012.02.002

Mouihate. A. (2014). TLR4-mediated brain inflammation halts neurogenesis: impact of hormonal replacement therapy. *Frontiers in Cellular Neuroscience*. https://doi.org/10.3389/fncel.2014.00146

Moussaoui, S. M., Le Prado, N., Bonici, B., Faucher, D. C., Cuiné, F., Laduron, P. M., & Garret, C. (1992). Distribution of neurokinin B in rat spinal cord and peripheral tissues: comparison with neurokinin A and substance P and effects of neonatal capsaicin treatment. *Neuroscience, 48*(4), 969–978. https://doi.org/10.1016/0306-4522(92)90285-a

Munhoz, C. D., Sorrells, S. F., Caso, J. R., Scavone, C., & Sapolsky, R. M. (2010). Glucocorticoids exacerbate lipopolysaccharide-induced signaling in the frontal cortex and hippocampus in a dose-dependent manner. *The Journal of Neuroscience, 30*(41), 13690–13698. https://doi.org/10.1523/JNEUROSCI.0303-09.2010

Murray, G. (2007). Diurnal mood variation in depression: A signal of disturbed circadian function? *Journal of Affective Disorders, 102*(1-3), 47–53. https://doi.org/10.1016/j.jad.2006.12.001

Nahvi, R. J., & Sabban, E. L. (2020). Sex differences in the neuropeptide Y system and implications for stress related disorders. *Biomolecules, 10*(9). https://doi.org/10.3390/biom10091248

Nakamura, M., & Nagamine, T. (2017). Neuroendocrine, autonomic, and metabolic responses to an orexin antagonist, suvorexant, in psychiatric patients with insomnia. *Innovations in Clinical Neuroscience, 14*(3-4), 30–37.

Namba, M. D., Leyrer-Jackson, J. M., Nagy, E. K., Olive, M. F., & Neisewander, J. L. (2021). Neuroimmune mechanisms as novel treatment targets for substance use disorders and associated comorbidities. *Frontiers in Neuroscience, 15*, 650785. https://doi.org/10.3389/fnins.2021.650785

Nandam, L. S., Brazel, M., Zhou, M., & Jhaveri, D. J. (2019). Cortisol and major depressive disorder-translating findings from humans to animal models and back. *Frontiers in Psychiatry, 10*, 974. https://doi.org/10.3389/fpsyt.2019.00974

Neufeld-Cohen, A., Tsoory, M. M., Evans, A. K., Getselter, D., Gil, S., Lowry, C. A., et al. (2010). A triple urocortin knockout mouse model reveals an essential role for urocortins in stress recovery. *Proceedings of the National Academy of Sciences of the United States of America, 107*(44), 19020–19025. https://doi.org/10.1073/pnas.1013761107

Ni, P., Tian, Y., Gu, X., Yang, L., Wei, J., Wang, Y., et al. (2019). Plasma neuropeptides as circulating biomarkers of multifactorial schizophrenia. *Comprehensive Psychiatry, 94*, 152114. https://doi.org/10.1016/j.comppsych.2019.152114

Niu, M., Wang, Y., Jia, Y., Wang, J., Zhong, S., Lin, J., et al. (2017). Common and specific abnormalities in cortical thickness in patients with major depressive and bipolar disorders. *eBioMedicine, 16*, 162–171. https://doi.org/10.1016/j.ebiom.2017.01.010

Noda, M., Furutani, Y., Takahashi, H., Toyosato, M., Hirose, T., Inayama, S., et al. (1982). Cloning and sequence analysis of cDNA for bovine adrenal preproenkephalin. *Nature, 295*(5846), 202–206. https://doi.org/10.1038/295202a0

Ogłodek, E. A., Szota, A. M., Moś, D. M., Araszkiewicz, A., & Szromek, A. R. (2015). Serum concentrations of chemokines (CCL-5 and CXCL-12), chemokine receptors (CCR-5 and CXCR-4), and IL-6 in patients with posttraumatic stress disorder and avoidant personality disorder. *Pharmacological Reports, 67*(6), 1251–1258. https://doi.org/10.1016/j.pharep.2015.05.023

Ogren, S. O., Razani, H., Elvander-Tottie, E., & Kehr, J. (2007). The neuropeptide galanin as an in vivo modulator of brain 5-HT1A receptors: Possible relevance for affective disorders. *Physiology & Behavior, 92*(1–2), 172–179. https://doi.org/10.1016/j.physbeh.2007.05.022

Oitzl, M. S., & de Kloet, E. R. (1992). Selective corticosteroid antagonists modulate specific aspects of spatial orientation learning. *Behavioral Neuroscience, 106*(1), 62–71. https://doi.org/10.1037/0735-7044.106.1.62

Orsolini, L., Pompili, S., Tempia Valenta, S., Salvi, V., & Volpe, U. (2022). C-reactive protein as a biomarker for major depressive disorder? *International Journal of Molecular Sciences, 23*(3). https://doi.org/10.3390/ijms23031616

Ortiz-Domínguez, A., Hernández, M. E., Berlanga, C., Gutiérrez-Mora, D., Moreno, J., Heinze, G., & Pavón, L. (2007). Immune variations in bipolar disorder: Phasic differences. *Bipolar Disorders, 9*(6), 596–602. https://doi.org/10.1111/j.1399-5618.2007.00493.x

Otsuka, M., & Yoshioka, K. (1993). Neurotransmitter functions of mammalian tachykinins. *Physiological Reviews, 73*(2), 229–308. https://doi.org/10.1152/physrev.1993.73.2.229

Owens, M. J., & Nemeroff, C. B. (1991). Physiology and pharmacology of corticotropin-releasing factor. *Pharmacological Reviews, 43*(4), 425–473.

Pace, T. W., Hu, F., & Miller, A. H. (2007). Cytokine-effects on glucocorticoid receptor function: relevance to glucocorticoid resistance and the pathophysiology and treatment of major depression. *Brain, Behavior, and Immunity, 21*(1), 9–19. https://doi.org/10.1016/j.bbi.2006.08.009

Pace, T. W., Wingenfeld, K., Schmidt, I., Meinlschmidt, G., Hellhammer, D. H., & Heim, C. M. (2012). Increased peripheral NF-κB pathway activity in women with childhood abuse-related posttraumatic stress disorder. *Brain, Behavior, and Immunity, 26*(1), 13–17. https://doi.org/10.1016/j.bbi.2011.07.232

Pagani, L., St Clair, P. A., Teshiba, T. M., Service, S. K., Fears, S. C., Araya, C., et al. (2016). Genetic contributions to circadian activity rhythm and sleep pattern phenotypes in pedigrees segregating for severe bipolar disorder. *Proceedings of the National Academy of Sciences of the United States of America, 113*(6), E754–E761. https://doi.org/10.1073/pnas.1513525113

Palta, P., Samuel, L. J., Miller, E. R., & Szanton, S. L. (2014). Depression and oxidative stress: Results from a meta-analysis of observational studies. *Psychosomatic Medicine, 76*(1), 12–19. https://doi.org/10.1097/PSY.0000000000000009

Pan, X., Wang, Z., Wu, X., Wen, S. W., & Liu, A. (2018). Salivary cortisol in post-traumatic stress disorder: A systematic review and meta-analysis. *BMC Psychiatry, 18*(1), 324. https://doi.org/10.1186/s12888-018-1910-9

Pariante, C. M., Dazzan, P., Danese, A., Morgan, K. D., Brudaglio, F., Morgan, C., et al. (2005). Increased pituitary volume in antipsychotic-free and antipsychotic-treated patients of the AEsop first-onset psychosis study. *Neuropsychopharmacology, 30*(10), 1923–1931. https://doi.org/10.1038/sj.npp.1300766

Passos, I. C., Vasconcelos-Moreno, M. P., Costa, L. G., Kunz, M., Brietzke, E., Quevedo, J., et al. (2015). Inflammatory markers in post-traumatic stress disorder: A systematic review, meta-analysis, and meta-regression. *Lancet Psychiatry, 2*(11), 1002–1012. https://doi.org/10.1016/S2215-0366(15)00309-0

Peng, W., Chen, Z., Yin, L., Jia, Z., & Gong, Q. (2016). Essential brain structural alterations in major depressive disorder: A voxel-wise meta-analysis on first episode, medication-naive patients. *Journal of Affective Disorders, 199*, 114–123. https://doi.org/10.1016/j.jad.2016.04.001

Perez, S. M., Boley, A., & Lodge, D. J. (2019). Region specific knockdown of Parvalbumin or Somatostatin produces neuronal and behavioral deficits consistent with those observed in schizophrenia. *Translational Psychiatry, 9*(1), 264. https://doi.org/10.1038/s41398-019-0603-6

Perkins, K. A., Coddington, S. B., Karelitz, J. L., Jetton, C., Scott, J. A., Wilson, A. S., & Lerman, C. (2009). Variability in initial nicotine sensitivity due to sex, history of other drug use, and parental smoking. *Drug and Alcohol Dependence, 99*(1-3), 47–57. https://doi.org/10.1016/j.drugalcdep.2008.06.017

Perroud, N., Dayer, A., Piguet, C., Nallet, A., Favre, S., Malafosse, A., & Aubry, J. M. (2014). Childhood maltreatment and methylation of the glucocorticoid receptor gene NR3C1 in bipolar disorder. *The British Journal of Psychiatry, 204*(1), 30–35. https://doi.org/10.1192/bjp.bp.112.120055

Pervanidou, P., Makris, G., Chrousos, G., & Agorastos, A. (2020). Early life stress and pediatric posttraumatic stress disorder. *Brain Sciences, 10*(3). https://doi.org/10.3390/brainsci10030169

Pinto, F. M., Almeida, T. A., Hernandez, M., Devillier, P., Advenier, C., & Candenas, M. L. (2004). mRNA expression of tachykinins and tachykinin receptors in different human tissues. *European Journal of Pharmacology, 494*(2–3), 233–239. https://doi.org/10.1016/j.ejphar.2004.05.016

Plotnikoff, N. P., Faith, R. E., Murgo, A. J., Herberman, R. B., & Good, R. A. (1997). Methionine enkephalin: A new cytokine—Human studies. *Clinical Immunology and Immunopathology, 82*(2), 93–101. https://doi.org/10.1006/clin.1996.4287

Porkka-Heiskanen, T., Kalinchuk, A., Alanko, L., Huhtaniemi, I., & Stenberg, D. (2004). Orexin A and B levels in the hypothalamus of female rats: The effects of the estrous cycle and age. *European Journal of Endocrinology, 150*(5), 737–742. https://doi.org/10.1530/eje.0.1500737

Pryce, C. R., & Fontana, A. (2017). Depression in autoimmune diseases. *Current Topics in Behavioral Neurosciences, 31*, 139–154. https://doi.org/10.1007/7854_2016_7

Raison, C. L., & Miller, A. H. (2001). The neuroimmunology of stress and depression. *Seminars in Clinical Neuropsychiatry, 6*(4), 277–294. https://doi.org/10.1053/scnp.2001.0060277

Raison, C. L., Capuron, L., & Miller, A. H. (2006). Cytokines sing the blues: Inflammation and the pathogenesis of depression. *Trends in Immunology, 27*(1), 24–31. https://doi.org/10.1016/j.it.2005.11.006

Rana, T., Behl, T., Sehgal, A., Singh, S., Sharma, N., Abdeen, A., et al. (2022). Exploring the role of neuropeptides in depression and anxiety. *Progress in Neuro-Psychopharmacology & Biological Psychiatry, 114*, 110478. https://doi.org/10.1016/j.pnpbp.2021.110478

Ranganathan, M., Braley, G., Pittman, B., Cooper, T., Perry, E., Krystal, J., & D'Souza, D. C. (2009). The effects of cannabinoids on serum cortisol and prolactin in humans. *Psychopharmacology, 203*(4), 737–744. https://doi.org/10.1007/s00213-008-1422-2

Reed, R. G., & Raison, C. L. (2016). Stress and the immune system. In C. Esser (Ed.), *Environmental influences on the immune system*. Springer.

Rein, T. (2016). FK506 binding protein 51 integrates pathways of adaptation: FKBP51 shapes the reactivity to environmental change. *Bioessays, 38*(9), 894–902. https://doi.org/10.1002/bies.201600050

Renner, V., & Petrowski, K. (2021). Stress-induced pro- and anti-inflammatory cytokine concentrations in PTSD and MD patients. *Psychoneuroendocrinology, 131*, 105513. https://doi.org/10.1016/j.psyneuen.2021.105513

Reul, J. M., & Holsboer, F. (2002). Corticotropin-releasing factor receptors 1 and 2 in anxiety and depression. *Current Opinion in Pharmacology, 2*(1), 23–33. https://doi.org/10.1016/s1471-4892(01)00117-5

Reul, J. M., Stec, I., Söder, M., & Holsboer, F. (1993). Chronic treatment of rats with the antidepressant amitriptyline attenuates the activity of the hypothalamic-pituitary-adrenocortical system. *Endocrinology, 133*(1), 312–320. https://doi.org/10.1210/endo.133.1.8391426

Reyes, T. M., Lewis, K., Perrin, M. H., Kunitake, K. S., Vaughan, J., Arias, C. A., et al. (2001). Urocortin II: A member of the corticotropin-releasing factor (CRF) neuropeptide family that is selectively bound by type 2 CRF receptors. *Proceedings of the National Academy of Sciences of the United States of America, 98*(5), 2843–2848. https://doi.org/10.1073/pnas.051626398

Robinson, S. L., & Thiele, T. E. (2017). The role of neuropeptide Y (NPY) in alcohol and drug abuse disorders. *International Review of Neurobiology, 136*, 177–197. https://doi.org/10.1016/bs.irn.2017.06.005

Rodríguez, B., Nani, J. V., Almeida, P. G. C., Brietzke, E., Lee, R. S., & Hayashi, M. A. F. (2020). Neuropeptides and oligopeptidases in schizophrenia. *Neuroscience and Biobehavioral Reviews, 108*, 679–693. https://doi.org/10.1016/j.neubiorev.2019.11.024

Roozendaal, B., McReynolds, J. R., Van der Zee, E. A., Lee, S., McGaugh, J. L., & McIntyre, C. K. (2009). Glucocorticoid effects on memory consolidation depend on functional interactions between the medial prefrontal cortex and basolateral amygdala. *The Journal of Neuroscience, 29*(45), 14299–14308. https://doi.org/10.1523/JNEUROSCI.3626-09.2009

Rosenblat, J. D., & McIntyre, R. S. (2017). Bipolar disorder and immune dysfunction: Epidemiological findings, proposed pathophysiology and clinical implications. *Brain Sciences, 7*(11). https://doi.org/10.3390/brainsci7110144

Rosenblat, J. D., Cha, D. S., Mansur, R. B., & McIntyre, R. S. (2014). Inflamed moods: A review of the interactions between inflammation and mood disorders. *Progress in Neuro-Psychopharmacology & Biological Psychiatry, 53*, 23–34. https://doi.org/10.1016/j.pnpbp.2014.01.013

Rowland, T., Perry, B. I., Upthegrove, R., Barnes, N., Chatterjee, J., Gallacher, D., & Marwaha, S. (2018). Neurotrophins, cytokines, oxidative stress mediators and mood state in bipolar disorder: Systematic review and meta-analyses. *The British Journal of Psychiatry, 213*(3), 514–525. https://doi.org/10.1192/bjp.2018.144

Ruiz, N. A. L., Del Ángel, D. S., Brizuela, N. O., Peraza, A. V., Olguín, H. J., Soto, M. P., & Guzmán, D. C. (2022). Inflammatory process and immune system in major depressive disorder. *The International Journal of Neuropsychopharmacology, 25*(1), 46–53. https://doi.org/10.1093/ijnp/pyab072

Saccani, F., Anselmi, L., Jaramillo, I., Bertoni, S., Barocelli, E., & Sternini, C. (2012). Protective role of μ opioid receptor activation in intestinal inflammation induced by mesenteric ischemia/reperfusion in mice. *Journal of Neuroscience Research, 90*(11), 2146–2153. https://doi.org/10.1002/jnr.23108

Sakai, K., Maeda, K., Chihara, K., & Kaneda, H. (1995). Increases in cortical neuropeptide Y and somatostatin concentrations following haloperidol-depot treatment in rats. *Neuropeptides, 29*(3), 157–161. https://doi.org/10.1016/0143-4179(95)90018-7

Salvador, A. F., de Lima, K. A., & Kipnis, J. (2021). Neuromodulation by the immune system: A focus on cytokines. *Nature Reviews. Immunology, 21*(8), 526–541. https://doi.org/10.1038/s41577-021-00508-z

Saper, C. B., Scammell, T. E., & Lu, J. (2005). Hypothalamic regulation of sleep and circadian rhythms. *Nature, 437*(7063), 1257–1263. https://doi.org/10.1038/nature04284

Schafer, M., Mousa, S. A., Zhang, Q., Carter, L., & Stein, C. (1996). Expression of corticotropin-releasing factor in inflamed tissue is required for intrinsic peripheral opioid analgesia. *Proceedings of the National Academy of Sciences of the United States of America, 93*(12), 6096–6100. https://doi.org/10.1073/pnas.93.12.6096

Schelling, G., Roozendaal, B., Krauseneck, T., Schmoelz, M., Quervain, D. E., & D., & Briegel, J. (2006). Efficacy of hydrocortisone in preventing posttraumatic stress disorder following critical illness and major surgery. *Annals of the New York Academy of Sciences, 1071*, 46–53. https://doi.org/10.1196/annals.1364.005

Schumacher, S., Niemeyer, H., Engel, S., Cwik, J. C., & Knaevelsrud, C. (2018). Psychotherapeutic treatment and HPA axis regulation in posttraumatic stress disorder: A systematic review and meta-analysis. *Psychoneuroendocrinology, 98*, 186–201. https://doi.org/10.1016/j.psyneuen.2018.08.006

Schumacher, S., Niemeyer, H., Engel, S., Cwik, J. C., Laufer, S., Klusmann, H., & Knaevelsrud, C. (2019). HPA axis regulation in posttraumatic stress disorder: A meta-analysis focusing on potential moderators. *Neuroscience and Biobehavioral Reviews, 100*, 35–57. https://doi.org/10.1016/j.neubiorev.2019.02.005

Schwarzer, C. (2009). 30 years of dynorphins—New insights on their functions in neuropsychiatric diseases. *Pharmacology & Therapeutics, 123*(3), 353–370. https://doi.org/10.1016/j.pharmthera.2009.05.006

Sgroi, S., & Tonini, R. (2018). Opioidergic modulation of striatal circuits, implications in Parkinson's disease and levodopa induced dyskinesia. *Frontiers in Neurology, 9*, 524. https://doi.org/10.3389/fneur.2018.00524

Shahraki, A. H., Tian, R., Zhang, C., Fregien, N. L., Bejarano, P., & Mirsaeidi, M. (2022). Anti-inflammatory properties of the alpha-melanocyte-stimulating hormone in models of granulomatous inflammation. *Lung, 200*(4), 463–472. https://doi.org/10.1007/s00408-022-00546-x

Shariq, A. S., Brietzke, E., Rosenblat, J. D., Barendra, V., Pan, Z., & McIntyre, R. S. (2018). Targeting cytokines in reduction of depressive symptoms: A comprehensive review. *Progress in Neuro-Psychopharmacology & Biological Psychiatry, 83*, 86–91. https://doi.org/10.1016/j.pnpbp.2018.01.003

Sharp, B. M., Roy, S., & Bidlack, J. M. (1998). Evidence for opioid receptors on cells involved in host defense and the immune system. *Journal of Neuroimmunology, 83*(1-2), 45–56.

Shen, Y., Fu, W. Y., Cheng, E. Y., Fu, A. K., & Ip, N. Y. (2013). Melanocortin-4 receptor regulates hippocampal synaptic plasticity through a protein kinase A-dependent mechanism. *The Journal of Neuroscience, 33*(2), 464–472. https://doi.org/10.1523/JNEUROSCI.3282-12.2013

Shen, Y., Tian, M., Zheng, Y., Gong, F., Fu, A. K. Y., & Ip, N. Y. (2016). Stimulation of the hippocampal POMC/MC4R circuit alleviates synaptic plasticity impairment in an Alzheimer's disease model. *Cell Reports, 17*(7), 1819–1831. https://doi.org/10.1016/j.celrep.2016.10.043

Shirayama, Y., Ishida, H., Iwata, M., Hazama, G. I., Kawahara, R., & Duman, R. S. (2004). Stress increases dynorphin immunoreactivity in limbic brain regions and dynorphin antagonism produces antidepressant-like effects. *Journal of Neurochemistry, 90*(5), 1258–1268. https://doi.org/10.1111/j.1471-4159.2004.02589.x

Şimşek, Ş., Yıldırım, V., Çim, A., & Kaya, S. (2016). Serum IL-4 and IL-10 levels correlate with the symptoms of the drug-naive adolescents with first episode, early onset schizophrenia. *Journal of Child and Adolescent Psychopharmacology, 26*(8), 721–726. https://doi.org/10.1089/cap.2015.0220

Singh, M. K., Chang, K. D., Chen, M. C., Kelley, R. G., Garrett, A., Mitsunaga, M. M., et al. (2012). Volumetric reductions in the subgenual anterior cingulate cortex in adolescents with bipolar I disorder. *Bipolar Disorders, 14*(6), 585–596. https://doi.org/10.1111/j.1399-5618.2012.01043.x

Sivukhina, E. V., Dolzhikov, A. A., Morozov, I. E., Jirikowski, G. F., & Grinevich, V. (2006). Effects of chronic alcoholic disease on magnocellular and parvocellular hypothalamic neurons in men. *Hormone and Metabolic Research, 38*(6), 382–390. https://doi.org/10.1055/s-2006-944522

Slavich, G. M., & Irwin, M. R. (2014). From stress to inflammation and major depressive disorder: a social signal transduction theory of depression. *Psychological Bulletin, 140*(3), 774–815. https://doi.org/10.1037/a0035302

Smith, S. M., & Vale, W. W. (2006). The role of the hypothalamic-pituitary-adrenal axis in neuroendocrine responses to stress. *Dialogues in Clinical Neuroscience, 8*(4), 383–395. https://doi.org/10.31887/DCNS.2006.8.4/ssmith

Smith, K. A., Fairburn, C. G., & Cowen, P. J. (1997). Relapse of depression after rapid depletion of tryptophan. *Lancet, 349*(9056), 915–919. https://doi.org/10.1016/s0140-6736(96)07044-4

Söderpalm, A., Nikolayev, L., & de Wit, H. (2003). Effects of stress on responses to methamphetamine in humans. *Psychopharmacology, 170*(2), 188–199. https://doi.org/10.1007/s00213-003-1536-5

Sorgdrager, F. J. H., Doornbos, B., Penninx, B. W. J. H., de Jonge, P., & Kema, I. P. (2017). The association between the hypothalamic pituitary adrenal axis and tryptophan metabolism in persons with recurrent major depressive disorder and healthy controls. *Journal of Affective Disorders, 222*, 32–39. https://doi.org/10.1016/j.jad.2017.06.052

Soria, V., González-Rodríguez, A., Huerta-Ramos, E., Usall, J., Cobo, J., Bioque, M., et al. (2018). Targeting hypothalamic-pituitary-adrenal axis hormones and sex steroids for improving cognition in major mood disorders and schizophrenia: A systematic review and narrative synthesis. *Psychoneuroendocrinology, 93*, 8–19. https://doi.org/10.1016/j.psyneuen.2018.04.012

Sparrow, A. M., Lowery-Gionta, E. G., Pleil, K. E., Li, C., Sprow, G. M., Cox, B. R., et al. (2012). Central neuropeptide Y modulates binge-like ethanol drinking in C57BL/6J mice via Y1 and Y2 receptors. *Neuropsychopharmacology, 37*(6), 1409–1421. https://doi.org/10.1038/npp.2011.327

Spivak, B., Shohat, B., Mester, R., Avraham, S., Gil-Ad, I., Bleich, A., et al. (1997). Elevated levels of serum interleukin-1 beta in combat-related posttraumatic stress disorder. *Biological Psychiatry, 42*(5), 345–348. https://doi.org/10.1016/S0006-3223(96)00375-7

Stein, C., Hassan, A. H., Przewłocki, R., Gramsch, C., Peter, K., & Herz, A. (1990). Opioids from immunocytes interact with receptors on sensory nerves to inhibit nociception in inflammation. *Proceedings of the National Academy of Sciences of the United States of America, 87*(15), 5935–5939. https://doi.org/10.1073/pnas.87.15.5935

Steinhoff, M. S., von Mentzer, B., Geppetti, P., Pothoulakis, C., & Bunnett, N. W. (2014). Tachykinins and their receptors: Contributions to physiological control and the mechanisms of disease. *Physiological Reviews, 94*(1), 265–301. https://doi.org/10.1152/physrev.00031.2013

Stellwagen, D., Beattie, E. C., Seo, J. Y., & Malenka, R. C. (2005). Differential regulation of AMPA receptor and GABA receptor trafficking by tumor necrosis factor-alpha. *The Journal of Neuroscience, 25*(12), 3219–3228. https://doi.org/10.1523/JNEUROSCI.4486-04.2005

Stephanou, A., Jessop, D. S., Knight, R. A., & Lightman, S. L. (1990). Corticotrophin-releasing factor-like immunoreactivity and mRNA in human leukocytes. *Brain, Behavior, and Immunity, 4*(1), 67–73. https://doi.org/10.1016/0889-1591(90)90007-d

Steudte-Schmiedgen, S., Kirschbaum, C., Alexander, N., & Stalder, T. (2016). An integrative model linking traumatization, cortisol dysregulation and posttraumatic stress disorder: Insight from recent hair cortisol findings. *Neuroscience and Biobehavioral Reviews, 69*, 124–135. https://doi.org/10.1016/j.neubiorev.2016.07.015

Stone, L., Girgenti, M., Krystal, J., & Duman, R. (2020). Neural transcriptomic alterations associated with neuropeptide Y and ghrelin in posttraumatic stress disorder. *Biological Psychiatry, 87*, S421–S422. https://doi.org/10.1016/j.biopsych.2020.02.1075

Streit, W. J., Sammons, N. W., Kuhns, A. J., & Sparks, D. L. (2004). Dystrophic microglia in the aging human brain. *Glia, 45*(2), 208–212. https://doi.org/10.1002/glia.10319

Susman, E. J. (2006). Psychobiology of persistent antisocial behavior: Stress, early vulnerabilities and the attenuation hypothesis. *Neuroscience and Biobehavioral Reviews, 30*(3), 376–389. https://doi.org/10.1016/j.neubiorev.2005.08.002

Takaesu, Y. (2018). Circadian rhythm in bipolar disorder: A review of the literature. *Psychiatry and Clinical Neurosciences, 72*(9), 673–682. https://doi.org/10.1111/pcn.12688

Teicher, M. H., Andersen, S. L., Polcari, A., Anderson, C. M., & Navalta, C. P. (2002). Developmental neurobiology of childhood stress and trauma. *The Psychiatric Clinics of North America, 25*(2), 397-426, vii-viii. https://doi.org/10.1016/s0193-953x(01)00003-x

Trickett, P. K., Noll, J. G., Susman, E. J., Shenk, C. E., & Putnam, F. W. (2010). Attenuation of cortisol across development for victims of sexual abuse. *Development and Psychopathology, 22*(1), 165–175. https://doi.org/10.1017/S0954579409990332

Trieu, B. H., Remmers, B. C., Toddes, C., Brandner, D. D., Lefevre, E. M., Kocharian, A., et al. (2022). Angiotensin-converting enzyme gates brain circuit-specific plasticity via an endogenous opioid. *Science, 375*(6585), 1177–1182. https://doi.org/10.1126/science.abl5130

Tsuchimine, S., Hattori, K., Ota, M., Hidese, S., Teraishi, T., Sasayama, D., et al. (2019). Reduced plasma orexin-A levels in patients with bipolar disorder. *Neuropsychiatric Disease and Treatment, 15*, 2221–2230. https://doi.org/10.2147/NDT.S209023

Tucker, P., Pfefferbaum, B., Jeon-Slaughter, H., Garton, T. S., & North, C. S. (2014). Extended mental health service utilization among survivors of the Oklahoma City bombing. *Psychiatric Services, 65*(4), 559–562. https://doi.org/10.1176/appi.ps.201200579

Tural, U., & Iosifescu, D. V. (2020). Neuropeptide Y in PTSD, MDD, and chronic stress: A systematic review and meta-analysis. *Journal of Neuroscience Research, 98*(5), 950–963. https://doi.org/10.1002/jnr.24589

Umeoka, E. H. L., van Leeuwen, J. M. C., Vinkers, C. H., & Joëls, M. (2021). The role of stress in bipolar disorder. *Current Topics in Behavioral Neurosciences, 48*, 21–39. https://doi.org/10.1007/7854_2020_151

Unschuld, P. G., Ising, M., Roeske, D., Erhardt, A., Specht, M., Kloiber, S., et al. (2010). Gender-specific association of galanin polymorphisms with HPA-axis dysregulation, symptom severity, and antidepressant treatment response. *Neuropsychopharmacology, 35*(7), 1583–1592. https://doi.org/10.1038/npp.2010.30

Upthegrove, R., Manzanares-Teson, N., & Barnes, N. M. (2014). Cytokine function in medication-naive first episode psychosis: A systematic review and meta-analysis. *Schizophrenia Research, 155*(1-3), 101–108. https://doi.org/10.1016/j.schres.2014.03.005

Valkanova, V., Ebmeier, K. P., & Allan, C. L. (2013). CRP, IL-6 and depression: A systematic review and meta-analysis of longitudinal studies. *Journal of Affective Disorders, 150*(3), 736–744. https://doi.org/10.1016/j.jad.2013.06.004

van den Berg, M. T., Wester, V. L., Vreeker, A., Koenders, M. A., Boks, M. P., van Rossum, E. F. C., & Spijker, A. T. (2020). Higher cortisol levels may proceed a manic episode and are related to disease severity in patients with bipolar disorder. *Psychoneuroendocrinology, 119*, 104658. https://doi.org/10.1016/j.psyneuen.2020.104658

Van Derveer, A. B., Bastos, G., Ferrell, A. D., Gallimore, C. G., Greene, M. L., Holmes, J. T., et al. (2021). A role for somatostatin-positive interneurons in neuro-oscillatory and information processing deficits in schizophrenia. *Schizophrenia Bulletin, 47*(5), 1385–1398. https://doi.org/10.1093/schbul/sbaa184

Vaughan, W. T., Sullivan, J. C., & Elmadjian, F. (1949). Immunity and schizophrenia; a survey of the ability of schizophrenic patients to develop an active immunity following the injection of pertussis vaccine. *Psychosomatic Medicine, 11*(6), 327–333. https://doi.org/10.1097/00006842-194911000-00001

Vazquez, M. I., Catalan-Dibene, J., & Zlotnik, A. (2015). B cells responses and cytokine production are regulated by their immune microenvironment. *Cytokine, 74*(2), 318–326. https://doi.org/10.1016/j.cyto.2015.02.007

Vieta, E., Gasto, C., Martinez de Osaba, M. J., Nieto, E., Canto, T. J., Otero, A., & Vallejo, J. (1997). Prediction of depressive relapse in remitted bipolar patients using corticotrophin-releasing hormone challenge test. *Acta Psychiatrica Scandinavica, 95*(3), 205–211. https://doi.org/10.1111/j.1600-0447.1997.tb09621.x

Viviani, D., Charlet, A., van den Burg, E., Robinet, C., Hurni, N., Abatis, M., et al. (2011). Oxytocin selectively gates fear responses through distinct outputs from the central amygdala. *Science, 333*(6038), 104–107. https://doi.org/10.1126/science.1201043

Walker, E., Mittal, V., & Tessner, K. (2008). Stress and the hypothalamic pituitary adrenal axis in the developmental course of schizophrenia. *Annual Review of Clinical Psychology, 4*, 189–216. https://doi.org/10.1146/annurev.clinpsy.4.022007.141248

Wamsley, J. K., Young, W. S., & Kuhar, M. J. (1980). Immunohistochemical localization of enkephalin in rat forebrain. *Brain Research, 190*(1), 153–174. https://doi.org/10.1016/0006-8993(80)91166-x

Wang, B., Chen, T., Xue, L., Wang, J., Jia, Y., Li, G., et al. (2019). Methamphetamine exacerbates neuroinflammatory response to lipopolysaccharide by activating dopamine D1-like receptors. *International Immunopharmacology, 73*, 1–9. https://doi.org/10.1016/j.intimp.2019.04.053

Wartchow, K. M., Cordeiro, R. C., & Scaini, G. (2023). Advances in the pathophysiology of bipolar disorder. *Current Opinion in Psychiatry, 36*(1), 20–27. https://doi.org/10.1097/YCO.0000000000000836

Watson, S., Gallagher, P., Porter, R. J., Smith, M. S., Herron, L. J., Bulmer, S., et al. (2012). A randomized trial to examine the effect of mifepristone on neuropsychological performance and mood in patients with bipolar depression. *Biological Psychiatry, 72*(11), 943–949. https://doi.org/10.1016/j.biopsych.2012.05.029

Weinberg, J., Sliwowska, J. H., Lan, N., & Hellemans, K. G. (2008). Prenatal alcohol exposure: Foetal programming, the hypothalamic-pituitary-adrenal axis and sex differences in outcome. *Journal of Neuroendocrinology, 20*(4), 470–488. https://doi.org/10.1111/j.1365-2826.2008.01669.x

Wemm, S. E., & Sinha, R. (2019). Drug-induced stress responses and addiction risk and relapse. *Neurobiology of Stress, 10*, 100148. https://doi.org/10.1016/j.ynstr.2019.100148

Werner, F. M., & Coveñas, R. (2010). Classical neurotransmitters and neuropeptides involved in major depression: A review. *The International Journal of Neuroscience, 120*(7), 455–470. https://doi.org/10.3109/00207454.2010.483651

Widerlöv, E., Lindström, L. H., Wahlestedt, C., & Ekman, R. (1988). Neuropeptide Y and peptide YY as possible cerebrospinal fluid markers for major depression and schizophrenia, respectively. *Journal of Psychiatric Research, 22*(1), 69–79. https://doi.org/10.1016/0022-3956(88)90030-1

Winsky-Sommerer, R., Boutrel, B., & de Lecea, L. (2005). Stress and arousal: The corticotrophin-releasing factor/hypocretin circuitry. *Molecular Neurobiology, 32*(3), 285–294. https://doi.org/10.1385/MN:32:3:285

Wochnik, G. M., Rüegg, J., Abel, G. A., Schmidt, U., Holsboer, F., & Rein, T. (2005). FK506-binding proteins 51 and 52 differentially regulate dynein interaction and nuclear translocation of the glucocorticoid receptor in mammalian cells. *The Journal of Biological Chemistry, 280*(6), 4609–4616. https://doi.org/10.1074/jbc.M407498200

Wybran, J., Appelboom, T., Famaey, J. P., & Govaerts, A. (1979). Suggestive evidence for receptors for morphine and methionine-enkephalin on normal human blood T lymphocytes. *Journal of Immunology, 123*(3), 1068–1070.

Wyrwoll, C. S., Holmes, M. C., & Seckl, J. R. (2011). 11β-hydroxysteroid dehydrogenases and the brain: From zero to hero, a decade of progress. *Frontiers in Neuroendocrinology, 32*(3), 265–286. https://doi.org/10.1016/j.yfrne.2010.12.001

Xu, Y. Y., Liang, J., & Xia, Q. R. (2017). Novel insights into the pharmacological effects of resveratrol on the management of depression: A short review. *Pharmazie, 72*(9), 499–502. https://doi.org/10.1691/ph.2017.7462

Yang, J. J., & Jiang, W. (2020). Immune biomarkers alterations in post-traumatic stress disorder: A systematic review and meta-analysis. *Journal of Affective Disorders, 268*, 39–46. https://doi.org/10.1016/j.jad.2020.02.044

Yang, R., Tuan, R. R. L., Hwang, F. J., Bloodgood, D. W., Kong, D., & Ding, J. B. (2023). Dichotomous regulation of striatal plasticity by dynorphin. *Molecular Psychiatry, 28*(1), 434–447. https://doi.org/10.1038/s41380-022-01885-0

Yehuda, R., & LeDoux, J. (2007). Response variation following trauma: A translational neuroscience approach to understanding PTSD. *Neuron, 56*(1), 19–32. https://doi.org/10.1016/j.neuron.2007.09.006

Yehuda, R., Teicher, M. H., Levengood, R. A., Trestman, R. L., & Siever, L. J. (1994). Circadian regulation of basal cortisol levels in posttraumatic stress disorder. *Annals of the New York Academy of Sciences, 746*, 378–380. https://doi.org/10.1111/j.1749-6632.1994.tb39260.x

Yildiz, E., Elboga, G., Uygun, M. C., Yilmaz, R., Selvi, Y., Ulusal, H., et al. (2022). Lower morning levels of cortisol and neuropeptides in blood samples from patients with bipolar disorder. *Journal of Affective Disorders Reports, 10*, 100406. https://doi.org/10.1016/j.jadr.2022.100406

Young, S. N. (1981). Mechanism of decline in rat brain 5-hydroxytryptamine after induction of liver tryptophan pyrrolase by hydrocortisone: roles of tryptophan catabolism and kynurenine synthesis. *British Journal of Pharmacology, 74*(3), 695–700. https://doi.org/10.1111/j.1476-5381.1981.tb10480.x

Young, L. J., Muns, S., Wang, Z., & Insel, T. R. (1997). Changes in oxytocin receptor mRNA in rat brain during pregnancy and the effects of estrogen and interleukin-6. *Journal of Neuroendocrinology, 9*(11), 859–865. https://doi.org/10.1046/j.1365-2826.1997.00654.x

Zellweger, M. J., Osterwalder, R. H., Langewitz, W., & Pfisterer, M. E. (2004). Coronary artery disease and depression. *European Heart Journal, 25*(1), 3–9. https://doi.org/10.1016/j.ehj.2003.09.009

Zhang, X. Y., Zhou, D. F., Zhang, P. Y., Wu, G. Y., Cao, L. Y., & Shen, Y. C. (2002). Elevated interleukin-2, interleukin-6 and interleukin-8 serum levels in neuroleptic-free schizophrenia: association with psychopathology. *Schizophrenia Research, 57*(2-3), 247–258. https://doi.org/10.1016/s0920-9964(01)00296-1

Zhang, Y., Berger, A., Milne, C. D., & Paige, C. J. (2006). Tachykinins in the immune system. *Current Drug Targets, 7*(8), 1011–1020. https://doi.org/10.2174/138945006778019363

Zhu, H., Wang, J., Li, J., & Li, S. (2011). Corticotropin-releasing factor family and its receptors: pro-inflammatory or anti-inflammatory targets in the periphery? *Inflammation Research, 60*(8), 715–721. https://doi.org/10.1007/s00011-011-0329-2

Zhu, F., Zhang, L., Liu, F., Wu, R., Guo, W., Ou, J., et al. (2018). Altered serum tumor necrosis factor and interleukin-1β in first-episode drug-naive and chronic schizophrenia. *Frontiers in Neuroscience, 12*, 296. https://doi.org/10.3389/fnins.2018.00296

Zimmerman, A., Halligan, S., Skeen, S., Morgan, B., Fraser, A., Fearon, P., & Tomlinson, M. (2020). PTSD symptoms and cortisol stress reactivity in adolescence: Findings from a high adversity cohort in South Africa. *Psychoneuroendocrinology, 121*, 104846. https://doi.org/10.1016/j.psyneuen.2020.104846

Zobel, A. W., Nickel, T., Sonntag, A., Uhr, M., Holsboer, F., & Ising, M. (2001). Cortisol response in the combined dexamethasone/CRH test as predictor of relapse in patients with remitted depression. A prospective study. *Journal of Psychiatric Research, 35*(2), 83–94. https://doi.org/10.1016/s0022-3956(01)00013-9

Zohar, J., Yahalom, H., Kozlovsky, N., Cwikel-Hamzany, S., Matar, M. A., Kaplan, Z., et al. (2011). High dose hydrocortisone immediately after trauma may alter the trajectory of PTSD: Interplay between clinical and animal studies. *European Neuropsychopharmacology, 21*(11), 796–809. https://doi.org/10.1016/j.euroneuro.2011.06.001

Index

H. E. Gendelman, T. Ikezu (eds.), *Neuroimmune Pharmacology and Therapeutics*,
https://doi.org/10.1007/978-3-031-68237-7

I

L

M

T

U

V

W

Z

GPSR Compliance

The European Union's (EU) General Product Safety Regulation (GPSR) is a set of rules that requires consumer products to be safe and our obligations to ensure this.

If you have any concerns about our products, you can contact us on ProductSafety@springernature.com

In case Publisher is established outside the EU, the EU authorized representative is:

Springer Nature Customer Service Center GmbH
Europaplatz 3
69115 Heidelberg, Germany

Batch number: 10756634

Printed by Printforce, the Netherlands